Mercury

The View after MESSENGER

Observations from the first spacecraft to orbit the planet Mercury have transformed our understanding of the origin and evolution of rocky planets. This volume is the definitive resource about Mercury for planetary scientists, from students to senior researchers. Topics treated in depth include Mercury's chemical composition; the structure of its crust, lithosphere, mantle, and core; Mercury's modern and ancient magnetic field; Mercury's geology, including the planet's major geologic units and their surface chemistry and mineralogy, its spectral reflectance characteristics, its craters and cratering history, its tectonic features and deformational history, its volcanic features and magmatic history, its distinctive hollows, and the frozen ices in its polar deposits; Mercury's exosphere and magnetosphere and the processes that govern their dynamics and their interaction with the solar wind and interplanetary magnetic field; the formation and large-scale evolution of the planet; and current plans and needed capabilities to explore Mercury further in the future.

SEAN C. SOLOMON is Director of the Lamont-Doherty Earth Observatory and William B. Ransford Professor of Earth and Planetary Science at Columbia University. He earlier served as Director of the Department of Terrestrial Magnetism at the Carnegie Institution of Washington and Professor of Geophysics at the Massachusetts Institute of Technology. He was the Principal Investigator for NASA's MESSENGER mission to Mercury from the initial mission concept in 1996 to the end of the project in 2018. He also served on the science teams for the Magellan mission to Venus, the Mars Global Surveyor mission, and the Gravity Recovery and Interior Laboratory mission to the Moon. A member of the U.S. National Academy of Sciences and the American Academy of Arts and Sciences and former President of the American Geophysical Union, Solomon in 2014 was awarded the National Medal of Science by President Barack Obama.

LARRY R. NITTLER conducts laboratory research on extraterrestrial materials and remote sensing observations of planets at the Carnegie Institution of Washington. He served on NASA's MESSENGER mission to Mercury as Participating Scientist from 2007 to 2012 and Deputy Principal Investigator from 2012 to 2018. He earlier participated in the Near Earth Asteroid Rendezvous, Stardust, and Genesis missions and is currently a science team member on the Japan Aerospace Exploration Agency's Hayabusa2 asteroid sample return mission and the BepiColombo mission to Mercury. He received the 2001 Alfred O. C. Nier Prize of the Meteoritical Society and was named Fellow of that society in 2010. Asteroid 5992 Nittler is named in his honor.

BRIAN J. ANDERSON is Principal Professional Staff Physicist at The Johns Hopkins University Applied Physics Laboratory, having served earlier as Magnetospheric Section supervisor and Space Physics Group supervisor. For MESSENGER he was Magnetometer Instrument Scientist from 1999 to 2009 and Deputy Project Scientist from 2007 to 2018 while also serving as Co-Investigator from 2009 to 2018. He was spacecraft magnetics lead and is on the science team of NASA's Magnetospheric Multiscale mission. He is the Principal Investigator of the National Science Foundation's Active Magnetosphere and Planetary Electrodynamics Response Experiment. His research includes the physics of magnetospheres, plasma wave–particle physics, and planetary magnetic fields.

Cambridge Planetary Science

Series Editors:

Fran Bagenal, David Jewitt, Carl Murray, Jim Bell, Ralph Lorenz, Francis Nimmo, Sara Russell

Books in the Series:

1. Jupiter: The Planet, Satellites and Magnetosphere†
   Edited by Bagenal, Dowling and McKinnon
   978-0-521-03545-3

2. Meteorites: A Petrologic, Chemical and Isotopic Synthesis†
   Hutchison
   978-0-521-03539-2

3. The Origin of Chondrules and Chondrites†
   Sears
   978-1-107-40285-0

4. Planetary Rings†
   Esposito
   978-1-107-40247-8

5. The Geology of Mars: Evidence from Earth-Based Analogs†
   Edited by Chapman
   978-0-521-20659-4

6. The Surface of Mars†
   Carr
   978-0-521-87201-0

7. Volcanism on Io: A Comparison with Earth†
   Davies
   978-0-521-85003-2

8. Mars: An Introduction to its Interior, Surface and Atmosphere†
   Barlow
   978-0-521-85226-5

9. The Martian Surface: Composition, Mineralogy and Physical Properties
   Edited by Bell
   978-0-521-86698-9

10. Planetary Crusts: Their Composition, Origin and Evolution†
    Taylor and McLennan
    978-0-521-14201-4

11. Planetary Tectonics†
    Edited by Watters and Schultz
    978-0-521-74992-3

12. Protoplanetary Dust: Astrophysical and Cosmochemical Perspectives†
    Edited by Apai and Lauretta
    978-0-521-51772-0

13. Planetary Surface Processes
    Melosh
    978-0-521-51418-7

14. Titan: Interior, Surface, Atmosphere and Space Environment
    Edited by Müller-Wodarg, Griffith, Lellouch and Cravens
    978-0-521-19992-6

15. Planetary Rings: A Post-Equinox View (Second edition)
    Esposito
    978-1-107-02882-1

16. Planetesimals: Early Differentiation and Consequences for Planets
    Edited by Elkins-Tanton and Weiss
    978-1-107-11848-5

17. Asteroids: Astronomical and Geological Bodies
    Burbine
    978-1-107-09684-4

18. The Atmosphere and Climate of Mars
    Edited by Haberle, Clancy, Forget, Smith and Zurek
    978-1-107-01618-7

19. Planetary Ring Systems
    Edited by Tiscareno and Murray
    978-1-107-11382-4

20. Saturn in the 21st Century
    Edited by Baines, Flasar, Krupp and Stallard
    978-1-107-10677-2

21. Mercury: The View after MESSENGER
    Edited by Solomon, Nittler and Anderson
    978-1-107-15445-2

† Reissued as a paperback

MERCURY
The View after MESSENGER

Edited by

SEAN C. SOLOMON
Lamont-Doherty Earth Observatory, Columbia University, New York, USA

LARRY R. NITTLER
Carnegie Institution of Washington, Washington, DC, USA

BRIAN J. ANDERSON
The Johns Hopkins University Applied Physics Laboratory, Laurel, Maryland, USA

CAMBRIDGE
UNIVERSITY PRESS

University Printing House, Cambridge CB2 8BS, United Kingdom

One Liberty Plaza, 20th Floor, New York, NY 10006, USA

477 Williamstown Road, Port Melbourne, VIC 3207, Australia

314–321, 3rd Floor, Plot 3, Splendor Forum, Jasola District Centre, New Delhi – 110025, India

79 Anson Road, #06–04/06, Singapore 079906

Cambridge University Press is part of the University of Cambridge.

It furthers the University's mission by disseminating knowledge in the pursuit of education, learning, and research at the highest international levels of excellence.

www.cambridge.org
Information on this title: www.cambridge.org/9781107154452
DOI: 10.1017/9781316650684

© Cambridge University Press 2018

This publication is in copyright. Subject to statutory exception and to the provisions of relevant collective licensing agreements, no reproduction of any part may take place without the written permission of Cambridge University Press.

First published 2018

The MESSENGER mission was sponsored by the U.S. Government (contract #s NAS5-97271/24 and NASW-00002).

Printed and bound in Great Britain by Clays Ltd, Elcograf S.p.A.

A catalogue record for this publication is available from the British Library.

Library of Congress Cataloging-in-Publication Data
Names: Solomon, Sean C., editor. | Nittler, Larry R., editor. | Anderson, Brian J., editor.
Title: Mercury : The view after MESSENGER / edited by Sean C. Solomon (Lamont-Doherty Earth Observatory, Columbia University, New York), Larry R. Nittler (Carnegie Institution of Washington, Washington, DC), Brian J. Anderson (The Johns Hopkins University, Applied Physics Laboratory, Laurel, Maryland).
Other titles: View after MESSENGER
Description: Cambridge : Cambridge University Press, [2018] | Includes bibliographical references.
Identifiers: LCCN 2018022383 | ISBN 9781107154452
Subjects: LCSH: Mercury (Planet) – Observations. | MESSENGER (Spacecraft)
Classification: LCC TL796.6.M47 M47 2018 | DDC 559.9/21–dc23
LC record available at https://lccn.loc.gov/2018022383

ISBN 978-1-107-15445-2 Hardback

Additional resources for this publication available at www.cambridge.org/mercury

Cambridge University Press has no responsibility for the persistence or accuracy of URLs for external or third-party internet websites referred to in this publication and does not guarantee that any content on such websites is, or will remain, accurate or appropriate.

DEDICATION

This book is dedicated to
Mario H. Acuña, Stanton J. Peale, and Jacob I. Trombka,
three original members of the MESSENGER science team who made contributions essential
to the success of the mission but left us before the journey ended.

CONTENTS

List of Contributors		*page* xi
Preface		xv
1	The MESSENGER Mission: Science and Implementation Overview SEAN C. SOLOMON AND BRIAN J. ANDERSON	1
2	The Chemical Composition of Mercury LARRY R. NITTLER, NANCY L. CHABOT, TIMOTHY L. GROVE, AND PATRICK N. PEPLOWSKI	30
3	Mercury's Crust and Lithosphere: Structure and Mechanics ROGER J. PHILLIPS, PAUL K. BYRNE, PETER B. JAMES, ERWAN MAZARICO, GREGORY A. NEUMANN, AND MARK E. PERRY	52
4	Mercury's Internal Structure JEAN-LUC MARGOT, STEVEN A. HAUCK, II, ERWAN MAZARICO, SEBASTIANO PADOVAN, AND STANTON J. PEALE	85
5	Mercury's Internal Magnetic Field CATHERINE L. JOHNSON, BRIAN J. ANDERSON, HAJE KORTH, ROGER J. PHILLIPS, AND LYDIA C. PHILPOTT	114
6	The Geologic History of Mercury BRETT W. DENEVI, CAROLYN M. ERNST, LOUISE M. PROCKTER, AND MARK S. ROBINSON	144
7	The Geochemical and Mineralogical Diversity of Mercury TIMOTHY J. MCCOY, PATRICK N. PEPLOWSKI, FRANCIS M. MCCUBBIN, AND SHOSHANA Z. WEIDER	176
8	Spectral Reflectance Constraints on the Composition and Evolution of Mercury's Surface SCOTT L. MURCHIE, RACHEL L. KLIMA, NOAM R. IZENBERG, DEBORAH L. DOMINGUE, DAVID T. BLEWETT, AND JÖRN HELBERT	191
9	Impact Cratering of Mercury CLARK R. CHAPMAN, DAVID M. H. BAKER, OLIVIER S. BARNOUIN, CALEB I. FASSETT, SIMONE MARCHI, WILLIAM J. MERLINE, LILLIAN R. OSTRACH, LOUISE M. PROCKTER, AND ROBERT G. STROM	217
10	The Tectonic Character of Mercury PAUL K. BYRNE, CHRISTIAN KLIMCZAK, AND A. M. CELÂL ŞENGÖR	249

11	The Volcanic Character of Mercury PAUL K. BYRNE, JENNIFER L. WHITTEN, CHRISTIAN KLIMCZAK, FRANCIS M. MCCUBBIN, AND LILLIAN R. OSTRACH	287
12	Mercury's Hollows DAVID T. BLEWETT, CAROLYN M. ERNST, SCOTT L. MURCHIE, AND FAITH VILAS	324
13	Mercury's Polar Deposits NANCY L. CHABOT, DAVID J. LAWRENCE, GREGORY A. NEUMANN, WILLIAM C. FELDMAN, AND DAVID A. PAIGE	346
14	Observations of Mercury's Exosphere: Composition and Structure WILLIAM E. MCCLINTOCK, TIMOTHY A. CASSIDY, AIMEE W. MERKEL, ROSEMARY M. KILLEN, MATTHEW H. BURGER, AND RONALD J. VERVACK, JR.	371
15	Understanding Mercury's Exosphere: Models Derived from MESSENGER Observations ROSEMARY M. KILLEN, MATTHEW H. BURGER, RONALD J. VERVACK, JR., AND TIMOTHY A. CASSIDY	407
16	Structure and Configuration of Mercury's Magnetosphere HAJE KORTH, BRIAN J. ANDERSON, CATHERINE L. JOHNSON, JAMES A. SLAVIN, JIM M. RAINES, AND THOMAS H. ZURBUCHEN	430
17	Mercury's Dynamic Magnetosphere JAMES A. SLAVIN, DANIEL N. BAKER, DANIEL J. GERSHMAN, GEORGE C. HO, SUZANNE M. IMBER, STAMATIOS M. KRIMIGIS, AND TORBJÖRN SUNDBERG	461
18	The Elusive Origin of Mercury DENTON S. EBEL AND SARAH T. STEWART	497
19	Mercury's Global Evolution STEVEN A. HAUCK, II, MATTHIAS GROTT, PAUL K. BYRNE, BRETT W. DENEVI, SABINE STANLEY, AND TIMOTHY J. MCCOY	516
20	Future Missions: Mercury after MESSENGER RALPH L. MCNUTT, JR., JOHANNES BENKHOFF, MASAKI FUJIMOTO, AND BRIAN J. ANDERSON	544
	Index	570
	Index of Place Names	582

CONTRIBUTORS

BRIAN J. ANDERSON
The Johns Hopkins University Applied Physics Laboratory, Laurel, MD 20723, USA

DAVID M. H. BAKER
Solar System Exploration Division, NASA Goddard Space Flight Center, Greenbelt, MD 20771, USA and Department of Earth, Environmental and Planetary Sciences, Brown University Providence, RI 02912, USA

DANIEL N. BAKER
Laboratory for Atmospheric and Space Physics, University of Colorado, Boulder, CO 80303, USA

OLIVIER S. BARNOUIN
The Johns Hopkins University Applied Physics Laboratory, Laurel, MD 20723, USA

JOHANNES BENKHOFF
European Space Research and Technology Centre, European Space Agency, 2201 AZ Noordwijk, The Netherlands

DAVID T. BLEWETT
The Johns Hopkins University Applied Physics Laboratory, Laurel, MD 20723, USA

MATTHEW H. BURGER
Space Telescope Science Institute, Baltimore, MD 21218, USA

PAUL K. BYRNE
Department of Marine, Earth, and Atmospheric Sciences, North Carolina State University, Raleigh, NC 27695, USA and Department of Terrestrial Magnetism, Carnegie Institution of Washington, Washington, DC 20015, USA

TIMOTHY A. CASSIDY
Laboratory for Atmospheric and Space Physics, University of Colorado, Boulder, CO 80303, USA

NANCY L. CHABOT
The Johns Hopkins University Applied Physics Laboratory, Laurel, MD 20723, USA

CLARK R. CHAPMAN
Southwest Research Institute, Boulder, CO 80302, USA

BRETT W. DENEVI
The Johns Hopkins University Applied Physics Laboratory, Laurel, MD 20723, USA

DEBORAH L. DOMINGUE
Planetary Science Institute, Tucson, AZ 85719, USA

DENTON S. EBEL
Department of Earth and Planetary Sciences, American Museum of Natural History, New York, NY 10024, USA

CAROLYN M. ERNST
The Johns Hopkins University Applied Physics Laboratory, Laurel, MD 20723, USA

CALEB I. FASSETT
NASA Marshall Space Flight Center, Huntsville, AL 35805, USA and Department of Astronomy, Mount Holyoke College, South Hadley, MA 01075, USA

WILLIAM C. FELDMAN
Planetary Science Institute, Tucson, AZ 85719, USA

MASAKI FUJIMOTO
Institute of Space and Astronautical Science, Japan Aerospace Exploration Agency, Sagamihara, Kanagawa, 252–5210, Japan

DANIEL J. GERSHMAN
Geospace Physics Laboratory, NASA Goddard Space Flight Center, Greenbelt, MD 20771, USA

MATTHIAS GROTT
Institute of Planetary Research, German Aerospace Center (DLR), 12489 Berlin, Germany

TIMOTHY L. GROVE
Department of Earth, Atmospheric and Planetary Sciences, Massachusetts Institute of Technology, Cambridge, MA 02139, USA

STEVEN A. HAUCK, II
Department of Earth, Environmental, and Planetary Sciences, Case Western Reserve University, Cleveland, OH 44106, USA

JÖRN HELBERT
Institute of Planetary Research, German Aerospace Center (DLR), 12489 Berlin, Germany

GEORGE C. HO
The Johns Hopkins University Applied Physics Laboratory, Laurel, MD 20723, USA

SUZANNE M. IMBER
Department of Physics and Astronomy, University of Leicester, Leicester, LE1 7RH, United Kingdom

NOAM R. IZENBERG
The Johns Hopkins University Applied Physics Laboratory, Laurel, MD 20723, USA

List of Contributors

PETER B. JAMES
Department of Geosciences, Baylor University, Waco, TX 76706, USA and Lunar and Planetary Institute, Universities Space Research Association, Houston, TX 77058, USA

CATHERINE L. JOHNSON
Department of Earth, Ocean and Atmospheric Sciences, University of British Columbia, Vancouver, BC V6T 1Z4, Canada and Planetary Science Institute, Tucson, AZ 85719, USA

ROSEMARY M. KILLEN
Solar System Exploration Division, NASA Goddard Space Flight Center, Greenbelt, MD 20771, USA

RACHEL L. KLIMA
The Johns Hopkins University Applied Physics Laboratory, Laurel, MD 20723, USA

CHRISTIAN KLIMCZAK
Department of Geology, University of Georgia, Athens, GA 30602, USA and Department of Terrestrial Magnetism, Carnegie Institution of Washington, Washington, DC 20015, USA

HAJE KORTH
The Johns Hopkins University Applied Physics Laboratory, Laurel, MD 20723, USA

STAMATIOS M. KRIMIGIS
The Johns Hopkins University Applied Physics Laboratory, Laurel, MD 20723, USA and Office of Space Research and Technology, Academy of Athens, Athens 10679, Greece

DAVID J. LAWRENCE
The Johns Hopkins University Applied Physics Laboratory, Laurel, MD 20723, USA

SIMONE MARCHI
Southwest Research Institute, Boulder, CO 80302, USA

JEAN-LUC MARGOT
Department of Earth, Planetary, and Space Sciences, University of California, Los Angeles, Los Angeles, CA 90095, USA

ERWAN MAZARICO
Solar System Exploration Division, NASA Goddard Space Flight Center, Greenbelt, MD 20771, USA

WILLIAM E. MCCLINTOCK
Laboratory for Atmospheric and Space Physics, University of Colorado, Boulder, CO 80303, USA

TIMOTHY J. MCCOY
Department of Mineral Sciences, National Museum of Natural History, Smithsonian Institution, Washington, DC 20560, USA

FRANCIS M. MCCUBBIN
NASA Johnson Space Center, Houston, TX 77058, USA

RALPH L. MCNUTT, JR.
The Johns Hopkins University Applied Physics Laboratory, Laurel, MD 20723, USA

AIMEE W. MERKEL
Laboratory for Atmospheric and Space Physics, University of Colorado, Boulder, CO 80303, USA

WILLIAM J. MERLINE
Southwest Research Institute, Boulder, CO 80302, USA

SCOTT L. MURCHIE
The Johns Hopkins University Applied Physics Laboratory, Laurel, MD 20723, USA

GREGORY A. NEUMANN
Solar System Exploration Division, NASA Goddard Space Flight Center, Greenbelt, MD 20771, USA

LARRY R. NITTLER
Department of Terrestrial Magnetism, Carnegie Institution of Washington, Washington, DC 20015, USA

LILLIAN R. OSTRACH
U.S. Geological Survey, Flagstaff, AZ 86001, USA and Solar System Exploration Division, NASA Goddard Space Flight Center, Greenbelt, MD 20771, USA

SEBASTIANO PADOVAN
Institute of Planetary Research, German Aerospace Center (DLR), 12489 Berlin, Germany

DAVID A. PAIGE
Department of Earth, Planetary, and Space Sciences, University of California, Los Angeles, Los Angeles, CA 90095, USA

STANTON J. PEALE (DECEASED)
Department of Physics, University of California, Santa Barbara, Santa Barbara, CA 93106, USA

PATRICK N. PEPLOWSKI
The Johns Hopkins University Applied Physics Laboratory, Laurel, MD 20723, USA

MARK E. PERRY
The Johns Hopkins University Applied Physics Laboratory, Laurel, MD 20723, USA

ROGER J. PHILLIPS
Department of Earth and Planetary Sciences and McDonnell Center for the Space Sciences, Washington University, St. Louis, MO 63130, USA

LYDIA C. PHILPOTT
Department of Earth, Ocean and Atmospheric Sciences, University of British Columbia, Vancouver, BC V6T 1Z4, Canada

LOUISE M. PROCKTER
Lunar and Planetary Institute, Universities Space Research Association, Houston, TX 77058, USA

JIM M. RAINES
Climate and Space Sciences and Engineering Department, University of Michigan, Ann Arbor, MI 48109, USA

MARK S. ROBINSON
School of Earth and Space Exploration, Arizona State University, Tempe, AZ 85287, USA

A. M. CELÂL ŞENGÖR
Department of Geology, Faculty of Mines and Eurasia Institute of Earth Sciences, Istanbul Technical University, Istanbul 34810, Turkey

JAMES A. SLAVIN
Climate and Space Sciences and Engineering Department, University of Michigan, Ann Arbor, MI 48109, USA

SEAN C. SOLOMON
Lamont-Doherty Earth Observatory, Columbia University, Palisades, NY 10964, USA

SABINE STANLEY
Department of Earth and Planetary Sciences, Johns Hopkins University, Baltimore, MD 21218, USA

SARAH T. STEWART
Department Earth and Planetary Sciences, University of California, Davis, Davis, CA 95616, USA

ROBERT G. STROM
Lunar and Planetary Laboratory, University of Arizona, Tucson, AZ 85721, USA

TORBJÖRN SUNDBERG
Queen Mary University of London, London, E1 4NS, United Kingdom

RONALD J. VERVACK, JR.
The Johns Hopkins University Applied Physics Laboratory, Laurel, MD 20723, USA

FAITH VILAS
Planetary Science Institute, Tucson, AZ 85719, USA

SHOSHANA Z. WEIDER
Department of Terrestrial Magnetism, Carnegie Institution of Washington, Washington, DC 20015, USA

JENNIFER L. WHITTEN
Center for Earth and Planetary Studies, National Air and Space Museum, Smithsonian Institution, Washington, DC 20013, USA

THOMAS H. ZURBUCHEN
Climate and Space Sciences and Engineering Department, University of Michigan, Ann Arbor, MI 48109, USA

PREFACE

Mercury, like Earth, is one of only four rocky planets in our solar system, yet the exploration of the innermost planet by spacecraft has lagged substantially behind that of Earth's two nearest neighbors, Venus and Mars. The first spacecraft to visit Venus was Mariner 2, which was developed by the National Aeronautics and Space Administration (NASA) and flew by Venus in December 1962. The first successful spacecraft flyby of Mars was by Mariner 4 in July 1965. The first spacecraft to orbit Mars was Mariner 9, which arrived at the red planet in November 1971. The first probe to orbit Venus was the Soviet Union's Venera 9 orbiter, which arrived at Venus in October 1975. Both planets have been visited dozens of times since those early missions by other spacecraft sent by multiple nations and space agencies.

In contrast, the first spacecraft encounter of Mercury was not until March 1974, when Mariner 10 completed the first of its three Mercury flybys. The third and final Mariner 10 flyby of Mercury was one year later in March 1975, and no spacecraft visited Mercury again for more than three decades. The spacecraft exploration of Mercury resumed when NASA's MErcury Surface, Space ENvironment, GEochemistry, and Ranging (MESSENGER) probe flew by Mercury three times in 2008–2009 and became the first spacecraft to orbit Mercury on 18 March 2011. MESSENGER operated in orbit about Mercury for more than four years, until 30 April 2015, and acquired the first global observations of Mercury's surface, interior, exosphere, magnetosphere, and heliospheric environment. The MESSENGER project and its science team continued to validate, archive, and analyze data acquired during the mission for more than two additional years, until the project formally ended on 30 September 2018.

The first spacecraft orbital mission to any planet, like Mariner 9 and the Venera 9 orbiter, enables many important discoveries and produces large new data sets. Collectively those first orbital data sets from a planet drive major increases in scientific understanding and raise multiple new scientific questions, not only about the target body but also about planetary and solar system processes more generally. So it has been with observations made by the MESSENGER spacecraft. The wealth of new data returned by MESSENGER and archived with NASA's Planetary Data System by the MESSENGER team continues to foster new investigations of this nearby yet remarkably distinctive sibling of Earth, and at the same time prompts new questions that expand the rationale for continued Mercury exploration.

This book is intended to synthesize the findings from the MESSENGER mission into a description of our current scientific understanding of Mercury. The book is timely, for two reasons. First, it was written after the end of data collection by MESSENGER, so that all of the measurements acquired over the course of the mission could be integrated and our markedly improved knowledge of Mercury could serve to update our understanding of the formation and evolution of the inner solar system's rocky planets. Second, the book was completed approximately eight years before the scheduled arrival of the next spacecraft at Mercury, the dual probes of the BepiColombo mission of the European Space Agency and the Japan Aerospace Exploration Agency.

The editors of this volume owe considerable thanks to many colleagues whose efforts contributed to the technical and scientific success of the MESSENGER mission. Among these individuals are members of the science team who, because of other responsibilities and interests, were not able to share in the writing for this book but in myriad other ways participated in the analysis and interpretation of observations from the mission. Hundreds of engineers, technicians, software developers, managers, and support personnel contributed to the successful design, construction, testing, launch, and operation of the MESSENGER spacecraft. Among those, eight warrant special thanks: the four individuals at the Johns Hopkins University Applied Physics Laboratory who served successively as MESSENGER Project Manager – Max R. Peterson, David G. Grant, Peter D. Bedini, and Helene L. Winters – and the four who served successively as MESSENGER's Mission Systems Engineer – Andrew G. Santo, James C. Leary, Eric J. Finnegan, and Daniel J. O'Shaughnessy. Each played a vital leadership role at a critical stage in the MESSENGER mission.

The editors are also indebted to the authors of the 20 chapters in this volume, most drawn from the MESSENGER science team but a few from outside the project who bring special expertise on a topic of importance to their chapter. Each of the chapters was reviewed not only by other members of the MESSENGER science team but also by an expert scientist from outside the project. The editors appreciate the thoughtful reviews of individual chapters by Erik Asphaug, Wolfgang Baumjohann, Doris Breuer, Masaki Fujimoto, Walter S. Kiefer, François Leblanc, H. Jay Melosh, Edwin J. Mierkiewicz, Stephen W. Parman, David A. Rothery, Christopher T. Russell, Richard A. Schultz, Matthew A. Siegler, Krista M. Soderlund, S. Alan Stern, David J. Stevenson, Jessica M. Sunshine, G. Jeffrey Taylor, Rebecca J. Thomas, and David A. Williams.

The editors are deeply grateful to Kimberly Schermerhorn for her substantial assistance with the preparation of all of the material for this book. We thank Nancy Chabot and Brett Denevi for their design of the maps located inside the front and back covers of the book; these maps provide a compact illustration of much of the data returned by MESSENGER as well as a ready means to locate many of the major features on Mercury's surface mentioned in the book's chapters. Brett also designed the image on the front cover and selected the MESSENGER images shown on the back cover. We thank Magda Saina for lending her graphical expertise to convert a number of the figures in this volume to publication quality.

Finally, we thank the editors at Cambridge University Press who worked with us from early discussions, to the writing of a formal book proposal, through the preparation of all of the chapters and supporting material, to copy editing and final production. We are particularly grateful for the sustained guidance of Lucy Edwards, Vince Higgs, and Esther Migueliz.

It is our hope that this volume will provide a standard reference on the planet Mercury for a number of years, at least until the next spacecraft after MESSENGER arrive to renew humankind's exploration of our solar system's innermost world.

Sean C. Solomon, Larry R. Nittler, and Brian J. Anderson

1

The MESSENGER Mission: Science and Implementation Overview

SEAN C. SOLOMON AND BRIAN J. ANDERSON

1.1 INTRODUCTION

Although a sibling of Earth, Venus, and Mars, the planet Mercury is an unusual member of the family (Solomon, 2003). Among the planets of our solar system, it is the smallest, at little more than 5% of an Earth mass, but its bulk density corrected for the effect of internal compression is the highest. Mercury's orbit is the most eccentric of the planets, and it is the only known solar system object in a 3:2 spin–orbit resonance, in which three sidereal days equal two periods of Mercury's revolution about the Sun. Mercury is the only inner planet other than Earth to host an internal magnetic field and an Earth-like magnetosphere capable of standing off the solar wind. The closest planet to the Sun, Mercury experiences a variation in surface temperature at the equator of 600°C over the course of a solar day, which because of Mercury's slow spin rate equals two Mercury years. The permanently shadowed floors of Mercury's high-latitude craters nonetheless are sufficiently cold to have trapped water ice and other frozen volatiles.

Thought to have been created by the same processes as the other inner planets and at the same early stage in the history of the solar system, Mercury with its unusual attributes has long held out the promise of deepening our understanding of how Earth and other Earth-like planets formed and evolved. Yet Mercury is not an easy object to study. Never separated from the Sun by more than 28° of arc when viewed from Earth, Mercury is forbidden as a target for the Hubble Space Telescope and other astronomical facilities because their optical systems would be severely damaged by exposure to direct sunlight. Located deep within the gravitational potential well of the Sun, Mercury has also long presented a challenge to spacecraft mission design. The first spacecraft to view Mercury at close range was Mariner 10, which after flying once by Venus encountered the innermost planet three times in 1974–1975. The encounters occurred nearly at Mercury's greatest distance from the Sun and were spaced approximately one Mercury solar day apart, so the same hemisphere of the planet was in sunlight at each flyby. Mariner 10 obtained images of 45% of the surface, discovered the planet's global magnetic field, assayed three neutral species (H, He, and O) in Mercury's tenuous atmosphere, and sampled the magnetic field and energetic charged particles in Mercury's dynamic magnetosphere (Dunne and Burgess, 1978).

After the Mariner 10 mission, the next logical step in the exploration of Mercury was widely viewed to be an orbiter mission (COMPLEX, 1978), and several notable discoveries by ground-based astronomers in the years since the Mariner 10 encounters (e.g., Potter and Morgan, 1985, 1986; Slade et al., 1992; Harmon and Slade, 1992) provided a wealth of new information about Mercury that whetted the appetite of the planetary science community for orbital observations. Nevertheless, substantial advances were needed in mission design, thermal engineering, and miniaturization of instruments and spacecraft subsystems before such a mission could be considered technically ready.

The MErcury Surface, Space ENvironment, GEochemistry, and Ranging (MESSENGER) mission to orbit Mercury was proposed under NASA's Discovery Program in 1996 and again in 1998 (Solomon et al., 2001; Gold et al., 2001; Santo et al., 2001) and was selected for flight in 1999. Development, construction, integration, and testing of the spacecraft and its instruments began in January 2000 and spanned the four and a half years leading to launch on 3 August 2004 (McNutt et al., 2006). MESSENGER completed gravity-assist flybys of Earth once, Venus twice, and Mercury three times (Figure 1.1) during a mission cruise phase that lasted 6.6 years. MESSENGER was inserted into orbit about Mercury on 18 March 2011 and conducted orbital observations of the innermost planet for more than four years, until 30 April 2015.

In this chapter we provide an overview of the MESSENGER mission from a historical perspective, including the mission's scientific objectives; the payload characteristics, data acquisition planning, and operational procedures adopted to achieve those objectives; and the scientific findings from flyby and orbital operations. We begin with summaries of the mission objectives, spacecraft, payload instruments, and orbit design. We then describe the procedures adopted to optimize the scientific return from the complex series of orbital data acquisition operations. We follow with an account of the primary mission, including the Mercury flybys and the first year of orbital observations. We then outline the rationale for and accomplishments of MESSENGER's first extended mission, conducted over the second year of orbital operations, and the second extended mission, conducted over the final two years of orbital operations. The second extended mission included a distinctive low-altitude campaign completed at the culmination of the mission. A concluding section briefly introduces the other chapters of this book.

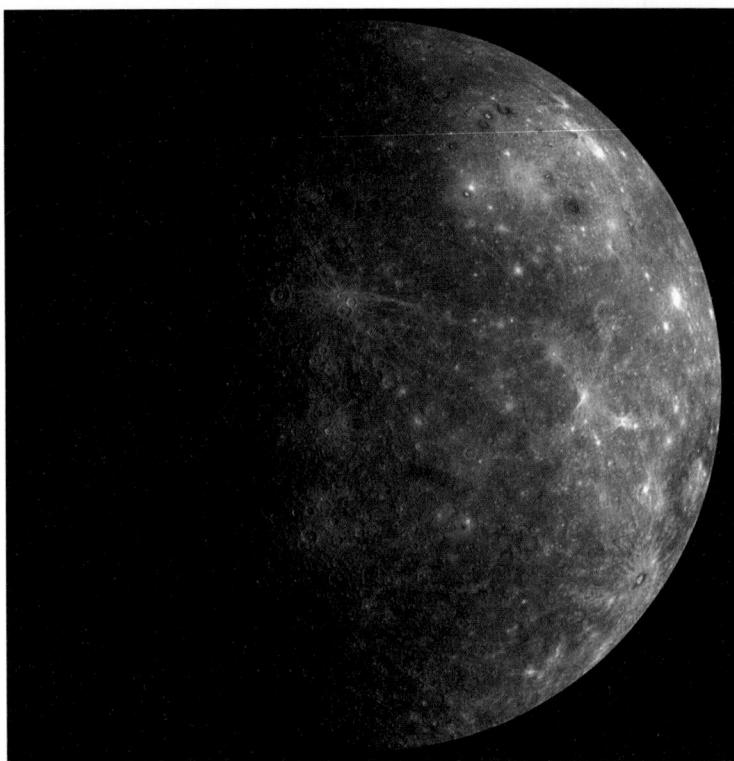

Figure 1.1. Image mosaic of Mercury acquired on departure from MESSENGER's first Mercury flyby on 14 January 2008. Mercury Dual Imaging System wide-angle camera images acquired through the narrow-band filters centered at 1000, 700, and 430 nm are projected in red, green, and blue in this color representation. Much of the area shown had not been imaged by Mariner 10.

1.2 MISSION OBJECTIVES, SPACECRAFT, PAYLOAD, AND ORBIT DESIGN

1.2.1 Key Scientific Questions

The MESSENGER mission was designed to address six key scientific questions. The questions were motivated by the knowledge of Mercury available at the time the mission was proposed, were capable of being substantially addressed by measurements that could be made from orbit, and would yield answers that would bear not only on the nature of Mercury but more generally on the origin and comparative evolution of the inner planets as a group. Those questions and a brief summary of the rationale for each were as follows.

1.2.1.1 What Planetary Formational Processes Led to the High Ratio of Metal to Silicate in Mercury?

The Mariner 10 spacecraft carried no elemental remote sensing instruments, so at the time the MESSENGER mission was proposed the single most important piece of information about the planet's bulk composition was its high uncompressed density, which implied that Mercury has an iron-rich core that occupies much higher fractions of the planet's mass and volume than do the cores of the other inner planets (e.g., Siegfried and Solomon, 1974). A variety of theories for the origin and early evolution of Mercury had been advanced to account for its high metal fraction, including formation from metal-enriched precursors resulting from either high-temperature fractionation or aerodynamical sorting in the solar nebula (e.g., Weidenschilling, 1978; Lewis, 1988) or removal of an initially larger silicate crust and mantle by evaporation or giant impact (e.g., Cameron, 1985; Wetherill, 1988; Benz et al., 1988). Those theories differed in their predictions for the bulk composition of the silicate fraction of the planet (e.g., Lewis, 1988), including the upper crust, which would be visible to geochemical remote sensing instruments on an orbiting spacecraft. Moreover, ground-based telescopic measurements of Mercury's surface reflectance showed few if any absorption features commonly seen in reflectance spectra of the Moon, Mars, and asteroids and attributable to the presence of ferrous iron in silicate minerals (e.g., Vilas, 1988), indicating both a low abundance of ferrous iron on Mercury's surface and the need to rely heavily on elemental remote sensing instruments to gain compositional information.

1.2.1.2 What Is the Geological History of Mercury?

Because of Mercury's size, intermediate between the Moon and Mars, as well as its high metal/silicate ratio, documenting the geological history of Mercury was viewed as crucial to understanding how terrestrial planet evolution depends on planet size and initial conditions. A broad geological history of Mercury had been developed from Mariner 10 images (e.g., Strom, 1979; Spudis and Guest, 1988), but the limited coverage and resolution of those images left many aspects of that history uncertain. Extensive plains units were documented by Mariner 10, and the youngest of those plains deposits were seen to be in stratigraphic positions similar to the volcanic lunar maria. Unlike the maria, however, the plains deposits on Mercury are not markedly lower in reflectance than the surrounding older terrain, and no volcanic landforms were visible at the resolution of Mariner 10 images, so both volcanic and impact ejecta processes for plains emplacement had been suggested (e.g., Strom et al., 1975; Wilhelms,

1976) and the importance of volcanism in Mercury's history was thus uncertain. Deformational features on Mercury were seen to be dominantly contractional, leading to the proposal that such features were the expression of global contraction resulting from interior cooling (e.g., Strom et al., 1975), although the restricted imaging coverage meant that the global contraction hypothesis remained untested over slightly more than a full hemisphere of the planet.

1.2.1.3 What Are the Nature and Origin of Mercury's Magnetic Field?

Measurements by Mariner 10 demonstrated that Mercury has an internal magnetic field (Ness et al., 1976) with a dipole component nearly orthogonal to the planet's orbital plane and an estimated moment near 300 nT R_M^3, where R_M is Mercury's mean radius (Connerney and Ness, 1988). Because external sources can dominate the total field measured at Mercury, and because of the limited sampling of the field during the two Mariner 10 flybys that penetrated Mercury's magnetosphere, the uncertainty in Mercury's dipole moment derived from Mariner 10 data was a factor of 2, and higher-order terms were linearly dependent and thus not resolvable (Connerney and Ness, 1988). A variety of mechanisms for producing Mercury's observed magnetic field had been proposed, including remanent or fossil fields in Mercury's crust and lithosphere (Stephenson, 1976; Srnka, 1976; Aharonson et al., 2004), hydromagnetic dynamos in a fluid outer core (e.g., Schubert et al., 1988; Stanley et al., 2005; Christensen, 2006), and a thermoelectric dynamo driven by temperature differences along the top of the core (Stevenson, 1987; Giampieri and Balogh, 2002). The different field generation models made different predictions regarding the geometry of the field, particularly for terms of higher order than the dipole term, and so measurements made from orbit about the planet were seen to be needed to distinguish among hypotheses.

1.2.1.4 What Are the Structure and State of Mercury's Core?

The size and physical state of Mercury's core are key to understanding the planet's bulk composition, thermal history, and magnetic field generation processes (Zuber et al., 2007). Peale (1976) realized that the existence and radius of a liquid outer core on Mercury can be determined by the measurement of Mercury's obliquity, the amplitude of its physical libration forced by variations in the torque exerted by the gravitational pull of the Sun over the planet's 88-day orbit period, and two quantities that define the shape of the planet's gravity field at spherical harmonic degree and order 2. The required coefficients in the spherical harmonic expansion of Mercury's gravity field had been estimated from radio tracking of the Mariner 10 flybys (Anderson et al., 1987) but not with high precision. All four quantities can be determined from measurements made by an orbiting spacecraft with sufficient precision to determine Mercury's polar moment of inertia and the moment of inertia of the planet's solid outer shell that participates in the 88-day libration (Peale, 1976; Peale et al., 2002), and from those quantities important aspects of Mercury's internal structure can be resolved. Mercury's obliquity and forced libration amplitude can also be measured from Earth-based radar observations, and such measurements were reported by Margot et al. (2007) before MESSENGER was inserted into orbit around Mercury, and then refined several years later (Margot et al., 2012). Although the measurements of libration amplitude and obliquity indicated that Mercury does indeed possess a fluid outer core (Margot et al., 2007), the uncertainties in Mercury's gravitational field coefficients at harmonic degree 2 dominated the uncertainty in the planet's moments of inertia. Radio tracking of an orbiting spacecraft was required to improve the determination of these key quantities.

1.2.1.5 What Are the Radar-Reflective Materials at Mercury's Poles?

The discovery in 1991 of radar-bright regions near Mercury's poles and the similarity of the radar reflectivity and polarization characteristics of such regions to those of icy satellites and the south residual polar cap of Mars led to the proposal that these areas host deposits of surface or near-surface water ice (Slade et al., 1992; Harmon and Slade, 1992). Subsequent radar imaging at improved resolution confirmed that the radar-bright deposits are confined to the floors of near-polar impact craters (e.g., Harmon et al., 2011). Because of Mercury's small obliquity, sufficiently deep craters are permanently shadowed and are predicted to be at temperatures at which water ice is stable for billions of years (Paige et al., 1992). Although a contribution from interior outgassing could not be excluded, impact volatilization of cometary and meteoritic material followed by transport of water molecules to polar cold traps was shown to provide sufficient polar ice to match the characteristics of the deposits (Moses et al., 1999).

Two alternative explanations for the radar-bright polar deposits of Mercury were nonetheless suggested. One was that the polar deposits are composed of elemental sulfur, on the grounds that sulfur would be stable in polar cold traps and the presence of sulfides in the regolith can account for a high disk-averaged index of refraction and low microwave opacity of surface materials (Sprague et al., 1995). The second was that the permanently shadowed portions of polar craters are radar-bright not because of trapped volatiles but because of either unusual surface roughness (Weidenschilling, 1998) or low dielectric loss (Starukhina, 2001) of near-surface silicates at extremely cold temperatures. Geochemical remote sensing measurements made from orbit around Mercury were recognized as able to distinguish among the competing proposals.

1.2.1.6 What Are the Important Volatile Species and Their Sources and Sinks on and near Mercury?

Mercury's atmosphere is a surface-bounded exosphere for which the composition and behavior are controlled by interactions with the magnetosphere and the surface. At the time the MESSENGER mission was under development, the atmosphere was known to contain at least six elements (H, He, O, Na, K, Ca). The Mariner 10 airglow spectrometer detected H and He and set an upper bound on O (Broadfoot et al., 1976), and ground-based spectroscopic observations led to the discovery of exospheric Na (Potter and Morgan, 1985), K (Potter and Morgan, 1986), and Ca (Bida et al., 2000). Exospheric H and He were thought to be dominated by solar wind ions neutralized by recombination at the surface, whereas proposed source processes for other exospheric species included diffusion from the planet's interior, evaporation, sputtering by photons and energetic ions, chemical sputtering by protons, and meteoroid impact and vaporization (e.g., Killen and Ip, 1999). That several of these processes play some role was suggested by the strong variations in exospheric characteristics observed as functions of local time, solar distance, and level of solar activity (e.g., Sprague et al., 1998; Hunten and Sprague, 2002; Leblanc and Johnson, 2003). It was long recognized that a spacecraft in orbit about Mercury can provide a range of opportunities for elucidating further the nature of the exosphere, through profiles of major exospheric neutral species versus time of day and solar distance and searches for new species (e.g., Domingue et al., 2007). In situ measurement of energetic and thermal plasma ions from orbit can also detect solar wind pickup ions that originated as exospheric neutral atoms (e.g., Koehn et al., 2002).

1.2.2 Scientific Objectives

The six key questions above led to a set of scientific objectives for the MESSENGER mission and in turn to a set of project requirements (Solomon et al., 2001), a suite of payload instruments (Gold et al., 2001), and a measurement strategy (Section 1.3). The scientific objectives for MESSENGER's primary mission are given in Table 1.1, and the project requirements for the primary mission are given in Table 1.2.

The objective to characterize the chemical composition of Mercury's surface led to a project requirement for global maps of major element composition at a resolution sufficient to discern the principal geological units and to distinguish material excavated and ejected by young impact craters from a possible veneer of cometary and meteoritic material. Information on surface mineralogy was also deemed important for this objective. The objective to determine the planet's geological history led to a project requirement for global monochrome imaging at a resolution of hundreds of meters or better, for topographic profiles across key geological features from altimetry or stereo, and for spectral measurements of major geologic units at spatial resolutions of several kilometers or better. The objective to characterize Mercury's magnetic field led to a project requirement for magnetometry, both near the planet and throughout the magnetosphere, as well as for energetic particle and plasma measurements so as to assist in the separation of external and internal fields. The objective to estimate the size and state of Mercury's core led to the project requirement for altimetric measurement of the amplitude of Mercury's physical libration as well as determination of the planet's obliquity and low-degree gravitational field. The objective to assay the volatile inventory at Mercury's poles led to the project requirement for ultraviolet spectrometry of the polar atmosphere and for gamma-ray and neutron spectrometry, imaging, and altimetry of polar-region craters. The objective to characterize the nature of Mercury's exosphere and magnetosphere led to the project requirement to identify all major neutral species in the exosphere and charged species in the magnetosphere.

1.2.3 Spacecraft

The design of the MESSENGER spacecraft (Figure 1.2) was driven largely by two requirements: to minimize mass and to survive the harsh thermal environment at Mercury (Santo et al., 2001; Leary et al., 2007). The largest launch vehicle available to the Discovery Program was the Delta II 7925-H, which could inject ~1100 kg into the required interplanetary trajectory. Because more than half of that total launch mass was needed for the propellant required to achieve the mission design, only 500 kg remained for the total spacecraft dry mass. To meet this constraint, the spacecraft structure was fabricated primarily with lightweight composite material and was fully integrated with a dual-mode propulsion system that

Table 1.1. *Scientific objectives for MESSENGER's primary mission.*

1. Determine the chemical composition of Mercury's surface.
2. Determine Mercury's geological history.
3. Determine the nature of Mercury's magnetic field.
4. Determine the size and state of Mercury's core.
5. Determine the volatile inventory at Mercury's poles.
6. Determine the nature of Mercury's exosphere and magnetosphere.

Table 1.2. *Project requirements for MESSENGER's primary mission.*

1. Provide major-element maps of Mercury to 10% relative uncertainty on the 1000-km scale and determine local composition and mineralogy at the ~20-km scale.
2a. Provide a global map with >90% coverage (monochrome) at 250-m average resolution and >80% of the planet imaged stereoscopically.
2b. Provide a global multispectral map at 2-km/pixel average resolution.
2c. Sample half of the northern hemisphere for topography at 1.5-m average height resolution.
3. Provide a multipole magnetic field model resolved through quadrupole terms with an uncertainty of less than ~20% in the dipole magnitude and direction.
4. Provide a global gravity field to degree and order 16 and determine the ratio of the solid-planet moment of inertia to the total moment of inertia to ~20% or better.
5. Identify the principal component of the radar-reflective material at Mercury's north pole.
6. Provide altitude profiles at 25-km resolution of the major neutral exospheric species and characterize the major ion-species energy distributions as functions of local time, Mercury heliocentric distance, and solar activity.

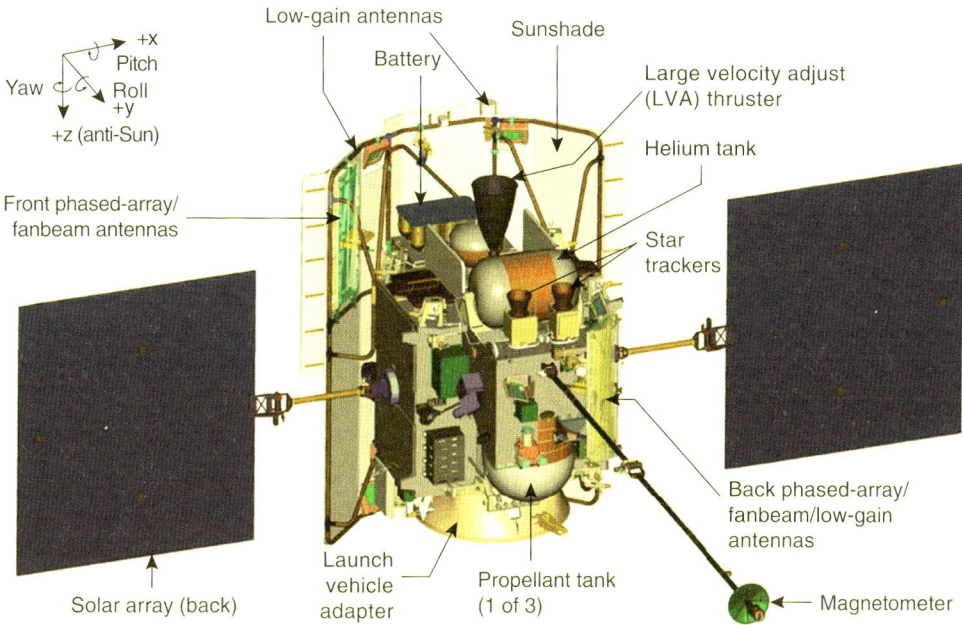

Figure 1.2. Engineering view of the MESSENGER spacecraft from behind the sunshade.

featured lightweight tanks for propellant (hydrazine), oxidizer (nitrous tetroxide), and pressurant (gaseous helium). The propulsion system included a total of 17 thrusters: a single large velocity adjustment bipropellant thruster; four 22-N monopropellant thrusters for thrust-vector steering during large spacecraft maneuvers and for trajectory-correction maneuvers; and 12 4.4-N monopropellant thrusters for attitude control, angular momentum management, and small trajectory-correction maneuvers.

A large number of mass-reduction measures were used in the development of the spacecraft. To avoid a cumbersome gimbaled antenna and the challenges associated with testing and operating it at high temperatures, an electronically steerable phased-array system was developed for the high-gain antenna. Used one at a time, each of two antennas – one on the spacecraft's Sun-facing side and one aft – could be steered about one axis while the spacecraft body rolled about a second axis to point the antenna toward Earth at any point in the mission. The phased-array antennas were complemented with two medium-gain fanbeam antennas and four low-gain antennas. Radio signals were transmitted to and received from the MESSENGER spacecraft at X-band frequencies (7.2-GHz uplink, 8.4-GHz downlink) by the 34-m and 70-m antennas at NASA's Deep Space Network stations in Goldstone, California; Madrid, Spain; and Canberra, Australia.

Mass was also conserved by limiting the number of spacecraft components that moved. With the lone exception of the imaging system (see next section), all science instruments were hard-mounted to the spacecraft. As a consequence, spacecraft attitude often had to be changed continuously in orbit about Mercury to permit the instruments to make their observations.

Spacecraft power was provided by two solar arrays (Figures 1.2 and 1.3), which could be articulated to manage array temperature, and by a battery during those orbits when the spacecraft was on Mercury's nightside and the Sun was eclipsed. In a fully redundant electronics system, a main processor performed all nominal spacecraft functions, while two other processors monitored spacecraft health and safety. The spacecraft attitude control system was three-axis stable and momentum biased and made use of four reaction wheels. Attitude knowledge was acquired through an inertial measurement unit, two star trackers, and a suite of Sun sensors as a backup to the primary attitude sensors.

Primarily passive thermal management techniques were used to minimize heating of spacecraft subsystems by the Sun and the dayside surface of Mercury. To protect the spacecraft from solar heating, all systems except the solar arrays were kept behind a ceramic-cloth sunshade that pointed toward the Sun. This approach simplified the design of the subsystems, which could be built with conventional electronics, but added a substantial constraint to the operation of the spacecraft. Throughout its time within the inner solar system and in orbit about Mercury, MESSENGER was constrained to maintain the orientation of the normal to the central sunshade panel in the sunward direction to within ±10° in Sun-relative elevation angle (pitch) and ±12° in Sun-relative azimuth (yaw) at all times.

1.2.4 Instrument Payload

The project requirements for MESSENGER's primary mission were met by a suite of seven scientific instruments plus the spacecraft communication system (Gold et al., 2001). There was a dual imaging system for wide and narrow fields of view, monochrome and color imaging, and stereo; gamma-ray, neutron, and X-ray spectrometers for surface chemical mapping; a magnetometer; a laser altimeter; a combined ultraviolet–visible and visible–near-infrared spectrometer to survey both exospheric species and surface mineralogy; and a combined energetic particle and plasma spectrometer to

Figure 1.3. View of the MESSENGER spacecraft during vibration testing at the Johns Hopkins University Applied Physics Laboratory. The solar arrays (mirrored surfaces) are stowed in their positions at the time of launch. Also visible are the Magnetometer boom (center), similarly in its stowed position, and thermal blankets (gold).

sample charged species in the magnetosphere (Figure 1.4). Brief descriptions of the payload instruments are as follows.

1.2.4.1 Mercury Dual Imaging System

The Mercury Dual Imaging System (MDIS) on the MESSENGER spacecraft (Hawkins et al., 2007), shown in Figure 1.4, consisted of a monochrome narrow-angle camera (NAC) and a multispectral wide-angle camera (WAC). The NAC was an off-axis reflector with a 1.5° field of view (FOV) and was co-aligned with the WAC, a four-element refractor with a 10.5° FOV and a 12-color filter wheel. The focal-plane electronics of each camera were identical and used a 1024 × 1024 charge-coupled-device detector. Only one camera operated at a time, a design that allowed them to share a common set of control electronics. The NAC and the WAC were mounted on a pivoting platform that provided a 90° field of regard, from 40° sunward to 50° anti-sunward from the spacecraft z-axis (Figure 1.2) – the boresight direction of most of MESSENGER's instruments. Onboard data compression provided capabilities for pixel binning, remapping of 12-bit data to 8 bits, and lossless or lossy compression. During MESSENGER's primary mission, four main MDIS data sets were planned: a monochrome global image mosaic at near-zero emission angles and moderate incidence angles, a stereo complement map at off-nadir geometry and near-identical lighting, multicolor images at low incidence angles, and targeted high-resolution images of key surface features. It was further planned that those data would be used to construct a global image base map, a digital terrain model, global maps of color properties, and mosaics of high-resolution image strips.

1.2.4.2 Gamma-Ray and Neutron Spectrometer

The Gamma-Ray and Neutron Spectrometer (GRNS) instrument (Figure 1.4) included separate Gamma-Ray Spectrometer (GRS) and Neutron Spectrometer (NS) sensors (Goldsten et al., 2007). The GRS detector was a mechanically cooled crystal of germanium, and the sensor detected gamma-ray emissions in the energy range 0.1–10 MeV and achieved an energy resolution of 3.5 keV full width at half maximum for ^{60}Co (1332 keV). Special construction techniques provided the necessary thermal isolation to

1.2 Objectives, Spacecraft, Payload, and Orbit Design

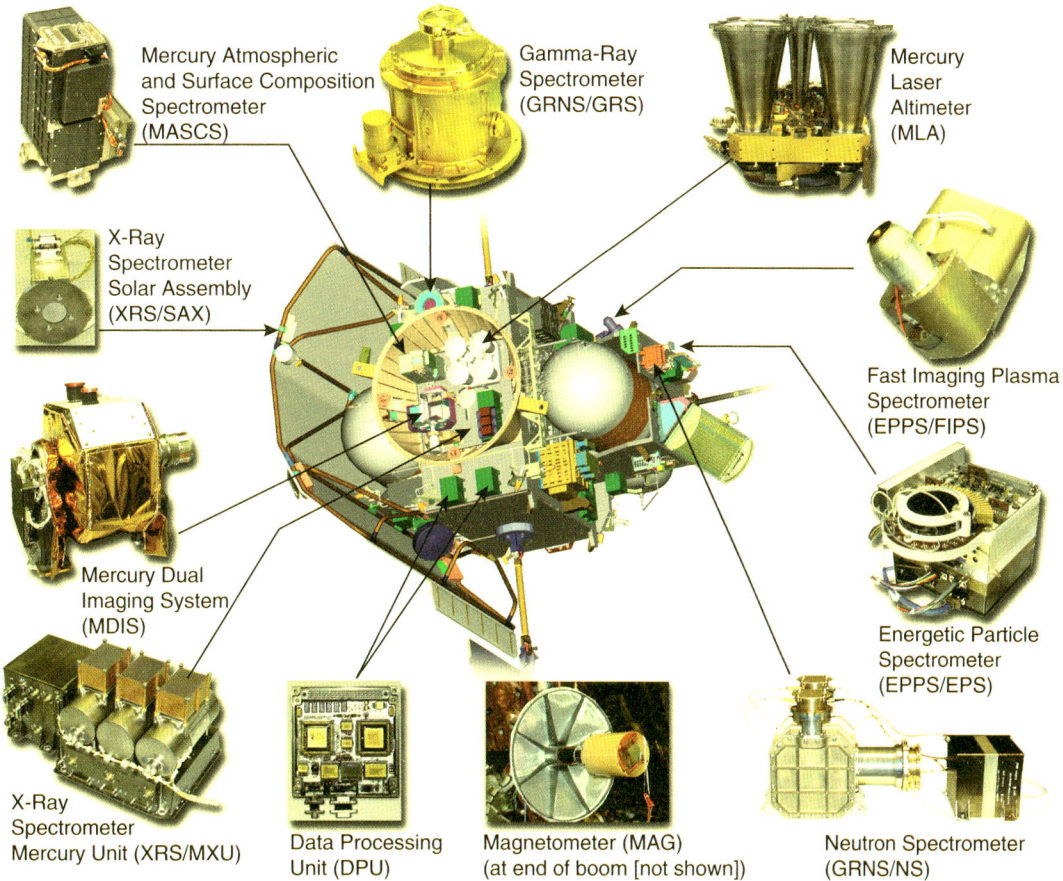

Figure 1.4. MESSENGER instruments and their locations on the spacecraft.

maintain the encapsulated detector at cryogenic temperatures (90 K) despite the high temperatures in Mercury's environment. The outer housing of the GRS sensor was equipped with an anticoincidence shield (ACS) to reduce the background from charged particles. The NS sensor consisted of a sandwich of three scintillation detectors working in concert to measure the flux of neutrons in three energy ranges from thermal to ~7 MeV.

1.2.4.3 X-Ray Spectrometer

The X-Ray Spectrometer (XRS) (Figure 1.4) measured the characteristic X-ray emissions induced on the surface of Mercury by the incident solar X-ray flux (Schlemm et al., 2007). The instrument detected the Kα lines for the elements Mg, Al, Si, S, Ca, Ti, Cr, Mn, and Fe. The planet-viewing sensor (Mercury X-ray Unit, MXU) consisted of three gas-filled proportional counters, one with a thin Mg foil over the entrance window, one with a thin Al foil over the entrance window, and one with no foil to separate the lower-energy lines from Mg, Al, and Si. The 12° field of view of the planet-viewing sensor allowed a spatial resolution that ranged from 42 km at 200-km altitude to 3200 km at 15,000-km altitude. A small Si-PIN detector (Solar Assembly for X-rays, SAX) mounted on the spacecraft sunshade (Leary et al., 2007) and directed sunward provided simultaneous measurement of the solar X-ray flux. The solar detector included a thermoelectric cooler that could also operate in a heater mode to anneal the sensor after radiation damage.

1.2.4.4 Magnetometer

MESSENGER's Magnetometer (MAG) was a low-noise, triaxial fluxgate instrument (Anderson et al., 2007). Its sensor was mounted on a 3.6-m-long boom that was directed generally anti-sunward (Figures 1.2 and 1.3). The instrument had both a coarse range, ±51,300 nT full scale (1.6-nT resolution), for preflight testing, and a fine range, ±1530 nT full scale (0.047-nT resolution), for operation near Mercury. A magnetic cleanliness program followed during the design and construction of the spacecraft minimized variable and static spacecraft-generated fields at the sensor. Analog signals from the three instrument axes were low-pass filtered (10-Hz cutoff) and sampled simultaneously by three 20-bit analog-to-digital converters every 50 ms. To accommodate variable telemetry rates, MAG provided 11 output rates from 0.01 s^{-1} to 20 s^{-1}. The instrument also provided continuous measurement of fluctuations by means of a digital 1–10-Hz bandpass filter. This fluctuation level was used to trigger high-time-resolution sampling in 8-min segments to record events of interest when continuous high-rate sampling was not possible.

1.2.4.5 Mercury Laser Altimeter

The Mercury Laser Altimeter (MLA) (Cavanaugh et al., 2007) (Figure 1.4) measured the round-trip time of flight of transmitted laser pulses reflected from the surface of Mercury which, in combination with the spacecraft orbit position and pointing data, gave a high-precision measurement of surface topography referenced to Mercury's center of mass. The laser transmitter was a diode-pumped Nd:YAG slab laser with passive Q-switching. The transmitter emitted 5-ns-wide pulses at an 8-Hz rate with 20 mJ of energy at a near-infrared wavelength of 1064 nm. The receiver consisted of four refractive telescopes and four equal-length optical fibers to couple the received optical signal onto a single silicon avalanche photodiode. The timing of laser pulses was measured with a set of time-to-digital converters and counters and a crystal oscillator operating at a frequency that was monitored regularly from Earth. MLA sampled the planet's surface to within a 1-m range error when the line-of-sight range to Mercury was less than 1500 km under spacecraft nadir pointing or the slant range was less than ~1000 km at off-nadir angles up to ~40°.

1.2.4.6 Mercury Atmospheric and Surface Composition Spectrometer

MESSENGER's Mercury Atmospheric and Surface Composition Spectrometer (MASCS) (McClintock and Lankton, 2007) consisted of a small Cassegrain telescope with 257-mm effective focal length and a 50-mm aperture that simultaneously fed an Ultraviolet and Visible Spectrometer (UVVS) and a Visible and Infrared Spectrograph (VIRS) (Figure 1.4). UVVS was a 125-mm-focal-length, scanning grating, Ebert–Fastie monochromator equipped with three photomultiplier tube detectors that covered far-ultraviolet (115–180 nm), middle-ultraviolet (160–320 nm), and visible (250–600 nm) wavelength ranges with an average spectral resolution of 0.6 nm. It was designed to measure profiles with altitude of known exospheric species, to search for previously undetected exospheric species, and to observe Mercury's surface in the far and middle ultraviolet at a spatial scale of 10 km or smaller. VIRS was a fixed concave grating spectrograph with a 210-mm focal length equipped with a beam splitter that simultaneously dispersed the spectrum onto a 512-element silicon visible-wavelength photodiode array (300–1050 nm) and a 256-element indium-gallium-arsenide infrared-wavelength photodiode array (850–1450 nm). The VIRS was designed to map surface reflectance with 5-nm spectral resolution in the wavelength range 300–1450 nm.

1.2.4.7 Energetic Particle and Plasma Spectrometer

The Energetic Particle and Plasma Spectrometer (EPPS) instrument on MESSENGER consisted of two sensors (Andrews et al., 2007), an Energetic Particle Spectrometer (EPS) and a Fast Imaging Plasma Spectrometer (FIPS) (Figure 1.4). The EPS was a hockey-puck-sized energy by time-of-flight spectrometer designed to measure in situ the energy, angular, and compositional distributions of the high-energy components of electrons (>20 keV) and ions (>5 keV/nucleon) near Mercury. The FIPS measured the energy, angular, and compositional distributions of the low-energy components of the ion distributions (<50 eV/charge to 20 keV/charge). The FIPS sensor featured an electrostatic analyzer system with a large (1.4 sr) instantaneous field of view.

1.2.4.8 Radio Science

The MESSENGER telecommunications subsystem was designed primarily to send commands to the spacecraft and to transmit to Earth both science measurements and information on the state of the spacecraft and instruments (Srinivasan et al., 2007). The subsystem doubled as a scientific tool by providing precise measurements of the spacecraft's velocity and range along the line of sight to Earth, information essential for spacecraft navigation and also for deriving Mercury's gravity field.

1.2.5 Orbit Design

The parameters selected for the MESSENGER orbit after the orbit insertion maneuver resulted from a complex trade-off of scientific objectives, spacecraft and instrument thermal design, communications and power constraints, and propellant budget (Santo et al., 2001). The original design for the initial orbit featured a periapsis altitude of 200 km, a periapsis latitude of 60°N, an inclination of 80° to the planet's equatorial plane, and a period of 12 h. The periapsis latitude and altitude, the high eccentricity of the orbit, and the phasing of the initial orbit relative to local time and Mercury true anomaly were all selected as part of the mission thermal design. The 12-h period was chosen to regularize the schedule of mission operations and permitted ample time for data downlink near apoapsis.

Orbit-correction maneuvers (OCMs) were planned for the orbital phase of the primary mission, because the gravitational pull of the Sun would raise periapsis altitude and latitude between successive orbits (McAdams et al., 2007). Such maneuvers were planned in pairs, with the first designed to lower periapsis altitude back to ~200 km and the second to adjust the orbit period after the first correction back to 12 h. The pairs of maneuvers were scheduled approximately one Mercury year apart in order to keep periapsis altitude below 500 km while meeting spacecraft sunshade pointing and science requirements.

1.3 MESSENGER'S SCIENCE DATA ACQUISITION PLANNING AND OPERATIONS

Planning for MESSENGER's scientific observations from orbit about Mercury required a novel approach to the design of payload operations and spacecraft attitude-control commanding. Experience with science planning for the Mercury flybys demonstrated the complex interplay between imaging and competing remote sensing observations, as well as with spacecraft operational constraints on pointing, power management, navigation, and achievable rates of change to spacecraft attitude. Planning for the flybys was conducted with conventional manual approaches to the design of observation and spacecraft command sequences with computational and visualization tools. Months of iterative, labor-intensive work were required to design, simulate, and review each encounter. The complexities of operations from orbit about Mercury, however, called for a marked change in the planning architecture, given that the orbital phase of the primary mission phase would be equivalent to two flybys every Earth day. To effect such a change, the observational requirements, spacecraft capabilities, and operational constraints had to be captured in carefully implemented software, which was then used together with orbit solutions in an automated search for optimal observation opportunities and spacecraft attitude, imaging pivot commanding, and instrument commanding. Those objectives were accomplished with a sophisticated, integrated suite of modules known as SciBox (Anderson et al., 2011b; Choo et al., 2014).

1.3.1 Science Observation Constraints

The observational opportunities at Mercury were highly constrained by MESSENGER's eccentric orbit and Mercury's low spin rate. A solar day on Mercury is approximately 176 Earth days, so during MESSENGER's year-long primary orbital mission there were only two opportunities to observe each longitude at a given solar illumination. In addition, the different science investigations had distinct and competing pointing requirements. For monochrome surface imaging, for instance, a specific range of solar incidence angles is optimal to reveal surface features while not obscuring terrain in shadow, whereas for color imaging near-normal solar incidence angles are best. Imaging plans also had to incorporate a favorable phase angle to minimize forward scattering of sunlight yet maintain surface resolution. Moreover, the choice of MDIS camera, wide-angle or narrow-angle, depended on altitude and viewing geometry as well as the need to balance resolution with the requirement to obtain as complete and overlapping imaging coverage as possible.

MESSENGER's other science investigations imposed still different requirements. Most of the instruments on the spacecraft's main instrument deck (Figure 1.4) had co-aligned fields of view and yielded optimal data for near-nadir viewing directions. In contrast, exospheric observations by the UVVS required turning the spacecraft so that the spacecraft z-axis, i.e., the normal to the instrument deck, pointed off the limb of the planet. The MLA observations yielded the highest signal-to-noise ratio for surfaces in darkness, whereas the XRS observations required that the surface be in daylight so as to be exposed to solar X-rays. The GRS observations were largely insensitive to surface illumination, but the NS observations were optimal for specific orientations of the NS detectors with respect to the spacecraft orbital velocity relative to Mercury. Finally, the EPS instrument yielded the most scientifically fruitful data when the magnetic field direction lay in the plane of the instrument field of view.

1.3.2 Spacecraft and Mission Operations Constraints

Over and above the scientific objectives and instrument observational constraints, ensuring the continued health of the spacecraft and payload demanded attention to spacecraft operations, communications, and navigation considerations. The constraints on spacecraft attitude were strictly enforced to maintain the orientation of the normal to the central sunshade panel in the direction of the Sun to within specified tolerances in the Sun-relative elevation angle and azimuth. Violation of these constraints would trigger autonomous spacecraft protection procedures and abort the science observation sequence, so the planning software imposed these constraints as hard limits on the commanded attitude. Communication passes for command uplink and data downlink and spacecraft angular momentum management were carefully planned and reserved for mission operations. Software tools were developed to allocate spacecraft resources, particularly the solid-state recorder (SSR), to track and predict the onboard data volume against the observation plan to ensure the return to Earth of all collected data. Orbit-adjustment maneuvers and other mission-critical activities were also strictly reserved for mission operations planning and treated as unavailable for science observations that required attitude commanding. Passive science data collection continued through communications operations.

1.3.3 Automated Science Opportunity Analyzer and Scheduler

The MESSENGER SciBox suite of software modules was designed to factor in the payload constraints and priorities for observation geometry and range within the constraints of orbit design and mission operations. The SciBox functional structure is shown schematically in Figure 1.5. Because a pivot about the spacecraft–Sun line was incorporated into the MDIS design, the imaging observations could be planned with this additional degree of freedom not available to the other instruments. This capability motivated an altitude-based hierarchy of science pointing priority, by which different instruments were assigned attitude control for specific ranges of altitude. Attitude control was assigned to MLA for all altitudes less than the ranging limit (~1500 km) to the surface for a nadir point on the planet's nightside. Otherwise, pointing control was assigned to XRS on the dayside if the allowed range of directions of the spacecraft z-axis intercepted the planet. The remaining time in which the planet was within the MDIS field of regard was assigned to MDIS control, and the remainder of the observing time was assigned to exosphere observations by MASCS. The science team also identified prioritized sets of specific targets on the planet for focused observation (e.g., high resolution, additional colors, greater pointing dwell time). These targets were assembled into a target database and were selected if unsubscribed opportunities were present. Each instrument was also assigned an allocation for data volume, and a nominal plan for altitude-dependent data collection was designed for each of the instruments other than MDIS.

This database and the mission design were ingested to derive a draft observation plan and predictions for SSR loading for the entire orbital mission. The spacecraft orbit (in the form of kernels in the NASA Spacecraft, Planet, Instrument, C-matrix, Events – or SPICE – toolkit), times reserved for mission operations, and constraints on spacecraft attitude slew rate and MDIS pivot rotation rate (captured as rules) were used by the Opportunity Analyzer in SciBox to identify all possible imaging opportunities. The MDIS imaging plan was then constructed by the Opportunity Analyzer, which identified all achievable imaging opportunities given the spacecraft attitude as constrained by MLA- and XRS-assigned attitude and the MDIS pivot. These opportunities were next evaluated against the desired properties for imaging and requirements for imaging overlap by the Optimizer module, resulting in the generation of an imaging plan and a spacecraft science attitude

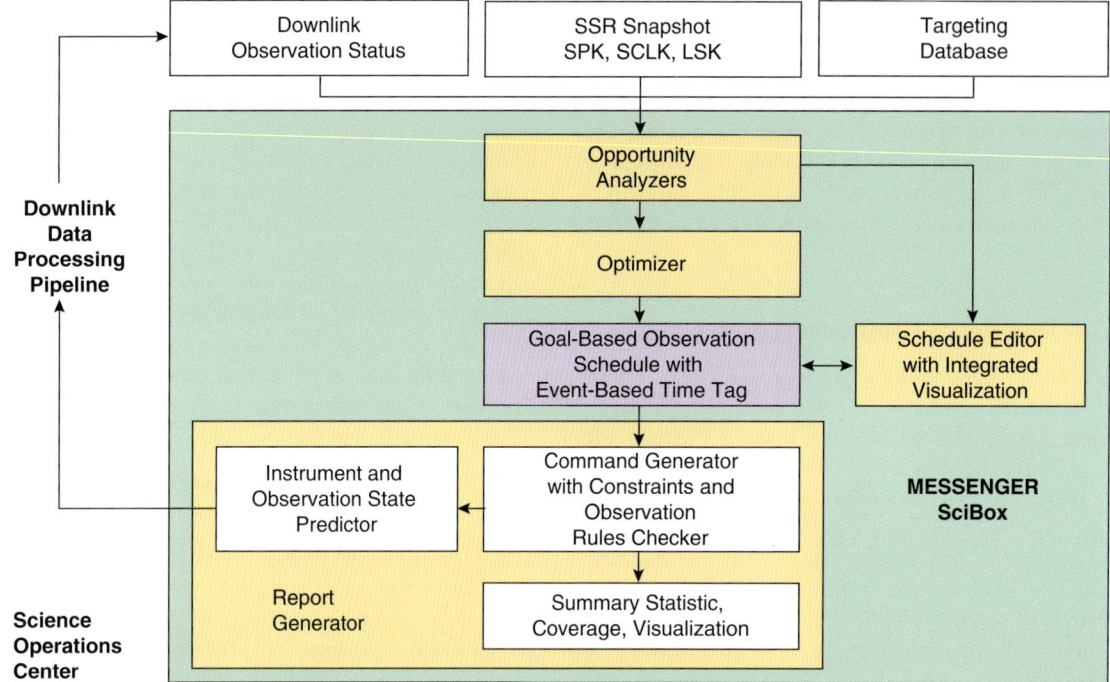

Figure 1.5. Functional schematic of the MESSENGER SciBox software suite (green box). Input data are indicated by white boxes at the top (downlink status, SSR data volume, attitude and orbit data in SPK, SCLK, and LSK SPICE kernels, and the targeting database); the main module elements are indicated by tan boxes; the key intermediate schedule product is indicated by the purple box; and key elements of the report generator are shown. The SciBox software was maintained and configured within the MESSENGER Science Operations Center, and the state predictor was used to update the SSR load predictions. SPK, SCLK, and LSK denote the Spacecraft and Planets Kernel, the Spacecraft Clock Kernel, and the Leapseconds Kernel, respectively.

plan. These plans were then checked against the spacecraft control constraints by the Rules Checker module, and draft spacecraft attitude and instrument commanding sequences were generated. The loading on the SSR was also evaluated and updated, including the expected downlink capacity for each communication pass. One key to ensuring accuracy of the planning against actual performance was that the times of commanding were keyed to orbit events rather than to absolute time. This procedure allowed the plan to transition smoothly from the orbit predictions far in the future to the immediate planned orbit using the latest orbit predictions to generate actual commands for the spacecraft.

The integrated plan was developed with an iterative approach by which successively more observations were included in the Opportunity Analyzer. Once it was demonstrated that the imaging goals could be met within the MLA and XRS constraints, the other pointed observations, including UVVS exosphere and surface targets, were included in the planning. In addition, the predictions for SSR loading were used to tailor the allowed instrument observation rates, and these revised data rates were included in refinements to the mission-long observation plan. Development and refinement of the modules continued throughout the orbital phase, including both extended missions, to track the additional science observation objectives.

1.3.4 Advance and Near-Term Science Planning

The integrated mission plan was used for both science planning and command generation (Berman et al., 2010). The plan was re-derived for each week of operations and updated with information acquired for imaging coverage, SSR loading, and any adjustments needed for instrument performance, operation, or spacecraft operations rules. This analysis constituted the Advance Science Planning process, which was the starting point for building the final command loads for the spacecraft. The command building process, known as Near-Term Science Planning, ran on a four-week cadence. A week's commands were generated four weeks ahead of execution on the spacecraft, and with each successive week the sequence was processed through different stages of review, quality assessment, and error checking. For actual spacecraft commanding, the SciBox suite included converters from the schedules to instrument operation and spacecraft attitude command requests in formats required by the mission operations scheduling and command-load development tools. These command requests were reviewed by the science and engineering teams for each instrument and then processed through the spacecraft command generator to verify compliance with all instrument and spacecraft rules. Each load was then run through the ground spacecraft simulator before being approved for upload and execution on the spacecraft.

1.3.5 Science Observation Performance

The performance of the observation planning during orbital operations resulted in imaging, mapping, and in situ surveys of the planet that met every project requirement for the

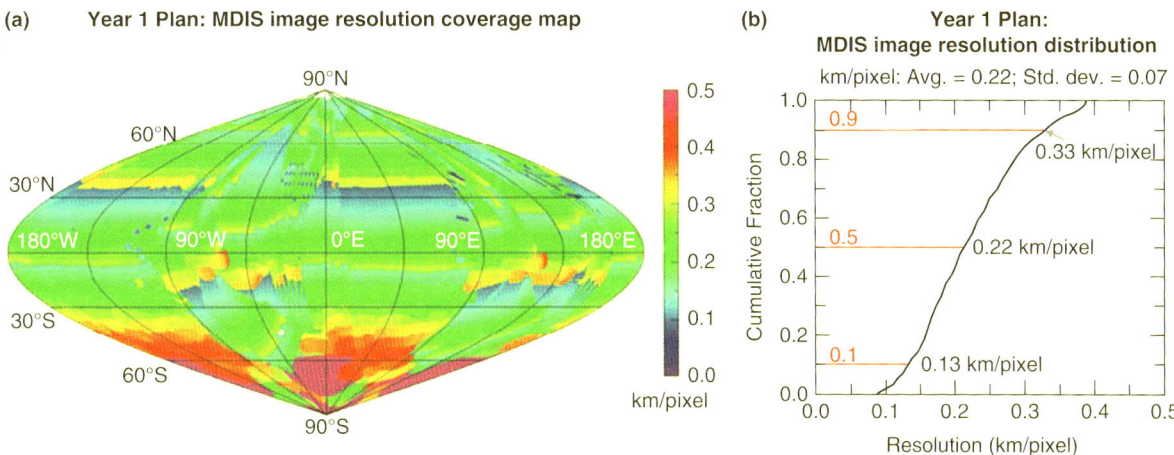

Figure 1.6. Example planning products for the first year of MESSENGER orbital observations, including (a) a map of the MDIS monochrome imaging resolution and (b) the cumulative distribution of image resolution. Products such as these were generated for each instrument to assess the characteristics of the planned observation plan against the observational requirements for each science investigation.

Figure 1.7. Four sequential views of the launch of the MESSENGER spacecraft on 3 August 2004.

mission. This record of mission success hinged critically on the capability of SciBox to design an entire yearlong imaging plan that integrated all of the science observations in a single overarching schedule. As an example, Figure 1.6 shows the planned resolution for MDIS monochrome mapping prior to orbit insertion in the form of a surface map, along with cumulative statistics. Transitions in the resolution over the surface correspond to transitions between the NAC and WAC and merging of imaging obtained on the ascending and descending legs of each orbit. Similar planning maps for image coverage, ensuring overlap to facilitate the construction of mosaics, as well as color imaging, XRS, MLA, VIRS, and GRS coverage, were all generated, and the mission performance met or exceeded all of these plans. Given the complexity of the observation plan and the need to build the entire observation plan at once, it is clear that the automated scheduling tool was essential to the acquisition of the data that allowed all of the advances achieved in our understanding of Mercury's characteristics.

1.4 MESSENGER'S PRIMARY MISSION

1.4.1 Overview of the Primary Mission

MESSENGER launched from Cape Canaveral Air Force Station, Florida, on 3 August 2004 (Figure 1.7). The cruise phase of the mission lasted 6.6 years and included six planetary flybys (McAdams et al., 2005, 2011). A gravity-assist flyby of Earth on 2 August 2005, approximately one year after launch, reduced the spacecraft's perihelion to 0.6 AU and moved the perihelion direction more than 60° closer to that of Mercury. The first of two flybys of Venus on 24 October 2006 increased the inclination of the spacecraft's orbit and reduced the orbit period. The second Venus flyby on 5 June 2007 lowered perihelion sufficiently to permit a Mercury flyby. Both Venus flybys moved the spacecraft's perihelion and aphelion closer to those of Mercury. A single loss of instrument functionality occurred during the primary mission: in April 2005 the high-voltage system on the time-of-flight portion of the EPS sensor failed,

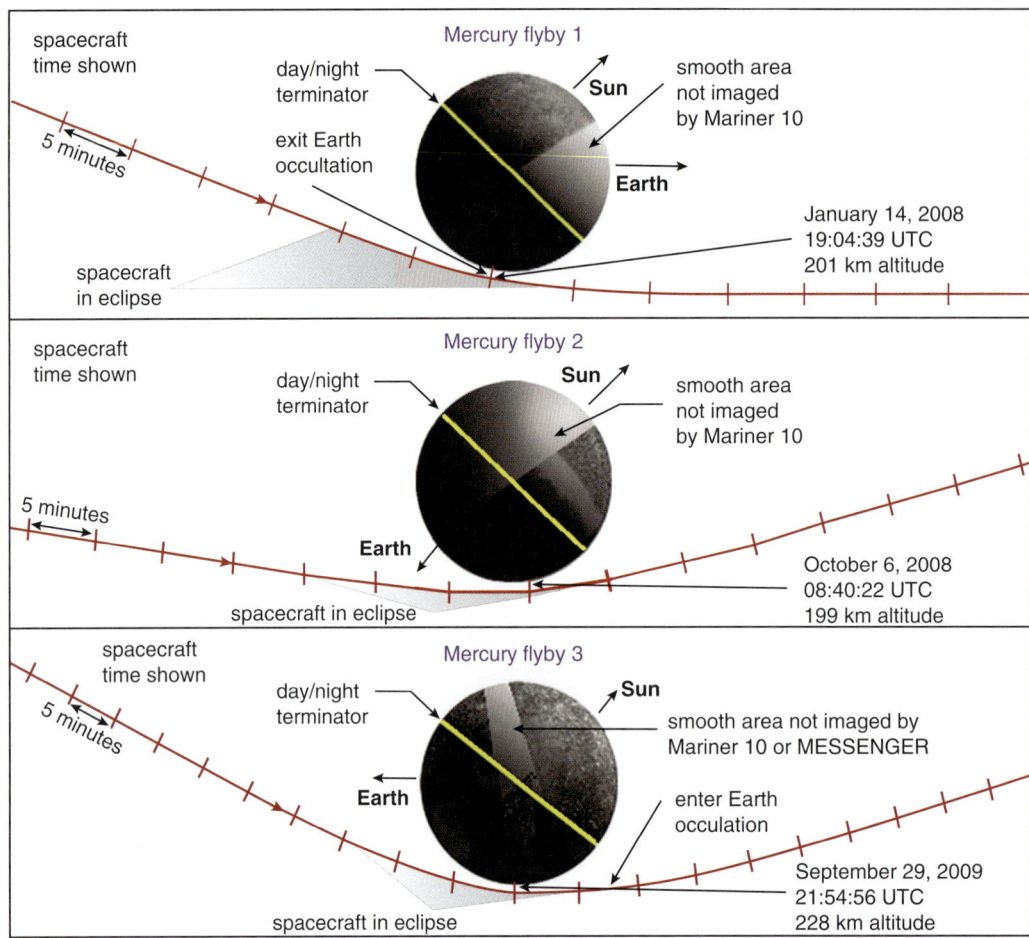

Figure 1.8. MESSENGER's Mercury flyby trajectories as viewed from above Mercury's north pole. Areas not imaged by Mariner 10 are shown in light gray. From McAdams et al. (2011).

and high voltages could not be restored despite repeated attempts over the next several months. The energy subsystem of the sensor (Andrews et al., 2007) was unaffected by this failure and operated until the end of the mission.

MESSENGER executed three flybys of Mercury on 14 January and 6 October 2008 and 29 September 2009. The three flybys, each followed about two months later by a large propulsive course-correction maneuver, completed the rotation of MESSENGER's orbit and changed the period of the orbit progressively closer to that of Mercury (McAdams et al., 2005). Each flyby followed a near-equatorial trajectory and involved closest approach on Mercury's nightside at ~200-km altitude (Figure 1.8). Operationally, the first two flybys proceeded flawlessly, but a safe-hold event triggered by a rising battery temperature shut off data acquisition midway through the third flyby. During the flybys, MESSENGER mapped nearly the entire planet in color, imaged most of the areas unseen by Mariner 10, completed initial measurements of the composition of Mercury's exosphere and neutral tail, and made initial characterizations of the structure and dynamics of Mercury's magnetosphere. Those three flybys returned the first new spacecraft data from Mercury in more than three decades (Figure 1.1). These data were invaluable to the planning of the yearlong orbital phase of MESSENGER's primary mission.

On 18 March 2011, the MESSENGER spacecraft was inserted into a highly eccentric, 12-h orbit about the planet Mercury. The orbit attained had an inclination of 82.5°, an initial periapsis altitude of ~200 km, an initial periapsis longitude of 60°N, and apoapsis at an altitude of ~15,200 km in the southern hemisphere (Figure 1.9).

During the primary mission, the periapsis altitude increased progressively and periapsis latitude drifted northward, both the effect of perturbations to the spacecraft trajectory by the gravitational pull of the Sun. A series of propulsive OCMs was designed to maintain the periapsis altitude within the approximate range 200–500 km (McAdams et al., 2012). OCM-1 and OCM-3 each lowered the periapsis to 200 km, but each maneuver also reduced the orbit period by ~15 min and so was followed by a smaller OCM that returned the period to ~12 h (Figure 1.10). Neither OCM-5 nor OCM-6 was followed by a maneuver to correct the orbit period.

1.4.2 Results from the Primary Mission

By the conclusion of MESSENGER's primary mission, all of the scientific objectives (Table 1.1) had been met, and all of the project requirements (Table 1.2) had been successfully accomplished. Scientific results from MESSENGER's primary mission substantially answered the six questions that had framed the mission.

1.4.2.1 Mercury's High Ratio of Metal to Silicate

Prior to the MESSENGER mission, most hypotheses put forward to explain Mercury's anomalously high core fraction invoked high-temperature processes that would have substantially depleted the planet's inventory of volatile elements. However, elemental measurements of Mercury's surface by MESSENGER's XRS and GRNS instruments during the orbital phase of the primary mission (Nittler et al., 2011; Peplowski et al., 2011; Evans et al., 2012) indicated that such moderately volatile elements as Na, K, and S are not depleted relative to other terrestrial planets. To the contrary, the surface S abundance is at least a factor of 10 higher than that of the surface of Earth or the Moon (Nittler et al., 2011). MESSENGER XRS and GRS measurements during the primary mission also showed that Mercury's surface material is low in iron (no more than ~4 wt% Fe) (Nittler et al., 2011; Evans et al., 2012). Collectively, these results are inconsistent with formation models calling for extended periods of high temperatures (e.g., evaporation in a hot solar nebula, formation from high-temperature condensates, or some giant impact scenarios) and suggest that Mercury's metal-rich, FeO-poor composition likely reflects chemically reduced precursor materials (Nittler et al., 2011), enriched in Fe metal by some aspect of the accretion process.

1.4.2.2 Mercury's Geological History

Global monochrome, color, and stereo images acquired during the Mercury flybys and from orbit revealed the presence of a range of landforms known to be associated with volcanism on other planets, and several lines of evidence suggested that the emplacement of Mercury's surface material has been dominated by volcanism. Indicators of volcanic resurfacing included extensive smooth plains that embay topographic lows, commonly with distinct reflectance and color properties; depressions that may be source vents (Head et al., 2008, 2011); a deficiency of large basins compared with the Moon, probably the result of volcanic burial (Fassett et al., 2012); and broad channels between plains deposits formed by sculpting of surrounding terrain, consistent with large-scale floods of highly fluid material (Byrne et al., 2013; Hurwitz et al., 2013). The similarity of observed features to fluvial landforms is consistent with formation by high-temperature lavas, as also suggested by the Mg-rich nature of Mercury's surface materials (Nittler et al., 2011). Laser altimetry by the MESSENGER spacecraft yielded a topographic model of the northern hemisphere of Mercury (Zuber et al., 2012), showing that the dynamic range of elevations is considerably smaller than those of Mars or the Moon. The most prominent topographic feature was found to be a broad lowland at high northern latitudes which hosts a large volcanic plains deposit covering ~6% of the planet (Head et al., 2011). Explosive volcanism was indicated by ~50 rimless depressions surrounded by diffuse, bright haloes that exhibit redder color

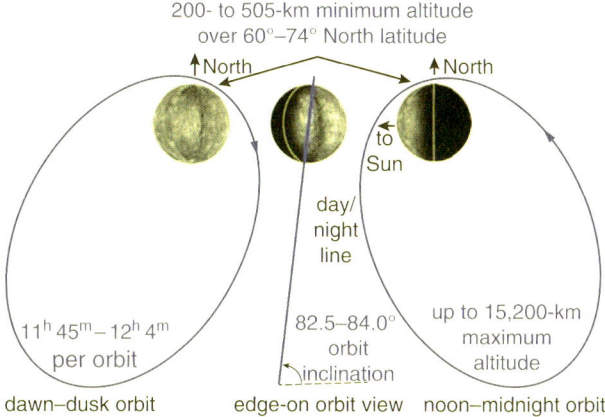

Figure 1.9. Three views of MESSENGER's 12-h orbit during the primary mission: a dawn–dusk orbit viewed from the Sun, an edge-on view, and a noon–midnight orbit viewed from a direction orthogonal to the planet–Sun line. From McAdams et al. (2012).

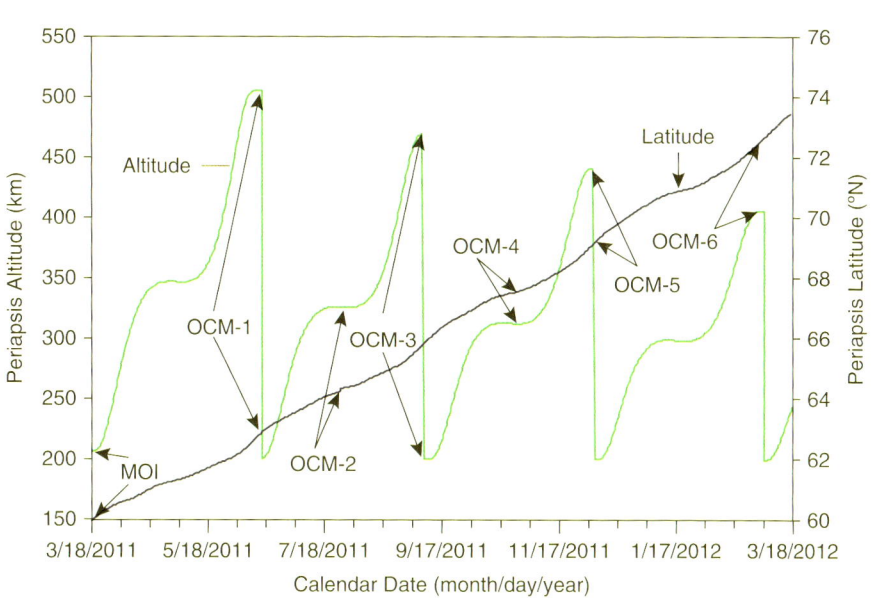

Figure 1.10. Progression of the altitude (green line) and latitude (blue line) of periapsis during MESSENGER's primary mission. The times of Mercury orbit insertion (MOI) and the six orbit-correction maneuvers (OCMs) during the primary mission are shown. From McAdams et al. (2012).

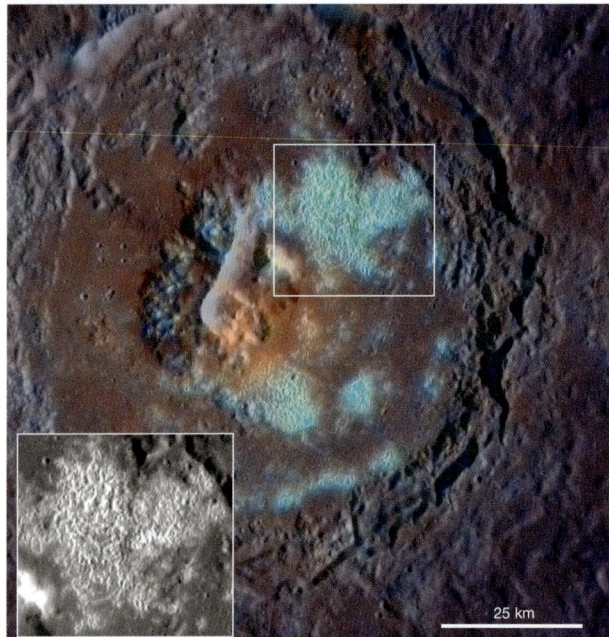

Figure 1.11. Hollows in Tyagaraja crater, 97 km in diameter. Bright areas shown in blue and with etched texture correspond to a high density of hollows (inset). The pit surrounded by reddish material in the center of the crater has been interpreted as a pyroclastic vent. From monochrome image EN0212327089 M, 111 m/pixel, with enhanced color from the eight-filter set EW0217266882I. From Blewett et al. (2011).

than plains materials, and are interpreted as pyroclastic deposits by analogy with such material on the Moon (Kerber et al., 2009; Goudge et al., 2014). Crater size–frequency distributions and stratigraphic analyses suggested that Mercury's smooth plains formation continued into the second half of Mercury's history on local scales, e.g., in the Rachmaninoff basin (Prockter et al., 2010).

One of the most surprising discoveries of the primary mission was the presence of "hollows," fresh-appearing, rimless depressions, commonly with high surface reflectance and often with bright haloes (Figure 1.11). Hollows appear concentrated in the low-reflectance material (LRM) color unit (Robinson et al., 2008; Denevi et al., 2009) within impact craters and basins (Blewett et al., 2011, 2013), and the fact that they are found within some of the freshest craters on Mercury suggests that their formation has been recent or even is ongoing. The host rocks are inferred to have been excavated from depth by impact (Robinson et al., 2008; Denevi et al., 2009), and likely formation mechanisms involve recent loss of volatiles through sublimation, space weathering, pyroclastic volcanism, or outgassing (Blewett et al., 2011, 2013).

1.4.2.3 Mercury's Magnetic Field

Orbital measurements revealed the structure of Mercury's internal magnetic field for the first time. On the basis of crossings of the magnetic equator, the internal field was shown to be consistent with that of a spin-aligned dipole with a moment of 195 ± 10 nT R_M^3; the tilt of the dipole from the spin axis is <0.8° (Anderson et al., 2011a, 2012; Johnson et al., 2012). One of the most surprising results of the MESSENGER mission is that the magnetic dipole is offset from the planet's equator by 479 ± 6 km, or ~0.2 R_M, so that the surface field is larger in magnitude by a factor of ~3 at the north pole than at the south pole, and the surface area of open magnetic flux in the southern hemisphere is larger by a factor of ~4 than in the northern hemisphere. The field geometry points to a core dynamo as the source of the field. Such an axially symmetric yet equatorially asymmetric dynamo is novel for the inner planets, and the low strength of multipolar terms higher than quadrupole is consistent with a deep source for the dynamo (Anderson et al., 2012). Magnetospheric measurements also revealed that Mercury's polar regions are important sources of Mercury's ionized exosphere (Zurbuchen et al., 2011). Further, bursts of energetic electrons were seen at a range of latitudes and times of day, implying that efficient acceleration mechanisms operate within Mercury's magnetosphere on a regular basis and produce electrons with energies up to hundreds of keV on timescales of seconds (Ho et al., 2011).

MESSENGER primary mission observations showed that Mercury's magnetosphere acts more effectively than anticipated to energize solar wind plasma and channel it to the surface. Magnetic reconnection between the planetary and solar wind magnetic fields at Mercury occurs with an intensity an order of magnitude greater than at Earth (Slavin et al., 2012a, b; DiBraccio et al., 2013), and shear instabilities at the magnetopause display similarly greater growth rates (Sundberg et al., 2012a). These interactions yield plasmas within 1000 km of the planetary surface with pressures often exceeding the magnetic pressure (Korth et al., 2011, 2012), leading to intense precipitation to the planetary surface (Winslow et al., 2012). The nearly ubiquitous occurrence of waves driven by ion-plasma instabilities indicates that non-thermal processes are central to plasma dynamics and precipitation behavior (Boardsen et al., 2012). Magnetic reconnection in the tail was also shown to be prevalent and intense, implying that plasmas are energized in the tail and convected convulsively planetward (Slavin et al., 2012a; Sundberg et al., 2012b).

1.4.2.4 Mercury's Core

MESSENGER primary mission data provided considerable insights into Mercury's interior structure. Radio tracking of the MESSENGER spacecraft provided a model of Mercury's gravity field (Smith et al., 2012); when combined with Earth-based measurements of Mercury's spin properties (Margot et al., 2012), the second-harmonic-degree coefficients in the gravity field yielded moments of inertia consistent with a core of ~2020-km radius (Hauck et al., 2013). The silicate mantle and crust together are no more than ~420 km thick, and in the northern hemisphere several large gravity anomalies, including candidate mass concentrations (mascons), exceed 100 mGal in amplitude (Smith et al., 2012; Hauck et al., 2013). From a model of a crust uniformly less dense than the mantle and laterally variable in thickness that fits the northern hemisphere topography and gravity field, Mercury's crust is thicker (50–80 km) at low northern latitudes and thinner (20–40 km) in the north polar region, and shows evidence for thinning beneath some impact basins such as Caloris (Zuber et al., 2012). A model for Mercury's radial density distribution consistent with these results includes a solid silicate crust and mantle overlying a liquid iron-rich outer core, with an overlying solid layer of iron sulfide and a solid inner core possible but not required (Smith et al., 2012; Hauck et al., 2013).

1.4.2.5 Mercury's Polar Deposits

Repeated imaging of Mercury's poles during MESSENGER's primary mission allowed the characterization of craters hosting radar-bright polar deposits, first identified two decades earlier from ground-based radar observations and postulated to consist of water ice (Slade et al., 1992; Harmon and Slade, 1992; Harmon et al., 2011). Mapping the areas of permanent and persistent shadow near each pole showed that nearly all such steadily shadowed regions at the highest latitudes host radar-bright material (Chabot et al., 2012, 2013). Small craters were shown to exhibit radar-bright material, as were craters that extend to latitudes equatorward of ±70°N. Thermal models that incorporate reflected sunlight and infrared radiation from the walls of idealized bowl-shaped craters (Vasavada et al., 1999) are inconsistent with the geologically long-term preservation of near-surface water ice in these small craters, supporting the inference that at least some of the polar deposits are geologically recent.

Neutron Spectrometer data collected during MESSENGER's primary mission indicated that Mercury's radar-bright polar deposits contain, on average, a hydrogen-rich layer more than tens of centimeters thick, generally covered by a surficial layer 10–30 cm thick that is less rich in hydrogen (Lawrence et al., 2013). Active measurements by the MLA of near-infrared (1064-nm wavelength) surface reflectance in permanently shadowed areas near Mercury's north pole revealed regions markedly darker and brighter than Mercury's average surface (Neumann et al., 2013). Both the MLA-dark and MLA-bright regions were shown to be collocated with areas of high radar backscatter in regions of persistent shadow. Correlation of observed reflectance with modeled surface and near-surface temperatures (Paige et al., 2013) indicated that the optically bright regions are consistent with the presence of surficial water ice, whereas MLA-dark regions have temperature structures consistent with water ice buried beneath an insulating surface layer of another volatile material, most likely complex organic deposits stable to somewhat higher temperatures than water ice. Impacts onto Mercury of comets or volatile-rich asteroids could have provided both the water ice and the dark, organic-rich material.

1.4.2.6 Mercury's Volatiles

As mentioned above, orbital data from MESSENGER's primary mission showed higher than expected surface abundances of volatile elements at the surface, including Na, K, and S. Correlation between Ca and S abundances (Nittler et al., 2011) suggested that at least some of the surface S is hosted by calcium sulfides. It was recognized by the end of the primary mission that volatile-rich materials, possibly including sulfides, play important roles in the formation of hollows (Blewett et al., 2011, 2013) and may have helped to drive the explosive volcanic eruptions that emplaced pyroclastic deposits (Kerber et al., 2009; Goudge et al., 2014).

Primary mission observations showed that Na, Ca, and Mg are the dominant species in Mercury's exosphere. Sodium is generally the most abundant and exhibits a two-component structure indicative of multiple source processes that supply Na with different energies (Killen et al., 2012). Calcium and Mg were seen to be less abundant overall and to show predominantly single-component altitude profiles reflective of a high-energy process. All three species show distinct variations in dayside near-surface densities: the Na abundance was seen to peak at local noon, whereas Ca showed a decreasing dawn-to-dusk gradient, and Mg was observed to be nearly isotropic (Burger et al., 2012; Merkel et al., 2012). The distinct distributions among these three species indicated that they are controlled by different source and transport mechanisms. Surveys conducted for other species yielded mostly upper limits. A weak O emission was detected above the subsolar point, and H, likely originating primarily from solar wind implantation, was routinely observed on the dayside and showed an altitude behavior similar to that observed during the Mariner 10 and MESSENGER flybys (Vervack et al., 2011).

1.5 MESSENGER'S FIRST EXTENDED MISSION

After MESSENGER successfully completed its primary mission on 18 March 2012, all spacecraft subsystems and payload instruments were healthy and sufficient propellant remained to continue orbital operations for at least an additional Earth year. Because a second Earth year of observations would provide a substantial advance in our understanding of Mercury beyond what was achieved as of the end of the primary mission, the MESSENGER team had earlier proposed and NASA had approved a first extended mission that lasted one Earth year, i.e., until 18 March 2013.

There were several overarching themes for MESSENGER's first extended mission which ensured that the second year of orbital operations would not be a simple continuation of the primary mission, including operation during a more active Sun, greater focus on observations at low spacecraft altitudes, and a greater variety of targeted observations. The extended mission permitted the first close-in observations of Mercury near a maximum in the solar cycle. A lower average altitude would be accomplished by reducing the period of the spacecraft orbit. The greater variety of targeted observations was enabled by the successful accomplishment of the global mapping objectives of the primary mission.

1.5.1 Objectives for the First Extended Mission

Six science questions framed the first extended mission. Each was motivated by discoveries made during the primary mission,

Table 1.3. *Scientific objectives for MESSENGER's first extended mission.*

1. Determine the morphological and compositional context of hollows and their relation to bright crater-floor deposits and pyroclastic vents.
2. Acquire targeted, high-resolution observations of volcanic materials of low impact crater density identified in the primary mission.
3. Document changes in long-wavelength topography versus geological time on Mercury from altimetric and complementary imaging measurements.
4. Characterize regions of enhanced exospheric density versus solar distance, proximity to geologic units, solar activity, and magnetospheric conditions.
5. Measure changes in exospheric neutrals and plasma ions as solar activity increases.
6. Infer sources and energization mechanism from the location, energy spectra, and temporal profiles of energetic electrons.

Table 1.4. *Project requirements for MESSENGER's first extended mission.*

1a. Image 70% of the planet in three colors at 600-m/pixel average spatial resolution.
1b. Acquire 100 sets of targeted images of hollows or pyroclastic vents at 60-m/pixel average spatial resolution.
1c. Acquire 20 targeted VIRS observations of hollows and pyroclastic vents at low solar incidence angle (i).
2. Acquire 30 sets of targeted images of young volcanic materials at 60-m/pixel average spatial resolution.
3a. Image 70% of the planet at 250-m/pixel average spatial resolution, targeting $i \sim 40°–65°$.
3b. Image 70% of the planet at 250-m/pixel average spatial resolution, targeting $i \sim 75°–85°$.
3c. Provide topographic profiles over 10 broadly elevated regions and the floors of 50 complex impact craters, including volcanically flooded craters.
4a. Survey dayside and nightside exosphere emissions at an average rate of once every third orbit.
4b. During dawn–dusk seasons, conduct repeated observations of exospheric emission over both poles to the maximum extent permitted by spacecraft pointing constraints.
4c. Conduct full-orbit, exosphere observation campaigns at equally spaced Mercury true anomalies over each of four Mercury years.
5. Measure the global distribution of planetary ions and the direction of plasma flow, within operational constraints.
6. Provide locations, energy spectra and pitch angles, and temporal profiles of energetic electrons across all magnetic longitudes in the northern hemisphere.

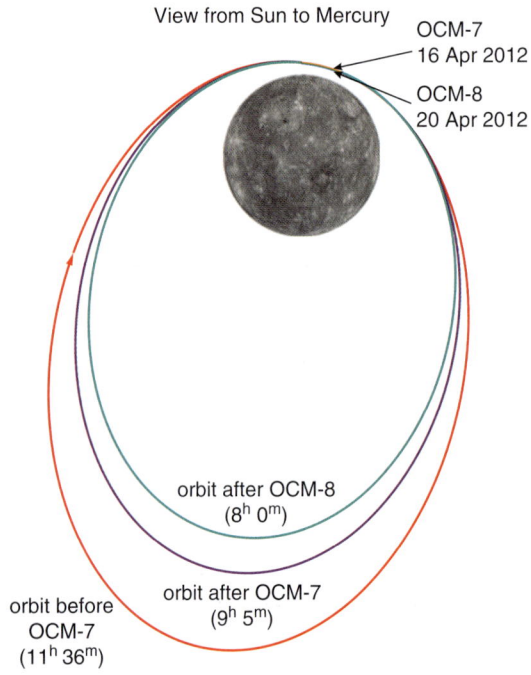

Figure 1.12. Orbit-correction maneuvers OCM-7 and OCM-8 in April 2012 changed MESSENGER's orbit period from just under 12 h to 8 h. North is up. From McAdams et al. (2012).

and collectively they addressed a broad range of coupled issues regarding Mercury's interior, surface, exosphere, and magnetosphere. Those six questions were as follows:

(1) What are the sources of surface volatiles on Mercury?
(2) How late into Mercury's history did volcanism persist?
(3) How did Mercury's long-wavelength topography change with time?
(4) What is the origin of localized regions of enhanced exospheric density at Mercury?
(5) How does the solar cycle affect Mercury's exosphere and volatile transport?
(6) What is the origin of Mercury's energetic electrons?

Those questions led to the set of scientific objectives for MESSENGER's first extended mission listed in Table 1.3. The project requirements corresponding to those objectives are listed in Table 1.4.

1.5.2 Results from the First Extended Mission

Less than five weeks after the start of the first extended mission, two OCMs four days apart in April 2012 reduced the period of MESSENGER's orbit from just under 12 h to 8 h (Figure 1.12). In its new orbit, the spacecraft spent more time per Earth day near Mercury's surface than during the primary mission. The periapsis altitude was ~280 km at the time of the two OCMs, which reduced the apoapsis altitude to ~10,300 km. Through most of the first extended mission, which saw no further OCMs, the periapsis altitude continued to increase progressively, and the periapsis latitude continued to drift northward. In early March 2013, when the periapsis altitude was ~450 km and the periapsis latitude was ~84°N, the changes to each quantity with successive orbits reversed sign, so that the periapsis moved progressively southward and downward thereafter.

All of the science objectives and project requirements for the first extended mission were achieved by the end of the second year of orbital operations, and substantial progress was made on the six science questions that provided the rationale for the mission extension.

1.5.2.1 Sources of Surface Volatiles

Targeted imaging of hollows during the first extended mission showed that well-developed hollows display a locally constant base level, suggesting either ablation of a layer having locally constant thickness or the development of a thermally insulating and mechanically resistant lag deposit that resists further volatile loss after reaching a given thickness (Blewett et al., 2013). The high abundance of S in Mercury's crust (Nittler et al., 2011; Weider et al., 2012), the instability of some sulfides at low pressure at Mercury's surface temperature (Helbert et al., 2013), and the concentration of hollows on sunward-facing slopes (Blewett et al., 2013) lent support to the hypothesis that hollows form by the ablation of sulfide-rich material within LRM deposits at Mercury's high daytime temperatures (Blewett et al., 2013). Analysis of orbital GRS observations yielded measurements of the surface abundance of Cl and indicated a chondritic Cl/K ratio and a higher abundance at high northern latitudes than nearer the equator, consistent with a role for Cl as a magmatic volatile in the eruption of the northern smooth plains (Evans et al., 2015).

1.5.2.2 History of Volcanism on Mercury

During MESSENGER's first extended mission, more detailed investigations were made of the composition and stratigraphy of Mercury's volcanic units, providing information about the volcanic contributions to the crust over time. XRS and GRS data showed that the northern smooth plains and Caloris interior plains have lower contents of Mg, Ca, and S and higher contents of Al and K (Weider et al., 2012; Peplowski et al., 2012) than do older intercrater plains and heavily cratered terrain, although K at low latitudes and near Mercury's hot poles may be continually removed by heating and redeposited at high latitudes (Peplowski et al., 2012). Only with the added observation time provided by the first extended mission did the coverage of high-energy XRS spectra reach the point at which regional variations in composition could begin to be mapped for other portions of the planet, in particular variations in Fe (Weider et al., 2014). Cratered and intercrater plains that predated the Caloris basin show some evidence for volcanic emplacement and may be older versions of the smooth plains (Denevi et al., 2009, 2013; Whitten et al., 2014). Low lava viscosities implied by flow features within the northern smooth plains are consistent with a composition intermediate between basaltic and ultramafic materials (Byrne et al., 2013; Hurwitz et al., 2013), as suggested by elemental abundance data.

Imaging during the primary and first extended MESSENGER mission revealed new details on the interplay between volcanism and tectonics on Mercury. Deformation within Mercury's large impact basins has been particularly complex and characterized by a diverse range of extensional and contractional features. The two largest basins, Rembrandt (Watters et al., 2009a) and Caloris (Byrne et al., 2012; Klimczak et al., 2013), each display complicated patterns of radial and concentric ridges and troughs, the age relations of which differ between the basins. Mechanisms that may have contributed to intrabasin deformation include loading of the basin interior by volcanic plains deposits and uplift of the basin floor by some combination of exterior loading and inward subsurface flow. Extensional deformation also occurred in buried, lava-filled basins and craters (Klimczak et al., 2012; Watters et al., 2012), probably as a result of cooling of the surficial lavas (Freed et al., 2012), and on at least one plateau, possibly a product of the relaxation of topographic relief.

1.5.2.3 Changes in Mercury's Long-Wavelength Topography

Global image mosaics of Mercury with lighting favorable to the characterization of morphology and stereo coverage plus MLA altimetry revealed a picture of Mercury's global tectonics far more complicated than the view immediately following the MESSENGER flybys, which emphasized lobate scarps and high-relief ridges as accommodators of global contraction (Watters et al., 2009b). Topography on Mercury is dominated not by impact basins, as on the Moon and Mars, but by broad rises (Preusker et al., 2011; Zuber et al., 2012). Some of these rises are superimposed on earlier volcanic flow features, occur within the otherwise low-relief northern smooth plains, and bow the floor of Caloris basin to elevations above the rim (Klimczak et al., 2013). On the northern smooth plains, outwardly tilting floors of volcanically infilled craters on the broad northern rise suggest that long-wavelength deformation postdated volcanism, and similar relations are seen on the long-wavelength topographic rises in Caloris and elsewhere (Balcerski et al., 2013). Some sets of lobate scarps bound broad rises that form monoclinal or anticlinal plateaus; such scarp systems are hypothesized to be outward-verging thrust faults in deformational assemblages that display similarities to terrestrial fold-and-thrust belts (Byrne et al., 2014).

1.5.2.4 Regions of Enhanced Exospheric Density

The evidence acquired during the primary mission for localized regions of enhanced exospheric density led to targeted observations during the first extended mission aimed at mapping their occurrence and understanding their origin. These observations revealed an exosphere in which the three most easily detected species – Na, Ca, and Mg – behave in ways not only different from one another but also at odds with hypotheses put forward to account for ground-based observations. The strong Ca enhancement in the dawn equatorial region, discovered during the MESSENGER flybys, showed persistence in both location and abundance (Burger et al., 2014) yet was seen to be composed of atoms too energetic to be derived from a strictly solar-release process, as the dawn location might suggest (Burger et al., 2012). An enhancement in Mg was seen near dawn local times, a phenomenon particularly notable near perihelion (Merkel et al., 2012, 2017).

In contrast to Ca and Mg, the Na exosphere exhibited less-marked localized enhancements, showing at most limited dawn–dusk asymmetry, contrary to many observations from the ground (Cassidy et al., 2015). Most surprising in the Na exosphere, however, was the general lack of short-term spatial and temporal variability in the MESSENGER observations, a result at odds with that seen in ground-based data (Killen et al., 2012). Whereas the Ca asymmetry would have been missed in ground-based observations owing to the limited geometry afforded from Earth, short-term variations in Na should not be as susceptible to differences in large-scale geometry. Instead, this difference between MESSENGER and ground-based observations suggests that the short-term variations originate almost completely in mid- to high-latitude dayside regions of the exosphere poorly probed by MESSENGER.

1.5.2.5 Effect of the Solar Cycle on Mercury's Neutral and Ionized Exosphere

Campaign observations spaced regularly through Mercury's orbit during the first extended mission, combined with daily measurements of the dayside and nightside exosphere during both the primary and first extended missions, revealed the overall behavior of the Na, Ca, and Mg exospheres over several Mercury years. All three species exhibited a relative persistence from year to year in the overall exospheric morphology, with seasonal variations in emission intensity in general agreement with that expected from variations in solar flux with Mercury true anomaly (Merkel et al., 2012, 2017; Burger et al., 2014; Cassidy et al., 2015). Contemporary magnetospheric measurements revealed a highly time-variable and spatially structured particle environment. Enhancements in planetary plasma ions were found in the dawn equatorial region, similar to neutral Ca enhancements, whereas on the nightside, observed asymmetries in planetary ions may be evidence of non-adiabatic behavior, expected but not previously observed (Raines et al., 2013). Despite the presence of localized enhancements in both the exosphere and magnetosphere, which suggest that feedback among these two systems and the surface is highly complex, the large-scale structure of the exosphere showed surprisingly little variation with changing solar conditions during the first two years of orbital observations. This finding was contrary to the expectations from current understanding of sputtering and other surface-interaction processes and the observed highly dynamic nature of the magnetosphere.

1.5.2.6 Mercury's Energetic Electrons

The existence of bursts of energetic electrons in Mercury's magnetosphere, a major discovery during the Mariner 10 flybys, was suggested by XRS signals seen during MESSENGER's Mercury flybys (Slavin et al., 2008) and confirmed almost as soon as MESSENGER began orbital science observations (Ho et al., 2011). Measurements during MESSENGER's primary and first extended mission revealed two groups of energetic electron events, one concentrated at northern high latitudes on the nightside and the other near the geographic equator at most local times (Ho et al., 2012). Not only were the two groups found at different spatial locations, but they also differed in energy distribution, with the high-latitude group tending to have energies in excess of 35 keV and the equatorial population having lower energies in the range 1–10 keV. Frequent observations of X-rays from Mercury's nightside surface, interpreted as X-ray emission induced by the interaction with ~1–10-keV electrons, indicated that the energetic electrons seen in orbit often precipitated onto Mercury's surface (Starr et al., 2012). Most of the observed energetic electron events displayed similar profiles of intensity versus time, with increases above background by up to three orders of magnitude within a few seconds (Ho et al., 2012). Although the spatial and large-scale temporal occurrence of these events was mapped, the unexpectedly irregular nature of these events, with the rapid rise in intensity and a velocity dispersion on timescales too short to be resolved by the EPS, hindered attempts to pin down the source or the acceleration mechanism. During the first extended mission, in June 2012, the cryocooler for the GRS sensor ceased to function at a time closely corresponding to its anticipated end of life. The ACS on the GRS nonetheless remained operational, and near the end of the first extended mission in February 2013 it was repurposed to measure energetic particles at a 10-ms sampling rate.

1.6 MESSENGER'S SECOND EXTENDED MISSION

MESSENGER's first extended mission raised new questions about Mercury that could be addressed only with new measurement campaigns. Given the healthy state of the spacecraft and instrument payload, an ample power margin, and remaining propellant as the end of the first extended mission drew near, the MESSENGER team proposed to NASA and the agency approved a second extended mission approximately two Earth years in duration. The questions that framed the second extended mission followed from discoveries made earlier in the mission or anticipated special aspects of either the timing of the observations or the geometry of MESSENGER's orbit. These questions addressed processes that have recently affected Mercury's surface, particularly at the locations of Mercury's hollows; the evolution of stress in Mercury's crust, and how that stress has been accommodated by a remarkably diverse set of tectonic landforms; changes in the composition of volcanic materials through geological time, and their implication for the evolution of magmatic source regions; the characteristics of volatile emplacement and sequestration in areas of permanent shadow in Mercury's north polar region; the consequences to Mercury's surface and neutral and ionized exosphere of the surface impact of ions and energetic electrons; the response of Mercury's exosphere and magnetosphere to continuing changes in solar activity; and the evolution of Mercury's crust and deeper interior as revealed by observations sensitive to variations over short horizontal scales.

A critical aspect of MESSENGER's second extended mission from the perspective of Mercury's exosphere, magnetosphere, and heliospheric environment was that the maximum in the solar cycle during which the mission had operated was predicted to occur during the first year of second extended mission operations, and the remainder of that year and all of the second year would capture the waning phase of the solar cycle. Solar disturbances were predicted to transition from coronal mass ejections (CMEs) up to and through solar maximum to high-speed streams during the declining phase of the cycle. It was recognized that the second extended mission therefore afforded a distinctive opportunity to characterize the response of Mercury's magnetosphere and exosphere to highly contrasting and intense forcing qualitatively different from that observed to date.

The second year of MESSENGER's second extended mission would also feature periapsis altitudes lower than at any earlier time in the mission, i.e., closer to Mercury's surface than any spacecraft had been before. Through the natural evolution of MESSENGER's orbit in response to the gravitational attraction of the Sun, together with an optimized set of OCMs conducted with MESSENGER's remaining propellant, it was planned that the spacecraft orbit during the final year would feature four separate campaigns of several days to one week each, during which the periapsis altitude would be nearly steady at 25 to 15 km. Such campaigns would provide opportunities to observe regions of Mercury at resolutions markedly superior to those yet attained, across the full suite of instruments. Observations would continue to extraordinarily low altitudes until the spacecraft finally impacted the planet at the end of mission operations.

1.6.1 Objectives for the Second Extended Mission

Seven science questions framed the second extended mission. Each was motivated by discoveries made during the primary and first extended missions, and collectively they addressed broad aspects of Mercury's characteristics, history, and interaction with the inner heliosphere. Those seven questions were as follows:

(1) What active and recent processes have affected Mercury's surface?
(2) How has the state of stress in Mercury's crust evolved over time?
(3) How have the compositions of volcanic materials on Mercury evolved over time?
(4) What are the characteristics of volatile emplacement and sequestration in Mercury's north polar region?
(5) What are the consequences of precipitating ions and energetic electrons at Mercury?
(6) How do Mercury's exosphere and magnetosphere respond to both extreme and stable solar wind conditions during solar maximum and the declining phase of the solar cycle?
(7) What novel insights into Mercury's thermal and crustal evolution can be obtained with high-resolution measurements from low altitudes?

Those questions led to the set of scientific objectives for MESSENGER's second extended mission listed in Table 1.5; the first number of each objective is tied to the corresponding science question. That the list of scientific objectives was longer than for the primary or first extended mission was a reflection of the maturation of our knowledge of the planet during the mission and the breadth and diversity of issues raised by the first two years of orbital observations. The project requirements corresponding to those objectives are listed in Table 1.6; many of the project requirements satisfied multiple science objectives.

1.6.2 Results from the Second Extended Mission

The design of the second extended mission combined optimum use of remaining propellant with spacecraft pointing strategies that balanced episodes of more intense heating of the spacecraft with opportunities for observations at lower altitudes than earlier in the mission (McAdams et al., 2014). The evolution of the altitude and latitude of periapsis during all but the final few weeks of the extended mission is illustrated in Figure 1.13. No OCMs were conducted during the first year of the second extended mission, and both the altitude and latitude of periapsis decreased progressively under the influence of the gravitational pull of the Sun. A series of four OCMs between June 2014 and December 2015 raised periapsis altitude and prolonged the mission duration.

During the final six weeks of the second extended mission, MESSENGER completed a low-altitude or "hover" campaign during which seven OCMs in March and April 2015 maintained the altitude at closest approach, relative to measured topography, between 5 and 37 km (McAdams et al., 2015). The campaign was unprecedented in several respects, including the short intervals between successive OCMs, the application of MLA altimetry data to validate trajectory solutions obtained by the project's mission design and navigation teams, and the use of propulsion system pressurant (helium gas) to impart thrust to the spacecraft during the final four OCMs once usable onboard hydrazine had been exhausted. After all usable pressurant as well as propellant had been consumed, no further OCMs were possible, and the spacecraft impacted Mercury's surface as expected on 30 April 2015. A plot of MESSENGER's closest-approach altitude during the hover campaign is shown in Figure 1.14.

All MESSENGER instruments continued to operate until the final transmission of data from the spacecraft, and by the end of MESSENGER's orbital operations all scientific objectives (Table 1.5) and project requirements (Table 1.6) for the second extended mission had been met. To illustrate this statement by example, we provide here an overview of some of the principal findings from the final two years of orbital observations, by scientific objective (Table 1.5). The full set of MESSENGER observations and their scientific implications for Mercury are described at greater length in the other chapters of this volume.

1.6.2.1 Recent Surface Processes

High-resolution images of hollows acquired at low elevations late in the second extended mission permitted measurements of the depths of hundreds of individual hollows; the results indicate an average depth of 24 ± 16 m and a range of values sufficiently narrow as to favor the hypothesis that hollows cease to increase in depth when a volatile-depleted lag deposit becomes sufficiently thick to protect the underlying surface (Blewett et al., 2016). Even the highest-resolution images reveal no superposed impact craters, implying that hollows are very young (Blewett et al., 2016). On the basis of the distribution of impact craters with high-reflectance ejecta, optical maturation or space weathering of surface material on Mercury is more rapid by a factor of as much as 4 than on the Moon (Braden and Robinson, 2013). Moreover, there are fewer optically immature craters per unit area on Mercury than on the Moon, indicating

Table 1.5. *Scientific objectives for MESSENGER's second extended mission.*

1.1 Investigate how hollows initiate and how they contribute to the exosphere.
1.2 Investigate how space weathering progressively modifies the optical properties of freshly exposed crustal materials.
1.3 Investigate how meteoritic materials contribute to the geochemistry of the surface.
2.1 Characterize how large-scale systems of lobate scarps spatially localize.
2.2 Determine whether there is evidence of recent contractional and extensional deformation.
2.3 Determine the crustal structure that is associated with contractional tectonics.
3.1 Investigate whether the northern plains are compositionally uniform.
3.2 Determine whether there are observable elemental and mineralogical differences between pyroclastic deposits and high-reflectance red plains.
3.3 Investigate the compositional relationship between low-reflectance blue plains and low-reflectance material.
3.4 Search for layers within plains units that could be volcanic flows.
3.5 Characterize the detailed nature of flow unit boundaries, and search for flow unit boundaries in the plains materials.
4.1 Characterize the morphology of small craters that host radar-bright material.
4.2 Determine which craters contain materials that are bright and dark at the wavelength (1064 nm) of the MLA and what physical features distinguish them.
4.3 Determine whether longitudinal variation of hydrogen concentrations within the north polar region is consistent with the distribution of radar-bright materials in permanently shadowed craters.
5.1 Determine fluences of protons, heavy ions, and electrons to the surface, and characterize how they vary with latitude and time.
5.2 Determine whether the signatures of particle precipitation to the surface are consistent with the inferred sources.
5.3 Determine what physical processes are revealed by the evolution of energetic electron events.
6.1 Characterize the nature of induced magnetic fields, and determine their effectiveness in controlling access of solar wind plasma to the surface.
6.2 Determine how the populations of heavy ions and protons in the cusps and the rest of the magnetosphere differ under extreme solar wind pressures.
6.3 Investigate how energetic electron events respond to increasing solar wind pressure or speed as well as prolonged stable solar wind, and determine whether exospheric density and distribution change under extreme conditions.
6.4 Characterize the time profiles of solar wind speed and density to which Mercury is exposed and how heliospheric pickup and suprathermal ions contribute to the exosphere.
6.5 Investigate how field-aligned currents close at low altitude.
7.1 Investigate how the lithosphere has evolved over time.
7.2 Characterize spatial variations in crustal thickness and density and determine the constraints these variations place on the history of crustal production.
7.3 Search for evidence of crustal magnetization, and evaluate the constraints that this evidence places on the evolution of the dynamo field.
7.4 Search for evidence of finite electrical conductivity of the mantle in induced magnetic field signatures and determine what constraints this evidence places on present mantle temperature structure.
7. 5 Discover the crustal geological characteristics and their variability with terrain types at small spatial scales not previously observable.

Table 1.6. *Project requirements for MESSENGER's second extended mission.*

1. Characterize faulted terrain by acquiring at least one of the following: (a) 20 NAC along-track stereo pairs or (b) 40 MLA topographic profiles.
2. Characterize fresh craters by acquiring at least one of the following: (a) 20 WAC 11-color image sets or (b) 20 NAC along-track stereo pairs.
3. Characterize hollows by acquiring (a) UVVS observations of exospheric species over eight clusters of hollows on three different dates and from two different viewing geometries per feature or (b) 20 along-track NAC stereo pairs and 20 11-color image sets each.
4. Characterize surface features at very high resolution by acquiring 750 NAC images at ≤10-m/pixel scale and 100 NAC images at ≤5-m/pixel scale
5. Search for color variations within the northern plains by acquiring 5-color MDIS images of 75% of the surface area north of 60°N at phase angles <60°.
6. Constrain the elemental composition of spectral end-member materials by acquiring targeted XRS spectra (a) from the large pyroclastic deposit northeast of Rachmaninoff and (b) of at least two different portions of low-reflectance blue plains exterior to the Caloris basin. For each target, acquire a minimum of 1000 s of spectral integration spread over at least five different orbits.

7. (a) Characterize MLA-bright and -dark materials by acquiring MLA ranging and reflectance data along portions of two orbits for which ground tracks cross each of 10 craters <20 km in diameter, and (b) characterize the north polar hydrogen distribution at high spatial resolution by acquiring NS measurements for 70% of the time that the spacecraft altitude is <150 km.
8. Characterize crustal structure at high resolution by acquiring Doppler tracking data for portions of 100 orbits at altitudes <100 km.
9. Characterize the structure of crustal magnetization at high resolution by acquiring MAG and FIPS observations along portions of 100 orbits at altitudes <50 km in the vicinity of the northern plains.
10. Characterize magnetospheric particle flows and pitch-angle distributions by acquiring a defined set of 970 EPPS measurements distributed across several different pointing scenarios.
11. Characterize the exospheric response to conditions during solar maximum and the declining phase of the solar cycle by acquiring a defined set of 5025 UVVS dayside and nightside observations, including searches for species with weaker resonant emissions.
12. Characterize the magnetospheric response to conditions during solar maximum and the declining phase of the solar cycle by acquiring MAG, EPPS, and NS/GRS observations for 75% of the time throughout the mission, including times at which the spacecraft altitude is <50 km.

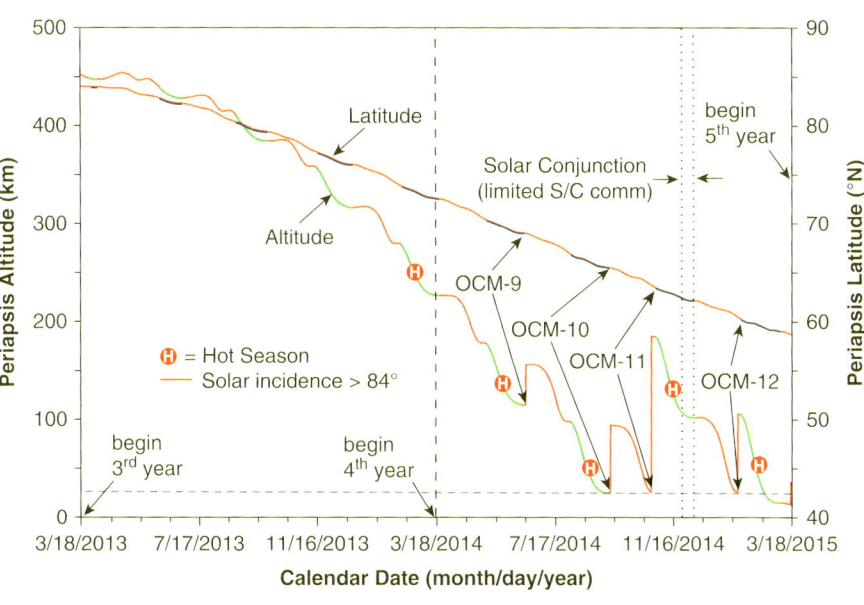

Figure 1.13. Evolution of the altitude and latitude of MESSENGER's periapsis during most of the second extended mission, from 18 March 2013 to 18 March 2015. Portions of the orbital observations when MESSENGER was in its "hot season" (near noon–midnight orbit configuration with periapsis on the dayside) are marked, as are periods when the solar incidence angle (measured from the vertical) along the dayside orbit track exceeded 84° and no imaging near periapsis was planned. A period of superior solar conjunction, when the spacecraft was on the opposite side of the Sun from Earth and thus communication with the spacecraft was limited, is indicated (S/C denotes spacecraft). From McAdams et al. (2014).

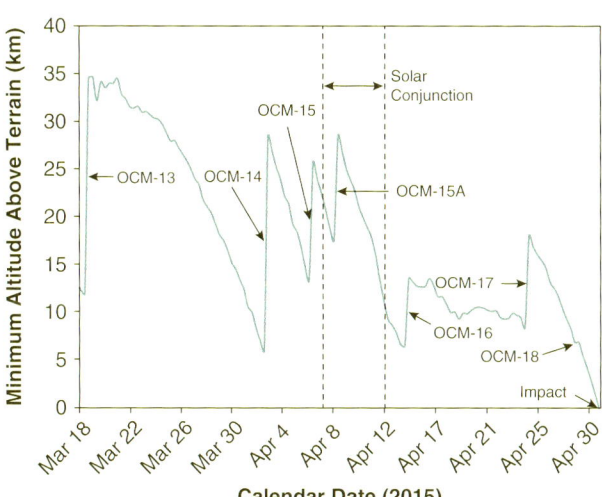

Figure 1.14. MESSENGER's altitude at closest approach, relative to measured topography, during the mission's hover campaign over the final six weeks of orbital operations. A period of superior solar conjunction, when the spacecraft was on the opposite side of the Sun from Earth, is indicated by vertical dashed lines. From McAdams et al. (2015).

that rayed craters on Mercury are younger on average than those on the Moon (Braden and Robinson, 2013).

1.6.2.2 Crustal Stress Field over Time

Images of Mercury's surface acquired from orbit by MESSENGER showed that the planet's global contraction involved a substantially greater number and variety of structures than previously recognized. The strain accommodated by identified tectonic features implies that Mercury contracted radially by as much as 7 km, well in excess of the 0.8–3 km previously reported from photogeological studies and resolving the long-standing discrepancy with the predictions of thermal history models (Byrne et al., 2014). Moreover, an additional 2 km of radial contraction may have been accommodated elastically prior to the development of widespread faulting (Klimczak, 2015). The distribution and orientation of tectonic features are consistent with scenarios in which tidal despinning accompanied the earliest phases of global contraction (Klimczak et al., 2015). That global contraction was underway within 1 Gyr of Mercury's formation is consistent with the cessation of widespread smooth plains formation around 3.5 Ga, given that lithospheric stresses characterized by horizontal compression would have inhibited the upward ascent of magma (Byrne et al., 2016). Crosscutting relations between lobate scarps and fresh craters indicate that contractional deformation on Mercury continued to geologically recent times (Banks et al., 2015).

1.6.2.3 Composition of Volcanic Materials over Time

The second extended mission doubled the opportunity to conduct elemental remote sensing of Mercury's surface over the primary and first extended missions. As the spatial resolution of such geochemical measurements improved, it became evident that Mercury's surface could be divided into approximately half a dozen geochemical terranes, each with distinctive compositional characteristics resolvable from several independent measurement types, including fluorescent X-rays (Weider et al., 2015), the flux of thermal neutrons (Peplowski et al., 2015), and the flux of fast neutrons (Lawrence et al., 2017). The chemical differences among terranes are broadly consistent with decreases in mantle potential temperature and degree of melting with time on Mercury (Namur et al., 2016).

1.6.2.4 Volatiles in the North Polar Region

During MESSENGER's second extended mission, images were acquired with the broadband clear filter on the MDIS WAC of persistently shadowed areas on the floors of impact craters at high northern latitudes (Chabot et al., 2014, 2016). On the floor of Prokofiev crater, a site previously identified on the basis of MLA reflectance measurements and thermal models as containing widespread surface water ice, the area in persistent shadow was seen to have a cratered texture that resembles the neighboring sunlit surface except for its uniformly higher reflectance, indicating that the surficial ice was emplaced after formation of the underlying craters (Chabot et al., 2014). In areas where water ice was inferred to be present from radar observations but covered by a thin layer of dark volatile material on the basis of MLA reflectance and thermal models, regions with uniformly low reflectance were seen to extend to the edges of the shadowed areas and terminate with sharp boundaries (Chabot et al., 2014). In images acquired during the low-altitude campaign late in the second extended mission, brightness variations across the low-reflectance deposits correlate with variations in the modeled biannual maximum surface temperature across the persistently shadowed regions, supporting the conclusion that multiple volatile organic compounds are present in addition to water ice (Chabot et al., 2016). Either a recent large impact by a comet or volatile-rich asteroid, or ongoing bombardment by volatile-rich micrometeoroids, could deliver water and volatile organic material to Mercury (Chabot et al., 2014, 2016).

1.6.2.5 Ion and Electron Precipitation

During MESSENGER's second extended mission, energetic electron events were characterized with data from multiple instruments, including measurements at high temporal resolution (10 ms) with the GRS ACS (Lawrence et al., 2015; Ho et al., 2016; Baker et al., 2016). The most energetic electron bursts detected by MESSENGER sensors appeared to be produced in the midnight sector of Mercury's magnetosphere, supporting the view that energetic electrons are accelerated in the near-tail region and are then injected onto closed magnetic field lines on the planetary nightside during substorm-like events (Baker et al., 2016). The electrons populate the plasma sheet and drift rapidly eastward toward the dawn and prenoon sectors, at times executing multiple complete drifts around the planet to form "quasi-trapped" populations (Lawrence et al., 2015; Ho et al., 2016; Baker et al., 2016). Observations of plasma in Mercury's magnetosphere show a north–south asymmetry on the nightside, with markedly lower fluxes at low altitudes in the northern hemisphere than at higher altitudes in the south on the same field line, an asymmetry consistent with particle loss to the southern hemisphere surface during bounce motion in Mercury's offset dipole magnetic field (Korth et al., 2014). Plasma measurements in Mercury's magnetospheric cusp show evidence of three processes: (1) direct inflow from the magnetosheath, (2) local production of planetary photoions and ions sputtered off the surface from solar wind impact that are then accelerated upward, and (3) flow of magnetosheath and magnetospheric plasma accelerated from dayside magnetic reconnection (Raines et al., 2014). During solar energetic particle events, FIPS measured fluxes of electrons at MeV energies equal to ~40% of their upstream values over Mercury's entire polar cap, indicating that space weathering of the surface by energetic electrons is not limited to the region of the cusp (Gershman et al., 2015).

1.6.2.6 Response of the Exosphere and Magnetosphere

MESSENGER magnetic field observations demonstrated the presence of electric currents that flow along magnetic lines of force toward and away from the planet above Mercury's northern hemisphere; such currents are analogous to Birkeland currents at Earth, but close not through an ionosphere but rather through the planet, radially through the low-conductivity outer silicate shell and laterally from dawn to dusk through more conductive material at depth (Anderson et al., 2014). Magnetic field observations over 15 Mercury years showed a small annual (88-day) variation in the planetary dipole moment, evidence that induced magnetic fields in Mercury's core act to oppose the decrease in subsolar magnetopause standoff distance with increasing solar wind ram pressure (Johnson et al., 2016). The shielding provided by induced currents is substantially offset, however, by the effects of dayside magnetic reconnection, which erodes magnetic flux from the dayside magnetosphere and can be particularly intense during extreme solar wind events (Imber et al., 2014; Slavin et al., 2014). Observations of frequent magnetic flux ropes in the cross-tail current sheet confirm the high rate of magnetic reconnection in that portion of the magnetosphere (DiBraccio et al., 2015).

1.6.2.7 New Insights from Low-Altitude Observations

The low periapsis altitudes during the second half of MESSENGER's second extended mission enabled a variety of discoveries. Magnetic field measurements obtained at altitudes less than 150 km demonstrated for the first time the presence of crustal magnetic fields inferred to have been acquired by at least 3.9–3.7 Ga (Johnson et al., 2015). A magnetic dynamo must therefore have operated in Mercury's core early in the planet's history. The low altitudes also permitted NS observations to be made of large expanses of low-reflectance material and of the largest identified pyroclastic deposit on Mercury, northeast of the Rachmaninoff basin. An increase in thermal neutron flux over LRM compared with surrounding terrain (Peplowski et al., 2016) coupled with the material's distinctive reflectance characteristics (Murchie et al., 2015) point to graphite as a major darkening agent on Mercury and in LRM in particular. The preferential location of hollows in LRM and the higher than average concentration of graphite in that material suggests that hollow formation may involve loss of carbon, e.g., by ion sputtering or conversion to methane by proton irradiation (Blewett et al., 2016). A decrease in the thermal neutron flux over the large pyroclastic deposit (Peplowski et al., 2016) together with targeted XRS observations and spectral reflectance measurements indicate that the deposit is depleted in S (relative to Ca and Si) and C compared with Mercury's average surface, consistent with oxidation of graphite and sulfides during magma ascent, via reaction with oxides in the magma or assimilated country rock, and the formation of S- and C-bearing volatile species (Weider et al., 2016).

1.7 CONCLUSIONS AND OVERVIEW OF OTHER CHAPTERS

The MESSENGER mission met or exceeded all of its scientific objectives across a broad spectrum of planetary science disciplines, as well as all of its project requirements. The spacecraft completed orbital operations at Mercury over a period that was a factor of 4 longer than originally planned, despite radiation and thermal hazards particular to Mercury's distance from the Sun. As a result, we have markedly deepened our understanding of Mercury, its interaction with the local heliospheric environment, and its role as a member of the family of inner solar system planets.

The state of that understanding, as of the end of the MESSENGER mission, is laid out in the chapters that follow. Chapters 2–4 are on the bulk properties of Mercury: the chemical composition, the structure of the crust and lithosphere, and the deeper interior structure, respectively. Chapter 5 summarizes our knowledge of Mercury's magnetic field. Chapters 6–13 address Mercury's geology: its major geological units, its variations in surface elemental chemistry and inferred mineralogy, its spectral reflectance characteristics and their variation, its impact craters and cratering history, its tectonic features and deformational history, its volcanic features and magmatic history, its hollows, and its polar deposits, respectively. Chapters 14 and 15 deal with observations of Mercury's exosphere and models of the physical processes that govern exospheric behavior, respectively. Chapters 16 and 17 summarize our understanding of the structure and dynamics, respectively, of Mercury's magnetosphere. Chapters 18 and 19 address the formation and large-scale evolution, respectively, of the planet. The final chapter of the book, Chapter 20, gives an overview of the future exploration of Mercury, from a mission now nearing launch to concepts for follow-on missions in the more distant future.

REFERENCES

Aharonson, O., Zuber, M. T. and Solomon, S. C. (2004). Crustal remanence in an internally magnetized non-uniform shell: A possible source for Mercury's magnetic field? *Earth Planet. Sci. Lett.*, **218**, 261–268.

Anderson, B. J., Acuña, M. H., Lohr, D. A., Scheifele, J., Raval, A., Korth, H. and Slavin, J. A. (2007). The Magnetometer instrument on MESSENGER. *Space Sci. Rev.*, **131**, 417–450.

Anderson, B. J., Johnson, C. L., Korth, H., Purucker, M. E., Winslow, R. M., Slavin, J. A., Solomon, S. C., McNutt, R. L., Jr., Raines, J. M. and Zurbuchen, T. H. (2011a). The global magnetic field of Mercury from MESSENGER orbital observations. *Science*, **333**, 1859–1862.

Anderson, B. J., Perry, M. E., Choo, T. H., Steele, R. J., Nguyen, L., Lucks, M., Prockter, L. M., McNutt, R. L., Jr. and Solomon, S. C. (2011b). MESSENGER science observation planning for orbital operations at Mercury. *Lunar Planet. Sci.*, **42**, abstract 1862.

Anderson, B. J., Johnson, C. L., Korth, H., Winslow, R. M., Borovsky, J. E., Purucker, M. E., Slavin, J. A., Solomon, S. C., Zuber, M. T. and McNutt, R. L., Jr. (2012). Low-degree structure in Mercury's planetary magnetic field. *J. Geophys. Res.*, **117**, E00L12, doi:10.1029/2012JE004159.

Anderson, B. J., Johnson, C. L., Korth, H., Slavin, J. A., Winslow, R. M., Phillips, R. J., McNutt, R. L., Jr. and Solomon, S. C. (2014). Steady-state field-aligned currents at Mercury. *Geophys. Res. Lett.*, **41**, 7444–7452.

Anderson, J. D., Colombo, G., Esposito, P. B., Lau, E. L. and Trager, G. B. (1987). The mass, gravity field, and ephemeris of Mercury. *Icarus*, **71**, 337–349.

Andrews, G. B., Zurbuchen, T. H., Mauk, B. H., Malcom, H., Fisk, L. A., Gloeckler, G., Ho, G. C., Kelley, J. S., Koehn, P. L., LeFevere, T. W., Livi, S. S., Lundgren, R. A. and Raines, J. M. (2007). The Energetic Particle and Plasma Spectrometer instrument on the MESSENGER spacecraft. *Space Sci. Rev.*, **131**, 523–556.

Baker, D. N., Dewey, R. M., Lawrence, D. J., Goldsten, J. O., Peplowski, P. N., Korth, H., Slavin, J. A., Krimigis, S. M., Anderson, B. J., Ho, G. C., McNutt, R. L., Jr., Raines, J. M., Schriver, D. and Solomon, S. C. (2016). Intense energetic electron flux enhancements in Mercury's magnetosphere: An integrated view with high-resolution observations from MESSENGER. *J. Geophys. Res. Space Physics*, **121**, 2171–2184.

Balcerski, J. A., Hauck, S. A., II, Sun, P., Klimczak, C., Byrne, P. K., Phillips, R. J. and Solomon, S. C. (2013). New constraints on timing and mechanisms of regional tectonism from Mercury's tilted craters. *Lunar Planet. Sci.*, **44**, abstract 2444.

Banks, M. E., Xiao, Z., Watters, T. R., Strom, R. G., Braden, S. E., Chapman, C. R., Solomon, S. C., Klimczak, C. and Byrne, P. K. (2015). Duration of activity on lobate-scarp thrust faults on Mercury. *J. Geophys. Res. Planets*, **120**, 1751–1762.

Benz, W., Slattery, W. L. and Cameron, A. G. W. (1988). Collisional stripping of Mercury's mantle. *Icarus*, **74**, 516–528.

Berman, A. F., Domingue, D. L., Holdridge, M. E., Choo, T. H., Steele, R. J. and Shelton, R. G. (2010). Testing and validation of orbital operations plans for the MESSENGER mission. In *Observatory Operations: Strategies, Processes, and Systems III*, ed. D. R. Silva, A. B. Peck and B. T. Soifer. *Proc. SPIE*, 7737, doi:10.1117/12.857107.

Bida, T. A., Killen, R. M. and Morgan, T. H. (2000). Discovery of calcium in Mercury's atmosphere. *Nature*, **404**, 159–161.

Blewett, D. T., Chabot, N. L., Denevi, B. W., Ernst, C. M., Head, J. W., Izenberg, N. R., Murchie, S. L., Solomon, S. C., Nittler, L. R., McCoy, T. J., Xiao, Z., Baker, D. M. H., Fassett, C. I., Braden, S. E., Oberst, J., Scholten, F., Preusker, F. and Hurwitz, D. M. (2011). Hollows on Mercury: MESSENGER evidence for geologically recent volatile-related activity. *Science*, **333**, 1856–1859.

Blewett, D. T., Vaughan, W. M., Xiao, Z., Chabot, N. L., Denevi, B. W., Ernst, C. M., Helbert, J., D'Amore, M., Maturilli, A., Head, J. W. and Solomon, S. C. (2013). Mercury's hollows: Constraints on formation and composition from analysis of geological setting and spectral reflectance. *J. Geophys. Res. Planets*, **118**, 1013–1032.

Blewett, D. T., Stadermann, A. C., Susorney, H. C., Ernst, C. M., Xiao, Z., Chabot, N. L., Denevi, B. W., Murchie, S. L., McCubbin, F. M., Kinczyk, M. J., Gillis-Davis, J. J. and Solomon, S. C. (2016). Analysis of MESSENGER high-resolution images of Mercury's hollows and implications for hollow formation. *J. Geophys. Res. Planets*, **121**, 1798–1813.

Boardsen, S. A., Slavin, J. A., Anderson, B. J., Korth, H., Schriver, D. and Solomon, S. C. (2012). Survey of coherent ~1 Hz waves in Mercury's inner magnetosphere. *J. Geophys. Res.*, **117**, A00M05, doi:10.1029/2012JA017822.

Braden, S. E. and Robinson, M. S. (2013). Relative rates of optical maturation of regolith on Mercury and the Moon. *J. Geophys. Res. Planets*, **118**, 1903–1914.

Broadfoot, A. L., Shemanski, D. E. and Kumar, S. (1976). Mariner 10: Mercury atmosphere. *Geophys. Res. Lett.*, **3**, 577–580.

Burger, M. H., Killen, R. M., McClintock, W. E., Vervack, R. J., Jr., Merkel, A. W., Sprague, A. L. and Sarantos, M. (2012). Modeling MESSENGER observations of calcium in Mercury's exosphere. *J. Geophys. Res.*, **117**, E00L11, doi:10.1029/2012JE004158.

Burger, M. H., Killen, R. M., McClintock, W. E., Merkel, A. W., Vervack, R. J., Jr., Cassidy, T. A. and Sarantos, M. (2014). Seasonal variations in Mercury's dayside calcium exosphere. *Icarus*, **238**, 51–58.

Byrne, P. K., Watters, T. R., Murchie, S. L., Klimczak, C., Solomon, S. C., Prockter, L. M. and Freed, A. M. (2012). A tectonic survey of the Caloris basin, Mercury. *Lunar Planet. Sci.*, **43**, abstract 1722.

Byrne, P. K., Klimczak, C., Williams, D. A., Hurwitz, D. M., Solomon, S. C., Head, J. W., Preusker, F. and Oberst, J. (2013). An assemblage of surface lava flow features on Mercury. *J. Geophys. Res. Planets*, **118**, 1303–1322.

Byrne, P. K., Klimczak, C., Şengör, A. M. C., Solomon, S. C., Watters, T. R. and Hauck, S. A., II (2014). Mercury's global contraction much greater than earlier estimates. *Nature Geosci.*, **7**, 301–307.

Byrne, P. K., Ostrach, L. R., Fassett, C. I., Chapman, C. R., Denevi, B. W., Evans, A. J., Klimczak, C., Banks, M. E., Head, J. W. and Solomon, S. C. (2016). Widespread effusive volcanism on Mercury likely ended by about 3.5 Ga. *Geophys. Res. Lett.*, **43**, 7408–7416.

Cameron, A. G. W. (1985). The partial volatilization of Mercury. *Icarus*, **64**, 285–294.

Cassidy, T. A., Merkel, A. W., Burger, M. H., Sarantos, M., Killen, R. M., McClintock, W. E. and Vervack, R. J., Jr. (2015). Mercury's seasonal sodium exosphere: MESSENGER orbital observations. *Icarus*, **248**, 547–559.

Cavanaugh, J. F., Smith, J. C., Sun, X., Bartels, A. E., Ramos-Izquierdo, L., Krebs, D. J., McGarry, J. F., Trunzo, R., Novo-Gradac, A. M., Britt, J. L., Karsh, J., Katz, R. B., Lukemire, A., Szymkiewicz, R., Berry, D. L., Swinski, J. P., Neumann, G. A., Zuber, M. T. and Smith, D. E. (2007). The Mercury Laser Altimeter instrument for the MESSENGER mission. *Space Sci. Rev.*, **131**, 451–480.

Chabot, N. L., Ernst, C. M., Denevi, B. W., Harmon, J. K., Murchie, S. L., Blewett, D. T., Solomon, S. C. and Zhong, E. D. (2012). Areas of permanent shadow in Mercury's south polar region ascertained by MESSENGER orbital imaging. *Geophys. Res. Lett.*, **39**, L09204, doi:10.1029/2012GL051526.

Chabot, N. L., Ernst, C. M., Harmon, J. K., Murchie, S. L., Solomon, S. C., Blewett, D. T. and Denevi, B. W. (2013). Craters hosting radar-bright deposits in Mercury's north polar region: Areas of persistent shadow determined from MESSENGER images. *J. Geophys. Res. Planets*, **118**, 26–36.

Chabot, N. L., Ernst, C. M., Denevi, B. W., Nair, H., Deutsch, A. N., Blewett, D. T., Murchie, S. L., Neumann, G. A., Mazarico, E., Paige, D. A., Harmon, J. K., Head, J. W. and Solomon, S. C. (2014). Images of surface volatiles in Mercury's polar craters acquired by the MESSENGER spacecraft. *Geology*, **12**, 1051–1064.

Chabot, N. L., Ernst, C. M., Paige, D. A., Nair, H., Denevi, B. W., Blewett, D. T., Murchie, S. L., Deutsch, A. N., Head, J. W. and Solomon, S. C. (2016). Imaging Mercury's polar deposits during MESSENGER's low-altitude campaign. *Geophys. Res. Lett.*, **43**, 9461–9468.

Choo, T. H., Murchie, S. L., Bedini, P. D., Steele, R. J., Skura, J. P., Nguyen, L., Nair, H., Lucks, M., Berman, A. F., McGovern, J. A. and Turner, F. S. (2014). SciBox: An end-to-end automated science planning and commanding system. *Acta Astronaut.*, **93**, 490–496.

Christensen, U. R. (2006). A deep dynamo generating Mercury's magnetic field. *Nature*, **444**, 1056–1058.

COMPLEX (Committee on Lunar and Planetary Exploration) (1978). *Strategy for Exploration of the Inner Planets: 1977–1987*. Washington, DC: National Research Council, 105 pp.

Connerney, J. E. P. and Ness, N. F. (1988). Mercury's magnetic field and interior. In *Mercury*, ed. F. Vilas, C. R. Chapman and M. S. Matthews. Tucson, AZ: University of Arizona Press, pp. 479–488.

Denevi, B. W., Robinson, M. S., Solomon, S. C., Murchie, S. L., Blewett, D. T., Domingue, D. L., McCoy, T. J., Ernst, C. M., Head, J. W., Watters, T. R. and Chabot, N. L. (2009). The evolution of Mercury's crust: A global perspective from MESSENGER. *Science*, **324**, 613–618.

Denevi, B. W., Ernst, C. M., Meyer, H. M., Robinson, M. S., Murchie, S. L., Whitten, J. L., Head, J. W., Watters, T. R., Solomon, S. C., Ostrach, L. R., Chapman, C. R., Byrne, P. K. and Peplowski, P. N. (2013). The distribution and origin of smooth plains on Mercury. *J. Geophys. Res. Planets*, **118**, 891–907.

DiBraccio, G. A., Slavin, J. A., Boardsen, S. A., Anderson, B. J., Korth, H., Zurbuchen, T. H., Raines, J. M., Baker, D. N., McNutt, R. L., Jr. and Solomon, S. C. (2013). MESSENGER observations of magnetopause structure and dynamics at Mercury. *J. Geophys. Res. Space Physics*, **118**, 997–1008.

DiBraccio, G. A., Slavin, J. A., Imber, S. M., Gershman, D. J., Raines, J. M., Jackman, C. M., Boardsen, S. A., Anderson, B. J., Korth, H., Zurbuchen, T. H., McNutt, R. L., Jr. and Solomon, S. C. (2015). MESSENGER observations of flux ropes in Mercury's magnetotail. *Planet. Space Sci.*, **115**, 77–89.

Domingue, D. L., Koehn, P. L., Killen, R. M., Sprague, A. L., Sarantos, M., Cheng, A. F., Bradley, E. T. and McClintock, W. E. (2007). Mercury's atmosphere: A surface-bounded exosphere. *Space Sci. Rev.*, **131**, 161–186.

Dunne, J. A. and Burgess E. (1978). *The Voyage of Mariner 10: Mission to Venus and Mercury*. Special Publication SP-424. Washington, DC: NASA Scientific and Technical Information Office.

Evans, L. G., Peplowski, P. N., Rhodes, E. A., Lawrence, D. J., McCoy, T. J., Nittler, L. R., Solomon, S. C., Sprague, A. L., Stockstill-Cahill, K. R., Starr, R. D., Weider, S. Z., Boynton, W. V. and

Hamara, D. K. (2012). Major-element abundances on the surface of Mercury: Results from the MESSENGER Gamma-Ray Spectrometer. *J. Geophys. Res.*, **117**, E00L07, doi:10.1029/2012JE004178.

Evans, L. G., Peplowski, P. N., McCubbin, F. M., McCoy, T. J., Nittler, L. R., Zolotov, M. Yu., Ebel, D. S., Lawrence, D. J., Starr, R. D., Weider, S. Z. and Solomon, S. C. (2015). Chlorine on the surface of Mercury: MESSENGER gamma-ray measurements and implications for the planet's formation and evolution. *Icarus*, **257**, 417–427.

Fassett, C. I., Head, J. W., Baker, D. M. H., Zuber, M. T., Smith, D. E., Neumann, G. A., Solomon, S. C., Strom, R. G., Chapman, C. R., Prockter, L. M., Phillips, R. J., Oberst, J. and Preusker, F. (2012). Large impact basins on Mercury: Global distribution, characteristics and modification history from MESSENGER orbital data. *J. Geophys. Res.*, **117**, E00L08, doi:10.1029/2012JE004154.

Freed, A. M., Blair, D. M., Watters, T. R., Klimczak, C., Byrne, P. K., Solomon, S. C., Zuber, M. T. and Melosh, H. J. (2012). On the origin of graben and ridges within and near volcanically buried craters and basins in Mercury's northern plains. *J. Geophys. Res.*, **117**, E00L06, doi:10.1029/2012JE004119.

Gershman, D. J., Raines, J. M., Slavin, J. A., Zurbuchen, T. H., Anderson, B. J., Korth, H., Ho, G. C., Boardsen, S. A., Cassidy, T. A., Walsh, B. M. and Solomon, S. C. (2015). MESSENGER observations of solar energetic electrons within Mercury's magnetosphere. *J. Geophys. Res. Space Physics*, **120**, 8559–8571.

Giampieri, G. and Balogh, A. (2002). Mercury's thermoelectric dynamo model revisited. *Planet. Space Sci.*, **50**, 757–762.

Gold, R. E., Solomon, S. C., McNutt, R. L., Jr., Santo, A. G., Abshire, J. B., Acuña, M. H., Afzal, R. S., Anderson, B. J., Andrews, G. B., Bedini, P. D., Cain, J., Cheng, A. F., Evans, L. G., Feldman, W. C., Follas, R. B., Gloeckler, G., Goldsten, J. O., Hawkins, S. E., III, Izenberg, N. R., Jaskulek, S. E., Ketchum, E. A., Lankton, M. R., Lohr, D. A., Mauk, B. H., McClintock, W. E., Murchie, S. L., Schlemm, C. E., II, Smith, D. E., Starr, R. D. and Zurbuchen, T. H. (2001). The MESSENGER mission to Mercury: Scientific payload. *Planet. Space Sci.*, **49**, 1467–1479.

Goldsten, J. O., Rhodes, E. A., Boynton, W. V., Feldman, W. C., Lawrence, D. J., Trombka, J. I., Smith, D. M., Evans, L. G., White, J., Madden, N. W., Berg, P. C., Murphy, G. A., Gurnee, R. S., Strohbehn, K., Williams, B. D., Schaefer, E. D., Monaco, C. A., Cork, C. P., Eckels, J. D., Miller, W. O., Burks, M. T., Hagler, L. B., Deteresa, S. J. and Witte, M. C. (2007). The MESSENGER Gamma-Ray and Neutron Spectrometer. *Space Sci. Rev.*, **131**, 339–391.

Goudge, T. A., Head, J. W., Kerber, L., Blewett, D. T., Denevi, B. W., Domingue, D. L., Gillis-Davis, J. J., Gwinner, K., Helbert, J., Holsclaw, G. M., Izenberg, N. R., Klima, R. L., McClintock, W. E., Murchie, S. L., Neumann, G. A., Smith, D. E., Strom, R. G., Xiao, Z., Zuber, M. T. and Solomon, S. C. (2014). Global inventory and characterization of pyroclastic deposits on Mercury: New insights into pyroclastic activity from MESSENGER orbital data. *J. Geophys. Res. Planets*, **119**, 635–658.

Harmon, J. K. and Slade, M. A. (1992). Radar mapping of Mercury: Full-disk images and polar anomalies. *Science*, **258**, 640–643.

Harmon, J. K., Slade, M. A. and Rice, M. S. (2011). Radar imagery of Mercury's putative polar ice: 1999–2005 Arecibo results. *Icarus*, **211**, 37–50.

Hauck, S. A., II, Margot, J.-L., Solomon, S. C., Phillips, R. J., Johnson, C. L., Lemoine, F. G., Mazarico, E., McCoy, T. J., Padovan, S., Peale, S. J., Perry, M. E., Smith, D. E. and Zuber, M. T. (2013). The curious case of Mercury's internal structure. *J. Geophys. Res. Planets*, **118**, 1204–1220.

Hawkins, S. E., III, Boldt, J. D., Darlington, E. H., Espiritu, R., Gold, R. E., Gotwols, B., Grey, M. P., Hash, C. D., Hayes, J. R., Jaskulek, S. E., Kardian, C. J., Keller, M. R., Malaret, E. R., Murchie, S. L., Murphy, P. K., Peacock, K., Prockter, L. M., Reiter, R. A., Robinson, M. S., Schaefer, E. D., Shelton, R. G., Sterner, R. E., II, Taylor, H. W., Watters, T. R. and Williams, B. D. (2007). The Mercury Dual Imaging System on the MESSENGER spacecraft. *Space Sci. Rev.*, **131**, 247–338.

Head, J. W., Murchie, S. L., Prockter, L. M., Robinson, M. S., Solomon, S. C., Strom, R. G., Chapman, C. R., Watters, T. R., McClintock, W. E., Blewett, D. T. and Gillis-Davis, J. J. (2008). Volcanism on Mercury: Evidence from the first MESSENGER flyby. *Science*, **321**, 69–72.

Head, J. W., Chapman, C. R., Strom, R. G., Fassett, C. I., Denevi, B. W., Blewett, D. T., Ernst, C. M., Watters, T. R., Solomon, S. C., Murchie, S. L., Prockter, L. M., Chabot, N. L., Gillis-Davis, J. J., Whitten, J. L., Goudge, T. A., Baker, D. M. H., Hurwitz, D. M., Ostrach, L. R., Xiao, Z., Merline, W. J., Kerber, L., Dickson, J. L., Oberst, J., Byrne, P. K., Klimczak, C. and Nittler, L. R. (2011). Flood volcanism in the northern high latitudes of Mercury revealed by MESSENGER. *Science*, **333**, 1853–1856.

Helbert, J., Maturilli, A. and D'Amore, M. (2013). Visible and near infrared reflectance spectra of thermally processed synthetic sulfide as a potential analog for the hollow forming materials on Mercury. *Earth Planet. Sci. Lett.*, **369–370**, 233–238.

Ho, G. C., Krimigis, S. M., Gold, R. E., Baker, D. N., Slavin, J. A., Anderson, B. J., Korth, H., Starr, R. D., Lawrence, D. J., McNutt, R. L., Jr. and Solomon, S. C. (2011). MESSENGER observations of transient bursts of energetic electrons in Mercury's magnetosphere. *Science*, **333**, 1866–1868.

Ho, G. C., Krimigis, S. M., Gold, R. E., Baker, D. N., Anderson, B. J., Korth, H., Slavin, J. A., McNutt, R. L., Jr., Winslow, R. M. and Solomon, S. C. (2012). Spatial distribution and spectral characteristics of energetic electrons in Mercury's magnetosphere. *J. Geophys. Res.*, **117**, A00M04, doi:10.1029/2012JA017983.

Ho, G. C., Starr, R. D., Krimigis, S. M., Vandegriff, J. D., Baker, D. N., Gold, R. E., Anderson, B. J., Korth, H., Schriver, D., McNutt, R. L., Jr. and Solomon, S. C. (2016). MESSENGER observations of suprathermal electrons in Mercury's magnetosphere. *Geophys. Res. Lett.*, **43**, 550–555.

Hunten, D. M. and Sprague, A. L. (2002). Diurnal variation of sodium and potassium at Mercury. *Meteorit. Planet. Sci.*, **37**, 1191–1195.

Hurwitz, D. M., Head, J. W., Byrne, P. K., Xiao, Z., Solomon, S. C., Zuber, M. T., Smith, D. E. and Neumann, G. A. (2013). Investigating the origin of candidate lava channels on Mercury with MESSENGER data: Theory and observations. *J. Geophys. Res. Planets*, **118**, 471–486.

Imber, S. M., Slavin, J. A., Boardsen, S. A., Anderson, B. J., Korth, H., McNutt, R. L., Jr. and Solomon, S. C. (2014). MESSENGER observations of large dayside flux transfer events: Do they drive Mercury's substorm cycle? *J. Geophys. Res. Space Physics*, **119**, 5613–5623.

Johnson, C. L., Purucker, M. E., Korth, H., Anderson, B. J., Winslow, R. M., Al Asad, M. M. H., Slavin, J. A., Alexeev, I., Phillips, R. J., Zuber, M. T. and Solomon, S. C. (2012). MESSENGER observations of Mercury's magnetic field structure. *J. Geophys. Res.*, **117**, E00L14, doi:10.1029/2012JE004217.

Johnson, C. L., Phillips, R. J., Purucker, M. E., Anderson, B. J., Byrne, P. K., Denevi, B. W., Feinberg, J. M., Hauck, S. A., II, Head, J. W., III, Korth, H., James, P. B., Mazarico, E., Neumann, G. A., Philpott, L. C., Siegler, M. A., Tsyganenko, N. A. and Solomon, S. C. (2015). Low-altitude magnetic field measurements by MESSENGER reveal Mercury's ancient crustal field. *Science*, **348**, 892–895.

Johnson, C. L., Philpott, L. C., Anderson, B. J., Korth, H., Hauck, S. A., II, Heyner, D., Phillips, R. J., Winslow, R. M. and Solomon, S. C.

(2016). MESSENGER observations of induced magnetic fields in Mercury's core. *Geophys. Res. Lett.*, **43**, 2436–2444.

Kerber, L., Head, J. W., Solomon, S. C., Murchie, S. L., Blewett, D. T. and Wilson, L. (2009). Explosive volcanic eruptions on Mercury: Eruption conditions, magma volatile content, and implications for interior volatile abundances. *Earth Planet. Sci. Lett.*, **285**, 263–271.

Killen, R. M. and Ip, W.-H. (1999). The surface-bounded atmospheres of Mercury and the Moon. *Rev. Geophys.*, **37**, 361–406.

Killen, R. M., Burger, M. H., Cassidy, T. A., Sarantos, M., Vervack, R. J., Jr., McClintock, W. E., Merkel, A. W., Sprague, A. L. and Solomon, S. C. (2012). Mercury's Na exosphere from MESSENGER data. *Bull. Amer. Astron. Soc.*, **44**, abstract 401.01.

Klimczak, C. (2015). Limits on the brittle strength of planetary lithospheres undergoing global contraction. *J. Geophys. Res. Planets*, **120**, 2135–2151.

Klimczak, C., Watters, T. R., Ernst, C. M., Freed, A. M., Byrne, P. K., Solomon, S. C., Blair, D. M. and Head, J. W. (2012). Deformation associated with ghost craters and basins in volcanic smooth plains on Mercury: Strain analysis and implications for plains evolution. *J. Geophys. Res.*, **117**, E00L03, doi:10.1029/2012JE004100.

Klimczak, C., Ernst, C. M., Byrne, P. K., Solomon, S. C., Watters, T. R., Murchie, S. L., Preusker, F. and Balcerski, J. A. (2013). Insights into the subsurface structure of the Caloris basin, Mercury, from assessments of mechanical layering and changes in long-wavelength topography. *J. Geophys. Res. Planets*, **118**, 2030–2044.

Klimczak, C., Byrne, P. K. and Solomon, S. C. (2015). A rock-mechanical assessment of Mercury's global tectonic fabric. *Earth Planet. Sci. Lett.*, **416**, 82–90.

Koehn, P. L., Zurbuchen, T. H., Gloeckler, G., Lundgren, R. A. and Fisk, L. A. (2002). Measuring the plasma environment at Mercury: The Fast Imaging Plasma Spectrometer. *Meteorit. Planet. Sci.*, **37**, 1173–1189.

Korth, H., Anderson, B. J., Raines, J. M., Slavin, J. A., Zurbuchen, T. H., Johnson, C. L., Purucker, M. E., Winslow, R. M., Solomon, S. C. and McNutt, R. L., Jr. (2011). Plasma pressure in Mercury's equatorial magnetosphere derived from MESSENGER Magnetometer observations. *Geophys. Res. Lett.*, **38**, L22201, doi:10.1029/2011GL049451.

Korth, H., Anderson, B. J., Johnson, C. L., Winslow, R. M., Slavin, J. A., Purucker, M. E., Solomon, S. C. and McNutt, R. L., Jr. (2012). Characteristics of the plasma distribution in Mercury's equatorial magnetosphere derived from MESSENGER Magnetometer observations. *J. Geophys. Res.*, **117**, A00M07, doi:10.1029/2012JA018052.

Korth, H., Anderson, B. J., Gershman, D. J., Raines, J. M., Slavin, J. A., Zurbuchen, T. H., Solomon, S. C. and McNutt, R. L., Jr. (2014). Plasma distribution in Mercury's magnetosphere derived from MESSENGER Magnetometer and Fast Imaging Plasma Spectrometer observations. *J. Geophys. Res. Space Physics*, **119**, 2917–2932.

Lawrence, D. J., Feldman, W. C., Goldsten, J. O., Maurice, S., Peplowski, P. N., Anderson, B. J., Bazell, D., McNutt, R. L., Jr., Nittler, L. R., Prettyman, T. H., Rodgers, D. J., Solomon, S. C. and Weider, S. Z. (2013). Evidence for water ice near Mercury's north pole from MESSENGER Neutron Spectrometer measurements. *Science*, **339**, 292–296.

Lawrence, D. J., Anderson, B. J., Baker, D. N., Feldman, W. C., Ho, G. C., Korth, H., McNutt, R. L., Jr., Peplowski, P. N., Solomon, S. C., Starr, R. D. Vandegriff, J. D. and Winslow, R. M. (2015). Comprehensive survey of energetic electron events in Mercury's magnetosphere with data from the MESSENGER Gamma-Ray and Neutron Spectrometer. *J. Geophys. Space Physics*, **120**, 2851–2876.

Lawrence, D. J., Peplowski, P. N., Beck, A. W., Feldman, W. C., Frank, E. A., McCoy, T. J., Nittler, L. R. and Solomon, S. C. (2017). Compositional terranes on Mercury: Information from fast neutrons. *Icarus*, **281**, 32–45.

Leary, J. C., Conde, R. F., Dakermanji, G., Engelbrecht, C. S., Ercol, C. J., Fielhauer, K. B., Grant, D. G., Hartka, T. J., Hill, T. A., Jaskulek, S. E., Mirantes, M. A., Mosher, L. E., Paul, M. V., Persons, D. F., Rodberg, E. H., Srinivasan, D. K., Vaughan, R. M. and Wiley, S. R. (2007). The MESSENGER spacecraft. *Space Sci. Rev.*, **131**, 187–217.

Leblanc, F. and Johnson, R. E. (2003). Mercury's sodium exosphere. *Icarus*, **164**, 261–281.

Lewis, J. S. (1988). Origin and composition of Mercury. In *Mercury*, ed. F. Vilas, C. R. Chapman, and M. S. Matthews. Tucson, AZ: University of Arizona Press, pp. 651–669.

Margot, J.-L., Peale, S. J., Jurgens, R. F., Slade, M. A. and Holin, I. V. (2007). Large longitude libration of Mercury reveals a molten core. *Science*, **316**, 710–714.

Margot, J.-L., Peale, S. J., Solomon, S. C., Hauck, S. A., II, Ghigo, F. D., Jurgens, R. F., Yseboodt, M., Giorgini, J. D., Padovan, S. and Campbell, D. B. (2012). Mercury's moment of inertia from spin and gravity data. *J. Geophys. Res.*, **117**, E00L09, doi:10.1029/2012JE004161.

McAdams, J. V., Dunham, D. W., Farquhar, R. W., Taylor, A. H. and Williams, B. G. (2005). Trajectory design and maneuver strategy for the MESSENGER mission to Mercury. *Spaceflight Mechanics 2005, Adv. Astronaut. Sci.*, **120**, Part II, 1185–1204.

McAdams, J. V., Farquhar, R. W., Taylor, A. H. and Williams, B. G. (2007). MESSENGER mission design and navigation. *Space Sci. Rev.*, **131**, 219–246.

McAdams, J. V., Moessner, D. P., Williams, K. E., Taylor, A. H., Page, B. R. and O'Shaughnessy, D. J. (2011). MESSENGER – Six primary maneuvers, six planetary flybys, and 6.6 years to Mercury orbit. *Astrodynamics 2011: Part III, Adv. Astronaut. Sci.*, **142**, 2191–2210.

McAdams, J. V., Solomon, S. C., Bedini, P. D., Finnegan, E. J., McNutt, R. L., Jr., Calloway, A. B., Moessner, D. P., Wilson, M. W., Gallagher, D. T., Ercol, C. J. and Flanigan, S. H. (2012). MESSENGER at Mercury: From orbit insertion to first extended mission. Presented at the *63rd International Astronautical Congress*, paper IAC-12-C1.5.6, 11 pp., Naples, Italy, 1–5 October.

McAdams, J. V., Bryan, C. G., Moessner, D. P., Page, B. R., Stanbridge, D. R. and Williams, K. E. (2014). Orbit design and navigation through the end of MESSENGER's extended mission at Mercury. *Space Flight Mechanics 2014: Part III, Adv. Astronaut. Sci.*, **152**, 2299–2318.

McAdams, J. V., Bryan, C. G., Bushman, S. S., Calloway, A. B., Carranza, E., Flanigan, S. H., Kirk, M. N., Korth, H., Moessner, D. P., O'Shaughnessy, D. J. and Williams, K. E. (2015). Engineering MESSENGER's grand finale at Mercury: The low-altitude hover campaign. *Astrodynamics Specialist Conference*, American Astronautical Society, paper AAS 15–634, 20 pp., Vail, CO., 9–13 August.

McClintock, W. E. and Lankton, M. R. (2007). The Mercury Atmospheric and Surface Composition Spectrometer for the MESSENGER mission. *Space Sci. Rev.*, **131**, 481–522.

McNutt, R. L., Jr., Solomon, S. C., Gold, R. E., Leary, J. C. and the MESSENGER team (2006). The MESSENGER mission to Mercury: Development history and early mission status. *Adv. Space Res.*, **38**, 564–571.

Merkel, A. W., McClintock, W. E., Sarantos, M., Cassidy, T. A., Vervack, R. J., Jr., Burger, M. H., Killen, R. M., Sprague, A. L. and Solomon, S. C. (2012). Seasonal variability and local time dependence of Mercury's dayside magnesium exosphere.

Presented at 2012 Fall Meeting, American Geophysical Union, abstract P33B-1929, San Francisco, CA, 3–7 December.

Merkel, A. W., Cassidy, T. A., Vervack, R. J., Jr., McClintock, W. E., Sarantos, M., Burger, M. H. and Killen, R. M. (2017). Seasonal variations of Mercury's magnesium dayside exosphere from MESSENGER observations. *Icarus*, **281**, 46–54.

Moses, J. I., Rawlins, K., Zahnle, K. and Dones, L. (1999). External sources of water for Mercury's putative ice deposits. *Icarus*, **137**, 197–221.

Murchie, S. L., Klima, R. L., Denevi, B. W., Ernst, C. M., Keller, M. R., Domingue, D. L., Blewett, D. T., Chabot, N. L., Hash, C. D., Malaret, E., Izenberg, N. R., Vilas, F., Nittler, L. R., Gillis-Davis, J. J., Head, J. W. and Solomon, S. C. (2015). Orbital multispectral mapping of Mercury with the MESSENGER Mercury Dual Imaging System: Evidence for the origins of plains units and low-reflectance material. *Icarus*, **254**, 287–305.

Namur, O., Collinet, M., Charlier, B., Grove, T. L., Holtz, F. and McCammon, C. (2016). Melting processes and mantle sources of lavas on Mercury. *Earth Planet. Sci. Lett.*, **439**, 117–128.

Ness, N. F., Behannon, K. W., Lepping, R. P. and Whang, Y. C. (1976). Observations of Mercury's magnetic field. *Icarus*, **28**, 479–488.

Neumann, G. A., Cavanaugh, J. F., Sun, X., Mazarico, E. M., Smith, D. E., Zuber, M. T., Mao, D., Paige, D. A., Solomon, S. C., Ernst, C. M. and Barnouin, O. S. (2013). Bright and dark polar deposits on Mercury: Evidence for surface volatiles. *Science*, **339**, 296–300.

Nittler, L. R., Starr, R. D., Weider, S. Z., McCoy, T. J., Boynton, W. V., Ebel, D. S., Ernst, C. M., Evans, L. G., Goldsten, J. O., Hamara, D. K., Lawrence, D. J., McNutt, R. L., Jr., Schlemm, C. E., II, Solomon, S. C. and Sprague, A. L. (2011). The major-element composition of Mercury's surface from MESSENGER X-ray spectrometry. *Science*, **333**, 1847–1851.

Paige, D. A., Wood, S. E. and Vasavada, A. R. (1992). The internal stability of water ice at the poles of Mercury. *Science*, **258**, 643–646.

Paige, D. A., Siegler, M. A., Harmon, J. K., Neumann, G. A., Mazarico, E. M., Smith, D. E., Zuber, M. T., Harju, E., Delitsky, M. L. and Solomon, S. C. (2013). Thermal stability of volatiles in the north polar region of Mercury. *Science*, **339**, 300–303.

Peale, S. J. (1976). Does Mercury have a molten core? *Nature*, **262**, 765–766.

Peale, S. J., Phillips, R. J., Solomon, S. C., Smith, D. E. and Zuber, M. T. (2002). A procedure for determining the nature of Mercury's core. *Meteorit. Planet. Sci.*, **37**, 1269–1283.

Peplowski, P. N., Evans, L. G., Hauck, S. A., II, McCoy, T. J., Boynton, W. V., Gillis-Davis, J. J., Ebel, D. S., Goldsten, J. O., Hamara, D. K., Lawrence, D. J., McNutt, R. L., Jr., Nittler, L. R., Solomon, S. C., Rhodes, E. A., Sprague, A. L., Starr, R. D. and Stockstill-Cahill, K. R. (2011). Radioactive elements on Mercury's surface from MESSENGER: Implications for the planet's formation and evolution. *Science*, **333**, 1850–1852.

Peplowski, P. N., Lawrence, D. J., Rhodes, E. A., Sprague, A. L., McCoy, T. J., Denevi, B. W., Evans, L. G., Head, J. W., Nittler, L. R., Solomon, S. C., Stockstill-Cahill, K. R. and Weider, S. Z. (2012). Variations in the abundances of potassium and thorium on the surface of Mercury: Results from the MESSENGER Gamma-Ray Spectrometer. *J. Geophys. Res.*, **117**, E00L04, doi:10.1029/2012JE004141.

Peplowski, P. N., Lawrence, D. J., Feldman, W. C., Goldsten, J. O., Bazell, D., Evans, L. G., Head, J. W., Nittler, L. R., Solomon, S. C. and Weider, S. Z. (2015). Geochemical terranes of Mercury's northern hemisphere as revealed by MESSENGER neutron measurements. *Icarus*, **253**, 346–353.

Peplowski, P. N., Klima, R. L., Lawrence, D. J., Ernst, C. M., Denevi, B. W., Frank, E. A., Goldsten, J. O., Murchie, S. L., Nittler, L. R. and Solomon, S. C. (2016). Remote sensing evidence for an ancient carbon-bearing crust on Mercury. *Nature Geosci.*, **9**, 273–276.

Potter, A. and Morgan, T. (1985). Discovery of sodium in the atmosphere of Mercury. *Science*, **229**, 651–653.

Potter, A. E. and Morgan, T. H. (1986). Potassium in the atmosphere of Mercury. *Icarus*, **67**, 336–340.

Preusker, F., Oberst, J., Head, J. W., Watters, T. R., Robinson, M. S., Zuber, M. T. and Solomon, S. C. (2011). Stereo topographic models of Mercury after three MESSENGER flybys. *Planet. Space Sci.*, **59**, 1910–1917.

Prockter, L. M., Ernst, C. M., Denevi, B. W., Chapman, C. R., Head, J. W., Fassett, C. I., Merline, W. J., Solomon, S. C., Watters, T. R., Strom, R. G., Cremonese, G., Marchi, S. and Massironi, M. (2010). Evidence for young volcanism on Mercury from the third MESSENGER flyby. *Science*, **329**, 668–671.

Raines, J. M., Gershman, D. J., Zurbuchen, T. H., Sarantos, M., Slavin, J. A., Gilbert, J. A., Korth, H., Anderson, B. J., Gloeckler, G., Krimigis, S. M., Baker, D. N., McNutt, R. L., Jr. and Solomon, S. C. (2013). Distribution and compositional variations of plasma ions in Mercury's space environment: The first three Mercury years of MESSENGER observations. *J. Geophys. Res. Space Physics*, **118**, 1604–1619.

Raines, J. M., Gershman, D. J., Slavin, J. A., Zurbuchen, T. H., Korth, H., Anderson, B. J. and Solomon, S. C. (2014). Structure and dynamics of Mercury's magnetospheric cusp: MESSENGER measurements of protons and planetary ions. *J. Geophys. Res. Space Physics*, **119**, 6587–6602.

Robinson, M. S., Murchie, S. L., Blewett, D. T., Domingue, D. L., Hawkins, S. E., III, Head, J. W., Holsclaw, G. M., McClintock, W. E., McCoy, T. J., McNutt, R. L., Jr., Prockter, L. M. Solomon, S. C. and T. R. Watters, T. R. (2008). Reflectance and color variations on Mercury: Regolith processes and compositional heterogeneity. *Science*, **321**, 66–69.

Santo, A. G., Gold, R. E., McNutt, R. L., Jr., Solomon, S. C., Ercol, C. J., Farquhar, R. W., Hartka, T. J., Jenkins, J. E., McAdams, J. V., Mosher, L. E., Persons, D. F., Artis, D. A., Bokulic, R. S., Conde, R. F., Dakermanji, G., Goss, M. E., Jr., Haley, D. R., Heeres, K. J., Maurer, R. H., Moore, R. C., Rodberg, E. H., Stern, T. G., Wiley, S. R., Williams, B. G., Yen, C. L. and Peterson, M. R. (2001). The MESSENGER mission to Mercury: Spacecraft and mission design. *Planet. Space Sci.*, **49**, 1481–1500.

Schlemm, C. E., II, Starr, R. D., Ho, G. C., Bechtold, K. E., Hamilton, S. A., Boldt, J. D., Boynton, W. V., Bradley, W., Fraeman, M. E., Gold, R. E., Goldsten, J. O., Hayes, J. R., Jaskulek, S. E., Rossano, E., Rumpf, R. A., Schaefer, E. D., Strohbehn, K., Shelton, R. G., Thompson, R. E., Trombka, J. I. and Williams, B. D. (2007). The X-Ray Spectrometer on the MESSENGER spacecraft. *Space Sci. Rev.*, **131**, 393–415.

Schubert, G., Ross, M. N., Stevenson, D. J. and Spohn, T. (1988). Mercury's thermal history and the generation of its magnetic field. In *Mercury*, ed. F. Vilas, C. R. Chapman and M. S. Matthews. Tucson, AZ: University of Arizona Press, pp. 429–460.

Siegfried, R. W., II and Solomon, S. C. (1974). Mercury: Internal structure and thermal evolution. *Icarus*, **23**, 192–205.

Slade, M. A., Butler, B. J. and Muhleman, D. O. (1992). Mercury radar imaging: Evidence for polar ice. *Science*, **258**, 635–640.

Slavin, J. A., Acuña, M. A., Anderson, B. J., Baker, D. N., Benna, M., Gloeckler, G., Gold, R. E., Ho, G. C., Killen, R. M., Korth, H., Krimigis, S. A., McNutt, R. L., Jr., Nittler, L. R., Raines, J. M., Schriver, D., Solomon, S. C., Starr, R. D., Trávníček, P. and Zurbuchen, T. H. (2008). Mercury's magnetosphere after MESSENGER's first flyby. *Science*, 321, 85–89.

Slavin, J. A., Anderson, B. J., Baker, D. N., Benna, M., Boardsen, S. A., Gold, R. E., Ho, G. C., Imber, S. M., Korth, H., Krimigis, S. M.,

McNutt, R. L., Jr., Raines, J. M., Sarantos, M., Schriver, D., Solomon, S. C., Trávníček, P. and Zurbuchen, T. H. (2012a). MESSENGER and Mariner 10 flyby observations of magnetotail structure and dynamics at Mercury. *J. Geophys. Res.*, **117**, A01215, doi:10.1029/2011JA016900.

Slavin, J. A., Imber, S. M., Boardsen, S. A., DiBraccio, G. A., Sundberg, T., Sarantos, M., Nieves-Chinchilla, T., Szabo, A., Anderson, B. J., Korth, H., Zurbuchen, T. H., Raines, J. M., Johnson, C. L., Winslow, R. M., Killen, R. M., McNutt, R. L., Jr. and Solomon, S. C. (2012b). MESSENGER observations of a flux-transfer-event shower at Mercury. *J. Geophys. Res.*, **117**, A00M06, doi:10.1029/2012JA017926.

Slavin, J. A., DiBraccio, G. A., Gershman, D. J., Imber, S. M., Poh, G. K., Raines, J. M. Zurbuchen, T. H., Jia, X., Baker, D. N., Glassmeier, K.-H., Livi, S. A., Boardsen, S. A., Cassidy, T. A., Sarantos, M., Sundberg, T., Masters, A., Johnson, C. L., Winslow, R. M., Anderson, B. J., Korth, H., McNutt, R. L., Jr. and Solomon, S. C. (2014). MESSENGER observations of Mercury's dayside magnetosphere under extreme solar wind conditions. *J. Geophys. Res. Space Physics*, **119**, 8087–8116.

Smith, D. E., Zuber, M. T., Phillips, R. J., Solomon, S. C., Hauck, S. A., II, Lemoine, F. G., Mazarico, E., Neumann, G. A., Peale, S. J., Margot, J.-L., Johnson, C. L., Torrence, M. H., Perry, M. E., Rowlands, D. D., Goossens, S., Head, J. W. and Taylor, A. H. (2012). Gravity field and internal structure of Mercury from MESSENGER. *Science*, **336**, 214–217.

Solomon, S. C. (2003). Mercury: The enigmatic innermost planet. *Earth Planet. Sci. Lett.*, **216**, 441–455.

Solomon, S. C., McNutt, R. L., Jr., Gold, R. E., Acuña, M. H., Baker, D. N., Boynton, W. V., Chapman, C. R., Cheng, A. F., Gloeckler, G., Head, J. W., III, Krimigis, S. M., McClintock, W. E., Murchie, S. L., Peale, S. J., Phillips, R. J., Robinson, M. S., Slavin, J. A., Smith, D. E., Strom, R. G., Trombka, J. I. and Zuber, M. T. (2001). The MESSENGER mission to Mercury: Scientific objectives and implementation. *Planet. Space Sci.*, **49**, 1445–1465.

Sprague, A. L., Hunten, D. M. and Lodders, K. (1995). Sulfur at Mercury, elemental at the poles and sulfides in the regolith. *Icarus*, **118**, 211–215.

Sprague, A. L., Schmitt, W. J. and Hill, R. E. (1998). Mercury: Sodium atmosphere enhancements, radar-bright spots, and visible surface features. *Icarus*, **136**, 60–68.

Spudis, P. D. and Guest, J. E. (1988). Stratigraphy and geologic history of Mercury. In *Mercury*, ed. F. Vilas, C. R. Chapman and M. S. Matthews. Tucson, AZ: University of Arizona Press, pp. 118–164.

Srinivasan, D. K., Perry, M. E., Fielhauer, K. B., Smith, D. E. and Zuber, M. T. (2007). The radio frequency subsystem and radio science on MESSENGER. *Space Sci. Rev.*, **131**, 557–571.

Srnka, L. J. (1976). Magnetic dipole moment of a spherical shell with TRM acquired in a field of internal origin. *Phys. Earth Planet. Inter.*, **11**, 184–190.

Stanley, S., Bloxham, J., Hutchison, W. E. and Zuber, M. T. (2005). Thin shell dynamo models consistent with Mercury's weak observed magnetic field. *Earth Planet. Sci. Lett.*, **234**, 27–38.

Starr, R. D., Schriver, D., Nittler, L. R., Weider, S. Z., Byrne, P. K., Ho, G. C., Rhodes, E. A., Schlemm, C. E., II, Solomon, S. C. and Trávníček, P. M. (2012). MESSENGER detection of electron-induced X-ray fluorescence from Mercury's surface. *J. Geophys. Res.*, **117**, E00L02, doi:10.1029/2012JE004118.

Starukhina, L. (2001). Water detection on atmosphereless celestial bodies: Alternative explanations of the observations. *J. Geophys. Res.*, **106**, 14701–14710.

Stephenson, A. (1976). Crustal remanence and the magnetic moment of Mercury. *Earth Planet. Sci. Lett.*, **28**, 454–458.

Stevenson, D. J. (1987). Mercury's magnetic field: A thermoelectric dynamo? *Earth Planet. Sci. Lett.*, **82**, 114–120.

Strom, R. G. (1979). Mercury: A post-Mariner 10 assessment. *Space Sci. Rev.*, **24**, 3–70.

Strom, R. G., Trask, N. J. and Guest, J. E. (1975). Tectonism and volcanism on Mercury, *J. Geophys. Res.*, **80**, 2478–2507.

Sundberg, T., Boardsen, S. A., Slavin, J. A., Anderson, B. J., Korth, H., Zurbuchen, T. H., Raines, J. M. and Solomon, S. C. (2012a). MESSENGER orbital observations of large-amplitude Kelvin-Helmholtz waves at Mercury's magnetopause. *J. Geophys. Res.*, **117**, A04216, doi:10.1029/2011JA017268.

Sundberg, T., Slavin, J. A., Boardsen, S. A., Anderson, B. J., Korth, H., Ho, G. C., Schriver, D., Uritsky, V. M., Zurbuchen, T. H., Raines, J. M., Baker, D. N., Krimigis, S. M., McNutt, R. L., Jr. and Solomon, S. C. (2012b). MESSENGER observations of dipolarization events in Mercury's magnetotail. *J. Geophys. Res.*, **117**, A00M03, doi:10.1029/2012JA017756.

Vilas, F. (1988). Surface composition of Mercury from reflectance spectrophotometry. In *Mercury*, ed. F. Vilas, C. R. Chapman and M. S. Matthews. Tucson, AZ: University of Arizona Press, pp. 59–76.

Vasavada, A. R., Paige, D. A. and Wood, S. E. (1999). Near-surface temperatures on Mercury and the Moon and the stability of polar ice deposits. *Icarus*, **141**, 179–193.

Vervack, R. J., Jr., McClintock, W. E., Killen, R. M., Sprague, A. L., Burger, M. H., Merkel, A. W. and Sarantos, M. (2011). MESSENGER searches for less abundant or weakly emitting species in Mercury's exosphere. Presented at 2011 Fall Meeting, American Geophysical Union, abstract P44A-02, San Francisco, CA, 5–9 December.

Watters, T. R., Head, J. W., Solomon, S. C., Robinson, M. S., Chapman, C. R., Denevi, B. W., Fassett, C. I., Murchie, S. L. and Strom, R. G. (2009a). Evolution of the Rembrandt impact basin on Mercury. *Science*, **324**, 618–621.

Watters, T. R., Solomon, S. C., Robinson, M. S., Head, J. W., André, S. L., Hauck, S. A., II and Murchie, S. L. (2009b). The tectonics of Mercury: The view after MESSENGER's first flyby. *Earth Planet. Sci. Lett.*, **285**, 283–296.

Watters, T. R., Solomon, S. C., Klimczak, C., Freed, A. M., Head, J. W., Ernst, C. M., Blair, D. M., Goudge, T. A. and Byrne, P. K. (2012). Extension and contraction within volcanically buried impact craters and basins on Mercury. *Geology*, **40**, 1123–1126.

Weidenschilling, S. J. (1978). Iron/silicate fractionation and the origin of Mercury. *Icarus*, **35**, 99–111.

Weidenschilling, S. J. (1998), Mercury's polar radar anomalies: Ice and/or cold rock? *Lunar Planet. Sci.*, **29**, abstract 1278.

Weider, S. Z., Nittler, L. R., Starr, R. D., McCoy, T. J., Stockstill-Cahill, K. R., Byrne, P. K., Denevi, B. W., Head, J. W. and Solomon, S. C. (2012). Chemical heterogeneity on Mercury's surface revealed by the MESSENGER X-Ray Spectrometer. *J. Geophys. Res.*, **117**, E00L05, doi:10.1029/2012JE004153.

Weider, S. Z., Nittler, L. R., Starr, R. D., McCoy, T. J. and Solomon, S. C. (2014). Variations in the abundance of iron on Mercury's surface from MESSENGER X-Ray Spectrometer observations. *Icarus*, **235**, 170–186.

Weider, S. Z., Nittler, L. R., Starr, R. D., Crapster-Pregont, E. J., Peplowski, P. N., Denevi, B. W., Head, J. W., Byrne, P. K., Hauck, S. A., II, Ebel, D. S. and S. C. Solomon (2015). Evidence of geochemical terranes on Mercury: Global mapping of major elements with MESSENGER's X-Ray Spectrometer. *Earth Planet. Sci. Lett.*, **416**, 109–120.

Weider, S. Z., Nittler, L. R., Murchie, S. L., Peplowski, P. N., McCoy, T. J., Kerber, L., Klimczak, C., Ernst, C. M., Goudge, T. A., Starr, R. D., Izenberg, N. R., Klima, R. L. and S. C. Solomon (2016). Evidence from MESSENGER for sulfur- and carbon-driven explosive volcanism on Mercury. *Geophys. Res. Lett.*, **43**, 3653–3661.

Wetherill, G. W. (1988). Accumulation of Mercury from planetesimals. In *Mercury*, ed. F. Vilas, C. R. Chapman and M. S. Matthews. Tucson, AZ: University of Arizona Press, pp. 670–691.

Whitten, J. L., Head, J. W., Denevi, B. W. and Solomon, S. C. (2014). Intercrater plains on Mercury: Insight into unit definition, characterization, and origin from MESSENGER datasets. *Icarus*, **241**, 97–113.

Wilhelms, D. E. (1976). Mercurian volcanism questioned. *Icarus*, **28**, 551–558.

Winslow, R. M., Johnson, C. L., Anderson, B. J., Korth, H., Slavin, J. A., Purucker, M. E. and Solomon, S. C. (2012). Observations of Mercury's northern cusp region with MESSENGER's Magnetometer. *Geophys. Res. Lett.*, **39**, L08112, doi:10.1029/2012GL051472.

Zuber, M. T., Aharonson, O., Aurnou, J. M., Cheng, A. F., Hauck, S. A., II, Heimpel, M. H., Neumann, G. A., Peale, S. J., Phillips, R. J., Smith, D. E., Solomon, S. C. and Stanley, S. (2007). The geophysics of Mercury: Current status and anticipated insights from the MESSENGER mission. *Space Sci. Rev.*, **131**, 105–132.

Zuber, M. T., Smith, D. E., Phillips, R. J., Solomon, S. C., Neumann, G. A., Hauck, S. A., II, Peale, S. J., Barnouin, O. S., Head, J. W., Johnson, C. L., Lemoine, F. G., Mazarico, E., Sun, X., Torrence, M. H., Freed, A. M., Klimczak, C., Margot, J.-L., Oberst, J., Perry, M. E., McNutt, R. L., Jr., Balcerski, J. A., Michel, N., Talpe, M. J. and Yang, D. (2012). Topography of the northern hemisphere of Mercury from MESSENGER laser altimetry. *Science*, **336**, 217–221.

Zurbuchen, T. H., Raines, J. M., Slavin, J. A., Gershman, D. J., Gilbert, J. A., Gloeckler, G., Anderson, B. J., Baker, D. N., Korth, H., Krimigis, S. M., Sarantos, M., Schriver, D., McNutt, R. L., Jr. and S. C. Solomon, S. C. (2011). MESSENGER observations of the spatial distribution of planetary ions near Mercury. *Science*, **333**, 1862–1865.

2

The Chemical Composition of Mercury

LARRY R. NITTLER, NANCY L. CHABOT, TIMOTHY L. GROVE, AND PATRICK N. PEPLOWSKI

2.1 INTRODUCTION

Chemical composition is a fundamental property of any planetary body. The bulk composition is determined by starting materials (the compositions of which are set by processes occurring in the Sun's protoplanetary disk) as well as secondary processes that may modify it, such as impacts with other bodies. Only materials at or near the surface of a planet, i.e., within its crust, are amenable to direct compositional measurements; interior (and hence bulk) compositional information must be inferred from indirect measurements and geophysical and/or geochemical considerations. The surface composition is influenced both by the bulk composition and geological processes such as planetary differentiation and volcanism as well as modification by impacts of other bodies. Prior to the MESSENGER mission, there were only a very few constraints on Mercury's elemental composition, namely the planet's anomalously high density, its surface reflectance properties, and the elements detected by ground-based astronomy in its weak exosphere.

Mercury's uncompressed (zero-pressure) density is 5.4 g/cm^3 (Anderson et al., 1987), much larger than those of the other terrestrial planets (e.g., 4.4 g/cm^3 for Earth and Venus). This high density clearly indicates a higher abundance of at least one heavy phase, of which metallic iron is by far the most plausible candidate given its high abundance in the bulk solar composition. Thus, it has long been recognized that Mercury must have a metallic iron-rich core that makes up a larger mass fraction than the cores of the other terrestrial planets (Siegfried and Solomon, 1974). The existence of a (partially molten) iron-rich core was also suggested by the Mariner 10 discovery that Mercury has an internal magnetic field (Chapter 5), though an active dynamo was not required to explain the limited magnetic field data returned by that spacecraft.

Despite the implied high ratio of iron to silicate in the bulk composition, ground-based observations of Mercury have long suggested relatively low amounts of FeO in surface silicates. For example, near-infrared spectral absorption features associated with FeO in the silicate minerals olivine and pyroxene are commonly seen in lunar and asteroidal reflectance spectra but are weak or absent in Mercury spectra. Pre-MESSENGER spectral measurements (summarized in Chapter 8) generally indicated that surface silicates contained no more than ~3–5 wt% FeO, and possibly much less. Comparison of mercurian spectra with lunar spectra led some researchers to conclude that Mercury's surface may be rich in anorthositic rocks (dominated by the plagioclase mineral anorthite, CaAl$_2$Si$_2$O$_8$), like the lunar highlands, and thus might have high Al and low Fe contents (McCord and Adams, 1972; Blewett et al., 1997).

Mariner 10 observations also revealed that Mercury is surrounded by a very thin atmosphere, a surface-based exosphere (Chapters 14 and 15); the spacecraft detected H, He, and possibly O via ultraviolet spectroscopy. Ground-based observations later indicated the presence of Na, K, and Ca (Potter and Morgan, 1985, 1986; Bida et al., 2000). Whereas the H and He most likely originate from the solar wind, the Na, K, and Ca were recognized as likely to be derived from surface materials, removed via one or more processes (e.g., sputtering), thus indicating that these elements are present at Mercury's surface, although micrometeorite material might also contribute to the exosphere. However, uncertainties regarding the partitioning between exogenous and indigenous sources for exosphere constituents meant that these observations could not be used to infer surface composition information.

Armed with these compositional constraints as well as evolving theoretical concepts of planet formation, pre-MESSENGER researchers suggested a wide range of possible bulk and/or surface compositions for Mercury (Goettel, 1988; Taylor and Scott, 2003; Chapter 18). By and large, the most popular models to explain the planet's anomalously high density (e.g., giant impacts, evaporation by an early active Sun) invoked high temperatures, so it was broadly expected that Mercury would be a volatile-depleted world (the observations of volatile Na and K in the planet's exosphere notwithstanding). However, some volatile-rich model compositions were considered (e.g., Morgan and Anders, 1980; Goettel, 1988), as were models involving metal-rich and/or highly chemically reduced meteorites (Wasson, 1988; Burbine et al., 2002; Taylor and Scott, 2003). Wurz et al. (2010) derived estimates of Mercury's surface mineralogical and elemental composition from a combination of reflectance spectroscopic and exosphere observations.

The advent of the MESSENGER mission marked the possibility of testing these models with direct compositional measurements of Mercury's surface. The MESSENGER payload included three sensors for determining surface elemental composition (Section 2.2): the X-Ray Spectrometer (XRS), the Gamma-Ray Spectrometer (GRS), and the Neutron Spectrometer (NS). The GRS and NS were integrated into a single instrument, but the sensors were independent in terms of placement on the spacecraft, functionality, and operation. During MESSENGER's flybys of Mercury in 2008–2009, compositional data over equatorial regions were obtained with both the GRS and NS. The GRS observations yielded the unsurprising discovery that silicon was present on Mercury's surface as well as upper limits on the abundances of Fe, Ti, K, and Th (Rhodes et al., 2011). The NS measurements indicated a surprisingly high level of neutron-absorbing elements (Lawrence et al., 2010) compared with expectations from modeled bulk compositions. By analogy with lunar neutron measurements and interpretation of MESSENGER Mercury flyby data (Denevi et al., 2009), the neutron data were interpreted as indicating relatively high surface abundances of Fe and Ti, postulated to be in the form of oxide minerals such as ilmenite ($FeTiO_3$). With the beginning of MESSENGER's orbital mission phase in March 2011, geochemical measurements of Mercury's surface with all three geochemical sensors began in earnest, and quickly revealed the planet to have a volatile-rich surface with a surprisingly high abundance of S and very low abundances of Fe and Ti (Nittler et al., 2011; Peplowski et al., 2011). This chapter summarizes the post-MESSENGER state of knowledge of Mercury's chemical composition, starting with the measured surface abundances of a wide range of elements and followed by estimates for the mantle and core compositions informed by the surface data and geochemical and petrological considerations. We conclude by putting this information together to estimate the bulk composition of Mercury and place it into the context of planet formation models. A much more detailed discussion of the origin of Mercury, constrained in part by the compositional data discussed here, can be found in Chapter 18. Implications of the measured surface composition and its heterogeneity for the geological history of Mercury are discussed throughout the book, especially in Chapters 6, 7, 8, 10, 11, 12, 18, and 19.

2.2 MESSENGER'S GEOCHEMICAL SENSORS

MESSENGER's three geochemical sensors, the XRS, GRS, and NS, all exploited the interactions of high-energy radiation with the surface to remotely obtain information about surface composition through fundamental principles of atomic and nuclear physics.

The XRS relied on the method of remote sensing X-ray fluorescence (XRF), whereby X-rays emitted by hot plasma and active regions low in the Sun's corona interact with atoms at an airless body's surface to induce emission of X-rays with energy characteristic of the fluorescing element. X-ray remote sensing is a surface technique, in that the detected X-rays come from within the top tens of micrometers of the planet's surface. Prior to MESSENGER, this method was successfully used to measure major-element surface compositions of the Moon (Adler et al., 1972a, b; Narendranath et al., 2011; Weider et al., 2012a) and the asteroid 433 Eros (Trombka et al., 2000; Nittler et al., 2001). The XRS, described in detail by Schlemm et al. (2007) and Starr et al. (2016), consisted of three collimated gas-filled proportional counter detectors to detect X-rays from Mercury and a silicon-based detector pointed at the Sun to simultaneously measure the solar X-ray spectrum (critical for converting the planetary spectra into elemental abundances). A "balanced-filter" approach (Adler et al., 1972b) was used to deconvolve Mg, Al, and Si signals, as these X-ray lines are not resolved by the proportional counters (Figure 2.1a). The 1–10-keV energy range of the instrument allowed measurement of the major and minor rock-forming elements Mg, Al, Si, S, Ca, Ti, Cr, Mn, and Fe. Because of MESSENGER's highly eccentric, near-polar orbit (Chapter 1), the spatial resolution of XRS measurements varied enormously by both latitude and time in the mission (due to the varying altitude of the spacecraft at periapsis), with measurement "footprint" sizes ranging from a few tens to hundreds of kilometers in the northern hemisphere and up to a few thousand kilometers over the south pole, observed when MESSENGER was at its farthest point in its orbit. Moreover, during typical (quiescent) solar conditions, the solar X-ray flux was sufficient to induce XRF from only Mg, Al, and Si; solar flares were required to enable measurements of heavier elements (Figure 2.1a). Since flares occur only sporadically and MESSENGER spent only a small fraction of each orbit over Mercury's northern hemisphere, the spatial coverage for the heavier elements in the northern hemisphere is incomplete. The methods used to analyze the XRS data and generate elemental abundance maps have been described in detail by Nittler et al. (2011) and Weider et al. (2012b, 2014, 2015).

The GRS detected gamma rays in the energy range 60 keV to 9 MeV, emitted both by the natural decay of radioactive isotopes and by nuclei through interactions with galactic cosmic rays (GCRs, i.e., high-energy particles, mostly protons, streaming through space and most likely originating in supernova explosions throughout the galaxy). The GCRs interact with near-surface materials (to depths of a few meters), liberating neutrons from atomic nuclei. These neutrons subsequently interact with other nuclei, placing them in higher-energy excited states that subsequently decay via the emission of gamma rays at element-characteristic energies (Figure 2.1b). Compared with the shallow origin of planetary X-rays, however, the detected gamma rays arise from the top tens of centimeters of the surface. The GRS (Goldsten et al., 2007) consisted of a mechanically cooled high-purity crystal of germanium to detect gamma rays and measure their energy, surrounded on three sides by a plastic anticoincidence shield (ACS). The ACS was used to discriminate background signals in the Ge detector from the gamma-ray signals of interest coming from Mercury. The mechanical cooler for the Ge crystal functioned nominally for about 14 months into the orbital mission before it failed, consistent with its expected lifetime, ending collection of gamma-ray data from Mercury early in MESSENGER's first extended mission (Chapter 1). Fortunately, a large amount of useful gamma-ray data was acquired prior to the failure of the cooler. The GRS data were used to characterize C, O, Na, Al, Si, K, S, Cl, Ca, Fe, Th, and U concentrations on Mercury's surface; results and analysis procedures have been described in detail by Peplowski et al. (2011, 2012a, b, 2014, 2015a) and Evans et al. (2012, 2015).

The NS characterized neutron emissions from Mercury in three broad energy bands: fast (~0.5 MeV to ~7 MeV), epithermal (~1 eV to ~500 keV), and thermal (~0.025 eV to ~1 eV). Fast neutrons are produced directly by GCR interactions with

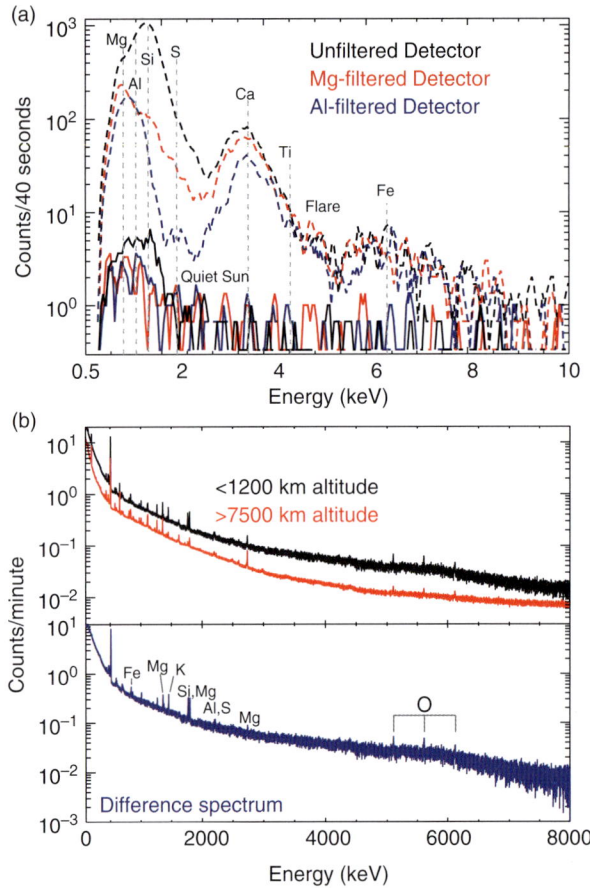

Figure 2.1. (a) Example X-ray spectra from Mercury recorded by the three planet-facing detectors on MESSENGER's XRS. Spectra acquired during typical quiet-Sun conditions (solid curves) show fluorescence from Mg, Al, and Si, whereas spectra during solar flares (dashed curves) have much higher signal-to-background ratios and show fluorescence from elements up to Fe. Adapted from Weider et al. (2015).
(b) Gamma-ray spectra measured by MESSENGER's GRS during the first two months of orbital operations. The high-altitude (>7500 km; red) spectrum samples instrument background, and the low-altitude (<1200 km; black) spectrum samples both background and signals from Mercury. The difference spectrum (blue) highlights the Mercury component. Several strong gamma-ray lines are labeled by element; the prominent line at 511 keV is due to electron–positron annihilation.

atomic nuclei, and the fast neutron flux measured by NS is proportional to the average atomic mass ($\langle A \rangle$) of the regolith. Epithermal neutrons are the result of downscattering (slowing down) of fast neutrons via inelastic scattering, a process that is most efficient in the presence of hydrogen. As a consequence, the epithermal neutron flux is highly sensitive to hydrogen, and by inference water, content. Finally, thermal neutron emissions are a balance between downscattering and neutron absorption, and the latter is sensitive to the bulk concentration of neutron-absorbing elements in the regolith. The NS consisted of three scintillator detectors: a central borated plastic detector that was sensitive to epithermal and fast neutrons and two Li-glass detectors that were sensitive to epithermal and thermal neutrons (Goldsten et al., 2007). MESSENGER NS data were used to

confirm the presence of large amounts of water ice in permanently shadowed polar impact craters (Lawrence et al., 2013; Chapter 13). Thermal neutron measurements, made late in the orbital mission when the spacecraft flew at very low periapsis altitudes (Chapter 1), were used to infer the presence of carbon in Mercury's regolith (Section 2.3.3; Peplowski et al., 2016). Apart from H and C (see Section 2.3.3), MESSENGER neutron measurements were not used to investigate individual elemental abundances on Mercury, and we focus in this chapter on XRS and GRS measurements of specific elements. However, neutron measurements acquired with both the NS and the GRS-ACS, which was re-purposed to provide measurements of neutrons complementary to those of the NS, have been used to derive important information on compositional variability across Mercury's surface, such as the definition of large-scale regions of distinct composition (Peplowski et al., 2015b; Lawrence et al., 2017; Chapter 7).

2.3 MERCURY'S SURFACE COMPOSITION

Mercury's average surface elemental composition, derived from XRS, GRS, and NS measurements, is summarized in Table 2.1. Select maps of elemental abundances or abundance ratios are shown in Figure 2.2. GRS measurements of radioactive elements (i.e., K, Th, and U) are directly quantifiable, so for these elements absolute abundances are reported. For most of the other measured elements, abundances are reported as ratios to that of silicon. This choice reflects the fact that elemental ratios are more easily measured by the XRS and GRS methods and ratioing reduces or eliminates some systematic uncertainties from the data. Moreover, Si tends to vary less than other major elements among typical rock types, so variations in a ratio such as Mg/Si tend to reflect primarily variations in Mg rather than Si. The GRS data support this statement, as Si gamma-ray emissions, and by extension Si concentrations, are observed to vary by just ~15% (the standard deviation of the measurements) across the surface (Peplowski et al., 2012a). For those element ratios measured by both XRS and GRS (Al/Si, Ca/Si, S/Si, and Fe/Si), there is generally good agreement in the average values determined by the two techniques, providing confidence in the results and suggesting that Mercury's surface composition is similar at depths of micrometers and tens of centimeters. Where there are differences outside one-standard-deviation errors (e.g., Fe/Si), these differences may reflect systematic errors in one or both of the techniques (e.g., Weider et al., 2014) or differences in sampling, due to depth or location. For example, GRS averages are over the northern hemisphere, whereas XRS maps for S/Si, Ca/Si, and Fe/Si are dominated by data in the southern hemisphere (Figure 2.2).

2.3.1 Volatile Elements

MESSENGER's geochemical observations definitively refuted the notion that the surface of Mercury is depleted in volatiles relative to the other terrestrial planets. Highly volatile species, including H and organics, are present but limited to Mercury's permanently shadowed regions in polar impact

Table 2.1. *Surface elemental composition of Mercury.*[a]

Element (ratio)	XRS	GRS[b]	NS[b]
K (ppm, average)		1288 ± 234	
K (ppm, range)		240–2500	
Th (ppm)		0.155 ± 0.054	
U (ppb)		90 ± 20	
K/Th		8000 ± 3200	
Mg/Si	0.436 (0.106)		
Al/Si	0.268 (0.048)	$0.29^{+0.05}_{-0.13}$	
S/Si	0.076 (0.019)	0.092 ± 0.015	
Ca/Si	0.165 (0.030)	0.24 ± 0.05	
Ti/Si	0.012 ± 0.001		
Cr/Si	0.006 ± 0.001		
Mn/Si	0.004 ± 0.001		
Fe/Si	0.053 (0.013)	0.077 ± 0.013	
Na/Si (average)		0.12 ± 0.01	
Na/Si (0–60°N)		0.107 ± 0.008	
Na/Si (80–90°N)		0.198 ± 0.030	
Cl/Si (average)		0.0057 ± 0.0010	
Cl/Si (0–60°N)		0.0049 ± 0.001	
Cl/Si (80–90°N)		0.014 ± 0.005	
O/Si		1.2 ± 0.1	
C (wt%)		1.4 ± 0.9	~1–4

[a] Ratios are by mass. Numbers in parentheses indicate the standard deviation (σ) of the XRS measurements, reflecting surface variability; ± symbol denotes the 1σ statistical uncertainty.

[b] GRS and NS data are from the northern hemisphere.
Data sources: Nittler et al. (2011, 2016); Peplowski et al. (2011, 2012b, 2014, 2015a, 2016); Evans et al. (2012, 2015); Weider et al. (2014, 2015); Frank et al. (2015); McCubbin et al. (2017)

craters (Chapter 13). Moderately volatile species, specifically Na, S, K, and Cl, are abundant and widespread (Nittler et al., 2011; Peplowski et al., 2011, 2012a, 2014; Weider et al., 2012b, 2015; Evans et al., 2015). The condensation temperatures (in kelvin) for these elements (Lodders and Fegley, 1998) – 970 (Na), 674 (S), 1000 (K), and 863 (Cl) – are sufficiently low to provide a sensitive test of the maximum temperatures experienced by Mercury and/or its precursor materials during planetary formation processes. Mercury's surface has elemental concentrations of Na, S, K, and Cl (Table 2.1, Figures 2.2 and 2.3) that are similar to those found on the surface of Mars, long regarded as the most volatile-rich planet in the inner solar system. Mercury's Na, S, and Cl abundances are higher than those observed on the volatile-depleted Moon by an order of magnitude or more. Mercury's K concentrations are similar to those found within lunar regions rich in KREEP (K, rare-earth elements, and P). However, lunar KREEP has elevated K levels as a result of the concentration of incompatible elements within the residues of fractional crystallization from a magma ocean, whereas on Mercury high K reflects a higher bulk volatile content, as discussed below.

Almost all elements detected on Mercury's surface exhibit clear spatial variability (Figure 2.2), and several "geochemical terranes" – regions with compositions distinct from their surroundings – have been defined (Peplowski et al., 2015b; Weider et al., 2015; Chapter 7). In principle, some of this heterogeneity for moderately volatile elements in Mercury's near-surface materials could arise from diffusion losses due to the extreme temperatures (up to ~700 K at places) experienced at the surface. Although the 50% condensation temperatures for the elements of interest (Na, S, K, Cl) are >670 K, diffusive loss can occur at substantially lower temperatures. For example, ~10-μm grains of Na-rich feldspars could lose sodium efficiently over a ~1-Gyr timescale at temperatures of ~400 K (Kasper, 1975; Giletti and Shanahan, 1997), and K loss occurs at a temperature of ~475 K. Peplowski et al. (2012a) demonstrated an inverse correlation between K concentrations and maximum near-surface temperature for regions experiencing temperatures >350 K, and they suggested the possibility that K was being lost from sufficiently warm (equatorial) regions and being redistributed to cooler regions or lost to space. However, Weider et al. (2015) noted a similar inverse correlation between K and non-volatile Mg, and they argued that the correlation between K and temperature is coincidental, on the grounds that Mg is not subject to thermal losses. Similarly, S concentrations are highly correlated with non-volatile Ca (Figure 2.4b), both being highest within Mercury's high-Mg region (Figure 2.2), an area at mid to equatorial latitudes (Chapter 7). Potassium, Na, and Cl all appear to have higher concentrations within Mercury's northern terrane (Peplowski et al., 2015b), which is generally associated with the large smooth plains unit that dominates Mercury's northernmost latitudes (Head et al., 2011; Chapter 6). These observations argue against thermal control of the spatial distribution of moderately volatile elements on Mercury, and instead, variability in K, Na, and Cl is attributed to differences in Mercury's geochemical terranes that result from their having been derived from compositionally distinct magmas, suggesting that Mercury's mantle is compositionally heterogeneous or that the parent magmas underwent distinct igneous evolutions (Section 2.4, Chapter 7).

While direct measurements have revealed that Mercury's surface is not depleted in moderately volatile elements, the question of Mercury's bulk volatile content is more complex. The absolute abundances of elements over a planet's surface can vary appreciably as a result of differences in melt generation and crustal emplacement, along with subsequent modification. As a result, the high concentrations of volatile elements observed in near-surface materials are not necessarily indicative of Mercury's bulk composition. The ratio of the moderately volatile lithophile element K relative to the refractory incompatible element Th is commonly adopted as a proxy for the bulk volatile content of a planet, as K/Th is thought to be preserved throughout igneous processing and so the value at the surface should reflect the bulk value. MESSENGER

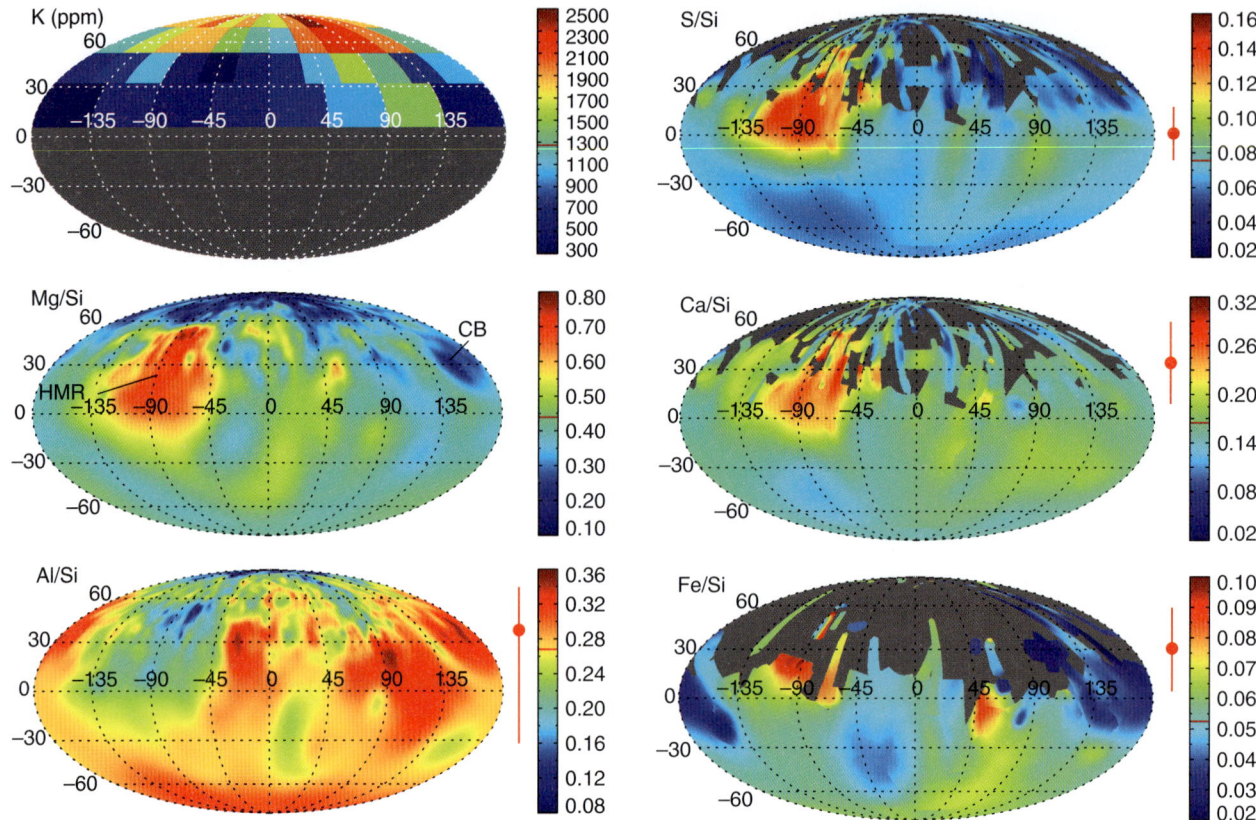

Figure 2.2. Elemental abundance and weight ratio maps for Mercury, derived from MESSENGER GRS (K; Peplowski et al., 2012a) and XRS (Weider et al., 2015; Nittler et al., 2016) data. Maps are shown in a Mollweide projection, centered on 0°N, 0°E. Red dots with one-standard-deviation errors are northern hemisphere averages from GRS (Table 2.1); red lines in color scales are area-weighted global averages of mapped data. HMR indicates the location of the high-Mg region, and CB is the Caloris impact basin.

revealed that Mercury's K/Th ratio is the highest observed in the inner solar system (Figure 2.3a), comparable to that of Mars and significantly higher than that for the volatile-depleted Moon (Peplowski et al., 2011, 2012a). This discovery argues against the simplistic expectation that volatile content in the solar system should vary inversely with solar distance (i.e., disk temperature). It also rules out many pre-MESSENGER theories for Mercury's formation that predicted a near-complete loss of volatiles from Mercury's surface and interior (Taylor and Scott, 2003; Chapter 18).

McCubbin et al. (2012) suggested that Mercury's high K/Th ratio may not be a reflection of its volatile content, but instead the result of sequestration of Th within Mercury's core. Specifically, they cited substantial uncertainties in elemental partitioning behavior under the highly reducing conditions inferred for Mercury (see Section 2.3.2). Although this issue is open, Mercury's Cl content can provide additional insights into Mercury's bulk volatile content. Like K, Cl is a moderately volatile lithophile element (Lodders, 2003), and as a result the Cl/K ratio at the surface is expected to reflect the bulk value. That Cl is more volatile than K makes it a test of even lower maximum temperatures than are probed by the K/Th ratio. Mercury's Cl/K value is similar to those of both carbonaceous chondrite meteorites (thought to represent unfractionated solar composition) and Mars, yet significantly higher than those of Earth and the Moon (Figure 2.3b; Evans et al., 2015). The subchondritic Cl/K values observed on Earth and the Moon are attributed to volatile loss resulting from protracted impact and accretion histories (Sharp and Draper, 2013). In contrast, Mars is thought to have formed rapidly, with limited subsequent removal of material by impacts (Dauphas and Pourmand, 2011). Evans et al. (2015) argued that the similarity between the Cl/K ratios of Mars and Mercury suggests that Mercury might have also had a rapid accretion and limited impact history.

Mercury's high bulk volatile contents have had a direct influence on the geological history of the surface of the planet. For example, pyroclastic volcanism was widespread on Mercury, with numerous source vents observed across the surface (Robinson and Lucey, 1997; Head et al., 2008; Kerber et al., 2009; Chapter 11). Pyroclastic volcanism requires one or more magmatic volatiles, and combined spectral, XRS, and NS data strongly indicate that these volatiles were S- and C-bearing (Weider et al., 2016; Chapters 8 and 11). There is also strong evidence for flood volcanism on Mercury (Head et al., 2011). Smooth plains units are overwhelmingly located within Mercury's northern hemisphere (Chapter 6), and the largest unit, the northern smooth plains, is generally collocated with the highest K concentrations found on the planet. This scenario mimics the Moon, where volcanic plains units (e.g., lunar maria) are concentrated on the lunar near side and are generally

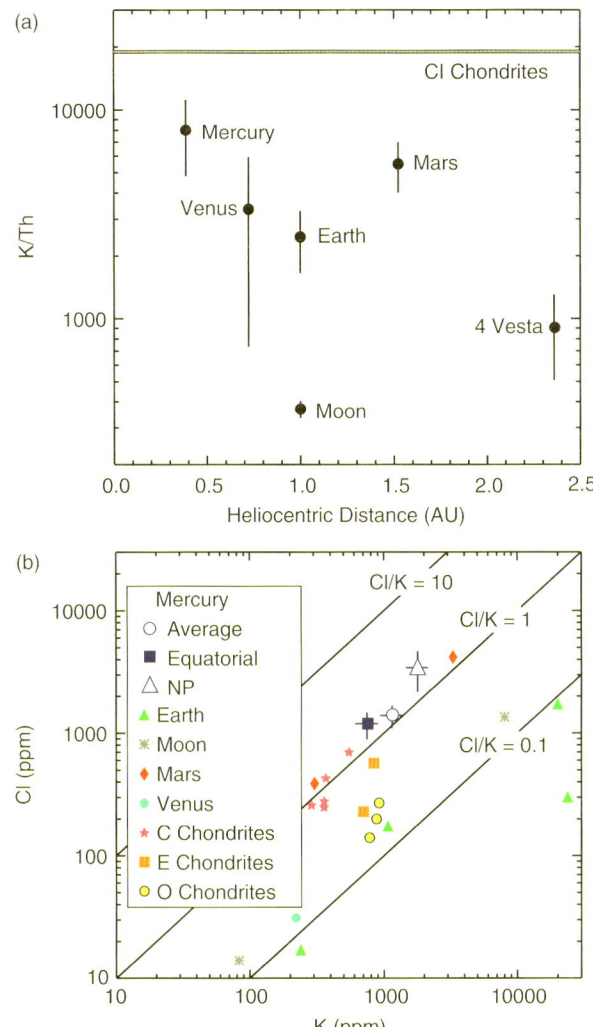
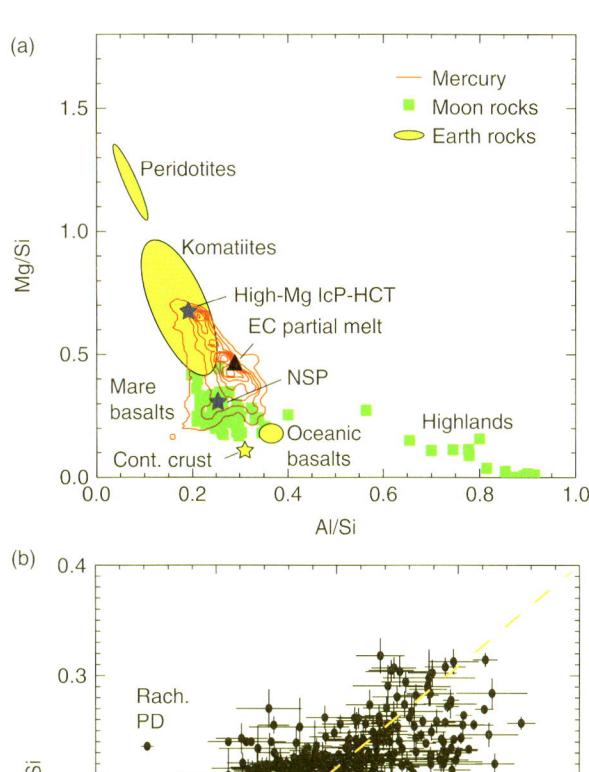
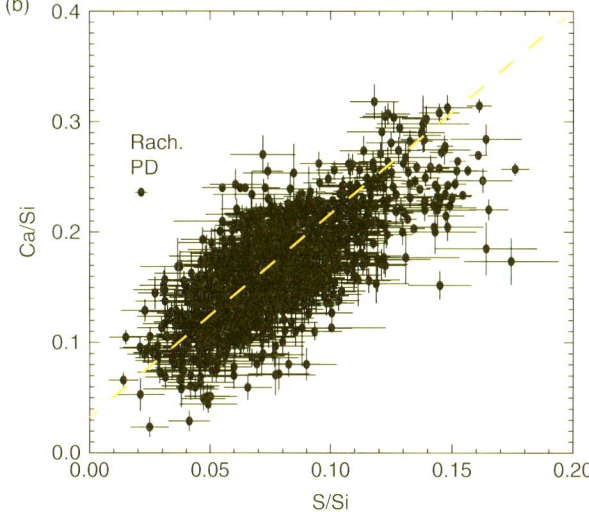

Figure 2.3. (a) K/Th mass ratios for inner solar system bodies as a function of heliocentric distance. Values and uncertainties are derived from Lodders and Fegley (1998), Prettyman et al. (2006), Taylor et al. (2006), Peplowski et al. (2012a), and Prettyman et al. (2015). (b) Mercury's surface Cl and K concentrations compared with those of other solar system objects. Adapted from Evans et al. (2015).

Figure 2.4. (a) Mg/Si versus Al/Si (elemental mass ratios) on Mercury compared with terrestrial and lunar rock compositions; after Nittler et al. (2011). Red contours indicate the distribution of Mercury values (from Figure 2.2), with the outer contour enclosing 99% of the data and other contours representing 5% increments. Blue stars indicate compositions modeled by Namur et al. (2016a); see Table 2.2. The black triangle is the composition of a 1425°C (29%) partial melt of the Indarch enstatite chondrite (McCoy et al., 1999). (b) Ca/Si versus S/Si (elemental weight ratios) for several hundred solar-flare XRS analyses with one-standard-deviation errors. Ca and S are highly correlated; the orange dashed line shows the best-fit line, which has a slope of 1.8. Rach. PD indicates a measurement of the pyroclastic volcanic deposit northeast of the Rachmaninoff impact basin. This deposit falls clearly off the correlation line, indicating loss of S. Adapted from Weider et al. (2016).

associated with high KREEP content. Although a KREEP-like material has not been identified on Mercury, the concentration of incompatible elements (K, Cl) in the northern terrane raises the possibility that regions with concentrated abundances of heat-producing radioactive elements such as K may have been a factor in powering Mercury's flood volcanism. In addition to their influence on volcanism, volatiles appear to play an ongoing key role in Mercury's recent geological evolution, in the formation of "hollows" (Blewett et al., 2011), shallow rimless depressions seen only on Mercury and discussed in detail in Chapter 12.

The pyroclastic deposits, some of which are as young as 1–3 Gyr (Goudge et al., 2014), and the even younger hollows provide evidence for recent and potentially ongoing release of volatiles from Mercury's interior. Relatively recently exposed volatiles would be subject to migration and cold trapping. Mercury's permanently shadowed regions (PSRs;

see Chapter 13) are known to host extensive deposits of water ice and possibly C-bearing organic compounds (Lawrence et al., 2013; Neumann et al., 2013; Paige et al., 2013; Chabot et al., 2014a). Impact gardening is expected to obscure PSR-sequestered volatiles with a burial rate of 0.4 cm/Myr (Crider and Killen, 2005); therefore, the observation of numerous exposed water ice deposits with sharp and presumably young boundaries leads to the possibility of recent, and perhaps ongoing, delivery of volatiles to the PSRs. It is possible that the polar deposits include some fraction of material derived from volatiles outgassed from Mercury's interior.

2.3.2 Non-Volatile Elements

XRS data acquired during the first few months of MESSENGER's orbital mission revealed Mercury's surface to have a major-element composition remarkably different from those of Earth and the Moon (Nittler et al., 2011). In addition to the surprising discovery that moderately volatile S is present at weight percent levels (Section 2.3.1, Table 2.1, Figure 2.2) compared with terrestrial crustal values of a few hundred parts per million (ppm), Mercury was found to have higher Mg/Si but lower Al/Si and Ca/Si ratios on average than the terrestrial or lunar crusts. Iron is present, but low, with an average abundance of ~1–2 wt% (Table 2.1) and a total observed range for Fe/Si of 0.02–0.1 (Weider et al., 2014). The relatively high Mg/Si and low Al/Si ratios of mercurian surface material (Figure 2.4a) rule out a plagioclase-dominated crust, as seen in the lunar highlands and previously suggested on the basis of spectral reflectance data (Section 2.1; Blewett et al., 1997). Although Mercury's surface composition was originally compared with that of terrestrial komatiites (Figure 2.4a; Nittler et al., 2011), due to the high Mg contents, subsequent petrologic modeling and experiments based on more complete elemental data sets indicated that the volcanic materials exposed on Mercury's surface are more similar to low-FeO terrestrial magnesian basalts and should be classified as norites (Stockstill-Cahill et al., 2012) or boninites (Vander Kaaden and McCubbin, 2016). As originally noted by Nittler et al. (2011), and confirmed by the full XRS data set for the entire mission, Ca/Si is highly correlated with S/Si (Figure 2.4b). Only the large pyroclastic deposit northeast of the Rachmaninoff impact basin has a composition clearly resolved from this correlation, and this departure is taken as evidence for loss of S-bearing volatiles during the explosive volcanism that generated the deposit (Section 2.3.1; Weider et al., 2016; Chapters 8 and 11).

Element maps (Figure 2.2) show that the surface of Mercury is remarkably heterogeneous chemically despite its relatively limited range of color and spectral properties (e.g., Chapter 8). Moreover, the geochemical terranes revealed by the elemental abundance maps are also seen, for example, in Mg/Si versus low-energy neutrons measured by the repurposed GRS-ACS (Section 2.2; Peplowski et al., 2015b) and in fast neutrons measured by the NS (Lawrence et al., 2017). The most chemically striking feature on the surface is a region of high Mg/Si, Ca/Si, S/Si, and Fe/Si, and low Al/Si situated at mid-northern to equatorial latitudes in the western hemisphere. In addition, the plains interior to Mercury's largest impact structure, the Caloris basin, are chemically distinct, with relatively high Al, but low Mg, Ca, S, Fe, and K abundances, compared with other smooth plains on the planet. The northern terrane, which includes much of the northern smooth plains (Head et al., 2011), is notably low in Mg, Al, Ca, and S, but enriched in the volatile elements K, Na, and Cl (Section 2.3.1). The presence of geochemical terranes on Mercury most likely points to melting of a heterogeneous mantle (Charlier et al., 2013; Weider et al., 2015; Namur et al., 2016a), as discussed in more detail in Chapter 7 (see also Section 2.4).

Nittler et al. (2011) noted that the major-element composition of Mercury's surface is similar, but not identical, to that found by partial melting of the highly reduced enstatite chondrite meteorites (McCoy et al., 1999; Burbine et al., 2002), and it is now recognized that the low Fe and high S contents of Mercury's surface are strong evidence of formation under highly chemically reducing conditions. There is decreasing incorporation of Fe and increasing incorporation of S into silicate melts as the availability of O goes down (Haughton et al., 1974; McCoy et al., 1999; Berthet et al., 2009; Namur et al., 2016b). The oxidation state of a magmatic system is often quantified in terms of oxygen fugacity (fO_2), which indicates the amount of O available to participate in chemical reactions and is commonly described on a logarithmic scale relative to an equilibrium reaction buffer. For example, the iron–wüstite buffer, denoted by IW, represents equilibrium in the reaction $Fe + \frac{1}{2}O_2 \rightarrow FeO$ (see Section 18.2). McCubbin et al. (2012) and Zolotov et al. (2013) used the MESSENGER S and Fe data to estimate that Mercury's interior has fO_2 some three to seven orders of magnitude below the iron–wüstite buffer (i.e., IW-3 to IW-7). More recently, Namur et al. (2016b) reported a large set of melting experiments under highly reducing conditions, and through comparison with the MESSENGER data argued that Mercury's interior has log fO_2 = IW-5.4 ± 0.4. These results indicate that Mercury is more reduced than any other planet and most known solar system materials, with important implications for its formation (Chapter 18). Some of the consequences of Mercury's low oxygen fugacity for mantle and core compositions are discussed in Sections 2.4 and 2.5, respectively.

The XRS measurements during solar flares also allowed detection of X-ray fluorescence from the minor elements Ti, Mn, and Cr. All three are clearly present on Mercury's surface, but in low abundances (<1 wt%; Nittler et al., 2011; Weider et al., 2014; Murchie et al., 2015). Average abundance ratios relative to Si are provided for these three elements in Table 2.1; because of their low abundances, these elements have larger statistical errors and are more susceptible to systematic uncertainties than the major elements measured by the XRS. The low abundance of Ti, along with the low bulk Fe abundance (see above), rules out the original interpretation of NS data acquired during MESSENGER's Mercury flybys (Section 2.1; Lawrence et al., 2010) as indicating substantial amounts of Fe–Ti oxides at the planet's surface. The high concentrations of thermal-neutron-absorbing elements observed in those measurements have subsequently been explained by the unexpectedly high Na and Cl content of Mercury's surface materials (Peplowski et al., 2015b).

The GRS has also provided measurements of the absolute concentrations of the naturally radioactive elements Th and U, which have mean northern hemisphere averages of 155 ± 54 and 90 ± 20 ppb, respectively. The decay of these elements, along with K (see Section 2.3.1), provides the primary source of long-lived heat generation for planetary interiors. Knowledge of K, Th, and U concentrations is therefore important for understanding the thermal and magmatic evolution of a planet. Directly relating the surface-measured concentrations to bulk values requires knowledge of the elemental partitioning behaviors of these elements, which are uncertain at the low fO_2 values predicted for Mercury. Nonetheless, the relative, time-varying concentrations of these elements can be characterized. This effort has shown that Mercury's initial heat production was higher than anticipated from pre-MESSENGER models of Mercury's composition, and that mantle heat production 4 Gyr ago was a few times larger than it is today (e.g., Peplowski et al., 2011; Tosi et al., 2013). These observations are consistent with the evidence for widespread volcanic activity that largely ended shortly after the end of the late heavy bombardment (Chapters 11 and 19). Mercury's Th/U ratio of about 2 is inconsistent with suggestions that U fractionated into Mercury's core (Malavergne et al., 2010). However, it has been suggested that the GRS-measured U concentrations may be overestimated as a consequence of radon transport and diffusion within Mercury's regolith (Meslin and Déprez, 2012), as U gamma rays originate from a number of U-decay products, including radon.

2.3.3 Oxygen and Carbon

Oxygen itself is the most abundant element in most silicate rocks, typically making up 40–50 wt%, but it is often difficult to measure quantitatively. Thus, for laboratory geochemical measurements, it is in many cases determined indirectly by assigning a certain amount of O to each major cation given knowledge of or assumptions about the oxidation state. Thus, for example, since Mg has an oxidation state of +2, Mg is assumed to be present in the chemical form MgO. Once the correct amount of O is assigned to every known element, the entire composition can be renormalized to 100%, including the O, or differences from 100% can be used to infer analytical errors, the presence of other elements, or incorrect assumptions regarding oxidation state. This procedure has been used both as part of the XRS data analysis procedures (Nittler et al., 2011) and in estimating bulk compositions of Mercury for petrologic modeling and experiments (Stockstill-Cahill et al., 2012; Charlier et al., 2013; Vander Kaaden and McCubbin, 2016; Vander Kaaden et al., 2017). However, O produces characteristic gamma rays and GRS data can thus be used to infer the surface O abundance (or the O/Si ratio) directly. Evans et al. (2012) reported a GRS-derived O/Si weight ratio of 1.4, lower than inferred from typical stoichiometry for measured cations (e.g., ~1.7; Lawrence et al., 2013). However, as the GRS methodology for O was not well calibrated at the time, Evans et al. (2012) cautioned against over-interpretation of their reported value. Subsequent refinement of the methodology for estimating O/Si yielded an even lower value of 1.2 ± 0.1 (McCubbin et al., 2017). Taken at face value, this anomalously low O/Si ratio would imply a very unusual surface mineralogy (Chapter 7). However, the low ratio may instead be due to secondary alteration of near-surface materials, for example by magmatic degassing in a process similar to industrial smelting (McCubbin et al., 2017), and hence may not be representative of the original composition of the lavas that formed the surface.

Mercury's surface is darker, on average, than that of the Moon (Robinson et al., 2008), and the darkest material on Mercury's surface (low-reflectance material or LRM) is associated with impact craters and their ejecta, suggesting that the reflectance-lowering component of Mercury's crust has an origin from some depth in the planet's crust (Denevi et al., 2009). On the Moon, low reflectance is largely driven by Fe- and Ti-bearing phases, but the discovery that surface abundances of Fe and Ti are low on Mercury (Section 2.3.2) ruled out the hypothesis that similar phases may be darkening that planet's surface. Murchie et al. (2015) carried out spectral modeling of MESSENGER reflectance data to show that C, in the form of fine-grained graphite, could match Mercury's global reflectance if present at ~1 wt% globally and ~5 wt% in the LRM. Peplowski et al. (2015a) developed a methodology to quantify C abundances from GRS spectra and found a mean northern hemisphere average C concentration of 1.4 ± 0.9 wt% (one-standard-deviation error). Adopting a three-standard-deviation threshold for positive identification led those authors to conclude that Mercury's surface has a mean C content of <4.1 wt%, consistent with the abundance needed to explain the planet's reflectance. The low altitudes reached by MESSENGER during its second extended mission allowed spatially resolved NS measurements of three deposits of LRM, all of which showed the neutron signature of enhanced C abundances (+1.1 to 3.1 wt%) relative to surrounding non-LRM materials (Peplowski et al., 2016), which have modeled C concentrations of ~1 wt%. These data, together with the spectral data of Murchie et al. (2015), convincingly demonstrate that weight-percent levels of C are present on Mercury. The association of higher C abundances with LRM, which is typically found within material excavated from depth by impact processes, indicates that the graphite is endogenic and may represent remnants of a primary graphite-rich crust formed by flotation in an early magma ocean and subsequently disrupted by impacts and buried by volcanism (Vander Kaaden and McCubbin, 2015) rather than solely by the addition to Mercury's surface of C from micrometeoroids (Bruck Syal et al., 2015). Moreover, although graphite itself is non-volatile, it is a potential source of volatile C-bearing compounds (e.g., through oxidation) which could contribute to pyroclastic volcanism and the formation of hollows (Section 2.3.1; Chapters 11, 12).

2.4 MERCURY'S MANTLE COMPOSITION

As discussed in the previous section, the MESSENGER mission provided the first quantitative information on the chemical compositions and on the compositional variability of the extensive volcanic materials that cover the surface of Mercury. Mercury's surface lavas are unlike any others observed in the solar system. They are very low in Fe (and essentially FeO-free), S-rich, and highly enriched in alkalis (Table 2.1). From crater size–frequency distributions it has been estimated that these volcanic deposits formed between 4.2 and 3.5 Ga (Marchi et al., 2013; Byrne et al., 2016; Chapters 6 and 9). The oldest volcanic rocks are found in the intercrater plains and heavily cratered terrain (IcP-HCT), and the youngest are found in smooth plains deposits, including the large expanse at high northern latitudes (Head et al., 2011; Chapter 6). A number of researchers have used petrologic modeling and experiments to infer candidate mineralogies and the melting conditions and mantle source compositions for the surface lavas (Stockstill-Cahill et al., 2012; Charlier et al., 2013; Namur et al., 2016a;

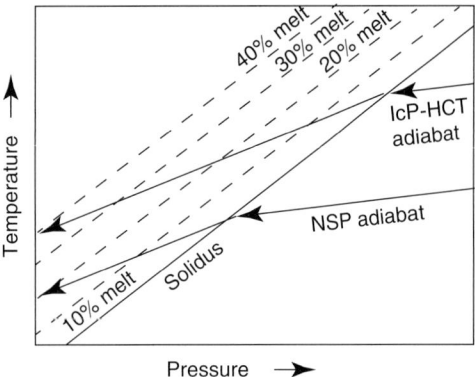

Figure 2.5. Schematic pressure–temperature diagram for partial melting of Mercury's mantle during adiabatic decompression. The IcP-HCT mantle adiabat is markedly hotter and crosses the solidus at a greater depth and higher temperature than that for the NSP. This higher temperature leads to a greater extent of melting. The solidus is the locus of pressure and temperature at which the mantle begins to partially melt. Dashed lines show the temperature at increasing extents of partial melting in 10% increments.

Vander Kaaden and McCubbin, 2016; Vander Kaaden et al., 2017). One approach to this methodology, discussed in detail in Chapter 7, relates various geochemical terranes to the overall geological history of Mercury. In this section, we use the recent results of Namur et al. (2016a) to estimate the bulk composition of Mercury's mantle.

Namur et al. (2016a) performed high-temperature (1320–1580°C), high-pressure (0.1–3 GPa) melting experiments under very reducing conditions (log fO_2 ~ IW-4 to IW-7; see Section 2.3.2) on synthetic analogs of the Mercury surface lavas. Compositions were chosen to correspond to the high-Mg portion of the IcP-HCT (Figure 2.2) and the low-Mg northern smooth plains (NSP). These compositions thus correspond to the "high Mg" and "northern terrane" compositions discussed in Chapter 7 and represent approximate end-members of Mercury's surface compositional range (Figure 2.4a). Namur et al. (2016a) mapped out a phase diagram for the selected compositions and identified multiple saturation points (points in pressure–temperature space at which two or more minerals are in equilibrium with the melt along the liquidus) involving forsterite + enstatite + liquid at 0.75 GPa and 1480°C for the high-Mg IcP-HCT composition and at 0.75 GPa and 1380°C for the low-Mg NSP composition. They also developed a thermodynamic model for the melting behavior of a lherzolitic source using experimental data from the system CMASN (CaO–MgO–Al$_2$O$_3$–SiO$_2$–Na$_2$O) and the MELTS/pMELTS algorithm (Ghiorso and Sack, 1995). The melting models predict an extent of melting (F = melt fraction) of 0.46 and 0.27 for the high-Mg IcP-HCT and NSP lavas, respectively.

The low multiple saturation point pressures indicate shallow depths of melt segregation (~60 km) within Mercury. The most straightforward mechanism for producing these high extents of melting would be batch adiabatic decompression melting. By this mechanism, a parcel of mantle ascends in a mantle plume or as a mobile blob (Figure 2.5), it cools slightly as its temperature follows the adiabatic gradient (0.3°C/km), and it begins to melt at the depth where the adiabat intersects the solidus. The primary control on melt production is the temperature of the upwelling mantle; this effect on melting is illustrated in Figure 2.5 for two different adiabatic paths. Magma is produced by the conversion of internal heat to heat of fusion. As a parcel of mantle continues to rise, melting continues and melt is extracted and transported upwards until a depth is reached at which the temperature of the parcel of mantle is lower than the solidus temperature for the residuum, i.e., the material remaining after melt is extracted. For Mercury lavas the melting rate is about 10 wt% per GPa change in pressure (or equivalent depth). For the high-Mg IcP-HCT lavas, melting would be predicted to begin at the base of Mercury's mantle near the core–mantle boundary at 400 km depth. For the NSP lavas, melting would commence at a depth of ~200 km. Therefore, the two end-member lavas sampled the mantle differently: one ascended from the base of the mantle, near the core–mantle boundary, and the second was the product of melting that started about halfway between the core–mantle boundary and the surface. In both situations, on the basis of multiple-saturation-point pressures, the batch melts produced by decompression melting segregated at the same depth (0.75 GPa, or ~60 km). The compositions of the solids in the residuum assemblage are from the Namur et al. (2016a) experiments, and they are similar in composition (Table 2.2). The proportions of enstatite and forsterite from the experiments are also similar for both compositions: 45 wt% forsterite and 55 wt% enstatite.

With the phase proportions and compositions of enstatite and forsterite from the Namur et al. (2016a) experiments, the original composition of the mantle from which the two melts were produced can be estimated for each element i from a mass balance equation:

$$W_i^{bulk} = FW_i^{melt} + (1-F)(\%Fo\ W_i^{Fo} + \%En\ W_i^{En})$$

where W_i^c is the weight fraction of element i in component c, F = melt fraction, %Fo = 0.45, the mass fraction of forsterite, and %En = 0.55, the mass fraction of enstatite in the residue.

Estimates of the original composition of the mantle sources of the high-Mg IcP-HCT and NSP lavas obtained from this mass balance calculation are given in Table 2.2 for elements considered by Namur et al. (2016a). The mantle compositions inferred from the two magma types are similar in SiO$_2$, Al$_2$O$_3$, and MgO contents, but they differ in CaO and Na$_2$O, suggesting that the mantle of Mercury is heterogeneous with respect to these elements. However, the sulfur content of the estimated silicate portion of Mercury's mantle has not been included in these mantle compositions. The high inferred CaO concentration of the high-Mg IcP-HCT lavas may be a consequence of the high S content of these lavas (4 wt%) and the possibility that Ca may be in part partitioned into an oldhamite (CaS) component (Section 2.3.2, Figure 2.4b). On the basis of the solubility of S in reduced silicate melts, Namur et al. (2016b) concluded that at Mercury's oxygen fugacity condition of IW-5.4 ± 0.4, the mantle of Mercury contains 7–11 wt% S.

Table 2.2. *Estimates of Mercury mantle silicate composition.*[a]

	SiO$_2$	TiO$_2$	Al$_2$O$_3$	FeO	MgO	CaO	Na$_2$O	K$_2$O
Mantle melts and residual minerals[b]								
NSP lava	58.7	0.40	13.8	0.04	13.9	5.81	7.0	0.20
IcP-HCT lava	52.7	0.40	8.79	0.04	27.8	7.27	2.75	0.08
Enstatite, NSP residuum	58.8	0.32	2.15	0.01	38.6	1.55	0.21	0
Enstatite, IcP residuum	59.3	0.07	0.65	0.01	37.2	1.44	0.07	0
Forsterite residuum	43.3			0.02	55.78	0.23		
Pre-melting mantle composition from mass balance								
NSP	53.67	0.24	4.57	0.02	36.89	2.26	1.97	0.05
IcP-HCT	51.98	0.21	4.24	0.03	37.64	3.84	1.29	0.04
Chondrite compositions								
Eagle (EL6)[c]	60.6	0.14	2.96	0	33.57	1.05	1.16	0.13
Eagle (EL6), Si reduced[d]	55.16	0.16	3.37	0	38.20	1.20	1.33	0.14
ALHA77295 (EH4)[c]	61.21	0.19	3.07	0	31.45	2.16	1.34	0.12
ALHA77295 (EH4), Si reduced[d]	51.98	0.21	3.50	0	35.83	2.46	1.52	0.14
Bencubbin (CB)[c]	61.55	0.25	2.47	0	32.58	2.00	1.02	0.13
Bencubbin (CB), Si reduced[d]	56.15	0.29	2.81	0	37.16	2.28	1.17	0.15

[a] Not including the possible presence of 7–11 wt% S (Namur et al., 2016b) or reduced Fe.
[b] From Namur et al. (2016a). NSP: northern smooth plains, IcP-HCT: intercrater plains and heavily cratered terrain; enstatite: MgSiO$_3$; forsterite: Mg$_2$SiO$_4$. Values in wt%.
[c] Silicate portion of analyses reported by Jarosewich (1990); Bencubbin is "Bencubbin II" analysis.
[d] Chondrite analyses from above, but with 20% of SiO$_2$ reduced to Si and assumed to have gone into the core.

Many researchers have suggested that enstatite chondrites and/or bencubbinites (E and CB chondrites, respectively) could be analogs to Mercury's precursor materials (Wasson, 1988; McCoy et al., 1999; Taylor and Scott, 2003; Brown and Elkins-Tanton, 2009; Malavergne et al., 2010; Nittler et al., 2011). These meteorites are reduced metal-rich rocks that share many geochemical characteristics with Mercury, including high S, Na, and K, but very little FeO in the silicates. The silicate portions of two E chondrites and a CB chondrite (Jarosewich, 1990) are shown in Table 2.2. They are higher in SiO$_2$ and lower in MgO than the Mercury mantle estimates inferred from the surface lavas. However, Si becomes more soluble in metallic melts under highly reducing conditions (Berthet et al., 2009; Malavergne et al., 2010; Chabot et al., 2014b; Namur et al., 2016b) and thus may be an important component of Mercury's core (Section 2.5). We therefore also removed 20% of the SiO$_2$ from the E and CB compositions, assuming that it was incorporated as Si metal into the core. When the chondrite compositions are renormalized for this reduction in SiO$_2$, the resulting silicate mantle has lower SiO$_2$ and higher MgO contents, much closer to those in the estimates of Mercury's mantle obtained from the surface lavas (Table 2.2). The compositions of these SiO$_2$-reduced chondrites and the estimates of the mantle source regions for the NSP and IcP-HCT lavas are compared in Figure 2.6. There is a remarkable similarity between the silicate residues and the estimated Mercury mantle compositions. The largest discrepancies are in the high CaO of the IcP-HCT mantle and the high Na$_2$O of the NSP mantle.

2.5 MERCURY'S CORE COMPOSITION

As has been known since the middle of the last century, Mercury's high density indicates the presence of a central metallic core that makes up a substantially larger fraction of the body than do the cores of the other terrestrial planets (Anderson et al., 1987). MESSENGER-derived estimates of Mercury's moments of inertia have been used to refine that value to a core mass fraction of ~69–77% (Hauck et al., 2013). Thus, Mercury's core has a substantial contribution to the bulk composition of the planet. However, determining the core composition of any planet is necessarily based on models, given that direct measurements are not possible. Even estimates of the composition of the best-characterized central metallic core in the solar system, the core of Earth, rely on models (e.g., Hillgren et al., 2000; McDonough, 2014; Li and Fei, 2014). One of the major constraints on these models comes from the fact that planetary cores make up a large fraction of their host planet, indicating that they must be composed of elements that are sufficiently abundant in the solar system to produce large cores. Nucleosynthesis processes result in Fe having much higher cosmochemical abundances than other heavy elements, and considerable seismic and laboratory evidence supports Fe as the dominant component of Earth's core (e.g., Jeanloz, 1990; Li and Fei, 2014). Nickel is less abundant than Fe but is present in iron meteorites, some of which are likely samples of disrupted cores of differentiated

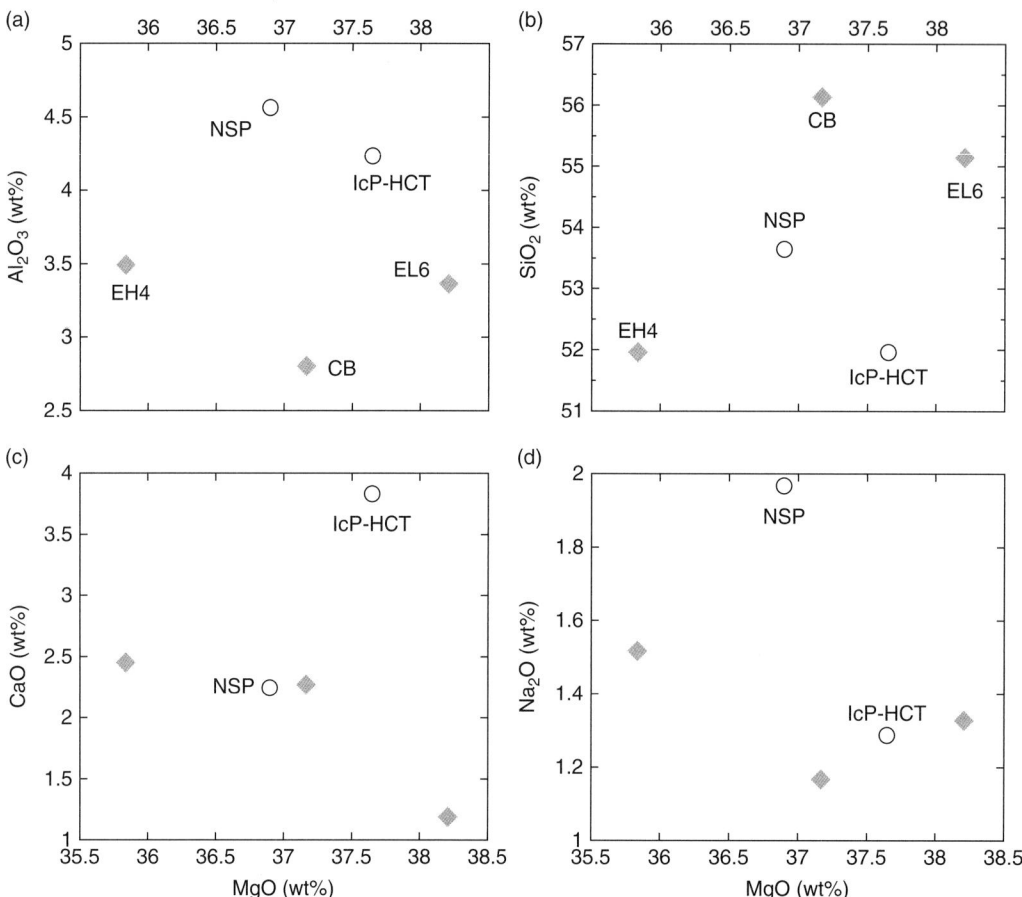

Figure 2.6. Estimated compositions (blue circles) of mantle sources of Mercury surface lavas from the high-Mg intercrater plains and heavily cratered terrain (IcP-HCT) and northern smooth plains (NSP) compared with compositions of the silicate portions of two enstatite and one Bencubbin-type chondrite with 20% of the SiO_2 removed and assumed to have been reduced to Si and sequestered in the core. Data from Table 2.2.

planetesimals, at concentrations that are generally ~5–15 wt% and can range up to even 50 wt% (Goldstein et al., 2009). Metal in the highly reduced aubrite meteorites, a potentially relevant analog given Mercury's reduced nature (Burbine et al., 2002), contains an average of 5 wt% Ni (McCoy and Bullock, 2017). A bulk compositional model of the Earth based on chondritic concentrations gives ~5 wt% Ni in Earth's core (McDonough, 2014), though given the similarities between Fe and Ni, more detailed constraints are limited (Li and Fei, 2014). By similar logic, Mercury's core is most likely composed of dominantly Fe with Ni perhaps at a level similar to its abundance in Earth, aubrites, or iron meteorites.

Mercury's core is not solid but rather at least the outer part is molten (Margot et al., 2007; Chapters 4 and 19), indicating on the basis of thermal models that Mercury's core is not pure Fe–Ni, since the temperatures reached within Mercury would have resulted in such a core having frozen solid by the present (Schubert et al., 1988). In contrast, phase relationships in Fe–Ni systems that also contain a "light-element" component, i.e., one or more elements lower in atomic mass than Fe and Ni, enable Fe–Ni alloys to remain as liquids to lower temperatures (e.g., Li and Fei, 2014). The observation that Mercury's outer core is presently molten indicates that it contains a light-element component, and Monte Carlo modeling has been used to show that a large range of light-element concentrations in Mercury's core can be consistent with the planet's radius, bulk density, and moment of inertia parameters (Hauck et al., 2013; Chapter 4). Thus, the identity and amount of the light-element component in Mercury's core are not highly constrained from measurements of the planet's bulk geophysical parameters, but they are important for understanding the planet's bulk composition, internal structure, and evolution, as well as the origin and history of its magnetic field (see, e.g., Chapters, 3, 4, 10, and 19).

Seismic data indicate that Earth's core also contains a light-element component in addition to Fe and Ni (e.g., Jeanloz, 1990). Potential major light elements that have been considered to account for the density of Earth's core are H, C, O, Si, and S (e.g., Hillgren et al., 2000), due to their cosmochemical abundances and their potential affinities to alloy with Fe–Ni. Additionally, iron meteorites contain sulfides and carbides (Buchwald, 1975), and iron meteorite trace-element trends support the presence of S during crystallization (e.g., Chabot and Haack, 2006), suggesting that S and C were incorporated into asteroidal cores and should be considered as potential light-element components in larger planetary cores. Even though the large majority of Earth's P is predicted to be in its core and iron meteorites contain phosphides, P has a lower abundance in the solar system, and only 0.2 wt% P is estimated to be in Earth's core (McDonough, 2014). Hence, P is not considered as a major light-element component. Applying a similar approach to evaluate the light-element component of Mercury's core, we consider here H, C, O, Si, and S.

Hydrogen is the lightest and most abundant element in the universe, and a small amount of H in the core by mass would result in a considerable amount of H when evaluated in atomic fractions. Current consideration of H in Earth's core involves incorporation of H into metal during core formation in a hydrous magma ocean or from reaction at Earth's core–mantle boundary (e.g., Fukai, 1984; Okuchi, 1997; Williams and Hemley, 2001; Terasaki et al., 2012). However, the pressure, temperature, and oxygen fugacity conditions discussed for the incorporation of H into Earth's core are much higher than could be experienced by Mercury currently, given its reduced nature and core–mantle boundary pressure of ~5.5 GPa (Hauck et al., 2013). On the other hand, one hypothesis to explain Mercury's large core fraction involves an impact between bodies comparable or larger in mass than Mercury (see Chapter 18); we discuss the potential implications of such a scenario in more detail after evaluating other potential light-element core constituents.

The measurement of ~1–5 wt% C associated with specific geological units on Mercury's surface (Section 2.3.3) indicates that C was likely available in Mercury's interior during the formation of Mercury's core. Carbon alloys strongly with Fe under a wide range of conditions, and experiments on metal–silicate mixtures with application to early Earth show that more C partitions into the metallic phase than into the silicate phase by a factor of ~200–5000 (Dasgupta et al., 2013; Chi et al., 2014). Thus, the identification of C within the silicate portion of Mercury could imply that Mercury's core contains a substantial amount of C (up to a few wt%, Lord et al., 2009). However, recent work (Li et al., 2015) has also shown that C partitions less strongly into the metallic phase at reducing conditions (log fO_2 = IW-4.7) and for Si-bearing metals (10 wt%). Such experiments still contain ~200 times more C in the metallic phase than in the silicate phase, but preliminary experiments at even more reducing conditions with higher Si contents in the metal suggest that the solubility of C in metal may decrease further (Li et al., 2016; Vander Kaaden et al., 2016), such that C may not be a major light-element component in Mercury's core if it formed under highly reducing conditions. One implication of scenarios with high amounts of C in both Mercury's silicate portion and metallic core is that it would imply a bulk composition for Mercury that is substantially more C-rich than the other terrestrial planets. Alternatively, the high levels of C measured on Mercury's surface may indicate that most C did not partition into Mercury's core and that Mercury experienced different conditions during core formation than those of Earth, perhaps extremely reducing. The proposed hypothesis of graphite as a stable phase in Mercury's early history (Vander Kaaden and McCubbin, 2015) could sequester much, but not all, of Mercury's C budget if conditions were such that Mercury's core was C saturated and graphite was stable during Mercury's core formation.

Oxygen is an abundant element in the solar system and is a key component of the silicate portions of all of the terrestrial planets. At elevated temperatures, the solubility of O in metal increases. The high-pressure and high-temperature conditions of an early magma ocean on Earth could result in ~2–8 wt% O in Earth's core and the subsequent reduction of Earth's mantle (Rubie et al., 2004, 2015; Siebert et al., 2013; Tsuno et al., 2013; Fischer et al., 2015). In contrast, Mars is thought to not have experienced sufficiently high temperatures during its core formation to result in O in the martian core (Rubie et al., 2004), though reactions at the core–mantle boundary have been proposed to produce ~3 wt% O in the core of Mars (Tsuno et al., 2011). Thus, a body of Mercury's current size would not be predicted to have a significant amount of O in the core.

Silicon is also an abundant and key element in the rocky portions of terrestrial planets, though under reducing conditions it also alloys strongly with Fe metal (Kilburn and Wood, 1997; Hillgren et al., 2000; Gessmann et al., 2001; Malavergne et al., 2004). Thus, the reduced nature of Mercury's surface, as indicated by its low apparent FeO content (Section 2.1), led to the suggestion that Si is a major component of Mercury's core even prior to orbital results from MESSENGER (Malavergne et al., 2010), and the surface compositional results discussed above in Section 2.3, especially the high S abundance, provided further evidence of Mercury's reduced nature and the possibility of abundant Si in its core (Hauck et al., 2013; Chabot et al., 2014b; Malavergne et al., 2014; Namur et al., 2016b). Specific predictions of the amount of Si in Mercury's core are assessed in more detail below.

As discussed in Section 2.3.3, the discovery that Mercury's surface has several wt% S led to estimates of the highly reduced nature of the planet (oxygen fugacity of IW-3 to IW-7; McCubbin et al., 2012; Zolotov et al., 2013), since the solubility of S in silicate is known to increase with decreasing oxygen fugacity (e.g., Haughton et al., 1974; Berthet et al., 2009). For log fO_2 levels lower than approximately IW-4, more S partitions into the silicate phase than the metallic phase (Kilburn and Wood, 1997; Berthet et al., 2009), though the precise partitioning behavior is also dependent on the pressure, temperature, and composition conditions (Boujibar et al., 2014; Namur et al., 2016b). Thus, the presence of high levels of S on Mercury's surface could indicate that Mercury's core formed under reducing conditions and that S is not a major component of the core. Considerations of the cosmochemical abundance and volatility of S led Dreibus and Palme (1996) to conclude that even though the majority of Earth's S is believed to be in the core, Earth's core cannot contain more than 1.7 wt% S. Consequently, if Mercury's silicate fraction and its core both contain wt% levels of S, the bulk composition of Mercury could be more S-rich than Earth.

Having introduced each potential light element, we now return to a consideration of the implications for Mercury's core composition if the modern planet is only a smaller remnant of a once larger body, such as predicted if Mercury's large core is a result of a large impact event. Proposed impact scenarios include a proto-Mercury that was a factor of 2–5 more massive than the present planet involved in an impact with either a body ~20–40% the mass of current Mercury (Benz et al., 2007) or one much larger, at 85% the mass of Earth, in a hit-and-run collision

(Asphaug and Reufer, 2014). All of these bodies are considered sufficiently large to have differentiated and formed central metallic cores prior to being involved in an impact event; the compositions of such cores are highly unconstrained and may have formed within bodies ranging in size from 20% of the mass of Mercury to 85% of the mass of Earth. During and following the impact event, the simulations of both Benz et al. (2007) and Asphaug and Reufer (2014) seem to show some level of mixing between the two cores to create Mercury's final core. The large-scale chemical evolution that would follow such a large impact event has not been examined. The energetics suggest that complete melting of the planet would have been possible, but the level of equilibration that would occur between a central metallic core largely present prior to the giant impact and the small fraction of silicate remaining is not clear, even if a global magma ocean was formed. Thus, if Mercury's large core is the result of a giant impact event, it may be even more challenging to constrain Mercury's core composition from measurements of its present silicate surface.

Though our current knowledge of Mercury cannot enable a unique identification of the core's composition, investigations that model the geophysical and geochemical evolution of the planet can place valuable constraints on abundances of light elements. From early MESSENGER measurements of Mercury's gravity field together with Earth-based measurements of Mercury's spin axis position and physical libration amplitude, Smith et al. (2012) suggested that a solid layer of FeS might be located at the top of the core, overlying a S- and Si-bearing fluid core. Subsequent revision to Mercury's obliquity (Margot et al., 2012) and modeling by Hauck et al. (2013) of Mercury's internal structure showed that a solid FeS layer was consistent with but not required by Mercury's geophysical parameters. Hauck et al. (2013) found numerous core composition solutions that ranged from 0–20 wt% S to 0–17 wt% Si and combinations of the two elements; extremely S-rich core compositions of >25 wt% were also examined but were not favored because of the high S/Fe ratio that would be implied for the planet in comparison with all other known solar system materials. Overall, Hauck et al. (2013) concluded that Mercury's core likely contains a substantial quantity of Si and/or S.

Taking a geochemical approach, Malavergne et al. (2010) examined the core composition that would result from the differentiation of a reduced, metal-rich (EH or CB, see Section 2.4) chondrite bulk composition. They concluded that Mercury's core likely contains at least 5 wt% Si and is composed of at least two distinct liquids, one Fe–Si with almost no S and another Fe–S with almost no Si. Following up on this work with the incorporation of MESSENGER's surface composition measurements, Malavergne et al. (2014) concluded that sulfides likely played a major role in Mercury's differentiation, both by creating a sulfide layer at the top of the core and by enabling sulfides to be stable within the silicate portion of the planet.

In an alternative geochemical approach, Chabot et al. (2014b) performed metal–silicate melting experiments to examine the combined partitioning behavior of both S and Si under a range of oxygen fugacity conditions. They found that metallic melts with a range of S and Si contents could be in equilibrium with a silicate melt with wt% levels of S, as measured by MESSENGER for Mercury's surface. The experiments illustrated that constraints could be placed on the specific combinations of S and Si that were consistent with equilibrium core formation in a magma ocean scenario. They concluded that Mercury's core likely contains Si, limiting the S content of the core to <20 wt%, and for core Si contents >10 wt% they predicted that <2 wt% S is also in the core.

Most recently, Namur et al. (2016b) experimentally examined the solubility of S in reduced silicate melts, using compositions and oxygen fugacity conditions of relevance to Mercury on the basis of MESSENGER results. Combining their extensive experimental results with those of previous studies, they developed a parameterization expression to model the S concentrations in Mercury's magmas, mantle, and core. They concluded that Mercury's core has <1.5 wt% S and is Si-bearing, and that a thin, <90-km-thick layer of either molten or solid FeS-rich material may be present at the core–mantle boundary, depending on Mercury's bulk S content.

Other potential constraints on the composition of Mercury's core may be derived from modeling the thermal evolution and associated global contraction of the planet (see also Chapter 10). The abundance of large scarps and other contractional geological features on Mercury's surface as measured from MESSENGER data indicates that the planet's radius has decreased by as much as 7 km since the late heavy bombardment (Byrne et al., 2014). Pre-MESSENGER models, based on a smaller estimate for the total global contraction and considering only S as a potential light element in the core, were used to make estimates of the core's composition and indicated $\gtrsim 6$ wt% S (Hauck et al., 2004; Grott et al., 2011). Similar models that include the possibility of Si in the core and account for Mercury's larger measured contraction have the potential to provide new constraints on Mercury's core composition.

Similarly, the presence of a dynamo-driven modern magnetic field (Ness et al., 1975; Anderson et al., 2011; Chapter 5) as well as evidence from crustal remanent magnetism for an ancient magnetic field (Johnson et al., 2015) also potentially provide constraints on the composition of Mercury's core. Previous studies have often considered only S as a candidate light element in Mercury's core (Riner et al., 2008; Rivoldini et al., 2009; Manglik et al., 2010; Vilim et al., 2010; Dumberry and Rivoldini, 2015), but future models that also incorporate Si would be worthwhile, especially given the much higher solubility of Si in solid Fe metal (Kuwayama and Hirose, 2004) and potential challenges for compositional convection in the core to drive a dynamo.

Overall, Mercury's molten outer core indicates the presence of a light-element component, but the identity and abundance of that light-element component is poorly constrained. Table 2.3 summarizes the results of recent studies and suggests a growing consensus since MESSENGER's measurements that Mercury's core is likely Si-bearing and perhaps contains S and C as well.

The light-element composition of Mercury's core and the conditions during core formation would also influence the concentrations of minor and trace elements in the core. In particular, understanding the concentrations of the heat-producing elements of K, U, and Th in Mercury's core is important, as their distribution has implications for the overall thermal

Table 2.3. *Estimated major element abundances in Mercury's core.*

	Fe (wt%)	Ni (wt%)	C (wt%)	Si (wt%)	S (wt%)
Typical iron meteorites[1]		~5–15			
Aubrite metal[2]		5			
Earth model[3]		5			
Model of graphite flotation crust[4]			>0		
Geophysical models[5]	~75–95	~5[a]		>0–17	0–20
Metal-rich chondrite differentiation model[6]	~75–89	~5[a]		>5	~1–15
Core formation experiments[7]	~75–83	~5[a]		>1	0–20
S solubility experiments[8]				>0	<1.5

[a] Value assumed to estimate the abundance range of Fe
Data sources: 1. Goldstein et al. (2009); 2. McCoy and Bullock (2017); 3. McDonough (2014); 4. Vander Kaaden and McCubbin (2015); 5. Hauck et al. (2013); 6. Malavergne et al. (2010); 7. Chabot et al. (2014b); 8. Namur et al. (2016b)

evolution of the planet (Chapter 17). On the basis of Mercury's reduced nature and experimental metal–silicate partitioning data, Malavergne et al. (2010) concluded that U could be an important heat-producing element in Mercury's core, whereas Th and K would not enter Mercury's core in significant amounts. However, in addition to oxygen fugacity, the S content of the phase can have a strong influence on the partitioning behavior of these elements, and if Mercury's core formed an FeS layer, such a layer could contain high concentrations of these heat-producing elements (McCubbin et al., 2012). If significant amounts of K, U, or Th are sequestered in Mercury's core, there also would be implications for interpreting the ratios of these elements measured by MESSENGER at Mercury's surface. Lastly, sulfides formed under reducing conditions in enstatite chondrite melting experiments contained elements that are more commonly lithophile, such as Mg, Ca, Cr, Mn, and Ti (McCoy et al., 1999; Berthet et al., 2009). Depending on the oxygen fugacity and partitioning of S during Mercury's core formation, Mercury's core could similarly contain minor amounts of elements not generally considered to be in planetary cores under more oxidizing conditions.

2.6 MERCURY'S BULK COMPOSITION

Here we use the estimated mantle and core compositions discussed in the previous sections to estimate the bulk composition of Mercury for key elements. In Section 2.4 we used petrologic results to estimate the compositions of mantle sources of two different measured surface compositions (Table 2.2). Since these two mantle estimates represent original compositions prior to crustal extraction, we take their average to be the bulk composition of Mercury's silicate shell, but we add 1 wt% Fe to match the surface abundance and 7 wt% S to match the mantle prediction of Namur et al. (2016b). From the geophysical modeling results of Hauck et al. (2013), we take Mercury's core radius to be 2020 km and its density to be 6980 kg m^{-3}, the silicate shell's density to be 3380 kg m^{-3}, and the planet's radius to be 2440 km. As discussed in detail in the previous section,

Mercury's core likely contains Si, and possibly S and C, but the precise composition is highly uncertain (Table 2.3), so we consider a range of possibilities. For simplicity, we assume that the core has a fixed abundance of 1.5 wt% S (Namur et al., 2016b), and we vary the Si abundance from 0 to 25 wt%. We also consider cases with and without a 100-km-thick layer of FeS at the base of the mantle.

Selected element ratios from the resulting bulk compositions are compared in Figure 2.7 to the measured surface and inferred original mantle ("Bulk Sil.") compositions and the compositions of E and CB chondrites, Earth, and the Sun. The long-recognized effect of Mercury's large core is apparent: Mercury's bulk composition is highly Fe-rich compared with the bulk solar composition and that of other terrestrial planets and chondrites. However, Mercury's high iron fraction has often been expressed as a high Fe/Si ratio. The discovery from MESSENGER data that Mercury is extremely chemically reduced and consequently may have substantial Si in its core shows that, in fact, bulk Mercury is not only enriched in Fe, but also in Si and possibly S, relative to other planets, and in principle could have a chondritic Fe/Si ratio, but very low ratios of other elements to Si (see, e.g., the 25 wt% core Si cases in Figure 2.7). These plots reinforce the decades-long view that Mercury's origin must entail some form of metal–silicate fractionation, though whether this occurred as a result of chaotic (e.g., giant impacts) or orderly (e.g., nebular metal–silicate fractionation) processes is still unknown, and answering this question will require substantial improvements in quantitative modeling of the chemical consequences of various proposed formation scenarios (see Chapter 18).

As discussed in Section 2.4, however, the estimated primitive mantle compositions are remarkably similar to the silicate compositions of E and CB chondrites if some Si is assumed to have been reduced to metallic form in the meteorite compositions. Indeed, for elements that don't partition into the core, Mercury's bulk composition is much closer to those of the other terrestrial planets; for example, Mercury has a chondritic-like Al/Mg ratio (Figure 2.7c) and Mars-like K/Th and K/Cl ratios (Figure 2.3). The presence at chondritic levels of elements such as Al, K, and

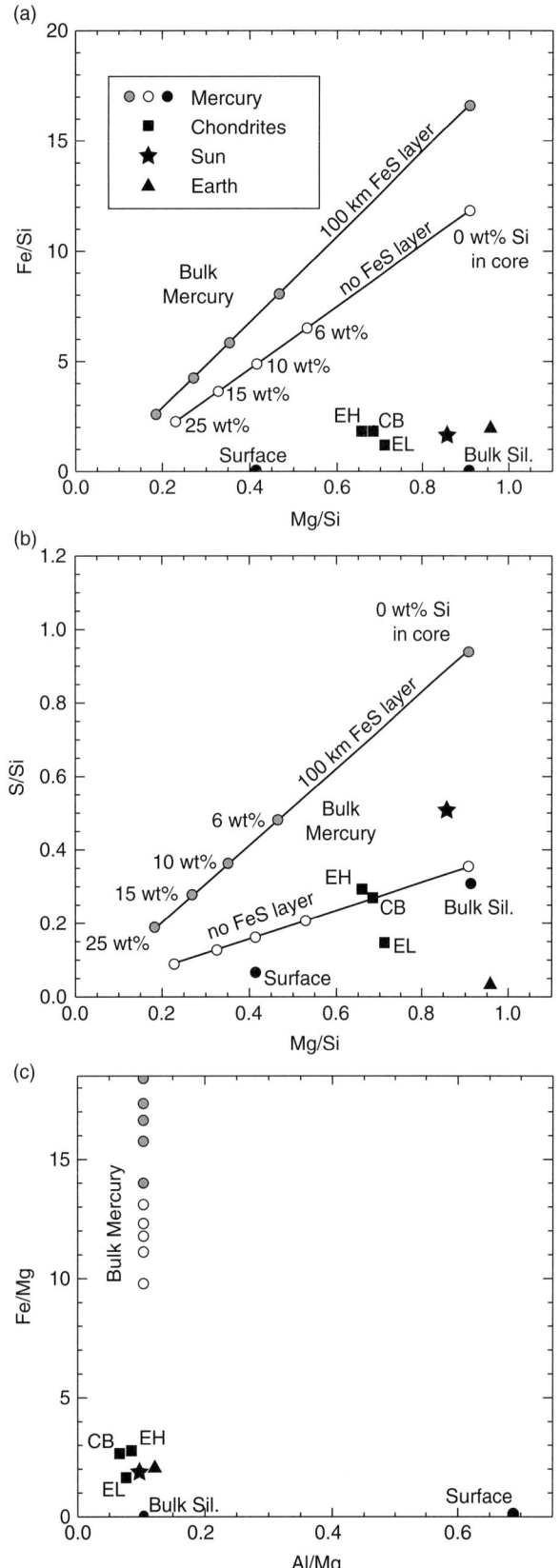

Na that preferentially partition into volcanic melts (e.g., surface Al/Mg ratio, Figure 2.6c) may provide constraints on Mercury origin scenarios. As discussed in Chapter 18, two types of impact models that have been discussed for Mercury's origin are a single giant impact stripping much of the silicate shell of a larger planet (e.g., Benz et al., 2007) and a hit-and-run scenario whereby a proto-Mercury impacts another body and transfers much of its silicate shell to the other planet before being scattered to the inner edge of the protoplanetary disk (e.g., Asphaug and Reufer, 2014). In the first scenario, the disrupted silicate material may be very well mixed before some of it re-accretes to the planet and thus may preserve the initial bulk volatile-rich composition (Section 18.4). However, if a proto-Mercury has already experienced a magma ocean stage and subsequent crystallization before being involved in a hit-and-run impact, incompatible elements may be enriched in the outer layers, and these may be preferentially lost. Mercury's present bulk silicate composition may thus argue against such an origin, but additional quantitative modeling is clearly needed.

The average of the Mercury mantle composition estimates derived in Section 2.4 is compared with the estimated compositions of the mantles of Earth (Hart and Zindler, 1986) and Mars (Dreibus and Wänke, 1985) on a plot of composition versus heliocentric distance in Figure 2.8. With the caveat that these mantle estimates are highly uncertain, especially for minor elements such as Ti, some characteristics of the silicate mantles are consistent with condensation temperatures that existed in the parts of the early solar nebula from which the bulk of material that accreted to form each planet originally condensed. There is a continual decrease in the mantle abundances of the refractory elements Ca, Al, and Ti from Mercury to Earth to Mars, which is consistent with Mercury having condensed in the hotter part of the solar nebula where CaO-, Al_2O_3-, and TiO_2-bearing minerals would initially have condensed. The continual increase in oxidized Fe from near zero for Mercury to a maximum of 17.5 wt% in the mantle of Mars is consistent with decreasing condensation temperatures and also with increasing fO_2 of the condensed minerals. However, other characteristics are much more enigmatic. Clearly there is an apparent decrease in SiO_2 with distance from the Sun, with Si-rich enstatite chondrite-like material most abundant in Mercury's mantle. Neither this nor the observation that Mercury is the most volatile-rich planet (e.g., Na) would be expected from classical condensation sequences as one moves to lower temperatures away from the Sun (e.g., Grossman, 1972). However, the low oxygen fugacity inferred for Mercury must play a key role. Ebel and Alexander (2011) showed that condensation from a system enriched in C-rich, water-poor dust analogous to

Figure 2.7. Bulk compositions of Mercury (open and gray-filled circles) derived from estimates of mantle and core compositions (Sections 2.4 and 2.5) and the physical parameters of Hauck et al. (2013) compared with compositions of meteorites (Table 2.2; from Jarosewich, 1990), Mercury surface and bulk silicate (black circles, Sections 2.3 and 2.4), Earth, and Sun. Core compositions include 1.5 wt% S, and the mantle composition corresponds to the average of the two mantle estimates in Table 2.2 with 1 wt% Fe and 7 wt% S added. Different bulk compositions correspond to different assumed Si concentrations in Mercury's core and different assumptions regarding internal structure, as indicated.

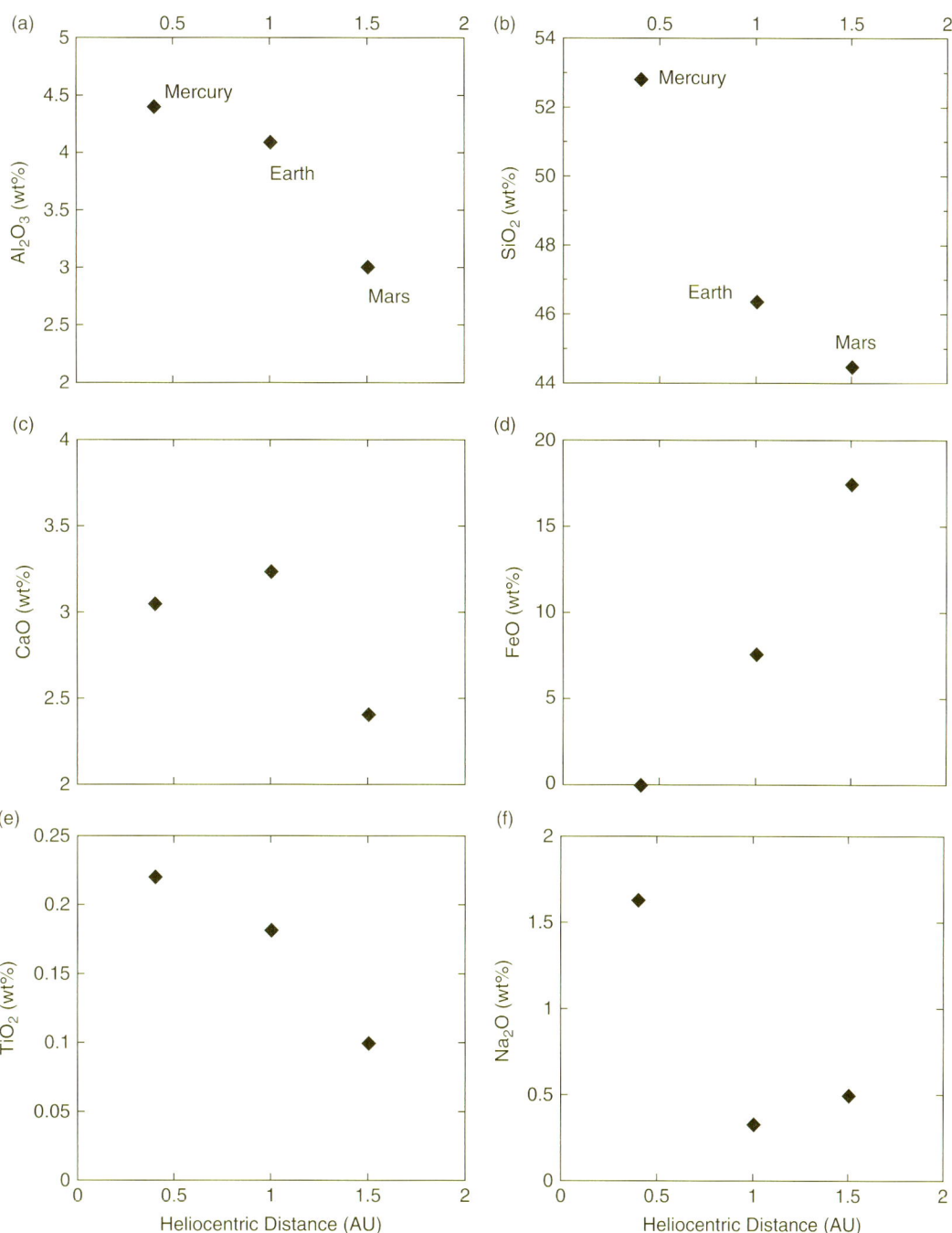

Figure 2.8. Comparison of Mercury's estimated mantle composition with those of Earth (Hart and Zindler, 1986) and Mars (Dreibus and Wänke, 1985), plotted versus heliocentric distance. Mercury compositions are averages of the two mantle estimates in Table 2.2 and Figure 2.6.

cometary dust particles can produce highly reduced, Fe-rich assemblages that could be precursors to E chondrites and Mercury. Moreover, under the inferred conditions, S, K, and Cl can all act as refractory elements (Ebel and Sack, 2013). Therefore, Mercury's volatile-rich and reduced silicate composition (and high overall Fe content in part) can be the result of condensation under highly reducing conditions perhaps influenced by exclusion of water ice from the feeding zone of Mercury's parental materials due to proximity to the Sun (Ebel and Alexander, 2011; Chapter 18).

2.7 OUTLOOK

After decades of argument over theories in the face of a severe paucity of geochemical data for Mercury (Section 2.1), MESSENGER and its geochemical instruments provided the first survey of the innermost planet's surface composition for a wide range of elements, as detailed in Section 2.3. These data clearly revealed Mercury to be a volatile-rich but highly chemically reduced world, the implications of which for planet formation and evolutionary processes in our solar system and in exoplanetary systems are still poorly understood. Although the fundamental scientific question of why Mercury is so metal rich is still unanswered, the importance of having data by which to test ideas and models cannot be overstated, and the MESSENGER compositional data set is rich indeed.

Scientific implications of the compositional data for Mercury's origin and geological history are discussed throughout this book (e.g., Chapters 7, 8, 10, 11, 12, 18, and 19), but the MESSENGER observations have opened a vast window for new questions and future work. Petrologic processes under highly reducing conditions are still not very well understood, but, largely motivated by MESSENGER results, are being addressed increasingly by experimental studies (McCoy et al., 1999; Berthet et al., 2009; Charlier et al., 2013; Chabot et al., 2014b; Namur et al., 2016a, b; Vander Kaaden and McCubbin, 2016). Such studies are needed to fully understand the origin of Mercury's geochemical terranes, the nature of its surface rocks, melting conditions in the planet's mantle, and element partitioning into Mercury's core. The discovery of abundant C on Mercury's surface (Section 2.3.3) indicates the need for experimental and theoretical work that includes this element. As discussed in detail in Chapter 18, a great deal of further theoretical work is needed to assess the array of proposed origin scenarios to account for Mercury's high metal-to-silicate ratio and especially to evaluate chemical consequences. The MESSENGER geochemical data set will be crucial to testing these scenarios. Finally, the availability of element abundance maps on Mercury's surface will be useful for better understanding Mercury's exosphere and the surface processes that supply it.

Looking beyond the MESSENGER data, the BepiColombo mission (Chapter 20) will carry a suite of geochemical instruments similar to those of MESSENGER (X-ray, gamma-ray, and neutron spectrometers) and will provide complementary – and for some areas improved – geochemical information about Mercury. BepiColombo's polar orbit will be much less eccentric than that of MESSENGER and thus provide southern hemisphere elemental measurements that were either not possible before (i.e., GRS measurements) or were made with poor spatial resolution (XRS). BepiColombo's thermal imaging spectrometer may provide crucial mineralogical data, not obtained with MESSENGER instruments, which can be combined with elemental abundance data to improve our understanding of the geological and space-weathering history of the surface.

REFERENCES

Adler, I., Trombka, J., Gerard, J., Lowman, P., Schmadebeck, R., Blodget, H., Eller, E., Yin, L., Lamothe, R., Gorenstein, P. and Bjorkholm, P. (1972a). Apollo 15 geochemical X-ray fluorescence experiment: Preliminary report. *Science*, **175**, 436–440.

Adler, I., Trombka, J., Gerard, J., Lowman, P., Schmadebeck, R., Blodget, H., Eller, E., Yin, L., Lamothe, R., Osswald, G., Gorenstein, P., Bjorkholm, P., Gursky, H. and Harris, B. (1972b). Apollo 16 geochemical X-ray fluorescence experiment: Preliminary report. *Science*, **177**, 256–259.

Anderson, B. J., Johnson, C. L., Korth, H., Purucker, M. E., Winslow, R. M., Slavin, J. A., Solomon, S. C., McNutt, R. L., Jr., Raines, J. M. and Zurbuchen, T. H. (2011). The global magnetic field of Mercury from MESSENGER orbital observations. *Science*, **333**, 1859–1862, doi:10.1126/science.1211001.

Anderson, J. D., Colombo, G., Espsitio, P. B., Lau, E. L. and Trager, G. B. (1987). The mass, gravity field, and ephemeris of Mercury. *Icarus*, **71**, 337–349.

Asphaug, E. and Reufer, A. (2014). Mercury and other iron-rich planetary bodies as relics of inefficient accretion. *Nature Geosci.*, **7**, 564–568, doi:10.1038/ngeo2189.

Benz, W., Anic, A., Horner, J. and Whitby, J. A. (2007). The origin of Mercury. *Space Sci. Rev.*, **132**, 189–202.

Berthet, S., Malavergne, V. and Righter, K. (2009). Melting of the Indarch meteorite (EH4 chondrite) at 1 GPa and variable oxygen fugacity: Implications for early planetary differentiation processes. *Geochim. Cosmochim. Acta*, **73**, 6402–6420, doi:10.1016/j.gca.2009.07.030.

Bida, T. A., Killen, R. M. and Morgan, T. H. (2000). Discovery of calcium in Mercury's atmosphere. *Nature*, **404**, 159–161.

Blewett, D. T., Lucey, P. G., Hawke, B. R., Ling, G. G. and Robinson, M. S. (1997). A comparison of Mercurian reflectance and spectral quantities with those of the Moon. *Icarus*, **129**, 217–231.

Blewett, D. T., Chabot, N. L., Denevi, B. W., Ernst, C. M., Head, J. W., Izenberg, N. R., Murchie, S. L., Solomon, S. C., Nittler, L. R., McCoy, T. J., Xiao, Z., Baker, D. M. H., Fassett, C. I., Braden, S. E., Oberst, J., Scholten, F., Preusker, F. and Hurwitz, D. M. (2011). Hollows on Mercury: MESSENGER evidence for geologically recent volatile-related activity. *Science*, **333**, 1856–1859, doi:10.1126/science.1211681.

Boujibar, A., Andrault, D., Bouhifd, M. A., Bolfan-Casanova, N., Devidal, J.-L. and Trcera, N. (2014). Metal-silicate partitioning of sulphur, new experimental and thermodynamic constraints on planetary accretion. *Earth Planet. Sci. Lett.*, **391**, 42–54.

Brown, S. M. and Elkins-Tanton, L. T. (2009). Compositions of Mercury's earliest crust from magma ocean models. *Earth Planet. Sci. Lett.*, **286**, 446–455.

Bruck Syal, M., Schultz, P. H. and Riner, M. A. (2015). Darkening of Mercury's surface by cometary carbon. *Nature Geosci*, **8**, 352–356, doi:10.1038/ngeo2397.

Buchwald, V. F. (1975). *Handbook of Iron Meteorites, Their History, Distribution, Composition, and Structure*. Berkeley, CA: University of California Press.

Burbine, T. H., McCoy, T. J., Nittler, L. R., Benedix, G. K., Cloutis, E. A. and Dickinson, T. L. (2002). Spectra of extremely reduced assemblages: Implications for Mercury. *Meteorit. Planet. Sci.*, **37**, 1233–1244.

Byrne, P. K., Klimczak, C., Şengör, A. M. C., Solomon, S. C., Watters, T. R. and Hauck, S. A., II (2014). Mercury's global contraction much greater than earlier estimates. *Nature Geosci.*, **7**, 301–307, doi:10.1038/ngeo2097.

Byrne, P. K., Ostrach, L. R., Fassett, C. I., Chapman, C. R., Denevi, B. W., Evans, A. J., Klimczak, C., Banks, M. E., Head, J. W. and Solomon, S. C. (2016). Widespread effusive volcanism on Mercury likely ended by about 3.5 Ga. *Geophys. Res. Lett.*, **43**, 7408–7416.

Chabot, N. L. and Haack, H. (2006). Evolution of asteroidal cores. In *Meteorites and the Early Solar System II*, ed. D. S. Lauretta and H. Y. McSween. Tucson, AZ: University of Arizona Press, pp. 747–771.

Chabot, N. L., Ernst, C. M., Denevi, B. W., Nair, H., Deutsch, A. N., Blewett, D. T., Murchie, S. L., Neumann, G. A., Mazarico, E., Paige, D. A., Harmon, J. K., Head, J. W. and Solomon, S. C. (2014a). Images of surface volatiles in Mercury's polar craters acquired by the MESSENGER spacecraft. *Geology*, **42**, 1051–1054, doi:10.1130/g35916.1.

Chabot, N. L., Wollack, E. A., Klima, R. L. and Minitti, M. E. (2014b). Experimental constraints on Mercury's core composition. *Earth Planet. Sci. Lett.*, **390**, 199–208.

Charlier, B., Grove, T. L. and Zuber, M. T. (2013). Phase equilibria of ultramafic compositions on Mercury and the origin of the compositional dichotomy. *Earth Planet. Sci. Lett.*, **363**, 50–60.

Chi, H., Dasgupta, R., Duncan, M. S. and Shimizu, N. (2014). Partitioning of carbon between Fe-rich alloy melt and silicate melt in a magma ocean: Implications for the abundance and origin of volatiles in Earth, Mars, and the Moon. *Geochim. Cosmochim. Acta*, **139**, 447–471.

Crider, D. and Killen, R. M. (2005). Burial rate of Mercury's polar volatile deposits. *Geophys. Res. Lett.*, **32**, L12201, doi:10.1029/2005GL022689.

Dasgupta, R., Chi, H., Shimizu, N., Buono, A. S. and Walker, D. (2013). Carbon solution and partitioning between metallic and silicate melts in a shallow magma ocean: Implications for the origin and distribution of terrestrial carbon. *Geochim. Cosmochim. Acta*, **102**, 191–212.

Dauphas, N. and Pourmand, A. (2011). Hf-W-Th evidence for rapid growth of Mars and its status as a planetary embryo. *Nature*, **473**, 489–492.

Denevi, B. W., Robinson, M. S., Solomon, S. C., Murchie, S. L., Blewett, D. T., Domingue, D. L., McCoy, T. J., Ernst, C. M., Head, J. W., Watters, T. R. and Chabot, N. L. (2009). The evolution of Mercury's crust: A global perspective from MESSENGER. *Science*, **324**, 613–618.

Dreibus, G. and Palme, H. (1996). Cosmochemical constraints on the sulfur content in the Earth's core. *Geochim. Cosmochim. Acta*, **60**, 1125–1130.

Dreibus, G. and Wänke, H. (1985). Mars, a volatile-rich planet. *Meteoritics*, **20**, 367–381.

Dumberry, M. and Rivoldini, A. (2015). Mercury's inner core size and core-crystallization regime. *Icarus*, **248**, 254–268.

Ebel, D. S. and Alexander, C. M. O'D. (2011). Equilibrium condensation from chondritic porous IDP enriched vapor: Implications for Mercury and enstatite chondrite origins. *Planet. Space Sci.*, **59**, 1888–1894, doi:10.1016/j.pss.2011.07.017.

Ebel, D. S. and Sack, R. O. (2013). Djerfisherite: Nebular source of refractory potassium. *Contrib. Mineral. Petrol.*, **166**, 923–934, doi:10.1007/s00410-013-0898-x.

Evans, L. G., Peplowski, P. N., Rhodes, E. A., Lawrence, D. J., McCoy, T. J., Nittler, L. R., Solomon, S. C., Sprague, A. L., Stockstill-Cahill, K. R., Starr, R. D., Weider, S. Z., Boynton, W. V., Hamara, D. K. and Goldsten, J. O. (2012). Major-element abundances on the surface of Mercury: Results from the MESSENGER Gamma-Ray Spectrometer. *J. Geophys. Res.*, **117**, E00L07, doi:10.1029/2012je004178.

Evans, L. G., Peplowski, P. N., McCubbin, F. M., McCoy, T. J., Nittler, L. R., Zolotov, M. Yu., Ebel, D. S., Lawrence, D. J., Starr, R. D., Weider, S. Z. and Solomon, S. C. (2015). Chlorine on the surface of Mercury: MESSENGER gamma-ray measurements and implications for the planet's formation and evolution. *Icarus*, **257**, 417–427.

Fischer, R. A., Nakajima, Y., Campbell, A. J., Frost, D. J., Harries, D., Langenhorst, F., Miyajima, N., Pollok, K. and Rubie, D. C. (2015). High pressure metal-silicate partitioning of Ni, Co, V, Cr, Si, and O. *Geochim. Cosmochim. Acta*, **167**, 177–194.

Frank, E. A., Nittler, L. R., Vorburger, A. H., Weider, S. Z., Starr, R. D. and Solomon, S. C. (2015). High-resolution measurements of Mercury's surface composition with the MESSENGER X-Ray Spectrometer. *Lunar Planet. Sci.*, **46**, abstract 1949.

Fukai, Y. (1984). The iron-water reaction and the evolution of the Earth. *Nature*, **308**, 174–175.

Gessmann, C. K., Wood, B. J., Rubie, D. C. and Kilburn, M. R. (2001). Solubility of silicon in liquid metal at high pressure: Implications for the composition of the Earth's core. *Earth Planet. Sci. Lett.*, **184**, 367–376.

Ghiorso, M. S. and Sack, R. O. (1995). Chemical mass transfer in magmatic processes IV. A revised and internally consistent thermodynamic model for the interpretation and extrapolation of liquid-solid equilibria in magmatic systems at elevated temperatures and pressures. *Contrib. Mineral. Petrol.*, **119**, 197–212.

Giletti, B. J. and Shanahan, T. M. (1997). Alkali diffusion in plagioclase feldspar. *Chemical Geology*, **139**, 3–20, doi:10.1016/S0009-2541(97)00026-0.

Goettel, K. A. (1988). Present bounds on the bulk composition of Mercury: Implications for planetary formation processes. In *Mercury*, ed. F. Vilas, C. R. Chapman and M. S. Matthews. Tucson, AZ: University of Arizona Press, pp. 613–621.

Goldstein, J. I., Scott, E. R. D. and Chabot, N. L. (2009). Iron meteorites: Crystallization, thermal history, parent bodies, and origin. *Chemie der Erde – Geochemistry*, **69**, 293–325, doi:10.1016/j.chemer.2009.01.002.

Goldsten, J. O., Rhodes, E. A., Boynton, W. V., Feldman, W. C., Lawrence, D. J., Trombka, J. I., Smith, D. M., Evans, L. G., White, J., Madden, N. W., Berg, P. C., Murphy, G. A., Gurnee, R. S., Strohbehn, K., Williams, B. D., Schaefer, E. D., Monaco, C. A., Cork, C. P., Del Eckels, J., Miller, W. O., Burks, M. T., Hagler, L. B., Deteresa, S. J. and Witte, M. C. (2007). The MESSENGER Gamma-Ray and Neutron Spectrometer. *Space Sci. Rev.*, **131**, 339–391.

Goudge, T. A., Head, J. W., Kerber, L., Blewett, D. T., Denevi, B. W., Domingue, D. L., Gillis-Davis, J. J., Gwinner, K., Helbert, J., Holsclaw, G. M., Izenberg, N. R., Klima, R. L., McClintock, W. E., Murchie, S. L., Neumann, G. A., Smith, D. E., Strom, R. G., Xiao, Z., Zuber, M. T. and Solomon, S. C. (2014). Global inventory and characterization of pyroclastic deposits on Mercury: New insights into pyroclastic activity from MESSENGER orbital data. *J. Geophys. Res. Planets*, **119**, 635–658.

Grossman, L. (1972). Condensation in the primitive solar nebula. *Geochim. Cosmochim. Acta*, **36**, 597–619.

Grott, M., Breuer, D. and Laneuville, M. (2011). Thermo-chemical evolution and global contraction of Mercury. *Earth Planet. Sci. Lett.*, **307**, 135–146.

Hart, S. R. and Zindler, A. (1986). In search of a bulk Earth composition. *Chemical Geology*, **57**, 247–267.

Hauck, S. A., II, Dombard, A. J., Phillips, R. J. and Solomon, S. C. (2004). Internal and tectonic evolution of Mercury. *Earth Planet. Sci. Lett.*, **222**, 713–728.

Hauck, S. A., II, Margot, J.-L., Solomon, S. C., Phillips, R. J., Johnson, C. L., Lemoine, F. G., Mazarico, E., McCoy, T. J., Padovan, S., Peale, S. J., Perry, M. E., Smith, D. E. and Zuber, M. T. (2013). The curious case of Mercury's internal structure. *J. Geophys. Res. Planets*, **118**, 1204–1220.

Haughton, D. R., Roeder, P. L. and Skinner, B. J. (1974). Solubility of sulfur in mafic magmas. *Economic Geology*, **69**, 451–467, doi:10.2113/gsecongeo.69.4.451.

Head, J. W., Murchie, S. L., Prockter, L. M., Robinson, M. S., Solomon, S. C., Strom, R. G., Chapman, C. R., Watters, T. R., McClintock, W. E., Blewett, D. T. and Gillis-Davis, J. J. (2008). Volcanism on Mercury: Evidence from the first MESSENGER flyby. *Science*, **321**, 69–72.

Head, J. W., Chapman, C. R., Strom, R. G., Fassett, C. I., Denevi, B. W., Blewett, D. T., Ernst, C. M., Watters, T. R., Solomon, S. C., Murchie, S. L., Prockter, L. M., Chabot, N. L., Gillis-Davis, J. J., Whitten, J. L., Goudge, T. A., Baker, D. M. H., Hurwitz, D. M., Ostrach, L. R., Xiao, Z., Merline, W. J., Kerber, L., Dickson, J. L., Oberst, J., Byrne, P. K., Klimczak, C. and Nittler, L. R. (2011).

Flood volcanism in the northern high latitudes of Mercury revealed by MESSENGER. *Science*, **333**, 1853–1856, doi:10.1126/science.1211997.

Hillgren, V. J., Gessmann, C. K., Li, J. and Righter, K. (2000). An experimental perspective on the light element in Earth's core. In *Origin of the Earth and Moon*, ed. R. M. Canup and K. Righter. Tucson, AZ: University of Arizona Press, pp. 245–263.

Jarosewich, E. (1990). Chemical analyses of meteorites: A compilation of stony and iron meteorite analyses. *Meteoritics*, **25**, 323–338.

Jeanloz, R. (1990). The nature of the Earth's core. *Annu. Rev. Earth Planet. Sci.*, **18**, 357–386, doi:10.1146/annurev.ea.18.050190.002041.

Johnson, C. L., Phillips, R. J., Purucker, M. E., Anderson, B. J., Byrne, P. K., Denevi, B. W., Feinberg, J. M., Hauck, S. A., Head, J. W., Korth, H., James, P. B., Mazarico, E., Neumann, G. A., Philpott, L. C., Siegler, M. A., Tsyganenko, N. A. and Solomon, S. C. (2015). Low-altitude magnetic field measurements by MESSENGER reveal Mercury's ancient crustal field. *Science*, **348**, 892–895, doi:10.1126/science.aaa8720.

Kasper, R. B. (1975). Cation and oxygen diffusion in albite. Ph.D. thesis, Brown University, Providence, RI, 143 pp.

Kerber, L., Head, J. W., Solomon, S. C., Murchie, S. L., Blewett, D. T. and Wilson, L. (2009). Explosive volcanic eruptions on Mercury: Eruption conditions, magma volatile content, and implications for interior volatile abundances. *Earth Planet. Sci. Lett.*, **285**, 263–271.

Kilburn, M. R. and Wood, B. J. (1997). Metal-silicate partitioning and the incompatibility of S and Si during core formation. *Earth Planet. Sci. Lett.*, **152**, 139–148.

Kuwayama, Y. and Hirose, K. (2004). Phase relations in the system Fe–FeSi at 21 GPa. *Amer. Mineral.*, **89**, 273–276.

Lawrence, D. J., Feldman, W. C., Goldsten, J. O., McCoy, T. J., Blewett, D. T., Boynton, W. V., Evans, L. G., Nittler, L. R., Rhodes, E. A. and Solomon, S. C. (2010). Identification and measurement of neutron-absorbing elements on Mercury's surface. *Icarus*, **209**, 195–209.

Lawrence, D. J., Feldman, W. C., Goldsten, J. O., Maurice, S., Peplowski, P. N., Anderson, B. J., Bazell, D., McNutt, R. L., Nittler, L. R., Prettyman, T. H., Rodgers, D. J., Solomon, S. C. and Weider, S. Z. (2013). Evidence for water ice near Mercury's north pole from MESSENGER Neutron Spectrometer measurements. *Science*, **339**, 292–296.

Lawrence, D. J., Peplowski, P. N., Beck, A. W., Feldman, W. C., Frank, E. A., McCoy, T. J., Nittler, L. R. and Solomon, S. C. (2017). Compositional terranes on Mercury: Information from fast neutrons. *Icarus*, **281**, 32–45, doi.org/10.1016/j.icarus.2016.07.018.

Li, J. and Fei, Y. (2014). Experimental constraints on core composition. In *The Mantle and Core*, ed. R. W. Carlson. *Treatise on Geochemistry*, 2nd edn, Vol. 3, ed. H. D. Holland and K. K. Turekian. Amsterdam, Oxford: Elsevier, pp. 527–557.

Li, Y., Dasgupta, R. and Tsuno, K. (2015). The effects of sulfur, silicon, water, and oxygen fugacity on carbon solubility and partitioning in Fe-rich alloy and silicate melt systems at 3 GPa and 1600 °C: Implications for core-mantle differentiation and degassing of magma oceans and reduced planetary mantles. *Earth Planet. Sci. Lett.*, **415**, 54–66.

Li, Y., Dasgupta, R., Tsuno, K., Monteleone, B. and Shimizu, N. (2016). Establishing the carbon and sulfur budget of the Earth's silicate reservoir by accretion and core formation process. *Lunar Planet. Sci.*, **47**, abstract 2486.

Lodders, K. (2003). Solar system abundances and condensation temperatures of the elements. *Astrophys. J.*, **591**, 1220–1247.

Lodders, K. and Fegley, B. (1998). *The Planetary Scientists's Companion*, New York: Oxford University Press.

Lord, O. T., Walter, M. J., Dasgupta, R., Walker, D. and Clark, S. M. (2009). Melting in the Fe–C system to 70 GPa. *Earth Planet. Sci. Lett.*, **284**, 157–167.

Malavergne, V., Siebert, J., Guyot, F., Gautron, L., Combes, R., Hammouda, T., Borensztajn, S., Frost, D. and Martinez, I. (2004). Si in the core? New high-pressure and high-temperature experimental data. *Geochim. Cosmochim. Acta*, **68**, 4201–4211.

Malavergne, V., Toplis, M. J., Berthet, S. and Jones, J. (2010). Highly reducing conditions during core formation on Mercury: Implications for internal structure and the origin of a magnetic field. *Icarus*, **206**, 199–209.

Malavergne, V., Cordier, P., Righter, K., Brunet, F., Zanda, B., Addad, A., Smith, T., Bureau, H., Surblé, S., Raepsaet, C., Charon, E. and Hewins, R. H. (2014). How Mercury can be the most reduced terrestrial planet and still store iron in its mantle. *Earth Planet. Sci. Lett.*, **394**, 186–197.

Manglik, A., Wicht, J. and Christensen, U. R. (2010). A dynamo model with double diffusive convection for Mercury's core. *Earth Planet. Sci. Lett.*, **289**, 619–628.

Marchi, S., Chapman, C. R., Fassett, C. I., Head, J. W., Bottke, W. F. and Strom, R. G. (2013). Global resurfacing of Mercury 4.0–4.1 billion years ago by heavy bombardment and volcanism. *Nature*, **499**, 59–61, doi:10.1038/nature12280.

Margot, J.-L., Peale, S. J., Jurgens, R. F., Slade, M. A. and Holin, I. V. (2007). Large longitude libration of Mercury reveals a molten core. *Science*, **316**, 710–714, doi:10.1126/science.1140514.

Margot, J.-L., Peale, S. J., Solomon, S. C., Hauck, S. A., II, Ghigo, F. D., Jurgens, R. F., Yseboodt, M., Giorgini, J. D., Padovan, S. and Campbell, D. B. (2012). Mercury's moment of inertia from spin and gravity data. *J. Geophys. Res. Planets*, **117**, doi:10.1029/2012JE004161.

McCord, T. B. and Adams, J. B. (1972). Mercury: Interpretation of optical observations. *Icarus*, **17**, 585–588.

McCoy, T. J. and Bullock, E. S. (2017). Differentiation under highly reducing conditions: New insights from enstatite meteorites and Mercury. In *Planetesimals: Early Differentiation and Consequences for Planets*, ed. L. T. Elkins-Tanton and B. P. Weiss. Cambridge: Cambridge University Press, pp. 71–91.

McCoy, T. J., Dickinson, T. L. and Lofgren, G. E. (1999). Partial melting of the Indarch (EH4) meteorite: A textural, chemical and phase relations view of melting and melt migration. *Meteorit. Planet. Sci.*, **34**, 735–746.

McCubbin, F. M., Riner, M. A., Vander Kaaden, K. E. and Burkemper, L. K. (2012). Is Mercury a volatile-rich planet? *Geophys. Res. Lett.*, **39**, L09202, doi:10.1029/2012GL051711.

McCubbin, F. M., Vander Kaaden, K. E., Peplowski, P. N., Bell, A. S., Nittler, L. R., Boyce, J. W., Evans, L. G., Keller, L. P., Elardo, S. M. and McCoy, T. J. (2017). A low O/Si ratio on the surface of Mercury: Evidence for silicon smelting? *J. Geophys. Res. Planets*, **122**, 2053–2076, doi:10.1002/2017JE005367

McDonough, W. F. (2014). Compositional model for the Earth's core. In *The Mantle and Core*, ed. R. W. Carlson. *Treatise on Geochemistry*, 2nd edn, Vol. 3, ed. H. D. Holland and K. K. Turekian. Amsterdam, Oxford: Elsevier, pp. 559–577.

Meslin, P.-Y. and Déprez, G. (2012). Radon exhalation as a possible explanation to the low Th/U ratio measured by MESSENGER GRS on Mercury. *Lunar Planet. Sci.*, **43**, abstract 2800.

Morgan, J. W. and Anders, E. (1980). Chemical composition of Earth, Venus, and Mercury. *Proc. Natl. Acad. Sci. USA*, **77**, 6973–6977.

Murchie, S. L., Klima, R. L., Denevi, B. W., Ernst, C. M., Keller, M. R., Blewett, D. T., Domingue, D. L., Chabot, N. L., Hash, C. D., Malaret, E., Izenberg, N. R., Vilas, F., Nittler, L. R., Gillis-Davis, J. J., Head, J. W. and Solomon, S. C. (2015). Orbital multispectral mapping of Mercury with the MESSENGER Mercury Dual Imaging System: Evidence for the origins of plains units and low-reflectance material. *Icarus*, **254**, 287–305.

Namur, O., Collinet, M., Charlier, B., Grove, T. L., Holtz, F. and McCammon, C. (2016a). Melting processes and mantle sources of lavas on Mercury. *Earth Planet. Sci. Lett.*, **439**, 117–128.

Namur, O., Charlier, B., Holtz, F., Cartier, C. and McCammon, C. (2016b). Sulfur solubility in reduced mafic silicate melts: Implications for the speciation and distribution of sulfur on Mercury. *Earth Planet. Sci. Lett.*, **448**, 102–114.

Narendranath, S., Athiray, P. S., Sreekumar, P., Kellett, B. J., Alha, L., Howe, C. J., Joy, K. H., Grande, M., Huovelin, J., Crawford, I. A., Unnikrishnan, U., Lalita, S., Subramaniam, S., Weider, S. Z., Nittler, L. R., Gasnault, O., Rothery, D., Fernandes, V. A., Bhandari, N., Goswami, J. N. and Wieczorek, M. A. (2011). Lunar X-ray fluorescence observations by the Chandrayaan-1 X-ray Spectrometer (C1XS): Results from the nearside southern highlands. *Icarus*, **214**, 53–66.

Ness, N. F., Behannon, K. W., Lepping, R. P. and Whang, Y. C. (1975). The magnetic field of Mercury, 1. *J. Geophys. Res.*, **80**, 2708–2716, doi:10.1029/JA080i019p02708.

Neumann, G. A., Cavanaugh, J. F., Sun, X., Mazarico, E. M., Smith, D. E., Zuber, M. T., Mao, D., Paige, D. A., Solomon, S. C., Ernst, C. M. and Barnouin, O. S. (2013). Bright and dark polar deposits on Mercury: Evidence for surface volatiles. *Science*, **339**, 296–300, doi:10.1126/science.1229764.

Nittler, L. R., Starr, R. D., Lim, L., McCoy, T. J., Burbine, T. H., Reedy, R. C., Trombka, J. I., Gorenstein, P., Squyres, S. W., Boynton, W. V., McClanahan, T. P., Bhangoo, J. S., Clark, P. E., Murphy, M. E. and Killen, R. (2001). X-ray fluorescence measurements of the surface elemental composition of asteroid 433 Eros. *Meteorit. Planet. Sci.*, **36**, 1673–1695.

Nittler, L. R., Starr, R. D., Weider, S. Z., McCoy, T. J., Boynton, W. V., Ebel, D. S., Ernst, C. M., Evans, L. G., Goldsten, J. O., Hamara, D. K., Lawrence, D. J., McNutt, R. L., Jr., Schlemm, C. E., II, Solomon, S. C. and Sprague, A. L. (2011). The major-element composition of Mercury's surface from MESSENGER X-ray spectrometry. *Science*, **333**, 1847–1850, doi:10.1126/science.1211567.

Nittler, L. R., Frank, E. A., Weider, S. Z., Crapster-Pregont, E., Vorburger, A., Starr, R. D. and Solomon, S. C. (2016). Global major-element maps of Mercury updated from four years of MESSENGER X-ray observations. *Lunar Planet. Sci.*, **47**, abstract 1237.

Okuchi, T. (1997). Hydrogen partitioning into molten iron at high pressure: Implications for Earth's core. *Science*, **278**, 1781–1784, doi:10.1126/science.278.5344.1781.

Paige, D. A., Siegler, M. A., Harmon, J. K., Neumann, G. A., Mazarico, E. M., Smith, D. E., Zuber, M. T., Harju, E., Delitsky, M. L. and Solomon, S. C. (2013). Thermal stability of volatiles in the north polar region of Mercury. *Science*, **339**, 300–303, doi:10.1126/science.1231106.

Peplowski, P. N., Evans, L. G., Hauck, S. A., McCoy, T. J., Boynton, W. V., Gillis-Davis, J. J., Ebel, D. S., Goldsten, J. O., Hamara, D. K., Lawrence, D. J., McNutt, R. L., Jr., Nittler, L. R., Solomon, S. C., Rhodes, E. A., Sprague, A. L., Starr, R. D. and Stockstill-Cahill, K. R. (2011). Radioactive elements on Mercury's surface from MESSENGER: Implications for the planet's formation and evolution. *Science*, **333**, 1850–1852, doi:10.1126/science.1211576.

Peplowski, P. N., Lawrence, D. J., Rhodes, E. A., Sprague, A. L., McCoy, T. J., Denevi, B. W., Evans, L. G., Head, J. W., Nittler, L. R., Solomon, S. C., Stockstill-Cahill, K. R. and Weider, S. Z. (2012a). Variations in the abundances of potassium and thorium on the surface of Mercury: Results from the MESSENGER Gamma-Ray Spectrometer. *J. Geophys. Res.*, **117**, E00L04, doi:10.1029/2012JE004141.

Peplowski, P. N., Rhodes, E. A., Hamara, D. K., Lawrence, D. J., Evans, L. G., Nittler, L. R. and Solomon, S. C. (2012b). Aluminum abundance on the surface of Mercury: Application of a new background-reduction technique for the analysis of gamma-ray spectroscopy data. *J. Geophys. Res.*, **117**, E00L10, doi:10.1029/2012JE004181.

Peplowski, P. N., Evans, L. G., Stockstill-Cahill, K. R., Lawrence, D. J., Goldsten, J. O., McCoy, T. J., Nittler, L. R., Solomon, S. C., Sprague, A. L., Starr, R. D. and Weider, S. Z. (2014). Enhanced sodium abundance in Mercury's north polar region revealed by the MESSENGER Gamma-Ray Spectrometer. *Icarus*, **228**, 86–95.

Peplowski, P. N., Lawrence, D. J., Evans, L. G., Klima, R. L., Blewett, D. T., Goldsten, J. O., Murchie, S. L., McCoy, T. J., Nittler, L. R., Solomon, S. C., Starr, R. D. and Weider, S. Z. (2015a). Constraints on the abundance of carbon in near-surface materials on Mercury: Results from the MESSENGER Gamma-Ray Spectrometer. *Planet. Space Sci.*, **108**, 98–107, doi:10.1016/j.pss.2015.01.008.

Peplowski, P. N., Lawrence, D. J., Feldman, W. C., Goldsten, J. O., Bazell, D., Evans, L. G., Head, J. W., Nittler, L. R., Solomon, S. C. and Weider, S. Z. (2015b). Geochemical terranes of Mercury's northern hemisphere as revealed by MESSENGER neutron measurements. *Icarus*, **253**, 346–363.

Peplowski, P. N., Klima, R. L., Lawrence, D. J., Ernst, C. M., Denevi, B. W., Frank, E. A., Goldsten, J. O., Murchie, S. L., Nittler, L. R. and Solomon, S. C. (2016). Remote sensing evidence for an ancient carbon-bearing crust on Mercury. *Nature Geosci.*, **9**, 273–276, doi:10.1038/ngeo2669.

Potter, A. and Morgan, T. (1985). Discovery of sodium in the atmosphere of Mercury. *Science*, **229**, 651–653.

Potter, A. E. and Morgan, T. H. (1986). Potassium in the atmosphere of Mercury. *Icarus*, **67**, 336–340.

Prettyman, T. H., Hagerty, J. J., Elphic, R. C., Feldman, W. C., Lawrence, D. J., McKinney, G. W. and Vaniman, D. T. (2006). Elemental composition of the lunar surface: Analysis of gamma ray spectroscopy data from Lunar Prospector. *J. Geophys. Res.*, **111**, E12007, doi:10.1029/2005JE002656.

Prettyman, T. H., Yamashita, N., Reedy, R. C., McSween, H. Y., Mittlefehldt, D. W., Hendricks, J. S. and Toplis, M. J. (2015). Concentrations of potassium and thorium within Vesta's regolith. *Icarus*, **259**, 39–52.

Rhodes, E. A., Evans, L. G., Nittler, L. R., Starr, R. D., Sprague, A. L., Lawrence, D. J., McCoy, T. J., Stockstill-Cahill, K. R., Goldsten, J. O., Peplowski, P. N., Boynton, W. V. and Solomon, S. C. (2011). Analysis of MESSENGER Gamma-Ray Spectrometer data from Mercury flybys. *Planet. Space Sci.*, **59**, 1829–1841, doi:10.1016/j.pss.2011.07.018.

Riner, M. A., Bina, C. R., Robinson, M. S. and Desch, S. J. (2008). Internal structure of Mercury: Implications of a molten core. *J. Geophys. Res.*, **113**, E08013, doi:10.1029/2007JE002993.

Rivoldini, A., Van Hoolst, T. and Verhoeven, O. (2009). The interior structure of Mercury and its core sulfur content. *Icarus*, **201**, 12–30.

Robinson, M. S. and Lucey, P. G. (1997). Recalibrated Mariner 10 color mosaics: Implications for mercurian volcanism. *Science*, **275**, 197–200.

Robinson, M. S., Murchie, S. L., Blewett, D. T., Domingue, D. L., Hawkins, S. E., Head, J. W., Holsclaw, G. M., McClintock, W. E., McCoy, T. J., McNutt, R. L., Jr., Prockter, L. M., Solomon, S. C. and Watters, T. R. (2008). Reflectance and color variations on Mercury: Regolith processes and compositional heterogeneity. *Science*, **321**, 66–69, doi:10.1126/science.1160080.

Rubie, D. C., Gessmann, C. K. and Frost, D. J. (2004). Partitioning of oxygen during core formation on the Earth and Mars. *Nature*, **429**, 58–61.

Rubie, D. C., Nimmo, F. and Melosh, H. J. (2015). Formation of the Earth's core. In *Evolution of the Earth*, ed. D. J. Stevenson. *Treatise on Geophysics*, 2nd edn, Vol. 9, ed. G. Schubert. Amsterdam, Oxford: Elsevier, pp. 43–79.

Schlemm, C. E., II, Starr, R. D., Ho, G. C., Bechtold, K. E., Benedict, S. A., Boldt, J. D., Boynton, W. V., Bradley, W., Fraeman, M. E., Gold, R. E., Goldsten, J. O., Hayes, J. R., Jaskulek, S. E., Rossano, E., Rumpf, R. A., Schaefer, E. D., Strohbehn, K., Shelton, R. G., Thompson, R. E., Trombka, J. I. and Williams, B. D. (2007). The X-Ray Spectrometer on the MESSENGER spacecraft. *Space Sci. Rev.*, **131**, 393–415.

Schubert, G., Ross, M. N., Stevenson, D. J. and Spohn, T. (1988). Mercury's thermal history and the generation of its magnetic field. In *Mercury*, ed. F. Vilas, C. R. Chapman and M. S. Matthews. Tucson, AZ: University of Arizona Press, pp. 429–460.

Sharp, Z. D. and Draper, D. S. (2013). The chlorine abundance of Earth: Implications for a habitable planet. *Earth Planet. Sci. Lett.*, **369**, 71–77.

Siebert, J., Badro, J., Antonangeli, D. and Ryerson, F. J. (2013). Terrestrial accretion under oxidizing conditions. *Science*, **339**, 1194–1197, doi:10.1126/science.1227923.

Siegfried, R. W., II and Solomon, S. C. (1974). Mercury: Internal structure and thermal evolution. *Icarus*, **23**, 192–205.

Smith, D. E., Zuber, M. T., Phillips, R. J., Solomon, S. C., Hauck, S. A., Lemoine, F. G., Mazarico, E., Neumann, G. A., Peale, S. J., Margot, J.-L., Johnson, C. L., Torrence, M. H., Perry, M. E., Rowlands, D. D., Goossens, S., Head, J. W. and Taylor, A. H. (2012). Gravity field and internal structure of Mercury from MESSENGER. *Science*, **336**, 214–217.

Starr, R. D., Schlemm, C. E., II, Ho, G. C., Nittler, L. R., Gold, R. E. and Solomon, S. C. (2016). Calibration of the MESSENGER X-Ray Spectrometer. *Planet. Space Sci.*, **122**, 13–25.

Stockstill-Cahill, K. R., McCoy, T. J., Nittler, L. R., Weider, S. Z. and Hauck, S. A., II (2012). Magnesium-rich crustal compositions on Mercury: Implications for magmatism from petrologic modeling. *J. Geophys. Res.*, **117**, E00L15, doi:10.1029/2012je004140.

Taylor, G. J. and Scott, E. R. D. (2003). Mercury. In *Meteorites, Comets, and Planets*, ed. A. M. Davis. *Treatise on Geochemistry*, Vol. 1, ed. H. D. Holland and K. K. Turekian. Oxford: Pergamon, pp. 477–485.

Taylor, G. J., Stopar, J. D., Boynton, W. V., Karunatillake, S., Keller, J. M., Brückner, J., Wänke, H., Dreibus, G., Kerry, K. E., Reedy, R. C., Evans, L. G., Starr, R. D., Martel, L. M. V., Squyres, S. W., Gasnault, O., Maurice, S., d'Uston, C., Englert, P., Dohm, J. M., Baker, V. R., Hamara, D., Janes, D., Sprague, A. L., Kim, K. J., Drake, D. M., McLennan, S. M. and Hahn, B. C. (2006). Variations in K/Th on Mars. *J. Geophys. Res.*, **111**, E03S06, doi:10.1029/2006JE002676.

Terasaki, H., Ohtani, E., Sakai, T., Kamada, S., Asanuma, H., Shibazaki, Y., Hirao, N., Sata, N., Ohishi, Y., Sakamaki, T., Suzuki, A. and Funakoshi, K.-i. (2012). Stability of Fe-Ni hydride after the reaction between Fe–Ni alloy and hydrous phase (δ-AlOOH) up to 1.2 Mbar: Possibility of H contribution to the core density deficit. *Phys. Earth Planet. Inter.*, **194–195**, 18–24, doi:10.1016/j.pepi.2012.01.002.

Tosi, N., Grott, M., Plesa, A. C. and Breuer, D. (2013). Thermochemical evolution of Mercury's interior. *J. Geophys. Res.*, **118**, 2474–2487, doi:10.1002/jgre.20168.

Trombka, J. I., Squyres, S. W., Bruckner, J., Boynton, W. V., Reedy, R. C., McCoy, T. J., Gorenstein, P., Evans, L. G., Arnold, J. R., Starr, R. D., Nittler, L. R., Murphy, M. E., Mikheeva, I., McNutt, R. L., Jr., McClanahan, T. P., McCartney, E., Goldsten, J. O., Gold, R. E., Floyd, S. R., Clark, P. E., Burbine, T. H., Bhangoo, J. S., Bailey, S. H. and Petaev, M. (2000). The elemental composition of asteroid 433 Eros: Results of the NEAR-Shoemaker X-ray spectrometer. *Science*, **289**, 2101–2105.

Tsuno, K., Frost, D. J. and Rubie, D. C. (2011). The effects of nickel and sulphur on the core-mantle partitioning of oxygen in Earth and Mars. *Phys. Earth Planet. Inter.*, **185**, 1–12.

Tsuno, K., Frost, D. J. and Rubie, D. C. (2013). Simultaneous partitioning of silicon and oxygen into the Earth's core during early Earth differentiation. *Geophys. Res. Lett.*, **40**, 66–71, doi:10.1029/2012GL054116.

Vander Kaaden, K. E. and McCubbin, F. M. (2015). Exotic crust formation on Mercury: Consequences of a shallow, FeO-poor mantle. *J. Geophys. Res. Planets*, **120**, 195–209.

Vander Kaaden, K. E. and McCubbin, F. M. (2016). The origin of boninites on Mercury: An experimental study of the northern volcanic plains lavas. *Geochim. Cosmochim. Acta*, **173**, 246–263, doi:10.1016/j.gca.2015.10.016.

Vander Kaaden, K. E., McCubbin, F. M., Ross, D. K., Rapp, J. F., Danielson, L. R., Keller, L. P. and Righter, K. (2016). Carbon solubility in Si-Fe-bearing metals during core formation on Mercury. *Lunar Planet. Sci.*, **47**, abstract 1474.

Vilim, R., Stanley, S. and Hauck, S. A., II (2010). Iron snow zones as a mechanism for generating Mercury's weak observed magnetic field. *J. Geophys. Res.*, **115**, E11003, doi:10.1029/2009JE003528.

Wasson, J. T. (1988). The building stones of the planets. In *Mercury*, ed. F. Vilas, C. R. Chapman and M. S. Matthews. Tucson, AZ: University of Arizona Press, pp. 622–650.

Weider, S. Z., Kellett, B. J., Swinyard, B. M., Crawford, I. A., Joy, K. H., Grande, M., Howe, C. J., Huovelin, J., Narendranath, S., Alha, L., Anand, M., Athiray, P. S., Bhandari, N., Carter, J. A., Cook, A. C., d'Uston, L. C., Fernandes, V. A., Gasnault, O., Goswami, J. N., Gow, J. P. D., Holland, A. D., Koschny, D., Lawrence, D. J., Maddison, B. J., Maurice, S., McKay, D. J., Okada, T., Pieters, C., Rothery, D. A., Russell, S. S., Shrivastava, A., Smith, D. R. and Wieczorek, M. (2012a). The Chandrayaan-1 X-ray Spectrometer: First results. *Planet. Space Sci.*, **60**, 217–228, doi:10.1016/j.pss.2011.08.014.

Weider, S. Z., Nittler, L. R., Starr, R. D., McCoy, T. J., Stockstill-Cahill, K. R., Byrne, P. K., Denevi, B. W., Head, J. W. and Solomon, S. C. (2012b). Chemical heterogeneity on Mercury's surface revealed by the MESSENGER X-Ray Spectrometer. *J. Geophys. Res.*, **117**, E00L05, doi:10.1029/2012je004153.

Weider, S. Z., Nittler, L. R., Starr, R. D., McCoy, T. J. and Solomon, S. C. (2014). Variations in the abundance of iron on Mercury's surface from MESSENGER X-Ray Spectrometer observations. *Icarus*, **235**, 170–186, doi:10.1016/j.icarus.2014.03.002.

Weider, S. Z., Nittler, L. R., Starr, R. D., Crapster-Pregont, E. J., Peplowski, P. N., Denevi, B. W., Head, J. W., Byrne, P. K., Hauck, S. A., II, Ebel, D. S. and Solomon, S. C. (2015). Evidence for geochemical terranes on Mercury: Global mapping of major elements with MESSENGER's X-Ray Spectrometer. *Earth Planet. Sci. Lett.*, **416**, 109–120.

Weider, S. Z., Nittler, L. R., Murchie, S. L., Peplowski, P. N., McCoy, T. J., Kerber, L., Klimczak, C., Ernst, C. M., Goudge, T. A., Starr, R. D., Izenberg, N. R., Klima, R. L. and Solomon, S. C. (2016).

Evidence from MESSENGER for sulfur- and carbon-driven explosive volcanism on Mercury. *Geophys. Res. Lett.*, **43**, 3653–3661.

Williams, Q. and Hemley, R. J. (2001). Hydrogen in the deep Earth. *Annu. Rev. Earth Planet. Sci.*, **29**, 365–418.

Wurz, P., Whitby, J. A., Rohner, U., Martín-Fernández, J. A., Lammer, H. and Kolb, C. (2010). Self-consistent modelling of Mercury's exosphere by sputtering, micro-meteorite impact and photon-stimulated desorption. *Planet. Space Sci.*, **58**, 1599–1616.

Zolotov, M. Yu., Sprague, A. L., Hauck, S. A., II, Nittler, L. R., Solomon, S. C. and Weider, S. Z. (2013). The redox state, FeO content, and origin of sulfur-rich magmas on Mercury. *J. Geophys. Res. Planets*, **118**, 138–146.

3

Mercury's Crust and Lithosphere: Structure and Mechanics

ROGER J. PHILLIPS, PAUL K. BYRNE, PETER B. JAMES, ERWAN MAZARICO,
GREGORY A. NEUMANN, AND MARK E. PERRY

3.1 INTRODUCTION

3.1.1 Chapter Overview

This chapter summarizes our current knowledge of the basic structure of the crust and lithosphere of Mercury, emphasizing the influence of MESSENGER observations. We begin by describing the data and analyses that provide an estimate of Mercury's shape. We follow with descriptions of the MESSENGER gravity field models, which relate directly to the global mass distribution down to and including the core and are a primary input to geophysical models and interpretation of shape. We show how gravity and shape together are used to obtain crustal thickness models and why these models are non-unique but nevertheless useful. The "static" crustal thickness model is derived by assuming that the observed gravity is the sum of contributions from just the shape and the crust–mantle boundary, and no compensation mechanism is explicitly invoked. We also review models that include some form of mechanical equilibrium, including the contribution of dynamic flow in the mantle. We discuss models with an elastic lithosphere (thickness T_e), which modulates the deformation of the crust–mantle boundary, and we review the rather limited number of estimates of T_e currently available for Mercury, with an emphasis on localized admittance techniques. We also compare large-scale tectonic landforms (or sets of such landforms) with regional crustal thickness variations, showing the utility of adding tectonic constraints to geophysical models of Mercury's interior.

We pay special attention to the spherical-harmonic-degree-2 (orders 0 and 2) components of the shape and geoid because of their historical significance in planetary geodesy, their relationship to the spin and orbital history of Mercury, and their anomalously high power relative to spherical harmonic terms of higher degree and order (i.e., terms that have shorter wavelengths). The emphasis here is on mechanisms that support the large non-hydrostatic component of the degree-2 shape, and we provide a comparison with the lunar case, which has been studied for over two centuries. We thoroughly review and critique the only current model that can account for Mercury's degree-2 shape and geoid. This model (Tosi et al., 2015) tests the hypothesis that these quantities can result from the temperature distribution in the lithosphere arising from the long-wavelength pattern of surface temperature associated with Mercury's 3:2 spin–orbit resonance and near-zero obliquity. We note that the use of degree-2 geoid information to constrain Mercury's basic radial structure (solid inner core, fluid outer core, and solid outer shell) is treated in Chapter 4.

The final section of this chapter is a summary that attempts to sort out what we know and what we do not know about the crust and lithosphere of Mercury, as well as the considerable information we might still obtain from the MESSENGER data sets.

3.1.2 A Word on Nomenclature and Other Matters

Throughout this chapter, the term "shape" refers to the radial distances from Mercury's center of mass to the surface less the radius of a reference sphere, which is usually set to 2440 km, slightly larger than the mean radius of 2439.4 km. The term "topography," in contrast, is the shape referenced to a geoid, a gravitational equipotential surface with a mean radius of 2440 km. In this chapter, we use "shape" almost exclusively; but in the community at large the terms "shape" and "topography" are sometimes used interchangeably or "shape" is delegated to the long-wavelength portion of topography (though "long" is often not defined). Here we lapse occasionally into using "topography" when describing local to regional features, though we are ignoring its definition, *sensu stricto*.

We use longitudes that are defined in a principal-axis framework (Mazarico et al., 2014).

Additionally, we use the term "compensated" in a qualitative sense to imply that the magnitude of the observed gravity signal is less than that expected from the shape alone. The description "fully compensated" or "isostasy" is reserved for the conditions of equal mass per unit area over some defined region as evaluated at an interior reference level.

3.1.3 Introducing Spherical Harmonics

Because spherical harmonics are used throughout this chapter, we provide a brief introduction to the topic here. Spherical harmonics (SH) are the natural set of basis functions to represent data on a sphere, just as the sine and cosine terms that make up a Fourier series are a common choice in the Cartesian coordinate system. Spherical harmonics are a complete set of orthogonal functions, and data distributed on a sphere can be represented with a linear series of SH functions (Wieczorek and Simons, 2005). For some latitude θ and longitude φ, a spherical harmonic, $\overline{Y}_{lm}(\theta, \varphi)$, of degree l and order m, is given by

$$\overline{Y}_{lm}(\theta, \varphi) = \overline{P}_{lm}(\sin \theta) \begin{bmatrix} \cos(m\varphi) \\ \sin(m\varphi) \end{bmatrix}, \tag{3.1}$$

where $\overline{P}_{lm}(\sin\theta)$ is an associated Legendre polynomial. Data X on and in a sphere can be represented by

$$X(r,\theta,\varphi) = \psi(r)\sum_{l=0}^{l_{max}}\sum_{m=0}^{l} f_l(r)\overline{P}_{lm}(\sin\theta)[\overline{C}_{lm}\cos(m\varphi) + \overline{S}_{lm}\sin(m\varphi)], \quad (3.2)$$

where $f_l(r)$ is a degree-dependent radial solution (for potential fields it is the continuation term) and $\psi(r)$ is a degree-and-order-independent function of the radius r. The solution coefficients $\overline{C}_{lm}$ and $\overline{S}_{lm}$ can be found by integrating the product of the data and the SH over the sphere for fixed r, though in practice with MESSENGER data, solutions for both the shape and geoid are obtained by underdetermined least-squares procedures. However, estimation of global basis function coefficients from partial data coverage can introduce undesirable correlations in their uncertainties. Spurious correlations of the coefficients are important, particularly at low degrees, because the coefficients have physical meaning (e.g., C_{20} represents polar flattening). The overbars indicate that the Legendre polynomials and thus the solution coefficients have been normalized to promote computational stability for expansions to high degree and order. Here we use the geodesy (or Kaula) normalization, where $\overline{Y}_{lm}^2(\theta,\varphi)$ integrated over the sphere yields 4π. The relationship between an un-normalized and a normalized coefficient is $C_{lm} = N_{lm}\overline{C}_{lm}$, where

$$N_{lm} = \{[(2-\delta_{0m})(2l+1)(l-m)!]/(l+m)!\}^{1/2} \quad (3.3)$$

and δ_{0m} is a Kronecker delta, and a parallel relation holds for S_{lm}. It follows from equation (3.2) that $C_{lm}P_{lm}(\sin\theta) = \overline{C}_{lm}\overline{P}_{lm}(\sin\theta)$. We use upper case ($C_{lm}$, S_{lm}) to represent gravity coefficients, lower case (c_{lm}, s_{lm}) to represent shape coefficients, and ($^T C_{lm}$, $^T S_{lm}$) to represent coefficients in a spherical harmonic expansion of Mercury's surface temperature, $T(\theta,\varphi) \equiv T_{\theta\varphi}$. Additionally, we use d_l to designate spherical harmonic degree l, and, unless we state otherwise, all orders m are included.

3.1.4 MESSENGER's Orbit

MESSENGER was inserted into orbit about Mercury on 18 March 2011 after 6.6 years of interplanetary cruise (Chapter 1). Primarily because of thermal constraints, the spacecraft was placed in an eccentric orbit with an initial inclination of 82.5°. The periapsis was initially at 200-km altitude and ~60°N latitude. During the one-year primary mission, the gravitational perturbations from the Sun raised the periapsis, particularly rapidly near the noon–midnight geometry, and six maneuvers were executed to maintain the periapsis altitude between 200 and 500 km. The gravitational perturbations also generally increased the periapsis latitude and the orbit inclination. For most of MESSENGER's instruments, the eccentric orbit caused data coverage and quality to vary with latitude. The thermal constraints on spacecraft attitude also produced longitudinal variations for most data sets.

Shortly after the first (Earth) year of orbital operation, the semi-major axis of MESSENGER's orbit was shortened such that the apoapsis was reduced from ~15,000 to ~10,000 km. This change lowered the orbital period to 8 h, providing more frequent low-altitude observations. The periapsis altitude was permitted to increase, reaching a maximum of 457 km with a periapsis latitude of 84.06°N on 7 February 2013. Subsequent to reaching these maxima, the strong secular effects of the gravitational pull of the Sun on the orbit decreased the periapsis altitude by as much as 90 km in ~44 days, limiting the duration of the MESSENGER mission. Through careful planning (McAdams et al., 2015), the remaining fuel was utilized to extend the mission and at the same time maximize the opportunities for conducting science at very low altitudes. From August 2014 to April 2015, periapses were typically <100 km and as low as ~5 km. The solar perturbations also caused the argument of periapsis to drift southward, eventually below the initial latitude of 60°N. These extremely low altitudes produced measurements with exceptionally high resolution for some of the investigations, including the planet's gravity field and internal magnetic field.

3.2 SHAPE OBSERVATIONS

There is no single set of observations that has sufficient coverage and quality to characterize the shape of Mercury. A combination of Earth-based radar observations (Harmon et al., 1986; Anderson et al., 1996), Mariner 10 images (Strom et al., 1975), and radio frequency (RF) occultations (Fjeldbo et al., 1976) of signals from Mariner 10 provided the first measurements of Mercury's radius and equatorial shape. On the basis of those results, the International Astronomical Union adopted a value in 1991 of 2439.7 km for Mercury's mean radius (Seidelmann et al., 2002). Given the considerable uncertainty in its shape, however, the MESSENGER project adopted a mean radius of 2440 km, with an uncertainty of 1 km, as a spherical datum for geophysical studies (Zuber et al., 2007, Table 1). Orbital observations by MESSENGER allow the shape to be determined by means of stereophotogrammetry (Oberst et al., 2011), stereophotoclinometry (Gaskell et al., 2008), limb profiles (Elgner et al., 2014), radio occultation (Perry et al., 2011), and laser altimetry, the most accurate of these measurements.

3.2.1 Mercury Laser Altimeter Observations of Mercury's Northern Hemisphere

The Mercury Laser Altimeter (MLA) (Cavanaugh et al., 2007) was designed to measure the shape of nearly the entire northern hemisphere via laser pulse time-of-flight data and spacecraft orbit position data, in an approach similar to that used with the Mars Orbiter Laser Altimeter (MOLA) at Mars (Zuber et al., 1992). The MLA was a single-beam laser altimeter with a beam divergence of 80 microradians; the beam illuminated a 20–100-m-diameter spot on the surface at nadir, depending on spacecraft altitude. The 8-Hz laser pulse rate provided a distance between along-track measurements of ~400 m. The MLA instrument was calibrated during cruise following an Earth flyby in 2005 (Smith et al., 2006); the calibration verified range precision at the sub-meter level and provided post-launch alignment corrections in weightless space of several

Table 3.1. *Altimetric data returned by the Mercury Laser Altimeter.*

Year	Orbits	Measurements
2008	2	7,959
2011	468	5,123,031
2012	774	6,283,515
2013	918	6,717,154
2014	812	6,530,210
2015	308	2,346,321
Total	3282	27,000,231

Note: The "orbits" in 2008 were Mercury flybys 1 and 2.

milliradians with respect to the pre-launch survey data. During altimetric operations, return pulses were detected on a single high-threshold trigger or on multiple low-threshold triggers that aided in separating ground returns from the noise.

The first two flybys of Mercury by MESSENGER in January and October 2008 provided the first direct altimetric measurements of Mercury's shape from space (Smith et al., 2010a). Each flyby measured the shape over ~80° of longitude on opposite hemispheres, within a narrow range of latitudes from 5°S to the equator. These flyby data confirmed with much smaller uncertainty the earlier results on the ellipticity and orientation of the equatorial shape (Anderson et al., 1996).

While in orbit about Mercury, the MLA operated during 80% of the orbital passes, making more than 26 million altimetry measurements (Table 3.1), primarily in the northern hemisphere because of MESSENGER's highly eccentric orbit and high northern periapsis, which left most of the southern hemisphere beyond MLA's maximum range. The boresight was nominally nadir-pointed for 20–45 min when MESSENGER was within 1500 km of the surface. Frequently, nadir pointing was preempted by other activities or prohibited by the spacecraft thermal constraints, which limited the viewing angles on many orbits. Moreover, the dawn–dusk orbits, which had the most favorable geometry, provided access only to portions of the planet, and the bulk of observations at low latitudes were off-nadir. The periapsis latitude and altitude were lowest at the end of the mission, providing access to lower latitudes, and MLA acquired ~75,000 measurements south of the equator and ~10,000 between 8°S and 15°S. The densest coverage was between 65°N and 84°N (Neumann et al., 2016).

The extended mission operations allowed for a denser-than-anticipated survey of the northern hemisphere, particularly north of 80°N, where the ~84° maximum inclination of the spacecraft orbit required off-nadir measurements. Although there are coverage gaps of >1° in longitude south of 60°N, redundant coverage within the polar region reduced most gaps to less than 1 km² at 78°–84°N latitude. A campaign of off-nadir observations partially filled in the higher latitudes on orbits during which the spacecraft could safely roll the MLA instrument deck to point northward. Oblique ranges, at slant angles of up to 50°, were less reliable than nadir observations but provided essential data within regions of persistent shadow.

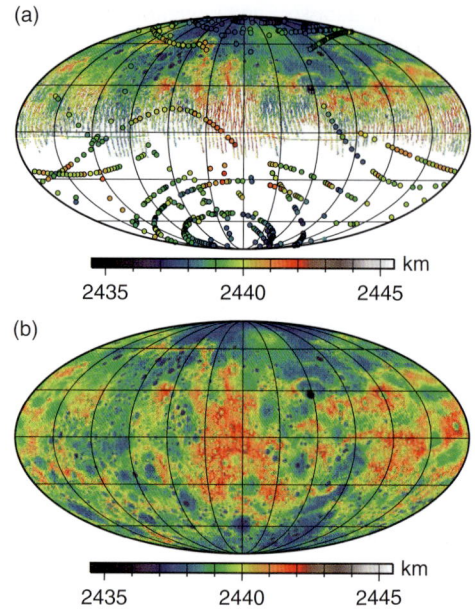

Figure 3.1. (a) Locations of radius measurements from RF occultations (circles) and MLA binned data (updated from Perry et al., 2015) and (b) the U.S. Geological Survey DEM constructed from MDIS WAC and NAC stereo images, with the bundle adjustment controlled by MLA in the northern hemisphere. Mollweide projections centered at 0°E longitude.

3.2.2 Radio Occultations Provide Shape Data for the Southern Hemisphere

Understanding global shape and differences in geological settings between hemispheres required southern hemisphere measurements that were not possible with MLA. The Radio Science investigation provided these measurements with high-resolution time series of radio signal diffraction cutoffs as MESSENGER was occulted from Earth by the Mercury limb (Perry et al., 2015). The occultation start and end times, recovered with 0.1-s accuracy or better by fitting edge-diffraction patterns to the RF power history, constrain the planetary radius perpendicular to the line of sight. The uncertainty in such radii is ~100–200 m for a typical geometry of MESSENGER, Mercury, and Earth. Occultation measurements preferentially sample the higher points along the RF path, resulting in upward bias. To correct for this bias and improve registration, an uncontrolled stereo elevation map of local terrain was used to locate the most prominent local topographic feature along the Earth–spacecraft RF path. The raw occultation measurement was then corrected to obtain a control point that represents the average elevation of the terrain surrounding the occultation location. The reliability of the southern hemisphere results was established by conducting the same analysis in the northern hemisphere and demonstrating that the northern occultations are consistent with MLA measurements.

Of the 557 occultations analyzed, 359 are in the southern hemisphere. Radio occultations are unaffected by spacecraft orientation or thermal distortions and place an important constraint on the low-degree shape. The occultation points are shown in Figure 3.1a. Large gaps in coverage remain in the southern hemisphere, but there are sufficient points to constrain the shape and its power spectral density up to degree and order 8. With correlations between the coefficients in the upper part of this SH band, the individual accuracy of higher-degree-and-order terms is reduced.

3.2.3 Image-Based Sources of Shape Information

Additional information was obtained from stereo imaging and limb scans by the Mercury Dual Imaging System (MDIS) wide-angle camera (WAC). Starting with the first two flybys (Oberst et al., 2010, 2011), these observations produced an estimate of mean radius and detailed, 1-km-resolution topographic maps of quadrangle-sized areas covering 30% of the surface. However, flyby geometry was not suitable to determine the overall shape, and the flattening of the poles was unresolved. Limb profile analysis by least-squares adjustment (Elgner et al., 2014) resolved the equatorial ellipticity and polar flattening, but the results were biased upwards by hundreds of meters when compared with the MLA and occultation data. This difference is not unexpected, as a limb image pixel spans several kilometers and, because of the curvature of Mercury's surface, an 800-m-high crater rim within ~40 km of the actual occultation point on the limb was sufficient to hide the surface.

Orbital observations included extensive stereo imaging in the southern hemisphere. Achieving the required 250-m resolution necessitated use of the MDIS narrow-angle camera (NAC). NAC images were merged with control networks derived from the WAC at lower resolutions. A bundle-adjusted control network consisting of 12,596,336 radii from the collocation of images acquired by both cameras provided a global digital elevation model (DEM) (Becker et al., 2012, 2016) that includes the southern hemisphere (Figure 3.1b). The stereo imaging analysis contains long-wavelength biases relative to MLA and RF occultations, possibly resulting from non-uniform illumination conditions, data density, or uncertainty in the NAC camera model (Neumann et al., 2016).

3.2.4 Shape Results and Spherical Harmonic Representations

In the first year of orbital operation, an MLA-based map of the northern hemisphere was obtained (Zuber et al., 2012). That map showed a ~10.4 km dynamic range of shape, from the deepest points in Prokofiev's innermost crater floor (85.48°N, 63.67°E; –6.26 km) to the northern rim of Raditladi crater (30.4°N, 116.9°E; +4.14 km), and a north polar radius of 2437.57 km. Figure 3.2a is a DEM of Mercury's northern hemisphere based entirely on MLA data (Neumann et al., 2016).

The spherical harmonic expansion for Mercury's shape is given by equation (3.2), with $\Psi(r) = f_l(r) = 1$ and with the coefficients in meters or kilometers. Both MLA and occultation data show a systematic flattening at the poles relative to the equator (Figure 3.3). To first order, this flattened shape can be approximated by an oblate spheroid with its short axis closely aligned with the rotation axis, and it can be represented by a subset of the d_2 spherical harmonic shape coefficients. However, poleward of 70°N and 70°S, the longitudinally

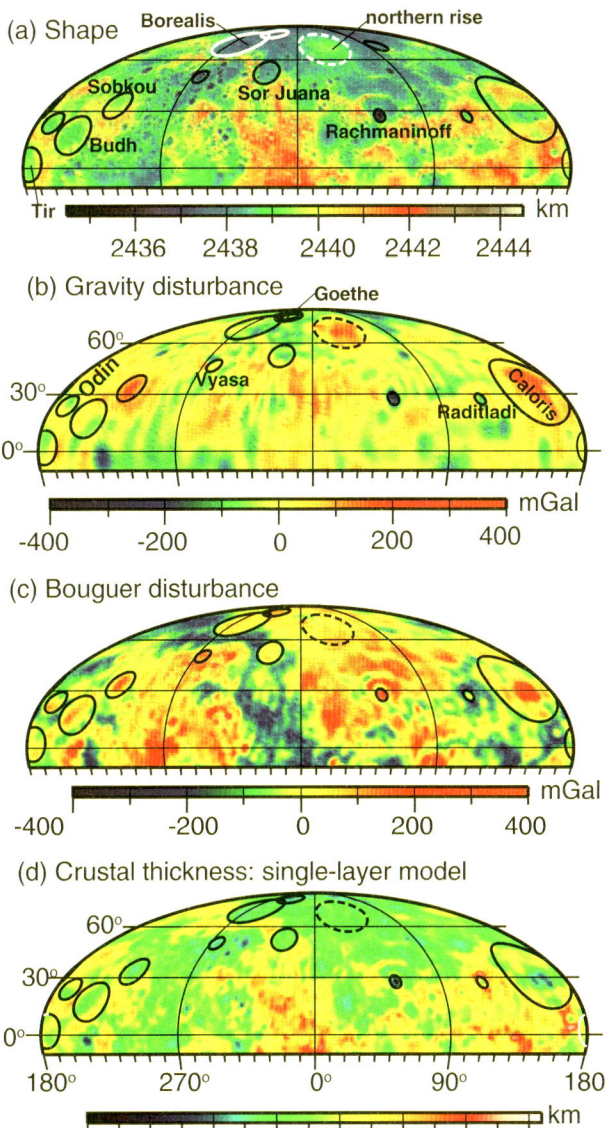

Figure 3.2. Mercury's shape, gravity field, Bouguer anomaly, and derived crustal thickness from MLA and gravity observations, expanded in spherical harmonics to degree (*l*) and order (*m*) 60. Craters and impact basins discussed in Section 3.6.5 are circled; the northern rise is outlined with dashed lines. (a) Shape from file gtmes_150_v05_sha.tab (Neumann et al., 2016). (b) Gravity disturbance from HgM007. The streaks are artifacts of MESSENGER's varying orbit (see Section 3.3.4). (c) Bouguer disturbance from MLA data and gravity analyses for a crustal density of 2900 kg/m³. (d) Crustal thickness derived for a single-layer model from the Bouguer anomaly and a crust–mantle density contrast of 200 kg/m³. Mollweide projections centered at 0°E, from 10°S to 90°N latitude; the projection distorts circles into ellipses.

averaged shape of Mercury from MLA and RF occultations is ~1 km deeper than the d_2 shape (Figure 3.3). A possible explanation for at least part of the northern depression is that the region contains two large impact basins (Borealis and Goethe), which host the widespread northern smooth plains (NSP), likely of volcanic origin (Head et al., 2011; Denevi et al., 2013). The crater Chao Meng-Fu (130-km diameter)

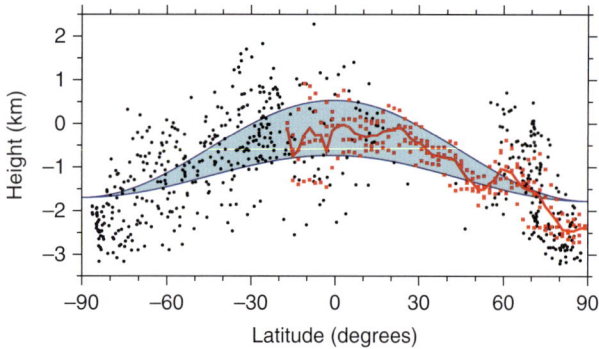

Figure 3.3. Occultation measurements of height (black circles) and longitudinal quadrant averages of MLA data (red squares) versus latitude, referenced to a sphere of radius $R_0 = 2440$ km. The blue background gives the maximum and minimum flattening of the shape from an oblate spheroid fit to the combined data. The red curve shows a longitudinal average of MLA data.

near the south pole may account for some of the southern discrepancy, but the origin of the anomalous depression is less certain than its northern counterpart. The d_2 shape is discussed in detail in Section 3.7; its origin was one of the major geophysical questions lingering at the end of the MESSENGER mission. The depressions that depart from the d_2 shape near the two poles also have a contribution from the d_4 terms of ~300 m, most of it poleward of 70°, and this effect is likely related to the influence that the long-wavelength insolation pattern has on the shape (Section 3.7).

The x, y, and z components (where x and y lie within Mercury's equatorial plane along longitudes 0° and 90°E, respectively, and z points northward along the spin axis) of the offset between the center of figure and the center of mass (COF–COM) are given by c_{11}, s_{11}, and c_{10}, respectively, and the estimated COF–COM offset vector, updated from Perry et al. (2015), is $[-14 \pm 40, 79 \pm 50, -100 \pm 50]$ m. The magnitude of this vector is small compared with the corresponding values for the Moon (Smith et al., 2010b), Earth (Melosh, 2011), and Mars (Smith et al., 1999). Values for the Moon and Earth likely result from hemispheric-scale variations in crustal thickness – the nearside–farside asymmetry for the former, and the Pacific Ocean basin and its antipodal, dominantly continental hemisphere for the latter. The Mars COF–COM z component reflects the thicker crust of the southern highlands relative to the northern lowlands, whereas the x and y components are influenced by the Tharsis rise and Hellas impact basin. Venus, like Mercury, has a small COF–COM offset (Bindschadler et al., 1994), suggesting that planetary size is not an important parameter, at least in a simple way, for influencing the magnitude of the COF–COM offset.

3.3 THE GRAVITY FIELD OF MERCURY

Planetary geodesy has a rich history of gravitational measurements, starting with the first detection of the lunar mass concentrations (mascons) beneath the Moon's large nearside impact basins, as determined from Lunar Orbiter tracking observations (Muller and Sjogren, 1968). The presence of these gravity anomalies greatly complicated mission planning for the Apollo Moon landings. Gravity features are derived from the perturbations they impart to spacecraft trajectories, which are generally observable by tracking spacecraft radio signals from ground stations or by satellite-to-satellite tracking (SST). Mapping resolutions achieved from planetary missions vary from high resolution at the Moon with SST of the dual Gravity Recovery and Interior Laboratory (GRAIL) spacecraft to low-resolution mapping of Europa from Galileo flybys, but even the lowest-order measurements play an important role in determining the largest-scale mass distributions and deep interior structure. At Mercury, MESSENGER's eccentric orbit and varying periapsis altitude resulted in a variable-resolution set of gravity measurements across the planet.

3.3.1 Spherical Harmonic Representation and Limitations

As with the shape, the gravitational potential, $\Delta V(r,\theta,\varphi)$, is typically expressed in terms of spherical harmonics with coefficients C_{lm} and S_{lm}. Equation (3.2) gives the expansion with $\Psi(r) = GM/r$ and $f_l(r) = (R_0/r)^l$, where G is the gravitational constant, R_0 is the reference radius, and M is the mass of Mercury.

Lower-degree terms describe the longer wavelengths of the gravity field, with C_{20} corresponding to the polar flattening of the field and C_{22} to its equatorial ellipticity. The spatial resolution of a field determined to degree l is $\pi R_0/l$ (half wavelength), and the number of necessary parameters, $l(l + 1)$, grows quadratically with resolution. Typically, gravity data sets used to estimate the C_{lm} and S_{lm} coefficients are not uniform spatially in terms of sampling, sensitivity, or both, and the introduction of a constraint is required. Although this regularization stabilizes the least-squares system used to find the coefficients, it still results in correlated terms. Thus, the gravitational field can often be reliably determined spatially, but the SH coefficients may not be well known individually.

3.3.2 Data and Geometry

A planet's gravity and mass distribution can be derived from spacecraft RF transmissions received by the Deep Space Network (DSN), NASA's network of Earth-based stations, which provides complete and continuous sky coverage of Mercury (Kegege et al., 2012). The RF data produce range and Doppler (range–rate) measurements that, after correction for a variety of geometric and environmental effects, relate to the line-of-sight distance and velocity of the spacecraft with respect to the ground station. These data allow for the precise reconstruction of the spacecraft trajectory, which depends on the planet's mass distribution. Discrepancies between the actual observations and the predictions of the force and measurement models are minimized through an iterative process, called "orbit determination," through which model parameters, such as the SH coefficients (C_{lm}, S_{lm}), are estimated and improved.

Although some spacecraft (such as Mars Global Surveyor and Cassini) have a high-fidelity frequency source to produce accurate timing for one-way, direct-to-Earth signals, spacecraft transmissions are usually referenced to a stable carrier signal generated at a ground station and transmitted to the spacecraft. The spacecraft re-transmits the received signal after frequency multiplication and telemetry modulation. These two-way measurements are of high quality but are affected by media delays twice, a situation that can affect their quality in certain geometries (Asmar and Armstrong, 2005; Srinivasan et al., 2007).

Tracking data from Mercury were first obtained by Mariner 10 at S-band (~2 GHz) frequencies during the spacecraft's three flybys in 1974–1975. From two of these flybys – one equatorial (~700 km closest-approach altitude) and one polar (~300 km closest-approach altitude) – a more accurate value of Mercury's mass was obtained along with the first estimates of C_{20} and C_{22} (Anderson et al., 1987). The uncertainties in these two coefficients were the dominant source of error in the estimation of the polar moment of inertia of Mercury until much better estimates were obtained from the orbital phase of the MESSENGER mission (see Chapter 4). The first MESSENGER tracking data from Mercury were obtained during its three flybys in 2008 and 2009.

The sensitivity of spacecraft velocity to subsurface mass distributions depends inversely on altitude. Although the additional constraint of Earth visibility to radio tracking limited the longitudinal and altitude coverage, the data collected at low altitudes in the last year of the MESSENGER mission proved invaluable to determining the shorter-wavelength, regional-scale gravity field. The sensitivity of tracking data to Mercury's orientation (pole position and spin rate) depended on periapsis latitude and was highest at the beginning and end of the mission, when the periapsis was at the lowest latitudes.

3.3.3 Methods and Data Analysis

MESSENGER tracking data were processed with the GEODYN II orbit determination and geodetic parameter estimation software, developed and maintained at the NASA Goddard Space Flight Center (Pavlis et al., 2013). GEODYN has been used in a number of geodetic and geophysical investigations in planetary science, from Clementine (Lemoine et al., 1997) to the Mars Reconnaissance Orbiter (Mazarico et al., 2008) and GRAIL (Zuber et al., 2013; Lemoine et al., 2013) missions. GEODYN uses detailed force calculations to model the observations for fitting and integrating the spacecraft trajectory (Lemoine et al., 2001; Mazarico et al., 2014). The measurement models include the various reference frames, the time system transformations, the shape of the spacecraft along with reflectivity coefficients of each of its surfaces, the ranges between spacecraft and tracking stations, and measurement corrections such as atmospheric delays and phase-center location. In addition to the central body and third-body gravitational forces, GEODYN models non-conservative accelerations such as those caused by solar radiation, planetary reflectance, and thermal radiation. The MESSENGER science team analyzed the tracking data to recover the gravity field, orientation, and ephemeris (Smith et al., 2012; Mazarico et al., 2014), and the navigation team, at KinetX, Inc., used the same data for maneuver design and for orbit prediction and reconstruction (Page et al., 2014).

The tracking data were first processed in short segments or "arcs" that were typically one day in length, with start and stop times selected to avoid propulsive maneuvers. In total, 1499 arcs were created to cover the orbital mission. Each arc was converged individually to achieve the best data fit. Only parameters affecting the data in this arc were adjusted, such as the spacecraft initial state and the radiation pressure coefficient, C_R, which is a scale factor that represents the effective, cross-sectional areas of the modeled spacecraft surfaces.

From the converged arcs, the partial derivatives of the tracking data measurements with respect to the variables were computed and transformed into a series of normal equation systems, which were then combined and inverted. The computed parameters included both arc parameters and global variables such as the SH coefficients (C_{lm} and S_{lm}) and the tidal Love number k_2. When solving for gravity expansions above degree and order 9, a regularization constraint such as the Kaula rule (Kaula, 1966) prevented the higher-degree coefficients from absorbing an excessive portion of the radio residuals. The Kaula rule provides weighting of the solution coefficients by the inverse of a variance K/l^2, where K is a constant. Although K was expected to be $\sim 4 \times 10^{-5}$ by scaling from other planetary gravity spectra, the lack of sensitivity in the southern hemisphere led to a better solution with $K \sim 1.25 \times 10^{-5}$.

As the mission progressed, this process was repeated, iterating the solution with each new set of data. Residual errors in the fits to the Doppler data were typically 0.2–0.3 mm/s, consistent with the intrinsic noise expected from solar plasma degradation of the radio data when the Sun–probe–Earth (SPE) angle was less than 20°. Range data were fit to a root-mean-squared (RMS) residual of ~5 m but required the use of biases when adjustments to Mercury's ephemeris were not incorporated into the analysis (Mazarico et al., 2014).

3.3.4 Gravity Results

The MESSENGER pre-orbit-insertion gravity recovery was relatively weak because of limited data and restricted geometry – two equatorial flybys (Smith et al., 2010a) – so the improvement over the Mariner 10 results (Anderson et al., 1987) was also limited. Analysis of the first six months of MESSENGER orbital data (Smith et al., 2012) yielded a substantial reduction in the uncertainties in the gravity coefficients C_{20} and C_{22}, leaving obliquity as the dominant contributor to the uncertainty in estimates of the polar moment of inertia and the ratio of the moment of the solid outer shell to that quantity (Margot et al., 2012; Hauck et al., 2013; Chapters 4 and 19).

Mazarico et al. (2014) obtained a solution based on nearly three Earth years of radio tracking data. Even though this analysis did not include the last 8 months of MESSENGER operations at Mercury, the solution, archived as HgM005, reflects well the state of understanding of Mercury's interior (see Chapter 4) before periapsis altitudes decreased below 200 km. Here, we present the results of a later gravity solution, HgM007, derived from the complete MESSENGER data set (Mazarico et al., 2016). The low-altitude coverage clearly shows short-wavelength variations in gravity in the northern hemisphere, but the sparse and heterogeneous altitude coverage produces variable resolution.

The HgM007 gravity field displayed in Figure 3.2b is formally termed the "gravity disturbance," $\delta g = g_n - \gamma_n$, the total gravity field (g_n) minus a reference field (γ_n). Both fields are usually calculated on the geoid, indicated by the subscript n. The geoid is expressed as a deviation from a simpler surface, the "reference ellipsoid." For Earth, the reference ellipsoid is biaxial (i.e., an oblate spheroid), so can be characterized by polar flattening terms (Section 3.7.1), and is tied to mean sea level. A spherical surface of radius R_0 (2440 km) with gravity $\gamma_0 = GM/R_0^2$ on that surface (0) is adequate for Mercury's reference gravity field, because the planet's slow spin rate leads to only a small polar flattening, about 2% of Earth's value. The reference field on the geoid is $\gamma_n \approx \gamma_0 + (\partial\gamma/\partial r)N$, where N is the geoid height and $\gamma(r) = GM/r^2$.

An alternative representation of the gravity field is given by the "gravity anomaly," $\Delta g = g_n - \gamma_0$, where the reference field is evaluated on the reference surface instead of the geoid. For either δg or Δg, the level of approximation involved in developing these relationships does not distinguish between the surfaces n and 0 in the SH evaluation of anomalous (non-reference) fields, and it is obviously more convenient to carry out the calculations on a sphere. The reference field given by $\gamma(r)$ is, however, sensitive to the choice of surface. The subtraction of γ_0 from g_n provides essentially a free-air correction of g_n from the geoid height, N, to the reference surface 0, with the correction given by $(\partial\gamma/\partial r)N$. Hence Δg has also been described as a "free-air gravity anomaly," g_{FA}. Because N is related linearly to the gravitational potential and the gravity field is the gradient of this potential, the correction is red-shifted, i.e., it is more effective at lower harmonic degrees. The net result is that the SH conversion of potential to gravity field introduces an $(l + 1)$ term in the calculation of g_n, which is reduced to $(l - 1)$ for Δg. Essentially, Δg is a high-pass-filtered version of δg. The difference in maps is subtle for Mercury, with a Δg map showing slightly better resolution and appearing slightly noisier than a δg map. This difference is less subtle for Mars, because the Tharsis rise produces large low-degree terms. Another consideration is that the gravity field resulting from the global shape associated with the COF–COM offset is not included in a Bouguer correction (Section 3.4.2) of Δg, thus affecting, for example, crustal thickness estimates, though for Mercury the error is small. This issue should not be confused with $C_{1m} = S_{1m} \equiv 0$ in a center-of-mass coordinate system.

The geoid does not have to be the surface on which gravity fields are calculated, and the reference ellipsoid is a valid choice. In fact, for gravity maps in this chapter, δg is calculated in this way with $\delta g = g_0 - \gamma_0$, so it has implicitly gone through a free-air correction and could be labeled as g_{FA}. The term "free air gravity anomaly" has been used loosely, sometimes without definition, in the planetary geophysics literature, and it is sometimes not stated whether δg or Δg is being employed. Here for gravity maps we will use the term "gravity disturbance," indicating that we calculated δg, with $(l + 1)$ in the SH expansion. Finally, we emphasize that estimation of the gravity coefficients C_{lm} and S_{lm} to produce a gravity field model does not depend on the considerations of δg versus Δg.

The range of δg variations is ~400 mGal, which is small compared with the ranges at the Moon or Mars (>2500 mGal and >3800 mGal, respectively). The small range of Mercury's gravity field may be due in part to the low global resolution of the field. Because of MESSENGER's eccentric orbit and high northern periapsis, the southern hemisphere is poorly resolved compared with high northern latitudes.

A useful means to ascertain the local resolution of the field is to compute from the covariance matrix the effective resolution of the field over the globe. Following Konopliv et al. (1999), we computed this SH "degree strength," $\ell_{strength}$. The degree strength is the intersection between the Kaula constraint and the spatially mapped uncertainty spectrum, i.e., the harmonic degree at which the signal-to-noise ratio (SNR) is unity. The resulting degree strength map (Figure 3.4a) is predominantly zonal, with a minimum value of ~10 at the south pole, the highest values (>70) north of 70°N, and a value of ~20 around the equator. The gravity fields of the Moon and Mars, if evaluated to degree and order 10, give ranges of ~250 mGal and ~1400 mGal, respectively, which are more comparable with that of Mercury. (To be sure, the Tharsis rise on Mars, as a single, large igneous complex, is responsible for the very large gravity field variations on that planet.) To illustrate the confidence in the spatial scale of the δg field actually resolved, at each point in a map grid we expanded the spherical harmonics of the HgM007 solution up to the estimated degree strength value for that locale [$\ell_{strength}(\theta, \varphi)$] instead of using the usual fixed value (ℓ_{max}) assigned to the entire grid (Figure 3.4b). In the full solution (Figure 3.2b) there are longitudinal streaks arising from variable altitude coverage. The streaks tend to disappear in the degree-strength-limited solution, and the remaining gravity disturbances are more suited to geophysical interpretation. Although only two possible mascons are directly identifiable at present (see Section 3.6.5), we expect stronger anomalies to be resolved

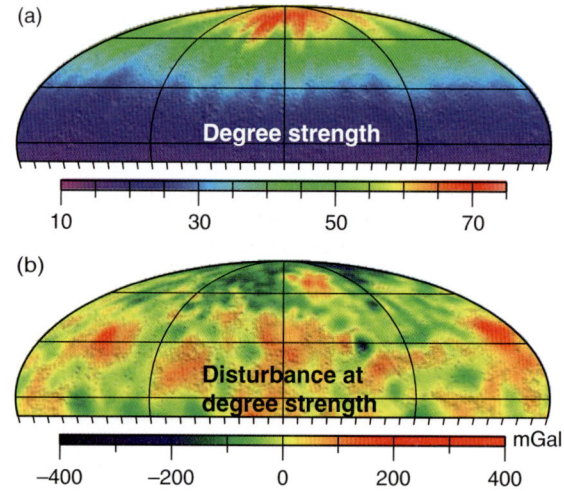

Figure 3.4. (a) Degree strength ($\ell_{strength}$) and (b) gravity disturbance obtained with the HgM007 gravity field. Each grid point in the gravity map was cut off at the local $\ell_{strength}$ in the spherical harmonic expansion. The large eccentricity and high northern periapsis of MESSENGER's orbit results in better resolution of the northern hemisphere. Mollweide projections centered at 0°E, from 10°S to 90°N.

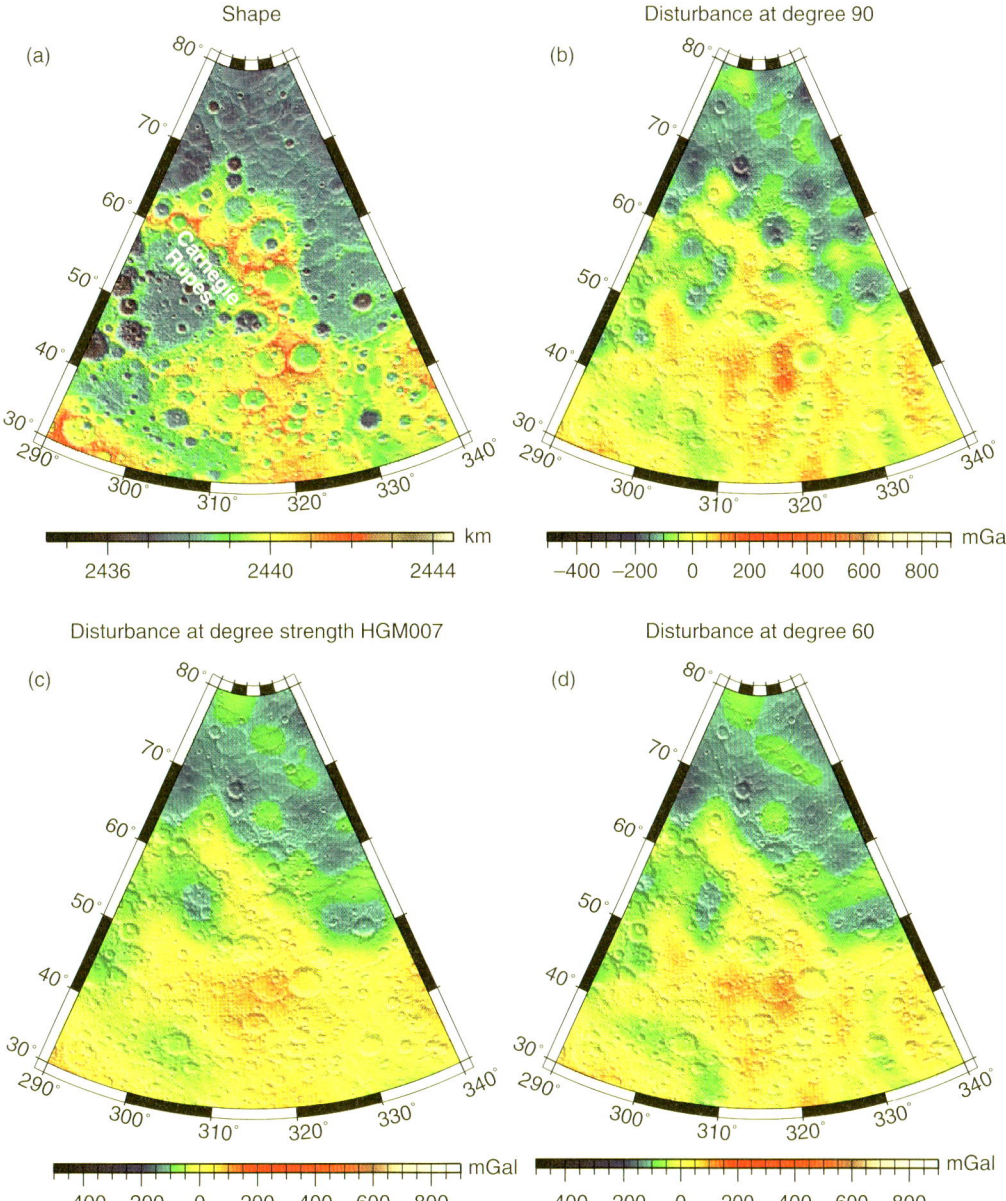

Figure 3.5. The region of Mercury surrounding Carnegie Rupes illustrates the varying correlation between (a) shape and (b)–(d) the gravity disturbance expanded to different choices of spherical harmonic degree and order, corresponding to different length scales. (b) An expansion to a higher degree ($l = m = 90$) than predicted by (c), the degree strength, shows improved correlation at shorter wavelengths. (d) An intermediate expansion (to $l = m = 60$) reduces some of the north–south gravity anomaly trends, which need to be carefully considered because of the uneven low-altitude coverage. Note that the $l = m = 90$ map does a superior job of resolving impact craters than the $l = m = 60$ map, but modeling the crater subsurface structure at $l = m = 90$ should be conducted with caution. Lambert azimuthal equal-area projections.

with lower-altitude data or with data from future spacecraft (Iess et al., 2009).

Expansions to degree and order 60 (e.g., Figure 3.2b) emphasize the fine-scale structure resolved at high northern latitudes, where low-altitude coverage obtained near the end of the mission resulted in degree strengths >75. The degree strength at latitudes south of 30°N is generally <25, however, and the features shown are less certain. In locations that were well tracked from low altitude, the degree strength can sometimes underestimate the true resolution. As shown in Figure 3.5, the δg anomalies show better correlation in some terrains (Figure 3.5a) when expanded to degree and order 90 (Figure 3.5b) rather than to the degree strength limit (Figure 3.5c) or to degree and order 60 (Figure 3.5d). In other locations and at finer length scales, the latest gravity solutions tend to show strong, short-wavelength, north–south anomaly trends that are the result of the non-uniform low-altitude coverage that provided very localized gravity information. Indeed, low-altitude tracks lying between high-altitude tracks tend to be difficult to ingest into a smooth gravity field, given the exponential altitude sensitivity (the continuation effect) to high degrees of the spherical harmonic coefficients, $f_l(r) = (R_0/r)^l$. The use of the degree strength is thus important, and the map of degree strength anomalies (Figure 3.5c) is informative when conducting analyses over a wide range of latitudes, helping to avoid over-interpretation of spurious, high-degree signals in the northern hemisphere.

3.4 FIRST-ORDER MODELS OF CRUSTAL THICKNESS

3.4.1 Introducing the Crust

A terrestrial planetary body's silicate shell is divided by composition into an overlying "crust" and an underlying "mantle." The former contains higher abundances of elements such as Si, Na, Ca, Al, and K that tend to be concentrated in lower-density silicate minerals, whereas the mantle has higher abundances of Mg and Fe (the latter particularly on terrestrial bodies other than Mercury) that tend to be concentrated in higher-density silicates. The development of a planetary crust is generally attributed to a limited number of processes, which we summarize here, and we refer the reader to Chapter 19 for an in-depth discussion.

Initial heating from accretion and metal–silicate differentiation likely produced a magma ocean in a forming terrestrial planet (and the Moon). Cooling led to fractional crystallization, and some of the minerals may have been positively buoyant in the remaining magma, floating upward to form a "primary" crust (also described as a "flotation" crust). This process is the likely origin of the Moon's plagioclase-rich crust (e.g., Shearer et al., 2006) and is a scenario that was proposed for Mercury (Brown and Elkins-Tanton, 2009) as a hypothesis to be tested with MESSENGER data. The results of laboratory experiments reported by Vander Kaaden and McCubbin (2015) indicated that plagioclase flotation was not likely to occur in Mercury's magma ocean and, in fact, the same statement can be applied to most other minerals in equilibrium with the magma. The low FeO and TiO_2 contents of Mercury's largely volcanic crust (Nittler et al., 2011; Evans et al., 2012) implies that mantle melts were similarly low in FeO and TiO_2 and this, along with a limited density range of the melts because of Mercury's thin mantle, led to the prediction of negative buoyancies for most crystallizing minerals and thus lack of development of a substantial flotation crust (Vander Kaaden and McCubbin, 2015). For carbon-rich bulk silicate compositions, graphite crystals are the exception; they have a positive buoyancy and can migrate upward through a low-FeO melt and form a primary crust. There is good evidence that such a flotation crust formed and its remnants are best exposed at the surface as the darkening agent in the spectral unit known as low-reflectance material (LRM) (Robinson et al., 2008; Denevi et al., 2009; Chapter 8). For the LRM darkening agent, carbon is the only element that will satisfy both the spectral reflectance at visible to near-infrared wavelengths as measured by the Mercury Dual Imaging System (MDIS) and the thermal neutron flux as measured by MESSENGER's Neutron Spectrometer (NS) (Peplowski et al., 2016).

The LRM deposits are associated with large impact craters and basins (Denevi et al., 2009; Ernst et al., 2010), suggesting that the primary occurrence of LRM is at mid-crustal levels. The thickness of LRM as a stratigraphic unit is also uncertain and depends on the bulk amount of carbon in the planet (Chapter 19). The LRM is the focus of the crustal carbon, where it is 1–3 wt% greater in abundance than the mean composition of the surrounding terrain (Peplowski et al., 2016). By impacts and magmatic disruption, the carbon apparently has been dispersed throughout much of the crust, which has a much lower reflectance than its lunar counterpart (Murchie et al., 2015).

New material added to an existing primary crust is termed "secondary" crust, which generally forms by the partial melting of solid material in the mantle and the transport of that melt to the shallow regions of the planet where it can erupt onto or intrude into earlier crust. Such melts can also mix with mantle and crustal material during ascent. One mechanism for generating partial melt is from the overturn of the gravitationally unstable crystallized products of magma ocean cooling (Brown and Elkins-Tanton, 2009), though it is not clear that gravitational instability would lead to overturn given the relatively thin and low-FeO mantle (Riner et al., 2009). Post-ocean radiogenic heating can also lead to partial melting in situ of the solid mantle. Early in Mercury's history, heat was probably removed from the mantle by solid-state convection (Tosi et al., 2013), and adiabatic decompression in the upwelling parts of the convective flow could have led to additional partial melting. The partial melts would likely remain buoyant during their entire upward transport through the mantle and solid crust (Vander Kaaden and McCubbin, 2015), maximizing the mass added to the crust. These melts would form or add to an existing crust, typically basaltic near the surface with coarser-grained equivalents formed by slower cooling at depth (Elkins-Tanton, 2012). This crust included contributions from the earlier graphite layer, which was substantially disrupted by impacts and magma assimilation. There is abundant evidence for volcanic eruption of iron-poor basalt and its higher-temperature counterparts (e.g., basaltic komatiites) at the surface of Mercury (Byrne et al., 2013; Denevi et al., 2013). Thermal models (Tosi et al., 2013) indicate that partial melts should have been generated through approximately the first half of Mercury's history. The combination of surface composition measurements indicating a diversity of rock types along with petrological arguments favor at least two chemically distinct source reservoirs for partial melts in the mantle and decompression melting as shallow as ~100 km depth (Charlier et al., 2013).

How do we know that a planet has a crust? On Earth, seismic observations show that at a depth that varies regionally there is a rapid increase in seismic P-wave velocity, from values measured for crustal rocks to values measured for ultramafic rocks, and this discontinuity has been identified as the crust–mantle boundary (the Mohorovičić discontinuity or "Moho"). A crust–mantle boundary (CrMB) was also detected on the Moon on the same basis with data from the Apollo seismic network (e.g., Khan et al., 2000). High-resolution lunar crustal thickness models (Wieczorek et al., 2013) are calculated from the gravity and shape fields described by spherical harmonics, and the results are "pinned" by the seismically determined crustal thickness on the central nearside, approximately within the triangular region for which the Apollo 12/14, 15, and 16 landing sites are at the vertices. Without a seismic anchor, geophysical approaches that establish the existence and structure of a crust on Mars, Mercury, and Venus must invoke a series of assumptions in order to estimate the crustal structure for each planet. All three planets have experienced large amounts of basaltic volcanism, so a crust at least partly derived from partial melting of the mantle certainly exists on each body, independent of any assumptions about crustal structure. Furthermore, as we discuss later, the combination of gravity field and shape constraints requires a strong density interface on Mercury, with a mean depth that is no greater than ~50 km (Padovan et al., 2015). The most obvious interpretation of this result is that it marks the crust–mantle boundary and a transition to denser rocks.

The geological and geochemical view of Mercury's crust described above is that of a complex amalgamation of graphite and basalt-like products from distinct mantle reservoirs. Significant reworking of the crust by impacts is likely required to account for the widespread distribution of LRM, inferred to have abundances of carbon substantially higher than average surface material. The usual geophysical approximation is far simpler: it is a crustal volume homogeneous in density with a sharp density contrast at the CrMB. Gravity and shape data allow solutions for variations in crustal thickness. The available gravity field models cannot resolve fine-scale crustal structure, and even if they could the results would be non-unique. For example, horizontal density interfaces will end up being interpreted as a single density interface positioned according to the depth-dependent weighted effects of the gravity signals from all horizontal interfaces, but this is probably a second-order effect on the position of the CrMB. Crustal densities and crust–mantle density contrasts are discussed in the next two subsections, and adopted values are obviously global averages that must be kept in mind when interpreting geophysical results on a regional scale.

3.4.2 Bouguer Correction, Crustal Density Choices, and Bouguer Disturbance

The gravity field δg may be modeled as the gravitational attraction of topographic relief (or shape), together with that of relief on deeper density interfaces and/or lateral variations of density within finite volumes at depth. Specifically, we obtain the "Bouguer disturbance" by subtracting from the observed gravity disturbance the contribution from shape (the "Bouguer correction") for a fixed upper crustal density, which must be assumed. Modeling local regions of high gravity–shape coherence in the NSP yielded a value of $\rho_{crust} = 2602 \pm 470$ kg/m^3 for crustal density (James et al., 2015a). Values as high as 3100 kg/m^3 have been assumed (Smith et al., 2012) on the basis of high-magnesium basaltic compositions. From XRS elemental abundance measurements, Padovan et al. (2015) estimated grain densities of 3014 kg/m^3 and 3082 kg/m^3 for heavily cratered and intercrater plains compositions, respectively. A lunar-like bulk porosity of 12% (Wieczorek et al., 2013) would make the density of the crust at least 2700 kg/m^3. Mercury's higher surface gravitational acceleration would be expected to reduce the porosity, and we adopt a value of 6%, which brings the bulk density for the Bouguer correction closer to 2900 kg/m^3, with an uncertainty in the range 200–300 kg/m^3.

The Bouguer disturbance (Figure 3.2c) was calculated with the SH coefficients from the HgM007 gravity field and the gtmes_150v05 shape model, both of which incorporate data through the end of the MESSENGER mission. This result was obtained by subtracting the SH coefficients of the gravitational potential of a finite-amplitude representation of the shape (Wieczorek and Phillips, 1998) from the observed gravitational potential represented by the HgM007 coefficients.

The Bouguer disturbance field provides direct insight into subsurface density structure without obfuscation from the strong shape signal. After first reviewing possible crust–mantle density contrasts, we assign the Bouguer δg entirely to relief on the CrMB in order to produce a global map of crustal thickness (Section 3.4.4).

3.4.3 Crust–Mantle Density Contrast

The implications of a large uncertainty in crustal density are less important than the density contrast, $\Delta\rho = \rho_{mantle} - \rho_{crust}$, at the CrMB, which affects, in nearly an inverse proportion, the crustal thickness variations required to explain the observed gravity field. Solutions are subject to the "minimum thickness constraint," which states that in a global solution, crustal thickness must everywhere be non-negative. This constraint places a lower bound on global mean crustal thickness, H_{cr}. Smith et al. (2012) assumed a small (200 kg/m^3) contrast, which implied a minimum H_{cr} of 50 km on the basis of preliminary gravity and shape results. Although it is possible that a larger density contrast would permit solutions with a thinner crust, we adopt this value for single-layer models discussed below. Mazarico et al. (2014) calculated an updated crustal map from the HgM005 model expanded to degree and order 50 for a mantle density of 3200 kg/m^3 and a 250 kg/m^3 density contrast. Padovan et al. (2015) assumed a density contrast of 400 ± 100 kg/m^3, similar to that employed in crustal models for the Moon and Mars. Mantle density at the base of the crust is not known, but from rotational dynamics (Chapter 4) the solid shell containing the crust and mantle was estimated by Smith et al. (2012) to have an average density of ~3650 ± 225 kg/m^3. Compositional stratification of the solid shell with a dense solid layer at the core–mantle boundary (CMB) (Smith et al., 2012; Hauck et al., 2013) has been proposed to reconcile the high density of the solid shell with lower mantle densities consistent with the low-iron, low-titanium source regions of volcanic material on Mercury's surface. An updated estimate of the solid shell density of ~3380 ± 200 kg/m^3 (Hauck et al., 2013) does not necessarily obviate the need for a high-density layer. Thermodynamically, it is plausible that the dense layer could be solid FeS positioned at the top of the core.

3.4.4 Crustal Thickness

The first estimate of Mercury's mean crustal thickness, H_{cr}, was obtained from the value of C_{22} derived from the tracking data acquired during the Mariner 10 flybys and from a c_{22} value estimated from Earth-based radar observations (Anderson et al., 1996). Simple mass balance of the c_{22} shape by variation in CrMB relief (i.e., Airy isostasy; see Section 3.5.2) yielded an H_{cr} of 203 ± 101 km. A ~200-km-thick crust would be difficult to reconcile with the average density of Mercury's solid outer shell if the mean densities of the crust and mantle are anywhere near their terrestrial counterparts. Moreover, given that the top of Mercury's fluid core is at only 420 km depth (Hauck et al., 2013; Chapter 19), it would be difficult to derive a crust of such thickness by partial melting of the mantle.

Through thermal modeling and analysis of lobate scarp fault structures inferred from scarp topographic profiles, Nimmo and Watters (2004) argued that H_{cr} is limited to an upper bound of 80–120 km; their investigation also recovered an estimate of the elastic lithosphere thickness, T_e, of 25–30 km at the time of fault activity (details are provided in Section 3.5.4.2). On the basis of data from two MESSENGER flybys of Mercury and from this elastic thickness result, Smith et al. (2010a) estimated an H_{cr} of 30 km. Their result was obtained by modeling the observed equatorial ellipticity of the shape and geoid, with the shape partially mass compensated by perturbations on the CrMB. The major difference between this result and the much larger crustal thickness value obtained by Anderson et al. (1996) is that the newer model incorporates an elastic lithosphere.

From MESSENGER orbital data, an H_{cr} estimate of 35 ± 18 km (Padovan et al., 2015) was obtained from modeling the observed geoid-to-shape ratio (often called the geoid-to-topography ratio, or GTR). Over the spectral range $l = 9–15$, the model is one of crustal isostasy, i.e., the surface relief and CrMB relief are in mass balance. The authors argued that large impact basins and areas of smooth plains may not be isostatically compensated, so these regions were excluded from their analysis. A model that describes mechanical equilibrium of surface relief usually contains an elastic lithosphere (see Section 3.5); the special case of $T_e = 0$ is an isostatic solution with a single variable, H_{cr}, which is appealing if the isostasy assumption is correct. We consider the implications of a non-zero T_e for the Padovan et al. (2015) solution in Section 3.5.2.

A crustal thickness model with $H_{cr} = 38$ km, which represents the downward continuation depth of the Bouguer δg in Figure 3.2c, is shown in Figure 3.2d. Note that the H_{cr} value of 38 km is assumed and is consistent with the Padovan et al. (2015) result. This "single-layer" approach uses relief on the CrMB to account for the entire Bouguer δg. Because downward continuation is unstable in the presence of noise and unmodeled lateral density variations, it is customary to apply a minimum-slope filter to an iterative solution in the spectral domain. As adopted for Figure 3.2d, the filter reduces to half the amplitude of the inferred crust–mantle relief at degree 50, smoothly tapering to extinction at degree 100. This approach suffers from non-uniqueness, but it serves to illustrate the main features of single-layer models and yields a minimum thickness of 6 km within Rachmaninoff basin and a maximum thickness of 66 km near Mercury's hot poles (see Section 3.7.5). The crust thins toward the north pole to <30 km beneath the NSP, but does not approach the minimum thickness constraint.

3.5 SHAPE COMPENSATION AND THE LITHOSPHERE

3.5.1 Introducing the Lithosphere

3.5.1.1 Rheology

Although "crust" is a term that describes the outermost compositional shell of a planet, planetary shells can also be defined in terms of rock strength. Under this convention, the outermost shell is termed the "lithosphere," and it has high strength, largely due to the relatively low temperatures near a planet's surface. The thickness and mechanical behavior of a lithosphere depend on the rheological properties of its crust and mantle components.

There are several categories of strength, and they produce different definitions of "lithosphere" (Phillips et al., 1997). In planetary studies, most common is the "elastic lithosphere," which is an elastic plate or shell overlying a strengthless interior and usually taken to be at least locally uniform in its rheological properties of Young's modulus, E, and Poisson's ratio, ν. The elastic lithosphere has the ability to support a load through flexural (i.e., bending) and membrane stresses, and its thickness, T_e, is a quantity that can be conveniently estimated from relations between observed gravity and shape, or from shape alone.

The rheology of a lithosphere, however, is far more complex than that of simple elastic behavior. In the shallow portions of the crust, rocks deform in a brittle fashion and can support stress up to a frictional yield stress, σ_f, which depends on pressure (depth) and beyond which stress is relieved by slip along pre-existing faults (Byerlee, 1978; Chapter 10). At greater depths, where temperatures are sufficient to enable relief of stress through viscous creep, rocks deform by ductile flow, and a viscous yield stress, σ_v, can be specified for a given flow law and strain rate. Between the regions of brittle and ductile deformation, the lithosphere may have an "elastic core." The lesser in magnitude of σ_f and σ_v, each a function of depth, defines a yield strength envelope (YSE: Brace and Kohlstedt, 1980; see Section 10.5 for a more thorough discussion). The brittle–ductile transition (BDT) is often taken to occur at the depth at which $\sigma_f = \sigma_v$. This sharp transition with increasing depth from faulting to viscous flow is an oversimplification, and there is most likely a transitional zone (often termed "semi-brittle") in which a number of deformation mechanisms (e.g., distributed shear) may operate (Kohlstedt et al., 1995; Kohlstedt and Mackwell, 2010). Furthermore, the difference between the viscous flow law constants for the compositionally distinct crust and mantle can lead to a complex YSE, with the possibility of a BDT in both the crust and upper mantle (see Figure 10.11). The use of a YSE essentially recasts the deformation of an elastic plate or shell into that of an elastic–plastic plate for which the bending moment depends on the brittle and ductile rheologies of the crust and upper mantle. The ductile behavior is strongly dependent on the temperature profile in the outer part of the planet as the plate or shell deforms in response to external and internal loads. McNutt (1984) showed that, for a specified plate curvature, the elastic–plastic bending moment could be equated to the bending moment of an equivalent elastic plate, which depends on T_e. This equivalence implies that an elastic lithosphere is in general just a mechanical stand-in for the complex rheology of an actual lithosphere and that T_e should be taken as the thickness of an "effective elastic lithosphere." Although dependent on many assumptions (e.g., ductile flow law parameters, crust and mantle densities, strain rate), it is possible to estimate heat flow into the base of the "mechanical lithosphere" of thickness T_m. This depth is the practical lower boundary of the YSE and is defined as the depth at which the highly temperature-dependent viscous creep stress falls below a threshold value, usually assumed to lie in the range 10–50 MPa and indicating that there is no long-term creep strength. Therefore, estimating T_e can provide important constraints on the thermal evolution of a planet. We are also interested in the "thermal lithosphere," the outer portion of a planet that transfers heat conductively, which for a convecting mantle is the outer thermal boundary layer.

Although there have not been widespread attempts to apply lithospheric concepts to Mercury, we note that Williams et al. (2011) constructed YSEs to show the effects of different surface temperatures resulting from Mercury's 3:2 spin–orbit resonance and near-zero obliquity (see Section 3.7.5). Additionally, Tosi et al. (2015) used a YSE formalism to validate the thickness of the elastic lithosphere required in a model of long-wavelength d_2 shape supported by solar-driven thermal anomalies. Nimmo and Watters (2004) and James et al. (2016) estimated elastic lithosphere thickness values that are considerably less than the Tosi et al. (2015) result (see Section 3.7.6).

3.5.1.2 Freezing T_e

Motivated largely by the promise of estimating the lithosphere's basal heat flux as a function of geological age, recovering the spatial and temporal behavior of T_e has been a vigorous research activity applied to the terrestrial planets (e.g., McGovern et al., 2002, 2004; Anderson and Smrekar, 2006; Grott and Breuer, 2008). A fundamental assumption in these endeavors is that T_e has been "frozen in," so that determinations of T_e correspond to the time that a deforming load was emplaced. This idea was first elucidated by McKenzie and Bowin (1976) from analysis of bathymetric and gravity data acquired on ship tracks traversing the Atlantic Ocean. Specifically, modeling the wavelength-dependent ratio of gravity to topography (see Sections 3.5.3 and 3.5.4) produced a T_e value for oceanic lithosphere that was smaller than estimates that were associated with the present era (e.g., flexural bending of the lithosphere at a subduction zone). From this result, McKenzie and Bowin (1976) speculated that the smaller value signaled a preservation of the response to loading of young, thin lithosphere found near spreading ridges. Watts (1978) analyzed gravity and topography from the Hawaiian–Emperor seamount chain and came to the same conclusion, albeit with a more detailed argument and with a knowledge of the ages of volcanic loads along the chain. Specifically, Watts (1978) found that local loading (e.g., seamounts) yields estimates of T_e that are proportional to the square root of plate age at the time the volcanic load was applied [see also Section 6.4 in the work of Watts (2001)]. The square-root dependence indicates that diffusive cooling of the plate controls the thickening rate of the elastic lithosphere, as expected. We emphasize that this methodology also requires assigning a geological age to the load, which for planetary studies is far less secure than for terrestrial oceanic lithosphere and is usually made from crater areal density measurements (Chapter 9). Such ambitious schemes can be used to constrain the thermal evolution of Mercury, but to date such analyses have been limited.

That the flexural deformation of a lithosphere in response to a load is "frozen in" as the lithosphere cools and thickens has been demonstrated in viscoelastic models (e.g., Lago and Cazenave, 1981) and elastoviscoplastic models (Albert and Phillips, 2000). The first model incorporates a Maxwell solid, defined by additive viscous and elastic strain rates; the second model adds a frictional sliding component to represent faulting (Byerlee, 1978). A mechanical lithosphere with a Maxwell rheology will generally behave as a viscous material at depth and as an elastic layer closer to the surface. The rheology boundary will migrate upward with time in a viscoelastic zone, thinning the elastic lithosphere (Courtney and Beaumont, 1983). This effect is quantified in terms of the ratio of dynamic viscosity (η) to elastic shear modulus (μ), or the Maxwell time, η/μ, marking the transition in time from elastic rheology to longer-term viscous rheology. Creep strain and elastic strain are equal at the Maxwell time.

Superposed on the thinning and often the dominant process of the two is the overall cooling and thickening of the mechanical lithosphere; this process serves to halt lithospheric thinning and stabilize T_e. Consider a positive load on an elastic lithosphere containing the CrMB and assume that the downward flexure of the lithosphere has achieved mechanical equilibrium. Now let the lithosphere start cooling. This is a transient regime; deviatoric stresses will decay only by driving the CrMB to a broader, shallower shape, which requires viscous flow of the mantle and crust in order to conform to the new equilibrium configuration. The process is controlled largely by the viscosity at the base of the elastic lithosphere and the activation energy of the viscous flow law in the mantle portion of the mechanical lithosphere (Courtney and Beaumont, 1983). If the viscosities are too high, the new equilibrium state will not be achieved; the stresses are "frozen in" and a T_e estimate may correspond to an age shortly after the time of loading. The outcome depends strongly on the cooling rate, and in planetary studies the value of T_e determined could be considerably younger than the age of the load as established by an analysis of the size–frequency distribution of superposed impact craters.

We have supplied detail here in a general sense to provide a background context when we return to this topic in Section 3.7.7.5, as it applies to thermoelastic stresses and the origin of Mercury's d_2 shape and geoid.

3.5.2 Support of Shape

In Section 3.4.4, we discussed the construction of a crustal thickness map that is based on the simplest of model assumptions, namely that the Bouguer disturbance is due entirely to the shape of the CrMB, with a specific result obtained by assigning a mean crustal thickness and the crust and mantle densities. Here we consider models that are in mechanical equilibrium under an assumed rheology, but we still insist that the model reproduce the observed shape and geoid. Such models usually start with a shape load, and we assume that the stresses therefrom have rearranged the interior density structure to achieve a mechanical equilibrium given the assumed rheology. The geoid signal from the model solution is compared with the observed geoid, and the model parameters are adjusted to improve the fit. An elastic lithosphere of thickness T_e is often an integral part of the model, and its presence contributes an additional level of non-uniqueness to the modeling efforts. When T_e approaches zero, the planetary interior is strengthless, and isostasy can be achieved if there are density interfaces or lateral variations of density in the interior. At shallow depths, shape can be compensated by lateral density variations in the crust such as those associated with porosity or temperature, or by vertical relief of the CrMB. We call the latter case "crustal compensation" (or "Airy compensation"); the former case is essentially "Pratt compensation" (Watts, 2001). At greater depths, shape can be compensated by thermal and compositional mantle heterogeneities, by an unevenly stratified mantle, or by relief on the CMB.

Alternatively, when T_e is not zero, bending and membrane stresses within a planet's lithosphere may prevent shape from reaching a fully compensated state. When stresses support some portion of a positive shape feature, that feature is "super-isostatic," since the internal mass deficit is insufficient to counterbalance locally the weight of the shape. Conversely, a shape with amplitudes smaller than the values necessary for compensation by an internal mass distribution is "sub-isostatic." In both cases, non-hydrostatic stresses in the lithosphere sustain the non-isostatic state and produce δg variations larger in magnitude than for the isostatic case.

The most straightforward inclusion of an elastic lithosphere involves its surface loading by the shape and a single interior density contrast at the CrMB, which has a mean depth set by H_{cr}. In equilibrium solutions for Mercury at long wavelengths (SH degree $\lesssim$10), the shape and CrMB are out of mass balance for E =100 GPa by 10%, 20%, and 30% for T_e values of 10 km, 20 km, and 30 km, respectively (Turcotte et al., 1981). Introduction of a non-zero T_e leads to a trade-off between T_e and H_{cr}, which is why Smith et al. (2010a) used an independent determination of T_e to come up with the 30-km solution for H_{cr}. In the T_e–H_{cr} trade-off, $T_e = 0$ gives an upper bound on H_{cr}, and that is perhaps how the Padovan et al. (2015) solution should be viewed, as we do not know with great accuracy the development history of Mercury's elastic lithosphere. We note that this upper-bound value of 35 km is close to the lower-bound value imposed by the minimum thickness constraint, so a value of H_{cr} in the range 30–40 km is consistent with both perspectives.

3.5.3 Inferences from Correlation and Admittance

Crustal thickness variations may account for much of Mercury's low-degree or long-wavelength shape (with the exception of the d_2 contribution; see Section 3.7), but the crust–mantle interface relief inferred from the Bouguer δg is largely non-isostatic. This departure from a state of crustal compensation likely results from a combination of at least three factors: (1) noise in the gravity data, (2) deep-seated dynamic compensation of shape, and (3) the support of shape by bending and membrane stresses. We review here the effects of data noise on spatial and spectral relationships between gravity and shape, because these effects must be understood before we can quantify the contributions of dynamic shape or elastic support to the observed fields.

Most simple compensation mechanisms, including crustal compensation, are associated with unitary correlation of gravity and shape. Under these conditions, with Airy isostatic compensation of surface loads, admittance gives an equivalent depth to the CrMB. The more common, non-unitary correlations of gravity and shape indicate the existence of multiple compensation mechanisms, noise in the gravity data set, or both. Examining the effects of noise, first, the observed gravity signal g_{obs} can be written as a combination of the true gravity signal g_{true} and data noise, I:

$$g_{obs} = g_{true} + I. \qquad (3.4)$$

These terms may be interpreted as spatial quantities [e.g., $g_{obs} = g_{obs}(r,\theta,\varphi)$]. Insofar as I is uncorrelated with shape, the observed admittance, Z_{obs}, is not biased by data noise:

$$Z_{obs} = \frac{\langle g_{obs} h\rangle}{\langle h^2\rangle} = \frac{\langle g_{true} h\rangle + \langle Ih\rangle}{\langle h^2\rangle} = \frac{\langle g_{true} h\rangle}{\langle h^2\rangle} = Z_{true}, \qquad (3.5)$$

where h denotes shape and the bracketed terms indicate the expectation of the enclosed quantity over the spatial domain. However, the correlation of observed gravity and shape is biased by the presence of gravity noise. If I is uncorrelated with the true gravity signal, g_{true}, it follows that the observed coherence (i.e., the correlation squared) between gravity and shape is biased downward by a degree-dependent factor:

$$\gamma_{obs} = \frac{\langle g_{obs} h\rangle}{\sqrt{\langle g_{obs}^2\rangle\langle h^2\rangle}} = \gamma_{true}\left(1 + \frac{\langle I^2\rangle}{\langle g_{true}^2\rangle}\right)^{-0.5}. \qquad (3.6)$$

The effects of a degree-dependent gravity noise on the gravity–shape correlation spectrum are quantified in Figure 3.6. The observed northern hemisphere correlation was found with a zonal Slepian taper (James et al., 2015b) having a bandwidth (the highest degree and order of the taper) of 2 and a localization radius of 90°. Although simple compensation mechanisms and elastic loading scenarios produce a unitary correlation of gravity and shape, the presence of noise in the data decreases the correlation at high degrees. The curves in Figure 3.6 correspond to data noise with a mean-squared amplitude proportional to $10^{l/100}$ and a true gravity signal with a mean-squared amplitude proportional to l^{-2}. In this case, the correlation curves for noisy data depend on a single parameter: the degree strength of the gravity field, $l_{strength}$. As defined earlier, this strength is the harmonic degree at which the SNR is equal to 1. Non-unitary global correlations at high harmonic degrees ($l \approx 45$–90) can be the result of data noise for a nominal degree strength of $l_{strength} = 30$, and northern hemisphere correlations at similar wavelengths can be the result of noise

3.5 Shape Compensation and the Lithosphere 65

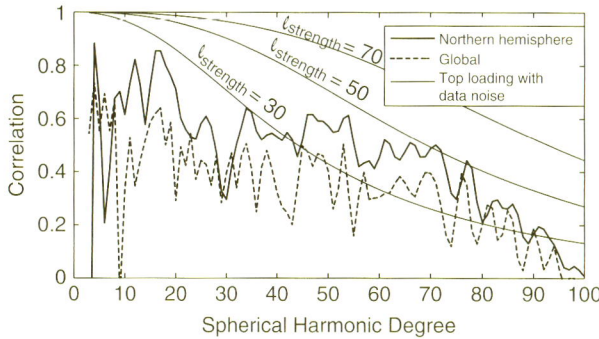

Figure 3.6. Observed gravity–shape correlation spectra for Mercury's northern hemisphere (localized with a Slepian taper) and the globe as a whole; thin solid lines indicate the theoretical correlations in the presence of data noise, given that the ratio of signal to error equals 1 at a prescribed degree strength, ℓ_{strength}.

with a degree strength of approximately $\ell_{\text{strength}} = 45$. Global correlation is less than the localized northern hemisphere correlation at most wavelengths, but this result is expected because of the lower data quality in the southern hemisphere and not likely caused by an actual difference between the hemispheres.

3.5.4 Estimating Elastic Thickness

3.5.4.1 Spectral Approach

Given that the nominal degree strength of the HgM007 gravity field exceeds $\ell_{\text{strength}} = 40$ in parts of the northern hemisphere, Figure 3.6 suggests that data noise cannot completely account for the departure of long-wavelength correlation of gravity and shape from unity. Another source of non-unitary correlation is the superposition of elastic loading. This loading can be from the top (e.g., where shape is produced by impact cratering or volcanic extrusion), the bottom (e.g., by stresses produced by thermally or chemically buoyant sub-lithospheric mantle), or a combination of the two. If we assume that top loading and bottom loading are uncorrelated with each other, we can calculate the resulting admittance and correlation spectra with a thin spherical shell approximation (Turcotte et al., 1981). Figure 3.7 shows the observed northern hemisphere spectra along with theoretical gravity–shape admittance and correlation spectra that arise from loading scenarios with varying amplitudes of uncorrelated top loading and bottom loading. The parameter F, which ranges from 0 to 1, indicates the fraction of total loading accommodated at the base of the lithosphere (McKenzie, 2003).

None of the theoretical correlation spectra plotted in Figure 3.7 successfully match the correlation spectrum at high harmonic degrees ($l > 50$) but, as explained above, the departure of the correlation spectrum from unity at high degrees likely results from the presence of data noise rather than an elastic loading scenario. The admittance spectrum at high degrees is also likely biased downward by the Kaula filtering used to produce the gravity field solution, so loading scenarios that fail to fit either the admittance or the correlation at high harmonic degrees should not necessarily be disregarded.

A loading model for Mercury should match both the observed admittance spectrum and the observed correlation spectrum. James et al. (2016) found best-fit parameter values for Mercury's lithosphere by employing a misfit function defined as the sum of RMS values for admittance misfit and correlation misfit. Normalized misfits between observed geophysical spectra and theoretical spectra are plotted in Figure 3.8. When solutions with misfits more than 25% larger than the minimum misfit are rejected, acceptable values of the loading parameter F fall in the range 0.2–0.6. For $F = 0.5$, indicating equal amplitudes of top loading and bottom loading, the best-fit elastic thickness is $T_e = 31 \pm 9$ km.

3.5.4.2 Other Estimates of T_e

The spectrally derived T_e values are consistent with the results of Nimmo and Watters (2004), who estimated that $T_e = 25$–30 km on the basis of an analysis of the penetration depth of a thrust fault beneath the Discovery Rupes monocline (a "lobate scarp"; see Chapter 10). The inferred maximum depth of faulting of 30–40 km (Watters et al., 2002; Zuber et al., 2010) was equated to the BDT, allowing construction of a yield strength envelope. Equating the resulting elastic–plastic bending moment (a function of rheology, temperature, and strain rate) to the elastic bending moment for a given curvature yields the estimate of T_e (McNutt, 1984). There are uncertainties in model parameters that broaden the T_e range, and " ... by far the largest uncertainty is in the curvature" (Nimmo and Watters, 2004). A factor of 10 increase or decrease in the curvature leads to a spread in T_e of ~10–60 km. As mentioned above, this analysis also yielded a crustal thickness constraint, namely an upper bound of 80–120 km. This inference was based on a crustal thermal model at ~4 Ga that (i) did not undergo basal melting and (ii) had a BDT depth as determined in their analyses.

Melosh and McKinnon (1988) estimated that at the time of the Caloris basin-forming impact, the elastic thickness there was in the range 75–125 km. Qualitatively, the radii of prominent topographic rings, which are likely associated with basin-concentric normal faults, increase with lithospheric thickness. To quantify the relationship for the Caloris basin will require numerical modeling that is able to predict the location and number of ring structures. Such a capability has only recently been applied to the Moon (e.g., Orientale basin; Johnson et al., 2016).

3.5.5 Role of Sub-Lithospheric Mantle: Two-Component Models

Non-unitary correlation of gravity and shape may result from compensation of shape at multiple depths. In practice, this scenario is sometimes quantified with a two-component model in which the crust–mantle interface and the deeper mantle both contribute to compensation (e.g., Banerdt, 1986; Herrick and Phillips, 1992). The shape and gravity contributions from the deeper compensation mechanism are properly given by dynamic flow kernels, which are solutions to steady-state Stokes flow of the mantle in a viscous, layered sphere (Hager and Clayton, 1989). Dynamic shape kernels are typically attenuated at short wavelengths, so dynamic flow is most likely to

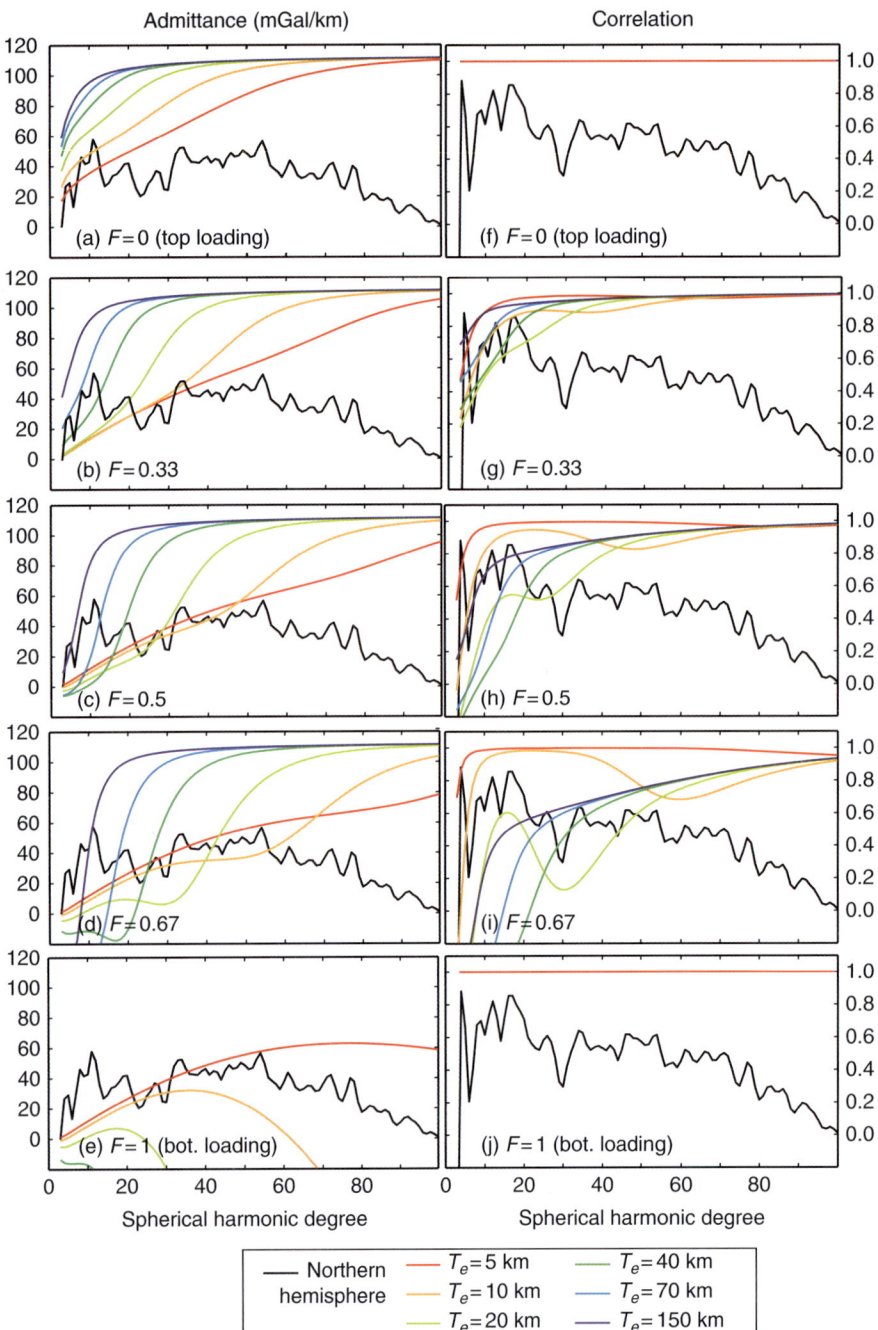

Figure 3.7. Admittance spectra (left column) and correlation spectra (right column) for combinations of top and bottom loading. Rows correspond to different values of the loading parameter F. Top loading and bottom loading are assumed to be uncorrelated, bottom loading is imposed on the crust–mantle interface at a depth of 40 km, and the Young's modulus is 100 GPa.

contribute to Mercury's shape and gravity at low spherical harmonic degrees. An inversion of Mercury's northern hemisphere gravity and shape (James et al., 2015b) for two-layer models produced maps of crustal thickness and the dynamic pressure from mantle mass anomalies (Figure 3.9). Dynamic pressure represents the upward pressure exerted on the planet's surface by buoyant anomalies at depth. This pressure scales linearly by a factor of approximately $\rho_{mantle} g_0$, where g_0 is the gravitational acceleration, and it is this pressure that supports the "dynamic shape." For the model shown in Figure 3.9, dynamic shape has a maximum value of 2.0 km in the vicinity of Caloris basin, and it has a minimum value of −1.6 km in a low-lying section of the NSP. The minimum thickness constraint indicates that H_{cr} must be greater than 38 km, slightly larger than in the single-layer solution.

Deep-seated dynamic flow in the presence of a weak elastic lithosphere produces large gravity/shape ratios (i.e., high admittance values), comparable with those due to top loading of an elastic lithosphere. It can be difficult to distinguish between deep compensation of shape and elastic support of shape, so we refer to shape supported by either of these mechanisms as "high-admittance shape." One way to distinguish between dynamic shape and elastically supported shape is to examine the prevalence of high-admittance shape at a

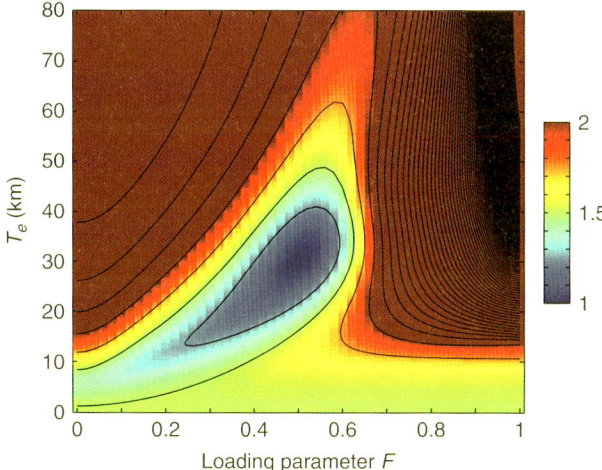

Figure 3.8. Misfit (normalized by the lowest value) between observed and theoretical admittance and correlation spectra for various values of T_e and the loading parameter, F. Contours of relative misfit are at intervals of 0.25.

range of wavelengths. James et al. (2015b) observed that a high-admittance compensation mechanism on Mercury would have relatively large amplitudes at long wavelengths ($l < 15$). This result is concordant with dynamic flow kernels associated with deep compensation of shape, which generally have a reduced shape expression for wavelengths comparable with, or shorter than, the compensation depth. It should be noted that the long-wavelength bias of high-admittance shape does not rule out elastic support of long-wavelength shape. However, the distribution of elastic loading would need to be spectrally red-shifted (i.e., more prevalent at longer wavelengths) relative to the spectral distribution of shape in order to reproduce the observed gravity and shape. At higher degrees ($l > 15$), high-admittance shape is no longer spectrally red-shifted. This pattern suggests that dynamic shape on Mercury predominates over elastically supported shape at low harmonic degrees, with the exception of d_2 (Section 3.7), and that elastically supported shape predominates at higher harmonic degrees.

3.6 REGIONAL-SCALE CRUST AND MANTLE FEATURES

3.6.1 Introduction

At a number of locations across Mercury, there is a strong correlation between gravity and shape and, in some of those places, a high admittance. Several such sites are also collocated with tectonic structures, principally those arising from crustal shortening (Chapter 10). An analysis of the degree of correlation and admittance at these sites, together with an assessment of the tectonics there (e.g., type, orientation, and relationship with shape), can provide a first-order insight into the crust or mantle processes that may have been responsible for the relationship between gravity and shape. This section examines a representative set of features but is far from a complete survey.

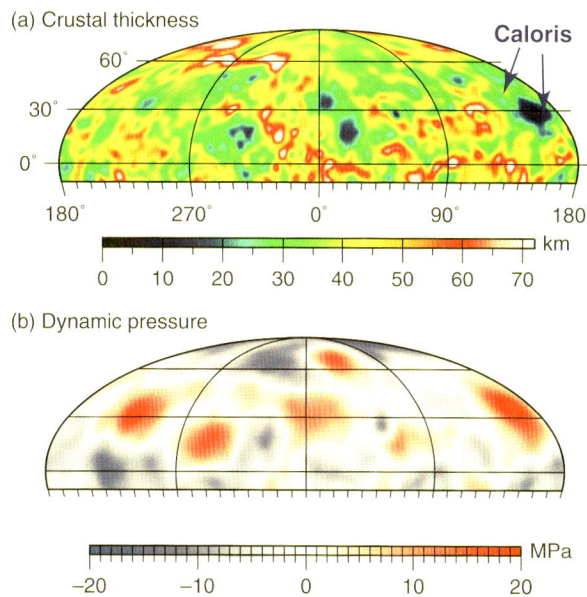

Figure 3.9. Two-component solutions to inversion of gravity and shape, including (a) crustal thickness and (b) dynamic pressure on Mercury's surface from mantle mass anomalies. Mollweide projections centered at 0°E, from 10°S to 90°N. Right arrow in (a) indicates an area of thin crust in the Caloris basin; left arrow marks the locale of moderately thick crust near the basin's northern boundary. Updated from James et al. (2015b).

3.6.2 Northern Rise

The northern rise is a domical swell approximately 1000 km wide and 1.5 km high situated at high northern latitudes in the largest single expanse of volcanic deposits on Mercury, the NSP (Klimczak et al., 2012; Zuber et al., 2012; see Chapter 10). The northern rise stands out as a substantial topographic high that is not underpinned by thickened crust. That structure is evidenced by the strongly positive δg (Figure 3.2b) but near-zero Bouguer δg (Figure 3.2c) and high geoid-to-shape ratios, all of which combine to indicate deeper sources of compensation or flexural support of the shape.

Long-wavelength gravity and shape are well correlated at the northern rise, raising the possibility that shape here is predominantly supported by a single mechanism. Localized admittances at the northern rise (80–100 mGal/km for $l = 10$) are approximately as large as those resulting from shape alone, suggesting that the northern rise is either supported by elastic top loading or compensated in the deep mantle. For a pure top-loading scenario, an elastic lithosphere thickness, T_e, greater than 70 km is required to fit the observed gravity and shape (Smith et al., 2012). Alternatively, dynamic flow must be driven by mass anomalies at depths of 250–400 km in order to fit the observations, and a mantle viscosity profile featuring a decrease in viscosity with depth is preferable to other viscosity structures (James et al., 2015b). Crustal thickness variations beneath and surrounding the northern rise are minimal (Figure 3.10) and thus provide little support to the shape.

For a top-loading scenario, James et al. (2014) determined that compression should occur at the rise, surrounded by a zone of extension. No extensional tectonic structures are centered on the rise, however (nor virtually anywhere on Mercury except at

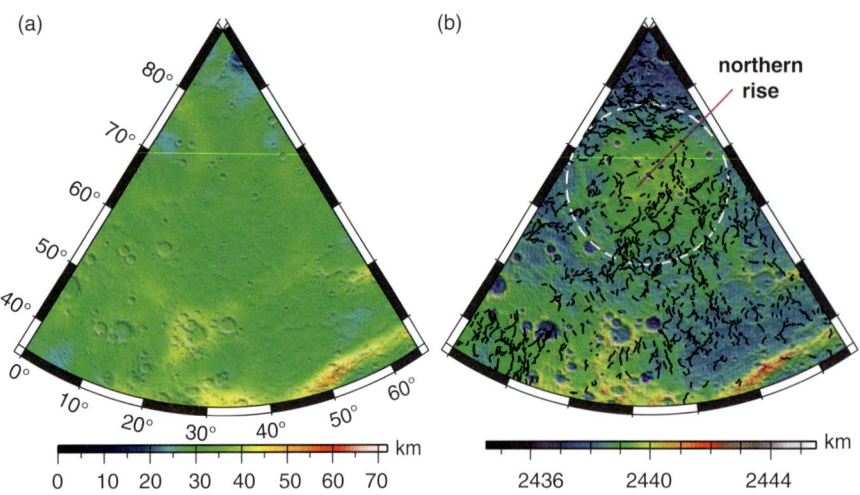

Figure 3.10. (a) Crustal thickness for a portion of the Hokusai (H-5) quadrangle of Mercury extending northward toward Borealis (H-1). (b) Shape for the same region. Black lines show surface traces of shortening structures in the northern volcanic plains mapped by Byrne et al. (2014), including those proximal to the northern rise. Crustal thickness calculated as in Figure 3.2d. Lambert azimuthal equal-area projections.

volcanically infilled craters; see Chapter 10). Moreover, the areal density of mapped shortening structures (Byrne et al., 2014), as a proxy for shortening strains, is no greater at the rise than elsewhere within the NSP (James et al., 2014, and Figure 3.10).

The timing of formation of the northern rise can be deduced by observations of associated impact and volcanic features. For example, MLA observations reveal that the floors of some volcanically flooded impact craters on the northern rise display tilts that generally match the local tilt of the long-wavelength shape in magnitude and direction (Zuber et al., 2012; Balcerski et al., 2013). Under the assumption that the top surface of the lavas inside these craters once followed a gravitational equipotential surface, the rise must have formed largely by uplift after the emplacement of the NSP at ~3.7 Ga (Head et al., 2011; Ostrach et al., 2015; Chapter 11). Notably, there is a population of younger, unfilled impact craters on the rise that show no tilt of their floors, implying that the growth of the northern rise was complete at some point prior to the formation of these unfilled craters (Balcerski et al., 2013).

These age relations present a difficulty for a top-loading scenario, which most likely calls for extrusive loading to be concurrent with the age of the smooth plains. Extensive near-subsurface volcanic intrusion could reconcile the observations of high admittance and tilted craters at the northern rise, provided that the intrusions are sufficiently shallow to overlie a majority of the rigid lithosphere. In this regard, Mercury's northern rise might be akin to the Marius Hills shield volcano complex on the Moon (Spudis, 1996). At 330 km in diameter and more than 2 km in relief (Spudis et al., 2013), however, this lunar complex differs in shape and size from the northern rise, and the Mercury feature lacks the distinctive volcanic cones and rilles that mark the Marius Hills as a major site of effusive volcanism (e.g., McCauley, 1967; Guest, 1971). Additionally, the major gravity anomaly high of the Marius Hills occupies only a small fraction of the physiographic area of this feature, whereas this fraction is close to unity for the northern rise.

Late-stage formation via uplift of the northern rise may instead be a product of deep-seated buoyant uplift. A relaxation model involving a perturbed FeS layer at the base of the mantle is able to reproduce a temporally delayed topographic uplift (James et al., 2015b). If the northern rise is at least 1 Gyr old, mantle viscosities in the FeS relaxation scenario would have to exceed 10^{23} Pa s, which is considerably more viscous than previously considered (e.g., Michel et al., 2013; Tosi et al., 2013). A simple dynamic shape scenario involving deep-seated mantle buoyancy can be sustained over geological time for lower mantle viscosities, but, unlike a layered relaxation mechanism, a deep-seated mantle anomaly associated with the melting that produced the NSP may not be able to account for the timing of the northern rise uplift or the mismatch between predicted and observed tectonic features. The origin of the northern rise thus remains an open question.

3.6.3 Caloris Basin

The northern rise is the clearest example of high-admittance shape associated with near-unitary correlation between gravity and shape, but similar high-admittance shape elsewhere on Mercury may be obscured by crustal thickness variations. For example, a two-component compensation model (Section 3.5.5) predicts the existence of thin crust and low mantle density under the Caloris impact basin (Figure 3.9). A low-density mantle anomaly in isolation would produce high admittances and unitary correlation, but for the Caloris basin, the net effect of crustal thickness variations superimposed on a high-admittance shape produces low correlations and low (or even negative) admittances.

Importantly, the Caloris basin also shows anomalous long-wavelength topographic undulations associated with thickened crust (Figure 3.11), and some portions of these plains have even been lifted above the basin rim (Preusker et al., 2011; Zuber et al., 2012; Klimczak et al., 2013; Chapter 10). In contrast to the northern domical rise, the topographic undulations in the Caloris plains are manifest as east–west-oriented, elongate highs and lows, with amplitudes of 2–3 km and wavelengths of 850–1100 km. These crests and troughs are also larger features than the northern rise and extend westward beyond the rim of the Caloris basin for almost 3000 km, where their wavelengths increase to ~1300 km (Figure 3.11). Nonetheless, as at the northern rise, the floors of some volcanically infilled

3.6 Regional-Scale Crust and Mantle Features

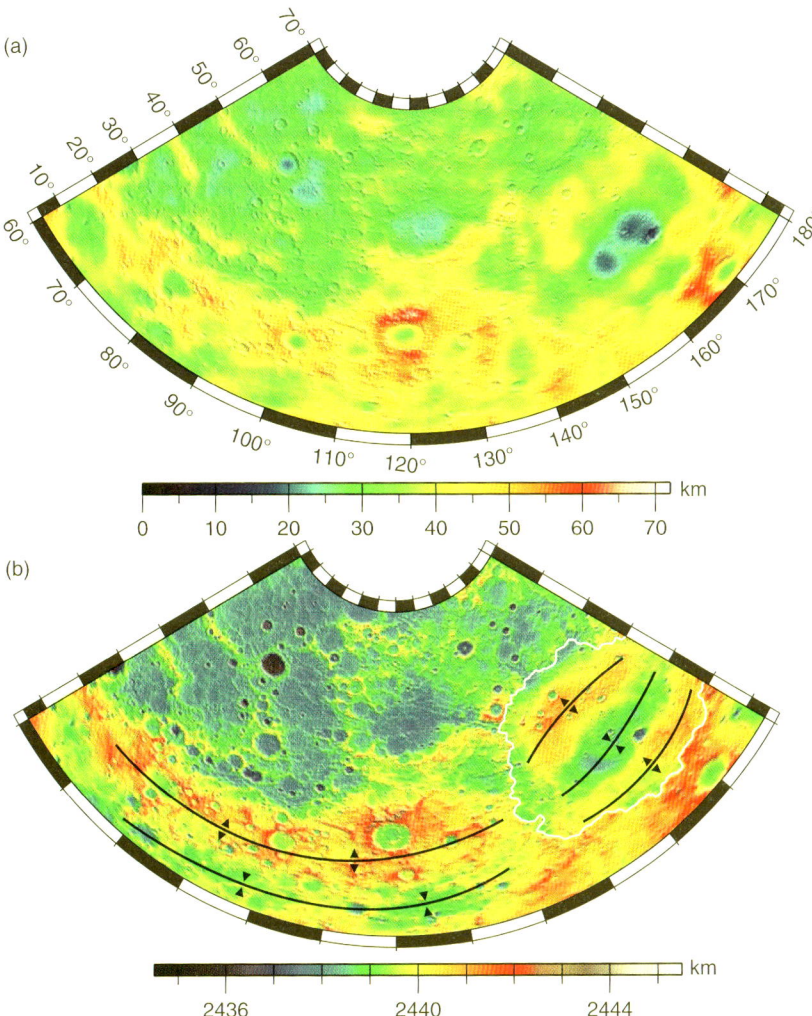

Figure 3.11. (a) Crustal thickness of the Raditladi (H-4) quadrangle of Mercury. (b) Shape of the same region. The Caloris basin is outlined in white. Black lines here denote axial traces of long-wavelength topographic undulations mapped by Byrne et al. (2014). Facing pairs of arrows correspond to troughs; opposite arrow pairs correspond to crests. Crustal thickness calculated as in Figure 3.2d. Lambert azimuthal equal-area projections.

craters display tilts similar in direction and magnitude to the local tilt of the long-wavelength shape of the Caloris plains (Balcerski et al., 2013). This pattern suggests a chronology similar to that of the northern rise, by which high-admittance shape postdates the emplacement of the Caloris volcanic plains but assumed its current form prior to its later record of cratering. Moreover, the earliest gravity signature in the Caloris basin may be the thin crust in the south, which could be indicative of a mascon (Smith et al., 2012).

Mantle convection on Mercury had been suggested to occur in the form of long roll patterns that might result in linear deformation of Mercury's long-wavelength shape (King, 2008). The limited thickness of Mercury's mantle indicated by MESSENGER observations challenges this hypothesis, however, as convection in the thin mantle would be expected to produce smaller features that may not have a roll-like morphology (Michel et al., 2013). Further, the amplitudes of modified long-wavelength shape are one to two orders of magnitude greater (Klimczak et al., 2013) than predicted by three-dimensional models of mantle convection.

Alternatively, these long-wavelength undulations may be a manifestation of the process that has governed the planet's tectonic history: global contraction. If so, then the troughs and crests within and proximal to the Caloris basin formed as a result of lithospheric buckling or folding (Dombard et al., 2001; Hauck et al., 2004) that was accompanied by crustal thickening. Elastic buckling stresses far exceed the strength of Mercury's lithosphere, but models of an elastic–plastic lithosphere allow for folding at stresses far below that threshold (McAdoo and Sandwell, 1985). Dombard et al. (2001) applied to Mercury a preliminary model of unstable deformation of a strong surface layer adjusted for folding. They found that large-scale changes in shape developed in response to global contraction-induced crustal shortening at wavelengths that are comparable with those measured in the Caloris basin (Chapter 10).

3.6.4 Fold-and-Thrust Belts

Crustal shortening structures occur across Mercury, and their global distribution is shown in Chapter 10 (Figure 10.3a). Although there is no globally coherent pattern of deformation similar to that predicted by early studies of tidal despinning (Melosh and McKinnon, 1988), there are regions where shortening structures form laterally contiguous, narrow bands several thousand kilometers long but only a few tens of kilometers wide. These bands have been recognized as fold-and-thrust

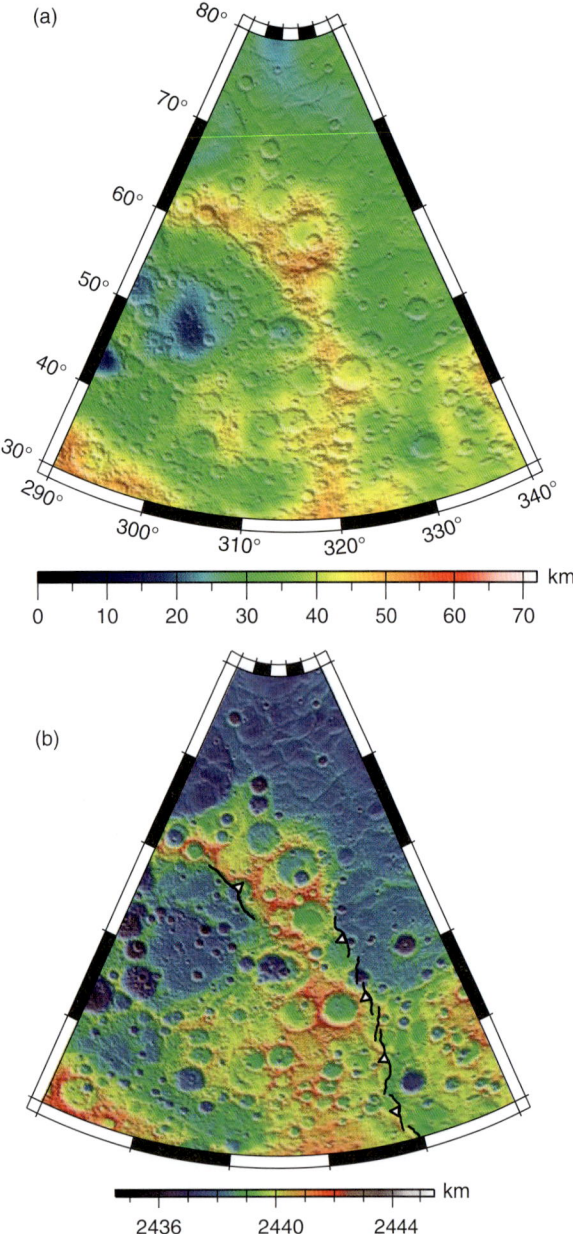

Figure 3.12. (a) Crustal thickness of the Victoria (H-25) quadrangle of Mercury. (b) Shape of the same region. Black lines show surface traces of crustal shortening structures mapped by Byrne et al. (2014); the white arrows indicate inferred down-dip directions of underlying faults. Crustal thickness calculated as in Figure 3.2d. Lambert azimuthal equal-area projections.

standing terrain for much of its length. This area of elevated terrain corresponds to a region of thicker crust, which is consistent with the dominant eastward vergence (see Chapter 10) of this fold belt. That is, many of the larger thrust-fault-related folds that make up this belt are, in cross section, inclined to the east, which indicates the direction of movement of the associated hanging-wall fault block. Accordingly, the underlying thrust fault dips to the west, under the portion of thickened crust.

This structural arrangement is seen for other areas of high-standing terrain that are bounded by large thrust faults, i.e., those that are hundreds of kilometers long and have accumulated many hundreds to several thousand meters of relief. Enterprise Rupes, which crosscuts the Rembrandt basin, is one such example. In the southern hemisphere are many other large thrust faults, for which crustal thickness estimates are not currently reliable. Nonetheless, large thrusts bordering areas of thickened crust may have effectively isolated those thicker crustal portions from neighboring blocks of thinner crust; with continued shortening deformation, these deep and long faults enable thicker crustal blocks to overthrust adjacent low-lying terrain, as is observed on Earth (e.g., Suess, 1909; Şengör et al., 1993). Some portions of Mercury's thicker crust may have developed because of shortening, through the formation of imbricate stacks, for example. Alternatively, it may be that other processes, such as bolide impacts, generated areas of thicker and thinner crust on the planet, and thrust faulting nucleated at the boundaries of such areas.

3.6.5 Geophysical Signatures of Mercury's Impact Basins

Aside from the thinned crust beneath the Caloris and Sobkou basins (Figure 3.2d), there are positive Bouguer δg signals associated with the Borealis, Budh, Goethe, Rachmaninoff, Raditladi, and Vyāsa basins, with Sor Juana Crater, and with Odin and Tir Planitiae. These positive δg values may indicate thinned crust, as is commonly the case for large impact features on the Moon. Whether that interpretation applies to Mercury, or the solution constraints cause an underestimation of gravity lows, remains to be determined. So, too, does the prospect of large-scale thrust faults mechanically bounding the elevated crust–mantle boundary within impact basins, as has been documented for the Moon's Crisium basin (Byrne et al., 2015). At the current level of resolution, the bull's-eye-shaped crustal annulus that surrounds central mantle uplift beneath lunar mascons is not seen on Mercury, but such annuli may not be well resolved by the ~130 km half-wavelength cutoff of our gravity inversion.

3.7 DEGREE-2 SHAPE AND GEOID

3.7.1 Why Isolate Degree 2?

Among a planet's first-order physical properties, beyond radius, mass, and internal magnetic field, are the properties of its geoid and shape ellipsoids. The ellipsoids are represented by spherical

belts (Byrne et al., 2014), similar to fold belts on Earth (e.g., Poblet and Lisle, 2011) and Venus (Burke et al., 1984; Zuber and Parmentier, 1995). Mercury's fold belts are composed of a combination of monoclines and asymmetric anticlines, structures that share a similar strike and sense of vergence (i.e., direction of tectonic transport: see Chapter 10).

One such example is found at mid-northern latitudes in Mercury's western hemisphere (Figure 3.12). Subtending over 40° of arc, this belt is aligned approximately north–south and demarcates the eastern margin of an area of high-harmonic coefficients of degree $l = 2$ and orders $m = 0, 1,$ and 2 (denoted "d_2" in aggregate). In a principal axis (PA) coordinate system (X, Y, Z), the $l = 2, m = 0$ terms (e.g., C_{20}) describe polar oblateness, axisymmetric about the Z axis. The $l = 2, m = 1$ terms (e.g., C_{21}, S_{21}) are identically zero for the geoid in a PA framework and for the shape give the orientation (tilt angle and its azimuth) of the ellipsoid z axis with respect to the Z axis. For the shape, the $l = 2, m = 2$ terms (c_{22}, s_{22}) are related to equatorial ellipticity and its rotation with respect to the X axis. For the geoid in a PA system, $S_{22} \equiv 0$, and there is no rotation; the minimum moment A lies along the X axis.

This d_2 information places constraints on the radial density and rheological structure of the interior (Chapter 4) and may also shed light on past mechanical, orbital, and spin states of the planet. In this section, we treat Mercury's d_2 shape and geoid coefficients separately from those at other degrees in order to understand better the planet's evolution. The d_2 terms for Earth, Mars, and the Moon have been studied intensively. What these coefficients tell us is not always clear; some of the interpretations have been contentious, and debates continue to this day. A useful approach to this problem is to measure the departure of a planetary body from hydrostatic equilibrium and to consider the thermal, mechanical, and dynamical evolution scenarios that may have led to this state.

A spinning planet that is density-stratified will achieve a state of hydrostatic equilibrium if internal shear stresses decay to zero. In this situation, the spherical harmonic gravity coefficient C_{20} completely defines the gravitational potential. Newton (see Chandrasekhar, 1987) was the first to recognize that equilibrium requires a spinning planet to take on the shape of an oblate spheroid, displaying a "rotational bulge" with the minor axis coincident with the spin axis. Hydrostatic equilibrium also displays decreasing oblateness of interior density boundaries with decreasing radial distance from a planet's center. Furthermore, the deformation of any density boundary depends on the deformation of all the other density boundaries in an effect known as "self-gravitation." The sum of hydrostatic and non-hydrostatic mass distributions determines a planet's principal moments of inertia ($A < B < C$).

Departures from the hydrostatic state of Earth and the other planets are best understood from the behavior of the geoid, which reflects the mass distribution of the entire planet. The oblateness of the geoid is characterized by the flattening, f, i.e., the fractional difference between the geoid's equatorial (a) and polar (c) radii: $f = (a - c)/a$. A first-order theory yields the observed flattening as

$$f = (3/2)J_2 + (1/2)m_e, \tag{3.7}$$

where $J_2 = -C_{20}$ and m_e is the ratio of centrifugal acceleration at the equator to the total gravitational acceleration and is proportional to the square of the angular speed of rotation. For Earth, departure from the hydrostatic state is so small that a second-order theory is necessary to describe the observed flattening accurately; it is given (Stacey and Davis, 2008) for degree 2 as

$$J_2 = \frac{2}{3}f\left(1 - \frac{f}{2}\right) - \frac{m_e}{3}\left(1 - \frac{3}{2}m_e - \frac{2}{7}f\right). \tag{3.8}$$

The first-order theory for hydrostatic flattening is

$$f_H = \frac{5}{2}m_e \Bigg/ \left\{1 + \left[\frac{5}{2}\left(1 - \frac{3}{2}\frac{C}{Ma^2}\right)\right]^2\right\}, \tag{3.9}$$

where M is the mass of the planet. The hydrostatic counterpart to equation (3.8) must be obtained numerically. The departure from the observed flattening for Earth is 0.5%. This small departure has a complex explanation and most likely involves the mass motion resulting from convection in the mantle and from rebound of the lithosphere following the end of the last glaciation (e.g., Chao, 2006).

Compared with the situation for Earth, the meager quantities of geophysical data available for the other terrestrial bodies would seem to make explanations for departures from the hydrostatic state very difficult to achieve. But Mars, the Moon, and Mercury are far from hydrostatic equilibrium, so the source of these departures may be readily modeled. (We do not discuss Venus because its moment of inertia is not known.)

Let us first consider Mars. The polar moment of inertia, C, of Mars can be determined from J_2 and from the precession of the spin pole (Folkner et al., 1997). The former had been well known from gravity models constructed with tracking data obtained from the Viking orbiters in the 1970s, and the latter was finally estimated 20 years later by combining Doppler and range data from the Viking and Pathfinder landers. Prior to that time, a hydrostatic assumption was used to obtain C, which was then corrected for non-hydrostatic effects. By far the most clever approach was that developed by Kaula (1979), who calculated the non-hydrostatic component of J_2 under an assumption that the mass distribution of the Tharsis rise was axisymmetric about the center of the rise. This assumption produced essentially the correct answer, confirmed nearly two decades later. The geoid of Mars is, to a good approximation, the sum of a hydrostatic Mars plus the Tharsis rise and the interior response to this massive load.

A planet or satellite is also subject to tidal forces and in hydrostatic equilibrium maintains a "tidal bulge" with a surface defined by the shape coefficient c_{22}, which gives the amplitude of equatorial ellipticity. The equilibrium distribution of density boundaries in this case is described by the addition of a C_{22} gravity field term, which is proportional to $A - B$. In the subsections that follow, C_{20} and C_{22} and their shape counterparts, c_{20} and c_{22}, play dominant roles in the narrative. The observed values of these four coefficients are very different from their hydrostatic counterparts for both the Moon and Mercury, and, among other topics, their values raise the contentious issue of "fossil bulges."

3.7.2 The Lunar Dilemma

For more than two centuries, it has been known that the lunar d_2 geoid is out of hydrostatic equilibrium with respect to its current rotational and tidal potentials (Laplace, 1878; Jeffreys, 1970). The cause of the larger-than-equilibrium magnitudes (Table 3.2) of the gravity coefficients associated with polar flattening (C_{20}) and equatorial ellipticity (C_{22}) has been debated for more than a century (Sedgwick, 1898). The main question is whether

Table 3.2. *Degree-2 shape and geoid properties for Mercury and the Moon.*

		Mercury	Moon
1	Shape rotation, φ*	15°	37°
2	Shape c-axis tilt[†]	4°	27°
3	Correlation coeff.	0.96	0.61
4[§]	J_2(Obs)/J_2(Eql)[‡]	58	21
5	C_{22}(Obs)/C_{22}(Eql)	73	8
6	j_2(Obs)/j_2(Eql)[∥]	309	54
7	c_{22}(Obs)/c_{22}(Eql)	412	9
8	J_2/C_{22} ratio[¶]	0.80	2.70
9	$(b-c)/(a-c)$ ratio[#,**]	0.76	1.91

Sources: Smith et al. (2010b); Zuber et al. (2013); Mazarico et al. (2014); Perry et al. (2015).
* Angle of rotation with respect to axes of the principal moments of inertia in the equatorial plane
† Tilt is with respect to the spin axis.
§ The Love numbers (k_2, h_2) used in equilibrium values (rows 4–7) are for a homogeneous mass distribution, which yields lower bounds on the coefficient ratios.
‡ Obs: observed value; Eql: equilibrium value.
∥ Lower-case coefficients (rows 6, 7, 9) refer to shape; upper-case coefficients refer to geoid.
¶ Ratio is $[J_2/C_{22}]_{Obs}/[J_2/C_{22}]_{Eql}$, where $J_2 = -C_{20}$.
# Ratio is $[(b-c)/(a-c)]_{Obs}/[(b-c)/(a-c)]_{Eql}$.
** The quantities in rows 8 and 9 do not depend on Love numbers.

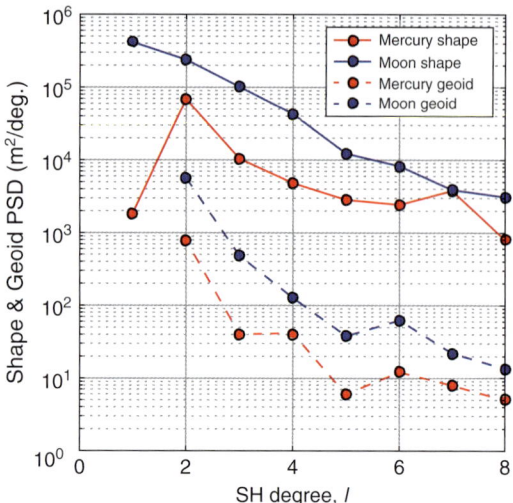

Figure 3.13. Power spectral density through degree 8 for the geoid and shape of Mercury and the Moon. Excess power in degree 2 (by extrapolation from higher degrees) is evident in both geoids and in Mercury's shape. Sources: Smith et al. (2010b), Zuber et al. (2013), Mazarico et al. (2014), Perry et al. (2015).

these bulges are rotational and tidal relics of an early Moon possessing a higher spin rate and smaller semi-major axis. A general renaissance in this topic began about a decade ago when Garrick-Bethell et al. (2006) showed that the Moon's non-hydrostatic shape can be explained if the Moon was locked into its shape while in a synchronous orbit with a semi-major axis equal to 22.9 Earth radii and eccentricity (e) = 0.49, or in a 3:2 spin–orbit resonance orbit with a semi-major axis equal to 24.8 Earth radii and e = 0.17. But it is a substantial challenge to find mechanisms that might maintain these "fossil" bulges as the Moon spun down and moved away from Earth. Meyer et al. (2010) showed that considerable melting from tidal heating would take place at eccentricities considered by Garrick-Bethell et al. (2006), and it was much more likely that the Moon's shape elastically deformed to follow the equilibrium values rather than storing gravitational potential energy in a fossil bulge. Garrick-Bethell et al. (2014) responded with a model that satisfies an impact-basin-corrected d_2 shape and geoid and rests on both crust building by tidal heating and the later development of a frozen tidal–rotational bulge. Qin et al. (2018) showed that bulge formation was controlled by the relative timing of lithosphere thickening and lunar orbit recession, and that acceptable model solutions indicated that lunar bulge formation was a geologically slow process lasting several hundred million years.

Garrick-Bethell et al. (2014) and Keane and Matsuyama (2014) both sought to reconstruct the pre-impact-basin geoid by separating the basin contributions from the global geoid data set. Garrick-Bethell et al. (2014) also estimated the pre-impact basin shape. A knowledge of the Moon's early d_2 geoid and shape should better constrain the Moon's early dynamical history. We can assume that the solutions found would mark the time when the lithosphere had just become sufficiently thick to preserve impact basins; this process should have contributed to taking the Moon out of an equilibrium state in addition to whatever relic bulge was maintained as the body spun down and its orbital semi-major axis was enlarged. Particularly relevant to Mercury is the orientation of the basin-corrected lunar shape relative to the basin-corrected lunar geoid.

3.7.3 So What About Mercury?

Now let us return to Mercury. Back-extrapolation from $l = 3$–8 to $l = 2$ for the geoid power spectral density (PSD) for Mercury indicates excess power at d_2 (Figure 3.13). The lunar spectrum also has a degree-2 excess, and this result has been used to argue that the Moon retains a fossil shape from an earlier and shorter Earth–Moon distance (Williams et al., 2001). The shape PSD for Mercury also indicates excess power in d_2 compared with other degrees (Figure 3.13).

The shapes of the Moon and Mercury also differ in their spin-axis alignment. Mercury's shape ellipsoid is very closely aligned with its spin (hence C) axis (it is off by 4°), whereas for the Moon the misalignment is 27° (Table 3.2, row 2). Correcting for lunar impact basins does not decrease this number (e.g., Garrick-Bethell et al., 2014). If the d_2 shape of Mercury is collocated with the principal axes because it is a relic of an early hydrostatic state, then either the formation of impact basins had little effect on the d_2 shape or there has been a continuation of processes that might have maintained the correlation. This situation is very different from that on the Moon,

3.7 Degree-2 Shape and Geoid   73

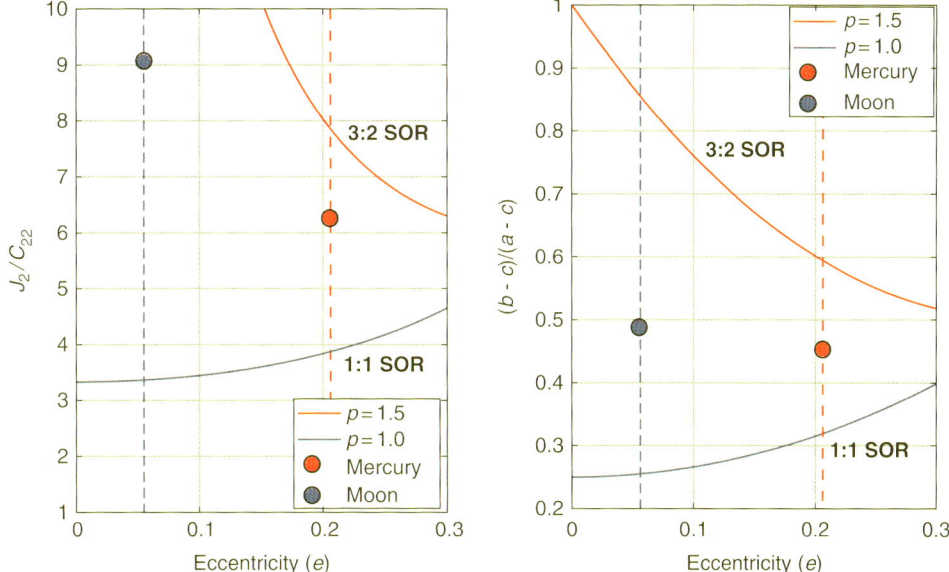

Figure 3.14. Estimates of the ratios J_2/C_{22} and $(b-c)/(a-c)$ for the geoid and shape ellipsoids, respectively, for the Moon (blue circles) and Mercury (red circles). Results are plotted at present orbital eccentricities (dashed lines). Theoretical curves for hydrostatic equilibrium are given by solid lines (Matsuyama and Nimmo, 2009). These ratios depend only on spin–orbit resonance state (p) and eccentricity (e) and, specifically, do not depend on the rheological structure of the interior.

where the South Pole–Aitken basin has a large influence on the d_2 shape (Keane and Matsuyama, 2014).

Quantitative departures from hydrostatic equilibrium for both Mercury and the Moon are listed in Table 3.2. The basic shape and geoid d_2 ratios of observed to hydrostatic values (Table 3.2, rows 4–7) suggest that Mercury is farther out of hydrostatic equilibrium than the Moon. Additionally, ratios of geoid and shape principal axes, J_2/C_{22} and $(b-c)/(a-c)$, respectively, are out of equilibrium for both bodies (Table 3.2, rows 8 and 9; Figure 3.14). These ratios do not depend on the rheology or structure of Mercury's interior. The high correlation (Table 3.2, row 3) between Mercury's d_2 shape and geoid, plus a value of 1/3 for the d_2 ratio of the observed geoid to that due to shape alone, indicate that partial compensation and lithospheric strength must each play a role in maintaining Mercury's d_2 shape, no matter its origin.

Could Mercury's d_2 shape be a fossil relic of a different orbit and spin state that existed early in Mercury's history? Matsuyama and Nimmo (2009) calculated solutions for the spherical harmonic gravity coefficients J_2 and C_{22} over the semi-major-axis–eccentricity (a_M–e) parameter space for various spin–orbit resonances in a manner similar to that of Garrick-Bethell et al. (2006). To match the Mariner 10 coefficient values requires $a_M < 0.1$ AU, an unrealistically low value. Repeating the exercise with MESSENGER values of J_2 and C_{22} (Mazarico et al., 2014) decreases the bound for a_M to only 0.05–0.06 AU (Perry et al., 2015).

Furthermore, Matsuyama and Nimmo (2009) showed that even if Mercury had been positioned at 0.1 AU, outward migration because of solar tides would have been negligible over the lifetime of the solar system, and Mercury would have been unable to reach its current orbit from the much earlier 0.1 AU orbit.

To sum up, Mercury is out of hydrostatic equilibrium to a much greater degree than the Moon, but fossil rotational and tidal bulges cannot provide a causative explanation for Mercury (Matsuyama and Nimmo, 2009), whereas they can for the Moon (Keane and Matsuyama, 2014; Garrick-Bethell et al., 2014). Unlike the Moon, Mercury's d_2 geoid and shape are highly correlated.

3.7.4 Simple Models to Help Understand the d_2 Shape and Geoid

In the past, Mercury's d_2 shape was out of hydrostatic equilibrium, stresses were generated in the interior, and the mechanical response of the planet was to relieve those stresses to the extent allowed by crustal and mantle rheology. That is, the dynamic response of the interior is to move toward balancing the mass associated with the non-hydrostatic shape. Mercury's d_2 geoid indicates that this compensation was not completely successful, but we note that even an isostatic state is not free of deviatoric stress (Phillips and Lambeck, 1980). In a planet, interior density boundaries can deform rapidly on geological timescales in response to large mass loads, relieving normal stresses but not necessarily shear stresses. We are most familiar with this concept when interpreting crustal thickness variations that are in approximate mass balance at some reference depth (e.g., Airy isostasy). To be sure, mechanical equilibrium can be achieved, including a flexural component, but this situation does not meet the definition, *sensu stricto*, of isostasy. In a more general sense, studying mechanisms of d_2 compensation provides insight into the thermomechanical evolution of Mercury.

The observed polar oblateness and equatorial ellipticity coefficients (C_{20} and C_{22}, respectively) of Mercury's geoid are each about one-third of the value expected from the shape alone, so

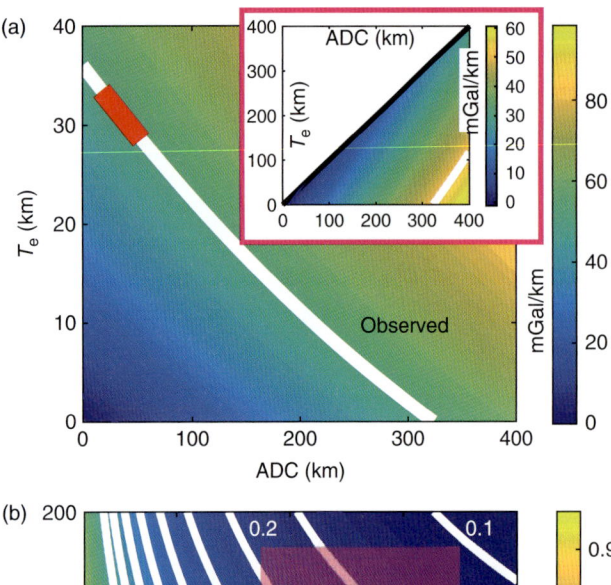

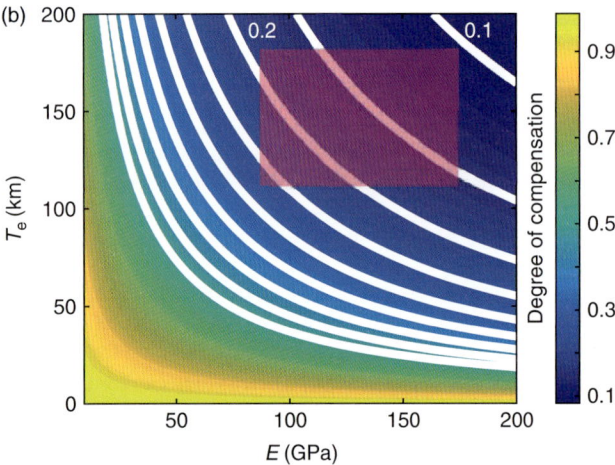

Figure 3.15. (a) Model of global admittance (gravity disturbance over shape) for top loading by the c_{20} shape component on an elastic lid of thickness T_e and Young's modulus of 50 GPa (James et al., 2015b). The shape load is compensated at an apparent depth of compensation (ADC), which for values less than about 50 km is likely to be the CrMB. The white curve is the observed value of the admittance, and the red portion corresponds to the mean crustal thickness range determined by Padovan et al. (2015). The inset shows the corresponding result for bottom loading for ADC ≥ T_e. (b) Contour plot of dimensionless "degree of compensation" (DC) (Turcotte et al., 1981) for Mercury as a function of Young's modulus (E) and elastic lithosphere thickness (T_e), for $l = 2$ and Poisson's ratio = 0.25. The shaded area provides information from Tosi et al. (2015): the T_e range corresponds to estimates obtained by matching output from their thermomechanical model to geoid and shape data, and the E range bounds the value assumed for the crust (left boundary) and for the mantle (right boundary). Most of the lithosphere is contributed by mantle material, so the DC will be less than about 0.2.

some degree of compensation has occurred. As discussed earlier, unlike the Moon, the d_2 shape c axis nearly coincides with the spin (C) axis, and most compensation mechanisms actually require such collocation. We note that C_{20} is large in the sense that it eliminates isostatic compensation dominated by crustal thickness variations in particular, and shallow density boundaries in general. Figure 3.15a contours the admittance coefficient, $Z_{20} = C_{20}/c_{20}$, for the loading of an elastic shell (James

et al., 2015b). The parameter axes are T_e and the apparent depth of compensation (ADC), which corresponds to a density boundary that, for depths less than about 50 km, would very likely be the CrMB. This model employs an elastic lid of thickness T_e and is based on the assumption that a top load, c_{20}, acts on the interior and is partially compensated at the ADC. The purely isostatic solution, where $T_e = 0$, places the density boundary at a depth of approximately 300 km for the observed admittance; this solution is indicated by the white line in Figure 3.15a, which traces the T_e–ADC combinations that produce the observed Z_{20}. It is unlikely that a dominant density boundary such as a phase change occurs at this depth in the mantle of Mercury (see Chapters 4 and 19). However, the ADC could be an equivalent depth for a vertical density distribution that produces the same gravity signal. It is also unlikely that a distributed density would reflect mantle convection, because the vertical-to-horizontal aspect ratio of convection has essentially no d_2 component (Michel et al., 2013). In summary, satisfying Mercury's large geoid term (C_{20}), as expressed by the admittance Z_{20}, requires either a deep source of mass compensation or a top-loaded elastic shell, and the latter is more likely. This situation is one of high-admittance shape (see Section 3.5.5).

Thus, for non-zero values of T_e, c_{20} is supported at least partially by elastic (and thermoelastic: Tosi et al., 2015; see also Section 3.7.7.1) membrane and flexural stresses; for mechanical equilibrium, a subsurface mass smaller in magnitude than and of opposite sign to the c_{20} mass results. There is a trade-off between T_e and ADC, and for an ADC equated to the crustal thickness range of Padovan et al. (2015), a T_e value of about 30 km is required (red line in Figure 3.15a). This value is close to that estimated by James et al. (2016).

Bottom loading of an elastic lithosphere leads to apparent depths of compensation larger than the isostatic value in order to offset the effects of an equilibrium subsurface mass greater in magnitude than that associated with c_{20} (Figure 3.15a inset). In addition to top and bottom loading, the density gradient within an elastic lithosphere influences both the shape and the geoid. The models discussed in the present section provide a framework for understanding more complex models.

3.7.5 The Role of the Temperature Poles

Because of Mercury's near-zero obliquity (Margot et al., 2012; Mazarico et al., 2014), there are substantial surface temperature differences (Figure 3.16) between the geographic (north, south) or spin poles and the equator. Because of Mercury's 3:2 spin–orbit resonance, there can be temperature differences (depending on the orbital eccentricity, e) of 100 K along the equator between the "hot" poles at 0°E and 180°E and the "cold" poles at 90°E and 270°E (Vasavada et al., 1999; Siegler et al., 2013; Figure 3.16). The reason for this variation is that the two longitudes (0°E and 180°E) of Mercury's axis of minimum principal moment of inertia (A) alternately face the Sun at successive perihelia, and the two longitudes (90°E and 270°E) of the intermediate principal axis (B) alternately face the Sun at successive aphelia. The net difference in insolation leads to a longitudinal dependence of average surface temperature, creating the "hot" and "cold" poles, with the coldest surface

3.7 Degree-2 Shape and Geoid

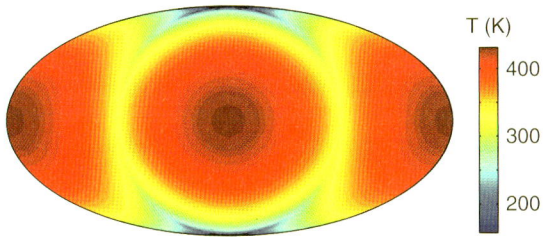

Figure 3.16. Global distribution of average temperature at 1-m depth (Siegler et al., 2013) for the present value of Mercury's eccentricity, 0.2056, and calculated with a spherical harmonic expansion to degree and order 80. Because of the low thermal conductivity of the shallow portions of the regolith, temperatures at 1-m depth are better boundary conditions for the interior temperature distribution than the actual surface temperatures. Mollweide projection with the equatorial hot poles, 0° and 180°E, at the center and map extremes, respectively.

temperatures at the north and south spin poles (not counting the effects of impact crater topography and shadowing). The spherical harmonic coefficients ($^TC_{lm}$, all $^TS_{lm} = 0$) of the resulting global surface temperature distribution, $T(\theta, \varphi) \equiv T_{0\varphi}$, are dominated by $^TC_{20}$ and $^TC_{22}$, with lesser contributions from the even-ordered d_4 terms ($m = 0, 2, 4$) and even smaller contributions from terms of higher even-numbered degree and order. Given the role that the principal axes play in controlling the surface temperature distribution, it is not surprising that the d_2 geoid and surface temperature distributions are highly correlated.

Williams et al. (2011) described the important consequences that these surface-temperature variations might have on the thermomechanical behavior of Mercury's subsurface. In particular, measurable variations in the thickness of the elastic and mechanical lithospheres should be expected. Phillips et al. (2014) noted that the d_2 shape and geoid correlation also means that the d_2 surface temperature and shape are correlated, and it is therefore a reasonable expectation that lateral variations in subsurface thermal buoyancy and thermal strain induced by the surface temperature distribution could be responsible for at least some of the d_2 shape.

Capture of Mercury into a 3:2 resonance is nearly certain at e between 0.2 and 0.41 (Makarov, 2012) and, once the spin state is captured, the resonance is very stable (Noyelles et al., 2014). Over the age of the solar system, Mercury's eccentricity covered a range from 0.0 to 0.4 (Correia and Laskar, 2004). The temperature difference between the hot and cold poles depends on e, and therefore so does the equatorial temperature ellipticity coefficient, $^TC_{22}$, which in fact shows a substantial dependence on eccentricity (Figure 3.17a).

The behavior of Mercury's eccentricity is chaotic (Correia and Laskar, 2004) on timescales $\gtrsim 1$ Gyr, but on timescales $\lesssim 100$ Myr, periodicities largely due to secular perturbations from the other terrestrial planets are superposed on this chaotic background. Quasi-resonances (Laskar, 1988) lead to dominant oscillation periods, P, which diffuse into the interior of a homogeneous planet according to a degree-l spherical Bessel function $j_l(kr)$, where $k = \sqrt{\pi/\kappa P}(1 + i)$, $i = \sqrt{-1}$, and κ is thermal diffusivity. The effects on the depth profile of the temperature difference between the hot pole and cold pole for a typical

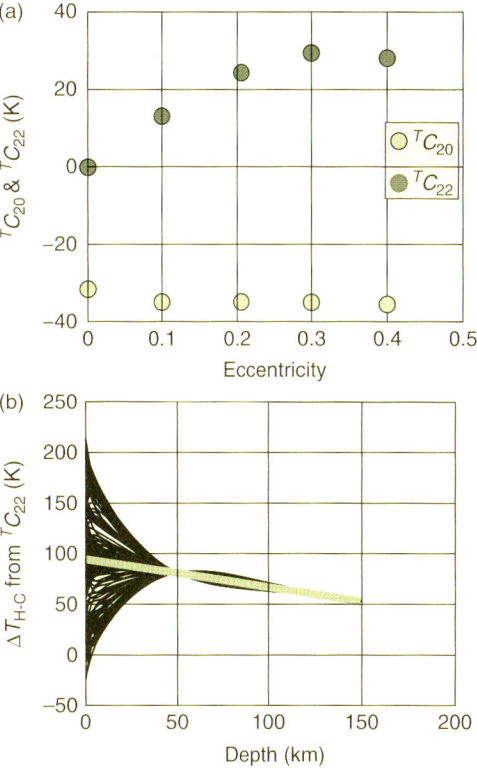

Figure 3.17. (a) Behavior of the normalized (2,0) and (2,2) coefficients of the spherical harmonic expansion of the surface temperature distribution for different eccentricities (e) of Mercury's orbit (Siegler et al., 2013). As expected, there is little change in $^TC_{20}$, which is affected by the difference between polar and average equatorial temperatures. The $^TC_{22}$ term reflects directly the changes in insolation between the hot and cold poles. At an eccentricity of 0.4, the Sun is large enough in the sky at perihelion to affect the insolation on the cold pole to the extent that the $^TC_{22}$ magnitude decreases relative to its 0.3 eccentricity value. (b) Envelope of possible conductive temperature profiles of the temperature anomaly between the hot pole and cold pole as determined by four dominant periodicities (~17 Myr, ~3 Myr, and two near 1 Myr) in e (Correia and Laskar, 2009; Wieczorek et al., 2011) plus an additional period of 100 Myr. The gray line is the mean value of e (taken as 0.2056). The black curves are the results of 200 Monte Carlo simulations of random phase differences between the various periodicities.

eccentricity time series (four spectral peaks in the period range 1–20 Myr plus a 100-Myr period with an amplitude of $e = 0.2$; Siegler et al., 2013) suggest that there could be considerable uncertainty in the $^TC_{22}$ temperature profile to depths approaching 50 km (Figure 3.17b), i.e., over much if not all of the crust. Given the stability of Mercury's near-zero obliquity (Yseboodt and Margot, 2006), a cautious approach to modeling insolation-driven thermal contributions to the shape and geoid would focus on $^TC_{20}$.

3.7.6 Tosi et al. (2015) Model: Basic Description

Tosi et al. (2015; hereafter "T15") provided a thermomechanical model of Mercury that showed a fairly successful outcome in a test of the hypothesis put forth by Phillips et al. (2014):

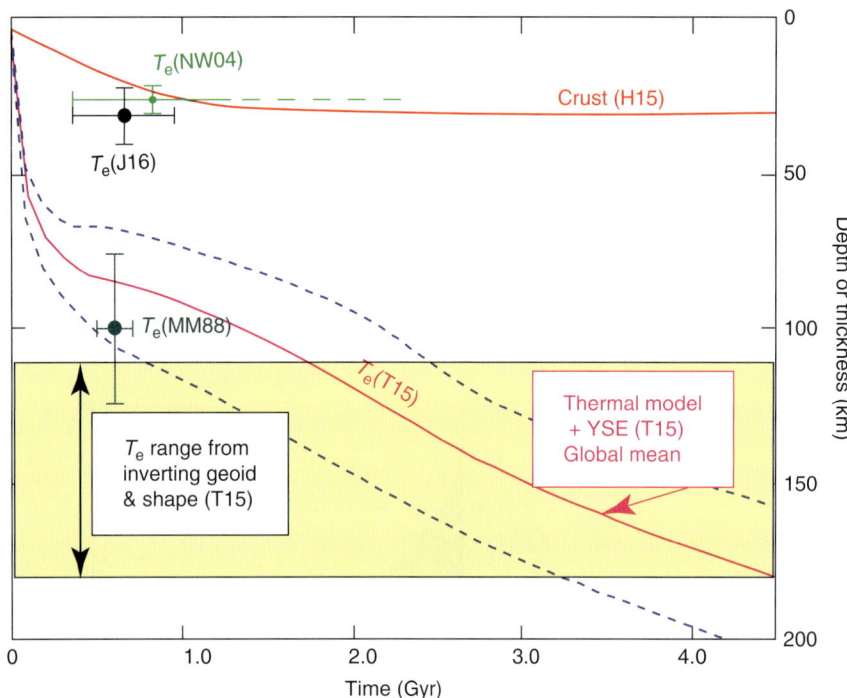

Figure 3.18. Determinations of T_e compared with models of crust and lithosphere evolution. The vertical extent of the yellow rectangle is the range in T_e estimated by Tosi et al. (2015) (T15) from inversion of geoid and shape data. In that model, the evolution of T_e is obtained by equating T_e to T_m (mechanical lithosphere thickness), the base of which is defined by a strongly temperature-dependent yield stress under ductile flow. Temperature is obtained from a three-dimensional thermal model. The magenta curve is the global mean value of T_e for a creep strain rate of 10^{-18} s^{-1}. The upper (lower) dotted blue curve is the minimum (maximum) T_e for a strain rate of 10^{-19} s^{-1} (10^{-17} s^{-1}). The red curve (H15) shows model-based crustal thickening (Hauck et al., 2015). Estimates of elastic thickness by Melosh and McKinnon (1988: MM88), Nimmo and Watters (2004: NW04), and James et al. (2016: J16) are also plotted. Adapted from James et al. (2016) and based partially on T15.

Mercury's non-hydrostatic d_2 shape and geoid result from the peculiar insolation pattern caused by the 3:2 spin–orbit resonance and near-zero obliquity. We refer to the d_2 shape and geoid, but the T15 modeling also included the smaller contribution of d_4, which is implicit in our discussion here. We describe the model in this subsection and review some of its associated issues in the next subsection. It may be that we are closer to understanding Mercury's non-hydrostatic d_2 shape and geoid than we are to their lunar counterparts, despite the two-century head start for the Moon.

Mercury's surface temperature distribution was used as a surface boundary condition for a three-dimensional thermal evolution model, which predicts that mantle heat transport was dominated by convection until about 1 Ga, after which time the mantle transported heat conductively. The density structure resulting from the temperature distribution in the thermal models was used to load a self-gravitating, compressible thermoelastic shell model with the same surface boundary condition. The deformation of the shell produces a model shape and geoid for which SH coefficients were compared with Mercury's observed geoid and shape SH coefficients associated with the degrees-2 and -4 surface temperature distribution. The best-fitting models (>95% variance reduction) yield a T_e of about 150 km, with a range of 110–180 km.

T15 posited that the present distribution of insolation-driven subsurface temperature anomalies is temporally part of a quasi-steady-state condition that was achieved following the last time that Mercury acquired a 3:2 spin–orbit resonance (SOR). Essentially, the assumption is that the 3:2 SOR thermal deformation, once established, does not adjust to a thickening lithosphere; it is "frozen in" (see Section 3.5.1.2), though T15 did not specifically use this terminology. So when in Mercury's history was the elastic lithosphere about 150 km thick? T15 addressed this question by making the assumption that the long-wavelength curvature in lithospheric deformation is small, which implies that the thicknesses of the elastic and mechanical lithospheres are nearly equal. As discussed in Section 3.5.1.1, T_m is defined by a threshold creep stress (15 MPa as given by T15) that is highly temperature dependent and can be predicted from the output of a thermal model. The result (Figure 3.18) is that the 150-km thickness is achieved about 1.5 Gyr into the evolution of Mercury, and the last 3:2 SOR reacquisition would have occurred approximately 0.5 Gyr earlier, or at about 3.5 Ga, to allow for the diffusion of the temperature perturbations into the interior.

3.7.7 Tosi et al. (2015) Model: Why Does It Work and What Are the Issues?

3.7.7.1 Why Does the Model Work?

There are three main contributions to the elastic shell deformation in the T15 model: (i) buoyancy forces associated with thermal density anomalies in the lithosphere generated by $T_{\Theta\varphi}$; (ii) thermal expansion and contraction associated with those anomalies; and (iii) radial displacement in the sub-lithosphere region, largely from thermal strain but with a minor contribution from CMB shape. The last contribution is likely the smallest of the three, partly because the temperature differences between the thermal poles decrease approximately linearly with depth and vanish on the CMB. In general, the geoid and shape signals are dominated by lithospheric processes, as indicated, for example, by the similarity in model predictions in the convective and conductive regimes.

The shape can be taken as a top load on an elastic lithosphere (Section 3.7.4), and we examine the resulting state of compensation with the T15 model parameters. The departure from mass balance is embodied in the dimensionless "degree of

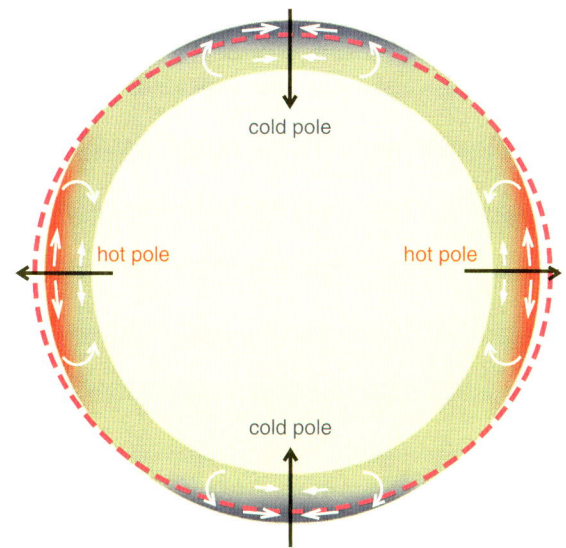

Figure 3.19. Schematic view of the elastic lithosphere component of the T15 model in Mercury's equatorial plane. Hot (cold) poles have surface temperatures higher (lower) than the planetary mean. The temperature anomaly lessens with depth, and the integrated buoyancy and radial strain lead to an elevation increase (decrease) in hot (cold) pole regions. Horizontal thermal strains are accommodated by upwarping (downwarping) in hot (cold) pole regions, and the thermoelastic stresses support at least half of the high-admittance shape (magenta dashed curve), which is an apparent top load.

compensation" (DC) introduced by Turcotte *et al.* (1981). Given the T15 best-fitting range of T_e, 110–180 km, and their Young's modulus for the mantle of 175 GPa, the degree of compensation for top loading by the shape is quite low, about 0.15 (Figure 3.15b). This result implies, *inter alia*, minor mass cancellation of the shape due to deformation of the CrMB in response to the load; this outcome is consistent with the relative insensitivity of T15 model fits to variation in H_{cr}. Although the DC results in Figure 3.15b are specifically for top loading, they do suggest, at least qualitatively, that internal buoyant loading within a lithosphere 150 km thick will have a muted response at the surface because of the membrane resistance of the elastic shell. The radially integrated thermal buoyancy across the entire silicate shell for a conduction solution is about 50% (with opposite sign) of the buoyancies stored in the shapes, $\rho_{crust} g_0 c_{20}$ and $\rho_{crust} g_0 c_{22}$. Membrane resistance means that the 50% fraction has to be taken as a generous upper bound of the buoyancy contribution to shape. With $T_e = 150$ km, radial thermal expansion beneath and mostly within the lithosphere can account for on the order of 10% of both the c_{20} and c_{22} terms, with the major contribution provided by the lithosphere. Buoyancy contributions likely contribute at about the same level. The sum of thermal buoyancy, compositional buoyancy (CrMB), and strain probably accounts for no more than half of the shape.

The remainder of the shape can be attributed to confined horizontal (θ, φ) thermoelastic strain. The T15 model produces a scenario that is geophysically comparable to top loading, in the sense that long-wavelength shape is supported partially by thermoelastic membrane and bending stresses in the lithospheric shell (Figure 3.19). The T15 mechanism does not require the emplacement of material at Mercury's surface or near subsurface. Rather, this process creates its own top load and is akin to warping, where long-wavelength shape is produced through uneven thermal expansion in a spherical shell and supported largely by thermoelastic stresses, as noted.

What about the geoid? We noted earlier that the observed d_2 geoid is about a third of that due to shape alone. To first order this ratio can be achieved largely with the density anomalies associated with the thermal buoyancy providing partial compensation of the shape, as the two masses have opposite signs. To a lesser extent, perturbation of the CrMB according to the 0.15 DC value also contributes to the compensation.

The model presented by T15 represents the only hypothesis advanced to date to explain the d_2 correlation of shape and surface temperature. The model also provides the necessary partially compensated d_2 surface load that reproduces both the shape and the geoid. So it looks as though solar energy in concert with the 3:2 SOR is responsible for Mercury's marked departure from a d_2 hydrostatic state, and this mechanism differs thoroughly from that responsible for the Moon's departure from a hydrostatic state. However, there are issues with the T15 model that are discussed in the next four subsections.

3.7.7.2 Late 3:2 SOR Acquisition?

Mercury should have entered a 3:2 SOR early in its history (Correia and Laskar, 2004, 2009; Makarov, 2012), but it is also possible that at later times impacts of sufficient energy temporarily disrupted the 3:2 SOR (Correia and Laskar, 2012; Noyelles et al., 2014), and it is the last of these events that set the thermal environment in the T15 model. However, a late acquisition of a 3:2 SOR is in conflict with geological observations. Tectonic mapping indicates that tidal spindown could have operated at the same time as, but did not postdate, global contraction from secular cooling (Klimczak et al., 2015); superposition relations show that tectonic deformation arising from global contraction was underway by about 3.6 Ga (see Section 10.8.1). Moreover, global contraction was likely operating for some time before then, because the tectonic manifestation of this process (i.e., faulting) was preceded by elastic deformation (Section 10.8.1). Thus, given that the strain pattern predicted by tidal spindown does not dominate Mercury's tectonic fabric, the acquisition of Mercury's current SOR must have occurred considerably before 3.6 Ga, a conclusion that is at odds with the T15 model.

3.7.7.3 Variations in Eccentricity

The Tosi et al. (2015) model neglects the consequences of the strong variations in Mercury's orbital eccentricity (Figure 3.17b). However, the quality of the model fit was judged by the reduction in the joint error variance of shape and geoid for all admissible d_2 and d_4 terms; this comparison is dominated by the two zonal ($m = 0$) terms, which are insensitive to eccentricity variations (Figure 3.17a). Nevertheless, the equatorial d_2 shape should be constantly changing, following orbital eccentricity changes with a diffusion time lag.

3.7.7.4 Relationship of T_e to T_m

The basic result of T15 depends on the assumption that $T_e = T_m$, i.e., there has been very little deformation at the top (via frictional sliding) and bottom (by viscous creep) of the mechanical lithosphere. Given that the model deformation is largely occurring at wavelengths corresponding to d_2 and to a lesser extent d_4, curvature should be small and setting $T_e = T_m$ is seemingly justified. At these same wavelengths, bending stresses are much smaller than membrane stresses. Whereas the bending stress profile passes from extensional to compressive with depth (or vice versa), the membrane stress profile is uniform with depth, leading to a different style of plate deformation. Furthermore, there is abundant evidence that long-wavelength strain has been localized (see Chapter 10), so curvature could have been sufficiently large locally to negate the idea that $T_e = T_m$. Additional processes such as impact cratering and volcanism can relieve lithospheric stresses over geological time. The net result is that the relationship between T_m and T_e is somewhat ambiguous.

The estimation of subsurface mechanical parameters from surface profiles over faults (Watters et al., 2002; Nimmo and Watters, 2004) had been limited with Mariner 10 data to a single structure, Discovery Rupes (see Section 3.5.4.2). With highly accurate profiles from MESSENGER MLA data, Peterson et al. (2017) extended this methodology to 31 fault structures. The elastic dislocation model that was employed to estimate fault parameters (dip, minimum and maximum fault depth) produced solutions at each of these 31 faults that were a good match to the MLA profiles. The maximum fault depth was equated to the BDT depth, and the temperature there, Ψ_{BDT}, was determined from the employed yield strength envelope. Ψ_{BDT} was linearly extrapolated to a depth at which the temperature generated the T15 threshold creep stress of 15 MPa to provide an estimate of T_m. The three regions studied (northern smooth plains, intercrater plains, and circum-Caloris plains) all have T_m mean values at the assumed ~4 Ga age that are distinctly smaller than the T15 thermal model result (Peterson et al., 2017). Differences in T_m estimates among the three regions may reflect differences in their volcanic histories. We do not compare the T_e estimates because they are based on diametrically opposed assumptions. The T15 approach was based on the assumption that stresses in the lithosphere do not reach the yield stress (i.e., the strength envelope) anywhere, and the Peterson et al. (2017) approach is predicated implicitly on the premise that the stresses are saturated at the yield stress everywhere in the lithosphere because the maximum depth of faulting is equated to the BDT depth. The mechanical thickness, T_m, is a stand-in for heat flux from Mercury's interior, so differences in the T_m results between T15 and Peterson et al. (2017) would at face value (and lateral heterogeneities aside) indicate distinctly different planets; but of course there is only one Mercury.

3.7.7.5 Age of Thermal Structure

Is the density configuration of the T15 model really frozen at ~3 Ga? Certainly, estimates of T_e as low as 30 km at ~4 Ga (e.g., James et al., 2016) imply, unsurprisingly, that stresses can be locked-in on Mercury. In the Tosi et al. (2015) model, thermoelastic stresses are largely responsible for elastic lithosphere deformation and the resulting shape. The corresponding geoid signal is generated mostly by the shape, with incomplete cancellation dominated by the density anomalies associated with the thermal buoyancy distribution.

As the elastic lithosphere cools and thickens, the thermoelastic strains change. Density boundaries, such as the CrMB, are "frozen" and do not adjust to accommodate the changing mechanical state because viscosities are too high. In contrast to the flexural problem (Section 3.5.1.2), thermoelastic processes do not require substantial viscous flow of the mantle to track the equilibrium stress state of the thickening lithosphere. That is, this state can be achieved largely by elastic strain, and by contrast the flexural problem also requires (implicitly) viscous strain, which is unavailable. Viscous strain is never seen explicitly in the classical flexural equilibrium solution because such strain is part of a transient phase. We argue that the T15 estimate of T_e and geoid is in fact germane to the current era on Mercury, but this issue can be resolved by engaging a viscoelastic model in place of the elastic model. The lack of a rheological memory would remove the requirement of a late final acquisition of the 3:2 SOR with a seemingly over-thick lithosphere. Note also that the notion that the equatorial shape ellipticity (c_{22}) tracks changes in orbital eccentricity (Section 3.7.7.3) is also predicated on an absence of rheological memory.

Assigning T_e to the present era is possibly more consistent with estimates of that quantity that are based on data, e.g., ~30 km at ~4 Ga (Figure 3.18). Extrapolating the 30-km value by following the same rate of elastic lithosphere thickening as that predicted by the T15 thermal model leads to a current value of about 125 km for T_e, which falls within the estimated range of T15. The thickening rate for T_e should be more reliable than specific numbers. Remaining unsolved is the discrepancy between estimates of T_e from gravity and shape data and estimates of T_e from thermal models, but reconsidering the assumption that $T_e = T_m$ would be a good place to start to resolve this difference.

3.7.8 An Alternative Model

We have already discussed another model (in Section 3.7.4 and shown in Figure 3.15a) that might depend on the surface temperature distribution. This is a top-loading model with an elastic lid and partial compensation at the CrMB. This model satisfies (i) the crustal thickness result of Padovan et al. (2015), (ii) the elastic thickness estimate of James et al. (2016), and (iii) the Z_{20} gravity/shape admittance. The challenge to this model is to identify the load implied by the c_{20} and c_{22} shape terms. One possibility is early crustal thickening that was sensitive to the insolation-driven thermal anomalies. Convection models (e.g., Tosi et al., 2015) show that temperature differences between the pole and the equator persist deep into the mantle and in a convection regime will generate more melt at the equator than at a spin pole (Figure 3.20). Under this scenario, melt would be brought upward into the crust by buoyancy of the melt itself, providing early crust that thickens toward the equator. Pressure-release partial melting of upward-advecting solid mantle material would add to the crustal volume (see Section 3.4.1). As

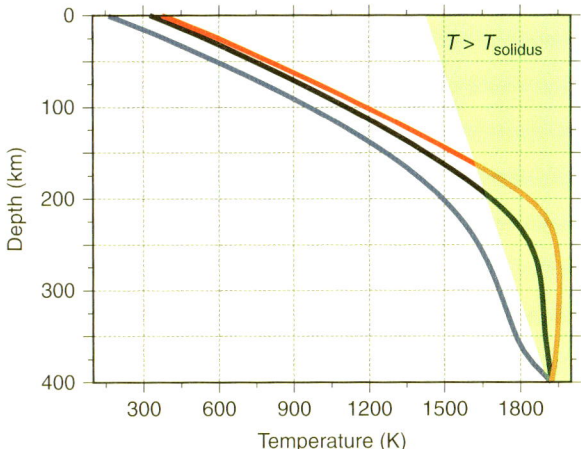

Figure 3.20. Solutions after 1 Gyr for the three-dimensional thermal evolution model of Tosi et al. (2015). Profiles are for the minimum (blue), average (black), and maximum (red) mantle temperatures; the differences are due largely to the insolation-driven surface temperature variations. The green region has temperatures above the solidus temperature, which is from Hirschmann (2000) and represents a quadratic fit to data from 63 sets of peridotite melting experiments.

noted earlier, partial melts would tend to remain buoyant as they traversed the mantle (Vander Kaaden and McCubbin, 2015), maximizing the mass difference between the equatorial and polar regions. The difficulty with this model may be that partial melting differences between the temperature poles are not large, and it may take an inordinate amount of time to build up the shape variations.

3.8 FINAL THOUGHTS

Following the end of the MESSENGER mission, we can evaluate our understanding of Mercury's crust and lithosphere from a geophysical perspective. We do so by asking a series of questions and supplying what we hope are objective answers.

1. How well do we know the shape of Mercury?

The answer depends on which is more important to a specific end user (see Section 3.2), accuracy or resolution. Stereo-derived digital elevation models have 250-m/pixel grids but suffer from long-wavelength biases. MLA grids in the northern hemisphere have high accuracy but lower resolution than the stereo. Spherical harmonic representations of shape beyond about degree 4 generally need to be localized to the northern hemisphere to be useful for modeling, though the global power spectral density can be used to about degree 8.

2. How well do we know the gravity field of Mercury?

The degree strength map (Figure 3.4) answers this question in full.

3. Does Mercury have a crust?

Yes, there is geochemical evidence for both primary and secondary crustal material (Section 3.4.1). Geophysical evidence is provided in the next answer.

4. If so, what is its mean thickness?

Estimates of mean crustal thickness are converging to a value less than 40–50 km (Section 3.4.4). More generally, there is good evidence for a strong density boundary within 50 km of the surface, and the most straightforward interpretation is that it is the crust–mantle boundary.

5. Is Mercury's shape isostatically compensated?

For the most part, the answer is no (Section 3.5.3). Crustal thickness variations derived from Bouguer δg signals are generally not isostatic. This result may be due, *inter alia*, to the contribution of dynamic pressure from mantle flow, or to the presence of a substantial elastic lithosphere ($T_e \geq 30$ km) early in Mercury's history.

6. Do we know the thickness of the elastic lithosphere and how it evolved with time?

The answer is no. Elastic lithosphere evolution can be linked to thermal model evolution if one is willing to equate T_e with T_m, the mechanical lithosphere thickness. Thermal models tend to produce thick lithospheres (e.g., >100 km at 4 Ga). Spectral analyses (admittance, correlation) yield T_e values equal to about one-third of the values from thermal modeling (Section 3.7.7.5). The assumption that $T_e = T_m$ is very likely too simplified and, furthermore, studies that use gravity and shape data to estimate T_e have been extremely limited to date.

7. What is the cause of the non-hydrostatic state of Mercury's second-degree shape and geoid (Section 3.7)?

A successful model appears to require a top load with substantial support by elastic membrane and flexural stresses, and variations in the load must be closely related to Mercury's principal axes. The insolation pattern resulting from Mercury's 3:2 spin–orbit resonance and near-zero obliquity, as a source of buoyancy anomalies and thermoelastic stresses in the lithosphere (Tosi et al., 2015), is a leading candidate. The top load in this case is a warping of the surface in response to spatially varying thermoelastic stresses. Among the challenges to this model is the timing of the acquisition of the subsurface temperature distribution that would create the necessary conditions in buoyancy and stress.

A second (and untested) mechanism is that the same insolation pattern causes variations in partial melting in the mantle that migrate upward to create crustal loads in phase with surface temperature (and thus with the principal axes).

8. What work is needed?

There are several routes for conducting additional work with MESSENGER data that would advance our understanding of the issues discussed in this chapter:

  a. Develop alternative methodologies for relating elastic thickness to thermal evolution models.
  b. Conduct hydrocode modeling of basin formation on Mercury in order to characterize lithospheric thickness as a function of basin age.
  c. Take advantage of the highest-resolution MESSENGER gravity data, acquired during the last year of the mission, to address problems at shorter length scales than before.

d. Investigate the origin and evolution of fold-and-thrust belts (Section 3.6.4) with an approach that integrates structural geology and geophysical modeling.
e. Look for tectonic evidence of changes in the equatorial shape in response to variations in orbital eccentricity.
f. Develop new models for the origin and support of the northern rise.

ACKNOWLEDGEMENTS

We thank Walter Kiefer, Nicola Tosi, and Sean Solomon for extremely constructive reviews. We also greatly appreciate the leadership of David Smith and Maria Zuber in the delivery, analysis, and interpretation of MESSENGER altimetry and gravity field measurements.

REFERENCES

Albert, R. A. and Phillips, R. J. (2000). Paleoflexure. *Geophys. Res. Lett.*, **27**, 2385–2388.

Anderson, F. S. and Smrekar, S. E. (2006). Global mapping of crustal and lithospheric thickness on Venus. *J. Geophys. Res.*, **111**, E08006, doi:10.1029/2004JE002395.

Anderson, J. D., Colombo, G., Esposito, P. B., Lau, E. L. and Trager, G. B. (1987). The mass, gravity field and ephemeris of Mercury. *Icarus*, **71**, 337–349.

Anderson, J. D., Jurgens, R. F., Lau, E. L., Slade M. A., III and Schubert, G. (1996). Shape and orientation of Mercury from radar ranging data. *Icarus*, **124**, 690–697.

Asmar, S. W. and Armstrong, J. W. (2005). Spacecraft Doppler tracking: Noise budget and accuracy achievable in precision radio science observations. *Radio Sci.*, **40**, RS2001, doi:10.1029/2004RS003101.

Balcerski, J. A., Hauck, S. A., II, Sun, P., Klimczak, C., Byrne, P. K., Phillips, R. J. and Solomon, S. C. (2013). New constraints on timing and mechanisms of regional tectonism from Mercury's tilted craters. *Lunar Planet. Sci.*, **44**, abstract 2444.

Banerdt, W. B. (1986). Support of long-wavelength loads on Venus and implications for internal structure. *J. Geophys. Res.*, **91**, 403–419.

Becker, K. J., Weller, L. A., Edmundson, K. L., Becker, T. L., Robinson, M. S., Enns, A. C. and Solomon, S. C. (2012). Global controlled mosaic of Mercury from MESSENGER orbital images. *Lunar Planet. Sci.*, **43**, abstract 2654.

Becker, K. J., Robinson, M. S., Becker, T. L., Weller, L. A., Edmundson, K. L., Neumann, G. A., Perry, M. E. and Solomon, S. C. (2016). First global digital elevation model of Mercury. *Lunar Planet. Sci.*, **47**, abstract 2959.

Bindschadler, D. L., Schubert, G. and Ford, P. G. (1994). Venus' center of figure–center of mass offset, *Icarus*, **111**, 417–432.

Brace, W. F. and Kohlstedt, D. L. (1980). Limits on lithospheric stress imposed by laboratory measurements. *J. Geophys. Res.*, **85**, 6248–6252.

Brown, S. M. and Elkins-Tanton, L. T. (2009). Compositions of Mercury's earliest crust from magma ocean models. *Earth Planet. Sci. Lett.*, **286**, 446–455, doi:10.1016/j.epsl.2009.07.010.

Burke, K. C., Şengör, A. M. C. and Francis, P. W. (1984). Maxwell Montes in Ishtar: A collisional plateau on Venus? *Lunar Planet. Sci.*, **15**, 104–105.

Byerlee, J. D. (1978). Friction of rocks. *Pure Appl. Geophys.*, **116**, 615–626.

Byrne, P. K., Klimczak, C., Williams, D. A., Hurwitz, D. M., Solomon, S. C., Head, J. W., Preusker, F. and Oberst, J. (2013). An assemblage of lava flow features on Mercury. *J. Geophys. Res. Planets*, **118**, 1303–1322.

Byrne, P. K., Klimczak, C., Şengör, A. M. C., Solomon, S. C., Watters, T. R. and Hauck II, S. A. (2014). Mercury's global contraction much greater than earlier estimates. *Nature Geosci.*, **7**, 301–307.

Byrne, P. K., Klimczak, C., McGovern, P. J., Mazarico, E., James, P. B., Neumann, G. A., Zuber, M. T. and Solomon, S. C. (2015). Deep-seated thrust faults bound the Mare Crisium lunar mascon. *Earth Planet. Sci. Lett.*, **427**, 183–190.

Cavanaugh, J. F., Smith, J. C., Sun, X., Bartels, A. E., Ramos-Izquierdo, L., Krebs, D. J., McGarry, J. F., Trunzo, R., Novo-Gradac, A.-M., Britt, J. L., Karsh, J. L., Katz, R. B., Lukemire, A. T., Symkiewicz, R., Berry, D. L., Swinski, J. P., Neumann, G. A., Zuber, M. T. and Smith, D. E. (2007). The Mercury Laser Altimeter instrument for the MESSENGER mission. *Space Sci. Rev.*, **131**, 451–480.

Chandrasekhar, S. (1987). *Ellipsoidal Figures of Equilibrium*. New York: Dover, 264 pp.

Chao, B. F. (2006). Earth's oblateness and its temporal variations. *Comptes Rendus Geoscience*, **338**, 1123–1129.

Charlier, B., Grove, T. L. and Zuber, M. T. (2013). Phase equilibria of ultramafic compositions on Mercury and the origin of the compositional dichotomy. *Earth Planet. Sci. Lett.*, **363**, 50–60.

Correia, A. C. M. and Laskar, J. (2004). Mercury's capture into the 3/2 spin-orbit resonance as a result of its chaotic dynamics. *Nature*, **429**, 848–850.

Correia, A. C. M. and Laskar, J. (2009). Mercury's capture into the 3/2 spin–orbit resonance including the effect of core–mantle friction. *Icarus*, **201**, 1–11.

Correia, A. C. M. and Laskar, J. (2012). Impact cratering on Mercury: Consequences for the spin evolution. *Astrophys. J.*, **751**, L43, 5 pp.

Courtney, R. C. and Beaumont, C. (1983). Thermally-activated creep and flexure of the oceanic lithosphere. *Nature*, **305**, 201–204, doi:10.1038/305201a0.

Denevi, B. W., Robinson, M. S., Solomon, S. C., Murchie, S. L., Blewett, D. T., Domingue, D. L., McCoy, T. J., Ernst, C. M., Head, J. W., Watters, T. R. and Chabot, N. L. (2009). The evolution of Mercury's crust: A global perspective from MESSENGER. *Science*, **324**, 613–618, doi:10.1126/science.1172226.

Denevi, B. W., Ernst, C. M., Meyer, H. M., Robinson, M. S., Murchie, S. L., Whitten, J. L., Head, J. W., Watters, T. R., Solomon, S. C., Ostrach, L. R., Chapman, C. R., Byrne, P. K., Klimczak, C. and Peplowski, P. N. (2013). The distribution and origin of smooth plains on Mercury. *J. Geophys. Res. Planets*, **118**, 891–907, doi:10.1002/jgre.20075.

Dombard, A. J., Hauck, S. A., II, Solomon, S. C. and Phillips, R. J. (2001). Potential for long-wavelength folding on Mercury. *Lunar Planet. Sci.*, **32**, abstract 2035.

Elgner, S., Stark, A., Oberst, J., Perry, M. E., Zuber, M. T., Robinson, M. S. and Solomon, S. C. (2014). Mercury's global shape and topography from MESSENGER limb images. *Planet. Space Sci.*, **103**, 299–308.

Elkins-Tanton, L. T. (2012). Magma oceans in the inner solar system. *Annu. Rev. Earth Planet. Sci.*, **40**, 113–139.

Ernst, C. M., Murchie, S. L., Barnouin, O. S., Robinson, M. S., Denevi, B. W., Blewett, D. T., Head, J. W., Izenberg, N. R., Solomon, S. C. and Roberts, J. H. (2010). Exposure of spectrally distinct material by impact craters on Mercury: Implications for global stratigraphy. *Icarus*, **209**, 210–223, doi:10.1016/j.icarus.2010.05.022.

Evans, L. G., Peplowski, P. N., Rhodes, E. A., Lawrence, D. J., McCoy, T. J., Nittler, L. R., Solomon, S. C., Sprague, A. L., Stockstill-Cahill, K. R., Starr, R. D., Weider, S. Z., Boynton, W. V., Hamara,

D. K. and Goldsten J. O. (2012). Major-element abundances on the surface of Mercury: Results from the MESSENGER Gamma-Ray Spectrometer. *J. Geophys. Res.*, **117**, E00L07, doi:10.1029/2012JE004178.

Fjeldbo, G.A., Kliore, A., Sweetnam, D., Esposito, P., Seidel, B. and Howard, T. (1976). The occultation of Mariner 10 by Mercury. *Icarus*, **27**, 439–444.

Folkner, W. M., Yoder, C. F., Yuan, D. N., Standish, E. M. and Preston, R. A. (1997). Interior structure and seasonal mass redistribution of Mars from radio tracking of Mars Pathfinder. *Science*, **278**, 1749–1752.

Garrick-Bethell, I., Wisdom, J. and Zuber, M. T. (2006). Evidence for a past high eccentricity lunar orbit. *Science*, **313**, 652–655.

Garrick-Bethell, I., Perera, V., Nimmo, F. and Zuber, M. T. (2014). The tidal-rotational shape of the Moon and evidence for polar wander. *Nature*, **512**, 181–184.

Gaskell, R. W., Barnouin-Jha, O. S., Scheeres, D. J., Konopliv, A. S., Mukia, T., Abe, S., Saito, J., Ishiguro, M., Kubota, T., Hashimoto, T., Kawaguhi, J., Yoshikawa, M. S., Kominato, T., Hirata, N. and Demura, H. (2008). Characterizing and navigating small bodies with imaging data. *Meteorit. Planet. Sci.*, **43**, 1049–1061.

Grott, M. and Breuer, D. (2008). The evolution of the Martian elastic lithosphere and implications for crustal and mantle rheology. *Icarus*, **193**, 503–515.

Guest, J. E. (1971). Centres of igneous activity in the maria. In *Geology and Physics of the Moon*, ed. G. Fielder. New York: Elsevier, pp. 41–53.

Hager, B. H. and Clayton, R. W. (1989). Constraints on the structure of mantle convection using seismic observations, flow models, and the geoid. In *Mantle Convection: Plate Tectonics and Global Dynamics*, ed. R. W. Peltier. New York: Gordon and Breach Science, pp. 657–763.

Harmon, J. K., Campbell, D. B., Bindschadler, D. L., Head, J. W. and Shapiro, I. I. (1986). Radar altimetry of Mercury: A preliminary analysis. *J. Geophys. Res.*, **91**, 385–401.

Hauck, S. A., II, Dombard, A. J., Phillips, R. J. and Solomon, S. C. (2004). Internal and tectonic evolution of Mercury. *Earth Planet. Sci. Lett.*, **222**, 713–728.

Hauck, S. A., II, Margot, J.-L., Solomon, S. C., Phillips, R. J., Johnson, C. L., Lemoine, F. G., Mazarico, E., McCoy, T. J., Padovan, S., Peale, S. J., Perry, M. E., Smith, D. E. and Zuber, M. T. (2013). The curious case of Mercury's internal structure. *J. Geophys. Res. Planets*, **118**, 1204–1220.

Hauck, S. A., II, Byrne, P. K., Denevi, B. W., Grott, M., McCoy, T. and Stanley, S. (2015). Mercury's global evolution: New views from MESSENGER. Presented at 2015 Fall Meeting, American Geophysical Union, abstract P53A-2105, San Francisco, CA, 14–18 December.

Head, J. W., Chapman, C. R., Strom, R. G., Fassett, C. I., Denevi, B. W., Blewett, D. T., Ernst, C. M., Watters, T. R., Solomon, S. C., Murchie, S. L., Prockter, L. M., Chabot, N. L., Gillis-Davis, J. J., Whitten, J. L., Goudge, T. A., Baker, D. M. H., Hurwitz, D. M., Ostrach, L. R., Xiao, Z., Merline, W. J., Kerber, L., Dickson, J. L., Oberst, J., Byrne, P. K., Klimczak, C. and Nittler, L. R. (2011). Flood volcanism in the northern high latitudes of Mercury revealed by MESSENGER. *Science*, **333**, 1853–1856.

Herrick, R. R. and Phillips, R. J. (1992). Geological correlations with the interior density structure of Venus. *J. Geophys. Res.*, **97**, 16,017–16,034.

Hirschmann, M. M. (2000). Mantle solidus: Experimental constraints and the effects of peridotite composition. *Geochem. Geophys. Geosyst.*, **1**, 1042, doi:10.1029/2000GC000070.

Iess, L., Asmar, S. and Tortora, P. (2009). MORE: An advanced tracking experiment for the exploration of Mercury with the mission BepiColombo. *Acta Astronautica*, **65**, 666–675.

James, P. B., Byrne, P. K., Solomon, S. C., Zuber, M. T. and Phillips, R. J. (2014). Surface strains associated with the evolution of Mercury's domical swells. Presented at 2014 Fall Meeting, American Geophysical Union, abstract P21C-393, San Francisco, CA, 15–19 December.

James, P. B., Mazarico, E., Genova, A., Smith, D. E., Neumann, G. A. and Solomon, S. C. (2015a). Mercury's lithospheric thickness and crustal density, as inferred from MESSENGER observations. Presented at 2015 Fall Meeting, American Geophysical Union, abstract P53A-2102, San Francisco, CA, 14–18 December.

James, P. B., Zuber, M. T., Phillips, R. J. and Solomon, S. C. (2015b). Support of long-wavelength topography on Mercury inferred from MESSENGER measurements of gravity and topography. *J. Geophys. Res. Planets*, **120**, 287–310.

James, P. B., Phillips, R. J., Grott, M., Hauck, S. A., II and Solomon, S. C. (2016). The thickness of Mercury's lithosphere inferred from MESSENGER gravity and topography. *Lunar Planet. Sci.*, **47**, abstract 1992.

Jeffreys, H. (1970). *The Earth, Its Origin, History and Physical Constitution*, 5th edn. Cambridge: Cambridge University Press.

Johnson, B. C., Blair, D. M., Collins, G. S., Melosh, H. J., Freed, A. M., Taylor, G. J., Head, J. W., Wieczorek, M. A., Andrews-Hanna, J. C., Nimmo, F., Keane, J. T., Miljković, K., Soderblom, J. M. and Zuber, M. T. (2016). Formation of the Orientale lunar multiring basin. *Science*, **354**, 441–444.

Kaula, W. M. (1966). *Theory of Satellite Geodesy*. Waltham, MA: Blaisdell.

Kaula, W. M. (1979). The moment of inertia of Mars. *Geophys. Res. Lett.*, **6**, 194–196.

Keane, J. T. and Matsuyama, I. (2014). Evidence for lunar true polar wander and a past low-eccentricity, synchronous lunar orbit. *Geophys. Res. Lett.*, **41**, 6610–6619.

Kegege, O., Fuentes, M., Meyer N. and Sil, A. (2012). Three-dimensional analysis of Deep Space Network antenna coverage. 2012 IEEE Aerospace Conference, 9 pp., Big Sky, Montana, 4–10 March, doi:10.1109/AERO.2012.6187124.

Khan, A., Mosegaard, K. and Rasmussen, K. L. (2000). A new seismic velocity model for the Moon from a Monte Carlo inversion of the Apollo lunar seismic data. *Geophys. Res. Lett.*, **27**, 1591–1594.

King, S. D. (2008). Pattern of lobate scarps on Mercury's surface reproduced by a model of mantle convection. *Nature Geosci.*, **1**, 229–232.

Klimczak, C., Watters, T. R., Ernst, C. M., Freed, A. M., Byrne, P. K., Solomon, S. C., Blair, D. M. and Head, J. W. (2012). Deformation associated with ghost craters and basins in volcanic smooth plains on Mercury: Strain analysis and implications for plains evolution. *J. Geophys. Res.*, **117**, E00L03, doi:10.1029/2012JE004100.

Klimczak, C., Ernst, C. M., Byrne, P. K., Solomon, S. C., Watters, T. R., Murchie, S. L., Preusker, F. and Balcerski, J. A. (2013). Insights into the subsurface structure of the Caloris basin, Mercury, from assessments of mechanical layering and changes in long-wavelength topography. *J. Geophys. Res. Planets*, **118**, 2030–2044.

Klimczak, C., Byrne, P. K. and Solomon, S. C. (2015). A rock-mechanical assessment of Mercury's global tectonic fabric. *Earth Planet. Sci. Lett.*, **416**, 82–90, doi:10.1016/j.epsl.2015.02.003.

Kohlstedt, D. L. and Mackwell, S. J. (2010). Strength and deformation of planetary lithospheres. In *Planetary Tectonics*, ed. T. R. Watters and R. A. Schultz. New York: Cambridge University Press, pp. 397–456.

Kohlstedt, D. L., Evans, B. and Mackwell, S. J. (1995). Strength of the lithosphere: Constraints imposed by laboratory experiments. *J. Geophys. Res.*, **100**, 17,587–17,602.

Konopliv, A. S., Banerdt, W. B. and Sjogren, W. L. (1999). Venus gravity: 180th degree and order model. *Icarus*, **139**, 3–18.

Lago, B. and Cazenave, A. (1981). State of stress in the oceanic lithosphere in response to loading. *Geophys. J. Roy. Astron. Soc.*, **64**, 785–799.

Laplace, P.-S. (1878). *Oeuvres Complètes de Laplace*. Paris: Gauthiers-Villars.

Laskar, J. (1988). Secular evolution of the Solar System over 10 million years. *Astron. Astrophys.*, **198**, 341–362.

Lemoine, F. G. R., Smith, D. E., Zuber, M. T., Neumann, G. A. and Rowlands, D. D. (1997). A 70th degree lunar gravity model (GLGM-2) from Clementine and other tracking data. *J. Geophys. Res.*, **102**, 16,339–16,359.

Lemoine, F. G., Smith, D. E., Rowlands, D. D., Zuber, M. T., Neumann, G. A., Chinn, D. S. and Pavlis, D. E. (2001). An improved solution of the gravity field of Mars (GMM-2B) from Mars Global Surveyor. *J. Geophys. Res.*, **106**, 23,359–23,376.

Lemoine, F. G., Goossens, S., Sabaka, T. J., Nicholas, J. B., Mazarico, E., Rowlands, D. D., Loomis, B. D., Chinn, D. S., Caprette, D. S., Neumann, G. A., Smith, D. E. and Zuber, M. T. (2013). High-degree gravity models from GRAIL primary mission data. *J. Geophys. Res. Planets*, **118**, 1676–1698.

Makarov, V. V. (2012). Conditions of passage and entrapment of terrestrial planets in spin-orbit resonances. *Astrophys. J.*, **752**, article 73, doi:10.1088/0004-637X/752/1/73.

Margot, J.-L., Peale, S. J., Solomon, S. C., Hauck, S. A., II, Ghigo, F. D., Jurgens, R. F., Yseboodt, M., Giorgini, J. D., Padovan, S. and Campbell, D. B. (2012). Mercury's moment of inertia from spin and gravity data. *J. Geophys. Res.*, **117**, E00L09, doi:10.1029/2012JE004161.

Matsuyama, I. and Nimmo, F. (2009). Gravity and tectonic patterns of Mercury: Effect of tidal deformation, spin-orbit resonance, nonzero eccentricity, despinning, and reorientation. *J. Geophys. Res.*, **114**, E01010, doi:10.1029/2008JE003252.

Mazarico, E., Zuber, M. T., Lemoine, F. G. and Smith, D. E. (2008). Observation of atmospheric tides in the Martian exosphere using Mars Reconnaissance Orbiter radio tracking data. *Geophys. Res. Lett.*, **35**, L09202, doi:10.1029/2008GL033388.

Mazarico, E., Genova, A., Goossens, S., Lemoine, F. G., Neumann, G. A., Zuber, M. T., Smith, D. E. and Solomon, S. C. (2014). The gravity field, orientation, and ephemeris of Mercury from MESSENGER observations after three years in orbit. *J. Geophys. Res. Planets*, **119**, 2417–2436.

Mazarico, E., Genova, A., Goossens, S., Lemoine, F. G., Smith, D. E., Zuber, M. T., Neumann, G. A. and Solomon, S. C. (2016). The gravity field of Mercury after MESSENGER. *Lunar Planet. Sci.*, **47**, abstract 2022.

McAdams, J. V., Bryan, C. G., Bushman, S. S., Calloway, A. B., Carranza, E., Flanigan, S. H., Kirk, M. N., Korth, H., Moessner, D. P., O'Shaughnessy, D. J. and Williams, K. E. (2015). Engineering MESSENGER's grand finale at Mercury: The low-altitude hover campaign. Astrodynamics Specialist Conference, American Astronautical Society, paper AAS 15–634, 20 pp., Vail, CO, 9–13 August.

McAdoo, D. C. and Sandwell, D. T. (1985). Folding of oceanic lithosphere. *J. Geophys. Res.*, **90**, 8563–8569.

McCauley, J. F. (1967). *Geologic Map of the Hevelius Region of the Moon*, Map I-1491, Miscellaneous Investigations Series. Denver, CO: U.S. Geological Survey.

McGovern, P. J., Solomon, S. C., Smith, D. E., Zuber, M. T., Simons, M., Wieczorek, M. A., Phillips, R. J., Neumann, G. A., Aharonson, O. and Head, J. W. (2002). Localized gravity/topography admittance and correlation spectra on Mars: Implications for regional and global evolution. *J. Geophys. Res.*, **107**, 5136, doi:10.1029/2002JE001854.

McGovern, P. J., Solomon, S. C., Smith, D. E., Zuber, M. T., Simons, M., Wieczorek, M. A., Phillips, R. J., Neumann, G. A., Aharonson, O. and Head, J. W. (2004). Correction to "Localized gravity/topography admittance and correlation spectra on Mars: Implications for regional and global evolution". *J. Geophys. Res.*, **109**, E07007, doi:10.1029/2004JE002286.

McKenzie, D. (2003). Estimating T_e in the presence of internal loads. *J. Geophys. Res.*, **108** (B9), 2438, doi:10.1029/2002JB001766.

McKenzie, D. and Bowin, C. (1976). The relationship between bathymetry and gravity in the Atlantic Ocean. *J. Geophys. Res.*, **81**, 1903–1915.

McNutt, M. K. (1984). Lithospheric flexure and thermal anomalies. *J. Geophys. Res.*, **89**, 11,180–11,194.

Melosh, H. J. (2011). *Planetary Surface Processes*, Cambridge Planetary Science Series, Cambridge: Cambridge University Press.

Melosh, H. J. and McKinnon, W. B. (1988). The tectonics of Mercury. In *Mercury*, ed. F. Vilas, C. R. Chapman and M. S. Matthews. Tucson, AZ: University of Arizona Press, pp. 374–400.

Meyer, J., Elkins-Tanton, L. and Wisdom, J. (2010). Coupled thermal–orbital evolution of the early Moon. *Icarus*, **208**, 1–10.

Michel, N. C., Hauck S. A., II, Solomon, S. C., Phillips, R. J., Roberts, J. H. and Zuber, M. T. (2013). Thermal evolution of Mercury as constrained by MESSENGER observations. *J. Geophys. Res. Planets*, **118**, 1033–1044.

Muller, P. M. and Sjogren, W. L. (1968). Mascons: Lunar mass concentrations. *Science*, **161**, 680–684.

Murchie, S. L., Klima, R. L., Denevi, B. W., Ernst, C. M., Keller, M. R., Domingue, D. L., Blewett, D. T., Chabot, N. L., Hash, C. D., Malaret, E., Izenberg, N. R., Vilas, F., Nittler, L. R., Gillis-Davis, J. J., Head, J. W. and Solomon, S. C. (2015). Orbital multispectral mapping of Mercury with the MESSENGER Mercury Dual Imaging System: Evidence for the origins of plains units and low-reflectance material. *Icarus*, **254**, 287–305, doi:10.1016/j.icarus.2015.03.027.

Neumann, G. A., Perry, M. E., Mazarico, E., Ernst, C. M., Zuber, M. T., Smith, D. E., Becker, K. J., Gaskell, R. E., Head, J. W., Robinson, M. S. and Solomon, S. C. (2016). Mercury shape model from laser altimetry and planetary comparisons. *Lunar Planet. Sci.*, **47**, abstract 2087.

Nimmo, F. and Watters, T. R. (2004). Depth of faulting on Mercury: Implications for heat flux and crustal and effective elastic thickness. *Geophys. Res. Lett.*, **31**, L02701, doi:10.1029/2003GL018847.

Nittler, L. R., Starr, R. D., Weider, S. Z., McCoy, T. J., Boynton, W. V., Ebel, D. S., Ernst, C. M., Evans, L. G., Goldsten, J. O., Hamara, D. K., Lawrence, D. J., McNutt, R. L., Jr., Schlemm, C. E., II, Solomon, S. C. and Sprague, A. L. (2011). The major-element composition of Mercury's surface from MESSENGER X-ray spectrometry. *Science*, **333**, 1847–1850, doi:10.1126/science.1211567.

Noyelles, B., Frouard, J., Makarov, V. V. and Efroimsky, M. (2014). Spin–orbit evolution of Mercury revisited. *Icarus*, **241**, 26–44.

Oberst, J., Preusker, F., Phillips, R. J., Watters, T. R., Head, J. W., Zuber, M. T. and Solomon, S. C. (2010). The morphology of Mercury's Caloris basin as seen in MESSENGER stereo topographic models. *Icarus*, **209**, 230–238.

Oberst, J., Elgner, S., Turner, F. S., Perry, M. E., Gaskell, R. W., Zuber, M. T., Robinson, M. S. and Solomon, S. C. (2011). Radius and limb topography of Mercury obtained from images acquired during the MESSENGER flybys. *Planet. Space Sci.*, **59**, 1918–1924.

Ostrach, L. R., Robinson, M. S., Whitten, J. L., Fassett, C. I., Strom, R. G., Head, J. W. and Solomon, S. C. (2015). Extent, age, and resurfacing history of the northern smooth plains on Mercury from MESSENGER observations. *Icarus*, **250**, 602–622.

Padovan, S., Wieczorek, M. A., Margot, J.-L., Tosi, N. and Solomon, S. C. (2015). Thickness of the crust of Mercury from geoid-to-topography ratios. *Geophys. Res. Lett.*, **42**, 1029–1038.

Page, B. R., Bryan, C. G., Williams, K. E., Taylor, A. H. and Williams, B. G. (2014). Tuning the MESSENGER state estimation filter for controlled descent to Mercury impact. Astrodynamics Specialist Conference, American Institute of Aeronautics and Astronautics/ American Astronautical Society, paper AIAA-2014–4129, 16 pp., San Diego, CA, 4–7 August.

Pavlis, D. E., Wimert, J. and McCarthy, J. J. (2013). *GEODYN II System Description, Volumes 1–5*, contractor report. Greenbelt, MD: SGT Inc.

Peplowski, P. N., Klima, R. L., Lawrence, D. J., Ernst, C. M., Denevi, B. W., Frank, E. A., Goldsten, J. O., Murchie, S. L., Nittler, L.R. and Solomon, S. C. (2016). Remote sensing evidence for an ancient carbon-bearing crust on Mercury. *Nature Geosci.*, **9**, 273–276, doi:10.1038/ngeo2669.

Perry, M. E., Kahan, D. S., Barnouin, O. S., Ernst, C. M., Solomon, S. C., Zuber, M. T., Smith, D. E., Phillips, R. J., Srinivasan, D. K., Oberst, J. and Asmar, S. W. (2011). Measurement of the radius of Mercury by radio occultation during the MESSENGER flybys. *Planet. Space Sci.*, **59**, 1925–1931.

Perry, M. E., Neumann, G. A., Phillips, R. J., Barnouin, O. S., Ernst, C. M., Kahan, D. S., Solomon, S. C., Zuber, M. T., Smith, D. E., Hauck, S. A., II, Peale, S. J., Margot, J.-L., Mazarico, E., Johnson, C. L., Gaskell, R. W., Roberts, J. H., McNutt, R. L., Jr. and Oberst, J. (2015). The low-degree shape of Mercury. *Geophys. Res. Lett.*, **42**, 6951–6958, doi:10.1002/2015GL065101.

Peterson, G. A., Johnson, C. L., Byrne, P. K., Phillips, R. J. and Neumann, G. A. (2017). Depth of faulting in Mercury's northern hemisphere from thrust fault morphology. *Lunar Planet. Sci.*, **48**, abstract 2315.

Phillips, R. J. and Lambeck, K. (1980). Gravity fields of the terrestrial planets: Long-wavelength anomalies and tectonics. *Rev. Geophys. Space Phys.*, **18**, 27–76.

Phillips, R. J., Johnson, C. L., Mackwell, S. J., Morgan, P., Sandwell, D. T. and Zuber, M. T. (1997). Lithospheric mechanics and dynamics of Venus. In *Venus II*, ed. S. W. Bougher, D. M. Hunten and R. J. Phillips. Tucson, AZ: University of Arizona Press, pp. 1163–1204.

Phillips, R. J., Johnson, C. L., Perry, M. E., Hauck, S. A., II, James, P. B., Mazarico, E., Lemoine, F. G., Neumann, G., Peale, S. J., Siegler, M. A., Smith, D. E., Solomon, S. C. and Zuber, M. T. (2014). Mercury's 2nd-degree shape and geoid: Lunar comparisons and thermal anomalies. *Lunar Planet. Sci.*, **45**, abstract 2634.

Poblet, J. and Lisle, R. J. (2011). Kinematic evolution and structural styles of fold-and-thrust belts. In *Kinematic Evolution and Structural Styles of Fold-and-Thrust Belts*, ed. J. Poblet and R. J. Lisle, Special Publication 349. London: Geological Society, pp. 1–24.

Preusker, F., Oberst, J., Head, J. W., Watters, T. R., Robinson, M. S., Zuber, M. T. and Solomon, S. C. (2011). Stereo topographic models of Mercury after three MESSENGER flybys. *Planet. Space Sci.*, **59**, 1910–1917, doi:10.1016/j.pss.2011.07.005.

Qin, C., Zhong, S. and Phillips, R. J. (2018). Formation of the lunar fossil bulges and its implication for the early Earth and Moon. *Geophys. Res. Lett.*, **45**, 1286–1296, doi:10.1002/2017GL076278.

Riner, M. A., Lucey, P. G., Desch, S. J. and McCubbin, F. M. (2009). Nature of opaque components on Mercury: Insights into a Mercurian magma ocean. *Geophys. Res. Lett.*, **36**, L02201, doi:10.1029/2008GL036128.

Robinson, M. S., Murchie, S. L., Blewett, D. T., Domingue, D. L., Hawkins, S. E., III, Head, J. W., Holsclaw, G. M., McClintock, W. E., McCoy, T. J., McNutt, R. L., Jr., Prockter, L. M., Solomon, S. C. and Watters T. R. (2008). Reflectance and color variations on Mercury: Regolith processes and compositional heterogeneity. *Science*, **321**, 66–69.

Sedgwick, W. F. (1898). On the figure of the Moon. *Messenger Math.*, **27**, 171–173.

Seidelmann, P. K., Abalakin, V. K., Bursa, M., Davies, M. E., de Bergh, C., Lieske, J. H., Oberst, J., Simon, J. L., Standish, E. M., Stooke, P. and Thomas, P. C. (2002). Report of the IAU/IAG Working Group on Cartographic Coordinates and Rotational Elements of the Planets and Satellites: 2000. *Celest. Mech. Dyn. Astron.*, **82**, 83–110.

Şengör, A. M. C., Natal'in, B. A. and Burtman, V. S. (1993). Evolution of the Altaid tectonic collage and Palaeozoic crustal growth in Eurasia. *Nature*, **364**, 299–307.

Shearer, C. K., Hess, P. C, Wieczorek, M. A., Pritchard, M. E., Parmentier, E. M., Borg, L. E., Longhi, J., Elkins-Tanton, L. T., Neal, C. R., Antonenko, I., Canup, R. M., Halliday, A. N., Grove, T. L., Hager, B. H., Lee, D.-C. and Wiechert, U. (2006). Thermal and magmatic evolution of the Moon. *Rev. Mineral. Geochem.*, **60**, 365–518.

Siegler, M. A., Bills, B. G. and Paige, D. A. (2013). Orbital eccentricity driven temperature variation at Mercury's poles. *J. Geophys. Res. Planets*, **118**, 930–937.

Smith, D. E., Zuber, M. T., Solomon, S. C., Phillips, R. J., Head, J. W., Garvin, J. B., Banerdt, W. B., Muhleman, D. O., Pettengill, G. H., Neumann, G. A., Lemoine, F. G., Abshire, J. B., Aharonson, O., Brown, C. D., Hauck, S. A., II, Ivanov, A. B., McGovern, P. J., Zwally, H. J. and Duxbury, T. C. (1999). The global topography of Mars and implications for surface evolution. *Science*, **284**, 1495–1503.

Smith, D. E., Zuber, M. T., Sun, X., Neumann, G. A., Cavanaugh, J. F., McGarry, J. F. and Zagwodzki, T. W. (2006). Two-way laser link over interplanetary distance. *Science*, **311**, 53.

Smith, D. E., Zuber, M. T., Phillips, R. J., Solomon, S. C., Neumann, G. A., Lemoine, F. G., Torrence, M., Peale, S. J., Margot, J.-L., Barmouin-Jha, O., Head, J. W. and Talpe, M. (2010a). The equatorial shape and gravity field of Mercury from MESSENGER flybys 1 and 2. *Icarus*, **209**, 247–255.

Smith, D. E., Zuber, M. T., Neumann, G. A. Lemoine, F. G., Mazarico, E., Torrence, M. H., McGarry, J. F., Rowlands, D. D., Head, J. W., Duxbury, T. H., Aharonson, O., Lucey, P. G., Robinson, M. S., Barnouin, O. S., Cavanaugh, J. F., Sun, X. L., Liva, P., Mao, D. D., Smith, J. C. and Bartles, A. E. (2010b). Initial observations from the Lunar Orbiter Laser Altimeter (LOLA). *Geophys. Res. Lett.*, **37**, L18204, doi:10.1029/2010GL043751.

Smith, D. E., Zuber, M. T., Phillips, R. J., Solomon, S. C., Hauck, S. A., II, Lemoine, F. G., Mazarico, E., Neumann, G. A., Peale, S. J., Margot, J. L., Johnson, C. L., Torrence, M. H., Perry, M. E., Rowlands, D. D., Goossens, S., Head, J. W. and Taylor, A. H. (2012). Gravity field and internal structure of Mercury from MESSENGER. *Science*, **336**, 214–217.

Spudis, P. D. (1996). *The Once and Future Moon*. Washington, DC: Smithsonian Institution Press, pp. 117–118,

Spudis, P. D., McGovern, P. J. and Kiefer, W. S. (2013). Large shield volcanoes on the Moon. *J. Geophys. Res. Planets*, **118**, 1063–1081.

Srinivasan, D. K., Perry, M. E., Fielhauer, K. B., Smith, D. E. and Zuber, M. T. (2007). The radio frequency subsystem and radio science on the MESSENGER mission. *Space Sci. Rev.*, **131**, 557–571.

Stacey, F. D. and Davis, P. M. (2008). *Physics of the Earth*, 4th edn. Cambridge: Cambridge University Press.

Strom, R. G., Trask, N. J. and Guest, J. E. (1975). Tectonism and volcanism on Mercury. *J. Geophys. Res.*, **80**, 2478–2507.

Suess, E. (1909). *Das Antlitz der Erde, Vol. III.2*, ed. E. Suess. Leipzig: F. Tempsky.

Tosi, N., Grott, M., Plesa, A. C. and Breuer, D. (2013). Thermochemical evolution of Mercury's interior. *J. Geophys. Res. Planets*, **118**, 2474–2487.

Tosi, N., Čadek, O., Běhounková, M., Káňová, M., Plesa, A. C., Grott, M., Breuer, D., Padovan, S. and Wieczorek, M. A. (2015). Mercury's low-degree geoid and topography controlled by insolation-driven elastic deformation. *Geophys. Res. Lett.*, **42**, 7327–7335.

Turcotte, D. L., Willemann, R. J., Haxby, W. F. and Norberry, J. (1981). Role of membrane stress in the support of planetary topography. *J. Geophys. Res.*, **86**, 3951–3959.

Vander Kaaden, K. E. and McCubbin F. M. (2015). Exotic crust formation on Mercury: Consequences of a shallow, FeO-poor mantle. *J. Geophys. Res. Planets*, **120**, 195–209.

Vasavada, A. R., Paige, D. A. and Wood, S. E. (1999). Near-surface temperatures on Mercury and the Moon and the stability of polar ice deposits. *Icarus*, **141**, 179–193.

Watters, T. R., Schultz, R. A., Robinson, M. S. and Cook, A. C. (2002). The mechanical and thermal structure of Mercury's early lithosphere. *Geophys. Res. Lett.*, **29**, 1542, doi:10.029/2001GL014308.

Watts, A. B. (1978). An analysis of isostasy in the world's oceans 1. Hawaiian-Emperor seamount chain. *J. Geophys. Res.*, **83**, 5989–6004.

Watts, A. B. (2001) *Isostasy and Flexure of the Lithosphere*. Cambridge: Cambridge University Press.

Wieczorek, M. A. and Phillips, R. J. (1998). Potential anomalies on a sphere: Applications to the thickness of the lunar crust. *J. Geophys. Res.*, **103**, 1715–1724.

Wieczorek, M. A. and Simons, F. J. (2005). Localized spectral analysis on the sphere. *Geophys. J. Int.*, **162**, 655–675.

Wieczorek, M. A., Correia, A. C. M., Le Feuvre, M., Laskar, J. and Rambaux, N. (2011). Mercury's spin–orbit resonance explained by initial retrograde and subsequent synchronous rotation. *Nature Geosci.*, **5**, 18–21.

Wieczorek, M. A., Neumann, G. A., Nimmo, F., Kiefer, W. S., Taylor, G. J., Melosh, H. J., Phillips, R. J., Solomon, S. C., Andrews-Hanna, J. C., Asmar, S. W., Konopliv, A. S., Lemoine, F. G., Smith, D. E., Watkins, M. M., Williams, J. G. and Zuber, M. T. (2013). The crust of the Moon as seen by GRAIL. *Science*, **339**, 671–675.

Williams, J. G., Boggs, D. H., Yoder, C. F., Ratcliff, J. T. and Dickey, J. O. (2001). Lunar rotational dissipation in solid body and molten core. *J. Geophys. Res.*, **106**, 27,933–27,968.

Williams, J.-P., Ruiz, J., Rosenburg, M. A., Aharonson, O. and Phillips, R. J. (2011). Insolation driven variations of Mercury's lithospheric strength. *J. Geophys. Res.*, **116**, E01008, doi:10.1029/2001JE003655.

Yseboodt, M. and Margot, J.-L. (2006). Evolution of Mercury's obliquity. *Icarus*, **181**, 327–337.

Zuber, M. T. and Parmentier, E. M. (1995). Formation of fold-and-thrust belts on Venus by thick-skinned deformation. *Nature*, **377**, 704–707.

Zuber, M. T., Smith, D. E., Solomon, S. C., Muhleman, D. O., Head, J. W., Garvin, J. B., Abshire, J. B. and Bufton, J. L. (1992). The Mars Observer Laser Altimeter investigation. *J. Geophys. Res.*, **97**, 7781–7797.

Zuber, M. T., Aharonson, O., Aurnou, J. M., Cheng, A. F., Hauck, S. A., II, Heimpel, M. H., Neumann, G. A., Peale, S. J., Phillips, R. J., Smith, D. E., Solomon, S. C. and Stanley, S. (2007). The geophysics of Mercury: Current status and anticipated insights from the MESSENGER mission. *Space Sci. Rev.*, **131**, 105–132.

Zuber, M. T., Montési, L. G. J., Farmer, G. T., Hauck, S. A., II, Ritzer, A., Phillips, R. J., Solomon, S. C., Smith, D. E., Talpe, M. J., Head, J. W., III, Neumann, G. A., Watters, T. R. and Johnson, C. L. (2010). Accommodation of lithospheric shortening on Mercury from altimetric profiles of ridges and lobate scarps measured during MESSENGER flybys 1 and 2. *Icarus*, **209**, 247–255.

Zuber, M. T., Smith, D. E., Phillips, R. J., Solomon, S. C., Neumann, G. A., Hauck, S. A., II, Peale, S. J., Barnouin, O. S., Head, J. W., Johnson, C. L., Lemoine, F. G., Mazarico, E., Sun, X., Torrence, M. H., Freed, A. M., Klimczak, C., Margot, J. L., Oberst, J., Perry, M. E., McNutt, R. L., Jr., Balcerski, J. A., Michel, N., Talpe, M. J. and Yang, D. (2012). Topography of the northern hemisphere of Mercury from MESSENGER laser altimetry. *Science*, **336**, 217–220.

Zuber, M. T., Smith, D. E., Watkins, M. M., Asmar, S. W., Konopliv, A. S., Lemoine, F. G., Melosh, H. J., Neumann, G. A., Phillips, R. J., Solomon, S. C., Wieczorek, M. A., Williams, J. G., Goossens, S. J., Kruizinga, G., Mazarico, E., Park, R. S. and Yuan, D. N. (2013). Gravity field of the Moon from the Gravity Recovery and Interior Laboratory (GRAIL) mission. *Science*, **339**, 668–671.

4

Mercury's Internal Structure

JEAN-LUC MARGOT, STEVEN A. HAUCK, II, ERWAN MAZARICO, SEBASTIANO PADOVAN, AND STANTON J. PEALE

4.1 INTRODUCTION

4.1.1 Importance of Planetary Interiors

We seek to understand the interior structures of planetary bodies because the interiors affect planetary properties and processes in several fundamental ways. First, a knowledge of the interior informs us about a planet's makeup and enables us to test hypotheses related to planet formation. Second, interior properties dictate the thermal evolution of planetary bodies and, consequently, the history of volcanism and tectonics on these bodies. Many geological features are the surface expression of processes that take place below the surface. Third, the structure of the interior and the nature of the interactions among inner core, outer core, and mantle have a profound influence on the evolution of the spin state and the response of the planet to external forces and torques. These processes dictate the planet's tectonic and insolation regimes and also affect its overall shape. Finally, interior properties control the generation of planetary magnetic fields, and, therefore, the development of magnetospheres.

Four of the six primary science objectives of the MESSENGER mission (Solomon et al., 2001; Chapter 1) rely on an understanding of the planet's interior structure. These four mission objectives pertain to the high density of Mercury, its geological history, the nature of its magnetic field, and the structure of its core.

4.1.2 Objectives

An ideal representation of a planetary interior would include the description of physical and chemical quantities at every location within the volume of the planetary body at every point in time. Here, we focus on a description of Mercury's interior at the current epoch. For a description of the evolution of the state of the planet over geological time, see Chapter 19. Because our ability to specify properties throughout the planetary volume is limited, we simplify the problem by assuming axial or spherical symmetry. Specifically, we seek self-consistent depth profiles of density, pressure, and temperature, informed by observational constraints (radius, mass, moment of inertia, composition). The solution requires the use of equations of state and assumptions about material properties, both guided by laboratory data. We compute the bulk modulus and thermal expansion coefficient as part of the estimation process, and we use the profiles to compute other rheological properties, such as viscosity and additional elastic moduli. Finally, we use our models to numerically evaluate the planet's tidal response and compare it with observational data. Our models of the interior structure are relevant to a wide range of problems, but Mercury's unusual insolation and thermal patterns violate our symmetry assumptions. These assumptions must be lifted for certain applications that require precise temperature distributions.

Our primary objective is to provide a family of simplified models of Mercury's interior that satisfy the currently available observational constraints. A secondary objective is to select, among these models, a recommended model that matches all available constraints. This model may be considered a preliminary reference Mercury model (PRMM), evoking a distant connection with its venerable Earth analog (Dziewonski and Anderson, 1981).

4.1.3 Available Observational Constraints

All of our knowledge about Mercury's interior comes from Earth-based observations, three Mariner 10 flybys, three MESSENGER flybys, and the four-year orbital phase of the MESSENGER mission. In the absence of seismological data, our information about the interior comes primarily from geodesy, the study of the gravity field, shape, and spin state of the planet, including solid-body tides. We also draw on constraints derived from the surface expression of global contraction and observations of surface composition, with the caveat that the composition at depth may be substantially different from that inferred for surface material. The structure of the magnetic field and its dynamo origin can also be used to inform interior models.

4.1.4 Outline

The primary observational constraints (Sections 4.2–4.4) are used to develop two- and three-layer structural models (Section 4.5). We then add compositional constraints (Section 4.6) and develop multi-layer models (Section 4.7). We examine the tidal response of the planet (Section 4.8) and the influence of an inner core (Section 4.9). We conclude with a discussion of a representative interior model (Section 4.10) and its implications (Section 4.11).

4.2 ROTATIONAL DYNAMICS

In his classic 1976 paper, Stanton J. Peale described the effects of a molten core on the dynamics of Mercury's rotation and proposed an ingenious method for measuring the size and state of the core (Peale, 1976). Most of our knowledge about Mercury's interior structure can be traced to Peale's ideas and to the powerful connection between dynamics and geophysics. We review aspects of Mercury's rotational dynamics that are relevant to determining its interior structure. Peale (1988) provided a more extensive review.

4.2.1 Spin–Orbit Resonance

Radar observations by Pettengill and Dyce (1965) revealed that the spin period of Mercury differs from its orbital period. To explain the radar results, Colombo (1965) correctly hypothesized that Mercury rotates on its spin axis three times for every two revolutions around the Sun. Mercury is the only known planetary body to exhibit a 3:2 spin–orbit resonance (Colombo, 1966; Goldreich and Peale, 1966).

4.2.2 Physical Librations

Peale's observational procedure allows the detection of a molten core by measuring deviations from the mean resonant spin rate of the planet. As Mercury follows its eccentric orbit, it experiences periodically reversing torques due to the gravitational influence of the Sun on the asymmetric shape of the planet. The torques affect the rotational angular momentum and cause small deviations of the spin frequency from its resonant value of 3/2 times the mean orbital frequency. The resulting oscillations in longitude are called physical librations, not to be confused with optical librations, which are the torque-free oscillations of the long axis of a uniformly spinning body about the line connecting it to a central body. Because the forcing and rotational response occur with a period $P \sim 88$ days dictated by Mercury's orbital motion, these librations have been referred to as forced librations. This terminology is not universally accepted (e.g., Bois, 1995) and loses meaning when the amount of angular momentum exchanged between spin and orbit is not negligible (e.g., Naidu and Margot, 2015). We will instead refer to these librations as 88-day librations, in part to distinguish them from librations with longer periods.

The amplitude ϕ_0 of the 88-day librations for a solid Mercury can be written as (Peale, 1972, 1988):

$$\phi_0 = \frac{3}{2}\frac{(B-A)}{C}\left(1 - 11e^2 + \frac{959}{48}e^4 + \cdots\right), \quad (4.1)$$

where $A < B < C$ are the principal moments of inertia and e is the orbital eccentricity, currently ~ 0.2056 (e.g., Stark et al., 2015b). This equation encapsulates the fact that the gravitational torques are proportional to the difference in equatorial moments of inertia $(B-A)$. The polar moment of inertia C appears in the denominator as it represents a measure of the resistance to changes in rotational motion. If the mantle is decoupled from a molten core that does not participate in the 88-day librations, then the moment of inertia in the denominator must be replaced by C_{m+cr}, the value appropriate for the mantle and crust. Peale (1976) noted that $C_{m+cr}/C \simeq 0.5$, suggesting that a measurement of the amplitude of the 88-day librations can be used to determine the state of the core if $(B-A)$ is known. This result holds over a wide range of core–mantle coupling behaviors (Peale et al., 2002; Rambaux et al., 2007).

4.2.3 Cassini State

Peale (1969, 1988) formulated general equations for the motion of the rotational axis of a triaxial body under the influence of gravitational torques. He wrote these equations

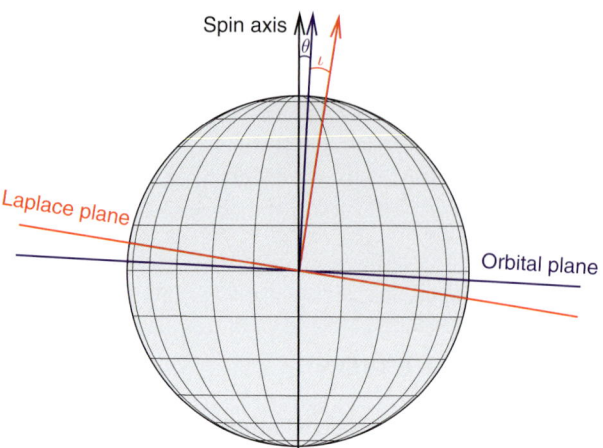

Figure 4.1. Geometry of Cassini state 1: the three vectors representing spin axis orientation (black), normal to the orbital plane (blue), and normal to the Laplace plane (red) remain coplanar as the orbit precesses around the Laplace plane with a $\sim 300{,}000$-year period. The inclination of Mercury's orbit with respect to the Laplace plane is represented by the angle ι, which is shown to scale. The tilt of Mercury's spin axis with respect to the orbit normal is the obliquity θ, which is shown with an exaggeration factor of 100 for clarity.

in the context of an orbit that precesses at a fixed rate around a reference plane called the *Laplace plane*, extending and refining earlier work by Colombo (1966). These equations generalize Cassini's laws and describe the dynamics of the Moon, Mercury, Galilean satellites, and other bodies. In the case of Mercury, the gravitational torques are due to the Sun, and the $\sim 300{,}000$-year precession of the orbit is due to the effect of external perturbers, primarily Jupiter, Venus, Saturn, and Earth.

On the basis of these theoretical calculations, Peale (1969, 1988) predicted that tidal evolution would carry Mercury to a Cassini state, in which the spin axis orientation, orbit normal, and normal to the Laplace plane remain coplanar (Figure 4.1). Specifically, he predicted that Mercury would reach Cassini state 1, with an obliquity near zero degrees. Numerical simulations (Bills and Comstock, 2005; Yseboodt and Margot, 2006; Peale, 2006; Bois and Rambaux, 2007) and analytical calculations (D'Hoedt and Lemaître, 2008) support these predictions.

In a Cassini state, the obliquity has evolved to a value where the spin precession period matches the orbit precession period (Gladman et al., 1996). Because the spin precession period and the gravitational torques depend on moment of inertia differences, there is a powerful relationship between the obliquity of a body in a Cassini state and its moments of inertia. Peale (1976, 1988) wrote:

$$K_1(\theta)\left(\frac{C-A}{C}\right) + K_2(\theta)\left(\frac{B-A}{C}\right) = K_3(\theta), \quad (4.2)$$

where K_1, K_2, K_3 are functions of the obliquity θ that involve the orbital eccentricity, inclination with respect to the Laplace plane, mean motion, spin rate, and precession rate. In this equation, the appropriate moment of inertia in the denominator is that of the entire planet, even if the core is molten, because it is hypothesized that the core follows the mantle on the ~300,000-year timescale of the orbital precession.

If we can confirm that Mercury is in a Cassini state, a measurement of the obliquity becomes extremely valuable: it provides a direct constraint on moment of inertia differences and, in combination with degree-2 gravity information, on the polar moment of inertia. A free precession of the spin axis about the Cassini state could, in principle, compromise the determination of the obliquity. However, such free precession would require a recent excitation because the corresponding damping timescale is ~10^5 yr (Peale, 2005).

4.2.4 Polar Moment of Inertia

Absent seismological data, the polar moment of inertia is arguably the most important quantity needed to quantify the interior structure of a planetary body. Peale (1976, 1988) showed that it is possible to measure the polar moment of inertia C by combining the obliquity with two quantities related to the gravity field. The gravity field of a body of mass M and radius R can be described with spherical harmonics (e.g., Kaula, 2000). The second-degree coefficients C_{20} and C_{22} in the spherical harmonic expansion are related to the moments of inertia, as follows:

$$C_{20} = -\frac{(C - (A+B)/2)}{MR^2}, \quad (4.3)$$

$$C_{22} = \frac{(B-A)}{4MR^2}. \quad (4.4)$$

Combining equations (4.2), (4.3), and (4.4), we find

$$\frac{C}{MR^2} = (-C_{20} + 2C_{22})\frac{K_1(\theta)}{K_3(\theta)} + 4C_{22}\frac{K_2(\theta)}{K_3(\theta)}, \quad (4.5)$$

which provides a direct relationship between the obliquity, gravity harmonics, and polar moment of inertia for bodies in Cassini state 1.

To complete Peale's argument, we determine the polar moment of inertia of the core, which can be done if the core is molten and does not participate in the 88-day librations. To do so, we write the identity

$$\frac{C_{m+cr}}{C} = \left(\frac{C_{m+cr}}{B-A}\right)\left(\frac{B-A}{MR^2}\right)\left(\frac{MR^2}{C}\right), \quad (4.6)$$

which yields the moment of inertia of the mantle and crust C_{m+cr} and, therefore, the moment of inertia of the core $C_c = C - C_{m+cr}$. Two spin-state quantities and two gravity quantities provide all the information necessary to determine these values. A measurement of the libration amplitude ϕ_0 provides a direct estimate of the first factor on the right-hand side of equation (4.6) via equation (4.1). A measurement of the gravitational harmonic C_{22} provides a direct estimate of the second factor. Measurements of the obliquity, C_{20}, and C_{22} yield an estimate of the third factor via equation (4.5).

The four quantities ϕ_0, θ, C_{20}, and C_{22} identified by Peale (1976, 1988) thus provide a powerful probe of the interior structure of the planet.

4.2.5 Orbital Precession

Implementing Peale's procedure requires precise knowledge of Mercury's orbital configuration. Whereas the mean motion and orbital eccentricity have been determined from centuries of observations, relatively little attention had been paid to the orientation of the Laplace plane and the orbital precession rate. Yseboodt and Margot (2006) used a Hamiltonian approach and numerical fits to ephemeris data to determine these ancillary quantities. They showed that the Laplace plane orientation varies due to planetary perturbations on ~10 kyr timescales, and they defined an *instantaneous Laplace plane* valid at the current epoch for the purpose of identifying the position of the Cassini state and interpreting spin-gravity data.

Yseboodt and Margot (2006) gave the coordinates of the normal to the instantaneous Laplace plane in ecliptic and equatorial coordinates at epoch J2000.0 as

$$\lambda_{inst} = 66.6°, \quad \beta_{inst} = 86.725°, \quad (4.7)$$

$$RA_{inst} = 273.72°, \quad DEC_{inst} = 69.53°, \quad (4.8)$$

where λ is ecliptic longitude, β is ecliptic latitude, RA is right ascension, and DEC is declination. The uncertainty in the determination is of order 1°, but the orientation of the narrow error ellipse is such that it can affect the interpretation of the spin-state data only at a level that is well below that due to measurement uncertainties.

The inclination of Mercury's orbit with respect to the instantaneous Laplace plane and the orbit precession rate about that plane at the current epoch are $\iota = 8.6°$ and $\dot{\Omega} = -0.110°$/century, respectively (Yseboodt and Margot, 2006). We will use both of these quantities to estimate Mercury's interior structure in Sections 4.5 and 4.7. Stark et al. (2015b) performed an independent analysis and confirmed the values of Yseboodt and Margot (2006), including the orientation of the instantaneous Laplace plane, the inclination ι, and the precession rate $\dot{\Omega}$. D'Hoedt et al. (2009) used a Hamiltonian approach and found an instantaneous Laplace plane orientation that differs from our preferred value by 1.4°.

4.3 GRAVITY CONSTRAINTS

4.3.1 Methods

We are interested in measuring the masses and sizes of planetary bodies because bulk density is a fundamental indicator of composition. In multi-planet systems, masses can be estimated by observing the effects of mutual orbital perturbations, manifested as variations in orbital elements or variations in transit times. Another common mass measurement technique is to determine the orbits of natural satellites.

The most precise mass estimates are obtained by radiometric tracking of a spacecraft while it is in close proximity to the body of interest, typically by using the onboard telecommunications system and a network of ground-based radio telescopes. The geodetic observations are then used to obtain a spherical harmonic expansion of the gravity field and to reconstruct the spacecraft trajectory with high fidelity. In addition to providing high-precision mass estimates, this technique enables the measurement of the spherical harmonic coefficients C_{20} and C_{22}, which provide important constraints on interior structure (Section 4.2.4).

In the following sections, we describe gravity results obtained from tracking the Mariner 10 spacecraft at a frequency of 2.3 GHz (S-band) during three flybys in 1974–1975 and the MESSENGER spacecraft at frequencies of 7.2 GHz uplink and 8.4 GHz downlink (X-band) during the flybys and orbital phase of the mission.

4.3.2 Mass and Density Results

The mass, size, and density of Mercury were known with remarkable precision prior to the exploration of the planet by spacecraft. After adding radar measurements to two centuries of optical observations, Ash et al. (1971) fit planetary ephemerides and determined Mercury's mass to 0.25% fractional uncertainty. They found a value of $6,025,000 \pm 15,000$ in inverse solar masses, i.e., $M = (3.300 \pm 0.008) \times 10^{23}$ kg, which is almost identical to the modern estimate. Using this measurement and the radar estimate of the average equatorial radius that was available at the time, $R = (2439 \pm 1)$ km, it was apparent that Mercury's bulk density was anomalously high, with $\rho = (5430 \pm 15)$ kg m^{-3}. On the basis of their density calculation, Ash et al. (1971) concluded that Mercury must be substantially richer in heavy elements than Earth. The pre-Mariner 10 estimates of mass, size, and density remain in excellent agreement with the MESSENGER results, but spacecraft data have enabled a reduction in uncertainties by a factor of ~50.

Howard et al. (1974) analyzed the tracking data from the first flyby of Mercury by Mariner 10 and obtained a gravitational parameter $GM = (2.2032 \pm 0.0002) \times 10^{13}$ m^{3}s^{-2}, where G is the gravitational constant. Analysis of data from all three Mariner 10 flybys yielded $GM = (2.203209 \pm 0.000091) \times 10^{13}$ m^{3}s^{-2} (Anderson et al., 1987). From more than three years of orbital tracking data of MESSENGER, Mazarico et al. (2014) obtained $GM = (2.203187080 \pm 0.000000086) \times 10^{13}$ m^{3}s^{-2}, estimated from a gravity field solution to degree and order 50. An independent analysis to degree and order 40 by Verma and Margot (2016) yielded $GM = (2.203187404 \pm 0.000000090) \times 10^{13}$ m^{3}s^{-2}. When translating the MESSENGER values to a mass estimate, the majority of the uncertainty comes from the 5×10^{-5} uncertainty in the gravitational constant. With $G = (6.67408 \pm 0.00031) \times 10^{-11}$ m^{3}kg^{-1}s^{-2} (Mohr et al., 2016), the current best estimate of the mass of Mercury is

$$M = (3.301110 \pm 0.00015) \times 10^{23} \text{ kg.} \quad (4.9)$$

From a combination of laser altimetry (Zuber et al., 2012) and radio occultation data, Perry et al. (2015) determined Mercury's average radius to be

$$R = (2439.36 \pm 0.02) \text{ km,} \quad (4.10)$$

although the stated radius uncertainty may be optimistic given the sparse sampling of the southern hemisphere. The corresponding bulk density is

$$\rho = (5429.30 \pm 0.28) \text{ kg m}^{-3}. \quad (4.11)$$

Mercury's bulk density is similar to that of Earth, $\rho_\oplus = 5514$ kg m^{-3}, despite the different sizes of the two bodies. The pressure P at the center of a homogeneous sphere scales as $P \propto \rho^2 R^2$, so materials in Earth's interior are more compressed (i.e., denser) than those in Mercury's interior. If we assume that both planets are made of a combination of a light component (i.e., silicates) and a heavy component (i.e., metals), we can infer from their similar densities and differing sizes that Mercury has a larger metallic component, as recognized by Ash et al. (1971).

4.3.3 C_{20} and C_{22} Results

The first measurements of the C_{20} and C_{22} gravity coefficients were obtained from Mariner 10 data recorded during one equatorial flyby with ~700 km minimum altitude and one polar flyby with ~300 km minimum altitude. Anderson et al. (1987) determined $C_{20} = (-6.0 \pm 2.0) \times 10^{-5}$ and $C_{22} = (1.0 \pm 0.5) \times 10^{-5}$. These values have large fractional uncertainties because there were only two favorable flybys, but the values are consistent with the most recent MESSENGER results (Mazarico et al., 2014; Verma and Margot, 2016). With the normalization that is commonly used in geodetic studies (Kaula, 2000, p. 7), the Mariner 10 values can also be expressed as $\overline{C}_{20} = C_{20}/\sqrt{5} = (-2.68 \pm 0.9) \times 10^{-5}$ and $\overline{C}_{22} = C_{22}/\sqrt{5/12} = (1.55 \pm 0.8) \times 10^{-5}$, where the overbar indicates normalized coefficients.

The next opportunity for measurements arose from the three MESSENGER flybys of Mercury in 2008–2009. However, the equatorial geometry of these flybys did not provide adequate leverage to measure C_{20} accurately. Because the Mariner 10 tracking data have been lost, it was not possible to perform a joint solution including both equatorial and polar flybys. For these reasons, Smith et al. (2010) cautioned that their recovery of $\overline{C}_{20} = (-0.86 \pm 0.30) \times 10^{-5}$ might not be reliable. However, the equatorial geometry was suitable for an accurate estimate of $\overline{C}_{22} = (1.26 \pm 0.12) \times 10^{-5}$.

Data acquired during the orbital phase of the MESSENGER mission provided significantly better sensitivity and lower uncertainties. Smith et al. (2012) analyzed the first six months of data (>300 orbits) and found $\overline{C}_{20} = (-2.25 \pm 0.01) \times 10^{-5}$ and $\overline{C}_{22} = (1.25 \pm 0.01) \times 10^{-5}$, where the error bars represent a calibrated uncertainty that is about 10 times the formal uncertainty of the fit. An independent analysis of the same data by Genova et al. (2013) confirmed these results. More recently, Mazarico et al. (2014) analyzed three years of data (2275 orbits) and estimated a gravity field solution to degree and order 50. This solution yielded an order-of-magnitude improvement in the calibrated uncertainties in C_{20} and C_{22}: $\overline{C}_{20} = (-2.2505 \pm 0.001) \times 10^{-5}$ and $\overline{C}_{22} = (1.2454 \pm 0.001) \times 10^{-5}$. An independent analysis by Verma and Margot (2016) confirmed these values to better than 0.4%.

The unnormalized quantities that we use in equations (4.3)–(4.6) are based on the Mazarico et al. (2014) values:

$C_{20} = (-5.0323 \pm 0.0022) \times 10^{-5}$ and $C_{22} = (0.8039 \pm 0.0006) \times 10^{-5}$. The $J_2/C_{22} = -C_{20}/C_{22}$ value of 6.26 is distinct from the equilibrium value of 7.86 for a body in a 3:2 spin–orbit resonance with the current value of the orbital eccentricity (Matsuyama and Nimmo, 2009), indicating that Mercury is not in hydrostatic equilibrium. It has been proposed that the values of the low-degree gravity coefficients can be explained by deep density anomalies induced by uneven insolation at the surface (Tosi et al., 2015).

4.3.4 Results for k_2

In addition to the static gravity field, Mazarico et al. (2014) also solved for the time-variable degree-2 potential that captures the tidal forcing due to the Sun. The tidal forcing is parameterized by the Love number k_2 (Section 4.8.1). Mazarico et al. (2014) obtained an estimate of $k_2 = 0.451 \pm 0.014$. However, because of potential mismodeling and systematic effects in the analysis, they could not rule out a wider range of values (0.43–0.50). The preferred value of Verma and Margot (2016) is $k_2 = 0.464 \pm 0.023$. They, too, encountered a wider range of best-fit values (0.420–0.465) in various trials. The weighted mean of these two estimates is $k_2 = 0.455 \pm 0.012$. These estimates are within the expected range from theoretical studies (Van Hoolst and Jacobs, 2003; Van Hoolst et al., 2007; Rivoldini et al., 2009) and from predictions of interior models informed by MESSENGER data and Earth-based radar data (Padovan et al., 2014).

4.4 SPIN-STATE CONSTRAINTS

Most of the quantities necessary to implement Peale's method of probing Mercury's interior were known when he wrote his paper in 1976. The mass, size, and density had been determined to <1% precision prior to the arrival of Mariner 10, the data from which confirmed and improved the ground-based estimates (Section 4.3). Values of the second-degree gravity coefficients C_{20} and C_{22} had also been determined, albeit with substantial uncertainties. In contrast, there were no satisfactory measurements of the spin state. Librations had not been detected, and the best spacecraft determination of the orientation of the rotation axis had a 50% error ellipse of ±2.6° by ±6.5° (Klaasen, 1976), about three orders of magnitude short of the required precision. Peale (1976) speculated that measurement of the obliquity and libration angles (θ and ϕ_0) would "almost certainly require rather sophisticated instrumentation on the surface of the planet." Fortunately, the measurements were obtained with Earth-based instruments as well as instruments aboard the MESSENGER orbiter.

4.4.1 Methods

Three observational methods have been used to measure Mercury's spin state: Earth-based radar observations, joint analysis of MESSENGER laser altimetry tracks and stereo-derived digital terrain models, and MESSENGER radio tracking observations. All three yielded estimates of Mercury's obliquity, but only the first two have yielded libration measurements so far. Another important distinction between these methods is that the first two measure the spin state of the rigid outer part of the planet, i.e., the lithosphere, whereas the gravity-based analyses are sensitive to the rotation of the entire planet.

The spin state of Mercury can be characterized to high precision with an Earth-based radar technique that relies on the theoretical ideas of Holin (1988, 1992). He showed that radar echoes from solid planets can display a high degree of correlation when observed by two receiving stations with appropriate positions in four-dimensional space–time. Normally each station observes a specific time history of fluctuations in the echo power (also known as *speckles*), and the signals recorded at separate antennas do not correlate. But during certain times on certain days of the year, the antennas become suitably aligned with the speckle trajectory, which is tied to the rotation of the observed planet (Figure 4.2). During these brief (~10–20 s) time

Figure 4.2. Radar echoes from Mercury sweep over the surface of the Earth. Diagrams show the trajectory of the speckles one hour before (left), during (center), and one hour after (right) the epoch of maximum correlation. Echoes from two receiver stations (red triangles) exhibit a strong correlation when the antennas are suitably aligned with the trajectory of the speckles (green dots shown with a 1-s time interval). From Margot et al. (2012).

intervals a cross-correlation of the two echo time series yields a high score at a certain value of the time lag (~5–10 s). The *epoch* at which the high correlation occurs provides a strong constraint on the orientation of the spin axis. The *time lag* at which the high correlation occurs provides a direct measurement of the spin rate. Margot et al. (2007, 2012) illuminated Mercury with monochromatic radiation (8560 MHz, 450 kW) from the Deep Space Network (DSN) 70-m antenna in Goldstone, California (DSS-14), and recorded the speckle patterns as they swept over two receiving stations (DSS-14 and the 100-m antenna in Green Bank, West Virginia). They obtained measurements of the instantaneous spin state of Mercury at 35 epochs between 2002 and 2012, from which they inferred both obliquity and libration angles.

Stark et al. (2015a) combined imaging (Hawkins et al., 2007) and laser altimetry (Cavanaugh et al., 2007) data obtained by MESSENGER during orbital operations to independently measure the spin state of Mercury. The basic idea was to produce digital terrain models (DTMs) from stereo analysis of the imaging data and to co-register the laser altimetry profiles to the DTMs (Stark et al., 2015c). During the co-registration step, a rotational model is adjusted in a way that minimizes the radial height differences between the two data sets. This adjustment enables the recovery of the spin axis orientation, which yields the value of the obliquity. It also enables the recovery of the amplitude of the physical librations because the laser profiles sample the topography of the surface at different phases of the libration cycle. In practice, Stark et al. (2015a) produced 165 individual gridded DTMs from thousands of images of the surface. Their DTMs cover ~50% of the northern hemisphere of Mercury with a grid spacing of 222 m/pixel, an effective horizontal resolution of 3.8 km, and an average height error of 60 m. For the co-registration step, they used 2325 laser profiles from three years of Mercury Laser Altimeter (MLA) observations. The laser altimetry data have a spacing between footprints that varied between 170 m and 440 m and a nominal ranging accuracy of 1 m.

The third method for estimating the spin state of Mercury is to adjust a rotational model of the planet during analysis of the radio tracking data (Section 4.3). Mazarico et al. (2014) and Verma and Margot (2016) analyzed three years of radio science data and produced estimates of the spin axis orientation. The detection of the physical librations with this technique is possible, but measuring the libration amplitude accurately remains challenging.

4.4.2 Obliquity Results

Analysis of the Earth-based radar data yielded an estimate of the obliquity $\theta = (2.042 \pm 0.08)$ arcminutes, where the adopted one-standard-deviation uncertainty corresponds to 5 arcseconds (Margot et al., 2012). Remarkably, the analysis of the spacecraft imaging and laser altimetry data, a completely independent data set, yielded an almost identical (0.6%) estimate of (2.029 ± 0.085) arcminutes, with similar uncertainties (Stark et al., 2015a). The weighted mean of these two estimates is $\theta = (2.036 \pm 0.058)$ arcminutes.

The best-fit spin axis orientation at epoch J2000.0 from analysis of the radar data is at equatorial coordinates (281.0103°, 61.4155°) and ecliptic coordinates (318.2352°, 82.9631°) in the corresponding J2000 frames (Margot et al., 2012). The MESSENGER DTM and laser altimetry results are within 0.8 arcseconds, at equatorial coordinates (281.0098°, 61.4156°) and ecliptic coordinates (318.2343°, 82.9633°) (Stark et al., 2015a).

Radio science tracking data can be used to estimate the orientation of the axis about which Mercury's gravity field rotates, which is not necessarily aligned with the axis about which the lithosphere rotates. Mazarico et al. (2014) and Verma and Margot (2016) used this technique and reported obliquities of (2.06 ± 0.16) and (1.88 ± 0.16) arcminutes, respectively. These results are consistent with those obtained by Margot et al. (2012) and Stark et al. (2015a), albeit with uncertainties that are twice as large (Figure 4.3).

Margot et al. (2007) provided observational evidence that Mercury is in or very near Cassini state 1, an important condition for the success of Peale's procedure. The current best-fit values place the radar-based and MESSENGER-based poles within 2.7 and 1.7 arcseconds of the Cassini state, respectively (Figure 4.3), confirming that Mercury closely follows the Cassini state. There are several possible interpretations for the imperfect agreement: (1) given the 5–6 arcsecond

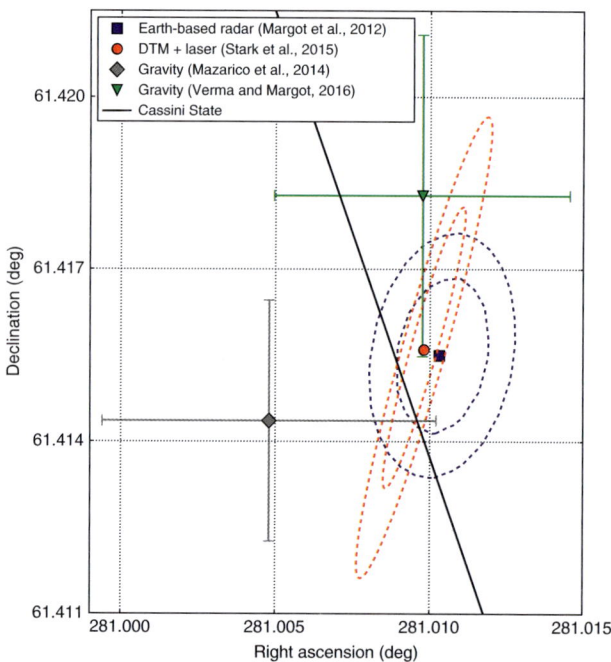

Figure 4.3. Orientation of the spin axis of Mercury obtained by three different techniques. The Earth-based radar results and the MESSENGER DTM and laser altimetry results are shown with contours representing the one- and two-standard-deviation uncertainty regions. The gravity results are shown with error bars representing the formal uncertainties of the fit multiplied by 10. The oblique line shows the predicted location of Cassini state 1 at epoch J2000.0 from the analysis of Yseboodt and Margot (2006). Points to the left and right of the line lead and lag the Cassini state, respectively.

uncertainty in spin axis orientation, Mercury may in fact be in the exact Cassini state; (2) Mercury may also be in the exact Cassini state if our knowledge of the location of that state is incorrect, which is possible because it is difficult to determine the exact Laplace pole orientation; (3) Mercury may lag the exact Cassini state by a few arcseconds; (4) Mercury may lead the exact Cassini state, although this seems less likely on the basis of the evidence at hand. Measurements of the offset between the spin axis orientation and the Cassini state location have been used to place bounds on energy dissipation due to solid-body tides and core–mantle interactions in the Moon (Yoder, 1981; Williams et al., 2001). However, the interpretation of an offset from the Cassini state at Mercury is complicated by the influence of various core–mantle coupling mechanisms (Peale et al., 2014) and the presence of an inner core (Peale et al., 2016).

4.4.3 Libration Results

Analysis of Earth-based radar observations obtained at 18 epochs between 2002 and 2006 yielded measurements of Mercury's instantaneous spin rate that revealed an obvious libration signature with a period of 88 days (Margot et al., 2007). From these data and the Mariner 10 estimate of C_{22} in equation (4.6), it was possible to show with 95% confidence that C_{m+cr}/C is smaller than unity. These results provided direct observational evidence that Mercury has a molten outer core (Margot et al., 2007). Measurements of Mercury's magnetic field prior to the radar observations had provided inconclusive suggestions about the nature of Mercury's core. A dynamo mechanism involving motion in an electrically conducting molten outer core was the preferred explanation (Ness et al., 1975; Stevenson, 1983), but alternative theories that did not require a liquid core, such as remanent magnetism in the crust, could not be ruled out (Stephenson, 1976; Aharonson et al., 2004).

Earth-based radar observations continued during the flyby and orbital phases of MESSENGER. By 2012, measurements at 35 epochs had been obtained (Figure 4.4). One can fit a libration model (Margot, 2009) to these data and derive the value of $(B-A)/C_{m+cr}$. Margot et al. (2012) found a value of $(B-A)/C_{m+cr} = (2.18 \pm 0.09) \times 10^{-4}$, which corresponds to a libration amplitude ϕ_0 of (38.5 ± 1.6) arcseconds, or a longitudinal displacement at the equator of 450 m.

Stark et al. (2015a) analyzed three years of MESSENGER DTM and laser altimetry data and found a libration amplitude of (38.9 ± 1.3) arcseconds, which corresponds to $(B-A)/C_{m+cr} = (2.206 \pm 0.074) \times 10^{-4}$. This estimate is in excellent agreement (1%) with the Earth-based radar value, giving confidence in the robustness of the results obtained by two independent techniques. The weighted means of these estimates are $(B-A)/C_{m+cr} = 2.196 \pm 0.057$ and $\phi_0 = (38.7 \pm 1.0)$ arcseconds.

4.4.4 Average Spin Rate

Questions remain about the precise spin behavior of Mercury, both in terms of its average spin rate and the presence of additional libration signatures. There are reasons to believe that longitudinal librations with periods of 2–20 yr exist, either because of planetary perturbations (Peale et al., 2007, 2009; Dufey et al., 2008; Yseboodt et al., 2010) or because of internal couplings and forcings (Dumberry, 2011; Veasey and Dumberry, 2011; Van Hoolst et al., 2012; Yseboodt et al., 2013; Koning and Dumberry, 2013; Dumberry et al., 2013). However, the addition of long-term libration components to the rotational model was not found to improve fits to the 2002–2012 radar data (Margot et al., 2012; Yseboodt et al., 2013). The duration of the MESSENGER data sets is not sufficiently long to detect a long-term libration signature, for which the primary period is expected to be ~12 yr. Therefore, Mazarico et al. (2014) and Stark et al. (2015a) did not attempt to fit for long-term librations. Instead, they obtained estimates of Mercury's average spin rate over the time span of the MESSENGER mission. Their estimates differ substantially from one another and from the expected mean resonant spin rate (Figure 4.5). One possible explanation for the discrepancy between theoretical and observational estimates is that the MESSENGER estimates are based on a 3- or 4-year period

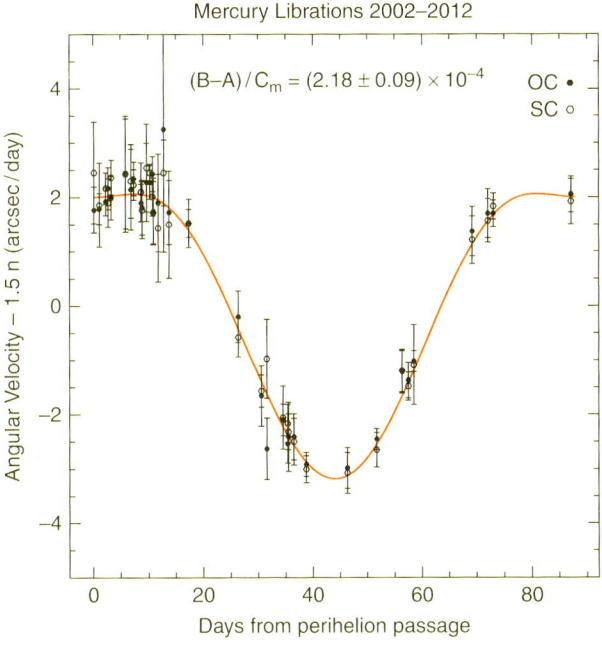

Figure 4.4. Mercury 88-day librations revealed by 35 instantaneous spin rate measurements obtained with Earth-based radar between 2002 and 2012. The vertical axis represents deviations of the angular velocity from the exact resonant rate of 3/2 times the mean orbital motion n. The measurements with their one-standard-deviation errors are shown in black. OC and SC represent measurements in two orthogonal polarizations (opposite-sense circular and same-sense circular, respectively). A numerical integration of the torque equation is shown in red. The flat top on the angular velocity curve near pericenter is due to the momentary retrograde motion of the Sun in the body-fixed frame and corresponding changes in the torque. The amplitude of the libration curve is determined by a one-parameter least-squares fit to the observations, which yields a value of $(B-A)/C_{m+cr} = (2.18 \pm 0.09) \times 10^{-4}$. From Margot et al. (2012).

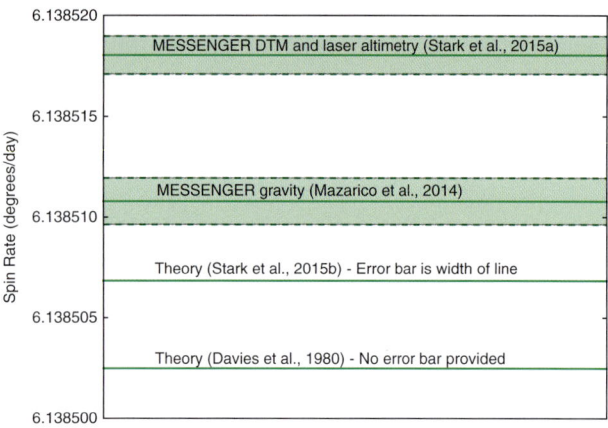

Figure 4.5. Theoretical and observational estimates of Mercury's mean resonant spin rate. The Davies et al. (1980) value was adopted in the latest report of the International Astronomical Union Working Group on Cartographic Coordinates and Rotational Elements (Archinal et al., 2011).

that represents only a small fraction of the long-term libration cycle.

4.5 TWO- AND THREE-LAYER STRUCTURAL MODELS

4.5.1 Governing Equations

The bulk density $\rho = M/V$ of a planetary body of mass M and volume V is an important indicator of composition, but it contains no information about the radial distribution of the material in the interior. Because we seek to calculate the radial density profile $\rho(r)$, we write expressions for the mass and bulk density of a spherically symmetric body of radius R that highlight the mass contributions from concentric spherical shells of width dr:

$$M = 4\pi \int_0^R \rho(r) r^2 dr, \quad (4.12)$$

$$\rho = \frac{3}{R^3} \int_0^R \rho(r) r^2 dr. \quad (4.13)$$

We write similar expressions for the polar moment of inertia C and its normalized value $\widetilde{C}$:

$$C = \frac{8\pi}{3} \int_0^R \rho(r) r^4 dr, \quad (4.14)$$

$$\widetilde{C} = \frac{C}{MR^2} = \frac{2}{\rho R^5} \int_0^R \rho(r) r^4 dr. \quad (4.15)$$

We first consider a two-layer model where a mantle with constant density ρ_m overlays a core with constant density ρ_c and radius R_c. In a gravitationally stable configuration, $\rho_c > \rho_m$. We use equations (4.13) and (4.15) to derive the analytical expressions for bulk density and normalized moment of inertia for this two-layer model:

$$\rho = \rho_c \alpha^3 + \rho_m (1 - \alpha^3), \quad (4.16)$$

$$\widetilde{C} = \frac{2}{5} \left[\frac{\rho_c}{\rho} \alpha^5 + \frac{\rho_m}{\rho} (1 - \alpha^5) \right], \quad (4.17)$$

where we have have used $\alpha = R_c/R$ for ease of notation. This system is underdetermined, because there are three unknowns (ρ_c, ρ_m, and R_c) and only two observables (ρ and $\widetilde{C}$). Even in the case of an oversimplified two-layer model, it is not possible to find a solution without making an additional assumption or securing an additional observable. For example, one could proceed by making an educated guess about the density of the mantle from measurements of the composition of the surface. A more rigorous approach is to obtain an additional observable that depends directly on the density of the mantle. We rely on Peale's procedure and the fact that Mercury is in a Cassini state (Section 4.4.2) to provide such an observable, the polar moment of inertia of the mantle plus crust as given by equation (4.6). For the two-layer model, this expression reduces to

$$\frac{C_{m+cr}}{C} = \frac{\rho_m (1 - \alpha^5)}{\rho_c \alpha^5 + \rho_m (1 - \alpha^5)}. \quad (4.18)$$

4.5.2 Moment of Inertia Results

Peale's formalism (Section 4.2.4) enabled a determination of Mercury's polar moment of inertia. Margot et al. (2012) combined measurements of the obliquity and librations with gravity data and found $\widetilde{C} = 0.346 \pm 0.014$. Stark et al. (2015a) also measured θ and ϕ_0, and found $\widetilde{C} = 0.346 \pm 0.011$. A uniform density sphere has $\widetilde{C} = 0.4$, and a body with a density profile that increases with depth has $\widetilde{C} < 0.4$. The Moon, with $\widetilde{C} \simeq 0.393$ (Williams et al., 1996), is nearly homogeneous, whereas the Earth, with $\widetilde{C} = 0.3307$ (Williams, 1994), has a substantial concentration of dense material near the center. Likewise, Mercury's $\widetilde{C}$ value suggests the presence of a dense metallic core.

The moment of inertia of Mercury's mantle and crust is also available from spin and gravity data (Equation 4.6). Margot et al. (2012) found $C_{m+cr}/C = 0.431 \pm 0.025$ and Stark et al. (2015a) found $C_{m+cr}/C = 0.421 \pm 0.021$.

Weighted means of the Margot et al. (2012) and Stark et al. (2015a) results provide the most reliable estimates to date of the moments of inertia. We find

$$\widetilde{C} = \frac{C}{MR^2} = 0.346 \pm 0.009, \quad (4.19)$$

$$\frac{C_{m+cr}}{C} = 0.425 \pm 0.016. \quad (4.20)$$

An error budget similar to that computed by Peale (1981, 1988) demonstrates that the dominant sources of uncertainties in the moment of inertia values can be attributed to spin quantities. Uncertainties arising from gravitational harmonics, tides, and orbital elements are at least an order of magnitude smaller (Noyelles and Lhotka, 2013; Baland et al., 2017). Further improvements to our knowledge of Mercury's moments of inertia therefore require better estimates of obliquity and libration amplitude. Such improved estimates may also enable a determination of the tidal quality factor Q (Baland et al., 2017).

4.5.3 Two-Layer Model Results

Using equations (4.16)–(4.18) and estimates of bulk density (4.11), $\widetilde{C}$ (4.19), and C_{m+cr}/C (4.20), we infer

$$R_c/R = 0.8209, \text{ i.e., } R_c = 2002 \text{ km}, \quad (4.21)$$

$$\rho_c/\rho = 1.3344, \text{ i.e., } \rho_c = 7245 \text{ kg m}^{-3}, \quad (4.22)$$

$$\rho_m/\rho = 0.5861, \text{ i.e., } \rho_m = 3182 \text{ kg m}^{-3}. \quad (4.23)$$

The results obtained with the two-layer model are within one standard deviation of the results of more elaborate, multi-layer models that take into account mineralogical, geochemical, and rheological constraints on the composition and physical properties of the interior (Hauck et al., 2013; Rivoldini and Van Hoolst, 2013; Section 4.7). Figure 4.6 illustrates the consistency of the two-layer solution (star) and of the multi-layer models of Hauck et al. (2013) (error bars). The two-layer model results are also consistent with results from multi-layer models that consider the total contraction of the planet (Knibbe and van Westrenen, 2015).

All points shown on Figure 4.6 are consistent with Mercury's bulk density ρ. Knowledge of the normalized moment of inertia $\widetilde{C}$ restricts acceptable models to a black, constant-$\widetilde{C}$ curve. The resulting degeneracy corresponds to the underdetermined system of equations (4.13) and (4.15). Knowledge of the moment of inertia of the mantle further restricts acceptable models to the blue curve. The intersection of the $\widetilde{C} = 0.346$ black curve (not shown) and of the $C_{m+cr}/C = 0.431$ blue curve yields the two-layer model solution.

Although three observables (ρ, $\widetilde{C}$, and C_{m+cr}/C) can be used to reliably estimate the parameters of a two-layer model (core size, core density, and mantle density), they provide no information about additional phenomena related to the origin, evolution, and present physical state of the planet (e.g., mineralogical composition of the mantle, composition of the core, presence of a solid inner core). Additional insight can be obtained with more elaborate three-layer and multi-layer models.

4.5.4 Three-Layer Models

We now consider a three-layer model with core, mantle, and crust of density ρ_{cr}. We express the core and mantle radii as fractions of the planetary radius, $\alpha = R_c/R$ and $\beta = R_m/R$. With this notation, we can write the bulk density, the polar moment of inertia, and the moment of inertia of the outer solid shell as follows:

$$\rho = \rho_c \alpha^3 + \rho_m (\beta^3 - \alpha^3) + \rho_{cr}(1 - \beta^3), \quad (4.24)$$

$$\widetilde{C} = \frac{2}{5}\left[\frac{\rho_c}{\rho}\alpha^5 + \frac{\rho_m}{\rho}(\beta^5 - \alpha^5) + \frac{\rho_{cr}}{\rho}(1 - \beta^5)\right], \quad (4.25)$$

$$\frac{C_{m+cr}}{C} = \frac{\rho_m(\beta^5 - \alpha^5) + \rho_c(1 - \beta^5)}{\rho_c \alpha^5 + \rho_m(\beta^5 - \alpha^5) + \rho_c(1 - \beta^5)}. \quad (4.26)$$

This system of equations has five unknowns and three observables. If we assume a crustal thickness value h_{cr} (i.e., β) and a crustal density value ρ_{cr}, the system of equations (4.24)–(4.26) can be solved. The thickness of the crust of Mercury has been estimated from the combined analysis of gravity and topography data (Mazarico et al., 2014; Padovan et al., 2015; James

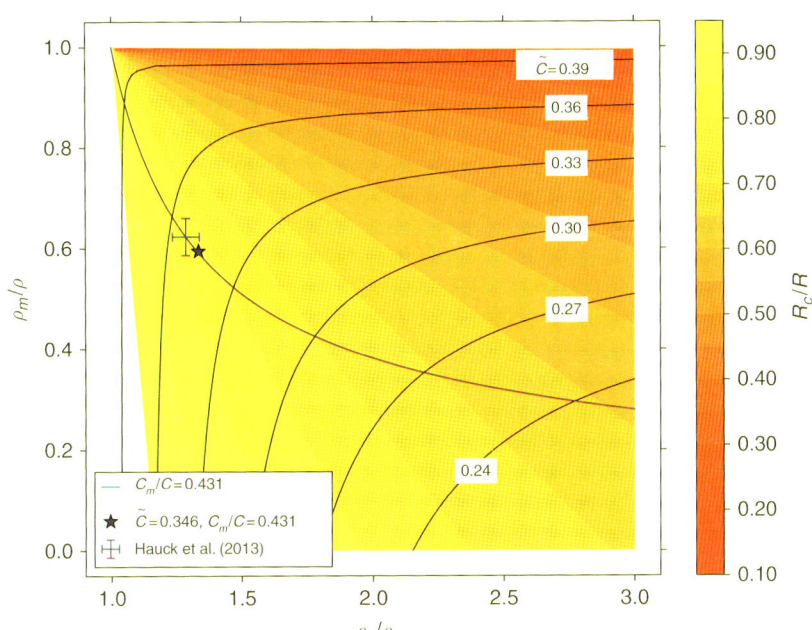

Figure 4.6. Mantle density versus core density showing the consistency of the two-layer model results (star) with those of more elaborate, multi-layer models (error bars). The position of the star corresponds to values of $\widetilde{C} = 0.346$ and $C_{m+cr}/C = 0.431$ (Margot et al., 2012). Error bars correspond to the one-standard-deviation intervals for ρ_c/ρ and ρ_m/ρ obtained by Hauck et al. (2013). The background color map indicates the value R_c/R in the two-layer model. Black curves illustrate models with various values of the normalized moment of inertia $\widetilde{C}$. The blue curve traces the locus of two-layer models with $C_{m+cr}/C = 0.431$.

et al., 2015). The density of the crust ρ_{cr} can be estimated from the measured composition of the surface of Mercury (e.g., Padovan et al., 2015).

We use the results of Padovan et al. (2015) and consider two end-member cases: a crust that is low-density and thin ($\rho_{cr} = 2700$ kg m^{-3}, $h_{cr} = 17$ km) and a crust that is high-density and thick ($\rho_{cr} = 3100$ kg m^{-3}, $h_{cr} = 53$ km). Compared with the two-layer model, the inferred radius of the core is almost unaffected by the inclusion of the crust, and the densities of the mantle and core change by less than 1%. This result can be explained by the small volume of the crust and the fact that its density is lower than that of the underlying layers. Consequently, the presence of the crust does not change the values of ρ, $\widetilde{C}$, and C_{m+cr}/C appreciably.

Another possible three-layer model includes a solid inner core, a liquid outer core, and a mantle. However, the composition of the core is not well constrained, and the system of equations (4.24)–(4.26) cannot be solved. To make further progress, we build multi-layer models (Section 4.7) that include additional, indirect constraints from the observed composition of the surface (Section 4.6) and from assumptions about interior properties guided by laboratory experiments. We then incorporate constraints that arise from the measurement of planetary tides (Section 4.8).

4.6 COMPOSITIONAL CONSTRAINTS

Measurements of the surface chemistry of Mercury by the MESSENGER spacecraft have provided important information on the composition of the interior (e.g., Chapter 2). Observations by the X-Ray Spectrometer (XRS) and Gamma-Ray and Neutron Spectrometer (GRNS) instruments have demonstrated that Mercury's surface has a low (<2.5 wt%) abundance of iron (Nittler et al., 2011; Evans et al., 2012; Weider et al., 2014; Chapter 2). This surface abundance, if also reflective of the mantle concentration of Fe (Robinson and Taylor, 2001), implies that the bulk density of the mantle is only modestly higher than those of the magnesium end-members of the likely major minerals, e.g., orthopyroxene enstatite with a density of 3200 kg m^{-3} (Smyth and McCormick, 1995). From the application of a normative mineralogy to the measured surface elemental abundances (Weider et al., 2015), Padovan et al. (2015) inferred grain densities for the crust of Mercury between 3000 and 3100 kg m^{-3}, a result driven primarily by the low Fe abundance. In addition to the low surface Fe abundance, Mercury has relatively large concentrations of sulfur in surface materials (Nittler et al., 2011; Chapter 2). When taken with the Fe observations, the measured S abundance of ~1.5–2.3 wt% in the crust implies strongly chemically reducing conditions (i.e., oxygen fugacities 2.6 to 7.3 log$_{10}$ units below the iron–wüstite buffer) in Mercury's interior during the partial melting that yielded these materials (Nittler et al., 2011; McCubbin et al., 2012; Zolotov et al., 2013). This inference is consistent with some pre-MESSENGER expectations (e.g., Wasson, 1988; Burbine et al., 2002; Malavergne et al., 2010). Two consequences of such reducing conditions are that, during global differentiation, S is more soluble in silicate melts that later crystallize as sulfides within the dominantly silicate material, and Si is more soluble in metallic Fe that segregates to the core. As a result, a wide range of core compositions has been considered when investigating Mercury's internal structure. The pressure, temperature, and compositional conditions relevant to Mercury's core have been tabulated by Rivoldini et al. (2009) and Hauck et al. (2013).

As Mercury's large bulk density has long implied, the planet has a large metallic core dominated by Fe that is likely alloyed with one or more lighter elements. Previous investigations focused on S as the major alloying element for Mercury's core (e.g., Stevenson et al., 1983; Schubert et al., 1988; Harder and Schubert, 2001; Van Hoolst and Jacobs, 2003; Hauck et al., 2007; Riner et al., 2008; Rivoldini et al., 2009; Dumberry and Rivoldini, 2015) because of its cosmochemical abundance and the greater availability of thermodynamic data. Sulfur has a strong effect on the density of Fe alloys, much greater than silicon or carbon for a given abundance. Additionally, S can lower the melting point of Fe alloys by hundreds of degrees, which is important for maintaining a liquid outer core, and it is relatively insoluble in solid Fe, the crystallizing phase in Fe-rich Fe–S systems. The latter property is important because it leads to a nearly pure Fe inner core and an outer core that is progressively enriched in S as a function of inner core growth.

For the most chemically reduced end-members of Mercury's inferred interior compositions, it is likely that Si is the primary, or sole, light alloying element in the metallic core. Alloys of Fe and Si have a markedly different behavior from Fe–S alloys in that they display a solid solution with a narrow phase loop, i.e., a narrow region between solidus and liquidus curves at high pressure (Kuwayama and Hirose, 2004). As a consequence, compositional differences between the potential solids and liquids in the core are much more limited, and thus density contrasts across the inner core boundary are smaller than for Fe–S core compositions. Silicon also has a smaller effect on the density and compressibility of Fe–Si alloys than does S on Fe–S alloys, with the consequence that more Si than S is required to achieve the same density reduction relative to pure Fe. Data on the equation of state of solid Fe–Si alloys are more plentiful than for liquid Fe–Si alloys, particularly at higher pressures, though the data are sufficient to construct models of Mercury's internal structure (Hauck et al., 2013). Due to the narrow phase loop and more limited melting point depression induced by Si in Fe alloys (e.g., Kuwayama and Hirose, 2004), inner core growth could be more extensive in Fe–Si systems than in S-bearing core alloys.

Over the range of inferred oxygen fugacities of 2.6–7.3 log$_{10}$ units below the iron–wüstite buffer for Mercury's interior, an alloy of Fe with both S and Si is likely in the core (Malavergne et al., 2010; Smith et al., 2012; Hauck et al., 2013; Namur et al., 2016b). Indeed, metal–silicate partitioning experiments motivated by the surface compositions measured by MESSENGER indicate that S and Si are likely both present in materials that make up Mercury's core (Chabot et al., 2014; Namur et al., 2016b). Unfortunately, data for the thermodynamic and thermoelastic properties of ternary alloys at high pressure are more limited than for their binary end-members. Experiments on the behavior of super-liquidus Fe–S–Si alloys have demonstrated large fields of two-liquid immiscibility (e.g., Sanloup and Fei, 2004; Morard and Katsura, 2010) with separate S-rich and Si-rich liquids at pressures relevant to Mercury's outermost core. Such immiscibility, if present in Mercury's core, would lead to a separation of phases with more S-rich liquids at the top of the core and Si-rich liquids deeper. In this situation, it is possible to assume end-member behavior in two separate compositional layers within the core and calculate properties separately for each layer (e.g., Smith et al., 2012; Hauck et al., 2013). However, liquid immiscibility in this system at higher pressures requires rather substantial amounts of both Si and S, which may or may not be appropriate. Experiments by Chabot et al. (2014) indicate a trade-off between Si and S in Mercury's metallic core that only minimally overlaps with current understanding of the Fe–S–Si liquid–liquid immiscibility phase field. Those results suggest that a mixture of Fe, S, and Si may be more likely. More recent work by Namur et al. (2016b), however, suggests that Mercury's core conditions may belong to the immiscible liquid field. In this case, Mercury's core may contain enough S for an FeS layer that is anywhere from negligibly thin to 90 km thick, depending on bulk S content of the planet. Regardless, the range of likely compositions for Mercury's core lies somewhere between an Fe–Si end-member and a (possibly segregated) mix of Fe, Si, and S.

4.7 MULTI-LAYER STRUCTURAL MODELS

We now wish to construct internal structure models with many layers in order to better match the gravity, spin state, and compositional constraints. We extend the approach of the two- and three-layer models (Section 4.5) to N-layer models with the goal of reproducing both discontinuous and continuous variations in density with depth. Such variations are expected on the basis of pressure-induced changes in the density of materials. For each material, an equation of state (EOS) describes the density as a function of pressure, temperature, and composition. Representing the effects of pressure variations inside Mercury's core requires an EOS, but the range of pressures expected across Mercury's thin silicate shell is relatively small. As a result, some models do not include an EOS for the silicate layer (Hauck et al., 2007, 2013; Smith et al., 2012; Dumberry and Rivoldini, 2015), although some models do (Harder and Schubert, 2001; Riner et al., 2008; Rivoldini et al., 2009; Rivoldini and Van Hoolst, 2013; Knibbe and van Westrenen, 2015). Multi-layer models provide an opportunity to reduce some of the non-uniqueness of simpler models through application of knowledge of the interior (e.g., potential core compositions) (Hauck et al., 2013; Rivoldini and Van Hoolst, 2013). They also enable investigations related to the structure of the core (Hauck et al., 2013; Dumberry and Rivoldini, 2015; Knibbe and van Westrenen, 2015) and the implications for the planet's thermal evolution and magnetic field generation.

4.7.1 Elements of the Model

Like two- and three-layer models, N-layer models consist of a series of layers defined by their composition and physical state. In contrast to simpler models, most of the geophysically defined layers in N-layer models are further subdivided into hundreds or thousands of sublayers. The sublayers provide for a smoother variation of density within the geophysically defined layers. Sublayer properties are functionally defined by the relevant EOS (Hauck et al., 2013; Rivoldini and Van Hoolst, 2013).

The basic internal organization of N-layer models is illustrated in Figure 4.7. The metallic core is divided into a solid inner core and a liquid outer core. Core densities vary according to the EOS. The solid outer portion of the planet is divided into one or more solid outer layers, most commonly with densities that are constant throughout their depth extent. Several models employ a traditional division of the solid outer shell into a crust and a mantle (Hauck et al., 2013; Rivoldini and Van Hoolst, 2013; Dumberry and Rivoldini, 2015; Knibbe and van Westrenen, 2015). Here, as did Hauck et al. (2013), we define up to three layers within the solid outermost portion of the planet: a basal layer at the bottom of the mantle, a mantle, and a silicate crust. The presence of a basal layer was suggested as a way to reconcile the low amounts of Fe observed at the planet's surface with the high bulk density of Mercury's outer solid shell inferred from spin and gravity data (Smith et al., 2012; Hauck et al., 2013). Evidence for deep compensation of domical swells on Mercury (James et al., 2015) also suggests that compositional variations deep within the solid outer shell are present, at least regionally.

4.7.2 Governing Equations

Any internal structure model for Mercury must be consistent with three quantities: the bulk density of the planet, the normalized moment of inertia $\tilde{C}$, and the fraction of the moment of inertia attributed to the librating, solid outer shell of the planet C_{m+cr}/C. This fraction is defined by

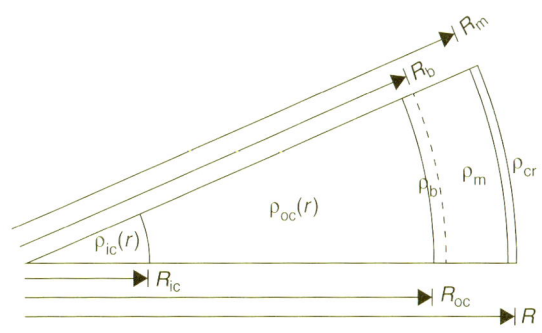

Figure 4.7. Schematic representation of the internal layers of Mercury in models with detailed sublayering aimed at capturing density variations due to changes in pressure, temperature, and composition with depth. Specific radii mark the transitions between layers, as follows: R_{ic} between solid inner core and liquid outer core, R_{oc} between liquid outer core and the solid outer shell of the planet, R_b between a compositionally distinct layer at the base of the mantle and the overlying mantle, and R_m between mantle and crust. The radius of the planet is R. The radially varying densities of the inner core and outer core are $\rho_{ic}(r)$ and $\rho_{oc}(r)$, respectively. The constant densities of any basal layer, mantle, and crust are ρ_b, ρ_m, and ρ_{cr}, respectively.

$$\frac{C_{m+cr}}{C} + \frac{C_c}{C} = 1, \quad (4.27)$$

where C_c/C is the fraction of the moment of inertia attributed to the core. The moment of inertia of the core C_c is calculated from equation (4.14) integrated from the center of the planet to the core–mantle boundary ($r = R_{oc}$ in Figure 4.7). The moment of inertia of the mantle plus crust C_{m+cr} can be determined from integration of equation (4.14) from $r = R_{oc}$ to $r = R$.

The EOSs that describe density variations with depth depend on the pressure and temperature of the materials. The pressure is a function of the overburden:

$$P(r) = \int_r^R \rho(x)g(x)dx, \quad (4.28)$$

and depends on the local gravity inside a sphere of radius r:

$$g(r) = \frac{G}{r^2}M(r) = \frac{G}{r^2}4\pi \int_0^r \rho(x)x^2 dx. \quad (4.29)$$

Equations (4.28) and (4.29) must be solved along with Equations (4.12) and (4.14) for the mass and polar moment of inertia of Mercury. Closing the set of four equations (4.12, 4.14, 4.28, 4.29), optionally augmented by equation (4.27), requires determination of the density as of a function of radius in the planet. Most models of Mercury's interior are based on a third-order Birch–Murnaghan EOS (Poirier, 2000):

$$P(r) = \frac{3K_0}{2}\left[\left(\frac{\rho(r)}{\rho_0}\right)^{\frac{7}{3}} - \left(\frac{\rho(r)}{\rho_0}\right)^{\frac{5}{3}}\right]$$
$$\times \left[1 + \frac{3}{4}(K'_0 - 4)\left\{\left(\frac{\rho(r)}{\rho_0}\right)^{\frac{2}{3}} - 1\right\}\right]$$
$$+ \alpha_0 K_0(T(r) - T_0), \quad (4.30)$$

where $T(r)$, T_0, ρ_0, K_0, K'_0, and α_0 are the local and reference temperatures, the reference density, the isothermal bulk modulus and its pressure derivative, and the reference volumetric coefficient of thermal expansion, respectively. The density, bulk moduli, and thermal expansivity are parameters for which ranges are determined from laboratory experiments and first-principles calculations. Values were given by, e.g., Hauck et al. (2013). The last term on the right of (4.30) relates to the increase in volume with increasing temperature.

The temperature as a function of radius can be determined for a conductive or convective mode of heat transfer. Most models for Mercury's core are based on the latter assumption. In the case of a thoroughly convective layer, the material is assumed to follow an adiabatic temperature gradient:

$$\frac{\partial T}{\partial P} = \frac{\alpha(T,P)T}{\rho(T,P)C_P}, \quad (4.31)$$

where α is the volume thermal expansion coefficient and C_P is the specific heat at constant pressure.

4.7.3 Methods

Investigations of Mercury's interior with N-layer models take the form of a basic parameter space study. The most fundamental parameter decision is the choice of core alloying elements because of their considerable influence on melting behavior (Section 4.6) and because the core occupies such a large fraction of the planet. The relative amounts of Fe and light elements are not known, so that broad ranges of possible core compositions tend to be considered. Indeed, Harder and Schubert (2001) considered all S contents from 0 wt% S (pure Fe) to 36.5 wt% S (pure FeS troilite). Most investigations in the post-MESSENGER era have used more limited compositional ranges. Other parameters considered include the thickness of the crust and the densities or density profiles of the crust and mantle.

The treatment of any crystallized solid layers within the metallic core represents another important modeling decision. Several models compare thermal gradients with an assumed, generally simplified, melting curve gradient for the core alloy (e.g., Rivoldini and Van Hoolst, 2013; Dumberry and Rivoldini, 2015). The intent is to develop a self-consistent prescription for the density structure of the core that includes the appropriate EOS for the regions of the core that are solid, liquid, or in the process of crystallizing from the top down (e.g., Dumberry and Rivoldini, 2015). This approach is most straightforward for Fe–S alloys because of their well-studied thermodynamic properties. However, these simplified phase diagrams tend to be based solely on eutectic compositions and do not account for mixing behavior that may be non-ideal (Chen et al., 2008). In addition, the melting relationships for Fe–Si and Fe–S–Si compositions are not well known. For these reasons, other studies consider the full range of possible solid inner core sizes (from zero to the entire core), irrespective of specific melting curves (Smith et al., 2012; Hauck et al., 2013).

With the constraints on Mercury's interior limited to the planetary radius, mass, and the moment of inertia parameters C/MR^2 and C_{m+cr}/C, knowledge of the planet's interior is necessarily non-unique. However, through a judicious set of assumptions regarding the composition of the interior and an exploration of parameter space, it is possible to place important constraints on Mercury's internal structure. Hauck et al. (2013) and Rivoldini and Van Hoolst (2013) employed Monte Carlo and Bayesian inversion approaches, respectively, in order to estimate the structure of Mercury's interior and to quantify the robustness of the most probable solution. One apparent difference in their approaches is that Hauck et al. (2013) included estimated uncertainties in the material parameters in the EOS of core material in addition to uncertainties in bulk density and moments of inertia, whereas Rivoldini and Van Hoolst (2013) included only the latter but considered depth-dependent density profiles for the mantle. Regardless of the details of the modeling and numerical approaches, several studies have converged on a common set of fundamental outcomes describing the internal structure of Mercury.

In assessing the agreement between interior models and observational constraints, we use a metric based on the fractional root mean square difference, defined as

$$\text{RMS} = \left[\frac{1}{2}\sum_{i=1}^{2}\left(\frac{O_i - C_i}{O_i}\right)^2\right]^{1/2}. \quad (4.32)$$

where O_i and C_i are observed and computed values, respectively, and the index i represents the two observables C/MR^2 and C_{m+cr}/C.

4.7.4 Results

Knowledge of the moment of inertia of a planet provides an integral measure of the distribution of density with radius. For Mercury, knowledge of the fraction of the polar moment of inertia due to the solid outer portion of the planet places further constraints on that density distribution. Still, taken together, the bulk density of the planet, C/MR^2, and C_{m+cr}/C represent a modest set of constraints on a body within which properties vary considerably with depth. As a result, N-layer models, which describe the internal density variation more precisely than the two- and three-layer models, are generally limited to describing a rather modest set of layers well. The most robust determinations include the bulk density of the solid, outermost planetary shell that overlies the liquid portion of the core, the bulk density of everything beneath that solid layer, and the location of the boundary between these two layers (Hauck et al., 2007, 2013; Smith et al., 2012; Rivoldini and Van Hoolst, 2013; Dumberry and Rivoldini, 2015). Although models based on the moments of inertia generally do not resolve the thickness of the crust or the density difference between the crust and mantle, studies of gravity and topography at higher-order harmonics do provide estimates of the crustal thickness and its regional variations (Smith et al., 2012; James et al., 2015; Padovan et al., 2015; Chapter 3).

The parameter of perhaps greatest interest regarding Mercury's interior is the location of the boundary between the liquid outer core and the solid outer shell. A similar answer is obtained with a wide variety of possible compositional models for Mercury's core: models with both more and less S than the Fe–S eutectic composition (Hauck et al., 2013; Rivoldini and Van Hoolst, 2013; Knibbe and van Westrenen, 2015), models that include Fe–Si alloys (Hauck et al., 2013), and models that include combinations of S, Si, and Fe (Hauck et al., 2013). Across all these models, the top of Mercury's liquid core has generally been estimated to be between 400 and 440 km beneath the surface with an estimated one-standard-deviation uncertainty of less than 10% of that value. Figure 4.8 illustrates a selection of results for the internal structure of Mercury with the Fe–Si core composition model results of Hauck et al. (2013). Interestingly, recent measurements of magnetic induction within Mercury are consistent with the top of the core being 400–440 km beneath the surface (Chapter 5).

The bulk densities of the material above and below the transition between the liquid core and outermost shell are also well established across a broad range of assumed core compositions and modeling approaches (e.g., Hauck et al., 2013; Rivoldini and Van Hoolst, 2013). The bulk density of the core material has been found to be distributed in the range 6750–7540 kg m^{-3}, with central values falling in the interval 6900–7300 kg m^{-3} and one-standard-deviation uncertainties of less than 5% of the central value (Hauck et al., 2013; Rivoldini and Van Hoolst, 2013). The bulk density of the solid outermost shell of Mercury is distributed in the range 3020–3580 kg m^{-3}, with central values falling in the interval 3200–3400 kg m^{-3} and one-standard-deviation uncertainties of approximately 6% of the central value.

One of the more intriguing proposals for the structure of Mercury's interior is the idea that a solid FeS layer could stably form at the core–mantle boundary. From a chemical standpoint, this layer originates in the core and resides at the top of the core. From a mechanical standpoint, however, a solid layer resides at the bottom of the mantle (Figure 4.7). The solid FeS layer hypothesis resulted from two observations. First, the chemically reducing conditions observed at the surface, if pertinent to the bulk of the planet, imply that Si will increasingly partition into the core with decreasing oxygen fugacity. At the pressures at Mercury's core–mantle boundary, Fe–S–Si liquids separate into two liquid phases over a broad range of compositions (Morard and Katsura, 2010). Hauck et al. (2013) estimated from the EOS of FeS in its high-pressure phase V that the solid was less dense than the residual liquid and could float rather than sink. Second, the best-fitting models (e.g., those with the lowest RMS values in Figure 4.8, but not necessarily with the highest histogram values) tend to have bulk densities for the solid outermost shell of Mercury that are larger than ~3200 kg m^{-3}, the approximate density expected for Fe-poor to Fe-absent mantle minerals such as forsterite and enstatite. For these reasons, Hauck et al. (2013) investigated the situation both with and without an FeS layer. However, the one-standard-deviation uncertainty in the outer shell bulk density is ~200 kg m^{-3} and permits a wide array of possible density configurations, with and without a solid FeS layer at the top of the core. Furthermore, calculations by Knibbe and van Westrenen (2015) question whether solid FeS is capable of floating at the top of the core, thus potentially preventing a substantial FeS layer from forming at the core–mantle boundary. Additional work on the EOS of solid FeS V at the appropriate conditions is warranted.

Experiments investigating the partitioning of S and Si between silicate and metallic melts for Mercury-like compositions (Chabot et al., 2014) have provided an opportunity to examine more closely the nature of the core–mantle boundary region. Figure 4.9 illustrates a comparison of the bulk core compositions of the internal structure models of Hauck et al. (2013) containing a possible solid FeS layer at the top of the core with the predicted ranges of core compositions compatible with MESSENGER geochemical observations of the surface (Chabot et al., 2014). Also shown are the limits on compositions in the Fe–S–Si system that display liquid–liquid immiscibility at the relevant pressures of 6 and 10 GPa. Compositions to the right of the immiscibility limit curves display immiscibility and are prone to phase separation at the given pressure. While the majority of core compositions in the Fe–S–Si models of Hauck et al. (2013) are consistent with the segregation of Fe–S-rich liquids at the top of Mercury's core, the general lack of overlap of recent geochemical predictions of possible core compositions with the immiscibility limits (Chabot et al., 2014) suggests that liquid–liquid phase separation may not be preferred. The further consequence, of course, is that the conditions for crystallization of an FeS phase at the top of the core appear less likely than the

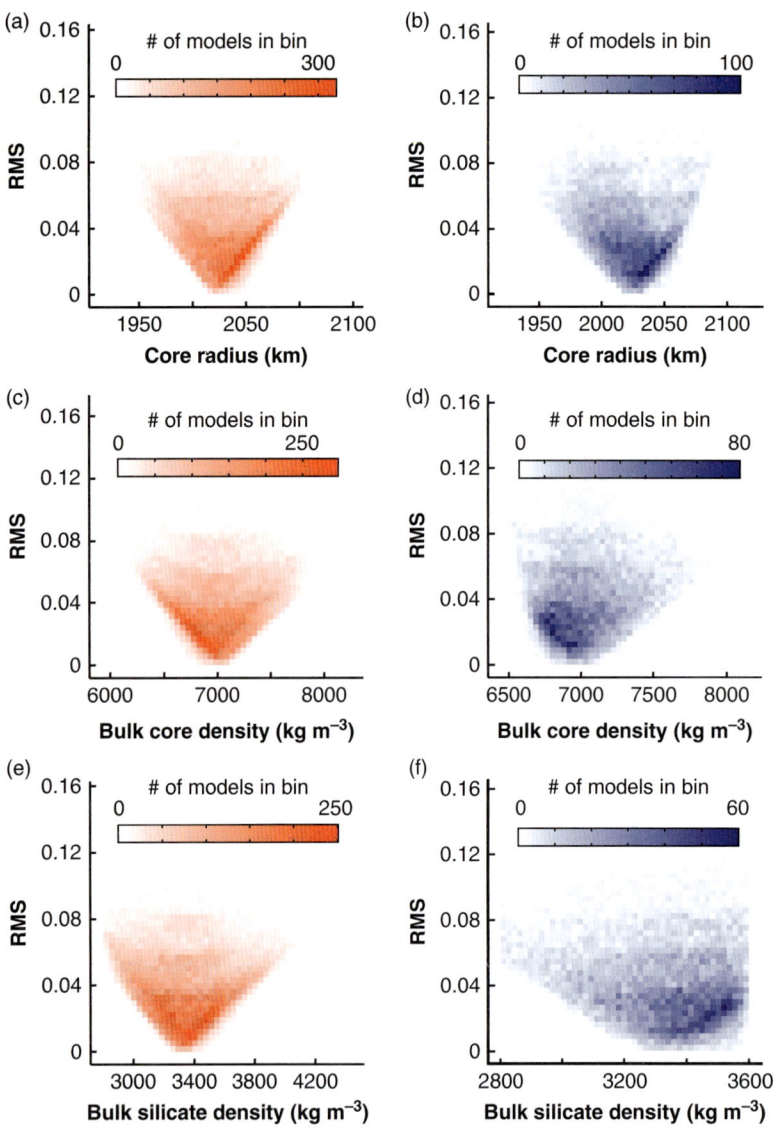

Figure 4.8. Two-dimensional histograms summarizing *N*-layer internal structure models of Mercury with Fe–Si core compositions based on the approach of Hauck et al. (2013) and current best estimates of C/MR^2 and C_{m+cr}/C (Section 4.5.2). The left column (a, c, and e) represents models that include a three-layer silicate shell with a crust, mantle, and denser solid layer at the base of the mantle. The right column (b, d, and f) represents models that include a two-layer silicate shell with a crust and mantle. Shown are the recoveries of the radius of the top of the liquid outer core (a and b), bulk density of the metallic core (c and d), and bulk density of the silicate, solid, outermost shell of the planet (e and f). The vertical axes show the goodness of fit expressed as a fractional root mean square difference; see equation (4.32). Si contents of the metallic core in these models vary from 0 to 17 wt% (Hauck et al., 2013).

immiscibility limits alone previously suggested. However, as is apparent from Figure 4.9, the preferred core compositions of Chabot et al. (2014) and the most probable models that match the density and moment of inertia parameters do not generally overlap. There are several possible explanations for the discrepancy. First, it may be that the surface abundance of S cannot yield reliable insights about core composition, either because the surface abundance is not representative of the planet's bulk silicate composition or because chemical equilibrium was not satisfied during core formation. Second, it is possible that a modeling approach not investigated so far is required, e.g., one involving a single, miscible Fe–S–Si liquid phase rather than two fully separated Fe–S and Fe–Si phases. Third, it is possible that the partitioning behavior observed at atmospheric pressure by Chabot et al. (2014) is not representative of core conditions. Indeed, a recent geochemical experimental study with differing silicate compositions and at slightly higher pressures (Namur et al., 2016b) suggests that the mantle may contain more S than the surface rocks. In that case, the bulk core S

content may be larger and the core conditions may belong to the immiscibility field. However, that conclusion and the thickness of any possible FeS layer depend strongly on Mercury's bulk S content.

Understanding the existence and size of an inner core on Mercury is a critical goal because an inner core influences several aspects of the planet's evolution, including magnetic field generation (Chapters 5 and 19), global contraction (Chapters 10 and 19), and rotational state (Section 4.9). However, the size of the inner core is difficult to quantify, for two reasons. First, the density contrast across the inner–outer core boundary is modest (e.g., Hauck et al., 2013; Rivoldini and Van Hoolst, 2013). Second, the inner core makes up only a small fraction of the mass and density distribution of the planet. Indeed, models with assumptions about the melting relationships of the core can typically place only upper limits on the radius of the inner core, and these upper limits are large. In models with core concentrations of S exceeding a few wt%, upper limits are ~1450 km,

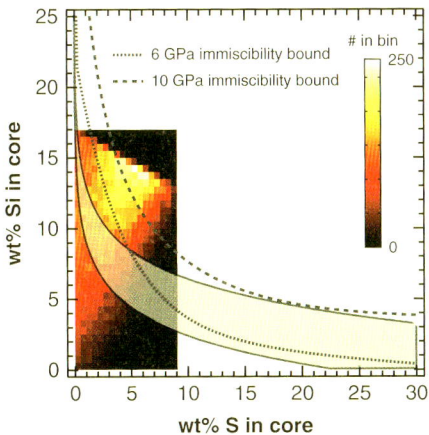

Figure 4.9. Representation of bulk core S and Si contents in a subset of the internal structure models of Hauck et al. (2013). All models shown have an Fe–S–Si core composition and a solid FeS layer at the base of the mantle, and only models that match the C/MR^2 and C_{m+cr}/C constraints (Section 4.5.2) are shown. The two-dimensional histogram indicates the relative number of successful models at each bulk core composition. Immiscibility limits in the Fe–S–Si system at two different pressures are shown by the dotted and dashed lines. Compositions to the right of these lines result in immiscible Fe–S-rich and Fe–Si-rich liquids at the indicated pressure. The gray region illustrates predicted bounds on core composition from metal–silicate partitioning experiments under the assumption that the S content at the surface of Mercury yields reliable constraints on core composition (Chabot et al., 2014). The lower and upper boundaries of the gray region denote the expected core compositions for representative surface S contents of 1 wt% and 4 wt%, respectively.

i.e., $R_{ic}/R \lesssim 0.6$ (Rivoldini and Van Hoolst, 2013; Dumberry and Rivoldini, 2015; Knibbe and van Westrenen, 2015). Upper limits as high as 1700–1800 km can be reached in models with low core concentrations of S (Rivoldini and Van Hoolst, 2013; Dumberry and Rivoldini, 2015; Knibbe and van Westrenen, 2015). Growth of an inner core to that size over the past ~4 billion years is likely incompatible with the amount of global contraction of the planet inferred from measurements of tectonic structures on the surface (Section 4.9, Chapter 19). Models without an assumed core melting relationship constraint do not place strong limits on the size of the solid inner core, although there is a slight preference for models with an inner core radius less than ~60% of the core radius or ~50% of the planetary radius (Hauck et al., 2013). Additional constraints on the size of the inner core are discussed in Section 4.9.

4.8 TIDAL RESPONSE

Additional insights about Mercury's interior structure can be gained by measuring the deformation that the planet experiences as a result of periodic tidal forces. These measurements are informative because the response of a planet to tides is a function of the density, rigidity (i.e., shear modulus), and viscosity of the subsurface materials. Tidal measurements have been used to support the hypothesis of a liquid core inside Venus (Konopliv and Yoder, 1996) and Mars (Yoder et al., 2003), and that of a global liquid ocean inside Titan (Iess et al., 2012). In principle, high-precision measurements of the tidal response can be used to rule out models that are otherwise compatible with the density and moment of inertia constraints (Section 4.7). When a global liquid layer is present, the tidal response is largely controlled by the strength and thickness of the outer solid shell (e.g., Moore and Schubert, 2000). Because Mercury has a molten outer core (Section 4.4) and because the thickness of the outer solid shell is known (Sections 4.5 and 4.7), tidal measurements enable investigations of the strength of the outer solid shell. This strength depends primarily on the mineralogy and thermal structure of the shell.

4.8.1 Tidal Potential Love Number k_2

The tidal perturbation generated by the Sun on Mercury simultaneously modifies the shape of the planet and the distribution of matter inside the planet. As a result of the redistribution of mass, solar tides also modify Mercury's gravitational field. From the standard expansion of the gravitational field in spherical harmonics (e.g., Kaula, 2000), the largest component of the tidal potential is a degree-2 component Φ_2 proportional to the mass of the Sun and with a long axis that is aligned with the Sun–Mercury line. The additional potential ϕ_{2t} resulting from the deformation of the planet in response to the tidal potential is parameterized by the tidal potential Love number k_2:

$$\phi_{2t} = k_2 \, \Phi_2. \tag{4.33}$$

The tidal component with the largest amplitude has a period $P_m = 87.9693$ days (Van Hoolst and Jacobs, 2003), corresponding to Mercury's orbital period. The Love number k_2 is a function of ω, $\rho(r)$, $\mu(r)$, and $\eta(r)$, where $\omega = 2\pi/P_m$ is the known forcing frequency and $\rho(r)$, $\mu(r)$, and $\eta(r)$ are the density, rigidity, and viscosity profiles. With the appropriate profiles, the Love number k_2 at the frequency ω can be calculated by solving the equations of motion inside the planet. These equations consist of three second-order equations that can be transformed into a system of six first-order linear differential equations in radius through a spherical harmonic decomposition in latitude and longitude (Alterman et al., 1959). We solve these equations with a slightly modified version of the propagator matrix method (e.g., Sabadini and Vermeersen, 2004), as described by Wolf (1994) and by Moore and Schubert (2000, 2003).

4.8.2 Rheological Models

The rheological response of solid materials is elastic, viscoelastic, or viscous, depending primarily on pressure, temperature, grain size, and timescale of the process under consideration. Other dependencies include melt fraction and water content. Earth's mantle has a quasi-elastic response on the short timescales associated with seismic waves and a fluid-like response on the long timescales of mantle convection.

The Maxwell rheological model is the simplest model that captures behavior on both short and long timescales. It is completely defined by two parameters, the unrelaxed (i.e., corresponding to an impulsive or infinite-frequency perturbation) rigidity μ_U and the dynamic viscosity η. The Maxwell time, defined as

$$\tau_M = \frac{\eta}{\mu_U}, \quad (4.34)$$

is a timescale that separates the elastic regime (forcing period $\ll \tau_M$) from the fluid regime (forcing period $\gg \tau_M$). This simple rheological model is sufficiently accurate to describe the crust of Mercury. The crust is cold and responds elastically ($\tau_{M,crust} = 10^5$ yr). We treat the liquid outer core as an inviscid fluid. We also use the Maxwell model to represent the rheology of the inner core, which, if present, has a negligible effect on the tidal response (Padovan et al., 2014). However, the Maxwell model fails to capture the response of the mantle at tidal frequencies (e.g., Efroimsky and Lainey, 2007; Nimmo et al., 2012), because it does not provide a good fit to laboratory and field data in the low-frequency seismological range.

We adopt the Andrade pseudo-period rheological model to estimate the response of Mercury's mantle to tidal forcing (Jackson et al., 2010; Padovan et al., 2014). In this model, the ratio of strain to stress, or inverse rigidity, is represented by a complex compliance. The expressions for the real (Re) and imaginary (Im) parts of the dynamic compliance in the Andrade model are (Jackson et al., 2010):

$$J_{Re} = J_U \left\{ 1 + \beta^* \Gamma(1+n) \omega_a^{-n} \cos\left(\frac{n\pi}{2}\right) \right\}, \quad (4.35)$$

$$J_{Im} = J_U \left\{ \beta^* \Gamma(1+n) \omega_a^{-n} \sin\left(\frac{n\pi}{2}\right) + \frac{1}{\omega_v \tau_M} \right\}. \quad (4.36)$$

J_U is the unrelaxed compliance, Γ is the gamma function, and n, $\beta^* = \beta/J_U$, and $\tau_M = \eta J_U$ are related to parameters appearing in the Andrade creep function $J(t) = J_U + \beta t^n + t/\eta$. The pressure ($P$), temperature ($T$), and grain size ($d$) dependencies are introduced through the pseudo-period master variable $X_B = 2\pi/\omega_{a,v}$:

$$X_B = T_0 \left(\frac{d}{d_{Ref}}\right)^{-m_{a,v}}$$
$$\times \exp\left[\left(\frac{-E}{R}\right)\left(\frac{1}{T} - \frac{1}{T_{Ref}}\right)\right]$$
$$\times \exp\left[\left(\frac{-V}{R}\right)\left(\frac{P}{T} - \frac{P_{Ref}}{T_{Ref}}\right)\right], \quad (4.37)$$

where T_0 is the period of the applied forcing (in this case the period of the primary tidal component), m_a (m_v) is the grain size exponent for anelastic (viscous) processes, and here R is the gas constant. The quantities P_{Ref}, T_{Ref}, and d_{Ref} indicate reference values (Table 4.1). The unrelaxed shear modulus $\mu_U = 1/J_U$ is itself dependent on pressure and temperature, which we characterize by a simple Taylor expansion truncated at linear terms: $\mu_U(P,T) = \mu_U^{Ref} + (\partial\mu/\partial P)(P - P_{Ref}) + (\partial\mu/\partial T)(T - T_{Ref})$. The frequency-dependent shear modulus $\mu(\omega)$, quality factor $Q(\omega)$, and viscosity $\eta(\omega)$ are all obtained from the dynamic compliance, as follows (Jackson et al., 2010; Padovan et al., 2014):

$$\mu(\omega) = [J_{Re}^2(\omega) + J_{Im}^2(\omega)]^{-1/2}, \quad (4.38)$$

$$Q(\omega) = \frac{J_{Re}(\omega)}{J_{Im}(\omega)}, \quad (4.39)$$

$$\eta(\omega) = \frac{1}{\omega_0 J_{Im}(\omega)}, \quad (4.40)$$

where $\omega_0 = 2\pi/T_0$. Our choice of model parameters is described in Table 4.1 and Section 4.8.3.

These choices of Andrade model parameter values (Table 4.1) are based on data obtained at periods smaller than 10^3 s (Jackson et al., 2010), whereas the main tide of Mercury has a period >10^6 s. The extrapolation to long timescales can be validated to some extent by two considerations. First, we verified that equation (4.40) yields viscosity values at long timescales (>10 Myr) that fall within the interval for convective viscosities commonly assumed in terrestrial mantle convection simulations (10^{20}–10^{23} Pa s). Second, we verified that, at timescales appropriate for glacial rebound on Earth (~10^4 yr), the predicted viscosity values (10^{20}–10^{21} Pa s) compare favorably with those inferred from geodynamical data (e.g., Kaufmann and Lambeck, 2000).

The choice of Andrade model parameter values (Table 4.1) is also based on laboratory data for olivine (Jackson et al., 2010), whereas we apply the model to a variety of mineralogies (Table 4.1). This extrapolation to other mineralogies is not strictly correct, especially for mantle models in which olivine is not the dominant phase. However, the Andrade model has been successfully applied to the description of dissipation in rocks, ices, and metals (e.g., Efroimsky, 2012, and references therein). The broad applicability of the model over a wide range of physical and chemical properties suggests that it can provide an adequate description of the rheology of silicate minerals.

Recent results of laboratory experiments and thermodynamic simulations based on Mercury surface compositions (Vander Kaaden and McCubbin, 2016; Namur et al., 2016a) suggest an olivine-rich source for both the northern smooth plains and the high-Mg region of the intercrater plains and heavily cratered terrain. These results are in accord with an olivine-dominated mineralogy for the mantle of Mercury and further support our model parameter choices.

4.8.3 Methods

We restrict our analysis to interior models that are compatible with the observed bulk density ρ, moment of inertia C, and moment of inertia of the solid outer shell C_{m+cr}. By design, the subset of interior models has distributions of ρ, C, and C_{m+cr} that are approximately Gaussian with means and standard deviations that match the nominal values of the observables and their one-standard-deviation errors. The mean density ρ has a Gaussian distribution with mean and standard deviation equal to 5430 kg m^{-3} and 10 kg m^{-3}, respectively. For C and C_{m+cr}, we choose Gaussian distributions with means and

Table 4.1. *Rheological models for the interior of Mercury.*

Layer	Model	Parameter	Definition	Value
Crust	Maxwell			
		μ_U	Unrelaxed rigidity	55 GPa
		η	Dynamic viscosity	10^{23} Pa s
Mantle	Andrade[a]			
		μ_U^{Ref}	Unrelaxed rigidity[b]	59–71 GPa
		T_b	Mantle basal temperature[c]	1600–1850 K
		n	Andrade creep coefficient	0.3
		β^*	Andrade creep parameter	0.02
		P_{Ref}	Reference pressure	0.2 GPa
		T_{Ref}	Reference temperature	1173 K
		d_{Ref}	Reference grain size	3.1 μm
		d	Grain size	1 mm–1 cm
		m_a, m_v	Grain size exponents	1.31, 3
		V	Activation volume	10^{-5} m^3 mol^{-1}
		E_B	Activation energy	303 kJ mol^{-1}
FeS	Andrade[d]			
Outer core	Inviscid fluid			
		μ_U	Unrelaxed rigidity	0 GPa
		η	Dynamic viscosity	0 Pa s
Inner core	Maxwell			
		μ_U	Unrelaxed rigidity	100 GPa
		η	Dynamic viscosity	10^{20} Pa s

Notes:
[a] The fixed parameters of the Andrade model are based on the results of Jackson (2010).
[b] The nominal value depends on the adopted mineralogy (Table 4.2).
[c] We report T_b because the relevant temperature in equation (4.37) is controlled by T_b.
[d] The FeS layer is assumed to have the same rheology as the base of the mantle.

standard deviations defined by the observed values and errors (Section 4.5.2).

We treat the interior of Mercury as a series of spherically symmetric, incompressible layers characterized by density, thickness, rigidity, and viscosity. We start with the density profiles calculated by Hauck et al. (2013), but we replace the ~1000 layers that characterize the core in these models with two homogeneous layers representing the solid inner core and the liquid outer core. This simplification is warranted because the tidal response is dominated by the presence of a liquid outer core and is largely independent of the detailed density structure of the core. It reduces the computational cost by about three orders of magnitude and introduces only a small (<2%) error in the estimated value of k_2. This error is smaller than the variations induced by the unknown mineralogy and thermal state of the mantle (Padovan et al., 2014).

For the core of Mercury, we focus on the Si-bearing subset of models analyzed by Hauck et al. (2013), because this subset is most consistent with the chemically reducing conditions inferred from surface materials (Section 4.6). We also consider the subset of models with a solid FeS V layer included at the base of the mantle (Hauck et al., 2013; Section 4.7). For the silicate mantle of Mercury, we consider six mineralogical models based on the works of Rivoldini et al. (2009) and Malavergne et al. (2010) (Table 4.2).

Our use of the Andrade model (Section 4.8.2) for the rheological properties of the mantle requires knowledge of the radial profiles of unrelaxed rigidity μ_U, temperature T, and pressure P in the outer solid shell. For each of the six mineralogical models, we compute a composite rigidity μ_U^{Ref} (Table 4.2) with Hill's expression (Watt et al., 1976) at $T = T_{Ref}$ and $P = P_{Ref}$. The pressure profile in the outer solid shell is obtained by evaluating the overburden pressure as a function of depth. The temperature in the mantle is computed by solving the static heat conduction equation with heat sources in spherical coordinates (e.g., Turcotte and Schubert, 2002) in the mantle and crust. For the crust, we adopted the surface value of the heat production rate $H_0 = 2.2 \times 10^{-11}$ W kg^{-1} (Peplowski et al., 2012). For the mantle, we used a value of $H_0/2.5$, which is compatible with the enrichment factor derived by Tosi et al. (2013). Temperature profiles are fairly insensitive to the value of the thermal conductivity of Mercury's silicate shell: we used a value of

Table 4.2. *Mineralogical models for the mantle of Mercury.*

Model	Grt	Opx	Cpx	Qtz	Spl	Pl	Mw	Ol	μ_U^{Ref} (GPa)
CB	–	66	4	22	4	4	–	–	59
EH	–	78	2	8	–	12	–	–	65
MA	23	32	15	–	–	–	–	30	69
TS	25	–	–	–	8	–	2	65	71
MC	15	50	9	–	–	–	–	26	68
EC	1	75	7	17	–	–	–	–	60

Notes: The adopted model mineralogies resemble those of enstatite chondrites (EC and EH), Bencubbin-like chondrites (CB), metal-rich chondrites (MC), a refractory-volatile model (TS), and a model based on fractionation processes in the solar nebula (MA). For details, see Malavergne et al. (2010; CB and EH), Morgan and Anders (1980; MA), Taylor and Scott (2003; TS and MC), and Wasson (1988; EC). Columns 2–9 give the mineralogical content in terms of the vol.% of its components, from Malavergne et al. (2010; CB, EH) and Rivoldini et al. (2009; MA, TS, MC, EC). Mineral abbreviations follow Siivola and Schmid (2007): Garnet (Grt), Orthopyroxene (Opx), Clinopyroxene (Cpx), Quartz (Qtz), Spinel (Spl), Plagioclase (Pl), Merwinite (Mw), Olivine (Ol). The composite rigidity μ_U^{Ref} is obtained with Hill's expression (Watt et al., 1976) at $T = T_{Ref}$ and $P = P_{Ref}$.

3.3 W m^{-1} K^{-1}, but confirmed that a value of 5 W m^{-1} K^{-1} yields essentially the same results. We establish two boundary conditions: the temperature at the surface of Mercury T_S and the temperature at the base of the mantle T_b. The latter provides the primary control on the temperature profile. T_S is set to 440 K, a value obtained with an equilibrium temperature calculation. Both Rivoldini and Van Hoolst (2013) and Tosi et al. (2013) indicate T_b values in the range 1600–1900 K. We define two end-member profiles: a cold-mantle profile with T_b and a hot-mantle profile with $T_b = 1850$ K. A larger value of T_b (e.g., 1900 K) would result in partial melting at the base of the mantle according to the peridotite solidus computed by Hirschmann (2000). We did not consider the presence of partial melting.

There is a scarcity of laboratory data for FeS V, which is the phase relevant at the pressure and temperature conditions at the bottom of the mantle of Mercury (Fei et al., 1995). We consider the effects of the FeS layer only in the cold-mantle case ($T_b = 1600$ K), because at higher temperatures the FeS would be liquid (see the phase diagram given by Fei et al., 1995). We model the rheological response of this layer by assuming that it has the same rheological properties as those at the base of the mantle. This assumption results in a lower bound on the k_2 estimates because we expect the viscosity of this layer to be lower than that of the silicate layer. The viscosity scales as the exponential of the inverse of the homologous temperature (i.e., the ratio of the temperature of the material to the solidus temperature) (e.g., Borch and Green, 1987). At $T = 1600$ K, the homologous temperature of the FeS V is larger than that of the silicates. In addition, the unrelaxed rigidity of FeS V is likely to be smaller than that for mantle material because the rigidity of troilite (or FeS I, the phase at standard pressure and temperature) is 31.5 GPa (Hofmeister and Mao, 2003).

We apply our calculations to five different models (nominal, cold and stiff, hot and weak, FeS-layer, and 1-mm grain size). Given the 1600–1850 K range for the basal mantle temperature and 59–71 GPa range for the unrelaxed rigidity of the mantle, we define a nominal model with $T_b = 1725$ K and $\mu_U = 65$ GPa. Changes in basal mantle temperature and unrelaxed rigidity have similar but opposite effects on the tidal response. Accordingly, we define two end-member models: a cold and stiff mantle model with $T_b = 1600$ K and $\mu_U = 71$ GPa and a hot and weak mantle model with $T_b = 1850$ K and $\mu_U = 59$ GPa. Our fourth model is a cold mantle model ($T_b = 1600$ K) with nominal rigidity ($\mu_U = 65$ GPa) and an FeS layer at the bottom of the mantle. In all of these four models, we use a nominal grain size $d = 1$ cm, a value compatible with the estimated grain size in the mantles of the Moon and Mars (Nimmo et al., 2012; Nimmo and Faul, 2013). Our fifth and last model is a variation of the nominal model in which we consider a grain size of $d = 1$ mm. Model parameters are summarized in Table 4.3.

Our procedure for evaluating the Love number k_2 and corresponding uncertainties is as follows. For each of the five cases described in Table 4.8.3, we use approximately 6×10^4 density profiles from the previously identified subsets of models from Hauck et al. (2013). For each profile, we construct an interior model and calculate the value of k_2. We then fit a Gaussian distribution to the ~6×10^4 calculated k_2 values, as was done by Padovan et al. (2014). We report the Love number and associated error as the mean and standard deviation of the Gaussian fit. Our values differ somewhat from those of Padovan et al. (2014) because we incorporated the most recent estimates of the moments of inertia in this work, from equations (4.19) and (4.20).

4.8.4 Results

Our Love number calculations for models with a molten outer core yield values $k_2 \simeq 0.5$. However, for models with a completely solid core, we found k_2 values that are approximately an order of magnitude smaller. Measurements of Mercury's tidal response (Section 4.3.4) therefore confirm the presence of a molten outer core.

Table 4.3. *Characteristics of five mantle models for the estimation of Mercury's tidal response.*

Model	μ_U, GPa	T_b, K	d, mm	FeS?
Nominal	65	1725	10	no
Cold and stiff	71	1600	10	no
Hot and weak	59	1850	10	no
FeS layer	65	1600	10	yes
1-mm grain size	65	1725	1	no

Notes: Model names correspond to those in Figure 4.10.

Our results also show that the tidal response is enhanced by higher mantle basal temperatures and by lower mantle rigidities (Figure 4.10).

The comparison of our calculated values with the k_2 value measured by Mazarico et al. (2014) indicates that the observed tidal signal is more compatible with cold, rigid mantle models (Figure 4.10). The k_2 value measured by Verma and Margot (2016) admits a wider range of models but still favors models with a cold and stiff mantle or a subset of the FeS-layer models. It is likely that models with an FeS layer at the bottom of the mantle and high mantle rigidity ($\mu_U = 71$ GPa) would also be compatible with k_2 measurements, but there are questions about the plausibility of such a layer (Knibbe and van Westrenen, 2015; Section 4.7.4).

The conclusion drawn from the modeling of the tidal Love number seems robust with respect to details of the thermal model. For instance, the inclusion of a surficial regolith layer with low thermal conductivity increases the temperature in the interior, which results in larger k_2 model values and further favors a cold and stiff mantle. The use of a higher solidus temperature and $T_b > 1850$ K would also strengthen the conclusion that Mercury's mantle is likely cold and stiff. Unfortunately, the robustness of the conclusion is undermined because of the large standard deviations associated with the modeled k_2 values and because the actual k_2 value may extend beyond the range given by the one-standard-deviation uncertainties. The overlap in simulated k_2 values for the five mantle models implies that even a more precise k_2 measurement would not be sufficient to identify a unique model at this time. However, a reduction in uncertainties of both the measured Love number and the moments of inertia would narrow the range of mantle models that are compatible with observations.

4.9 INFLUENCE OF SOLID INNER CORE

Torques between layers in Mercury's interior can influence the spin state. Peale et al. (2014) derived the behavior of Mercury's spin axis orientation under the influence of a variety of core–mantle torques, including gravitational, tidal, magnetic, topographic, viscous, and pressure torques. They showed that tidal torques are small in comparison with magnetic and topographic torques, which are themselves small compared with viscous torques. These dissipative torques would drive the mantle spin

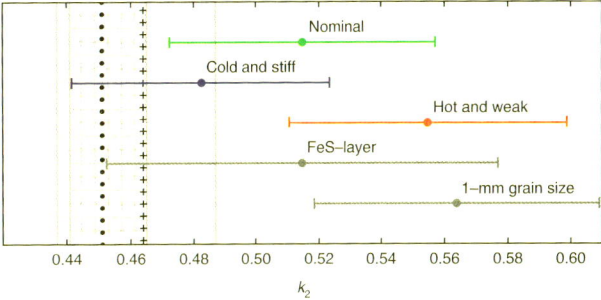

Figure 4.10. Calculated values of the tidal Love number k_2 for five models constructed under different assumptions about the rheological properties and physical structure of the outer solid shell of Mercury (Table 4.3). The vertical lines and hatch patterns represent two independent determinations of k_2 and associated one-standard-deviation uncertainties measured by radio tracking of the MESSENGER spacecraft. The vertical dot symbols correspond to the results of Mazarico et al. (2014), and the plus symbols correspond to the results of Verma and Margot (2016).

away from the Cassini state if it were not for the action of a pressure torque between the outer core and the mantle. The pressure torque is due to fluid pressure at the core–mantle boundary, which is not spherically symmetric because of its hydrostatic, approximately ellipsoidal shape. In the absence of an inner core, the pressure torque would dominate the spin axis evolution and drive the mantle spin close to the Cassini state position.

Peale et al. (2016) considered the additional torques due to an inner core. Their theoretical formalism is general and applicable to other planets, including Earth. The shape of the inner core is distorted by the non-radial gravitational field, and a gravitational torque between the inner core and mantle develops. The relationship between the observed obliquity and the moment of inertia (Equation 4.5), which is based on solar torques, must be modified to account for this additional torque. If the inner core is small ($R_{ic}/R < 0.35$), the mantle spin follows the Cassini state orientation sufficiently closely that the moment of inertia determination is not compromised. However, if the inner core radius exceeds 35% of the planetary radius, the additional torque would drag the mantle spin away from the Cassini state by an amount that exceeds the current observational uncertainty of 5 arcseconds, and the polar moment of inertia would have to be reevaluated. In the presence of an inner core, the obliquity of the mantle spin axis corresponds to a smaller polar moment of inertia than that inferred from the situation with no inner core. This change in the value of the moment of inertia can be evaluated for a variety of interior models by tracking the evolution of the spin under the action of all relevant torques and enforcing the requirement that the mantle spin axis orientation remains within the uncertainty region of the radar observations. Peale et al. (2016) performed this calculation for a variety of inner core sizes and inner core densities. They found that the required adjustment to the value of the moment of inertia increases with both inner core density and inner core size. For an inner core density of 9300 kg m^{-3}, they found corrected values of $C/MR^2 = 0.346, 0.343, 0.330, 0.327,$

and 0.323 for inner core sizes of $R_{ic}/R = 0.0, 0.3, 0.4, 0.5$, and 0.6, respectively (Peale et al., 2016). Because 85% of the best-fit interior models (Section 4.10) have inner core densities below 9300 kg m^{-3}, the corrections identified by Peale et al. (2016) likely represent upper bounds on any necessary adjustment to the moment of inertia due to the presence of an inner core.

Because of the possible impact of an inner core on the determination of Mercury's moment of inertia (Dumberry et al., 2013; Peale et al., 2016), it is important to place bounds on the size of the inner core. We review six lines of evidence: (1) Peale et al. (2016) found that in models with inner cores larger than $R_{ic}/R = 0.3$, the inferred mantle densities were only ~3000 kg m^{-3}. Such low mantle densities are difficult to explain because they are lower than those of materials that likely dominate Mercury's Fe-poor interior, such as Mg-rich olivine and Mg-rich orthopyroxene, which have densities of 3200–3300 kg m^{-3}. If the density information is a reliable indicator, the calculations of Peale et al. (2016) suggest an inner core size $R_{ic}/R \leq 0.3$. (2) A similar conclusion can be reached by examining the distribution of internal structure models. We find that 65% of models that provide the best fit to existing observations (Section 4.10) have a small inner core ($R_{ic}/R < 0.35$). (3) Independent constraints on inner core size arise from the fact that planetary contraction is due in part to inner core solidification. The observed planetary contraction of 7 km (Byrne et al., 2014; Chapter 10) suggests that the inner core size does not exceed 800–1000 km, i.e., $R_{ic}/R \lesssim 0.4$ (Grott et al., 2011; Dumberry and Rivoldini, 2015). Knibbe and van Westrenen (2015) found upper bounds as large as $R_{ic}/R \lesssim 0.7$ for certain values of model parameters, but they did not consider the effects of mantle contraction, which may account for about half of the planetary contraction (Tosi et al., 2013). (4) Simulations of Mercury's magnetic field provide another indicator about the size of the inner core. Dynamo models that can reproduce the observed features of Mercury's magnetic field (Cao et al., 2014) favor small inner cores ($R_{ic}/R_{oc} < 0.5$, i.e., $R_{ic}/R < 0.4$). (5) Dumberry and Rivoldini (2015) further argued that, in some situations, the dynamics of snow formation in the fluid core would place an upper limit on the inner core radius of 650 km ($R_{ic}/R < 0.27$). (6) Finally, several authors have noted that a large inner core ($R_{ic}/R > 0.4$) would produce detectable signatures in the librations of the planet (Veasey and Dumberry, 2011; Dumberry, 2011; Van Hoolst et al., 2012), but such signatures have not been detected to date. There is considerable interest in improving measurements of the longitudinal librations in an attempt to place bounds on the size of Mercury's inner core (Veasey and Dumberry, 2011; Dumberry, 2011; Van Hoolst et al., 2012), although it is not clear that the precision of the current measurement techniques would enable a detection of the inner core signature.

To summarize, there is some circumstantial evidence that Mercury's inner core is small ($R_{ic}/R \lesssim 0.35$) and that the existing estimate of $C/MR^2 = 0.346 \pm 0.009$ remains valid. However, no direct measurements of the inner core size exist, which reduces our confidence in the knowledge of Mercury's moment of inertia. Improved measurements of the librations or direct measurements of the inner core size will be required to eliminate the uncertainty. One approach would be to deploy seismometers on the surface and measure seismic signals triggered by tides, internal activity, explosive charges, or impacts.

4.10 REPRESENTATIVE MODEL

The observational evidence from spin, tidal, and compositional observations, summarized in Table 4.4, favors a Mercury interior model with a core composition dominated by Fe–Si and with a small or no solid FeS layer. Therefore, models in which the core is treated as an Fe–Si end-member are likely representative of Mercury's interior.

Our preferred models include bounds on crustal thickness and density. Analyses of gravity-to-topography ratios suggest an average crustal thickness of 35 ± 18 km (Padovan et al., 2015) and >38 km (James et al., 2015). We combine these bounds into a preferred crustal thickness in the range 35–53 km (Table 4.4). The grain density of crustal material can be determined from a normative mineralogy, which itself is guided by observations of elemental abundances at the surface of Mercury (Weider et al., 2014). With this approach, Padovan et al. (2015) obtained grain densities of 3014 kg m^{-3} and 3082 kg m^{-3} for the northern smooth plains and for heavily cratered terrain and intercrater plains, respectively. If we take into account porosity values of up to 12% as observed on the Moon (Wieczorek et al., 2013), our preferred crustal densities are in the range 2700–3100 kg m^{-3} (Table 4.4).

We updated the analysis of Hauck et al. (2013) to conform to the radius and density values listed in Table 4.4. In addition, we specified an initial crustal thickness in the range 0–70 km, a crustal density in the range 2700–3100 kg m^{-3}, and a core Si content in the range 0–17 wt%. This analysis yielded 1,016,236 Fe–Si interior models with considerable scatter in structural properties. From these models, one can extract a random sample of models for which the distributions of C/MR^2 and C_{m+cr}/C values match the observed values and corresponding one-standard-deviation uncertainties (Table 4.4). We further restricted the set of preferred models to those that provide the closest agreement to the observed values of C/MR^2 and C_{m+cr}/C. All 1479 models in this subset have RMS < 0.005, where the RMS metric is described by equation (4.32). These 1479 best-fit models constitute a family of representative models that can be used to illustrate the remaining scatter in the values of Mercury's internal structure parameters (Table 4.5). Among the subset of models that provide the closest match to observational data, the radius of Mercury's core, $R_{oc} = 2024 \pm 9$ km, is determined with <0.5% precision and represents 83% of the radius of the planet.

We describe an example among the 1479 models in some detail (Table 4.5 and Figure 4.11). This model is representative in the sense that its structural properties match Mercury's mass, radius, and moments of inertia, as well as our preferred bounds on crustal thickness and density. However, we emphasize that Mercury's inner core properties are unknown. The inner core properties of the chosen model are therefore illustrative and not representative. We also emphasize that our chosen model is no

Table 4.4. *Summary of observational constraints used for the calculation of internal structure models.*

Parameter	Symbol	Value	Uncertainty	Unit
Mass	M	3.301110	0.00015	10^{23} kg
Radius	R	2439.36	0.02	km
Density	ρ	5429.30	0.28	kg m^{-3}
Gravity spherical harmonic	C_{20}	−5.0323	0.0022	10^{-5}
Gravity spherical harmonic	C_{22}	0.8039	0.0006	10^{-5}
Tidal Love number	k_2	0.455	0.012	
Obliquity	θ	2.036	0.058	arcminutes
Amplitude of longitude librations	ϕ_0	38.7	1.0	arcseconds
Moment of inertia factor	C/MR^2	0.346	0.009	
Moment of inertia of mantle and crust	C_{m+cr}/C	0.425	0.016	
Crustal thickness	h_{cr}	~35–53		km
Crustal density	ρ_{cr}	~2700–3100		kg m^{-3}

Notes: The first eight values are direct measurements. The remaining four values are derived quantities that rely on a variety of assumptions. These assumptions, described below, are justified considering the data obtained to date and our knowledge of terrestrial planets. However, additional data are required to fully verify the validity of some of these assumptions. Moment of inertia assumptions: (1) Mercury is in Cassini state 1, (2) core does not follow mantle on the 88-day timescale of longitude librations, (3) core does follow mantle on the 300 000-year timescale of orbital precession, (4) $R_{ic}/R < 0.35$. Crustal thickness assumptions: (1) filtering of gravity and topography data is effective in isolating the crustal signal, (2) compensation of topography is well approximated by Airy isostasy. Crustal density assumptions: (1) elemental abundances derived from X-ray fluorescence measurements sampling the uppermost 100 μm of the surface are applicable to the entire crust, (2) normative mineralogy derived from elemental abundances correctly captures crustal minerals, (3) porosity of the crust does not exceed 12%.

better than any other model that fits the observational data. The model does have desirable structural properties, and, as such, it may be useful for a variety of modeling tasks. We refer to this model as the Preliminary Reference Mercury Model (PRMM).

In PRMM, Mercury's mass is divided among inner core (0.5%), outer core (73.4%), mantle (23.5%), and crust (2.5%). The central pressure is 35.77 GPa, and the pressure at the core–mantle boundary is 5.29 GPa. Table 4.6 lists the parameters that we used to construct PRMM. PRMM values are available at https://escholarship.org/uc/item/6xk8m1wx.

PRMM was constructed with the benefit of Earth-based and MESSENGER observations that were not available in earlier modeling efforts. Salient differences between PRMM and pre-MESSENGER models include narrower ranges of admissible structural parameter values compared with the ranges considered by Harder and Schubert (2001), Van Hoolst and Jacobs (2003), and Riner et al. (2008) and a core size that is substantially larger than the core sizes assumed by Siegfried and Solomon (1974; 1660–1900 km), Stevenson et al. (1983; 1840–1900 km), Spohn et al. (2001; 1860 ± 80 km), and Breuer et al. (2007; 1900 km).

4.11 IMPLICATIONS

4.11.1 Thermal Evolution

An accurate understanding of Mercury's thermal evolution requires knowledge of the internal structure, because interior properties dictate the processes and boundary conditions that have governed the evolution. The ~400 km thickness of the silicate mantle and crust has wide-ranging implications. The thickness of this layer is a fundamental control on both the vigor and ultimately the longevity of mantle convection (e.g., Michel et al., 2013; Tosi et al., 2013; Chapter 19). The vigor of the convection is described by the Rayleigh number, which is the ratio of buoyancy forces to viscous forces in a fluid and scales as the cube of the thickness of the layer (e.g., Schubert et al., 2001). Pre-MESSENGER models typically invoked a mantle thickness of ~600 km and therefore overestimated the vigor of the convection by a factor of a few. The MESSENGER-derived value enables more accurate calculations. In particular, the thin mantle implies that convection in Mercury's mantle has been less vigorous than previously thought and may have completely ceased if the Rayleigh number fell below the critical value for convection. A detailed analysis of Mercury's thermal evolution is given in Chapter 19.

4.11.2 Surface Geology

Volcanism is intimately tied to mantle convection because decompression melting is the primary source of magma in terrestrial planets. Mercury's crust, which is the product of perhaps the most efficient crustal extraction among the inner planets (James et al., 2015; Padovan et al., 2015), was dominantly generated early in the planet's history when radiogenic heat production was higher (Chapter 19). Mercury's thin mantle limits the amount of heat transfer because of the reduced vigor

Table 4.5. *Statistical properties of interior structure model parameters and corresponding PRMM values.*

Parameter	Minimum	1st quartile	Median	3rd quartile	Maximum	Mean	Std. dev.	PRMM
C/MR^2	0.34430	0.34523	0.34596	0.34670	0.34771	0.34597	0.00089	0.34573
C_{m+cr}/C	0.42294	0.42418	0.42496	0.42578	0.42712	0.42497	0.00102	0.42482
R_{ic}	0.01877	310.780	623.280	1003.60	1790.82	666.577	420	369.433
R_{oc}	2009.31	2016.69	2021.30	2029.62	2062.56	2023.66	9.09	2015.48
R_m	2369.37	2385.60	2401.37	2419.32	2439.35	2402.61	19.9	2401.20
ρ_{ic}	7368.25	8295.31	8659.58	8991.33	10 214.90	8652.52	488	8215.62
ρ_{oc}	5937.29	6775.76	7010.49	7087.14	7187.97	6909.98	237	7109.73
ρ_m	3206.19	3288.90	3333.75	3388.10	3593.18	3343.35	71.8	3278.98
ρ_{cr}	2700.28	2807.00	2898.57	3006.28	3099.78	2903.03	116	2979.19
ρ_{ic+oc}	6671.42	6976.74	7053.32	7102.67	7190.40	7034.32	88.3	7116.54
ρ_{m+cr}	3198.01	3255.43	3286.49	3327.32	3531.21	3295.84	53.0	3247.21
ρ	5428.34	5429.11	5429.30	5429.52	5430.53	5429.32	0.31	5429.66
M_{ic}	2.588×10^8	1.101×10^{21}	8.962×10^{21}	3.582×10^{22}	1.773×10^{23}	2.288×10^{22}	2.95×10^{22}	1.735×10^{21}
M_{oc}	6.728×10^{22}	2.084×10^{23}	2.351×10^{23}	2.428×10^{23}	2.446×10^{23}	2.213×10^{23}	2.93×10^{22}	2.423×10^{23}
M_m	6.964×10^{22}	7.464×10^{22}	7.789×10^{22}	8.152×10^{22}	8.631×10^{22}	7.813×10^{22}	4.15×10^{21}	7.771×10^{22}
M_{cr}	1.998×10^{18}	4.319×10^{21}	8.020×10^{21}	1.147×10^{22}	1.567×10^{22}	7.822×10^{21}	4.21×10^{21}	8.368×10^{21}
M_{ic+oc}	2.432×10^{23}	2.439×10^{23}	2.441×10^{23}	2.444×10^{23}	2.454×10^{23}	2.442×10^{23}	3.95×10^{20}	2.441×10^{23}
M_{m+cr}	8.484×10^{22}	8.583×10^{22}	8.611×10^{22}	8.639×10^{22}	8.702×10^{22}	8.609×10^{22}	3.92×10^{20}	8.622×10^{22}
M	3.301×10^{23}	3.301×10^{23}	3.301×10^{23}	3.301×10^{23}	3.302×10^{23}	3.301×10^{23}	1.93×10^{19}	3.301×10^{23}

Notes: Statistical properties of structural parameters of 1479 best-fit models (see text) extracted from about a million models of Mercury's interior generated with the method of Hauck et al. (2013). All of these models incorporate an Fe–Si core composition but no solid FeS layer. Masses, radii, and densities are expressed in kg, km, and kg m^{-3}, respectively. Symbols are defined in Figure 4.7. The last column describes a representative model, PRMM, with desirable structural properties. Values for the inner core in PRMM are illustrative only.

Table 4.6. *Parameters used to construct PRMM.*

Parameter	Symbol	Value	Units
Mass fraction of Si (below R_{oc})	χ_{Si}	12.83	wt%
Liquid Fe reference density	$\rho_{0,Fe,l}$	6471.29	kg m^{-3}
Liquid Fe coefficient of thermal expansion	$\alpha_{0,Fe,l}$	9.2×10^{-5}	K^{-1}
Liquid Fe bulk modulus	$K_{0,Fe,l}$	115.47	GPa
Liquid Fe pressure derivative of bulk modulus	$K'_{0,Fe,l}$	4.93	
Solid γ Fe reference density	$\rho_{0,Fe,s}$	7381.34	kg m^{-3}
Solid γ Fe coefficient of thermal expansion	$\alpha_{0,Fe,s}$	6.4×10^{-5}	K^{-1}
Solid γ Fe bulk modulus	$K_{0,Fe,s}$	190.73	GPa
Solid γ Fe pressure derivative of bulk modulus	$K'_{0,Fe,s}$	5.62	

Notes: Temperature-dependent parameters are calculated for the value of the temperature at the core–mantle boundary $T_{cmb} = 1945$ K.

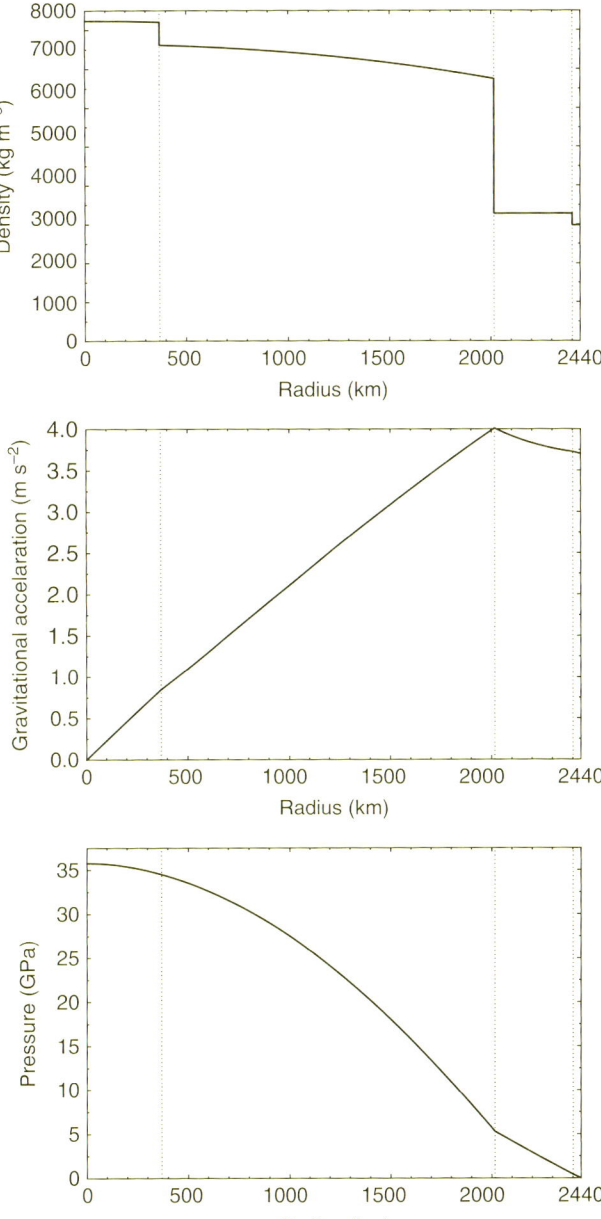

Figure 4.11. Illustration of the density, gravitational acceleration, and pressure corresponding to a model of Mercury's interior that closely matches the mass, radius, and moments of inertia of Mercury (PRMM). This model also matches our preferred bounds on crustal thickness and density (Table 4.4). It incorporates an Fe–Si core composition but has no solid FeS layer. Inner core properties are merely illustrative and not representative. Vertical dotted lines indicate transitions with increasing radius between inner and outer core, core and mantle, and mantle and crust.

of convection and a possible transition to conduction (Section 4.11.1). The reduced heat transfer lowers the amount of volcanism, cooling, and ensuing global contraction, all of which affect the geological evolution of the surface. In particular, the reduced heat transfer hypothesis is consistent with observations of limited volcanism in the past ~3.5 billion years and an amount of radial contraction accommodated by thrust faulting of no more than 7 km (Byrne et al., 2014; Chapters 10 and 19). Tectonic patterns observed at the surface may be due to the interplay of tidal despinning and global contraction (Chapter 10). Surface composition is also affected by mantle thickness, because the horizontal scale of convection cells is similar to the thickness of the convecting layer. Investigations of the source regions of surface volcanic material indicate at least two separate sources (Charlier et al., 2013), consistent with limited mixing of the mantle due to the small horizontal scales and limited vigor of convection (Chapter 19).

4.11.3 Capture in 3:2 Resonance

Mercury's distinctive 3:2 spin–orbit resonance was established at least in part because of Mercury's internal structure. The structure of the interior and the nature of the interactions among inner core, outer core, and mantle have a profound influence on the evolution of the spin state and the response of the planet to external forces and torques. These processes dictate the overall tectonic and insolation regimes, which, in turn, have wide-ranging implications for a variety of questions related to Mercury's shape, surface geology, thermal regime, and even the presence of polar ice deposits.

The history of Mercury's spin–orbit configurations has been markedly affected by the presence of a liquid core. It has been suggested that increased energy dissipation at a core–mantle interface would have led to near-certain capture in specific spin–orbit resonances (Goldreich and Peale, 1967; Counselman and Shapiro, 1970; Peale, 1988), although some models indicate 100% capture probability in the 2:1 resonance (Peale and Boss, 1977), which would prevent evolution to the current configuration. A solution to this problem was found by Correia and Laskar (2004), who showed that chaotic variations in orbital eccentricity destabilize most spin–orbit resonances and ultimately lead to a 55% capture probability in the 3:2 resonance. After spin state observations revealed Mercury's liquid core, Correia and Laskar (2009) added core–mantle friction to their model. They found capture probabilities of 32% (2:1), 26% (3:2), and 22% (5:2). While capture in the 3:2 spin–orbit configuration is not the most probable, the specific outcome depends on the particular realization of orbital eccentricity evolution that Mercury experienced. The capture probability can be increased either if Mercury's eccentricity reached very low values (Correia and Laskar, 2009) or if Mercury started in a retrograde spin configuration and became locked in a synchronous state that was later destabilized by large impacts (Wieczorek et al., 2012). Core–mantle friction also affects Mercury's obliquity evolution, which itself can affect resonance capture probabilities (Correia and Laskar, 2010).

The capture probability results depend on the choice of the tidal torque formulation, which often relies on assumptions of constant time lag or constant phase lag. Models that incorporate a different formulation based on a Darwin–Kaula expansion of the tidal torque yield different capture probabilities (Makarov, 2012; Noyelles et al., 2014) from models that rely on a formulation with constant time lag or constant phase lag. The model of Makarov (2012) predicts 100% capture probability in the 3:2 resonance but does not include orbital eccentricity variations. The model of Noyelles et al. (2014) predicts capture in a 2:1 or higher resonance unless Mercury was captured in the 3:2 resonance early in its evolution, i.e., before differentiation was complete. However, Correia and Laskar (2012) argued that large collisions destabilized all spin–orbit resonances experienced early in Mercury's history and that orbital eccentricity evolution dictated the final outcome. According to Correia and Laskar (2012), the most probable outcome (~50%) is capture in the 3:2 resonance, regardless of the details of the tidal formulation, core–mantle friction formulation, or collisional history.

Estimates for the timing of capture in the 3:2 resonance range from very early (i.e., before differentiation was complete, Noyelles et al., 2014) to very late (i.e., 10^9 yr after formation, Tosi et al., 2015).

4.11.4 Magnetic Field Generation

Knowledge of Mercury's internal structure played a key role in solving a long-standing puzzle related to the origin of the magnetic field. The field that was detected by Mariner 10 (Ness et al., 1974) appeared to have an orientation similar to that of the spin axis. For many years, a dynamo mechanism involving motion in an electrically conducting molten outer core was the preferred explanation for the origin of the field (Ness et al., 1975; Stevenson, 1983), but alternative theories that do not require a currently liquid core, such as remanent magnetism in the crust, could not be ruled out (Stephenson, 1976; Aharonson et al., 2004). Because an active dynamo was not the only possible mechanism for producing the observed field, the detection of the magnetic field left the nature of Mercury's core uncertain. The unambiguous dynamical evidence provided by libration measurements (Section 4.4.3) indicated that Mercury's outer librating shell is decoupled from the deep interior and that Mercury's outer core must be molten. Because a liquid core is a necessary condition for dynamo action, the case for a currently active dynamo was strengthened by the spin state observations. Magnetic field observations from MESSENGER's first two flybys could not be unambiguously attributed to a dynamo mechanism (Anderson et al., 2008, 2010). After orbital insertion, however, the case for a deep dynamo gradually became incontrovertible (Anderson et al., 2012; Chapter 5).

Stevenson (1983, 2010) showed that the existence of convection in a partially molten core, rather than the vigor of that convection, is the primary determinant of dynamo action. He estimated that a fluid layer thickness of order 100 km or more is required for sustaining convection by compositional buoyancy in Mercury. Given the ~2000 km radius of the fluid outer core determined by Mercury's moments of inertia, a convecting layer of sufficient depth can be easily accommodated. If it were not, the signature of an enormous inner core would be detectable (Section 4.9). The lack of information about the size of Mercury's inner core prevents a thorough investigation of the working of the dynamo responsible for Mercury's magnetic field. Measurement of the inner core size is therefore an important goal for future investigations. Detailed discussions of Mercury's magnetic field and models for the generation of that field over the history of the planet are given in Chapters 5 and 19.

4.12 CONCLUSIONS

We have reviewed Mercury's rotational dynamics (Section 4.2) and showed how gravity (Section 4.3) and spin (Section 4.4) observations can provide powerful bounds on Mercury's internal structure. We discussed the results of two- and three-layer structural models (Sections 4.5), which provide a good approximation to the results of more complex models.

Additional constraints derived from compositional studies (Section 4.6) enable the development of multi-layer models, which admit a wide range of solutions (Section 4.7). To further constrain the range of possible models, we calculated the tidal response of the planet and compared it with observations of the k_2 Love number (Section 4.8). We examined the influence of an inner core on the spin state and the determination of the moment of inertia (Section 4.9), and we presented circumstantial evidence for a small inner core.

We have described the statistical properties of 1479 interior models that provide the best fit to the moment of inertia data. We also described a Preliminary Reference Mercury Model that incorporates all existing constraints, including constraints on crustal density and thickness (Section 4.10). The description of radial profiles of density, gravitational acceleration, and pressure will prove useful for a variety of modeling tasks.

We have discussed the wide-ranging implications of Mercury's internal structure on its thermal evolution, surface geology, capture in its distinctive spin–orbit resonance, and magnetic field generation (Section 4.11).

Peale (1976)'s ingenious procedure to determine the size and state of Mercury's core permeates this work. His insight allowed us to quantify the properties of Mercury's core such that, at the time of this writing, we know more about the core of Mercury than that of any planet other than Earth.

Additional observations are necessary to place bounds on the size of Mercury's inner core, either by improved measurements of longitudinal librations or seismological observations. The BepiColombo mission (Novara, 2002, Balogh and Giampieri, 2002; Jehn et al., 2004; Balogh et al., 2007; Benkhoff et al., 2010; Pfyffer et al., 2011; Cicalò and Milani, 2012; Chapter 20) is expected to improve our knowledge of Mercury's internal structure substantially, as would a future lander mission (Wu et al., 1995).

REFERENCES

Aharonson, O., Zuber, M. T. and Solomon, S. C. (2004). Crustal remanence in an internally magnetized non-uniform shell: A possible source for Mercury's magnetic field? *Earth Planet. Sci. Lett.*, **218**, 261–268, doi:10.1016/S0012-821X(03)00682-4.

Alterman, Z., Jarosch, H. and Pekeris, C. L. (1959). Oscillations of the Earth. *Proc. R. Soc. A*, **252**, 80–95, doi:10.1098/rspa.1959.0138.

Anderson, B. J., Acidla, M. H., Korth, H., Purucker, M. E., Johnson, C. L., Slavin, J. A., Solomon, S. C. and McNutt, R. L., Jr. (2008). The structure of Mercury's magnetic field from MESSENGER's first flyby. *Science*, **321**, 82–85, doi:10.1126/science.1159081.

Anderson, B. J., Acuna, M. H., Korth, H., Slavin, J. A., Uno, H., Johnson, C. L., Purucker, M. E., Solomon, S. C., Raines, J. M., Zurbuchen, T. H., Gloeckler, G. and McNutt, R. L., Jr. (2010). The magnetic field of Mercury. *Space Sci. Rev.*, **152**, 307–339, doi:10.1007/978-1-4419-5901-0_10.

Anderson, B. J., Johnson, C. L., Korth, H., Winslow, R. M., Borovsky, J. E., Purucker, M. E., Slavin, J. A., Solomon, S. C., Zuber, M. T. and McNutt, R. L., Jr. (2012). Low-degree structure in Mercury's planetary magnetic field. *J. Geophys. Res.*, **117**, E00L12, doi:10.1029/2012JE004159.

Anderson, J. D., Colombo, G., Esposito, P. B., Lau, E. L. and Trager, G. B. (1987). The mass, gravity field, and ephemeris of Mercury. *Icarus*, **71**, 337–349, doi:10.1016/0019-1035(87)90033-9.

Archinal, B. A., A'Hearn, M. F., Bowell, E., Conrad, A., Consolmagno, G. J., Courtin, R., Fukushima, T., Hestroffer, D., Hilton, J. L., Krasinsky, G. A., Neumann, G., Oberst, J., Seidelmann, P. K., Stooke, P., Tholen, D. J., Thomas, P. C. and Williams, I. P. (2011). Report of the IAU Working Group on Cartographic Coordinates and Rotational Elements: 2009. *Celest. Mech. Dyn. Astron.*, **109**, 101–135, doi:10.1007/s10569-010-9320-4.

Ash, M. E., Shapiro, I. I. and Smith, W. B. (1971). The system of planetary masses. *Science*, **174**, 551–556, doi:10.1126/science.174.4009.551.

Baland, R.-M., Yseboodt, M., Rivoldini, A. and Van Hoolst, T. (2017). Obliquity of Mercury: Influence of the precession of the pericenter and of tides. *Icarus*, **291**, 136–159, doi:10.1016/j.icarus.2017.03.020.

Balogh, A. and Giampieri, G. (2002). Mercury: The planet and its orbit. *Rep. Prog. Phys.*, **65**, 529–560, doi:10.1088/0034-4885/65/4/202.

Balogh, A., Grard, R., Solomon, S. C., Schulz, R., Langevin, Y., Kasaba, Y. and Fujimoto, M. (2007). Missions to Mercury. *Space Sci. Rev.*, **132**, 611–645, doi:10.1007/978-0-387-77539-5_16.

Benkhoff, J., van Casteren, J., Hayakawa, H., Fujimoto, M., Laakso, H., Novara, M., Ferri, P., Middleton, H. R. and Ziethe, R. (2010). BepiColombo – Comprehensive exploration of Mercury: Mission overview and science goals. *Planet. Space Sci.*, **58**, 2–20, doi:10.1016/j.pss.2009.09.020.

Bills, B. G. and Comstock, R. L. (2005). Forced obliquity variations of Mercury. *J. Geophys. Res.*, **110**, E04006, doi:10.1029/2003JE002116.

Bois, E. (1995). Proposed terminology for a general classification of rotational swing motions of the celestial solid bodies. *Astron. Astrophys.*, **296**, 850–857.

Bois, E. and Rambaux, N. (2007). On the oscillations in Mercury's obliquity. *Icarus*, **192**, 308–317, doi:10.1016/j.icarus.2007.07.015.

Borch, R. S. and Green, H. W., II (1987). Dependence of creep in olivine on homologous temperature and its implications for flow in the mantle. *Nature*, **330**, 345–348, doi:10.1038/330345a0.

Breuer, D., Hauck, S. A., II, Buske, M., Pauer, M. and Spohn, T. (2007). Interior evolution of Mercury. *Space Sci. Rev.*, **132**, 229–260, doi:10.1007/978-0-387-77539-5_4.

Burbine, T. H., McCoy, T. J., Nittler, L. R., Benedix, G. K., Cloutis, E. A. and Dickinson, T. L. (2002). Spectra of extremely reduced assemblages: Implications for Mercury. *Meteorit. Planet. Sci.*, **37**, 1233–1244, doi:10.111141945-5100.2002.tb00892.x.

Byrne, P. K., Klimczak, C., Şengör, A. M. C., Solomon, S. C., Watters, T. R. and Hauck, S. A., II. (2014). Mercury's global contraction much greater than earlier estimates. *Nature Geosci.*, **7**, 301–307, doi:10.1038/ngeo2097.

Cao, H., Aurnou, J. M., Wicht, J., Dietrich, W., Soderlund, K. M. and Russell, C. T. (2014). A dynamo explanation for Mercury's anomalous magnetic field. *Geophys. Res. Lett.*, **41**, 4127–4134 doi:10.1002/2014GL060196.

Cavanaugh, J. F., Smith, J. C., Sun, X., Bartels, A. E., RamosIzquierdo, L., Krebs, D. J., McGarry, J. F., Trunzo, R., NovoGradac, A. M., Britt, J. L., Karsh, J., Katz, R. B., Lukemire, A. T., Szymkiewicz, R., Berry, D. L., Swinski, J. P., Neumann, G. A., Zuber, M. T. and Smith, D. E. (2007). The Mercury Laser Altimeter instrument for the MESSENGER mission. *Space Sci. Rev.*, **131**, 451–479, doi:10.1007/s11214-007-9273-4.

Chabot, N. L., Wollack, E. A., Klima, R. L. and Minitti, M. E. (2014). Experimental constraints on Mercury's core composition. *Earth Planet. Sci. Lett.*, **390**, 199–208, doi:10.1016/j.epsl.2014.01.004.

Charlier, B., Grove, T. L. and Zuber, M. T. (2013). Phase equilibria of ultramafic compositions on Mercury and the origin of the compositional dichotomy. *Earth Planet. Sci. Lett.*, **363**, 50–60, doi:10.1016/j.epsl.2012.12.021.

Chen, B., Li, J. and Hauck, S. A., II. (2008). Non-ideal liquidus curve in the Fe–S system and Mercury's snowing core. *Geophys. Res. Lett.*, **35**, L07201, doi:10.1029/2008GL033311.

Cicalò, S. and Milani, A. (2012). Determination of the rotation of Mercury from satellite gravimetry. *Mon. Not. Roy. Astron. Soc.*, **427**, 468–482, doi:10.1111/j.1365-2966.2012.21919.x.

Colombo, G. (1965). Rotational period of the planet Mercury. *Nature*, **208**, 575, doi:10.1038/208575a0.

Colombo, G. (1966). Cassini's second and third laws. *Astron. J.*, **71**, 891–896, doi:10.1007/978-94-010-3529-3_2.

Correia, A. C. M. and Laskar, J. (2004). Mercury's capture into the 3/2 spin-orbit resonance as a result of its chaotic dynamics. *Nature*, **429**, 848–850, doi:10.1038/nature02609.

Correia, A. C. M. and Laskar, J. (2009). Mercury's capture into the 3/2 spin-orbit resonance including the effect of core-mantle friction. *Icarus*, **201**, 1–11, doi:10.1016/j.icarus.2008.12.034.

Correia, A. C. M. and Laskar, J. (2010). Long-term evolution of the spin of Mercury. I. Effect of the obliquity and core–mantle friction. *Icarus*, **205**, 338–355, doi:10.1016/j.icarus.2009.08.006.

Correia, A. C. M. and Laskar, J. (2012). Impact cratering on Mercury: Consequences for the spin evolution. *Astrophys. J.*, **751**, L43, doi:10.1088/2041-8205/751/2/L43.

Counselman, C. C., III and Shapiro, I. I. (1970). Spin-orbit resonance of Mercury. *Symp. Math.*, **3**, 121–169.

Davies, M. F., Abalakin, V. K., Duncombe, R. L., Masursky, H., Morando, B., Owen, T. C., Seidelmann, P. K., Sinclair, A. T., Wilkins, G. A. and Cross, C. A. (1980). Report of the IAU Working Group on Cartographic Coordinates and Rotational Elements of the Planets and Satellites. *Celest. Mech.*, **22**, 205–230, doi:10.1007/BF01229508.

D'Hoedt, S. and Lemaître, A. (2008). Planetary long periodic terms in Mercury's rotation: A two dimensional adiabatic approach. *Celest. Mech. Dyn. Astron.*, **101**, 127–139, doi:10.1007/s10569-007-9115-4.

D'Hoedt, S., Noyelles, B., Dufey, J. and Lemaitre, A. (2009). Determination of an instantaneous Laplace plane for Mercury's rotation. *Adv. Space Res.*, **44**, 597–603, doi:10.1016/j.asr.2009.05.008.

Dufey, J., Lemaître, A. and Rambaux, N. (2008). Planetary perturbations on Mercury's libration in longitude. *Celest. Mech. Dyn. Astron.*, **101**, 141–157, doi:10.1007/s10569-008-9143-8.

Dumberry, M. (2011). The free librations of Mercury and the size of its inner core. *Geophys. Res. Lett.*, **38**, L16202, doi:10.1029/2011GL048277.

Dumberry, M. and Rivoldini, A. (2015). Mercury's inner core size and core-crystallization regime. *Icarus*, **248**, 254–268, doi:10.1016/j.icarus.2014.10.038.

Dumberry, M., Rivoldini, A., Van Hoolst, T. and Yseboodt, M. (2013). The role of Mercury's core density structure on its longitudinal librations. *Icarus*, **225**, 62–74, doi:10.1016/j.icarus.2013.03.001.

Dziewonski, A. M. and Anderson, D. L. (1981). Preliminary reference Earth model. *Phys. Earth Planet. Inter.*, **25**, 297–356, doi:10.1016/0031-9201(81)90046-7.

Efroimsky, M. (2012). Bodily tides near spin-orbit resonances. *Celest. Mech. Dyn. Astron.*, **112**, 283–330, doi:10.1007/s10569-011-9397-4.

Efroimsky, M. and Lainey, V. (2007). Physics of bodily tides in terrestrial planets and the appropriate scales of dynamical evolution. *J. Geophys. Res.*, **112**, E12003, doi:10.1029/2007JE002908.

Evans, L. G., Peplowski, P. N., Rhodes, E. A., Lawrence, D. J., McCoy, T. J., Nittler, L. R., Solomon, S. C., Sprague, A. L., Stockstill-Cahill, K. R., Starr, R. D., Weider, S. Z., Boynton, W. V., Hamara, D. K. and Goldsten, J. O. (2012). Major-element abundances on the surface of Mercury: Results from the MESSENGER Gamma-Ray Spectrometer. *J. Geophys. Res.*, **117**, EOOL07, doi:10.1029/2012JE004178.

Fei, Y., Prewitt, C. T., Mao, H.-K. and Bertka, C. M. (1995). Structure and density of FeS at high pressure and high temperature and the internal structure of Mars. *Science*, **268**, 1892–1894, doi:10.1126/science.268.5219.1892.

Genova, A., Iess, L. and Marabucci, M. (2013). Mercury's gravity field from the first six months of MESSENGER data. *Planet. Space Sci.*, **81**, 55–64, doi:10.1016/j.pss.2013.02.006.

Gladman, B., Dane Quinn, D., Nicholson, P. and Rand, R. (1996). Synchronous locking of tidally evolving satellites. *Icarus*, **122**, 166–192, doi:10.1006/icar.1996.0117.

Goldreich, P. and Peale, S. (1966). Spin-orbit coupling in the solar system. *Astron. J.*, **71**, 425–438, doi:10.1086/109947.

Goldreich, P. and Peale, S. (1967). Spin-orbit coupling in the solar system. II. The resonant rotation of Venus. *Astron. J.*, **72**, 662–668, doi:10.1086/110289.

Grott, M., Breuer, D. and Laneuville, M. (2011). Thermo-chemical evolution and global contraction of Mercury. *Earth Planet. Sci. Lett.*, **307**, 135–146, doi:10.1016/j.epsl.2011.04.040.

Harder, H. and Schubert, G. (2001). Sulfur in Mercury's core? *Icarus*, **151**, 118–122, doi:10.1006/icar.2001.6586.

Hauck, S. A., II, Solomon, S. C. and Smith, D. A. (2007). Predicted recovery of Mercury's internal structure by MESSENGER. *Geophys. Res. Lett.*, **34**, L18201, doi:10.1029/2007GL030793.

Hauck, S. A., II, Margot, J. L., Solomon, S. C., Phillips, R. J., Johnson, C. L., Lemoine, F. G., Mazarico, E., McCoy, T. J., Padovan, S., Peale, S. J., Perry, M. E., Smith, D. E. and Zuber, M. T. (2013). The curious case of Mercury's internal structure. *J. Geophys. Res. Planets*, **118**, 1204–1220, doi:10.1002/jgre.20091.

Hawkins, S. E., III, Boldt, J. D., Darlington, E. H., Espiritu, R., Gold, R. E., Gotwols, B., Grey, M. P., Hash, C. D., Hayes, J. R., Jaskulek, S. E., Kardian, C. J., Keller, M. R., Malaret, E. R., Murchie, S. L., Murphy, P. K., Peacock, K., Prockter, L. M., Reiter, R. A., Robinson, M. S., Schaefer, E. D., Shelton, R. G., Sterner, R. E., II, Taylor, H. W., Watters, T. R. and Williams, B. D. (2007). The Mercury Dual Imaging System on the MESSENGER spacecraft. *Space Sci. Rev.*, **131**, 247–338, doi:10.1007/s11214-007-9266-3.

Hirschmann, M. M. (2000). Mantle solidus: Experimental constraints and the effects of peridotite composition. *Geochem. Geophys. Geosyst.*, **1**, 1042–1068, doi:10.1029/2000GC000070.

Hofmeister, A. M. and Mao, H. K. (2003). Pressure derivatives of shear and bulk moduli from the thermal Gruneisen parameter and volume-pressure data. *Geochim. Cosmochim. Acta*, **67**, 1207–1227, doi:10.1016/S0016-7037(02)01289-9.

Holin, I. V. (1988). Space-time coherence of a signal diffusely scattered by an arbitrarily moving surface in the case of monochromatic sounding. *Izvestiya Vysshikh Uchebnykh Zavedenii, Radiofizika*, **31**(5), 515–518.

Holin, I. V. (1992). Accuracy of body-rotation-parameter measurement with monochromatic illumination and two-element reception. *Izvestiya Vysshikh Uchebnykh Zavedenii, Radiofizika*, **35**(5), 433–439, doi:10.1007/BF01038312.

Howard, H. T., Tyler, G. L., Esposito, P. B., Anderson, J. D., Reasenberg, R. D., Shapiro, I. I., Fjeldbo, G., Kliore, A. J., Levy, G. S., Brunn, D. L., Dickinson, R., Edelson, R. E., Martin, W. L., Postal, R. B., Seidel, B., Sesplaukis, T. T., Shirley, D. L., Stelzried, C. T., Sweetnam, D. N., Wood, G. E. and Zygielbaum, A. I. (1974). Mercury: Results on mass, radius, ionosphere, and atmosphere from Mariner 10 dual-frequency radio signals. *Science*, **185**, 179–180, doi:10.1126/science.185.4146.179.

Iess, L., Jacobson, R. A., Ducci, M., Stevenson, D. J., Lunine, J. I., Armstrong, J. W., Asmar, S. W., Racioppa, P., Rappaport, N. J. and Tortora, P. (2012). The tides of Titan. *Science*, **337**, 457–459, doi:10.1126/science.1219631.

Jackson, I., Faul, U. H., Suetsugu, D., Bina, C., Inoue, T. and Jellinek, M. (2010). Grainsize-sensitive viscoelastic relaxation in olivine: Towards a robust laboratory-based model for seismological application. *Phys. Earth Planet. Inter.*, **183**, 151–163, doi:10.1016/j.pepi.2010.09.005.

James, P. B., Zuber, M. T., Phillips, R. J. and Solomon, S. C. (2015). Support of long-wavelength topography on Mercury inferred from MESSENGER measurements of gravity and topography. *J. Geophys. Res. Planets*, **120**, 287–310, doi:10.1002/2014JE004713.

Jehn, R., Corral, C. and Giampieri, G. (2004). Estimating Mercury's 88-day libration amplitude from orbit. *Planet. Space Sci.*, **52**, 727–732, doi:10.1016/j.pss.2003.12.012.

Kaufmann, G. and Lambeck, K. (2000). Mantle dynamics, postglacial rebound and the radial viscosity profile. *Phys. Earth Planet. Inter.*, **121**, 301–324, doi:10.1016/S0031-9201(00)00174-6.

Kaula, W. M. (2000). *Theory of Satellite Geodesy: Applications of Satellites to Geodesy*. Mineola, NY: Dover Publications.

Klaasen, K. P. (1976). Mercury's rotation axis and period. *Icarus*, **28**, 469–478, doi:10.1016/0019-1035(76)90120-2.

Knibbe, J. S. and van Westrenen, W. (2015). The interior configuration of planet Mercury constrained by moment of inertia and planetary contraction. *J. Geophys. Res. Planets*, **120**, 1904–1923, doi:10.1002/2015JE004908.

Koning, A. and Dumberry, M. (2013). Internal forcing of Mercury's long period free librations. *Icarus*, **223**, 40–47, doi:10.1016/j.icarus.2012.11.022.

Konopliv, A. S. and Yoder, C. F. (1996). Venusian k_2 tidal Love number from Magellan and PVO tracking data. *Geophys. Res. Lett.*, **23**, 1857–1860, doi:10.1029/96GL01589.

Kuwayama, Y. and Hirose, K. (2004). Phase relations in the system Fe–FeSi at 21 GPa. *Amer. Mineral.*, **89**, 273–276, doi:10.2138/am-2004-2-303.

Makarov, V. V. (2012). Conditions of passage and entrapment of terrestrial planets in spin-orbit resonances. *Astrophys. J.*, **752**, 73–80, doi:10.1088/0004-637X/752/1/73.

Malavergne, V., Toplis, M. J., Berthet, S. and Jones, J. (2010). Highly reducing conditions during core formation on Mercury: Implications for internal structure and the origin of a magnetic field. *Icarus*, **206**, 199–209, doi:10.1016/j.icarus.2009.09.001.

Margot, J. L. (2009). A Mercury orientation model including nonzero obliquity and librations. *Celest. Mech. Dyn. Astron.*, **105**, 329–336, doi:10.1007/s10569-009-9234-1.

Margot, J. L., Peale, S. J., Jurgens, R. F., Slade, M. A. and Holin, I. V. (2007). Large longitude libration of Mercury reveals a molten core. *Science*, **316**, 710–714, doi:10.1126/science.1140514.

Margot, J. L., Peale, S. J., Solomon, S. C., Hauck, S. A., II, Ghigo, F. D., Jurgens, R. F., Yseboodt, M., Giorgini, J. D., Padovan, S. and Campbell, D. B. (2012). Mercury's moment of inertia from spin and gravity data. *J. Geophys. Res.*, **117**, EOOL09, doi:10.1029/2012JE004161.

Matsuyama, I. and Nimmo, F. (2009). Gravity and tectonic patterns of Mercury: Effect of tidal deformation, spin-orbit resonance, non-zero eccentricity, despinning, and reorientation. *J. Geophys. Res.*, **114**, E01010, doi:10.1029/2008JE003252.

Mazarico, E., Genova, A., Goossens, S., Lemoine, F. G., Neumann, G. A., Zuber, M. T., Smith, D. E. and Solomon, S. C. (2014). The gravity field, orientation, and ephemeris of Mercury from MESSENGER observations after three years in orbit. *J. Geophys. Res. Planets*, **119**, 2417–2436, doi:10.1002/2014JE004675.

McCubbin, F. M., Riner, M. A., Vander Kaaden, K. E. and Burkemper, L. K. (2012). Is Mercury a volatile-rich planet? *Geophys. Res. Lett.*, **39**, L09202, doi:10.1029/2012GL051711.

Michel, N. C., Hauck, S. A., II, Solomon, S. C., Phillips, R. J., Roberts, J. H. and Zuber, M. T. (2013). Thermal evolution of Mercury as constrained by MESSENGER observations. *J. Geophys. Res. Planets*, **118**, 1033–1044, doi:10.1002/jgre.20049.

Mohr, P. J., Newell, D. B. and Taylor, B. N. (2016). CO-DATA recommended values of the fundamental physical constants: 2014. *Rev. Mod. Phys.*, **88**(3), 035009, doi:10.1103/RevModPhys.88.035009.

Moore, W. B. and Schubert, G. (2000). Note: The tidal response of Europa. *Icarus*, **147**, 317–319, doi:10.1006/icar.2000.6460.

Moore, W. B. and Schubert, G. (2003). The tidal response of Ganymede and Callisto with and without liquid water oceans. *Icarus*, **166**, 223–226, doi:10.1016/j.icarus.2003.07.001.

Morard, G. and Katsura, T. (2010). Pressure-temperature cartography of Fe–S–Si immiscible system. *Geochim. Cosmochim. Acta*, **74**, 3659–3667, doi:10.1016/j.gca.2010.03.025.

Morgan, J. W. and Anders, E. (1980). Chemical composition of Earth, Venus, and Mercury. *Proc. Natl. Acad. Sci.*, **77**, 6973–6977, doi:10.1073/pnas.77.12.6973.

Naidu, S. P. and Margot, J. L. (2015). Near-Earth asteroid satellite spins under spin-orbit coupling. *Astron. J.*, **149**, 80–90, doi:10.1088/0004-6256/149/2/80.

Namur, O., Collinet, M., Charlier, B., Grove, T. L., Holtz, F. and McCammon, C. (2016a). Melting processes and mantle sources of lavas on Mercury. *Earth Planet. Sci. Lett.*, **439**, 117–128, doi:10.1016/j.epsl.2016.01.030.

Namur, O., Charlier, B., Holtz, F., Cartier, C. and McCammon, C. (2016b). Sulfur solubility in reduced mafic silicate melts: Implications for the speciation and distribution of sulfur on Mercury. *Earth Planet. Sci. Lett.*, **448**, 102–114, doi:10.1016/j.epsl.2016.05.024.

Ness, N. F., Behannon, K. W., Lepping, R. P., Whang, Y. C. and Schatten, K. H. (1974). Magnetic field observations near Mercury: Preliminary results from Mariner 10. *Science*, **185**, 151–160, doi:10.1126/science.185.4146.151.

Ness, N. F., Behannon, K. W., Lepping, R. P. and Whang, Y. C. (1975). The magnetic field of Mercury. I. *J. Geophys. Res.*, **80**, 2708–2716, doi:10.1017/S1539299600002562.

Nimmo, F. and Faul, U. H. (2013). Dissipation at tidal and seismic frequencies in a melt-free, anhydrous Mars. *J. Geophys. Res. Planets*, **118**, 2558–2569, doi:10.1002/2013JE004499.

Nimmo, F., Faul, U. H. and Garnero, E. J. (2012). Dissipation at tidal and seismic frequencies in a melt-free Moon. *J. Geophys. Res.*, **117**, E09005, doi:10.1029/2012JE004160.

Nittler, L. R., Starr, R. D., Weider, S. Z., McCoy, T. J., Boynton, W. V., Ebel, D. S., Ernst, C. M., Evans, L. G., Goldsten, J. O., Hamara, D. K., Lawrence, D. J., McNutt, R. L., Jr., Schlemm, C. E., Solomon, S. C. and Sprague, A. L. (2011). The major-element composition of Mercury's surface from MESSENGER X-ray spectrometry. *Science*, **333**, 1847–1850, doi:10.1126/science.1211567.

Novara, M. (2002). The BepiColombo ESA cornerstone mission to Mercury. *Acta Astronaut.*, **51**, 387–395, doi:10.1016/S0094-5765(02)00065-6.

Noyelles, B. and Lhotka, C. (2013). The influence of orbital dynamics, shape and tides on the obliquity of Mercury. *Adv. Space Res.*, **52**, 2085–2101, doi:10.1016/j.asr.2013.09.024.

Noyelles, B., Frouard, J., Makarov, V. V. and Efroimsky, M. (2014). Spin-orbit evolution of Mercury revisited. *Icarus*, **241**, 26–44, doi:10.1016/j.icarus.2014.05.045.

Padovan, S., Margot, J. L., Hauck, S. A., II, Moore, B. and Solomon, S. C. (2014). The tides of Mercury and possible implications for its interior structure. *J. Geophys. Res. Planets*, **119**, 850–866, doi:10.1002/2013JE004459.

Padovan, S., Wieczorek, M. A., Margot, J. L., Tosi, N. and Solomon, S. C. (2015). Thickness of the crust of Mercury from geoid-to-topography ratios. *Geophys. Res. Lett.*, **42**, 1029–1038, doi:10.1002/2014GL062487.

Peale, S. J. (1969). Generalized Cassini's laws. *Astron. J.*, **74**, 483–489, doi:10.1086/110825.

Peale, S. J. (1972). Determination of parameters related to the interior of Mercury. *Icarus*, **17**, 168–173, doi:10.1016/0019–1035(72)90052–8.

Peale, S. J. (1976). Does Mercury have a molten core? *Nature*, **262**, 765–766, doi:10.1038/262765a0.

Peale, S. J. (1981). Measurement accuracies required for the determination of a Mercurian liquid core. *Icarus*, **48**, 143–145, doi:10.1016/0019–1035(81)90160–3.

Peale, S. J. (1988). The rotational dynamics of Mercury and the state of its core. In *Mercury*, ed. F. Vilas, C. R. Chapman and M. S. Matthews. Tucson, AZ: University of Arizona Press, pp. 461–493.

Peale, S. J. (2005). The free precession and libration of Mercury. *Icarus*, **178**, 4–18, doi:10.1016/j.icarus.2005.03.017.

Peale, S. J. (2006). The proximity of Mercury's spin to Cassini state 1 from adiabatic invariance. *Icarus*, **181**, 338–347, doi:10.1016/j.icarus.2005.10.006.

Peale, S. J. and Boss, A. P. (1977). A spin-orbit constraint on the viscosity of a Mercurian liquid core. *J. Geophys. Res.*, **82**, 743–749, doi:10.1029/JB082i005p00743.

Peale, S. J., Phillips, R. J., Solomon, S. C., Smith, D. E. and Zuber, M. T. (2002). A procedure for determining the nature of Mercury's core. *Meteorit. Planet. Sci.*, **37**, 1269–1283, doi:10.1111/j.1945-5100.2002.tb00895.x.

Peale, S. J., Yseboodt, M. and Margot, J. L. (2007). Long-period forcing of Mercury's libration in longitude. *Icarus*, **187**, 365–373, doi:10.1016/j.icarus.2006.10.028.

Peale, S. J., Margot, J. L. and Yseboodt, M. (2009). Resonant forcing of Mercury's libration in longitude. *Icarus*, **199**, 1–8, doi:10.1016/j.icarus.2008.09.002.

Peale, S. J., Margot, J. L., Hauck, S. A., II and Solomon, S. C. (2014). Effect of core–mantle and tidal torques on Mercury's spin axis orientation. *Icarus*, **231**, 206–220, doi:10.1016/j.icarus.2013.12.007.

Peale, S. J., Margot, J. L., Hauck, S. A., II and Solomon, S. C. (2016). Consequences of a solid inner core on Mercury's spin configuration. *Icarus*, **264**, 443–455, doi:10.1016/j.icarus.2015.09.024.

Peplowski, P. N., Lawrence, D. J., Rhodes, E. A., Sprague, A. L., McCoy, T. J., Denevi, B. W., Evans, L. G., Head, J. W., Nittler, L. R., Solomon, S. C., Stockstill-Cahill, K. R. and Weider, S. Z. (2012). Variations in the abundances of potassium and thorium on the surface of Mercury: Results from the MESSENGER Gamma-Ray Spectrometer. *J. Geophys. Res.*, **117**, E00L04, doi:10.1029/2012JE004141.

Perry, M. E., Neumann, G. A., Phillips, R. J., Barnouin, O. S., Ernst, C. M., Kahan, D. S., Solomon, S. C., Zuber, M. T., Smith, D. E., Hauck, S. A., II, Peale, S. J., Margot, J. L., Mazarico, E., Johnson, C. L., Gaskell, R. W., Roberts, J. H., McNutt, R. L., Jr. and Oberst, J. (2015). The low-degree shape of Mercury. *Geophys. Res. Lett.*, **42**, 6951–6958, doi:10.1002/2015GL065101.

Pettengill, G. H. and Dyce, R. B. (1965). A radar determination of the rotation of the planet Mercury. *Nature*, **206**, 1240, doi:10.1038/2061240a0.

Poirier, J. P. (2000). *Introduction to the Physics of the Earth*, 2nd edn. Cambridge: Cambridge University Press.

Pfyffer, G., Van Hoolst, T. and Dehant, V. (2011). Librations and obliquity of Mercury from the BepiColombo radio-science and camera experiments. *Planet. Space Sci.*, **59**, 848–861, doi:10.1016/j.pss.2011.03.017.

Rambaux, N., Van Hoolst, T., Dehant, V. and Bois, E. (2007). Inertial core–mantle coupling and libration of Mercury. *Astron. Astrophys.*, **468**, 711–719, doi:10.1051/0004-6361:20053974.

Riner, M. A., Bina, C. R., Robinson, M. S. and Desch, S. J. (2008). Internal structure of Mercury: Implications of a molten core. *J. Geophys. Res.*, **113**, E08013, doi:10.1029/2007JE002993.

Rivoldini, A. and Van Hoolst, T. (2013). The interior structure of Mercury constrained by the low-degree gravity field and the rotation of Mercury. *Earth Planet. Sci. Lett.*, **377**, 62–72, doi:10.1016/j.epsl.2013.07.021.

Rivoldini, A., Van Hoolst, T. and Verhoeven, O. (2009). The interior structure of Mercury and its core sulfur content. *Icarus*, **201**, 12–30, doi:10.1016/j.icarus.2008.12.020.

Robinson, M. S. and Taylor, G. J. (2001). Ferrous oxide in Mercury's crust and mantle. *Meteorit. Planet. Sci.*, **36**, 841–847, doi:10.1111/j.1945-5100.2001.tb01921.x.

Sabadini, R. and Vermeersen, B. (2004). *Global Dynamics of the Earth: Applications of Normal Mode Relaxation Theory to Solid-Earth Geophysics*. Dordrecht, the Netherlands: Kluwer Academic Publishers.

Sanloup, C. and Fei, Y. (2004). Closure of the Fe–S–Si liquid miscibility gap at high pressure. *Phys. Earth Planet. Inter.*, **147**, 57–65, doi:10.1016/j.pepi.2004.06.008.

Schubert, G., Ross, M. N., Stevenson, D. J. and Spohn, T. (1988). Mercury's thermal history and the generation of its magnetic field. In *Mercury*, ed. F. Vilas, C. R. Chapman and M. S. Matthews. Tucson, AZ: University of Arizona Press, pp. 429–460.

Schubert, G., Turcotte, D. L. and Olson, P. (2001). *Mantle Convection in the Earth and Planets*. Cambridge: Cambridge University Press.

Siegfried, R. W., II and Solomon, S. C. (1974). Mercury: Internal structure and thermal evolution. *Icarus*, **23**, 192–205, doi:10.1016/0019-1035(74)90005-0.

Siivola, J. and Schmid, R. (2007). List of mineral abbreviations: Recommendations by the IUGS Subcommission on the Systematics of Metamorphic Rocks. Electronic Source: http://www.bgs.ac.uk/scmr/docs/papers/paper_12.pdf.

Smith, D. E., Zuber, M. T., Phillips, R. J., Solomon, S. C., Neumann, G. A., Lemoine, F. G., Peale, S. J., Margot, J. L., Torrence, M. H., Talpe, M. J., Head, J. W., Hauck, S. A., II, Johnson, C. L., Perry, M. E., Barnouin, O. S., McNutt, R. L., Jr. and Oberst, J. (2010). The equatorial shape and gravity field of Mercury from MESSENGER flybys 1 and 2. *Icarus*, **209**, 88–100, doi:10.1016/j.icarus.2010.04.007.

Smith, D. E., Zuber, M. T., Phillips, R. J., Solomon, S. C., Hauck, S. A., II, Lemoine, F. G., Mazarico, E., Neumann, G. A., Peale, S. J., Margot, J. L., Johnson, C. L., Torrence, M. H., Perry, M. E., Rowlands, D. D., Goossens, S., Head, J. W. and Taylor, A. H. (2012). Gravity field and internal structure of Mercury from MESSENGER. *Science*, **336**, 214–217, doi:10.1126/science.1218809.

Smyth, J. R. and McCormick, T. C. (1995). Crystallographic data for minerals. In *Mineral Physics and Crystallography: A Handbook of Physical Constants*, ed. T. J. Ahrens. Washington, DC: American Geophysical Union, pp. 1–17.

Solomon, S. C., McNutt, R. L., Jr., Gold, R. E., Acuña, M. H., Baker, D. N., Boynton, W. V., Chapman, C. R., Cheng, A. F., Gloeckler, G., Head, J. W., III, Krimigis, S. M., McClintock, W. E., Murchie, S. L., Peale, S. J., Phillips, R. J., Robinson, M. S., Slavin, J. A., Smith, D. E., Strom, R. G., Trombka, J. I. and Zuber, M. T. (2001). The MESSENGER mission to Mercury: Scientific objectives and implementation. *Planet. Space Sci.*, **49**, 1445–1465, doi:10.1016/S0032-0633(01)00085-X.

Spohn, T., Sohl, F., Wieczerkowski, K. and Conzelmann, V. (2001). The interior structure of Mercury: What we know, what we expect from BepiColombo. *Planet. Space Sci.*, **49**, 1561–1570, doi:10.1016/S0032-0633(01)00093-9.

Stark, A., Oberst, J., Preusker, F., Peale, S. J., Margot, J. L., Phillips, R. J., Neumann, G. A., Smith, D. E., Zuber, M. T. and Solomon, S. C. (2015a). First MESSENGER orbital observations of Mercury's librations. *Geophys. Res. Lett.*, **42**, 7881–7889, doi:10.1002/2015GL065152.

Stark, A., Oberst, J. and Hussmann, H. (2015b). Mercury's resonant rotation from secular orbital elements. *Celest. Mech. Dyn. Astron.*, **123**, 263–277, doi:10.1007/s10569-015-9633-4.

Stark, A., Oberst, J., Preusker, F., Gwinner, K., Peale, S. J., Margot, J. L., Phillips, R. J., Zuber, M. T. and Solomon, S. C. (2015c). Mercury's rotational parameters from MESSENGER image and laser altimeter data: A feasibility study. *Planet. Space Sci.*, **117**, 64–72, doi:10.1016/j.pss.2015.05.006.

Stephenson, A. (1976). Crustal remanence and the magnetic moment of Mercury. *Earth Planet. Sci. Lett.*, **28**, 454–458, doi:10.1016/0012-821X(76)90206-5.

Stevenson, D. J. (1983). Planetary magnetic fields. *Rep. Prog. Phys.*, **46**, 555–620, doi:10.1016/50012-821X(02)01126-3.

Stevenson, D. J. (2010). Planetary magnetic fields: Achievements and prospects. *Space Sci. Rev.*, **152**, 651–664, doi:10.1007/978-1-4419-5901-0_20.

Stevenson, D. J., Spohn, T. and Schubert, G. (1983). Magnetism and thermal evolution of the terrestrial planets. *Icarus*, **54**, 466–489, doi:10.1016/0019-1035(83)90241-5.

Taylor, G. J. and Scott, E. R. D. (2003). Mercury. In *Treatise on Geochemistry*, ed. H. D. Holland and K. K. Turekian. Oxford: Pergamon, pp. 477–485.

Tosi, N., Grott, M., Plesa, A.-C. and Breuer, D. (2013). Thermochemical evolution of Mercury's interior. *J. Geophys. Res. Planets*, **118**, 2474–2487, doi:10.1002/jgre.20168.

Tosi, N., Čadek, O., Běhounková, M., Káňová, M., Plesa, A.-C., Grott, M., Breuer, D., Padovan, S. and Wieczorek, M. A. (2015). Mercury's low-degree geoid and topography controlled by insolation-driven elastic deformation. *Geophys. Res. Lett.*, **42**, 7327–7335, doi:10.1002/2015GL065314.

Turcotte, D. L. and Schubert, G. (2002). *Geodynamics*, 2nd edn. Cambridge: Cambridge University Press.

Van Hoolst, T. and Jacobs, C. (2003). Mercury's tides and interior structure. *J. Geophys. Res.*, **108**, 5121–5136, doi:10.1029/2003JE002126.

Van Hoolst, T., Sohl, F., Holin, I., Verhoeven, O., Dehant, V. and Spohn, T. (2007). Mercury's interior structure, rotation, and tides. *Space Sci. Rev.*, **132**, 203–227, doi:10.1007/s11214-007-9202-6.

Van Hoolst, T., Rivoldini, A., Baland, R.-M. and Yseboodt, M. (2012). The effect of tides and an inner core on the forced longitudinal libration of Mercury. *Earth Planet. Sci. Lett.*, **333**, 83–90, doi:10.1016/j.epsl.2012.04.014.

Vander Kaaden, K. E. and McCubbin, F. M. (2016). The origin of boninites on Mercury: An experimental study of the northern volcanic plains lavas. *Geochim. Cosmochim. Acta*, **173**, 246–263, doi:10.1016/j.gca.2015.10.016.

Veasey, M. and Dumberry, M. (2011). The influence of Mercury's inner core on its physical libration. *Icarus*, **214**, 265–274, doi:10.1016/j.icarus.2011.04.025.

Verma, A. K. and Margot, J. L. (2016). Mercury's gravity, tides, and spin from MESSENGER radio science data. *J. Geophys. Res. Planets*, **121**, 1627–1640, doi:10.1002/2016JE005037.

Wasson, J. T. (1988). The building stones of the planets. In *Mercury*, ed. F. Vilas, C. R. Chapman and M. S. Matthews. Tucson, AZ: University of Arizona Press, pp. 622–650.

Watt, J. P., Davies, G. F. and O'Connell, R. J. (1976). The elastic properties of composite materials. *Rev. Geophys. Space Phys.*, **14**, 541–563, doi:10.1029/RG014i004p00541.

Weider, S. Z., Nittler, L. R., Starr, R. D., McCoy, T. J. and Solomon, S. C. (2014). Variations in the abundance of iron on Mercury's surface from MESSENGER X-Ray Spectrometer observations. *Icarus*, **235**, 170–186, doi:10.1016/j.icarus.2014.03.002.

Weider, S. Z., Nittler, L. R., Starr, R. D., Crapster-Pregont, E. J., Peplowski, P. N., Denevi, B. W., Head, J. W., Byrne, P. K., Hauck, S. A., II, Ebel, D. S. and Solomon, S. C. (2015). Evidence for geochemical terranes on Mercury: Global mapping of major elements with MESSENGER's X-Ray Spectrometer. *Earth Planet. Sci. Lett.*, **416**, 109–120, doi:10.1016/j.epsl.2015.01.023.

Wieczorek, M. A., Correia, A. C. M., Le Feuvre, M., Laskar, J. and Rambaux, N. (2012). Mercury's spin–orbit resonance explained by initial retrograde and subsequent synchronous rotation. *Nature Geosci.*, **5**, 18–21, doi:10.1038/ngeo1350.

Wieczorek, M. A., Neumann, G. A., Nimmo, F., Kiefer, W. S., Taylor, G. J., Melosh, H. J., Phillips, R. J., Solomon, S. C., Andrews-Hanna, J. C., Asmar, S. W., Konopliv, A. S., Lemoine, F. G., Smith, D. E., Watkins, M. M., Williams, J. G. and Zuber, M. T. (2013). The crust of the Moon as seen by GRAIL. *Science*, **339**, 671–675, doi:10.1126/science.1231530.

Williams, J. G. (1994). Contributions to the Earth's obliquity rate, precession, and nutation. *Astron. J.*, **108**, 711–724, doi:10.1086/117108.

Williams, J. G., Newhall, X. X. and Dickey, J. O. (1996). Lunar moments, tides, orientation, and coordinate frames. *Planet. Space Sci.*, **44**, 1077–1080, doi:10.1016/0032-0633(95)00154-9.

Williams, J. G., Boggs, D. H., Yoder, C. F., Ratcliff, J. T. and Dickey, J. O. (2001). Lunar rotational dissipation in solid body and molten core. *J. Geophys. Res.*, **106**, 27933–27968, doi:10.1029/2000JE001396.

Wolf, D. (1994). Lamé's problem of gravitational viscoelasticity: The isochemical, incompressible planet. *Geophys. J. Int.*, **116**, 321–348, doi:10.1111/j.1365-246X.1994.tb01801.x.

Wu, X., Bender, P. L. and Rosborough, G. W. (1995). Probing the interior structure of Mercury from an orbiter plus single lander. *J. Geophys. Res.*, **100**, 1515–1525, doi:10.1029/94JE02833.

Yoder, C. F. (1981). The free librations of a dissipative moon. *Phil. Trans. R. Soc. London A*, **303**, 327–338, doi:10.1098/rsta.1981.0206.

Yoder, C. F., Konopliv, A. S., Yuan, D. N., Standish, E. M. and Folkner, W. M. (2003). Fluid core size of Mars from detection of the solar tide. *Science*, **300**, 299–303, doi:10.1126/science.1079645.

Yseboodt, M. and Margot, J. L. (2006). Evolution of Mercury's obliquity. *Icarus*, **181**, 327–337, doi:10.1016/j.icarus.2005.11.024.

Yseboodt, M., Margot, J. L. and Peale, S. J. (2010). Analytical model of the long-period forced longitude librations of Mercury. *Icarus*, **207**, 536–544, doi:10.1016/j.icarus.2009.12.020.

Yseboodt, M., Rivoldini, A., Van Hoolst, T. and Dumberry, M. (2013). Influence of an inner core on the long-period forced librations of Mercury. *Icarus*, **226**, 41–51, doi:10.1016/j.icarus.2013.05.011.

Zolotov, M. Yu., Sprague, A. L., Hauck, S. A., II, Nittler, L. R., Solomon, S. C. and Weider, S. Z. (2013). The redox state, FeO content, and origin of sulfur-rich magmas on Mercury. *J. Geophys. Res. Planets*, **118**, 138–146, doi:10.1029/2012JE004274.

Zuber, M. T., Smith, D. E., Phillips, R. J., Solomon, S. C., Neumann, G. A., Hauck, S. A., II, Peale, S. J., Barnouin, O. S., Head, J. W., Johnson, C. L., Lemoine, F. G., Mazarico, E., Sun, X., Torrence, M. H., Freed, A. M., Klimczak, C., Margot, J. L., Oberst, J., Perry, M. E., McNutt, R. L., Jr., Balcerski, J. A., Michel, N., Talpe, M. J. and Yang, D. (2012). Topography of the northern hemisphere of Mercury from MESSENGER laser altimetry. *Science*, **336**, 217–220, doi:10.1126/science.1218805.

5

Mercury's Internal Magnetic Field

CATHERINE L. JOHNSON, BRIAN J. ANDERSON, HAJE KORTH, ROGER J. PHILLIPS,
AND LYDIA C. PHILPOTT

5.1 INTRODUCTION

The MESSENGER mission to Mercury (Chapter 1) has enabled the global characterization of Mercury's magnetic field, provided insights into the planet's present and past interior states, and completed the first-order picture of the magnetic fields of the planets in the inner solar system. A planetary magnetic field provides key information on a planet's interior structure and thermal evolution and governs the nature of the planet's interaction with the solar wind. Global-scale magnetic fields can be generated in electrically conducting fluid regions within a differentiated body, and the existence of such fields is thus a powerful diagnostic of the interior state. For the terrestrial planets, global magnetic fields provide indirect evidence for an at least partly fluid iron–nickel-rich core and of a present-day thermal state that supports a self-sustaining dynamo. On geological timescales, the dynamo history can be recorded in magnetized rocks of the crust and/or the upper mantle. The net magnetization reflects the combined effects of the field in which an original magnetic remanence was acquired, the magnetic mineralogy, and the physical and chemical conditions to which the rocks were subsequently subjected. Magnetization can be measured directly in the laboratory or detected remotely via the magnetic fields it generates.

Global planetary magnetic fields interact with the magnetized solar wind, resulting in a region in which the planetary field is confined, the *magnetosphere*, bordered by the *magnetopause*, a boundary carrying an electric current and defined by the balance of planetary field pressure and solar wind dynamic pressure (Chapter 16). The geometry of, and time variations in, the magnetopause current layer provide information on the planet's internal field geometry and strength and on how the magnetosphere responds to changes in solar wind conditions. The electric currents along the magnetopause and current systems within the magnetosphere create magnetic fields (Ampere's law), and the time variations of such fields induce secondary electromagnetic fields (Faraday's law) in a planet's interior. These induced fields can be used to probe internal electrical conductivity structure, which in turn depends largely on composition, volatile content, and temperature of a planet's crust and mantle, as well as the radius of its electrically conducting iron-rich core. Currents that result from the magnetic connection between the interplanetary magnetic field and the planetary field are of particular interest. At Mercury, this interaction generates an electromotive potential in the dawn-to-dusk direction that is conveyed along planetary magnetic field lines to the planet's surface at latitudes poleward of 50° in both hemispheres. If there is finite electrical conductivity between the surface footpoints at which the potential is applied, electric currents flow from high altitudes at the magnetopause down to the surface along magnetic lines of force and close through the planet. These field-aligned or Birkeland currents (e.g., Baumjohann and Treumann, 1996) thus also provide information on the planet's interior electrical conductivity.

Earth is the canonical example of a planet with a global dipolar field driven by thermochemical convection in its fluid outer core. A few essential ingredients are required for planetary dynamos to operate. First, a highly electrically conductive fluid layer (e.g., the inferred iron–nickel outer core for inner solar system planets) must be present in the planet's interior. Second, a mechanism that can drive sufficiently vigorous convection to generate a self-sustaining magnetic field must exist, i.e., the timescale for magnetic field generation by advection of the electrically conductive fluid must be much shorter than the timescale for that magnetic field to diffuse away. Third, planetary rotation plays an essential role in organizing the fluid motions and the resulting magnetic field, such that the latter has a coherent, large-scale structure (Proudman, 1916; Taylor, 1917; Busse, 2002). Thus, given a metallic core that is at least partly liquid, the question is whether the convection is sufficiently vigorous to sustain a dynamo. Core convection is driven by thermal and/or chemical buoyancy. The super-adiabatic radial thermal gradients across the fluid core necessary for thermal convection can arise from secular cooling of the planet, the presence within the core of radioactive elements such as potassium, and the latent heat of freezing of the inner core (Gubbins, 1977; Nimmo, 2015). However, thermal convection in Earth's core by itself is not sufficiently vigorous at present to permit a dynamo (Gubbins et al., 2003; Nimmo, 2015). If the liquid core freezes and forms a solid inner core, light elements can be excluded, setting up an unstable density gradient in the liquid layer. This chemical buoyancy source can also drive convection, and some combination of thermal and chemical convection can sustain a dynamo. However, on Earth, an early dynamo was driven by thermal convection (e.g., Labrosse, 2003; Aubert et al., 2010); the solid inner core did not exist (Labrosse et al., 2001), and there was a high heat flux across the core–mantle boundary. Presently, chemical convection dominates the Earth's core, and this situation likely holds for Mercury.

In a general sense, the history of a planet's core dynamo field is intimately tied to the thermal evolution of the planet and the growth history of the inner core (e.g., Aubert et al., 2010). Additionally, chemical buoyancy sources can arise at any place in the core where the temperature is at the solidus and drive convection either by the release of light elements upward or by the precipitation of iron (Vilim et al., 2010). Numerical dynamo models further suggest that the strength and geometry of the field, in particular regional-scale fields and the temporal variability in the field, are determined by the combined effects of the geometry of the convecting portion of the fluid core (i.e., the radius of the outer core shell and the presence or absence of stably stratified layers), the spatial pattern of heat flow through the core–mantle boundary, and the distribution of buoyancy sources in the core (see reviews by Stanley and Glatzmaier, 2010; Wicht and Tilgner, 2010).

For Earth's field, the strength, geometry, and variation of the global-scale field over historical timescales yield constraints on the dynamo process and have been characterized with satellite, marine, and land-based observations (e.g., Bloxham et al., 1989). The longevity (Tarduno et al., 2014, 2015), dominantly dipolar nature (Opdyke and Henry, 1969), and variations in strength and geometry of the field over thousand-year to billion-year timescales (e.g., Aubert et al., 2010; Johnson and McFadden, 2015) have been established from records of remanent magnetization in rock samples. Satellite measurements have also been used to study crustal magnetization (Langel and Hinze, 1998) and to probe interior electrical conductivity structure (see review by Olsen, 1999).

For other inner solar system bodies, orbiting spacecraft provide the primary, or only, source of magnetic field data. Spacecraft observations of Venus, Mars, and the Moon indicate that neither of Earth's nearest neighbors nor its Moon possesses a present dynamo field. One of the most notable discoveries of the Mars Global Surveyor mission was that Mars has crustal magnetic fields (Acuña et al., 1999; Connerney et al., 1999). These magnetic anomalies have generally been interpreted to result from thermal remanent magnetization (TRM) acquired in a dynamo field prior to ~4.1 Ga (Lillis et al., 2008; Robbins et al., 2013), although a later timing or longer duration (Schubert et al., 2000; Milbury et al., 2012) of the dynamo have also been proposed, as has acquisition of remanence via chemical alteration (Chassefière et al., 2013). Laboratory measurements of martian meteorites have provided constraints on magnetic mineralogies capable of carrying the inferred strong magnetizations (Kletetschka et al., 2000; Gattacceca and Rochette, 2004; Dunlop and Arkani-Hamed, 2005). In contrast, Venus shows no evidence for remanent crustal magnetization. However, an ancient dynamo field cannot be ruled out, because the relatively young surface age suggests substantial volcanic resurfacing of the planet within the past ~1 Gyr (Phillips et al., 1992; Schaber et al., 1992; McKinnon et al., 1997; Herrick and Rumpf, 2011), presenting a challenge for the (thermal) survival of any magnetization acquired prior to that time. The high (~455°C) surface temperature presents additional challenges to preservation of remanence for some magnetic minerals. Apollo samples from, and spacecraft and surface magnetic field measurements of, the Moon have demonstrated that some but not all lunar rocks are magnetized. The sample record is difficult to interpret because of the effects of shock (e.g., Gattacceca et al., 2010a, b). However, several lines of evidence point to a dynamo that operated either intermittently or continuously from ~4.2 Ga to ~3.3 Ga, with a surface field strength comparable to that of Earth (Cisowski et al., 1983; Fuller and Cisowski, 1987; Garrick-Bethell et al., 2009; Hood, 2011; Cournède et al., 2012; Wieczorek et al., 2012; Weiss and Tikoo, 2014). Alternative interpretations invoke transient fields associated with impacts and remanence acquired during shock rather than TRM (e.g., Hood and Huang, 1991; Hood and Artemieva, 2008; Hood et al., 1979, 2013).

Mercury's magnetic field was discovered from Mariner 10 measurements during the first flyby (M10-I) of the planet in March 1974 (Ness et al., 1974). The large distance (~700 km) of closest approach, the equatorial trajectory, and the substantial field variability on the outbound portion of the flyby (Figure 5.1), however, meant that an internal origin for the field was uncertain. The third Mariner 10 flyby (M10-III), with a closest approach distance of 327 km at a geographic latitude of 68°N (Figure 5.1), confirmed the existence of an intrinsic magnetic field (Ness et al., 1975). Magnetic field data from M10-I and M10-III indicated a dominantly dipolar internal magnetic field, tilted approximately 10–20° from the planet's rotation axis, with an equivalent dipole moment of 136–350 nT R_M^3, where R_M is Mercury's radius (2440 km) (Ness et al., 1976; Lepping et al., 1979; Connerney and Ness, 1988). The range of dipole moment estimates arose from uncertainties in fields contributed by external current systems (i.e., current sources above the planetary surface) and from the non-dipole contributions of the internal field (Ness et al., 1975; Connerney and Ness, 1988), which could not be distinguished given the limited geometry of the available observations. Importantly, the results from all analyses indicated an internal field much weaker than Earth's.

The origin of Mercury's magnetic field was a puzzle because it seemed to be too weak to be a dynamo field but too strong to be a remanent field. Solar-wind induction was considered after M10-I (e.g., Herbert et al., 1976), but the predicted field strengths were much lower than observed during M10-III. An active dynamo on Mercury was not expected because calculations implied early solidification of an iron core (e.g., Solomon, 1976; Schubert et al., 1988). Subsequent thermal evolution models indicated that a liquid outer core could exist at present, depending on the concentration of light elements such as sulfur (Schubert et al., 1988; Conzelmann and Spohn, 1999; Hauck et al., 2004; Williams et al., 2007; Grott et al., 2011; Tosi et al., 2013). Later Earth-based radar measurements of the planet's obliquity and the amplitude of its forced librations provided observational evidence for a liquid outer core (Margot et al., 2007). However, a dynamo field was still regarded as problematic, because energetics and force balance arguments suggested that Mercury's field should be at least two orders of magnitude stronger than observed (Stevenson, 2003). A variety of dynamo models have since been proposed that can yield weak fields by treating the thickness of the liquid outer core as a free parameter (Stanley et al., 2005; Heimpel et al., 2005; Takahashi and Matsushima, 2006) or by invoking the presence of one or more stably stratified layers (Christensen, 2006; Christensen and Wicht, 2008; Manglik et al., 2010), precipitation of solid iron within one or two intervals of radius (Vilim et al., 2010), or feedback

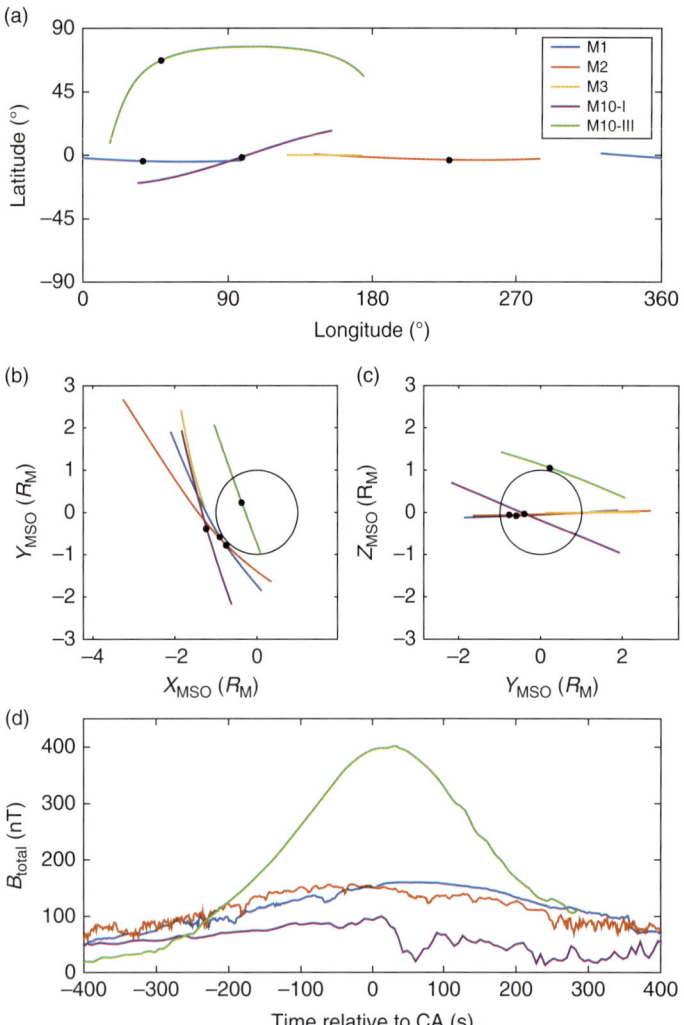

Figure 5.1. Observation geometry of, and magnetic field strength measured on, the first (M10-I) and third (M10-III) Mariner 10 and the first (M1), second (M2), and third (M3) MESSENGER flybys of Mercury. Spacecraft trajectories are shown (a) in Mercury body-fixed (MBF) coordinates and (b, c) in Mercury solar orbital (MSO) coordinates. See Section 5.2 for definitions of coordinate systems. Black circles mark the point of closest approach (CA) on flybys M10-I, M10-III, M1, and M2. No data were recorded at CA on M3. (d) Total field strength versus time relative to the time of closest approach. CA altitudes above the surface of the planet were 705 km (M10-I), 327 km (M10-III), 201 km (M1), and 199 km (M2). Peak field strengths on each flyby were 98 nT (M10-I), 400 nT (M10-III), 159 nT (M1), and 158 nT (M2).

between the magnetospheric and core dynamo fields (Grosser et al., 2004; Glassmeier et al., 2007a, b; Heyner et al., 2011). A thermoelectric dynamo has also been proposed (Stevenson, 1987; Giampieri and Balogh, 2002).

Remanent crustal magnetic fields were proposed to explain the Mariner 10 magnetic field data (Stephenson, 1976; Srnka, 1976), but such an explanation was not initially favored because either strong magnetizations or a large magnetized-layer thickness would be required to explain the observed signals. However, a crustal origin was reconsidered after data for Mars revealed that crustal fields an order of magnitude stronger than on Earth are possible (Acuña et al., 1999), albeit for quite different magnetic mineralogies than might be expected at Mercury (Dunlop and Arkani-Hamed, 2005). Spatial variations in solar insolation on Mercury could give rise to long-wavelength variations in the thickness of a magnetized layer (Aharonson et al., 2004) that, in turn, predict specific long-wavelength longitudinal structure in the resulting magnetic field.

Irrespective of the source of the global magnetic field, the combined effects of the weak intrinsic field and a higher solar-wind density at Mercury than that at Earth suggested a magnetosphere a factor of 7–8 times smaller than its terrestrial counterpart (Slavin and Holzer, 1979; Russell et al., 1988).

Consequently, Mercury occupies a much larger fraction of its magnetosphere than does Earth. The magnetosphere is the region enclosed by the magnetopause current layer that constitutes the boundary between the planetary magnetic field and the magnetic field of the shocked solar wind in the magnetosheath (see Figure 16.1). The magnetopause currents are a major source of external magnetic fields near Mercury. Magnetopause crossings from the M10 flybys were used to calculate the equivalent subsolar distance, R_{SS}, of the magnetopause from the dipole origin, and R_{SS} was estimated to be 1.5 R_M to 1.9 R_M. Because Mercury's magnetosphere is small and because the core radius is ~80% of the planetary radius, the external fields have substantial contributions to the total field at spacecraft altitudes, at the planetary surface, and even at the core–mantle boundary (CMB). These contributions must be quantified to establish reliable descriptions of Mercury's internal field.

Hints that the magnetosphere is also highly dynamic were suggested by the differing magnetic field signatures during the inbound and outbound trajectories of M10-I (Russell et al., 1988). Variations in solar-wind conditions drive changes in the magnetosphere size and morphology. For example, R_{SS} is governed to first order by a pressure balance between the solar-wind ram pressure, P_{ram}, and the magnetic field pressure just inside the magnetopause. Thus, the magnetopause approaches or recedes from the planet under higher or lower P_{ram} conditions, respectively. An open question after M10-I and M10-III was whether the dayside magnetopause could reach the planetary surface, exposing the entire dayside to the solar wind. Contrasting predictions were made regarding the extent to which fields arising from currents induced at the top of Mercury's large metallic core could help stand off the solar wind versus the extent to which the magnetopause position would be "eroded" by magnetic reconnection and moved closer to the planet (Hood and Schubert, 1979; Slavin and Holzer, 1979).

The magnetic field of the innermost planet provides key connections that link the planet's interior and surface with the solar wind environment. A primary goal of the MESSENGER mission was to characterize the nature of Mercury's magnetic field and to understand its interaction with the solar wind. Questions that MESSENGER data have been able to address directly, and on which we focus in this chapter, are: (1) What is the origin of Mercury's global-scale field? (2) If a core dynamo field is present, what are the spatial structure and strength of that field, and is there any evidence for secular (time) variation since the Mariner 10 flybys? (3) Do time-varying external fields induce fields at the top of Mercury's core that have observable signals at spacecraft altitudes and, if so, can these, and/or any other magnetic field signatures, be used to constrain Mercury's interior conductivity structure? (4) Is there evidence for magnetization of crustal and/or upper mantle rocks at Mercury and, if so, what constraints does this finding place on magnetic mineralogy and on the magnetic field history of the planet?

We next outline MESSENGER magnetic field observations and techniques that can be used to estimate contributions to the observations from fields of internal origin (Section 5.2). Critically important to all aspects of internal field studies at Mercury is the non-trivial issue of identifying and separating fields of external origin. In the case of Mercury, some currents driven by external processes flow near or even beneath the surface of the planet. Insofar as these currents are driven by the interaction of the planet and its magnetic field with the solar wind, we touch on aspects of the average state of and time dependence in the magnetosphere. A comprehensive synthesis of magnetospheric structure and dynamics is given in Chapters 16 and 17. We summarize briefly the evidence for a core dynamo field provided by MESSENGER flyby observations, and we review the core field structure inferred from observations acquired during the MESSENGER orbital mission and bounds on secular variation since the time of the Mariner 10 flybys (Section 5.3). We discuss constraints on interior electrical conductivity provided by induced fields and field-aligned currents (Section 5.4). In Section 5.5 we focus on magnetic fields arising from magnetized rocks, using data from the last few months of the MESSENGER mission, when spacecraft periapsis altitudes were below 100 km (Chapter 1). For brevity, we use "crustal magnetization" and "crustal fields" to refer to magnetizations of crustal/upper mantle rocks and the fields that may result from them, and we lay out the rationale for the view that most of the remanence is carried in the crust. Finally, we discuss the implications of the results presented in this chapter for Mercury's interior structure and evolution and suggest avenues for future work.

5.2 MESSENGER MAGNETIC FIELD OBSERVATIONS AND APPROACHES FOR ANALYZING INTERNAL FIELDS

The MESSENGER spacecraft was launched on 3 August 2004, made three flybys of the innermost planet between January 2008 and September 2009, and was inserted into orbit about Mercury on 18 March 2011 (Chapter 1). Observations were made in orbit around Mercury for more than four Earth years (17 Mercury years), and the spacecraft impacted the planet on 30 April 2015. Magnetic field observations inside Mercury's magnetosphere were acquired by the Magnetometer (MAG) at a rate of 20 vector samples per second (Anderson et al., 2007).

5.2.1 Coordinate Systems

Three coordinate systems are relevant to magnetic fields at Mercury. With the exception of induced fields, internally generated fields are described in the Mercury body-fixed (MBF) frame. In this planetocentric coordinate system, +X points toward the prime meridian on Mercury, +Z points northward along the rotation axis, and +Y lies in the equatorial plane, perpendicular to +X and +Z, and completes the right-handed system. Interactions of the planet with the solar wind are often described in the local-time, planetocentric Mercury solar orbital (MSO) coordinate system, in which +X is sunward, +Y is duskward, and +Z is normal to the orbital plane and positive northward, completing the right-handed system. Because of Mercury's small obliquity (Margot et al., 2007, 2012), the MBF and MSO +Z axes are nearly identical. The Mercury solar magnetospheric (MSM) coordinate system captures the interaction of the dipole field with the solar wind. This system has the same axis directions as MSO but is shifted northward by Z_0, where Z_0 is the mean offset of the dipole magnetic field from the planetary center (Section 5.3). The MSO and MSM positions for observations of Mercury's magnetopause and field inside the magnetosphere were corrected for solar wind aberration (Paschmann and Daly, 1998), using Mercury's instantaneous orbital speed and a mean solar wind speed taken from ENLIL simulations (Odstrcil, 2003).

5.2.2 Geometry of MESSENGER MAG Observations

All three MESSENGER flybys were equatorial, with closest approach on the nightside of the planet (Figure 5.1). MAG data were acquired continuously during the first two flybys, M1 and M2, and the MBF locations of closest approach were 4.5°S, 37.7°E, and 3.4°S, 228.9°E at altitudes of 201 km and 199 km, respectively. No data suitable for internal field modeling were acquired during the third flyby, M3, because of a spacecraft safe-hold anomaly.

During its first Earth year of orbital operations, MESSENGER's orbital period was 12 h, periapsis altitudes were between 200 km and 505 km at 60–73°N, and apoapsis altitudes were ~15,000 km. At the beginning of the second year of operations, the orbital period was decreased to 8 h, and periapsis altitude increased steadily from 200 km to a maximum of 450 km in March 2013 and then decreased steadily to 113 km in June 2014. Between June 2014 and April 2015, multiple orbit-correction maneuvers were conducted. These maneuvers enabled two brief distinct periods of observations at periapsis altitudes as low as 25 km in September and October 2014, and an extended interval in 2015 with periapsis altitudes below 60-km altitude (Chapter 1).

The time spent by the spacecraft inside the magnetosphere during each orbit ranged from 1 to 2 h. MAG measurements sampled the northern magnetosphere at altitudes less than ~1000 km and the nightside southern lobe of the magnetotail to distances of just under 5 R_M (Figure 5.2). Observations spanning all body-fixed longitudes and local times were obtained every 59 and 88 days, respectively.

5.2.3 Field Modeling Approaches

5.2.3.1 Spherical Harmonic Descriptions: MESSENGER Data Limitations

Vector magnetic fields measured in a source-free region can be described by the gradient of a scalar potential, V, due to internal (V_{int}) and external (V_{ext}) sources, expressed by a spherical harmonic expansion:

$$\begin{aligned} V &= V_{int} + V_{ext} \\ &= R \sum_{l=1}^{L_{int}} \sum_{m=0}^{l} (g_l^m \cos m\phi + h_l^m \sin m\phi) \left(\frac{R}{r}\right)^{l+1} P_l^m(\cos\theta) \\ &+ R \sum_{l=1}^{L_{ext}} \sum_{m=0}^{l} (q_l^m \cos m\phi + s_l^m \sin m\phi) \left(\frac{r}{R}\right)^{l} P_l^m(\cos\theta), \end{aligned} \quad (5.1)$$

where (r, θ, ϕ) are radius, colatitude, and longitude; R is the reference radius (typically the radius of the planet); and P_l^m are the associated Legendre functions of spherical harmonic degree, l, and order, m. The expansion describes contributions to the observed field of different spatial scales or wavelength, λ, where $l \sim 2\pi R/\lambda$. L_{int} and L_{ext} are the maximum spherical harmonic degrees of the internal and external field expansions, respectively, or equivalently the shortest wavelengths in the expansion, and the Gauss coefficients (g_l^m, h_l^m) and (q_l^m, s_l^m) describe the amplitude of each spherical harmonic term. A rule of thumb in potential field studies is that the minimum resolvable wavelength in the internal field, λ_{min}, scales with the altitude of the spacecraft observations, h, such that the range of λ_{min} is ~$h/2$ to ~h. Fields that are symmetric about the planet's rotation axis have no dependence on longitude and are described by the $m = 0$ terms (the zonal or axial terms). The $l = 1$ terms in the internal field describe the field due to a planetocentric dipole. The terms g_1^0, g_2^0, and g_3^0 are known as the axial dipole, axial quadrupole, and axial octupole terms, respectively. Terms even in $l - m$ are symmetric about the equator, and terms odd in $l - m$ are antisymmetric about the equator.

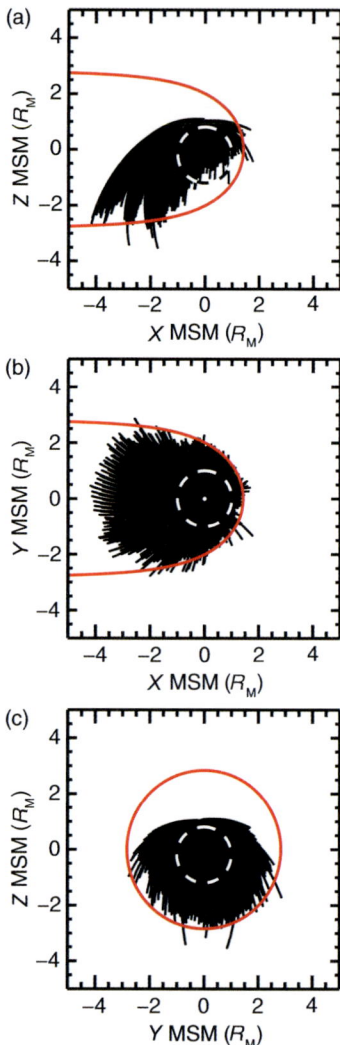

Figure 5.2. Magnetic field data coverage for the first seven Mercury years of Mercury orbital observations. Observation locations are shown projected onto the MSM (a) X–Z, (b) X–Y, and (c) Y–Z planes. Orbit tracks are shown for portions of the orbit inside the observed magnetopause (Chapter 16) on each magnetospheric transit. The average location of the magnetopause is shown in red, and the outline of the planet is shown with the dashed line. Figure modified from Korth et al. (2015).

For uncorrelated errors in the observations, the data vector, **d**, comprising the vector field observations at the spacecraft locations, can be written

$$\mathbf{d} = \mathbf{G}\,\mathbf{x} + \mathbf{e}. \quad (5.2)$$

In a spherical harmonic description of the field, the model vector **x** comprises the Gauss coefficients (g_l^m, h_l^m, q_l^m, s_l^m), the matrix **G** encompasses the spatial distribution of the observations, and **e** is the error vector. The inverse problem is usually solved by finding the model, **x**, that minimizes the misfit of the model predictions to the observations in a least-squares sense. This basic approach (albeit with modifications and more complexity in the implementation) has been used to analyze spacecraft magnetic field data at Earth, Mars, and

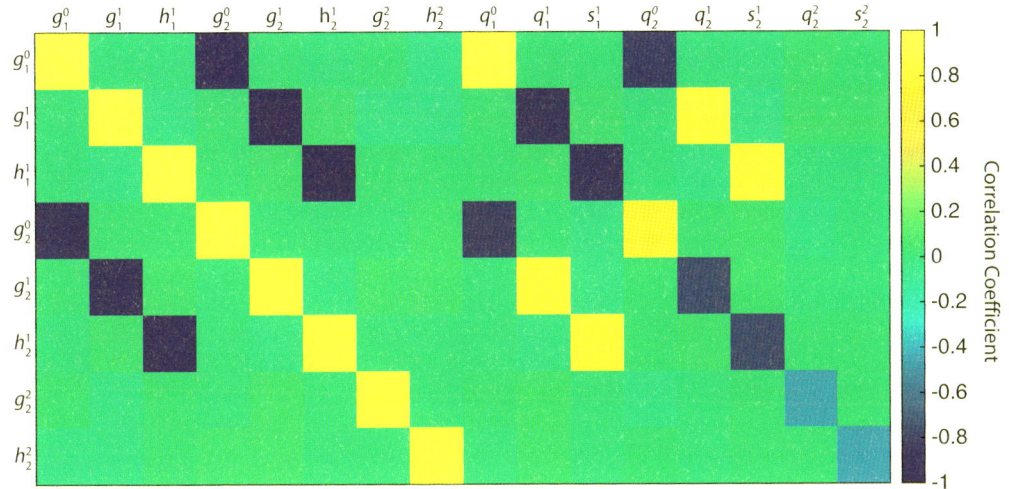

Figure 5.3. Correlations among terms in a classic spherical harmonic separation of internal and external fields that result from MESSENGER's orbit geometry. The correlation matrix was estimated from the first two Mercury years of orbital data. Spherical harmonic coefficients of degree l and order m in the internal field are denoted by g_l^m and h_l^m; those in the external field are denoted by q_l^m and s_l^m. Terms up to degree and order 2 are shown.

the outer planets (e.g., Olsen et al., 2010; Connerney, 2015; Hulot et al., 2015).

Mercury's small magnetospheric dimensions mean that external fields generated by electric currents on the magnetopause and in the cross-tail (magnetotail) current sheet contribute substantial signals to the field measured by MESSENGER everywhere along its trajectory inside the magnetosphere, and these signals must be considered in any interpretation. MESSENGER's altitude varied systematically with latitude, and spherical harmonic terms estimated via a least-squares solution of equations (5.1) and (5.2) are correlated (Figure 5.3) in several ways. First, zonal structure in the internal field and external field co-varies along the spacecraft orbit, and so internal and external $m = 0$ terms are strongly correlated (e.g., $g_1^0 - q_1^0$ and $g_1^0 - q_2^0$ correlations). Second, there are correlations among non-zonal internal and external terms, e.g., day–night structure in the external field (q_2^1) is correlated with the internal g_1^1 and g_2^1 terms. Third, internal terms that are symmetric ($l - m$ even) or antisymmetric ($l - m$ odd) about the equator are correlated. Specifically, g_1^0 and g_2^0 are strongly anticorrelated, as was observed in the Mariner 10 data (e.g., Connerney and Ness, 1988) and as was anticipated for MESSENGER's orbital mission (Korth et al., 2004). Thus, aside from more fundamental issues regarding modeling of the external field (see below), the orbit geometry presented challenges to establishing even the long-wavelength structure of the internal field.

We note two crucial ways in which the spherical harmonic formalism does not apply to Mercury. First, the external currents extend from the magnetopause to the surface, so the volume of the magnetosphere is not source-free, violating the basic requirement for equation (5.1). Indeed, MESSENGER's orbit passed through the magnetopause, magnetotail, and field-aligned currents and detected plasma present in the magnetosphere even very close to the planet (Chapter 16 and references therein). Second, the presumed separation of internal and external potentials by negative and positive powers of r, respectively, does not hold for Mercury because the signals from field-aligned currents increase with decreasing altitude. Modifications to equation (5.1) are possible to describe the toroidal fields generated by currents in the observation region (e.g., Olsen et al., 2010) but are not well suited to MESSENGER's orbit geometry.

5.2.3.2 Empirical Models for External Sources

Although physics-based simulations of the interaction of the solar wind with the planetary field have been conducted (e.g., Kabin et al., 2000; Trávníček et al., 2007; Benna et al., 2010) and could, in principle, be used to derive external field corrections, they are not practical for extensive parameter searches or inversions involving large quantities of data. An alternative approach to deriving external field estimates involves empirically parameterized models for the magnetospheric current sources. There is a rich history of the development of such models for Earth (e.g., Alekseev and Shabansky, 1972; Tsyganenko, 1995), and they have also been applied to Mercury (Alexeev et al., 2008, 2010; Johnson et al., 2012; Korth et al., 2015; Chapter 16). In such models, the internal field is represented by a dipole field, and the geometry and strength of the tail current sheet and the geometry of the magnetopause boundary are specified. The magnetopause fields are then computed under the assumption that the normal component of the magnetic field across the magnetopause is zero. Analytical solutions exist for specific magnetopause geometries (e.g., Alekseev and Shabansky, 1972), and numerical solutions can be computed for more general and more likely magnetopause shapes (Tsyganenko, 1995). MESSENGER's orbit allowed determination of the magnetopause boundary and magnetotail current sheet geometries (Winslow et al., 2013; Chapter 16), allowing the development of magnetospheric models for Mercury (Alexeev et al., 2010; Johnson et al., 2012; Korth et al., 2015). For details, see Chapter 16 and references therein. These empirical models are used to derive quantitative estimates for the external field. Models to date invoke a dipolar internal field, although this assumption is not strictly required (see further discussion in Section 5.3.2), and do not include contributions from field-aligned currents (Chapter 16).

5.2.3.3 Inverse Techniques and Localization Approaches

The development of magnetospheric models has been key to establishing the first-order dipole structure in Mercury's global internal field (Section 5.3). However, characterization of the general non-dipole structure of the core field and/or crustal field requires other approaches. MESSENGER's orbit geometry results in strongly varying spatial resolution in potential field models (see also Chapter 3). The challenge is evident by considering a typical least-squares solution to equation (5.1), in which the spherical harmonic expansion is limited to internal terms and truncated. Structure in the resulting model is restricted by the truncation degree, L_{int}, imposed. If L_{int} is set to capture the shortest wavelengths resolvable near periapsis, the model will have unrealistic structure in regions of lower spatial resolution. An alternative approach, used extensively in geomagnetism (e.g., Bloxham et al., 1989; Parker, 1994), is to permit power in terms of higher spherical harmonic degree and order but to impose a smoothness constraint (known as damping or regularization) to avoid unrealistic structure. These inverse techniques allow the downward continuation of noisy observations that have substantial variations in spatial resolution. Global basis functions can also be replaced by functions that have local support, such as spherical cap harmonics (Haines, 1985), the recently developed spatially localized vector spherical harmonics (Plattner and Simons, 2015), or the more well-established equivalent-source techniques (e.g., Mayhew 1979; von Frese et al., 1981; Kother et al., 2015). Localization approaches can be computationally advantageous because they allow regional solutions, and they can reduce the level of regularization needed to stabilize the inversion.

One localization approach that we discuss in this chapter is the equivalent-source dipole technique. The magnetic field measured at a given location results from the net contributions of model dipolar sources located at different distances from the observation location. The scalar potential due to an individual source is given by

$$V_{int} = -\frac{\mu_0}{4\pi} \mathbf{m_d} \cdot \nabla \frac{1}{s} \quad (5.3)$$

where $\mathbf{m_d}$ is the magnetic moment of a dipole located at (r_d, θ_d, ϕ_d), and the distance s between the observation location (r, θ, ϕ) and the source is given by

$$s = \sqrt{r_d^2 + r^2 - 2r_d r \left(\cos\theta \cos\theta_d + \sin\theta \sin\theta_d \cos(\phi - \phi_d) \right)}. \quad (5.4)$$

The dipole sources are placed a priori at one or more specified depths below the planetary surface, and the dipole mesh design and spacing are guided by the data distribution and desired model resolution. The moments, $\mathbf{m_d}$, are estimated by solving the linear system in equation (5.2), with or without regularization, where the model vector $\mathbf{x}$ comprises the three components of $\mathbf{m_d}$ for every dipole source. For crustal fields, the dipole moments, $\mathbf{m_d}$, can be converted into an equivalent magnetization, $\mathbf{M}$, for an assumed thickness of the magnetized layer. For core fields, the dipole moments have no physical meaning; they simply provide a mathematical description of the magnetic field.

5.2.3.4 Magnetospheric Activity Considerations

At Earth, indices that represent the level of magnetic activity inside the magnetosphere (e.g., Mayaud, 1980; Parkinson, 1983) have been of great use in identifying quiet times for geomagnetic field modeling. Although Mercury's magnetosphere exhibits phenomena analogous to dynamics at Earth (Chapter 17), the variability in magnetic field strength is more rapid and intense relative to the planetary magnetic field than at Earth. Techniques that are unbiased with respect to orbit geometry and heliocentric distance were developed to quantify the level of magnetospheric disturbance, allowing observations to be ranked by the level of activity (Anderson et al., 2013). The magnetic field variability in three period bands spanning 0.1 s to 300 s was used to construct an activity index that ranges from 0 to 100. An important point is that signatures of magnetospheric activity are not restricted to high-frequency variations. For a spacecraft periapsis speed of ~4 km s^{-1}, 300 s corresponds to signals with along-track wavelengths of ~1200 km or ~30° in latitude. The activity index has allowed the selection of quiet orbits for several internal field studies.

5.3 CORE DYNAMO FIELD

5.3.1 MESSENGER Flybys: Evidence for a Dynamo Origin of the Field

The MESSENGER flybys provided the first opportunity since the 1970s to measure Mercury's magnetic field. The consistent peak field amplitudes and the decrease in field strength with increasing spacecraft altitude observed during M1 and M2 (Figure 5.1d), combined with data from the M10-III flybys, confirmed the presence of an intrinsic magnetic field and definitively established its dominantly dipolar geometry (Anderson et al., 2008, 2010; Uno et al., 2009b). The inferred dipole moment was 174–250 nT R_M^3 (Table 1 of Anderson et al., 2010). Moments at the lower end of the range were obtained if external fields were not considered and if spherical harmonic structure in the internal field above degree 1 was permitted, reflecting trade-offs in estimating the internal and external fields similar to those shown in Figure 5.3. External fields were treated either via spherical harmonic separation (with full recognition of the caveats described in Section 5.2), or by scaling the magnetopause and magnetotail currents in the Earth-based empirical model of Tsyganenko (1995) and then subtracting the resulting external field estimate from the data before conducting internal field analyses (Anderson et al., 2008, 2010; Uno, 2009; Uno et al., 2009). The similarity in the longest-wavelength fields from M1 and M2 suggested that the dipole field is axisymmetric in the MBF frame; dipole tilts were estimated to be less than 5° (Uno et al., 2009; Anderson et al., 2010). Crustal fields, particularly long-wavelength signatures of spatial variations in magnetized layer thickness resulting from geographical variations in surface insolation (Aharonson et al., 2004), were not observed (Purucker et al., 2009; Uno et al., 2009).

Collectively, these observations established the dynamo origin of the global-scale internal field. Furthermore, under the presumption that external fields could be estimated and subtracted from the observations, simulations by Uno et al. (2009) predicted that regularized inversion approaches applied to the orbital data could successfully recover northern hemisphere structure in the internal field (if present) to spherical harmonic degree and order 10. A limitation of studies with the flyby data was that external field treatments were inadequate, as evidenced by correlated residuals (Uno, 2009; Uno et al., 2009; Anderson et al., 2010). It was clear that a magnetospheric model specific to Mercury was needed, but that development of such a model would require systematic observations of the magnetopause and magnetotail from orbit.

5.3.2 Orbital Data: Global Core Field Structure

Important products of the analysis of MESSENGER orbital magnetic field data were empirical magnetospheric models for Mercury (Chapter 16). In their more general form, such models can include time dependence in the external fields, driven by solar-wind conditions, and can be used to establish the global-scale time-averaged internal field and to investigate induction currents in the planetary interior driven by time-varying external fields.

In the first generation of magnetospheric models the magnetopause was prescribed as a paraboloid of revolution (Alexeev et al., 2010; Johnson et al., 2012) with a time-averaged value for R_{SS} of 1.45 R_M derived from three Mercury years of MESSENGER magnetopause crossings (Winslow et al., 2013). Data from the entire mission yielded a minor update to this value, R_{SS} = 1.43 R_M. The paraboloid geometry provides a good approximation to the magnetopause shape in the near-planet region but is incorrect downtail (Figure 16.19). The magnetospheric model of Korth et al. (2015), hereafter KT14, involved a magnetopause shape fit to the observations. In both magnetospheric models, the internal field was specified by the moment, tilt, azimuth, and offset from the planetary center of the planetary dipole, but the KT14 model allows arbitrary internal magnetic field sources and magnetopause shape.

5.3.2.1 MAG Observations: The Offset Axial Dipole (OAD) Description

The magnetic field strength inside the magnetosphere during the second Mercury year of observations increased with increasing latitude, suggestive of a dipole field (Figure 5.4). After accounting for variations in spacecraft altitude, there are no obvious large-scale departures from axisymmetry in the field. It was recognized (Anderson et al., 2011) that knowledge of the magnetic equator position could resolve the trade-off between equatorially symmetric and equatorially antisymmetric structure in the field imposed by the orbit geometry, at least at the lowest spherical harmonic degrees. The normal to the plane of the magnetic equator specifies the dipole tilt and azimuth. Furthermore, the field from a dipole at an origin offset from the planetary center can be expressed as an equivalent spherical harmonic expansion in the planetocentric frame (Bartels, 1936).

Motivated by the apparent axisymmetry in the field, the magnetic field components were transformed into a cylindrical coordinate system with the Z-axis aligned with the Z_{MSO} axis. For an approximately axially aligned dipole moment, the north–south position of the zero crossing of the cylindrical radial

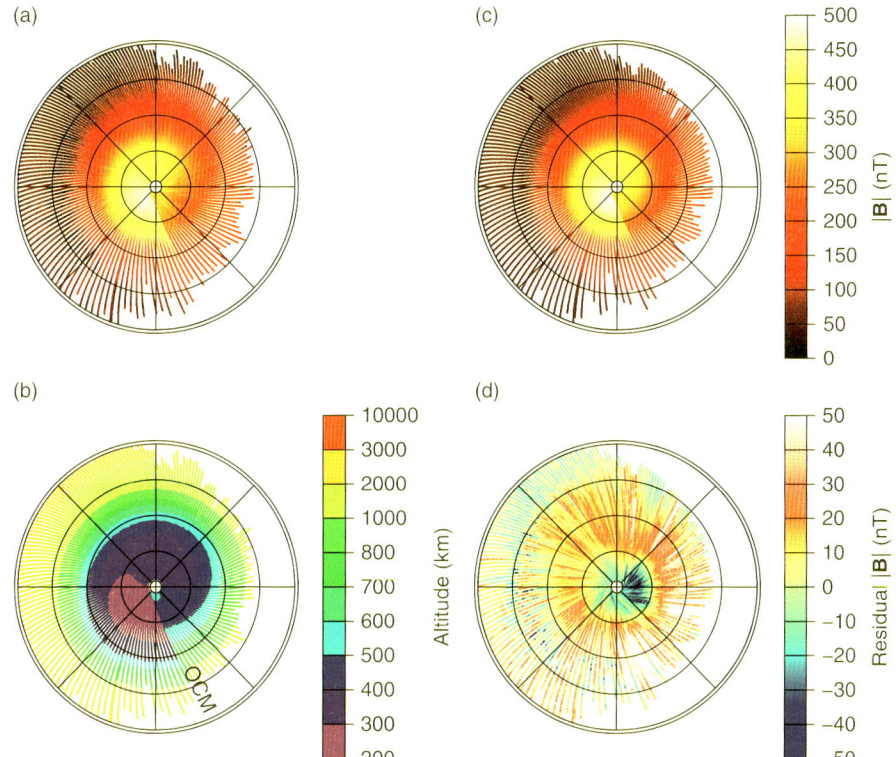

Figure 5.4. Stereographic projections in aberrated MSO coordinates of (a) magnetic field magnitude, B, (b) spacecraft altitude for descending tracks from the second Mercury year of observations (20 June – 15 September 2011), (c) predicted magnetic field magnitude from the magnetospheric model of Johnson et al. (2012), and (d) the corresponding residuals, where $|\mathbf{B}_{resid}|$ = $|\mathbf{B}_{data}| - |\mathbf{B}_{model}|$. Magnetic local noon is to the right. Grid lines are every 30° in latitude (30°S to north pole) and every 3 h in local time. Data and model predictions are shown within the model magnetosphere, or within the actual magnetosphere if the observed magnetopause boundaries fell within the model magnetopause boundary. The altitude scale is non-linear; note the large variations in spacecraft altitude. The increase in periapsis altitude during a Mercury year is evident, and OCM marks an orbit-correction maneuver to lower periapsis altitude to ~200 km. Figure modified from Johnson et al. (2012).

component, $\rho = \sqrt{X_{MSO}^2 + Y_{MSO}^2}$, of the field, the $B_\rho = 0$ point, gives the average offset of the magnetic equator from the geographic equator (see Figure 2 of Anderson et al., 2011). The Z_{MSO} location of the $B_\rho = 0$ crossing was found on an orbit-by-orbit basis by fitting a quadratic function to the B_ρ–Z_{MSO} data in a region near the geographic equator (Anderson et al., 2011, 2012). The average Z_{MSO} equator position was then calculated for the orbits considered in each study and the confidence intervals estimated. Data now available for the entire mission allow an update of these analyses and some refinements to the earlier algorithm. Figure 5.5 shows an example magnetic equator crossing. A cubic spline, constrained to be monotonically decreasing, was fit to the B_ρ–Z_{MSO} data inside the magnetopause in a 15° sliding latitude window. The window was moved from 55°S to 55°N in steps of 1° to avoid any bias in the choice of latitude interval. The equator Z position was then found from the $B_\rho = 0$ crossing of the spline fit. For any latitude window where the $B_\rho = 0$ crossing was found, the coefficient of determination, R^2, was calculated. The latitude window and corresponding spline fit that resulted in the highest R^2 value was used to estimate the equator position for the crossing.

Observed magnetic equator positions for the entire orbital mission are shown in Figure 5.6a. In the analysis of the equator position only equator-crossing fits with $R^2 > 0.7$ and at least 5 min inside the magnetopause boundary were included. Significant variation in the equator position is observed with increasing distance from the planet on the nightside and may result from tilting of the magnetotail in response to north–south excursions in solar wind velocity (Anderson et al., 2012). A mean magnetic equator position of $Z_0 = 484$ km was found

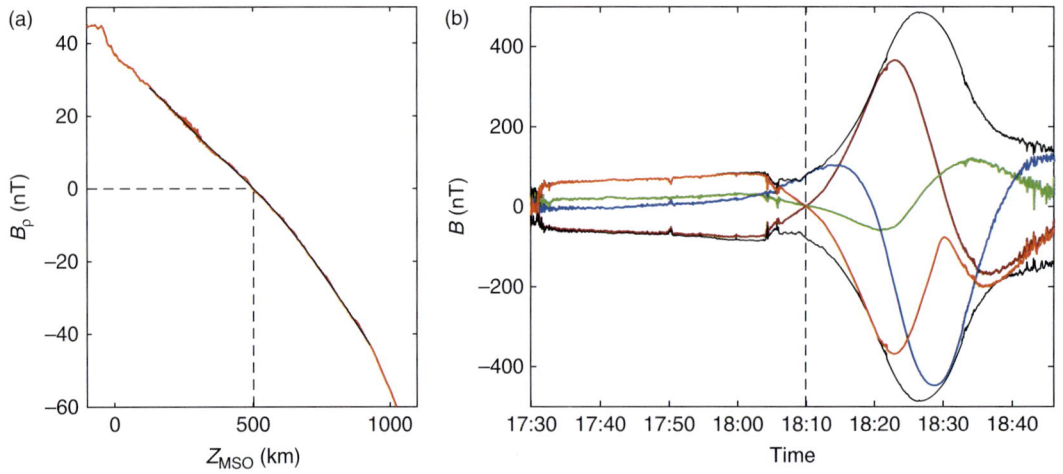

Figure 5.5. Example of magnetic equator crossing. (a) B_ρ (cylindrical radial component of the field, see Section 5.3.2.1) versus Z_{MSO} for MESSENGER orbit 3000 (20 April 2014) plotted at a 1-s sampling interval (orange curve) with monotonically decreasing cubic spline fit (black) to data in a 15° latitude window centered on 9.5°N. Dashed lines indicate the Z_{MSO} position of the magnetic equator, identified as the $B_\rho = 0$ crossing. In this case, the value of Z_{MSO} when $B_\rho = 0$ is $Z_{\rho 0} = 502 \pm 2$ km. (b) Complete MESSENGER transit inside the magnetosphere on this orbit showing the B_X (red), B_Y (green), and B_Z (blue) field components in MSO coordinates, the cylindrical B_ρ component (orange), and the total field magnitude and its negative (black lines) versus coordinated universal time (HH:MM). Vertical dashed line indicates the time of the magnetic equator crossing. Other examples can be seen in Figure 3 of Anderson et al. (2012).

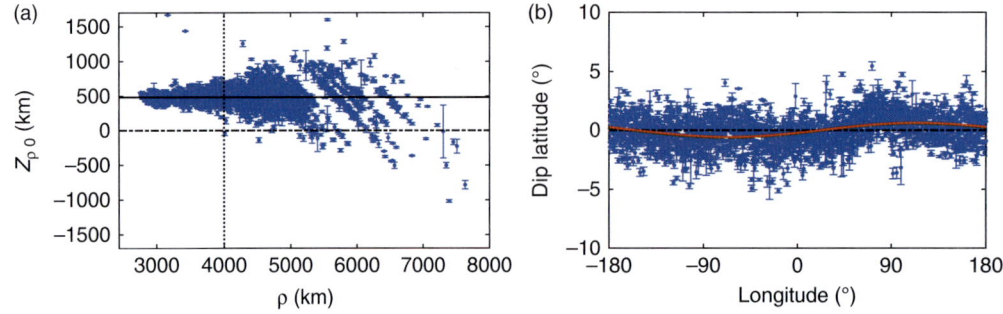

Figure 5.6. Observed magnetic equator crossings having a coefficient of determination criterion $R^2 > 0.7$ and at least 5 min inside the magnetopause. (a) Positions of observed instantaneous magnetic equator, $Z_{\rho 0}$, versus $\rho = \sqrt{X_{MSO}^2 + Y_{MSO}^2}$, during the orbital mission (filled blue circles). Error bars indicate bootstrap estimates (from 1000 bootstrap samples) of three standard deviations in the equator position. The average equator position, Z_0, of 484 km for crossings with $\rho < 4000$ km (vertical dotted line) is indicated by the solid black line. (b) Dip equator latitude versus MBF longitude (filled blue circles). Error bars indicate bootstrap estimates of three standard deviations in the dip latitude. Only equator crossings with $\rho < 4000$ km are included; two outlying points are not shown. A best-fit sine curve to the data, constrained to have a period of 360° and a mean dip latitude of zero, is shown in red.

for crossings close to the planet ($\rho < 4000$ km), with an uncertainty (three standard errors of the mean) of ±4 km.

To estimate the tilt of the dipole relative to the rotation axis, the magnetic field and spacecraft position data were transformed to a body-fixed coordinate system, in which the origin is offset north along the Z-direction by Z_0. Following Anderson et al. (2011), the latitude of the magnetic dip equator ($B_r = 0$) was then estimated. A small but non-zero tilt of 0.6° ± 0.1° was found, with a pole longitude of 114°. However, the tilt signal is substantially smaller than orbit-to-orbit variations in the equator position (Figure 5.6b). The latter have been confirmed to have no dependence on magnetic activity. In addition, structure is seen around 70°E, the origin of which (even whether it is internal) is currently unknown.

Thus, MESSENGER orbital data indicate that to first order Mercury's internal field can be described by the field of a dipole with moment $\mathbf{m}_{dip}$ aligned with the planet's rotation axis and offset northward along the rotation axis by Z_0. [Note that we use $\mathbf{m}_{dip}$ to denote the global dipole moment, to distinguish it from the localized sources, $\mathbf{m}_d$, in equation (5.4).] We refer to this geometry as the offset axial dipole (OAD) description of the field. The relationship between the Gauss coefficients in the body-fixed frame and a dipole with moment $\mathbf{m}_{dip}$ directed with a polar angle θ_m and azimuth ϕ_m, offset along the rotation axis by Z_0, are given in Table 5.1. The large Z_0 results in substantial $m = 0$ terms in the body-fixed spherical harmonic expansion up to and including the coefficient, g_4^0. In the absence of any other non-dipolar structure in the field, only the $m = 1$ non-axial terms are non-zero. For the OAD description, i.e., $\theta_m = 0$, the $m = 1$ terms are also zero.

Anderson et al. (2012) examined whether alternative spherical harmonic field models could match the magnetic equator results and the vector field observations inside the magnetosphere. For example, an internal field comprising g_1^0 and g_3^0 terms would yield three magnetic equators close to the planet: one at the planetary equator and one each at low latitudes in the northern and southern hemispheres. Tests were performed to investigate whether such fields, combined with MESSENGER's orbit, could yield detection of only the northernmost equator. The magnetic equator crossings both close to the planet and in the far tail were found to be incompatible with the predictions for such fields, and alternative low-spherical-harmonic-degree axial fields also did not match the field strength inside the magnetosphere. Furthermore, plasma observations (Chapter 16) provide independent evidence that the inferred magnetic equator offset best organizes the plasma distributions.

As shown in Table 5.1, the magnetic equator offset and tilt describe the global-scale internal structure for a given dipole moment. A grid search was conducted to determine the best-fit dipole moment to the observations in a root-mean-square (RMS) sense using three Mercury years of MESSENGER orbital data and the paraboloid magnetospheric model (Johnson et al., 2012). All available orbits were included, as this study preceded the development of an activity index (Section 5.2.3.4; Anderson et al., 2013). A best-fit dipole moment of 190 nT R_M^3 was found, with a minimum misfit of 18.9 nT (Figure 5.7). The RMS misfit is within 20% of this minimum value for dipole moments within ±10 nT R_M^3 of 190 nT R_M^3. The result was confirmed with seven Mercury years of observations and the KT14 model. Lower RMS misfits are obtained for magnetospherically quiet orbits, but the best-fit moment value is unchanged.

Comparison of the predicted model field magnitudes (Figure 5.4c) with the corresponding MAG observations (Figure 5.4a) demonstrates that the combined magnetopause, magnetotail, and OAD fields predict most of the signal in the observations. Residuals (data minus model) are typically less than 50 nT in magnitude (Figure 5.4d and Figure 16.20), compared with a maximum signal in the data of ~500 nT. Notably, external fields contribute up to ~40% of the observed

Table 5.1. *Relationships between Gauss coefficients in planet-centered (unprimed quantities) and axially displaced (primed quantities) coordinates in terms of the axial offset, Z_0, and offset magnetic dipole $\mathbf{m}_{dip} = (m_{dip}, \theta_m, \phi_m)$.*

Degree	Planetocentric coefficients	Offset coefficients
1	$g_1^0 = g_1^{0'}$	$g_1^{0'} = m_{dip} \cos\theta_m$
	$g_1^1 = g_1^{1'}$	$g_1^{1'} = m_{dip} \sin\theta_m \cos\phi_m$
	$h_1^1 = h_1^{1'}$	$h_1^{1'} = m_{dip} \sin\theta_m \sin\phi_m$
2	$g_2^0 = 2Z_0 g_1^{0'}$	n/a
	$g_2^1 = g_1^{1'} Z_0/\sqrt{3}$	
	$h_2^1 = h_1^{1'} Z_0/\sqrt{3}$	
	$g_2^2 = h_2^2 = 0$	
3	$g_3^0 = 3Z_0^2 g_1^{0'}$	n/a
	$g_3^1 = g_1^{1'} Z_0^2/\sqrt{6}$	
	$h_3^1 = h_1^{1'} Z_0^2/\sqrt{6}$	
	$g_3^m = h_3^m = 0$ for $m > 1$	
4	$g_4^0 = 4Z_0^3 g_1^{0'}$	n/a

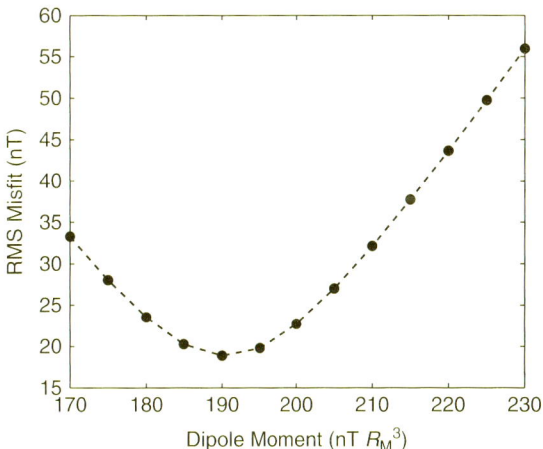

Figure 5.7. RMS misfit between predicted and observed magnetic field magnitude, B, as a function of dipole moment. The RMS misfit was computed using observations at MBF latitudes north of 30°N. The best-fit dipole moment is 190 nT R_M^3. Figure modified from Johnson et al. (2012).

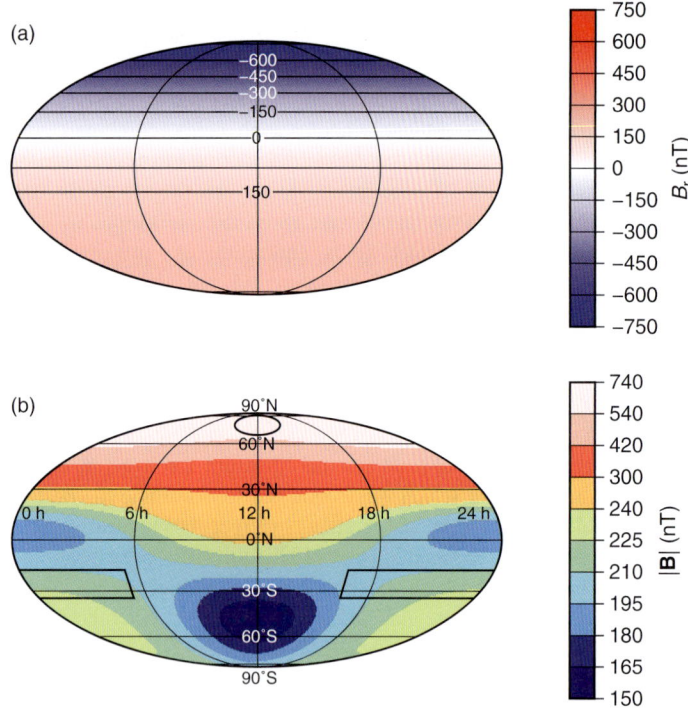

Figure 5.8. Mercury's global magnetic field structure shown in Mollweide projection. (a) Radial component of the internal field, B_r, shown at the planetary surface in the body-fixed frame, centered on 0° longitude. (b) Field strength B for the total magnetospheric field (internal field plus magnetopause and magnetotail fields) at the planetary surface in the MSO frame, centered on local noon. Black lines mark the approximate outline of the two regions for which estimates of the surface field strength have been made with proton reflectometry (see Section 5.3.2.2). Note the non-linear color scale in (b).

signal along MESSENGER's trajectory inside the magnetosphere and have substantial contributions to the field at the planetary surface and even at the top of the core (see Section 5.4.1).

Mercury's global-scale field is summarized in Figure 5.8. The radial component of the internal field, B_r, shown in MBF coordinates, demonstrates the equatorial asymmetry and the strong axisymmetry in the field (Figure 5.8a). The equatorial asymmetry is also seen in the total field strength, B, from the combined contributions of the OAD, magnetopause, and magnetotail fields shown in the MSO frame at the planetary surface (Figure 5.8b). The planet-centered spherical harmonic description of the OAD is found from Table 5.1 and the best-fit dipole moment of 190 nT R_M^3, which yields $g_1^0 = -190$ nT, $g_2^0 = -75$ nT, $g_3^0 = -22$ nT, and $g_4^0 = -6$ nT. We discuss the implications of this core field structure in Section 5.7.

5.3.2.2 Proton Reflectometry: Constraints on Surface Field Strength

Electron reflectometry (ER) has been used to sense remotely the magnetic field strength at the surface of the Moon (Lin et al., 1998) and at 185-km altitude at Mars (Lillis et al., 2008). The ER technique depends on the magnetic mirroring effect, that is, the reflection of charged particles by convergent magnetic field lines. For an airless body, electrons that would mirror below the surface are lost, and the flux of reflected electrons exhibits a sharp drop at the pitch angle (the angle between the particle velocity vector and the local magnetic field) corresponding to mirroring at the surface. The in situ magnetic field together with the pitch angle of the last reflected electrons (the cut-off pitch angle) indicates the surface magnetic field strength. This technique, in principle, provides an estimate of surface magnetic field that is not subject to the issues of spatial resolution and downward continuation found in classical magnetic field studies. For an orbit such as MESSENGER's, the technique could offer a higher-spatial-resolution view of the surface field at low and southern latitudes than is available from the MAG data. Winslow et al. (2014) adapted the ER technique to protons detected by the Fast Imaging Plasma Spectrometer (FIPS) (Andrews et al., 2007; Zurbuchen et al., 2011; Chapters 16 and 17) to estimate Mercury's surface magnetic field strength.

A challenge in proton reflectometry with MESSENGER data was that at any given time the incident and reflected proton populations were not observed simultaneously because of the limited, ~1.4π sr, FIPS field of view. Thus, pitch angle distributions from times when the incident population was observed were combined with those obtained when the reflected population was visible. Winslow et al. (2014) focused on two regions where the highest proton counts were detected by FIPS: Mercury's northern magnetospheric cusp and the low-latitude (0°–30°S) region of the nightside (Figure 5.8b).

Estimates of the surface magnetic field strength were 412 ± 98 nT for the northern cusp and 113^{+87}_{-61} nT for the low-latitude nightside. These values are markedly less than the magnetic field strength predicted by the magnetospheric model of Johnson et al. (2012). Such a difference is expected, however, because plasma generates a diamagnetic field, which reduces the surface field below the vacuum model prediction (Korth et al., 2011, 2012; Winslow et al., 2012). After accounting for estimates of the diamagnetic effect, the surface field strength in the northern cusp (498 ± 99 nT) and low-latitude southern hemisphere (157^{+87}_{-61} nT) showed fair agreement with the paraboloid model predictions. Furthermore, the ratio between the surface magnetic field strength in the northern cusp and that in

5.3 Core Dynamo Field   125

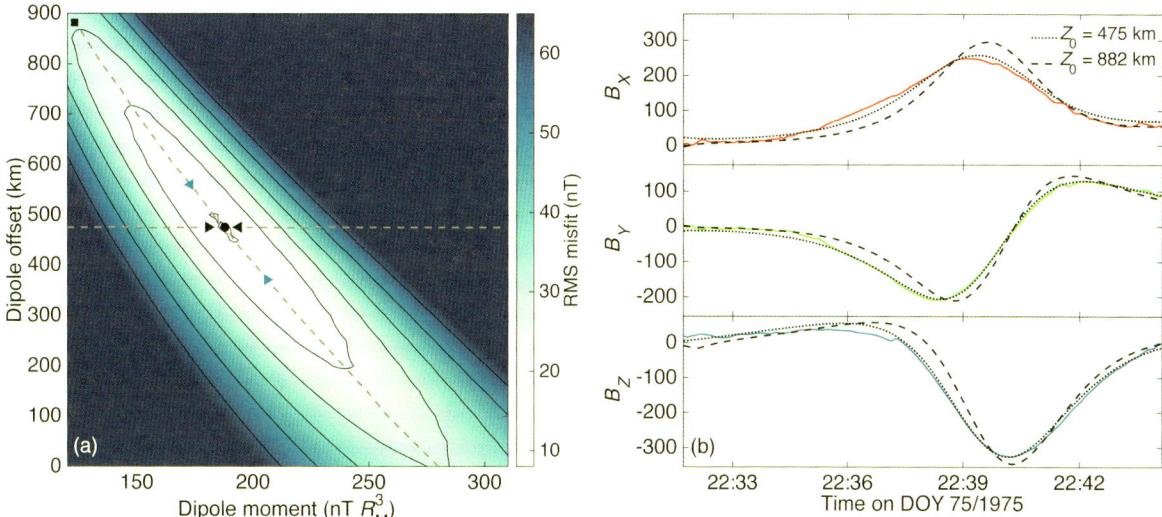

Figure 5.9. (a) RMS misfit between model and M10-III data obtained from a grid search over dipole moment and offset of the dipole from the geographic equator. Contours indicate misfits from 10 nT to 60 nT in 10-nT steps. The parameter pairs for the best-fit offset of $Z_0 = 475$ km and the M10-III magnetic equator offset of 882 km are marked by the filled black circle and filled square, respectively. The diagonal dashed line shows the trade-off between best-fit dipole offset and dipole moment; blue triangles correspond to parameter pairs that yield an RMS misfit 25% above the minimum value. The horizontal dashed line shows the location of an RMS misfit profile corresponding to the best-fit dipole offset; a 25% increase above the minimum misfit would yield dipole moments of 181 and 194 nT R_M^3, shown by the black triangles. (b) Magnetic field observations from M10-III inside the magnetopause (solid lines) and best-fit models to the data for dipole offsets of 475 km and 882 km. Figure modified from Philpott et al. (2014).

the southern hemisphere was found to be $3.2^{+3.0}_{-1.6}$, in agreement with the expected value of 3.3 for an offset dipole field alone and distinguishable from the expected ratio of 1.5 for a centered dipole field at the latitudes sampled. Proton-reflection magnetometry thus provided independent confirmation of the equatorial asymmetry in, and OAD representation of, Mercury's magnetic field.

5.3.3 Constraints on Secular Variation

In addition to the present average field structure, the secular (or temporal) variation, and in particular its spatial power spectrum, provides a constraint on core dynamo models (see Section 5.6). Techniques developed for the analysis of MESSENGER orbital data were applied to the M10-III data to search for evidence of secular variation in the planetary field (Philpott et al., 2014), specifically to investigate any changes in the dipole moment or dipole offset.

The instantaneous position of the magnetic equator in the M10-III data was identified and compared with the range of values seen in the MESSENGER orbital data. The M10-III data yielded an instantaneous equator position of $Z_{\rho 0} = 882 \pm 35$ km, significantly northward of the average position calculated from the MESSENGER orbital data, but not atypical of instantaneous crossings from MESSENGER orbits with trajectories most similar to that of the M10-III flyby. As the M10-III magnetic equator crossing was not necessarily a good indication of the average dipole offset in 1975, the paraboloid magnetospheric model of Johnson et al. (2012) was used to conduct a grid search over both the dipole offset, Z_0, and the magnitude of the dipole moment, $m_{\rm dip}$, to find the internal field model that best fit the M10-III data. External fields during the M10-III flyby were prescribed using the M10-III observations of the magnetopause and MESSENGER average values for the tail fields (details were given by Philpott et al., 2014).

The best-fit model to the data has an RMS misfit of 9.6 nT, a dipole offset, Z_0, of 475 km, and a dipole moment of $m_{\rm dip} = 188$ nT R_M^3 (Figure 5.9). As previously reported (Connerney and Ness, 1988; Korth et al., 2004), the M10-III trajectory results in a trade-off between the dipole offset and dipole moment (Figure 5.9a). For example, a 25% increase in RMS misfit corresponds to a change in the ($m_{\rm dip}$, Z_0) parameters from (173 nT R_M^3, 560 km) to (206 nT R_M^3, 370 km). For a given dipole offset, the misfit minimum is better defined. The model with a dipole offset given by the observed instantaneous equator crossing position, $Z_{\rho 0} = 882$ km, is clearly not a good fit to the observed planetary field (Figure 5.9b).

The best-fit model yields no resolvable evidence for secular variation in either the dipole moment or its offset since the time of the Mariner 10 flybys, and Philpott et al. (2014) also concluded that there is no evidence for significant secular variation in Mercury's dipole tilt. The trade-off between dipole moment and offset provides bounds on secular variation. Limits on the low-degree axial spherical harmonic coefficients g_1^0, g_2^0, and g_3^0 during M10-III were estimated to be (−206, −62, −14) nT and (−173, −79, −27) nT, compared with the mean values from the MESSENGER orbital data of (−190, −75, −22) nT (Anderson et al., 2012). Thus, changes in the dipole moment (or the g_1^0 coefficient) of less than 10% may be accompanied by changes in the degree-2 and degree-3 axial terms of up to 16% and

35%, respectively, yielding substantial changes in the slope of the low-degree spatial power spectrum (see discussion in Section 5.6).

5.3.4 Beyond the OAD: Residual Signatures of Internal Origin?

The models of Johnson et al. (2012) and Korth et al. (2015) successfully predict ~90% of the signal in the MAG observations, demonstrating the overall structure of the global-scale magnetospheric and internal fields. The remaining or residual signals can comprise systematic signatures of internal or external origin that are present in the natural system but not captured by the models, or they could be random fluctuations, e.g., driven by orbit-to-orbit variations in solar wind conditions.

Residuals to the paraboloid model for the first four Mercury sidereal days show signals up to 50 nT amplitude in the MBF B_θ and B_ϕ components that are not consistent from one sidereal day to the next (Figure 5.10). Positive signatures in B_r are seen at latitudes north of 60°N. At least part of this signal repeats at similar MBF longitudes every third sidereal day, rather than every sidereal day, consistent with signatures that are organized in the MSO frame and the associated aliasing into the MBF frame from Mercury's 3:2 spin–orbit resonance. This pattern was confirmed by examining the residuals in the MSO frame (Johnson et al., 2012) and verified with data spanning seven Mercury years and the KT14 model (Korth et al., 2015). Furthermore, residual amplitudes correlate with magnetospheric activity, even at long, $O(1000$ km$)$, wavelengths (see Figure 16.20, and orbit-to-orbit amplitude variations in Figure 5.10). These signals are dominantly of external origin and result from at least two different effects. First, diamagnetic depressions of the field by local plasma pressure are particularly strong in the plasma sheet at the magnetic equator, even within 0.5 R_M altitude (Korth et al., 2012, 2014), and in the northern magnetospheric cusp (Winslow et al., 2012, 2014; Raines et al., 2013). Second, the large signals in the horizontal field components (notable in B_ϕ in Figure 5.10) reflect field-aligned or Birkeland currents (Anderson et al., 2014). The fields due to Birkeland currents increase with decreasing altitude, are present to the lowest spacecraft altitudes, and close in the planetary interior (Anderson et al., 2014, 2016). Finally, we note that residuals to the paraboloid and KT14 magnetospheric models are very similar in their large-scale structure (Johnson et al., 2012; Korth et al., 2015; Chapter 16), indicating that they reflect real signals in the field.

The large-scale quasi-steady external residual fields present severe challenges to identifying remaining fields of internal origin, in particular core field structure with wavelengths corresponding to spherical harmonic degrees 2–10 that, if present, should be resolvable over the northern hemisphere (Uno et al., 2009). The possibility that Birkeland currents close through the planet is particularly problematic because even signals observed at very low altitude that appear to have a source within the planet in the formalism of equation (5.1) may actually arise from currents driven by the solar wind interaction with the magnetosphere. Fortunately, the amplitudes of these signals are also highly correlated with magnetic activity (Anderson et al., 2014, 2016), and this relation suggests that establishing core field structure beyond the OAD may be possible with data from magnetically quiet conditions. We return to further discussion of this topic in Section 5.6. Shorter-wavelength, steady internal fields can be identified because of their stronger altitude dependence than fields from Birkeland currents and because the quasi-steady external contributions are observed to have less power at these wavelengths. This situation has enabled the detection of crustal fields (Section 5.5). However, the large amplitudes and time dependence of the magnetospheric fields have permitted two distinct investigations of interior electrical conductivity, which we discuss next.

5.4 INTERIOR ELECTRICAL CONDUCTIVITY STRUCTURE

An open question for Mercury prior to the MESSENGER mission was whether changes in the magnetopause driven by solar wind ram pressure (P_{ram}) could induce currents in the planetary interior that, in turn, would produce detectable fields at or above the planetary surface (Hood and Schubert, 1979; Suess and Goldstein, 1979; Grosser et al., 2004; Glassmeier et al., 2007b). Induction was initially of interest because it increases the solar wind dynamic pressure needed to compress the dayside magnetosphere to the surface of the planet. This effect could inhibit direct bombardment of the dayside surface by the solar wind and limit surface space weathering and exospheric species production by charged particle bombardment (Leblanc and Johnson 2003, 2010; Wurz et al., 2010; Jia et al., 2015). However, magnetopause reconnection and erosion of the dayside magnetospheric field can offset or even exceed the effects of induction (Slavin and Holzer, 1979; DiBraccio et al., 2013; Slavin et al., 2014), and so it was unknown whether induction would be observable.

Induced fields are sensitive to the interior electrical conductivity structure of the planet. For an electrical conductivity model consisting of an inner conducting sphere, of conductivity σ_0 (~10^6 S/m), and a silicate shell, of conductivity σ_m (~0) (Grosser et al., 2004), induction in the silicate shell is negligible relative to that in the conducting sphere at periods longer than 1–2 h. The induced currents are confined to a very thin shell at the top of the conducting sphere, and MESSENGER data have enabled magnetic "sounding" of the core–mantle boundary from induction (Johnson et al., 2016a). A complementary opportunity to probe the electrical conductivity properties of the silicate part of the planet has been provided by the discovery of field-aligned currents at Mercury (Anderson et al., 2014). Constraints on the spatial distribution, dynamics, and origin of the currents are described in Chapter 16; the implications for conductivity structure (Anderson et al., 2016) are summarized here.

5.4.1 Induced Fields in Mercury's Core

In the presence of induction, an increase in P_{ram} will compress the magnetosphere, i.e., decrease the magnetopause subsolar distance, R_{SS}, and induce currents in the core that increase the planetary field to oppose the decrease in R_{SS}. Conversely, a decrease in P_{ram} leads to a decrease in the planetary field to

5.4 *Interior Electrical Conductivity Structure* 127

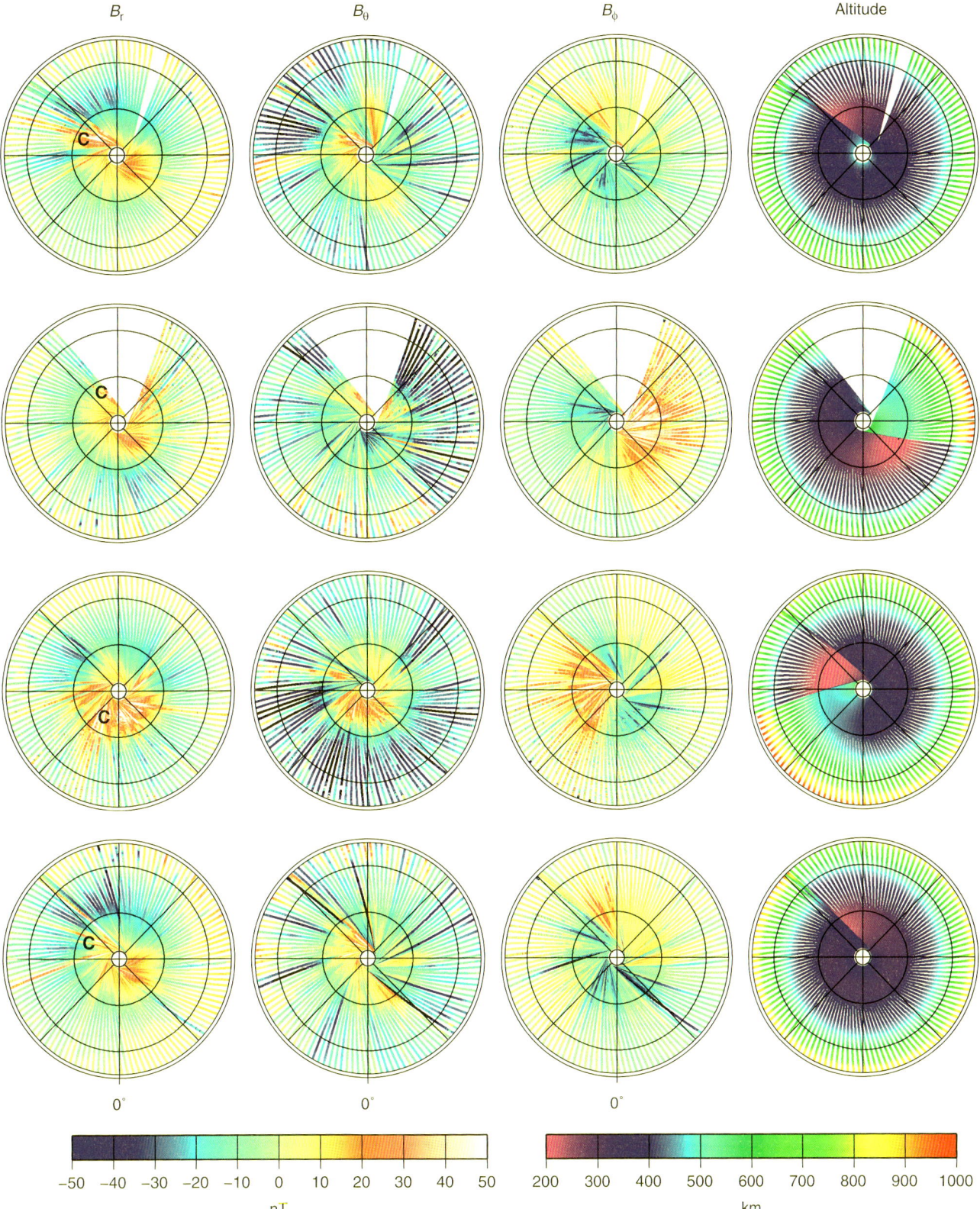

Figure 5.10. Residuals in MBF format for the first four coverages (1 Mercury day, 59 Earth days, each) of body-fixed longitude (rows top to bottom). Columns left to right show B_r, B_θ, and B_ϕ residuals in nT (bottom left color bar) and spacecraft altitude in km (bottom right color bar). Structure in the spacecraft altitude figures reflects the drift of periapsis altitude to higher elevations and the periapsis-lowering maneuvers that occurred every 88 days. 0° MBF longitude is toward the bottom of each figure; figures show latitudes only northward of 15°N; grid lines are at 30°N, 60°N, and at 45° longitude intervals. The region labeled "C" denotes the magnetospheric cusp. Figure modified from Johnson et al. (2012).

oppose the increase in R_{SS}. R_{SS} values inferred from three Mercury years of MAG observations suggested that induction occurs at Mercury, but the limited data set and absence of observational constraints on P_{ram} (Winslow et al., 2013) could not exclude the case of no induction. MESSENGER data spanning 15 Mercury years were subsequently used to investigate the presence of an annual induction signal (Johnson et al., 2016a). Because P_{ram} varies as r_h^{-2}, where r_h is heliocentric distance, direct estimates of P_{ram} are not required, and in the absence of induction R_{SS} will vary as $r_h^{1/3}$. If induction occurs, R_{SS} will vary as r_h^b, where the exponent b is less than 1/3. In addition, the change in the planetary field due to induction can be investigated from observations inside the magnetosphere. The largest induction signals will be in the axial dipole term, g_1^0, in the internal field: a relative increase in the dipole moment will occur at perihelion, and a decrease will occur at aphelion.

The annual induction signal was investigated using orbits for which the magnetospheric activity index (Anderson et al., 2013) was less than 33, i.e., quiet orbits, because R_{SS} values show a substantial dependence on magnetospheric activity at all r_h (Figure 5.11a). A least-squares fit to $R_{SS} = a\,r_h^b$ yielded an exponent $b = 0.29 \pm 0.03$, where the uncertainty gives the 95% confidence limits on b (Figure 5.11b). Thus, the value for b is significantly different from the value of 1/3 expected in the absence of induction. The inferred R_{SS}–r_h relation corresponds to a change in dipole moment of 9.5 nT R_M^3 between perihelion and aphelion, with 95% confidence limits of 3.8 and 15.2 nT R_M^3.

Two subsets of quiet orbits near perihelion and aphelion (colored symbols, Figure 5.11b) were used to investigate the best-fit dipole moment to the MAG data inside the magnetosphere given the KT14 model and the fitting approach described in Section 5.3.1. The best-fit dipole moment at aphelion was found to be 7.5 nT R_M^3 greater than that at perihelion (Figure 5.11c), consistent with the presence of an induction signal and in good agreement with the mean change in dipole moment inferred from the R_{SS}–r_h relation.

For the two-layer electrical conductivity model described above (Grosser et al., 2004), the transfer function between the induced and the inducing spherical harmonic coefficients in the planetocentric local time frame at a given period and given spherical harmonic degree, l, is given by $l/(l+1)(R_{COND}/R_M)^{(l+1)}$ (Rikitake, 1966; Grosser et al., 2004). Typically, the radius of the conducting sphere, R_{COND}, is assumed to be the radius of the core, R_C. Prior to the MESSENGER mission, a commonly used value for R_C was 1800 km, with a permitted range from ~1760 km to even 2400 km (Harder and Schubert, 2001; Hauck et al., 2007). MESSENGER observations of Mercury's gravity field, in combination with Earth-based observations of Mercury's rotational state, indicate $R_C = 2020$ km $\pm$ 30 km (Hauck et al., 2013; Chapters 4 and 19). The possibility of a solid FeS layer at the base of the silicate mantle has been suggested (Hauck et al., 2013), although conditions under which such a layer could occur are restrictive (Chabot et al., 2014). If present, such a layer may have an electrical conductivity much greater than that of the silicate rocks above it; for such a scenario, R_{COND} would correspond to the radius of the top of this layer rather than R_C.

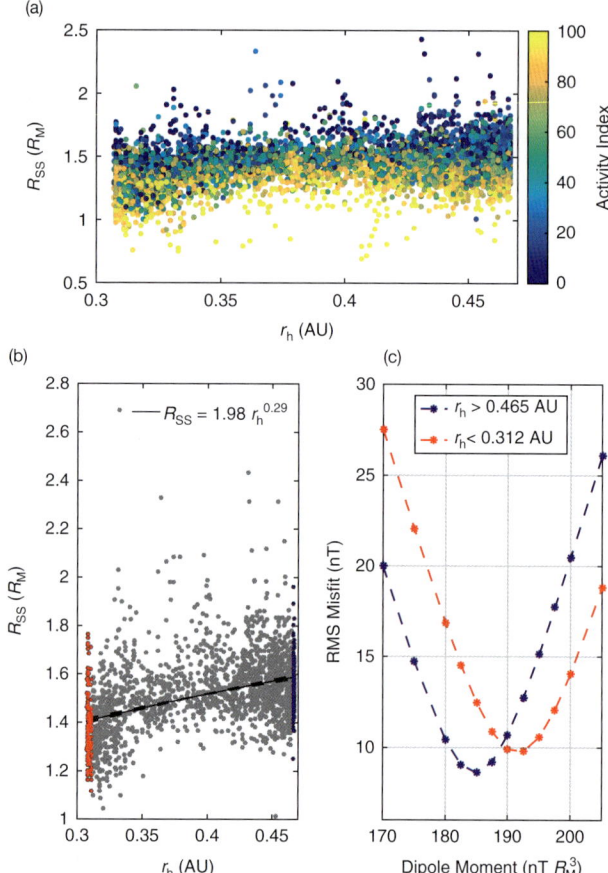

Figure 5.11. (a) Subsolar magnetopause distance, R_{SS}, versus heliocentric distance, r_h, computed for magnetopause crossings spanning the period 24 March 2011 – 31 October 2014 (15 Mercury years) and color-coded by activity index (Section 5.2.3.4). One crossing for which $R_{SS} > 2.5\,R_M$ is not shown. (b) R_{SS} versus r_h for 1943 quiet orbits (see text) along with the best-fit power law $R_{SS} = 1.98\,r_h^{0.29}$ (black solid line) and 95% confidence limits (black dashed lines). Red and blue dots denote R_{SS} values for 76 and 93 orbits within 0.005 AU of perihelion and 0.002 AU of aphelion, respectively. (c) RMS misfit of predicted B_r to observed B_r inside the magnetosphere as a function of dipole moment for orbits near perihelion (red) and aphelion (blue). From Johnson et al. (2016a).

The inducing field geometry was computed for a given change in R_{SS} by subtracting spherical harmonic expansions of the KT14 magnetopause fields computed at the initial and final R_{SS} values. The inducing field is dominated by the q_1^0 and q_2^1 terms, reflecting the equator-to-pole and day–night structure in the magnetopause field. The predicted induced terms (g_1^0, g_2^1) are shown in Figure 5.12 for the pre-MESSENGER range of R_{COND} estimates. MESSENGER's orbit geometry yielded uncertainty in the knowledge of the tail field, and the resulting shielding field and two inducing field models (Figure 5.12, see details given by Johnson et al., 2016a) were examined to account for this uncertainty. The predicted g_1^0 induction signals for $R_{COND} = 2020$ km and the two inducing field models are 7.5 and 9.0 nT (Figure 5.12).

Thus, MESSENGER observations of Mercury's magnetopause position and of the field inside the magnetosphere show evidence for an annual induction signal. The average values obtained from the two approaches show excellent agreement with the predicted values for $R_{COND} = R_C = 2020$ km (Hauck et al., 2013), and the pre-MESSENGER canonical value, $R_C \sim 1800$ km, is not favored. The annual induction signal thus constrains the radius of the top of the highly electrically conductive region, providing an assessment of Mercury's core radius that is independent of traditional geodetic approaches (Hauck et al., 2013).

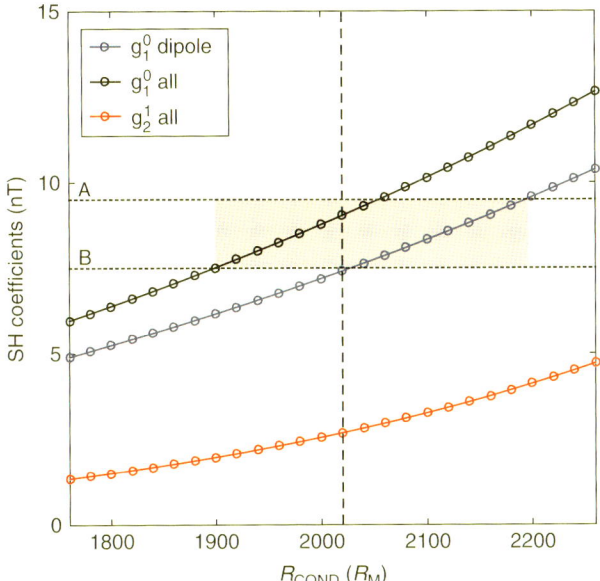

Figure 5.12. The predicted annual induction signal for spherical harmonic (SH) terms g_1^0 and g_2^1, resulting from a change in R_{SS} from 1.4 R_M to 1.6 R_M, as a function of R_{COND}. Coefficients g_1^0 (black) and g_2^1 (red) are computed from the full external field (the tail field and both the dipole and tail shielding fields) with the KT14 model (Korth et al., 2015). The induced g_1^0 term (blue) for a magnetopause field that shields only the planetary dipole field is also shown. The dashed vertical line denotes R_{COND} = 2020 km. Dotted horizontal lines are the mean values for the annual g_1^0 induction signal inferred from $R_{SS} - r_h$ (labeled "A") and the field inside the magnetosphere (labeled "B"). The purple shaded region indicates the range of R_{COND} consistent with the observed average dipole induction signal and the predicted values from the KT14 model. Figure modified from Johnson et al. (2016a).

5.4.2 Implications of Birkeland Current Closure

The observation of fields from Birkeland currents raised the question of where such currents close and suggested that if the currents close through the planet, they may provide an indirect means to probe the electrical conductivity of the crust and mantle. Since different candidate materials are expected to exhibit substantially different electrical conductivities at a given pressure and temperature (Verhoeven et al., 2009), the discovery of steady-state Birkeland currents at Mercury provides an indirect constraint on composition and temperature in the silicate part of the planet. The effective electrical conductance (conductivity integrated over the conductive path length) consistent with the observed currents and the electric potential applied to Mercury's magnetosphere is ~ 1 S (Anderson et al., 2014; Chapter 16). This value is too high for either exospheric or regolith conductances (see Anderson et al., 2014, and references therein), so Anderson et al. (2014) evaluated a simple spherical shell conductivity model (Janhunen and Kallio, 2004) to assess whether closure of Birkeland currents through the planet is consistent with the range of plausible electrical conductivities for the crust and mantle. Under the assumption that all of the current closes radially through a thin shell to the much more highly conducting core and then laterally across the top of the core, one can relate the resulting solutions to a conductivity profile, using as a boundary condition the inferred surface current density of the Birkeland currents. For an exponentially increasing conductivity with depth, $\sigma(d) = \sigma_s \exp(d/\lambda_\sigma)$, where $d = R_M - r$ and R_M is Mercury's radius, σ_s is the surface value of the electrical conductivity, and λ_σ is a characteristic scale length for the increase in electrical conductivity with depth, the solutions for the closure of Birkeland currents imply $\sigma_s/\lambda_\sigma \sim (5.0 \pm 2.4) \times 10^{-11}$ S m^{-2}. An estimate of 5 km for λ_σ gives $\sigma_s \sim 3 \times 10^{-8}$ S m^{-1}, consistent with expectations for a wide range of candidate mineralogies for Mercury (Verhoeven et al., 2009).

Current closure solutions for an arbitrary variation of conductivity with depth were obtained by Anderson et al. (2016). A radial conductivity profile of the form shown in Figure 5.13a was considered, for depths to the CMB of 370 to 470 km, consistent with values for Mercury's core radius (Hauck et al., 2013; Johnson et al., 2016a; Chapter 4). Just above the core, the conductivity was assigned a maximum value for the mantle, σ_{max}, of 0.01 to 1.0 S m^{-1}. The exponential increase in σ with depth described

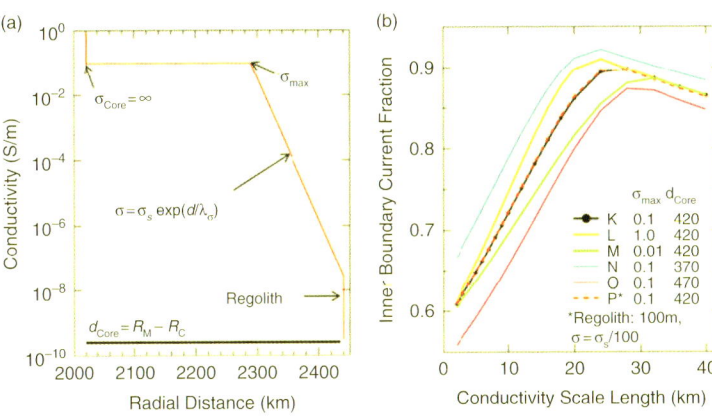

Figure 5.13. (a) Model radial conductivity profile showing the variation of conductivity in the regolith, increasing to σ_{max} exponentially with depth, d, and extending to the outer radius of the core, R_C, as indicated by the horizontal black bar, and (b) fraction of total current closing through the core for a range of conductivity profiles with different σ_{max} and d_{Core}. From Anderson et al. (2016).

above was used, where σ_s was the value either at the surface or just below an assumed 100-m-thick regolith with $\sigma = 0.01\sigma_s$. Solutions for the electrostatic potential and current density distribution were obtained for $\sigma_s = 10^{-8}$ S m^{-1} and λ_σ between 1 and 40 km, given the measured Birkeland current density distributions mapped to the surface. Figure 5.13 shows the form of the conductivity profile and the fraction of current closing to the core for a range of cases. For all cases, even those with very steeply increasing conductivity with depth, more than half of the current closes through the core, and the results are insensitive to the presence of a low-conductance regolith. The current flowing to the core is maximized for λ_σ near 20 to 30 km. The reason is that as λ_σ varies from ~1 km to 20 km the current to the core increases as the lateral conductance through the crust and mantle increases, but for λ_σ greater than 30 km, the resistance to the core increases relative to the lateral conductance through the crust and mantle. The key result is that the ratio between the potential and the scale length is invariant, so that $\sigma_s/\lambda_\sigma \sim$ constant holds, even for these complex conductivity profiles, with the same value as obtained for the shell model. Fundamentally, more than 90% of the potential drop occurs in the outer few scale lengths, and how the current closes at greater depth has almost no influence on the potential, so the effective resistance is set by the series resistance of the upper few scale lengths, as argued by Anderson et al. (2014). The insensitivity of the result to closure at depth implies that measurement of the electric potential together with the currents could be used to constrain the electrical conductivity of the near-surface material.

5.5 CRUSTAL FIELDS

Low-altitude observations made near the end of MESSENGER's orbital mission provided an unprecedented opportunity to investigate short-wavelength and/or low-amplitude internal fields of either core or crustal origin. As discussed in Section 5.3.4, after subtraction of the major magnetospheric fields and the OAD, the remaining signals are dominated by external contributions that can vary on an orbit-to-orbit basis in their frequency content and by more than 10 nT in amplitude (Figure 5.10). These fields dominate even at the lowest periapsis altitudes, indicating that fields resulting from crustal magnetization, if present, are very weak. Thus, detection of crustal fields requires additional data processing, followed by verification that any remaining signals are of internal origin. The initial identification of crustal fields was made from two periods of very-low-altitude observations (periapsis altitudes close to 25 km) in September and October 2014, over two distinct regions of Mercury's northern hemisphere (Johnson et al., 2015). Additional observations obtained from January until April 2015 enabled mapping of crustal fields over much of the northern hemisphere (Johnson et al., 2016b).

5.5.1 Identification of Crustal Fields in Low-Altitude Observations

Crustal fields were identified via a two-step process (Johnson et al., 2015). First, predictions of the KT14 model (Korth et al., 2015) for the major magnetospheric fields and the OAD were subtracted from the observations on an orbit-by-orbit basis. Because models for the remaining external contributions were not available, the long-wavelength residual fields (Figure 5.14a) were removed with a high-pass filter to better reveal short-wavelength fields of possible internal origin. The filter was designed to ensure that fields with spatial scales greater than ~1900 km were removed and those with spatial scales less than about 520 km were retained

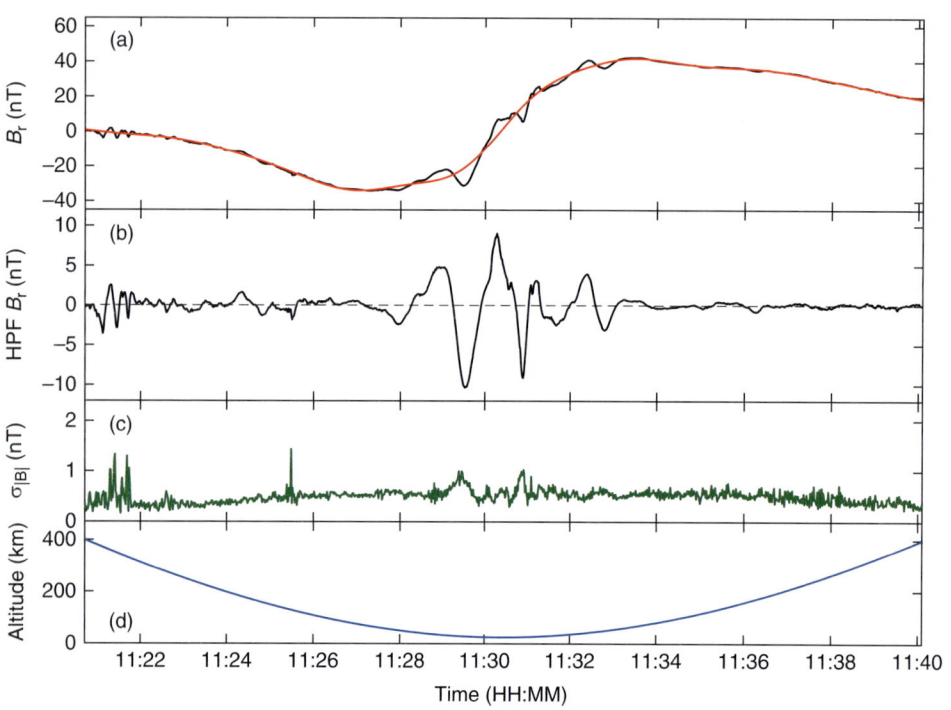

Figure 5.14. Magnetic field observations from 8 September 2014 to show filtering used to isolate short-wavelength signals on an orbit-by-orbit basis. The orbit track crosses the Suisei Planitia region shown in Figure 5.17. (a) MBF radial component of the field, B_r (black), after subtraction of the KT14 model-predicted field; and the low-pass filtered signal (red). (b) High-pass-filtered (HPF) signal showing the near-periapsis signals. (c) High-frequency (> 1 Hz) variability in the total field, $\sigma_{|B|}$, a measure of the external field noise remaining in the HPF signals. (d) Spacecraft altitude. Periapsis altitude was 25 km; 100 s corresponds to a horizontal scale of ~385 km at periapsis. Modified from Johnson et al. (2015).

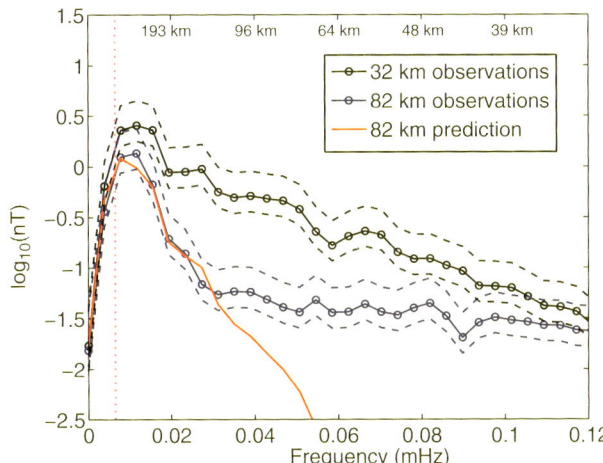

Figure 5.15. Amplitude spectra for HPF B_r for orbits over Suisei Planitia at average altitudes of 32 km (black) and 82 km (blue). The dashed lines are 95% confidence limits. Predicted amplitudes from upward continuation of the spectrum observed at 32-km altitude to 82-km altitude are in red. Equivalent wavelengths are marked inside the top axis. The sharp roll-off in the spectrum at frequencies below 0.0075 Hz (magenta dotted line) reflects the roll-off in the HPF between 0.0075 Hz and 0.002 Hz. From Johnson et al. (2015).

5.5.2 Geographical Distribution of Crustal Fields

Operational constraints resulted in the acquisition of most observations below ~100-km altitude at local times between 1600 and 0800 (through 0000) with excellent geographical coverage (Figure 5.16a). This timing was advantageous because magnetic field data in this local time interval were typically quieter than dayside and dawn–dusk data, and so weak crustal signals could be clearly identified (Figure 5.16b). The strongest fields occur in the region 120°E to 210°E, a region that includes both smooth and heavily cratered terrain to the north and northwest of the Caloris impact basin. Weaker signals characterize heavily cratered terrain in other regions and the northern volcanic plains.

In this data set, 21 orbits had periapsis altitudes less than ~10 km, and the corresponding magnetic field profiles are of interest because they afford the best opportunity to detect larger-amplitude, shorter-wavelength signals. These orbits were found to exhibit a variety of signals. Some show no substantial signal even at very low altitudes, e.g., over the heavily cratered Victoria and Carnegie Rupes regions (e.g., Figure 5.16b) and portions of orbits 4097–4100 (Figure 5.17). HPF signals are seen in association with some, but not all, craters. For example, for the tracks shown in Figure 5.17, spacecraft altitudes were sufficiently low to resolve signals over Turgenev crater and unnamed crater C1 if present, yet no such signals were observed. In contrast, Strindberg basin (centered at 53°N, 194°E, 187-km diameter) shows clear signals associated with the crater rim and the peak ring within the basin interior, and very weak signals are associated with the rim and ejecta of the basin on the northern rim of the Shakespeare basin on MESSENGER's last complete orbit (Figure 5.17). Many signals are not associated with obvious surface features, such as the peak HPF B_r signals of −20 nT and −30 nT at 56.5°N, −142.7°E, 5-km altitude and 61.2°N, −142.7°E, 6-km altitude on orbits 4102 and 4103, respectively. The high-frequency signals on the last three orbits are indicative of plasma rather than crustal fields. Notably, even at these very low spacecraft altitudes, the shortest wavelengths in the crustal fields are ~40 km.

5.5.3 Magnetization Models and Altitude-Normalized Field Maps

Systematic variations in spacecraft altitude with latitude and the multiple periapsis-raising orbit-correction maneuvers (Figure 5.16a) make it difficult to compare observations taken over different regions. The equivalent source dipole technique (Section 5.2.3.3) has been used to invert for magnetization distributions and to produce altitude-normalized magnetic field maps. At the time of writing, these studies are in their early stages, but comprise an investigation of the entire low-altitude data set (Johnson et al., 2016b) as well as independent investigations focusing on the Suisei Planitia and the Caloris and Sobku basins (Hood, 2015, 2016). The main differences among the modeling approaches are (a) the a-priori data filtering, (b) the model dipole mesh design, and (c) the inversion approach. Johnson et al. (2016b) used the high-pass filter technique described above with a 20-km-spacing icosahedral grid, allowing wavelengths from ~40 km to at least 520 km to be

(Figure 5.14b; for details see Johnson et al., 2015). After filtering, higher-frequency external signals due to waves and local plasmas remained but could be identified by enhancements in the variability in the total field, $\sigma_{|B|}$, at frequencies above 1 Hz. On portions of many orbits with spacecraft altitudes below ~100 km, the high-pass-filtered (HPF) data showed smoothly varying signals (Figure 5.14b) with amplitudes that were low (a few to ~10 nT) but more than three times that of $\sigma_{|B|}$ (Figure 5.14c). Tests indicated that it may be possible to extend the frequency range of the high-pass filter to retain all signals with wavelengths less than ~770 km (Johnson et al., 2015). However, the more conservative filter choice was made to ensure little or no contamination by external contributions, with the trade-off that the amplitude of the resulting crustal field signal could be underestimated by ~10–15%.

The first identification of these signals was made from data over Suisei Planitia (~50–70°N, 200–230°E) taken at an average altitude of 32 km in September 2014 (Johnson et al., 2015). Repeated observations over the region in January 2015 indicated weaker signals at an average altitude of 85 km. Along-track amplitude spectra of the data at the two different average altitudes were calculated (Figure 5.15). The amplitude of a signal of wavelength λ of shallow internal origin should attenuate between two altitudes z and $z + \Delta z$ as $\exp(-2\pi\Delta z/\lambda)$, an effect known as upward continuation. As shown in Figure 5.15, the predicted amplitudes from upward continuation of the spectrum observed at 32-km altitude to 82-km altitude ($\Delta z = 50$ km) match the 82-km-altitude observations for frequencies below 0.03 Hz (wavelengths longer than ~125 km). These results confirm an internal origin for the source(s) of the observed HPF signals and also indicate that the source is shallower than the CMB.

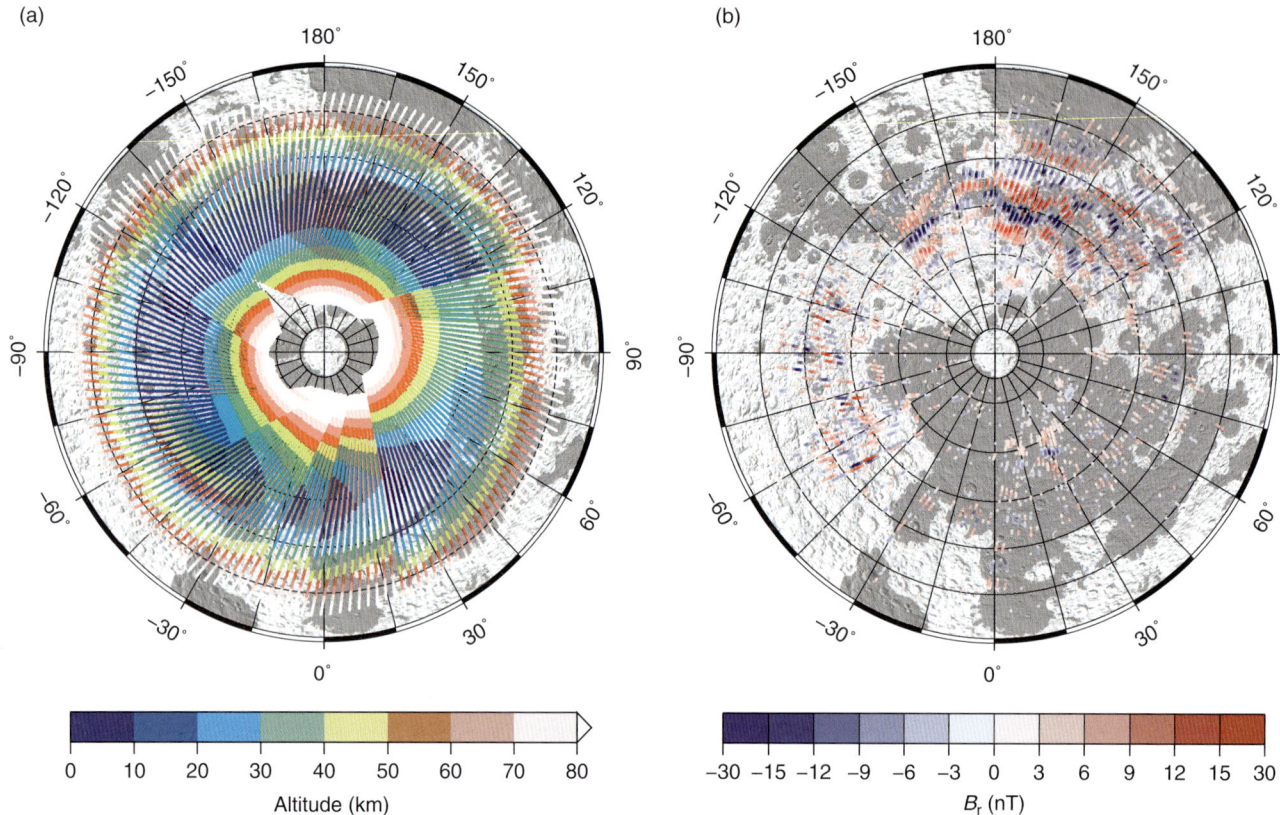

Figure 5.16. Lambert azimuthal equal area projections from 32°N to 90°N. (a) Spacecraft altitudes along orbit segments for which the spacecraft altitude was less than 100 km, and for which local times were between 1600 and 0800. Portions of the tracks with altitudes from 80 to 100 km are in white. Additional daytime coverage between 0800 and 1600 was obtained in the longitude band ~310°E to 230°E. Background image shows shaded relief with smooth plains (gray) and intercrater plains (white) (Denevi et al., 2013). Abrupt changes in altitude on adjacent orbits are seen in several locations and resulted from OCMs conducted to extend the low-altitude phase of the mission as long as possible (Chapter 1). (b) Map of high-pass-filtered radial field (B_r). Signals are shown only if the magnitude of B_r is at least 1 nT and if the signal-to-noise ratio is at least 4. Grid lines every 10° of latitude and 15° of longitude. Figure modified from Johnson et al. (2016b).

modeled. Hood (2015, 2016) used a dipole spacing of ~43 km, and the data processing was more restrictive, allowing wavelengths only between ~85 and 215 km to be modeled. The magnetization direction was constrained in all studies because of the non-uniqueness inherent in such inversions. Johnson et al. (2016b) also used regularization in the solution of equation (5.2) to stabilize the inversion and downward continuation of the field, as described in Section 5.2, and modeled the full low-altitude data set, performing suites of inversions with subsets of the data (e.g., B_r data only versus the full vector field, nightside data versus all local times) to check for consistency among the solutions. As a result, the magnetization model, as well as altitude-normalized maps, can be examined over much of the northern hemisphere. We summarize the results of that approach here. The magnetization model (Figure 5.18) was used to calculate a map of B_r at 20-km altitude (Figure 5.19).

The strongest magnetizations (0.1–0.4 A m^{-1} for a 10-km-thick layer) and magnetic fields and the largest coherence length scales are spatially associated with the Caloris and circum-Caloris region (Hood, 2015, 2016; Johnson et al., 2016b), as shown in Figures 5.18 and 5.19. Elsewhere, magnetization and magnetic field amplitudes are weaker and exhibit shorter coherence length scales. In general, weak signals are seen in association with the intercrater plains in the western hemisphere and with the northern smooth plains. Magnetization contrasts are sometimes, but not always, found in association with craters (Johnson et al., 2016b), and magnetization magnitudes are compatible with the theoretical lower bounds derived from the earliest low-altitude observations over Suisei Planitia (Johnson et al., 2015).

5.6 DISCUSSION AND FUTURE DIRECTIONS

5.6.1 Core Field Structure and Modern Dynamo

MESSENGER observations have confirmed that Mercury's core field is dominantly dipolar but quite unlike the dipolar planetary fields elsewhere in the solar system in its combination of weak strength, axisymmetry, and equatorial asymmetry. Spherical harmonic magnetic field models are often summarized via the spatial power spectrum given by

5.6 Discussion and Future Directions

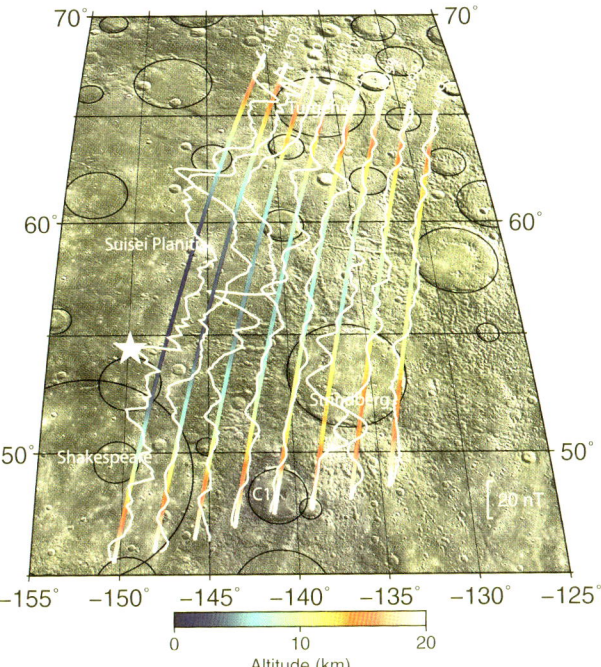

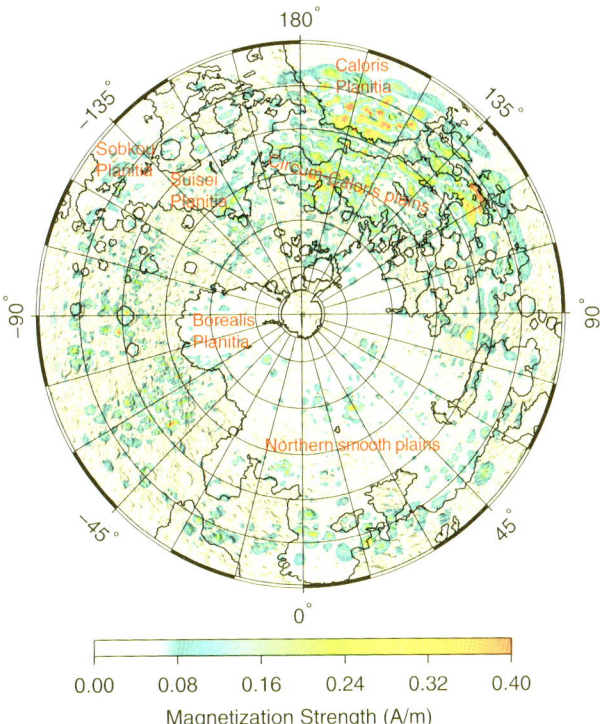

Figure 5.17. HPF B_r in nT (white lines) projected along the orbit tracks for the last eight complete orbits of the MESSENGER mission. Orbits have periapsis altitudes below 10 km, and track segments are color coded by altitude above the surface of the planet and limited to altitudes less than 20 km. Positive B_r plots to the right of each track and negative to the left, with signal amplitude given by the scale bar. Underlying image is a Mercury Dual Imaging System (MDIS) mosaic. Mollweide projection. The white star indicates the best estimate of MESSENGER's impact location. The HPF B_r signals are discussed in the text. Periapsis local time varied from ~3:30 pm on orbit 4097 to ~2 pm on orbit 4104. The last three orbits show substantially increased high-frequency signal associated with plasma in the afternoon cusp region and inside the entire dayside magnetosphere. Shakespeare basin, Suisei Planitia, Strindberg basin, Turgenev crater, and unnamed crater C1 are labeled in white. Orbit numbers are given in black at the northern end of each orbit track.

Figure 5.18. Magnetization strength for a 10-km-thick magnetized layer, computed from an equivalent-source dipole model with sources at a depth of 20 km. The weakest magnetizations are not shown. The low-altitude data distribution yields little to no sensitivity to crustal magnetization south of ~35°N and north of ~80°N. The model was obtained from an inversion of B_r data spanning local times 1600–0800 using Tikhonov regularization and corresponds to the corner of the L–curve (see, e.g., Aster et al., 2013). The reduction in variance was over 95% with a final root-mean-square misfit to the data of just under 0.7 nT. Lambert azimuthal equal area projections from 32° to 90° N. Background image shows shaded relief with smooth plains regions outlined by the black contour (Denevi et al., 2013).

$$R_l(r) = (l+1)\left(\frac{R_p}{r}\right)^{2l+4} \sum_{m=0}^{m=l} (g_l^{m2} + h_l^{m2}), \quad (5.5)$$

where R_p is the planetary radius. The secular variation spectrum is similarly defined, but with the spherical harmonic coefficients in equation (5.5) replaced by their time derivatives. The spherical harmonic power spectrum for Mercury's OAD field and Earth's IGRF-12 field, scaled to Mercury, are shown in Figure 5.20a. Earth's field has anomalously high power at $l = 1$, but for $l = 2$–13, the power spectrum is white (i.e., flat) at the CMB (Backus et al., 1996). Mercury's spectrum falls off more steeply than Earth's for $l = 2$–4, even when only the zonal part of Earth's core field is considered, at both the planetary surface and the CMB. In fact, Mercury's core field spectrum would be white at a radius of 1000 km or ~ $R_C/2$. Although there is no a priori reason to suspect that Mercury's core field spectrum should be the same as that of Earth, if it is even close to white at the top of the source region, then the source region is very deep. As discussed in Section 5.1, and in Chapter 19, a common aspect of many dynamo models that have had success in predicting a weak field with minimal non-zonal structure is the presence of a stable (non-convecting) layer at the top of the core (e.g., Christensen, 2006; Christensen and Wicht, 2008; Tian et al., 2015). In these models the top of the dynamo source region corresponds to the base of the stably stratified layer. The above discussion on the radius to the top of the source region suggests that the stably stratified layer could be up to ~0.5 R_C in thickness, a result suggested by some dynamo studies (e.g., Tian et al., 2015). In such a situation, a further consequence is that the inner core radius, R_{IC}, must be substantially less than 0.5 R_C. There are currently no independent determinations of R_{IC}, but recent work suggests that R_{IC} is less than 0.5 R_C and perhaps less than 0.33 R_C (Chapters 4, 19, and references therein).

The asymmetry in the field about the geographic equator implied by the OAD description is substantial and has implications for the spatial extent of the magnetospheric cusp and the difference in the proximity of the magnetopause boundary to the planetary surface between the southern and the northern hemispheres. As described below, from a dynamo perspective the

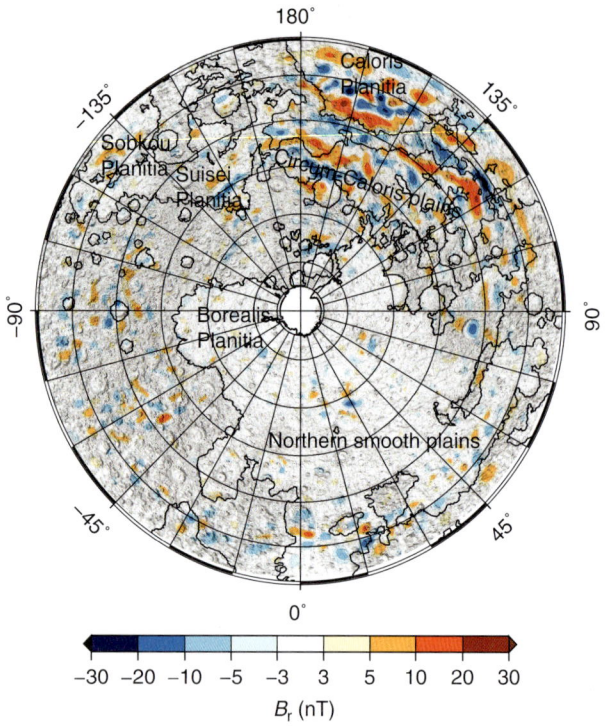

Figure 5.19. Map of B_r at 20-km altitude predicted by the magnetization model shown in Figure 5.18. Background image shows shaded relief with smooth plains regions outlined by the black contour (Denevi et al., 2013). Lambert azimuthal equal area projections from 32° to 90° N. Figure modified from Johnson et al. (2016b).

field structure is challenging to explain. The spatial power spectrum is currently unconstrained beyond the OAD description, and while simple low-degree-and-order alternatives to the OAD description have been ruled out (Anderson et al., 2012), a systematic study of allowable alternative representations of the field has not been conducted. In addition, as described earlier, the existence of non-zonal and regional-scale core field structure has proven difficult to establish because of the large contributions from external fields, in particular those due to Birkeland currents. One attempt has been made to derive core field models from the vector magnetic field data using equivalent source dipoles (Oliviera et al., 2015) and the first solar day of MESSENGER data. That study confirmed the dominantly axisymmetric nature of the internal field but lacked any treatment of even the major magnetospheric fields. Future progress in establishing internal field structure from MESSENGER data must account for not only the major magnetospheric fields but also residual external fields, in particular those due to Birkeland currents, and requires localized inversion approaches for the regional-scale internal field (Section 5.2.3.3). The BepiColombo mission (Chapter 20) will make substantive advances in understanding Mercury's core field because it will make measurements of Mercury's low-latitude and southern hemisphere field at altitudes of ~480–1500 km.

The proton reflectometry study described in Section 5.3.2.3 was also extended in an attempt to map non-OAD structure in the internal field (Winslow, 2014). Because of the limited FIPS viewing geometry and long integration times, however, sufficiently high proton counts to allow investigations of regional field structure in the MBF frame were available only in the latitude band 70–80°N (i.e., in the northern cusp region). The

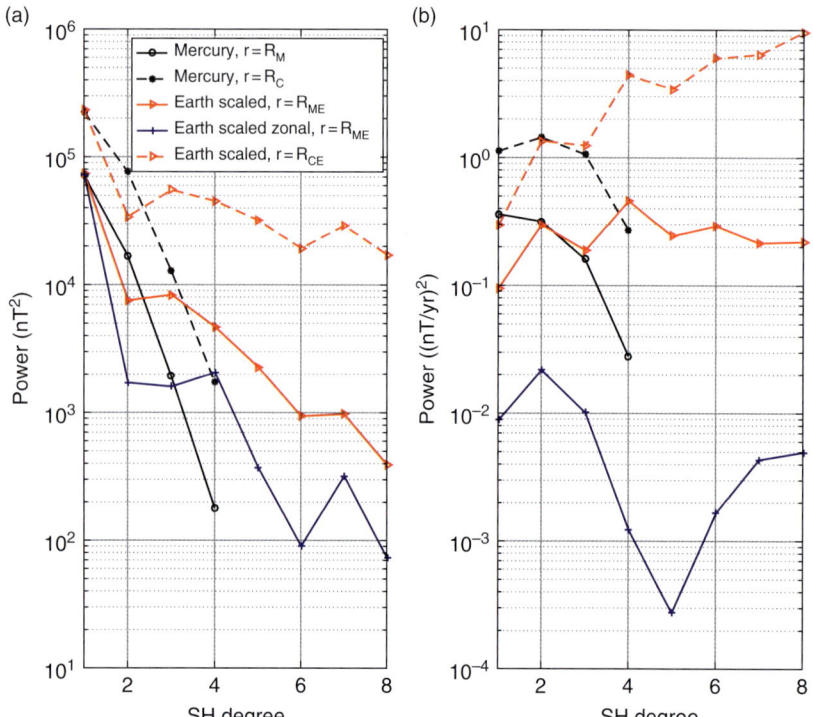

Figure 5.20. (a) Spatial power spectra for the OAD model for Mercury's core field estimated at Mercury's surface (black solid line, radius $r = R_M$) and CMB (black dashed line, $r = R_C$) and the IGRF-12 (Thébault et al., 2015) model for Earth's field (red) and Earth's zonal field only (blue). Because the OAD model is by definition a zonal field, the IGRF-12 spherical harmonic coefficients are scaled such that the power in the axial dipole term at Earth's CMB (radius R_{CE}) is the same as the power in Mercury's field at R_C. The scaled Earth model is shown downward continued to R_{CE} and also to a radius R_{ME}, equivalent to the distance of Mercury's surface above Mercury's CMB, such that $R_{ME}/R_{CE} = R_M/R_C$. (b) Spatial power spectra for secular variation. The spectra for Mercury are upper bounds from the results of Philpott et al. (2014). The IGRF-12 spherical harmonic coefficients are scaled as in (a) and plotted at $r = R_{ME}$ and $r = R_{CE}$.

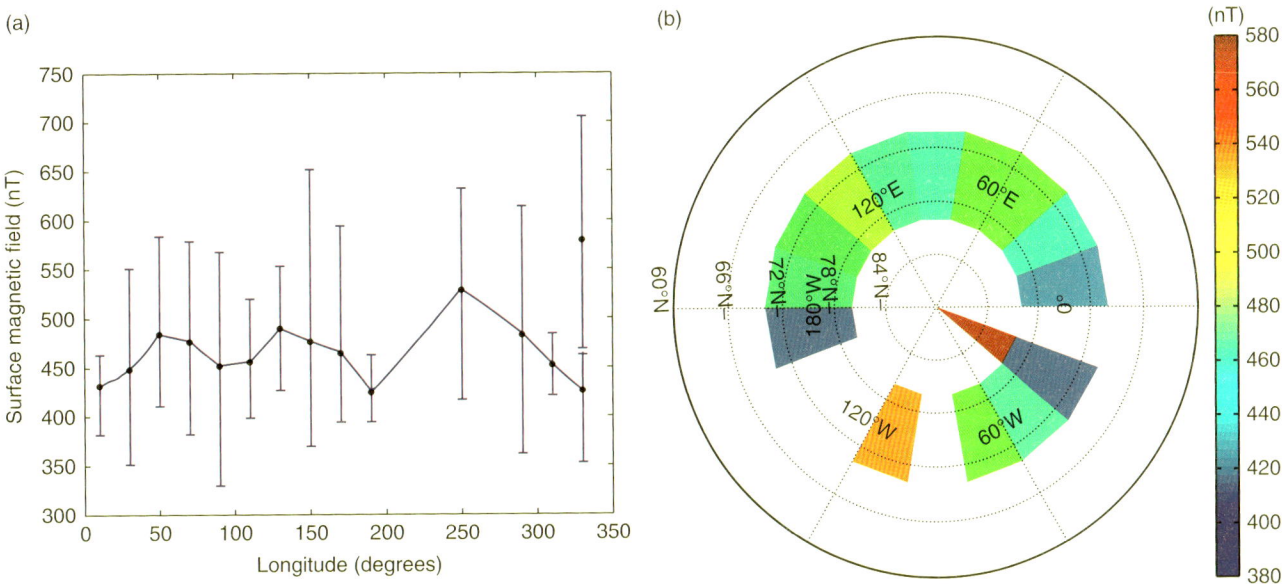

Figure 5.21. Surface magnetic field strength evaluated from proton reflectometry (Winslow, 2014) in 20° longitude bins over the latitude band 70–80°N and in one longitude bin from 80 to 90°N. Surface field strength values have been corrected for the diamagnetic field. (a) Average surface field strength with uncertainties plotted versus MBF longitude. See Winslow et al. (2014) for estimates of uncertainties. The 80–90°N estimate is the higher estimate in the bin centered on 330°E. (b) Average surface field strength in nT plotted in the geographical bins analyzed, shown in MBF stereographic projection as viewed from above the north pole. Modified from Winslow (2014).

inferred surface field strengths and their uncertainties are shown in Figure 5.21. These include internal (core and crust) and external contributions (similar in each geographic bin because they are all in the cusp region) to the field, but they have been corrected for the diamagnetic effects of plasma. In the one bin for which estimates in two latitude bands were available, an increase in field strength with latitude is seen as expected. The results show that within uncertainties no regional structure in the surface field strength was detectable at 70–80°N.

The secular variation result of Philpott et al. (2014) suggests that the Mariner 10 and MESSENGER data do not require any secular variation in the OAD field over the past 40 years, but an estimate of the upper bound on secular variation compatible with the observations was also determined (see Figure 5.9). Secular variation power spectra provide important constraints on dynamo models that are inherently dynamic. For Earth, secular variation spectra are deduced by estimating annual changes in the global magnetic field from satellite and observatory measurements (e.g., Thébault et al., 2015). Such measurements are not available for Mercury, but the bounds on changes in the global field between the times of the Mariner 10 and MESSENGER observations (Philpott et al., 2014) can be used to compute an upper limit on the power spectrum of the secular variation. This upper bound is compared with that in Earth's present field in Figure 5.20b. Interestingly, the upper bound on Mercury's zonal secular variation spectrum for $l = 2–4$ has a similar shape to that of Earth's zonal secular variation, but is approximately an order of magnitude larger. At Earth, the power in the non-zonal secular variation at any given spherical harmonic degree is 10–1000 times larger than the power in the zonal terms. The same could be true at Mercury in the dynamo region, but need not be observed at the surface (or even the CMB), e.g.,

because of the presence of a stably stratified layer. A more appropriate timescale for the discussion of secular variation may be the magnetic diffusion time, $\tau = L^2/(\pi^2 \eta)$, where η is the magnetic diffusivity of the core fluid (~ 1 m^2 s^{-1}), and L is the scale length of the core dynamo region, typically taken as the thickness of the convecting shell. The discussion above suggests that L for Mercury could be ~ 1000 km or less, about half that of L for Earth. This value yields a diffusion time about four times less and power in the secular variation spectrum at least an order of magnitude greater for Mercury than for Earth. Thus, scaled for the different diffusion times of Earth and Mercury's cores, the upper bound on Mercury's secular variation power spectrum is similar to Earth's. Taken together, the secular variation power spectrum (Figure 5.19b) and spatial power spectrum (Figure 5.19a) can be used to compute correlation timescales, τ_l, at each spherical harmonic degree (Hulot et al., 2015). The correlation times, $\tau_1 = 450$ yr, $\tau_2 = 230$ yr, $\tau_3 = 110$ yr, and $\tau_4 = 80$ yr for Mercury, are minimum values because the secular variation spectrum is an upper bound. However, they are still of interest because structure in each spherical harmonic degree need not be preserved over periods longer than several correlation times. Thus, whereas the OAD description of the internal field is valid for the present and for 1974–1975 when Mariner 10 visited Mercury, whether it is a steady feature of the planet's field, required on timescales of several thousand years and longer, is unknown.

The core field structure and its secular variation provide important constraints on models for magnetic field generation at present. As described in the introduction to this chapter, substantial effort has been devoted to explaining the weak strength of Mercury's core field; more recent efforts have focused on attempting to explain the strength, equatorial asymmetry, and strong axisymmetry in the field (Cao et al., 2014; Tian et al., 2015; Chapter 19). Many models include a stably stratified layer at the top of the core that acts to reduce the field strength and dampen secular variation. Such a stably stratified layer may be present as a result of differences in the light element concentrations and the different pressure regime of Mercury's outer core relative to that of Earth. These differences may also affect the distribution of compositional buoyancy sources within the core (Vilim et al., 2010; Cao et al., 2014) that result from, e.g., multiple iron precipitation (iron snow) regions (Chen et al., 2008). The geographical pattern and amplitude of the CMB heat flux play important roles in driving fluid flow and magnetic field generation in the core, and both $l = 1$ and $l = 2$ zonal heat flux variations have been invoked to produce dynamo fields with equatorial asymmetry (Cao et al., 2014; Tian et al., 2015). As discussed in Chapter 19 (Section 19.6.2), any CMB heat flux pattern must be consistent with other geological and geophysical constraints for the structure and evolution of Mercury's mantle and crust. Furthermore, the evolution of Mercury's core over time will have affected the distribution of thermal and compositional buoyancy sources within the core, and so Mercury's ancient (see Section 5.6.3) and modern dynamos need not have been similar (Soderlund and Schubert, 2016).

5.6.2 Electrical Conductivity Structure

MESSENGER observations have allowed some assessment of the electrical conductivity structure of Mercury's mantle and electromagnetic sounding of the CMB. However, detailed studies of interior conductivity structure have not been possible because of MESSENGER's eccentric orbit, which precluded characterization of the southern hemisphere magnetic field and could not provide simultaneous measurements of magnetospheric driving and response. The latter limitation means that the competing effects of erosion and induction cannot be fully separated on an orbit-by-orbit basis, in particular during times of extreme events when the induction signal should be larger. Refinement of the estimate of R_{COND} by Johnson et al. (2016a) and of interior conductivity structure more generally requires improved knowledge of the inducing field geometry and information at periods other than one Mercury year. The BepiColombo mission is expected to make substantial advances in this regard through dual-spacecraft measurements of tail loading and its inductive response (Heyner et al., 2016).

5.6.3 Crustal Magnetization: Origin, Magnetic Mineralogy, and Ancient Field Strength

Crustal magnetization, **M**, reflects the combined effects of the strength of the magnetizing field, **B**, and the bulk magnetic properties of the crust, given by its susceptibility, χ. An important question for Mercury is whether the magnetization was acquired in an ancient or modern field. For magnetizations induced in the present field, $M = M_{ind} = \chi_{ind} B_{present}/\mu_0$, where χ_{ind} is the low-field susceptibility, μ_0 is the permeability of free space, and $B_{present}$ denotes the present field strength. The latter is known, and M_{ind} can be calculated for the susceptibilities of relevant magnetic minerals. For a thermoremanent magnetization (TRM) acquired in an unknown ancient field, $B_{ancient}$, $M_{TRM} = \chi_{TRM} B_{ancient}/\mu_0$. The field $B_{ancient}$ is unknown but can be calculated for a range of candidate mineralogies. A further trade-off occurs because the measured crustal fields reflect the product of magnetization and magnetized layer thickness, and so inferred magnetizations depend on the assumed layer thickness.

The susceptibilities, χ_{ind} and χ_{TRM}, of Mercury's crust are unknown, as they depend on the magnetic minerals present and on their relative volumetric abundances. The chemically reduced characteristics of Mercury's surface materials (see Chapter 2 and references therein) suggest that iron metal, iron alloys, and iron sulfides are possible magnetic carriers. The values of χ_{ind} and χ_{TRM} for pyrrhotite, iron metal, and high-iron (EH) and low-iron (EL) enstatite chondrites were scaled for volume fractions of the magnetic carrier consistent with the 1.5–2 wt% average iron content inferred from MESSENGER X-ray fluorescence observations (Johnson et al., 2015, 2016b). To explain the magnetizations inferred from the observed crustal fields entirely as an induced magnetization for an EH magnetic mineralogy requires layer thicknesses in excess of 100–200 km in the regions of strongest magnetization, under the assumptions that χ_{ind} is independent of temperature and that all the iron is partitioned into magnetic phases. Neither such a large layer thickness nor the required iron partitioning is plausible. Pyrrhotite would require layer thicknesses almost an order of magnitude greater. Johnson et al. (2015) also considered viscous remanence (VRM) acquired during prolonged exposure of the magnetic minerals to the planetary field, and concluded that it, too, was unlikely to fully explain the observations.

Thus, although it is likely that induced magnetization and perhaps some contribution from VRM are present, the strongest crustal fields, in particular those associated with the Caloris region, require a thermoremanent contribution. Figure 5.22 shows the trade-off between χ_{TRM}, $B_{ancient}$, and layer thickness if the fields are entirely the result of TRM. Ancient surface field strengths that lie between values comparable to those from Mercury's current dynamo and Earth-like values are most likely, given the possible magnetic minerals in Mercury's crust (Johnson et al., 2015). Two lines of evidence point toward very weak magnetizations at the shallowest depths. First, even at the lowest spacecraft altitudes, signals are of low amplitude and have minimum wavelengths of ~40 km (Section 5.5.2). Second, preliminary investigations of source depths with the equivalent-source dipole model suggest that the data are better matched by sources at depths of 10–30 km (Johnson et al., 2016b). Although the latter are rather coarsely weighted mean depths of the vertically distributed magnetization, together with models for Mercury's crustal thickness (see Chapter 3 and references therein) they suggest that most of the magnetization lies within the crust and that there are at most weak contributions from sources in the shallowest (<10 km) crust and in the mantle.

The observations that the strongest magnetizations occur in and around the Caloris basin, including the Suisei Planitia region (Johnson et al., 2015), suggest an age for the remanence at least that of the circum-Caloris and Caloris interior smooth plains (Johnson et al., 2015), and possibly as old as the Caloris basin itself. Thus, the crustal field observations point to a dynamo operating at 3.9–3.7 Ga (Johnson et al., 2015; Chapters 6 and 9).

An obvious question is whether TRM can be acquired and retained for a large fraction of the age of the planet given the thermal conditions in Mercury's crust today and in the past. The planet's orbital eccentricity, 3:2 spin–orbit resonance, and near-zero obliquity lead to geographical variations in the average daily surface temperature (Figure 5.23; see Siegler et al., 2013). The corresponding depth, Z_C, to the Curie temperature, T_C, for a magnetic mineral can be calculated given the average temperature gradient in the crust. For present conditions, Z_C is greater than ~40 km everywhere, even for low-T_C minerals such as pyrrhotite (T_C = 320°C). For ancient conditions, an upper bound on the surface temperature distribution can be estimated from the maximum average daily surface temperature (see Figure 5.23) predicted for a range of Mercury's orbital eccentricities from 0 to 0.4 (Correia and Laskar, 2009). With these and estimates for ancient thermal gradients (Williams et al., 2007; Grott et al., 2011; Tosi et al., 2013), Z_C at ~4 Ga is greater than 14 km and 59 km for pyrrhotite and iron, respectively, minerals that span the likely range of T_C for candidate magnetic carriers in Mercury's crust. These results imply that acquisition and subsequent preservation of an ancient crustal remanence by magnetic minerals with T_c values of at least 320°C are consistent with thermal models (Johnson et al., 2015).

The spatial distribution of crustal magnetization, in particular the overall weaker magnetizations associated with the intercrater plains and northern smooth plains compared with the Caloris region, likely reflects the combined effects of (a) the dynamo field history, (b) the distribution of magnetic minerals in the crust, (c) the relative contributions of induced and remanent magnetizations, and (d) the effects of cratering and magmatism subsequent to acquisition of any ancient remanence. The detection of remanent fields places important (and restrictive) constraints on the interior thermal and compositional history (Chapter 19). Results to date indicate that a dynamo field was required at ~3.9–3.7 Ga, with a strength comparable to, or up to 100 times stronger than, Mercury's present field. The direction of the ancient field is not resolvable because of the inherent non-uniqueness in the inversion for magnetization distributions.

Further understanding of Mercury's crustal magnetic field requires several avenues of investigation. First, the current state of knowledge of the magnetic properties of plausible magnetic carriers compatible with petrologic constraints and thermal conditions is limited (Strauss et al., 2016). For example, magnetic carriers such as carbides have not heretofore been considered, but may be possible given recent results on the presence of carbon in Mercury's crust (Chapters 2, 7, 19, and references therein), and the dependence of magnetic properties of candidate minerals on temperature, grain size, and

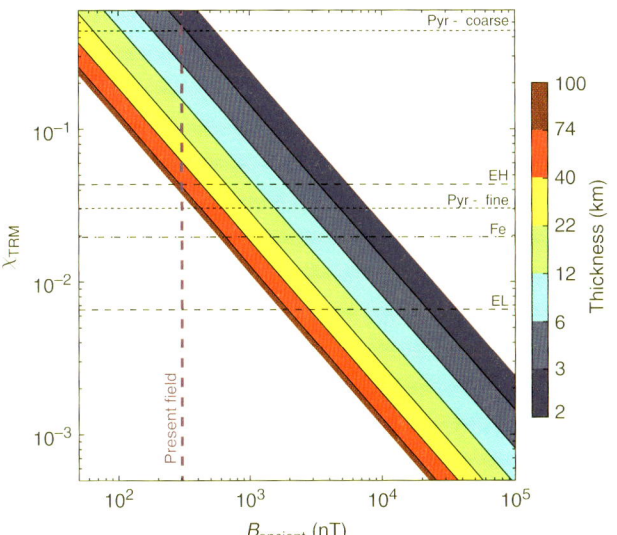

Figure 5.22. Magnetic susceptibility (χ_{TRM}) and magnetizing field strength ($B_{ancient}$) required to produce the observed peak HPF magnetic field strength over Suisei Planitia. Susceptibilities for pyrrhotite for grain sizes ranging from 5 to 250 μm (black dotted lines), multidomain iron (black dashed-dot line), and high-iron (EH) and low-iron (EL) enstatite chondrites (black dashed lines) are shown. The values are scaled for volume fractions of the magnetic carrier consistent with the average iron content inferred from MESSENGER observations. The surface field strength for the present internal dipole (vertical dashed line) is that for the average magnetic latitude of the region (46°N). From Johnson et al. (2015).

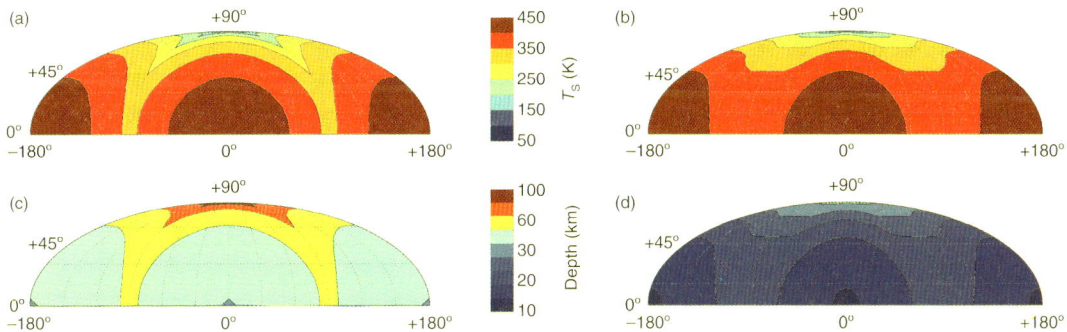

Figure 5.23. Average daily surface temperature, T_S, calculated at 1-m depth (i.e., beneath the penetration depth of the diurnal temperature variations) calculated (a) for the present eccentricity (e = 0.2056), (b) from the maximum T_S at any location for eccentricities ranging from 0.0 to 0.4, corresponding to an upper bound on T_S for an unknown ancient eccentricity. Corresponding depth to the Curie temperature for pyrrhotite (T_C = 320°C) for (c) the T_S in (a) and an average thermal gradient of 4 K km^{-1}, and (d) the T_S in (b) and an average thermal gradient of 10 K km^{-1}. For a given T_S the depth is inversely proportional to dT/dz. The depth to the Curie temperature for Fe (T_C = 770°C) would be greater than that for pyrrhotite by 112 km for the conditions in (c) and by 45 km for the conditions in (d).

microstructures should be more fully explored. Second, dynamo models compatible with thermal evolution scenarios for Mercury are needed (Soderlund and Schubert, 2016) to explore early global fields compatible with MESSENGER observational constraints. Third, the global distribution of crustal magnetization can be ascertained only with new orbital measurements made below 100-km altitude at latitudes south of ~30°N. Such observations are not planned as part of the BepiColombo baseline mission, but should the orbit of the Mercury Planetary Orbiter spacecraft eventually decay, like that of the MESSENGER spacecraft, new measurements of Mercury's internal field could be made at very low altitudes in the future.

5.7 CONCLUDING REMARKS

Results from the MESSENGER mission have markedly advanced our understanding of the magnetic field of the innermost planet. Prior to MESSENGER, little was known about Mercury's internal magnetic field other than that its surface strength is about 100 times weaker than Earth's. Similarly, little was known about Mercury's magnetosphere other than that, when scaled by planetary radius, it is ~7–8 times smaller than its terrestrial counterpart. Data from MESSENGER have established that the global planetary field results from a core dynamo and that it is dominantly dipolar, weak, highly axisymmetric, and highly equatorially asymmetric. These combined characteristics render Mercury's field unique among the dynamo fields of planets in our solar system. MESSENGER orbital observations provide an estimate of the spherical harmonic power spectrum of the core field and its secular variation estimated over a ~40-yr time interval that are, in turn, important constraints on core dynamo models.

The absence of a substantial atmosphere, the close proximity to the Sun, the weak internal field, and large fractional core radius (~0.83) result in a magnetosphere that is not only highly dynamic but is also coupled to the interior. Time variations in the magnetospheric fields driven by changes in solar-wind ram pressure induce electric currents in the planetary interior. MESSENGER observations have revealed the fields that result from such currents, providing a confirmation of Mercury's core radius that is independent of traditional geodetic techniques. Furthermore, the identification of field-aligned currents at Mercury has been accompanied by the discovery that these currents close through the planetary interior, consistent with the range of possible electrical conductivity structures of the silicate mantle and crust.

The unprecedented data set afforded by MESSENGER observations at altitudes below 100 km, combined with the extensive groundwork laid in understanding the large-scale structure and dynamics of the magnetosphere, has led to the discovery of crustal magnetic fields with peak amplitudes of ~30 nT measured a few to ~15 km above the planetary surface. Present data on plausible magnetic mineralogies, together with calculations of ancient thermal conditions, indicate that these crustal fields were at least partly acquired in an ancient field, at least ~3.7 Ga, with a strength that could have been greater than the present field strength by a factor of as much as 100. The clear correlation of the strongest crustal fields with the Caloris and circum-Caloris region suggests that the thermal, compositional, and structural effects of this basin have played a major role in the geographical distribution, and possibly history, of crustal magnetization on Mercury.

REFERENCES

Acuña, M. H., Connerney, J. E. P., Ness, N. F., Lin, R. P., Mitchell, D., Carlson, C. W., McFadden, J., Anderson, K. A., Rème, H., Mazelle, C., Vignes, D., Wasilewski, P. and Cloutier, P. (1999). Global distribution of crustal magnetization discovered by the Mars Global Surveyor MAG/ER experiment. *Science*, **284**, 790–793, doi:10.1126/science.284.5415.790.

Aharonson, O., Zuber, M. T. and Solomon, S. C. (2004). Crustal remanence in an internally magnetized non-uniform shell: A possible source for Mercury's magnetic field? *Earth Planet. Sci. Lett.*, **218**, 261–268, doi:10.1016/S0012-821X(03)00682-4.

Alekseev, I. I. and Shabansky, V. P. (1972). A model of a magnetic field in the geomagnetosphere. *Planet. Space. Sci.*, **20**, 117–113, doi:10.1016/0032-0633(72)90146-8.

Alexeev, I. I., Belenkaya, E. S., Bobrovnikov, S. Y., Slavin, J. A. and Sarantos, M. (2008). Paraboloid model of Mercury's magnetosphere. *J. Geophys. Res.*, **113**, A12210, doi:10.1029/2008JA013368.

Alexeev, I. I., Belenkaya, E. S., Slavin, J. A., Korth, H., Anderson, B. J., Baker, D. N., Boardsen, S. A., Johnson, C. L., Purucker, M. E., Sarantos, M. and Solomon, S. C. (2010). Mercury's magnetospheric magnetic field after the first two MESSENGER flybys. *Icarus*, **209**, 23–39, doi:10.1016/j.icarus.2010.01.024.

Anderson, B. J., Acuña, M. H., Lohr, D. A., Scheifele, J., Raval, A., Korth, H. and Slavin, J. A. (2007). The Magnetometer instrument on MESSENGER. *Space Sci. Rev.*, **131**, 417–450, doi:10.1007/s11214-007-9246-7.

Anderson, B. J., Acuña, M. H., Korth, H., Purucker, M. E., Johnson, C. L., Slavin, J. A., Solomon, S. C. and McNutt, R. L., Jr. (2008). The structure of Mercury's magnetic field from MESSENGER's first flyby. *Science*, **321**, 82–85, doi:10.1126/science.1159081.

Anderson, B. J., Acuña, M. H., Korth, H., Slavin, J. A., Uno, H., Johnson, C. L., Purucker, M. E., Solomon, S. C., Raines, J. M., Zurbuchen, T. H., Gloeckler, G. and McNutt, R. L., Jr. (2010). The magnetic field of Mercury. *Space Sci. Rev.*, **152**, 307–339, doi:10.1007/s11214-009-9544-3.

Anderson, B. J., Johnson, C. L., Korth, H., Purucker, M. E., Winslow, R. M., Slavin, J. A., Solomon, S. C., McNutt, R. L., Jr., Raines, J. M. and Zurbuchen T. H. (2011). The global magnetic field of Mercury from MESSENGER orbital observations. *Science*, **333**, 1859–1862, doi:10.1126/science.1211001.

Anderson, B. J., Johnson, C. L., Korth, H., Winslow, R. M., Borovsky, J. E., Purucker, M. E., Slavin, J. A., Solomon, S. C., Zuber, M. T. and McNutt, R. L., Jr. (2012). Low-degree structure in Mercury's planetary magnetic field. *J. Geophys. Res.*, **117**, E00L12, doi:10.1029/2012JE004159.

Anderson, B. J., Johnson, C. L. and Korth, H. (2013). A magnetic disturbance index for Mercury's magnetic field derived from MESSENGER Magnetometer data. *Geochem. Geophys. Geosyst.*, **14**, 3875–3886, doi:10.1002/ggge.20242.

Anderson, B. J., Johnson, C. L., Korth, H., Slavin, J. A., Winslow, R. M., Phillips, R. J., McNutt, R. L., Jr. and Solomon, S. C. (2014). Steady-state field-aligned currents at Mercury. *Geophys. Res. Lett.*, **41**, 7444–7452, doi:10.1002/2014GL061677.

Anderson, B. J., Korth, H., Johnson, C. L., Phillips, R. J., Philpott, L. C. and Solomon, S. C. (2016). Closure of Birkeland currents at

Mercury: Constraints on the electrical conductivity of the crust and mantle. *Lunar Planet. Sci.*, **47**, abstract 1243.

Andrews, B. G., Zurbuchen, T. H., Mauk, B. H., Malcom, H., Fisk, L. A., Gloeckler, G., Ho, G. C., Kelley, J. S., Koehn, P. L., LeFevere, T. W., Livi, S. S., Lundgren, R. A. and Raines, J. M. (2007). The Energetic Particle and Plasma Spectrometer instrument on the MESSENGER spacecraft. *Space Sci. Rev.*, **131**, 523–556, doi:10.1007/s11214-007-9272-5.

Aster, R. C., Borchers, B. and Thurber, C. H. (2013). *Parameter Estimation and Inverse Problems*, 2nd edn. Oxford: Academic Press.

Aubert, J., Johnson, C. L. and Tarduno, J. A. (2010). Observations and models of the long-term evolution of Earth's magnetic field. *Space Sci. Rev.*, **155**, 337–370, doi:10.1007/s11214-010-9684-5.

Backus, G., Parker, R. and Constable, C. (1996). *Foundations of Geomagnetism*. Cambridge: Cambridge University Press.

Bartels, J. (1936). The eccentric dipole approximating the Earth's magnetic field. *Terr. Magn. Atmos. Electr.*, **41**, 225–250, doi:10.1029/TE041i003p00225.

Baumjohann, W. and Treumann, R. A. (1996), *Basic Space Plasma Physics*. London: Imperial College Press.

Benna, M., Anderson, B. J., Baker, D. N., Boardsen, S. A., Gloeckler, G., Gold, R. E., Ho, G. C., Killen, R. M., Korth, H., Krimigis, S. M., Purucker, M. E., McNutt, R. L., Jr., Raines, J. M., McClintock, W. E., Sarantos, M., Slavin, J. A., Solomon, S. C. and Zurbuchen, T. H. (2010). Modeling of the magnetosphere of Mercury at the time of the first MESSENGER flyby. *Icarus*, **209**, 3–10, doi:10.1016/j.icarus.2009.11.036.

Bloxham J., Gubbins D. and Jackson A. (1989). Geomagnetic secular variation. *Phil. Trans. Roy. Soc. London A*, **329**, 415–502, doi:10.1098/rsta.1989.0087.

Busse, F. H. (2002). Convective flows in rapidly rotating spheres and their dynamo action. *Phys. Fluids*, **14**, 1301–1314, doi:10.1063/1.1455626.

Cao, H., Aurnou, J. M., Wicht, J., Dietrich, W., Soderlund, K. M. and Russell, C. T. (2014). A dynamo explanation for Mercury's anomalous magnetic field. *Geophys. Res. Lett.*, **41**, 2014GL060196, doi:10.1002/2014gl060196.

Chabot, N. L., Wollack, E. A., Klima, R. L. and Minitti, M. E. (2014). Experimental constraints on Mercury's core composition, *Icarus*, **390**, 199–208, doi:10.1016/j.epsl.2014.01.004.

Chassefière, E., Langlais, B., Quesnel, Y. and Leblanc, F. (2013). The fate of early Mars' lost water: The role of serpentinization. *J. Geophys. Res. Planets*, **118**, 1123–1134, doi:10.1002/jgre.20089.

Chen, B., Li, J. and Hauck, S. A., II (2008). Non-ideal liquidus curve in the Fe–S system and Mercury's snowing core. *Geophys. Res. Lett.*, **35**, L07201, doi:10.1029/2008gl033311.

Christensen, U. R. (2006). A deep dynamo generating Mercury's magnetic field. *Nature*, **444**, 1056–1058, doi:10.1038/nature05342.

Christensen, U. R. and Wicht, J. (2008). Models of magnetic field generation in partly stable planetary cores: Applications to Mercury and Saturn. *Icarus*, **196**, 16–34, doi:10.1016/j.icarus.2008.02.013.

Cisowski, S. M., Collinson, D. W., Runcorn, S. K., Stephenson, A. and Fuller, M. (1983). A review of lunar paleointensity data and implications for the origin of lunar magnetism. *J. Geophys. Res.*, **88**, A691–A704.

Connerney, J. E. P. (2015). Planetary magnetism. In *Physics of Terrestrial Planets and Moons*, ed. T. Spohn. *Treatise on Geophysics*, 2nd edn, Vol. 10, ed. G. Schubert. Amsterdam, Oxford: Elsevier, pp. 195–237, doi:10.1016/B978-0-444-53802-4.00171-8.

Connerney, J. E. P. and Ness, N. F. (1988). Mercury's magnetic field and interior. In *Mercury*, ed. F. Vilas, C. R. Chapman and M. S. Matthews. Tucson, AZ: University of Arizona Press, pp. 494–513.

Connerney, J. E. P., Acuña, M. H., Wasilewski, P. J., Ness, N. F., Rème, H., Mazelle, C., Vignes, D., Lin, R. P., Mitchell, D. L. and Cloutier, P. A. (1999). Magnetic lineations in the ancient crust of Mars. *Science*, **284**, 794–798, doi:10.1126/science.284.5415.794.

Conzelmann, V. and Spohn, T. (1999). New thermal evolution models suggesting a hot, partially molten Mercurian interior. *Bull. Amer. Astron. Soc.*, **31**, 31st DPS Meeting Program, Padua, Italy, abstract 18.02.

Correia, A. C. M. and Laskar, J. (2009). Mercury's capture into the 3/2 spin–orbit resonance including the effect of core–mantle friction. *Icarus*, **201**, 1–11.

Cournède, C., Gattacceca, J. and Rochette, P. (2012). Magnetic study of large Apollo samples: Possible evidence for an ancient centered dipolar field on the Moon. *Earth Planet. Sci. Lett.*, **331**, 31–42, doi:10.1016/j.epsl.2012.03.004.

DiBraccio, G. A., Slavin, J. A., Boardsen, S. A., Anderson, B. J., Korth, H., Zuburchen, T. H., Raines, J. M., Baker, D. N., McNutt, R. L., Jr. and Solomon, S. C. (2013). MESSENGER observations of magnetopause structure and dynamics at Mercury. *J. Geophys. Res. Space Physics*, **118**, 997–1008, doi:10.1002/jgra.50123.

Dunlop, D. J. and Arkani-Hamed, J. (2005). Magnetic minerals in the Martian crust. *J. Geophys. Res.*, **110**, E12S04, doi:10.1029/2005JE002404.

Fuller, M. and Cisowski, S. M. (1987). Lunar paleomagnetism. In *Geomagnetism*, Vol. 2, ed. J. A. Jacobs. Orlando, FL: Academic Press, pp. 307–455.

Garrick-Bethell, I., Weiss, B. P., Shuster, D. L. and Buz, J. (2009). Early lunar magnetism. *Science*, **323**, 356–359, doi:10.1126/science.1166804.

Gattacceca, J. and Rochette, P. (2004). Toward a robust normalized magnetic paleointensity method applied to meteorites. *Earth Planet. Sci. Lett.*, **227**, 377–393, doi:10.1016/j.epsl.2004.09.013.

Gattacceca, J., Boustie, M., Hood, L., Cuq-Lelandais, J.-P., Fuller, M., Bezaeva, N. S., de Resseguier, T. and Berthe, L. (2010a). Can the lunar crust be magnetized by shock: Experimental groundtruth. *Earth Planet. Sci. Lett.*, **299**, 42–53, doi:10.1016/j.epsl.2010.08.011.

Gattacceca, J., Boustie, M., Lima, E. A., Weiss, B. P., de Resseguier, T. and Cuq-Lelandais, J.-P. (2010b). Unraveling the simultaneous shock magnetization and demagnetization of rocks. *Phys. Earth Planet. Inter.*, **182**, 42–49, doi:10.1016/j.pepi.2010.06.009.

Giampieri, G. and Balogh, A. (2002). Mercury's thermoelectric dynamo model revisited. *Planet. Space Sci.*, **50**, 757–762, doi:10.1016/S0032-0633(02)00020-X.

Glassmeier, K.-H., Auster, H.-U. and Motschmann, U. (2007a). A feedback dynamo generating Mercury's magnetic field. *Geophys. Res. Lett.*, **34**, L22201, doi:10.1029/2007GL031662.

Glassmeier, K.-H., Grosser, J., Auster, U., Constantinescu, D., Narita, Y. and Stellmach, S. (2007b). Electromagnetic induction effects and dynamo action in the Hermean system. *Space Sci. Rev.*, **132**, 511–527, doi:10.1007/s11214-007-9244-9.

Grosser, J., Glassmeier, K.-H. and Stadelmann, A. (2004). Induced magnetic field effects at planet Mercury. *Planet. Space Sci.*, **52**, 1251–1260, doi:10.1016/j.pss.2004.08.005.

Grott, M., Breuer, D. and Laneuville, M. (2011). Thermo-chemical evolution and global contraction of Mercury. *Earth. Planet. Sci. Lett.* **307**, 135–146, doi:10.1016/j.epsl.2011.04.040.

Gubbins, D. (1977). Energetics of the Earth's core. *J. Geophys.*, **43**, 453–464.

Gubbins, D., Alfè, D., Masters, G., Price, G. D. and Gillan, M. J. (2003). Can the Earth's dynamo run on heat alone? *Geophys. J. Int.*, **155**, 609–622, doi:10.1046/j.1365-246X.2003.02064.x.

Haines, G. V. (1985). Spherical harmonic cap analysis. *J. Geophys. Res.*, **90**, 2583–2591, doi:10.1029/JB090iB03p02583.

Harder, H. and Schubert, G. (2001). Sulfur in Mercury's core? *Icarus*, **151**, 118–122, doi:10.1006/icar.2001.6586.

Hauck, S. A., II, Dombard, A. J., Phillips, R. J. and Solomon, S. C. (2004). Internal and tectonic evolution of Mercury. *Earth Planet. Sci. Lett.*, **222**, 713–728, doi:10.1016/j.epsl.2004.03.037.

Hauck, S. A., II, Solomon, S. C. and Smith, D. A. (2007). Predicted recovery of Mercury's internal structure by MESSENGER. *Geophys. Res. Lett.*, **34**, L18201, doi:10.1029/2007GL030793.

Hauck, S. A., II, Margot, J.-L., Solomon, S. C., Phillips, R. J., Johnson, C. L., Lemoine, F. G., Mazarico, E., McCoy, T. J., Padovan, S., Peale, S. J., Perry, M. E., Smith, D. E. and Zuber, M. T. (2013). The curious case of Mercury's internal structure. *J. Geophys. Res. Planets*, **118**, 1204–1220, doi:10.1002/jgre.20091.

Heimpel, M., Aurnou, J. M., Al-Shamali, F. M. and Gomez Perez, N. (2005). A numerical study of dynamo action as a function of spherical shell geometry. *Earth Planet. Sci. Lett.*, **236**, 542–557, doi:10.1016/j.epsl.2005.04.032.

Herbert, F., Wiskerchen, M., Sonnet, C. P. and Chao, J. K. (1976). Solar wind induction in Mercury: Constraints on the formation of a magnetosphere. *Icarus*, **28**, 489–500, doi:10.1016/0019-1035(76)90122-6.

Herrick, R. R. and Rumpf, M. E. (2011). Postimpact modification by volcanic or tectonic processes as the rule, not the exception, for Venusian craters. *J. Geophys. Res.*, **116**, E02004, doi:10.1029/2010JE003722.

Heyner, D., Wicht, J., Gomez-Perez, N., Schmitt, D., Auster, H.-U. and Glassmeier, K.-H. (2011). Evidence from numerical experiments for a feedback dynamo generating Mercury's magnetic field. *Science*, **334**, 1690–1693, doi:10.1126/science.1207290.

Heyner, D., Nabert, C., Liebert, E. and Glassmeier, K.-H. (2016). Concerning reconnection–induction balance at the magnetopause of Mercury, *J. Geophys. Res. Space Physics*, **121**, 2935–2961, doi:10.1002/2015JA021484.

Hood, L. L. (2011). Central magnetic anomalies of Nectarian-aged lunar impact basins: Probable evidence for an early core dynamo. *Icarus*, **211**, 1109–1128, doi:10.1016/j.icarus.2010.08.012.

Hood, L. L. (2015). Initial mapping of Mercury's crustal magnetic field: Relationship to the Caloris impact basin. *Geophys. Res. Lett.*, **42**, 10,565–10,572, doi:10.1002/2015GL066451.

Hood, L. L. (2016). Magnetic anomalies concentrated near and within Mercury's impact basins: Early mapping and interpretation. *J. Geophys. Res. Planets*, **121**, 1016–1025, doi:10.1002/2016JE005048.

Hood, L. L. and Artemieva, N. A. (2008). Antipodal effects of lunar basin-forming impacts: Initial 3D simulations and comparisons with observations. *Icarus*, **193**, 485–502, doi:10.1016/j.icarus.2007.08.023.

Hood, L. L. and Huang, Z. (1991). Formation of magnetic anomalies antipodal to lunar impact basins: Two-dimensional model calculations. *J. Geophys. Res.*, **96**, 9837–9846, doi:10.1029/91JB00308.

Hood, L. L. and Schubert, G. (1979). Inhibition of solar wind impingement on Mercury by planetary induction currents. *J. Geophys. Res.*, **84**, 2641–2647, doi:10.1029/JA084iA06p02641.

Hood, L. L., Coleman, P. J., Jr. and Wilhelms, D. E. (1979). The Moon: Sources of the crustal magnetic anomalies. *Science*, **204**, 53–57, doi:10.1126/science.204.4388.53.

Hood, L. L., Richmond, N. C. and Spudis, P. D. (2013). Origin of strong lunar magnetic anomalies: Further mapping and examinations of LROC imagery in regions antipodal to young large impact basins. *J. Geophys. Res. Planets*, **118**, 1265–1284, doi:10.1002/jgre.20078.

Hulot, G., Sabaka, T. J., Olsen, N. and Fournier, A. (2015). The present and future geomagnetic field. In *Geomagnetism*, ed. M. Kono. *Treatise on Geophysics*, 2nd edn, Vol. 5, ed. G. Schubert. Amsterdam, Oxford: Elsevier, pp. 33–78, doi:10.1016/B978-0-444-53802-4.00096-8.

Janhunen, P. and Kallio, E. (2004). Surface conductivity of Mercury provides current closure and may affect magnetospheric symmetry. *Ann. Geophys.*, **22**, 1829–1837, doi:10.5194/angeo-22-1829-2004.

Jia, X., Slavin, J. A., Gombosi, T. I., Daldorff, L. K. S., Toth, G. and Holst, B. (2015). Global MHD simulations of Mercury's magnetosphere with coupled planetary interior: Induction effect of the planetary conducting core on the global interaction. *J. Geophys. Res. Space Physics*, **120**, 4763–4775, doi:10.1002/2015JA021143.

Johnson, C. L. and McFadden, P. L. (2015). Time-averaged field and paleosecular variation. In *Geomagnetism*, ed. M. Kono. *Treatise on Geophysics*, 2nd edn, Vol. 5, ed. G. Schubert. Amsterdam, Oxford: Elsevier, pp. 385–417, doi:10.1016/B978-0-444-53802-4.00105-6.

Johnson, C. L., Purucker, M. E., Korth, H., Anderson, B. J., Winslow, R. M., Al Asad, M. M. H., Slavin, J. A., Alexeev, I. I., Phillips, R. J., Zuber, M. T. and Solomon, S. C. (2012). MESSENGER observations of Mercury's magnetic field structure. *J. Geophys. Res.*, **117**, E00L14, doi:10.1029/2012JE004217.

Johnson, C. L., Phillips, R. J., Purucker, M. E., Anderson, B. J., Byrne, P. K., Denevi, B. W., Feinberg, J. M., Hauck, S. A., II, Head, J. W., III, Korth, H., James, P. B., Mazarico, E., Neumann, G. A., Philpott, L. C., Siegler, M. A., Tsyganenko, N. A. and Solomon, S. C. (2015). Low-altitude magnetic field measurements by MESSENGER reveal Mercury's ancient crustal field. *Science*, **348**, 892–895, doi:10.1126/science.aaa8720.

Johnson, C. L., Philpott, L. C., Anderson, B. J., Korth, H., Hauck, S. A., II, Heyner, D., Phillips, R. J., Winslow, R. M. and Solomon, S. C. (2016a). MESSENGER observations of induced magnetic fields in Mercury's core. *Geophys. Res. Lett.*, **43**, 2436–2444, doi:10.1002/2015GL067370.

Johnson, C. L., Phillips, R. J., Philpott, L. C., Anderson, B. J., Byrne, P. K., Denevi, B. W., Fan, K., Feinberg, J. M., Hauck, S. A., II, Head, J. W., III, Korth, H., Mazarico, E., Neumann, G. A., Purucker, M. E., Strauss, B. M. and Solomon, S. C. (2016b). Mercury's lithospheric magnetic field. *Lunar Planet. Sci.*, **47**, abstract 1391.

Kabin, K., Gombosi, T. I., DeZeeuw, D. L. and Powell, K. G. (2000). Interaction of Mercury with the solar wind. *Icarus*, **143**, 397–406, doi:10.1006/icar.1999.6252.

Kletetschka, G., Wasilewski, P. J. and Taylor, P. T. (2000). Mineralogy of the sources for magnetic anomalies on Mars. *Meteorit. Planet. Sci.*, **35**, 895–899, doi:10.1111/j.1945-5100.2000.tb01478.x.

Korth, H., Anderson, B. J., Acuña, M. H., Slavin, J. A., Tsyganenko, N. A., Solomon, S. C. and McNutt, R. L., Jr. (2004). Determination of the properties of Mercury's magnetic field by the MESSENGER mission. *Planet. Space. Sci.*, **52**, 733–746, doi:10.1016/j.pss.2003.12.008.

Korth, H., Anderson, B. J., Raines, J. M., Slavin, J. A., Zurbuchen, T. H., Johnson, C. L., Purucker, M. E., Winslow, R. M., Solomon, S. C. and McNutt, R. L., Jr. (2011). Plasma pressure in Mercury's equatorial magnetosphere derived from MESSENGER Magnetometer observations. *Geophys. Res. Lett.*, **38**, L22201, doi:10.1029/2011GL049451.

Korth, H., Anderson, B. J., Johnson, C. L., Winslow, R. M., Slavin, J. A., Purucker, M. E., Solomon, S. C. and McNutt, R. L., Jr. (2012). Characteristics of the plasma distribution in Mercury's equatorial magnetosphere derived from MESSENGER Magnetometer observations. *J. Geophys. Res.*, **117**, A00M07, doi:10.1029/2012JA018052.

Korth, H., Anderson, B. J., Gershman, D. J., Raines, J. M., Slavin, J. A., Zurbuchen, T. H., Solomon, S. C. and McNutt, R. L., Jr. (2014). Plasma distribution in Mercury's magnetosphere derived from MESSENGER Magnetometer and Fast Imaging Plasma Spectrometer observations. *J. Geophys. Res. Space Physics*, **119**, 2917–2932, doi:10.1002/2013JA019567.

Korth, H., Tsyganenko, N. A., Johnson, C. L., Philpott, L. C., Anderson, B. J., Al Asad, M. M., Solomon, S. C. and McNutt, R. L., Jr. (2015). Modular model for Mercury's magnetospheric magnetic field confined within the average observed magnetopause. *J. Geophys. Res. Space Physics*, **120**, 4503–4518, doi:10.1002/2015JA021022.

Kother, L., Hammer, M. D., Finlay, C. C. and Olsen, N. (2015). An equivalent source method for modelling the global lithospheric magnetic field. *Geophys. J. Int.*, **203**, 553–566, doi:10.1093/gji/ggv317.

Labrosse, S. (2003). Thermal and magnetic evolution of the Earth's core. *Phys. Earth Planet. Inter.*, **140**, 127–143, doi:10.1016/j.pepi.2003.07.006.

Labrosse, S., Poirier, J.-P. and Le Mouël, J.-L. (2001). The age of the inner core. *Earth Planet. Sci. Lett.*, **190**, 111–123, doi:10.1016/j.pepi.2003.07.006.

Langel, R. A. and Hinze, W. J. (1998). *The Magnetic Field of the Earth's Lithosphere*. Cambridge: Cambridge University Press, 445 pp.

Leblanc, F. and Johnson, R. E. (2003). Mercury's sodium exosphere. *Icarus*, **164**, 261–281, doi:10.1016/S0019-1035(03)00147-7.

Leblanc, F. and Johnson, R. E. (2010). Mercury exosphere I. Global circulation model of its sodium component. *Icarus*, **209**, 280–300, doi:10.1016/j.icarus.2010.04.020.

Lepping, R. P., Ness, N. F. and Behannon, K. W. (1979). *Summary of Mariner 10 Magnetic Field and Trajectory Data for Mercury I and III Encounters*, NASA-TM 80600. Greenbelt, MD: NASA Goddard Space Flight Center.

Lillis, R. J., Frey, H. V. and Manga, M. (2008). Rapid decrease in martian crustal magnetization in the Noachian era: Implications for the dynamo and climate of early Mars. *Geophys. Res. Lett.*, **35**, L14203, doi:10.1029/2008GL034338.

Lin, R. P., Mitchell, D. L., Curtis, D. W., Anderson, K. A., Carlson, C. W., McFadden, J., Acuña, M. H., Hood, L. L. and Binder A. (1998). Lunar surface magnetic fields and their interaction with the solar wind: Results from Lunar Prospector. *Science*, **281**, 1480–1484, doi:10.1126/science.281.5382.1480.

Manglik, A., Wicht, J. and Christensen, U. R. (2010). A dynamo model with double diffusive convection for Mercury's core. *Earth Planet. Sci. Lett.*, **289**, 619–628, doi:10.1016/j.epsl.2009.12.007.

Margot, J. L., Peale, S. J., Jurgens, R. F., Slade, M. A. and Holin, I. V. (2007). Large longitude libration of Mercury reveals a molten core. *Science*, **316**, 710–714, doi:10.1126/science.1140514.

Margot, J. L., Peale, S. J., Solomon, S. C., Hauck, S. A., II, Ghigo, F. D., Jurgens, R. F., Yseboodt, M., Giorgini, J. D., Padovan, S. and Campbell, D. B. (2012). Mercury's moment of inertia from spin and gravity data. *J. Geophys. Res.*, **117**, E00L09, doi:10.1029/2012JE004161.

Mayaud, P. N. (1980). *Derivation, Meaning, and Use of Geomagnetic Indices*. Geophysical Monograph 22. Washington, DC: American Geophysical Union, doi:10.1029/GM022.

Mayhew, M. A. (1979). Inversion of satellite magnetic anomaly data. *J. Geophys.*, **45**, 119–128.

McKinnon, W. B., Zahnle, K. J., Ivanov, B. A. and Melosh, H. J. (1997). Cratering on Venus: Models and observations. In *Venus II: Geology, Geophysics, Atmosphere, and Solar Wind Environment*, ed. S. W. Bougher, D. M. Hunten and R. J. Phillips. Tucson, AZ: University of Arizona Press, pp. 969–1014.

Milbury, C., Schubert, G., Raymond, C. A., Smrekar, S. E. and Langlais, B. (2012). The history of Mars' dynamo as revealed by modeling magnetic anomalies near Tyrrhenus Mons and Syrtis Major. *J. Geophys. Res.*, **117**, E10007, doi:10.1029/2012JE004099.

Ness, N. F., Behannon, K. W., Lepping, R. P., Whang, Y. C. and Schatten, K. H. (1974). Magnetic field observations near Mercury: Preliminary results from Mariner 10. *Science*, **185**, 151–160, doi:10.1126/science.185.4146.151.

Ness, N. F., Behannon, K. W., Lepping, R. P. and Whang, Y. C. (1975). Magnetic field of Mercury confirmed. *Nature*, **255**, 204–205, doi:10.1038/255204a0.

Ness, N. F., Behannon, K. W., Lepping, R. P. and Whang, Y. C. (1976). Observations of Mercury's magnetic field. *Icarus*, **28**, 479–488, doi:10.1016/0019-1035(76)90121-4.

Nimmo, F. (2015). Energetics of the core. In *Core Dynamics*, ed. P. Olsen. *Treatise on Geophysics*, 2nd edn, Vol. 8, ed. G. Schubert. Amsterdam, Oxford: Elsevier, pp. 27–55, doi:10.1016/B978-0-444-53802-4.00139-1.

Odstrcil, D. (2003). Modeling 3-D solar wind structure. *Adv. Space Res.*, **32**, 497–506, doi:10.1016/S0273-1177(03)00332-6.

Oliveira, J. S., Langlais, B., Pais, M. A. and Amit, H. (2015), A modified Equivalent Source Dipole method to model partially distributed magnetic field measurements, with application to Mercury. *J. Geophys. Res. Planets*, **120**, 1075–1094, doi:10.1002/2014JE004734.

Olsen, N. (1999). Induction studies with satellite data. *Surv. Geophys.*, **20**, 309–340, doi:10.1023/A:1006611303582.

Olsen, N., Glassmeier, K.-H. and Jia, X. (2010). Separation of the magnetic field into external and internal parts. *Space Sci. Rev.*, **152**, 135–157, doi:10.1007/s11214-009-9563-0.

Opdyke, N. D. and Henry, K. W. (1969) A test of the dipole hypothesis. *Earth Planet. Sci. Lett.*, **6**, 139–151, doi:10.1016/0012-821X(69)90132-0.

Parker, R. L. (1994). *Geophysical Inverse Theory*. Princeton, NJ: Princeton University Press.

Parkinson, W. D. (1983). *Introduction to Geomagnetism*. Edinburgh, UK: Scottish Academic Press.

Paschmann, G. and Daly, P. W. (eds.) (1998). *Analysis Methods for Multi-spacecraft Data*. ISSI Scientific Report Series, 1. Noordwijk, the Netherlands: European Space Agency.

Phillips, R. J., Raubertas, R. F., Arvidson, R. E., Sarkar, I. C., Herrick, R. R., Izenberg, N. and Grimm, R. E. (1992). Impact craters and Venus resurfacing history. *J. Geophys. Res.*, **97**, 15,923–15,948, doi:10.1029/92JE01696.

Philpott, L. C., Johnson, C. L., Winslow, R. M., Anderson, B. J., Korth, H., Purucker, M. E. and Solomon, S. C. (2014). Constraints on the secular variation of Mercury's magnetic field from the combined analysis of MESSENGER and Mariner 10 data. *Geophys. Res. Lett.*, **41**, 6627–6634, doi:10.1002/2014GL061401.

Plattner, A. and Simons, F. J. (2015). High resolution local magnetic field models for the martian south pole from Mars Global Surveyor data. *J. Geophys. Res. Planets*, **120**, 1543–1566, doi:10.1002/2015JE004869.

Proudman, J. (1916). On the motion of solids in a liquid possessing vorticity. *Proc. R. Soc. London A*, **92**, 408–424.

Purucker, M. E., Sabaka, T. J., Solomon, S. C., Anderson, B. J., Korth, H., Zuber, M. T. and Neumann, G. A. (2009). Mercury's internal magnetic field: Constraints on large- and small-scale fields of crustal origin. *Earth Planet. Sci. Lett.*, **285**, 340–346, doi:10.1016/j.epsl.2008.12.017.

Raines, J. M., Gershman, D. J., Zurbuchen, T. H., Sarantos, M., Slavin, J. A., Gilbert, J. A., Korth, H., Anderson, B. J., Gloeckler, G., Krimigis, S. M., Baker, D. N., McNutt, R. L., Jr. and Solomon, S. C. (2013). Distribution and compositional variations of plasma ions in Mercury's space environment: The first three Mercury years of MESSENGER observations. *J. Geophys. Res. Space Physics*, **118**, 1604–1619, doi:10.1029/2012ja018073.

Rikitake, T. (1966). *Electromagnetism and the Earth's Interior, Developments in Solid Earth Geophysics*, Vol. 2. Amsterdam: Elsevier.

Robbins, S. J., Hynek, B. M., Lillis, R. J. and Bottke, W. F. (2013). Large impact crater histories of Mars: The effect of different model crater age techniques. *Icarus*, **225**, 173–184, doi:10.1016/j.icarus.2013.03.019.

Russell, C. T., Baker, D. N. and Slavin, J. A. (1988). The magnetosphere of Mercury. In *Mercury*, ed. F. Vilas, C. R. Chapman and M. S. Matthews. Tucson, AZ: University of Arizona Press, pp. 514–561.

Schaber, G. G., Strom, R. G., Moore, H. J., Soderblom, L. A., Kirk, R. L., Chadwick, D. J. Dawson, D. D., Gaddis, L. R., Boyce, J. M. and Russell, J. (1992). Geology and distribution of impact craters on Venus: What are they telling us? *J. Geophys. Res.*, **97**, 13,257–13,302, doi:10.1029/92JE01246.

Schubert, G., Ross, M. N., Stevenson, D. J. and Spohn, T. (1988). Mercury's thermal history and the generation of its magnetic field. In *Mercury*, ed. F. Vilas, C. R. Chapman and M. S. Matthews. Tucson, AZ: University of Arizona Press, pp. 429–460.

Schubert, G., Russell, C. T. and Moore, W. B. (2000). Geophysics: Timing of the martian dynamo. *Nature*, **408**, 666–667, doi:10.1038/35047163.

Siegler, M. A., Bills, B. G. and Paige, D. A. (2013). Orbital eccentricity driven temperature variation at Mercury's poles. *J. Geophys. Res. Planets*, **118**, 930–937.

Slavin, J. A. and Holzer, R. E. (1979). The effect of erosion on the solar wind stand-off distance at Mercury. *J. Geophys. Res.*, **84**, 2076–2082, doi:10.1029/JA084iA05p02076.

Slavin, J. A., DiBraccio, G. A., Gershman, D. J., Imber, S. M., Poh, G. K., Raines, J. M., Zurbuchen, T. H., Jia, X. Z., Baker, D. N., Glassmeier, K. H., Livi, S. A., Boardsen, S. A., Cassidy, T. A., Sarantos, M., Sundberg, T., Masters, A., Johnson, C. L., Winslow, R. M., Anderson, B. J., Korth, H., McNutt, R. L., Jr. and Solomon, S. C. (2014). MESSENGER observations of Mercury's dayside magnetosphere under extreme solar wind conditions. *J. Geophys. Res. Space Physics*, **119**, 8087–8116, doi:10.1002/2014ja020319.

Soderlund, K.-M. and Schubert, G. (2016). Evolution of Mercury's core dynamo. *Lunar Planet. Sci.*, **47**, abstract 2262.

Solomon, S. C. (1976). Some aspects of core formation in Mercury. *Icarus*, **28**, 509–521, doi:10.1016/0019-1035(76)90124-X.

Srnka, L. J. (1976). Magnetic dipole moment of a spherical shell with TRM acquired in a field of internal origin. *Phys. Earth Planet. Inter.*, **11**, 184–190, doi:10.1016/0031-9201(76)90062-5.

Stanley, S. and Glatzmaier, G. (2010). Dynamo models for planets other than Earth. *Space Sci. Rev.*, **152**, 617–649, doi:10.1007/s11214-009-9573-y.

Stanley, S., Bloxham, J., Hutchison, W. E. and Zuber, M. T. (2005). Thin shell dynamo models consistent with Mercury's weak observed magnetic field. *Earth Planet. Sci. Lett.*, **234**, 27–38, doi:10.1016/j.epsl.2005.02.040.

Stephenson, A. (1976). Crustal remanence and the magnetic moment of Mercury. *Earth Planet. Sci. Lett.*, **28**, 454–458, doi:10.1016/0012-821X(76)90206-5.

Stevenson, D. J. (1987). Mercury's magnetic field: A thermoelectric dynamo? *Earth Planet. Sci. Lett.*, **82**, 114–120, doi:10.1016/0012-821X(87)90111-7.

Stevenson, D. J. (2003). Planetary magnetic fields. *Earth Planet. Sci. Lett.*, **208**, 1–11, doi:10.1016/S0012-821X(02)01126-3.

Strauss, B. E., Feinberg, J. M. and Johnson, C. L. (2016). Magnetic mineralogy of the Mercurian lithosphere, *J. Geophys. Res. Planets*, **121**, 2225–2238, doi:10.1002/2016JE005054.

Suess, S. T. and Goldstein, B. E. (1979). Compression of the hermaean magnetosphere by the solar wind. *J. Geophys. Res.*, **84**, 3306–3312, doi:10.1029/JA084iA07p03306.

Takahashi, F. and Matsushima, M. (2006). Dipolar and non-dipolar dynamos in thin spherical shell geometry with implications for the magnetic field of Mercury. *Geophys. Res. Lett.*, **33**, L10202, doi:10.1029/2006GL025792.

Tarduno, J. A., Blackman, E. G. and Mamajek, E. E. (2014). Detecting the oldest geodynamo and attendant shielding from the solar wind: Implications for habitability. *Phys. Earth Planet. Inter.*, **233**, 68–87, doi:10.1016/j.pepi.2014.05.007.

Tarduno, J. A., Cottrell, R. D., Davis, W. J., Nimmo. F., Bono. R. K. (2015). A Hadean to Paleoarchaean geodynamo recorded by single zircon crystals. *Science*, **349**, 521–524, doi:10.1126/science.aaa9114.

Taylor, G. (1917). Motion of solids in fluids when the flow is not irrotational. *Proc. Roy. Soc. London A*, **93**, 99–113, doi:10.1098/rspa.1917.0007.

Tian, Z., Zuber, M. T. and Stanley, S. (2015). Magnetic field modeling for Mercury using dynamo models with a stable layer and laterally variable heat flux. *Icarus*, **260**, 263–268, doi:10.1016/j.icarus.2015.07.019.

Thébault, E., Finlay, C. C, Beggan, C. D., Alken, P., Aubert, J., Barrois, O., Bertrand, F., Bondar, T., Boness, A., Brocco, L., Canet, E., Chambodut, A., Chulliat, A., Coïsson, P., Civet, F., Du, A., Fournier, A, Fratter, I., Gillet, N., Hamilton, B., Hamoudi, M., Hulot, G., Jager, T., Korte, M., Kuang, W., Lalanne, X., Langlais, B., Léger, J.-M., Lesur, V., Lowes, F. J., Macmillan, S., Mandea, M., Manoj, C., Maus, S., Olsen, N., Petrov, V., Ridley, V., Rother, M., Sabaka, T. J., Saturnino, D., Schachtschneider, R., Sirol, O., Tangborn, A., Thomson, A., Tøffner-Clausen, L., Vigneron, P., Wardinski, I. and Zvereva, T. (2015). International Geomagnetic Reference Field: The 12th generation. *Earth Planets Space*, **67**, 19 pp., doi:10.1186/s40623-015-0228-9.

Tosi, N., Grott, M., Plesa, A.-C. and Breuer, D. (2013). Thermochemical evolution of Mercury's interior. *J. Geophys. Res. Planets*, **118**, 2474–2487, doi:10.1002/jgre.20168.

Trávníček, P. M., Hellinger, P. and Schriver, D. (2007). Structure of Mercury's magnetosphere for different pressure of the solar wind: Three-dimensional hybrid simulations. *Geophys. Res. Lett.*, **34**, L05104, doi:10.1029/2006GL028518.

Tsyganenko, N. A. (1995). Modeling the Earth's magnetospheric magnetic field confined within a realistic magnetopause. *J. Geophys. Res.*, **100**, 5599–5612, doi:10.1029/94JA03193.

Uno, H. (2009). New constraints on Mercury's internal magnetic field. M.Sc. thesis, The University of British Columbia, Vancouver, BC, Canada.

Uno, H., Johnson, C. L., Anderson, B. J., Korth, H. and Solomon, S. C. (2009). Modeling Mercury's internal magnetic field with smooth inversions. *Earth Planet. Sci. Lett.*, **285**, 328–339, doi:10.1016/j.epsl.2009.02.032.

Verhoeven, O., Tarits, P., Vacher, P., Rivoldini, A. and Van Hoolst, T. (2009). Composition and formation of Mercury: Constraints from future electrical conductivity measurements. *Planet. Space Sci.*, **57**, 296–305, doi:10.1016/j.pss.2008.11.015.

Vilim, R., Stanley, S. and Hauck, S. A., II (2010). Iron snow zones as a mechanism for generating Mercury's weak observed magnetic field. *J. Geophys. Res.*, **115**, E11003, doi:10.1029/2009JE003528.

von Frese, R. R. B., Hinze, W. J. and Braile, L. W. (1981). Spherical earth gravity and magnetic anomaly analysis by equivalent point source inversion. *Earth. Planet. Sci. Lett.*, **53**, 69–83, doi:10.1016/0012-821X(81)90027-3.

Weiss, B. P. and Tikoo, S. M. (2014). The lunar dynamo. *Science*, **346**, 1198, doi:10.1126/science.1246753.

Wicht, J. and Tilgner, A. (2010). Theory and modeling of planetary dynamos. *Space Sci. Rev.*, **152**, 501–542, doi:10.1007/s11214-010-9638-y.

Wieczorek, M. A., Weiss, B. P. and Stewart, S. T. (2012). An impactor origin for lunar magnetic anomalies. *Science*, **335**, 1212–1215, doi:10.1126/science.1214773.

Williams, J.-P., Aharonson, O. and Nimmo, F. (2007). Powering Mercury's dynamo. *Geophys. Res. Lett.* **34**, L21201, doi:10.1029/2007GL031164.

Winslow, R. M. (2014). Investigation of Mercury's magnetospheric and surface magnetic fields. Ph.D. thesis, The University of British Columbia, Vancouver, BC, Canada.

Winslow, R. M., Johnson, C. L., Anderson, B. J., Korth, H., Slavin, J. A., Purucker, M. E. and Solomon, S. C. (2012). Observations of Mercury's northern cusp region with MESSENGER's Magnetometer. *Geophys. Res. Lett.*, **39**, L08112, doi:10.1029/2012GL051472.

Winslow, R. M., Anderson, B. J., Johnson, C. L., Slavin, J. A., Korth, H., Purucker, M. E., Baker, D. N. and Solomon, S. C. (2013). Mercury's magnetopause and bow shock from MESSENGER Magnetometer observations. *J. Geophys. Res. Space Physics*, **118**, 2213–2227, doi:10.1002/jgra.50237.

Winslow, R. M., Johnson, C. L., Anderson, B. J., Gershman, D. J., Raines, J. M., Lillis, R. J., Korth, H., Slavin, J. A., Solomon, S. C., Zurbuchen, T. H. and Zuber, M. T. (2014). Mercury's surface magnetic field determined from proton-reflection magnetometry. *Geophys. Res. Lett.*, **41**, 4463–4470, doi:10.1002/2014GL060258.

Wurz, P., Whitby, J. A., Rohner, U., Martín-Fernández, J. A., Lammer, H. and Kolb, C. (2010). Self-consistent modelling of Mercury's exosphere by sputtering, micro-meteorite impact and photon-stimulated desorption. *Planet. Space Sci.*, **58**, 1599–1616, doi:10.1016/j.pss.2010.08.003.

Zurbuchen, T. H., Raines, J. M., Slavin, J. A., Gershman, D. J., Gilbert, J. A., Gloeckler, G., Anderson, B. J., Baker, D. N., Korth, H., Krimigis, S. M., Sarantos, M., Schriver, D., McNutt, R. L., Jr. and Solomon, S. C. (2011). MESSENGER observations of the spatial distribution of planetary ions near Mercury. *Science*, **333**, 1862–1865, doi:10.1126/science.1211302.

6

The Geologic History of Mercury

BRETT W. DENEVI, CAROLYN M. ERNST, LOUISE M. PROCKTER, AND MARK S. ROBINSON

6.1 INTRODUCTION

Our understanding of Mercury's geology, first brought into focus by Mariner 10, is now maturing into a holistic view of the history of the innermost planet. It is not an overstatement to say that our knowledge of Mercury has been revolutionized by MESSENGER's three flybys and four years of orbital observations. The combination of global imaging at incidence angles favorable for viewing surface morphology; global multispectral imaging; high-resolution targeted monochrome and color observations of sites of high scientific interest; topographic information from laser altimetry and stereo imaging; compositional information from X-ray, neutron, and gamma-ray spectroscopy; and measurement of the gravity and magnetic fields and geodetic parameters provide a multi-faceted picture of the planet. From this broad perspective, we now have answers to many longstanding questions about Mercury's history. However, some of the questions still unanswered are remarkably similar to those that were posed in the decades after Mariner 10 imaged approximately half of the planet. Completion of the MESSENGER mission provides an opportunity to assess Mercury's major geologic units, the crustal stratigraphy, and their implications for the formation and evolution of the planet's crust.

6.2 IMAGE DATA

Images from MESSENGER were obtained with the Mercury Dual Imaging System (MDIS), which included a narrow-angle camera (NAC, 1.5° field of view) with a central wavelength of 750 nm (full-width half-maximum of 50 nm) and a wide-angle camera (WAC, 10.5° field of view) with 11 narrow-band filters (centered near 430–1020 nm) and a clear filter (used mainly for images of stars and permanently shadowed regions) (Hawkins et al., 2007, 2009). Five major monochrome imaging campaigns were completed during MESSENGER's orbital mission (Chabot et al., 2016): (1) a moderate-incidence map, with images acquired at an incidence angle (illumination angle measured from the surface normal) as close to 68° as possible and emission angle (viewing angle measured from the surface normal) minimized; (2) a low-incidence map with both incidence and emission angles minimized; (3) a stereo campaign designed to complement the low- and moderate-incidence maps with off-nadir imaging at higher emission angles; (4) a morphology map with incidence angles as close to 80° as possible and illumination from the east; and (5) a complementary morphology map with illumination from the west. The resulting mosaics each have an average pixel scale of ~200 m (e.g., images acquired for the moderate-incidence map have an average pixel scale of 192 m, with a standard deviation of 65 m), despite the wide range of altitudes from which the images were acquired as a result of MESSENGER's highly eccentric orbit (initially ~200-km altitude at high northern periapsis, ~15,000-km altitude at high southern apoapsis, and inclination of 82.5°). The near-uniform pixel scale for these products was achieved by combining NAC imaging in the southern hemisphere with WAC imaging (750-nm filter) in the north, and applying selective on-chip pixel binning (required by limitations to onboard data storage) where the pixel scale was smaller than the requirement by more than a factor of 2.

Three major color imaging campaigns were also conducted from orbit: (1) an 8-color map with global coverage at a pixel scale of ~1 km (Figure 6.1); (2) a higher-resolution (~300 m/pixel) 3-color map of latitudes northward of ~40°S; and (3) a 5-color map of regions north of 60°N acquired with off-nadir geometry to minimize phase angle (defined as the angle at the surface between the direction to the Sun and the camera boresight direction). A common method to highlight color variations using MDIS data is a color composite created through a principal component analysis, where the second principal component is viewed in the red channel, the first principal component in the green channel, and a ratio of the image with the 430-nm filter to that with the 1000-nm filter in blue (e.g., Robinson et al., 2008; see Chapter 8 for further description of MESSENGER color products).

In addition to global and regional imaging campaigns, targeted color images with pixel scales as small as 40 m and monochrome images with pixel scales as small as several meters were acquired. Together with compositional information (e.g., Nittler et al., 2011; Peplowski et al., 2012, 2015; Weider et al., 2015), topography from the Mercury Laser Altimeter (Zuber et al., 2012), and image-based digital terrain models (e.g., Becker et al., 2016), these data sets enabled global geologic mapping. In the following sections we describe Mercury's major crustal units, and we outline the sequence of events that governed the evolution of Mercury's surface from the planet's formation to the present.

6.3 THE MAJOR GEOLOGIC TERRAINS

Though based on a view of less than half of the planet with much less capable instrumentation, Mercury's major terrain types defined on the basis of Mariner 10 observations (Murray et al.,

6.3 The Major Geologic Terrains

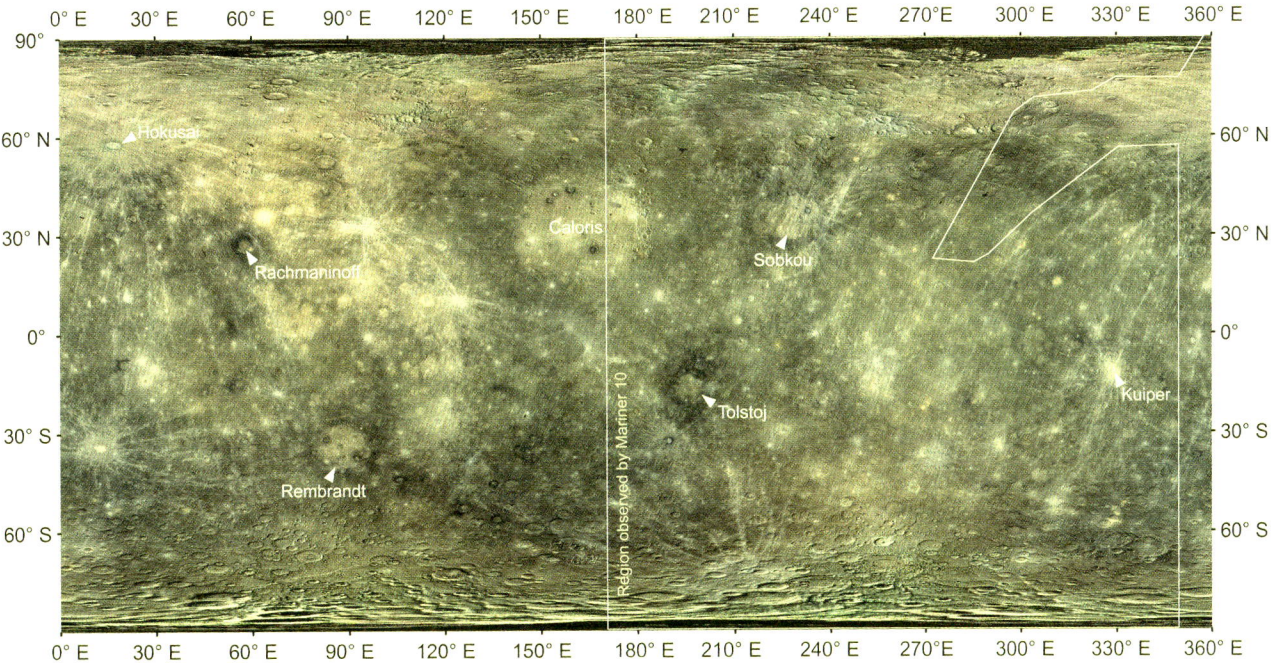

Figure 6.1. Global map of Mercury from the MDIS WAC. Images acquired at 1000, 750, and 430 nm are shown in red, green, and blue, respectively. The region viewed during the Mariner 10 flybys (~170–350°E) is outlined in white.

1975; Trask and Guest, 1975; Spudis and Guest, 1988) have not required substantial alteration after the acquisition of MESSENGER data. These broad-scale units included plains with a range of crater densities (intercrater to smooth) as well as impact craters in a variety of states of degradation and units associated with the basin formation process. What has changed is our understanding of their distribution and origin, and the insights gained from compositional measurements, global multispectral observations, topographic information, and geophysical models of interior structure and crustal thickness.

Mercury's geology was originally explored and units were defined from monochrome images (Murray et al., 1974; Trask and Guest, 1975; Spudis and Guest, 1988). Mariner 10 also acquired color images at 355 and 575 nm with a vidicon detector, though the calibration was challenging and studies at the time suggested there was little correlation between color boundaries and geologic boundaries (Hapke et al., 1980; Rava and Hapke, 1987). Recalibration of the Mariner 10 color images (Robinson and Lucey, 1997; Robinson and Taylor, 2001) and subsequent MDIS observations (Murchie et al., 2008, 2015; Robinson et al., 2008; Denevi et al., 2009), in contrast, showed many examples where color boundaries follow morphologic boundaries, such as at the edges of smooth plains, and revealed important color-stratigraphic relationships. However, we can now also see regions where geology, color, and composition are not correlated (Weider et al., 2012; Murchie et al., 2015; Peplowski et al., 2015). The lack of a strong, lunar-like link between color and morphology appears to be a result of the fact that much of Mercury may have been resurfaced by lavas with a limited range of compositions (little to no ferrous iron in silicate minerals) and thus also a limited range of spectral properties. The range of color properties for each terrain type is briefly described in the following sections

(see also Chapter 8). Mercury's measured surface composition, which is Mg-rich and Ca-, Al-, and Fe-poor compared with typical lunar and terrestrial materials (Nittler et al., 2011; Weider et al., 2012, 2015) (see Chapter 2), is consistent with a range of lithologies from magnesian basalts to basaltic komatiites (Stockstill-Cahill et al., 2012; Charlier et al., 2013), and Mg-rich trachyandesites or alkali-rich boninites (Vander Kaaden and McCubbin, 2016).

An additional key observation for understanding Mercury's geology is that its average reflectance is low and unimodal (Denevi and Robinson, 2008; Blewett et al., 2009; Murchie et al., 2015), with an average value ~15% lower than the lunar nearside (Warell, 2004; Robinson et al., 2008). Materials of especially low reflectance (up to 30% below average for Mercury) appear to be predominantly concentrated at depth and exposed on the surface where excavated by impact events (Robinson et al., 2008; Denevi et al., 2009; Ernst et al., 2010, 2015; Klima et al., 2016). The range of reflectance in different terrains, from low-reflectance material (LRM) to high-reflectance red plains (HRP), is thought to be controlled by the presence of a darkening agent, interpreted to be carbon in the form of graphite on the basis of spectral reflectance and thermal neutron measurements (Murchie et al., 2015; Peplowski et al., 2016) (see Chapter 8).

6.3.1 Intercrater Plains

The intercrater plains are found between large craters and are gently rolling terrain with a high abundance of craters 5–10 km in diameter (Trask and Guest, 1975) (Figure 6.2). In this context "large craters" are typically those 30–50 km in diameter and larger (e.g., Prockter et al., 2016), and the diameter cutoff above which craters in this size range are treated as a separate unit

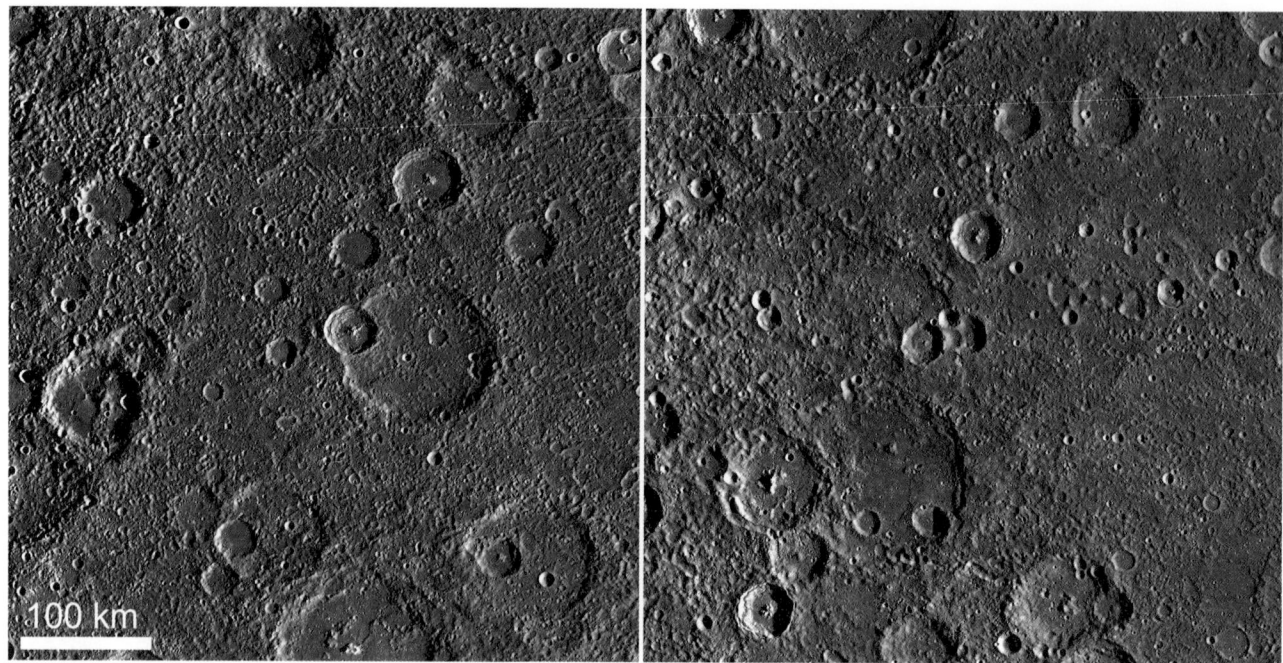

Figure 6.2. Examples of intercrater plains from MDIS monochrome high-incidence maps. Area at left is centered at 26.3°N, 284.5°E; area at right is centered at 11.8°S, 35.6°E. Scale bar applies to both scenes.

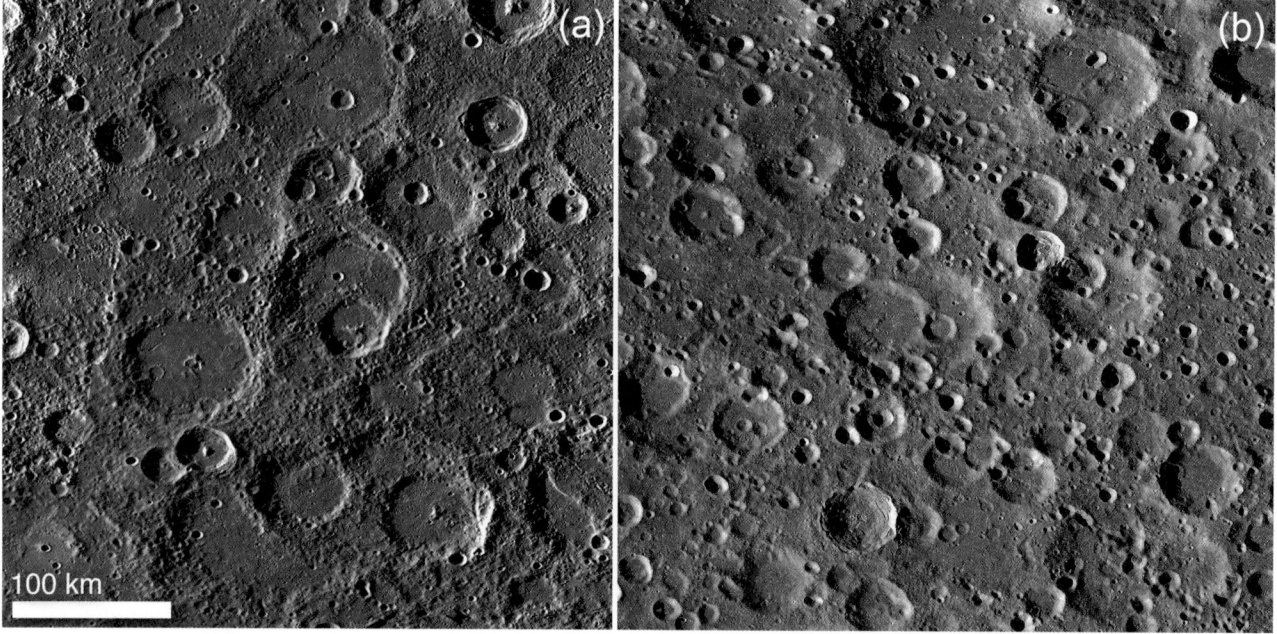

Figure 6.3. The most heavily cratered regions on (a) Mercury (37.6°S, 165.9°E, from MDIS high-incidence mosaic) and (b) the Moon (35.7°N, 215.5°E, from Lunar Reconnaissance Orbiter Camera (LROC) WAC mosaic). Compared with the lunar highlands, Mercury has a similar density of large craters, but patchy resurfacing of topographic lows has removed many smaller craters. Where unaffected by relatively recent resurfacing, much of the surface of Mercury is rougher than that of the Moon because of the influence of secondary craters. Scale bar applies to both scenes.

from the intercrater plains is related solely to the scale of geologic mapping. Indeed, most of these craters superpose the intercrater plains, which have been substantially modified by impact craters across all size ranges (Trask and Guest, 1975; Strom, 1977; Spudis and Guest, 1988; Strom et al., 2008, 2011; Whitten et al., 2014). The high abundance of superposed secondary craters was noted as the "principal characteristic" of the intercrater plains (Trask and Guest, 1975), and the observation that the intercrater plains are "nearly saturated with visually indistinct secondary craters" (Schaber and McCauley, 1980) is confirmed in higher-resolution MDIS images (Figures 6.2 and 6.3a). The heavy modification by these secondary craters, which leads to increased roughness at the kilometer scale (Susorney et al., 2016), is not seen in highland terrain on the Moon (Figure 6.3b) and reflects the larger sizes of secondary craters on Mercury than on the Moon due to Mercury's higher surface gravitational acceleration and the especially high velocities at which primary impactors strike Mercury (Gault et al., 1975; Schultz and Singer, 1980; Xiao et al., 2014) (see Chapter 9).

On average, intercrater plains are modestly lower in reflectance than smooth plains (see below) and display a lower spectral slope from near-ultraviolet to near-infrared wavelengths, but there is substantial overlap in the spectral properties of the two terrains (Denevi et al., 2009; Murchie et al., 2015). Spatially resolved compositional information from MESSENGER's X-Ray Spectrometer (Nittler et al., 2011; Weider et al., 2012, 2015) and neutron measurements from the anticoincidence shield on MESSENGER's Gamma-Ray Spectrometer (Peplowski et al., 2015), available for the northern hemisphere, also show that the intercrater plains and smooth plains commonly have similar compositions. Although it was once suspected that the intercrater plains may be similar in composition to the lunar highlands on the basis of their smooth visible to near-infrared reflectance spectra (e.g., Adams and McCord, 1977; Blewett et al., 1997, 2007), their overall reflectance is substantially lower than that of the lunar highlands (Denevi and Robinson, 2008; Braden and Robinson, 2013). MESSENGER compositional measurements also show that the intercrater plains cannot be anorthositic like the lunar highlands but are instead more similar to low-Fe basalts with low Al/Si and Ca/Si and high Mg/Si (Nittler et al., 2011; Weider et al., 2012) (see Chapter 7). One broad region of intercrater plains that appears to have a composition distinct from any measured smooth plains is an area high in magnesium (Mg/Si >~0.6) located at ~30°N, 270°E (Weider et al., 2015) (see Section 6.4.2.1 and Chapter 7).

The intercrater plains are the most areally extensive unit on Mercury and are interpreted to be the oldest regional terrain type (Trask and Guest, 1975; Spudis and Guest, 1988; Prockter et al., 2016). But despite the fact that the intercrater plains have been heavily modified by impacts, a critical observation is that the density of superposed craters is substantially lower than that of the lunar highlands at crater diameters greater than ~20 km (Figures 6.4 and 6.5; see also Chapter 9). Mercury's relatively low $N(20)$ values, where $N(D)$ is the number of craters with diameter $\geq D$ (in km) per million km^2, was well established for terrain viewed by Mariner 10 (e.g., Murray et al., 1975; Strom, 1977) and is now confirmed as a global phenomenon (Fassett et al., 2011; Strom et al., 2011; Marchi et al., 2013). This result is in contrast to the estimate that the cratering rate is approximately a factor of 3 higher on Mercury than on the Moon (Le Feuvre and Wieczorek, 2008, 2011; Greenstreet et al., 2012; Marchi et al., 2013). On the basis of these comparatively low crater densities, it is inferred that Mercury's oldest surfaces, along with a substantial portion of the earliest population of craters, are no longer extant due to erosion and resurfacing (e.g., Fassett et al., 2011; Strom et al., 2011; Marchi et al., 2013); the intercrater plains are the result of that resurfacing.

Two interpretations for the origin of the intercrater plains and the nature of this resurfacing have been proposed: an impact origin and a volcanic origin, both of which can lead to burial of pre-existing impact craters and produce plains deposits. Under the first scenario, large impact events produced ejecta that traveled radially outward in a large curtain and formed ground-hugging debris surges, during which primary ejecta mixed with local material in a process known as ballistic erosion and sedimentation (Howard et al., 1974; Oberbeck et al., 1975). Close to each basin, craters were largely obliterated within the continuous ejecta deposit (for example, note the low crater density surrounding the lunar Orientale basin, 19°S, 266°E, in Figures 6.4 and 6.5). Farther from the basin, deposits of mixed primary and secondary debris tended to collect in topographic lows and formed smooth deposits that have relatively sharp contacts with the surrounding older terrain (Howard et al., 1974; Oberbeck et al., 1975). Because of the large fraction of local material incorporated into the deposits, plains far from their related impact event are not expected to have strong compositional or color contrasts with the terrain that surrounds them. Understanding of this process was one of the major results from the 1972 Apollo 16 mission to the lunar Cayley plains, which were found to be ejecta from the Imbrium basin impact event (Eggleton and Schaber, 1972), in contrast to the previously widely accepted interpretation that they were of volcanic origin (Wilhelms and McCauley, 1971; Milton, 1972; Trask and McCauley, 1972). Thus, the burden of proof shifted, and to positively identify plains as volcanic, an origin as basin ejecta had to be ruled out. On the basis of this reasoning, Wilhelms (1976) and Oberbeck et al. (1977) made the case for an impact origin for Mercury's plains.

Could impact basin ejecta alone be the cause of Mercury's global resurfacing and the formation of the intercrater plains? Wilhelms (1976) argued that the large volume of intercrater plains could be explained if Mercury had more impact basins than the Moon. The observed density of basins >300-km diameter, however, is lower on Mercury than the Moon (Fassett et al., 2012; see also Chapter 9), though it is almost certain that there is a population of basins that is no longer recognizable. Retention of the topographic signature of basins has likely been affected by viscous relaxation of relief, and the long-wavelength deformation of the crust that distorts even young basins such as Caloris (Oberst et al., 2010; Zuber et al., 2012; James et al., 2015) testifies to the likelihood that the most ancient features are difficult to discern. We can nonetheless recognize some extremely degraded basins that have been almost entirely resurfaced by large expanses of intercrater plains (see Section 6.4.2). The higher impact velocities that Mercury experiences lead to a larger volume of melt for a given size impact (e.g., Gault et al., 1975) and a more fluid-like behavior for its ejecta, meaning that a larger fraction of the surface is likely to be affected per basin [although this effect would be mitigated by the more restricted ballistic range of ejecta on Mercury (Gault et al., 1975)]. However, the vast areal extents of some intercrater plains units that have distinct color properties and thicknesses comparable with those of volcanic smooth plains deposits are inconsistent with an impact origin and suggest that volcanism contributed a substantial fraction of the intercrater plains (Section 6.4.2).

The global view of Mercury's crater density demands that resurfacing was extensive. However, there are some regions where the difference between the crater populations of Mercury and the Moon is minimal for diameters ≥65 km

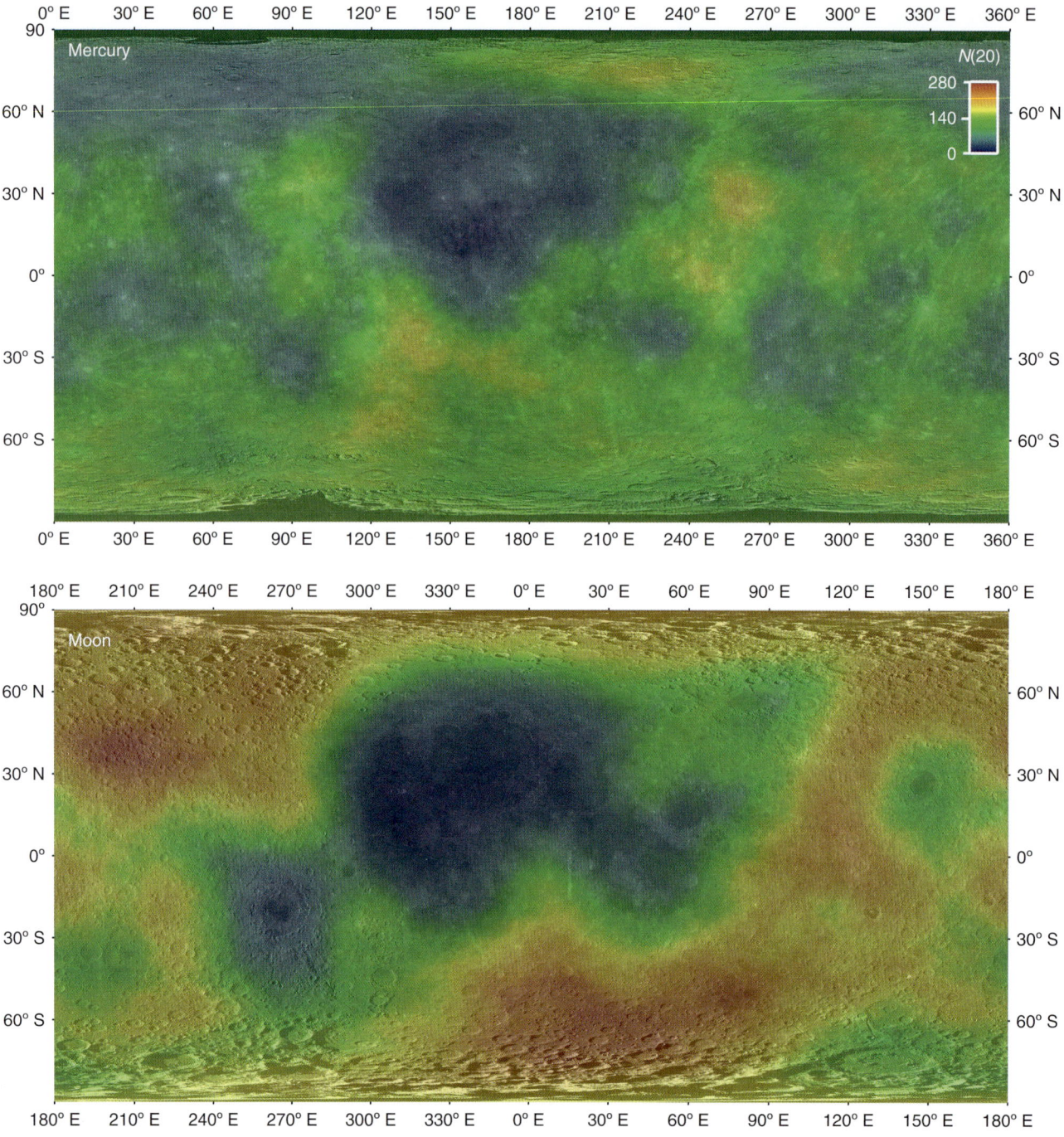

Figure 6.4. Areal density of craters ≥20 km in diameter per million square kilometers, calculated within a moving circular neighborhood 500 km in radius centered on each pixel (22.5-km/pixel scale), for Mercury and the Moon, following the methodology of Fassett et al. (2011). The color scale is the same for both maps, and crater density is shown at 50% transparency over an MDIS mosaic for Mercury and an LROC WAC mosaic for the Moon. Crater density has been derived from the compilations of Fassett et al. (2011) and Head et al. (2010).

(Figure 6.5), despite a clear contrast at smaller diameters (Figure 6.4). The largest contiguous region on Mercury with $N(65)$ values that are similar to those of the lunar highlands is located in the southern hemisphere at ~150°E. Evidence for partial resurfacing of this area seen in images of the region includes patchy deposits of smooth plains filling crater floors and other topographic lows (Figure 6.3a), deposits that are likely responsible for the removal of small craters and the low $N(20)$ values. Areas that did not experience complete resurfacing may represent the oldest crustal materials still exposed and thus are key to understanding the origin of Mercury's early crust. The geology of this region is further examined in Section 6.4.1.

6.3 The Major Geologic Terrains

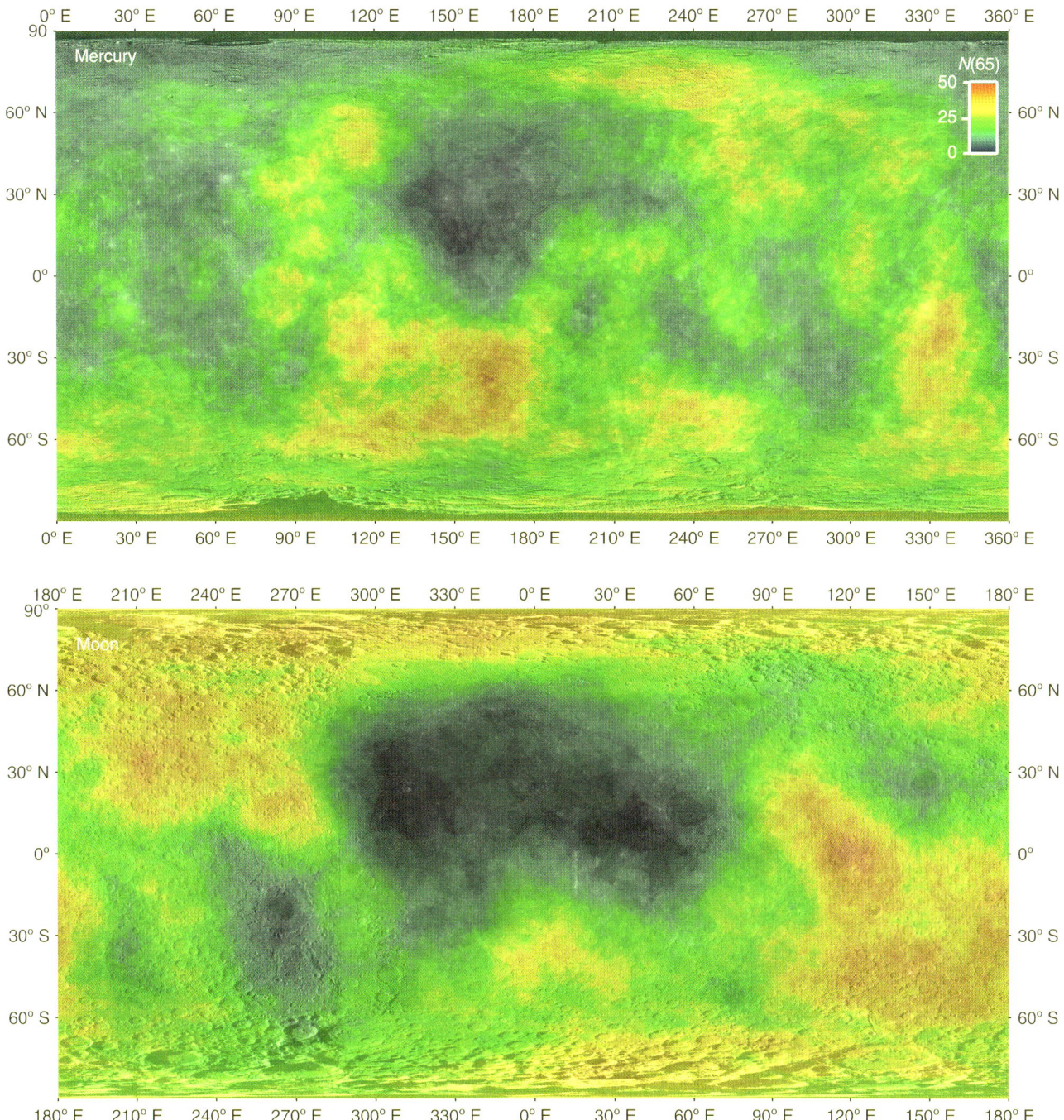

Figure 6.5. Areal density of craters ≥65 km in diameter per million square kilometers, calculated within a moving circular neighborhood 500 km in radius centered on each pixel (22.5-km/pixel scale), for Mercury and the Moon. The color scale is the same for both maps, and crater density is shown at 50% transparency over an MDIS mosaic for Mercury and an LROC WAC mosaic for the Moon. Crater density has been derived from the compilations of Fassett et al. (2011) and Head et al. (2010).

6.3.2 Smooth Plains

Mercury's smooth plains are sparsely cratered deposits that display sharp boundaries with adjacent terrain and are level to gently sloped over baselines of 100–200 km (Figure 6.6) (Trask and Guest, 1975; Denevi et al., 2013a; Prockter et al., 2016). The smooth plains show evidence in some areas for contractional deformation, expressed as wrinkle ridges and scarps, but extensional features have also been observed on plains within several basins and in association with buried ("ghost") craters (Watters et al., 2009a, c; Klimczak et al., 2012; Byrne et al., 2013b, 2014) (Chapter 10). The density of features resulting from tectonic deformation within smooth plains varies greatly, from essentially absent in the smallest deposits to an enormous complexity of overlapping features in

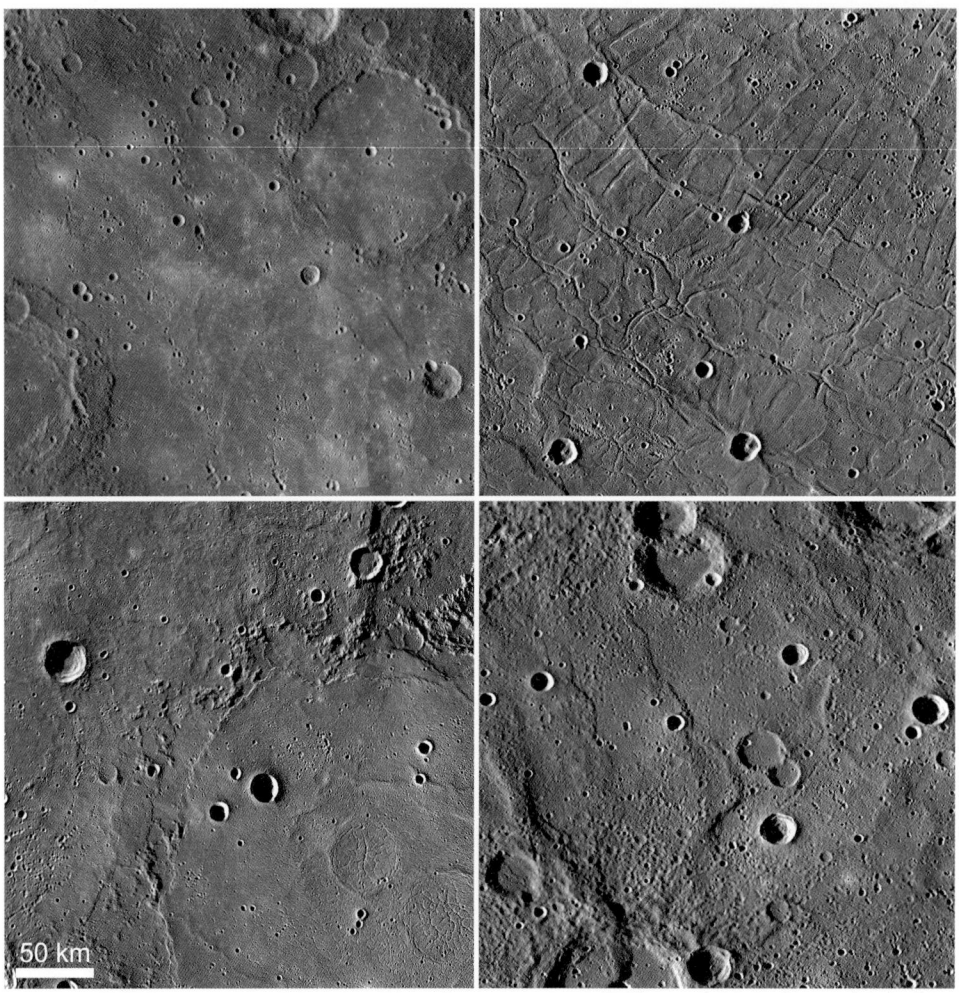

Figure 6.6. Examples of mercurian smooth plains. Top left: Aparāngi Planitia (scene centered at 12.4°N, 63.5°E). Top right: smooth plains within the Caloris basin that have been heavily modified by wrinkle ridges and graben (22.4°N, 157.2°E). Bottom left: smooth plains within and around Goethe basin. Wrinkle ridges localized by buried ("ghost") crater rims are visible at lower right; graben are also found within these buried craters (80.8°N, 298.0°E). Bottom right: one of the most heavily cratered regions to be classified as smooth plains (30.8°N, 332.1°E). Scale bar applies equally to each panel.

the Caloris basin (Figure 6.6) (Strom et al., 1975; Byrne et al., 2013b). Long-wavelength deformation (regional tilt) also appears to have affected some smooth plains deposits, including those within the Caloris basin and the northern smooth plains, after their formation (Oberst et al., 2010; Zuber et al., 2012; James et al., 2015) and without substantially altering their morphology (see Chapter 10).

Estimates of the thicknesses of smooth plains have been made for several large units. The Caloris interior plains are the thickest measured smooth plains unit, with thicknesses of 2.5–4 km (Ernst et al., 2010, 2015; Klimczak et al., 2013). The thicknesses of smooth plains units inside other large basins tend to range from ~1.5 to 3 km, with lower values observed toward the basin edges (André et al., 2005; Watters et al., 2009b). Caloris exterior plains are thinner than those filling the basin, with typical thickness estimates ranging from 0.5 to 2 km (Fassett et al., 2009); those surrounding the Rembrandt basin are also relatively thin, at 0.2–0.4 km (Whitten and Head, 2015). Thickness measurements of the northern smooth plains generally range from 1 to 2 km (Head et al., 2011; Klimczak et al., 2012; Ostrach et al., 2015), with multiple phases of flooding indicated. A few craters ~35–45 km in diameter in the northern smooth plains expose LRM in their central peaks; given the depth of excavation of such material (Ernst et al., 2010), this compositional difference indicates a maximum thickness of the plains in this area of 3–4 km. In general, the major smooth plains units range from 0.5 to 4 km in thickness where estimated across the planet.

In some Mariner-10-era geologic maps of Mercury, the term "intermediate plains" was used to describe a category of plains with higher crater densities and morphologic boundaries that were more gradational than those of the smooth plains but did not reach the threshold for intercrater plains (Schaber and McCauley, 1980; Guest and Greeley, 1983; McGill and King, 1983; Grolier and Boyce, 1984; Spudis and Prosser, 1984; Trask and Dzurisin, 1984; King and Scott, 1990; Strom et al., 1990). Characterization of the crater size–frequency distribution for regions mapped as smooth and intercrater plains shows a wide

Table 6.1. *Values of crater density, N(20), for Mercury shown in Figure 6.7.*

Name/Location	Type	Latitude (°N)	Longitude (°E)	N(20)
East of Gibran	Intercrater plains	35	252	162 ± 35[c]
North of Thoreau	Intercrater plains	13	227	162 ± 35[c]
Near Simonides	Intercrater plains	−30	314	131 ± 32[c]
Near Jokai	Intercrater plains	74	211	115 ± 30[c]
Calder–Hodgkins interior plains	Intercrater plains	16	23	94 ± 9[b]
Tolstoj basin	Basin material	−16	195	93 ± 15[f]
Near Vincente	Intercrater plains	−56	229	92 ± 27[c]
Utaridi	Smooth plains	−64	84	65 ± 16[e]
Near Sholem Aleichem	Intercrater plains	52	261	62 ± 22[c]
East of Tolstoj	Intercrater plains	−16	209	62 ± 22[c]
Rembrandt basin	Basin material	−34	88	58 ± 16[a]
Rembrandt exterior low-reflectance plains	Basin material	−19	101	53 ± 11[d]
Rembrandt interior plains	Smooth plains	−33	87	45 ± 12[b]
Near Barma	Smooth plains	−48	201	44 ± 17[e]
Caloris basin	Basin material	31	161	41 ± 9[f]
Lugus Planitia	Smooth plains	−7	261	35 ± 17[e]
Tir Planitia	Smooth plains	−9	169	32 ± 9[b]
Beethoven interior plains	Smooth plains	−22	233	32 ± 12[b], 31 ± 13[e]
Caloris Planitia	Smooth plains	31	163	29 ± 4[f]
Papsukkal Planitia	Smooth plains	−14	89	25 ± 18[e], 11 ± 11[d]
Northern plains (Borealis Planitia)	Smooth plains			23 ± 2[g]
West of Caloris	Smooth plains	19	135	25 ± 8[b]
Turms Planitia	Smooth plains	−30	6	22 ± 11[e]
Aparāngi Planitia	Smooth plains	5	72	17 ± 7[b], 14 ± 6[e]
Tolstoj interior plains	Smooth plains	−16	196	15 ± 15[e]
Sihtu Planitia	Smooth plains	−3	305	10 ± 10[b]
Near Pushkin	Smooth plains	−60	358	n/a[e]

[a]Watters et al. (2009b); [b]Denevi et al. (2013a); [c]Whitten et al. (2014); [d]Whitten and Head (2015); [e]Byrne et al. (2016); [f]Ernst et al. (2017); [g]Ostrach et al. (2015). Unit types are from Prockter et al. (2016).

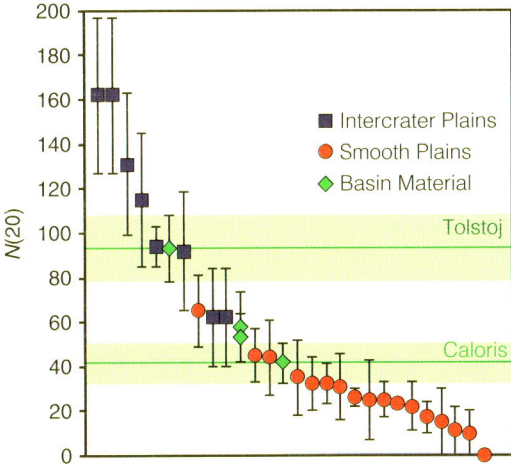

Figure 6.7. *N*(20) values for intercrater plains, smooth plains, and selected basin materials (Table 6.1). Values are plotted in order of decreasing crater density. After Whitten et al. (2014). See Chapter 9 for details and caveats involved in interpreting crater density in terms of relative age.

range of crater densities (Figure 6.7, Table 6.1) (Spudis and Guest, 1988; Denevi et al., 2013a; Whitten et al., 2014; Byrne et al., 2016). There is no strict division between smooth and intercrater plains on the basis of crater density, though at *N*(20) values below ~50 all plains have been classified as smooth, and at *N*(20) values above ~80 all plains have been classified as intercrater (Figure 6.7). Whitten et al. (2014) noted that some intercrater plains have relatively low densities of primary craters (*N*(20) values of ~60) but have been so modified by secondary craters from proximal impact events that they are classified as intercrater plains. They also found that, in spite of the fact that there is a continuum of crater densities, intercrater plains can be readily distinguished from smooth plains largely on the basis of their altered texture due to secondary craters, and thus they recommended against continued use of the term intermediate plains. Subsequent mapping has distinguished only smooth and intercrater plains (Prockter et al., 2016).

After the Mariner 10 flybys, questions concerning the origin of the smooth plains were the same as those for the intercrater plains, namely which deposits were volcanic and which impact-produced, and how do we distinguish between those two modes of formation? Many workers favored an effusive volcanic

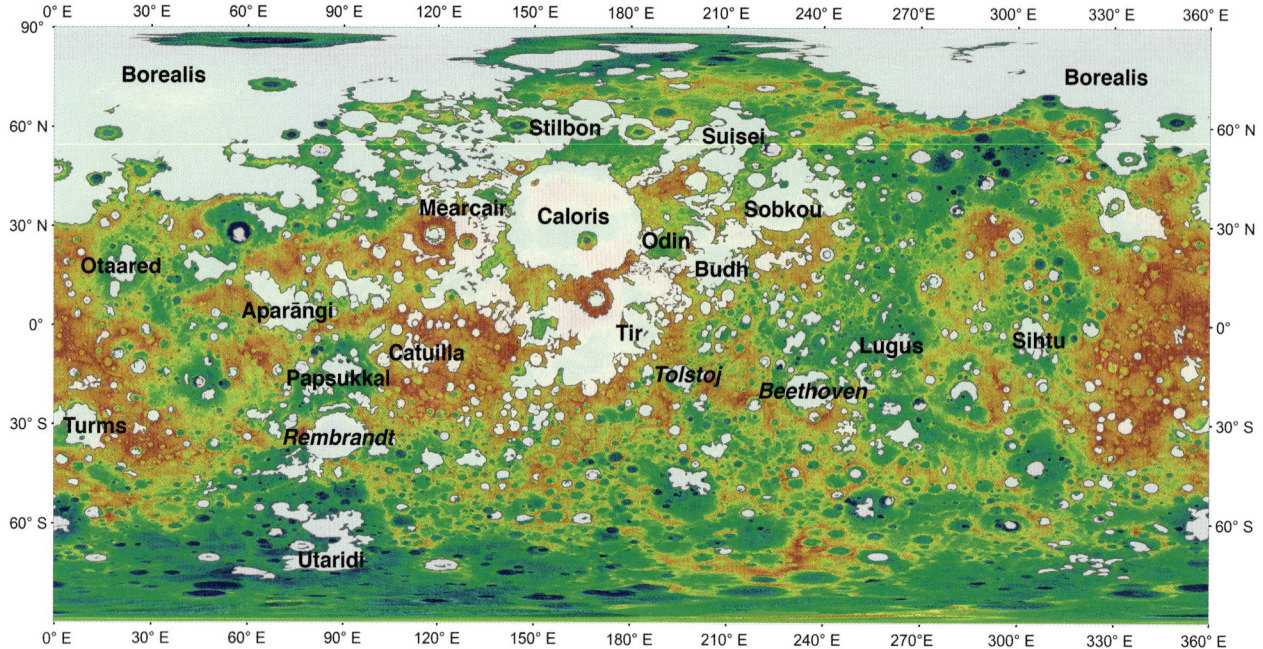

Figure 6.8. The distribution of smooth plains (Denevi et al., 2013a) shown over a digital elevation map of Mercury, ranging from a radius of 2434.018 km (blue) to a radius of 2443.897 km (red) (Becker et al., 2016). Named planitiae (and one planum, Catuilla) are labeled, and large, unnamed smooth plains deposits within named basins are indicated in italics.

origin, analogous to the lunar maria, for most smooth plains. The morphologic similarities with the lunar maria, the large areas of smooth plains, and the fact that smooth plains occur in many basins across the planet, including those substantially older than Caloris, are consistent with volcanism (Strom et al., 1975; Trask and Guest, 1975). However, an impact origin could not be ruled out because of (1) the proximity of many of the smooth plains to the Caloris basin (Trask and Guest, 1975; Wilhelms, 1976), (2) the lack of any indication of compositional contrasts between the smooth plains and their surroundings, and (3) the lack of any identified volcanic landforms. It is now possible to address each of these points.

6.3.2.1 Distribution of Smooth Plains

Mariner 10 viewed terrain from ~170° to 350°E during its three flybys of Mercury (Figure 6.1), and illumination conditions ideal for assessing surface morphology (high incidence angles) held only for the poles, the eastern portion of the Caloris basin, and a region of intercrater plains on the opposite hemisphere. Within these regions, the majority of smooth plains were in close proximity to the Caloris basin. Thus, in the limited terrain viewed by Mariner 10, the geographic connection between the Caloris basin and smooth plains, and the inability to determine the full distribution of impact basins in relation to the observed smooth plains, left open the possibility that *all* plains were of impact origin (e.g., Wilhelms, 1976). With the global picture provided by MESSENGER, it is now known from the distribution of smooth plains that such units cannot all have been emplaced as basin ejecta. Over one-quarter of the surface is covered by smooth plains (Denevi et al., 2013a), and large, contiguous smooth plains deposits are found far from the Caloris basin and, in some cases, far from any basin (Head et al., 2008, 2009a, 2011; Denevi et al., 2009, 2013a), including the extensive northern smooth plains (Borealis Planitia) that show no clear relationship to any one basin (Head et al., 2011).

And yet, the global distribution of smooth plains does confirm the outsize role of the Caloris basin. The Caloris interior plains (2% of the surface) and the circum-Caloris plains (7%) together make up approximately one-third of all smooth plains, and they contribute to an asymmetric distribution of smooth plains on Mercury (Figure 6.8) (Denevi et al., 2013a). Excluding these and the northern smooth plains (an additional 9% of the surface), the remaining smooth plains deposits are distributed essentially equally between the northern and southern hemispheres. Support for a volcanic origin is found for the northern smooth plains and the Caloris interior plains (Murchie et al., 2008; Head et al., 2011), but the circum-Caloris plains defy simple categorization (see Section 6.3.3). Small patches of smooth plains with little color contrast relative to their surroundings are some of the most likely to be of impact origin. An assessment of the distribution of this type of plains deposit in relation to impact basins has not yet been completed and could aid in evaluating their origin (Meyer et al., 2016). On the other hand, many such patches could be volcanic and similar in nature to the patches of mare basalt that fill topographic lows in the Australe region of the Moon (Whitford-Stark, 1979; Lawrence et al., 2013).

6.3.2.2 Color of Smooth Plains

Initial analysis of Mariner 10 vidicon images acquired at 355- and 575-nm wavelengths for limited portions of the planet found little correlation between color boundaries and the smooth plains on Mercury (Hapke et al., 1975, 1980; Rava

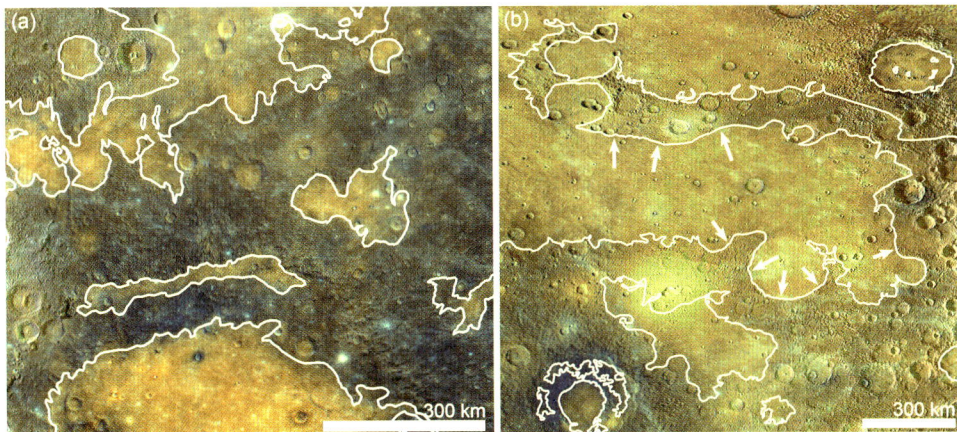

Figure 6.9. (a) Example of color boundaries that follow morphologic boundaries. Portion of Rembrandt basin (bottom of image) and Papsukkal Planitia (top), scene centered at 21°S, 89°E. (b) Example of region where no strong color differences are found along a boundary between smooth plains and intercrater plains. Arrows indicate portions of the boundary along which color differences are minimal. Portion of the northern volcanic plains, scene centered at 43°N, 63°E. Both examples are shown with enhanced color (second principal component, first principal component, and 430-nm/1000-nm reflectance ratio in red, green, and blue, respectively) over a monochrome map that highlights terrain differences. White lines indicate boundaries of smooth plains; smooth plains are the brighter and redder regions in (a), and the arrows in (b) point to the boundaries from inside the smooth plains.

and Hapke, 1987). Subsequently, recalibration of the data to remove vidicon blemishes and correct radiometric errors revealed color contrasts at the boundaries of some smooth plains deposits (Robinson and Lucey, 1997; Robinson and Taylor, 2001). These color differences were interpreted in terms of a difference in the content of an opaque mineral [now thought to be graphite (Murchie et al., 2015; Peplowski et al., 2016)]. Because lunar plains emplaced as impact ejecta generally have compositions similar to their surroundings, the color contrast of some mercurian smooth plains provides support for a volcanic origin (Robinson and Lucey, 1997). Global color imaging from MESSENGER provides many more instances of color boundaries following the contacts between smooth plains and surrounding terrain (e.g., Figure 6.9a) (Murchie et al., 2008; Robinson et al., 2008; Denevi et al., 2009, 2013a), and differences in elemental abundances from their surroundings have been confirmed for Caloris Planitia and portions of the northern smooth plains (Nittler et al., 2011; Weider et al., 2012, 2015; Peplowski et al., 2015).

However, as with the initial observations from Mariner 10, there are examples in MESSENGER data of boundaries between smooth and intercrater plains where no color differences are observed (e.g., Figure 6.9b). Moreover, variations in elemental composition do not always track morphologic boundaries (Murchie et al., 2015; Peplowski et al., 2015; Weider et al., 2015), nor do color and compositional measurements always follow the same boundaries (see the discussion by Murchie et al. (2015) and in Chapters 7 and 8). The expectation that older, more heavily cratered units should be distinct in color and composition is one based on analogy with the Moon, where the ancient anorthositic crust provides an unmistakable compositional contrast with basaltic deposits produced by later mare volcanism. If many of Mercury's intercrater and smooth plains both formed in the same manner, via effusive volcanism, any differences in composition are likely to be much smaller and need not correlate with differences in age.

6.3.2.3 Volcanic Landforms

The case for a volcanic origin of smooth plains after the Mariner 10 flybys was also hindered by the lack of any definitive examples of volcanic landforms in Mariner 10 images, though it was noted that the typical resolution of ~1 km/pixel might preclude their recognition (Wilhelms, 1976; Milkovich et al., 2002). Resolution was indeed the limiting factor, as a host of volcanic landforms have been identified in images having higher resolution (Figure 6.10; see Chapter 11). These include pit craters and pyroclastic deposits (Head et al., 2008; Murchie et al., 2008; Gillis-Davis et al., 2009; Kerber et al., 2009, 2011; Goudge et al., 2014; Rothery et al., 2014; Thomas et al., 2014b, c), features interpreted as evidence for intrusive volcanism (Head et al., 2009b), and channels hosting streamlined landforms interpreted to have formed by lava flow erosion (Byrne et al., 2013a; Hurwitz et al., 2013). Rimless depressions and the diffuse deposits of high-reflectance material having a steep spectral slope across visible to near-infrared wavelengths that often surround them were interpreted as pyroclastic vents and deposits and are by far the most numerous candidate volcanic features on Mercury. Approximately 90% of these features are found in impact craters (Goudge et al., 2014). Smooth plains do not typically have associated pyroclastic deposits except at the margins of some large deposits, including the Caloris interior plains, the smooth plains within the Tolstoj basin, and portions of the northern smooth plains (Denevi et al., 2013a; Goudge et al., 2014). This geographic relationship mirrors that seen for pyroclastic deposits that cluster around the margins of the lunar maria and may imply a genetic relationship between the plains and pyroclastic deposits (Head and Wilson, 1979; Gaddis et al., 1985, 2003; Weitz et al., 1998; Goudge et al., 2014). The lack of volcanic vents within the interiors of smooth plains deposits may be the result of burial of such features by volcanic deposits that collect in topographic lows and for some units are postulated to have erupted at high effusion rates and in large volumes

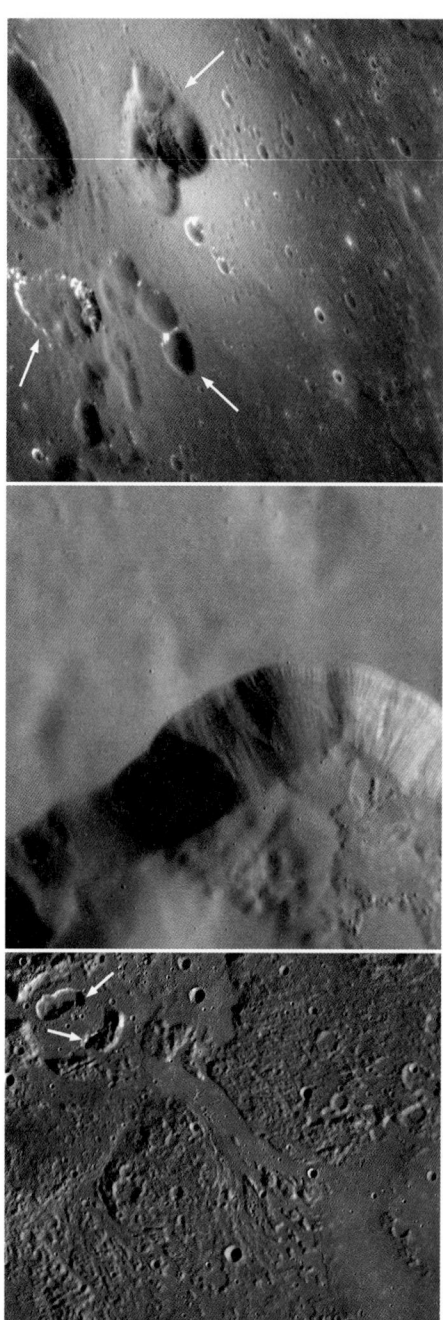

Figure 6.10. Examples of volcanic landforms on Mercury. Top: volcanic vents (indicated by arrows) within the Caloris basin, surrounded by high-reflectance pyroclastic deposits. Large irregular depression at top center has been interpreted as a low-relief shield volcano (Head et al., 2008). MDIS image EN0261627279M; scene is 105 km across and centered at 21.96°N, 146.35°E. Middle: High-resolution view of a portion of the volcanic vent located to the northeast of Rachmaninoff basin (36.07°N, 63.83°E; scene is 26.5 km across). The smooth texture to the north of the vent is characteristic of a deposit of fine pyroclastic glass. MDIS Image EN1003843866M. Bottom: Rimless pit craters (indicated by arrows) thought to mark the sources of lavas that flowed to the south and east, eroding the surface and leaving streamlined features (Byrne et al., 2013a; Hurwitz et al., 2013). Portion of MDIS monochrome mosaic centered at 57.38°N, 113.92°E; scene is 200 km across.

(Head et al., 2011). Several candidate flow fronts have been suggested within smooth plains deposits, though distinguishing these from tectonic features is difficult (Head et al., 2009a, 2011; Chapter 11).

6.3.3 Craters and Basins

In addition to the density of impact craters on a surface, the range of degradation states of these craters can also be used to understand the relative ages of geologic units and to make comparisons among solar system bodies (e.g., Trask, 1967; Chapman, 1968; Wood et al., 1977; McCauley et al., 1981). The classification system for crater degradation on Mercury (Wood et al., 1977; McCauley et al., 1981) was recently updated with MESSENGER data (Kinczyk et al., 2016) to emphasize the most prominent attributes of each crater class and to streamline the classification scheme as a whole (Figure 6.11). The five crater classes, in the order youngest to oldest, also correspond to the five divisions of Mercury's time–stratigraphic system (Section 6.4).

The youngest craters (class 1) have crisp rims, terraces, and central peaks or peak rings, distinct contacts between wall and floor units, and floors with hummocky material or smooth plains. They have well-defined, radially textured ejecta deposits and continuous fields of relatively crisp secondary craters. The defining characteristic of class 1 craters is that they exhibit high-reflectance proximal (or continuous) ejecta deposits and distal rays, although Kinczyk et al. (2016) proposed a subclass that includes craters with high-reflectance proximal ejecta but no obvious rays. This subclass is likely transitional between class 1 and class 2, which includes craters that have relatively fresh rims and slightly degraded wall terraces but no high-reflectance ejecta or rays. Class 2 craters also have distinct contacts between their interior walls and floors, and the floor may contain hummocky deposits or be partially filled with smooth plains. Radially textured continuous ejecta deposits are common, as are clearly defined fields of secondary craters. Central peaks and rings are crisp, and there is a low density of superposed craters.

Impact features of intermediate age (class 3) have continuous, rounded rims with a degraded appearance (i.e., they lack the sharp, angular edge of the rims of class 1 and 2 craters). Wall terraces generally show evidence of slumping, and the floor–wall boundary is commonly indistinct but still identifiable. The floor of the crater may be filled with smooth plains material or have a hummocky texture, which is likely the exposed original floor. If central peaks or rings are visible, they tend to be subdued relative to those of class 2 craters. Ejecta may be continuous to discontinuous, and some chains or clusters of secondaries may still be visible. A low to moderate density of superposed craters is found.

Class 4 craters and basins also exhibit rims that are rounded and degraded and can be continuous to discontinuous. Some remnant terraces are visible in larger examples of class 4 impact features; central peaks and peak rings are rare, and crater interiors may be filled with smooth plains material. A discontinuous ejecta deposit may still be visible, and the density of superposed craters is moderate. The oldest recognizable impact features on Mercury (class 5) are characterized

Figure 6.11. Examples of crater classes showing a progression in relative degradation state from youngest (class 1) to oldest (class 5). Inset shows the high-reflectance rays that characterize class 1 craters.

Figure 6.12. The Caloris basin in orthographic projection centered at 32°N, 162°E. Grey: smooth plains; yellow: Odin Formation; pink: Van Eyck Formation; green: Caloris Montes; blue: Nervo Formation and Caloris rim materials. Units from Denevi et al. (2013a), Goosmann et al. (2016), and Prockter et al. (2016).

by discontinuous, highly degraded rims. Class 5 craters and basins lack terraces or wall structures, have no central peaks or peak rings regardless of their size, and have no detectable ejecta deposits. These impact features are commonly filled with smooth plains deposits and may have a moderate to high density of superposed craters.

The largest well-preserved impact feature on Mercury, the Caloris basin (class 3), has distinct basin-related deposits that were mapped as discrete formations that make up the Caloris group (Figure 6.12). The Nervo Formation is composed of smooth material draped within the massifs of the Caloris Montes (the Caloris rim) and was suggested to have originated as impact melt (Spudis and Guest, 1988) or fallback ejecta (McCauley et al., 1981) from the Caloris event. MESSENGER observations show that the Nervo Formation is present in several locations along the Caloris rim (Goosmann et al., 2016). The Van Eyck Formation contains radial lineations and secondary craters (McCauley et al., 1981) and is distributed broadly around the basin (Fassett et al., 2009). Also known as "Caloris sculpture," the Van Eyck Formation includes ridges and troughs 5–30 km wide, some of which likely formed as secondary crater chains (Fassett et al., 2009).

The Odin Formation is defined by the presence of distinctive kilometer-scale knobs within areas of smooth plains (McCauley et al., 1981; Fassett et al., 2009), the type location of which is Odin Planitia, located to the east of the Caloris basin (Figure 6.13). The Odin Formation is found among smooth plains that surround the basin and often has a gradational transition to knobless smooth plains (Fassett et al., 2009; Denevi et al., 2013a). The density of knobs also appears to show a moderate decrease with distance from the basin rim (Fassett et al., 2009; Ackiss et al., 2015). Clear morphologic analogs to the Odin Formation are found on the Moon in the knobby member of the Montes Rook Formation, found both inside and outside the main Cordillera ring of the Orientale basin, and the Alpes Formation found more extensively around the Imbrium basin (Figure 6.13) (McCauley et al., 1981; Denevi et al., 2013a); both formations have been interpreted as mixtures of impact melt and blocky basin ejecta (Page, 1970; Head, 1974; Scott et al., 1977).

A reanalysis of the crater size–frequency distribution conducted from orbital images of the rim of the Caloris basin (Caloris Montes and Nervo Formation; Ernst et al., 2017) confirms previous measurements made with Mariner 10 and MESSENGER flyby images (Strom et al., 2008; Fassett et al., 2009) that show a difference in crater density among the basin rim, the interior plains, and the Odin Formation, including Odin Planitia (Figure 6.14). The crater size–frequency distribution for the Odin Formation appears to suffer from poor statistics resulting from a relatively small area: the region contains no craters larger than 41 km in diameter, and there is a paucity of craters having diameters between 13 and 26 km, which results in a lower overall crater density than the Caloris rim. Previous work has suggested that the Odin Formation is likely Caloris ejecta (Schaber and McCauley, 1980; McCauley et al., 1981; Guest and Greely, 1983; Spudis and Guest 1988; Denevi et al., 2013a), and if so the variation in crater density between Odin Planitia and the basin rim is likely due to poor

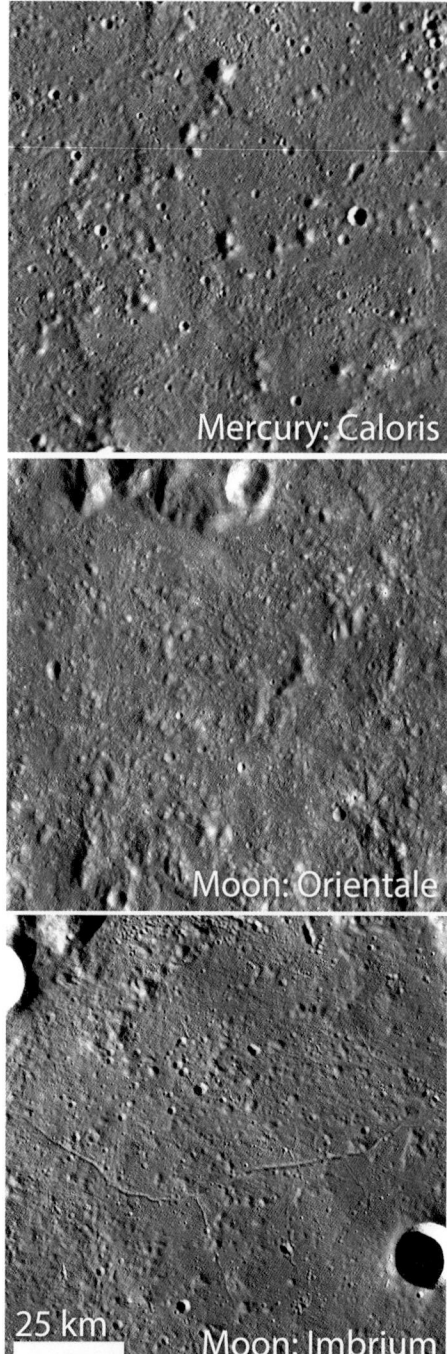

Figure 6.13. A comparison of the Odin Formation to the east of the Caloris basin (32.5°N, 197.3°E) with the Montes Rook Formation of the lunar Orientale basin (6.9°S, 265.1°E) and the Alpes Formation of the lunar Imbrium basin (65.7°N, 4.1°E), after Denevi et al. (2013a). Lunar examples are from an LROC WAC mosaic with a pixel scale of 100 m. The scale bar applies to all panels.

statistics or differences in target properties or crater retention. Fassett et al. (2009) favored an impact origin and subsequent volcanic resurfacing of the Odin Formation, and others also suggested partial embayment by smooth plains (e.g., Schaber and McCauley, 1980; Guest and Greely, 1983; Spudis and Guest, 1988). However, resurfacing events typically preserve

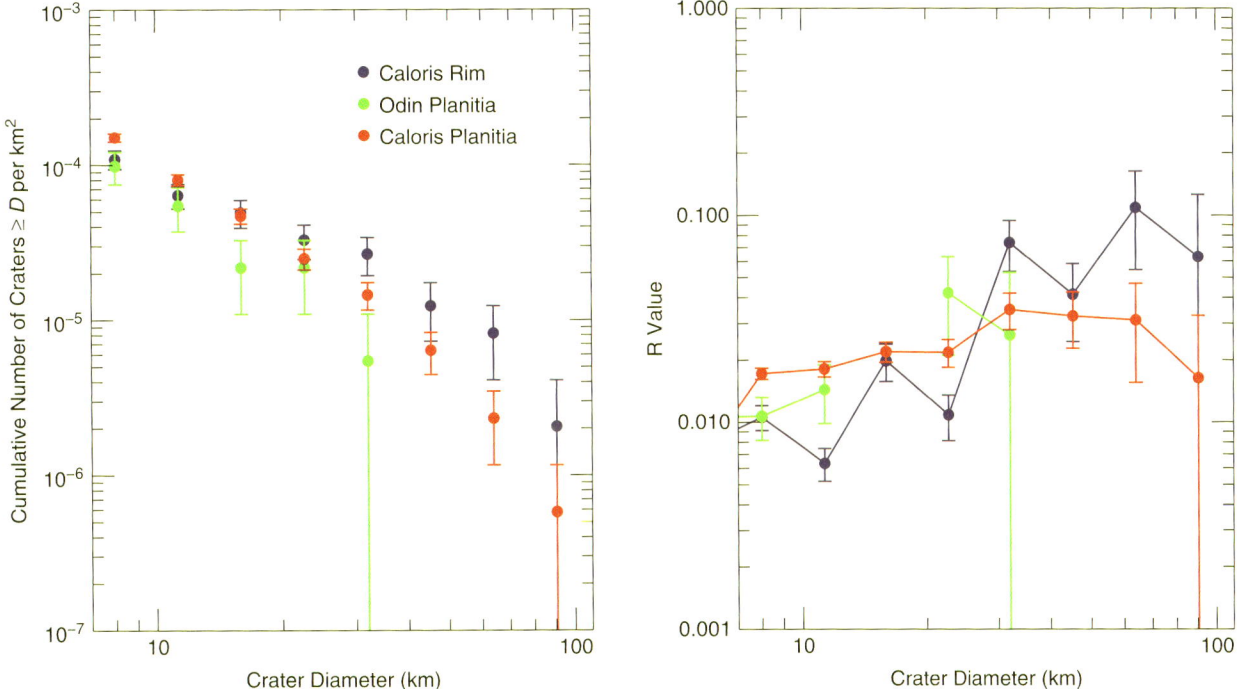

Figure 6.14. Crater size–frequency distribution for the Caloris rim and Caloris Planitia (complete for craters ≥10 km in diameter; Ernst et al., 2017), and a portion of the Odin Formation to the east of Caloris (complete for craters ≥1 km in diameter; Denevi et al., 2013a). Cumulative histogram and R-plot (see Chapter 9) follow conventions of the Crater Analysis Techniques Working Group (1978).

large craters at the expense of smaller ones, whereas here the difference is greatest for craters greater than ~20 km in diameter (Figure 6.14). Given the relatively small areas for determining crater density, ambiguities in defining the extent of the Caloris rim (Ernst et al., 2017), and morphologic similarities to the ejecta deposits of lunar basins (e.g., Figure 6.13), we cannot dismiss the possibility of a genetic relationship between the Odin Formation and the Caloris basin.

Many portions of the plains surrounding Caloris do not share the morphologic attributes of Odin Planitia and instead are either entirely smooth or contain only isolated knobs. These areas may represent locations where impact melt and ejecta deposits from Caloris were resurfaced by later volcanism, leaving at most isolated kipukas of ejecta behind (Fassett et al., 2009; Denevi et al., 2013a). However, it is also possible that some of these plains, especially those that do not have color properties distinct from those of their surroundings, are of impact origin.

The Rembrandt basin has morphologic analogs to the formations defined for the Caloris basin (Watters et al., 2009b; Whitten and Head, 2015). However, most large impact basins on Mercury are substantially degraded to the point that only portions of the rim and sculpture, analogous to the Caloris Montes and Van Eyck Formation, can be discerned. Hilly and lineated terrain has been identified antipodal to the Caloris basin (Murray et al., 1974; Blewett et al., 2010) and may have formed as a result of the convergence of large-amplitude seismic waves (Schultz and Gault, 1975; Hughes et al., 1977; Lü et al., 2011) or ejecta emplacement (Moore et al., 1974) from the Caloris impact event. No such terrain has been identified antipodal to other basins on Mercury, however, either because it did not form or because it has been obscured by subsequent impact events and smooth plains formation (Blewett et al., 2010).

6.4 TIME–STRATIGRAPHIC SYSTEM

Mercury's time–stratigraphic system (Spudis, 1985; Spudis and Guest, 1988) is based on the equivalent system established for the Moon (Shoemaker and Hackman, 1962; Wilhelms, 1987) and relates the terrains described in the preceding section to their relative ages and provides estimates of their absolute ages. From oldest to youngest, the time–stratigraphic system includes the pre-Tolstojan, Tolstojan, Calorian, Mansurian, and Kuiperian Systems (Figure 6.15, Table 6.2). New work, detailed in this section, has resulted in revised estimates of the absolute ages of the base of each system (Banks et al., 2016; Ernst et al., 2017). For reference, ages from Spudis and Guest (1988) are also included in Figure 6.15 and Table 6.2. These age estimates were based on the qualitative assumption that geologic units on Mercury and the Moon with similar crater densities have similar ages, now known to be incorrect (see Chapter 9), but are included because, as the canonical values for many years, they have had a large influence on interpretations of the timing of many events in Mercury's history. The Calorian Period in particular may have been substantially longer, and the Mansurian and Kuiperian

158 The Geologic History of Mercury

Periods shorter, than the original estimates, as described below.

Table 6.2. *Model ages for the base of each time–stratigraphic system (Ga) for Mercury.*

Model	Tolstojan	Calorian	Mansurian	Kuiperian
Spudis and Guest (1988)	3.9–4.0	3.9	3.0–3.5	1.0
Marchi et al. (2009)	3.9	3.8	1.7	0.28
Le Feuvre and Wieczorek (2011)	3.7	3.5	0.85	0.13

Notes: The original age estimates of Spudis and Guest (1988), which were based on comparison with the equivalent lunar time–stratigraphic system, are provided for context. Ages for the Mansurian and Kuiperian are from Banks et al. (2016); those for the Tolstojan and Calorian are from Ernst et al. (2017).

The boundaries that mark the start of the Tolstojan and Calorian Systems are defined by the formation of the Tolstoj and Caloris basins, respectively. The Tolstoj basin is substantially degraded, but its ejecta deposit, the Goya Formation, is still clearly visible and is defined by hummocky to lineated terrain (Spudis, 1985). In addition, the Goya Formation is a type example of LRM, with reflectance values that are among the lowest on the planet (Denevi et al., 2009; Murchie et al., 2015), making its extent relatively easy to define (Figure 6.16a). The $N(20)$ values for the Goya Formation and the boundary of the pre-Tolstojan/Tolstojan are 93 ± 15. Absolute ages estimated from the crater production models of Le Feuvre and Wieczorek (2011) and Marchi et al. (2009) suggest that the basin formed at 3.9–3.7 Ga (Table 6.2, Figure 6.15). For all system boundaries, the Le Feuvre and Wieczorek model results in the youngest age estimates (Table 6.2). Comparing ages for Tolstoj and Caloris within a single model (Table 6.2) suggests that the duration of the

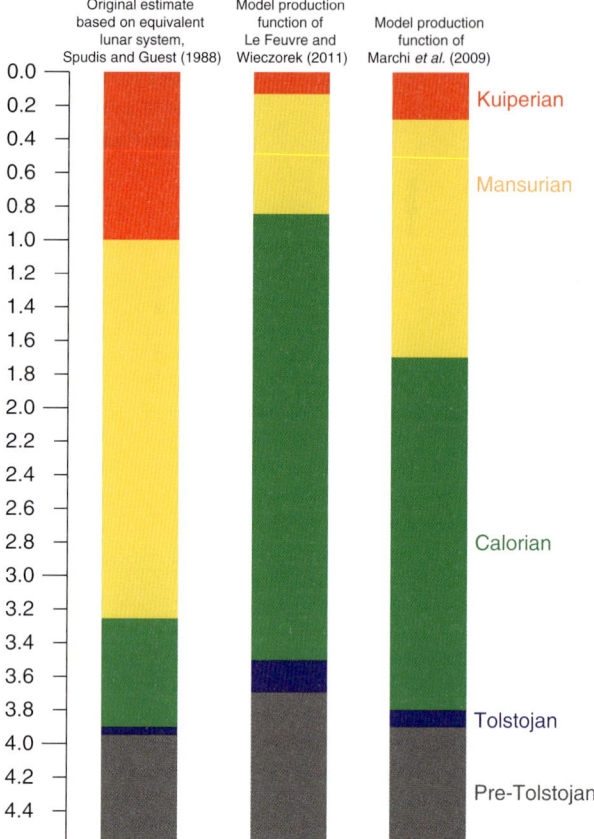

Figure 6.15. Estimates for the absolute ages (Ga) of the boundaries of Mercury's time–stratigraphic system, from Banks et al. (2016) and Ernst et al. (2017), calculated from two models for crater production (Marchi et al., 2009; Le Feuvre and Wieczorek, 2011). Also shown are approximate age estimates from Spudis and Guest (1988), which were extrapolated from the equivalent lunar time–stratigraphic system (where a range of estimates was given for a boundary (Table 6.2), the average was used here).

Tolstojan System was approximately 100–200 Myr. Thirty-five basins (diameter >300 km; Table 6.3) and 712 craters >50 km in diameter that formed prior to Tolstoj (class 5) are identified on

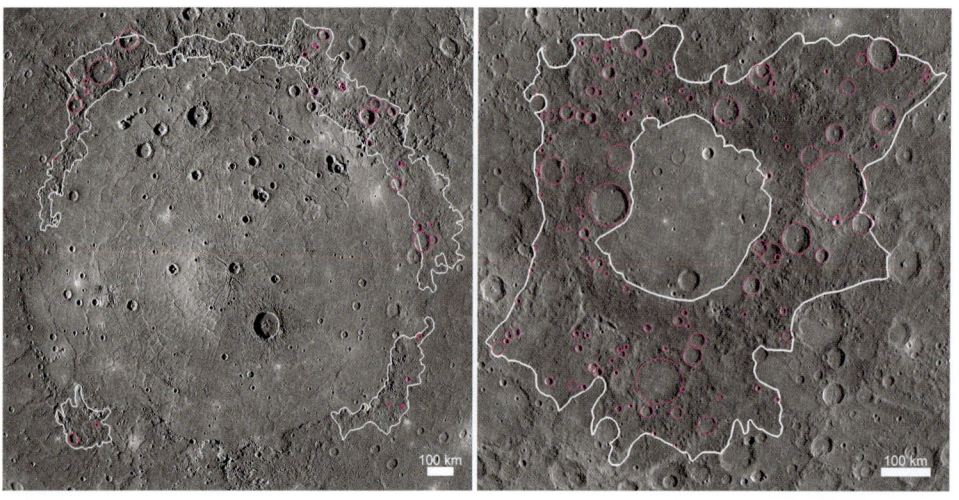

Figure 6.16. (a) The approximate extent of the ejecta deposit (Goya Formation, outlined in white) of the Tolstoj basin and 110 superposed craters ≥10 km in diameter (pink). (b) Portions of the rim materials of the Caloris basin (outlined in white) and 59 superposed craters ≥10 km in diameter (pink).

Table 6.3. *Basins >300 km in diameter as updated by Prockter et al. (2016) from Fassett et al. (2012).*

System	Name	Latitude (°N)	Longitude (°E)	Diameter (km)
Calorian	Rembrandt	−32.9	87.8	716
	Caloris	31.1	156.6	1532
Tolstojan	Homer	−1.2	323.3	307
	Vyāsa	49.7	275.4	310
	Raphael	−20.3	283.9	320
	Unnamed	56.6	252.3	329
	Unnamed	−3.2	315.8	356
	Dostoevskij	−44.6	276.8	394
	Aneirin	−27.4	356.8	435
	Sanai	−13.4	353.4	460
	Tolstoj	−16.4	194.9	491
	Beethoven	−20.8	236.1	633
Pre-Tolstojan	Unnamed	−50.8	92.5	302
	Unnamed	−16.2	160.7	307
	Unnamed	−25.0	261.2	308
	Unnamed	28.9	246.2	308
	Unnamed	6.5	134.8	309
	Goethe	81.5	305.7	319
	Unnamed	−44.8	217.3	339
	Unnamed	−17.0	337.0	340
	Shakespeare	48.9	207.7	357
	Unnamed	−29.8	153.2	360
	Unnamed	55.7	349.4	369
	Unnamed	−45.8	266.1	389
	Unnamed	28.1	201.6	389
	Unnamed	−26.6	218.0	392
	Unnamed	−2.6	303.9	393
	Unnamed	−39.0	258.6	420
	Unnamed	−10.3	102.7	452
	Unnamed	−73.1	149.9	456
	Unnamed	−55.3	256.1	517
	Unnamed	3.4	75.9	549
	Unnamed	52.4	331.7	583
	Unnamed	−59.6	358.2	615
	Unnamed	17.5	207.5	678
	Unnamed	1.4	151.7	682
	Unnamed	−52.2	197.4	692
	Unnamed	−30.1	6.0	712
	Unnamed	−7.6	21.6	735
	Unnamed	45.2	43.3	766
	Unnamed	36.5	3.6	772
	Sobkou	33.4	227.0	773
	Unnamed	70.9	279.0	776
	Unnamed	−42.8	309.0	830
	Unnamed	−24.4	285.1	887
	Lennon–Picasso	−18.4	45.3	1450
	Calder–Hodgkins	17.1	21.7	1460

the basis of their degradation state (as described in Section 2.3) but not strictly by the relative density of superposed craters (Prockter et al., 2016). Ten basins and 887 craters >50 km in diameter are identified as Tolstojan in age (class 4, Table 6.3).

The Caloris impact event clearly had a major effect on Mercury's history (Figure 6.12) and marks a natural boundary in time; the deposits of the Caloris Group (McCauley et al., 1981; Fassett et al., 2009) are assumed to have formed at one time. As with the Tolstoj basin, clear stratigraphic relations are not always available, but they can be inferred from superposed crater density. Given the lack of complete consensus for the origin of some of the circum-Caloris deposits, we adopt the measured $N(20)$ values of 41 ± 9 (Ernst et al., 2017) for the Caloris rim materials (Figure 6.16b) as representative of the Calorian boundary. Model ages for the Caloris basin range from 3.8 to 3.5 Ga (Table 6.2). Among basins >300 km in diameter, only Caloris and Rembrandt are found to be Calorian in age (class 3), and 417 craters >50 km in diameter formed during the Calorian Period (Prockter et al., 2016).

The Mansurian and Kuiperian Systems are more loosely defined. The Mansurian System includes all class 2 craters: those with only moderately degraded ejecta deposits within which fine-scale features can still be observed but for which no rays or high-reflectance ejecta are visible (Figure 6.11) (Spudis and Guest, 1988). The type example is Mansur crater, which superposes the Van Eyck Formation of the Caloris Group. The definition of the Mansurian System is equivalent to the lunar Eratosthenian System, and the age of its base was originally estimated to be approximately 3.5–3.0 Ga (Spudis and Guest, 1988). Braden and Robinson (2013) found that the density of Mansurian craters ≥10 km in diameter is ~1.2 greater than that of Eratosthenian craters, though the expected impact flux is a factor of ~3 greater for Mercury. However, they noted that the increased impact flux on Mercury also degrades craters more rapidly, and they proposed that craters thus transition from characteristics that define the Mansurian System to those that correspond to the Calorian System in a shorter time. A more rapid degradation is also consistent with the larger and highly erosive population of secondary craters on Mercury, and suggests that the age of the base of Mansurian System should be younger than its lunar equivalent. From the combined population of Mansurian and Kuiperian craters, model ages for the base of the Mansurian System are indeed found to be substantially lower than the original estimate of Spudis and Guest (1988) but show a large range, from 1.7 Ga to 850 Ma (Banks et al., 2016) (Table 6.2, Figure 6.15). Of craters >50 km in diameter, 206 have been identified as Mansurian in age (class 2) (Prockter et al., 2016). No basins >300 km in diameter formed during the Mansurian Period.

The Kuiperian System includes all rayed craters on Mercury, typified by Kuiper crater (Figure 6.11). As with the Mansurian System, there are arguments as to why the base of the Kuiperian System [originally assumed to be ~1 Ga (Spudis and Guest, 1988)] is likely to be substantially younger than its lunar equivalent, the Copernican (Denevi and Robinson, 2008; Braden and Robinson, 2013). Space weathering, the process by which the freshly exposed material in ejecta and crater rays matures until it is indistinguishable from the background, may be more rapid on Mercury than on the Moon by a factor of ~4 (Braden and Robinson, 2013). This difference is due to Mercury's lesser

Table 6.4. *Kuiperian craters >50 km in diameter (Prockter et al., 2016).*

Name	Latitude (°N)	Longitude (°E)	Diameter (km)
Erté	27.50	242.72	50.6
Degas	37.08	232.76	52.3
Unnamed	2.97	312.66	54.2
Futabatei	−16.10	276.60	55.3
Kuiper	−11.33	328.75	62.0
Spitteler	−69.17	299.83	65.8
de Graft	22.05	2.00	66.8
Seuss	7.69	33.25	67.3
Balzac	10.63	215.36	68.0
Bashō	−32.36	189.66	71.8
Debussy	−33.95	12.66	80.8
Hokusai	57.78	16.84	97.1
Tyagaraja	3.90	211.22	98.0
Amaral	−26.48	117.90	101.4
Bartók	−29.26	224.94	107.1

solar distance, where it is subjected to an intense thermal and radiation environment and experiences a high flux of impacting meteoroids and micrometeoroids, resulting in more melt and vapor produced on Mercury for a given size impact and a faster gardening rate than on the Moon (Cintala, 1992; Borin et al., 2009; Domingue et al., 2014) (see Chapter 8). From a catalog of all rayed craters >20 km in diameter, Banks et al. (2016) found ages for the base of the Kuiperian System that range from 280 to 130 Ma (Table 6.2, Figure 6.15). Fifteen Kuiperian craters (class 1) >50 km in diameter have been identified globally (Table 6.4) (Prockter et al., 2016).

In the following sections we track the geologic evolution of Mercury through time. Simplified geologic maps of Mercury at different points in its history, based on the work of Prockter et al. (2016), are shown in Figure 6.17.

6.4.1 Pre-Tolstojan: Evidence for the Nature of Mercury's Early Crust

As described in Section 6.3.1, no large expanse of surface on Mercury is as heavily cratered as the lunar highlands, as measured by the $N(20)$ values (Figure 6.4), leading to the conclusion

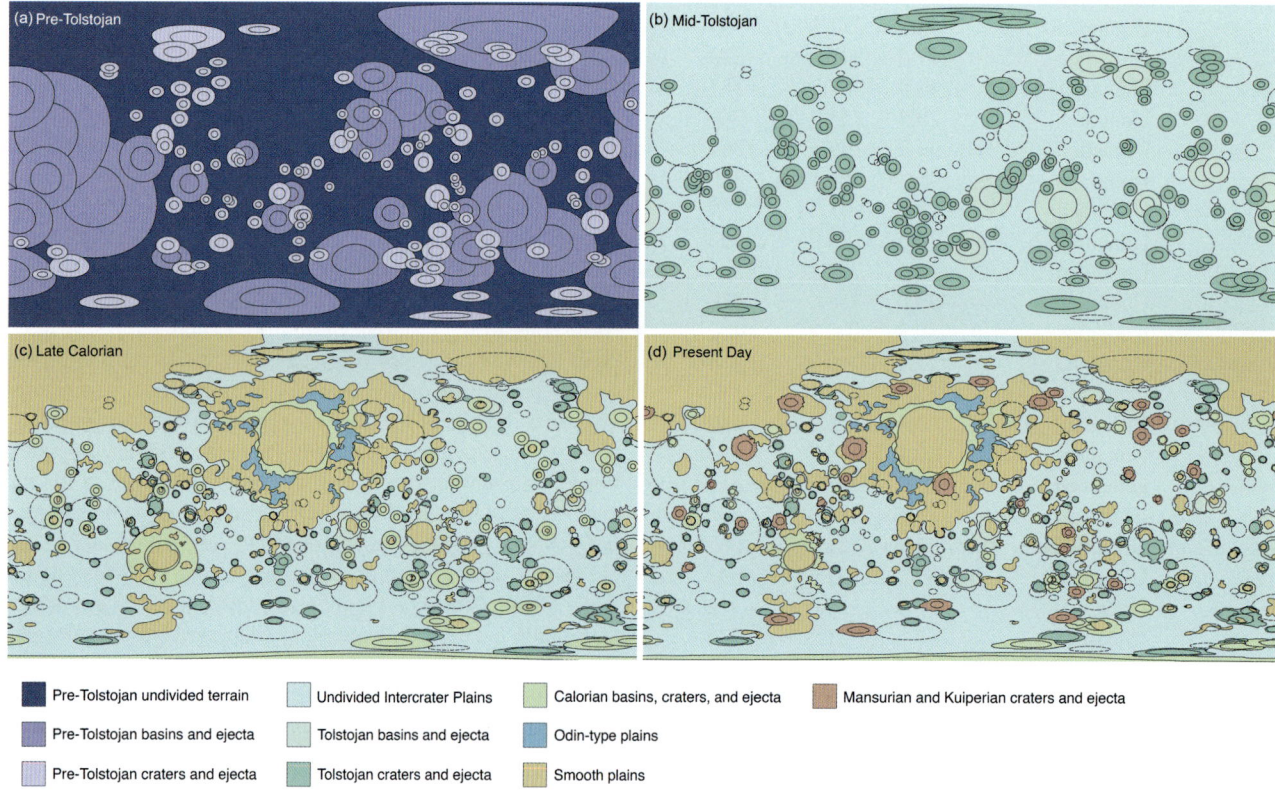

Figure 6.17. Mercury's geologic history, from the geologic map of Prockter et al. (2016). (a) Pre-Tolstojan. Undivided pre-Tolstojan terrain is interpreted as a mixture of early crust and products of volcanic and impact processes. Pre-Tolstojan basins and craters >100 km in diameter are indicated; the approximate original extent of their ejecta (assumed to be one crater radius beyond the rim) is also shown. (b) Mid-Tolstojan. Dashed lines indicate pre-Tolstojan craters and basins, which have been largely resurfaced. Undivided intercrater plains include both pre-Tolstojan and Tolstojan terrain. Tolstojan basins and ejecta and craters >100 km in diameter and their ejecta are indicated. (c) Late Calorian. Smooth plains and Calorian craters and basins and their ejecta are shown. (d) Mansurian–Kuiperian. Mansurian and Kuiperian craters >100 km in diameter are shown. Each map is a simple cylindrical projection from 0 to 360°E, 90°S to 90°N.

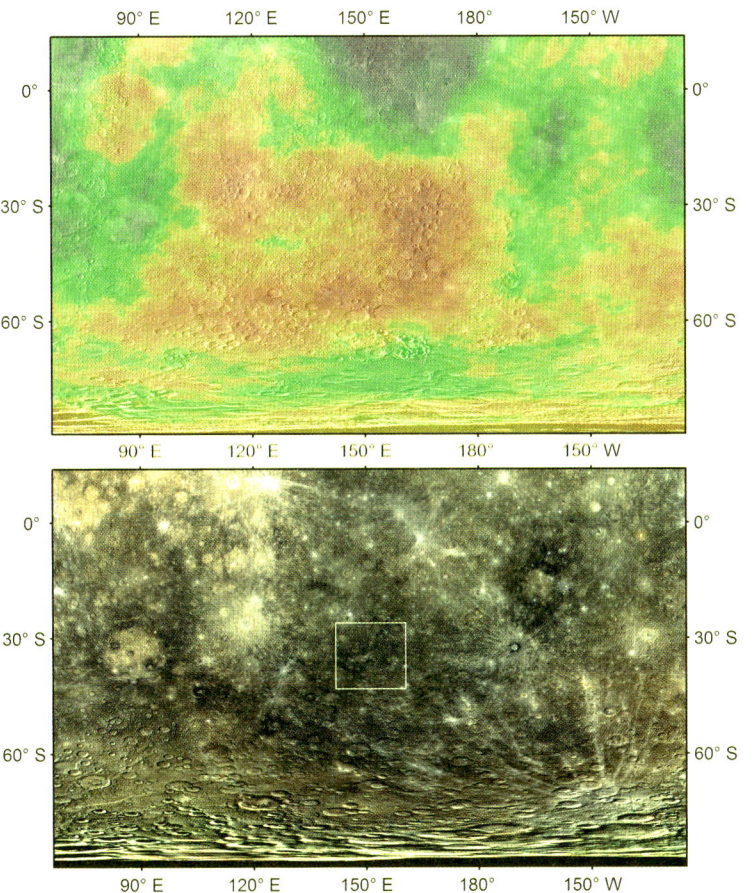

Figure 6.18. Top: $N(65)$ values for Mercury, from zero (dark blue) to >40 (red) craters per million square kilometers superposed on an MDIS monochrome map. Bottom: Image mosaic of the same region in color; images acquired with filters at 1000, 750, and 430 nm are shown in red, green, and blue, respectively. The white box indicates the location of Figure 6.19.

that Mercury was globally resurfaced (Strom et al., 2011; Fassett et al., 2011; Marchi et al., 2013). The most heavily cratered terrain, identified on the basis of its $N(25)$ value, was modeled to have an age of ~4.1–4.0 Ga (Marchi et al., 2013). However, resurfacing may have been less effective at removing craters of larger diameters. At least two broad regions have $N(65)$ values (up to 46 craters/10^6 km^2) that are similar to values for much of the lunar highlands and approach those of the most heavily cratered lunar highlands (up to 50–56 craters/10^6 km^2) (Figure 6.5), and several other areas have patchy exposures with $N(65)$ values nearly as high, though these patches may not all have differences in their crater populations that are statistically significant. The similarities in the density of large craters suggests that these areas on Mercury and the Moon may have reached saturation equilibrium (Hartmann, 1984; Fassett et al., 2011), despite the fact that the $N(20)$ values in these areas are substantially below those of the lunar highlands (Figure 6.4). Saturation equilibrium describes the situation in which the density of craters on a surface has reached a steady state because, on average, the formation of a new crater results in the destruction of an existing crater of the same diameter by some combination of crater excavation, ejecta deposition, and erosion from secondary cratering (Hartmann, 1984; see Chapter 9). If the crater population has reached saturation equilibrium in some areas of Mercury, the regional geologic unit formed at an earlier date than can be estimated from its crater size–frequency distribution.

We examined an area south of the Caloris basin that represents the broadest expanse of terrain with the highest $N(65)$ value; the area generally corresponds to a regional expanse of LRM (Denevi et al., 2009; Klima et al., 2016) and to the southern heavily cratered terrain identified by Marchi et al. (2013) (Figure 6.18). Evidence for partial resurfacing is observed in images of the region: patchy deposits of smooth plains fill crater floors and other topographic lows (Figure 6.19), and these deposits are likely responsible for the removal of small craters and the low $N(20)$ values. Their sporadic distribution and distance from both Caloris and Rembrandt (3–4 basin radii from the rim) suggests that these smooth plains could have originated as impact ejecta, in a manner similar to the smooth plains around the lunar Orientale basin (Meyer et al., 2016). If this is the case, their color contrast (moderately higher in reflectance and steeper spectral slope) with surrounding terrain indicates they likely contain substantial primary basin ejecta. Alternatively, these smooth plains may be volcanic in origin. Within the smooth plains, some craters exposed high-reflectance, spectrally red material, but craters elsewhere exposed material with reflectance values that are either similar to or lower than the background terrain (Figure 6.19).

The stratigraphy of areas outside the patchy smooth plains deposits, as suggested by the observation that material excavated from depth is either comparable to or lower in reflectance than the surrounding terrain, is in contrast to other regions where both low- and high-reflectance materials were excavated from depth (see Section 6.4.2). These observations,

Figure 6.19. Monochrome (left) and enhanced color (as defined in Figure 6.9) overlaid on monochrome (right) views of the area indicated by the square in Figure 6.18. The region experienced partial resurfacing by patches of smooth plains materials (white outlines at right) that are moderately redder than their surroundings. Some craters that formed within these smooth plains exposed red material; outside of the smooth plains crater deposits are LRM (dark blue). The right panel has been stretched to enhance subtle color differences (see Figure 6.18).

(a) Pre-Tolstojan Terrain

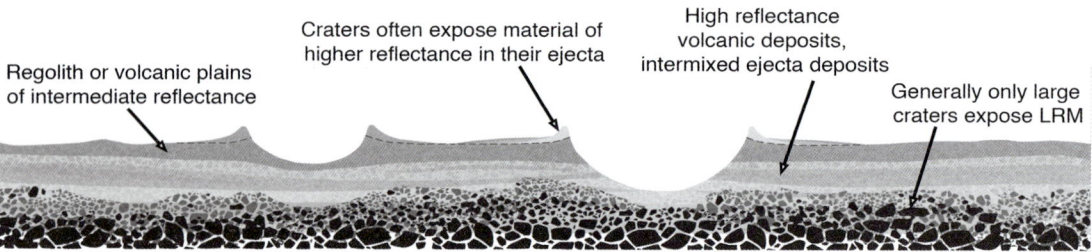

(b) Pre-Tolstojan to Tolstojan Intercrater Plains

(c) Smooth plains within Caloris

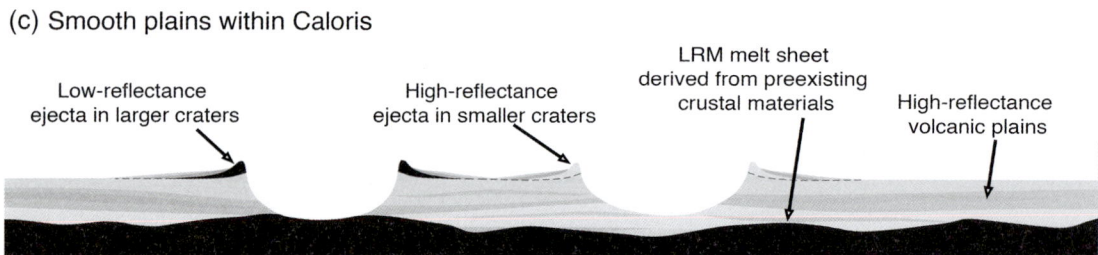

Figure 6.20. (a) Notional cross section of a region that experienced minimal resurfacing, as seen in Figure 6.18. (b) Intercrater plains as in Figures 6.22 and 6.23. (c) Smooth plains, such as in the Caloris basin. Illustration by Kenneth Moscati and Susan Selkirk.

combined with the high $N(65)$ value, suggest that this region may be one where early crust escaped wholesale resurfacing and remains exposed at the surface, albeit in a heavily gardened form (Figure 6.20). The LRM seen concentrated in crater ejecta may have at one time made up the surface in this area but is now found beneath a mixed regolith with a reflectance that has been modestly raised by mixing of distant material from large impacts, particularly by basin-sized

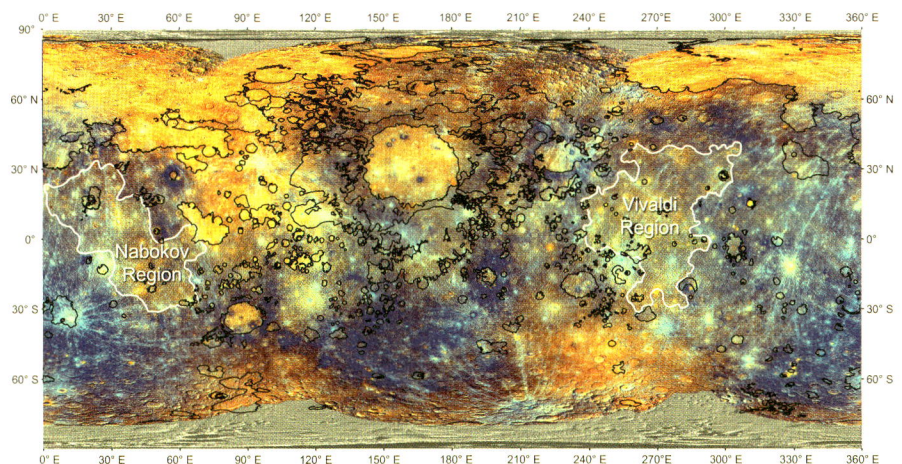

Figure 6.21. Enhanced color view of Mercury. Smooth plains (outlined in black) and the two regions of intercrater plains (outlined in white) discussed in Section 6.4.2 are shown. Colors are the same as those in Figure 6.9.

events. In addition to the two adjacent Calorian basins, eight basins are found near the margins of this region. Here an approximate analogy may be the lunar highlands, where pure anorthosite (thought to represent the original flotation crust that formed in the cooling magma ocean) is found beneath a more mafic megaregolith of mixed local material and basin ejecta and is exposed mainly within large craters and basins (e.g., Hawke et al., 2003). However, on Mercury, at least in this region, the early crust had low reflectance. As LRM has been interpreted to be rich in graphite on the basis of its spectral reflectance and neutron absorption properties (Murchie et al., 2015; Peplowski et al., 2016), this scenario is consistent with the idea that Mercury had a magma ocean that produced a graphite flotation crust (Vander Kaaden and McCubbin, 2015) (see Section 6.4).

6.4.2 Pre-Tolstojan to Tolstojan: Resurfacing by the Intercrater Plains

The formation of the most extensive geologic unit on Mercury, the intercrater plains, began in the pre-Tolstojan and continued into the Tolstojan (Figure 6.7). We have not definitively subdivided the intercrater plains into regional units; they may include a variety of materials that do not all share a common origin. Further characterization of the intercrater plains is warranted, but here we detail two regional examples near Vivaldi and Nabokov craters (Figure 6.21). The first is pre-Tolstojan in age with stratigraphic relationships consistent with thick (3–4 km), relatively high-reflectance deposits overlying LRM; the second is Tolstojan in age and has a complicated stratigraphy whereby relatively high-reflectance deposits overlie two basins, each nearly as large as the Caloris basin, and craters expose both high-reflectance deposits and LRM.

6.4.2.1 Vivaldi Region: Pre-Tolstojan Intercrater Plains

We identified a broad region (~1000 × 2000 km), on the basis of its color properties (Figure 6.22), that is centered at ~10°N, 270°E, near Vivaldi crater. Throughout this region, the intercrater plains have moderate reflectance, and most impact craters with distinct ejecta deposits (Calorian–Kuiperian) exposed high-reflectance red material, defined by a steeper spectral slope and elevated reflectance. High-reflectance red material is spectrally equivalent to HRP, and examples of such material exposed by impact craters examined by Ernst et al. (2010) were thought to have been derived from an older generation of buried volcanic plains. Where a spectrally distinct material is observed within a crater (ejecta or central peak), craters 20–100 km in diameter typically exposed only high-reflectance red material. Only a single crater between 20 and 45 km in diameter exposed LRM (compared with 48 craters in this size range that exposed high-reflectance red materials), and only seven craters 45–100 km in diameter exposed LRM (compared with 20 craters that exposed high-reflectance red materials) (Figure 6.22). The abundance of craters 20–100 km in diameter that exposed high-reflectance red material and the dearth of LRM excavated by craters 20–45 km in diameter suggest a minimum thickness of 3–4 km for a stratigraphic layer of high-reflectance red material, on the basis of the estimated depths of excavation and uplift inferred with the methodology of Ernst et al. (2010). The margins of this region are approximate and mark a transition to more frequent exposures of LRM within craters and lower overall reflectance and shallower spectral slope (Figure 6.22). However, even in the terrain beyond the outlined Vivaldi region, many craters exposed high-reflectance red material (white circles, Figure 6.22).

Our proposed stratigraphy of the Vivaldi region is consistent with a thin deposit of plains material of moderate reflectance covering a deposit of high-reflectance red material that is generally 3–4 km thick and in turn overlies LRM (Figure 6.20b), though there is likely substantial local heterogeneity as well as the possibility of additional compositional variations with depth at finer vertical scales. We interpret the high-reflectance red material as a volcanic deposit that is pre-Tolstojan in age and originally had spectral properties similar to those of the HRP; the range of thicknesses of this deposit is similar to that of many volcanic smooth plains deposits (Section 6.3.2). The unit encompasses terrain with a range of $N(20)$ values, and pre-Tolstojan craters are the most numerous in the northern region, suggesting that the deposit did not form in a single volcanic episode. Smooth plains deposits in the area are sparse and limited in extent, consistent with cessation of volcanic activity

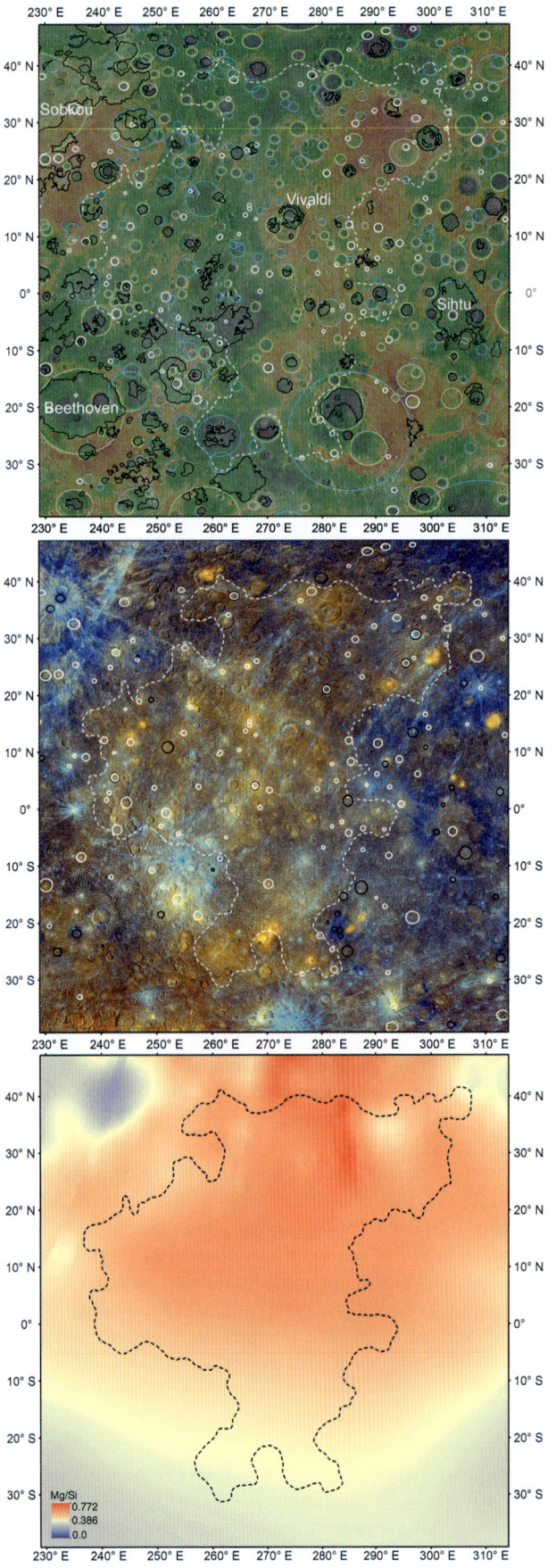

relatively early in this area (Figure 6.22). The region is also notable for its lack of basins >300 km in diameter, further evidence for substantial resurfacing (Fassett et al., 2012). Some of the surrounding terrain also appears to have been affected by volcanic resurfacing, given the numerous craters that expose high-reflectance red material (Figure 6.22). The volcanic deposits in these areas may have been thinner, resulting in more impact mixing with the underlying LRM and a lower overall reflectance, or they may have been resurfaced by plains of intermediate reflectance. The surrounding regions were also affected by the ejecta from Beethoven basin and the basins containing the Sihtu and Sobkou Planitia. In the Sihtu Planitia region, the basin appears to postdate an episode of resurfacing by high-reflectance red material, as craters on the margins of its ejecta deposits expose high-reflectance red material from beneath material of lower reflectance (Denevi et al., 2009; Ernst et al., 2010).

The Vivaldi region also largely coincides with the high-magnesium region identified by Weider et al. (2015) from XRS measurements. Those workers suggested that the distinctive composition of the region may have been the result of an ancient impact basin, more than a factor of 1.5 larger in diameter than Caloris, which experienced extensive volcanic resurfacing with high-degree partial melts of a distinct mantle source. No morphologic evidence is found for the presence of such a basin, though lack of an extant surface expression may be a consequence of the hypothesized volcanic resurfacing (Weider et al., 2015). The stratigraphic evidence here is consistent with volcanic resurfacing, but it does not add a compelling argument for the presence of a large, ancient basin. The thinning of the high-reflectance intercrater plains deposits toward the margins, with more craters exposing LRM at the edges, is similar to the color–stratigraphic relationships observed within the Caloris basin (Ernst et al., 2015). The numerous craters and basins that exposed LRM in the surrounding region (Figure 6.22) could have excavated material from within the ejecta deposits of the hypothesized basin. These relations, however, can be equally well explained without invoking the presence of such a basin.

6.4.2.2 Nabokov Region: Tolstojan Intercrater Plains

An area of intercrater plains centered at ~0°N, 40°E, near Nabokov crater, also provides an interesting study of intercrater plains stratigraphy (Figure 6.23). This area includes several proposed large basins. The first, Calder–Hodgkins, was identified as a "probable" basin (b30) by Fassett et al. (2012), with a diameter of 1390 km, though we measured a diameter of 1460 km (Table 6.3). Evidence for its presence includes preserved portions of a rim and basin sculpture, localized tectonic features, vents associated with the rim, thinned crust, and

Figure 6.22. Summary of excavation of spectrally distinct material by craters in the Vivaldi region. Dashed outline approximately demarks a region of intercrater plains. Top: Topography shown over MDIS monochrome mosaic. Pre-Tolstojan (blue) and Tolstojan (green) craters and basins are shown. Smooth plains are outlined in black. Middle: Enhanced color (as defined in Figure 6.9); white circles indicate craters between 20 and 100 km in diameter that exposed red material, and black circles indicate craters of the same size range that exposed LRM. Bottom: Mg/Si ratio derived from XRS (Weider et al., 2015).

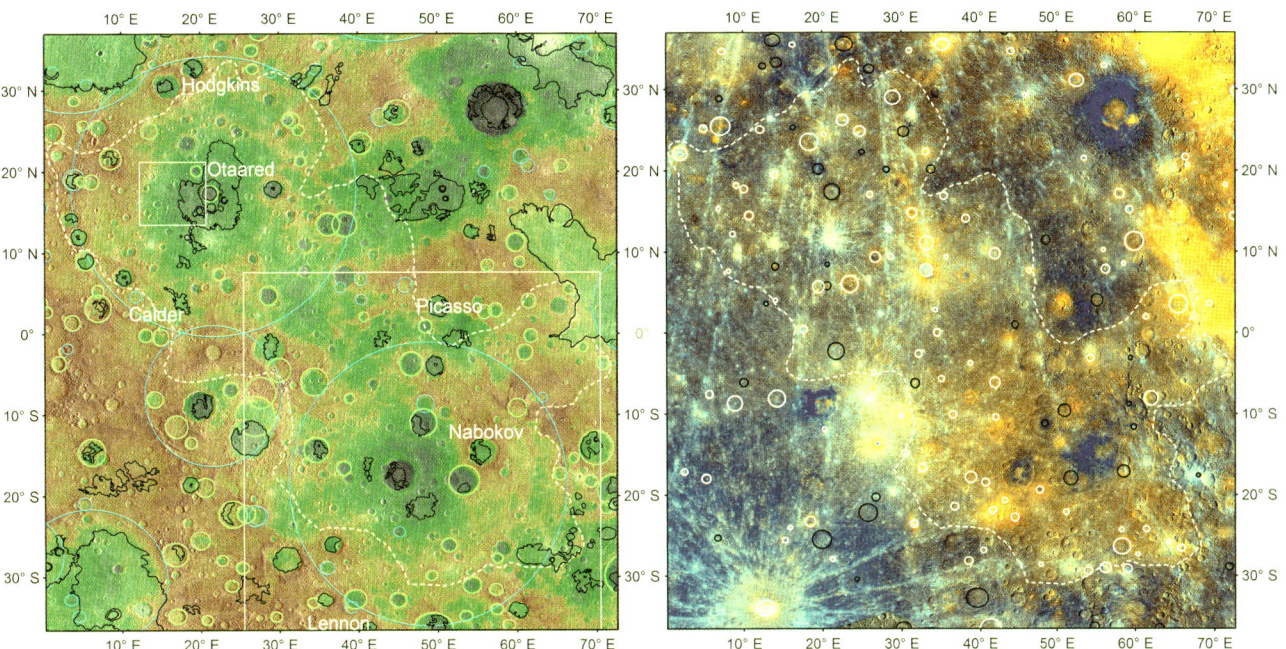

Figure 6.23. Summary of excavation of spectrally distinct material by craters in the Nabokov region. Left: Topography shown over MDIS monochrome mosaic. Pre-Tolstojan (blue circles) and Tolstojan (green circles) craters and basins are shown. Smooth plains are outlined in black. White boxes show the locations of Figures 6.24 (smaller box) and 6.25 (larger box). Right: Enhanced color (as defined in Figure 6.9); white circles indicate craters between 20 and 100 km in diameter that exposed red material, and black circles indicate craters of the same size range that exposed LRM.

Figure 6.24. Smooth (sp, Otaared Planitia) and intercrater (icp) plains near the center of the Calder–Hodgkins basin (location given in Figure 6.23).

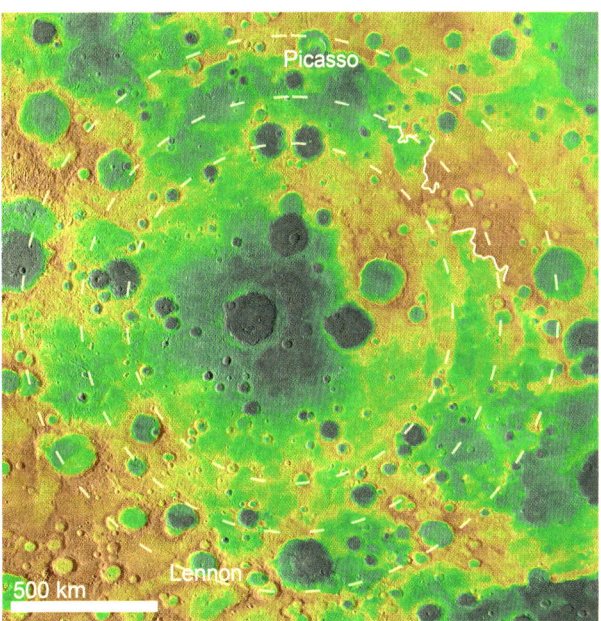

Figure 6.25. Lennon–Picasso basin shown over a digital elevation model (Becker et al., 2016) that spans 7 km in elevation from blue to red. The solid white lines indicate scarps interpreted to be remnants of the basin rim, which approximately follows the middle dashed line (1446 km in diameter). The inner and outer dashed lines follow tectonic features. Location given in Figure 6.23.

interior smooth plains (Denevi et al., 2013b). Within the center of the basin is a region of smooth plains (Otaared Planitia) ~400 km across. These smooth plains embay intercrater plains (Figure 6.24), which cover the remainder of the basin floor and portions of the rim and exterior, extending to the southeast. A second ~1450-km-diameter basin, centered to the southeast of the Calder–Hodgkins basin at 18°S, 48°E, was described as

"suggested but unverified" (basin b56) by Fassett et al. (2012). We find evidence for the presence of this basin as well, and we include it in our list of pre-Tolstojan basins (Lennon–Picasso,

Table 6.3). An arc of massifs that we interpret as a portion of the eastern basin rim (Figure 6.25) rises more than 3.5 km above Mercury's mean radius and includes the highest point on the planet (Becker et al., 2016). The remainder of the rim is not readily discernable, as it was degraded as a result of subsequent cratering events (particularly a ~750-km-diameter basin to the northwest) and intercrater plains resurfacing. As with the Calder–Hodgkins basin, the locations of tectonic features appear to have been influenced by the presence of the basin, and concentric wrinkle ridges found both interior and exterior to the basin may mark a portion of the western rim and a possible outer ring (Figure 6.25).

Throughout the region, craters expose both high-reflectance red material (Ernst et al., 2010) and LRM (Figure 6.23), but high-reflectance red material is associated with craters 20–45 km in diameter nearly twice as often. At smaller diameters, craters that expose LRM appear to be more common near the margins of the proposed basin rims. Near the center of the Lennon–Picasso basin, the smallest crater to expose LRM in its central peak is 34 km in diameter. At the sites of some nested craters that excavated LRM, smaller craters exposed LRM from the floor or ejecta of a larger crater. The estimated depths of excavation and uplift of these craters show that the thickness of the layer of high-reflectance red material varies from ~1 km to >4 km. Similar to the Vivaldi region described above, the margins are approximate and based on the color properties of the intercrater plains, which are overall higher in reflectance and have a redder spectral slope than surrounding terrain. Numerous craters exterior to the region also exposed high-reflectance red material in their ejecta, including Kuiper crater to the southwest (Figure 6.23).

The $N(20)$ value for the northern part of this region is 94 ± 9 (Denevi et al., 2013b), equivalent to that for the Tolstoj basin, and there is a distinct paucity of craters classified as pre-Tolstojan superposing the intercrater plains (just seven with diameters >50 km, Figure 6.23). We interpret this area as a region of Tolstojan-age intercrater plains of volcanic origin and with spectral properties similar to the HRP that flooded much of the Calder–Hodgkins and Lennon–Picasso basins and portions of their exteriors and are buried beneath volcanic plains of intermediate reflectance or a mixed regolith. The LRM exposed from beneath the intercrater plains by craters in the two basins may represent basin floor material, as suspected for LRM within Caloris (Ernst et al., 2015).

The two regions of intercrater plains described here are provided as examples of how volcanism influenced Mercury's crust in the pre-Tolstojan and Tolstojan periods, while at the same time large impact events punctuated the history of broad regions. The combination of these two processes early in Mercury's history resulted in a complicated stratigraphy of interbedded volcanic and impact deposits. Detailed geologic mapping of the intercrater plains will be critical to developing a full picture of the evolution of the crust during the pre-Tolstojan and Tolstojan, though it is clear from these areas and evidence described in Section 6.3.1 that volcanic resurfacing was extensive during this time.

6.4.3 Calorian–Mansurian: Smooth Plains Formation and Pyroclastic Volcanism

As the impact flux dwindled following the late heavy bombardment, volcanism continued into the Calorian and is recognized mainly in the form of smooth plains. The density of superposed craters indicates that smooth plains formation actually began in the Tolstojan and that there is a gradational transition from intercrater to smooth plains, though the majority of smooth plains are Calorian in age (Figure 6.7).

As described in Section 6.3.2, the distribution of smooth plains is hemispherically asymmetrical, with the largest occurrences found predominantly in the northern hemisphere and near the Caloris basin (Figure 6.8). It is likely that this asymmetry is one of timing, with early effusive volcanism (now observed as the intercrater plains) occurring broadly across Mercury, and continuing into the Calorian Period (as smooth plains) in a more limited distribution (Denevi et al., 2013a). However, it is still an interesting question as to why volcanism persisted longer in some regions than in others. In some cases, there is a clear relation to basin formation, as many of the largest smooth plains deposits are found within or surrounding basins. The formation of impact basins may promote or enable local volcanism by removing overburden pressure, uplifting isotherms at depth, fracturing the lithosphere to create conduits to the surface, and depositing impact energy as heat in the underlying mantle (e.g., Arkani-Hamed, 1973; Elkins-Tanton et al., 2004; Roberts and Barnouin, 2012; Klimczak, 2015).

However, the northern smooth plains are not evidently related to any single impact event. Potassium abundance is elevated in the north (Peplowski et al., 2012), which could indicate a relationship between the duration of volcanism in some regions and the concentrations of heat-producing elements, such as in the lunar KREEP (K, rare-earth elements, and P) terrane (e.g., Jolliff et al., 2000). Alternatively, the enhanced concentration of potassium at high latitudes may be the result of diurnal heating and mobilization of this volatile element at lower latitudes rather than a measure of the mantle source region for the northern smooth plains (Peplowski et al., 2012). However, an anticorrelation between potassium and non-volatile magnesium at high latitudes (Weider et al., 2015) argues against this possibility. The lack of similar elemental data for the southern hemisphere currently precludes any further assessment of the relation between smooth plains and the distribution of potassium.

On the whole, it appears that the majority of volcanism ended on Mercury by ~3.7–3.5 Ga (Denevi et al., 2013a; Ostrach et al., 2015; Byrne et al., 2016), though there are large uncertainties in deriving absolute ages for features on Mercury (see Figure 6.15 and Chapter 9). Despite crater densities that vary by up to a factor of 5, model ages for the largest smooth plains deposits fall in the range 3.8–3.5 Ga because of the rapidly declining impactor flux during that time. The cessation of volcanism was likely the combined result of cooling and consequent decreases in the amount of partial melting of the mantle (Michel et al., 2013; Tosi et al., 2013) and global lithospheric contraction that inhibited the ascent of magmas to the surface (Solomon, 1978; Byrne et al., 2014, 2016). Effusive volcanism may have continued to times as recent as 1 Ga in small deposits contained wholly within relatively young impact features such as the peak-ring basin Rachmaninoff (Prockter et al., 2010; Marchi et al., 2011). Thus, effusive volcanism appears to have ceased entirely by the early Mansurian Period; intrusive magmatism may have continued beyond the duration of most smooth plains formation but may also have been influenced by global contraction (Head et al., 2009b) (see Chapter 11). In contrast to the compelling case for the volcanic origin of Rachmaninoff's smooth plains on the basis of embayment relations, contrasting spectral properties, and crater density differences inside and outside of its peak ring, most small smooth plains contained within relatively young craters are not spectrally distinct from their surroundings, do not embay craters superposed on the impact features, and are likely to have originated as impact melt.

Explosive volcanism appears to have continued to times more recent than the times of emplacement of most smooth plains units. Pits and pitted terrain with surrounding deposits of the very-high-reflectance red unit are found in some 150 locations across the planet, many located in Mansurian craters (Goudge et al., 2014; Thomas et al., 2014b, c) (Chapter 11). Four candidate pyroclastic deposits have been identified in late Mansurian to Kuiperian craters (Thomas et al., 2014b) and thus may be as young as several hundred million years old (Figure 6.15).

6.4.4 Kuiperian: Craters and Hollows

Though some small fraction of pyroclastic volcanism may have continued into the Kuiperian Period (Thomas et al., 2014b), and the presence of small, young scarps suggests that tectonic deformation driven by global contraction likely has continued to the present (Banks et al., 2015; Watters et al., 2015), the last several hundred million years are most notable for the formation of impact craters and hollows. Hollows are shallow, irregular rimless depressions commonly surrounded by high-reflectance haloes that range in horizontal dimension from tens of meters to several kilometers (Blewett et al., 2011, 2013; Thomas et al., 2014a; Chapter 12) and were first identified as high-reflectance deposits in Mariner 10 images (Dzurisin, 1977). They are typically found in impact craters or basins that expose LRM, and they are thought to form by loss of volatile materials from minerals such as sulfides (MgS, CaS) or graphite, which weakens the local material and causes scarp retreat (see Chapter 12). Hollows may have formed throughout Mercury's history, but their small sizes mean they are relatively rapidly degraded and thus only Kuiperian hollows remain.

The population of Kuiperian impact craters was described in Section 6.3, and here we note several important geologic characteristics that are the most readily visible in such fresh craters on Mercury. First, impact melt appears to be more abundant on Mercury than on the Moon, flooding more of the crater floor (Ostrach et al., 2012), and sometimes occurring as melt ponds outside craters (Beach et al., 2012) (see Chapter 9). Some Mariner 10 quadrangle maps included "very smooth plains" that were interpreted mainly as impact melt (Schaber and McCauley, 1980; Guest and Greeley, 1983; Spudis and Prosser, 1984; Trask and Dzurisin, 1984; Strom et al., 1990), though more recent work has not distinguished such a unit from smooth plains (Denevi et al., 2013a; Prockter et al., 2016). The large volumes of impact melt are consistent with predictions based on the higher average impact velocity for Mercury

Figure 6.26. The rays of Hokusai crater (top, 97 km in diameter) stretch across much of Mercury.

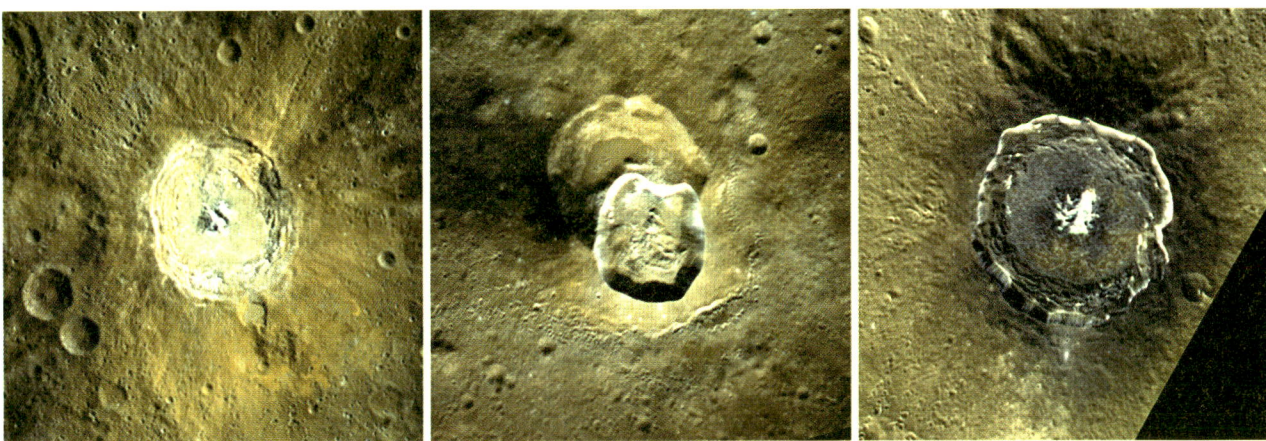

Figure 6.27. Kuiperian craters that expose spectrally distinct material. Kuiper crater (left) is 62 km in diameter, Ailey crater (center) is 20 km in diameter, and Degas crater (right) is 51 km in diameter. Images acquired through filters at 1000, 750, and 430 nm are shown in red, green, and blue, respectively.

than for other locations in the solar system. Second, though there are relatively few large Kuiperian craters (Table 6.4), they have an outsized influence on Mercury's surface. The rays of Hokusai, for instance, cross essentially an entire hemisphere of Mercury (Figure 6.26) (Ernst et al., 2016; Xiao et al., 2016), and those of craters such as Kuiper and Debussy mask a substantial portion of the surrounding terrain. The presence of immature material in crater rays increases not only surface reflectance, but also the roughness as indicated by 12.6-cm radar backscatter (Neish et al., 2013). The roughness of mercurian rays exceeds that of lunar crater rays, likely the result of larger secondary craters in mercurian rays that lead to more decimeter-size clasts or exposure of rocky material in crater walls (Neish et al., 2013). Finally, Kuiperian craters also excavated subsurface material of a variety of spectral properties. For example, compare Kuiper crater, which exposed high-reflectance red material from beneath a diffuse region of LRM, with Degas, which exposed LRM, or Ailey, which contains both LRM and high-reflectance red material in its ejecta (Figure 6.27). These instances not only point to the complicated stratigraphy of Mercury, but also provide the best opportunities to examine the properties of materials that have experienced minimal space weathering (Murchie et al., 2015; Chapter 8).

6.5 CRUSTAL FORMATION AND EVOLUTION

Understanding the nature of LRM and plains deposits is critical to determining the origin of Mercury's crust. Developing a coherent story that accounts for the origin of these two key units within the context of Mercury's crustal formation and evolution is now possible by combining geologic mapping, compositional information, and geochemical models. Verifying the accuracy of that story is another matter and will likely require in situ geochemical measurements or the study of meteorites from Mercury or documented returned samples, but that a narrative is emerging is progress.

Resurfacing on Mercury was so extensive that evaluating the composition and formation of the earliest crust is difficult. However, the global view of $N(65)$ values does point to some areas where early crustal materials may be exposed at the surface, albeit in a heavily gardened form. The geology and stratigraphy of the most extensive such region (Section 6.4.1) is consistent with an early LRM-rich crust exposed from beneath a more mixed regolith. As the LRM is thought to be enriched in graphite relative to average surface material (Murchie et al., 2015; Peplowski et al., 2016), these observations are compatible with geochemical models which suggest that a cooling magma ocean on Mercury may have developed an early graphite flotation crust (Vander Kaaden and McCubbin, 2015). Interestingly, because of the low iron concentrations in Mercury's silicate fraction, graphite may have been the only buoyant mineral in such a magma ocean (Brown and Elkins-Tanton, 2009; Vander Kaaden and McCubbin, 2015) and would thus be the only candidate mineral to form a flotation crust. However, the thickness of such a flotation crust is likely to have been small, <100 m for carbon concentrations similar to estimates for the mantle of Earth or Mars. In an extreme case, a graphite flotation crust thickness of up to ~1–20 km would have been possible if Mercury's mantle contained carbon in abundances comparable with those of carbonaceous chondrites (Vander Kaaden and McCubbin, 2015) and no carbon is sequestered in Mercury's core. The thickness of LRM deposits is not well constrained but at least locally can be >7 km (Ernst et al., 2010, 2015), and LRM is thought to contain only <5 wt% graphite (Murchie et al., 2015). The low-weight fraction of the graphite and the thickness of the low-reflectance material at depth suggests that, if it existed, the graphite flotation crust must have been mixed into the early forming crust as a result of multiple magmatic and impact events, and that only remnants of this mixed, brecciated material remain.

Adiabatic melting during cumulate overturn was a possible source of large-scale volcanism (e.g., Brown and Elkins-Tanton, 2009) that formed the first stable silicate crust. The ratio of early mantle-derived melt to graphite flotation crust must have been high, to account for the low-weight fraction of carbon in Mercury's modern crust, even in LRM. This situation contrasts with that on the Moon, where the plagioclase flotation crust is thought to constitute a large fraction of the modern highland crust. The relation of LRM to this early silicate crust on Mercury is not clear. With existing observations, it may not be possible to determine whether LRM formed via flotation (crystal–liquid fractionation in a magma ocean), as a product of early volcanism (adiabatic melting of overturned cumulates), or both, because these early crustal materials were substantially modified by later magmatism and impact disruption and mixing. Though its composition and mode of formation is not well understood, an early stable crust formed and was able to support the large-scale emplacement of younger magmas and the formation of impact basins. We refer to this material, in shorthand, as the "early crust."

If graphite were concentrated in the early crust, subsequent volcanic material derived from partial melting of the mantle would be generally higher in reflectance than the early crust, as is observed for the intercrater and smooth plains (Robinson et al., 2008; Denevi et al., 2009, 2013a; Murchie et al., 2015). And because LRM is often exposed from depths of several kilometers (Ernst et al., 2010, 2015), and crater densities indicate that little terrain escaped wholesale resurfacing, this scenario implies substantial volcanism subsequent to the development of LRM as

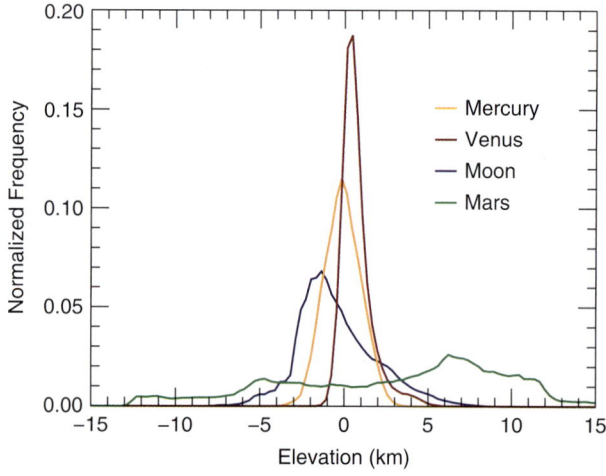

Figure 6.28. Distribution of elevations on Mercury (Becker et al., 2016), Venus (McNamee et al., 1993), the Moon (Scholten et al., 2012), and Mars (Smith et al., 2001).

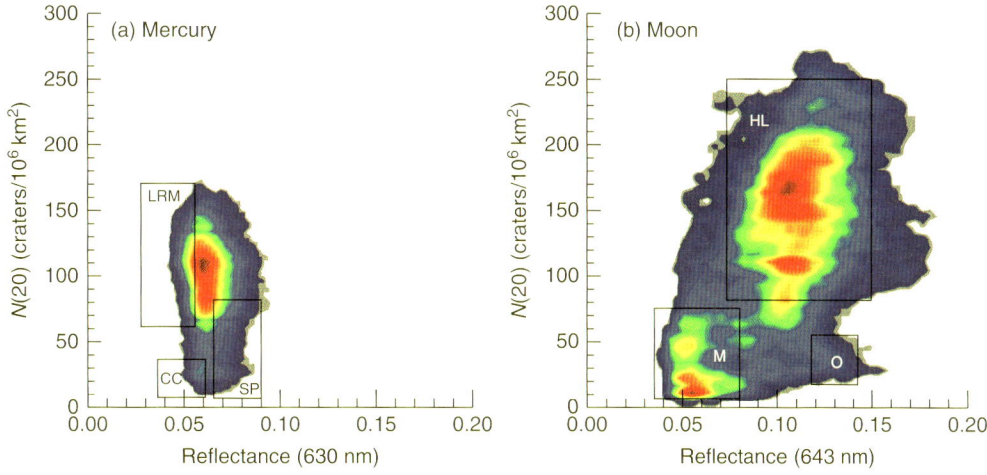

Figure 6.29. Two-dimensional histogram of reflectance versus crater density across Mercury and the Moon; colors indicate frequency (blue is low, red is high). (a) Mercury. Boxes indicate the approximate extent of LRM, circum-Caloris plains (CC), and smooth plains (SP). (b) The Moon. Boxes indicate the approximate extent of the highlands (HL), maria (M), and Orientale ejecta (O).

widespread crustal material. Thus, we interpret the emplacement history of the crust to include the development of an early, low-reflectance, graphite-rich surface by some combination of flotation and volcanism, and the continuation of widespread volcanic activity for which the contribution to crustal formation was largest in the pre-Tolstojan (intercrater plains) and continued during the Tolstojan (intercrater and smooth plains) but had largely ceased by the end of the Calorian Period (smooth plains). This sequence of events is consistent with earlier inferences that much of Mercury's crust formed through eruptions of magmas over an extended period of time (Denevi et al., 2009), perhaps the first billion years of Mercury's history.

Future efforts to expand regional studies like those in Section 6.4.1 and 6.4.2 will be essential to understanding the formation of the intercrater plains and the overall importance of volcanism in Mercury's early evolution; intercrater plains are currently "undivided" terrain (Figure 6.17) that remains after smooth plains and crater materials have been mapped. However, several lines of evidence are consistent with the majority of intercrater plains being volcanic in origin: (1) their low crater density (Figures 6.4 and 6.5), (2) the overlap in composition and color between smooth and intercrater plains (Denevi et al., 2013a; Murchie et al., 2015), (3) a range of compositions consistent with partial melting of the mantle (Chapter 7), and (4) a limited range of surface elevations (Figure 6.28), consistent with volcanic infilling of topographic lows.

The early history of Mercury appears to have been much different from that of the Moon, where the bulk of the crust is thought to have formed through a relatively simple flotation process and effusive volcanic activity was more limited. Mercury's narrow range of compositions and overall low abundance of iron likely meant that magmas on Mercury were not inhibited from reaching the surface by a low-density crust, as they were on the Moon. In fact, because of Mercury's composition and thin mantle, partial melts from the mantle would be buoyant at all depths, from the core–mantle boundary to the surface (Vander Kaaden and McCubbin, 2015). Thus, magmas on Mercury were less likely to stall, and a greater fraction of mantle partial melt that formed may have erupted on Mercury's surface (Vander Kaaden and McCubbin, 2015), until interior cooling and global contraction worked to inhibit magma ascent through the lithosphere.

These differences are highlighted by the relative reflectance and crater density distributions of the Moon and Mercury. In contrast to the relatively narrow range of reflectance and crater density seen for Mercury because of its extensive volcanism, the Moon shows a clear bimodal distribution between the heavily cratered highlands and less cratered volcanic maria (Figure 6.29). Mercury's high-reflectance plains are a small tail of higher reflectance and low $N(20)$ values on the distribution of these quantities; smoothing of the crater densities over neighborhoods of 500-km radius obscures many small units (Figure 6.4). Note also the wide spread of $N(20)$ values within the lunar highlands and the tail of high-reflectance, low crater density values that results from impact resurfacing of the ancient highlands (Figure 6.29b). Areas with low reflectance and low crater densities on Mercury include the circum-Caloris plains; determining whether any young regions of low reflectance are of volcanic origin or include impact deposits is important to evaluating the origin of these and older crustal units. Further work to understand Mercury's color–stratigraphic relationships will aid in deciphering the global picture apart from regional effects of impacts (e.g., Rivera-Valentin and Barr, 2014) and volcanism. But the overall picture is clear: unlike the Moon, Mercury does not a have a well-preserved flotation crust with a composition that contrasts strongly with mantle-derived magmas.

Though the formation of Mercury's early crust appears to have been by processes different from those on the Moon, Mercury may have followed an early evolutionary track similar to other terrestrial planets. No major flotation crust is likely to have formed on Mars, and neither would a quenched "stagnant lid" have been stable or preserved (Elkins-Tanton et al., 2005). Instead, magma ocean solidification would have been from the bottom up, leading to overturn as a result of gravitational instability, and basaltic volcanism fueled by adiabatic decompression melting likely formed the earliest martian crust (Elkins-Tanton et al., 2005). We have no record of the earliest crust of Venus, though volcanism has clearly been of crucial importance in its evolution, and Earth's original crust appears to have been lost, but it may also have been basaltic (e.g., Kamber, 2007).

Planetary crusts are sometimes divided into two stages: primary crusts, which formed within a short time interval as a consequence of planetary differentiation, and secondary crusts, which formed by later partial melting of the mantle over a more extended period (Taylor, 1989; Taylor and McLennan, 2010). This view is strongly shaped by our understanding of the Moon. For Mercury, as with Mars, however, the difference between primary and secondary crust may be a distinction without much practical meaning. Given the general lack of buoyant minerals in a cooling magma ocean (apart from graphite), it is likely that melting during cumulate overturn would have resulted in the essentially global eruption of lavas that would have formed most of the early crust (Brown and Elkins-Tanton, 2009). Thus, magmatic activity likely commenced early in Mercury's history and appears to have been so widespread that much of the crust was built by repeated eruptions and intrusions of magmas produced by varying degrees of partial melting of the mantle.

Future work to discriminate among the intercrater plains and to understand the modes of occurrence of the LRM will be key in evaluating further the nature of Mercury's crust. And in the long term, in situ geochemical analysis of LRM to determine its detailed elemental composition and petrology will be critical to understanding the process of crustal formation. An ancient, LRM-rich terrain would make an attractive landing site for a future Mercury lander mission.

REFERENCES

Ackiss, S. E., Buczkowski, D. L., Ernst, C. M., McBeck, J. A. and Seelos, K. D. (2015). Knob heights within circum-Caloris geologic units on Mercury: Interpretations of the geologic history of the region. *Earth Planet. Sci. Lett.*, **430**, 542–550, doi:10.1016/j.epsl.2015.08.003.

Adams, J. B. and McCord, T. B. (1977). Mercury: Evidence for an anorthositic crust from reflectance spectra. *Bull. Amer. Astron. Soc.*, **9**, 457.

André, S. L., Watters, T. R. and Robinson, M. S. (2005). The long wavelength topography of Beethoven and Tolstoj basins, Mercury. *Geophys. Res. Lett.*, **32**, L21202, doi:10.1029/2005GL023627.

Arkani-Hamed, J. (1973). On the thermal history of the Moon. *Moon*, **6**, 380–383.

Banks, M. E., Xiao, Z., Watters, T. R., Strom, R. G., Braden, S. E., Chapman, C. R., Solomon, S. C., Klimczak, C. and Byrne, P. K. (2015). Duration of activity on lobate-scarp thrust faults on Mercury. *J. Geophys. Res. Planets*, **120**, 1751–1762, doi:10.1002/2015JE004828.

Banks, M. E., Xiao, Z., Braden, S. E., Marchi, S., Chapman, C. R., Barlow, N. G. and Fassett, C. I. (2016). Revised age constraints for Mercury's Kuiperian and Mansurian systems. *Lunar Planet. Sci.*, **47**, abstract 2943.

Beach, M. J., Head, J. W., Ostrach, L. R., Robinson, M. S., Denevi, B. W. and Solomon, S. C. (2012). The influence of pre-existing topography on the distribution of impact melt on Mercury. *Lunar Planet. Sci.*, **43**, abstract 1335.

Becker, K. J., Robinson, M. S., Becker, T. L., Weller, L. A., Edmundson, K. L., Neumann, G. A., Perry, M. E. and Solomon, S. C. (2016). First global digital elevation model of Mercury. *Lunar Planet. Sci.*, **47**, abstract 2959.

Blewett, D. T., Lucey, P. G. and Hawke, B. R. (1997). A comparison of mercurian spectral reflectance and spectral quantities with those of the Moon. *Icarus*, **129**, 217–231.

Blewett, D. T., Hawke, B. R., Lucey, P. G. and Robinson, M. S. (2007). A Mariner 10 color study of mercurian craters. *J. Geophys. Res.*, **112**, E02005, doi:10.1029/2006JE002713.

Blewett, D. T., Robinson, M. S., Denevi, B. W., Gillis-Davis, J. J., Head, J. W., Solomon, S. C., Holsclaw, G. M. and McClintock, W. E. (2009). Multispectral images of Mercury from the first MESSENGER flyby: Analysis of global and regional color trends. *Earth Planet. Sci. Lett.*, **285**, 272–282.

Blewett, D. T., Denevi, B. W., Robinson, M. S., Ernst, C. M., Purucker, M. E. and Solomon, S. C. (2010). The apparent lack of lunar-like swirls on Mercury: Implications for the formation of lunar swirls and for the agent of space weathering. *Icarus*, **209**, 239–246, doi:10.1016/j.icarus.2010.03.008.

Blewett, D. T., Chabot, N. L., Denevi, B. W., Ernst, C. M., Head, J. W., Izenberg, N. R., Murchie, S. L., Solomon, S. C., Nittler, L. A., McCoy, T. J., Xiao, Z., Baker, D. M. H., Fassett, C. I., Braden, S. E., Oberst, J., Scholten, F., Preusker, F. and Hurwitz, D. M. (2011). Hollows on Mercury: MESSENGER evidence for geologically recent volatile-related activity. *Science*, **333**, 1856–1859.

Blewett, D. T., Vaughan, W. M., Xiao, Z., Chabot, N. L., Denevi, B. W., Ernst, C. M., Helbert, J., D'Amore, M., Maturilli, A., Head, J. W. and Solomon, S. C. (2013). Mercury's hollows: Constraints on formation and composition from analysis of geological setting and spectral reflectance. *J. Geophys. Res. Planets*, **118**, 1013–1032, doi:10.1029/2012JE004174.

Borin, P., Cremonese, G., Marzari, F., Bruno, M. and Marchi, S. (2009). Statistical analysis of micrometeoroids flux on Mercury. *Astron. Astrophys.*, **503**, 259–264, doi:10.1051/0004-6361/200912080.

Braden, S. E. and Robinson, M. S. (2013). Relative rates of optical maturation of regolith on Mercury and the Moon. *J. Geophys. Res. Planets*, **118**, 1903–1914, doi:10.1002/jgre.20143.

Brown, S. M. and Elkins-Tanton, L. T. (2009). Compositions of Mercury's earliest crust from magma ocean models. *Earth Planet. Sci. Lett.*, **286**, 446–455, doi:10.1016/j.epsl.2009.07.010.

Byrne, P. K., Klimczak, C., Williams, D. A., Hurwitz, D. M., Solomon, S. C., Head, J. W., Preusker, F. and Oberst, J. (2013a). An assemblage of lava flow features on Mercury. *J. Geophys. Res. Planets*, **118**, 1303–1322, doi:10.1002/jgre.20052.

Byrne, P. K., Klimczak, C., Blair, D. M., Ferrari, S., Solomon, S. C., Freed, A. M., Watters, T. R. and Murchie, S. L. (2013b). Tectonic complexity within volcanically infilled craters and basins on Mercury. *Lunar Planet. Sci.*, **44**, abstract 1261.

Byrne, P. K., Klimczak, C., Şengör, A. M. C., Solomon, S. C., Watters, T. R. and Hauck, S. A., II (2014). Mercury's global contraction much greater than earlier estimates. *Nature Geosci.*, **7**, 301–307, doi:10.1038/ngeo2097.

Byrne, P. K., Ostrach, L. R., Fassett, C. I., Chapman, C. R., Denevi, B. W., Evans, A. J., Klimczak, C., Banks, M. E., Head, J. W. and Solomon, S. C. (2016). Widespread effusive volcanism on Mercury likely ended by about 3.5 Ga. *Geophys. Res. Lett.*, **43**, 7408–7416, doi:10.1002/2016GL069412.

Chabot, N. L., Denevi, B. W., Murchie, S. L., Hash, C. D., Ernst, C. M., Blewett, D. T., Nair, H., Laslo, N. R. and Solomon, S. C. (2016). Mapping Mercury: Global imaging strategy and products from the MESSENGER mission. *Lunar Planet. Sci.*, **47**, abstract 1256.

Chapman, C. R. (1968). Interpretation of the diameter–frequency relation for lunar craters photographed by Rangers VII, VIII, and IX. *Icarus*, **8**, 1–22, doi:10.1016/0019-1035(68)90058-4.

Charlier, B., Grove, T. L. and Zuber, M. T. (2013). Phase equilibria of ultramafic compositions on Mercury and the origin of the compositional dichotomy. *Earth Planet. Sci. Lett.*, **363**, 50–60, doi:10.1016/j.epsl.2012.12.021.

Cintala, M. J. (1992). Impact-induced thermal effects in the lunar and mercurian regoliths. *J. Geophys. Res.*, **97**, 947–973.

Crater Analysis Techniques Working Group (1978). Standard techniques for presentation and analysis of crater size–frequency data. *Icarus*, **37**, 467–474.

Denevi, B. W. and Robinson, M. S. (2008). Mercury's albedo from Mariner 10: Implications for the presence of ferrous iron. *Icarus*, **197**, 239–246, doi:10.1016/j.icarus.2008.04.021.

Denevi, B. W., Robinson, M. S., Solomon, S. C., Murchie, S. L., Blewett, D. T., Domingue, D. L., McCoy, T. J., Ernst, C. M., Head, J. W., Watters, T. R. and Chabot, N. L. (2009). The evolution of Mercury's crust: A global perspective from MESSENGER. *Science*, **324**, 613–618.

Denevi, B. W., Ernst, C. M., Meyer, H. M., Robinson, M. S., Murchie, S. L., Whitten, J. L., Head, J. W., Watters, T. R., Solomon, S. C., Ostrach, L. R., Chapman, C. R., Byrne, P. K., Klimczak, C. and Peplowski, P. N. (2013a). The distribution and origin of smooth plains on Mercury. *J. Geophys. Res. Planets*, **118**, 891–907, doi:10.1002/jgre.20075.

Denevi, B. W., Ernst, C. M., Whitten, J. L., Head, J. W., Murchie, S. L., Watters, T. R., Byrne, P. K., Blewett, D. T., Solomon, S. C. and Fassett, C. I. (2013b). The volcanic origin of a region of intercrater plains on Mercury. *Lunar Planet. Sci.*, **44**, abstract 1218.

Domingue, D. L., Chapman, C. R., Killen, R. M., Zurbuchen, T. H., Gilbert, J. A., Sarantos, M., Benna, M., Slavin, J. A., Schriver, D., Trávníček, P. M., Orlando, T. M., Sprague, A. L., Blewett, D. T., Gillis-Davis, J. J., Feldman, W. C., Lawrence, D. J., Ho, G. C., Ebel, D. S., Nittler, L. R., Vilas, F., Pieters, C. M., Solomon, S. C., Johnson, C. L., Winslow, R. M., Helbert, J., Peplowski, P. N., Weider, S. Z., Mouawad, N., Izenberg, N. R. and McClintock, W. E. (2014). Mercury's weather-beaten surface: Understanding Mercury in the context of lunar and asteroidal space weathering studies. *Space Sci. Rev.*, **181**, 121–214, doi:10.1007/s11214-014-0039-5.

Dzurisin, D. (1977). Mercurian bright patches: Evidence for chemical alteration of surface material? *Geophys. Res. Lett.*, **4**, 383–396.

Eggleton, R. E. and Schaber, G. G. (1972). Cayley Formation interpreted as basin ejecta. In *Apollo 16 Preliminary Science Report*, Special Publication SP-315. Washington, DC: National Aeronautics and Space Administration, pp. 29-7–29-16.

Elkins-Tanton, L. T., Hager, B. H. and Grove, T. L. (2004). Magmatic effects of the lunar late heavy bombardment. *Earth Planet. Sci. Lett.*, **222**, 17–27, doi:10.1016/j.epsl.2004.02.017.

Elkins-Tanton, L. T., Hess, P. C. and Parmentier, E. M. (2005). Possible formation of ancient crust on Mars through magma ocean processes. *J. Geophys. Res.*, **100**, E12S01, doi:10.1029/2005JE002480.

Ernst, C. M., Murchie, S. L., Barnouin, O. S., Robinson, M. S., Denevi, B. W., Blewett, D. T., Head, J. W., Izenberg, N. R., Solomon, S. C. and Roberts, J. H. (2010). Exposure of spectrally distinct material by impact craters on Mercury: Implications for global stratigraphy. *Icarus*, **209**, 210–223, doi:10.1016/j.icarus.2010.05.022.

Ernst, C. M., Denevi, B. W., Barnouin, O. S., Klimczak, C., Chabot, N. L., Head, J. W., Murchie, S. L., Neumann, G. A., Prockter, L. M., Robinson, M. S., Solomon, S. C. and Watters, T. R. (2015). Stratigraphy of the Caloris basin, Mercury: Implications for volcanic history and basin impact melt. *Icarus*, **250**, 413–429, doi:10.1016/j.icarus.2014.11.003.

Ernst, C. M., Chabot, N. L. and Barnouin, O. S. (2016). Examining the potential contribution of the Hokusai impact to water ice on Mercury. *Lunar Planet. Sci.*, **47**, abstract 1374.

Ernst, C. M., Denevi, B. W. and Ostrach, L. R. (2017). Updated absolute age estimates for the Tolstoj and Caloris basins, Mercury. *Lunar Planet. Sci.*, **48**, abstract 2934.

Fassett, C. I., Head, J. W., Blewett, D. T., Chapman, C. R., Dickson, J. L., Murchie, S. L., Solomon, S. C. and Watters, T. R. (2009). Caloris impact basin: Exterior geomorphology, stratigraphy, morphometry, radial sculpture, and smooth plains deposits. *Earth Planet. Sci. Lett.*, **285**, 297–308.

Fassett, C. I., Kadish, S. J., Head, J. W., Solomon, S. C. and Strom, R. G. (2011). The global population of large craters on Mercury and comparisons with the Moon. *Geophys. Res. Lett.*, **38**, L10202, doi:10.1029/2011GL047294.

Fassett, C. I., Head, J. W., Baker, D. M. H., Zuber, M. T., Smith, D. E., Neumann, G. A., Solomon, S. C., Klimczak, C., Strom, R. G., Chapman, C. R., Prockter, L. M., Phillips, R. J., Oberst, J. and Preusker, F. (2012). Large impact basins on Mercury: Global distribution, characteristics, and modification history from MESSENGER orbital data. *J. Geophys. Res.*, **117**, E00L08, doi:10.1029/2012JE004154.

Gaddis, L. R., Pieters, C. M. and Hawke, B. R. (1985). Remote-sensing of lunar pyroclastic mantling deposits. *Icarus*, **61**, 461–489.

Gaddis, L. R., Staid, M. I., Tyburczy, J. and Hawke, B. R. (2003). Compositional analyses of lunar pyroclastic deposits. *Icarus*, **161**, 262–280.

Gault, D. E., Guest, J. E., Murray, J. B., Dzurisin, D. and Malin, M. C. (1975). Some comparisons of impact craters on Mercury and the Moon. *J. Geophys. Res.*, **80**, 2444–2460.

Gillis-Davis, J. J., Blewett, D. T., Gaskell, R. W., Denevi, B. W., Robinson, M. S., Strom, R. G., Solomon, S. C. and Sprague, A. L. (2009). Pit-floor craters on Mercury: Evidence of near-surface igneous activity. *Earth Planet. Sci. Lett.*, **285**, 243–250, doi:10.1016/j.epsl.2009.05.023.

Goosmann, E., Buczkowski, D. L., Ernst, C. M., Denevi, B. W. and Kinczyk, M. J. (2016). Geologic map of the Caloris basin, Mercury. *Lunar Planet. Sci.*, **47**, abstract 1254.

Goudge, T. A., Head, J. W., Kerber, L., Blewett, D. T., Denevi, B. W., Domingue, D. L., Gillis-Davis, J. J., Gwinner, K., Helbert, J., Holsclaw, G. M., Izenberg, N. R., Klima, R. L., McClintock, W. E., Murchie, S. L., Neumann, G. A., Smith, D. E., Strom, R. G., Xiao, Z., Zuber, M. T. and Solomon, S. C. (2014). Global inventory and characterization of pyroclastic deposits on Mercury: New insights into pyroclastic activity from MESSENGER orbital data. *J. Geophys. Res. Planets*, **119**, 635–658, doi:10.1002/2013JE004480.

Greenstreet, S., Ngo, H. and Gladman, B. (2012). The orbital distribution of near-Earth objects inside Earth's orbit. *Icarus*, **217**, 355–366, doi:10.1016/j.icarus.2011.11.010.

Grolier, M. J. and Boyce, J. M. (1984). *Geologic Map of the Borealis Region of Mercury*, Map I-1660, Miscellaneous Investigations Series. Reston, VA: U.S. Geological Survey.

Guest, J. E. and Greeley, R. (1983). *Geologic Map of the Shakespeare Quadrangle of Mercury*, Map I-1408, Miscellaneous Investigations Series. Reston, VA: U.S. Geological Survey.

Hapke, B., Danielson, G. E., Klaasen, K. and Wilson, L. (1975). Photometric observations of Mercury from Mariner 10. *J. Geophys. Res.*, **80**, 2431–2443.

Hapke, B., Christman, C., Rava, B. and Mosher, J. (1980). A color-ratio map of Mercury. *Proc. Lunar Planet. Sci. Conf.*, **11**, 817–821.

Hartmann, W. K. (1984). Does crater "saturation equilibrium" occur in the solar system? *Icarus*, **60**, 56–74.

Hawke, B. R., Peterson, C. A., Blewett, D. T., Bussey, D. B. J., Lucey, P. G., Taylor, G. J. and Spudis, P. D. (2003). Distribution and modes of occurrence of lunar anorthosites. *J. Geophys. Res.*, **108** (E6), 5050, doi:10.1029/2002JE001890.

Hawkins, S. E., III, Boldt, J. D., Darlington, E. H., Espiritu, R., Gold, R. E., Gotwols, B., Grey, M. P., Hash, C. D., Hayes, J. R., Jaskulek, S. E., Kardian, C. J., Keller, M. R., Malaret, E. R., Murchie, S. L., Murphy, P. K., Peacock, K., Prockter, L. M., Reiter, R. A., Robinson, M. S., Schaefer, E. D., Shelton, R. G., Sterner, R. E., II, Taylor, H. W., Watters, T. R. and Williams, B. D. (2007). The Mercury Dual Imaging System on the MESSENGER spacecraft. *Space Sci. Rev.*, **131**, 247–338.

Hawkins, S. E., III, Murchie, S. L., Becker, K. J., Selby, C. M., Turner, F. S., Noble, M. W., Chabot, N. L., Choo, T. H., Darlington, E. H., Denevi, B. W., Domingue, D. L., Ernst, C. M., Holsclaw, G. M., Laslo, N. R., McClintock, W. E., Prockter, L. M., Robinson, M. S., Solomon, S. C. and Sterner, R. E., II (2009). In-flight performance of MESSENGER's Mercury Dual Imaging System. In *Instruments and Methods for Astrobiology and Planetary Missions XII*, ed. R. B. Hoover, G. V. Levin, A. Y. Rozanov and K. D. Retherford. *Proc. SPIE*, Vol. 7441. Bellingham, WA: SPIE, 12 pp., doi:10.1117/12.826370.

Head, J. W. (1974). Orientale multi-ringed basin interior and implications for the petrogenesis of lunar highland samples. *Moon*, **11**, 327–356.

Head, J. W. and Wilson, L. (1979). Alphonsus-type dark halo craters: Morphology, morphometry and eruption conditions. *Proc. Lunar Planet. Sci. Conf.*, **10**, 2861–2897.

Head, J. W., Murchie, S. L., Prockter, L. M., Robinson, M. S., Solomon, S. C., Strom, R. G., Chapman, C. R., Watters, T. R., McClintock, W. E., Blewett, D. T. and Gillis-Davis, J. J. (2008). Volcanism on Mercury: Evidence from the first MESSENGER flyby. *Science*, **321**, 69–72.

Head, J. W., Murchie, S. L., Prockter, L. M., Solomon, S. C., Chapman, C. R., Strom, R. G., Watters, T. R., Blewett, D. T., Gillis-Davis, J. J., Fassett, C. I., Dickson, J. L., Morgan, G. A. and Kerber, L. (2009a). Volcanism on Mercury: Evidence from the first MESSENGER flyby for extrusive and explosive activity and the volcanic origin of plains. *Earth Planet. Sci. Lett.*, **285**, 227–242, doi:10.1016/j.epsl.2009.03.007.

Head, J. W., Murchie, S. L., Prockter, L. M., Solomon, S. C., Strom, R. G., Chapman, C. R., Watters, T. R., Blewett, D. T., Gillis-Davis, J. J., Fassett, C. I., Dickson, J. L., Hurwitz, D. M. and Ostrach, L. R. (2009b). Evidence for intrusive activity on Mercury from the first MESSENGER flyby. *Earth Planet. Sci. Lett.*, **285**, 251–262, doi:10.1016/j.epsl.2009.03.008.

Head, J. W., Fassett, C. I., Kadish, S. J., Smith, D. E., Zuber, M. T., Neumann, G. A. and Mazarico, E. (2010). Global distribution of large lunar craters: Implications for resurfacing and impactor populations. *Science*, **329**, 1504–1507.

Head, J. W., Chapman, C. R., Strom, R. G., Fassett, C. I., Denevi, B. W., Blewett, D. T., Ernst, C. M., Watters, T. R., Solomon, S. C., Murchie, S. L., Prockter, L. M., Chabot, N. L., Gillis-Davis, J. J., Whitten, J. L., Goudge, T. A., Baker, D. M. H., Hurwitz, D. M., Ostrach, L. R., Xiao, Z., Merline, W. J., Kerber, L., Dickson, J. L., Oberst, J., Byrne, P. K., Klimczak, C. and Nittler, L. R. (2011). Flood volcanism in the northern high latitudes of Mercury revealed by MESSENGER. *Science*, **333**, 1853–1856, doi:10.1126/science.1211997.

Howard, K. A., Wilhelms, D. E. and Scott, D. H. (1974). Lunar basin formation and highland stratigraphy. *Rev. Geophys. Space Physics*, **12**, 309–327.

Hughes, H. G., App, F. N. and McGetchin, T. R. (1977). Global seismic effects of basin-forming impacts. *Phys. Earth Planet. Inter.*, **15**, 251–263.

Hurwitz, D. M., Head, J. W., Byrne, P. K., Xiao, Z., Solomon, S. C., Zuber, M. T., Smith, D. E. and Neumann, G. A. (2013). Investigating the origin of candidate lava channels on Mercury with MESSENGER data: Theory and observations. *J. Geophys. Res. Planets*, **118**, 471–486, doi:10.1029/2012JE004103.

James, P. B., Zuber, M. T., Phillips, R. J. and Solomon, S. C. (2015). Support of long-wavelength topography on Mercury inferred from MESSENGER measurements of gravity and topography. *J. Geophys. Res. Planets*, **120**, 287–310, doi:10.1002/2014JE004713.

Jolliff, B. L., Gillis, J. J., Haskin, L. A. and Korotev, R. L. (2000). Major lunar crustal terranes: Surface expressions and crust–mantle origins. *J. Geophys. Res.*, **105**, 4197–4216.

Kamber, B. S. (2007). The enigma of the terrestrial protocrust: Evidence of its former existence and the importance of its complete disappearance. In *The Earth's Oldest Rocks*, ed. M. van Kranendonk, R. H. Smithies and V. C. Bennett. Amsterdam: Elsevier, pp. 75–89.

Kerber, L., Head, J. W., Solomon, S. C., Murchie, S. L., Blewett, D. T. and Wilson, L. (2009). Explosive volcanic eruptions on Mercury: Eruption conditions, magma volatile content, and implications for interior volatile abundances. *Earth Planet. Sci. Lett.*, **285**, 263–271, doi:10.1016/j.epsl.2009.04.037.

Kerber, L., Head, J. W., Blewett, D. T., Solomon, S. C., Wilson, L., Murchie, S. L., Robinson, M. S., Denevi, B. W. and Domingue, D. L. (2011). The global distribution of pyroclastic deposits on Mercury: The view from MESSENGER flybys 1–3. *Planet. Space Sci.*, **59**, 1895–1909, doi:10.1016/j.pss.2011.03.020.

Kinczyk, M. J., Prockter, L. M., Chapman, C. R. and Susorney, H. C. (2016). A morphologic evaluation of crater degradation on Mercury: Revisiting crater classification with MESSENGER data. *Lunar Planet. Sci.*, **47**, abstract 1573.

King, J. S. and Scott, D. H. (1990). *Geologic Map of the Beethoven Quadrangle of Mercury*, Map I-2048, Miscellaneous Investigations Series. Reston, VA: U.S. Geological Survey.

Klima, R. L., Blewett, D. T., Denevi, B. W., Ernst, C. M., Frank, E. A., Head, J. W., Izenberg, N. R., Murchie, S. L., Nittler, L. R., Peplowski, P. N. and Solomon, S. C. (2016). Global distribution and spectral properties of low-reflectance material on Mercury. *Lunar Planet. Sci.*, **47**, abstract 1195.

Klimczak, C. (2015). Limits on the brittle strength of planetary lithospheres undergoing global contraction. *J. Geophys. Res. Planets*, **120**, 2135–2151, doi:10.1002/2015JE004851.

Klimczak, C., Watters, T. R., Ernst, C. M., Freed, A. M., Byrne, P. K., Solomon, S. C., Blair, D. M. and Head, J. W. (2012). Deformation associated with ghost craters and basins in volcanic smooth plains on Mercury: Strain analysis and implications for plains evolution. *J. Geophys. Res.*, **117**, E00L03, doi:10.1029/2012JE004100.

Klimczak, C., Ernst, C. M., Byrne, P. K., Solomon, S. C., Watters, T. R., Murchie, S. L., Preusker, F. and Balcerski, J. A. (2013). Insights into the subsurface structure of the Caloris basin, Mercury, from assessments of mechanical layering and changes in long-wavelength topography. *J. Geophys. Res. Planets*, **118**, 2030–2044, doi:10.1002/jgre.20157.

Lawrence, S. J., Stopar, J. D., Robinson, M. S., Hawke, B. R., Jolliff, B. L. and Giguere, T. A. (2013). Mare deposits in the Australe region: Extent, topography, and stratigraphy. *Lunar Planet. Sci.*, **44**, abstract 2671.

Le Feuvre, M. and Wieczorek, M. A. (2008). Nonuniform cratering of the terrestrial planets. *Icarus*, **197**, 291–306.

Le Feuvre, M. and Wieczorek, M. A. (2011). Nonuniform cratering of the Moon and a revised crater chronology of the inner Solar System. *Icarus*, **214**, 1–20, doi:10.1016/j.icarus.2011.03.010.

Lü, J., Sun, Y., Toksöz, M. N., Zheng, Y. and Zuber, M. T. (2011). Seismic effects of the Caloris basin impact, Mercury. *Planet. Space Sci.*, **59**, 1981–1991, doi:10.1016/j.pss.2011.07.013.

Marchi, S., Mottola, S., Cremonese, G., Massironi, M. and Martellato, E. (2009). A new chronology for the Moon and Mercury. *Astron. J.*, **137**, 4936–4948, doi:10.1088/0004-6256/137/6/4936.

Marchi, S., Massironi, M., Cremonese, G., Martellato, E., Giacomini, L. and Prockter, L. (2011). The effects of the target material properties and layering on the crater chronology: The case of Raditladi and Rachmaninoff basins on Mercury. *Planet. Space Sci.*, **59**, 1968–1980, doi:10.1016/j.pss.2011.06.007.

Marchi, S., Chapman, C. R., Fassett, C. I., Head, J. W., Bottke, W. F. and Strom, R. G. (2013). Global resurfacing of Mercury 4.0–4.1 billion years ago by heavy bombardment and volcanism. *Nature*, **499**, 59–61, doi:10.1038/nature12280.

McCauley, J. F., Guest, J. E., Schaber, G. G., Trask, N. J. and Greeley, R. (1981). Stratigraphy of the Caloris basin, Mercury. *Icarus*, **47**, 184–202.

McGill, G. E. and King, E. A. (1983). *Geologic Map of the Victoria (H-2) Quadrangle of Mercury*, Map I-1409, Miscellaneous Investigations Series. Reston, VA: U.S. Geological Survey.

McNamee, J. B., Borderies, N. J. and Sjogren, W. L. (1993). Venus: Global gravity and topography. *J. Geophys. Res.*, **98**, 9113–9128.

Meyer, H. M., Denevi, B. W., Boyd, A. K. and Robinson, M. S. (2016). The distribution and origin of lunar light plains around Orientale basin. *Icarus*, **273**, 135–145, doi:10.1016/j.icarus.2016.02.014.

Michel, N. C., Hauck, S. A., II, Solomon, S. C., Phillips, R. J., Roberts, J. H. and Zuber, M. T. (2013). Thermal evolution of Mercury as constrained by MESSENGER observations. *J. Geophys. Res. Planets*, **118**, 1033–1044, doi:10.1002/jgre.20049.

Milkovich, S. M., Head, J. W. and Wilson, L. (2002). Identification of mercurian volcanism: Resolution effects and implications for MESSENGER. *Meteorit. Planet. Sci.*, **37**, 1209–1222.

Milton, D. J. (1972). *Geologic Map of the Descartes Region of the Moon: Apollo 16 Pre-mission Map*, Map I-748, Miscellaneous Investigations Series. Reston, VA: U.S. Geological Survey.

Moore, H. J., Hodges, C. A. and Scott, D. H. (1974). Multiring basins: Illustrated by Orientale and associated features. *Proc. Lunar Sci. Conf.*, **5**, 71–100.

Murchie, S. L., Watters, T. R., Robinson, M. S., Head, J. W., Strom, R. G., Chapman, C. R., Solomon, S. C., McClintock, W. E., Prockter, L. M., Domingue, D. L. and Blewett, D. T. (2008). Geology of the Caloris basin, Mercury: A view from MESSENGER. *Science*, **321**, 73–76.

Murchie, S. L., Klima, R. L., Denevi, B. W., Ernst, C. M., Keller, M. R., Domingue, D. L., Blewett, D. T., Chabot, N. L., Hash, C. D., Malaret, E., Izenberg, N. R., Vilas, F., Nittler, L. R., Gillis-Davis, J. J., Head, J. W. and Solomon, S. C. (2015). Orbital multispectral mapping of Mercury with the MESSENGER Mercury Dual Imaging System: Evidence for the origins of plains units and low-reflectance material. *Icarus*, **254**, 287–305, doi:10.1016/j.icarus.2015.03.027.

Murray, B. C., Belton, M. J. S., Danielson, G. E., Davies, M. E., Gault, D. E., Hapke, B., O'Leary, B., Strom, R. G., Suomi, V. and Trask, N. (1974). Mercury's surface: Preliminary description and interpretation from Mariner 10 pictures. *Science*, **185**, 169–179, doi:10.1126/science.185.4146.169.

Murray, B. C., Strom, R. G., Trask, N. J. and Gault, D. E. (1975). Surface history of Mercury: Implications for terrestrial planets. *J. Geophys. Res.*, **80**, 2508–2514.

Neish, C. D., Blewett, D. T., Harmon, J. K., Coman, E. I., Cahill, J. T. S. and Ernst, C. M. (2013). A comparison of rayed craters on the Moon and Mercury. *J. Geophys. Res. Planets*, **118**, 2247–2261, doi:10.1002/jgre.20166.

Nittler, L. R., Starr, R. D., Weider, S. Z., McCoy, T. J., Boynton, W. V., Ebel, D. S., Ernst, C. M., Evans, L. G., Goldsten, J. O., Hamara, D. K., Lawrence, D. J., McNutt, R. L., Schlemm, C. E., Solomon, S. C. and Sprague, A. L. (2011). The major-element composition of Mercury's surface from MESSENGER X-ray spectrometry. *Science*, **333**, 1847–1850, doi:10.1126/science.1211567.

Oberbeck, V. R., Hörz, F., Morrison, R. H., Quaide, W. L. and Gault, D. E. (1975). On the origin of the lunar smooth plains. *Moon*, **12**, 19–54.

Oberbeck, V. R., Quaide, W. L., Arvidson, R. E. and Aggarwal, H. R. (1977). Comparative studies of lunar, Martian, and Mercurian craters and plains. *J. Geophys. Res.*, **82**, 1687–1698.

Oberst, J., Preusker, F., Phillips, R. J., Watters, T. R., Head, J. W., Zuber, M. T. and Solomon, S. C. (2010). The morphology of Mercury's Caloris basin as seen in MESSENGER stereo topographic models. *Icarus*, **209**, 230–238, doi:10.1016/j.icarus.2010.03.009.

Ostrach, L. R., Robinson, M. S. and Denevi, B. W. (2012). Distribution of impact melt on Mercury and the Moon. *Lunar Planet. Sci.*, **43**, abstract 1113.

Ostrach, L. R., Robinson, M. S., Whitten, J. L., Fassett, C. I., Strom, R. G., Head, J. W. and Solomon, S. C. (2015). Extent, age, and resurfacing history of the northern smooth plains on Mercury from MESSENGER observations. *Icarus*, **250**, 602–622, doi:10.1016/j.icarus.2014.11.010.

Page, N. J. (1970). *Geologic Map of the Cassini Quadrangle of the Moon*, Map I-666, Miscellaneous Investigations Series. Reston, VA: U.S. Geological Survey.

Peplowski, P. N., Lawrence, D. J., Rhodes, E. A., Sprague, A. L., McCoy, T. J., Denevi, B. W., Evans, L. G., Head, J. W., Nittler, L. R., Solomon, S. C., Stockstill-Cahill, K. R. and Weider, S. Z. (2012). Variations in the abundances of potassium and thorium on the surface of Mercury: Results from the MESSENGER Gamma-Ray Spectrometer. *J. Geophys. Res.*, **117**, E00L04, doi:10.1029/2012JE004141.

Peplowski, P. N., Lawrence, D. J., Feldman, W. C., Goldsten, J. O., Bazell, D., Evans, L. G., Head, J. W., Nittler, L. R., Solomon, S. C. and Weider, S. Z. (2015). Geochemical terranes of Mercury's northern hemisphere as revealed by MESSENGER neutron measurements. *Icarus*, **253**, 346–363, doi:10.1016/j.icarus.2015.02.002.

Peplowski, P. N., Klima, R. L., Lawrence, D. J., Ernst, C. M., Denevi, B. W., Frank, E. A., Goldsten, J. O., Murchie, S. L., Nittler, L. R. and Solomon, S. C. (2016). Remote sensing evidence for an ancient carbon-bearing crust on Mercury. *Nature Geosci.*, **9**, 273–276, doi:10.1038/NGEO2669.

Prockter, L. M., Ernst, C. M., Denevi, B. W., Chapman, C. R., Head, J. W., Fassett, C. I., Merline, W. J., Solomon, S. C., Watters, T. R., Strom, R. G., Cremonese, G., Marchi, S. and Massironi, M. (2010). Evidence for young volcanism on Mercury from the third MESSENGER flyby. *Science*, **329**, 668–671.

Prockter, L. M., Kinczyk, M. J., Byrne, P. K., Denevi, B. W., Head, J. W., Fassett, C. I., Whitten, J. L., Thomas, R. J., Buczkowski, D. L., Hynek, B. M., Ostrach, L. R., Blewett, D. T., Ernst, C. M. and the MESSENGER Mapping Group (2016). The first global geological map of Mercury. *Lunar Planet. Sci.*, **47**, abstract 1245.

Rava, B. and Hapke, B. (1987). An analysis of the Mariner 10 color ratio map of Mercury. *Icarus*, **71**, 397–429.

Rivera-Valentin, E. G. and Barr, A. C. (2014). Impact-induced compositional variations on Mercury. *Earth Planet. Sci. Lett.*, **391**, 234–242, doi:10.1016/j.epsl.2014.02.003.

Roberts, J. H. and Barnouin, O. S. (2012). The effect of the Caloris impact on the mantle dynamics and volcanism of Mercury. *J. Geophys. Res.*, **117**, E02007, doi:10.1029/2011JE003876.

Robinson, M. S. and Lucey, P. G. (1997). Recalibrated Mariner 10 color mosaics: Implications for mercurian volcanism. *Science*, **275**, 197–200.

Robinson, M. S. and Taylor, G. J. (2001). Ferrous oxide in Mercury's crust and mantle. *Meteorit. Planet. Sci.*, **36**, 841–847, doi:10.1111/j.1945-5100.2001.tb01921.x.

Robinson, M. S., Murchie, S. L., Blewett, D. T., Domingue, D. L., Hawkins, S. E., III, Head, J. W., Holsclaw, G. M., McClintock, W. E., McCoy, T. J., McNutt, R. L., Prockter, L. M., Solomon, S. C. and Watters, T. R. (2008). Reflectance and color variations on Mercury: Regolith processes and compositional heterogeneity. *Science*, **321**, 66–69.

Rothery, D. A., Thomas, R. J. and Kerber, L. (2014). Prolonged eruptive history of a compound volcano on Mercury: Volcanic and tectonic implications. *Earth Planet. Sci. Lett.*, **385**, 59–67, doi:10.1016/j.epsl.2013.10.023.

Schaber, G. G. and McCauley, J. F. (1980). *Geologic Map of the Tolstoj Quadrangle of Mercury*, Map I-1199, Miscellaneous Investigations Series. Reston, VA: U.S. Geological Survey.

Scholten, F., Oberst, J., Matz, K.-D., Roatsch, T., Wählisch, M., Speyerer E. J. and Robinson, M. S. (2012). GLD100: The near-global lunar 100 m raster DTM from LROC WAC stereo image data. *J. Geophys. Res.*, **117**, E00H17, doi:10.1029/2011JE003926.

Schultz, P. H. and Gault, D. E. (1975). Seismic effects from major basin formations on the Moon and Mercury. *Moon*, **12**, 159–177.

Schultz, P. H. and Singer, J. (1980). A comparison of secondary craters on the Moon, Mercury and Mars. *Proc. Lunar Planet. Sci. Conf.*, **11**, 2243–2259.

Scott, D. H., McCauley, J. F. and West, M. N. (1977). *Geologic Map of the West Side of the Moon*, Map I-1034, Miscellaneous Investigations Series. Reston, VA: U.S. Geological Survey.

Shoemaker, E. M. and Hackman, R. J. (1962). Stratigraphic basis for a lunar time scale. In *The Moon*, ed. Z. Kopal and S. K. Miklhalov. London: Academic Press, pp. 289–300.

Smith, D. E., Zuber, M. T., Frey, H. V., Garvin, J. B., Head, J. W., Muhleman, D. O., Pettengill, G. H., Phillips, R. J., Solomon, S. C., Zwally, H. J., Banerdt, W. B., Duxbury, T. C., Golombek, M. P., Lemoine, F. G., Neumann, G. A., Rowlands, D. D., Aharonson, O., Ford, P. G., Ivanov, A. B., Johnson, C. L., McGovern, P. J., Abshire, J. B., Afzal, R. S. and Sun, X. (2001). Mars Orbiter Laser Altimeter: Experiment summary after the first year of global mapping of Mars. *J. Geophys. Res.*, **106**, 23,689–23,722, doi:10.1029/2000JE001364.

Solomon, S. C. (1978). On volcanism and thermal tectonics on one-plate planets. *Geophys. Res. Lett.*, **5**, 461–464, doi:10.1029/GL005i006p00461.

Spudis, P. D. (1985). A Mercurian chronostratigraphic classification. In *Reports of Planetary Geology and Geophysics Program – 1984*, Technical Memorandum 87563. Washington, DC: NASA, pp. 595–597.

Spudis, P. D. and Guest, J. E. (1988). Stratigraphy and geologic history of Mercury. In *Mercury*, ed. F. Vilas, C. R. Chapman and M. S. Matthews. Tucson, AZ: University of Arizona Press, pp. 118–164.

Spudis, P. D. and Prosser, J. G. (1984). *Geologic Map of the Michelangelo Quadrangle of Mercury*, Map I-1659, Miscellaneous Investigations Series. Reston, VA: U.S. Geological Survey.

Stockstill-Cahill, K. R., McCoy, T. J., Nittler, L. R., Weider, S. Z. and Hauck, S. A., II (2012). Magnesium-rich crustal compositions on Mercury: Implications for magmatism from petrologic modeling. *J. Geophys. Res.*, **117**, E00L15, doi:10.1029/2012JE004140.

Strom, R. G. (1977). Origin and relative age of lunar and mercurian intercrater plains. *Phys. Earth Planet. Inter.*, **15**, 156–172.

Strom, R. G., Trask, N. J. and Guest, J. E. (1975). Tectonism and volcanism on Mercury. *J. Geophys. Res.*, **80**, 2478–2507.

Strom, R. G., Malin, M. C. and Leake, M. A. (1990). *Geologic Map of the Bach Region of Mercury*, Map I-2015, Miscellaneous Investigations Series. Reston, VA: U.S. Geological Survey.

Strom, R. G., Chapman, C. R., Merline, W. J., Solomon, S. C. and Head, J. W. (2008). Mercury cratering record viewed from MESSENGER's first flyby. *Science*, **321**, 79–81.

Strom, R. G., Banks, M. E., Chapman, C. R., Fassett, C. I., Forde, J. A., Head, J. W., III, Merline, W. J., Prockter, L. M. and Solomon, S. C. (2011). Mercury crater statistics from MESSENGER flybys: Implications for stratigraphy and resurfacing history. *Planet. Space Sci.*, **59**, 1960–1967, doi:10.1016/j.pss.2011.03.018.

Susorney, H. C., Barnouin, O. S. and Ernst, C. M. (2016). The distribution of surface roughness in and around complex craters on Mercury. *Lunar Planet. Sci.*, **47**, abstract 1705.

Taylor, S. R. (1989). Growth of planetary crust. *Tectonophysics*, **161**, 147–156.

Taylor, S. R. and McLennan, S. M. (2010). *Planetary Crusts: Their Composition, Origin and Evolution*. Cambridge: Cambridge University Press.

Thomas, R. J., Rothery, D. A., Conway, S. J. and Anand, M. (2014a). Hollows on Mercury: Materials and mechanisms involved in their formation. *Icarus*, **229**, 221–235, doi:10.1016/j.icarus.2013.11.018.

Thomas, R. J., Rothery, D. A., Conway, S. J. and Anand, M. (2014b). Long-lived explosive volcanism on Mercury. *Geophys. Res. Lett.*, **41**, 6084–6092, doi:10.1002/2014GL061224.

Thomas, R. J., Rothery, D. A., Conway, S. J. and Anand, M. (2014c). Mechanisms of explosive volcanism on Mercury: Implications from its global distribution and morphology. *J. Geophys. Res. Planets*, **119**, 2239–2254, doi:10.1002/2014JE004692.

Tosi, N., Grott, M., Plesa, A.-C. and Breuer, D. (2013). Thermochemical evolution of Mercury's interior. *J. Geophys. Res. Planets*, **118**, 2474–2487, doi:10.1002/jgre.20168.

Trask, N. J. (1967). Distribution of Lunar craters according to morphology from Ranger VIII and IX photographs. *Icarus*, **6**, 270–276, doi:10.1016/0019-1035(67)90023-1.

Trask, N. J. and Dzurisin, D. (1984). *Geologic Map of the Discovery Quadrangle of Mercury*, Map I-1658 (H11), Miscellaneous Investigations Series. Reston, VA: U.S. Geological Survey.

Trask, N. J. and Guest, J. E. (1975). Preliminary geologic terrain map of Mercury. *J. Geophys. Res.*, **80**, 2461–2477.

Trask, N. J. and McCauley, J. F. (1972). Differentiation and volcanism in the lunar highlands: Photogeologic evidence and Apollo 16 implications. *Earth Planet. Sci. Lett.*, **14**, 201–206.

Vander Kaaden, K. E. and McCubbin, F. M. (2015). Exotic crust formation on Mercury: Consequences of a shallow, FeO-poor mantle. *J. Geophys. Res. Planets*, **120**, 195–209, doi:10.1002/2014JE004733.

Vander Kaaden, K. E. and McCubbin, F. M. (2016). The origin of boninites on Mercury: An experimental study of the northern volcanic plains lavas. *Geochim. Cosmochim. Acta*, **173**, 246–263, doi:10.1016/j.gca.2015.10.016.

Warell, J. (2004). Properties of the Hermean regolith: IV. Photometric parameters of Mercury and the Moon contrasted with Hapke modelling. *Icarus*, **167**, 271–286.

Watters, T. R., Murchie, S. L., Robinson, M. S., Solomon, S. C., Denevi, B. W., André, S. L. and Head, J. W. (2009a). Emplacement and tectonic deformation of smooth plains in the Caloris basin, Mercury. *Earth Planet. Sci. Lett.*, **285**, 309–319, doi:10.1016/j.epsl.2009.03.040.

Watters, T. R., Head, J. W., Solomon, S. C., Robinson, M. S., Chapman, C. R., Denevi, B. W., Fassett, C. I., Murchie, S. L. and Strom, R. G. (2009b). Evolution of the Rembrandt impact basin on Mercury. *Science*, **324**, 618–621, doi:10.1126/science.1172109.

Watters, T. R., Solomon, S. C., Robinson, M. S., Head, J. W., André, S. L., Hauck, S. A., II and Murchie, S. L. (2009c). The tectonics of Mercury: The view after MESSENGER's first flyby. *Earth Planet. Sci. Lett.*, **285**, 283–296, doi:10.1016/j.epsl.2009.01.025.

Watters, T. R., Solomon, S. C., Daud, K., Banks, M. E., Selvans, M. M., Robinson, M. S., Murchie, S. L., Chabot, N. L., Denevi, B. W., Ernst, C. M., Chapman, C. R., Fassett, C. I., Klimczak, C., Byrne, P. K. and Blewett, D. T. (2015). Small thrust fault scarps on Mercury revealed in low-altitude MESSENGER images. *Lunar Planet. Sci.*, **46**, abstract 2240.

Weider, S. Z., Nittler, L. A., Starr, R. D., McCoy, T. J., Stockstill-Cahill, K. R., Byrne, P. K., Denevi, B. W., Head, J. W. and Solomon, S. C. (2012). Chemical heterogeneity on Mercury's surface revealed by the MESSENGER X-Ray Spectrometer. *J. Geophys. Res.*, **117**, E00L05, doi:10.1029/2012JE004153.

Weider, S. Z., Nittler, L. R., Starr, R. D., Crapster-Pregont, E. J., Peplowski, P. N., Denevi, B. W., Head, J. W., Byrne, P. K., Hauck, S. A., Ebel, D. S. and Solomon, S. C. (2015). Evidence for geochemical terranes on Mercury: Global mapping of major elements with MESSENGER's X-Ray Spectrometer. *Earth Planet. Sci. Lett.*, **416**, 109–120, doi:10.1016/j.epsl.2015.01.023.

Weitz, C. M., Head, J. W. and Pieters, C. M. (1998). Lunar regional dark mantle deposits: Geologic, multispectral, and modeling studies. *J. Geophys. Res.*, **103**, 22725–22759.

Whitford-Stark, J. L. (1979). Charting the southern seas: The evolution of the lunar Mare Australe. *Proc. Lunar Planet. Sci. Conf.*, **10**, 2975–2994.

Whitten, J. L. and Head, J. W. (2015). Rembrandt impact basin: Distinguishing between volcanic and impact-produced plains on Mercury. *Icarus*, **258**, 350–365, doi:10.1016/j.icarus.2015.06.022.

Whitten, J. L., Head, J. W., Denevi, B. W. and Solomon, S. C. (2014). Intercrater plains on Mercury: Insights into unit definition, characterization, and origin from MESSENGER datasets. *Icarus*, **241**, 97–113, doi:10.1016/j.icarus.2014.06.013.

Wilhelms, D. E. (1976). Mercurian volcanism questioned. *Icarus*, **28**, 551–558.

Wilhelms, D. E. (1987). *The Geologic History of the Moon*. Professional Paper 1348. Denver, CO: U.S. Geological Survey.

Wilhelms, D. E. and McCauley, J. F. (1971). *Geologic Map of the Near Side of the Moon*, Map I-703, Miscellaneous Investigations Series. Reston, VA: U.S. Geological Survey.

Wood, C. A., Head, J. W. and Cintala, M. J. (1977). Degradation trends of mercurian craters and correlation with the Moon. *Proc. Lunar Sci. Conf.*, **8**, 3503–3520.

Xiao, Z., Strom, R. G., Chapman, C. R., Head, J. W., Klimczak, C., Ostrach, L. R., Helbert, J. and D'Incecco, P. (2014). Comparisons of fresh complex impact craters on Mercury and the Moon: Implications for controlling factors in impact excavation processes. *Icarus*, **228**, 260–275, doi:10.1016/j.icarus.2013.10.002.

Xiao, Z., Prieur, N. C. and Werner, S. C. (2016). The self-secondary crater population of the Hokusai crater on Mercury. *Geophys. Res. Lett.*, **43**, 7424–7432, doi:10.1002/2016GL069868.

Zuber, M. T., Smith, D. E., Phillips, R. J., Solomon, S. C., Neumann, G. A., Hauck, S. A., Peale, S. J., Barnouin, O. S., Head, J. W., Johnson, C. L., Lemoine, F. G., Mazarico, E., Sun, X., Torrence, M. H., Freed, A. M., Klimczak, C., Margot, J.-L., Oberst, J., Perry, M. E., McNutt, R. L., Jr., Balcerski, J. A., Michel, N., Talpe, M. J. and Yang, D. (2012). Topography of the northern hemisphere of Mercury from MESSENGER laser altimetry. *Science*, **336**, 217–220.

7

The Geochemical and Mineralogical Diversity of Mercury

TIMOTHY J. MCCOY, PATRICK N. PEPLOWSKI, FRANCIS M. MCCUBBIN, AND SHOSHANA Z. WEIDER

7.1 INTRODUCTION

The MESSENGER spacecraft (Solomon et al., 2008) carried two remote sensing instruments to measure the chemical composition of Mercury's surface: the X-Ray Spectrometer (XRS) and the Gamma-Ray and Neutron Spectrometer (GRNS), the latter with separate Gamma-Ray Spectrometer (GRS) and Neutron Spectrometer (NS) sensors (Chapter 1). These instruments provided spatially resolved compositional data for a variety of major, minor, and trace elements. This information included abundance estimates for U, K, and Th, as well as elemental ratios (relative to Si by weight) for Na, Mg, Al, S, Cl, Ca, Ti, Cr, Mn, Fe, C, and O. Early orbital XRS measurements (Nittler et al., 2011) with large (<50 to >3000 km) "footprint" areas indicated that the surface of Mercury exhibits a high Mg/Si ratio (0.33–0.67), i.e., intermediate between terrestrial oceanic or lunar mare basalts and highly magnesian komatiites, lower Al/Si and Ca/Si ratios than typical terrestrial or lunar basalts (Lodders and Fegley, 1998; Papike et al., 1998), and Fe/Si ratios of 0.03–0.15 that correspond to an upper limit of ~4 wt% Fe. Most surprisingly, high S/Si ratios (0.05–0.15) suggested surface abundances of S (typically a volatile element) up to ~4 wt%.

The low Fe and high S concentrations indicate that Mercury formed under highly reducing conditions, between 3 and 7 $\log_{10}$ units below the iron–wüstite (IW) buffer (McCubbin et al., 2012; Zolotov et al., 2013; Namur et al., 2016; Chapters 2, 18). Observations from the MESSENGER GRS are in broad agreement with XRS data for S, Ca, Al, and Fe (Evans et al., 2012) and also provide evidence against a volatile-depleted composition for Mercury. Peplowski et al. (2011) first reported average surface compositions for the moderately volatile element K of 1150 ± 220 ppm (since updated to 1288 ± 234 ppm; Peplowski et al., 2012), as well as for the refractory elements Th (220 ± 60 ppb) and U (90 ± 20 ppb) at latitudes northward of ~20°S. Mercury's K/Th ratio (5200 ± 1800) is comparable with those of other terrestrial planets but is much higher than that of the volatile-depleted lunar crust (~360) (Taylor et al., 2006). Furthermore, Mercury exhibits a surface K/Cl ratio similar to that of primitive chondritic meteorites (Lodders and Fegley, 1998), which also points to its volatile-rich nature (Evans et al., 2015). In addition, GRS data reveal a northern hemisphere average Na/Si content of 0.12 (Evans et al., 2012), but a significantly higher Na/Si value of 0.2 at high northern latitudes (Peplowski et al., 2014).

This global view of an Mg-rich, Fe-poor crust that formed under reducing conditions, but that is not depleted in volatiles, remains robust in light of more extensive data sets (e.g., Weider et al., 2014, 2015). The more recent X-ray, gamma-ray, and neutron measurements, however, indicate substantial variation in geochemical compositions and the presence of distinct geochemical terranes on the surface of Mercury. Mapping of geochemical terranes has previously proven to be a powerful tool in planetary sciences (e.g., Davis and Spudis, 1985; Jolliff et al., 2000). For instance, geochemical terranes can provide insights into the lateral, vertical, and temporal variation of geochemical compositions and, by extension, insights into the processes of initial differentiation, later volcanism, and impact processes that shaped the planet's crust. Of particular importance is the correlation between geochemistry and other characteristics, including morphology, spectral features, and crustal structure. These correlations can provide additional information regarding the extent and boundaries of a geochemical terrane, as well as its potential origin.

The number of geochemical terranes on Mercury that have thus far been defined has varied in the literature. In addition to the average composition of the planet, Peplowski et al. (2015) favored three terranes on the basis of neutron absorption and Mg/Si values, whereas Weider et al. (2015) defined six on the basis of Mg/Si versus Al/Si, and Vander Kaaden et al. (2017) defined nine. Lawrence et al. (2017) also added one terrane to the inventory of Peplowski et al. (2015). Peplowski et al. (2015) recognized – on the basis of measurements of high-energy neutrons – different boundaries for one of the units of Weider et al. (2015). In this chapter, it is our aim to define a robust set of geochemical terranes for Mercury. For each terrane that we define, we discuss the distinguishing features, the geochemistry, the derived mineralogy, and the possible origin. By comparing these terranes, we obtain new insights into the processes that formed Mercury's heterogeneous crust.

7.2 DATA AND METHODS

It is important to establish clear definitions when assigning geochemical terranes, and we have thus considered three essential features. First, the region must be spatially continuous. Spatially separated areas of comparable composition do not make up a distinct terrane. Second, the terrane must be geochemically distinct from the average crustal composition of Mercury. We determined the average composition by considering the planet's southern hemisphere, for which the spatial resolution of geochemical measurements is poor. Last, we require that a terrane be spatially extensive. Each of the terranes we discuss is greater than 1000 km in its minimum horizontal dimension. As a counterexample, we note that the largest of Mercury's pyroclastic deposits, NE Rachmaninoff, has an S/Si ratio that is clearly different from the global average (Weider et al., 2016), but we do not define this area as a distinct geochemical terrane because of its limited spatial extent and similarity in other geochemical parameters to the average crustal composition.

For our study of Mercury's surface geochemical diversity, we used the average oxygen abundance (Evans et al., 2012) and latitude-dependent Na, Cl, and K (Peplowski et al., 2012, 2014; Evans et al., 2015) from gamma-ray data. In addition, "global" average compositions for Ti, Cr, and Mn (mainly for regions in the southern hemisphere) were taken from Weider et al. (2014). The remaining abundances (for Mg, Al, S, Ca, and Fe) were derived from regional averages of XRS data (Weider et al., 2015) for the different terranes. We also used bulk compositional parameters in our terrane identification and classification. These parameters included the total macroscopic neutron absorption cross section (Σ_a) and average atomic mass ($\langle A \rangle$), values for which were adopted from Peplowski et al. (2015) and Lawrence et al. (2017), respectively. We did not include U and Th in our investigation, as they are not major rock-forming elements.

To convert elemental ratios to chemical compositions, we followed the methodology of Vander Kaaden et al. (2017) and Lawrence et al. (2017), i.e., we treated the Si abundance as a free parameter and adjusted it so that the sum of all elements (consistent with their elemental ratios relative to Si) is equal to 100% by weight. Although the compositions of some rock-forming elements (e.g., P) are unknown, these elements likely occur in at most minor concentrations, and their absence in our calculations should not markedly alter our conclusions. The resultant compositions for our geochemical terranes are given in Table 7.1. For each chemical composition, we also used the MELTS program of Gualda and Ghiorso (2015) to calculate the liquidus temperature and the method of Shaw (1972) to calculate a magma viscosity. Our calculations of the liquidus temperature and viscosity do not incorporate the effect of S, which can depress melting temperatures by up to 10°C per wt%.

While the geochemical composition of these terranes is well constrained, the mineralogy is largely unconstrained, yet such information provides important insights into the nature of the materials that make up the surface of Mercury. In the absence of spatially resolved spectroscopy of the type to be obtained by the Mercury Radiometer and Thermal Infrared Spectrometer on BepiColombo (Benkhoff et al., 2010) or equilibrium crystallization experiments on the compositions derived for the terranes, we used a modified CIPW normative calculation to derive the mineralogy of our terranes. Named for the four authors who first proposed the concept (Cross, Iddings, Pirsson, and Washington; Cross et al., 1903), a normative mineralogy is a hypothetical mineral assemblage calculated from the bulk composition of a rock. In this calculation, it is assumed that the composition represents a single igneous unit that crystallized according to the expected partitioning of elements among phases. Although the assumption of a single igneous unit is invalid for some of the terranes (see below), and the CIPW calculation is designed for rocks that crystallized under terrestrial oxygen fugacity conditions, normative mineralogies can be useful for comparing different regions on Mercury.

Table 7.1. *Chemical composition (in wt%) and derived mineralogy (in modal abundance) of geochemical terranes on Mercury.*

	Southern hemisphere	Northern	Low-Fast	High-Magnesium	Caloris Interior Plains
O	39.65	42.27	41.13	37.21	41.31
Na	2.83	5.74	2.94	2.66	2.95
Mg	12.44	7.55	12.34	16.48	9.15
Al	7.79	6.04	7.05	5.32	9.44
Si	28.32	30.19	29.38	26.58	29.51
S	2.07	2.11	1.76	2.92	1.77
Cl	0.14	0.45	0.24	0.13	0.15
K	0.13	0.20	0.15	0.10	0.08
Ca	4.55	4.23	3.82	5.58	4.43
Ti	0.34	0.36	0.35	0.32	0.35
Cr	0.14	0.15	0.15	0.13	0.15
Mn	0.11	0.12	0.12	0.11	0.12
Fe	1.48	0.60	0.59	2.44	0.59
Total	100.02	100.01	100.01	99.98	100.00
T_{liq} (°C)	1460	1350	1476	1542	1365
η (Pa s)	3.0	37.8	3.6	0.4	28.5
Plagioclase	**50.4**	**55.7**	**48.2**	**37.4**	**57.7**
Or	0.8	1.3	1.0	0.6	0.5
Ab	29.2	54.3	30.7	27.0	30.7
An	20.4	0.0	16.5	9.8	26.5
Pyroxene	**37.3**	**29.0**	**47.5**	**26.2**	**32.8**
Di	5.4	20.0	4.8	17.8	0.0
Hy	31.9	9.0	42.7	8.4	32.8
Olivine	**7.5**	**7.5**	**0.0**	**29.5**	**0.0**
Sulfides	**4.7**	**4.3**	**3.7**	**6.8**	**3.7**
Accessory	None	None	Qz	None	Qz, Cor

Mineral abbreviations: Or: Orthoclase, Ab: Albite, An: Anorthite, Di: Diopside, Hy: Hypersthene, Qz: Quartz, Cor: Corundum.

To accommodate this calculation for Mercury compositions and conditions (i.e., oxygen fugacity fO_2 of 2.6 to 7.3 $\log_{10}$ units below the iron–wüstite buffer; McCubbin et al., 2012; Zolotov et al., 2013; Chapters 2, 18), it was therefore necessary to partition cations to form sulfides. We first assigned the minor elements Fe, Mn, Cr, and Ti to sulfides, following the experimental results of Vander Kaaden and McCubbin (2016). All geochemical terranes had sufficient sulfur abundances that none of these transition metals occur as oxides. In the absence of redox-sensitive elements, particularly iron, in the CIPW normative calculation, the resulting mineralogy should be largely insensitive to the differences in oxygen fugacity between Mercury compositions and those for which the CIPW normative calculation was originally designed. Data from enstatite

Table 7.2. *Cations expressed as oxides and normalized to 100% after removal of sulfur and cations bound as sulfides.*

	Southern hemisphere	Northern	Low-Fast	High Magnesium	Caloris Interior Plains
SiO_2	57.38	63.94	59.71	54.34	59.86
Al_2O_3	13.95	11.30	12.66	9.61	16.92
CaO	5.77	5.37	4.75	7.07	5.54
MgO	19.01	11.36	18.74	25.28	13.68
Na_2O	3.62	7.65	3.76	3.43	3.77
K_2O	0.15	0.24	0.17	0.11	0.09
Cl	0.13	0.15	0.22	0.13	0.14
Total	100.01	100.01	100.01	99.97	100.00
Mg	0.34	0.63	0.45	0.54	0.45
Ca	0.19	0.35	0.25	0.3	0.25
Fe	1.48	0.61	0.59	2.44	0.59
Cr	0.14	0.15	0.15	0.13	0.15
Ti	0.34	0.36	0.35	0.32	0.35
Mn	0.11	0.12	0.12	0.11	0.12
S	2.07	2.12	0.59	2.92	0.59

Notes: All figures are wt%. The oxide abundances shown were used in the CIPW calculations.

chondrites, which formed under similarly highly reducing conditions as Mercury, also indicate that these elements are chalcophilic (Keil, 1989). The remaining S (i.e., that which is not accommodated by these cations) was combined with Mg and Ca in a 75:25 molar proportion, as determined for natural and experimental samples by Stockstill-Cahill et al. (2012). The abundances of the cations bound with sulfur are given in Table 7.2. Following these sulfide calculations, we renormalized the elements to 100% and recalculated the cations as oxides (because the CIPW normative calculation requires oxides rather than elements) using standard oxidation states (e.g., Si^{4+}, Al^{3+}, Mg^{2+}). We finally renormalized these oxide abundances to 100%. This procedure yielded CIPW norms using only abundances for SiO_2, Al_2O_3, MgO, CaO, Na_2O, and K_2O. These oxide abundances are given in Table 7.2. This recalculation, however, produced an inconsistency with the measured compositions. By calculating oxides in this manner, the O/Si ratios we derived range from 1.6 to 1.8. In contrast, the average GRS-measured O/Si ratio for Mercury is 1.40±0.03 (Evans et al., 2012). The resulting difference in oxygen abundance from the elemental (measured) and oxide (derived) compositions thus ranges from 5.9 to 12.2 wt% O. Furthermore, in our calculations we utilized only averages for the elemental ratios, but we varied the elemental ratios in the calculation (using the derived standard deviations) to test the robustness of our derived minor phases. We discuss the results of varying the elemental ratios, within uncertainties, in the following sections.

An alternative to the CIPW normative calculation (i.e., using oxides) for the derivation of mineralogy would be to assume that one or more of the elements occur as metals rather than oxides.

The only major cations remaining after the sulfide calculations are Si, Mg, Al, and Ca. Among these, only Si is known to occur in the metallic phase in highly reduced enstatite-rich meteorites (Keil, 1989). In addition, the relative temperature–fO_2 dependence of the metal–oxide buffers for these four elements suggests that Si metal would be the most likely to form. The measured O/Si ratio would thus require approximately 9 wt% metallic Si on the surface of Mercury. We suggest instead that the measured O/Si ratio reflects reduction of Mercury's gardened near-surface materials, perhaps through the formation of silicides as a result of space weathering (Anand et al., 2004), and that the original composition is best reflected by the oxides calculated using the standard oxidation states. The calculated oxide compositions are presented in a total alkalis versus silica (TAS) diagram, on which the regions for high-Mg compositions after Le Bas and Streckeisen (1991) and Le Bas (2000) are overlain. After calculation of the CIPW norms, we renormalized the derived mineral abundances to 100% (minus the sulfide abundances) to determine the final normative mineralogy. We present the silicate mineral abundances in standard International Union of Geological Sciences classification schemes for plutonic rocks. Although these classifications often imply specific origins for terrestrial rocks, rocks of similar composition might form through different petrogenetic processes on other planetary bodies. As strictly classificational tools, these schemes serve the useful function of allowing comparisons between geochemical terranes and planetary bodies. Although we have chosen to classify the compositions using plutonic rock classification schemes, extrusive equivalents of these rocks are also likely to exist and may even dominate on the surface of Mercury.

7.3 GEOCHEMICAL TERRANES

In addition to Mercury's average surface composition (here taken from the southern hemisphere), we recognize four distinct geochemical terranes, for which we use the following nomenclature: (i) the Northern Terrane, (ii) the Caloris Interior Plains Terrane, (iii) the High-Magnesium Terrane, and (iv) the Low-Fast Terrane. In this section we describe the main features of these four terranes. We discuss their geographic distribution (see Figure 7.1) and the features that allowed their recognition, and we provide a brief comparison with other morphological or spectral data sets with which they can be discerned. We also describe the distinguishing geochemical features (Figure 7.2), the derived mineralogy, and the resultant geochemical and mineralogical classification of each terrane (Figures 7.3 and 7.4).

7.3.1 Southern Hemisphere (Mercury Average)

Mercury's southern hemisphere comprises multiple units of both volcanic and impact origin (Chapter 6). Geologic mapping, however, reveals that smooth volcanic plains cover a significantly smaller area than in the northern hemisphere (Denevi et al., 2013). The southern hemisphere is instead dominated by intercrater plains, a geomorphological unit generally associated with a composition that is similar to an average of

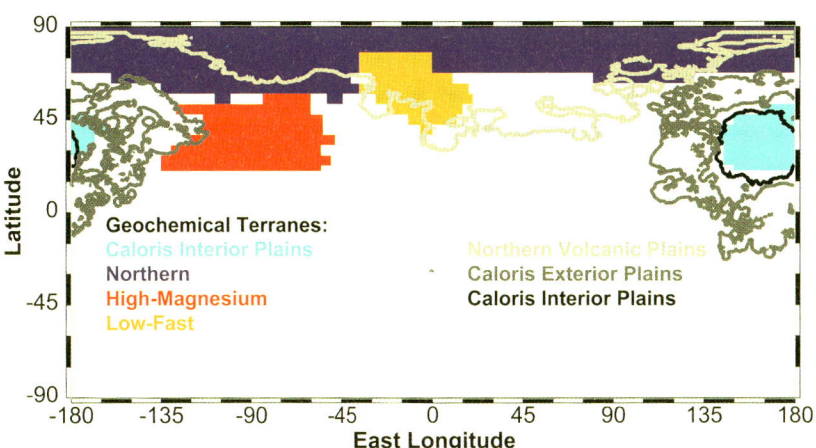

Figure 7.1. Location of the four geochemical terranes in Mercury's northern hemisphere defined in this paper. Boundaries of major smooth plains, as mapped by Denevi et al. (2013), are also shown.

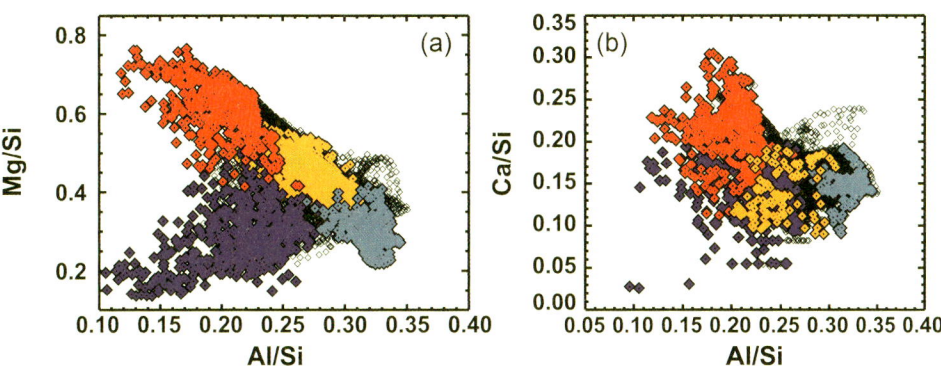

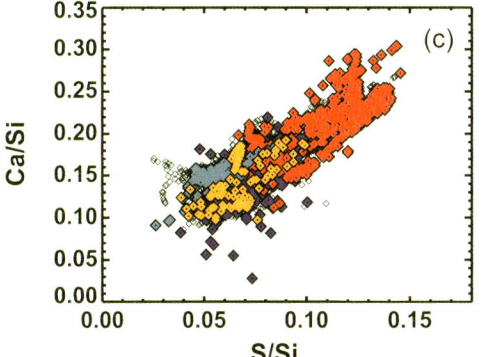

Figure 7.2. Plots of (a) Mg/Si versus Al/Si, (b) Ca/Si versus Al/Si, and (c) Ca/Si versus S/Si. Data are colored according to the corresponding geochemical terrane (refer to Figure 7.1). Abundance ratios were derived from the MESSENGER XRS elemental maps (Weider et al., 2015) and re-binned into equal-area pixels, 5° × 5° at the equator.

Mercury's surface (Peplowski et al., 2015). XRS measurements of the southern hemisphere therefore serve as a close approximation to the average surface composition of Mercury. MESSENGER's highly eccentric orbit (an apoapsis of ~10,000–15,000 km and a periapsis of 200–500 km, except during the low-altitude campaign late in the mission; see Chapter 1) meant that for the majority of each orbit the XRS observed the southern hemisphere, albeit with poor spatial resolution. The best coverage for Mg/Si, Al/Si, S/Si, Ca/Si, and Fe/Si elemental ratios is thus available for the southern part of the planet. If data at higher spatial resolution were available for the southern hemisphere, it is possible that additional geochemical terranes would be apparent. MESSENGER's eccentric orbit also meant that GRS-derived elemental ratios are unavailable south of ~20°S latitude (Peplowski et al., 2011). We therefore adopted O, Na, and Cl ratios for the southern hemisphere that are typical of equatorial to mid-latitude northern hemisphere measurements. We also adopted the global average values of Weider et al. (2014) for Ti, Cr, and Mn, relative to Si. With the exception of Na, these elements (i.e., Cl, Ti, Cr, and Mn) occur at minor to trace abundances. The uncertainties connected with their non-measurement should not, therefore, substantially alter our conclusions.

The southern hemisphere is characterized by high Mg/Si, S/Si, and inferred Na/Si, as well as low Al/Si and Ca/Si compared with terrestrial oceanic basalts (originally noted by Nittler et al., 2011, and Peplowski et al., 2011). Chemically, the southern hemisphere lies on the boundary between basaltic andesite

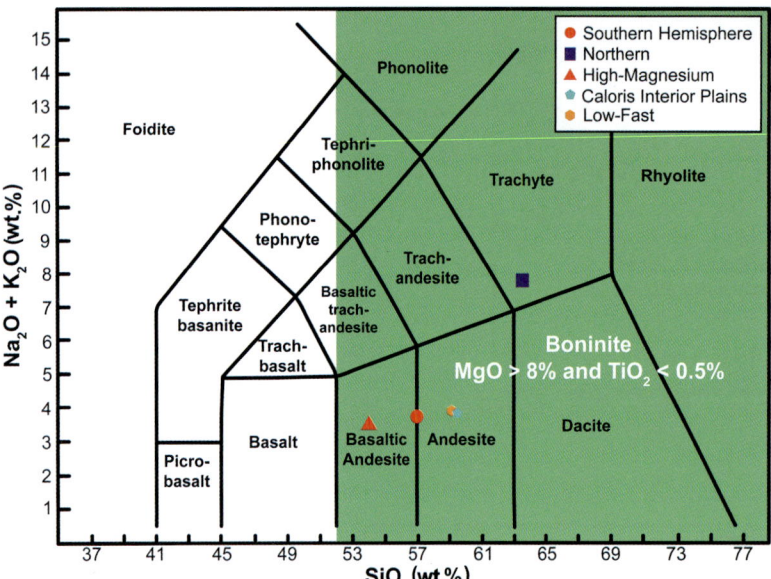

Figure 7.3. Total alkalis versus silicate (TAS) diagram (for the classification of extrusive igneous rocks) showing the normalized, sulfide-free compositions of the geochemical terranes. The field for high-Mg rocks (Le Bas, 2000) is overlaid in green.

and andesite in a TAS diagram (Figure 7.3), although its high normalized MgO concentration (19.0 wt%) means that it can also be classified as a boninite. A melt of this composition has a liquidus temperature of 1460°C (an upper limit that excludes the influence of S) and a viscosity of 3.0 Pa s.

Mineralogically, the modified CIPW norm indicates that the southern hemisphere is dominated by major plagioclase (50.4 wt%; $Ab_{58}An_{40}Or_2$ by mode) and pyroxene (37.3%, primarily orthopyroxene), with lesser olivine (7.5 wt%) and sulfides (4.7 wt%). No other phases are calculated from our CIPW normative calculation. The sulfide we derive is complex, with a modal composition of $Fe_{0.46}Mg_{0.25}Ti_{0.12}Ca_{0.08}Cr_{0.05}Mn_{0.04}S$. In the plutonic rock classification system (Figure 7.4), the average southern hemisphere plots in the olivine gabbro (norite) field in a pyroxene–olivine–plagioclase (px–ol–pl) ternary and as a norite in the orthopyroxene–clinopyroxene–plagioclase (opx–cpx–pl) ternary. The southern hemisphere is thus best described as an olivine norite.

7.3.2 Northern Terrane

The first discussion of geochemically distinct units on Mercury's surface came from early XRS orbital results (Nittler et al., 2011; Weider et al., 2012). Weider et al. (2012) demonstrated that, on average, the northern plains differ from the surrounding, older intercrater plains and heavily cratered terrain. The northern smooth plains (NSP) are dominantly of volcanic origin and cover more than ~7% of Mercury's surface area (Head et al., 2011; Denevi et al., 2013), spanning a 5.59 × 10^6 km² region (Ostrach et al., 2015). The NSP is less cratered than its surroundings, exhibits the "high-reflectance red plains" (HRP) spectral signature typical of several other smooth plains deposits, and is likely the product of flood volcanism (Head et al., 2011; Denevi et. al., 2013; Ostrach et al., 2015). The NSP exhibit lower Mg/Si, S/Si, Ca/Si, and higher Al/Si ratios than the older surrounding regions. Further support for a distinct geochemical unit at high northern latitudes was provided by GRS measurements of volatile elements, including low-spatial-resolution measurements of K (Peplowski et al., 2012). Additional, latitudinally resolved measurements of Na/Si and Cl/Si reveal significantly higher (by a factor of ~2) ratios poleward of 80°N than for mid-northern latitudes (Peplowski et al., 2014; Evans et al., 2015). Although the region of Na enrichment corresponds to portions of the NSP, the continuous volcanic deposits extend to lower latitudes. Indeed, a model of the latitude-dependent Na/Si variation in which Na was assumed to be concentrated in and uniformly distributed across volcanic plains units, including the NSP and the Caloris interior plains, provided a poor fit to the observed data (Peplowski et al., 2014). In contrast, the best-fit Na model was only partially consistent with the spatial extent of the NSP, but was a good match to the high-K region and subsequently identified low-Mg/Si areas. These observations, along with mismatches from X-ray spectroscopy (e.g., Mg/Si on the western edge of NSP), provide evidence that compositional and geomorphological units do not coincide.

Surface reflectance of major units on Mercury varies by a factor of 1.7 at 750 nm, with high-reflectance red plains up to 20% above the global mean and low-reflectance blue plains 15% below the global mean (Denevi et al., 2009; Chapter 8), although spatially minor red pyroclastic units and hollows exhibit nearly twice the reflectance of the global mean. Despite the mismatch between compositional and geomorphological units, that the Northern Terrane is an area of HRP suggests that the Northern Terrane hosts lower than average concentrations of Mercury's darkening agent. The nature of this opaque phase has been debated extensively, although it is generally thought to have a compositional or mineralogical origin (e.g., Lawrence et al., 2010). The most recent results – from multispectral imaging (Murchie et al., 2015) and neutron spectroscopy (Peplowski et al., 2016) – indicate that carbon (in the form of graphite) causes much of

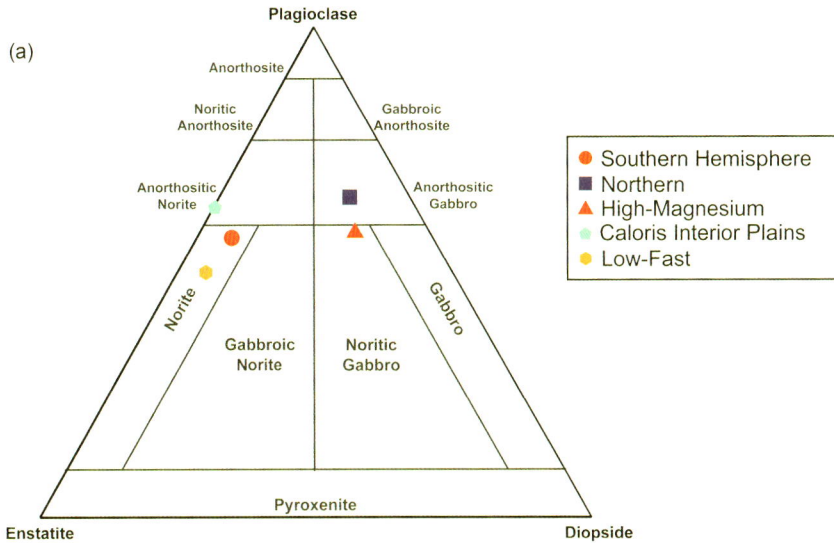

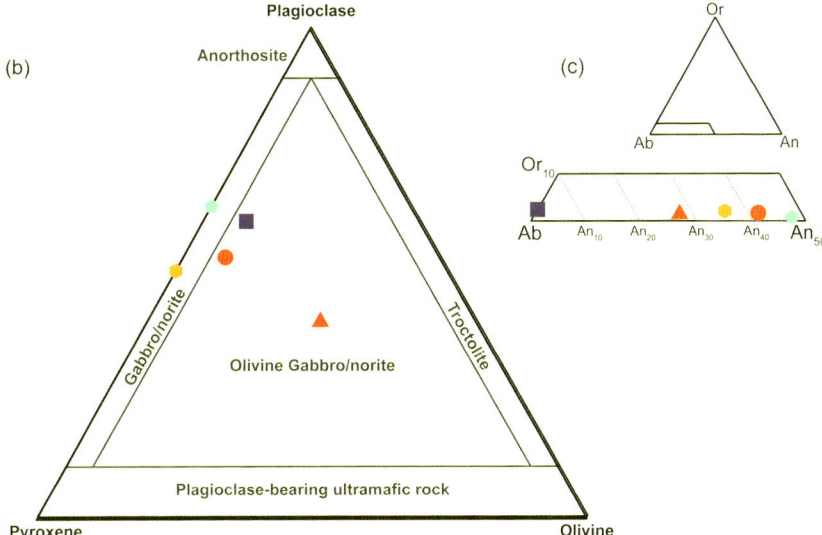

Figure 7.4. International Union of Geological Sciences rock classification diagrams for plutonic rocks, showing the silicate mineralogies of the geochemical terranes, derived from modified CIPW normative calculations. The mineralogical compositions are shown on (a) a plagioclase–pyroxene–olivine (Pl–Px–Ol) ternary diagram and (b) a plagioclase–orthopyroxene–clinopyroxene (Pl–Opx–Cpx) ternary diagram. The mineral compositions for feldspars (where Ab is albite, An is anorthite, Or is orthoclase) are also illustrated in (c). The albitic (Na-rich) corner of the feldspar ternary is shown as an inset to (c).

the darkening of Mercury's surface (Chapter 8). The HRP may contain 1–4 wt% less C than low-reflectance material.

Recognizing the apparent mismatch between the spatial extent of the northern plains and that of the geochemical signatures, Peplowski et al. (2015) defined the spatial extent of the "Northern Geochemical Terrane" (i.e., the Northern Terrane in Figure 7.1) as centered on the pole and having an outer boundary approximately coinciding with 65°N, giving an area of $\sim 3.6 \times 10^6$ km^2. A $\pm 5°$ uncertainty on the boundary yields an area uncertainty of $\pm 1.5 \times 10^6$ km^2. In contemporaneous XRS-based work, Weider et al. (2015) found that the NSP are not compositionally homogeneous and divided them into the "low-Mg northern plains" and the "intermediate-Mg northern plains" (see the section below on Low-Fast Terrane for a discussion of the intermediate-Mg northern plains). The edges of the XRS-derived units were arbitrarily assigned as the geomorphologically mapped (Denevi et al., 2013) boundaries of the NSP. Despite this assignment, the spatial extent of the low-Mg northern plains was generally consistent with the high-K region and the modeled high-Na region.

The Northern Terrane that we define has a relatively low Mg/Si ratio, as well as moderate to low Al/Si, Ca/Si, and S/Si ratios (Table 7.1, Figure 7.2). This composition corresponds to markedly low Mg and Al concentrations (and high Si and O contents). The most notable geochemical feature of this terrane is the substantial enrichment in the volatile elements Na, K, and Cl compared with the mean northern hemisphere concentration. Chemically, the Northern Terrane lies within the trachyte field of a TAS diagram (Figure 7.3), although we caution that this label does not imply a formation mechanism similar to that for trachytes on Earth. We also note that the high normalized-MgO concentration (11.4 wt%) of this terrane means that it can also be classified as a boninite (Figure 7.3), as also suggested by Vander Kaaden and McCubbin (2016) for their northern volcanic plains unit (although their composition plotted in the trachyandesite field). A melt of this composition has a liquidus temperature (1350°C) that is more than 100°C lower than that for Mercury's average surface composition and a viscosity (37.8 Pa s) that is more than an order of magnitude higher (see Section 7.4).

Compared with the average surface composition for Mercury, the Northern Terrane is slightly enriched in plagioclase (55.7 versus 50.4%) and depleted in pyroxene (29.0 versus 37.3%), but with similar abundances of olivine and sulfides. We find that the plagioclase is markedly more albitic ($Ab_{98}Or_2$) than the average for Mercury, which is consistent with the higher Na concentration, whereas the pyroxene has considerably more normative diopside ($(Mg,Ca)SiO_3$) than hypersthene ($MgSiO_3$) (Table 7.1). The sulfide composition we derive ($Mg_{0.45}Fe_{0.18}Ca_{0.15}Ti_{0.13}Cr_{0.05}Mn_{0.04}S$) is richer in Mg and Ca than that for the southern hemisphere. This terrane's material is best classified as an olivine-bearing anorthositic gabbro, if considered as a plutonic rock (Figure 7.4).

The markedly high Na concentration (5.75 wt%) in the Northern Terrane presents challenges for the CIPW normative calculation. If we use the one-standard-deviation result for the Al/Si ratio (0.24) instead of the average (0.20), the plagioclase abundance increases from 55.7 to 66.1 wt% and has a corresponding composition of $Ab_{95}An_3Or_2$. Without this adjustment, the CIPW normative calculation yields acmite ($NaFe^{3+}Si_2O_6$), despite the absence of oxidized iron in the normative calculation, and a corresponding negative abundance of magnetite (–3.4 wt%). Alternatively, some of the Na could be the result of thermal redistribution. Evidence for such a possibility includes the inverse correlation between the maximum surface temperature and potassium concentration (Peplowski et al., 2012), as well as the similarity in latitudinal profiles of Na and K (Peplowski et al., 2014), suggesting redistribution of volatile Na and K. Weider et al. (2015) noted, however, that K and Mg are anticorrelated. Given that Mg (as a major element) cannot be thermally redistributed, Na and K are unlikely to have been redistributed by this process. Furthermore, the Northern Terrane was originally defined on the basis of Mg, Al, and fast neutrons (i.e., with energies of 0.5 to ~10 MeV), and thus cannot be a thermal feature.

7.3.3 Caloris Interior Plains Terrane

The Caloris basin is Mercury's largest well-preserved impact feature, with an estimated diameter of 1420–1550 km (Murchie et al., 2008; Ernst et al., 2015), and the basin floor hosts one of our recognized geochemical terranes – the Caloris Interior Plains Terrane (CIP). Whereas most of Mercury's geochemical terranes have a moderate-to-poor spatial correlation with geologic units and spectral data, the boundary of the CIP is well correlated with the inner basin wall, as well as with the HRP spectral unit (Denevi et al., 2009) observed within the basin's interior. Although the CIP does extend somewhat beyond the inner wall in the northeastern quadrant, this is also the case for the HRP unit. The association of the CIP and the HRP suggests that the CIP, like the Northern Terrane, also has depleted levels of Mercury's darkening agent, likely graphite (Chapter 8). The basin's interior is filled with a smooth plains unit, which is thought to be the result of extensive flood volcanism that occurred sometime after basin formation (Murchie et al., 2008; Denevi et al., 2013; Chapter 6). Evidence for flood volcanism includes the presence of embayed craters on the basin floor, as well as multiple pyroclastic vents and associated deposits along portions of the basin rim (Murchie et al., 2008). On the basis of the depths of excavation of spectrally different material by younger craters within the basin, these smooth plains are consistently found to be at least 2.5 km thick, and in places have a thickness of up to 3.5 km (Ernst et al., 2015).

Early indications of the Caloris interior's distinct composition was provided by lower Mg/Si and higher Al/Si ratios compared with the intercrater plains and heavily cratered terrain (i.e., representing an average surface composition) (Weider et al., 2012) and low K concentration (Peplowski et al., 2012). The distinct composition of the Caloris interior plains was more clearly illustrated in the subsequent work of Weider et al. (2015) and Peplowski et al. (2015), who used the GRS-derived Σ_a and XRS-derived Mg/Si measurements to define a geochemical terrane there. The CIP clearly stands out from other northern hemisphere regions because of its low Σ_a values ($<48.5 \times 10^{-4}$ cm^2/g) – the lowest reported on the surface – and its low Mg/Si (<0.38) and high Al/Si (>~0.3) values.

The most distinctive geochemical features of the CIP are its low Mg and high Al concentrations (Figure 7.2, Table 7.1). In fact, lower Mg concentrations are observed only within the Northern Terrane. The CIP also exhibits the highest Al concentration of any of our recognized terranes. In the low-Mg TAS diagram (Figure 7.3), the CIP falls firmly within the andesite field, but it can also be classified as a boninite because of its normalized-Mg content (13.7 wt% MgO). The CIP composition has a liquidus temperature (1365°C) and a viscosity (28.5 Pa s) similar to those of the Northern Terrane.

Compared with the average Mercury composition, the CIP is enriched in plagioclase and markedly depleted in olivine, with similar pyroxene and sulfide abundances (Figure 7.4, Table 7.1). The plagioclase composition we derive is slightly more calcic ($An_{46}Ab_{53}Or_1$) than average for Mercury, and the pyroxene is entirely hypersthene. Ground-based mid-infrared observations of Caloris led to suggestions that the terrane contains clinopyroxene, K-feldspar, Na-plagioclase, as well as minor amounts of Mg-rich orthopyroxene (Sprague et al., 2009). Although the suggestion of Na-plagioclase appears to be correct, we find that the pyroxene component is consistent with enstatite rather than with diopside. The sulfides we derive from the CIPW normative calculation are Mg-dominant ($Mg_{0.39}Fe_{0.22}Ti_{0.16}Ca_{0.13}Cr_{0.06}Mn_{0.04}S$) and the calculation also yields – in addition to the minerals listed in Table 7.1 – 5.1 wt% quartz and 0.5 wt% corundum. The corundum finding is not robust, however, given the uncertainties in the Al/Si ratio. In contrast, we are confident in the quartz result, although we note that the abundance is small and the combined one-standard-deviation uncertainties in the Al/Si, Mg/Si, and Ca/Si ratios are sufficient to allow an olivine-normative composition. Given this uncertainty, we plot the CIP in the pyroxene–olivine–plagioclase classification ternary (Figure 7.4b) at 0% olivine, which provides a gabbro or norite classification. In addition, the absence of diopside and significant plagioclase abundance suggests that the material should be classified as an anorthositic norite.

7.3.4 High-Magnesium Terrane

The first strong evidence for a compositional dichotomy within Mercury's intercrater plains and highly cratered terrain (i.e., areas not classified as smooth plains) was present in the XRS data of Weider et al. (2012). The data for these regions exhibited two clusters: one with moderate Mg/Si (~0.45) and Al/Si (~0.3) ratios, and another with high Mg/Si (~0.75) and low Al/Si (~0.15) ratios. The second of these clusters included data from the "high-Mg region" that was later delineated by Weider et al. (2015). This region exhibits Mercury's highest Mg/Si ratios as well as low Al/Si, and we refer to it here as the High-Magnesium Terrane. Centered at ~30°N, −90°E, the High-Magnesium Terrane has a diameter of >3000 km and an area of >5 × 10^6 km^2, making it the largest of our geochemical terranes. In addition, the XRS data set (Weider et al., 2015) shows that this terrane has relatively high S/Si, Ca/Si, and Fe/Si ratios (Figure 7.2, Chapter 2). It has also been noted (Weider et al., 2015) that the High-Magnesium Terrane is an area of relatively high crater density, low elevation, and low crustal thickness (although there is considerable variability across the region). Peplowski et al. (2015) further observed that the High-Magnesium Terrane corresponds to an area of high neutron absorption. The western and northern boundaries of this terrane also closely match the borders of the Caloris exterior plains (which are substantially younger) and, to a lesser extent, the NSP (Figure 7.1). Interestingly, relatively high absorption levels for fast neutrons have been observed for the High-Magnesium Terrane and in a similar-sized area (incompletely sampled) centered at about ~10°S, 90°E (Lawrence et al., 2017). This result suggests that the second region's bulk composition produces a higher $\langle A \rangle$ value than the average for Mercury. No comparable high Mg/Si area, however, is observed in the global XRS data set at the site of the eastern high-fast-neutron region.

Among the elements for which spatially resolved measurements have been made, the High-Magnesium Terrane is distinct in almost every aspect. It has the highest Mg/Si, Ca/Si, and S/Si ratios, as well as the lowest Al/Si ratio of any of our terranes (Figure 7.2). These extreme elemental ratios result in a composition that is low in Si, O, and Al, and high in Mg, Ca, and S (Table 7.1). The low Si content of the High-Magnesium Terrane is notionally consistent with the Si gamma-ray map of Peplowski et al. (2012). With a normalized, sulfide-free, bulk SiO_2 concentration of 54.4 wt%, the High-Magnesium Terrane is the most mafic of the terranes. It thus falls as a basaltic andesite in the TAS diagram (Figure 7.3), although the exceptionally high normalized-MgO concentration (25.3 wt%) clearly places it as a boninite. The High-Magnesium Terrane has the highest liquidus temperature (1542°C) and lowest melt viscosity (0.4 Pa s) of any geochemical unit on Mercury.

The unusual chemical composition of the High-Magnesium Terrane yields a CIPW normative mineralogy that is enriched in diopside and has, by far, the highest enrichment of olivine, relative to the average Mercury surface (Figure 7.4, Table 7.1). The plagioclase composition ($An_{26}Ab_{72}Or_2$) is slightly more sodic than the southern hemisphere, probably because of substantial incorporation of Ca into diopside. The High-Magnesium Terrane also has the highest abundance of sulfides (6.8 wt%). As for the southern hemisphere average composition, but unlike the other northern hemisphere terranes, these sulfides are Fe-dominant ($Fe_{0.52}Mg_{0.26}Ca_{0.09}Ti_{0.08}Cr_{0.03}Mn_{0.02}S$). We do not obtain any other phases from the CIPW normative calculation, and we therefore mineralogically classify this terrane's rock as an olivine gabbro.

7.3.5 Low-Fast Terrane

As described above, Mercury's NSP do not constitute a geochemically homogenous unit. Lawrence et al. (2017) reported that an area – contained largely within, but not corresponding precisely with, the lower latitudes of the NSP – exhibits relatively low count rates for fast neutrons. Fast neutron measurements are primarily sensitive to the $\langle A \rangle$ value of near-surface materials (e.g., Gasnault et al., 2000) and thus bulk composition. These measurements are particularly sensitive to the relative proportions of low-mass (e.g., O, Mg, Si, and Al) and high-mass (e.g., Ca, Ti, and Fe) elements. Lawrence et al. (2017) used the name "Low-Fast Terrane" for this region, which we also adopt.

Centered at ~60°N, 10°W (Figure 7.1), the Low-Fast Terrane adjoins the Northern Terrane and corresponds in some places to local Mg, S, and Ca lows (Lawrence et al., 2017). The Low-Fast Terrane, however, is chemically distinct from the Northern Terrane, as the former exhibits lower Na/Si and Ca/Si ratios, as well as a higher Mg/Si ratio. We note that the Low-Fast Terrane corresponds closely to the region for which there is a spatial mismatch between the high-Na and high-K measurements and the NSP unit. The differences in Na and Mg are most statistically significant between the Northern and Low-Fast Terranes. In contrast, there are mostly only minimal compositional differences between the Low-Fast Terrane and the southern hemisphere. Although the Low-Fast Terrane has a higher Cl content, as well as lower Ca (the lowest Ca content of all the terranes), all the other elemental concentrations are similar to the surface average. On the elemental ratio plots (Figure 7.2), the Low-Fast Terrane tends to cluster close to the center of all the composition distributions.

Unsurprisingly, given their similar compositions, the Low-Fast Terrane and the southern hemisphere plot near each other in the TAS diagram (Figure 7.3), i.e., in the low-Mg andesite field. As for the other terranes, the Low-Fast Terrane can also be classified as a boninite because of its high sulfide-free normalized-MgO concentration (18.7 wt%). The liquidus temperature (1476°C) and melt viscosity (3.6 Pa s) of the Low-Fast Terrane are similar to average values for Mercury. Although the CIP and Low-Fast Terrane plot close together on the TAS diagram, these terranes differ markedly in Mg and Al.

Mineralogically, the Low-Fast Terrane is dominated by major plagioclase (48.2 wt%; $Ab_{64}An_{34}Or_2$) and pyroxene (primarily orthopyroxene), with lesser amounts of sulfides (3.7 wt%). The CIPW norm also contains 0.4 wt% quartz, although uncertainties in Al/Si alone are sufficient to yield an olivine-normative composition. Thus, we do not consider the occurrence of quartz as robust and plot the Low-Fast Terrane at 0% olivine in the pyroxene–olivine–plagioclase classification ternary (Figure 7.4b). We also obtain a complex sulfide composition, i.e., $Mg_{0.39}Fe_{0.22}Ti_{0.16}Ca_{0.13}Cr_{0.06}Mn_{0.04}S$. In the plutonic rock classification system (Figure 7.4), the Low-Fast Terrane plots in the norite field.

7.4 DISCUSSION

7.4.1 Similarities among the Terranes

Despite their geochemical and inferred mineralogical differences, the geochemical terranes we have defined share a number of common features. For instance, Mercury's geochemical terranes exhibit universally higher Mg/Si and S/Si and lower Al/Si, Ca/Si, and Fe/Si ratios than typical terrestrial oceanic basalts (Lodders and Fegley, 1998). The low Fe and high S concentrations suggest that all of the terranes formed under highly reducing conditions. The high Mg/Si and S/Si ratios, as well as the low Al/Si and Fe/Si ratios, are also similar to the compositions of melts produced from single-stage partial melting of enstatite chondrites (McCoy et al., 1999; Burbine et al., 2002; Berthet et al., 2009), which have been proposed as a candidate composition for Mercury's precursor materials (Wasson, 1988). Geochemical, relative age, and thermal considerations, however, provide evidence against an undifferentiated chondritic mantle as the source region of the melts that erupted to Mercury's surface. The observed high Na concentrations of the average surface composition of Mercury, relative to an enstatite chondrite, would require low-degree partial melts (<35%; Berthet et al., 2009). The high Na concentration of the Northern Terrane appears inconsistent with even low-degree partial melts. In contrast, high-degree partial melting (>50%) would be required to produce the Mg-rich, Al-poor compositions (McCoy et al., 1999; Berthet et al., 2009).

Our terrane compositions all yield liquidus temperatures (1350–1542°C, an upper limit that does not account for 2–3 wt% S) that are greater than for most terrestrial ocean floor basalts (consistent with the higher Mg concentrations). The melt viscosities we have calculated, which range over two orders of magnitude (0.4–37.8 Pa s), are substantially lower than for typical terrestrial basaltic magmas that are not enriched in Mg (>100 Pa s; Basaltic Volcanism Study Project, 1981). As a result, extrusive magmas on Mercury would have produced thin, laterally extensive deposits, which is consistent with the evidence for flood volcanism in the northern smooth plains and the Caloris interior plains.

Mineralogically, all of Mercury's terranes are rich in plagioclase, which makes up 37–58% of the normative mineralogies. Indeed, the CIP and Northern Terrane compositions are sufficiently feldspathic to be classified as anorthositic norite/gabbro. Stockstill-Cahill et al. (2012) predicted that the expected plagioclase-rich nature of the Na-rich NSP would produce a higher-reflectance surface, unless a darkening agent was present. The NSP are generally associated with HRP, Mercury's highest-reflectance spectral unit. We note, however, that a darkening phase, likely graphite, is required globally (including within the HRP) to account for Mercury's low reflectance (see Section 7.4.2).

7.4.2 Constraints on an Early Magma Ocean

It has been argued (Schubert et al., 1988) that the heat from Mercury's accretion would have been sufficient to melt the entire planet. In addition, Mercury's large core (e.g., Hauck et al., 2013) indicates that high degrees of melting must have occurred to allow efficient metal segregation and core formation. This level of melting is thus likely to have produced a magma ocean (Brown and Elkins-Tanton, 2009; Riner et al., 2009; Charlier et al., 2013; Vander Kaaden and McCubbin, 2015, 2016). The nature and products of this magma ocean's solidification depend critically on the bulk composition of the silicate shell of Mercury. In particular, the FeO concentration of the magma ocean would have critically affected the formation of a potential flotation crust and mantle overturn that occurred after solidification (Brown and Elkins-Tanton, 2009). With the hindsight provided by the MESSENGER data, the most relevant of the Brown and Elkins-Tanton (2009) models seems to be the composition of CB chondrites adjusted to low FeO concentrations. In such models, a differentiating magma ocean produces a basal layer of dunite (100% olivine), overlain by layers of harzburgite (70% olivine, 30% orthopyroxene) and garnet-bearing wehrlite (60% clinopyroxene, 30% olivine, 10% garnet), and capped by gabbro (50% plagioclase, 40% clinopyroxene, 10% opaques). Thicknesses of these units were modeled under the pre-MESSENGER estimate of a 600-km-thick silicate shell, whereas current estimates for the silicate shell thickness are closer to 400 km (Hauck et al., 2013; Chapters 4, 19). Given a core–mantle boundary depth of 420 km and a pressure at that depth of ~5.7 GPa (Hauck et al., 2013), as well as a linear increase in pressure with depth, pressure increases 1 GPa for each ~75 km (Chapter 4). With this updated silicate shell thickness, and similar thickness proportions for the layers to those of Brown and Elkins-Tanton (2009), we find that the gabbro and garnet-bearing wehrlite layers would be ~85 km thick, the harzburgite layer ~80 km thick, and the underlying dunite layer ~150 km thick.

Brown and Elkins-Tanton (2009) also showed that with higher modeled bulk FeO concentrations, an enrichment of FeO in the melt during solidification would have been produced. Ultimately, a magma of sufficient density would have been created so that plagioclase was buoyant and could form a flotation crust, akin to that of the Moon. In addition, even modest FeO concentrations (0.2 wt% FeO) could produce density contrasts that would have caused overturn in the solidified silicate shell (Brown and Elkins-Tanton, 2009). MESSENGER observations, however, indicate that Mercury's mantle has a very low FeO content (e.g., Murchie et al., 2015) and much (or all) of the observed Fe may be complexed with S (Zolotov et al., 2013; Weider et al., 2014). With such a low concentration of FeO in the crystallizing magma ocean, no flotation of any silicate phase could occur. Vander Kaaden and McCubbin (2015) subsequently examined the issue of mineral flotation for compositions that are more appropriate to Mercury. Their results indicate that graphite is the only phase that could have formed a flotation crust. This crust could have a thickness ranging from 1 m to 20 km, depending on the bulk C concentration of the magma ocean. The low total FeO of Mercury's mantle means that there are only minor density contrasts through the mantle, and gravitational overturn of the solidified magma ocean is thus unlikely to have occurred (Vander Kaaden and McCubbin, 2015). Vander Kaaden and McCubbin (2016) noted that the concentration of incompatible elements near the surface of Mercury could be a permanent geochemical feature of the planet, after the magma ocean phase. If a solidifying magma ocean produced layers broadly similar to those modeled for the low-FeO CB chondrite case (Brown and Elkins-Tanton, 2009), as well as the graphite flotation crust proposed by Vander Kaaden and McCubbin (2015), the distinct geochemical terranes might be evidence that the initially layered structure was sampled by impact excavation.

Charlier et al. (2013) addressed the possibility that Mercury retains a quenched crust comparable in composition to the bulk magma ocean. However, this notion was dismissed because the MESSENGER-derived chemical compositions available to them at the time did not resemble any of the posited bulk magma ocean compositions. This conclusion remains robust with the recognition of additional geochemical diversity on the surface of Mercury. In contrast, impact excavation of a buried graphite flotation crust (and subsequent deposition of C-rich material on the planet's surface) is consistent with MESSENGER observations. Modeling of reflectance spectra for Mercury's low-reflectance material (LRM; Denevi et al., 2009) suggests that the addition of ~5 wt% graphite could account for the darkness of LRM, and a graphite content of <1 wt% could yield the average reflectance spectrum of the northern hemisphere (Murchie et al., 2015). Low-altitude Neutron Spectrometer measurements over LRM deposits suggest that they have C concentrations that are 1–4 wt% higher than surrounding, higher-reflectance regions (Peplowski et al., 2016; Chapter 8). These measurements, along with the association of LRM and impact-excavated materials, strongly suggest that LRM contains at least a remnant of the postulated graphite flotation crust, possibly impact-mixed and gardened with later volcanic materials. In this aspect, some of the spectral and geochemical variation on the surface of Mercury may result directly from sampling original layers of the stratified solidification products of a cooling magma ocean.

Sampling the deeper-seated layers of a stratified mantle might also be possible through impact excavation or melting. Although the CIP were clearly formed through volcanism (Murchie et al., 2008; Chapter 11), impact melting and/or impact-induced volcanism may have occurred as a result of the Caloris impact (Roberts and Barnouin, 2012). The Caloris basin, with a rim-to-rim diameter of 1420 km, would have had a transient crater diameter of ~730 km, an excavation depth of ~73 km, and a melting depth of 220 km (Ernst et al., 2015). An impact-melt sheet 3–15 km thick underlies the smooth plains deposits in Caloris (Ernst et al., 2015). By scaling the Brown and Elkins-Tanton (2009) CB chondrite post-solidification magma ocean to a silicate shell thickness of 400 km, we estimate that excavation would have reached downward to near the boundary between the gabbro and garnet-bearing wehrlite layers, with melting well into the harzburgite layer. Impact melts produced directly from these layers should have been clinopyroxene-rich if the model of Brown and Elkins-Tanton (2009) is applicable to Mercury. The paucity of clinopyroxene in the CIP (Table 7.1) suggests that little of this impact melt is exposed at the surface and refutes the idea of sampling of magma ocean layers as the sole contributor to surface geochemical heterogeneity.

The High-Magnesium Terrane poses a more interesting possibility for impact sampling of the deepest layers of a stratified, solidified magma ocean. On the basis of its generally low elevations and thinner-than-average crust, Weider et al. (2015) suggested that this terrane may be an ancient giant impact basin. This postulated basin (defined by an ellipse of ~3700 × 3000 km; Weider et al., 2015) would be more than twice the diameter of Caloris (1420 km; Ernst et al., 2015). By scaling the relationships given by Ernst et al. (2015), the transient crater would thus be ~1500 km in diameter, with an excavation depth of ~150 km. The associated melting would have reached the base of Mercury's silicate shell, and the impact-melt sheet might have ranged from 5 to 30 km in thickness. As noted by Weider et al. (2015), the Mg-rich nature of this region may have resulted from removal of the upper, incompatible-element-enriched portions of the mantle. A signature of these incompatible-element-enriched layers remains in the High-Magnesium Terrane in the concentrations of Ca, Al, and S, as well as the corresponding modest-to-high abundances of sodic plagioclase and diopside. Despite retaining some signature of the incompatible-element-enriched portions of the mantle, the High-Magnesium Terrane is the most mafic of the terranes (Figure 7.3) and has an exceptionally high abundance of olivine (29.5 wt%) compared with the other terranes (≤7.5 wt%). It also has a very high liquidus temperature (~1540°C; Table 7.1).

Charlier et al. (2013) studied a Mg-rich composition, similar to that of the High-Magnesium Terrane, and argued that it was produced by partial melting of a lherzolitic source (i.e., similar to that formed at moderate depths in a solidifying magma ocean). Alternatively, the High-Magnesium Terrane may include a component of a mafic, olivine-rich impact melt formed from the basal layers of the magma ocean. In this model, the High-Magnesium Terrane would be a composite of impact-melted basal layers of the magma ocean and contributions from melting of both deeper-seated and shallower material. Frank et al. (2017) examined the possibility of a mantle contribution to the High-Magnesium Terrane. They argued that the extensive resurfacing needed to obscure a basin of this magnitude, and the lack of radial variation in composition (as expected of impact ejecta), is evidence against an impact origin for the High-Magnesium Terrane. With the exception of the heterogeneous distribution of carbon, the geochemical variation on the surface of Mercury is unlikely to result from sampling of a stratified magma ocean alone.

7.4.3 Explaining Crustal Heterogeneity

With the paucity of evidence for direct lateral or vertical sampling of a heterogeneous solidified magma ocean, the crust of Mercury may be largely comprised of mantle-derived melts. Indeed, geological observations provide compelling evidence that Mercury's crust has a largely volcanic origin (e.g., Head et al., 2011; Byrne et al., 2013; Denevi et al., 2013; Whitten et al., 2014; Chapter 11). Even without overturn of a gravitationally unstable mantle, melting of the mantle may have resulted from the decay of radioactive elements and/or adiabatic decompression melting driven by thermal convection. As noted above, MESSENGER data indicate that Mercury's silicate shell is about 400 km thick (Hauck et al., 2013; Chapters 4 and 19) and thus mantle convection is unlikely to be vigorous (because the dimensionless number governing the instability of a layer to convection varies as the cube of the layer thickness). Even with such a thin silicate layer, convection is likely to have continued to about 1.5–1 Ga, and may even continue today (Michel et al., 2013; Chapter 19). Michel et al. (2013) modeled the temperature of the mantle at 3.9 Ga – the approximate age of the CIP and the NSP – and found that 30–50% partial melting (depending on the assumed heat production from radioactive elements) of the mid-to-upper mantle would have been possible.

Three of the geochemical terranes (Northern, Low-Fast, CIP) are clearly volcanic in origin and formed at ~3.8–3.7 Ga as inferred from the cratering record (Chapter 9), times near the cessation of the late heavy bombardment (Head et al., 2011; Denevi et al., 2013). Despite having approximately the same age and common volcanic origin, these terranes exhibit substantial geochemical differences. These differences are especially noteworthy for Na, Al, Mg, and Ca abundances, which give rise to our derived differences in plagioclase abundance and the relative abundance of orthopyroxene and clinopyroxene. This geochemical diversity could result from different degrees of partial melting of a common mantle reservoir, from fractional crystallization of a common magma, or from melting of a vertically and/or laterally heterogeneous mantle.

The possibility of time- and temperature-dependent melting of a common mantle reservoir was suggested by Weider et al. (2012). On the basis of the lower Mg concentration of the NSP, they suggested that its source was more evolved (i.e., Mg-depleted) and cooler than that which produced the older, more cratered regions of Mercury. A cooling mantle that produced less magnesian and cooler melts with time is broadly consistent with the composition and liquidus temperatures of the Northern and CIP terranes (Table 7.1). Although Weider et al. (2012) rejected the idea of melting of a common reservoir in the mantle, they argued that the more heavily cratered regions of Mercury's surface derived from relatively early and high-temperature melts, at high degrees of melting. They also proposed that the lower-degree partial melts of the NSP (and CIP) emanated from a separate mantle source, at a later period and cooler temperatures. Weider et al. (2015), however, later rejected this simple time- and temperature-dependent melting model because the expanded XRS data set showed that Mercury's oldest regions span the planet's entire dynamic range of Mg/Si, and that there is substantial variation in Mg/Si for the younger regions. Weider et al. (2015) instead suggested that lateral and vertical heterogeneity in the mantle might better account for the observed compositional variations.

Fractional crystallization of a common magma can also produce variations in composition of volcanic deposits. In a set of experiments, Charlier et al. (2013) examined Na-free compositions that are similar to our southern hemisphere and High-Magnesium Terrane compositions. Charlier et al. (2013) found that the presence or absence of clinopyroxene in their two compositions leads to contrasting differentiation paths and precludes their being related through differentiation accompanying cooling and crystallization, at either high or low pressure. The relative, experimentally derived liquidus temperatures of the two compositions further preclude fractional melting as an explanation for the different components (the clinopyroxene-bearing component should have a lower temperature than the clinopyroxene-free component, but this outcome is not seen in the experimental data). Namur et al. (2016) conducted an updated investigation by considering the geochemical terranes of Weider et al. (2015). These authors reached the same conclusions as Charlier et al. (2013), but they considered the High-Magnesium Terrane as one composition and all the smooth plains regions (i.e., our CIP, Northern, and Low-Fast terranes) as a distinct composition.

Charlier et al. (2013) proposed that mineralogically distinct source regions are required to produce Mercury's different surface compositions, i.e., a clinopyroxene-bearing lherzolite source was required for the high-Mg compositions and a clinopyroxene-free harzburgite source was necessary for the low-Mg regions. Partial melting of the stratified products of a magma ocean, at pressures of <1 GPa, was invoked as a possible origin of the different source regions and surface compositions. Namur et al. (2016) considered the effect of Na and S on the pressure and temperature at which the forsterite–enstatite–melt multiple saturation point (MSP) occurs. They concluded that the smooth plains regions – including our Northern, Low-Fast, and CIP terranes – required a harzburgitic source at shallow depth and a lherzolitic source at greater depth (>1.5 GPa) for the High-Magnesium Terrane. Vander Kaaden and McCubbin (2015) reached a similar conclusion regarding the origin of the NSP. They found that the most likely scenario for the formation of this terrane was high-degree partial melting at shallow depth (<1.4 GPa), leaving only olivine in the source. They were unable, however, to rule out lower-degree partial melting at the olivine – orthopyroxene–melt MSP. The suggestion (Namur et al., 2016) of shallower harzburgite is in contrast to the model of Brown and Elkins-Tanton (2009), in which the wehrlite (lherzolite) is found at shallower depths and the harzburgite is at greater depth. In both cases, pressures for the MSP range from 1 to 2 GPa (which corresponds to a depth of ~80–150 km), consistent with the depth of the wehrlite layer (~85–170 km) in the Brown and Elkins-Tanton (2009) model, and slightly above the harzburgite layer. Vander Kaaden and McCubbin (2015) also noted that there is an inconsistency in requiring an olivine-dominated source at shallow depth when magma ocean crystallization models suggest a pyroxene-rich upper mantle. A possible resolution to this mismatch is convective motion of the mantle prior to melting. Such motions could have brought the hotter harzburgite layer to shallower depth, with associated adiabatic decompression melting.

7.5 SUMMARY

Despite considerable uncertainties, we can construct a notional model for the evolution of Mercury's mantle and thus offer a notional explanation for the origin of the geochemical and mineralogical diversity (i.e., geochemical terranes) at the surface of the planet. Our proposed model (Figure 7.5) may be summarized as follows:

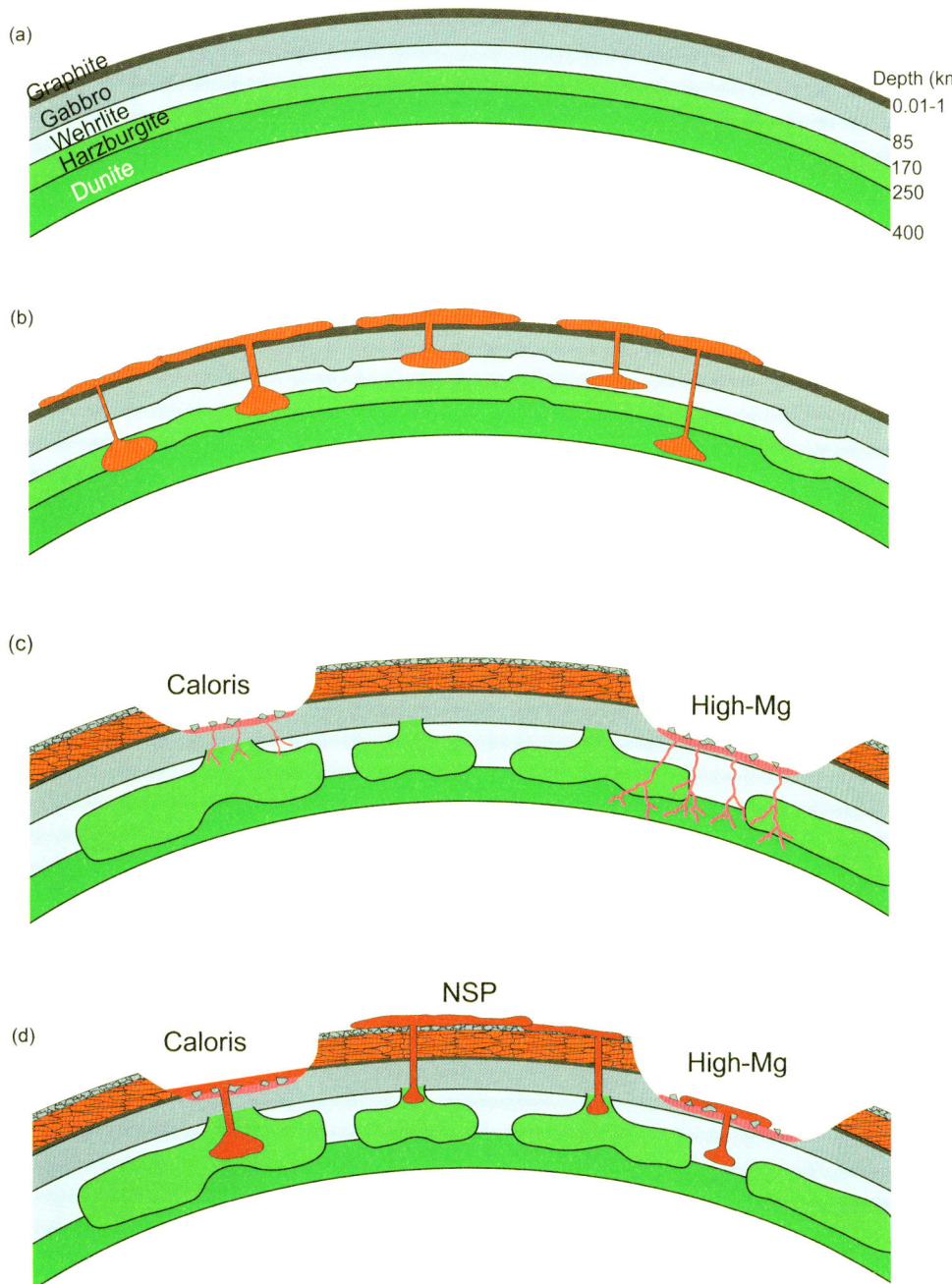

Figure 7.5. Notional view of the evolution of Mercury's mantle and crust. (a) A global magma ocean crystallizes basal ultramafic materials with a graphite flotation crust. Mineralogy and thickness of layers from the Brown and Elkins-Tanton (2009) model for melting of a CB chondrite, with low FeO, adapted to a 400-km-thick silicate shell. (b) Partial melting occurs throughout the mantle, forming the crust. Melts are buoyant throughout the mantle, and eruptive products form thin, laterally extensive volcanic deposits. (c) Basin-scale impacts excavate through the crust to the upper mantle, forming mafic to ultramafic impact melt sheets and depositing crustal and upper mantle materials on the crust, including graphite. (d) Partial melting of a convecting, chemically heterogeneous mantle forms smooth plains deposits in Caloris and the northern smooth plains and partially covers earlier deposits in the High-Magnesium Terrane.

(1) Accretional heating, combined with heat released during core formation, melted the silicate shell of Mercury and produced a global magma ocean ~400 km thick. Within this magma ocean, stratification developed during crystallization, with ultramafic material at the base grading to increasingly incompatible-element-enriched material near the surface. The only mineral sufficiently buoyant to form a flotation crust was graphite. With an exceptionally low FeO concentration, the solidified magma ocean likely did not experience any overturn (Figure 7.5a).

(2) Following magma ocean solidification, a second stage of differentiation began with the formation of partial melts (the result of adiabatic decompression melting during mantle convection and/or heating from the decay of radioactive elements). These high-Mg, high-temperature partial melts were exceptionally fluid and on eruption produced thin, laterally extensive units. This stage marked the onset of the formation of the crust as a layer distinct from the underlying mantle (i.e., formed during magma ocean solidification). The nominally gabbroic nature of the uppermost mantle layers, however, may diminish the chemical distinction between the upper mantle and crust. This boundary might be best marked by the occurrence of the graphite flotation crust. Owing to the low density of the melts and the thin mantle, melts from all mantle depths were buoyant and able to rise to the surface (Vander Kaaden and McCubbin, 2015). Given the incompatible-element-enriched nature of the average surface (i.e., southern hemisphere, Table 7.1), however, it seems likely that most of the crust originated as partial melts from the upper mantle, i.e., from late-stage crystallization products of the cooling magma ocean, rather than from the ultramafic basal layer of the mantle (Figure 7.5b).

(3) Although impacts occurred throughout the period of crustal formation, the largest (i.e., basin-forming) impacts may have greatly influenced the geochemical composition of the surface we observe today. The Caloris basin (and the postulated basin in the High-Magnesium Terrane) excavated through the crust and well into the upper layers of the mantle. Impact-melt sheets at the bottom of the postulated basin in the High-Magnesium Terrane may have sampled materials from the deepest layers of the mantle. By the impact-basin scenario for the origin of this terrane, therefore, it exhibits high-Mg concentrations and associated high abundances of ultramafic minerals (e.g., olivine and orthopyroxene). Impact excavations might also have deposited material from the upper mantle, including the graphite-rich flotation crust, as ejecta and impact melt at the top of the crust. Given its large size and lack of obvious geomorphic evidence for extensive flood basalts, the High-Magnesium Terrane may offer the best-preserved samples of these impact-derived products (Figure 7.5c).

(4) In contrast, smooth plains deposits in the CIP and the NSP originated from high-temperature and fluid lava flows. These lavas created laterally extensive flood basalts that efficiently covered any pre-existing layers. Post-impact volcanism of smooth plains deposits in Caloris and the NSP appears to be related to partial melting of a harzburgite layer in the mantle, perhaps as a result of convection-driven adiabatic decompression melting. Volcanism in the High-Magnesium Terrane was likely the result of partial melting of a lherzolite layer in the mantle. The existence of these two distinct source compositions suggests that stirring by mantle convection was insufficient to have homogenized the mantle by ~3.8–3.5 Ga when the CIP deposits formed (Chapter 6). These erupting melts may have also assimilated upper mantle and crustal materials (e.g., the graphite-rich flotation crust), leading to crystallization of graphite in the erupting melts and darkening of the surface.

REFERENCES

Anand, M., Taylor, L. A., Nazarov, M. A., Shu, J., Mao, H.-K. and Hemley, R. (2004). Space weathering on airless planetary bodies: Clues from the lunar mineral hapkeite. *Proc. Natl. Acad. Sci.*, **101**, 6847–6851.

Basaltic Volcanism Study Project (1981). *Basaltic Volcanism on the Terrestrial Planets*. New York: Pergamon, 1286 pp.

Benkhoff, J., van Casteren, J., Hayakawa, H., Fujimoto, M., Laaksso, H., Novara, M., Ferri, P., Middleton, H. R. and Ziethe, R. (2010). BepiColombo – Comprehensive exploration of Mercury: Mission overview and science goals. *Planet. Space Sci.*, **58**, 2–20.

Berthet, S., Malavergne, V. and Righter, K. (2009). Melting of the Indarch meteorite (EH4 chondrite) at 1 GPa and variable oxygen fugacity: Implications for early planetary differentiation processes. *Geochim. Cosmochim. Acta*, **73**, 6402–6420.

Brown, S. M. and Elkins-Tanton, L. T. (2009). Compositions of Mercury's earliest crust from magma ocean models. *Earth Planet. Sci. Lett.*, **286**, 446–455.

Burbine, T. H., McCoy, T. J., Nittler, L. R., Benedix, G. K., Cloutis, E. A. and Dickinson, T. L. (2002). Spectra of extremely reduced assemblages: Implications for Mercury. *Meteorit. Planet. Sci.*, **37**, 1233–1244.

Byrne, P. K., Klimczak, C., Williams, D. A., Hurwitz, D. M., Solomon, S. C., Head, J. W., Preusker, F. and Oberst, J. (2013). An assemblage of lava flow features on Mercury. *J. Geophys. Res. Planets*, **118**, 1303–1322, doi:10.1002/jgre.20052.

Charlier, B., Grove, T. L. and Zuber, M. T. (2013). Phase equilibria of ultramafic compositions on Mercury and the origin of the compositional dichotomy. *Earth. Planet. Sci. Lett.*, **363**, 50–60.

Cross, W., Iddings, J. P., Pirsson, L. V. and Washington, H. S. (1903). *Quantitative Classification of Igneous Rocks*. Chicago, IL: University of Chicago Press, 315 pp.

Davis, P. A. and Spudis, P. D. (1985). Petrologic province map of the lunar highlands derived from orbital geochemical data. *Proc. 16th Lunar Planet. Sci. Conf., J. Geophys. Res.*, **90**, D61–D74.

Denevi, B. W., Robinson, M. S., Solomon, S. C., Murchie, S. L., Blewett, D. T., Domingue, D. L., McCoy, T. J., Ernst, C. M., Head, J. W., Watters, T. R. and Chabot, N. L. (2009). The evolution of Mercury's crust: A global perspective from MESSENGER. *Science*, **324**, 613–618.

Denevi, B. W., Ernst, C. M., Meyer, H. M., Robinson, M. S., Murchie, S. L., Whitten, J. L., Head, J. W., Watters, T. R., Solomon, S. C., Ostrach, L. R., Chapman, C. R., Byrne, P. K., Klimczak, C. and Peplowski, P. N. (2013). The distribution and origin of smooth plains on Mercury. *Icarus*, **118**, 891–907.

Ernst, C. M., Denevi, B. W., Barnouin, O. S., Klimczak, C., Chabot, N. L., Head, J. W., Murchie, S. L., Neumann, G. A., Prockter, L. M., Robinson, M. S., Solomon, S. C. and Watters, T. R. (2015).

Stratigraphy of the Caloris basin, Mercury: Implications for volcanic history and basin impact melt. *Icarus*, **250**, 413–429.

Evans, L. G., Peplowski, P. N., Rhodes, E. A., Lawrence, D. J., McCoy, T. J., Nittler, L. R., Solomon, S. C., Sprague, A. L., Stockstill-Cahill, K. R., Starr, R. D., Weider, S. Z., Boynton, W. V., Hamara, D. K. and Goldsten, J. O. (2012). Major-element abundances on the surface of Mercury: Results from the MESSENGER Gamma-Ray Spectrometer. *J. Geophys. Res.*, **117**, E00L07, doi:10.1029/2012JE004178.

Evans, L. G., Peplowski, P. N., McCubbin, F. M., McCoy, T. J., Nittler, L. R., Zolotov, M. Yu., Ebel, D. S., Lawrence, D. J., Starr, R. D., Weider, S. Z. and Solomon, S. C. (2015). Chlorine on the surface of Mercury: MESSENGER gamma-ray measurements and implications for the planet's formation and evolution. *Icarus*, **257**, 417–427, doi:10.1016/j.icarus.2015.04.039.

Frank, E. A., Potter, R. W. K., Abramov, O., James, P. B., Klima, R. L., Mojzsis, S. J. and Nittler, L. R. (2017). Evaluating an impact origin for Mercury's high-magnesium region. *J. Geophys. Res. Planets*, **122**, 614–632, doi:10.1002/2016JE005244.

Gasnault, O., d'Uston, C., Feldman, W. C. and Maurice, S. (2000). Lunar fast neutron leakage flux calculation and its elemental abundance dependence. *J. Geophys. Res.*, **105**, 4263–4263.

Gualda, G. A. R. and Ghiorso, M. S. (2015). MELTS_Excel: A Microsoft Excel-based MELTS interface for research and teaching of magma properties and evolution. *Geochem. Geophys. Geosyst.*, **16**, 315–324.

Hauck, S. A., II, Margot, J.-L., Solomon, S. C., Phillips, R. J., Johnson, C. L., Lemoine, F. G., Mazarico, E., McCoy, T. J., Padovan, S., Peale, S. J., Perry, M. E., Smith, D. E. and Zuber, M. T. (2013). The curious case of Mercury's internal structure. *J. Geophys. Res. Planets*, **118**, 1204–1220, doi:10.1002/jgre.20091.

Head, J. W., Chapman, C. R., Strom, R. G., Fassett, C. I., Denevi, B. W, Blewett, D. T., Ernst, C. M., Watters, T. R., Solomon, S. C., Murchie, S. L., Prockter, L. M., Chabot, N. L., Gillis-Davis, J. J., Whitten, J. L., Goudge, T. A., Baker, D. M. H., Hurwitz, D. M., Ostrach, L. R., Xiao, Z., Merline, W. J., Kerber, L., Dickson, J. L., Oberst, J., Byrne, P. K., Klimczak, C. and Nittler L. R. (2011). Flood volcanism in the northern high latitudes of Mercury revealed by MESSENGER. *Science*, **333**, 1853–1856.

Jolliff, B. L., Gillis, J. J., Haskin, L. A., Korotev, R. L. and Wieczorek, M. A. (2000). Major lunar crustal terranes: Surface expressions and crust–mantle origins. *J. Geophys. Res.*, **105**, 4197–4216.

Keil, K. (1989). Enstatite meteorites and their parent bodies. *Meteoritics*, **24**, 195–208.

Lawrence, D. J., Feldman, W. C., Goldsten, J. O., McCoy, T. J., Blewett, D. T., Boynton, W. V., Evans, L. G., Nittler, L. R., Rhodes, E. A. and Solomon, S. C. (2010). Identification and measurement of neutron-absorbing elements on Mercury's surface. *Icarus*, **209**, 195–209.

Lawrence, D. J., Peplowski, P. N., Beck, A. W., Feldman, W. C., Frank, E. A., McCoy, T. J., Nittler L. R. and Solomon S. C. (2017). Compositional terranes on Mercury: New information from fast neutrons. *Icarus*, **281**, 32–45.

Le Bas, M. J. (2000). IUGS reclassification of the high-Mg and picritic volcanic rocks. *J. Petrol.*, **41**, 1467–1470.

Le Bas, M. J. and Streckeisen, A. L. (1991). The IUGS systematics of igneous rocks. *J. Geol. Soc. London*, **148**, 825–833.

Lodders, L. and Fegley, B. (1998). *The Planetary Scientist's Companion*. New York: Oxford University Press.

McCoy, T. J., Dickinson, T. L. and Lofgren, G. E. (1999). Partial melting of the Indarch (EH4) meteorite: A textural, chemical, and phase relations view of melting and melt migration. *Meteorit. Planet. Sci.*, **34**, 735–746.

McCubbin, F. M., Riner, M. A., Vander Kaaden, K. E. and Burkemper, L. K. (2012). Is Mercury a volatile-rich planet? *Geophys. Res. Lett.*, **39**, L09202, doi:10.1029/2012GL051711.

Michel, N. C., Hauck, S. A., II, Solomon, S. C., Phillips, R. J., Roberts, J. H. and Zuber, M. T. (2013). Thermal evolution of Mercury as constrained by MESSENGER observations. *J. Geophys. Res. Planets*, **118**, 1033–1044.

Murchie, S. L., Watters, T. R., Robinson, M. S., Head, J. W., Strom, R. G., Chapman, C. R., Solomon, S. C., McClintock, W. E., Prockter, L. M., Domingue, D. L. and Blewett, D. T. (2008). Geology of the Caloris basin, Mercury: A view from MESSENGER. *Science*, **321**, 73–76.

Murchie, S. L., Klima, R. L., Denevi, B. W., Ernst, C. M., Keller, M. R., Domingue, D. L., Blewett, D. T., Chabot, N. L., Hash, C. D., Malaret, E., Izenberg, N. R., Vilas, F., Nittler, L. R., Gillis-Davis, J. J., Head, J. W. and Solomon, S. C. (2015). Orbital multispectral mapping of Mercury with the MESSENGER Mercury Dual Imagining System: Evidence for the origins of plains units and low-reflectance material. *Icarus*, **254**, 287–305.

Namur, O., Collinet, M., Charlier, B., Grove, T. L., Holt, F. and McCammon, C. (2016). Melting processes and mantle sources of lavas on Mercury. *Earth Planet. Sci. Lett.*, **439**, 117–128.

Nittler, L. R., Starr, R. D., Weider, S. Z., McCoy, T. J., Boynton, W. V, Ebel, D. S., Ernst, C. M., Evans, L. G., Goldsten, J. O., Hamara, D. K., Lawrence, D. J., McNutt, R. L., Jr., Schlemm, C. E., Solomon, S. C. and Sprague, A. L. (2011). The major-element composition of Mercury's surface from MESSENGER X-ray spectrometry. *Science*, **333**, 1847–1850, doi:10.1126/science.1211567.

Ostrach, L. R., Robinson, M. S., Whitten, J. L., Fassett, C. I., Strom, R. G., Head, J. W. and Solomon, S. C. (2015). Extent, age, and resurfacing history of the northern smooth plains on Mercury from MESSENGER observations. *Icarus*, **250**, 602–622.

Papike, J. J., Ryder, G. and Shearer, C. K. (1998). Lunar samples. In *Planetary Materials*, ed. J. J. Papike. *Reviews in Mineralogy*, **36**. Washington, DC: Mineralogical Society of America, pp. 5-1–5-234.

Peplowski, P. N., Evans, L. G., Hauck, S. A., II, McCoy, T. J., Boynton, W. V., Gillis-Davis, J. J., Ebel, D. S., Goldsten, J. O., Hamara, D. K., Lawrence, D. J., McNutt, R. L., Jr., Nittler, L. R., Solomon, S. C., Rhodes, E. A., Sprague, A. L., Starr, R. D. and Stockstill-Cahill, K. R. (2011). Radioactive elements on Mercury's surface from MESSENGER: Implications for the planet's formation and evolution. *Science*, **333**, 1850–1852.

Peplowski, P. N., Lawrence, D. J., Rhodes, E. A., Sprague, A. L., McCoy, T. J., Denevi, B. W., Evans, L. G., Head, J. W., Nittler, L. R., Solomon, S. C., Stockstill-Cahill, K. R. and Weider, S. Z. (2012). Variations in the abundances of potassium and thorium on the surface of Mercury: Results from the MESSENGER Gamma-Ray Spectrometer. *J. Geophys. Res.*, **117**, E00L04, doi:10.1029/2012JE004141.

Peplowski, P. N., Evans, L. G., Stockstill-Cahill, K. R., Lawrence, D. J., Goldsten, J. O., McCoy, T. J., Nittler, L. R., Solomon, S. C., Sprague, A. L., Starr, R. D. and Weider, S. Z. (2014). Enhanced sodium abundance in Mercury's north polar region revealed by the MESSENGER Gamma-Ray Spectrometer. *Icarus*, **228**, 86–95.

Peplowski, P. N., Lawrence, D. J., Feldman, W. C., Goldsten, J. O., Bazell, D., Evans, L. G., Head, J. W., Nittler, L. R., Solomon, S. C. and Weider, S. Z. (2015). Geochemical terranes of Mercury's northern hemisphere as revealed by MESSENGER neutron measurements. *Icarus*, **253**, 346–363.

Peplowski, P. N., Klima, R. L., Lawrence, D. J., Ernst, C. M., Denevi, B. W., Goldsten, J. O., Murchie, S. L., Nittler, L. R. and Solomon, S. C. (2016). Remote sensing evidence for an ancient carbon-bearing crust on Mercury. *Nature Geosci.*, **9**, 273–276.

Riner, M. A., Lucey, P. G., Desch, S. J. and McCubbin, F. M. (2009). Nature of opaque components on Mercury: Insights into a

Mercurian magma ocean. *Geophys. Res. Lett.*, **36**, L02201, doi:10.1029/2008GL036128.

Roberts, J. H. and Barnouin, O. S. (2012). The effect of the Caloris impact on the mantle dynamics and volcanism of Mercury. *J. Geophys. Res.*, **117**, E02007, doi:10.1029/2011JE003876.

Schubert, G., Ross, M. N., Stevenson, D. J. and Spohn, T. (1988). Mercury's thermal history and the generation of its magnetic field. In *Mercury*, ed. F. Vilas, C. R. Chapman and M. S. Matthews. Tucson, AZ: University of Arizona Press, pp. 622–650.

Shaw, H. R. (1972). Viscosities of magmatic silicate liquids: An empirical method of prediction. *Amer. J. Sci.*, **272**, 870–893.

Solomon, S. C., McNutt, R. L., Jr., Watters, T. R., Lawrence, D. J., Feldman, W. C., Head, J. W., Krimigis, S. M., Murchie, S. L., Phillips, R. J., Slavin, J. A. and Zuber, M. T. (2008). Return to Mercury: A global perspective on MESSENGER's first Mercury flyby. *Science*, **321**, 59–62, doi:10.1126/science.1159706.

Sprague, A. L., Donaldson Hanna, K. L., Kozlowski, R. W. H., Helbert, J., Maturilli, A., Warell, J. B. and Hora J. L. (2009). Spectral emissivity measurements of Mercury's surface indicate Mg- and Ca-rich mineralogy, K-spar, Na-rich plagioclase, rutile, with possible perovskite, and garnet. *Planet. Space Sci.*, **57**, 364–383.

Stockstill-Cahill, K. R., McCoy, T. J., Nittler, L. R., Weider, S. Z. and Hauck, S. A., II (2012). Magnesium-rich crustal compositions on Mercury: Implications for magmatism from petrologic modeling. *J. Geophys. Res.*, **117**, E00L15, doi:10.1029/2012JE004140.

Taylor, G. J., Boynton, W., Brückner, J., Wänke, H., Dreibus, G., Kerry, K., Keller, J., Reedy, R., Evans, L., Starr, R., Squyres, S., Karunatillake, S., Gasnault, O., Maurice, S., d'Uston, C., Englert, P., Dohm, J., Baker, V., Hamara, D., James, D., Sprague, A., Kim, K. and Drake, D. (2006). Bulk composition and early differentiation of Mars. *J. Geophys. Res.*, **111**, E03S10, doi:10.1029/2005JE002645.

Vander Kaaden, K. E. and McCubbin, F. M. (2015). Exotic crust formation on Mercury: Cosequences of a shallow, FeO-poor mantle. *J. Geophys. Res. Planets*, **120**, 195–209.

Vander Kaaden, K. E. and McCubbin, F. M. (2016). The origin of boninites on Mercury: An experimental study of the northern volcanic plains lavas. *Geochim. Cosmochim. Acta*, **173**, 246–263.

Vander Kaaden, K. E., McCubbin, F. M., Nittler, L. R., Peplowksi, P. N., Weider S. Z., Frank, E. A. and McCoy, T. J. (2017). Geochemistry, mineralogy, and petrology of boninitic and komatiitic rocks on the mercurian surface: Insights into the mercurian mantle. *Icarus*, **285**, 155–168, doi:10.1016/j.icarus.2016.11.041.

Wasson, J. T. (1988), The building stones of the planets. In *Mercury*, ed. F. Vilas, C. R. Chapman and M. S. Matthews. Tucson, AZ: University of Arizona Press, pp. 622–650.

Weider, S. Z., Nittler, L. R., Starr, R. D., McCoy, T. J., Stockstill-Cahill, K. R., Byrne, P. K., Denevi, B. W., Head, J. W. and Solomon, S. C. (2012). Chemical heterogeneity on Mercury's surface revealed by the MESSENGER X-Ray Spectrometer. *J. Geophys. Res.*, **117**, E00L05, doi:10.1029/2012JE004153.

Weider, S. Z., Nittler, L. R., Starr, R. D., McCoy, T. J. and Solomon, S. C. (2014). Variations in the abundance of iron on Mercury's surface from MESSENGER X-Ray Spectrometer observations. *Icarus*, **235**, 170–186.

Weider, S. Z., Nittler, L. R., Starr, R. D., Crapster-Pregont, E. J., Peplowski, P. N., Denevi, B. W., Head, J. W., Byrne, P. K., Hauck, S. A., II, Ebel, D. S. and Solomon, S. C. (2015). Evidence for geochemical terranes on Mercury: Global mapping of major elements with MESSENGER's X-Ray Spectrometer. *Earth Planet. Sci. Lett.*, **416**, 109–120, doi:10.1016/j.epsl.2015.01.023.

Weider, S. Z., Nittler, L. R., Murchie, S. L., Peplowski, P. N., Ernst, C. M., McCoy, T. J., Goudge, T. A., Kerber, L., Starr, R. D., Izenberg, N. R., Klima, R. L., Head, J. W., Denevi, B. W. and Solomon, S. C. (2016). Evidence from MESSENGER for sulfur- and carbon-driven explosive volcanism on Mercury. *Geophys. Res. Lett.*, **43**, 3653–3661.

Whitten, J. L., Head, J. W., Denevi, B. W. and Solomon, S. C. (2014). Intercrater plains on Mercury: Insight into unit definition, characterization, and origin from MESSENGER datasets. *Icarus*, **241**, 97–113.

Zolotov, M. Yu., Sprague, A. L., Hauck, S. A., II, Nittler, L. R., Solomon, S. C. and Weider, S. Z. (2013). The redox state, FeO content, and origin of sulfur-rich magmas on Mercury. *J. Geophys. Res. Planets*, **118**, 138–146.

8

Spectral Reflectance Constraints on the Composition and Evolution of Mercury's Surface

SCOTT L. MURCHIE, RACHEL L. KLIMA, NOAM R. IZENBERG, DEBORAH L. DOMINGUE,
DAVID T. BLEWETT, AND JÖRN HELBERT

8.1 INTRODUCTION

Spectral reflectance has long been a key tool for investigating the composition and geology of Mercury. In the 1970s, the imaging system on Mariner 10 provided the first moderate-resolution views of the surface, revealing the heavily cratered plains and color differences associated with different geologic units. Over the next four decades, ground-based measurements revealed a featureless spectrum lacking evidence for absorptions due to iron in silicate minerals. The most comprehensive measurements, covering ultraviolet (UV) through short-wave infrared (SWIR) wavelengths, were acquired by the multispectral Mercury Dual Imaging System (MDIS) wide-angle camera (WAC) (Hawkins et al., 2007, 2009) and the Mercury Atmospheric and Surface Composition Spectrometer (MASCS) Visible and Infrared Spectrograph (VIRS) and Ultraviolet and Visible Spectrometer (UVVS) (McClintock and Lankton, 2007) on the MESSENGER spacecraft (Solomon et al., 2001). In this chapter, we review and synthesize key findings from analyses of the MDIS and MASCS spectral measurements of Mercury's surface, in the context of results from Earth-based telescopic studies and from the MESSENGER Gamma-Ray and Neutron Spectrometer (GRNS; Goldsten et al., 2007) and X-Ray Spectrometer (XRS) (Schlemm et al., 2007). Key questions we focus upon include: (a) How does the reflectance of Mercury compare with that of the Moon, and what does this comparison indicate about Mercury's composition? (b) Which photometric models perform best at mosaicking MDIS and MASCS/VIRS data, and what do photometric model parameters indicate about regolith properties? (c) What do spatial variations in photometric properties indicate about variations among selected geologic formations? (d) What are the major spectral heterogeneities on Mercury, and which underlying sources of spectral variability are responsible for them? (e) How do variations in spectral reflectance among geologic units constrain the composition, structure, and history of Mercury's crust? (f) How do spectral reflectance variations constrain geologic processes, including space weathering and pyroclastic volcanism?

Compared with the surfaces of other differentiated silicate bodies, Mercury's surface materials lack the 1-µm crystal-field absorption due to ferrous iron in silicate yet are unusually low in reflectance. Spectral modeling suggests that the most likely darkening phase is carbon as graphite, and its occurrence is confirmed by data acquired by MESSENGER's Neutron Spectrometer. Control of reflectance by this relatively minor opaque phase, rather than by the abundance of iron in silicates as on the Moon, prevents the correlation of spectral reflectance and major element composition that is observed on the Moon. Variations in reflectance and color nevertheless serve as markers for the structure of the upper crust, revealing that at least 5 km of volcanic plains overlie low-reflectance material (LRM). The one definitive absorption due to an oxidized transition metal, an oxygen-metal charge transfer (OMCT) band in the UV observed in bright, pyroclastic deposits, may originate by oxidation of the darkening carbon as well as sulfide phases, reducing enough iron from the ferrous to metallic state to unsaturate the OMCT band. The implied content of ferrous iron in crustal silicates is bounded by the presence of this feature and the lack of a 1-µm feature to be between 0.1 and 1 wt%.

8.2 WHAT CAN SPECTRAL REFLECTANCE REVEAL ABOUT MERCURY?

From mid-UV through SWIR wavelengths (0.2–2.5 µm), reflection of sunlight by a silicate planetary regolith is controlled by two major factors: the partitioning of radiation that interacts with the surface between absorbance and reflectance, and the scattering of reflected light in different directions. The behavior of each factor provides information on the composition and texture of the surface.

Key variables that affect the balance of absorbance and reflectance include the amount and oxidation state of transition metals, principally iron, in silicates and oxides; the presence of opaque components, including ilmenite and carbon-bearing phases; and the degree of space weathering, that is, the modification of the optical properties of the surface by sustained interaction with the space environment. Ferrous iron in minerals creates crystal-field absorptions near 0.85–1.05-µm wavelength that are particularly important in distinguishing key Fe-bearing phases, including oxides, olivine, pyroxene, and glasses. The presence of certain anions causes additional vibrational absorptions, for example, an absorption from OH near 1.4 µm. Opaque minerals typically darken and flatten the reflectance spectrum of a mixture in which they occur. Space weathering occurs on the Moon and near-Earth asteroids and is widely thought to occur on Mercury, by the formation of sub-microscopic amorphous rims on regolith grains, which contain nanometer-scale inclusions of Fe or FeS formed from impact-generated vapor and solar wind sputtering (e.g., Hapke et al., 1975; Gaffey et al.,

1993; Pieters, 1993; Hapke, 2001; Domingue et al., 2014). Space weathering on those bodies has the effect of darkening and reddening (i.e., steepening the slope of reflectance versus wavelength of) silicate spectra at visible to SWIR wavelengths (McCord and Adams, 1972a, b; Fischer and Pieters, 1994) and brightening and bluing (i.e., reducing the slope of) the spectrum in the UV (Hendrix and Vilas, 2006; Hendrix et al., 2012). Both opaque minerals and space weathering also tend to subdue mineralogic absorptions.

The manner in which a surface scatters reflected light in different directions is known as the surface's photometric behavior. Three key angles define the scattering geometry: incidence angle (between incoming sunlight and the surface normal), emergence angle (between outgoing scattered light and the surface normal), and phase angle (between incoming and outgoing rays). The first widely used models of directional scattering, developed by Hapke (1981, 1984, 1986, 2002), predict effects of particle shapes and transparency, packing, and roughness at the scale of grains or larger. In general, as incidence and phase angles increase, reflectance decreases because of the reduction of insolation per unit of surface area and because of the appearance of shadows cast by surface roughness elements. At small phase angles, $<\sim 5°–20°$, reflectance spikes within an "opposition surge," due to disappearance of shadows from the perspective of the observer and the appearance of coherent backscattering by relatively transparent particles (e.g., Shkuratov et al., 1999; Hapke, 2002). However, laboratory studies have shown that relating the magnitude of directional scattering to fundamental physical properties of regolith grains remains highly uncertain (Shepard and Helfenstein, 2007; Helfenstein and Shepard, 2011; Shkuratov et al., 2012a). Models other than those of Hapke, for example those of Kaasalainen and Shkuratov (Kaasalainen et al., 2001; Shkuratov et al., 2011; Schröder et al., 2013), are inherently empirical but may more accurately predict directional scattering under certain scattering geometries (Domingue et al., 2016).

Whichever modeling approach is adopted, two surface attributes derived from photometric modeling are nevertheless informative about regolith composition and texture: albedo, and the relative magnitudes of scattering at forward (away from the Sun) and backward (back toward the Sun) geometries. Albedo differs from reflectance in that it is referenced to a predefined geometry rather than the one measured, such as reflectance where all three photometric angles have a value of 0° (normal albedo). Other definitions of albedo were reviewed by Hapke (1993). Normal albedo includes effects of the opposition surge, a scattering geometry that is not observed in many Mercury data sets including those collected by MESSENGER. Instead, this chapter mostly references "standardized reflectance," which is reflectance extrapolated to the common laboratory measurement geometry of 0° incidence angle, 30° emergence angle, and 30° phase angle, outside the opposition surge. Differences in standardized reflectance result from the same compositional factors that influence albedo, especially the presence of opaque phases, absorption features, and sub-microscopic iron. In contrast, the forward- or backward-scattering nature of the surface is thought to arise from grain-scale transparency and structure: grains that are transparent with relatively little internal scattering are dominantly forward-scattering, whereas opaque grains or grains with relatively abundant internal scatterers are more backward-scattering (Domingue et al., 2016).

8.2.1 Mercury in the Context of the Moon

Understanding what Mercury's spectral reflectance can reveal about surface composition benefits from consideration within the context of spectral reflectance properties of other silicate planetary surfaces. Prior to MESSENGER, the Moon had widely been assumed to foreshadow Mercury's properties. On the Moon, there are three primary causes for variations in spectral reflectance: (a) the modal abundance of iron-containing pyroxene, which varies between the basaltic maria and anorthositic highlands, resulting in a lower standardized reflectance and stronger 1-μm absorption in the maria (Adams and McCord, 1970; Pieters, 1993); (b) variations in UV to visible standardized reflectance and spectral slope due to differences in the abundance of the opaque mineral ilmenite ($FeTiO_3$) (Charette et al., 1974; Lucey et al., 1995, 1998; Blewett et al., 1997a); and (c) "optical maturation" (darkening and reddening in the visible to SWIR) of regolith by space weathering (e.g., Fischer and Pieters, 1994; Lucey et al., 1995, 1998, 2000a, b; Blewett et al., 1997a; Hapke, 2001). In hindsight, the lunar model underestimates the role of opaque phases and space weathering on Mercury, and overestimates the effects of the major element composition.

8.2.2 Mercury's Spectral Reflectance as Known Before MESSENGER

A number of Earth-based telescopic observers prior to MESSENGER investigated Mercury's UV-SWIR spectral reflectance to assess mineralogy and the iron content of surface material. The spectrum is red-sloped with no 1-μm iron crystal-field absorption detectable above the level of noise in the data (e.g., Vilas and McCord, 1976; McCord and Clark, 1979; Vilas et al., 1984). In contrast, on the Moon, despite the action of space weathering, a 1-μm absorption is preserved in both mature mare and most mature highland materials. Most spectral studies of Mercury concluded that the lack of a 1-μm feature indicates that the amount of Fe in surface silicates is <2–5 wt% (McCord and Adams, 1972a; Hapke, 1977; McCord and Clark, 1979; Blewett et al., 1997b; Warell and Blewett, 2004; Warell et al., 2006). The lack of a 1-μm feature required pre-MESSENGER mineralogical interpretations to be made mainly on the basis of estimates of albedo at UV-SWIR wavelengths and mineral signatures from mid- to thermal-infrared emission spectra (Vilas and McCord, 1976; Vilas et al., 1984; Sprague et al., 1994, 1995, 2002; Emery et al., 1998; Warell et al., 2006). Although most hypothesized compositions were low in iron, the proposed silicate mineralogies ranged from plagioclase-rich through pyroxene-rich.

Spectral reflectance measurements of Mercury by Mariner 10 and MESSENGER have helped to revolutionize the understanding of Mercury's surface composition and stratigraphy. The Mariner 10 imaging system acquired images covering about 40% of Mercury's surface through 0.355- and 0.575-μm wavelength filters using a vidicon detector. At this wavelength range, spectral slope and standardized reflectance can be used to estimate content of opaque minerals and differences in optical maturity with the methods of Lucey et al. (1998, 2000b). However, the wavelength range measured by the Mariner 10 vidicon does not cover the 1-μm absorption, precluding identification of regions with differing content of iron in silicate phases. Also, difficulties in calibrating Mariner 10 vidicon images resulted in initial studies focusing on qualitative comparisons of selected regions (e.g., Hapke et al., 1975; Rava and Hapke, 1987). In the 1990s, extensive work was done to reprocess and perform new calibrations of Mariner 10 images of both the Moon (Robinson et al., 1992) and Mercury (Robinson and Lucey, 1997). Analysis of the recalibrated data yielded key findings about Mercury's surface: that there are two primary spectral trends, one involving variations in opaque content and the other involving optical maturity; that spatial variations in color serve as stratigraphic markers that support a volcanic origin for some intrabasin plains on the basis of their distinct color; and that craters and basins exhume from depth materials with both higher and lower opaque mineral contents, indicating vertical heterogeneity in composition (Robinson and Lucey, 1997; Robinson and Taylor, 2001; Blewett et al., 2007).

8.2.3 Mercury's Spectral Reflectance Measured by MESSENGER

The MESSENGER spacecraft (Solomon et al., 2001) measured Mercury's spectral reflectance from UV through SWIR wavelengths using three instruments. The MDIS/WAC (Hawkins et al., 2007, 2009) sampled the spectrum through 11 discrete filters centered at 0.43–1.02 μm, with a pixel instantaneous field of view (IFOV) of 179 μrad and a 10.5° × 10.5° field of view (FOV). MDIS was mounted on a pivot, enabling nadir pointing and data to be collected from Mercury orbit at low emergence and phase angles, with a goal of obtaining images at the lowest possible solar incidence angle at each latitude. MASCS/VIRS (McClintock and Lankton, 2007) was a point spectrometer with a 400-μrad (0.023°) IFOV sampling the wavelength range 0.3–1.44 μm at 5 nm channel^{-1}. VIRS was fixed to the spacecraft and constrained by spacecraft-pointing restrictions to view the surface at higher phase angles of 78° to 102°. MASCS/UVVS (McClintock and Lankton, 2007) was a separate point spectrometer with distinct modes for measuring the surface and atmosphere. The surface mode had a 0.05° × 0.04° IFOV with a far-UV detector sampling the wavelength range 0.115–0.19 μm at 0.3 nm channel^{-1}, a mid-UV detector sampling 0.16–0.32 μm at 0.7 nm channel^{-1}, and a near-UV to visible channel sampling 0.25–0.6 μm at 0.6 nm channel^{-1}.

The three flybys of Mercury by the MESSENGER spacecraft in 2008–2009 provided >90% coverage of the surface by MDIS/WAC with 11-color multispectral imaging, but only regional coverage at the low to moderate incidence and emergence angles preferred for spectral measurements. Moreover, the flybys provided only sparse measurements of representative surface units by MASCS/VIRS and very limited surface measurements by MASCS/UVVS. The flyby multispectral imaging provided important evidence that Mercury's surface contains widespread, smooth volcanic plains (Head et al., 2008, 2009; Denevi et al., 2009) that vary in spectral character from lower standardized reflectance and a less red spectral slope through higher standardized reflectance with a redder slope. There is a continuous range of variation, with standardized reflectance at 0.75 μm ranging from ~0.8 to 1.25 that of the average surface (Denevi et al., 2009). Flyby color imaging revealed all of the major spectral units that were later characterized in more detail from orbit (Robinson et al., 2008). Both MDIS and MASCS/VIRS data confirmed the general absence of a 1-μm crystal field absorption, and thus a low ferrous iron content in silicates (McClintock et al., 2008; Robinson et al., 2008; Blewett et al., 2009), to the limits of the instruments' spatial resolutions and coverage.

During operations in orbit about Mercury from March 2011 through April 2015, three MDIS multispectral mapping campaigns were conducted (Denevi et al., 2018): a global 8-color map at ~1000 m pixel^{-1} for global characterization of surface units (shown in a false-color representation in Figure 8.1a), a 3-color map of northern and equatorial latitudes at a finer spatial sampling of ~400 m pixel^{-1} to improve knowledge of stratigraphic relations, and a 5-color map of the northern smooth plains (for discussion of this feature, see Chapter 6) at a uniform phase angle to search for subtle spectral units within the plains. In addition, 11-color image sets were specifically targeted at hollows (for discussion of this class of feature, see Chapter 12), selected impact craters, and pyroclastic deposits to improve spectral resolution. Most published analyses of orbital multispectral data to date have used the 8-color map, which was constructed in an effort to cover the planet at the lowest incidence angles available at a given latitude and at a low emergence angle (Figures 8.1b and 8.1c) (e.g., Denevi et al., 2013a; Domingue et al., 2015, 2016; Murchie et al., 2015). The 8-color map's spectral sampling every ~110 nm, on average, can be used to detect and map opaque phases, differences in optical maturity, or Fe^{2+}-bearing olivine, pyroxene, or glasses if Fe^{2+} is present in sufficient quantity. The version shown in Figure 8.1 was described by Denevi et al. (2018) and has an updated calibration and a more accurate photometric normalization than the versions described by Domingue et al. (2015) and Murchie et al. (2015), enabling quantitative comparisons of nearly all portions of the map.

Globally distributed mapping with MASCS/VIRS was accomplished by measuring spectral reflectance profiles along the track of the instrument FOV formed by the spacecraft's north–south motion, with spectra covering all wavelengths simultaneously collected typically at a 1-s cadence. Pointing perpendicular to the tracks in each orbit was offset to distribute the spectral profiles of the surface as evenly as possible (Figure 8.2a). This strategy attained the densest practical coverage for a point spectrometer but nevertheless left discontinuities across track, as well as along track at northern latitudes where orbital altitudes were typically lower and spacecraft velocities higher. In addition, specific features of interest were targeted either with single spectral profiles or small raster scans. The restricted MASCS/VIRS phase angle range of 78°–102° resulted in either incidence or emergence angle having a high value if the other had a low value (Figures 8.2b and 8.2c). Results of MASCS/VIRS global mapping have been described by Izenberg et al.

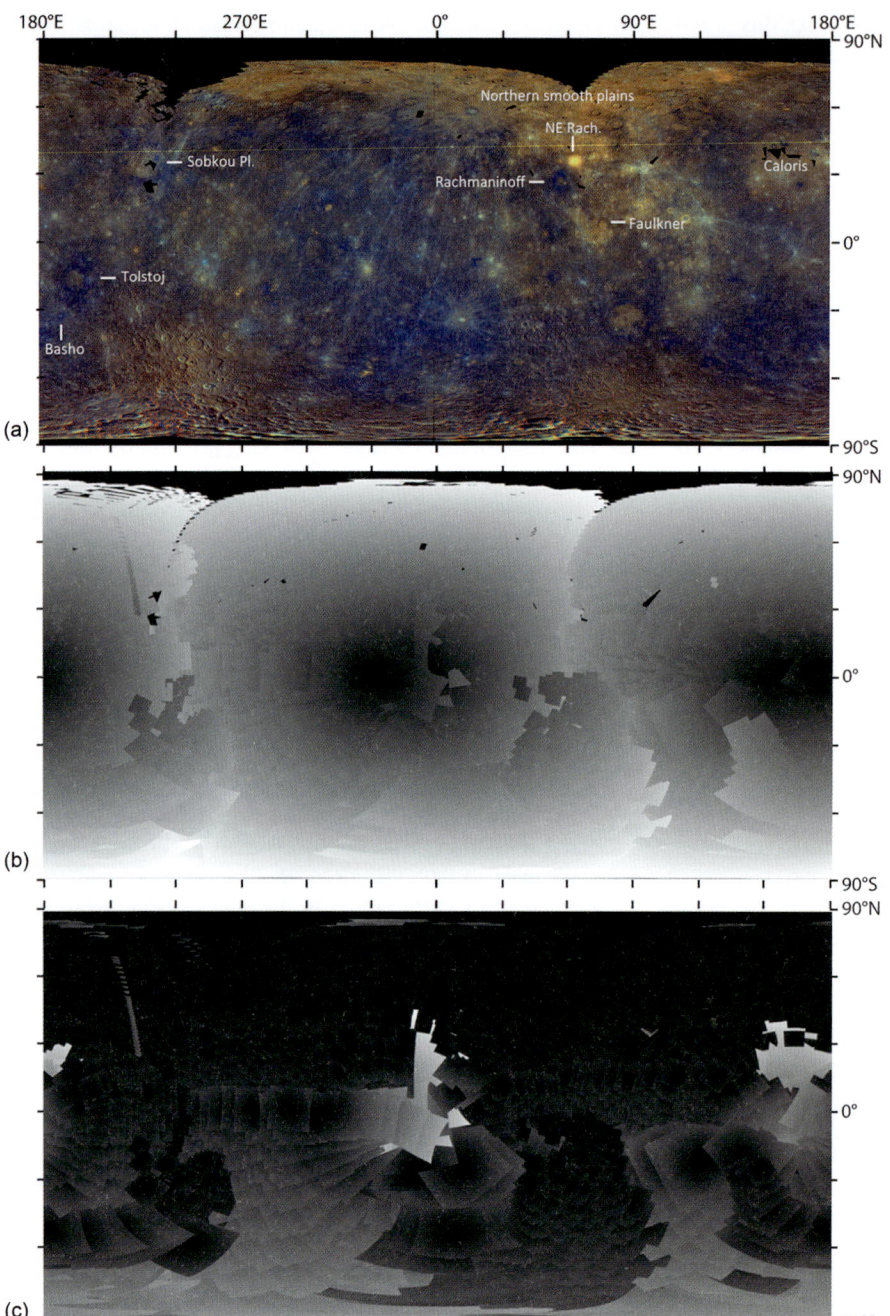

Figure 8.1. Synoptic view of MDIS/WAC normalized spectral reflectance of Mercury, showing parameterized spectral variations and illumination and viewing geometries in simple cylindrical projection. (a) Red (R), green (G), blue (B) image with R = principal component 2 (PC2, see Section 8.5.2), stretched from 0.01 to 0.045; G = principal component 1 (PC1), stretched from 0.072 to 0.372; and B = 0.43-μm/0.99-μm reflectance ratio, stretched from 0.349 to 0.717. Key geographic features are indicated. (b) Incidence angle at which the measurements were taken, stretched 0° (black) to 90° (white). (c) Emergence angle at which the measurements were taken, stretched 0° (black) to 80° (white).

(2014). Several additional studies have examined spectral units revealed in these data and how they compare with results from MDIS (e.g., Helbert et al., 2012; D'Amore et al., 2012, 2013, 2014).

MASCS/UVVS measurements were acquired from Mercury orbit separately from VIRS spectra, because of inherent design differences between the two spectrometers. The UVVS scanned its grating to build wavelength coverage, so that measurements of differing wavelengths were non-simultaneous. To ensure that all wavelengths sampled the same region of the surface, while the grating scanned through wavelengths, the MESSENGER spacecraft pointed and tracked a fixed point on the surface. The spacecraft angular velocity required to do so restricted coverage by high-quality UVVS spectra to those equatorial and southern latitudes that were observed from higher orbital altitudes (Figure 8.3). Most sampled spots form a grid that covers a variety of spectral and morphologic units, but some discrete features were targeted. To date, from MASCS/UVVS surface measurements, only analyses of the mid-UV channel have been reported (Maxwell et al., 2016).

To a high degree, the MDIS/WAC and MASCS/VIRS data sets reveal the same spatial distribution of major spectral units and are complementary. MDIS imaging benefits from continuity of coverage, spatial resolution, and favorable and more continuous viewing geometry (compare Figures 8.1b, 8.1c with 8.2b, 8.2c). MASCS/VIRS spectra benefit from higher spectral resolution, simultaneous coverage at all wavelengths, and greater wavelength range, particularly the UV coverage of the wing of the

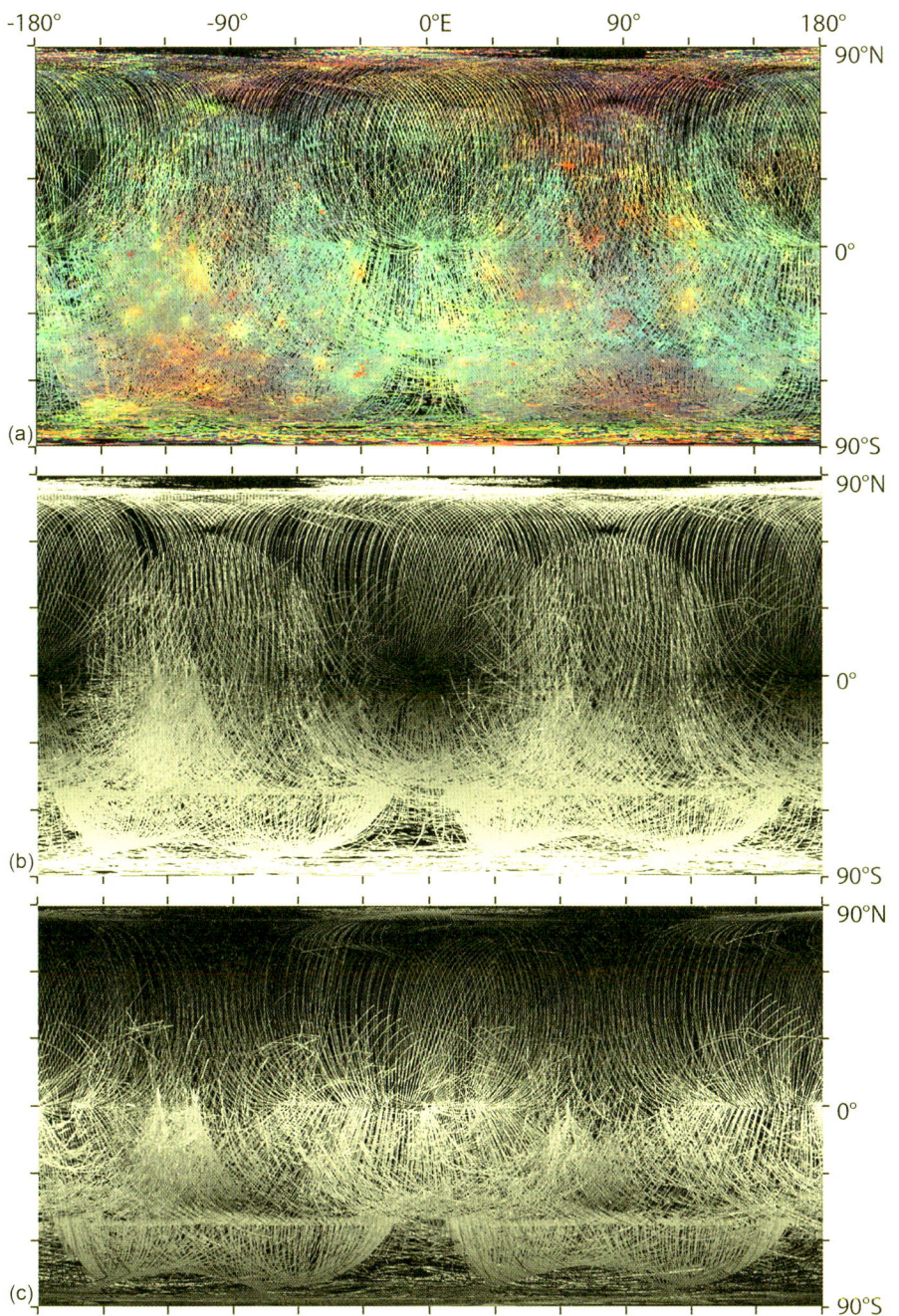

Figure 8.2. Synoptic view of MASCS/VIRS spectral reflectance of Mercury showing parameterized spectral variations and illumination and viewing geometries in simple cylindrical projection. (a) RGB image with R = reflectance at 0.575 μm, stretched from 0.03 to 0.10; G = 0.415-μm/0.75-μm reflectance ratio, stretched from 0.48 to 0.65; and B = 0.31-μm/0.39-μm reflectance ratio, stretched from 0.57 to 0.71. (b) Incidence angle at which the measurements were taken, stretched 0° (black) to 90° (white). (c) Emergence angle at which the measurements were taken, stretched 0° (black) to 90° (white).

OMCT band, which is sensitive to even traces of Fe^{2+} in silicates. The MASCS/UVVS data measure the core of the OMCT band.

8.3 REFLECTANCE OF MERCURY COMPARED WITH THAT OF OTHER BODIES

Some aspects of Mercury's surface composition may be inferred by comparing its estimated normal albedo with those of other planetary bodies at 0.55–0.60 μm, for which there are measurements by modern ground-based and spacecraft imaging systems covering a sufficient range of photometric angles to estimate normal albedo (Table 8.1). Mercury's normal albedo is intermediate to those of surfaces optically dominated by ferrous-iron-containing silicates (e.g., 4 Vesta, 433 Eros, the Moon) and those of low-albedo C- and D-type asteroids. C- and D-type asteroids are low in albedo, but C-types are spectrally neutral, displaying little slope at visible wavelengths, whereas D-types are strongly red-sloped. Both asteroid types are thought to be darkened by opaque phases, including magnetite and carbon-containing phases such as organic matter and graphite, which are present in carbonaceous chondrite meteorites (e.g., Cloutis et al., 2011). Eros is an S-type asteroid composed of olivine, pyroxene, and metal, with an ordinary-chondrite-like

Table 8.1. *Comparison of the normal albedo of Mercury with that of other planetary bodies.*

Body	Wavelength, μm	Normal albedo	Description	Reference
253 Mathilde	0.55	0.047±0.005	Undifferentiated, C-type	Clark et al. (1999)
Deimos	0.55	0.068±0.007	Undifferentiated, D-type	Thomas et al. (1996)
Phobos	0.55	0.071±0.012	Undifferentiated, D-type	Simonelli et al. (1998)
Mercury	0.559	0.091	Differentiated, komatiite to andesite	Domingue et al. (2015)*
Moon (avg. nearside)	0.603	0.14	Differentiated, basalt+anorthosite	Shkuratov et al. (2011)
4 Vesta	0.554	0.27±0.03	Differentiated, basalt	Li et al. (2013)
433 Eros	0.55	0.29±0.02	Undifferentiated, S-type	Domingue et al. (2002)

Notes: Bodies listed are those with silicate surfaces, for which normal albedo has been determined with data from spacecraft encounters. Bodies are ordered by increasing albedo.

* Obtained with Kaasalainen–Shkuratov model 3 described by the authors.

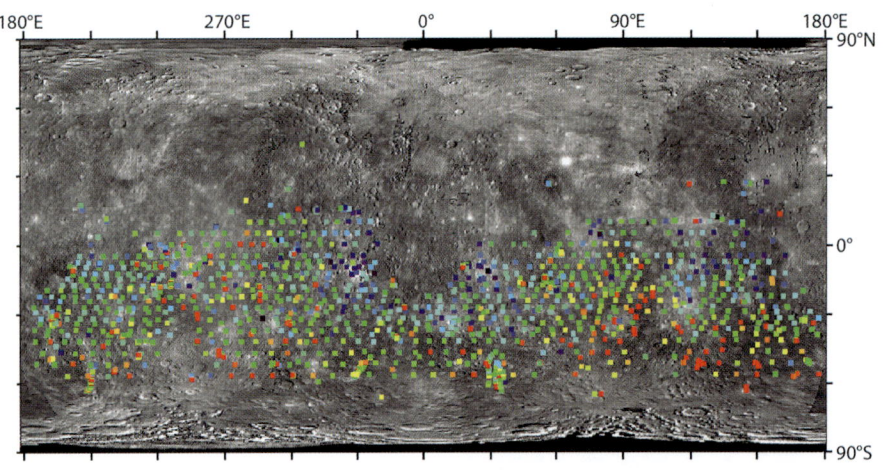

Figure 8.3. Synoptic view of MASCS/UVVS grid of spectral reflectance measurements of Mercury in simple cylindrical projection. Each measurement was taken so that observations at all wavelengths were of the same geographic location. The color scale represents the ratio of reflectances at 0.22 μm and 0.29 μm, on a rainbow scale where violet to red represent values from 0.25 to 0.45. The base is MDIS reflectance at 0.75 μm standardized to incidence angle $i = 30°$, emergence angle $e = 0°$, and phase angle $g = 30°$.

composition (Peplowski et al., 2015b). Vesta has a similar bulk composition to Eros but is differentiated and, like the Moon, has an igneous surface (Russell et al., 2012).

On average, the normal albedo of Mercury is lower than that of the lunar nearside (Denevi and Robinson, 2008) and approximates that of the maria, except without spectral evidence that ferrous iron in silicates darkens the surface (Domingue et al., 2014). This comparison indicates the occurrence of one or more pervasive opaque phases on Mercury that are either absent from or much less abundant on the Moon. With Mariner 10 data, Denevi and Robinson (2008) compared average reflectance properties of the two bodies as well as fresh crater ray material least affected by space weathering. The contrast between fresh and space-weathered regolith is greater on Mercury than on the Moon, suggesting more intense space weathering on Mercury. In addition, Mercury is ~15% lower in standardized reflectance than the lunar nearside, and fresh crater ray material is ~35% lower in standardized reflectance on Mercury than in the lunar highlands. Both results indicate that, in the absence of Fe^{2+} in silicates, there must be a greater abundance of opaque minerals in mercurian surface material. Braden and Robinson (2013) reexamined the comparison of lunar and mercurian reflectances with MDIS orbital measurements. They found that mercurian fresh crater ray material is even darker than recognized by Denevi and Robinson (2008), with comparable lunar highland crater rays ~1.9 times brighter. Also, MDIS observations show that there are fewer immature craters per unit area on Mercury than on the Moon. Given the greater impact rate at Mercury (~5.5 times greater impactor flux; Cintala, 1992), this difference suggests both a younger average age for mercurian than lunar rayed craters, and that space weathering occurs ~4 times more rapidly on Mercury.

In addition to the difference in average standardized reflectance, the frequency distribution of standardized reflectance differs between Mercury and the Moon. Mariner 10 imaging showed that, in contrast to the bimodal distribution of lunar reflectance, with the two modes representing maria and highlands, Mercury exhibits a single mode (Blewett et al., 2007; Denevi and Robinson, 2008). This result was confirmed and explored in more detail with MESSENGER imaging (Blewett et al., 2009). Besides the differing modality, reflectance contrast is lower on Mercury. Whereas the lunar highlands standardized reflectance mode is ~2.0 times greater than the mare mode, the brightest major spectral unit on Mercury (high-reflectance red plains, or HRP; see Section 8.5.1) is only about 1.6 times brighter than the darkest major unit (LRM) (Blewett et al., 2009; Denevi et al., 2009).

8.4 PHOTOMETRIC PROPERTIES OF MERCURY'S REGOLITH

8.4.1 Character and Objectives of Photometric Models

Photometric models are employed to predict the variations in reflectance of a planetary regolith that occur with changes in incidence, emergence, and phase angles. Typically the models are developed by fitting measurements acquired at multiple geometries and solving for the model parameters that best describe the set of measurements. One major purpose of this type of modeling is to provide a photometric normalization of image and spectral data acquired at differing geometries to a common illumination and observation geometry. This normalization is essential for the construction of mosaics in which variations in reflectance ideally are the result only of inherent differences in surface properties (e.g., Figures 8.1a and 8.2a). Such normalization enables comparison of spectral properties of one region with those of another, even if observations of the regions were illuminated and/or measured at different geometries. It also enables the comparison of spectral reflectance observations with laboratory measurements. For the latter reason, a standard normalization geometry is the common laboratory measurement geometry of incidence angle $i = 30°$, emergence angle $e = 0°$, and phase angle $g = 30°$.

Another goal of photometric modeling is to derive information about regolith texture and scattering properties, and to compare these properties among geologic units and among solar system objects. Textural and scattering properties are linked to the physical structure of the regolith and the particles comprising it that arise from processes forming the regolith and material properties of the source rock formations. Besides photometric modeling, other photometric analysis techniques such as phase-ratio analysis may identify regions with distinct regolith texture. In phase-ratio analysis, images of the same region acquired at different phase angles (differing typically by >10°) are co-registered and ratioed, to identify materials with different scattering characteristics.

Two classes of photometric models that have been widely applied to spacecraft imaging data are those of Hapke (1981, 1984, 1986, 1993, 2002, 2008, 2012a), whose models are based on geometric optics, and those of Kaasalainen and Shkuratov (abbreviated as "KS" below; Kaasalainen et al., 2001; Shkuratov et al., 2011; Schröder et al., 2013), in which the dependence of reflectance on phase angle is explicitly decoupled from the dependence on incidence and emergence angles. Each class of models has its own set of parameters intended to represent different surface characteristics. Although qualitative relations have been noted between Hapke model parameters and sample characteristics measured in the laboratory, definitive correlations have not been established (Shepard and Helfenstein, 2007; Shkuratov et al., 2007; Helfenstein and Shepard, 2011; Souchon et al., 2011). Similar tests between the KS model parameters and sample characteristics measured in the laboratory have not been conducted.

To interpret the results of photometric models, it is important to consider the length scale to which the models are sensitive. The composition and texture of planetary regoliths vary on multiple spatial scales, from geologic units (meters to kilometers) to soil particles (micrometers to millimeters) (Shkuratov et al., 2011). Photometric models with parameters tied to specific regolith properties are thus mathematically complicated, making derivations of unique solutions difficult, especially when attempting to do so with limited observations. Therefore, broad qualitative comparisons can be made with the least uncertainty. In addition, models typically have difficulty separating effects of distinct physical characteristics, such as roughness and particle size, on photometric properties.

In addition to its mapping and targeted observation campaigns, MDIS acquired repeated observations of three regions of Mercury at a wide range of photometric geometries, to provide data from which to derive the photometric model needed to create global mosaics from images obtained during seven distinct campaigns (Denevi et al., 2018; the 8-, 5-, and 3-color mapping campaigns, plus campaigns to acquire monochrome maps at four distinct lighting conditions: moderate incidence angle averaging near ~50°, high incidence angle averaging near 65°, and very high incidence angle averaging near 80° illuminated separately from the east and west). The three sets of photometric observations also provided a basis for qualitatively comparing scattering properties of Mercury's surface with those of the Moon and several asteroids. The photometric observations were acquired near Beethoven basin (8°–50°S, 200°–270°E), Rembrandt basin (17°–44°S, 55°–99°E), and Matabei crater (3°–40°S, 296°–346°E). At the first two locations, off-nadir spacecraft pointing was used to maximize the range of incidence and emergence angles, with relatively little sampling of the near-nadir geometries predominantly used for the mapping campaigns. In contrast, the Matabei crater region was sampled as part of the 8-color mapping campaign at a large variety of incidence angles, but predominantly at low emergence angles. All three regions were used in the photometric analyses of Domingue et al. (2016) discussed below, from which the photometric normalization used to create global mosaics was derived. Photometric measurements were modeled using both the Hapke and KS approaches in order to identify the normalization that left the least residuals. Model results were compared and contrasted to examine regolith characteristics.

Both classes of models describe the scattering properties comparably well in areas where incidence angle is less than 40° and emergence angle is greater than 40°, and in areas where incidence angle is 40° to 60° and emergence angle is less than 40°. In regions where there are large contrasts in emergence angle between adjacent images, the KS model provided more consistent corrections to standardized reflectance, and hence photometric normalization, with smaller residuals (Figure 8.4). The end-of-mission MDIS map products delivered to the NASA Planetary Data System (PDS; Denevi et al., 2018) utilize the KS model for photometric normalization to $i = 30°$, $e = 0°$, $g = 30°$, because contrasting emergence angles are the most common discontinuity in photometric geometry. (Earlier versions of MDIS global maps in earlier PDS deliveries used the Hapke model. To minimize residuals, differing map products used differing sets of model parameters, a procedure that was physically problematic but necessary practically.)

198 Spectral Reflectance Constraints

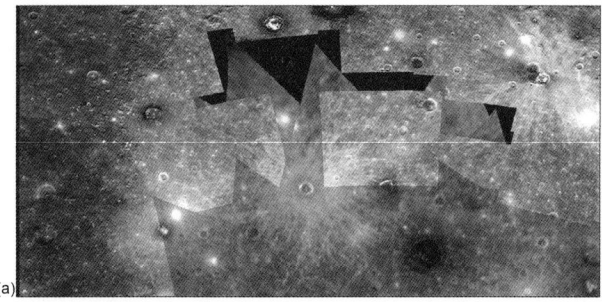
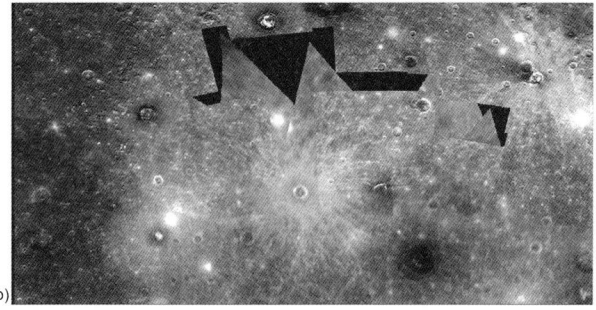

Figure 8.4. These MDIS image mosaics show 0.99-μm reflectance of the interior of Caloris basin taken during the 8-color mapping campaign, normalized to $i = 30°$, $e = 0°$, $g = 30°$ using (a) the Hapke model and (b) the Kaasalainen–Shkuratov model. Seams are most notable between component images acquired with large differences in emergence angle, and residuals appear larger in the Hapke normalization. Both mosaics are stretched to the same dynamic range.

The end-of-mission MASCS/VIRS and MASCS/UVVS products used a relatively simpler empirical photometric normalization based on the Lommel–Seeliger function, in which reflectance is proportional to cos i/(cos i + cos e) (Lumme and Bowell, 1981), and a phase function (reflectance as a function of phase angle) $p(g) = (0.0576 - 0.000352g)$ (Izenberg and Ward, 2016; Izenberg et al., 2016). Both data sets were normalized to $i = 45°$, $e = 45°$, $g = 90°$, which is inside the range of geometries measured by MASCS.

8.4.2 Average Regolith Scattering Properties from Photometric Models

Photometric modeling of MDIS orbital measurements supports qualitative statements regarding characteristics of particles making up the top layer of Mercury's regolith. Hapke model parameters can be divided into those that are related to individual particle scattering properties and those that are related to interparticle properties (porosity and roughness). Mercury's low value for one of the individual particle parameters, single-scattering albedo (the fraction of light scattered in a single bounce from a regolith particle), is driven by the overall low reflectance, indicating a material highly absorbing of UV–SWIR light. Comparison with the higher single-scattering albedo for the Moon (Sato et al., 2014) indicates that the surface of Mercury is more absorbing than either the average lunar surface or the lunar highlands. Mercury is as absorbing as the lunar maria at wavelengths <0.57 μm, but more absorbing at longer wavelengths. Such behavior is consistent with inferences from Mercury's lower normal albedo and lower standardized reflectance of fresh crater rays on Mercury than on the Moon, supporting a greater content of opaque phases in Mercury's regolith than in lunar regolith (Denevi and Robinson, 2008; Robinson et al., 2008; Braden and Robinson, 2013). Relationships between parameters from the Hapke model and laboratory scattering measurements also suggest that Mercury's regolith is unusually backscattering (McGuire and Hapke, 1995; Hapke, 2012b; Domingue et al., 2016). Such behavior could result from regolith particles with a high density of internal scatterers, for example, glass with abundant inclusions of opaque and other minerals. Such a regolith texture would be consistent with the presence of surface material with relatively abundant opaque minerals that has experienced extensive impact melting (Domingue et al., 2014, 2016).

During MESSENGER's Mercury flybys, there was a campaign of disk-integrated MDIS observations, in which single images covered all of Mercury's disk to measure integrated brightness, to construct the global phase curve of Mercury. The images covered moderate to large phase angles that are not measurable in Earth-based observations (Domingue et al., 2010, 2016). Low phase angles spanning the opposition surge (<5°–20°) were not measured by MESSENGER but are available from ground-based measurements (Mallama et al., 2002).

Modeling the combined suite of spacecraft and ground-based observations provides some insights into Mercury's globally averaged regolith textural properties. The width and magnitude of the opposition surge are thought to be affected by both the porosity and particle size distribution of the uppermost regolith. Comparison of Hapke modeling results with results from similar analyses of the Moon and asteroids suggests that the surface porosity and grain size characteristics on Mercury are different from those on these bodies (Domingue et al., 2010, 2016). If the particle size distribution within regoliths of all these objects is similar, then Mercury probably has a more porous surface. Alternatively, if the porosity within the regoliths of these objects is similar, then the size range of particles is probably smaller within Mercury's regolith.

In Hapke models, the shape of the phase curve at moderate phase angles is strongly affected by roughness of the surface at the scales at which most shadows are cast. Fractal analysis suggests that the dominant length scale in lunar-like regolith is of order hundreds of micrometers to a millimeter, that is, the size of shadows cast by individual regolith grains (Shepard and Campbell, 1998; Helfenstein and Shepard, 1999). The dominance of roughness at that length scale is supported by analysis of Apollo close-up stereo images (Helfenstein and Shepard, 1999; Goguen et al., 2010). Photometric modeling results for Mercury suggest that at this scale Mercury's regolith is on average smoother than that of the Moon. A surface smoother than that on the Moon could be consistent with greater impact melting, as suggested above to help explain the relatively backscattering nature of the surface materials.

8.4.3 Information from Spatial Variations in Photometry

Spatial variations in the photometric properties of the mercurian regolith provide evidence for differences in particle size and/or grain-scale roughness among geologic units. The technique of phase-ratio analysis compares relative reflectances within co-registered image sets taken at different phase angles and has been applied to several locations on the lunar surface to identify differences in regolith texture. These include the Apollo, Luna, and Surveyor landing sites (Kreslavsky and Shkuratov, 2003; Kaydash et al., 2011, 2012, 2014; Kaydash and Shkuratov, 2012, 2014; Shkuratov et al., 2013; Clegg et al., 2014), impact craters (Kaydash et al., 2012), impact melts (Shkuratov et al., 2012b), and haloes and rays associated with young impact features (Kaydash et al., 2014).

Blewett et al. (2014) used phase-ratio analysis to investigate differences in regolith texture among mercurian hollows, pyroclastic deposits, impact melt, and their surrounding geologic units. Hollows consist of irregular, flat-floored depressions tens of meters deep that commonly are surrounded by haloes tens to a few hundred meters in radial extent that are higher in reflectance and bluer than their substrate and occur in clusters up to tens of kilometers in size (Blewett et al., 2011, 2013; Xiao et al., 2013; Thomas et al., 2014a; see also Chapter 12). The depressions themselves have been interpreted to result from removal of a subsurface volatile-bearing phase by heating or radiation sputtering (Blewett et al., 2013); ballistic emplacement of the material in the haloes has been suggested (Xiao et al., 2013). Pyroclastic deposits have been described by Kerber et al. (2009, 2011, 2014), Goudge et al. (2014), and Thomas et al. (2014b, c); see also Chapter 11. Pyroclastic deposits surround central depressions interpreted as vents that occur singly or in a cluster. The largest vent, northeast of the Rachmaninoff impact basin (NE Rachmaninoff), has a mean diameter of 30 km (Goudge et al., 2014) and a depth of 4 km (Thomas et al., 2014b). Typical pyroclastic deposits are relatively high in reflectance, have a relatively red spectral slope, and are ~20 km in radius (Kerber et al., 2011); however, the NE Rachmaninoff deposit reaches a maximum radius of 130 km (Thomas et al., 2014b). Some vents exhibit raised rims and surrounding deposits tens of meters in depth, consistent with ballistic emplacement of a pyroclastic blanket (Thomas et al., 2014b, c). Phase-ratio images of hollows and pyroclastic deposits constructed by Blewett et al. (2014) are shown in Figures 8.5b and 8.5d. In these phase-ratio images, reflectance at a lower phase angle is divided by reflectance at the higher phase angle. Low phase-ratio values, which are observed

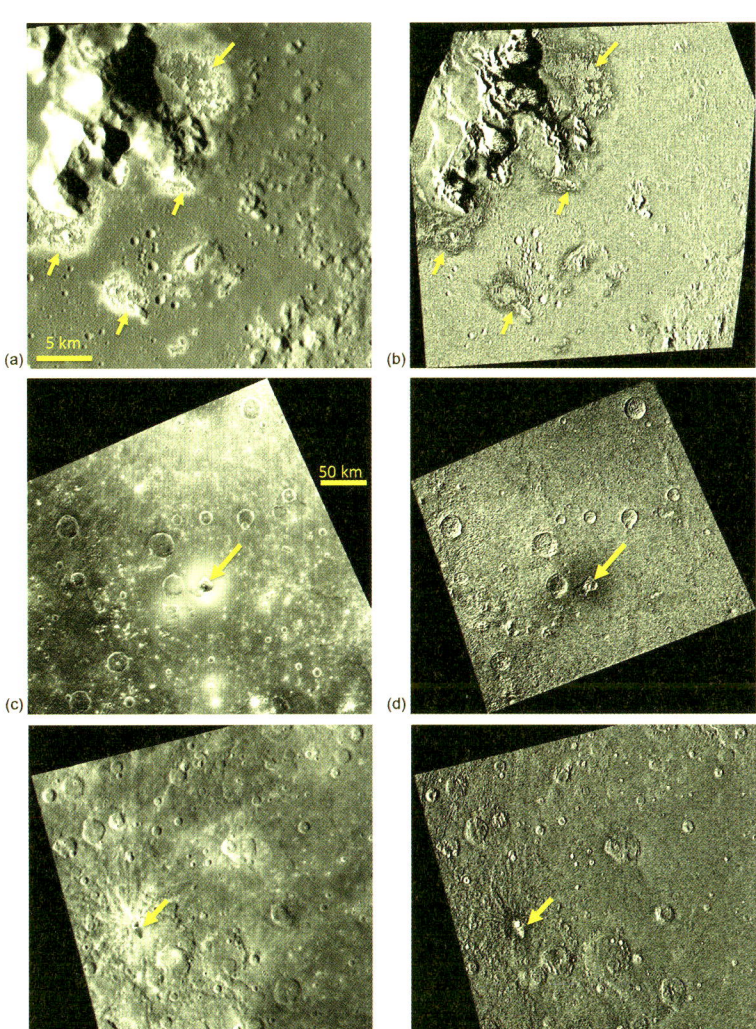

Figure 8.5. MDIS 0.75-μm normalized reflectance images (a, c, e) and phase-ratio images (b, d, f) of features discussed in the text. (a, b) Haloes of bright material surrounding hollows (arrows) that formed along the peak ring of Eminescu basin, located near 10.7°N, 245.7°E. The phase ratio was constructed from an image taken at 46° phase angle in the numerator and one taken at 70° phase angle in the denominator, and is stretched from 0.80 (black) to 1.25 (white). (c, d) Pyroclastic deposit (arrow) in the southwestern part of Caloris basin, near 22.0°N, 146.1°E. The phase angles of the numerator and denominator in the ratio image are 29° and 44°, and the ratio image is stretched from 1.023 (black) to 1.124 (white). (e, f) Dark tongue of impact melt (arrow) extending to the south from the rayed crater Waters, located near 9.0°S, 254.6°E. The phase angles of the numerator and denominator for the ratio image are 45° and 78°, and the ratio image is stretched from 0.880 (black) to 1.125 (white).

in the haloes of both types of features, correspond to a shallower slope of the phase function, which implies smoother surfaces or smaller particles. Either difference is consistent with a blanket of fine-grained material having been draped on top of surrounding, rougher regolith. In contrast, impact melt that extends as a lobate tongue from the Kuiperian-age Waters crater (Figure 8.5f; see Chapter 6 for an explanation of Mercury's stratigraphic system) displays a higher phase ratio than surrounding regolith, consistent with a rougher surface (Blewett et al., 2014).

8.5 MERCURY'S SPECTRAL PROPERTIES AND SPECTRAL VARIABILITY

8.5.1 Spectral Units and Relationship to Morphologic Features

The surface of Mercury is dominated by three geomorphic units: smooth plains, intercrater plains, and the rims, walls, and ejecta of impact craters and basins (Trask and Guest, 1975; Denevi et al., 2013a, b; Whitten et al., 2014). Intercrater plains have been described as comparatively level to smooth plains on a tens-of-kilometers scale, but with a higher density of superposed, mostly secondary craters than on smooth plains (Trask and Guest, 1975). Whitten et al. (2014) concluded that an "intermediate plains" geomorphic unit identified initially in Mariner 10 images is indistinct from intercrater plains and transitional to smooth plains.

MDIS multispectral imaging and MASCS/VIRS spectral measurements show that variations in standardized reflectance and spectral slope divide the major geomorphic units into four spatially extensive spectral units, listed in Table 8.2. Three additional minor spectral units (Table 8.2) correspond to specific classes of endogenic and impact features. All of these units were defined originally by Robinson et al. (2008) and Denevi et al. (2009) and later modified by Murchie et al. (2015). Three of the four extensive units form smooth and intercrater plains, and the fourth mostly represents ejecta of relatively well-preserved impact basins. Type spectra of these units are shown in Figure 8.6, and type locations of the units are given in Table 8.2. The three plains units are low-reflectance blue plains (LBP), intermediate plains (IP), and high-reflectance red plains (HRP). Examples of all three spectral units occur in both the smooth and intercrater plains. The two largest expanses of smooth plains that are definitively endogenic in origin – the northern smooth plains (Head et al., 2009) and Caloris interior plains (Murchie et al., 2008; Watters et al., 2009; Ernst et al., 2015) – are dominated by HRP (Denevi et al., 2013a; Murchie et al., 2015). The fourth unit, low-reflectance material or LRM (Robinson et al., 2008), is darker than LBP. LRM is concentrated on the floors, rims, and ejecta of impact craters and basins, and also occurs as widespread patches in heavily cratered regions at southern latitudes, as discussed by Klima et al. (2018). A broad, shallow, absorption-like upward curvature in the spectrum of LRM, centered near 0.6 μm, is particularly evident when the spectrum is ratioed to spectrally bland HRP (Figure 8.6b, c). This curvature originally led to an initial interpretation that LRM may contain a large fraction of ilmenite (which has such a spectral feature; Figure 8.7) or other Ti-rich phases as on the Moon (Robinson et al., 2008; Denevi et al., 2009). Subsequently, after the average content of Ti on Mercury's surface was found to be too low to account for large amounts of such phases (Nittler et al., 2011; Weider et al., 2012, 2014; Evans et al., 2012; Murchie et al., 2015), the 0.6-μm feature was reinterpreted as possibly originating from graphite (which has a spectral feature similar to that in ilmenite; Figure 8.7), or from a mixture of grain sizes of metallic iron (which can have similar spectral curvature), or some combination of the two (Murchie et al., 2015). Section 8.6.2 summarizes how MESSENGER elemental abundance measurements support graphite as the source of this spectral feature.

Three additional spectral units with limited geographic distributions, all comparatively high in standardized reflectance, are distinct from the spectral units that dominate the plains (Figure 8.6). (1) Fresh crater materials are brighter and less red than plains or LRM, consistent with their having experienced less space weathering (Robinson et al., 2008; Blewett et al., 2009). However, as mentioned above, they are 30–50% lower in standardized reflectance than fresh crater materials in the lunar highlands, suggesting that an indigenous mercurian opaque component is present in most or all crustal materials (Robinson et al., 2008; Denevi and Robinson, 2008; Braden and Robinson, 2013). (2) A "red unit" consisting of pyroclastic deposits (Robinson et al., 2008; Goudge et al., 2014) is brighter but only slightly redder than HRP. However, the red unit (as well as hollows formed on it) has a much more prominent falloff in reflectance below 0.4 μm than do other materials, consistent with a better-defined oxygen–metal charge transfer band (Figure 8.6c). The OMCT band results from an oxidized transition metal in silicates. Given measured elemental abundances on Mercury, the OMCT band most likely results from a small concentration of Fe^{2+} (McClintock et al., 2008; Izenberg et al., 2014; Goudge et al., 2014). The abundance of Fe^{2+} required to create the observed moderate OMCT band, possibly 0.1 wt% (Cloutis et al., 2008), is too low to create a 1-μm crystal field absorption detectable above the noise and residuals in MASCS and MDIS data (Klima et al., 2011). (3) Bright materials within and forming haloes around the central depressions of hollows (Blewett et al., 2011) are comparable in reflectance to most surrounding materials at 1 μm but are less red and brighter at shorter wavelengths. Subtle absorption features seen in hollows materials at Dominici and Hopper craters indicate the presence of MgS or graphite (see Section 8.6.3; Vilas et al., 2016). Most hollows are closely associated with LRM (Robinson et al., 2008; Blewett et al., 2011, 2013), but others also occur in the red unit.

The strategy for acquiring MASCS/UVVS spectra resulted in sampling of a limited number of locations compared with the MDIS/WAC and MASCS/VIRS data sets. However, spectra were obtained for representative locations in most spectral units (Table 8.3), and these spectra are compared in Figure 8.8 with MASCS/VIRS spectra of the same spots. UVVS spectra corroborate the differences in relative strengths of a UV turndown into the OMCT band found in VIRS data. The red unit exhibits the strongest turndown and a relatively well-defined band minimum; HRP, IP, and LBP lack comparably well-

Table 8.2. *Definition and description of major and minor spectral units discussed in this chapter.*

Unit name	Abbreviation	Typical values from MDIS		Description and characteristics	
		Reflectance at 0.75 μm, $i = 30°$, $e = 0°$, $g = 30°$	430-nm/1000-nm reflectance ratio, relative to type HRP	Type area	Morphology
Low-reflectance material	LRM	0.05	1.20–1.25	Rachmaninoff 29.4°N, 58.7°E	Hummocky to lineated crater and basin ejecta; crater rims and central peaks
Low-reflectance blue plains	LBP	0.06	1.15	Circum-Caloris 11.9°N, 141.7°E	Typically intercrater plains; limited occurrences in circum-Caloris smooth plains where unit was originally defined
Intermediate plains	IP	0.07	1.05–1.10	Rudaki plains −1.5°N, 304.4°E	Smooth and intercrater plains; also in crater rims, ejecta within other plains units
High-reflectance red plains	HRP	0.08	1.0	Northern smooth plains 54.2°N, 63.6°E	Smooth and intercrater plains; also in crater rims, ejecta within other plains units
Red unit	—	0.08–0.14	1.0	NE Rachmaninoff 36.0°N, 62.6°E	Diffuse haloes surrounding deep, scallop-rimmed depressions (interpreted as pyroclastic vents)
Fresh crater material	—	0.08–0.13	>1 to 1.4	Unnamed crater −28.87°N, 3.95°E	Interiors, ejecta, and rays of Kuiperian-age craters
Bright hollows (in red unit) Bright hollows (in LRM)	—	0.08–0.14	>1 to 1.4	Tyagaraja 4.2°N, 211.8°E Basho −32.90°N 189.57°E	Diffuse low-relief haloes surrounding shallow, flat-floored depressions (interpreted as erosional)

Notes: Nomenclature follows Murchie et al. (2015). Content is compiled from that reference, Blewett et al. (2013), Goudge et al. (2014), and this chapter. Murchie et al. (2015) proposed that plains spectral units be prefaced by descriptors "smooth" or "intercrater" to describe morphology distinctly from spectral properties.

defined band minima (Maxwell et al., 2016). Mid-UV reflectance in LRM is comparable to or higher than in IP or LBP, an important observation for spectrally modeling the opaque phase present in LRM (Section 8.6.2).

None of the resolved spectral units exhibit an identifiable 1-μm crystal-field absorption attributable to Fe^{2+} in silicates (Izenberg et al., 2014; Murchie et al., 2015). Pre-MESSENGER searches for a 1-μm crystal field absorption examined mostly mature regolith over large areas, where space weathering would partly obscure a 1-μm band. MESSENGER's XRS and Gamma-Ray Spectrometer (GRS) investigations indicate a low total iron abundance, with a global range of ~1–2 wt% (1.3–2.6 wt% expressed as FeO) (Nittler et al., 2011; Weider et al., 2012, 2014) and an average abundance over the northern hemisphere of 1.9 ± 0.3 wt% (Evans et al., 2012). The strongest 1-μm signature of Fe^{2+}, if incorporated into silicates, is expected to occur in fresh crater materials. Klima et al. (2013) examined over 720 fresh mercurian craters and found no detectable 1-μm absorption in any of the features. In comparison, spectra of fresh lunar highlands craters and laboratory spectra of olivines and pyroxenes with iron contents as small as 1% have a 1-μm absorption several percent or greater in depth (Figure 8.9). From these observations and geochemical modeling, Klima et al. (2013) and Zolotov et al. (2013) concluded that FeO in mercurian silicates must be <1 wt%, and that the majority of the iron probably exists as free metal or sulfides.

8.5.2 Primary and Secondary Spectral Trends

Mercury's spectral units occupy two major spectral trends, each defined by variations in spectral slope and standardized reflectance at UV to SWIR wavelengths (Figure 8.10a). These trends

202 *Spectral Reflectance Constraints*

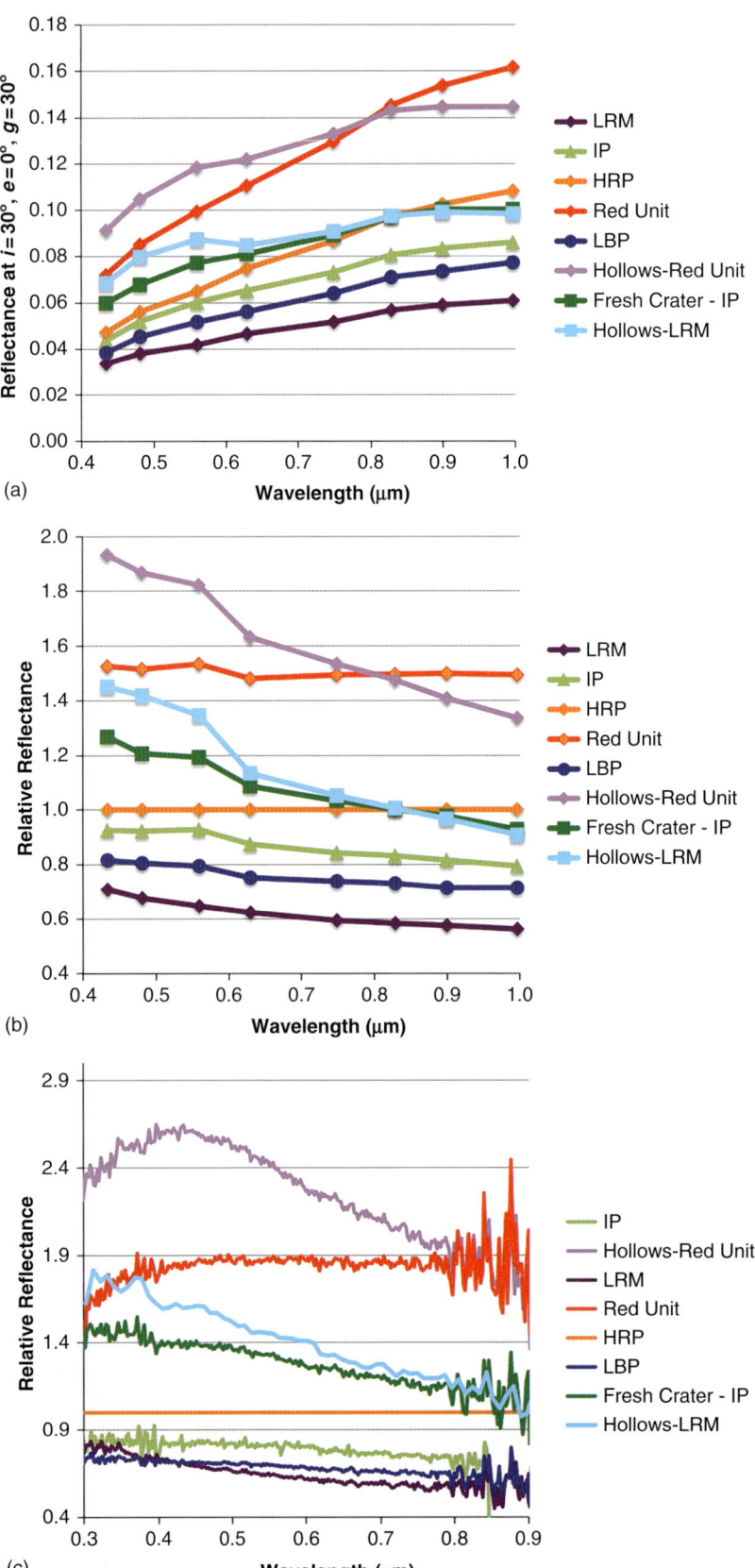

Figure 8.6. Type spectra of major and minor spectral units measured by MASCS/VIRS and MDIS/WAC. Unit nomenclature follows the conventions of Murchie et al. (2015). Locations of the type areas are given in Table 8.2. (a) MDIS spectra corrected to $i = 30°$, $e = 0°$, $g = 30°$. (b) The same spectra ratioed to HRP to highlight differences. (c) MASCS/VIRS spectra ratioed to HRP to highlight differences. Note: The standard photometric correction for MASCS/VIRS data is to $i = 45°$, $e = 45°$, $g = 90°$, which is within the range of photometric angles actually sampled by the data set; those data are not routinely shown on the same scale as MDIS spectra normalized to $i = 30°$, $e = 0°$, $g = 30°$.

8.5 Spectral Properties and Spectral Variability

Table 8.3. *Locations of representative MASCS/UVVS spectra.*

Unit name	Abbreviation	Representative area
Low-reflectance material	LRM	Tolstoj ejecta −12.5°N, 200.2°E
Low-reflectance blue plains	LBP	SW of Tolstoj −19.5°N, 188.7°E
Intermediate plains	IP	Between Picasso, Firdousi, and Nabokov −1.8°N, 61.2°E
High-reflectance red plains	HRP	Rembrandt interior plains −37.1°N, 89.7°E
Red unit	–	In crater 500 km NNE of Joplin −27.4°N, 30.9°E
Hollows in red unit	–	Tyagaraja −4.0°N, 211.6°E

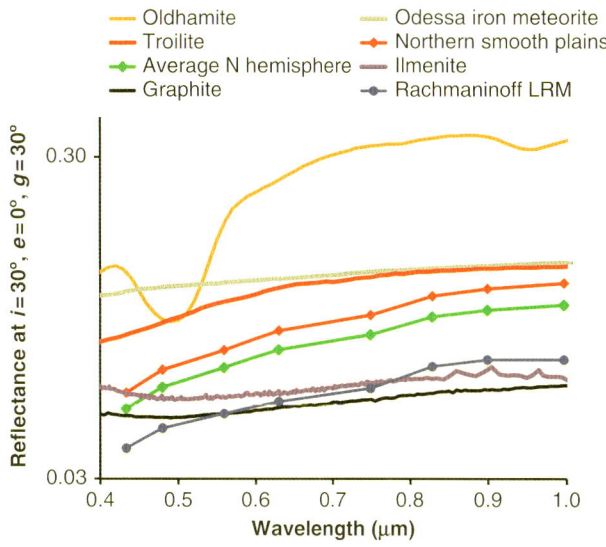

Figure 8.7. Spectra of candidate darkening phases on Mercury (lines without symbols) compared with MDIS spectra of HRP in the northern smooth plains, LRM surrounding Rachmaninoff basin, and the average northern hemisphere of Mercury (lines with symbols) that have been normalized to the laboratory measurement geometry of $i = 30°$, $e = 0°$, $g = 30°$. Reflectance is plotted on a logarithmic scale to facilitate comparison of the spectral shapes of brighter and darker materials.

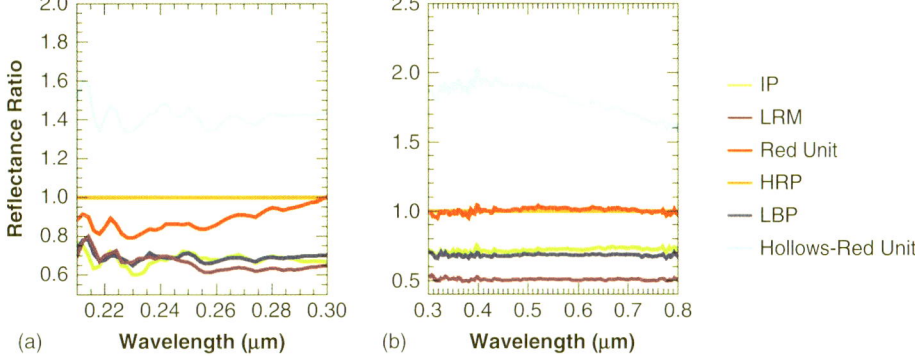

Figure 8.8. Representative spectra of all major and most minor spectral units measured by MASCS/UVVS and MASCS/VIRS. Unit nomenclature follows the conventions of Murchie et al. (2015). Locations of the areas for which spectra are shown are given in Table 8.3. Spectra are corrected to $i = 45°$, $e = 45°$, $g = 90°$, and ratioed to the spectrum of HRP as in Figure 8.6 to highlight differences. (a) UVVS spectra. (b) VIRS spectra of the same locations. The locations of these spectra are not type locations, so spectral contrast is typically less than in Figure 8.6, which shows type locations. Note also that the spots sampled by the two spectrometers were measured at different times and geometries, so data may disagree at the overlap wavelength of 0.3 µm.

were revealed with the assistance of principal component analysis, or PCA (Davis, 1973), of the 8-color data. PCA is a technique in which multidimensional – such as spectral – data are transformed into the coordinate system of the eigenvectors of the data set. The first principal component, or PC1, corresponds to the first eigenvector and represents overall variations in standardized reflectance. The primary spectral trend, defined by PC2, is between the end-members LRM (having a low value of PC2) and the red unit (having a high value of PC2), with LBP, IP, and HRP intermediate to the end-members and becoming respectively more like the red unit (Murchie et al., 2015; Klima et al., 2016). In the 0.4–1.0 µm wavelength range, the trend is between a lower standardized reflectance and shallower spectral slope in LRM, and a higher standardized reflectance and slightly steeper spectral slope in the red unit. Regions of LRM with the most well-defined 0.6-µm absorption-like feature have the lowest PC2 values (Murchie et al., 2015; Klima et al., 2016). In MASCS/VIRS data, a similar trend is defined by UV spectral properties, specifically the 0.31-µm/0.39-µm UV reflectance ratio, which parameterizes the UV turndown into the OMCT band: lower values represent a stronger band, and higher values a weaker band (Izenberg et al., 2014). The red unit has the lowest UV reflectance ratio and the most well-defined OMCT band, and LRM has a high UV reflectance ratio and less well-defined OMCT band.

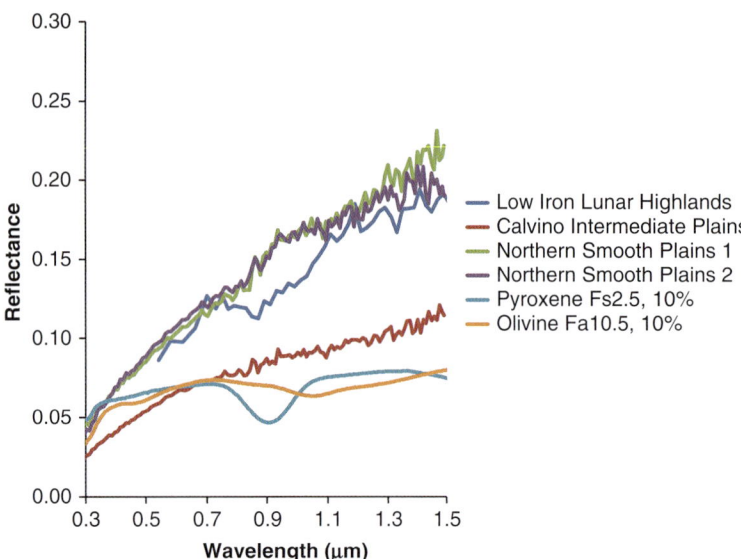

Figure 8.9. Representative spectra of fresh craters formed in HRP in the northern smooth plains and IP near Calvino on Mercury, compared with a Moon Mineralogy Mapper (Green et al., 2011) spectrum of low-iron lunar highlands and laboratory spectra of low-iron olivine and pyroxene. The laboratory spectra have been scaled in reflectance to overlie the spectrum of IP near Calvino on the short-wavelength wing of the 1-μm absorption. "Fs2.5" and "Fa10.5" refer to the percentages of the Fe end-members ferrosilite and fayalite, respectively, in the pyroxene and olivine solid solution series with Mg end-members; the two spectra are plotted at 10% of their absolute value.

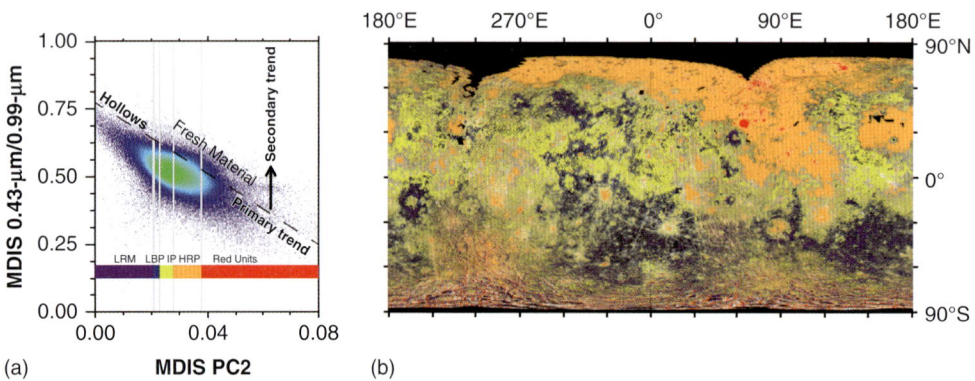

Figure 8.10. (a) Plot of MDIS PC2 versus 0.43-μm/0.99-μm reflectance ratio, color coded by density of points and with coordinates given by the values of PC2 and ratio of reflectances measured at 0.43 μm and 0.99 μm. Points are color-coded with a rainbow color scale; warmer colors have more points in the global map. Approximate PC2 boundaries of the major spectral units forming the primary trend are shown. The PC2 values in spectral units are different from those listed by Murchie et al. (2015) as a result of updated data calibration. The primary and secondary spectral trends are indicated; fresh crater materials and hollows are offset vertically from the primary trend and plot above the dashed line. (b) Approximate geographic distribution of spectral units forming the primary trend, shown on a base map of MDIS standardized reflectance at 0.75 μm.

The secondary spectral trend is characterized by increasing reflectance and a progressively less red ("bluer") spectral slope, with the spectral slope defined in MDIS data by higher values of the 0.43-μm/0.99-μm reflectance ratio (Murchie et al., 2015), and in MASCS/VIRS data by a higher 0.415-μm/0.75-μm reflectance ratio (Izenberg et al., 2014). For the most part, higher reflectance ratios correspond with optically immature fresh crater materials, and low reflectance ratios with mature, space-weathered materials. In Figure 8.10a, less mature materials in any given spectral unit are offset positive vertically in the plot; those that plot above the dashed line correspond to the optically freshest materials. Most hollows have low values of PC2 in MDIS data, like LRM, and are thus undistinguished from LRM by PC2 despite their higher reflectance. However, hollows are offset from LRM along the direction of the secondary spectral trend (Murchie et al., 2015). The overlap of PC2 values for most hollows and LRM is consistent with the findings by Blewett et al. (2013) and Thomas et al. (2014a) that most bright hollows materials are closely associated spatially with LRM. Izenberg et al. (2015) noted that hollows also occur in the red unit, and those that do resemble the red unit in having a low UV 0.31-μm/0.39-μm reflectance ratio and a stronger turndown into the OMCT band (compare hollows spectra in Figure 8.6). The spectral contrast with the red unit for these hollows (higher reflectance, bluer spectral slope at 0.4–1.0 μm) is comparable to the contrast between hollows formed in LRM and the LRM itself. On this basis, Izenberg et al. (2015) suggested that bright hollows materials may be fresher, less-space-weathered equivalents of their substrate, whether the red unit or LRM. Presumably, fresher materials were exposed by disturbance of the regolith when volatile materials were withdrawn to form the central depressions in the hollows (Blewett et al., 2011, 2013; Vaughan et al., 2012; Thomas et al., 2014a; Vilas et al., 2016). Alternately, in some cases for the red unit, MASCS/VIRS data may actually sample an areal mixture of red unit and distinct hollows material.

8.5.3 Spatial Distributions of Units

The spatial distribution of materials in the spectral units along the primary trend has been described by Klima et al. (2018) and is shown in Figure 8.10b. This spectral unit classification is based solely on the value of PC2 and thus is independent of superimposed effects of space weathering. Materials with high values of PC2 or low values of the MASCS/VIRS 0.31-μm/ 0.39-μm reflectance ratio (i.e., HRP and the red unit) are unevenly distributed across Mercury's surface. The two largest continuous areas of HRP are the northern smooth plains (Head et al., 2011) and plains interior to the Caloris basin (Murchie et al., 2008; Watters et al., 2009; Ernst et al., 2015). Those are two of the three largest occurrences of smooth plains mapped by Denevi et al. (2013a). The remaining large occurrence of smooth plains, the Caloris exterior plains, was interpreted by Denevi et al. (2013a) as likely to include basin ejecta as well as volcanic material, although the size–frequency distribution of impact craters on the Caloris exterior plains indicates a crater retention age resolvably younger than the basin (Fassett et al., 2009) (but see also Chapter 6). Other patches of smooth HRP include interior plains of the Rachmaninoff, Rembrandt, and Tolstoj impact basins, and scattered patches of smooth plains within and around other smaller basins, including several west of Caloris, most conspicuously Faulkner. There are scattered exposures of HRP material throughout the IP within impact crater rims, walls, and ejecta, which are interpreted as HRP generally buried by overlying volcanic IP but excavated locally by the impacts (Ernst et al., 2010). In addition, large areas of HRP material form intercrater plains at southern mid latitudes from 220°E to 260°E longitude, and in northern equatorial to mid latitudes between Rachmaninoff and Caloris basins from 60°E to 120°E longitude. Most of Mercury's surface consists of IP, which forms both smooth and intercrater plains. LBP forms patches of smooth plains southwest of Caloris, where it was originally identified (Denevi et al., 2009), but the major occurrences are large patches of intercrater plains over the region 60° S–50°N, 290°–350°E, within and to the south and east of the "high-Mg" region identified from XRS measurements (Weider et al., 2015). LRM occurs in rims and ejecta of impact craters and basins, most conspicuously Rachmaninoff, Tolstoj, and Rembrandt. There are also large annuli of LRM defining the ejecta around a number of highly degraded and infilled impact basins such as Sobkou basin (Fassett et al., 2012) and several others discussed in Section 8.6. Large, relatively diffuse exposures of LRM composed of numerous small patches occur at southern mid latitudes over the longitude ranges 310°E–30°E and 120°E–180°E, in areas with the highest density of large craters, which may represent Mercury's oldest preserved crust (Klima et al., 2018; see also Chapter 6).

8.5.4 Heterogeneity within Spectral Units

Two of the spectral units exhibit heterogeneous spectral properties that indicate differences either in their composition or in their extent of space weathering. First, LRM having comparable standardized reflectance and spectral slope is exposed in craters and basins ranging in age from rayed Kuiperian craters through degraded ejecta of Tolstojan basins (Murchie et al., 2015). Some younger examples, such as Kuiperian ejecta of the rayed crater Bashō and Mansurian ejecta of Rachmaninoff, have a broad, deep 0.6-μm absorption-like feature seen in ratioed MDIS and MASCS/VIRS spectra, whereas in Calorian and older materials this feature is weaker or absent (Figure 5 of Murchie et al., 2015). Murchie et al. (2015) suggested that freshly exposed LRM exhibits this feature, which over time becomes obscured by the effects of space weathering, i.e., by amorphous coatings with nanoscale grain inclusions, possibly Fe^0 or sulfides (Domingue et al., 2014). Alternatively, LRM may have regional differences in opaque phases responsible for darkening it, and not all of the opaque minerals have the 0.6-μm feature. Murchie et al. (2015) used spectral mixture modeling to simulate LRM by mixing spectra of candidate low-reflectance phases with spectra of mature northern plains, under the assumption that the primary spectral trend arises from variations in the content of an opaque phase and that space weathering affects different units similarly. They found that among candidate opaque phases – Mg or Ca sulfides (Weider et al., 2012; Blewett et al., 2013; Helbert et al., 2013), troilite, ilmenite, graphite (Robinson et al., 2008), and metallic iron occurring as a mix of nanophase (nanometer-sized) to microphase (micrometer-sized) grains (Lucey and Riner, 2011; Gillis-Davis et al., 2013) – only graphite or ilmenite, or a mixture of Fe^0 or possibly sulfides having different grain sizes including nanophase grains, is sufficiently dark to impart the low reflectance of LRM (Figure 8.7). Ilmenite is excluded by the low elemental abundance of Ti, leaving graphite or a mixture of microphase and nanophase iron or FeS by the process of elimination.

Second, the red unit exhibits variation in the strength of its UV turndown with shorter wavelengths, a feature that has been attributed to the wing of an OMCT band. For a given low Fe^{2+} content, an OMCT band is stronger and more easily detected than a 1-μm crystal-field absorption, and the OMCT band reaches ~20% depth, such as that observed in the red unit for an Fe^{2+} content in plagioclase of only ~0.1 wt% Fe (Cloutis et al., 2008). The UV turndown is very weak or absent on average on Mercury's surface, but within the red unit the strength of the UV turndown increases with increasing reflectance (Goudge et al., 2014). Izenberg et al. (2014) and Goudge et al. (2014) suggested that the correlation between reflectance and UV turndown may originate from obscuration of the OMCT band by an opaque phase or phases in darker materials. A problem with this hypothesis is that material outside that red unit that is comparable in UV reflectance lacks a UV turndown (Figure 8.6c). In Section 8.6.4 below, we show that a variety of MESSENGER data support an alternate hypothesis, proposed by Izenberg et al. (2014), that the content of oxidized iron is lower in the red unit than in other materials, and the OMCT band is detectable there because elsewhere on Mercury it is saturated and thus not evident.

Figure 8.11. Unnamed basin containing the 43-km-diameter crater Calvino, shown in coverage acquired during MESSENGER's second Mercury flyby. This is the highest-resolution multispectral image of this location. (a) Standardized reflectance at 0.75 μm. (b) RGB composite with red = PC2, green = PC1, and blue equal to the ratio of reflectance at 0.43 μm to that at 0.99 μm. HRP is orange, IP brown, and LRM blue. Calvino, indicated by the arrow in (b) and notable for its orange rim material, is centered at 3.9°S, 304°E.

8.6 IMPLICATIONS FOR CRUSTAL STRUCTURE AND GEOLOGIC PROCESSES

Mercury's spectral reflectance properties provide insights into four key aspects of mercurian geology: (a) the structure of the upper crust, (b) formation of LRM in the earliest crust, (c) modern processes that form hollows, and (d) processes that drove pyroclastic volcanism. These insights come from considering spectral variations in the context of geomorphology, crustal elemental composition, and Mercury's oxidation state. MESSENGER XRS and GRS data (Chapters 2 and 7) show that, overall, Mercury's surface rocks range in composition from low-Fe, Mg-rich basalts (Stockstill-Cahill et al., 2012) to komatiites (Charlier et al., 2013). Elemental composition varies regionally: the most magnesium-rich regions are interpreted to be komatiitic or boninitic basalts, and most plains are basalt to basaltic andesite (Vander Kaaden et al., 2017). HRP in Caloris and the northern plains is more siliceous, with compositions of andesite to dacite. Sulfur is abundant, with a global range of ~1–4 wt% S (Nittler et al., 2011; Weider et al., 2012) and a northern hemisphere average of 2.3 ± 0.4 wt% S (Evans et al., 2012). Geochemical modeling indicates that the low abundance of Fe and high abundance of S are consistent with a highly reducing environment, with oxygen fugacity values (Chapters 2 and 18) ranging from 4.5 to 7.3 $\log_{10}$ units below the iron–wüstite buffer. Under these conditions, sulfide–silicate melt equilibria predict that iron will be present in sulfide or metal instead of in silicates, and that magnesium and calcium should both be partitioned between silicate and sulfide phases (Zolotov et al., 2013).

8.6.1 Structure of the Upper Crust

Spectral variations among deposits that form smooth and intercrater plains act as tracers with which to reconstruct the structure of Mercury's upper crust. Figure 8.11 shows a clear example of plains filling an unnamed basin superposed by the 43-km-diameter crater Calvino. There, LRM was excavated from depth by the impact that formed the basin rim and floor materials. Subsequently, at least two episodes of volcanism largely filled the basin and embayed craters superposed on the basin floor, with HRP emplaced first and IP second. Finally, the plains were excavated by Calvino crater. The underlying HRP was exposed in the crater wall, and deeper LRM from the basin floor was exposed in the crater's central peak. Where craters such as Calvino expose subsurface layers, impact scaling relations allow bounds to be placed on the depth of excavation of ejecta and central peak material and thus on the thicknesses of different layers in the target area. With that approach, Ernst et al. (2010) examined several regions imaged multispectrally by MDIS during MESSENGER's first two flybys of Mercury. They concluded that much of the upper 5 km of Mercury's crust consists of layers of HRP, IP, and LBP. LRM forms the bottom of the stratigraphic column in many locations, on the grounds that it is not penetrated by well-preserved craters or basins. By applying similar methods to the Caloris interior plains, Ernst et al. (2015) inferred that >2.5 km of HRP buries basin floor material consisting of LRM, and that LRM extends to at least 11 km depth. Using this same approach, Denevi et al. (2013b) examined a highly infilled Caloris-sized basin ("b30" of Fassett et al., 2012) and inferred that >500 m of intercrater IP buries at least 3.5 kilometers of older HRP in that location. Modeling of overlapping ejecta of multiple impacts suggests that, to account for the observed areal extent of LRM, the source depth from which it originates is on average about 30 km, but that subsurface LRM is laterally heterogeneous (Rivera-Valentin and Barr, 2014).

8.6.2 Composition and Formation of LRM

The stratigraphy, spectral reflectance properties, and elemental composition of LRM provide evidence for the formation of a primary flotation crust on Mercury that is very different from that on the Moon. The position of LRM at the bottom of the recoverable portion of Mercury's stratigraphic column indicates that its darkening agent is intrinsic to the crust and is not a comparatively recent veneer as has been suggested by others (e.g., Bruck Syal et al., 2015). Given the low abundance of Ti on

Mercury both globally (Nittler et al., 2011; Weider et al., 2012, 2014; Evans et al., 2012) and locally in the LRM exposed by Rachmaninoff basin (Murchie et al., 2015), the only opaque phases plausibly present at abundances sufficiently high to account for the low reflectance of LRM are graphite, a mixture of nanophase and microphase metallic iron or iron sulfide, or some combination. Modeling of UV spectral reflectance further supports the presence of carbon. Treating the opaque phase(s) in LRM as a mixture of microphase and nanophase iron and carbon, originating from either igneous processes or space weathering, Trang et al. (2016) found that iron alone would produce a lower UV reflectance than is observed. In contrast, an iron–carbon mixture can accurately reproduce the UV spectrum of LRM. Murchie et al. (2015) summarized three possible origins for a layer of a carbon- and/or iron-bearing LRM excavated from depth: a primary graphite flotation crust predicted by geochemical modeling of an early magma ocean (Vander Kaaden and McCubbin, 2015) which was mixed with magmatic, volcanic, and perhaps upper mantle material by large impacts; a late-accreting carbonaceous veneer (Wänke, 1981; Wänke and Dreibus, 1994); or a layer that was shock-darkened during and before the early part of the late heavy bombardment (Gillis-Davis et al., 2013). These hypotheses make distinct predictions about abundances of carbon and iron in LRM compared with plains units: C-enriched, C- and Fe-enriched, or broadly similar, respectively.

The iron and carbon contents of LRM were bounded by measurements by MESSENGER's XRS and Neutron Spectrometer (NS) acquired at low altitudes late in the mission (Peplowski et al., 2016). Three regions of LRM were measured, associated with the 100-km-diameter crater Akutagawa (48.2°N, 219.1°E), the 195-km-diameter Sholem-Aleichem basin (50.9°N, 269.7°E), and overlapping ejecta of the Borealis and Derzhavin-Sor Juana basins at approximately 50°N, 300°E. None of these regions is as dark as the end-member Rachmaninoff LRM modeled optically by Murchie et al. (2015). Carbon and iron are detectable with neutron spectroscopy because carbon is a thermal neutron scatterer and manifests its presence as an increase in the thermal neutron flux, whereas iron is a neutron absorber and manifests its presence as a decrease in the thermal neutron flux. The NS detected thermal neutron enhancements highly correlated with all low-altitude overflights of the three LRM occurrences, consistent with an increase in the carbon content of 1–3 wt% above the average abundance outside the LRM. Iron content was determined independently by the XRS for one of these occurrences of LRM (the last) and confirmed to be low (Peplowski et al., 2016). UV–SWIR spectral modeling of both LRM and average surface material (Murchie et al., 2015) predicted an average carbon content in the northern hemisphere of ~1 wt% and up to ~5 wt% in end-member LRM in Rachmaninoff. The modest carbon enhancement in the non-end-member LRM measured by neutron spectroscopy is consistent with the optical estimates, and with the northern hemisphere average carbon content of 1.4±0.9 wt% estimated from gamma-ray spectroscopy as described by Peplowski et al. (2015a).

The widespread presence of ~1–2 wt% concentrations of carbon on Mercury sharply contrasts with the 0.001–0.1 wt% observed in surface materials on Earth, Moon, Mars, and Vesta (Lodders and Fegley, 1998). Gradual infall of carbon-rich (15–25 wt%) cometary dust has been suggested as a source for Mercury's darkening phase (Bruck Syal et al., 2015). Such a process should be a solar-system-wide effect, yet there is approximately an order-of-magnitude difference between the C concentration predicted by this process (800–1500 ppm) and that observed (20–200 ppm) on the Moon (Haskin and Warren, 1991). These results together with the stratigraphy of LRM strongly support the hypothesis that LRM is endogenous material. The results also suggest that graphite or another form of carbon accounts for much of the opaque content that imparts Mercury's low standardized reflectance. Peplowski et al. (2016) proposed that LRM incorporated portions of the primary graphite flotation crust of Mercury (Vander Kaaden and McCubbin, 2015) mixed with younger volcanic and intrusive materials, either by impacts or by assimilation into ascending magmas.

8.6.3 Modern Processes that Form Hollows

Two issues regarding hollows are the nature of the bright halo materials surrounding the topographic depressions, and the volatile phase or phases sublimated to form the hollows' depressions; see also Chapter 12. Several spectral reflectance results discussed above constrain the nature of the halo materials: (a) their relatively forward-scattering photometric behavior, consistent with a smooth blanket of fine particles (Blewett et al., 2014); (b) the correspondence of the UV spectral properties of haloes with those of the substrate (Izenberg et al., 2015); and (c) that the spectral contrast of hollows with their substrates is greater than or similar to the contrast between fresh crater materials and their substrates, i.e., a higher reflectance and bluer spectral slope indicative of a lesser degree of space weathering. A greater spectral contrast between hollows and their substrates could result simply from an extremely young exposure age of the hollows' bright haloes, compared with a greater age for ejecta from fresh rayed craters that has allowed space weathering to progress further. Overall, these results are consistent with bright halo material having formed by emplacement of a blanket of fine-grained, optically fresh regolith particles that were mobilized by decomposition of a volatile material. If the particles were ejected from the central depressions of the hollows, as the haloes' distribution along rims suggests, ejection velocities to attain distances of tens of meters to 100 m from the rim may have been meters per second, for a 45° ejection angle from the vertical. A moderately energetic process that launched halo-forming particles is implied (Xiao et al., 2013); simple diffusion of slowly volatilized phases through granular regolith is insufficiently energetic to explain particle trajectories (Blewett et al., 2016). Alternative mechanisms for hollow formation include deposition of a chemical decomposition product from a volatile mineral (Blewett et al., 2016).

Identifying the volatile phase removed from hollows' central depressions is more problematic, because it should have been partially to completely removed by the hollows-forming process. Hollows occur preferentially in LRM, on sunward facing slopes (Blewett et al., 2011, 2013), and near the mercurian hot-pole longitudes (Thomas et al., 2014a);

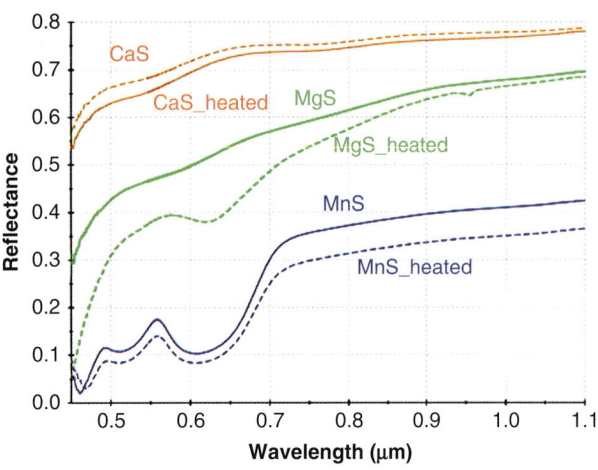

Figure 8.12. Reflectance spectra of magnesium sulfide (MgS), calcium sulfide (CaS), and manganese sulfide (MnS) before and after heating (indicated by "heated"). Adapted from Helbert et al. (2013).

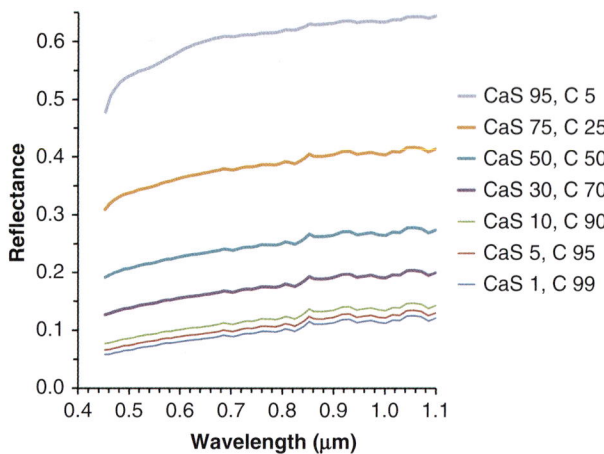

Figure 8.13. Synthetic spectra from mixture modeling of laboratory CaS and a VIRS spectrum of Caloris plains material. The spectrum of CaS, from Helbert et al. (2013), is of a sample for which reflectance was decreased by heating to Mercury surface temperature. Adapted from Izenberg et al. (2014).

these relations suggest formation under the action of solar radiation involving a phase abundant in LRM. A candidate for the phase being removed is one or more sulfides (Blewett et al., 2011, 2013). Sulfides may segregate from cooling mercurian lavas, forming a flotation crust that becomes a hollows-forming layer (Vaughan et al., 2012). Laboratory experiments in vacuum show that at mercurian temperatures manganese, calcium, and magnesium sulfides can decompose energetically (Helbert et al., 2013). During MESSENGER's second extended mission, a survey was conducted with MASCS/UVVS in its atmospheric mode to look for photo-emissions from gas or ions emanating from hollows (Prockter et al., 2013). That experiment found no detectable emanations; however, the most likely atom to be released by decomposition of sulfides, S, has a low probability of emission and thus of measurement compared with other species (Killen et al., 2009). These sulfide phases also have characteristic absorptions at 0.5–0.7 µm that are potentially detectable with reflectance spectroscopy (Figure 8.12). Calcium sulfide, suggested to be present on Mercury because of the correlation of the abundances of Ca and S, darkens after exposure to mercurian daytime temperatures. MgS undergoes a reddening and a weakening of its characteristic absorption (Helbert et al., 2013). However, the degree of darkening demonstrated in the laboratory would still leave these sulfides much brighter than typical mercurian regolith. Izenberg et al. (2014) conducted a global search for 0.5–0.7-µm absorptions due to sulfides in MASCS/VIRS data and found no exposures several kilometers in size or larger having an absorption with a depth a factor of 5 greater than the level of noise. To quantify the implied abundance limits, they modeled a mixture of mature HRP in Caloris interior plains with CaS that had been heated to mercurian temperatures. They found that because mercurian surface materials are so dark, and would attenuate absorptions in the much brighter sulfides, ~75 wt% CaS might remain obscured in data with a signal-to-noise ratio like that of MASCS/VIRS data (Figure 8.13). Vilas et al. (2016) conducted a similar search with MDIS data, which are capable of spatially resolving smaller exposures. They found small patches of fresh-appearing bright hollows material that appear to have an absorption near 0.65 µm, consistent with MgS (see Figure 12.19). Such a feature also appears in hollows formed in LRM comprising the rim of Bashō crater, shown in Figure 8.6. This result could suggest that sulfides are the phase removed from hollows, but that some of the material survives, or alternatively that the phase removed is also low in density and segregates together with sulfides. Another possible interpretation of the feature attributed by Vilas et al. (2016) to sulfides is that it originates from the same 0.6-µm feature as that attributed to graphite in LRM, which is stronger in hollows because the material is optically fresher.

Blewett et al. (2015) identified graphite in LRM as an alternative possible parent of a volatile phase, the removal of which could form hollows; see also Chapter 12. Whereas graphite is itself refractory, it may react with solar protons to form CH_4, which would be gaseous under mercurian conditions. Although slow effluence of methane from solar-irradiated regolith would not be sufficiently energetic to lift regolith particles, perhaps methane buildup and explosive release from graphite inclusions could do so.

8.6.4 Processes that Drove Pyroclastic Volcanism

Mercury's red unit is distinct both in its spectral properties and in its style of emplacement. Most of Mercury's volcanic materials are plains-forming and interpreted to have been emplaced as low-viscosity fluids in flood volcanic eruptions at high effusion rates (Head et al., 2008, 2009, 2011). In contrast, the red unit occurs in diffuse-edged blankets typically ~20 km and as much as 130 km in radius and are interpreted as pyroclastic deposits emplaced during explosive volcanic eruptions. The known population of such deposits has grown as the coverage, resolution, and calibration accuracy of released MDIS data have improved. Kerber et al. (2011) recognized 40 candidate deposits on the basis of the presence of both a central pit or pits and a surrounding halo of the red unit; Goudge et al. (2014) recognized 51; Kerber et al. (2014) recognized 137; and Thomas et al.

(2014b) recognized 150. Kerber et al. (2009, 2011) and Thomas et al. (2014b) considered the content of exsolved volatiles required to account for the measured extent of the haloes as a result of ballistic emplacement during eruption. Kerber et al. (2011) estimated that for volatiles that are well mixed into erupting magma, at the gravitational acceleration near Mercury's surface, and for a typical deposit ~20 km in radius, the areal extent of the halo would require exsolution of 0.5 wt% volatiles if that volatile were CO. The required content of a well-mixed volatile is inversely proportional to its molecular weight; there are a variety of C- and S- containing volatiles that could possibly have driven eruptions under chemically reducing conditions appropriate to Mercury's interior, including CO, CS_2, COS, and S_2 (Zolotov, 2011), and different contents of each would be required. Under slightly different assumptions, Thomas et al. (2014b) estimated a typical volatile content of 1.1 wt% if the volatile consisted entirely of CO, and a content of up to 10 wt% for the largest deposit surrounding the vent northeast of Rachmaninoff basin. Exsolved volatiles may have been concentrated at the top of a magma chamber prior to eruption, in which case a lesser overall volatile content would have been required to account for the radius of the halo.

The elemental composition of the NE Rachmaninoff pyroclastic deposit is constrained by a targeted XRS measurement and by NS measurements acquired during the low-altitude campaign near the end of the MESSENGER mission; see Chapter 1. To acquire the XRS measurement, when possible during the last two years of its mission the MESSENGER spacecraft was rotated to keep the instrument's field of view centered on NE Rachmaninoff for periods of up to tens of minutes. The goal was to increase the chance of observing the site during solar flares, and hence to obtain high-quality measurements of elemental abundances. On 14 December 2013, a large solar flare occurred during a targeted XRS observation. It revealed a composition with Mg/Si, Al/Si, Ca/Si, and Fe/Si ratios typical of IP, but a depleted S/Si ratio (Weider et al., 2016), equivalent to a S depletion of ~1.7 wt% given the observed global correlation between Ca/Si and S/Si (Figure 2.4) and a typical SiO_2 abundance of 51 wt% (Vander Kaaden et al., 2017). The NS analysis of Peplowski et al. (2016) described above also included multiple overflights of NE Rachmaninoff. With an analysis comparable to that performed for LRM, Peplowski et al. (2016) found a depletion of thermal neutrons consistent with a depletion of 1 wt% C relative to typical IP. That C depletion is equivalent to all or nearly all of the average northern hemisphere abundance of graphite estimated from spectral modeling by Murchie et al. (2015) and gamma-ray spectroscopy by Peplowski et al. (2015a).

As described by Weider et al. (2016) and summarized below, for NE Rachmaninoff the combination of MDIS and MASCS/VIRS spectral reflectance data, XRS and NS data, and measurements of the pyroclastic halo size from MDIS together provide constraints that point toward the source and amount of exsolved volatiles. These constraints include: (a) higher standardized reflectance than other mercurian surface materials, consistent with a depletion of the opaque phase, which available evidence indicates is carbon as graphite; (b) a stronger turndown of reflectance into UV wavelengths than for other materials with comparable UV reflectance; (c) depletion of ~1.7 wt% S compared with otherwise chemically analogous materials; (d) depletion of ~1 wt% C compared with otherwise chemically analogous materials; and (e) a spatial extent requiring up to but probably less than 10 wt% exsolved volatiles if the volatile was CO (Kerber et al., 2011; Weider et al., 2016).

Geochemical modeling makes predictions that are testable with these measurements. Zolotov (2011) proposed that highly reduced mercurian magmas could have been slightly oxidized by oxides such as FeO and SiO_2 assimilated into rising magma from country rock. Those oxides would have acted as an oxygen donor to oxidize graphite and sulfides to C- and S-containing volatiles, the exsolution of which would then drive explosive eruption. Suitable country rock exists: typical crustal rocks contain at least ~50 wt% SiO_2 and some are as SiO_2-rich (up to ~63%) as dacite (Vander Kaaden et al., 2017). Although the absence of a 1-μm crystal-field absorption limits the content of Fe^{2+} in silicates to <1 wt%, more than a few tenths of a percent Fe^{2+} creates an OMCT band so strong that it saturates, and there is no obvious turndown of reflectance into the UV; thus materials outside the red unit could have ~0.3–1 wt% Fe^{2+} available to oxidize graphite and sulfides. Furthermore, units of varied composition are vertically heterogeneous in Mercury's crust to at least 5 km and possibly as much as 30 km depth (Ernst et al., 2010, 2015; Denevi et al., 2013b; Rivera-Valentin and Barr, 2014) and thus available to be incorporated into ascending magma or temporarily stationary magma bodies.

The evidence described above is consistent with the idea that oxidation of C and S supplied volatile gases that drove pyroclastic volcanism (Weider et al., 2016). The red unit's strongest-observed UV turndown on Mercury is consistent with an Fe^{2+} content being lowered to ~0.1 wt% by reduction to Fe^0; the higher reflectance is consistent with carbon having been oxidized and escaped; and the depleted C and S contents measured by the NS and XRS are consistent with graphite and sulfides having been oxidized and escaped. In addition, typical amounts of S and C available for oxidation are of the order of the exsolved volatile content required to account for observed halo sizes, especially if volatiles were concentrated at the tops of magma chambers. For a typical pyroclastic halo 20 km in radius, depletion from a typical surface composition of 1.7 wt% S and 1 wt% C, as suggested at NE Rachmaninoff, could yield, for example, 1.7 wt% S_2 plus 2.3 wt% CO, which are equivalent to 3.1 wt% CO. This figure is 3–6 times greater than the equivalent volatile content of 0.5–1.1 wt% CO estimated by Kerber et al. (2011) and Thomas et al. (2014b) that would be required for emplacement of typical pyroclastic haloes. As an extreme case, oxidation of magma with the highest inferred contents of S and C indicated by the XRS and NS, 4 wt% S (Nittler et al., 2011; Weider et al., 2012) and 4 wt% C (Peplowski et al., 2016), could yield, for example, 4 wt% S_2 plus 9.3 wt% CO, which are equivalent to 11.1 wt% CO. This CO content would be adequate to explain the large halo of NE Rachmaninoff even without volatile segregation.

In summary, typical pyroclastic haloes could have been emplaced by incomplete oxidation of typical contents of graphite and sulfides, and emplacement of extremely large haloes such as that of NE Rachmaninoff could have involved concentration of volatiles in the upper parts of the magma bodies, more complete oxidation, and/or larger contents of C and S. In any case, evidence summarized above supports the hypothesis that pyroclastic volcanism on Mercury was driven by oxidation of C and S that removed these elements from refractory phases into eruption-driving volatile phases.

8.6.5 Relationship of Spectral Properties and Chemical Composition

Despite the insights that Mercury's spectral reflectance properties provide on the four geologic issues discussed above, one issue that is poorly constrained by spectral reflectance is the spatial variation in major element composition. For example, the two large northern hemisphere deposits of HRP, the northern smooth plains and Caloris interior plains, are lower in Fe, Mg, Ca, and S and higher in Al than nearby terrain (Weider et al., 2012, 2014, 2015; Chapters 2 and 7). However, none of these elements exhibits a correlation with reflectance globally. Areas with high Al other than the two large HRP deposits are low in reflectance (Weider et al., 2015). And even though large areas of HRP are low in magnesium, the "high-Mg region" of intercrater plains is undistinguished by its spectral reflectance (Weider et al., 2015).

Recall that three major factors affecting spectral reflectance are the types and abundances of silicates containing large amounts of Fe^{2+}, the presence and abundance of opaque phases, and the extent of space weathering. The Moon has a surface composition dominated by two very different lithologies, high-reflectance, high-Al, low-Fe anorthosite, and low-reflectance, high-Fe basalt. Reflectance is strongly correlated with Fe and Al abundances, once corrections are made for the effects of space weathering (Lucey et al., 1995, 1998; Blewett et al., 1997a). In this context, the reason for Mercury's lack of correlation between spectral reflectance and major element composition is clear: Mercury lacks Fe^{2+} in crustal silicates that is both abundant and exhibits large spatial variations. Olivine, pyroxene, and feldspar, which on the Moon differ strongly in spectral reflectance, are all bright and spectrally featureless on Mercury because they are low in Fe^{2+}. The dominant spectral trend, between HRP and LRM, is instead the result of spatial variations in the content of a relatively minor opaque phase (the second factor controlling color and reflectance). Mercury's secondary spectral trend, brightening and a less red spectral slope in fresh crater and bright hollows materials, results from spatial differences in the third factor controlling color and reflectance, space weathering.

8.7 SYNTHESIS

The ultraviolet through short-wavelength infrared spectrum of Mercury is red-sloped and relatively featureless. Unlike the lunar surface, Mercury's surface lacks crystal-field absorptions due to transition metals in silicate, principally iron. However some mercurian materials exhibit either of two weak features at these wavelengths, an OMCT band at <0.3 μm and a broad, shallow feature at 0.6 μm. Additionally, there are variations in overall reflectance, spectral slope, and directionality of scattering. These spectral reflectance properties, taken in the context of elemental abundance measurements, provide several insights into the composition and evolution of Mercury's crust:

(1) Mercury's surface is lower in reflectance than the surface of the Moon because of the presence of opaque phases. At least in LRM, and probably globally, the major opaque phase is inferred to be graphite on the basis of low reflectance, a broad absorption near 0.6 μm, UV spectral modeling, and low-altitude neutron spectroscopy.

(2) Without large variations in ferrous iron in silicates to create differences in reflectance and spectral slope such as those between the lunar maria and highlands, spatial variations in spectral reflectance result instead from two major variables: variations in the content of opaque phases, which form a continuum between LRM and the red unit, and the extent of space weathering, which results in differences in overall reflectance and spectral continuum.

(3) Spatial variations in spectral reflectance serve as stratigraphic markers that enable reconstruction of crustal structure. The dominant surface units are volcanic plains, which display subtle color differences. Exposures of buried plains by impact features reveal that volcanic materials constitute at least the uppermost 5 km thickness of the typical crust. The plains overlie LRM that probably originated by impact and magmatic mixing of a primary, graphite flotation crust with overlying and underlying rock units.

(4) On average, Mercury's surface is more backscattering than that of the Moon and possibly smoother at very small scales. Such a surface might have resulted from a generally higher velocity of impacting bolides and consequently higher degrees of impact melting, with formation of glassy grains having abundant inclusions. Locally, bright hollows materials and pyroclastic deposits forming the red spectral unit are more forward scattering, possibly indicating that they consist of a blanket formed of finer particles and/or have smoother surfaces.

(5) Spectral properties of minor bright units provide evidence for their origins. The pyroclastic red unit has a UV absorption edge absent from other materials, interpreted to result from an OMCT band in material with a very low, ~0.1 wt%, content of Fe^{2+}. This feature is consistent with predictions of geochemical models for explosive volcanic eruptions that invoke oxidation of graphite and sulfides during magma ascent by reduction of FeO and SiO_2, in the process unsaturating the OMCT band and thus strengthening the UV absorption edge. Consumption of graphite and sulfides would brighten the pyroclastic material at visible wavelengths and create exsolved C- and S-bearing volatiles to drive eruptions. In contrast, the haloes surrounding hollows are brighter than their substrates but have a bluer spectral slope. In the UV, the haloes exhibit a UV turndown where the substrate does but otherwise are bland. These contrasts between hollow haloes and their substrates are consistent with exposure of less space-weathered material by hollows-forming processes.

REFERENCES

Adams, J. B. and McCord, T. B. (1970). Remote sensing of lunar surface mineralogy: Implications from visible and near-infrared reflectivity of Apollo 11 samples. *Proc. Apollo 11 Lunar Science Conf.*, **3**, 1937–1945.

Blewett, D. T., Lucey, P. G., Hawke, B. R. and Jolliff, B. L. (1997a). Clementine images of the lunar sample-return stations: Refinement of FeO and TiO_2 mapping techniques. *J. Geophys. Res.*, **102**, 16319–16325.

Blewett, D. T., Lucey, P. G., Hawke, B. R., Ling, G. G. and Robinson, M. S. (1997b). A comparison of mercurian reflectance and spectral quantities with those of the Moon. *Icarus*, **129**, 217–231.

Blewett, D. T., Hawke, B. R., Lucey, P. G. and Robinson, M. S. (2007). A Mariner 10 color study of Mercurian craters. *J. Geophys. Res.*, **112**, E02005, doi:10.1029/2006JE002713.

Blewett, D. T., Robinson, M. S., Denevi, B. W., Gillis-Davis, J. J., Head, J. W., Solomon, S. C., Holsclaw, G. M. and McClintock, W. E. (2009). Multispectral images of Mercury from the first MESSENGER flyby: Analysis of global and regional color trends. *Earth Planet. Sci. Lett.*, **285**, 272–282.

Blewett, D. T., Chabot, N. L., Denevi, B. W., Ernst, C. M., Head, J. W., Izenberg, N. R., Murchie, S. L., Solomon, S. C., Nittler, L. R., McCoy, T. J., Xiao, Z., Baker, D. M. H., Fassett, C. I., Braden, S. E., Oberst, J., Scholten, F., Preusker, F. and Hurwitz, D. M. (2011). Hollows on Mercury: Evidence from MESSENGER for geologically recent volatile-related activity. *Science*, **333**, 1856–1859.

Blewett, D. T., Vaughan, W. V., Xiao, Z., Chabot, N. L., Denevi, B. W., Ernst, C. M., Helbert, J., D'Amore, M., Maturilli, A., Head, J. W. and Solomon, S. C. (2013). Mercury's hollows: Constraints on formation and composition from analysis of geological setting and spectral reflectance. *J. Geophys. Res. Planets*, **118**, 1013–1032.

Blewett, D. T., Levy, C. L., Chabot, N. L., Denevi, B. W., Ernst, C. M. and Murchie, S. L (2014). Phase-ratio images of the surface of Mercury: Evidence for differences in sub-resolution texture. *Icarus*, **242**, 142–148.

Blewett, D. T., Stadermann, A., Susorney, H. C., Ernst, C. M., Xiao, Z., Chabot, N. L., Denevi, B. W., Murchie, S. L., McCubbin, F. M., Kinczyk, M. J., Gillis-Davis, J. J. and and Solomon, S. C. (2016). Analysis of MESSENGER high-resolution images of Mercury's hollows and implications for hollow formation. *J. Geophys. Res. Planets*, **121**, 1798–1813, doi:10.1002/2016JE005070.

Braden, S. E. and Robinson, M. S. (2013). Relative rates of optical maturation of regolith on Mercury and the Moon. *J. Geophys. Res. Planets*, **118**, 1903–1914.

Bruck Syal, M., Schultz, P. H. and Riner, M. A. (2015). Darkening of Mercury's surface by cometary carbon. *Nature Geosci.*, **8**, 352–356.

Charette, M. P., McCord, T. B., Pieters, C. M. and Adams, J. B. (1974). Application of remote spectral reflectance measurements to lunar geology classification and determination of titanium content of lunar soils. *J. Geophys. Res.*, **79**, 1605–1613.

Charlier, B., Grove, T. L. and Zuber, M. T. (2013). Phase equilibria of ultramafic compositions on Mercury and the origin of the compositional dichotomy. *Earth Planet. Sci. Lett.*, **363**, 50–60.

Cintala, M. J. (1992). Impact-induced thermal effects in the lunar and mercurian regoliths. *J. Geophys. Res.*, **97**, 947–973.

Clark, B. E., Veverka, J., Helfenstein, P., Thomas, P. C., Bell, J. F., Harch, A., Robinson, M. S., Murchie, S. L., McFadden, L. A. and Chapman, C. R. (1999). NEAR photometry of asteroid 253 Mathilde. *Icarus*, **140**, 53–65.

Clegg, R. N., Jolliff, B. L., Robinson, M. S., Hapke, B. W. and Plescia, J. B. (2014). Effects of rocket exhaust on lunar soil reflectance properties. *Icarus*, **227**, 176–194.

Cloutis, E. A., McCormack, K. A., Bell, J. F., Hendrix, A. R., Bailey, D. T., Craig, M. A., Mertzman, S. A., Robinson, M. S. and Riner, M. A. (2008). Ultraviolet spectral reflectance properties of common planetary minerals. *Icarus*, **197**, 321–347.

Cloutis, E. A., Hudon, P., Hiroi, T., Gaffey, M. J. and Mann, P. (2011). Spectral reflectance properties of carbonaceous chondrites: 2. CM chondrites. *Icarus*, **216**, 309–346.

D'Amore, M., Helbert, J., Maturilli, A., Head, J. W., Sprague, A. L., Izenberg, N. R., Holsclaw, G. M., McClintock, W. E., Vilas, F. and Solomon, S. C. (2012). Global classification of MESSENGER spectral reflectance data and a detailed look at Rudaki plains. *Lunar Planet. Sci.*, **43**, abstract 1413.

D'Amore, M., Helbert, J., Holsclaw, G. M., Izenberg, N. R., McClintock, W. E., Head, J. W. and Solomon, S. C. (2013). Exploiting the Mercury surface reflectance spectroscopy dataset from MESSENGER: Making sense of three million spectra. *Lunar Planet. Sci.*, **44**, abstract 1900.

D'Amore, M., Helbert, J., Ferrari, S., Maturilli, A., Nittler, L. R., Domingue, D. L., Weider, S. Z., Starr, R. D., Crapster-Pregont, E. J., Ebel, D. S. and Solomon, S. C. (2014). Unsupervised classification of Mercury's visible–near-infrared reflectance spectra: Comparison with major element compositions. *Lunar Planet. Sci.*, **45**, abstract 1073.

Davis, J. C. (1973). *Statistics and Data Analysis in Geology*. New York: John Wiley & Sons, 550 pp.

Denevi, B. W. and Robinson, M. S. (2008). Mercury's albedo from Mariner 10: Implications for the presence of ferrous iron. *Icarus*, **197**, 239–246.

Denevi, B. W., Robinson, M. S., Solomon, S. C., Murchie, S. L., Blewett, D. T., Domingue, D. L., McCoy, T. J., Ernst, C. M., Head, J. W., Watters, T. R. and Chabot, N. L. (2009). The evolution of Mercury's crust: A global perspective from MESSENGER. *Science*, **324**, 613–618.

Denevi, B. W., Ernst, C. M., Meyer, H. M., Robinson, M. S., Murchie, S. L., Whitten, J. L., Head, J. W., Watters, T. R., Solomon, S. C., Ostrach, L. R., Chapman, C. R., Byrne, P. K., Klimczak, C. and Peplowski P. N. (2013a). The distribution and origin of smooth plains on Mercury. *J. Geophys. Res. Planets*, **118**, 891–907.

Denevi, B. W., Ernst, C. M., Whitten, J. L., Head, J. W., Murchie, S. L., Watters, T. R., Byrne, P. K., Blewett, D. T., Solomon, S. C. and Fassett C. I. (2013b). The volcanic origin of a region of intercrater plains on Mercury. *Lunar Planet. Sci.*, **44**, abstract 1218.

Denevi, B. W., Chabot, N. L., Murchie, S. L., Becker, K. J., Blewett, D. T., Domingue, D. I., Ernst, C. M., Hash, C. D., Hawkins, S. E., III, Keller, M. R., Laslo, N. R., Nair, H., Robinson, M. S., Seelos, F. P., Stephens, G. K., Turner, F. S. and Solomon, S. C. (2018). Calibration, projection, and final image products of MESSENGER's Mercury Dual Imaging System. *Space Sci. Rev.*, **214**, 2.

Domingue, D. L., Robinson, M., Carcich, B., Joseph, J., Thomas, P. and Clark, B. E. (2002). Disk-integrated photometry of 433 Eros. *Icarus*, **155**, 205–219.

Domingue, D. L., Vilas, F., Holsclaw, G. M., Warell, J., Izenberg, N. R., Murchie, S. L., Denevi, B. W., Blewett, D. T., McClintock, W. E., Anderson, B. J. and Sarantos, M. (2010). Whole-disk spectrophotometric properties of Mercury: Synthesis of MESSENGER and ground-based observations. *Icarus*, **209**, 101–124.

Domingue, D. L., Chapman, C. R., Killen, R. M., Zurbuchen, T. II., Gilbert, J. A., Sarantos, M., Benna, M., Slavin, J. A., Schriver, D., Travnicek, P. M., Orlando, T. M., Sprague, A. L., Blewett, D. T., Gillis-Davis, J. J., Feldman, W. C., Lawrence, D. J., Ho, G. C.,

Ebel, D. S., Nittler, L. R., Vilas, F., Pieters, C. M., Solomon, S. C., Johnson, C. L., Winslow, R. M., Helbert, J., Peplowski, P. N., Weider, S. Z., Izenberg, N. R. and McClintock, W. E. (2014). Mercury's weather-beaten surface: Understanding Mercury in the context of lunar and asteroidal space weathering studies. *Space Sci. Rev.*, **181**, 121–214.

Domingue, D. L., Murchie, S. L., Denevi, B. W., Ernst, C. M. and Chabot, N. L. (2015). Mercury's global color mosaic: An update from MESSENGER's orbital observations. *Icarus*, **257**, 477–488.

Domingue, D. L., Denevi, B. W., Murchie, S. L. and Hash, C. (2016). Application of multiple photometric models to disk-resolved measurements of Mercury's surface: Insights into Mercury's regolith characteristics. *Icarus*, **268**, 172–203.

Emery, J. P., Sprague, A. L., Witteborn, F. C., Colwell, J. E., Kozlowski, R. W. H. and Wooden, D. H. (1998). Mercury: Thermal modeling and mid-infrared (5–12 µm) observations. *Icarus*, **136**, 104–123.

Ernst, C. M., Murchie, S. L., Barnouin-Jha, O. S., Robinson, M. S., Denevi, B. W., Blewett, D. T., Head, J. W., Izenberg, N. R. and Solomon, S. C. (2010). Exposure of spectrally distinct material by impact craters on Mercury: Implications for global stratigraphy. *Icarus*, **209**, 210–223.

Ernst, C. M., Denevi, B. W., Barnouin, O. S., Klimczak, C., Chabot, N. L., Head, J. W., Murchie, S. L., Neumann, G. A., Prockter, L. M., Robinson, M. S., Solomon, S. C. and Watters, T. R. (2015). Stratigraphy of the Caloris basin, Mercury: Implications for volcanic history and basin impact melt. *Icarus*, **250**, 413–429.

Evans, L. G., Peplowski, P. N., Rhodes, E. A., Lawrence, D. J., McCoy, T. J., Nittler, L. R., Solomon, S. C., Sprague, A. L., Stockstill-Cahill, K. R., Starr, R. D., Weider, S. Z., Boynton, W. V. and Hamara, D. K. (2012). Major-element abundances on the surface of Mercury: Results from the MESSENGER Gamma-Ray Spectrometer. *J. Geophys. Res.*, **117**, E00L07, doi:10.1029/2012JE004178.

Fassett, C. I., Head, J. W., Blewett, D. T., Chapman, C. R., Dickson, J. L., Murchie, S. L., Solomon, S. C. and Watters, T. R. (2009), Caloris impact basin: Exterior geomorphology, stratigraphy, morphometry, radial sculpture, and smooth plains deposits. *Earth Planet. Sci. Lett.*, **285**, 297–308, doi:10.1016/j.epsl.2009.05.022.

Fassett, C. I., Head, J. W., Baker, D. M. H., Zuber, M. T., Smith, D. E., Neumann, G. A., Solomon, S. C., Klimczak, C., Strom, R. G., Chapman, C. R., Prockter, L. M., Phillips, R. J., Oberst, J. and Preusker, F. (2012). Large impact basins on Mercury: Global distribution, characteristics, and modification history from MESSENGER orbital data. *J. Geophys. Res.*, **117**, E00L08, doi:10.1029/2012JE004154.

Fischer, E. M. and Pieters, C. M. (1994). Remote determination of exposure degree and iron concentration of lunar soils using VIS-NIR spectroscopic methods. *Icarus*, **111**, 475–488.

Gaffey, S. J., McFadden, L. A., Nash, D. and Pieters, C. M. (1993). Ultraviolet, visible, and near-infrared reflectance spectroscopy: Laboratory spectra of geologic materials. In *Remote Geochemical Analysis: Elemental and Mineralogic Composition*, ed. C. M. Pieters and P. A. J. Englert. Cambridge: Cambridge University Press, pp. 43–78.

Gillis-Davis, J. J., van Niekerk, D., Scott, E. R. D., McCubbin, F. M. and Blewett, D. T. (2013). Impact darkening: A possible mechanism to explain why Mercury is spectrally dark and featureless. Presented at 2013 Fall Meeting, American Geophysical Union, abstract P11A-07, San Francisco, CA, 9–13 December.

Goguen, J. D., Stone, T. C., Kieffer, H. H. and Buratti, B. J. (2010). A new look at photometry of the Moon, *Icarus*, **208**, 548–557.

Goldsten, J. O., Rhodes, E. A., Boynton, W. V., Feldman, W. C., Lawrence, D. J., Trombka, J. I., Smith, D. M., Evans, L. G., White, J., Madden, N. W., Berg, P. C., Murphy, G. A., Gurnee, R. S., Strohbehn, K., Williams, B. D., Schaefer, E. D., Monaco, C. A., Cork, C. P., Eckels, J. D., Miller, W. O., Burks, M. T., Hagler, L. B., DeTeresa, S. J. and Witte, M. C. (2007). The MESSENGER Gamma-Ray and Neutron Spectrometer. *Space Sci. Rev.*, **131**, 339–391.

Goudge, T. A., Head, J. W., Kerber, L., Blewett, D. T., Denevi, B. W., Domingue, D. L., Gillis-Davis, J. J., Gwinner, K., Helbert, J., Holsclaw, G. M., Izenberg, N. R., Klima, R. L., McClintock, W. E., Murchie, S. L., Neumann, G. A., Smith, D. E., Strom, R. G., Xiao, Z., Zuber, M. T. and Solomon, S. C. (2014). Global inventory and characterization of pyroclastic deposits on Mercury: New insights into pyroclastic activity from MESSENGER orbital data. *J. Geophys. Res. Planets*, **119**, 635–658.

Green, R. O., Pieters, C., Mouroulis, P., Eastwood, M., Boardman, J., Glavich, T., Isaacson, P., Annadurai, M., Besse, S., Barr, D., Buratti, B., Cate, D., Chatterjee, A., Clark, R., Cheek, L., Combe, J., Dhingra, D., Essandoh, V., Geier, S., Goswami, J. N., Green, R., Haemmerle, V., Head, J., Hovland, L., Hyman, S., Klima, R., Koch, T., Kramer, G., Kumar, A. S. K., Lee, K., Lundeen, S., Malaret, E., McCord, T., McLaughlin, S., Mustard, J., Nettles, J., Petro, N., Plourde, K., Racho, C., Rodriquez, J., Runyon, C., Sellar, G., Smith, C., Sobel, H., Staid, M., Sunshine, J., Taylor, L., Thaisen, K., Tompkins, S., Tseng, H., Vane, G., Varanasi, P., White, M. and Wilson, D. (2011). The Moon Mineralogy Mapper (M3) imaging spectrometer for lunar science: Instrument description, calibration, on-orbit measurements, science data calibration and on-orbit validation. *J. Geophys. Res.*, **116**, E00G19, doi:10.1029/2011JE003797.

Hapke, B. (1977). Interpretations of optical observations of Mercury and the Moon, *Phys. Earth Planet. Inter.*, **15**, 264–274.

Hapke, B. (1981). Bidirectional reflectance spectroscopy: 1. Theory. *J. Geophys. Res.*, **86**, 3039–3054.

Hapke, B. (1984). Bidirectional reflectance spectroscopy, 3: Correction for macroscopic roughness. *Icarus*, **59**, 41–59.

Hapke, B. (1986). Bidirectional reflectance spectroscopy, 4: The extinction coefficient and opposition effect. *Icarus*, **67**, 264–280.

Hapke, B. (1993). *Theory of Reflectance and Emittance Spectroscopy*. Cambridge: Cambridge University Press, 455 pp.

Hapke, B. (2001). Space weathering from Mercury to the asteroid belt. *J. Geophys. Res.*, **106**, 10,039–10,073.

Hapke, B. (2002). Bidirectional reflectance spectroscopy, 5: The coherent backscatter opposition effect and anisotropic scattering. *Icarus*, **157**, 523–534.

Hapke, B. (2008). Bidirectional reflectance spectroscopy. 6. Effects of porosity. *Icarus*, **195**, 918–926.

Hapke, B. (2012a). *Theory of Reflectance and Emittance Spectroscopy*, 2nd edn. Cambridge: Cambridge University Press, 513 pp.

Hapke, B., (2012b). Bidirectional reflectance spectroscopy, 7. The single particle phase function hockey stick relation. *Icarus*, **221**, 1079–1083.

Hapke, B., Danielson, G. E., Klaasen, K. and Wilson, L. (1975). Photometric observations of Mercury from Mariner 10. *J. Geophys. Res.*, **80**, 2431–2443.

Haskin, L. and Warren, P. (1991). Lunar chemistry. In *A User's Guide to the Moon*, ed. G. H. Heiken, D. T. Vaniman and B. M. French. New York: Cambridge University Press, pp. 357–474.

Hawkins, S. E., III, Boldt, J. D., Darlington, E. H., Espiritu, R., Gold, R. E., Gotwols, B., Grey, M. P., Hash, C. D., Hayes, J. R., Jaskulek, S. E., Kardian, C. J., Jr., Keller, M. R., Malaret, E. R., Murchie, S. L., Murphy, P. K., Peacock, K., Prockter, L. M., Reiter, R. A., Robinson, M. S., Schaefer, E. D., Shelton, R. G., Sterner, R. E., II, Taylor, H. W., Watters, T. R. and Williams, B. D. (2007). The Mercury Dual Imaging System on the MESSENGER spacecraft. *Space Sci. Rev.*, **131**, 247–338.

Hawkins, S. E., III, Murchie, S. L., Becker, K. J., Selby, C. M., Turner, F. S., Noble, M. W., Chabot, N. L., Choo, T. H., Darlington, E. H., Denevi, B. W., Domingue, D. L., Ernst, C. M., Holsclaw, G. M., Laslo, N. H., McClintock, W. E., Prockter, L. M., Robinson, M. S., Solomon, S. C. and Sterner, R. E. (2009). In-flight performance of MESSENGER's Mercury Dual Imaging System. In *Instruments and Methods for Astrobiology and Planetary Missions XII*, SPIE Proceedings, Vol. 7441, ed. R. B. Hoover, G. V. Levin, A. Y. Rozanov and K. D. Retherford. Paper 74410Z, 12 pp.

Head, J. W., Murchie, S. L., Prockter, L. M., Robinson, M. S., Solomon, S. C., Strom, R. G., Chapman, C. R., Watters, T. R., McClintock, W. E., Blewett, D. T. and Gillis-Davis, J. J. (2008). Volcanism on Mercury: Evidence from the first MESSENGER flyby. *Science*, **321**, 69–72.

Head, J. W., Murchie, S. L., Prockter, L. M., Solomon, S. C., Chapman, C. R., Strom, R. G., Watters, T. R., Blewett, D. T., Gillis-Davis, J. J., Fassett, C. I., Dickson, J. L., Morgan, G. A. and Kerber, L. (2009). Volcanism on Mercury: Evidence from the first MESSENGER flyby for extrusive and explosive activity and the volcanic origin of plains. *Earth Planet. Sci. Lett.*, **285**, 227–242.

Head, J. W., Chapman, C. R., Strom, R. G., Fassett, C. I., Denevi, B. W., Blewett, D. T., Ernst, C. M., Watters, T. R., Solomon, S. C., Murchie, S. L., Prockter, L. M., Chabot, N. L., Gillis-Davis, J. J., Whitten, J. L., Goudge, T. A., Baker, D. M. H., Hurwitz, D. M., Ostrach, L. R., Xiao, Z., Merline, W. J., Kerber, L., Dickson, J. L., Oberst, J., Byrne, P. K., Klimczak, C. and Nittler, L. R. (2011). Flood volcanism in the high northern latitudes of Mercury revealed by MESSENGER. *Science*, **333**, 1853–1856.

Helbert, J., D'Amore, M., Izenberg, N. R., Domingue, D. L., Head, J. W., D'Incecco, P., Maturilli, A., Holsclaw, G. M., McClintock, W. E. and Solomon, S. C. (2012). Surface units on Mercury defined by unsupervised classification analysis of MESSENGER spectral reflectance data from the first year in orbit. Presented at 2012 Fall Meeting, American Geophysical Union, abstract P33B-1939, San Francisco, CA, 10–14 December.

Helbert, J., Maturilli, A. and D'Amore, M. (2013). Visible and near-infrared reflectance spectra of thermally processed synthetic sulfides as a potential analog for the hollow forming materials on Mercury. *Earth Planet. Sci. Lett.*, **369**, 233–238.

Helfenstein, P. and Shepard, M. K. (1999). Submillimeter-scale topography of the lunar regolith. *Icarus*, **141**, 107–131.

Helfenstein, P. and Shepard, M. K. (2011). Testing the Hapke photometric model: Improved inversion and the porosity correction. *Icarus*, **215**, 83–100.

Hendrix, A. R. and Vilas, F. (2006). The effects of space weathering at UV wavelengths: S-class asteroids. *Astron. J.*, **132**, 1396–1404.

Hendrix, A. R., Retherford, K. D., Gladstone, G. R., Hurley, D. M., Feldman, P. D., Egan, A. F., Kaufmann, D. E., Miles, P. F., Parker, J. W., Horvath, D., Rojas, P. M., Versteeg, M. H., Davis, M. W., Greathouse, T. K., Mukherjee, J., Steffl, A. J., Pryor, W. R. and Stern, S. A. (2012). The lunar far-UV albedo: Indicator of hydration and weathering. *J. Geophys. Res.*, **117**, E12001, doi:10.1029/2012JE004252.

Izenberg, N. R. and Ward, J. G. (2016). MESSENGER MASCS VIRS Calibrated Data Record, Derived Data Record, and Derived Analysis Product Software Interface Specification. Available online, http://pds-geosciences.wustl.edu/messenger/mess-e_v_h-mascs-3-virs-cdr-caldata-v1/messmas_2001/document/virs_cdr_ddr_dap_sis.pdf.

Izenberg, N. R., Klima, R. L., Murchie, S. L., Blewett, D. T., Holsclaw, G. M., McClintock, W. E., Malaret, E., Mauceri, C., Vilas, F., Sprague, A. L., Helbert, J., Domingue, D. L., Head, J. W., III, Goudge, T. A., Solomon, S. C., Hibbitts, C. A. and Dyar, M. D. (2014). The low-iron, reduced surface of Mercury as seen in spectral reflectance by MESSENGER. *Icarus*, **228**, 364–374.

Izenberg, N. R., Thomas, R. J., Blewett, D. T. and Nittler, L. R. (2015). Are there compositionally different types of hollows on Mercury? *Lunar Planet. Sci.*, **46**, abstract 1344.

Izenberg, N. R., Domingue, D. L., Holsclaw, G. M. and McClintock, W. E. (2016). Photometric normalization error in MASCS pipeline: Effects and remediation. Available online, http://pds-geosciences.wustl.edu/messenger/mess-e_v_h-mascs-3-virs-cdr-caldata-v1/messmas_2001/document/mascs_photometric_error.pdf.

Kaasalainen, M., Torppa, J. and Muinonen K. (2001). Optimization methods for asteroid lightcurve inversion. *Icarus*, **153**, 37–51.

Kaydash, V. G. and Shkuratov, Yu. G. (2012). Structural disturbances of the lunar surface caused by spacecraft. *Solar Syst. Res.*, **46**, 108–118.

Kaydash, V. G. and Shkuratov, Yu. G. (2014). Structural disturbances of the lunar surface near the Lunokhod-1 spacecraft landing site. *Solar Syst. Res.*, **48**, 167–175.

Kaydash, V., Shkuratov, Yu., Korokhin, V. and Videen, G. (2011). Photometric anomalies in the Apollo landing sites as seen from the Lunar Reconnaissance Orbiter. *Icarus*, **211**, 89–96.

Kaydash, V., Shkuratov, Yu. and Videen, G. (2012). Phase-ratio imagery as a planetary remote-sensing tool. *J. Quant. Spectrosc. Radiat. Trans.*, **113**, 2601–2607.

Kaydash, V., Shkuratov, Yu. and Videen, G. (2014). Dark halos and rays of young lunar craters: A new insight into interpretation. *Icarus*, **231**, 22–33.

Kerber, L., Head, J. W., Solomon, S. C., Murchie, S. L., Blewett, D. T. and Wilson, L. (2009). Explosive volcanic eruptions on Mercury: Eruption conditions, magma volatile content, and implications for interior volatile abundances. *Earth Planet. Sci. Lett.*, **285**, 263–271.

Kerber, L., Head, J. W., Blewett, D. T., Solomon, S. C., Wilson, L., Murchie, S. L., Robinson, M. S., Denevi, B. W. and Domingue, D. L. (2011). The global distribution of pyroclastic deposits on Mercury: The view from MESSENGER flybys 1–3. *Planet. Space Sci.*, **59**, 1895–1909.

Kerber, L., Besse, S., Head, J. W., Blewett, D. T., Goudge, T. A. and Jussieu, P. (2014). The global distribution of pyroclastic deposits on Mercury: The view from orbit. *Lunar Planet. Sci.*, **45**, abstract 2862.

Killen, R., Shemansky, D. and Mouawad, N. (2009). Expected emission from Mercury's exospheric species, and their ultraviolet–visible signatures. *Astrophys. J. Supp.*, **181**, 351–359.

Klima, R. L., Dyar, M. D. and Pieters, C. M. (2011). Near-infrared spectra of clinopyroxenes: Effects of calcium content and crystal structure. *Meteorit. Planet. Sci.*, **42**, 235–253.

Klima, R. L., Izenberg, N. R., Murchie, S. L., Meyer, H. M., Stockstill-Cahill, K. R., Blewett, D. T., D'Amore, M., Denevi, B. W., Ernst, C. M., Helbert, J., McCoy, T. J., Sprague, A. L., Vilas, F. and Weider, S. Z. (2013). Constraining the ferrous iron content of minerals in Mercury's crust. *Lunar Planet. Sci.*, **44**, abstract 1602.

Klima, R. L., Denevi, B. W., Ernst, C. M., Murchie, S. L. and Peplowski, P. N. (2018). Global distribution and spectral properties of low-reflectance material on Mercury. *Geophys. Res. Lett.*, **45**, 2945–2953.

Kreslavsky, M. A. and Shkuratov, Yu. G. (2003). Photometric anomalies of the lunar surface: Results from Clementine data. *J. Geophys. Res.*, **108**, 5015, doi:10.1029/2002JE001937, E3.

Li, J.-Y., Le Corre, L., Schröder, S. E., Reddy, V., Denevi, B. W., Buratti, B. J., Mottola, S., Hoffmann, M., Gutierrez-Marques, P., Nathues, A., Russell, C. T. and Raymond, C. A. (2013). Global photometric properties of asteroid (4) Vesta observed with Dawn Framing Camera. *Icarus*, **226**, 1252–1274.

Lodders, K. and Fegley, B. (1998). *The Planetary Scientist's Companion*. New York: Oxford University Press.

Lucey, P. G. and Riner, M. A. (2011). The optical effects of small iron particles that darken but do not redden: Evidence of intense space weathering on Mercury. *Icarus*, **212**, 451–462.

Lucey, P. G., Taylor, G. J. and Malaret, E. (1995). Abundance and distribution of iron on the Moon. *Science*, **268**, 1150–1153.

Lucey, P. G., Blewett D. T. and Hawke, B. R. (1998). Mapping the FeO and TiO_2 content of the lunar surface with multispectral imaging. *J. Geophys. Res.*, **103**, 3679–3699.

Lucey, P. G., Blewett, D. T. and Jolliff, B. L. (2000a). Lunar iron and titanium abundance algorithms based on final processing of Clementine UVVIS data. *J. Geophys. Res.*, **105**, 20,297–20,306.

Lucey, P. G., Blewett, D. T., Taylor, G. J. and Hawke, B. R. (2000b). Imaging of lunar surface maturity. *J. Geophys. Res.*, **105**, 20,377–20,386.

Lumme, K. and Bowell, E. (1981). Radiative transfer in the surfaces of atmosphereless bodies. I. Theory. *Astron. J.*, **86**, 1694–1704.

Mallama, A., Wang. D. and Howard, R. A. (2002). Photometry of Mercury from SOHO/LASCO and Earth: The phase function from 2 to 170°. *Icarus*, **155**, 253–264.

Maxwell, R. E., Izenberg, N. R. and Holsclaw, G. M. (2016). Implications for iron and carbon in Mercury surface materials from ultraviolet reflectance. *Lunar Planet. Sci.*, **47**, abstract 1606.

McClintock, W. E. and Lankton, M. R. (2007). The Mercury Atmospheric and Surface Composition Spectrometer for the MESSENGER mission. *Space Sci. Rev.*, **131**, 481–522.

McClintock, W. E., Izenberg, N. R., Holsclaw, G. M., Blewett, D. T., Domingue, D. L., Head, J. W., III, Helbert, J., McCoy, T. J., Murchie, S. L., Robinson, M. S., Solomon, S. C., Sprague, A. L. and Vilas, F. (2008). Spectroscopic observations of Mercury's surface reflectance during MESSENGER's first Mercury flyby. *Science*, 321, 62–65, doi:10.1126/science.1159933.

McCord, T. B. and Adams, J. B. (1972a). Mercury: Surface composition from the reflection spectrum. *Science*, **178**, 745–747.

McCord, T. B. and Adams, J. B. (1972b). Mercury: Interpretation of optical observations. *Icarus*, **17**, 585–588.

McCord, T. B. and Clark, R. N. (1979). The Mercury soil: Presence of Fe^{2+}. *J. Geophys. Res.*, **84**, 7664–7668.

McGuire, A. and Hapke, B. (1995). An experimental study of light scattering by large irregular particles. *Icarus*, **113**, 134–155.

Murchie, S. L., Watters, T. R., Robinson, M. S., Head, J. W., Strom, R. G., Chapman, C. R., Solomon, S. C, McClintock, W. E., Prockter, L. M., Domingue, D. L. and Blewett, D. T. (2008). Geology of the Caloris basin, Mercury: A view from MESSENGER. *Science*, **321**, 73–76.

Murchie, S. L., Klima, R. L., Denevi, B. W., Ernst, C. M., Keller, M. R., Domingue, D. L., Blewett, D. T., Chabot, N. L., Hash, C. D., Malaret, E., Izenberg, N. R., Vilas, F., Nittler, L. R., Gillis-Davis, J. J., Head, J. W. and Solomon, S. C. (2015). Orbital multispectral mapping of Mercury with the MESSENGER Mercury Dual Imaging System: Evidence for the origins of plains units and low-reflectance material. *Icarus*, **254**, 287–305.

Nittler, L. R., Starr, R. D., Weider, S. Z., McCoy, T. J., Boynton, W. V., Ebel, D. S., Ernst, C. M., Evans, L. G., Goldsten, J. O., Hamara, D. K., Lawrence, D. J., McNutt, R. L., Jr., Schlemm, C. E., II, Solomon, S. C. and Sprague, A. L. (2011). The major-element composition of Mercury's surface from MESSENGER X-ray spectrometry. *Science*, **333**, 1847–1850.

Peplowski, P. N., Lawrence, D. J., Evans, L. G., Klima, R. L., Blewett, D. T., Goldsten, J. O., Murchie, S. L., McCoy, T. J., Nittler, L. R., Solomon, S. C., Starr, R. D. and Weider, S. Z. (2015a). Constraints on the abundance of carbon in near-surface materials on Mercury: Results from the MESSENGER Gamma-Ray Spectrometer. *Planet. Space Sci.*, **108**, 98–107.

Peplowski, P. N., Bazell, D., Evans, L. G., Goldsten, J. O., Lawrence, D. J. and Nittler, L. R. (2015b). Hydrogen and major element concentrations on 433 Eros: Evidence for an L- or LL-chondrite-like surface composition. *Meteorit. Planet. Sci.*, **50**, 353–367.

Peplowski, P. N., Klima, R. L., Lawrence, D. J., Ernst, C. M., Denevi, B. W., Frank, E. A., Goldsten, J. O., Murchie, S. L., Nittler, L. R. and Solomon, S. C. (2016). Remote sensing evidence for an ancient carbon-bearing crust on Mercury. *Nature Geosci.*, 9, 273–276, doi:10.1038/ngeo2669.

Pieters, C. M. (1993). Compositional diversity and stratigraphy of the lunar crust derived from reflectance spectroscopy. In *Remote Geochemical Analysis: Elemental and Mineralogic Composition*, ed. C. M. Pieters and P. A. J. Englert. Cambridge: Cambridge University Press, pp. 309–340.

Prockter, L. M., Murchie, S. L., Solomon, S. C., Nittler, L. R., McNutt, R. L., Jr., Chabot, N. L., Lawrence, D. J., Evans, L. G., Johnson, C. L., Phillips, R. J., Vervack, R. J., Jr., Korth, H., Perry, M. E., Bedini, P. D. and Winters, H. L. (2013). MESSENGER's second extended mission: Exploring Mercury's dynamic magnetosphere and complex surface at unprecedented scales. *Lunar Planet. Sci.*, **44**, abstract 2907.

Rava, B. and Hapke, B. (1987). An analysis of the Mariner 10 color ratio map of Mercury. *Icarus*, **71**, 397–429.

Rivera-Valentin, E. G. and Barr, A. C. (2014). Impact-induced compositional variations on Mercury. *Earth Planet. Sci. Lett.*, **391**, 234–242.

Robinson, M. S. and Lucey, P. G. (1997). Recalibrated Mariner 10 color mosaics: Implications for mercurian volcanism. *Science*, **275**, 197–200.

Robinson, M. S. and Taylor, G. J. (2001). Ferrous oxide in Mercury's crust and mantle. *Meteorit. Planet. Sci.*, **36**, 841–847.

Robinson, M. S., Hawke, B. R., Lucey, P. G. and Smith, G. A. (1992). Mariner 10 multispectral images of the eastern limb and farside of the Moon. *J. Geophys. Res.*, **97**, 18,265–18,274.

Robinson, M. S., Murchie, S. L., Blewett, D. T., Domingue, D. L., Hawkins, S. E., III, Head, J. W., Holsclaw, G. M., McClintock, W. E., McCoy, T. J., McNutt, R. L., Jr., Prockter, L. M., Solomon, S. C. and Watters T. R. (2008). Reflectance and color variations on Mercury: Regolith processes and compositional heterogeneity. *Science*, **321**, 66–69.

Russell, C. T., Raymond, C. A., Coradini, A., McSween, H. Y., Zuber, M. T., Nathues, A., De Sanctis, M. C., Jaumann, R., Konopliv, A. S., Preusker, F., Asmar, S. W., Park, R. S., Gaskell, R., Keller, H. U., Mottola, S., Roatsch, T., Scully, J. E. C., Smith, D. E., Tricarico, P., Toplis, M. J., Christensen, U. R., Feldman, W. C., Lawrence, D. J., McCoy, T. J., Prettyman, T. H., Reedy, R. C., Sykes, M. E. and Titus, T. N. (2012). Dawn at Vesta: Testing the protoplanetary paradigm. *Science*, **336**, 684–686.

Sato, H., Robinson, M. S., Hapke, B., Denevi, B. W. and Boyd, A. K. (2014). Resolved Hapke parameter maps of the Moon. *J. Geophys. Res. Planets*, **119**, 1775–1805.

Schlemm, C. E., II, Starr, R. D., Ho, G. C., Bechtold, K. E., Hamilton, S. A., Boldt, J. D., Boynton, W. V., Bradley, W., Fraeman, M. E., Gold, R. E., Goldsten, J. O., Hayes, J. R., Jaskulek, S. E., Rossano, E., Rumpf, R. A., Schaefer, E. D., Strohbehn, K., Shelton, R. G., Thompson, R. E., Trombka, J. I. and Williams, B. D. (2007). The X-Ray Spectrometer on the MESSENGER spacecraft. *Space Sci. Rev.*, **131**, 393–415.

Schröder, S. E., Mottola, S., Keller, H. U., Raymond, C. A. and Russell, C. T. (2013). Resolved photometry of Vesta reveals physical properties of crater regolith. *Planet. Space Sci.*, **85**, 198–213.

Shepard, M. K. and Campbell, R. (1998). Shadows on a planetary surface and implications for photometric roughness. *Icarus*, **134**, 279–291.

Shepard, M. K. and Helfenstein, P. (2007). A test of the Hapke photometric model, *J. Geophys. Res.*, **112**, E03001, doi:10.1029/2005JE002625.

Shkuratov, Yu., Starukhina, L., Hoffmann, H. and Arnold, G. (1999). A model of spectral albedo of particulate surfaces: Implication to optical properties of the Moon. *Icarus*, **137**, 235–246.

Shkuratov, Yu., Bondarenko, S., Kaydash, V., Videen, G., Munos, O. and Volten, H. (2007). Photometry and polarimetry of particulate surfaces and aerosol particles over a wide range of phase angles. *J. Quant. Spectrosc. Radiat. Trans.*, **106**, 487–508.

Shkuratov Yu., Kaydash, V., Korokhin, V., Velikodsky, Y., Opanasenko, N. and Videen, G. (2011). Optical measurements of the Moon as a tool to study its surface. *Planet. Space Sci.*, **59**, 1326–1371.

Shkuratov, Yu., Kaydash, V., Korokhin, V., Velikodsky, Y., Petrov, D., Zubko, E., Stankevich, D. and Videen, G. (2012a). A critical assessment of the Hapke photometric model. *J. Quant. Spectrosc. Radiat. Trans.*, **113**, 2431–2456.

Shkuratov, Yu., Kaydash, V. and Videen, G. (2012b). The lunar crater Giordano Bruno as seen with optical roughness imagery. *Icarus*, **218**, 525–533.

Shkuratov, Yu., Kaydash, V., Sysolyatina, X., Razim, A. and Videen, G. (2013). Lunar surface traces of engine jets of Soviet sample return probes: The enigma of the Luna-23 and Luna-24 landing sites. *Planet. Space Sci.*, **75**, 28–36.

Simonelli, D. P., Wisz, M., Switala, A., Adinolfi, D., Veverka, J., Thomas, P. C. and Helfenstein, P. (1998). Photometric properties of Phobos surface materials from Viking images. *Icarus*, **131**, 52–77.

Solomon, S. C., McNutt, R. L., Jr., Gold, R. E., Acuña, M. H., Baker, D. N., Boynton, W. V., Chapman, C. R., Cheng, A. F., Gloeckler, G., Head, J. W., III, Krimigis, S. M., McClintock, W. E., Murchie, S. L., Peale, S. J., Phillips, R. J., Robinson, M. S., Slavin, J. A., Smith, D. E., Strom, R. G., Trombka, J. I. and Zuber, M. T. (2001). The MESSENGER mission to Mercury: Scientific objectives and implementation. *Planet. Space Sci.*, **49**, 1445–1465.

Souchon, A. L., Pinet, P. C., Chevrel, S. D., Daydou, Y. H., Baratoux, D., Kurita, K., Shepard, M. K. and Helfenstein, P. (2011). An experimental study of Hapke's modeling of natural granular surface samples. *Icarus*, **215**, 313–331.

Sprague, A. L., Kozlowski, R. W. H., Witteborn, F. C., Cruikshank, D. P. and Wooden, D. H. (1994). Mercury: Evidence for anorthosite and basalt from mid-infrared (7.3–13.5 micrometers) spectroscopy. *Icarus*, **109**, 156–167.

Sprague, A. L., Hunten, D. M. and Lodders, K. (1995). Sulfur at Mercury, elemental at the poles and sulfides in the regolith. *Icarus*, **118**, 211–215.

Sprague, A. L., Emery, J. P., Donaldson, K. L., Russell, R. W., Lynch, D. K. and Mazuk, A. L. (2002). Mercury: Mid-infrared (3–13.5 μm) observations show heterogeneous composition, presence of intermediate and basic soil types, and pyroxene. *Meteorit. Planet. Sci.*, **37**, 1255–1268.

Stockstill-Cahill, K. R., McCoy, T. J., Nittler, L. R., Weider, S. Z. and Hauck, S.A., II (2012). Magnesium-rich crustal compositions on Mercury: Implications for magmatism from petrologic modeling. *J. Geophys. Res.*, **117**, E00L15, doi:10.1029/2012JE004140.

Thomas, P.C., Adinolfi, D., Helfenstein, P., Simonelli, D. and Veverka, J. (1996). The surface of Deimos: Contribution of materials and processes to its unique appearance. *Icarus*, **123**, 536–556.

Thomas, R. J., Rothery, D. A., Conway, S. J. and Anand, M. (2014a). Hollows on Mercury: Materials and mechanisms involved in their formation. *Icarus*, **229**, 221–235.

Thomas, R. J., Rothery, D. A., Conway, S. J. and Anand, M. (2014b). Mechanisms of explosive volcanism on Mercury: Implications from its global distribution and morphology. *J. Geophys. Res. Planets*, **119**, 2239–2254.

Thomas, R. J., Rothery, D. A., Conway, S. J. and Anand, M. (2014c). Long-lived explosive volcanism on Mercury. *Geophys. Res. Lett.*, **41**, 6084–6092.

Trang, D., Lucey, P. G. and Izenberg, N. R. (2016). Mapping of submicroscopic carbon and iron on Mercury with radiative transfer modeling of MESSENGER VIRS reflectance spectra. *Lunar Planet. Sci.*, **47**, abstract 1396.

Trask, N. J. and Guest, J. E. (1975). Preliminary geologic terrain map of Mercury. *J. Geophys. Res.*, **80**, 2461–2477.

Vander Kaaden, K. E. and McCubbin, F. M. (2015). Exotic crust formation on Mercury: Consequences of a shallow, FeO-poor mantle. *J. Geophys. Res. Planets*, **120**, 195–209.

Vander Kaaden, K. E., McCubbin, F. M., Nittler, L. R., Peplowski, P. N., Weider, S. Z., Frank, E. A. and McCoy, T. J. (2017). Geochemistry, mineralogy, and petrology of boninitic and komatiitic rocks on the mercurian surface: Insights into the mercurian mantle. *Icarus*, **285**, 155–168.

Vaughan, W. M., Helbert, J., Blewett, D. T., Head, J. W., Murchie, S. L., Gwinner, K., McCoy, T. J. and Solomon, S. C. (2012). Hollow-forming layers in impact craters on Mercury: Massive sulfide deposits formed by impact melt differentiation? *Lunar Planet Sci.*, **43**, abstract 1187.

Vilas, F. and McCord, T. B. (1976). Mercury: Spectral reflectance measurements (0.33–1.06 μm) 1974/75. *Icarus*, **28**, 593–599.

Vilas, F., Leake, M. A. and Mendell, W. W. (1984). The dependence of reflectance spectra of Mercury on surface terrain. *Icarus*, **59**, 60–68.

Vilas, F., Domingue, D. L., Helbert, J., D'Amore, M., Maturilli, A., Klima, R. L., Stockstill-Cahill, K. R., Murchie, S. L., Izenberg, N. R., Blewett, D. T., Vaughan, W. M. and Head, J. W. (2016). Mineralogical indicators of Mercury's hollows composition in MESSENGER color observations. *Geophys. Res. Lett.*, **43**, 1450–1456, doi:10.1002/2015GL067515.

Wänke, H. (1981). Constitution of terrestrial planets. *Phil. Trans. Roy. Soc. London A*, **303**, 287–302.

Wänke, H. and Dreibus G. (1994). Water abundance and accretion history of terrestrial planets. In *Papers Presented to the Conference on Deep Earth and Planetary Volatiles*. Houston, TX: Lunar and Planetary Institute, p. 46.

Warell, J. and Blewett, D. T. (2004). Properties of the hermean regolith: V. New optical reflectance spectra, comparison with lunar anorthosites, and mineralogical modeling. *Icarus*, **168**, 257–276.

Warell, J., Sprague, A. L., Emery, J. P., Kozlowski, R. W. H. and Long, A. (2006). The 0.7–5.3 μm spectra of Mercury and the Moon: Evidence for high-Ca pyroxene on Mercury. *Icarus*, **180**, 281–291.

Watters, T. R., Murchie, S. L., Robinson, M. S., Solomon, S. C., Denevi, B. W., André, S. L. and Head, J. W. (2009). Emplacement and tectonic deformation of smooth plains in the Caloris basin, Mercury. *Earth Planet. Sci. Lett.*, **285**, 309–319.

Weider, S. Z., Nittler, L. R., Starr, R. D., McCoy, T. J., Stockstill-Cahill, K. R., Byrne, P. K., Denevi, B. W., Head, J. W. and Solomon, S. C. (2012). Chemical heterogeneity on Mercury's surface revealed by the MESSENGER X-Ray Spectrometer. *J. Geophys. Res.*, **117**, E00L05, doi:10.1029/2012JE004153.

Weider, S. Z., Nittler, L. R., Starr, R. D., McCoy, T. J. and Solomon, S. C. (2014). Variations in the abundance of iron on Mercury's surface from MESSENGER X-Ray Spectrometer observations. *Icarus*, **235**, 170–186.

Weider, S. Z., Nittler, L. R., Starr, R. D., Crapster-Pregont, E. J., Peplowski, P. N., Denevi, B. W., Head, J. W., Byrne, P. K., Hauck, S. A., II, Ebel, D. S. and Solomon, S. C. (2015). Evidence for geochemical terranes on Mercury: Global mapping of major elements with MESSENGER's X-Ray Spectrometer. *Earth Planet. Sci. Lett.*, **416**, 109–120.

Weider, S. Z., Nittler, L. R., Murchie, S. L., Peplowski, P. N., McCoy, T. J., Kerber, L., Klimczak, C., Ernst, C. M., Goudge, T. A., Starr, R. D., Izenberg, N. R., Klima, R. L. and Solomon, S. C. (2016). Evidence from MESSENGER for sulfur- and carbon-driven explosive volcanism on Mercury. *Geophys. Res. Lett.*, **43**, 3653–3661, doi:10.1002/2016GL068325.

Whitten, J. L., Head, J. W., Denevi, B. W. and Solomon, S. C. (2014). Intercrater plains on Mercury: Insights into unit definition, characterization, and origin from MESSENGER datasets. *Icarus*, **241**, 97–113.

Xiao, Z., Strom, R. G., Blewett, D. T., Byrne, P. K., Solomon, S. C., Murchie, S. L., Sprague, A. L., Domingue, D. L. and Helbert, J. (2013). Dark spots on Mercury: A distinctive low-reflectance material and its relation to hollows. *J. Geophys. Res. Planets*, **118**, 1752–1765,

Zolotov, M. Yu. (2011). On the chemistry of mantle and magmatic volatiles on Mercury. *Icarus*, **212**, 24–41.

Zolotov, M. Yu., Sprague, A. L., Hauck, S. A., II, Nittler, L. R., Solomon, S. C. and Weider, S. Z. (2013). The redox state, FeO content, and origin of sulfur-rich magmas on Mercury. *J. Geophys. Res. Planets*, **118**, 138–146.

9

Impact Cratering of Mercury

CLARK R. CHAPMAN, DAVID M. H. BAKER, OLIVIER S. BARNOUIN, CALEB I. FASSETT, SIMONE MARCHI, WILLIAM J. MERLINE, LILLIAN R. OSTRACH, LOUISE M. PROCKTER, AND ROBERT G. STROM

9.1 INTRODUCTION

Impact cratering has been a major geological process that has strongly affected the crust and surface of Mercury throughout its history. Whereas giant impacts during the first tens of millions of years of solar system history presumably influenced the planet's overall geophysical and cosmochemical character, the topographical expression of Mercury's impact history during the first half-billion years has been destroyed by subsequent impacts and volcanism. Yet enormous, early basin-forming impacts and continuing impact cratering to the present have, in competition with volcanic plains emplacement and tectonics, shaped the planet that the MESSENGER spacecraft imaged and measured during its flybys and orbital mission.

Mercury's surface looked superficially lunar-like in the images acquired by Mariner 10 during its three flybys in 1974–1975, with the Caloris basin centered advantageously on the terminator with good lighting. But a closer look at Mercury's craters revealed distinct differences from those on other worlds, undoubtedly affected by the planet's appreciable surface gravitational acceleration and especially the unusually high velocities of asteroids and comets that have impacted the planet. Secondary crater fields were noted as especially prominent on Mercury, and many craters were deformed by near-ubiquitous lobate scarps. Now, with high-resolution MESSENGER images, we see additionally that Mercury-specific processes, such as the formation of hollows (Chapter 12), have helped to degrade crater topography with time.

In this chapter, we address measurements of Mercury's impact craters and basins and summarize interpretations of cratering in terms of the planet's evolving geological history.

9.2 BASINS AND EARLY IMPACTS

9.2.1 Very Early Bombardment of Mercury

Mercury, like other terrestrial planets, formed during an accretionary phase during which innumerable small planetesimals evolved to a much smaller number of planetary-sized bodies. This process could have spanned from a few million years (as for Mars) to several tens of million years (as for Earth) after condensation of the first solids. The last stage of planetary accretion was characterized by fewer, yet likely more energetic, collisions. This stage is well documented by the record of oldest impact basins on Mars and the Moon, and indirectly by the terrestrial mantle enrichment in highly siderophile elements (HSEs) thought to have been accreted after the Moon-forming giant collision (e.g., Walker, 2009).

It has been suggested that a late hit-and-run collision, stripping away proto-Mercury's outer layers, could have been responsible for the planet's high ratio of metal to silicate (Wetherill, 1988; Benz et al., 1988; Stewart et al., 2016), though a challenge to this idea is that much of the ejecta from such a collision may have re-accreted. The timing and magnitude of such a large collision or series of collisions remains unconstrained, and alternative explanations have also been suggested (see Chapter 18). The lack of samples from Mercury, however, prevents us from drawing firm conclusions about the earliest collisional evolution of Mercury, which notoriously has been regarded an end-member planet, often ignored or poorly investigated with terrestrial planet formation models.

The earliest asteroidal bombardment after the differentiation of Mercury's core could have delivered a substantial inventory of HSEs to the planet's silicate fraction, analogous to what is found on Earth, the Moon, and Mars. However, Mercury's relatively thin silicate shell might have allowed HSEs to merge with the core so that they would not be retained in the silicates. A much higher average impact speed at Mercury could have resulted in net erosion of the planet, rather than accretion (Raymond et al., 2013), although Rivera-Valentin and Barr (2014) considered that late-veneer compositional heterogeneities might remain visible today.

The earliest evolution of Mercury continues to elude us (but see Chapter 18). Some approximate insights can be gained from an extrapolation of the nearest well-studied bodies, Earth and the Moon. Estimates based on a lunar bombardment flux (Marchi et al., 2009, 2013; Le Feuvre and Wieczorek, 2011) have shown that, currently, the crater production rate on Mercury per unit area is about three times that for the Moon. (Crater production rates differ among target bodies because of numbers of projectiles, their velocities, the target body's surface gravitational acceleration, and other factors discussed below in Section 9.5.2.) Thus, all else being equal, the observed ~43 basins >300 km in diameter on the Moon (Neumann et al., 2015) would translate to ~250 basins of >300-km diameter on Mercury, given that the latter body has approximately double the surface area. This total is a lower limit since it comprises basins formed *after* the formation of the Moon and solidification of its crust, roughly 4.4 Ga. With its larger surface area and higher impact velocity than the Moon, earlier basins could have formed on Mercury that were larger than the largest lunar basin, the 2500-km-diameter South Pole–Aitken basin.

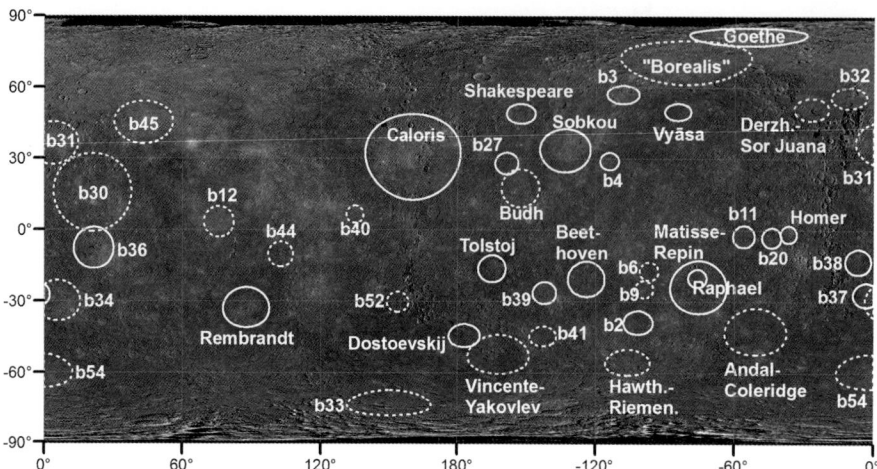

Figure 9.1. Distribution of large ($D \geq 300$ km) impact basins on Mercury classified as certain (solid outline) or probable (dashed). From Fassett et al. (2012).

9.2.2 Global Distribution of Basins, Relationships to Plains

Mercury has only 46 "certain and probable" basins having a diameter D from 300 to 1550 km (Figure 9.1) plus 41 additional "suggested but unverified" basins 320 to ~2000 km in diameter visible in MESSENGER image and topography data (Fassett et al., 2012). [It is possible that the high-magnesium region covering ~15% of Mercury's surface – see Chapter 2 – reflects a giant, early impact (Weider et al., 2015; Frank et al., 2016, 2017), but that region is not considered a basin here.] Their numbers and areas cover a modest fraction of Mercury's surface, far undersaturated at these diameters (Fassett et al., 2012; Marchi et al., 2013). It may be that on the Moon basins never completely saturated the surface subsequent to solidification of the lunar crust, as suggested by Gravity Recovery and Interior Laboratory data (Neumann et al., 2015; Soderblom et al., 2015). But on the basis of Mercury's much higher cratering rate, it is more likely that its surface was saturated by basins during the same time interval and that subsequent plains formation, tectonics, and erosion by later, smaller impacts have destroyed most early basins so that those visible today are comparatively few.

More recent volcanism has also modified, and in some cases erased, basins; smooth plains are found within and surrounding most of the known basins (Fassett et al., 2012; see also Figure 9.2). Although some plains may be solidified impact melt (e.g., Whitten and Head, 2015), a volcanic origin for most such smooth plains is supported by the fact that many are relatively sparsely cratered compared with the basins they superpose, requiring that substantial time passed between basin formation and plains formation. Indeed, formation of many smooth plains may be genetically linked to basin formation itself, as the impacts may have pervasively fractured the crust, relieved any pre-existing horizontal compressional stress in the lithosphere, and imparted a stress state that enabled the ascent of magma (Roberts and Barnouin, 2012; Marchi et al., 2013; see also Chapter 11).

One of the most intriguing aspects of the geographic distribution of probable and certain basins mapped by Fassett et al. (2012) is that they are not uniformly distributed (Figure 9.1). For example, the eastern hemisphere has less than half the number of mapped basins seen in the western hemisphere. If the processes forming and erasing basins were uniform and random, this dichotomy would be expected to arise only ~1% of the time. Three hypotheses might explain this observation. First, a large spatial difference in basin formation probabilities would be expected if Mercury were once in a 1:1 spin–orbit resonance (synchronous rotation) before being captured into its present 3:2 state (Wieczorek et al., 2012). However, Noyelles et al. (2014) reexamined the capture of Mercury into its present 3:2 resonance with a model that included the effects of tides, and they argued that 3:2 capture likely occurred quickly, perhaps within 10–20 Myr, far faster than would explain the basin distribution. Second, major asymmetries in basin distribution could mainly reflect differences in the intensity of resurfacing; smooth plains volcanism on Mercury was spatially non-uniform (Denevi et al., 2013; Chapters 6 and 11), and complete erasure of basins may have been non-uniform as well. A third explanation is that hemispheric to global-scale differences in target properties may have played a role. Miljković et al. (2013) showed that temperature variations strongly influenced the final diameters of large basins on the Moon, thus accounting for asymmetries in the lunar basin distribution between the nearside and farside. Mercury's basin distribution might have been similarly affected if it also experienced large-scale lateral temperature variations.

9.2.3 Morphometry and Topography of Basins, Protobasins, and Peak Rings

9.2.3.1 The Transition from Crater to Basin

Impact structures on planetary bodies exhibit a spectrum of morphologies as they increase in size from small bowl-shaped craters to giant multi-ring basins. The transition from impact craters to basins with increasing diameter is marked morphologically by two or more concentric ring structures (Figure 9.3). Specifically, complex craters possessing central peaks transition to peak-ring basins, which possess both a rim crest and an interior ring of peaks. Other transitional morphologies occur, including protobasins, which have both a central peak and a peak ring (Figure 9.3). MESSENGER image and altimetry data

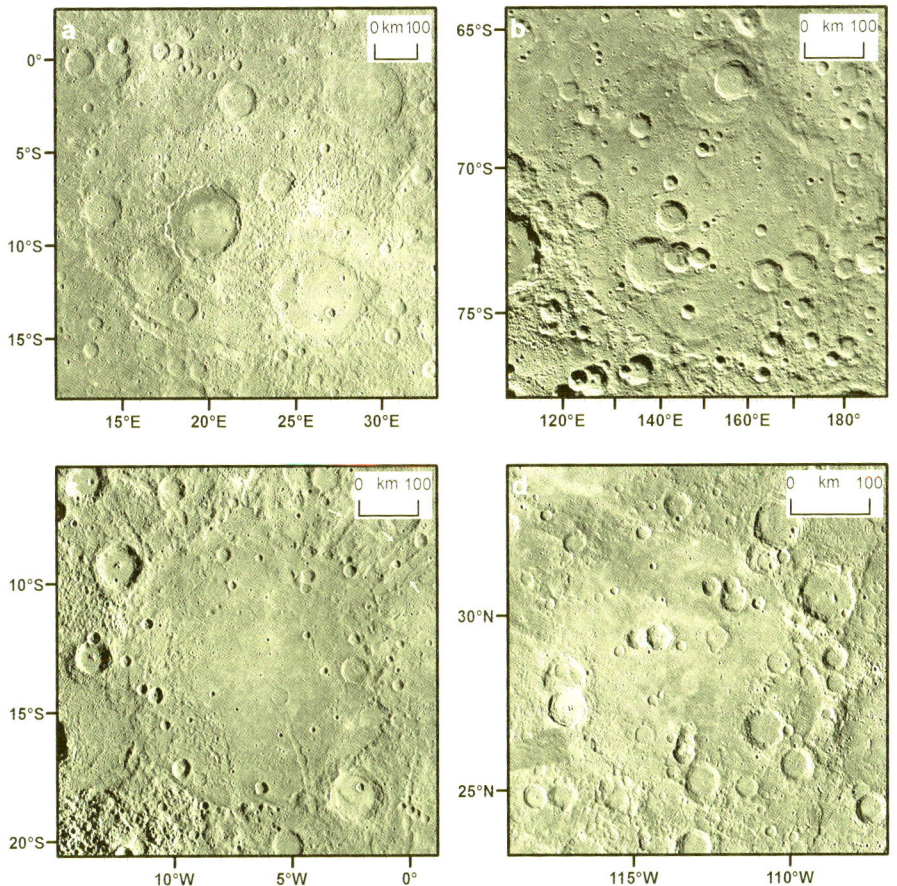

Figure 9.2. Examples of four large impact basins on Mercury. (a) A 730-km-diameter basin (b36) centered at 7.6°S, 21.6°E. The basin's interior is degraded and superposed by three peak-ring basins, including the fresh peak-ring basin Derain (~8°S, 20°E). (b) A 470-km-diameter basin (b33) centered at 72.9°S, 149.9°E. This degraded basin is superposed by numerous younger craters, but its southwestern rim remains prominent. (c) A 470-km-diameter basin (b38) centered at 13.4°S, 6.6°E. The basin has prominent radial troughs to its northeast (white arrows), interpreted as sculpture related to basin ejecta, and is floored by smooth plains. (d) A 310-km-diameter basin (b4) centered at 28.9°N, 113.8°W. Basin designations are from Figure 9.1.

have permitted global reassessment and updating of previous catalogs of complex craters, protobasins, and peak-ring basins on Mercury (e.g., Pike, 1988), including new measurements of their diameters and morphometric parameters (Fassett et al., 2011; Baker et al., 2011a; Baker and Head, 2013).

MESSENGER data have confirmed that Mercury has the most peak-ring basins and protobasins of any solar system body: 110 and 70 in total number, respectively; in contrast, the Moon has only 17 preserved peak-ring basins and three protobasins (Baker and Head, 2013). This difference is due in part to Mercury's larger surface area, but also to the smaller transition diameter from complex craters to basins on Mercury and Mercury's higher cratering rate. Peak-ring basins have a diameter range from 84 to 320 km on Mercury (Baker and Head, 2013). The onset diameter of peak-ring basins on Mercury with increasing diameter, averaging 109^{+23}_{-19} km, occurs at a smaller diameter than on the Moon (~206 km) and is larger than on Venus (42^{+10}_{-8} km). These onset diameters on the terrestrial planets show an inverse dependence on planetary surface gravitational acceleration, a relation that may also be modulated by differences in mean impact velocity (Pike, 1988; Baker et al., 2011b). Interior central peak and peak-ring structures on Mercury are also much better preserved than on the Moon, as evidenced by the larger fraction of fresh, less-degraded craters and basins possessing these interior structures for a given crater diameter range on Mercury (Baker and Head, 2013).

The morphometric trends of peak-ring basins have specific characteristics. MESSENGER's Mercury Laser Altimeter (MLA) profiles reveal peak-ring topographic signatures that are distinct from those of smaller complex craters. The floors of peak-ring basins are higher in elevation in an annulus between the base of the basin wall and the peak ring than towards the center. A peak ring is typically marked by a scarp that is steeper on its inward face than its outward face, a geometry that serves to bound the deeper cavity-like center of the basin. Deepening of the center of the basin can result from vertical contraction during cooling of a central melt sheet (e.g., Vaughan et al., 2013) or from contraction and subsidence following emplacement of volcanic fill (Blair et al., 2013). In contrast, floors of complex craters are relatively flat, with small-scale hummocks and with a center that rises to form the central peak. Compared with the Moon, the ratios of floor area to entire crater area are larger for both complex craters and peak-ring basins on Mercury. This trend has been interpreted to have resulted from the higher impact melt production and retention for impact events on Mercury compared with the Moon (Ostrach et al., 2012; Baker and Head, 2013). The ratios of depth to diameter for the largest complex craters and smallest peak-ring basins on Mercury are similar at values near 0.03; these values then decrease with increasing rim-crest diameter to ratios less than 0.02.

A plot of peak-ring diameter versus rim-crest diameter (Figure 9.4) reveals a trend that is distinct from that observed for

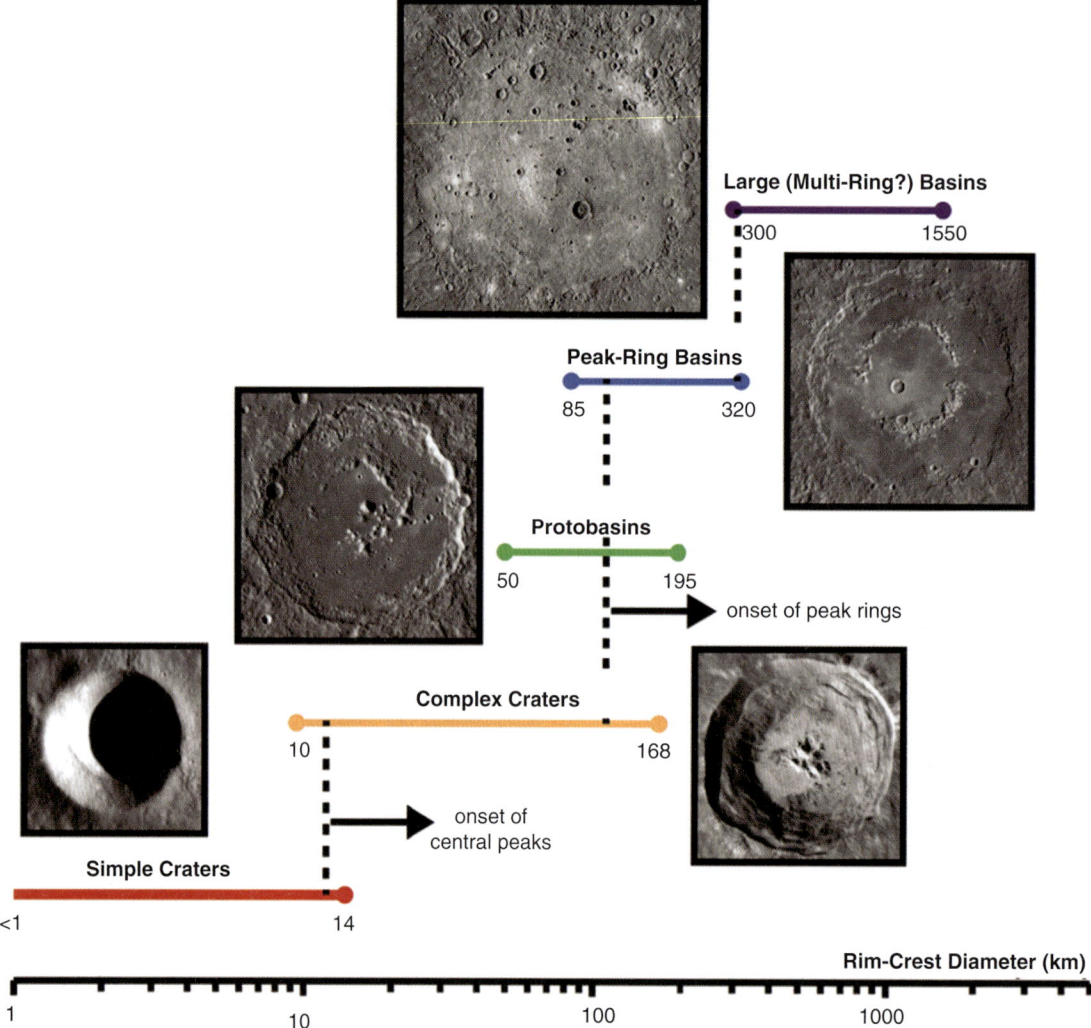

Figure 9.3. The variation of impact crater morphology with feature size on Mercury. Simple craters dominate at small diameters up to ~14 km (Pike, 1988). The transition from simple to complex craters near 12-km diameter is marked by rim-wall terraces and central peaks. Complex craters transition with increasing diameter to peak-ring basins near 100-km diameter; protobasins are intermediate types, with both a central peak and a peak ring. Near 300-km diameter, peak-ring basins transition to large impact basins that reach sizes up to 1550 km (Caloris basin) and can possess multiple concentric rings. Images of each crater or basin are mosaics of MESSENGER Mercury Dual Imaging System (MDIS) narrow-angle (NAC) and wide-angle camera (WAC) images: Simple crater (unnamed, 10-km diameter, centered at 40.2°N, 26.5°E); complex crater (Fonteyn, 29 km, 32.85°N, 95.7°E); protobasin (Velázquez, 123 km, 37.7°N, 304.4°E); peak-ring basin (Raditladi, 263 km, 27°N, 119°E); large basin (Caloris, 1550 km, 31°N, 160°E).

central-peak diameters, suggesting that the transition from complex crater to peak-ring basin is abrupt, likely requiring a process threshold to be reached before peak rings are formed (Baker et al., 2016). Such a process might be (e.g., Collins et al., 2002) that central peaks of large craters become gravitationally unstable and then collapse downward and outward to form peak rings. This idea is also supported by the observation for both Mercury and the Moon that central peak heights and areas increase continuously up to the transition to peak-ring basins (Baker and Head, 2013). However, between diameters 84 km and 168 km on Mercury, multiple interior uplift morphologies for a given crater size can occur (Figure 9.3). This overlap in crater morphology, which does not occur on the Moon, may be related to the relatively small peak-ring onset diameter and the broader range of impact velocities occurring at Mercury, which may have heavily affected the mechanics and character of floor uplift during crater formation, including modifying impact melt volumes and maximum heights of central uplifts.

9.2.3.2 Transition from Peak-Ring Basin to Multi-Ring Basin

Above about 500–600-km diameter, lunar peak-ring basins generally give way to multi-ring basins with three or more concentric rings, which are the largest impact structures on terrestrial bodies (e.g., Wilhelms, 1987) (Figure 9.3). However, large impact basins with three or more concentric rings are relatively rare on Mercury. In contrast to previous catalogs (e.g., Pike and Spudis, 1987), the vast majority

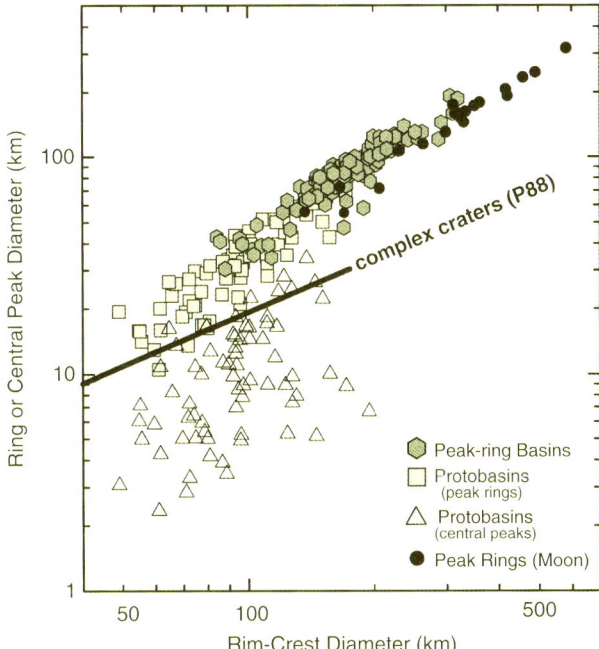

Figure 9.4. Trends in peak-ring or central-peak diameter versus rim-crest diameter for peak-ring basins, protobasins, and complex craters on Mercury and the Moon. Peak-ring basin and protobasin data are from Baker and Head (2013). The trend for complex crater central peaks on Mercury is from Pike (1988) or P88 ($D_{pk} = 0.44 D_r^{0.82}$, where D_{pk} is the central peak diameter and D_r is the rim-crest diameter, both in km). Peak rings follow a trend that is distinct from that for central peaks.

(>75%) of certain and probable basins identified by Fassett et al. (2012) have only one prominent ring, with little evidence for additional rings. Why Mercury lacks true multi-ring basins may be that such features fail to form for some reason, relax away, or are preferentially degraded by post-impact volcanism and nearby or superposed impacts. Most cataloged basins are highly degraded and infilled with volcanic plains or ejecta, suggesting that post-impact processes are especially important factors in erasing basin topography. However, the paucity of multiple rings in the largest basins is in stark contrast to the abundance of smaller, well-preserved peak-ring basins on Mercury. We have yet to understand this apparent size dependence of the appearance of basin interior ring structures, as revealed by MESSENGER.

9.2.4 Graben, Scarps, and Ridges in Basins

Mercury's dominant form of surface deformation documented by tectonic landforms is contractional, with extension almost entirely confined to locations within impact basins >200-km diameter (Murchie et al., 2008; Watters et al., 2009b; Prockter et al., 2009, 2010, 2012; see also Chapter 10 and discussion of ghost craters below). Impact structures as small as the 116-km diameter Abedin (61.7°N, 349.5°E) may show troughs within smooth plains on their floors that may be related to the cooling and contraction of solidified impact melt (e.g., Xiao et al., 2014b); such troughs also occur in mid-sized lunar craters (Howard and Wilshire, 1975; Schultz, 1976). Some wrinkle ridges are found within smooth plains on the floors of larger basins, such as Rachmaninoff and Mozart, and lobate scarps may deform basins long after their formation, as in Rembrandt (Watters et al., 2012). The complexity of deformation appears to increase with basin size (see Chapter 10), with the largest basins (e.g., Caloris and Rembrandt) having experienced multiple episodes of volcanic infilling interspersed with tectonic deformation (Murchie et al., 2008; Fassett et al., 2009; Watters et al., 2009a, c).

The prominent family of radial troughs, Pantheon Fossae, in Caloris (e.g., Figure 10.9) appears to have formed from flexural uplift of the basin center, possibly in response to inward flow of the lower crust (Watters et al., 2005) or the volcanic emplacement of the circum-Caloris smooth plains (Freed et al., 2009). It has also been proposed that the troughs (interpreted as graben) formed as the surface expression of dike propagation (Head et al., 2008), but a mechanical analysis of the graben characteristics does not favor this explanation (Klimczak et al., 2010). The suggestion that formation of the large crater Apollodorus near the center of the system was responsible for the radial features (Freed et al., 2009) has generally been discounted in favor of the coincidental formation of the crater at this location (Klimczak et al., 2010). None of these formation mechanisms provide stress orientations that can account for circumferential graben such as those in Rachmaninoff or Raditladi, which instead may result from variations in the thickness of the youngest of a set of cooling volcanic layers (Freed et al., 2012). Buried basin rings or central depressions can strongly localize radial extensional stresses within these layers and favor circumferential graben formation (Figure 9.5b) (Blair et al., 2013). Graben in mixed orientations within basins may result when rapidly emplaced, thick lava flows cool and thermally contract, producing horizontal extensional stresses (Freed et al., 2012). Models of cooling volcanic layers predict extension at the very center of a basin, but because extensional tectonic features are uncommon in such areas it has been postulated that such structures have been covered with later, thin volcanic flows. Ridge formation within basins may be the result in part of global contraction associated with the cooling of the planet's interior (Freed et al., 2012; Blair et al., 2013), and lithospheric flexure in response to loading by volcanic plains emplacement may have also contributed (Melosh and McKinnon, 1988; Kennedy et al., 2008; Watters et al., 2009b). Detailed discussions of the mechanisms that could have formed the tectonic structures interior to impact basins are given in Chapter 10.

9.2.5 The Geology of Individual Basins

High-resolution imaging and spectral reflectance data from MESSENGER have enabled studies of the detailed geology of specific basins on Mercury, including their complex deformational histories, later modification by smooth plains emplacement, and interior deformation by troughs and wrinkle ridges.

9.2.5.1 Caloris Basin

The relatively young Caloris basin, the largest unambiguous impact basin on Mercury, has a diameter equal to more than

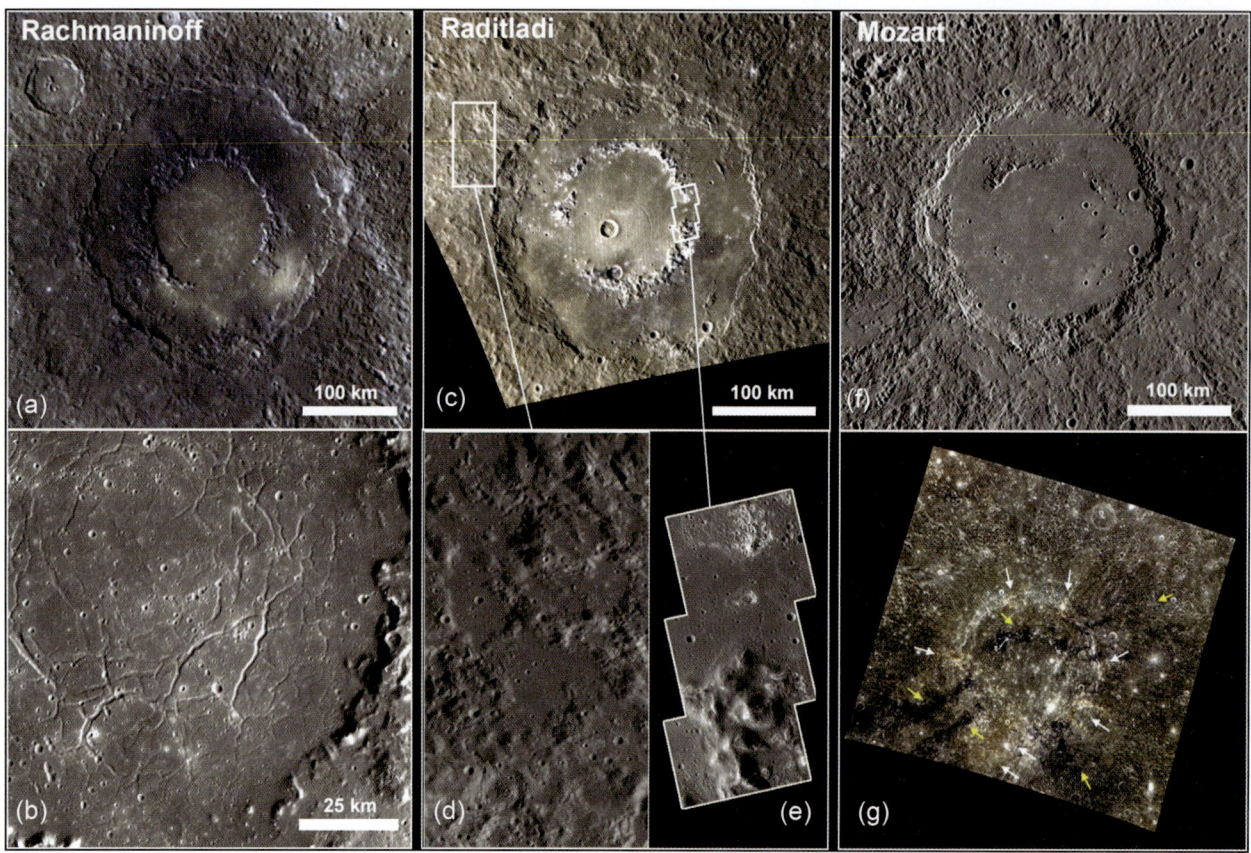

Figure 9.5. Various morphologies of three of the largest, most geologically complex peak-ring basins on Mercury. (a) Enhanced color view (second and first principal components and 430-nm/1000-nm ratio in red, green, and blue, respectively) of Rachmaninoff (292-km diameter), which has volcanic plains within its peak ring that are much younger than the time of basin formation (Prockter et al., 2010). (b) A view of the basin floor shows that these plains have undergone extensive fracturing, forming a polygonal circumferential pattern of troughs. (c) Enhanced color image of Raditladi (263-km diameter), one of the youngest basins on Mercury. (d) Pockets of solidified impact melt are on top of the basin's continuous ejecta deposit. (e) The peak ring has high-reflectance "hollows" material. (f) Monochrome view of Mozart (236-km diameter), which sits atop Caloris' southern exterior plains. Most of the central peak ring has been overlain by volcanic plains. (g) Enhanced color image of Mozart showing low-reflectance material (LRM) excavated in both the peak ring and as streaks within the basin ejecta (yellow arrows). A reddish unit, likely excavated high-reflectance red plains (HRP) material, is also visible in linear patches across the rim, especially to the south (white arrows).

60% of Mercury's radius (Figure 6.12). Thus, it is one of the most important geological features on Mercury and has been the subject of many studies since it was discovered in Mariner 10 flyby images (which revealed only its eastern third).

MESSENGER has provided much new data about Caloris, clarifying the basin's size, stratigraphy, tectonics, relationship to volcanism, and influence on its surroundings. Caloris is larger than originally inferred from Mariner 10 data, with a maximum rim-to-rim diameter of ~1550 km (Murchie et al., 2008). It is irregularly shaped, with a larger east–west than north–south extent (Fassett et al., 2009), and has an ellipticity of ~1.16, consistent with other large impact basins in the solar system having diameters that are a significant fraction of their planet's radius (Andrews-Hanna and Zuber, 2010).

Caloris has been substantially modified by volcanic activity, and its interior plains are volcanic (e.g., Murchie et al., 2008). The plains are spectrally distinct from both the basin's rim and the plains surrounding the basin (e.g., Robinson et al., 2008). Whereas initial measurements found a much lower crater density on the plains than on the rim of the basin itself (Fassett et al., 2009), newer data suggest that the crater densities on the interior and exterior plains are within the large uncertainty in the crater density on the rim (Ernst et al., 2017). In addition, large craters have excavated spectrally distinct material from beneath the interior plains (Murchie et al., 2008; Ernst et al., 2010, 2015), demonstrating that those plains are a discrete unit of perhaps as much as ~2.5–4 km thickness. Additional basin-associated volcanic features include several volcanic pit complexes, which appear concentrated around the circumference of Caloris near the inner edge of its rim and are sources of pyroclastic deposits (Head et al., 2009).

It is attractive to think of Caloris as analogous to Imbrium or Orientale on the Moon (e.g., McCauley, 1977; McCauley et al., 1981) since, like these lunar basins, Caloris is relatively young, is partially buried by volcanic plains, and has recognizable ejecta, but there are significant differences. The largest difference is that Caloris is deformed by both extensional and compressional tectonic features to a much greater extent than are large, young lunar basins. As discussed above, its central plains are deformed by an extensive system of radial graben, Pantheon

Fossae, as well as numerous wrinkle ridges (Strom et al., 1975; Murchie et al., 2008; Watters et al., 2009c), which crosscut each other in a complex manner.

Stereo imaging of Caloris and orbital MLA data show that the northern half of the interior of Caloris is topographically higher than its rim (Oberst et al., 2010; Zuber et al., 2012). There is little to no correlation between the long-wavelength undulations that contribute to this unusual topography and tectonic features on the interior plains or the interpreted thickness of the plains (Klimczak et al., 2013; Ernst et al., 2015). Moreover, many of the craters on the plains are tilted by the long-wavelength rise on the basin floor (Klimczak et al., 2013), which requires that the topographic rise formed well after the basin and its interior volcanic plains. Since there are other regions of long-wavelength undulation in Mercury's topography (see Chapter 10), this topographic modification of Caloris is probably not related directly to the basin itself.

9.2.5.2 Rembrandt Basin

The second largest well-preserved impact structure on Mercury is Rembrandt (Figure 9.6), which at 715-km diameter is about half the size of Caloris. Situated at 33°S, 88°E, Rembrandt is one of the youngest basins on Mercury, comparable in age to Caloris on the basis of the areal density of craters >20-km diameter along its rim. Radially lineated ejecta deposits are well preserved, especially to the north and northeast of the basin rim (Figure 9.6a, b).

Detailed mapping of Rembrandt by Watters et al. (2009a) showed that the basin is partially flooded with volcanic plains and its floor exhibits a "wheel-and-spoke" pattern of basin-radial and basin-concentric wrinkle ridges and graben. Much of the basin's interior is covered by smooth plains (Figure 9.6c). Although larger basins are expected to have relatively larger proportions of impact melt to volcanic fill in their interiors (e.g., Grieve and Cintala, 1992), a volcanic origin for the interior plains is indicated by their high relative reflectance (Whitten and Head, 2015), analogous to high-reflectance red plains (HRP) interpreted elsewhere to be of volcanic origin (Robinson et al., 2008; Denevi et al., 2009), embayment relations, and partially flooded impact craters (Figure 9.6d) that imply a prolonged period of plains formation. Watters et al. (2009a) estimated that the thickest smooth plains – at least ~2 km in thickness – are near the center of the basin, similar to estimates of at least 2.5 km for the fill in Caloris (Ernst et al., 2015).

Rembrandt's radial and concentric ridges are 1–10 km wide and extend up to 180 km in length, whereas the troughs are 1–3 km in width and considerably shorter. Concentric ridges form an almost complete ring within the basin, with a diameter of ~375 km, and the majority of the radial wrinkle ridges and troughs are found within this ring.

Stratigraphic relations indicate a multistage history of infilling by volcanic plains, interspersed with overlapping phases of contractional and extensional deformation (Figure 9.6e and f) (Watters et al., 2009a). Crosscutting relations indicate that much of the contractional deformation within the smooth plains generally preceded extensional deformation, although there is some overlap in timing. The driving stresses for this deformation are discussed above and in Chapter 10.

The most recent deformation episode is shown by a crosscutting lobate scarp, evidently the surface expression of a major thrust fault (Watters et al., 2009a; Ferrari et al., 2014), which cuts Rembrandt's rim and extends almost 400 km across the basin floor, offsetting two impact craters ~60 km in diameter (Figure 9.6g). This scarp likely formed primarily as a result of the global contraction that accompanied cooling of the planet's interior (see Chapter 10).

9.2.5.3 Rachmaninoff Basin

The 292-km-diameter Rachmaninoff peak-ring basin, at 28°N, 58°E, is surrounded by a continuous ejecta deposit and a few secondary crater chains but has no visible rays (Figure 9.5a). It has a partial third ring to the south and southwest of the main rim, extending across an arc of ~120°, which may have resulted from the impact process that formed Rachmaninoff or may be the remnant of an older, heavily degraded basin that was further obscured by the formation of Rachmaninoff.

Rachmaninoff's rim appears relatively crisp but has undergone some modification, with slumping and terrace formation. The inner peak-ring is a discontinuous, slightly elongated ring that measures ~135 km north to south and has a color similar to that of the rim and terraces. A distinct, smooth, reddish, HRP unit is largely confined within the peak ring (Figure 9.5a), appearing to be thicker at its southern extent and completely overlying the southern part of the peak ring. (By "reddish" we mean displaying a reflectance that increases more strongly with

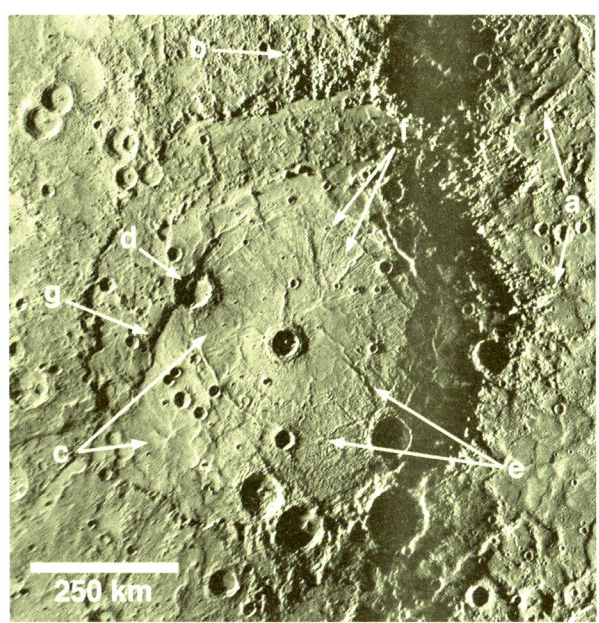

Figure 9.6. The 715-km-diameter Rembrandt basin displays a variety of morphological and tectonic features, including (a, b) radial ejecta deposits, (c) smooth plains infilling most of the interior and (d) embaying some craters, (e) radial and circumferential wrinkle ridges, (f) radial troughs, and (g) lobate scarp. Illumination is from the northwest in the left half of the mosaic and from the east in the right half.

wavelength from the near-ultraviolet to the near-infrared than that for the average surface.) On the basis of crater densities and its distinct color, this unit is probably volcanic, likely emplaced long after the basin formed and representing some of the youngest volcanism on Mercury (Prockter et al., 2010; Marchi et al., 2011; Chapman et al., 2012).

The inner plains are crosscut by troughs and ridges, arranged in a broadly circumferential pattern (Figure 9.5b) (Prockter et al., 2009; Blair et al., 2013), similar to those found in Raditladi (discussed below). Smooth plains lying in an annulus between the peak ring and the basin rim, in places embaying the rim terraces, tend to be lower in reflectance (Figure 9.5a) than the interior plains. The reflectance characteristics are consistent with those of low-reflectance material (LRM) observed elsewhere on Mercury as both diffuse deposits and concentrated in central peaks and the ejecta deposits of craters and basins such as Tolstoj and Rembrandt (Robinson et al., 2008; Denevi et al., 2009; Whitten and Head, 2015).

Rachmaninoff's continuous ejecta deposit has the same intermediate bluish color as the plains farther beyond (Figure 9.5a). Several pockets of smooth plains on top of the continuous ejecta, especially southwest of the basin, have been interpreted to be solidified impact melt (Prockter et al., 2010; Whitten and Head, 2015), asymmetrically distributed because of an oblique impact that formed the basin, with the projectile coming from the northeast (e.g., Howard and Wilshire, 1975; Hawke and Head, 1977a).

Rachmaninoff contains the deepest depression on Mercury, 5.8 km below the mean northern hemisphere radius at the center of the basin (Zuber et al., 2012; Becker et al., 2016; see Chapter 3) and with a measured rim-to-floor depth of $4.76^{+0.12}_{-0.14}$ km (Baker and Head, 2013). Despite the fact that volcanic plains were emplaced within its center, those plains still lie at markedly lower elevations than the rest of the basin, and the basin's rim and continuous ejecta deposit lie between about 0.5 and 1.5 km below mean radius. Possibly, as suggested by its partial third ring, the impact that formed Rachmaninoff struck the already depressed interior of an older basin.

9.2.5.4 Raditladi Basin

Centered at 27°N, 119°E, the 263-km-diameter Raditladi basin has a continuous ejecta deposit and few superposed craters (Figure 9.5c), so it must be relatively young. Numerous secondary clusters and chains are arranged radially around the basin, extending from the outer edges of the continuous ejecta deposit to as much as 500 km from the basin center. The density of craters a few kilometers in diameter on Raditladi's floor is more than an order of magnitude lower than that of the Caloris exterior plains immediately to the east and a factor of ~4 less than on the floor of Rachmaninoff, implying that Raditladi is extremely young, possibly as young as the last several hundred million years (Strom et al., 2008; Marchi et al., 2011; Chapman et al., 2012; see Section 9.5.2.2).

Raditladi's basin rim is relatively well preserved, although its walls have been modified and in places exhibit pronounced terraces (Mancinelli et al., 2014). A distinctive peak ring, ~125 km in diameter, is slightly offset from the basin center to the north and is elongated north to south. High-resolution images show that parts of the peak ring are made up of hollows – shallow, pitted terrain ranging from a few tens of meters to kilometers in extent, with high-reflectance interiors and diffuse bright haloes (Figure 9.5e) (Blewett et al., 2011, 2013; Chapter 12). Hollows may have formed because of loss of volatiles from materials that were excavated by the impact and have since been degraded at the surface by sublimation and/or sputtering (Thomas et al., 2014). Craters 100–200 m in diameter at the foot of Raditladi's peak ring have since been buried by talus, apparently created by the hollow-forming process, leading Blewett et al. (2013) to suggest that hollow formation at Raditladi is relatively recent or even ongoing.

Plains units within Raditladi's interior lack distinct boundaries, unlike those within Rachmaninoff. Exterior to the peak ring, the plains have relatively low reflectance and are bluer than other plains within the basin (Figure 9.5c). On the basin floor, some LRM deposits are seen in the southern portions of the annulus between the peak ring and basin rim and have been excavated by a small crater just inside the basin's peak ring to the south. The LRM deposits are lower in reflectance than the reddish interior plains and ejecta blanket and are interpreted as basement material excavated and redeposited during the basin-forming impact (Prockter et al., 2009).

Much of the basin floor is covered with HRP material that embays the rim, terrace, and peak-ring units and is reddish except where overlain by bright rays from nearby young craters (Figure 9.5c). These plains may be either volcanic or solidified impact melt. Plains with similar reddish colors are found in small discrete patches ranging from a few to a few tens of kilometers in horizontal extent, distributed within local topographic lows on top of Raditladi's terraces and continuous ejecta deposit (Figure 9.5d). Their location suggests that they are solidified impact melt that was deposited during the basin-formation event (Prockter et al., 2009; Mancinelli et al., 2014). A network of subcircular troughs, interpreted as graben, lie within Raditladi's peak ring close to the basin center, arranged in a concentric pattern ~70 km in diameter (Blair et al., 2013). These features are broadly similar to those found in Rachmaninoff and Mozart.

Topographic data from laser ranging (Cavanaugh et al., 2007) and stereo imaging (Preusker et al., 2011) show that Raditladi was formed on a regional high and has a rim-to-floor depth of $2.9^{+0.3}_{-0.6}$ km (Baker and Head, 2013). Mancinelli et al. (2014) observed that the relief is much greater on the western and northwestern ejecta deposits than elsewhere, which they attributed to an oblique impact or to pre-impact topographic relief striking southwest–northeast.

9.2.5.5 Mozart Basin

The ~236-km-diameter Mozart peak-ring basin (Figure 9.5f) is centered at 8°N, 169.5°E, and is superposed on the southern circum-Caloris smooth plains, with ejecta deposits overlying the Caloris Montes (the Caloris basin rim). Mozart has a sharp rim scarp, a well-preserved secondary field, and a low density of superposed craters (Schaber and McCauley, 1980), so it is relatively young, but age determinations have been ambiguous. Compared with Rachmaninoff and Raditladi, Mozart's interior is more heavily infilled with volcanic material and has only a few geological units that can be discriminated.

Mozart's rim wall has terraces, generally indistinguishable in color or relative reflectance from surrounding terrain (Figure 9.5g). Two arcuate mountainous segments near Mozart's center and some isolated hummocks likely are remnants of an inner peak ring (Baker et al., 2011b; Baker and Head, 2013). If intact, the peak ring would measure ~120 km in diameter. The peak-ring remnants differ spectrally from the surrounding terrain and are much lower in reflectance. Much of Mozart's interior is filled with smooth plains that have embayed the peak ring, rim terrace, and other interior units. These plains have intermediate spectral traits broadly similar to those of Mozart's continuous ejecta deposit and the plains immediately surrounding the basin (Figure 9.5g). As with Raditladi and Rachmaninoff, small patches of reddish smooth plains, up to several tens of kilometers in extent, are seen in topographic lows on top of the continuous ejecta, probably solidified impact melt from the basin's formation.

Mozart's inner plains contain narrow extensional troughs, up to ~1.5 km wide and ~3–25 km long, loosely circumferential to the basin center (Blair et al., 2013). The troughs are generally found in the central and northeast parts of the basin, bounded by the peak-ring massifs, but are absent to the southwest where there is little trace of the peak ring. Some troughs appear to be shallower toward areas where they are no longer visible, suggesting infilling by later volcanic flows. Near the basin center are some subtle undulating ridges, approximately 10–20 km long and a few kilometers wide, which do not appear to have interacted with the nearby troughs although they lie in close proximity.

Some bright, reddish material is associated with the southern portion of Mozart's rim and also appears in linear patches across parts of the rim (Figure 9.5g). Such deposits are interpreted to be pre-existing HRP material excavated by the impact. Prominent linear streaks, up to 20 km wide and 120 km long, with spectrally distinct low reflectance (Figure 9.5g), extend from the inner basin rim and overlie parts of the basin terraces, rim, and continuous ejecta deposit to the south. Similar LRM is also found along the northern remnants of Mozart's peak ring and was probably excavated and uplifted from deep within the basin during its formation, forming the peak ring and some late-stage ejecta.

Mozart is shallow compared with Raditladi and Rachmaninoff; on the basis of stereo topographic models, its floor lies between 0.5 and 1 km below the reference radius for the surrounding region (Preusker et al., 2011). The lowest regions are along the missing peak ring on the basin's western side and just outside the peak-ring trace to the east.

9.3 GEOLOGY OF IMPACT CRATERS

9.3.1 Primary Craters: Introduction

Craters on Mercury were formed by countless impacts by asteroids, large and small, by occasional comets, and conceivably by earlier populations of planetesimals (e.g., "vulcanoids," which are hypothesized early asteroids, now extinct, on orbits inside that of Mercury; see Section 9.6.1). Comets currently are a very small fraction of impacting small bodies in the inner solar system; possibly they were important in early epochs, though recent research suggests otherwise (Rickman et al., 2017). In this chapter we consider only asteroids, some of which were once comets but are now inactive. Each explosive impact by an asteroid generates ejecta, which can travel far beyond the primary crater, forming secondary craters. Such secondaries may be clearly distinguishable from primaries on the basis of their morphology and spatial relationships, but such attributes may be less obvious for those formed by the highest-velocity ejecta far from their primary craters. There are other processes on Mercury that create topographic forms resembling impact craters (e.g., volcanic vents, hollows), but those features can usually be distinguished from impact craters by their morphology. We discuss the continuum of impact crater sizes from those barely resolved in the highest-resolution images (some meters across) to those a few hundred kilometers in diameter (many larger ones are technically basins) (Figure 9.3).

Once formed, craters undergo degradation by several processes, which may eventually obliterate the craters so that they become unrecognizable. Naturally, smaller craters are more quickly degraded and obliterated, although some processes preferentially work to highly degrade the largest, oldest craters (e.g., viscous or viscoelastic relaxation). At the smallest sizes, certainly below 1-km diameter, the combination of primary and secondary craters can saturate the older surfaces and create regolith, analogous to that formed on the lunar maria. Such regolith may control surficial thermal processes on Mercury, affect downslope mass wasting on slopes, and affect the retention of volatiles in permanently shaded craters.

It is plausible that the upper crust of Mercury has been converted into a megaregolith – blocky or fractured bedrock produced by the cumulative effects of numerous impactors that created craters of order 100 km in diameter – that now extends many kilometers in depth. However, compared with the Moon, it is possible that very voluminous magmatic and volcanic processes competed with the late heavy bombardment (LHB), ~4 Ga, when most large craters and basins were formed, greatly changing the character of the crust from that of an idealized megaregolith.

Here we describe the different kinds of craters, explore measurements of their traits, and discuss the evolution of Mercury's surface under cratering bombardment. Next, size–frequency distributions of craters are presented and analyzed, and finally we address the absolute chronology inferred for the evolution of Mercury's surface.

9.3.2 Distinguishing among Primary, Secondary, and Other Types of Craters

Recent, fresh primary craters are generally easy to recognize, especially larger ones. They are explosion craters and have classic morphologies well described in the literature from studies of craters on the Moon and other solid-surfaced bodies (e.g., Melosh, 1989). Primary impact craters $\lesssim 10$ km in diameter on Mercury are formed with a classic bowl shape, except for rare cases of highly oblique impacts. Craters $\gtrsim 15$ km in diameter exhibit the classic shape of complex craters (see below). Such recent craters commonly display an ejecta blanket and a higher-reflectance halo and ray system, with strings and clusters of secondary craters immersed in the rays.

Secondary craters formed near a primary crater are often subdued and shallow, typically clustered or in chains, ill-shapen, and sometimes exhibiting chevron- or herringbone-like patterns due to interaction with other ejecta or the partially fluidized regolith into which the ejecta are falling. The impact velocities are too low to form idealized explosion craters. But the farther away a cluster or block of ejecta falls, the higher the impact velocity, the less likely to be close to other ejecta, and the more likely to form a crater with an initial morphology more closely resembling that of small primary craters. In other planetary contexts, such distant craters may be recognized as secondaries from non-obvious but still measurable spatial clustering (Bierhaus et al., 2005) or by other (e.g., photometric) similarities (e.g., the secondary crater system formed by the martian crater Zunil; McEwen et al., 2005). Commonly, researchers separately classify (or even exclude) so-called "obvious secondaries" but necessarily fail to identify the less obvious secondaries. This is an issue for Mercury's secondaries, which are larger and more common than on other bodies.

Endogenic craters (i.e., those not formed by impact) are uncommon on most planetary surfaces, including Mercury, and are usually readily identified. Volcanic vents on Mercury are typically not circular in shape, may lie atop a topographic mound, and may have special reflectance traits (Head et al., 2009). The smaller hollows might individually resemble a tiny, perhaps eroded, primary or secondary impact crater, but they occur in obvious clusters in very restricted regions and typically have unusually high reflectance (see Chapter 12), so the chances of confusion are minimal. The most ambiguous cases on Mercury concern crater chains. Some are obviously chains of secondary craters, but others have spatial relationships with troughs and graben and are likely not impact craters. Most uncertain are some that are radial to primary craters and may reflect mass wasting into radial cracks but do not look particularly different from secondary chains.

Inevitably, craters of any origin degrade morphologically with time, and it becomes more difficult to identify their origin with any certainty. But statistical characteristics (e.g., spatial relationships, size–frequency attributes) may be applied in many cases.

9.3.3 Morphology of Simple and Complex Craters on Mercury

During a crater's formation, its morphology is primarily influenced by the magnitude of the surface gravitational acceleration (g) of the target body (e.g., Gault et al., 1975; Pike, 1980) but also by the physical attributes (strength, porosity, layering) of the target material (e.g., Fulmer and Roberts, 1963; Pike, 1988; Senft and Stewart, 2007; Kalynn et al., 2013) and the density and velocity of the projectile (e.g., Cintala et al., 1977; Gault and Wedekind, 1978; Schultz, 1988; Holsapple, 1993; Hermalyn and Schultz, 2011). On bodies without a significant atmosphere, crater morphology is influenced after formation by erosion by nearby younger craters (Gault 1970; Wood et al., 1977), infill by volcanic activity, tectonic activity, or relaxation of topography by interior flow (Mohit et al., 2009).

Impact crater morphology, specifically ratios of depth (d) or rim height (h) to diameter (D) and the transition diameter (D_t) from simple to complex shape, can provide inferences on the strength of the surface and subsurface. Further, comparison of the dimensions of fresh, crisp craters to non-fresh ones (as well as to embayed and filled ones) can constrain the efficiency of subsequent impact erosion and volcanic processes in influencing surface evolution. Finally, changes in crater shapes due to horizontal compression or extension can quantify tectonic strain.

Ejecta blanket morphology can also yield insights about surface properties and processes. Whereas for most atmosphereless bodies, ejecta deposits are well explained by ballistic deposition, some lunar ejecta deposits may have flowed (see below). Mercury may elucidate such processes on the Moon and Mars, given (a) its somewhat similar surface properties and environments to those on the Moon but dissimilar surface gravitational acceleration and impact velocities, and (b) the similar surface gravitational acceleration of Mercury and Mars but dissimilar surface environments and impact velocities.

9.3.4 Ratio of Crater Depth to Diameter

Prior to MESSENGER, measurements of d and h for craters on Mercury were derived with traditional photoclinometry and shadow-length techniques (e.g., Pike, 1988) from Mariner 10 imaging. They revealed (e.g., Malin and Dzurisin, 1977) that, as on the Moon and Mars, Mercury's craters become progressively more complex with increasing diameter, from simple craters to complex craters, protobasins, and basin morphologies, as discussed earlier (Figure 9.3). The depths of simple craters were found to be like those for the Moon and Mars, but those for larger complex craters were seen to be greater for Mercury. D_t was found to be near $10.3^{+4.4}_{-3.1}$ km, as expected by extrapolation between D_t for the Moon and the Earth as a function of g. Surprisingly, this value differs from both recent (5.6 ± 2.3 km; Robbins and Hynek, 2012) and past (5.8 ± 2.3 km; Pike, 1980) estimates of D_t for Mars, despite the similarity in g. These estimates include the transition of all morphological attributes (e.g., flat floors, terracing, rim-height transition) from simple to complex craters.

MESSENGER laser altimetry and imaging data circumvented some of the pitfalls encountered in analyzing Mariner 10 data and have yielded new crater morphology data for Mercury. Barnouin et al. (2012) combined shadow measurements from images and MLA data from MESSENGER's first two Mercury flybys (M1 and M2) and largely confirmed the older data (Pike, 1988), particularly for the freshest simple craters in equatorial regions. They also found that d values for fairly fresh large complex craters plot along the lower bound for depths from Mariner 10 data. The study also confirmed that the more degraded craters defined by conventional morphology classes (e.g., Trask, 1967) are shallower than fresher craters of the same diameter. From MLA crater elevation profiles acquired during the first year of MESSENGER orbital operations, Talpe et al. (2012) showed that the data were largely consistent with the results of Barnouin et al. (2012), but Talpe et al. also compared the depths of craters in different types of plains units and found that those in heavily cratered plains were usually shallower than those in smooth plains at a given diameter, reflecting the more degraded state of typical older craters on older terrain and possibly an admixture of shallower secondary craters. Moreover, in contrast with earlier studies, Talpe et al. (2012) found no statistical difference between d for craters

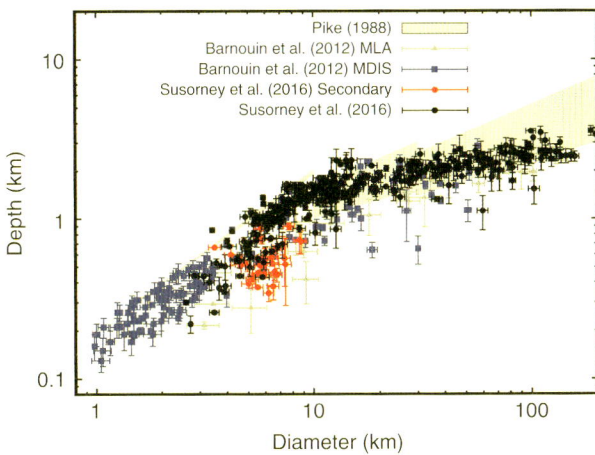

Figure 9.7. Depth versus diameter for craters on Mercury (modified from Barnouin et al., 2012; Susorney et al., 2016). The Susorney et al. (2016) results used MLA data, whereas Barnouin et al. (2012) used shadow measurements from MDIS and depth measurements from MLA. Gray shaded lines show Mariner 10 results from Pike (1988).

that hosted radar-bright deposits (indicating likely ice deposits; see Chapter 13) and d for other craters investigated.

MESSENGER Mercury flyby data indicated a value for D_t of about 12 km (Barnouin et al., 2012), whereas the analysis of MLA profiles acquired from orbit by Talpe et al. (2012) indicated $D_t \sim 8$ km. Both studies used only the intersection of power laws fit to d versus D for simple and complex craters to calculate D_t and did not use the more complete approach of Pike (1980, 1988). The results may differ because of inclusion of shallower secondary craters by Talpe et al. (2012), which could change the power laws. From MLA data collected during the entire MESSENGER mission, Susorney et al. (2016) measured d, h, wall-width (w), and peak or peak-ring height (r), and they used MDIS images with guidance from MLA to estimate D. They also assessed crater degradation state as well as regional terrain type and data on the shapes of secondary craters, which are prolific in the northern smooth plains (NSP), and they considered all morphological attributes typically used to determine D_t (e.g., Pike, 1988), which they found to be 11.7 ± 1.2 km. Besides confirming other MESSENGER results (Figure 9.7), the group found that secondary craters on average are shallower than similar sized primaries, and no significant differences were found in d values between fresh craters in heavily cratered plains and those in smooth plains. The latter finding is consistent with the Talpe et al. (2012) result, which included both degraded and fresh craters.

The lack of a dependence of d on terrain type for fresh craters is a bit surprising since lunar craters and simple martian craters show such a dependence (Baldwin, 1963, 1965; Pike, 1974, 1980; Boyce and Garbeil, 2007; Stewart and Valiant, 2006; Robbins and Hynek, 2012). Possibly the smaller age difference between the smooth and cratered plains on Mercury compared with that between the lunar maria and highlands would mean that megaregolith properties would not differ much and would display smaller differences in diameter scaling. The likely volcanic origin of both the intercrater and smooth plains (e.g., Trask and Guest, 1975; Whitten et al., 2014; Spudis and Guest, 1988; Denevi et al., 2013; see Chapter 6) may also have reduced differences in terrain properties and resulting crater depths. Alternatively, the much wider range in impact velocities at Mercury compared with other planets may have overwhelmed other factors, leading to similar d values across all terrains. Because terrain effects are more likely to be noticed when D is small, for which target properties are more important relative to g, more detailed studies of small crater morphologies on Mercury are warranted.

The D_t result of 11.7 ± 1.2 km reported by Susorney et al. (2016) for Mercury is consistent with earlier results of Pike (1988) in that it is different from that on Mars ($D_t = 5.6 \pm 1.2$ km; Robbins and Hynek, 2012), despite similar values of g. Variations in target strength are not likely responsible, given the similarity in d/D for simple craters on Mars and Mercury. The higher-velocity impacts on Mercury could have produced deeper final craters (Schultz, 1988; Barnouin et al., 2011) because they reduce the coupling time between a projectile and a target as the shock travels more quickly through the projectile. Experimental impacts in low-friction granular targets (Barnouin-Jha et al., 2007) yield shallow transient craters for high velocities but slightly deeper final craters. Numerical investigations suggest that such a process might occur until a critical impact velocity is reached (Jutzi and Michel, 2014; Bray and Schenk, 2015), above which all projectile velocity effects are lost. Beyond deeper final craters on Mercury relative to Mars, the transition to complex craters might be expected to occur at a different D on Mercury as well. But the similar d values for simple craters on Mars and Mercury emphasize the need for more experimental and numerical studies to understand the interplay between target strength and impact velocity on crater shapes and resulting value of D_t.

9.3.5 Ejecta Morphology

Mariner 10 studies (Gault et al., 1975; Cintala, 1979) showed that ejecta deposits on Mercury for simple and complex craters are similar to those for corresponding features on the Moon and are presumably just the product of ballistic sedimentation (Oberbeck, 1975). The radial sequence from the crater rim begins with a hummocky inner ejecta blanket, then progresses radially outward to the development of radial features, a field of secondaries, and finally rays. The distances are more compressed toward the crater rim on Mercury because of its higher g.

Quantitative, high-resolution studies of ejecta deposits on Mercury have not yet been conducted. However, qualitative assessments are consistent with the Mariner 10 results for many craters (e.g., Figure 9.8). The inner edge of the hummocky terrain is closer to the crater rim and the radial zone is far less extensive than on the Moon, for craters of the same size. The region where secondary cratering begins also is closer to the crater but is broader than on the Moon. Preliminary assessments of surface roughness near impact craters show that the secondary cratering zone is a main contributor and affects Mercury's surface topography at lateral scales of 1 km (Kreslavsky et al., 2014; Susorney et al., 2015).

Some ejecta facies appear to have flowed (Xiao and Komatsu, 2013; Barnouin et al., 2015); such ejecta deposits have broad,

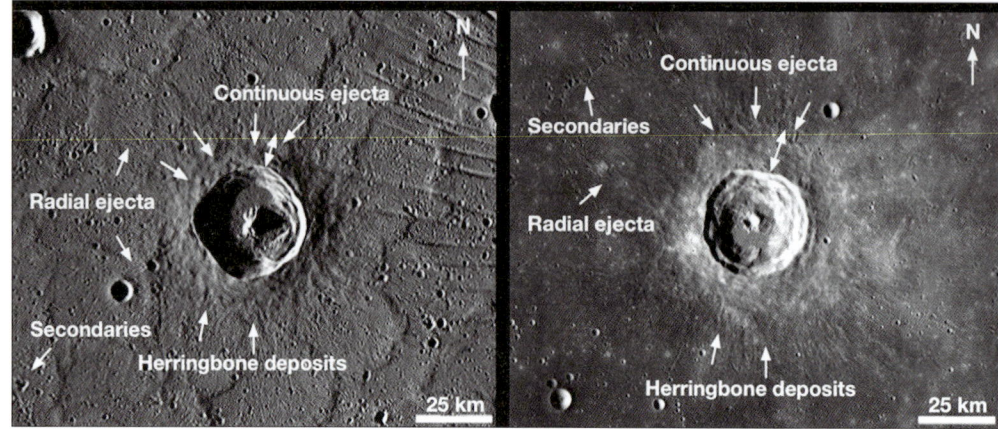

Figure 9.8. Comparison of 36-km-diameter crater Cunningham on Mercury (left; 30.4°N, 202.9°E) with 34-km-diameter lunar crater Timocharis (right; 26.7°N, 246.9°E). Note the shorter extent of the near-rim continuous ejecta deposit on Mercury and the sharper and less extensive herringbone pattern along its edge. Timocharis possesses a more diffuse and wider herringbone zone. Note also that secondaries for Cunningham seen throughout its ejecta field are typically larger than secondaries of Timocharis at the same radial distance from the crater center.

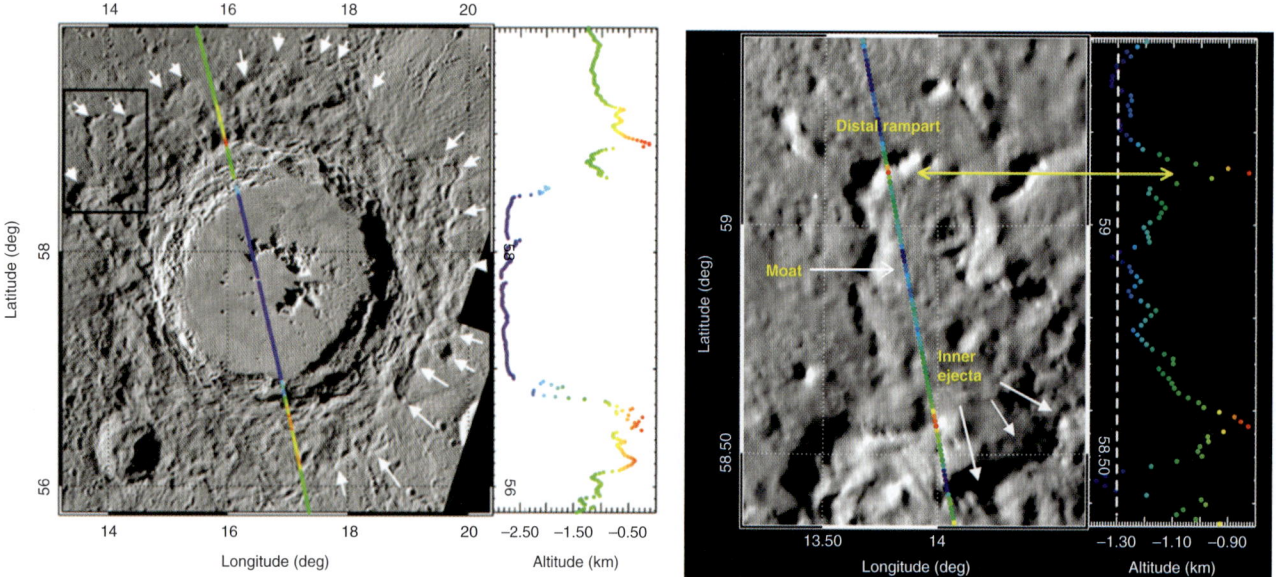

Figure 9.9. (Left) Hokusai crater with an MLA profile, color-coded by altitude, along the indicated line. The black box outlines the region shown on the right. White arrows indicate distal ejecta ramparts. (Right) A distal rampart is separated from inner ejecta by a topographic moat. Ejecta in the moat are minimally thick, as indicated by elevation relative to the height of surrounding terrain (white dashed line).

low-curvature distal rises, resembling landslides (Barnouin-Jha et al., 2005), which MLA data reveal are quite thick (500–800 m in the case of a 60-km-diameter crater at 45°N, 288°E). These more common landslide-like flows form on pre-existing slopes >5° and seem to result from outward slumping of the crater rim, like those at Tsiolkovsky basin on the Moon (Guest and Murray, 1969). The morphology of these ejecta flows differs from that of fluidized ejecta (single-layered ejecta) deposits on Mars (Barnouin-Jha et al., 2005), which have a region of thin ejecta separating a sharp distal rampart from a near-rim thick ejecta deposit.

The unique, very young crater Hokusai (Figure 9.9) does have a fluidized ejecta morphology more akin to single-layer ejecta morphology. Unlike the landslide-like ejecta, this facies is formed on a very slight (<1°) regional slope. Furthermore, MLA data show that the thickness of the Hokusai ejecta deposit does not decrease with radial distance from the crater rim by a power-law function with a −3 exponent typically associated with ballistically emplaced ejecta (e.g., McGetchin et al., 1973) that have not flowed extensively across the surface. The surface roughness pattern around Hokusai is distinctive, with ropey near-rim ejecta and a less well-pronounced secondary field relative to similar-sized craters Abedin and Stieglitz, which are also in the NSP (Susorney et al., 2015). Surrounding about two-thirds of the crater, the ropey ejecta are bounded by a distal, terminal rampart, which is generally separated by a topographic moat from the continuous ejecta and can be ~500 m in height. Although the Hokusai rampart is high relative to those on Mars (~200 m for similar-sized craters) and does not run out as far (<1 crater radius in contrast to >2 crater radii on Mars), it is otherwise morphologically identical to single-layer martian ejecta facies (Barnouin-Jha et al., 2005). The much more numerous melt pools seen in Hokusai's ejecta facies relative to Abedin's may account for the origin of the fluidized ejecta, which could have intermixed with solid ejecta, allowing it to flow easily and then stop to form a rampart. This mechanism would be a mercurian analog to the hypothesis that water mainly contributed to single-layer ejecta on Mars (e.g., Carr et al., 1977; Greeley et al., 1980).

9.3.6 Special Types of Craters and Associated Features

9.3.6.1 Elliptical and Polygonal Craters

Elliptical impact craters, perhaps 2–4% of Mercury's primaries (Collins et al., 2011), result from highly oblique impacts (Gault and Wedekind, 1978) (Figure 9.10). Of course, smaller secondary and endogenic craters may more commonly exhibit elliptical shapes in map view.

Polygonal impact craters have rims partly composed of at least two straight segments (Öhman et al., 2005) and form when the growth or modification of a crater is influenced by discrete strength heterogeneities (e.g., faults) within a target (e.g., Watters et al., 2017, and references therein). On Mercury, Weihs et al. (2015) found 33 polygonal impact craters among 291 that are >12 km in diameter, ~11% of the population. This total is consistent with estimates that 10–15% of craters are polygonal (Öhman et al., 2005) and with the finding of Wood et al. (1977) from the limited Mariner 10 data set that 16% of craters are quasi-polygonal. Weihs et al. (2015) found that the diameters of polygonal craters range from 65 km to 240 km and average ~120 km, consistent with the idea that polygonal impact craters tend to be mid-sized complex craters (Pohn and Offield, 1970; Öhman et al., 2005). Weihs et al. (2015) reported regional differences in the percentages of craters that are polygonal.

Figure 9.10. Sveinsdóttir, at 2.8°S, 100.3°E, is one of the largest elliptical impact craters on Mercury at 220 km × 120 km.

9.3.6.2 Ghost or Buried Craters

In addition to obvious impact craters superposed on volcanic smooth plains units (e.g., the NSP and the Caloris interior and exterior plains), some partially to fully embayed craters are seen in these plains units (Head et al., 2008, 2009, 2011; Watters et al., 2009c, 2012; Klimczak et al., 2012; Ostrach et al., 2015). Buried, or "ghost," craters are identified by arcuate tectonic features resembling wrinkle ridges (or occasionally graben), often forming rings, that are thought to have nucleated above the rims of buried impact craters (e.g., Watters, 1993). These are subtle topographic features, and many buried craters must have been entirely covered over and their signatures erased by the emplacement of thick volcanic or crater ejecta deposits.

Within the NSP, ghost craters with diameters as small as 4 km and as large as several hundred kilometers (e.g., Goethe basin, ~317-km diameter) are reliably observed. Unambiguous identification of relict, partially to completely buried craters is aided by images at high solar-incidence angle, which typically show long shadows that emphasize the subtle topography of tectonic features marking the rims of buried craters. However, even with such images, buried craters ≤25–30 km in diameter are more difficult to discern because of the complexity of tectonic structures within the NSP (Byrne et al., 2013; Ostrach et al., 2015; Chapter 10).

Buried craters are categorized as one of three morphological types (Klimczak et al., 2012; Watters et al., 2012). Type-1 buried craters display a wrinkle-ridge ring and span a wide range of diameters (Figure 9.11a). Type-2 buried craters contain arcuate wrinkle ridges that may not form a full ring and also contain interior graben (Figure 9.11b). Type-2 are generally large buried craters or basins, none being less than 40 km in diameter. A rare third category of buried craters (Type-3) consists of features with minimal or absent wrinkle-ridge rings; instead, a graben ring outlines the rim of the buried crater (Figure 9.11c). Type-3 buried craters range from 10 to ~50 km in diameter and are closer to the boundaries of the NSP than type-1 or type-2 buried craters. Buried craters enable estimates to be made of the thickness (and thus volume) of volcanic smooth plains. Given that an impact crater was fresh when it was buried by volcanic material above its rim, morphometric relations (rim height above surrounding terrain and d to D ratio; Pike, 1988; Barnouin et al., 2012) can be used to estimate pre-flooding rim heights and thus the minimum thickness of material needed to cover the crater.

9.3.6.3 Rayed Craters and Crater Chains

Young craters with bright rays are prominent on Mercury (Figure 9.12). Rays can extend up to 4500 km from the parent crater (Izenberg et al., 2009), as for Hokusai (Xiao and Werner, 2015). Although the cratering rate on Mercury is about a factor of 3 higher than on the Moon, there are a factor of ~2 fewer rayed craters per unit area on Mercury (Braden and Robinson, 2013), implying that Mercury's regolith matures faster than lunar regolith (Denevi and Robinson, 2008) because of Mercury's higher solar wind and meteoroid bombardment fluxes and higher temperature. Neish et al. (2013) showed that mercurian crater rays appear bright not only in optical images but also at radar

230   *Impact Cratering of Mercury*

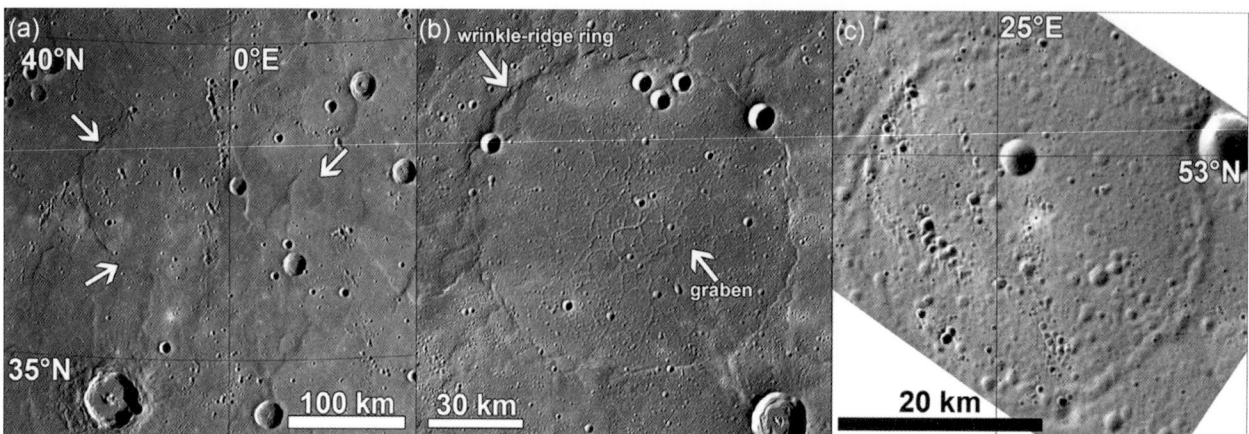

Figure 9.11. Three morphological types of buried craters on Mercury: (a) Type-1 craters exhibit an arcuate wrinkle-ridge boundary or a wrinkle-ridge ring; two examples are noted by arrows. Type-1 craters are the most commonly observed type of buried craters. Image centered at ~37.1°N, 1.4°E. (b) Type-2 craters have an arcuate wrinkle-ridge boundary and interior graben; an example of this type is at ~60.2°N, 36.5°E. (c) Type-3 craters have a graben ring instead of a wrinkle-ridge ring, and no wrinkle ridges or additional graben are associated with the buried crater; an example is centered at ~52.8°N, 24.8°E. (a) and (c) modified from Klimczak et al. (2012).

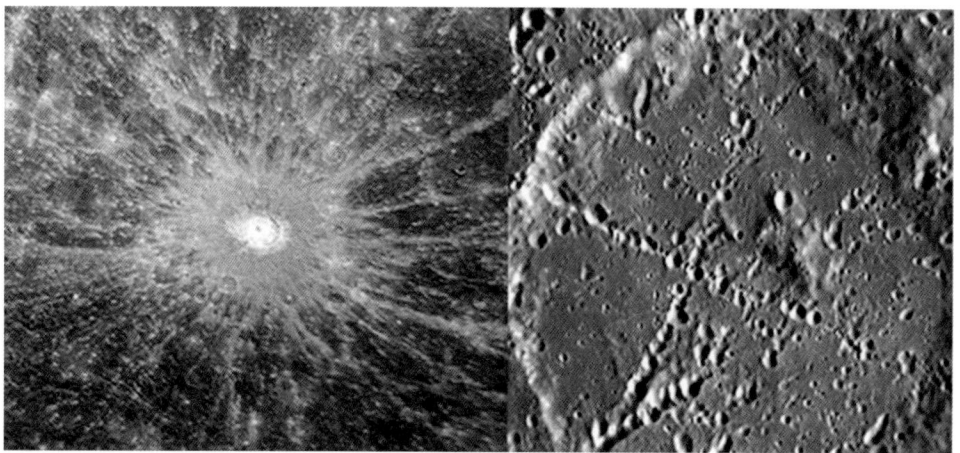

Figure 9.12. (Left) The 85-km-diameter Debussy crater, one of the geologically youngest large craters on Mercury, shows an extensive system of bright rays. (Right) The floor of the ~134-km-diameter Stevenson crater is crisscrossed by prominent crater chains, most resulting from ejecta from two primary impacts outside this view.

wavelengths, because of centimeter- to decimeter-sized scatterers, i.e., rocks and boulders deposited as ejecta and excavated by dense concentrations of secondary craters. Radar polarization ratios and optical maturity permit the assignment of relative ages for rayed craters; Debussy (Figure 9.12), Amaral, and Kuiper are among the youngest (Neish et al., 2013).

Crater chains on Mercury are formed from ejecta thrown out by an impact. Such craters may display characteristic herringbone patterns pointing back toward the crater of origin. Crater chains can be found thousands of kilometers from their source crater. Individual craters within a chain commonly have irregular rims and may be elongated because of the shallow ejection and impact angles of the ejecta (Figure 9.12b). Secondary crater chains are usually associated with younger impact structures, presumably because such relatively small-scale features have been degraded and erased over time.

9.3.6.4 Identification of Solidified Impact Melt

Although Apollo-era lunar images revealed "ponds" interpreted as solidified impact melt (Shoemaker et al., 1968; El-Baz, 1972; Howard, 1972; Howard and Wilshire, 1975; Hawke and Head, 1977a, b) and several smooth-ponded deposits were identified in mercurian craters imaged by Mariner 10, analyses of impact melt on Mercury awaited the higher-resolution images from varying viewing geometries afforded by MESSENGER's orbital mission (Beach et al., 2012; D'Incecco et al., 2012; Ostrach et al., 2012).

Ostrach et al. (2012) identified solidified impact melt in Kuiperian- and Mansurian-aged craters (see Chapter 6 for definitions of these stratigraphic periods) on the basis of three primary factors (Figure 9.13): (1) interior pond morphology (smooth texture, few superposed craters, and a distinct

9.4 Crater Size–Frequency Distributions and Statistics

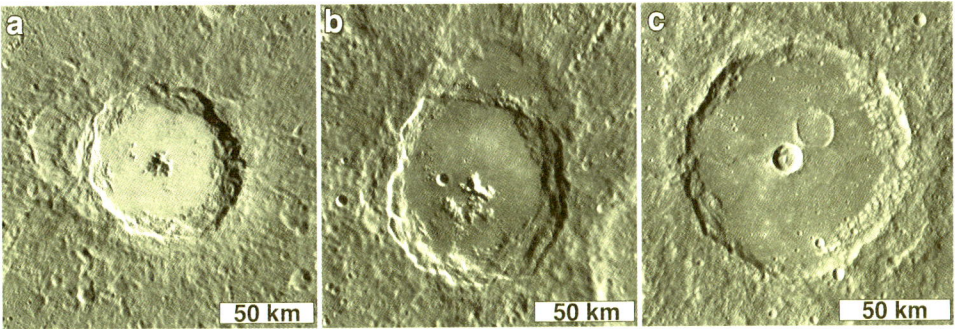

Figure 9.13. Representative craters illustrating criteria for distinguishing solidified impact melt materials from volcanic materials. (a) Debussy crater (33.95°S, 12.5°E, 81-km diameter) and (b) Sibelius crater (49.5°S, 214.6°E, 94-km diameter) are relatively young impact craters containing smooth interior deposits with well-defined contacts on the crater floors and hosting exterior ponds or flows and perched ponds on crater wall terraces, which are primary indicators of materials interpreted to be solidified impact melt on Mercury. Additionally, solidified impact melt may be distinguished from volcanic materials within impact craters by an absence of embayed or filled craters within smooth floor deposits, since impact melt is emplaced soon after initial impact crater formation. (c) Machaut crater (2.05°S, 277.6°E, 104-km diameter) is an example of a volcanically modified impact crater, where volcanic materials flooded the crater interior and embayed pre-existing features (arrow denotes an embayed crater).

fill–wall contact); (2) the presence of exterior flows, ponds, and veneers; and (3) an absence of craters embayed by a flow unit, which would otherwise require volcanism long after the crater formed (e.g., Strom et al., 1975; Wilhelms, 1976). A complicating factor for Mercury compared with the Moon is that the reflectances of volcanic and impact-generated plains on Mercury are similar (e.g., Robinson et al., 2008; Denevi et al., 2009); in contrast, solidified impact melts on the Moon are typically distinct from darker basaltic volcanic infill.

At high resolution, material interpreted to be solidified impact melt appears smooth and fresh, with minimal erosion and few superposed craters. Any linear and polygonally arranged surface cracks on ponds are interpreted as cooling cracks (e.g., Howard and Wilshire, 1975; Schultz, 1976; Xiao et al., 2014b). Impact melt ponds, veneers, or flows exterior to the parent crater postdate the surrounding terrain since they were expelled from the crater and then flowed into their current positions immediately following ejecta emplacement (e.g., Osinski et al., 2011).

9.4 CRATER SIZE–FREQUENCY DISTRIBUTIONS AND STATISTICS

9.4.1 Methodology and R-Plots

A primary tool that has been used to study crater populations on solar system bodies is the size–frequency distribution (SFD), i.e., the distribution of areal density by crater diameter, which typically is approximated by a power law and is shown on a log–log plot. Several methods used for plotting such data and deriving fits to the SFDs have been used by different researchers; most were defined by the Crater Analysis Techniques Working Group (1979). The fundamental plot is the differential SFD, which plots, against diameter D, the number of craters counted in a diameter interval per square kilometer divided by the width of the interval in kilometers. (The plotted slope of a power-law fit to such differential data is the power-law exponent.) Such a plot is inconvenient in several ways (e.g., it has a very steep slope and often has large error bars), so some researchers plot cumulative frequency data (number of craters greater than or equal to diameter D per square kilometer). That approach, which yields a shallower slope by about 1 unit, can be misleading and hides important features of SFDs (Chapman and Haefner, 1967; Chapman, 2015). Also, error bars are calculated incorrectly (they are too small) for cumulative plots by some commonly used tools such as Craterstats (2016; versions created prior to June 2016) in the sense that the statistical significance of changes in curve shape cannot be ascertained from the error bars of adjoining cumulative numbers since they are not independent. Other improved methodologies have been proposed (Robbins et al., 2018) but have not yet been widely used. Therefore, we choose to use the relative plot or R-plot, in which a differential SFD is divided by D^{-3}. The R-plot has the intuitive property that the vertical frequency axis plots the spatial density (R is relative density) of craters with respect to a horizontal line (at a value of order 1) that represents crater saturation.

In this section we present R-plots of Mercury crater SFDs for different geological units and different types of craters. In general, we do not show data believed to be incomplete at the small-diameter end of the diameter range being plotted (e.g., because they approach the resolution limit of the image). We usually show one-standard-deviation error bars derived from Poisson $\sqrt{N}$ statistics, which do not account for the variety of systematic errors that may be present. Readers are cautioned that such SFDs are subject to several uncertainties, as discussed by Robbins et al. (2014). We also note that one-standard-deviation errors, commonly used in this field, do not constrain statistical significance as tightly as two- or three-standard-deviation errors commonly used in other fields.

Another useful measure of crater density is $N(D)$, which is defined as the cumulative number of craters with diameter $\geq D$ per 10^6 km^2, where D is in kilometers. This measure, with $\sqrt{N}$ errors, provides a statistically robust estimate of crater density, independent of size, which can be interpreted in terms of age, provided that the shape of the SFD over this diameter range matches the production function (PF), i.e., the function that describes the distribution of the rate of formation of impact craters by size.

9.4.2 Interpretation of Features in R-Plots

The cratering record on Mercury shows two primary crater populations roughly similar to those on the Moon and Mars. We thus assume in what follows, consistent with our understanding of the dynamics of small bodies in the inner solar system, that the Moon and terrestrial planets have all been cratered by essentially the same populations of small-body impactors. The major variable affecting the scaling between impactor and crater sizes is mean impact velocity, which differs greatly from Mercury to Mars. The term "Population 1" has been used to denote craters on the older, more heavily cratered terrains on which most craters formed during the late heavy bombardment, approximately from 4.1 to 3.9 Ga. The term "Population 2" denotes the craters that dominate on the younger, lightly cratered plains and primarily formed after the end of the LHB on the Moon, Venus, Mars, and Mercury (Strom et al., 2005, 2015). Figure 9.14 is an R-plot of the SFDs for several crater populations on terrestrial bodies. The lunar highlands SFD at the top illustrates Population 1, and the three SFDs toward the bottom, below $R = 0.01$, have the typical horizontal slope (though different densities) for Population 2. Population 1 on the Moon has a complex SFD, characterized on an R-plot by a differential SFD with −2.2 exponent at D less than about 50 km, a nearly horizontal part (−3 exponent) between 50 and 100 km in diameter, and a part that slopes downward to the right (−4 exponent) from $D \approx 100$ km to 300 km. The SFD also may slope upward from ~400 to 1000 km, although statistics are poor for such large features. Population 2 has an SFD that generally follows a −3 differential SFD exponent, i.e., it is a nearly horizontal line on an R-plot. The crater density of Population 1 typically exceeds that of Population 2 by at least an order of magnitude for $D > 10$ km.

In interbody comparisons of crater SFDs, one must recognize that asteroids currently impact these bodies at different mean velocities, about 42 km/s at Mercury, 19 km/s at the Moon, and 12 km/s at Mars. This difference results, on average, in larger craters on Mercury than on the Moon or Mars for a given impactor size, so for Mercury the crater curve of Population 1 is shifted to the right (toward larger diameters) of the lunar SFD, and that for Mars is shifted to the left (toward smaller diameters). These velocity corrections have not been applied to R-plots of SFDs in this chapter.

On Mercury the shape of the SFD for craters on heavily cratered terrains appears steeper at smaller diameters (<50 km) than for comparable terrains on the Moon. This difference was originally ascribed (Strom, 1977) to loss of many smaller craters on Mercury by voluminous effusive volcanism during the formation of the intercrater plains. Whereas the degradation of small, older craters has also occurred on the Moon and particularly on Mars by a variety of processes, the formation of intercrater plains was especially efficacious in obliterating smaller craters on Mercury. The amount of crater loss and the steepness of the SFD depends on the degree of flooding by intercrater plains; the greater the flooding, the steeper the curve (Strom et al., 2011). One of the plots in Figure 9.15 (labeled "Mercury Flooded") shows the SFD for a very densely cratered area that has been partially flooded by intercrater plains. At $D > 25$ km the SFD is similar in shape and spatial density to that of the lunar highlands, but at $D < 20$ km craters have evidently been erased by intercrater plains emplacement, causing the SFD to bend down sharply.

SFDs for Mercury's heavily cratered terrain, NSP, the smooth plains exterior to Caloris, and Class 1 craters are compared with those for the lunar highlands in Figure 9.15. (Class 1 craters are defined as having a pristine morphology, a well-defined continuous ejecta blanket, and fresh secondary craters on the surrounding terrain; it is doubtful that Class 1 represents the same range of ages at all sizes, but the class is an approximation for "recent" craters.) The shape of the SFDs for smooth plains on Mercury is the same as that for Mercury's Class 1 craters and relatively recent lunar Copernican and Eratosthenian craters (see Figure 9.14). The SFD for the heavily cratered terrains with interspersed intercrater plains is similar to that for the lunar highlands but has a lower crater density because of the greater loss of craters on Mercury by erosional and depositional processes, including volcanic emplacement of the intercrater plains. The upturn in the SFD at diameters near 10 km and smaller is due to secondary craters, as discussed

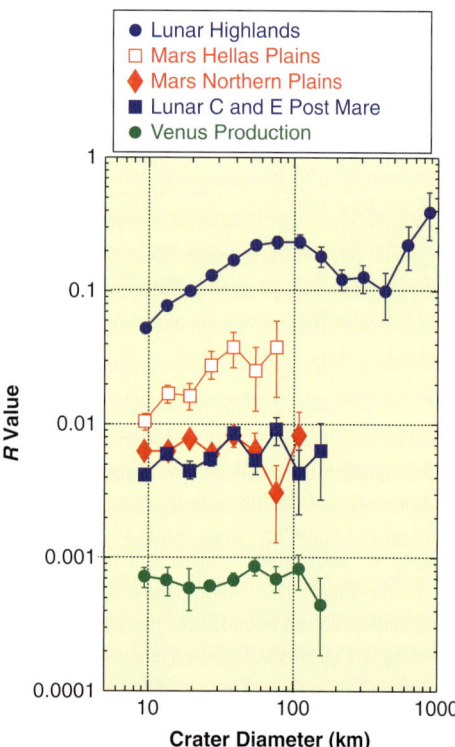

Figure 9.14. These R-plots summarize the inner solar system cratering record for $D > 10$ km. The representative SFDs show the two distinctly different crater populations. Those above $R \sim 0.01$ have the complex shape of Population 1, whereas the lower plots show the nearly horizontal trend of Population 2. The "Lunar C and E" craters are post-mare Copernican and Eratosthenian in age, and the "Venus Production" is a composite of all craters and multiple craters. Modified from Strom et al. (2015).

9.4 Crater Size–Frequency Distributions and Statistics 233

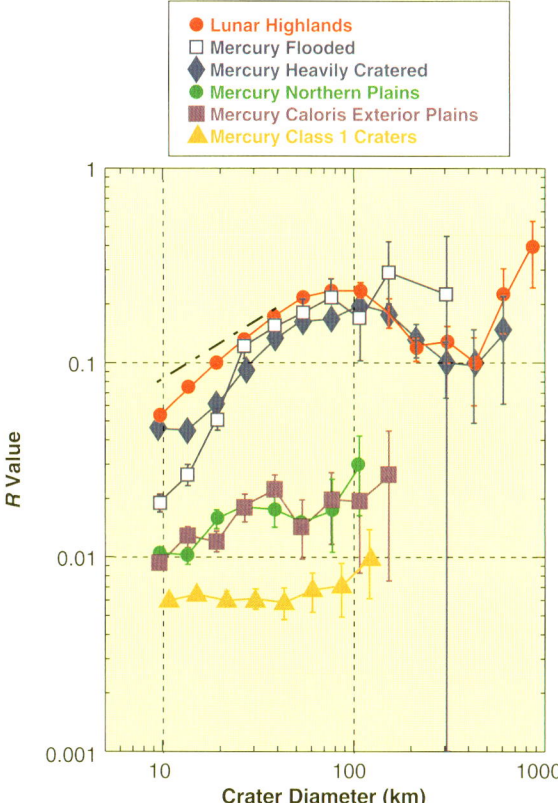

Figure 9.15. R-plot of impact crater SFDs for Mercury's heavily cratered terrain, the NSP (from Ostrach et al., 2015), Caloris exterior plains, and Class 1 craters compared with that for the lunar highlands. Also shown (Mercury Flooded) is the SFD for a heavily cratered area where emplacement of intercrater plains substantially obliterated craters <25 km in diameter. The black dash-dot curve is a hypothetical SFD at diameters <50 km discussed in the text.

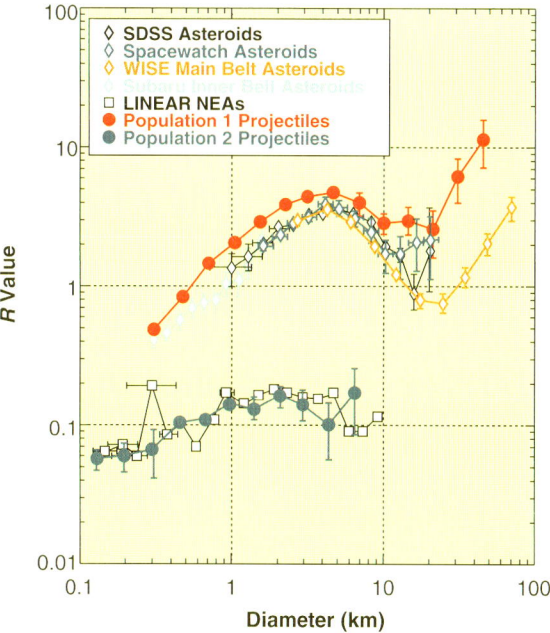

Figure 9.16. The SFDs of the impactors derived from the crater SFDs for Populations 1 and 2 compared with those of MBAs and NEAs. The Subaru asteroid data are from the inner belt, whereas the Spacewatch, Sloan Digital Sky Survey (SDSS), and NEOWISE data are from the entire main belt; the NEA data are from LINEAR (sources are cited in the text). Impactors derived from the lunar highlands SFD represent Population 1, and impactors derived from the young Mars plains SFD represent Population 2. The absolute values of R are arbitrary in this plot, so the astronomical data can be slid vertically to allow easy comparison with the projectiles derived from the crater data.

below. SFDs for the two plains units have the same crater density, indicating that volcanism ended in those regions at about the same time, but they have an SFD slope intermediate between those of Populations 1 and 2, indicating that in addition to Population 2 there was a modest admixture of the earlier Population 1 (Strom et al., 2008, 2011). Therefore, these plains must have been emplaced near the end of the LHB, after the Caloris impact, at a time when the LHB impact rate had fallen to a level such that subsequent cratering by Population 2 dominated the late Population 1 cratering. The fresh Class 1 craters counted in equatorial areas have a horizontal SFD and a low density characteristic of "pure" Population 2.

9.4.3 Relationship between Crater and Impactor SFDs

From a comparison of the SFDs of main belt and near-Earth asteroids with those of Population 1 and 2 craters, Strom et al. (2005) proposed that Population 1 impactors were main belt asteroids (MBAs) and Population 2 impactors were near-Earth asteroids (NEAs). They related the size of an impact crater to the size of the impactor by a Pi-group scaling law (Croft, 1985; Schmidt and Housen, 1987; Melosh, 1989; Collins et al., 2005). Figure 9.16 is an R-plot comparing the two asteroid populations with the inferred Population 1 and 2 projectiles. The NEA data are based on bias-corrected NEA diameters from the Lincoln Near-Earth Asteroid Research (LINEAR) survey (Stuart and Binzel, 2004). The MBA data are derived from four published data sets: (1) Spacewatch (Jedicke and Metcalfe, 1998); (2) Sloan Digital Sky Survey (SDSS; Ivezić et al., 2001); (3) Wide-field Infrared Survey Explorer (NEOWISE) first release data (Masiero et al., 2011); and (4) Subaru Main Belt Asteroid Survey (Yoshida et al., 2003). The "red" asteroid data from the SDSS were used, and from the other data sets only MBA data from the inner part of the main asteroid belt were used. Asteroid observations have continued during the years since NEOWISE, but the qualitative features in Figure 9.16 remain unchanged.

The reason the SFDs for the MBAs and NEAs differ somewhat is the subject of current research, but size dependence of processes such as the thermal Yarkovsky effect (due to asymmetric absorption and re-radiation of sunlight by a spinning asteroid), which help remove NEAs from the main belt by moving them toward resonances, may be responsible. The projectile populations early in solar system history and those causing the LHB, which itself is still under debate about its timing, magnitude, and origin, are also the subjects of continuing research. Debated topics relevant to the projectile population SFDs include different scenarios for giant planet migration, such as "jumping Jupiter," scenarios involving just the tail-end of accretion but no LHB, arguments that the impactor SFD has never changed, and inferences from new lunar and martian data that the MBA population may not be a good match to the older cratering record at all diameters (e.g., Marchi et al., 2013; Minton et al., 2015b; Neumann et al., 2015; Roig and Nesvorný, 2015; Bottke et al., 2017; Evans et al., 2017; Morbidelli et al., 2017; Orgel et al., 2017; Werner, 2017).

For Population 1, the dynamical ejection event from the asteroid belt inferred to have caused the LHB apparently was one that largely preserved the MBA size distribution, such as a planetary gravitational event that caused the migration of secular resonances, particularly the 2:1, 3:1, and ν_6 resonances, through the asteroid belt, ejecting asteroids into terrestrial-planet-crossing orbits (Strom et al., 2005, 2015). Other planetary gravitational interaction scenarios, such as the "Nice model" (Tsiganis et al., 2005) and other scenarios cited above, continue to be researched.

9.4.4 Saturation Equilibrium

An interesting question about Mercury's early bombardment history is whether the most heavily cratered terrains on the planet have reached "empirical saturation equilibrium." The concept of such an equilibrium was introduced by Hartmann (1984), who noticed that heavily cratered terrains on most solid-surfaced bodies in the solar system tend to have the same spatial densities to within a modest factor (very approximately $R = 0.3$ on an R-plot), a density well below some theoretical, geometrical saturation densities discussed in the literature, some of which even exceed $R = 3$ (Gault, 1970; Marcus, 1970; Woronow, 1977). While a simple algorithm can lay down a high density of recognizable circles, actual cratering involves the creation of ejecta blankets and secondary crater fields, seismic shaking that can induce downslope mass-wasting, and other effects that degrade the topography of pre-existing craters well beyond the rim of the latest crater, greatly reducing the spatial density of visible craters no matter how many times the surface has been struck. Compounding the problem of defining an equilibrium density is that the recognizability of highly degraded craters on the verge of totally disappearing is highly subjective and depends on factors that are difficult to control (e.g., solar illumination angle, potential differences in target properties). Recent numerical simulations (e.g., Cratered Terrain Evolution Model: Richardson, 2009; Minton et al., 2015a) capture more of the details of real-world cratering than earlier calculations, but it remains uncertain whether a particular cratered surface can be said to be in empirical saturation equilibrium from the spatial density (R value) alone. As new images are taken of solar system bodies (e.g., Ceres, Mimas, and Pluto), debates continue about whether empirical saturation has been reached in individual cases, but the issue probably has little relevance to Mercury, as we now discuss.

It has been widely thought that the most heavily cratered regions on the Moon are in saturation equilibrium, although such a conclusion has been disputed by Woronow (1977) and more recently by Marchi et al. (2012). Marchi et al. (2013) showed that the most heavily cratered terrains on Mercury (which they termed the northern heavily cratered terrains, or NHCT, and the southern heavily cratered terrains, or SHCT) have spatial densities similar to lunar post-Nectarian terrains, with $N(20) \approx 170$–180, which in turn have a somewhat lower

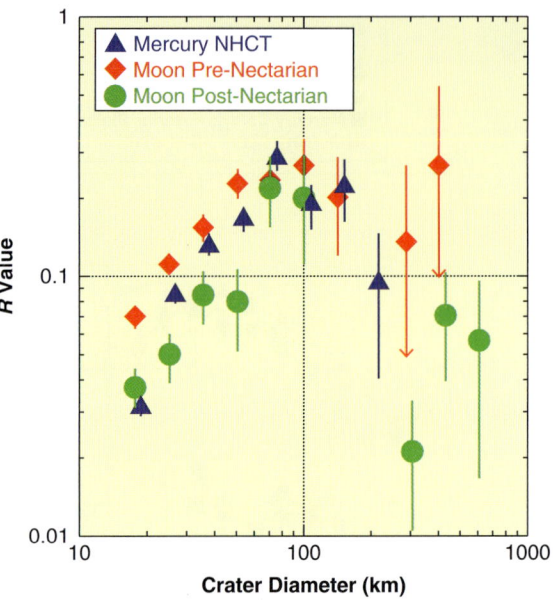

Figure 9.17. R-plot for heavily cratered terrains on Mercury and the Moon. Blue triangles are for Mercury's NHCT, red diamonds for lunar pre-Nectarian (most heavily cratered) terrains, and green circles for lunar post-Nectarian terrains. From Marchi et al. (2013).

spatial density than the most heavily cratered lunar terrains, with $N(20) > 200$ (Fassett et al., 2011); see Figure 9.17. Since the cratering rate for Mercury is approximately a factor of 3 greater than the lunar rate, Mercury must have been saturated by craters and basins even if the most heavily cratered lunar highlands were not quite saturated. Thus, the fact that Mercury's NHCT and SHCT are even less cratered means that the earliest craters and basins still visible on Mercury must have formed in competition not only with near-saturation cratering but also with additional crater-obliteration processes, including the enhanced secondary cratering on Mercury and, especially, the voluminous effusive volcanism that formed the intercrater plains. The degree to which that volcanism, which would have preferentially degraded and erased smaller craters, has modified the underlying production function of the impactors is unclear. Marchi et al. (2013) argued that the SFD of visible craters in the NHCT is similar to their model PF derived from the Moon and extrapolated to Mercury, suggesting minimal size-dependent erasure of craters. On the other hand, many cratered regions of Mercury have long been known to be depleted in craters smaller than 20–50 km in diameter, compared with the Moon, as discussed above. The dominant conclusion remains, as first proposed by Leake (1979), that the earliest landforms still visible on Mercury date from a period of combined heavy impact bombardment and emplacement of volcanic intercrater plains.

9.4.5 Secondary Crater SFDs and Transition from Primary SFDs

Secondary impact craters are very widely distributed and dominate the small crater population on many planetary surfaces.

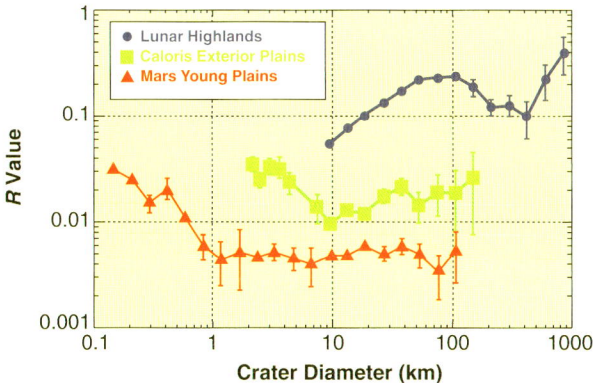

Figure 9.18. R-plots for Mercury's Caloris exterior plains and young plains on Mars, down to small crater sizes where secondary craters are abundant. Upturns due to secondaries at small diameters can be seen near 1-km diameter for Mars but near 10 km for Mercury. A schematic SFD for lunar highlands is also shown. From Strom et al. (2015).

Robbins and Hynek (2011) have shown that secondaries on Mars can affect crater age-dating unless they can be unambiguously distinguished from primaries or unless crater counts are restricted to larger crater sizes at which secondaries are very rare. Xiao and Strom (2012) showed that secondaries dominate the small crater ($D < 1$ km) population on both young and old lunar surfaces. The size and spatial distribution of secondary craters generally depend on the size of the primary impact crater, the impact velocity, and the surface gravitational acceleration of the planet or satellite (e.g., Xiao et al., 2014a). SFDs for secondary craters generally exhibit a very steep differential slope (about -4 power law). On Mercury, secondaries from a given-size primary crater are considerably larger than on other planets (Strom et al., 2011, 2015), perhaps due to the higher impact velocities on Mercury than on other planets (Xiao et al., 2014a), although there are other explanations (McEwen et al., 2017). Figure 9.18 shows three R-plots in which the dominant presence of secondaries is evident by the sharp difference in slope at small diameters. This upturn occurs for craters smaller than $D \sim 1$ km on Mars and the Moon but smaller than 10 km for Mercury. It may be that some especially young, small units are minimally contaminated by secondary craters, but these must be evaluated individually, including analysis of whether the units are near relatively large, recent, secondary-producing primary craters.

9.4.6 Small Crater SFDs, Regolith, and Rates of Topographic Degradation

Mercury's cratering record permits us to assess the ages of terrains and the rates of degradation of topography and resurfacing at different horizontal and vertical scales. Large craters, those with diameters in excess of tens of kilometers, form and are resurfaced at very different rates from small craters (diameters of tens of meters) forming in the regolith. Because the Moon has been intensively studied and its impact flux reasonably well calibrated by radiometric dates of returned samples, we consider the Moon in evaluating Mercury's resurfacing rates. Abundant post-Apollo literature on the evolution of the lunar regolith has been supplemented recently by analysis of high-resolution Lunar Reconnaissance Orbiter Camera images. To first order, we might expect that saturation of Mercury's surface many times over by the same steeply sloped power-law SFD of small craters (primaries plus secondaries) yields an analogous surficial regolith.

Primarily because of the greater crater production rate on Mercury, there are important differences in resurfacing rates between Mercury and the Moon by craters tens of meters to a few kilometers in scale, the formation of which can degrade the topography of older and much larger craters and topographic features. It has been recognized since the 1960s that the high flux of meteoroids puts craters with $D \leq 100$–300 m into saturation equilibrium on the lunar maria (e.g., Wilhelms, 1987), consistent with the growth of regolith ~ 5–10 m in thickness. Crater SFDs for mercurian smooth plains measured from MESSENGER's final year of highest-resolution images (see Chapter 1) suggest that they, too, have reached saturation (Fassett and Crowley, 2016). These images reveal that the crossover diameter between craters in production and in saturation equilibrium is at a larger diameter on Mercury than on the Moon: $D \sim 1$–1.5 km (see Figure 9.19) versus $D \sim 200$ m, respectively (e.g., Hartmann and Gaskell, 1997). Three factors can account for this difference: (1) the higher impactor flux at Mercury, (2) the greater efficiency of secondary crater production on Mercury, and (3) a probable older age for these smooth plains than for typical lunar maria. So the smooth plains regolith must have grown more quickly than lunar regolith and reached at least ~ 15–30 m in depth, consistent with other observations of the smooth plains, including crater morphologies and surface roughness (Kreslavsky and Head, 2015). It has been suggested that the rate of topographic degradation on Mercury could be twice that on the Moon (Fassett and Crowley, 2016; Fassett et al., 2017).

9.4.7 Interpreting Anomalous or Inconsistent Crater SFDs

Some crater SFDs, as observed on Mercury and other bodies, show bumps, wiggles, and other unexpected features. So far, we have considered two elements that contribute to an observed crater SFD: (a) primary craters with diameters that result from scaling the corresponding SFD of small bodies, chiefly asteroids with a very small admixture of comets (we showed earlier that these SFDs changed shape with time around the end of the LHB); and (b) secondary craters produced by ejecta from the primary craters formed during the impact explosions. Generally, the SFD for primary craters has the shape of a production function, such as those of Neukum (1983) or Marchi et al. (2009). Although these production functions can be approximated by power laws over certain diameter ranges, they are actually smoothly curving functions as depicted on an R-plot. Population 1 has greater curvature than Population 2 (see Figure 9.14). The SFDs for secondary craters generally have exponents 1 to 2 units more negative ("steeper") than the SFD for Population 2 for diameters several times smaller than the largest secondary, rolling over to a shallower slope at still smaller diameters.

As discussed briefly in Section 9.4.2, another kind of deviation exists for some observed crater SFDs compared with the primary PF: the shallower sloped portion of Population 1 is expressed somewhat differently on Mars and Mercury than on the Moon, for craters about 10–50 km in diameter (Figure 9.15). On Mercury, this feature has been ascribed to preferential loss of smaller craters by the formation of the volcanic intercrater plains. A wider variety of atmospheric, glacial, and geological processes on Mars has resulted in an even greater loss of smaller craters. It is possible that smaller lunar craters have also been preferentially erased by later impacts and mass wasting, in which case the true Population 1 PF might deviate from the observed SFD for the Moon, perhaps following a relation similar to the black dash-dot curve in Figure 9.15. If it deviated considerably, then it would become inconsistent with the SFD for main belt asteroids (see Figure 9.16). But the main belt SFD varies in shape for different semi-major axes and families (Cibulková et al., 2014) and, as a result of collisional evolution, it is uncertain how closely its form during the LHB resembled the modern SFD.

There are other reasons why an observed crater SFD might depart from a straightforward scaling of an asteroidal SFD onto a planetary surface, with superposition of secondaries. As demonstrated by Quaide and Oberbeck (1968) and more recently by Massironi et al. (2009), differences in the strength of layers in the target deposits can cause kinks, even if the PF in that size range is a simple power law (i.e., a straight line on a log-log plot). Also, if secondary or endogenic craters are inadequately removed from the data, they may superimpose bumps and wiggles, reflecting their own different SFDs.

Very commonly, kinks in observed SFDs for solar system bodies have been interpreted in terms of "resurfacing episodes," in which a process such as volcanic flooding may degrade craters of certain sizes and completely obliterate still smaller ones, yielding a fresh surface on which the PF is then subsequently expressed (e.g., Chapman et al., 1970; Neukum and Horn, 1976). This scenario can yield a distinct "kink," an offset, or even a dip in an R-plot of the SFD over a narrow range of diameters, and the portions of the SFD at lesser and greater diameters correspond to different model ages (e.g., Hiesinger et al., 2002; Hartmann et al., 2008; Williams et al., 2008). A complex resurfacing history may yield a crater SFD without obvious kinks but perhaps a different slope from the PF (e.g., Michael and Neukum, 2010).

9.5 MERCURY'S CHRONOLOGY

9.5.1 Crater Spatial Densities: Stratigraphy and Relative Ages

Data on the spatial densities of craters on different geological units have long been used to assess relative ages of such units on Mercury (e.g., Spudis and Guest, 1988, and references therein). Crater densities are determined from $N(D)$, where D is often 1, 10, or 20 km (e.g., Neukum, 1983). Such measures of relative crater density permit determination of relative ages of generic geologic units (e.g., smooth plains, heavily cratered terrain, intercrater plains) or of more

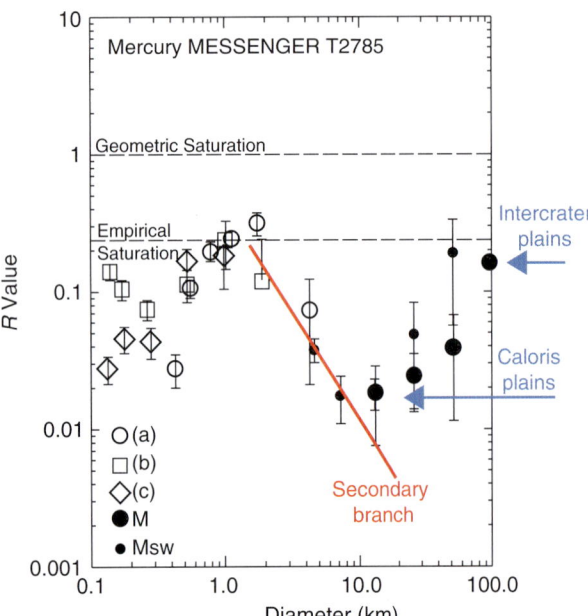

Figure 9.19. (Top panel) Representative part of Mercury's northern smooth plains (NSP) (near 70°N, 321°E), illustrating crater measurements in a full MDIS NAC frame (a). Measurements to much smaller diameters (not shown) were made in parts (b) and (c). (Bottom panel) R-plot shows that the "secondary branch" of the SFD reaches the empirical saturation line near $D = 1.5$ km. (Here we define "geometrical saturation" as the unrealistically high spatial density of craters for which $R = 1$.) The open circles, squares, and diamonds are counts from the frame for the target area (T2758) shown in the top panel, including part (b), which contains a cluster of secondary craters, and part (c), which does not. The solid points are counts for larger craters from an NSP mosaic in this region (M) and from the southwestern quadrant of the mosaic (Msw) that contains the target area. Arrows indicate spatial densities for intercrater plains and Caloris exterior plains. The steeply sloping secondary branch reaches the empirical saturation level at D values between 1 and 1.5 km. Modified from Chapman et al. (2011).

specifically defined units that are statistically distinguishable from each other (e.g., Denevi et al., 2013; Whitten et al., 2014; Ostrach et al., 2015; see Chapter 6). Additionally, SFD plots provide a graphical assessment of relative ages for crater populations measured for different geologic units. Relatively older geologic units have greater crater densities (plot higher) than relatively younger geologic units. To be valid, however, the SFDs from different units must have the same *shape* to within errors; otherwise $N(D)$ data are inconsistent with the fundamental assumption that the same primary production function describes the population of impactors that struck the surface for different durations.

Another assumption in relative age determination is that scaling of crater diameters from projectile diameters (given a mean impact velocity) is everywhere the same. But this assumption may not be valid if there are lateral variations in target strength for different units, or if strength varies with depth differently in different areas. It is certainly possible that strength varies laterally on Mercury, despite the fact that Mercury lacks the marked compositional differences evident on the Moon between the feldspathic highlands and basaltic maria. And there must be vertical variations in target strength with depth, from the porous surficial regolith through the megaregolith to bedrock in the lower crust, and those variations might be functions of surface age or lithology. Unfortunately, even on the Moon, effective target strengths are poorly known at the spatial scales of craters hundreds of meters to tens or hundreds of kilometers in diameter, and strengths are even more speculative for Mercury. Although it is difficult to correct for target strengths, it must be kept in mind that crater density differences at a particular size between two localities could be affected by differences in strength in addition to differences in age (Marchi et al., 2009; Kirchoff et al., 2015).

To be useful, it is necessary that counts be made within geological units that are homogeneous in relevant properties. One definition of a homogeneous unit is a region that has everywhere the same crater density and the same SFD, within errors. But usually units are additionally defined on the basis of homogeneity in other observable properties, such as spectral character, roughness, or elemental composition. Definition of unit boundaries is often rather subjective, and intercomparisons of data from different researchers may be difficult if the unit boundaries differ.

On a large scale, different geological units are indicated by variations around Mercury's globe in the density of craters (Fassett et al., 2011; Marchi et al., 2013). In a global map of density of craters ≥20 km in diameter (Fassett et al., 2011), the terrain within and immediately surrounding the Caloris basin is distinct from many other regions on Mercury, exhibiting an $N(20)$ value of ~25. This low-density region corresponds well with the Caloris interior and exterior smooth plains units. Regions of higher crater density are also evident. From MESSENGER-derived $N(25)$ values, two regions with the highest crater density on Mercury were identified and interpreted to reflect the most heavily cratered terrain surviving on the planet (Marchi et al., 2013). There are interesting differences in crater density maps at $N(50)$ and other diameters (see Chapter 6 for details and discussion).

Interpretation of potential geologic units derived from such maps is not always straightforward (e.g., Ostrach et al., 2015). Crater density maps are derived with a moving neighborhood technique that typically determines the number of craters within a circular region defined about each output cell (i.e., each pixel of the resulting map). Varying the neighborhood radius alters the spatial structure observed in the density map; small neighborhood sizes emphasize local (and possibly statistical) variations, whereas larger neighborhood sizes may smooth over real variations. By selecting a neighborhood radius that includes >30 craters per average neighborhood (to provide statistically robust sampling; Silverman, 1986; Davis, 2002), the resulting density map can be judged likely to indicate resolvable geologic units with different relative ages.

Units determined by crater densities may not correspond precisely with geologic unit boundaries determined on the basis of morphology or color, for several reasons. The moving neighborhood often yields fuzzy boundaries. Sharp color or reflectance boundaries may seem preferable, but they can be products of only surficial differences. Thus, careful interpretation of unit boundaries is important. The crater density across the NSP, for instance, is similar to that of a randomly distributed population, which indicates that differently aged sub-units within those plains are not resolved and that the last resurfacing event likely encompassed the full or nearly the full spatial extent of the NSP (Ostrach et al., 2015).

Crater density measurements on Mercury's plains are discussed and tabulated in Chapter 6. In the next section, we tabulate some representative crater densities, associate them with R values in R-plots for specific diameter ranges, and assign absolute ages. Other crater density values (e.g., from Chapter 6) can be interpolated from the absolute age scale.

9.5.2 Absolute Crater Chronology

Once relative ages are established, an important goal is to convert relative to absolute ages so that we may determine whether major features and units are ancient or youthful and relate the geological history of Mercury to its geophysical evolution. The most straightforward approach is to extrapolate Mercury's cratering chronology from that for the Moon, given that the ages of several cratered units on the Moon have been linked to radiometric dates of returned lunar samples. The small-body populations (e.g., NEAs) in the inner solar system are well characterized and their dynamical behavior well understood, from which it may be inferred that essentially the same population of objects now strikes both Mercury and the Moon, though at different rates and velocities. It is less certain, but still likely, that the source population (SFDs and dynamical properties) of impactors striking the Moon prior to and during the LHB was similar for Mercury (we consider possible Mercury-specific impactors in Section 9.6). Production functions for lunar cratering have been developed and extrapolated to Mercury (Marchi et al., 2009; Le Feuvre and Wieczorek, 2011).

To interpret the ancient crater SFD on Mercury from the lunar chronology, one must account for the different impact velocities between Mercury and the Moon as well as gravitational focusing and other factors that affect crater scaling relationships. On the basis of the current Moon-crossing and Mercury-crossing asteroid populations (Bottke et al., 2002), more craters in the diameter range 20–300 km should form on Mercury than on the Moon by a factor of 3 to 3.5 at a given diameter. Marchi et al. (2013) adopted a factor of 3 (valid at 20-km diameter), the value we have adopted in this chapter, consistent with the independent estimate of Le Feuvre and Wieczorek (2011).

A chronology often used for past discussions of the geological history of Mercury was that of Neukum et al. (2001), for which the cratering rate on Mercury was taken to be a factor of only 1.1 greater than that on the Moon. Whereas debate may continue on the contribution of secondary craters, differences in the shape of the Neukum et al. (2001) PF from other production functions, and the lunar chronology function, the Neukum et al. (2001) chronology for Mercury is now considered obsolete, for two primary reasons (for a detailed discussion, see the Supplementary Information by Marchi et al., 2013). First, the older NEA catalog used underestimated by more than a factor of 2 what we now know to be the ratio of Mercury-crossing to Moon-crossing NEAs. Second, Neukum et al. (2001) adopted 32.2 km/s as the mean impact velocity of NEAs at Mercury, whereas modern calculations put that figure at ~42 km/s. A very active program of telescopic searches for NEAs began in the late 1990s, and NEA discoveries continue at a rapid pace, so we may expect these factors to be further refined in the future. Therefore, in this chapter we do not use the Neukum et al. (2001) chronology, but rather we illustrate the range of modern perspectives concerning Mercury's absolute chronology by applying the Marchi et al. (2009, 2013) and Le Feuvre and Wieczorek (2011) chronologies, with uncertainties indicated by the differences between them.

9.5.2.1 Age of Mercury's Oldest Remaining Terrains

Marchi et al. (2013) studied two regions on Mercury with the highest $N(25)$ values, concentrating on the northern of these two heavily cratered terrains (NHCT). They applied their main belt asteroid PF to the observed craters >20 km in diameter. This PF has provided suitable fits to older Population 1 units on Mercury, the Moon, and Mars, and it fits the NHCT data quite well. Furthermore, they adopted a recently revised early lunar chronology (Morbidelli et al., 2012) for ages older than 3 Ga (Figure 9.20). Two alternatives are shown for this earliest chronology, a time of declining bombardment by planetesimals left over from terrestrial planet formation.

Beginning at about 4.1 Ga (in the range ~4.2–4.0 Ga) there was a spike (the LHB), of somewhat uncertain magnitude and duration in the bombardment rate, suggested to be the result of asteroids ejected from the primordial asteroid belt by sweeping resonances during giant planet migration (Strom et al., 2005, 2015), and the rate then declined over at least the subsequent 0.6 Gyr (Bottke et al., 2012). This spike is manifested in Figure 9.20 by the break in slope at 4.1 Ga. Figure 9.20 also shows the expected cumulative cratering flux for Mercury, scaled by the factor of 3 discussed above. The results show that the NHCT has an age of about 4.1–4.0 Ga, several hundred million years younger than the most ancient lunar terrains if they date from 4.4 Ga, as interpreted by Morbidelli et al. (2012), or 4.4–4.3 Ga as inferred by Elkins-Tanton (2012). These ages are based on the assumption (which was shown in Section

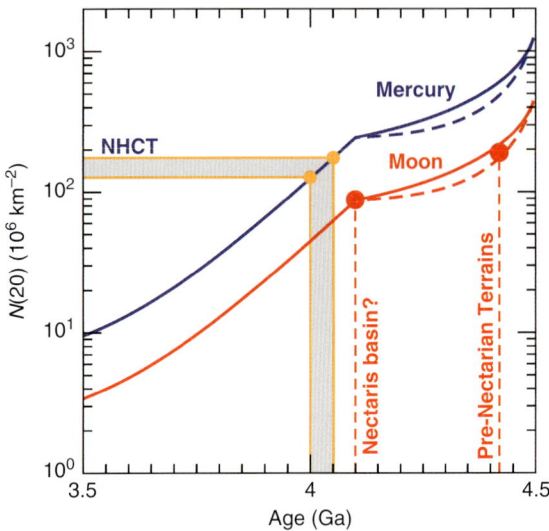

Figure 9.20. Cumulative crater density versus age for Mercury and the Moon (see text). The gray bands indicate a range of crater densities observed within the NHCT and their corresponding ages. The dashed lines are alternative early chronologies but do not affect post-Nectarian ages on the Moon or their counterparts on Mercury. Modified from Marchi et al. (2013).

9.4.4 as likely to hold) that these old terrains have not reached empirical saturation. Even if the heavily cratered terrains on the Moon and Mercury were in empirical saturation, however, the age difference between the lunar and mercurian terrains would be reduced but Mercury's NHCT crater retention age would still be post-Nectarian because of Mercury's greater cratering rate.

9.5.2.2 Ages of Smooth Plains and Implied History of Volcanism

Here we use a similar approach to estimate absolute ages for several other representative younger terrains on Mercury, on the basis of the lunar chronology of Marchi et al. (2009). The analysis makes use of Marchi et al.'s (2009) PF, which is appropriate for modern NEAs and is distinct and slightly different from the PF for main belt asteroids. We have selected several representative crater SFD data sets to serve as absolute age calibration points in Mercury's post-4.0 Ga geological evolution. In particular, in Figure 9.21 we show fits of the Marchi et al. (2009) PF to crater SFDs that represent the beginnings of the Mansurian and Kuiperian periods, as evaluated by Banks et al. (2017).

We also show a fit of the Marchi et al. (2009) PF to the crater population of the NSP, which bear an uncertain relation to the base of the Calorian Period, defined by the time of formation of the Caloris basin (see Chapter 6). Despite earlier indications (Fassett et al., 2009) that the Caloris basin rim materials were much older than the Caloris interior and exterior plains, updated crater counts suggest that, while $N(20)$ numbers are greater on the rim, the basin's age cannot be statistically distinguished from the age of the associated plains (Ernst et al., 2017), because of the large uncertainty in the crater SFD for the rim. Additionally, it has been shown by Ostrach et al. (2015) that the SFD for the NSP is homogeneous across the

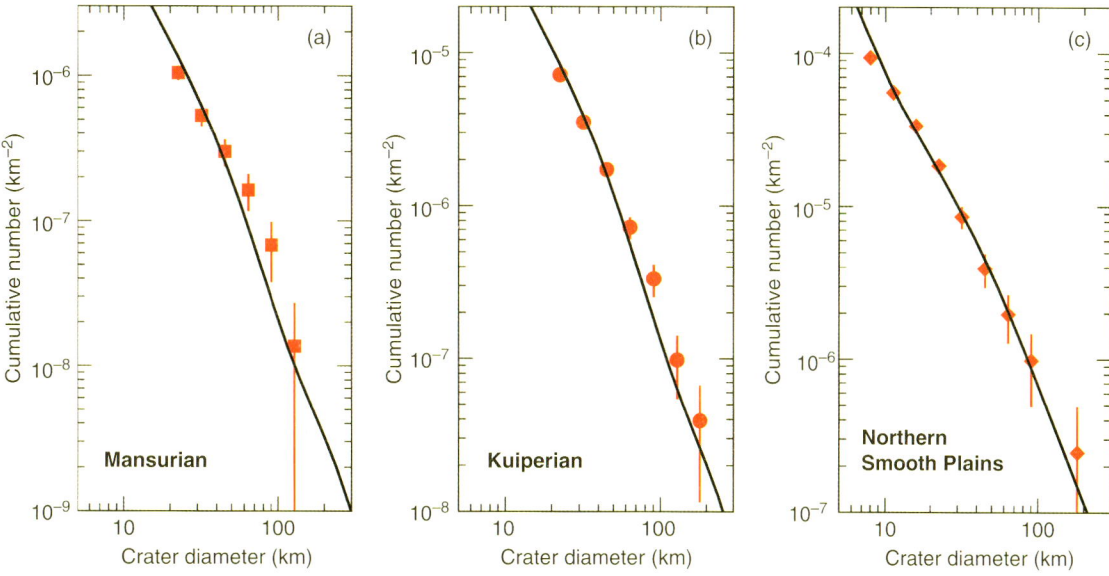

Figure 9.21. Fits of the Marchi et al. (2009) PF to cumulative numbers of craters that formed since the onset of the (a) Mansurian (1.7 ± 0.2 Ga) and (b) Kuiperian (280 ± 60 Ma) periods (Banks et al., 2017) and (c) for the NSP (3.7 ± 0.2 Ga; see text). (The plotted and cited uncertainties are simplified and probably too small, because cumulative counts in different diameter bins are treated as independent.)

broad expanse of plains and is statistically indistinguishable from the SFDs for the Caloris interior and exterior plains (Strom et al., 2011; updated interior plains counts by Ernst et al., 2017, are similar). Therefore, the NSP could define a lower bound on the age for the base of the Calorian Period. But the base of the Calorian could be older by as much as ~200 Myr, reflecting the statistical uncertainty in the Caloris rim crater counts and the marginally significantly higher $N(20)$ values for the rim compared with the Ostrach et al. (2015) values for the NSP.

For these fits, a hard-rock scaling law of Holsapple and Housen (2007) was applied with an intermediate value for the target strength of 2×10^7 dyne cm^{-2} (Banks et al., 2017). For the larger sub-area of NSP and craters measured in that area as defined by Ostrach et al. (2015; "NSP1"), the Marchi et al. (2009) PFs for NEAs and for main belt asteroids were fit, resulting in a nominal age for the NSP of 3.4 Ga for both PFs. Possible variations in target strength change the ages very little (e.g., by 50 Myr). The baseline PF assumption (NEA SFD) yields a poorer fit to the data than the main belt asteroid fit. Of course, although the NEA SFD represents today's population of impactors, the SFD may have varied over the last several billion years as asteroids collided and were catastrophically disrupted. As noted above, the SFD was different (Population 1) during the LHB from what it has been more recently (Population 2). During the critical period from 3.9 to 3.4 Ga, moreover, there may have been contamination of Population 1 by an inward extension of the asteroid belt called the E-belt (Bottke et al., 2012).

One approach to an estimate for the absolute age for the NSP is to follow Marchi et al. (2013) (see curves in Figure 9.20) but apply the improved NSP crater data of Ostrach et al. (2015), which show $N(20) = 24 \pm 2$, yielding an age of 3.7 Ga. The difference between 3.4 Ga (determined from the baseline SFD) and 3.7 Ga [determined from $N(20)$] is well within the associated uncertainty of about 200 Myr. We select 3.7 Ga for what follows.

There have been a number of determinations of relative ages for geological units on Mercury, derived either from measurements such as $N(20)$ or directly from R-plots of spatial densities of craters over a relevant diameter range (Table 9.1). Values in Table 9.1 are very approximate, determined on the basis of data shown in this chapter, tabulated in Table 6.1, and from the literature. A longer list of areas distinguished by $N(20)$ value is given in Table 6.1. We may apply the absolute ages just discussed to calibrate the relative ages in both tables and interpolate to determine absolute ages for other units.

Caveats must be emphasized concerning both the relative and absolute ages shown in Table 9.1. The relative ages are typically uncertain by at least 50%, for several reasons. First, some of the studies reported have considered craters with $D < 10$ km and may be dominated by secondaries; even $N(10)$ values may be contaminated by secondaries. Furthermore, statistics may be poor for densities based only on larger craters likely to exclude secondaries, especially for small or more recent units; indeed, reliable statistics are nearly non-existent for the Raditladi and Rachmaninoff plains. Finally, there are inherent uncertainties, especially between different researchers, in identifying and measuring craters (Robbins et al., 2014). The absolute ages may additionally be uncertain by perhaps a factor of 2 (for ages more recent than the LHB) for reasons discussed above: target strengths are uncertain and may vary, the adopted PFs have uncertainties [e.g., the PFs of Marchi et al. (2009) and Le Feuvre and Wieczorek (2011) differ], and the lunar chronology is unevenly constrained, with particularly large uncertainties in the middle of lunar history.

Table 9.1. *Absolute ages of geological units and periods on Mercury.*

R	N(10)	N(20)	Age (Ga)	Location
0.02	100	60	3.8	Floor of Rembrandt
0.02	60	25	3.7	Northern smooth plains
0.02	60	25	3.7	Southern smooth plains
0.006	30	15	3.1	Plains inside Tolstoj
0.007	60	10	1.7	Base of Mansurian
0.001	5	1.5	0.3	Base of Kuiperian
0.0003	3	–	0.2	Raditladi rim and inner plains
0.0003	3?	–	0.2	Rachmaninoff inner plains

Notes: Values are very approximate and intended only to show a representative correspondence among R values, $N(10)$ and $N(20)$ values, and inferred age. The "southern smooth plains" are those analyzed by Byrne et al. (2016). From data shown in Figure 9.19, Table 6.1, Prockter et al. (2010), Chapman et al. (2012), Byrne et al. (2016), and Banks et al. (2017). $N(10)$ values for Raditladi and Rachmaninoff are extrapolated from measurements at smaller diameters.

9.6 SEARCHES FOR POTENTIAL MERCURY IMPACTORS: VULCANOIDS AND SATELLITES

Leake et al. (1987) proposed that a population of small bodies having orbits inside the orbit of Mercury, called "vulcanoids," might have preferentially cratered Mercury during early epochs, perhaps continuing to the present time. Such a contribution of additional impactors would invalidate assumptions made above that the same population of impactors has struck the Moon and Mercury throughout the portion of solar system history preserved in the cratering record. The orbital and collisional evolution of vulcanoids was treated theoretically by Evans and Tabachnik (1999), Vokrouhlický et al. (2000), and Stern and Durda (2000). Upper limits have been set by several ground-based and space-based searches for vulcanoids (Leake et al., 1987; Durda et al., 2000; Schumacher and Gay, 2001; Steffl et al., 2013), all of which were hampered by the close angular proximity of any such bodies to the Sun. Although theoretical arguments have been made against the presence of satellites of Mercury (e.g., Donnison, 1978), and none have been found, such bodies might also have impacted Mercury's surface in the past, motivating searches for smaller satellites.

Theory suggests that a dynamically stable region for any vulcanoids is ~0.08–0.21 AU from the Sun (Mercury, with its eccentric orbit, ranges between 0.31 AU and 0.47 AU from the Sun). Closer to the Sun, even large iron bodies would evaporate over solar system history (Lebofsky, 1975; Campins et al., 1996). Beyond 0.21 AU, Mercury perturbations would destabilize vulcanoids on short timescales (Evans and Tabachnik, 1999). If vulcanoids ever existed, they could have cratered Mercury's surface after the LHB (while minimally affecting the cratering record of the other terrestrial planets) and thus may have skewed Mercury's geological chronology, making its surface appear older than it is. Satellites might exist anywhere within Mercury's rather compressed Hill sphere, the volume of space within which Mercury dominates the gravitational attraction of any potential satellite. If found, satellites would have implications for Mercury's formation and, in addition, could point to the existence of additional earlier satellites with orbits that may have decayed, impacting Mercury and potentially forming some of the basins discussed in this chapter.

MESSENGER's delivery of cameras close to both the Sun and Mercury provided an opportunity to search for such small bodies, which are difficult to detect from Earth. The searches and results have been described by Merline et al. (2016). While observations from near Mercury had a huge advantage (compared with Earth-based observations) in the apparent brightness of the targets (a factor of 5 closer for vulcanoids, equivalent to a 3.5-magnitude boost, and even better for satellites), the MDIS wide-angle camera (WAC) had only an 8-mm aperture, so the resulting limiting magnitude in a single 10-s (maximum) exposure was only about visual magnitude $V = 8$.

9.6.1 Vulcanoids

Prior to the MESSENGER survey, the upper limit on vulcanoid diameters was about 60 km (limit of $V = 8$ from Earth) from a search made with the Solar and Heliospheric Observatory spacecraft by Durda et al. (2000). Bodies smaller than about 1 km are not expected because of the Yarkovsky effect (Vokrouhlický et al., 2000). The original goal of the MESSENGER search was to reduce the size limit to at least 15-km diameter. Vulcanoid searches were conducted at six opportunities during the interplanetary cruise phase of the mission, prior to orbit insertion (Chapter 1), and covered 46% of the potential volume of vulcanoids. Cruise-phase searches achieved a completeness of ~95% and ~50% at 10 and 6.5 km in diameter, respectively. No definitive evidence of vulcanoids was found. Simultaneous with the MESSENGER effort, a search using the pair of Solar Terrestrial Relations Observatory spacecraft, employing occulting disks to block the Sun, found no vulcanoids >5.7 km in diameter (three-standard-deviation uncertainty, Steffl et al., 2013). To push to fainter limits, two additional MESSENGER searches were made from orbit about Mercury. Relative to the cruise-phase searches, more images of restricted portions of the vulcanoid region were co-added to search for fainter objects, at the expense of the previously wider coverage of the vulcanoid region. Each search covered 5% of the potential vulcanoid volume with the limit pushed to nearly 100% completeness at 6.5-km diameter and 90% completeness at 4 km. Again, no vulcanoids were found.

9.6.2 Satellites of Mercury

Despite a headline-producing but false detection of a Mercury moon during the Mariner 10 flybys of Mercury in 1974–1975 (Sullivan, 1974; Tucson Daily Citizen, 1974), data from those encounters eventually put size limits at 5 km for satellites inside 30 Mercury radii (R_M) (Murray et al., 1974). A ground-based search from the observatory on La Palma in the Canary Islands

(Warell and Karlsson, 2007) was claimed to be complete for ranges $(19–73)R_M$ at 1.6-km diameter, and partially complete over $(6–19)R_M$ at 0.5-km diameter. The Hill sphere for Mercury, within which satellites might be found, extends only to about 75 R_M, the smallest for any planet and even smaller than that for Ceres, because of Mercury's eccentric orbit and proximity to the Sun. A planned search by MESSENGER within the full Hill sphere down to 1-km diameter, during the third and final Mercury flyby (M3), was lost due to a spacecraft safe-hold anomaly (Chapter 1). So satellite searches were restricted to the orbital mission, during which any satellites would have been much closer to the spacecraft, but a complete search from "inside looking out" over 4π sr would have been prohibitive in terms of number of images and spacecraft resources. Four separate satellite searches were made, with a combined coverage of about 20% of the orbital phase space. No satellites were found at 90% completeness at diameters of about 10 m at 2 R_M and 100 m at 75 R_M.

9.7 CONCLUSIONS

Although the searches to date and theoretical studies have not completely ruled out the past presence of vulcanoids or satellites of Mercury, we regard their presence as very unlikely. The assumptions in this chapter that the customary small-body populations have been cratering both the Moon and Mercury thus seem to represent a good working hypothesis. Unfortunately, serious uncertainties affect the assessment of both relative and absolute cratering ages for different geological units on Mercury. Nonetheless, we can conclude that Mercury's history prior to 4.1–4.0 Ga is hidden by pervasive cratering and volcanism. Mercury's subsequent geological activity, like volcanism and tectonism, can be mapped into the absolute cratering chronology described in this chapter. The period of major volcanic smooth plains emplacement ended rather soon after the last great basin (Caloris) formed on Mercury (Byrne et al., 2016) and did not extend, as it did in a limited way on the Moon, by more than a few hundreds of millions of years (apart from the floors of one or two younger small basins; Prockter et al., 2010). Tectonic activity due to Mercury's global contraction began after the LHB and has gradually slowed until the present (Banks et al., 2015; Crane and Klimczak, 2017). Since the cratering rate on Mercury has been much greater than on the Moon, some geological processes, such as hollows formation, were active during the very recent Kuiperian period and probably are still continuing.

Impact basin formation early in Mercury's history was important in creating major topography and perhaps contributing to widespread early volcanism, at least since 4.1 Ga. However, the heavy bombardment and volcanism has erased evidence of Mercury's earlier geological history, probably explaining why basins are far undersaturated. For reasons not well understood, peak-ring basins are very prevalent on Mercury, but even the largest show minimal evidence of multi-ring morphology. The most recent, localized volcanism on Mercury is within a few small, relatively recent basins.

REFERENCES

Andrews-Hanna, J. C. and Zuber, M. T. (2010). Elliptical craters and basins on the terrestrial planets. In *Large Meteorite Impacts and Planetary Evolution IV*, ed. R. L. Gibson and W. U. Reimold. Special Paper 465. Denver, CO: Geological Society of America, pp. 1–13, doi:10.1130/2010.2465(01).

Baker, D. M. H. and Head, J. W. (2013). New morphometric measurements of craters and basins on Mercury and the Moon from MESSENGER and LRO altimetry and image data: An observational framework for evaluating models of peak-ring basin formation. *Planet. Space Sci.*, **86**, 91–116, doi:10.1016/j.pss.2013.07.003.

Baker, D. M. H, Head, J. W., Fassett, C. I., Kadish, S. J., Smith, D. E., Zuber, M. T. and Neumann, G. A. (2011a). The transition from complex crater to peak-ring basin on the Moon: New observations from the Lunar Orbiter Laser Altimeter (LOLA) instrument. *Icarus*, **214**, 377–393, doi:10.1016/j.icarus.2011.05.030.

Baker, D. M. H., Head, J. W., Schon, S. C., Ernst, C. M., Prockter, L. M., Murchie, S. L., Denevi, B. W., Solomon, S. C. and Strom, R. G. (2011b). The transition from complex crater to peak-ring basin on Mercury: New observations from MESSENGER flyby data and constraints on basin-formation models. *Planet. Space Sci.*, **59**, 1932–1948, doi:10.1016/j.pss.2011.05.010.

Baker, D. M. H., Head, J. W., Collins, G. S. and Potter, R. W. K. (2016). The formation of peak-ring basins: Working hypotheses and path forward in using observations to constrain models of impact-basin formation. *Icarus*, **273**, 146–163, doi:10.1016/j.icarus.2015.11.033.

Baldwin, R. B. (1963). *The Measure of the Moon*. Chicago, IL: University of Chicago Press, 488 pp.

Baldwin, R. B. (1965). The crater diameter–depth relationship from Ranger VII photographs. *Astron. J.*, **70**, 545–547.

Banks, M. E., Xiao, Z., Watters, T. R., Strom, R. G., Braden, S. E., Chapman, C. R., Solomon, S. C., Klimczak, C. and Byrne, P. K. (2015). Duration of activity on lobate-scarp thrust faults on Mercury. *J. Geophys. Res. Planets*, **120**, 1751–1762.

Banks, M. E., Xiao, Z., Braden, S. E., Barlow, N. G., Chapman, C. R., Fassett, C. I. and Marchi, S. (2017). Revised constraints on absolute age limits for Mercury's Kuiperian and Mansurian stratigraphic systems. *J. Geophys. Res. Planets*, **122**, 1010–1020, doi:10.1002/2016JE005254.

Barnouin, O. S., Ernst, C. M., Heinick, J. T., Sugita, S., Cintala, M. J., Crawford, D. A. and T. Matsui, T. (2011). Experimental results investigating the impact velocity effects on crater growth and the transient crater diameter-to-depth ratio. *Lunar Planet. Sci.*, **42**, abstract 2258.

Barnouin, O. S., Zuber, M. T., Smith, D. E., Neumann, G. A., Herrick, R. R., Chappelow, J. E., Murchie, S. L. and Prockter, L. M. (2012). The morphology of craters on Mercury: Results from MESSENGER flybys. *Icarus*, **219**, 414–427, doi:10.1016/j.icarus.2012.02.029.

Barnouin, O. S., Ernst, C. M. and Susorney, H. C. M. (2015). The remarkable Hokusai Crater, Mercury. *Lunar Planet. Sci.*, **46**, abstract 2672.

Barnouin-Jha, O. S., Baloga, S. and Glaze, L. (2005). Comparing landslides to fluidized crater ejecta on Mars. *J. Geophys. Res.*, **110**, E04010, doi:10.1029/2003JE002214.

Barnouin-Jha, O. S., Yamamoto, S., Toriumi, T., Sugita, S. and Matsui, T. (2007). Non-intrusive measurements of crater growth. *Icarus*, **188**, 506–521, doi:10.1016/j.icarus.2007.01.009.

Beach, M. J., Head, J. W., Ostrach, L. R., Robinson, M. S., Denevi, B. W. and Solomon, S. C. (2012). The influence of pre-existing topography on the distribution of impact melt on Mercury. *Lunar Planet. Sci.*, **43**, abstract 1335.

Becker, K. J., Robinson, M. S., Becker, T. L., Weller, L. A., Edmundson, K. L., Neumann, G. A., Perry, M. E. and Solomon, S. C. (2016). First global digital elevation model of Mercury. *Lunar Planet. Sci.*, **47**, abstract 2959.

Benz, W., Slattery, W. L. and Cameron, A. G. W. (1988). Collisional stripping of Mercury's mantle. *Icarus*, **74**, 516–528.

Bierhaus, E. B., Chapman, C. R. and Merline, W. J. (2005). Secondary craters on Europa and implications for cratered surfaces. *Nature*, **437**, 1125–1127, doi:10.1038/nature04069.

Blair, D. M., Freed, A. M., Byrne, P. K., Klimczak, C., Prockter, L. M., Ernst, C. M., Solomon, S. C., Melosh, H. J. and Zuber, M. T. (2013). The origin of graben and ridges in Rachmaninoff, Raditladi, and Mozart basins, Mercury. *J. Geophys. Res. Planets*, **118**, 47–58, doi:10.1029/2012JE004198.

Blewett, D. T., Chabot, N. L., Denevi, B. W., Ernst, C. M., Head, J. W., Izenberg, N. R., Murchie, S. L., Solomon, S. C., Nittler, L. R., McCoy, T. J., Xiao, Z., Baker, D. M. H., Fassett, C. I., Braden, S. E., Oberst, J., Scholten, F., Preusker F. and Hurwitz, D. M. (2011). Hollows on Mercury: Evidence for geologically recent volatile-related activity. *Science*, **333**, 1856–1859, doi:10.1126/science.1211681.

Blewett, D. T., Vaughan, W. M., Xiao, Z., Chabot, N. L., Denevi, B. W., Ernst, C. M., Helbert, J., D'Amore, M., Maturilli, A., Head, J. W. and Solomon, S. C. (2013). Mercury's hollows: Constraints on formation and composition from analysis of geological setting and spectral reflectance. *J. Geophys. Res. Planets*, **118**, 1013–1032, doi:10.1029/2012JE04174.

Bottke, W. F., Morbidelli, A., Jedicke, R., Petit, J.-M., Levison, H. F., Michel, P. and Metcalfe, T. S. (2002). Debiased orbital and absolute magnitude distribution of the near-Earth objects. *Icarus*, **156**, 399–433.

Bottke, W. F., Vokrouhlický, D., Minton, D., Nesvorný, D., Morbidelli, A., Brasser, R., Simonson, B. and Levison, H. F. (2012). An Archaean heavy bombardment from a destabilized extension of the asteroid belt. *Nature*, **485**, 78–81.

Bottke, W. F., Nesvorný, D., Roig, F., Marchi, S. and Vokrouhlický, D. (2017). Evidence for two impacting populations in the early bombardment of Mars and the Moon. *Lunar Planet. Sci.*, **48**, abstract 2572.

Boyce, J. M. and Garbeil, H. (2007). Geometric relationships of pristine Martian complex impact craters, and their implications to Mars geologic history. *Geophys. Res. Lett.*, **34**, L16201, doi:10.1029/2007GL029731.

Braden, S. E. and Robinson, M. S. (2013). Relative rates of optical maturation of regolith on Mercury and the Moon, *J. Geophys. Res. Planets*, **118**, 1903–1914, doi:10.1002/jgre.20143.

Bray, V. J. and Schenk, P. M. (2015). Pristine impact crater morphology on Pluto: Expectations for New Horizons. *Icarus*, **246**, 156–164.

Byrne, P. K., Klimczak, C., Williams, D. A., Hurwitz, D. M., Solomon, S. C., Head, J. W., Preusker, F. and Oberst, J. (2013). An assemblage of lava flow features on Mercury. *J. Geophys. Res. Planets*, **118**, 1303–1322, doi:10.1002/jgre.20052.

Byrne, P. K., Ostrach, L. R., Fassett, C. I., Chapman, C. R., Denevi, B. W., Evans, A. J., Klimczak, C., Banks, M. E., Head, J. W. and Solomon, S. C. (2016). Widespread effusive volcanism on Mercury likely ended by about 3.5 Ga. *Geophys. Res. Lett.*, **43**, 7408–7416.

Campins, H., Davis, D. R., Weidenschilling, S. J. and Magee, M. (1996). Searching for vulcanoids. In *Completing the Inventory of the Solar System*, ed. T. W. Rettig and J. M. Hahn. ASP Conference Series, Vol. 107. San Francisco, CA: Astronomical Society of the Pacific, pp. 85–96.

Carr, M. H., Crumpler, L. S. and Cutts, J. A. (1977). Martian impact craters and emplacement of ejecta by surface flow. *J. Geophys. Res.*, **82**, 4055–4065.

Cavanaugh, J. F., Smith, J. C., Sun, X., Bartels, A. E., Ramos-Izquierdo, L., Krebs, D. J., McGarry, J. F., Trunzo, R., Novo-Gradac, A.-M., Britt, J. L., Karsh, J. L., Katz, R. B., Lukemire, A. T., Symkiewicz, R., Berry, D. L., Swinski, J. P., Neumann, G. A., Zuber, M. T. and Smith, D. E. (2007). The Mercury Laser Altimeter instrument for the MESSENGER mission. *Space Sci. Rev.*, **131**, 451–479, doi:10.1007/s11214-007-9273-4.

Chapman, C. R. (2015). A critique of methods for analysis of crater size–frequency distributions. In *Workshop on Issues in Crater Studies and the Dating of Planetary Surfaces*. Contribution 1841. Houston, TX: Lunar and Planetary Institute, abstract 9039.

Chapman, C. R. and Haefner, R. R. (1967). A critique of methods for analysis of the diameter–frequency relation for craters with special application to the Moon. *J. Geophys. Res.*, **72**, 549–557.

Chapman, C. R., Mosher, J. A. and Simmons, G. (1970). Lunar cratering and erosion from Orbiter 5 photographs. *J. Geophys. Res.*, **75**, 1445–1466.

Chapman, C. R., Merline, W. J., Ostrach, L. R., Xiao, Z., Solomon, S. C. and Head, J. W. (2011). Small craters (secondaries) on Mercury's northern plains. *EPSC-DPS Joint Meeting Abstracts and Program*, **6**, abstract EPSC-DPS2011–1497.

Chapman, C. R., Merline, W. J., Marchi, S., Prockter, L. M., Fassett, C. I., Head, J. W., Solomon, S. C. and Xiao, Z. (2012). The young inner plains of Mercury's Rachmaninoff basin reconsidered. *Lunar Planet. Sci.*, **43**, abstract 1607.

Cibulková, H., Brož, M. and Benavidez, P. G., (2014). A six-part collisional model of the main asteroid belt. *Icarus*, **241**, 358–372, doi:10.1016/j.icarus.2014.07.016.

Cintala, M. J. (1979). Mercurian crater rim heights and some interplanetary comparisons. *Proc. Lunar Planet. Sci. Conf.*, **10**, 2635–2650.

Cintala, M. J., Wood, C. A. and Head, J. W. (1977). The effects of target characteristics on fresh crater morphology: Preliminary results for the Moon and Mercury. *Proc. Lunar Sci. Conf.*, **8**, 3409–3425.

Collins, G. S., Melosh, H. J., Morgan, J. V. and Warner, M. R. (2002). Hydrocode simulations of Chicxulub crater collapse and peak-ring formation. *Icarus*, **157**, 24–33.

Collins, G. S., Melosh, H. J. and Marcus, R. A. (2005). Earth Impact Effects Program: A Web-based computer program for calculating the regional environmental consequences of a meteoroid impact on Earth. *Meteorit. Planet. Sci.*, **40**, 817–840, doi:10.1111/j.1945-5100.2005.tb00157.x.

Collins, G. S., Elbeshausen, D., Davison, T. M., Robbins, S. J. and Hynek, B. M. (2011). The size–frequency distribution of elliptical impact craters. *Earth Planet. Sci. Lett.*, **310**, 1–8.

Crane, K. T. and Klimczak, C. (2017). Timing and rate of global contraction on Mercury. *Geophys. Res. Lett.*, **44**, 3082–3089, doi:10.1002/2017GL072711.

Crater Analysis Techniques Working Group (1979). Standard techniques for presentation and analysis of crater size–frequency data. *Icarus*, **37**, 467–474.

Craterstats (2016). www.geo.fu-berlin.de/en/geol/fachrichtungen/planet/software/index.html.

Croft, S. K. (1985). The scaling of complex craters. *J. Geophys. Res.*, **90**, Suppl., C828–C842.

Davis, J. C. (2002). *Statistics and Data Analysis in Geology*, 3rd edn. New York: John Wiley and Sons, 638 pp.

Denevi, B. W. and Robinson, M. S. (2008). Mercury's albedo from Mariner 10: Implications for the presence of ferrous iron. *Icarus*, **197**, 239–246, doi:10.1016/j.icarus.2008.04.021.

Denevi, B. W., Robinson, M. S., Solomon, S. C., Murchie, S. L., Blewett, D. T., Domingue, D. L., McCoy, T. J., Ernst, C. M., Head, J. W., Watters, T. R. and Chabot, N. L. (2009). The evolution of Mercury's crust: A global perspective from MESSENGER. *Science*, **324**, 613–618, doi:10.1126/science.1172226.

Denevi, B. W., Ernst, C. M., Meyer, H. M., Robinson, M. S., Murchie, S. L., Whitten, J. L., Head, J. W., Watters, T. R., Solomon, S. C., Ostrach, L. R., Chapman, C. R., Byrne, P. K., Klimczak, C. and Peplowski, P. N. (2013). The distribution and origin of smooth plains on Mercury. *J. Geophys. Res. Planets*, **118**, 891–907, doi:10.1002/jgre.20075.

D'Incecco, P., Helbert, J., Head, J. W., D'Amore, M., Maturilli, A., Izenberg, N. R., Holsclaw, G. M., Domingue, D. L., McClintock, W. E. and Solomon, S. C. (2012). Kuiper crater on Mercury: An opportunity to study recent surface weathering trends with MESSENGER. *Lunar Planet. Sci.*, **43**, abstract 1815.

Donnison, J. R. (1978). The escape of natural satellites from Mercury and Venus. *Astrophys. Space Sci.*, **59**, 499–501.

Durda, D. D., Stern, S. A., Colwell, W. B., Parker, J. W., Levison, H. F. and Hassler, D. M. (2000). A new observational search for vulcanoids in SOHO/LASCO coronagraph images. *Icarus*, **148**, 312–315, doi:10.1006/icar.2000.6520.

Elkins-Tanton, L. T. (2012). Magma oceans in the inner solar system. *Annu. Rev. Earth Planet. Sci.*, **40**, 113–139.

El-Baz, F. (1972). King crater and its environs. In *Apollo 16 Preliminary Science Report*, Special Publication SP-315. Washington, DC: National Aeronautics and Space Administration, pp. 29-62–29-70.

Ernst, C. M., Murchie, S. L., Barnouin, O. S., Robinson, M. S., Denevi, B. W., Blewett, D. T., Head, J. W., Izenberg, N. R., Solomon, S. C. and Roberts, J. H. (2010). Exposure of spectrally distinct material by impact craters on Mercury: Implications for global stratigraphy. *Icarus*, **209**, 210–223.

Ernst, C. M., Denevi, B. W., Barnouin, O. S., Klimczak, C., Chabot, N. L., Head, J. W., Murchie, S. L., Neumann, G. A., Prockter, L. M., Robinson, M. S., Solomon, S. C. and Watters, T. R. (2015). Stratigraphy of the Caloris basin, Mercury: Implications for volcanic history and basin impact melt. *Icarus*, **250**, 413–429, doi:10.1016/j.icarus.2014.11.003.

Ernst, C. M., Denevi, B. W. and Ostrach, L. R. (2017). Updated absolute age estimates for the Tolstoj and Caloris basins, Mercury. *Lunar Planet. Sci.*, **48**, abstract 2934.

Evans, A. J., Andrews-Hanna, J. C., Soderblom, J. M., Solomon, S. C. and Zuber, M. T. (2017). Insights into early lunar chronology from GRAIL data. *Lunar Planet. Sci.*, **48**, abstract 1276.

Evans, N. W. and Tabachnik, S. (1999). Possible long-lived asteroid belts in the inner solar system. *Nature*, **399**, 41–43.

Fassett, C. I. and Crowley, M. C. (2016). High-resolution stereo digital terrain models of Mercury: Crater degradation and morphometry. *Lunar Planet. Sci.*, **47**, abstract 1046.

Fassett, C. I., Head, J. W., Blewett, D. T., Chapman, C. R., Dickson, J. L., Murchie, S. L., Solomon, S. C. and Watters, T. R. (2009). Caloris impact basin: Exterior geomorphology, stratigraphy, morphometry, radial sculpture, and smooth plains deposits. *Earth Planet. Sci. Lett.*, **285**, 297–308.

Fassett, C. I., Kadish, S. J., Head, J. W., Solomon, S. C. and Strom, R. G. (2011). The global population of large craters on Mercury and comparison with the Moon. *Geophys. Res. Lett.*, **38**, L10202, doi:10.1029/2011GL047294.

Fassett, C. I., Head, J. W., Baker, D. M. H., Zuber, M. T., Smith, D. E., Neumann, G. A., Solomon, S. C., Klimczak, C., Strom, R. G., Chapman, C. R., Prockter, L. M., Phillips, R. J., Oberst, J. and Preusker, F. (2012). Large impact basins on Mercury: Global distribution, characteristics, and modification history from MESSENGER orbital data. *J. Geophys. Res.*, **117**, E00L08, doi:10.1029/2012JE004154.

Fassett, C. I., Crowley, M. C., Leight, C., Dyar, M. D., Minton, D. A., Hirabayashi, M., Thomson, B. J. and Watters, W. A. (2017). Evidence for rapid topographic evolution and crater degradation on Mercury from simple crater morphometry. *Geophys. Res. Lett.*, **44**, 5326–5335, doi:10.1002/2017GL073769.

Ferrari, S., Massironi, M., Marchi, S., Byrne, P. K., Klimczak, C., Martellato, E. and Cremonese, G. (2014). Age relationships of the Rembrandt basin and Enterprise Rupes, Mercury. In *Volcanism and Tectonism Across the Inner Solar System*, ed. T. Platz, M. Massironi, P. K. Byrne and H. Hiesinger, Special Publication 401. London: Geological Society, pp. 159–172, doi:10.1144/SP401.20.

Frank, E. A., Potter, R. W. K., Abramov, O., Mojzsis, S. and Nittler, L. R. (2016). Investigations into the origin of Mercury's high-magnesium region. *Lunar Planet. Sci.*, **47**, abstract 1270.

Frank, E. A., Potter, R. W. K., Abramov, O., James, P. B., Klima, R. L., Mojzsis, S. J. and Nittler, L. R. (2017). Evaluating an impact origin for Mercury's high-magnesium region. *J. Geophys. Res. Planets*, **122**, 614–632, doi:10.1002/2016JE005244.

Freed, A. M., Solomon, S. C., Watters, T. R., Phillips, R. J. and Zuber, M. T. (2009). Could Pantheon Fossae be the result of the Apollodorus crater-forming impact within the Caloris basin, Mercury? *Earth Planet. Sci. Lett.*, **285**, 320–327.

Freed, A. M., Blair, D. M., Watters, T. R., Klimczak, C., Byrne, P. K., Solomon, S. C., Zuber, M. T. and Melosh, H. J. (2012). On the origin of graben and ridges within and near volcanically buried craters and basins in Mercury's northern plains. *J. Geophys. Res.*, **117**, E00L06, doi:10.1029/2012JE004119.

Fulmer, C. V. and Roberts, W. A. (1963). Rock induration and crater shape. *Icarus*, **2**, 452–465.

Gault, D. E. (1970). Saturation and equilibrium conditions for impact cratering on the lunar surface: Criteria and implications. *Radio Science*, **5**, 273–291.

Gault, D. E. and Wedekind, J. A. (1978). Experimental studies of oblique impact. *Proc. Lunar Planet. Sci. Conf.*, **9**, 3843–3875.

Gault, D. E., Guest, J. E., Murray, J. B., Dzurisin, D. and Malin, M. C. (1975). Some comparisons of impact craters on Mercury and the Moon. *J. Geophys. Res.*, **80**, 2444–2460, doi:10.1029/JB080i017p02444.

Greeley, R., Fink, J., Gault, D. E., Snyder, D. B., Guest, J. E. and Schultz, P. H. (1980). Impact cratering in viscous targets: Laboratory experiments. *Proc. Lunar Planet. Sci. Conf.*, **11**, 2075–2097.

Grieve, R. A. F. and Cintala, M. J. (1992). An analysis of differential impact melt-crater scaling and implications for the terrestrial cratering record. *Meteoritics*, **27**, 526–538.

Guest, J. E. and Murray, J. B. (1969). Nature and origin of Tsiolkovsky crater, lunar farside. *Planet. Space Sci.*, **17**, 121–141.

Hartmann, W. K. (1984). Does crater "saturation equilibrium" occur in the solar system? *Icarus*, **60**, 56–74, doi:10.1016/0019-1035(84)90138-6.

Hartmann, W. K. and Gaskell, R. W. (1997). Planetary cratering 2: Studies of saturation equilibrium. *Meteorit. Planet. Sci.*, **32**, 109–121, doi:10.1111/j.1945-5100.1997.tb01246.x.

Hartmann, W. K., Neukum, G. and Werner, S. (2008). Confirmation and utilization of the "production function" size–frequency distributions of Martian impact craters. *Geophys. Res. Lett.*, **35**, L02205, doi:10.1029/2007GL031557.

Hawke, B. R. and Head, J. W. (1977a). Impact melt on lunar crater rims. In *Impact and Explosion Cratering: Planetary and Terrestrial Implications*, ed. D. J. Roddy, R. O. Pepin and R. B. Merrill. New York: Pergamon Press, pp. 815–841.

Hawke, B. R. and Head, J. W. (1977b). Impact melt in lunar crater interiors. *Lunar Sci.*, **8**, 415–416.

Head, J. W., Murchie, S. L., Prockter, L. M., Robinson, M. S., Solomon, S. C., Strom, R. G., Chapman, C. R., Watters, T. R., McClintock, W. E., Blewett, D. T. and Gillis-Davis, J. J. (2008). Volcanism on Mercury: Evidence from the first MESSENGER flyby. *Science*, **321**, 69–72, doi:10.1126/science.1159256.

Head, J. W., Murchie, S. L., Prockter, L. M., Solomon, S. C., Chapman, C. R., Strom, R. G., Watters, T. R., Blewett, D. T., Gillis-Davis, J. J., Fassett, C. I., Dickson, J. L., Morgan, G. A. and Kerber, L. (2009). Volcanism on Mercury: Evidence from the first MESSENGER flyby for extrusive and explosive activity and the volcanic origin of plains. *Earth Planet. Sci. Lett.*, **285**, 227–242, doi:10.1016/j.epsl.2009.03.007.

Head, J. W., Chapman, C. R., Strom, R. G., Fassett, C. I., Denevi, B. W., Blewett, D. T., Ernst, C. M., Watters, T. R., Solomon, S. C., Murchie, S. L., Prockter, L. M., Chabot, N. L., Gillis-Davis, J. J., Whitten, J. L., Goudge, T. A., Baker, D. M. H., Hurwitz, D. M., Ostrach, L. R., Xiao, Z., Merline, W. J., Kerber, L. A., Dickson, J. L., Oberst, J., Byrne, P. K., Klimczak, C. and Nittler, L. R. (2011). Flood volcanism in the northern high latitudes of Mercury revealed by MESSENGER. *Science*, **333**, 1853–1856, doi:10.1126/science.1211997.

Hermalyn, B. and Schultz, P. H. (2011). Time-resolved studies of hypervelocity vertical impacts into porous particulate targets: Effects of projectile density on early-time coupling and crater growth. *Icarus*, **216**, 269–279.

Hiesinger, H., Head, J. W., III, Wolf, U., Jaumann, R. and Neukum, G. (2002). Lunar mare basalt flow units: Thicknesses determined from crater size–frequency distributions. *Geophys. Res. Lett.*, **29**, 1248, doi:10.1029/2002GL014847.

Holsapple, K. A., (1993). The scaling of impact processes in planetary sciences. *Annu. Rev. Earth Planet. Sci.*, **21**, 333–373.

Holsapple K. A. and Housen, K. R. (2007). A crater and its ejecta: An interpretation of deep impact. *Icarus*, **191**, 586–597, doi:10.1016/j.icarus.2006.08.035.

Howard, K. A. (1972). Ejecta blankets of large craters exemplified by King crater. In *Apollo 16 Preliminary Science Report*, Special Publication SP-315. Washington, DC: National Aeronautics and Space Administration, pp. 29-70–29-77.

Howard, K. A. and Wilshire, H. G. (1975). Flows of impact melt at lunar craters. *J. Res. U.S. Geol. Surv.*, **3**, 237–251.

Ivezić, Ž., Tabachnik, S., Rafikov, R., Lupton, R. H., Quinn, T., Hammergren, M., Eyer, L., Chu, J., Armstrong, J. C., Fan, X., Finlator, K., Geballe, T. R., Gunn, J. E., Hennessy, G. S., Knapp, G. R., Leggett, S. K., Munn, J. A., Pier, J. R., Rockosi, C. M., Schneider, D. P., Strauss, M. A., Yanny, B., Brinkmann, J., Csabai, I., Hindsley, R. B., Kent, S., Lamb, D. Q., Margon, B., McKay, T. A., Smith, J. A., Waddel, P. and York, D. G. (2001). Solar system objects observed in the Sloan Digital Sky Survey commissioning data. *Astron. J.*, **122**, 2749–2784, doi:10.1086/323452.

Izenberg, N. R., Blewett, D. T., McNutt, R. L., Chabot, N. L., Chapman, C. R., Denevi, B. W., Robinson, M. S., Prockter, L. M. and Murchie, S. L. (2009). MESSENGER views of crater rays on Mercury. *Lunar Planet. Sci.*, **40**, abstract 1676.

Jedicke, R. and Metcalfe, T. S. (1998). The orbital and absolute magnitude distributions of main belt asteroids. *Icarus*, **131**, 245–260, doi:10.1006/icar.1997.5876.

Jutzi, M. and Michel, P. (2014). Hypervelocity impacts on asteroids and momentum transfer. I. Numerical simulations using porous targets. *Icarus*, **229**, 247–253.

Kalynn, J., Johnson, C. L., Osinski, G. R. and Barnouin, O. (2013). Topographic characterization of lunar complex craters. *Geophys. Res. Lett.*, **40**, 38–42, doi:10.1029/2012GL053608.

Kennedy, P. J., Freed, A. M. and Solomon, S. C. (2008). Mechanisms of faulting in and around Caloris basin, Mercury, *J. Geophys. Res.*, **113**, E08004, doi:10.1029/2007JE002992.

Kirchoff, M. R., Marchi, S. and Wunnemann, K. (2015). The effects of terrain properties on determining crater model ages on lunar surfaces. *Lunar Planet. Sci.*, **46**, abstract 2121.

Klimczak, C., Schultz, R. A. and Nahm, A. L. (2010). Evaluation of the origin hypotheses of Pantheon Fosse, central Caloris basin, Mercury. *Icarus*, **209**, 262–270, doi:10.1016/j.icarus.2010.04.014.

Klimczak, C., Watters, T. R., Ernst, C. M., Freed, A. M., Byrne, P. K., Solomon, S. C., Blair, D. M. and Head, J. W. (2012). Deformation associated with ghost craters and basins in volcanic smooth plains on Mercury: Strain analysis and implications for plains evolution. *J. Geophys. Res.*, **117**, E00L03, doi:10.1029/2012JE004100.

Klimczak, C., Ernst, C. M., Byrne, P. K., Solomon, S. C., Watters, T. R., Murchie, S. L., Preusker, F. and Balcerski, J. A. (2013). Insights into the subsurface structure of the Caloris basin, Mercury, from assessments of mechanical layering and changes in long-wavelength topography. *J. Geophys. Res. Planets*, **118**, 2030–2044, doi:10.1002/jgre.20157.

Kreslavsky, M. A. and Head, J. W. (2015). A thicker regolith on Mercury. *Lunar Planet. Sci.*, **46**, abstract 1246.

Kreslavsky, M. A., Head, J. W., Neumann, G. A., Zuber, M. T. and Smith, D. E. (2014). Kilometer-scale topographic roughness of Mercury: Correlation with geologic features and units. *Geophys. Res. Lett.*, **41**, 8245–8251, doi:10.1002/2014GL062162.

Leake, M. A. (1979). The intercrater plains of Mercury. *Lunar Planet. Sci.*, **10**, 710–712.

Leake, M. A., Chapman, C. R., Weidenschilling, S. J., Davis, D. R. and Greenberg, R. (1987). The chronology of Mercury's geological and geophysical evolution: The vulcanoid hypothesis. *Icarus*, **71**, 350–375.

Lebofsky, L. A. (1975). Stability of frosts in the solar system. *Icarus*, **25**, 205–217, doi:10.1016/0019-1035(75)90020-2.

Le Feuvre, M. and Wieczorek, M. A. (2011). Nonuniform cratering of the Moon and a revised crater chronology of the inner solar system. *Icarus*, **214**, 1–20, doi:10.1016/j.icarus.2011.03.010.

Malin, M. C. and Dzurisin, D. (1977). Landform degradation on Mercury, the Moon, and Mars: Evidence from crater depth/diameter relationships. *J. Geophys. Res.*, **82**, 376–388.

Mancinelli, P., Minelli, F., Mondini, A., Pauselli, C. and Federico, C. (2014). A downscaling approach for geological characterization of the Raditladi basin of Mercury. In *Volcanism and Tectonism Across the Inner Solar System*, ed. T. Platz, M. Massironi, P. K. Byrne and H. Hiesinger, Special Publication 401. London: Geological Society, pp. 57–75, doi:10.1144/SP401.10.

Marchi, S., Mottola, S., Cremonese, G., Massironi, M. and Martello, E. (2009). A new chronology for the Moon and Mercury. *Astron. J.*, **137**, 4936–4948, doi:10.1088/0004-6256/137/6/4936.

Marchi S., Massironi, M., Cremonese, G., Marellato, E., Giacomini, L. and Prockter, L. (2011). The effects of the target material properties and layering on the crater chronology: The case of Raditladi and Rachmaninoff basins on Mercury. *Planet. Space Sci.*, **59**, 1968–1980.

Marchi, S., McSween, H. Y., O'Brien, D. P., Schenk, P., De Sanctis, M. C., Gaskell, R., Jaumann, R., Mottola, S., Preusker, F., Raymond, C. A., Roatsch, T. and Russell, C. T. (2012). The violent collisional history of asteroid 4 Vesta. *Science*, **336**, 690–694.

Marchi, S., Chapman, C. R., Fassett, C. I., Head, J. W., Bottke, W. F. and Strom, R. G. (2013). Global resurfacing of Mercury 4.0–4.1 billion years ago by heavy bombardment and volcanism. *Nature*, **499**, 59–61, doi:10.1038/nature12280.

Marchi, S., Bottke, W. F., Cohen, B. A., Wünnemann, K., Kring, D. A., McSween, H. Y., de Sanctis, M. C., O'Brien, D. P., Schenk, P.,

Raymond, C. A. and Russell, C. T. (2013). High-velocity collisions from the lunar cataclysm recorded in asteroidal meteorites. *Nature Geosci.*, **6**, 303–307.

Marcus, A. H. (1970). Comparison of equilibrium size distributions for lunar craters. *J. Geophys. Res.*, **75**, 4977–4984, doi:10.1029/JB075i026p04977.

Masiero, J. R., Mainzer, A. K., Grav, T., Bauer, J. M., Cutri, R. M., Dailey, J., Eisenhardt, P. R. M., McMillan, R. S., Spahr, T. B., Skrutskie, M. F., Tholen, D., Walker, R. G., Wright, E. L., DeBaun, E., Elsbury, D., Gautier, T., IV, Gomillion, S. and Wilkins, A. (2011). Main belt asteroids with WISE/NEOWISE. I. Preliminary albedos and diameters. *Astrophys. J.*, **741**, 68, 20 pp., doi:10.1088/0004-637X/741/2/68.

Massironi, M., Cremonese, G., Marchi, S., Martellato, E., Mottola, S. and Wagner, R. J. (2009). Mercury's geochronology revised by applying Model Production Function to Mariner 10 data: Geological implications. *Geophys. Res. Lett.*, **36**, L21204, doi:10.1029/2009GL040353.

McCauley, J. F. (1977). Orientale and Caloris. *Phys. Earth Planet. Inter.*, **15**, 220–250, doi:10.1016/0031-9201(77)90033-4.

McCauley, J. F., Guest, J. E., Schaber, G. G., Trask, N. J. and Greeley, R. (1981). Stratigraphy of the Caloris basin, Mercury. *Icarus*, **47**, 184–202.

McEwen, A. S., Preblich, B. S., Turtle, E. P., Artemieva, N. A., Golombek, M. P., Hurst, M., Kirk, R. L., Burr, D. M. and Christensen, P. R. (2005). The rayed crater Zunil and interpretations of small impact craters on Mars. *Icarus*, **176**, 351–381.

McEwen, A. S., Robbins, S. J. and Bierhaus, E. B. (2017). Why are there many more large secondary craters on Mercury than on the Moon or Mars? *Lunar Planet. Sci.*, **48**, abstract 2028.

McGetchin, T. R., Settle, M. and Head, J. W. (1973). Radial thickness variation in impact crater ejecta: Implications for lunar basin deposits. *Earth Planet. Sci. Lett.*, **20**, 226–236, doi:10.1016/0012-821X(73)90162-3.

Melosh, H. J. (1989). *Impact Cratering: A Geologic Process*. London: Oxford University Press, 253 pp.

Melosh, H. J. and W. B. McKinnon (1988). The tectonics of Mercury. In *Mercury*, ed. F. Vilas, C. R. Chapman and M. S. Matthews. Tucson, AZ: University of Arizona Press, pp. 374–400.

Merline, W. J., Chapman, C. R., Tamblyn, P. M., Nair, H., Chabot, N. L., Enke, B. L. and Solomon, S. C. (2016). Search for vulcanoids and Mercury satellites from MESSENGER. *Lunar Planet. Sci.*, **47**, abstract 2765.

Michael, G. G. and Neukum, G. (2010). Planetary surface dating from crater size–frequency distribution measurements: Partial resurfacing events and statistical age uncertainty. *Earth Planet. Sci. Lett.*, **294**, 223–229, doi:10.1016/j.epsl.2009.12.041.

Miljković, K., Wieczorek, M. A., Collins, G. S., Laneuville, M., Neumann, G. A., Melosh, H. J., Solomon, S. C., Phillips, R. J., Smith, D. E. and Zuber, M. T. (2013). Asymmetric distribution of lunar impact basins caused by variations in target properties. *Science*, **342**, 724–726, doi:10.1126/science.1243224.

Minton, D. A., Richardson, J. E. and Fassett, C. I. (2015a). Testing crater counting assumptions with the Cratered Terrain Evolution Model. In *Workshop on Issues in Crater Studies and the Dating of Planetary Surfaces*, Contribution No. 1841. Houston, TX: Lunar and Planetary Institute, abstract 9042.

Minton, D. A., Richardson, J. E. and Fassett, C. I. (2015b). Re-examining the main asteroid belt as the primary source of ancient lunar craters. *Icarus*, **247**, 172–190, doi:10.1016/j.icarus.2014.10.018.

Mohit, P. S., Johnson, C. L., Barnouin-Jha, O., Zuber, M. T. and Solomon, S. C. (2009). Shallow basins on Mercury: Evidence of relaxation? *Earth Planet. Sci. Lett.*, **285**, 355–363.

Morbidelli, A., Marchi, S., Bottke, W. F. and Kring, D. A. (2012). A sawtooth-like timeline for the first billion years of lunar bombardment. *Earth Planet. Sci. Lett.*, **355–356**, 144–151.

Morbidelli, A., Nesvorný, D., Laurenz, V., Marchi, S., Rubie, D. C., Elkins-Tanton, L. and Jacobson, S. A. (2017). The lunar Late Heavy Bombardment as a tail-end of planet accretion. *Lunar Planet. Sci.*, **48**, abstract 2298.

Murchie, S. L., Watters, T. R., Robinson, M. S., Head, J. W., Strom, R. G., Chapman, C. R., Solomon, S. C., McClintock, W. E., Prockter, L. M., Domingue, D. L. and Blewett, D. T. (2008). Geology of the Caloris basin, Mercury: A view from MESSENGER. *Science*, **321**, 73–76, doi:10.1126/science.1159261.

Murray, B. C., Belton, M. J. S., Danielson, G. E., Davies, M. E., Gault, D. E., Hapke, B., O'Leary, B., Strom, R. G., Suomi, V. and Trask, N. (1974). Mercury's surface: Preliminary description and interpretation from Mariner 10 pictures. *Science*, **185**, 169–179.

Neish, C. D., Blewett, D. T., Harmon, J. K., Coman, E. I., Cahill, J. T. S. and Ernst, C. M. (2013). A comparison of rayed craters on the Moon and Mercury. *J. Geophys. Res. Planets*, **118**, 2247–2261, doi:10.1002/jgre.20166.

Neukum, G. (1983). Meteoritenbombardement und Datierung Planetarer Oberflächen. Habilitation dissertation for faculty membership, University of Munich.

Neukum, G. and Horn, P. (1976). Effects of lava flows on lunar crater populations. *Moon*, **15**, 205–222, doi:10.1007/BF00562238.

Neukum, G., Ivanov, B. A. and Hartmann, W. K. (2001). Cratering records in the inner solar system in relation to the lunar reference system. *Space Sci. Rev.*, **96**, 55–86.

Neumann, G. A., Zuber, M. T., Wieczorek, M. A., Head, J. W., Baker, D. M. H., Solomon, S. C., Smith, D. E., Lemoine, F. G., Mazarico, E., Sabaka, T. J., Goossens, S. J., Melosh, H. J., Phillips, R. J., Asmar, S. W., Konopliv, A. S., Williams, J. G., Sori, M. M., Soderblom, J. M., Miljković, K., Andrews-Hanna, J. C., Nimmo, F. and Kiefer, W. S. (2015). Lunar impact basins revealed by Gravity Recovery and Interior Laboratory measurements. *Science Advances*, **1**, e1500852, doi:10.1126/sciadv.1500852.

Noyelles, B., Frouard, J., Makarov, V. V. and Efroimsky, M. (2014). Spin–orbit evolution of Mercury revisited. *Icarus*, **241**, 26–44, doi:10.1016/j.icarus.2014.05.045.

Oberbeck, V. R. (1975). The role of ballistic erosion and sedimentation in lunar stratigraphy. *Rev. Geophys. Space Phys.*, **13**, 337–362.

Oberst, J., Preusker, F., Phillips, R. J., Watters, T. R., Head, J. W., Zuber, M. T. and Solomon, S. C. (2010). The morphology of Mercury's Caloris basin as seen in MESSENGER stereo topographic models. *Icarus*, **209**, 230–238, doi:10.1016/j.icarus.2010.03.009.

Öhman, T., Aittola, M., Kostama, V.-P. and Raitala, J., (2005). The preliminary analysis of polygonal impact craters within greater Hellas region, Mars. In *Impact Tectonics*, ed. C. Koeberl and H. Henkel. Berlin: Springer-Verlag, pp. 131–160.

Orgel, C., Michael, G. G. and Freie, T. (2017). Ancient bombardment of the inner solar system: Reinvestigation of the "fingerprints" of different impactor populations on the lunar surface. *Lunar Planet. Sci.*, **48**, abstract 1033.

Osinski, G. R., Tornabene, L. L. and Grieve, R. A. F. (2011). Impact ejecta emplacement on terrestrial planets. *Earth Planet. Sci. Lett.*, **310**, 167–181, doi:10.1016/j.epsl.2011.08.012.

Ostrach, L. R., Robinson, M. S. and Denevi, B. W. (2012). Distribution of impact melt on Mercury and the Moon. *Lunar Planet. Sci.*, **43**, abstract 1113.

Ostrach, L. R., Robinson, M. S., Whitten, J. L., Fassett, C. I., Strom, R. G., Head, J. W. and Solomon, S. C. (2015). Extent, age, and resurfacing history of the northern smooth plains on Mercury from MESSENGER observations. *Icarus*, **250**, 602–622, doi:10.1016/j.icarus.2014.11.010.

Pike, R. J. (1974). Depth/diameter relations of fresh lunar craters: Revision from spacecraft data. *Geophys. Res. Lett.*, **1**, 291–294.

Pike, R. J. (1980). Control of crater morphology by gravity and target type: Mars, Earth, Moon. *Proc. Lunar Planet. Sci. Conf.*, **11**, 2159–2189.

Pike, R. J. (1988). Geomorphology of impact craters on Mercury. In *Mercury*, ed. F. Vilas, C. R. Chapman and M. S. Matthews. Tucson, AZ: University of Arizona Press, pp. 165–273.

Pike, R. J. and Spudis, P. D. (1987). Basin-ring spacing on the moon, Mercury, and Mars. *Earth Moon Planets*, **39**, 129–194, doi:10.1007/BF00054060.

Pohn, H. A. and Offield, T. W. (1970). Lunar crater morphology and relative age determination of lunar geologic units. Part 1: Classification. Part 2: Applications. In *Geological Survey Research 1970*, Professional Paper 700-C. Denver, CO: U.S. Geological Survey, pp. C153–C162.

Preusker, F., Oberst, J., Head, J. W., Watters, T. R., Robinson, M. S., Zuber, M. T. and Solomon, S. C. (2011). Stereo topographic models of Mercury after three MESSENGER flybys. *Planet. Space Sci.*, **59**, 1910–1917.

Prockter, L. M., Watters, T. R., Chapman, C. R., Denevi, B. W., Head, J. W., III, Solomon, S. C., Murchie, S. L., Barnouin-Jha, O. S., Robinson, M. S., Blewett, D. T., Gillis-Davis, J. and Gaskell, R. W. (2009). The curious case of Raditladi basin. *Lunar Planet. Sci.*, **40**, abstract 1758.

Prockter, L. M., Ernst, C. M., Denevi, B. W., Chapman, C. R., Head, J. W., Fassett, C. I., Merline, W. J., Solomon, S. C., Watters, T. R., Strom, R. G., Cremonese, G., Marchi, S. and Massironi, M. (2010). Evidence for young volcanism on Mercury from the third MESSENGER flyby. *Science*, **329**, 668–671, doi:10.1126/science.1188186.

Prockter, L. M., Murchie, S. L., Ernst, C. M., Baker, D. M. H., Byrne, P. K., Head, J. W., Watters, T. R., Denevi, B. W., Chapman, C. R. and Solomon, S. C. (2012). The geology of medium-sized basins on Mercury: Implications for surface processes and evolution. *Lunar Planet. Sci.*, **43**, abstract 1326.

Quaide, W. L. and Oberbeck, V. R. (1968). Thickness determinations of the lunar surface layer from lunar impact craters. *J. Geophys. Res.*, **73**, 5247–5270.

Raymond, S. N., Schlichting, H. E., Hersant, F. and Selsis, F. (2013). Dynamical and collisional constraints on a stochastic late veneer on the terrestrial planets. *Icarus*, **226**, 671–681.

Richardson, J. E. (2009). Cratering saturation and equilibrium: A new model looks at an old problem. *Icarus*, **204**, 697–715, doi:10.1016/j.icarus.2009.07.029.

Rickman, H., Wiśniowski, T., Gabryszewski, R., Wajer, P., Wójcikowski, K., Szutowicz, S., Valsecchi, G. B. and Morbidelli, A. (2017). Cometary impact rates on the Moon and planets during the late heavy bombardment. *Astron. Astrophys.*, **598**, A67, 15 pp., doi:10.1051/0004-6361/201629376.

Rivera-Valentin, E. G. and Barr, A. C. (2014). Estimating the size of late veneer impactors from impact-induced mixing on Mercury. *Astrophys. J.*, **782**, L8, doi:10.1088/2041-8205/782/1/L8.

Robbins, S. J. and Hynek, B. M. (2011). Distant secondary craters from Lyot crater, Mars, and implications for surface ages of planetary bodies. *Geophys. Res. Lett.*, **38**, L05201, doi:10.1029/2010GL046450.

Robbins, S. J. and Hynek, B. M. (2012). A new global database of Mars impact craters ≥1 km, 2. Global crater properties and regional variations of the simple-to-complex transition diameter. *J. Geophys. Res.*, **117**, E06001, doi:101.1029/2011JE003967.

Robbins, S. J., Antonenko, I., Kirchoff, M. R., Chapman, C. R., Fassett, C. I., Herrick, R. R., Singer, K., Zanetti, M., Lehan, C., Huang, D. and Gay, P. L. (2014). The variability of crater identification among expert and community crater analysts. *Icarus*, **234**, 109–131, doi:10.1016/j.icarus.2014.02.022.

Robbins, S. J., Riggs, J., Weaver, B. P., Bierhaus, E. B., Chapman, C. R., Kirchoff, M. R., Singer, K. N. and Gaddis, L. R. (2018). Revised recommended methods for analyzing crater size–frequency distributions. *Meteorit. Planet. Sci.*, **53**, 583–637.

Roberts, J. H. and Barnouin, O. S. (2012). The effect of the Caloris impact on the mantle dynamics and volcanism of Mercury. *J. Geophys. Res.*, **117**, E02007, doi:10.1029/2011JE003876.

Robinson, M. S., Murchie, S. L., Blewett, D. T., Domingue, D. L., Hawkins, S. E., Head, J. W., Holsclaw, G. M., McClintock, W. E., McCoy, T. J., McNutt, R. L., Jr., Prockter, L. M., Solomon, S. C. and Watters, T. R. (2008). Reflectance and color variations on Mercury: Regolith processes and compositional heterogeneity. *Science*, **321**, 66–69, doi:10.1126/science.1160080.

Roig, F. and Nesvorný, D. (2015). The evolution of asteroids in the jumping-Jupiter migration model. *Astron. J.*, **150**, article 186, 15 pp.

Schaber, G. and McCauley, J. F. (1980). *Geologic Map of the Tolstoj (H-8) Quadrangle of Mercury*, Map I-1199, Miscellaneous Investigations Series. Denver, CO: U.S. Geological Survey.

Schmidt, R. M. and Housen, K. R. (1987). Some recent advances in the scaling of impact and explosion cratering. *Int. J. Impact Eng.*, **5**, 543–560.

Schultz, P. H. (1976). *Moon Morphology: An Interpretation Based on Lunar Orbiter Photography*. Austin, TX: University of Texas Press, 626 pp.

Schultz, P. H. (1988). Cratering on Mercury: A relook. In *Mercury*, ed. F. Vilas, C. R. Chapman and M. S. Matthews. Tucson, AZ: University of Arizona Press, pp. 274–335.

Schumacher, G. and Gay, J. (2001). An attempt to detect vulcanoids with SOHO/LASCO images. I. Scale relativity and quantization of the solar system. *Astron. Astrophys.*, **368**, 1108–1114.

Senft, L. E. and Stewart, S. T. (2007). Modeling impact cratering in layered surfaces. *J. Geophys. Res.*, **112**, E11002, doi:10.1029/2007JE002894.

Shoemaker, E. M., Batson, R. M., Holt, H. E., Morris, E. C., Rennilson, J. J. and Whitaker, E. A. (1968). Television observations from Surveyor VII. In *Surveyor VII: A Preliminary Report*, Special Publication SP-173. Washington, DC: National Aeronautics and Space Administration, pp. 13–81.

Silverman, B. W. (1986). *Density Estimation for Statistics and Data Analysis*. New York: Chapman and Hall, 176 pp.

Soderblom, J. M., Evans, A. J., Johnson, B. C., Melosh, H. J., Miljković, K., Phillips, R. J., Andrews-Hanna, J. C., Bierson, C. J., Head, J. W., Milbury, C., Neumann, G. A., Nimmo, F., Smith, D. E., Solomon, S. C., Sori, M. M., Wieczorek, M. A. and Zuber, M. T. (2015). The fractured Moon: Production and saturation of porosity in the lunar highlands from impact cratering. *Geophys. Res. Lett.*, **42**, 6939–6944, doi:10.1002/2015GL065022.

Spudis, P. D. and Guest, J. E. (1988). Stratigraphy and geologic history of Mercury. In *Mercury*, ed. F. Vilas, C. R. Chapman and M. S. Matthews. Tucson, AZ: University of Arizona Press, pp. 118–164.

Steffl, A. J., Cunningham, N. J., Shinn, A. B., Durda, D. D. and Stern, S. A. (2013). A search for vulcanoids with the STEREO Heliospheric Imager. *Icarus*, **223**, 48–56, doi:10.1016/j.icarus.2012.11.031.

Stern, S. A. and Durda, D. D. (2000). Collisional evolution in the vulcanoid region: Implications for present-day population constraints. *Icarus*, **143**, 360–370.

Stewart, S. T. and Valiant, G. J. (2006). Martian subsurface properties and crater formation processes inferred from fresh impact crater geometries. *Lunar Planet. Sci.*, **37**, abstract 2427.

Stewart, S. T., Lock, S. J., Petaev, M. I., Jacobsen, S. B., Sarid, G., Leinhardt, Z. M., Mukhopadhyay, S. and Humayun, M. (2016). Mercury impact origin hypothesis survives the volatile crisis: Implications for terrestrial planet formation. *Lunar Planet. Sci.*, **47**, abstract 2954.

Strom, R. G. (1977). Origin and relative age of lunar and Mercurian intercrater plains. *Phys. Earth Planet. Inter.*, **15**, 156–172, doi:10.1016/0031-9201(77)90028-0.

Strom, R. G., Murray, B. C., Belton, M. J. S., Danielson, G. E., Davies, M. E., Gault, D. E., Hapke, B. W., O'Leary, B., Trask, N. J. and Guest, J. E. (1975). Preliminary imaging results from the second Mercury encounter. *J. Geophys. Res.*, **80**, 2345–2356.

Strom, R. G., Malhotra, R., Ito, T., Yoshida, F. and Kring, D. A. (2005). The origin of planetary impactors in the inner solar system. *Science*, **309**, 1847–1850.

Strom, R. G., Chapman, C. R., Merline, W. J., Solomon, S. C. and Head, J. W. (2008). Mercury cratering record viewed from MESSENGER's first flyby. *Science*, **321**, 79–81, doi:10.1126/science.1159317.

Strom, R. G., Banks, M. E., Chapman, C. R., Fassett, C. I., Forde, J. A., Head, J. W., Merline, W. J., Prockter, L. M. and Solomon, S. C. (2011). Mercury crater statistics from MESSENGER flybys: Implications for stratigraphy and resurfacing history. *Planet. Space Sci.*, **59**, 1960–1967, doi:10.1016/j.pss.2011.03.018.

Strom, R. G., Malhotra, R., Xiao, Z.-Y., Ito, T., Yoshida, F. and Ostrach, L. R. (2015). The inner solar system cratering record and the evolution of impactor populations. *Res. Astron. Astrophys.*, **15**, 407–434, doi:10.1088/1674-4527/15/3/009.

Stuart, J. S. and Binzel, R. P. (2004). Bias-corrected population, size distribution, and impact hazard for the near-Earth objects. *Icarus*, **170**, 295–311, doi:10.1016/j.icarus.2004.03.018.

Sullivan, W. (1974). A possible moon of Mercury is detected. *New York Times*, 1 April 1974, p. 65.

Susorney, H. C. M., Barnouin, O. S. and Ernst, C. M. (2015). The surface roughness of Mercury: Investigating the effects of impact cratering, volcanism, and tectonics. *Lunar Planet. Sci.*, **46**, abstract 2088.

Susorney, H. C. M., Barnouin, O. S., Ernst, C. M. and Johnson, C. L. (2016). Morphometry of impact craters on Mercury from MESSENGER altimetry and imaging. *Icarus*, **271**, 180–193.

Talpe, M. J., Zuber, M. T., Yang, D., Neumann, G. A., Solomon, S. C., Mazarico, E. and Vilas, F. (2012). Characterization of the morphometry of impact craters hosting polar deposits in Mercury's north polar region. *J. Geophys. Res.*, **117**, E00L13, doi:10.1029/2012JE004155.

Thomas, R. J., Rothery, D. A., Conway, S. J. and Anand, M. (2014). Hollows on Mercury: Materials and mechanisms involved in their formation. *Icarus*, **229**, 221–235.

Trask, N. J. (1967). Distribution of lunar craters according to morphology from Ranger VIII and IX photographs. *Icarus*, **6**, 270–276.

Trask, N. J. and Guest, J. E. (1975). Preliminary geologic terrain map of Mercury. *J. Geophys. Res.*, **80**, 2461–2477.

Tsiganis, K., Gomes, R., Morbidelli, A. and Levison, H. F. (2005). Origin of the orbital architecture of the giant planets of the solar system. *Nature*, **435**, 459–461.

Tucson Daily Citizen (1974). Arizona man names new moon "Charley." 1 April 1974, p. 4.

Vaughan, W. M., Head, J. W., Wilson, L. and Hess, P. C. (2013). Geology and petrology of enormous volumes of impact melt on the Moon: A case study of the Orientale basin impact melt sea. *Icarus*, **223**, 749–765, doi:10.1016/j.icarus.2013.01.017.

Vokrouhlický, D., Farinella, P. and Bottke, W. F. (2000). The depletion of the putative vulcanoid population via the Yarkovsky effect. *Icarus*, **148**, 147–152.

Walker, R. J. (2009). Highly siderophile elements in the Earth, Moon and Mars: Update and implications for planetary accretion and differentiation. *Chemie der Erde – Geochemistry*, **69**, 101–125.

Warell, J. and Karlsson, O. (2007). A search for natural satellites of Mercury. *Planet. Space Sci.*, **55**, 2037–2041, doi:10.1016/j.pss.2007.06.004.

Watters, T. R. (1993). Compressional tectonism on Mars. *J. Geophys. Res.*, **98**, 17,049–17,060, doi:10.1029/93JE01138.

Watters, T. R., Nimmo, F. and Robinson, M. S. (2005). Extensional troughs in the Caloris basin of Mercury: Evidence of lateral crustal flow. *Geology*, **33**, 669–672.

Watters, T. R., Head, J. W., Solomon, S. C., Robinson, M. S., Chapman, C. R., Denevi, B. W., Fassett, C. I., Murchie, S. L. and Strom, R. G. (2009a). Evolution of the Rembrandt impact basin on Mercury. *Science*, **324**, 618–621, doi:10.1126/science.1172109.

Watters, T. R., Solomon, S. C., Robinson, M. S., Head, J. W., André, S. L., Hauck, S. A., II and Murchie, S. L. (2009b). The tectonics of Mercury: The view after MESSENGER's first flyby. *Earth Planet. Sci. Lett.*, **285**, 283–296.

Watters, T. R., Murchie, S. L., Robinson, M. S., Solomon, S. C., Denevi, B. W., André, S. L. and Head, J. W. (2009c). Emplacement and tectonic deformation of smooth plains in the Caloris basin, Mercury. *Earth Planet. Sci. Lett.*, **285**, 309–319, doi:10.1016/j.epsl.2009.03.040.

Watters, T. R., Solomon, S. C., Klimczak, C., Freed, A. M., Head, J. W., Ernst, C. M., Blair, D. M., Goudge, T. A. and Byrne, P. K. (2012). Extension and contraction within volcanically buried impact craters and basins on Mercury. *Geology*, **40**, 1123–1126, doi:10.1130/G33725.1.

Watters, W. A., Geiger, L. M., Fendrock, M., Gibson, R. and Hundal, C. B. (2017). The role of strength defects in shaping impact crater planforms. *Icarus*, **286**, 15–34.

Weider, S. Z., Nittler, L. R., Starr, R. D., Crapster-Pregont, E. J., Peplowski, P. N., Denevi, B. W., Head, J. W., Byrne, P. K., Hauck, S. A., Ebel, D. S. and Solomon, S. C. (2015). Evidence for geochemical terranes on Mercury: Global mapping of major elements with MESSENGER's X-Ray Spectrometer. *Earth Planet. Sci. Lett.*, **416**, 109–120.

Weihs, G. T., Leitner, J. J. and Firneis, M. G. (2015). Polygonal impact craters on Mercury. *Planet. Space Sci.*, **111**, 77–82.

Werner, S. C. (2017). Could Mars have witnessed giant planet migration? *Lunar Planet. Sci.*, **48**, abstract 1856.

Wetherill, G. W. (1988). Accumulation of Mercury from planetesimals. In *Mercury*, ed. F. Vilas, C. R. Chapman and M. S. Matthews. Tucson, AZ: University of Arizona Press, pp. 670–691.

Whitten, J. L. and Head, J. W. (2015). Rembrandt impact basin: Distinguishing between volcanic and impact-produced plains on Mercury. *Icarus*, **258**, 350–365, doi:10.1016/j.icarus.2015.06.022.

Whitten, J. L., Head, J. W., Denevi, B. W. and Solomon, S. C. (2014). Intercrater plains on Mercury: Insights into unit definition, characterization, and origin from MESSENGER datasets. *Icarus*, **241**, 97–113.

Wieczorek, M. A., Correia, A. C. M., Le Feuvre, M., Laskar, J. and Rambaux, N. (2012). Mercury's spin–orbit resonance explained by initial retrograde and subsequent synchronous rotation. *Nature Geosci.*, **5**, 18–21, doi:10.1038/ngeo1350.

Wilhelms, D. E. (1976). Mercurian volcanism questioned. *Icarus*, **28**, 551–558.

Wilhelms, D. E. (1987). *The Geologic History of the Moon*. Professional Paper 1348. Denver, CO: U.S. Geological Survey.

Williams, D. A., Greeley, R., Werner, S. C., Michael G., Crown, D. A., Neukum, G. and Raitala, J. (2008). Tyrrhena Patera: Geologic history derived from *Mars Express* High Resolution Stereo Camera. *J. Geophys. Res.*, **113**, E11005, doi:10.1029/2008JE003104.

Wood, C. A., Head, J. W. and Cintala M. J. (1977). Crater degradation on Mercury and the Moon: Clues to surface evolution. *Proc. Lunar Sci. Conf.*, **8**, 3503–3520.

Woronow, A. (1977). Crater saturation and equilibrium: A Monte Carlo simulation. *J. Geophys. Res.*, **82**, 2447–2456, doi:10.1029/JB082i017p02447.

Xiao, Z. and Komatsu, G. (2013). Impact craters with ejecta flows and central pits on Mercury. *Planet. Space Sci.*, **82–83**, 62–78.

Xiao, Z. and Strom, R. G. (2012). Problems determining relative and absolute ages using the small crater population. *Icarus*, **220**, 254–267.

Xiao, Z. and Werner, S. C. (2015). Size–frequency distribution of crater populations in equilibrium on the Moon. *J. Geophys. Res. Planets*, **120**, 2277–2292, doi:10.1002/2015JE004860.

Xiao, Z., Strom, R. G., Chapman, C. R., Head, J. W., Klimczak, C., Ostrach, L. R., Helbert, J. and D'Incecco, P. (2014a). Comparisons of fresh complex impact craters on Mercury and the Moon: Implications for controlling factors in impact excavation processes. *Icarus*, **228**, 260–275, doi:10.1016/j.icarus.2013.10.002.

Xiao, Z., Zeng, Z., Li, Z., Blair, D. M. and Xiao, L. (2014b). Cooling fractures in impact melt deposits on the Moon and Mercury: Implications for cooling solely by thermal radiation. *J. Geophys. Res. Planets*, **119**, 1496–1515, doi:10.1002/2013JE004560.

Yoshida, F., Nakamura, T., Watanabe, J., Kinoshita, D., Yamamoto, N. and Fuse, T. (2003). Size and spatial distributions of sub-km main-belt asteroids. *Publ. Astron. Soc. Jpn.*, **55**, 701–715, doi:10.1093/pasj/55.3.701.

Zuber, M. T., Smith, D. E., Phillips, R. J., Solomon, S. C., Neumann, G. A., Hauck, S. A., II, Peale, S. J., Barnouin, O. S., Head, J. W., Johnson, C. L., Lemoine, F. G., Mazarico, E., Sun, X., Torrence, M. H., Freed, A. M., Klimczak, C., Margot, J.-L., Oberst, J., Perry, M. E., McNutt, R. L., Jr., Balcerski, J. A., Michel, N., Talpe, M. J. and Yang, D. (2012). Topography of the northern hemisphere of Mercury from MESSENGER laser altimetry. *Science*, **336**, 217–220, doi:10.1126/science.1218805.

10

The Tectonic Character of Mercury

PAUL K. BYRNE, CHRISTIAN KLIMCZAK, AND A. M. CELÂL ŞENGÖR

10.1 INTRODUCTION

From its three flybys in 1974–1975, Mariner 10 told us that Mercury is a tectonic world. That is, the planet has experienced a long and complicated history of tectonic deformation, recorded by its preserved landforms. As the study of tectonics naturally intersects with volcanology, chemistry, interior structure, and thermal evolution, understanding the tectonic character of a world – the nature, distribution, and formational histories of its tectonic landforms, and their spatial and temporal relationship to the interior – is a crucial means by which to comprehend more fully the geological history of that body.

In this chapter, then, we seek to tie together the various strands of observational and analytical studies of Mercury's tectonic character conducted since the first Mercury flyby of the MESSENGER mission. First, we introduce Mercury's tectonics as understood after the Mariner 10 flybys, and we outline some of the outstanding questions raised by that mission, indeed questions that helped frame the MESSENGER project. In Section 10.2, we begin to discuss Mercury's tectonic character – the types and distributions of the most widespread record of Mercury's tectonic deformation: its population of shortening structures. In the subsequent section, we discuss the planet's spatially limited but important set of extensional structures. In Section 10.4, we briefly review a set of systematic long-wavelength modifications to Mercury's topography that remain unexplained but that may have been tectonically driven. In Section 10.5, we discuss the extent of our understanding of the structure and mechanical behavior of Mercury's lithosphere – the mechanically strong outer layer of the planet (Chapter 3). The mechanisms for tectonic deformation, chief among them global contraction but also those that appear to operate solely within volcanically flooded impact features, are then discussed in Section 10.6. Other mechanisms known to drive tectonic activity are briefly visited in Section 10.7.

In Section 10.8, we explore the other major aspect of Mercury's tectonics – *when* deformation took place – as we attempt to describe at least in broad terms the tectonic history of the planet. The influence that the planet's tectonic properties and evolution have played on its volcanism is then addressed in Section 10.9. In the final section, we list some major questions regarding Mercury's tectonics that remain open, and suggest how they might yet be answered.

10.1.1 Observations of Tectonic Features by Mariner 10

Images returned by the Mariner 10 spacecraft of about ~45% of Mercury's surface show a landscape battered by impact bombardment, its terrains discriminated largely by the level of cratering. Notably, evidence of tectonic deformation was readily identified across the entire portion of the planet surface imaged by that spacecraft (Murray et al., 1974; Strom et al., 1975).

The greatest degree of structural complexity was observed in Caloris Planitia, an expansive deposit of smooth plains interpreted to be volcanic and situated within the largest well-preserved impact basin on the planet, with an east–west diameter of some 1640 km (Byrne et al., 2013a; Chapter 9). Although little more than a third of the spatial extent of these plains was imaged, they were seen to include lineations and graben-like structures, interpreted to represent extensional strain, as well as a complex pattern of ridges regarded as evidence of crustal shortening (Murray et al., 1974; Strom et al., 1975). Vertical motion was proposed to account for both types of landform, with the graben and ridges within the Caloris basin attributed to uplift from isostatic readjustment and subsidence from volcanic loading, respectively (Strom et al., 1975; Dzurisin, 1978).

A large population of escarpments identified outside the Caloris basin were also inferred to be shortening structures (e.g., Strom et al., 1975). In fact, it was apparent from the outset that, for the entire portion of Mercury viewed by Mariner 10, by far the most widespread form of tectonic deformation accommodated horizontal shortening (e.g., Trask and Guest, 1975). Thrust-fault-related landforms are so abundant that Strom et al. (1975) concluded that their formation mechanism must be global in nature.

The style of crustal tectonics on a planet is intimately tied to the thermal evolution of the planetary interior (Solomon, 1977). The widespread occurrence of shortening structures therefore implies a thermal history characterized by global contraction, in which the planet's interior cools through time and the body experiences a net reduction in volume (Solomon, 1978) (Section 10.6.1). Global contraction was invoked as a causal mechanism for those shortening structures observed across Mercury outside the smooth plains units (Murray et al., 1974; Strom et al., 1975). On the basis of their length, height (as a measure for vertical displacement), and a range of fault dip angles, Strom et al. (1975) calculated that the scarps observed over about 24% of Mercury's surface, extrapolated to a global population of shortening structures, represent a reduction in planetary radius of 1–2 km.

This value is a key parameter for thermal evolution models of Mercury, which address, among other aspects, the bulk abundances of heat-producing elements in the planet's silicate portion, the nature of mantle convection through time, and the history of cooling and present-day structure of the planet's large metallic core, the source of its internal magnetic field (e.g., Hauck et al., 2004; Grott et al., 2011; Michel et al., 2013; Tosi et al., 2013, 2015; Chapters 4 and 19). Accordingly, much of the study of Mercury's tectonic evolution has focused on the extent to which the planet has contracted, as recorded by its assemblage of shortening structures. Following Strom et al. (1975), studies of planetary radius change from Mariner 10 photogeology consistently reported estimates of ~1–2 km (Watters et al., 1998, 2009a; Watters and Nimmo, 2010) – substantially less than the ~5–10 km predicted by interior thermal history models of the planet (e.g., Solomon, 1977; Schubert et al., 1988; Dombard and Hauck, 2008) (Section 10.6.1). This discrepancy between photogeological observation and model prediction persisted long after the Mariner 10 mission.

10.1.2 Post-Mariner 10 Questions

The findings from Mariner 10 raised far more questions than they answered, strongly whetting the scientific community's appetite for more Mercury science. Indeed, the gaps in our understanding of Mercury's geological history served to frame several of the driving science questions for the MESSENGER mission (Chapter 1). As they pertained to the innermost planet's tectonics, those questions called for a full characterization of the spatial and temporal distribution of tectonic landforms and a determination of their contribution to planetary radius change.

Specifically, did expressions of Mercury's tectonics occur as widely across the hemisphere not viewed by Mariner 10 as across the imaged hemisphere? Relatedly, was the tectonic deformation observed in the eastern portion of the Caloris basin floor representative of these plains as a whole? Aside from the interior of the Caloris basin, where else (if anywhere) was extension preserved on Mercury? When did the planet's tectonic landforms start to develop, when did they cease to be active, and what drove their formation? To what extent do the planet's shortening landforms reflect the effects of global contraction rather than some other process or processes? And by how much did Mercury *actually* contract as a result of secular interior cooling?

A key science objective of MESSENGER's primary mission was to study the geological history of Mercury (Chapter 1). That history encompasses geological, geochemical, and geophysical processes, of which tectonic deformation is an important part. Mercury's tectonics have not only helped shape the planet surface but have likely played a controlling role in its volcanic history (Section 10.9 and Chapter 11) – and have in turn been influenced at least in part by the planet's record of impact bombardment (Section 10.6.4 and Chapter 9). Numerous measurements of the planet's properties have given insight into its tectonics, including its shape, topography, crustal thickness, surface composition, and even its interior structure.

10.1.3 Mercury Tectonics with MESSENGER Data

Absent an ability to visit them in the field, the most effective means of assessing tectonic landforms is with remotely sensed photographic and topographic data. The Mariner 10 mission facilitated early studies of Mercury's tectonics by means of its Television Photography Experiment (Murray et al., 1974). This instrument returned photographic images of tectonic landforms, from which their morphology and distribution could be determined (Trask and Guest, 1975). Topographic data were obtained from photoclinometry (Strom et al., 1975) and, in some cases, from stereophotogrammetry (Spudis and Guest, 1988).

The MESSENGER spacecraft carried a far more capable instrument payload (Solomon et al., 2008; Chapter 1) and returned global coverage of Mercury's surface at resolutions far higher than those of Mariner 10 data. Many MESSENGER-based tectonic studies have used the global mosaic base maps derived from the Mercury Dual Imaging System (MDIS) (Hawkins et al., 2007). Consisting of image data from the wide- and narrow-angle cameras (WAC and NAC, respectively), and featuring mean resolutions of 250 m/pixel, several global base maps with a variety of solar incidence and illumination azimuths were produced (Chabot et al., 2016), facilitating regional to global tectonic surveys (e.g., Rothery and Massironi, 2010; Di Achille et al., 2012; Byrne et al., 2014). More detailed structural analyses can be accomplished with thousands of high-resolution MDIS NAC images (both targeted and opportunistically acquired), which have typical resolutions of tens of meters per pixel trending to meters per pixel for those acquired toward the end of MESSENGER's low-altitude campaign (Chapter 1).

Although it did not carry a dedicated stereo imaging system, repeated passes by MESSENGER over the surface enabled the production of digital elevation models (DEMs) with stereophotogrammetric techniques. Elevation data for substantial portions of Mercury were calculated at resolutions at or below ~1 km/pixel (e.g., Oberst et al., 2010; Preusker et al., 2011), and DEMs have been created for selected regions at higher resolutions (e.g., Fassett and Crowley, 2016). A global DEM, at a resolution of 665 m/pixel, was also generated from the control network derived from the development of the global image base maps (Becker et al., 2016). Notably, topographic measurements of Mercury's northern hemisphere were also acquired by MESSENGER's Mercury Laser Altimeter (Cavanaugh et al., 2007; Zuber et al., 2012); individual laser altimetric profiles were then combined to form an interpolated DEM of the northern hemisphere, also at a resolution of 665 m/pixel. These regional and global elevation data products in particular have been of considerable use for thrust fault displacement–length scaling analyses (e.g., Byrne et al., 2014; Section 10.6.1), as well for characterizing the relationship between tectonics and topography (Section 10.2.1).

10.2 SHORTENING STRUCTURES ON MERCURY

Perhaps the most prominent product of Mercury's tectonic history is its population of landforms thought to have formed by brittle deformation. Like all planetary surfaces on which tectonic processes other than plate tectonics have acted, these landforms lend themselves to classification as either primarily shortening or lengthening (i.e., extensional) in nature. Consistent with Mariner 10 findings, MESSENGER observations have shown that lithospheric shortening is the dominant form of tectonic deformation on Mercury and has occurred globally. In this section, we first review the types of shortening structures documented across Mercury, drawing comparisons where possible with cognate structures on Earth. Next, we review the spatial distributions of shortening structures on the innermost planet, before presenting a kinematic analysis of shortening-related deformation on Mercury, once more informed by centuries of exploration of tectonic deformation on our own planet.

10.2.1 Types of Shortening Structures

Throughout the literature regarding Mercury's shortening tectonics, terms such as "wrinkle ridge," high-relief ridge," and "lobate scarp" have been commonly used (e.g., Strom et al., 1975; Dzurisin, 1978; Melosh and McKinnon, 1988; Watters et al., 1998, 2004, 2009a, 2015; Watters and Nimmo, 2010; Egea-González et al., 2012). [Indeed, Dzurisin (1978) developed a classification scheme consisting of at least six discrete categories into which positive-relief landforms were grouped.] On the basis of Mariner 10 data, Mercury's wrinkle ridges were observed to resemble those in the lunar maria – often manifest as broad, steep-sided but low-relief arches symmetric in cross section, variously with or without crenulated crests (Strom et al., 1975; Dzurisin, 1978); numerous examples appear flat-topped. Qualitatively, wrinkle ridges are shorter, narrower, and possess less relief than other types of shortening-related landforms. A characteristic property of wrinkle ridges is the substantial morphological variation along strike, with changes to width, height, and number of sides often observed (Figure 10.1a). As suggested by their name, high-relief ridges tend to possess greater relief than wrinkle ridges, but they were also noted to be generally symmetric in cross section (Watters et al., 2001). Although considerable along-strike changes in size and shape are less commonly seen for high-relief ridges, some variation in width and height occurs (Figure 10.1b). In contrast to the generally symmetrical cross sections of wrinkle- and high-relief ridges, Strom et al. (1975) described lobate scarps as highly asymmetric in transverse view, with steep slopes on one side and gentle backslopes on the other; in map view they have relatively steep and long escarpments that typically show a broadly lobate outline (Figure 10.1c). Some lobate scarps, such as Beagle Rupes, have strongly arcuate forms, however (Rothery and Massironi, 2010) (Figure 10.1d). Many examples show smaller, subordinate scarps along their leading edges; such accessory scarps are visible, for example, where the larger structures in Figures 10.1c and 10.1d cut through the smooth-floored Duccio and Sveinsdóttir craters, respectively.

Despite differences in morphology between these landform types, in all cases these features have been interpreted as being a manifestation of lithospheric shortening, representing some combination of thrust faulting and/or folding (e.g., Strom et al., 1975; Dzurisin, 1978; Melosh and McKinnon, 1988; Watters et al., 2004; Watters and Nimmo, 2010). Comparable assessments have been made of morphologically similar landforms on other worlds (e.g., Colton et al., 1972; Lucchitta, 1976; Schultz, 1985; Mueller and Golombek, 2004).

Tectonic deformational features on Earth tell us that lobate scarps, as upthrust volumes of rock, are likely the folded portions of hanging walls atop thrust faults (i.e., low-angle faults formed in compression) (Figure 10.2). With the nomenclature developed for tectonics on Earth, then, lobate scarps are fault-propagation- or fault-bend folds, which together may be classed as "fault displacement-gradient folds" (Wickham, 1995). On the basis of shape alone, lobate scarps are asymmetric hanging-wall anticlines or monoclines (arch- or step-like folds, respectively). This interpretation is consistent with observations of narrow graben along the crests of scarps on several worlds (Section 10.3), which likely reflect outer-arc (or extrados) extension along the fold hinge lines (Figure 10.2). Moreover, the East Kaibab monocline in southwestern Utah, United States, matches the geometry of exemplar lobate scarps on Mercury, Mars, and the Moon once its eroded volume is restored (Byrne et al., 2016a). The subordinate lobate scarps along the leading edges of larger counterparts (Strom et al., 1975) may be thrust duplexes – imbricate (overlapping) stacks of smaller fault-bound slices of upthrust rock regularly seen within shortening systems on Earth (e.g., Butler, 1982). The asymmetry of lobate scarps therefore reflects the vergence of the hanging-wall anticlinal fold: the landform can be said to "verge" in the direction the steeper scarp faces – the direction the fold is inclined – which in turn indicates the direction of tectonic transport of the hanging-wall block along the underlying fault (e.g., Byrne et al., 2014) (Figure 10.2).

Landforms proposed as analogs to wrinkle ridges have been described at numerous sites on Earth, including Algeria, Australia, the Solomon Islands, and the United States (e.g., Plescia and Golombek, 1986; Petterson et al., 1997; Last et al., 2012), and in each case they feature the folding of rocks over thrust faults. Although several wrinkle ridge properties, including the orientation and depth of the causative faults and the kinematics of fault-related folding, remain to be fully characterized (e.g., Watters, 1991; Golombek et al., 1991; Plescia, 1991, 1993; Schultz and Tanaka, 1994; Zuber, 1995; Schultz, 2000; Mueller and Golombek, 2004), these landforms have also been considered as forms of fault displacement-gradient folds (Plescia and Golombek, 1986). The steep escarpments that bound wrinkle ridges can themselves be regarded as monoclines (albeit at scales smaller than those characterizing lobate scarps) such that where they occur, two opposite-facing ridge-bounding scarps are essentially paired monoclines, similar to those observed in the Southern Rocky Mountains, United States (Tweto, 1975), Libya (Fodor et al., 2005), and the United Kingdom (Woodcock and Soper, 2006). Accordingly, where wrinkle ridges have steep sides, they, too, can be said to have vergence.

To our knowledge, no examples on Earth of high-relief ridges have yet been documented. This may not be a function of geology, however, but rather may reflect a limitation in the use of morphology as a discriminator for landform classification. High-relief ridges have been suggested to differ from lobate scarps only in that the dip angle of their underlying fault is greater than that of a lobate scarp fault (Watters and Nimmo, 2010), and so these landforms may occupy two points on a single morphological continuum. Under that interpretation, a high-relief ridge is simply a hanging-wall anticline over a

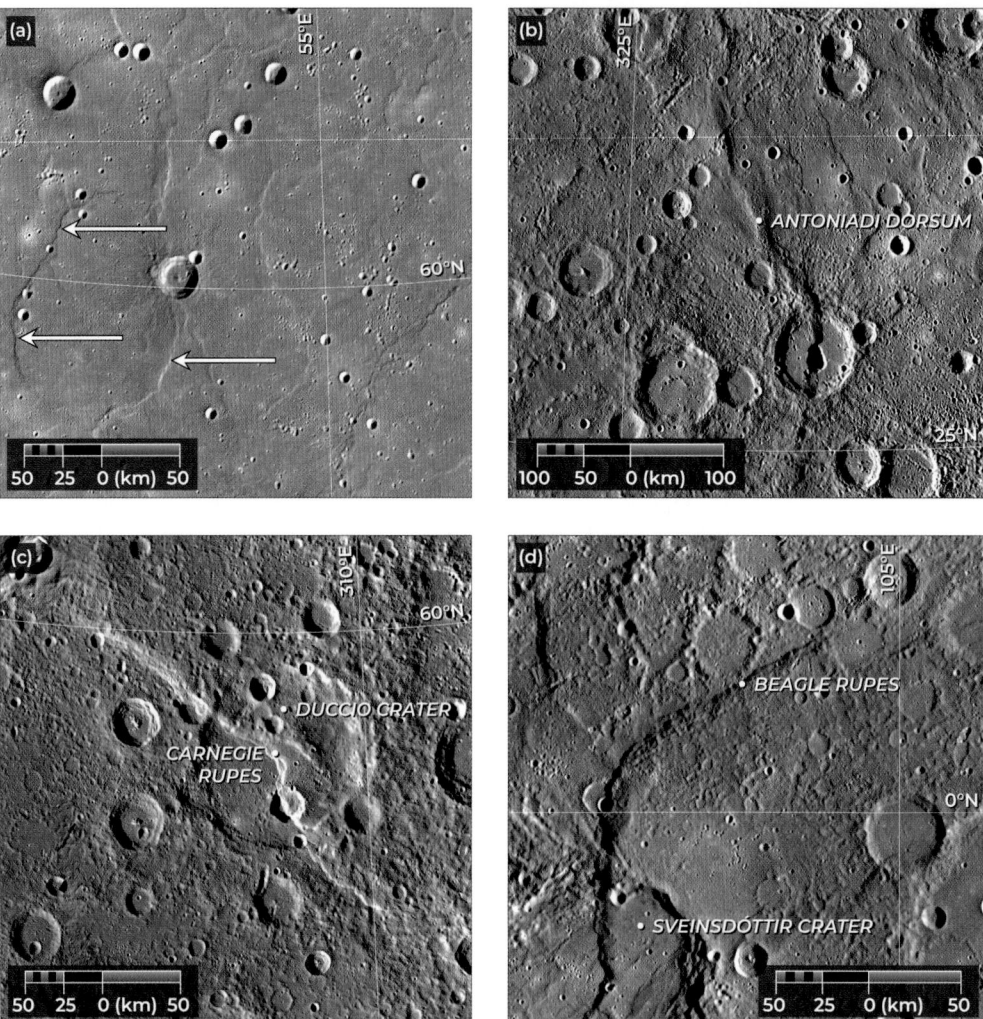

Figure 10.1. Examples of shortening landforms on Mercury. (a) Smooth plains structures in Mercury's northern plains (the majority of which are "wrinkle ridges"). Note the localization of shortening landforms at the rims of buried craters (shown by white arrows). (b) Antoniadi Dorsum, one of the comparatively rare class of tectonic landform historically termed a "high-relief ridge"; this example is 360 km in length. (c) Carnegie Rupes, an example of a relatively linear monocline (a "lobate scarp"); this feature, almost 270 km long, cuts through the Duccio crater (133 km in diameter). (d) Another monocline, Beagle Rupes, noted for its pronounced arcuate shape in map view. About half of the 630-km-long scarp is shown. Azimuthal equidistant projections, centered as follows: (a) 60.3°N, 52.9°E; (b) 29.0°N, 329.5°E; (c) 58.5°N, 306.7°E; and (d) 2.0°S, 103.0°E. These and all subsequent images in this chapter are taken from the global morphology base map (Chabot et al., 2016).

relatively steep reverse fault, with the symmetry of the fold precluding a definitive identification of vergence and thus the associated fault's slip direction.

Despite the published work comparing extraterrestrial shortening landforms with those on Earth, however, the terms lobate scarp, high-relief ridge, and wrinkle ridge – which are *never* used for terran tectonics – persist in the planetary tectonic literature and render it opaque. Yet the variations in shape of these landforms on Mercury is such that these terms of classification fail us in many cases. For example, although lobate scarps can generally be described as having a single steep escarpment with wrinkle ridges possessing two scarp-like sides (Figures 10.1c and 10.1d), this description does not always hold. The portion of Enterprise Rupes located outside the Rembrandt basin (Watters et al., 2009b) and one of Mercury's largest systems of lobate scarps, has both a front- and a backscarp, and in that respect resembles an enormous wrinkle ridge. Conversely, many wrinkle ridges on Mercury display only a single scarp, whereas others change along their length from single- to double-sided. These morphological distinctions are further challenged in cases where one landform type transitions into another. Lobate scarps can change to high-relief ridges along strike, for instance, a phenomenon illustrated in Figure 10.1b. As noted above, this change may simply reflect a variation in dip angle of the underlying fault (e.g., Watters and Nimmo, 2010). Additionally, several shortening structures terminate as lobate scarps at the western margin of the Caloris basin, continuing along the same strike as wrinkle ridges within those plains. A similar situation is observed on the Moon, where a lobate scarp transitions to a wrinkle ridge at the northern boundary between Mare Serenitatis and the surrounding lunar highlands (Masursky et al., 1978).

Table 10.1. *The number and cumulative length of shortening structures by class on Mercury (after Byrne et al., 2014).*

Structure class	Number of structures	% of total number	Cumulative length (m)	% of total length
Smooth plains	3751	63.2	2.07×10^8	49.8
Cratered plains	1831	30.9	1.64×10^8	39.5
Crater-related	252	4.2	2.01×10^7	4.8
High-terrain-bounding	100	1.7	2.45×10^7	5.9
Total	**5934**	**100.0**	**4.16×10^8**	**100.0**

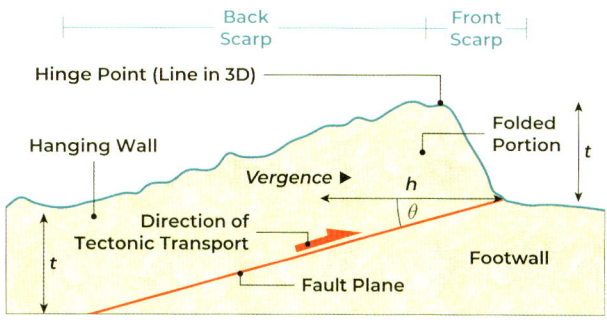

Figure 10.2. A schematic cross section of a lobate scarp, a type of thrust-fault-related landform. A thrust fault (red line) separates the upthrust volume of rock (i.e., the hanging wall) from the volume of rock that does not move (i.e., the footwall). The upper portion of the hanging wall is likely an asymmetric fold, which constitutes the lobate scarp itself. The direction that the steeper front scarp faces is the direction of vergence of the fold, which in turn gives the direction of tectonic transport of the hanging wall (red arrow). The hinge point is located at the crest of the fold; the hinge line extends perpendicular to the plane of view. When making estimates of horizontal shortening accommodated by such a structure, we assume that the heave, h, of the thrust fault is related to its throw, t, by the fault dip angle, θ (Section 10.6.1).

Therefore, a classification scheme for shortening structures based on morphology, such as that used historically for Mercury, works only in a general way but cannot capture the broad variation in geometry of these landforms. Geological setting, however, provides a clear and objective means by which to describe Mercury's shortening structures (Byrne et al., 2014) and is the basis on which we present those structures in this chapter (see Figure 10.3a). With this approach, shortening structures are classified by the primary terrain type in which they occur: "smooth plains" (Denevi et al., 2013) or "cratered plains," a term we use here that encompasses both the intercrater plains and heavily cratered terrain units described from Mariner 10 images (Trask and Guest, 1975; Chapter 6). Most such mapped landforms are therefore categorized here as "smooth plains structures" or "cratered plains structures." The remaining landforms are either spatially associated with impact craters and so are termed "crater-related structures" or border broad areas of substantially elevated terrain and are catalogued as "high-terrain-bounding structures." Topographic data, such as the DEM derived by Becker et al. (2016) shown in Figure 10.3b, provide the basis for this last classification. Several statistical measures for these structures, including number and cumulative length as fractions of the entire population of shortening landforms on Mercury, are given in Table 10.1.

Smooth plains structures (light blue lines in Figure 10.3a) are those hosted within Mercury's eponymous plains units. Smooth plains on the innermost planet are sparsely cratered relative to other terrain, are gently rolling to essentially level, and generally have clear boundaries (Trask and Guest, 1975). They occupy some 27% of Mercury's surface and occur at all longitudes, nearly all latitudes, and primarily low elevations (Denevi et al., 2013; Chapter 6). At least two-thirds of these deposits are interpreted as volcanic (Denevi et al., 2013; Chapter 11). The majority of thrust-fault-related landforms within smooth plains would traditionally have been classified as wrinkle ridges, but this class also includes some monoclines and asymmetric hanging-wall anticlines (i.e., lobate scarps) (Byrne et al., 2014). Smooth plains structures represent ~63% of all mapped structures but only ~50% of cumulative structure length (Table 10.1). Over 1500 structures (~40% of this population) are located in the vast Borealis Planitia, the "northern smooth plains" (NSP) of Head et al. (2011); the remaining structures are situated within the extensive plains encircling the Caloris basin or in impact-feature-hosted smooth plains within more heavily cratered terrain (Denevi et al., 2013).

Cratered plains structures include those located in both the intercrater plains and heavily cratered terrain units described from images returned by the Mariner 10 spacecraft (Strom et al., 1975; Trask and Guest, 1975) (purple lines in Figure 10.3a). These units are older than the smooth plains and may have formed through voluminous volcanism during or at the end of the late heavy bombardment (LHB) of the solar system (Chapters 9 and 11), though at least some portion may have been emplaced as basin ejecta (Chapter 9). Excluding those landforms within isolated pockets of impact-related smooth plains, shortening structures in intercrater and heavily cratered plains are almost exclusively monoclines and asymmetric anticlines (Byrne et al., 2014). They constitute ~31% of all mapped structures and ~40% of cumulative structure length (Table 10.1).

Byrne et al. (2014) identified 252 arcuate or near-circular basin- and crater-related structures (teal lines in Figure 10.3a), which demarcate volcanically filled or buried impact features (typically marking the presence of a "ghost crater": Section 10.6.4). Although most structures of this class occur within smooth plains, with 62% in the NSP alone, a subset consists of large monoclines within impact basins that follow, and verge toward, the basin perimeter (e.g., Fegan et al., 2017). Indeed, the longest single structure mapped by Byrne et al. (2014) defines part of the third-largest impact feature recognized on Mercury to date, the 950-km-diameter Matisse–Repin basin (Spudis and Guest, 1988). In its entirety this structure class represents ~4% and ~5% of all surveyed shortening landforms and of cumulative structure length, respectively (Table 10.1).

The fourth structure classification described by Byrne et al. (2014) consists of high-terrain-bounding structures, a category that includes 100 monoclines and asymmetric anticlines that,

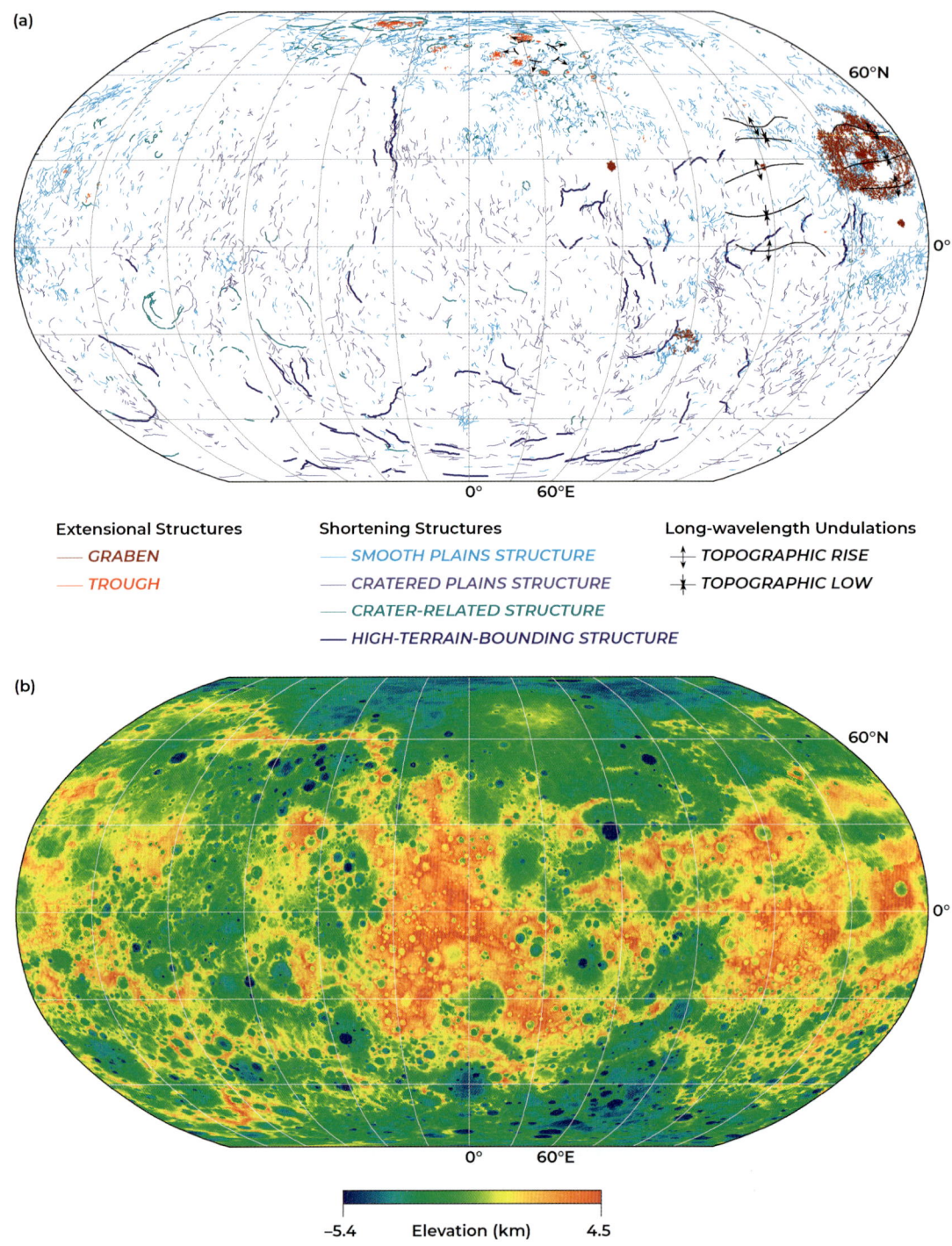

Figure 10.3. The shortening and extensional tectonic landforms of Mercury. (a) Shortening structures are shown in cool colors; extensional landforms are shown in warm colors. Long-wavelength topographic undulations within and proximal to the Caloris basin, as well as in the NSP, are shown with black lines; arrows denote downslope directions. The global population of shortening structures is from Byrne et al. (2014); extensional and shortening landforms in Rembrandt basin are from Ferrari et al. (2015). The troughs mapped in the NSP and the circum-Caloris plains are from Klimczak et al. (2012). (b) The global topography of Mercury, derived from a digital elevation model (Becker et al., 2016). Both maps are in a Robinson projection centered at 0°E; the graticule is in 30° increments in latitude and longitude.

individually or as systems of structures, lie along the margins of and thus partially delineate some portions of high-standing terrain (thick dark blue lines in Figure 10.3a). These landforms have a dominant sense of vergence toward adjacent lows and possess some of the greatest structural relief recorded on Mercury. For example, parts of Enterprise Rupes rise 3 km above the surrounding plains. These thrust-fault-related landforms together represent ~2% of the total structure count but possess a disproportionately large ~6% of the total cumulative structure length (Table 10.1).

Importantly, there is no evidence for widespread strike-slip deformation on Mercury, as is observed at Earth's conservative plate margins and in escape tectonics settings on Earth and Venus. However, numerous shortening landforms show elements of strike-slip or oblique motion. For example, Rothery and Massironi (2010) investigated in detail the morphology of Beagle Rupes (Figure 10.1d) and interpreted the trapezoidal shape of this landform as a system composed of a frontal monocline and two lateral ramps showing evidence of transpression (i.e., combined orthogonal shortening and strike-slip deformation). (A ramp is a steep and usually short segment of a thrust fault as it climbs to a higher stratigraphic level.) Given the westward vergence of the frontal monocline, Rothery and Massironi (2010) inferred that the fault system must have accommodated oblique slip with right-lateral sense along the northern ramp and left-lateral sense along the southern ramp. These authors further concluded that, on the basis of its very prominent polygonal map pattern shape, the Beagle Rupes system is underlain by listric (curved) thrust faults that root to a basal décollement – a horizontal detachment surface – that extends several hundred kilometers to the east.

Other strike-slip kinematic indicators have been documented on Mercury. Massironi et al. (2015) identified oblique-shear kinematics along monoclines and anticlines in the form of lateral ramps, strike-slip duplexes, and restraining bends, as well as pop-up and pull-apart structures and en échelon (staggered or overlapping) folds. Oblique or lateral ramps associated with major thrust systems were reported, including at the distal ends of Enterprise Rupes, at the southern termination of Blossom Rupes, along Belgica Rupes, at the southern end of La Dauphine Rupes, and at the ends of Paramour Rupes (Galluzzi et al., 2015; Massironi et al., 2015). Additional examples of oblique fault slip were identified on the basis of changes in circularity and topography of faulted impact craters (Galluzzi et al., 2015). More such structures will likely be recognized on Mercury, as studies of the planet's thrust-fault-related landforms have so far focused predominantly only on the pure shortening they accommodate, with little detailed analysis of any components of oblique slip so often observed in intraplate shortening systems on Earth (e.g., Cunningham et al., 1996; Norris and Cooper, 1997; Liu et al., 2012).

10.2.2 Distribution of Shortening Structures

A fundamental consequence of global contraction is a global stress field that is horizontally isotropic, which in the absence of other sources of stress or heterogeneities in lithospheric strength predicts a global, evenly distributed population of shortening structures with no preferred orientations (e.g., Solomon, 1977). Early observations of Mercury did not show evenly distributed thrust-fault-related landforms, but resurfacing by impact bombardment and volcanism, together with suboptimal illumination conditions and image resolution, hampered efforts to characterize fully the distribution of shortening structures across the hemisphere imaged by Mariner 10 (Strom et al., 1975).

Nonetheless, from Mariner 10 observations, Dzurisin (1978) identified a "tectonic grid" on Mercury, which was investigated further (Melosh and Dzurisin, 1978; Melosh and McKinnon, 1988; Thomas et al., 1988). This grid was attributed to the effects of tidal despinning of Mercury from an original rotation rate substantially higher than that of today (e.g., Melosh and McKinnon, 1988) (Sections 10.6.2 and 10.6.3). However, although mapping with MESSENGER data has shown that thrust faulting is not uniformly distributed on the planet – the NSP, for example, represent just ~6% of the planet's surface but host a disproportionately large number of shortening landforms – there is *no* definitive evidence of a globally coherent lithospheric fracture pattern that survived the LHB (Byrne et al., 2014), such as that predicted to have been influenced by tidal despinning.

Of note, shortening landforms appear concentrated along quasi-longitudinal bands at approximately 0°E, 90°E, 180°E, and 270°E. At these locations, the densities of mapped structures per 10° × 10° bins are greatest (Figure 10.4). These longitudes, however, correspond to portions of the global photomosaic base maps from which the structure map in Figure 10.3a was produced that were obtained with the highest values of incidence angle, *i*, measured from the surface normal (Byrne et al., 2014). Such images, taken when the Sun was low in the

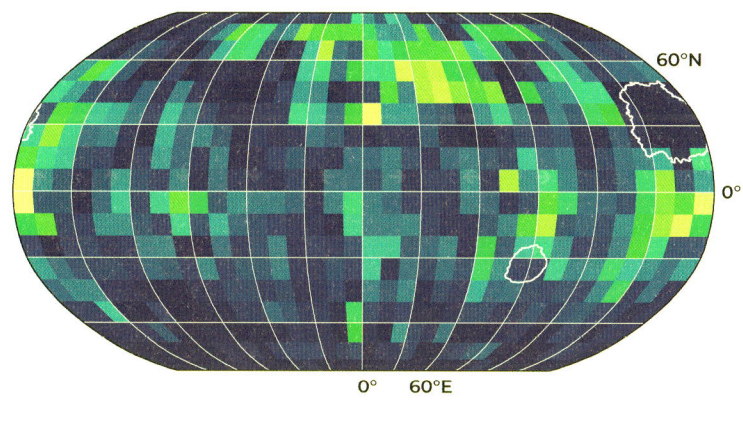

Figure 10.4. The spatial density of the shortening landforms shown in Figure 10.3a, color-coded by the number of discrete structures per 10° × 10° bin. These data are adapted from Byrne et al. (2014) but exclude the structures inside the Caloris and Rembrandt impact basins (outlined in white). The map is in a Robinson projection centered at 0°E; the graticule is in 30° increments in latitude and longitude.

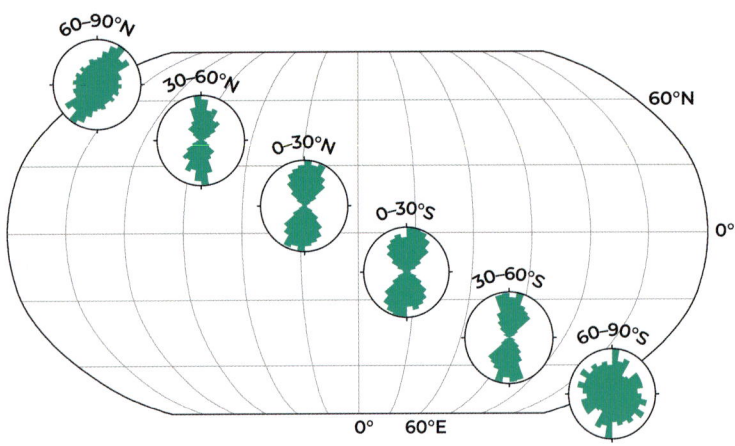

Figure 10.5. The distribution of shortening structures in Figure 10.3a, grouped into 30° latitudinal bands (with northern and southern hemisphere structures shown on the left and right, respectively). These data are adapted from Byrne et al. (2014) (and exclude structures in the Caloris and Rembrandt basins). The map is in a Robinson projection centered at 0°E; the graticule is in 30° increments in latitude and longitude.

sky, feature long shadows and so facilitate the ready identification and mapping of morphologically subtle tectonic landforms; regions imaged at lower incidence angles thus have fewer mapped structures.

Further, the orientations of structures between 60°N and 60°S show a predominance of north–south trends (Figure 10.5), consistent with the illumination of linear landforms from solar azimuth directions (measured clockwise) near ~90° and ~270° due to Mercury's very low obliquity (~2 arcmin). Structures north of 60°N, including a portion of the densely distributed landforms in the NSP, show some clustering at southwest–northeast trends but are not as strongly oriented as those at mid latitudes. South of 60°S, structures display no preferred orientation, but these latitudes host the lowest number of mapped shortening-related landforms per area (Figure 10.5).

Even the earliest workers studying Mercury noted a roughly north–south trend to tectonic structures on Mercury and suggested that lighting conditions might be at least in part responsible (Strom et al., 1975). The sensitivity to solar illumination of the identification of tectonic landforms was also noted by other workers (e.g., Melosh and McKinnon, 1988; Thomas et al., 1988). Although Byrne et al. (2014) demonstrated that shortening landforms visible in the image data are not uniformly distributed across Mercury, as was predicted for global contraction (e.g., Solomon, 1977) and by early models of tidal despinning (Melosh and McKinnon, 1988) (Section 10.6.3), the longitudinal bands of increased structure density shown in Figure 10.4 are likely a function principally of lighting geometry and not geology. The mapping of some basin-related structures is also likely influenced by an illumination bias (Fegan et al., 2017). Moreover, other processes are capable of producing a pattern of shortening strain that deviates from spatial homogeneity. Large impact basins act as zones of weakness in the lithosphere and concentrate shortening strains at their peripheries (e.g., Watters et al., 2001), leading to the class of landforms Byrne et al. (2014) termed crater-related structures (Section 10.2.1). Moreover, large basins and their extensive ejecta deposits are responsible for large areas devoid of thrust-fault-related landforms where illumination conditions are otherwise amenable to tectonic mapping, as is observed at the Vivaldi impact basin, in the planet's western hemisphere.

Yet there *is* evidence of systematic, regional-scale shortening on Mercury that may be geological in nature. In several places, groups of thrust-fault-related landforms form laterally contiguous, narrow bands of considerable length. One such system extends for some 1700 km (over 40° of arc) across Mercury's northern hemisphere and includes Victoria and Endeavour Rupes and Antoniadi Dorsum (Figure 10.6). Many of its constituent landforms are cratered plains structures, but those that comprise the northernmost third of this system border an area of high-standing terrain to the west and verge eastward onto the adjacent smooth plains. This sense of vergence is echoed by the other high-terrain-bounding structures along its length. An even longer system, 1800 km long, extends from 19°N, 55°E, to 23°S, 61°E, and also has a dominant westward vergence.

These systems have been recognized as fold-and-thrust belts (FTBs) (Byrne et al., 2014), counterparts to fold belts on Earth (e.g., Poblet and Lisle, 2011) and Venus (e.g., Burke et al., 1984). In their simplest terms, FTBs are linear, regionally contiguous sets of thrust faults with associated hanging-wall anticlines that have a single predominant sense of vergence. Many FTBs on Earth feature an extensive décollement at depth (McClay, 1992; Roeder, 2009). Such a structural arrangement may not commonly exist for Mercury's FTBs, but it has been inferred to underlie individual arcuate monoclines on the planet (e.g., Rothery and Massironi, 2010) (Section 10.2.1).

10.2.3 Kinematics of Shortening Structures

The contrast in morphology and density of tectonic structures hosted by Mercury's younger smooth plains and the older cratered plains (compare Figure 10.1a with Figures 10.1b–10.1d) could reflect differences in rheological and structural fabric characteristics between the two terrain types. For example, MESSENGER X-Ray Spectrometer (XRS) and Gamma-Ray Spectrometer (GRS) measurements of elemental abundances indicate that the surface composition of at least some heavily cratered terrain is more magnesian but less feldspathic than the low-iron basalt-like NSP (Weider et al., 2012; Chapter 7). With their internal structure rendered more complex by sustained impacts and volcanic resurfacing (e.g., Denevi et al., 2009; Ernst et al., 2010), the cratered plains material may thus have isotropic textures on the scale of tens to hundreds of kilometers

10.2 Shortening Structures on Mercury

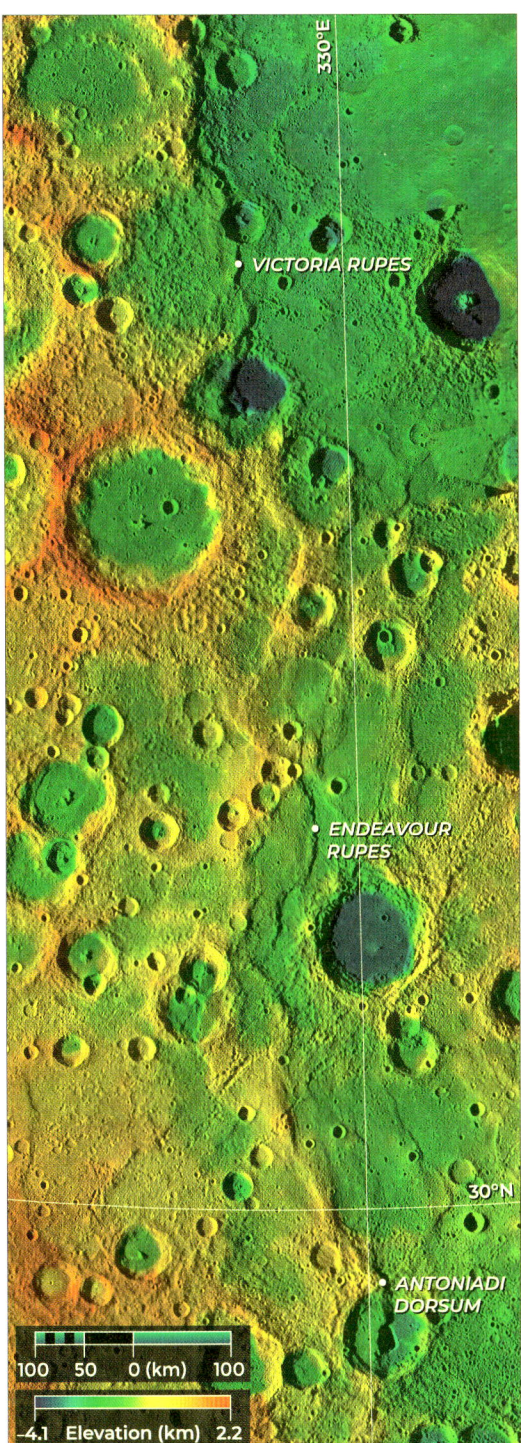

Figure 10.6. A fold-and-thrust belt on Mercury. This example is ~1700 km long and consists of an aligned assemblage of monoclines and asymmetric and symmetric anticlines. Named parts of this system include Victoria and Endeavour Rupes and Antoniadi Dorsum (Figure 10.1b). The colors correspond to elevation, derived here and in subsequent figures from the global DEM of Becker et al. (2016). Azimuthal equidistant projection, centered at 40.0°N, 328.0°E.

that facilitated the development of thick-skinned, large-scale fault and fold systems, which root to (bottom out at) depths of tens of kilometers (and possibly to deep basal décollements)

(Figure 10.7), consistent with the results of forward modeling studies of topographic profiles across several monoclines and anticlines (Watters et al., 1998).

In contrast, the smooth plains likely feature strong vertical variations in mechanical properties inherent to layered volcanic strata (e.g., Jerram and Widdowson, 2005; Section 11.4), which promote detachments and shallow tectonic deformation (Freed et al., 2012). Smooth plains structures occur consistently within stratified units on Mars and, although there is debate as to whether they are thick- or thin-skinned in nature (e.g., Watters, 1991; Zuber, 1995; Mangold et al., 1998; Schultz, 2000; Golombek et al., 2001; Montési and Zuber, 2003; Mueller and Golombek, 2004), fits of topographic profiles from elastic dislocation models to observed profiles indicate that the thrust faults beneath these landforms root to shallow crustal levels (Watters, 2004). Given the similarity in setting and morphology between smooth plains structures on Mars and Mercury, landforms on the latter body are probably thin-skinned, their faults rooting to shallow décollements – likely some mix of the interface between the smooth plains and the underlying regolith-covered basement and interbeds within the plains deposits themselves (e.g., Jerram and Widdowson, 2005) (Figure 10.7). Similar concentric folding of basalt flows above décollements has been described on Earth, with those in the Yakima fold belt that straddles the Washington–Oregon border (Last et al., 2012) and the Malaita anticlines at the Solomon Islands–Ontong Java collisional front (Petterson et al., 1997) among the best-known examples.

Where shortening landforms demarcate buried impact craters in smooth plains (Figure 10.1a), such structures were likely formed by the concentration of compressive stresses above crater rims (Watters et al., 2009a). Conversely, outward-verging monoclines and anticlines within impact basin rims may represent partitioning of shortening strain between the basin and its interior smooth fill, either as a result of a difference in elastic moduli between the fill and basin floor material or, as for the smooth plains structures, by the rooting of thrust faults into a décollement between the two deposits (e.g., Fegan et al., 2017). At some sites where shortening structures and impact craters interact, syntaxis is observed (Figure 10.8). This phenomenon occurs when the hanging wall of a thrust fault encounters an impediment, such as a crater's central peak, and the continued propagation of the unobstructed portions of the fold form a distinct bend in the strike of the shortening structure, as is famously observed in the Himalaya (Suess, 1883).

Regions of high- and low-standing terrain on Mercury do not appear to correlate spatially with free-air gravity anomalies over the planet's northern hemisphere, a result indicating that topography is largely isostatically compensated, presumably by variations in crustal thickness (e.g., Smith et al., 2012; James et al., 2015; Chapter 3). Those tectonic structures that border high-standing terrain may therefore have served to isolate thicker crustal blocks from neighboring, thinner portions of the crust, localizing substantial shortening along their length and accumulating considerable strain by penetrating far into the lithosphere. With continued shortening, thicker crustal blocks overthrust adjacent low-lying terrain, as on Earth (Suess, 1909; Şengör, 1993). Moreover, where smooth plains structures are located adjacent to, and share a dominant strike with,

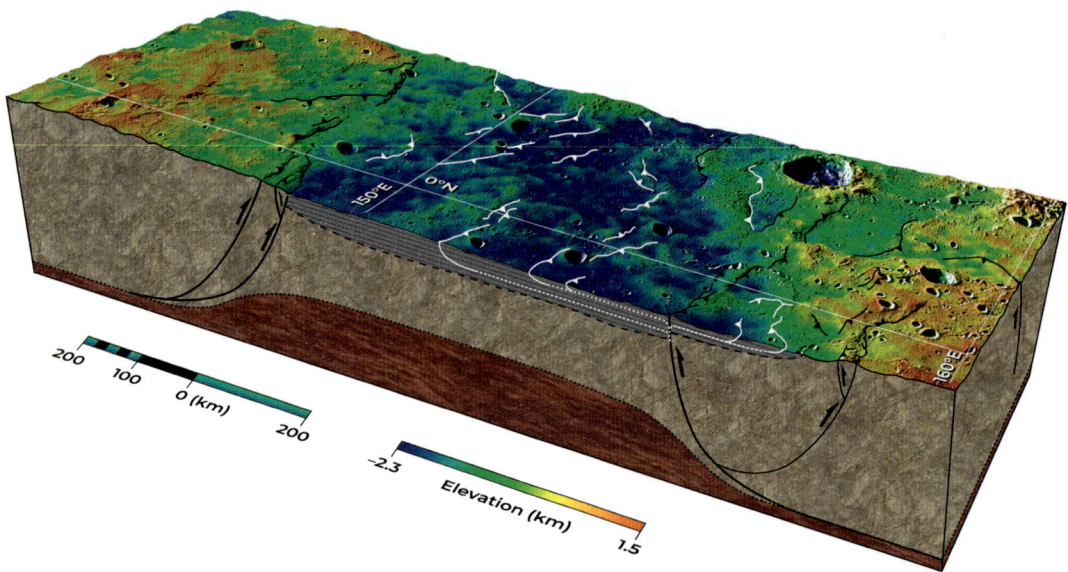

Figure 10.7. A schematic cross section showing the possible kinematic relations of thin- and thick-skinned shortening structures on Mercury. Here, thick crustal blocks are deformed and bounded by deep-seated "germanotype" structures (black lines), whereas low-lying smooth plains (dominantly consisting of ponded lavas) are shortened by thin-skinned "alpinotype" faults (white lines). These shallow structures may root into the deeper, larger faults, as observed widely on Earth. The black arrows indicate the direction of tectonic transport along the larger faults, which are shown here to bottom out at the crust–mantle boundary (dotted line). In cross-section, smooth plains material is shown in grey, older crustal material in light brown, and mantle material in dark brown. Features shown at depth are not to scale.

highland-bounding and cratered plains structures, the smaller "thin-skinned" structures may root into the deeper faults – a structural arrangement seen in the Alps (Trümpy, 1980), the European Hercynides (Bois et al., 1986), the Appalachian–Ouachita orogen (Hatcher, 1989), the Himalaya (McDougall et al., 1993), the U.S. Rocky Mountains (Stille, 1940), and the Andes (Roeder, 1988) (Figure 10.7). (Under this interpretation, the thin-skinned shortening landforms are cognates of Stille's (1920) "alpinotype" structures, connected mechanically to the larger, deep-seated "germanotype" thrust faults.)

10.3 EXTENSIONAL STRUCTURES

Although thrust-fault-related landforms are by far the most abundant type of tectonic structure on Mercury, extensional deformation was also recognized in images acquired by the Mariner 10 spacecraft, albeit spatially restricted to the Caloris basin. Within these plains, Strom et al. (1975) noted a pattern of deformation unlike any seen elsewhere in the solar system, with a complex system of ridges and "fractures" arrayed throughout the observed portion of the basin floor. Noting their morphological similarity to graben and tension fractures on the Moon, Strom et al. (1975) considered these flat-floored, graben-like features to be extensional structures, an interpretation adopted by subsequent workers (e.g., Dzurisin, 1978; Melosh and McKinnon, 1988). (The term "fracture" to denote extensional landforms alone has since been abandoned, correctly, since shortening structures also fall under this general terminological description (Schultz and Fossen, 2008) but display a reverse sense of slip.)

Further evidence for extensional deformation on Mercury was not forthcoming until the MESSENGER spacecraft returned images of the planet. Observations of the entirety of the Caloris basin revealed substantially greater structural complexity than had been seen by Mariner 10, including a vast network of radially oriented graben emanating from the basin center named Pantheon Fossae (Murchie et al., 2008; Watters et al., 2009c) (Section 10.6.4). These landforms were interpreted as graben, a structure formed where two parallel normal faults that are antithetic (dipping toward one another) bound a down-dropped block. This interpretation is consistent with their relatively straight walls, flat floors, and in some instances tapered ends, which correspond to slip displacements trending to zero at the fault tips. Moreover, there are numerous examples of normal fault segmentation and linkage, as evinced by ramps in overlapping stepover regions (relay structures) and by abrupt bends (jogs) in the fault traces (Klimczak et al., 2012, 2013a) (Figure 10.9a).

Perhaps more importantly, MESSENGER observations revealed evidence for extension at numerous other sites across the planet. For example, sets of basin-radial and -concentric graben have been documented in the volcanic plains interior to the Rembrandt basin (Watters et al., 2009b; Ferrari et al., 2015); smaller assemblages of graben have been noted in several mid-sized basins, including the Mozart, Rachmaninoff, and Raditladi basins (Prockter et al., 2009, 2010, 2011; Blair et al., 2013); and sets of graben multiple orientations have been described within numerous volcanically infilled craters (Freed et al., 2012; Klimczak et al., 2012; Watters et al., 2012) (Figure 10.9b). We describe these structures further in Section 10.6.4.

As a consequence of the progressive lowering of its periapsis altitude between orbit-correction maneuvers during its second extended mission (Chapter 1), the MESSENGER spacecraft was able to image the surface of Mercury at higher resolution than earlier in the mission. In consequence, MESSENGER

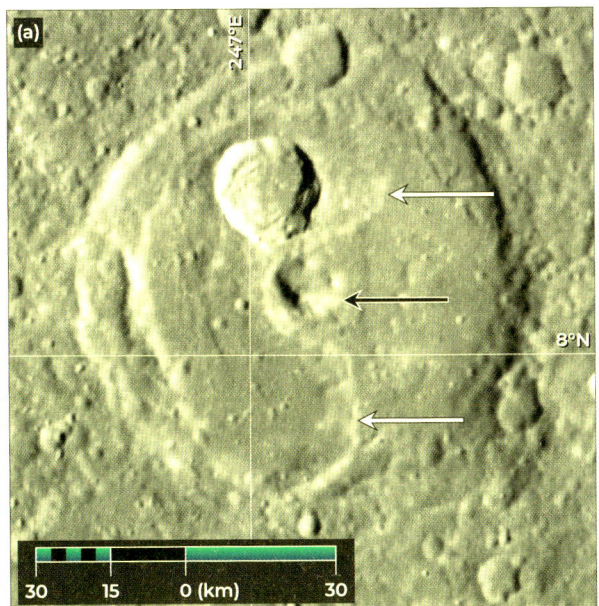

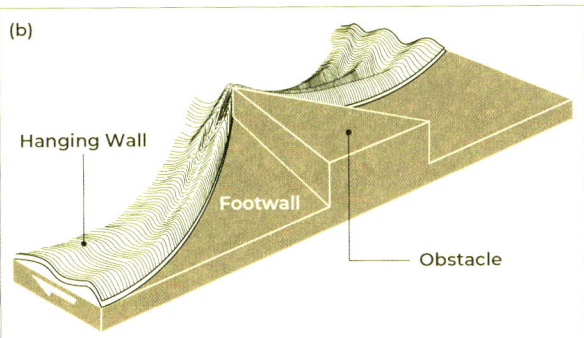

Figure 10.8. An example of syntaxis on Mercury. (a) A large shortening structure crosses an unnamed 80-km-diameter impact crater; where the fold meets the crater's central peak (shown by a black arrow) it is impeded, such that the entire shortening structure develops a pronounced bend in its leading edge (white arrows). Azimuthal equidistant projection, centered at 8.3°N, 247.3°E. (b) A schematic sketch of this phenomenon, showing the sharp bend in the hanging wall of a detachment fault upon encountering an obstacle.

imaged a set of narrow, trough-like landforms in the smooth plains units of relatively well-preserved craters, including Degas crater (Figure 10.9c). These troughs likely correspond to opening-mode fractures (also termed "joints") and small graben, formed in the same manner as discussed below for larger graben, in thermally contracting impact melt and volcanic deposits (Freed et al., 2012; Klimczak et al., 2012; Watters et al., 2012) (Section 10.6.4).

Notably, extensional strains have been seen in only one structural setting other than within smooth plains units: as narrow graben sets along the crests of large shortening landforms (Banks et al., 2015). Similar structures have been recorded in comparable settings on other worlds (e.g., Plescia and Golombek, 1986; Mueller and Golombek, 2004). These narrow extensional landforms are strain-compatible with, and provide insight into the kinematics of, thrust faulting on Mercury (Section 10.2.1 and Figure 10.18). Another situation where extensional deformation is associated with shortening structures is opposite large syntaxes of arcs (Wilson, 1954; Jacobs, 1959). No such deformation has yet been observed on Mercury, but whether this is because such extension does not exist there, or because it has yet to be resolved with image data, remains unclear.

10.4 LONG-WAVELENGTH TOPOGRAPHIC CHANGES

The record of brittle tectonic activity on Mercury was apparent from the earliest photographs returned to Earth by the Mariner 10 mission, and the study of these landforms has naturally tended to dominate investigations of the innermost planet's tectonic history. Interestingly, however, there is another class of landform on Mercury that has been recognized only with MESSENGER data, and then only with topographic measurements: long-wavelength "warps" in the planet's topography that have virtually no other surficial signature. We include a brief description and discussion of those landforms here, but the tectonic process through which they originated is by no means certain. Nonetheless, these warps are spatially collocated with tectonic landforms, and so their presence, geometry, and likely timing are worth briefly discussing here.

DEMs derived from stereophotogrammetric processing of MDIS flyby images showed long-wavelength variations in topography along the floor of the Caloris basin (Oberst et al., 2010). Orbital observations from the Mercury Laser Altimeter confirmed these topographic variations within the basin to be real and not artifacts of the DEM, and, further, revealed several similar long-wavelength undulations elsewhere across Mercury's northern hemisphere (Zuber et al., 2012), including exterior to the basin (Klimczak et al., 2013a) (Figure 10.3b). These variations are apparent as east–west-oriented, elongate topographic highs and lows, far greater in length than in amplitude, and can be described as nearly sinusoidal in cross section (Klimczak et al., 2013a) (Figure 10.10). Within the Caloris basin, the wavelengths of these topographic undulations range from 850 km to up to 1120 km, with amplitudes of up to 2.5–3 km (corresponding to values of horizontal shortening strain, ε, of only $\sim 10^{-5}$) (Klimczak et al., 2013a). Surprisingly, some areas of the basin floor have even been vertically displaced to now lie above the basin rim (Zuber et al., 2012; Figure 10.10). In addition, the flat portions of the floors of many impact craters, including craters that are volcanically infilled and host floors that presumably once followed a gravitational equipotential surface, are tilted in the same direction as the downslope trend of the long-wavelength topography (Balcerski et al., 2012; Zuber et al., 2012; see also the Atget crater in Figure 10.10).

Tilted crater floors and their relationships with long-wavelength topography can be observed at numerous other sites across Mercury's northern hemisphere (Balcerski et al., 2012), perhaps most prominently in association with a broad rise in the NSP some 1000 km across that stands ~1.5 km above the surrounding terrain (Zuber et al., 2012; Klimczak et al., 2012)

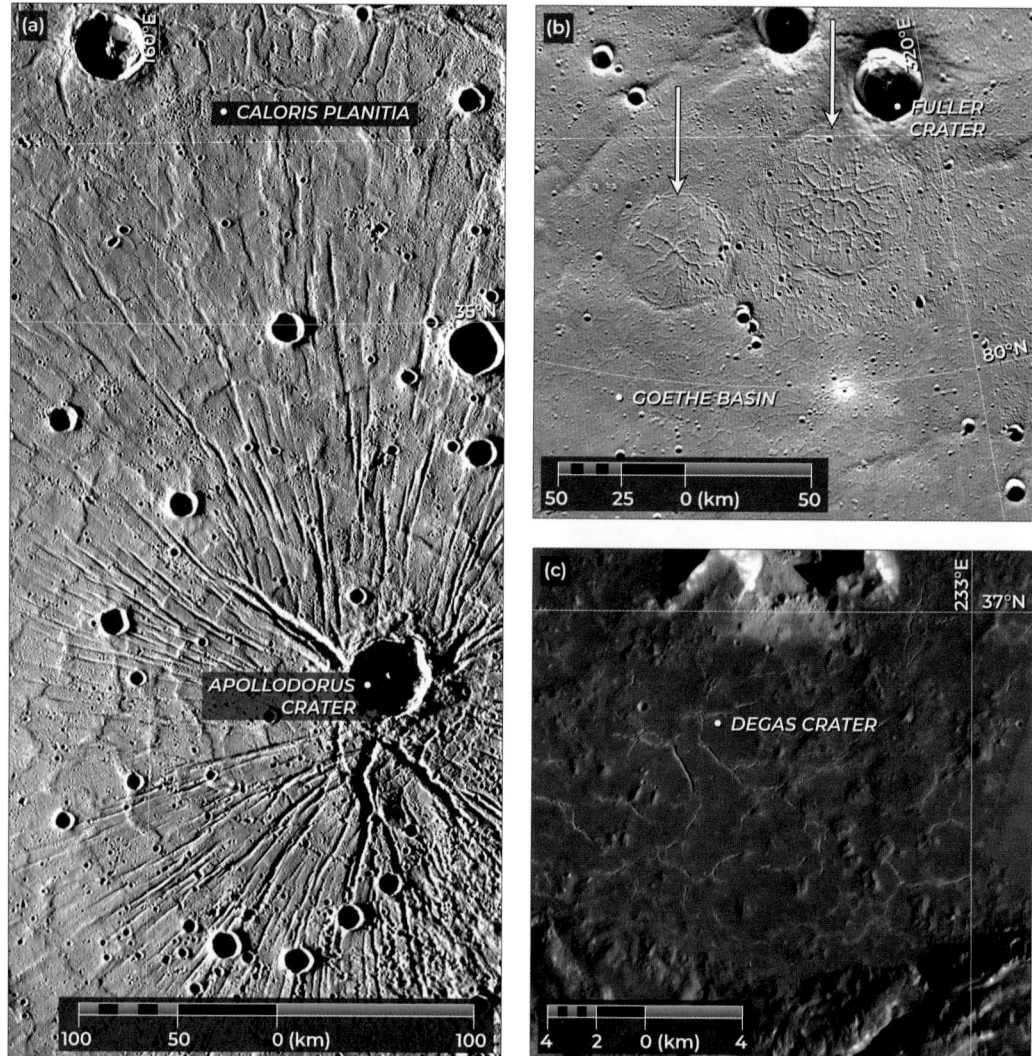

Figure 10.9. Examples of extensional landforms on Mercury. (a) A portion of Pantheon Fossae, the radial population of graben within the Caloris basin. (b) Graben with multiple orientations inside two ghost craters (white arrows), themselves situated within the 317-km-diameter Goethe basin. Note several additional graben immediately south of the larger of the two ghost craters. (c) Structures interpreted to be large joints and very small graben that formed within the impact melt deposit of the 55-km-diameter Degas crater. The crater's central peak and wall terraces are visible at the top and bottom of the image, respectively. Azimuthal equidistant projections, centered as follows: (a) 30.0°N, 161.0°E; (b) 81.0°N, 309.0°E; and (c) 60.3°N, 52.9°E. Images from the global morphology base map (Chabot et al., 2016).

(Figure 10.3b). On its flanks, the floors of both fresh and infilled craters tilt away from the center of that rise at angles similar to those of the flank slopes themselves (Klimczak et al., 2012). Undulations to the west of the Caloris basin have wavelengths and amplitudes of approximately 1300 km and 3 km, respectively (Klimczak et al., 2013a). Further, several of Mercury's small population of valles (impact-sculpted troughs shaped into flat-floored channels by lava: Section 11.2.1), situated at the periphery of the NSP, also show broad rises orthogonal to their long axes (Byrne et al., 2013b). These portions of elevated topography have wavelengths of several hundred kilometers and amplitudes of ~1 km; these changes in relief far exceed the local downhill gradient (Byrne et al., 2013b). Notably, the undulations that cross the valles strike approximately parallel to their larger counterparts near and within the Caloris basin, and all of these linear crests and troughs are approximately circumferential to the northern rise (Figure 10.3a).

10.5 MERCURY'S LITHOSPHERE

Collating observations of tectonic landforms with measurements of Mercury's topography and gravity field provides information on the planet's interior structure. The structure, properties, and evolution of the interior of Mercury are examined at length elsewhere in this volume (Chapters 3, 4, and 19), but here we discuss the insights afforded by tectonic landforms into the planet's brittle and ductile lithospheric regimes (Sections 10.5.1 and 10.5.2, respectively), followed by a discussion of the likely structure of the lithosphere (Section 10.5.3).

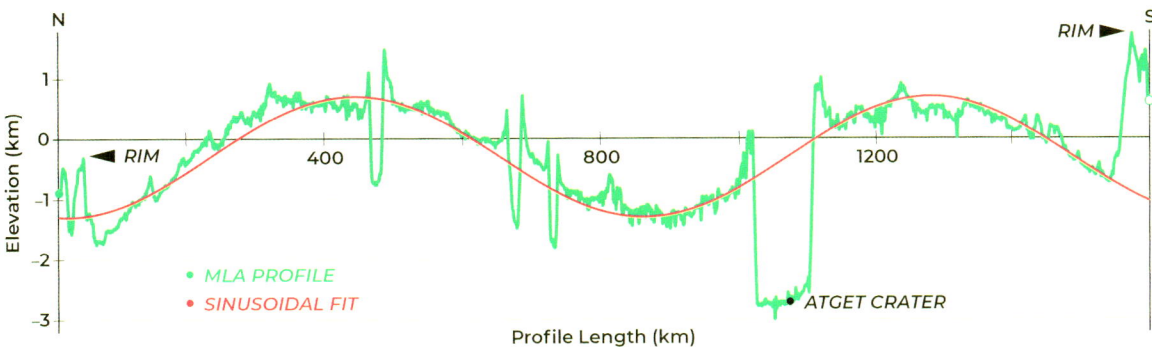

Figure 10.10. A Mercury Laser Altimeter profile (shown in teal) across the central portion of the Caloris basin. The basin floor shows sinusoidal topographic undulations approximated with a wavelength of 850 km and an amplitude of 2.5 km (shown in pink). The profile, which has a vertical exaggeration of ~100:1, shows both the tilted floor of the Atget crater and where the highest elevations of the basin floor exceed those of the northern basin rim.

For a general overview of planetary lithospheres, including that of Mercury, see Chapter 3.

10.5.1 Mercury's Brittle Regime

As for any solid-surface body in the solar system, Mercury's lithosphere makes up its strong, deformable outer shell, as defined by its mechanical properties (Chapter 3). The lithosphere consists of brittle and ductile regimes, under which deformation is accommodated by localized fracturing processes and by distributed plastic flow, respectively (e.g., Kohlstedt and Mackwell, 2010). The brittle regime occupies the relatively cold, upper part of the lithosphere. Deformation here is manifest as fractures, with two major fracture modes observed across the surface of the planet: mode-I or tensile fractures (joints: Section 10.3 and Figure 10.9a) and mode-II and -III or shear fractures (faults: Figures 10.1, 10.9b, and 10.9c). Tensile and shear fractures form only when tectonic stresses meet the respective strengths of the volume of rock that is subject to these stresses. Therefore, a characterization of the magnitude, direction, and orientation of stresses is crucial for understanding brittle rock behavior in Mercury's lithosphere.

Several geophysical processes have been invoked to produce regional-to-global fracture patterns under Mercury's brittle regime. Stress models for these processes have been derived for global contraction (e.g., Melosh and McKinnon, 1988), tidal despinning (e.g., Melosh, 1977), thermal expansion or contraction (e.g., Turcotte, 1983), polar wander (e.g., Melosh, 1980), and mantle convection (e.g., Solomatov and Moresi, 2000), and then compared with the observed pattern of tectonic landforms.

In addition, a variety of non-tectonic processes introduce discontinuities and tensile fractures into Mercury's lithosphere, such as the cooling of emplaced volcanic units and impact melts (e.g., Freed et al., 2012; Klimczak et al., 2012; Watters et al., 2012; Figure 10.9a), impact damage (Melosh, 1984; Ahrens and Rubin, 1993; Xia and Ahrens, 2001; Collins et al., 2004), and igneous intrusion. The depth and degree of fracturing of Mercury's lithosphere is poorly determined, but first-order estimates inferred from deep mines (Bieniawski, 1989) and boreholes (Emmermann and Lauterjung, 1997) on Earth, as well as from seismic (Dainty et al., 1974; Toksöz et al., 1974) and gravity data analyses for the Moon (Wieczorek et al., 2013), suggest that Mercury's lithosphere may be moderately to heavily fractured to depths of ~15 km (Klimczak, 2015). Brittle deformation within Earth's lithosphere can extend to depths of tens of kilometers and so, accounting for the effect of gravitational acceleration on the penetration depth of fractures (e.g., Klimczak, 2015; Heap et al., 2017), quakes on Mercury at depths considerably greater than 15 km are likely.

Most stress models considered to date for the processes listed above have been purely elastic, and so the stress magnitudes derived from these models depend on the elastic rock properties, e.g., Poisson's ratio (v) and Young's modulus (E) (or the closely related shear modulus, G). Poisson's ratio, a measure of the shortening in the direction of an applied load relative to the lengthening in the direction perpendicular to that load, is often taken to be 0.25 ± 0.05 in basalts (Schultz, 1995). Young's modulus describes the resistance of a material to elastic deformation (essentially its longitudinal "stiffness") and, for intact basaltic rock samples, laboratory measurements yield values of $E \approx 70\text{--}100$ GPa (Schultz, 1993).

The use of Young's modulus in elastic stress models, however, is appropriate only for rock samples intact at the hand-sample scale (Walsh, 1965; Kulhawy, 1975; Segall, 1984; Kachanov, 1992; Schultz, 1996). At larger scales, no rock volume is intact, and so stresses in Mercury's lithosphere are more appropriately modeled by incorporating some degree of lithospheric fracturing (e.g., Klimczak, 2015; Klimczak et al., 2015). The resistance to elastic deformation of a fractured volume of rock, termed the in situ modulus of deformation (E^*, Bieniawski, 1989), is relatively insensitive to rock type, with several empirical studies establishing some relationship to the degree of fracturing (see the summary by Hoek and Diederichs (2006) for more information). For low degrees of fracturing within Mercury's lithosphere, the deformation modulus $E^* \approx 40\text{--}70$ GPa and, for moderate degrees of fracturing, a deformation modulus of $E^* \approx 5\text{--}15$ GPa is appropriate (e.g., Klimczak et al., 2015).

Faults are, by far, the most prominent and largest fractures in Mercury's lithosphere (Section 10.2) (Figures 10.1, 10.3, and 10.9). The formation and growth of a given fault population is dependent on the brittle strength of the lithosphere, which is a

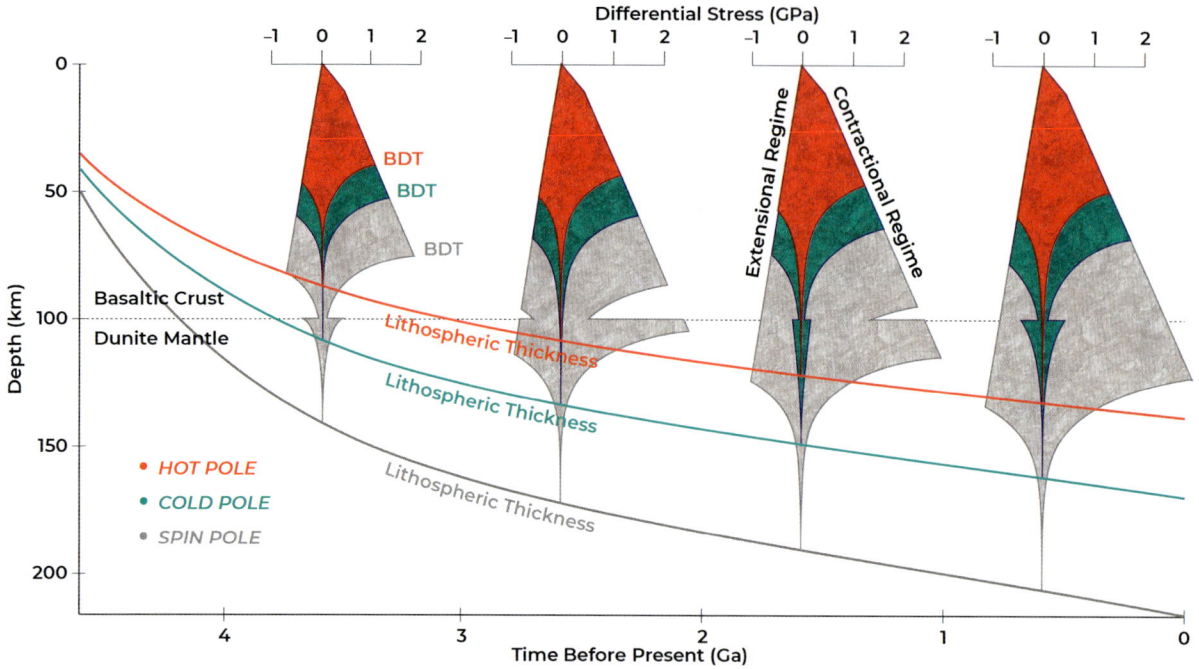

Figure 10.11. One possible scenario for the evolution of lithospheric strength and thickness for Mercury's different thermal environments, shown for the equatorial hot (red) and cold (teal) poles as well as the spin poles (gray) for both extensional and contractional tectonic regimes. The brittle regime is represented with Byerlee's law (after Klimczak, 2015); the lithospheric thickness and ductile regime are modified after the models of Williams et al. (2011). Lithospheric strength envelopes are shown in 1 Ga increments after Mercury's formation, for a horizontal strain rate of 10^{-19} s^{-1} and for ductile flow laws appropriate to a dry basaltic crust and a dry dunite mantle (with their interface shown here for an exemplar depth of 100 km). Note that brittle strength is independent of rock type, but the ductile portions of the strength envelopes vary with the flow law of the mineral that limits the rate of deformation. The brittle–ductile transition (BDT) is defined as the depth at which the brittle and ductile lithospheric strength envelopes intersect.

measure of the amount of stress that can be withstood by the lithosphere without permanent deformation; this value is dominated by the frictional properties of the rock volume. Under the assumption that Mercury's lithosphere contains fractures of all orientations, stresses within the lithosphere must overcome the frictional resistance to sliding for rock failure to occur along a fracture plane optimally oriented for such sliding (e.g., Byerlee, 1978). Friction between fracture planes has been shown to be independent of rock type, fracture surface roughness, and slip rates (e.g., Byerlee, 1968, 1978; Barton, 1976). The coefficient of friction, μ, for Mercury's lithosphere is likely therefore to be consistent with the narrow range of values of $0.6 < \mu < 1.0$ observed for faults on Earth.

Importantly, we must have an understanding of how a rock will respond to an applied stress, for which we use failure criteria: if the stress applied to a given rock volume meets such a criterion, we assume that failure occurs. Failure criteria previously used to characterize the brittle strength of Mercury's lithosphere include Byerlee's law (e.g., Zuber et al., 2010; Williams et al., 2011; Klimczak, 2015), the Coulomb criterion (Klimczak, 2015), and the Hoek–Brown criterion (Klimczak, 2015; Klimczak et al., 2015). These criteria allow for the assessment of the resistance of fractures to frictional sliding for an increasingly complex set of boundary conditions, such as the incorporation of rock cohesion (the Coulomb and Hoek–Brown criteria) and accounting for various degrees of pre-existing fracturing in the lithosphere (the Hoek–Brown criterion).

In their simplest form, lithospheric strength envelopes (the change of strength with depth) calculated with Byerlee's law in the brittle regime describe a linear dependence on depth of the maximum stresses permissible in the lithosphere for both intermediate- and high-pressure conditions (Brace and Kohlstedt, 1980; Kohlstedt and Mackwell, 2010) (Figure 10.11). Differential stresses demarcate the limits of stable conditions and so indicate when faulting – frictional sliding on pre-existing zones of weakness – will be triggered (Figure 10.11). These stresses are defined as the difference between the lithospheric strength and the reference state of stress, which is usually taken as the overburden pressure, or "lithospheric state of stress": the weight of the rock column at a given depth below the surface. Note that, in Figure 10.11, conditions for frictional sliding are given for entirely compressive stress states; tensile stresses instead trigger opening-mode deformation and thus the formation of joints [for details see Klimczak et al. (2015)].

A horizontal, tectonically governed stress component in a lithosphere, σ_H, that is greater than the stress component acting vertically, σ_V (where $\sigma_H > \sigma_V > 0$, with compressive stress taken to be positive), yields positive stress differences. These stresses, in turn, define a contractional tectonic regime under which thrust faults form (e.g., Anderson, 1951; Engelder, 1992) if the stress differences in the lithosphere reach their maximum permissible values (Figure 10.11). The planet-wide presence of thrust-fault-related landforms on Mercury (Figure 10.3a) indicates that stresses reached the compressive strength of the lithosphere, and thus that such a stress state was globally dominant. This conclusion supports the early inferences from Mariner 10 data that Mercury experienced lithospheric shortening on a global scale (e.g., Strom et al., 1975).

Negative stress differences (i.e., $\sigma_H < \sigma_V$, where $\sigma_V > \sigma_H > 0$) define an extensional tectonic regime (Figure 10.11) that allows for the formation of normal faults and graben (e.g., Anderson, 1951; Engelder, 1992) and facilitates volcanism by providing favorable conditions for the vertical ascent of magma and the formation of rift zones and dikes (Section 11.7.2). The asymmetry between the two sides of the brittle portion of a strength envelope (Figure 10.11) illustrates that a lithosphere under extension is much weaker than one under a contractional regime, and so it is easier to overcome the resistance to frictional sliding to form normal faults. However, joints, normal faults, and graben on Mercury are spatially restricted to relatively late effusive volcanic deposits (Figure 10.3a; Section 10.3), indicating that any extensional stress states within Mercury's lithosphere for which evidence remains today have been present only regionally.

10.5.2 Mercury's Ductile Regime

At depths where temperatures are sufficiently high to activate plastic microdeformation mechanisms, such as dislocation glide or diffusion creep (e.g., Passchier and Trouw, 2006), rocks behave in a ductile manner via plastic flow (e.g., Kohlstedt and Mackwell, 2010). Ductile strength is strongly dependent on temperature, viscosity (and thus mineralogy), and strain rate (e.g., Kohlstedt et al., 1995); the respective strength envelope is typically assumed to follow a decrease with depth governed by a non-linear (e.g., power-law) relation between stress and strain rate (Figure 10.11). Ductile-strength models are sensitive to geothermal gradient and most often are constructed for a constant strain rate, a uniform lateral distribution of deformation, and the premise that the strength of the entire lithosphere under the ductile regime is governed by a single mineralogy (usually that which displays the weakest rheology). Mercury's ductile strength has typically been modeled with the rheology of a dry (anhydrous) basalt (Zuber et al., 2010; Williams et al., 2011; Egea-González et al., 2012) or of dry anorthosite in the crust (Nimmo and Watters, 2004; Williams et al., 2011) and dry dunite in the mantle (Nimmo and Watters, 2004; Zuber et al., 2010; Williams et al., 2011). The depth to the crust–mantle interface within Mercury has variously been taken to be less than 140 km (Nimmo and Watters, 2004), or about 50 km (Zuber et al., 2010), or around 100 km (Zuber et al., 2010; Williams et al., 2011). Analysis of MESSENGER topography and gravity measurements place the crust–mantle boundary in Mercury's northern hemisphere at ~35 km depth (Padovan et al., 2015).

Different assumptions for flow laws and crust–mantle boundary depths can lead either to a small strength contrast between the crust and mantle or to a lower crustal weak zone that serves to mechanically decouple the crust and mantle (e.g., Williams et al., 2011). For a scenario under which the crust and mantle are coupled and have a modest difference in strength, tectonic deformation at the surface will likely also be coupled to strain throughout the entire lithosphere and so may directly reflect mantle processes (e.g., Kohlstedt and Mackwell, 2010). In turn, brittle deformation initiated at the surface might root into broader shear zones when reaching the ductile regime (e.g., Cowie et al., 2013). It may be that strain localization by plastic microdeformation mechanisms in Mercury's ductile regime leads to dynamic recrystallization (whereby new mineral grains grow in response to stress rather than chemical potential) and thus a heterogeneous lateral distribution of deformation in Mercury's lower crust and upper mantle. Excavation of this portion of Mercury's lithosphere (e.g., by impact) could, under this scenario, reveal a range of structural rock types (e.g., Şengör and Sakınç, 2001) characterized by their deformation history (rather than lithology), including mylonites, cataclasites, and pseudotachylites.

10.5.3 Lithospheric Structure

With increasing depth and temperature, rock behavior transitions from brittle to semi-brittle to ductile. This transition is commonly termed the brittle–ductile transition (BDT). The depth of the BDT is generally taken to be the depth at which the brittle and ductile strengths of the lithosphere are equal and thus by the locus of where the brittle and ductile strength envelopes intersect (Figure 10.11). In such strength models, the BDT is manifest as an abrupt change between the two regimes. In reality, though, the transition from fully brittle to fully ductile behavior probably occurs more gradually with depth across a semi-brittle or plastic zone in which microdeformation mechanisms that allow for some relaxation of stress become activated at points of stress concentrations (e.g., Kohlstedt et al., 1995). Moreover, given that the depth of the BDT on Earth varies spatially as a function of rock type, geothermal gradient, and other factors (e.g., Gettings, 1988), similar spatial variations are likely to occur in Mercury's lithosphere.

The topography of a fault-related landform can be used to infer the geometry of the underlying fault itself (Klimczak, 2014; Byrne et al., 2015), which in turn can place estimates on the depth of the seismogenic layer, the surface heat flow, and the depth of the BDT at the time of faulting. Analysis of topographic profiles across several thrust-fault-related landforms, including Discovery Rupes (Watters et al., 2002; Nimmo and Watters, 2004), a group of three unnamed monoclines in the equatorial region (Ritzer et al., 2010), and a set of four structures including Santa Maria Rupes (Egea-González et al., 2012), provided preliminary estimates of such parameters for Mercury. In each case, the maximum depths to which these faults were found to have propagated were between 30 and 40 km (Watters et al., 2002; Ritzer et al., 2010; Egea-González et al., 2012), giving a candidate depth to the BDT in that range (e.g., Nimmo and Watters, 2004; Williams et al., 2011). It is important to note, however, that faults may root into zones of mechanical weakness in the lithosphere that do not correspond to the BDT itself, and also that the BDT has likely grown deeper since those faults developed (Figure 10.11) – and so this depth range is but a minimum for present-day conditions.

Models for the evolution of Mercury's lithosphere that include the effect of latitudinal and longitudinal variations in surface temperature (Vasavada et al., 1999) on lithospheric strength and thickness over time (Figure 10.11) indicate that substantial differences in lithospheric strength arise when lateral variations in Mercury's thermal structure are considered (Williams et al., 2011). After the end of the LHB, when most thrust-fault-related landforms observed today are believed to have formed (Banks et al., 2015) (Section 10.8), differences of up to 15 km in lithospheric thickness developed in Mercury's equatorial region between hot- and cold-pole latitudes, and even greater lithospheric thicknesses developed at the spin poles (Figure 10.11). Williams and co-workers (2011) suggested that the differences in lithospheric strength and thickness may be expressed in the localization, orientation, and depth of deformation of Mercury's tectonic structures.

10.6 TECTONIC PROCESSES

The tectonics of Mercury have been dominated by the process of global contraction from secular interior cooling, but other tectonic mechanisms have shaped the planet as well. Although extensional structures are rare, where they occur they are abundant, and their origin must be explained against a backdrop of planetary contraction. Further, not all of the shortening structures on Mercury necessarily trace their formation to global contraction. And other, planetary-scale processes may have operated throughout Mercury's history, perhaps not in a dominating role like global contraction but to an extent sufficient to shape at least in part the distribution and character of tectonic deformation we see today.

In Section 10.6.1, we first discuss the causes, effects, and means to estimate the amount of global contraction. We then review the process of tidal despinning (Section 10.6.2), before appraising the effects of both processes acting together (Section 10.6.3). Finally, in Section 10.6.4, we discuss in detail the processes that drove extensional deformation within myriad volcanically flooded impact craters and basins on Mercury.

10.6.1 Global Contraction

As we noted in Section 10.1.1, Strom et al. (1975) invoked global contraction [possibly arising from phase changes in the core (Murray et al., 1974)] to account for the widespread distribution of shortening structures they observed. Secular cooling of a planetary body will lead to a reduction in volume of the interior relative to the lithosphere, causing an increase in horizontal compressive stresses and a consequent failure mode favoring large-scale thrust faulting of that lithosphere (Élie de Beaumont, 1829, 1852; Solomon, 1978). In contrast to other global-scale processes (Sections 10.6.2 and 10.7), global contraction is not predicted to produce a distinctive pattern of faulting, per se. Instead, absent the superposition of other sources of stress or a pre-existing structural fabric, the resultant compressive stresses are horizontally isotropic, and so the ensuing shortening structures should be distributed evenly across the surface without preferred orientations (e.g., Solomon, 1977) – much like on the surface of a shrinking apple, an analogy first proposed for Earth by Ægidius Romanus in the thirteenth century (Dana, 1863).

Both Mariner 10- and MESSENGER-based observations indicated that Mercury's shortening tectonics are not homogeneously distributed across its surface (e.g., Strom et al., 1975; Melosh and McKinnon, 1988; Byrne et al., 2014) (Figure 10.3a). The degree to which the distribution of crustal shortening deviates from that predicted solely from global contraction simply as a result of lighting bias (Section 10.2.2) rather than because of other contributory processes (Section 10.7) has yet to be determined. Even so, global contraction remains the most plausible mechanism with which to account for the planet-wide population of shortening structures mapped on Mercury (e.g., Byrne et al., 2014), and those structures can in turn provide bounds on the amount by which the planet has decreased in volume.

For example, from their assessment of Mercury's population of shortening structures visible with Mariner 10 data, Strom et al. (1975) estimated that the planet has experienced a reduction in radius of 1–2 km since a time near the end of the LHB (Section 10.1.1). Later studies of Mercury's tectonics from Mariner 10 and early MESSENGER observations offered similar estimates (Watters et al., 1998, 2009a; Watters and Nimmo, 2010), bracketed by values from as low as ~0.8 km (Watters et al., 2015) to as much as 3.6 km (Di Achille et al., 2012). Of course, all of these values were derived from the observed thrust-fault-related landforms alone and did not account for lithospheric strength, nor that some portion of the compressive stresses imposed on a lithosphere by global contraction can be accommodated before the frictional resistance to sliding on lithospheric faults is overcome (Section 10.5.1). Importantly, this additional component of radius change amounts to ~0.4–2.1 km, is preserved in the lithosphere even after thrust faulting is initiated (Klimczak, 2015), and so must be added to any radius change estimates made from mapped shortening structures alone (Section 10.8.1).

In any case, such mapping-based estimates were consistently below those predicted from thermal history models of Mercury, which generally returned radius change values of ~5–10 km (Solomon, 1977; Schubert et al., 1988; Hauck et al., 2004; Dombard and Hauck, 2008). Only when considering a restricted set of model parameters (including "extraordinary compositions"), or invoking hidden shortening strain on Mercury, could the constraint of 1–2 km of radius reduction since the LHB be even approximately satisfied (e.g., Dombard and Hauck, 2008).

With their global survey of mapped structures, Byrne et al. (2014) produced revised estimates for the accumulated decrease in Mercury's radius using fault displacement and length data. Several earlier studies employed this technique (e.g., Watters et al., 1998, 2009a; Watters and Nimmo, 2010; Di Achille et al., 2012), whereby a linear scaling relation (set by a proportionality constant γ) between maximum fault displacement (D_{max}) and fault length (L) is established for a subset of a total fault population; this scaling ratio is then extrapolated to calculate the strain accommodated by the entire fault population from their mapped lengths alone (Scholz and Cowie, 1990; Cowie et al., 1993). Byrne et al. (2014) determined, for 216 monoclines

Table 10.2. *The decrease in radius of Mercury since the LHB, inferred from mapped shortening structures and calculated with the displacement–length scaling method (after Byrne et al., 2014).*

Fault plane dip angle	25°	30°	35°
Derived value of γ	9.6×10^{-3}	8.1×10^{-3}	7.1×10^{-3}
Radius change (km) from all (5934) structures			
Surface area of Mercury (7.48×10^{13} m^2)	6.9	5.6	4.6
Surface area excluding the Caloris and Rembrandt interior plains (7.27×10^{13} m^2)	7.1	5.7	4.7
Radius change (km) from all (2183) structures in areas outside of smooth plains			
Surface area of Mercury (7.48×10^{13} m^2)	4.6	3.7	3.1
Surface area excluding smooth plains and the Caloris and Rembrandt interior plains (5.46×10^{13} m^2)	6.3	5.1	4.2

and anticlines across Mercury for which the lengths had already been measured, the maximum relief of those landforms. This value is assumed to correspond, in the absence of subaerial erosion, to the vertical component (i.e., the throw, t) of the underlying fault (Figure 10.2). These workers then calculated D_{max}/L ratios (i.e., γ) by finding the best-fit linear regression scaling statistic γ (Clark and Cox, 1996) for values of fault dip angle, θ, of 25°, 30°, and 35°, values used in previous such analyses (e.g., Watters and Nimmo, 2010; Di Achille et al., 2012). By computing the surface lengths of all mapped thrust faults (i.e., Figure 10.3a), Byrne et al. (2014) found their representative displacements with the appropriate scaling relation for an assumed dip angle. Cumulative shortening strains and corresponding changes to Mercury's radius were then obtained for the entire surface area of the planet, for that portion of the surface that excludes the Caloris and Rembrandt interior plains, and for that portion that excludes all smooth plains units (Byrne et al., 2014); these results are given in Table 10.2.

The areas of the Caloris and Rembrandt interior plains were omitted as part of this analysis because, although the plains host substantial structural complexity (Section 10.6.4), the extent to which those structures can be attributed to global contraction rather than basin-related processes is not well understood (Byrne et al., 2014). The areal shortening strain accommodated by all mapped structures outside those plains corresponds to a change in planetary radius of 7.1 km, 5.7 km, and 4.7 km for $\theta =$ 25°, 30°, and 35°, respectively (Byrne et al., 2014) (Table 10.2). Although smooth plains structures have been attributed to global contraction in some studies (e.g., Freed et al., 2012), they have also at least partially been ascribed to load-induced flexure and subsidence (e.g., Melosh and McKinnon, 1988; Watters et al., 2005) (Section 10.6.4), and so have been left out of other estimates of global contraction (e.g., Watters et al., 2013). Byrne et al. (2014) therefore also calculated values of radial contraction from areal shortening strains excluding any contributions from smooth plains units. Such an analysis requires that no smooth plains units on Mercury experienced the effects of global contraction, which is geologically implausible since there are no large structures at the peripheries of these units into which shortening strains have obviously been partitioned. Moreover, there is evidence for the formation of a relatively young population of monoclines on Mercury (Section 10.8.1), and thrust fault reactivation has been observed in the Caloris basin (Section 10.8.2); both types of observation are consistent with sustained global contraction into the geologically recent. Nonetheless, when the corresponding surface area of smooth plains [~27% of the total planetary surface: Denevi et al. (2013)] was disregarded, together with that of the Caloris and Rembrandt interior plains, the change in planetary radius since the LHB for $\theta =$ 25°, 30°, and 35° is 6.3 km, 5.1 km, and 4.2 km, respectively (Byrne et al., 2014) (Table 10.2).

Even accounting for various combinations of smooth and cratered plains structures, these values are all substantially greater than those from earlier photogeological studies. Previous reports of 0.8–2 km of radial shortening (Strom et al., 1975; Watters et al., 1998; Watters and Nimmo, 2010) were derived from analyses of the 45% of Mercury's surface imaged by Mariner 10. Yet the arrival of MESSENGER at Mercury did not herald the end of underreported radial shortening: a displacement–length scaling analysis with MDIS data considered only a subset of the monoclines and symmetric anticlines on Mercury and returned a value of ~1.0 km for $\theta =$ 30° (Watters et al., 2015). Even the highest previously reported value for radius change (3 km for $\theta =$ 30°) was based on an analysis of but one-fifth of Mercury's surface imaged by MESSENGER that was then extrapolated to the entire planet (Di Achille et al., 2012), and that study adopted a D_{max}/L ratio derived by Watters and Nimmo (2010) from only eight monoclines imaged by Mariner 10.

In comparison, the D_{max}–L analysis of Byrne et al. (2014), obtained with a scaling relation derived from 27 times as many structures as previous analyses and applied to the entire population of shortening structures on Mercury, gives a value for radial contraction of 5.7 km (5.1 km if smooth plains structures are excluded) for $\theta =$ 30° (Table 10.2). Even these estimates of radius change accommodated by shortening landforms are likely to underestimate the actual extent of Mercury's global contraction since the LHB, in part because of limitations on illumination geometry in many areas of the base map used in that study (Byrne et al., 2014). Further, the history of Mercury's global contraction as recorded in its shortening structures is incomplete, since the planet experienced an initial reduction in radius without concomitant brittle deformation (see the discussion above, as well as Section 10.8.1).

As an independent check on this analysis, Byrne et al. (2014) also summed, along eight great circles, individual estimates of horizontal shortening across the 216 structures described above. Again, under the assumption that the relief of a given landform corresponds to the vertical component or throw, t, of the offset of the underlying fault, the horizontal component (the heave, h) of the fault offset along the great circle was calculated from the relation $h = t \tan^{-1} \theta$ (Figure 10.2). (For structures for which the

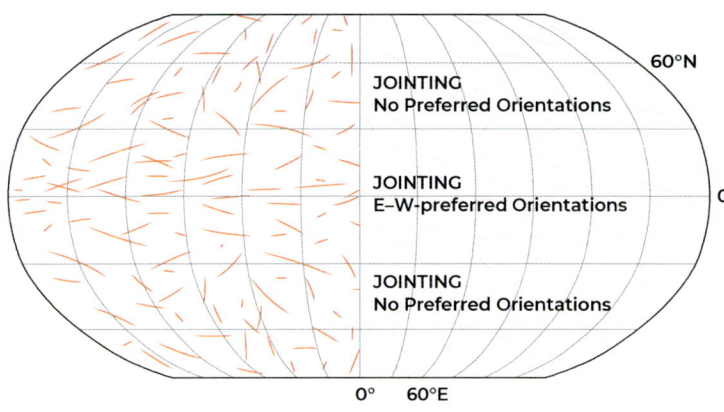

Figure 10.12. The global pattern of joints predicted to result from tidal despinning alone (after Klimczak et al., 2015). In the equatorial regions, east–west-oriented jointing is expected to form, in contrast to randomly oriented joints near the poles. The map is in a Robinson projection centered at 0°E; the graticule is in 30° increments of latitude and longitude.

strike direction was not orthogonal to a great circle, apparent dip, θ_a, was determined from θ with the relation $\theta_a = \tan^{-1}(\tan\theta \sin\psi)$, where ψ is the angle between the structure's strike direction and the great circle.) The change in planetary radius, ΔR, accommodated along each great circle was then calculated by $\Delta R = h_{\text{CUMULATIVE}} (2\pi)^{-1}$. For assumed dip angles once more of 25°, 30°, and 35° for the thrust faults underlying each shortening landform, this method returned a change in planetary radius averaged over the eight great circles of 5.5 km, 4.4 km, and 3.7 km, respectively (Byrne et al., 2014).

These results provide an important constraint for understanding the thermal history of Mercury's interior (Chapter 19). MESSENGER orbital measurements of the relative surface abundances of K, U, and Th (the major heat-producing elements in planetary interiors) indicate larger fractions of K and U and their comparatively shorter-lived isotopes (Peplowski et al., 2011; Chapter 2) than had been used in earlier thermal evolution studies (e.g., Hauck et al., 2004). It is therefore likely that Mercury experienced a greater change in heat production since the LHB (and so has cooled more) than earlier models had suggested (Chapter 19), further exacerbating the mismatch between theoretical findings and previous photogeological observations. Moreover, the finding of changes in Mercury's radius of up to a factor of 7 greater than previous results resolved a nearly four-decades-old paradox – which arose in the first place because of incomplete mapping – by making consistent the history of heat production and loss and the accumulated global contraction of Mercury (Solomon, 1977; Schubert et al., 1988; Dombard et al., 2008; Byrne et al., 2014).

10.6.2 Tidal Despinning

Mercury is locked in a 3:2 spin–orbit resonance, in which it rotates in a prograde manner precisely three times about its spin axis for every two orbits about the Sun. Mercury's current rotational period is approximately 59 Earth days, but its initial rotational period may have been as short as 20 Earth hours, as its spin angular momentum would then have compared well with those of most other planets (Kaula, 1968; Burns, 1975). Importantly, a decrease in spin rate by tidal torques would have been accompanied by the relaxation of an equatorial bulge (Burns, 1976; Melosh, 1977; Melosh and McKinnon, 1988). The stresses resulting from the relaxation of such a tidal bulge were predicted to have been sufficiently large to have pervasively fractured the planet's lithosphere, producing a distinctive "despinning" pattern (Melosh, 1977; Melosh and Dzurisin, 1978; Melosh and McKinnon, 1988). This pattern was interpreted to include an equatorial province of north–south-oriented thrust faults, a zone of strike-slip faults in the mid latitudes, and a region of east–west-oriented normal faulting in the polar regions. More recent modeling that incorporated variations in lithospheric thickness resolved a despinning stress distribution that was interpreted to produce an equatorial set of strike-slip faults and a population of east–west-oriented normal faults at the poles (Beuthe, 2010).

Stress magnitudes and orientations arising from tidal spindown alone were later assessed with failure criteria (Section 10.5.1) and were found to result in a fracture pattern composed only of a set of surficial joints (Klimczak et al., 2015) (Figure 10.12). This work showed that, regardless of the degree of preexisting fracturing and corresponding strength of the lithosphere, the presence of tensile stresses at all latitudes predicts jointing across Mercury's surface. Moreover, for low to moderate degrees of lithospheric fracturing, the stress magnitudes were found to be insufficient to promote frictional sliding at depth, and so despinning alone would have produced no global faulting pattern at all (Klimczak et al., 2015).

10.6.3 Global Contraction with Tidal Despinning

Although Mariner 10 observations of the distribution and orientation of thrust-fault-related landforms on Mercury suggested that tidal despinning had played at least some role in their formation, early studies determined that despinning alone was not sufficient to explain all such observations (Dzurisin, 1978; Melosh and Dzurisin, 1978). MESSENGER observations of a global set of thrust faults but a general lack of normal and strike-slip faults (Byrne et al., 2014; Figure 10.3a), in contrast, indicate that a tectonic pattern from tidal despinning was either not preserved, was overprinted, or was strongly affected by global contraction, depending on whether the despinning process predated or overlapped (at least in part) with global contraction (Klimczak et al., 2015).

Indeed, global stress distributions were modeled for a scenario under which despinning and contraction acted in concert (Pechmann and Melosh, 1979; Dombard and Hauck, 2008;

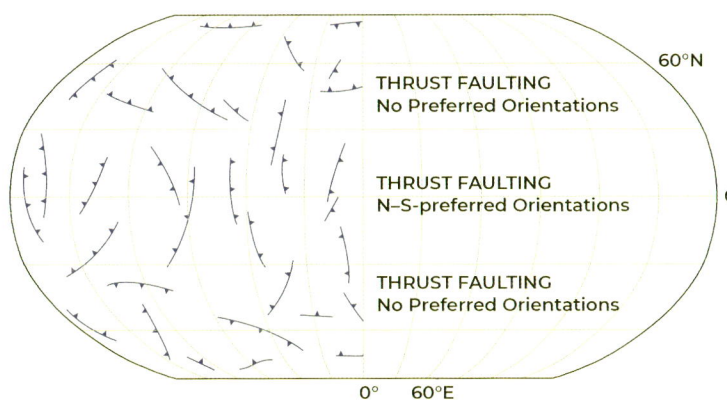

Figure 10.13. The global pattern of thrust faults on Mercury predicted to result from the overlapping of tidal despinning and global contraction. North–south-oriented thrust faults are predicted to develop at low latitudes, with randomly oriented thrust faults forming at the poles. The map is in a Robinson projection centered at 0°E; the graticule is in 30° increments in latitude and longitude. After Klimczak et al. (2015).

Beuthe, 2010). In all cases, fault patterns resulting from combinations of stresses from global contraction and tidal despinning were predicted to produce a global population of thrust faults with preferred north–south orientations at low to mid latitudes, although differences in the interpretation of the stress pattern led to variations in the predicted faulting style and/or orientation at higher latitudes (e.g., Pechmann and Melosh, 1979; Beuthe, 2010; Klimczak et al., 2015).

The tectonic maps of Mercury derived from MESSENGER orbital data (e.g., Byrne et al., 2014; Figure 10.3a) enable model predictions for various despinning and radius change scenarios to be comprehensively compared with tectonic observations from across the entire planet (Klimczak et al., 2015). For example, both Byrne et al. (2014) and Watters et al. (2015) reported a preponderance of north–south thrust fault orientations at equatorial and mid latitudes on the planet. At higher latitudes, however, a range of thrust fault orientations was found by Byrne et al. (2014), whereas Watters et al. (2015) reported a predominantly east–west-oriented set of thrust faults. This difference in interpretation of the tectonic fabric on Mercury likely results again from the use of different mapping criteria: Watters et al. (2015) chose to record only those thrust-fault-related landforms with several hundred meters of relief and more than 50 km in length, whereas Byrne et al. (2014) mapped all classes of shortening landforms on all terrains at individual feature lengths as short as ~10 km.

Nonetheless, both sets of observations agree with the predicted roughly north–south orientation of thrust faults at low to mid latitudes predicted by stress models for a combination of tidal despinning and global contraction. Of note, Watters et al. (2015) suggested that the east–west thrust fault pattern they recognized at higher latitudes is consistent with the interpreted fault pattern of an equal contribution of stresses from despinning and contraction by Beuthe (2010). In contrast, the more complete assessment of stresses with failure criteria (Klimczak et al., 2015) revealed that the small variation in horizontal stresses at and near the poles arising from despinning permits the formation of a wide variety of thrust fault orientations at the polar regions (Figure 10.13), in agreement with the mapping by Byrne et al. (2014).

Neither the onset nor the duration of despinning has been well characterized to date (Chapter 4), and it may be that despinning was a sufficiently early and short-lived process either to have had no substantive effect on Mercury's tectonic fabric or to have overlapped not with global contraction but with a potential earlier phase of global expansion instead (Solomon and Chaiken, 1976; Chapter 19). Under such a scenario, the pattern of jointing predicted to have formed from despinning alone (Figure 10.12) would have further developed into graben and ultimately rift zones with those same orientations. Conversely, if tidal despinning did influence the development of global contraction-induced faults on Mercury, then either the despinning process began later and/or operated for longer than has been assumed before (e.g., Peale, 1988) or an early tectonic pattern generated long before global contraction-induced thrust faulting began somehow survived intense later volcanic resurfacing and impact bombardment (Section 10.8.1).

Importantly, this discussion is based on studies of the effect on Mercury's tectonics of the deceleration by tidal forces of an initially faster rate of prograde rotation. However, a planetary body's rotation rate can be altered by variations in orbital eccentricity (Correia and Laskar, 2004) as well as by momentum imparted by large impacts (e.g., Melosh, 1975; Lissauer, 1985; Wieczorek and Le Feuvre, 2009), such that the rate can either decrease *or* increase. On the basis of Mercury's cratering record, Correia and Laskar (2012) and Wieczorek et al. (2012) found evidence that the planet may once have been in synchronous rotation before transitioning to its current 3:2 spin–orbit resonance as a result of spin-up by a large impact event. Moreover, if its spin rate following such an impact was greater than that of the 3:2 resonance, tidal forces may have decelerated the planet, whereas a lower post-impact spin rate could have been tidally accelerated to the 3:2 resonance during times of high orbital eccentricities (Wieczorek et al., 2012). And the Caloris impact event, which likely occurred at about the same time as the onset of global contraction (Section 10.8.1), probably increased Mercury's rate of rotation (Wieczorek and Zuber, 2001) even to those of resonances higher than 4:1 (Wieczorek et al., 2012).

Conversely, Noyelles et al. (2014) argued that Mercury acquired its 3:2 spin–orbit resonance very early, over a timescale of as short as 20 Myr and possibly as a homogeneous body that had yet to undergo core–mantle segregation (i.e., differentiation). Noyelles et al. (2014) also disputed the possibility that Mercury was ever in a synchronous rotation rate. Later, Knibbe and van Westrenen (2016) concluded that Mercury may have had one or more higher-order resonances prior to acquiring its present spin rate, with the secular evolution of its spin–orbit

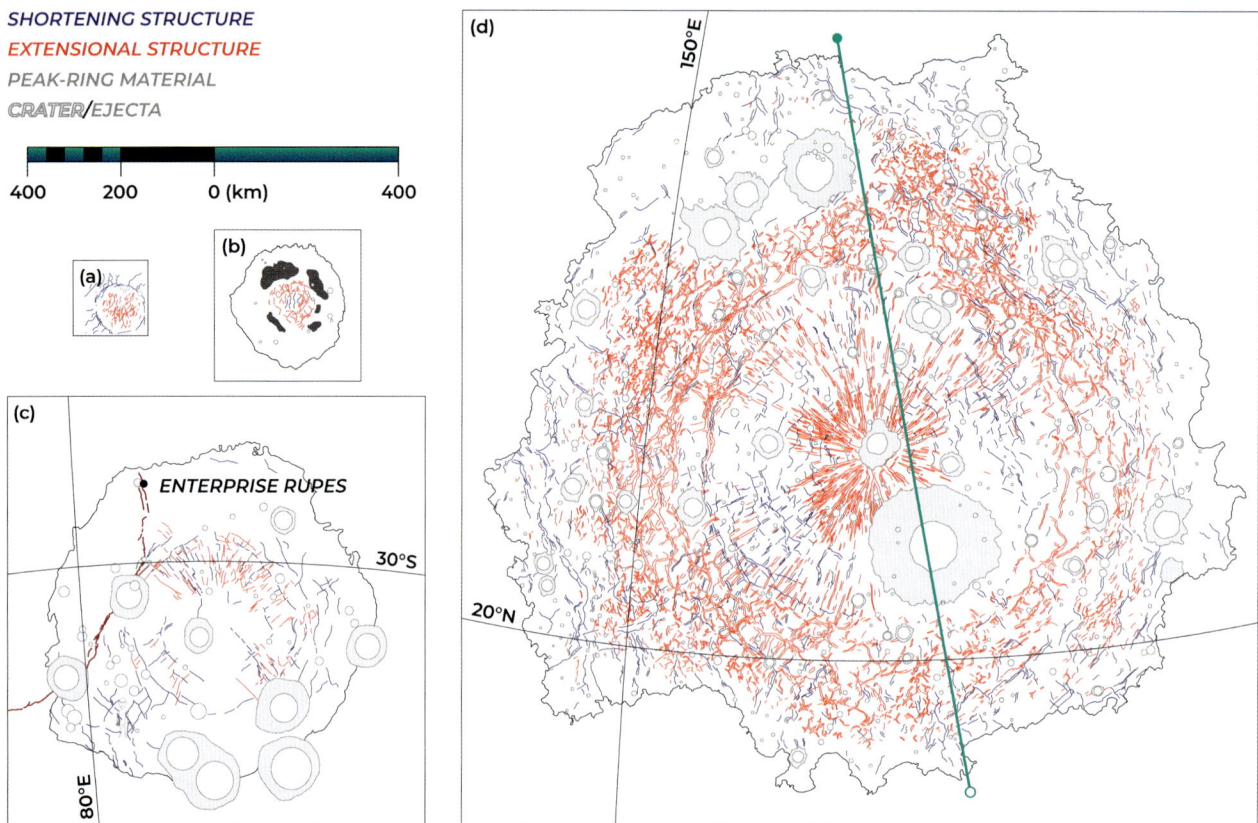

Figure 10.14. Extensional and shortening structures within flooded impact features. (a) An unnamed ghost crater in the NSP. (b) The Mozart basin. (c) The Rembrandt basin, the second-largest well-preserved impact structure on Mercury. (d) The Caloris basin, the largest well-preserved impact structure on the planet and host to one of the largest smooth plains deposits on Mercury. Extensional landforms are shown in red; shortening landforms are shown in blue. The teal line through the basin corresponds to the location of the topographic profile shown in Figure 10.9. Each impact structure is shown at the same scale. Azimuthal equidistant projections centered at: (a) 60.3°N, 36.7°E; (b) 7.8°N, 169.6°E; (c) 33.5°S, 88°E; and (d) 30°N, 161°E.

resonance a function of both its impact history and its chaotic orbital eccentricity (Correia and Laskar, 2009). Whatever the evolution of Mercury's spin rate, then, the possibility exists that the planet has been subject to accelerations in spin rate both positive and negative and more than once – and so models in which its shortening structures develop against a backdrop of simple tidal spindown have probably not explored sufficiently the complex interaction between tectonics and celestial mechanics. This interaction remains one of the outstanding questions in the temporal development of Mercury's tectonic structures (Section 10.10.1).

10.6.4 Basin Tectonics

As we have shown, the dominant form of tectonic deformation on Mercury involves shortening, and extensional tectonic features are almost entirely restricted to impact craters and basins that host volcanic smooth plains (Murchie et al., 2008; Watters et al., 2009a; Blair et al., 2013). However, extensional and shortening tectonic landforms within those impact features vary enormously in structural complexity – from sets of graben that describe polygonal patterns in craters tens of kilometers in diameter to collocated basin-radial, basin-circumferential, and basin-oblique graben and shortening structures within the Caloris basin, which could reflect a combination of local- and global-scale processes.

To illustrate this increase in complexity with increasing crater diameter, we show in Figure 10.14 the *tectonic* map patterns of four exemplar volcanically infilled impact features on Mercury: a 100-km-diameter ghost crater (Figure 10.14a), the 240-km-diameter Mozart basin (Figure 10.14b), the Rembrandt basin, some 720 km in diameter (Figure 10.14c), and the titanic Caloris basin (Figure 10.14d).

Ghost craters are impact features that have been partially to entirely covered by lavas (Chapter 11). Where such burial is incomplete, these craters can still be recognized by their rims. Where burial is total, however, ghost craters are nearly always identified by tectonic landforms localized by the buried rim. For the example shown in Figure 10.14a, situated in the NSP, the buried crater rim is delineated by a ring of shortening structures ~100 km across (individual segments of this ring are several to tens of kilometers in length and hundreds of meters to kilometers in width, respectively). The interior of the crater has been deformed by a number of graben with no strongly preferred orientations; these structures are several kilometers long and hundreds of meters in rim-to-rim width. Notably, the intersections of these graben render the interior crater fill as a set of polygonal blocks (Figure 10.9b). The graben extend from the crater center to ~80% of its radius. Shortening structures in the surrounding plains terminate at the crater's periphery and do not penetrate its interior. This structural arrangement is characteristic of graben-hosting ghost craters in general, which primarily occur in the NSP and circum-Caloris smooth plains (e.g., Klimczak et al., 2012). Some variations can occur: in several instances, graben superpose the shortening structures that ring ghost craters, and in others it is only graben that delineate buried craters, with shortening structures entirely absent.

At a larger scale, several of Mercury's mid-sized basins feature both shortening and extensional landforms within their volcanic infill. For example, the Mozart basin contains both graben and shortening structures, although both types of tectonic deformation occur only in the area bounded by its peak ring (Blair et al., 2013) (Figure 10.14b). Basin-circumferential graben form an annulus spanning a distance of ~20–40% of the basin's radius from and concentric to its center; individually these structures measure kilometers to tens of kilometers in length and several hundred meters to about a kilometer in rim-to-rim width (Blair et al., 2013). From this annulus to the peak ring, located radially at about 50% of the basin's radius, graben are found in a mix of basin-radial, -circumferential, and -oblique orientations; these structures are kilometers long and hundreds of meters wide. Thrust-fault-related landforms, kilometers wide and tens of kilometers long, occur within the annular graben and show no preferred orientations. Tectonic structures of any kind are fewer in number in the southwestern sector of basin infill within Mozart's peak ring. Notably, this general pattern of deformation is also observed in the smooth plains units situated within the peak rings of the similarly sized Raditladi and Rachmaninoff basins (Blair et al., 2013).

The largest well-preserved impact feature in Mercury's southern hemisphere, Rembrandt basin, is riven along its northwestern interior by the northeast–southwest-trending, 820-km-long fault array termed Enterprise Rupes (Figure 10.14c). Apart from this enormous structure, however, the basin floor is heavily deformed by extensional and shortening structures (Watters et al., 2009b; Ferrari et al., 2015). Collocated basin-radial sets of graben and shortening structures form a fan-like pattern centered on the basin outward from around 20% of its radius, such that the radial graben in Rembrandt resemble a less-developed version of Pantheon Fossae in the center of the Caloris basin (Section 10.3). Both sets of radial structures in Rembrandt basin have similar dimensions, from tens of to about one hundred kilometers in length, and hundreds of meters in width. At a radial distance from the basin center of ~50% of the basin radius, these structures are spatially bound by circumferential thrust faults and graben of similar widths but shorter segment lengths. Local clusters of thrust-fault-related landforms without preferred orientations lie in the southwestern and eastern portions of the basin interior beyond about half the basin radius from its center (Ferrari et al., 2015); these structures are tens of kilometers long and ~1–2 km across.

The Caloris basin also hosts the greatest structural complexity on the planet (Figures 10.3a and 10.14d). The pronounced, basin-radial set of graben, Pantheon Fossae (Murchie et al., 2008), is the dominant tectonic assemblage in the basin, originates from a point near the center, and extends to ~55% of the basin's radius. These graben range in length from ~5 km to 110 km and vary from ~1 km to 8 km in rim-to-rim width (Murchie et al., 2008). Pantheon Fossae is bound by circumferential graben that form a near-complete annulus from ~45% to 55% of the basin radius; this annulus is tens of kilometers across, but its constituent graben are hundreds of meters to a few kilometers wide. Its outward extension is manifest by basin-oblique graben that define a complex polygonal map pattern and steadily decrease in width, depth, and length toward the basin margin; near the basin periphery, these structures are but a few kilometers in length and several hundred meters wide. The most abundant types of shortening structures are basin-circumferential, and extend from close to the basin center to about 70% of its radius. Radially oriented thrust faults also occur within this zone but are less common than their basin-concentric counterparts. Beyond ~70% of the basin radius, thrust-fault-related landforms show no strongly preferred orientations and so also form a polygonal pattern that becomes less prominent toward the basin rim. There is less variation in size of the shortening structures in the Caloris basin, with most segments tens of kilometers long and up to several kilometers across.

Almost every tectonically deformed impact structure on Mercury is characterized by a scarcity of definitive crosscutting relations, challenging recognition of a developmental sequence for attendant structures (Section 10.8). Where shortening and extensional landforms spatially coincide within ghost craters, their superposition relations are often unclear (Watters et al., 2012); they do not coincide in Mozart at all (Blair et al., 2013). Graben appear to superpose, and thus postdate, thrust-fault-related structures in the Caloris and Rembrandt interior plains (e.g., Murchie et al., 2008; Ferrari et al., 2015; Cunje and Ghent, 2016), but no clear dip- or strike-slip offsets are observed in either basin.

However, finite-element modeling results for ghost craters (Freed et al., 2012) and mid-sized basins such as Mozart (Blair et al., 2013) provide some insight into the complex tectonic deformation recorded in Mercury's large, volcanically filled impact basins. For example, the thermal contraction of thick, rapidly emplaced lava flows (Chapter 11) produces horizontal tensile stresses that favor the formation of mixed-orientation joints which, if allowed to grow sufficiently, can develop into graben. This mechanism also plausibly accounts for the myriad graben and joints observed within both volcanic and impact melt deposits across Mercury (Section 10.3 and Figure 10.3a). Moreover, models of thermally contracting lavas show that a buried crater rim or ring serves to strongly concentrate radial tensile stresses in the overlying unit, thus facilitating the formation of circumferential graben (Blair et al., 2013). In contrast, some combination of shortening in response to cooling and contraction of the planet's interior and flexure due to vertical loading – as has been proposed for lavas inside the summit caldera complex of Mars' Olympus Mons volcano, for example (Zuber and Mouginis-Mark, 1992) – is likely responsible for the formation of thrust-fault-related landforms at a variety of scales on Mercury, including the shortening structures in volcanically infilled impact features (Watters et al., 2009b; Freed et al., 2012; Blair et al., 2013). The truncation of shortening landforms in the

NSP at the peripheries of thrust-bound ghost craters may reflect partitioning of shortening strain into existing structures, in effect shielding the (ghost) crater interior from a compressive stress state and so permitting the formation of joints and graben therein.

Earlier workers suggested that the prominent fossae within the Caloris basin may have formed as the surface expression of dike propagation (Head et al., 2008, 2009) – and indeed, this system of extension bears a resemblance to radial graben sets on Venus interpreted to be the result of upward flexure of magma chambers within mechanically layered lithospheres (Le Corvec et al., 2015). The strain associated with Pantheon Fossae, however, does not compare well with measured strains of radial dike systems on Earth (Klimczak et al., 2010). Other studies explored the idea that Pantheon Fossae may be the result of a bolide impact to the center of a flexurally uplifted basin center (Freed et al., 2009), with either the volcanic emplacement of the smooth plains exterior to the basin (Melosh and McKinnon, 1988; Kennedy et al., 2008) or the inward flow of the lower crust (Watters et al., 2005) responsible for that uplift. Of note, the radial graben of the Rembrandt interior plains are similar to those of the Caloris basin, but there are no extensive plains exterior to Rembrandt (Denevi et al., 2013), and the prospect of lower crustal or mantle flow operating within Mercury, particularly at scales below those of the diameter of the Caloris basin, is unlikely given the planet's relatively thin silicate shell (Chapter 19). However, the strains represented by Pantheon Fossae correspond to an uplift on the order of 10 km (Klimczak et al., 2010), which is not observed for the Caloris basin, and so the means by which those fossae formed remains unknown.

In a manner similar to those within flooded craters and smaller basins, graben of mixed orientation in the Caloris and Rembrandt plains may reflect near-isotropic horizontal extension of rapidly emplaced lavas (Blair et al., 2013). Under this scenario, the shoaling of graben floors, and overall reduction in graben size, with increasing distance from basin center likely indicates corresponding reductions in plains thickness, and thus original basin depth, toward the rim (Freed et al., 2012; Klimczak et al., 2013a). Further, if the pronounced circumferential graben within both of these plains units reflect extension above a buried basin ring, then the impact structures in which they are hosted may be multi-ring basins – a class of impact feature not yet clearly documented for Mercury (Fassett et al., 2012).

Finally, although the shortening structures within these basins may reflect some amount of subsidence (e.g., Melosh and McKinnon, 1988), the extent to which this mechanism has operated is likely not substantial. Moreover, it may be difficult to differentiate crustal shortening as a result of volcanic loading from that caused by global contraction, especially given the pervasiveness and longevity of the latter process (Section 10.8.1). Indeed, there is evidence that shortening continued within the Caloris basin for some amount of geological time after its network of extensional structures formed (Section 10.8.2). Nonetheless, the differences in orientation of thrust faults in the Caloris and Rembrandt interior plains from basin-circumferential or -radial to -oblique could reflect a transition of the responsible stress field from strongly basin-shape-influenced to horizontally isotropic with increasing distance from the basin center.

10.7 OTHER DEFORMATIONAL PROCESSES

Although Mercury's tectonic character has been dominated by global contraction, with a possible contribution from tidal despinning, studies of Mercury's tectonic history have also considered other processes to help account for observations made first by Mariner 10 and later by MESSENGER. Such candidate processes include mantle convection (Section 10.7.1), lithospheric folding (Section 10.7.2), planetary reorientation by processes internal and/or external to Mercury (Section 10.7.3), and some combination thereof. Several of these mechanisms were suggested before MESSENGER observations of the innermost planet, however, and as a result are less tractable today than when first proposed. Nonetheless, some of the processes discussed below may have contributed at least in part to the present-day tectonic character of Mercury.

10.7.1 Mantle Convection

Convective motion within Mercury's silicate shell has been suggested as a means to control or at least influence tectonic deformation. For example, Watters et al. (2004) proposed that mantle downwelling might thicken Mercury's crust and so localize compression in a manner similar to intraplate downwelling on Earth (Neil and Houseman, 1999). In addition, from three-dimensional numerical simulations of mantle convection within Mercury, King (2008) had suggested that the linear roll structures in his models might concentrate the roughly north–south-oriented shortening strains at low latitudes mapped with Mariner 10 data (Watters et al., 2004) (Section 10.2.2).

Mantle downwelling as a mechanism for developing elongate zones of crustal shortening, however, must be reconciled with indications from MESSENGER that the thickness of Mercury's silicate shell is substantially less than values accepted earlier. The simulations of King (2008), for instance, incorporated a mantle thickness of 600 km, whereas MESSENGER measurements of Mercury's gravity field, together with determinations of the planet's obliquity and libration amplitude, place the core–mantle boundary at a depth of about 420 km (Hauck et al., 2013). Although recent thermal evolution models with a core–mantle boundary at this depth are permissive of convection in Mercury's mantle (Michel et al., 2013; Tosi et al., 2013), and convection continues to the present for a subset of these models, this process would likely occur as discrete cells of up- and downwelling having horizontal dimensions comparable with the thickness of the mantle rather than as long rolls (Chapter 19).

Therefore, although it is possible that mantle downwelling has played some role in the localization of shortening strains on Mercury's surface, this process was likely not responsible for the apparent roughly north–south fabric identified in the mapping of the planet's thrust-fault-related landforms (Byrne et al., 2014; Watters et al., 2015) – if indeed this pattern is real in the first place and not a function of a lighting bias. On the other hand, mantle up- and downwelling may have contributed to regional variations in crustal thickness, which in turn could influence the surface manifestation of shortening strains on

Mercury. Yet because the vigor and longevity of the convection of Mercury's mantle is sensitive to a range of parameters and conditions (e.g., the possibility of compositional stratification or lateral heterogeneity) that are not yet fully characterized (Chapter 19), the contribution of mantle convection to near-surface compressive stresses, and so to the formation, distribution, and orientation of Mercury's shortening structures, is uncertain.

10.7.2 Lithospheric Folding

The process or set of processes that has substantially modified Mercury's topography at large length scales has yet to be determined, although the origin of these undulations is addressed in more detail in Chapter 3. Nonetheless, where they are collocated, tectonics may provide some insight into the nature and origin of Mercury's anomalous topographic lows and highs.

For example, high values of admittance (i.e., the ratio of gravity to topography in the wavenumber domain) at low spherical harmonic degree and order indicate that the topographic rise within the NSP is either supported by buoyancy near the base of the mantle or elastically supported by lithospheric strength (James et al., 2014). This latter scenario predicts compression centered at the rise and extension at its periphery. Yet the areal density of mapped shortening structures (Byrne et al., 2014) (Figure 10.3a), if a proxy for horizontal strain across the NSP, shows no greater density of structures on the long-wavelength northern rise than in the surrounding plains (James et al., 2014). For the lithosphere to have supported the rise, it would have to have a thickness of more than 100 km since the time that the rise was formed (James et al., 2014).

It has also been proposed that mantle dynamic processes might result in long-wavelength modification to Mercury's topography (King, 2008). To first order, the linear undulations within and adjacent to the Caloris basin, for example, are consistent with the long roll-like convection patterns predicted by some three-dimensional models of convection within Mercury's mantle (King, 2008). As noted above, however, convection cells within Mercury are not likely to follow such a pattern (Chapter 19), and the amplitude of dynamic topography predicted by convection models that display these patterns is one to two orders of magnitude (King, 2008) below that of the observed undulations (Klimczak et al., 2013a) (Section 10.4).

The long-wavelength warps may be a manifestation of Mercury's global contraction that formed as a result of lithospheric buckling or folding (Dombard et al., 2001; Hauck et al., 2004; Solomon et al., 2012). Although calculated elastic buckling stresses far exceed the strength of the lithosphere, models simulating elastic–plastic behavior of the lithosphere allow for folding much below the expected threshold of elastic buckling stresses (McAdoo and Sandwell, 1985). In an effort to reconcile estimates of planetary radius change from earlier, incomplete mapping studies with those predicted from thermal evolution models (Section 10.6.1), Dombard et al. (2001) explored the possibility that hidden shortening strain was manifest on Mercury in the form of long-wavelength, low-amplitude folding.

When Dombard et al. (2001) conducted their study there was no evidence that such folding existed on Mercury. The dimensions of the long-wavelength undulations that were ultimately discovered on the planet yield shortening strains ε of only $\sim 10^{-5}$ (Section 10.4), far less than those of brittle structures and so representing an extremely small contribution to Mercury's decrease in radius. Even so, Dombard et al. (2001) found that in a model of unstable deformation of a strong surface layer (Fletcher and Hallet, 1983), adjusted for folding, long-wavelength changes in topography did develop in response to global contraction-induced crustal shortening at wavelengths comparable with those measured on Mercury. If related to global contraction, these undulations may correspond to the long-abandoned geosyncline and geanticline hypothesis for Earth, developed when the tectonics of our own world was thought to be dominated by vertical crustal motion driven by global contraction (Élie de Beaumont, 1852; Dana, 1873).

10.7.3 Planetary Reorientation

A body in which mass is unequally distributed may undergo reorientation so that its axis of greatest moment of inertia aligns with its spin axis. This reorientation can yield stresses sufficient to fracture the body's outer rigid shell, with the resultant strain recorded as tectonic landforms (e.g., Melosh, 1977), both extensional and shortening in nature. Polar wander on Earth can be driven by changes in mass distribution from plate motions (e.g., Duncan and Richards, 1991), but on one-plate planetary bodies polar wander may arise from surface loading (as with the Tharsis Rise on Mars: Melosh, 1980; Zuber and Smith, 1997; Zhong, 2009) as well as from variations in interior density (Nimmo and Pappalardo, 2006) or thermal structure (Roberts and Nimmo, 2008).

Beyond inducing changes in spin rate (Section 10.6.3), large impacts can also drive changes in mass distribution (Nimmo and Matsuyama, 2007; Karimi and Dombard, 2014; Matsuyama et al., 2014) and thus may also force the reorientation of one-plate bodies. This mechanism has been proposed for numerous bodies in the solar system: for example, the Moon (Wieczorek and Le Feuvre, 2009), Mars (Kuang et al., 2014), and Uranus (Morbidelli et al., 2012) may all have experienced substantial reorientation from large hypervelocity impacts. At the greatest scales, such collisions have been invoked to account for the formation of the Moon (e.g., Asphaug, 2014) and the large mass fraction of Mercury's core (e.g., Benz et al., 1988; Chapter 18).

The prospect of planetary reorientation influenced by the gravity anomaly associated with the Caloris basin was investigated by Matsuyama and Nimmo (2009). These authors suggested that a combination of despinning, reorientation, and global contraction generated a stress field consistent with Mariner 10 observations of Mercury's tectonics. However, this predicted stress field does not compare as well with MESSENGER-derived mapping observations, and the relative timing of despinning, contraction, and reorientation (should it have occurred) is as yet poorly characterized (Section 10.6.3).

Interestingly, the NSP on Mercury occupy a broad region of low gravitational potential and thin crust and are so termed because these plains occur at and near the planet's present north pole. Although there is no morphological evidence that points to a single, giant basin in which these plains lie (Head et al., 2011), polar flattening is about the same at both poles (Chapter 3), and there is no southern counterpart to the ~1–2-km-thick NSP. If the northern gravity low and corresponding thin crust correspond to an ancient basin for which surficial evidence no longer survives, then this basin may have caused Mercury to reorient early in its history. Nonetheless, at present the largest impact basin recognized on Mercury is Caloris, and the effects of any early global-scale tectonic process may well have been overprinted by the global contraction that followed (Section 10.6.3).

10.8 TIMING OF DEFORMATION

A full characterization of Mercury's tectonics cannot be complete without an understanding of the planet's history of deformation. Chief among the means we can use to establish this history is the law of superposition, by which the uppermost unit of a geological sequence is interpreted as the youngest, and vice versa. This approach has been refined for Earth over the past several centuries and generally works well in the mapping of geological units on other worlds. However, unlike temporally discrete events such as bolide strikes or volcanic eruptions, faults can be reactivated or can accommodate sustained slip over time, making it far more difficult to place the initiation or cessation of faulting into a stratigraphic column. To that end, remotely sensed observations of a fault superposing a given geological unit indicate that the *most recent* tectonic activity occurred after the unit was emplaced, but tell us nothing of when the fault first formed. Nonetheless, superposition relations between faults and their related landforms and Mercury's geological units, together with crater areal density measurements (Chapter 9), give us a first-order insight into the tectonic evolution of the innermost planet.

Here, we review first the evidence for when shortening deformation occurred on Mercury (Section 10.8.1). In Section 10.8.2, we do the same for extensional deformation. In Section 10.8.3, we discuss the likely timing of the long-wavelength modifications of Mercury's topography. We finish with an assessment of the implications of these findings for understanding the planet's thermal history (Section 10.8.4).

10.8.1 Shortening Deformation

Observations from MESSENGER data indicate that thrust-fault-related landforms interpreted to have formed from global contraction deform all major surface units present on Mercury, which broadly places the relative timing of deformation by these landforms as having continued at least until after the emplacement of the units they superpose. Thrust-fault-related landforms are abundant in Mercury's older intercrater terrain (e.g., Byrne et al., 2014; Chapter 6), but they also deform the younger smooth plains units (e.g., Watters et al., 2009b; Byrne et al., 2014; Ferrari et al., 2015) (Figure 10.3a). With the exception of a single, ambiguous superposition relationship (Watters et al., 2009a), the lack of embayment relations between thrust faults and plains units has been taken as evidence that the bulk of global contraction-induced deformation did not occur until the youngest major plains units were in place (e.g., Strom et al., 1975; Melosh and McKinnon, 1988; Spudis and Guest, 1988; Solomon et al., 2008; Banks et al., 2015). Absolute model ages derived from crater size–frequency distribution (SFD) measurements indicate that the emplacement of the youngest expanses of smooth plains units was largely complete by about 3.5 Ga (Strom et al., 2008, 2011; Fassett et al., 2009; Head et al., 2011; Denevi et al., 2013; Marchi et al., 2013; Ferrari et al., 2015; Ostrach et al., 2015; Byrne et al., 2016b; see also Chapters 6, 9, and 11), which therefore indirectly signifies the point after which the majority of shortening strain was accommodated on Mercury.

This finding is borne out by the results of several regional studies. For example, through mapping shortening structures and other lineations in Apārangi Planitia, situated at low latitudes in Mercury's eastern hemisphere, López et al. (2015) identified two stages of deformation. These authors interpreted low-relief thrust-fault-related landforms to have formed prior to the emplacement of the smooth plains units in the region; these smooth plains were emplaced at around 3.7 Ga (Byrne et al., 2016b). In contrast, a set of topographically more pronounced thrust-fault-related landforms was inferred by López et al. (2015) to have formed entirely after the emplacement of the units they deform. The orientation of this set was also found to be strain incompatible with, and so must have formed at a different time than, the low-relief landforms. Crosscutting relations and crater SFD measurements of the Rembrandt impact basin, its interior fill, and the Enterprise Rupes thrust fault system (Figure 10.14c) gave model ages for major tectonic activity along the system of between 3.8 and 3.6 Ga (Ferrari et al., 2015). And Giacomini et al. (2015) found, with SFD-derived model ages, that activity along Blossom Rupes occurred at around 3.7–3.5 Ga.

Although crater statistics can provide a general overview of the timing of shortening in a given region affected by thrust faulting, superposition relationships of thrust-fault-related landforms with individual impact craters can contribute toward a planet-wide understanding of the onset and duration of global contraction-induced thrust faulting (Banks et al., 2015). The state of degradation of a crater can be linked to one of Mercury's time–stratigraphic periods (e.g., Pohn and Offield, 1970; Trask, 1971, 1975; Moore et al., 1980) that, from oldest to youngest, include pre-Tolstojan, Tolstojan, Calorian, Mansurian, and Kuiperian (Chapter 6). Although qualitative, some age information can be gleaned regarding the relative timing of thrust faulting at a given location if a crater for which degradation state can be determined superposes, or is superposed by, a shortening structure.

For example, Banks et al. (2015) found that of a number of craters with degradation states consistent with the pre-Tolstojan and Tolstojan periods are all crosscut by shortening landforms, indicating that the thrust faulting visible today postdates the formation of these craters. Of the craters with degradation states consistent with the Calorian period that these authors investigated, the majority had been deformed by thrust faults. However, a few of the craters determined to be of this period do not appear to be deformed by the thrusts they superpose (Figure 10.15). Under the assumption that these shortening landforms are the result of global contraction, this process was

10.8 Timing of Deformation 273

Figure 10.15. A 200-km-long unnamed thrust-fault-related landform (white arrows) superposed by a heavily degraded 130-km-diameter Calorian crater (black arrows). Note that the crater's ejecta superposes and mutes the surface expression of the proximal portion of the thrust fault. The image is a mosaic composed of MESSENGER images EN0220316553, EN0220319665, EN0220321873, EN0220404657, EN0220407308, EN0220490886, and EN0220493578. Azimuthal equidistant projection, centered at 42.0°S, 19.0°E.

Figure 10.16. The 51-km-diameter Mansurian crater Martial, which is crosscut by an unnamed thrust-fault-related landform (white arrows). The image is a portion of MESSENGER image EW0213416030G. Azimuthal equidistant projection, centered at 68.0°N, 182.0°E.

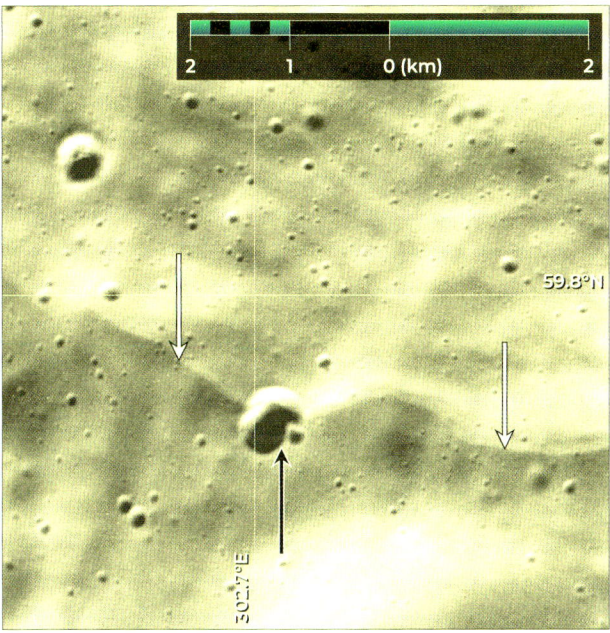

Figure 10.17. A small thrust-fault-related landform (white arrows) appearing to crosscut a relatively fresh 500-m-diameter crater (black arrow). This thrust is adjacent to, and has the same strike as, Carnegie Rupes (Figure 10.1c). The image is a portion of MESSENGER image EN1036136378M. Azimuthal equidistant projection, centered at 59.8°N, 302.7°E.

underway during or even prior to the Calorian – which, in turn, implies at least some overlap between the waning phase of flood basalt emplacement and global contraction.

Banks et al. (2015) also assessed superposition relations between thrust faults and craters for which degradation states correspond to the Mansurian period, and they found that a majority of these craters are undeformed. Yet a few such craters *are* deformed (Figure 10.16), indicating that global contraction continued to drive thrust faulting during this period. Of note, none of the Kuiperian craters investigated by Banks et al. (2015) was found to be unequivocally deformed by thrusts – although these authors identified other evidence that crustal shortening continued into the Kuiperian.

This evidence is manifest in part as a population of small monoclines, at least an order of magnitude smaller than those previously recognized, some of which crosscut craters only 1–2 km in diameter for which the state of degradation (or lack thereof) implies that they correspond to the Kuiperian period (Figure 10.17). These small shortening landforms are typically only tens of meters in relief and but a few kilometers in length (Watters et al., 2016). Although several of these small structures superpose craters, the rarity of this type of crosscutting relationship is likely a function of monocline size and the relatively low cratering rate in the Kuiperian. In addition, narrow graben have been identified along the crests of larger shortening structures (Section 10.2.1) (Figure 10.18); these extensional landforms are consistent with the interpretation of Mercury's thrust-fault-related landforms as fault displacement-gradient folds (Wickham, 1995), with the graben corresponding to outer-arc extension along the fold hinge axes

(Section 10.2.1). In any case, given assumed cratering rates on Mercury (Le Feuvre and Wieczorek, 2011; Marchi et al., 2013), landform degradation even in the Kuiperian is expected to be geologically rapid, such that small shortening and extensional structures are not likely to survive for extended periods of time (Watters et al., 2016). That these landforms are visible today therefore indicates that thrust faulting has operated on Mercury into the geologically recent. To date, there is no evident explanation for why new structures should form from global contraction when a worldwide network of faults was already present. Nonetheless, that global contraction continues today is not surprising, since the presence of a magnetic field at Mercury and the amplitude of the planet's forced libration (Chapter 4) indicate that its outer core is still molten and so the planet must continue to experience secular cooling.

Taken together, these observations show that globally distributed thrust faulting was underway during or at the end of the Calorian period and continued into the Kuiperian. Importantly, we cannot ascertain with certainty whether thrust faulting commenced in the Calorian or instead craters formed in this period were deformed by thrusts that were already there and simply experienced slip events subsequent to impact. Given that the oldest surface units on Mercury date to ~4.1 Ga (Marchi et al., 2013; Chapter 9), it may be that earlier volcanism and impact bombardment removed entirely any pre-Tolstojan or Tolstojan shortening structures. If so, then those superposed by Calorian craters are representative of the oldest population of preserved thrust-fault-related landforms on the planet.

Even so, a rock-mechanical assessment of the brittle strength of Mercury's lithosphere suggests that an appreciable amount of radial decrease – between 400 m and 2.1 km – is necessary before stresses become sufficiently large to overcome frictional resistance to sliding along pre-existing discontinuities in the lithosphere (Klimczak, 2015) (Section 10.6.1 and Figure 10.19). This finding requires that some considerable time must have passed between the start of secular cooling and the formation of shortening landforms, indicating that even the earliest reliable evidence of thrust faulting in the Calorian must postdate the onset of global contraction. Moreover, the calculated reduction in Mercury's radius of up to 2.1 km necessary to trigger slip along thrust faults also implies that the rate of global contraction was likely greater early in the planet's geological history (Klimczak, 2015) and, for a radius change of 3.1–7.1 km since the Calorian (Table 10.1), must have continued at a very slow average rate of radius reduction of $(1.1-2.5) \times 10^{-20}$ s^{-1} (1.1–2.5 μm per Earth year) after thrust faulting began (Figure 10.19).

Figure 10.18. A set of small, sub-parallel graben along the crest of Calypso Rupes (white arrows). The image is a mosaic composed of MESSENGER images EN0249987811M and EN0250016605M. Azimuthal equidistant projection, centered at 19.6°N, 45.7°E.

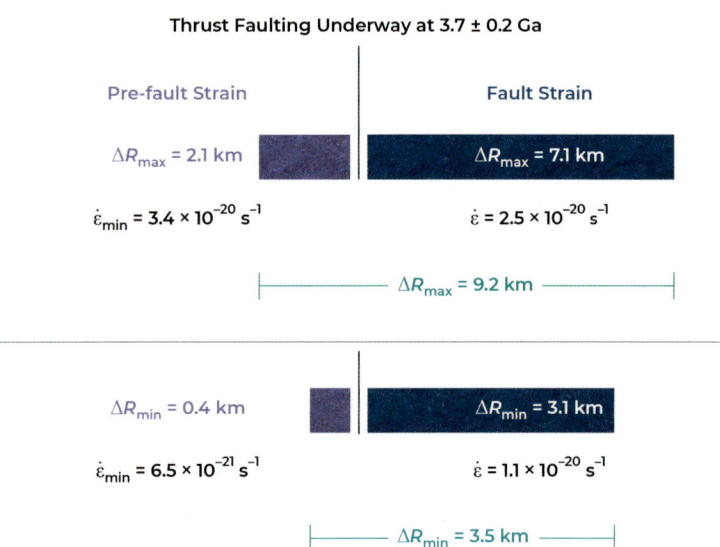

Figure 10.19. Maximum and minimum amounts of total radius change ΔR from the different components of global contraction. Mercury's lithosphere accommodated a change in radius prior to, as well as after, the onset of thrust faulting. The resultant rates of contractional strain, $\dot{\varepsilon}$, are given for a scenario under which brittle failure was underway by 3.7 ± 0.2 Ga.

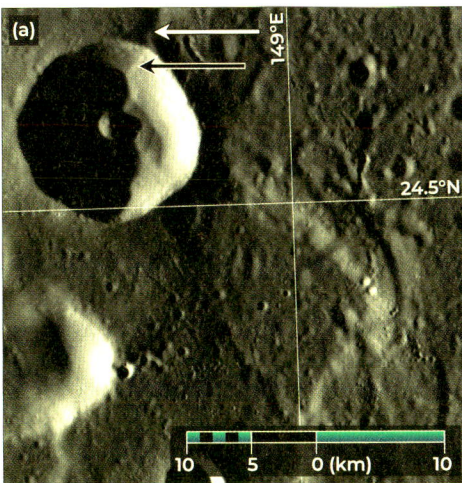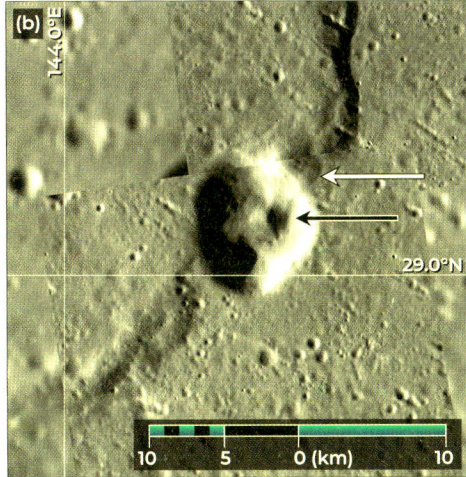

Figure 10.20. Examples of craters within the Caloris basin that have been modified by thrust faulting. (a) An unnamed crater 14 km in diameter, the northeastern portion of which is crosscut (black arrow) by a thrust-fault-related landform (white arrow). The image is a portion of MESSENGER image EN1015368165M. (b) Another unnamed crater, 8 km in diameter, cut (black arrow) by a northeast–southwest-trending shortening structure (white arrow). The image is a mosaic composed of MESSENGER images EN0250998103M and EN0251055707M. Azimuthal equidistant projections, centered at 24.5°N, 145.0°E and 29.0°N, 144.5°E, respectively.

10.8.2 Extensional Deformation

Where they are observed, joints and graben on Mercury are thought to represent the effects of cooling and thermal contraction of the units in which they are found (e.g., Freed et al., 2012; Klimczak et al., 2012; Watters et al., 2012; Blair et al., 2013) (Section 10.6.4). Given that the timescales over which these lavas (and impact melt) would likely have cooled (Freed et al., 2012), these fractures must therefore have formed soon after the emplacement of their host rocks. If so, then the joints that have so far been recognized in impact melt deposits on the floors of relatively well-preserved impact features, such as Hokusai, Abedin, or Degas craters (Figure 10.9c), must themselves be relatively young.

Crosscutting relations between extensional tectonic features and impact craters support this interpretation. For example, the graben identified in the NSP, which are located exclusively within ghost craters (e.g., Klimczak et al., 2012; Figures 10.3a and 10.9b), do not appear to modify any superposing craters, consistent with the inference that these structures formed soon after their host rocks were emplaced. Absolute model ages for the NSP are ~3.7–3.9 Gyr (Head et al., 2011; Denevi et al., 2013; Marchi et al., 2013; Ostrach et al., 2015), a range of ages that must therefore also correspond to the formation times of these graben. Likewise, the graben in the Caloris basin do not superpose any of the craters on those plains (Klimczak et al., 2013a), and so they must be approximately the same age as those ~3.7–3.9-Gyr-old basalts (Strom et al., 2008, 2011; Fassett et al., 2009; Denevi et al., 2013; Marchi et al., 2013). The graben mapped within the Rachmaninoff, Raditladi, and Mozart basins (e.g., Figure 10.14b) are therefore also about the same age as their volcanic host rocks (e.g., Blair et al., 2013). Interestingly, the lavas within the inner peak ring of the Raditladi basin may be as young as about 1 Gyr (Prockter et al., 2010); if so, so too are the graben therein.

The Rembrandt and Caloris interior plains units boast a complex history of deformation (Figures 10.14c, 10.14d, and Section 10.6.4), and in numerous instances, even where tectonic structures crosscut one another, timing relations are ambiguous. Nevertheless, Cunje and Ghent (2016) compiled a series of relative timing relationships for several examples of graben-shortening-landform intersections in the Caloris basin. These authors found that tectonic deformation of these plains began with the formation of shortening structures, which then partially overlapped with the development of basin-radial graben (i.e., Pantheon Fossae) (Sections 10.3 and 10.6.4). Some time thereafter, when thrust faulting within the plains ceased but the radial graben were still forming, basin-concentric graben developed.

However, although neither basin-concentric nor basin-radial graben modify any craters (e.g., Klimczak et al., 2013a), several craters have been deformed by thrust-fault-related landforms (Figure 10.20). This observation does not necessarily require that shortening entirely postdated extension in the Caloris basin; instead, whatever mechanism was responsible for initiating thrust faulting may have been superseded by global contraction after basin-scale processes were complete. If so, then the uncertainty regarding the sequence of deformation within the Caloris basin illustrates once more that fault reactivation can frustrate efforts to determine a definitive geological history for a heavily deformed region.

10.8.3 Long-Wavelength Deformation

One of the key arguments against Mercury's long-wavelength undulations being constructional (i.e., volcanic) in origin is the presence of superposed craters with once-level floors that are now tilted in direction and slope comparable with the local topography of the undulations (Section 10.4). This observation also provides some insight into the timing of this enigmatic modification to Mercury's topography. For example, in the

Caloris basin there are many tilted craters that superpose but are not crosscut either by extensional or shortening tectonic landforms (Klimczak et al., 2013a). Given the density of structures within the plains (Figure 10.14d), and therefore the likelihood of one type of landform crossing another, it follows then that the modification to topography here occurred after plains emplacement and the formation of most, if not all, brittle tectonic structures.

This inference is supported by the pattern of tectonic deformation in the Caloris basin (Section 10.6.4 and Figure 10.14), which is profoundly influenced by the near-circularity of the basin shape. Even though the northern portion of the Caloris basin interior rises more than 1000 m above the northern basin rim, the circular arrangement of ridges and graben neither there nor in the southern portion shows any influence from this elevated topography. In contrast, for example, there is a systematic pattern of radial and concentric faults centered on (and that characterize) domical rises associated with magma chamber tumescence (Marti et al., 1994; Walter and Troll, 2001). This apparent lack of influence of the elevated topography on the orientations of graben and ridges is consistent with the inference that the brittle tectonic deformation in this area occurred prior to the establishment of the regional (warped) topography.

Moreover, the degradation states of those craters that are tilted within the Caloris basin suggest that they were formed well after the Calorian period; accordingly, the long-wavelength undulations developed relatively late in Mercury's geological history (Balcerski et al., 2013). For example, the floor of the Atget crater, a morphologically fresh impact structure within the basin, shows substantial deviation from horizontal (Figure 10.10). Craters with such degradation states are thought to correspond to the Mansurian period, which requires that the long-wavelength modification to the Caloris basin occurred in (or after) that same period.

Similarly, the northern rise hosts ghost craters with floors that tilt systematically away from the topographic high (Section 10.4), but craters superposing the rise that are not volcanically filled show less or no correlation in floor tilt with the long-wavelength topography (Klimczak et al., 2012; Balcerski et al., 2012). Modification of the region's topography thus began after plains emplacement and continued for some time thereafter, slowing and then ceasing as fresh craters continued to accumulate (Balcerski et al., 2013). Notably, some of the tilted but unfilled craters superpose, but are not modified by, tectonic landforms (both shortening and extensional), indicating that the growth of the undulation continued to times later than the periods of crustal shortening and thermal contraction of the NSP lavas. And since the amplitudes of the rises within Mercury's valles far exceed the local downhill gradient, the modification in these locations of long-wavelength topography must have occurred after the valles were formed and so, again, were not constructional in origin (Byrne et al., 2013b).

10.8.4 Implications for Mercury's Thermal History

The thermal history and evolution of Mercury must satisfy a number of geological constraints that are informed by the observations described in this chapter (and summarized in Figure 10.19) and elsewhere in this volume (Chapter 19). For example, the total decrease in the radius of Mercury from secular cooling since the LHB amounts to 3.5–9.2 km and consists of two components. The first includes a 3.1–7.1 km radius change accommodated by thrust faulting (Byrne et al., 2014), and the second component accounts for 0.4–2.1 km of radius change (Klimczak, 2015) (Section 10.6.1) accommodated elastically and occurring sometime during or before the Calorian period, generally before the onset of thrust faulting (Banks et al., 2015).

A change in planetary radius of 3.1 to 7.1 km since the onset of widespread thrust faulting requires an average rate of radius reduction of $(1.1–2.5) \times 10^{-20}$ s^{-1} throughout Mercury's geological past. The radius change calculated to have occurred prior to the start of thrust faulting (averaged over the entire geological history of Mercury prior to the Calorian period) yields a minimum contraction rate of 6.5×10^{-21} s^{-1} to 3.4×10^{-20} s^{-1}. If global contraction commenced a few hundred million years after Mercury's formation, however, as the majority of thermal models predict, then values for the initial contraction rate would have been considerably greater, exceeding the average rate of contraction since the formation of thrust faults. To first order, this finding indicates that the rate of Mercury's global contraction has slowed over time.

10.9 TECTONISM AND VOLCANISM

In addition to tectonic activity, volcanism is a major planetary process that has shaped the surfaces of numerous solar system bodies, including the Moon, Mars, Venus, Io, and our own world. Mercury, too, records a detailed history of volcanic activity, albeit one largely constrained to the first ~20% of the planet's history (Chapters 6 and 11). As with those other bodies, the histories of volcanism and tectonic deformation on Mercury are intertwined – yet perhaps one is controlled by the other more so there than anywhere else in the solar system. In this section, we first review the prospect for widespread effusive volcanism on a planet undergoing global contraction (Section 10.9.1), and then appraise the utility of faults as conduits for the ascent of magma (Section 10.9.2).

10.9.1 Flood Volcanism on a Contracting Planet

Under a tectonic regime dominated by global contraction, the least compressive stresses act vertically and are governed by the overburden (the weight of the overlying rock volume: Section 10.5.1); the greatest compressive stresses act in the horizontal plane. Such a stress field is compatible with the formation of thrust faults, but also inhibits the vertical ascent of magma (e.g., Glazner, 1991; Hamilton, 1995; Watanabe et al., 1999), which is much more likely to propagate laterally. This stress state is thus not readily conducive to widespread effusive volcanism (e.g., Solomon, 1978; Marrett and Emmerman, 1992), and so volcanic flooding is not expected to occur after the onset of global contraction (Byrne et al., 2016b). An obvious inference, then, is that the volcanic smooth plains units on Mercury must have been mostly emplaced prior to the onset of global contraction.

Thrust fault activity has been found to have temporally overlapped, at least to some extent, with the emplacement of some of the relatively late flood lavas (Banks et al., 2015) (Section 10.8). Nonetheless, it has long been noted that many volcanic smooth plains deposits on Mercury are situated within older impact basins, including Caloris Planitia and those volcanic units in Beethoven, Rembrandt, and Tolstoj basins (e.g., Strom et al., 1975; Fassett et al., 2012). Many smaller deposits across the planet (Denevi et al., 2013), at least some of which are likely volcanic (Prockter et al., 2010; Marchi et al., 2011), are similarly situated in older impact craters.

This collocation of many of the youngest effusive volcanic units on Mercury with impact structures is consistent with predictions for a planet undergoing contraction from secular interior cooling (Solomon, 1978). The impact process removes overburden, resets prevailing stresses, and either entirely destroys (and thus forms anew) the pre-existing lithosphere or substantially fractures and weakens it, so that conditions amenable to magma ascent are in place locally. A large impact also converts kinetic energy to heat in the crust and mantle that may trigger or enhance the production of partial melt, even some time after the impact event (Roberts and Barnouin, 2012; Chapter 19). Impact structures are likely therefore to be prime (if not necessarily the *only*) sites for late-stage effusive volcanic eruptions under a tectonic regime otherwise generally unfavorable to extrusive activity.

10.9.2 Tectonic Structures as Magma Conduits

Despite its anticipated effects on effusive volcanism, global contraction has not prevented explosive volcanic activity on Mercury. Irregular depressions that lack raised rims, are morphologically dissimilar to impact craters, and are often surrounded by diffuse high-reflectance deposits have been interpreted as volcanic vents that have served as sites of explosive eruptions (Head et al., 2008; Kerber et al., 2009, 2011; Goudge et al., 2014; Thomas et al., 2014) (Section 11.2). Several of these vents superpose and thus postdate Caloris Planitia (Head et al., 2008), and at least one example (that which cuts the wall terraces within the 26-km-diameter Kuniyoshi crater) may be as young as ~1 Gyr (Thomas et al., 2014). Volatiles provide the primary driving force for explosive volcanic eruptions (e.g., Cashman, 2004), and volatile-rich melts are more buoyant and therefore more capable of ascent and eruption than effusive magmas of otherwise comparable composition.

The majority of pyroclastic deposits on Mercury are collocated with sites of pre-existing weakness in the lithosphere (Chapters 11 and 19) (Figure 10.21), including the heavily fractured central peaks, peak rings, and rims of impact craters and along the traces of thrust-fault-related landforms (e.g., Kerber et al., 2011; Klimczak et al., 2013a; Goudge et al., 2014; Habermann and Klimczak, 2015). Given that global contraction was at least underway in the Calorian period, explosive volcanism on Mercury has also operated under a regime of global contraction. This stress state therefore accounts for the spatial association between sites of

Figure 10.21. A group of north–south-trending thrust-fault-related landforms that crosscut craters, including the 89-km-diameter Glinka "pit-floor crater" in the upper part of this image. The pit inside Glinka is surrounded by a relatively bright "halo," interpreted to be a pyroclastic deposit. The crater in the lower part of the image is that shown in Figure 10.7. The image is a three-color mosaic composed of MESSENGER images EW0242128483G, EW0242128487F, and EW0242128491I; red: 996-nm, green: 749-nm, and blue: 433-nm wavelength. Azimuthal equidistant projection, centered at 11.0°N, 248.0°E.

pyroclastic activity and faults and fractures that, especially if critically stressed, may have acted as conduits for the upward migration of volatile-rich magmas (Klimczak et al., 2013b) (Section 11.7).

Accordingly, that few to no pyroclastic vents have been observed in smooth plains units, e.g., in the NSP, suggests that there are no deep-seated faults in such locations along which volatile-rich magma can ascend. There are, however, several vents and pyroclastic deposits along the interior margin of the Caloris basin, and a few large vents are present on the margins of the smooth plains within the Tolstoj and Rembrandt basins. The presence and spatial distribution of these vents may reflect a stress state favorable to magma ascent and eruption similar to that calculated for regions at or near the rims of large mare-filled basins on the Moon (McGovern and Litherland, 2011).

10.10 FURTHER WORK

Despite providing a remarkable advance in our understanding of Mercury's tectonic character and history, the results of the MESSENGER mission challenge us to ask yet more questions about the innermost planet. Here, we discuss briefly some of those questions, and why it is important to resolve them (Section 10.10.1). We then propose a number of future observations that might enable us to answer these questions, either by missions currently planned for flight or by missions for which architectures have yet to be developed (Section 10.10.2).

10.10.1 Outstanding Tectonic Questions

With Mercury's tectonic character so dominated by global contraction, it is prudent to continue to explore this fundamental process. Interior cooling and its consequences have played controlling roles in the planet's thermal, volcanic, and tectonic evolution (Chapters 11 and 19), but a complete understanding of this control is possible only once we characterize fully the history of global contraction for Mercury.

For example, we still do not know when this process started, in part because Mercury's record of tectonic deformation may date only from the end of the LHB (Section 10.8.1), and in part because thermal evolution models can describe multiple cooling scenarios of varying onset and duration. Similarly, the history of recent global contraction is not well understood. Given that the planet's extant magnetic field (Chapter 5) requires a molten and convecting core, heat must flow from the core to the mantle (Chapter 19), and so Mercury must still be cooling and contracting. Relatively small and well-preserved landforms interpreted to represent crustal shortening (Figures 10.17 and 10.18) attest to thrust faulting, and therefore likely global contraction, having occurred within the past several hundred million years (Section 10.8.1). However, although the rate of contraction has likely decreased through time (Section 10.8.4), we do not yet know for certain if the stresses from global contraction today remain sufficient to overcome the frictional strength of Mercury's lithosphere. Accordingly, exploring how the strain rate from contraction has varied through time will improve constraints on thermal models of the planet's interior evolution (e.g., Michel et al., 2013; Tosi et al., 2013).

Understanding the contribution to tectonic deformation of Mercury's surface from processes other than global contraction remains an important task, too. As assessment of the areal strain recorded by shortening structures in the intercrater plains versus that for smooth plains structures may indicate whether stresses from vertical loading have contributed to the formation of the latter set of landforms (e.g., Melosh and McKinnon, 1988; see also Section 10.6.4). Similarly, the effects of solar tidal stresses and thermal stresses (e.g., Beuthe, 2010; Williams et al., 2011) from Mercury's spin–orbit resonance on its lithosphere and on its global fault pattern remain to be fully investigated. And, of course, determining the mechanism(s) responsible for modifying Mercury's topography at long wavelengths is important for understanding whether these enigmatic troughs and rises reflect a component of global contraction, and whether such features might exist on other worlds. Characterizing the planet-wide distribution of the long-wavelength undulations, including establishing whether they are present in the southern hemisphere, will be key to this understanding.

The influence, if any, of changes in Mercury's spin rate on the planet's tectonics also remains to be fully characterized. For scenarios under which tidal spindown occurred early in the planet's history and spanned some hundreds of millions of years, or started during or even after the LHB but spanned a much shorter duration, or was affected by a change in Mercury's rate of rotation during or after the LHB by one or more large impacts and/or variations in orbital eccentricity, it is not difficult to imagine that despinning overlapped temporally with global contraction. If, on the other hand, the change in Mercury's spin rate was a short-lived process and/or operated only before the LHB, then for this mechanism to have influenced the tectonic pattern seen on the planet today the structures it formed must have been sufficiently deep-seated to survive resurfacing from the impact bombardment and effusive volcanism that was responsible for erasing the planet's surface features older than ~4.1 Ga (Marchi et al., 2013; Chapter 9).

Determining whether the perceived north–south fabric of Mercury's shortening structures is an artifact of solar illumination, with artificially illuminated DEMs, for example, is also an important objective, for the answer will either update existing models for the planet's tectonic and thermal evolution or call for the formulation of new ones.

Mercury's tectonic landforms themselves warrant further analysis. For example, the distinction in type of shortening structure we give here (e.g., smooth plains structure, cratered plains structure) is motivated by the observation that such landforms often share morphological characteristics that defy a straightforward classification as wrinkle ridge or lobate scarp (Section 10.2.1). The broad variety in form of these structures has not been quantified for Mercury, and so as yet there is no *systematically* robust classification scheme for the planet's inventory of thrust-fault-related landforms. Such an analysis would also provide key structural information such as fault linkage, frequency and extent of thrust duplexes at the leading edges of shortening structures, and the geometry and depth of penetration of the underlying fault surfaces. These last parameters are especially important, as the fault dip angles used in estimates of global contraction-induced radius change (i.e., $\theta = 25°–35°$) (Section 10.6.1) are greater than dip angles found for select thrust faults on Mercury (Galluzzi et al., 2015), as well as for deep-seated thrusts beneath the Mare Crisium basin on the Moon (Byrne et al., 2015). The use of fault dip angles as low as $7°$ (Galluzzi et al., 2015) would increase estimates of the amount of Mercury's radial shortening accommodated by brittle structures beyond even the highest published values to date (Byrne et al., 2014). Moreover, the structural arrangement on the Moon, whereby large thrust faults bound an elevated portion of the crust–mantle boundary beneath large impact basins (Byrne et al., 2015), may apply to Mercury but has yet to be investigated.

Finally, the initial state of global contraction is thought to be characterized geometrically by extension at the surface (e.g., Delamétherie, 1795a; Dana, 1873; Ampferer, 1923), with extensional structures propagating to some depth (de Buffon, 1788; Delamétherie, 1795b). As contraction progresses, the outermost layers would be put into compression, separated from layers continuing to extend below by a surface of no strain that deepens with time (e.g., Wilson, 1954; Jeffreys, 1976) – much as thermally contracting lavas behave (e.g., Blair et al., 2013), albeit at a far greater scale. Moreover, some thermal evolution models for Mercury predict an early phase of global expansion as a function of mantle differentiation (e.g., Grott et al., 2011; Tosi et al., 2013). Such incipient and/or large-scale extension has not been substantially investigated for Mercury, and any resulting deformation probably preceded the emplacement of even the oldest surface now preserved on the planet (Marchi et al., 2013). Yet high-resolution gravity gradiometry data for the Moon have revealed a global system of deep-seated features with no surface manifestation that are interpreted to be ancient, giant dikes formed in an early stage of lunar expansion (Andrews-Hanna et al., 2013). The acquisition of similar high-resolution gravity field data for Mercury could establish whether such buried structures are present within the planet. The identification of such structures would provide compelling support for an early period of contraction-induced rifting or even planetary expansion, with those rifts facilitating the rapid and widespread eruption of flood basalts onto the surface of the planet (e.g., Whitten et al., 2014; Evans et al., 2015; Chapter 11).

10.10.2 Future Observations

The wealth of data sets returned by the MESSENGER spacecraft has provided a detailed and integrated insight into the character and history of the innermost planet, from which we can identify further observations needed to address the questions above, among many others. Such observations would help frame the science campaigns of future missions to Mercury, such as the joint European Space Agency–Japan Aerospace Exploration Agency BepiColombo mission (e.g., Benkhoff et al., 2010), scheduled to launch in 2018, and may even inspire later missions to the innermost planet (Chapter 20).

For example, much remains to be done in characterizing the tectonics of Mercury's southern hemisphere. High-resolution targeted images of shortening landforms there would complement similar data for northern hemisphere structures, as enabled by MESSENGER's periapsis at high northern latitudes (Chapter 1). Similarly, high-resolution measurements of the topography and gravity field and the development of crustal thickness models for the southern hemisphere would provide a geophysical context for photogeological observations of the tectonics there. Such new data would also act as a basis for comparing fault displacement-gradient fold morphology, fault penetration depth, and displacement–length scaling of the structures in both hemispheres, as well as information on the depth and degree of fracturing of the planet's lithosphere in general.

But what of yet more ambitious exploration efforts? Interferometric synthetic aperture radar (InSAR), for example, has been used to identify and characterize ongoing tectonic deformation on Earth (e.g., Bürgmann et al., 2000). InSAR measurements from orbital assets at Mercury could be used to search for evidence of active surface change on the planet, including fault slip – a crucial observation for understanding the rate of global contraction and whether tectonic deformation is active on the innermost planet today. A seismic station, or better yet a seismic network, would yield unprecedented views of Mercury's interior. Such a mission would contribute not only to the search for active faulting, but could also yield critical information on the structure of the lithosphere and whether Mercury possesses a solid inner core (Chapter 19). Linking measurements of Mercury's present interior structure with observations of almost four billion years of tectonic deformation would considerably expand our understanding not only of that planet's thermal, tectonic, and volcanic histories, but also of silicate bodies throughout the solar system. Moreover, with the increasing number of terrestrial planets identified in extrasolar planetary systems, including ones as small as Mercury (e.g., Barclay et al., 2013), our solar system's innermost planet may come to serve as a case study with which to understand the global cooling and contractional histories of rocky, one-plate planets in general.

The objectives of future missions to Mercury, from BepiColombo forward, will reflect and build on the findings of the MESSENGER mission. Our understanding of the planet's tectonic character has never been more comprehensive, but there is much left to learn. What we can state with certainty for now, however, is that MESSENGER has given us compelling reasons to continue to investigate enigmatic Mercury, and that it is worth going back.

REFERENCES

Ahrens, T. J. and Rubin, A. M. (1993). Impact-induced tensional failure in rock. *J. Geophys. Res.*, **98**, 1185–1203.

Ampferer, A. (1923). Beiträge zur Auflösung der Mechanik der Alpen. *Jahrb. Geol. Bundesanst.*, **76**, 125–151.

Anderson, E. M. (1951). *The Dynamics of Faulting and Dyke Formation with Applications to Britain*, 2nd edn. Edinburgh, UK: Oliver and Boyd.

Andrews-Hanna, J. C., Asmar, S. W., Head, J. W., Kiefer, W. S., Konopliv, A. S., Lemoine, F. G., Matsuyama, I., Mazarico, E., McGovern P. J., Melosh, H. J., Neumann, G. A., Nimmo, F., Phillips, R. J., Smith, D. E., Solomon, S. C., Taylor, J., Wieczorek, M. A., Williams, J. G. and Zuber M. T. (2013). Ancient igneous intrusions and early expansion of the Moon revealed by GRAIL gravity gradiometry. *Science*, **339**, 675–678.

Asphaug, E. (2014). Impact origin of the Moon? *Annu. Rev. Earth Planet. Sci.*, **42**, 551–578.

Balcerski, J. A., Hauck, S. A., II and Sun, P. (2012). Tilted crater floors: Recording the history of Mercury's long-wavelength deformation. *Lunar Planet. Sci.*, **43**, abstract 1850.

Balcerski, J. A., Hauck, S. A., II and Sun, P. (2013). New constraints on timing and mechanisms of regional tectonism from Mercury's tilted craters. *Lunar Planet. Sci.*, **44**, abstract 2444.

Banks, M. E., Xiao, Z., Watters, T. R., Strom, R. G., Braden, S. E., Chapman, C. R., Solomon, S. C., Klimczak, C. and Byrne, P. K. (2015). Duration of activity on lobate-scarp thrust faults on Mercury. *J. Geophys. Res. Planets*, **120**, 1751–1762.

Barclay, T., Rowe, J. F., Lissauer, J. J., Huber, D., Fressin, F., Howell, S. B., Bryson, S. T., Chaplin, W. J., Désert, J. M., Lopez, E. D., Marcy, G. W., Mullally, F., Ragozzine, D., Torres, G., Adams, E. R., Agol, E., Barrado, D., Basu, S., Bedding, T. R., Buchhave, L. A., Charbonneau, D., Christiansen, J. L., Christensen-Dalsgaard,

J., Ciardi, D., Cochran, W. D., Dupree, A. K., Elsworth, Y., Everett, M., Fischer, D. A., Ford, E. B., Fortney, J. J., Geary, J. C., Haas, M. R., Handberg, R., Hekker, S., Henze, C. E., Horch, E., Howard, A. W., Hunter, R. C., Isaacson, H., Jenkins, J. M., Karoff, C., Kawaler, S. D., Kjeldsen, H., Klaus, T. C., Latham, D. W., Li, J., Lillo-Box, J., Lund, M. N., Lundkvist, M., Metcalfe, T. S., Miglio, A., Morris, R. L., Quintana, E. V., Stello, D., Smith, J. C., Still, M. and Thompson, S. E. (2013). A sub-Mercury-sized exoplanet. *Nature*, **494**, 452–454.

Barton, N. (1976). Rock mechanics review: The shear strength of rock and rock joints. *Int. J. Rock Mech. Min. Sci.*, **13**, 255–279.

Becker, K. J., Robinson, M. S., Becker, T. L., Weller, L. A., Edmundson, K. L., Neumann, G. A., Perry, M. E. and Solomon, S. C. (2016). First global digital elevation model of Mercury. *Lunar Planet. Sci.*, **47**, abstract 2959.

Benkhoff, J., van Casteren, J., Hayakawa, H., Fujimoto, M., Laakso, H., Novara, M., Ferri, P., Middleton, H. R. and Ziethe, R. (2010). BepiColombo – Comprehensive exploration of Mercury: Mission overview and science goals. *Planet. Space Sci.*, **58**, 2–20.

Benz, W., Slattery, W. L. and Cameron, A. G. W. (1988). Collisional stripping of Mercury's mantle. *Icarus*, **74**, 516–528.

Beuthe, M. (2010). East–west faults due to planetary contraction. *Icarus*, **209**, 795–817.

Bieniawski, Z. T. (1989). *Engineering Rock Mass Classifications*. New York: Wiley.

Blair, D. M., Freed, A. M., Byrne, P. K., Klimczak, C., Prockter, L. M., Ernst, C. M., Solomon, S. C., Melosh, H. J. and Zuber, M. T. (2013). The origin of graben and ridges in Rachmaninoff, Raditladi, and Mozart basins, Mercury. *J. Geophys. Res. Planets*, **118**, 47–58.

Bois, C., Cazes, M., Damotte, B., Galdéano, A., Hirn, A., Mascle, A., Matte, P., Raoult, J. F. and Torreilles, G. (1986). Deep seismic profiling of the crust in northern France: The Ecors project. In *Reflection Seismology: A Global Perspective*, eds. M. Barazangi and L. Brown, Geodynamics Series, Vol. 13. Washington, DC: American Geophysical Union, pp. 21–29.

Brace, W. F. and Kohlstedt, D. L. (1980). Limits on lithospheric stress imposed by laboratory experiments. *J. Geophys. Res.*, **85**, 6248–6252.

Bürgmann, R., Rosen, P. A. and Fielding, E. J. (2000). Synthetic aperture radar interferometry to measure Earth's surface topography and its deformation. *Annu. Rev. Earth Planet. Sci.*, **28**, 169–209.

Burke, K. C., Şengör, A. M. C. and Francis, P. W. (1984.) Maxwell Montes in Ishtar: A collisional plateau on Venus? *Lunar Planet. Sci.*, **15**, 104–105.

Burns, J. A. (1975). The angular momenta of solar system bodies: Implications for asteroid strengths. *Icarus*, **25**, 545–554.

Burns, J. A. (1976). Consequences of the tidal slowing of Mercury. *Icarus*, **28**, 453–458.

Butler, R. W. H. (1982). The terminology of structures in thrust belts. *J. Struct. Geol.*, **4**, 239–245.

Byerlee, J. D. (1968). Brittle–ductile transition in rocks. *J. Geophys. Res.*, **73**, 4741–4750.

Byerlee, J. D. (1978). Friction of rocks. *Pure Appl. Geophys.*, **116**, 615–626.

Byrne, P. K., Klimczak, C., Blair, D. M., Ferrari, S., Solomon, S. C., Freed, A. M., Watters, T. R. and Murchie, S. L. (2013a). Tectonic complexity within volcanically infilled craters and basins on Mercury. *Lunar Planet. Sci.*, **44**, abstract 1261.

Byrne, P. K., Klimczak, C., Williams, D. A., Hurwitz, D. M., Solomon, S. C., Head, J. W., Preusker, F. and Oberst, J. (2013b). An assemblage of lava flow features on Mercury. *J. Geophys. Res. Planets*, **118**, 1303–1322.

Byrne, P. K., Klimczak, C., Şengör, A. M. C., Solomon, S. C., Watters, T. R. and Hauck, S. A., II (2014). Mercury's global contraction much greater than earlier estimates. *Nature Geosci.*, **7**, 301–307.

Byrne, P. K., Klimczak, C., McGovern, P. J., Mazarico, E., James, P. B., Neumann, G. A., Zuber, M. T. and Solomon, S. C. (2015). Deep-seated thrust faults bound the Mare Crisium lunar mascon. *Earth Planet. Sci. Lett.*, **427**, 183–190.

Byrne, P. K., Klimczak, C. and LaFond, J. K. (2016a). The East Kaibab monocline: A Terran lobate scarp? *Lunar Planet. Sci.*, **47**, abstract 1022.

Byrne, P. K., Ostrach, L. R., Fassett, C. I., Chapman, C. R., Denevi, B. W., Evans, A. J., Klimczak, C., Banks, M. E., Head, J. W. and Solomon, S. C. (2016b). Widespread effusive volcanism on Mercury likely ended by about 3.5 Ga. *Geophys. Res. Lett.*, **43**, 7408–7416.

Cashman, K. V. (2004). Volatile controls on magma ascent and eruption. In *The State of the Planet: Frontiers and Challenges in Geophysics*, ed. R. S. J. Sparks and C. J. Hawkesworth. Washington, DC: American Geophysical Union, pp. 109–124.

Cavanaugh, J. F., Smith, J. C., Sun, X., Bartels, A. E., Ramos-Izquierdo, L., Krebs, D. J., McGarry, J. F., Trunzo, R., Novo-Gradac, A. M., Britt, J. L., Karsh, J., Katz, R. B., Lukemire, A. T., Szymkiewicz, R., Berry, D. L., Swinski, J. P., Neumann, G. A., Zuber, M. T. and Smith, D. E. (2007). The Mercury Laser Altimeter instrument for the MESSENGER mission. *Space Sci. Rev.*, **131**, 451–479.

Chabot, N. L., Denevi, B. W., Murchie, S. L., Hash, C. D., Ernst, C. M., Blewett, D. T., Nair, H., Laslo, N. R. and Solomon, S. C. (2016). Mapping Mercury: Global imaging strategy and products from the MESSENGER mission. *Lunar Planet. Sci.*, **47**, abstract 1256.

Clark, R. M. and Cox, S. (1996). A modern regression approach to determining fault displacement–length scaling relationships. *J. Struct. Geol.*, **18**, 147–152.

Collins, G. S., Melosh, H. J. and Ivanov, B. A. (2004). Modeling damage and deformation in impact simulations. *Meteorit. Planet. Sci.*, **39**, 217–231.

Colton, G. W., Howard, K. A. and Moore, H. J. (1972). Mare ridges and arches in southern Oceanus Procellarum. In *Apollo 16 Preliminary Science Report*, Special Publication SP-315. Washington, DC: NASA, pp. 29-90–29-93.

Correia, A. C. M. and Laskar, J. (2004). Mercury's capture into the 3/2 spin–orbit resonance as a result of its chaotic dynamics. *Nature*, **429**, 848–850.

Correia, A. C. M. and Laskar, J. (2009). Mercury's capture into the 3/2 spin–orbit resonance including the effect of core–mantle friction. *Icarus*, **201**, 1–11.

Correia, A. C. M. and Laskar, J. (2012). Impact cratering on Mercury: Consequences for the spin evolution. *Astrophys. J. Lett.*, **751**, L43, 5 pp.

Cowie, P. A., Scholz, C. H., Edwards, M. and Malinverno, A. (1993). Fault strain and seismic coupling on mid-ocean ridges. *J. Geophys. Res.*, **98**, 17,911–17,920.

Cowie, P. A., Scholz, C. H., Roberts, G. P., Faure Walker, J. P. and Steer, P. (2013). Viscous roots of active seismogenic faults revealed by geologic slip rate variations. *Nature Geosci.*, **6**, 1036–1040.

Cunje, A. B. and Ghent, R. R. (2016). Caloris basin, Mercury: History of deformation from an analysis of tectonic landforms. *Icarus*, **268**, 131–144.

Cunningham, W. D., Windley, B. F., Dorjnamjaa, D., Badamgarov, G. and Saandar, M. (1996). Late Cenozoic transpression in south-western Mongolia and the Gobi Altai–Tien Shan connection. *Earth Planet. Sci. Lett.*, **140**, 67–82.

Dainty, A. M., Toksöz, M. N., Anderson, K. R., Pines, P. J., Nakamura, Y. and Latham, G. (1974). Seismic scattering and shallow structure of the Moon in Oceanus Procellarum. *Moon*, **9**, 11–29.

Dana, J. D. (1863). *Manual of Geology: Treating of the Principles of the Science with Special Reference to American Geological History, for the Use of Colleges, Academies and Schools of Science*. Philadelphia, PA: Theodore Bliss & Co., 798 pp.

Dana, J. D. (1873). On some results of the Earth's contraction from cooling, including a discussion of the origin of mountains and the nature of the Earth's interior. *Amer. J. Sci.*, **5**, 423–443.

Delamétherie, J.-C. (1795a). *Théorie de la Terre*, tome premier. Paris: Chez Maradan, 422 pp.

Delamétherie, J.-C. (1795b). *Théorie de la Terre*, tome second. Paris: Chez Maradan, 456 pp.

de Buffon, C. (1778). *Histoire Naturelle Générale et Particulière*. Paris: Imprimerie Royale, 615 pp.

Denevi, B. W., Robinson, M. S., Solomon, S. C., Murchie, S. L., Blewett, D. T., Domingue, D. L., McCoy, T. J., Ernst, C. M., Head, J. W., III, Watters T. R. and Chabot, N. L. (2009). The evolution of Mercury's crust: A global perspective from MESSENGER. *Science*, **324**, 613–618.

Denevi, B. W., Ernst, C. M., Meyer, H. M., Robinson, M. S., Murchie, S. L., Whitten, J. L., Head, J. W., Watters, T. R., Solomon, S. C., Ostrach, L. R., Chapman, C. R., Byrne, P. K., Klimczak, C. and Peplowski, P. N. (2013). The distribution and origin of smooth plains on Mercury. *J. Geophys. Res. Planets*, **118**, 891–907.

Di Achille, G., Popa, C., Massironi, M., Epifani, E. M., Zusi, M., Cremonese, G. and Palumbo, P. (2012). Mercury's radius change estimates revisited using MESSENGER data. *Icarus*, **221**, 456–460.

Dombard, A. J. and Hauck, S. A., II (2008). Despinning plus global contraction and the orientation of lobate scarps on Mercury: Predictions for MESSENGER. *Icarus*, **198**, 274–276.

Dombard, A. J., Hauck, S. A., II, Solomon, S. C. and Phillips, R. J. (2001). Potential for long-wavelength folding on Mercury. *Lunar Planet. Sci.*, **32**, abstract 2035.

Duncan, R. A. and Richards, M. A. (1991). Hotspots, mantle plumes, flood basalts, and true polar wander. *Rev. Geophys.*, **29**, 31–50.

Dzurisin, D. (1978). The tectonic and volcanic history of Mercury as inferred from studies of scarps, ridges, troughs, and other lineaments. *J. Geophys. Res.*, **83**, 4883–4906.

Egea-González, I., Ruiz, J., Fernández, C., Williams, J.-P., Márquez, Á. and Lara, L. M. (2012). Depth of faulting and ancient heat flows in the Kuiper region of Mercury from lobate scarp topography. *Planet. Space Sci.*, **60**, 193–198.

Élie de Beaumont, L. (1829). Faits pour Servir a l'Histoire des Montagnes de l'Oisans. *Mémoires de la Société d'Histoire Naturelle de Paris*, 1–32.

Élie de Beaumont, L. (1852). *Notice sur les Systèmes de Montagnes*, III. Paris: P. Bertrand, pp. 1069–1543.

Emmermann, R. and Lauterjung, J. (1997). The German continental deep drilling program KTB: Overview and major results. *J. Geophys. Res.*, **102**, 18,179–18,201.

Engelder, T. (1992). *Stress Regimes in the Lithosphere*. Princeton, NJ: Princeton University Press.

Ernst, C. M., Murchie, S. L., Barnouin, O. S., Robinson, M. S., Denevi, B. W., Blewett, D. T., Head, J. W., Izenberg, N. R., Solomon, S. C. and Roberts, J. H. (2010). Exposure of spectrally distinct material by impact craters on Mercury: Implications for global stratigraphy. *Icarus*, **209**, 210–223.

Evans, A. J., Brown, S. M. and Solomon, S. C. (2015). Characteristics of early mantle convection and melting on Mercury. *Lunar Planet. Sci.*, **46**, abstract 2414.

Fassett, C. I. and Crowley, M. C. (2016). High-resolution stereo digital terrain models of Mercury: Crater degradation and morphometry. *Lunar Planet. Sci.*, **47**, abstract 1046.

Fassett, C. I., Head, J. W., Blewett, D. T., Chapman, C. R., Dickson, J. L., Murchie, S. L., Solomon, S. C. and Watters, T. R. (2009). Caloris impact basin: Exterior geomorphology, stratigraphy, morphometry, radial sculpture, and smooth plains deposits. *Earth Planet. Sci. Lett.*, **285**, 297–308.

Fassett, C. I., Head, J. W., Baker, D. M. H., Zuber, M. T., Smith, D. E., Neumann, G. A., Solomon, S. C., Klimczak, C., Strom, R. G., Chapman, C. R., Prockter, L. M., Phillips, R. J., Oberst, J. and Preusker, F. (2012). Large impact basins on Mercury: Global distribution, characteristics, and modification history from MESSENGER orbital data. *J. Geophys. Res.*, **117**, E00L08.

Fegan, E. R., Rothery, D. A., Marchi, S., Massironi, M., Conway, S. J. and Anand, M. (2017). Late movement of basin-edge lobate scarps on Mercury. *Icarus*, **288**, 226–234.

Ferrari, S., Massironi, M., Marchi, S., Byrne, P. K., Klimczak, C., Martellato, E. and Cremonese, G. (2015). Age relationships of the Rembrandt basin and Enterprise Rupes, Mercury. In *Volcanism and Tectonism Across the Solar System*, ed. T. Platz, M. Massironi, P. K. Byrne and H. Hiesinger, Special Publication 401. London: Geological Society, pp. 159–172.

Fletcher, R. C. and Hallet B. (1983). Unstable extension of the lithosphere: A mechanical model for basin-and-range structure. *J. Geophys. Res.*, **88**, 7457–7466.

Fodor, L., Turki, S. M., Dalub, H. and Al Gerbi, A. (2005). Fault-related folds and along-dip segmentation of breaching faults: Syn-diagenetic deformation in the south-western Sirt basin, Libya. *Terra Nova*, **17**, 121–128.

Freed, A. M., Solomon, S. C., Watters, T. R., Phillips, R. J. and Zuber, M. T. (2009). Could Pantheon Fossae be the result of the Apollodorus crater-forming impact within the Caloris basin, Mercury? *Earth Planet. Sci. Lett.*, **285**, 320–327.

Freed, A. M., Blair, D. M., Watters, T. R., Klimczak, C., Byrne, P. K., Solomon, S. C., Zuber, M. T. and Melosh, H. J. (2012). On the origin of graben and ridges within and near volcanically buried craters and basins in Mercury's northern plains. *J. Geophys. Res.*, **117**, E00L06, doi:10.1029/2012JE004119.

Galluzzi, V., Di Achille, G., Ferranti, L., Popa, C. and Palumbo, P. (2015). Faulted craters as indicators for thrust motions on Mercury. In *Volcanism and Tectonism Across the Solar System*, ed. T. Platz, M. Massironi, P. K. Byrne and H. Heisinger, Special Publication 401. London: Geological Society, pp. 313–326.

Gettings, M. E. (1988). Variation of depth to the brittle-ductile transition due to cooling of a midcrustal intrusion. *Geophys. Res. Lett.*, **15**, 213–216.

Giacomini, L., Massironi, M., Marchi, S., Fassett, C. I., Di Achille, G. and Cremonese, G. (2015). Age dating of an extensive thrust system on Mercury: Implications for the planet's thermal evolution. In *Volcanism and Tectonism Across the Solar System*, ed. T. Platz, M. Massironi, P. K. Byrne and H. Heisinger, Special Publication 401. London: Geological Society, pp. 291–312.

Glazner, A. F. (1991). Plutonism, oblique subduction, and continental growth: An example from the Mesozoic of California. *Geology*, **19**, 784–786.

Golombek, M. P., Plescia, J. B. and Franklin, B. J. (1991). Faulting and folding in the formation of planetary wrinkle ridges. In *Proceedings of Lunar and Planetary Science*, **21**. Houston, TX: Lunar and Planetary Institute, pp. 679–693.

Golombek, M. P., Anderson, F. S. and Zuber, M. T. (2001). Martian wrinkle ridge topography: Evidence for subsurface faults from MOLA. *J. Geophys. Res.*, **106**, 23,811–23,821.

Goudge, T. A., Head, J. W., Kerber, L., Blewett, D. T., Denevi, B. W., Domingue, D. L., Gillis-Davis, J. J., Gwinner, K., Helbert, J., Holsclaw, G. M., Izenberg, N. R., Klima, R. L., McClintock, W. E., Murchie, S. L., Neumann, G. A., Smith, D. E., Strom, R. G., Xiao, Z., Zuber, M. T. and Solomon, S. C. (2014). Global inventory and characterization of pyroclastic deposits on Mercury: New insights into pyroclastic activity from MESSENGER orbital data. *J. Geophys. Res. Planets*, **119**, 635–658.

Grott, M., Breuer, D. and Laneuville, M. (2011). Thermo-chemical evolution and global contraction of Mercury. *Earth Planet. Sci. Lett.*, **307**, 135–146.

Habermann M. A. and Klimczak, C. (2015). Tectonic controls of pyroclastic volcanism on Mercury. Presented at 2018 Fall Meeting, American Geophysical Union, abstract P53A-2101, San Francisco, CA, 14–18 December.

Hamilton, W. B. (1995). Subduction systems and magmatism. In *Volcanism Associated with Extension at Consuming Plate Margins*, ed. J. L. Smellie, Special Publication 81. London: Geological Society, pp. 3–28.

Hatcher, R. D., Jr. (1989). Tectonic synthesis of the U.S. Appalachians. In *The Appalachian-Ouachita Orogen in the United States*, ed. R. D. Hatcher, Jr., W. A. Thomas and G. W. Viele. Boulder, CO: Geological Society of America, pp. 511–536.

Hauck, S. A., II, Dombard, A. J., Phillips, R. J. and Solomon, S. C. (2004). Internal and tectonic evolution of Mercury. *Earth Planet. Sci. Lett.*, **222**, 713–728.

Hauck, S. A., II, Margot, J.-L., Solomon, S. C., Phillips, R. J., Johnson, C. L., Lemoine, F. G., Mazarico, E., McCoy, T. J., Padovan, S., Peale, S. J., Perry, M. E., Smith, D. E. and Zuber, M. T. (2013). The curious case of Mercury's internal structure. *J. Geophys. Res. Planets*, **118**, 1204–1220.

Hawkins, S. E., III, Boldt, J. D., Darlington, E. H., Espiritu, R., Gold, R. E., Gotwols, B., Grey, M. P., Hash, C. D., Hayes, J. R., Jaskulek, S. E., Kardian, C. J., Jr., Keller, M. R., Malaret, E. R., Murchie, S. L., Murphy, P. K., Peacock, K., Prockter, L. M., Reiter, R. A., Robinson, M. S., Schaefer, E. D., Shelton, R. G., Sterner, R. E., II, Taylor, H. W., Watters, T. R. and Williams, B. D. (2007). The Mercury Dual Imaging System on the MESSENGER spacecraft. *Space Sci. Rev.*, **131**, 247–338.

Head, J. W., Murchie, S. L., Prockter, L. M., Robinson, M. S., Solomon, S. C., Strom, R. G., Chapman, C. R., Watters, T. R., McClintock, W. E., Blewett, D. T. and Gillis-Davis, J. J. (2008). Volcanism on Mercury: Evidence from the first MESSENGER flyby. *Science*, **321**, 69–72.

Head, J. W., Murchie, S. L., Prockter, L. M., Solomon, S. C., Strom, R. G., Chapman, C. R., Watters, T. R., Blewett, D. T., Gillis-Davis, J. J., Fassett, C. I., Dickson, J. L., Hurwitz, D. M. and Ostrach, L. R. (2009). Evidence for intrusive activity on Mercury from the first MESSENGER flyby. *Earth Planet. Sci. Lett.*, **285**, 251–262.

Head, J. W., Chapman, C. R., Strom, R. G., Fassett, C. I., Denevi, B. W., Blewett, D. T., Ernst, C. M., Watters, T. R., Solomon, S. C., Murchie, S. L., Prockter, L. M., Chabot, N. L., Gillis-Davis, J. J., Whitten, J. L., Goudge, T. A., Baker, D. M. H., Hurwitz, D. M., Ostrach, L. R., Xiao, Z., Merline, W. J., Kerber, L., Dickson, J. L., Oberst, J., Byrne, P. K., Klimczak, C. and Nittler, L. R. (2011). Flood volcanism in the northern high latitudes of Mercury revealed by MESSENGER. *Science*, **333**, 1853–1856.

Heap, M. J., Byrne, P. K. and Mikhail, S. (2017). Low surface gravitational acceleration of Mars results in a thick and weak lithosphere: Implications for topography, volcanism, and hydrology. *Icarus*, **281**, 103–114.

Hoek, E. and Diederichs, M. S. (2006). Empirical estimation of rock mass modulus. *Int. J. Rock Mech. Min. Sci.*, **43**, 203–215.

Jacobs, J. A., Russell, R. D. and Wilson, J. T. (1959). *Physics and Geology*. New York: McGraw-Hill, 424 pp.

James, P. B., Byrne, P. K., Solomon, S. C., Zuber, M. T. and Phillips, R. J. (2014). Surface strains associated with the evolution of Mercury's domical swells. Presented at 2014 Fall Meeting, American Geophysical Union, abstract P21C-3939, San Francisco, CA, 15–19 December.

James, P. B., Zuber, M. T., Phillips, R. J. and Solomon, S. C. (2015). Support of long-wavelength topography on Mercury inferred from MESSENGER measurements of gravity and topography. *J. Geophys. Res. Planets*, **120**, 287–310.

Jeffreys, H. (1976). *The Earth: Its Origin, History and Physical Constitution*, 6th edn. Cambridge: Cambridge University Press, 574 pp.

Jerram, D. A. and Widdowson, M. (2005). The anatomy of continental flood basalt provinces: Geological constraints on the processes and products of flood volcanism. *Lithos*, **79**, 385–405.

Kachanov, M. (1992). Effective elastic properties of cracked solids: Critical review of some basic concepts. *Appl. Mech. Rev.*, **45**, 304–335.

Karimi, M. and Dombard, A. J. (2014). A study regarding the possibility of true polar wander on the asteroid Vesta. Presented at 2014 Fall Meeting, Amererican Geophysical Union, abstract P43C-4003, San Francisco, CA, 15–19 December.

Kaula, W. M. (1968). *An Introduction to Planetary Physics: The Terrestrial Planets*. New York: John Wiley.

Kennedy, P. J., Freed, A. M. and Solomon, S. C. (2008). Mechanisms of faulting in and around Caloris basin, Mercury. *J. Geophys. Res.*, **113**, E08004, doi:10.1029/2007JE002992.

Kerber, L., Head, J. W., Solomon, S. C., Murchie, S. L., Blewett, D. T. and Wilson, L. (2009). Explosive volcanic eruptions on Mercury: Eruption conditions, magma volatile content, and implications for interior volatile abundances. *Earth Planet. Sci. Lett.*, **285**, 263–271.

Kerber, L., Head, J. W., Blewett, D. T., Solomon, S. C., Wilson, L., Murchie, S. L., Robinson, M. S., Denevi, B. W. and Domingue, D. L. (2011). The global distribution of pyroclastic deposits on Mercury: The view from MESSENGER flybys 1–3. *Planet. Space Sci.*, **59**, 1895–1909.

King, S. D. (2008). Pattern of lobate scarps on Mercury's surface reproduced by a model of mantle convection. *Nature Geosci.*, **1**, 229–232.

Klimczak, C. (2014). Geomorphology of lunar grabens requires igneous dikes at depth. *Geology*, **42**, 963–966.

Klimczak, C. (2015). Limits on the brittle strength of planetary lithospheres undergoing global contraction. *J. Geophys. Res. Planets*, **120**, 2135–2151.

Klimczak, C., Schultz, R. A. and Nahm, A. L. (2010). Evaluation of the origin hypotheses of Pantheon Fossae, central Caloris basin, Mercury. *Icarus*, **209**, 262–270.

Klimczak, C., Watters, T. R., Ernst, C. M., Freed, A. M., Byrne, P. K., Solomon, S. C., Blair, D. M. and Head, J. W. (2012). Deformation associated with ghost craters and basins in volcanic smooth plains on Mercury: Strain analysis and implications for plains evolution. *J. Geophys. Res.*, **117**, E00L03, doi:10.1029/2012JE004100.

Klimczak, C., Ernst, C. M., Byrne, P. K., Solomon, S. C., Watters, T. R., Murchie, S. L., Preusker, F. and Balcerski, J. A. (2013a). Insights into the subsurface structure of the Caloris basin, Mercury, from assessments of mechanical layering and changes in long-wavelength topography. *J. Geophys. Res. Planets*, **118**, 2030–2044.

Klimczak, C., Byrne, P. K., Solomon, S. C., Nimmo, F., Watters, T. R., Denevi, B. W., Ernst, C. M. and Banks, M. E. (2013b). The role of thrust faults as conduits for volatiles on Mercury. *Lunar Planet. Sci.*, **44**, abstract 1390.

Klimczak, C., Byrne, P. K. and Solomon, S. C. (2015). A rock-mechanical assessment of Mercury's global tectonic fabric. *Earth Planet. Sci. Lett.*, **416**, 82–90.

Knibbe, J. S. and van Westrenen, W. (2016). Mercury's past rotation and cratering distribution. *Lunar Planet. Sci.*, **47**, abstract 1445.

Kohlstedt, D. L. and Mackwell, S. J. (2010). Strength and deformation of planetary lithospheres. In *Planetary Tectonics*, ed. T. R. Watters and R. A. Schultz. New York: Cambridge University Press, pp 397–456.

Kohlstedt, D. L., Evans, B. and Mackwell, S. J. (1995). Strength of the lithosphere: Constraints imposed by laboratory experiments. *J. Geophys. Res.*, **100**, 17,587–17,602.

Kuang, W., Jiang, W., Roberts, J. and Frey, H. V. (2014). Could giant basin-forming impacts have killed Martian dynamo? *Geophys. Res. Lett.*, **41**, 8006–8012.

Kulhawy, F. H. (1975). Stress deformation properties of rock and rock discontinuities. *Eng. Geol.*, **9**, 327–350.

Last, G. V., Winsor, K. and Unwin, S. G. (2012). *A Summary of Information on the Behavior of the Yakima Fold Belt as a Structural Entity*. Topical Report, Richland, WA: Pacific Northwest National Laboratory, 82 pp.

Le Corvec, N., McGovern, P. J., Grosfils, E. B. and Galanga, G. (2015). Effects of crustal-scale mechanical layering on magma chamber failure and magma propagation within the Venusian lithosphere. *J. Geophys. Res. Planets*, **120**, 1279–1297.

Le Feuvre, M. and Wieczorek, M. A. (2011). Nonuniform cratering of the Moon and a revised crater chronology of the inner Solar System. *Icarus*, **214**, 1–20.

Lissauer, J. J. (1985). Can cometary bombardment disrupt synchronous rotation of planetary satellites? *J. Geophys. Res.*, **90**, 11,289–11,293.

Liu, L., Li, S., Dai, L., Suo, Y., Liu, B., Zhang, G., Wang, Y. and Liu, E. (2012). Geometry and timing of Mesozoic deformation in the western part of the Xuefeng Tectonic Belt, South China: Implications for intra-continental deformation. *J. Asian Earth Sci.*, **49**, 330–338.

Lopez, V., Ruiz, J. and Vázquez, A. (2015). Evidence for two stages of compressive deformation in a buried basin of Mercury. *Icarus*, **254**, 18–23.

Lucchitta, B. K. (1976). Mare ridges and related highland scarps: Results of vertical tectonism. *Proc. Lunar Sci. Conf.*, **7**, 2761–2782.

Mangold, N., Allemand, P. and Thomas, P. G. (1998). Wrinkle ridges of Mars: Structural analysis and evidence for shallow deformation controlled by ice-rich décollements. *Planet. Space. Sci.*, **46**, 345–356.

Marchi, S., Massironi, M., Cremonese, G., Martellato, E., Giacomini, L. and Prockter, L. M. (2011). The effects of the target material properties and layering on the crater chronology: The case of Raditladi and Rachmaninoff basins on Mercury. *Planet. Space Sci.*, **59**, 1968–1980.

Marchi, S., Chapman, C. R., Fassett, C. I., Head, J. W., Bottke, W. F. and Strom, R. G. (2013). Global resurfacing of Mercury 4.0–4.1 billion years ago by heavy bombardment and volcanism. *Nature*, **499**, 59–61.

Marrett, R. and Emerman, S. H. (1992). The relations between faulting and mafic magmatism in the Altiplano-Puna plateau (central Andes). *Earth Planet. Sci. Lett.*, **112**, 53–59.

Marti, J., Ablay, G. J., Redshaw, L. T. and Sparks, R. S. J. (1994). Experimental studies of collapse calderas. *J. Geol. Soc. London*, **151**, 919–929.

Massironi, M., Di Achille, G., Rothery, D. A., Galluzzi, V., Giacomini, L., Ferrari, S., Zusi, M., Cremonese, G. and Palumbo P. (2015). Lateral ramps and strike-slip kinematics on Mercury. In *Volcanism and Tectonism Across the Solar System*, ed. T. Platz, M. Massironi, P. K. Byrne and H. Heisinger, Special Publication 401. London: Geological Society, pp. 269–290.

Masursky, H., Colton, G. W. and El-Baz, F. (1978). *Apollo over the Moon: A View from Orbit*, Special Publication SP-362. Washington, DC: NASA.

Matsuyama, I. and Nimmo, F. (2009). Gravity and tectonic patterns of Mercury: Effect of tidal deformation, spin-orbit resonance, non-zero eccentricity, despinning, and reorientation. *J. Geophys. Res.*, **114**, E01010, doi:10.1029/2008JE003252.

Matsuyama, I., Nimmo, F. and Mitrovica, J. X. (2014). Planetary reorientation. *Annu. Rev. Earth Planet. Sci.*, **42**, 605–634.

McAdoo, D. C. and Sandwell, D. T. (1985). Folding of oceanic lithosphere. *J. Geophys. Res.*, **90**, 8563–8569.

McClay, K. R. (1992). *Thrust Tectonics*. London: Chapman and Hall.

McDougall, J. W., Hussain, A. and Yeats, R. S. (1993). The Main Boundary Thrust and propagation of deformation into the foreland fold-and-thrust belt in northern Pakistan near the Indus River. In *Himalayan Tectonics*, ed. P. J. Treloar and M. P. Searle, Special Publication 74. London: Geological Society, pp. 581–588.

McGovern, P. J. and Litherland, M. M. (2011). Lithospheric stress and basaltic magma ascent on the Moon, with implications for large volcanic provinces and edifices. *Lunar Planet. Sci.*, **42**, abstract 2587.

Melosh, H. J. (1975). Large impact craters and the Moon's orientation. *Earth. Planet. Sci. Lett.*, **26**, 353–360.

Melosh, H. J. (1977). Global tectonics of a despun planet. *Icarus*, **31**, 221–243.

Melosh, H. J. (1980). Tectonic patterns on a tidally distorted planet. *Icarus*, **43**, 334–337.

Melosh, H. J. (1984). Impact ejection, spallation, and the origin of meteorites. *Icarus*, **59**, 234–260.

Melosh, H. J. and Dzurisin, D. (1978). Mercurian global tectonics: A consequence of tidal despinning? *Icarus*, **35**, 227–236.

Melosh, H. J. and McKinnon, W. B., (1988). The tectonics of Mercury. In *Mercury*, ed. F. Vilas, C. R. Chapman and M. S. Matthews. Tucson, AZ: University of Arizona Press, pp. 374–400.

Michel, N. C., Hauck, S. A., II, Solomon, S. C., Phillips, R. J., Roberts, J. H. and Zuber, M. T. (2013). Thermal evolution of Mercury as constrained by MESSENGER observations. *J. Geophys. Res. Planets*, **118**, 1033–1044.

Montési, L. G. J. and Zuber, M. T (2003). Clues to the lithospheric structure of Mars from wrinkle ridge sets and localization instability. *J. Geophys. Res.*, **108**, E65048, doi:10.1029/2002JE001974.

Moore, H. J., Boyce, J. M. and Hahn, D. A. (1980). Small impact craters in the lunar regolith: Their morphologies, relative ages, and rates of formation. *Moon Planets*, **23**, 231–252.

Morbidelli, A., Tsiganis, K., Batygin, K., Crida, A. and Gomes, R. (2012). Explaining why the uranian satellites have equatorial prograde orbits despite the large planetary obliquity. *Icarus*, **219**, 737–740.

Mueller, K. and Golombek, M. (2004). Compressional structures on Mars. *Annu. Rev. Earth Planet. Sci.*, **32**, 435–464.

Murchie, S. L., Watters, T. R., Robinson, M. S., Head, J. W., Strom, R. G., Chapman, C. R., Solomon, S. C., McClintock, W. E., Prockter, L. M., Domingue, D. L. and Blewett, D. T. (2008). Geology of the Caloris basin, Mercury: A view from MESSENGER. *Science*, **321**, 73–76.

Murray, B. C., Belton, M. J. S., Danielson, G. E., Davies, M. E., Gault, D. E., Hapke, B., O'Leary, B., Strom, R. G., Suomi, V. and Trask, N. (1974). Mercury's surface: Preliminary description and interpretation from Mariner 10 pictures. *Science*, **185**, 169–179.

Neil, E. A. and Houseman, G. A. (1999). Rayleigh–Taylor instability of the upper mantle and its role in intraplate orogeny. *Geophys. J. Int.*, **138**, 89–107.

Nimmo, F. and Watters, T. R. (2004). Depth of faulting on Mercury: Implications for heat flux and crustal and effective elastic thickness. *Geophys. Res. Lett.*, **31**, L02701, doi:10.1029/2003GL018847.

Nimmo, F. and Matsuyama, I. (2007). Reorientation of icy satellites by impact basins. *Geophys. Res. Lett.*, **34**, L19203, doi:10.1029/2007GL030798.

Nimmo, F. and Pappalardo, R. T. (2006). Diapir-induced reorientation of Saturn's moon Enceladus. *Nature*, **441**, 614–616.

Norris, R. J. and Cooper, A. F. (1997). Erosional control on the structural evolution of a transpressional thrust complex on the Alpine fault, New Zealand. *J. Struct. Geol.*, **19**, 1323–1342.

Noyelles, B., Frouard, J., Makarov, V. V. and Efroimsky, M. (2014). Spin–orbit evolution of Mercury revisited. *Icarus*, **241**, 26–44.

Oberst, J., Preusker, F., Phillips, R. J., Watters, T. R., Head, J. W., Zuber, M. T. and Solomon, S. C. (2010). The morphology of Mercury's Caloris basin as seen in MESSENGER stereo topographic models. *Icarus*, **209**, 230–238.

Ostrach, L. R., Robinson, M. S., Whitten, J. L., Fassett, C. I., Strom, R. G., Head, J. W. and Solomon, S. C. (2015). Extent, age, and resurfacing history of the northern smooth plains on Mercury from MESSENGER observations. *Icarus*, **250**, 602–622.

Padovan, S., Wieczorek, M. A., Margot, J.-L., Tosi, N. and Solomon, S. C. (2015). Thickness of the crust of Mercury from geoid-to-topography ratios. *Geophys. Res. Lett.*, **42**, 1029–1038,

Passchier, C. W. and Trouw, R. A. J. (2006). *Microtectonics*. New York: Springer Verlag.

Peale, S. J. (1988). The rotational dynamics of Mercury and the state of its core. In *Mercury*, ed. F. Vilas, C. R. Chapman and M. S. Matthews. Tucson, AZ: University of Arizona Press, pp. 461–493.

Pechmann, J. B. and Melosh, H. J. (1979). Global fracture patterns of a despun planet: Application to Mercury. *Icarus*, **38**, 243–250.

Peplowski, P. N., Evans, L. G., Hauck, S. A., II, McCoy, T. J., Boynton, W. V., Gillis-Davis, J. J., Ebel, D. S., Golsten, J. O., Hamara, D. K., Lawrence, D. J., McNutt, R. L., Nittler, L. R., Solomon, S. C., Rhodes, E. A., Sprague, A. L., Starr, R. D. and Stockstill-Cahill, K. R. (2011). Radioactive elements on Mercury's surface from MESSENGER: Implications for the planet's formation and evolution. *Science*, **333**, 1850–1852.

Petterson, M. G., Neal, C. R., Mahoney, J. J., Kroenke, L. W., Saunders, A. D., Babbs, T. L., Duncan, R. A., Tolia, D. and McGrail, B. (1997). Structure and deformation of north and central Malaita, Solomon Islands: Tectonic implications for the Ontong Java Plateau–Solomon arc collision, and for the fate of oceanic plateaus. *Tectonophysics*, **283**, 1–33.

Plescia, J. B. (1991). Wrinkle ridges in Lunae Planum Mars: Implications for shortening and strain. *Geophys. Res. Lett.*, **18**, 913–916.

Plescia, J. B. (1993). Wrinkle ridges of Arcadia Planitia, Mars. *J. Geophys. Res.*, **98**, 15,049–15,059.

Plescia, J. B. and Golombek, M. P. (1986). Origin of planetary wrinkle ridges based on the study of terrestrial analogs. *Geol. Soc. Amer. Bull.*, **97**, 1289–1299.

Poblet, J. and Lisle, R. J. (2011). Kinematic evolution and structural styles of fold-and-thrust belts. In *Kinematic Evolution and Structural Styles of Fold-and-Thrust Belts*, ed. J. Poblet and R. J. Lisle, Special Publication 349. London: Geological Society, pp.1–24.

Pohn, H. A. and Offield, T. W. (1970). Lunar crater morphology and relative age determination of lunar geologic units. Part 1: Classification. Part 2: Applications. In *Geological Survey Research 1970*, Professional Paper 69–209. Denver, CO: U.S. Geological Survey, 35 pp.

Preusker, F., Oberst, J., Head, J. W., Watters, T. R., Robinson, M. S., Zuber, M. T. and Solomon, S. C. (2011). Stereo topographic models of Mercury after three MESSENGER flybys. *Planet. Space Sci.*, **59**, 1910–1917.

Prockter, L. M., Watters, T. R., Chapman, C. R., Denevi, B. W., Head, J. W., Solomon, S. C., Murchie, S. L., Barnouin, O. S., Robinson, M. S., Blewett, D. T., Gillis-Davis, J. J. and Gaskell, R. W. (2009). The curious case of Raditladi basin. *Lunar Planet. Sci.*, **40**, abstract 1758.

Prockter, L. M., Ernst, C. M., Denevi, B. W., Chapman, C. R., Head, J. W., Fassett, C. I., Merline, W. J., Solomon, S. C., Watters, T. R., Strom, R. G., Cremonese, G., Marchi, S. and Massironi, M. (2010). Evidence for young volcanism on Mercury from the third MESSENGER flyby. *Science*, **329**, 668–671.

Prockter, L. M., Baker, D. M. H., Head, J. W., Murchie, S. L., Ernst, C. M., Chapman, C. R., Denevi, B. W., Solomon, S. C., Watters, T. R. and Massironi, M. (2011). The geology of medium-sized impact basins on Mercury. *Abstracts with Programs*, **43**, paper 142-11. Boulder, CO: Geological Society of America.

Ritzer, J. A., Hauck, S. A., II and Barnouin, O. S. (2010). Mechanical structure of Mercury's lithosphere from MESSENGER observations of lobate scarps. *Lunar Planet. Sci.*, **41**, abstract 2122.

Roberts, J. H. and Barnouin, O. S. (2012). The effect of the Caloris impact on the mantle dynamics and volcanism of Mercury. *J. Geophys. Res.*, **117**, E02007.

Roberts, J. H. and Nimmo, F. (2008). Near-surface heating on Enceladus and the south polar thermal anomaly. *Geophys. Res. Lett.*, **35**, L09201, doi:10.1029/2011JE003876.

Roeder, D. (1988). Andean-age structure of Eastern Cordillera (Province of La Paz, Bolivia). *Tectonics*, **7**, 23–39.

Roeder, D. (2009). *American and Tethyan Fold-Thrust Belts*. Berlin: Gebrüder Borntraeger, 168 pp.

Rothery, D. A. and Massironi, M. (2010). Beagle Rupes: Evidence for a basal decollement of regional extent in Mercury's lithosphere. *Icarus*, **209**, 256–261.

Scholz, C. H. and Cowie, P. A. (1990). Determination of total strain from faulting using slip measurements. *Nature*, **346**, 837–839.

Schubert, G., Ross, M. N., Stevenson, D. J. and Spohn, T. (1988). Mercury's thermal history and the generation of its magnetic field. In *Mercury*, ed. F. Vilas, C. R. Chapman and M. S. Matthews. Tucson, AZ: University of Arizona Press, pp. 429–460.

Schultz, R. A. (1985). Assessment of global and regional tectonic models for faulting in the ancient terrains of Mars. *J. Geophys. Res.*, **90**, 7849–7860.

Schultz, R. A. (1993). Brittle strength of basaltic rock masses with applications to Venus. *J. Geophys. Res.*, **98**, 10,883–10,895.

Schultz, R. A. (1995). Limits on strength and deformation properties of jointed basaltic rock masses. *Rock. Mech. Rock. Eng.*, **28**, 1–15.

Schultz, R. A. (1996). Relative scale and the strength and deformability of rock masses. *J. Struct. Geol.*, **18**, 1139–1149.

Schultz, R. A. (2000). Localization of bedding plane slip and backthrust faults above blind thrust faults: Keys to wrinkle ridge structure. *J. Geophys. Res.*, **105**, 12,035–12,052.

Schultz, R. A. and Fossen, H. (2008). Terminology for structural discontinuities. *Amer. Assoc. Petrol. Geol. Bull.*, **92**, 853–867.

Schultz, R. A. and Tanaka, K. L. (1994). Lithospheric-scale buckling and thrust structures on Mars: The Coprates rise and south Tharsis ridge belt. *J. Geophys. Res.*, **99**, 8371–8385.

Segall, P. (1984). Formation and growth of extensional fracture sets. *Geol. Soc. Amer. Bull.*, **95**, 454–462.

Şengör, A. M. C. and Sakınç, M. (2001). Structural rocks: Stratigraphic implications. In *Paradoxes in Geology*, ed. U. Briegel and W. J. Xiao. Amsterdam: Elsevier.

Şengör, A. M. C., Natal'in, B. A. and Burtman, V. S. (1993). Evolution of the Altaid tectonic collage and Palaeozoic crustal growth in Eurasia. *Nature*, **364**, 299–307.

Smith, D. E., Zuber, M. T., Phillips, R. J., Solomon, S. C., Hauck, S. A., Lemoine, F. G., Mazarico, E., Neumann, G. A., Peale, S. J., Margot, J. L., Johnson, C. L., Torrence, M. H., Perry, M. E., Rowlands, D. D., Goossens, S., Head, J. W. and Taylor, A. H. (2012). Gravity field and internal structure of Mercury from MESSENGER. *Science*, **336**, 214–217.

Solomatov, V. S. and Moresi, L. N. (2000). Scaling of time-dependent stagnant lid convection: Application to small-scale convection on Earth and other terrestrial planets. *J. Geophys. Res.*, **105**, 21,795–21,817.

Solomon, S. C. (1977). The relationship between crustal tectonics and internal evolution in the Moon and Mercury. *Phys. Earth Planet. Inter.*, **15**, 135–145.

Solomon, S. C. (1978). On volcanism and thermal tectonics on one-plate planets. *Geophys. Res. Lett.*, **5**, 461–464.

Solomon, S. C. and Chaiken, J. (1976). Thermal expansion and thermal stress in the Moon and terrestrial planets: Clues to early thermal history. *Proc. Lunar Sci. Conf.*, **7**, 3229–3243.

Solomon, S. C., McNutt, R. L., Jr., Watters, T. R., Lawrence, D. J., Feldman, W. C., Head, J. W., Krimigis, S. M., Murchie, S. L., Phillips, R. J., Slavin, J. A. and Zuber, M. T. (2008). Return to Mercury: A global perspective on MESSENGER's first Mercury flyby. *Science*, **321**, 59–62.

Solomon, S. C., Klimczak, C., Byrne, P. K., Hauck, S. A., II, Balcerski, J. A., Dombard, A. J., Zuber, M. T., Smith, D. E., Phillips, R. J., Head, J. W. and Watters, T. R. (2012). Long-wavelength topographic change on Mercury: Evidence and mechanisms. *Lunar Planet. Sci.*, **43**, abstract 1578.

Spudis, P. D. and Guest, J. E. (1988). Stratigraphy and geologic history of Mercury. In *Mercury*, ed. F. Vilas, C. R. Chapman and M. S. Matthews. Tucson, AZ: University of Arizona Press, pp. 118–164.

Stille, H. (1920). Über Alter und Art der Phasen variszischer Gebirgsbildung. *Nachr. Ges. Wiss. Göttingen*, 1–7.

Stille, H. (1940). Einführung in den Bau Amerikas. In *Einführung in den Bau Amerikas*. Berlin: Gebrüder Borntraeger.

Strom, R. G., Trask, N. J. and Guest, J. E. (1975). Tectonism and volcanism on Mercury. *J. Geophys. Res.*, **80**, 2478–2507.

Strom, R. G., Chapman, C. R., Merline, W. J., Solomon, S. C. and Head, J. W. (2008). Mercury cratering record viewed from MESSENGER's first flyby. *Science*, **321**, 79–81.

Strom, R. G., Banks, M. E., Chapman, C. R., Fassett, C. I., Forde, J. A., Head, J. W., Merline, W. J., Prockter, L. M. and Solomon, S. C. (2011). Mercury crater statistics from MESSENGER flybys: Implications for stratigraphy and resurfacing history. *Planet. Space Sci.*, **59**, 1960–1967.

Suess, E. (1883). In *Das Antlitz der Erde, Vol. I*, ed. E. Suess. Leipzig: F. Tempsky.

Suess, E. (1909). In *Das Antlitz der Erde, Vol. III.2*, ed. E. Suess. Leipzig: F. Tempsky.

Thomas, P. G., Masson, P. and Fleitout, L. (1988). Tectonic history of Mercury. In *Mercury*, ed. F. Vilas, C. R. Chapman and M. S. Matthews, Tucson, AZ: University of Arizona Press, pp. 401–428.

Thomas, R. J., Rothery, D. A., Conway, S. J. and Anand, M. (2014). Long-lived explosive volcanism on Mercury. *Geophys. Res. Lett.*, **41**, 6084–6092.

Toksöz, M. N., Dainty, A. M., Solomon, S. C. and Anderson, K. R. (1974). Structure of the Moon. *Rev. Geophys. Space Phys.*, **12**, 539–567.

Tosi, N., Grott, M., Plesa, A. C. and Breuer, D. (2013). Thermochemical evolution of Mercury's interior. *J. Geophys. Res. Planets*, **118**, 2474–2487.

Tosi, N., Čadek, O., Běhounková, M., Káňová, M., Plesa, A. C., Grott, M., Breuer, D., Padovan, S. and Wieczorek, M. A. (2015). Mercury's low-degree geoid and topography controlled by insolation-driven elastic deformation. *Geophys. Res. Lett.*, **42**, 7327–7335.

Trask, N. J. (1971). Geologic comparison of mare materials in the lunar equatorial belt, including Apollo 11 and Apollo 12 landing sites. In *Geophysical Survey Research 1971*, Professional Paper 750-D. Denver, CO: U.S. Geological Survey, pp. 138–144.

Trask, N. J. (1975). Cratering history of the heavily cratered terrain on Mercury. *Proc. Int. Colloq. Planet. Geol., Geol. Rom.*, **15**, 471–476.

Trask, N. J. and Guest, J. E. (1975). Preliminary geologic terrain map of Mercury. *J. Geophys. Res.*, **80**, 2461–2477.

Trümpy, R. (1980). *Geology of Switzerland, A Guide-book: An Outline of the Geology of Switzerland*, ed. R. Trümpy. Bern: Schweizerische Geologische Kommission, 334 pp.

Turcotte, D. L. (1983). Thermal stresses in planetary elastic lithospheres. *J. Geophys. Res.*, **88**, A585–A587.

Tweto, O. (1975). Laramide (Late Cretaceous–Early Tertiary) Orogeny in the Southern Rocky Mountains. In *Cenozoic History of the Southern Rocky Mountains*, ed. B. F. Curtis. Memoirs, Vol. 144. Boulder, CO: Geological Society of America, pp. 1–44.

Vasavada, A. R., Paige, D. A. and Wood, S. E. (1999). Near-surface temperatures on Mercury and the Moon and the stability of polar ice deposits. *Icarus*, **141**, 179–193.

Walsh, J. B. (1965). The effect of cracks on the uniaxial compression of rocks. *J. Geophys. Res.*, **70**, 399–411.

Walter, T. R. and Troll, V. R. (2001). Formation of caldera periphery faults: An experimental study. *Bull. Volcanol.*, **63**, 191–203.

Watanabe, T., Koyaguchi, T. and Seno, T. (1999). Tectonic stress controls on ascent and emplacement of magmas. *J. Volcanol. Geotherm. Res.*, **91**, 65–78.

Watters, T. R. (1991). Origin of periodically spaced wrinkle ridges on the Tharsis Plateau of Mars. *J. Geophys. Res.*, **96**, 15,599–15,616.

Watters, T. R. (2004). Elastic dislocation modeling of wrinkle ridges on Mars. *Icarus*, **171**, 284–294.

Watters, T. R. and Nimmo, F. (2010). The tectonics of Mercury. In *Planetary Tectonics*, ed. T. R. Watters and R. A. Schultz. New York: Cambridge University Press, pp. 15–80.

Watters, T. R., Robinson, M. S. and Cook, A. C. (1998). Topography of lobate scarps on Mercury: New constraints on the planet's contraction. *Geology*, **26**, 991–994.

Watters, T. R., Cook, A. C. and Robinson, M. S. (2001). Large-scale lobate scarps in the southern hemisphere of Mercury. *Planet. Space Sci.*, **49**, 1523–1530.

Watters, T. R., Schultz, R. A., Robinson, M. S. and Cook, A.C. (2002). The mechanical and thermal structure of Mercury's early lithosphere. *Geophys. Res. Lett.*, **29**, 1542, doi:10.1029/2001GL014308.

Watters, T. R., Robinson, M. S., Bina, C. R. and Spudis, P. D. (2004). Thrust faults and the global contraction of Mercury. *Geophys. Res. Lett.*, **31**, L04701, doi:10.1029/2003GL019171.

Watters, T. R., Nimmo, F. and Robinson, M. S. (2005). Extensional troughs in the Caloris basin of Mercury: Evidence of lateral crustal flow. *Geology*, **33**, 669–672.

Watters, T. R., Solomon, S. C., Robinson, M. S., Head, J. W., André, S. L., Hauck, S. A., II and Murchie, S. L. (2009a). The tectonics of Mercury: The view after MESSENGER's first flyby. *Earth Planet. Sci. Lett.*, **285**, 283–296.

Watters, T. R., Head, J. W., Solomon, S. C., Robinson, M. S., Chapman, C. R., Denevi, B. W., Fassett, C. I., Murchie, S. L. and Strom, R. G. (2009b). Evolution of the Rembrandt impact basin on Mercury. *Science*, **324**, 618–621.

Watters, T. R., Murchie, S. L., Robinson, M. S., Solomon, S. C., Denevi, B. W., André, S. L. and Head, J. W. (2009c).

Emplacement and tectonic deformation of smooth plains in the Caloris basin, Mercury. *Earth Planet. Sci. Lett.*, **285**, 309–319.

Watters, T. R., Solomon, S. C., Klimczak, C., Freed, A. M., Head, J. W., Ernst, C. M., Blair, D. M., Goudge, T. A. and Byrne, P. K. (2012). Extension and contraction within volcanically buried impact craters and basins on Mercury. *Geology*, **40**, 1123–1126.

Watters, T. R., Selvans, M. M., Banks, M. E., Hauck, S. A., II, Becker, K. J. and Robinson, M. S. (2015). Distribution of large-scale contractional tectonic landforms on Mercury: Implications for the origin of global stresses. *Geophys. Res. Lett.*, **42**, 3755–3763.

Watters, T. R., Daud, K., Banks, M. E., Selvans, M. M., Chapman, C. R. and Ernst, C. M. (2016). Recent tectonic activity on Mercury revealed by small thrust fault scarps. *Nature Geosci.*, **9**, 743–747.

Weider, S. Z., Nittler, L. R., Starr, R. D., McCoy, T. J., Stockstill-Cahill, K. R., Byrne, P. K., Denevi, B. W., Head, J. W. and Solomon, S. C. (2012). Chemical heterogeneity on Mercury's surface revealed by the MESSENGER X-Ray Spectrometer. *J. Geophys. Res.*, **117**, E00L05.

Whitten, J. L., Head, J. W., Denevi, B. W. and Solomon, S. C. (2014). Intercrater plains on Mercury: Insights into unit definition, characterization, and origin from MESSENGER datasets. *Icarus*, **241**, 97–113.

Wickham, J. (1995). Fault displacement-gradient folds and the structure at Lost Hills, California (U.S.A.). *J. Struct. Geol.*, **17**, 1293–1302.

Wieczorek, M. A. and Le Feuvre, M. (2009). Did a large impact reorient the Moon? *Icarus*, **200**, 358–366.

Wieczorek, M. A. and Zuber, M. T. (2001). A Serenitatis origin for the Imbrium grooves and South Pole–Aitken thorium anomaly. *J. Geophys. Res.*, **106**, 27,856–27,864.

Wieczorek, M. A., Correia, A. C. M., Le Feuvre, M., Laskar, J. and Rambaux, N. (2012). Mercury's spin–orbit resonance explained by initial retrograde and subsequent synchronous rotation. *Nature Geosci.*, **5**, 18–21.

Wieczorek, M. A., Neumann, G. A., Nimmo, F., Kiefer, W. S., Taylor, G. J., Melosh, H. J., Phillips, R. J., Solomon, S. C., Andrews-Hanna, J. C., Asmar, S. W., Konopliv, A. S., Lemoine, F. G., Smith, D. E., Watkins, M. M., Williams, J. G. and Zuber, M. T. (2013). The crust of the Moon as seen by GRAIL. *Science*, **339**, 671–675.

Williams, J.-P., Ruiz, J., Rosenburg, M. A., Aharonson, O. and Phillips, R. J. (2011). Insolation driven variations of Mercury's lithospheric strength. *J. Geophys. Res.*, **116**, E01008, doi:10.1029/2010JE003655.

Wilson, J. T. (1954). The development and structure of the crust. In *The Earth as a Planet: The Solar System II*, ed. G. P. Kuiper. Chicago, IL: University of Chicago Press, pp. 138–214.

Woodcock, N. H. and Soper, N. J. (2006). The Acadian Orogeny: The mid-Devonian phase of deformation that formed slate belts in England and Wales. In *The Geology of England and Wales*, ed. P. J. Brenchley and P. F. Rawson. London: Geological Society, pp. 131–146.

Xia, K. and Ahrens, T. J. (2001). Impact induced damage beneath craters. *Geophys. Res. Lett.*, **28**, 3525–3527.

Zhong, S. (2009). Migration of Tharsis volcanism on Mars caused by differential rotation of the lithosphere. *Nature Geosci.*, **2**, 19–23.

Zuber, M. T. (1995). Wrinkle ridges, reverse faulting, and the depth penetration of lithospheric strain in Lunae Planum, Mars. *Icarus*, **114**, 80–92.

Zuber, M. T. and Mouginis-Mark, P. J. (1992). Caldera subsidence and magma chamber depth of the Olympus Mons volcano, Mars. *J. Geophys. Res.*, **97**, 18,295–18,307.

Zuber, M. T. and Smith, D. E. (1997). Mars without Tharsis. *J. Geophys. Res.*, **102**, 28,673–28,685.

Zuber, M. T., Montési, L. G. J., Farmer, G. T., Hauck, S. A., II, Ritzer, J. A., Phillips, R. J., Solomon, S. C., Smith, D. E., Talpe, M. J., Head, J. W., III, Neumann, G. A., Watters, T. R. and Johnson, C. L. (2010). Accommodation of lithospheric shortening on Mercury from altimetric profiles of ridges and lobate scarps measured during MESSENGER flybys 1 and 2. *Icarus*, **209**, 247–255.

Zuber, M. T., Smith, D. E., Phillips, R. J., Solomon, S. C., Neumann, G. A., Hauck, S. A., II, Peale, S. J., Barnouin, O. S., Head, J. W., Johnson, C. L., Lemoine, F. G., Mazarico, E., Sun, X., Torrence, M. H., Freed, A. M., Margot, J. L., Oberst, J., Perry, M. E., McNutt, R. L., Jr., Balcerski, J. A., Michel, N., Talpe, M. J. and Yang, D. (2012). Topography of the northern hemisphere of Mercury from MESSENGER laser altimetry. *Science*, **336**, 217–220.

11

The Volcanic Character of Mercury

PAUL K. BYRNE, JENNIFER L. WHITTEN, CHRISTIAN KLIMCZAK, FRANCIS M. MCCUBBIN,
AND LILLIAN R. OSTRACH

11.1 INTRODUCTION

From its three flybys in 1974–1975, Mariner 10 raised the tantalizing prospect that Mercury is a volcanic world. That is, the planet has experienced a geological history that included partial melting of the interior and the ascent of melt to, and eruption upon, the surface. As the study of volcanism naturally intersects with chemistry, tectonics, interior structure, and thermal evolution, understanding the volcanic character of a world – the nature, distribution, and formational histories of its volcanic landforms and products – is a crucial means by which to comprehend more fully the geological history of that body.

In this chapter, then, we seek to tie together the various strands of observational and analytical studies of Mercury's volcanic character conducted since the first Mercury flyby of the MESSENGER mission. First, we introduce the topic of mercurian volcanism as it was understood after the Mariner 10 flybys, and we outline some of the outstanding questions raised by that mission, indeed questions that helped frame the MESSENGER project. In Section 11.2, we begin to discuss Mercury's volcanic character, first in terms of effusive volcanism (which encompasses lava plains, erosional landforms, and spectral characteristics), next in regards to the planet's explosive volcanic activity, and then from the perspective of intrusive magmatism. In the subsequent section, we discuss the planet's ancient yet spatially expansive intercrater plains and visit the prospect that they, too, are volcanic. In Section 11.4, we combine the observations of and inferences for Mercury's smooth and intercrater plains to propose a model for the planet's crustal stratigraphy. We then discuss the extent of our understanding of the petrology of the surface materials on Mercury, including compositions and lithologies, mineral assemblages, physicochemical properties, and volatile contents (Section 11.5).

In Section 11.6, we explore the other major aspect of Mercury's volcanic character – *when* such activity took place – as we attempt to describe at least in broad terms the history of effusive and explosive volcanism on the planet. The influence that the planet's tectonic properties and evolution have played on volcanism is then addressed in Section 11.7. In the final section, we list some major questions regarding Mercury's volcanic character that remain open and suggest how they might yet be answered.

11.1.1 Observations of Candidate Volcanic Features by Mariner 10

Notably, no landforms unequivocally recognizable as volcanic were observed on Mercury from Mariner 10 data (Strom et al., 1975; Trask and Guest, 1975). Nonetheless, two primary types of plains material *were* readily identified, of which one was the most widespread terrain on the planet and the other morphologically similar to the lunar maria. That spatially dominant unit, the so-called "intercrater plains" of Trask and Guest (1975), was suggested by Strom et al. (1975) to be volcanic on the basis of its great volume and because there were no obvious source impact basins from which most of these plains could have emanated as ejecta. The relatively less cratered and so likely younger "smooth plains" were observed to be ponded in and to have embayed pre-existing lows, and to have ridge-like landforms akin to the shortening structures observed within lunar maria. In at least one instance, smooth plains within a large impact basin were seen to host partially or nearly entirely infilled craters (termed "ghost craters"), indicating that, because those craters must have formed after the basin floor had cooled and solidified, these plains could not have been the direct result of the impact that formed the basin in which they were situated (Strom et al., 1975). As a result, early observations of Mercury hinted at a volcanic origin for at least some of the planet's smooth plains deposits (e.g., Murray et al., 1974; Strom et al., 1975; Trask and Guest, 1975; Trask and Strom, 1976; Spudis and Guest, 1988).

Yet this consensus was not universally shared. Samples returned from the Apollo 16 mission indicated that expansive plains proximal to large impact basins on the Moon were emplaced not as lava flows but as fluidized impact breccias (Eggleton and Schaber, 1972). Wilhelms (1976) regarded the smooth plains adjacent to the Caloris impact basin on Mercury (the largest preserved impact structure on the planet) as morphologically similar to those impact-derived "lunar light plains" and he thus concluded that at least some mercurian smooth plains were not volcanic. His view was supported by the minor difference in reflectance between the plains adjacent to Caloris and the surrounding terrain, in contrast to the strong difference in reflectance between the dark, basaltic lunar maria and the light, anorthositic highland rocks on the Moon (Hapke et al., 1975), and by the fact that no craters younger than the Caloris basin but older than the plains around the basin had been found (Wilhelms, 1976).

Long after the Mariner 10 mission ended, images returned by the spacecraft were reprocessed with an improved calibration, and a resolvable spectral distinction between a smooth plains deposit near the Rudaki crater and its surrounding terrain was identified (Robinson and Lucey, 1997). This distinction implied a compositional difference between these plains and their surroundings, consistent with – if not definitively supporting – a volcanic origin for those plains. Later recalibration of Mariner 10 clear-filter data also revealed a higher reflectance for at least some smooth plains units on Mercury than the average for the planet, further supporting a compositional contrast between these younger plains and the earlier crust (Denevi and Robinson, 2008). But Mariner 10 observations alone were not sufficient to definitively resolve the role of volcanism in the formation of Mercury's smooth plains, and so the question of whether volcanism had occurred on the innermost planet remained open.

11.1.2 Post-Mariner 10 Questions

The findings from Mariner 10 raised far more questions than they answered, strongly whetting the scientific community's appetite for more Mercury science. Indeed, the gaps in our understanding of Mercury's geological history served to frame several of the driving scientific questions for the MESSENGER mission (Chapter 1). As they pertained to volcanism, those questions included definitively determining whether volcanic activity had taken place on Mercury in the first place, and then fully characterizing the spatial and temporal distribution of volcanic landforms and their link to the thermal evolution of the planet.

Specifically, were the smooth plains seen across the hemisphere imaged by Mariner 10 unequivocally volcanic? If so, how were they distributed globally, when were they first emplaced, and how late into the planet's geological history did plains volcanism extend? Were there any shield volcanoes on Mercury, as per its inner solar system siblings, or was plains volcanism the sole manifestation of this process? Was evidence for intrusive activity preserved on the surface? Were Mercury's lavas compositional cognates of those on Earth, Mars, Venus, and the Moon? What was the nature of the older plains units on the planet? And what was the effect upon Mercury's eruptive record of global contraction, as suggested by its considerable inventory of shortening tectonic landforms?

A key science objective of MESSENGER's primary mission was to study the geological history of Mercury (Chapter 1). That history encompasses geological, geochemical, and geophysical processes, of which volcanic activity is an important part. Mercury's history of effusive and explosive volcanism has not only helped shape the planet surface, but has both influenced and been influenced by the planet's record of impact bombardment (Chapter 9) and the tectonic history of Mercury (Chapter 10). Numerous measurements of the planet's properties give insight into its volcanic character, including its surface textures, topography, spectral and compositional properties, and even its interior structure.

11.1.3 Observations of Mercury Volcanic Features with MESSENGER Data

Short of in situ field observations, the most effective means of investigating a planet's volcanic landforms and products is with remotely sensed photographic, topographic, and spectral data. The Mariner 10 mission facilitated early studies of Mercury's surface by means of its Television Photography Experiment (Murray et al., 1974). This instrument returned photographic images of Mercury's plains units from which their morphology and distribution could be determined (Trask and Guest, 1975) and their prospective volcanic nature assessed (Strom et al., 1975; Robinson and Lucey, 1997). Topographic data were obtained from photoclinometry (Strom et al., 1975) and, in some cases, from stereophotogrammetry (Spudis and Guest, 1988; André et al., 2005). (Mariner 10 did not carry any spectrometers for elemental remote sensing of the planet's surface materials.)

The MESSENGER spacecraft carried a far more capable instrument payload (Solomon et al., 2008; Chapter 1) and returned global coverage of Mercury's surface at resolutions far higher than those of Mariner 10 data. Many MESSENGER-based volcanic studies have used the global mosaic base maps derived from the Mercury Dual Imaging System (MDIS) (Hawkins et al., 2007). Consisting of wide- and narrow-angle camera (WAC and NAC, respectively) image data and featuring mean resolutions of 250 m/pixel, several global base maps with a variety of solar incidence and illumination azimuths have been produced (Chabot et al., 2016), and these maps have enabled regional to global mapping surveys, particularly of volcanic landforms (e.g., Head et al., 2008, 2009a, 2011; Denevi et al., 2009, 2013; Byrne et al., 2013; Ostrach et al., 2015; Prockter et al., 2016). Such surveys can be enhanced with thousands of high-resolution MDIS NAC images (both targeted and opportunistically acquired), which typically have resolutions of tens of meters per pixel (trending to meters per pixel for those acquired toward the end of MESSENGER's low-altitude campaign, described in Chapter 1). These high-resolution data have assisted in the identification, for example, of depressions associated with volcanic channels (Byrne et al., 2013; Section 11.2.1).

Although MESSENGER did not carry a dedicated stereo imaging system, repeated passes by the spacecraft over the surface enabled the production of digital elevation models (DEMs) with stereophotogrammetric techniques. Elevation data for substantial portions of Mercury were calculated at resolutions at or below ~1 km/pixel (e.g., Oberst et al., 2010; Preusker et al., 2011), and DEMs have been created for selected regions at higher resolutions (e.g., Fassett, 2016). A global DEM, at a resolution of 665 m/pixel, was also generated from the control network derived from the development of the global image base maps (Becker et al., 2016). Notably, topographic measurements of Mercury's northern hemisphere were also acquired by MESSENGER's Mercury Laser Altimeter (Cavanaugh et al., 2007; Zuber et al., 2012); individual laser altimetric profiles were then combined to form an interpolated DEM of the northern hemisphere, also at a resolution of 665 m/pixel. Elevation data (from all sources) have been useful in characterizing the depths of landforms such as channels, depressions, and pyroclastic vents, as well as facilitating estimates of depths of volcanic fill within impact structures (e.g., Section 11.2).

Finally, the MESSENGER spacecraft was also equipped with two instruments capable of measuring the chemical makeup of Mercury's surface, including an X-Ray Spectrometer (XRS) (Schlemm et al., 2007) and a Gamma-Ray and Neutron Spectrometer (GRNS), with independent Gamma-Ray Spectrometer (GRS) and Neutron Spectrometer (NS) sensors (Goldsten et al., 2007). The spatial resolution of individual elemental measurements varied as a function of spacecraft altitude and integration time but was generally superior at northern latitudes (e.g., Nittler et al., 2011). These instruments have provided crucial insight into the composition of the planet's surface units, which has in turn enabled us to characterize to first order the lithologies present on Mercury (Chapters 2 and 7).

11.2 THE PHYSICAL VOLCANOLOGY OF MERCURY

The most prominent manifestations of volcanism on Mercury are its expansive flood lava deposits, and so it follows that our discussion of Mercury's volcanic landforms should start with those emplaced by effusive eruptions. We first review evidence for the view that the majority of the planet's smooth plains are volcanic, and we then consider major examples of ponded lavas and how rapidly emplaced, voluminous flows have shaped the landscape on Mercury. We then turn to those landforms thought to have arisen from explosive volcanism, before presenting what limited evidence exists for intrusive magmatic activity having occurred on the innermost planet. (We visit the other dominant surface unit present on Mercury, the intercrater plains, in Section 11.3.)

First, though, a note on nomenclature. It has become customary in planetary science to devise names for landforms and other surface features solely on the basis of their appearance from telescopic views or spacecraft; names that might imply a given genetic origin are largely avoided. Without direct sampling or other in situ measurements, this is a sound approach: to apply a name to a landform that invokes a given process could prejudice later studies of that type of landform. In time, consensus can result in a genetic name gaining widespread use, e.g., "impact crater." Most surface features, however, tend to retain their original name, even if a plausible formational mechanism or property has been identified. And so it is for smooth plains on Mercury: there is strong evidence that most of these plains deposits are volcanic, but the term "volcanic plains" tends not to be used. (Of note, there are numerous relatively small deposits of smooth plains on Mercury for which a volcanic origin is debatable or even unlikely.) Nonetheless, throughout this chapter we use terms such as "flood basalts" and "ponded lavas" as synonyms for "volcanic smooth plains." (Evidence that these lavas in fact are basaltic is presented in Section 11.5.) We also use another term for these deposits that has gained widespread use for analogous deposits on Earth: "large igneous province" (LIP).

Large igneous provinces are regions on Earth where voluminous outpourings of dominantly mafic lavas have been preserved (e.g., Coffin and Eldholm, 1994). LIPs tend to develop first as low-volume eruptions that then transition into a main phase of repeated emplacement of large-volume, expansive tabular flows and flow fields from spatially restricted vents and fissures; these flows build a thick stratigraphy of lavas relatively quickly, before the waning volcanic flux reduces flow volume and leads to fewer, more broadly distributed eruptive sites (e.g., Jerram and Widdowson, 2005). A key characteristic of a LIP on Earth is an effusive flux far greater than that of the background rate of effusion from normal convergent and divergent plate boundary processes. The best-preserved examples are Mesozoic and Cenozoic in age, but LIPs seem to have formed throughout most of Earth history. The largest examples – accorded an "A" rating by the Large Igneous Provinces Commission (www.largeigneou sprovinces.org) – are greater than 100,000 km^2 in area (or more than 100,000 km^3 in volume, where known) and were emplaced over geologically short timescales (substantially less than ~50 Myr) (Jerram and Widdowson, 2005; Bryan and Ernst, 2008). We discuss later in this chapter indications that at least some of Mercury's major flood basalts were emplaced rapidly, and certainly several such deposits are far greater than 100,000 km^2 in area (Tables 11.1 and 11.2). The role (if any) of mantle plumes – sites of convecting upwelling – in the eruption of these basalts, the precise duration of emplacement, and the volumetric ratio of intrusive to extrusive material are unknown, and so arguably these deposits are not true large igneous provinces in the sense that this term is used for Earth (e.g., Bryan and Ernst, 2008; Ernst et al., 2013). However, as high-volume, short-duration mafic eruptions on planetary surfaces (Bryan et al., 2010), the largest volcanic smooth plains units on Mercury must surely qualify as LIPs, and on the basis of their volume and rapidity of emplacement, these provinces on Mercury may even provide new insight into the genesis of flood basalt eruptions on other planetary bodies, including Earth.

Table 11.1. *Reported N(10) values for the largest flood basalt provinces on Mercury.*

Site	Area (km^2)*	N(10)	Reference(s)
Caloris Planitia	1.7×10^6	80 ± 7, 75 ± 7	Denevi et al. (2013), Fassett et al. (2009)
Plains exterior to Caloris basin	6.2×10^5	91 ± 15	Denevi et al. (2013)
Northern smooth plains (NSP)	5.6×10^6	67 ± 4	Ostrach et al. (2015)
Rembrandt interior plains	2.9×10^5	103 ± 19, 110 ± 23	Denevi et al. (2013), Whitten et al. (2014)
Rudaki	9.8×10^4	51 ± 23	Denevi et al. (2013)
Intercrater plains		154 ± 34 → 370 ± 53	Whitten et al. (2014)

*Area values are from Denevi et al. (2013) except for that for the NSP, which is from Ostrach et al. (2015). Note that the areal extent of the plains exterior to the Caloris basin given here is a minimum.

Table 11.2. *The location, count area, crater spatial density, and model age for nine additional large igneous provinces on Mercury (after Byrne et al., 2016).*

Site	Area (km^2)	N(4)*	N(10)*	N(20)*	N(4)	N(10)
				Model Age (Gyr)**		
Alver/Disney	2.6×10^5	517 ± 44	145 ± 23	65 ± 16	3.7	3.8
Aneirin	1.8×10^5	311 ± 42	72 ± 20	22 ± 11	3.7	3.7
Barma	1.6×10^5	446 ± 53	138 ± 29	44 ± 17	3.7	3.8
Beethoven	2.0×10^5	595 ± 55	92 ± 22	31 ± 13	3.8	3.7
Faulkner	3.6×10^5	391 ± 34	34 ± 10	14 ± 6	3.7	3.2
Philoxenus/Machaut	1.2×10^5	365 ± 56	130 ± 34	35 ± 17	3.7	3.8
Pushkin	6.4×10^4	327 ± 71	47 ± 27	n/a***	3.7	3.0
Steichen/Ruysch	8.0×10^4	314 ± 63	75 ± 31	25 ± 18	3.7	3.6
Tolstoj	6.9×10^4	234 ± 58	29 ± 21	15 ± 15	3.5	2.6

* The confidence intervals given here are ± one standard deviation, taken to equal the square root of the number of craters counted, normalized to an area of 10^6 km^2 (e.g., Crater Analysis Techniques Working Group, 1979; Fassett et al., 2009; Section 11.5.1).
** These ages are derived by Poisson timing analysis (Michael et al., 2016) for the porous-target scaling of Le Feuvre and Wieczorek (2011) applied to the crater data reported by Byrne et al. (2016).
*** This site has no superposed craters 20 km or greater in diameter.

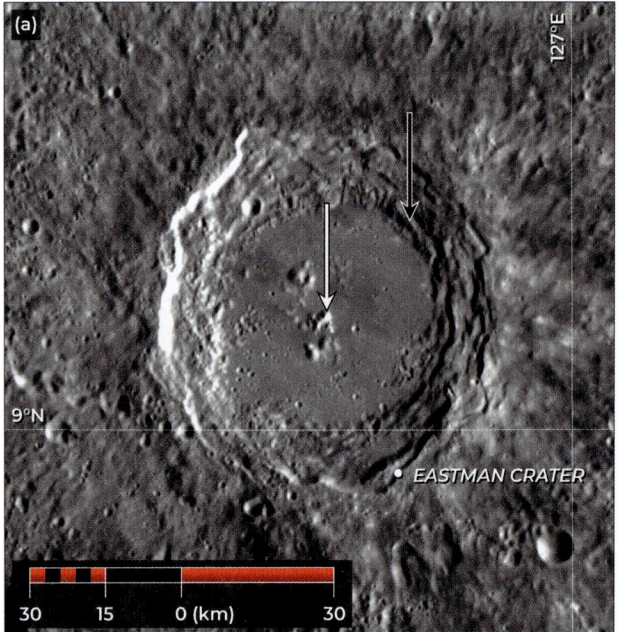
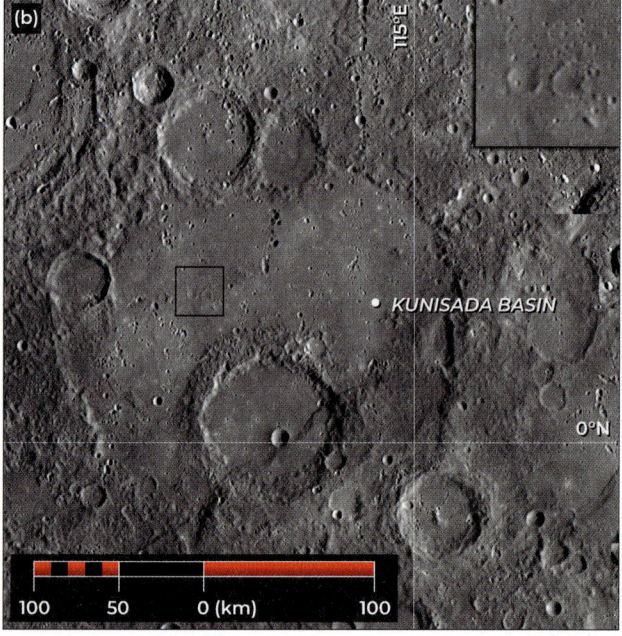

Figure 11.1. Evidence for flood volcanism on Mercury, from the comparison of (a) the 67-km-diameter Eastman crater and (b) the 240-km-diameter Kunisada basin (after Head et al., 2008). The former impact structure is substantially deeper than its larger counterpart; the crater's central peak (white arrow) and terraced walls (black arrow) are readily apparent. In contrast, neither wall terracing nor a central peak or peak ring is visible within Kunisada, although the basin hosts several ghost craters on its floor (inset), denoting at least one phase of flood volcanism for the basin. These and all subsequent images in this chapter are taken from the global morphology base map (Chabot et al., 2016). The images are in an azimuthal equidistant projection centered at (a) 9.4°N, 125.8°E and (b) 1.8°N, 112.6°E.

11.2.1 Effusive Volcanism

The first definitive evidence for volcanism on Mercury was obtained from flyby data returned by the MESSENGER spacecraft in 2008. In addition to evidence of explosive volcanism (Section 11.2.2), numerous instances were documented where large impact structures and adjacent terrains appeared to have been flooded with lava (Head et al., 2008; Murchie et al., 2008; Robinson et al., 2008). This interpretation was based in part on the difference in morphology between impact craters and basins that host smooth plains units and those that do not; for example, Head et al. (2008) compared the appearance of the 67-km-diameter Eastman crater with that of the 240-km-diameter Kunisada basin (Figure 11.1). The former impact structure possesses a rough, flat floor (likely covered by ponded impact melt), a central peak, and terraced walls, whereas, with about 1 km from floor to rim, the Kunisada basin is anomalously shallow, with evidence for neither a central peak or peak ring nor interior terracing.

Further, there are several younger craters within Kunisada for which the proximal ejecta deposits are embayed or buried by the smooth plains that occupy almost the entire basin interior. Together with at least two ghost craters on its floor (Figure 11.1), these morphological and superposition relations are that the smooth plains within the basin are volcanic, emplaced by at least one phase of effusive, flood-mode activity. Of note, Head et al. (2008) estimated that the Kunisada basin would be ~5 km deep if devoid of a smooth plains infill (Pike, 1988); even a depth of ~3 km would, on the basis of volume, qualify the lava inside this basin as a large igneous province on Earth.

Notably absent from any investigations of Mercury's volcanic character to date is the recognition of individual flow units on the planet: there has been no reported observation of discrete volcanic deposits that might plausibly correspond to pāhoehoe or 'a'ā lava flows (Dutton, 1884), either individually or grading from one to the other, nor of associated flow features such as leveed or brecciated margins, partially collapsed lava tubes, or chains of rootless vents. Most lava flows on Earth tend to be on the order of tens of meters across, meters to tens of meters thick, and perhaps a few hundred to a few thousand meters long (e.g., Walker, 1973); such dimensions would challenge (if not entirely preclude) the identification of individual lava flow features on Mercury from most MDIS data. Nonetheless, some terran flows have reached substantially greater lengths (Chester et al., 1985; Thordarson et al., 1996; Cashman et al., 1998; Stephenson et al., 1998; Riker et al., 2009), and pāhoehoe lava flows on Venus, Mars, the Moon, and Io can be several hundred to thousands of kilometers long (e.g., Wilson and Head, 1983; Stofan et al., 1998; Zimbelman, 1998). By comparison with other terrestrial worlds, then, MESSENGER ought to have returned images of lava flows on Mercury, or at least the striations formed by stacks of flows commonly observed in extraterrestrial flow fields (e.g., Head et al., 1991; Bleacher et al., 2007). However, sustained impact bombardment and deposition of ejecta may have together resulted in the muting or removal entirely of lava flow morphologies on the planet. The average impact velocity at Mercury is more than twice that at the Moon, for example (Le Feuvre and Wieczorek, 2011), and so the greater destructive force of impacts on Mercury could account for the dearth of resolvable flows on Mercury compared with the Moon, as well as for the manifestation of flood basalts on Mercury only as expansive and relatively smoothly textured regions.

Nonetheless, the presence of ghost craters and embayment relations within smooth plains makes a compelling case that these deposits are volcanic, and spectral contrast can provide additional supporting evidence for a volcanic origin. Fluidized impact ejecta, by definition, incorporate a large fraction of local material during emplacement (Oberbeck, 1975), and thus ought to show little compositional difference from underlying terrain. Spectrally distinct plains were first recognized on Mercury by Robinson and Lucey (1997) with recalibrated Mariner 10 color data, implying that effusive volcanism had taken place on the planet (Section 11.1.1). Subsequent MESSENGER imaging showed that many smooth plains deposits interpreted to be volcanic have color properties distinct from their surroundings (Figure 11.2). Such deposits are often higher in reflectance and

Figure 11.2. An example of how smooth plains regarded as volcanic differ spectrally and in reflectance from older intercrater plains (after Head et al., 2009a). (a) Monochrome image of smooth plains situated within the 176-km-diameter Dali basin (lower left) and an unnamed 120-km-diameter basin (upper right), both of which are located to the northwest of the Caloris basin. (b) Those same plains seen with an MDIS WAC composite false-color mosaic in which images taken with filters at 430, 750, and 1000 nm are projected in blue, green, and red, respectively. In this view, the difference in color between the smooth and intercrater plains is clear. Both images are in an azimuthal equidistant projection centered at 47.2°N, 121.5°E.

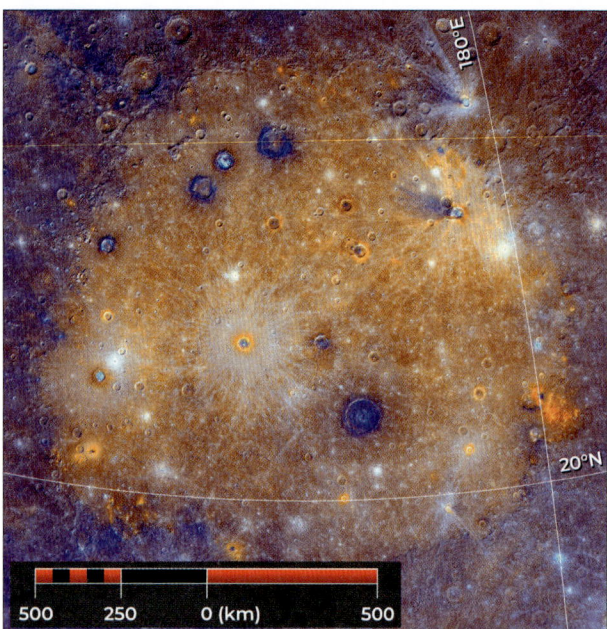

Figure 11.3. An enhanced-color mosaic of the flood-basalt province that makes up the Caloris Planitia, situated within the Caloris basin (after Ernst et al., 2015). The image is a composite of MDIS WAC "enhanced color" images, where PC2 (principal component 2), PC1, and relative visible color (430-nm/1000-nm ratio) are shown in the red, green, and blue image planes, respectively. The mosaic is in an azimuthal equidistant projection centered at 32.7°N, 161.7°E.

have a steeper slope of reflectance versus wavelength (Murchie et al., 2008, 2015; Robinson et al., 2008; Blewett et al., 2009; Denevi et al., 2009), although at least one low-reflectance deposit (an example south of Calypso Rupes) is likely volcanic (Denevi et al., 2013).

On the other hand, because the rocks on Mercury's surface have an extremely low ferrous iron content (e.g., Vilas, 1985; Izenberg et al., 2014; Section 11.5), and some of the largest compositional variation is among elements such as magnesium (Weider et al., 2012, 2015) that have little effect on spectral shape, not all smooth plains are spectrally distinct from surrounding terrain (e.g., Murchie et al., 2015). Additionally, most of the planet's crust is likely volcanic in origin (Section 11.4), and so a compositional contrast between smooth plains and older terrain need not be present. Nonetheless, some of the largest effusive deposits do have distinct spectral and compositional signatures and have even been classified as individual geochemical terranes (Peplowski et al., 2015; Weider et al., 2015; Chapter 7; Section 11.5.1).

Among the largest volumes of flood basalts on Mercury are those that occupy almost the entirety of the titanic Caloris basin (Table 11.1). These interior plains were named Caloris Planitia in 1976 after their eastern third was imaged by Mariner 10 (Strom et al., 1975) and were imaged in full by MESSENGER during its first flyby of the innermost planet (Murchie et al., 2008) (Figure 11.3). The Caloris Planitia are higher in normalized reflectance and have a redder spectral continuum (i.e., a more steeply sloped curve of spectral reflectance versus wavelength, from the ultraviolet to the near-infrared) than the rim of the basin, implying that they are compositionally distinct from the disrupted, pre-existing terrain and are thus volcanic, a conclusion independently reached on the basis of GRS and XRS data. The Caloris Planitia show a remarkable level of tectonic deformation (Section 10.6.4) which, together with their general spectral homogeneity, hampers the detection of individual flow units within the basin and thus the development of an emplacement sequence for these plains. Nevertheless, this spectral homogeneity argues against the presence of major sub-units within the Caloris Planitia, implying that these voluminous lavas may have been emplaced geologically rapidly (and so constitute a large igneous province).

The smooth plains surrounding the Caloris basin likely represent another LIP, although ambiguity persists as to whether they are entirely volcanic. The presence of ghost craters to the south and east of Caloris (Tir and Budh Planitiae, respectively) implies that these parts of the circum-Caloris plains are composed of flood basalts (Denevi et al., 2013). Yet other parts of the plains adjacent to Caloris have knobby and hummocky textures and are known as Odin-type plains (Trask and Guest, 1975; McCauley et al., 1981; Guest and Greeley, 1983), which closely resemble impact ejecta deposits associated with some lunar basins (Wilhelms, 1987). Curiously, regional crater size–frequency distributions (SFDs: Chapter 9) do not definitively demonstrate that Odin-type plains were emplaced as part of the Caloris impact event (Fassett et al., 2009), a finding that seemingly contradicts stratigraphic evidence that discrete portions of this hummocky terrain were embayed by smooth plains (Denevi et al., 2013; Chapter 6). Considerable uncertainty in crater statistics, local differences in target material strength, and spatially variable secondary cratering may together account for this apparent contradiction. Conversely, an as yet unrecognized process may have rendered volcanic plains hummocky, in which case the circum-Caloris plains could be fully volcanic (Denevi et al., 2013). So, although it appears that some fraction of the circum-Caloris smooth plains are impact related, it is likely that the majority of these plains are volcanic (Rothery et al., 2017) and therefore represent another major deposit of effusively emplaced, ponded lavas on Mercury.

The largest contiguous area of smooth plains deposits on the planet is that of the "northern smooth plains" (NSP), an especially broad expanse of flood basalts at mid-to-high northern latitudes (Figure 11.4). The locations of this and the two deposits associated with the Caloris basin mean that most smooth plains units lie in Mercury's northern hemisphere. Although portions were glimpsed by Mariner 10 (e.g., Strom et al., 1975; Trask and Guest, 1975), the NSP were first documented as a single smooth plains unit on the basis of early MESSENGER orbital image data (Head et al., 2011). Covering almost 7% of the planet's surface (Ostrach et al., 2015) (Table 11.1), the NSP are almost entirely homogeneous in color and generally spectrally distinctive from neighboring terrain. Whereas the Caloris Planitia are delimited by the rim of the basin in which they occur, the margins of the NSP are defined, as for those of the circum-Caloris plains, by a major contrast in texture with the surrounding intercrater plains. The NSP are replete with ghost craters, a key indication that this deposit is volcanic, although there is little by way of other volcanological landforms in the region (Head et al., 2011;

11.2 The Physical Volcanology of Mercury 293

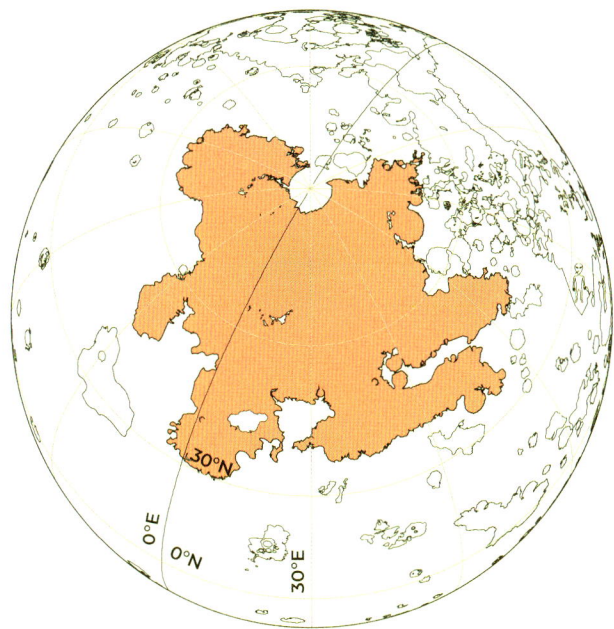

Figure 11.4. Situated at mid-to-high latitudes, the northern smooth plains occupy some 7% of Mercury's surface; they are shown here in light red in an orthographic map centered at 65.0°N, 30.0°E. The NSP outline is part of the global survey of smooth plains by Denevi et al. (2013).

Kreslavsky et al., 2016). Comparisons of the depths of ghost craters and basins with estimates of their unfilled volumes from crater scaling laws (Pike, 1988; Barnouin et al., 2012) indicate that the NSP are regionally 700–1800-m thick; for a surface area of about 5.6×10^6, the aggregate volume of the NSP is thus between 4×10^6 and 10^7 km^3 (Ostrach et al., 2015). Such a volume renders the NSP comparable to the combined volume of lunar mare lavas (Head and Wilson, 1992) and is an amount far greater than the extrusive components of several major Phanerozoic LIPs, including the Columbia River Basalts, the Deccan Traps, and the North Atlantic Volcanic Province (Coffin and Eldholm, 1994, and references therein).

Importantly, the population of craters superposed on the NSP does not show statistically resolvable variations in spatial density (Ostrach et al., 2015), so there are no obvious component sub-units of resolvably distinct age within this region. The inescapable conclusion, then, is that the NSP were emplaced in a single phase of flood volcanism – albeit one that may have lasted several tens of millions of years and likely consisted of multiple episodes of eruption (Ostrach et al., 2015). This finding is not only similar to that for the Caloris Planitia (and at least some portion of the circum-Caloris plains) but is consistent with predictions by thermochemical evolution models of voluminous partial melting and volcanic resurfacing – especially early in the planet's history (A. J. Evans et al., 2015; Byrne et al., 2016), an era of volcanism that may have been enhanced by pressure-release melting during overturn of the solidified products of a cooled magma ocean (Brown and Elkins-Tanton, 2009). Early MESSENGER flyby observations informed models of how volcanism might operate on Mercury, which also predicted large-scale flood volcanism: on the basis of lobate landforms interpreted to be lava flow fronts, Wilson and Head (2008) concluded that Mercury's lavas were emplaced during high-effusion-rate, large-volume eruptions that were fed by dikes from sources in the crust and mantle.

That huge volumes of partial melt were emplaced rapidly onto the surface now seems incontrovertible, but caution must be taken when considering lobate landforms as candidate lava flow fronts. Many of these lobate features, which are common in the NSP (Head et al., 2011), may in fact be tectonic in nature. The thermal history of Mercury is dominated by interior cooling and global contraction, which has resulted in the formation of thousands of crustal shortening landforms across the planet's surface (Strom et al., 1975; Byrne et al., 2014; Chapter 10). The tectonic landforms are likely folds atop thrust faults (Section 10.2), and although they bear some morphological similarity to inflated lava flow lobes, theirs is an entirely different mechanism of formation. Head et al. (2011) noted numerous instances of what appear to be flow fronts following and embaying pre-existing terrain, having apparently been constrained by earlier-formed impact structures, and following local topographic gradients. Yet these landforms closely resemble smooth-plains-type shortening structures observed across the planet (Byrne et al., 2013, 2014) (Figure 11.5). Both types of feature are generally linear in planform with minor, sinuous variations in trend at small scales, both can be accompanied by parallel scarps that face away from the main lobate feature, and neither type is morphologically like the complex, digitate fronts of lava flows observed on other terrestrial worlds (e.g., Wilson and Head, 1983; Stofan et al., 1998; Zimbelman, 1998). Moreover, superposition relations interpreted as flow fronts abutting against crater walls (Head et al., 2011) mirror observations of much larger lobate landforms situated within and parallel to the perimeters of volcanically infilled impact structures (such as Duyfken Rupes inside the Beethoven basin: Byrne et al., 2014). The presence of craters truncated by these larger lobate features shows them to be tectonic, likely formed by the partitioning of shortening strain into the basin-filling basalts along a detachment surface between the volcanic fill and the basin floor (e.g., Fegan et al., 2017; Section 10.2.3). The landforms interpreted as flow fronts in the NSP are tens to hundreds of meters thick (e.g., Wilson and Head, 2008), well within the bounds of lava flows observed on Earth (Walker, 1973) where the surface gravitational acceleration is nearly a factor of 3 greater than that of Mercury, and comparable in size to smooth-plains-type shortening structures (Byrne et al., 2014). Topographic relief is therefore not a useful discriminator of volcanic and tectonic landforms – but identifying a means for robustly differentiating volcanic flow fronts from tectonic shortening structures remains an outstanding challenge in studies of Mercury's geological character (Section 10.10.1).

Early MESSENGER flyby observations facilitated the recognition of lavas inside and adjacent to the Caloris basin, as well as other major deposits such as those within the previously unseen Rembrandt basin (Watters et al., 2009; Whitten et al., 2014; Ferrari et al., 2015; Table 11.1), but orbital data permitted the *global* distribution of smooth plains units on Mercury to be comprehensively characterized (Denevi et al., 2013). We now know that smooth plains occupy some 27% of the planet's surface, are dominantly volcanic on the basis of embayment

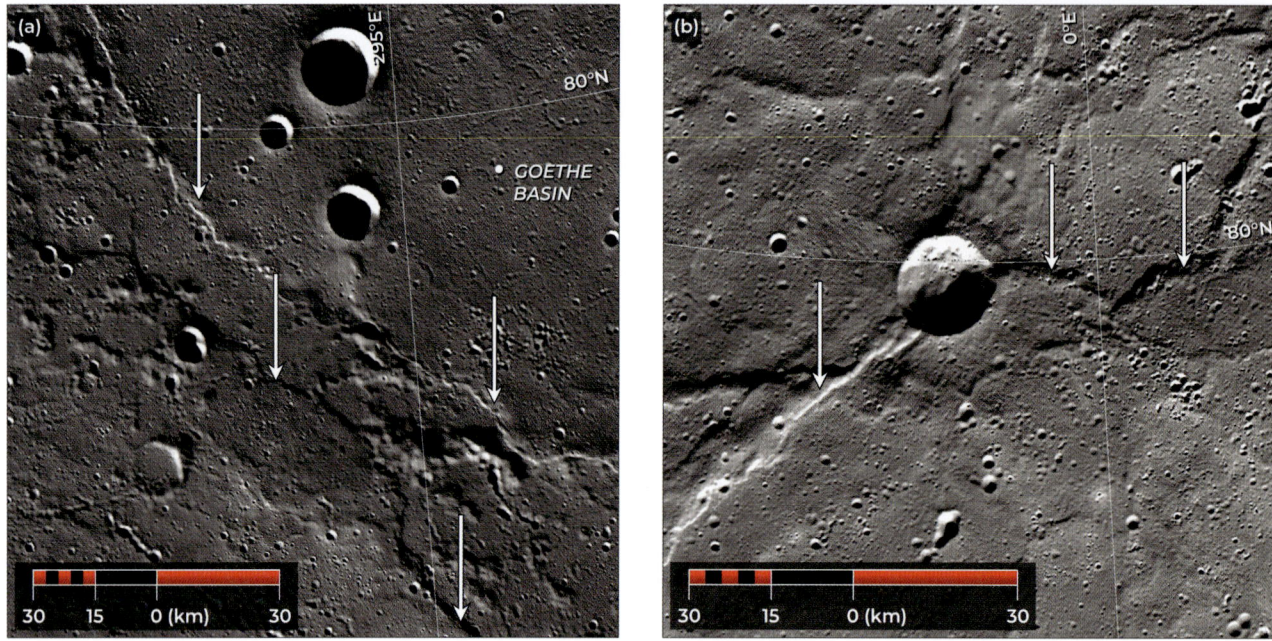

Figure 11.5. Examples of landforms (a) proposed to be flow fronts and (b) interpreted as tectonic shortening structures. Both types of landform share morphological characteristics, including broadly linear forms in plan view and undulating scarp faces (white arrows). The image in (a) shows the same landforms as Figure 3e of Head et al. (2011); both are in azimuthal equidistant projections centered at 79.0°N, 290.0°E and 80.0°N, 356.5°E, respectively.

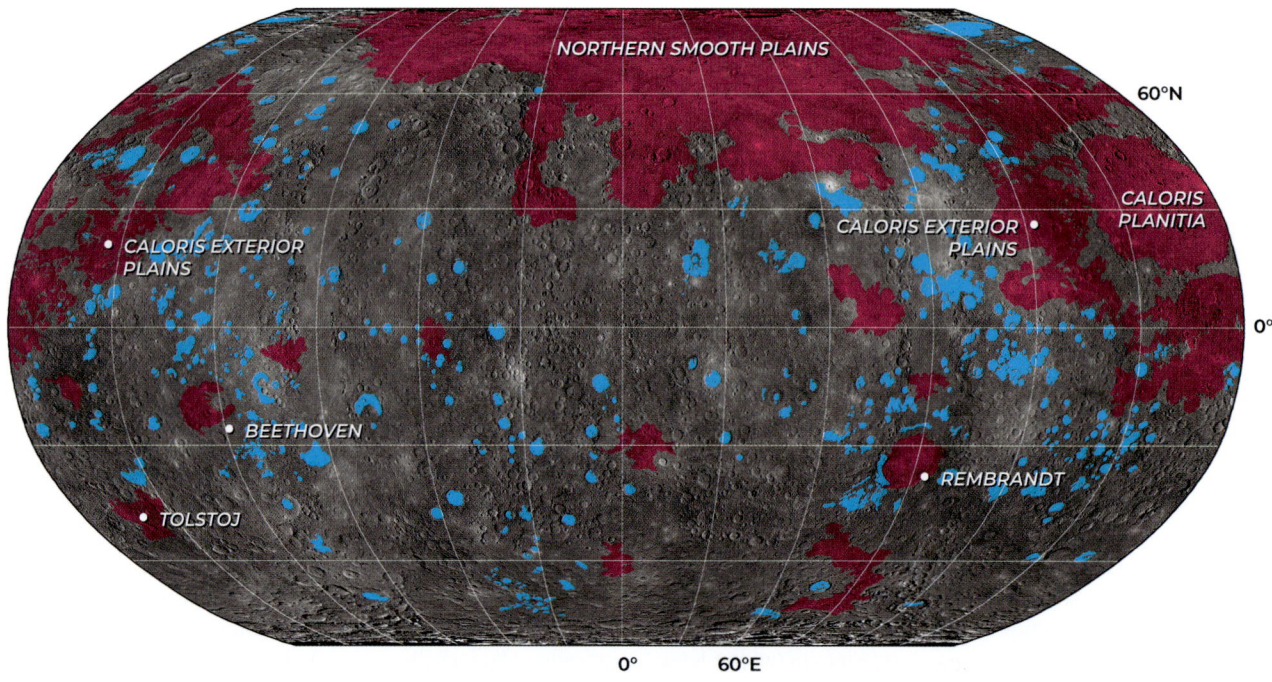

Figure 11.6. The distribution of smooth plains on Mercury (after Denevi et al., 2013); key deposits are labeled. Crater size–frequency distributions have been measured for the deposits shown in magenta, all of which are likely volcanic (including at least some portion of the Caloris exterior plains). The deposits shown in light blue have yet to be so assessed. The map is in a Robinson projection centered at 0°E; the graticule is in 30° increments in latitude and longitude.

relations, spectral properties, and crater size–frequency distributions, and are heavily concentrated in the northern hemisphere (Figure 11.6). At least the dozen largest such deposits are almost certainly volcanic (Byrne et al., 2016; Tables 11.1 and 11.2), affirming that the primary manifestation of volcanism on the innermost planet is expansive flood basalts. Many of the smaller (though still substantial) smooth plains deposits are also probably volcanic, although at the time of writing they have not

11.2 The Physical Volcanology of Mercury 295

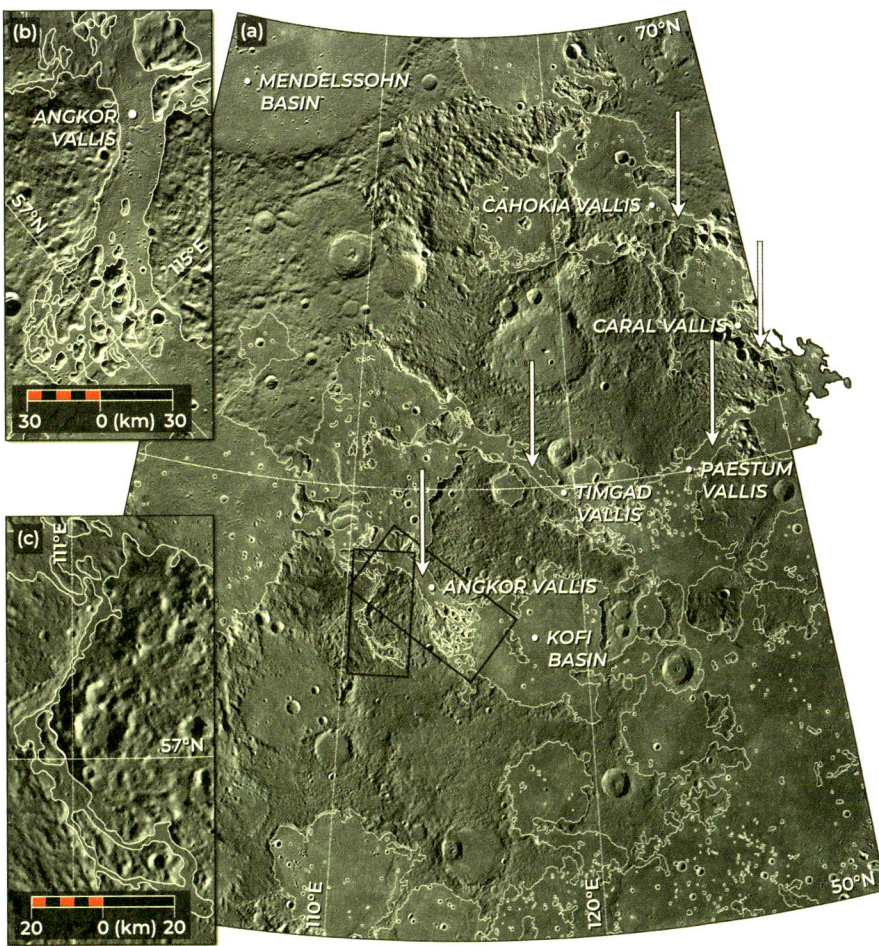

Figure 11.7. Mercury's enigmatic valles. (a) The region in which the valles (marked with white arrows) occur is bounded by the 50°N and 70°N parallels and the 90°E and 120°E meridians. The contiguous smooth plains in this region are outlined in white (after Byrne et al., 2013). The image is an orthographic projection centered at 60°N, 115°E. (b) An exemplar broad channel, Angkor Vallis, which is characterized by a relatively flat floor, straight edges, and streamlined erosional remnants. (c) A narrower channel, situated at the northern terminus of Angkor Vallis (after Hurwitz et al., 2013). Both (b) and (c) are in azimuthal equidistant projections, centered at 57.4°N, 114.1°E and 56.9°N, 111°E, respectively.

been adjudged as such; spectral characteristics are not a reliable means for distinguishing *every* instance of lava from impact melt on Mercury and so, for these smaller units, morphological and superpositional characteristics alone must be used to determine whether they are volcanic.

It is a curious thing that several of the biggest LIPs on Mercury, including the Caloris Planitia and the flood basalts in the Beethoven, Rembrandt, and Tolstoj basins, are situated within pre-existing impact structures (e.g., Strom et al., 1975; Murchie et al., 2008; Watters et al., 2009; Fassett et al., 2012) (Figure 11.6). In fact, there appear to be few exceptions to this collocation of effusive volcanic deposits and craters and basins globally (Prockter et al., 2010; Denevi et al., 2013). [Interestingly, although there is no morphological evidence that points to a single, giant basin in which the NSP lie (Head et al., 2011), polar flattening is about the same at both poles (Chapter 3), there is no southern counterpart to these ~1–2-km-thick plains, and the NSP occupy a broad region of low gravitational potential and thin crust. Were the NSP ponded in an ancient impact basin, the hemispherical dichotomy in volcanic smooth plains distribution on Mercury could be explained by its stochastic impact history.] Collectively, these observations are consistent with predictions for a planet undergoing contraction from secular interior cooling (Solomon, 1978). The importance to volcanism of impacts on Earth has been controversial (e.g., Ivanov and Melosh, 2003; Glickson, 2004; Elkins-Tanton and Hager, 2005), but it may be that the impact process is critical to the transport of magma to the surface of a planet undergoing global contraction (Section 11.7.2).

A major challenge associated with understanding the development of large igneous provinces is the burial of source vents and obvious flow-related features by successive lava flow fields (e.g., Jerram and Widdowson, 2005). Such burial has added to the paucity of primary volcanic landforms on Mercury (Strom et al., 1975; Head et al., 2008) and renders the determination of flow directions, unit thicknesses, and detailed emplacement histories difficult. Yet a distinctive assemblage of channels in Mercury's northern hemisphere provides some insight into how lava has shaped the surface of Mercury, via both the burial and erosion of pre-existing terrain (Head et al., 2011; Byrne et al., 2013; Hurwitz et al., 2013). Situated to the southeastern margin of the NSP, five broad, linear channels or valles (Latin for "valleys") interconnect a number of volcanically flooded impact basins and craters (Figure 11.7a). The valles are substantially longer than they are wide and are characterized by generally smooth floors and steep sides (Figure 11.7b). No stratigraphic layering, lineations, or similar structural detail is visible along the vallis walls in high-resolution images, but at least one vallis features terraces along its margins, and several display a furrowed or ridged texture along their floors. This region also hosts five additional channel-like landforms, although these features are more sinuous and more narrow (with respect to their length) than their broader counterparts (Figure 11.7c); a further five such sinuous channels were reported in the planet's southern hemisphere (Hurwitz et al., 2013).

Although the region as a whole is replete with kīpukas – portions of pre-existing terrain that escaped subsequent burial and so now appear as "islands" – those located within the broad channels are of particular interest. Many of these vallis-hosted kīpukas are elongate and aligned approximately parallel to the host channel's long axis; some have a lenticular planform. As with the vallis walls, MDIS NAC images do not reveal any stratigraphic or linear details along the sides of vallis-hosted islands.

The valles are morphologically similar to wide outflow channel systems observed on Mars (e.g., Baker and Kochel, 1978) and Venus (e.g., Head et al., 1991), which are characterized by low channel sinuosity, streamlined bedforms, terraced margins, and longitudinal grooves (Baker and Kochel, 1979; Baker et al., 1992; Leverington, 2007). Liquid water is not stable at the surface temperature and pressure on Mercury and so cannot be responsible for their formation; it is also unlikely that impact melt, lacking the necessary flux, volume, and spatial distribution, managed to erode these distinctive landforms. Instead, their proximity to the NSP and circum-Caloris plains strongly implies that the flow of lava itself was responsible for creating the valles.

Lavas mechanically erode substrata by physically degrading and removing material; thermal erosion is accomplished by the ablation and melting of country rock (e.g., Williams et al., 2001a; Leverington, 2007). Consolidated substrata, such as solidified basalts, resist mechanical erosion to a greater extent than less cohesive material (e.g., impact regolith), whereas hotter surfaces are thermally eroded in less time than that required for cooler material (Hurwitz et al., 2010). Hurwitz et al. (2013) calculated and discussed possible erosion rates for one of the valles, under the assumption that the present depth of the channel represents the maximum extent of incision. They concluded that thermal erosion was likely to have been the dominant regime in which it formed, with the time taken to incise the landform differing by an order of magnitude (30–300 days) depending on the lava viscosity and channel slope. In addition, the ability of lava to shape Mercury's valles would have been augmented by the presence of any pre-existing linear depressions. Fassett et al. (2009) described a set of troughs and grooves radial to the Caloris basin within the geological unit named the Van Eyck Formation (McCauley et al., 1981), which is regarded as terrain sculpted by the Caloris basin impact on the basis of its similarity to sets of radial furrows around large lunar basins. All valles have been mapped as part of the Van Eyck Formation, and all but one are radial to Caloris.

These observations, then, provide support for the formation of the valles by lavas that first flooded, and then modified the shape of, pre-existing topographic depressions (Byrne et al., 2013). Other radial troughs in the Van Eyck Formation of similar lengths and distances to Caloris feature scalloped margins in contrast to the regular edges of the valles, and they are not smooth-floored, do not feature elongate kīpukas aligned parallel to the trough axes, and do not form an interconnected network of smooth plains. Relatedly, the five narrow channels are morphologically similar to sinuous rilles documented on the Moon (e.g., Schubert et al., 1970; Hurwitz et al., 2012) and Mars (e.g., Carr, 1974; Byrne et al., 2012), as well as the "canali" on Venus (Komatsu et al., 1992, 1993). Although there is some debate as to the fluid agent responsible for martian rille formation (Bleacher et al., 2010; Murray et al., 2010), sinuous rilles on the Moon are generally thought to have formed by thermal erosion of the surface by high-effusion-rate lava flows (Hulme, 1973; Mouginis-Mark et al., 1984; Williams et al., 2000; Hurwitz et al., 2012), with pooling and subsequent drainage of lava a contributing factor, at least in the case of Vallis Schroteri (Garry and Bleacher, 2011). Effusive lavas (albeit of more exotic compositions) have been suggested as responsible for venusian canali (e.g., Baker et al., 1992). In the absence of pre-existing linear depressions, such as those of impact origin inferred for the valles, and with no evidence of any tectonic control over their course, the thermal and mechanical erosion of substrata by effusive flows across the hummocky intercrater plains is the leading mechanism by which Mercury's narrow channels formed (Hurwitz et al., 2012, 2013).

There is a curious change in terrain texture proximal to the valles: some portions of the surrounding intercrater plains show a marked softening in topographic expression, with the characteristic hummocky texture grading evenly to smooth terrain with no clear boundary between the two (Figure 11.8). This grading is accompanied in places by linear topographic highs that resemble the furrows along the vallis floors, and its areal extent implies covering to some depth by lavas (with this depth a function of flow volume and pre-existing topography). Under this interpretation, the inundation of the plains surrounding the valles indicates that at least some volume of erupted lava was not constrained to the channels but instead flowed overland in a flood lava mode, perhaps because of temporary increases in eruptive flux. Such flood volcanism is consistent with the preferred mechanism for the emplacement of the NSP as a whole (Head et al., 2011). The linear patterns visible in parts of the graded terrain strongly suggest erosion by vast sheets of rapidly emplaced lavas, via the removal of some small mounds and knolls and the aggradation of lava behind other such features. That the furrows occur close to the margins of the valles but then disappear could reflect a reduction in the erosive capacity of the lavas as they moved progressively beyond the confines of the broad channels.

The presence of the valles so close to the northern smooth plains raises the prospect that the channels and the NSP are causally related. The inundation of the plains surrounding the valles attests to some measure of overland flood volcanism, like that assumed for the NSP, but of insufficient volume for complete burial. As flood volcanism is likely to be supply-limited rather than cooling-limited (Head et al., 2009a), the lavas that shaped the region in which the valles lie could have been emplaced toward the end, and so reflect the waning supply (cf. Jerram and Widdowson, 2005), of the eruptive phase responsible for the NSP themselves. There are no examples of broad channels recognized anywhere else on Mercury at present but, if this interpretation is correct, then the valles are distinctive only in that they are preserved

Figure 11.8. An example of an area where the intercrater plains surrounding the valles display a change in texture toward a more muted topographic expression (white arrows). Some small craters superposed on intercrater plains material appear partially flooded without being obviously connected to the contiguous smooth plains units (black arrow). Solid white lines demarcate boundaries between smooth and intercrater plains (after Byrne et al., 2013). The image is in an azimuthal equidistant projection, centered at 60.8°N, 120.9°E.

today. Similar landforms may have commonly prefaced total burial by flood basalts emplaced in an impact-sculpted and -cratered region, but for most such features no evidence remains after the eruptive phase was complete. The geological character of this region of Mercury, then, may serve as the basis for understanding the development of large igneous provinces across the planet in general.

11.2.2 Explosive Volcanism

With observations of candidate effusive volcanic deposits on Mercury from Mariner 10, and the propensity for explosive eruptions during flood volcanism on Earth (e.g., Kamo et al., 2003; McClintock and White, 2006), efforts were made to look for evidence from Mariner 10 images of explosive activity on Mercury, too. Indeed, Rava and Hapke (1987) and Robinson and Lucey (1997) identified diffuse, relatively high-reflectance material in the Homer and Lermontov craters as possible pyroclastic deposits. Yet it was only after the first MESSENGER encounter with Mercury that it became apparent that the body had not only experienced pyroclastic eruptions but that its volatile content was substantially greater than had been thought. Situated within the southern margin of the Caloris Planitia, several irregular depressions that lack raised rims but are surrounded by relatively bright haloes were documented (Head et al., 2008) (Figure 11.9). These depressions differ from the regular, near-circular shapes of primary or secondary impact craters and, through comparison with similar landforms on the Moon (e.g., Lucchitta and Schmitt, 1974), were interpreted as volcanic vents (Head et al., 2008; Murchie et al., 2008; Robinson et al., 2008).

The bright deposits are distinctive because of their spectral and textural properties. They have a higher reflectance than surrounding terrain and a distinctly redder spectral slope than is typical for Mercury (Blewett et al., 2009; Kerber et al., 2009, 2011; Goudge et al., 2014; Thomas et al., 2014a). These deposits also appear to mantle the underlying landscape, becoming progressively more diffuse with increasing distance from the vents with which they are associated (Figure 11.9). Together, these morphological and spectral characteristics strongly suggest that the bright haloes are unconsolidated pyroclastic deposits, i.e., tephra, emplaced by explosive eruptions from the irregular depressions. MESSENGER image data do not resolve individual blocks or bombs within these deposits; although they may be present, it is possible that most of the pyroclastic material on Mercury consists of coarse and fine ash.

A minimum eruption velocity of 300 m/s was calculated by Kerber et al. (2009) for the deposit surrounding a large vent in the Caloris Planitia (Figure 11.9), on the basis of the maximum radial range of deposits from the central vent, and volatile contents as great as 0.36–1.35% were inferred. Such a volatile fraction is greater than had been assumed for Mercury prior to MESSENGER and is comparable to those in basalts erupted at mid-ocean ridges and hotspots on Earth (Gerlach, 1986; Kerber et al., 2009). The vent itself is situated on a broad dome (Head et al., 2008); MESSENGER orbital data have shown that this feature is a very-low-angle shield volcano with flank slopes of only 0.2°, and that the vent consists of at least nine overlapping depressions, such that the entire structure is actually a compound volcano (Rothery et al., 2014). Notably, this is the only constructional landform definitively recognized on the

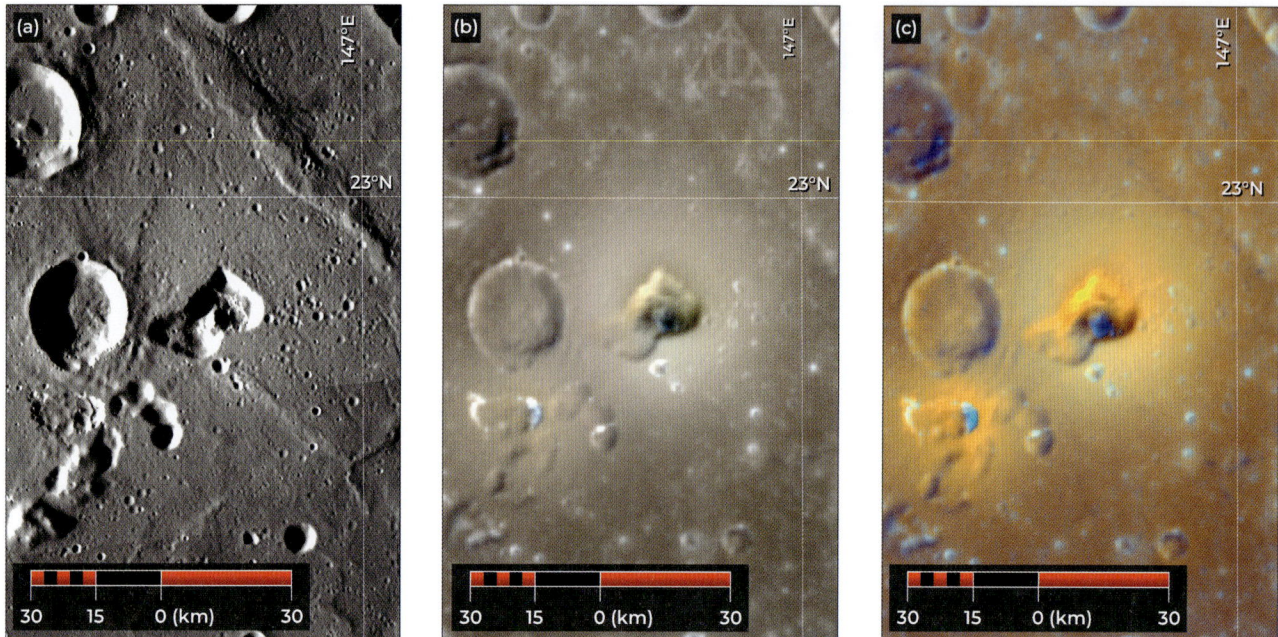

Figure 11.9. A large, irregularly shaped depression in the southwestern interior of the Caloris basin, interpreted to be a volcanic vent, surrounded by a relatively bright halo of pyroclastic material (after Head et al., 2008). The vent is shown (a) with mosaicked MDIS NAC images, (b) with the MDIS WAC composite false-color mosaic, and (c) with the MDIS WAC composite enhanced color mosaic. The images are in azimuthal equidistant projections centered at 22.3°N, 146°E.

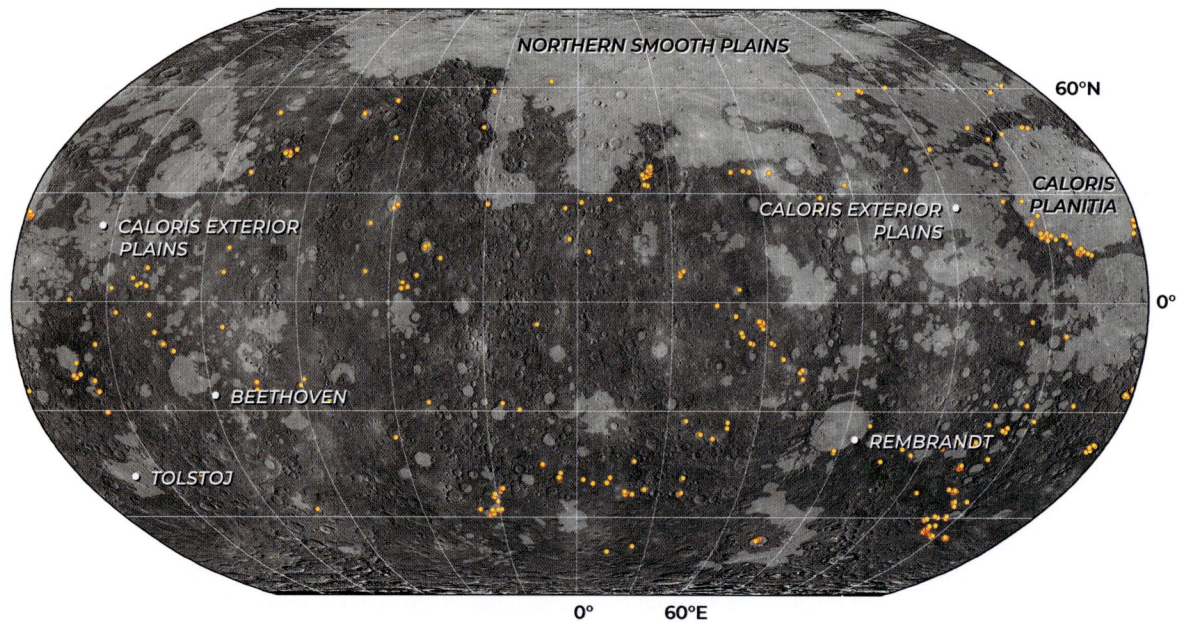

Figure 11.10. The distribution across Mercury of depressions surrounded by relatively bright, spectrally red deposits that are interpreted to be pyroclastic in nature (yellow points with red outlines), after Thomas et al. (2014a). The smooth plains mapped by Denevi et al. (2013) are shown in light gray. Note the profusion of deposits along the inner perimeter of the Caloris basin, and the dearth of pyroclastic vents within most smooth plains deposits. The map is in a Robinson projection centered at 0°E; the graticule is in 30° increments in latitude and longitude.

innermost planet, although other such candidate volcanoes have been proposed (e.g., Wright et al., 2017).

Subsequent surveys have shown that pyroclastic vents (often identified from the spectral properties of their deposits) occur widely across Mercury (Kerber et al., 2011; Goudge et al., 2014) (Figure 11.10); at least 150 have been reported planetwide (Thomas et al., 2014a). The vast majority are situated within impact craters and/or along large thrust faults (Goudge et al., 2014; Thomas et al., 2014a; Section 10.9.2), are often irregular in planform shape, and in some instances show evidence for their having coalesced from earlier, smaller depressions. They typically have depths of ~1–2 km (Kerber et al., 2011; Goudge et al., 2014; Rothery et al., 2014). Numerous vents have been documented along the interior margins of large impact structures: in addition to those within the Caloris basin (Head et al., 2008; Murchie et al., 2008), large vents are present on the margins of the flood basalts within the Tolstoj and Rembrandt basins.

The presence and spatial distribution of these vents is likely to be the result of a stress state favorable to magma ascent and eruption similar to that calculated for regions at or near the rims of large mare-filled basins on the Moon (McGovern and Litherland, 2011). Further, Rothery et al. (2014) suggested that normal faults associated with the Pantheon Fossae system in the Caloris Planitia may have structurally influenced the eruptive history of the large vent in the southwest of the basin. More generally, that few to no pyroclastic vents have been observed in volcanic smooth plains, e.g., in the NSP, suggests that there are no deep-seated faults there along which volatile-rich magma can ascend. These observations are consistent with sites of relatively late explosive volcanism (Section 11.6.2) developing in, and being strongly influenced by, a prevailing stress state governed by global contraction (Section 11.7.2).

Several other depressions have been identified on Mercury in association with the system of valles near the NSP (Byrne et al., 2013; Hurwitz et al., 2013). These landforms have scalloped, irregular outlines and flat floors, and one kidney-shaped example appears to have a smaller, circular pit nested within it; none has a raised rim, but instead all have rounded and smooth margins. Notably, there is no attendant rise, bright halo, or spectral contrast between any of these depressions and its surroundings, and so if these structures were source vents, any erupted material was likely not pyroclastic in nature but effusive. Yet, although they are superposed upon the surrounding smooth plains units, in no cases do resolvable lava flows issue from the depressions themselves (Byrne et al., 2013). The presence of the smaller, circular pit within the larger kidney-shaped depression resembles nesting of calderas as is observed on Mars, for example, within the summit caldera complexes of the Tharsis Montes (e.g., Crumpler and Aubele, 1978; Byrne et al., 2012). No associated structural evidence of caldera collapse (such as a peripheral fault zone) is observed around this or any other depression in the area, though such coherent structures are not necessary for the collapse of overlying strata into subsurface voids. Under a scenario in which these depressions formed by the lateral movement of magma through subsurface lava tubes, then, their flat floors would have formed by the infill of country rock that slumped into void space after withdrawal. The depressions may therefore be essentially very large collapse pits or calderas, similar to "pit-floor craters" described by Gillis-Davis et al. (2009) that do not feature associated pyroclastic haloes (Section 11.2.3), and thus may be more closely related to intrusive activity on Mercury than explosive volcanism.

In any case, the spectral properties of Mercury's pyroclastic deposits can provide useful insight into their composition. For example, the most distinctive spectral feature of these deposits is a downturn at wavelengths shorter than ~400 nm, which may reflect an oxygen–metal charge transfer (OMCT) band centered at 200–300 nm (Goudge et al., 2014; Izenberg et al., 2014). This transfer of charge between oxygen and metal is extremely effective at absorbing ultraviolet light, even when very low abundances of transition metals such as iron are present. The OMCT band observed in spectra of pyroclastic deposits on Mercury may be the result of lower than usual FeO contents (e.g., <0.1 wt% FeO), whereas the band is saturated at slightly higher abundances of FeO (~1 wt% FeO) elsewhere on Mercury (Goudge et al., 2014; Weider et al., 2016; Chapter 8).

Additional geochemical insight is available for the largest pyroclastic deposit on Mercury, the high-reflectance unit >150 km in diameter that surrounds a depression ~38 km × 20 km in horizontal extent to the northeast of the Rachmaninoff basin. A lower thermal neutron count rate for this deposit has been attributed to a C abundance 1–2 wt% lower than its surroundings (Peplowski et al., 2016), and XRS measurements show that the deposit is substantially depleted in S compared with units of otherwise similar composition (Weider et al., 2016). Given that the oxidation of S and C through reactions with oxides such as FeO and SiO_2 can provide the necessary volatiles to drive explosive eruptions (Zolotov, 2011), the low abundances of these elements in this pyroclastic deposit have been interpreted as evidence for the oxidation of graphite and sulfides and their ensuing loss as species such as SO_2 and CO (Weider et al., 2016). This oxidation and loss would account for the OMCT band signature seen for pyroclastic material on Mercury in general (as the reduction of FeO to Fe would undersaturate the OMCT band), as well as for the characteristically high reflectance of pyroclastic deposits (as low-reflectance graphite would be consumed before being erupted) (Weider et al., 2016). However, comprehensive geochemical measurements were not acquired for other sites of explosive activity on Mercury, and so for now it is not clear if the Rachmaninoff deposit is characteristic of pyroclastic material on the planet in general.

11.2.3 Evidence for Intrusive Activity

All of the landforms we have described so far occur on the surface and are thus the consequence of extrusive activity. But what of intrusive activity? The ratio of intrusive to extrusive igneous activity by volume on Earth globally is 8:1, with estimated ratios of 5:1 for oceanic crust and 10:1 for continental crust (Crisp, 1984); the ratio is likely even higher for Mars (Black and Manga, 2016) and the Moon (Head and Wilson, 1992). The densities of magmas on Mercury, however, are considerably lower than that of the surrounding crust (Vander Kaaden and McCubbin, 2015), promoting the ready ascent and eruption onto the surface of melts (Section 11.5.3). On the other hand, much of Mercury's history has been dominated by global contraction (Chapter 10), which limits the ability for magma to rise through the crust (Section 11.6). The intrusive to extrusive ratio for Mercury is therefore unknown, but even for a ratio as low as 1:1 there would have been a substantial volume of igneous material within Mercury that was never erupted.

Even so, evidence for magmatic activity in planetary crusts is often equivocal, because intrusive bodies may be too small and/or too deep for reliable detection with geophysical remote sensing from spacecraft (e.g., gravity and magnetic anomalies). Equally, the surface expressions of intrusions may be sufficiently subtle so as to be difficult to differentiate from those

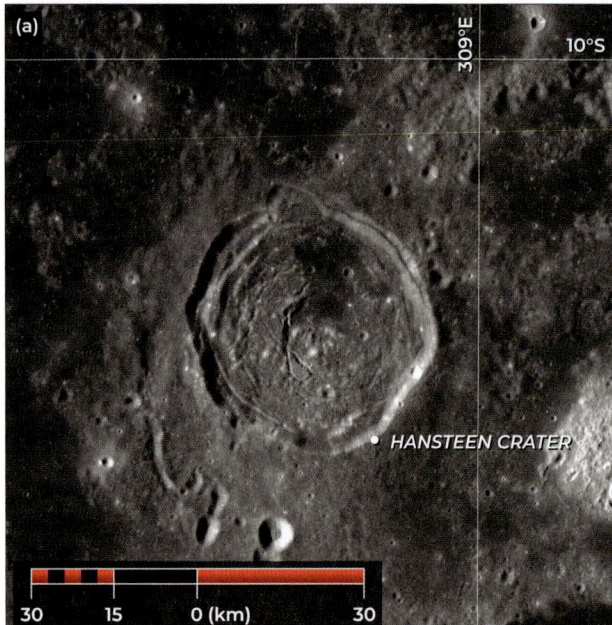

Figure 11.11. Floor-fractured craters on the Moon and Mercury. (a) The 45-km-diameter floor-fractured crater Hansteen on the Moon, located at 11.5°S, 308.0°E. (b) A ~40-km-diameter crater on Mercury that hosts small fractures in the western portion of its floor, situated at about 6.5°N, 100.5°E. Both images are shown here at the same scale and in azimuthal equidistant projections.

caused by other surface processes and/or may have been obscured by later geological activity. Where they are preserved, however, the surficial signature of a shallow magmatic intrusion may be manifest as positive topographic relief (i.e., doming: Johnson and Pollard, 1973; Wöhler et al., 2009), as sites of eruption or collapse such as vents and pits (which may also be aligned linearly, e.g., Okubo and Martel, 1998), or as fractures and faults (e.g., Rubin, 1992).

There is scant evidence for elevated topography on Mercury that can be attributed to intrusive bodies. On the Moon, so-called "floor-fractured" craters (Schultz, 1976; Wichman and Schultz, 1996; Jozwiak et al., 2015) are characterized by irregular, radial, and concentric patterns of troughs on their floors (Figure 11.11a). These tectonic landforms are attributed to uplift of the crater floors by shallow, relatively small laccoliths (i.e., lens-shaped horizontal intrusions) (Jozwiak et al., 2012, 2015). Whereas more than 170 floor-fractured craters have been identified on the Moon, with the largest population concentrated around the lunar maria, only one candidate floor-fractured crater has been described on Mercury (Figure 11.11b) (Head et al., 2008, 2009b). The floor of this 35-km-diameter crater is irregular, and its western portion appears elevated; four fractures are oriented radially to this local topographic high. Should this crater be a mercurian counterpart to the lunar floor-fractured craters, then its floor is also likely underlain by a shallow laccolith (Head et al., 2008, 2009b).

Mercury also hosts numerous examples of craters that feature pits on their floors; these "pit-floor craters" range in diameter from 55 to 130 km and have arcuate, elliptical, or near-circular pits without raised rims at or near their centers (Gillis-Davis et al., 2009) (Figure 11.12). The major axes of these pits are

Figure 11.12. Two examples of "pit-floor craters" on Mercury: the 60-km-diameter Beckett crater (white arrow) and an unnamed crater 55 km in diameter (black arrow). This image, modified from Gillis-Davis et al. (2009), is in an azimuthal equidistant projection centered at 41°S, 113°E.

~10–40 km in length, and the pits may represent sites of collapse where the eruption or lateral withdrawal of magma occurred (Gillis-Davis et al., 2009). Indeed, some of these pits are surrounded by a "halo" of material spectrally distinct from the rest of the crater floor that has been interpreted as a pyroclastic deposit (e.g., Kerber et al., 2009; Section 11.2.2). The irregular depressions observed in association with some of Mercury's valles (Section 11.2.1) may have a similar collapse origin, although none of those sites features a candidate pyroclastic deposit. In any case, candidate collapse structures are relatively rare on the planet, and where they do occur they are overwhelmingly situated near the centers of impact craters. Mercury shows no evidence for pit crater chains, which might correspond to dikes stalled at shallow crustal levels such as have been proposed for Mars (e.g., Mège et al., 2003) or the Moon (e.g., Wilson et al., 2011), possibly because of the buoyancy of mercurian melts (Section 11.5.3).

Stalled or surface-breaking (near-) vertical intrusions (i.e., dikes) are frequently accompanied by graben (Pollard et al., 1983; Mastin and Pollard, 1988). These tectonic landforms are characterized by a linear topographic low caused by the down-dropping of a crustal block along two parallel, bounding antithetic normal faults (e.g., Schultz et al., 2010; Chapter 10). On the Moon, such landforms are associated with pit crater chains or pyroclastic deposits (e.g., Wilson et al., 2011; Klimczak, 2014) and occur within impact basins or around the lunar maria (e.g., Solomon and Head, 1979; Golombek and McGill, 1983; Klimczak, 2014). On Mars, some graben are associated with pit crater chains (Wyrick et al., 2004), and others form large complexes radial to the large volcanoes on the Tharsis Rise, and even to the rise itself (e.g., Plescia, 1991). Further, subtle topographic rises indicative of the opening of a dike at depth have been found in association with both lunar and martian graben (Schultz et al., 2004; Klimczak, 2014). Together, these observations imply that many graben on both the Moon and Mars are underlain by dikes (e.g., Head and Wilson, 1993; Ernst et al., 2001; Scott et al., 2002).

On Mercury, almost all graben are situated within impact structures that have a substantial volcanic infill or impact melt deposit (Chapter 10). Such impact craters and basins include those volcanically flooded examples in the NSP, relatively fresh craters such as Hokusai and Degas, the peak-ring basins Mozart, Raditladi, and Rachmaninoff, and the two largest preserved impact basins on the planet, Rembrandt and Caloris (Figures 10.14a–10.14d). Within the Caloris basin, in particular, the enigmatic radial system of graben labeled Pantheon Fossae has been proposed as a radial dike swarm on account of its morphological similarity to landforms termed novae on Venus (e.g., Head et al., 2008, 2009b; Basilevsky et al., 2011). However, no lava flows or pyroclastic deposits have been observed anywhere within this graben system, and none of these landforms exhibits definitive dike-related topography (Klimczak and Byrne, 2013). Moreover, an areal strain analysis of Pantheon Fossae found no evidence for a dike swarm radiating from a pressurized magma chamber (Klimczak et al., 2010), as suggested for the graben system by Head et al. (2009b). Elsewhere on Mercury, a sill-like intrusion was suggested to have formed the concentric graben documented in the volcanic fill inside the peak ring of the Raditladi basin (Head et al., 2009b); numerical models subsequently showed, however, that such graben can arise solely from the thermal contraction of pooled volcanic flows, so no laccolith is required to account for these tectonic structures (Blair et al., 2013).

11.3 THE INTERCRATER PLAINS OF MERCURY

If smooth plains form one of two dominant types of surface unit on Mercury, then the intercrater plains constitute the other. That most smooth plains units are volcanic is unequivocal, but the same cannot be said for the intercrater plains. Nonetheless, there is evidence suggesting that *some* portions of the intercrater plains have a volcanic origin, and so the unit as a whole warrants inclusion in this chapter.

Intercrater plains were one of the fundamental geological units identified with Mariner 10 data (Trask and Guest, 1975) and were described as consisting of uneven plains materials texturally distinct from smooth plains and crater-related units. The morphological definition of Mercury's intercrater plains has not meaningfully changed since they were first described, but the data returned by MESSENGER provide a wealth of new insights with which our understanding of the composition and formation of these plains has been considerably improved. Here, we discuss first the appearance and then the distribution of this globally important unit, before reviewing evidence for a volcanic origin for the intercrater plains.

11.3.1 Appearance

Intercrater plains have a rough, hummocky texture and a high density of small, superposed craters <15 km in diameter (Trask and Guest, 1975; Strom, 1977; Schaber and McCauley, 1980; Leake, 1981); these plains are substantially more rugged in appearance than Mercury's smooth plains units (Figure 11.13). The intercrater plains occur on both low- and high-standing terrain and do not have a clear relation to topography (Whitten et al., 2014). Compared with the lunar highlands – another example of an areally dominant, rugged geological unit on a terrestrial body – the intercrater plains are substantially rougher at length scales <2 km but are smoother at length scales >2 km (Fa et al., 2016). The roughness characteristics of the intercrater plains have been attributed in part to the infilling of many larger primary impact craters (Fa et al., 2016).

It is the superposed population of small craters that gives the plains their distinctive hummocky texture (Trask and Guest, 1975; Schaber and McCauley, 1980). The crater size–frequency distributions of this population show a pronounced upturn with decreasing crater diameter at small (i.e., <10 km) diameters (Strom, 1977; Strom et al., 2008, 2011; Whitten et al., 2014). This upturn denotes a greater than expected number of impact craters at these diameters; a population solely of primary craters would not deviate from the expected negative correlation between crater diameter and frequency (Crater Analysis Techniques Working Group, 1979). These superposed craters are therefore interpreted as largely the result of secondary impacts. In some parts of the intercrater plains, secondary craters are obvious: discrete clusters or chains of craters are resolvable, as are sharp crater rims (Figures 11.13a, 11.13c, and 11.13d). In other places, there are few to no readily identifiable secondary crater rims, just a heavily textured landscape (Figures 11.13b and 11.13c). This appearance is enhanced by variations in the degradation states of secondary craters, which implies a sustained and prolonged history of secondary impact cratering.

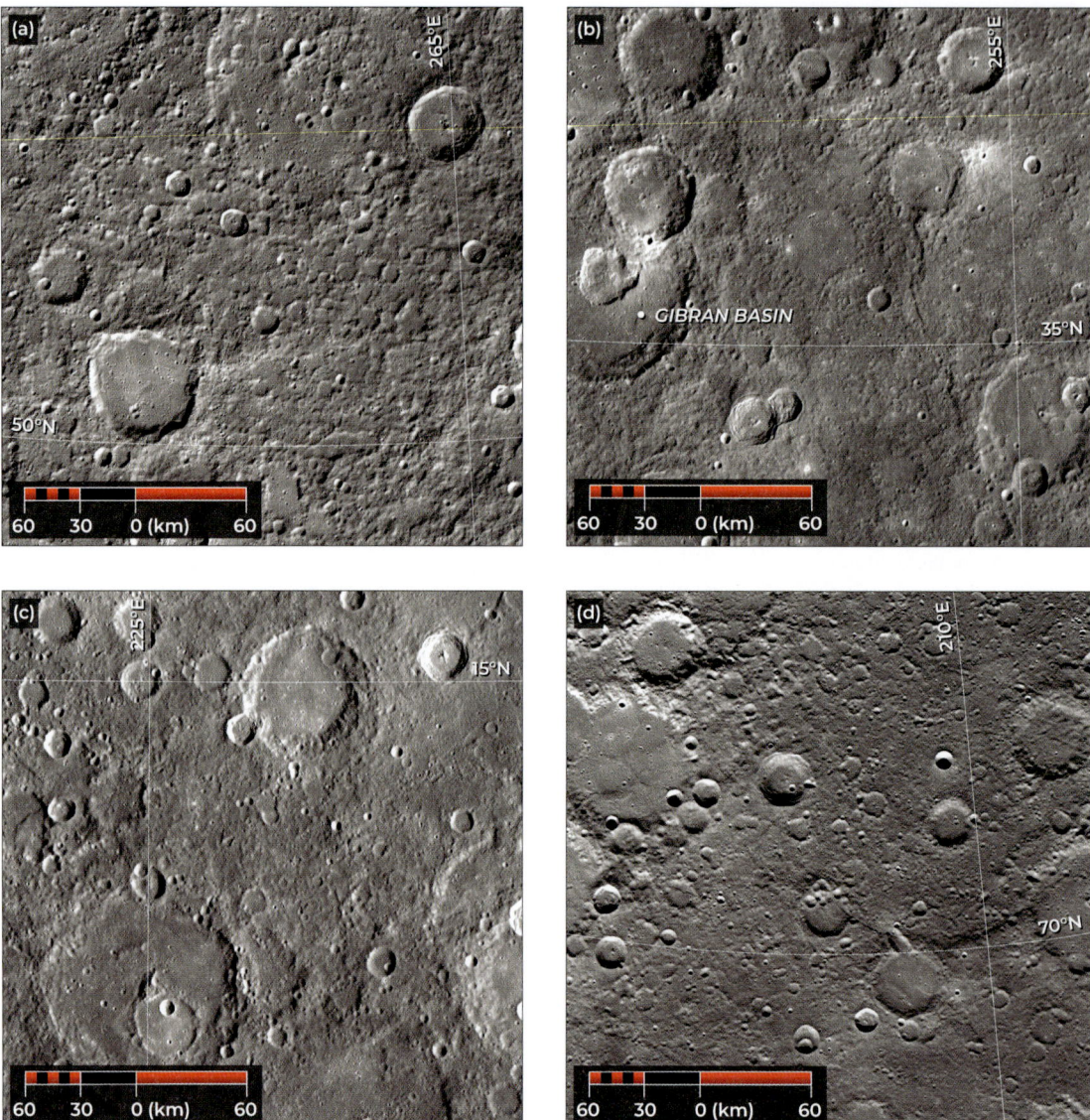

Figure 11.13. Different morphologies observed across the intercrater plains. (a) A region of intercrater plains with distinct, if slightly degraded, secondary craters. (b) Intercrater plains with a hummocky texture interpreted to result from highly degraded secondary craters. (c) Here, a portion of the intercrater plains hosts both a background hummocky texture and a set of relatively fresh superposed secondary craters. (d) Another portion of intercrater plains with slightly degraded secondary craters. The areas shown here are from Whitten et al. (2014); all images are at the same scale and in azimuthal equidistant projections, centered at (a) 52.0°N, 261.0°E, (b) 36.0°N, 252.0°E, (c) 13.0°N, 227.0°E, and (d) 71.0°N, 204.0°E.

11.3.2 Distribution

Numerous regional and global geological maps have shown that intercrater plains are the most areally extensive geologic unit on the planet (Trask and Guest, 1975; Schaber and McCauley, 1980; DeHon et al., 1981; Guest and Greeley, 1983; McGill and King, 1983; Grolier and Boyce, 1984; Spudis and Prosser, 1984; Trask and Dzurisin, 1984; King and Scott, 1990; Strom et al., 1990; Whitten et al., 2014; Prockter et al., 2016; Chapter 6) (Figure 11.14). Indeed, Trask and Guest (1975) classified primary craters ≥30 km in diameter as part of a "heavily cratered terrain" unit superposed on the intercrater plains. Under this interpretation, then, any parts of Mercury's surface not mapped as smooth plains, heavily cratered terrain, or units associated with a given impact structure are, by definition, part of the intercrater plains.

Several workers have attempted to subdivide the intercrater plains on the basis of surface roughness, MDIS color properties, or crater statistics, but with little success (e.g., Trask and Guest, 1975; Schaber and McCauley, 1980; Denevi et al., 2009; Whitten et al., 2014). Each of these properties is spatially variable, and no consistent metrics by which geological contacts can be drawn reliably or systematically have so far been identified. Nonetheless, the spatial distribution of intercrater plains can be approximated with areal crater density measurements, e.g., through the comparison of specific density values for given intercrater plains deposits (e.g., Whitten et al., 2014) with global areal crater density maps (e.g., Fassett et al., 2011; see also the supplementary information for Marchi et al., 2013). The spatial distributions of relatively high areal crater density values match well those sites mapped as intercrater plains. This result is

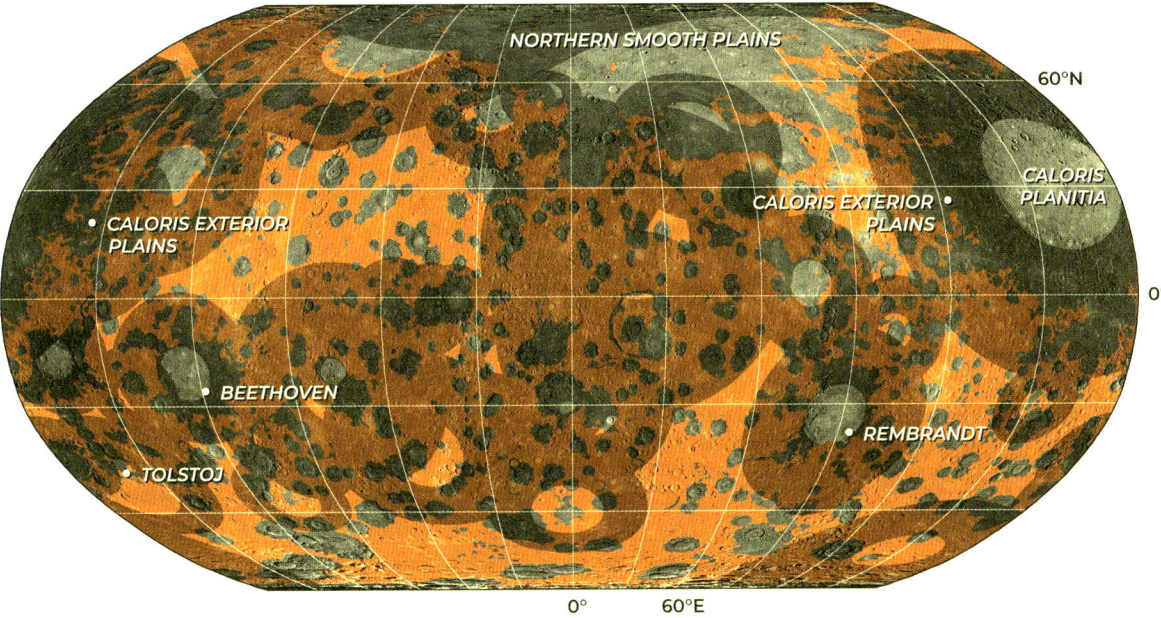

Figure 11.14. The global distribution of intercrater plains across Mercury (orange, after Prockter et al., 2016) compared with the maximum extent of impact melt from basins >300 km in diameter (shown by dark gray haloes that extend approximately one basin diameter from the rim crest). For many of the smaller basins, the impact melt would likely occur only on the basin rim as small, areally discontinuous deposits (e.g., Whitten and Head, 2015). These estimates of maximum extent of basin impact melt deposits, coupled with the discontinuous distribution of the impact melt, cannot account for the present distribution of the intercrater plains. The map is in a Robinson projection centered at 0°E; the graticule is in 30° increments in latitude and longitude.

unsurprising, given the assumption that impact cratering is the leading mechanism responsible for giving intercrater plains their distinctive characteristics.

11.3.3 Evidence for a Volcanic Origin

Given their vast spatial extent (especially if they also underlie the less heavily cratered units), there are only two plausible geological processes that can be invoked for the formation of the intercrater plains: emplacement as fluidized ejecta (Wilhelms, 1976; Oberbeck et al., 1977) or as widespread volcanic flows (Murray et al., 1975; Malin, 1976; Strom, 1977; Kiefer and Murray, 1987; Spudis and Guest, 1988; Strom et al., 2011). The prospect of an early global resurfacing event on Mercury by one or both of these processes is supported by crater statistics, which show a planet-wide deficit of impact craters 20–128 km in diameter (Strom et al., 1977; Fassett et al., 2011; Chapter 9). Moreover, the oldest surviving parts of Mercury's surface have been dated as ~4.1 Gyr old (Marchi et al., 2013), indicating that a substantial portion of the planet's history is no longer preserved at the surface.

The predicted extent of impact-related resurfacing by the ejecta deposits of preserved basins and primary impact craters (Gault et al., 1975) does not correlate with the mapped distribution of intercrater plains across the surface (Figure 11.14). To account for this discrepancy, many large impact events would have had to have occured within a geologically short period of time and to have produced thick fluidized ejecta deposits that completely buried most pre-existing impact craters. However, the preserved large impact crater population (Fassett et al., 2012) could not have produced near-global ejecta deposits; as a result, it is unlikely that impact processes alone can account for such large-scale resurfacing of Mercury.

Several observations of intercrater plains first made with Mariner 10 data – including their areal extent, crater size–frequency distributions, and stratigraphic relations with the heavily cratered terrain – support the interpretation that the majority of intercrater plains deposits are volcanic (e.g., Murray et al., 1975; Trask and Guest, 1975; Strom, 1977). Analyses with MESSENGER data have provided more support for the volcanic emplacement of intercrater plains, strengthening the interpretation of widespread and dominantly volcanic resurfacing of Mercury early in the planet's history (e.g., Denevi et al., 2009; Fassett et al., 2011; Strom et al., 2011; Marchi et al., 2013; Whitten et al., 2014; Murchie et al., 2015; Byrne et al., 2016; Chapter 6).

A picture has therefore emerged of intercrater plains forming as a result of the modification of formerly smooth volcanic plains by subsequent impact cratering (e.g., Denevi et al., 2009; Whitten et al., 2014) (Figure 11.15). This sequence can be observed even for smooth plains units: secondary craters from the Gaudí and Steiglitz craters (81 and 100 km in diameter, respectively), situated in the NSP, have disrupted the surrounding terrain, resulting in a texture not unlike that of the intercrater plains (Whitten et al., 2014) (Figures 11.15a and 11.15b). Such modification of the landscape is even more apparent at larger scales, where the superposed secondary crater populations from the Ahmad Baba and Strindberg basins (126 and 189 km in diameter, respectively) have created an areally extensive intercrater plains deposit (Whitten et al., 2014) (Figures 11.15c and 11.15d).

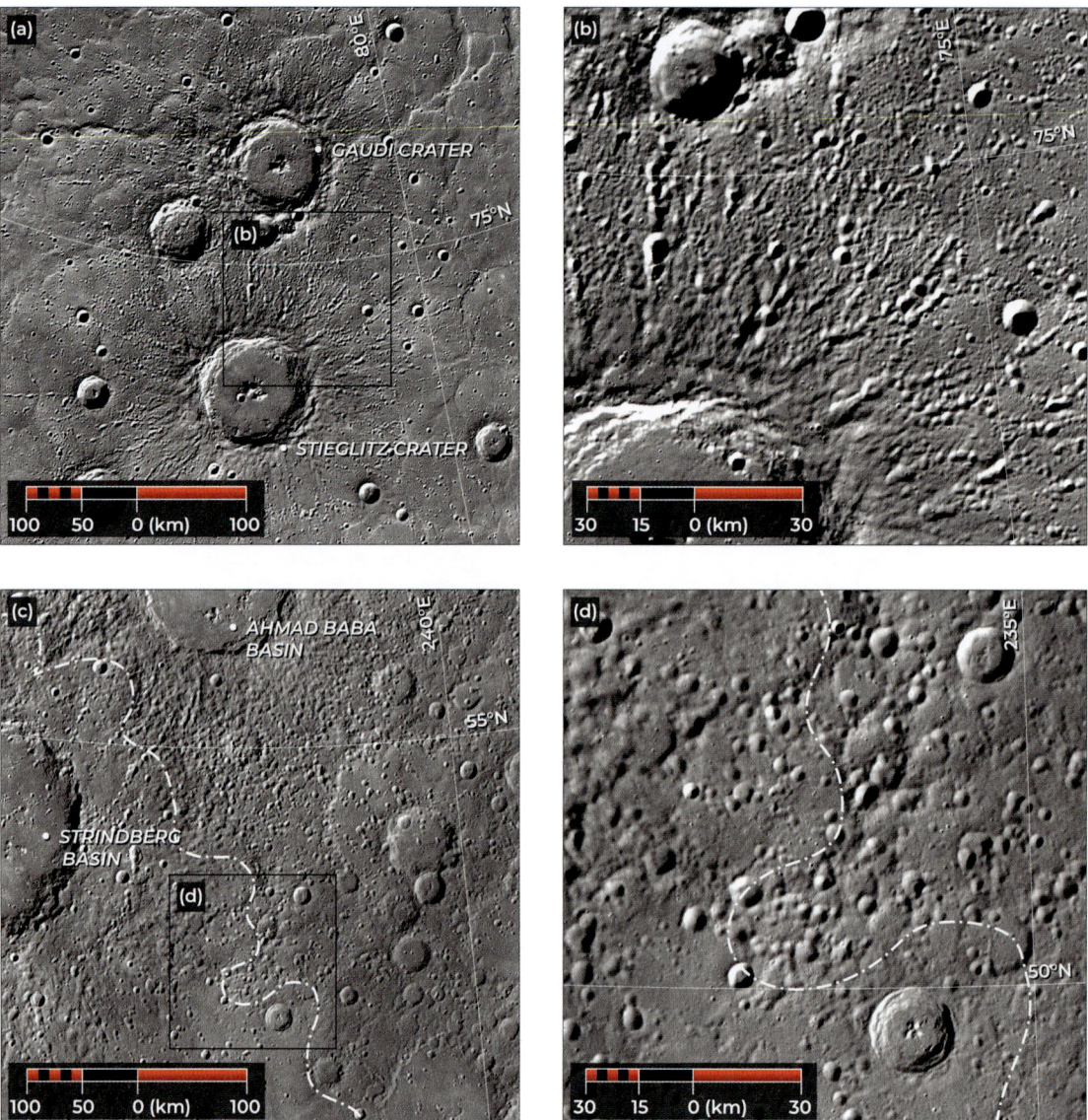

Figure 11.15. The transition from smooth plains to intercrater plains via secondary impact cratering. (a) A region of the northern smooth plains near the relatively young craters Gaudí (top) and Stieglitz (bottom) that illustrates how the overlap of only two populations of secondary impact craters can alter the morphology of smooth plains to resemble that of intercrater plains. The black box shows the location of (b), a closer view of the overlapping sets of secondary craters. (c) Another example of how overlapping secondary crater populations, from the Strindberg (left) and Ahmad Baba (top) basins northwest of Sobkou Planitia, create an intercrater-plains-like surface texture. The black box shows the location of (d); the dashed lines in (c) and (d) denote the boundary of smooth plains mapped by Denevi et al. (2009). The areas shown here are from Whitten et al. (2014); (a) is shown here at the same scale as (c) and (b) is shown at the same scale as (d). All images are in azimuthal equidistant projections, centered at 75.0°N, 68.0°E [for (a) and (b)] and 53.8°N, 232.5°E [for (c) and (d)].

That MDIS color properties of the intercrater plains resemble those of some smooth plains (Murchie et al., 2015) is consistent with the interpretation that the serial superposition of secondary craters transforms smooth plains into intercrater plains. This interpretation is bolstered by numerous observations of gradational contacts between smooth and intercrater plains units, such as those at the northwest boundary of Sobkou Planitia (Whitten et al., 2014). Moreover, the lowest areal crater density values for the intercrater plains are, within error, about the same as the greatest such values for the smooth plains (Whitten et al., 2014; Byrne et al., 2016; Section 11.6.1).

The origin of intercrater plains through the rapid emplacement of enormous volumes of lava mirrors the formation mechanism we infer for the two most areally extensive smooth plains deposits on the planet, the NSP and the plains interior to the Caloris basin (Section 11.2.1). Such voluminous, high-eruption-rate volcanism is supported by thermochemical evolution models, especially within the planet's early history (Brown and Elkins-Tanton, 2009; A. J. Evans et al., 2015; Section 11.2.1). It thus becomes easy to conceive of such widespread effusive volcanism as the driving mechanism by which the material that would later develop into the intercrater plains was formed.

Repeated episodes of volcanic flooding are observed in numerous smooth plains deposits, where superposition relations (including variations in the depths of fill of ghost craters) indicate multiple phases of volcanic resurfacing (e.g., Head et al., 2008, 2011; Ostrach et al., 2015; Byrne et al., 2016). Serial flooding events presumably also formed the intercrater plains, an assumption supported by superposition relations and crater statistics (Malin 1976; Fassett et al., 2011; Marchi et al., 2013). It seems likely, then, that the global unit we map as intercrater plains is largely volcanic in nature, was not emplaced as a single deposit in one event, but represents a complex history of primarily voluminous effusive volcanic flows that were subsequently modified by impacts, predominantly during the early history of Mercury.

11.4 CRUSTAL STRATIGRAPHY

By considering the geological observations of both the intercrater plains and the flood basalts discussed in Section 11.2, and informed by fieldwork in LIPs on Earth, it is possible to develop a model for the internal stratigraphy of Mercury's crust. Taylor (1989) put forward a model for three types of planetary crust, in which primary crust forms after accretional heating, secondary crust is produced by partial melting of the mantle, and tertiary crust is formed by the processing of secondary crust. Continental crust on Earth is (to date) the only known example of a definitively tertiary planetary crust; terran oceanic crust and the lavas that dominate the venusian, martian, and ionian surfaces make up at least the upper portions of the secondary crusts on these bodies; and the anorthositic lunar highlands are the type example of a primary crust, the product of crystallization and ascent of buoyant minerals in a cooling magma ocean (e.g., Taylor and McLennan, 2010). Arguably, the effusively emplaced lavas that constitute Mercury's LIPs represent secondary crust; under the assumption that they, too, are volcanic, so do the intercrater plains (Section 11.3).

Geochemical modeling suggests that in any magma ocean present on Mercury very early in its history, graphite would have been the only major mineral that would have been buoyant and thus the only candidate component of a flotation crust (Vander Kaaden and McCubbin, 2015). Remnants of this crust may be manifest as the low-reflectance material (LRM) that has been observed across the planet (Denevi et al., 2009; Klima et al., 2015); the darkening material that renders this spectral unit low in reflectance is likely graphitic carbon, on the basis of combined spectral and thermal neutron measurements (Murchie et al., 2015; Peplowski et al., 2016; Chapters 6 and 8). LRM appears to be largely concentrated at depth and exposed only by large impact events. If the graphite in this material once formed a surficial primary crust, it was subsequently buried in places by volcanic deposits many kilometers in thickness and substantially mixed (and diluted) through magmatism and impacts (Section 6.5). Importantly, many large impact events do *not* appear to have exposed LRM (Ernst et al., 2010), suggesting that no uniform global stratum of LRM exists on Mercury.

Given the general lack of buoyant minerals (apart from graphite) in a cooling mercurian magma ocean, it is likely that partial melting during cumulate overturn would have resulted in an essentially global eruption of lavas that, in turn, formed the early portion of the secondary crust (Brown and Elkins-Tanton, 2009). Under this scenario, volcanism likely commenced early in Mercury's history and was sustained, at least for a time, by the heat-producing elements in the mantle (e.g., Peplowski et al., 2011). With the majority of the planet's flood basalts emplaced within the first billion years (Strom et al., 2008; Head et al., 2011; Ostrach et al., 2015; Byrne et al., 2016; Section 11.5.1), and the first 400 Myr of surface history lost through resurfacing (Marchi et al., 2013; Section 11.3.3), Mercury's secondary crust appears to have been built by repeated eruptions and intrusions of magma produced by varying degrees of partial melting in the mantle (Section 6.5).

Geophysical measurements suggest that the crust of Mercury is, on average, at least about 40 km thick (Chapter 3), indicating that the planet has experienced the most efficient extraction of crustal material from the mantle of any of the terrestrial planets (James et al., 2015; Padovan et al., 2015). From geological observations, we now know that much of that material consists of mafic and ultramafic lavas emplaced in a flood mode, with the NSP the largest late-stage instance of such resurfacing. A cross section of Mercury's crust would therefore show tens of kilometers of stacked lava flows, with complex thickening and thinning of individual units together with on- and off-lap stacking patterns; these flows would be interspersed by pyroclastic and volcanoclastic deposits, layers of impact-induced regolith, and possibly the remains of the planet's flotation crust, which together would constitute weathering horizons, erosion surfaces, and paleosols (Jerram and Widdowson, 2005) (Figure 11.16). The crust would also be intruded by diapirs, plutons, and batholiths (progressively larger intrusive bodies) as well as by laccoliths, sills, and dikes at volumes proportionate to that amount of material extruded at the time of their intrusion (Crisp, 1984) – although this proportion is likely much less for Mercury than for the other terrestrial worlds (Vander Kaaden and McCubbin, 2015; Sections 11.2.3 and 11.5.3). The interplay of the regional heterogeneity of heat-producing elements (and thus melt production, intrusion, and eruption) and a sustained and prolonged history of impact bombardment has surely rendered the interior structure of most of Mercury's crust complex and spatially highly variable. From eroded and incised flood basalt provinces on Earth, we might assume that the internal architecture of Mercury's preserved large igneous provinces (i.e., today's major volcanic smooth plains deposits) is more straightforward, consisting of areally expansive tabular lava flows both fed and intruded by dikes and sills (Coffin and Eldholm, 1994; Ernst et al., 2013); impact bombardment has presumably not yet destroyed the majority of the internal stratigraphy of these deposits.

11.5 PETROLOGY

The MESSENGER mission has enabled the first quantitative characterization of the geochemistry of the surface of the

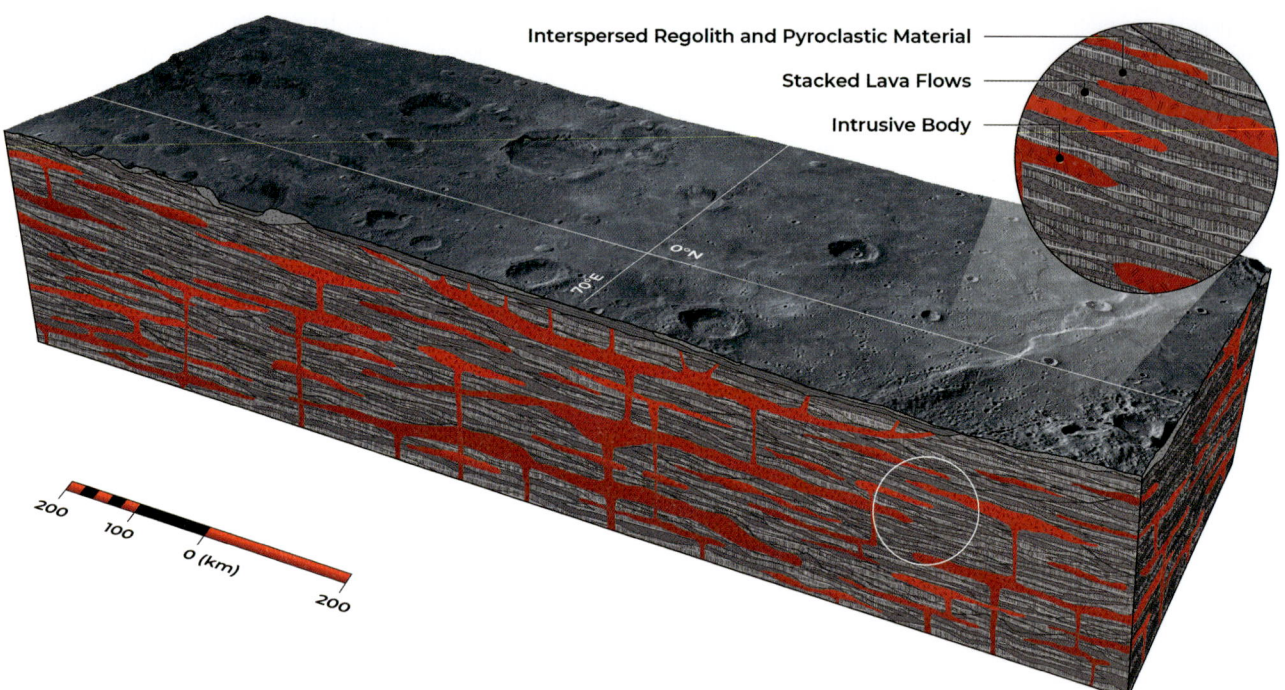

Figure 11.16. A schematic cross section of the possible internal stratigraphy of Mercury's crust. Here, the crust is composed of kilometers of stacked lava flows, pyroclastic and volcanoclastic material, and impact regolith. This extrusive material is accompanied by intrusive bodies such as diapirs, plutons, and batholiths (although at a ratio likely much lower than that for other terrestrial worlds; the ratio shown here is approximately 1:1). Smooth plains deposits probably retain much of their original internal stratigraphy. Features shown at depth are not to scale.

innermost planet, establishing Mercury's status as a geochemical end-member among the terrestrial planets. The materials that make up the planet's surface were originally volcanic (see Sections 11.2 and 11.3), and hence they are essentially igneous rocks that have undergone some amount of space weathering and mechanical and thermal processing by impact cratering (Section 11.4), similar to what has been documented for the surfaces of other airless bodies such as the Moon and the asteroids. Despite being igneous in nature, however, the materials at the surface of Mercury lack diagnostic spectral absorption features that can be used to determine definitively its mineralogical composition (and thus, broadly, its "petrology") because of the paucity of FeO-bearing minerals (McCord and Clark, 1979; Vilas, 1988).

Even so, the observations returned by MESSENGER's XRS and GRNS instruments have revealed substantial geochemical diversity across the surface (e.g., Lawrence et al., 2010, 2017; Nittler et al., 2011; Peplowski et al., 2011, 2016; L. G. Evans et al., 2012, 2015; Weider et al., 2012, 2016; Vander Kaaden et al., 2017; Chapters 2 and 7). These data have, in turn, been used to infer the compositions of Mercury's surface materials from mineralogical modeling and petrological experiments on mercurian geochemical analogs (e.g., Brown and Elkins-Tanton, 2009; Stockstill-Cahill et al., 2012; Charlier et al., 2013; Parman et al., 2014, 2016; Sehlke and Whittington, 2015; Vander Kaaden and McCubbin, 2015, 2016; Namur et al., 2016; Vander Kaaden et al., 2017). In this section, we describe the current state of knowledge of mercurian petrology, derived from many of those modeling and experimental efforts, as it pertains first to the types of rock on the planet (Section 11.5.1), then to those rocks' likely mineral assemblages (Section 11.5.2), and finally to the physical and chemical properties (11.5.3) and volatile contents (11.5.4) of mercurian lavas.

11.5.1 Surface Compositions and Rock Types

The surface of Mercury can be divided into as few as four or as many as nine distinct geochemical regions on the basis of MESSENGER data (Weider et al., 2015; Vander Kaaden et al., 2017; Figures 7.1 and 7.2). These regions and their respective inferred geochemistry are discussed in detail elsewhere, particularly by Vander Kaaden et al. (2017) and in Chapter 7. Here, we focus on the lithologies (i.e., rock types) that are inferred to make up these geochemical regions, from which we can gain insight into the physical properties of the lavas that form much of the surface of Mercury. Petrology can be classified on the basis of bulk composition, following the International Union of Geological Sciences (IUGS) classification protocols for igneous rocks (Le Bas, 2000, and references therein; Le Maitre et al., 2002). Of course, these protocols enable classifications based solely on the chemical composition of regions on Mercury's surface; they do not, a priori, indicate specific geological processes or settings.

Among the most important factors to consider when applying the IUGS protocols to the classification of rocks solely on the basis of chemical composition is the abundance of alkalis (Na_2O and K_2O), silica (SiO_2), and magnesium (MgO). When the abundance of MgO in rocks is below 8 wt%, a total alkalis versus silica (TAS) diagram is used to differentiate among various lithologies (Le Bas, 2000). The TAS diagram has been used to classify many rock types on Earth and Mars (Le Bas, 2000; Le Maitre et al., 2002; McSween et al., 2006, 2009; Nekvasil et al., 2007; Stolper et al., 2013; Santos et al., 2015). However, at the scale of observations that can be conducted for Mercury from orbit, all of the surface compositions range from 12 to 25 wt% MgO (Vander Kaaden et al., 2017). Consequently, rocks on Mercury must be classified with the high-Mg classification diagram (Chapter 7). This classification scheme indicates that all of the rocks on Mercury's surface are either boninites or komatiites, Mg-rich rocks with more than or less than 52 wt% SiO_2, respectively. The entire range of silica on the surface of Mercury is 49–59 wt% (Vander Kaaden et al., 2017). As noted by these workers, however, the mercurian surface is more alkali-rich than boninites and komatiites on Earth, and so Mercury's rocks actually fall outside of the range for the naming protocols. Adopting the nomenclature of Vander Kaaden et al. (2017) to describe them, then, the dominant lithologies on Mercury are alkali-rich komatiites and alkali-rich boninites.

11.5.2 Mineralogy as Inferred from Modeling and Experiments

Absent any diagnostic spectral absorption features (McCord and Clark, 1979; Vilas, 1988), we must infer the mineralogy of Mercury's surface materials from geochemical data. Such inferences have been made with several methods, including CIPW normative calculations (Cross et al., 1902), crystallization experiments of candidate lava compositions, and computational modeling of mercurian lava crystallization (Stockstill-Cahill et al., 2012; Charlier et al., 2013; Namur et al., 2016; Vander Kaaden and McCubbin, 2016). Although each geochemical region may not be represented by a single lithology or a specific mineral assemblage that can be easily deduced by bulk normative mineralogical calculations or experiments, these methods are nonetheless useful for obtaining first-order estimates of the mineralogy of Mercury's surface.

Although the prevalence of boninitic and komatiitic rocks on Mercury represents an exotic set of lithologies compared with those typical of the surfaces of Earth, the Moon, and Mars, the inferred mineralogies of these rocks are in keeping with what are considered to be typical for rocks on terrestrial worlds. All of the normative calculations and petrological experiments and modeling indicate that the surface of Mercury is comprised of plagioclase, pyroxene, and olivine, with minor amounts of quartz (or other crystalline forms of silica) (Stockstill-Cahill et al., 2012; Charlier et al., 2013; Namur et al., 2016; Vander Kaaden and McCubbin, 2016; Vander Kaaden et al., 2017; Chapter 7). Where Mercury seems to differ from its rocky counterparts is in its plagioclase: this phase is much more albitic (i.e., sodium-rich) on Mercury than the typical plagioclase reported in terran, martian, or lunar basalts (Vander Kaaden and McCubbin, 2016). Further, the ferromagnesian silicate phases on Mercury (i.e., pyroxene and olivine) are essentially Mg end-members. Interestingly, elevated Mg abundances in ferromagnesian silicates are typical of the most primitive melt compositions on Earth, and the most albitic plagioclase is associated with the most evolved rock compositions. The peculiar coupling of these two phases, although both typical rock-forming minerals, allows for the existence of exotic alkali-rich boninites and komatiites on Mercury unlike those that have been observed on other planetary bodies.

There have also been efforts to characterize what accessory minerals may be present on Mercury. The planet's surface has elevated S but low Fe abundances (Nittler et al., 2011; Figure 2.2) and so probably hosts some exotic sulfides. Experimental and modeling efforts have shown that sulfides on Mercury likely consist of multicomponent phases that include Fe, Cr, Mn, Ti, Ca, and Mg (Vander Kaaden and McCubbin, 2016; Vander Kaaden et al., 2017; Chapter 7). Mercury also has Cl abundances at its surface that are above detection limits (0.14 ± 0.03 wt%: Evans, L. G. et al., 2015); the Cl is likely hosted by sulfides such as djerfisherite, within metal chloride salts, or within the mineral sodalite (L. G. Evans et al., 2015). And, intriguingly, abundances of C of 1.4 ± 0.9 wt% have been detected on the surface of Mercury (Murchie et al., 2015; Peplowski et al., 2015, 2016). On the basis of experimental and modeling efforts, Vander Kaaden and McCubbin (2015) inferred that this carbon could represent the remnants of the primary graphite flotation crust thought to have formed during the early stages of magma ocean crystallization on Mercury (Section 11.4; Chapters 6 and 7).

11.5.3 Physicochemical Properties of Mercurian Lavas

The discovery of laterally extensive volcanic deposits on Mercury that appear to have been emplaced rapidly (e.g., Head et al., 2011) posed immediate questions as to how they were formed. Early MESSENGER geochemical measurements suggested that at least some of these lavas might have compositions approaching that of komatiites. It was not clear, therefore, whether these voluminous volcanic deposits were simply the result of high effusion rates and large magma volumes (i.e., flood volcanism), or if exceptionally low-viscosity melts also played a part.

At the time of writing, there are no known samples of Mercury's surface materials in any meteorite collection worldwide; it is also unlikely that a sample return mission will be dispatched to the innermost planet for the foreseeable future. We must therefore infer the physiochemical properties of mercurian lavas from the modeling of and experimentation with analog materials, as informed by MESSENGER observations. For example, computational modeling results suggest that the Mg-rich compositions of Mercury's lithologies would have lower viscosities than their terran counterparts, because of the higher temperatures required to melt rocks rich in magnesium (Stockstill-Cahill et al., 2012).

However, results from experiments on the dependence of mercurian lava viscosities on strain rate show that, for a broad range of likely compositions, these lavas have a higher viscosity than Hawaiian basalt for a given temperature (Sehlke and Whittington, 2015). We show, in Figure 11.17, lava viscosity as a function of temperature for four representative types of mercurian lava, as well as a number of tholeiitic and komatiitic lavas on Earth, the Moon, Mars, and Io, calculated with the algorithm of Shaw et al. (1972). Interestingly, a lava composition corresponding to Mercury's intercrater plains (the "intermediate terrane" of Vander Kaaden et al., 2017) matches most

308 The Volcanic Character of Mercury

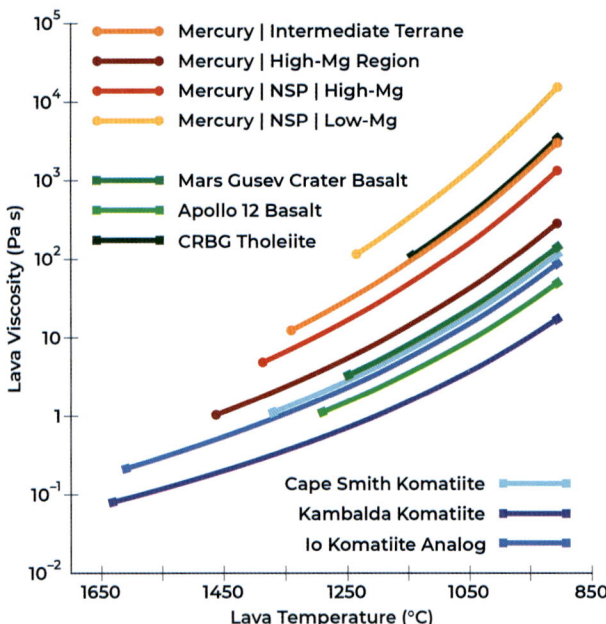

Figure 11.17. Lava dynamic viscosity versus temperature. Values are shown for four discrete geochemical terranes identified on Mercury: the intermediate terrane, the high-magnesium region, and two portions of the northern smooth plains with relatively high and low magnesium content, respectively (Weider et al., 2016; Vander Kaaden et al., 2017). Shown for comparison are values for a martian basalt (McSween et al., 2004), a low-Ti mare basalt sample returned by the Apollo 12 mission (Walker et al., 1976), and a Columbia River Basalt Group (CRBG) tholeiite (Murase and McBirney, 1973), as well as a sample of a Cape Smith komatiite (19% MgO: Barnes et al., 1982), a komatiite from Kambalda, Australia (32% MgO: Williams et al., 1999), and an Io-analog komatiite from Commondale, South Africa (31% MgO: Williams et al., 2001b). Of note, the dynamic viscosity of lava corresponding to the intermediate terrane on Mercury, which largely consists of intercrater plains (Vander Kaaden et al., 2017), most closely matches that of the CRBG tholeiitic lava.

closely the behavior of a tholeiitic basalt from the Columbia River Basalts Group (Murase and McBirney, 1973).

Importantly, Sehlke and Whittington (2015) also reported that the rheological properties of lavas on Mercury are broadly similar to those of terran basalts, which implies that vast magma volumes and high effusion rates – and not exotic compositions – are likely the cause of the laterally extensive smooth plains on Mercury. Nonetheless, despite similarities in their rheological properties, the melt densities of lavas on Mercury are probably much lower, and within a narrower range, than those for basaltic lavas on Earth (2.51–2.58 g/cm^3 versus 2.64–2.83 g/cm^3, respectively); these differences have been attributed to the paucity of FeO in the bulk silicate portion of Mercury (Vander Kaaden and McCubbin, 2015).

11.5.4 Oxygen Fugacity and Magmatic Volatiles

Magmatic volatiles – for example, hydrogen, carbon, nitrogen, and sulfur (and their oxides) – influence both the physicochemical properties and eruption styles of magmas (e.g., Section 11.2.2). This influence is, in turn, a function of the melt-phase and vapor-phase speciation of the volatiles. To characterize the speciation of volatile elements in a system, it is important to understand the oxygen fugacity (fO_2: Chapters 2 and 18) in that system, because many magmatic volatiles change speciation as a function of oxygen fugacity (Ehlmann et al., 2016). MESSENGER XRS data show that the surface of Mercury has an elevated abundance of S but an exceptionally low abundance of Fe, from which the oxygen fugacity of magmas on Mercury can be estimated. These measurements suggest that mercurian melts have an exceedingly low fO_2 (i.e., 2.6 to 7.3 $\log_{10}$ units below the iron–wüstite buffer: Chapter 2), a value that is much lower than for any other terrestrial planetary body. The next lowest calculated fO_2 value, for example, is for the Moon, at 1 $\log_{10}$ unit below the iron–wüstite buffer (McCubbin et al., 2012; Zolotov et al., 2013).

On Earth, the most influential volatile species in magmatic systems are H_2O and CO_2 (Hirschmann, 2006; Dasgupta and Hirschmann, 2010). Both of these phases, however, are unstable at mercurian fO_2 values. Instead, the C- and H-phases on Mercury are likely dominated by H_2, CH_4, graphite, carbides, and CO. Nitrogen is typically a gas in the form of N_2 under terran conditions, but on Mercury this element is likely to occur as nitrides or as other reduced forms. Similarly, sulfur is probably present on Mercury only as a sulfide, rather than sulfate.

Intriguingly, the effect of the low oxygen fugacity on Mercury on the physical properties and eruptive styles of magmas may be different from what has been encountered elsewhere in the solar system. There is, at present, a paucity of experimental data pertinent to Mercury melt compositions and fO_2 values; nonetheless, the experimental data that are available for magmatic volatiles under reducing conditions support the conclusion that low fO_2 has a profound effect on volatile speciation (e.g., Mysen et al., 2009; Mysen and Fogel, 2010; Hirschmann et al., 2012; Ardia et al., 2013; McCubbin et al., 2014; Vander Kaaden and McCubbin, 2016). There is little doubt that volatiles have influenced the eruptive behavior of magmas on Mercury: the pyroclastic deposits identified on the planet are a primary example of this outcome (e.g., Kerber et al., 2011; Thomas et al., 2014a, b). We can but speculate as to the identities of the full array of volatiles that drove those explosive eruptions, but previous studies have implicated H-, C-, and S-bearing species (Kerber et al., 2009; Weider et al., 2016). Additional experiments, under highly reducing conditions, are needed to assess fully the effects of volatiles in melt compositions relevant to Mercury, as well as the identities of the vapor species that enhanced the eruptions of magma to the surface of a planet undergoing global contraction (Chapter 10).

11.6 TIMING OF VOLCANISM

A full assessment of Mercury's volcanic character cannot be complete without understanding the planet's history of effusive and explosive activity. Chief among the means we can use to establish this history is the law of superposition, by which the uppermost unit of a geological sequence is generally interpreted as the youngest. This approach has been refined for Earth over the last few centuries and generally works well in the mapping of geological units on other worlds. Further, unlike tectonic landforms, which can be reactivated so that superposition relations tell us only when the last deformation event occurred, volcanic eruptions are temporally discrete events and so represent useful chronostratigraphic markers when one is compiling the geological history of a region, or even an entire planet. Together with areal crater density measurements (Chapter 9), then, the superposition relations between volcanic landforms or deposits and surrounding terrain give us a first-order insight into the volcanic evolution of the innermost planet.

In Section 11.6.1, we review first what crosscutting observations tell us of the emplacement history of Mercury's effusive volcanic units, and then discuss what areal crater density measurements and absolute models ages can add to this story. In Section 11.6.2, we do the same for the planet's explosive volcanic products, addressing first what we infer from superposition relations, and then from areal crater density data. We finish with an assessment of the implications of these findings for understanding the thermal history of Mercury (Section 11.6.3).

11.6.1 Effusive Volcanism

Superposition relations represent one of the means by which Mercury's smooth plains units are identified: as either impact melt or lava flows, these units by definition postdate the surrounding landscape and are in part delineated on the basis of that superposition. Moreover, the interpretation of many smooth plains deposits as volcanic arises from observations of their ponding in topographic lows, embaying pre-existing terrain, varying spatially in compositions and color properties, and hosting partially to almost entirely infilled craters (e.g., Head et al., 2008, 2011; Denevi et al., 2013; Byrne et al., 2016; Section 11.2.1).

There is abundant evidence of superposition and embayment relations within the largest volcanic province on the planet, the northern smooth plains. Head et al. (2011) identified two crater populations there: superposed primary impact craters ("post-plains craters") and ghost craters. Ostrach et al. (2015) used stratigraphic embayment relations within the ghost crater population in the NSP to argue for at least two episodes of major resurfacing of the region over a relatively short geological time frame (on the order of 100 Myr or less); Ferrari et al. (2015) reached similar conclusions for the emplacement duration of smooth plains within the Rembrandt basin. There are embayment relations similar to those documented for the NSP in ghost crater populations within numerous other large igneous provinces on Mercury, including the smooth plains within the Beethoven, Tolstoj, and Aneirin basins and surrounding the Faulkner, Barma, and Rudaki craters (Denevi et al., 2013; Byrne et al., 2016), suggesting that serial phases of effusive activity are the rule, not the exception, for Mercury's volcanic smooth plains deposits. We might reasonably infer a similar eruption history for the planet's volcanic intercrater plains (Section 11.3.3).

Figure 11.18. Possible evidence for spatially limited volcanic resurfacing within the Caloris basin. Here, tectonic structures (folds and graben) are absent in what appears to be a local topographic low (marked with white arrows), and a 4-km-diameter crater appears partially infilled with smooth plains material (black arrow). The image is in an azimuthal equidistant projection centered at 22.5°N, 171.5°E.

No ghost craters have been definitively resolved within the Caloris interior plains, even though these units are demonstrably volcanic. The reason for the apparent dearth of pre-existing craters within this LIP is unclear, although it is possible that the tectonic complexity within the basin has obscured any preceding craters (Ernst et al., 2015; Section 10.6.4). It may also be that insufficient time passed between the formation of the basin and the emplacement of the Caloris Planitia for a record of craters to accumulate there (Klimczak et al., 2013; Ernst et al., 2015). Such a scenario would be enhanced by elevated subsurface temperatures following the impact (Roberts and Barnouin, 2012), preventing the material constituting the basin floor from cooling to its elastic blocking temperature (i.e., forming a lithosphere) until after the interior plains lavas were erupted. Nonetheless, there is some evidence of localized resurfacing within the Caloris basin, where tectonic structures are absent and relatively small craters appear partially infilled (Figure 11.18). The lack of reliable crosscutting relations at such sites of suspected later volcanism, however, coupled with there being too few craters for the reliable discrimination of unit boundaries, makes this evidence for resurfacing circumstantial at best.

Of course, the superposition of impact craters upon volcanic units provides an additional means to characterize the timing of volcanism on Mercury, via the assessment of areal crater density measurements and crater size–frequency distributions. In principle, the more heavily cratered a given unit is, the older it is likely to be, because it is presumed to have been exposed to incident asteroids and comets for longer than a less heavily cratered unit (see Chapter 9 for a more comprehensive overview of this topic). The cumulative number of impact craters superposed on a region of interest can be given as $N(D)$, where N is the number of craters of diameter greater than or equal to D (in km) per unit area (usually 10^6 km^2; e.g., Ostrach et al., 2015). These values are often quoted with an uncertainty, which is taken as the square root of the number of craters per diameter bin per 10^6 km^2 (Crater Analysis Techniques Working Group, 1979). This approach has the benefit of allowing comparisons of crater populations across different studies and regions of different sizes.

Early MESSENGER observations indicated that three of the largest igneous provinces on Mercury, the NSP, the Caloris Planitia, and the plains exterior to Caloris, have similar crater SFDs (Fassett et al., 2009, 2011; Head et al., 2011; Strom et al., 2011) (Table 11.1). Subsequent studies of areal crater densities across the planet showed that numerous other (if smaller) LIPs have similar values, including smooth plains near Rudaki crater, plains inside the Rembrandt, Beethoven, and Tolstoj basins, and plains adjacent to the Rachmaninoff basin (Denevi et al., 2013; Whitten et al., 2014; Ferrari et al., 2015; Byrne et al., 2016) (Table 11.2). For example, Byrne et al. (2016) found a span of N(10) values for nine volcanic smooth plains deposits of 29 ± 21 to 145 ± 23, which encompasses fully the range of corresponding values for the Caloris interior and exterior plains, the Rembrandt interior plains, and the NSP. Interestingly, as noted in Section 11.3.3, the greatest areal density measurement for craters ≥10 km in diameter these authors reported, for the plains between the Alver and Disney craters (145 ± 23), is close to the lowest $N(10)$ value of 154 ± 34 reported by Whitten et al. (2014) for a representative intercrater plains unit to the east of Tolstoj. The overlap in uncertainty estimates for most major volcanic smooth plains units therefore indicates that their *relative* ages are statistically indistinguishable.

But what of *absolute* age? Without physical rock samples of these units with which to perform radiometric age dating, we can rely only on estimates of absolute ages derived from modeled impact crater fluxes (Chapter 9). At present, there are two widely accepted model production functions (PFs) for impact cratering on Mercury, each of which builds upon older such models that are scaled from the lunar impact flux by incorporating revised crater scaling laws, target-specific properties, and improved estimates of impactor populations (Marchi et al., 2005, 2009, 2011; Le Feuvre and Wieczorek, 2011; Chapter 9). Crater size–frequency analyses have shown that the NSP and the Caloris and Rembrandt interior plains were emplaced at around 3.7 Ga (Fassett et al., 2009; Head et al., 2011; Strom et al., 2011; Denevi et al., 2013; Ferrari et al., 2015; Ostrach et al., 2015) for many of the published chronology models for Mercury (e.g., Marchi et al., 2009), including the model of Le Feuvre and Wieczorek (2011), which incorporates the effects of a porous megaregolith. The areal densities of impact craters for other LIPs on Mercury imply that they have ages comparable to those of these largest deposits, and the PFs for Mercury support this inference. Byrne et al. (2016) found that no major volcanic smooth plains units on the planet were emplaced after about 3.5 Ga for ages calculated with either the Marchi et al. (2009) or Le Feuvre and Wieczorek (2011) model production functions (Table 11.2). (In their study, Byrne et al. (2016) applied the term "major" to any volcanic deposit greater than about 1×10^5 km^2 in area.) Taken together, then, morphological observations, crater size–frequency distribution data, and model age results suggest that widespread effusive volcanism on Mercury, a key process by which the planet's secondary crust was likely developed (Section 11.4), ended within approximately the first billion years after planet formation. Such a volcanic history contrasts strongly with the protracted (i.e., multi-billion-year) records of major effusive volcanism on Earth (e.g., Bryan et al., 2010), Mars (e.g., Carr and Head, 2010), and Venus (e.g., Ivanov and Head, 2013) – but may mirror that of the Moon, where later volcanism has been of relatively little volume (e.g., Hiesinger et al., 2003).

Importantly, the cessation of large-scale effusive volcanism on Mercury did not spell the end of volcanic activity on the planet as a whole. At least two other smooth plains units have been identified as potentially less than ~1 Gyr old, far younger than the areally extensive deposits we discuss above. However, both of these much younger smooth plains units – each ~1.4 × 10^4 km^2 in area – are situated within the peak rings of medium-sized impact basins: the 290-km-diameter Rachmaninoff basin (Prockter et al., 2010) and the 265-km-diameter Raditladi basin (Strom et al., 2008; Marchi et al., 2011). Although there is strong evidence for a volcanic origin for the plains in the Rachmaninoff basin, it is possible that the deposit within the Raditladi basin, which may be only hundreds of millions years old, is ponded impact melt and not volcanic at all (Prockter et al., 2010; Marchi et al., 2011).

As discussed in Section 11.3, the majority of intercrater plains on Mercury are impact-modified, effusively emplaced volcanic rocks, some of which are among the oldest materials on the planet's surface. We can say as much because of the morphology of these plains, their superposition relations with other geological units, and particularly their crater statistics: the highest areal crater density values are found for some of the most rugged portions of the intercrater plains (e.g., Marchi et al., 2013; Whitten et al., 2014). That the highest areal crater densities for smooth plains overlap with the lowest such values for intercrater plains further reinforces the notion that the former evolve into the latter with sustained impact bombardment (Section 11.3.3).

A comparison of representative areal crater density values for intercrater plains (Whitten et al., 2014) with those for smooth plains (e.g., Denevi et al., 2013) and impact basins (Fassett et al., 2009; Whitten and Head, 2015) shows that most intercrater plains formed prior to the Caloris and Rembrandt impact events. This finding thus provides for those intercrater plains a context within Mercury's time–stratigraphic sequence (e.g., Pohn and Offield, 1970; Trask, 1971, 1975; Spudis and Guest, 1988) that, from oldest to youngest, includes the pre-Tolstojan, Tolstojan, Calorian, Mansurian, and Kuiperian periods (Chapter 6). In preceding the formation of the Caloris impact basin, which defines the base of the eponymous time–stratigraphic period, all of the intercrater plains were thus formed before the Calorian, i.e., in the Tolstojan and/or Pre-Tolstojan periods (Spudis and Guest, 1988; Chapter 6). Since the intercrater plains were themselves once smooth plains, the phases of effusive volcanism in which they were emplaced must also have been Tolstojan and/or Pre-Tolstojan. Crater statistics for the most heavily cratered terrain on Mercury suggest, under the assumption that the cratering record is not saturated (see Chapter 9), that the oldest portions of the surface date to about 4.1–4.0 Ga (Marchi et al., 2013), placing an upper limit on the age of *preserved* volcanic material on the planet. There is presumably, then, a considerable volume of effusively emplaced volcanic rock beneath the present intercrater and smooth plains that extends back to the onset of secondary crust production on Mercury (Section 11.4), but for which no surficial evidence remains.

11.6.2 Explosive Volcanism

Observations of superposition relations also provide a means to assess the relative timing of pyroclastic activity. For example, the pyroclastic vents superposing Caloris Planitia identified from MESSENGER flyby data (Head et al., 2008, 2009a; Murchie et al., 2008) must postdate the emplacement of those plains. Goudge et al. (2014) noted numerous crosscutting relations between pyroclastic deposits and other landforms on Mercury, including a secondary crater chain from the Hokusai impact crater that superposes the deposit in the Praxiteles crater 2600 km away. These authors also reported explosive volcanic deposits that appeared to be crosscut by shortening landforms (e.g., along Victoria Rupes) and described numerous examples of pyroclastic material predating the formation of hollows (Goudge et al., 2014). Rothery et al. (2014) argued, on the basis of crosscutting vents, for repeated eruptive events at the large pyroclastic shield in Caloris Planitia (Figure 11.9). And Thomas et al. (2014a) found at least one pyroclastic vent cutting the wall terraces within the 26-km-diameter Kuniyoshi crater, which is interpreted as having formed in the Kuiperian period, i.e., within the past billion years, and perhaps as recently as 280 Ma (Banks et al., 2017; Chapter 6).

The majority of pyroclastic deposits are found within large impact craters or basins (Section 11.2.2), and so assigning relative (and, in some cases, absolute) ages to the host crater can in turn provide information on the relative (and, in some cases, absolute) timing of pyroclastic activity (e.g., Goudge et al., 2014; Thomas et al., 2014a). One way to estimate ages of impact structures is on the basis of their degradation state, which is generally regarded as a proxy for relative age and so is tied to the time–stratigraphic sequence for Mercury (e.g., McCauley et al., 1981; Spudis and Guest, 1988; Kinczyk et al., 2016; Prockter et al., 2016; Chapter 6). If a given crater or basin is attributed to the Mansurian period, for instance, then a pyroclastic deposit within that basin could have been emplaced no earlier than the Mansurian.

By this reasoning, Goudge et al. (2014) reported evidence for explosive volcanism in an approximately equal number of Mansurian, Calorian, and Tolstojan craters. In contrast, Thomas et al. (2014a) noted in their survey that the majority of pyroclastic units on Mercury are situated within Calorian impact structures, with some deposits in Mansurian craters and only a few equivocal examples in Kuiperian craters. Of course, the determination of the age of a crater or basin from its morphology is subjective and qualitative; there is as yet no reliable means for quantitatively characterizing a crater's age from its degradation state, and additional processes such as proximity weathering (Spudis and Guest, 1988; Kinczyk et al., 2016) add further uncertainty to efforts to associate a crater with a given time–stratigraphic period. Even so, Goudge et al. (2014) and Thomas et al. (2014a) both concluded that, unlike large-scale effusive volcanism, pyroclastic activity was not limited to the first billion years or so after planet formation but rather operated throughout much of Mercury's history.

Several complications arise when attempting to apply crater SFDs to the age dating of pyroclastic deposits. The surficial nature of, and the mantling of underlying terrain by, explosive volcanic material makes determining whether an observed crater superposes or precedes a deposit difficult. The unconsolidated nature of pyroclastic ash can also have substantial effects on the size and preservation of impact craters (e.g., Lucchitta and Schmitt, 1974; van der Bogert et al., 2016). Additionally, the spatial extent of pyroclastic deposits is typically small, which limits the measurement area from which a meaningful SFD can be derived. This technique (and by extension, the derivation of absolute model ages) requires a count area sufficiently large to contain a statistically robust number of craters. This constraint is compounded by the need to exclude craters below the diameter at which secondary craters come to dominate; on Mercury, this threshold may be as great as 10 km (e.g., Strom et al., 2008; Chapter 9). Nevertheless, Thomas et al. (2014a) calculated absolute model ages from crater SFDs for five pyroclastic deposits, finding that pyroclastic activity occurred at ~3.9–3.3 Ga, although these ages were determined with the now-obsolete PF of Neukum et al. (2001). Thomas and co-workers also applied the Marchi et al. (2009) PF to one deposit, which returned an absolute model age of 3.7 Ga.

11.6.3 Implications for Mercury's Thermal History

The thermal history and evolution of Mercury must comport with a number of geological constraints that are informed by the observations described in this chapter and elsewhere in this volume (e.g., Chapters 10 and 19). For example, the cessation of widespread plains volcanism at about 3.5 Ga may reflect a waning magma supply as a consequence of secular interior cooling, as suggested by thermochemical evolution models for Mercury (e.g., Michel et al., 2013; Tosi et al., 2013; A. J. Evans et al., 2015). This cooling would have had another major effect on Mercury, which in turn would have played a controlling role in its volcanic character: the phenomenon of global contraction.

The surface of Mercury hosts thousands of tectonic landforms, including lobate scarps and wrinkle ridges, which are interpreted to have formed in response to horizontal shortening that resulted from a reduction in planetary volume from interior cooling (e.g., Strom et al., 1975). Together, these landforms have accommodated a decrease in planetary radius by as much as 7 km (Byrne et al., 2014) and, coupled with elastic shortening strain, indicate that Mercury may have contracted by up to 9 km (Chapter 10). Under a tectonic regime governed by global contraction, the least compressive stresses act vertically and are governed by the overburden, whereas the most compressive stresses act in the horizontal plane, such that vertical ascent of magma, and thus widespread effusive volcanism, is inhibited. Widespread effusive volcanism on Mercury is therefore not expected to have occured after the onset of global contraction (we discuss this point further in the next section).

Superposition relations between impact craters and shortening structures on Mercury indicate that global contraction and effusive volcanism overlapped at least for a time (e.g., Strom et al., 1975; Banks et al., 2015; Byrne et al., 2016). The oldest craters that superpose segments of lobate scarps are estimated to have formed after the end of the late heavy bombardment (LHB), during the Calorian period – an assessment made on the basis of those craters' degradation states (Spudis and Guest, 1988; Banks et al., 2015). Observations of Calorian craters superposing scarps require, by definition, that brittle failure of Mercury's surface at least regionally was underway at some point after the formation of the Caloris basin, which would subsequently host a large igneous province.

Indeed, many of the largest volcanic smooth plains units on Mercury, including the Caloris Planitia and those LIPs in Beethoven, Rembrandt, and Tolstoj, are situated within older impact basins and craters (e.g., Strom et al., 1975; Fassett et al., 2012). So, too, are many smaller areas of smooth plains across the planet (Denevi et al., 2013), at least some of which are likely volcanic (Prockter et al., 2010; Marchi et al., 2011). This collocation of many of the youngest effusive volcanic units on Mercury with impact structures is consistent with predictions for a planet undergoing contraction from secular interior cooling, as the impact process could create transient, localized sites in which effusive volcanism can occur (Section 11.7.2). Impact structures may therefore be prime sites for late-stage eruptions under a global tectonic regime otherwise generally unfavorable to extrusive activity (e.g., Head and Wilson, 1992). The size and location of the (relatively) young smooth plains deposits in the Rachmaninoff and Raditladi basins are consistent with this view, whereby plains volcanism is limited in volume and collocated with areas of pre-existing weakness (i.e., impact structures). It may be that many more of the small smooth plains units situated in impact craters and basins across Mercury (Denevi et al., 2013) are composed of ponded lava flows, but no such deposit is greater than about 6×10^4 km^2 in area (Byrne et al., 2016) and so the conclusion that global contraction precipitated the end of widespread effusive volcanism on Mercury seems robust.

Yet it is clear that the onset of global contraction did not similarly inhibit explosive volcanic activity on Mercury. Volatiles provide the primary driving force for explosive volcanic eruptions (e.g., Cashman, 2004; Kerber et al., 2009), and volatile-rich melts are more capable of ascent and eruption than effusive magmas, particularly if those magmas are themselves more buoyant than those on Earth in the current epoch (Section 11.5.3). Nonetheless, the majority of pyroclastic deposits on Mercury are collocated with sites of pre-existing weakness in the lithosphere, including the heavily fractured central peaks, peak rings, and rims of impact craters and basins (e.g., Kerber et al., 2011), as well as along the surface traces of thrust faults thought to underlie lobate scarps (e.g., Goudge et al., 2014). Fractures and faults, especially if critically stressed (e.g., Barton et al., 1995), may act as conduits for the upward migration of volatile-rich magmas (e.g., Klimczak et al., 2013), and so these spatial relations indicate that the history of explosive volcanism on Mercury has also been influenced, at least in part, by global contraction.

11.7 VOLCANISM AND TECTONISM

In addition to volcanic activity, tectonism is a major planetary process and has shaped the surfaces of numerous solar system bodies, including the Moon, Mars, Venus, Io, and our own world. Mercury, too, records a detailed history of tectonic activity, which has played a large role in its geological evolution (Chapters 6 and 10). As with those other bodies, the histories of tectonism and volcanic activity on Mercury are intertwined – yet perhaps one is controlled by the other more so on that body than anywhere else in the solar system. In this section, we first review the means by which magma can ascend to the surface on Mercury (Section 11.7.1), and we then discuss the prospect for widespread effusive volcanism on a planet undergoing global contraction (Section 11.7.2).

11.7.1 Magma Ascent through Mercury's Lithosphere

For volcanism to occur, magmas must migrate vertically through the lithosphere and reach the surface, and so the mechanisms of magma ascent influence the locations, styles, and even timing of eruptive activity on the body. Magma ascent through the ductile portion of the lithosphere (see Chapters 3 and 10) has not been discussed extensively for Mercury, but likely involves the rise of diapirs such as that proposed for Earth (e.g., Bateman, 1984; Rubin, 1993; Miller and Paterson, 1999; Burov et al., 2003). Under this scenario, magma rises through and so affects the surrounding country rock by viscous flow.

The ascent of a positively buoyant magma through the brittle portion of Mercury's lithosphere has been considered qualitatively in the literature (Solomon, 1978; Wilson and Head, 2008; Head et al., 2009b). Here, magmas either actively intrude the lithosphere by creating their own accommodation space (i.e., hydrofracturing) or exploit pre-existing weaknesses within the rock as pathways to the surface (e.g., Rubin, 1993). The extent to which ascending magmas actively or passively intrude has yet to be fully investigated for Mercury, although melts there are more buoyant, and so likely ascend more readily, than terran magma (Section 11.5.3). In any case, the utilization of pre-existing fractures requires much lower magma pressure than that needed to actively fracture the lithosphere, and so passive intrusion may be the dominant mode of ascent for positively buoyant magmas on Mercury (Klimczak et al., 2013). The use of existing fractures also offers a means by which magma can ascend through a lithosphere under a net horizontal compressive stress state, such as that arising from global contraction (see below).

11.7.2 Volcanism on a Contracting Planet

It has long been recognized that magma ascent through the brittle regime of a planetary body's lithosphere must be closely

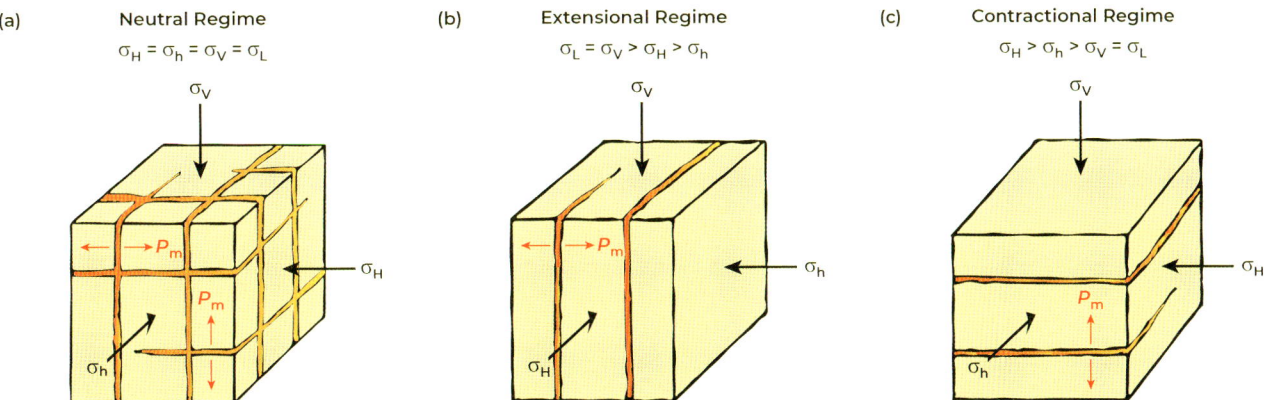

Figure 11.19. The orientations of magma-filled fractures under different tectonic regimes. (a) Under a neutral tectonic regime, vertical and horizontal magma migration is possible as long as the magma pressure (P_m) reaches the tensile strength of the host rocks and remains sufficiently large to act against the lithostatic stress (σ_L). (b) Under an extensional tectonic regime, magma is capable of ascending vertically, as the least compressive stresses act horizontally and so allow for the opening of vertically oriented fractures. (c) Under a contractional tectonic regime, vertical magma ascent is suppressed, as the vertical stress component is the least compressive and so facilitates the opening of a horizontal fracture when magma pressures reach the strength of the host rocks.

tied to the prevailing tectonic regime (e.g., Solomon, 1978). We can describe three primary such regimes: neutral, extensional, and contractional. The behavior of magma under each of these regimes is controlled by the relative magnitudes of one vertical and two horizontal stress components (here taken to be positive under compression). Under a neutral tectonic regime, the vertical (σ_V) and horizontal stresses (σ_H and σ_h, where $\sigma_H > \sigma_h$) are equal to, and so governed by, the lithostatic stress, σ_L. In contrast, under extensional and contractional regimes the vertical stress corresponds to the overburden stress (i.e., the weight of the overlying rock at any given depth), but tectonic processes act to decrease or increase the horizontal stresses.

Magma may migrate in any direction under a neutral tectonic regime (i.e., $\sigma_H = \sigma_h = \sigma_V = \sigma_L$) and is therefore equally capable of forming dikes or sills – as long as the magma pressure (P_m) reaches the tensile strength of the host rock so as to actively drive fracturing, or remains sufficiently large to act against the lithostatic stress and thus keep pre-existing fractures open (Figure 11.19a). Under an extensional tectonic regime, horizontal compressive stresses (where $\sigma_L = \sigma_V > \sigma_H > \sigma_h$) are of sufficiently low magnitude for positively buoyant magma to open (and keep open) vertical fractures for use as conduits through which they can ascend (e.g., Solomon, 1977); the magma, as dikes, can then rise to the surface (Figure 11.19b). Such a tectonic regime therefore readily promotes magma ascent and effusive and explosive volcanism. In contrast, under a contractional tectonic regime, the ascent of positively buoyant magma through the lithosphere is suppressed, as large horizontal compressive stresses prevent the opening up of fractures that might otherwise be utilized by dikes (Hamilton, 1995; Watanabe et al., 1999). Vertical stresses are lower than the horizontal stress components ($\sigma_H > \sigma_h > \sigma_V = \sigma_L$), such that horizontal fractures are opened and kept open by magma before stresses can reach the levels necessary to facilitate its vertical migration. Thus, magma preferentially migrates horizontally

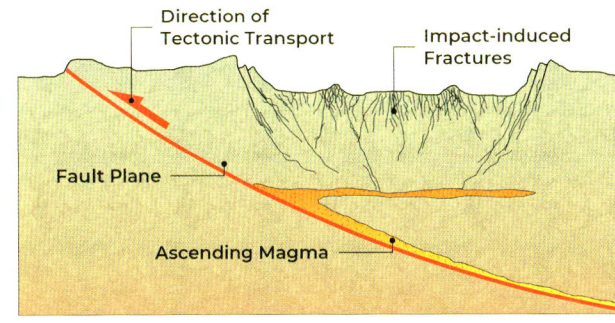

Figure 11.20. A schematic view of how magma may ascend along pre-existing weaknesses in the lithosphere, such as deep-seated thrust faults (the direction of tectonic transport is given by the red arrow) and fractures associated with impact craters and basins.

under contractional tectonic regimes, producing laccoliths, lopoliths, and sills over the formation of any vertically oriented intrusions (Hamilton, 1995; Watanabe et al., 1999; Sibson, 2003) (Figure 11.19c).

Globally, the lithosphere of Mercury has been under a contractional tectonic regime for most of solar system history (e.g., Strom et al., 1975; Byrne et al., 2014), driven by the sustained cooling of the planet's interior (Solomon, 1977; Chapter 10). With the onset of global contraction, widespread effusive volcanism ceased (Byrne et al., 2016; Section 11.5.1). However, explosive volcanism was likely much longer lived (Goudge et al., 2014; Thomas et al., 2014a; Section 11.6.2), overlapping substantially with global contraction and thus indicating that magma ascent was not fully suppressed by the prevailing tectonic regime. As noted above, the vast majority of pyroclastic vents and their associated deposits are situated at or near major weaknesses in the lithosphere, such as thrust faults or the rims, central peaks, and peak rings of impact craters and basins (Chapter 10). These spatial relations support the view that magma ascent through Mercury's lithosphere, at least since the start of global contraction, has occurred primarily through pre-existing fractures (Klimczak et al., 2013; Habermann and Klimczak, 2015; Section 10.9.2) (Figure 11.20).

It may be that a neutral or even extensional tectonic regime existed on Mercury prior to global contraction, especially during any early phase of planetary expansion (e.g., Solomon and Chaiken, 1976). Rifting, the ready emplacement of dikes, and consequent voluminous effusive volcanism (especially with very high degrees of partial melting because of elevated interior temperatures) may well have accompanied such a period, but aside from the presumed volcanic nature of the intercrater plains (Section 11.3.3), there is no evidence yet detected in Mercury's geological record of rift zones or widespread dike intrusion. High-resolution gravity gradiometry data for the Moon returned by the Gravity Recovery and Interior Laboratory (GRAIL) mission (e.g., Zuber et al., 2013) revealed a system of deep-seated structures with no surface manifestation that have been interpreted as ancient, giant dikes formed in an early stage of lunar expansion (Andrews-Hanna et al., 2013). No corresponding structures have yet been detected on Mercury, and thermal evolution models do not unequivocally predict such an expansionary phase early in the planet's history (Chapter 19).

Spatially limited neutral and even extensional tectonic regimes may have been temporarily present on Mercury after global contraction began, however. For example, neutral tectonic regimes could have existed on Mercury in and around impact craters and basins shortly after formation, as the impact process removes overburden, substantially fracturing or entirely destroying the pre-existing lithosphere, and so in effect "resets" the prevailing stress regime (Klimczak, 2015; Byrne et al., 2016; Section 10.9.1). Impact structures might thus be sites where volcanism was facilitated by a local, neutral stress regime even against a background of pervasive global contraction, accounting for why many large impact basins on Mercury are filled with thick volcanic plains units, i.e., the Caloris, Rembrandt, and Beethoven basins (Spudis and Guest, 1988; Murchie et al., 2008; Denevi et al., 2009, 2013; Watters et al., 2009; Whitten et al., 2014; Ferrari et al., 2015; Sections 10.6.4 and 11.2.1), why numerous peak-ring basins contain lavas (Prockter et al., 2010; Byrne et al., 2016), and why most pyroclastic vents are situated within impact craters (Klimczak et al., 2013; Thomas et al., 2014a, b; Section 11.2.2). Moreover, a local extensional regime is predicted to arise when impact melt deposits and lavas, such as those within buried impact craters in the NSP or that fill the Caloris basin, cool at a faster rate than the surrounding rocks (Freed et al., 2012; Blair et al., 2013). These lavas frequently host extensional tectonic landforms such as graben, which cannot form in a prevailing contractional regime; superposition relations indicate that many of these graben are situated within craters that formed after global contraction was underway (Section 10.8.1).

11.8 FURTHER WORK

Although they have enabled a remarkable advancement in our understanding of Mercury's volcanic character and history, the results of the MESSENGER mission challenge us to ask yet more questions of the innermost planet. Here, we review briefly some of those questions, and why they are important to resolve (Section 11.8.1). We then suggest the types of future investigations that might enable us to answer these questions, either by the joint European Space Agency–Japan Aerospace Exploration Agency BepiColombo mission (Benkhoff et al., 2010), scheduled to launch in 2018, or by missions for which architectures have yet to be developed (Section 11.8.2).

11.8.1 Outstanding Questions for Volcanism on Mercury

The recognition that the two dominant surface units on Mercury, the intercrater plains and the smooth plains, share a similar mode of emplacement provides strong evidence for voluminous flood-mode volcanism early in the planet's history (Section 11.3). Evidence of serial effusive volcanism preserved in some smooth plains deposits today hints at similar processes for the formerly smooth intercrater plains, although geological boundaries along which these older plains can be divided are only starting to be recognized (Chapter 6). Yet presumably, even with major plains volcanism on Mercury as late as around the end of the LHB (Head et al., 2011; Ostrach et al., 2015; Section 11.6), the portion of the planet covered by intercrater plains today was not resurfaced in a single event – a supposition supported by spatial variations in areal crater density, color variations, and superposition relations across and within this entire unit (Ernst et al., 2010; Whitten et al., 2014; Chapter 6).

Given the inference, then, that the deposits identified today as intercrater plains were emplaced sequentially, has any record of that emplacement history been preserved on Mercury? It is unlikely that discrete landforms such as flow boundaries in the intercrater plains have survived subsequent bombardment, but it may be that portions of these plains can be distinguished, and an eruption sequence developed, on the basis of crater size–frequency distribution, color, or other properties. The recognition of such discrete portions would provide better constraints on estimates of effusive volcanic fluxes early in Mercury's history, with corresponding insight into the rate of melt production and the thermal evolution of this planet in particular, and possibly of terrestrial worlds in general. Similarly, establishing a means to distinguish between fluidized ejecta deposits and flood basalts would assist in determining the relative contribution of each of these materials in the early, widespread resurfacing of the planet (Section 11.3.3).

A major outstanding question relates to the composition of Mercury's rocks, most of which are volcanic. Geochemical measurements from the MESSENGER XRS and GRS indicate that the dominant rock types on the planet are alkali-rich komatiites and boninites (Section 11.5.1). Yet the resolution of these measurements varies considerably across Mercury, and so there may be local differences in surface composition that cannot be resolved with available data, especially in the planet's southern hemisphere. Moreover, these instruments sense composition only within the top ~100 μm (XRS) to the top tens of centimeters (GRS) of the crust (e.g., Schlemm et al., 2007; Weider et al., 2016), and so the compositions identified at the surface and their horizontal variations may not necessarily correspond to those at depth. And, at present, there is no independent means by which we can verify these compositions at the rock scale or their petrological or mineralogical interpretations: there are no robust candidates for samples of the innermost planet in any of the world's meteorite collections (although candidates have been proposed, e.g., Irving et al., 2013). As a result, the precise compositions (and compositional variability) of the rocks on the surface of Mercury remain to be fully characterized.

This issue is more acute for the unobserved intrusive equivalents to the surface lithologies that may be present; alkali-rich komatiites and boninites, for example, suggest the presence of presumably equally alkali-rich rocks (such as olivine adcumulates and sanukitoids: e.g., Shirey and Hanson, 1984) at depth. Other processes may have worked to alter the compositions of Mercury's subsurface lithologies after they were formed, including metasomatism (the chemical alteration of a rock by hydrothermal fluids, resulting in basic hornfels in the case of Mercury, for instance) and contact metamorphism from subsequent proximal intrusions, as well as shock metamorphism from nearby impact events. Affirming that these processes have operated on Mercury, however, is all but impossible with orbital data alone. Equally intangible is the character of Mercury's very earliest volcanism: what were conditions like at the onset of secondary crust formation? Was there a substantial if transient atmosphere at Mercury from early volcanic outgassing? Is there a record in the planet's crustal stratigraphy of a change in composition from deep to shallow levels that reflects progressively lower degrees of partial melting as interior cooling took hold?

Other outstanding issues persist that are perhaps more amenable to progress, including the origin of several of the planet's volcanic landforms. For example, how did the depressions observed in proximity to the network of valles in the northern hemisphere, and other pits without raised rims and no obvious pyroclastic signature, form? The role of magma eruption versus subsurface lateral withdrawal in the creation of these structures remains an open question, the answer to which might provide more information on why large-scale caldera complexes, even dating from before the onset of global contraction, are not observed on Mercury. Another question is whether the valles are simply examples of flooded impact-sculpted terrain that happened to escape burial, with other examples hidden from view by subsequent volcanism. More broadly, what of the spatial collocation between large igneous provinces and large impact structures? Superposition and areal crater density measurements show that, in most cases, the formation of a basin predates by some geologically considerable amount of time the emplacement of any smooth plains deposits it may host. Nonetheless, it is telling that many large basins show little or no evidence of their infilling lavas having flowed into the basin from the surrounding terrain. This arrangement is also common for smaller basins and large craters, although for these examples it is less clear how many host voluminous basalts instead of impact melt deposits.

It is possible that at least some of the landforms identified with early MESSENGER observations as lava margins (e.g., Head et al., 2009a, 2011) are in fact tectonic landforms. It is important to resolve which (if any) of these lobate landforms are definitively volcanic, and whether any such volcanic landform has experienced lava inflation. In doing so, we may have the means to estimate lava properties such as yield strength from measurements of these flow lobe dimensions (e.g., Moore et al., 1978; Wilson and Head, 2008; Chevrel et al., 2013; Sehlke and Whittington, 2015). On that note, the initial state of global contraction is characterized by extension at the surface (e.g., Dana, 1873; Ampferer, 1923) and so, as Mercury's early lithosphere cooled initially at a rate faster than the interior, it must have experienced tensile stresses in a manner similar to the thermal contraction of ponded lavas (e.g., Freed et al., 2012; Blair et al., 2013). Such incipient extension has not been substantially investigated for Mercury, and any such deformation probably preceded the emplacement of even the oldest surface now preserved on the planet (Marchi et al., 2013). Yet an early phase of contraction-induced rifting, in which those rifts facilitated the rapid and widespread eruption of material onto the surface, is consistent with the growing body of evidence that Mercury's early history featured widespread effusive volcanism (Sections 11.2 and 11.3).

11.8.2 Future Investigations

The wealth of data returned by MESSENGER has provided a detailed and integrated insight into the volcanic character of the innermost planet. Of course, like Mariner 10, MESSENGER has left us with more questions than it has provided answers – and so we must identify the additional observations, measurements, and analyses needed to address the questions we include above, among many others. By doing so, we can help frame the science campaigns of future missions, including the BepiColombo mission (Benkhoff et al., 2010), as well as later missions to Mercury (Chapter 20).

For example, more remains to be done in characterizing the planet's intercrater plains, including their compositions, relationship with impact structures, and emplacement histories. High-resolution targeted monochrome and color images of these plains, especially in the southern hemisphere, would complement existing data from MESSENGER and may help to distinguish impact melt from ponded lavas and even resolve different flow units. Such new data, together with high-resolution topographic observations, would also assist in developing a set of criteria for differentiating flow lobes from shortening structures. Similarly, high-resolution geochemical measurements of Mercury's surface units, particularly of heavier elements resolved by MESSENGER's XRS only during solar flares (e.g., S, Ca, Ti, and Fe: Nittler et al., 2011; Weider et al., 2012), would enhance our ability to characterize the compositions of rocks on the surface of Mercury. And the acquisition of high-resolution gravity field data for the entire planet with a GRAIL-like mission could establish whether deep-seated dikes, such as those inferred for the Moon (Andrews-Hanna et al., 2013), are present within Mercury; the identification of such buried structures would provide compelling support for an early period of global expansion with concomitant rapid, voluminous volcanism.

Among the most ambitious future exploration efforts we can consider is in situ analysis of the surface rocks of Mercury – or even the return to Earth of samples from the innermost planet. Landing a spacecraft on Mercury would be a formidable challenge (e.g., Hauck et al., 2010), and returning material to Earth would be even more so. Yet our understanding of the chemistry of the planet's surface, the physicochemical properties of its lavas, and the history of melt production would be advanced enormously by a landed mission to Mercury. The recognition of a Mercury meteorite would represent a substantial step toward this understanding, and could even help inform the design of such a mission.

The objectives of future missions to Mercury, from BepiColombo forward, will reflect and build on the findings of the MESSENGER mission. Our understanding of the planet's volcanic character has never been more comprehensive, but there is much left to learn. With the increasing number of terrestrial planets identified in extrasolar planetary systems (e.g., Barclay et al., 2013), Mercury may come to serve as a case study with which to understand the thermal and volcanic histories of small, metal-rich, one-plate planets in general. The question, then, is not *whether* we should return to Mercury – but *when* will we go? And what more will we find?

REFERENCES

Ampferer, A. (1923). Beiträge zur Auflösung der Mechanik der Alpen. *Jahrb. Geol. Bundesanst.*, **76**, 125–151.

André, S. L., Watters, T. R. and Robinson, M. S. (2005) The long wavelength topography of Beethoven and Tolstoj basins, Mercury. *Geophys. Res. Lett.*, **32**, L21202, doi:10.1029/2005GL023627.

Andrews-Hanna, J. C., Asmar, S. W., Head, J. W., Kiefer, W. S., Konopliv, A. S., Lemoine, F. G., Matsuyama, I., Mazarico, E., McGovern, P. J., Melosh, H. J., Neumann, G. A., Nimmo, F., Phillips, R. J., Smith, D. E., Solomon, S. C., Taylor, J., Wieczorek, M. A., Williams, J. G. and Zuber M. T. (2013). Ancient igneous intrusions and early expansion of the Moon revealed by GRAIL gravity gradiometry. *Science*, **339**, 675–678.

Ardia, P., Hirschmann, M. M., Withers, A. C. and Stanley, B. D. (2013). Solubility of CH_4 in a synthetic basaltic melt, with applications to atmosphere–magma ocean-core partitioning of volatiles and to the evolution of the Martian atmosphere. *Geochim. Cosmochim. Acta*, **114**, 52–71.

Baker, V. R. and Kochel, R. C. (1978). Morphometry of streamlined forms in terrestrial and martian channels. *Proc. Lunar Planet. Sci. Conf.*, **9**, 3193–3203.

Baker, V. R. and Kochel, R. C. (1979). Martian channel morphology: Maja and Kasei Valles. *J. Geophys. Res.*, **84**, 7961–7983.

Baker, V. R., Komatsu, G., Parker, T. J., Gulick, V. C., Kargel, J. S. and Lewis, J. S. (1992). Channels and valleys on Venus: Preliminary analysis of Magellan data. *J. Geophys. Res.*, **97**, 13,421–13,444.

Banks, M. E., Xiao, Z., Watters, T. R., Strom, R. G., Braden, S. E., Chapman, C. R., Solomon, S. C., Klimczak, C. and Byrne, P. K. (2015). Duration of activity on lobate-scarp thrust faults on Mercury. *J. Geophys. Res. Planets*, **120**, 1751–1762.

Banks, M. E., Xiao, Z., Braden, S. E., Barlow, N. G., Chapman, C. R., Fassett, C. I. and Marchi, S. S. (2017). Revised constraints on absolute age limits for Mercury's Kuiperian and Mansurian stratigraphic systems. *J. Geophys. Res. Planets*, **122**, 1010–1020, doi:10.1002/2016JE005254.

Barclay, T., Rowe, J. F., Lissauer, J. J., Huber, D., Fressin, F., Howell, S. B., Bryson, S. T., Chaplin, W. J., Désert, J. M., Lopez, E. D., Marcy, G. W., Mullally, F., Ragozzine, D., Torres, G., Adams, E. R., Agol, E., Barrado, D., Basu, S., Bedding, T. R., Buchhave, L. A., Charbonneau, D., Christiansen, J. L., Christensen-Dalsgaard, J., Ciardi, D., Cochran, W. D., Dupree, A. K., Elsworth, Y., Everett, M., Fischer, D. A., Ford, E. B., Fortney, J. J., Geary, J. C., Haas, M. R., Handberg, R., Hekker, S., Henze, C. E., Horch, E., Howard, A. W., Hunter, R. C., Isaacson, H., Jenkins, J. M., Karoff, C., Kawaler, S. D., Kjeldsen, H., Klaus, T. C., Latham, D. W., Li, J., Lillo-Box, J., Lund, M. N., Lundkvist, M., Metcalfe, T. S., Miglio, A., Morris, R. L., Quintana, E. V., Stello, D., Smith, J. C., Still, M. and Thompson, S. E. (2013). A sub-Mercury-sized exoplanet. *Nature*, **494**, 452–454.

Barnes, S. J., Coats, C. J. and Naldrett, A. J. (1982). Petrogenesis of a Proterozoic nickel-sulfide-komatiite association; the Katiniq Sill, Ungava, Quebec. *Econ. Geol.*, **77**, 413–429.

Barnouin, O. S., Zuber, M. T., Smith, D. E., Neumann, G. A., Herrick, R. R., Chappelow, J. E., Murchie, S. L. and Prockter, L. M. (2012). The morphology of craters on Mercury: Results from MESSENGER flybys. *Icarus*, **219**, 414–427.

Barton, C. A., Zoback, M. D. and Moos, D. (1995). Fluid flow along potentially active faults in crystalline rock. *Geology*, **23**, 683–686.

Basilevsky, A. T., Head, J. W., Fassett, C. I. and Michael, G. (2011). History of tectonic deformation in the interior plains of the Caloris basin, Mercury. *Solar Syst. Res.*, **45**, 471–497.

Bateman, R. (1984). On the role of diapirism in the segregation, ascent and final emplacement of granitoid magmas. *Tectonophysics*, **110**, 211–231.

Becker, K. J., Robinson, M. S., Becker, T. L., Weller, L. A., Edmundson, K. L., Neumann, G. A., Perry, M. E. and Solomon, S. C. (2016). First global digital elevation model of Mercury. *Lunar Planet. Sci.*, **47**, abstract 2959.

Benkhoff, J., van Casteren, J., Hayakawa, H., Fujimoto, M., Laakso, H., Novara, M., Ferri, P., Middleton, H. R. and Ziethe, R. (2010). BepiColombo – Comprehensive exploration of Mercury: Mission overview and science goals. *Planet. Space Sci.*, **58**, 2–20.

Black, B. A. and Manga, M. (2016). The eruptibility of magmas at Tharsis and Syrtis Major on Mars. *J. Geophys. Res. Planets*, **121**, 944–964.

Blair, D. M., Freed, A. M., Byrne, P. K., Klimczak, C., Prockter, L. M., Ernst, C. M., Solomon, S. C., Melosh, H. J. and Zuber, M. T. (2013). The origin of graben and ridges in Rachmaninoff, Raditladi, and Mozart basins, Mercury. *J. Geophys. Res. Planets*, **118**, 47–58.

Bleacher, J. E., Greeley, R., Williams, D. A., Cave, S. R. and Neukum, G. (2007). Trends in effusive style at the Tharsis Montes, Mars, and implications for the development of the Tharsis province. *J. Geophys. Res.*, **112**, E09005, doi:10.1029/2006JE002873.

Bleacher, J. E., de Wet, A. P., Garry, W. B., Zimbelman, J. R. and Trumble, M. E. (2010). Volcanic or fluvial: Comparison of an Ascraeus Mons, Mars, braided and sinuous channel with features of the 1859 Mauna Loa flow and Mare Imbrium flows. *Lunar Planet. Sci.*, **41**, abstract 1612.

Blewett, D. T., Robinson, M. S., Denevi, B. W., Gillis-Davis, J. J., Head, J. W., Solomon, S. C., Holsclaw, G. M. and McClintock W. E. (2009). Multispectral images of Mercury from the first MESSENGER flyby: Analysis of global and regional color trends. *Earth Planet. Sci. Lett.*, **285**, 272–282.

Brown, S. M. and Elkins-Tanton, L. T. (2009). Compositions of Mercury's earliest crust from magma ocean models. *Earth Planet. Sci. Lett.*, **286**, 446–455.

Bryan, S. E. and Ernst, R. E. (2008). Revised definition of large igneous provinces (LIPs). *Earth Sci. Rev.*, **86**, 175–202.

Bryan, S. E., Peate, I. U., Peate, D. W., Self, S., Jerram, D. A., Mawby, M. R., Marsh, J. S. and Miller, J. A. (2010). The largest volcanic eruptions on Earth. *Earth Sci. Rev.*, **102**, 207–229.

Burov, E., Jaupart, C. and Guillou-Frottier, L. (2003). Ascent and emplacement of buoyant magma bodies in brittle-ductile upper crust. *J. Geophys. Res.*, **108** (B4), 2177, doi:10.1029/2002JB001904.

Byrne, P. K., van Wyk de Vries, B., Murray, J. B. and Troll, V. R. (2012). A volcanotectonic survey of Ascraeus Mons, Mars. *J. Geophys. Res.*, **117**, E01004, doi:10.1029/2011JE003825.

Byrne, P. K., Klimczak, C., Williams, D. A., Hurwitz, D. M., Solomon, S. C., Head, J. W., Preusker, F. and Oberst, J. (2013). An assemblage of lava flow features on Mercury. *J. Geophys. Res. Planets*, **118**, 1303–1322.

Byrne, P. K., Klimczak, C., Şengör, A. M. C., Solomon, S. C., Watters, T. R. and Hauck, S. A., II (2014). Mercury's global contraction much greater than earlier estimates. *Nature Geosci.*, **7**, 301–307.

Byrne, P. K., Ostrach, L. R., Fassett, C. I., Chapman, C. R., Denevi, B. W., Evans, A. J., Klimczak, C., Banks, M. E., Head, J. W. and Solomon, S. C. (2016). Widespread effusive volcanism on Mercury likely ended by about 3.5 Ga. *Geophys. Res. Lett.*, **43**, 7408–7416.

Carr, M. H. (1974). The role of lava erosion in the formation of lunar rilles and Martian channels. *Icarus*, **22**, 1–23.

Carr, M. H. and Head, J. W. (2010). Geologic history of Mars. *Earth Planet. Sci. Lett.*, **294**, 185–203.

Cashman, K. V. (2004). Volatile controls on magma ascent and eruption. In *The State of the Planet: Frontiers and Challenges in Geophysics*, ed. R. S. J. Sparks and C. J. Hawkesworth. Washington, DC: American Geophysical Union, pp. 109–124.

Cashman, K. V., Pinkerton, H. and Stephenson, J. (1998). Introduction to special section: Long lava flows. *J. Geophys. Res.*, **103**, 27,281–27,289.

Cavanaugh, J. F., Smith, J. C., Sun, X., Bartels, A. E., Ramos-Izquierdo, L., Krebs, D. J., McGarry, J. F., Trunzo, R., Novo-Gradac, A. M., Britt, J. L., Karsh, J., Katz, R. B., Lukemire, A. T., Szymkiewicz, R., Berry, D. L., Swinski, J. P., Neumann, G. A., Zuber, M. T. and Smith, D. E. (2007). The Mercury Laser Altimeter instrument for the MESSENGER mission. *Space Sci. Rev.*, **131**, 451–479.

Chabot, N. L., Denevi, B. W., Murchie, S. L., Hash, C. D., Ernst, C. M., Blewett, D. T., Nair, H., Laslo, N. R. and Solomon, S. C. (2016). Mapping Mercury: Global imaging strategy and products from the MESSENGER mission. *Lunar Planet. Sci.*, **47**, abstract 1256.

Charlier, B., Grove, T. L. and Zuber, M. T. (2013). Phase equilibria of ultramafic compositions on Mercury and the origin of the compositional dichotomy. *Earth Planet. Sci. Lett.*, **363**, 50–60.

Chester, D. K., Duncan, A. M., Guest, J. E. and Kilburn, C. R. J. (1985). *Mount Etna*. New York, Chapman and Hall.

Chevrel, M. O., Platz, T., Hauber, E., Baratoux, D., Lavallée, Y. and Dingwell, D. B. (2013). Lava flow rheology: A comparison of morphological and petrological methods. *Earth Planet. Sci. Lett.*, **384**, 109–120.

Coffin, M. F. and Eldholm, O. (1994). Large igneous provinces: Crustal structure, dimensions, and external consequences. *Rev. Geophys.*, **32**, 1–36.

Crater Analysis Techniques Working Group (1979). Standard techniques for presentation and analysis of crater size–frequency data. *Icarus*, **37**, 467–474.

Crisp, J. A. (1984). Rates of magma emplacement and volcanic output. *J. Volcanol. Geotherm. Res.*, **20**, 177–211.

Cross, W., Iddings, J. P., Pirsson, L. V. and Washington, H. S. (1902). A quantitative chemico-mineralogical classification and nomenclature of igneous rocks. *J. Geol.*, **10**, 555–690.

Crumpler, L. S. and Aubele, J. C. (1978). Structural evolution of Arsia Mons, Pavonis Mons, and Ascreus Mons: Tharsis region of Mars. *Icarus*, **34**, 496–511.

Dana, J. D. (1873). On some results of the Earth's contraction from cooling, including a discussion of the origin of mountains and the nature of the Earth's interior. *Amer. J. Sci.*, **5**, 423–443.

Dasgupta, R. and Hirschmann, M. M. (2010). The deep carbon cycle and melting in Earth's interior. *Earth Planet. Sci. Lett.*, **298**, 1–13.

DeHon, R. A., Scott, D. H. and Underwood, J. R. (1981). *Geologic Map of the Kuiper (H-6) Quadrangle of Mercury*, Map I-1233. Denver, CO: U.S. Geological Survey.

Denevi, B. W. and Robinson, M. S. (2008). Mercury's albedo from Mariner 10: Implications for the presence of ferrous iron. *Icarus*, **197**, 239–246.

Denevi, B. W., Robinson, M. S., Solomon, S. C., Murchie, S. L., Blewett, D. T., Domingue, D. L., McCoy, T. J., Ernst, C. M., Head, J. W., Watters, T. R. and Chabot, N. L. (2009). The evolution of Mercury's crust: A global perspective from MESSENGER. *Science*, **324**, 613–618.

Denevi, B. W., Ernst, C. M., Meyer, H. M., Robinson, M. S., Murchie, S. L., Whitten, J. L., Head, J. W., Watters, T. R., Solomon, S. C., Ostrach, L. R., Chapman, C. R., Byrne, P. K., Klimczak, C. and Peplowski, P. N. (2013). The distribution and origin of smooth plains on Mercury. *J. Geophys. Res. Planets*, **118**, 891–907.

Dutton, C. E. (1884). Hawaiian Volcanoes. *Annual Report U.S. Geological Survey*, **4**, 81–219.

Eggleton, R. E. and Schaber, G. G. (1972). Cayley Formation interpreted as basin ejecta. In *Apollo 16 Preliminary Science Report*, Special Publication SP-315. Washington, DC: National Aeronautics and Space Administration, pp. 29-7–29-16.

Ehlmann, B. L., Anderson, F. S., Andrews-Hanna, J. C., Catling, D. C., Christensen, P. R., Cohen, B. A., Dressing, C. D., Edwards, C. S., Elkins-Tanton, L. T., Farley, K. A., Fassett, C. I., Fischer, W. W., Fraeman, A. A., Golombek, M. P., Hamilton, V. E., Hayes, A. G., Herd, C. D. K., Horgan, B., Hu, R., Jakosky, B. M., Johnson, J. R., Kasting, J. F., Kerber, L., Kinch, K. M., Kite, E. S., Knutson, H. A., Lunine, J. I., Mahaffy, P. R., Mangold, N., McCubbin, F. M., Mustard, J. F., Niles, P. B., Quantin-Nataf, C., Rice, M. S., Stack, K. M., Stevenson, D. J., Stewart, S. T., Toplis, M. J., Usui, T., Weiss, B. P., Werner, S. C., Wordsworth, R. D., Wray, J. J., Yingst, R. A., Yung, Y. L. and Zahnle, K. J. (2016). The sustainability of habitability on terrestrial planets: Insights, questions, and needed measurements from Mars for understanding the evolution of Earth-like worlds. *J. Geophys. Res. Planets*, **121**, 1927–1961.

Elkins-Tanton, L. T. and Hager, H. B. (2005). Giant meteoroid impacts can cause volcanism. *Earth Planet. Sci. Lett.*, **239**, 219–232.

Ernst, C. M., Murchie, S. L., Barnouin, O. S., Robinson, M. S., Denevi, B. W., Blewett, D. T., Head, J. W., Izenberg, N. R., Solomon, S. C. and Roberts, J. H. (2010). Exposure of spectrally distinct material by impact craters on Mercury: Implications for global stratigraphy. *Icarus*, **209**, 210–223.

Ernst, C. M., Denevi, B. W., Barnouin, O. S., Klimczak, C., Chabot, N. L., Head, J. W., Murchie, S. L., Neumann, G. A., Prockter, L. M., Robinson, M. S., Solomon, S. C. and Watters, T. R. (2015). Stratigraphy of the Caloris basin, Mercury: Implications for volcanic history and basin impact melt. *Icarus*, **250**, 413–429.

Ernst, R. E., Grosfils, E. B. and Mège, D. (2001). Giant dike swarms: Earth, Venus, and Mars. *Annu. Rev. Earth Planet. Sci.*, **29**, 489–534.

Ernst, R. E., Bleeker, W., Söderlund, U. and Kerr, A. C. (2013). Large igneous provinces and supercontinents: Toward completing the plate tectonic revolution. *Lithos*, **174**, 1–14.

Evans, A. J., Brown, S. M. and Solomon, S. C. (2015). Characteristics of early mantle convection and melting on Mercury. *Lunar. Planet. Sci.*, **46**, abstract 2414.

Evans, L. G., Peplowski, P. N., Rhodes, E. A., Lawrence, D. J., McCoy, T. J., Nittler, L. R., Solomon, S. C., Sprague, A. L., Stockstill-Cahill, K. R., Starr, R. D., Weider, S. Z., Boynton, W. V., Hamara, D. K. and Goldsten, J. O. (2012). Major-element abundances on the surface of Mercury: Results from the MESSENGER Gamma-Ray Spectrometer. *J. Geophys. Res.*, **117**, E00L07, doi:10.1029/2012JE004178.

Evans, L. G., Peplowski, P. N., McCubbin, F. M., McCoy, T. J., Nittler, L. R., Zolotov, M. Yu., Ebel, D. S., Lawrence, D. J., Starr, R. D., Weider, S. Z. and Solomon, S. C. (2015). Chlorine on the surface of Mercury: MESSENGER gamma-ray measurements and implications for the planet's formation and evolution. *Icarus*, **257**, 417–427.

Fa, W., Cai, Y., Xiao, Z. and Tian, W. (2016). Topographic roughness of the northern high latitudes of Mercury from MESSENGER laser altimeter data. *Geophys. Res. Lett.*, **43**, 3078–3087.

Fassett, C. I. (2016). Ames stereo pipeline-derived digital terrain models of Mercury from MESSENGER stereo imaging. *Planet. Space Sci.*, **134**, 19–28.

Fassett, C. I., Head, J. W., Blewett, D. T., Chapman, C. R., Dickson, J. L., Murchie, S. L., Solomon, S. C. and Watters, T. R. (2009). Caloris impact basin: Exterior geomorphology, stratigraphy, morphometry, radial sculpture, and smooth plains deposits. *Earth Planet. Sci. Lett.*, **285**, 297–308.

Fassett, C. I., Kadish, S. J., Head, J. W., Solomon, S. C. and Strom, R. G. (2011). The global population of large craters on Mercury and comparison with the Moon. *Geophys. Res. Lett.*, **38**, L10202, doi: 10.1029/2011GL047294.

Fassett, C. I., Head, J. W., Baker, D. M. H., Zuber, M. T., Smith, D. E., Neumann, G. A., Solomon, S. C., Klimczak, C., Strom, R. G., Chapman, C. R., Prockter, L. M., Phillips, R. J., Oberst, J. and Preusker, F. (2012). Large impact basins on Mercury: Global distribution, characteristics, and modification history from MESSENGER orbital data. *J. Geophys. Res.*, **117**, E00L08, doi: 10.1029/2012JE004154.

Fegan, E. R., Rothery, D. A., Marchi, S., Massironi, M., Conway, S. J. and Anand, M. (2017). Late movement of basin-edge lobate scarps on Mercury. *Icarus*, **288**, 226–234.

Ferrari, S., Massironi, M., Marchi, S., Byrne, P. K., Klimczak, C., Martellato, E. and Cremonese, G. (2015). Age relationships of the Rembrandt basin and Enterprise Rupes, Mercury. In *Volcanism and Tectonism Across the Solar System*, ed. T. Platz, M. Massironi, P. K. Byrne and H. Heisinger, Special Publication 401. London: Geological Society, pp. 159–172.

Freed, A. M., Blair, D. M., Watters, T. R., Klimczak, C., Byrne, P. K., Solomon, S. C., Zuber, M. T. and Melosh, H. J. (2012). On the origin of graben and ridges within and near volcanically buried craters and basins in Mercury's northern plains. *J. Geophys. Res.*, **117**, E00L06, doi:10.1029/2012JE004119.

Garry, W. B. and Bleacher, J. E. (2011). Emplacement scenarios for Vallis Schröteri, Aristarchus Plateau, the Moon. In *Recent Advances and Current Research Issues in Lunar Stratigraphy*, ed. W. A. Ambrose and D. A. Williams, Special Paper 477. Boulder, CO: Geological Society of America, pp. 77–93.

Gault, D. E., Guest, J. E., Murray, J. B., Dzurisin, D. and Malin, M.C. (1975). Some comparisons of impact craters on Mercury and the Moon. *J. Geophys. Res.*, **80**, 2444–2460.

Gerlach, T. M. (1986). Exsolution of H_2O, CO_2, and S during eruptive episodes at Kilauea volcano, Hawaii. *J. Geophys. Res.*, **91**, 12,177–12,185.

Gillis-Davis, J. J., Blewett, D. T., Gaskell, R. W., Denevi, B. W., Robinson, M. S., Strom, R. G., Solomon, S. C. and Sprague, A. L. (2009). Pit-floor craters on Mercury: Evidence of near-surface igneous activity. *Earth Planet. Sci. Lett.*, **285**, 243–250.

Glickson, A. Y. (2004). Impacts do not initiate volcanic eruptions: Eruptions close to the crater: Comment and reply. *Geology*, **32**, e48.

Goldsten, J. O., Rhodes, E. A., Boynton, W. V., Feldman, W. C., Lawrence, D. J., Trombka, J. I., Smith, D. M., Evans, L. G., White, J., Madden, N. W., Berg, P. C., Murphy, G. A., Gurnee, R. S., Strohbehn, K., Williams, B. D., Schaefer, E. D., Monaco, C. A., Cork, C. P., Del Eckels, J., Miller, W. O., Burks, M. T., Hagler, L. B., Deteresa, S. J. and Witte, M. C. (2007). The MESSENGER Gamma-Ray and Neutron Spectrometer. *Space Sci. Rev.*, **131**, 339–391.

Golombek, M. P. and McGill, G. E. (1983). Grabens, basin tectonics, and the maximum total expansion of the Moon. *J. Geophys. Res.*, **88**, 3563–3578.

Goudge, T. A., Head, J. W., Kerber, L., Blewett, D. T., Denevi, B. W., Domingue, D. L., Gillis-Davis, J. J., Gwinner, K., Helbert, J., Holsclaw, G. M., Izenberg, N. R., Klima, R. L., McClintock, W. E., Murchie, S. L., Neumann, G. A., Smith, D. E., Strom, R. G., Xiao, Z., Zuber, M. T. and Solomon, S. C. (2014). Global inventory and characterization of pyroclastic deposits on Mercury: New insights into pyroclastic activity from MESSENGER orbital data. *J. Geophys. Res. Planets*, **119**, 635–658.

Grolier, M. J. and Boyce, J. M. (1984). *Geologic Map of the Borealis Region (H-1) of Mercury*, Map I-1660. Denver, CO: U.S. Geological Survey.

Guest, J. E. and Greeley, R. (1983). *Geologic Map of the Shakespeare (H-3) Quadrangle of Mercury*, Map I-1408. Miscellaneous Investigations Service, Denver, CO: U.S. Geological Survey.

Habermann M. A. and Klimczak, C. (2015). Tectonic controls of pyroclastic volcanism on Mercury. Presented at 2015 Fall Meeting American Geophysical Union, abstract P53A–2101, San Francisco, CA, 14-18 December.

Hapke, B., Danielson, G. E., Klaasen, K. and Wilson, L. (1975). Photometric observations of Mercury from Mariner 10. *J. Geophys. Res.*, **80**, 2431–2443.

Hamilton, W. B. (1995). Subduction systems and magmatism. In *Volcanism Associated with Extension at Consuming Plate Margins*, ed. J. L. Smellie, Special Publication 81. London: Geological Society, pp. 3–28.

Hauck, S. A., II, Eng, D. A. and Tahu, G. J. (2010). *Mercury Lander Mission Concept Study*. Washington, DC: National Aeronautics and Space Administration.

Hawkins, S. E., III, Boldt, J. D., Darlington, E. H., Espiritu, R., Gold, R. E., Gotwols, B., Grey, M. P., Hash, C. D., Hayes, J. R., Jaskulek, S. E., Kardian, C. J., Jr., Keller, M. R., Malaret, E. R., Murchie, S. L., Murphy, P. K., Peacock, K., Prockter, L. M., Reiter, R. A., Robinson, M. S., Schaefer, E. D., Shelton, R. G., Sterner, R. E., II, Taylor, H. W., Watters, T. R. and Williams, B. D. (2007). The Mercury Dual Imaging System on the MESSENGER spacecraft. *Space Sci. Rev.*, **131**, 247–338.

Head, J. W. and Wilson, L. (1992). Lunar mare volcanism: Stratigraphy, eruption conditions, and the evolution of secondary crusts. *Geochim. Cosmochim. Acta*, **56**, 2155–2175.

Head, J. W. and Wilson, L. (1993). Lunar graben formation due to near-surface deformation accompanying dike emplacement. *Planet. Space Sci.*, **41**, 719–727.

Head, J. W., Campbell, D. B., Elachi, C., Guest, J. E., McKenzie, D. P., Saunders, R. S., Schaber, G. G. and Schubert, G. (1991). Venus volcanism: Initial analysis from Magellan data. *Science*, **252**, 276–288.

Head, J. W., Murchie, S. L., Prockter, L. M., Robinson, M. S., Solomon, S. C., Strom, R. G., Chapman, C. R., Watters, T. R., McClintock, W. E., Blewett, D. T. and Gillis-Davis, J. J. (2008). Volcanism on Mercury: Evidence from the first MESSENGER flyby. *Science*, **321**, 69–72.

Head, J. W., Murchie, S. L., Prockter, L. M., Solomon, S. C., Chapman, C. R., Strom, R. G., Watters, T. R., Blewett, D. T., Gillis-Davis, J. J. and Fassett, C. I. (2009a). Volcanism on Mercury: Evidence from the first MESSENGER flyby for extrusive and explosive activity and the volcanic origin of plains. *Earth Planet. Sci. Lett.*, **285**, 227–242.

Head, J. W., Murchie, S. L., Prockter, L. M., Solomon, S. C., Strom, R. G., Chapman, C. R., Watters, T. R., Blewett, D. T., Gillis-Davis, J. J., Fassett, C. I., Dickson, J. L., Hurwitz, D. M. and Ostrach, L. R.

(2009b). Evidence for intrusive activity on Mercury from the first MESSENGER flyby. *Earth Planet. Sci. Lett.*, **285**, 251–262.

Head, J. W., Chapman, C. R., Strom, R. G., Fassett, C. I., Denevi, B. W., Blewett, D. T., Ernst, C. M., Watters, T. R., Solomon, S. C., Murchie, S. L., Prockter, L. M., Chabot, N. L., Gillis-Davis, J. J., Whitten, J. L., Goudge, T. A., Baker, D. M. H., Hurwitz, D. M., Ostrach, L. R., Xiao, Z., Merline, W. J., Kerber, L., Dickson, J. L., Oberst, J., Byrne, P. K., Klimczak, C. and Nittler, L. R. (2011). Flood volcanism in the northern high latitudes of Mercury revealed by MESSENGER. *Science*, **333**, 1853–1856.

Hiesinger, J., Head, J. W., Wolf, U., Jaumann, R. and Neukum, G. (2003). Ages and stratigraphy of mare basalts in Oceanus Procellarum, Mare Nubium, Mare Cognitum, and Mare Insularum. *J. Geophys. Res.*, **108** (E7), 5065, doi:10.1029/2002JE001985.

Hirschmann, M. M. (2006). Water, melting, and the deep Earth H_2O cycle. *Annu. Rev. Earth Planet. Sci.*, **34**, 629–653.

Hirschmann, M. M., Withers, A. C., Ardia, P. and Foley, N. T. (2012). Solubility of molecular hydrogen in silicate melts and consequences for volatile evolution of terrestrial planets. *Earth Planet. Sci. Lett.*, **345**, 38–48.

Hulme, G. (1973). Turbulent lava flow and the formation of lunar sinuous rilles. *Mod. Geol.*, **4**, 107–117.

Hurwitz, D. M., Fassett, C. I., Head, J. W. and Wilson, L. (2010). Formation of an eroded lava channel within an Elysium Planitia impact crater: Distinguishing between a mechanical and thermal origin. *Icarus*, **210**, 626–634.

Hurwitz, D. M., Head, J. W., Wilson, L. and Hiesinger, H. (2012). Origin of lunar sinuous rilles: Modeling effects of gravity, surface slope, and lava composition on erosion rates during the formation of Rima Prinz. *J. Geophys. Res.*, **117**, E00H14, doi:10.1029/2011JE004000.

Hurwitz, D. M., Head, J. W., Byrne, P. K., Xiao, Z., Solomon, S. C., Zuber, M. T., Smith, D. E. and Neumann, G. A. (2013). Investigating the origin of candidate lava channels on Mercury with MESSENGER data: Theory and observations. *J. Geophys. Res. Planets*, **118**, 471–486.

Irving, A. J., Kuehner, S. M., Bunch, T. E., Ziegler, K., Chen, G., Herd, C. D. K., Conrey, R. M. and Ralew, S. (2013). Ungrouped mafic achondrite Northwest Africa 7325: A reduced, iron-poor cumulate olivine gabbro from a differentiated planetary parent body. *Lunar Planet. Sci.*, **44**, abstract 2164.

Ivanov, B. A. and Melosh, H. J. (2003). Impacts do not initiate volcanic eruptions: Eruptions close to the crater. *Geology*, **31**, 869–872.

Ivanov, M. A. and Head, J. W. (2013). The history of volcanism on Venus. *Planet. Space Sci.*, **84**, 66–92.

Izenberg, N. R., Klima, R. L., Murchie, S. L., Blewett, D. T., Holsclaw, G. M., McClintock, W. E., Malaret, E. Mauceri, C., Vilas, F., Sprague, A. L., Helbert, J., Domingue, D. L., Head, J. W., Goudge, T. A., Solomon, S. C., Hibbitts, C. A. and Dyar, M. D. (2014). The low-iron, reduced surface of Mercury as seen in spectral reflectance by MESSENGER. *Icarus*, **228**, 364–374.

James, P. B., Zuber, M. T., Phillips, R. J. and Solomon, S. C. (2015). Support of long-wavelength topography on Mercury inferred from MESSENGER measurements of gravity and topography. *J. Geophys. Res. Planets*, **120**, 287–310.

Jerram, D. A. and Widdowson, M. (2005). The anatomy of continental flood basalt provinces: Geological constraints on the processes and products of flood volcanism. *Lithos*, **79**, 385–405.

Johnson, A. M. and Pollard, D. D. (1973). Mechanics of growth of some laccolithic intrusions in the Henry Mountains, Utah: I. Field observations, Gilbert's model, physical properties and flow of the magma. *Tectonophysics*, **18**, 261–309.

Jozwiak, L. M., Head, J. W., Zuber, M. T., Smith, D. E. and Neumann, G. A. (2012). Lunar floor-fractured craters: Classification, distribution, origin and implications for magmatism and shallow crustal structure. *J. Geophys. Res.*, **117**, E11005, doi:10.1029/2012JE004134.

Jozwiak, L. M., Head, J. W. and Wilson, L. (2015). Lunar floor-fractured craters as magmatic intrusions: Geometry, modes of emplacement, associated tectonic and volcanic features, and implications for gravity anomalies. *Icarus*, **248**, 424–447.

Kamo, S. L., Czamanske, G. K., Amelin, Y., Fedorenko, V. A., David, D. W. and Trofimov, V. R. (2003). Rapid eruption of Siberian flood-volcanic rocks and evidence for coincidence with the Permian-Triassic boundary and mass extinction at 251 Ma. *Earth Planet. Sci. Lett.*, **214**, 75–91.

Kerber, L., Head, J. W., Solomon, S. C., Murchie, S. L., Blewett, D. T. and Wilson, L. (2009). Explosive volcanic eruptions on Mercury: Eruption conditions, magma volatile content, and implications for interior volatile abundances. *Earth Planet. Sci. Lett.*, **285**, 263–271.

Kerber, L., Head, J. W., Blewett, D. T., Solomon, S. C., Wilson, L., Murchie, S. L., Robinson, M. S., Denevi, B. W. and Domingue, D. L. (2011). The global distribution of pyroclastic deposits on Mercury: The view from MESSENGER flybys 1–3. *Planet. Space Sci.*, **59**, 1895–1909.

Kiefer, W. S. and Murray, B. C. (1987). The formation of Mercury's smooth plains. *Icarus*, **72**, 477–491.

Kinczyk, M. J., Prockter, L. M., Chapman, C. R. and Susorney, H. C. M. (2016). A morphological evaluation of crater degradation on Mercury: Revisiting crater classification with MESSENGER data. *Lunar Planet. Sci.*, **47**, abstract 1573.

King, J. S. and Scott, D. H. (1990). *Geologic Map of the Beethoven (H-7) Quadrangle of Mercury*, Map I-2048. Denver, CO: U.S. Geological Survey.

Klima, R. L., Denevi, B. W., Ernst, C. M., Izenberg, C. M., Murchie, S. L., Peplowski, P. N. and Solomon, S. C. (2015). Global distribution and spectral properties of low-reflectance material on Mercury. Presented at 2015 Fall Meeting, American Geophysical Union, abstract P53A–2094, San Francisco, CA, 14-18 December.

Klimczak, C. (2014). Geomorphology of lunar grabens requires igneous dikes at depth. *Geology*, **42**, 963–966.

Klimczak, C. (2015). Limits on the brittle strength of planetary lithospheres undergoing global contraction. *J. Geophys. Res. Planets*, 120, 2135–2151.

Klimczak, C. and Byrne, P. K. (2013). The prospect of diking on the Moon and Mercury. Presented at 2013 Fall Meeting, American Geophysical Union, abstract P23B–03, San Francisco, CA, 9–13 December.

Klimczak, C., Schultz, R. A. and Nahm, A. L. (2010). Evaluation of the origin hypotheses of Pantheon Fossae, central Caloris basin, Mercury. *Icarus*, **209**, 262–270.

Klimczak, C., Byrne, P. K., Solomon, S. C., Nimmo, F., Watters, T. R., Denevi, B. W., Ernst, C. M. and Banks, M. E. (2013). The role of thrust faults as conduits for volatiles on Mercury. *Lunar Planet. Sci.*, **44**, abstract 1390.

Komatsu, G., Kargel, J. S. and Baker, V. R. (1992). Canali-type channels on Venus: Some genetic constraints. *Geophys. Res. Lett.*, **19**, 1415–1418.

Komatsu, G., Baker, V. R. and Gulick, V. C. (1993). Venusian channels and valleys: Distribution and volcanological implications. *Icarus*, **102**, 1–25.

Kreslavsky, M. A., Head, J. W., Neumann, G. A., Zuber, M. T. and Smith, D. E. (2016). Features of the northern smooth plains on Mercury revealed by detrended MLA topography: Comparison with the Moon. *Lunar Planet. Sci.*, **47**, abstract 1333.

Lawrence, D. J., Feldman, W. C., Goldsten, J. O., McCoy, T. J., Blewett, D. T., Boynton, W. V., Evans, L. G., Nittler, L. R., Rhodes, E. A. and Solomon, S. C. (2010). Identification and measurement of neutron-absorbing elements on Mercury's surface. *Icarus*, **209**, 195–209.

Lawrence, D. J., Peplowski, P. N., Beck, A. W., Feldman, W. C., Frank, E. A., McCoy, T. J., Nittler, L. R., Chabot, N. L., Ernst, C. M. and Solomon, S.C. (2017). Compositional terranes on Mercury: Information from fast neutrons. *Icarus*, **281**, 32–45.

Le Bas, M. J. (2000). IUGS reclassification of the high-Mg and picritic volcanic rocks. *J. Petrol.*, **41**, 1467–1470.

Le Feuvre, M. and Wieczorek, M. A. (2011). Nonuniform cratering of the Moon and a revised crater chronology of the inner Solar System. *Icarus*, **214**, 1–20.

Le Maitre, R. W., Streckeisen, A., Zanettin, B., Le Bas, M. J., Bonin, B., Bateman, P., Bellieni, G., Dudek, A., Efremova, S., Keller, J., Lameyre, J., Sabine, P. A., Schmid, R., Sorensen, H. and Woolley, A. R. (2002). *Igneous Rocks: A Classification and Glossary of Terms*. New York: Cambridge University Press, 256 pp.

Leake, M. A. (1981). The intercrater plains of Mercury and the Moon: Their nature, origin, and role in terrestrial planet evolution. Ph.D. thesis, University of Arizona, Tucson, AZ.

Leverington, D. W. (2007). Was the Mangala Valles system incised by volcanic flows? *J. Geophys. Res.*, **112**, E11005, doi:10.1029/2007JE002896.

Lucchitta, B. K. and Schmitt, H. H. (1974). Orange material in the Sulpicius Gallus Formation at the southwestern edge of Mare Serenitatis. *Proc. Lunar Sci. Conf.*, **5**, 223–234.

Malin, M. C. (1976). Observations of intercrater plains on Mercury. *Geophys. Res. Lett.*, **3**, 581–584.

Marchi, S., Morbidelli, A. and Cremonese, G. (2005). Flux of meteoroid impacts on Mercury. *Astron. Astrophys.*, **431**, 1123–1127.

Marchi, S., Mottola, S., Cremonese, G., Massironi, M. and Martellato, E. (2009). A new chronology for the Moon and Mercury. *Astron. J.*, **137**, 4936–4948.

Marchi, S., Massironi, M., Cremonese, G., Martellato, E., Giacomini, L. and Prockter, L. M. (2011). The effects of the target material properties and layering on the crater chronology: The case of Raditladi and Rachmaninoff basins on Mercury. *Planet. Space Sci.*, **59**, 1968–1980.

Marchi, S., Chapman, C. R., Fassett, C. I., Head, J. W., Bottke, W. F. and Strom, R. G. (2013). Global resurfacing of Mercury 4.0–4.1 billion years ago by heavy bombardment and volcanism. *Nature*, **499**, 59–61.

Mastin, L. G. and Pollard, D. D. (1988). Surface deformation and shallow dike intrusion processes at Inyo Craters, Long Valley, California. *J. Geophys. Res.*, **93**, 13,221–13,235.

McCauley, J. F., Guest, J. E., Schaber, G. G., Trask, N. J. and Greeley, R. (1981). Stratigraphy of the Caloris basin, Mercury. *Icarus*, **47**, 184–202.

McClintock, M. and White, D. L. (2006). Large phreatomagmatic vent complex at Coombs Hills, Antarctica: Wet, explosive initiation of flood basalt volcanism in the Ferrar-Karoo LIP. *Bull. Volcanol.*, **68**, 215–239.

McCord, T. B. and Clark, R. N. (1979). The Mercury soil: Presence of Fe^{2+}. *J. Geophys. Res.*, **84**, 7664–7668.

McCubbin, F. M., Riner, M. A., Vander Kaaden, K. E. and Burkemper, L. K. (2012). Is Mercury a volatile-rich planet? *Geophys. Res. Lett.*, **39**, L09202, doi:10.1029/2012GL051711.

McCubbin, F. M., Sverjensky, D. A., Steele, A. and Mysen, B. O. (2014). In-situ characterization of oxalic acid breakdown at elevated P and T: Implications for organic C-O-H fluid sources in petrologic experiments. *Amer. Mineral.*, **99**, 2258–2271.

McGill, G. E. and King, E. A. (1983). *Geologic Map of the Victoria (H-2) Quadrangle of Mercury*, Map I-1409. Denver, CO: U.S. Geological Survey.

McGovern, P. J. and Litherland, M. M. (2011). Lithospheric stress and basaltic magma ascent on the Moon, with implications for large volcanic provinces and edifices. *Lunar Planet. Sci.*, **42**, abstract 2587.

McSween, H. Y., Arvidson, R. E., Bell, J. F., Blaney, D., Cabrol, N. A., Christensen, P. R., Clark, B. C., Crisp, J. A., Crumpler, L. S., Des Marais, D. J., Farmer, J. D., Gellert, R., Ghosh, A., Gorevan, S., Graff, T., Grant, J., Haskin, L. A., Herkenhoff, K. E., Johnson, J. R., Jolliff, B. L., Klingelhoefer, G., Knudson, A. T., McLennan, S., Milam, K. A., Moersch, J. E., Morris, R. V., Rieder, R., Ruff, S. W., de Souza, P. A., Squyres, S. W., Wänke, H., Wang, A., Wyatt, M. B., Yen, A. and Zipfel, J. (2004). Basaltic rocks analyzed by the Spirit rover in Gusev crater. *Science*, **305**, 842–845.

McSween, H. Y., Ruff, S. W., Morris, R. V., Bell, J. F., Herkenhoff, K. E., Gellert, R., Stockstill, K. R., Tornabene, L. L., Squyres, S. W., Crisp, J. A., Christensen, P. R., McCoy, T. J., Mittlefehldt, D. W. and Schmidt, M. (2006). Alkaline volcanic rocks from the Columbia Hills, Gusev crater, Mars. *J. Geophys. Res.*, **111**, E09S91, doi:10.1029/2006JE002698.

McSween, H. Y., Taylor, G. J. and Wyatt, M. B. (2009) Elemental composition of the Martian crust. *Science*, **324**, 736–739.

Mège, D., Cook, A. C., Lagabrielle, Y., Garel, E. and Cormier, M.-H. (2003). Volcanic rifting at Martian grabens. *J. Geophys. Res.*, **108** (E5), 5044, doi:10.1029/2002JE001852.

Michael, G. G., Kneissl, T. and Neesemann, A. (2016). Planetary surface dating from crater size–frequency distribution measurements: Poisson timing analysis. *Icarus*, **277**, 279–285.

Michel, N. C., Hauck, S. A., II, Solomon, S. C., Phillips, R. J., Roberts, J. H. and Zuber, M. T. (2013). Thermal evolution of Mercury as constrained by MESSENGER observations. *J. Geophys. Res. Planets*, **118**, 1033–1044.

Miller, R. B. and Paterson, S. R. (1999). In defense of magmatic diapirs. *J. Struct. Geol.*, **21**, 1161–1173.

Moore, H. J., Arthur, D. W. G. and Schaber, G. G. (1978). Yield strengths of flows on the earth, Mars, and moon. *Proc. Lunar Sci. Conf.*, **3**, 3351–3378.

Mouginis-Mark, P. J., Wilson, L., Head, J. W., Brown, S. R., Hall, J. L. and Sullivan, K. D. (1984). Elysium Planitia, Mars: Regional geology, volcanology, and evidence for volcano-ground ice interactions. *Earth Moon Planets*, **30**, 149–173.

Murase, T. and McBirney, A. R. (1973). Properties of some common igneous rocks and their melts at high temperatures. *Geol. Soc. Amer. Bull.*, **84**, 3563–3592.

Murchie, S. L., Watters, T. R., Robinson, M. S., Head, J. W., Strom, R. G., Chapman, C. R., Solomon, C. R., McClintock, W. E., Prockter, L. M., Domingue, D. L. and Blewett, D. T. (2008). Geology of the Caloris basin, Mercury: A new view from MESSENGER. *Science*, **321**, 73–76.

Murchie, S. L., Klima, R. L., Denevi, B. W., Ernst, C. M., Keller, M. R., Domingue, D. L., Blewett, D. T., Chabot, N. L., Hash, C. D., Malaret, E., Izenberg, N. R., Vilas, F., Nittler, L. R., Gillis-Davis, J. J., Head, J. W. and Solomon, S. C. (2015). Orbital multispectral mapping of Mercury with the MESSENGER Mercury Dual Imaging System: Evidence for the origins of plains units and low-reflectance material. *Icarus*, **254**, 287–305.

Murray, B. C., Belton, M. J. S., Danielson, G. E., Davies, M. E., Gault, D. E., Hapke, B., O'Leary, B., Strom, R. G., Suomi, V. and Trask, N. (1974). Mercury's surface: Preliminary description and interpretation from Mariner 10 pictures. *Science*, **185**, 169–179.

Murray, B. C., Strom, R. G., Trask, N. J. and Gault, D. E. (1975). Surface history of Mercury: Implications for terrestrial planets. *J. Geophys. Res.*, **80**, 2508–2514.

Murray, J. B., van Wyk de Vries, B., Marquez, A., Williams, D. A., Byrne, P. K., Muller, J.-P. and Kim, J.-R. (2010). Late-stage water eruptions from Ascraeus Mons volcano, Mars: Implications for its structure and history. *Earth Planet. Sci. Lett.*, **294**, 479–491.

Mysen, B. O. and Fogel, M. L. (2010). Nitrogen and hydrogen isotope compositions and solubility in silicate melts in equilibrium with reduced (N plus H)-bearing fluids at high pressure and temperature: Effects of melt structure. *Amer. Mineral.*, **95**, 987–999.

Mysen, B. O., Fogel, M. L., Morrill, P. L. and Cody, G. D. (2009). Solution behavior of reduced C-O-H volatiles in silicate melts at high pressure and temperature. *Geochim. Cosmochim. Acta*, **73**, 1696–1710.

Namur, O., Collinet, M., Charlier, B., Grove, T. L., Holtz, F. and McCammon, C. (2016). Melting processes and mantle sources of surface lavas on Mercury. *Earth Planet. Sci. Lett.*, **439**, 117–128.

Nekvasil, H., Filiberto, J., McCubbin, F. M. and Lindsley, D. H. (2007). Alkalic parental magmas for the chassignites? *Meteorit. Planet. Sci.*, **42**, 979–992.

Neukum, G., Ivanov, B. A. and Hartmann, W. K. (2001). Cratering records in the inner solar system in relation to the lunar reference system. *Space Sci. Rev.*, **96**, 55–86.

Nittler, L. R., Starr, R. D., Weider, S. Z., McCoy, T. J., Boynton, W. V., Ebel, D. S., Ernst, C. M., Evans, L. G., Goldsten, J. O., Hamara, D. K., Lawrence, D. J., McNutt, R. L., Schlemm, C. E., Solomon, S. C. and Sprague, A. L. (2011). The major-element composition of Mercury's surface from MESSENGER X-ray spectrometry. *Science*, **333**, 1847–1850.

Oberbeck, V. R. (1975). The role of ballistic erosion and sedimentation in lunar stratigraphy. *Rev. Geophys. Space Phys.*, **13**, 337–362.

Oberbeck, V. R., Quaide, W. L., Arvidson, R. E. and Aggarwal, H. R. (1977). Comparative studies of lunar, Martian, and Mercurian craters and plains. *J. Geophys. Res.*, **82**, 1687–1698.

Oberst, J., Preusker, F., Phillips, R. J., Watters, T. R., Head, J. W., Zuber, M. T. and Solomon, S. C. (2010). The morphology of Mercury's Caloris basin as seen in MESSENGER stereo topographic models. *Icarus*, **209**, 230–238.

Okubo, C. H. and Martel, S. J. (1998). Pit crater formation on Kilauea volcano, Hawaii. *J. Volcanol. Geotherm. Res.*, **86**, 1–18.

Ostrach, L. R., Robinson, M. S., Whitten, J. L., Fassett, C. I., Strom, R. G., Head, J. W. and Solomon, S. C. (2015). Extent, age, and resurfacing history of the northern smooth plains on Mercury from MESSENGER observations. *Icarus*, **250**, 602–622.

Padovan, S., Wieczorek, M. A., Margot, J.-L., Tosi, N. and Solomon, S. C. (2015). Thickness of the crust of Mercury from geoid-to-topography ratios. *Geophys. Res. Lett.*, **42**, 1029–1038.

Parman, S. W., O'Brien, H. P., Vaughn, W. M. and Head, J. W. (2014). Experimental constraints on melting conditions in Mercury. *Lunar Planet. Sci.*, **45**, abstract 2367.

Parman, S. W., Parmentier, E. M. and Wang, S. (2016). Crystallization of Mercury's sulfur-rich magma oceans. *Lunar Planet. Sci.*, **47**, abstract 2990.

Peplowski, P. N., Evans, L. G., Hauck, S. A., McCoy, T. J., Boynton, W. V., Gillis-Davis, J. J., Ebel, D. S., Goldsten, J. O., Hamara, D. K., Lawrence, D. J., McNutt, R. L., Nittler, L. R., Solomon, S. C., Rhodes, E. A., Sprague, A. L., Starr, R. D. and Stockstill-Cahill, K. R. (2011). Radioactive elements on Mercury's surface from MESSENGER: Implications for the planet's formation and evolution. *Science*, **333**, 1850–1852.

Peplowski, P. N., Lawrence, D. J., Feldman, W. C., Goldsten, J. O., Bazell, D., Evans, L. G., Head, J. W., Nittler, L. R., Solomon, S. C. and Weider, S. Z. (2015). Geochemical terranes of Mercury's northern hemisphere as revealed by MESSENGER neutron measurements. *Icarus*, **253**, 346–353.

Peplowski, P. N., Klima, R. L., Lawrence, D. J., Ernst, C. M., Denevi, B. W., Frank, E. A., Goldsten, J. O., Murchie, S. L., Nittler, L R. and Solomon, S. C. (2016). Remote sensing evidence for an ancient carbon-bearing crust on Mercury. *Nature Geosci.*, **9**, 273–276.

Pike, R. J. (1988). Geomorphology of impact craters on Mercury. In *Mercury*, ed. F. Vilas, C. R. Chapman and M. S. Matthews, Tucson, AZ: University of Arizona Press, pp.165–273.

Plescia, J. B. (1991). Wrinkle ridges in Lunae Planum, Mars: Implications for shortening and strain. *Geophys. Res. Lett.*, **18**, 913–916.

Pohn, H. A. and Offield, T. W. (1970). Lunar crater morphology and relative age determination of lunar geologic units. Part 1: Classification. In *Geological Survey Research 1970*, Professional Paper 69–209. Denver, CO: U.S. Geological Survey, 35 pp.

Pollard, D. D., Delaney, P. T., Duffield, W. A., Endo, E. T. and Okamura, A. T. (1983). Surface deformation in volcanic rift zones. *Tectonophysics*, **94**, 541–584.

Preusker, F., Oberst, J., Head, J. W., Watters, T. R., Robinson, M. S., Zuber, M. T. and Solomon, S. C. (2011). Stereo topographic models of Mercury after three MESSENGER flybys. *Planet. Space Sci.*, **59**, 1910–1917.

Prockter, L. M., Ernst, C. M., Denevi, B. W., Chapman, C. R., Head, J. W., Fassett, C. I., Merline, W. J., Solomon, S. C., Watters, T. R., Strom, R. G., Cremonese, G., Marchi, S. and Massironi, M. (2010). Evidence for young volcanism on Mercury from the third MESSENGER flyby. *Science*, **329**, 668–671.

Prockter, L. M., Kinczyk, M. J., Byrne, P. K., Denevi, B. W., Head, J. W., Fassett, C. I., Whitten, J. L., Thomas, R. J., Buczkowski, D. L., Hynek, B. M., Ostrach, L. R., Blewett, D. T., Ernst, C. M. and the MESSENGER Mapping Group (2016). The first global geological map of Mercury. *Lunar Planet. Sci.*, **47**, abstract 1245.

Rava, B. and Hapke, B. (1987). An analysis of the Mariner 10 color ratio map of Mercury. *Icarus*, **71**, 397–429.

Riker, J. M., Cashman, K. V., Kauahikaua, J. P. and Montierth, C. M. (2009). The length of channelized lava flows: Insight from the 1859 eruption of Mauna Loa Volcano, Hawai'i. *J. Volcanol. Geotherm. Res.*, **183**, 139–156.

Roberts, J. H. and Barnouin, O. S. (2012). The effect of the Caloris impact on the mantle dynamics and volcanism of Mercury. *J. Geophys. Res.*, **117**, E02007, doi:10.1029/2011JE003876.

Robinson, M. S. and Lucey, P. G. (1997). Recalibrated Mariner 10 color mosaics: Implications for mercurian volcanism. *Science*, **275**, 197–200.

Robinson, M. S., Murchie, S. L., Blewett, D. T., Domingue, D. L., Hawkins, S. E., III, Head, J. W., Holsclaw, G. M., McClintock, W. E., McCoy, T. J., McNutt, R. L., Jr., Prockter, L. M., Solomon, S. C. and Watters, T. R. (2008). Reflectance and color variations on Mercury: Regolith processes and compositional heterogeneity. *Science*, **321**, 66–69.

Rothery, D. A., Thomas, R. J. and Kerber, L. (2014). Prolonged eruptive history of a compound volcano on Mercury: Volcanic and tectonic implications. *Earth Planet. Sci. Lett.*, **385**, 59–67.

Rothery, D. A., Mancinelli, P., Guzzetta, L. and Wright, J. (2017). Mercury's Caloris basin: Continuity between the interior and exterior plains. *J. Geophys. Res. Planets*, **122**, 560–576.

Rubin, A. M. (1992). Dike-induced faulting and graben subsidence in volcanic rift zones. *J. Geophys. Res.*, **97**, 1839–1858.

Rubin, A. M. (1993). Dikes vs. diapirs in viscoelastic rock. *Earth Planet. Sci. Lett.*, **117**, 653–670.

Santos, A. R., Agee, C. B., McCubbin, F. M., Shearer, C. K., Burger, P. V., Tartese, R. and Anand, M. (2015). Petrology of igneous clasts in Northwest Africa 7034: Implications for the petrologic diversity of the martian crust. *Geochim. Cosmochim. Acta*, **157**, 56–85.

Schaber, G. G. and McCauley, J. F. (1980). *Geologic Map of the Tolstoj (H-8) Quadrangle of Mercury*, Map I-1199. Denver, CO: U.S. Geological Survey.

Schlemm, C. E., II, Starr, R. D., Ho, G. C., Bechtold, K. E., Hamilton, S. A., Boldt, J. D., Boynton, W. V., Bradley, W., Fraemen, M. E., Gold, R. E., Goldsten, J. O., Hayes, J. R., Jaskulek, S. E., Rossano, E., Rumpf, R. A., Schaefer, E. D., Strohbehn, K., Shelton, R. G., Thompson, R. E., Trombka, J. I. and Williams, B. D. (2007) The X-Ray Spectrometer on the MESSENGER spacecraft. *Space Sci. Rev.*, **131**, 393–415.

Schubert, G., Lingenfelter, R. E. and Peale, S. J. (1970). The morphology, distribution, and origin of lunar sinuous rilles. *Rev. Geophys.*, **8**, 199–224.

Schultz, P. H. (1976). Floor-fractured lunar craters. *Moon*, **15**, 241–273.

Schultz, R. A., Okubo, C. H., Goudy, C. L. and Wilkins, S. J. (2004). Igneous dikes on Mars revealed by Mars Orbiter Laser Altimeter topography. *Geology*, **32**, 889–892.

Schultz, R. A., Hauber, E., Kattenhorn, S. A., Okubo, C. H. and Watters, T. R. (2010). Interpretation and analysis of planetary structures. *J. Struct. Geol.*, **32**, 855–875.

Scott, E. D., Wilson, L. and Head, J. W. (2002). Emplacement of giant radial dikes in the northern Tharsis region of Mars. *J. Geophys. Res.*, **107** (E4), 5019, doi:10.1029/2000JE001431.

Sehlke, A. and Whittington, A. G. (2015). Rheology of lava flows on Mercury: An analog experimental study. *J. Geophys. Res. Planets*, **120**, 1924–1955.

Shaw, H. R. (1972). Viscosities of magmatic silicate liquids: An empirical method of prediction. *Amer. J. Sci.*, **272**, 870–893.

Shirey, S. B. and Hanson, G. N. (1984) Mantle-derived Archaean monzodiorites and trachyandesites. *Nature*, **310**, 222–224.

Sibson, R. H. (2003). Brittle-failure controls on maximum sustainable overpressure in different tectonic regimes. *Amer. Assoc. Petrol. Geol. Bull.*, **87**, 901–908.

Solomon, S. C. (1977). The relationship between crustal tectonics and internal evolution in the Moon and Mercury. *Phys. Earth Planet. Inter.*, **15**, 135–145.

Solomon, S. C. (1978). On volcanism and thermal tectonics on one-plate planets. *Geophys. Res. Lett.*, **5**, 461–464.

Solomon, S. C. and Chaiken, J. (1976). Thermal expansion and thermal stress in the Moon and terrestrial planets: Clues to early thermal history. *Proc. Lunar Sci. Conf.*, **7**, 3229–3243.

Solomon, S. C. and Head, J. W. (1979). Vertical movement in mare basins: Relation to mare emplacement, basin tectonics, and lunar thermal history. *J. Geophys. Res.*, **84**, 1667–1682.

Solomon, S. C., McNutt, R. L., Watters, T. R., Lawrence, D. J., Feldman, W. C., Head, J. W., Krimigis, S. M., Murchie, S. L., Phillips, R. J., Slavin, J. A. and Zuber, M. T. (2008). Return to Mercury: A global perspective on MESSENGER's first Mercury flyby. *Science*, **321**, 59–62.

Spudis, P. D. and Guest, J. E. (1988). Stratigraphy and geologic history of Mercury. In *Mercury*, ed. F. Vilas, C. R. Chapman and M. S. Matthews, Tucson, AZ: University of Arizona Press, pp. 118–164.

Spudis, P. D. and Prosser, J. G. (1984). *Geologic Map of the Michaelangelo (H-12) Quadrangle of Mercury*, Map I-1659. Denver, CO: U.S. Geological Survey.

Stephenson, P. J., Burch-Johnson, A. T., Whitehead, R. W. and Stanton, D. (1998). Three long lava flows in north Queensland. *J. Geophys. Res.*, **103**, 27,359–27,370.

Stockstill-Cahill, K. R., McCoy, T. J., Nittler, L. R., Weider, S. Z. and Hauck, S.A., II (2012). Magnesium-rich crustal compositions on Mercury: Implications for magmatism from petrologic modeling. *J. Geophys. Res.*, **117**, E00L15, doi:10.1029/2012JE004140.

Stofan, E. R., Guest, J. E., Anderson, S. W. and Smrekar, S. E. (1998). Development of planetary lava flow fields. *Lunar Planet. Sci.*, **29**, abstract 1099.

Stolper, E. M., Baker, M. B., Newcombe, M. E., Schmidt, M. E., Treiman, A. H., Cousin, A., Dyar, M. D., Fisk, M. R., Gellert, R., King, P. L., Leshin, L., Maurice, S., McLennan, S. M., Minitti, M. E., Perrett, G., Rowland, S., Sautter, V., Wiens, R. C. and MSL Team (2013). The petrochemistry of Jake_M: A Martian mugearite. *Science*, **341**, 1239463.

Strom, R.G. (1977). Origin and relative age of lunar and mercurian intercrater plains. *Phys. Earth Planet. Inter.*, **15**, 156–172.

Strom, R. G., Trask, N. J. and Guest, J. E. (1975). Tectonism and volcanism on Mercury. *J. Geophys. Res.*, **80**, 2478–2507.

Strom, R. G., Malin, M. C. and Leake, M. A. (1990). *Geologic Map of the Bach (H-15) Quadrangle of Mercury*, Map I-2015. Denver, CO: U.S. Geological Survey.

Strom, R. G., Chapman, C. R., Merline, W. J., Solomon, S. C. and Head, J. W. (2008). Mercury cratering record viewed from MESSENGER's first flyby. *Science*, **321**, 79–81.

Strom, R. G., Banks, M. E., Chapman, C. R., Fassett, C. I., Forde, J. A., Head, J. W., Merline, W. J., Prockter, L. M. and Solomon, S. C. (2011). Mercury crater statistics from MESSENGER flybys: Implications for stratigraphy and resurfacing history. *Planet. Space Sci.*, **59**, 1960–1967.

Taylor, S. R. (1989). Growth of planetary crust. *Tectonophysics*, **161**, 147–156.

Taylor, S. R. and McLennan, S. M. (2010). *Planetary Crusts: Their Composition, Origin and Evolution*. Cambridge Planetary Science. Cambridge: Cambridge University Press.

Thomas, R. J., Rothery, D. A., Conway, S. J. and Anand, M. (2014a). Long-lived explosive volcanism on Mercury. *Geophys. Res. Lett.*, **41**, 6084–6092.

Thomas, R. J., Rothery, D. A., Conway, S. J. and Anand, M. (2014b). Mechanisms of explosive volcanism on Mercury: Implications from its global distribution and morphology. *J. Geophys. Res. Planets*, **119**, 2239–2254.

Thordarson, T., Self, S., Oskarsson, N. and Hulsebosch, T. (1996). Sulfur, chlorine and fluorine degassing and atmospheric loading by the 1783–1784 AD Laki (Skaftar fires) eruption in Iceland. *Bull. Volcanol.*, **58**, 205–225.

Tosi, N., Grott, M., Plesa, A. C. and Breuer, D. (2013). Thermochemical evolution of Mercury's interior. *J. Geophys. Res. Planets*, **118**, 2474–2487.

Trask, N. J. (1971). Geologic comparison of mare materials in the lunar equatorial belt, including Apollo 11 and Apollo 12 landing sites. In *Geological Survey Research 1971*, Professional Paper 750-D. Denver, CO: U.S. Geological Survey, pp. 138–144.

Trask, N. J. (1975). Cratering history of the heavily cratered terrain on Mercury. *Proc. Int. Colloq. Planet. Geol., Geol. Rom.*, **15**, 471–476.

Trask, N. J. and Strom, R. G. (1976) Additional evidence of Mercurian volcanism. *Icarus*, **28**, 559–563.

Trask, N. J. and Dzurisin, D. (1984). *Geologic Map of the Discovery (H-11) Quadrangle of Mercury*, Map I-1658. Denver, CO: U.S. Geological Survey.

Trask, N. J. and Guest, J. E. (1975). Preliminary geologic terrain map of Mercury. *J. Geophys. Res.*, **80**, 2461–2477.

van der Bogert, C. H., Gaddis, L., Hiesinger, H., Ivanov, M., Jolliff, B. Mahanti, P. and Paskert, J. H. (2016). Revisiting the CSFDs of the Taurus Littrow dark mantle deposit: Implications for age determinations of pyroclastic deposits. *Lunar Planet. Sci.*, **47**, abstract 1616.

Vander Kaaden, K. E. and McCubbin, F. M. (2015). Exotic crust formation on Mercury: Consequences of a shallow, FeO-poor mantle. *J. Geophys. Res. Planets*, **120**, 195–209.

Vander Kaaden, K. E. and McCubbin, F. M. (2016) The origin of boninites on Mercury: An experimental study of the northern volcanic plains lavas. *Geochim. Cosmochim. Acta*, **173**, 246–263.

Vander Kaaden, K. E., McCubbin, F. M., Nittler, L. R., Peplowski, P. N., Weider, S. Z., Frank, E. A. and McCoy, T. J. (2017). Geochemistry, mineralogy, and petrology of boninitic and komatiitic rocks on the mercurian surface: Insights into the mercurian mantle. *Icarus*, **285**, 155–168.

Vilas, F. (1985). Mercury: Absence of crystalline Fe^{2+} in the regolith. *Icarus*, 64, 133–138.

Vilas, F. (1988). Surface composition of Mercury from reflectance spectrophotometry. In *Mercury*, ed. F. Vilas, C. R. Chapman and M. S. Matthews. Tucson, AZ: University of Arizona Press, pp. 59–76.

Walker, D., Kirkpatrick, R. J., Longhi, J. and Hays, J. F. (1976). Crystallization history of lunar picritic basalt sample 12002: Phase-equilibria and cooling-rate studies. *Geol. Soc. Amer. Bull.*, **87**, 646–656.

Walker, G. P. L. (1973). Lengths of lava flows. *Phil. Trans. Roy. Soc. London A*, **274**, 107–118.

Watanabe, T., Koyaguchi, T. and Seno, T. (1999). Tectonic stress controls on ascent and emplacement of magmas. *J. Volcanol. Geotherm. Res.*, **91**, 65–78.

Watters, T. R., Head, J. W., Solomon, S. C., Robinson, M. S., Chapman, C. R., Denevi, B. W., Fassett, C. I., Murchie, S. L. and Strom, R. G. (2009). Evolution of the Rembrandt impact basin on Mercury. *Science*, **324**, 618–621.

Weider, S. Z., Nittler, L. R., Starr, R. D., McCoy, T. J., Stockstill-Cahill, K. R., Byrne, P. K., Denevi, B. W., Head, J. W. and Solomon, S. C. (2012). Chemical heterogeneity on Mercury's surface revealed by the MESSENGER X-Ray Spectrometer. *J. Geophys. Res.*, **117**, E00L05, doi:10.1029/2012JE004153.

Weider, S. Z., Nittler, L. R., Starr, R. D., Crapster-Pregont, E. J., Peplowski, P. N., Denevi, B. W., Head, J. W., Byrne, P. K., Hauck, S. A., II, Ebel, D. S. and Solomon, S. C. (2015). Evidence for geochemical terranes on Mercury: Global mapping of major elements with MESSENGER's X-Ray Spectrometer. *Earth Planet. Sci. Lett.*, **416**, 109–120.

Weider, S. Z., Nittler, L. R., Murchie, S. L., Peplowski, P. N., McCoy, T. J., Kerber, L., Klimczak, C., Ernst, C. M., Goudge, T. A., Starr, R. D., Izenberg, N. R., Klima, R. L. and Solomon, S. C. (2016). Evidence from MESSENGER for sulfur- and carbon-driven explosive volcanism on Mercury. *Geophys. Res. Lett.*, **43**, 3653–3661.

Whitten, J. L. and Head, J. W. (2015). Rembrandt impact basin: Distinguishing between volcanic and impact-produced plains on Mercury. *Icarus*, **258**, 350–365.

Whitten, J. L., Head, J. W., Denevi, B. W. and Solomon, S. C. (2014). Intercrater plains on Mercury: Insights into unit definition, characterization, and origin from MESSENGER datasets. *Icarus*, **241**, 97–113.

Wichman, R. W. and Schultz, P. H. (1996). Crater-centered laccoliths on the Moon: Modeling intrusion depth and magmatic pressure at the crater Taruntius. *Icarus*, **122**, 193–199.

Wilhelms, D. E. (1976). Mercurian volcanism questioned. *Icarus*, **28**, 551–558.

Wilhelms, D. E. (1987). *The Geologic History of the Moon*. Professional Paper 1348. Denver, CO: U.S. Geological Survey.

Williams, D. A., Kerr, R. C. and Lesher, C. M. (1999). Thermal and fluid dynamics of komatiitic lavas associated with magmatic Ni-Cu-(PGE) sulphide deposits. In *Dynamic Processes in Magmatic Ore Deposits and their Application in Mineral Exploration*, ed. R. R. Keays, C. M. Lesher, P. C. Lightfoor and C. E. Farrow. Short Course, Geological Association Canada, **13**, 367–412.

Williams, D. A., Fagents, S. A. and Greeley, R. (2000). A reassessment of the emplacement and erosional potential of turbulent, low-viscosity lavas on the Moon. *J. Geophys. Res.*, **105**, 20,189–20,205.

Williams, D. A., Kerr, R. C., Lesher, C. M. and Barnes, S. J. (2001a). Analytical/numerical modeling of komatiite lava emplacement and thermal erosion at Perseverance, Western Australia. *J. Volcanol. Geotherm. Res.*, **110**, 27–55.

Williams, D. A., Greeley, R., Lopes, R. M. and Davies, A. G. (2001b). Evaluation of sulfur flow emplacement on Io from Galileo data and numerical modeling. *J. Geophys. Res.*, **106**, 33,161–33,174.

Wilson, L. and Head, J. W. (1983). A comparison of eruption processes on Earth, Moon, Mars, Io and Venus. *Nature*, **302**, 663–669.

Wilson, L. and Head, J. W. (2008). Volcanism on Mercury: A new model for the history of magma ascent and eruption. *Geophys. Res. Lett.*, **35**, L23205, doi:10.1029/2008GL035620.

Wilson, L., Hawke, B. R., Giguere, T. A. and Petrycki, E. R. (2011). An igneous origin for Rima Hyginus and Hyginus crater on the Moon. *Icarus*, **215**, 584–595.

Wöhler, C., Lena, R. and the Geologic Lunar Research Group (2009). Lunar intrusive domes: Morphometric analysis and laccolith modeling. *Icarus*, **204**, 381–398.

Wright, J., Rothery, D. A., Balme, M. R. and Conway, S. J. (2017). Volcanic shields on Mercury identified at last? *Lunar Planet. Sci.*, **48**, abstract 1871.

Wyrick, D. Y., Ferill, D. A., Morris, A. P., Colton, S. L. and Sims, D. W. (2004). Distribution, morphology, and origins of Martian pit crater chains. *J. Geophys. Res.*, **109**, E06005, doi:10.1029/2004JE002240.

Zimbelman, J. R. (1998). Emplacement of long lava flows on planetary surfaces. *J. Geophys. Res.*, **103**, 27,503–27,516.

Zolotov, M. Yu. (2011). On the chemistry of mantle and magmatic volatiles on Mercury. *Icarus*, **212**, 24–41.

Zolotov, M. Yu., Sprague, A. L., Hauck, S. A., II, Nittler, L. R., Solomon, S. C. and Weider, S. Z. (2013). The redox state, FeO content, and origin of sulfur-rich magmas on Mercury. *J. Geophys. Res. Planets*, **118**, 138–146.

Zuber, M. T., Smith, D. E., Phillips, R. J., Solomon, S. C., Neumann, G. A., Hauck, S. A., Peale, S. J., Barnouin, O. S., Head, J. W., Johnson, C. L., Lemoine, F. G., Mazarico, E., Sun, X., Torrence, M. H., Freed, A. M., Margot, J. L., Oberst, J., Perry, M. E., McNutt, R. L., Balcerski, J. A., Michel, N., Talpe, M. J. and Yang, D. (2012). Topography of the northern hemisphere of Mercury from MESSENGER laser altimetry. *Science*, **336**, 217–220.

Zuber, M. T., Smith, D. E., Watkins, M. M., Asmar, S. W., Konopliv, A. S., Lemoine, F. G., Melosh, H. J., Neumann, G. A., Phillips, R. J., Solomon, S. C., Wieczorek, M. A., Williams, J. G., Goossens, S. J., Kruizinga, G., Mazarico, E., Park, R. S. and Yuan, D.-N. (2013). Gravity field of the Moon from the Gravity Recovery and Interior Laboratory (GRAIL) mission. *Science*, **339**, 668–671.

12

Mercury's Hollows

DAVID T. BLEWETT, CAROLYN M. ERNST, SCOTT L. MURCHIE, AND FAITH VILAS

12.1 INTRODUCTION

Hollows are a landform on Mercury that has no close counterpart on other airless silicate surfaces. They are small, shallow, irregular depressions that lack rims and have flat floors. Evidence suggests that hollows form by volatile loss, and that they are very young relative to Mercury's surface as a whole. This chapter reviews the morphology, distribution, and spectral and compositional properties of hollows. We also consider analog features on other planetary bodies and the processes by which hollows are formed. The discussion begins with the history of spacecraft observations of these interesting and unusual features.

Images of Mercury returned by Mariner 10 during its three flybys of the innermost planet in 1974 and 1975 revealed an ancient surface which to first order is similar to that of the Moon, dominated by impact basins, craters, and bright crater rays (e.g., Murray et al., 1974; Strom, 1984). Several authors described craters that had high-reflectance floors (Hapke et al., 1975; Dzurisin, 1977; Schultz, 1977). In a few areas where color-ratio images were available, the bright patches proved to have anomalous color (Dzurisin, 1977; Schultz, 1977; Rava and Hapke, 1987; Robinson and Lucey, 1997; Robinson and Taylor, 2001; Blewett et al., 2007). An example of an impact crater with high reflectance and anomalous "red" color (a low ultraviolet/orange reflectance ratio) is Lermontov basin (166-km diameter, centered at 15.2°N, 311.1°E). Schultz (1977) and Rava and Hapke (1987) suggested that the reddish material in Lermontov could have been emplaced by pyroclastic activity. This conjecture was later confirmed by MESSENGER observations (e.g., Murchie et al., 2008; Robinson et al., 2008; Head et al., 2008, 2009; Kerber et al., 2009, 2011; Goudge et al., 2014; Chapter 11). Whereas all surface materials on Mercury have an increasing spectral reflectance with increasing wavelength (a "red" spectral slope) from the visible to the near infrared (Chapter 8), spectral slopes that are less steep than the global average are termed "blue." Refer to Figure 8.6 for examples of reflectance spectra for areas of Mercury's surface that are relatively red and blue.

Dzurisin (1977) noted that the craters Balzac, Tyagaraja, Theophanes, and Zeami have high-reflectance patches with relatively blue color (high ultraviolet/orange reflectance ratio) in their interiors, and the interior of the crater now named Hopper (35 km, 12.4°S, 304.1°E) was identified as having very high reflectance. In fact, Hapke et al. (1975) determined that Hopper (crater 48 in their Table 2) has an overall reflectance equal to that of the famously bright, rayed crater Kuiper. Schultz (1977) mentioned the unusual morphology of the bright patches in Balzac, Tyagaraja, and Zeami.

Researchers in the Mariner 10 era were in most cases limited to interpretation of photographic prints, and hence were likely hindered by the inability to adjust the contrast stretch of the images with which they were working. The floor of Hopper is highly saturated in the *Atlas of Mercury* (Figure 6-D of Davies et al., 1978) (Figure 12.1). The bright area on the floor is clearly unlike the floor of any lunar crater. Dzurisin (1977) hypothesized that the bright, blue patches could have resulted from endogenic physical or chemical alteration, perhaps related to impact-induced fractures. Schultz (1977) mentioned that the highly reflective, blue materials could be products of alteration by shock processes.

The bright patches on Mercury remained only a minor curiosity for more than 30 years until new images of Mercury were returned by MESSENGER during its three flybys in 2008 and 2009. Monochrome and 11-color images collected by the Mercury Dual Imaging System (MDIS) (Hawkins et al., 2007) led to the discovery of additional examples of impact craters with high-reflectance, relatively blue floor materials (Robinson et al., 2008; Blewett et al., 2009, 2010). Called bright crater-floor deposits (BCFDs) by the MESSENGER team, these surfaces were shown to have reflectance and color properties that made them outliers relative to the global spectral trend from the freshest, least-weathered materials to typical mature regolith (see Figure 8.6). Thus it is possible that the BCFDs owe their unusual spectral properties to differences in composition and/or physical state that are distinct from the differences that characterize most optically fresh crater materials. In terms of the geological setting of the deposits, the MESSENGER flyby images indicated that bright, blue material was typically associated with impact craters, e.g., on the peak-ring mountains of some basins, such as Eminescu and Raditladi, and the floors of some craters, such as Sander and Balanchine. The bright materials on the floors of craters Kertész and de Graft (neither imaged by Mariner 10) appeared to have lobate or flow-like boundaries, leading to the inference that they could be related to impact melt deposits (Blewett et al., 2010).

The mosaics of Mercury's illuminated hemisphere that were collected during the MESSENGER flybys (Solomon et al., 2008) typically had pixel scales of several hundred meters for images from the MDIS monochrome narrow-angle camera (NAC), and a few kilometers in multispectral images from the MDIS wide-angle camera (WAC). This spatial resolution limited the ability to resolve morphological details and further interpret the nature of BCFDs. Systematic mapping of the planet

12.1 Introduction

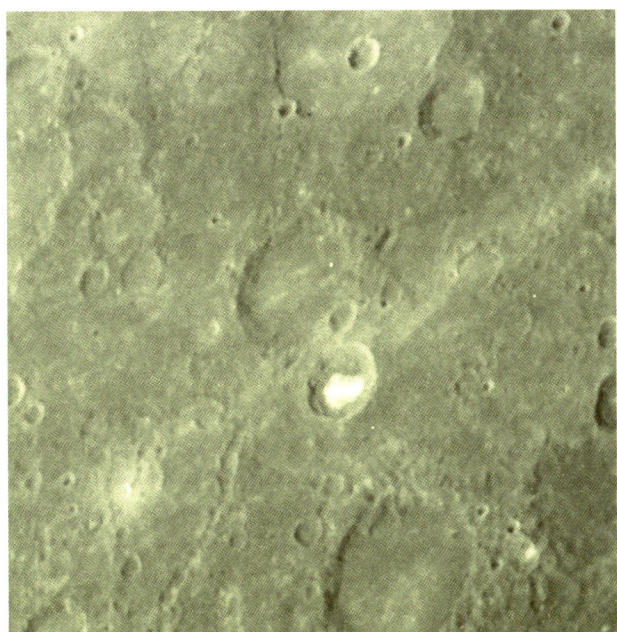

Figure 12.1. The 35-km-diameter crater Hopper, in a Mariner 10 image from that spacecraft's second Mercury flyby. Hopper's floor is occupied by high-reflectance material. The name Hopper was adopted by the International Astronomical Union in 2012. The crater is centered at ~12.4°S, 304.1°E. Portion of clear filter image 0166637 with pixel dimension ~600 m, obtained from the Mariner 10 image archive website maintained by M. S. Robinson at Arizona State University.

Figure 12.2. MESSENGER enhanced-color image of Hopper crater. Principal component analysis was performed on a three-band image cube. The color composite has principal component 2 (PC2) in the red channel, PC1 in the green channel, and the 433-nm/996-nm reflectance ratio in the blue channel. The inset is the 749-nm image with the contrast stretch adjusted to better show the details of the high-reflectance area of the floor. The 749-nm reflectance at a spot on the bright crater floor deposit (BCFD, tip of orange arrow) is 0.133; that at a location outside the southeastern rim (tip of white arrow) is 0.059. Derived from MDIS images EW1051081483I (997 nm), EW1051081479G (749 nm), and EW1051081475F (433 nm), equirectangular projection, 202 m/pixel, center at ~12.6°S, 304.0°E.

began once the spacecraft was inserted into orbit about Mercury in March 2011, with campaigns to produce a global monochrome base map at 250 m/pixel under lighting conditions favorable for morphological interpretation and a color base map (using eight of the WAC's 11 color filters) at 1000 m/pixel. Targeted images and those obtained when the spacecraft's highly eccentric orbit took it closer to the surface (over the northern hemisphere) have better spatial resolution. Figure 12.2 illustrates the appearance of Hopper at ~202 m/pixel, showing features almost three times smaller than the Mariner 10 image in Figure 12.1. Reflectance values for locations shown in Figure 12.2 indicate that the BCFD within Hopper is more than a factor of 2 brighter than the surrounding surface. The enhanced colors in Figure 12.2 reveal the relatively blue character of the BCFD.

As the first year of MESSENGER's orbital operations progressed, higher-resolution images with pixel scales at or better than a few tens of meters were returned for more of the surface. From the flyby data, BCFDs had been targeted for high-resolution observations. It was soon realized that all the areas corresponding to BCFDs shared a characteristic morphology: they are composed of groups of small (a few tens of meters to a few kilometers across), shallow, rimless depressions with rounded, irregular outlines (Blewett et al., 2011a). The floors are flat, as opposed to having bowl or V-shapes. These depressions typically exhibit high-reflectance interiors and haloes, though in some cases the haloes or bright interiors are not present. In a number of areas, the depressions occur in clusters or have coalesced to give the surface an "etched" appearance. Figure 12.3 is a targeted NAC monochrome image for Hopper at

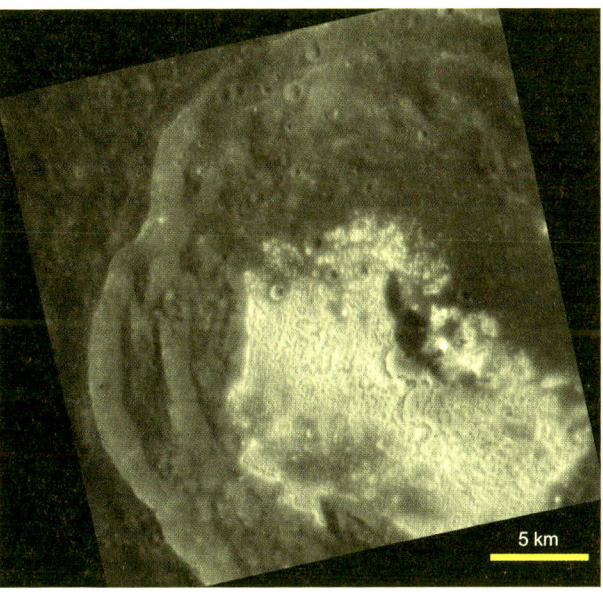

Figure 12.3. The interior of Hopper crater. MDIS image EN1066258636M, 26 m/pixel, in orthographic projection centered at 12.4°S, 303.9°E. Much of the crater floor and parts of the central peak are covered with hollows, shallow irregular depressions having high-reflectance interiors and haloes.

26 m/pixel, illustrating the unusual morphology of its floor. It was also found that the strange, bright depressions occurred in a variety of locations, including on crater central peaks (Figure 12.3), rims, walls, and ejecta deposits. Thus the term "bright crater floor deposits" was seen to be incomplete.

The International Astronomical Union (IAU) is in charge of naming surface features on the planets and has a set of approved Latin names for different feature types. The MESSENGER science team discussed the issue of what to call these newly discovered features. Several questions arose. Should an appropriate existing term be identified and applied to these small depressions on Mercury? Or should a new name be invented and proposed to the IAU for adoption?

An alternative option was to use an informal name. This possibility has precedent on the Moon. For example, the term "lunar swirl," applied to features such as Reiner Gamma, is accepted and widely used in the lunar science community (e.g., Blewett et al., 2011b), but there is no official IAU term for these features, and individual swirls are not given official IAU names according to a theme. The same is true of features such as the lunar Mairan domes or the Marius hills, which have not been assigned a special feature type and do not have official names. In the process of considering the problem, it was noted that there is an IAU descriptor that would be a good fit for the Mercury features: *cavus* (plural *cavi*) is used for "hollows, irregular steep-sided depressions usually in arrays or clusters." The term *cavi* is used on some icy satellites and in a few places on Mars. However, the decision was made to not seek approval for use of this term. The main reason was that the features in question are quite common on Mercury and are overwhelmingly associated with impact craters and basins. No benefit would come from assigning individual names to all of them. It was deemed preferable to refer to the features by the name of the impact structure with which they are associated. However, the word "hollows" from the IAU definition of *cavi* was attractive, and team members began to use hollows as shorthand for "irregular, rimless, flat-floored, shallow depressions that often have high-reflectance haloes and interiors and characteristic blue spectral slope." The term hollows is also helpful to distinguish these features from other non-impact depressions on Mercury that differ in their morphology and are likely to be formed by different processes, such as vents related to explosive (pyroclastic) volcanism (Murchie et al., 2008; Head et al., 2008, 2009; Kerber et al., 2009, 2011; Goudge et al., 2014; Thomas et al., 2014b, c) and pits that may result from collapse of a near-surface magma chamber (Gillis-Davis et al., 2009). Pyroclastic vents and volcanic collapse pits tend to have sloping floors and gently sloped walls, whereas hollows have flat floors and steeply sloped walls (Thomas et al., 2014a).

In this chapter we first review what is known about the nature of hollows, including their morphology and morphometry, the geologic settings in which they are found, their distribution around the planet, the composition of the units that host hollows, and their spectral and textural properties. We then discuss the evidence that hollows are very young compared with the majority of Mercury's surface, and we consider the means by and conditions under which hollows and the associated bright haloes may form, taking instruction from analog landforms on other planetary bodies (including features on icy surfaces).

12.2 MORPHOLOGY AND DISTRIBUTION OF HOLLOWS

12.2.1 Styles and Geologic Setting

Hollows are found predominantly within impact craters and basins (Blewett et al., 2011a, 2013; Thomas et al., 2014a). Floors of complex craters may host a few isolated hollows or small clusters (e.g., Abu Nuwas) or be covered with extensive fields of coalesced hollows, as at Hopper (Figure 12.3), Sander, de Graft, and Kertész. In a number of craters, hollows are found around a portion of the floor's outer perimeter where the floor meets the crater wall (Suess, Warhol; see Figure 12.4) or the inner perimeter where the floor meets central peaks (Warhol, Kyosai). Hollows can occur on the walls, terraces, rim (Dominici), and ejecta (Xiao Zhao, Figure 12.5) of craters. Prominent hollows cover the central peaks of craters such as Eminescu and Degas and the peak rings of impact basins such as Vivaldi and Raditladi. Figure 12.6 shows small hollows distributed on the terraces and central peaks of the 75-km-diameter Boznańska crater.

Relatively high-reflectance material also occurs as layers that are visible in the walls of simple craters (Figure 12.6) or exposed in the walls of volcanic vents, such as the pit crater on the floor of Scarlatti basin and in the vent northeast of Rachmaninoff (Figure 12.7). When images with sufficient spatial resolution are available, it can be seen that erosion of the layer of bright material has produced indentations or depressions in the wall and that bright material is being shed downslope.

Occasionally, isolated hollows are seen on plains or rolling terrain that are not obviously related to an impact structure, as shown in Figure 12.8. Cases such as this may represent the formation of hollows on the relict rims, peaks, or ejecta of ancient impact structures. Thomas et al. (2014a) identified two large, dispersed groups of such isolated hollows to the northwest of the Caloris basin. Some small isolated hollows are surrounded by low-reflectance surface haloes called "dark spots" by Xiao et al. (2013). The dark spots are thin and have diffuse edges (Figure 12.9), and they have the lowest values of reflectance that have been measured on Mercury.

12.2.2 Sizes, Shapes, Depths

Hollows are rimless depressions with irregular, rounded outlines. They range in diameter from several tens of meters to several kilometers, though the larger ones are likely to have formed by the merger of smaller hollows. Thomas et al. (2014a) determined that hollows cover 0.08% of the part of Mercury's surface for which images at better than 180 m/pixel were available at the time, which was 96% of the planet's surface. That work was completed prior to the end of the mission; the area covered at better than 180 m/pixel at the end of the mission is 98%.

The floors of hollows are generally flat, although bumps or mounds are present that could be remnants of the former surface. In craters such as Hopper (Figure 12.3) and Kertész, hollows appear to have a uniform depth over a wide area. Depth measurements of hollows were reported by Blewett et al. (2011a, 2016), Vaughan et al. (2012), Thomas et al.

12.2 Morphology and Distribution of Hollows 327

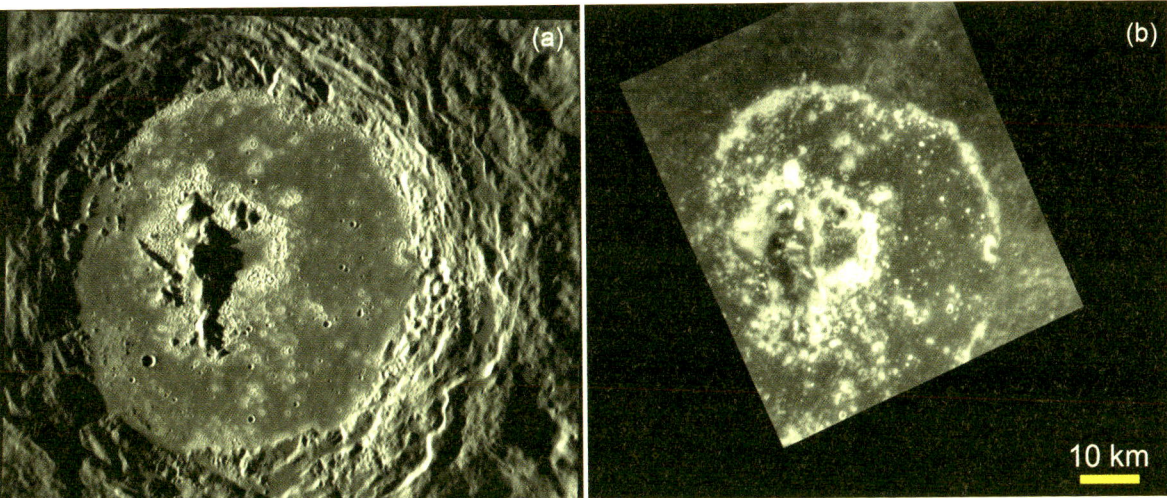

Figure 12.4. Hollows within Warhol crater. (a) Image at 79° incidence angle emphasizes texture and topography (EN1022253846M). (b) Image at a small incidence angle (2.6°) is dominated by reflectance differences (EN1035068908M). Hollows occur around the perimeter of the floor and surrounding the central peaks. Both images are in the same orthographic map projection: center latitude 2.6°S, center longitude 354.1°E, 78 m/pixel.

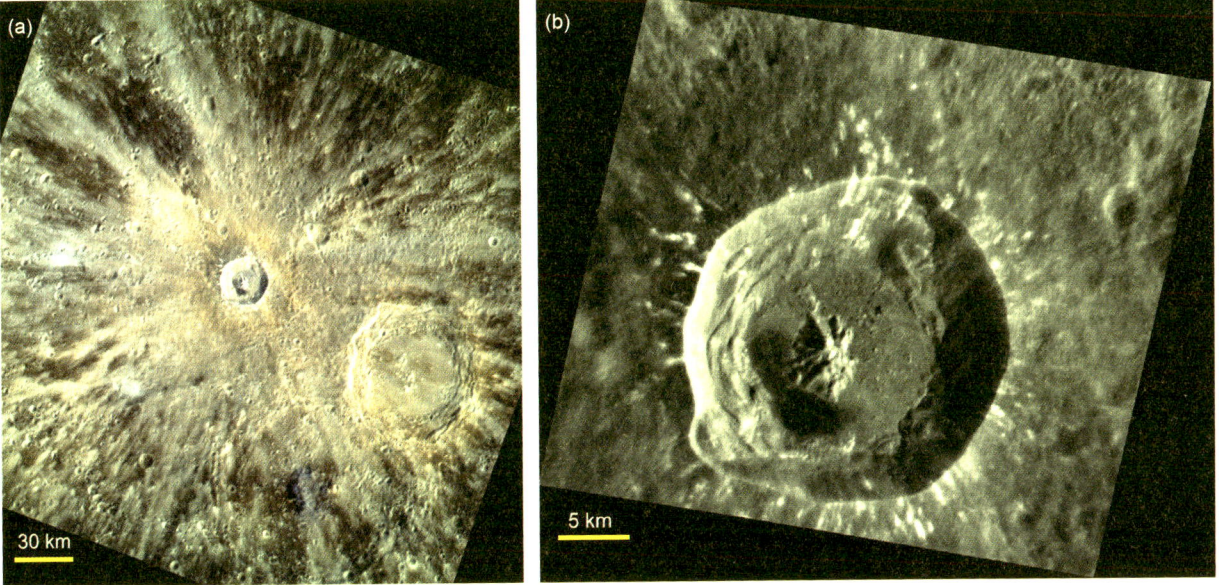

Figure 12.5. Crater Xiao Zhao, 24 km in diameter. The crater has an extensive system of bright rays and was judged by Braden and Robinson (2013) to be one of the youngest rayed craters on Mercury. (a) Color composite of EW1013176470I (997 nm), EW1013176462G (749 nm), and EW1013176458F (430 nm) displayed as red–green–blue (RGB). Image centered at 10.3°N, 123.9°E, 280 m/pixel. (b) High-reflectance streaks consistent with the presence of hollows occur on the central peak, walls, rim, and ejecta. Targeted NAC image EN1028472584M, centered at 10.6°N, 123.9°E, 35 m/pixel.

(2014a), and Fassett (2016). As a result of the small size of hollows relative to uncertainties in the location of measurements by MESSENGER's Mercury Laser Altimeter (MLA), few MLA profiles of hollows have been reported. Therefore, depth determinations have been made primarily by measuring the lengths of shadows cast by the brink, in combination with knowledge of the Sun's azimuth and angular elevation above the horizon when an image was collected. Blewett et al. (2011a) found a depth of 44 m for hollows on the floor of the Raditladi basin, and Vaughan et al. (2012) determined that hollows on the floor of Kertész are ~30-m deep. Thomas et al. (2014a) made 108 depth measurements of 27 clusters of hollows and found depths ranging from 5 to 98 m; the average value they obtained is 47 m, with a standard deviation of 21 m. Fassett (2016) determined the depths of 125 hollows that were resolved in digital elevation models constructed from image stereo pairs and found an average depth of 37 m. Blewett et al. (2016) made the most comprehensive set of shadow-length measurements to date using 565 images with pixel scales <20 m (the average was 13.1 m/pixel). The images were located at latitudes between ~3°S and 78°N, with a global spread in longitude. The resulting depth measurements for 2518 individual hollows (Figure 12.10) yielded a mean depth of 24 m, with a standard deviation of 16 m.

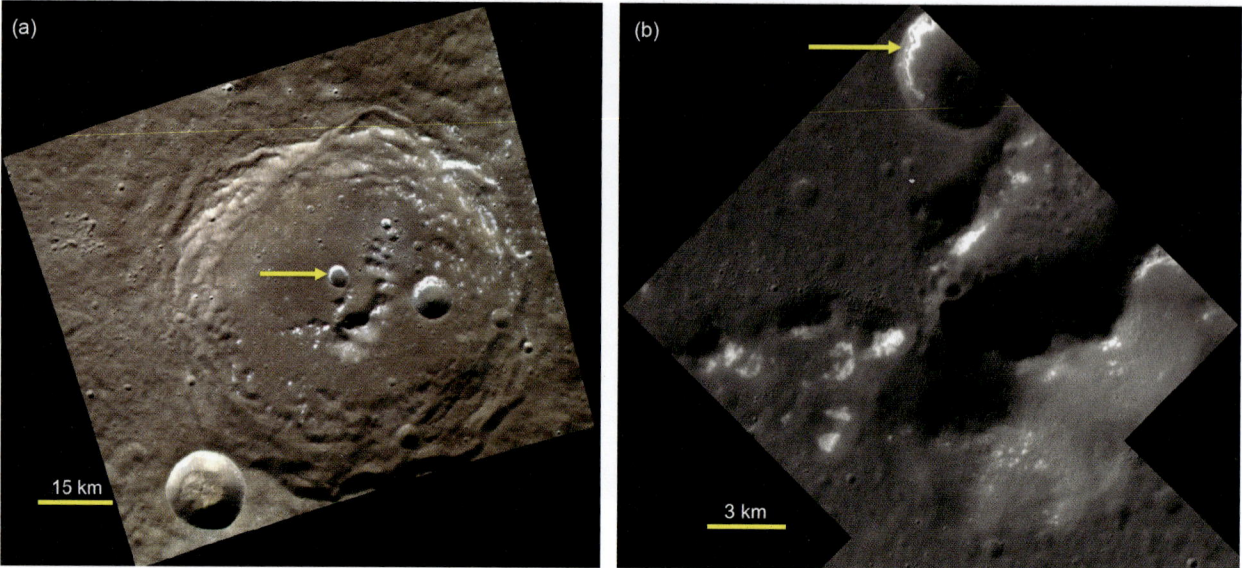

Figure 12.6. Hollows on the walls, terraces, and central peaks of Boznańska crater (75-km diameter). (a) WAC image at 108 m/pixel with 997 nm (EW1005457480I), 749 nm (EW1005457472G), and 430 nm (EW1005457476F) as RGB. The arrow indicates the small bowl-shaped crater at the top of (b). (b) High-resolution image shows a layer of hollows-forming material in the upper wall of the small crater. NAC ride-along image EN1063406030M, 15 m/pixel. North is toward the top.

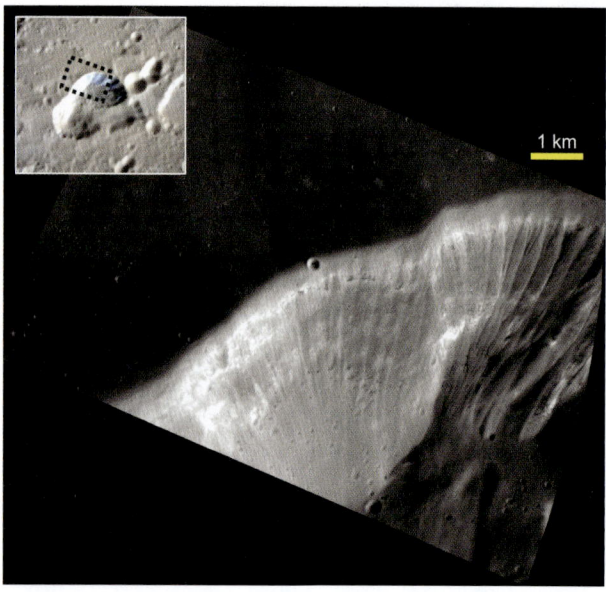

Figure 12.7. Hollows and high-reflectance layers in the wall of a large volcanic vent northeast of Rachmaninoff and west of Copland crater. The inset shows the entire vent, which is ~50 km in the long dimension and 4 km in maximum depth. The dashed box indicates the area of the high-resolution targeted NAC image (EN1059620367M, ~7 m/pixel), centered near 36.1°N, 64.0°E.

12.2.3 Geographic Distribution and Compositional Affinities

Catalogs of the locations of hollows have been published by Blewett et al. (2013, 2016) and Thomas et al. (2014a, 2016) (Figure 12.11). Thomas et al. (2014a) also assessed the association of hollows with other features. MESSENGER's orbit was highly eccentric, with initial periapsis altitudes of several hundred kilometers over high northern latitudes (Chapter 1). The apoapsis altitude over the southern hemisphere ranged from ~15,000 km during the first Earth year of orbital operations to ~10,000 km during the final three years in orbit. Therefore, there is a latitudinal asymmetry in the number of high-resolution images that are available across the planet. Because hollows are small features, high-resolution images are needed in order to identify them. Consequently only 14 of the 445 clusters of hollows listed by Thomas et al. (2014a) are located south of 50°S (Figure 12.11). When normalized to the fraction of the surface imaged at pixel scales <180 m, the greatest occurrence of identified hollows is in the latitude bin 0° to 30°N (Thomas et al., 2014a). At high northern latitudes, large solar incidence angles (measured from the surface normal) can make it difficult to detect hollows from their high reflectance. Further, extensive shadows, particularly in crater interiors, may prevent identification of hollows if they are present. Smooth plains dominate the northern polar region of Mercury (Head et al., 2011). As discussed below, hollows are rarely found to form within a substrate of smooth plains material. Therefore the lower abundance of hollows poleward of ~50°N can be attributed partly to lack of a favorable substrate.

Mercury's 3:2 spin–orbit resonance causes longitudes 0° and 180°E to experience noon on successive perihelia. These "hot-pole" longitudes are thus subjected to greater levels of solar radiation and reach higher surface temperatures than longitudes closer to 90° and 270°E. If higher temperatures or higher solar flux promote the loss of material and lead to the formation or enlargement of hollows, then a greater extent of hollowed terrain should be found near hot-pole longitudes. Thomas et al. (2014a) corrected for the proportion of area occupied by smooth plains and found a weak tendency for hollows to be

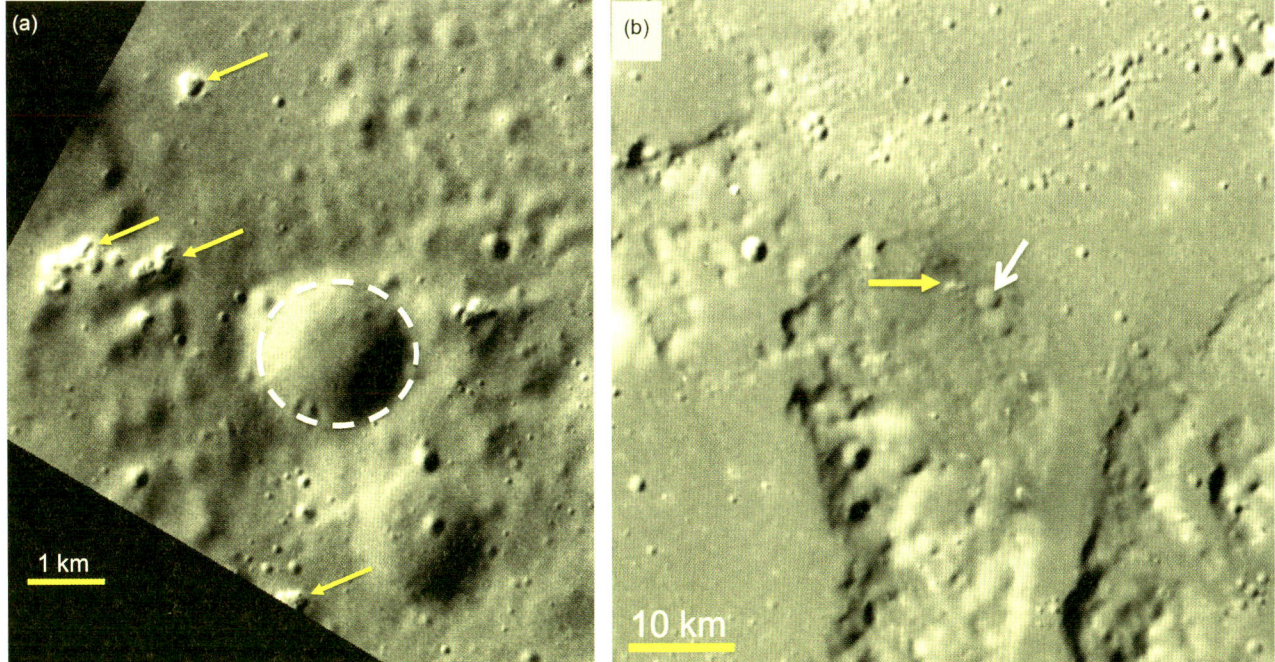

Figure 12.8. (a) Isolated hollows (arrows) near 57.6°N, 124.3°E. Portion of NAC ride-along image EN1071214280M, 17.9 m/pixel. Dashed circle indicates a small crater marked with a white arrow in the context image shown in (b). (b) Portion of image EW1025420872G, providing regional context for the isolated hollows. Yellow arrow points to the westernmost cluster of hollows shown in (a).

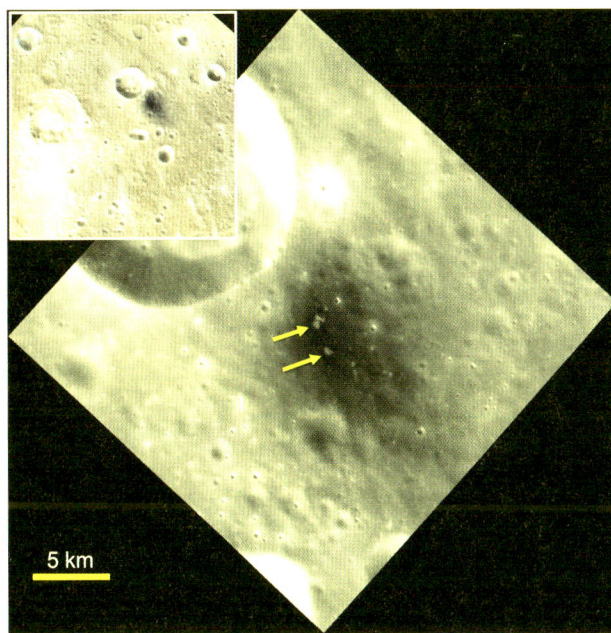

Figure 12.9. One of the dark spots identified by Xiao et al. (2013). Small hollows (arrows) are found within a low-reflectance patch with diffuse edges. Targeted NAC image EN0240207569M, 27 m/pixel, centered at 39.51°N, 25.90°E. Inset is a targeted three-color WAC image that illustrates the low reflectance and blue color of the dark spot. EW0240165835I (997 nm), EW0240165831G (749 nm), and EW0240165827F (430 nm) as RGB.

more abundant around hot-pole longitudes (Figure 12.11), implying that the formation of hollows is somewhat favored by high temperatures or greater exposure to solar flux. Superimposed on the hot-pole–cold-pole trend, the greatest areal extent of hollows was found by Thomas et al. (2014a) to lie in the non-hot-pole longitude bin 300°–320°E. Therefore it is likely that an additional compositional or geological factor also contributes to the formation of hollows.

On the basis of a few examples, Blewett et al. (2011a, 2013) suggested that hollows form preferentially on Sun-facing slopes, consistent with the idea that high temperatures or high solar flux could promote loss of a volatile-bearing phase. For example, in the crater Boznańska (latitude 59.4°N, 319.3°E) shown in Figure 12.6, hollows appear to occur predominantly on south-facing slopes. Thomas et al. (2014a) performed a quantitative analysis of the orientation of slopes on which hollows are found. Statistics were limited, partly because in many images only Sun-facing slopes were illuminated, and images are not available with lighting on the opposite face in order to determine if hollows are present there. However, in the small proportion of hollows for which a preferred slope aspect could be determined, Thomas et al. (2014a) reported that a strong correlation with Sun-facing slopes exists.

Blewett et al. (2011a, 2013) noted that hollows were most commonly found in material with a lower reflectance than the global average. One of Mercury's major color units, called low-reflectance material (LRM) (Chapters 6 and 8), is characterized by reflectance and spectral slope at visible to near-infrared wavelengths that are less than the planetary averages

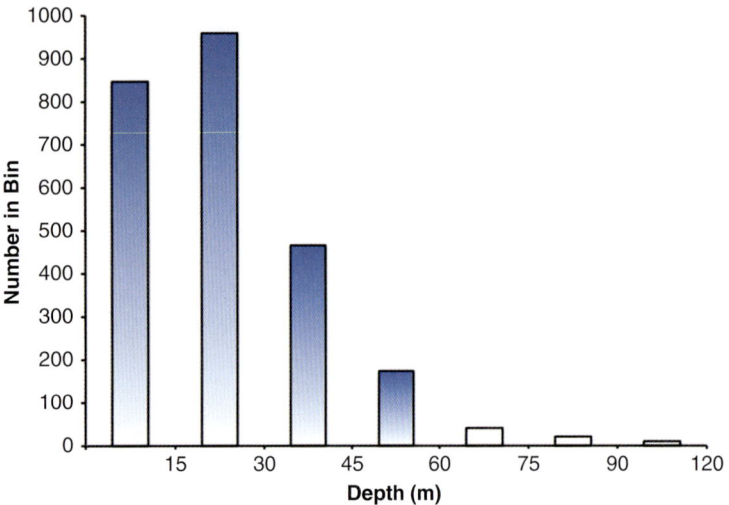

Figure 12.10. Depths determined by shadow-length measurements of 2518 hollows in 565 MDIS images (Blewett et al., 2016).

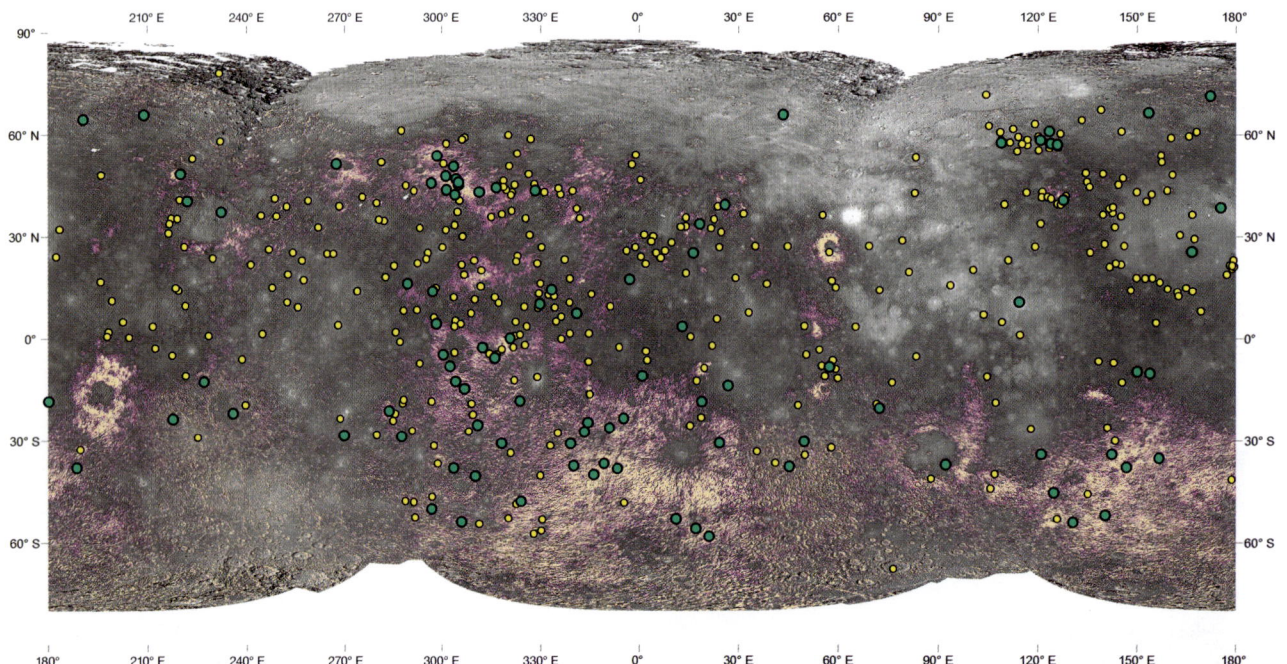

Figure 12.11. Map of hollows locations on Mercury. Peach-colored areas are LRM, and purple areas are LBP, mapped on the basis of reflectance and color properties (Klima et al., 2016). Large green dots are hollows that plot within LRM or LBP. Small yellow dots are hollows located in neither LRM nor LBP at the scale of the global color images used for mapping, although higher-resolution images reveal that many of these occurrences are found within localized outcrops of LRM or LBP. The base map is principal component 2 (PC2) derived from global eight-color MDIS imaging. Low values of PC2 correspond to LRM; red materials have high values of PC2. Locations are from Thomas et al. (2014a). The map is centered at 0° longitude. The hot-pole longitudes are 0° and 180°.

(Robinson et al., 2008; Blewett et al., 2009; Denevi et al., 2009; Murchie et al., 2015). Both LRM and hollows have spectral slopes that are relatively "blue." Mapping by Blewett et al. (2013) and Thomas et al. (2014a) showed that the overwhelming majority of hollows are associated with LRM, low-reflectance blue plains (LBP) (Denevi et al., 2009; Chapter 8), or dark spots (Xiao et al., 2013). Maps of LRM and LBP show that LBP are especially prevalent from 300° to 320°E (Klima et al., 2016; Figure 12.11), the longitude range in which Thomas et al. (2014a) noted abundant hollows.

In 38 cases examined by Thomas et al. (2014a) where hollows occur within high-reflectance red plains (HRP) (Robinson et al., 2008; Denevi et al., 2009), LRM is present locally at 37 of them. This spatial association suggests that

12.2 Morphology and Distribution of Hollows

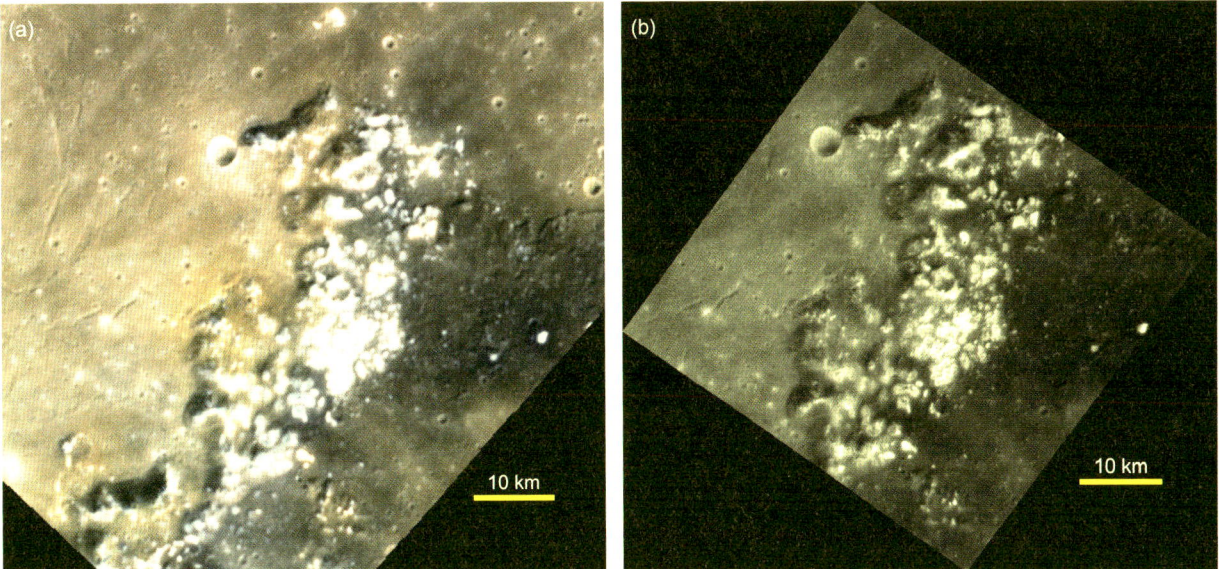

Figure 12.12. Interior of the Raditladi basin, illustrating the contact between the HRP plains to the left and the LRM to the right. Hollows occur within the LRM on the basin floor and the LRM that forms the mountains of the central peak ring. (a) Color image is a WAC composite with images EW1013234506I (997 nm), EW1013234498G (749 nm), and EW1013234494F (430 nm) as RGB, 186 m/pixel. (b) Targeted NAC image EN0233984961M, 47 m/pixel. Both images are in orthographic projection centered at 26.6°N, 120.5°E.

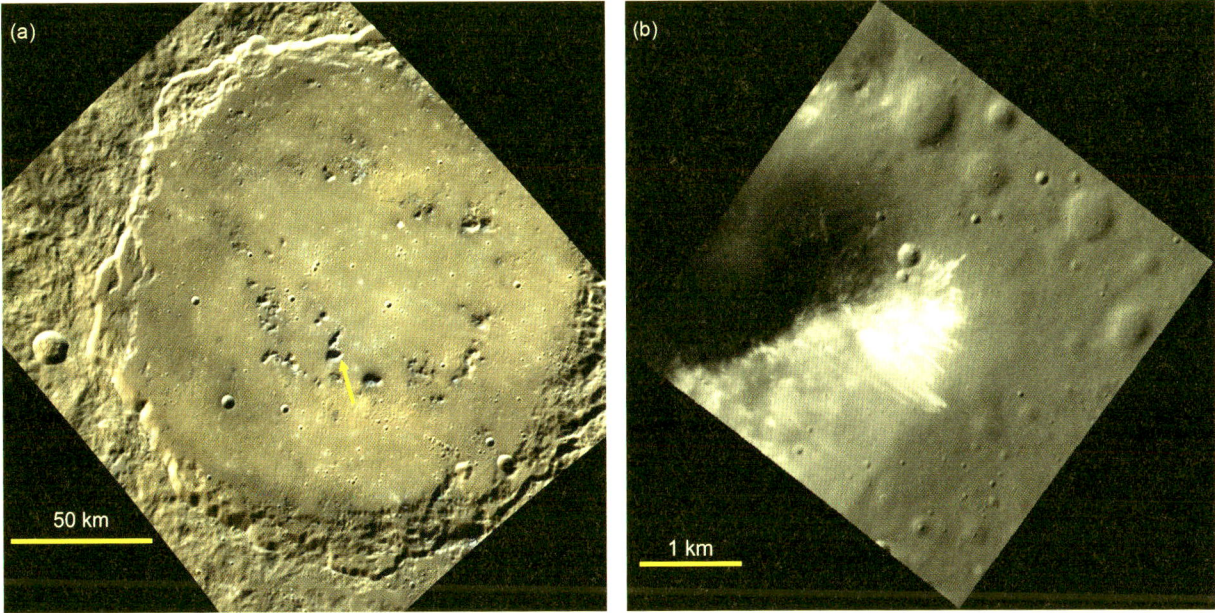

Figure 12.13. Hollows forming on a peak-ring mountain composed of LRM within the 200-km-diameter Rustaveli basin. (a) Color composite context image with EW0219478808I (997 nm), EW0219478804G (749 nm), and EW0219478800F (430 nm) as RGB, 187 m/pixel. Equirectangular projection centered at 52.4°N, 81.8°E. Arrow points to the hollows-covered peak in (b). Hollows are absent on the HRP plains filling the basin. (b) Targeted NAC image shows delicate streamers of bright material that extend downslope from hollows at higher elevations on the peak. EN1044173928M, 8 m/pixel, centered at 51.9°N, 82.7°E.

hollows are forming in small surface exposures of LRM surrounded by HRP, or in an LRM substrate that is only thinly covered by HRP. Examples illustrating the affinity of hollows for LRM are shown in Figures 12.12 and 12.13.

An association between hollows and pyroclastic deposits was discussed by Blewett et al. (2011a, 2013). Prominent examples include the floors of Lermontov basin and Tyagaraja crater, which contain hollows, vents, and red pyroclastic deposits. Thomas et al. (2014a) found that 77% of volcanic vents within impact craters have nearby hollows, whereas only 47% of vents outside craters do. Thomas et al. (2014a) also described "spectrally red pitted ground," which

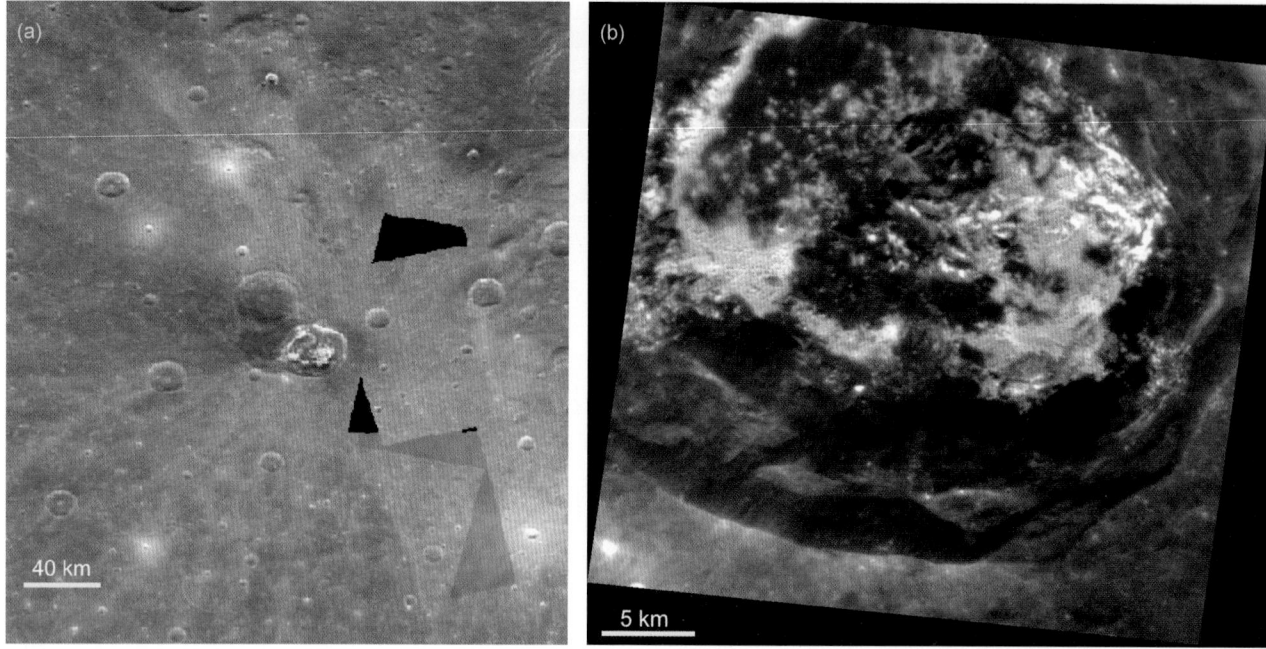

Figure 12.14. The rayed crater Balanchine, 41 km in diameter. (a) Context image showing color contrasts and Balanchine's rays. Dark polygonal areas are image artifacts. Extracted from the MDIS three-color base map (997 nm, 749 nm, and 430 nm as RGB). (b) High-Sun (40° incidence angle, 40° phase angle) image of the southern portion of the crater. Extensive hollows are present on the floor, peaks, and walls. Targeted NAC image EN1024844936M, 28 m/pixel, centered at 38.3°N, 175.6°E.

consists of shallow, irregular depressions with depths similar to hollows but that are floored by reddish material. The red pitted ground also differs from hollows in that the margins of the depressions are more subdued than those of hollows. The 24 areas of spectrally red pitted ground identified by Thomas et al. (2014a) are mostly located northwest of Caloris near the convergence of Timgad and Paestum Valles, at ~59°N, 122°E. All of the pitted-ground locations have hollows within 50 km.

In the case of Lermontov, which appears to have formed in LRM, it could be that the presence of the reddish pyroclastic blanket on top of the LRM floor has acted to limit formation of hollows by protecting the LRM from external hollow-forming agents (e.g., solar heating, space weathering; see Section 12.6.1). Alternatively, in some cases heating by volcanic materials could have promoted loss of volatiles (see discussion in Section 12.6.1), or hollows may form in pyroclastic deposits in which a volcanic volatile distinct from the volatiles in LRM had condensed and subsequently was removed.

12.3 EVIDENCE FOR YOUNG AGES

A number of lines of evidence point to hollows as being among the youngest non-impact features on Mercury.

12.3.1 Occurrence within Rayed Craters

The youngest time–stratigraphic system on Mercury is the Kuiperian (Spudis and Guest, 1988; Chapter 6). Units in the Kuiperian period are related to impact craters that have crisp morphology and possess rays, analogous to Copernican craters on the Moon. Under the assumption of a lunar-like impact flux, the base of the Kuiperian had been estimated at approximately 1 Ga (Spudis and Guest, 1988; cf. Hawke et al., 2004). However, on the basis of an updated impactor flux at Mercury, Xiao et al. (2012) determined that crater size–frequency distributions of all Mercury craters with distinct rays indicate an average model age of $\lesssim 270$ Myr. Braden and Robinson (2013) inventoried the number of rayed craters per unit area on the Moon and Mercury and found that there are approximately twice as many Copernican craters per unit area on the Moon as Kuiperian craters per unit area on Mercury. This difference can be attributed to a higher rate of optical maturation on Mercury (estimated to be a factor of 4 faster than on the Moon), leading to more rapid loss of rays by processes that darken the surface. Braden and Robinson (2013) also concluded, on the basis of the difference in density of rayed craters and impact flux, that the base of the Kuiperian system is two to four times younger than the base of the Copernican system. Thus, on average, rayed craters on Mercury are younger than rayed craters on the Moon.

Rayed craters on Mercury that contain prominent hollows include Ailey, Balanchine (Figures 12.14 and 12.15), Bashō, Cunningham, Degas, and Dominici. Kuiper crater has small hollows on its floor (Xiao et al., 2013) and likely on its central peak and walls. Xiao Zhao (Figure 12.5) is one of the youngest rayed craters on Mercury (Braden and Robinson, 2013). It has high-reflectance markings that are likely to be hollows.

12.3.2 Crisp Morphology and Fine Structures

Hollows are small features, with lateral extents as small as tens of meters and depths that average ~24 m (Section 12.2.2). They have a fresh appearance and do not show evidence of having been subjected to extensive erosion by impacts. Delicate features such as the feathered edges of the bright haloes of many hollows and the thin, bright streamers that appear to be related to formation of hollows on the central peak of Rustaveli (Figure 12.13) would not be preserved for a long period of time given the impactor flux that affects Mercury, which is estimated to be approximately a factor of 3 greater than that at the Moon (Marchi et al., 2013).

12.3.3 Lack of Superposed Craters

Available images indicate that hollows lack superposed craters at the resolution of the data. Blewett et al. (2013) presented images of the Raditladi basin, where downslope movement of material that is likely related to formation of hollows on top of the central peak ring has formed a talus that partially covers craters ~100–200 m in diameter. Perhaps the best-quality high-resolution image of hollows obtained by MESSENGER is shown in Figure 12.16. The hollows here have remarkably flat floors, with small bumps that may be partially devolatilized material. The transition from the flat floor to the steep, straight walls is sharp, occurring over a distance of just a pixel or two. No impact craters are visible on the floors. Since the image is at 3 m/pixel, any craters on the floors must be smaller than ~10 m in diameter. The area exterior to the hollows is lightly cratered and has a smooth, subdued appearance, perhaps as a result of mantling by material related to formation of the hollows. The smallest discernable craters in the area surrounding the hollows are ~10 m in diameter.

The stratigraphic relationships revealed by consideration of small impact craters suggest that hollows are very young in geological terms, and that some hollows may be actively forming today. Related to the relative age of hollows and the craters that contain them, Thomas et al. (2014a) looked at the proportion of crater floors that are covered by hollows as a function of size and the degradation state of the craters. If hollows began to grow immediately after the formation of the crater, then older

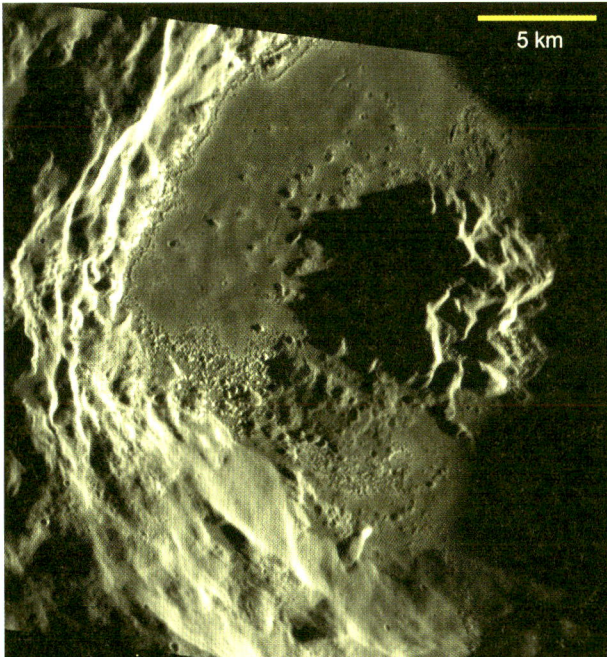

Figure 12.15. Texture of hollows on the floor of Balanchine crater seen in a low-Sun (81° incidence angle, 131° phase angle) image (compare with Figure 12.14). Targeted NAC EN1022827014M, 24 m/pixel, image center 38.4°N, 175.2°E.

 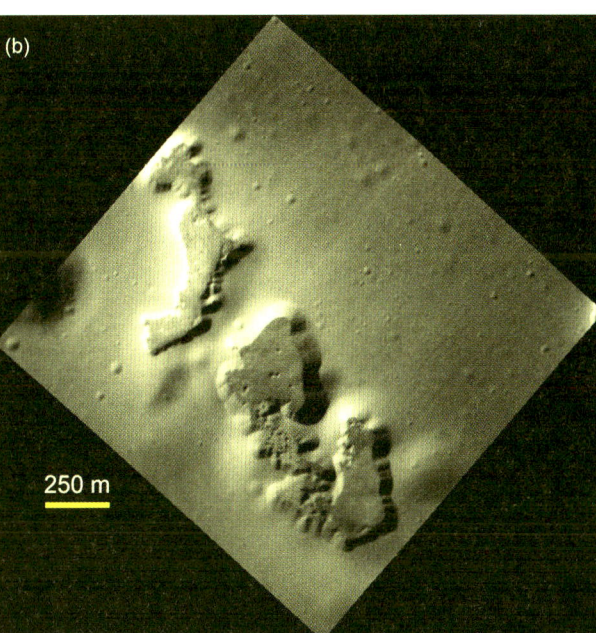

Figure 12.16. Hollows in small outcrops of LRM on the wall of the degraded Sholem Aleichem basin. (a) Color composite image of the eastern portion of Sholem Aleichem. The bright spot indicated by the arrow is the small cluster of hollows in (b). Images EW0241878939I (997 nm), EW0241878931G (749 nm), and EW0241878935F (430 nm) displayed as RGB, 149 m/pixel, centered near 50.8°N, 272.3°E. (b) High-resolution (3 m/pixel) NAC ride-along image EN1051631967M, centered at 52.0°N, 272.0°E.

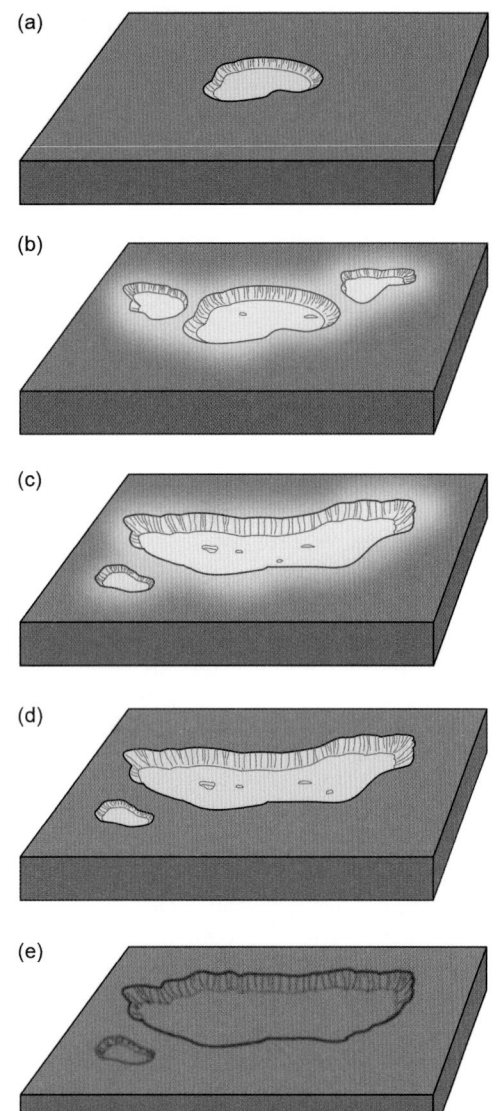

Figure 12.17. Illustration of the inferred sequence in the development of hollows. (a) Initiation of a small hollow. (b) The hollow has enlarged, and additional hollows have begun to grow. The interiors are bright, and the depressions are surrounded by bright haloes. Small bumps on the floor may consist of volatile-depleted remnant material. (c) The three hollows in (b) have enlarged and merged, and an additional small hollow has begun to grow. (d) Activity is waning; the bright haloes have disappeared. (e) Activity has ceased; the interiors have the same reflectance as the surroundings, and impact erosion has begun to modify the landforms of the hollows.

craters (those with a greater degradation-state classification) would be expected to have a higher fraction of their floors covered with hollows. Thomas et al. (2014a) found no clear increase in the area covered by hollows in older craters. Therefore, in many cases hollows must be substantially younger than the crater or basin surfaces in which they are forming (e.g., Figures 12.6, 12.8a, 12.12, 12.13, and 12.16). Thus, an important question is the nature of the trigger that initiated growth of hollows at times more recent than the formation of the hosting impact crater. The topic of initiation is addressed in Section 12.6.3.

12.3.4 Inferred Age Sequence and Formation Rates

Blewett et al. (2011a, 2013) proposed a development sequence for hollows by which those with high-reflectance interiors and haloes are actively forming, those with only high-reflectance interiors have waning activity, and those with interiors and exteriors that match the background are no longer growing. Xiao et al. (2013) suggested that the earliest stage of growth begins with the appearance of a hollow within a dark spot, followed by growth of the hollow, fading of the dark spot, and development of the bright halo. This sequence is depicted in Figure 12.17.

Features with morphologies consistent with old, degraded hollows are not common in the available images. Figure 12.18a shows examples of fresh and waning hollows, with some instances where growth may have slowed or ceased. Figure 12.18b illustrates flat-floored depressions that have a softened appearance and lack reflectance contrasts with the surroundings; these are candidates for hollows that stopped growing and have been overtaken by impact erosion and mixing. The rarity of old hollows like these suggests that hollows, as shallow features of small lateral extent, are relatively short lived and are erased relatively quickly once they are no longer active. However, observational bias also plays a role, because old hollows lack the distinctive high reflectance and blue color that drew MESSENGER team members to specifically target younger hollows for high-resolution imaging opportunities.

The rate at which hollows enlarge can be estimated if the age of the hollows or the crater containing them is available. Blewett et al. (2011a) used the ~1 Ga age estimate for the Raditladi basin (Strom et al., 2008) as an upper limit on the age of the hollows on the basin's floor. Under the assumption that the hollows enlarge by radial scarp retreat, the ~137-m characteristic radius of the Raditladi floor hollows implies that they grow at a minimum of 0.14 μm/year (1 cm per ~70,000 yr), or faster if the hollows are younger than the basin.

Another estimate of the rate at which hollows form can be made from observations of a rayed (Kuiperian) crater. Banks et al. (2016) placed the base of the Kuiperian at 300 Ma. Rayed crater Balanchine (Figure 12.14) has extensive floor hollows. The low-Sun image in Figure 12.15 shows the morphology of the hollows more clearly than does the high-Sun image (Figure 12.14b). On the western margin of the floor, hollows appear to have formed by scarp retreat from the crater wall by a distance of ~300 m (Blewett et al., 2016). Under the assumption that Balanchine's age is 300 Ma and that the hollows began to form immediately after the Balanchine impact, the rate of lateral enlargement was 1 cm per 10,000 Earth years. This growth rate is for steady enlargement, but it could be that growth of hollows is episodic and occurs at rates faster than the average values estimated here. To place these rates into context, the erosion of kilogram-sized rocks on the lunar surface by micrometeoroid bombardment is estimated to proceed at 1 cm per 10^7 yr (Ashworth, 1977). Malin (1987) measured the average aeolian abrasion rates of different rock types in Antarctica to be between 1 cm per 700 yr and 1 cm per 100 yr.

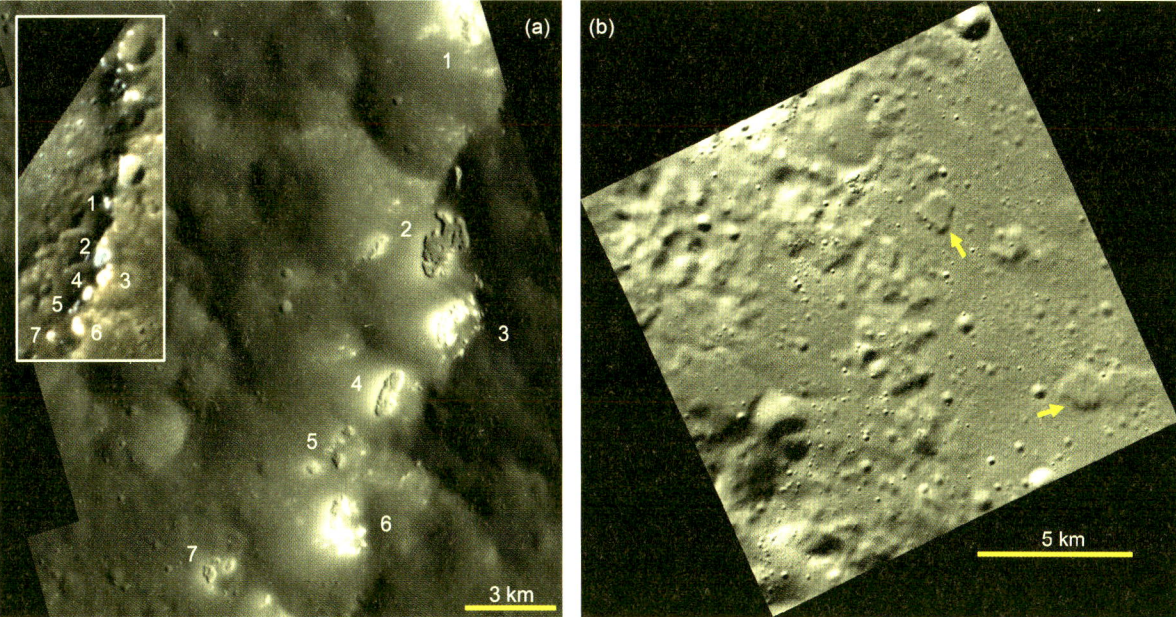

Figure 12.18. (a) The peak ring of Aksakov shows several stages of hollows evolution. Some hollows have bright interiors and bright haloes. Other hollows (numbered 2, 5, 7) have less prominent or no high-reflectance haloes; some may have ceased enlargement and begun to erode away. Main image is a mosaic of EN0213851669M and EN0213851674M, 15 m/pixel, centered at ~34.4°N, 282.1°E. The color composite (inset) illustrates the relatively blue color of the hollows and is composed of portions of EW1005860272I (997 nm), EW1005860264G (749 nm), and EW1005860268F (433 nm) as RGB. (b) Rimless, flat-floored depressions with softened morphology that may be old hollows (arrows). The image is located within the inner ring of a degraded two-ring basin, in an area described by Thomas et al. (2014) as "spectrally red pitted ground." Targeted NAC image EN0249182311M, 15 m/pixel, centered at 62.9°N, 118.3°E.

12.4 REMOTE SENSING PROPERTIES AND COMPOSITIONAL DATA

12.4.1 Texture and Roughness of Hollows

Earth-based radar images of Mercury, obtained prior to MESSENGER's arrival, show surface features on both the half of the planet viewed by Mariner 10 and the other half (Harmon 2007, 2008; Harmon et al., 2007). Harmon et al. (2007) presented S-band (12.6 cm) images of craters that are now known to contain hollows, including Tyagaraja, Zeami, Balzac, Theophanes, Eminescu, and Cunningham. The radar return is primarily sensitive to wavelength-scale roughness. The radar images often show bright rims and ray systems. Spatial resolution is low (~5 km/pixel), and at this scale no anomalous returns that could be associated with the presence of hollows are obvious.

Information on surface texture can also be obtained from analysis of the change in a surface's reflectance as a function of the viewing and illumination geometry. The phase angle is the angle subtended at the surface between the direction to the Sun and the direction to the detector (camera). Most surfaces become more reflective at smaller phase angles as a result of shadow hiding (e.g., Hapke, 2012). As the phase angle decreases, rougher surfaces tend to brighten more than smoother surfaces because of their greater shadow-hiding effect. Images that were collected at different phase angles can be co-registered in order to make a phase-ratio image. In such a ratio image, differences in the ratio value can be attributed to textural contrasts or differences in scattering behavior (e.g., Kaydash et al., 2012). Blewett et al. (2014) used phase-ratio analysis to show that the high-reflectance haloes of the hollows on the floor of Eminescu crater have phase behavior consistent with the presence of material that is smoother or finer-grained than the nearby normal impact-generated regolith (see Figure 8.5). Thomas et al. (2016) conducted more detailed phase-ratio analysis of several hollows locations and concluded that the flat floors of hollows are rougher textured or have larger particle sizes than the haloes and background regolith. Such clues to the particle size or sub-resolution texture of the hollows are important when considering mechanisms for their formation (see Section 12.6).

12.4.2 Color and Spectral Properties

MESSENGER obtained a variety of remote sensing data for hollows. These include monochrome and multispectral images (in 3, 8, or 11 colors) from MDIS (Murchie et al., 2015), and reflectance spectra obtained by the Visible and Infrared Spectrograph (VIRS) component of the Mercury Atmospheric and Surface Composition Spectrometer (MASCS) (McClintock and Lankton, 2007; Izenberg et al., 2014).

Enhanced-color and color-composite images (e.g., Figures 12.2 and 12.12) reveal the relative blue color of the hollows. The shape of the reflectance spectrum for hollows and other Mercury surfaces in the visible to near-infrared is shown by the MDIS eight-color spectra presented in Figure 8.6. All the spectra are characterized by reflectance that increases toward longer wavelengths (i.e., a red spectral slope), although this increase is less pronounced in the hollows, leading to their color being described as "blue." The hollows spectra in

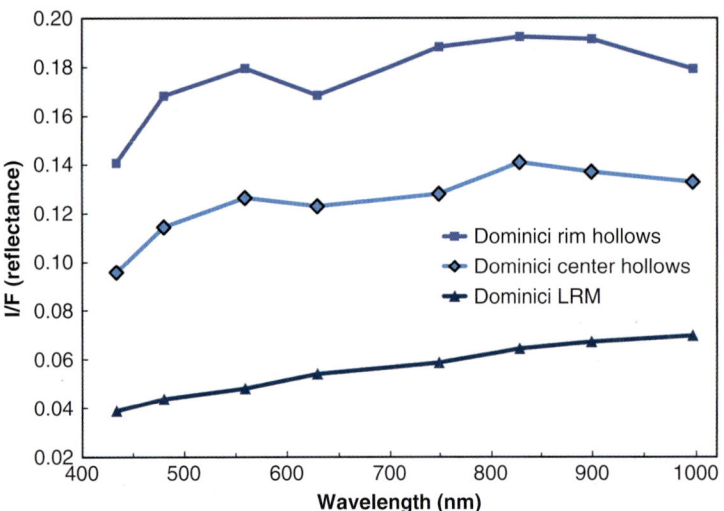

Figure 12.19. Spectra for locations in and around Dominici crater, made by averaging spectra extracted from five or six separate MDIS WAC image cubes that covered the crater (Vilas et al., 2016). The spectra for the hollows on the rim and in the crater show evidence of an absorption feature that decreases reflectance in the 629- and 749-nm filters. The average spectrum for LRM near Dominici does not exhibit this absorption feature.

Figure 8.6b have been divided by the spectrum of an HRP reference area, in order to highlight spectral features in LRM and pyroclastic materials. The relative hollows spectra have a negative slope, emphasizing that they depart from the spectral character of the great majority of Mercury's surface.

Reflectance spectra of most materials on Mercury are red-sloped and featureless. The major exceptions are some occurrences of LRM, which show a broad absorption near 600 nm that Murchie et al. (2015) attributed possibly to graphite, and many of the pyroclastic materials, which exhibit the long-wavelength edge of an ultraviolet oxygen–metal charge-transfer band at wavelengths less than 400 nm (Goudge et al., 2014; Izenberg et al., 2014). In addition, there is some evidence for a shallow absorption feature near 600 nm in spectra for a small number of hollows locations (Vilas et al., 2016). Calibration and interpretation of MDIS WAC color spectra are complicated by corrections for the temperature-dependent sensitivity of the WAC detector array, correction for the removal of scattered light, limitations of the photometric normalization, and changes in instrument response over the course of the orbital mission (Keller et al., 2013; Murchie et al., 2015). Spectra for the bright haloes surrounding hollows formed in LRM on the rims and in the centers of Dominici and Hopper craters, obtained early in the mission prior to known changes in instrument response, exhibit decreases in reflectance in the 629- and 749-nm filters (Figure 12.19) that may be consistent with absorption by magnesium-rich sulfide minerals (Vilas et al., 2016; cf. Burbine et al., 2002). Alternatively, as suggested in Chapter 8, the absorption may be the same as that found in much of the LRM, except that it is deeper because of the optical freshness of the hollows material.

In both MASCS and MDIS data, bright material in and surrounding hollows exhibits spectral properties comparable to those of the brightest fresh craters. In MASCS VIRS data, hollows have high reflectance and shallow spectral slopes from the visible to near-infrared (MASCS parameters 575 nm reflectance and ratio of reflectance at 415 nm to that at 750 nm). These characteristics are similar to those of immature, freshly exposed material, suggesting that hollows are also young in terms of exposure to the agents of space weathering (Izenberg et al., 2014). From principal component analysis of MDIS data, hollows are undistinguished by the second principal component (PC2), which defines the major spectral trend between LRM and HRP. Rather, as with MASCS data, they are distinguished by high reflectance and spectral slope, as measured with MDIS data by the 430-nm/997-nm reflectance ratio (Murchie et al., 2015). Interestingly, where clusters of hollows are resolvable by MASCS at ultraviolet wavelengths, their spectral properties may be inherited from those of their substrate. Hollows formed on pyroclastic materials appear to exhibit the same oxygen–metal charge-transfer band as that typical of pyroclastic deposits (Goudge et al., 2014; Izenberg et al., 2014).

12.4.3 Elemental Composition

The spatial resolution of elemental-composition maps from the MESSENGER X-Ray Spectrometer (XRS) (Schlemm et al., 2007), Gamma-Ray Spectrometer (GRS), and Neutron Spectrometer (NS) (Goldsten et al., 2007) measurements is too coarse to resolve small features such as hollows. However, targeted observations with the Ultraviolet and Visible Spectrometer (UVVS) channel of MASCS (McClintock and Lankton, 2007) that attempted to detect exospheric species associated with the largest fields of hollows were made late in the mission when the spacecraft was at low altitudes over the northern hemisphere.

Although the MESSENGER elemental composition sensors were not able to make specific measurements confined to hollows, compositional data are available for the LRM, the global color unit in which hollows are found (Chapters 6 and 8). The LRM appears to have a magnesium content that is higher than the planet's average. Globally, calcium is correlated with sulfur (Weider et al., 2012); the highest calcium and sulfur values are found in the intercrater plains and heavily cratered terrain (IcP-HCT), which includes the LRM. In the IcP-HCT, magnesium is correlated with sulfur (Weider et al., 2012). Thus, it appears that LRM is enriched in these three elements relative to the planetary average. Calcium and magnesium sulfides are therefore likely to be present. Sulfides are also expected on the basis of petrologic modeling under chemically reducing conditions similar to those thought to correspond to Mercury's surface materials (e.g., Burbine et al., 2002; Weider et al., 2012). Note that the sulfur content of Mercury's surface, which averages ~2 wt% and has a maximum of ~4 wt%, is surprisingly high, far greater than that found in the Moon or the bulk silicate Earth (Nittler et al., 2011; Chapter 2).

Murchie et al. (2015) noted that on Mercury the correlation between color and elemental composition is generally weak, unlike for the Moon with its clear color and compositional dichotomy between maria and highlands. The difference is that on Mercury the abundance of ferrous iron in silicates is so low as to have no discernable influence on the observed reflectance. Instead, reflectance and color are dominated by minor darkening agents. Murchie et al. (2015) reviewed evidence for the identity of the component(s) responsible for causing the low reflectance and relatively flat spectral slope of the LRM. Candidates include dark, opaque phases such as Fe- or Ti-bearing oxides (e.g., Denevi et al., 2009; Riner et al., 2009), metallic iron in the microphase and nanophase size range (Lucey and Riner, 2011), microphase or nanophase opaque sulfide minerals (Blewett et al., 2013), and the carbon allotrope graphite (Vander Kaaden and McCubbin, 2015). Only graphite or a mix of nano- and microphase iron was found to be sufficiently low in reflectance to be the darkening agent in LRM (Murchie et al., 2015). Low-altitude MESSENGER Neutron Spectrometer data (Peplowski et al., 2016) indicate that carbon is more abundant by 1–3 wt% in LRM than in non-LRM areas. The background level in non-LRM has been estimated to be near 1 wt% with three independent techniques, including mixture modeling of MDIS spectral data (Murchie et al., 2015), Gamma-Ray Spectrometer measurements of the average carbon content of the northern hemisphere (Peplowski et al., 2015), and the depletion observed in Mercury's largest high-reflectance deposit, the pyroclastic material northeast of Rachmaninoff basin (Peplowski et al., 2016). The total estimated carbon content in LRM is thus 2–4 wt%.

Given that hollows occur almost exclusively in the LRM, then some characteristic of the LRM must be favorable to formation of hollows relative to other terrains on Mercury. Because LRM is enriched in magnesium, calcium, and sulfur, it is plausible that phases containing one or more of these three elements could be involved in the formation of hollows. In addition, LRM contains a greater abundance of carbon, probably as graphite, than the global average, accounting for its low reflectance. Therefore, graphite could be the component responsible for formation of hollows, or one or more hollow-forming phases could be geochemically associated with carbon.

12.5 PLANETARY ANALOGS

12.5.1 Negative-Relief Features

A variety of processes can form irregular rimless depressions on planetary surfaces. Existing evidence strongly suggests that hollows are neither secondary impact craters nor volcanic pits. Neither are hollows closely similar to lunar irregular mare patches.

12.5.1.1 Secondary Impact Craters

Chunks or clusters of rock that are ejected at less than a body's escape velocity by a crater-forming impact will re-impact the surface and produce secondary craters. Secondaries formed by overlapping, low-angle, or low-speed ejecta may be shallow and have irregular outlines. However, these types of secondaries are usually found in chains and clusters and within bright, diffuse ray segments. Hollows, by virtue of their flat floors, steep walls, anomalous color and texture, and association with a particular color unit (LRM), are not secondary craters. Furthermore, secondary craters have a random distribution across the surface, whereas hollows are strongly associated with crater and basin interiors.

12.5.1.2 Volcanic Pits

Several types of negative-relief features related to volcanism or magmatism have been identified on Mercury. Approximately 150 pyroclastic vents with surrounding bright, red deposits have been mapped (Kerber et al., 2009, 2011; Goudge et al., 2014; Thomas et al., 2014b, c; Chapter 11). The distinctive red color contrasts with the relatively blue color of hollows. The largest of these, e.g., in southwestern Caloris and northeast of Rachmaninoff, are as much as several tens of kilometers across and several kilometers deep (Murchie et al., 2008; Head et al., 2008, 2009; Chapter 11).

Pit-floor craters (Gillis-Davis et al., 2009) on Mercury are impact craters that contain rimless, steep-sided, irregularly shaped depressions. These pits are large (~20–40 km in their longest dimension), far larger than hollows, and some have associated reddish deposits that may be eruptive materials. The leading hypothesis for formation of such pits is collapse following withdrawal of magma from a near-surface chamber after an explosive eruption.

12.5.1.3 Irregular Mare Patches

A class of features on the Moon, known as irregular mare patches (IMPs) and typified by Ina (also known as the "D caldera": Whitaker, 1972; El Baz and Worden, 1972; Strain and El Baz, 1980), bears a resemblance in planform to hollows on Mercury. IMPs range in horizontal dimension from ~100 m to 5 km, though the average size is 485 m (Braden et al., 2014). IMPs are likely to be young, on the basis of their sharp meter-scale morphology and few superposed impact craters (Braden et al., 2014; see also Schultz et al., 2006). They occur in low-lying portions of the lunar maria and unlike hollows are not associated with impact crater floors, walls, peaks, ejecta, or topographic highs that could be old crater or basin rims.

Within IMPs, a smooth upper unit lies on top of a lower, rougher unit that often has higher reflectance than the smooth unit, although IMPs do not have bright haloes as do hollows. The margins of the smooth unit are frequently convex-upward (see images and topographic profile in the supplemental material of Braden et al., 2014), consistent with the idea that the smooth

Figure 12.20. Mars Reconnaissance Orbiter High Resolution Imaging Science Experiment image of "Swiss-cheese" terrain near the south pole of Mars. The rounded, flat-floored depressions likely form by sublimation of CO_2 ice. Portion of image ESP_012271_0940, ~25 cm/pixel, centered near 85.919°S, 272.556°E. Illumination is from the left.

unit consists of mare basalt flow lobes. The curving, convex-upward profile of the smooth unit margins contrasts with the straight walls of hollows that are seen where permitted by the spatial resolution of MESSENGER images (e.g., Figure 12.16b). The IMP lower units in most cases look like kipukas, i.e., areas that have been surrounded, but not covered by, lobate smooth-unit flows. By this interpretation IMPs are constructional features, in contrast to the erosional nature of hollows. Therefore, the available evidence suggests that IMPs may be features related to special conditions of mare lava emplacement (Garry et al., 2012; Braden et al., 2014; Qiao et al., 2017, 2018).

12.5.2 Icy Analogs

Certain features of icy surfaces share morphological similarities with Mercury's hollows (Blewett et al., 2011a). Rounded, shallow, flat-floored depressions are found in the remnant ice cap at the south pole of Mars (e.g., Sharp, 1973; Thomas et al., 2000; Malin et al., 2001). These depressions (Figure 12.20), known collectively as "Swiss cheese terrain," are tens to hundreds of meters in their longest dimension and occur as isolated individual features as well as in large fields where many of the depressions have coalesced. Temperature measurements and the rate at which the polar depressions are seen to enlarge by scarp retreat (~1 m per Earth year) point to sublimation of CO_2 ice as the means of formation (Byrne and Ingersoll, 2003). Kargel (2013) noted similarities between Mercury hollows and martian thermokarst features where water ice has sublimed from ice-rich dust (e.g., Dundas et al., 2015).

The icy surfaces of outer-planet satellites also have landforms that resemble hollows in some aspects. Scarp-bounded, flat-floored depressions with arcuate outlines occur within the "cantaloupe terrain" and other portions of Triton's northern hemisphere (Smith et al., 1989; Moore and Spencer, 1990). This geomorphology (Figure 12.21) appears to be the result of scarp recession caused by collapse in response to sublimation of

Figure 12.21. Voyager 2 view of scarp-bounded depressions on Neptune's moon Triton. The scarps have irregular outlines and sharp brinks and likely formed by loss of a volatile phase such as N_2 or CH_4. Portion of image C1139533, center at ~39°N, 11°E, with north toward the top. See Moore et al. (1996).

a volatile component. Examples of similar terrain may occur on Europa and Ganymede (Moore et al., 1996). Sublimation-driven surface modification has also been proposed for Callisto (e.g., Moore et al., 1999; Howard and Moore, 2008; White et al., 2016).

From their survey of the Mars and icy-satellite examples cited above, Moore et al. (1996) concluded that "sublimation degradation" is a solar-system-wide geological process. They attributed the irregular, scarp-bounded depressions on icy surfaces to a "manifestation of exogenic degradation due to the loss of a volatile rock-forming matrix or cement." We find this concept to be an attractive hypothesis for the formation of hollows, although of course the nature of the volatile phase(s) on Mercury must be very different from the ices involved in loss and collapse in the martian poles and outer-planet satellites. Section 12.6 discusses possible materials and loss mechanisms on Mercury.

12.6 CONCEPTS FOR FORMATION OF HOLLOWS

Key open issues related to the formation of hollows are the growth sequence, the identity of the volatile phase or phases that are being lost, the process or processes driving the loss, the source of the volatiles, and the origin of the bright interiors and haloes.

12.6.1 Formation Sequence

The formation sequence envisioned here involves vertical growth downward to an approximately constant depth, followed by lateral enlargement by scarp retreat. Figure 12.17 illustrates the steps – initiation, loss, crumbling, enlargement in depth and diameter, emplacement of halo, continued diameter growth through scarp retreat, downslope movement from walls and lag deposition, waning of activity, and quiescence.

The description of hollows morphology in Section 12.2.2 noted the relatively similar depths that have been measured for hollows found across the planet (a few tens of meters). Blewett et al. (2013) proposed two possible controls on the depths of hollows: exhaustion of the volatile-bearing layer (i.e., the thickness of the unit hosting the hollows), or development of a lag deposit that protects the substrate from further losses. The LRM and LBP units that host hollows are in many places much thicker than the depths of the hollows (Denevi et al., 2009; Ernst et al., 2010), suggesting that it is not the thickness of the hollows-hosting layer that limits the depths. This observation favors a scenario in which a lag provides thermal or physical armor against the agents driving loss.

12.6.2 Identity of the Volatile-Bearing Phase(s) and Loss Mechanisms

As mentioned in Section 12.4.3, MESSENGER geochemical data suggest that the LRM unit in which hollows are found is richer than average in the elements magnesium, calcium, sulfur, and carbon. Loss of phases containing these volatile elements (e.g., MgS, CaS, graphite) might thus be responsible for the formation of hollows. In addition to the high daytime temperatures on Mercury, there are other factors in the environment that could cause sublimation or destruction of volatile-bearing phases. These include space-weathering processes such as ultraviolet photolysis, heating and vaporization by micrometeoroid impacts, and sputtering induced by ion bombardment (see Domingue et al., 2014, for a review of space weathering on Mercury), as well as sublimation by heating on contact with or in proximity to volcanic materials or impact melt.

Several lines of evidence suggest that loss of sulfide minerals could take place on the surfaces of airless solar system bodies. Helbert et al. (2013) reported that synthetic MgS and CaS decomposed when heated to temperatures of ~500°C (773 K), well below the melting points of these compounds. These workers suggested that on Mercury the S would escape at low energy, whereas the Mg and Ca would remain behind in atomic form. Photodesorption could then move Mg and Ca into the exosphere. Bennett et al. (2016) performed experiments in which the photon-stimulated desorption of neutral calcium was induced by irradiation of CaS powder with near-ultraviolet (355 nm) laser light. Calcium and Mg are prominent exospheric species (McClintock et al., 2009; Vervack et al., 2010; Chapters 14 and 15). Unlike Ca and Mg, S has an extremely low probability for resonantly scattering sunlight. Therefore, atomic S in the exosphere is unlikely to have been detected by MESSENGER's MASCS instrument, even when present in relatively large abundances, unless with a substantially long stare at a fixed altitude. Such stares over the larger fields of hollows were attempted with the UVVS, but they did not reveal enhancements of Mg or Ca, or any evidence for S (R. J. Vervack, Jr., personal communication, 2015).

Iron sulfide (the mineral troilite, FeS) cannot exist in great abundance on Mercury's surface because of the low abundance (1–2 wt%) of iron measured by MESSENGER's X-Ray Spectrometer and Gamma-Ray Spectrometer (Nittler et al., 2011; Evans et al., 2012; Weider et al., 2012, 2015). However, the surface of asteroid 433 Eros was found by the Near Earth Asteroid Rendezvous spacecraft to have a sulfur deficit compared with ordinary chondrite meteorites (Nittler et al., 2001). Destruction of FeS by space weathering (micrometeoroid and solar-wind ion bombardment) and escape of the sulfur could be responsible for this difference (Killen, 2003; Kracher and Sears, 2005; Loeffler et al., 2008). If FeS is being destroyed on Eros, then perhaps CaS and/or MgS are similarly being lost in the more intense space-weathering environment of Mercury.

As noted above (Section 12.2.3), hollows form preferentially on equator-facing slopes (Blewett et al., 2011a, 2013; Thomas et al., 2014a), implying that maximal solar heating or exposure to solar radiation could be driving loss of volatile-bearing phases such as sulfides. In addition to solar heating, heating from impact melts or volcanism could promote formation of hollows. Heating of volatile-bearing rocks by contact with or in close proximity to impact melt could account for examples where hollows have formed along crater floor perimeters and around the base of central peak mountains, as illustrated in Section 12.2.1 (e.g., Figure 12.4). Blewett et al. (2011a, 2013) described examples where hollows are found in close association with pyroclastic deposits, as in Tyagaraja crater and Praxiteles, Lermontov, and Scarlatti basins. Heat from intrusions or erupted materials could drive sublimation and formation of hollows. Spectrally red pitted ground (see Section 12.2.3 and Figure 12.18b), exemplified by features in Rachmaninoff basin, may form in a manner similar to hollows when lava covered a volatile-bearing substrate, causing release and escape of the volatile component (Thomas et al., 2014a).

Graphite is a refractory material often used for crucibles to hold molten metal. The melting temperature of graphite is over 4000 K (Savvatimskiy, 2005), far higher than the ~700 K daytime temperature of Mercury's surface. However, two loss mechanisms for carbon can be envisioned (Blewett et al., 2016). First, graphite is likely to be sputtered more efficiently than silicates under ion bombardment, leading to loss of carbon. Second, it is possible that irradiation of graphite-bearing rocks (such as the LRM, see Section 12.4.3 above) by solar-wind or magnetospheric protons could lead to production of methane (CH_4) (e.g., Chen et al., 1999, 2001), and thus to disruption of the surface, collapse following volume loss, and scarp retreat. Destruction of the graphite darkening agent might cause the remaining material to have higher reflectance than the original mineral assemblage, potentially accounting for the bright interiors of hollows. However, Thomas et al. (2016) argued that the volatile-bearing phase removed from hollows is redder than the host material (LRM), favoring calcium or magnesium sulfides as the lost phase.

12.6.3 The Availability of Volatiles and Localization of Hollows

A question related to the identity of the volatile-bearing phase(s) involved in formation of hollows is the distribution of the volatiles. That is, have volatiles been especially concentrated in areas where hollows form, or are the volatiles present as a pervasive rock-forming phase? If the volatiles are pervasive, why do hollows form only in certain locations?

Several means by which spatial concentration of volatiles could occur have been proposed. Blewett et al. (2011a, 2013) envisioned a scenario by which magmatic volatiles from volcanic eruptions (e.g., CO, CO_2, H_2O, SO_2) condensed within the subsurface or on the cold nighttime surface. Results from MESSENGER geochemical remote sensing have demonstrated that the large pyroclastic deposit northeast of the Rachmaninoff basin is depleted in sulfur and carbon relative to its surroundings, suggesting that the explosive eruption was driven by S- and C-bearing volatile compounds (Weider et al., 2016). If such volatiles were sequestered when buried by continued emplacement of lava flows or pyroclastic deposits, the volatiles could later be exposed when the cap rock was breached by impact cratering. The volatiles themselves, or alteration products produced by contact of the volatiles with the country rock, might then be lost in the harsh environment of the surface. Proximity to active faults or fracturing by young craters can expose volatile-bearing material or provide access of volatiles to the surface (Thomas et al., 2014a).

Spatial concentration of sulfides has been proposed, in two related variants. Vaughan et al. (2012) recognized that, given that Mercury's surface silicate composition is similar to that of komatiites, sulfides would be immiscible in silicate melt. Thus, impact melts could differentiate to produce a substantial flotation layer of sulfide, of a thickness consistent with the uniform depth of hollows observed in craters such as Kertész. Insight from industrial steel production led Helbert et al. (2013) to note that S binds with Mg and Ca in molten ore to form MgS and CaS in the slag that floats on the liquid metal. These authors suggested that rising silicate melts that encounter S-rich materials within Mercury's crust could experience reactions that produce sulfides, and that the sulfide slag might form a layer on top of erupted lavas. Hollows could then form by removal of this sulfide-rich surface material.

The spatial-accumulation scenarios are attractive in some specific cases, such as the extensive hollows on the floors of Sander and Kertész, where differentiation and flotation of a slag-like sulfide layer seems plausible. However, many of the geologic settings where hollows are found (e.g., on central peaks, crater walls, ejecta) would appear to be unlikely locations for accumulation of lava or impact melt followed by creation of a hollow-forming layer by differentiation and flotation. Thus, the existence of a pervasive rock-forming phase that contains volatiles and is susceptible to loss when exposed at the surface of Mercury may be a more generally applicable hypothesis for the formation of hollows than explanations that involve spatial concentration of volatiles. However, it could be that both the presence of a pervasive phase and spatial concentration have been responsible for formation of hollows in different geologic settings. For example, hollows in LRM could form by loss of a pervasive phase, whereas hollows associated with pyroclastic deposits could be related to concentrations of magmatic volatiles, and hollows on some crater floors may have resulted from the loss of a phase concentrated by differentiation of the impact melt.

In the case of hollows not associated with impact melt pools or volcanic eruptive products (e.g., Figures 12.6, 12.7, 12.8, 12.12, and 12.13; Section 12.2.3), growth of the hollows presumably involves loss of a phase that is present throughout the rocks, and at concentrations large enough that macroscopic effects on the landscape result. If the volatile-bearing phase is inherent to the LRM and LBP, it must be explained why hollows are not present everywhere that these units are exposed at the surface. Blewett et al. (2016) proposed that hollows grow in depth until a lag deposit of devolatilized material develops to a thickness sufficient to protect the volatile-bearing substrate from the agents that cause loss of the volatiles. Lateral growth can continue via scarp retreat, with loss taking place from the walls. Thus, the regolith that covers the LRM and LBP may be devolatilized, implying that material underneath may be richer in volatiles than indicated by orbital sensing. If the lag is disrupted, for example by small impacts, then growth may be restarted at a location that had been dormant. Mass wasting on steep topographic slopes would inhibit accumulation of a lag. This reasoning could account for the common occurrence of hollows on steeply sloped locations such as crater walls (Figure 12.6), vent walls (Figure 12.7), and crater central peaks (Figures 12.3, 12.4, 12.5, 12.6, and 12.13).

Given the current state of knowledge, the leading candidate phases that are susceptible to loss are CaS, MgS, and graphite, all of which may be abundant in and may contribute to the low reflectance of the LRM and LBP, the color unit in which hollows form (see Sections 12.4.3 and 12.6.2).

12.6.4 Origin of Bright Interiors and Haloes

As discussed in Section 12.3, a developmental sequence of hollows has been inferred (Figure 12.17), i.e., those with high-reflectance interiors and haloes are active, those with only bright interiors have waning activity, and those for which the interiors match their surroundings have ceased growing. Here we consider the cause of the high-reflectance interiors and haloes.

Blewett et al. (2013) listed four potential causes for the (relatively) high reflectance associated with hollows: the presence or exposure of compositionally different alteration products, presence of vapor-deposited coatings, destruction of a darkening agent, or physical differences (e.g., smaller particle size, or special texture or scattering behavior). Spectral differences suggest a fifth potential cause, exposure of optically fresh regolith particles less affected by space weathering. These candidate causes apply to the interiors of the hollows. For the haloes, the list would seem to be more limited. For example, vaporized material generated in the interior could be deposited on the surroundings, creating the high-reflectance haloes. This vapor-deposited material could also have contrasting physical characteristics that produce high reflectance.

The hollow-forming process may involve generation of dust, and thermal fracturing (e.g., Molaro and Byrne, 2012) could enhance crumbling of remnant matrix and aid in exposure of fresh volatile-bearing material. The observation of talus slopes that appear to be smooth and fine-grained at the scale of the images (a few tens of meters per pixel) at the base of hollows-bearing peak-ring mountains, e.g., in Raditladi (Figure 7 of Blewett et al., 2013), supports the idea that creation of fine-grained material is involved in the formation of hollows. The phase-ratio image analysis of Blewett et al. (2014) indicates that the bright haloes around hollows in Eminescu crater have a photometric behavior consistent with finer particle sizes or scattering characteristics that differ from those in non-hollows background areas (Figure 8.5). The fine-grained material remaining after loss of the volatile phase would be fresh and might have a relatively high reflectance as a consequence of having less exposure to darkening by space-weathering processes.

Calculations based on the physics of dust movement on comets (e.g., Whipple, 1951; Meech and Svoren, 2004) show that expected fluxes of sublimating gas are not capable of lofting dust grains in Mercury's high surface gravitational acceleration (Blewett et al., 2016), so simple sublimation is unlikely to be driving deposition of dust in the bright haloes. Electrostatic levitation and motion of dust, which are of interest on the Moon (e.g., Colwell et al., 2007), could have a role in dispersing dust generated within the hollows onto the surroundings, producing the bright haloes. Other means by which the bright haloes could develop include recondensation of less-volatile components liberated during formation of the hollows, or physical modification or chemical alteration of the surface by re-deposited sublimation products.

12.7 CONCLUSIONS

Hollows are a landform unique to Mercury, with no close equivalent on other planetary bodies. However, clues to understanding hollows come from consideration of morphological features associated with ice-bearing surfaces on Mars and icy satellites, and of processes leading to loss of sulfur from ordinary chondrite asteroids. Hollows are a geological manifestation of the planet's high volatile content, consistent with the relatively high surface abundances of potassium, sodium, sulfur, chlorine, and sulfur discovered by MESSENGER, and with the magmatic gas contents needed to drive volcanic eruptions that produced the pyroclastic deposits observed on Mercury.

On the basis of currently available evidence, the following conclusions regarding hollows may be drawn: (1) Hollows are among the youngest non-impact features on Mercury. (2) Hollows form when sublimation or destruction of a volatile-bearing phase weakens the host rock, causing crumbling, collapse, and scarp retreat. The phase susceptible to loss may be calcium sulfide and/or magnesium sulfide or graphite. (3) Solar heating, exposure to solar ultraviolet radiation, exposure to the solar wind, sputtering by magnetospheric ions, and micrometeoroid bombardment may all play roles in driving loss of the volatile component. (4) The depth to which hollows grow may be controlled by accumulation of a lag deposit that protects against further volatile loss. (5) The volatile-bearing phase that is lost appears to be a pervasive component of the host rock (LRM or LBP), but in some cases the hollow-forming phase may have been concentrated by volcanic processes or differentiation of melts.

REFERENCES

Ashworth, D. G. (1977). Lunar and planetary impact erosion. In *Cosmic Dust*, ed. J. A. M. McDonnell. Hoboken, NJ: Wiley, pp. 427–526.

Banks, M. E., Xiao, Z., Braden, S. E., Marchi, S., Chapman, C. R., Barlow, N. G. and Fassett, C. I. (2016). Revised age constraints for Mercury's Kuiperian and Mansurian systems. *Lunar Planet. Sci.*, **47**, abstract 2943.

Bennett, C. J., McLain, J. L., Sarantos, M., Gann, R. D., DeSimone, A. and Orlando, T. M. (2016). Investigating potential sources of Mercury's exospheric calcium: Photon-stimulated desorption of calcium sulfide. *J. Geophys. Res. Planets*, **121**, 137–146, doi:10.1002/2015JE004966.

Blewett, D. T., Hawke, B. R., Lucey, P. G. and Robinson, M. S. (2007). A Mariner 10 color study of Mercurian craters. *J. Geophys. Res.*, 112, E02005, doi:10.1029/2006JE002713.

Blewett, D. T., Robinson, M. S., Denevi, B. W., Gillis-Davis, J. J., Head, J. W., Solomon, S. C., Holsclaw, G. M. and McClintock, W. E. (2009). Multispectral images of Mercury from the first MESSENGER flyby: Analysis of global and regional color trends. *Earth Planet. Sci. Lett.*, **285**, 272–282.

Blewett, D. T., Denevi, B. W., Robinson, M. S., Ernst, C. M., Purucker, M. E. and Solomon, S. C. (2010). The apparent lack of lunar-like swirls on Mercury: Implications for the formation of lunar swirls and for the agent of space weathering. *Icarus*, **209**, 239–246.

Blewett, D. T., Chabot, N. L., Denevi, B. W., Ernst, C. M., Head, J. W., Izenberg, N. R., Murchie, S. L., Solomon, S. C., Nittler, L. R., McCoy, T. J., Xiao, Z., Baker, D. M. H., Fassett, C. I., Braden, S. E., Oberst, J., Scholten, F., Preusker, F. and Hurwitz, D. M. (2011a). Hollows on Mercury: MESSENGER evidence for geologically recent volatile-related activity. *Science*, **333**, 1856–1859, doi:10.1126/science.1211681.

Blewett, D. T., Coman, E. I., Hawke, B. R., Gillis-Davis, J. J., Purucker, M. E. and Hughes, C. G. (2011b). Lunar swirls: Examining crustal magnetic anomalies and space weathering trends. *J. Geophys. Res.*, **116**, E02002, doi:10.1029/2010JE004656.

Blewett, D. T., Vaughan, W. M., Xiao, Z., Chabot, N. L., Denevi, B. W., Ernst, C. M., Helbert, J., D'Amore, M., Maturilli, A., Head, J. W. and Solomon, S. C. (2013). Mercury's hollows: Constraints on formation and composition from analysis of geological setting and spectral reflectance. *J. Geophys. Res. Planets*, **118**, 1013–1032, doi:10.1029/2012JE004174.

Blewett, D. T., Levy, C. L., Chabot, N. L., Denevi, B. W., Ernst, C. M. and Murchie, S. L. (2014). Phase-ratio images of the surface of Mercury: Evidence for differences in sub-resolution texture. *Icarus*, **242**, 142–148.

Blewett, D. T., Stadermann, A. C., Susorney, H. C., Ernst, C. M., Xiao, Z., Chabot, N. L., Denevi, B. W., Murchie, S. L., McCubbin, F. M., Kinczyk, M. J., Gillis-Davis, J. J. and Solomon, S. C. (2016). Analysis of MESSENGER high-resolution images of Mercury's hollows and implications for hollow formation. *J. Geophys. Res. Planets*, **121**, 1798–1813, doi:10.1002/2016JE005070.

Braden, S. E. and Robinson, M. S. (2013). Relative rates of optical maturation of regolith on Mercury and the Moon. *J. Geophys. Res. Planets*, **118**, 1903–1914, doi:10.1002/jgre.20143.

Braden, S. E., Stopar, J. D., Robinson, M. S., Lawrence, S. J., van der Bogert, C. H. and Hiesinger, H. (2014). Evidence for basaltic volcanism on the Moon within the past 100 million years. *Nature Geosci.*, **7**, 787–791, doi:10.1038/ngeo2252.

Burbine, T. H., McCoy, T. J., Nittler, L. R., Benedix, G. K., Cloutis, A. and Dickinson, T. L. (2002). Spectra of extremely reduced assemblages: Implications for Mercury. *Meteorit. Planet. Sci.*, **37**, 1233–1244.

Byrne, S. and Ingersoll, A. P. (2003). A sublimation model for martian south polar ice features. *Science*, 299, 1051–1053.

Chen, A. Y. K., Davis, J. W. and Haasz, A. A. (1999). Chemical erosion of graphite under simultaneous O^+ and H^+ irradiation. *J. Nucl. Mater.*, **266–269**, 399–405.

Chen, A. Y. K., Davis, J. W. and Haasz, A. A. (2001). Methane formation in graphite and boron-doped graphite under simultaneous O^+ and H^+ irradiation. *J. Nucl. Mater.*, **290–293**, 61–65.

Colwell, J. E., Batiste, S., Horányi, M., Robertson, S. and Sture, S. (2007). Lunar surface: Dust dynamics and regolith mechanics. *Rev. Geophys.*, **45**, RG2006, doi:10.1029/2005RG000184.

Davies, M. E., Dwornik, S. E., Gault, D. E. and Strom, R. G. (1978). *Atlas of Mercury*, Special Publication SP-423. Washington, DC: National Aeronautics and Space Administration.

Denevi, B. W., Robinson, M. S., Solomon, S. C., Murchie, S. L., Blewett, D. T., Domingue, D. L., McCoy, T. J., Ernst, C. M., Head, J. W., Watters, T. R. and Chabot, N. L. (2009). The evolution of Mercury's crust: A global perspective from MESSENGER. *Science*, **324**, 613–618, doi:10.1126/science.1172226.

Domingue, D. L., Chapman, C. R., Killen, R. M., Zurbuchen, T. H., Gilbert, J. A., Sarantos, M., Benna, M., Slavin, J. A., Schriver, D., Trávníček, P. M., Orlando, T. M., Sprague, A. L., Blewett, D. T., Gillis-Davis, J. J., Feldman, W. C., Lawrence, D. J., Ho, G. C., Ebel, D. S., Nittler, L. R., Vilas, F., Pieters, C. M., Solomon, S. C., Johnson, C. L., Winslow, R. M., Helbert, J., Peplowski, P. N., Weider, S. Z., Mouawad, N., Izenberg, N. R. and McClintock, W. E. (2014). Mercury's weather-beaten surface: Understanding Mercury in the context of lunar and asteroidal space weathering studies. *Space Sci. Rev.*, **181**, 121–214, doi:10.1007/s11214-014-0039-5.

Dundas, C. M., Byrne, S. and McEwen, A. S. (2015). Modeling the development of martian sublimation thermokarst landforms. *Icarus*, **262**, 154–169.

Dzurisin, D. (1977). Mercurian bright patches: Evidence for physio-chemical alteration of surface material? *Geophys. Res. Lett.*, **4**, 383–386, doi:10.1029/GL004i010p00383.

El Baz, F. and Worden, A. W. (1972). Orbital-science photography, Part A: Visual observations from lunar orbit. In *Apollo 15 Preliminary Science Report*, Special Publication SP-289. Washington, DC: National Aeronautics and Space Administration, pp. 25-1–25-25.

Ernst, C. M., Murchie, S. L., Barnouin, O. S., Robinson, M. S., Denevi, B. W., Blewett, D. T., Head, J. W., Izenberg, N. R., Solomon, S. C. and Roberts, J. H. (2010). Exposure of spectrally distinct material by impact craters on Mercury: Implications for global stratigraphy. *Icarus*, **209**, 210–223.

Evans, L. G., Peplowski, P. N., Rhodes, E. A., Lawrence, D. J., McCoy, T. J., Nittler, L. R., Solomon, S. C., Sprague, A. L., Stockstill-Cahill, K. R., Starr, R. D., Weider, S. Z., Boynton, W. V., Hamara, D. K. and Goldsten, J. O. (2012). Major-element abundances on the surface of Mercury: Results from the MESSENGER Gamma-Ray Spectrometer. *J. Geophys. Res.*, **117**, E00L07, doi:10.1029/2012JE004178.

Fassett, C. I. (2016). Ames stereo pipeline-derived digital terrain models of Mercury from MESSENGER stereo imaging. *Planet. Space Sci.*, **134**, 19–28.

Garry, W. B., Robinson, M. S., Zimbelman, J. R., Bleacher, J. E., Hawke, B. R., Crumpler, L. S., Braden, S. E. and Sato, H. (2012). The origin of Ina: Evidence for inflated lava flows on the Moon. *J. Geophys. Res.*, **117**, E00H31, doi:10.1029/2011JE003981.

Gillis-Davis, J. J., Blewett, D. T., Gaskell, R. W., Denevi, B. W., Robinson, M. S., Strom, R. G., Solomon, S. C. and Sprague, A. L. (2009). Pit-floor craters on Mercury: Evidence of near-surface igneous activity. *Earth Planet. Sci. Lett.*, **285**, 243–250, doi:10.1016/j.epsl.2009.05.023.

Goldsten, J. O., Rhodes, E. A., Boynton, W. V., Feldman, W. C., Lawrence, D. J., Trombka, J. I., Smith, D. M., Evans, L. G., White, J., Madden, N. W., Berg, P. C., Murphy, G. A., Gurnee, R. S., Strohbehn, K., Williams, B. D., Schaefer, E. D., Monaco, C. A., Cork, C. P., Del Eckels, J., Miller, W. O., Burks, M. T., Hagler, L. B., DeTeresa, S. J. and Witte, M. C. (2007). The MESSENGER Gamma-Ray and Neutron Spectrometer. *Space Sci. Rev.*, **131**, 339–391.

Goudge, T. A., Head, J. W., Kerber, L., Blewett, D. T., Denevi, B. W., Domingue, D. L., Gillis-Davis, J. J., Gwinner, K., Helbert, J., Holsclaw, G. M., Izenberg, N. R., Klima, R. L., McClintock, W. E., Murchie, S. L., Neumann, G. A., Smith, D. E., Strom, R. G., Xiao, Z., Zuber, M. T. and Solomon, S. C. (2014). Global inventory and characterization of pyroclastic deposits on Mercury: New insights into pyroclastic activity from MESSENGER orbital data. *J. Geophys. Res. Planets*, **119**, 635–658, doi:10.1002/2013JE004480.

Hapke, B., Danielson, G. E., Jr., Klaasen, K. and Wilson, L. (1975). Photometric observations of Mercury from Mariner 10. *J. Geophys. Res.*, 80, 2431–2443.

Hapke, B. (2012). *Theory of Reflectance and Emittance Spectroscopy*, 2nd edn. New York: Cambridge University Press.

Harmon, J. K. (2007). Radar imaging of Mercury. *Space Sci. Rev.*, 132, 307–349.

Harmon, J. K. (2008). Radar imagery of the southern Caloris region, Mercury. *Icarus*, 196, 298–301.

Harmon, J. K., Slade, M. A., Butler, B. J., Head, J. W., III, Rice, M. S. and Campbell, D. B. (2007). Mercury: Radar images of the equatorial and midlatitude zones. *Icarus*, **187**, 374–405, doi:10.1016/j.icarus.2006.09.026.

Hawke, B. R., Blewett, D. T., Lucey, P. G., Smith, G. A., Bell, J. F., Campbell, B. A. and Robinson, M. S. (2004). The origin of lunar crater rays. *Icarus*, **170**, 1–16, doi:10.1016/j.icarus.2004.02.013.

Hawkins, S. E., III, Boldt, J. D., Darlington, E. H., Espiritu, R., Gold, R. E., Gotwols, B., Grey, M. P., Hash, C. D., Hayes, J. R., Jaskulek, S. E., Kardian, C. J., Jr., Keller, M. R., Malaret, E. R., Murchie, S. L., Murphy, P. K., Peacock, K., Prockter, L. M., Reiter, R. A., Robinson, M. S., Schaefer, E. D., Shelton, R. G., Sterner, R. E., II, Taylor, H. W., Watters, T. R. and Williams, B. D. (2007). The Mercury Dual Imaging System on the MESSENGER spacecraft. *Space Sci. Rev.*, **131**, 247–338.

Head, J. W., Murchie, S. L., Prockter, L. M., Robinson, M. S., Solomon, S. C., Strom, R. G., Chapman, C. R., Watters, T. R., McClintock, W. E., Blewett, D. T. and Gillis-Davis, J. J. (2008). Volcanism on Mercury: Evidence from the first MESSENGER flyby. *Science*, **321**, 69–72, doi:10.1126/science.1159256.

Head, J. W., Murchie, S. L., Prockter, L. M., Solomon, S. C., Chapman, C. R., Strom, R. G., Watters, T. R., Blewett, D. T., Gillis-Davis, J. J., Fassett, C. I., Dickson, J. L., Morgan, G. A. and Kerber, L. (2009). Volcanism on Mercury: Evidence from the first MESSENGER flyby for extrusive and explosive activity and the volcanic origin of plains. *Earth Planet. Sci. Lett.*, **285**, 227–242, doi:10.1016/j.epsl.2009.03.007.

Head, J. W., Chapman, C. R., Strom, R. G., Fassett, C. I., Denevi, B. W., Blewett, D. T., Ernst, C. M., Watters, T. R., Solomon, S. C., Murchie, S. L., Prockter, L. M., Chabot, N. L., Gillis-Davis, J. J., Whitten, J. L., Goudge, T. A., Baker, D. M. H., Hurwitz, D. M., Ostrach, L. R., Xiao, Z., Merline, W. J., Kerber, L., Dickson, J. L., Oberst, J., Byrne, P. K., Klimczak, C. and Nittler, L. R. (2011). Flood volcanism in the northern high latitudes of Mercury revealed by MESSENGER. *Science*, **333**, 1853–1856.

Helbert, J., Maturilli, A. and D'Amore, M. (2013). Visible and near-infrared reflectance spectra of thermally processed synthetic sulfide as a potential analog for the hollow forming materials on Mercury. *Earth Planet. Sci. Lett.*, **369–370**, 233–238.

Howard, A. D. and Moore, J. M. (2008). Sublimation-driven erosion on Callisto: A landform simulation model test. *Geophys. Res. Lett.*, **35**, L03203, doi:10.1029/2007GL032618.

Izenberg, N. R., Klima, R. L., Murchie, S. L., Blewett, D. T., Holsclaw, G. M., McClintock, W. E., Malaret, E., Mauceri, C., Vilas, F., Sprague, A. L., Helbert, J., Domingue, D. L., Head, J. W., Goudge, T. A., Solomon, S. C., Hibbitts, C. A. and Dyar, M. D. (2014). The low-iron, reduced surface of Mercury as seen in spectral reflectance by MESSENGER. *Icarus*, **228**, 364–374.

Kargel, J. S. (2013). Mercury's hollows: Chalcogenide pyrothermokarst analog of thermokarst on Earth, Mars, and Titan. *Lunar Planet. Sci.*, **44**, abstract 2840.

Kaydash, V., Shkuratov, Y. and Videen, G. (2012). Phase-ratio imagery as a planetary remote-sensing tool. *J. Quant. Spectrosc. Radiat. Transf.*, **113**, 2601–2607.

Keller, M. R., Ernst, C. M., Denevi, B. W., Murchie, S. L., Chabot, N. L., Becker, K. J., Hash, C. D., Domingue, D. L. and Sterner, R. E., II (2013). Time-dependent calibration of MESSENGER's wide-angle camera following a contamination event. *Lunar Planet. Sci.*, **44**, abstract 2489.

Kerber, L., Head, J. W., Solomon, S. C., Murchie, S. L., Blewett, D. T. and Wilson, L. (2009). Explosive volcanic eruptions on Mercury: Eruption conditions, magma volatile content, and implications for interior volatile abundances. *Earth Planet. Sci. Lett.*, **285**, 263–271, doi:10.1016/j.epsl.2009.04.037.

Kerber, L., Head, J. W., Blewett, D. T., Solomon, S. C., Wilson, L., Murchie, S. L., Robinson, M. S., Denevi, B. W. and Domingue, D. L. (2011). The global distribution of pyroclastic deposits on Mercury: The view from MESSENGER flybys 1–3. *Planet. Space Sci.*, **59**, 1895–1909, doi:10.1016/j.pss.2011.03.020.

Killen, R. M. (2003). Depletion of sulfur on the surface of the asteroid Eros and the Moon. *Meteorit. Planet. Sci.*, **38**, 383–388.

Klima, R. L., Blewett, D. T., Denevi, B. W., Ernst, C. M., Frank, E. A., Head, J. W., III, Izenberg, N. R., Murchie, S. L., Nittler, L. R., Peplowski, P. N. and Solomon, S. C. (2016). Global distribution and spectral properties of low-reflectance material on Mercury. *Lunar Planet. Sci.*, **47**, abstract 1195.

Kracher, A. and Sears, D. W. G. (2005). Space weathering and the low sulfur abundance of Eros. *Icarus*, **174**, 36–45, doi:10.1016/j.icarus.2004.10.010.

Loeffler, M. J., Dukes, C. A., Chang, W. Y., McFadden, L. A. and Baragiola, R. A. (2008). Laboratory simulations of sulfur depletion at Eros. *Icarus*, **195**, 622–629, doi:10.1016/j.icarus.2008.02.002.

Lucey, P. G. and Riner, M. A. (2011). The optical effects of small iron particles that darken but do not redden: Evidence of intense space weathering on Mercury. *Icarus*, **212**, 451–462, doi:10.1016/j.icarus.2011.01.022.

Malin, M. C. (1987). Abrasion in ice-free areas of southern Victoria Land, Antarctica. *Antarct. J. U.S.*, **22**, 38.

Malin, M. C., Caplinger, M. A. and Davis, S. D. (2001). Observational evidence for an active surface reservoir of solid carbon dioxide on Mars. *Science*, **294**, 2146–2148.

Marchi, S., Chapman, C. R., Fassett, C. I., Head, J. W., Bottke, W. F. and Strom, R. G. (2013). Global resurfacing of Mercury 4.0–4.1 billion years ago by heavy bombardment and volcanism. *Nature*, **499**, 59–61.

McClintock, W. E. and Lankton, M. R. (2007). The Mercury Atmospheric and Surface Composition Spectrometer for the MESSENGER mission. *Space Sci. Rev.*, **131**, 481–522.

McClintock, W. E., Vervack, Jr., R. J., Bradley, E. T., Killen, R. M., Mouawad, N., Sprague, A. L., Burger, M. H., Solomon, S. C. and Izenberg, N. R. (2009). MESSENGER observations of Mercury's exosphere: Detection of magnesium and distribution of constituents. *Science*, **324**, 610–613.

Meech, K. and Svoren, J. (2004). Using cometary activity to trace the physical and chemical evolution of cometary nuclei. In *Comets II*, ed. M. C. Festou, H. Uwe Keller and H. A. Weaver. Tucson, AZ: University of Arizona Press, pp. 317–335.

Molaro, J. and Byrne, S. (2012). Rates of temperature change of airless landscapes and implications for thermal stress weathering. *J. Geophys. Res.*, **117**, E10011, doi:10.1029/2012JE004138.

Moore, J. M. and Spencer, J. R. (1990). Koyaanismuuyaw: The hypothesis of a perennially dichotomous Triton. *Geophys. Res. Lett.*, **17**, 1757–1760.

Moore, J. M., Mellon, M. T. and Zent, A. P. (1996). Mass wasting and ground collapse in terrains of volatile-rich deposits as a Solar System-wide geological process: The pre-Galileo view. *Icarus*, **122**, 63–78.

Moore, J. M., Asphaug, E., Morrison, D., Spencer, J. R., Chapman, C. R., Bierhaus, B., Sullivan, R. J., Chuang, F. C., Klemaszewski, J. E., Greeley, R., Bender, K. C., Geissler, P. E., Helfenstein, P. and Pilcher, C. B. (1999). Mass movement and landform degradation on the icy Galilean satellites: Results of the Galileo nominal mission. *Icarus*, **140**, 294–312.

Murchie, S. L., Watters, T. R., Robinson, M. S., Head, J. W., Strom, R. G., Chapman, C. R., Solomon, S. C., McClintock, W. E., Prockter, L. M., Domingue, D. L. and Blewett, D. T. (2008). Geology of the Caloris basin, Mercury: A view from MESSENGER. *Science*, **321**, 73–76.

Murchie, S. L., Klima, R. L., Denevi, B. W., Ernst, C. M., Keller, M. R., Domingue, D. L., Blewett, D. T., Chabot, N. L., Hash, C. D., Malaret, E., Izenberg, N. R., Vilas, F., Nittler, L. R., Gillis-Davis, J. J., Head, J. W. and Solomon, S. C. (2015). Orbital multispectral mapping of Mercury with the MESSENGER Mercury Dual Imaging System: Evidence for the origins of plains units and low-reflectance material. *Icarus*, **254**, 287–305.

Murray, B. C., Belton, M. J. S., Danielson, G. E., Davies, M. E., Gault, D. E., Hapke, B., O'Leary, B., Strom, R. G., Suomi, V. and Trask, N. (1974). Mercury's surface: Preliminary description and interpretation from Mariner 10 pictures. *Science*, **185**, 169–179.

Nittler, L. R., Starr, R. D., Lim, L., McCoy, T. J., Burbine, T. H., Reedy, R. C., Trombka, J. I., Gorenstein, P., Squyres, S. W., Boynton, W. V., McClanahan, T. P., Bhangoo, J. S., Clark, P. E., Murphy, M. E. and Killen, R. (2001). X-ray fluorescence measurements of the surface elemental composition of Asteroid 433 Eros. *Meteorit. Planet. Sci.*, **36**, 1673–1695.

Nittler, L. R., Starr, R. D., Weider, S. Z., McCoy, T. J., Boynton, W. V., Ebel, D. S., Ernst, C. M., Evans, L. G., Goldsten, J. O., Hamara, D. K., Lawrence, D. J., McNutt, R. L., Jr., Schlemm, C. E., II, Solomon, S. C. and Sprague, A. L. (2011). The major-element composition of Mercury's surface from MESSENGER X-ray spectrometry. *Science*, **333**, 1847–1850.

Peplowski, P. N., Lawrence, D. J., Evans, L. G., Klima, R. L., Blewett, D. T., Goldsten, J. O., Murchie, S. L., McCoy, T. J., Nittler, L. R., Solomon, S. C., Starr, R. D. and Weider, S. Z. (2015). Constraints on the abundance of carbon in near-surface materials on Mercury: Results from the MESSENGER Gamma-Ray Spectrometer. *Planet. Space Sci.*, **108**, 99–107, doi:10.1016/j.pss.2015.01.008.

Peplowski, P. N., Klima, R. L., Lawrence, D. J., Ernst, C. M., Denevi, B. W., Frank, E. A., Goldsten, J. O., Murchie, S. L., Nittler, L. R. and Solomon, S. C. (2016). Remote sensing evidence for an ancient carbon-bearing crust on Mercury. *Nature Geosci.*, **9**, 273–276, doi:10.1038/NGEO2669.

Qiao, L., Head, J., Wilson, L., Xiao, L., Kreslavsky, M. and Dufek, J. (2017). Ina pit crater on the Moon: Extrusion of waning-stage lava lake magmatic foam results in extremely young crater retention ages. *Geology*, **45**, 455–458.

Qiao, L., Head, J. W., Xiao, L., Wilson, L. and Dufek, J. D. (2018). The role of substrate characteristics in producing anomalously young crater retention ages in volcanic deposits on the Moon: Morphology, topography, subresolution roughness, and mode of emplacement of the Sosigenes lunar irregular mare patch. *Meteorit. Planet. Sci*, **53**, 778–812.

Rava, B. and Hapke, B. (1987). An analysis of the Mariner 10 color ratio map of Mercury. *Icarus*, **71**, 397–429, doi:10.1016/j.bbr.2011.03.031.

Riner, M. A., Lucey, P. G., Desch, S. J. and McCubbin, F. M. (2009). Nature of opaque components on Mercury: Insights into a mercurian magma ocean. *Geophys. Res. Lett.*, **36**, L02201, doi:10.1029/2008GL036128.

Robinson, M. S. and Lucey, P. G. (1997). Recalibrated Mariner 10 color mosaics: Implications for mercurian volcanism. *Science*, **275**, 197–200, doi:10.1126/science.275.5297.197.

Robinson, M. S. and Taylor, G. J. (2001). Ferrous oxide in Mercury's crust and mantle. *Meteorit. Planet. Sci*, **36**, 841–847.

Robinson, M. S., Murchie, S. L., Blewett, D. T., Domingue, D. L., Hawkins, S. E., III, Head, J. W., Holsclaw, G. M., McClintock, W. E., McCoy, T. J., McNutt, R. L., Jr., Prockter, L. M., Solomon, S. C. and Watters, T. R. (2008). Reflectance and color variations on Mercury: Regolith processes and compositional heterogeneity. *Science*, **321**, 66–69, doi:10.1126/science.1160080.

Savvatimskiy, A. I. (2005). Measurements of the melting point of graphite and the properties of liquid carbon (a review for 1963–2003). *Carbon*, **43**, 1115–1142.

Schlemm, C. E., II, Starr, R. D., Ho, G. C., Bechtold, K. E., Hamilton, S. A., Boldt, J. D., Boynton, W. V., Bradley, W., Fraeman, M. E., Gold, R. E., Goldsten, J. O., Hayes, J. R., Jaskulek, S. E., Rossano, E., Rumpf, R. A., Schaefer, E. D., Strohbehn, K., Shelton, R. G., Thompson, R. E., Trombka, J. I. and Williams, B. D. (2007). The X-Ray Spectrometer on the MESSENGER spacecraft. *Space Sci. Rev.*, **131**, 393–415.

Schultz, P. H. (1977). Endogenic modification of impact craters on Mercury. *Phys. Earth Planet. Inter.*, **15**, 202–219, doi:10.1016/j.bbr.2011.03.031.

Schultz, P. H., Staid, M. I. and Pieters, C. M. (2006). Lunar activity from recent gas release. *Nature*, **444**, 184–186.

Sharp, R. P. (1973). Mars: South polar pits and etched terrain. *J. Geophys. Res*, **78**, 4222–4230.

Smith, B. A., Soderblom, L. A., Banfield, D., Barnet, C., Basilevsky, A. T., Beebe, R. F., Bollinger, K., Boyce, J. M., Brahic, A., Briggs, G. A., Brown, R. H., Chyba, C., Collins, S. A., Colvin, T., Cook, A. F., II, Crisp, D., Croft, S. K., Cruikshank, D., Cuzzi, J. N., Danielson, G. E., Davies, M. E., De Jong, E., Dones, L., Godfrey, D., Goguen, J., Grenier, I., Haemmerle, V. R., Hammel, H., Hansen, C. J., Helfenstein, C. P., Howell, C., Hunt, G. E., Ingersoll, A. P., Johnson, T. V., Kargel, J., Kirk, R., Kuehn, D. I., Limaye, S., Masursky, H., McEwen, A., Morrison, D., Owen, T., Owen, W., Pollack, J. B., Porco, C. C., Rages, K., Rogers, P., Rudy, D., Sagan, C., Schwartz, J., Shoemaker, E. M., Showalter, M., Sicardy, B., Simonelli, D., Spencer, J., Sromovsky, L. A., Stoker, C., Strom, R. G., Suomi, V. E., Synott, S. P., Terrile, R. J., Thomas, P., Thompson, W. R., Verbiscer, A. and Veverka, J. (1989). Voyager 2 at Neptune: Imaging science results. *Science*, **246**, 1422–1449.

Solomon, S. C., McNutt, R. L., Jr., Watters, T. R., Lawrence, D. J., Feldman, W. C., Head, J. W., Krimigis, S. M., Murchie, S. L., Phillips, R. J., Slavin, J. A. and Zuber, M. T. (2008). Return to Mercury: A global perspective on MESSENGER's first Mercury flyby. *Science*, **321**, 59–62.

Spudis, P. D. and Guest, J. E. (1988). Stratigraphy and geologic history of Mercury. In *Mercury*, ed. F. Vilas, C. R. Chapman and M. S. Matthews. Tucson, AZ: University of Arizona Press, pp. 118–164.

Strain, P. and El Baz, F. (1980). The geology and morphology of Ina. *Proc. Lunar Planet. Sci. Conf.*, **11**, 2437–2446.

Strom, R. G. (1984). Mercury. In *The Geology of the Terrestrial Planets*, ed. M. H. Carr., Special Publication SP-469. Washington, DC: National Aeronautics and Space Administration.

Strom, R. G., Chapman, C. R., Merline, W. J., Solomon, S. C. and Head, J. W. (2008). Mercury cratering record viewed from MESSENGER's first flyby. *Science*, **321**, 79–81, doi:10.1126/science.1159317.

Thomas, P. C., Malin, M. C., Edgett, K. S., Carr, M. H., Hartmann, W. K., Ingersoll, A. P., James, P. B., Soderblom, L. A., Veverka, J. and Sullivan, R. (2000). North–south geological differences between the residual polar caps on Mars. *Nature*, **404**, 161–164.

Thomas, R. J., Rothery, D. A., Conway, S. J. and Anand, M. (2014a). Hollows on Mercury: Materials and mechanisms involved in their formation. *Icarus*, **229**, 221–235, doi:10.1016/j.icarus.2013.11.018.

Thomas, R. J., Rothery, D. A., Conway, S. J. and Anand, M. (2014b). Long-lived explosive volcanism on Mercury. *Geophys. Res. Lett.*, **44**, 6084–6092, doi:10.1002/2014GL061224.

Thomas, R. J., Rothery, D. A., Conway, S. J. and Anand, M. (2014c). Mechanisms of explosive volcanism on Mercury: Implications from its global distribution and morphology. *J. Geophys. Res. Planets*, **119**, 2239–2254, doi:10.1002/2014JE004692.

Thomas, R. J., Hynek, B. M., Rothery, D. A. and Conway, S. J. (2016), Mercury's low-reflectance material: Constraints from hollows. *Icarus*, **277**, 455–465.

Vander Kaaden, K. E. and McCubbin, F. M. (2015). Exotic crust formation on Mercury: Consequences of a shallow, FeO-poor mantle. *J. Geophys. Res. Planets*, **120**, 195–209, doi:10.1002/2014JE004733.

Vaughan, W. M., Helbert, J., Blewett, D. T., Head, J. W., Murchie, S. L., Gwinner, K., McCoy, T. J. and Solomon, S. C. (2012). Hollow-forming layers in impact craters on Mercury: Massive sulfide deposits formed by impact melt differentiation? *Lunar Planet. Sci.*, **43**, abstract 1187.

Vervack, R. J., Jr., McClintock, W. E., Killen, R. M., Sprague, A. L., Anderson, B. J., Burger, M. H., Bradley, E. T., Mouawad, N., Solomon, S. C. and Izenberg, N. R. (2010). Mercury's complex exosphere: Results from MESSENGER's third flyby. *Science*, **329**, 672–675.

Vilas, F., Domingue, D. L., Helbert, J., D'Amore, M., Maturilli, A., Klima, R. L., Stockstill-Cahill, K. R., Murchie, S.L., Izenberg, N. R., Blewett, D. T., Vaughan, W. M. and Head, J. W. (2016), Mineralogical indicators of Mercury's hollows composition in MESSENGER color observations, *Geophys. Res. Lett.*, **43**, 1450–1456, doi:10.1002/2015GL067515.

Weider, S. Z., Nittler, L. R., Starr, R. D., McCoy, T. J., Stockstill-Cahill, K. R., Byrne, P. K., Denevi, B. W., Head, J. W. and Solomon, S. C.

(2012). Chemical heterogeneity on Mercury's surface revealed by the MESSENGER X-Ray Spectrometer. *J. Geophys. Res.*, **117**, E00L05, doi:10.1029/2012JE004153.

Weider, S. Z., Nittler, L. R., Starr, R. D., Crapster-Pregont, E. J., Peplowski, P. N., Denevi, B. W., Head, J. W., Byrne, P. K., Hauck, S. A., Ebel, D. S. and Solomon, S. C. (2015). Evidence for geochemical terranes on Mercury: Global mapping of major elements with MESSENGER's X-Ray Spectrometer. *Earth Planet. Sci. Lett.*, **416**, 109–120, doi:10.1016/j.epsl.2015.01.023.

Weider, S. Z., Nittler, L. R., Murchie, S. L., Peplowski, P. N., McCoy, T. J., Kerber, L., Klimczak, C., Ernst, C. M., Goudge, T. M., Starr, R. D., Izenberg, N. R., Klima, R. L. and Solomon, S. C. (2016). Evidence from MESSENGER for sulfur- and carbon-driven explosive volcanism on Mercury. *Geophys. Res. Lett.*, **43**, 3653–3661, doi:10.1002/2016GL068325.

Whipple, F. L. (1951). A comet model. II. Physical relations for comets and meteors. *Astrophys. J.*, **113**, 464–474.

Whitaker, E. A. (1972). Orbital-science photography, Part N: An unusual mare feature. In *Apollo 15 Preliminary Science Report*, Special Publication SP-289. Washington, DC: National Aeronautics and Space Administration, pp. 25-84–25-85.

White, O. L., Umurhan, O. M., Moore, J. M. and Howard, A. D. (2016). Modeling of ice pinnacle formation on Callisto. *J. Geophys. Res. Planets*, **121**, 21–45.

Xiao, Z., Strom, R. G., Blewett, D. T., Chapman, C. R., Denevi, B. W., Head, J. W., Fassett, C. I., Braden, S. E., Gwinner, K., Solomon, S. C., Murchie, S. L., Watters, T. R. and Banks, M. E. (2012). The youngest geologic terrains on Mercury. *Lunar Planet. Sci.*, **43**, abstract 2143.

Xiao, Z., Strom, R. G., Blewett, D. T., Byrne, P. K., Solomon, S. C., Murchie, S. L., Sprague, A. L., Domingue, D. L. and Helbert, J. (2013). Dark spots on Mercury: A distinctive low-reflectance material and its relation to hollows. *J. Geophys. Res. Planets*, **118**, 1752–1765.

13

Mercury's Polar Deposits

NANCY L. CHABOT, DAVID J. LAWRENCE, GREGORY A. NEUMANN, WILLIAM C. FELDMAN, AND DAVID A. PAIGE

13.1 INTRODUCTION

Two and a half decades ago, the discovery of Mercury's polar deposits provided the first evidence for the possibility of water ice on the solar system's innermost planet. Radar observations in 1991 with the Goldstone 70-m radio antenna as the transmitter and the Very Large Array as the receiving instrument revealed a highly reflective region near Mercury's north pole with a high same-sense to opposite-sense circular polarization ratio (Slade et al., 1992). The high reflectivity and circular polarization ratio of the observed radar-bright feature resembled the distinctive radar characteristics of the icy Galilean satellites (Campbell et al., 1978; Ostro et al., 1980; Hapke, 1990; Hapke and Blewett, 1991) and at the south polar cap of Mars (Muhleman et al., 1991), and so were interpreted as evidence for ice in Mercury's north polar region. The north polar radar-bright feature was subsequently identified in Arecibo observations, confirming the Goldstone discovery, and soon afterwards Arecibo observations led to the discovery of a south polar radar-bright feature as well (Harmon and Slade, 1992). In particular, the south polar radar-bright feature was estimated to lie within the large impact crater Chao Meng-Fu, suggesting that the radar-bright material favored the shadowed, low-temperature environment of this crater's interior, a setting consistent with the presence of long-lived water ice on Mercury. Evidence for the presence of water ice was also supported by thermal model calculations, which showed that the maximum temperatures in permanently shadowed regions within high-latitude craters on Mercury could provide thermal environments where water ice is stable for billions of years (Paige et al., 1992; Ingersoll et al., 1992). Analysis of the radar data further indicated that the water ice deposits are at least several meters in depth and that the ice layer itself must be very pure, with less than about 5% silicates by volume, leading to the suggestion that the ice was deposited on Mercury over a relatively short period of time (Butler et al., 1993).

Water ice within permanently shadowed cold traps had previously been proposed on theoretical grounds for the Moon (Urey, 1952; Watson et al., 1961) and also for Mercury (Thomas, 1974), but the Earth-based radar measurements of Mercury provided the first observational evidence to support the presence of such deposits. In contrast, Earth-based radar observations of the Moon did not reveal similar polar deposits (Stacy et al., 1997). Subsequent Earth-based radar measurements greatly improved the spatial resolution of the Mercury radar images, resolving many individual radar-bright spots near both of Mercury's poles and confirming the high radar reflectivity and circular polarization ratio of the features (Harmon et al., 1994, 2001, 2011; Harcke, 2005; Harmon, 2007). Many radar-bright deposits were identified as lying within specific impact craters imaged by Mariner 10 in 1974–1975, but many other radar-bright features could not be mapped to craters or other geologic features because Mariner 10 acquired images covering only slightly less than half of Mercury's polar regions (Davies et al., 1978).

In parallel with the acquisition of radar images of Mercury's polar regions at increasingly improved resolution, more detailed modeling refined the quantitative understanding of the thermal environments of candidate host craters. Large permanently shadowed craters near the poles, such as Chao Meng-Fu, were concluded to have maximum surface temperatures conducive to the long-term stability of water ice at the surface (Salvail and Fanale, 1994; Vasavada et al., 1999). In contrast, the many radar-bright host craters located more than about 10° in latitude from the pole were found to have maximum surface temperatures that exceeded 110 K, and under such conditions water ice at the surface would be rapidly lost to thermal sublimation (Vasavada et al., 1999). However, burial of the deposits below a layer tens of centimeters thick was found to insulate water ice deposits to the point that they could remain stable for billions of years (Vasavada et al., 1999). Radar observations acquired at different wavelengths also were suggestive of a thin layer covering the deposits (Harmon et al., 2011). Although a thin layer of overlying regolith could enable water ice deposits to be thermally stable, how the deposits were buried sufficiently rapidly to prevent the complete loss of water ice at expected rates of impact gardening of the regolith was seen to be problematic (Harmon et al., 2001). An alternate suggestion was that a slightly dirty ice deposit might sublimate until its contaminant load formed a thin insulating lag deposit at the surface (Vasavada et al., 1999).

Small, simple craters on Mercury, which have diameters generally <10 km, posed a separate thermal challenge. Thermal models suggested that such craters could not host thermally stable water ice if located more than 2° in latitude from the poles, even if such deposits were buried by an insulating layer (Vasavada et al., 1999), yet Earth-based radar observations identified small craters farther from the pole that host radar-bright deposits (Harmon et al., 2001). Additionally, radar-bright features were discovered at latitudes as far from the pole as 67°N (Harmon et al., 2001), locations at which interior conditions were thought to present potentially challenging thermal environments for the stability of water ice, though the geologic setting of the majority of these lower-latitude deposits could not be identified as they were in the hemisphere of the planet not imaged by Mariner 10.

Overall, Arecibo observations showed that Mercury's radar-bright deposits are extensive, covering ~12,500 km^2 in the north polar region, with comparable deposits also in the south (Harmon et al., 2011), motivating questions about the formation and evolution of the material if it consists mostly of water ice. Simulations showed that water ice could migrate to cold traps at Mercury's poles with ~10% efficiency (Butler, 1997) and suggested that asteroids, comets, or interplanetary dust could have provided sufficient water to account for Mercury's radar-bright deposits, though impacts by a few large comets or asteroids were favored (Moses et al., 1999). From models of burial by regolith gardening, it was concluded that Mercury's polar deposits must have been emplaced geologically recently, within the last 50 Myr, in order that such deposits not be more deeply buried (Crider and Killen, 2005). Materials other than water ice were also suggested for the radar-bright deposits, including sulfur (Sprague et al., 1995) and silicates with dielectric properties altered at very low temperatures (Starukhina, 2001). The stark contrast between Mercury's extensive radar-bright deposits and the lack of any similar deposits on the Moon also raised a question. Thus, Earth-based radar observations and Mariner 10 images provided provocative but limited evidence to constrain the nature of Mercury's radar-bright polar deposits.

One of the six major science questions that motivated the MESSENGER mission (Chapter 1) was "What are the radar-reflective materials at Mercury's poles?" In the course of the mission's more than four Earth years of operations in orbit about Mercury, MESSENGER acquired multiple data sets to address this question. In this chapter, we review MESSENGER's observations of Mercury's polar deposits and discuss the resulting implications.

13.2 MESSENGER OBSERVATIONS OF MERCURY'S POLAR DEPOSITS

13.2.1 Mapping Results and Illumination Conditions

Whereas MESSENGER's three flybys of Mercury were equatorial and did not provide new views of Mercury's polar regions, upon entering orbit about Mercury, MESSENGER provided complete imaging of Mercury's surface with its Mercury Dual Imaging System (MDIS) (Hawkins et al., 2007). This global imaging enabled the first complete identification of the geologic features that host radar-bright deposits.

One of the earliest studies completed in orbit regarding Mercury's polar regions was a mapping of regions of permanent shadow near Mercury's south pole (Chabot et al., 2012). The MESSENGER spacecraft had a highly eccentric orbit about Mercury, which during the mission's first year of orbital operations had a minimum altitude of ~200 km in Mercury's northern hemisphere and a maximum altitude of 15,000 km in the southern hemisphere (Chapter 1). At the higher altitudes, MDIS's wide-angle camera (WAC) was able to capture the entire sunlit region from Mercury's south pole to a latitude of ~73°S in a single image. Such imaging was repeated approximately every other Earth day, for an entire Mercury solar day (176 Earth days), producing a data set of 89 images that captured Mercury's south polar region at a spatial scale of 1.7 km/pixel under all illumination conditions. By thresholding each image into sunlit and shadowed regions, an illumination map of Mercury's south polar region was created, as shown in Figure 13.1.

Many regions of permanent shadow from 73°S to the south pole are evident in Figure 13.1, covering a total area of ~43,000 km^2. It is Mercury's low obliquity of 2.04 arcminutes (0.034°) (Margot et al., 2012) that accounts for the large area of terrain in shadow during a solar day. Moreover, regions in shadow in a single solar day remain in shadow indefinitely because the obliquity is stabilized by Mercury's 3:2 spin–orbit resonance, and these shadowed regions likely date to the times of formation of their host craters given that Mercury was probably captured into its current spin–orbit resonance early in solar system history (Peale, 1988).

The highest-resolution Earth-based radar image obtained of Mercury's south polar region is also shown in Figure 13.1. These data were acquired on 24–25 March 2005 at the Arecibo Observatory in S-band (12.6-cm wavelength) with a range resolution of 1.5 km (Harmon et al., 2011). All of the Arecibo radar-bright features either collocate with the regions identified as permanently shadowed or are within a few kilometers of an area of permanent shadow, consistent with the registration and resolution limitations of the MDIS and Arecibo data sets (Chabot et al., 2012). As originally identified by Harmon and Slade (1992), the ~180-km-diameter crater Chao Meng-Fu hosts an extensive radar-bright deposit, which is both the largest radar-bright region and the largest permanently shadowed area on the entire planet.

Goldstone X-band (3.5-cm wavelength) radar observations have also been acquired of Mercury's south polar region, with a range resolution of 6 km (Harcke, 2005). Though the Goldstone data are lower in resolution than the Arecibo observations, they were acquired from a viewing direction nearly opposite to that of the 2005 Arecibo image and were used to catalog radar-bright features beyond the Arecibo radar horizon of the 2005 observations. All of the Goldstone radar-bright features also correspond to areas identified as in permanent shadow from the MDIS data set (Chabot et al., 2012).

Given MESSENGER's highly eccentric orbit, images acquired of Mercury's north polar region were at considerably higher resolution than those of the south polar region, allowing detailed identification of geologic features that host radar-bright deposits. However, a single MDIS WAC image could not capture the entire sunlit northern polar region to enable mapping of permanently shadowed regions, as was completed for Mercury's south polar region. Instead, from ~6,500 images of Mercury's north polar region that were acquired during MESSENGER's one-year primary orbital mission, regions of "persistent shadow" – meaning surfaces in shadow during all images acquired – were identified (Chabot et al., 2013). Even from this limited one-year data set, radar-bright features in Mercury's north polar region (Harmon et al., 2011) were found to be associated with locations persistently shadowed in MDIS images. Subsequently, with over 16,000 images from MESSENGER's full orbital mission, Deutsch et al. (2016)

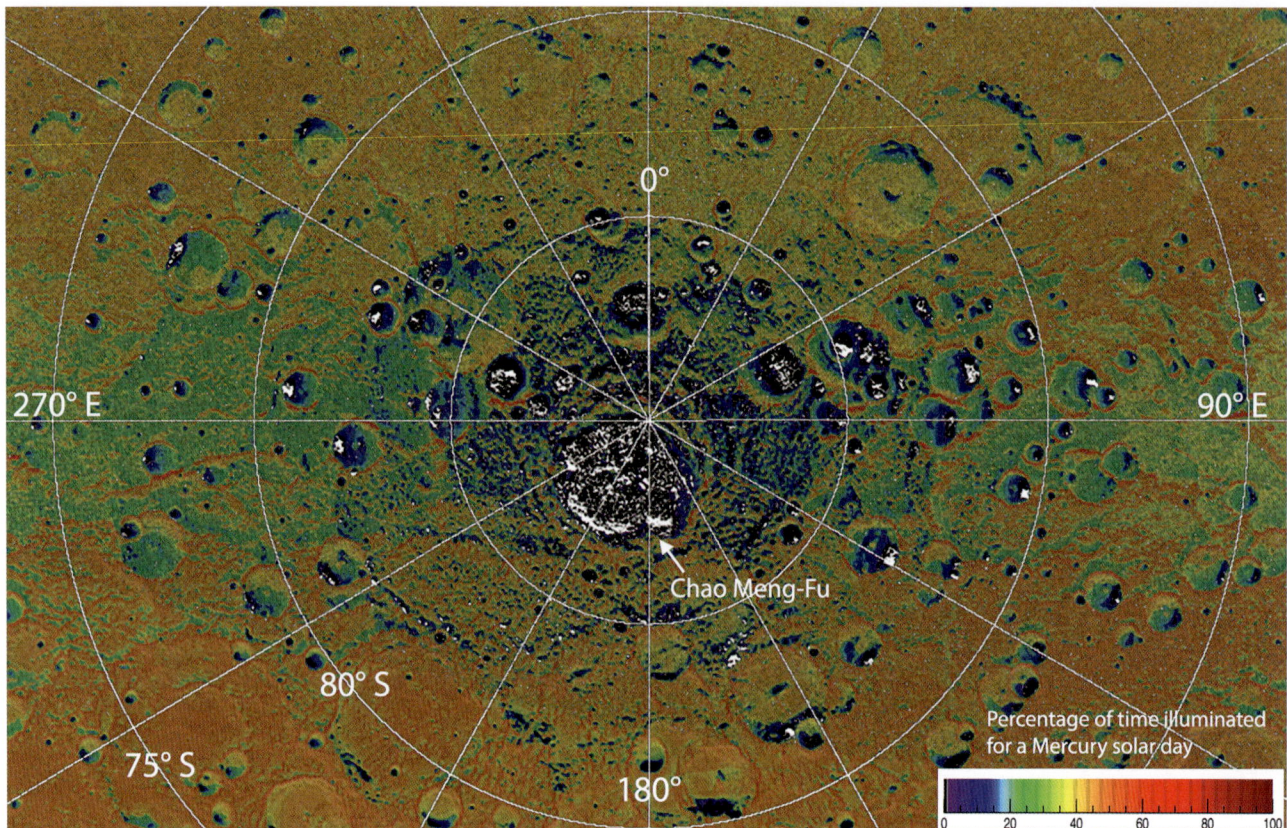

Figure 13.1. Mercury's south polar region colored by the percentage of illumination time (Chabot et al., 2012), with areas of high radar backscatter in the highest-resolution Arecibo radar image (Harmon et al., 2011) overlain in white.

further confirmed the association of radar-bright and persistently shadowed regions and substantially improved the identification of persistently shadowed regions within 4° of the north pole.

Additionally, the lower spacecraft altitude enabled topographic measurements by MESSENGER's Mercury Laser Altimeter (MLA) (Cavanaugh et al., 2007) at northern high latitudes that were not possible at comparable latitudes in the south. The solar illumination at Mercury's north polar region was modeled with MLA topographic data in a manner similar to that used to investigate permanently shadowed regions on the Moon (Mazarico et al., 2011), and regions of permanent shadow were identified from insolation models independent of the imaging data (Mazarico et al., 2014; Deutsch et al., 2016). Figure 13.2 shows the regions modeled as permanently shadowed with an MLA-based elevation model obtained from data acquired over MESSENGER's full orbital mission. There is good agreement between regions mapped as permanently shadowed by MLA and persistently shadowed by MDIS (Deutsch et al., 2016).

Also shown in Figure 13.2 is the highest-resolution radar image yet produced of Mercury's north polar region, with a range resolution of 1.5 km (Harmon et al., 2011). The radar image was assembled from multiple Arecibo observations obtained at S-band from 1999 to 2005, covering a wide range of viewing geometries that serve to reduce regions in radar shadow or beyond the radar horizon, in contrast to the south pole radar image of Figure 13.1 that was obtained from a single observing run. As seen in Figure 13.2, all of the well-defined radar-bright features near Mercury's north pole are associated with regions of permanent shadow.

Although all of the radar-bright features near both of Mercury's poles are associated with permanently shadowed regions, not all permanently shadowed regions also host radar-bright features. For water ice deposits, permanent shadow is a necessary but not sufficient condition to host such material, as the temperature within permanently shadowed regions can still be too high for the long-term stability of water ice, depending on the host crater's shape, latitude, longitude, and other factors (Paige et al., 1992; Vasavada et al., 1999). However, for permanently shadowed regions within complex craters located within 10° of the pole, the thermal environment is expected to be generally conducive to the long-term stability of surface or buried water ice (Vasavada et al., 1999). On Mercury, the transition with increasing diameter between simple and complex crater morphologies for primary craters occurs at a diameter of ~10 km (Pike, 1988; Barnouin et al., 2012; Susorney et al., 2016). For craters >10 km in diameter within 10° of Mercury's north pole that also contain regions of permanent shadow, 65% host radar-bright deposits (Deutsch et al., 2016). Similarly, in the south polar region, most, but not all, shadowed craters host radar-bright deposits (Chabot et al., 2012). In the north polar region, for example, Burke (29-km diameter) and Sapkota (27-km diameter) craters, labeled in Figure 13.2, have interiors that are almost in complete shadow, yet neither hosts an extensive radar-bright deposit, in contrast to similarly sized

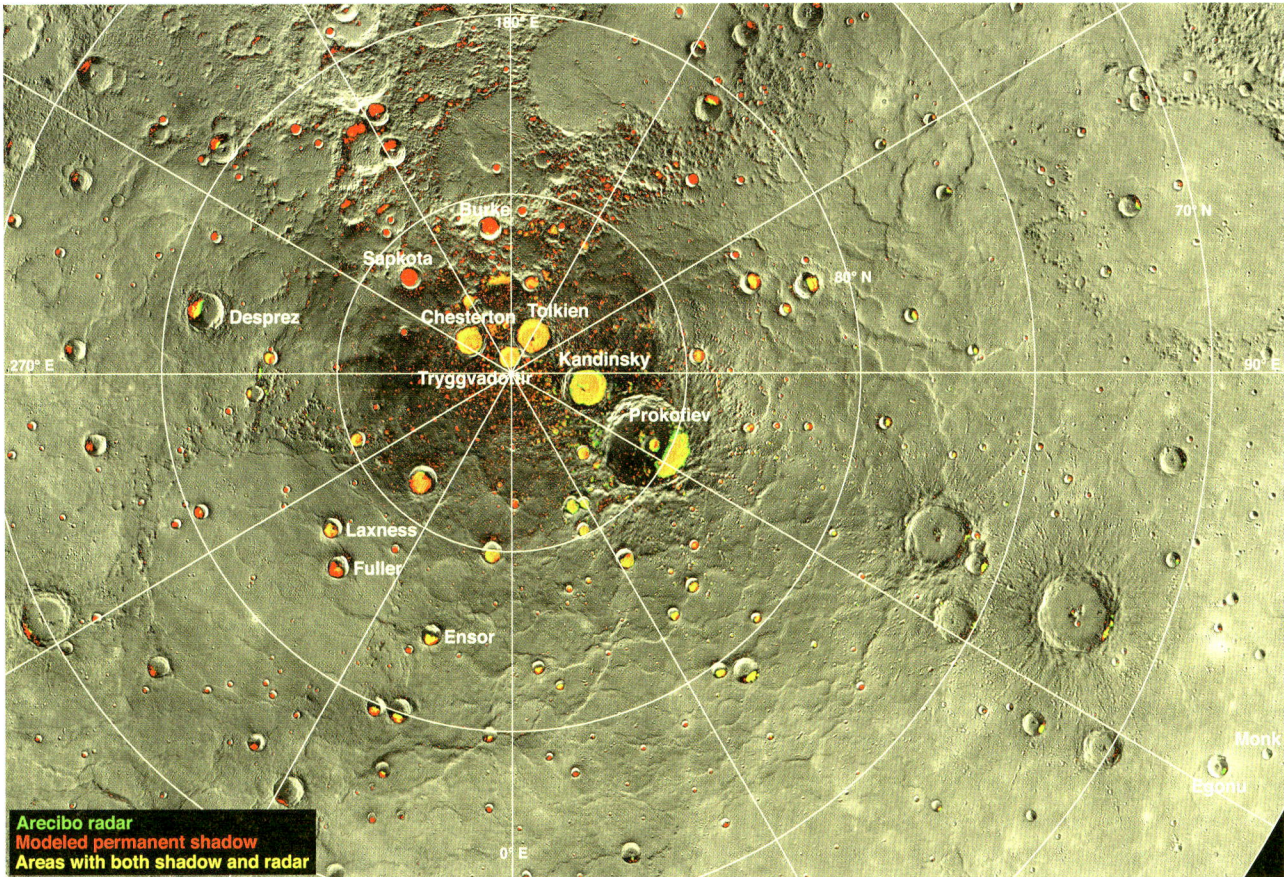

Figure 13.2. A MESSENGER mosaic of Mercury's north polar region, overlain with the highest-resolution Arecibo radar image in green (Harmon et al., 2011) and with a map of regions of permanent shadow in red (Deutsch et al., 2016). Regions that are both in permanent shadow and radar-bright are colored yellow.

neighboring craters. Thus, Burke, Sapkota, and other craters may lack radar-bright deposits, indicating that not all of Mercury's available cold-traps are currently occupied by water ice. Alternatively, such deposits may be present but were not observed during the Arecibo measurements or are buried too deep to be detected. As discussed by Deutsch et al. (2016), consideration of the Arecibo radar viewing opportunities does not support the view that the lack of extensive radar-bright deposits in these craters is a result of limited radar coverage, though the observed longitudinal distribution of permanently shadowed craters that lack radar-bright deposits is not correlated with thermal conditions, possibly suggesting some sort of viewing or detection limitation with the radar observations.

Because of Mercury's 3:2 spin–orbit resonance, the "hot pole" longitudes of 0°E and 180°E always experience local noon at Mercury's perihelion, resulting in a ~130 K higher temperature at the equator than experienced for local noon at the "cold pole" longitudes of 90°E and 270°E (Vasavada et al., 1999). For locations farther than 10° from the poles, there is a preference for radar-bright features to be located along the cold-pole longitudes, as noted from the radar observations (Harmon et al., 2001, 2011). MESSENGER observations have confirmed that for regions >10° latitude from the poles, the radar-bright deposits do not uniformly occupy all permanently shadowed regions but rather show a preference for longitudes near the 90° E and 270°E cold-pole longitudes (Chabot et al., 2012, 2013). At these longitudes, radar-bright features have been identified in craters at latitudes a considerable distance from the poles; in the south, the most equatorward radar-bright deposit is at about 74° S (Figure 13.1), and in the north (Figure 13.2) Egonu (67°N) and Monk (66°N) are the lowest-latitude craters to host radar-bright deposits.

Thermal models prior to MESSENGER measurements indicated that the interiors of simple bowl-shaped craters experience considerably warmer temperatures than flatter-floored complex craters because of increased indirect heating from scattering off crater walls (Vasavada et al., 1999). From a model of an idealized bowl-shaped simple crater, <10 km in diameter with a depth-to-diameter ratio of 5:1, Vasavada et al. (1999) concluded that simple craters on Mercury cannot host long-lived water ice except if located within 2° latitude of the pole. MESSENGER images have shown that there are numerous radar-bright deposits within small craters <10 km in diameter located considerably farther from the poles than 2° in both the north and south polar regions (Chabot et al., 2012, 2013; Ernst et al., 2014; Deutsch et al., 2016), in addition to the few noted previously by Harmon et al. (2001) from Mariner 10 images. In particular, many of the small craters surrounding Prokofiev host radar-bright deposits, and secondary craters on Mercury are generally shallower than primary craters (Susorney et al., 2016). Assessments of the morphology of these small craters from MLA measurements indicate that they are shallower than was assumed for the thermal model of Vasavada et al. (1999) (Ernst et al., 2014), which could affect the thermal stability of water ice in their shadowed interiors.

The thickness of the radar-bright deposits is poorly constrained. A minimum thickness of several meters is implied by the absence of a clear drop in radar cross section with increasing wavelength for observations at 3.5-cm, 12.6-cm and 70-cm wavelengths, indicating that the scattering layer is many wavelengths thick at all three wavelengths (Black et al., 2010). A maximum thickness of a few hundred meters has been estimated from statistical similarities between the depths and diameters of craters that host radar-bright deposits and those that do not, as determined by MLA measurements (Talpe et al., 2012). From a pre-MESSENGER study of 12 craters that host radar-bright deposits, Vilas et al. (2005) concluded that these craters were shallower than normal, but MESSENGER MLA measurements have not supported that result (Talpe et al., 2012), likely illustrating the greater accuracy of laser altimetry compared with topographic information derived from Mariner 10 imaging data. From a digital elevation model (DEM) produced from MLA measurements to compare the topography of six craters that host radar-bright deposits and six craters that do not, Eke et al. (2017) estimated a maximum thickness of the radar-bright deposits of 150 m, with a typical excess height associated with the radar-bright deposits of 50 ± 35 m (one standard deviation, or 1σ). From individual MLA tracks that cross the boundaries of two radar-bright deposits, Susorney et al. (2017) did not detect any statistically significant change in topography associated with the radar-bright deposits in either Prokofiev or Desprez craters, reporting deposit thickness limits of 33 ± 60 m and 45 ± 38 m, respectively (1σ uncertainties). Susorney et al. (2017) concluded that the natural topographic variability of the crater floors, e.g., from younger and smaller impact craters, boulders, and variations in impact melt thickness, may be on the same scale or larger than the thickness of the radar-bright deposits. Additionally, from a characterization of the depths and diameters of nine small impact craters imaged within radar-bright regions, Deutsch et al. (2017b) estimated that the radar-bright deposits in those regions may have a maximum thickness of 24–95 m if the small craters predate the emplacement of the radar-bright deposits. Overall, the fact that these studies have not been able to measure directly the thickness of the radar-bright deposits but rather have placed increasingly smaller limits on the maximum thickness suggest that the thickness of the radar-bright deposits may be toward the lower end of the current estimated range, e.g., tens of meters or even less.

13.2.2 Neutron Spectrometer Measurements

To test the hypothesis that Mercury's polar deposits are composed mostly of water, the MESSENGER spacecraft was equipped with a Neutron Spectrometer (NS) as part of the Gamma-Ray and Neutron Spectrometer instrument (Goldsten et al., 2007). Planetary neutron spectroscopy is a robust technique for measuring hydrogen concentrations on airless or nearly airless planetary bodies (Prettyman, 2007), and the high hydrogen abundance in water ice relative to other candidate radar reflective materials allows neutron spectroscopy to constrain the radar-bright material (Feldman et al., 1997). This technique involves measurement of the neutrons created by nuclear spallation reactions when galactic cosmic rays (GCRs) strike a planetary surface. These neutrons have a broad spectrum of kinetic energy E_n that is typically divided into three energy ranges: thermal ($E_n < 0.4$ eV), epithermal (0.4 eV $< E_n <$ 0.5 MeV), and fast ($E_n > 0.5$ MeV). A hydrogen atom has nearly the same mass as a neutron, which allows a highly efficient momentum transfer from the spallation neutrons, resulting in strong neutron moderation or downscatter in energy. This effect causes the number of epithermal neutrons to be depressed, so that variation of the epithermal count rate with location can be used as a sensitive indicator of the presence of hydrogen in planetary materials.

In contrast to techniques that measure surficial properties, such as spectral reflectance or X-ray spectroscopy, neutron spectroscopy is sensitive to the composition of planetary materials to a depth of tens of centimeters. Planetary epithermal neutrons were first measured at the Moon with the Lunar Prospector mission, and their fluxes permitted the conclusive identification of hydrogen enhancements at both lunar poles in the vicinity of permanently shadowed craters (Feldman et al., 1998). Neutron spectroscopy has also been used to measure and map hydrogen concentrations at Mars (Feldman et al., 2002) and the asteroid Vesta (Prettyman et al., 2012).

For Mercury, neutron spectroscopy provides a means to characterize the polar deposits that is complementary to other techniques in that it measures the bulk composition, in contrast to surface reflective (Section 13.2.3) or morphologic (Section 13.2.5) properties. If the radar-bright deposits are dominantly composed of hydrogen, then the polar regions should have an epithermal neutron flux that is lower than non-polar regions that have little to no hydrogen (Feldman et al., 1997). If, to the contrary, the radar-bright regions are dominantly composed of sulfur, as suggested by Sprague et al. (1995), then Mercury's polar regions should have a decreased thermal-neutron flux compared with non-polar regions but very little change in epithermal neutrons. Finally, if non-compositional effects, such as the alteration of the dielectric properties of silicate materials by unusually cold temperatures (Starukhina et al., 2001), are responsible for the radar-bright properties, then there should be no difference in neutron flux between polar and non-polar regions.

Although the neutron signatures for these different hypotheses are, in principle, clear and unambiguous, the actual neutron measurements proved to be challenging owing to several aspects of the MESSENGER mission. Planetary neutron measurements are generally count-rate limited, and therefore optimum measurements require specific measurement parameters. These parameters include close proximity to the target of interest (in this case the north polar radar-bright deposits), constant and/or regularly changing viewing geometry with a clear field of view to the target to enable simple and well-behaved viewing-geometry corrections, and multi-day to multi-week accumulation times to provide measurements with high statistical precision. Because MESSENGER operated at Mercury in an eccentric near-polar orbit with a constantly changing altitude and limited time near the north pole, and because of the multiple types of measurements conducted by the MESSENGER spacecraft, the viewing geometry for the NS sensor was highly variable. None of the measurement parameters was fully optimized for NS observations. As a consequence, substantial analysis and data reduction were required to successfully obtain MESSENGER's north polar neutron measurements. From pre-orbit-insertion predictions of the orbital mission ephemeris, Lawrence et al. (2011) estimated that if the northern permanently shadowed regions contained extensive amounts of water ice, epithermal neutrons would show a count-rate decrease of 4% or less poleward of 60°–70°N compared with count rates at lower latitudes. The challenging aspect of this measurement was that this small signal variation had to be identified on top of systematic and non-compositional variations of between 200% and 300%. In addition, because of the spacecraft's high altitude (>200 km) above the radar-bright deposits, Lawrence et al. (2011) concluded that none of the radar-bright deposits would be individually spatially resolved, but rather the latitudinal variation of epithermal neutrons would be the primary signature of a polar hydrogen enhancement.

A full analysis of epithermal and fast neutron data from the first 10 months of the MESSENGER orbital mission was presented by Lawrence et al. (2013). The problem of detecting a <4% signal within the >200% non-compositional variations was solved by carrying out a quantitative simulation of the NS neutron-detection performance. This simulation accounted for the planetary neutron creation and transport with the Monte Carlo particle transport code MCNPX (Pelowitz et al., 2005). The neutron transport from the planet to the spacecraft was calculated from the analytic expressions of Feldman et al. (1989). Finally, the angular- and energy-dependent response of the NS was modeled with an MCNPX-based geometry model of the full MESSENGER spacecraft. Further details of these simulations were given by Lawrence et al. (2011, 2013).

The measured epithermal neutron count-rate data were compared with simulated rates for two end-member cases of hydrogen concentrations within the polar deposits identified by Earth-based radar: 100 wt% water-equivalent hydrogen (WEH) and 0 wt% WEH. This comparison is shown in Figure 13.3a, where the measured and simulated values are plotted versus latitude. The monotonic equator-to-pole variation is due mostly to the latitude-dependent radial neutron-velocity Doppler effect (Feldman and Drake, 1986). This effect is removed (Figure 13.3b) by normalizing both the 100 wt% WEH simulation results and the measurements to the no-water simulation values. The dominant remaining signal in the measured epithermal neutron data is a relative count-rate decrease versus latitude poleward of approximately latitude 70°N. The magnitude and latitude profile of this variation closely matches that of the simulated count rate for a thick, surficial layer of 100 wt% water ice at all of the radar-bright regions identified from radar observations (Harmon et al., 2011).

Given the highest-latitude value as the maximum polar signal, the epithermal neutron data show a measured polar signal of 0.976 ± 0.0025 (two standard deviations, or 2σ), relative to an equatorial neutron signal of 1. After accounting for remaining systematic variations due to background counts from energetic electron events and possible subsurface temperature variations,

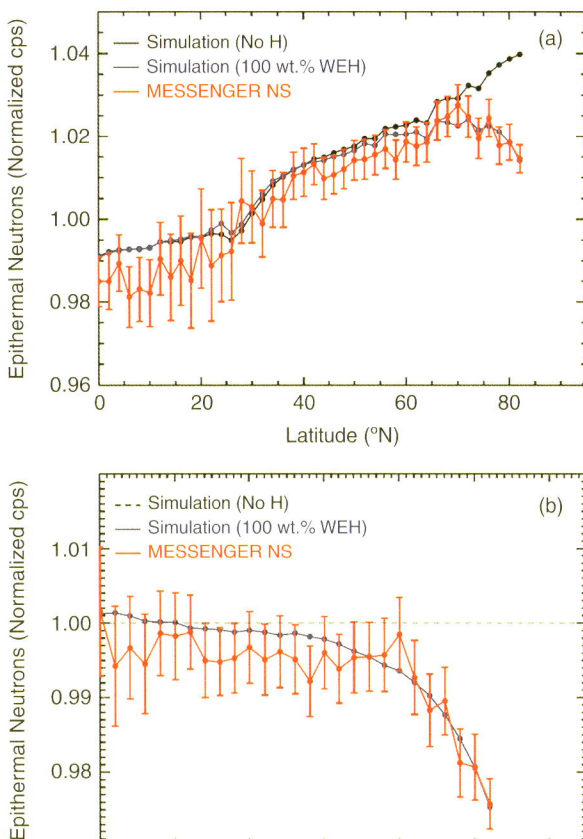

Figure 13.3. (a) Measured (red) and simulated (black, blue) epithermal neutron count rates averaged over 2°-wide latitude bins and plotted as a function of latitude. All corrections except for a radial Doppler effect have been applied to the data. Counts are normalized to the mean count rate (~60 counts per second, or cps) at an altitude of 400 km. Simulated count rates are shown for the cases of no hydrogen (black) and a thick layer of 100 wt% water-equivalent hydrogen (WEH) (blue) in the radar-bright regions. The error bars denote twice the measured standard deviation of the mean in each latitude bin. (b) Simulated and measured epithermal neutron count rates after correcting for the radial Doppler effect, which is accomplished by normalizing to the simulation with no hydrogen. From Lawrence et al. (2013).

Lawrence et al. (2013) concluded that the epithermal neutron data provide strong evidence that Mercury's north polar radar-bright regions contain high concentrations of hydrogen, consistent with the presence of water ice. If it is assumed that the water ice is located within the radar-bright regions as a single thick layer, then the epithermal neutron data are consistent with the presence of up to 100 wt% WEH within these regions. That the 2σ uncertainty of the measurements extends to a slightly larger signal than is given for the hydrogen-rich simulation indicates that the epithermal neutron data are consistent with (but do not require) a larger total area than is specified by regions identified as radar-bright by Harmon et al. (2011).

The NS also measured fast neutrons, which depend on hydrogen concentration, but with less sensitivity and diagnostic specificity than epithermal neutrons. Fast neutrons have a different sensitivity to hydrogen-rich material buried under tens of centimeters of hydrogen-poor material than epithermal neutrons, and the combination of epithermal and fast neutron measurements can constrain the burial depth of a hydrogen-rich layer under a hydrogen-poor layer (Feldman et al., 2007; Miller et al., 2014). When the epithermal and fast neutron measurements are combined, the inferred polar hydrogen enhancements are consistent with an average two-layer stratigraphy in which the hydrogen concentration in the upper layer is 0–25 wt% WEH, the hydrogen concentration in the lower layer is 12–100 wt% WEH, and the effective surface density of the upper layer is 12–35 g/cm^2 (Lawrence et al., 2013). If a typical planetary regolith density of 1.5 g/cm^3 is assumed (Carrier et al., 1991), this surface density corresponds to a physical thickness of 8–23 cm.

The results from radar and neutron observations give a consistent picture of the polar deposits. Specifically, the identification from neutron data of large concentrations of hydrogen within the radar-bright regions strongly supports the idea that the high radar backscatter of the polar deposits is the result of nearly pure water ice (Harmon et al., 2011). Multi-wavelength radar data also support the interpretation that the water-rich layer, on average, is buried beneath an insulating layer of ~10 cm thickness (Harmon et al., 2011). This thickness falls within the range 8–23 cm inferred from the neutron data. Finally, because the neutron simulations used the same locations of radar-bright features as MESSENGER illumination studies (Chabot et al., 2013; Deutsch et al., 2016; also Section 13.2.1), the combined results provide a self-consistent basis for interpreting the locations of hydrogen, permanent shadow, and radar-bright deposits.

13.2.3 Laser Reflectance Measurements

MESSENGER's Mercury Laser Altimeter (MLA) (Cavanaugh et al., 2007) obtained the first topographic measurements of Mercury's surface (Chapter 3), which contributed to identifying regions of permanent shadow near Mercury's north pole, as discussed in Section 13.2.1. Additionally, MLA measured the reflectance of the surface at the laser wavelength of 1064 nm, providing key new insights into Mercury's polar deposits.

The single-beam time-of-flight MLA ranging system measured the surface elevation at 8 Hz (~400-m intervals) as well as zero-phase reflectance at 1064-nm wavelength during much of the orbital mission. Brief (~6 ns) laser pulses with ~20–100-m-diameter footprints were reflected from the surface and collimated by four telescopes into a single detector photodiode, converted to an electronic signal, and amplified. As little as 0.1 fJ of detected return signal energy sufficed to make a range measurement. The laser output, range, and incidence and emission angles all affect the reflectance intensity, and to determine the surface reflectance all of these factors must be considered. The energy of the return pulse as a fraction of the outgoing laser pulse energy provides the reflectance, which is obtained by solving the lidar link equation:

$$E_{rx} = E_{tx}\eta_r \frac{A_{rx}}{R^2}\frac{r_s}{\pi}, \tag{13.1}$$

where E_{rx} is the received signal pulse energy, E_{tx} is the transmitted laser pulse energy, η_r is the receiver optics transmission at the laser wavelength, A_{rx} is the receiver telescope aperture area, R is the distance from the MESSENGER spacecraft to the illuminated surface, and r_s is the target surface reflectivity (relative to a Lambertian sphere, i.e., an ideal, diffusely reflecting surface). A narrow, 0.3-nm-wide transmission window rejects almost all background solar radiation. The energy of the outgoing pulse was measured directly after attenuation through the laser output turning mirror. In ground testing, the laser energy varied as a function of temperature from ~14 to 22 mJ and was recorded by the energy monitor with ~10% precision. The return pulse energy was inferred from the differential pulse width of detector signal triggers measured simultaneously at low- and high-voltage thresholds (Sun and Neumann, 2015). Although less precise than a direct measurement, the dual-threshold detection scheme maximized the range sensitivity and provided an estimate of pulse spreading as well. The detector electronics incorporated programmable gain and threshold settings that allowed operation nominally between 200 and 1500 km range. Under optimal conditions, the energy measurement had a 25% precision at one standard deviation. Ground calibration of detector parameters provided a conversion of electrical signal to energy. The MLA measurements could be made in complete darkness as well as sunlight and were virtually unaffected by background solar flux, but required near-nadir incidence to produce a suitable pulse. Off-nadir measurements generally involved ranging to longer distances, which resulted in the laser footprint being wider, its radiance being fainter, and the higher incidence angle combined with the slope of the terrain spreading the return signal energy over a longer interval than the response time associated with the detector circuit. Thus, off-nadir observations often resulted in signals too low to trigger both thresholds. Moreover, the steepest poleward-facing slopes are hidden from MLA measurements, and overall the MLA coverage was limited by spacecraft geometry. Detection of ground returns over steep slopes was aided by a matched filter design for the receiver, allowing detection of surface returns at high angles but without energy measurements. Over the steepest and darkest surfaces, the very few dual-threshold returns indicated a very low reflectance.

The north polar region was densely sampled by MLA at nadir incidence from altitudes of ~250 to 460 km over the course of four years and more than 4000 orbits. A map of the surface reflectance (Neumann et al., 2013; Deutsch et al., 2017a) in regions of high density and good signal determined from MESSENGER's full orbital data set is shown in Figure 13.4, which can be compared with the image mosaic in Figure 13.2. Reflectance measurements north of 84°N required off-nadir observations, due to the spacecraft's orbital inclination, and hence these measurements are noisier than the nadir observations at lower latitudes. Along with the reflectance measured off-nadir, the sampling at latitude >84°N was also sparse, resulting in an overall degradation of the data quality and an inability to determine precisely the reflectance contrast in this northernmost region.

Whereas Mercury's sunlit regolith in the north polar region has an average surface reflectivity value r_s of 0.17, it is apparent

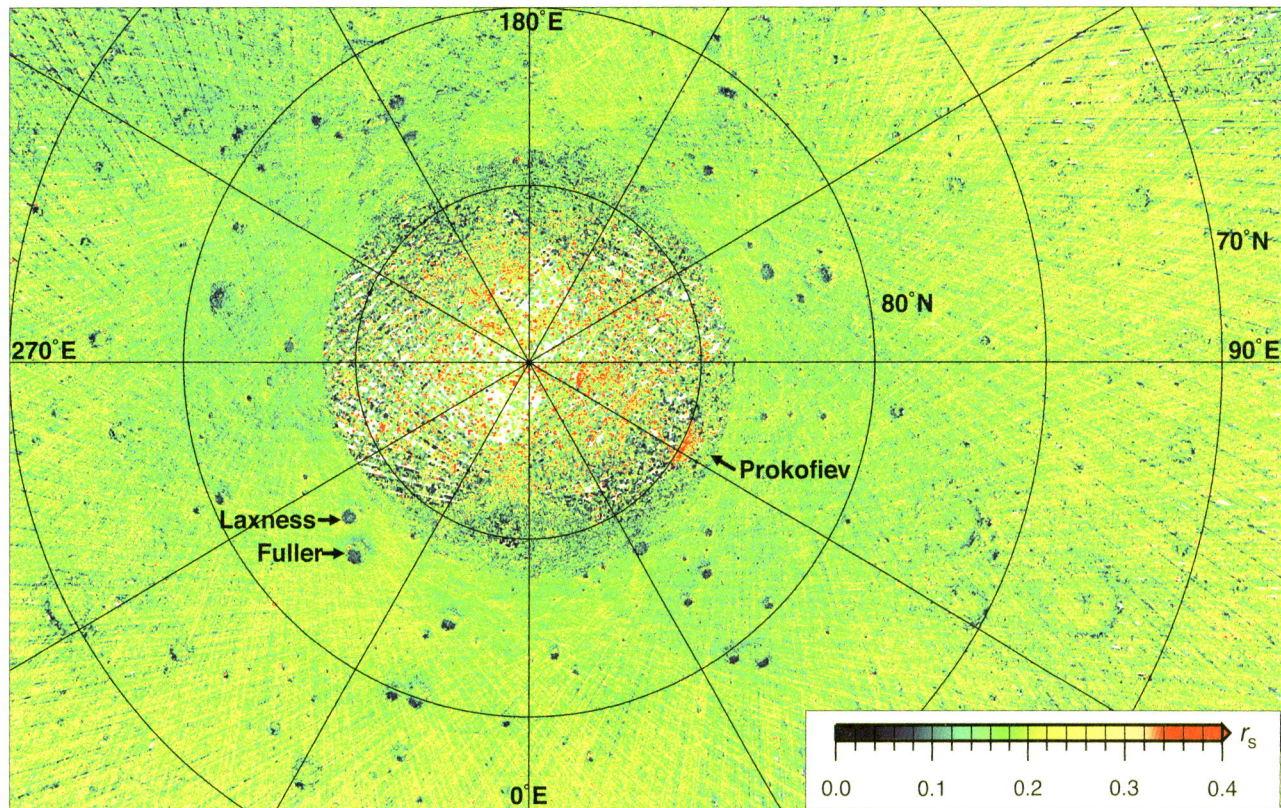

Figure 13.4. The bidirectional reflectance (r_s) of Mercury's north polar region at 1064-nm wavelength (Neumann et al., 2013; Deutsch et al., 2017a), median binned at 0.4-km resolution using nearest-neighbor interpolation. Lower reflectance values within many permanently shadowed craters, such as Laxness and Fuller, are evident, as are higher-reflectance regions, such as those within Prokofiev. The MLA reflectance measurements in regions poleward of Prokofiev were acquired at off-nadir viewing geometries and were more limited because of the spacecraft's orbital inclination.

from Figure 13.4 that the large majority of permanently shadowed surfaces within polar craters contain material with considerably lower reflectance ($r_s < 0.1$). The low-reflectance surfaces within the craters are preferentially located on poleward-facing slopes and are not found in regions that experience direct solar illumination. The lowest reflectance values represent only an upper bound, because such surfaces attenuate or extinguish laser returns to levels below the sensitivity of the detector. Neumann et al. (2013) noted that, where coverage was available, all radar-bright regions at latitudes <84°N have low-reflectance surfaces, a finding confirmed by subsequent MESSENGER MLA observations from the entire orbital mission.

Examples of craters with low-reflectance shadowed surfaces are shown in Figure 13.5 for Laxness and Fuller. Portions of these flat-floored, central-peak craters (Figure 13.5a) have regions of permanent shadow (Figure 13.5b) and high radar reflectivity (Figure 13.5c), and in these regions the reflectance is lower than in the surrounding illuminated areas (Figure 13.5d). The location of the low-reflectance boundary is not distinctly defined, but the low-reflectance surface includes the radar-bright region in Figure 13.5c and is generally consistent with the region of permanent shadow inferred from topography.

Low-reflectance surfaces are also found in many small, simple craters where pre-MESSENGER thermal models predicted that water ice could not be stable at the surface or near subsurface (Vasavada et al., 1999). The 6-km-diameter, bowl-shaped crater near Laxness, labeled "L4" in Figure 13.5, is one such example. The poleward-facing interiors of many similarly sized craters, even some at latitudes south of 70°N, also have very low-reflectance signals as measured by MLA. Low-reflectance surfaces are found in permanently shadowed regions other than crater interiors, such as on the poleward-facing scarp at 83°N, 270°E, which crosses craters Qiu Ying and Bechet. Additionally, scattered areas on the northernmost flanks of some craters also have lower reflectance values in the MLA map, as is seen for Laxness and Fuller in Figure 13.5. These areas do not host large regions of permanent shadow, but it is possible that small shadowed regions, with dimensions below the spatial resolution of current shadow maps (Mazarico et al., 2014; Deutsch et al., 2016) or Arecibo observations (Harmon et al., 2011), are present, perhaps as a result of surface roughness and the limited sunlight that illuminates this location. In fact, a gradual poleward darkening of the surface from ~72°N to ~85°N in the MLA reflectance map has been attributed to the presence of low-reflectance volatile deposits in "micro-scale cold traps," which are at spatial scales below those resolved by MLA (Paige et al., 2014; Neumann et al., 2017). Modeling suggests that surface roughness can result in an areal fraction for micro-scale cold traps of up to ~20% for Mercury's north polar region (Rubanenko et al., 2017).

In contrast to the low-reflectance surfaces identified within the majority of permanently shadowed craters, the permanently

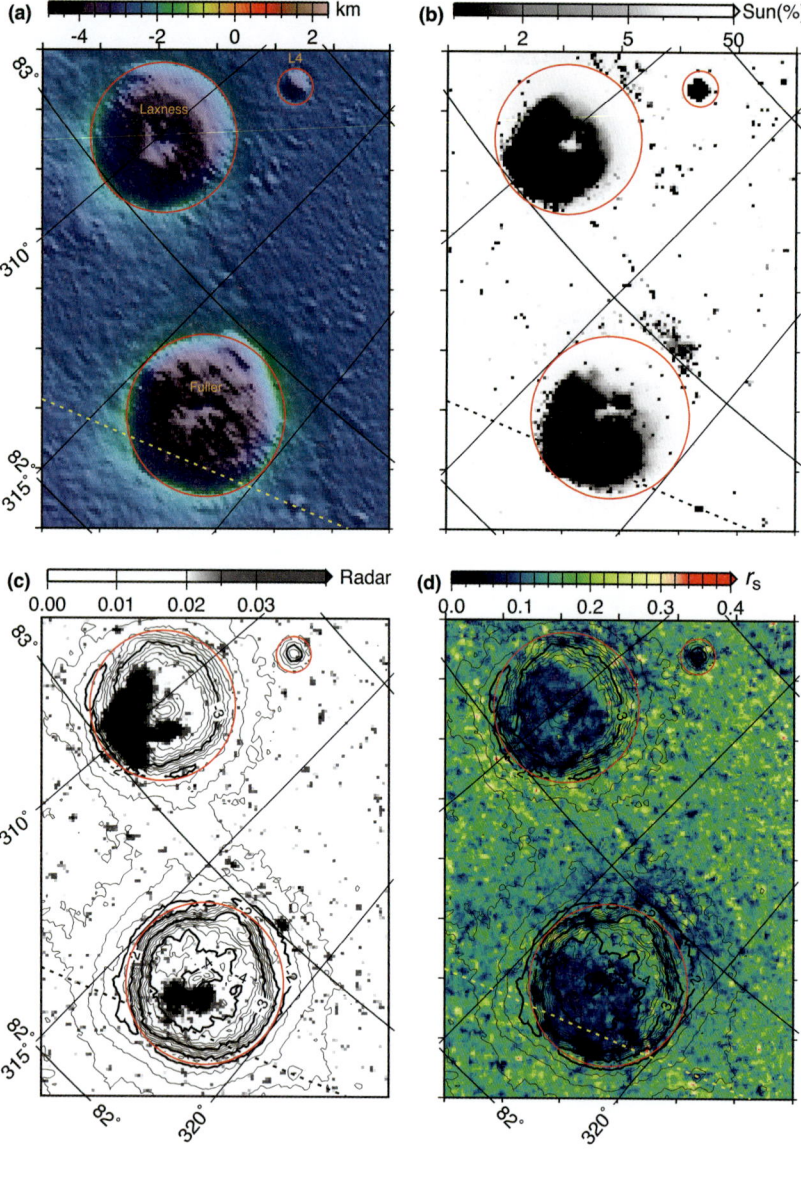

Figure 13.5. (a) MLA topography of complex craters Laxness (26-km diameter) and Fuller (27-km diameter) and a 6-km-diameter crater northeast of Laxness denoted L4. Red circles denote rim crests. Polar stereographic projection of height relative to a 2440-km-radius spherical datum. (b) Percentage of time illuminated by sunlight determined from MLA topography (Mazarico et al., 2014; Deutsch et al., 2016). (c) Radar reflectivity (Harmon et al., 2011). Contours of topography are shown at 0.2-km intervals. (d) MLA reflectance (r_s). Dashed lines show the location of the profile through Fuller in Figure 13.6.

shadowed interior of the 112-km-diameter Prokofiev crater contains an anomalously high-reflectance surface, as seen by the red area in Figure 13.4. During MESSENGER's second year of orbital operations, several MLA profiles were obtained with off-nadir targeting at modest incidence angles (<10°) through Prokofiev, and these profiles showed areas of enhanced reflectance. The precise boundary of the high-reflectance region is not well resolved, but the average reflectance in this region is $r_s = 0.3$, unusually high for Mercury, and is interpreted as indicating a surface exposure of water ice (Neumann et al., 2013). The northernmost large craters, Kandinsky, Tolkien, Chesterton, and Tryggvadóttir (Figure 13.2), which have permanently shadowed interiors conducive to stable surface water ice (Paige et al., 2013; Section 13.2.4), also display evidence for high-reflectance surfaces (Deutsch et al., 2017a) but from more limited and scattered MLA reflectance data than Prokofiev, as seen in Figure 13.4. Additionally, four small but resolvable areas 2–5 km in diameter within 8° of Mercury's north pole have been shown to exhibit clusters of high MLA reflectance values, consistent with the presence of surface ice exposed in small cold traps in these locations (Deutsch et al., 2017a). Overall, the MLA reflectance map shows an increase in the latitudinally averaged reflectance value from ~85°N to Mercury's north pole, interpreted to be the result of increasing amounts of surface water ice present in micro-scale cold traps (Neumann et al., 2017).

MLA topography and reflectance measurements for the low-reflectance surface observed in Fuller crater and the high-reflectance surface detected in Prokofiev crater are shown in Figure 13.6. As shown in Figure 13.6a for Fuller, the ~2-km-deep crater wall on the left is too steep (>27° slope) for successful reflectance measurements, but the interior has reflectance values lower than the surrounding plains by a factor of at least 2. For Prokofiev (Figure 13.6b), the steeper walls of the crater have low reflectance, but the floor has much higher reflectance than the surroundings.

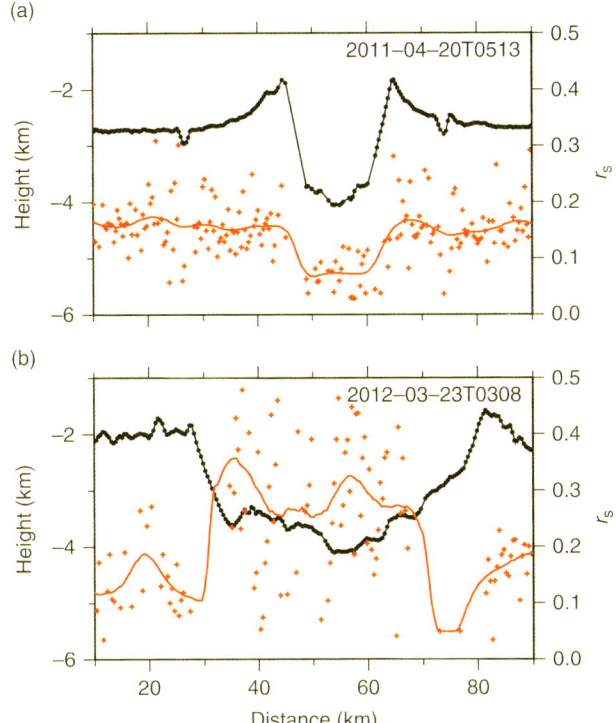

Figure 13.6. Profiles of MLA topography (black, labels at upper right), individual MLA reflectance measurements (red symbols), and reflectance averaged over a 12-km-wide cosine window (red curve). Surface height relative to a 2440-km-radius spherical datum is plotted with a 10:1 vertical exaggeration relative to the distance. (a) For Fuller crater, the profile location is shown in Figure 13.5. (b) For Prokofiev crater, the profile is taken along the 85°N latitude line.

13.2.4 Thermal Modeling Results

Measurements from MLA enabled major advances in constructing thermal models of Mercury's surface and near surface from the measured topography (Chapter 3). Previous thermal modeling work had necessarily relied on idealized crater shapes (Paige et al., 1992; Ingersoll et al., 1992; Salvail and Fanale, 1994; Vasavada et al., 1999). MESSENGER's MLA measurements led to the production of a digital elevation model that was used for thermal model calculations of Mercury's north polar region (Paige et al., 2013). The thermal model results were obtained from three-dimensional ray-tracing calculations that yielded surface and near-surface temperatures, as described by Paige et al. (2013), and were similar to thermal models determined earlier for the Moon (Paige et al., 2010). The resulting biannual maximum surface temperatures and biannual average temperatures at 2-cm depth are shown in Figure 13.7. Biannual maxima and averages are shown because it takes Mercury two orbits about the Sun to complete the full thermal cycle experienced by the planet. The results in Figure 13.7 are similar to those published by Paige et al. (2013) but generated from an MLA-derived DEM with more complete coverage than was available previously.

The thermal model results indicate that the maximum surface temperatures in large craters at high latitudes are sufficiently low to permit water ice deposits to be stable at the surface for geologically long intervals (Figure 13.7c). The results suggest that substantial water ice may be expected on the surfaces of the interiors of craters Kandinsky, Tolkien, Chesterton, Tryggvadóttir, and Prokofiev (Figure 13.2), consistent with the observation of higher surface reflectance values measured by MLA in these permanently shadowed regions (Figure 13.4). However, the results also show that the majority of permanently shadowed craters near Mercury's north pole experience biannual maximum surface temperatures that are greater than ~110 K, values too high for water ice to be stable at the surface (Figure 13.7c).

On Mercury, biannual average temperatures can be interpreted as close approximations to the nearly constant subsurface temperatures that exist below the penetration depths of the diurnal temperature wave, which is tens of centimeters for a lunar-like regolith (Paige et al., 2013). Many permanently shadowed craters in Mercury's north polar region have average temperatures conducive to the thermal stability of subsurface water ice (Figure 13.7c), and these locations correlate with regions that are also radar-bright (Figure 13.7d). Conversely, regions with average temperatures higher than ~110 K are not observed to host substantial radar-bright deposits (Paige et al., 2013). This result suggests that the radar-bright deposits are dominantly composed of a volatile species that is not thermally stable at temperatures higher than 110 K, providing strong evidence for water ice as the major component of Mercury's polar deposits.

Additionally, in these locations conducive to the thermal stability of subsurface water ice, MLA measurements consistently have shown low-reflectance surfaces (Figure 13.4). Comparison of the MLA reflectance measurements with the thermal model results indicate that low-reflectance surfaces are absent in regions with biannual average temperatures greater than 210 K and biannual maximum temperatures greater than 300 K, leading to the conclusion that the distribution of low-reflectance substances must be controlled by the presence of volatile compounds that are not thermally stable above these temperatures (Paige et al., 2013). Combining the evidence for subsurface water ice with that for a low-reflectance layer of a higher volatility temperature, Paige et al. (2013) concluded that the low-reflectance deposits form as sublimation lags, eventually insulating the water ice below, similar to the earlier suggestion of Vasavada et al. (1999) that the ice deposits might sublimate until their contaminant load formed thin insulating layers. In such a scenario, a source of both water ice and low-reflectance volatiles is required, both of which migrate to the polar cold traps, leading to the suggestion that the low-reflectance volatiles, stable only below ~300 K, are organic-rich compounds derived from asteroidal or cometary impacts (Paige et al., 2013). Figure 13.8 (Zhang and Paige, 2009, 2010) illustrates that a range of organic compounds found in primitive meteorites (Botta and Bada, 2002), such as aromatic hydrocarbons, linear amides, and carboxylic acids, are potential compounds to consider for the identity of the low-reflectance layer covering most of Mercury's polar deposits.

One important issue to address with future thermal models is that of small simple craters. On the basis of pre-MESSENGER thermal models, derived for idealized bowl-shaped craters, it was concluded that simple craters on Mercury (<10-km

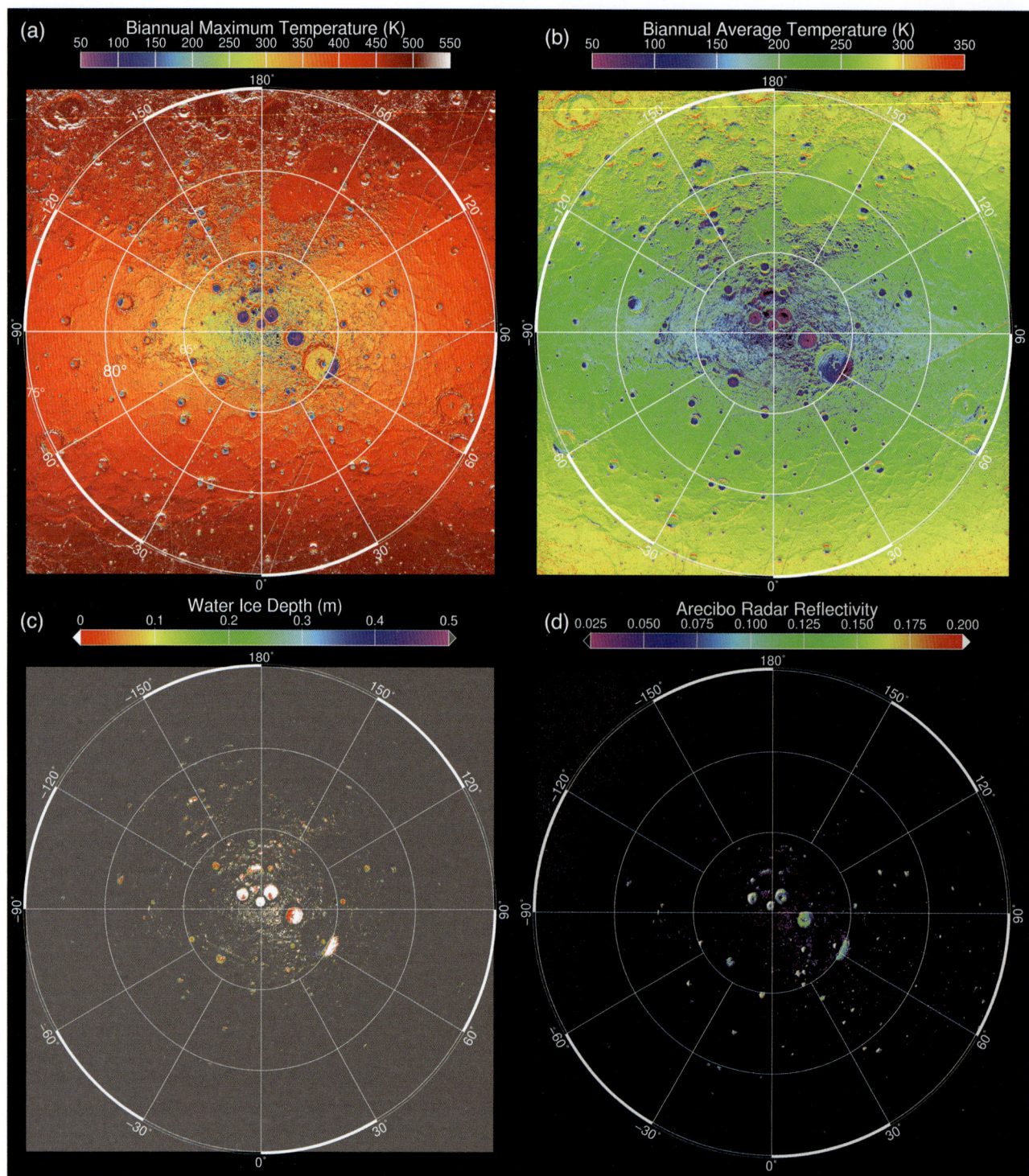

Figure 13.7. Results from thermal model calculations (Paige et al., 2013) for (a) biannual maximum surface temperatures and (b) biannual average surface temperatures at 2-cm depth for Mercury's north polar region. (c) Calculated depths at which water ice would be lost to sublimation at a rate of less than 1 kg m^{-2} per billion years given an insulating cover of lunar-like regolith; white regions indicate locations where water ice is thermally stable at the surface, and colored regions show the minimum depths at which water ice must be buried below the surface to be thermally stable. (d) Arecibo radar image from Harmon et al. (2011), to which the figure in (c) bears strong resemblance.

diameter) could not host long-lived water ice except if located within 2° of the pole (Vasavada et al., 1999). Yet mapping of radar-bright host craters (Section 13.2.1) indicates that many small craters contain radar-bright deposits, and MLA measurements show these deposits to have low-reflectance surfaces (Figure 13.5, Section 13.2.3). Such craters are shallower (Ernst et al., 2014) than was assumed for the earlier thermal models. Thermal models based on MESSENGER

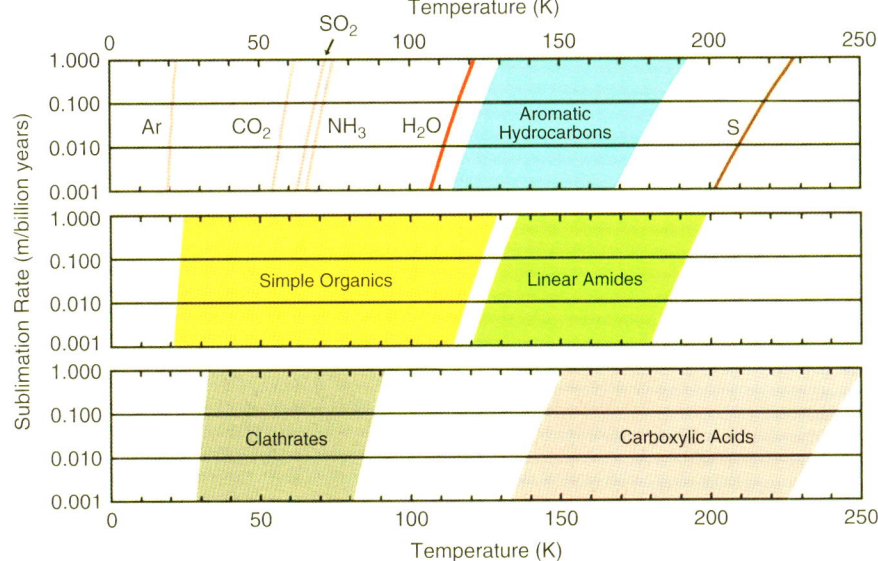

Figure 13.8. Shaded regions represent the range of vacuum evaporation rates calculated for representative organic compounds measured in primitive meteorites. Most simple organic compounds have volatility temperatures lower than that of water, but more complex organic compounds with higher molecular masses are stable to higher temperatures than water, consistent with the maximum surface temperatures calculated for Mercury's low-reflectance polar deposits. From Zhang and Paige (2009, 2010).

measurements of the topography of these small simple craters can yield insight into whether such craters can provide long-term stable environments for water ice or whether Mercury's water ice must be relatively young. Extending thermal models to cover the potentially challenging thermal environments experienced by the lowest-latitude craters that host radar-bright material, such as Egonu and Monk (Figure 13.2), could also provide insights into water ice stability on Mercury.

Currently, no thermal model calculations that include Mercury's surface topography have been performed for Mercury's south polar region, since MLA measurements were limited largely to Mercury's northern hemisphere. Studies with MDIS-derived DEMs, and suitable assumptions to deal with permanently shadowed surfaces within such DEMs, may yield thermal modeling results for Mercury's south polar region and enable comparison with the north polar results.

13.2.5 Direct Imaging of Polar Deposits

Directly imaging the permanently shadowed surfaces of radar-bright deposits was not a measurement objective of MESSENGER's one-year primary mission nor a required measurement for the MDIS instrument. However, as the spacecraft continued to operate and the mission was given extensions (Chapter 1), new measurement objectives were added, and one such addition was a campaign to image the surfaces of Mercury's polar deposits. Such observations are challenging, as the permanently shadowed surface is illuminated only by very low amounts of sunlight scattered off nearby terrain. Similar imaging had successfully revealed the permanently shadowed surfaces inside lunar craters (Haruyama et al., 2008; Speyerer and Robinson, 2013), indicating that attempts by MESSENGER might be successful. The MDIS WAC was equipped with 11 narrow-band filters (4–20-nm bandpasses) and one broadband clear filter (600-nm bandpass) centered at 700 nm (Hawkins et al., 2007). Designed with the main purpose of obtaining calibration images of stars, the WAC broadband filter yielded images that quickly saturated when viewing Mercury's sunlit surface but provided high sensitivity to the low light levels necessary to image shadowed regions. To reveal details within shadowed areas, such regions had to be targeted on the edge of the WAC charge-coupled device (CCD) that was read out first, so that the low signal would not be compromised by saturation effects from the surrounding saturated terrain. With this approach, MESSENGER was able to image the radar-bright surfaces within a number of north polar craters; the spacecraft's highly eccentric orbit, with an altitude about 20 times higher above the south polar region than the north, did not enable a similar WAC broadband imaging campaign for Mercury's south polar radar-bright deposits.

The largest crater that hosts extensive radar-bright deposits in Mercury's north polar region is Prokofiev (112 km in diameter), the shadowed interior of which, as discussed above, exhibits evidence for surficial water ice, on the basis of high MLA reflectance measurements (Neumann et al., 2013) and thermal model predictions for water ice stability (Paige et al., 2013). Two WAC broadband images revealed that a portion of Prokofiev's floor has a higher reflectance than the surrounding surface (Chabot et al., 2014), as seen in Figure 13.9. The higher-reflectance surface is located in the portion of Prokofiev that hosts the large radar-bright deposit, consistent with surface water ice at this location. In the WAC broadband images, the higher-reflectance region has only 4–5% greater relative reflectance than the immediately neighboring surface, but the illumination conditions during these images are complicated: grazing sunlight at high solar incidence angles scatters off multiple surfaces, contributing to the combined illumination on each portion of the scene. Given the complex illumination conditions, the determination of absolute reflectance values from the WAC broadband images would require highly detailed modeling of the multiply scattered sunlight incident on each surface. In contrast, MLA measurements were obtained at a phase angle of zero and provided a more direct measurement of absolute reflectance values for regions within Prokofiev.

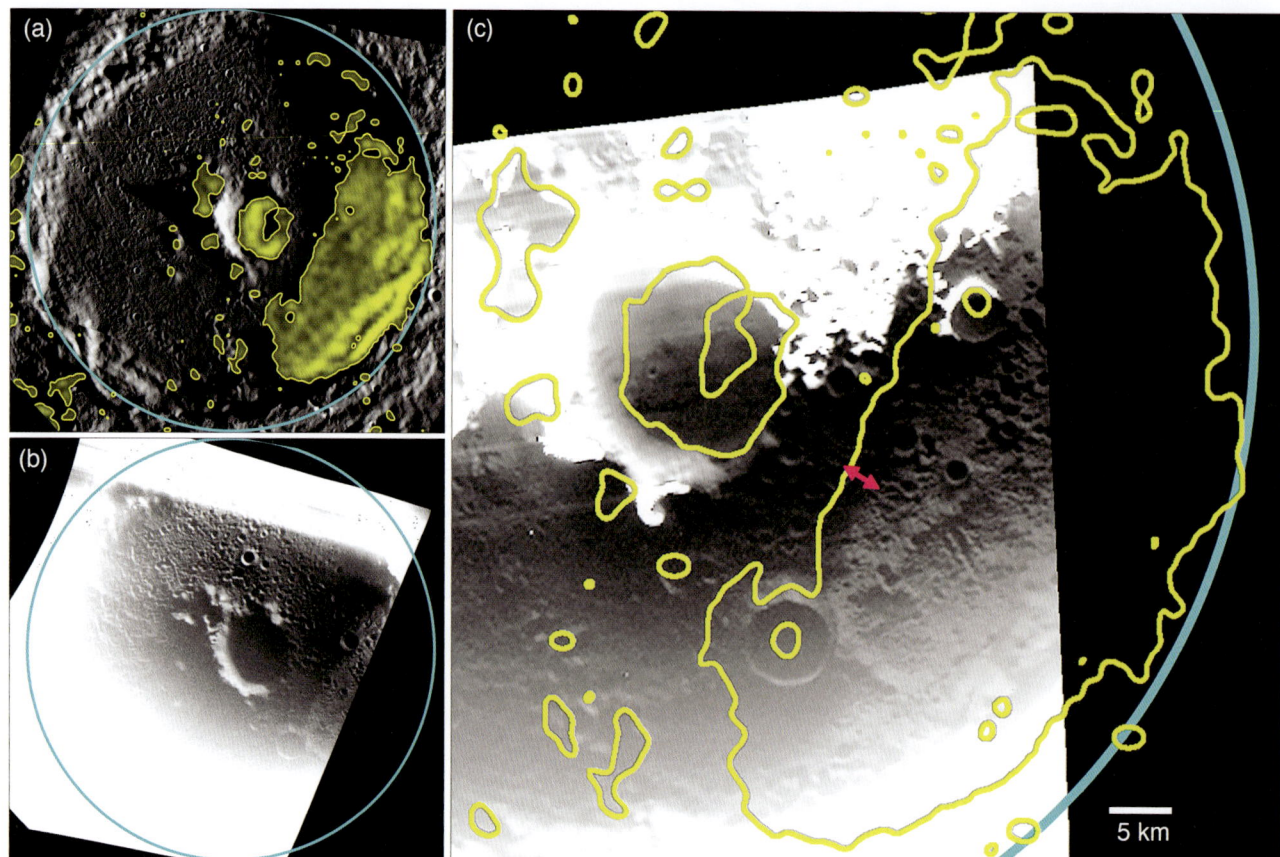

Figure 13.9. Prokofiev (112-km diameter) crater, with the rim outlined in cyan. All images are in stereographic projection about the north pole, with 180°E to the top. (a) Arecibo radar data are shown in yellow (Harmon et al., 2011); variations in radar backscatter across the permanently shadowed portion of the crater floor are related to topography. Two WAC broadband images, (b) EW1020581108B and (c) EW1019169411B, for which sunlit portions are saturated, reveal a higher-reflectance surface for the permanently shadowed, radar-bright region (Chabot et al., 2014). The pink arrows denote the approximately 3-km offset between the outer boundaries of the radar-bright region and the higher-reflectance surface.

Such measurements indicate that the geometric albedo within the radar-bright region of Prokofiev is approximately a factor of 2 higher than Mercury's average surface value at the near-infrared wavelength of the laser (Neumann et al., 2013).

The WAC broadband images also reveal the surface texture of the radar-bright region in Prokofiev. At the 85 m/pixel scale of the images, the surface texture appears similar to that seen on the neighboring portions of the crater floor that regularly receive direct solar illumination. This result indicates that if the higher-reflectance region is water ice exposed at the surface, its thickness does not noticeably affect the texture of the surface at this scale. In particular, the water ice is not sufficiently thick to make the covered surface appear smoother than the uncovered surface at 85 m/pixel. Small craters dominate the surface within Prokofiev, and yet the higher-reflectance region is fairly uniform in its surface reflectance properties. The images do not show evidence that higher-reflectance material is covered in local regions by ejecta from any of the small craters widely distributed across the floor of Prokofiev. This observation suggests that the higher-reflectance material in Prokofiev was emplaced after the small craters formed.

The higher-reflectance region in Prokofiev identified in the WAC broadband images is located within but is slightly smaller than the radar-bright region, as seen in Figure 13.9c. The higher-reflectance surface is offset inward by ~3 km relative to the boundary of the radar-bright deposit, and the intervening ~3-km-wide region has a relative reflectance that is not distinguishable from the portions of Prokofiev's floor that receive direct solar illumination. Studies of mare–highland contacts on the Moon indicate that impact gardening has achieved lateral transport over a scale of ~4–5 km (Li and Mustard, 2000), a scale comparable to the difference in boundary position between the higher-reflectance and the radar-bright regions in Prokofiev. Lateral mixing at lunar mare–highland contacts has operated over billions of years. It is possible that the higher frequency and generally higher velocity of impacts at Mercury relative to the Moon (Cintala, 1992; Borin et al., 2009, 2016) or a sufficiently thin deposit within Prokofiev could yield similar lateral transport effects on a shorter timescale. In general, if the ~3-km offset boundary between the higher-reflectance and radar-bright regions in Prokofiev is the result of lateral mixing having covered the outermost edge of surface ice with regolith, the emplacement of the water ice occurred sufficiently far in the past to allow the effects of lateral transport to develop undisturbed. Alternatively, the thermal environment in Prokofiev may support a large area of surface ice surrounded by a marginal zone of subsurface ice, which is a general pattern often predicted in thermal models (Paige et al., 2010, 2013). Up to now, thermal models have focused on global studies of Mercury's entire north polar region (Paige et al., 2013), and the precise region of permanent shadow within Prokofiev has not been determined with sufficient accuracy to address this offset region (Deutsch et al., 2016). A higher-resolution, localized study of the topography and thermal conditions in the Prokofiev region, at a scale approaching that of the WAC broadband images, would provide information key to interpreting the 3-km offset observed between the higher-reflectance and radar-bright regions.

WAC broadband images were also obtained of the radar-bright, permanently shadowed interiors of the four largest craters nearest Mercury's north pole: Chesterton, Kandinsky, Tolkien, and Tryggvadóttir (Figure 13.2). Although thermal models predict that these four craters, like Prokofiev, are capable of sustaining long-lived water ice at the surface (Figure 13.7c), the WAC broadband images showed that the floors of these craters do not look atypical for Mercury, and, in contrast to Prokofiev, no clear regions of differing reflectance properties were noted (Chabot et al., 2014). However, unlike Prokofiev, the entire floors of these high-latitude craters are in permanent shadow, so clear reflectance boundaries across their floors are not expected. Thus, the current analysis of the WAC broadband images of these four craters neither supports nor negates the possibility that water ice is exposed at the surface in these locations.

In contrast to Prokofiev and the few other large craters nearest Mercury's north pole, the polar deposits in the large majority of host craters have low-reflectance surfaces as measured by MLA at 1064 nm (Neumann et al., 2013) and are predicted to sustain long-lived water ice only if covered by an insulating layer a few tens of centimeters in thickness (Paige et al., 2013). As discussed above, this evidence has led to the interpretation that the insulating layer is a lag deposit composed of low-reflectance, cold-trapped, organic-rich volatile compounds stable at temperatures somewhat higher than water ice (Figure 13.8). WAC broadband images were acquired for many of these lower-latitude craters and, in total, reveal numerous examples of radar-bright locations with low-reflectance surfaces (Chabot et al., 2014, 2016). In the images, the low-reflectance surfaces are ~20% lower in relative reflectance than the nearby crater floor, but, as discussed for Prokofiev, the complex illumination conditions during these images complicates any determination of absolute reflectance values. The zero-phase-angle MLA reflectance measurements are better suited to provide this information and indicate that the low-reflectance deposits are approximately a factor of 2 lower in reflectance than Mercury's average surface (Neumann et al., 2013).

The best images of low-reflectance, radar-bright deposits were obtained during the final year of MESSENGER's orbital operations, when data were acquired at lower altitudes than in prior years (Chapter 1). The low altitudes enabled data sets of progressively higher spatial resolution for many instruments, including WAC broadband imaging within Mercury's shadowed polar craters. The low-altitude imaging opportunities were limited in area, however, and did not enable higher-resolution imaging within Prokofiev or other high-latitude craters. However, WAC broadband images were acquired of lower-latitude craters at resolutions as good as 24 m/pixel (Chabot et al., 2016), with examples shown in Figure 13.10.

A striking feature observed in the WAC broadband images of the low-reflectance deposits is the sharpness of the boundaries. These well-defined boundaries of the low-reflectance deposits align extremely closely with the boundaries of the permanently shadowed regions. In many cases, the boundaries of the radar-bright deposits are also similar, but in some locations, such as at Fuller crater, Arecibo radar data do not indicate an extensive radar-bright signal in the crater (Figure 13.2), but the crater shows a sizable low-reflectance deposit (Figure 13.10a). Though an absence of ice in such craters is one possibility, viewing or detection limitations of the radar observations can also affect the completeness of the identification of radar-bright features in Arecibo images. In fact, all of the 35 distinct craters with permanently shadowed surfaces revealed during the low-altitude imaging campaign show low-reflectance surfaces with well-defined boundaries (Chabot et al., 2016). This observation supports the suggestion that all of Mercury's available cold traps are occupied by volatiles. The well-defined boundaries of the low-reflectance deposits are not confined to topographic lows. To the contrary, the low-reflectance boundaries shown in Figure 13.10 extend up crater walls to just below the rim, notwithstanding that elsewhere on Mercury relatively bright material can be exposed on crater walls by slumping. Transects of relative surface brightness across the boundaries of the low-reflectance deposits show a transition zone approximately 400-m wide with intermediate brightness values (Chabot et al., 2016). This transition zone could be due to mixing between the low-reflectance compounds and regolith, or it could be due to variations in the amount of sunlit and permanently shadowed areas in this region on a scale smaller than the spatial resolution of the images.

If impact gardening of a low-reflectance deposit tens of centimeters thick exposed or thermally disturbed an underlying layer of water ice, any water ice exposed at the surface would quickly sublimate (1 m in 1 Myr at 130 K, and 1 m in 1000 yr at 150 K; Vasavada et al., 1999; Paige et al., 2013). While rapidly restoring the disturbed region to a stable configuration, new lag deposits of low-reflectance compounds would likely be formed, and this ongoing process could continually maintain well-defined boundaries. Such a process would not be active in colder permanently shadowed regions where water ice is stable at the surface, such as in Prokofiev, and hence it might account for the difference between the 3-km offset of the boundaries between the higher-reflectance region and the radar-bright material in Prokofiev and the sharp boundaries of the low-reflectance deposits in other craters. Regardless of the precise process, the well-defined boundaries of the low-reflectance deposits, even when imaged at resolutions of tens of meters, support the conclusion that the low-reflectance deposits are geologically young relative to the timescale for lateral mixing by impacts. This inference points either to delivery of volatiles to Mercury in the geologically recent past or to an ongoing process that restores the deposits and maintains sharp boundaries.

From the first WAC broadband images to resolve the low-reflectance deposits, Chabot et al. (2014) concluded that the reflectance of the deposits appeared uniform rather than patchy. However, subsequent higher-resolution WAC broadband images revealed that the low-reflectance deposits do display variations in brightness, such as seen in the example images of

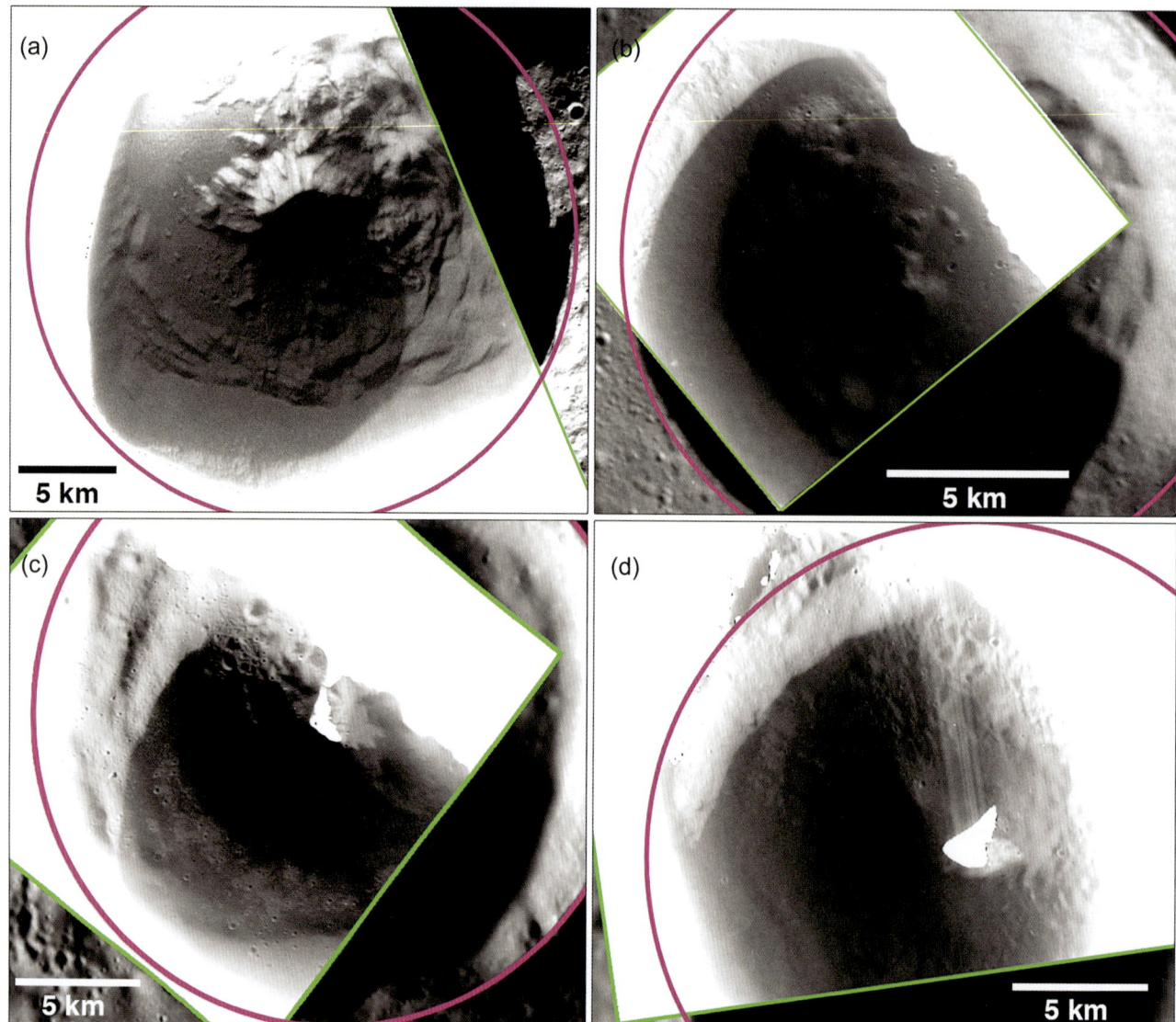

Figure 13.10. MDIS WAC broadband images, outlined in green, of Mercury's low-reflectance polar deposits (Chabot et al., 2016). Crater rims are outlined in magenta. All images are in stereographic projection about the north pole, with 180°E to the top. (a) Fuller crater (27-km diameter, 82.63°N, 317.35°E, EW1047206595B). (b) Unnamed crater at 80.30°N, 293.47°E (18-km diameter, EW1068017709B). (c) Ensor crater (25-km diameter, 82.32°N, 342.47°E, EW1051458815B). (d) Laxness crater (26-km diameter, 83.27°N, 309.96°E, EW1052529039B).

Ensor and Laxness in Figures 13.10c and 13.10d. In some craters, these brightness variations are correlated with variations in the modeled biannual maximum surface temperature, suggesting that the brightness variations are thermally controlled (Chabot et al., 2016). Control of surface reflectance variations by temperature could be the result of multiple volatile species that differ in reflectance and are stable to different maximum temperatures, implying the presence of multiple low-reflectance volatile compounds within Mercury's polar deposits. In contrast, in other craters, some of the small-scale brightness variations look as if they could be related to small impact craters that have disturbed the surfaces of the low-reflectance regions (Chabot et al., 2016), and such an observation may be inconsistent with these deposits being geologically young, as suggested by their well-defined boundaries. Higher-resolution localized topography and thermal modeling studies of these craters could provide additional insights for interpreting the evolution of the low-reflectance deposits.

13.3 IMPLICATIONS

13.3.1 Evaluation of Pre-MESSENGER Hypotheses

Measurements and observations by multiple MESSENGER instruments, specifically NS, MLA, and MDIS, have greatly increased our knowledge about Mercury's polar deposits. These new data sets, combined with the Earth-based radar observations and thermal modeling results, enable an evaluation of previously proposed hypotheses for the composition of the polar deposits.

One proposed hypothesis was that Mercury's polar deposits are composed largely of elemental sulfur (Sprague et al., 1995). Sprague et al. (1995) preferred sulfur rather than water ice as the material in Mercury's polar deposits because of its stability to slightly higher temperatures and because there were good reasons to expect a sufficient abundance of sulfur at Mercury. Indeed, once MESSENGER entered orbit about Mercury and obtained compositional measurements of the surface, a high level of 1–4 wt% sulfur was discovered (Nittler et al., 2011). Although sulfur is now confirmed to be present on Mercury's surface, there is no quantitative study showing that a layer of elemental sulfur would create the high reflectivity and circular polarization ratio measured by the Earth-based radar observations, such as modeled by Black et al. (2001) for water ice on the Galilean satellites. Further, the complete sublimation of a 1-m-thick deposit of elemental sulfur over 1 Gyr would require a surface temperature of ~220 K, which is considerably higher than the corresponding ~110 K temperature for water ice (Vasavada et al., 1999). Whereas the higher stability temperature could support long-lived deposits even in the higher temperatures experienced within small simple craters (Vasavada et al., 1999), it also supports a polar cap of elemental sulfur exposed at the surface within 1° latitude of the pole and buried beneath the surface within 4° latitude of the pole (Butler, 1997; Vasavada et al., 1999). Neither Earth-based radar observations nor MESSENGER data sets support the existence of such an exposed or buried polar cap at either Mercury's north or south poles, however. In particular, the radar-bright regions identified from the highest-resolution radar images (Harmon et al., 2011) collocate with permanently shadowed regions even very near the pole (Figures 13.1 and 13.2), suggesting that the volatiles that compose the polar deposits require the colder thermal environments provided by the permanently shadowed regions. Additionally, MESSENGER's discovery of high-reflectance surfaces within craters that thermally can support long-lived water ice at the surface but low-reflectance surfaces for craters whose temperatures require water ice to be buried to be stable (Figures 13.4 and 13.7) indicates a transition temperature for the volatile material that is consistent with that of water ice but not with the higher stability temperature of sulfur. Lastly, MESSENGER's detection of enhanced hydrogen in Mercury's north polar region (Figure 13.3) is strong evidence that water ice is the dominant volatile in Mercury's polar deposits; polar deposits of sulfur provide no explanation for the enhanced polar hydrogen measured by MESSENGER.

Another hypothesis proposed for Mercury's radar-bright materials is that the high-radar-backscatter signals originate from the altered dielectric properties of silicates by very low temperatures (Starukhina, 2001). Although this hypothesis is consistent with the fact that the radar-bright materials are found in locations that experience the lowest temperatures on Mercury, it is not the best explanation for other observations. For example, the permanently shadowed regions of the Moon, which are even colder than those on Mercury (Paige et al., 2010), lack similar strong radar-backscatter signals, though this difference has been suggested as due to the different silicate compositions of the two bodies (Starukhina, 2001). However, the altered-dielectric-properties hypothesis does not offer an explanation for the high-reflectance and low-reflectance surfaces measured in the radar-bright regions by the MLA and MDIS instruments. Finally, enhanced polar hydrogen abundances are not predicted by the altered-dielectric-properties hypothesis, as this hypothesis predicts no compositional difference between neighboring radar-bright and non-radar-bright surfaces.

The original hypothesis put forward to explain the first radar detections (Slade et al., 1992; Harmon and Slade, 1992; Butler et al., 1993) was that Mercury's polar deposits are dominantly composed of water ice, and multiple MESSENGER observations now provide further support for this interpretation. The enhanced hydrogen abundance measured in Mercury's north polar region by MESSENGER's NS is strong evidence for the presence of water ice (Lawrence et al., 2013). The determination that radar-bright deposits in both Mercury's north and south polar regions lie within regions of permanent shadow (Chabot et al., 2012, 2013; Mazarico et al., 2014; Deutsch et al., 2016) is consistent with the prediction of the water ice hypothesis. Thermal modeling with the MESSENGER-determined topography of Mercury's north polar region shows that the permanently shadowed regions have temperatures conducive to the long-lived stability of water ice deposits, either at the surface or in the near subsurface (Paige et al., 2013). Additionally, in regions where thermal models predict that surface water ice would be stable, higher reflectance surfaces are observed by both active near-infrared reflectance measurements by MLA (Neumann et al., 2013; Deutsch et al., 2017a) and visible-wavelength imaging by MDIS (Chabot et al., 2014) and are interpreted as evidence for water ice exposed at the surface. In regions where the thermal models predict that water ice is stable only if buried a few tens of centimeters below the surface, the same measurements reveal low-reflectance surfaces, interpreted to be a thin covering of volatile, organic-rich material that developed as a sublimation lag deposit and that insulates the water ice beneath it. Thus, the body of evidence from Earth-based radar and MESSENGER observations provide complementary and compelling evidence to answer MESSENGER's major science question: "What are the radar-reflective materials at Mercury's poles?" Mercury's polar deposits are dominantly composed of water ice.

13.3.2 Constraints on the Source of Mercury's Polar Water Ice Deposits

MESSENGER's observations of Mercury's polar deposits provide more than just evidence that the deposits are composed of water ice. Data from MESSENGER also yield new insights into the source, and hence the age and evolution, of water ice on Mercury and the history of volatiles in the inner solar system. Below are four key constraints, followed by discussion of the implications and consideration of the possible sources of Mercury's polar water ice deposits.

13.3.2.1 Constraints

(1) Mercury's Polar Deposits Are Dominantly but Not Solely Water Ice

One of the key new insights obtained from MESSENGER data is that although Mercury's polar deposits are dominantly water ice, they also contain other volatile materials. Although there are no measurements of the composition of these other volatiles, they are thought to be the key constituents of the low-reflectance insulating layer observed in permanently shadowed regions that have surface temperatures greater than ~110 K, where surface water ice is not stable. The surfaces of this material have very low reflectance values, approximately half that of Mercury's normal regolith surface as measured by MLA at 1064 nm (Neumann et al., 2013), and surface brightness variations suggest the presence of multiple low-reflectance compounds in the deposits (Chabot et al., 2016). The low-reflectance surfaces are located only in permanently shadowed regions, suggesting that the low-reflectance layer is not stable in environments with maximum temperatures exceeding ~300 K. These combined observations support the hypothesis that the low-reflectance layers are composed of carbon-rich materials that have low reflectance values, and, while volatile, are stable to higher temperatures than water ice (Paige et al., 2013). Organic compounds vary widely in their volatilities, but those that are stable to sublimation at higher temperatures than water ice are complex and have high molecular weights, in contrast with most simple organic compounds, which are more volatile than water (Figure 13.8; Zhang and Paige, 2009, 2010). Such complex organic molecules are found in primitive meteorites (Botta and Bada, 2002), leading Zhang and Paige (2009, 2010) to suggest that if such materials were found in the polar cold traps on Mercury or the Moon, an impact origin for their source would be implied. The formation of higher-order organic compounds in Mercury's polar deposits through the chemical processing of simple ices by energetic magnetospheric particles, GCRs, and other sources has also been suggested (Crites et al., 2013; Delitsky et al., 2017). Paige et al. (2013) proposed that the low-reflectance layers could have formed rapidly after an impact on Mercury delivered both water and organic compounds to cold traps in the same permanently shaded locations. Subsequent evolution of these cold traps would have favored the sublimation of near-surface water ice, leaving a surface lag of complex organic material having higher volatility temperature and lower reflectance. Simulations of large impact events suggest that most of this organic material would be destroyed during the impact, but some fraction may survive, depending on impact conditions (Pierazzo and Chyba, 1999, 2006), though such simulations have not yet been specifically conducted for Mercury.

The presence of volatile materials other than water ice is an important constraint on the source of Mercury's polar deposits. If Mercury's water were predominantly formed by interaction of surface regolith with protons from the solar wind, such interactions would produce OH or H_2O but the production of other volatiles is not predicted (McCord et al., 2011).

Outgassing of volatile materials from Mercury's interior is also an option for the source of Mercury's polar deposits. MESSENGER found considerable evidence for past volcanic activity on Mercury, including features interpreted as vents surrounded by pyroclastic deposits interpreted to be the product of explosive volcanic eruptions driven by exsolution of volatiles in ascending magma (Chapter 11). Additionally, MESSENGER determined that Mercury is not depleted in volatile elements, with both sulfur and carbon measured on the surface (Chapter 2). Graphite and amorphous carbon are not volatile but rather stable at temperatures across Mercury's surface (Bruck Syal et al., 2015), whereas the volatility temperatures of CO_2 and SO_2 are lower than that of H_2O, as are those of simple C-bearing molecules (Figure 13.8; Zhang and Paige, 2009, 2010). Hence, such materials cannot account for the low-reflectance surfaces observed on Mercury's polar deposits. However, elemental S is volatile but also stable to a higher temperature, as noted by Sprague et al. (1995) and shown on Figure 13.8, generally consistent with the biannual maximum surface temperatures experienced by the low-reflectance deposits.

Although indirect evidence supports the inference that the low-reflectance volatiles are organic compounds, determining the chemical and molecular composition of the low-reflectance volatile material would resolve this question and provide critical information on the source and evolution of Mercury's water ice deposits.

(2) Mercury's Polar Deposits Have Distinct Surface Reflectance Values and Sharp Boundaries

Another key new observation from MESSENGER about Mercury's polar deposits is that the surfaces of the deposits have distinctive reflectance characteristics that differ from the rest of Mercury's surface. The existence of these reflectance differences, and the fact that regolith gardening or other processes have not destroyed them, provides a constraint on the age of the deposits. Crider and Killen (2005) modeled the burial of water ice on Mercury through regolith emplacement and concluded that 20 cm of regolith would cover the water ice deposits in <50 Myr if water ice is lost from Mercury's surface by meteoroid impacts. Whereas migration simulations suggest that 98% of water molecules in polar cold traps are likely to recondense in those traps if disturbed (Butler et al., 1993), rendering the specific age estimate by Crider and Killen (2005) uncertain, MESSENGER observations indicate that Mercury's polar deposits are not dominantly buried by regolith but rather by material lower in reflectance than normal regolith, and the low-reflectance deposits show well-defined boundaries (Figure 13.10). The preservation of these distinctive reflectance properties and sharp boundaries indicates a geologically young age for the deposits.

The purity of the water ice exposed on the floor of Prokofiev is not known, but laboratory studies indicate that, depending on the ice particle size and viewing conditions, ice concentrations less than ~50% do not provide strong effects on the reflectance at visible to near-infrared wavelengths (Yoldi et al., 2015). Thus, since MLA zero-phase-angle observations and MDIS high-phase-angle imaging were both able to detect the higher-reflectance surface within Prokofiev, the majority of the surface must be composed of exposed water ice, placing a limit on the silicate regolith mixed into the region. Additionally, the temperatures within Prokofiev are sufficiently low that water molecules have very little diffusive mobility, and hence a diffusion process cannot account for the high-reflectance surface (Paige et al., 2013).

Other processes that would alter the water ice deposits on Mercury have also been proposed, including destruction by Lyman alpha photodissociation (Morgan and Shemansky,

1991), organic synthesis within ice bombarded by GCRs and solar energetic particles (Crites et al., 2013; Delitsky et al., 2017), and chemical reactions induced by magnetospheric charged particles funneled onto Mercury's surface at high latitudes that could produce organic compounds and dark refractory materials in the ice (Delitsky et al., 2017). Of these potential energy inputs that might initiate chemical reactions in Mercury's polar ices to yield new heavier-molecular-weight organic products, Delistky et al. (2017) concluded that magnetospheric particle energy deposition is two or more orders of magnitude greater than each of the other potential energy sources. However, the fact that Mercury has polar deposits with both higher-reflectance surfaces, such as in Prokofiev (Figure 13.9), and lower-reflectance surfaces, such as shown in Figure 13.10, argues against the idea that the surface reflectance of the deposits is produced dominantly by a weathering or alteration process, which would be expected to affect all of Mercury's polar deposits in a relatively uniform manner. Instead, the well-preserved surface reflectance properties, with both high- and low-reflectance surfaces, and the sharp boundaries of the polar deposits suggest that they are geologically young. The implication for the source of the water ice is that the material was either recently delivered to the planet or is actively refreshed and maintained, either from external sources or local redistribution.

(3) Mercury's Polar Deposits Are Relatively Pure and Extensive

The conclusion that Mercury's polar deposits are relatively pure water ice is not a new inference, but it remains an important constraint on the source, age, and evolution of water ice on Mercury. Radar observations of materials with a high circular polarization ratio and high radar reflectivity are indicative of icy materials on other solar system objects. Such deposits produce coherent backscatter effects from the essentially transparent nature of ice at radar wavelengths that results in multiple volume scattering with little radar energy absorption (Hapke, 1990; Hapke and Blewett, 1991). The volume scattering occurs from variations in density, which can be caused by silicate inclusions or fractures and voids. However, calculations indicate that to match the observed radar properties from Mercury's north polar region, even a small volume fraction of silicates mixed with the ice would contribute substantial absorption, leading to the conclusion that the ice in Mercury's polar deposits must be nearly pure, estimated at less than ~5% silicates by volume (Butler et al., 1993). The high radar reflectivity and circular polarization ratio have been observed at radar wavelengths of 3.5, 12.6, and 70 cm. If many multiple scatterings are required to yield the bright radar returns, then the scattering layer must be many wavelengths thick, implying that the ice has a minimum thickness of several meters (Black et al., 2010). However, the maximum possible thickness of Mercury's water ice deposits is poorly constrained by radar or MESSENGER observations and has been estimated to range from a few hundreds of meters to tens of meters or less (Section 13.2.1).

That Mercury's extensive water ice is relatively pure has also been interpreted to suggest that the water ice was likely deposited over a relatively short time period, because silicate particles would otherwise be present in larger amounts within the deposits (Butler et al., 1993). An episodic source, such as the impact of a large comet or asteroid, is favored by such an argument, rather than a more nearly continuous source for Mercury's water ice. More continuous sources, such as sustained planetary outgassing or micrometeoroid delivery, are less likely to build up relatively pure water ice deposits, given that regolith gardening processes operate concurrently. However, the ability to produce pure water ice deposits would depend on the specific rates for all processes involved in the evolution of the deposits. Specific models that examine quantitatively the purity of water ice deposits that could be produced by micrometeoroid impacts or planetary outgassing on Mercury have yet to be developed. Improved constraints on the thickness of the water ice deposits and their purity could also provide limits on the total volume of material delivered and the timing of its emplacement.

(4) Polar Deposits Similar to Those on Mercury Are Not Seen on the Moon

The Moon, like Mercury, has regions of permanent shadow near its poles (Bussey et al., 1999, 2005, 2010; Noda et al., 2008; Mazarico et al., 2011; Speyerer and Robinson, 2013), and, also as on Mercury, these regions are sufficiently cold to host thermally stable water ice (Paige et al., 2010). However, in contrast to Mercury, Earth-based radar observations of the Moon do not show extensive radar-bright deposits in these shadowed regions (Stacy et al., 1997; B. A. Campbell et al., 2003; D. B. Campbell et al., 2006). Spacecraft radar observations, although suggestive of the potential for patchy occurrences of lunar ice, have not shown evidence for lunar water ice deposits in quantities comparable to the deposits in the polar regions of Mercury (Nozette et al., 1996; Simpson and Tyler, 1999; Spudis et al., 2010; Neish et al., 2011; Thomson et al., 2012). Neutron spectrometer measurements indicate enhanced hydrogen in the polar regions of the Moon, but the inferred concentrations are substantially lower than those observed at Mercury (Feldman et al., 1998, 2000, 2001; Lawrence et al., 2006; Teodoro et al., 2010; Mitrofanov et al., 2010; Basilevsky et al., 2012; Miller et al., 2012). Similarly, water vapor detected in the ejecta plume from the Lunar Crater Observation and Sensing Satellite (LCROSS) suggested $5.6 \pm 2.9\%$ by mass for water ice in the lunar regolith at the permanently shadowed impact site within Cabeus crater (Colaprete et al., 2010), a weight fraction considerably lower than Mercury's relatively pure water ice deposits. While it is possible that higher concentrations of water ice are buried in the lunar subsurface, as a result of the downward migration of water ice from impact gardening, temperature oscillations, or changing thermal environments during the Moon's orbital evolution (Schorghofer and Taylor, 2007; Siegler et al., 2011, 2015, 2016), similar measurements conducted at both bodies indicate that Mercury's polar regions contain substantially more water ice than the polar regions on the Moon.

Furthermore, whereas laser reflectance measurements at 1064 nm for Mercury have revealed both low-reflectance and high-reflectance surfaces for the polar deposits, similar observations at the Moon have shown permanently shadowed surfaces with reflectance values only ~10% higher than average, indicative of modest amounts of water frost or a reduction in the effectiveness of space weathering (Lucey et al., 2014). No permanently shadowed regions with low-reflectance surfaces at 1064 nm have been seen on the Moon. Visible-wavelength imaging of permanently shadowed lunar craters has not revealed surfaces with anomalously high- or low-reflectance surfaces (Haruyama et al., 2008; Speyerer and Robinson, 2013), such as those seen in comparable images of Mercury's polar deposits. Observations at far-ultraviolet wavelengths of permanently shadowed lunar craters indicate redder, low-reflectance surfaces, consistent with high porosity and ~1–2 wt% water frost (Gladstone et al., 2012).

Although the Moon presently shows evidence for water ice in its polar regions, the form of the deposits is qualitatively different from that of polar deposits on Mercury (Lawrence, 2017). Thus, there is a fundamental difference between the volatile inventories of Mercury and the Moon, and resolving this enigma is central to advancing our understanding of the evolution of volatiles across the inner solar system.

13.3.2.2 Implications for the Source

The proposed sources for Mercury's water ice can be divided into four major categories: interactions with the solar wind, outgassing from Mercury's interior, delivery by micrometeoroids, or delivery by one or more large impact events. Table 13.1 summarizes some implications of the observational constraints discussed in the previous section for each of these proposed sources. It is possible, of course, that a combination of processes has acted together to deliver water to Mercury's polar regions, and it is also possible that different processes have provided the dominant source of water ice on Mercury and the Moon.

Solar wind interactions that could create water molecules are expected on both Mercury and the Moon, and Mercury's proximity to the Sun and the stronger interplanetary magnetic field could result in differences between the two bodies. However, solar wind interactions face other difficulties as the dominant source of Mercury's polar deposits. In particular, the production of other volatile materials to create the low-reflectance deposits observed on Mercury is not predicted. Additionally, an ongoing and steady production of water molecules by solar wind interactions, and their migration to the polar cold traps, has not been shown to be consistent with the extensive and relatively pure nature of Mercury's polar water ice deposits.

Mercury and the Moon have different surface chemical compositions and volcanic histories. Different histories of planetary outgassing could therefore be envisioned for the two bodies, potentially providing different amounts of water and other volatiles available to be cold-trapped in the polar regions. From MESSENGER, we now know that Mercury is more enhanced in volatiles than previously expected (Chapter 2), that volcanism, including pyroclastic eruptions that released magmatic volatiles, was a widespread process in shaping the planet (Chapter 11), and that some volatile loss may indeed still be occurring on Mercury at present, as evidenced by hollows (Chapter 12). The sharp boundaries and distinct surface reflectance of the polar deposits would require planetary outgassing within Mercury's recent past to maintain these characteristics. The rate of accumulation of water ice in Mercury's polar cold traps from outgassing would have to be considerably higher than that of regolith gardening and other surface modification processes to produce the high purity of water ice in Mercury's polar deposits. Whereas Mercury's surface composition is now

Table 13.1. *Implications for possible sources of Mercury's water ice from observational constraints of the planet's polar deposits.*

Source	Constraints on Mercury's polar water ice deposits			
	Not solely water ice	Distinct surface reflectances with sharp boundaries	Relatively pure water ice	Not seen on the Moon
Solar wind interactions	Not predicted	Ongoing process may maintain reflectances, but low-reflectance surfaces are not predicted	Regular and steady process not expected to produce pure deposits	Solar wind interactions occur on both but may manifest differently on Mercury and the Moon
Planetary outgassing	Low-reflectance deposits are volatiles from volcanic eruptions	Would need to be active at present or in recent past to explain surface reflectances	Rate of accumulation from outgassing would have to be faster than regolith mixing	Different compositions and volcanic histories can result in different outgassing
Micrometeoroids	Low-reflectance deposits are organic compounds and volatiles from micrometeoroids	Ongoing process may maintain surface reflectances and sharp boundaries	Regular and steady process not expected to produce pure deposits	Flux of micrometeoroids on Mercury is much larger than on the Moon
Large impact events	Low-reflectance deposits are organic compounds and volatiles from comets or asteroids	Occurred geologically recently to account for the sharp boundaries and surface reflectances	Episodic process can produce pure deposits by emplacement of large volumes of ice at one time	Mercury experienced a recent large impact event; the Moon has not but may have in the past

known to include volatiles, measurements of surface elemental abundances indicate that the planet's chemical composition is highly reduced (Chapter 2). An examination of the nature of possible magmatic volatiles that are expected given Mercury's chemically reduced conditions suggests that its interior is depleted in H and H_2O, and that a combination of C-, S-, Cl-, and Ni-bearing gases drove its pyroclastic activity (Zolotov, 2011). If interior is depleted in water, planetary outgassing would not be likely to produce the broad distribution and dominantly water ice composition of polar deposits that are observed.

Differences between the populations of regularly impacting micrometeoroids could also deliver different amounts of water to Mercury and the Moon, accounting for the different volatile inventories in the polar regions of the two bodies. With an analytical approach, Cintala (1992) calculated that the micrometeoroid flux is 5.5 times greater at Mercury than at the Moon. More recently, with a numerical approach, Borin et al. (2009, 2016) concluded that substantially more micrometeoroids impact Mercury than the Moon, with a flux that is ~58 times higher than that reported by Cintala (1992). Borin et al. (2009, 2016) assumed that the source of micrometeoroids was the main asteroid belt, whereas dynamical modeling work by Nesvorný et al. (2010) suggests that >90% of inner solar system micrometeoroids arise from Jupiter-family comets. This different source for the micrometeoroids would require the numerically modeled impact rate at Mercury to be scaled (Borin et al., 2009); one estimate gives the flux at Mercury as ~39 times greater than that at the Moon (Bruck Syal et al., 2015). However, cometary particles would also each contain greater amounts of water for delivery to Mercury. Monte Carlo models indicate that 20–50% of water molecules emplaced anywhere on the lunar surface will migrate to lunar cold traps, whereas Mercury's cold traps would capture 5–15% of such molecules (Butler, 1997). From simulations that account for the impacting populations, impact velocity, impact angle, and material properties, Bruck Syal and Schultz (2015) concluded that micrometeoroids provide >99% of Mercury's delivered water whereas water delivery to the Moon is dominated at >95% by asteroids. However, although a steady and ongoing process such as micrometeoroid bombardment can be envisioned to regularly refresh the surfaces of Mercury's polar deposits, perhaps creating the distinctive reflectances and sharp boundaries observed, it may be challenging for this process to build up extensive water ice deposits that have the purity determined on Mercury. Thus, while there is general agreement that substantially more micrometeoroids impact Mercury than the Moon, quantitatively robust models are needed to understand the delivery of water to Mercury by micrometeoroids, the subsequent evolution of that water, and the rate of its accumulation in polar cold traps relative to other processes, as well as to investigate whether micrometeoroids are a dominant source for the delivery of water to either object.

Alternatively, an episodic source, such as the impact of one or several large comets or volatile-rich asteroids, could yield the differences in the water ice concentrations observed in the polar regions of Mercury and the Moon. Monte Carlo simulations by Moses et al. (1999) indicated that a small number of massive objects might have delivered most of Mercury's water inventory, with Jupiter-family comets in particular having the potential to deliver large amounts of water to the planet. If a small number of large impacts were the dominant source for water ice on Mercury or the Moon, then differences between the inventories of these two bodies would be expected, depending on how long ago the most recent such impact occurred on each object. In such a scenario, Mercury's presently extensive water ice deposits could indicate that a large impact delivered water to the planet in the relatively recent past, whereas the Moon may have experienced such events in the more distant past, perhaps as much as billions of years ago (Siegler et al., 2015, 2016). The differences in characteristics of water ice deposits among different lunar polar cold traps has been suggested as evidence for an episodic rather than steady-state source (Spudis et al., 2010; Miller et al., 2014). A relatively recent event, such as the impact of a large comet or asteroid in the recent past, provides a natural explanation for Mercury's relatively pure water ice deposits with their sharp boundaries and their distinctive surface reflectances arising from low-reflectance, organic-rich volatile compounds delivered by the impactor at the same time as the water ice.

A major question for a large comet or asteroid impact source is whether the water ice can survive the energetic impact event. Modeling the volatile retention from cometary impacts on the Moon, Ong et al. (2010) found that impacts at low velocities (~5 km/s) retain nearly all of the water but at high velocities (~60 km/s) nearly all of the water is lost. Overall, Ong et al. (2010) concluded that asteroids provide six times more water to the Moon than comets and that such delivery could match the measured lunar concentrations of water ice. However, impacts on Mercury have a higher average impact velocity and span a larger overall range of impact velocities (Le Feuvre and Wieczorek, 2008; Marchi et al., 2009), which could decrease water retention during the impact event, though the larger surface gravitational acceleration could increase retention relative to the Moon. On the basis of a comet-impact simulation on the Moon that began with the impact event and followed the water until it was lost or deposited in a lunar cold trap, Stewart et al. (2011) concluded that ~0.1% of the cometary water was retained on the Moon in polar cold traps. They also concluded that migration of water molecules by hopping on the lunar surface, such as modeled by Butler (1997), is not the most important process for the transport and deposition that occurs during a large impact, given the density of the initial gas in the comet impact and the potential for the creation of a transient atmosphere (Stewart et al., 2011). Impact simulations at 45°N and 45°S latitudes predicted trapping of water molecules near both the lunar north and south poles (Stewart et al., 2011), suggesting that a large impact event could be responsible for water ice being transported to both polar regions. Subsequent numerical simulations indicated that a transient, impact-generated lunar atmosphere could be sufficiently dense that its lower layers are shielded from photodestruction, prolonging the lifetime of water molecules and allowing greater amounts of water to reach lunar polar cold traps (Prem et al., 2015). Similar modeling of a large impact event on Mercury would provide valuable insight into the possibility that Mercury's water ice was delivered by a large impact, especially considering the differences in the gravitational acceleration, impact velocity, rotation rate, and thermal environment between the Moon and Mercury.

Given that Mercury's total water ice abundance is estimated at ~10^{16}–10^{18} g (Moses et al., 1999; Lawrence et al., 2013) and that estimates for the amount of impacting water captured in polar cold traps range from ~0.1% (for a cometary impact on the Moon rather than Mercury) (Stewart et al., 2011) to ~1% (for Monte Carlo modeling of Jupiter-family comets impacting Mercury followed by random-hop migration) (Butler, 1997; Moses et al., 1999), a single impacting object would have to contain 10^{18}–10^{21} g of water. Given that a comet of pure water ice without porosity provides a lower limit to the size of such an impacting object, an impactor diameter of at least ~10–100 km would be needed. The diameter of the crater associated with such an impact would be approximately an order of magnitude larger than that of the impacting object and would depend on impact conditions (e.g., Collins et al., 2005). There are considerable uncertainties associated with this simple estimate, but if a single impact event delivered all of the water ice observed in Mercury's polar deposits, such an impact should have produced a sizable crater. The 96-km-diameter crater Hokusai is distinctive, as it is one of Mercury's youngest large craters and its formation generated the planet's longest set of rays, which extend for thousands of kilometers (Xiao et al., 2016). The size of Hokusai crater is on the lower end of estimates for the source crater of Mercury's water ice deposits (Ernst et al., 2016), but size alone is insufficient to rule out Hokusai as a source-crater candidate because of the uncertainties currently associated with estimates for the retention of volatiles on Mercury during impact events. In general, if the sources of Mercury's water ice deposits were large, recent impact events, investigations to identify the corresponding crater or craters should be pursued.

In summary, a recent large impact event can most easily account for the observed constraints on Mercury's polar deposits, but there is much work left to be done to examine the proposed source hypotheses further, given the new results from the MESSENGER mission.

13.4 FUTURE EXPLORATION

Results from the MESSENGER mission, combined with previous Earth-based radar observations, provide multiple lines of evidence that Mercury's polar deposits are composed of substantial amounts of water ice. Moreover, the characteristics of the water ice deposits, such as the presence of low-reflectance volatiles, the distinctive surface reflectances, the relative purity of the water ice, and the contrast with the polar regions on the Moon, provide constraints on the source of the water, and hence on its age and evolution. Future research can provide important insights into the origin and evolution of water ice on Mercury and the delivery of volatiles throughout the inner solar system.

Further analyses of MESSENGER data are needed. These include, but are not limited to, high-resolution studies to explore specific water-ice-bearing craters, analyses to constrain the depth and thickness of the water ice deposits, investigations of small craters that host water ice deposits, and models of the reflectance properties of the deposits in relation to their purity and surface composition. The MESSENGER results also provide a clear rationale for modeling efforts to simulate the wide-ranging processes that have been hypothesized as the source of Mercury's water ice. Additional Earth-based radar observations could provide insight into the similarities and differences between individual deposits and improve overall coverage, especially in the south polar region.

BepiColombo, a joint mission of the European Space Agency and the Japan Aerospace Exploration Agency (Benkhoff et al., 2010), which is currently scheduled to launch in 2018 and to insert two spacecraft into orbit about Mercury in 2025, has the opportunity to provide new observations of Mercury's water ice deposits (Chapter 20). The majority of MESSENGER's observations of Mercury's polar deposits were limited to the north polar region because of the spacecraft's highly eccentric orbit and high northern periapsis, whereas BepiColombo's Mercury Planetary Orbiter (MPO) will have a near-equatorial periapsis and should pass at similar altitudes over both of Mercury's poles. Depending on the final altitude of MPO, BepiColombo could provide the first neutron spectrometry, laser altimetry, and visible-wavelength imaging of the shadowed surfaces of the water ice deposits in Mercury's south polar region. Such measurements would be complementary to similar measurements made by MESSENGER for the north polar region. Additionally, the Mercury Radiometer and Thermal Infrared Spectrometer (MERTIS) instrument on BepiColombo's MPO includes an infrared spectrometer that will operate in the wavelength range 7–14 μm and a radiometer that will operate in the wavelength range 7–40 μm, and there are plans to operate in a special "polar mode" to enable measurements of Mercury's polar deposits (Hiesinger et al., 2010). No such measurements were made by MESSENGER, so MERTIS has the potential to provide new compositional and thermal measurements of Mercury's polar deposits. In particular, MERTIS spectral measurements may constrain the composition of the low-reflectance layer that covers the majority of Mercury's polar deposits. Such information could test hypotheses for the source of Mercury's water ice and would have implications for the distribution of volatiles in the solar system.

Future exploration of Mercury by spacecraft after BepiColombo that are focused on understanding Mercury's water ice deposits or the inventory of inner solar system volatiles may use landed payloads, providing sensitive chemical and isotopic measurements not possible from remote orbital observations. Resolving the differences between Mercury's extensive water ice deposits and the polar volatile deposits measured on the Moon remains a fundamental question tied to the overarching goal to understand the sources and evolution of water in the inner solar system. In situ measurements of water on both the Moon and Mercury will be key to resolving these questions and gaining this larger understanding, including an understanding of the sources of water on Earth.

REFERENCES

Barnouin, O. S., Zuber, M. T., Smith, D. E., Neumann, G. A., Herrick, R. R., Chappelow, J. E., Murchie, S. L. and Prockter, L. M. (2012). The morphology of craters on Mercury: Results from MESSENGER flybys. *Icarus*, **219**, 414–427.

Basilevsky, A. T., Abdrakhimov, A. M. and Dorofeeva, V. A. (2012). Water and other volatiles on the Moon: A review. *Solar Syst. Res.*, **46**, 89–107.

Benkhoff, J., van Casteren, J., Hayakawa, H., Fujimoto, M., Laakso, H., Novara, M., Ferri, P., Middleton, H. R. and Ziethe, R. (2010). BepiColombo – Comprehensive exploration of Mercury: Mission overview and science goals. *Planet. Space Sci.*, **58**, 2–20.

Black, G. J., Campbell, D. B. and Nicholson, P. D. (2001). Icy Galilean satellites: Modeling radar reflectivities as coherent backscatter effect. *Icarus*, **151**, 167–180.

Black, G. J., Campbell, D. B. and Harmon, J. K. (2010). Radar measurements of Mercury's north pole at 70 cm wavelength. *Icarus*, **209**, 224–229, doi:10.1016/j.icarus.2009.10.009.

Borin, P., Cremonese, G., Marzari, F., Bruno, M. and Marchi, S. (2009). Statistical analysis of micrometeoroids flux on Mercury. *Astron. Astrophys.*, **503**, 259–264, doi:10.1051/0004-6361/200912080.

Borin, P., Cremonese, G., Bruno, M. and Marzari F. (2016). Asymmetries in the dust flux at Mercury. *Icarus*, **264**, 220–226.

Botta, O. and Bada, J. L. (2002). Extraterrestrial organic compounds in meteorites. *Surv. Geophys.*, **23**, 411–467.

Bruck Syal, M. and Schultz, P. H. (2015). Impact delivery of water at the Moon and Mercury. *Lunar Planet. Sci.*, **46**, abstract 1680.

Bruck Syal, M., Schultz, P. H. and Riner, M. A. (2015). Darkening of Mercury's surface by cometary carbon. *Nature Geosci.*, **8**, 352–356, doi:10.1038/NGE02397.

Bussey, D. B. J., Spudis, P. D. and Robinson, M. S. (1999). Illumination conditions at the lunar south pole. *Geophys. Res. Lett.*, **26**, 1187–1190.

Bussey, D. B. J., Fristad, K. E., Schenk, P. M., Robinson, M. S. and Spudis, P. D. (2005). Constant illumination at the lunar north pole. *Nature*, **434**, 842.

Bussey, D. B. J., McGovern, J. A., Spudis, P. D., Neish, C. D., Noda, H., Ishihara, Y. and Sorensen, S.-A. (2010). Illumination conditions of the south pole of the Moon derived using Kaguya topography. *Icarus*, **208**, 558–564.

Butler, B. J. (1997). The migration of volatiles on the surfaces of Mercury and the Moon. *J. Geophys. Res.*, **102**, 19,283–19,291.

Butler, B. J., Muhleman, D. O. and Slade, M. A. (1993). Mercury: Full-disk radar images and the detection and stability of ice at the north pole. *J. Geophys. Res.*, **98**, 15,003–15,023.

Campbell, B. A., Campbell, D. B., Chandler, J. F., Hine, A. A., Nolan, M. C. and Perillat, P. J. (2003). Radar imaging of the lunar poles. *Nature*, **426**, 137–148.

Campbell, D. B., Chandler, J. F., Ostro, S. J., Pettengill, G. H. and Shapiro, I. I. (1978). Galilean satellites: 1976 radar results. *Icarus*, **34**, 254–267.

Campbell, D. B., Campbell, B. A., Carter, L. M., Margot, J.-L. and Stacy, N. J. S. (2006). No evidence for thick deposits of ice at the lunar south pole. *Nature*, **443**, 835–837.

Carrier, W. D., III, Olhoeft, G. R. and Mendell W. (1991). Physical properties of the lunar surface. In *Lunar Sourcebook: A User's Guide to the Moon*, ed. G. Heiken, D. Vaniman and B. M. French. Cambridge: Cambridge University Press, pp. 475–594.

Cavanaugh, J. F., Smith, J. C., Sun, X., Bartels, A. E., Ramos-Izquierdo, L., Krebs, D. J., McGarry, J. F., Trunzo, R., Novo-Gradac, A. M., Britt, J. L., Karsh, J., Katz, R. B., Lukemire, A. T., Szymkiewicz, R., Berry, D. L., Swinski, J. P., Neumann, G. A., Zuber, M. T. and Smith, D. E. (2007). The Mercury Laser Altimeter instrument for the MESSENGER mission. *Space Sci. Rev.*, **131**, 451–480.

Chabot, N. L., Ernst, C. M., Denevi, B. W., Harmon, J. K., Murchie, S. L., Blewett, D. T., Solomon, S. C. and Zhong, E. D. (2012). Areas of permanent shadow in Mercury's south polar region ascertained by MESSENGER orbital imaging. *Geophys. Res. Lett.*, **39**, L09204, doi:10.1029/2012GL051526.

Chabot, N. L., Ernst, C. M., Harmon, J. K., Murchie, S. L., Solomon, S. C., Blewett, D. T. and Denevi, B. W. (2013). Craters hosting radar-bright deposits in Mercury's north polar region: Areas of persistent shadow determined from MESSENGER images. *J. Geophys. Res. Planets*, **118**, 26–36.

Chabot, N. L., Ernst, C. M., Denevi, B. W., Nair, H., Deutsch, A. N., Blewett, D. T., Murchie, S. L., Neumann, G. A., Mazarico, E., Paige, D. A., Harmon, J. K., Head, J. W. and Solomon, S. C. (2014). Images of surface volatiles in Mercury's polar craters acquired by the MESSENGER spacecraft. *Geology*, **42**, 1051–1054.

Chabot, N. L., Ernst, C. M., Paige, D. A., Nair H., Denevi, B. W., Blewett, D. T., Murchie, S. L., Deutsch, A. N., Head, J. W. and Solomon, S. C. (2016). Imaging Mercury's polar deposits during MESSENGER's low-altitude campaign. *Geophys. Res. Lett.*, **43**, 9461–9468.

Cintala, M. J. (1992). Impact-induced thermal effects in the lunar and mercurian regoliths. *J. Geophys. Res.*, **97**, 947–973.

Colaprete, A., Schultz, P., Heldmann, J., Wooden, D., Shirley, M., Ennico, K., Hermalyn, B., Marshall, W., Ricco, A., Elphic, R. C., Goldstein, D., Summy, D., Bart, G. D., Asphaug, E., Korycansky, D., Landis, D. and Sollitt, L. (2010). Detection of water in the LCROSS ejecta plume. *Science*, **330**, 463–468.

Collins, G. S., Melosh, H. J. and Marcus, R. A. (2005). Earth impact effects program: A web-based computer program for calculating the regional environmental consequences of a meteoroid impact on Earth. *Meteorit. Planet. Sci.*, **40**, 817–840.

Crider, D. and Killen, R. M. (2005). Burial rate of Mercury's polar volatile deposits. *Geophys. Res. Lett.*, **32**, L12201, doi:10.1029/2005GL022689.

Crites, S. T., Lucey, P. G. and Lawrence, D. J. (2013). Proton flux and radiation dose from galactic cosmic rays in the lunar regolith and implications for organic synthesis at the poles of the Moon and Mercury. *Icarus*, **226**, 1192–1200.

Davies, M. E., Dwornik, S. E., Gault, D. E. and Strom, R. G. (1978). *Atlas of Mercury*, Special Publication SP-423. Washington, DC: National Aeronautics and Space Administration.

Delitsky, M. L., Paige, D. A., Siegler, M. A., Harju, E. R., Schriver, D., Johnson, R. E. and Travnicek, P. (2017). Ices on Mercury: Chemistry of volatiles in permanently cold areas of Mercury's north polar region. *Icarus*, **281**, 19–31.

Deutsch, A. N., Chabot, N. L., Mazarico, E., Ernst, C. M., Head, J. W., Neumann, G. A. and Solomon, S. C. (2016). Comparison of areas in shadow in the north polar region of Mercury from imaging and altimetry, with implications for polar ice deposits. *Icarus*, **280**, 158–171.

Deutsch, A. N., Neumann, G. A. and Head, J. W. (2017a). New evidence for surface water ice in small-scale cold traps and in three large craters at the north polar region of Mercury from the Mercury Laser Altimeter. *Geophys. Res. Lett.*, **44**, 9233–9241, doi:10.1002/2017GL074723.

Deutsch, A. N., Head, J. W., Neumann, G. A. and Chabot, N. L. (2017b). Constraining the depth of polar ice deposits and evolution of cold traps on Mercury with small craters in permanently shadowed regions. *Lunar Planet. Sci.*, **48**, abstract 1634.

Eke, V. R., Lawrence, D. J. and Teodoro, L. F. A. (2017). How thick are Mercury's polar water deposits? *Icarus*, **284**, 407–415.

Ernst, C. M., Chabot, N. L., Susorney, H. C., Barnouin, O. S., Harmon, J. K. and Paige, D. A. (2014). Exploring the morphology of simple craters that host polar deposits on Mercury: Implications for the source and stability of water ice. *Lunar Planet. Sci.*, **45**, abstract 1238.

Ernst, C. M., Chabot, N. L. and Barnouin, O. S. (2016). Examining the potential contributions of the Hokusai impact to water ice on Mercury. *Lunar Planet. Sci.*, **47**, abstract 1374.

Feldman, W. C. and Drake, D. M. (1986). A Doppler filter technique to measure the hydrogen content of planetary surfaces. *Nucl. Instrum. Methods A*, **245**, 182–190.

Feldman, W. C., Drake, D. M., O'Dell, R. D., Brinkley, F. W., Jr. and Anderson, R. C. (1989). Gravitational effects on planetary neutron flux spectra. *J. Geophys. Res.*, **94**, 513–525.

Feldman, W. C., Barraclough, B. L., Hansen, C. J. and Sprague, A. L. (1997). The neutron signature of Mercury's volatile polar deposits. *J. Geophys. Res.*, **102**, 25,565–25,574.

Feldman, W. C., Maurice, S., Binder, A. B., Barraclough, B. L., Elphic, R. C. and Lawrence, D. J. (1998). Fluxes of fast and epithermal neutrons from Lunar Prospector: Evidence for water ice at the lunar poles. *Science*, **281**, 1496–500.

Feldman, W. C., Lawrence, D. J., Elphic, R. C., Barraclough, B. L., Maurice, S., Genetay, I. and Binder, A. B. (2000). Polar hydrogen deposits on the Moon. *J. Geophys. Res.*, **105**, 4175–4195.

Feldman, W. C., Maurice, D., Lawrence, D. J., Little, R. C., Lawson, S. L., Gasnault, O., Wiens, R. C., Barraclough, B. L., Elphic, R. C., Prettyman, T. H., Steinberg, J. T. and Binder, A. B. (2001). Evidence for water ice near the lunar poles. *J. Geophys. Res.*, **106**, 23,231–23,251, doi:10.1029/2000JE001444.

Feldman, W. C., Boynton, W. V., Tokar, R. L., Prettyman, T. H., Gasnault, O., Squyres, S. W., Elphic, R. C., Lawrence, D. J., Lawson, S. L., Maurice, S., McKinnet, G. W., Moore, K. R. and Reedy, R. C. (2002). Global distribution of neutrons from Mars: Results from Mars Odyssey, *Science*, **297**, 75–78.

Feldman, W. C., Mellon, M. T., Gasnault, O., Diez, B., Elphic, R. C., Hagerty, J. J., Lawrence, D. J., Maurice, S. and Prettyman, T. H. (2007). Vertical distribution of hydrogen at high northern latitudes on Mars: The Mars Odyssey Neutron Spectrometer. *Geophys. Res. Lett.*, **34**, L05201, doi:10.1029/2006GL028530.

Gladstone, G. R., Retherford, K. D., Egan, A. F., Kaufmann, D. E., Miles, P. F., Parker, J. W., Horvath, D., Rojas, P. M., Versteeg, M. H., Davis, M. W., Greathouse, T. K., Slater, D. C., Mukherjee, J., Steffl, A. J., Feldman, P. D., Hurley, D. M., Pryor, W. R., Hendrix, A. R., Mazarico, E. and Stern, S. A. (2012). Far-ultraviolet reflectance properties of the Moon's permanently shadowed regions. *J. Geophys. Res.*, **117**, E00H04, doi:10.1029/2011JE003913.

Goldsten, J. O., Rhodes, E. A., Boynton, W. V., Feldman, W. C., Lawrence, D. J., Trombka, J. I., Smith, D. M., Evans, L. G., White, J., Madden, N. W., Berg, P. C., Murphy, G. A., Gurnee, R. S., Strohbehn, K., Williams, B. D., Schaefer, E. D., Monaco, C. A., Cork, C. P., Del Eckels, J., Miller, W. O., Burks, M. T., Hagler, L. B., DeTeresa, S. J. and Witte, M. C. (2007). The MESSENGER Gamma-Ray and Neutron Spectrometer. *Space Sci. Rev.*, **131**, 339–391.

Hapke, B. (1990). Coherent backscatter and the radar characteristics of outer planet satellites. *Icarus*, **88**, 407–17.

Hapke, B. and Blewett, D. (1991). Coherent backscatter model for the unusual radar reflectivity of icy satellites. *Nature*, **352**, 46–47.

Harcke, L. J. (2005). Radar imaging of solar system ices. Ph.D. thesis, Stanford University, Stanford, CA, 201 pp.

Harmon, J. K. (2007). Radar imaging of Mercury. *Space Sci. Rev.*, **132**, 307–349.

Harmon, J. K. and Slade, M. A. (1992). Radar mapping of Mercury: Full-disk images and polar anomalies. *Science*, **258**, 640–643.

Harmon, J. K., Perillat, P. J. and Slade, M. A. (2001). High-resolution radar imaging of Mercury's north pole. *Icarus*, **149**, 1–15.

Harmon, J. K., Slade, M. A., Vélez, R. A., Crespo, A., Dryer, M. J. and Johnson, J. M. (1994). Radar mapping of Mercury's polar anomalies. *Nature*, **369**, 213–215.

Harmon, J. K., Slade, M. A. and Rice, M. S. (2011). Radar imagery of Mercury's putative polar ice: 1999–2005 Arecibo results. *Icarus*, **211**, 37–50.

Haruyama, J., Ohtake, M., Matsunaga, T., Morota, T., Honda, C., Yokota, Y., Pieters, C. M., Hara, S., Hioki, K., Saiki, K., Miyamoto, H., Iwasaki, A., Abe, M., Ogawa, Y., Takeda, H., Shirao, M., Yamaji, A. and Josset, J. L. (2008). Lack of exposed ice inside lunar south pole Shackleton crater. *Science*, **322**, 938–939, doi:10.1126/science.1164020.

Hawkins, S. E., III, Boldt, J. D., Darlington, E. H., Espiritu, R., Gold, R. E., Gotwols, B., Grey, M. P., Hash, C. D., Hayes, J. R., Jaskulek, S. E., Kardian, C. J., Jr., Keller, M. R., Malaret, E. R., Murchie, S. L., Murphy, P. K., Peacock, K., Prockter, L. M., Reiter, R. A., Robinson, M. S., Schaefer, E. D., Shelton, R. G., Sterner, R. E., II, Taylor, H. W., Watters, T. R. and Williams, B. D. (2007). The Mercury Dual Imaging System on the MESSENGER spacecraft. *Space Sci. Rev.*, **131**, 247–338.

Hiesinger, H., Helbert, J. and MERTIS Co-I Team (2010). The Mercury Radiometer and Thermal Infrared Spectrometer (MERTIS) for the BepiColombo mission. *Planet. Space Sci.*, **58**, 144–165.

Ingersoll, A. P., Svitek, T. and Murray, B. C. (1992). Stability of polar frosts in spherical bowl-shaped craters on Moon, Mercury, and Mars. *Icarus*, **100**, 40–47.

Lawrence, D. J. (2017). A tale of two poles: Toward understanding the presence, distribution, and origin of volatiles at the polar regions of the Moon and Mercury. *J. Geophys. Res. Planets*, **122**, 21–52.

Lawrence, D. J., Feldman, W. C., Elphic, R. C., Hagerty, J. J., Maurice, S., McKinney, G. W. and Prettyman, T. H. (2006). Improved modeling of Lunar Prospector neutron spectrometer data: Implications for hydrogen deposits at the lunar poles. *J. Geophys. Res.*, **111**, E08001, doi:10.1029/2005JE002637.

Lawrence, D. J., Harmon, J. K., Feldman, W. C., Goldsten, J. O., Paige, D. A., Peplowski, P. N., Rhodes, E. A., Selby, C. M. and Solomon S. C. (2011). Predictions of MESSENGER Neutron Spectrometer measurements of Mercury's north polar region. *Planet. Space Sci.*, **59**, 1665–1669.

Lawrence, D. J., Feldman, W. C., Goldsten, J. O., Maurice, S., Peplowski, P. N., Anderson, B. J., Bazell, D., McNutt, R. L., Nittler, L. R., Prettyman, T. H., Rodgers, D. J., Solomon, S. C. and Weider, S. Z. (2013). Evidence for water ice near Mercury's north pole from MESSENGER Neutron Spectrometer measurements. *Science*, **339**, 292–296, doi:10.1126/science.1229953.

Le Feuvre, M. and Wieczorek, M. A. (2008). Nonuniform cratering of the terrestrial planets. *Icarus*, **197**, 291–306.

Li, L. and Mustard, J. F. (2000). Compositional gradients across mare-highland contacts: Importance and geological implication of lateral transport. *J. Geophys. Res.*, **105**, 20,431–20,450.

Lucey, P. G., Neumann, G. A., Riner, M. A., Mazarico, E., Smith, D. E., Zuber, M. T., Paige, D. A., Bussey, D. B., Cahill, J. T., McGovern, A., Isaacson, P., Corley, L. M., Torrence, M. H., Melosh, H. J., Head, J. W. and Song, E. (2014). The global albedo of the Moon at 1064 nm from LOLA. *J. Geophys. Res. Planets*, **119**, 1665–1679, doi:10.1002/2013JE004592.

Marchi, S., Mottola, S., Cremonese, G., Massironi, M. and Martellato, E. (2009). A new chronology for the Moon and Mercury. *Astron. J.*, **137**, 4936–4948.

Margot, J.-L., Peale, S. J., Solomon, S. C., Hauck, S. A., II, Ghigo, F. D., Jurgens, R. F., Yseboodt, M., Giorgini, J. D., Padovan, S. and Campbell, D. B. (2012). Mercury's moment of inertia from spin and gravity data. *J. Geophys. Res.*, **117**, E00L09, doi:10.1029/2012JE004161.

Mazarico, E., Neumann, G. A., Smith, D. E., Zuber, M. T. and Torrence, M. H. (2011). Illuminations conditions of the lunar polar regions using LOLA topography. *Icarus*, **211**, 1066–1081.

Mazarico, E., Nicholas, J. B., Neumann, G. A., Smith, D. E. and Zuber, M. T. (2014). Illumination conditions at the poles of the Moon and Mercury, and application to data analysis. *Lunar Planet. Sci.*, **45**, abstract 1867.

McCord, T. B., Taylor, L. A., Combe, J.-P., Kramer, G., Pieters, C. M., Sunshine, J. M. and Clark, R. N. (2011). Sources and physical processes responsible for OH/H_2O in the lunar soil as revealed by the Moon Mineralogy Mapper (M^3). *J. Geophys. Res.*, **116**, E00G05, doi:10.1029/2010JE003711.

Miller, R. S., Nerurkar, G. and Lawrence, D. J. (2012). Enhanced hydrogen at the lunar poles: New insights from the detection of epithermal and fast neutron signatures. *J. Geophys. Res.*, **117**, E11007, doi:10.1029/2012JE004112.

Miller, R. S., Lawrence, D. J. and Hurley, D. M. (2014). Identification of surface hydrogen enhancements within the Moon's Shackleton crater. *Icarus*, **233**, 229–232.

Mitrofanov, I. G., Sanin, A. B., Boynton, W. V., Chin, G., Garvin, J. B., Golovin, D., Evans, L. G., Harshman, K., Kozyrev, A. S., Litvak, M. L., Malakhov, A., Mazarico, E., McClanahan, T., Milikh, G., Mokrousov, M., Nandikotkur, G., Neumann, G. A., Nuzhdin, I., Sagdeev, R., Shevchenko, V., Shvetsov, V., Smith, D. E., Starr, R., Tretyakov, V. I., Trombka, J., Usikov, D., Varenikov, A., Vostrukhin, A. and Zuber, M. T. (2010). Hydrogen mapping of the lunar south pole using the LRO neutron detector experiment LEND. *Science*, **330**, 483–486.

Morgan, T. H. and Shemansky, D. C. (1991). Limits to the lunar atmosphere. *J. Geophys. Res.*, 96, 1351–1367.

Moses, J. I., Rawlins, K., Zahnle, K. and Dones, L. (1999). External sources of water for Mercury's putative ice deposits. *Icarus*, **137**, 197–221.

Muhleman, D. O., Butler, B. J., Grossman, A. W. and Slade, M. A. (1991). Radar images of Mars. *Science*, **253**, 1508–1513.

Neish, C. D., Bussey, D. B. J., Spudis, P., Marshall, W., Thomson, B. J., Patterson, G. W. and Carter, L. M. (2011). The nature of lunar volatiles as revealed by Mini-RF observations of the LCROSS impact site. *J. Geophys. Res.*, **116**, E01005, doi:10.1029/2010JE003647.

Nesvorný, D., Jenniskens, P., Levison, H. F., Bottke, W. F., Vokrouhlický, D. and Gounelle, M. (2010). Cometary origin of the zodiacal cloud and carbonaceous micrometeorites. Implications for host debris disks. *Astrophys. J.*, **713**, 816–836, doi:10.1088/0004-637X/713/2/816.

Neumann, G. A., Cavanaugh, J. F., Sun, X., Mazarico, E., Smith, D. E., Zuber, M. T., Mao, D., Paige, D. A., Solomon, S. C., Ernst, C. M. and Barnouin, O. S. (2013). Bright and dark polar deposits on Mercury: Evidence for surface volatiles. *Science*, **339**, 296–300.

Neumann, G. A., Sun, X., Mazarico, E., Deutsch, A. N., Head, J. W., Paige, D. A., Rubanenko, L. and Susorney, H. C. M. (2017). Latitudinal variation in Mercury's reflectance from the Mercury Laser Altimeter. *Lunar Planet. Sci.*, **48**, abstract 2660.

Nittler, L. R., Starr, R. D., Weider, S. Z., McCoy, T. J., Boynton, W. V., Ebel, D. S., Ernst, C. M., Evans, L. G., Goldsten, J. O., Hamara, D. K., Lawrence, D. J., McNutt, R. L., Jr., Schlemm, C. E., II, Solomon, S. C. and Sprague, A. L. (2011). The major-element composition of Mercury's surface from MESSENGER X-ray spectrometry. *Science*, **333**, 1847–1850.

Noda, H., Araki, H., Goossens, S., Ishihara, Y., Matsumoto, K., Tazawa, S., Kawano, N. and Sasaki, S. (2008). Illumination conditions at the lunar polar regions by KAGUYA (SELENE) laser altimeter. *Geophys. Res. Lett.*, **35**, L24203, doi:10.1029/2008GL035692.

Nozette, S., Lichtenberg, C. L., Spudis, P., Bonner, R., Ort, W., Malaret, E., Robinson, M. and Shoemaker, E. M. (1996). The Clementine bistatic radar experiment. *Science*, **274**, 1495–1498.

Ong, L., Asphaug, E. I., Korycansky, D. and Coker, R. F. (2010). Volatile retention from cometary impacts on the Moon. *Icarus*, **207**, 578–589.

Ostro, S. J., Campbell, D. B., Pettengill, G. H. and Shapiro, I. I. (1980). Radar observations of the icy Galilean satellites. *Icarus*, **44**, 431–440.

Paige, D. A., Wood, S. E. and Vasavada, A. R. (1992). The thermal stability of water ice at the poles of Mercury. *Science*, **258**, 643–646.

Paige, D. A., Siegler, M. A., Zhang, J. A., Hayne, P. O., Foote, E. J., Bennett, K. A., Vasavada, A. R., Greenhagen, B. T., Schofield, J. T., McCleese, D. J., Foote, M. C., DeJong, E., Bills, B. G., Hartford, W., Murray, B. C., Allen, C. C., Snook, K., Soderblom, L. A., Calcutt, S., Taylor, F. W., Bowles, N. E., Bandfield, J. L., Elphic, R., Ghent, R., Glotch, T. D., Wyatt, M. B. and Lucey, P. G. (2010). Diviner lunar radiometer observations of cold traps in the Moon's south polar region. *Science*, **330**, 479–482.

Paige, D. A., Siegler, M. A., Harmon, J. K., Neumann, G. A., Mazarico, E. M., Smith, D. E., Zuber, M. T., Harju, E., Delitsky, M. L. and Solomon, S. C. (2013). Thermal stability of volatiles in the north polar region of Mercury. *Science*, **339**, 300–303.

Paige, D. A., Hayne, D. A., Siegler, M. A., Smith, D. E., Zuber, M. T., Neumann, G. A., Mazarico, E. M., Denevi, B. W. and Solomon, S. C. (2014). Dark surface deposits in the north polar region of Mercury: Evidence for widespread small-scale volatile cold traps. *Lunar Planet. Sci.*, **45**, abstract 2501.

Peale, S. J. (1988). The rotational dynamics of Mercury and the state of its core. In *Mercury*, ed. F. Vilas, C. R. Chapman and M. S. Matthews. Tucson, AZ: University of Arizona Press, pp. 461–493.

Pelowitz, D. B. (ed.) (2005). *MCNPX User's Manual, Version 2.5.0.* Report LA-UR-94–1817. Los Alamos, NM: Los Alamos National Laboratory.

Pierazzo, E. and Chyba, C. F. (1999). Amino acid survival in large cometary impacts. *Meteorit. Planet. Sci.*, **34**, 909–918.

Pierazzo, E. and Chyba, C. F. (2006). Impact delivery of prebiotic organic matter to planetary surfaces. In *Comets and the Origin and Evolution of Life*, 2nd edn, ed. P. J. Thomas, R. D. Hicks, C. F. Chyba and C. P. McKay. Advances in Astrobiology and Biogeophysics. Berlin: Springer-Verlag, pp. 137–168, doi:10.1007/10903490_5.

Pike, R. J. (1988). Geomorphology of impact craters on Mercury. In *Mercury*, ed. F. Vilas, C. R. Chapman and M. S. Matthews. Tucson, AZ: University of Arizona Press, pp. 165–273.

Prem, P., Artemieva, N. A., Goldstein, D. B., Varghese, P. L. and Trafton, L. M. (2015). Transport of water in a transient impact-generated lunar atmosphere, *Icarus*, **255**, 148–158, doi:10.1016/j.icarus.2014.10.017.

Prettyman, T. H. (2007). Remote chemical sensing using nuclear spectroscopy. In *Encyclopedia of the Solar System*, 2nd edn, ed. L. A. McFadden, P. R. Weissman and T. V. Johnson. San Diego, CA: Academic Press, pp. 765–786.

Prettyman, T. H., Mittlefehldt, D. W., Yamashita, N., Lawrence, D. J., Beck, A. W., Feldman, W. C., McCoy, T. J., McSween, H. Y., Toplis, M. J., Titus, T. N., Tricarico, P., Reedy, R. C., Hendricks, J. S., Forni, O., Le Corre, L., Li, J.-Y., Mizzon, H., Reddy, V., Raymond, C. A. and Russell, C. T. (2012). Elemental mapping by Dawn reveals exogenic H in Vesta's regolith. *Science*, **338**, 242–246.

Rubanenko, L., Mazarico, E., Neumann, G. A. and Paige, D. A. (2017). Evidence for surface and subsurface ice inside micro cold-traps on Mercury's north pole. *Lunar Planet. Sci.*, **48**, abstract 1461.

Salvail, J. R. and Fanale, F. P. (1994). Near-surface ice on Mercury and the Moon: A topographic thermal model. *Icarus*, **111**, 441–455.

Schorghofer, N. and Taylor, G. J. (2007). Subsurface migration of H$_2$O at lunar cold traps. *J. Geophys. Res.*, **112**, E02010, doi:10.1029/2006JE002779.

Siegler, M. A., Bills, B. G. and Paige, D. A. (2011). Effects of orbital evolution on lunar ice stability. *J. Geophys. Res.*, **116**, E03010, doi:10.1029/2010JE003652.

Siegler, M., Paige, D., Williams, J.-P. and Bills, B. (2015). Evolution of lunar polar ice stability. *Icarus*, **255**, 78–87.

Siegler, M. A., Miller, R. S., Keane, J. T., Paige, D. A., Matsuyama, I., Lawrence, D. J., Crotts, A. and Poston, M. J. (2016). Lunar true polar wander inferred from polar hydrogen. *Nature*, **531**, 480–484.

Simpson, R. A. and Tyler, G. L. (1999). Reanalysis of Clementine bistatic radar data from the lunar South Pole. *J. Geophys. Res.*, **104**, 3845–3862.

Slade, M. A., Butler, B. J. and Muhleman, D. O. (1992). Mercury radar imaging: Evidence for polar ice. *Science*, **258**, 635–640.

Speyerer, E. J. and Robinson, M. S. (2013). Persistently illuminated regions at the lunar poles: Ideal sites for future exploration. *Icarus*, **222**, 122–136.

Sprague, A. L., Hunten, D. M. and Lodders, K. (1995). Sulfur at Mercury, elemental at the poles and sulfides in the regolith. *Icarus*, **118**, 211–215.

Spudis, P. D., Bussey, D. B. J., Baloga, S. M., Butler, B. J., Carl, D., Carter, L. M., Chakraborty, M., Elphic, R. C., Gillis-Davis, J. J., Goswami, J. N., Heggy, E., Hillyard, M., Jensen, R., Kirk, R. L., LaVallee, D., McKerracher, P., Neish, C. D., Nozette, S., Nylund, S., Palsetia, M., Patterson, W., Robinson, M. S., Raney, R. K., Schulze, R. C., Sequeira, H., Skura, J., Thompson, T. W., Thomson, B. J., Ustinov, E. A. and Winters, H. L. (2010). Initial results for the north pole of the Moon from Mini-SAR, Chandrayaan-1 mission. *Geophys. Res. Lett.*, **37**, L06204, doi:10.1029/2009GL042259.

Stacy, N. J. S., Campbell, D. B. and Ford, P. G. (1997). Arecibo radar mapping of the lunar poles: A search for ice deposits. *Science*, **276**, 1527–1530.

Starukhina, L. (2001). Water detection on atmosphereless celestial bodies: Alternative explanations of the observations. *J. Geophys. Res.*, **106**, 14,701–14,710.

Stewart, B. D., Pierazzo, E., Goldstein, D. B., Varghese, P. L. and Trafton, L. M. (2011). Simulations of a comet impact on the Moon and associated ice deposition in polar cold traps. *Icarus*, **215**, 1–16.

Sun, X. and Neumann, G. A. (2015). Calibration of the Mercury Laser Altimeter on the MESSENGER spacecraft. *IEEE Trans. Geosci. Remote Sensing*, **53**, 2860–2874.

Susorney, H. C. M., Barnouin, O. S., Ernst, C. M. and Johnson, C. L. (2016). Morphometry of impact craters on Mercury from MESSENGER altimetry and imaging. *Icarus*, **271**, 180–193.

Susorney, H. C. M., James, P. B., Chabot, N. L., Ernst, C. M., Mazarico, E. M. and Neumann, G. A. (2017). Measuring the thickness of Mercury's polar water ice deposits using the Mercury Laser Altimeter. *Lunar Planet. Sci.*, **48**, abstract 2059.

Talpe, M. J., Zuber, M. T., Yang, D., Neumann, G. A., Solomon, S. C., Mazarico, E. and Vilas, F. (2012). Characterization of the morphometry of impact craters hosting polar deposits in Mercury's north polar region. *J. Geophys. Res.*, **117**, E00L13, doi:10.1029/2012je004155.

Teodoro, L. F. A., Eke, V. R. and Elphic R. C. (2010). Spatial distribution of lunar polar hydrogen deposits after KAGUYA (SELENE). *Geophys. Res. Lett.*, **37**, L12201, doi:10.1029/2010GL042889.

Thomas, G. E. (1974). Mercury: Does its atmosphere contain water? *Science*, **183**, 1197–1198.

Thomson, B. J., Bussey, D. B. J., Neish, C. D., Cahill, J. T. S., Heggy, E., Kirk, R. L., Patterson, G. W., Raney, R. K., Spudis, P. D., Thompson, T. W. and Ustinov, E. A. (2012). An upper limit for ice in Shackleton crater as revealed by LRO Mini-RF orbital radar. *Geophys. Res. Lett.*, **39**, L14201, doi:10.1029/2012GL052119.

Urey, H. C. (1952). *The Planets: Their Origin and Development*. New Haven, CT: Yale University Press, 245 pp.

Vasavada, A. R., Paige, D. A. and Wood, S. E. (1999). Near-surface temperatures on Mercury and the Moon and the stability of polar ice deposits. *Icarus*, **141**, 179–193.

Vilas, F., Cobain, P. S., Barlow, N. G. and Lederer, S. M. (2005). How much material do the radar-bright craters at the Mercurian poles contain? *Planet. Space Sci.*, **53**, 1496–1500.

Watson, K. B., Murray, C. and Brown, H. (1961). The behavior of volatiles on the lunar surface. *J. Geophys. Res.*, **66**, 3033–3045.

Xiao, Z., Prieur, N. C. and Werner, S. C. (2016). The self-secondary crater population of the Hokusai crater on Mercury. *Geophys. Res. Lett.*, **43**, 7424–7432, doi:10.1002/2016GL069868.

Yoldi, Z., Pommerol, A., Jost, B., Poch, O., Gouman, J. and Thomas, N. (2015). VIS-NIR reflectance of water ice/regolith analogue mixtures and implications for the detectability of ice mixed within planetary regoliths. *Geophys. Res. Lett.*, **42**, 6205–6212, doi:10.1002/2015GL064780.

Zhang, J. A. and Paige, D. A. (2009). Cold-trapped organic compounds at the poles of the Moon and Mercury: Implications for origins. *Geophys. Res. Lett.*, **36**, L16203, doi:10.1029/2009GL038614.

Zhang, J. A. and Paige, D. A. (2010). Correction to "Cold-trapped organic compounds at the poles of the Moon and Mercury: Implications for origins." *Geophys. Res. Lett.*, **37**, L03203, doi:10.1029/2009GL041806.

Zolotov, M. Yu. (2011). On the chemistry of mantle and magmatic volatiles on Mercury. *Icarus*, **212**, 24–41.

14

Observations of Mercury's Exosphere: Composition and Structure

WILLIAM E. MCCLINTOCK, TIMOTHY A. CASSIDY, AIMEE W. MERKEL, ROSEMARY M. KILLEN, MATTHEW H. BURGER, AND RONALD J. VERVACK, JR.

14.1 INTRODUCTION

Mercury is surrounded by a tenuous exosphere in which particles travel on ballistic trajectories under the influence of a combination of gravity and solar radiation pressure. The densities are so small that the surface forms the exobase and particles in the exosphere are more likely to collide with it rather than with each other. For a planet with a more substantial collision-dominated atmosphere, a population of particles that enters from below the exobase supplies the exosphere. In contrast, Mercury's exosphere is supplied both by incoming sources, including the solar wind (hydrogen and helium), micrometeoroids (dust) and meteoroids, and comets and by particles released from the surface through a variety of processes that include sputtering by solar wind ions, desorption by solar photons and electrons, impacts by micrometeoroids, and thermal desorption of surface materials. These source processes are balanced by loss processes, which include impact with and sticking to the surface, Jeans (or thermal) escape, ionization followed by transport along magnetic field lines, and acceleration by solar radiation pressure to escape velocity.

Ground-based attempts to detect an atmosphere around Mercury before Mariner 10 first visited the planet in 1974 were unsuccessful and led only to increasingly tight upper limits, culminating in a limiting value for surface atmospheric pressure of 0.015 Pascal (Pa) determined by Fink et al. (1974).

During the three flybys of Mercury by Mariner 10 between March 1974 and March 1975, the spacecraft's Ultraviolet Spectrometer made robust measurements of hydrogen and helium and made a tentative detection of oxygen that was described as "unconfirmed" (Broadfoot et al., 1976). These observations were followed a decade later by discoveries with Earth-based telescopes of exospheric sodium and potassium (Potter and Morgan, 1985, 1986). Subsequently, Bida et al. (2000) reported the discovery of exospheric calcium from observations with the High Resolution Echelle Spectrometer (HIRES) at the Keck I telescope.

During the three flybys of Mercury by the MESSENGER spacecraft, measurements by the Mercury Atmospheric and Surface Composition Spectrometer (MASCS) instrument (McClintock and Lankton, 2007) yielded detections of exospheric magnesium (McClintock et al., 2009) and ionized calcium (Vervack et al., 2010). Other exospheric species have been reported from both ground-based observers and additional MESSENGER measurements. Doressoundiram et al. (2009) made a provisional detection of aluminum. Bida and Killen (2016) reported measurements of aluminum and iron (three-standard-deviation detections). Most recently, Vervack et al. (2016) reported the discovery of manganese and definitive measurements of aluminum and ionized calcium from MASCS data. Thus, the total inventory of confirmed exospheric neutral species now includes H, He, Na, K, Ca, Mg, Al, Fe, and Mn.

Since its discovery in 1985, sodium has been the most studied species from both ground-based observatories and from space. As described in Section 14.3.2, several groups have imaged emission from the sodium exosphere around the planet and in an extended anti-sunward tail (Potter et al., 2002a; Baumgardner et al., 2008). These studies provide insight into the behavior of the sodium exosphere on timescales as short as a few hours (e.g., Leblanc et al., 2009). Schleicher et al. (2004) and Potter et al. (2013) also observed sodium in absorption against the Sun during Mercury solar transits. These observations are complementary to those made by MESSENGER from orbit around the planet that characterized the seasonal and annual behavior of the sodium exosphere (Cassidy et al., 2015, 2016).

Potassium has also been observed from the ground by a number of researchers (Potter et al., 2002a; Doressoundiram et al., 2010; Killen et al., 2010). Calcium is much less studied from the ground than sodium or potassium, with only one group having reported observations (Bida et al., 2000). Emissions from hydrogen, helium, and magnesium occur at ultraviolet wavelengths that are absorbed by the Earth's atmosphere, so only space-based instruments can observe these constituents.

This chapter summarizes both ground-based and space-based observations that have been made of Mercury's exosphere, from its initial discovery by Mariner 10 through the MESSENGER mission, focusing on work published after the pre-MESSENGER reviews by Killen et al. (2007) and Domingue et al. (2007). In order to place the observations in context, the chapter begins with a brief summary of Chamberlain's theory of exospheres, which provides a framework for discussing the observations. That introduction is followed by a description of the techniques that various researchers have employed to observe Mercury's exosphere and then by a detailed discussion of the results for each of the known species. Models of source and loss processes derived from the MESSENGER observations are discussed in Chapter 15.

14.2 CHAMBERLAIN'S THEORY

The theory of exospheres developed by Chamberlain (1963) is often used to describe the composition and temperature of Mercury's exosphere. Chamberlain's formulation begins with the assumption that the collisonless exosphere is supplied from the lower collisional atmosphere through a boundary layer, termed the exobase, by gas that has a Maxwellian velocity distribution at a single temperature. In contrast, Mercury's surface is the lower boundary of its exosphere, which is populated by many components with different velocity distributions. Surface sticking, ionization followed by transport along magnetic field lines, Jeans escape, and acceleration by solar radiation pressure are known to deplete Mercury's exosphere, whereas the only loss process in Chamberlain's original formulation was Jeans escape. Solar radiation pressure was later added by Bishop and Chamberlain (1989) to extend the general formulation. Although a Chamberlain exosphere is spherically symmetric and Mercury's actual exosphere does not have such symmetry, a Chamberlain model nonetheless provides a simple, analytic description that is routinely used to provide insight into the exospheric processes at work on Mercury.

A classical Chamberlain exosphere contains three populations of particles: those that travel on ballistic orbits that reenter the atmosphere through the exobase, those that are in quasi-trapped satellite orbits, and those that are escaping. Because there is no atmospheric reservoir below its exobase, a surface-bounded exosphere such as Mercury's has only the ballistic and escaping components, and its density, n, as a function of the radial distance from the center of the planet, r, is approximated by

$$n = \zeta n_0 e^{-(U-U_0)/kT}, \qquad (14.1)$$

$$U = -GMm/r. \qquad (14.2)$$

In equations (14.1) and (14.2), n_0 is the surface density, U is the gravitational potential energy of a particle of mass m, U_0 is the potential energy at the surface, k is Boltzmann's constant, T is the absolute temperature, G is the gravitational constant, and M is the mass of Mercury. The factor ζ (Chamberlain, 1963) accounts for the fraction of the initial isotropic Maxwellian distribution that is actually present at a given altitude. A value of $\zeta = 1$, which is appropriate for a cool gas deep within a gravitational well (Chamberlain and Hunten, 1987), is an appropriate assumption for particles that are on ballistic orbits that intersect the exobase (or the surface, in the case of Mercury). This approximation begins to fail, however, as particles reach escape energy.

Equations (14.1) and (14.2) are valid for an exosphere in which only gravity acts on the particles. However, some atoms in Mercury's exosphere are accelerated by non-negligible solar radiation pressure. When an atom absorbs a photon, it is accelerated in the direction of the photon because photons possess both energy and momentum. An atom that emits a photon is also accelerated, but in a direction opposite to that of the emitted photon. Because sunlight incident on Mercury's exosphere is nearly unidirectional, and reemission is nearly isotropic, the effect of many absorption–emission events is a net anti-sunward acceleration. This effect can be included in the model by modifying U to include a combination of both gravity and radiation pressure (Bishop and Chamberlain, 1989):

$$U = -GMm/r + mbr\cos\theta. \qquad (14.3)$$

Here b is the net acceleration by photon pressure, and θ is the solar zenith angle (the angle between the Mercury–Sun line and the local radius vector from the planet center).

14.3 OBSERVATIONAL TECHNIQUES

Although the Fast Imaging Plasma Spectrometer (FIPS) component of the Energetic Particle and Plasma Spectrometer instrument on the MESSENGER spacecraft (Andrews et al., 2007) measured concentrations of sodium and oxygen ions around Mercury in situ (Raines et al., 2013), the neutral exosphere has been studied from Earth and space only with telescopes, either to sense remotely the intensity of sunlight resonantly scattered by its constituents or to observe sunlight absorbed by the exosphere as the planet passes in front of the Sun during a transit.

14.3.1 Remote Sensing Observations

Resonant scattering occurs when an atom (or molecule) in its lowest energy state absorbs a photon and transitions to an excited state. The atom promptly returns to the ground state by emitting a photon of equal energy. In the case of solar radiation that is resonantly scattered in an exosphere in which densities are vanishingly small, most of the photons that are absorbed are removed from the incident beam and are reemitted isotropically. Small asymmetries that arise for some transitions (Chamberlain, 1961) are usually neglected. For an optically thin medium, the resonantly scattered radiance (the term "emission" is used interchangeably in this Chapter), I, measured in photons cm^{-2} steradian^{-1} s^{-1}, is related to the total number of emitters along a line of sight by

$$4\pi I = gN. \qquad (14.4)$$

In equation (14.4), g (often referred to as the g-value) is the photon scattering coefficient in photons atom^{-1} s^{-1}, and N is the line-of-sight column density in atoms cm^{-2} (Hunten et al., 1956). The value of g is proportional to the absorption cross section of the atomic transition in cm^2 and to the solar irradiance incident on the exosphere at the rest wavelength of the transition in photons cm^{-2} s^{-1}. Atmospheric scientists often express radiance in rayleighs, where 1 rayleigh (R) is 10^6 photons emitted into 4π steradians in 1 s:

$$4\pi R = gN/10^6. \qquad (14.5)$$

Solar radiation reflected from the planet's surface also illuminates the exosphere, but its contribution to the observed exospheric intensity is small and is usually neglected except for the case of multiple scattering (Killen, 2006). In principle, electron-impact excitation can also contribute to the observed emission,

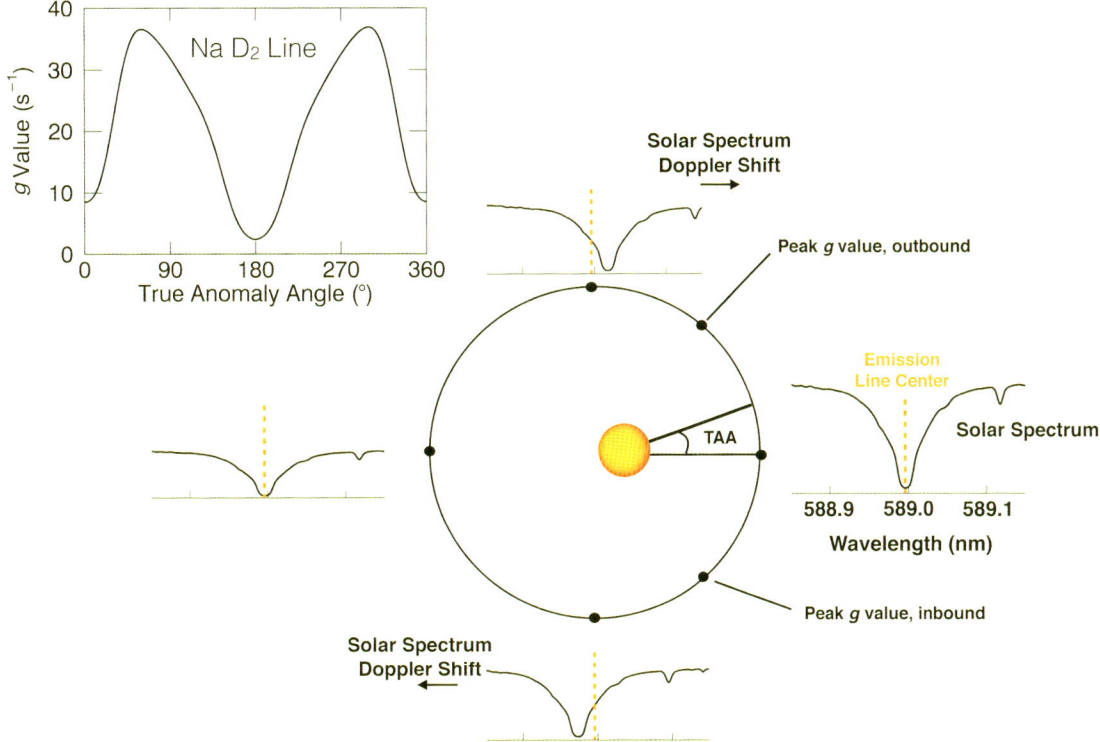

Figure 14.1. Mercury's distance from the Sun traces out an ellipse (central figure) when plotted as a function of true anomaly angle (TAA), the counterclockwise angular distance from perihelion when viewed from the Sun's north celestial pole. The solar spectrum around the sodium D_2 resonance transition is shown at the right. It exhibits a deep photospheric absorption feature (Fraunhofer line) near the center (rest wavelength) of the transition (dashed yellow line). This spectrum impinges upon atoms in Mercury's exosphere at perihelion, where the heliocentric radial velocity of the planet is zero. As Mercury orbits the Sun on the outbound leg of its orbit, its radial component of velocity begins to increase, causing the solar spectrum to shift to longer wavelengths. The shift reaches a maximum near TAA = 90° and then declines to zero at aphelion. On the inbound leg of Mercury's orbit the solar spectrum is shifted to shorter wavelengths, reaching a maximum shift near TAA = 270°. In addition to being Doppler-shifted, the magnitude of the solar irradiance varies with distance from the Sun, such that the magnitude of the solar irradiance at periapsis is a factor of 2.3 greater than that at apoapsis. The variation in g-value as a function of TAA, which is shown in the upper left panel, results from both changing Doppler shift and changing solar distance.

but the electron density in Mercury's magnetosphere is too low for it to be significant. Chamberlain (1963) described a variety of approximations for integrating densities in equation (14.1) along a line of sight to obtain values of N, although in practice this integral can be performed numerically.

Owing to the eccentricity of the planet's orbit, the heliocentric radial velocity of the planet varies up to ± 10 km s^{-1} during a Mercury year, causing the solar spectrum to be Doppler shifted relative to the exosphere. This shift results in substantial variation in g-values, because the solar spectrum in the vicinity of resonance transitions of important exospheric species is modulated by the presence of absorption features, known as Fraunhofer lines, that form in the Sun's relatively cool photosphere. This effect is illustrated in Figure 14.1 for one of the resonance lines of sodium (D_2, see Section 14.3.2), which has a rest wavelength of 588.995 nm in air. Changes in g-value cause the observed sodium intensity emitted by the exosphere to vary with an annual cycle, brighter near Mercury true anomaly angles (TAA) 65° and 295° but fainter near TAA = 0° and TAA = 180°. Larger g-values also lead to larger average radiation forces, so that atoms are more strongly accelerated anti-sunward near TAA = 65° and TAA = 295° than at other times during the Mercury year. The g-value, which depends linearly on the incident solar irradiance, is also modulated by Mercury's changing distance from the Sun.

When a planet passes between the Sun and Earth, atmospheric constituents above its limb scatter the solar radiation away from its original path, creating an absorption feature in the solar spectrum. The area of an absorption feature is related to the line-of-sight column density by

$$wE_{\text{Sun}} = gN, \tag{14.6}$$

where w is the equivalent width of the absorption feature measured in nm and E_{Sun} is the solar irradiance at the wavelength of the feature, measured in photons cm^{-2} nm^{-1} s^{-1} (Potter et al., 2013). The "equivalent width" is the width the absorption feature would have if it were completely absorbing.

14.3.2 Ground-Based Techniques

Most ground-based observers employ high-resolution spectrographs to measure Mercury's spectrum. High resolution is necessary to remove backgrounds from the exospheric signal that arise from sunlight reflected by the surface of the planet and from sunlight scattered by Earth's atmosphere. The right half of

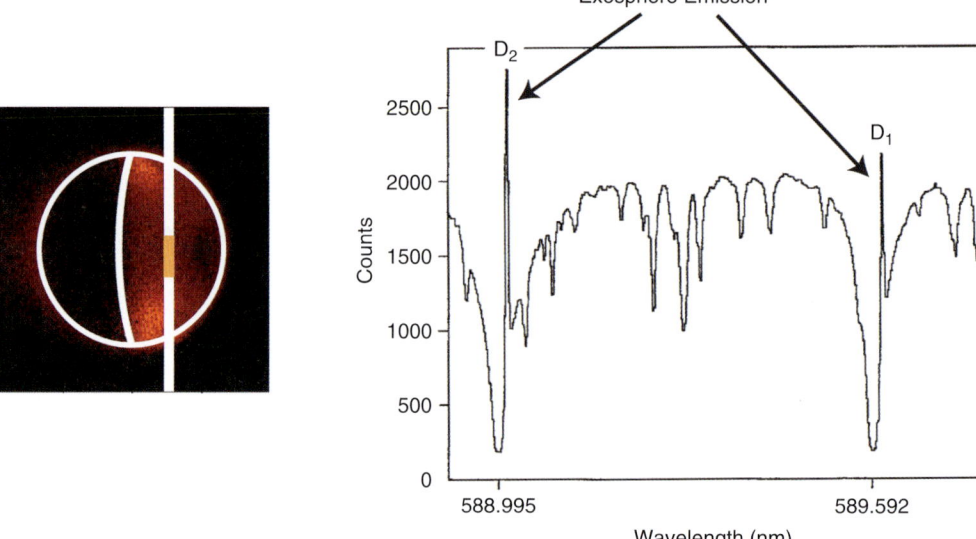

Figure 14.2. Ground-based observers employ telescopes equipped with high-resolution spectrographs to separate exosphere emission from sunlight reflected from Mercury's surface. As shown here, the telescope images the planet onto the spectrograph entrance slit (left). The spectrograph disperses the light into a spectral-spatial image that is projected onto a two-dimensional detector in which the spectrum on each row originates from a unique location along the slit. (As an illustration, the spectrum on the right originates from the center of the orange band along the slit). Exosphere sodium emission appears at narrow peaks above the continuous reflected spectrum and is extracted from each row to construct a slice of the full disk image. The entire image is built up by slewing the telescope to scan the slit across the disk. In a second approach, a telescope is used to image the planet onto the input of an image slicer. The slicer reformats the two-dimensional image and projects it onto the entrance slit of a spectrograph as a one-dimensional line image. This technique produces an image without the need to slew the telescope.

Figure 14.2 illustrates a typical high-resolution spectrum (Sprague et al., 1997) of the surface exosphere that covers wavelengths near the resonance lines of sodium, which are located at air wavelengths of 588.995 nm and 589.592 nm and are historically referred to as D_2 and D_1, respectively. At a resolving power of $\lambda/\Delta\lambda \sim 160{,}000$, where λ is wavelength and $\Delta\lambda$ is the spectral resolution, the exospheric components are clearly evident as a pair of discrete, narrow peaks superimposed on the broad, continuous light reflected from the surface. Equation (14.4) relates the emission in these peaks to the total column density, in atoms cm^{-2}, along the line of sight from the observer to the planet's surface.

Two approaches are commonly used to produce images from high-resolution spectra. In the first, as illustrated in Figure 14.2, the telescope images Mercury directly onto the entrance slit of the spectrograph, and the resulting spectra are analyzed to extract the exosphere intensities along the slit. A two-dimensional image is built up by slewing the telescope to move the target image across the slit in a series of discrete steps and recording a line image at each step. This process is sometimes referred to as "push-broom" imaging. The second employs an optical image slicer (Pierce, 1965). In this approach, the telescope forms an image of Mercury on the slicer, which rearranges the image to fit in the narrow entrance slit of an imaging spectrograph. As a result, two-dimensional areas much wider than the slit can be observed instantaneously. The advantage of this approach is that spectra from multiple points over most (or in some cases all) of Mercury's disk can be obtained simultaneously and without the need to stitch multiple line images together (e.g., Potter and Morgan, 1990). Fiber optic bundles can also be used instead of image slicers to rearrange a two-dimensional target image and project it onto the slit of an imaging spectrograph (Doressoundiram et al., 2010). Typical integration times required to obtain a high-quality image are instrument dependent and are in the range of several minutes to an hour (e.g., Leblanc et al., 2009).

Two additional techniques have been used to study Mercury's exosphere from Earth. First, a small telescope equipped with a narrow interference filter and an occulting disk that blocks the direct image of the planet's surface has been used to measure resonantly scattered sunlight from sodium atoms that are sufficiently far from the planet that reflected sunlight from the disk is negligible (Baumgardner et al., 2008; Schmidt et al., 2012). Second, two groups (Schleicher et al., 2004; Potter et al., 2013) have used telescopes equipped with high-spectral-resolution Fabry–Perot interferometers to measure the absorption of sunlight by sodium atoms in Mercury's exosphere during solar transit. During these observations Mercury appears as a dark disk superimposed on the image of the Sun. Whereas the disk of the planet completely blocks sunlight, sodium in the exosphere surrounding the disk only partly absorbs it. Rather than appearing as excess emission superimposed on a solar continuum, these spectra exhibit excess absorption in the nominal solar photospheric profile resulting from the partial extinction of sunlight passing through the exosphere.

14.3.3 Space-Based Techniques

Two spacecraft equipped with small telescope spectrometers have visited Mercury. These were Mariner 10 with its Ultraviolet Spectrometer (UVS) (Broadfoot et al., 1974 and 1976) and MESSENGER with its MASCS Ultraviolet and

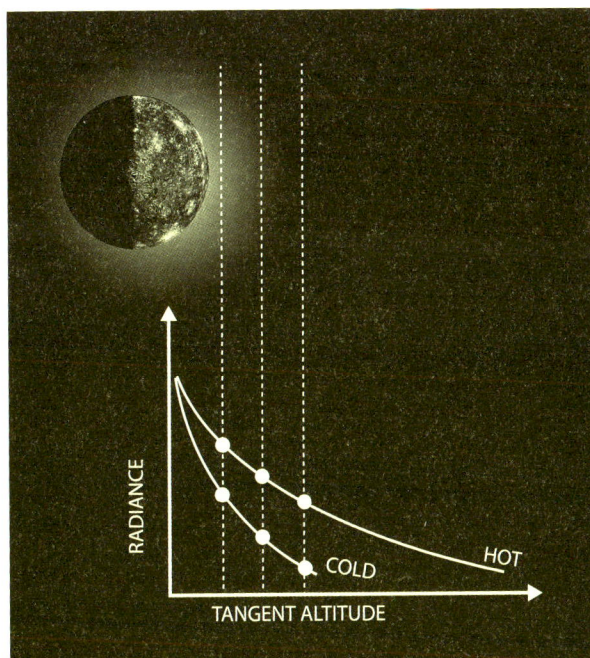

Figure 14.3. Small telescope spectrometers on spacecraft near Mercury use limb-scan viewing to study the exosphere. In this approach, the telescope images the exosphere, which is shown as a faint gray glow surrounding the planet, onto the spectrometer entrance in order to view lines of sight that are tangent to the limb of the planet. Different lines of sight probe different altitudes, enabling the measurement of exospheric column density as a function of altitude. The slope of the radiance profile is indicative of its temperature, with a warmer exosphere producing a shallower slope.

Visible Spectrometer (UVVS) (McClintock and Lankton, 2007). The spectral passbands of these instruments were approximately 0.5–1 nm. This bandwidth was too large for these instruments to detect the exosphere while viewing the sunlit surface of the planet because the intensity of continuous reflected sunlight integrated over a ~1 nm passband is much larger than that of exospheric emission lines, which have full widths that are typically 0.005 nm or less. Instead they relied on limb-scan viewing, in which a telescope views the exosphere with the entrance slit of a spectrograph along a line of sight that passes above the surface at a given altitude. Figure 14.3 illustrates the technique for two exospheres with identical surface densities at two different temperatures. These are sampled along three lines of sight through Mercury's exosphere. For each line of sight, there is a vector from the center of the planet that is perpendicular to it. The point at which that vector intersects the surface is referred to as the tangent point, and the distance from the surface to the line of sight along that vector is referred to as the minimum ray height or tangent altitude. Limb emission profiles, which are plots of radiance versus tangent altitude, are assembled from the line-of-sight radiances. (The terms limb scan and limb profile are often used interchangeably. In this chapter, we use "limb scan" to identify the observation – the motion of the telescope field of view – and "limb profile" to identify the measured geophysical quantity as a function of tangent altitude.)

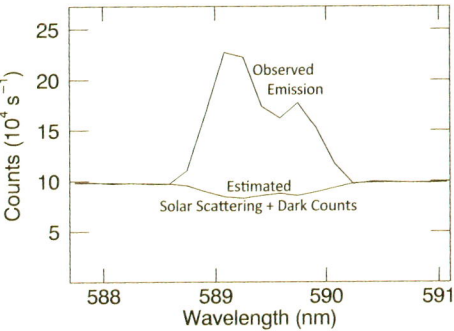

Figure 14.4. A sodium emission spectrum observed at 100 km tangent altitude near the subsolar point by the MESSENGER UVVS, which had a spectral resolution of ~0.6 nm. The graph shows the total observed counts per second recorded by one of the instrument's photomultiplier tubes as a function of wavelength. The background due to the combination of sunlight reflected from the surface and scattered into the telescope and detector dark counts is subtracted to extract the exospheric D_1 and D_2 emission lines at 589.0 and 589.6 nm, respectively. These lines appear as two broad profiles that have an approximate 2:1 intensity ratio.

Equation (14.4) relates the spectrograph signal to the total column density, in atoms cm^{-2}, along the line of sight from the observer through the exosphere. If a Chamberlain exosphere is assumed, the vertical column density, N, is related to the exosphere density at the tangent point, n_h, and the distance from the center of the planet, r, by

$$N = n_h \sqrt{2\pi r/H}, \qquad (14.7)$$

$$H = n \Big/ \left(\frac{dn}{dr}\right) = kT/(GMm/r^2 + mb\cos\theta), \qquad (14.8)$$

where H is the scale height, or the e-folding distance, and is defined as the ratio of density to the rate of density change at altitude h. Although an exosphere is not hydrostatic, the definition of H for an exosphere with $b = 0$ is analogous to that for a plane-parallel hydrostatic atmosphere in which H is the kinetic energy of a particle divided by the gravitational force acting on it.

Space-based observations provide altitude information that is not available from Earth, because they do not include backgrounds caused by light from the bright atmosphere above the illuminated disk that is scattered along the slit by the telescope and spectrograph optics. A typical dayside sodium spectrum acquired by the MESSENGER UVVS is shown in Figure 14.4. It covers the same emission lines as those shown in Figure 14.2 but at much lower spectral resolution. The total emission (measured in either R or kR) is obtained by subtracting the estimated background (dark counts and surface reflectance), integrating the resulting signal, and multiplying by a laboratory-measured calibration factor (McClintock and Lankton, 2007). The removal of surface-reflected sunlight from the MESSENGER calcium and magnesium emissions, which are less intense than those from sodium, was described by Burger et al. (2010) and Merkel et al. (2017), respectively.

14.3.3.1 Flyby Observations

Both the Mariner 10 UVS and the MESSENGER UVVS observed Mercury's exosphere during flybys of the planet. The

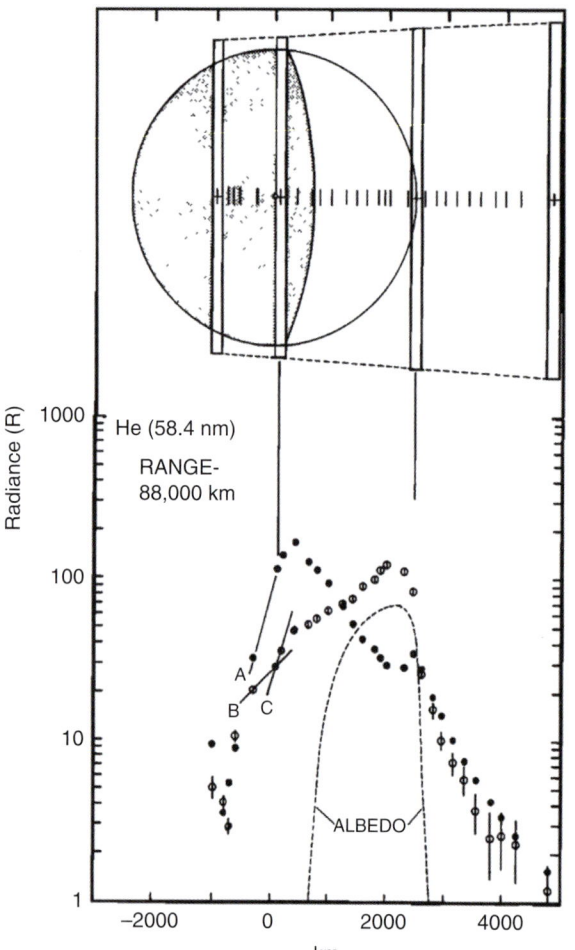
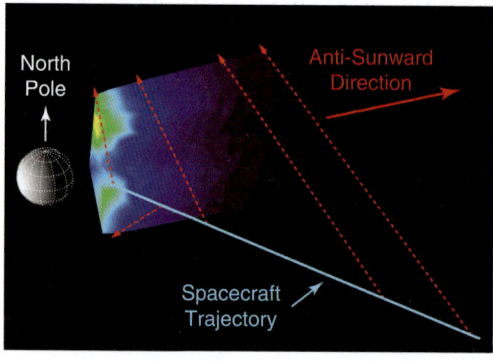

Figure 14.5. (Left) The Mariner 10 UVS observing geometry and a profile of helium (58.4 nm) emission versus altitude obtained at a distance of 88,000 km, during the third encounter of the spacecraft with the planet on 16 March 1975. Open circles with vertical errors represent the observed radiance, which includes emission from the exosphere and reflection from the surface (albedo). Filled circles are radiance computed from a model. Lines labeled A and C illustrate model scale heights at their respective locations. The line labeled B represents the observed scale height at its location. There are significant differences in both emission and scale height between the observations and the model. Adapted from Broadfoot et al. (1976). (Right) The UVVS observation geometry and a reconstructed image of emission from the sodium resonance lines during the first MESSENGER flyby (M1) on 14 January 2008. Adapted from McClintock et al. (2008).

left panel of Figure 14.5 illustrates the Mariner 10 UVS observing geometry and an altitude profile for the helium 58.4 nm line obtained at a distance of 88,000 km during the third encounter of the spacecraft with the planet on 16 March 1975. These observations were made by pointing the spacecraft and allowing the UVS field of view to drift across the disk and into the exosphere. In this measurement the spacecraft was oriented so that the spectrograph slit was projected north–south and swept out the entire disk as the spacecraft flew sunward past the planet. The UVS had no spatial resolution in the north–south direction, and the one-dimensional profiles were produced by summing the signal from all the light entering the spectrometer slit. From these observations and Monte Carlo models, Broadfoot et al. (1976) derived surface densities and altitude distributions for hydrogen and helium. They reported an upper limit for oxygen emission as well as upper limits on selected molecular species derived from a solar occultation experiment. Their results are discussed in further detail in Section 14.4.4.

The MESSENGER encounters were much closer to the planet than those executed by Mariner 10. The right panel in Figure 14.5 illustrates the UVVS observation geometry during the first MESSENGER flyby (M1), which took place on 14 January 2008. As the spacecraft approached the planet from its anti-sunward dusk side, it oriented the UVVS slit approximately perpendicular to the Sun–Mercury line and executed a series of partial roll maneuvers that swept the slit north–south during approach. This geometry enabled the UVVS to construct images of the sodium radiances behind the planet. Similar tail observations were executed in order to construct images of sodium, calcium, and magnesium during the second and third flybys (M2 and M3), which occurred on 6 October 2008 and 29 September 2009, respectively (e.g., Vervack et al., 2010).

As the spacecraft entered the shadow of Mercury near closest approach to the planet during M1, it executed a 180° roll (the so-called "fantail" observations) that swept the UVVS line-of-sight from dawnward to north to duskward before

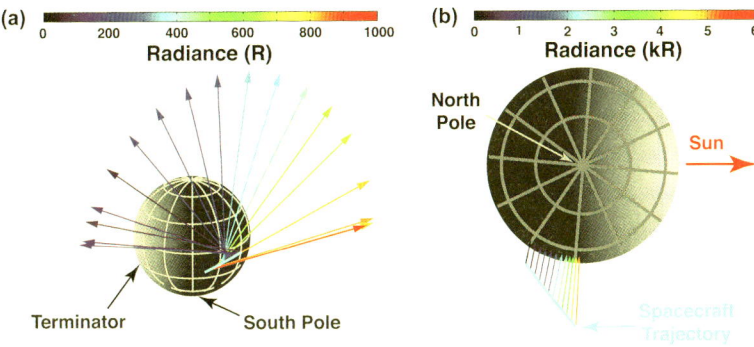

Figure 14.6. Viewing geometry and observed radiances from calcium: (a) near-tail and (b) near-terminator observations during MESSENGER's first flyby. In (a), UVVS line-of-sight vectors are color coded according to the observed radiances. In (b), line-of-sight vectors are seen as the spacecraft emerges from Mercury's shadow. The observed radiances here correspond to columns of illuminated calcium atoms between MESSENGER and Mercury's surface. Adapted from McClintock et al. (2008).

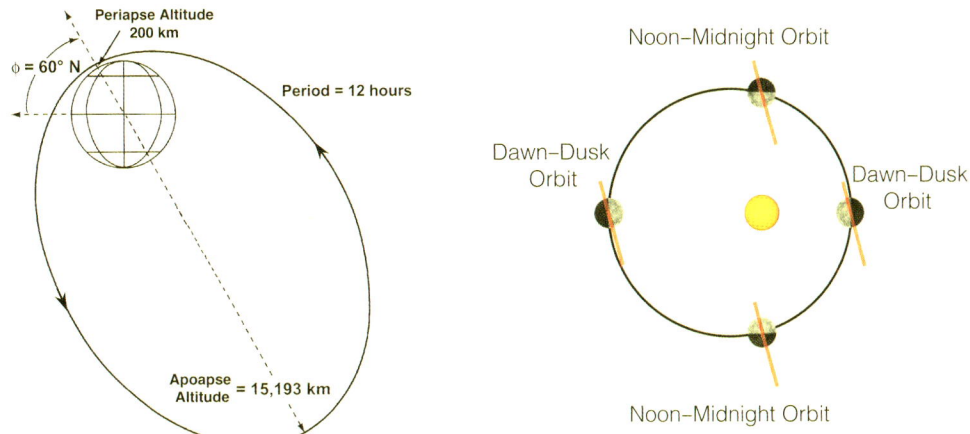

Figure 14.7. (Left) MESSENGER's initial 12-h orbit about Mercury. Most UVVS observations of the exosphere occurred near apoapsis when the instrument had the widest coverage in local time. (Right) The orbit plane of MESSENGER through one Mercury year. "Dawn–dusk" and "noon–midnight" refer to times when the MESSENGER orbit was more or less aligned with the terminator and Sun–Mercury lines, respectively. This varying orientation placed restrictions on the spacecraft attitude that limited local time coverage, particularly near TAA = 0° (perihelion) and TAA = 180° (aphelion), where observations were possible only near dawn and dusk.

exiting on the dawn side, as illustrated in Figure 14.6a. As MESSENGER emerged from shadow, the UVVS pointed toward the surface, measuring emission over the pre-dawn and dawn terminator regions (Figure 14.6b). The UVVS obtained similar observations during M2. During M3 the initial line-of-sight was pointed south, and the spacecraft rolled it through dawn toward north as part of a planned 270° maneuver. However, the spacecraft entered safe mode midway through the flyby, and data that would have been acquired during the last 90° were lost. Fantail observations were performed very infrequently during orbital operations.

14.3.3.2 Orbital Observations

MESSENGER was inserted into orbit around Mercury on 18 March 2011. The initial orbit, shown in the left panel of Figure 14.7, had a 12-h period, an inclination of 82.5°, an apoapsis of 15,200 km, and a periapsis of 200 km. After four Mercury years of operations, the MESSENGER flight team reduced apoapsis to ~10,000 km, which reduced the period to 8 h. The UVVS performed most of its exospheric observations near apoapsis, from which it could view the largest range of local times.

The right panel of Figure 14.7 shows the orientation of MESSENGER's orbital plane as a function of TAA just after orbit insertion. MESSENGER was equipped with a sunshade that shielded the spacecraft from illumination by the Sun through most of the cruise phase and the entire orbital phase of the mission, and the requirement that the normal to the sunshade's central panel remain within 12° of the Sun–Mercury line limited the local-time coverage of the dayside exosphere over the course of a Mercury year. In particular, only local times near dawn and dusk could be sampled when Mercury was near perihelion (TAA ~ 0°) and aphelion (TAA ~ 180°). During the rest of the Mercury year, covering TAA ranges 20°–160° and 225°–340°, the spacecraft could execute small scan maneuvers about axes close to the Sun–Mercury line, enabling UVVS to make observations above the dayside limb. The most complete local-time coverage was obtained when the orbit orientation aligned more or less parallel to the Sun–Mercury line near TAA 90° and 270°.

Typical viewing geometries for the two primary observation sequences executed by UVVS on a regular basis are illustrated in Figure 14.8. In the illustration the spacecraft orbital plane was close to the Sun–Mercury line, allowing UVVS to perform dayside limb scans at up to seven local times, beginning near 0600 and ending near 1800 (Figures 14.8a and 14.8b). On the nightside, the

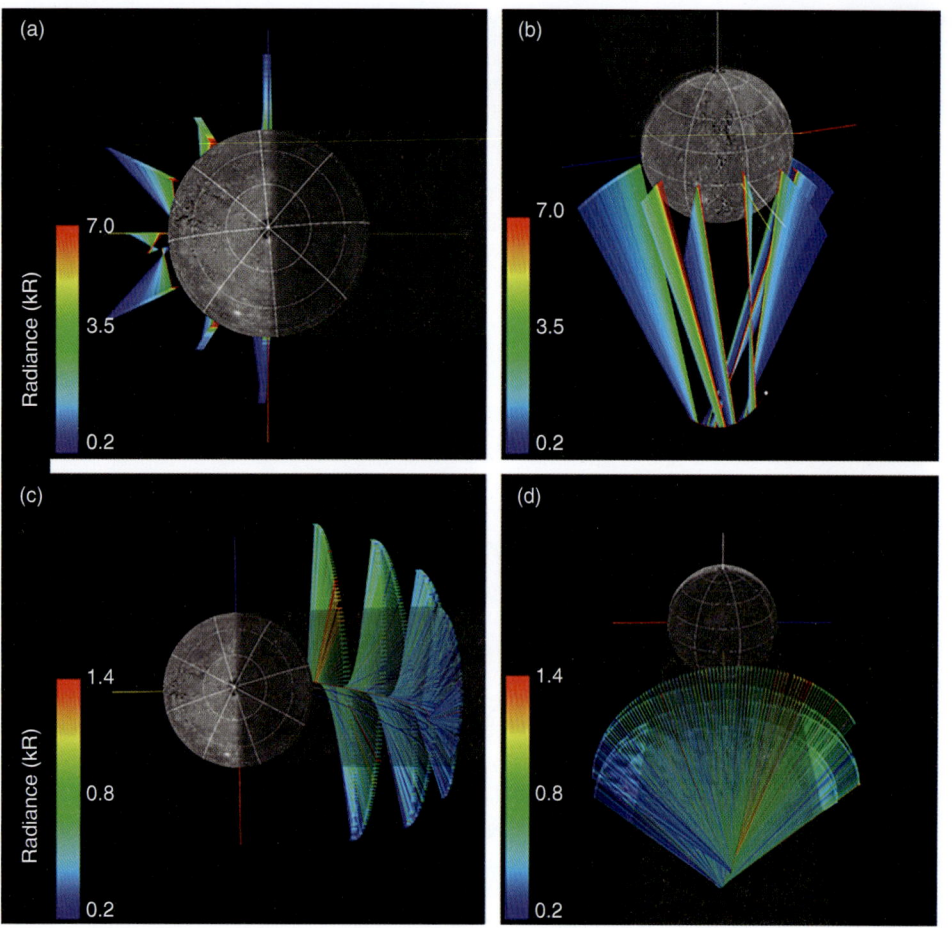

Figure 14.8. Examples of typical UVVS observing sequences used to sample the exosphere on the day and night sides of Mercury. The white, blue, red, and yellow lines radiating from the planet center represent Mercury's rotation axis, dawn and dusk terminators, and the subsolar point, respectively. In this example, the color bar represents magnesium emission in kilorayleighs, increasing linearly from blue to red. Dayside limb scans are shown (a) looking down on the north pole and (b) from a view of the dayside. Nightside tail sweeps are illustrated (c) looking down on the north pole and (d) looking toward the Sun from the nightside of the planet.

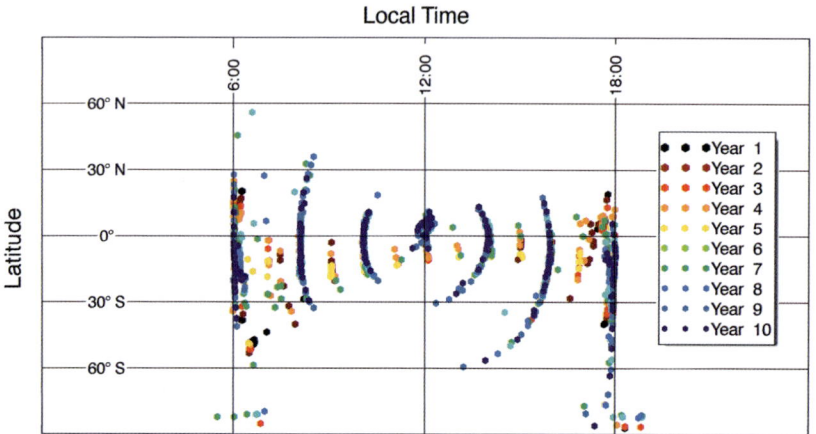

Figure 14.9. UVVS limb scan coverage. Each point indicates the local time and latitude of a limb scan's tangent point at minimum tangent altitude. Each symbol color represents a different Mercury year during MESSENGER's orbital mission. Adapted from Cassidy et al. (2015).

spacecraft executed short back-and-forth roll maneuvers (referred to as tail sweeps) to scan the field of view in the dawn–dusk direction. In this way, UVVS was able to map out the anti-sunward region of the exosphere, as shown in Figures 14.8c and 14.8d.

Although observations above the south pole were possible when the spacecraft was near TAA 0° and 180°, the tangent points of the UVVS limb scans were mostly limited to the equatorial region of Mercury, due to the combination of spacecraft orbit geometry and sunshade pointing constraints. Figure 14.9, adapted from Cassidy et al. (2015), shows the locations of the tangent points for sodium dayside limb scans acquired during the first 10 Mercury years of the mission. No routine limb scans were made over the planet's magnetospheric cusps or above the north pole.

14.4 OBSERVATIONAL RESULTS

The known inventory of Mercury's exosphere as documented by remote observations consists of nine neutral species and one ionized species. Of these, sodium, potassium, calcium, and

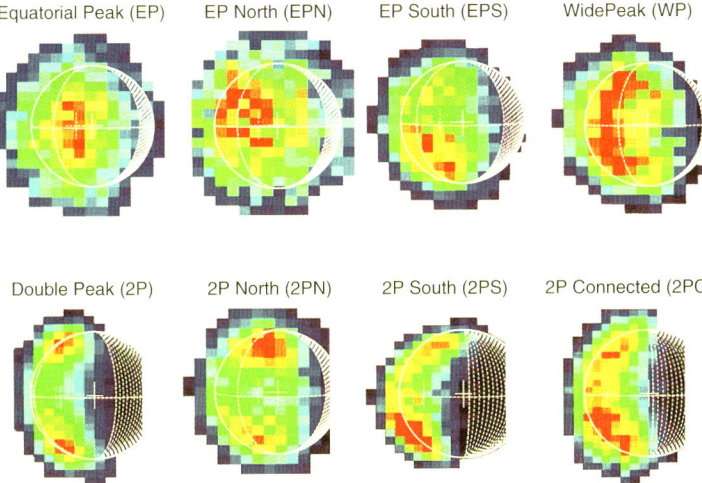

Figure 14.10. Common patterns of exospheric sodium emission seen during an extended set of images of exospheric sodium collected between 2009 and 2013 with the THEMIS telescope. Colors represent radiance on a linear rainbow scale from blue to red. Adapted from Mangano et al. (2015).

magnesium have been well studied (sodium and calcium from both the ground and space, potassium from the ground, and magnesium from space). MESSENGER observed hydrogen routinely but not systematically. Aluminum and iron have been provisionally detected from the ground on a few occasions, with three-standard-deviation upper limits reported for ionized calcium, whereas aluminum, manganese, and ionized calcium have been clearly detected by MESSENGER but over a limited range of TAA values. Only Mariner 10 observed helium, as the helium emission wavelength was outside the range of MESSENGER's instruments. MESSENGER did not confirm Mariner 10's tentative detection of oxygen, but observations of ionized oxyzen by FIPS (Raines et al., 2013) suggest that some neutral oxygen is present in the exosphere.

14.4.1 Sodium

14.4.1.1 Ground-Based Observations

Mercury's sodium exosphere was first observed by Potter and Morgan (1985), who measured the emission with a high-resolution echelle spectrograph at the McDonald Observatory. Subsequent observations have been made with a variety of techniques to measure the speed distribution, flow speed, spatial distribution, and temporal variability of the sodium exosphere. This section focuses on work published after the pre-MESSENGER literature reviews by Killen et al. (2007) and Domingue et al. (2007).

Spatial Distribution

Starting with Potter and Morgan (1990), who imaged Mercury's sodium emissions with image slicers at the McMath–Pierce Solar Telescope, a number of observers have studied the spatial distribution of sodium around Mercury. Some of the first disk-resolved observations showed high-latitude enhancements in sodium emission, leading to the hypothesis that ion bombardment at Mercury's cusps and open field lines liberates sodium atoms via a variety of processes. These include ion sputtering (McGrath et al., 1986), the ejection of neutral atoms by ion-initiated collision cascades; chemical sputtering (Potter, 1995), a chemical reaction in the regolith with incoming protons that ejects sodium; and ion-enhanced diffusion (Killen et al., 2004; Sarantos et al., 2008), a process that aids the diffusion of sodium out of regolith grains to where other processes such as photo-desorption can eject the sodium.

Another possibility is that the high-latitude enhancements are due to an observational bias (Potter et al., 2006) that may disappear when line-of-sight effects are removed from observations (Mouawad et al., 2011). Regardless of the cause, these enhancements have been seen by many observers who used all of the methods described in Section 14.3.2. When present, the enhancement in emission is typically tens of percent greater than the emission at lower latitudes. Borland and Taylor (2007) pointed out that enhancement of this magnitude can appear more pronounced in publication figures due to the use of "rainbow" false-color scales that employ a non-linear stretch, which emphasizes small differences between values.

Mangano et al. (2015) categorized the observed emission patterns in a large data set taken by the Télescope Héliographique pour l'Etude du Magnétisme et des Instabilités Solaires (THEMIS) solar telescope with push-broom imaging by a long-slit spectrograph (Figure 14.10), as described in Section 14.3.2 and illustrated in Figure 14.2. The most common emission pattern had distinct emission peaks at high latitudes in both the northern and southern hemispheres. This is the "2P" pattern in Figure 14.10. Other common configurations included the equatorial peak ("EP"), featuring emission peaks at low latitudes near the subsolar point and several variations on the high-latitude emission pattern.

One feature of the high-latitude enhancements is their variability, documented on the timescale of days in earlier papers (Potter et al., 1999; Potter and Morgan, 1990). More recently, shorter-term (hourly) variability was reported in observations with the THEMIS telescope (Leblanc et al., 2006, 2008, 2009, 2013; Doressoundiram et al., 2010; Mangano et al., 2009, 2013, 2015). Figure 14.11a shows an example adapted from Mangano et al. (2013); from one day of observations, they reported changes in emission that they attributed to magnetospheric dynamics (Figure 14.11b). These changes were on the order of 10% at both the equator and mid-latitude cusp regions. Unlike in earlier studies of episodic variability, Mangano et al. (2013) corrected

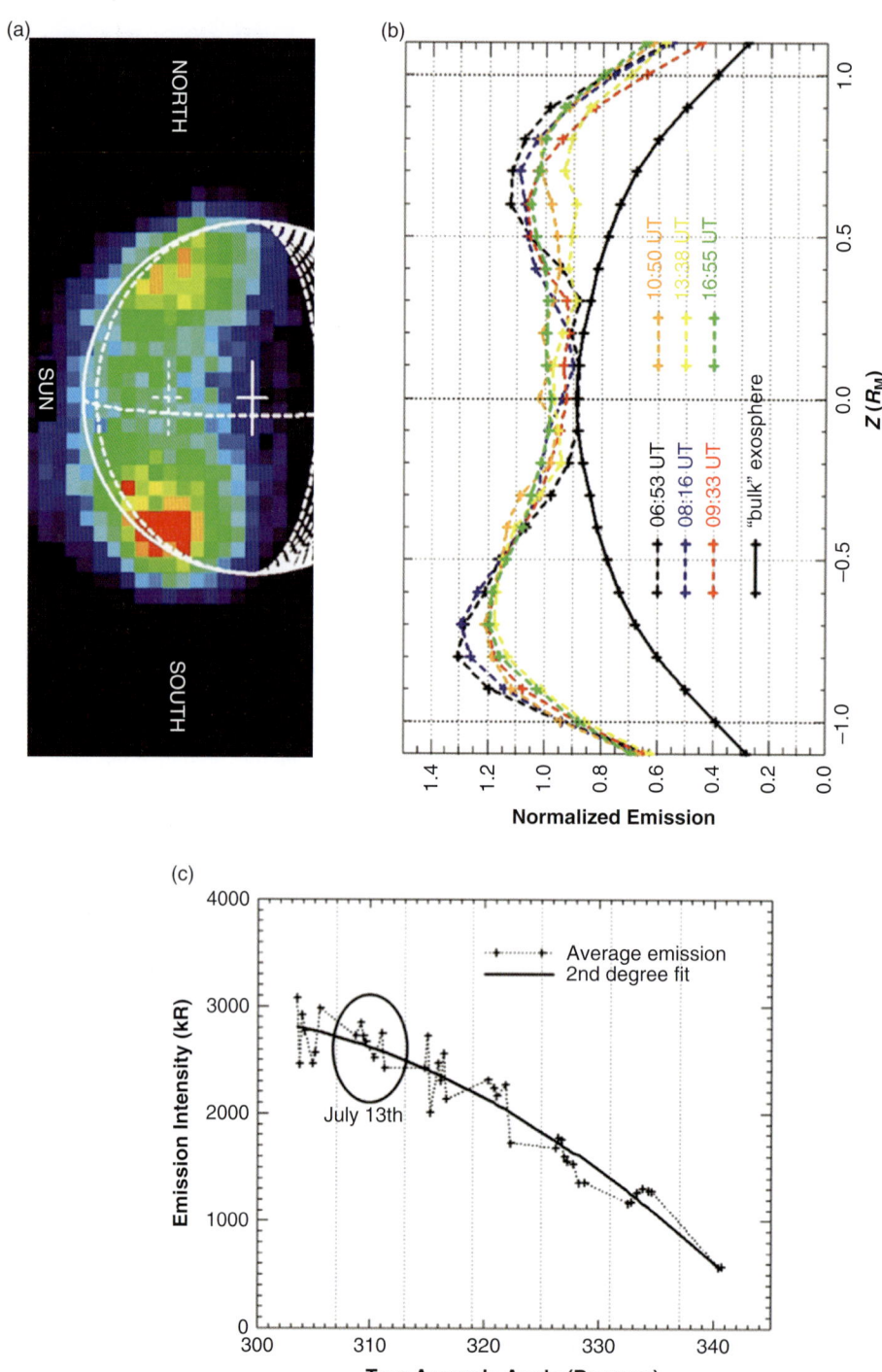

Figure 14.11. THEMIS solar telescope observations of mid-latitude enhancements in the sodium exosphere. (a) A ground-based observation of Mercury's dayside sodium exosphere. (b) Emission along the north–south axis for several observations, showing variations over the course of ~10 h. (c) Seasonal variations that were used to normalize the emissions in (b). Adapted from Mangano et al. (2013).

for longer-term seasonal changes in sodium brightnesses (Figure 14.11c) that were assumed to be independent of short-term magnetospheric activity. They did not, however, discuss the effects of atmospheric seeing and opacity, both of which can contribute to the observed variability.

Mangano et al. (2015) attempted to confirm their hypothesis that magnetospheric dynamics contribute to short-term emission variability by searching for correlations between ground-based sodium observations and interplanetary magnetic field (IMF) measurements made by MESSENGER's Magnetometer (Anderson et al., 2007). They concluded that there is a correlation between the magnitude of the IMF southward component and the presence of mid-latitude emission peaks in a subset of their observations. A southward IMF encourages magnetic reconnection and auroral activity at Earth and so might plausibly result in higher ion fluxes to Mercury's surface (Baker et al., 2013), but the Mangano et al. (2015) conclusion is at variance with the finding from MESSENGER observations that the IMF direction is not as strong a forcer of solar wind–magnetosphere coupling as it is at Earth (DiBraccio et al., 2013).

Mangano et al. (2015) also found that southern peaks in emission were no brighter, on average, than those in the northern hemisphere. Likewise, Potter et al. (2006) applied a statistical test to show that excess emission is equally likely to be observed in the northern or southern hemisphere. This lack of north–south asymmetry is surprising in light of the MESSENGER discovery of a northward offset to the dipole magnetic field (Anderson et al., 2011), which results in the southern hemisphere receiving a factor of ~4 more solar wind ions than the northern hemisphere (Winslow et al., 2012, 2014).

Contrary to the conclusions of Mangano et al. (2015) and Potter et al. (2006), Schmidt (2013) reported that there is a larger source of sodium in the southern hemisphere on the basis of observations of Mercury's anti-sunward sodium "tail" (discussed in more detail below). Evidence for this asymmetry includes several observations showing excess northern tail emission (Baumgardner et al., 2008; Potter and Killen, 2008). With a Monte Carlo simulation, Schmidt showed that a northern enhancement can arise from a southern hemisphere source because sodium atoms switch hemispheres as they are pushed anti-sunward by radiation pressure, at least when they are ejected from the surface within a certain energy range. Schmidt concluded that a northern tail enhancement can be explained by a process in which solar wind ions impinging on the southern nightside hemisphere at low latitudes (Winslow et al., 2014) liberate sodium for later photodesorption at dawn in a manner similar to the mechanism proposed by Mura et al. (2009). Schmidt also cited an apparent asymmetry in the tail observed during the first MESSENGER flyby (McClintock et al., 2008), although Burger et al. (2010) concluded that this asymmetry was most likely an observational artifact and that sodium source rates from the two hemispheres were approximately equal during the flyby. Observations during Mercury transits of the Sun (Schleicher et al., 2004; Yoshikawa et al., 2008; Potter et al., 2013), discussed below, have shown no evidence of a large sodium source in the southern hemisphere. Schleicher et al. (2004) found approximately equal sodium column densities above the two poles, whereas Potter et al. (2013) found that the sodium column density was 50–100% larger above the north pole.

There have also been many studies of the distribution of sodium emission by local time (east–west). Several groups, including Sprague et al. (1997), reported that, on average, sodium emissions are brighter in the morning than in the afternoon. They attributed this pattern to the thermal desorption of sodium frozen on Mercury's nightside surface. Transit observations made at TAA = 149° by Schleicher et al. (2004) also exhibited larger sodium emission above the dawn than the dusk terminator, although Potter et al. (2013) found a dawn–dusk column density ratio close to 1 in their transit observation made at TAA = 328°. In contrast, Potter et al. (2009) claimed the detection of sodium desorption near the dawn terminator in one observation of line-of-sight Doppler shifts. They interpreted

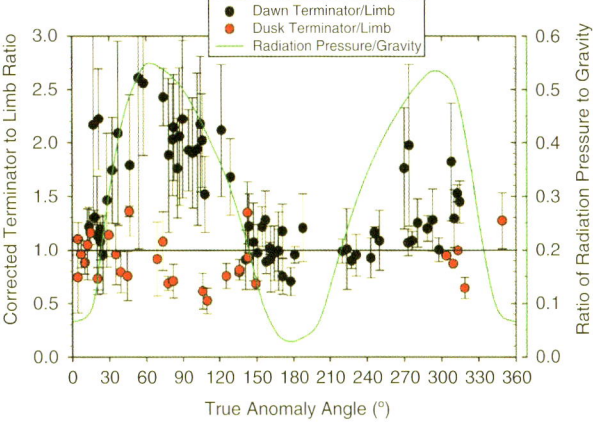

Figure 14.12. Variation with true anomaly angle of the ratio of sodium emission at Mercury's terminator to that at the limb, from observations at the McMath–Pierce Solar Telescope. For the dawn terminator data, ratio values greater than unity appear at true anomaly angles up to about 140°, and again at angles greater than about 250°. When the dusk terminator is in view, the ratio is generally less than unity, with little or no variation with true anomaly angle or radiation acceleration. The green curve is the ratio of radiation acceleration to gravitational acceleration. In these data, values of the terminator-to-limb ratio greater than unity are seen to be located in regions of maximum radiation acceleration, and the smallest values are seen to be located in regions of minimum radiation acceleration. Comparable MESSENGER observations are shown in Figure 14.24. Adapted from Potter et al. (2006).

their data as consistent with an upward flow of sodium from the surface during Mercury's early morning.

Potter et al. (2006) also surveyed the distribution of sodium emission with local time as a function of Mercury's TAA. Their results, which are summarized in Figure 14.12, indicated that emission near the dawn terminator was greater than emission from the limb (which typically corresponded to midday) during the outbound leg of Mercury's orbit, up to TAA ~ 140°, and during Mercury's inbound leg beginning near TAA ~ 250°. In contrast, their observed ratio for the dusk terminator was near unity for all observed true anomaly angles, which suggests that there is a deficit of sodium at dusk. They concluded that this apparent dawn enhancement is evidence of thermal desorption when the terminator is in motion (for much of the year the terminator is nearly motionless). At these times, sodium adsorbed on the cold nightside could be thermally desorbed. Potter et al. (2006) also pointed out that terminator-to-limb ratios greater than unity occur at times of greatest radiation pressure, whereas the smallest values occur at times of minimum pressure. They argued that radiation pressure enhances dawn emission by transporting sodium desorbed in the morning toward the terminator and into shadow.

MESSENGER observations, discussed below, also show that pre-noon local times are, on average, brighter than post-noon and that this pattern is a consequence of variations in solar radiation pressure and terminator velocity throughout the Mercury year. Those data show no evidence of thermal desorption and no sustained dawn–dusk asymmetry.

Sodium Velocity Distribution

Interest in the sodium flow arises from the early modeling work of Ip (1986) and Smyth (1986), who showed that sodium D-line

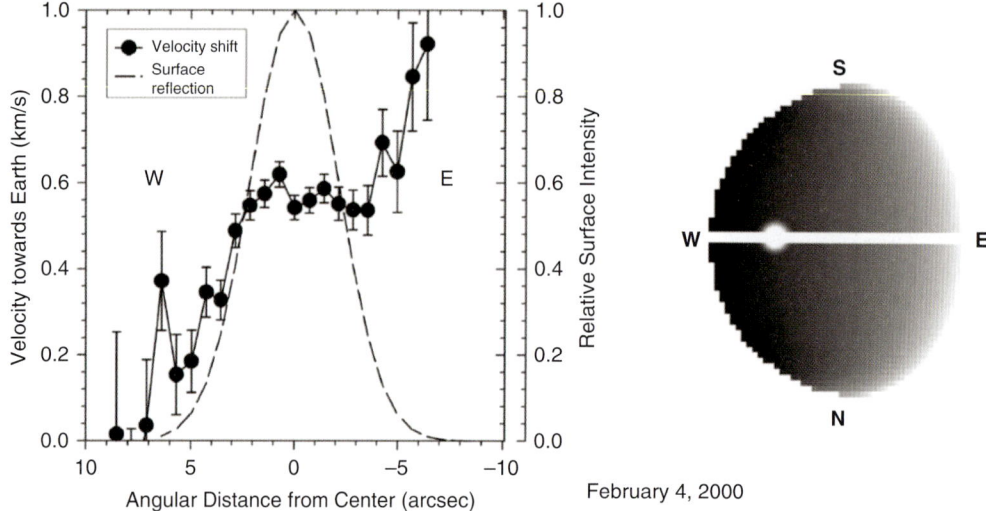

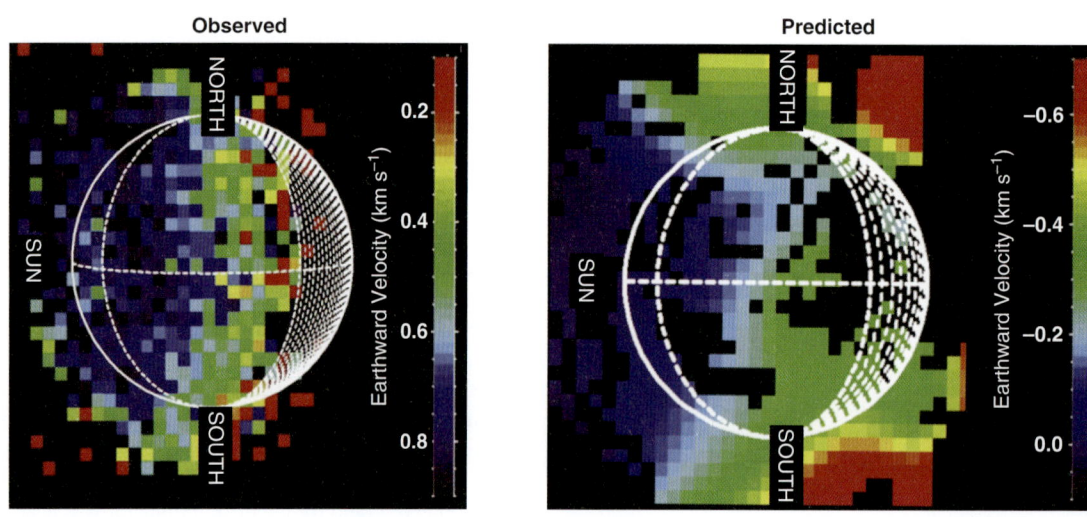

Figure 14.13. Observations of bulk sodium flow speeds in Mercury's exosphere. (a) East–west profile of Doppler shifts in sodium emission with the McMath–Pierce Solar Telescope. A zero value indicates rest with respect to Mercury's surface. The geometry of the observation is shown at right, with the subsolar point indicated by the bright circle. Adapted from Potter et al. (2009). (b) Doppler shifts with the THEMIS telescope compared with the shifts predicted by a Monte Carlo model. The observed line-of-sight velocities are comparable in magnitude with those of Potter et al., but the model-predicted Doppler shifts differ by about 1 km s^{-1}. Adapted from Leblanc et al. (2013).

resonant scattering of solar photons results in radiation pressure acceleration that, during certain times of the Mercury year, is nearly as large as surface gravitational acceleration (see also Figure 14.1). Unfortunately, there are relatively few reliable measurements of the sodium bulk-flow velocity distribution on the dayside of Mercury. The Doppler shifts involved are small, typically less than 1 km s^{-1}, and are difficult to measure. This flow is so small that in their discovery paper Potter and Morgan (1985) reported that the sodium exosphere was at rest with respect to Mercury.

Leblanc et al. (2008) published the first map of bulk sodium flow on Mercury's dayside, and their work was followed by that of Potter et al. (2009) and Leblanc et al. (2009, 2013). Figure 14.13 shows Doppler shift observations taken with high-resolution spectrographs from Potter et al. (2009) and Leblanc et al. (2013). Both found line-of-sight (Earthward) motion between 0 and 1 km s^{-1} relative to Mercury's surface, and both reported that the line-of-sight velocity increased away from the subsolar point. Leblanc et al. (2008) reported that the velocity continued to increase all the way to the terminator, whereas Potter et al. (2009) found that the velocity usually has a maximum between the subsolar point and the terminator.

Though the two sets of observations have many similarities, the two papers had different interpretations. Potter et al. (2009) concluded that they had observed sodium flowing from the subsolar point to the terminators as a result of acceleration by radiation pressure. Leblanc et al. (2013), in contrast, rejected that interpretation because their Monte Carlo model (Leblanc

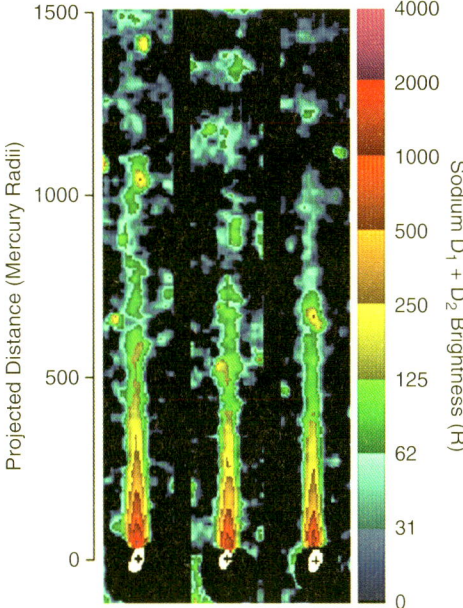

Figure 14.14. (Left) Sodium absorption observed during Mercury's 2003 solar transit (TAA 147) with a Fabry–Perot spectrograph in the Vacuum Tower Telescope at Izaña, Tenerife. Adapted from Schleicher et al. (2004). (Right) Similar observation taken during a 2006 transit (TAA 323) with the Interferometric Bidimensional Spectrometer at the Dunn Solar Telescope. Adapted from Potter et al. (2013). The equivalent width of the absorption is indicated by the bottom color bar.

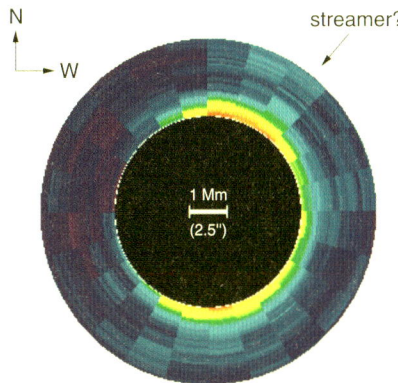
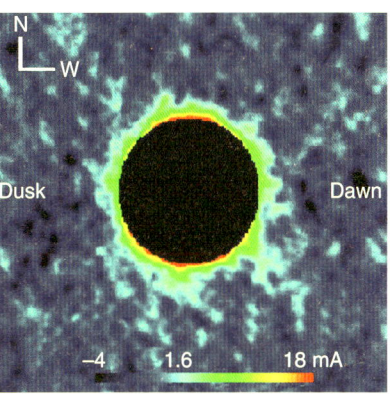

Figure 14.15. (Left) Coronagraph observation of Mercury's sodium tail taken at the Boston University station at McDonald Observatory on 30 April 2009 with a narrowband filter centered on the sodium D lines. Image adapted from Schmidt et al. (2012). (Right) Processed sodium tail observations from the same facility, taken in May–June 2007. The Sun is in the downward direction, and the white area indicates the uncertainty in Mercury's position.

and Johnson, 2010) predicted that the exosphere should be, on average, at rest with respect to Mercury. Although the patterns of flow were similar between observation and model, the magnitude of these Doppler shifts differed by about 1 km s^{-1}. Leblanc et al. (2013) concluded that their observed Doppler shifts were miscalibrated, i.e., their measurements of sodium Doppler shift over Mercury's dayside were consistently in error by about 1 km s^{-1}. The alternative is that a well-established Monte Carlo model failed to reproduce basic features of sodium transport.

Terminator Observations

Two groups (Schleicher et al., 2004; Potter et al., 2013) measured sodium absorption above Mercury's limb, as described in Section 14.3.2. These observations permitted measurements of the sodium scale height, Doppler shift, and energy distribution. Both observations are shown in Figure 14.14. Similar to the Schleicher et al. (2004) observations, Potter et al. (2013) observed enhanced absorption near the poles, but in contrast to the findings of Schleicher et al. (2004) they did not observe a dawn–dusk asymmetry. These workers reached broadly similar conclusions about the sodium speed distribution: scale heights are consistent with a temperature on the order of 1000 K (with large variations), and the Doppler width is consistent with that temperature combined with a bulk anti-sunward flow on the order of 1 km s^{-1}.

Sodium Tail Observations

Ip (1986) and Smyth (1986) predicted that radiation pressure acceleration as a result of resonant scattering of solar photons would be sufficiently strong to accelerate sodium atoms to escape velocity and so form a long anti-sunward tail. Such a tail was first successfully observed by Potter et al. (2002a) using the McMath Pierce image slicer. Later, Baumgardner et al. (2008) and Schmidt et al. (2010) observed the tail using a narrow-band interference filter at the McDonald Observatory – a first for Mercury's exosphere and ideal for the distant sodium tail, where scattered sunlight from the surface is negligible. The Baumgardner et al. and Schmidt et al. observations were made with a wide-field coronagraph that occulted Mercury's disk, allowing long exposures that could detect sodium from ~20 to ~1000 R_M anti-sunward of Mercury (where R_M is Mercury's radius), as illustrated in Figure 14.15. Schmidt et al. (2012) modeled these observations and concluded that the tail is supplied primarily by low-energy sources such as impact vaporization (approximated in the model as a 3000 K Maxwellian distribution) and/or photon-stimulated desorption (approximated as a 1500 K Maxwellian distribution). Additionally, they concluded that there must also be a weaker, high-energy source component in order to match the observed widths of north–south cross-tail profiles. They suggested that this source is supplied by the high-energy "tail" of the energy distribution for photon-stimulated desorption observed by Johnson et al. (2002).

Potter and Killen (2008) observed the emission and Doppler shift of sodium atoms in the tail from ~4000 km to ~46,000 km anti-sunward of Mercury's center (Figure 14.16). Unlike the dayside Doppler shift observations discussed above, there is little ambiguity in estimating sodium velocity in the distant tail. The direction of the flow is known to be approximately anti-sunward, and the magnitude of the Doppler shift is many kilometers per second, much larger than the Doppler shifts on the dayside. From the variability of the tail with TAA, particularly the apparent disappearance of the tail below a particular threshold value of radiation acceleration, Potter and Killen concluded, contrary to the conclusion of Schmidt et al. (2012), that the distant tail is primarily supplied by a high-energy source such as sputtering.

14.4.1.2 MESSENGER Observations

During MESSENGER's Mercury flybys, the UVVS observed the sodium exosphere by obtaining profiles of the tail and performing "fantail" observations during a spacecraft roll in Mercury's shadow (McClintock et al., 2008, 2009; Vervack et al., 2010). An image constructed from the M1 observations of the sodium tail is given in Figure 14.5, and images constructed from M2 and M3 observations are shown in Figure 14.17. A key feature of these images is a brightening of the exosphere above the poles in all three observations. The weak emission during M3 (TAA = 331°) relative to that observed during M1 (TAA = 285°) and M2 (TAA = 293°) was at least partly the result of lower solar radiation pressure near perihelion (Figure 14.1).

Sodium fantail radiances for M2 and M3 are shown in Figure 14.18a along with two models of the M2 observations constructed by Burger et al. (2010). These comparisons indicate that the M2 observations are better fit by a high-latitude dayside source than a uniform source.

MESSENGER also measured limb profiles of sodium emission versus altitude above the polar regions during M3 (Vervack et al., 2010), which are displayed in Figure 14.19a. These observations revealed a two-component structure to the sodium exosphere consistent with that observed routinely during MESSENGER's orbital phase (see below). Fits to exponential functions yielded near-equal intensities for both poles and profiles with e-folding distances of ~200 km for altitudes less than 800 km and ~500 km for higher altitudes.

Coincident with the M1 flyby, observations of Mercury's dayside were obtained with the McMath Pierce image slicer. Mouawad et al. (2011) used both spacecraft and ground-based observations as constraints for a Monte Carlo exosphere model. They concluded that the primary sodium source had an energy distribution similar to a 1000 K Maxwellian, possibly with a high-energy tail, in order to match simultaneously the dayside

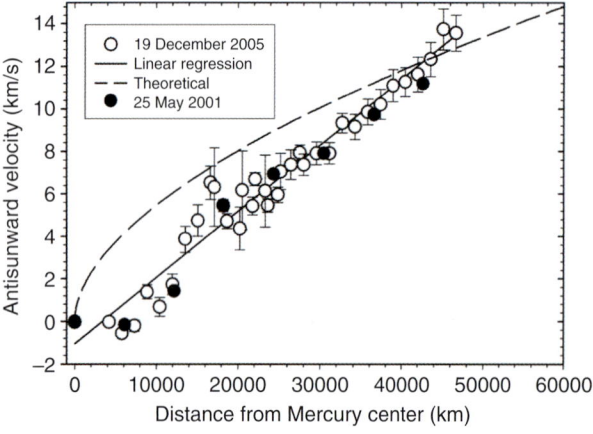

Figure 14.16. Sodium tail velocities derived from the Doppler shift observed with the high-resolution echelle stellar spectrograph at the McMath–Pierce Solar Telescope. Adapted from Potter and Killen (2008).

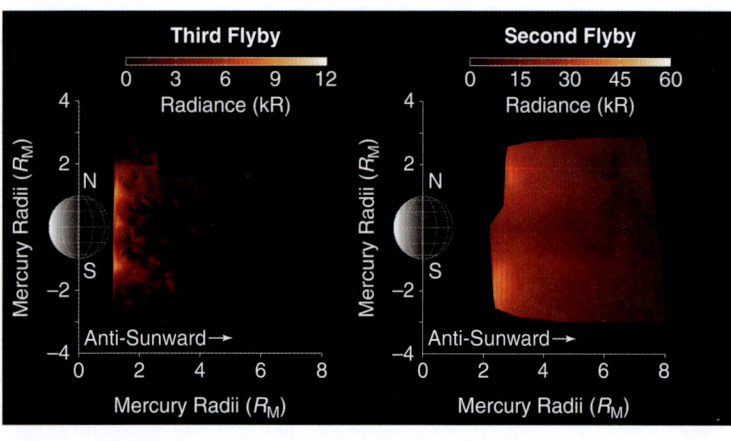

Figure 14.17. Comparison of observations from the third (left) and second (right) MESSENGER flybys of Mercury, which occurred at TAA = 331° and TAA = 293°, respectively. The markedly lower sodium tail emission during M3 is the result of relatively small solar radiation pressure during M3. The images show observed column emissions projected onto the plane containing the Sun–Mercury line and Mercury's spin axis, interpolated to fill in unobserved regions. The region (1–2) R_M anti-sunward from the planet was not sampled during M2.

14.4 Observational Results

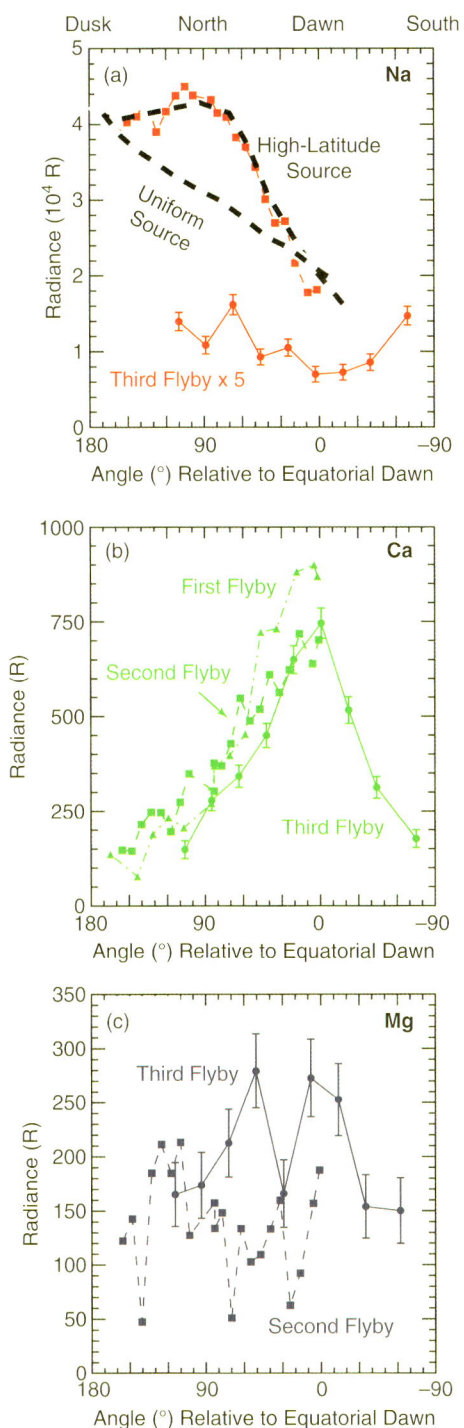

Figure 14.18. Fantail observations of (a) sodium, (b) calcium, and (c) magnesium in Mercury's tail taken during MESSENGER's Mercury flybys. Each panel shows observed radiance as a function of angle relative to equatorial dawn (0°) for the second and third flybys. Data from the third flyby are shown with one-standard-deviation uncertainties; uncertainties in the observations from the first and second flybys are slightly smaller owing to longer integration times but are omitted for clarity. Predictions of two models of the sodium observations from the second flyby, including a uniform source and a high-latitude northern-hemisphere source, are shown in (a) with black dashed lines (Burger et al., 2010). The sodium emissions during the third flyby were much lower because of the low sodium g-value at that time. By contrast, calcium observations (b) show a clear dawn–dusk asymmetry, with dawn (0°) brighter than dusk (180°). Adapted from Vervack et al. (2010) and Burger et al. (2010).

and tail data. They also obtained an upper limit on an impact vaporization source (assumed to be at 5000 K). This upper limit was consistent with the impact vapor production predicted by Morgan et al. (1988) but two orders of magnitude smaller than that of Borin et al. (2010).

Observations of Mercury's sodium exosphere from orbit were described by Cassidy et al. (2015), who focused on dayside limb scans (Figures 14.3 and 14.8) above the equatorial region (approximately 30°S to 30°N, Figure 14.9) and similar scans above the south polar terminator. Figure 14.20, taken from Cassidy et al. (2015), shows the total sodium emission as a function of tangent altitude for 1400 local time and TAA ~ 112°. These are typical sodium dayside limb profiles that illustrate features seen throughout the mission. There are two distinct temperature components (Vervack et al., 2010), a relatively dense and cold component with a small scale height (~100 km) and a tenuous hot component seen at the highest altitudes probed by the limb scans (up to 4000 km).

Cassidy et al. (2015) extracted information from limb profiles using the model of Bishop and Chamberlain (1989), which includes the effects of radiation acceleration (Section 14.2). This modeling permitted estimates of temperature and near-surface density as functions of TAA, as shown in Figure 14.21 for limb scans taken near the subsolar point. These results were obtained for the relatively cold and dense component that comprises the bulk of the exosphere. The derived temperature of this colder component is relatively constant at about 1200 K, which is warmer than the surface but much colder than predicted by many candidate source processes, suggesting that the most likely source process is photon-stimulated desorption. Cassidy et al. (2015) were not able to constrain the temperature of the hot component from limb profiles. They showed that models with temperatures in the range 5000–20,000 K fit the hot component equally well.

Ground-based observers typically use Doppler broadening to infer temperatures, which is not possible with the relatively low spectral resolution of MASCS measurements. Temperatures inferred in this way are, with one exception (Killen et al., 1999), much larger than the ~1200 K inferred by MASCS limb scan data (see summary by Cassidy et al., 2015). However measurements of the exosphere's scale height during solar transits (Schleicher et al., 2004; Potter et al., 2013) found values comparable with those measured by MASCS (Figure 14.21).

Thermalization, the thermal equilibration of atoms with a surface, is seen in laboratory experiments with sodium atoms adsorbed on a variety of substrates (e.g., Yakshinskiy and Madey, 2005) and is also a common assumption underlying Monte Carlo models of Mercury's exosphere, such as those of Smyth and Marconi (1995), Mura et al. (2009), and Leblanc and Johnson (2010). Cassidy et al. (2015) searched their data for a colder, thermalized component. Their demonstration of what limb profiles would look like with the addition of a thermalized component is given in Figure 14.22 for two different local times: early morning, during which thermal desorption might release sodium atoms adsorbed on the nightside; and at the subsolar point, where the surface is warmest. They added a surface-temperature component to the fitted model, giving it the same vertical column density as the warmer component. The surface temperature used was typical for that local time (Yan et al., 2006). The top panel of Figure 14.22 (early morning) shows that an abundant thermalized component would be

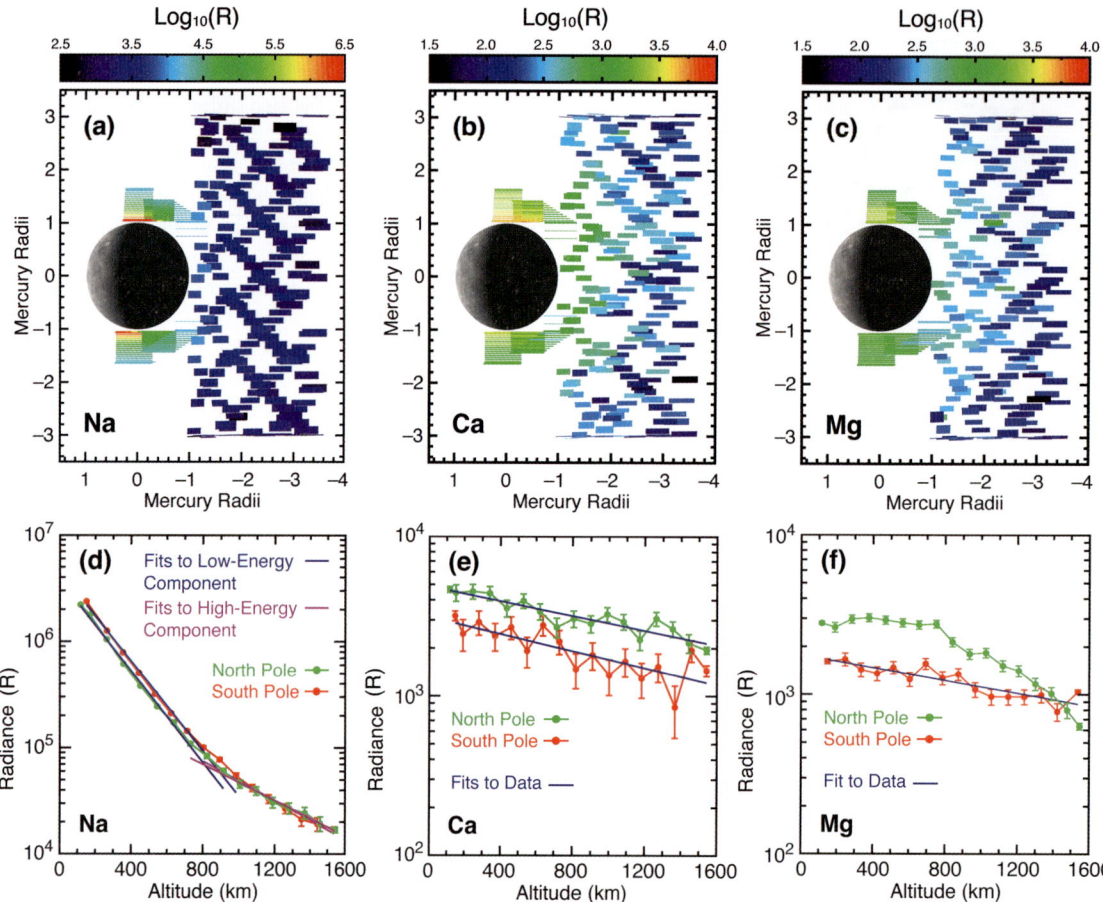

Figure 14.19. (a)–(c) Observations of sodium, calcium, and magnesium emission in the polar, nightside, and tail regions of the planet during MESSENGER's third flyby. Each observation is represented by the projection of the instrument's rectangular field of view onto the plane defined by the Sun–Mercury line and the spin axis of Mercury. The horizontal dimension of the filled rectangles indicates the length of the slit as projected onto the plane and becomes smaller as the spacecraft approaches Mercury; the vertical extent of the rectangles indicates the range over which the UVVS slit moved during the integration time (1.5–2.0 s), with shorter distances corresponding to slower rates of slewing by the spacecraft. Each observation is thus an average of the emission over the indicated region. Owing to the location of the spacecraft during these observations, the lines of sight are not perpendicular to the plane but are directed sunward by 5–10°. (d)–(f) Profiles of sodium, calcium, and magnesium emission with altitude over the north and south poles of the planet (shown with one-standard-deviation uncertainties). Exponential fits to the data indicating the general behavior of each species are also shown. The lone exception is magnesium over the north pole, which cannot be fit with such a simple model, as discussed in the text.

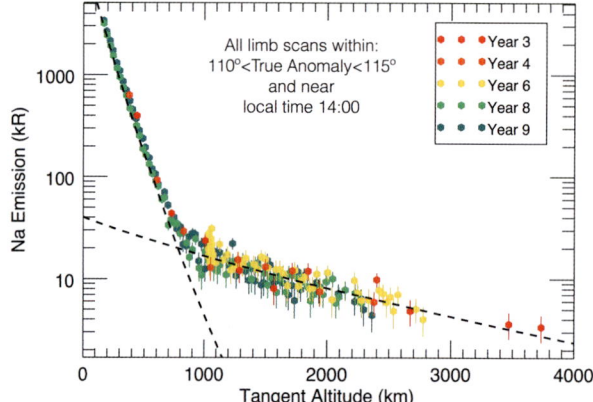

Figure 14.20. MESSENGER limb emission profiles of sodium in Mercury's exosphere taken from orbit (adapted from Cassidy et al., 2015) and associated Chamberlain models (dashed lines). This figure includes every limb emission profile within short ranges of true anomaly angles and local time (as labeled). Repeatable observation geometry during the mission permitted comparison among Mercury years. As is typical for dayside sodium limb emission profiles, there are two components at different temperatures; these profiles are well approximated by Chamberlain models at 1230 K and 20,000 K.

14.4 Observational Results   387

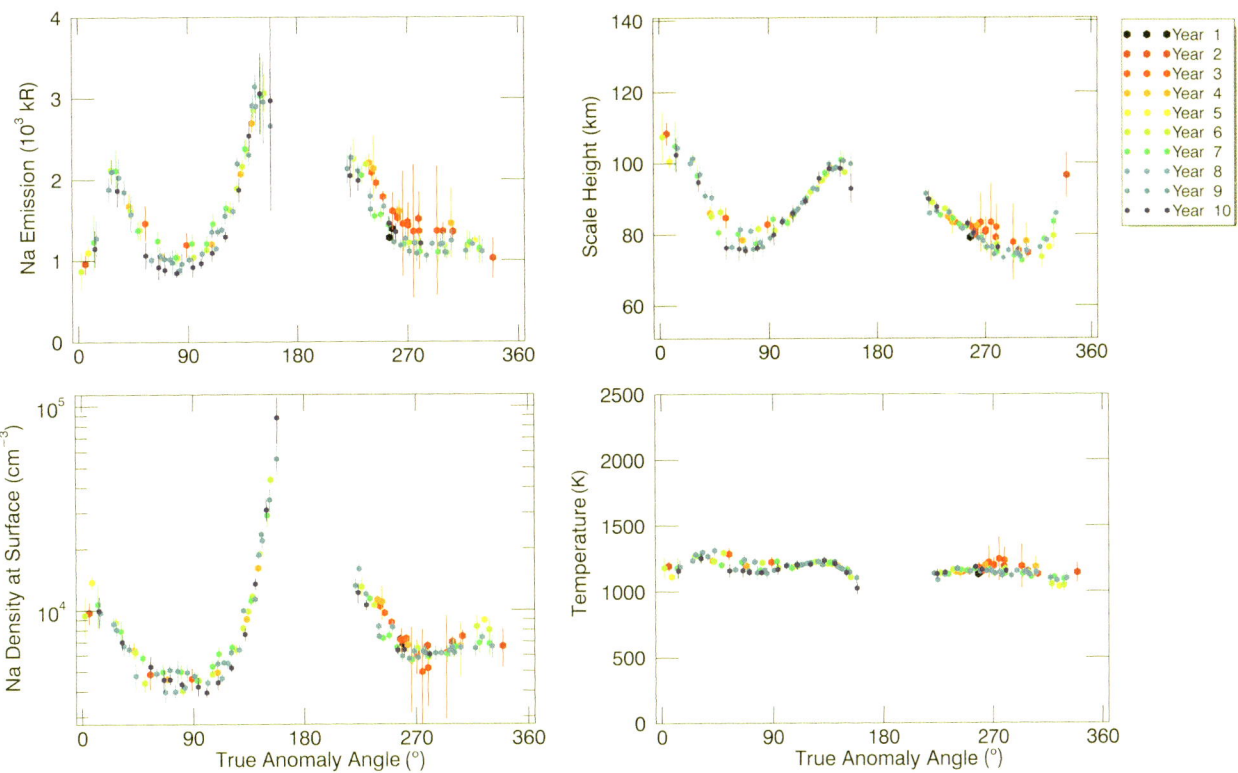

Figure 14.21. Properties of Mercury's sodium exosphere near the subsolar point as a function of true anomaly angle. (Top left) Radiance at 300-km altitude obtained by interpolation of the limb profiles. (Top right) Scale height of the dominant colder component. (Bottom left) Surface density. (Bottom right) Temperature of the colder component. Adapted from Cassidy et al. (2015).

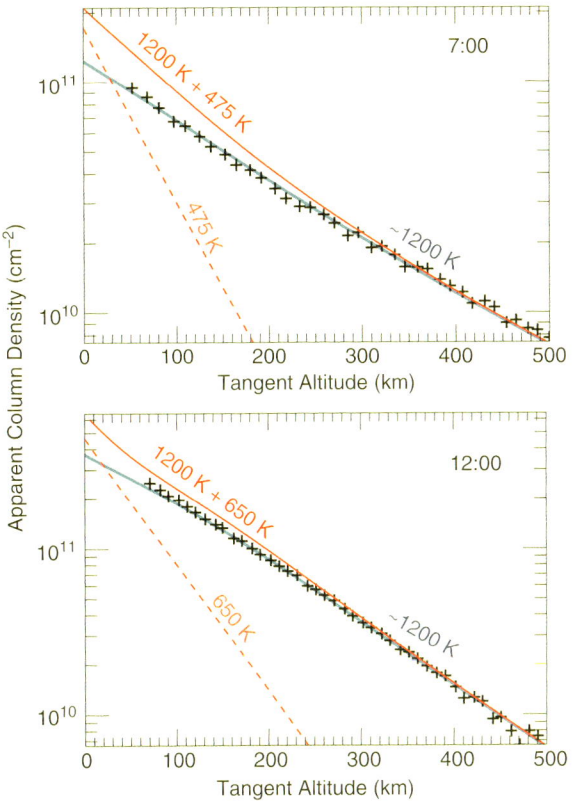

Figure 14.22. Cassidy et al. (2015) found no evidence for a substantial cold sodium exosphere that would indicate accommodation with the surface temperature. Examples of their search for a surface-temperature component are shown for early morning (07:00 local time) and the subsolar point (12:00). The solid red line shows a notional limb emission profile near the surface with the addition of a surface-temperature component added to the observed 1200 K component. For this example, it was assumed that the surface-temperature component has the same vertical column density as the 1200 K component.

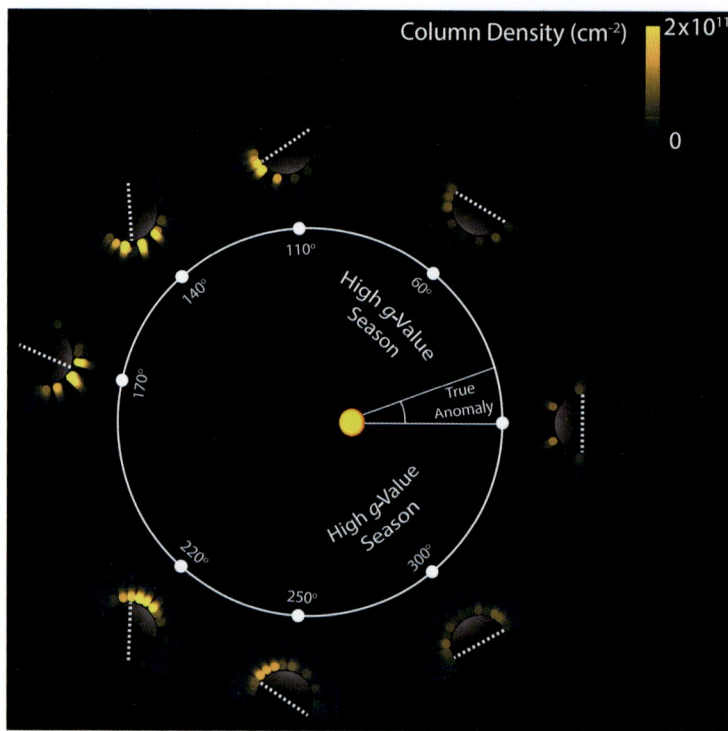

Figure 14.23. The distribution of dayside sodium column density over the course of one Mercury year as seen from above the north pole. Sodium emission is enhanced above one of the two "cold pole" longitudes (connected by dotted lines). As that surface rotates into the sunlight, it becomes a strong localized source of sodium. There is no persistent dawn–dusk asymmetry. Whereas dawn is brighter than dusk on the outbound leg from TAA = 60° to TAA = 180°, dusk is brighter than dawn on the inbound leg from TAA = 180° to TAA = 300°. Near perihelion the emission is approximately uniform over the dayside. To produce this figure, MESSENGER UVVS observations with low-latitude line-of-sight tangent points (<30° latitude) were projected onto the equatorial plane.

obvious but was not evident in the data. At the subsolar point (bottom panel), the difference was much more subtle. Although Cassidy et al. (2015) did not rule out the presence of a thermal component, they concluded that it is constrained to be much less abundant than the ~1200 K component in the dayside limb profiles.

Thermal desorption of sodium at dawn or in the morning has been one of the mechanisms invoked in the literature to explain dawn–dusk asymmetries in some of the ground-based observations of sodium emission (e.g., Leblanc and Johnson, 2003, 2010; Peplowski et al., 2014), as discussed above. MESSENGER observations show that changes in the dawn-to-dusk emission ratio do not necessarily imply a thermal desorption source but are rather the consequence of seasonal variations in solar radiation pressure and Mercury's orbital angular rate. Figure 14.23 illustrates the sodium column densities observed by MESSENGER throughout a Mercury year. Over the TAA range ~270° to ~90°, centered on perihelion, the planet's rotation rate is nearly equal to its orbital angular rate. During this time the terminators are nearly stationary and sit near the "cold poles" (geographical longitudes −90° and 90° E). At these times the sodium g-value reaches its maximum and anti-sunward transport is most robust (see Figure 14.1). As a consequence, the cold poles collect sodium atoms that migrate from the dayside to the nightside in response to solar radiation pressure. As Mercury's distance from the Sun increases, the planet's orbital angular rate slows and the dawn cold pole rotates into sunlight, where the sodium sequestered on the surface is released into the exosphere. The release-enhanced emission over the cold pole declines slowly as Mercury progresses in its orbit, and it continues as the pole rotates toward the evening terminator. Thus, the morning is substantially brighter than the afternoon for most of the outbound portion of Mercury's orbit, and the afternoon is slightly brighter than morning during the inbound portion of Mercury's orbit (Cassidy et al., 2016).

This result is consistent with the ground-based observations shown in Figure 14.12. In Figure 14.24 the radiances observed by MESSENGER in the morning (local times 0600–1000) and afternoon (1400–1800) are ratioed to those at midday (1000–1400). Morning and afternoon correspond approximately to what Potter et al. (2006) refer to as the "dawn" and "dusk" terminators, as ground-based observations taken near the terminators will necessarily include emissions from much of the dayside as a result of smear from imperfect atmospheric seeing. The "limb" in the observations of Potter et al. (2006) was usually near midday. As in Figure 14.12, MESSENGER observations show a strong morning enhancement for much of the outbound leg (TAA < 180°) and no morning–afternoon asymmetry near perihelion. Unlike Figure 14.12, MESSENGER observations show that the afternoon is brighter than the morning for much of each outbound leg (180° < TAA < 300°). This TAA range is where Figure 14.12 has a gap in "dusk" observations.

A remarkable feature of the MESSENGER sodium limb profiles is their annual repeatability. For a given true anomaly and local time, observations change little over many Mercury years. It is unlikely that ion bombardment from the ever-changing space weather environment (Slavin et al., 2014) could produce the observed pattern. Although MESSENGER orbit and pointing constraints precluded low-altitude (<1000 km above the surface) limb scans directly above the cusp regions (Figure 14.8), the observations shown in Figure 14.21 were taken over portions of the surface that are bombarded by ions on the nightside (Winslow et al., 2014). No clear signature of this process can be seen in the MESSENGER data.

14.4.2 Potassium

Potassium in Mercury's exosphere can be observed with ground-based telescopes via emissions from its D lines located at 766.49-nm and 769.90-nm wavelength. It has been observed less frequently than sodium because of its lower abundance and because large Doppler shifts are required to avoid contamination from telluric molecular oxygen lines. Figure 14.25 compares sodium and potassium spectra obtained on 17 January 2008 at the McMath–Pierce Solar Telescope at the Kitt Peak National Observatory (Killen et al., 2010). Whereas sodium emission dominates the signal at 588.918 nm, the potassium signal is a weak feature superimposed on a much brighter spectrum of sunlight reflected from the planet's surface. Nonetheless, disk-averaged potassium data were published for 18 dates in 1988, 1990, 1998, and 1999 (Potter et al., 2002b). These show that the potassium column density varied over the range $(0.7–4) \times 10^9$ cm^{-2}. The sodium/potassium ratio was observed to be highly variable, between ~30 and ~140, and was highly correlated to Mercury sub-observer longitude and weakly correlated to radiation pressure acceleration.

Examples of disk-resolved observations of sodium and potassium are shown in Figure 14.26. The left panel shows disk images obtained by Killen et al. (2010) on 17 January 2008 near the time of the first MESSENGER flyby, and the right panel shows sodium/potassium ratios observed at several positions by Doressoundiram et al. (2010) on 17 and 19 June 2006. Killen et al. (2010) reported that the sodium/potassium ratio peaked at high latitudes on the dayside, at a value on the order of 100, and that the equatorial ratio was smaller than this value, at about 30. The sodium/potassium ratio varied with altitude above the limb, as expected, given that the sodium scale height is expected to far exceed that of heavier potassium. Their southern hemisphere potassium column density was 1.3×10^9 cm^{-2}, and their average potassium zenith column in the north, 8×10^8 cm^{-2}, was close to the detection limit of 6.8×10^8 cm^{-2}. Doressoundiram et al. (2010) also reported sodium/potassium column density ratios that varied over Mercury's surface, ranging from ~60 near the equator to ~400 at high latitudes on the dayside and up to 160 near the pole on the nightside. Their potassium column densities were reported to be between 6×10^8 and 1×10^{10} cm^{-2}.

The spectral range of the MESSENGER UVVS instrument was limited to wavelengths less than 610 nm, which is below the strong potassium D-line observations described above. Attempts to detect the potassium emission lines located at 404.48 nm and 404.52 nm, which have g-values that are

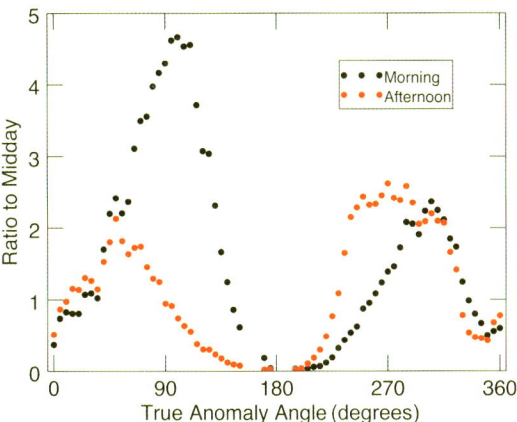

Figure 14.24. Morning–afternoon asymmetries in MESSENGER observations of exospheric sodium. These quantities, the ratio of morning or afternoon radiance to that at midday, are approximately comparable to the terminator-to-limb ratios shown in Figure 14.12. Both sets of observations show a pronounced morning enhancement during the outbound leg of Mercury's orbit, whereas the afternoon is brighter than the morning during the inbound leg, where Figure 14.12 has a gap in observations.

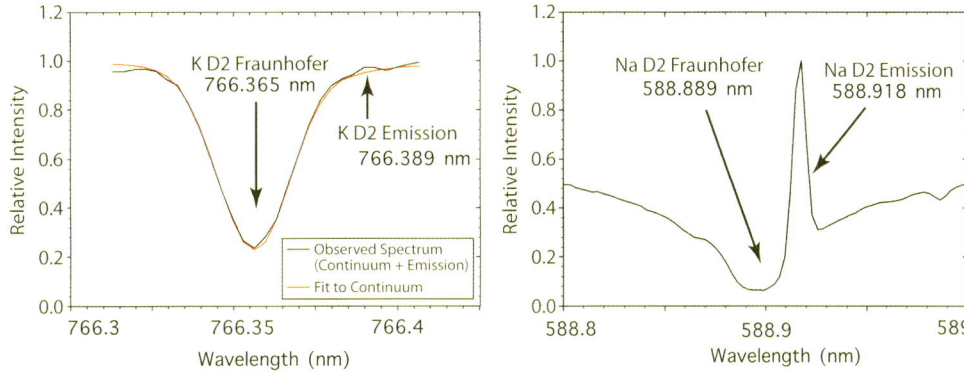

Figure 14.25. (Left) Observations of exospheric potassium on Mercury acquired at the McMath–Pierce Solar Telescope. The wavelength scale is given in Earth's rest frame, which is Doppler shifted by -0.101 nm with respect to Mercury's rest frame. (Right) Sodium D_2 line emission, also plotted in Earth's reference frame for comparison, shows the relative difficulty of observing potassium. Adapted from Killen et al. (2010).

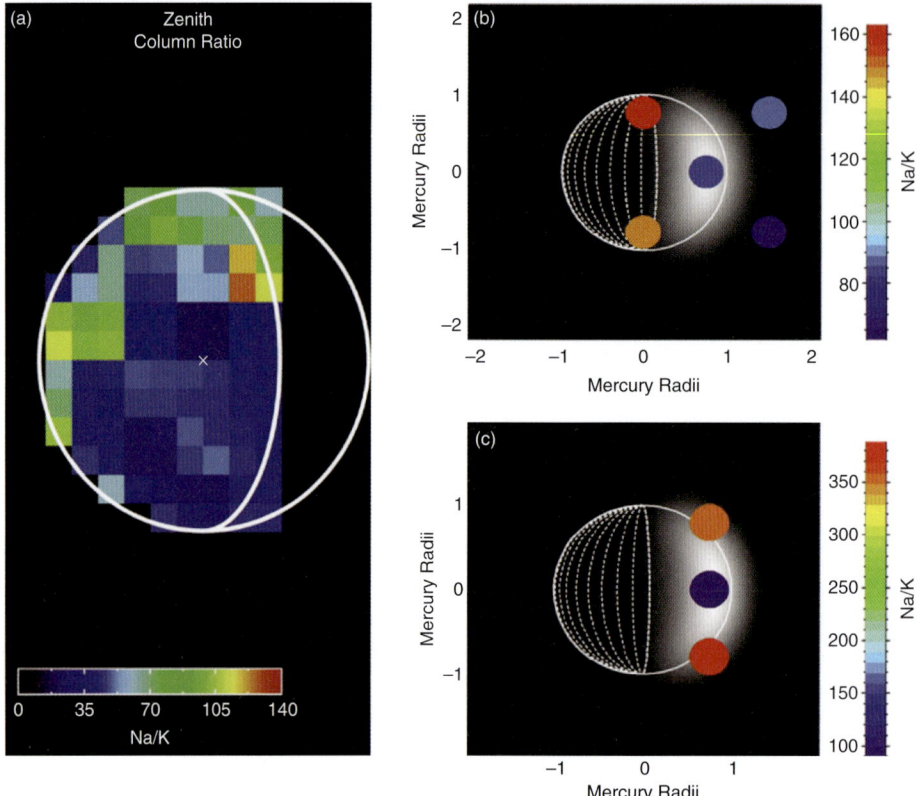

Figure 14.26. (a) Line-of-sight sodium/potassium ratio measured with the McMath–Pierce Solar Telescope. The ratio in the southern hemisphere (~36), which is close to the noise-equivalent column, is close to the smallest ratio found in the data sets considered by Potter et al. (2002a), whereas the ratio in the northern hemisphere (~74) is near the mid-value found previously (Killen et al., 2010). (b) Sodium/potassium ratios measured by Doressoundiram et al. (2010) from ~60 near the equator to ~400 at high latitudes on the dayside and up to 160 near the pole on the nightside.

approximately 1000 times smaller than those for the D lines, were not successful.

14.4.3 Calcium

14.4.3.1 Ground-Based Observations

Mercury's calcium exosphere was discovered with ground-based observations at the Keck Observatory by Bida et al. (2000); see Figure 14.27. From the observed line widths, it was clear that calcium is hot, with a temperature ~12,000 K. It was also determined that the calcium exosphere is much more tenuous than the sodium exosphere; the estimated calcium column density was more than three orders of magnitude lower than the sodium column density (e.g., Potter and Morgan, 1985). The observations described a calcium exosphere that varies both temporally and spatially. The spatial distribution of calcium was not well determined; emission was seen over the poles but was not observed over the subsolar point because scattering from the surface made detection over this point unreliable. Bida et al. (2000) considered three sources for the observed neutral calcium: photon-stimulated desorption, micrometeoroid impact vaporization, and ion sputtering, preferring the last because of the observed high energy of the calcium and the enhanced emission over the poles.

Killen et al. (2005) partially mapped the calcium around Mercury using observations acquired at the Keck Observatory between 1997 and 2002. A gradual increase in radiance was seen during this period, although this increase can partly be explained by variations in the g-value, which increased over time for the nights analyzed. However, a small increase in calcium column density was observed over time, which may be consistent with the seasonal variability in calcium source rate found from MESSENGER data by Burger et al. (2014) and discussed below. Killen et al. (2005) also confirmed the presence of energetic calcium through measurements of the emission line width (consistent with temperatures between 12,000 K and 20,000 K) and the Doppler shift of the emission from the expected wavelength. The emission was found to be blue-shifted by motion toward the observer at speeds of a few kilometers per second. As only the dawn hemisphere was observed by Killen et al. (2005), it was not clear whether this high-velocity calcium was present at dusk as well. Because of the high energy of the ejected calcium, the preferred ejection process was either ion sputtering or the photodissociation of calcium-bearing molecules (such as calcium oxide) resulting from micrometeoroid impact vapor.

14.4.3.2 MESSENGER Observations

MESSENGER observed calcium emission from Mercury's exosphere during all three flybys and on a near-daily basis once in orbit around the planet. During the first MESSENGER flyby, UVVS observed calcium in two regions. First, calcium was seen

14.4 Observational Results 391

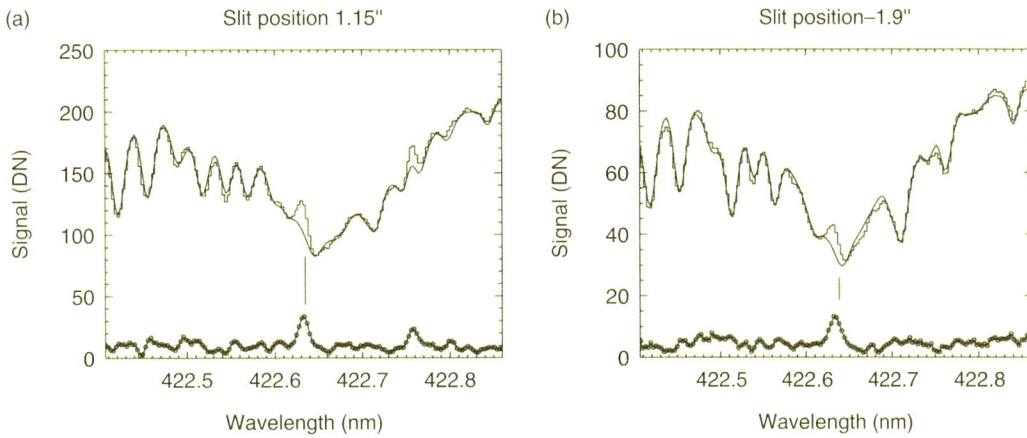

Figure 14.27. Samples of calcium emission spectra obtained at the W. M. Keck Observatory by Bida et al. (2000) on (a) 7 July 1998 and (b) 19 July 1998. The histogram plots show the observed spectra, and the solid lines are the fitted sky continuum. Differencing these two reveals the exosphere emission shown by the circle-symbol plots. The expected position of the Doppler-shifted calcium emission line is indicated by a vertical line. DN denotes digital number. Adapted from Bida et al. (2000).

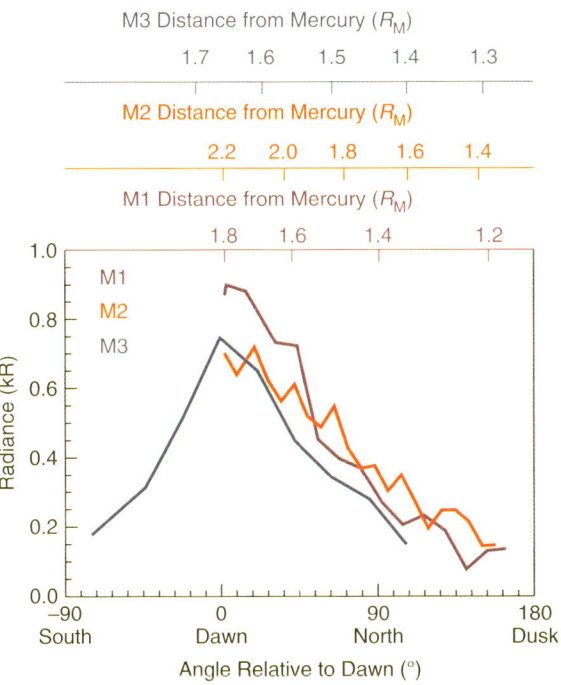

Figure 14.28. Calcium fantail measurements from each of the MESSENGER flybys of Mercury. The abscissa gives the direction of the MASCS boresight relative to the equatorial dawn direction, with positive values indicating that UVVS was pointing north of the equatorial plane and negative values indicating southward pointing. The top axes give the radial distance of MESSENGER from Mercury's center. These observations presented the first evidence that calcium is concentrated toward dawn.

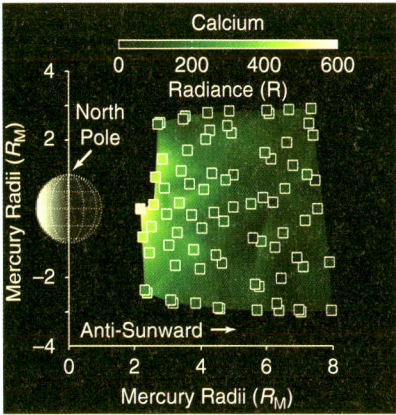

Figure 14.29. Observations of calcium emission from Mercury's tail acquired during M2. Individual observations (white squares) are overlaid on an image generated by projecting the observed line of sight onto the plane containing the spin axis and the Sun–Mercury line. Adapted from McClintock et al. (2009).

during the fantail observations (Section 14.3.3.1), which occurred in Mercury's shadow near closest approach to the planet. As MESSENGER emerged from shadow, UVVS pointed toward the surface, measuring emission over the pre-dawn and dawn terminator regions where the sodium/calcium ratio was observed to increase from 10 at the end of the fantail to 50 just before the UVVS line-of-sight crossed the terminator (McClintock et al., 2008). Similar fantail observations were made during the second and third flybys (McClintock et al., 2009; Vervack et al., 2010). Comparisons of the fantail measurements from the three flybys are shown in the middle panel of Figure 14.18 and in Figure 14.28. Flybys M2 and M3 also included observations in the tail over the nightside of the planet (Figure 14.29). During M3, UVVS also measured profiles of calcium emission versus altitude above the north and south poles (Figure 14.19b and 14.19e) which exhibited a single temperature component. Vervack et al. (2010) made exponential fits to these profiles and reported e-folding values of 1878 km in the north and 1621 km in the south as well as a north/south intensity ratio of 1.6 at the surface. Analysis by Burger et al. (2012) produced similar e-folding values (1840 ± 140 km and 1716 ± 200 km) and surface intensity ratio (1.4).

The flyby observations revealed several key features of the calcium exosphere, which indicated substantial differences from the sodium exosphere. First, the large scale heights over the poles (~1800 km) indicated an energetic calcium source, consistent with the ground-based observations. The tail observations are strikingly different from the sodium tail observations (Figures 14.5 and 14.17). Whereas sodium shows the brightest emission north and south of the equator, calcium was brightest in the equatorial region. Also, the calcium radiances measured in the tail region were approximately stable between the two flybys whereas sodium showed a large seasonal variation because of differences in the radiation pressure. Finally, the fantail observations indicated a large dawn–dusk asymmetry. During all three flybys, the peak radiance was measured when UVVS pointed toward dawn and decreased as the spacecraft rolled through north toward dusk. Burger et al. (2012) modeled this dawn–dusk asymmetry with a source located entirely on the dawn hemisphere.

A more complete view of the exosphere was derived from the orbital observations, which spanned a wide range of observing geometries, as described above. Burger et al. (2014) summarized observations of dayside limb scans from MESSENGER's primary and first extended missions. These data confirmed the dawn enhancement in the calcium emission. Figure 14.30 presents a visual representation of the limb profiles from three orbits at different true anomalies. Small-scale variations in the images are not real but are created by statistical variations in the radiances of the individual limb profiles. However, the large-scale trends seen in the data (in particular the dawn–dusk asymmetry) are real. Burger et al. (2014) cautioned that the images are intended as a qualitative visualization of the data, and the interpolated values should not be used in quantitative analysis.

Burger et al. (2014) also discovered a strong seasonal variation in the calcium emission. Figure 14.31 presents the results of their fits of exponential functions to limb profiles using the function

$$4\pi I(A, L) = 4\pi I_0(L) e^{-A/H} \quad (14.9)$$

where $4\pi I_0$ is the radiance at the surface at local time L, A is the altitude above the surface, and H is the e-folding width. Burger et al. (2014) pointed out that H is not a true scale height [equation (14.8)] because the e-folding width is strongly affected by ionization, and the calcium ionization rate is less than 1 h (Huebner and Mukherjee, 2015). Therefore, much of the calcium ejected from the surface is ionized within the first several thousand kilometers of leaving the surface. Several features are apparent from these plots. First, there is a strong seasonal variation in the surface radiance at dawn (Figure 14.31a). Second, there is very little year-to-year variability in the surface radiance, indicating that the exosphere is repeatable from one year to the next. Third, the e-folding width is approximately constant over a Mercury year, although the results are somewhat noisy (Figure 14.31b). There may be a small seasonal variation, with H peaking at aphelion and being minimized at perihelion, consistent with the inverse-square variation in calcium ionization rate with distance from the Sun. Much of the variation in the surface radiance can be explained by variations in the g-value. Figure 14.31c shows the approximate surface column density derived from $N = 10^9 \, 4\pi I/g$, where $4\pi I$ has units of kR, g is the g-value of an atom at rest relative to Mercury, and the column density at the surface, N, has units of cm^{-2}. Use of this equation introduces a systematic error because of the assumption that the atoms are at rest relative to Mercury. As discussed above, however, it is known that calcium is ejected energetically from Mercury. Killen et al. (2005) found temperatures up to 20,000 K; the models of Burger et al. (2012, 2014) suggest T values of ~50,000–70,000 K. The thermal speeds associated with these temperatures are 3–5 km s^{-1}. Speeds this high can result in significant differences in the g-value of such an atom from that for one at rest relative to Mercury, particularly at the subsolar point where atoms have a large component of their velocity along the Sun–Mercury line. This systematic error has the greatest impact on the plot of N in Figure 14.31c at perihelion and aphelion because the Mercury g-value is minimized and both positive and negative deviations in radial velocity tend to increase the true g-value. Use of the Mercury-at-rest g-value therefore underestimates the true g-value, and the derived value of N_0 is too large. This effect yields an overestimate of the peak at perihelion seen in Figure 14.31c and creates a pseudo-peak at aphelion. A correction for this systematic error would be both a strong function of local time and model dependent (it depends on both the assumed temperature and the spatial distribution of the source), and therefore it is not straightforward to remove the g-value bias in the data. A more complete understanding of the seasonal variation in the calcium data requires a model that takes into account the motion of calcium atoms relative to Mercury.

Release from the surface followed by molecular photodissociation, first proposed by Killen et al. (2005), is often suggested as a process that can lead to the observed high temperatures for calcium. More recently, Killen and Hahn (2015) and Christou et al. (2015) showed that the calcium source rate observed by MESSENGER is consistent with micrometeoroid impact vaporization by dust supplied by the interplanetary dust disk and by comet 2P/Encke, respectively (Chapter 15).

Killen et al. (2016) ruled out photodissociation as the source of the energetic calcium: new results from the Burger et al. (2014) model showed that part of the calcium source is in the region behind the terminator, where photons do not have access. Instead, Killen suggested that the energetic calcium can be produced directly by impact vaporization for sufficiently energetic impactors.

14.4.4 Magnesium

Magnesium was first detected by the UVVS during MESSENGER's second flyby of Mercury (M2) on 6 October 2008 (McClintock et al., 2009). Figure 14.32 shows the magnesium image of the nightside tail region that was acquired by UVVS during approach. The magnesium image is more consistent with a uniform distribution than either sodium (Figure 14.17) or calcium (Figure 14.29). Fantail observations of magnesium, shown in Figure 14.18c, also support a nearly uniform distribution during both M2 and M3.

During M3, UVVS also measured profiles of magnesium emission versus altitude above the north and south poles

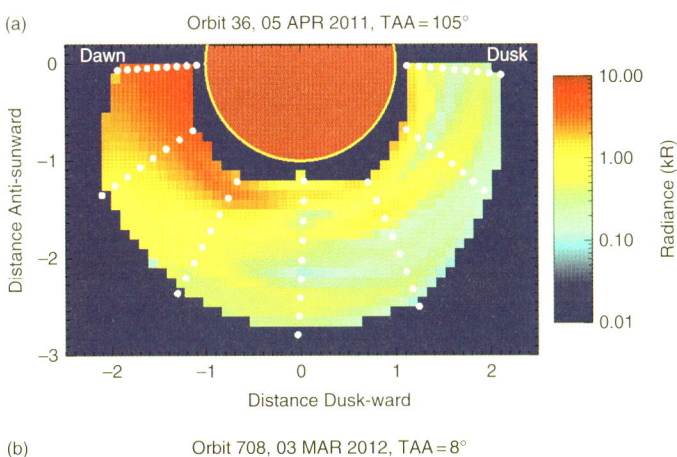

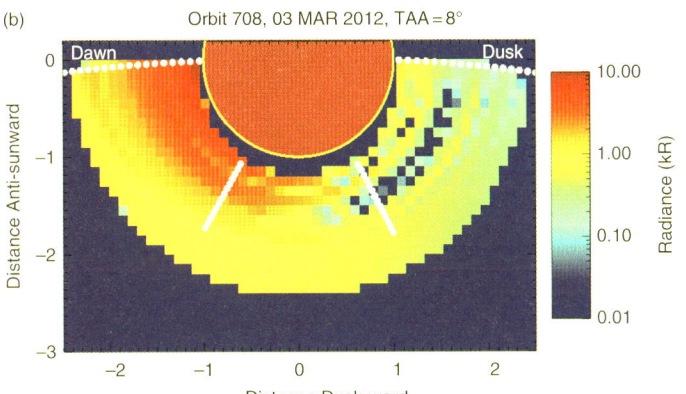

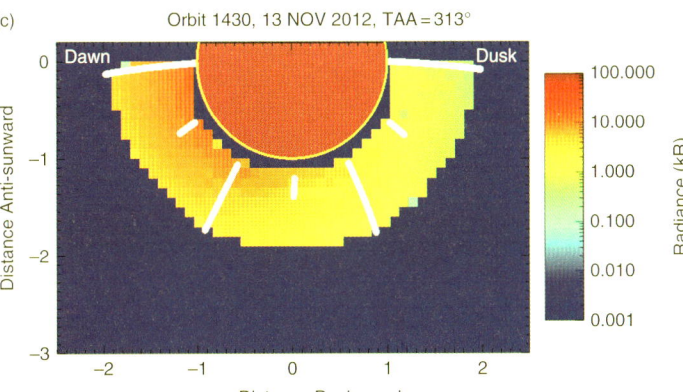

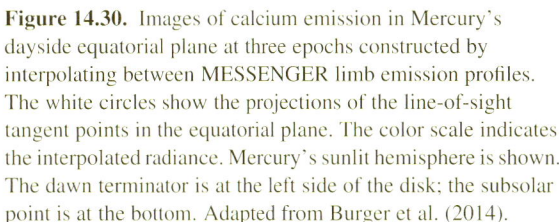

Figure 14.30. Images of calcium emission in Mercury's dayside equatorial plane at three epochs constructed by interpolating between MESSENGER limb emission profiles. The white circles show the projections of the line-of-sight tangent points in the equatorial plane. The color scale indicates the interpolated radiance. Mercury's sunlit hemisphere is shown. The dawn terminator is at the left side of the disk; the subsolar point is at the bottom. Adapted from Burger et al. (2014).

(Figures 14.19c and 14.19f). Vervack et al. (2010) fit an exponential function to the southern profile and reported an e-folding value of 2160 km. In the north, magnesium exhibited higher radiances, indicating larger release rates. Moreover, the magnesium profile above the north pole is strikingly different from a simple exponential. Vervack et al. (2010) suggested that this profile is indicative of additional source processes at work.

Chamberlain models fit to the flyby tail data (Killen et al., 2010; Sarantos et al., 2011) showed that the inferred magnesium density distribution could not be derived with a single source process. Rather, the combination of a hot source >20,000 K and a cooler source <5000 K was needed to adequately describe the magnesium distribution in the near and very far tail (Sarantos et al., 2011).

During the MESSENGER orbital phase, dayside limb scans of magnesium were observed at a systematic cadence. Merkel et al. (2017) analyzed a subset of these data from March 2013 to April 2015. Figure 14.33 illustrates their results for a typical sequence acquired on 12 August 2014 when Mercury's TAA was 77° (and when MESSENGER was in nearly a noon–midnight orbit).

The central image in Figure 14.33 summarizes the results, whereas separate plots represent the limb emission profiles and uncertainties at each local time. Maximum emission level decreased as the local time progressed toward dusk until the signal was near the noise level by local time 1800. The colored center figure represents the seven limb profiles projected onto the equatorial plane (viewed from the north) and interpolated to

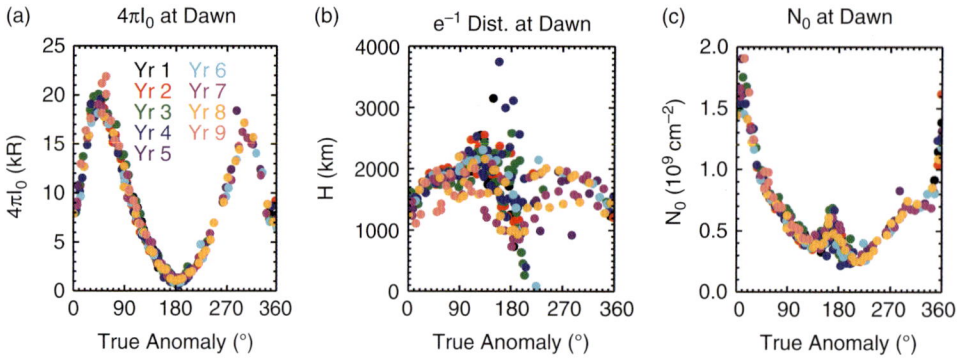

Figure 14.31. (a) Calcium 422.6-nm emission at the surface at Mercury dawn determined from exponential fits to the radial limb profiles. Different Mercury years are indicated by different colors. (b) The e-folding distance at dawn determined from the exponential fits. (c) Apparent tangent column density at dawn. A systematic error that creates the apparent peak near aphelion has been introduced by the assumption that all atoms are at rest relative to Mercury. Adapted from Burger et al. (2014).

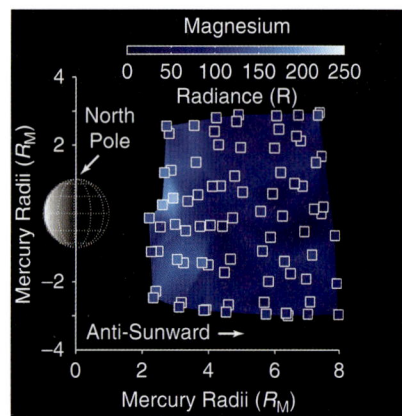

Figure 14.32. Emission from magnesium in Mercury's tail observed during M2. Individual observations (white squares) are overlaid on an image generated by projecting the observed line of sight onto the plane containing the spin axis and the Sun–Mercury line. Adapted from McClintock et al. (2009).

show the variation over local time, and it clearly shows an enhancement in emission in the morning.

Merkel et al. (2017) also investigated magnesium emission and column density as functions of local time and season (i.e., Mercury true anomaly). Figure 14.34a illustrates the distribution of dayside emission at a fixed altitude of 500 km versus true anomaly angle and local time. The data, which include observations acquired during approximately seven Mercury years, exhibit a prominent emission enhancement (maximum ~2.5 kR) on the dawn hemisphere near perihelion. Figure 14.34b illustrates the distinct seasonal pattern of the morning (0600–1000) enhancement in emission. Although there is some scatter, the seasonal pattern has a strong year-to-year repeatability. Figure 14.34c shows the distribution of column density at an altitude of 500 km versus true anomaly and local time. The seasonal variability has changed from that exhibited in the emission data, but the morning enhancement is maintained. Figure 14.34d shows that the peak in seasonal variability has shifted toward the inbound leg, which is related to the removal of the variation of the g-value over Mercury's orbit. The morning dayside magnesium column density at 500 km ranged between 1×10^9 cm^{-2} and 8×10^9 cm^{-2}.

Merkel et al. (2017) fit Chamberlain models to the limb profile data in order to estimate the temperature and near-surface density of the magnesium exosphere. Example fits are shown in Figure 14.35. These examples represent the characteristic range of temperatures retrieved from data acquired during seven Mercury years. Retrieved temperature, temperature uncertainty, and χ^2 error for the fits are reported for each example. The blue dashed lines represent Chamberlain models with three different fixed temperature sources (5000, 10,000, and 20,000 K) as guides to compare with the data. Note that there is not much difference in slope between profiles for 10,000 K and 20,000 K. Because the hottest temperatures are difficult to constrain (Burger et al., 2014; Cassidy et al., 2015), Merkel et al. (2017) set an upper limit of 20,000 K. Although the flyby analysis (Sarantos et al., 2011) indicated the need for two distinct temperatures to characterize the data adequately, these examples show that the orbital-phase dayside limb profiles can be fit well with a only a single component.

The distributions of near-surface density and temperature versus Mercury's true anomaly angle and local time retrieved by Merkel et al. (2017) for the magnesium exosphere are shown in Figures 14.36a–14.36d. Figures 14.36b and 14.36d illustrate the seasonal variation of density and temperature for data taken between 0600 and 1000. Retrieved temperatures range between 3500 K and 20,000 K, with the average near ~6000 K at a small uncertainty. For ~15% of the time the temperatures are >10,000 K and typically occur near the terminators (e.g., Figure 14.35d). At the dawn terminator, the density and temperature solutions are robust; however, the data near 1800 local time are noisy and therefore less reliable. There is a small population of solutions that could not be constrained and are assigned the 20,000 K temperature limit. The majority of the retrieved temperatures are lower than 10,000 K and seem to be evenly distributed around all local times. There is little seasonal variation in temperature aside from a population of low temperatures near 90° TAA.

The distribution of the magnesium production rate versus true anomaly and local time is shown in Figures 14.36e and 14.36f. The production of magnesium is concentrated in the

14.4 Observational Results 395

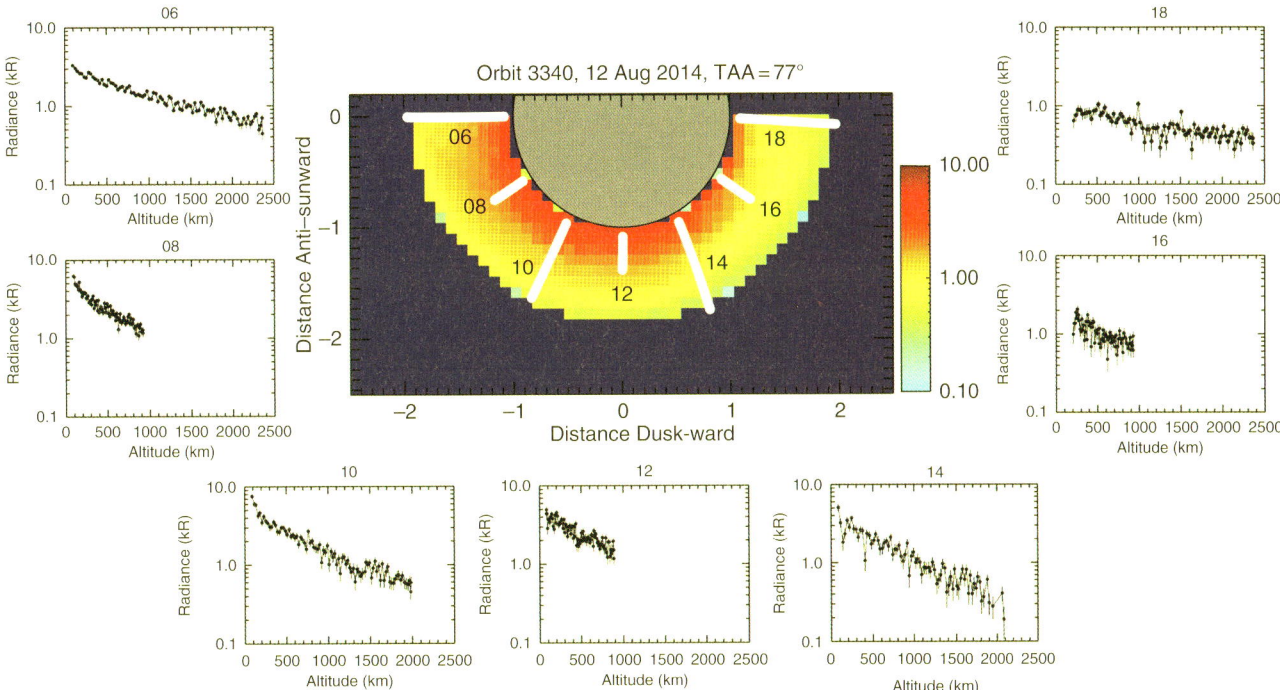

Figure 14.33. Example emission profiles for one orbit of the magnesium dayside limb-scan observations from MESSENGER. Seven local times were typically observed every third orbit throughout the orbit phase. The emission, in kR, is shown versus altitude at each local time on 12 August 2014, along with uncertainties. The color plot represents the limb emission profiles interpolated onto the equatorial plane (viewed from the north) and shows the local time variation over this orbit. The white lines represent the local time location and altitude range of each profile.

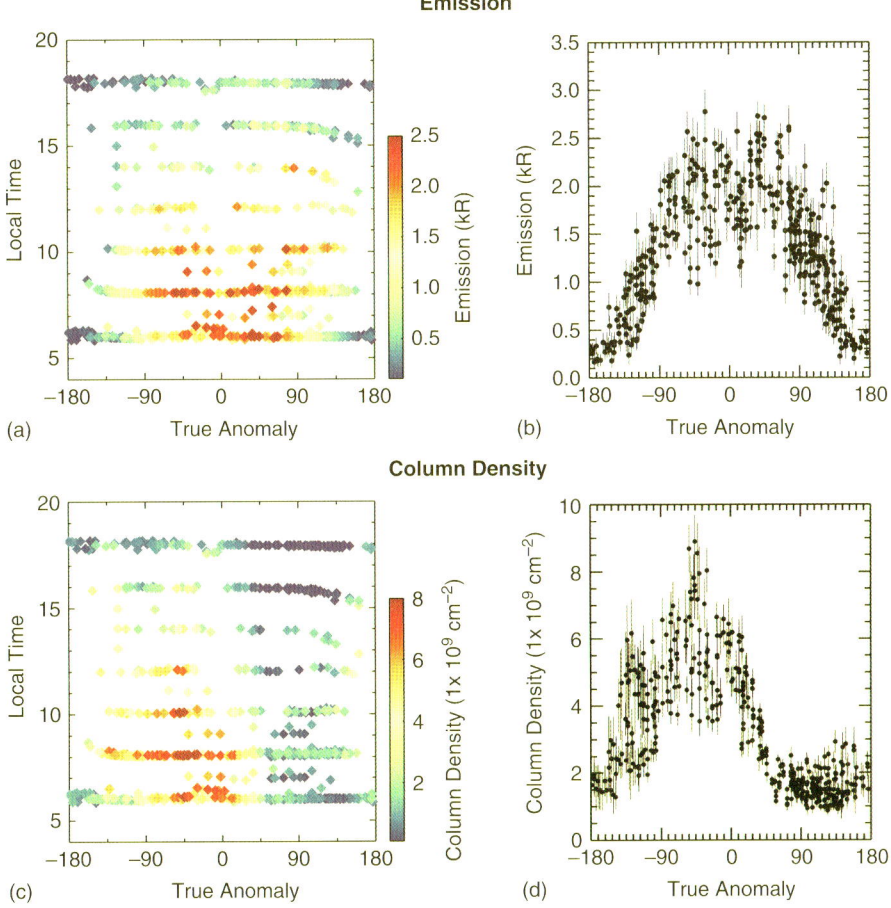

Figure 14.34. (a) and (c) Magnesium emission and column density, respectively, at 500-km altitude on the dayside as functions of Mercury true anomaly and local time (hours). The data encompass approximately seven Mercury years (March 2013–April 2015). (b) and (d) Plots of magnesium emission and column density, respectively, at local times 0600–1000 versus Mercury true anomaly illustrate the seasonal variation of the morning enhancement.

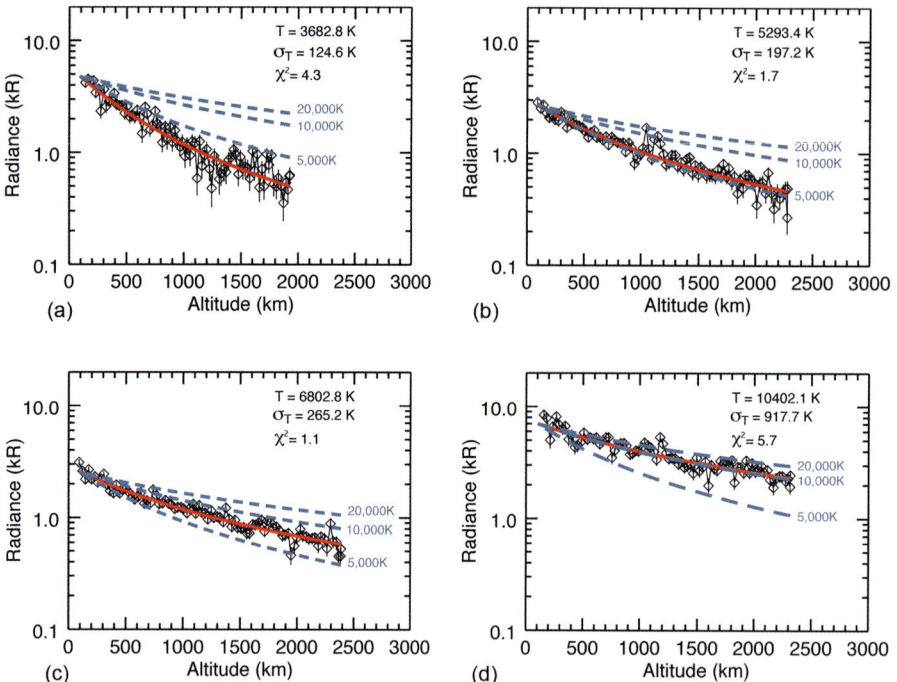

Figure 14.35. Examples of Chamberlain models fit to magnesium dayside limb profiles. The black diamonds are the emission data with one-standard-deviation uncertainties. The blue dashed lines represent Chamberlain models at three different fixed temperatures as guides to compare with the data. The red line gives the best fit in each case. The retrieved temperature, the temperature uncertainty σ_T, and the χ^2 error estimate are included in the right-hand corner of each panel.

morning and predominantly on the inbound leg of Mercury's orbit. Production is mostly influenced by surface density, as seen by the similarity between the seasonal variation in the two quantities shown in Figures 14.36b and 14.36e. The production of magnesium is enhanced in the morning, with the largest peak on the inbound leg (315°) and a smaller peak on the outbound leg (~140°) with rates ranging from 2×10^5 cm^{-2} s^{-1} to 8×10^5 cm^{-2} s^{-1}.

Whereas the analyses of MESSENGER's flyby data of the magnesium tail provided evidence for two sources – a lower-energy source with temperature < 5000 K and a higher energy source with temperature > 20,000 K (Sarantos et al., 2011) – Merkel et al. (2017) concluded that the observed enhanced production rate in the morning and mean temperatures near ~6000 K are most consistent with impact vaporization as the source of the dayside magnesium exosphere. However, they noted that a relatively high-energy source ($T > 10,000$ K) was retrieved 15% of the time near the dawn terminator, which is in agreement with the higher-energy source inferred from the MESSENGER flyby tail data (Killen et al., 2010; Sarantos et al., 2011). A hotter source at the dawn terminator is reminiscent of the hot dawn-enhanced calcium distribution described above. Contrary to the situation for calcium, there is no evidence so far for a comet stream as a possible source process, as discussed by Killen and Hahn (2015). However, Merkel et al. (2018) provided evidence that the production and distribution of magnesium in the exosphere is related to the distribution of magnesium on Mercury's surface.

14.4.5 Additional Species

Searches for additional species in Mercury's exosphere have been conducted from the ground, by Mariner 10, and by MESSENGER. These species are more difficult to observe than species such as sodium and calcium, because they either emit more weakly or are less abundant in Mercury's exosphere. Nevertheless, several such species have been detected in Mercury's exosphere and are discussed in this section.

14.4.5.1 Hydrogen

Mariner 10 Observations

Hydrogen in Mercury's exosphere was first detected by the UVS on Mariner 10 (Broadfoot et al., 1974; Shemansky and Broadfoot, 1977). These observations revealed two interesting features in Mercury's hydrogen exosphere distribution, which are illustrated in Figure 14.37. The first was a "bump" near 200-km altitude that may have been an instrumental artifact (Hunten et al., 1988). The second was the two-component nature of the distribution. A "warm" component had a scale height associated with dayside temperatures (420 K), whereas the other component was consistent with nightside temperatures (110 K) (Shemansky and Broadfoot, 1977). Unfortunately, the geometry of the Mariner 10 observations did not allow for measurement of the hydrogen profile at points other than normal to the subsolar longitude, so little else was learned about the distribution of hydrogen about the planet from that mission.

MESSENGER Observations

Measurements of the hydrogen distribution over the subsolar point during the first two MESSENGER flybys of Mercury, reported by Vervack et al. (2010), are compared with the most extended Mariner 10 hydrogen profile in Figure 14.38. In general, the UVVS and Mariner 10 profiles are quite similar in structure, but the UVVS radiances are uniformly higher than the Mariner 10 values. These overall higher hydrogen radiances

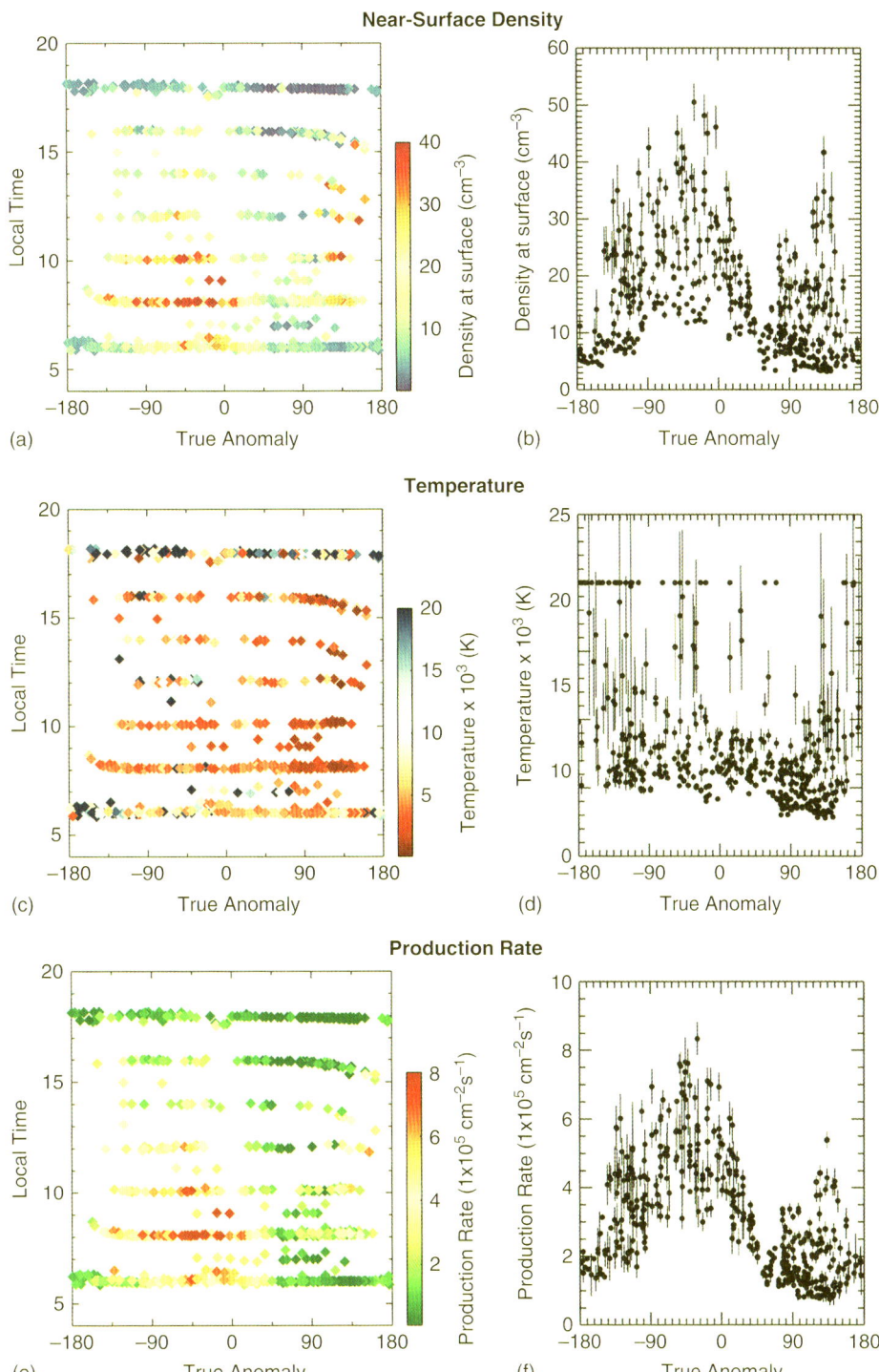

Figure 14.36. (a) and (b) Magnesium near-surface density, (c) and (d) temperature, and (e) and (f) production rate as functions of Mercury true anomaly and local time (hours). The data encompass approximately seven Mercury years (March 2013–April 2015). The right panels show the seasonal variation of the results for morning local times (0600–1000).

observed by UVVS relative to the Mariner 10 data may have several causes, including differences in viewing geometry (the Mariner 10 UVS slit subtended a much larger range of latitudes than did the UVVS), instrument calibration, incident solar flux, removal of the background interplanetary hydrogen component, and exospheric hydrogen density. Adopting a range of 100–300 rayleighs as the maximum profile radiance results in a near-surface line-of-sight column density between $\sim 8 \times 10^9$ cm^{-2} and $\sim 2.4 \times 10^{10}$ cm^{-2} for $g \sim 1.25 \times 10^{-2}$ photons atom^{-1} s^{-1} (Killen et al., 2009).

Vervack et al. (2009) fit a range of Chamberlain models to the UVVS data (Figure 14.38). Their resulting values of temperature and surface density, n_0, are shown for each flyby. Vervack et al. (2009) also noted that relatively large radiance values were measured during M2 at high latitude in the altitude range between 2500 and 3500 km. They postulated that these could have resulted from larger hydrogen densities there, possibly as a result of magnetospheric influences. Alternatively, they argued that the equatorial region may be depleted in hydrogen because of greater desorption near the subsolar point.

The flyby data were not suitable for searching for the cold component observed by Mariner 10. Instead, Vervack et al. (2011) used profiles measured during the orbital phase of the MESSENGER mission, which have much better spatial resolution, to further investigate this feature of the Mariner 10 distribution. Figure 14.39 illustrates the better altitude resolution of the orbital profiles, and it suggests that there may in fact be a "cold" component to the distribution. Vervack et al. (2011) pointed out that sunlight scattered into the UVVS telescope from the planet's surface can mimic a cold component, and any possible artifact resulting from this effect had not been removed from their data. Thus the existence of a cold component has not been conclusively verified with MESSENGER observations. To date, no evidence has been found in the UVVS data that matches the bump in the Mariner 10 profiles.

14.4.5.2 Helium

Helium is the exospheric species best measured by Mariner 10. Numerous scans across the disk and above the limb were carried out for the helium 58.4-nm emission, an example of which is shown in the left panel of Figure 14.5. Attempts to model the distribution (left panel), which led to predictions of an antisolar/subsolar density ratio near 200 depending on the source velocity (e.g., Smith et al., 1978), failed to match the observed value of 50 inferred from the terminator region.

Also measured was the profile of emission above the subsolar point (Figure 14.40). Unlike the case for hydrogen, the helium distribution appears to correspond to a single component exhibiting dayside temperatures. The near-surface line-of-sight column density inferred from the maximum 100 rayleigh signal is $\sim 2.5 \times 10^{12}$ cm^{-2} for $g \sim 4 \times 10^{-5}$ photons atom^{-1} s^{-1} (Killen et al., 2009). UVVS did not cover the wavelengths of helium emission. However, models of the He$^+$ distribution measured by FIPS may eventually provide constraints on the distribution of neutral He.

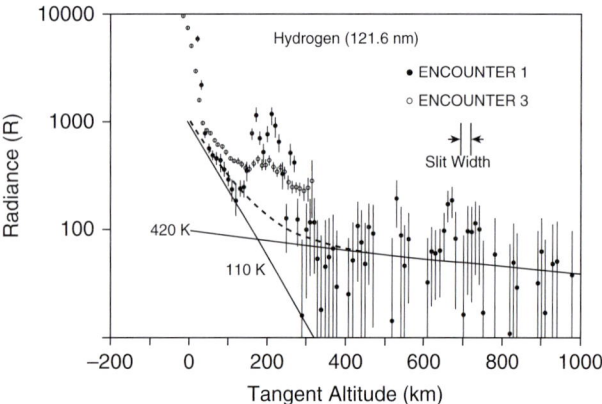

Figure 14.37. Profile of the H Lyman α emission measured normal to the subsolar point during two of the Mariner 10 flybys. The profile is fit with a two-component model, one with a temperature close to the dayside temperature (420 K) and one with a temperature close to that of the nightside (110 K). The bump near 200 km is a mystery. Adapted from Hunten et al. (1988).

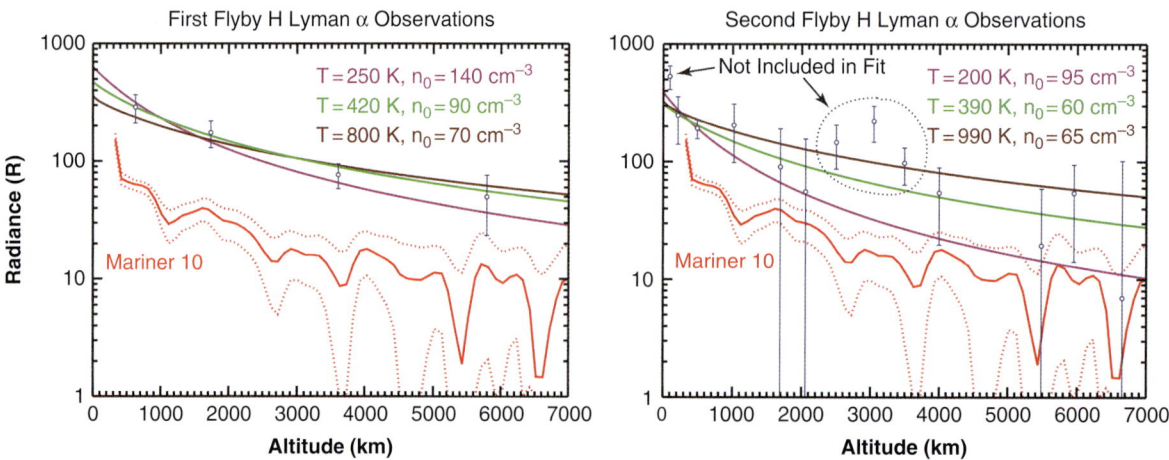

Figure 14.38. Hydrogen Lyman α radiances (corrected for background interplanetary H emission) measured by the MESSENGER UVVS (Vervack et al., 2009) compared with radiances obtained during the first Mariner 10 flyby (Broadfoot et al., 1976). The UVVS data have been binned in altitude (averages over 0–1000 km, 1000–2500 km, 2500–4500 km, and >4500 km for M1 and averages centered on multiples of 500 km for M2). Chamberlain models fit to the data are shown for both M1 and M2. The green curves are "nominal" models. For M2, this is the best fit to the data. The sparseness of the altitude sampling and large bin sizes for M1 rendered the fits unreliable, and a 420 K model is shown as representative. The magenta and brown curves are visually estimated limiting cases. Three points between 2500- and 3500-km altitude were excluded from the fit for M2 because they were obtained during times when the spacecraft pointed to high latitudes for operational reasons.

14.4 Observational Results 399

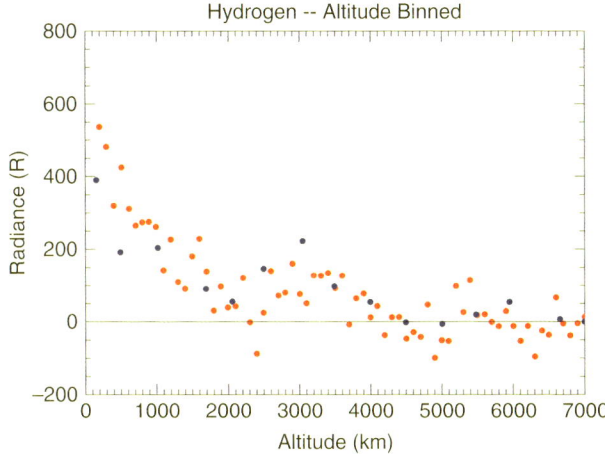

Figure 14.39. Comparison of the H distribution measured during MESSENGER's second Mercury flyby and binned in altitude to multiples of 500 km (blue circles) with a typical H profile measured over the subsolar point during the MESSENGER orbital phase (red circles). The orbital profile has been binned into 100-km-wide altitude bins.

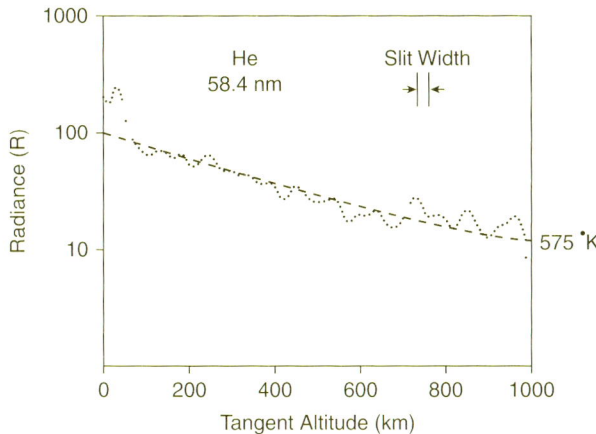

Figure 14.40. Altitude profile of helium 58.4 nm emission above the subsolar point from Mariner 10. Dots indicate measured radiance values, and the dashed line is a single-temperature Monte Carlo model. Adapted from Broadfoot et al. (1976).

14.4.5.3 Oxygen

During the third Mariner 10 flyby, a detection of 130.4-nm oxygen emission was reported (Broadfoot et al., 1976). However, high noise levels precluded measurement of a profile, and the actual level of the detection was uncertain (63–200 rayleighs) (Broadfoot et al., 1976; Hunten et al., 1988). The inferred Mariner 10 oxygen column density ($\sim 3 \times 10^{11}$ cm^{-2}) in the exosphere is comparable to that of sodium ($\sim 2 \times 10^{11}$ cm^{-2}).

Searches for oxygen 130.4-nm emission with UVVS were regularly conducted throughout the orbital phase, but no convincing detection was found. Vervack et al. (2016) reported that levels of oxygen emission consistent with those reported for Mariner 10 would have easily been seen by UVVS. A typical example of a dayside spectrum, taken from Vervack et al. (2016), is illustrated in Figure 14.41.

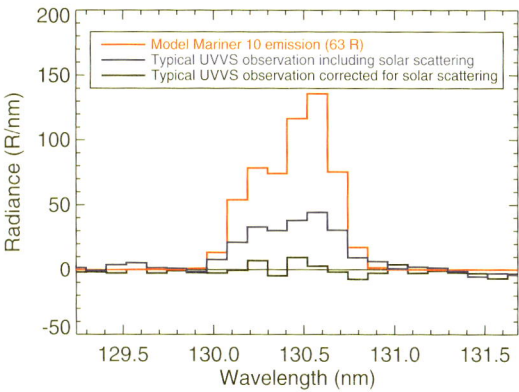

Figure 14.41. Example of a typical MESSENGER observation of the O 130.4-nm emission region. The blue line denotes the observed spectrum, which includes a solar scattered component from the surface. The black line is this spectrum after removal of the solar scattering. For this spectrum, the one-standard-deviation upper limit for emission is 2 R. The red line is a model spectrum of 130.4-nm emission corresponding to 63 R, which is the lower bound of the Mariner 10 detections. Thus, the UVVS upper limits are typically a factor of 30 less than the reported Mariner 10 detection.

They estimated that the one-standard-deviation upper limit for oxygen emission in the example observation is 2 rayleighs, which corresponds to a tangent column density of 2×10^{10} cm^{-2} and is an order of magnitude less than a typical sodium value.

14.4.5.4 Aluminum

From observations of the 396.152-nm emission line for aluminum acquired at the New Technology Telescope at La Silla, Chile, Doressoundiram et al. (2009) reported an upper limit for aluminum in Mercury's exosphere of 7.8×10^9 cm^{-2} (but they referred to this limit as a "tentative detection"); their observations are shown in Figure 14.42. Observations of aluminum made with the HIRES spectrometer on the Keck I telescope were reported by Bida and Killen (2011, 2016), who claimed measurements of the aluminum 396.152-nm emission at a four-to-five-standard-deviation level during runs in 2008 and 2013. Bida and Killen (2016) reported tangent columns of $(2.5–5.1) \times 10^7$ cm^{-2} over altitudes 860–2100 km, significantly lower than the figure reported by Doressoundiram et al. (2009). The scale height of the aluminum gas is consistent with a kinetic temperature of 4800–6900 K.

Searches for minor species conducted by the MESSENGER UVVS during dayside limb scans and nightside tail sweeps were unsuccessful. Nonetheless, aluminum (as well as ionized calcium and manganese; see below) was detected in the pre-dawn nightside region of the planet (0200–0500) over the TAA range 0°–70° during ride-along observations with FIPS (Vervack et al., 2016) that used the specialized viewing geometry illustrated in Figure 14.43. Similar geometries at other times in the orbital phase of MESSENGER did not reveal these species, so there may be substantial variability in their distributions both spatially and/or temporally.

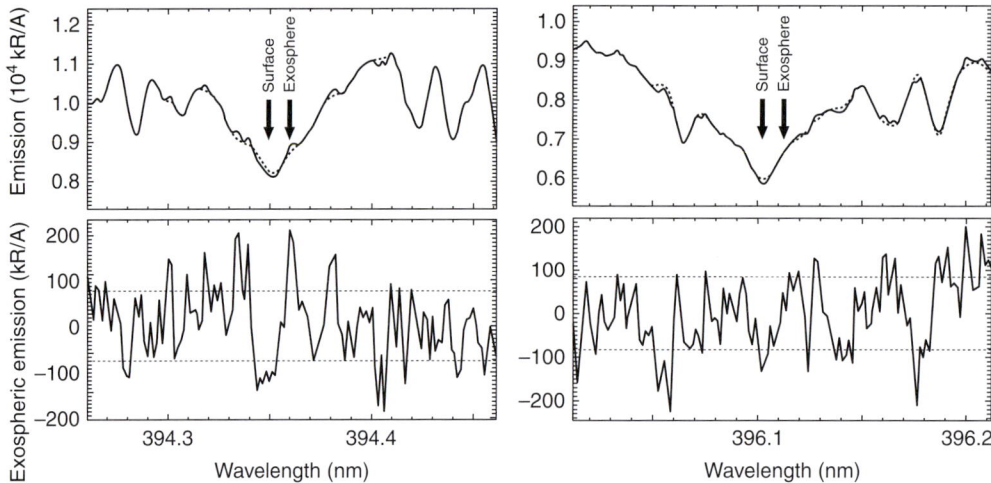

Figure 14.42. Spectra of the Al 394.4 nm (left upper panel) and 396.1 nm (right upper panel) resonant emission lines as measured in Mercury's exosphere. The positions of the exospheric line and of the solar absorption line reflected at Mercury's surface as expected from the planetary ephemeris are also indicated. A solar spectrum has been reduced in resolution, scaled to the measured spectra, and plotted as a dotted line. Left and right lower panels are differences between the measured spectra and the modified solar spectra in the panels above. The dashed horizontal lines on the lower panels indicate the one-standard-deviation level. From Doressoundiram et al. (2009).

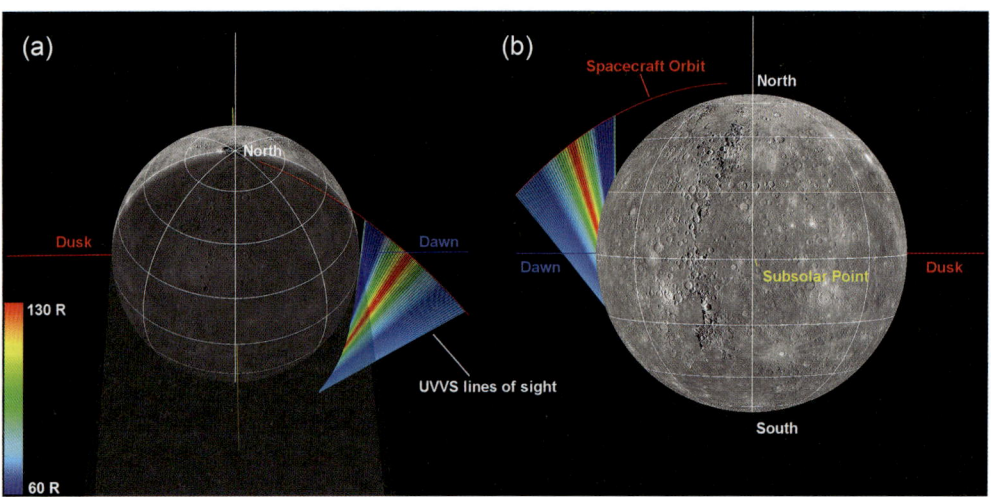

Figure 14.43. MESSENGER observation geometry for the aluminum, ionized calcium, and manganese observations. North is up in these panels, and the observations were made on the dawn side of the planet. The red arc represents the spacecraft orbit, and the spacecraft motion was from north to south (moving away from the planet). The colored lines from the spacecraft positions represent the UVVS line of sight for each spectrum, with the color scale indicating the radiance of the ionized calcium 393.48-nm line. These lines terminate at the line-of-sight tangent point for each observation. The left panel illustrates that the line of sight started in a north–south orientation (looking south) and then rotated during the observational sequence to end up pointing in a direction that is more oriented in the tail direction. The right panel illustrates that a portion of the lines of sight were in Mercury's shadow during most of the observation.

An example of one such observation displaying emissions from two lines of aluminum and two lines of ionized calcium (discussed below) is shown in Figure 14.44. The high signal-to-noise ratio of the UVVS measurements and the observation of two lines provide definitive confirmation of the presence of aluminum in Mercury's exosphere. Figure 14.45 shows the radiance in the two aluminum lines observed by UVVS as a function of tangent altitude. The Chamberlain formalism cannot be reliably applied to these data because the UVVS line of sight was in constant motion owing to spacecraft maneuvers so that it skirted Mercury's shadow rather than drifting in altitude. Vervack et al. (2016) converted their average radiance value to column density to provide an approximate estimate of 7.7×10^7 cm^{-2}, in reasonable agreement with the line-of-sight column densities reported by Bida and Killen (2016).

14.4 Observational Results

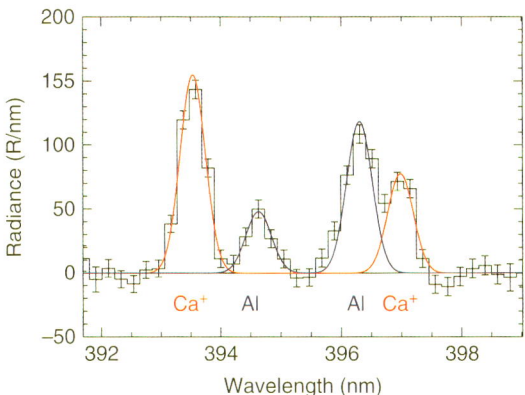

Figure 14.44. A MESSENGER spectrum of an observation of aluminum and ionized calcium with one-standard-deviation errors. The background-subtracted spectrum has been averaged over the altitude range 200–700 km to enhance the signal-to-noise ratio. That the two lines of each species are in the ratios expected provides clear evidence that these species have been detected.

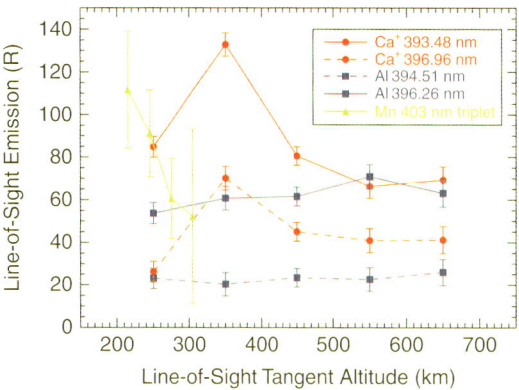

Figure 14.45. Profiles of aluminum, ionized calcium, and manganese radiance as functions of tangent altitude from one observation sequence. The different curves for aluminum and ionized calcium are for the two lines detected. The profile for manganese is the sum over the three lines of the triplet.

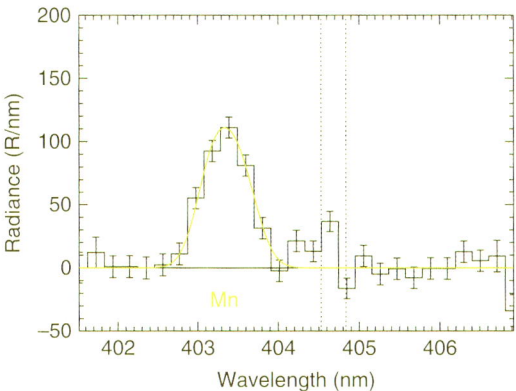

Figure 14.46. A MESSENGER spectrum for one of the observations of manganese, which has been averaged over the altitude range 200–320 km and background-subtracted, is shown with one-standard-deviation error bars. The three lines of the triplet are not individually resolved but sum to produce a "line" that is wider than the standard UVVS line shape. The evidence for potassium emission near 404–405 nm (vertical dotted lines) is not conclusive.

14.4.5.5 Ionized Calcium

Emission from ionized calcium was first detected in Mercury's exosphere during the third MESSENGER flyby (Vervack et al., 2010). The ions were observed at approximately $(1-2) R_M$ tailward of the planet and mostly near the equatorial plane. Because velocities for ionized calcium are generally much larger than those for neutral calcium and the line-of-sight column density estimates for ionized calcium and neutral calcium $(1-2) R_M$ downtail were of the same order of magnitude, it was determined that local production of ionized calcium from calcium was unlikely to yield the observed ionized calcium distribution. Instead, Vervack et al. (2010) concluded that a combination of magnetospheric convection and centrifugal acceleration, channeling ionized calcium produced elsewhere into the tail region, was required to account for the distribution (Gershman et al., 2014).

Vervack et al. (2016) also observed two lines of ionized calcium during the same observations from which aluminum was detected by UVVS (Figure 14.44). These later observations confirmed the presence of detectable ionized calcium emission from Mercury's exosphere. Figure 14.45 shows the radiance profiles for the ionized calcium lines as a function of tangent altitude. These profiles exhibit different altitude distributions from those of aluminum over the same observation geometries, suggesting differences in their source and loss processes. Vervack et al. (2016) reported an average line-of-sight column density of 3.1×10^7 cm^{-2} for the viewing geometry shown in Figure 14.43.

Although the wavelengths of the two ionized calcium lines detected by UVVS can be observed from the ground, attempts to do so have thus far yielded only upper limits. Bida and Killen (2016) reported three-standard-deviation upper limits near Mercury's equatorial antisolar limb, from which a column density limit of 3.9×10^6 cm^{-2} at 1630-km altitude was derived.

14.4.5.6 Manganese

Manganese was an unexpected constituent of the exosphere that was also detected during near-dawn observations reported by Vervack et al. (2016). It was added to the UVVS search list because it has a high neutron absorption cross section, so its presence on the surface could influence the interpretation of MESSENGER Gamma-Ray and Neutron Spectrometer data (Evans et al., 2012). MESSENGER X-Ray Spectrometer data also revealed Mn to be present on Mercury's surface (Nittler et al., 2011; Chapter 2). Figure 14.46 shows an average spectrum of the manganese triplet near 403 nm. This robust detection is clear evidence for manganese in Mercury's exosphere. The radiance profile as a function of altitude, shown in Figure 14.45, exhibits distinct differences from the profiles for aluminum and ionized calcium. These differences suggest that different processes control manganese, aluminum, and ionized calcium in Mercury's exosphere (Chapter 15). Vervack et al. (2016) reported a value of 4.9×10^7 cm^{-2} for the average

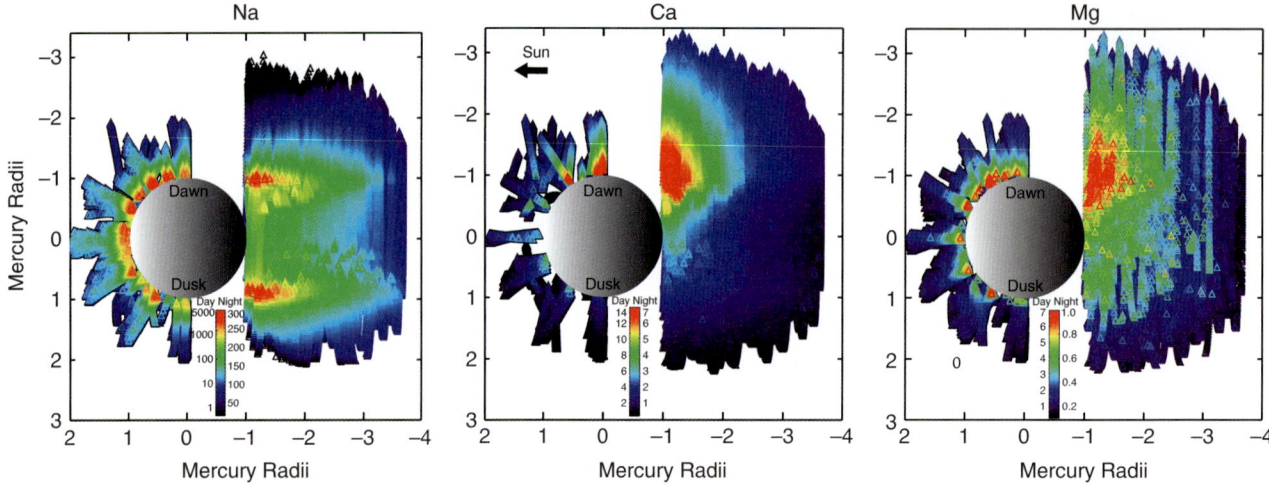

Figure 14.47. Geometrical representation of emission (in kR) from sodium, calcium, and magnesium during the orbital phase of the MESSENGER mission. These plots illustrate the two types of systematic observations (as shown in Figure 14.8) used to measure these constituents on Mercury's dayside and nightside. The color scaling is specific to each species, and scaling for daytime observations is different from that for nighttime observations. Each species has a different region of localized enhanced brightness.

column density with a viewing geometry similar to that shown in Figure 14.43.

14.4.5.7 Other Species

A number of other species have been sought in Mercury's exosphere. Doressoundiram et al. (2009) reported upper limits for iron, lithium, and silicon of 1×10^{11} cm^{-2}, 4×10^{7} cm^{-2}, and 5×10^{10} cm^{-2}, respectively, on the basis of averages of observations acquired during the nights of 30–31 October 2005. Bida and Killen (2016) reported a single three-standard-deviation detection of iron from HIRES observations at the Keck I Observatory in 2009. With the same simple model as that employed for their aluminum observations, they found a zenith column density of 8.2×10^{8} cm^{-2}. The scale height of the iron gas is consistent with a kinetic temperature of 5000–13,000 K. Although sought on numerous occasions, including in concert with the observations that yielded detections of aluminum and ionized calcium, iron was not observed by UVVS.

14.5 SUMMARY

Mercury's exosphere was discovered in 1974 when the UVS experiment on the Mariner 10 spacecraft detected hydrogen and helium about the planet. This result was followed by the first measurements of Mercury's sodium and potassium exospheres by Potter and Morgan (1985, 1986) and by the detection of calcium by Bida et al. (2000). Several groups have imaged the sodium exosphere with ground-based telescopes equipped with high-resolution spectrographs. These images have revealed a dynamic exosphere with localized enhancements in intensity (by tens of percent) at high latitudes that appear and disappear on timescales of hours. These variations are thought to result from the impact of solar-wind ions that penetrate Mercury's magnetosphere at its cusps. However, attempts to correlate such variations with interplanetary magnetic field measurements at Mercury made by the MESSENGER Magnetometer have been inconclusive (Section 14.4.1.1). Although ground-based observers can observe localized changes (analogous to weather patterns on Earth) in global-scale images of Mercury's exosphere, Mercury orbits too close to the Sun to allow ground-based synoptic studies of exospheric behavior as a function of TAA or to allow simultaneous observations of both terminators. Moreover, detailed ground-based studies of Mercury's exosphere have been limited primarily to sodium. Only a few potassium images have been acquired, and calcium observations are limited to a small number of long-slit spectra acquired at the Keck Observatory.

During its orbital phase, MESSENGER provided the first high-spatial-resolution synoptic measurements of Mercury's sodium, calcium, and magnesium exospheres. These observations revealed three distinct spatial distributions, each with its own characteristic temperature structure. Figure 14.47 illustrates the spatial distributions of sodium, calcium, and magnesium about the planet made by superimposing all the observations of these species made between March 2014 and March 2015 in the TAA range 0°–90°. Whereas sodium is more or less uniformly distributed over the dayside and symmetric about Mercury's shadow at night, calcium emission exhibits a strong dawn–dusk asymmetry. The magnesium distribution is similar to that of calcium, except that the dayside emission is distributed over a larger range of local time (0600–1000) and the dawn–dusk contrast is less pronounced.

MESSENGER's orbit allowed the UVVS to monitor the seasonal variation in Mercury's exosphere (analogous to seasonal climate variations on Earth). From nearly 12 Mercury years in orbit, UVVS measurements of intensity distributions of sodium and calcium as functions of TAA remained constant within the errors of the measurements (Figures 14.20 and 14.28). This constancy was also true for magnesium, but the data were limited to only the last seven years in orbit (Figure 14.31). Individual temperature distributions for each species showed little variation with TAA and were also stable from year to year. Whereas sodium temperature values were near 1200 K, suggesting that photon-stimulated desorption is a dominant surface release process (Cassidy et al., 2015), and magnesium temperatures of 6000 K are consistent with impact vaporization, calcium's more extreme temperature, near 50,000 K, suggests that more complex processes contribute to the calcium exosphere.

In general, MESSENGER sodium measurements are consistent with and complementary to ground-based observations. Although there is no evidence for short time variations in the MESSENGER data, this result is most likely due to limitations on the observation geometry imposed by the spacecraft orbit and pointing constraints. These constraints generally restricted observations to occur near orbit apoapsis (Figure 14.8) and limited the latitude of the tangent point to lie within ±30° of the equator (Figure 14.9). In this configuration, the line of sight passed above the magnetospheric cusps at very high altitudes, where the local densities are vanishingly small. Further, UVVS observed a small number of discrete local times with a cadence of 12 h before apoapsis was lowered and MESSENGER's orbital period was reduced to 8 h. In both orbits, changes in observed intensity were aliased with changes in viewing geometry. On the other hand, the lack of a cold sodium component (Figure 14.22) agues that thermal desorption, thought to be an important release mechanism on the basis of pre-MESSENGER work (Killen et al., 2007; Leblanc and Johnson, 2010; Mura et al., 2009), is not an important surface-release process.

Several weakly emitting species have now been identified about Mercury. Tentative ground-based identifications of aluminum have been confirmed by MESSENGER observations. Manganese and ionized calcium have also been discovered by MESSENGER. The limited number of detections suggests that their spatial distributions and dependence on TAA are different from those of routinely observed species and are possibly highly time-variable, but there are too few measurements to make definitive conclusions. Both Mariner 10 and MESSENGER observed hydrogen about Mercury. Whereas both investigations measured similar temperature distributions, the inferred hydrogen column densities by MESSENGER are about a factor of 4 larger than those by Mariner 10. MESSENGER did not confirm the tentative identification of oxygen by Mariner 10.

The total inventory of confirmed exospheric neutral species is nine. Of these, helium and sodium dominate, exhibiting near-surface line-of-sight column densities in the range 10^{11} cm^{-2}–10^{12} cm^{-2}. Magnesium and hydrogen are about a factor of 10 less abundant, potassium and calcium are about a factor of 100 less abundant, and the minor species aluminum and manganese are less abundant by factors near 10,000. The MESSENGER upper limit for oxygen places it with magnesium and hydrogen; however, this limit is not particularly strong because oxygen resonance lines occur near 130 nm and have very small g-values. This situation also holds for sulfur, an element that is abundant in Mercury's surface material (Chapter 2). Placing the observations of ionized species of calcium (UVVS) and sodium and oxygen (FIPS) in context with the neutral species is not straightforward because the transport of ions is dominated by the morphology of the planet's magnetic field.

The combination of modern ground-based observing tools coupled with the first orbital observations have markedly increased our understanding of Mercury's exospheric composition and structure relative to the flyby observations made by Mariner 10, which detected only those species delivered to the planet by the solar wind. Advances in ground-based observing techniques, particularly improvements in spatial resolution, which is currently limited by atmospheric seeing, will certainly provide additional insights into the workings of the system. The next breakthrough will occur when the BepiColombo Mercury Planetary Orbiter and Mercury Magnetospheric Orbiter spacecraft (Benkhoff et al., 2010, Chapter 20) are inserted into orbit around the planet, an event currently scheduled for 2025. On those spacecraft will be a number of instruments designed to make detailed studies of the exosphere. These include the following investigations:

- Probing of Hermean Exosphere by Ultraviolet Spectroscopy (PHEBUS), an ultraviolet spectrometer covering the wavelength range 55–315 nm;
- Search for Exosphere Refilling and Emitted Neutral Abundances (SERENA), a suite of in situ neutral and ionized particle detectors;
- Mercury Sodium Atmospheric Spectral Imager (MSASI), a Fabry–Perot imaging system designed to produce full-disk images of the sodium exosphere;
- Mercury Dust Monitor (MDM), which will provide measurements of dust impacts to the planet's surface; and
- Mercury Plasma Particle Experiment (MPPE), which will provide measurements of exospheric ions.

BepiColombo will provide the first near-global high-resolution images of Mercury's sodium exosphere, a comprehensive inventory of exospheric species (both ionized and neutral), and key measurements of the inputs to Mercury's surface from the space environment. These observations will substantially advance our understanding of Mercury's coupled surface–exosphere–space environment system.

REFERENCES

Anderson, B. J., Acuña, M. H., Lohr, D. A., Scheifele, J., Raval, A., Korth, H. and Slavin, J. A. (2007). The Magnetometer instrument on MESSENGER. *Space Sci. Rev.*, **131**, 417–450, doi:10.1007/s11214-007-9246-7.

Anderson, B. J., Johnson, C. L., Korth, H., Purucker, M. E., Winslow, R. M., Slavin, J. A., Solomon, S. C., McNutt, R. L., Jr., Raines, J. M. and Zurbuchen, T. H. (2011). The global magnetic field of Mercury from MESSENGER orbital observations. *Science*, **333**, 1859–1862, doi:10.1126/science.1211001.

Andrews, G. B., Zurbuchen, T. H., Mauk, B. H., Malcom, H., Fisk, L. A., Gloeckler, G., Ho, G. C., Kelley, J. S., Koehn, P. L., Lefevere, T. W., Livi, S. S., Lundgren, R. A. and Raines, J. M. (2007). The Energetic Particle and Plasma Spectrometer instrument on the MESSENGER spacecraft. *Space Sci. Rev.*, **131**, 523–556, doi:10.1007/s11214-007-9272-5.

Baker, D. N., Poh, G., Odstrcil, D., Arge, C. N., Benna, M., Johnson, C. L., Korth, H., Gershman, D. J., Ho, G. C., McClintock, W. E., Cassidy, T. A., Merkel, A., Raines, J. M., Schriver, D., Slavin, J. A., Solomon, S. C., Travnicek, P. M., Winslow, R. M. and

Zurbuchen, T. H. (2013). Solar wind forcing at Mercury: WSA-ENLIL model results. *J. Geophys. Res. Space Physics*, **118**, 45–57, doi:10.1029/2012JA018064.

Baumgardner, J., Wilson, J. K. and Mendillo, M. (2008). Imaging the sources and full extent of the sodium tail of the planet Mercury. *Geophys. Res. Lett.*, **35**, L03201, doi:10.1029/2007GL032337.

Benkhoff, J., van Casteren, J., Hayakawa, H., Fujimoto, M., Laakso, H., Novara, M., Ferri, P., Middleton, H. R. and Ziethe, R. (2010). BepiColombo – Comprehensive exploration of Mercury: Mission overview and science goals. *Planet. Space Sci.*, **58**, 2–20, doi:10.1016/j.pss.2009.09.020.

Bida, T. A. and Killen, R. M. (2011). Observations of Al, Fe, and Ca^+ in Mercury's exosphere. *EPSC-DPS Joint Meeting Abstracts and Program*, **6**, abstract EPSC-DPS2011-1621. European Planetary Science Congress – Division for Planetary Sciences Joint Meeting, Nantes, France, 2–7 October. Available at http://adsabs.harvard.edu/abs/2011epsc.conf.1621B.

Bida, T. A. and Killen, R. M. (2016). Observations of the minor species Al, Fe, and Ca^+ in Mercury's exosphere. *Icarus*, **268**, 32–36, doi:10.1016/j.icarus.2016.10.019.

Bida, T. A., Killen, R. M. and Morgan, T. H. (2000). Discovery of calcium in Mercury's atmosphere. *Nature*, **404**, 159–161, doi:10.1038/35004521.

Bishop, J. and Chamberlain, J. W. (1989). Radiation pressure dynamics in planetary exospheres: A "natural" framework. *Icarus*, **81**, 145–163, doi:10.1016/0019-1035(89)90131-0.

Borin, P., Bruno, M., Cremonese, G. and Marzari, F. (2010). Estimate of the neutral atoms' contribution to the Mercury exosphere caused by a new flux of micrometeoroids. *Astron. Astrophys.*, **517**, A89, doi:10.1051/0004-6361/201014312.

Borland, D. and Taylor, R. M. (2007). Rainbow color map (still) considered harmful. *IEEE Comput. Graph. Appl.*, **27**, 14–17.

Broadfoot, A. L., Kumar, S., Belton, M. J. S. and McElroy, M. B. (1974). Mercury's atmosphere from Mariner 10: Preliminary results. *Science*, **185**, 166–169, doi:10.1126/science.185.4146.166.

Broadfoot, A. L., Shemansky, D. E. and Kumar, S. (1976). Mariner 10: Mercury atmosphere. *Geophys. Res. Lett.*, **3**, 577–580, doi:10.1029/GL003i010p00577.

Burger, M. H., Killen, R. M., Vervack, R. J., Jr., Bradley, E. T., McClintock, W. E., Sarantos, M., Benna, M. and Mouawad, N. (2010). Monte Carlo modeling of sodium in Mercury's exosphere during the first two MESSENGER flybys. *Icarus*, **209**, 63–74, doi:10.1016/j.icarus.2010.05.007.

Burger, M. H., Killen, R. M., McClintock, W. E., Vervack, R. J., Jr., Merkel, A. W., Sprague, A. L. and Sarantos, M. (2012). Modeling MESSENGER observations of calcium in Mercury's exosphere. *J. Geophys. Res.*, **117**, E00L11, doi:10.1029/2012JE004158.

Burger, M. H., Killen, R. M., McClintock, W. E., Merkel, A. W., Vervack, R. J., Cassidy, T. A. and Sarantos, M. (2014). Seasonal variations in Mercury's dayside calcium exosphere. *Icarus*, **238**, 51–58, doi:10.1016/j.icarus.2014.04.049.

Cassidy, T. A., Merkel, A. W., Burger, M. H., Sarantos, M., Killen, R. M., McClintock, W. E. and Vervack, R. J., Jr. (2015). Mercury's seasonal sodium exosphere: MESSENGER orbital observations. *Icarus*, **248**, 547–559, doi:10.1016/j.icarus.2014.10.037.

Cassidy, T. A., McClintock, W. E., Killen, R. M., Sarantos, M., Merkel, A. W., Vervack, R. J., Jr. and Burger, M. H. (2016). A cold-pole enhancement in Mercury's sodium exosphere. *Geophys. Res. Lett.*, **43**, 11,121–11,128, doi:10.1002/2016GL071071.

Chamberlain, J. W. (1961). *Physics of the Aurora and Airglow*. New York: Academic Press. Available at: http://onlinelibrary.wiley.com/doi/10.1002/9781118668047.fmatter/summary.

Chamberlain, J. W. (1963). Planetary coronae and atmospheric evaporation. *Planet. Space Sci.*, **11**, 901–960, doi:10.1016/0032-0633(63)90122-3.

Chamberlain, J. W. and Hunten, D. M. (1987). *Theory of Planetary Atmospheres. An Introduction to Their Physics and Chemistry*. International Geophysics Series, Vol. 36. Orlando, FL: Academic Press.

Christou, A. A., Killen, R. M. and Burger, M. H. (2015). The meteoroid stream of comet Encke at Mercury: Implications for MErcury Surface, Space ENvironment, GEochemistry, and Ranging observations of the exosphere. *Geophys. Res. Lett.*, **42**, 7311–7318, doi:10.1002/2015GL065361.

DiBraccio, G. A., Slavin, J. A., Boardsen, S. A., Anderson, B. J., Korth, H., Zurbuchen, T. H., Raines, J. M., Baker, D. N., McNutt, R. L., Jr. and Solomon, S. C. (2013). MESSENGER observations of magnetopause structure and dynamics at Mercury. *J. Geophys. Res. Space Physics*, **118**, 997–1008, doi:10.1002/jgra.50123.

Domingue, D. L., Koehn, P. L., Killen, R. M., Sprague, A. L., Sarantos, M., Cheng, A. F., Bradley, E. T. and McClintock, W. E. (2007). Mercury's atmosphere: A surface-bounded exosphere. *Space Sci. Rev.*, **131**, 161–186, doi:10.1007/s11214-007-9260-9.

Doressoundiram, A., Leblanc, F., Foellmi, C. and Erard, S. (2009). Metallic species in Mercury's exosphere: EMMI/New Technology Telescope observations. *Astron. J.*, **137**, 3859–3863, doi:10.1088/0004-6256/137/4/3859.

Doressoundiram, A., Leblanc, F., Foellmi, C., Gicquel, A., Cremonese, G., Donati, J.-F. and Veillet, C. (2010). Spatial variations of the sodium/potassium ratio in Mercury's exosphere uncovered by high-resolution spectroscopy. *Icarus*, **207**, 1–8, doi:10.1016/j.icarus.2009.11.020.

Evans, L. G., Peplowski, P. N., Rhodes, E. A., Lawrence, D. J., McCoy, T. J., Nittler, L. R., Solomon, S. C., Sprague, A. L., Stockstill-Cahill, K. R., Starr, R. D., Weider, S. Z., Boynton, W. V., Hamara, D. K. and Goldsten, J. O. (2012). Major-element abundances on the surface of Mercury: Results from the MESSENGER Gamma-Ray Spectrometer. *J. Geophys. Res.*, **117**, E00L07, doi:10.1029/2012JE004178.

Fink, U., Larson, H. P. and Poppen, R. F. (1974). A new upper limit for an atmosphere of CO_2, CO on Mercury. *Astrophys. J.*, **187**, 407–416, doi:10.1086/152647.

Gershman, D. J., Slavin, J. A., Raines, J. M., Zurbuchen, T. H., Anderson, B. J., Korth, H., Baker, D. N. and Solomon, S. C. (2014). Ion kinetic properties in Mercury's pre-midnight plasma sheet. *Geophys. Res. Lett.*, **41**, 5740–5747, doi:10.1002/2014GL060468.

Huebner, W. F. and Mukherjee, J. (2015). Photoionization and photodissociation rates in solar and blackbody radiation fields. *Planet. Space Sci.*, **106**, 11–45, doi:10.1016/j.pss.2014.11.022.

Hunten, D. M., Roach, F. E. and Chamberlain, J. W. (1956). A photometric unit for the airglow and aurora. *J. Atmos. Terr. Phys.*, **8**, 345–346.

Hunten, D. M., Shemansky, D. E. and Morgan, T. H. (1988). The Mercury atmosphere. In *Mercury*, ed. F. Vilas, C. R. Chapman and M. S. Matthews. Tucson, AZ: University of Arizona Press, pp. 562–612.

Ip, W. H. (1986). The sodium exosphere and magnetosphere of Mercury. *Geophys. Res. Lett.*, **13**, 423–426, doi:10.1029/GL013i005p00423.

Johnson, R. E., Leblanc, F., Yakshinskiy, B. V. and Madey, T. E. (2002). Energy distributions for desorption of sodium and potassium from ice: The Na/K ratio at Europa. *Icarus*, **156**, 136–142, doi:10.1006/icar.2001.6763.

Killen, R. M. (2006). Curve-of-growth model for sodium D2 emission at Mercury. *Publ. Astron. Soc. Pac.*, **118**, 1344–1350, doi:10.1086/508070.

Killen, R. M. (2016). Pathways for energization of Ca in Mercury's exosphere. *Icarus*, **268**, 32–36, doi:10.1016/j.icarus.2015.12.035.

Killen, R. M. and Hahn, J. M. (2015). Impact vaporization as a possible source of Mercury's calcium exosphere. *Icarus*, **250**, 230–237, doi:10.1016/j.icarus.2014.11.035.

Killen, R. M., Potter, A., Fitzsimmons, A. and Morgan, T. H. (1999). Sodium D2 line profiles: Clues to the temperature structure of Mercury's exosphere. *Planet. Space Sci.*, **47**, 1449–1458, doi:10.1016/S0032-0633(99)00071–9.

Killen, R. M., Sarantos, M., Potter, A. E. and Reiff, P. (2004). Source rates and ion recycling rates for Na and K in Mercury's atmosphere. *Icarus*, **171**, 1–19, doi:10.1016/j.icarus.2004.04.007.

Killen, R. M., Bida, T. A. and Morgan, T. H. (2005). The calcium exosphere of Mercury. *Icarus*, **173**, 300–311, doi:10.1016/j.icarus.2004.08.022.

Killen, R., Cremonese, G., Lammer, H., Orsini, S., Potter, A. E., Sprague, A. L., Wurz, P., Khodachenko, M. L., Lichtenegger, H. I. M., Milillo, A. and Mura, A. (2007). Processes that promote and deplete the exosphere of Mercury. *Space Sci. Rev.*, **132**, 433–509, doi:10.1007/s11214-007–9232-0.

Killen, R. M., Mouawad, N. and Shemansky, D. E. (2009). Expected emission from Mercury's exospheric species, and their ultraviolet-visible signatures. *Astrophys. J. Suppl. Ser.*, **181**, 351–359.

Killen, R. M., Potter, A. E., Vervack, R. J., Bradley, E. T., McClintock, W. E., Anderson, C. M. and Burger, M. H. (2010). Observations of metallic species in Mercury's exosphere. *Icarus*, **209**, 75–87, doi:10.1016/j.icarus.2010.02.018.

Leblanc, F. and Johnson, R. E. (2003). Mercury's sodium exosphere. *Icarus*, **164**, 261–281, doi:10.1016/S0019-1035(03)00147–7.

Leblanc, F. and Johnson, R. E. (2010). Mercury exosphere I. Global circulation model of its sodium component. *Icarus*, **209**, 280–300, doi:10.1016/j.icarus.2010.04.020.

Leblanc, F., Barbieri, C., Cremonese, G., Verani, S., Cosentino, R., Mendillo, M., Sprague, A. and Hunten, D. (2006). Observations of Mercury's exosphere: Spatial distributions and variations of its Na component during August 8, 9 and 10, 2003. *Icarus*, **185**, 395–402, doi:10.1016/j.icarus.2006.08.006.

Leblanc, F., Doressoundiram, A., Schneider, N., Mangano, V., López Ariste, A., Lemen, C., Gelly, B., Barbieri, C. and Cremonese, G. (2008). High latitude peaks in Mercury's sodium exosphere: Spectral signature using THEMIS solar telescope. *Geophys. Res. Lett.*, **35**, L18204, doi:10.1029/2008GL035322.

Leblanc, F., Doressoundiram, A., Schneider, N. M., Massetti, S., Wedlund, M., Lopez Ariste, A., Barbieri, C., Mangano, V. and Cremonese, G. (2009). Short-term variations of Mercury's Na exosphere observed with very high spectral resolution. *Geophys. Res. Lett.*, **36**, L07201, doi:10.1029/2009GL038089.

Leblanc, F., Chaufray, J. Y., Doressoundiram, A., Berthelier, J. J., Mangano, V., Lopez-Ariste, A. and Borin, P. (2013). Mercury exosphere. III: Energetic characterization of its sodium component. *Icarus*, **223**, 963–974, doi:10.1016/j.icarus.2012.08.025.

Mangano, V., Leblanc, F., Barbieri, C., Massetti, S., Milillo, A., Cremonese, G. and Grava, C. (2009). Detection of a southern peak in Mercury's sodium exosphere with the TNG in 2005. *Icarus*, **201**, 424–431, doi:10.1016/j.icarus.2009.01.016.

Mangano, V., Massetti, S., Milillo, A., Mura, A., Orsini, S. and Leblanc, F. (2013). Dynamical evolution of sodium anisotropies in the exosphere of Mercury. *Planet. Space Sci.*, **82–83**, 1–10, doi:10.1016/j.pss.2013.03.002.

Mangano, V., Massetti, S., Milillo, A., Plainaki, C., Orsini, S., Rispoli, R. and Leblanc, F. (2015). THEMIS Na exosphere observations of Mercury and their correlation with in-situ magnetic field measurements by MESSENGER. *Planet. Space Sci.*, **115**, 102–109, doi:10.1016/j.pss.2015.04.001.

McClintock, W. E. and Lankton, M. R. (2007). The Mercury Atmospheric and Surface Composition Spectrometer for the MESSENGER mission. *Space Sci. Rev.*, **131**, 481–521, doi:10.1007/s11214-007–9264-5.

McClintock, W. E., Bradley, E. T., Vervack, R. J., Jr., Killen, R. M., Sprague, A. L., Izenberg, N. R. and Solomon, S. C. (2008). Mercury's exosphere: Observations during MESSENGER's first Mercury flyby. *Science*, **321**, 92–94.

McClintock, W. E., Vervack, R. J., Bradley, E. T., Killen, R. M., Mouawad, N., Sprague, A. L., Burger, M. H., Solomon, S. C. and Izenberg, N. R. (2009). MESSENGER observations of Mercury's exosphere: Detection of magnesium and distribution of constituents. *Science*, **324**, 610–613, doi:10.1126/science.1172525.

McGrath, M. A., Johnson, R. E. and Lanzerotti, L. J. (1986). Sputtering of sodium on the planet Mercury. *Nature*, **323**, 694–696, doi:10.1038/323694a0.

Merkel, A. W., Cassidy, T. A., Vervack, R. J., McClintock, W. E., Sarantos, M., Burger, M. H. and Killen, R. M. (2017). Seasonal variations of Mercury's magnesium dayside exosphere from MESSENGER observations. *Icarus*, **281**, 46–54, doi:10.1016/j.icarus.2016.08.032.

Merkel, A. W., Vervack, R. J., Jr., Killen, R. M., Cassidy, T. A., McClintock, W. E., Nittler, L. R. and Burger, M. H. (2018). Evidence connecting Mercury's magnesium exosphere to its magnesium-rich surface terrane. *Geophys. Res. Lett.*, **45**, 6790-6797 doi:10.1029/2018GL078407.

Morgan, T. H., Zook, H. A. and Potter, A. E. (1988). Impact-driven supply of sodium and potassium to the atmosphere of Mercury. *Icarus*, **75**, 156–170, doi:10.1016/0019–1035(88)90134–0.

Mouawad, N., Burger, M. H., Killen, R. M., Potter, A. E., McClintock, W. E., Vervack, R. J., Jr., Bradley, E. T., Benna, M. and Naidu, S. (2011). Constraints on Mercury's Na exosphere: Combined MESSENGER and ground-based data. *Icarus*, **211**, 21–36, doi:10.1016/j.icarus.2010.10.019.

Mura, A., Wurz, P., Lichtenegger, H. I. M., Schleicher, H., Lammer, H., Delcourt, D., Milillo, A., Orsini, S., Massetti, S. and Khodachenko, M. L. (2009). The sodium exosphere of Mercury: Comparison between observations during Mercury's transit and model results. *Icarus*, **200**, 1–11, doi:10.1016/j.icarus.2008.11.014.

Nittler, L. R., Starr, R. D., Weider, S. Z., McCoy, T. J., Boynton, W. V., Ebel, D. S., Ernst, C. M., Evans, L. G., Goldsten, J. O., Hamara, D. K., Lawrence, D. J., McNutt, R. L., Schlemm, C. E., Solomon, S. C. and Sprague, A. L. (2011). The major-element composition of Mercury's surface from MESSENGER X-ray spectrometry. *Science*, **333**, 1847–1850, doi:10.1126/science.1211567.

Peplowski, P. N., Evans, L. G., Stockstill-Cahill, K. R., Lawrence, D. J., Goldsten, J. O., McCoy, T. J., Nittler, L. R., Solomon, S. C., Sprague, A. L., Starr, R. D. and Weider, S. Z. (2014). Enhanced sodium abundance in Mercury's north polar region revealed by the MESSENGER Gamma-Ray Spectrometer. *Icarus*, **228**, 86–95.

Pierce, A. K. (1965). Construction of a Bowen image slicer. *Publ. Astron. Soc. Pac.*, **77**, 216–217, doi:10.1086/128199.

Potter, A. E. (1995). Chemical sputtering could produce sodium vapor and ice on Mercury. *Geophys. Res. Lett.*, **22**, 3289–3292, doi:10.1029/95GL03181.

Potter, A. E. and Killen, R. M. (2008). Observations of the sodium tail of Mercury. *Icarus*, **194**, 1–12, doi:10.1016/j.icarus.2007.09.023.

Potter, A. E. and Morgan, T. H. (1985). Discovery of sodium in the atmosphere of Mercury. *Science*, **229**, 651–653, doi:10.1126/science.229.4714.651.

Potter, A. E. and Morgan, T. H. (1986). Potassium in the atmosphere of Mercury. *Icarus*, **67**, 336–340, doi:10.1016/0019–1035(86)90113–2.

Potter, A. E. and Morgan, T. H. (1990). Evidence for magnetospheric effects on the sodium atmosphere of Mercury. *Science*, **248**, 835–838, doi:10.1126/science.248.4957.835.

Potter, A. E., Killen, R. M. and Morgan, T. H. (1999). Rapid changes in the sodium exosphere of Mercury. *Planet. Space Sci.*, **47**, 1441–1448, doi:10.1016/S0032-0633(99)00070-7.

Potter, A. E., Killen, R. M. and Morgan, T. H. (2002a). The sodium tail of Mercury. *Meteorit. Planet. Sci.*, **37**, 1165–1172, doi:10.1111/j.1945-5100.2002.tb00886.x.

Potter, A. E., Anderson, C. M., Killen, R. M. and Morgan, T. H. (2002b). Ratio of sodium to potassium in the Mercury exosphere. *J. Geophys. Res.*, **107**, 5040, doi:10.1029/2000JE001493.

Potter, A. E., Killen, R. M. and Sarantos, M. (2006). Spatial distribution of sodium on Mercury. *Icarus*, **181**, 1–12, doi:10.1016/j.icarus.2005.10.026.

Potter, A. E., Morgan, T. H. and Killen, R. M. (2009). Sodium winds on Mercury. *Icarus*, **204**, 355–367, doi:10.1016/j.icarus.2009.06.028.

Potter, A. E., Killen, R. M., Reardon, K. P. and Bida, T. A. (2013). Observation of neutral sodium above Mercury during the transit of November 8, 2006. *Icarus*, **226**, 172–185, doi:10.1016/j.icarus.2013.05.029.

Raines, J. M., Gershman, D. J., Zurbuchen, T. H., Sarantos, M., Slavin, J. A., Gilbert, J. A., Korth, H., Anderson, B. J., Gloeckler, G., Krimigis, S. M., Baker, D. N., McNutt, R. L. and Solomon, S. C. (2013). Distribution and compositional variations of plasma ions in Mercury's space environment: The first three Mercury years of MESSENGER observations. *J. Geophys. Res. Space Physics*, **118**, 1604–1619, doi:10.1029/2012JA018073.

Sarantos, M., Killen, R. M., Sharma, A. S. and Slavin, J. A. (2008). Influence of plasma ions on source rates for the lunar exosphere during passage through the Earth's magnetosphere. *Geophys. Res. Lett.*, **35**, L04105, doi:10.1029/2007GL032310.

Sarantos, M., Killen, R. M., McClintock, W. E., Todd Bradley, E., Vervack, R. J., Benna, M. and Slavin, J. A. (2011). Limits to Mercury's magnesium exosphere from MESSENGER second flyby observations. *Planet. Space Sci.*, **59**, 1992–2003, doi:10.1016/j.pss.2011.05.002.

Schleicher, H., Wiedemann, G., Wohl, H., Berkefeld, T. and Soltau, D. (2004). Detection of neutral sodium above Mercury during the transit on 2003 May 7. *Astron. Astrophys.*, **425**, 1119–1124, doi:10.1051/0004-6361:20040477.

Schmidt, C. A. (2013). Monte Carlo modeling of north–south asymmetries in Mercury's sodium exosphere. *J. Geophys. Res. Space Physics*, **118**, 4564–4571, doi:10.1002/jgra.50396.

Schmidt, C. A., Wilson, J. K., Baumgardner, J. and Mendillo, M. (2010). Orbital effects on Mercury's escaping sodium exosphere. *Icarus*, **207**, 9–16, doi:10.1016/j.icarus.2009.10.017.

Schmidt, C. A., Baumgardner, J., Mendillo, M. and Wilson, J. K. (2012). Escape rates and variability constraints for high-energy sodium sources at Mercury. *J. Geophys. Res.*, **117**, A03301, doi:10.1029/2011JA017217.

Shemansky, D. E. and Broadfoot, A. L. (1977). Interaction of the surfaces of the moon and Mercury with their exospheric atmospheres. *Rev. Geophys. Space Phys.*, **15**, 491–499, doi:10.1029/RG015i004p00491.

Slavin, J. A., DiBraccio, G. A., Gershman, D. J., Imber, S. M., Poh, G. K., Raines, J. M., Zurbuchen, T. H., Jia, X., Baker, D. N., Glassmeier, K.-H., Livi, S. A., Boardsen, S. A., Cassidy, T. A., Sarantos, M., Sundberg, T., Masters, A., Johnson, C. L., Winslow, R. M., Anderson, B. J., Korth, H., McNutt, R. L. and Solomon, S. C. (2014). MESSENGER observations of Mercury's dayside magnetosphere under extreme solar wind conditions. *J. Geophys. Res. Space Physics*, **119**, 8087–8116, doi:10.1002/2014JA020319.

Smith, G. R., Shemansky, D. E., Broadfoot, A. L. and Wallace, L. (1978). Monte Carlo modeling of exospheric bodies: Mercury. *J. Geophys. Res.*, **83**, 3783–3790, doi:10.1029/JA083iA08p03783.

Smyth, W. H. (1986). Nature and variability of Mercury's sodium atmosphere. *Nature*, **323**, 696–699, doi:10.1038/323696a0.

Smyth, W. H. and Marconi, M. L. (1995). Theoretical overview and modeling of the sodium and potassium atmospheres of Mercury. *Astrophys. J.*, **441**, 839–864, doi:10.1086/175407.

Sprague, A. L., Kozlowski, R. W. H., Hunten, D. M., Schneider, N. M., Domingue, D. L., Wells, W. K., Schmitt, W. and Fink, U. (1997). Distribution and abundance of sodium in Mercury's atmosphere, 1985–1988. *Icarus*, **129**, 506–527, doi:10.1006/icar.1997.5784.

Vervack, R. J., Jr., McClintock, W. E., Bradley, E. T., Killen, R. M., Sprague, A. L., Mouawad, N., Izenberg, N. R., Kochte, M. C. and Lankton, M. R. (2009). MESSENGER observations of Mercury's exosphere: Discoveries and surprises from the first two flybys. *Lunar Planet. Sci.*, **40**, abstract 2220.

Vervack, R. J., Jr., McClintock, W. E., Killen, R. M., Sprague, A. L., Anderson, B. J., Burger, M. H., Bradley, E. T., Mouawad, N., Solomon, S. C. and Izenberg, N. R. (2010). Mercury's complex exosphere: Results from MESSENGER's third flyby. *Science*, **329**, 672–675, doi:10.1126/science.1188572.

Vervack, R. J., Jr., Killen, R. M., Sprague, A. L., Burger, M. H., Merkel, A. W. and Sarantos, M. (2011). Early MESSENGER results for less abundant or weakly emitting species in Mercury's exosphere. *EPSC-DPS Joint Meeting Abstracts and Program*, **6**, abstract EPSC-DPS2011-1131. European Planetary Science Congress – Division for Planetary Sciences Joint Meeting, Nantes, France, 2–7 October. Available at http://adsabs.harvard.edu/abs/2011epsc.conf.1131V.

Vervack, R. J., Jr., Killen, R. M., McClintock, W. E., Merkel, A. W., Burger, M. H., Cassidy, T. A. and Sarantos, M. (2016). New discoveries from MESSENGER and insights into Mercury's exosphere. *Geophys. Res. Lett.*, **43**, 11,545–11,551, doi:10.1002/2016GL071284.

Winslow, R. M., Johnson, C. L., Anderson, B. J., Korth, H., Slavin, J. A., Purucker, M. E. and Solomon, S. C. (2012). Observations of Mercury's northern cusp region with MESSENGER's Magnetometer. *Geophys. Res. Lett.*, **39**, L08112, doi:10.1029/2012GL051472.

Winslow, R. M., Johnson, C. L., Anderson, B. J., Gershman, D. J., Raines, J. M., Lillis, R. J., Korth, H., Slavin, J. A., Solomon, S. C., Zurbuchen, T. H. and Zuber, M. T. (2014). Mercury's surface magnetic field determined from proton-reflection magnetometry. *Geophys. Res. Lett.*, **41**, 4463–4470, doi:10.1002/2014GL060258.

Yakshinskiy, B. V. and Madey, T. E. (2005). Temperature-dependent DIET of alkalis from SiO_2 films: Comparison with a lunar sample. *Surf. Sci.*, **593**, 202–209, doi:10.1016/j.susc.2005.06.062.

Yan, N., Chassefire, E., Leblanc, F. and Sarkissian, A. (2006). Thermal model of Mercury's surface and subsurface: Impact of subsurface physical heterogeneities on the surface temperature. *Adv. Space Res.*, **38**, 583–588, doi:10.1016/j.asr.2005.11.010.

Yoshikawa, I., Ono, J., Yoshioka, K., Murakami, G., Ezawa, F., Kameda, S. and Ueno, S. (2008). Observation of Mercury's sodium exosphere during the transit on November 9, 2006. *Planet. Space Sci.*, 56, 1676–1680.

15

Understanding Mercury's Exosphere: Models Derived from MESSENGER Observations

ROSEMARY M. KILLEN, MATTHEW H. BURGER, RONALD J. VERVACK, JR., AND TIMOTHY A. CASSIDY

15.1 INTRODUCTION

Mercury's exosphere is quite complex, and a number of models have been developed to explain the observations presented in Chapter 14. The first models were based upon a few simple assumptions and primarily explored the dynamics of sodium atoms pushed anti-sunward by radiation pressure (e.g., Ip, 1986; Smyth and Marconi, 1995a, b). More recently, these early models have been superseded by simulations with an increasing number of interdependent processes (e.g., Leblanc and Johnson, 2003; Mura et al., 2009; Leblanc and Johnson, 2010; Burger et al., 2010, 2012, 2014). In this chapter, we briefly summarize the various source and loss processes before describing the published exosphere models, first for the three species observed almost continuously during the MESSENGER mission (Na, Mg, and Ca) by the Ultraviolet and Visible Spectrometer (UVVS) channel of the Mercury Atmospheric and Surface Composition Spectrometer (MASCS), and then more briefly for other species that have been observed or for which new upper limits have been derived.

Although a number of processes have been known for many years to be likely source and loss progenitors for surface-bounded exospheres, the complexity of the processes and the nature of line-of-sight observations require sophisticated and rigorous modeling of these various processes to discern their relative importance for each of the observed species populating the exosphere. In the next section we discuss five source processes: thermal desorption, impact vaporization, photon-stimulated desorption, ion sputtering, and chemical sputtering. Thermal desorption is a function of surface temperature, which varies not only with solar zenith angle but also with heliocentric distance. In addition, thermal desorption is a function of the binding energy of the atom with the surface, which can be a complicated function of the space weathering history of the surface grains and the fractional coverage of the adsorbate. Impact vaporization is a process that produces ejecta at a high temperature, which decreases with decreasing impact angle relative to the horizontal (Schultz, 1996). This temperature can be several thousand kelvin for silicates, but the precise temperature is not well determined. The vapor released as a function of impactor mass is a steep function of impact velocity and also depends on the relative densities of impactor and target and their compositions. Meteoroid and cometary impact is the most important process in regolith gardening, the pulverization and overturn of the upper crust. Large impacts excavate deeper layers, thereby exposing fresh material to the surface, but it is the small micrometeoroids that contribute ejecta, melt, and vapor on short timescales, including daily to yearly timescales. Photon-stimulated desorption (PSD) is important for the volatile elements, notably for the alkalis Na and K, and is a function of solar insolation and also of adsorbate coverage. PSD-derived vapor is of a much higher temperature than that from thermal desorption but is cooler than impact-derived vapor. Ion sputtering releases very energetic vapor, but the yield and velocity distribution of these ejecta are somewhat controversial. Sputtering yields depend strongly on the impacting ion species and its energy and impact angle, and will also depend on the grain size distribution, space weathering history, and composition of the regolith. Chemical sputtering relates to the production of ejecta through chemical reactions on the surface. Although it has been suggested as a source for Na and for water, little work has been done to determine rates for these chemical processes.

The relative importance of these processes can be constrained given the spatial and temporal distributions of the exospheric species to the extent that their source rates, velocity distributions, and forcing parameters are known. These topics are discussed in the following section.

15.2 OVERVIEW OF SOURCE AND LOSS PROCESSES

15.2.1 Source Processes

15.2.1.1 Thermal Desorption

Thermal desorption (or thermal evaporation) is the release of adsorbed atoms from a surface via heating. The process is related to the binding energy of the atom on the surface and to the vibrational frequency of this bound atom, such that the rate of thermal desorption is given by

$$R_{TD} = \nu_{TD}\, C \exp\left(-\frac{U}{kT_S}\right), \qquad (15.1)$$

where ν_{TD}, the vibrational frequency of the atom on the surface, is commonly approximated as the ratio of the spacing between adsorption sites to the mean thermal speed of an adsorbed atom. U is the binding energy with the surface, k is the Boltzmann constant, C is the concentration of the desorbing species on the

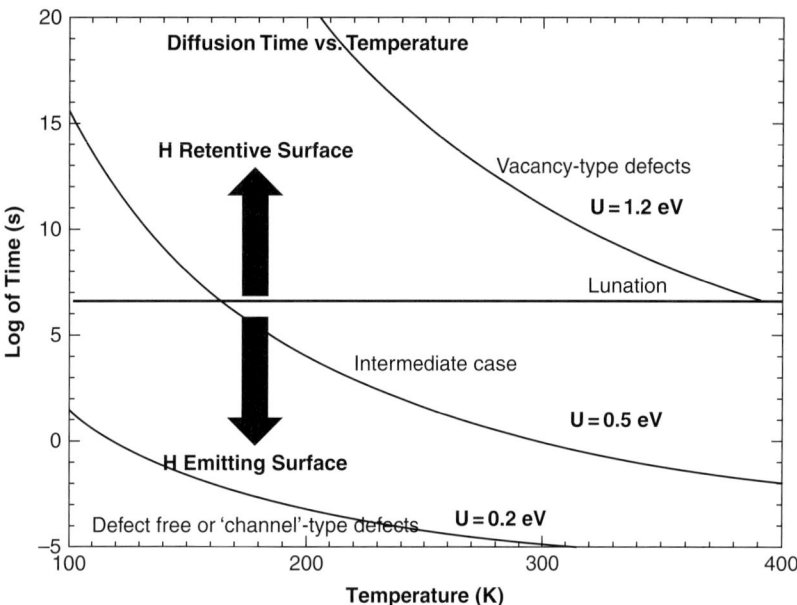

Figure 15.1. Diffusion time as a function of temperature for a family of binding energies, U, for H implanted at a depth of 100 nm onto a crystal with and without channel-type defects. This figure is illustrative of the effect of binding energy on release of volatiles, whereby an intermediate binding energy will result in highly temperature-dependent release or release dependent on plasma bombardment (solar wind on the dayside or magnetospheric effects on the nightside). High binding energy results in retentive surfaces, whereas low binding energy results in an emitting surface. From Farrell et al. (2015), with permission.

surface, and T_S is the surface temperature (e.g., Hunten et al., 1988). The vibrational frequency is often set at 10^{13} s^{-1} (Yakshinskiy et al., 2000), although Killen et al. (2007) argued that it might be several orders of magnitude smaller, following the arguments of Holmlid (2006). Leblanc and Johnson (2003) considered a Gaussian distribution of binding sites between 1.4 and 2.7 eV with a most probable value of 1.85 eV. After a number of vibrations on the order of $\exp(U/kT)$, where T is the absolute temperature, the atom is likely to escape the potential well. Thus the amount of time the atom spends in the adsorbed state is minuscule for small adsorption energies. Any process that contributes adsorbed atoms to the surface will contribute to a surface reservoir. The retention time on the surface is an exponential function of binding energy. Hunten and Sprague (1997, 2002) suggested that thermal desorption competes with PSD and sputtering, depleting the Na atoms in the surface layers of grains.

For sodium, thermal desorption produces atoms at only 0.03–0.05 eV energy (Yakshinskiy et al., 2000) compared with the escape energy of 2.07 eV. The energy distribution of atoms ejected by thermal desorption follows a Maxwell–Boltzmann flux distribution given by

$$f(E, \theta) = 2 \cos\theta \, \frac{E}{(kT_s)^2} \exp\left(\frac{-E}{kT_s}\right), \quad (15.2)$$

where E is the energy of the ejecta and θ is the angle between the velocity vector of the ejecta and the normal to the surface.

No evidence for thermal desorption has been found for any of the species observed by the MESSENGER UVVS (e.g., Burger et al., 2012; Cassidy et al., 2015). This result is surprising in light of the predictions that thermal desorption should dominate near the subsolar point (Mura et al., 2009) and at aphelion (Leblanc et al., 2010). A possible explanation for the lack of observation of a thermal component is the observation that binding sites shift to higher energies after prolonged bombardment of the surface by 1-keV He$^+$ (Madey et al., 1998; Yakshinskiy et al., 2000) and, by analogy, by space weathering of an exposed surface. After one hour of bombardment by 1-keV He$^+$ ions, the thermal desorption of Na from SiO$_2$ shifts to temperatures near 1000 K (Madey et al., 1998). Bonding at defect sites (which increase with ion bombardment) is more important for wide-band-gap oxides (such as MgO) than for narrow-band-gap oxides (insulators such at TiO$_2$). Madey et al. (1998) also reported that the binding energy of alkalis increases with lower fractional coverage (<1 mono-layer). At higher coverage (>0.1 mono-layer), alkali bonding is influenced by cohesive interactions between the adsorbed atoms, which may decrease the adsorbate–substrate bond (Madey et al., 1998). The fractional coverage of atoms on surfaces affects thermal vaporization in such a way that atoms are less likely to desorb from surfaces when there is less than a mono-layer of coverage (Sneh et al., 1996).

If U [see equation (15.1)] increases with surface temperature, with lower fractional coverage, or with exposure to solar photons or ions, then the thermal desorption rate does not increase as fast with solar insolation as would be the case if U remained constant. Because the surface composition of Mercury is highly non-uniform, thermal desorption may be dependent not only on local solar time but also on the underlying mineralogy. Thus, thermal desorption is extremely complex and difficult to model accurately. Whether a surface is retentive or emitting of a particular volatile constituent depends strongly on the binding energy and temperature (Figure 15.1).

Thermally desorbed atoms ejected from the surface follow a Maxwellian flux distribution of the form

$$f(v) \propto v^3 \exp(-v^2/v_{\text{th}}^2), \quad (15.3)$$

where v is the neutral atom's velocity, $v_{\text{th}} = 2kT_S/m$ is the mean thermal speed, and m is the mass of the ejected neutral atom. The energy distribution of thermally desorbed Na atoms at 700 K is shown by the magenta line in Figure 15.2.

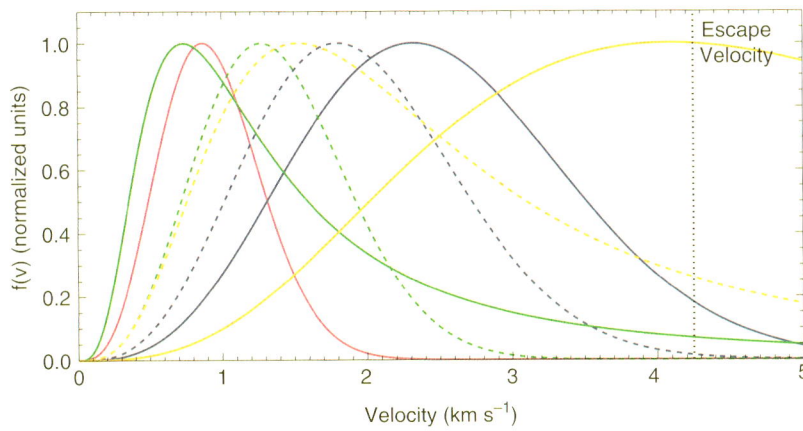

Figure 15.2. Velocity distributions for the major source processes expected for exospheric species: (1) magenta: Maxwellian at 700 K (thermal desorption); (2) dashed blue: Maxwellian at 3000 K (impact vaporization); (3) solid blue: Maxwellian at 5000 K (impact vaporization); (4) solid green: Weibull distribution used by Burger et al. (2010) for PSD; (5) dashed green: Maxwellian at 1500 K used by Leblanc et al. (2003) for PSD; (6) solid orange: sputter distribution with $U = 2$ eV used by Burger et al. (2010); (7) dashed broken orange: sputtering with $U = 0.28$ eV used by Leblanc et al. (2003).

The effectiveness of thermal desorption is influenced by surface diffusion between sites with different desorption energies (Farrell et al., 2015). The rate of transport of volatiles from grain interiors to the surface is governed by diffusion, grain size, grain composition, presence and thickness of a glassy rim, and grain lifetime at the regolith surface (Killen et al., 2004, 2007). For example, atoms diffuse to the surface of smaller grains more quickly than from larger grains, and thus small grains are depleted more rapidly in volatiles via thermal desorption (Killen et al., 2004). However, small grains also serve as a more efficient repository for adsorption of volatiles owing to their larger surface-to-volume ratio. Because the activation energy for diffusion is about one-tenth of the binding energy, surface diffusion – a mechanism whereby atoms can move along a surface – can result in atoms accessing deeper binding sites, where they are more stable against thermal vaporization, or becoming cold trapped underneath surface grains.

15.2.1.2 Impact Vaporization

Impact vaporization is the most universal process that promotes atomic species into the exosphere in that it is energetic enough to eject all species and for impact velocities > 5 km/s will probably vaporize all species rather than a select few. Mangano et al. (2007) concluded that micrometeoroid impact vaporization is the sole process continuously refilling the exosphere for Mg, Al, Si, S, Ca, and O. Impact vaporization has been variously described as a minor source for Na, delivering only a small percentage of the total source rate to Mercury's exosphere (Smyth and Marconi, 1995a, b), or a major source (Morgan et al., 1988; Borin et al., 2010), and so its importance remains somewhat uncertain. It is believed that impact vaporization is either the dominant or the exclusive process ejecting Ca, a refractory element, into the exosphere (Killen and Hahn, 2015). The moderately volatile species Mg shows temporal and spatial variation in its distribution and exospheric temperature suggestive of impact vaporization, as discussed below. Impacting particles of small sizes (<100 μm) constantly rain onto Mercury's surface at a mean velocity of ~20 km/s (Cintala, 1992), churning the regolith and vaporizing the surface. Larger meteoroids impact more sporadically, but with higher mean velocity and a double-peaked velocity distribution with peaks near 30 and 40 km/s (Mangano et al., 2007; Marchi et al., 2005). Many minor species that are refractory during vaporization of silicates in vacuum are highly volatile during hypervelocity impacts owing to the high temperatures and pressures (Gerasimov et al., 1999).

Impact events probe to a depth of several diameters of the impacting body and are therefore important in terms of supply of fresh grains to the surface, a process referred to as regolith gardening. This process is important in exposing fresh, unweathered material to the surface. Because meteoroid impacts probe much deeper than any process other than venting, and because the energy density of the process is very high, the exospheric products of this emission process most closely represent the surface composition as a whole. Mangano et al. (2007) performed simulations to analyze the effects in terms of the gaseous cloud produced by impacts of objects in the range 1 cm to 1 m. Particularly noticeable is the case of 10-cm meteoroids for which the enhancement, depending on the species considered, varies from one to four orders of magnitude over the mean exospheric background values.

Larger meteoroids will impact Mercury but at an unknown rate. Marchi et al. (2005) provided the distribution of impact probability as a function of impactor radius, up to objects of 100 m in radius. In particular, meteoroid impactors coming from the main asteroid belt are expected to impact on Mercury as well. The contribution by these larger meteoroids to the global mercurian exosphere is negligible on average; nevertheless, their impact is expected to produce strong, localized, but temporary increases in the exospheric density, enriched in material coming from deeper layers (Mangano et al., 2007). Mangano et al. (2007) calculated that a 1-m meteoroid impacting Mercury's dayside will produce an enhancement in the Na exosphere visible over the background at 400-km altitude for 800 s, and for 6000 s for the O exosphere. The nightside enhancements last longer relative to the background because the scale height of the background exosphere is lower. Although the impact frequency of such objects at Mercury is not negligible relative to the duration of the MESSENGER orbital mission phase (more than four years), they would be difficult to detect given their short lifetimes and localized effects.

Regardless of the size of the impactor, the initial ejecta from an impact will be high-temperature vapor (~3500–5000 K). This ejecta will quickly be followed by liquid + vapor at a slightly lower temperature (2500 K). The vapor ejected by cometary impacts may be hotter and include a large ionized component (Hornung et al., 2000; Kurosawa et al., 2010), followed quickly by thermalized vapor. In general, impact vapor may be much cooler than previously supposed because of rapid cooling. Energy distributions for Na produced by impact vaporization, defined by the Maxwellian flux speed distribution in equation (15.2), are shown in Figure 15.2 for vapor at $T = 3000$ K (dashed blue line) and $T = 5000$ K (solid blue line).

Soil grains melted by impact may subsequently release vapor as a result of the extreme temperatures (>2000°C) reached by some melts, giving off dissociated molecules, ions, and atoms into the vapor (Keller and McKay, 1997). This process places a portion of the silicate melt into vapor. Some of the vapor condenses on the rims of all soil grains, largely as amorphous glass rinds from 50 to 100 nm thick (McCord et al., 2011). Some of the charged ions neutralize, creating energetic atoms, such as the observed hot Ca and Mg (Killen, 2016).

Killen and Hahn (2015) showed that the source rate for calcium in the exosphere of Mercury can be produced by impact vaporization of interplanetary dust, fine grained micrometeoritic material which is believed to preferentially impinge on the dawn-side hemisphere – the ram direction (e.g., Janches et al., 2006). Dust ejection at the Moon was shown to be markedly concentrated in the ram direction by the results of the Lunar Atmosphere and Dust Environment Explorer (LADEE) Lunar Dust Experiment (Horanyi et al., 2015, 2016), lending further evidence that micrometeoroid dust impact is strongly asymmetric. A permanent, asymmetric dust cloud peaking near the dawn terminator was observed at the Moon and attributed to impacts by high-speed dust particles of cometary origin (Horanyi et al., 2015). This distribution is similar to the distribution of Ca in Mercury's exosphere. It is not known whether such a dust cloud exists at Mercury.

Seasonal variations seen in the planet-wide Ca emission rate (Burger et al., 2014) can be explained as Mercury traverses the interplanetary dust disk, given the relative inclination of the dust disk and Mercury's orbital plane of about 3.5°. One exception is near Mercury true anomaly angle (TAA) 25°–30°, where it has been shown that additional cometary dust from comet 2P/Encke is likely to encounter Mercury (Christou et al., 2015; Killen and Hahn, 2015).

The observed extreme temperature of the Ca exosphere (>50,000 K) (Burger et al., 2014) is more difficult to explain than the source rate. Killen (2016) attempted to outline possible processes that could produce these extreme temperatures, but more laboratory work is desirable to determine the excess energies of the various processes proposed.

15.2.1.3 Photon-Stimulated Desorption

Photon-stimulated desorption was first suggested as a source process for the Na exosphere of Mercury by McGrath and Johnson (1986). Photon-stimulated desorption results from an electron transfer induced by photon bombardment of the surface with energies greater than the threshold value of 3 to 4 eV. Fractional mono-layers of Na deposited on oxide surfaces adsorb as Na^+ (whereas multi-layers contain metallic Na). A photon can induce an electron transfer to an unoccupied Na^+ 3s level, neutralizing the adsorbed Na^+. This state is highly repulsive, but the Na atom thus created can either re-adsorb by charge transfer from the substrate or desorb from the surface, whichever process is faster. PSD acts only on adsorbed Na because photons do not penetrate the bulk of the solid.

A number of experiments have been performed to measure the cross sections for photon-stimulated desorption, electron-stimulated desorption, and ion-induced desorption for alkali elements (Madey et al., 1998; Yakshinskiy and Madey, 2004, 2005). The PSD cross section for Na desorption from SiO_2 films was measured by Yakshinskiy and Madey (2000) to be $\sim 3 \times 10^{-20}$ cm^2 for 5 eV photons. Burger et al. (2010) used a PSD cross section of 3×10^{-21} cm^2, taking into account porosity and temperature effects, and they set the photon-limited PSD rate to 2.7×10^8 cm^{-2} s^{-1} at the subsolar point during the first and second MESSENGER flybys. They argued that the PSD rate is everywhere diffusion-limited, consistent with the results of Killen et al. (2004). Burger al. (2010) found that a diffusion rate of 10^6–10^7 cm^{-2} s^{-1} is consistent with the data, depending on the sticking coefficient used, although this rate was increased by ion precipitation, which heats the surface and creates defects in the regolith lattice that allow diffusion to proceed more rapidly.

It was subsequently shown (Yakshinskiy and Madey, 2004) that the desorption cross section has an exponential dependence on the binding energy of a transient atom. If there are multiple adsorption sites on the surface for which the lifetimes are different, then the relative concentration in the higher-energy site increases. This behavior gives rise to PSD-inactive and PSD-active sites similar to the retentive and emitting surfaces shown in Figure 15.1. Either increased temperature or bombardment by high-energy ions or electrons can produce such a migration to high-energy sites, both of which are prevalent at Mercury. Sodium desorbs more easily from a wide-band-gap insulator such as glass than from a crystalline surface (Yakshinskiy and Madey, 2004). Glassy rims are expected to coat the surfaces of most grains on the surface of Mercury (Keller and McKay, 1997).

Mura (2012) pointed out that the loss rate of sodium particles from Mercury due to PSD release is a function of both TAA (heliocentric distance and velocity) and the assumed source velocity distribution. The energy distribution of these particles has been modeled with a Maxwell–Boltzmann flux distribution for a temperature of ~1200–1500 K (e.g., Leblanc and Johnson 2003),

$$f(E) = \frac{E}{kT} e^{(-E/kT)}, \qquad (15.4)$$

and with a Weibull function (Johnson et al., 2002),

$$f(E) = x(1+x) \frac{E\beta^x}{(E+\beta)^{2+x}}, \qquad (15.5)$$

where E is the energy of the emitted particle, T is the temperature of the distribution, x is a free parameter (assumed to be

0.7 by Mura), and β is the characteristic energy of the Weibull distribution. Burger and co-workers (Burger et al., 2010, 2012, 2014; Mouawad et al., 2011) also used a Weibull distribution with $x = 0.7$ but with β equal to the binding energy $U = 0.052$ eV and with the leading term, $x(1 + x)$, replaced with a normalization constant. The "fast" PSD velocity distribution of Burger et al. was taken from Johnson et al. (2002), who measured electron-stimulated desorption (ESD) of Na from amorphous ice. The Maxwellian velocity distribution at 1200 K was based on measurements by Yakshinskiy and Madey (1999) of Na ejected from SiO_2 by ESD. Yakshinskiy and Madey (2004) showed that the peak velocity of Na desorbing from a lunar sample by PSD is ~800 m/s (~900 K), less than that observed for desorption of Na from a SiO film (Yakshinskiy and Madey, 2000). A Maxwellian distribution for the velocity of Na ejected at 1500 K (dashed green line) and the Weibull distribution (solid green line) are shown in Figure 15.2.

Modeled fits to the rates of Na, K, and Mg ejected into the exosphere by PSD depend strongly on the assumed sticking probability and thermal accommodation coefficient. This is because if the atom sticks upon re-impact with the surface then a higher ejection rate is required to maintain the observed exosphere than would be needed if the atom bounces. Similarly, the mix of different processes derived from the models depends critically on the assumed thermal accommodation. If the thermal accommodation is assumed to be zero, or the sticking probability is unity, as in the models of Cassidy et al. (2015), the conclusion will be that PSD is the only ejection process because the exosphere is observed to have a characteristic temperature of ~1200 K, consistent with the presumed PSD velocity distribution. However, the sticking probability for Na measured by Yakshinskiy and Madey (2005) is 0.2–0.3 in the temperature range appropriate for Mercury's dayside. If the thermal accommodation coefficient is assumed to be 0.2 for Na, an admixture of at least 20% of impact vapor at a temperature of ~3000 K is required to maintain the observed exospheric temperature and also to populate the Na tail (Burger et al., 2010; Schmidt, 2012). With the smaller sticking probability, a smaller desorption cross section is required than for the sticking coefficient of unity assumed by Sarantos et al. (2012), because the atoms are reemitted. These reemitted atoms are not assumed to be "new" atoms in the models of Burger et al. (2012, 2014). According to Yakshinskiy and Madey (2005), potassium has a higher probability than Na of losing energy to substrate phonons and becoming trapped. The lack of temperature dependence in the sticking for K is an indication of a very efficient kinetic energy transfer to the substrate (i.e., K should be thermally accommodated to the surface) (Yakshinskiy and Madey, 2005). Presumably, the reverse is true for Na: a strong temperature dependence in sticking indicates that Na should not be thermally accommodated. This presumption seems to be confirmed by the observations: MESSENGER UVVS did not observe a thermal component in the Na exosphere.

15.2.1.4 Ion Sputtering

Ion sputtering results from the impinging of an ion of mass m_1 onto a surface; if the impact energy (E_i) is high enough, a new particle of mass m_2 may be extracted. Solar-wind protons (H^+) strike the lunar soil grains with enough energy to penetrate to depths of 5–10 nm (Starukhina and Shkuratov, 2000; Johnson, 1990). Grain rims are typically 50–100 nm thick (Noble et al., 2005), so the protons penetrate only the rim, not the crystalline grain, if the grains are coated with glass. Thus protons preferentially sputter from glassy rims, which may not have compositions consistent with the bulk regolith. Plagioclase rims on lunar grains analyzed by Keller and McKay (1997) are depleted in Al by 50% and depleted in Ca by 80% relative to their host grains; the orthopyroxene rim is depleted of Mg by 80% and enriched in O (presumed to be bound to H). Excess oxygen found in rims analyzed by Keller and McKay is approximately 12–15% by number in the rims of both orthopyroxene and plagioclase grains, presumably bound to H in the form of hydroxyl. On the other hand, these rims are enriched in Si, S, Fe, Al, and Ti relative to their host grains. The amorphous rims on cristobalite grains show compositions nearly identical to those of the host grains. The wide range in the chemical properties of amorphous grains relative to their host grains on the Moon means that most likely only impact vaporization produces ejecta reflective of the bulk composition of the regolith, and the sources from different regions must reflect different compositional terrains.

For light ions (e.g., H^+), ion sputtering is a two-step process: backscattering of the ion over a surface target, and ejection of a second surface atom by the backscattered ion; in most cases, the ejected particle is neutral (Hofer, 1991). The velocity function for neutral ejecta peaks at few eV (Sigmund, 1969; Sieveka and Johnson, 1984) and can be empirically reproduced by the following function:

$$f_s(E_e, T_m) = c_n \frac{E_e}{(E_e + E_b)^3} \left[1 - \left(\frac{E_e + E_b}{T_m} \right)^{1/2} \right], \quad (15.6)$$

where E_b is the surface binding energy of the atomic species extracted, E_e the energy of the emitted particles, and c_n the normalization constant. For Na sputtered by H, T_m is given by (e.g., Mura et al., 2009)

$$T_m = E_i \frac{4 m_H m_{Na}}{(m_H + m_{Na})^2}. \quad (15.7)$$

In this context E_b is the chemisorption energy and E_i is the energy of the incoming proton. Although the binding energy for Na has sometimes been taken to be ~2 eV (McGrath et al., 1986), Yakshinskiy et al. (2000) reported that their measurements are consistent with multiple binding sites between 1.4 and 2.7 eV, with a most probable value of 1.85 eV. Consequently, Leblanc and Johnson (2003) used a Gaussian distribution for E_b with the most probable value equal to 1.85 eV. The sputtered Na energy distribution with $U = 2$ eV is shown in Figure 15.2 (solid orange line). For heavy ions, the ion impact direction does not affect the angular distribution of ejecta, which is a $\cos^n(\alpha)$ function, where α is the angle from the surface normal direction and n is usually between 1 and 2 (note that this function has never been measured on powdered surfaces). For light ions, the angular distribution is related to the ion impact direction and exhibits a maximum close to the mirror angle, where the angle of ejection equals the angle of incidence along the incoming direction of the ion. The resulting neutral differential flux is

$$\frac{d\Phi_n}{dE_e} = cY \int_{E_{min}}^{E_{max}} \frac{d\Phi_i}{dE_i} f_s(E_e, E_i) dE_i, \qquad (15.8)$$

where Y is the process yield, c is the surface relative abundance of the atomic species considered, Φ_i is the ion flux, f_s is the distribution function of ejection energy, E_e is the energy of the emitted particle and is related to E_b, the surface binding energy (Mura et al., 2007), and E_i is the energy of the incoming ion. The yield reduction due to regolith porosity (Cassidy and Johnson, 2005) is expected to be about one-third and is not considered in equation (15.8). Also not considered in equation (15.8) are the energy dependence of the yield and the angular distribution of emitted particles.

A fraction of atoms and ions sputtered from the surface are ejected in excited states, which will decay by radiative transitions. These excited states decay within millimeters of the surface and are therefore not taken into account in analysis of remote observations of exospheric emission (Dzioba and Kelly, 1980).

15.2.1.5 Chemical Sputtering

Chemical sputtering relates to the release of atoms and molecules from regolith grains during chemical reactions between implanted solar-wind or magnetospheric ions (predominantly implanted protons) and regolith material. The radiolytic processes leading to chemical sputtering consist of several steps: implantation of reactive ions, followed by chemical reactions with target atoms or molecules, and, finally, desorption of the reaction products (Roth, 1983).

Chemical sputtering may produce and remove elemental sodium (Potter, 1995), hydrogen, hydroxyl, and water (e.g., Crider and Vondrak, 2003) from the surface to the exosphere. Mura et al. (2009) argued that chemical sputtering does not eject Na atoms directly but liberates them for another process, such as PSD, to eject them from the surface. In this sense, chemical sputtering is similar to the ion-enhanced diffusion proposed by Burger et al. (2010) to explain the increase in PSD rate in regions open to solar-wind ion precipitation. This process deserves further study.

15.2.1.6 Gas–Surface Interaction

All processes discussed here are affected by gas–surface interactions, which are functions of surface composition and space weathering effects. As discussed earlier, thermal accommodation, sticking probabilities, crystal defects and binding sites, and grain structures such as agglutinates and glassy rims all contribute. Interactions with the surface are described in terms of adsorption and sticking, which we will discuss here.

Sticking is a term that relates to the fraction of atoms impinging onto a surface that are retained. When a surface is sufficiently cold, the sticking fraction can be approximated by unity, so that every atom hitting the surface is retained, either by chemisorption (bonding with the surface), or physisorption by van der Waals interactions. With the exception of H and He, returning particles essentially stick with a residence time that depends on the surface temperature and the availability of deep sites (Smith and Kay, 1997; Yakshinskiy et al., 2000). In laboratory studies a distinction can be made between direct reflection, sticking–migration–desorption, and sticking and becoming bound in a deep well (Smith and Kay, 1997). In modeling, one typically uses a net sticking coefficient, S, which combines the probability of physisorption and the probability of finding a binding site. Yakshinskiy et al. (2000) gave the net sticking probability for Na on a silicate: $S \sim 0.5$ at 250 K, decreasing to 0.2 at 500 K. However, as demonstrated for water on lunar materials, the sticking coefficient and adsorption probabilities are highly dependent not only on composition but also on the physical state of the surface and the weathering history (Poston et al., 2015). Scanning electron microscope (SEM) images of typical lunar agglutinates extracted from Apollo soil samples show that the glassy surface is extensively coated with small, fine-grained soil fragments. These images demonstrate the irregular shapes and delicate structures common to agglutinates. Visible on the surface of these agglutinates are regions of glassy, fragment-free surfaces adjacent to fragment-laden surfaces. The texture of the grain surface, composed of a coating of tightly welded fine-fragment material, will affect desorption and sequestration (Domingue et al., 2014). The surface area of the grains is very important in determining the adsorptive efficiency, so small grains can be highly adsorptive.

Adsorption is a term that encompasses both physisorption and chemisorption, which are in fact very different interactions. As mentioned above, physisorption involves the van der Waals potential, thus is electrostatic in nature and related to the creation of induced dipole moments at the surface. A physisorbed atom can be desorbed by thermally exciting internal states or a surface state (Madey et al., 2002). Exciting a surface state can lead to ejection of an atom or molecule, whereas excitation of the internal states leads to ejection of an atom only if it is pointing outward from the surface.

If the atom is not reflected, any atom returning to the surface likely remains in a physisorbed state a very short time before finding a binding site through surface diffusion. That is, the returning atoms or molecules initially become weakly adsorbed on the surface. In this state they can either migrate along the surface of a grain until they find a deep adsorption site (characterized by chemisorption) or they can desorb thermally. In order to desorb, an activation energy greater than the difference between the bound potential and the surface potential must be available. The activation energy can be obtained by a thermal or photon process.

Once an atom finds a deep potential well it becomes chemisorbed, characterized by chemical bonding. The depth of the bonding sites depends on the particular physical state of the surface. Chemisorption sites extend over a wide range with a distribution of adsorption energies characterized by a Weibull distribution. This phenomenon has been studied extensively for water by temperature-programmed desorption (TPD) (e.g., Poston et al., 2015). An important finding is that the adsorption energy – or more properly the distribution of binding sites – is highly dependent on the composition of the surface, the grain sizes, and the weathering history. For instance, mature anorthositic lunar soil has a distribution of binding sites for water that peaks at about 0.7 eV and extends to 1.5 eV, whereas sub-mature low-titanium basalt has binding energies extending only to about 0.6 eV.

Description of chemisorption in terms of the usual energy–distance curve is an oversimplification because more coordinates are needed. Binding energies depend not only also on the coordinate along the surface (e.g., top site versus hollow site) but also on the interatomic distance in the precursor molecule, its orientation with respect to the surface, and the individual coordinates of the dissociation products. Thus, the problem of obtaining the relevant energies for the chemisorption process is extremely complicated.

An additional factor is surface charging. Dayside excitation by solar ultraviolet (UV) and X-rays causes the photoemission of electrons from surface grains, creating a positive potential (Farrell et al., 2007). The surfaces of grains can be oppositely charged by photon and electron fluxes (Jurac et al., 1995) affecting the amount of adsorbate available for desorption. Proton bombardment can result in positive charging and, therefore, inward sodium diffusion, whereas low-energy electrons can cause negative charging and, therefore, outward diffusion of sodium (Madey et al., 2002). The magnitude of the effect is at present poorly constrained by laboratory measurements.

15.2.1.7 Synopsis of Source Processes

Despite continued measurements needed for accurate models (e.g., the PSD cross sections), the actual relative mix of source processes has remained ambiguous, partly because of uncertainties in the velocity distributions, thermal accommodation, and sticking coefficients of the atoms when they interact with the surface. Estimates of removal by PSD of Na and K made by Wurz et al. (2010) indicated that this process dominates by about three orders of magnitude over either micrometeoroid impact vaporization or physical sputtering. Other simulation studies, however, have suggested that impact vaporization is a significant contributor, ranging from about 20% of that for PSD for Na (Burger et al., 2010; Mouawad et al., 2011) to all the exospheric Na produced via impact vaporization (Morgan et al., 1988; Borin et al., 2010). The Na production rate due to impact vaporization required to populate the tail in the models of Schmidt et al. (2012) – 1.8×10^6 atoms cm^{-2} s^{-1} for a 5000 K velocity distribution – is matched by the impact vaporization rate given by Burger et al. (2010) scaled to the Na wt% measured by MESSENGER.

15.2.2 Loss Processes

Neutral species in Mercury's exosphere are lost by collisions with the surface, gravitational escape (i.e., reaching the Hill sphere), or photoionization, the process by which solar photons remove electrons from neutral atoms and create pickup ions that are governed by electrodynamic forces rather than gravitational forces. About 85% of photoionized species are estimated to be entrained in the solar wind and lost from Mercury (e.g., Leblanc et al., 2003). Radiation pressure can accelerate atoms to above the escape velocity that otherwise would not escape. These processes are discussed below.

15.2.2.1 Radiation Pressure Effects

The constituents in Mercury's exosphere are subjected to radiation acceleration which pushes neutral species anti-sunward, forming a comet-like tail behind Mercury. Radiation acceleration is a consequence of the resonant-scattering emission process. Solar photons with a single incident direction are absorbed at the resonant wavelength (e.g., the Na D_1 and D_2 lines at 589.7 nm and 589.1 nm in the Na atom's rest frame) and nearly instantly reemitted approximately isotropically. Because photons have finite momentum, the atom experiences an impulse equal to the momentum difference between the incident and scattered photons. The magnitude of the radiation acceleration depends on the solar flux at the Doppler-shifted wavelength in the atom's rest frame. Because of deep Fraunhofer absorption lines in the solar spectrum, the photon flux experienced by an atom is a strong function of its radial velocity relative to the Sun. The magnitude of the radiation acceleration $a_{\rm rad}$ was given approximately by Smyth (1983):

$$a_{\rm rad} = \sum_i \frac{h}{m\lambda_i} g_i, \qquad (15.9)$$

where h is Planck's constant, m is the mass of the scattering atom, λ is the resonant wavelength, and g is the g-value of the transition (the product of the photon flux at the transition and the scattering probability per atom). The acceleration is the sum over all resonant transitions, i, but in practice only the strongest transitions need to be considered. The g-values for important transitions of species predicted to be in Mercury's exosphere were compiled by Killen et al. (2009).

The magnitude of the radiation acceleration as a function of Mercury true anomaly angle is shown for Na, Ca, and Mg atoms at rest relative to Mercury in Figure 15.3. Radiation acceleration has the greatest effect on the motion of Na atoms and is not much of a factor when determining the motion of Mg atoms because of weak absorptions. For Ca, the trajectories are

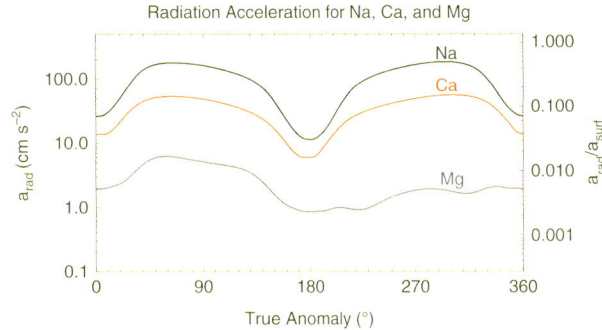

Figure 15.3. Acceleration due to radiation pressure is shown in absolute units (cm s^{-2}) for Na (black), Ca (red), and Mg (blue) at rest relative to Mercury (left axis) and as a fraction of gravitational acceleration at the surface (right axis). The magnitude of radiation acceleration ($a_{\rm rad}$) depends on the true anomaly because of Mercury's changing heliocentric distance and the variation in Doppler shift of the relevant transitions. Deep Fraunhofer features contribute to the strongly varying radiation pressure of Na and Ca versus true anomaly. Acceleration due to radiation pressure for Na approaches 50% of surface gravitational acceleration and remains high except near aphelion and perihelion.

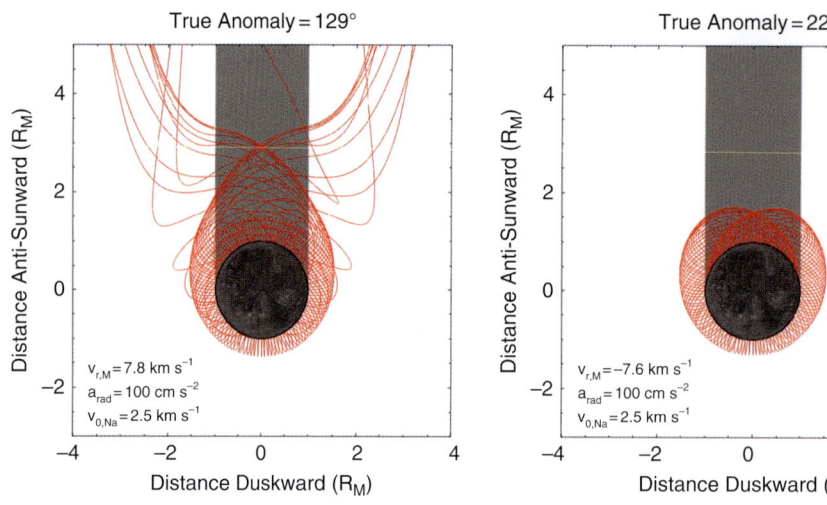

Figure 15.4. Comparison of Na trajectories ejected from Mercury at two true anomalies when the initial radiation acceleration is the same but the direction of Mercury's radial motion relative to the Sun is different. The Sun is downward in this representation, and Mercury's shadow is shown by the gray region extending upward. Na atoms experience no radiation acceleration while in Mercury's shadow. Na atoms are ejected radially outward from Mercury with an initial speed with respect to Mercury of $v_{0,Na} = 2.5$ km s^{-1}. The radiation acceleration for each case is 100 cm s^{-2}, but the left panel is during Mercury's outbound phase (TAA = 129°) with Mercury's distance from the Sun increasing (radial velocity $v_{r,M} = 7.8$ km s^{-1}), and the left panel is during the inbound phase (TAA = 229°) with Mercury's distance decreasing ($v_{r,M} = -7.6$ km s^{-1}).

affected, although the short photoionization lifetime (discussed below) results in a large fraction of Ca atoms being photoionized before radiation acceleration effects become significant.

Potter et al. (2007) showed that radiation acceleration affects the trajectories of Na atoms differently, for the same initial magnitude of radiation acceleration, at different parts of Mercury's orbit. On the outbound portion of Mercury's orbit (0° < TAA < 180°, when Mercury's distance from the Sun is increasing), there is positive feedback such that the radiation acceleration of Na atoms increases as their heliocentric radial velocity increases. On the inbound portion of the orbit where heliocentric velocity is negative (180° < TAA < 360°), the opposite occurs: there is negative feedback that retards the heliocentric acceleration of the Na atoms. These feedback effects occur because of the deep solar Fraunhofer lines that dominate the shape of the solar spectrum around the wavelengths of Na D transitions. In general, a_{rad} increases with the magnitude of the radial velocity toward or away from the Sun. During the outbound leg, Na atoms at rest with respect to Mercury have a net positive radial velocity relative to the Sun (i.e., they are moving away from the Sun). Radiation acceleration is in the same direction as the atom's initial motion, so the radial velocity increases, which in turn increases the radiation acceleration, resulting in positive feedback. During the inbound leg, Na atoms at rest with respect to the planet are moving with a net speed toward the Sun (a negative radial velocity). The radiation acceleration is directed in the opposite direction so the atom's speed toward the Sun decreases, decreasing the radiation acceleration and setting up negative feedback. This pattern is illustrated in Figure 15.4, which shows the trajectories of Na atoms at times when a_{rad} for an atom at rest relative to Mercury is the same but Mercury's radial velocity relative to the Sun is in opposite directions. It can be seen that it is easier for atoms to escape down the tail on the outbound leg (left panel) than the inbound leg (right panel) for atoms ejected well below Mercury's escape velocity (4.25 km s^{-1}).

15.2.2.2 Photoionization

Because photoionization rates depend on the incident solar flux at Mercury, they are a strong function of Mercury true anomaly angle and slightly dependent on the solar cycle. Photoionization lifetimes (the inverse of the photoionization rate) for Na, Ca, and Mg over the course of a Mercury year are given in Figure 15.5; these lifetimes are based on rates at 1 AU presented by Huebner et al. (1992) and Huebner and Mukherjee (2015). It can be seen that Mg has the longest lifetime, up to 100 h, whereas Ca has the shortest, with a lifetime under an hour.

There are uncertainties in the exact photoionization lifetime for each species. Huebner et al. (1992) presented two significantly different values for the photoionization rates of Na based on experimental (red line in Figure 15.5a) and theoretical (black line) cross sections, with a preference toward the theoretical cross section. Although both sets of cross sections have been used in Mercury models (e.g., Leblanc et al., 2003; Burger et al., 2010), calculations by Combi et al. (1997) and observations of the comet Hale–Bopp Na tail by Cremonese et al. (1997) were both consistent with the theoretical value published by Huebner et al. (1992) [see Killen et al. (2007) for further discussion]. More recently, Huebner and Mukherjee (2015) published an updated photoionization rate for Na using cross sections from two databases from Verner and coworkers (Verner et al., 1993, 1996; Verner and Yakovlev, 1995) and TOPbase (http://cdsweb.u-strasbg.fr/topbase/xsections.html). The new cross sections imply a ~20% adjustment in the Na ionization rate compared with the theoretical value given by Huebner et al. (1992).

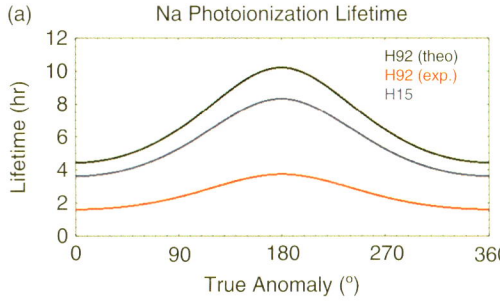
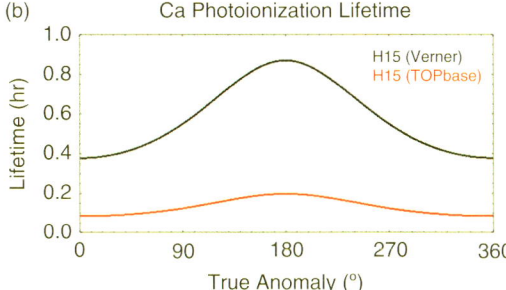
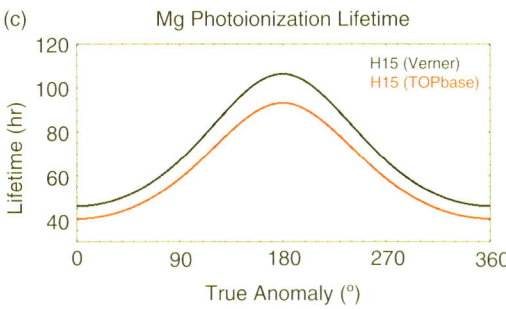

Figure 15.5. Photoionization lifetimes as a function of true anomaly angle for (a) Na, (b) Ca, and (c) Mg. References: H92: Huebner et al. (1992); H15: Huebner and Mukherjee (2015). For Na, Huebner et al. (1992) gave theoretical and experimental values for the cross section. Huebner and Mukherjee (2015) gave cross sections from two groups: Verner and co-workers (Verner et al., 1993, 1996; Verner and Yakovlev, 1995), and TOPbase. See text for further discussion.

The Ca photoionization rate is even more uncertain. From a pre-publication personal communication from W. Huebner to R. M. Killen, Burger et al. (2012, 2014) used the ionization rate associated with Verner and co-workers in Huebner and Mukherjee (2015) (black line in Figure 15.5b). Huebner and Mukherjee (2015) favored the value from TOPbase (red) as Verner and co-workers had smoothed over resonances near the threshold energy. This TOPbase ionization rate is 4.4 times the Verner rate. Had Burger et al. (2014) used this more recent value, they would have required a significantly higher temperature for the Ca than the 70,000 K their model required, possibly as high as >150,000 K. Unfortunately, the uncertainties in the cross sections are unclear from Huebner et al. (1992). The Mg ionization rate also shows a discrepancy between the two databases used by Huebner and Mukherjee (2015), although it is only ~15%.

15.2.2.3 Kinetic Escape

An atom will escape from the gravity well of a planetary body when its velocity exceeds the escape velocity:

$$v_{\rm esc} = \left(\frac{2GM}{r}\right)^{1/2}, \qquad (15.10)$$

where G is the gravitational constant, M is the mass of the planet, and r is the radial distance from the planet center. Because the escape velocity decreases as the distance from the planet center increases, the rate of Jeans escape (atom cm^{-2} s^{-1}), the escape from a planetary atmosphere by thermal evaporation, depends on the exobase distance. Jeans escape is derived for thermal escape from a planetary atmosphere and is thus not strictly applicable to a surface-bounded exosphere, but the formulation is sometimes used as an approximation. The velocity distribution is assumed to be Maxwellian. The Jeans flux is given by

$$F_{\rm Jeans}(r) = \frac{N(r)v_{\rm m}}{2\pi^{1/2}} e^{-\lambda}(\lambda + 1), \qquad (15.11)$$

where $v_{\rm m}$ is the most probable velocity of a Maxwellian distribution,

$$v_{\rm m} = \left(\frac{2kT}{m}\right)^{1/2}, \qquad (15.12)$$

T is the temperature, m is the atomic mass, $N(r)$ is the number density at radius r, and λ is given by

$$\lambda = \frac{v_{\rm esc}^2}{v_{\rm m}^2} = \frac{GMm}{rkT}, \qquad (15.13)$$

and is generally referred to as the escape parameter (essentially the gravitational potential energy in units of kT). It is useful to look at gravitational escape as a function of planetary mass, temperature, and mass of the atomic species as shown in Figure 15.6.

Even for water, the fraction of released vapor that escapes at 3000 K is less than 25% on the initial trajectory. Therefore, for metals (Na, Mg), escape or permanent loss is expected to result from photoionization or some process that produces a very hot vapor, such as sputtering or the high-velocity tail of the impact vapor. For Na, radiation pressure enhances escape at some true anomaly angles, but radiation pressure is insignificant for Mg. Nevertheless escaping Mg has been observed at Mercury, which necessitates an energetic source process. The extreme temperature of the Ca exosphere (Burger et al., 2014) implies that the Ca is escaping; however, there may be an unobserved molecular component that condenses back to the surface (Killen, 2015).

15.3 MESSENGER OBSERVATIONS AND MODELS

15.3.1 Sodium

The sodium tail – the component of the exosphere that is escaping anti-sunward primarily due to radiation-pressure-induced acceleration and first imaged by Potter et al. (2002) – was imaged by UVVS during each of MESSENGER's three

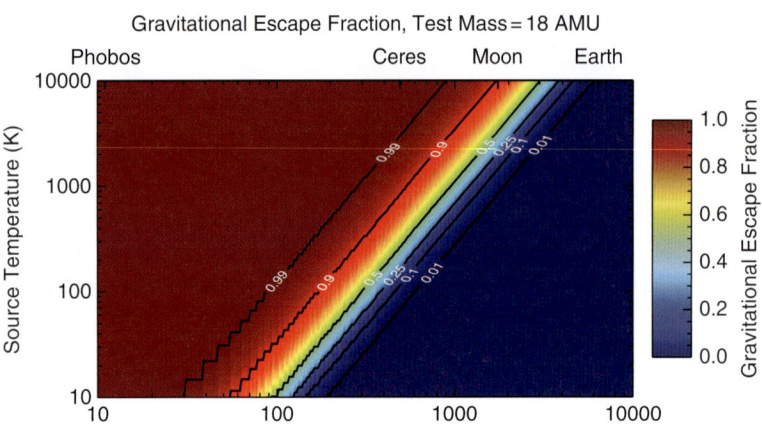
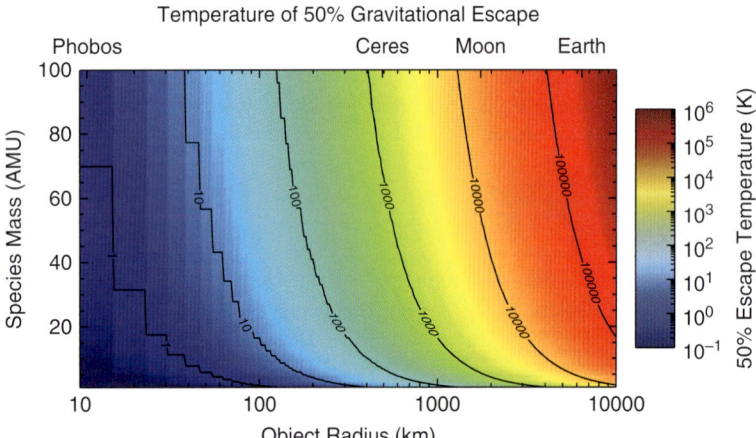
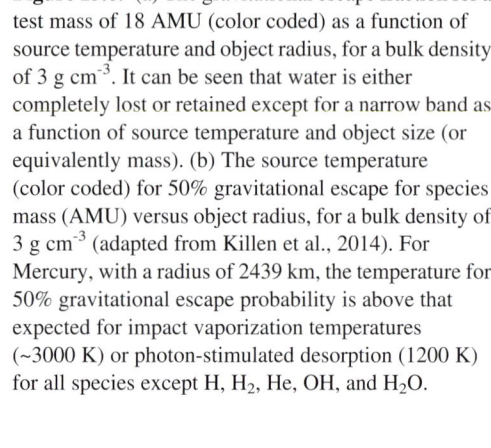

Figure 15.6. (a) The gravitational escape fraction for a test mass of 18 AMU (color coded) as a function of source temperature and object radius, for a bulk density of 3 g cm^{-3}. It can be seen that water is either completely lost or retained except for a narrow band as a function of source temperature and object size (or equivalently mass). (b) The source temperature (color coded) for 50% gravitational escape for species mass (AMU) versus object radius, for a bulk density of 3 g cm^{-3} (adapted from Killen et al., 2014). For Mercury, with a radius of 2439 km, the temperature for 50% gravitational escape probability is above that expected for impact vaporization temperatures (~3000 K) or photon-stimulated desorption (1200 K) for all species except H, H_2, He, OH, and H_2O.

Mercury flybys. The tail exhibited a nearly uniform north–south emission strength during the first and second flybys (Burger et al., 2010). Scaled to the same levels, the tail was practically non-existent during the third flyby. Although the radiation pressure for an atom at rest with respect to Mercury is near its peak for TAA 292° and only beginning to decrease for TAA 326° (see Figure 15.3), there is a negative feedback in the radiation pressure for true anomaly angles >180° such that the effective radiation pressure at TAA 326° is quite weak (e.g., Potter et al., 2007; Cassidy et al., 2015). UVVS observations of the sodium tail confirm the seasonal variation expected from the radiation pressure (Figure 15.7).

McClintock et al. (2009) used a Monte Carlo model (see Burger et al., 2010) to demonstrate that the Na observed during the second MESSENGER flyby was not consistent with a spherically symmetric source of Na but instead showed a northern enhancement. On the third flyby there was no apparent northern enhancement in Na relative to east or west, and only a very slight excess of Na over the south pole relative to the north pole (Vervack et al., 2010). This pattern is in contrast to the large excesses in both Ca and Mg over the north pole relative to the south pole during the third flyby, indicating very different source processes. High-latitude enhancement in the Na surface content (Peplowski et al., 2014) must be considered in future simulations of the Na exosphere, especially for models at high latitudes for which the Na content of the regolith is double that at the equator.

The observations during the orbital phase of MESSENGER were of several general types, termed dayside limb scans, nightside tail sweeps, ride-alongs, and stares. Because of orbital constraints, most of the limb scans have tangent points near the equator or at low latitudes (Figure 14.9). Notable exceptions are the limb scans approximately perpendicular to the spin axis at the south pole. In spite of limited spatial coverage, the UVVS provided unprecedented spatial resolution and observation cadence for more than 16 Mercury years of near-daily observations.

The UVVS observations analyzed by Cassidy et al. (2015) showed year-to-year repeatability: at a given local time and TAA the emissions were nearly identical from one Mercury year to the next. Cassidy et al. (2015) interpreted the UVVS Na limb scan data with a simple model to estimate the temperature and density of the near-surface exosphere within approximately 50–1500 km of the surface. The model accounts for the effects of radiation acceleration and includes single scattering with a uniform phase function. Cassidy et al. (2015) derived a temperature of ~1200 K with some local-time variation (Figure 15.8), including observations over the south polar region. They found that the tangent column density at 300 km above the subsolar point was a factor of 3 higher at aphelion than at perihelion. This difference may be accounted for by the fact that radiation pressure reduces the scale height at perihelion, so this measurement does not give a direct comparison with total abundance. Ground-based observations taken over

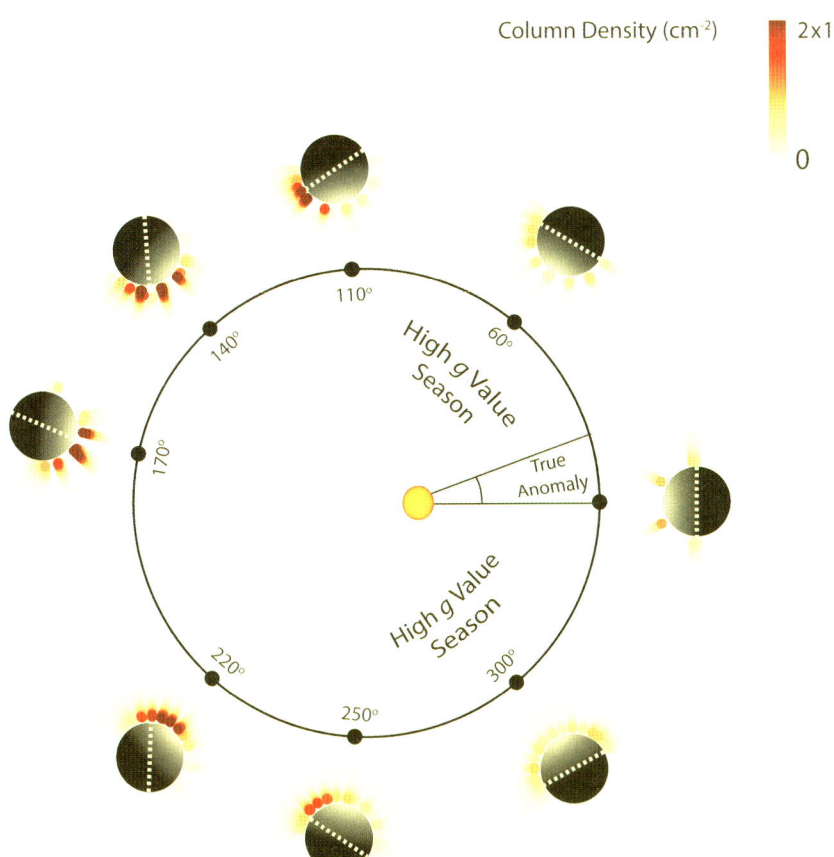

Figure 15.7. Illustration of "seasonal" variations in Mercury's exospheric emission. Observations of Mercury's sodium tail were taken at the indicated true anomalies. Compare with Figure 15.3 for radiation pressure. Note that the shadow region anti-sunward of Mercury prevents resonant emission, so the "streamers" are not related to density. The emission scales as the inverse square of heliocentric distance, so the column abundance scales as the product of the emission and the squared heliocentric distance, a factor of 2.3 increase from aphelion to perihelion.

a period of seven years reported by Potter et al. (2007) showed on average about the same disk-averaged column abundance at aphelion and perihelion, but with a spread of a factor of 3 at aphelion and a much smaller spread at perihelion. These results suggest that photon-stimulated desorption is the primary source process. However, a high-energy process must also be present to produce both seasonal and episodically variable effects seen especially in ground-based data, and to populate the tail (Schmidt, 2012). Vaporization by micrometeoroid impact has been shown to be seasonally repeatable (Killen and Hahn, 2015) and thus represents a seasonal high-energy source that can potentially populate the tail. Impacts by larger meteoroids are rare, and the effects are short-lived (Mangano et al., 2007). Sputtering would produce a high-energy, episodically variable component, but observations of ion sputtering by the UVVS instrument were precluded by instrument noise produced by high-energy charged particles, especially during solar energetic particle (SEP) events.

Given that impact vaporization is almost certainly the most important source of the calcium exosphere (see Section 15.3.2; Killen and Hahn, 2015), it must play a role in imparting Na to the exosphere as well. The Na content, as well as that of most of the other species seen in the exosphere, is spatially variable in the surface grains (Peplowski et al., 2014), ranging from 2.8 wt% at low northern latitudes (0°–15° N) to 4.9 wt% at high northern latitudes (80°–90°N). Given an average Na surface concentration of 2.8 wt%, and the impact vaporization rate from Cintala (1992), the Na impact vaporization rate should be 1.0×10^6 cm^{-2} s^{-1} at aphelion and 2.4×10^6 cm^{-2} s^{-1} at perihelion, corresponding to global rates of 7.5×10^{23} s^{-1} at aphelion to 1.8×10^{24} s^{-1} at perihelion. These values represent 4% of the rate needed to populate the exosphere for a one-bounce model (i.e., the atoms stick on re-contact with the surface) or all of the required rate if the lifetime is the photoionization lifetime (i.e., the atoms do not stick on re-contact with the surface). Assumptions concerning the gas–surface interaction are therefore critical to conclusions concerning the importance of the various source processes. In addition, pre-MESSENGER models all used lunar composition and their results must be scaled. For instance, Mouawad et al. (2011) assumed a Na fraction of 0.5 wt% (i.e., a lunar composition) in deriving a Na impact vaporization rate of 3.5×10^5 cm^{-2} s^{-1}, but they gave an upper limit on the impact vaporization rate from observational data of 2.1×10^6 cm^{-2} s^{-1}, consistent with the impact vaporization rate that would be derived if scaled to MESSENGER surface composition measurements. If the total PSD desorption rate is 3.5×10^{24} s^{-1} (Burger et al., 2010) and the total impact vaporization rate is 1.8×10^{24} s^{-1}, then impact vaporization represents about 35% of the total ejection of Na at Mercury rather than the 2% estimated by Mouawad et al. (2011) on the basis of an assumed lunar composition. If this is the case, then some thermal accommodation must be taking place to maintain the exosphere at 1200–1400 K.

A PSD desorption rate of 3.5×10^{24} s^{-1} was derived by Burger et al. (2010) from the PSD cross section of

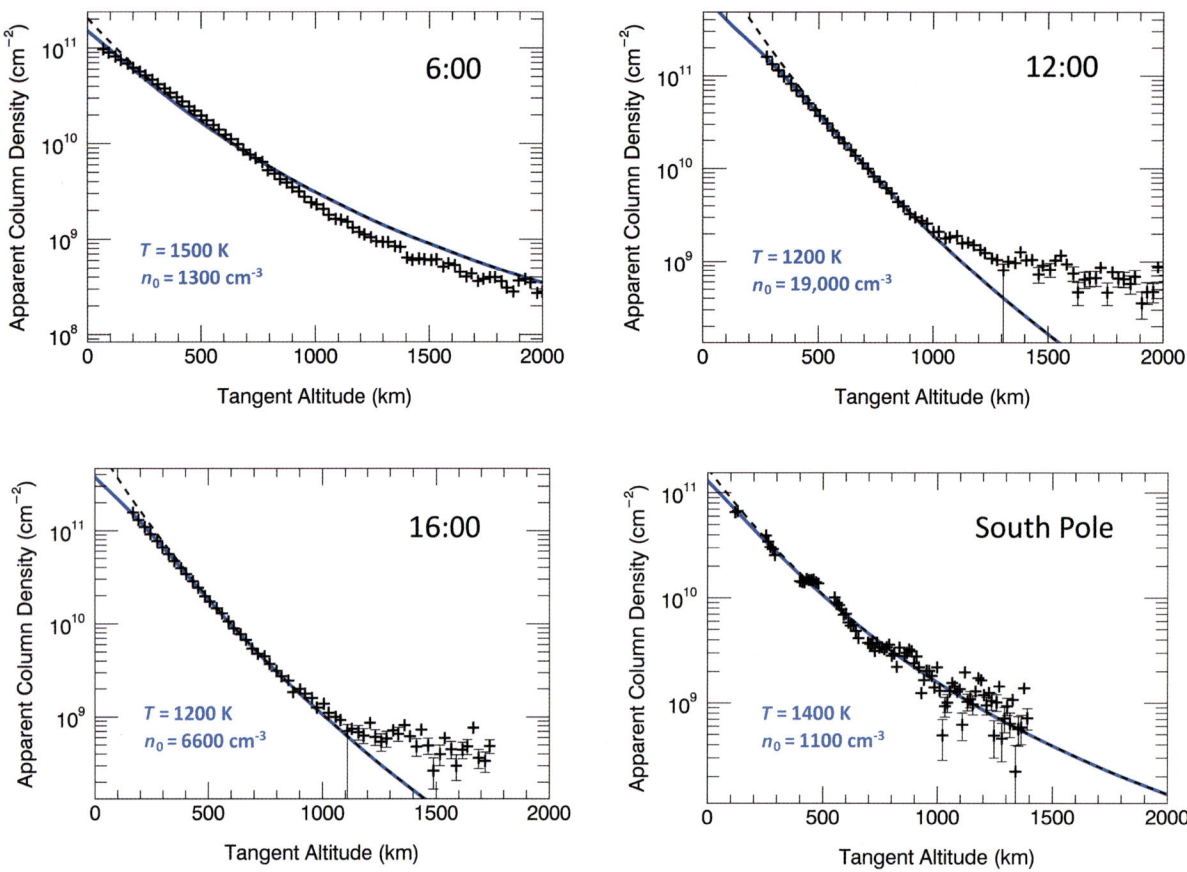

Figure 15.8. Examples of fits to near-equatorial dayside limb scans (local times indicated) and the south pole observations for Na (Cassidy et al., 2015). Only the low-altitude component of the profile was fit. Data are represented by crosses, and the fit is shown by a black dashed line. The blue line is a Chamberlain fit with the optical depth correction. The resulting temperature and surface density used for the fits are indicated by blue text. Observations were taken on 6 June 2012, except for the south pole observation, which took place on 17 October 2011.

Yakshinskiy and Madey (1999), 3×10^{-21} cm^2, and a total surface number density of 7.5×10^{14} cm^{-2} for the first two MESSENGER flybys at a heliocentric distance of 0.35 AU. However, Burger et al. (2010) assumed a lunar abundance of Na, 0.5 wt%, and a surface density of 7.5×10^{14} cm^{-2}. Scaling the PSD source rate to the measured Na wt% would imply a PSD desorption rate of $\sim 1.9 \times 10^{25}$ s^{-1}.

The simulation of Schmidt (2012) yielded a total source rate from impact vaporization in the range 1.8×10^6 to 3.6×10^6 cm^{-2} s^{-1} Na atoms in order to obtain the Na escape rate required to match the tail observations, given that the escape is due to an impact vaporization source. Because the Burger (2010) impact vaporization rate was derived from first principles for a lunar composition (Na wt% = 0.5), scaling that figure by a factor of 5.6 to the Na mass fraction observed by MESSENGER (Na wt% = 2.8) would give an impact vaporization rate of 2×10^6 Na atoms cm^{-2} s^{-1}, consistent with the Schmidt (2012) rate derived by matching the observed Na escaping down the tail.

Mura (2012) found that for a Weibull distribution of velocities (their "fast" distribution) for PSD, up to one-third of the sodium particles escape if the characteristic energy β = 0.086 eV, whereas for the Maxwellian case the loss rate is smaller by a factor of 3 but still accounts for 10% of the source.

Thus, the high-velocity portion of the PSD source distribution can populate Mercury's tail if a high-energy tail to the distribution is assumed. Schmidt et al. (2012) also performed three-dimensional time-dependent modeling of Mercury's extended sodium tail, considering the effects of orbital motion, gas–surface interaction, variable source rates, and spatially non-uniform distributions of the sources (see Section 15.2.1). They concluded that either a combination of a slow impact vaporization source (3000 K) and a "fast" PSD source or a combination of a slow PSD source and a fast (5000 K) impact vaporization source can result in a ~20% loss of the released Na atoms down the tail, depending on orbital phase. They estimated that a loss rate of $\sim 10^{24}$ Na atoms s^{-1} is required to populate the tail on average, and that a sputter source can supply at most 25% of this required rate except in exceptional circumstances.

If the average loss rate of Na atoms is 3.5×10^{23} Na atoms s^{-1}, as derived by Leblanc and Johnson (2003), this average represents a loss rate of 26% of the impact vaporization source. Schmidt et al. (2012) estimated that about 15% of the atoms derived from a slow impact vaporization source or 30% of those from a fast impact vaporization source would escape. Thus the derived loss rate is closer to the fast impact vaporization loss rate than the slow rate, but a combination of impact vaporization and PSD would also fit the observations.

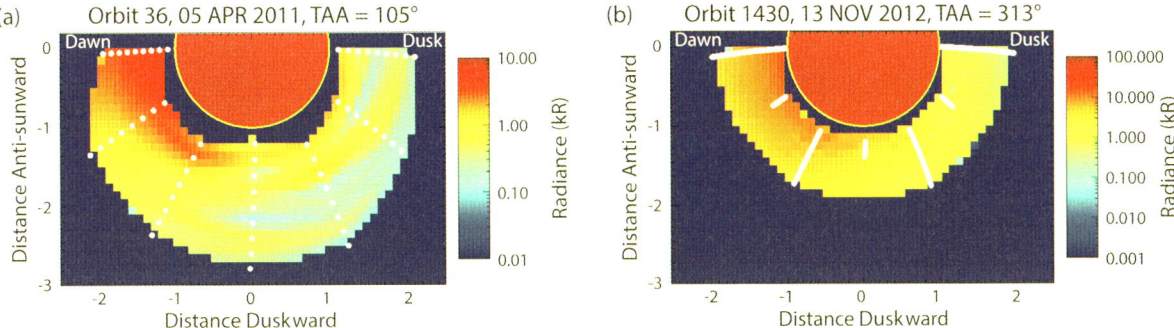

Figure 15.9. Simulated images of Ca emission in Mercury's dayside equatorial plane at two locations in Mercury's orbit were produced by interpolating between observations obtained at the white points approximately perpendicular to the plane of these images (Burger et al., 2014). In each image, dawn is to the left and the Sun is down. The white points represent positions where the UVVS line of sight crosses the equatorial plane; the altitude sampling is higher in the right panel. The images reflect large-scale local-time variations (small-scale variations in the images are not real). Although the magnitude of the emission varies with Mercury true anomaly, Ca is always brightest in the dawn hemisphere, usually, but not always, peaking at dawn.

The robust nature of the year-to-year repeatability and nearly constant near-surface exospheric temperature were unexpected. These findings are surprising in light of the published ground-based observations, which have suggested that sodium is ejected from the surface by a complex mixture of processes. However, because of the geometrical limitations of the UVVS limb scans, with a tangent point almost always at low latitudes on the dayside, a global model is precluded from these data.

15.3.2 Calcium

15.3.2.1 Observations

Prior to MESSENGER, ground-based observations identified Ca mainly over the poles and in the anti-sunward direction (Killen et al., 2005). Significant Ca emission (>2 kR) was seen up to two planetary radii behind the planet. The lines of sight for ground-based observations are limited in their ability to separate source regions in the east–west (i.e., dawn–dusk) direction, and scattering from the surface precluded observations near the dayside. During MESSENGER's Mercury flybys, UVVS observations were able to isolate dawn from dusk, and it was seen that the source of Ca is highly concentrated in the dawn equatorial region (Figure 15.9) (McClintock et al., 2009; Vervack et al., 2010; Burger et al., 2012).

The temperature was also determined to be extremely high on the basis of the scale height of Ca observed above Mercury's poles and by the fact that Ca extends into the tail region despite the short photoionization lifetime (<1 h) (McClintock et al., 2009). Once MESSENGER was in orbit, UVVS began near daily (i.e., at least one orbit per Earth day) observations of Ca. Although its abundance in the exosphere is small compared with those of Na, H, and Mg, Ca is easily detected by UVVS owing to a combination of calcium's high scattering probability at a wavelength where the solar flux is large and the high sensitivity of the instrument at that wavelength.

The calcium in Mercury's exosphere appears to originate from a region near the dawn equatorial point, but the source may move toward mid-morning at some true anomaly angles. Burger et al. (2012) studied the UVVS Ca data from the MESSENGER flybys of Mercury and the first seven orbits of the orbital science phase of the mission. Burger et al. (2014) extended this study to include dayside limb scans from MESSENGER's primary mission (18 March 2011 to 17 March 2012) and first extended mission (18 March 2012 to 17 March 2013). The Ca observations are well fit by a source centered at dawn that drops off exponentially with an e-folding radius of ~50° of arc. The source appeared to be fixed in local time and was not tied to the surface geology. The models required the atomic Ca source to be extremely energetic: the best fit was found to be ~70,000 K under the assumption that it has a thermal flux distribution (a thermal speed of 5.4 km s^{-1}), or a velocity greater than the escape velocity (4.25 km s^{-1}) if it is characterized by a Gaussian distribution. Burger et al. (2014) found a strong seasonal dependence in the source rate (Figure 15.10), although there was little year-to-year variability. The Ca source rate is greatest at a Mercury true anomaly angle of 20°, shortly after perihelion (TAA = 0°), and reaches a minimum just after aphelion (TAA = 180°).

15.3.2.2 Models

Wurz and Lammer (2003) used a Monte Carlo code that treated thermal release, particle sputtering, photon-stimulated desorption, and micrometeoroid impact to develop Mercury exospheric models for a variety of atomic and molecular species. They assumed that Ca is sputtered from the surface because of the large line width measured by Bida et al. (2000), consistent with a thermal temperature of 12,000 K. Killen et al. (2005) suggested that Ca is vaporized in the form of molecules by micrometeoroid impact and is subsequently dissociated in an exothermic process, releasing additional binding energy.

A Monte Carlo model of Mercury's exosphere was successfully developed to model UVVS observations of Na (Burger et al., 2010; Mouawad et al., 2011) and later was extended to Ca (Burger et al., 2012, 2014). The model follows the evolution of atomic and molecular species ejected from Mercury's surface or produced above the surface by molecular dissociation. Source processes were characterized by the initial spatial and energy distributions of test particles; a fifth-order, variable-step-size Runge–Kutta integrator (Press et al., 2007) was used to solve

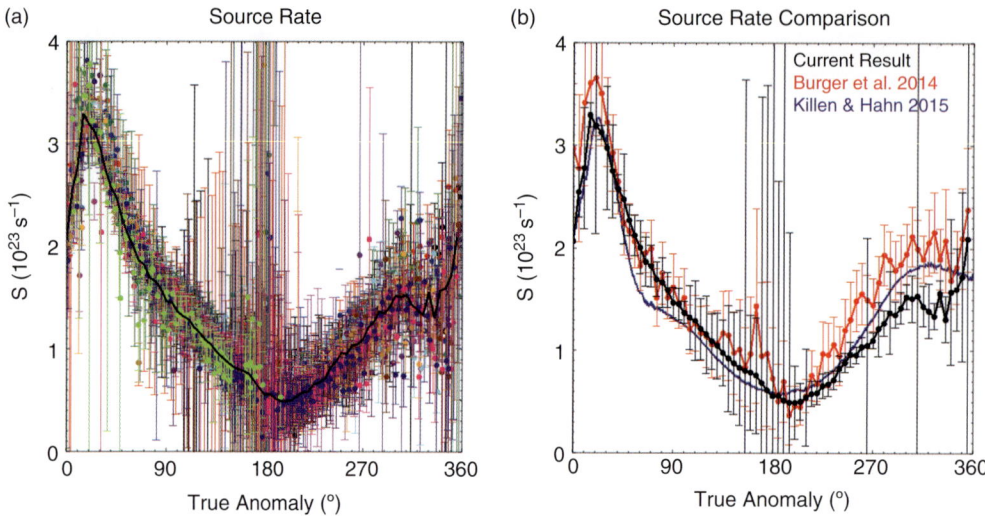

Figure 15.10. (a) Derived source rate S of calcium for a source temperature of 70,000 K and an exponential fall off from the dawn equatorial point with an e-folding distance of 50° of arc along Mercury's surface (Christou et al., 2015). Each point represents a model of a single MESSENGER orbit; 2375 orbits from the first 16 Mercury years of the MESSENGER mission were used. The black line shows the mean source rate in bins of width 5° in true anomaly. (b) Comparison of the mean source rate (black) with that derived by Burger et al. (2014) using only the limb scans from the first nine Mercury years (red) and with the micrometeoroid flux determined by Killen and Hahn (2015) scaled to match the data from Burger et al. (2014) (blue).

the equations of motion for these particles under the forces of gravity from Mercury and radiation acceleration caused by the isotropic scattering of solar photons incident from a single direction. In practice, either a constant-step-size Runge–Kutta integrator or the more computationally intensive, but also more accurate and numerically stable, variable-step-size integrator was used.

Because the g-value (emission probability) depends on the solar flux absorbed by the atom, it is a function of the distance from and radial velocity of the atom relative to the Sun. The Burger et al. (2012, 2014) model determines the g-value at each integration step using the instantaneous radial velocity relative to the Sun as computed by the integrator in order to compute radiation pressure acceleration. The g-values for Ca at 272.2 nm, 422.8 nm, and 456.7 nm were calculated as a function of radial velocity by Killen et al. (2009) (>99% of the radiation acceleration is due to the 422.8 nm line). A linear interpolation between tabulated values was used to calculate the g-value at the heliocentric velocity of the atom. Atomic species were tracked by the integrator until they were lost by photoionization, gravitational escape outside the region of interest (~10 R_M), or sticking to the surface. Charge exchange and electron-impact ionization do not contribute significantly to the loss because of the low plasma densities. Similarly, the only emission mechanism considered was resonant scattering of sunlight. However, the ability to include plasma effects on loss and emission was incorporated in the model. It was assumed that Ca atoms that return to the surface stick with 100% efficiency, although the sticking efficiency and thermal accommodation coefficients are adjustable parameters in the model (Burger et al., 2010; Mouawad et al., 2011).

Burger et al. (2012, 2014) assumed Maxwellian flux distributions as a means of estimating the mean energy of the outward-directed Ca. Burger et al. (2014) also demonstrated that Gaussian flux distributions are consistent with the data when the mean velocity is greater than Mercury's escape velocity. This distribution approximates a dissociation source: dissociating Ca-bearing molecules give the fresh neutral Ca atom an energy boost from the excess energy of the dissociating photon, electron, or ion (e.g., Killen, 2015, 2016).

If atomic Ca is produced from the dissociation of Ca-bearing molecules near the surface, the initial speed distribution will be approximately Gaussian, in the form

$$f_v = e^{-(v-v_p)^2/2\eta^2}, \quad (15.14)$$

where v_p is the most probable speed and η is the width of the distribution.

Electron-stimulated desorption (ESD) has also been suggested as a possible source process for Mercury's exosphere (McLain et al., 2011; Schriver et al., 2011). Burger et al. (2012, 2014) found that ESD is an unlikely Ca source, however, owing to the year-to-year stability of the Ca exosphere compared with the solar-wind interaction with Mercury's magnetosphere and the lack of evidence for precipitating electrons at dawn. Killen (2016) concluded that electron impact dissociation (EID) cannot be responsible for the energization of the Ca due to the small cross section for EID.

The atomic calcium in Mercury's exosphere is quickly ionized by solar UV radiation. Burger et al. (2012, 2014) assumed that the photoionization lifetime varied between 23 min at perihelion and 52 min at aphelion, although there is a large uncertainty in these values, as discussed in Section 15.2.2.2.

Burger et al. (2012, 2014) took a data-fitting approach in their study of Ca observations by MASCS. The goal of these studies was to simulate the observed data from the MESSENGER flybys and orbital phase to determine the spatial and energy distributions of the escaping Ca. Other authors (e.g., Killen and Hahn, 2015; Christou et al., 2015; Killen, 2015, 2016) considered the physical mechanisms responsible for producing the modeled source distributions, concentrating on impact vaporization. A process-driven procedure can be used to determine the expected distributions that MESSENGER would observe for a given process.

Given that the high latitudes were not well observed by the UVVS instrument, the complete mix of Ca source mechanisms is currently somewhat uncertain. Ion sputtering or electron-stimulated desorption were considered unimportant (Burger et al., 2012, 2014) because the magnetosphere is more variable than the data, and ion precipitation is not expected to peak at dawn (Kallio et al., 2008; Benna et al., 2010). Pfleger et al. (2015) computed three-dimensional exosphere models for sputtering by solar-wind H^+ and He^{++} and concluded that solar-wind-sputtered Ca could provide a minor population with respect to the MESSENGER observations. The average dayside number density of Ca produced by solar-wind sputtering was estimated to be less than 1 cm^{-3}, which would probably not be measured by the UVVS instrument but would not be insignificant relative to the 1–4 cm^{-3} peak Ca density reported by Burger et al. (2014). Pfleger et al. (2015) concluded that a dayside Ca surface density comparable to the long-term observations could be expected from extreme solar events. Unfortunately, the UVVS instrument either went into safe hold or was overwhelmed by noise during extreme solar wind events and could not measure the response of the exosphere to these solar events.

The most promising hypothesis for the production of the Ca exosphere is that the primary source is micrometeoroid impact vaporization, which produces Ca-bearing molecules or ions that quickly dissociate to produce energetic Ca atoms (Killen, 2015). This idea was first proposed by Killen et al. (2005) and was considered by Burger et al. (2014). There are several features that make this proposal attractive. First, recent radar observations and models of micrometeoroids at Earth show that there is a strong dawn enhancement in the impactor flux (Janches et al., 2006; Pifko et al., 2013). Horanyi et al. (2015) reported a dawn–dusk asymmetry in dust at the Moon from LADEE dust observations. They indicated that the dust flux at dawn can be up to a factor of 6 larger than at dusk, in contrast to estimates of a factor of 3 dawn-to-dusk ratio of dust influx at Earth (Pifko et al., 2013).

Calcium-bearing molecules are more likely to be produced in impact vapor plumes than atomic Ca (Berezhnoy and Klumov, 2008), and the molecules expected to be produced (CaO, CaOH, and/or $Ca(OH)_2$, depending on the temperature of the vapor plume) quickly dissociate to release atomic Ca (Berezhnoy and Klumov, 2008; Berezhnoy, 2013). Killen (2015) estimated that the most likely mechanism for creating high-energy atomic Ca is dissociative ionization of precursor molecules. Killen and Hahn (2015) showed that the total Ca source rate is consistent with the dust flux expected as Mercury traverses the interplanetary dust disk and, in addition, intersects a cometary dust stream near true anomaly 25° (Figure 15.11), possibly associated with comet 2P/Encke (Christou et al., 2015). They did not estimate probable dawn–dusk asymmetries.

The exospheric Ca source is unlikely to be adsorbed Ca-bearing material on the nightside that thermally desorbs as it enters sunlight, because the models of Burger et al. (2012) required a pre-dawn Ca source and because the temperature required to vaporize Ca is over 3000 K.

Pfleger et al. (2015) concluded that sputtering could be responsible for an important fraction of the Ca exosphere, slightly below detectability by the UVVS instrument. This finding might explain the Ca observed over Mercury's poles in ground-based spectra (e.g., Killen et al., 2005) and possibly the enhancement of Ca over the poles during the third MESSENGER flyby (Vervack et al., 2010).

15.3.3 Magnesium

The discovery measurements of magnesium (Mg) obtained during MESSENGER's second Mercury flyby were modeled to constrain the source and loss processes for this species (Killen et al., 2010; Sarantos et al., 2011). These measurements provided a unique opportunity for constraining the portion of the exosphere produced by energetic processes, because they probed ejecta at very large downtail distances that could not be probed again after MESSENGER's orbit insertion. The flyby data were matched to Chamberlain models (Killen et al., 2010; Sarantos et al., 2011) and non-uniform sputtering models (Sarantos et al., 2011) in which transport was computed from Liouville's theorem (i.e., gravity forces were included but not radiation pressure or losses to photoionization). Sarantos et al. (2011) found that the distribution of magnesium with tail distance is suggestive of possibly two energetic ejection processes, because the superposition of a hot source ($T_{HOT} \geq 20,000$ K) with a cooler source ($T_{COOL} \leq 5000$ K) improved the description of emission detected in the near and far tail (Figure 15.12). The more energetic component populating the tail required higher rates by a factor of at least 5 than ion sputtering could provide under the solar-wind conditions prevailing during the flyby (Sarantos et al., 2011). This conclusion is subject to some uncertainty, however, both from modeling access of the solar wind to the surface and from the fact that the g-value was used for atoms at rest with respect to Mercury.

Dayside limb scans and near-tail measurements during MESSENGER's orbital phase provide evidence for a non-uniform Mg source. Analysis of individual limb scans with Chamberlain fits provides convincing evidence for a dawn–dusk asymmetry with peak dayside abundances at local times of 8–10 h (Merkel et al., 2017). Given only limb scan data, the necessity for two temperature components on the dayside is unclear, as these data can be fit with a single source at 5000 K on most days. The source of exospheric Mg appears to originate near dawn or in the mid-morning (Figure 15.13).

The main conclusion from Sarantos et al. (2011) was that the total amount of magnesium at altitudes exceeding 100 km is consistent with predictions from impact vaporization models if micrometeoroid impacts eject both Mg atoms and Mg-bearing molecules (e.g., MgO, MgS) with molecular dissociation lifetimes of no more than 2 min (Figure 15.14). It is conceivable that reactions taking place during a fireball expansion result in at least half of the magnesium being bound in molecules (Berezhnoy and Klumov, 2008). Longer dissociation lifetimes would require a mix of hot atoms from dissociating molecules and fast atoms from sputtering. Observations taken at low

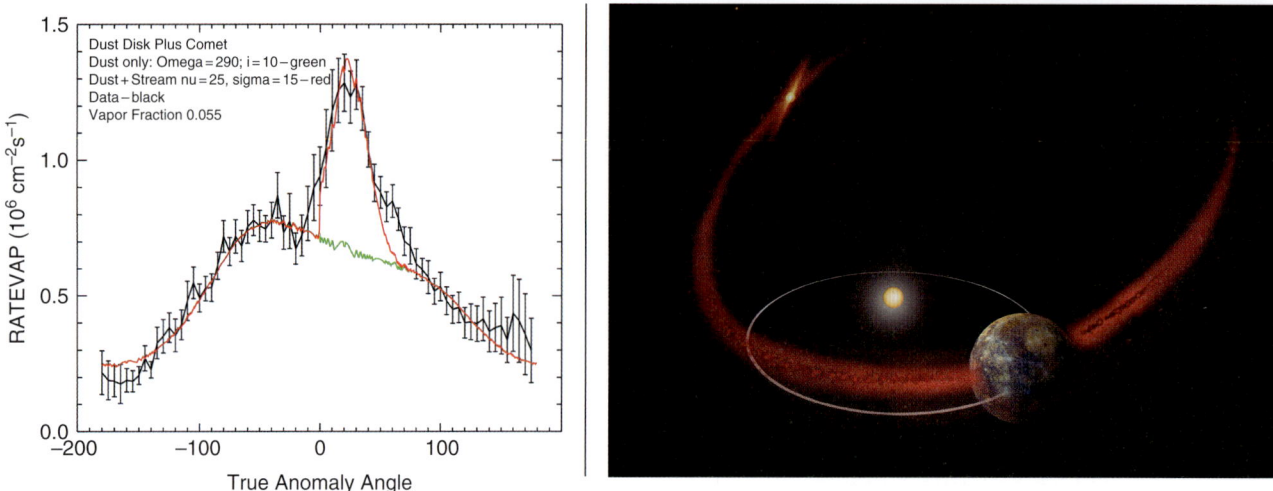

Figure 15.11. (Left) Derived source rate of Ca from the first nine Mercury years of UVVS observations plotted as a function of Mercury true anomaly angle, along with the modeled hot component of impact vaporization by the flux from interplanetary dust disk onto Mercury (green) with the addition of dust from the comet 2P/Encke centered near 25° (red) (Killen and Hahn, 2015). Away from 0–70° TAA the red line follows the dust-disk-only curve. (Right) Artist's conception of comet dust impacting Mercury (not to scale).

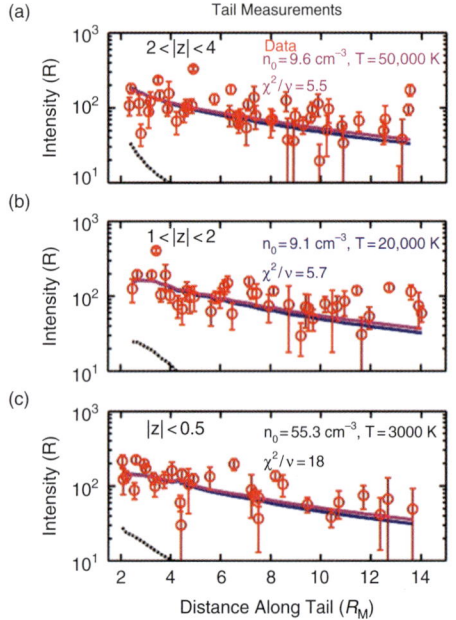

Figure 15.12. Magnesium 285-nm cross-tail line-of-sight radiances (red circles) measured down the tail to 14 Mercury radii (R_M) are compared at vertical distances from the equator of (a) (2–4) R_M (b) (1–2) R_M, and (c) (−0.5–0.5) R_M. Models based on source temperatures 50,000 K (magenta) and 20,000 K (blue) are indistinguishable, and both fit the data. The model prediction for a 3000 K source lies well below the data for all vertical distances. From Sarantos et al. (2011).

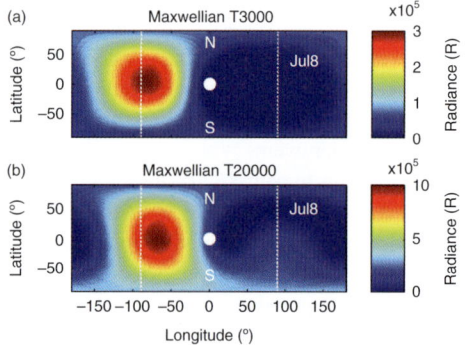

Figure 15.13. The derived Mg flux from the surface (atoms cm^{-2} s^{-1}) on 8 July 2012, under the assumption of a Maxwellian velocity distribution for released ejecta. The data are consistent with a warm component at 3000 K covering the dawn hemisphere and a hot component at about 20,000 K in the morning quadrant. The white dot denotes the subsolar point, and the dawn terminator is at −90° longitude (vertical dotted line to the left). Other fits are possible, including a single source at 5000 K. Adapted from Sarantos et al. (2012).

spacecraft altitudes just nightward of the dawn terminator region and before crossing to the dayside could be interpreted as indicative of thermally accommodated particles (Sarantos et al., 2011). Data from the third MESSENGER flyby, which could provide equally important constraints for hot processes, have not been modeled. Both Mg and Ca exhibit a very hot component, but that for Ca (~70,000 K) (Burger et al., 2014) is more energetic than that for Mg (~20,000 K). However, this temperature was derived under the assumption of a short photoionization lifetime. Both Ca and Mg require some process in addition to impact vaporization to provide the additional energy, and their observed energies are consistent with the dissociation of a precursor molecule (e.g., Killen, 2015) or sputtering. Presumably a sputtered component should be seen at high latitudes (Pfleger et al., 2015), in contrast to the observed distributions peaking at the equator (Sarantos et al., 2012). There is no evidence for a 3000–5000 K component for the Ca exosphere as there is for Mg.

Sarantos et al. (2012) fit limb scans and tail data from the first three Mercury years after MESSENGER's orbit insertion by employing a Monte Carlo model that included the effects of photoionization, radiation pressure, and velocity-dependent

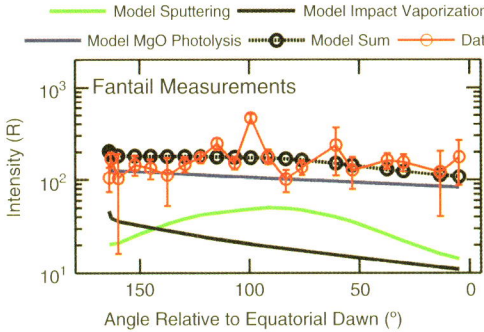

Figure 15.14. Fantail measurements of the Mg emission during the second MESSENGER flyby (McClintock et al., 2009) are best fit with a model of MgO dissociation (Sarantos et al., 2011). The addition of smaller contributions from impact vaporization and sputtering yields a better total fit to the details of the observations (red circles with error bars).

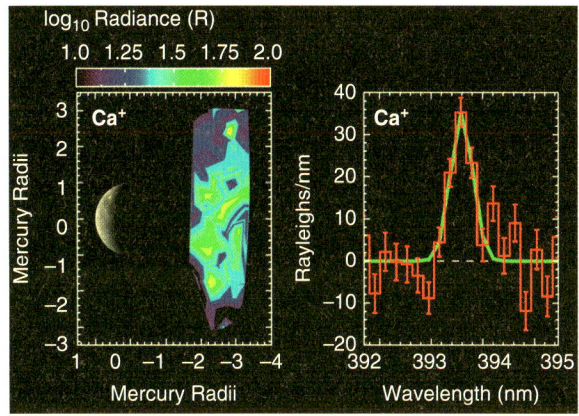

Figure 15.15. (Left) This pseudo-image of the Ca$^+$ emission detected during the third MESSENGER flyby was generated by projecting the observed column emissions onto a plane containing the Sun–Mercury line and the planet's spin axis and interpolating to fill in unobserved regions. (Right) The red spectrum represents the average of all the Ca$^+$ emission-line observations (one-standard-deviation error bars are shown) during the third flyby; the green line is a Gaussian fit to the average Ca$^+$ line. Adapted from Vervack et al. (2010).

g-values. They used combinations of ejection processes with Maxwellian and sputtered initial distributions. The three-dimensional output from each tested source process was discretized in a way that enabled searches for non-uniform ejection in a systematic manner. A Monte Carlo method was used to eject test particles within each of 2000 surface patches for each assumed source temperature. These particles, which were traced until photoionization or contact with the surface, represent the mapping of a "unit flux" moving from one surface element into the three-dimensional (3D) volume. Output from the modeled surface elements was saved separately, and a linear combination of the modeled intensities from each was fit to the emission data using a penalized least-squares regression method. The retrieved unknown variables are the fluxes leaving each surface patch. Such a method was used to estimate the best spatial release pattern that fit the data each day under the assumption of a given velocity distribution function for released ejecta.

15.3.4 Ionized Calcium

Ionized calcium (Ca$^+$) was observed close to the equatorial plane in a relatively small region approximately (1–2) R_M tailward of the planet on the third MESSENGER flyby (Vervack et al., 2010), as shown in Figure 15.15. Similar line-of-sight column densities for Ca$^+$ and Ca were measured, an unlikely phenomenon if Ca$^+$ is produced locally by ionization of Ca because of large differences in their respective observed velocities (Ca, several km s^{-1}; Ca$^+$ up to hundreds of km s^{-1}). The observations occurred tailward of the magnetospheric X-line while the spacecraft was inbound and lend support for the view that a magnetospheric convection pattern led to a concentration of Ca$^+$ in the region behind the X-line before the Ca$^+$ ions were ejected down the tail. UVVS intensities compared favorably with those reported by Bida and Killen (2017) from observations at the Keck 1 telescope of a region (1.7–2.0) R_M from planet center (Section 14.4.5.5). This observation corresponds to about a factor of 5 greater than the three-standard-deviation abundance measured on 15 May 2008 and 3 May 2009 (Bida and Killen, 2011, 2017). Bida and Killen (2017) obtained an upper limit to the Ca$^+$ column equal to 3.9×10^6 cm^{-2} at approximately the same altitude (1630 km above the limb) as the UVVS observation.

The lifetime of Ca against photoionization is short, ~1500 s (Huebner and Mukherjee, 2015), so most exospheric Ca$^+$ ions are formed within 2.5 R_M and the Ca$^+$ pickup process occurs within Mercury's magnetosphere. The Ca$^+$ produced over a relatively large volume might be concentrated by magnetospheric convection into the near-tail equatorial region. Vervack et al. (2010) estimated that although Ca$^+$ would fill the entire width of the magnetotail, only about 65% of the Ca$^+$ was in sunlight where it would have been observable by the UVVS.

15.3.5 Weakly Emitting or Less Abundant Species

Although it was expected that emission from several weaker or less abundant species such as Al, Fe, and Ca$^+$ would be detected with some regularity in Mercury's exosphere, such an expectation was not met. The MESSENGER UVVS observations in particular are interesting in that no emission from Al, Ca$^+$, or Mn was detected during the orbital phase until late in the mission (the final Earth year in orbit) (Vervack et al., 2011, 2015, 2016). These species have been elusive both from the ground and from MESSENGER observations, pointing to a high level of variability. Even more interesting is that all of the detections of these species have been confined to a particular spatial and temporal pattern. Spatially, all the detections of these three species by UVVS during the orbital phase of MESSENGER occurred in the pre-dawn nightside region (local times around 2–5 h). Also, they were detected only during the outbound leg of Mercury's orbit, between true anomaly angles of 0° and 70° (Vervack et al., 2016). Aluminum was detected from ground-based observations with the Keck I telescope at TAA 103° and 117°, farther from perihelion but still on the outbound leg of the orbit (Bida and Killen, 2017).

The location and timing of these detections is highly suggestive of a connection to the comet 2P/Encke dust stream that was proposed by Killen and Hahn (2015) to explain the spike in the Ca emission over similar true anomaly angles (see Section 15.3.2.2). Dust from Encke impacting the dawn side of the planet may have led to an enhancement in these species to levels that were detectable by UVVS. It is also possible that some of the material may be cometary in origin. The marked difference in the observed profile for Mn (Figure 14.45) compared with those for Al and Ca^+ may indicate that Mn perhaps derives from the cometary dust rather than the surface of Mercury itself.

The Encke stream may also explain why these species were not detected by UVVS earlier in the mission, despite observations at the same general location and time. Due to the eccentricity of Mercury's orbit, Mercury's terminator moves backward (toward the nightside) from a TAA of about 340° to TAA 20°. A delayed source as a result of this rock-back of the terminator pre-perihelion would exhibit the same source rate every Mercury year, counter to the observations. Comet streams at Earth are responsible for meteor showers, and occasionally there are larger "clumps" of dust that lead to more spectacular meteor storms. Such behavior is not uncommon with the Taurids at Earth, which also derive from comet Encke (Jenniskens, 2006). A similar change in the amount of dust that is impacting Mercury in a given year (or years) could account for the onset of detection of these species by UVVS. Furthermore, it may not be coincidental that Encke had a very close passage to Mercury in 2013, and these detections began three Mercury years later.

15.4 CONCLUSIONS AND UNANSWERED QUESTIONS

Observations of exospheric Ca and Mg show that their respective exospheres are very hot, much hotter than can be explained by impact vaporization. Although sputtering can certainly produce vapor that is as hot as that measured for these species, Mercury's magnetosphere generally shields the surface from direct penetration by solar wind except at the cusps (Raines et al., 2013). It is unlikely, on the grounds of both measured sputtering yields and inferred ion fluxes to the surface, that the Ca and Mg exospheres are wholly produced by a sputter source. Killen et al. (2005) suggested that the hot vapor could be produced by dissociation of a molecular precursor. Subsequently Berezhnoy (2013) calculated the equilibrium fraction of various atomic and molecular species produced by impact vaporization and concluded that CaO or $Ca(OH)_2$ would be the most likely calcium-bearing species in the fireball at the quenching temperatures of greater or less than ~3750 K, respectively. Killen (2015) looked at the likely energy of various dissociation mechanisms and concluded that dissociative ionization would be the most likely candidate to produce escaping Ca. There has been no work on the production of hot Mg, which appears to have either a mixture of 3000 K and 20,000 K components, or a single 5000 K component (Sarantos et al., 2012; Merkel et al., 2017). Further analysis of the UVVS Mg observations is required to determine the true energy distribution for Mg and the likely source processes. Further work is also needed to determine the effect of high-mass and high-energy ions that may penetrate to the surface more widely.

The enhanced abundance of Ca seen following perihelion, at a TAA of about 25°–30°, has been attributed to enhanced impact vaporization by a meteoroid shower due to comet 2P/Encke, which is known to cross Mercury's orbit (Killen and Hahn, 2015; Christou et al., 2015). Observations of other species, such as Al, Mn, and Ca^+, may also be enhanced by this cometary dust stream's impact on Mercury. The primarily dawnward location of the Mg source suggests that hot Mg atoms could be produced by the same physical process as the hot Ca. In order for micrometeoroid impacts to be responsible for producing gaseous Mg and Ca around Mercury, models of micrometeoroid precipitation onto Mercury's surface must account for the production of molecules with a dawn–dusk asymmetry. It is likely that the Mg and Ca sources centered near dawn are correlated with impacting dust peaking in Mercury's ram direction, as seen at the Moon (Szalay and Horanyi, 2016). Because the ram direction moves with TAA (Figure 15.16), the dawn sources may shift to slightly pre- or post-dawn locations over the course of a Mercury year. The combined source fluxes of Mg and Ca inferred from orbital phase data locally approach 2×10^6 atoms $cm^{-2} s^{-1}$, consistent with the flyby results, and may be provided by impacts as previously surmised.

Despite targeting observations at wavelengths near the oxygen 130.4 nm triplet for over 16 Mercury years, no detection was found in the UVVS observations. However, the analysis is difficult owing to scattered solar O emission from the dayside. If the impact vapor produces oxides rather than atomic oxygen and if the oxide is dissociated, an extremely hot O corona would be produced. A very tenuous oxygen corona would by its nature be difficult to observe. However, MESSENGER's Fast Imaging Plasma Spectrometer observed a group of ions, including possible constituents O, OH, and H_2O. If a mass spectrometer with higher mass resolution were flown on a future mission, then it would be possible to separate these components and to determine the true oxygen abundance at the spacecraft altitude.

UVVS observations of sodium did not uncover conclusive evidence for a sputtered component, in contrast to expectations from ground-based observations (e.g., Potter et al., 2006; Mangano et al., 2015; Pfleger et al., 2015). One possible explanation is that most of the data over the dayside, and almost all of the data that were analyzed by Cassidy et al. (2015), were taken at low latitudes. UVVS limb scans were most often taken tangent to the equatorial limb while looking from south to north; thus, they would have missed a high-latitude source near the cusp region for the most part. Future observations that fill this gap in coverage would be worthwhile.

High northern latitudes should be targeted in the future to determine whether meteoroid showers expected to result from comets Bradfield (at TAA 130°) and Tempel–Tuttle lead to enhancements in the region of impact. Concurrent measurement of the dust flux onto Mercury, and especially its spatial distribution with respect to leading and trailing hemispheres and expected meteoroid streams, is highly desirable.

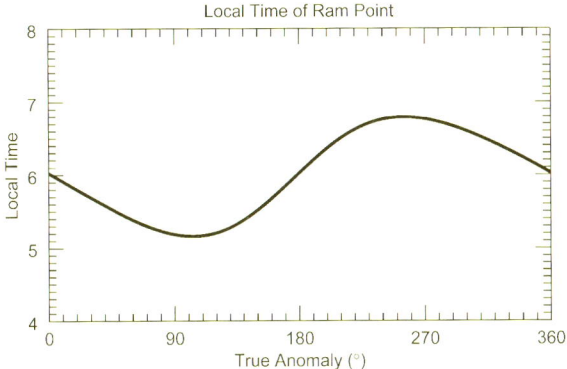

Figure 15.16. The ram point (local time of the orbital velocity vector) moves as a function of true anomaly angle because of the non-uniform relationship between orbital velocity and rotational velocity as a result of Mercury's eccentric orbit.

Mercury's surface composition is highly non-uniform (Weider et al., 2015). A correlation has been found between the local surface composition and the Mg exosphere (Merkel et al., 2018). Further investigations should seek correlations between the exosphere and underlying surface, especially between regions for which there is a large difference in composition.

Although there have been ground-based detections of potassium in Mercury's exosphere from observations of the strong 766.3 nm D_2 line, this wavelength was outside the range of UVVS. Searches for K with UVVS had to use the weaker lines at 404.5 and 404.8 nm (with a combined g-value of <0.07). To date, no K has been detected by UVVS. The Na/K ratio in Mercury's exosphere has been reported to be quite high, but this result may be due to a difference in the spatial distribution of the two species. Doressoundiram et al. (2010) reported Na/K in Mercury's exosphere to be between 80 and 400. Killen et al. (2010) reported Na/K between 22 and 49 at times when the signal-to-noise ratio of the K observation was >5. The large variations in the measured ratios are not surprising in light of the large variation in the surficial K abundance, reported by Peplowski et al. (2012) to be between 300 and 2400 ppm, whereas the Na wt% varies by only about a factor of ~2 from 2.6 wt% in the equatorial regions to 5 wt% at high northern latitudes (Peplowski et al., 2014).

A factor of approximately 2.3 was found between the surface Na/Si weight ratios in the northern polar and equatorial regions, respectively, from MESSENGER Gamma-Ray Spectrometer (GRS) data (Peplowski et al., 2014). Peplowski et al. sought to determine whether the sodium could have been redistributed through desorption in equatorial regions and redeposition and cold trapping near the poles. If redistribution is not an important process, then the composition of the polar regions may represent the original Na abundance in that region. Invoking the assumption that the Na content is lowest in regions that experience maximum near-surface temperatures above 400 K and from temperature maps at the surface and at a depth of 7 cm (Vasavada et al., 1999), Peplowski et al. (2014) concluded that the measured Na distribution supports the hypothesis that the equatorial Na abundances are consistent with depletion via thermal modification similar to that proposed for K by Peplowski et al. (2012), who found a correlation between the distribution of measured surficial K abundances and models of the maximum temperature. Cassidy et al. (2016), on the other hand, found a correlation between Na abundance in the dawn exosphere and the equatorial temperature at Mercury's hot and cold poles. Other species, notably Mg and Ca, have highly heterogeneous distributions of surface composition (Weider et al., 2015).

REFERENCES

Benna, M., Anderson, B. J., Baker, D. N., Boardsen, S. A., Gloeckler, G., Gold, R. E., Ho, G. C., Killen, R. M., Korth, H., Krimigis, S. M., Purucker, M. E., McNutt, R. L., Raines, J. M., McClintock, W. E., Sarantos, M., Slavin, J. A., Solomon, S. C. and Zurbuchen, T. H. (2010). Modeling of the magnetosphere of Mercury at the time of the first MESSENGER flyby, *Icarus*, **209**, 3–10, doi:10.1016/j.icarus.2009.11.036.

Berezhnoy, A. A. (2013). Chemistry of impact events on the Moon. *Icarus*, **226**, 205–211, doi:10.1016/j.icarus.2013.05.030.

Berezhnoy, A. A. and Klumov, B. A. (2008). Impacts as sources of the exosphere on Mercury. *Icarus*, **195**, 511–522, doi:10.1016/j.icarus.2008.01.005.

Bida, T. A. and Killen, R. M. (2011). Observations of Al, Fe and Ca$^+$ in Mercury's exosphere. *EPSC-DPS Joint Meeting Abstracts and Program*, **6**, EPSC-DPS2011-1621. European Planetary Science Congress–Division for Planetary Sciences Joint Meeting, Nantes, France, October 2–7.

Bida, T. A. and Killen, R. M. (2017). Observations of the minor species Al and Fe in Mercury's exosphere. *Icarus*, **289**, 227–238, doi:10.1016/j.icarus.2016.10.019.

Bida, T. A., Killen, R. M. and Morgan, T. H. (2000). Discovery of calcium in Mercury's atmosphere. *Nature*, **404**, 159–161.

Borin, P., Bruno, M., Cremonese, G. and Marzari, F. (2010). Estimate of the neutral atoms' contribution to the Mercury exosphere caused by a new flux of micrometeoroids. *Astron. Astrophys.*, **517**, A89, doi:10.1051/0004-6361/201014312.

Burger, M. H., Killen, R. M., Vervack, R. J., Jr., Bradley, E. T., McClintock, W. E., Sarantos, M., Benna, M. and Mouawad, N. (2010). Monte Carlo modeling of sodium in Mercury's exosphere during the first two MESSENGER flybys. *Icarus*, **209**, 63–74.

Burger, M. H., Killen, R. M., McClintock, W. E., Vervack, R. J., Jr., Merkel, A. W., Sprague, A. L. and Sarantos, M. (2012). Modeling MESSENGER observations of calcium in Mercury's exosphere. *J. Geophys. Res.*, **117**, E0L11B, doi:10.1029/2012JE004158.

Burger, M. H., Killen, R. M., McClintock, W. E., Merkel, A. W., Vervack, R. J., Jr., Cassidy, T. A. and Sarantos, M. (2014). Seasonal variations in Mercury's dayside calcium exosphere. *Icarus*, **238**, 51–58, doi:10.1016/j.icarus.2014.04.049.

Cassidy, T. A. and Johnson, R. E. (2005). Monte Carlo model of sputtering and other ejection processes within a regolith. *Icarus*, **176**, 499–507.

Cassidy, T. A., Merkel, A. W., Burger, M. H., Sarantos, M., Killen, R. M., McClintock, W. E. and Vervack, R. J., Jr. (2015). Mercury's seasonal sodium exosphere: MESSENGER orbital observations. *Icarus*, **248**, 547–559, doi:10.1016/j.icarus.2014.10.037.

Cassidy, T. A., McClintock, W. E., Killen, R. M., Sarantos, M., Merkel, A. W., Vervack, R. J., Jr. and Burger, M. H. (2016). A cold-pole enhancement in Mercury's sodium exosphere. *Geophys. Res. Lett.*, **43**, 11,121–11,128.

Christou, A. A., Killen, R. M. and Burger, M. H. (2015). The meteoroid stream of comet Encke at Mercury: Implications for MErcury Surface, Space ENvironment, GEochemistry, and Ranging observations of the exosphere. *Geophys. Res. Lett.*, **42**, 7311–7318, doi:10.1002/2015GL065361.

Cintala, M. J. (1992). Impact-induced thermal effects in the lunar and Mercurian regoliths. *J. Geophys. Res.*, **97**, 947–973.

Combi, M. R., DiSanti, M. A. and Fink, U. (1997). The spatial distribution of gaseous atomic sodium in the comae of comets: Evidence for direct nucleus and extended plasma sources. *Icarus*, **130**, 336–354, doi:10.1006/icar.1997.5832.

Cremonese, G. and Verani, S. (1997). High resolution observations of the sodium emission from the Moon. *Adv. Space Res.*, **19**, 1561–1569, doi:10.1016/S0273-1177(97)00369-4.

Crider, D. H. and Vondrak, R. R. (2003). Space weathering effects on lunar cold trap deposits. *J. Geophys. Res.*, **108** (E7), 5079, doi:10.1029/2002JE002030.

Domingue, D. L., Chapman, C. R., Killen, R. M., Zurbuchen, T. H., Gilbert, J. A., Sarantos, M., Benna, M., Slavin, J. A., Schriver, D., Trávníček, P. M., Orlando, T. M., Sprague, A. L., Blewett, D. T., Gillis-Davis, J. J., Feldman, W. C., Lawrence, D. J., Ho, G. C., Ebel, D. S., Nittler, L. R., Vilas, F., Pieters, C. M., Solomon, S. C., Johnson, C. L., Winslow, R. M., Helbert, J., Peplowski, P. N., Weider, S. Z., Mouawad, N., Izenberg, N. R. and McClintock, W. E. (2014). Mercury's weather-beaten surface: Understanding Mercury in the context of lunar and asteroidal space weathering studies. *Space Sci. Rev.*, **181**, 121–214, doi:10.1007/s11214-014-0039-5.

Doressoundiram, A., Leblanc, F., Foellmi, C., Gicquel, A., Cremonese, G., Donati, J.-F. and Veillet, C. (2010). Spatial variations of the sodium/potassium ratio in Mercury's exosphere uncovered by high-resolution spectroscopy. *Icarus*, **207**, 1–8, doi:10.1016/j.icarus.2009.11.020.

Dzioba, S. and Kelly, R. (1980). On the kinetic energies of sputtered excited particles II. Theory and applications to group IIA fluorides. *Surf. Sci.*, **100**, 119–134.

Farrell, W. M., Hurley, D. M. and Zimmerman, M. I. (2015). Solar wind implantation into lunar regolith: Hydrogen retention in a surface with defects. *Icarus*, **255**, 116–126, doi:10.1016/j.icarus.2014.09.014.

Farrell, W. M., Stubbs, T. J., Vondrak, R. R., Delory, G. T. and Halekas, J. S. (2007). Complex electric fields near the lunar terminator: The near-surface wake and accelerated dust. *Geophys. Res. Lett.*, **34**, L14201, doi:10.1029/2007GL029312.

Gerasimov, M. V., Dikov, Yu. P., Yakolev, O. I., Wlotzka, F. and Distler, V. V. (1999). The volatility of microelements during impact vaporization of meteorites. *Meteorit. Planet. Sci.*, **34**, Suppl., A42.

Hofer, W. O. (1991). Angular, energy, and mass distribution of sputtered particles. In *Sputtering by Particle Bombardment III*, ed. R. Behrisch and K. Wittmaack, Topics in Applied Physics, Vol. 64. Berlin: Springer-Verlag, pp. 15–90.

Holmlid, L. (2006). The alkali metal atmospheres on the Moon and Mercury: Explaining the stable exospheres by heavy Rydberg matter clusters. *Planet. Space Sci.*, **54**, 101–112, doi:10.1016/j.pss.2005.10.005.

Horanyi, M., Szalay, J. R., Kempf, S., Schmidt, J., Grun, E., Srama, R. and Sternovsky, Z. (2015). A permanent, asymmetric dust cloud around the Moon. *Nature*, **522**, 324–326, doi:10.1038/nature14479.

Horanyi, M., Szalay, J. R., Grun, E., Glenar, D., Wang, X. and Zakharov, A. (2016). The dust environment of the Moon. *New Views of the Moon*, abstract 6005. Houston, TX: Lunar and Planetary Institute.

Hornung, K., Malama, Y. G. and Kestenboim, K. S. (2000). Impact vaporization and ionization of cosmic dust particles. *Astrophys. Space Sci.*, **274**, 355–363.

Huebner, W. F. and Mukherjee, J. (2015). Photoionization and photodissociation rates in solar and blackbody radiation fields. *Planet. Space Sci.*, **106**, 11–45, doi:10.1016/j.pss.2014.11.022.

Huebner, W. F., Keady, J. J. and Lyon, S. P. (1992). Solar photo rates for planetary atmospheres and atmospheric pollutants. *Astrophys. Space Sci.*, **195**, 1–294.

Hunten, D. M. and Sprague, A. L. (1997). Origin and character of the lunar and mercurian atmospheres. *Adv. Space Res.*, **19**, 1551–1560, doi:10.1111/j.1945-5100.2002.tb00888.x.

Hunten, D. M. and Sprague, A. L. (2002). Diurnal variation of sodium and potassium at Mercury. *Meteorit. Planet. Sci.*, **37**, 1191–1195, doi:10.1111/j.1945-5100.2002.tb00888.x.

Hunten, D. M., Morgan, T. H. and Shemansky, D. E. (1988). The Mercury atmosphere. In *Mercury*, ed. F. Vilas, C. R. Chapman and M. S. Matthews. Tucson, AZ: University of Arizona Press, pp. 562–612.

Ip, W.-H. (1986). The sodium exosphere and magnetosphere of Mercury. *Geophys. Res. Lett.*, **13**, 423–426.

Janches, D., Heinselman, C. J., Chau, J. L., Chandran, A. and Woodman, R. (2006). Modeling the global micrometeor input function in the upper atmosphere observed by high power and large aperture radars. *J. Geophys. Res.*, **111**, A07317, doi:10.1029/2006JA011628.

Jenniskens, P. (2006). *Meteor Showers and Their Parent Comets*. Cambridge: Cambridge University Press, 752 pp.

Johnson, R. E. (1990). *Energetic Charged-Particle Interactions with Atmospheres and Surfaces*. Berlin: Springer.

Johnson, R. E., Leblanc, F., Yakshinskiy, B. V. and Madey, T. E. (2002). Energy distributions for desorption of sodium and potassium from ice: The Na/K ratio at Europa. *Icarus*, **156**, 136–142.

Jurac, S., Baragiola, R. A., Johnson, R. E. and Sittler, E. C. (1995). Charging of ice grains by low-energy plasmas: Application to Saturn's E-ring. *J. Geophys. Res.*, **100**, 14,821–14,831.

Kallio, E, Wurz, P., Killen, R. M., McKenna-Lawlor, S., Milillo, A., Mura, A., Massetti, S., Orsini, S., Lammer, H., Janhunen, P. and Ip, W.-H. (2008). On the impact of multiply charged heavy solar wind ions on the surface of Mercury, the Moon and Ceres. *Planet. Space Sci.*, **56**, 1506–1516.

Keller, L. P. and McKay, D. S. (1997). The nature and origin of rims on lunar soil grains. *Geochim. Cosmochim. Acta*, **61**, 2331–2341, doi:10.1016/S0016-7037(97)00085-9.

Killen, R. M. (2015). Processes producing the extremely hot Ca and Mg exospheres at Mercury. Presented at 2015 Fall Meeting, American Geophysical Union, abstract P53A–2090, San Francisco, CA, 14–18 December.

Killen, R. M. (2016). Pathways for energization of Ca and Mg in Mercury's exosphere. *Icarus*, **268**, 32–36, doi:10.1016/j.icarus.2015.12.035.

Killen, R. M. and Hahn, J. M. (2015). Impact vaporization as a possible source of Mercury's calcium exosphere. *Icarus*, **250**, 230–237, doi:10.1016/j.icarus.2014.11.035.

Killen, R. M., Sarantos, M., Potter, A. E. and Reiff, P. (2004). Source rates and ion recycling rates for Na and K in Mercury's atmosphere. *Icarus*, **171**, 1–19.

Killen, R. M., Bida, T. A. and Morgan, T. H. (2005). The calcium exosphere of Mercury, *Icarus*, **173**, 300–311.

Killen, R. M., Cremonese, G., Lammer, H., Orsini, S., Potter, A. E., Sprague, A. L., Wurz, P., Khodachenko, M. L., Lichtenegger, H. I. M., Milillo, A. and Mura, A. (2007).

Processes that promote and deplete the exosphere of Mercury. *Space Sci. Rev.*, **132**, 433–509, doi:10.1007/s11214-007-9232-0.

Killen, R. M., Shemansky, D. E. and Mouawad, N. (2009). Expected emission from Mercury's exospheric species, and their UV-visible signatures. *Astrophys. J. Suppl. Ser.*, **181**, 351–359.

Killen, R. M., Potter, A. E., Vervack, R. J., Jr., Bradley, E. T., McClintock, W. E., Anderson, C. M. and Burger, M. H. (2010). Observations of metallic species in Mercury's exosphere. *Icarus*, **209**, 75–87, doi:10.1016/j.icarus.2010.02.018.

Killen, R. M., Burger, M. H., Hurley, D. M., Sarantos, M. and Farrell, W. M. (2014). Exospheres from asteroids to planets. In *Asteroids, Comets, Meteors: Book of Abstracts, Helsinki, Finland, 2014*. Helsinki: University of Helsinki, p. 285.

Kurosawa, K., Sugita, S., Kadono, T., Shigemori, K., Hironaka, Y., Ozaki, N., Shiroshita, A., Cho, Y., Sakaiya, T., Fujioka, S., Tachibana, S., Vinci, T., Kodama, R. and Matsui, T. (2010). Roles of shock-induced ionization due to > 10 km/s impacts on evolution of silicate vapor clouds. *Lunar Planet. Sci.*, **41**, abstract 1785.

Leblanc, F. and Johnson, R. E. (2003). Mercury's sodium exosphere. *Icarus*, **164**, 261–281, doi:10.1016/S0019-1035(03)00147-7.

Leblanc, F. and Johnson, R. E. (2010). Mercury exosphere I. Global circulation model of its sodium component. *Icarus*, **209**, 280–300.

Leblanc, F., Delcourt, D. and Johnson, R. E. (2003). Mercury's sodium exosphere: Magnetospheric ion recycling. *J. Geophys. Res.*, **108** (E12), 5136, doi:10.1029/2003JE002151.

Madey, T. E., Yakshinskiy, B. V. and Ageev, V. N. (1998). Desorption of alkali atoms and ions from oxide surfaces: Relevance to origins of Na and K in atmospheres of Mercury and the Moon. *J. Geophys. Res.*, **103**, 5873–5887.

Madey, T. E., Johnson, R. E. and Orlando, T. M. (2002). Far-out surface science: Radiation-induced surface processes in the Solar System. *Surf. Sci.*, **500**, 838–858.

Mangano, V., Milillo, A., Mura, A., Orsini, S., DeAngelis, E., DiLellis, A. M. and Wurz, P. (2007). The contribution of impulsive meteoritic impact vapourization to the Hermean exosphere. *Planet. Space Sci.*, **55**, 1541–1556.

Mangano, V., Massetti, S., Milillo, A., Plainaki, C., Orsini, S., Rispoli, R. and Leblanc, F. (2015). THEMIS Na exosphere observations of Mercury and their correlation with in-situ magnetic field measurements by MESSENGER. *Planet. Space Sci.*, **115**, 102–109, doi:10.1016/j.pss.2015.04.001.

Marchi, S., Morbidelli, A. and Cremonese, G. (2005). Flux of meteoroid impacts on Mercury. *Astron. Astrophys.*, **431**, 1123–1127.

McClintock, W. E., Vervack, R. J., Jr., Bradley, E. T., Killen, R. M., Mouawad, N., Sprague, A. L., Burger, M. H., Solomon, S. C. and Izenberg, N. R. (2009). Mercury's exosphere during MESSENGER's second flyby: Detection of magnesium and distinct distributions of neutral species. *Science*, **324**, 610–613.

McCord, T. B., Taylor, L. A., Combe, J.-P., Kramer, G., Pieters, C. M., Sunshine, J. M. and Clark, R. N. (2011). Sources and physical processes responsible for OH/H$_2$O in the lunar soil as revealed by the Moon Mineralogy Mapper (M^3). *J. Geophys. Res.*, **116**, E00G05, doi:10.1029/2010JE003711.

McGrath, M. A., Johnson, R. E. and Lanzerotti, L. J. (1986). Sputtering of sodium on the planet Mercury. *Nature*, **323**, 694–696.

McLain, J. L., Sprague, A. L., Grieves, G. A., Schriver, D., Trávníček, P. M. and Orlando, T. M. (2011). Electron-stimulated desorption of silicates: A potential source for ions in Mercury's space environment. *J. Geophys. Res.*, **116**, E03007, doi:10.1029/2010JE003714.

Merkel, A. W., Cassidy, T. A., Vervack, R. J., Jr., McClintock, W. E., Sarantos, M., Burger, M. H. and Killen, R. M. (2017). Seasonal variations of Mercury's magnesium dayside exosphere from MESSENGER observations. *Icarus*, **281**, 46–54, doi:10.1016/j.icarus.2016.08.032.

Merkel, A. W., Vervack, R. J., Jr., Killen, R. M., Cassidy, T. A., McClintock, W. E., Nittler, L. R. and Burger, M. H. (2018). Evidence connecting Mercury's magnesium exosphere to its magnesium-rich surface terrane. *Geophys. Res. Lett.*, **45**, 6790–6797, doi:10.1029/2018GL078407.

Morgan, T. H., Zook, H. and Potter, A. E. (1988). Impact-driven supply of sodium and potassium to the atmosphere of Mercury. *Icarus*, **75**, 156–170.

Mouawad, N., Burger, M. H., Killen, R. M., Potter, A. E., McClintock, W. E., Vervack, R. J., Bradley, E. T., Benna, M. and Naidu, S. (2011). Constraints on Mercury's Na exosphere: Combined MESSENGER and ground-based data. *Icarus*, **211**, 21–36, doi:10.1016/j.icarus.2010.10.019.

Mura, A. (2012). Loss rates and time scales for sodium at Mercury. *Planet. Space Sci.*, **63–64**, 2–7, doi:10.1016/j.pss.2011.08.012.

Mura, A., Milillo, A., Orsini, S. and Massetti, S. (2007). Numerical and analytical model of Mercury's exosphere: Dependence on surface and external conditions. *Planet. Space Sci.*, **55**, 1569–1583.

Mura, A., Wurz, P., Lichtenegger, H. I. M., Schleicher, H., Lammer, H., Delcourt, D., Milillo, A., Orsini, S., Massetti, S. and Khodachenko, M. L. (2009). The sodium exosphere of Mercury: Comparison between observations during Mercury's transit and model results. *Icarus*, **200**, 1–11, doi:10.1016/j.icarus.2008.11.014

Noble, S. K., Keller, L. P. and Pieters, C. (2005). Evidence of space weathering in regolith breccias I: Lunar regolith breccias. *Meteorit. Planet. Sci.*, **40**, 397–408.

Peplowski, P. N., Lawrence, D. J., Rhodes, E. A., Sprague, A. L., McCoy, T. J., Denevi, B. W., Evans, L. G., Head, J. W., Nittler, L. R., Solomon, S. C., Stockstill-Cahill, K. R. and Weider, S. Z. (2012). Variations in the abundances of potassium and thorium on the surface of Mercury: Results from the MESSENGER Gamma-Ray Spectrometer. *J. Geophys. Res.*, **117**, E00L04, doi:10.1029/2012JE004141.

Peplowski, P. N., Evans, L. G., Stockstill-Cahill, K. R., Lawrence, D. J., Goldsten, J. O., McCoy, T. J., Nittler, L. R., Solomon, S. C., Sprague, A. L., Starr, R. D. and Weider, S. Z. (2014). Enhanced sodium abundance in Mercury's north polar region revealed by the MESSENGER Gamma-Ray Spectrometer. *Icarus*, **228**, 86–95, doi:10.1016/j.icarus.2013.09.007.

Pfleger, M., Lichtenegger, H. I. M., Wurz, P., Lammer, H., Kallio, E., Alho, M., Mura, A., McKenna-Lawler, S. and Martin-Fernandez, J. A. (2015). 3D-modeling of Mercury's solar wind sputtered surface-exosphere environment. *Planet. Space Sci.*, **115**, 90–101, doi:10.1016/j.pss.2015.04.016.

Pifko, S., Janches, D., Close, S., Sparks, J., Nakamura, T. and Nesvorny, D. (2013). The Meteoroid Input Function and predictions of mid-latitude meteor observations by the MU radar. *Icarus*, **223**, 444–459, doi:10.1016/j.icarus.2012.12.014.

Poston, M. J., Grieves, G. A., Aleksandrov, A. B., Hibbitts, C. A., Dyar, M. D. and Orlando, T. M. (2015). Temperature programmed desorption studies of water interactions with Apollo lunar samples 12001 and 72501. *Icarus*, **255**, 24–29, doi:10.1016/j.icarus.2014.09.049

Potter, A. E. (1995). Chemical sputtering could produce sodium vapor and ice on Mercury. *Geophys. Res. Lett.*, **22**, 3289–3292.

Potter, A. E., Killen, R. M. and Morgan, T. H. (2002). The sodium tail of Mercury. *Meteorit. Planet. Sci.*, 37, 1165–1172.

Potter, A. E., Killen, R. M. and Sarantos, M. (2006). Spatial distribution of sodium on Mercury. *Icarus*, 181, 1–12, doi:10.1016/j.icarus.2005.10.026.

Potter, A. E., Killen, R. M. and Morgan, T. H. (2007). Solar radiation acceleration effects on Mercury's sodium emission. *Icarus*, **186**, 571–580, doi:10.1016/j.icarus.2006.09.025.

Press, W. H., Teukolsky, S. A., Vetterling, W. T. and Flannery, B. P. (2007). *Numerical Recipes: The Art of Scientific Computing*, 3rd edn. Cambridge: Cambridge University Press.

Raines, J. M., Gershman, D. J., Zurbuchen, T. H., Sarantos, M., Slavin, J. A., Gilbert, J. A., Korth, H., Anderson, B. J., Gloeckler, G., Krimigis, S. M., McNutt, R. L., Jr. and Solomon, S. C. (2013). Distribution and compositional variations of plasma ions in Mercury's space environment: The first three Mercury years of MESSENGER observations. *J. Geophys. Res. Space Physics*, **118**, 1604–1619, doi:10.1029/2012JA018073.

Roth, J. (1983). Chemical sputtering. In *Sputtering by Particle Bombardment II*, ed. R. Behrish, Topics in Applied Physics, Vol. 52. Berlin: Springer-Verlag, pp. 91–146, doi:10.1007/3-540-12593-0_3.

Sarantos, M., Killen, R. M., McClintock, W. E., Bradley, E. T., Vervack, R. J., Jr., Benna, M. and Slavin, J. A. (2011). Limits to Mercury's magnesium exosphere from MESSENGER second flyby observations. *Planet. Space Sci.*, **59**, 1992–2003, doi:10.1016/j.pss.2011.05.002.

Sarantos, M., Killen, R. M., McClintock, W. E., Vervack, R. J., Jr., Merkel, A. W., Burger, M. H., Cassidy, T. A., Slavin, J. A., Sprague, A. L. and Solomon, S. C. (2012). Mercury's Mg exosphere from MESSENGER data. *EPSC Abstracts*, **7**, abstract EPSC2012-707-1. European Planetary Science Congress, Madrid, Spain, 23–28 September.

Schmidt, C. A., Baumgardner, J., Mendillo, M. and Wilson, J. K. (2012). Escape rates and variability constraints for high-energy sodium sources at Mercury. *J. Geophys. Res.*, **117**, A03301, doi:10.1029/2011JA017217.

Schriver, D., Trávníček, P., Ashour-Abdalla, M., Richard, R. L., Hellinger, P., Slavin, J. A., Anderson, B. J., Baker, D. N., Benna, M., Boardsen, S. A., Gold, R. E., Ho, G. C., Korth, H., Krimigis, S. M., McClintock, W. E., McLain, J. L., Orlando, T. M., Sarantos, M., Sprague, A. L. and Starr, R. D. (2011). Electron transport and precipitation at Mercury during the MESSENGER flybys: Implications for electron-stimulated desorption. *Planet. Space Sci.*, **59**, 2026–2036, doi:10.1016/j.pss.2011.03.008.

Schultz, P. (1996). Effect of impact angle on vaporization. *J. Geophys. Res.*, **101**, 21,117–21,136, doi:10.1029/96JE02266.

Sieveka, E. M. and Johnson, R. E. (1984). Ejection of atoms and molecules from Io by plasma-ion impact. *Astrophys. J.*, **287**, 418–426, doi:10.1086/162701.

Sigmund, P. (1969). Theory of sputtering I: Sputtering yields of amorphous and polycrystalline targets. *Phys. Rev.*, **184**, 383–416.

Smith, R. S. and Kay, B. D. (1997). Adsorption, desorption, and crystallization kinetics in nanoscale water films. *Recent Res. Dev. Phys. Chem.*, **1**, 209–219.

Smyth, W. H. (1983). Io's sodium cloud: Explanation of the east–west asymmetries. II. *Astrophys. J.*, **264**, 708–725.

Smyth, W. H. and Marconi, M. L. (1995a). Theoretical overview and modeling of the sodium and potassium atmospheres of the Moon. *Astrophys. J.*, **443**, 371–392.

Smyth, W. H. and Marconi, M. L. (1995b). Theoretical overview and modeling of the sodium and potassium atmospheres of Mercury. *Astrophys. J.*, **441**, 839–864.

Sneh, O., Cameron, M. A. and George, S. M. (1996). Adsorption and desorption kinetics of H_2O on a fully hydroxylated SiO_2 surface. *Surf. Sci.*, **364**, 61–78.

Starukhina, L. V. and Shkuratov, Y. G. (2000). The lunar poles: Water ice or chemically trapped hydrogen?, *Icarus*, **147**, 585–587, doi:10.1006/icar.2000.6476.

Szalay, J. R. and Horanyi, M. (2016). Annual variation and synodic modulation of the sporadic meteoroid flux to the Moon. *Geophys Res. Lett.*, 42, 10,580–10,584, doi:10.1002/2015GL066908.

Vasavada, A. R., Paige, D. A. and Wood, S. E. (1999). Near-surface temperatures on Mercury and the Moon and the stability of polar ice deposits. *Icarus*, **141**, 179–193, doi:10.1006/icar.1999.6175.

Verner, D. A. and Yakovlev, D. G. (1995). Analytic FITS for partial photoionization cross sections. *Astron. Astrophys. Suppl. Ser.*, **109**, 125–135.

Verner, D. A., Yakovlev, D. G., Band, I. M. and Trzhaskovskaya, M. B. (1993). Subshell photoionization cross sections and ionization energies of atoms and ions from He to Zn. *Atom. Data Nucl. Data Tables*, **55**, 233–280.

Verner, D. A., Ferland, G. J., Korista, K. T. and Yakovlev, D. G. (1996). Atomic data for astrophysics. II. New analytic FITS for photoionization cross sections of atoms and ions. *Astrophys. J.*, **465**, 487–498.

Vervack, R. J., Jr., McClintock, W. E., Killen, R. M., Sprague, A. L., Anderson, B. J., Burger, M. H., Bradley, E. T., Mouawad, N., Solomon, S. C. and Izenberg, N. R. (2010). Mercury's complex exosphere: Results from MESSENGER's third flyby. *Science*, **329**, 672–675.

Vervack, R. J., Jr., McClintock, W. E., Killen, R. M., Sprague, A. L., Burger, M. H., Merkel, A. W. and Sarantos, M. (2011). Early MESSENGER results for less abundant or weakly emitting species in Mercury's exosphere. *EPSC-DPS Joint Meeting Abstracts and Program*, **6**, abstract EPSC-DPS2011-1131. European Planetary Science Congress–Division for Planetary Sciences Joint Meeting, Nantes, France, 2–7 October.

Vervack, R. J., Jr., McClintock, W. E., Killen, R. M., Merkel, A. W., Burger, M. H., Sarantos, M. and Cassidy, T. A. (2015). Mercury's exosphere: New detections, discoveries, and insights. *Abstracts, 47th Division for Planetary Sciences Annual Meeting*, abstract 107.01. National Harbor, MD, 8–13 November, pp. 19–20.

Vervack, R. J., Jr., Killen, R. M., McClintock, W. E., Merkel, A. W., Burger, M. H., Cassidy, T. A. and Sarantos, M. (2016). New discoveries from MESSENGER and insights into Mercury's exosphere. *Geophys. Res. Lett.*, **43**, 11,545–11,551.

Weider, S. Z., Nittler, L. R., Starr, R. D., Crapster-Pregont, E. J., Peplowski, P. N., Denevi, B. W., Head, J. W., Byrne, P. K., Hauck, S. A., Ebel, D. S. and Solomon, S. C. (2015). Evidence for geochemical terranes on Mercury: Global mapping of major elements with MESSENGER's X-Ray Spectrometer. *Earth Planet. Sci. Lett.*, **416**, 109–120, doi:10.1016/j.epsl.2015.01.023.

Wurz, P. and Lammer, H. (2003). Monte-Carlo simulation of Mercury's exosphere. *Icarus*, **164**, 1–13.

Wurz, P., Whitby, J. A., Rohner, U., Martin-Fernandez, J. A., Lammer, H. and Kolb, C. (2010). Self-consistent modelling of

Mercury's exosphere by sputtering, micro-meteorite impact and photon-stimulated desorption. *Planet. Space Sci.*, **58**, 1599–1616, doi:10.1016/j.pss.2010.08.003.

Yakshinskiy, B. V. and Madey, T. E. (1999). Photon-stimulated desorption as a substantial source of sodium in the lunar atmosphere. *Nature*, **400**, 642–644.

Yakshinskiy, B. V. and Madey, T. E. (2000). Desorption induced by electronic transitions of Na from SiO_2: Relevance to tenuous planetary atmospheres. *Surf. Sci.*, **451**, 160–165.

Yakshinskiy, B. V. and Madey, T. E. (2004). Photon-stimulated desorption of Na from a lunar sample: Temperature-dependent effects. *Icarus*, **168**, 53–59.

Yakshinskiy, B. V. and Madey, T. E. (2005). Temperature-dependent DIET of alkalis from SiO_2 films: Comparison with a lunar sample. *Surf. Sci.*, **593**, 202–209.

Yakshinskiy, B. V., Madey, T. E. and Ageev, V. N. (2000). Thermal desorption of sodium atoms from thin SiO_2 films. *Surface Rev. Lett.*, **7**, 75–87.

16

Structure and Configuration of Mercury's Magnetosphere

HAJE KORTH, BRIAN J. ANDERSON, CATHERINE L. JOHNSON, JAMES A. SLAVIN, JIM M. RAINES,
AND THOMAS H. ZURBUCHEN

16.1 INTRODUCTION

Of the terrestrial planets, only Mercury and Earth possess global magnetic fields and, in consequence, magnetospheres. Mercury's magnetosphere is substantially different from Earth's in a number of key respects. In the inner heliosphere, the solar wind subjects Mercury's magnetosphere to a much higher ram pressure, 10 to 30 nPa, than the ~2 nPa at Earth, and the magnitude of the interplanetary magnetic field (IMF) is much higher, ~30 nT, than at Earth (~5 nT). Also, the IMF is predominantly radially (sunward or anti-sunward) directed at Mercury, so that its bow-shock normal is more likely to be quasi-parallel to the solar wind flow than that of Earth. Moreover, the lower Alfvén Mach number – the ratio of plasma speed to Alfvén speed – leads to a somewhat weaker bow shock and lower magnetosheath plasma β – the ratio of the plasma thermal pressure to the magnetic pressure – at Mercury. These environmental differences have profound consequences for both the structure and dynamics of Mercury's magnetosphere. In addition, the relatively small planetary moment implies that Mercury's magnetosphere is much smaller than Earth's, so the characteristic timescales for convection and wave transits through Mercury's magnetosphere are nearly two orders of magnitude shorter than at Earth. The volume fraction of Mercury's magnetosphere occupied by the planet itself is about a factor of 500 larger at Mercury than at Earth, implying that plasma dynamics within Mercury's magnetosphere is substantially different and reflects the direct interaction with the planetary surface. Furthermore, owing to Mercury's large iron core, the effects of magnetic induction within the core on the magnetosphere are clearly evident at Mercury. Finally, Mercury possesses no permanent atmosphere of appreciable density and, hence, supports no ionosphere, so the inner boundary of the magnetosphere is fundamentally different from Earth's. Given the breadth of contrasts between Earth's and Mercury's magnetospheres, the quantitative comparison of the two systems affords a critical test of our understanding of the physics of planetary magnetospheres.

The first in situ observations of Mercury and its space environment made four decades ago by the Mariner 10 spacecraft revealed that the innermost planet has a magnetic field that is sufficiently strong to stand off the solar wind and form a magnetosphere. Mariner 10 executed three flybys of Mercury, two of which passed through the magnetosphere – one near the equatorial plane and one approximately over the northern pole. The magnetic field data acquired during these flybys revealed the presence of a weak, global-scale planetary magnetic field that was represented by a dipole with a southward-directed planetary moment, like at Earth, but with a surface field weaker by a factor of ~1000 (Ness et al., 1974, 1975). Mercury's internal field produces a magnetosphere, albeit much smaller than Earth's, with a magnetotail, a magnetopause, and a bow shock, as illustrated in Figure 16.1. The boundary enveloping the planetary magnetic field lines and separating them from the IMF lines is the magnetopause. The magnetosphere is an obstacle to the supersonic flow of the solar wind, and consequently a bow shock wave forms upstream of the magnetopause. At the dayside magnetopause, magnetic reconnection opens magnetic flux that is subsequently transported anti-sunward by the solar wind to form a quasi-cylindrical structure on the nightside of the planet, with northern and southern lobes that contain magnetic flux linked to the northern and southern polar regions of the planet, respectively (Russell et al., 1988). The plasma sheet, the layer of plasma separating the northern and southern lobes of the magnetotail, exhibits a plasma β value that is greater than unity. It is composed of closed magnetic flux closer to the planet but open magnetic flux farther downstream, beyond the distance at which the magnetic flux from the two lobes reconnects (Slavin et al., 2007).

Estimates of Mercury's planetary dipole moment were made from the Mariner 10 observations under the assumption that the dipole is centered on the planet. The non-dipole structure was only loosely constrained because the limited spatial distribution of the observations led to high correlations among the spherical harmonic coefficients describing the model field (Connerney and Ness, 1988; Chapter 5). In the absence of more extensive observations of the magnetosphere and its boundaries, our understanding of the magnetospheric structure remained largely conceptual, and models for Mercury's magnetospheric field were either highly simplified (Whang, 1977; Grosser et al., 2004) or scaled-Earth models (Jackson and Beard, 1977; Luhmann et al., 1998; Korth et al., 2004).

Mariner 10 also obtained observations of Mercury's plasma environment by measuring low-energy (~100 eV) electrons in the magnetosheath and electrons with energies up to the instrument limit of 1 keV in the plasma sheet and its boundary layer (Ogilvie et al., 1974). The electron observations indicated that the solar wind is the primary source for the plasma sheet. No charged particle observations were obtained in the energy range 1–100 keV, and the population of energetic particles with energies >100 keV (Simpson et al., 1974) may have been overestimated because of an instrumental effect (Armstrong et al.,

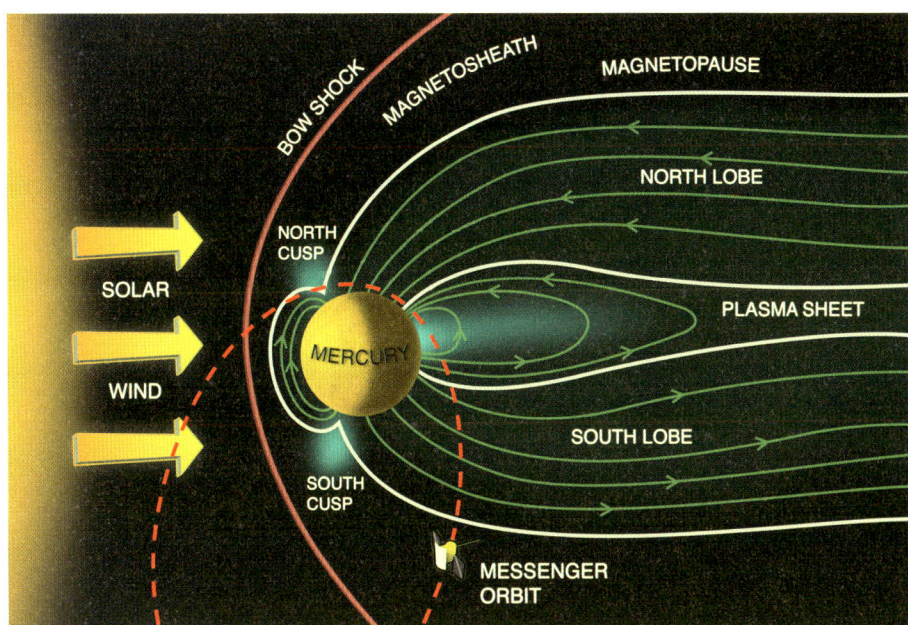

Figure 16.1. Schematic view of Mercury's magnetosphere with boundaries, major regions, magnetic field orientation (green curves and arrows), and MESSENGER orbit (dashed red curve) identified. Adapted from Zurbuchen et al. (2011).

1975). Unfortunately, the plasma ion instrument failed in flight, so Mariner 10 was unable to return information about ions at Mercury.

Many new insights into Mercury's magnetosphere were enabled by data returned from the MESSENGER spacecraft. MESSENGER completed three equatorial flybys of Mercury during the 6.6-year mission cruise phase and conducted orbital observations for more than four Earth years (more than 16 Mercury years) (see Chapter 1). The extensive magnetic field and particle observations accumulated in orbit allowed detailed characterization of the structure and configuration of Mercury's magnetosphere. MESSENGER magnetic field observations were used to determine definitively the orientation, intensity, and location of the internal planetary magnetic moment by first fitting the magnetic equator crossing points (Anderson et al., 2011b) to yield the position and orientation of an equivalent planetary magnetic dipole relative to the body center and then using these constraints in the evaluation of the magnetic moment (Johnson et al., 2012). (See also Chapter 5 for an extensive discussion of Mercury's internally generated field.) The result is that the dipole is aligned to within $0.6° \pm 0.1°$ of the planetary rotation axis and is offset 484 ± 4 km northward, along the rotation axis, and with these specifications an internal dipole moment of 190 ± 10 nT R_M^3, where R_M is Mercury's mean radius (2439.4 km), was determined (Chapter 5). These specifications for the internally generated magnetic field formed the basis for subsequent exploration of Mercury's magnetosphere.

MESSENGER observations were used to establish the configuration of the magnetopause, bow shock, cross-tail current sheet, and field-aligned or Birkeland currents. Plasma observations were used to determine the distribution and composition of plasma in the magnetosphere. This suite of analyses revealed processes unique to Mercury among magnetospheres in our solar system.

In this chapter, we review our understanding of the geometry and dominant physical processes of Mercury's magnetosphere inferred from MESSENGER data. We first review the solar wind environment in the inner solar system, because such an overview provides the context that governs the magnetospheric geometry and because the solar wind is the source of most of the plasma in Mercury's magnetosphere. We consider the shape and location of the magnetospheric boundaries and discuss the fundamental regions and configuration of the magnetosphere. We then describe the magnetospheric current systems and present state-of-the-art models of the magnetospheric magnetic field that combine mathematical descriptions of most of these current systems and the planetary dipole. We conclude by considering the plasma environment in Mercury's magnetosphere, discussing the sources and losses of plasma and describing the processes by which plasma is transported and heated.

16.2 MERCURY'S SOLAR WIND ENVIRONMENT

The solar wind near Mercury is composed of fast and slow solar wind, and these components are identified by their distinct origins, composition, and dynamic properties. The fast solar wind at Mercury's heliocentric latitudes originates in so-called coronal holes from which outflowing plasma expands into space on solar magnetic field lines (Zurbuchen, 2007). It has a typical velocity of ~700 km s^{-1} and tends to be time-stationary in composition; the plasma has substantial non-thermal He and heavy-ion components that have velocities exceeding those of the protons by about the local Alfvén speed, $v_A = \sqrt{B^2/(\mu_0 \rho)}$, where B is the magnetic field magnitude, μ_0 is the permeability of free space, and ρ is the plasma mass density (von Steiger et al., 1995, 2000). At Mercury's heliocentric distance, the nonthermal He and heavy ions flow up to ~150 km s^{-1} faster than the protons. The ion temperatures in the fast solar wind are not equal but are proportional to the ion mass, leading to heavy-ion temperatures of up to ~10^6 K (Marsch et al., 1982). MESSENGER observations showed that nearly 40% of the thermal pressure and up to 20% of the momentum flux is carried by alpha particles and heavy ions (Gershman et al., 2012). The dynamic pressure of the fast solar wind is higher and the Mach number lower than expected from adiabatic propagation of the much more thermalized and equilibrated solar wind observed near Earth.

The slow solar wind originates at coronal hole boundaries (Neugebauer et al., 1998), in active regions (Neugebauer et al., 2002), and in coronal streamers (Gosling et al., 1981). In the latter, the solar magnetic field is nominally perpendicular to the outward flow, resulting in a lower speed (v_{sw} ~ 350 km s^{-1}), higher density, and greater thermalization (Zurbuchen, 2007). In contrast to the fast solar wind, the density, flow speed, number flux, and temperature of the slow solar wind stream are highly variable on timescales of hours to days (Gosling, 1997). The momentum and Mach number in the slow wind are dominated by protons, and different ion species have similar temperatures and flow speeds (von Steiger and Zurbuchen, 2006). The slow solar wind also hosts the heliospheric current sheet that separates the northern and southern magnetic polarity regions of the heliosphere (Smith, 2001).

16.3 SHAPE AND LOCATION OF MAGNETOSPHERIC BOUNDARIES

The boundaries of the magnetosphere are the bow shock and the magnetopause. At the bow shock, the plasma and magnetic field are compressed as the particles' bulk flow speed is substantially reduced (Spreiter et al., 1966a). The flow energy is converted to thermal energy so that the plasma in the magnetosheath located immediately downstream of the shock is denser and hotter than in the solar wind. The location and shape of the bow shock are controlled by the solar wind momentum, solar wind Mach number, and, to a lesser extent, the IMF direction (Fairfield et al., 2001). The magnetopause position is determined to first order by the balance of the dynamic pressure exerted by the solar wind ram flow, which is partially converted to thermal pressure at the bow shock, and the magnetic pressure of the planetary field (Schield, 1969).

The bow shock and magnetopause locations vary substantially in response to dynamics arising from boundary waves and reconnection (Chapter 17). Reconnection of interplanetary and planetary magnetic field lines can modify the bow shock and magnetopause locations by eroding the dayside magnetopause and loading the magnetosphere with magnetic flux on the field lines opened by reconnection, in turn leading to flaring of the magnetopause. Because Mercury can be embedded in either the fast or slow solar wind, the degree of compression of the magnetosphere and imposed solar wind plasma varies markedly with time. This variability, together with that introduced by the influence of the IMF orientation and magnitude on the boundaries, leads to correspondingly large variations in the magnetopause and bow shock locations. The extensive observations provided by the orbital phase of the MESSENGER mission were therefore important to constraining the average boundary positions.

The MESSENGER spacecraft crossed the bow shock and magnetopause twice on every orbit, allowing a quantitative characterization of their location and shape throughout the range of heliocentric distances spanned by Mercury's orbit. Both the bow shock and the magnetopause have been mapped in detail, using MESSENGER observations to document their morphology and provide insight into the physical processes that lead to their formation and influence their dynamics. The bow shock and magnetopause shape were characterized by Winslow et al. (2013) from magnetic field observations during three Mercury years that spanned a broad range of solar wind and IMF conditions. Figure 16.2 shows an example for a magnetosphere transit from the dayside to the nightside. Observations are shown in the Mercury solar orbital (MSO) coordinate system, in which +X points toward the Sun, +Y is the direction opposite to the planet's orbital motion and orthogonal to +X (toward dusk), and +Z completes the right-handed system. The inbound (outbound) bow shock encounter is marked by a sharp increase (decrease) in B, which was most readily observed when the subsolar shock normal was quasi-perpendicular to the IMF, as was the case during the interval shown. In addition, there was a pronounced increase in the 1–10-Hz bandpass amplitude, denoted B_{AC} (Anderson et al., 2007). For quasi-parallel shocks, the bow shock was often quite broad and was conservatively bracketed in time by noting the outermost and innermost excursions in B.

Magnetopause crossings were most reliably identified by the rotation of the magnetic field across the magnetopause current layer. The example in Figure 16.2 illustrates a high-shear magnetopause marked by the rotation signature in B_Y and B_Z. The magnetopause also exhibited an increase in B_{AC}, as was almost always the case even when the magnetic shear was smaller. Because shear angles between the magnetospheric and magnetosheath magnetic fields were often <45°, the peak in B_{AC} and/or a change in the character of the <1-Hz field fluctuations at the boundary served as alternate identifiers of the magnetopause when the field rotation was not evident. Nearly all passes of MESSENGER across the bow shock and magnetopause exhibited multiple crossings, attributed to motions of the boundaries. Because it was not practical to note every crossing of the bow shock or magnetopause over the spacecraft, only the innermost and outermost boundaries were noted.

The average location and shape of the bow shock and the magnetopause were determined from parametric fits to the observed mean of the inner and outer boundary locations under the assumption – later tested and verified (Winslow et al., 2013) – that the boundaries are to first order figures of revolution in Mercury solar magnetospheric (MSM) coordinates. The MSM coordinate system has the same axis directions as the MSO frame, but the MSM origin is offset northward from the MSO origin by $Z_0 = 479$ km [an earlier estimate for the dipole offset (Anderson et al., 2012) used for many of the calculations discussed here]. The sampling distribution versus angle from the MSM X-axis was non-uniform, so to ensure uniform weighting of the boundary crossings, the probability distributions of the mean crossings were constructed as a function of angle from the +X_{MSM} axis, and fits were also made to the most probable crossing distance in each direction (Winslow et al., 2013).

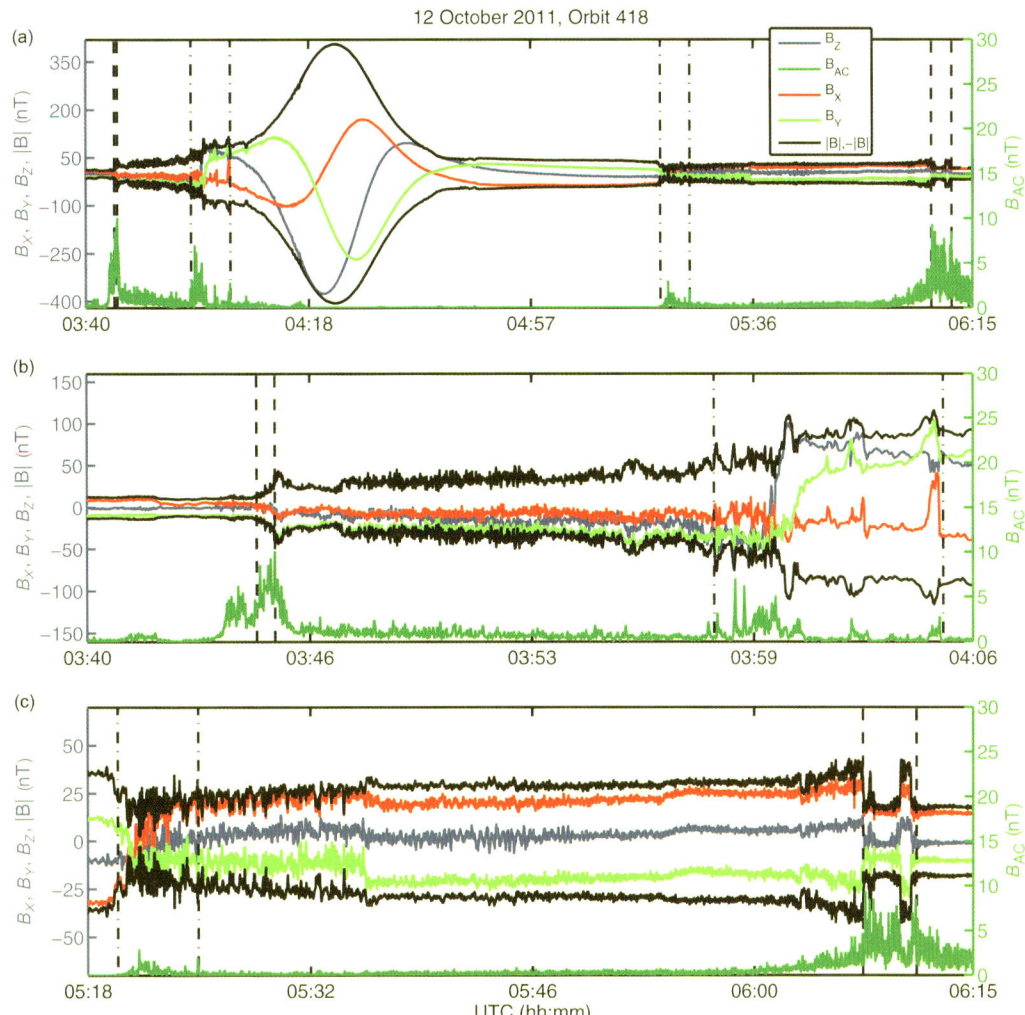

Figure 16.2. (a) MESSENGER Magnetometer data in MSO coordinates for the first magnetospheric transit on 12 October 2011 (orbit 418). Shown are the magnetic field components B_X (red), B_Y (light green), B_Z (blue), and B and $-B$ (black), with the scale given on the left axes, as well as the 1–10-Hz bandpass amplitude, B_{AC} (dark green), with the scale given on the right axes. Vertical lines denote the times of the inner and outer crossings of the bow shock (dashed) and magnetopause (dot-dashed). (b) Close-up view of the inbound portion of the orbit. (c) Close-up view of the outbound portion of the orbit. Times here and in other figures are in Coordinated Universal Time (UTC). Adapted from Winslow et al. (2013).

The bow shock was fit by a conic section (Slavin et al., 2009b) of the form

$$\sqrt{(X-X_0)^2 + \rho^2} = \frac{p\varepsilon}{1+\varepsilon \cos\theta}, \quad (16.1)$$

where $\rho = \sqrt{Y^2 + Z^2}$ is the distance from the MSM x-axis, the focus of the conic section lies at X_0 along the line through the planetary dipole that parallels the X-axis at X_0, and p and ε are the focal and eccentricity parameters, respectively. The observed shock crossings are shown in Figure 16.3a, and the probability distributions are shown in Figure 16.3c. The best-fit parameters are $X_0 = 0.5\ R_M$, $p = 2.75\ R_M$, and $\varepsilon = 1.04$. The extrapolated nose distance from the dipole origin is 1.90 R_M. Winslow et al. (2013) showed that the bow shock moves toward the planet during times of high-solar-wind Alfvén Mach number, $M_A = v_{sw}/v_A$, but, unlike at Earth, M_A does not affect the flaring of the bow shock on average, so the shape does not change. Corrected for M_A, the best-fit parameters were $X_0 =$ 0.5 R_M, $p = 2.9\ R_M$, and $\varepsilon = 1.02$, which yield a solar wind standoff distance of 1.96 R_M.

The observed magnetopause crossings and their uncertainties are shown in Figure 16.3b, and the corresponding probability distributions are shown in Figure 16.3d. The time-averaged magnetopause is best fit using the functional form proposed by Shue et al. (1997):

$$R = \sqrt{X^2 + \rho^2} = R_{SS}\left(\frac{2}{1+\cos\theta}\right)^{\alpha}, \quad (16.2)$$

where R is the distance from the dipole center, $\theta = \tan^{-1}(\rho/X)$, and α is a flaring parameter that governs whether the magnetotail is closed ($\alpha < 0.5$) or open ($\alpha \geq 0.5$). The best-fit magnetopause has a subsolar standoff distance of $R_{SS} = 1.45\ R_M$, has a flaring of $\alpha = 0.5$, and becomes cylindrical at a small, ~(2–3) R_M, downstream distance, where it has a radius of ~2.7 R_M. The resulting average subsolar magnetosheath thickness of 0.45 R_M is consistent with that predicted

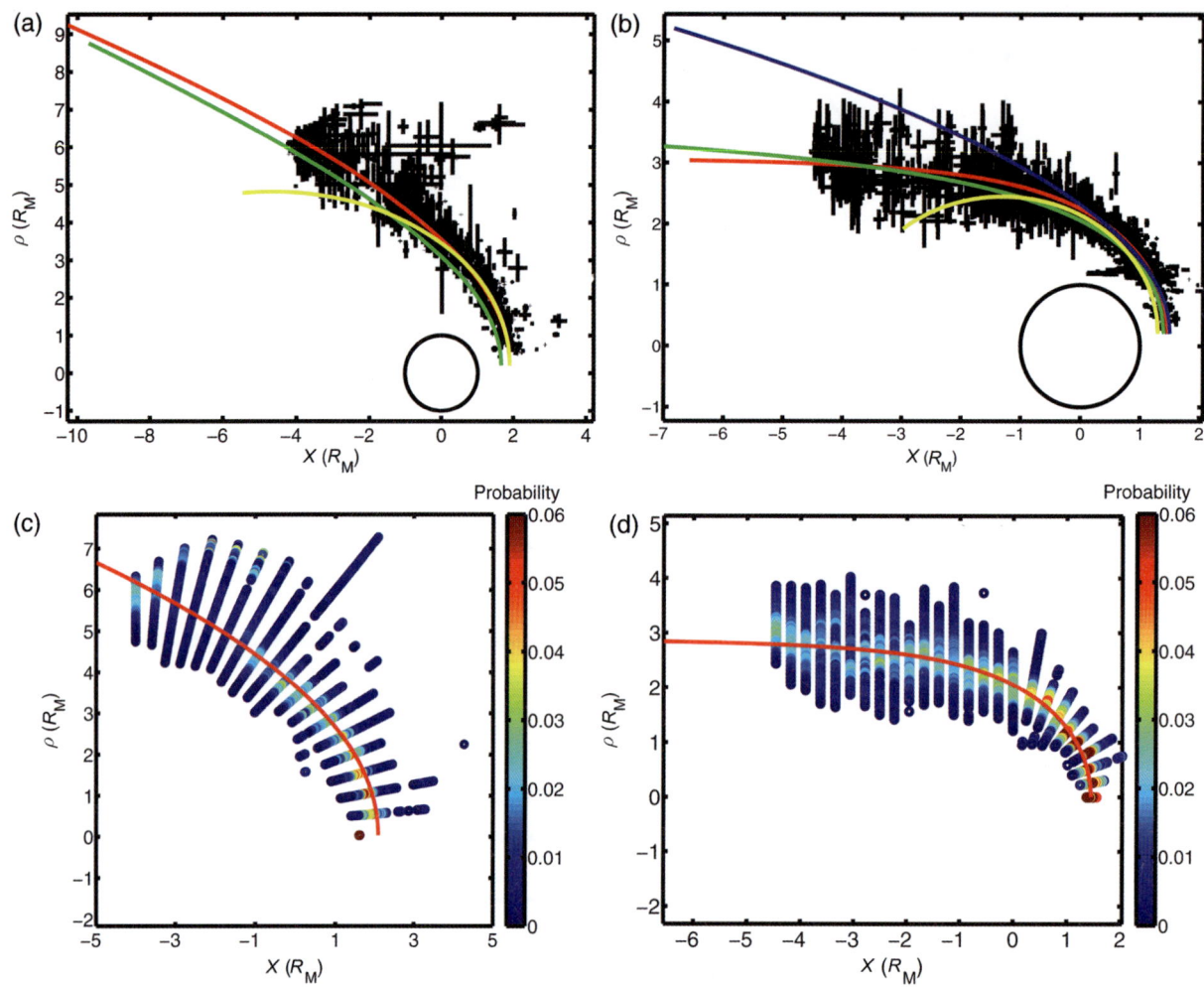

Figure 16.3. Midpoints between the inner and outer crossing positions of (a) the bow shock and (b) the magnetopause, and probability distributions computed at different angular distances from the $+X_{MSM}$ direction for (c) the bow shock and (d) the magnetopause. The error bars in (a) and (b) show the distance between the inner and outer crossings. Red curves show the parametric fits to the crossing probability distributions of (c) and (d). In (a) and (b), the green and yellow curves represent models from previous studies based on Mariner 10 observations by Slavin et al. (2009b) and Russell (1977), respectively. In (b), the blue curve is the best-fit paraboloid. Adapted from Winslow et al. (2013).

by global magnetohydrodynamic and hybrid models (e.g., Benna et al., 2010; Müller et al., 2012). An updated model fit with magnetopause crossings over seven Mercury years yielded $\alpha = 0.5$ and $R_{SS} = 1.42\ R_M$ (Korth et al., 2015), and these values have been confirmed with crossings from the entire mission. The best-fit paraboloid model was also computed, as given by

$$X(\rho) = -\left(\frac{\gamma^2 + 1}{4R_{SS}}\right)\rho^2 + R_{SS}, \qquad (16.3)$$

where $R_{SS} = 1.5\ R_M$ and $\gamma = 1$. The standoff distance, calculated from the first three Mercury years of data, was found to vary with solar wind ram pressure, p_{ram}, in nPa as $R_{SS} = (2.15 \pm 0.10)p_{ram}^{[(-1/6.75)\pm 0.024]}$ (Winslow et al., 2013). The p_{ram} estimates were obtained from simulations with the ENLIL model (Odstrcil, 2003) and mainly reflected p_{ram} variations associated with changes in heliocentric distance during one Mercury year. The difference of the exponent from the value of $-1/6$ expected for simple pressure balance may result from the effect of induction in Mercury's conductive interior (Johnson et al., 2016; Chapter 5), which reinforces the dayside magnetic field configuration during intervals of higher than average p_{ram} and decreases the dayside field during intervals of lower than average p_{ram}. The bow shock moves planetward during times of higher p_{ram}, but, unlike at Earth, p_{ram} does not appear to affect the flaring of the bow shock.

16.4 REGIONS AND CONFIGURATION OF THE MAGNETOSPHERE

16.4.1 Magnetotail Configuration

Mariner 10 found that the large-scale structure of Mercury's magnetotail is similar to that of Earth's and of other magnetospheres created by interaction of the solar wind with a dipolar planetary magnetic field oriented largely perpendicular to the solar wind velocity (Russell et al., 1988). This result was confirmed by the MESSENGER flyby observations (Slavin et al., 2008), and an example of the magnetic field measured during a traversal of the plasma sheet during the orbital phase of the mission is displayed in Figure 16.4. The two lobes are identified

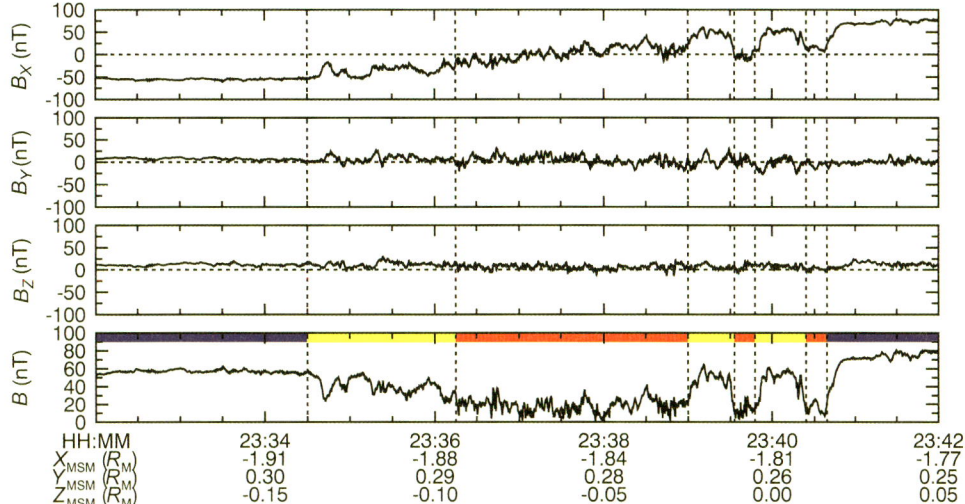

Figure 16.4. The magnetic field in MSM coordinates measured by MESSENGER on 3 February 2013 during a traversal from the south magnetic tail lobe (blue horizontal bar at left), through the plasma sheet (red bars) and plasma-sheet boundary layer (yellow bars), into the northern tail lobe (blue bar at right).

by the strong magnetic field, the low level of fluctuations, and the field orientations away from the Sun in the southern hemisphere and toward the Sun in the northern hemisphere.

The intensity of the magnetic field in the lobes is determined by the inward pressure exerted by the solar wind dynamic pressure (magnetosheath thermal pressure) on the flaring magnetopause (Slavin et al., 1985, 2012). The weakness of the magnetic field in the central plasma sheet, as well as the boundary layer at the outer edge of the plasma sheet, is due to the presence of higher-density, hot plasma. The direction of the magnetic field in this region is predominantly northward, i.e., $B_Z > 0$ (see Section 16.4.2). As at Earth, the cross-tail current sheet across which the magnetic field in the two lobes reverses is embedded in the plasma sheet (Slavin et al., 1985).

During the orbital phase of the mission, MESSENGER sampled the structure of Mercury's magnetotail from a downstream distance of ~(1.25–3.25) R_M. The magnetic field strength, B, in Mercury's lobes and plasma sheet follows a power-law decrease with increasing anti-sunward distance $|X|$, $B \sim |X|^G$, with G varying from −5.4 for northward IMF to −1.6 for southward IMF (Slavin et al., 2012). The slower decrease for southward IMF is attributed to the tail field being more stretched when dayside reconnection and the rate of magnetic flux transfer are greatest. The average length of Mercury's magnetotail derived from models of the rate of magnetic flux transfer between the dayside and nightside magnetosphere is ~150 R_M (Milan and Slavin, 2011). Statistical properties of the north–south component of the magnetic field in the plasma sheet have been evaluated from 333 plasma-sheet crossings identified during the orbital phase of the mission from 23 March 2011 to 30 April 2015 (Poh et al., 2015), and the results are shown in Figure 16.5. The left-hand panel shows distributions of B_Z for different downstream distances. The results indicate that, when averaged across the full width of the plasma sheet, B_Z is positive and the plasma-sheet magnetic field is closed, at least out to $X \sim -2.6\ R_M$. However, the mean northward magnetic field steadily decreases with increasing distance. The right-hand panels of Figure 16.5 display the mean B_Z in four downstream distance ranges as a function of Y. Closest to Mercury, B_Z exhibits a maximum near midnight (i.e., $Y = 0$), which is fit with a quadratic function (shown in red in Figure 16.5). The magnitude of B_Z decreases monotonically from the dawn side to the dusk side of the plasma sheet. The reason for this variation is not clear, but it suggests that the cross-tail current sheet should be thickest, and more stable against reconnection, at dawn and thinnest and most likely to reconnect on the dusk side (Poh et al., 2015). Interestingly, this result indicates that the cross-tail current sheet has its greatest north–south thickness on the side of the tail where the dayside low-latitude boundary layer is also widest in the east–west direction (Liljeblad et al., 2015).

In the magnetotail, field lines of the northern and southern lobes reconnect in the magnetic equatorial plane to form closed field lines. Magnetotail reconnection occurs along a line referred to as the X-line, named because the magnetic field topology resembles that letter at the reconnection site. This line is also termed the neutral line because the magnetic field strength vanishes at these locations. The mean distance of the distant X-line in Mercury's magnetotail is remarkably close to the planet. The right-hand panels of Figure 16.5 show that the B_Z magnetic field approaches zero near the center of the tail (i.e., local midnight) at $X \sim -2.3\ R_M$ to $-2.6\ R_M$. These results strongly suggest that the mean distance at which reconnection X-lines form in Mercury's plasma sheet is near local midnight, at $X \sim -2.5\ R_M$ to $-3.0\ R_M$ (Poh et al., 2015). This location is much closer to the planet than at Earth, where the distant X-line is found at $X \sim -100\ R_E$ to $-140\ R_E$ and R_E = 6371 km is Earth's mean radius (Slavin et al., 1985). Planetward of the X-line, the magnetic field in the plasma sheet has a northward component to the magnetic field, and the convection (i.e., charged particle drift due to the dawn-to-dusk electric field imposed by the solar wind across the magnetosphere) is sunward. Downstream of that distance, the magnetic field in the plasma sheet has a southward magnetic field component, and the convection is anti-sunward as at Earth (e.g., Slavin et al., 1985). That this X-line is close to the planet is supported by analyses of plasmoid formation in Mercury's tail during disturbed intervals that show

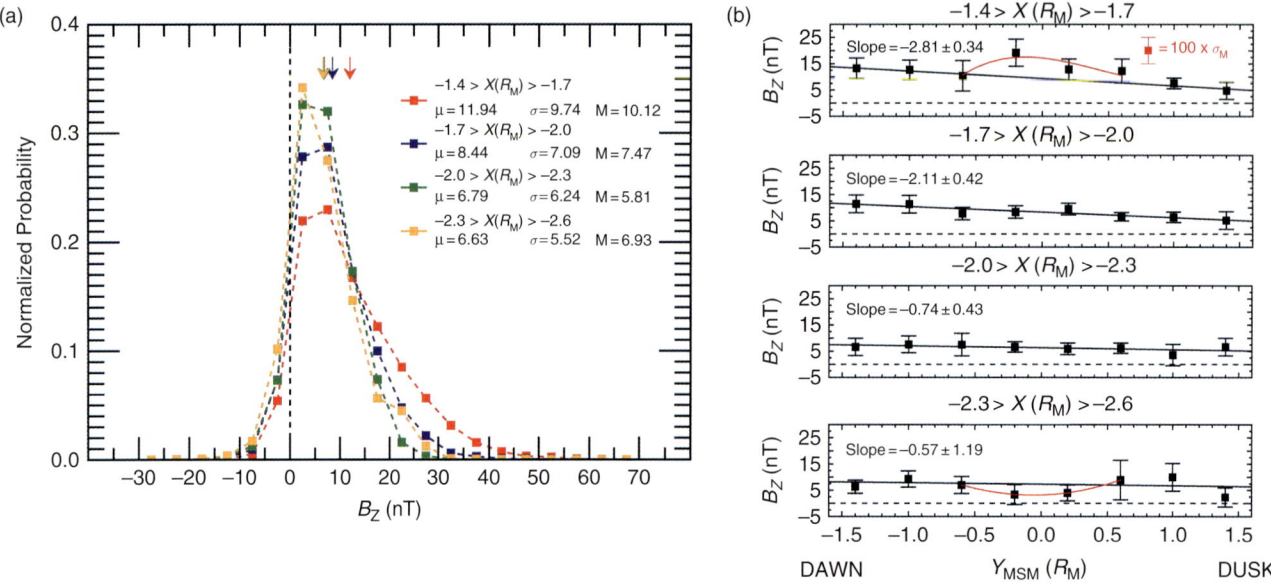

Figure 16.5. (a) Distribution of the B_Z component in MSM coordinates of the plasma sheet magnetic field at four 0.3-R_M-wide ranges from $X = -1.4$ R_M to -2.6 R_M. Mean (μ), median (M), and standard deviation (σ) for each range are provided. (b) The average B_Z magnetic field component as a function of the cross-tail location, -1.6 $R_M < Y < 1.6$ R_M, for four downstream distances. For each distance range, the standard error of the mean, σ_M, is displayed after being multiplied by 100, and the slope of a linear fit to the data is given. Quadratic fits to the strong B_Z in the top panel (closest to Mercury) and the weak B_Z in the bottom panel (farthest from Mercury) are displayed in red.

the near-Mercury neutral line forming at distances of $X \sim -1.5$ R_M to -2.5 R_M (Slavin et al., 2009a, 2010, 2012; DiBraccio et al., 2015b).

16.4.2 Plasma Sheet

The plasma sheet hosts the majority of the plasma in Mercury's magnetosphere. The transport of flux into the tail and toward the plasma sheet in the lobes corresponds to a dawn-to-dusk electric field, **E**, and this electric field transports plasma on open magnetic flux tubes in the tail lobes via drift in the $\mathbf{E} \times \mathbf{B}$ direction. MESSENGER acquired plasma observations with the Fast Imaging Plasma Spectrometer (FIPS) (Andrews et al., 2007), which measured protons and heavy-ion species in the energy range from 46 eV e^{-1} to 13 keV e^{-1}, where e is the electron charge, with a 1.4π sr field of view.

FIPS observations during MESSENGER's first Mercury flyby on 15 January 2008 showed that Mercury's plasma sheet is Earth-like with respect to proton densities, pressures, and plasma β (Raines et al., 2011). Near the planet, the plasma sheet extends to low altitudes just poleward of the boundary between open and closed magnetic field lines located at mid latitudes (Korth et al., 2014; Sun et al., 2015). The sources of the plasma in the plasma sheet are the solar wind and the planet, and charged particles populate this region via the processes described in Section 16.6.3. The average distribution functions for both solar wind and planetary ions in Mercury's pre-midnight plasma sheet are well described by hot Maxwell–Boltzmann distributions, and the plasma bulk properties are shown in Figure 16.6 (Gershman et al., 2014). Densities and temperatures of the H^+-dominated plasma sheet are in the ranges ~1–10 cm^{-3} and ~5–30 MK, respectively, and the plasma sheet maintains a thermal pressure of ~1 nPa. The plasma-sheet density decreases with increasing solar wind velocity, v_{sw}, whereas the temperature increases with v_{sw}.

The average bulk properties of other ion species in the plasma sheet relative to H^+ ($n = 7.8$ cm^{-3}, $T = 9.3$ MK) are shown in Figure 16.7. The dominant planetary ion species are Na^+-group ions [mass per charge (m/q) ratio = 21–30], which exhibit number densities ~10% of those of H^+ on average, followed by He^{2+} and O^+-group ions (m/q = 16–20) with number densities 3.5% and 1.5%, respectively. The average temperature of the planetary ion species is a factor of ~1.5 larger than that of the protons observed nearby. These values imply that planetary ions could contribute ~15% to the plasma thermal pressure and ~50% to the mass density in the nightside plasma sheet. Whereas solar wind ions (i.e., H^+, He^{2+}, O^{6+}) show mass-proportional temperatures, the temperatures of planetary ions are approximately equal. The latter characteristic may be additional evidence of non-adiabatic particle motion in Mercury's magnetosphere, because it is consistent with the ion motion induced by a potential drop, rather than an $\mathbf{E} \times \mathbf{B}$ drift, as would be expected for energetic heavy ions gyrating with a large radius in a weak magnetic field (see Section 16.6.2).

The spatial distribution of plasma-sheet protons in the magnetic equatorial plane has been determined from FIPS observations and inferred independently from Magnetometer data. Although MESSENGER's orbit line of apsides is substantially inclined with respect to Mercury's equatorial plane, the distribution of plasma at the magnetic equator may be obtained by mapping observations along magnetic field lines. Statistical mapping of the plasma-sheet proton flux (Figure 16.8a) reveals the existence of a plasma enhancement within a toroidal section, which is centered at the magnetic equator near local midnight and extends on the nightside from dusk to dawn (Korth et al., 2014). An enhanced plasma population is also found near the

16.4 Regions and Configuration of the Magnetosphere 437

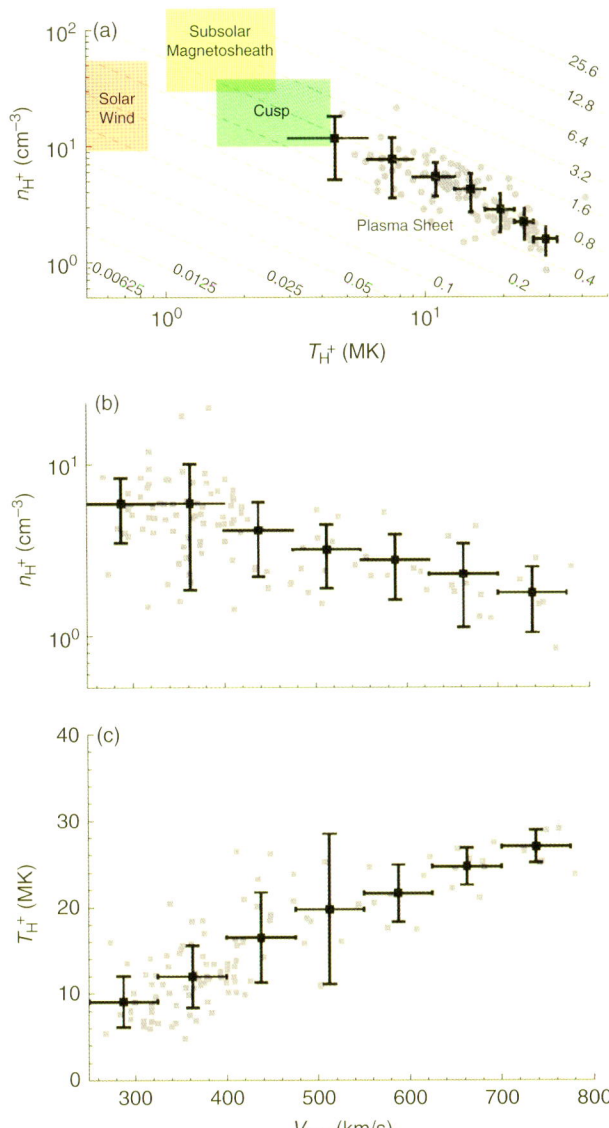

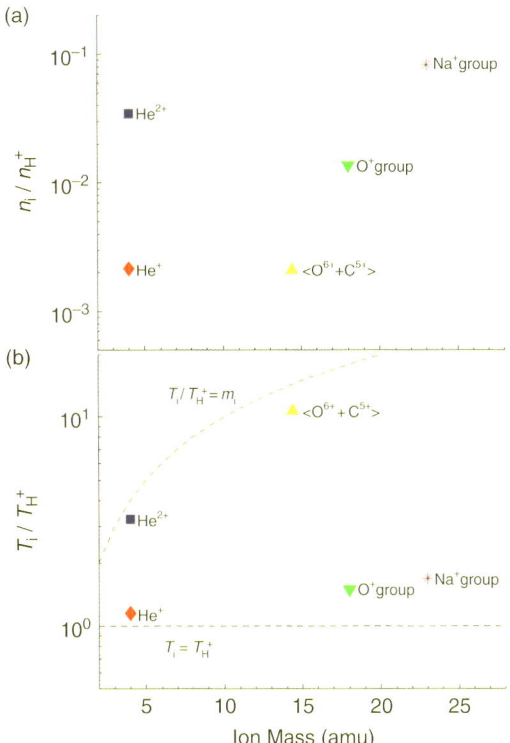

Figure 16.6. (a) Orbit-averaged density and temperature of H$^+$ in Mercury's pre-midnight/dusk-side plasma sheet at heliocentric distance $R \approx 0.35$ AU for 113 orbits. Lines of constant pressure (in nPa) are dashed. Ranges of H$^+$ density and temperature observed elsewhere in Mercury's magnetosphere are indicated with red, yellow, and green boxes for the solar wind (Gershman et al., 2012; Baker et al., 2013), subsolar magnetosheath (Gershman et al., 2013), and northern magnetospheric cusp (Raines et al., 2014), respectively. (b) Plasma-sheet density versus upstream solar wind speed. (c) Plasma temperature versus upstream solar wind speed. Black squares denote bin averages, vertical error bars denote the standard deviation of values in each bin, and horizontal error bars correspond to the bin size. Adapted from Gershman et al. (2014).

Figure 16.7. Average (a) density n_i and (b) temperature T_i of ion species relative to H$^+$. Dashed lines corresponding to $T_i = T_{H^+}$ and $T_i/T_{H^+} = m_i$ are also shown. Adapted from Gershman et al. (2014).

magnetopause flanks, indicating entry of magnetosheath plasma into the low-latitude boundary layer of the magnetosphere. The plasma bulk properties are derived from the measured fluxes (Gershman et al., 2012), and regions showing higher fluxes also exhibit higher plasma pressures (Korth et al., 2014).

Consistent with the distribution of protons, the average densities of all planetary ions within 30° of the planetary equator are depressed at the subsolar point relative to the dawn and dusk terminators (Raines et al., 2013). The effect is largest for Na$^+$-group ions, which are 49% lower in density at the subsolar point than at the terminators. The observations show that dense plasma does not form a closed distribution around the planet, likely because of the dynamic solar wind and IMF conditions, which prevent the formation of drift paths that close around the planet.

Enhancements of the plasma pressure in the plasma sheet were independently determined from diamagnetic depressions of the background magnetic field (Korth et al., 2012). These decreases in the magnetic field magnitude arise from reductions in magnetic pressure in the presence of an increase in plasma thermal pressure to maintain total pressure balance. The magnetic-field technique complements observations from FIPS because the FIPS instrument observes only 35% of the full solid angle. Using the paraboloid magnetic field model described in Section 16.5.3 to represent the average magnetospheric magnetic field, Korth et al. (2012) inferred the plasma pressure enhancements from the magnetic field perturbations with respect to that baseline. The resulting distribution of the plasma pressure enhancement (Figure 16.8b) qualitatively reproduces that of the proton flux in the inner magnetosphere (Figure 16.8a), where the pressures are large and localized. The magnitude of the average magnetic pressure deficit normalized for heliocentric distance is 1.45 nPa and exhibits a weak, 0.05-nPa h^{-1}, dusk-to-dawn gradient with local time. The magnitudes of the pressure agree with estimates derived from plasma observations on average but can deviate for individual events by factors of up to ~3 (Korth et al., 2014).

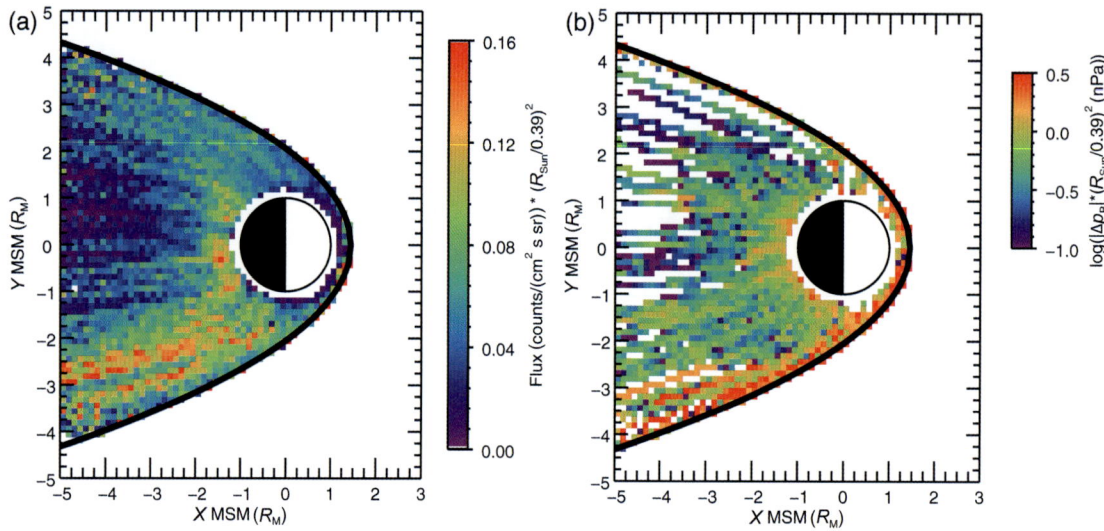

Figure 16.8. Distributions of (a) the mean proton flux observed by FIPS and (b) the mean magnetic pressure deficit determined from Magnetometer data mapped to Mercury's magnetic equatorial plane and normalized to a heliocentric distance of 0.39 AU. The circle denotes the planet, the Sun is to the right, and the magnetopause of the magnetic field model is represented by the solid black curves. Adapted from Korth et al. (2014).

16.4.3 Cusp

In a vacuum representation, the cusps are singularities in the magnetic field immediately inside the magnetopause. The magnetic field vanishes at the cusps, and all magnetic field lines immediately inside the current layer thread the cusps (Olson, 1984). The Chapman–Ferraro currents (Chapman and Ferraro, 1930, 1931) that form the dayside magnetopause flow around the cusps, and for Mercury the current loops are counterclockwise in the north and clockwise in the south when viewed from the Sun (Olson, 1984). In Mercury's magnetosphere, the cusps are regions of plasma exchange between the magnetosheath and the magnetosphere, for two reasons. First, because the magnetic field is weak near the cusps, the magnetic pressure cannot stand off the thermal pressure of the magnetosheath, so magnetosheath plasma has ready access to magnetospheric field lines. The cusps are therefore regions of high thermal pressure and high magnetosheath densities near and even within the magnetopause. Second, because magnetic fields on the dayside magnetopause map to the cusps, fields that are reconnected with the magnetosheath field on the dayside map near the cusp and its projection to the planetary surface. Reconnection at the dayside magnetopause therefore injects plasma into the cusps and drives convection of magnetic flux through the cusps and into the polar caps, which are the regions threaded by field lines connected to the tail lobes. Open field lines allow access of plasma from both magnetosheath and planetary sources, so that the plasma population in the vicinity of the cusp is a mixture of these plasmas. In addition, because Mercury's atmosphere is tenuous, the cusp plasma can interact directly with the surface, causing space weathering and providing a source of sputtered neutral atoms and ions to the exosphere and magnetosphere.

The location of Mercury's northern cusp at dayside high latitudes was first studied statistically by Winslow et al. (2012) with MESSENGER data acquired during the first and second Mercury years in orbit. Using the method summarized in Section 16.4.2, these authors inferred plasma pressure enhancements in the cusp from diamagnetic depressions in the magnetic field magnitude. The cusp was found on average to span a region extending 11° in latitude and 4.5 h in local time at spacecraft altitudes (Figure 16.9). The bounds of the northern cusp are 55.8°N and 83.6°N MSO latitude and 7.2 h and 15.9 h local time, and the cusp is approximately symmetric about noon (Figure 16.9). Because the MESSENGER orbit was eccentric and periapsis during this phase of the mission was on the descending latitude portion of the orbit, the cusp was encountered at lower altitudes on the descending than on the ascending orbit leg. Consistent with the expected shift in cusp latitude closer to the magnetopause, the high-altitude cusp was observed on average a few degrees equatorward of that seen at lower altitude.

Winslow et al. (2014) independently determined the extent of the northern cusp using proton reflectometry to measure the proton loss cone indicating persistent ion precipitation to the surface. Tracing the locations of FIPS observations in the cusp taken from 7 June 2011 to 7 June 2012 along magnetic field lines to the surface with a paraboloid magnetic field model (Section 16.5.3) showed that the cusp is centered on noon at 76.4°N latitude and extends 15.6° in latitude and 7.5 h in local time. The cusp location coincides with a region where high fluxes of planetary Na^+-group and O^+-group ions peaking near noon and 60°N latitude are observed (Zurbuchen et al., 2011). Because of the northward offset of the planetary dipole, the extent of the cusp in the southern hemisphere is expected to be larger than in the north. However, as a result of MESSENGER's highly eccentric orbit, the spacecraft was located outside the

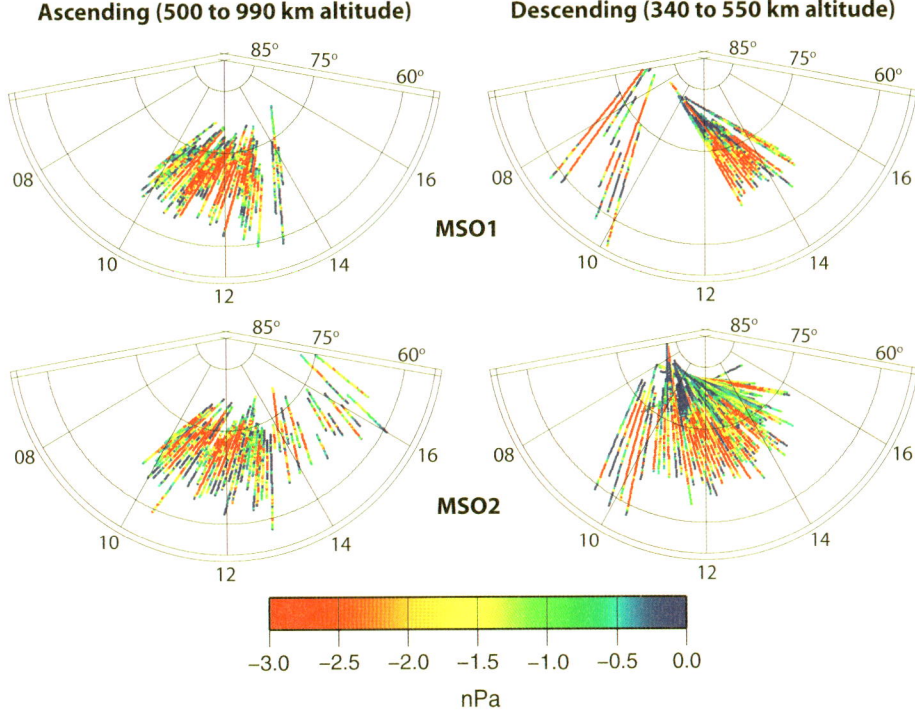

Figure 16.9. Stereographic projections of the pressure deficit along profiles across the cusp shown in MSO coordinates corrected for aberration of the magnetosphere resulting from Mercury's finite orbital velocity. During portions of MESSENGER's first Mercury year in orbit (MSO1), the Magnetometer was off when the spacecraft experienced long eclipses or was close to the planet, resulting in gaps in data coverage (between ~10 h and ~12 h local time) for the descending tracks. Complete coverage was obtained during MESSENGER's second Mercury year in orbit (MSO2). Projections span local times from 6.7 h to 17.3 h and latitudes 55°N to the pole. The color bar is saturated so that observed, but localized, pressure deficits greater in magnitude than −3 nPa are shown in red. Adapted from Winslow et al. (2012).

magnetosphere when traversing the dayside high-latitude region in the southern hemisphere, so the southern cusp could not be mapped with the techniques mentioned above.

The plasma populations of the northern cusp were observed by the FIPS sensor. On most orbits that crossed the cusp, MESSENGER traversed the cusp at altitudes ranging from 200 to 600 km, whereas the sampling altitude decreased to as low as 11 km during the final phase of the mission (Chapter 1). Consistent with the location of the cusp identified above (Figure 16.9), overall lower fluxes of cusp ions were observed during times when the spacecraft was in a dawn–dusk orbit. The FIPS observations showed that protons are the dominant ion species in the cusp, followed by alpha particles (He^{2+}) and Na^+-group ions (Raines et al., 2014). Other planetary ions, e.g., O^+-group ions and He^+, were also observed but with lower fluxes. The distribution of Na^+-group ions, shown in Figure 16.10, exhibits a peak in the flux within the northern cusp.

Protons enter the cusp from the solar wind directly from the dayside reconnection region or drift toward the planet along newly reconnected field lines as they convect across the polar cap. Some protons have sufficient energy parallel to the magnetic field to overcome the mirror force of the planetary magnetic field (which increases in strength toward the surface) and impact the surface. The remaining protons mirror and flow upward along the field lines away from the planet and into the magnetotail. Precipitating energetic protons impacting at the surface can lead to ion and neutral sputtering, a process that results in release of

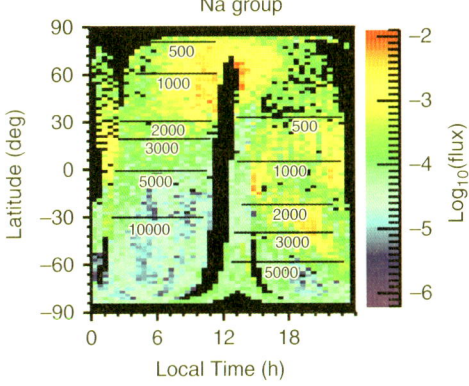

Figure 16.10. Distribution of Na^+-group ions as a function of latitude and local time sampled along the MESSENGER trajectory during 130 orbits in 2011. Horizontal lines denote spacecraft altitude in km. Adapted from Zurbuchen et al. (2011).

neutral atoms (~90–99% by number density) and ions (~1–10% by number density) into the exosphere and the magnetosphere (Killen et al., 2007). Similarly, precipitating solar wind electrons contribute to the exospheric and magnetospheric populations through electron-stimulated desorption. Evidence of the precipitation process leading to depletion of flux in the upward direction is shown in Figure 16.11 (Raines et al., 2014). Figure 16.11a

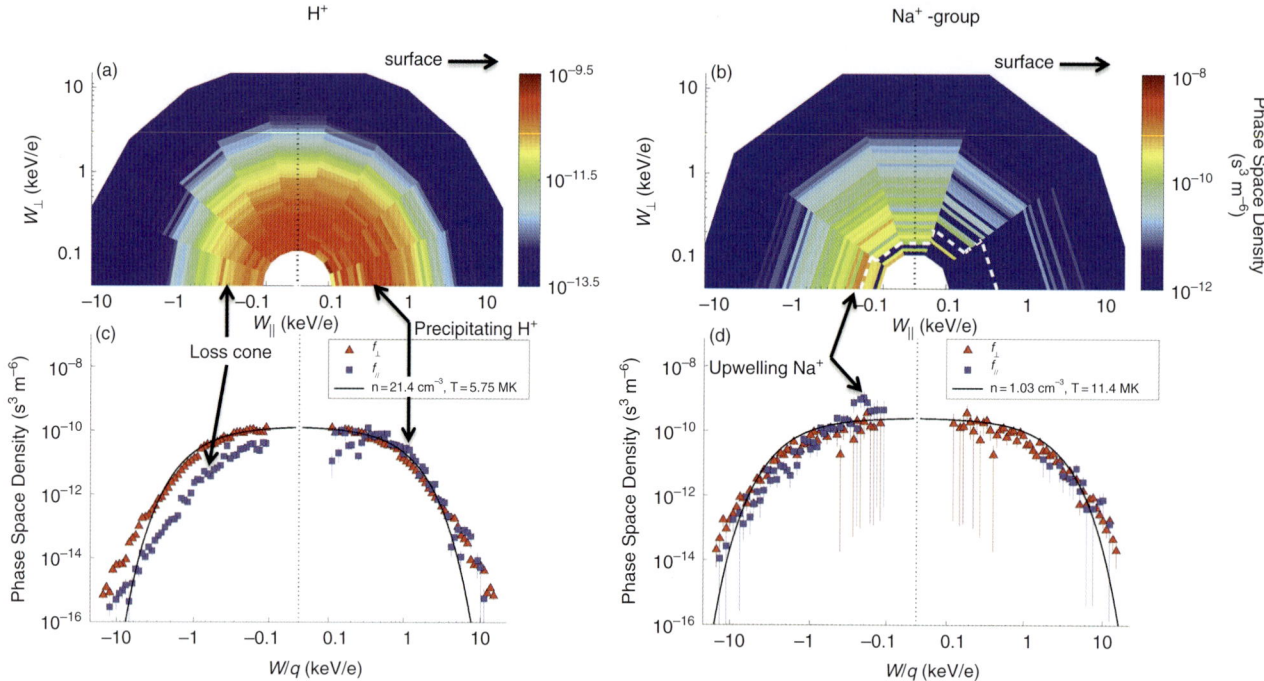

Figure 16.11. (Top) Pitch-angle distributions for (a) protons and (b) Na$^+$-group ions summed over 77 selected cusp crossings resolved by energy parallel, $W_\parallel$, and perpendicular, $W_\perp$, to the local magnetic field. Positive (negative) $W_\parallel$ values correspond to precipitating (upwelling) particles. (Bottom) Phase-space density versus energy per charge W/q for directions parallel ($f_\parallel$, blue squares to the right of the dotted vertical line), antiparallel ($-f_\parallel$, blue squares to the left of the dotted vertical line), and perpendicular ($f_\perp$, red triangles) to the magnetic field for (c) protons and (d) Na$^+$-group ions. The black curves indicate the best-fit Maxwell–Boltzmann distributions, and uncertainties in the counting statistics are shown as vertical bars. Adapted from Raines et al. (2014).

shows the proton distribution as a function of the pitch angle, $\alpha = \tan^{-1}(\sqrt{W_\perp/W_\parallel})$, where $W_\parallel$ and $W_\perp$ are the particle energies parallel and perpendicular to the magnetic field, respectively. Positive (negative) $W_\parallel$ values correspond to precipitating (upwelling) particles. Figure 16.11a shows that the precipitating proton flux is larger than that of the upwelling population, implying that protons are lost to the surface. This loss is also evident in the distribution of the phase-space density (Figure 16.11c) and is labeled as the loss cone. Loss processes are discussed in more detail in Section 16.6.3.

The energy of the Na$^+$-group ions in the cusp typically ranges from about 800 eV to a few keV, with an average of 2.7 keV and a detected maximum of 13 keV limited by the measurement range of the FIPS sensor. Planetary ions produced locally in the cusp initially have much lower energies, 0.1–10 eV, depending on their source (Raines et al., 2014, and references therein), implying that some mechanism accelerates the Na$^+$-group ions to their observed energies. Several mechanisms that have been shown to operate at Earth, e.g., the "cleft ion fountain" (Lockwood et al., 1985) and wave–particle interactions (Ashour-Abdalla et al., 1981), may also be acting at Mercury but are expected to yield energies of only tens of eV, much lower than observed. Raines et al. (2014) proposed that the keV-energy Na$^+$-group ions in the cusp are created when neutral Na atoms, which have drifted beyond the magnetopause boundary, are photoionized and accelerated as they are picked up in the convection of newly reconnected magnetic field lines, which flow anti-sunward over the polar cap. The local Alfvén speed, hundreds of kilometers per second, is consistent with acceleration of Na$^+$-group ions to the observed energies because, for a Na$^+$ ion, an energy of 2.7 keV corresponds to a speed of 210 km s^{-1}.

A population of Na$^+$-group ions moving upward away from the planet was also observed in the cusp (Figures 16.11b and 16.11d, Raines et al., 2014) and provides the first evidence for the predicted effect of ion precipitation and interaction with surface material (Killen et al., 2007). The upwelling component is visible in the energy-resolved pitch-angle distribution (Figure 16.11b) as a narrow band of enhanced flux in the 160° to 180° pitch-angle bin (the bin closest to the $-W_\parallel$ direction) and as a bump in the phase space density in the one-dimensional anti-sunward cut (Figure 16.11d). The energy of these upwelling ions ranges from 100 to 300 eV, implying that they were accelerated by factors of 10–100 after generation at the surface.

16.4.4 Dayside Boundary Layer

The first MESSENGER flyby of Mercury revealed a striking transition near the morning magnetopause during which the magnetic field intensity decreased by nearly a factor of 2 without a significant change in orientation, as shown in Figure 16.12 (Anderson et al., 2011a). This transition occurred within 200–300 km of the magnetopause and was attributed by Anderson et al. (2011a) to a boundary layer (BL in Figure 16.12) of plasma of solar wind origin just inside the magnetopause, consistent with results from hybrid simulations of Mercury's magnetosphere (Trávníček et al., 2007). The mechanism by which this plasma is transported across the magnetopause is not

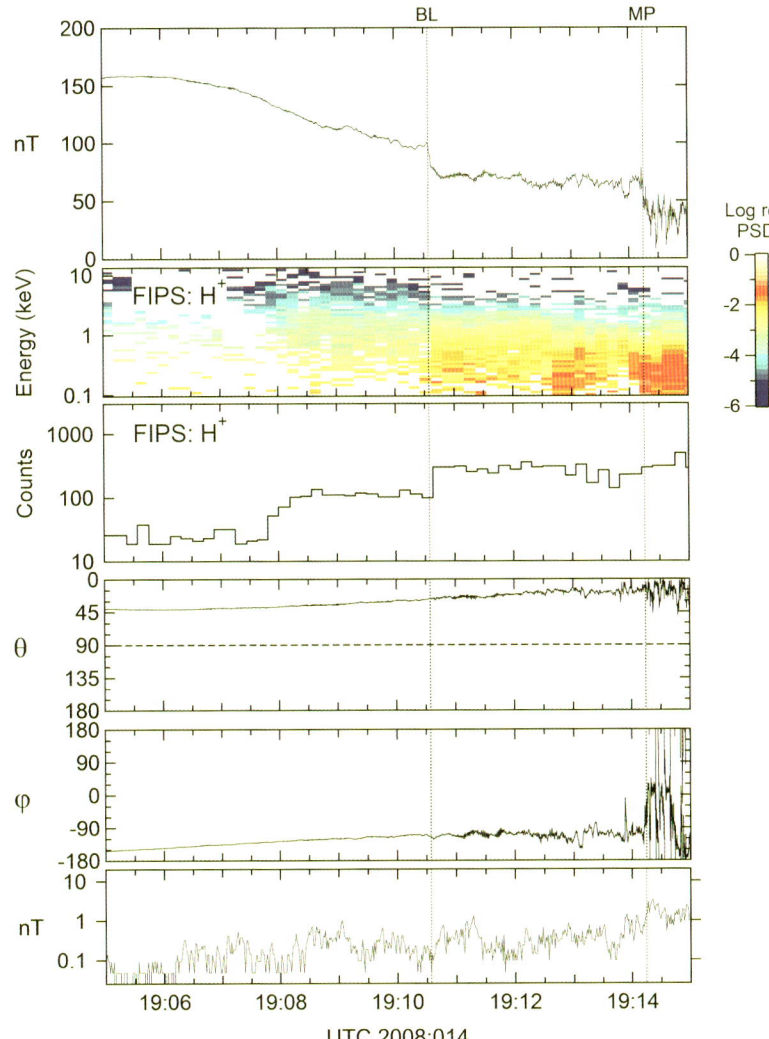

Figure 16.12. Magnetic field and proton data for the outbound magnetosphere crossing of MESSENGER's first Mercury flyby. The inner edge of the boundary layer and the magnetopause are labeled BL and MP, respectively. From top to bottom, the panels show the magnetic field magnitude; the proton phase-space density (PSD) relative to the maximum in the interval; the proton counts in each 8-s integration; the polar (θ) and azimuthal (φ) direction angles of the magnetic field, where $\theta = 0°$ is northward and $\varphi = 0°$ is sunward; and the amplitude of the magnetic fluctuations in the passband 1 Hz to 10 Hz. Adapted from Anderson et al. (2011a).

completely understood, but transport in the cusp region via finite-gyroradius effects may play a role (Section 16.6.3). Alternatively, Slavin et al. (2008) noted that the thickness of the layer corresponds approximately to the gyroradius of Na$^+$ picked up in the solar wind, and they proposed that the layer is due to the Na$^+$ pressure. If this is the case, then the boundary layer should appear in the morning for southward IMF and in the afternoon for northward IMF (Anderson et al., 2011a). Subsequent analysis of FIPS and Magnetometer data from Mercury orbit (Liljeblad et al., 2015) confirmed the presence of a dayside boundary layer at morning local times with a thickness commensurate with the initial estimates. However, Liljeblad et al. (2015) showed that the boundary layer is persistently present in the morning independent of the IMF north–south polarity, implying that Na$^+$ pickup ions do not govern its formation. Additionally, the local-time asymmetry in the boundary layer, which is present in the morning but absent for local times after noon, may contribute to the prevalence on the dusk flank of Kelvin–Helmholtz (KH) waves (e.g., Liljeblad et al., 2014) arising from plasma flow shears across the magnetopause boundary, because the KH instability threshold is lower for thinner boundary layers and hence for greater velocity shear.

16.4.5 Plasma Depletion Layer

The magnetosphere is directly impacted by the shocked solar wind in the magnetosheath (Figure 16.1) rather than by the solar wind itself, and this distinction is important for the dynamics of Mercury's magnetosphere. Of particular significance is the formation of a plasma depletion layer (PDL), with decreased plasma density and increased magnetosheath magnetic field relative to the plasma immediately downstream of the shock (Zwan and Wolf, 1976; Denton and Lyon, 1996). The physical processes downstream of the bow shock that determine whether a PDL will form are governed by the solar wind environment, so we review the salient features of the solar wind at Mercury that distinguish it from nominal conditions at Earth.

To date the most complete observations of the solar wind and IMF at Mercury orbit were made by the Helios spacecraft (Marsch et al., 1982). Some of the derived solar wind parameters that are most important for Mercury's magnetosphere are illustrated in Figure 16.13 (Sarantos and Slavin, 2009). At 1 AU, the sonic and Alfvénic Mach numbers are generally in the range ~7–10, which places Earth, and the more distant planets, in the hypersonic and hyper-Alfvénic flow regime (Marsch et al., 1982). For these upstream conditions, the bow-shock jump conditions are near their

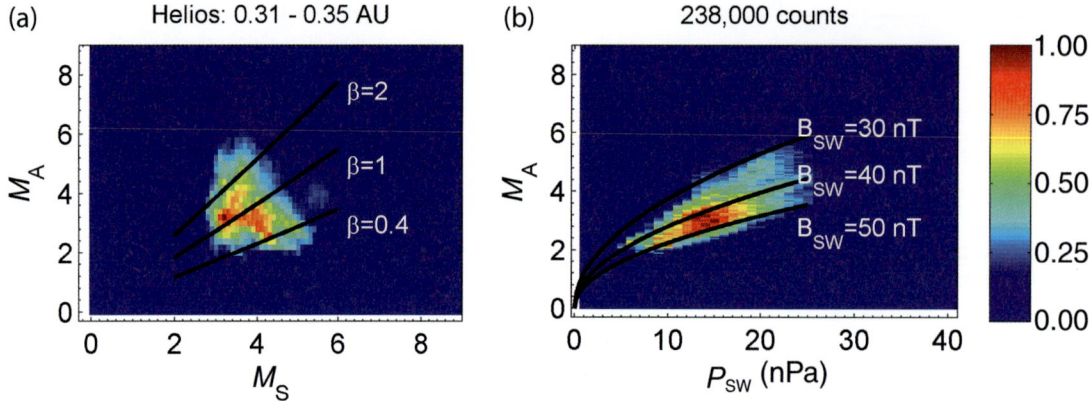

Figure 16.13. Relative probability of occurrence of solar wind properties observed by Helios at heliocentric distances near Mercury perihelion (0.31–0.35 AU). (a) The relationship between the Alfvén and sonic Mach numbers, M_A and M_S, respectively, governs the strength of the bow shock and conditions in the magnetosheath. Lines of constant plasma β are indicated. (b) Variation of M_A with dynamic pressure, p_{sw}. Lines of constant IMF magnitude B_{SW} are indicated. Adapted from Sarantos and Slavin (2009).

asymptotic limit (e.g., Spreiter et al., 1966b) and relatively insensitive to changing Mach number. In contrast, the Mach numbers at Mercury are much lower, ~2–5, as shown in Figure 16.13. The upstream conditions at Mercury therefore fall into the low-Mach-number regime in which the bow-shock jump conditions and its standoff distance are quite sensitive to changes in the solar wind and IMF. The six black lines shown in Figure 16.13 indicate relatively low solar wind β and strong IMF magnetic fields at Mercury's orbit. Figure 16.13b displays the positive correlation between M_A and P_{sw}. Whereas typical solar wind dynamic pressures near Earth are ~1–2 nPa, Helios observed a peak in P_{sw} in the ~10–20-nPa range near Mercury. The orbital phase of the MESSENGER mission was marked by generally weak solar activity, and Baker et al. (2013) used ENLIL solar wind simulations (Toth and Odstrcil, 1996; Odstrcil, 2003) to infer the solar wind conditions during the MESSENGER orbital mission phase. They determined typical dynamic pressure values of ~5–15 nPa, placing the solar wind conditions at Mercury in the low-M_A regime.

One of the most important effects of the low Alfvénic Mach numbers at Mercury is the formation of a PDL. The development of PDLs just upstream of planetary "obstacles" (i.e., a magnetopause or ionopause) was predicted from magnetohydrodynamic theory (Zwan and Wolf, 1976) and confirmed at Earth (Anderson et al., 1991; Fuselier et al., 1991; Phan et al., 1994). The effect of high or low Alfvénic Mach number on conditions in the magnetosheath is illustrated in Figure 16.14 (Gershman et al., 2013). As shown, the region of sub-Alfvénic flow expands greatly in the right-hand panel where the upstream Alfvén Mach number is low (Gershman et al., 2013). In fact, the degree of depletion as measured by plasma β and the extent of the PDL grow as the inverse square of M_A (Zwan and Wolf, 1976).

Gershman et al. (2013) conducted a detailed examination of the degree of depletion and the thickness of the PDL at Mercury as functions of M_A and plasma β. Figure 16.15 illustrates their results for the PDL thickness as a function of the ratio between β just outside the magnetopause in the magnetosheath, β_{MP}, to β just downstream of the bow shock, β_{MS}. Although there is substantial scatter, the thickness of the PDL increases with decreasing β_{MP}/β_{MS}, in qualitative agreement with the

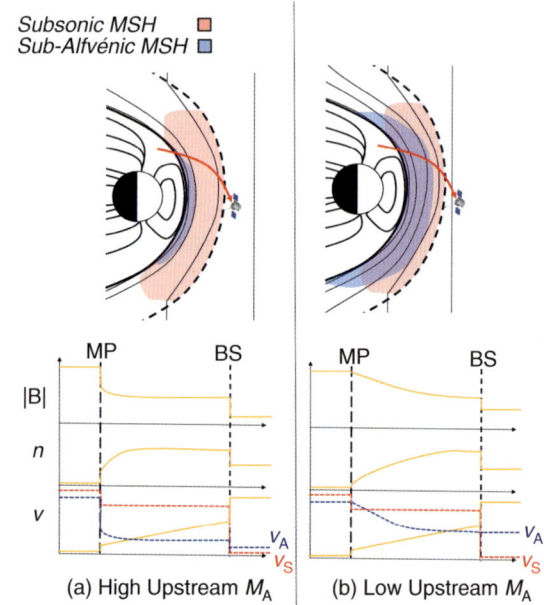

Figure 16.14. Comparison of schematic radial profiles in plasma speed and density and magnetic field intensity from the magnetosphere to the magnetosheath and the upstream solar wind for (a) high-solar-wind M_A and (b) low-solar-wind M_A; a portion of a typical MESSENGER orbit is shown as a red curve ending in a symbol of the spacecraft. With decreasing M_A, a larger fraction of the subsolar subsonic magnetosheath (red shading) shows $v < v_A$, i.e., is sub-Alfvénic, as indicated by the blue shaded regions in the cross sections. In addition, a thicker region of magnetic flux pileup is evident by an increase in B and a decrease in plasma density, n. The Alfvén speed (v_A) and sound speed (v_S) are shown as dashed blue and red lines, respectively, and the magnetopause and bow shock are labeled MP and BS, respectively. Adapted from Gershman et al. (2013).

predictions of Zwan and Wolf (1976). The characteristic length scale for the depletion layer at Mercury is 335 ± 49 km or ~ 0.1 R_{SS}.

As discussed in Chapter 17, one important consequence of a PDL is that the jump in magnetic field intensity across the

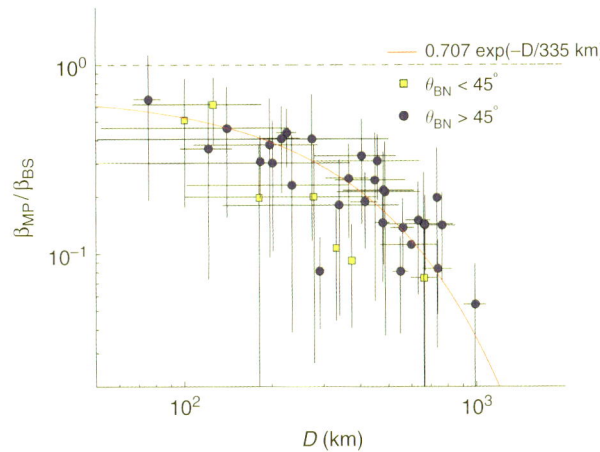

Figure 16.15. Depletion ratio β_{MP}/β_{BS} as a function of measured PDL thickness, D, for all orbits with $\beta_{MP}/\beta_{BS} < 1/\sqrt{2}$, regardless of upstream conditions. Events are classified by the shock angle, θ_{BN}, between the magnetic field and the shock normal direction. Those associated with quasi-parallel, $\theta_{BN} < 45°$, and quasi-perpendicular, $\theta_{BN} > 45°$, shocks are shown as yellow squares and blue circles, respectively. A best-fit exponential relationship (red line) is shown to match the data well. Adapted from Gershman et al. (2013).

magnetopause is less than the jump without plasma depletion (Phan et al., 1996; Anderson et al., 1997a). This pattern is particularly true during the impact of a coronal mass ejection (CME) onto the magnetosphere when the upstream Alfvénic Mach number at Mercury can approach unity (Sarantos and Slavin, 2009). Analyses of such impacts at Mercury by Slavin et al. (2014) indicated very high reconnection rates at the dayside magnetopause and extremely high numbers of flux transfer events, deep magnetospheric cusps because of intense plasma injections, and large numbers of flux ropes in the plasma sheet, even when the angle between the magnetosheath and magnetospheric magnetic fields at the magnetopause was substantially less than 90°. Further, DiBraccio et al. (2013) found that the magnetopause reconnection rate at Mercury is relatively insensitive to IMF orientation but increases as β_{MP} decreases. Such a dependence, but over a smaller range of plasma β, has been observed at Earth (Scurry et al., 1994; Farrugia et al., 1995). For extreme conditions at Earth, Anderson et al. (1997a) showed that the reconnection rate should vary as the square of M_A and suggested that the relative increase in magnetic field in the PDL facilitates near-subsolar reconnection for a broad range of magnetosheath field orientations relative to the magnetospheric field if component reconnection occurs (Sonnerup, 1984). In this way, it appears that the occurrence of subsolar reconnection for nearly all IMF orientations is likely a consequence of the prevalence of a PDL at Mercury.

16.5 CURRENT SYSTEMS AND MAGNETIC FIELD MODELS

16.5.1 Cross-Tail Current

The stretching of the magnetic field in the magnetotail (Figure 16.1) implies a dawn-to-dusk current centered at the magnetic equator. Orbits from the first three Mercury years of

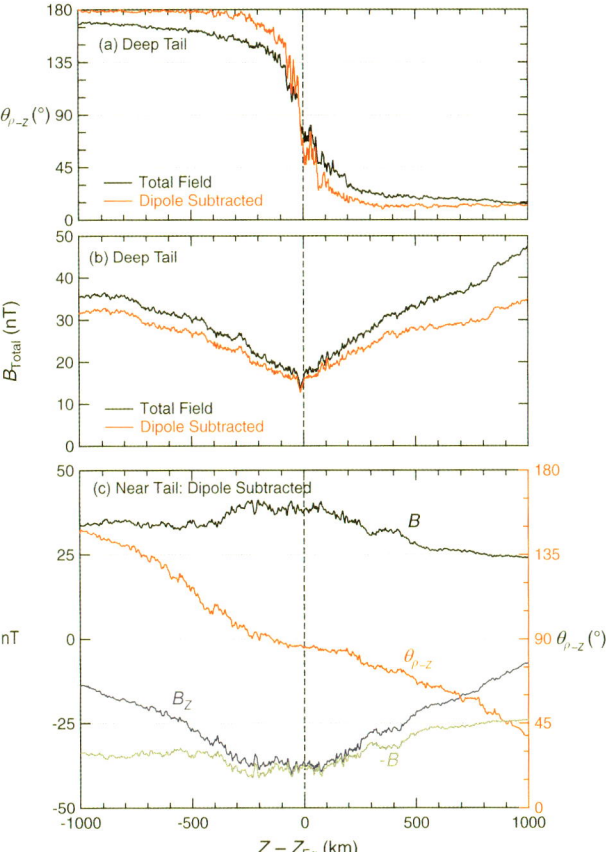

Figure 16.16. Stacks of measurements from 79 deep-tail current-sheet crossings with (red) and without (black) the dipole field removed for (a) tilt, $\theta = \cos^{-1}(B_\rho/B_{\rho Z})$ in degrees, where $B_{\rho Z}$ is the magnitude of **B** projected into the ρ–Z plane, and (b) field magnitude (B) in nT. Each orbit is aligned on its equator crossing, Z_{Eq}, before stacking. (c) Stacks for 47 near-tail orbits (see text for selection criteria) showing B, B_Z, and θ after removal of the dipole field. Adapted from Johnson et al. (2012).

observations with magnetic equator crossings within 3 h of local midnight were used to constrain the current-sheet half-thickness, D_D, in the deep tail and the distance from the planetary spin axis to the inner edge of the current sheet, R_2, in the near tail (Johnson et al., 2012). For the deep tail, the current sheet was indicated by a rotation in the field direction from dominantly anti-sunward in the southern tail lobe to dominantly sunward in the northern tail lobe. A mean thickness for the current sheet in this region was obtained from a superposed epoch analysis. The magnetic equator crossing for each orbit was assigned as the reference time, and the tilt of the field in the ρ–Z plane was calculated from $\theta_{\rho-Z} = \cos^{-1}(B_\rho/B_{\rho Z})$, where $B_{\rho Z}$ is the magnitude of **B** projected onto the ρ–Z plane; $\theta_{\rho-Z}$ values close to 0° and 180° indicate anti-sunward and sunward field directions, respectively. The angles $\theta_{\rho-Z}$ and the magnetic field magnitudes for the 79 selected deep-tail orbits were averaged, and the results are shown in Figures 16.16a and 16.16b. The rotation of the field direction from the southern to the northern lobe is clear (black curve), and after removal of the dipole field (red curve) the field is almost purely anti-sunward in the southern lobe and sunward in the northern lobe. The depression in field magnitude associated with the plasma sheet (Korth et al., 2011) is centered on the field reversal, although the magnetic depression is broader than the field rotation, indicating that the plasma sheet is thicker than the current sheet on average. The field rotation is 95% complete within 140 km (0.09 R_M) of the current-sheet center, and this observation was used to define the current-sheet half-thickness in the far tail, $D_D = 0.09 \, R_M$.

Orbits that sampled the near-tail region provide information on how close, on average, the current sheet comes to the planet (R_2). These near-tail trajectories generally traversed the equator planetward of the cross-tail current. Depressions in the field magnitude, indicating spacecraft encounters with the plasma sheet, occurred on some but not all crossings of the near-tail region. Near-tail orbits with stronger plasma-sheet signatures were selected by evaluating the minimum value of the magnetic field strength near the equator crossing. Orbits with the strongest plasma-sheet signatures and deepest magnetic field minima should be those that passed closest to the current sheet. Superposed epoch averages were obtained for the 25% of the orbits with the lowest minimum field magnitudes (Figure 16.16c). The vector dipole field was subtracted to assess the field properties from external currents only, and averages of the residual field magnitude (B), Z-component (B_Z), and polar angle ($\theta_{\rho-Z}$) were taken. The near-tail field, even for these orbits, was quite different from the far-tail field. First, $\theta_{\rho-Z}$ came only within ~30° of the 180° or 0° direction, indicating that the orbits were, on average, planetward of the current-sheet and tail lobes. Second, the magnitude of the external field (after removal of the dipole field) increased rather than decreased near the equator crossing and was almost entirely in the $-B_Z$ direction, i.e., southward, at the equator. This behavior indicates that the external field in this region was dominated by the fringing field of the cross-tail current planetward of the tail current sheet. Thus, these orbits passed close to but not through the current sheet. Therefore, a lower bound on R_2 of 1.41 R_M was estimated from the mean radial distance ($\sqrt{X^2 + Y^2}$) to these equator crossings.

16.5.2 Birkeland Currents

The interaction of the solar wind and IMF with Mercury's magnetic field drives the magnetic convection cycle, also known as the Dungey cycle (Dungey, 1963; Slavin et al., 2007), and signatures of magnetic reconnection at the magnetopause, in the cusp, and in the magnetotail all confirm this process at Mercury. This convection implies the imposition of an electrical potential on the open magnetic flux of Mercury's magnetosphere, which is estimated to be in the range 15–30 kV (DiBraccio et al., 2013, 2015b). At Earth, the Dungey cycle is responsible for the field-aligned or Birkeland current system that conveys stress between the ionosphere and magnetosphere, and closure of these currents drives Joule dissipation (Cowley, 2000; Richmond and Thayer, 2000). The average Birkeland currents at Earth for southward IMF consist of two concentric upward/downward pairs of currents appearing approximately in arcs at constant magnetic latitude centered at dawn and dusk. The poleward pair of currents, denoted Region 1, is upward at dusk and downward at dawn, whereas the second pair, denoted Region 2, is within ~5° latitude of Region 1 and has the opposite polarity, downward at dusk and upward at dawn (Iijima and Potemra, 1976; Anderson et al., 2008). Whether Mercury supports steady-state Birkeland currents without a conducting ionosphere was not known, and a variety of suggestions had been made for their configuration and closure at Mercury (Glassmeier, 2000; Ip and Kopp, 2004; Janhunen and Kallio, 2004).

MESSENGER observations revealed that Birkeland currents corresponding to the terrestrial Region 1 polarity are present at Mercury (Anderson et al., 2014). To identify the signals of these currents, the magnetic residuals within Mercury's magnetosphere over the northern hemisphere were calculated by removing a model field, B_m, that includes both an internal field, B_{int}, represented as an axially aligned, offset dipole (Anderson et al., 2012; Johnson et al., 2012), and an external field, B_{ext}, accounting for magnetopause and magnetotail currents (Korth et al., 2015; and Section 16.5.3). Writing the total model field as $B_m = B_{int} + B_{ext}$ and the observed magnetic field as B_{obs}, the residuals are $\delta B = B_{obs} - B_m$.

The δB vector components perpendicular to B_m projected onto the MSO X–Y plane as viewed from above the north pole and plotted along the orbit trajectory are shown for two sets of three sequential orbits in Figure 16.17 (Anderson et al., 2014). The sets of orbits are ~90 days apart, corresponding to slightly more than one Mercury year and ~1.5 planetary spin periods. In the upper and lower sets of orbits, the planetary orientation is nearly opposite: relative to its orientation in the upper set of orbits, the planet rotated ~200° counterclockwise for the bottom sets of orbits. The residuals poleward of 60°N are consistently sunward regardless of the planet's orientation, and their magnitude varies by a factor of 2 or more from one orbit to the next. These features indicate that the signals reflect currents of external origin that vary in intensity on timescales comparable with or shorter than the MESSENGER orbit period. The signals were interpreted by Anderson et al. (2014) as fields associated with Birkeland currents flowing between altitudes above and below the spacecraft: an upward current in the evening and a downward current in the morning, corresponding to the poleward Region 1 currents documented at Earth.

To derive field-aligned current densities, the portion of δB parallel to B_m was subtracted to obtain the transverse residual, $\delta B_\perp$ (Anderson et al., 2014). These data were averaged to obtain maps of $\delta B_\perp$ for each year of orbit operations, binned by magnetic disturbance level (Anderson et al., 2013). The field-aligned current density threading the mean "orbit surface" for each year was calculated from the curl of $\delta B_\perp$ and mapped to the planetary surface along B_m to obtain the surface radial current density, j_{rS}. Figure 16.18 shows maps of j_{rS} versus latitude and local time for low, moderate, and high levels of magnetic disturbance for the first three years of orbit operations obtained from data acquired during the descending orbit segments (Anderson et al., 2016). Results for ascending orbit segments are essentially the same. The increase of current with activity is clear, as is the basic structure of an upward current in the dusk sector and a downward current in the dawn sector. The total average currents range from just under 20 kA for quiet conditions to nearly 40 kA during disturbed conditions (Anderson et

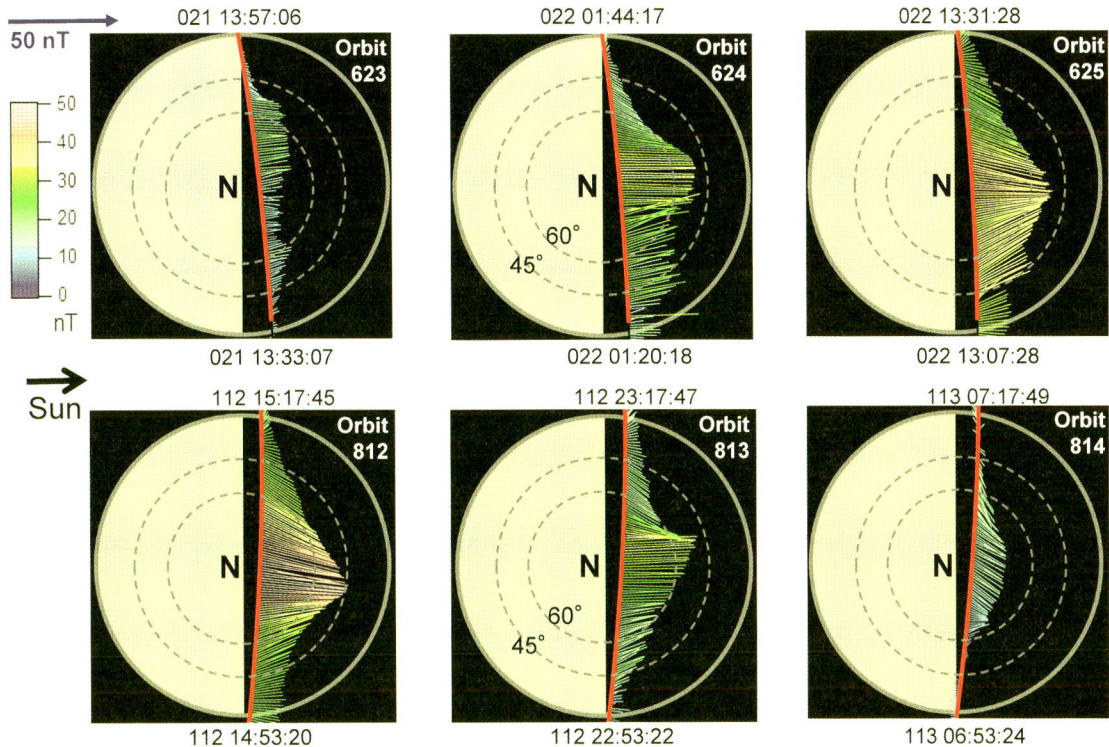

Figure 16.17. Magnetic perturbations recorded by MESSENGER as viewed from above Mercury's north pole with the Sun to the right during 21–22 January 2012 (upper panels) and 21–22 April 2012 (lower panels). The spacecraft trajectory is shown in red below 1000-km altitude. Magnetic residuals perpendicular to the total model field and projected onto the X–Y plane, $\delta \mathbf{B}_{XY}$, are plotted at 12-s intervals and shown by colored lines originating at the observation point. The directions correspond to the $\delta \mathbf{B}_{XY}$ direction, and the color and length indicate $|\delta \mathbf{B}_{XY}|$ (see color bar and blue reference arrow at upper left). Start and end times are given by day of year and UTC. Adapted from Anderson et al. (2014).

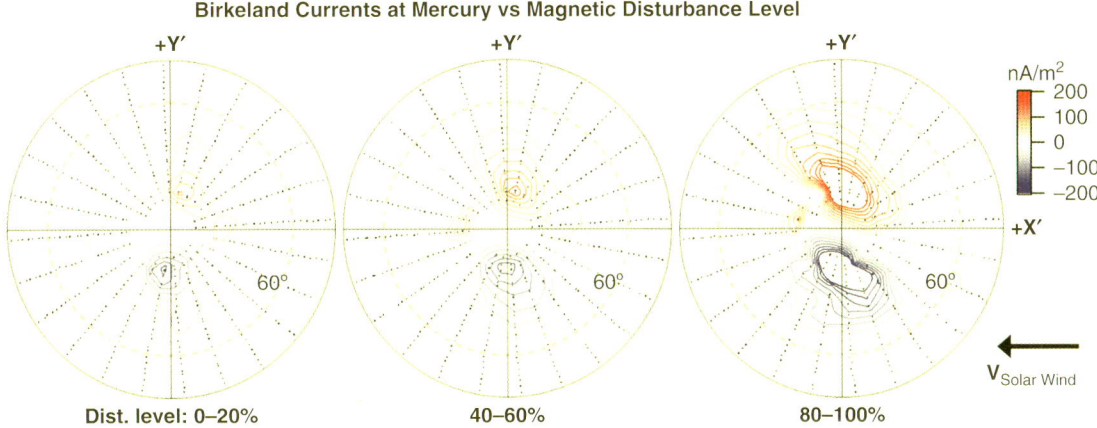

Figure 16.18. Average Birkeland current densities in aberrated MSO coordinates determined from descending orbit segments of MESSENGER Magnetometer data from 23 March 2011 through 31 March 2014 and mapped to the planetary surface. Upward currents are indicated in red and downward currents in blue, with contours every 20 nA m^{-2} and current densities as indicated by the color bar. From left to right, panels show current density for the quietest 20% of orbits, middle disturbance level (40–60%), and most magnetically disturbed orbits (80–100%), where the disturbance level is the magnetic activity index for Mercury of Anderson et al. (2013). From Anderson et al. (2016).

al., 2014), a factor of ~100 lower than the Birkeland currents at Earth (Richmond and Thayer, 2000). Significantly, there is no evidence of a Region 2 current system at Mercury, consistent with the expectation that Mercury's magnetosphere does not support a ring current plasma population, which is central to the Region 2 currents at Earth (e.g., Slavin et al., 2007).

Although the convection potential applied to Mercury's magnetosphere, ~30 kV, is lower than that at Earth, which is typically ~100 kV during active conditions, the decrement is only a factor of 3, whereas the currents are weaker by a factor of 100, indicating a much greater electrical resistance to their closure at Mercury. Anderson et al. (2014) used a simple single spherical shell model for interior electrical conductivity to show that the observed current distribution and estimated total potential are consistent with current closure through the planet, as proposed by Janhunen and Kallio (2004) for a nominal conductivity structure (e.g., Verhoeven et al., 2009). Anderson et al. (2016) extended this work to allow a general radially varying conductivity and analyzed a range of possible conductivity–depth profiles to show that the initial result holds generally, indicating that the current most likely does close through the planet and between 50% and 90% of the current closes through the core. In addition to core induction effects on the magnetosphere (Johnson et al., 2016), this coupling between Mercury and its magnetosphere is unique in the solar system in that the solid planet itself is an integral part of the magnetospheric electrodynamic system.

16.5.3 Magnetic Field Models

A consequence of Mercury's weak internal magnetic field is that external fields generated by magnetospheric current systems contribute substantially to the observed magnetic field throughout the small magnetosphere. Accurate modeling of the magnetospheric magnetic field is therefore important to analyzing planetary fields arising from the internal dynamo, crustal magnetization, and induction effects. Because current systems are present throughout the magnetosphere, spatially distributed observations are required to characterize them. MESSENGER's sampling of the magnetic field in the inner magnetosphere was sufficiently dense to allow the magnetic field at spacecraft altitudes less than ~800 km to be predicted to better than 10% accuracy (Johnson et al., 2012; Korth et al., 2015). Two models were developed over the course of the MESSENGER mission, with the first derived early in the orbital mission phase to enable other analyses and the latter taking advantage of the full mission data.

The first magnetospheric model developed was the paraboloid model (Alexeev et al., 2010; Johnson et al., 2012). The paraboloid model represents the magnetospheric magnetic field as the sum of the contributions from sources internal and external to the planet. It includes the internal magnetic field of the offset dipole, $\mathbf{B}_{int}$, and external magnetic fields, $\mathbf{B}_t$ and $\mathbf{B}_{cf}$, generated by cross-tail and magnetopause currents, respectively. The total magnetic field,

$$\mathbf{B}_m = \mathbf{B}_{int} + \mathbf{B}_t + \mathbf{B}_{cf}, \quad (16.4)$$

is confined within a magnetopause prescribed by a paraboloid of revolution. All except one of the parameters describing the model can be established directly from MESSENGER observations. The parameters describe the magnetopause shape, geometry of the cross-tail current sheet, magnetic flux in the magnetotail, and dipole location and orientation. In this model, the magnetopause is represented by a paraboloid [Section 16.3, equation (16.3)]. The cross-tail current sheet of the model is parameterized by its thickness, $0.09\ R_M$, and the location of its inner edge, $1.41\ R_M$ (Section 16.5.1). The magnetic flux, $F = B\pi r_{MP}^2/2 = 2.6$ MWb, was estimated from the average magnetic field magnitude, B, in the magnetotail and the radius, r_{MP}, to the Shue et al. (1997) magnetopause (Section 16.3). Finally, the properties of Mercury's axisymmetric internal dipole field, which is offset by $0.196\ R_M$ to the north (Anderson et al., 2011b, 2012), are described in Chapter 5. Using the above parameters as a priori constraints in the paraboloid model, the dipole moment, $\mu = 190$ nT R_M^3, was obtained by minimizing the root mean square (rms) value of the residual magnetic field,

$$\overline{\delta B} = \sqrt{\sum_{i=1}^{N}\left[(B_X - B_{m,X})^2 + (B_Y - B_{m,Y})^2 + (B_Z - B_{m,Z})^2\right]/N},$$

$$(16.5)$$

between the components B_i of the magnetic field observed within the magnetosphere and the components $B_{m,i}$ of the model field obtained for the set of N observations (Johnson et al., 2012). Magnetic field lines of this model are shown as blue dashed lines in Figure 16.19, and the gray solid and dashed ellipses indicate the observational sampling of the MESSENGER orbit.

The most serious limitation of the paraboloid model is the shape of the model magnetopause (Figure 16.19), which deviates substantially from the observed magnetopause and flares too much with anti-sunward distance from the planet. The result is that the predicted contribution from magnetopause currents is too small at high northern latitudes. Less serious but also problematic is the sharp inner edge of the cross-tail current sheet in this model, which yields magnetic islands at which magnetic field lines loop back on themselves rather than map to the planet. The rigid analytical formalism of the paraboloid model allows neither modifications to resolve these discrepancies nor

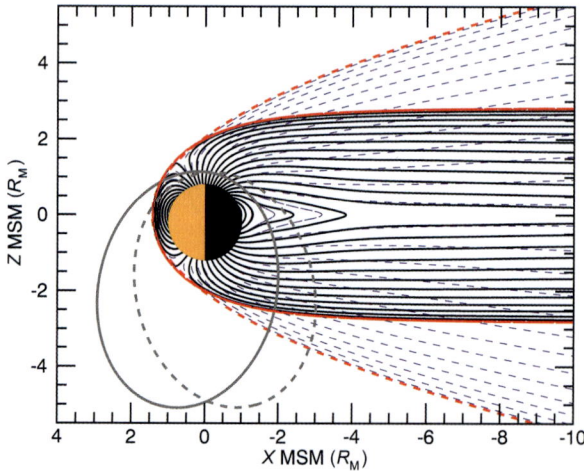

Figure 16.19. Magnetic field lines of the paraboloid (dashed blue) and KT14 (solid black) models in the MSM X–Z plane confined within the observed average magnetopause (red), modeled after Shue et al. (1997) using best-fit parameters determined from MESSENGER Magnetometer observations. The planet is shown as a circle with dayside and nightside in orange and black, respectively, and MESSENGER orbits 2723 and 2822 are shown as solid and dashed gray lines, respectively. The model magnetopause for the paraboloid model is indicated by the dashed red line.

extensions to include additional current systems, such as the Birkeland currents discussed above.

The extensive orbital observations provided the basis to model Mercury's magnetospheric magnetic field while allowing for an arbitrary magnetopause shape, a representation of the magnetic field that is continuous in space, in particular in the near-tail region on the nightside, and use of a modular, expandable model framework. The development of this model, termed KT14 (Korth et al., 2015), followed the data-based approach to generate magnetic shielding fields used for Earth's magnetosphere (e.g., Tsyganenko, 2013). In the absence of reconnection, these shielding fields eliminate the normal component of magnetic field, B_n, at the magnetopause and separate interplanetary and planetary magnetic field lines. The numerical approach to minimizing B_n can be applied to determine the shielding field and, hence, the magnetopause currents for an arbitrary magnetopause shape. The shielding field is defined as

$$\mathbf{B}_{cf} = -\nabla U, \qquad (16.6)$$

where the scalar potential, U, is represented by Cartesian harmonic basis functions in the MSM frame:

$$U = \sum_{i,k=1}^{N} a_{ik} \exp\left[\sqrt{p_i^2 + p_k^2}\, X\right] \cos(p_i Y) \sin(p_k Z). \qquad (16.7)$$

The N^2 linear coefficients a_{ik} and N non-linear coefficients p_i are obtained by minimizing the RMS residual of the magnetic field component normal to the boundary surface:

$$\sigma = \sqrt{\sum_{j=1}^{M} [(\mathbf{B}_j - \nabla U) \cdot \mathbf{n}]/M}, \qquad (16.8)$$

where M is the number of data points at the boundary surface. The boundary surface is given by the Shue et al. (1997) magnetopause with best-fit parameterization (Winslow et al., 2013) (Section 16.3). The shielding field is defined separately for each magnetospheric source field, thus ensuring modular extensibility of the model.

To provide a spatially continuous representation, the magnetic field of the cross-tail current sheet in the KT14 model is composed of two distinct contributions. In the far magnetotail, $\gtrsim 10\ R_M$ anti-sunward, the current density at the magnetic equator is held constant with radial distance but decreases to the north and south and vanishes at the northern and southern edges of the current sheet. Closer to the planet on the nightside, $\sim (1.5–5)\ R_M$ from Mercury's center, the current density is higher, corresponding to higher magnetic field strength in the lobes. The resulting current distribution is represented with a disk-shaped current sheet (Tsyganenko and Peredo, 1994) in the near magnetotail and a sheet-shaped current farther tailward. The disk current, I, is expressed as a piecewise continuous function of radial distance, $\rho = \sqrt{X^2 + Y^2}$:

$$\begin{aligned}
I(\rho) &= 0 & \rho &< \rho_1, \\
I(\rho) &= I_m \sin^2\left(\frac{\pi}{2}\frac{\rho - \rho_1}{\rho_2 - \rho_1}\right) & \rho_1 &< \rho < \rho_2, \\
I(\rho) &= I_m \exp\left(\frac{\rho_2 - \rho}{L}\right) & \rho &> \rho_2.
\end{aligned} \qquad (16.9)$$

In equation (16.9), I_m is the peak current, ρ_1 and ρ_2 are the cylindrical radial distances to the inner edge of the current sheet and the peak current, respectively, and the e-folding scale L imposes a decrease of the current for large ρ. This form provides a smooth inner edge that avoids magnetic islands. Consistent with the paraboloid model, the inner edge of the current sheet is $\rho_1 = 1.41\ R_M$, and $\rho_2 = 1.56\ R_M$ and $L = 1.43\ R_M$ were chosen a posteriori to yield a gradient near the inner edge as steep as could be fit using the procedure below and a decay to $I_m/10$ at a radial distance of $5\ R_M$. These geometrical approximations were introduced to avoid an unphysically sharp inner edge of the cross-tail current. A magnetic vector potential, $\mathbf{A}$, was then sought, so that the magnetic field of the disk current,

$$\mathbf{B}_d = t_1 \nabla \times \mathbf{A}, \qquad (16.10)$$

corresponds to that of the current represented by equation (16.9), where the amplitude parameter t_1 is proportional to the strength of the cross-tail current. For distances $>5\ R_M$ tailward, the current disk merges into a sheet current having a vector potential of the form $\mathbf{A} = (0, A_Y, 0)$, where

$$A_Y = -2\, t_2 \ln \cosh\left(\frac{z}{d}\right), \qquad (16.11)$$

and where t_2 is the current amplitude, d is the half-thickness of the current sheet, and z is the distance to the current sheet. The disk and the sheet currents are both centered on the magnetic equator and widen infinitely toward the dayside magnetosphere, where the cross-tail current vanishes.

As in the paraboloid model, the magnetic field is composed of the sum of internal and external contributions. The KT14 model parameters for the dipole moment and offset and the nominal current-sheet half-thickness are identical to those of the paraboloid model. The amplitude parameters of the disk and sheet currents, t_1 and t_2, respectively, were obtained by minimization of the RMS residual of the magnetic field observed within the magnetosphere and corrected for aberration. The expansion of the tail current-sheet thickness in MSO X and Y directions was set empirically to match the boundary between open and closed magnetic field lines inferred from plasma observations (Korth et al., 2014). The model was developed from a data set acquired over seven Mercury years with an updated value of the subsolar standoff distance of $R_{SS} = 1.42\ R_M$ (Winslow et al., 2013). The magnetic field lines of this model are shown in Figure 16.19, with solid traces indicating the confinement within the observed magnetopause shape.

The KT14 model RMS residual is $\overline{\delta B} = 24.8$ nT, and the resulting model satisfies $\nabla \cdot \mathbf{B} = 0$ in the magnetosphere. Figure 16.20 shows the spatial distributions of magnetic field residuals in the radial (a and d), southward (b and e), and eastward (c and f) components with respect to the KT14 model. For the analysis period, the ascending (a–c) and descending (d–f) orbit legs correspond to higher- and lower-altitude observations, respectively.

The magnetic field residuals exhibit several systematic features. The overall structures of the residuals obtained from the KT14 and paraboloid models were found to be very similar (Korth et al., 2015), reflecting in part the limitations of the MESSENGER data distribution in prescribing the nightside external fields. The largest

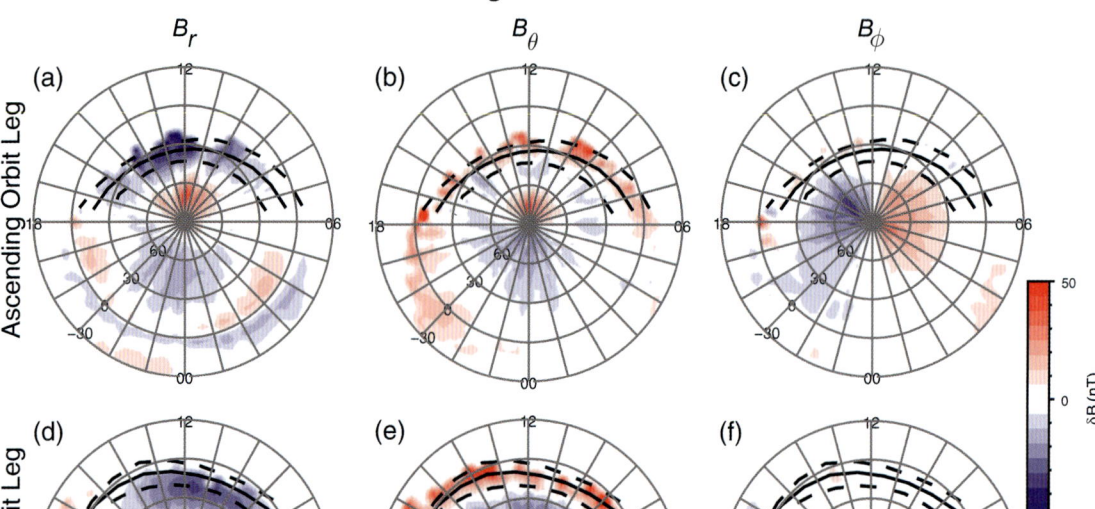

Figure 16.20. Magnetic field residuals with respect to the KT14 model in the (a and d) radial, (b and e) southward, and (c and f) eastward directions for ascending (a–c) and descending (d–f) orbit legs averaged over the period 24 March 2011 to 23 November 2012. Grid lines are labeled in aberrated local time and MSM latitude, and the color bar denotes the magnitude of the residuals. In each panel, the latitude range over which MESSENGER crossed the magnetopause on the dayside is delineated by the dashed lines, and the average crossing latitude is indicated by the solid line. Adapted from Korth et al. (2015).

KT14 residuals are in δB_r and δB_θ near the dayside magnetopause and may originate from a dayside boundary layer (Section 16.4.4). Similar residuals associated with the presence of plasmas are observed in the northern cusp and in the plasma sheet near the magnetic equator. Because neither the KT14 model nor the paraboloid model include the diamagnetic effects of the plasma populations, these models cannot account for local magnetic field variations associated with the maintenance of total pressure balance in these regions. Second, large-scale residuals in δB_ϕ, which are directed eastward at dawn and westward at dusk, are found on both the ascending and descending orbit legs and extend from ~30°N to the north pole. Their distribution is consistent with the existence of steady field-aligned Birkeland currents (Section 16.5.2). Third, residuals in the downward ($-\delta B_r$) and poleward ($-\delta B_\theta$) directions are observed in a partial ring structure extending between latitudes 30°N and 60°N; the structure is especially pronounced on the descending orbit legs. These residuals indicate additional magnetic field sources not represented by either model.

16.6 PLASMA

16.6.1 Sources

Mercury is embedded in the most dense and most dynamic solar wind of any planet in our solar system, and MESSENGER provided critical insights regarding both the dynamics of solar plasma acceleration and the effects of these unique solar wind conditions on Mercury's magnetosphere. At the Sun, energy transfer from the solar magnetic field accelerates solar plasma within the magnetically dominant solar atmosphere, the solar corona, to produce the solar wind that fills the entire heliosphere. The solar wind outflow reaches supersonic and super-Alfvénic speeds at heliocentric radial distances between (2–10) R_S and (10–20) R_S (where R_S = 695,700 km is the nominal solar radius), respectively. In situ observations in the inner heliosphere by MESSENGER provided evidence that this heating process is both transient in time and highly spatially dependent, especially near Mercury's orbit (Gershman et al., 2012).

With the exception of He ions, the fast solar wind associated with the solar poles and coronal holes is nearly photospheric in ion composition. Slow and transient streams from CMEs are enriched relative to the photospheric composition in ions with a first ionization potential (FIP) <10 eV and also in Ne (von Steiger et al., 2000). Overall, the abundance of He^{2+} relative to that of H^+ is depleted in the solar wind when compared with the photospheric composition, with an average He^{2+}/H^+ ratio of ~3% compared with the solar value of 9%. All ions from the Sun, with the exception of those from rare events associated with CMEs, reflect electron temperatures of the corona of >1 MK. CME-associated plasmas contain both the ions with the hottest electron temperatures (up to 3 MK) and also those with the coolest temperatures, associated with erupting solar filaments (<0.5 MK), keeping He singly ionized and providing additions of heavy ions with much lower charge states than usual.

In addition to solar wind ions, heliospheric pickup ions are also part of Mercury's space environment. These pickup ions originate from neutral particles from both galactic and interplanetary sources and can be ionized near the Sun and picked up in the expanding magnetic field of the solar wind (Gloeckler et al., 2000; Gershman et al., 2013). Because of gravitational focusing of the interstellar neutral atoms at the Sun in the antiparallel direction of the Sun's movement through the galaxy, this component is time dependent at Mercury.

The main source of plasma in Mercury's magnetosphere is the solar wind, as is evident from the dominance of solar wind protons in number density in most regions of the magnetosphere and under most conditions. Nonetheless, heavy ions released from the surface have also been observed with plasma pressures comparable with the proton pressure. Charged particles are produced either directly at the surface or indirectly from ionization of neutral atoms removed from the surface. Exospheric neutrals are produced by impact vaporization, photon- and electron-stimulated desorption, and surface sputtering processes and are subsequently ionized, e.g., by interaction with photons or electrons (see Chapter 15). For low-FIP elements (e.g., Na, Mg, Al), photoionization is most likely the dominant process, but other ionization processes can also contribute (Leblanc and Johnson, 2003, 2010). The impact of energetic ions onto the surface can also release ions directly from the surface, although ions represent ≤10% of the sputtering products (Killen et al., 2007). The production of these species is highly dependent on surface composition and, for all processes except photon-stimulated desorption, on incident particle atomic mass, energy, and flux. Models of these processes have been developed (Lammer et al., 2003) and used to predict the composition of the global exosphere (e.g., Leblanc and Johnson, 2003, 2010; Wurz et al., 2010).

These models have been used along with photoionization rates to estimate ion production rates, but none yet incorporate the ion component that comes from surface processes directly. Also, computation of ion source rates requires consideration of exospheric and magnetospheric transport (Sarantos and Slavin, 2009).

Ionization rates have not been well constrained because of the lack of observations of electron distribution functions and because of the challenges to accounting for transport processes at higher energies. Newly created photoions have essentially the same energy as their parent neutral atom, peaking at no more than a few eV (e.g., Burger et al., 2012; Cassidy et al., 2015). During most of the MESSENGER mission, the minimum energy for detection by the FIPS sensor was 46 eV, so direct measurement of these ions was impossible. Without reliable in situ observations, photoionization rates at Mercury must be obtained by scaling estimates from 1 AU, which exist for many exospheric species (e.g., He, O, Na, S, Ca, K) (Huebner et al., 1992; Killen et al., 2005, 2007), to the inner heliosphere. Exospheric ionization processes and production rates are described in Chapters 14 and 15.

16.6.2 Composition, Properties, and Distribution

16.6.2.1 Ions

The overwhelming majority of ions observed in Mercury's magnetosphere are protons from the solar wind. In addition, alpha particles (He^{2+}) and highly ionized heavy ions (C^{5+}, O^{6+}), which also originate from solar sources, are typically observed collocated with protons, as well as He^+ pickup ions generated in the magnetosphere (Gershman et al., 2013). Solar wind entry into the magnetosphere primarily results from reconnection at the dayside magnetopause, but particles also enter the magnetosphere directly, e.g., at flanks via the KH instability (Sundberg et al., 2010). Figure 16.21 shows the proton flux measured on the descending legs of orbits over a 10-month period mapped onto a grid of local time and invariant latitude, which describes the field line of each observation

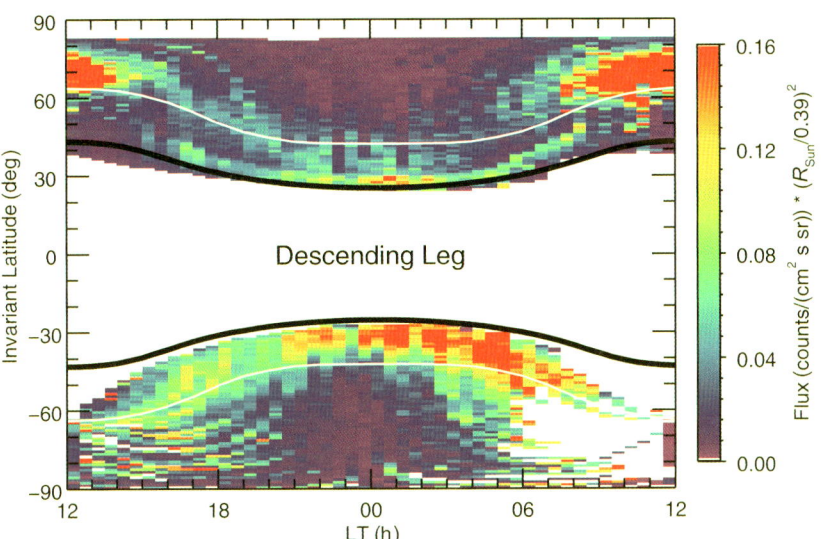

Figure 16.21. Distribution of the mean proton flux with respect to invariant latitude and local time observed on the descending leg of each orbit during the period 11 April 2011 to 12 February 2012, normalized to Mercury's mean heliospheric distance to account for seasonal variations in solar-wind environment. The black lines outline the region of closed model field lines near the equator not sampled by MESSENGER. The boundaries between open and closed field lines in the best-fit paraboloid magnetic field model are represented by white lines. Adapted from Korth et al. (2014).

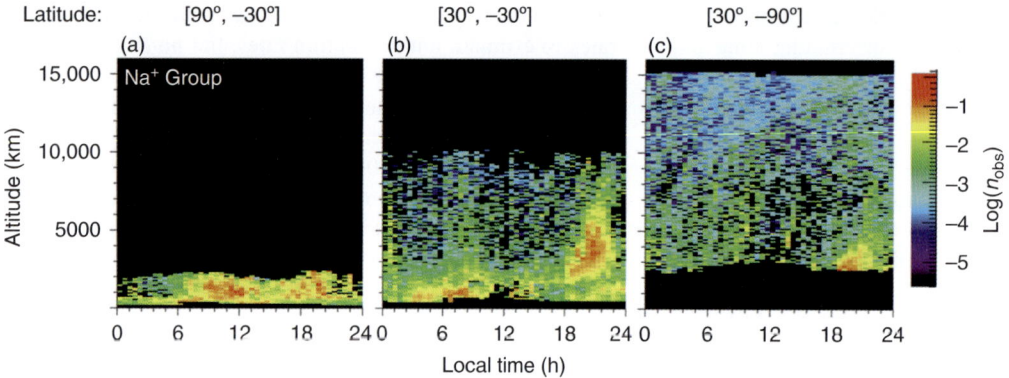

Figure 16.22. Distribution of Na+-group ions as a function of local time and altitude for the latitude ranges (a) +90° to −30°, (b) +30° to −30°, and (c) +30° to −90°N. The color black indicates regions without observations, and distinct plasma populations are indicated by colors. Adapted from Raines et al. (2013).

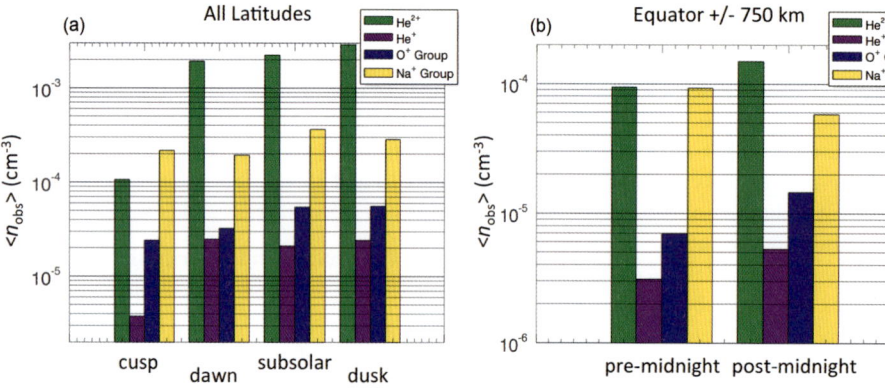

Figure 16.23. Plasma composition in Mercury's space environment averaged over the period 25 March 2011 to 31 December 2011. (a) Dayside regions averaged over 3 h in local time and centered on the indicated location. (b) Plasma-sheet composition within 3 h pre- and post-midnight and within 750 km of the geographic equator. Adapted from Raines et al. (2013).

and thus the symmetry with respect to the magnetic equator (Korth et al., 2014). The highest concentrations of protons are found in the northern cusp at high dayside latitudes and near the magnetic equator in the nightside plasma sheet, but protons and alpha particles are present throughout the magnetosphere, albeit with reduced densities. Although solar wind plasma is observed on closed field lines on the dayside, the fluxes in this region are very low on average (Figure 16.8a), indicating that Mercury does not have a persistent, trapped equatorial plasma population as is found at Earth (Korth et al., 2014). These observations are consistent with results from numerical simulations of Mercury's magnetosphere with hybrid kinetic codes (e.g., Kallio and Janhunen, 2003a; Trávníček et al., 2007, 2010; Wang et al., 2010; Müller et al., 2012) and magnetohydrodynamic (MHD) codes (e.g., Kabin et al., 2000; Benna et al., 2010; Jia et al., 2015).

Ions of planetary origin are also present in Mercury's magnetosphere. Among the planetary ions, Na+-group ions are the dominant magnetospheric ion species (Zurbuchen et al., 2011). The second- and third-most abundant ions are O+-group and He+ ions, respectively, which have average abundances equal to 16% and 6.7%, respectively, of those of the Na+ group (Raines et al., 2013). Additional ion species having $m/q > 30$, such as S+, and doubly charged ions in the range $4.5 \leq m/q \leq 12$ were detected in the plasma sheet at a radial distance between 1.6 R_M and 2.2 R_M during MESSENGER's first flyby of Mercury (Zurbuchen et al., 2008). The distribution of planetary ions varies substantially by region, as shown in Figure 16.22. Enhancements of these ion species are evident at both dawn and dusk, although the variations with altitude are different.

Figure 16.23 compares the abundances of major ion species in distinct locations within the magnetosphere.

Notable findings are that, in the northern cusp, the abundance of Na+-group ions exceeds that of solar wind alpha particles by a factor of 2 on average (Figure 16.23a), and, in the central plasma sheet (Figure 16.23b), planetary ions can contribute up to 50% of the mass density and 15% of the plasma thermal pressure (Gershman et al., 2014). Compared with the post-midnight sector, Na+-group ions in the pre-midnight plasma sheet, with typical energies of a few keV, are substantially enhanced over a broad altitude range, ~1500–6000 km (Figure 16.22, plasma population 4) (Raines et al., 2013; Gershman et al., 2014). This observation is consistent with test particle simulations by Delcourt (2013), which showed that Na+ and O+ ions are preferentially transported into this altitude range on the pre-midnight side of the central plasma sheet by the mechanism detailed in Section 16.6.3.

There also appears to be a dependence of ion abundance on solar wind conditions as manifested in a seasonal enhancement in both Na+-group and O+-group ions (Raines et al., 2013). Both species show enhancements by a factor of more than 2 for measurements acquired at Mercury true anomalies of 120° and 315°. The extrema in ion abundance do not strictly correlate with Mercury's heliocentric distance, which is known to modulate this planet's solar wind environment (Korth et al., 2012), so an as yet unidentified physical process must govern the seasonal variation of these ion populations. He+ ions do not show such enhancement, presumably because these ions are partially sourced by the solar wind.

The enhancement of Na$^+$-group ions in the dawn-side magnetosphere (Figure 16.22b, plasma population 3) is collocated with a persistent source of high-energy neutral Ca observed by MESSENGER's Ultraviolet and Visible Spectrometer (McClintock and Lankton, 2007) in the exosphere (Burger et al., 2012), although a direct link with this source has not been established. Similarly, seasonal enhancements in the ion population noted above do not correlate with those of the exospheric neutral atoms, which exhibit much lower variability (Cassidy et al., 2015). To examine the relationship between the neutral and ionized components, the *e*-folding heights of planetary ions were compared with the scale heights of exospheric neutral atoms, and the former were typically found to be much larger (Raines et al., 2013). The *e*-folding distance of Na$^+$-group ions is a factor of 5–10 greater than the Na scale height (Cassidy et al., 2015). This difference implies that unlike Na, which is cool and gravitationally bound to the surface, the Na$^+$ ions are more energetic and may escape to high altitudes, e.g., into the magnetosphere.

16.6.2.2 Electrons

Magnetospheric electrons with energies >35 keV were routinely observed by the Energetic Particle Spectrometer (EPS) (Andrews et al., 2007), beginning shortly after Mercury orbit insertion (Ho et al., 2011b). The energetic electrons were registered as bursts lasting from seconds to hours with typical energies up to 100 keV, although energies >200 keV were measured occasionally. The events were most often observed close to the planet in the northern hemisphere from the magnetic equator to the pole, but they were present over a broad range of local times (Ho et al., 2012). The pitch-angle distributions of the electrons suggest that the bulk of the population does not execute complete drift paths around the planet (Ho et al., 2011b) and thus does not form long-lived radiation belts such as those found at Earth. As discussed in Section 16.6.3, low fluxes of ~200 keV electrons detectable by the Gamma-Ray Spectrometer (GRS) sensor (Goldsten et al., 2007) did on occasion exhibit periodicities consistent with the azimuthal drift time of electrons around the planet (drift echoes) but with lifetimes of not more than approximately five complete drifts around the planet (Baker et al., 2016). Interestingly, no energetic ions with energies >25 keV were detected above the EPS detector intensity threshold even though 1-keV ions are prevalent in Mercury's magnetosphere (Ho et al., 2011b).

The energy range of the EPS sensor was designed to sense energies observed by Mariner 10 (>35 keV). Fortuitously, three other sensors of the MESSENGER payload – the X-Ray Spectrometer (XRS) (Schlemm et al., 2007), the Neutron Spectrometer (NS) (Goldsten et al., 2007), and the GRS – all returned signals from orbit about Mercury attributable to energetic electrons with energies that EPS was not designed to measure. The XRS instrument observed photons resulting from low-energy (~10 keV) electrons impinging on its detectors approximately uniformly throughout the magnetosphere (Ho et al., 2011a). Although the energy spectra could not be measured directly by XRS, simulation of the XRS detector response indicates an energy spectrum that peaks in the range 0.7–1.0 keV and has a flux consistent with that observed by EPS at 45 keV. The spatial distribution of the electrons inferred from XRS observations is shown in Figure 16.24 (Ho et al., 2016). The events spanned all local times, with the highest concentration near dawn and dusk, and, similarly to the proton distribution (Figure 16.21), were clustered within a narrow latitude band located at higher latitudes on the dayside than on the nightside.

The GRS and NS detectors responded to bremsstrahlung photons produced when energetic electrons impacted materials nearby and were found to be substantially more sensitive to this

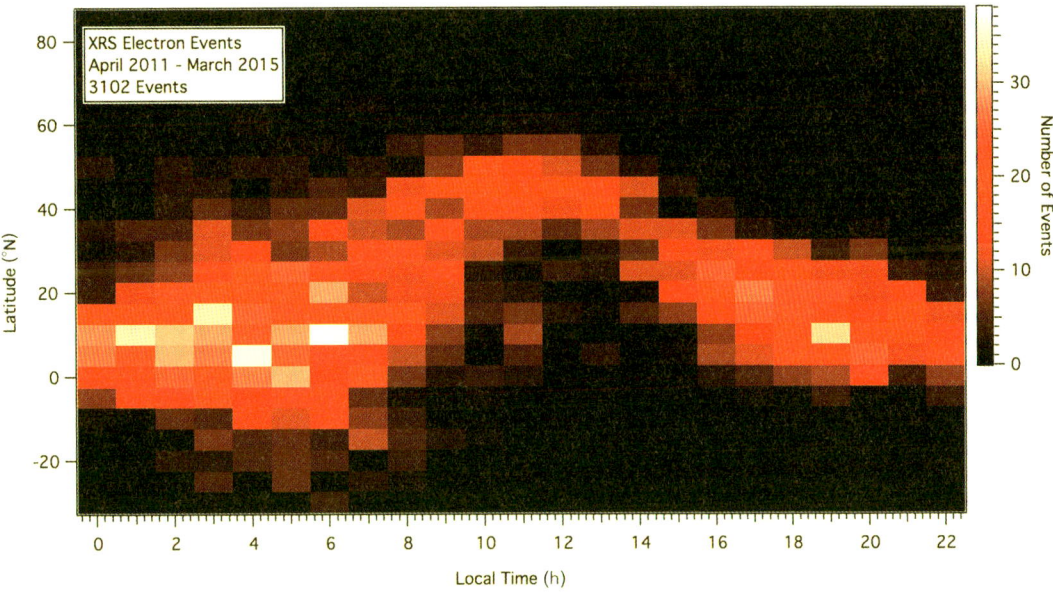

Figure 16.24. Distribution of suprathermal electron events detected by the MESSENGER X-Ray Spectrometer (XRS) organized by latitude and local time. Adapted from Ho et al. (2016).

population than the EPS instrument, owing to their comparatively large geometric factors for responding to electron impacts. These sensors regularly detected energetic electrons with energies from tens to hundreds of keV, mostly within the magnetosphere and predominantly on closed magnetic field lines (Lawrence et al., 2015). A subset of these energetic electron events showed periodicities in their occurrence rates of a few minutes, consistent with Mercury's Dungey cycle period (Section 16.6.3), indicating that they may be related to reconnection or magnetospheric convection.

16.6.3 Transport and Heating

Interaction of Mercury's planetary magnetic field with the solar wind and IMF gives rise to circulation of magnetic flux (Dungey cycle, Section 16.5.2) and the corresponding convection electric field, **E**, and plasma circulation drifts (e.g., Hughes, 1996). This circulation was directly observed by MESSENGER (DiBraccio et al., 2015a) in the plasma mantle, a persistent layer of tailward-flowing magnetosheath-like plasma inside of and adjacent to the magnetopause (Rosenbauer et al., 1975), and is similar to that observed at Earth but is one to two orders of magnitude faster, i.e., with cycle times of minutes instead of hours at Mercury (Slavin et al., 2009a, 2010). In the region of the magnetosphere dominated by the dipole field, ions and electrons gyrate about magnetic field lines of force and bounce along magnetic field lines between the northern and southern hemisphere while they also drift from the nightside to the dayside. The actual drift trajectories depend on the strength of the convection electric field and on the particle's energy, charge, and pitch angle. Because of Mercury's weak planetary field, the region of quasi-dipolar magnetic field configuration is very close to the planet, within 1 R_M, and everywhere else the magnetic field gradient and curvature drifts are negligible relative to convective $\mathbf{E} \times \mathbf{B}$ drift motion. Within ~1 R_M altitude, the gradient-curvature drift might be important and, in principle, a very narrow annulus near the magnetic equator close to the planet might support drift motions that close on themselves around the planet.

The character of charged particle motion in electromagnetic fields depends on the ratio between the gradient scale of the magnetic field and the particle gyroradius, $r_g = m v_\perp / (|q| B) = \sqrt{2 m W_\perp} / (|q| B)$, where m is the mass, $v_\perp$ is the particle velocity perpendicular to the magnetic field, $W_\perp = m v_\perp^2 / 2$, and q is the particle charge. When the magnetic field varies on scales much greater than r_g, particles conserve the magnetic moment of their gyromotion, known as the first adiabatic invariant, $\mu = (m v_\perp) / (2B)$, and the particle transport can be described in the guiding center approximation, by which one considers the drift motion only of the gyrocenter subject to the $\mathbf{E} \times \mathbf{B}$ and gradient-curvature drifts. In the magnetotail, the minimum radius of curvature of the magnetic field on a given field line, R_c, occurs at the cross-tail current sheet within the plasma sheet (Sections 16.4.2 and 16.5.1). This location is also where the magnetic field is a minimum and r_g is a maximum. For this reason, the magnetotail equator is a localized region of non-adiabatic motion within which the guiding center treatment breaks down, and one must evaluate particle motions from the Lorentz force explicitly.

Non-adiabatic behavior is most important for the heavy ions, e.g., Na$^+$, because, for a fixed energy, the gyroradius is proportional to $\sqrt{m}$. For a given field line, the parameter $\kappa = \sqrt{R_c / r_{g,max}}$, where $r_{g,max}$ is the maximum gyroradius on the field line, is a useful measure of non-adiabatic behavior (Büchner and Zelenyi, 1989). The motion is generally adiabatic for $\kappa > 3$, whereas for $\kappa < 1$ particles experience meandering motions about the field minimum (Speiser, 1965; Delcourt and Martin, 1994). For intermediate κ, particles have their magnetic moments quasi-randomly altered with each crossing of the magnetic equator, a process referred to as μ-scattering (Birmingham, 1984; Anderson et al., 1997b; Delcourt et al., 2003).

For Mercury, the guiding center drift approximation is violated throughout most of the magnetosphere, and stochastic processes must be considered in plasma-sheet transport and heating (Korth et al., 2012). Korth et al. (2011) showed that plasma-sheet protons with energies in the range 0.5–5 keV have $1 < \kappa < 3$, and their motion is non-adiabatic throughout much of the nightside. Heavy ions are more strongly non-adiabatic because their gyroradii are larger by a factor of ~20 than those of protons, so that Na$^+$ motions will be fully chaotic with $\kappa < 1$. Furthermore, non-adiabatic transport extends closer to the surface for higher ion energies. For 5-keV protons, the first adiabatic invariant fails to be conserved even within 500-km altitude at midnight, implying that wave–particle interactions are not required to scatter protons into the loss cone because μ-scattering occurs simply by virtue of non-adiabatic motions (e.g., Anderson et al., 1997b). By contrast, electrons should remain adiabatic even to fairly high energies, e.g., hundreds of keV.

At Mercury, bombardment of the surface by charged particles is expected to yield low-energy heavy ions of planetary origin (O$^+$, Na$^+$, or even Ca$^+$) near the surface. Remarkably, a low-energy, <1 eV, ion may be abruptly energized parallel to the magnetic field and transported from Mercury's surface into the magnetosphere by curvature-related centrifugal acceleration. The centrifugal force occurs as the magnetic field line about which a particle gyrates is transported over the polar cap to the nightside by the $\mathbf{E} \times \mathbf{B}$ drift of the Dungey convection. If the cross-polar-cap potential drop, $\Delta\phi$, reaches ~10 keV, the centrifugal acceleration overcomes the gravitational force, and the ion escapes into the magnetotail and may gain substantial energy in the neutral sheet as it moves parallel to **E** (Delcourt et al., 2003).

Convectively driven centrifugal acceleration has been modeled extensively with both a guiding-center approximation as well as a Lorentz-force treatment of single-particle motion in an analytical magnetic field (Delcourt et al., 2003, 2012; Delcourt, 2013). In addition, such acceleration has been reproduced in test-particle calculations in self-consistent electromagnetic fields from MHD simulations (Sarantos et al., 2007) and in ion transport dynamics in hybrid simulations (Trávníček et al., 2007). Typical results are illustrated in Figure 16.25, which shows the drift path and energy gain of 1-eV Na$^+$ ions originating in the cusp for three values of the imposed potential, $\Delta\phi$. For $\Delta\phi = 1$ kV (gray), the centrifugal force is insufficient for escape into the magnetotail to occur, and the ions ultimately return to

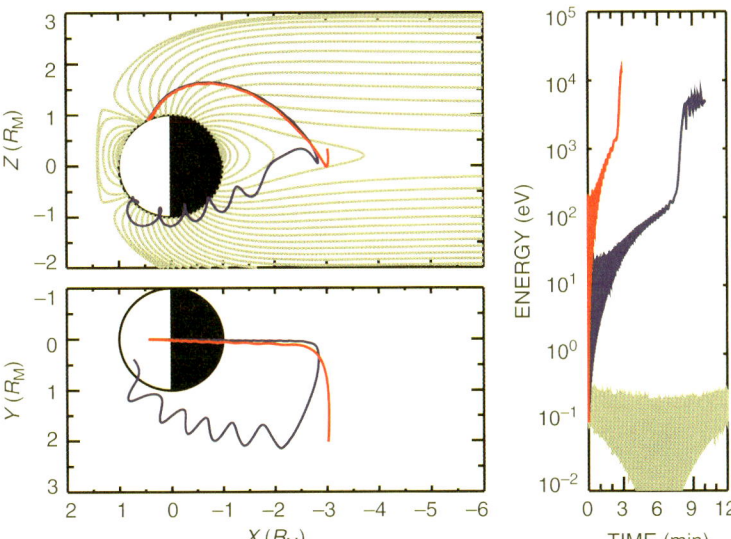

Figure 16.25. Model trajectories of Na$^+$ ions. (Left) Projections in the MSM (top) X–Z and (bottom) X–Y planes. (Right) Kinetic energy versus time for ions launched from 65°N latitude on the noon meridian. The trajectories correspond to cross-polar-cap potential drops of 1 kV (gray), 10 kV (blue), and 30 kV (red). Adapted from Delcourt (2013).

the surface. For $\Delta\phi \geq 10$ kV (red, blue), Na$^+$ ions are transported on similar trajectories into the magnetic equatorial plane at a downtail distance of ~3 R_M, where their paths diverge when they encounter the cross-tail current sheet and their motions become strongly non-adiabatic. During the transport, the ions are energized in a two-step process, gaining a small percentage of $q\Delta\phi$ in energy while traversing the magnetotail lobe and up to 50% of $q\Delta\phi$ in the central plasma sheet (Delcourt, 2013). Depending on the magnitude of $\Delta\phi$, low-energy planetary ions may be energized up to 10 keV (Figure 16.25, right panel).

The centrifugal acceleration process should be more effective in the northern than in the southern hemisphere. Because the northward offset of the planetary dipole moment yields weaker magnetic fields in the southern hemisphere, ions originating in the southern hemisphere do not attain the same energy as ions launched from the same latitude in the northern hemisphere. Moreover, the inner magnetosphere is preferentially populated by heavy ions originating at the northern hemisphere surface (Delcourt, 2013), because ions originating from the southern cusp reach the magnetotail center farther downtail, at distances of ~4 R_M, and are quickly lost to the magnetopause.

Direct confirmation of the centrifugal acceleration process would require simultaneous observations of the electric field and three-dimensional observations of Na$^+$. Although these observations have not been made, both the spatial distribution and energy spectra of Na$^+$ observed within Mercury's magnetosphere are consistent with the theoretical predictions. The trajectory apex of Na$^+$ ions in the simulation varied from 1500 to 6000 km as the initial latitude at noon local time ranged from 85°N to 60°N. This range of altitudes is comparable with that of the density enhancements observed by MESSENGER in the inner magnetosphere (Figure 16.22). The preponderance of >5-keV Na$^+$ ions in the plasma sheet, with a strong bias to the evening side as observed (Figure 16.22), is additional evidence for this process because such an evening bias is a prediction of the mechanism.

In the magnetotail, plasma drifts from the lobes into the magnetic equatorial plane to form the plasma sheet.

Planetward of the magnetotail reconnection site, the return convection transports plasma sunward toward the planet. At Earth, the magnetosphere is ~10 times larger than the planet and the region within ~10 R_E is quasi-dipolar, so that gradient curvature drifts grow to dominate the $\mathbf{E} \times \mathbf{B}$ convection as plasma convects Earthward from the tail. As a consequence, plasma is diverted around the Earth in the distance range 5 R_E to 8 R_E. In addition, Earth's rotation rate is 59 times greater than Mercury's, so the corotation electric field is much stronger at Earth and dominates the drift for particles with energies of ~1 keV and lower within ~6 R_E. At Mercury, the cross-tail current is very close to the planetary surface, the region of quasi-dipolar field is within <1 R_M of the surface, and the corotation drift is negligible; thus, the azimuthal drifts are very small, allowing the plasma sheet to approach to within <1 R_M of the surface.

The heating of ions during the convection cycle depends on their source and the transport mechanism(s) to which they are subjected. Planetary ions are approximately equally energized during transport, as indicated in Figure 16.7b by the similar temperatures of Na$^+$-group, O$^+$-group, and He$^+$ ions in the plasma sheet. This similarity suggests that acceleration is independent of m/q, as expected for large gyroradius ions in the plasma sheet (Speiser, 1965) and as illustrated in Figure 16.25. In contrast, the solar wind ions measured by MESSENGER have temperatures that are approximately mass-proportional relative to H$^+$ in the central plasma sheet (Figure 16.7b). This result is consistent with a mechanism that accelerates or heats ions to the same velocity (Fuselier and Lewis, 2011) as that predicted either by the Sweet–Parker model of magnetic reconnection (Vasyliunas, 1975) or by turbulent heating (Chandran et al., 2010). The measurements strongly indicate that solar wind ions are largely energized by reconnection at the dayside magnetopause (and the resulting convection) and do not experience substantial energization within the plasma sheet during steady-state conditions. Other transport and heating phenomena, such as sunward (planetward) flows of plasma in the magnetotail, are more closely tied to transient events, such as magnetotail reconnection and magnetospheric substorms (Chapter 17).

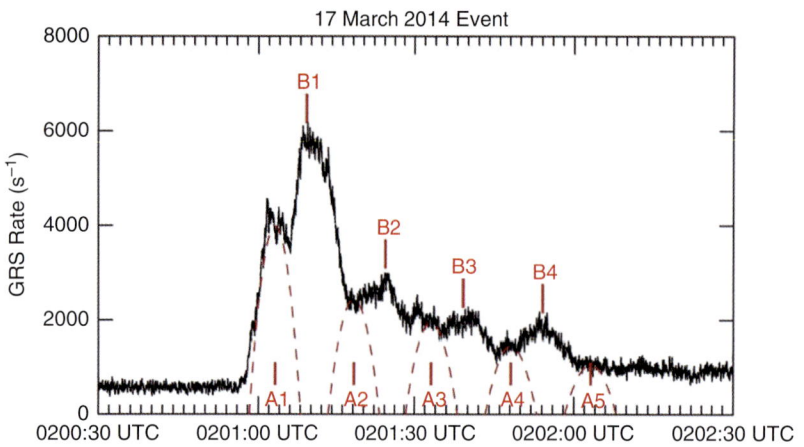

Figure 16.26. GRS count rate (black) versus time showing electron drift echoes from two separate injections of energetic electrons onto closed magnetic field lines. The recurring observations of the injection populations A and B at 15-s intervals are marked in red. Adapted from Baker et al. (2016).

Transport signatures of energetic electrons were also observed by MESSENGER. Baker et al. (2016) reported energetic particle injections likely associated with magnetospheric substorms. Figure 16.26 shows indirect observations of electrons by the GRS when MESSENGER was located near dawn. The data exhibit an abrupt two-stage increase in the GRS count rate associated with the injection of two distinct populations of energetic electrons, labeled A and B, onto closed magnetic field lines. After the initial increase, and embedded in the subsequent decay of the signal, several peaks spaced regularly at 15-s intervals were observed in the count rate. These peaks were interpreted to be drift echoes of 200-keV electrons observed as they drift repeatedly around the planet on closed drift paths. This result confirms that, for electrons, the particle motion can be treated as adiabatic. The signal decays as the particles are scattered into the loss cone during the drift. Closer to the injection location near midnight, less-dispersed energetic electron distributions were indicated by more time-symmetric peaks in the GRS count rate. These observations, together with evidence of dayside magnetopause reconnection (DiBraccio et al., 2013) and flux rope and plasmoid formation on the nightside (DiBraccio et al., 2015b), suggest that Earth-like magnetospheric substorms do occur at Mercury.

16.6.4 Losses

Charged particles are lost from Mercury's magnetosphere either by precipitation to the planetary surface or by escape through the magnetopause or down the magnetotail. The relative importance of precipitation and escape depends primarily on the mass, energy, and pitch angle of the charged particle and on the topology of the magnetic field. Particle mass and energy determine the ion gyroradius, and ions with gyroradii comparable with or larger than Mercury's radius are susceptible to loss through collision with the planet or escape through the magnetopause while gyrating about the local magnetic field direction. In addition, charged particles with small pitch angles may be lost through impact onto the planet surface while bouncing along closed field lines between hemispheres. The resulting precipitation occurs primarily at mid latitudes and produces a distinctive loss-cone distribution. Finally, ions on open field lines rooted at the planet at high latitudes tend to be lost to space. This heliospheric loss is a primary loss mechanism for

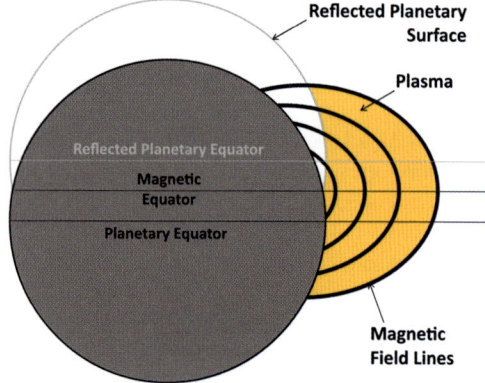

Figure 16.27. Location of a plasma void at low altitudes in the northern hemisphere of the planet (filled gray circle), with the planetary and magnetic equators marked by black lines. For conceptual understanding of the asymmetric particle access to the surface (orange shading), the planetary surface and equator are also shown reflected or mirrored about the magnetic equator (in light gray). The heavy black lines indicate a set of closed field lines originating from a spin-axis-aligned dipole offset northward as illustrated. Adapted from Korth et al. (2014).

exospheric ions, especially those with large mass, which are not effectively confined by Mercury's magnetic field.

Mercury's offset magnetic field controls the precipitation of plasma to the surface in a way that is unique in the solar system (Korth et al., 2014). Precipitation occurs when charged particles impact the surface as they bounce between the mirror points in the northern and southern hemispheres. The mirror points on a magnetic field line are defined in terms of their magnetic field strength, $B_m = B/\sin^2 \alpha$, which depends on the local magnetic field strength, B, and the particle pitch angle, α. On any closed field line, bounce motion is supported for pitch angles with B_m smaller than the magnetic field at the surface, B_s, whereas charged particles with pitch angles having $B_m > B_s$ would mirror below the surface and are lost to the planet. Because of the northward offset of the planetary dipole, the surface magnetic field on any given closed field line is larger in the northern hemisphere than it is in the south. The magnetospheric plasma is thus lost preferentially to the southern hemisphere. The orange-shaded region in Figure 16.27 illustrates the plasma access to low altitudes, which extends to the surface in the southern hemisphere. In the northern hemisphere, field-line access is limited to those portions where the magnetic field strength is lower than that at the southern hemisphere surface, as indicated by the reflection of the planetary surface about the magnetic equator (light gray circle). The portions of field lines in the northern hemisphere that lie below this surface are populated only by particles for which the mirror points lie below the surface in the southern hemisphere. Because the trapped population cannot access this region, a plasma void exists on closed field lines at low to mid northern altitudes.

Observational evidence for such asymmetric precipitation is provided by the proton flux distribution in Figure 16.21. In that figure, the closed-field-line region is located equatorward of the white lines, which represent the boundaries along which the field-line topology changes from open to closed, as inferred from the paraboloid model. Field lines near the magnetic equator were not intersected by the MESSENGER orbit, yielding a gap in the observations (poleward limits are indicated by the black lines in the figure). Within the observed closed-field-line region there is a pronounced north–south asymmetry in the nightside flux distribution, which indicates that the fluxes in the southern hemisphere are much larger than those seen on the same field line in the north. The southern hemisphere fluxes are also distributed over a wider latitude range than in the north, where there is a sharp decrease near ~30°N at midnight that increases with latitude toward dawn and dusk. This cutoff in the flux coincides with the modeled location where the spacecraft travels from south to north into the plasma void region (Figure 16.27). The north–south asymmetry in the distribution of the flux is thus a direct consequence of preferential precipitation of particles onto the southern hemisphere surface expected for a northward-offset magnetic field. This loss process produces a substantial loss cone in the plasma-sheet proton distribution (Winslow et al., 2014), which, in turn, acts to promote local ion plasma-wave instabilities (e.g., Boardsen et al., 2012; and Chapter 17).

The computation of precipitation rates generally requires detailed knowledge of the magnetospheric process controlling plasma sources, transport, and losses. In the cusps, however, such estimates are more straightforward because of the more direct access of plasma to the surface. Massetti et al. (2003) modeled the precipitation of cusp protons to the surface with a semi-empirical scaled-Earth magnetic field model defining the open area of the magnetospheric cusps under varying solar wind conditions. For a planet-centered and spin-axis-aligned internal magnetic field, the mean proton precipitation flux into the cusp was estimated at 4.1×10^8 cm^{-2} s^{-1} for southward B_Z. Similar analysis with a hybrid kinetic-fluid model predicted proton precipitation fluxes of up to 10^9 cm^{-2} s^{-1} in this region for a range of typical solar wind and IMF conditions (Kallio and Janhunen, 2003b). More accurate assessments of the precipitation flux in the cusps were enabled by MESSENGER plasma observations. Under an adiabatic approximation, precipitation fluxes for protons were estimated from FIPS observations during several hundred cusp crossings (Raines et al., 2014). These fluxes were found to be highly variable, ranging from 10^4 to 10^8 cm^{-2} s^{-1}, with most of the variability apparently due to changing solar wind conditions. Winslow et al. (2012) also estimated proton precipitation rates with a simple model of a circular cusp of diameter equal to the observed cusp crossing and found these rates to vary similarly, 10^{21} to 10^{24} s^{-1} over an area of 5×10^{11} m^2 (Winslow et al., 2012). Finally, the precipitation flux was estimated from an average-size loss cone determined from MESSENGER observations under the assumption that protons are scattered into the loss cone by pitch angle diffusion (Winslow et al., 2014). For average proton densities and temperatures, the precipitation flux into the northern cusp amounts to 3.7×10^8 cm^{-2} s^{-1}, a rate that is somewhat higher than those observed directly. The latter method may also be applied to the southern cusp, where the average precipitation flux of 4.4×10^7 cm^{-2} s^{-1} is about an order of magnitude lower.

16.7 CONCLUSIONS

MESSENGER has obtained a wealth of data on Mercury's magnetosphere and solar wind environment, and these observations have led to a number of discoveries that establish the uniqueness of Mercury's magnetosphere in our solar system. This uniqueness is reflected in the following characteristics. First, because Mercury's magnetic moment is weaker by a factor of 1000 than that of Earth and because the size of the magnetosphere scales with the magnitude of the planetary field, the linear dimensions of Mercury's magnetosphere are smaller by a factor of 20 than Earth's and among the smallest in our solar system. Second, the dynamic pressure exerted by the solar wind is much stronger in the inner heliosphere than at 1 AU, to the point that the planet barely stands off the dayside magnetopause. The locations of the bow shock and magnetopause boundary are also very dynamic over short timescales because of large variability in the solar wind environment near the Sun and on longer timescales because of variations in Mercury's heliocentric distance along its eccentric orbit. Third, the proximity of the magnetopause to the planetary surface implies substantial and highly variable contributions to the magnetic field from external current systems even at low altitudes and results in chaotic motion of charged particles throughout much of Mercury's magnetosphere, including near the surface. Fourth, the offset internal magnetic field yields a surface magnetic field strength in the northern hemisphere that is about twice as large as that in the south, and the effective difference in polar cap area implies a hemispheric asymmetry in particle precipitation and associated space weathering. The electrodynamic coupling of the magnetosphere to low altitudes occurs via field-aligned currents, which, in the absence of a conducting ionosphere at Mercury, evidently close through the planet.

The small magnetosphere of Mercury is therefore very rich in magnetospheric and plasma physics, as evidenced by the unique range of plasma sources, variety of loss processes, intense electrodynamics, and surface–magnetosphere interactions. Mercury is particularly distinctive in the solar system with regard to the intimate electrodynamic coupling between the solid planet and the magnetosphere, which makes this system, although small on a physical scale relative to other accessible magnetospheres, unusually important as a proving ground for our understanding of the physics of planetary magnetospheres in general.

REFERENCES

Alexeev, I. I., Belenkaya, E. S., Slavin, J. A., Korth, H., Anderson, B. J., Baker, D. N., Boardsen, S. A., Johnson, C. L., Purucker, M. E., Sarantos, M. and Solomon, S. C. (2010). Mercury's magnetospheric magnetic field after the first two MESSENGER flybys. *Icarus*, **209**, 23–39, doi:10.1016/j.icarus.2010.01.024.

Anderson, B. J., Fuselier, S. A. and Murr, D. (1991). Electromagnetic ion cyclotron waves observed in the plasma depletion layer. *Geophys. Res. Lett.*, **18**, 1955–1958, doi:10.1029/91gl02238.

Anderson, B. J., Phan, T. D. and Fuselier, S. A. (1997a). Relationships between plasma depletion and subsolar reconnection. *J. Geophys. Res.*, **102**, 9531–9542, doi:10.1029/97ja00173.

Anderson, B. J., Decker, R. B., Paschalidis, N. P. and Sarris, T. (1997b). Onset of nonadiabatic particle motion in the near-Earth magnetotail. *J. Geophys. Res.*, **102**, 17,553–17,569, doi:10.1029/97ja00798.

Anderson, B. J., Acuña, M. H., Lohr, D. A., Scheifele, J., Raval, A., Korth, H. and Slavin, J. A. (2007). The Magnetometer instrument on MESSENGER. *Space Sci. Rev.*, **131**, 417–450, doi:10.1007/s11214-007-9246-7.

Anderson, B. J., Korth, H., Waters, C. L., Green, D. L. and Stauning, P. (2008). Statistical Birkeland current distributions from magnetic field observations by the Iridium constellation. *Ann. Geophys.*, **26**, 671–687.

Anderson, B. J., Slavin, J. A., Korth, H., Boardsen, S. A., Zurbuchen, T. H., Raines, J. M., Gloeckler, G., McNutt, R. L., Jr. and Solomon, S. C. (2011a). The dayside magnetospheric boundary layer at Mercury. *Planet. Space Sci.*, **59**, 2037–2050, doi:10.1016/j.pss.2011.01.010.

Anderson, B. J., Johnson, C. L., Korth, H., Purucker, M. E., Winslow, R. M., Slavin, J. A., Solomon, S. C., McNutt, R. L., Jr., Raines, J. M. and Zurbuchen, T. H. (2011b). The global magnetic field of Mercury from MESSENGER orbital observations. *Science*, **333**, 1859–1862, doi:10.1126/science.1211001.

Anderson, B. J., Johnson, C. L., Korth, H., Winslow, R. M., Borovsky, J. E., Purucker, M. E., Slavin, J. A., Solomon, S. C., Zuber, M. T. and McNutt, R. L., Jr. (2012). Low-degree structure in Mercury's planetary magnetic field. *J. Geophys. Res.*, **117**, E00L12, doi:10.1029/2012JE004159.

Anderson, B. J., Johnson, C. L. and Korth, H. (2013). A magnetic disturbance index for Mercury's magnetic field derived from MESSENGER Magnetometer data. *Geochem. Geophys. Geosyst.*, **14**, 3875–3886, doi:10.1002/ggge.20242.

Anderson, B. J., Johnson, C. L., Korth, H., Slavin, J. A., Winslow, R. M., Phillips, R. J., Solomon, S. C. and McNutt, R. L., Jr. (2014). Steady-state field-aligned currents at Mercury. *Geophys. Res. Lett.*, **41**, 7444–7452, doi:10.1002/2014GL061677.

Anderson, B. J., Korth, H., Johnson, C. L., Phillips, R. J., Philpott, L. C. and Solomon, S. C. (2016). Closure of Birkeland currents at Mercury: Constraints on the electrical conductivity of the crust and mantle. *Lunar Planet. Sci.*, **47**, abstract 1243.

Andrews, G. B., Zurbuchen, T. H., Mauk, B. H., Malcom, H., Fisk, L. A., Gloeckler, G., Ho, G. C., Kelley, J. S., Koehn, P. L., LeFevere, T. W., Livi, S. S., Lundgren, R. A. and Raines, J. M. (2007). The Energetic Particle and Plasma Spectrometer instrument on the MESSENGER spacecraft. *Space Sci. Rev.*, **131**, 523–556, doi:10.1007/s11214-007-9272-5.

Armstrong, T. P., Krimigis, S. M. and Lanzerotti, L. J. (1975). Reinterpretation of reported energetic particle fluxes in the vicinity of Mercury. *J. Geophys. Res.*, **80**, 4015–4017, doi:10.1029/Ja080i028p04015.

Ashour-Abdalla, M., Okuda, H. and Cheng, C. Z. (1981). Acceleration of heavy ions on auroral field lines. *Geophys. Res. Lett.*, **8**, 795–798, doi:10.1029/Gl008i007p00795.

Baker, D. N., Poh, G., Odstrcil, D., Arge, C. N., Benna, M., Johnson, C. L., Korth, H., Gershman, D. J., Ho, G. C., McClintock, W. E., Cassidy, T. A., Merkel, A., Raines, J. M., Schriver, D., Slavin, J. A., Solomon, S. C., Trávníček, P. M., Winslow, R. M. and Zurbuchen, T. H. (2013). Solar wind forcing at Mercury: WSA-ENLIL model results. *J. Geophys. Res. Space Physics*, **118**, 45–57, doi:10.1029/2012ja018064.

Baker, D. N., Dewey, R. M., Lawrence, D. J., Goldsten, J. O., Korth, H., Slavin, J. A., Krimigis, S. M., Anderson, B. J., Ho, G. C., McNutt, R. L., Jr., Raines, J. M., Schriver, D. and Solomon, S. C. (2016). Energetic electron flux enhancements in Mercury's magnetosphere: An integrated view with high-resolution observations from MESSENGER. *J. Geophys. Res. Space Physics*, **121**, 2171–2184.

Benna, M., Anderson, B. J., Baker, D. N., Boardsen, S. A., Gloeckler, G., Gold, R. E., Ho, G. C., Killen, R. M., Korth, H., Krimigis, S. M., Purucker, M. E., McNutt, R. L., Jr., Raines, J. M., McClintock, W. E., Sarantos, M., Slavin, J. A., Solomon, S. C. and Zurbuchen, T. H. (2010). Modeling of the magnetosphere of Mercury at the time of the first MESSENGER flyby. *Icarus*, **209**, 3–10, doi:10.1016/j.icarus.2009.11.036.

Birmingham, T. J. (1984). Pitch angle diffusion in the Jovian magnetodisc. *J. Geophys. Res.*, **89**, 2699–2707, doi:10.1029/Ja089ia05p02699.

Boardsen, S. A., Slavin, J. A., Anderson, B. J., Korth, H., Schriver, D. and Solomon, S. C. (2012). Survey of coherent ~1 Hz waves in Mercury's inner magnetosphere from MESSENGER observations. *J. Geophys. Res.*, **117**, A00M05, doi:10.1029/2012JA017822.

Büchner, J. and Zelenyi, L. M. (1989). Regular and chaotic charged-particle motion in magnetotail-like field reversals, 1. Basic theory of trapped motion. *J. Geophys. Res.*, **94**, 11,821–11,842, doi:10.1029/Ja094ia09p11821.

Burger, M. H., Killen, R. M., McClintock, W. E., Vervack, R. J., Jr., Merkel, A. W., Sprague, A. L. and Sarantos, M. (2012). Modeling MESSENGER observations of calcium in Mercury's exosphere. *J. Geophys. Res.*, **117**, E00L11, doi:10.1029/2012je004158.

Cassidy, T. A., Merkel, A. W., Burger, M. H., Sarantos, M., Killen, R. M., McClintock, W. E. and Vervack, R. J., Jr. (2015). Mercury's seasonal sodium exosphere: MESSENGER orbital observations. *Icarus*, **248**, 547–559, doi:10.1016/j.icarus.2014.10.037.

Chandran, B. D. G., Li, B., Rogers, B. N., Quataert, E. and Germaschewski, K. (2010). Perpendicular ion heating by low-frequency Alfvén-wave turbulence in the solar wind. *Astrophys. J.*, **720**, 503–515, doi:10.1088/0004-637X/720/1/503.

Chapman, S. and Ferraro, V. C. A. (1930). A new theory of magnetic storms. *Nature*, **126**, 129–130, doi:10.1038/126129a0.

Chapman, S. and Ferraro, V. C. A. (1931). A new theory of magnetic storms. *Terr. Mag.*, **36**, 171–186, doi:10.1029/TE036i003p00171.

Connerney, J. E. P. and Ness, N. F. (1988). Mercury's magnetic field and interior. In *Mercury*, ed. F. Vilas, C. R. Chapman and M. S. Matthews. Tucson, AZ: University of Arizona Press, pp. 494–513.

Cowley, S. W. H. (2000). Magnetosphere–ionosphere interactions: A tutorial review. In *Magnetospheric Current Systems*, ed. S. Ohtani, R. Fujii, M. Hesse and R. L. Lysak. Geophysical Monograph 118. Washington, DC: American Geophysical Union, pp. 91–106.

Delcourt, D. C. (2013). On the supply of heavy planetary material to the magnetotail of Mercury. *Ann. Geophys.*, **31**, 1673–1679, doi:10.5194/angeo-31-1673-2013.

Delcourt, D. C. and Martin, R. F. (1994). Application of the centrifugal impulse model to particle motion in the near-Earth magnetotail. *J. Geophys. Res.*, **99**, 23,583–23,590, doi:10.1029/94ja01845.

Delcourt, D. C., Grimald, S., Leblanc, F., Berthelier, J. J., Millilo, A., Mura, A., Orsini, S. and Moore, T. E. (2003). A quantitative model of the planetary Na$^+$ contribution to Mercury's magnetosphere. *Ann. Geophys.*, **21**, 1723–1736.

Delcourt, D. C., Seki, K., Terada, N. and Moore, T. E. (2012). Centrifugally stimulated exospheric ion escape at Mercury. *Geophys. Res. Lett.*, **39**, L22105, doi:10.1029/2012gl054085.

Denton, R. E. and Lyon, J. G. (1996). Density depletion in an anisotropic magnetosheath. *Geophys. Res. Lett.*, **23**, 2891–2894, doi:10.1029/96gl01590.

DiBraccio, G. A., Slavin, J. A., Boardsen, S. A., Anderson, B. J., Korth, H., Zurbuchen, T. H., Raines, J. M., Baker, D. N., McNutt, R. L., Jr. and Solomon, S. C. (2013). MESSENGER observations of magnetopause structure and dynamics at Mercury. *J. Geophys. Res. Space Physics*, **118**, 997–1008, doi:10.1002/jgra.50123.

DiBraccio, G. A., Slavin, J. A., Raines, J. M., Gershman, D. J., Tracy, P. J., Boardsen, S. A., Zurbuchen, T. H., Anderson, B. J., Korth, H., McNutt, R. L., Jr. and Solomon, S. C. (2015a). First observations of Mercury's plasma mantle as seen by MESSENGER. *Geophys. Res. Lett.*, **42**, 9666–9675, doi:10.1002/2015GL065805.

DiBraccio, G. A., Slavin, J. A., Imber, S. M., Gershman, D. J., Raines, J. M., Jackman, C. M., Boardsen, S. A., Anderson, B. J., Korth, H., Zurbuchen, T. H., McNutt, R. L., Jr. and Solomon, S. C. (2015b). MESSENGER observations of flux ropes in Mercury's magnetotail. *Planet. Space Sci.*, **115**, 77–89, doi:10.1016/j.pss.2014.12.016.

Dungey, J. W. (1963). Interactions of solar plasma with the geomagnetic field. *Planet. Space Sci.*, **10**, 233–237, doi:10.1016/0032-0633(63)90020-5.

Fairfield, D. H., Cairns, I. H., Desch, M. D., Szabo, A., Lazarus, A. J. and Aellig, M. R. (2001). The location of low Mach number bow shocks at Earth. *J. Geophys. Res.*, **106**, 25,361–25,376, doi:10.1029/2000ja000252.

Farrugia, C. J., Erkaev, N. V., Biernat, H. K. and Burlaga, L. F. (1995). Anomalous magnetosheath properties during Earth passage of an interplanetary magnetic cloud. *J. Geophys. Res.*, **100**, 19,245–19,257, doi:10.1029/95ja01080.

Fuselier, S. A. and Lewis, W. S. (2011). Properties of near-Earth magnetic reconnection from in-situ observations. *Space Sci. Rev.*, **160**, 95–121, doi:10.1007/s11214-011-9820-x.

Fuselier, S. A., Klumpar, D. M., Shelley, E. G., Anderson, B. J. and Coates, A. J. (1991). He^{2+} and H^+ dynamics in the subsolar magnetosheath and plasma depletion layer. *J. Geophys. Res.*, **96**, 21,095–21,104, doi:10.1029/91ja02145.

Gershman, D. J., Zurbuchen, T. H., Fisk, L. A., Gilbert, J. A., Raines, J. M., Anderson, B. J., Smith, C. W., Korth, H. and Solomon, S. C. (2012). Solar wind alpha particles and heavy ions in the inner heliosphere observed with MESSENGER. *J. Geophys. Res.*, **117**, A00M02, doi:10.1029/2012ja017829.

Gershman, D. J., Slavin, J. A., Raines, J. M., Zurbuchen, T. H., Anderson, B. J., Korth, H., Baker, D. N. and Solomon, S. C. (2013). Magnetic flux pileup and plasma depletion in Mercury's subsolar magnetosheath. *J. Geophys. Res. Space Physics*, **118**, 7181–7199, doi:10.1002/2013ja019244.

Gershman, D. J., Slavin, J. A., Raines, J. M., Zurbuchen, T. H., Anderson, B. J., Korth, H., Baker, D. N. and Solomon, S. C. (2014). Ion kinetic properties in Mercury's pre-midnight plasma sheet. *Geophys. Res. Lett.*, **41**, 5740–5747, doi:10.1002/2014gl060468.

Glassmeier, K. H. (2000). Currents in Mercury's magnetosphere. In *Magnetospheric Current Systems*, ed. S. Ohtani, R. Fujii, M. Hesse and R. L. Lysak. Geophysical Monograph 118. Washington, DC: American Geophysical Union, pp. 371–380.

Gloeckler, G., Fisk, L. A., Zurbuchen, T. H. and Schwadron, N. A. (2000). Sources, injection and acceleration of heliospheric ion populations. In *Acceleration and Transport of Energetic Particles Observed in the Heliosphere*, ed. R. A. Mewaldt, M. Miller, J. R. Jokipii, M. A. Lee, T. H. Zurbuchen and E. Mobius. New York: AIP Publishing, pp. 221–228.

Goldsten, J. O., Rhodes, E. A., Boynton, W. V., Feldman, W. C., Lawrence, D. J., Trombka, J. I., Smith, D. M., Evans, L. G., White, J., Madden, N. W., Berg, P. C., Murphy, G. A., Gurnee, R. S., Strohbehn, K., Williams, B. D., Schaefer, E. D., Monaco, C. A., Cork, C. P., Del Eckels, J., Miller, W. O., Burks, M. T., Hagler, L. B., DeTeresa, S. J. and Witte, M. C. (2007). The MESSENGER Gamma-Ray and Neutron Spectrometer. *Space Sci. Rev.*, **131**, 339–391, doi:10.1007/s11214-007-9262-7.

Gosling, J. T. (1997). Physical nature of the low-speed solar wind. In *Robotic Exploration Close to the Sun: Scientific Basis*, ed. S. R. Habbal. New York: AIP Publishing, pp. 17–24.

Gosling, J. T., Borrini, G., Asbridge, J. R., Bame, S. J., Feldman, W. C. and Hansen, R. T. (1981). Coronal streamers in the solar wind at 1 AU, *J. Geophys. Res.*, **86**, 5438–5448, doi:10.1029/Ja086ia07p05438.

Grosser, J., Glassmeier, K. H. and Stadelmann, A. (2004). Induced magnetic field effects at planet Mercury. *Planet. Space Sci.*, **52**, 1251–1260, doi:10.1016/j.pss.2004.08.005.

Ho, G. C., Starr, R. D., Gold, R. E., Krimigis, S. M., Slavin, J. A., Baker, D. N., Anderson, B. J., McNutt, R. L., Jr., Nittler, L. R. and Solomon, S. C. (2011a). Observations of suprathermal electrons in Mercury's magnetosphere during the three MESSENGER flybys. *Planet. Space Sci.*, **59**, 2016–2025, doi:10.1016/j.pss.2011.01.011.

Ho, G. C., Krimigis, S. M., Gold, R. E., Baker, D. N., Slavin, J. A., Anderson, B. J., Korth, H., Starr, R. D., Lawrence, D. J., McNutt, R. L., Jr. and Solomon, S. C. (2011b). MESSENGER observations of transient bursts of energetic electrons in Mercury's magnetosphere. *Science*, **333**, 1865–1868, doi:10.1126/science.1211141.

Ho, G. C., Krimigis, S. M., Gold, R. E., Baker, D. N., Anderson, B. J., Korth, H., Slavin, J. A., McNutt, R. L., Jr., Winslow, R. M. and Solomon, S. C. (2012). Spatial distribution and spectral characteristics of energetic electrons in Mercury's magnetosphere. *J. Geophys. Res.*, **117**, A00M04, doi:10.1029/2012JA017983.

Ho, G. C., Starr, R. D., Krimigis, S. M., Vandegriff, J. D., Baker, D. N., Gold, R. E., Anderson, B. J., Korth, H., Schriver, D., McNutt, R. L., Jr. and Solomon, S. C. (2016). MESSENGER observations of suprathermal electrons in Mercury's magnetosphere. *Geophys. Res. Lett.*, **43**, 550–555, doi:10.1002/2015GL066850.

Huebner, W. F., Keady, J. J. and Lyon, S. P. (1992). Solar photo rates for planetary atmospheres and atmospheric pollutants. *Astrophys. Space Sci.*, **195**, 1–294, doi:10.1007/Bf00644558.

Hughes, W. J. (1996). The magnetopause, magnetotail and magnetic reconnection. In *Introduction to Space Physics*, ed. M. G. Kivelson and C. T. Russell. New York: Cambridge University Press, pp. 227–287.

Iijima, T. and Potemra, T. A. (1976). The amplitude distribution of field-aligned currents at northern high latitudes observed by Triad. *J. Geophys. Res.*, **81**, 2165–2174, doi:10.1029/Ja081i013p02165.

Ip, W. H. and Kopp, A. (2004). Mercury's Birkeland current system. *Adv. Space Res.*, **33**, 2172–2175, doi:10.1016/s0273-1177(03)00444-7.

Jackson, D. J. and Beard, D. B. (1977). Magnetic field of Mercury. *J. Geophys. Res.*, **82**, 2828–2836, doi:10.1029/Ja082i019p02828.

Janhunen, P. and Kallio, E. (2004). Surface conductivity of Mercury provides current closure and may affect magnetospheric symmetry. *Ann. Geophys.*, **22**, 1829–1837.

Jia, X. Z., Slavin, J. A., Gombosi, T. I., Daldorff, L. K. S., Toth, G. and van der Holst, B. (2015). Global MHD simulations of Mercury's magnetosphere with coupled planetary interior: Induction effect of the planetary conducting core on the global interaction. *J. Geophys. Res. Space Physics*, **120**, 4763–4775, doi:10.1002/2015JA021143.

Johnson, C. L., Purucker, M. E., Korth, H., Anderson, B. J., Winslow, R. M., Al Asad, M. M. H., Slavin, J. A., Alexeev, I. I., Phillips, R.

J., Zuber, M. T. and Solomon, S. C. (2012). MESSENGER observations of Mercury's magnetic field structure. *J. Geophys. Res.*, **117**, E00L14, doi:10.1029/2012JE004217.

Johnson, C. L., Philpott, L. C., Anderson, B. J., Korth, H., Hauck, S. A., II, Heyner, D., Phillips, R. J., Winslow, R. M. and Solomon, S. C. (2016). MESSENGER observations of induced magnetic fields in Mercury's core. *Geophys. Res. Lett.*, **43**, 2436–2444, doi:10.1002/2015GL067370.

Kabin, K., Gombosi, T. I., DeZeeuw, D. L. and Powell, K. G. (2000). Interaction of Mercury with the solar wind. *Icarus*, **143**, 397–406, doi:10.1006/icar.1999.6252.

Kallio, E. and Janhunen, P. (2003a). Modelling the solar wind interaction with Mercury by a quasi-neutral hybrid model. *Ann. Geophys.*, **21**, 2133–2145.

Kallio, E. and Janhunen, P. (2003b). Solar wind and magnetospheric ion impact on Mercury's surface. *Geophys. Res. Lett.*, **30**, 1877, doi:10.1029/2003gl017842.

Killen, R. M., Bida, T. A. and Morgan, T. H. (2005). The calcium exosphere of Mercury. *Icarus*, **173**, 300–311, doi:10.1016/j.icarus.2004.08.022.

Killen, R. M., Cremonese, G., Lammer, H., Orsini, S., Potter, A. E., Sprague, A. L., Wurz, P., Khodachenko, M. L., Lichtenegger, H. I. M., Milillo, A. and Mura, A. (2007). Processes that promote and deplete the exosphere of Mercury. *Space Sci. Rev.*, **132**, 433–509, doi:10.1007/s11214-007-9232-0.

Korth, H., Anderson, B. J., Acuña, M. H., Slavin, J. A., Tsyganenko, N. A., Solomon, S. C. and McNutt, R. L., Jr. (2004). Determination of the properties of Mercury's magnetic field by the MESSENGER mission. *Planet. Space Sci.*, **52**, 733–746, doi:10.1016/j.pss.2003.12.008.

Korth, H., Anderson, B. J., Raines, J. M., Slavin, J. A., Zurbuchen, T. H., Johnson, C. L., Purucker, M. E., Winslow, R. M., Solomon, S. C. and McNutt, R. L., Jr. (2011). Plasma pressure in Mercury's equatorial magnetosphere derived from MESSENGER Magnetometer observations. *Geophys. Res. Lett.*, **38**, L22201, doi:10.1029/2011GL049451.

Korth, H., Anderson, B. J., Johnson, C. L., Winslow, R. M., Slavin, J. A., Purucker, M. E., Solomon, S. C. and McNutt, R. L., Jr. (2012). Characteristics of the plasma distribution in Mercury's equatorial magnetosphere derived from MESSENGER Magnetometer observations. *J. Geophys. Res.*, **117**, A00M07, doi:10.1029/2012JA018052.

Korth, H., Anderson, B. J., Gershman, D. J., Raines, J. M., Slavin, J. A., Zurbuchen, T. H., Solomon, S. C. and McNutt, R. L., Jr. (2014). Plasma distribution in Mercury's magnetosphere derived from MESSENGER Magnetometer and Fast Imaging Plasma Spectrometer observations. *J. Geophys. Res. Space Physics*, **119**, 2917–2932, doi:10.1002/2013JA019567.

Korth, H., Tsyganenko, N. A., Johnson, C. L., Philpott, L. C., Anderson, B. J., Al Asad, M. M., Solomon, S. C. and McNutt, R. L., Jr. (2015). Modular model for Mercury's magnetospheric magnetic field confined within the average observed magnetopause. *J. Geophys. Res. Space Physics*, **120**, 4503–4518, doi:10.1002/2015JA021022.

Lammer, H., Wurz, P., Patel, M. R., Killen, R., Kolb, C., Massetti, S., Orsini, S. and Milillo, A. (2003). The variability of Mercury's exosphere by particle and radiation induced surface release processes. *Icarus*, **166**, 238–247, doi:10.1016/j.icarus.2003.08.012.

Lawrence, D. J., Anderson, B. J., Baker, D. N., Feldman, W. C., Ho, G. C., Korth, H., McNutt, R. L., Jr., Peplowski, P. N., Solomon, S. C., Starr, R. D., Vandegriff, J. D. and Winslow, R. M. (2015). Comprehensive survey of energetic electron events in Mercury's magnetosphere with data from the MESSENGER Gamma-Ray and Neutron Spectrometer. *J. Geophys. Res. Space Physics*, **120**, 2851–2876, doi:10.1002/2014JA020792.

Leblanc, F. and Johnson, R. E. (2003). Mercury's sodium exosphere. *Icarus*, **164**, 261–281, doi:10.1016/S0019-1035(03)00147-7.

Leblanc, F. and Johnson, R. E. (2010). Mercury exosphere, I. Global circulation model of its sodium component. *Icarus*, **209**, 280–300, doi:10.1016/j.icarus.2010.04.020.

Liljeblad, E., Sundberg, T., Karlsson, T. and Kullen, A. (2014). Statistical investigation of Kelvin–Helmholtz waves at the magnetopause of Mercury. *J. Geophys. Res. Space Physics*, **119**, 9670–9683, doi:10.1002/2014JA020614.

Liljeblad, E., Karlsson, T., Raines, J. M., Slavin, J. A., Kullen, A., Sundberg, T. and Zurbuchen, T. H. (2015). MESSENGER observations of the dayside low-latitude boundary layer in Mercury's magnetosphere. *J. Geophys. Res. Space Physics*, **120**, 8387–8400, doi:10.1002/2015JA021662.

Lockwood, M., Chandler, M. O., Horwitz, J. L., Waite, J. H., Moore, T. E. and Chappell, C. R. (1985). The cleft ion fountain. *J. Geophys. Res.*, **90**, 9736–9748, doi:10.1029/Ja090ia10p09736.

Luhmann, J. G., Russell, C. T. and Tsyganenko, N. A. (1998). Disturbances in Mercury's magnetosphere: Are the Mariner 10 "substorms" simply driven? *J. Geophys. Res.*, **103**, 9113–9119, doi:10.1029/97ja03667.

Marsch, E., Muhlhauser, K. H., Rosenbauer, H., Schwenn, R. and Neubauer, F. M. (1982). Solar wind helium ions: Observations of the Helios solar probes between 0.3 and 1 AU. *J. Geophys. Res.*, **87**, 35–51, doi:10.1029/Ja087ia01p00035.

Massetti, S., Orsini, S., Milillo, A., Mura, A., De Angelis, E., Lammer, H. and Wurz, P. (2003). Mapping of the cusp plasma precipitation on the surface of Mercury. *Icarus*, **166**, 229–237, doi:10.1016/j.icarus.2003.08.005.

McClintock, W. E. and Lankton, M. R. (2007). The Mercury Atmospheric and Surface Composition Spectrometer for the MESSENGER mission. *Space Sci. Rev.*, **131**, 481–521, doi:10.1007/s11214-007-9264-5.

Milan, S. E. and Slavin, J. A. (2011). An assessment of the length and variability of Mercury's magnetotail. *Planet. Space Sci.*, **59**, 2058–2065, doi:10.1016/j.pss.2011.05.007.

Müller, J., Simon, S., Wang, Y. C., Motschmann, U., Heyner, D., Schüle, J., Ip, W. H., Kleindienst, G. and Pringle, G. J. (2012). Origin of Mercury's double magnetopause: 3D hybrid simulation study with A.I.K.E.F. *Icarus*, **218**, 666–687, doi:10.1016/j.icarus.2011.12.028.

Ness, N. F., Behannon, K. W., Lepping, R. P., Whang, Y. C. and Schatten, K. H. (1974). Magnetic field observations near Mercury: Preliminary results from Mariner 10. *Science*, **185**, 151–160, doi:10.1126/science.185.4146.151.

Ness, N. F., Behannon, K. W., Lepping, R. P. and Whang, Y. C. (1975). Magnetic field of Mercury confirmed. *Nature*, **255**, 204–205, doi:10.1038/255204a0.

Neugebauer, M., Forsyth, R. J., Galvin, A. B., Harvey, K. L., Hoeksema, J. T., Lazarus, A. J., Lepping, R. P., Linker, J. A., Mikic, Z., Steinberg, J. T., von Steiger, R., Wang, Y. M. and Wimmer-Schweingruber, R. F. (1998). Spatial structure of the solar wind and comparisons with solar data and models. *J. Geophys. Res.*, **103**, 14,587–14,599, doi:10.1029/98ja00798.

Neugebauer, M., Liewer, P. C., Smith, E. J., Skoug, R. M. and Zurbuchen, T. H. (2002). Sources of the solar wind at solar activity maximum. *J. Geophys. Res.*, **107**, 1488, doi:10.1029/Ja000306.

Odstrcil, D. (2003). Modeling 3-D solar wind structure, *Adv. Space Res.*, **32**, 497–506, doi:10.1016/S0273-1177(03)00332-6.

Ogilvie, K. W., Scudder, J. D., Hartle, R. E., Siscoe, G. L., Bridge, H. S., Lazarus, A. J., Asbridge, J. R., Bame, S. J. and Yeates, C. M.

(1974). Observations at Mercury encounter by the plasma science experiment on Mariner 10. *Science*, **185**, 145–151, doi:10.1126/science.185.4146.145.

Olson, W. P. (1984). Introduction to the topology of magnetospheric current systems. In *Magnetospheric Currents*, ed. T. A. Potemra. Geophysical Monograph 28. Washington, DC: American Geophysical Union, pp. 49–62.

Phan, T. D., Paschmann, G., Baumjohann, W., Sckopke, N. and Luhr, H. (1994). The magnetosheath region adjacent to the dayside magnetopause: AMPTE/IRM observations. *J. Geophys. Res.*, **99**, 121–141, doi:10.1029/93ja02444.

Phan, T. D., Larson, D. E., Lin, R. P., McFadden, J. P., Anderson, K. A., Carlson, C. W., Ergun, R. E., Ashford, S. M., McCarthy, M. P., Parks, G. K., Reme, H., Bosqued, J. M., D'Uston, C., Wenzel, K. P., Sanderson, T. R. and Szabo, A. (1996). The subsolar magnetosheath and magnetopause for high solar wind ram pressure: WIND observations. *Geophys. Res. Lett.*, **23**, 1279–1282, doi:10.1029/96gl00845.

Poh, G., Slavin, J. A., Jia, X., Raines, J. M. and Gershman, D. J. (2015). MESSENGER observations of reconnection ion diffusion region structure at Mercury. Presented at 12th Annual Asia Oceania Geosciences Society Conference, abstract PS05-D5-PM2-P-015, Singapore, 2–7 August.

Raines, J. M., Slavin, J. A., Zurbuchen, T. H., Gloeckler, G., Anderson, B. J., Baker, D. N., Korth, H., Krimigis, S. M. and McNutt, R. L., Jr. (2011). MESSENGER observations of the plasma environment near Mercury. *Planet. Space Sci.*, **59**, 2004–2015, doi:10.1016/j.pss.2011.02.004.

Raines, J. M., Gershman, D. J., Zurbuchen, T. H., Sarantos, M., Slavin, J. A., Gilbert, J. A., Korth, H., Anderson, B. J., Gloeckler, G., Krimigis, S. M., Baker, D. N., McNutt, R. L., Jr. and Solomon, S. C. (2013). Distribution and compositional variations of plasma ions in Mercury's space environment: The first three Mercury years of MESSENGER observations. *J. Geophys. Res. Space Physics*, **118**, 1604–1619, doi:10.1029/2012ja018073.

Raines, J. M., Gershman, D. J., Slavin, J. A., Zurbuchen, T. H., Korth, H., Anderson, B. J. and Solomon, S. C. (2014). Structure and dynamics of Mercury's magnetospheric cusp: MESSENGER measurements of protons and planetary ions. *J. Geophys. Res. Space Physics*, **119**, 6587–6602, doi:10.1002/2014ja020120.

Richmond, A. D. and Thayer, J. P. (2000). Ionospheric electrodynamics: A tutorial. In *Magnetospheric Current Systems*, ed. S. Ohtani, R. Fujii, M. Hesse and R. L. Lysak. Geophysical Monograph 118. Washington, DC: American Geophysical Union, pp. 131–146.

Rosenbauer, H., Grünwaldt, H., Montgomery, M. D., Paschmann, G. and Sckopke, N. (1975). Heos 2 plasma observations in the distant polar magnetosphere: The plasma mantle. *J. Geophys. Res.*, **80**, 2723–2737, doi:10.1029/Ja080i019p02723.

Russell, C. T. (1977). On the relative locations of bow shocks of the terrestrial planets. *Geophys. Res. Lett.*, **4**, 387–390, doi:10.1029/Gl004i010p00387.

Russell, C. T., Baker, D. N. and Slavin, J. A. (1988). The magnetosphere of Mercury. In *Mercury*, ed. F. Vilas, C. R. Chapman and M. S. Matthews. Tucson, AZ: University of Arizona Press, pp. 514–561.

Sarantos, M. and Slavin, J. A. (2009). On the possible formation of Alfvén wings at Mercury during encounters with coronal mass ejections. *Geophys. Res. Lett.*, **36**, L04107, doi:10.1029/2008gl036747.

Sarantos, M., Killen, R. M. and Kim, D. (2007). Predicting the long-term solar wind ion-sputtering source at Mercury. *Planet. Space Sci.*, **55**, 1584–1595, doi:10.1016/j.pss.2006.10.011.

Schield, M. A. (1969). Pressure balance between solar wind and magnetosphere. *J. Geophys. Res.*, 74, 1275–1286.

Schlemm, C. E., II, Starr, R. D., Ho, G. C., Bechtold, K. E., Hamilton, S. A., Boldt, J. D., Boynton, W. V., Bradley, W., Fraeman, M. E., Gold, R. E., Goldsten, J. O., Hayes, J. R., Jaskulek, S. E., Rossano, E., Rumpf, R. A., Schaefer, E. D., Strohbehn, K., Shelton, R. G., Thompson, R. E., Trombka, J. I. and Williams, B. D. (2007). The X-Ray Spectrometer on the MESSENGER spacecraft. *Space Sci. Rev.*, **131**, 393–415, doi:10.1007/s11214-007-9248-5.

Scurry, L., Russell, C. T. and Gosling, J. T. (1994). Geomagnetic activity and the beta dependence of the dayside reconnection rate. *J. Geophys. Res.*, **99**, 14,811–14,814, doi:10.1029/94JA00794.

Shue, J. H., Chao, J. K., Fu, H. C., Russell, C. T., Song, P., Khurana, K. K. and Singer, H. J. (1997). A new functional form to study the solar wind control of the magnetopause size and shape. *J. Geophys. Res.*, **102**, 9497–9511, doi:10.1029/97ja00196.

Simpson, J. A., Eraker, J. H., Lamport, J. E. and Walpole, P. H. (1974). Electrons and protons accelerated in Mercury's magnetic field. *Science*, **185**, 160–166, doi:10.1126/science.185.4146.160.

Slavin, J. A., Smith, E. J., Sibeck, D. G., Baker, D. N., Zwickl, R. D. and Akasofu, S. I. (1985). An ISEE-3 study of average and substorm conditions in the distant magnetotail. *J. Geophys. Res.*, **90**, 875–895, doi:10.1029/Ja090ia11p10875.

Slavin, J. A., Krimigis, S. M., Acuña, M. H., Anderson, B. J., Baker, D. N., Koehn, P. L., Korth, H., Livi, S., Mauk, B. H., Solomon, S. C. and Zurbuchen, T. H. (2007). MESSENGER: Exploring Mercury's magnetosphere. *Space Sci. Rev.*, **131**, 133–160, doi:10.1007/s11214-007-9154-x.

Slavin, J. A., Acuna, M. H., Anderson, B. J., Baker, D. N., Benna, M., Gloeckler, G., Gold, R. E., Ho, G. C., Killen, R. M., Korth, H., Krimigis, S. M., McNutt, R. L., Jr., Nittler, L. R., Raines, J. M., Schriver, D., Solomon, S. C., Starr, R. D., Travnicek, P. and Zurbuchen T. H. (2008). Mercury's magnetosphere after MESSENGER's first flyby. *Science*, **321**, 85–89, doi:10.1126/science.1159040.

Slavin, J. A., Acuña, M. H., Anderson, B. J., Baker, D. N., Benna, M., Boardsen, S. A., Gloeckler, G., Gold, R. E., Ho, G. C., Korth, H., Krimigis, S. M., McNutt, R. L., Jr., Raines, J. M., Sarantos, M., Schriver, D., Solomon, S. C., Trávníček, P. and Zurbuchen, T. H. (2009a). MESSENGER observations of magnetic reconnection in Mercury's magnetosphere. *Science*, **324**, 606–610, doi:10.1126/science.1172011.

Slavin, J. A., Anderson, B. J., Zurbuchen, T. H., Baker, D. N., Krimigis, S. M., Acuña, M. H., Benna, M., Boardsen, S. A., Gloeckler, G., Gold, R. E., Ho, G. C., Korth, H., McNutt, R. L., Jr., Raines, J. M., Sarantos, M., Schriver, D., Solomon, S. C. and Trávníček, P. (2009b). MESSENGER observations of Mercury's magnetosphere during northward IMF. *Geophys. Res. Lett.*, **36**, L02101, doi:10.1029/2008GL036158.

Slavin, J. A., Anderson, B. J., Baker, D. N., Benna, M., Boardsen, S. A., Gloeckler, G., Gold, R. E., Ho, G. C., Korth, H., Krimigis, S. M., McNutt, R. L., Jr., Nittler, L. R., Raines, J. M., Sarantos, M., Schriver, D., Solomon, S. C., Starr, R. D., Trávníček, P. M. and Zurbuchen, T. H. (2010). MESSENGER observations of extreme loading and unloading of Mercury's magnetic tail. *Science*, **329**, 665–668, doi:10.1126/science.1188067.

Slavin, J. A., Anderson, B. J., Baker, D. N., Benna, M., Boardsen, S. A., Gold, R. E., Ho, G. C., Imber, S. M., Korth, H., Krimigis, S. M., McNutt, R. L., Jr., Raines, J. M., Sarantos, M., Schriver, D., Solomon, S. C., Trávníček, P. and Zurbuchen, T. H. (2012). MESSENGER and Mariner 10 flyby observations of magnetotail structure and dynamics at Mercury. *J. Geophys. Res.*, **117**, A01215, doi:10.1029/2011JA016900.

Slavin, J. A., DiBraccio, G. A., Gershman, D. J., Imber, S. M., Poh, G. K., Raines, J. M., Zurbuchen, T. H., Jia, X. Z., Baker, D. N., Glassmeier, K. H., Livi, S. A., Boardsen, S. A., Cassidy, T. A.,

Sarantos, M., Sundberg, T., Masters, A., Johnson, C. L., Winslow, R. M., Anderson, B. J., Korth, H., McNutt, R. L., Jr. and Solomon, S. C. (2014). MESSENGER observations of Mercury's dayside magnetosphere under extreme solar wind conditions. *J. Geophys. Res. Space Physics*, **119**, 8087–8116, doi:10.1002/2014ja020319.

Smith, E. J. (2001). The heliospheric current sheet. *J. Geophys. Res.*, **106**, 15,819–15,831, doi:10.1029/2000ja000120.

Sonnerup, B. U. Ö. (1984). Magnetic field reconnection at the magnetopause: An overview. In *Magnetic Reconnection in Space and Laboratory Plasmas*, ed. E. W. Hones. Geophysical Monograph 30. Washington, DC: American Geophysical Union, pp. 92–103.

Speiser, T. W. (1965). Particle trajectories in model current sheets: 1. Analytical solutions. *J. Geophys. Res.*, **70**, 4219–4226, doi:10.1029/Jz070i017p04219.

Spreiter, J. R., Summers, A. L. and Alksne, A. Y. (1966a). Hydromagnetic flow around the magnetosphere. *Planet. Space Sci.*, **14**, 223–253, doi:10.1016/0032-0633(66)90124-3.

Spreiter, J. R., Alksne, A. Y. and Abraham-Shrauner, B. (1966b). Theoretical proton velocity distributions in the flow around the magnetosphere. *Planet. Space Sci.*, 14, 1207–1220.

Sun, W. J., Slavin, J. A., Fu, S. Y., Raines, J. M., Zong, Q. G., Imber, S. M., Shi, Q. Q., Yao, Z. H., Poh, G., Gershman, D. J., Pu, Z. Y., Sundberg, T., Anderson, B. J., Korth, H. and Baker, D. N. (2015). MESSENGER observations of magnetospheric substorm activity in Mercury's near magnetotail. *Geophys. Res. Lett.*, **42**, 3692–3699, doi:10.1002/2015gl064052.

Sundberg, T., Boardsen, S. A., Slavin, J. A., Blomberg, L. G. and Korth, H. (2010). The Kelvin–Helmholtz instability at Mercury: An assessment. *Planet. Space Sci.*, **58**, 1434–1441, doi:10.1016/j.pss.2010.06.008.

Toth, G. and Odstrcil, D. (1996). Comparison of some flux corrected transport and total variation diminishing numerical schemes for hydrodynamic and magnetohydrodynamic problems. *J. Comput. Phys.*, **128**, 82–100, doi:10.1006/jcph.1996.0197.

Trávníček, P. M., Hellinger, P. and Schriver, D. (2007). Structure of Mercury's magnetosphere for different pressure of the solar wind: Three dimensional hybrid simulations. *Geophys. Res. Lett.*, **34**, L05104, doi:10.1029/2006GL028518.

Trávníček, P. M., Schriver, D., Hellinger, P., Herčík, D., Anderson, B. J., Sarantos, M. and Slavin, J. A. (2010). Mercury's magnetosphere–solar wind interaction for northward and southward interplanetary magnetic field: Hybrid simulation results. *Icarus*, **209**, 11–22, doi:10.1016/j.icarus.2010.01.008.

Tsyganenko, N. A. (2013). Data-based modelling of the Earth's dynamic magnetosphere: A review. *Ann. Geophys.*, **31**, 1745–1772, doi:10.5194/angeo-31-1745-2013.

Tsyganenko, N. A. and Peredo, M. (1994). Analytical models of the magnetic field of disk-shaped current sheets. *J. Geophys. Res.*, **99**, 199–205, doi:10.1029/93ja02768.

Vasyliunas, V. M. (1975). Theoretical models of magnetic field line merging, 1. *Rev. Geophys.*, **13**, 303–336, doi:10.1029/RG013i001p00303.

Verhoeven, O., Tarits, P., Vacher, P., Rivoldini, A. and Van Hoolst, T. (2009). Composition and formation of Mercury: Constraints from future electrical conductivity measurements. *Planet. Space Sci.*, **57**, 296–305, doi:10.1016/j.pss.2008.11.015.

von Steiger, R. and Zurbuchen, T. H. (2006). Kinetic properties of heavy solar wind ions from Ulysses-SWICS. *Geophys. Res. Lett.*, **33**, L09103, doi:10.1029/2005gl024998.

von Steiger, R., Geiss, J., Gloeckler, G. and Galvin, A. B. (1995). Kinetic properties of heavy ions in the solar wind from SWICS/Ulysses. *Space Sci. Rev.*, 72, 71–76.

von Steiger, R., Schwadron, N. A., Fisk, L. A., Geiss, J., Gloeckler, G., Hefti, S., Wilken, B., Wimmer-Schweingruber, R. F. and Zurbuchen, T. H. (2000). Composition of quasi-stationary solar wind flows from Ulysses/Solar Wind Ion Composition spectrometer. *J. Geophys. Res.*, **105**, 27,217–27,238, doi:10.1029/1999ja000358.

Wang, Y. C., Mueller, J., Motschmann, U. and Ip, W. H. (2010). A hybrid simulation of Mercury's magnetosphere for the MESSENGER encounters in year 2008. *Icarus*, **209**, 46–52, doi:10.1016/j.icarus.2010.05.020.

Whang, Y. C. (1977). Magnetospheric magnetic field of Mercury. *J. Geophys. Res.*, **82**, 1024–1030, doi:10.1029/Ja082i007p01024.

Winslow, R. M., Johnson, C. L., Anderson, B. J., Korth, H., Slavin, J. A., Purucker, M. E. and Solomon, S. C. (2012). Observations of Mercury's northern cusp region with MESSENGER's Magnetometer. *Geophys. Res. Lett.*, **39**, L08112, doi:10.1029/2012GL051472.

Winslow, R. M., Anderson, B. J., Johnson, C. L., Slavin, J. A., Korth, H., Purucker, M. E., Baker, D. N. and Solomon, S. C. (2013). Mercury's magnetopause and bow shock from MESSENGER Magnetometer observations. *J. Geophys. Res. Space Physics*, **118**, 2213–2227, doi:10.1002/jgra.50237.

Winslow, R. M., Johnson, C. L., Anderson, B. J., Gershman, D. J., Raines, J. M., Lillis, R. J., Korth, H., Slavin, J. A., Solomon, S. C., Zurbuchen, T. H. and Zuber, M. T. (2014). Mercury's surface magnetic field determined from proton-reflection magnetometry. *Geophys. Res. Lett.*, **41**, 4463–4470, doi:10.1002/2014gl060258.

Wurz, P., Whitby, J. A., Rohner, U., Martin-Fernandez, J. A., Lammer, H. and Kolb, C. (2010). Self-consistent modelling of Mercury's exosphere by sputtering, micro-meteorite impact and photon-stimulated desorption. *Planet. Space Sci.*, **58**, 1599–1616, doi:10.1016/j.pss.2010.08.003.

Zurbuchen, T. H. (2007). A new view of the coupling of the Sun and the heliosphere. *Ann. Rev. Astron. Astrophys.*, **45**, 297–338, doi:10.1146/annurev.astro.45.010807.154030.

Zurbuchen, T. H., Raines, J. M., Gloeckler, G., Krimigis, S. M., Slavin, J. A., Koehn, P. L., Killen, R. M., Sprague, A. L., McNutt, R. L., Jr. and Solomon, S. C. (2008). MESSENGER observations of the composition of Mercury's ionized exosphere and plasma environment. *Science*, **321**, 90–92, doi:10.1126/science.1159314.

Zurbuchen, T. H., Raines, J. M., Slavin, J. A., Gershman, D. J., Gilbert, J. A., Gloeckler, G., Anderson, B. J., Baker, D. N., Korth, H., Krimigis, S. M., Sarantos, M., Schriver, D., McNutt, R. L., Jr. and Solomon, S. C. (2011). MESSENGER observations of the spatial distribution of planetary ions near Mercury. *Science*, **333**, 1862–1865, doi:10.1126/science.1211302.

Zwan, B. J. and Wolf, R. A. (1976). Depletion of solar-wind plasma near a planetary boundary. *J. Geophys. Res.*, **81**, 1636–1648, doi:10.1029/JA081i010p01636.

17

Mercury's Dynamic Magnetosphere

JAMES A. SLAVIN, DANIEL N. BAKER, DANIEL J. GERSHMAN, GEORGE C. HO, SUZANNE M. IMBER, STAMATIOS M. KRIMIGIS, AND TORBJÖRN SUNDBERG

17.1 INTRODUCTION

The Mariner 10 flybys of Mercury on 29 March 1974 and 16 March 1975 provided our first close-up glimpse of the innermost planet and its magnetosphere (Ness, 1979). The weak planetary magnetic field, lack of an ionosphere, bursts of energetic particles, and proximity to the Sun sparked great speculation as to how Mercury's magnetosphere might be different from that of Earth (Russell et al., 1988; Slavin, 2004; Vasyliunas, 2004; Baumjohann et al., 2006; Fujimoto et al., 2007; Slavin et al., 2007; Sundberg and Slavin, 2015; Raines et al., 2015).

The MErcury Surface, Space ENvironment, GEochemistry, and Ranging (MESSENGER) spacecraft made three close (~200 km) equatorial flybys of Mercury on 14 January 2008, 6 October 2008, and 29 September 2009, two of which are displayed in Figure 17.1. Insertion into its initial orbit with a ~12-h period, 82.5° inclination, and high eccentricity (~200 × 15,000-km altitude) took place on 18 March 2011 (Solomon et al., 2007). Apoapsis was lowered to ~10,000-km altitude by two maneuvers on 16 and 20 April 2012, which reduced the orbital period to ~8 h. After completing a final low-altitude campaign that saw the spacecraft return measurements from altitudes as low as ~5 km, the MESSENGER spacecraft impacted Mercury's surface on 30 April 2015.

The MESSENGER Magnetometer (MAG) measured the ambient magnetic field at a cadence of 20 s^{-1} (Anderson et al., 2007). The MAG measurements from MESSENGER's first flyby confirmed that Mercury's magnetic field is largely dipolar, with the same polarity as Earth's field. It has a moment of ~190 nT R_M^3, where R_M is Mercury's radius, 2440 km, and the dipole is aligned with the planetary rotation axis to within ~1° (Anderson et al., 2008a). Magnetic field observations collected during the second flyby and the orbital phase of the mission refined the dipole value and determined a northward offset of ~0.2 R_M (Alexeev et al., 2008, 2010; Anderson et al., 2011, 2012; Johnson et al., 2012).

As described in Chapter 16, and illustrated in Figure 17.1, the solar wind interaction with Mercury's dipole creates a miniature magnetosphere, structurally quite similar to Earth's. The mean distance from the internal dipole to the nose of Mercury's magnetopause is ~(1.4–1.5) R_M (Slavin et al., 2009b; Winslow et al., 2013; Zhong et al., 2015a), and the tail diameter is ~5 R_M under average solar wind conditions (Slavin et al., 2012a; Winslow et al., 2013). Numerical simulations of the solar wind interaction with Mercury's offset dipole magnetic field support these MESSENGER results (Kabin et al., 2000; Janhunen and Kallio, 2004; Kidder et al., 2008; Trávníček et al., 2009, 2010; Benna et al., 2010; Wang et al., 2010; Muller et al., 2012; Varela et al., 2015; Jia et al., 2015).

In addition to the magnetic field, MESSENGER measured the energy per charge and composition of ions up to 13 keV/e with 10-s time resolution using its Fast Imaging Plasma Spectrometer (FIPS) (Andrews et al., 2007). The FIPS measurements showed that Mercury's magnetosphere and the surrounding space environment are permeated by planetary ions derived from the photoionization of Mercury's exosphere (Zurbuchen et al., 2008, 2011; Sarantos et al., 2009; Korth et al., 2011b; Raines et al., 2013; Gershman et al., 2013). As shown in Figure 17.1, Mercury's exosphere, and the resulting population of planetary ions, is observed in the magnetosheath and throughout the magnetosphere (Raines et al., 2011, 2013; Schriver et al., 2011a,b; Zurbuchen et al., 2011; Gershman et al., 2013; DiBraccio et al., 2015b). FIPS was also highly sensitive to energetic electrons with energies in the MeV range that penetrated the instrument during solar energetic particle (SEP) events. These energetic electrons produced counts in the FIPS detectors that were recorded with ~150-ms time resolution. These electrons have been used as field-line tracers to separate magnetic fields that are connected to the solar wind from those with both ends rooted in the planet (Gershman et al., 2015b).

MESSENGER's Energetic Particle Spectrometer (EPS) measured energetic electrons from ~35 keV to ~1 MeV with a temporal resolution of 3 s (Andrews et al., 2007). The X-Ray Spectrometer (XRS) (Starr et al., 2012) and the Gamma-Ray and Neutron Spectrometer (GRNS) (Goldsten et al., 2007) sensor systems were also sensitive to electrons with energies from several keV up to a few hundred keV (Lawrence et al., 2015; Baker et al., 2016; Ho et al., 2016). Measurements from these MESSENGER instruments have determined that the energetic particle bursts detected by Mariner 10 (Simpson et al., 1974; Christon et al., 1979) are composed of energetic electrons with energies from tens of keV up to several hundred keV (Ho et al., 2011, 2012; Lawrence et al., 2015; Baker et al., 2016).

Whereas Chapter 16 concentrated on the structure and configuration of Mercury's magnetosphere, our focus here is on its remarkable dynamics and the intense solar wind forcing that drives it. To frame this review of MESSENGER's most important results we list below some key sets of questions drawn from the scientific literature regarding magnetospheric dynamics at Mercury. However, the complete answers for most of them go well beyond the MESSENGER mission-level objectives and

(a) 14 January 2008, Northward IMF

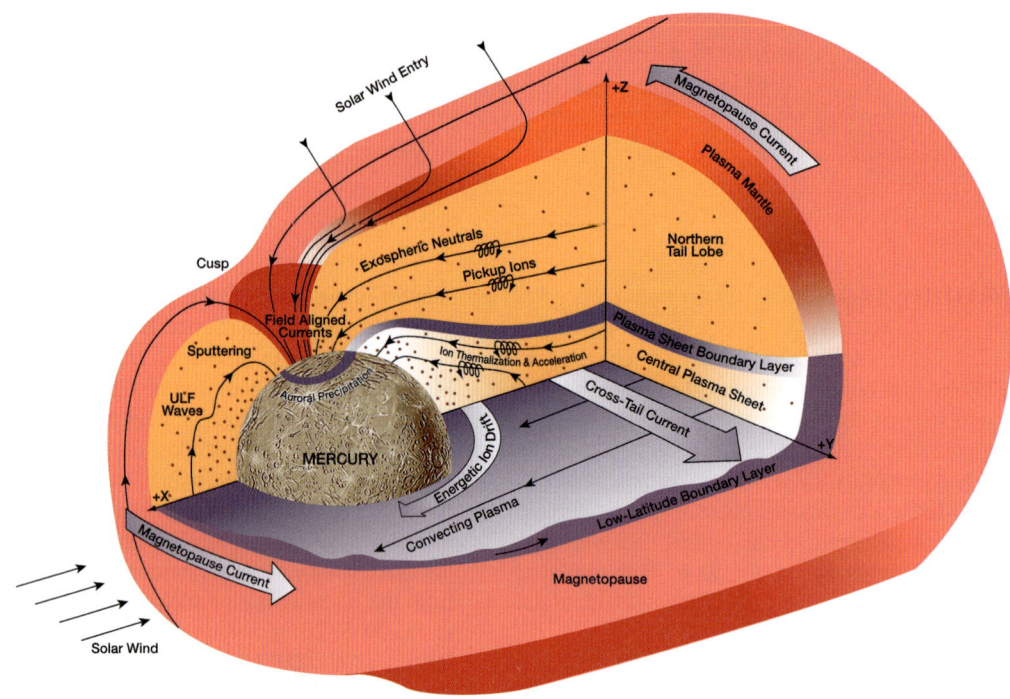

(b) 6 October 2008, Southward IMF

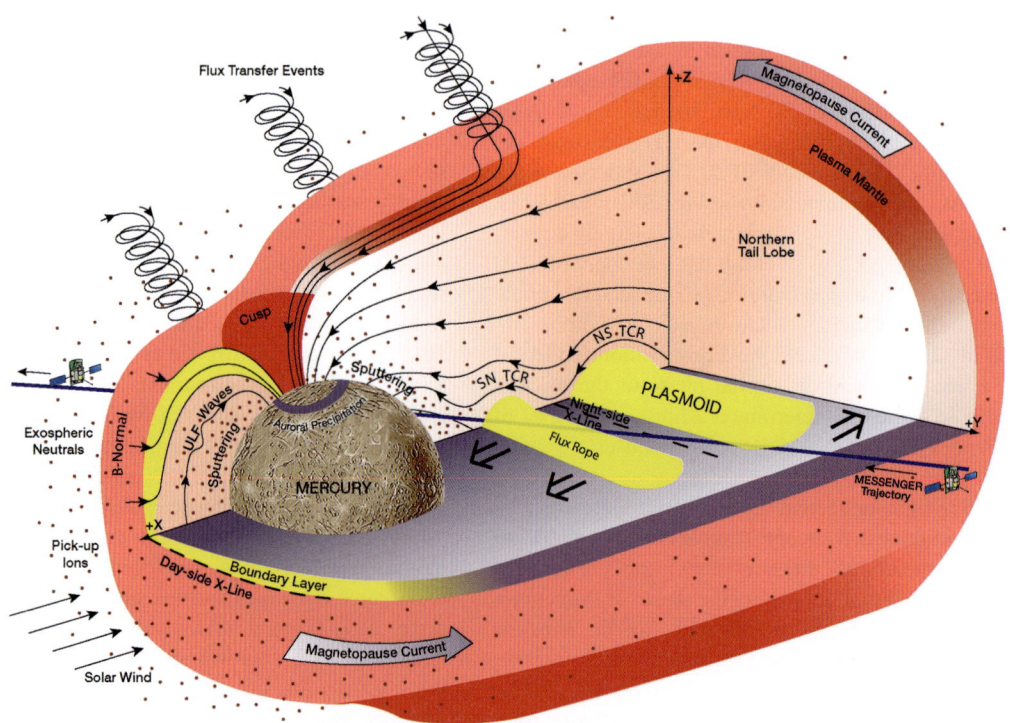

Figure 17.1. (a) Quiet magnetospheric conditions were found by MESSENGER during its first Mercury flyby. From Slavin et al. (2008). (b) During MESSENGER's second Mercury flyby, the magnetosphere was very active. From Slavin et al. (2009a).

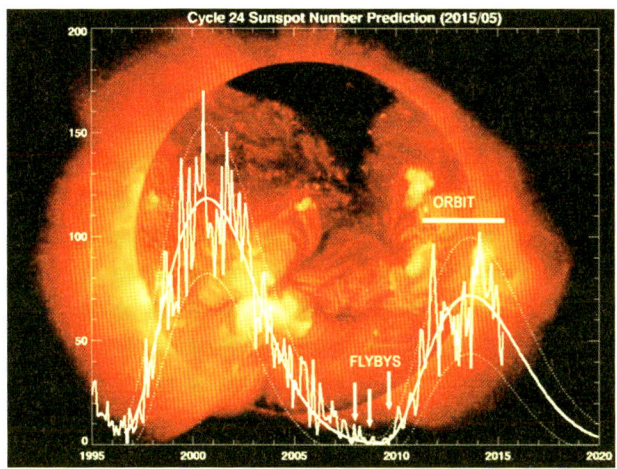

Figure 17.2. Sunspot numbers for solar cycles 23 and 24. The times of MESSENGER's Mercury flybys are marked by downward arrows and the orbital phase of the mission by a horizontal bar. Image provided courtesy of D. Hathaway at NASA Ames Research Center and the Yohkoh Soft X-ray Telescope Consortium.

will have to await new measurements to be collected by the upcoming dual-spacecraft BepiColombo mission (Benkhoff et al., 2010).

Question set 1: How do magnetosheath conditions at Mercury differ from what is found at the other planets?

Question set 2: How do conditions in Mercury's magnetosheath contribute to the dynamic nature of its magnetosphere? How does magnetopause reconnection at Mercury differ from that at Earth? Are flux transfer events (FTEs) a major driver of magnetospheric convection at Mercury?

Question set 3: Does reconnection ever erode the dayside magnetosphere to the point that the subsolar region of the surface is exposed to direct solar wind impact? To what extent do induction currents driven in Mercury's interior limit the solar wind flux to the surface? Do FTEs contribute significantly to the solar wind flux reaching the surface?

Question set 4: What effects do heavy planetary ions have on Mercury's magnetosphere?

Question set 5: Does Mercury's magnetotail store and dissipate magnetic energy in a manner similar to substorms at Earth? How is the process affected by the lack of an ionosphere and the expected high electrical resistivity of the crust?

Question set 6: How does Mercury's magnetosphere accelerate energetic electrons up to hundreds of keV?

17.2 MERCURY'S PLACE IN THE HELIOSPHERE

As depicted in Figure 17.2 (Baker et al., 2013), the three MESSENGER flybys occurred very close to the minimum in solar activity between sunspot cycles 23 and 24.

The MESSENGER orbital phase (2011–2015) spanned the rising portion of cycle 24 and most of its peak. The ability of FIPS to measure solar wind properties was strongly impacted by the need to locate the sensor behind the spacecraft's sunshade because of Mercury's challenging thermal environment (Gershman et al., 2012). However, the MAG measurements of the interplanetary magnetic field (IMF) (Korth et al., 2011a) and numerical solar wind modeling (Baker et al., 2009, 2011, 2013) have confirmed the expected mixture of slow and fast solar wind punctuated by coronal mass ejections (CMEs) over the course of the mission (Slavin et al., 2014; Dewey et al., 2015). Typical solar wind dynamic pressures at 1 AU are ~1–2 nPa. Solar wind measurements at Mercury's heliocentric distance (~0.31–0.47 AU) were made by the Helios spacecraft in the 1980s. Those measurements indicated solar wind pressure in the range ~10–20 nPa (Marsch et al., 1982), similar to values inferred from radial scaling of measurements at 1 AU, numerical simulations of solar wind evolution, and the MESSENGER data (Slavin and Holzer, 1981; Baker et al., 2013; Winslow et al., 2013; Dewey et al., 2015).

Mercury is unique in the solar system with respect to its low-Alfvénic-Mach-number ($M_A = B/(\mu_0 \rho)^{1/2}$) interplanetary environment – where B is IMF intensity, μ_0 is the permeability in vacuum, and ρ is the solar wind mass density – and all that this implies for its magnetosphere. Solar wind M_A at 1 AU typically ranges from ~7 to 10 and increases linearly with radial distance from the Sun. For this reason the solar wind flow about Earth and the more distant planets is said to be hyper-Alfvénic. In this regime, the plasma and magnetic field compression at the bow shock are near their asymptotic limits and relatively insensitive to further increases in Mach number (Spreiter et al., 1966). In contrast, M_A at Mercury is much lower, ~2 to 5 (Slavin and Holzer, 1981; Sarantos et al., 2007; Gershman et al., 2013). In this regime, bow shock and magnetosheath properties are very sensitive to upstream conditions.

17.3 MAGNETOSPHERIC DYNAMICS

Magnetic reconnection at the dayside magnetopause splices together interplanetary and planetary magnetic fields to create new flux tubes that have one end embedded in the fast antisunward-directed solar wind and the other rooted in the planet. In this manner the magnetopause current sheet that separates the interplanetary and magnetospheric magnetic fields is transformed from an impenetrable tangential discontinuity (Figure 17.1a) into a rotational discontinuity (Figure 17.1b) across which the solar wind enters the magnetosphere (Dungey, 1961). Reconnection at the dayside magnetopause transfers magnetic flux from the closed-field-line region in the forward magnetosphere into the open-field-line lobe regions of the magnetotail. The return circulation is driven by reconnection in the cross-tail current sheet. The complete circulation of magnetic flux and plasma to the tail and back again is termed the Dungey cycle. The speed of the internal plasma convection and the scale of the system determine the time necessary to complete the Dungey cycle in a terrestrial-type magnetosphere. Mercury's magnetosphere is small compared with that of Earth, and its internal convection speed is expected to be high because of Mercury's lack of an ionosphere and, therefore, weak "line-tying" effects (Coroniti and Kennel, 1973; Hill et al., 1976). For this reason the several-hour Dungey cycle time observed at Earth (e.g., Borovsky et al., 1993; Tanskanen, 2009) is reduced to only a few minutes for Mercury's magnetosphere (Siscoe et al., 1975; Slavin et al., 2010a). Hence, Mercury's magnetosphere can move from quiescent to disturbed states, as illustrated in Figures 17.1a and 17.1b, in response to changing solar wind conditions in just a few minutes.

At Earth a large magnetic shear across the magnetopause due to a southward interplanetary magnetic field is necessary before intense magnetic reconnection can occur (Paschmann et al., 1986; Sonnerup et al., 1990). Therefore, Earth's magnetosphere has been described as a "half-wave rectifier" with maximum (minimum) reconnection and energy transfer to the magnetosphere for southward (northward) IMF. The low-M_A solar wind interaction with Mercury's magnetosphere results in a low-β magnetosheath, where β is the ratio of plasma thermal pressure to magnetic pressure, and the development of a strong plasma depletion layer (PDL) adjacent to the magnetopause (Gershman et al., 2013). Reconnection at Mercury is therefore substantially different from that at Earth in that it occurs between magnetic fields that are often comparable in magnitude. Reconnection between magnetic fields of similar magnitude is said to be "symmetric." Under such conditions reconnection can occur for all non-zero magnetic shear angles and at higher rates for a given magnetic shear angle than under asymmetric conditions (Sonnerup, 1974; DiBraccio et al., 2013). In contrast with Mercury, asymmetric magnetopause reconnection is the most common situation at Earth and the outer planets (Masters et al., 2012; Eastwood et al., 2013).

Since the initial Mariner 10 discovery of Mercury's weak dipolar magnetic field (Ness et al., 1974; Ness, 1979), there has been great interest in determining how often the solar wind compresses and/or erodes the altitude of the forward magnetopause to within a few proton gyroradii of the surface. However, direct solar wind impact also occurs whenever magnetospheric magnetic fields are opened by reconnection at the magnetopause. Initially these open field lines all map to the magnetospheric cusps. As the ends of the field lines embedded in the solar wind are carried downstream, this magnetic flux is added to the lobes of the tail, where the flux tubes convect toward the cross-tail current sheet. Solar wind particles with a component of their velocities parallel to the magnetic field and directed inward toward the planet will impact the surface unless they mirror first. As a result, the magnetospheric cusps are believed to be primary sites for solar wind impact on Mercury's surface (Massetti et al., 2003; Sarantos et al., 2007). The direct impact of solar wind protons not only sputters neutral atoms and ions into Mercury's environment (e.g., Pfleger et al., 2015), it also contributes to the "space weathering" of the surface (Domingue et al., 2014).

Numerous studies of the distribution of solar wind pressure have all shown that simple Earth-like compression of the forward magnetopause will place the magnetopause close to or beneath the surface no more than a small percentage of the time (e.g., Siscoe and Christopher, 1975; Zhong et al., 2015b; Johnson et al., 2016). However, Slavin and Holzer (1979) argued, on the basis of models developed for Earth, that the rate of reconnection at Mercury would be substantially greater due to the high Alfvénic speed (i.e., low-M_A) conditions in the inner solar system. They scaled the effects of reconnection at Earth to Mercury and found that the magnetopause should frequently be eroded to very low altitudes, where direct solar wind impact would take place. However, shortly after these results were reported, Hood and Schubert (1979) and Suess and Goldstein (1979) presented models of the effects of induction currents driven within Mercury's interior by solar wind pressure enhancements. Their results indicated that induction would "stiffen" the planetary magnetic field to the point at which it would not be possible to compress the magnetopause to the surface even for the highest solar wind pressures anticipated at Mercury. Further, induction increases the amount of magnetic flux in the dayside magnetosphere that is available to reconnect with the IMF and must be transferred to the tail before the magnetopause may be eroded to the surface. The possible effects of magnetic reconnection were not included in the Hood and Schubert (1979) and Suess and Goldstein (1979) studies, just as the effects of induction-generated magnetic fields were not included by Slavin and Holzer (1979).

At Earth the Dungey cycle does not operate in a steady predictable manner by which the time profile of reconnection in the magnetotail can be directly determined by the time history of the reconnection that took place earlier at the dayside magnetopause. Instead, the magnetotail is observed to store magnetic flux for variable amounts of time (Borovsky et al., 1993). This storage interval ends with the onset of intense reconnection in the tail, which soon drives the system to a lower energy state. All of the processes in Earth's space environment that participate in this magnetic energy storage and release make up what is termed the "magnetospheric substorm" (McPherron et al., 1973; Hones et al., 1977; Baker et al., 1996). The tail lobes are "loaded" with magnetic flux during the growth phase of the substorm and unloaded during the expansion phase. The onset of reconnection in the magnetotail is thought to be due to the tail lobe magnetic fields reaching unsustainable levels, the growth of an instability in the coupling between the ionosphere and the magnetotail, or the "triggering" of tail reconnection by perturbations in the solar wind such as rapid changes in the IMF or solar wind pressure (see McPherron, 1991). In particular, the role of the ionosphere in closing the field-aligned currents that link it to the magnetosphere has led to the suggestion that changes in electrical conductivity at the foot of the open flux tubes, possibly as a result of charged particle precipitation, may be central to the occurrence of substorms in Earth's magnetosphere (Akasofu, 1964; Kan et al., 1988). However, as we describe below, MESSENGER observations suggest that substorms also occur at Mercury despite its lack of an ionosphere and the presence of an electrically resistive regolith (Hill et al., 1976).

17.3.1 Magnetopause Reconnection

An interplanetary shock wave passed over MESSENGER at 05:20 UTC on 23 November 2011. The shock wave was followed by the arrival of a CME around 07:30 UTC. The WSA-

17.3 Magnetospheric Dynamics 465

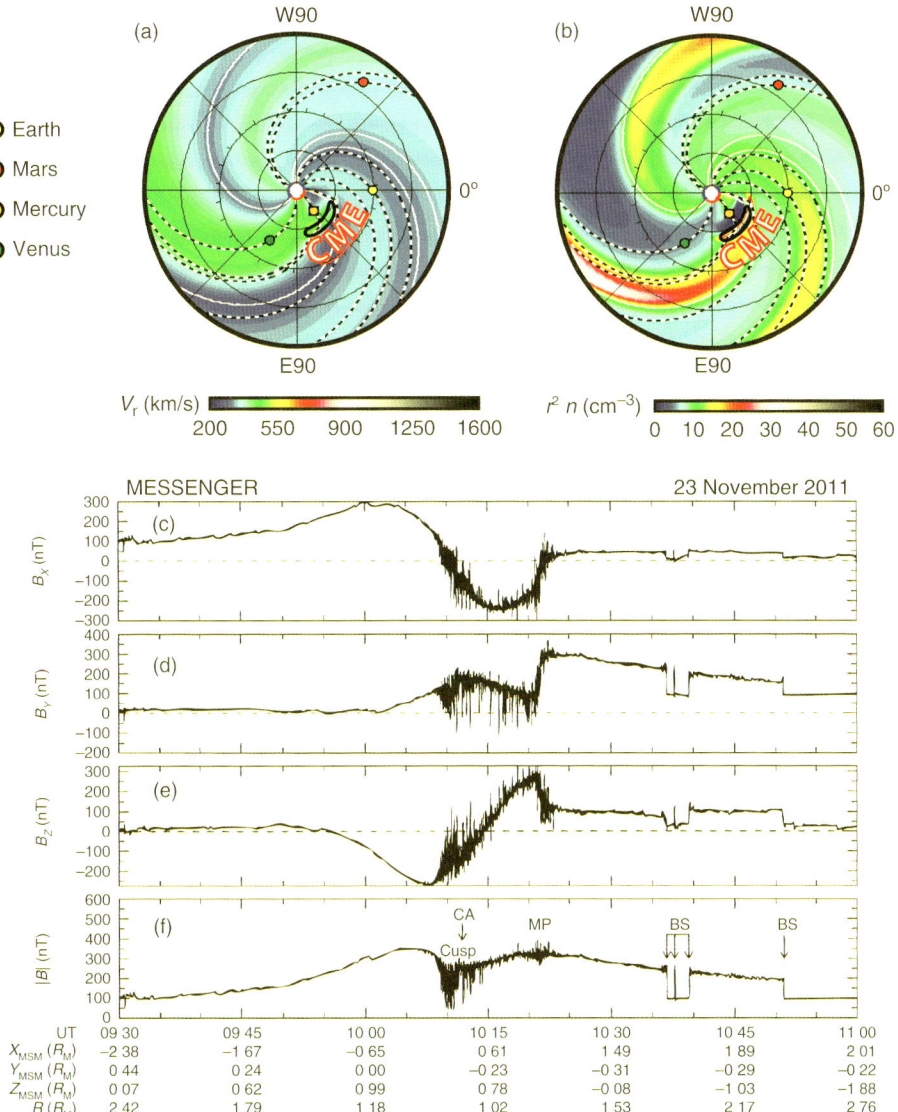

Figure 17.3. (a) WSA-ENLIL model of the radial solar wind speed and (b) the product of solar wind density and the square of distance from the Sun in AU for 23 November 2011 at 24:00 UTC. Sample interplanetary magnetic field lines are depicted as white lines, which are dashed when they pass near planets. The center of a coronal mass ejection impacted Mercury and its magnetosphere. (c)–(f) MESSENGER magnetic field measurements taken during a periapsis pass on 23 November 2011 when the magnetosphere was interacting with a CME are displayed in MSM coordinates. The northern cusp, closest approach (CA), magnetopause (MP), and bow shock (BS) are labeled. From Slavin et al. (2014).

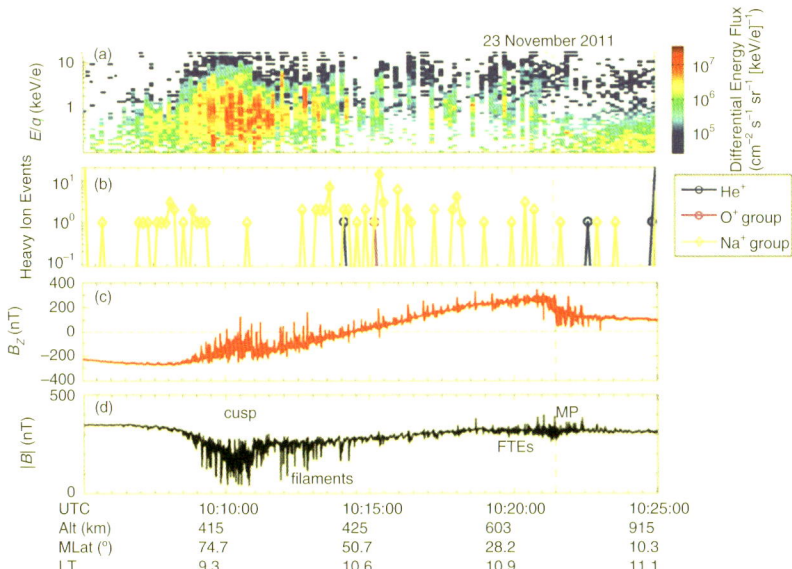

Figure 17.4. An example of symmetric reconnection at Mercury's magnetopause on 23 November 2011. (a) Proton differential energy flux versus energy per charge (E/q). (b) Heavy ion counts in three bins: He$^+$, O$^+$ group ($m/q = 14$–20), and Na$^+$ group ($m/q = 21$–30). (c) B_Z in MSM coordinates. (d) Total magnetic field intensity. The locations of the cusp and magnetopause (MP) are marked, as are the intervals dominated by cusp plasma filaments and, closer to the magnetopause, FTEs. Note the nearly equal magnetic field intensity of magnetospheric and magnetosheath sides of the magnetopause current sheet. Altitude (Alt), magnetic latitude (MLat), and local time (LT) are displayed at the bottom. From Slavin et al. (2014).

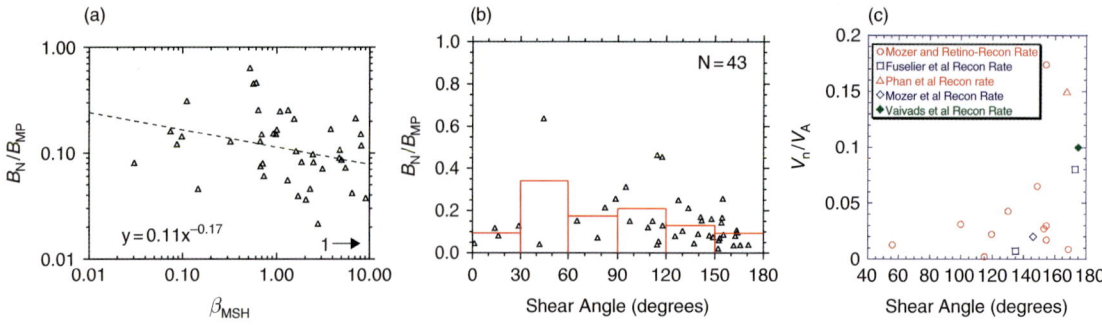

Figure 17.5. Magnetopause reconnection rate, B_N/B_{MP}, at Mercury as a function of (a) magnetosheath plasma β and (b) magnetic shear angle θ, from DiBraccio et al. (2013), and (c) reconnection rate at Earth, V_n/V_A, as a function of magnetic shear angle, from Fuselier and Lewis (2011).

ENLIL magnetohydrodynamic solar wind simulation (Toth and Odstrcil, 1996; Odstrcil et al., 2004; Baker et al., 2009, 2013) has been used to provide global context for this event (Slavin et al., 2014). The initial conditions for WSA-ENLIL are taken from the Wang–Sheeley–Arge (WSA) model of the coronal magnetic field derived from ground-based observations of the photospheric magnetic field gathered over a solar rotation (Arge et al., 2004). Coronal mass ejections are included by means of the "cone model" which launches the CME at a distance from the Sun of 21.5 solar radii. At this distance the CME is assumed to be a spherical parcel of plasma, but it evolves in response to the surrounding solar wind in which it is embedded as it is carried outward. Equatorial views of the solar wind radial velocity, V_r, and density, n, are presented in Figures 17.3a and 17.3b. The plasma density has been multiplied by the square of radial distance (in units of AU) from the center of the Sun to detrend for the decrease in density with increasing distance. These views are snapshots at 24:00 UTC on 23 November 2011 after the CME passed over Mercury. The solar wind velocity snapshot (Figure 17.3a) shows that the CME was near the trailing edge of a higher-speed stream. The solar wind density snapshot (Figure 17.3b) depicts the expected compression signatures as the higher-speed stream overtakes the slower solar wind. The CME does not stand out in the velocity display because its speed is only ~450 km/s, but it is very clear in the density display because of the strong compression signature.

The magnetic field measured by MESSENGER over a 90-min interval beginning at 09:30 UTC on 23 November 2011 is displayed in Figures 17.3c–17.3f in Mercury solar magnetospheric (MSM) coordinates. This coordinate system is centered on Mercury's magnetic dipole, which is shifted northward from the center of Mercury by 0.2 R_M (Alexeev et al., 2008, 2010; Anderson et al., 2011, 2012). In that system, X is directed from Mercury's dipole to the center of the Sun, Z is perpendicular to Mercury's orbital plane and points northward, and Y completes the right-handed system and is positive in a direction opposite to Mercury's orbital motion. Primes indicate that the axes have been aberrated using a mean solar wind speed and the actual orbital speed of the planet, so that X' is antiparallel to the mean solar wind flow direction. The sequence begins with MESSENGER in the magnetotail just north of the plasma sheet. After passing over the north polar region, it began moving equatorward and reached its closest approach (CA) in a very broad magnetospheric cusp centered on ~10:11 UTC. Note the large-amplitude magnetic perturbations centered on the cusp that produced brief reductions in the field intensity by over 90%. MESSENGER then continued to move southward as its altitude increased until the magnetopause was crossed (Slavin et al., 2014). Short-duration increases and decreases in the magnetic field were present throughout this dayside pass, and they maximized near the magnetopause. MESSENGER then traversed the magnetosheath and experienced multiple crossings of the bow shock before remaining in the solar wind. The outward motion of the bow shock appears to have been the result of a decreasing upstream Mach number. The decreasing Mach number is reflected in the declining magnetic field jumps across each successive shock crossing. Low solar-wind Mach number conditions are especially common during CMEs (e.g., Farrugia et al., 1995; Lavraud and Borovsky, 2008; Sarantos and Slavin, 2009).

A closer view of the dayside magnetosphere during this CME event in the MAG and FIPS data is displayed in Figure 17.4 (Slavin et al., 2014). As MESSENGER moved equatorward it traversed a broad magnetospheric cusp within which the total field was highly depressed and variable compared with what was observed under more typical conditions (see Winslow et al., 2012; Raines et al., 2014). FIPS proton differential energy flux measured during this periapsis pass is displayed in Figure 17.4a. Consistent with the earlier investigations (Zurbuchen et al., 2008; Raines et al., 2014), high fluxes of protons with energies from ~100 eV to ~3 keV were present during this cusp pass. The count rates of Na$^+$-group ions, defined as having mass per charge $m/q = 21$–30 amu/q (Raines et al., 2013), are displayed in Figure 17.4b. These heavy planetary ions were observed across this entire dayside pass. Brief enhancements in the proton flux near 1 keV were seen from the poleward edge of the cusp to the magnetopause.

Starting just poleward of the cusp and continuing through the high-altitude dayside magnetosphere, the magnetic field exhibited large-amplitude perturbations in response to a phenomenon we term cusp plasma filaments and FTEs. At higher latitudes these single-energy-scan proton flux enhancements were generally correlated with the narrow cusp filaments, identified by short-lived decreases in the magnetic field strength; whereas at lower latitudes these peaks in the proton flux were frequently coincident with the enhanced fields observed during FTEs. The spacecraft then continued southward and increased in altitude until it crossed the magnetopause at 10:21:18 UTC, at

17.3 Magnetospheric Dynamics 467

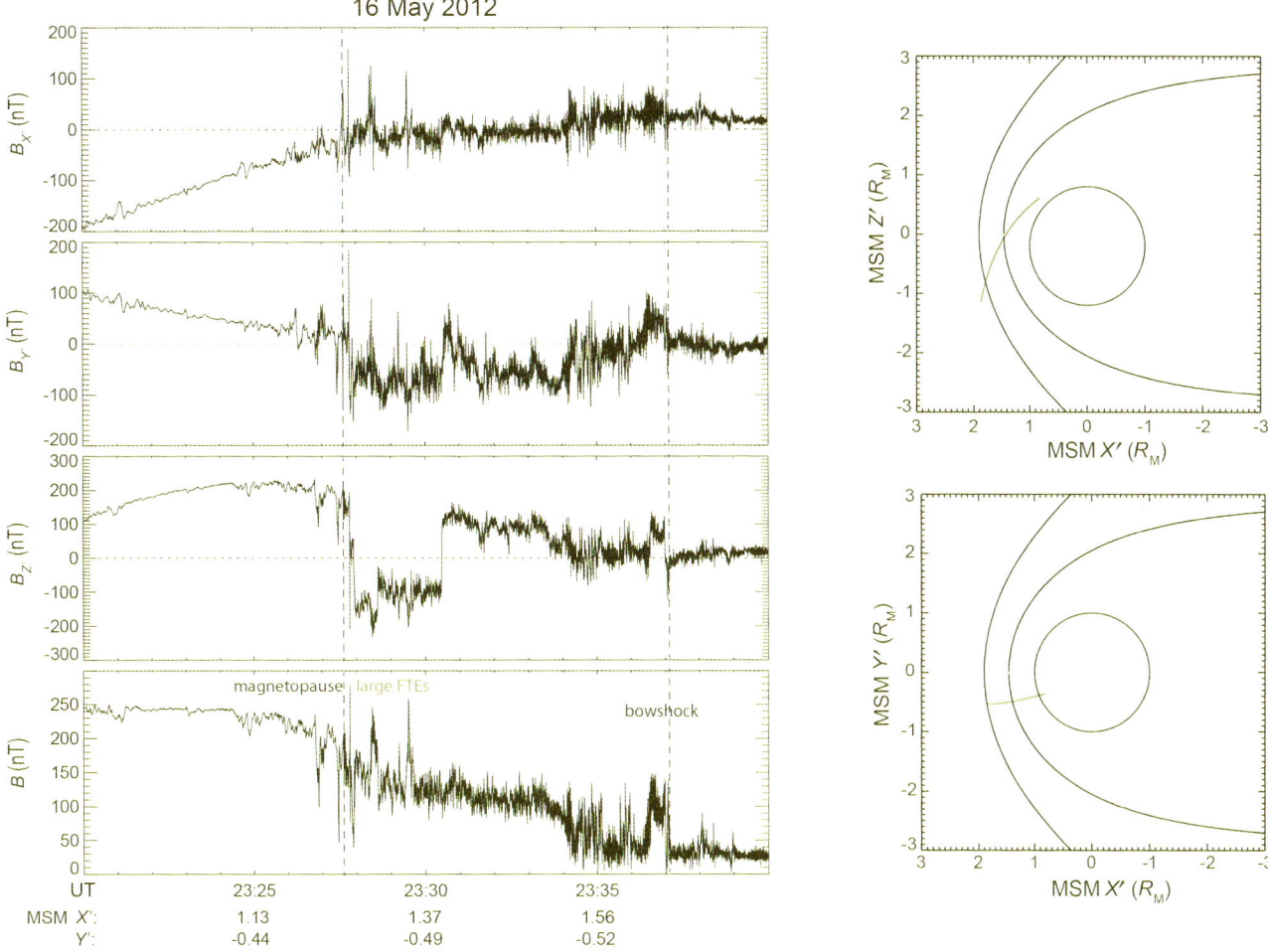

Figure 17.6. Magnetic field measurements in aberrated MSM coordinates during a 40-min interval on 16 May 2012. (Left) From top to bottom, the panels show $B_{X'}$, $B_{Y'}$, $B_{Z'}$, and B. Vertical dashed lines mark the average locations of the magnetopause and the bow shock. (Right) The spacecraft trajectory during this period projected onto the MSM X'–Z' and X'–Y' planes, along with average model locations for the magnetopause and bow shock. From Imber et al. (2014).

a local time (LT) of ~11:00, magnetic latitude (MLat) of ~25°, and altitude (Alt) of ~700 km (Slavin et al., 2014).

The low upstream Mach number during this dayside pass is reflected in the near-constant magnetic field intensity across the magnetopause, ~319 nT, indicative of a strong plasma depletion layer in the adjacent magnetopause (Zwan and Wolf, 1976; Gershman et al., 2013). The location of the magnetopause is identified by the rotation in the magnetic field (see the change in B_Z marked with a vertical dashed line). Analysis of the magnetopause current sheet by Slavin et al. (2014) indicated that the dynamic pressure of the solar wind at the time of the crossing was ~51 nPa. Under the assumption of quasi-isotropy in the spacecraft frame, moment-based calculations of density and a single isotropic temperature using the FIPS proton measurements just upstream of the magnetopause yield values of 15.4 cm^{-3} and 2.62 MK, respectively. The plasma β calculated from these MESSENGER data is only ~2×10^{-3}, as expected from the constant magnetic field magnitude across the magnetopause (Figure 17.4).

The rate of reconnection in single-fluid magnetohydrodynamics is measured by one of two equivalent methods. The first measure of the reconnection rate is the ratio of the flow speed into the diffusion region, V_n, to the outflow region speed, which equals the local Alfvén speed, V_A. The second measure is the ratio of the magnetic field normal at the magnetopause, B_N, to the magnetic field intensity just inside the magnetosphere, B_{MP} (see Fuselier and Lewis, 2011; Klimas et al., 2015). As already described, FIPS could not be configured to measure three-dimensional plasma velocities inside the magnetosphere. For this reason, the determination of reconnection rate from plasma flow speed is not possible with the MESSENGER observations. However, when the magnetic field across the magnetopause can be rotated into well-determined boundary-normal coordinates, the average normal field component over the width of the magnetopause current sheet may be computed. In the boundary crossing on 23 November 2011 the normal directions were well determined, and a value of $B_N = -31.9$ nT was measured. The dimensionless reconnection rate, B_N/B_{MP}, computed in this

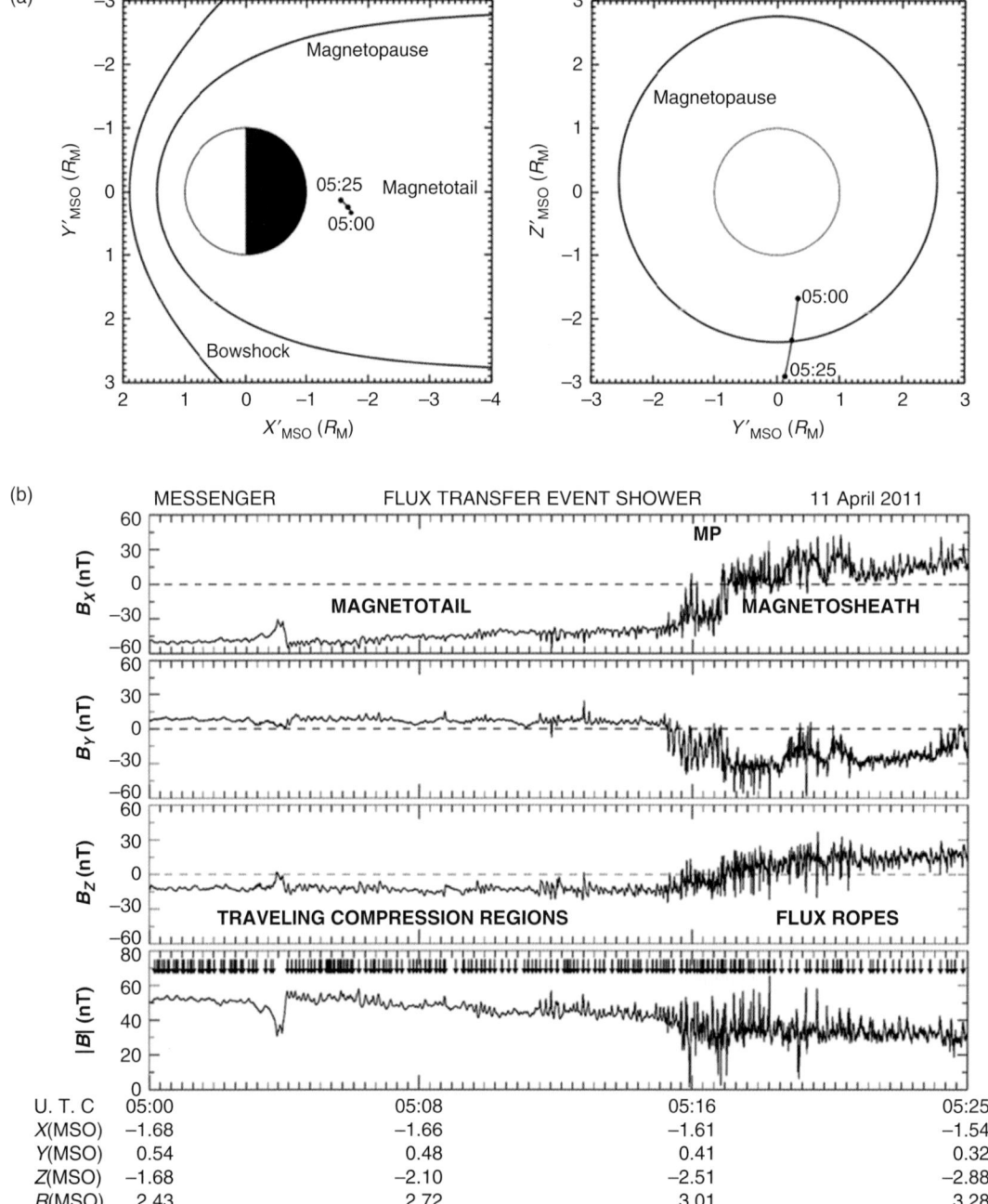

Figure 17.7. (a) MESSENGER trajectory (in red) from 05:00 to 05:25 UTC on 11 April 2011, projected onto the aberrated MSO X'–Y' and Y'–Z' planes. Note that the bow shock and magnetopause surfaces are shifted northward by 0.20 R_M to match the northward offset in Mercury's internal magnetic dipole. (b) Magnetic field measurements taken during this interval span the outer portion of the southern lobe of Mercury's magnetotail, the magnetopause, and the nearby magnetosheath. The many small vertical arrows in the fourth panel mark 97 TCRs inside the magnetotail and 66 FTE-type flux ropes in the adjacent magnetosheath. From Slavin et al. (2012b).

manner is 0.10, with an estimated uncertainty of 10 to 30% (DiBraccio et al., 2013; Slavin et al., 2014).

This result for the magnetopause reconnection rate is representative of the values found for a large ensemble of magnetopause crossings at Mercury by DiBraccio et al. (2013). They determined reconnection rates from 0.02 to 0.8 with a mean B_N/B_{MP} of ~0.15, as shown in Figures 17.5a and 17.5b. These reconnection events differ from those measured at Earth in at least two important respects. First, reconnection rates at Earth are typically only 0.01–0.05, as displayed in Figure 17.5c (Mozer and Retino, 2007; Fuselier and Lewis, 2011). Second, reconnection at Mercury was observed to occur for all magnetopause crossings with non-zero magnetic shear angles, as shown in Figure 17.5b (DiBraccio et al., 2013), where the shear angle is defined as the angular rotation of the magnetic field from the magnetosheath into the magnetosphere. While the reconnection rates at Mercury determined by DiBraccio et al. (2013) show little dependence on the magnetic shear angle across the

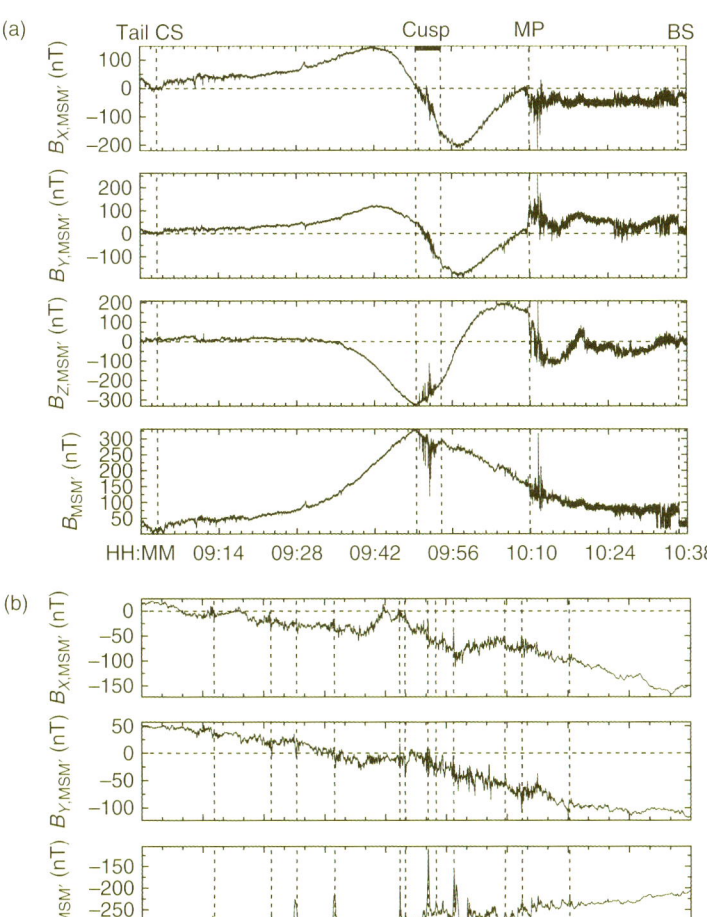

Figure 17.8. (a) Magnetic field measurements of filamentary activities in the cusp acquired by MESSENGER on 20 May 2011, in aberrated MSM coordinates. (b) Close-up view of magnetic field measurements in the cusp region, delimited by the vertical dashed lines in (a). In (b) each cusp filament is marked with a vertical dashed line. Note that the minor time increment in (a) is 2 minutes. From Poh et al. (2016).

magnetopause (Figure 17.5b), there is a clear inverse relationship with plasma β just upstream of the magnetopause (Figure 17.5a), similar to that observed at Earth (Scurry et al., 1994).

17.3.2 Flux Transfer Events and Cusp Filaments

As described above, reconnection alters the structure of the magnetopause current sheet to that of a rotational discontinuity with a non-zero magnetic field normal to the magnetopause (Sonnerup, 1990). However, some of the magnetic flux that has reconnected with the IMF is incorporated into flux ropes called flux transfer events (Russell and Elphic, 1978; Rijnbeek et al., 1984). As illustrated in Figure 17.1b (Slavin et al., 2009a), the outer layers of FTEs have a helical topology that winds magnetic flux about an axial magnetic field near the center of the structure. The axial "core" field in these FTE-type flux ropes is compressed by the tension in those outer layers. For this reason the magnetic field intensity maximizes at the center of the structure unless strong gradients in the plasma pressure are present. Flux ropes are believed to form as a result of simultaneous reconnection at two or more X-lines (i.e., localized regions of high electrical dissipation in space where the magnetic fields "break" and their topology can change) at the magnetopause (Lee and Fu, 1985; Zong et al., 2005; Raeder, 2006; Fear et al., 2007; Hasegawa et al., 2010; Trenchi et al., 2011). If the spacecraft passes close to a flux rope but fails to penetrate the flux rope proper, then the compression of the surrounding, draped magnetic flux may be detected as a traveling compression region (TCR) (Slavin et al., 1984; Liu et al., 2008). Gershman et al. (2016) used FIPS observations of highly energetic electrons during SEP events to demonstrate that FTE flux tubes are indeed connected to the interplanetary magnetic field, as illustrated in Figure 17.1b. The FTEs are then carried tailward by the solar wind flow in the magnetosheath and added to the magnetotail (Cowley and Owen, 1989; Cooling et al., 2001).

Flux transfer events were first detected at Mercury from Mariner 10 data (Russell and Walker, 1985). Slavin et al. (2009a, b) conducted the initial investigations of FTEs in the MESSENGER measurements. They called attention to their very large amplitudes, i.e., core fields up to several hundred nT, and very short durations of only a few seconds. The MESSENGER observations are especially remarkable in that very large numbers of FTEs are present at most magnetopause crossings (e.g., Slavin et al., 2010b, 2014). Imber et al. (2014) conducted a statistical study of the largest-amplitude dayside FTEs and found that their average duration is 2.5 s and that their typical core field strength is ~200–300 nT, which is substantially higher than the typical magnetic field intensity just inside the magnetopause. By comparison, the duration of an FTE at Earth is typically 1–2 min, and the characteristic amplitude is a few tens of nT (e.g., Kuo et al., 1995; Sanny et al., 1998). Imber et al. (2014) determined the orientation of each MESSENGER FTE and its magnetic flux content with a minimum variance analysis and a force-free flux rope model (Lepping et al., 1995, 1996). The average flux content of these large-amplitude FTEs at Mercury was found to be 0.06 MWb. Given their short durations, a typical FTE at Mercury will produce a transient increase in the cross-polar-cap potential of ~25 kV, which represents approximately a doubling of cross-magnetospheric electric potentials determined by Slavin et al. (2009a, 2010b) and DiBraccio et al. (2013, 2015b). Hence, the impact of FTEs on Mercury's magnetosphere appears to be quite large.

An example of a magnetopause crossing in the magnetic field observations on 16 May 2012 is shown in Figure 17.6 (Imber et al., 2014). The right panels show MESSENGER's trajectory in solar wind aberrated MSM coordinates (X', Y', Z') during a 20-min interval, along with average magnetopause and bow shock surfaces. The corresponding magnetic field measurements by MESSENGER are plotted on the left. Two FTEs can clearly be identified as short-duration ~250-nT peaks in the total magnetic field strength observed just after the spacecraft crossed the magnetopause and entered the magnetosheath. The increase in the magnetic field intensity combined with the smooth rotation of the magnetic field component normal to the magnetopause confirms that the spacecraft traversed the central portion of the flux rope (Imber et al., 2014).

During a 25-min-long interval on 11 April 2012, MESSENGER observed a total of 163 TCRs and FTEs as it crossed the magnetopause and entered the magnetosheath, as displayed in Figure 17.7 (Slavin et al., 2012b). The location of the spacecraft is displayed in solar wind aberrated Mercury Solar Orbiter (MSO) coordinates. The MSO system differs from the MSM system only in that its origin is at the center of Mercury rather than on the internal magnetic dipole. The FTE and TCR signatures had an average duration of 2–3 s and a repetition rate of approximately once per 10 s. The interval presented in Figure 17.7 is an example of solar wind forcing that results in quasi-periodic reconnection signatures that have been termed "FTE showers" (Slavin et al., 2012b). Multiple FTEs are observed on nearly all of MESSENGER's encounters with Mercury's magnetopause both on the dayside and downtail, suggesting that the solar wind is constantly driving dynamics and magnetic flux transport through Mercury's magnetosphere. Consistent with this high level of FTE activity, a well-developed plasma mantle appears to be a quasi-permanent attribute of the high-latitude tail lobes (DiBraccio et al., 2015b).

Imber et al. (2014) combined estimates of the average FTE flux content (0.06 MWb) with a repetition rate of 1 every 10 s to conclude that FTEs are able to transport ~1 MWb of magnetic flux into the magnetotail over a timescale of 3 min. This flux constitutes ~30% of the total transport required to cycle all of the magnetic flux in the tail during a large substorm at Mercury. This fraction is significantly higher than the estimated ~2% of flux transport contributed by FTEs during a typical substorm at Earth (Huang et al., 2009). The primary implication is that bursts of multiple X-line dayside reconnection events are very common at Mercury and that FTEs are a major influence on its magnetospheric dynamics. Global magnetohydrodynamic (MHD) simulations of Ganymede's magnetosphere (Jia et al., 2010) also produce frequent, quasi-periodic FTEs. This result is important because Ganymede's surface magnetic field, its small magnetospheric dimensions, and the low-M_A (<1) of the jovian magnetospheric plasma flow in which it is immersed are closer to Mercury's situation than that of the other planets. In fact, it has been suggested that the relative size and frequency of FTEs may be linked to Alfvénic Mach number, with more frequent events occurring during low-M_A conditions (Jia et al., 2010).

The uppermost panels of Figure 17.8 present 1 h of MESSENGER magnetic field measurements from 20 May 2011 (Poh et al., 2016). As shown in the expanded view in the lower panels, there are 12 sharp transient decreases in magnetic field strength clearly visible during the cusp encounter. These structures were first reported by Slavin et al. (2014) in their analysis of the effect of CMEs on Mercury's magnetosphere. They termed them "cusp filaments." Minimum variance analysis of each filament has shown that majority of the filaments are quasi-cylindrical or slightly flattened magnetic flux tubes that are aligned with the ambient dayside magnetic field (Poh et al., 2016). FIPS measurements indicate that the plasma in each filament has energies similar to the magnetosheath plasma with pitch angles near 90°, where the pitch angle is the angle between the particle velocity vector and the local magnetic field (Poh et al., 2016).

It is believed that the cusp filaments are the low-altitude extensions of FTEs transiting the dayside magnetopause as they move poleward into the cusp (Slavin et al., 2014; Poh et al., 2016). Localized, transient reconnection produces FTEs with one end connected to the magnetosheath field lines and the other to planetary field lines. This impulsive reconnection accelerates and directs the magnetosheath plasma into these newly opened flux tubes and downward toward the surface. As the magnetic field becomes stronger at lower altitudes, the plasma slows and gains perpendicular energy. The local increase in perpendicular plasma pressure is balanced by a corresponding diamagnetic decrease in the local magnetic pressure to produce the filaments shown in Figure 17.8. Further analysis of the low-altitude-campaign results by Poh et al. (2016) indicates that these filaments can be observed down to very low altitudes (<20 km). This result suggests that at least some of the plasma inside each filament precipitates onto Mercury's surface. Poh et al. (2016) estimated the mean hemispheric precipitation rate aggregated over the estimated number of filaments present at a given moment was ~$(2.70 \pm 0.09) \times 10^{25}$ s^{-1}, which is comparable to the average total precipitation in Mercury's northern cusp inferred by Winslow et al. (2012). The implication is that these filaments, much like the FTEs to which they are believed to be connected, make a major contribution to the total solar wind flux impacting Mercury's surface.

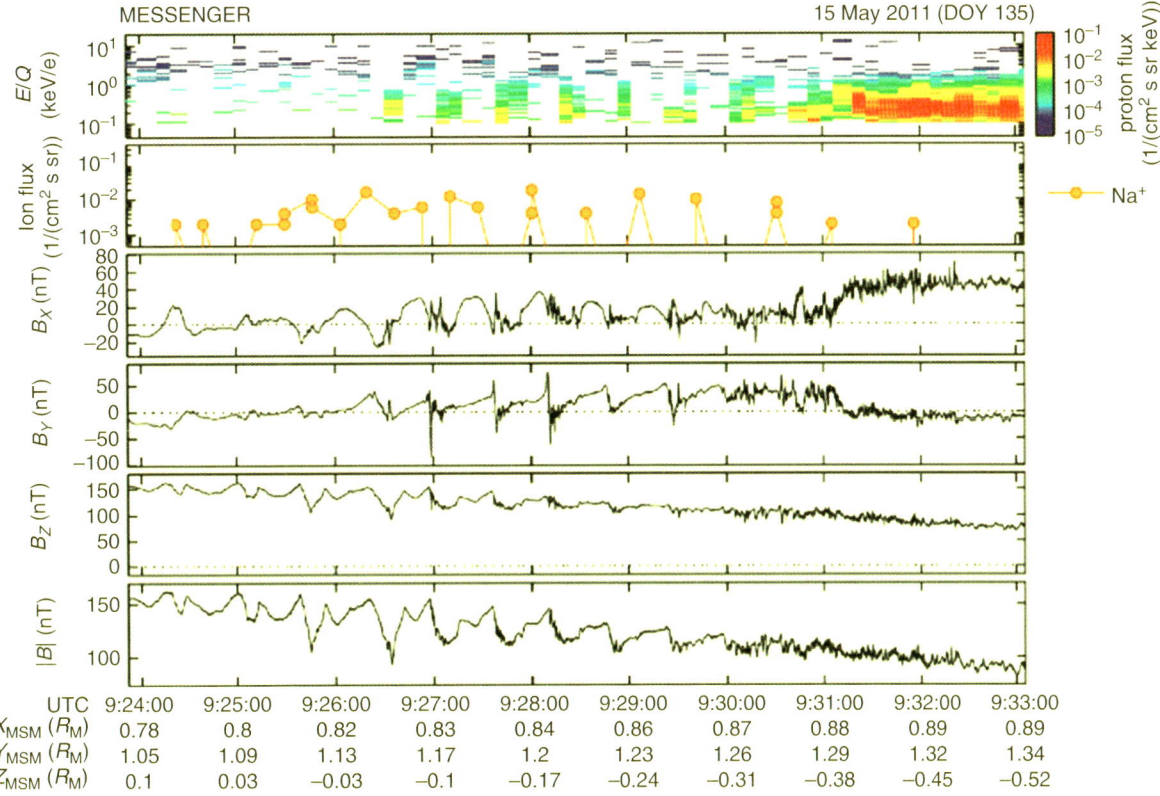

Figure 17.9. A Kelvin–Helmholtz wave train observed in the post-noon region of the magnetopause. The figure shows the FIPS proton and sodium ion flux (top two panels), the MSM X, Y, and Z components of the magnetic field (panels 3–5), and the magnetic field magnitude (bottom panel). A clear ~40-s wave pattern is seen in the magnetic field data, which also corresponds to periodic bursts in the proton flux measurements. These bursts are indicative of magnetosheath plasma entering into the magnetosphere, mediated by the KH waves. From Sundberg et al. (2012a).

17.3.3 Kelvin–Helmholtz Waves

MESSENGER measurements have shown that fully developed Kelvin–Helmholtz (KH) waves are common at Mercury's magnetopause, especially on the dusk side of the magnetosphere (Sundberg et al., 2012a). These waves, which are generated by the KH instability, are surface waves that can develop at boundaries separating two fluids with different streaming velocities, for example, ocean waves driven by wind blowing over water. The waves form from an initial perturbation of the boundary that develops into large-scale boundary waves and, eventually, rolled-up vortices. Such vortices are of particular importance for magnetospheric dynamics, as they can lead to an interchange of magnetospheric and solar wind plasma across the magnetopause and contribute to the large-scale plasma convection cycles within the magnetosphere.

Kelvin–Helmholtz waves were first detected at Mercury's magnetopause following MESSENGER's third flyby of the planet (Boardsen et al., 2010). Subsequent measurements have shown that they are ubiquitous at Mercury's post-noon and dusk-side magnetopause when the IMF is northward, whereas they are only sporadically present on the dawn side and during southward IMF (Sundberg et al., 2012a; Liljeblad et al., 2015a). The wave amplitudes are often on the order of 100 nT or more in the dayside region, and the wave periods observed are in the range 5–40 s. A typical example of a KH wave is shown in Figure 17.9. As many of these waves are observed in regions where the magnetosheath flow velocity is expected to be subsonic, such measurements indicate higher instability growth rates than are typically observed at Earth. This difference may be due to the limited wave energy dissipation at Mercury's highly resistive regolith surface (Sundberg et al., 2012a).

The strong dusk-side bias in the observations (Sundberg et al., 2012a) is most commonly attributed to kinetic effects because of the large gyroradii of the protons and sodium ions in the boundary layer relative to the overall width of the shear layer. There are differences in the growth rates on the two sides, enhancing and/or suppressing certain wavelengths (e.g., Nagano et al., 1979; Glassmeier and Espley, 2006; Sundberg et al., 2010) due to the opposite vorticity of the KH waves on the dawn- and dusk-side flanks, which results in the ions co-rotating with the waves on the dawn side and counter-rotating at dusk. Modifications of the ion gyroradius by the convective electric fields may also lead to a broadened shear layer on the dawn side and a reduced shear layer at dusk, making the dusk side more prone to KH wave growth (Nakamura et al., 2010; Paral and Rankin, 2013). External factors may also influence the growth rates, such as the structure of the low-latitude boundary layer, where the thicker boundary layer observed on the dawn side may inhibit KH formation (Sundberg and Slavin, 2015; Liljeblad et al., 2015b). Furthermore, the density of the large-gyroradius, keV planetary ions at Mercury exhibits strong asymmetries with respect to both dusk and dawn and night and day (Raines et al., 2013; Gershman et al., 2015a).

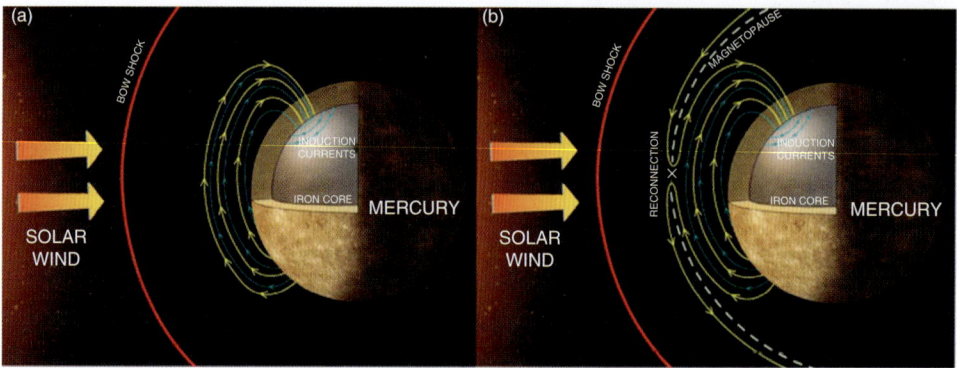

Figure 17.10. (a) Increases in solar wind pressure drive induction currents (green loops) at the top of Mercury's large iron-rich core. The sense of these currents is to oppose the compression of the intrinsic magnetic field (yellow) by generating additional magnetic flux (green field lines) which, when added to the intrinsic flux, acts to balance the increased solar wind pressure. (b) Magnetic reconnection between the interplanetary magnetic field and the intrinsic planetary magnetic field opposes the effectiveness of induction by removing magnetic flux from the dayside magnetosphere and transporting it to the tail. From Slavin et al. (2014).

Not only may finite gyroradius effects influence the wave evolution in the growth stage, but it is also possible for a resonant interaction to take place between fully developed KH waves and the energetic sodium ions on the dusk-side flank of the magnetosphere (Gershman et al., 2015a). These sodium ions may contribute ~10–30% of the total plasma pressure in the pre-midnight plasma sheet, as shown by Gershman et al. (2014). Gershman et al. (2015a) also found nightside KH waves at the Na^+ gyrofrequency, indicating that multi-scale processes were acting within the waves. Conversely, dayside KH vortices, in which there were few hot (keV) planetary ions, were found with a nearly constant wave period. The detailed physics of this wave–particle interaction has since been examined by Gingell et al. (2015), who showed using local hybrid simulations that the counter-rotating sodium ions can become entrained in the KH vortex pattern and thereby enhance the wave power at frequencies close to the Na^+ gyrofrequency.

There are some indications that the KH waves observed at Mercury also mediate a transfer of solar wind plasma and energy into the magnetosphere. For example, the FIPS proton measurements in Figure 17.9 show a clear periodic signature of magnetosheath-like ions inside the magnetopause proper (indicated by the step change in B_X and B_Y at 09:31:00 UTC). This observation provides evidence that the KH waves are transporting magnetosheath protons across the magnetopause. However, definite proof requires a clear observation of magnetosheath plasma on magnetospheric field lines, something that would be difficult to verify using only the MESSENGER measurements. It is also clear that KH waves at the magnetopause can couple to ultra-low-frequency (ULF) pulsations in the magnetosphere at the same frequency (Sundberg et al., 2012a). This result means that solar wind energy is being transmitted into the system through wave coupling, and that the magnetopause KH waves may have a global impact on wave dynamics in the magnetosphere.

17.3.4 Magnetopause Compression, Reconnection-Driven Erosion, and Induction in Mercury's Interior

With its ~2000-km-radius, iron-rich metallic core (Smith et al., 2012), Mercury has the highest uncompressed density of all the planetary bodies in our solar system. Because of the high electrical conductivity of iron, it is expected that a change in the external magnetic field would take ~10^4–10^5 years to diffuse to the center of the planet (Glassmeier, 2000). The short timescales for changes in solar wind speed and density, i.e., minutes to hours, strongly suggest that Mercury reacts as a perfectly electrically conducting sphere to variations in solar wind pressure. In Figure 17.10 the magnetic field lines generated by the dynamo in the fluid core are shown in yellow. Increases in solar wind pressure compress these magnetic field lines. However, any change in the magnetic field normal to the surface of the core will generate currents according to Faraday's law. These electric currents circulating in a thin layer at the top of the core will flow in such a manner as to oppose the change in the normal magnetic field. The net result is that any compression of Mercury's magnetic field will be opposed by the addition of dipolar magnetic field lines (in green) generated by the induction currents (green loops) (Hood and Schubert, 1979; Suess and Goldstein, 1979; Glassmeier et al., 2007). The induced currents should be similar in magnitude about the magnetic equator, but the magnetopause is closer to the planet in the southern hemisphere. For this reason the induction is expected to be somewhat greater south of the magnetic equator (e.g., Johnson et al., 2016). However, magnetic reconnection between the interplanetary magnetic field and the intrinsic planetary magnetic field will negate some or all of the effects of the induction currents by removing magnetic flux from the dayside magnetosphere and transporting it to the tail, as shown in Figure 17.10b (Slavin and Holzer, 1979; Heyner et al., 2016).

It is for this reason that Slavin et al. (2014) analyzed the subsolar magnetopause crossings on 23 November 2011, 8 May 2012, and 11 May 2012, for which solar wind dynamic pressure was about three to five times the typical values. The high pressures were the result of CME impacts for the first two events and a high-speed stream for the third. The upstream IMF orientations for these events produced magnetic shear angles across the magnetopause from ~27° to 166°. Slavin et al. (2014) mapped these magnetopause crossings to the

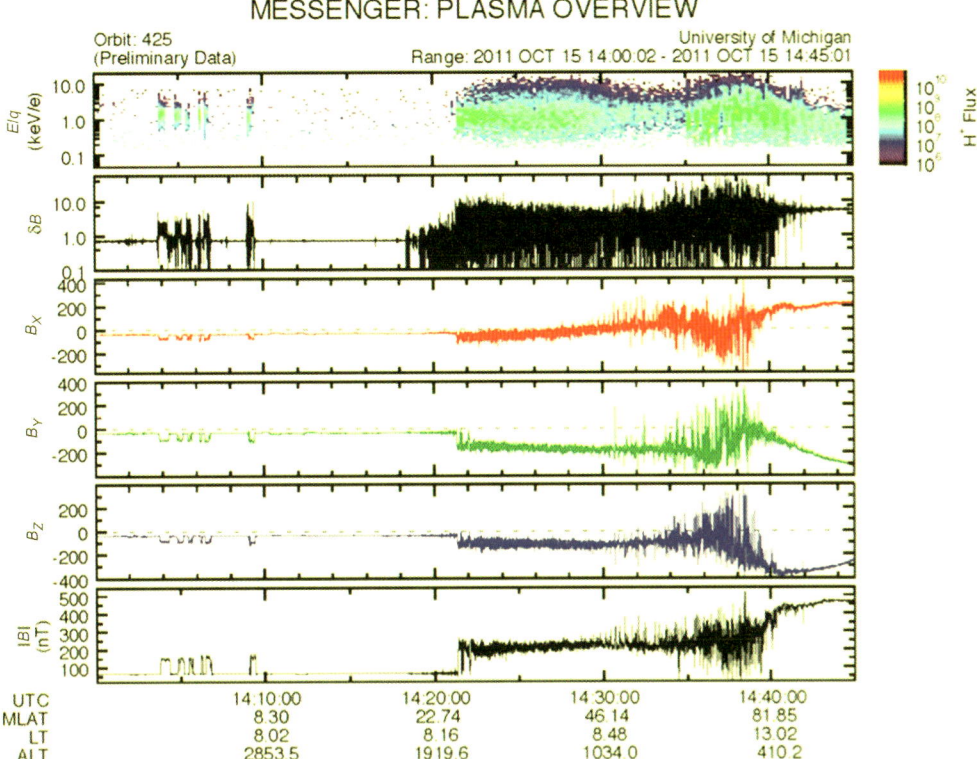

Figure 17.11. Example of a disappearing dayside magnetosphere event. The FIPS proton energy spectra are displayed in the top panel, and the magnetic field is shown in MSM coordinates in the lower panels. These measurements indicate that MESSENGER never entered the dayside magnetosphere. Instead, magnetosheath-like plasma and magnetic fields were observed until ~14:35 UTC, when the spacecraft entered the nightside polar magnetosphere.

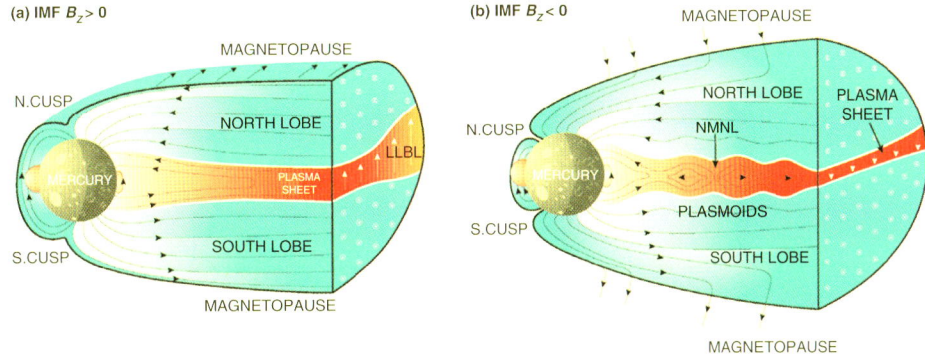

Figure 17.12. (a) Mercury's magnetosphere in its ground state with minimal magnetic field normal to the magnetopause, weak tail magnetic field, a thick plasma sheet, no near-Mercury reconnection X-lines, and a well-developed low-latitude boundary layer. (b) Substorm onset in a highly stressed magnetosphere with large magnetic fields normal to the magnetopause, a strongly loaded tail, a thinned plasma sheet, multiple near-Mercury X-lines, and plasmoids about to be ejected. NMNL denotes near-Mercury neutral line. From Slavin et al. (2012a).

nose of the magnetopause by assuming a mean magnetopause shape (Winslow et al., 2013). The altitude of the nose of the magnetopause for these events, (1.03–1.13) R_M, and the inferred solar wind dynamic pressure, 44–65 nPa, were then compared against the predictions of the induction models (Glassmeier et al., 2007). The result was that these high-solar-wind-pressure MESSENGER magnetopause crossings occurred at much lower altitudes than predicted by the induction models. Slavin et al. (2014) concluded that the most likely reason for the very low altitudes of the forward magnetopause during these events was the effect of reconnection and the resulting transfer of magnetic flux into the tail, as originally predicted by Slavin and Holzer (1979). On the basis of their analysis, it appears that direct solar wind impact, especially in the southern hemisphere where the intrinsic magnetic field is weaker, may be common during extreme solar wind events.

Another investigation of the effect of solar wind pressure on the altitude of the subsolar magnetopause and the implications of the induction currents for the compressibility of Mercury's magnetosphere was carried out by Zhong et al. (2015b). They examined the

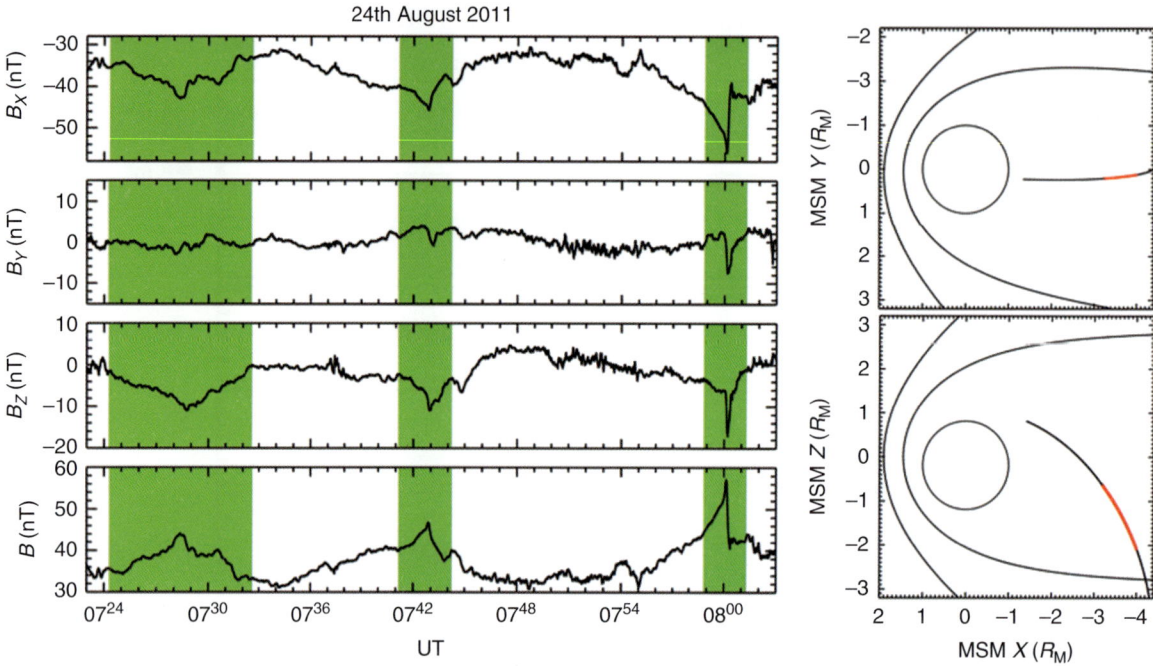

Figure 17.13. (Left) Magnetic field measurements are displayed in MSM coordinates for an interval when MESSENGER was located near the center of the southern tail lobe. Note the three several-minute-long periods shaded in green when the magnetic field intensity first increased and then decreased (tail lobe loading and unloading events). (Right) The spacecraft trajectory during this tail pass relative to projections on the MSM X–Y and X–Z planes of mean bow shock and magnetopause surfaces, with the intervals of interest shown in red.

effect of distance from the Sun on the compressibility of the magnetosphere to learn more about the effects of reconnection and induction on magnetopause standoff distance, using a database of 2826 magnetopause crossings (Zhong et al., 2015a). At perihelion, solar wind pressure and M_A are expected to be higher and lower, respectively, than at aphelion. The analysis of Zhong et al. (2015b) found the expected trends, but the overall dimensions of Mercury's magnetosphere changed substantially from the simple sixth-root dependence on solar wind pressure expected for a dipolar magnetic field (Spreiter et al., 1966). They interpreted their results as indicating that dayside reconnection counterbalanced the effects of induction in Mercury's interior at perihelion, because the effects of reconnection are stronger for the lower solar wind M_A found nearer to the Sun. However, at aphelion, where solar wind pressure is lower and M_A is higher, Zhong et al. (2015b) found that induction effects tend to be dominant.

MESSENGER also experienced dayside passes where the magnetopause was sufficiently close to Mercury's surface that MESSENGER never entered the magnetosphere despite reaching altitudes below 400 km (i.e., 0.16 R_M). Such events were first reported in the MESSENGER data by Middleton et al. (2014) and Zhong et al. (2015a), with the former terming them "disappearing dayside magnetosphere" events. We present an additional example of such an event on 15 February 2011 in Figure 17.11. The event was associated with a CME and an estimated peak solar wind dynamic pressure of ~37 nPa on the basis of WSA-ENLIL (L. Mays, personal communication, 2014). As shown, the 30-min-long interval began with MESSENGER upstream of the bow shock and measuring a ~60 nT interplanetary magnetic field oriented largely southward. Large numbers of FTEs are evident in the magnetic field observations as the several-second-long enhancements in the total field and bipolar variations in the other components, starting around 14:30 UTC near an altitude of ~1000 km. In fact, the strongest FTE core magnetic field intensities exceeded 500 nT at ~14:38 UTC and ~473-km altitude. These FTE fields are approximately twice the background magnetosheath magnetic field. The FIPS proton dynamic spectra in the top panel of Figure 17.11 show that the magnetosheath plasma continued to be observed and its temperature increased as MESSENGER moved northward and lower in altitude. However, MESSENGER never entered the dayside magnetosphere, where strong, low-variance magnetic fields with $B_Z > 0$ would have been observed. Further, the strong decrease in proton flux expected when crossing into the magnetosphere was not observed (DiBraccio et al., 2013; Liljeblad et al., 2015b). Instead, MESSENGER continued to observe magnetosheath plasma until it traversed the cusp region and entered the high-latitude polar magnetosphere at an altitude of ~400 km at 13:53 local time and ~82°N magnetic latitude around 14:40 UTC. The underlying reasons for these dayside disappearing-magnetosphere passes are still unclear, but they appear to be associated with high solar wind dynamic pressure and large shear across the magnetopause (i.e., IMF $B_Z < 0$). The fact that FTEs continued to be generated and to move northward to join the cusp indicates ongoing reconnection between the IMF and dayside planetary magnetic fields, but the interaction must have taken place within a few hundred kilometers, or less than 10 proton gyroradii, of the surface. This scale, combined with the large numbers of FTEs, suggests that during these events the solar wind plasma must have had direct access to most or possibly the entire sunlit surface of Mercury along open flux tubes connecting the solar wind to the surface.

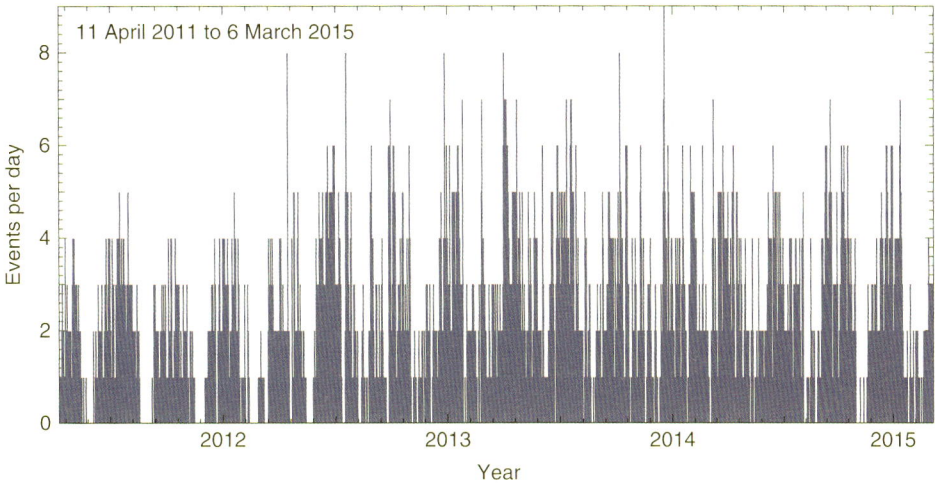

Figure 17.14. Rate of low-energy (i.e., keV) electron events per day as determined from the MESSENGER XRS from April 2011 to March 2015. There was an increase in the rate in April 2012 because the spacecraft orbital period was changed from 12 h to 8 h. On average there are about two events per orbit. From Ho et al. (2016).

17.3.5 Magnetotail Loading and Unloading

On the basis of experience at Earth, it is expected that an isolated magnetospheric substorm at Mercury would begin with the magnetotail in its relaxed state, with minimal magnetic flux in the lobes and flaring of the magnetopause, as illustrated in Figure 17.12a. This situation may be considered the "ground state" for Mercury's magnetosphere, which exists only during an extended interval of minimal dayside reconnection. All planetary magnetic field lines in the near-Mercury environment are contained within the magnetopause in this state, and reconnection signatures are absent or at very low levels. Figure 17.12b depicts Mercury's magnetotail in a much higher energy state, which, by analogy to Earth, we label as being near the end of the energy storage or growth phase and close to the onset of the substorm expansion phase (McPherron et al., 1973). Dayside reconnection has opened planetary magnetic fields to the solar wind, and these field lines are being dragged anti-sunward. As a result of this transport, the open flux content of the lobes is substantially higher than in the ground state. In turn, this change results in increased flaring of the magnetopause and enhanced pressure throughout the tail. The mechanism by which the cross-tail current sheet thins is still not well understood (Kuznetsova et al., 2007; Winglee et al., 2009; Raeder et al., 2010). This configuration is not stable against reconnection because of the reduction in the northward magnetic field component in the thinned cross-tail current layer. Once the current sheet thins to electron inertial scales, the magnetic energy stored in the tail is released by reconnection between the magnetic fields of the tail lobes. At Earth this process begins with the current sheet fragmenting into flux ropes, as shown in Figure 17.12b, followed by the reconnection of open magnetic flux, which unloads the lobes and returns closed field lines to the dayside to complete the flux transfer cycle. The fast sunward and tailward reconnection-driven flows emanating from the X-lines drive many other substorm phenomena, including dipolarization fronts, the formation of a substorm current wedge, energetic particle injections, particle precipitation along the boundary between open and closed magnetic field lines to produce auroras in ovals centered on the Earth's magnetic poles, and plasmoid ejection (Baumjohann et al., 1991, 1999; Angelopoulos et al., 1992; Slavin et al., 1993, 2003; Nagai et al., 1998).

A comprehensive analysis of tail loading and unloading in the MESSENGER observations has yet to be carried out, but Figure 17.13 shows examples of this phenomenon. Between 07:23 and 08:03 UTC on 24 August 2011, MESSENGER was near local midnight in the southern lobe of the tail (see Figure 17.13b). The magnetic field measured during this time period is displayed in Figure 17.13a, in MSM coordinates. Three loading–unloading events are shaded in green, during which the B_Z component decreased as the field magnitude rose (loading), followed by an increase in B_Z as the magnitude fell (unloading). The elevation of the magnetic field to the equatorial plane shows that the amplitude of the flaring variation for these events was ~10°–20°. In this manner the magnetic field flared farther away from the central axis of the tail when loaded with additional flux and flared less when it was unloaded and closer to its ground state. The total change in the tail field intensity for the three events shaded in green ranges from ~10% to 50%. At Earth the amplitude of the lobe magnetic field during the substorm loading–unloading cycle is much lower, typically ~10–20% (Baker et al., 1996; Milan et al., 2004; Huang et al., 2009). The duration of the loading–unloading events in Mercury's tail, ~2–6 min, is very close to the Dungey convection cycle times inferred from measurements of the dayside magnetopause reconnection rate (Slavin et al., 2009a; DiBraccio et al., 2013).

17.3.6 Plasma Sheet Dynamics

The FIPS ion measurements in the nightside magnetosphere have shown plasma sheet H^+ densities and temperatures of ~1–10 cm^3 and ~ 0.4–2.5 keV, respectively (Raines et al., 2013; Korth et al., 2011b, 2014; Gershman et al., 2014). The dominant planetary ion, Na^+, has number densities about

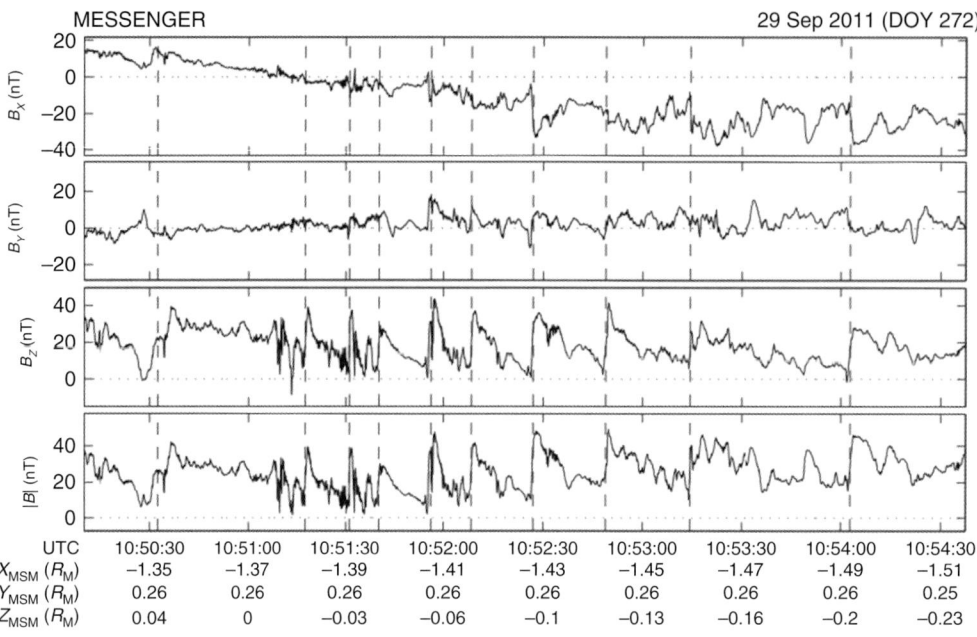

Figure 17.15. A series of dipolarization fronts identified from full-resolution (20 vectors/s) magnetic field observations in Mercury's near tail, displayed here in MSM coordinates. Each dipolarization front, marked by a step-like increase in the B_Z component, is denoted by a vertical dashed line. From Sundberg et al. (2012b).

10% that of H$^+$, with increased fluxes observed in the pre-midnight plasma sheet (Raines et al., 2013; Gershman et al., 2014). Ions naturally experience curvature and gradient drift westward as they convect sunward. For planetary ions originating at high dayside latitudes, the small radius of curvature of Mercury's magnetic field generates strong centrifugal forces that enhance the westward drift of these ions and accelerate them to keV energies as they move in the direction of the dawn-to-dusk cross-magnetospheric electric field (Delcourt et al., 2002, 2012). It is these accelerated ions that are believed to be the dominant source of keV Na$^+$ in the plasma sheet (Delcourt et al., 2003, 2010; Delcourt, 2013; Raines et al., 2013; Gershman et al., 2014).

Plasma electrons in the ~1–10 keV energy range were detected by MESSENGER using the self-fluorescence they excited within the XRS instrument (Starr et al., 2012; Ho et al., 2016). Figure 17.14 shows the number of low-energy (1–10 keV) electron events detected per day on the XRS between 10 April 2011 and 6 March 2015 (Ho et al., 2016). The number of events detected rose from two to three events per day when MESSENGER's orbital period was reduced from 12 h to 8 h in April 2012. They were detected at all local times, but with the highest concentration in the dawn region, consistent with the expected eastward curvature and gradient drift of electrons in a dipole magnetic field. The XRS electrons and the FIPS ions exhibit many similarities in terms of their detection during essentially all passes through the equatorial plasma sheet and its low-altitude/high-latitude extensions (i.e., the "horns" of the plasma sheet) (Gershman et al., 2014; Ho et al., 2016). The ratio of plasma ion temperature to electron temperature, T_i/T_e, in Earth's magnetosphere is normally in the range ~5–10 (Slavin et al., 1985; Baumjohann et al., 1990). If the plasma in Mercury's magnetosphere has the same T_i/T_e ratio as at Earth, then the FIPS ion temperature measurements imply that the plasma sheet electrons will have temperatures of only ~0.1–0.5 keV. These values are much smaller than the temperatures implied by the XRS-measured energies of ~1–10 keV. Accordingly, it appears unlikely that the electrons measured by XRS are thermal plasma-sheet electrons. Rather, as suggested by Ho et al. (2016), the XRS electron events probably constitute the low-energy tail of the highly energetic electron bursts observed by EPS and GRNS, as will be discussed in Section 17.4.

17.3.7 Near-Tail Dipolarization

A dipolarization front is a thin current sheet at the leading edge of a region of dipolar flux tubes being transported sunward by a fast reconnection-driven flow in the near tail (Ohtani et al., 2004). It is identified by a sudden increase in the B_Z component of the magnetic field, which signifies the rapid reconfiguration from a stretched, tail-like configuration to a more dipolar field. These reconfigurations are driven by sunward Alfvénic flows, termed bursty bulk flows (Angelopoulos et al., 1992), emanating from reconnection X-lines in the near-Mercury magnetotail (e.g., Kidder et al., 2008; Jia et al., 2015). The entire interval during which the near-tail magnetic field has this more dipolar magnetic field configuration is termed a dipolarization event (Baumjohann et al., 1999). At Earth, dipolarization is associated with the development of a system of field-aligned currents (FACs), termed the substorm current wedge (SCW), which couples the near-tail plasma sheet to the auroral ionosphere (McPherron et al., 1973). The processes contributing field-aligned current to the SCW include: (1) the inertial current generated during the braking of the sunward-directed bursty bulk flows and (2) the diversion of bursty bulk flow away

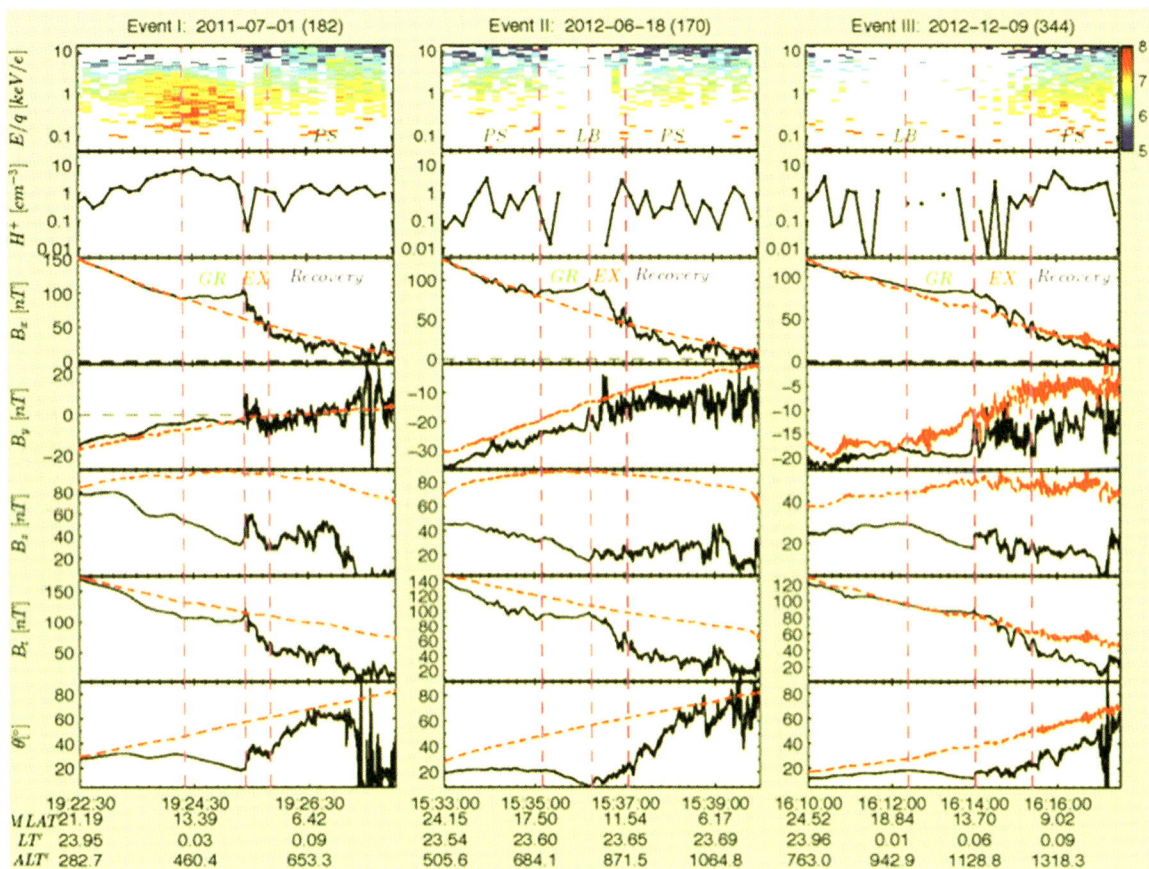

Figure 17.16. MESSENGER observations of three magnetospheric substorms at Mercury. FIPS H$^+$ E/q spectra and density are displayed in the top two rows of panels. The three components, total intensity, and elevation angle of the vector magnetic field (20 samples/s) are displayed in MSM coordinates in the bottom four rows of panels. Red dashed lines in each panel display the magnetic field for adjacent orbits when there was no evidence of substorm activity. The first, second, and third vertical dashed lines from the left indicate the start times of the substorm growth (GR), expansion (EX), and recovery phases for each event. From Sun et al. (2015a).

from the region of highest pressure near midnight, which creates rotation in the flow (Kepko et al., 2015). It must further be recognized that the substorm dipolarization events at Earth are thought to be the aggregate effect of multiple bursty bulk flows impacting on the inner magnetosphere during the substorm expansion phase. Multi-point observations of the magnetotail during substorms at Earth have shown close temporal relationships between reconnection-driven bursty bulk flows, dipolarization in the near tail, and plasmoid ejection (Slavin et al., 2002; Angelopoulos et al., 2008). These observations are well reproduced at Earth by global MHD simulations (e.g., Sitnov et al., 2009; Ashour-Abdalla et al., 2015).

Mariner 10 provided the first tantalizing evidence that Mercury may be host to Earth-type substorms. Two possible dipolarization events associated with energetic charged particle injections were reported from measurements made during this spacecraft's first Mercury flyby (Eraker and Simpson, 1986; Baker et al., 1986; Christon et al., 1987). Indeed, these observations led Baker et al. (1987) to propose that Mercury may possess an auroral oval created by the heating of the surface when charged particles injected during substorms precipitate as they drift about the planet. Sundberg et al. (2012b) conducted the first examination of dipolarization events with MESSENGER magnetic field measurements. As displayed in Figure 17.15 the dipolarization events analyzed by Sundberg et al. (2012b) were short-lived, lasting on the order of 10 s, and frequently reached amplitudes of ~50 nT, or a factor of ~5 higher than the background magnetic field. Further, these events were often quasi-periodic, with recurrence periods from a few seconds up to a minute (Sundberg et al., 2012b).

Sun et al. (2015b) also identified dipolarization events in the MESSENGER plasma and magnetic field measurements. The selection of events differed from that of the Sundberg et al. (2012b) study in that passes were chosen on the basis of clear substorm growth and expansion phase signatures being present in the MESSENGER magnetic field data around the time of the sudden enhancement in B_Z. The three substorm events considered in the Sun et al. (2015b) study are displayed in Figure 17.16. The first and second rows contain the proton energy per charge, E/q, spectra and proton density measured by FIPS (Raines et al., 2013). Before the time indicated by the first vertical dashed line in the figure, the X component of the magnetic field, B_X, was nearly the same as for the non-substorm orbits just before or after all three events. After that time the magnetic field in the near tail began to increase as measured by the B_X component. This point is identified as the

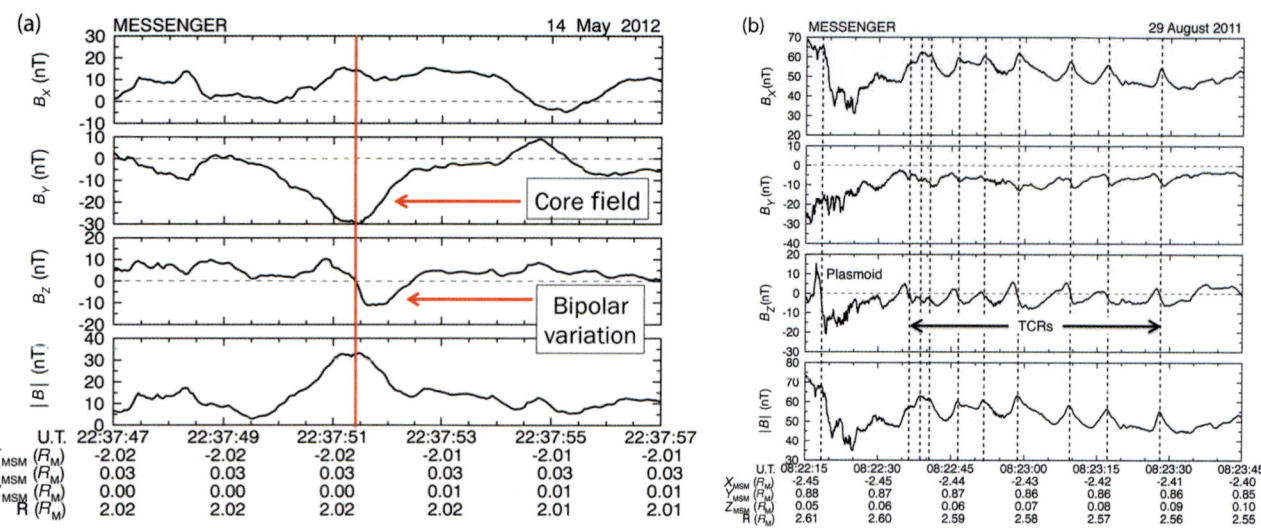

Figure 17.17. (a) A single plasmoid-type flux rope observed with MESSENGER magnetic field measurements. From DiBraccio et al. (2015a). (b) A chain event consisting of one plasmoid-type flux rope and nine TCRs moving tailward. From Raines et al. (2015).

beginning of the substorm growth, or loading, phase (Sun et al., 2015a). As MESSENGER moved toward the equatorial plane the B_X component of the magnetic field was expected to decrease, as shown by the non-substorm example in the figure, due to the diamagnetic effect of the plasma in the plasma sheet and the decrease in the intrinsic planetary magnetic field with decreasing latitude. However, for the substorm passes of Sun et al. (2015a), B_X remained constant or increased slightly while the magnetic elevation angles decreased for all events, indicating that the magnetic field lines continued to be stretched down the tail. Further, the FIPS measurements show that for two of the three passes, i.e., events I and II, MESSENGER was in the outer plasma sheet at the beginning of the growth phase but quickly found itself in the high-latitude, very-low-β lobe region as the plasma sheet thinned in response to the loading of the tail lobes with magnetic flux. All three substorm growth phases ended with clear dipolarization events (second vertical line shown for each pass) indicated by sharp increases in B_Z and the appearance of plasma-sheet particles in the FIPS observations. The measured B_X at the end of the growth phase for the three events was ~80%, ~50%, and ~30% larger than that of non-substorm observations, consistent with the tail loading–unloading measurements discussed earlier. It is worth noting that other possible dipolarization fronts were observed in events II and III. These may be the signatures of a series of planetward plasma flows (Sun et al., 2015a). The growth phase lasted ~58 s, ~1 min 2 s, and ~1 min 35 s for events I, II, and III, respectively. The transition from the expansion phase to the recovery phase, again marked with a vertical dashed line, is the end of the unloading interval inferred from the relaxation of B_X to the non-substorm trend line. The duration of the expansion phase in these substorms ranged from ~30 s to ~1 min. The total duration of the substorms was ~3–4 min, which is comparable to the determinations of the Dungey cycle time at Mercury by MESSENGER (Slavin et al., 2010a).

Perhaps the most important difference between the dipolarization events studied by Sundberg et al. (2012b) and by Sun et al. (2015a) is that the latter were longer-lived, i.e., tens of seconds to ~1 min compared with ~1–10 s for the former. As discussed by Kepko et al. (2015), the existence of long-duration dipolarization events relative to the Alfvén transit time between the boundary between open and closed field lines and the equatorial near tail, estimated to be a few seconds (e.g., Slavin et al., 2004, 2007), is significant because it would imply the existence of at least a weak substorm current wedge supported by the continued braking and diversion of at least a portion of the sunward reconnection flow bursts (Lyatsky et al., 2010).

Mercury's lack of an ionosphere and the expected high electrical resistivity of its regolith had been thought to preclude the development of long-lived field-aligned currents. However, Anderson et al. (2014) carried out an analysis of the MESSENGER magnetic field measurements and, after careful subtraction of the planetary field contributions, demonstrated the existence and measured the magnitude of Region 1 (R1) currents, which, just as at Earth, flow at the boundary between the open magnetic flux in the high-latitude polar caps and the closed magnetic flux at lower latitudes. They are driven by the first half of the Dungey cycle, which transfers magnetic flux from the dayside magnetosphere into the magnetotail. The R1 system is the most intense of the FAC systems at Earth, with a total current of ~(0.5–1.5) × 10^6 A (Iijima and Potemra, 1976; Anderson et al., 2008b). Anderson et al. (2014) found that the R1 FACs at Mercury have a total current of ~30 kA when the magnetosphere is strongly driven by dayside reconnection. Given the ~30 kV cross-magnetosphere electric potential derived from MESSENGER's observations of intense reconnection at the magnetopause (Slavin et al., 2009a, 2010b; DiBraccio et al., 2013), the integrated conductance of the R1 closure path is, therefore, ~1 S (Anderson et al., 2014). While these field-aligned currents are substantially weaker than at Earth for equivalent cross-magnetospheric potentials, the relatively high resistivity of Mercury's outer silicate shell reduces but does not stop FACs from flowing radially through that shell to close across Mercury's highly conductive iron-rich core ~400 km below the surface (Janhunen and Kallio, 2004; Anderson et al.,

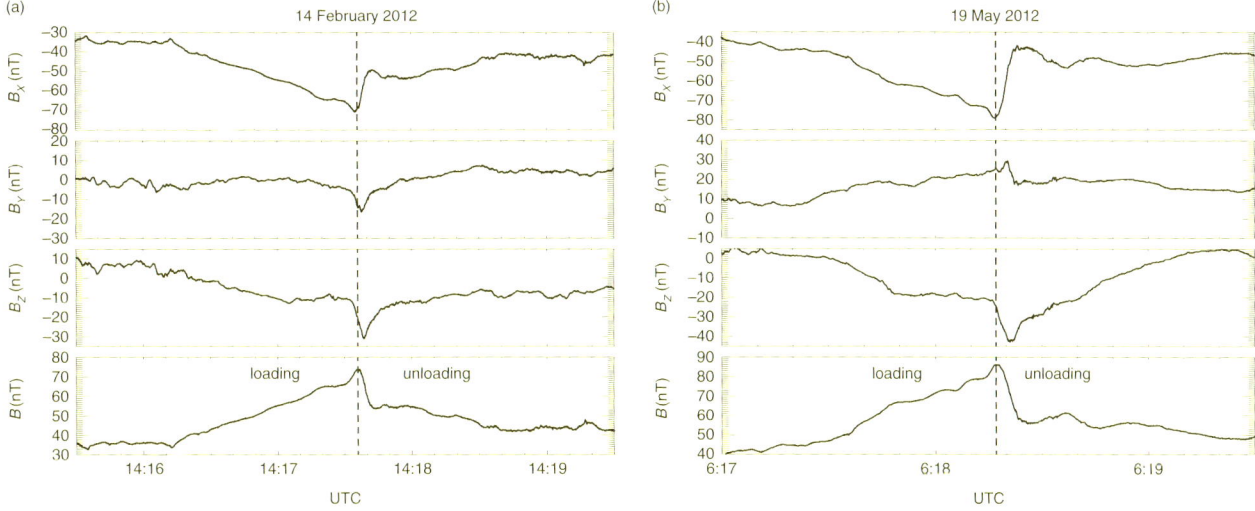

Figure 17.18. Two examples of large-amplitude magnetotail loading and unloading events, each ~2–3 min in duration, as seen in magnetic field observations (MSM coordinates). (a) Event of 14 February 2012 and (b) event of 19 May 2012. Note the single tailward-moving TCR that marks the transition from loading to unloading in each event.

2014). For this reason it is perhaps not surprising that the magnetic field dipolarization observed in the near tail appears consistent with a weak substorm current wedge. Indeed, the closure of Region 1 and possibly SCW field-aligned currents across Mercury's highly electrically conducting iron-rich core may be the ultimate explanation for why Earth-like substorm processes are observed by MESSENGER despite the lack of an electrically conducting ionosphere.

17.3.8 Magnetotail Flux Ropes and Traveling Compression Regions

A fundamental aspect of the reconnection process is the filamentation of the cross-tail current layer into magnetic islands (Hones et al., 1984; Hesse and Kivelson, 1998). Similar to the situation with FTEs at the magnetopause, spacecraft that do not penetrate into the flux rope proper often observe the draping and compression of the lobe magnetic field in the form of a traveling compression region. At Earth, plasmoids and TCRs are highly correlated with the onset of magnetospheric substorms (Moldwin and Hughes, 1992; Slavin et al., 1992; Taguchi et al., 1998). Many flux ropes can be formed during a given reconnection event, with some being carried sunward and others tailward by the fast Alfvénic jetting of plasma away from reconnection X-lines (Slavin et al., 2003). Indeed, analyses of MESSENGER measurements during the initial Mercury flybys revealed the presence of sunward- and anti-sunward-moving plasmoids and TCRs in Mercury's magnetotail (Slavin et al., 2009a, 2012a).

A close-up of a flux rope as seen in MESSENGER's magnetic field observations is shown in Figure 17.17a (DiBraccio et al., 2015a). In a statistical survey of 49 similar flux ropes, encountered between 1.7 R_M and 2.8 R_M down the tail, the mean diameter was found to be ~345 km, or ~0.14 R_M (DiBraccio et al., 2015a). This diameter is comparable to a proton gyroradius in the plasma sheet, or ~380 km. The events in this survey demonstrated that the magnetic structures of flux ropes at Mercury are similar to those observed at Earth, but with timescales that are 40 times smaller.

The magnetic field observations in Figure 17.17b were collected just after MESSENGER entered the south lobe of the magnetotail ~3.2 R_M downstream of the planet during the second Mercury flyby on 6 October 2008. At 06:12:37 UTC the spacecraft encountered a flux rope. It is identified by the ~1.5-s-long, large-amplitude, north-then-south B_Z perturbation followed by a ~10-s interval of magnetic field with a southward orientation and higher-frequency fluctuations typical of the plasma sheet (Slavin et al., 2009a). MESSENGER entered the northern lobe of the tail immediately after its encounter with the flux rope. At that point a series of nine TCRs, which are characterized by ~1–2-s-long north-then-south B_Z perturbations and enhancements of the total magnetic field of ~10–15%, were observed (Slavin et al., 2009a, 2012a).

MESSENGER did not have the capability to measure the plasma flow during these events, but the mean ejection speed for plasmoid-type flux ropes in Earth's near tail is ~500–600 km s^{-1} (Ieda et al., 1998; Slavin et al., 2003). If we assume a speed of 500 km s^{-1} for these flux-rope and TCR events at Mercury, then the average diameter of these structures at Mercury is ~500 km, or 0.2 R_M. This diameter compares with the ~(1–3) R_E flux rope diameters in the near tail of Earth (Slavin et al., 2003), where R_E is Earth's radius. Given the factor of ~8 scaling between the dimensions of these two magnetospheres, the diameters of flux ropes at Mercury and Earth appear to take up similar volumes relative to the size of these respective magnetospheres. It should also be noted that "chains" of plasmoids and TCRs, such as are displayed in Figure 17.17b, are also common at Earth (Slavin et al., 2005; Imber et al., 2011). What is still not understood is whether these chains form near simultaneously as a result of tearing-mode reconnection at multiple X-lines (Schindler, 1974) or are the result of periodic episodes of reconnection at a smaller number of X-lines. Interestingly, the mean interval of 9 s between the plasmoid and TCR events in Figure 17.17b is very close to the ~8–10-s spacing between flux transfer events observed at Mercury by Slavin et al. (2012b).

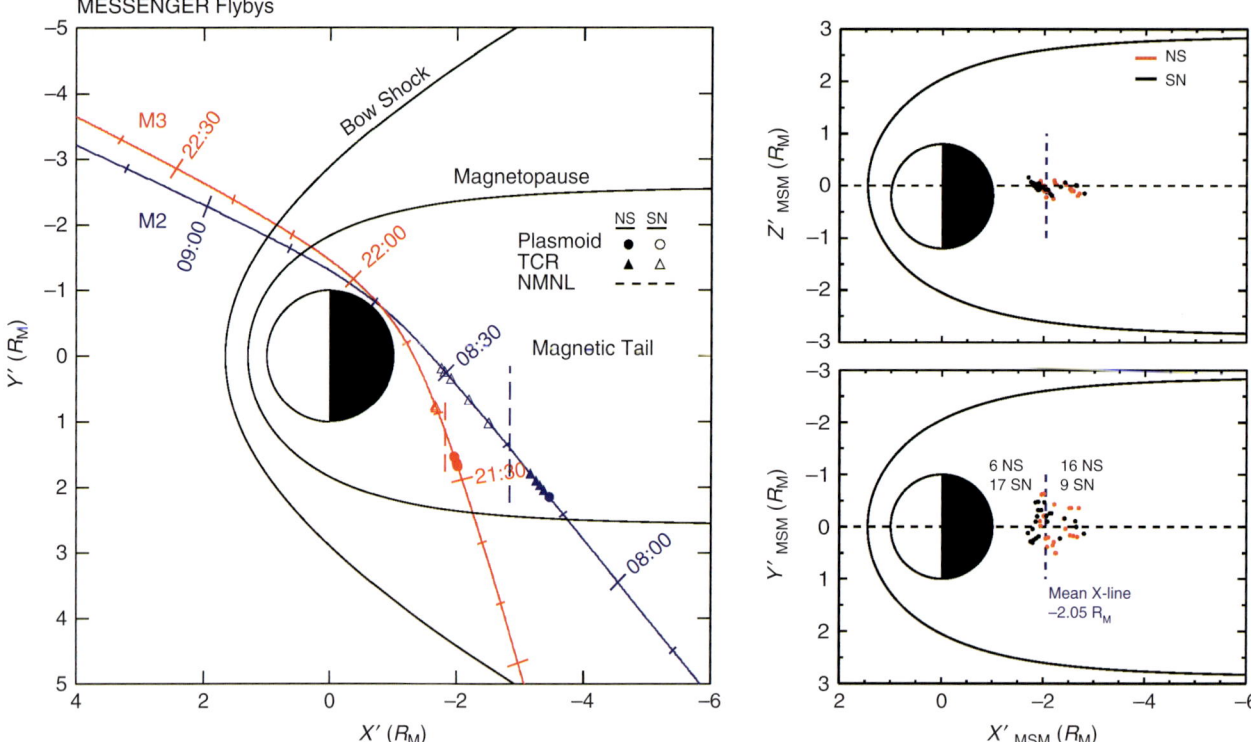

Figure 17.19. Near-Mercury neutral line (i.e., reconnection X-line) positions derived from plasmoid and TCR observations. (Left) Positions during the Mercury flybys. In this figure, NS denotes north-then-south variations in B_Z, and SN denotes south-then-north variations. From Slavin et al. (2012a). (Right) Positions from the orbital phase of the mission. From DiBraccio et al. (2015a).

At Earth the ejection of flux ropes and the observation of the TCRs that they drive are highly correlated with a peak in tail loading and the onset of the substorm expansion phase (Moldwin and Hughes, 1992; Slavin et al., 1992, 2003). Figures 17.18a and 17.18b show examples of this same behavior at Mercury on 14 February 2012 and 19 May 2012, respectively. In each case MESSENGER was located in a central part of the southern lobe of the tail. Both intervals contain a clear ~2-min-long, large-amplitude loading–unloading event. The total increase in the lobe magnetic field from the start to the peak was ~100%, similar to what was observed during the third Mercury flyby (Slavin et al., 2010b). A single TCR was then observed in both cases, coinciding with the initiation of the unloading phase. An important feature of the TCRs observed during these two loading–unloading events was the unipolar nature of the B_Z component of the magnetic field in contrast with the more common bipolar B_Z variation (e.g., Figure 17.17a). This signature is commonly observed in Earth's tail lobes when the observing spacecraft is directly over a stationary, growing flux rope (Slavin et al., 1990; Taguchi et al., 1998). The first half of the compression signature in this type of TCR is not due to the motion of the underlying flux rope, but rather the increase in its diameter as magnetic flux is added to the structure by reconnection. The sudden southward enhancement in B_Z marks the onset of fast reconnection, which ejects the plasmoid tailward while simultaneously unloading the tail lobes (see additional examples at Earth given by Slavin et al., 1992, 2002, 2005).

Flux ropes and TCRs may also be used to infer the mean location of X-lines in planetary magnetotails by identifying the downtail distance where the flux ropes and TCRs transition from being dominated by south-then-north B_Z variations, indicating planetward motion of the flux ropes, to north-then-south events, which move tailward (e.g., Slavin et al., 2003; Imber et al., 2011). Slavin et al. (2012a) applied this approach to observations from the second and third Mercury flybys and found that the mean X-line locations for these intervals were $X \sim -2\,R_M$ and $-3\,R_M$, as shown in Figure 17.19a. The spatial distribution of planetward and tailward flux ropes in the DiBraccio et al. (2015a) study of observations from orbit about Mercury indicated a mean X-line location at $X_{MSM} \sim -2\,R_M$, as indicated in Figure 17.19b.

17.3.9 Plasma Waves

There is an abundance of wave activity in Mercury's magnetosphere, including intense waves generated upstream of the bow shock, in the magnetosheath, and within the magnetosphere. Because of the small size of the system, there is a high degree of coupling across these regions, resulting in frequent wave activity within the magnetosphere that is driven by magnetosheath and/or bow shock phenomena. Additionally, without the ionosphere as a source for the magnetospheric plasma population, the solar wind protons dominate the thermal properties of the magnetotail ions. As this keV-plasma population is hotter than typically is the case for the eV plasma of ionospheric origin at Earth, the high temperature limits the cold plasma modes, making the characteristic wave modes in Mercury's magnetotail

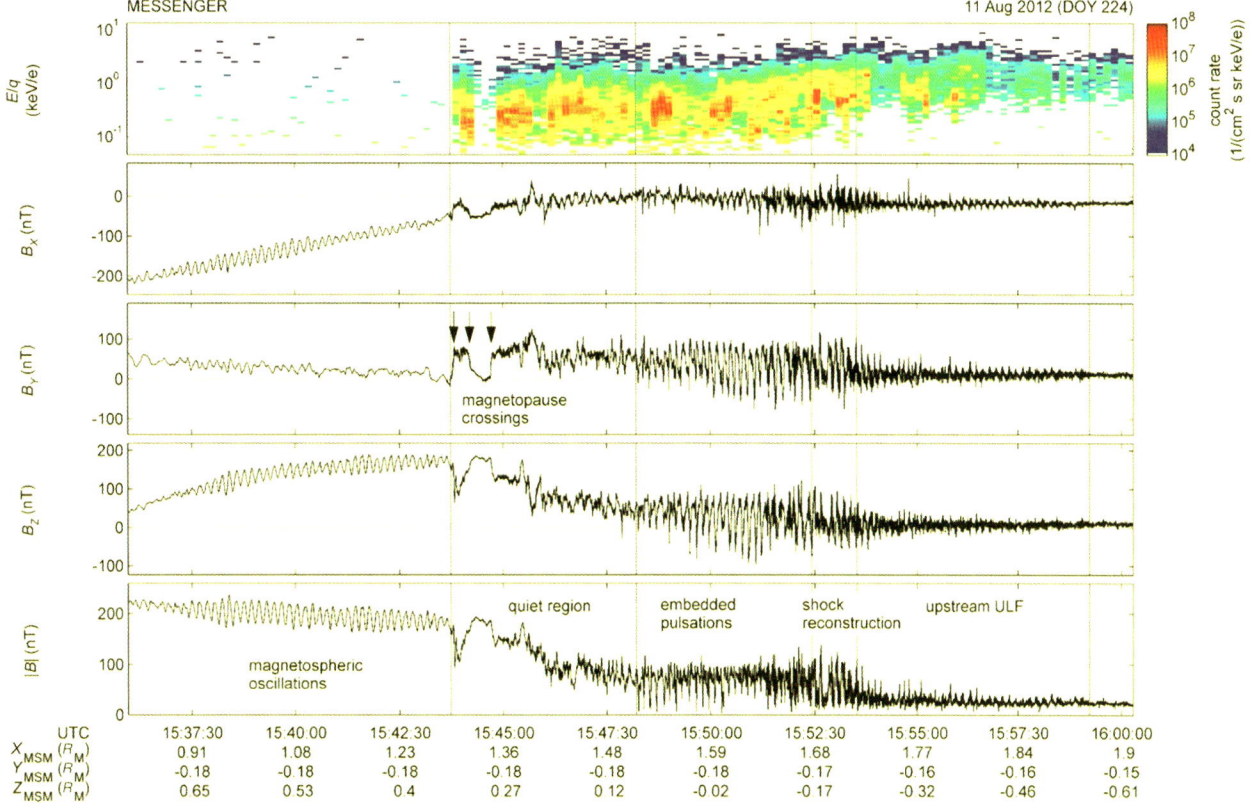

Figure 17.20. Example of a periodic reformation of the quasi-parallel bow shock as seen in the FIPS proton flux (top panel) and magnetic field (four lower panels, MSO coordinates). Such bow-shock reformation generates large-amplitude ULF waves within the magnetosheath and the magnetosphere. From Sundberg et al. (2013).

substantially different from those at Earth (Boardsen et al., 2015). Kinetic-scale fluctuations also play a major role in Mercury's magnetosphere; the turbulence spectrum is primarily in the ion-kinetic regime, strongly influenced by finite gyroradius effects; fluid-like fluctuations play only a minor part in the energy cascade (Uritsky et al., 2011).

The bow shock is one of the most important sources of wave activity in Mercury's space environment. As already discussed, the Alfvén speed in the inner solar system is higher because of the stronger magnetic field relative to the plasma density, and for this reason Mercury's bow shock is generally weaker than that at Earth. Mercury's shock structure rarely shows any pronounced "foot" regions or overshoots, phenomena that are associated with strong ion thermalization at the shock front. In addition, the small scale of the system and the bow shock's small radius of curvature reduce the extent of the ion and electron foreshock (Le et al., 2013). For this reason, the longer-period upstream waves rarely have time to grow, which constrains the upstream wave field primarily to the higher-frequency domain; longer-wavelength pulsations, typically at ~0.3 Hz, are observed only sporadically. The most common waves found upstream of the shock are whistler waves at ~2 Hz, similar in their characteristics to the typical 1 Hz waves observed at Earth. Short-lived wave bursts at ~0.8 Hz superimposed on the whistler wave trains have also been observed (Le et al., 2013). This wave mode had not previously been detected, and it has no counterpart at the terrestrial shock. Hot-flow-anomaly-like events have also been detected at the bow shock, but these events typically show signs of being in the early stages of their evolution, before the compressional regions surrounding the core have been fully established (Uritsky et al., 2014).

The magnetosheath wave population is dominated by electromagnetic ion-cyclotron (EMIC) waves (Fairfield and Behannon, 1976; Sundberg et al., 2015). These waves are generated by perpendicular temperature anisotropies in the proton population downstream of the shock, and they are found primarily behind the quasi-perpendicular section of the bow shock where the upstream magnetic field is perpendicular to the bow-shock normal. Coherent EMIC waves are observed at ~50% of the quasi-perpendicular crossings, and they are often strongly Doppler-shifted to frequencies above the local proton cyclotron frequency. The magnetosheath downstream of the quasi-parallel shock is typically much more turbulent as a result of the dynamic behavior of the local bow shock. Although the shock transition at the quasi-parallel shock often is turbulent in nature, observations by MESSENGER have shown that, given stable upstream conditions, there may be an inherent reformation frequency associated with the shock, leading to a periodic growth of the upstream low-frequency waves, which in turn generate new shock fronts (Sundberg et al., 2013). This reformation also leads to the periodic injection of large-scale low-magnetic-field structures into the magnetosheath. These structures are probably populated by partly unheated solar wind plasma (Sundberg et al., 2013). The structures can survive deep into the magnetosheath, and

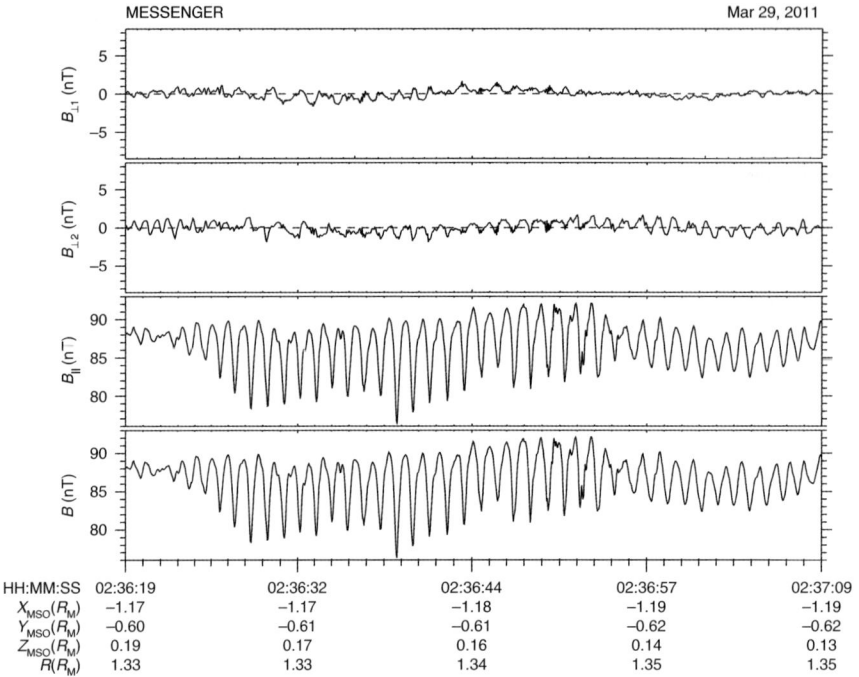

Figure 17.21. ULF waves observed near the magnetic equator in Mercury's magnetotail in magnetic field measurements rotated into coordinates parallel and perpendicular to the local magnetic field during this interval. The oscillations are strongly compressional and have a peak-to-peak amplitude near 10 nT. The wave frequency in this case is slightly below that of the local ion cyclotron frequency. From Boardsen et al. (2012).

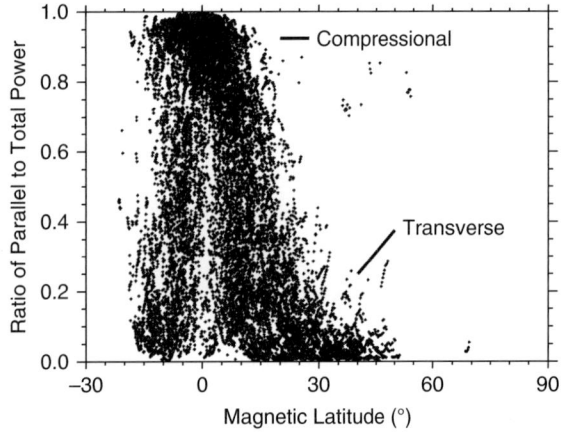

Figure 17.22. Occurrence statistics of magnetospheric ULF waves near the ion cyclotron frequency. The figure shows the ratio of power parallel to the local magnetic field to total power as a function of magnetic latitude. Compressional waves are clustered near the magnetic equator, whereas transverse waves are observed primarily at higher latitudes. From Boardsen et al. (2012).

they drive substantial low-frequency (~0.1 Hz) wave activity in the magnetosphere, presumably due to a cyclic variation in the magnetosheath dynamic pressure. Figure 17.20 shows one such example, a stable upstream wave period that was transmitted through the shock transition layer and into the magnetosheath, thereby driving waves nearly within the entire crossing of the dayside magnetosphere.

The quasi-parallel shock also commonly generates high-amplitude ULF waves in the downstream medium. Such waves will dominate the magnetic field signatures, with amplitudes up to 100 nT peak to peak. These waves are strongest near the perfectly parallel shock, and they can persist through large parts of the magnetosheath. They are most likely generated by upstream waves, which can be transmitted into the downstream medium at intermediate-Mach-number shocks (Krauss-Varban and Omidi, 1991; Sundberg et al., 2015) and/or bow-shock reformation cycles. Three-dimensional kinetic simulations of Mercury's magnetosphere also show shock transitions that are similar to those found at Earth. Long-wavelength pulsations dominate the region upstream of the quasi-parallel shock, and the magnetosheath shows clear signs of ion-anisotropy-driven instabilities, with both ion cyclotron and mirror waves downstream of the quasi-perpendicular shock (Trávníček et al., 2009). It is likely that mirror waves develop at times in Mercury's magnetosheath, but they are much less common than the ion-cyclotron waves, and no definitive observations of mirror waves have yet been documented (Sundberg et al., 2015). The difficulty in detecting mirror waves may be due partly to the short duration of each wave encounter, the high level of background turbulence in the magnetosheath, and the long integration time of the ion measurements.

Mercury's magnetosphere is full of wave activity. The inner magnetosphere, in particular the nightside region, is commonly host to narrow-band harmonic waves with a fundamental frequency near the proton cyclotron frequency (Russell, 1989; Boardsen et al., 2009a, b, 2012). Figure 17.21 shows one typical example, reported by Boardsen et al. (2012), and Figure 17.22 shows the statistical properties of these waves. It is apparent from this work that the bulk of the wave observations are clustered around the magnetic equator, and ~75% of the waves

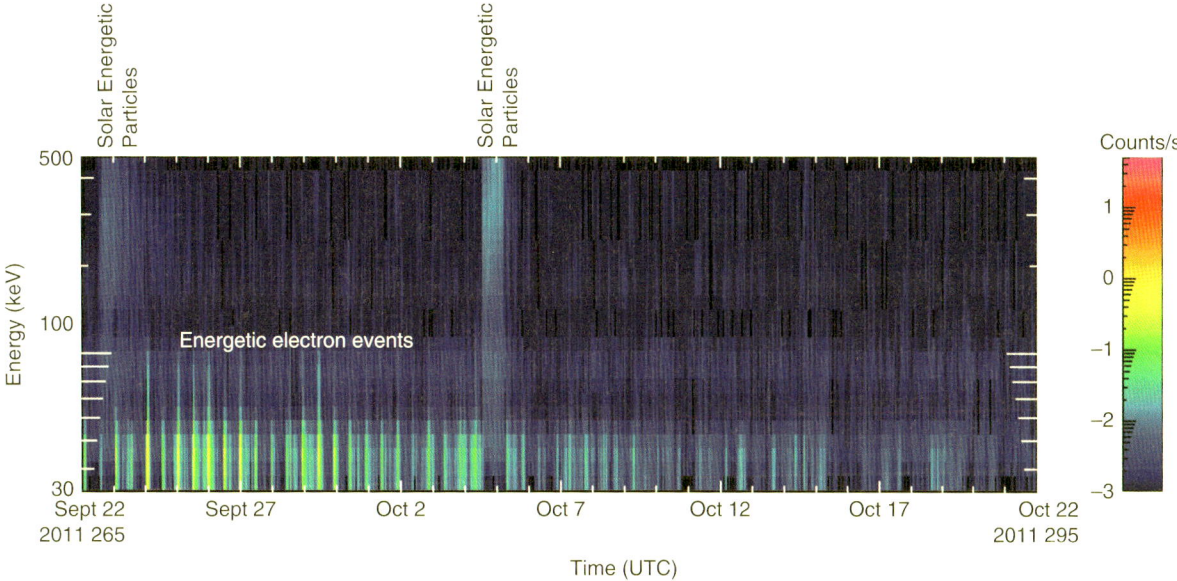

Figure 17.23. Energy spectrogram for the electron events observed by EPS from 22 September 2011 to 22 October 2011. The energetic electron events appear as increases in intensity in the energy range 30 to 60 keV (in orange and light blue) and generally fade into the instrument background at energies above 100 keV. Event intensities, even at the lower energies, gradually dropped close to the instrument background level on 16 October 2011. From Ho et al. (2012).

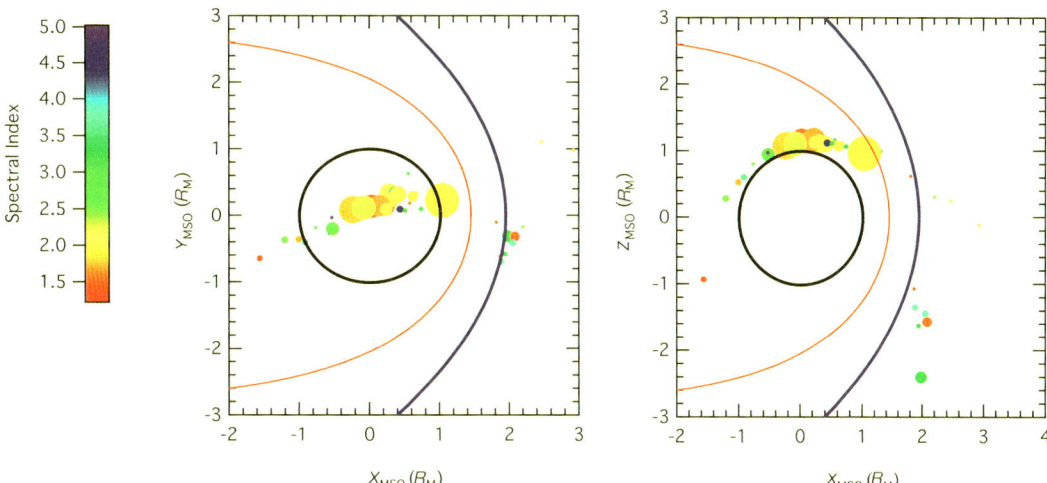

Figure 17.24. The locations of the 51 most intense electron events detected by EPS (>200 particles cm^{-2} s^{-1} sr^{-1} keV^{-1} at 45 keV) during the first year of MESSENGER observations are projected onto two orthogonal planes in MSO coordinates. The circles denote the location of MESSENGER at the time of the peak intensity during each event. Circle size increases with event intensity, and the color represents the power-law index fit to the energy spectrum. The average magnetopause and bow shock surfaces are shown in red and blue, respectively; the planetary outline is shown in black. From Ho et al. (2012).

are compressionally dominant. Transverse-dominant waves are also observed, but these are typically seen farther from the magnetic equator. The wave power maximizes at the equator and peaks in the occurrence frequency in the pre-midnight and post-midnight sectors, with a clear predominance for dusk-side waves (Boardsen et al., 2012). Boardsen et al. (2015) interpreted these characteristics as a form of ion-Bernstein-mode wave. As these waves propagate back and forth across the magnetic equator, the polarization and transmission properties change with plasma β, and the waves cycle between two different branches of the instability, one with high compression at the magnetic equator, and one for which the compression maximum occurs off, and symmetric to, the equatorial plane (Boardsen et al., 2015). These properties agree well with what is seen in the observations. In addition to the narrow-band harmonic waves, there is also evidence for quasi-periodic waves at frequencies of ~0.1 Hz, i.e., below the typical ion cyclotron frequency. These waves are likely externally driven, as they have been observed in association with both Kelvin–Helmholtz activity on the magnetopause and large-amplitude magnetosheath waves downstream of the quasi-parallel bow shock (Sundberg et al., 2012a, 2013).

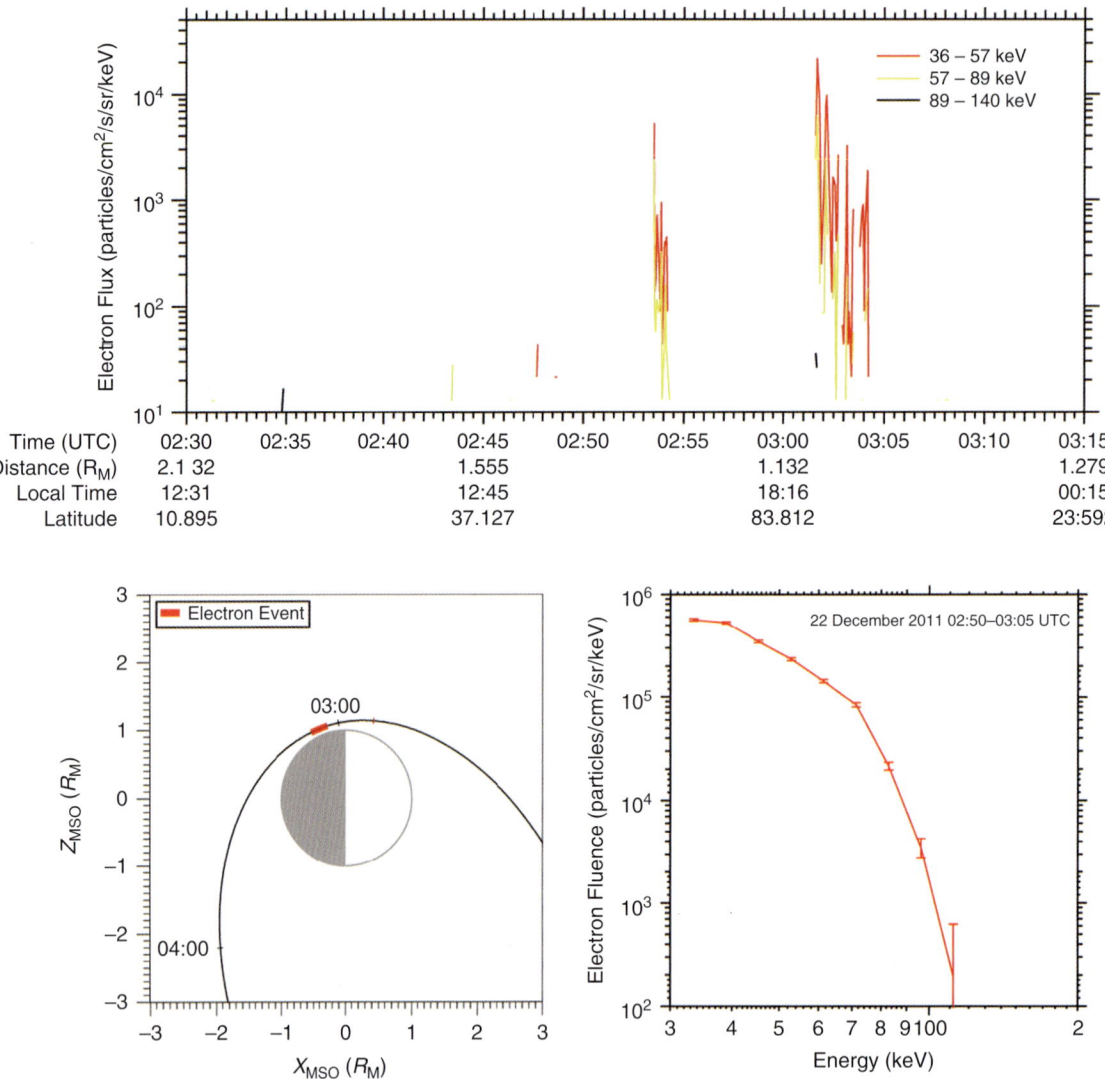

Figure 17.25. The intensity versus time, event-averaged energy spectrum, and location for a pair of two closely spaced high-latitude energetic electron bursts on 22 December 2011. From Ho et al. (2012).

17.4 ENERGETIC ELECTRON BURSTS

After Mercury's global planetary magnetic field, arguably the most surprising Mariner 10 mission discovery was that this small magnetosphere is a source of intense bursts of energetic (>35 keV) charged particles (Simpson et al., 1974; Eraker and Simpson, 1986). Unfortunately, instrumental effects made an unambiguous determination of species, flux, and energy spectrum for the Mariner 10 events impossible (Armstrong et al., 1975; Christon et al., 1979). For this reason, definitive measurement of the properties and acceleration processes for the energetic particles in Mercury's magnetosphere has long been a priority for the planetary magnetosphere community (Sundberg and Slavin, 2015; Seki et al., 2015; Raines et al., 2015).

Data from the MESSENGER EPS have shown that these energetic particle bursts are composed entirely of electrons (Ho et al., 2011). EPS made measurements of these electrons from ~30 to 300 keV energy during its 3-s scans. The durations of the energetic electron bursts ranged from several minutes to nearly an hour, and they tended to be centered on the time of the spacecraft's closest approach to the planet. The energy, E, of these electrons sometimes exceeded 200 keV, but usually the energy distributions exhibited a cutoff near $E = 100$ keV. However, no ions with energies >35 keV were detected by EPS anywhere in Mercury's magnetosphere, and no evidence of a stably trapped high-energy charged particle population was found. This latter result was not a surprise, because Mercury's magnetic field is weak, and the dayside magnetosphere is too small, to support the stably trapped radiation belts found at Earth (e.g., Shriver et al., 2011a; Walsh et al., 2013).

An energy–time spectrogram for the EPS electron measurements covering one month early in the orbital phase of the MESSENGER mission is shown in Figure 17.23 (Ho et al., 2012). Electron bursts were detected during each of the 12-h orbits from 22 September to 22 October 2011. Two SEP events (on 22 September and 4 October) occurred during this interval, and they are identified by their characteristic high energies (>100 keV) and several-day durations. Some of the events reported by Ho et al. (2012) occurred near the magnetic equator,

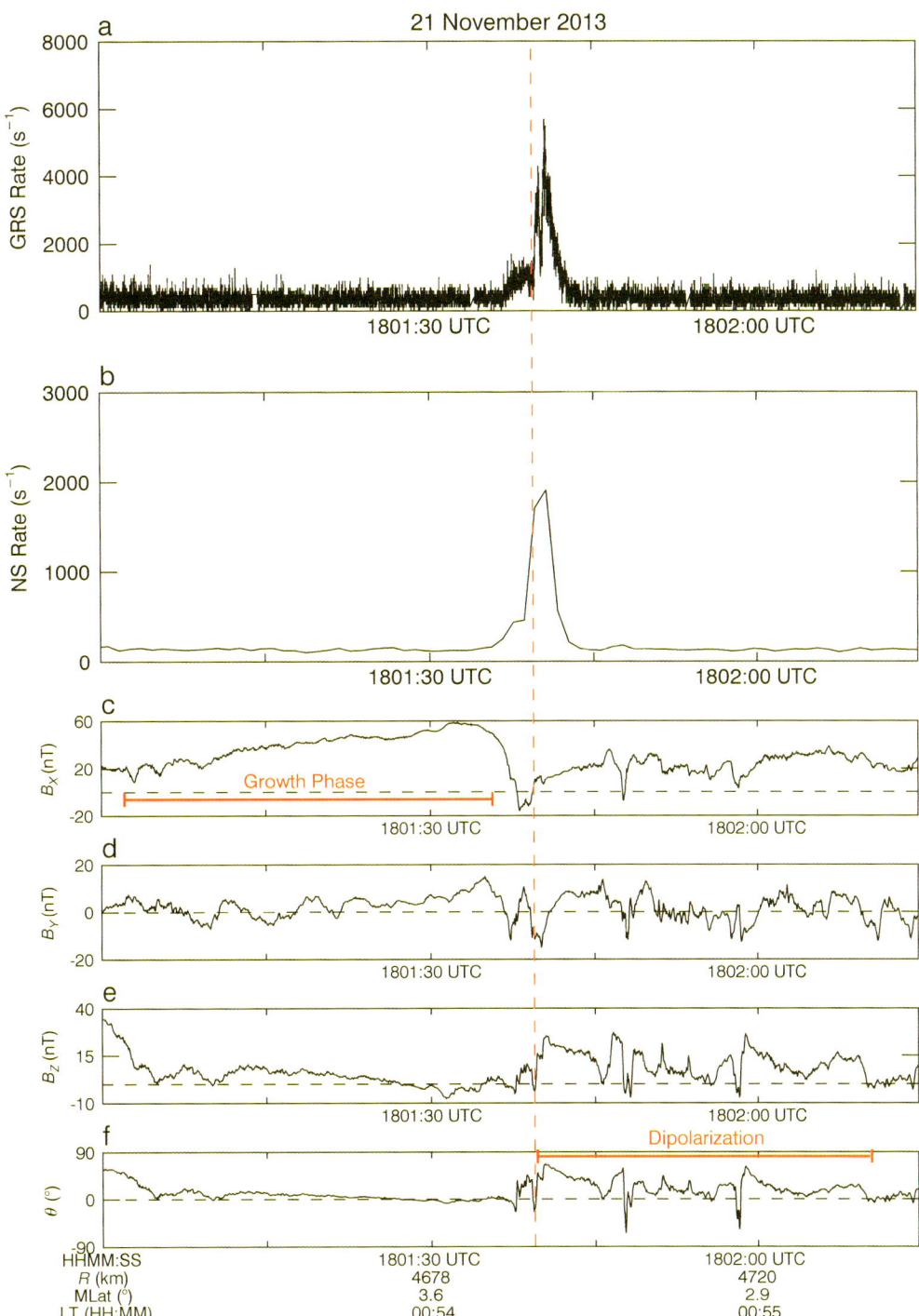

Figure 17.26. An electron burst event near local midnight on 21 November 2013 displayed in (a) GRS, (b) NS, and (c)–(f) Magnetometer measurements in MSM coordinates. A ~1-minute-long growth phase denotes loading of the tail, and a sudden increase in the B_Z magnetic field coordinate (vertical dashed line) marks the onset of a dipolarization event in the near tail. From Baker et al. (2016).

but most were concentrated at high northern latitudes, near spacecraft periapsis. As shown in Figure 17.24, these electron events appeared mainly around local midnight near the magnetic equator and at high latitudes near the magnetospheric cusp region on open as well as closed magnetic field lines.

Two examples of the energetic electron bursts measured by EPS observed on 22 December 2011 are displayed in Figure 17.25 (Ho et al., 2012). Typically, these events were brief, lasting a few seconds to a few minutes. They have pancake pitch angle distributions and power-law energy spectra for energies above ~50 keV. The first event in Figure 17.25 was detected at ~02:55 UTC and lasted ~1 min. The second burst began at 03:01 UTC and continued for about 3 min. Multiple events are often seen during a single orbit, as in this example. The energy spectrum shown bends over at energies below ~50 keV, and it is well fit with a kappa function. However, these events are usually characterized by power law indices of ~1.5–4.5 (Ho et al., 2012).

The Gamma-Ray and Neutron Spectrometer (GRNS), with separate Gamma-Ray Spectrometer (GRS) and Neutron Spectrometer (NS) sensors, was sensitive to electrons with $E >$ 50 keV, and it made measurements with temporal resolutions from seconds to 10 milliseconds (Goldsten et al., 2007). Lawrence et al. (2015) conducted a survey of energetic electron burst properties as observed at Mercury with the NS sensor covering the period from orbit insertion through 31 December 2013. The NS had a much larger geometric factor than EPS and a time resolution of 1 s or 20 s. It responded primarily to incident electrons in the energy range ~20–40 keV. The Lawrence et al. (2015) study identified 2711 electron events and surveyed their temporal, spatial, and spectral behavior. The duration of the events ranged from tens of seconds to nearly 20 min, and the events were classified as "bursty" (large amplitude variation with time during an event) or "smooth." Almost all events were detected inside Mercury's magnetosphere on closed field lines. The bursty events were observed most frequently near dawn, whereas the smooth events were detected most frequently in the midnight sector at lower latitudes. Some of the NS events exhibited periodicities from hundreds of seconds to tens of milliseconds. Lawrence et al. (2015) attributed the short-period variations to particle dynamics such as north–south bouncing within dipolar magnetic flux tubes, whereas the longer few-minute variations were interpreted as substorm injection events in Mercury's magnetic tail, similar to the Mariner 10 burst events (e.g., Siscoe et al., 1975; Baker et al., 1986; Christon et al., 1987; Ip, 1987; Delcourt et al., 2005).

The GRNS gamma-ray sensor was cryocooled and measured gamma rays in the energy range ~50 keV to 10 MeV (Goldsten et al., 2007). It was used primarily for Mercury surface composition measurements. The GRS operated in this mode until June 2012 when the cryocooler failed after reaching its expected lifetime. The GRS sensor incorporated a borated plastic anticoincidence shield (ACS) surrounding its germanium detector. The ACS was sensitive to electrons with energies from ~50 keV to several hundred keV. Beginning on 25 February 2013, the germanium detector's telemetry was re-allocated to the ACS system so that it could provide near-continuous measurements of energetic electrons with 10-ms time resolution. Baker et al. (2016), building on the initial analyses of the high-resolution GRS data by Lawrence et al. (2015), conducted a detailed examination of intense electron bursts in the high-time-resolution ACS data collected from 1 March 2013 to October 2014. Energetic electrons measured with FIPS during SEP events were also observed in closed-field regions of the magnetotail and were consistent with the average spatial distribution of events from GRS (Gershman et al., 2015b).

The acceleration of electrons at Earth is often closely associated with the formation of reconnection X-lines in the cross-tail current sheet (see the review by Birn et al., 2012). The magnetic flux loading and unloading of the tail lobes, the dipolarization of the plasma sheet magnetic field as it piles up at the inner edge of the tail, and the formation of flux ropes in the cross-tail current layer at Earth are all frequently observed at Mercury. Inductive electric fields resulting from the rapid reconfiguration of the magnetic field at X-lines, and Fermi acceleration in flux ropes as they contract, can readily accelerate electrons to energies of hundreds of keV on very short timescales (Sarris et al., 1976; Baker and Stone, 1977; Richardson et al., 1996; Øieroset et al., 2002; Hoshino, 2005; Drake et al., 2006; Ashour-Abdalla et al., 2011; Birn et al., 2012). Given the high frequency and intensity of reconnection at Mercury (Slavin et al., 2009a, 2010b; DiBraccio et al., 2013, 2015a), these processes are the most likely acceleration mechanisms for energetic electrons at Mercury.

An example of one of the energetic (~100–200 keV) electron injection events in the GRS and NS data studied by Baker et al. (2016) is displayed in Figure 17.26. Sharp enhancements in NS and GRS count rates occurred at ~18:01:36 UTC on 21 November 2013. The MESSENGER spacecraft was located at a planetocentric distance of 1.90 R_M in the post-midnight region at ~00:54 local time. Figure 17.26a shows that there was a pulse of energetic electrons with complex structure lasting until ~18:01:45 UTC. The magnetic field associated with this electron event showed "tail-like" stretching (Sundberg et al., 2012b; Sun et al., 2015a) prior to the particle injection, with the B_X component strengthening and the B_Z component diminishing between ~18:01:00 and 18:01:35 UTC. At the time of the energetic electron flux injection (18:01:36 UTC), the B_X component decreased and the B_Z component increased. The elevation angle, θ, of the magnetic field to the magnetic equator is shown in the bottom panel. Its sudden increase marks the arrival of a dipolarization front, and a pulse of energetic electrons was observed as in the Mariner 10 events (Baker et al., 1986; Christon et al., 1987; Delcourt et al., 2005).

Baker et al. (2016) found that the most intense energetic electron bursts detected by MESSENGER appeared to be produced in the midnight sector of Mercury's magnetosphere. The data show that the accelerated electrons were frequently observed on closed, rapidly reconfiguring magnetic field lines during substorm-like events. Further, the high-time-resolution GRS ACS (10-ms sampling) electron data show that injected electrons can in some instances complete one or more drifts around the planet on closed, quasi-trapped paths, creating Earth-like drift-echo events (Baker et al., 1996; Schriver et al., 2011a). However, it is also important to bear in mind the compactness of Mercury's dayside magnetosphere. In most cases, it is likely that substorm-injected electrons will not be able to execute even one complete azimuthal drift around the planet before being lost. In fact, relatively few instances of electron burst events in the dusk or pre-midnight sectors were found (Baker et al., 2016), consistent with most of the injected electron bursts not having executed complete drifts about Mercury. Overall, the EPS and GRNS data clearly indicate that Earth-like substorm energetic electron injection events take place at Mercury (Ho et al., 2012; Lawrence et al., 2015; Baker et al., 1996). Indeed, these results support the hypothesis proposed by Baker et al. (1987) that Mercury might possess an auroral oval analog whereby an annular region of the surface centered on the boundary between open and closed magnetic field lines is warmed by the precipitation of energetic particles injected during substorms. In point of fact, Lindsay et al. (2016) recently used MESSENGER's XRS measurements to construct a statistical image of X-ray fluorescence emissions from the surface mostly resulting from the precipitation of magnetospheric electrons with energies of ~1–10 keV. These emissions originate on the nightside of Mercury near the boundary between open and closed magnetic field lines at ~50°N and ~20°S, and they resemble the auroral ovals observed at Earth and the other planets possessing intrinsic magnetic fields and dense neutral atmospheres (Lindsay et al., 2016).

17.5 MESSENGER'S ANSWERS

To summarize the new insights that MESSENGER has provided into the dynamics of Mercury's magnetosphere, we return to the science questions listed in the introduction. For each question set, brief answers are provided here, to the extent that they can be confidently inferred from the MESSENGER data. We then close by considering, very briefly, how the observations from the upcoming BepiColombo mission to Mercury are likely to lead to major new advances in our understanding of Mercury's magnetosphere.

Question set 1: How do magnetosheath conditions at Mercury differ from what is found at the other planets?

Answer: The strong interplanetary magnetic fields in the inner heliosphere result in low M_A, especially during coronal mass ejections. Under these conditions the bow shock is weak, the plasma β in the inner magnetosheath is very low, and a thick plasma depletion layer forms adjacent to the dayside magnetopause.

Question set 2: How do these conditions in Mercury's magnetosheath contribute to the dynamic nature of its magnetosphere? How does magnetopause reconnection at Mercury differ from that at Earth? Are flux transfer events (FTEs) a major driver of magnetospheric convection at Mercury?

Answers: The intensity of the magnetic fields on the two sides of the magnetopause current layer tend to be comparable at Mercury due to the formation of strong plasma depletion layers. Under these conditions reconnection is said to be "symmetric," in contrast with the typical situation at Earth where the field intensity inside the magnetosphere is usually much stronger than in the magnetosheath. As expected for symmetric conditions, reconnection at Mercury occurs for all non-zero magnetic shear angles and appears to be several times faster than at Earth. Flux transfer events occur much more frequently, and they carry relatively more magnetic flux than at Earth. At Mercury they are estimated to account for a third of the total magnetic flux transferred from the dayside magnetosphere into the magnetotail and a similar fraction of the dawn-to-dusk magnetospheric electric field.

Question set 3: Does reconnection ever erode the dayside magnetosphere to the point that the subsolar region of the surface is exposed to direct solar wind impact? To what extent do induction currents driven in Mercury's interior limit the solar wind flux to the surface? Do FTEs contribute significantly to the solar wind flux reaching the surface?

Answers: During some CME events the magnetopause has been observed to be eroded and/or compressed to altitudes of <0.2 R_M in the subsolar region, enabling direct solar wind impact. However, in other cases the effects of reconnection-driven erosion and compression-driven induction currents in Mercury's interior appear to balance each other. The solar wind also reaches the surface near the equatorward edge of the magnetospheric cusps via parallel flows along newly reconnected flux tubes. These reconnection-driven plasma injections may be the source of the cusp filaments in the MESSENGER observations. The solar wind flux channeled to the surface in these filaments approximately equals the total flux to the surface estimated for the large-scale magnetospheric cusps.

Question set 4: What effects do heavy planetary ions have on Mercury's magnetosphere?

Answers: A clear dusk-side asymmetry in the occurrence of Kelvin–Helmholtz boundary waves at Mercury appears due to the large gyroradii of sodium and other heavy ions in the magnetopause boundary layer. These large-gyroradius, keV-energy planetary ions are seen in greater numbers and higher energies on the dusk side of the plasma sheet, where the KH boundary waves are most prevalent. In this region resonant interactions between the KH waves and the energetic sodium ions may occur.

Question set 5: Does Mercury's magnetotail store and dissipate magnetic energy in a manner similar to substorms at Earth? How is the process affected by the lack of an ionosphere and the expected high electrical resistivity of the crust?

Answers: Many of the phenomena observed during substorms at Earth are also present in the MESSENGER observations at Mercury. During the substorm growth phase, the magnetic flux content of Mercury's tail is observed to increase as the plasma sheet thins. Expansion-phase onset is closely associated with the ejection of plasmoids down the tail, just as at Earth. Dipolarization fronts and energetic particle injections are also seen at this point in the inner magnetosphere as the plasma sheet is observed to thicken. The lack of an ionosphere and the presence of a resistive regolith is expected to limit the intensity of the field-aligned currents that form the substorm current wedge. The extent to which the pileup of dipolarized magnetic flux just sunward of the plasma sheet slows and diverts flow bursts in the plasma sheet is not clear, but it may be sufficient to form a steady-state substorm current wedge.

Question set 6: How does Mercury's magnetosphere accelerate energetic electrons up to hundreds of keV?

Answers: MESSENGER observed frequent electron acceleration events with energies up to several hundred keV. However, there is no durable trapping of these energetic electrons. Hence, Mercury has the only intrinsic-field planetary magnetosphere with no "Van Allen" radiation belts. Reconnection, including X-line formation, dipolarization fronts, and flux rope formation, is known to accelerate electrons to these energies at Earth. For these reasons the episodes of intense reconnection observed by MESSENGER are the most likely cause of the energetic electron bursts at Mercury.

17.6 FUTURE INVESTIGATIONS OF MERCURY'S MAGNETOSPHERE

Mariner 10's many discoveries during its three flybys of Mercury exceeded all expectations for a reconnaissance mission. Building upon this foundation, MESSENGER's exploration of this remarkable planet and its dynamic magnetosphere have led to many new discoveries and a global perspective on Mercury, its exosphere, and its magnetosphere as a coupled system. As detailed in this monograph, much has been learned, but in acquiring this knowledge the limits to our understanding have also been revealed.

The upcoming BepiColombo mission (Benkhoff et al., 2010; Chapter 20) will initiate an era of intensive study that promises to greatly advance our knowledge of Mercury and planetary science as a whole. From the magnetospheric standpoint, these advances will come about as a result of at least four major new types of measurements to be collected by BepiColombo: (1) The mission's dual spacecraft will provide the first direct measurements of the response of Mercury's magnetosphere and near-space environment to dynamic changes in the solar wind. These multi-point measurements will enable scientific discoveries ranging from the origins of Mercury's sodium-dominated exosphere to the properties of the interior inferred from magnetic induction driven by solar wind transients, and, possibly, internal magnetospheric dynamics. (2) BepiColombo's magnetic and electric fields instrumentation will measure the role of plasma-wave–charged-particle resonances, kinetic-scale instabilities, and energy transfer via field-aligned currents and waves in magnetospheric dynamics and energy transfer. (3) The mission's charged particle instruments will provide the high-time-resolution distribution function measurements for electrons and individual ion species necessary for understanding magnetic reconnection, energetic particle acceleration processes, and the role of kinetic effects in such phenomena as Kelvin–Helmholtz boundary waves and reconnection at the magnetopause. (4) BepiColombo's neutral mass spectrometer measurements of a broad range of exospheric constituents for a wide range of solar wind and magnetospheric conditions will provide a foundation for understanding how neutral atoms and molecules are promoted into the exosphere and the role of these processes in coupling the exosphere to the magnetosphere.

ACKNOWLEDGEMENTS

This chapter is dedicated to George L. Siscoe and Vytenis M. Vasyliunas for their pioneering research and leadership in the comparative study of planetary magnetospheres. We thank everyone who contributed to the success of the MESSENGER mission. The authors also gratefully acknowledge conversations concerning this review with B. J. Anderson, T. A. Cassidy, G. A. DiBraccio, J. M. Jasinski, X. Jia, R. L. McNutt, Jr., G. Poh, J. M. Raines, and W.-J. Sun.

REFERENCES

Akasofu, S.-I. (1964). The development of the auroral substorm. *Planet. Space Sci.*, **12**, 273–282.

Alexeev, I. I., Belenkaya, E. S., Bobrovnikov, S. Y., Slavin, J. A. and Sarantos, M. (2008). Paraboloid model of Mercury's magnetosphere. *J. Geophys. Res.*, **113**, A12210, doi:10.1029/2008JA013368.

Alexeev, I. I., Belenkaya, E. S., Slavin, J. A., Korth, H., Anderson, B. J., Baker, D. N., Boardsen, S. A., Johnson, C. L., Purucker, M. E., Sarantos, M. and Solomon, S. C. (2010). Mercury's magnetospheric magnetic field after the first two MESSENGER flybys. *Icarus*, **209**, 23–39, doi:10.1016/j.icarus.2010.01.024.

Anderson, B. J., Acuña, M. H., Lohr, D. A., Scheifele, J., Raval, A., Korth, H. and Slavin, J. A. (2007). The Magnetometer instrument on MESSENGER. *Space Sci. Rev.*, **131**, 417–540, doi:10.1007/s11214-007-9246-7.

Anderson, B. J., Acuña, M. H., Korth, H., Purucker, M. E., Johnson, C. L., Slavin J. A., Solomon, S. C. and McNutt, R. L., Jr. (2008a). The structure of Mercury's magnetic field from MESSENGER's first flyby. *Science*, **321**, 82–85, doi:10.1126/science.1159081.

Anderson B. J., Korth, H., Waters, C. L., Green, D. L. and Stauning, P. (2008b). Statistical Birkeland current distributions from magnetic field observations by the Iridium constellation. *Ann. Geophys.*, **26**, 671–687, doi:10.5194/angeo-26-671-2008.

Anderson, B. J., Johnson, C. L., Korth, H., Purucker, M. E., Winslow, R. M., Slavin, J. A., Solomon, S. C., McNutt, R. L., Jr., Raines, J. M. and Zurbuchen, T. H. (2011). The global magnetic field of Mercury from MESSENGER orbital observations. *Science*, **333**, 1859–1862, doi:10.1126/science.1211001.

Anderson, B. J., Johnson, C. L., Korth, H., Winslow, R. M., Borovsky, J. E., Purucker, M. E., Slavin, J. A., Solomon, S. C., Zuber, M. T. and McNutt, R. L., Jr. (2012). Low-degree structure in Mercury's planetary magnetic field. *J. Geophys. Res.*, **117**, E00L12, doi:10.1029/2012JE004159.

Anderson, B. J., Johnson, C. L., Korth, H., Slavin, J. A., Winslow, R. M., Phillips, R. J., McNutt, R. L., Jr. and Solomon, S. C. (2014). Steady-state field-aligned currents at Mercury. *Geophys. Res. Lett.*, **41**, 7444–7452, doi:10.1002/2014GL061677.

Andrews, G. B., Zurbuchen, T. H., Mauk, B. H., Malcom, H., Fisk, L. A., Gloeckler, G., Ho, G. C., Kelley, J. S., Koehn, P. L., LeFevere, T. W., Livi, S. S., Lundgren, R. A. and Raines, J. M. (2007). The Energetic Particle and Plasma Spectrometer instrument on the MESSENGER spacecraft. *Space Sci. Rev.*, **131**, 523–556, doi:10.1007/s11214-007-9272-5.

Angelopoulos, V., Baumjohann, W., Kennel, C. F., Coroniti, F. V., Kivelson, M. G., Pellat, R., Walker, R. J., Lühr, H. and Paschmann, G. (1992). Bursty bulk flows in the inner central plasma sheet. *J. Geophys. Res.*, **97**, 4027–4039, doi:10.1029/91JA02701.

Angelopoulos, V., McFadden, J. P., Larson, D., Carlson, C. W., Mende, S. B., Frey, H., Phan, T., Sibeck, D. G., Glassmeier, K.-H., Auster, U., Donovan, E., Mann, I. R., Rae, I. J., Russell, C. T., Runov, A., Xhou, X. and Kepko, L. (2008). Tail reconnection triggering substorm onset. *Science*, **321**, 931–935, doi:10.1126/Science.1160495.

Arge, C. N., Luhmann, J. G., Odstrcil, D., Shriver, C. J. and Li, Y. (2004). Stream structure and coronal sources of the solar wind during the May 12th, 1997 CME. *J. Atmos. Solar Terr. Phys.*, **66**, 1295–1309.

Armstrong, T. P., Krimigis, S. M. and Lanzerotti, L. J. (1975). A reinterpretation of the reported energetic particle fluxes in the vicinity of Mercury. *J. Geophys. Res.*, **80**, 4015–4017, doi:10.1029/JA080i028p04015.

Ashour-Abdalla, M., El-Alaoui, M., Goldstein, M. L., Zhou, M., Schriver, D., Richard, R., Walker, R., Kivelson, M. G. and Hwang, K. J. (2011). Observations and simulations of non-local acceleration of electrons in magnetotail magnetic reconnection events. *Nature Phys.*, **7**, 360–365, doi:10.1038/nphys1903.

Ashour-Abdalla, M., Lapenta, G., Walker, R. J., El-Alaoui, M. and Liang, H. (2015). Multiscale study of electron energization during unsteady reconnection events. *J. Geophys. Res. Space Physics*, **120**, 4784–4799, doi:10.1002/2014JA020316.

Baker, D. N. and Stone, E. C. (1977). Observations of energetic electrons ($E \gtrsim 200$ keV) in the Earth's magnetotail: Plasma sheet and fireball observations. *J. Geophys. Res.*, **82**, 1532–1546, doi:10.1029/JA082i010p01532.

Baker, D. N., Simpson, J. A. and Eraker, J. H. (1986). A model of impulsive acceleration and transport of energetic particles in Mercury's magnetosphere. *J. Geophys. Res.*, **91**, 8742–8748, doi:10.1029/JA091iA08p08742.

Baker, D. N., Borovsky, J. E., Burns, J. O., Gisler, G. R. and Zeilik, M. (1987). Possible calorimetric effects at Mercury due to solar wind-magnetosphere interactions. *J. Geophys. Res.*, **92**, 4707–4712, doi:10.1029/JA092iA05p04707.

Baker, D. N., Pulkkinen, T. I., Angelopoulos, V., Baumjohann, W. and McPherron, R. L. (1996). Neutral line model of substorms: Past results and present view. *J. Geophys. Res.*, **101**, 12,975–13,010, doi:10.1029/95JA03753.

Baker, D. N., Odstrcil, D., Anderson, B. J., Arge, C. N., Benna, M., Gloeckler, G., Raines, J. M., Schriver, D., Slavin, J. A., Solomon, S. C., Killen, R. M. and Zurbuchen, T. H. (2009). Space environment of Mercury at the time of the first MESSENGER flyby: Solar wind and interplanetary magnetic field modeling of upstream conditions. *J. Geophys. Res.*, **114**, A10101, doi:10.1029/2009JA014287.

Baker, D. N., Anderson, B. J., Arge, C. N., Benna, M., Gloeckler, G., Odstrcil, D., Korth, H., Mayer, L., Raines, J. M., Schriver, D., Slavin, J. A., Solomon, S. C., Trávníček, P. and Zurbuchen, T. H. (2011). The space environment of Mercury at the times of the second and third MESSENGER flyby. *Planet. Space Sci.*, **59**, 2066–2074, doi:10.1016/j.pss.2011.01.018.

Baker, D. N., Poh, G. K., Odstrcil, D., Arge, C. N., Benna, M., Johnson, C. L., Korth, H., Gershman, D. J., Ho, G. C., McClintock, W. E., Cassidy, T. A., Merkel, A., Raines, J. M., Schriver, D., Slavin, J. A., Solomon, S. C., Trávníček, P. M., Winslow, R. M. and Zurbuchen, T. H. (2013). Solar wind forcing at Mercury: WSA-ENLIL model results. *J. Geophys. Res. Space Physics*, **118**, 45–57, doi:10.1029/2012JA018064.

Baker, D. N., Dewey, R. M., Lawrence, D. J., Goldsten, J. O., Peplowski, P. N., Korth, H., Slavin, J. A., Krimigis, S. M., Anderson, B. J., Ho, G. C., McNutt, R. L., Jr., Raines, J. M., Schriver, D. and Solomon, S. C. (2016). Intense energetic electron flux enhancements in Mercury's magnetosphere: An integrated view with high-resolution observations from MESSENGER. *J. Geophys. Res. Space Physics*, **121**, 2171–2184, doi:10.1002/2015JA021778.

Baumjohann, W., Paschmann, G. and Lühr, H. (1990). Characteristics of high-speed ion flows in the plasma sheet. *J. Geophys. Res.*, **95**, 3801–3809.

Baumjohann, W., Paschmann, G., Nagai, T. and Lühr, H. (1991). Superposed epoch analysis of the substorm plasma sheet. *J. Geophys. Res.*, **96**, 11,605–11,608, doi:10.1029/91JA00775.1991.

Baumjohann, W., Hesse, M., Kokubun, S., Mukai, T., Nagai, T. and Petrukovich, A. A. (1999). Substorm dipolarization and recovery. *J. Geophys. Res.*, **104**, 24,995–25,000, doi:10.1029/1999JA900282.

Baumjohann, W., Matsuoka, A., Glassmeier, K. H., Russell, C. T., Nagai, T., Hoshino, M., Nakagawa, T., Balogh, A., Slavin, J. A., Nakamura, R. and Magnes, W. (2006). The magnetosphere of Mercury and its solar wind environment: Open issues and scientific questions. *Adv. Space Res.*, **38**, 604–609, doi:10.1016/j.asr.2005.05.117.

Benkhoff, J., van Casteren, J., Hayakawa, H., Fujimoto, M., Laakso, H., Novara, M., Ferri, P., Middleton, H. R. and Ziethe, R. (2010). BepiColombo – Comprehensive exploration of Mercury: Mission overview and science goals. *Planet. Space Sci.*, **58**, 2–20.

Benna, M., Anderson, B. J., Baker, D. N., Boardsen, S. A., Gloeckler, G., Gold, R. E., Ho, G. C., Killen, R. M., Korth, H., Krimigis, S. M., Purucker, M. E., McNutt, R. L., Jr., Raines, J. M., McClintock, W. E., Sarantos, M., Slavin, J. A., Solomon, S. C. and Zurbuchen, T. H. (2010). Modeling of the magnetosphere of Mercury at the time of the first MESSENGER flyby. *Icarus*, **209**, 3–10, doi:10.1016/j.icarus.2009.11.036.

Birn, J., Artemyev, A. V., Baker, D. N., Echim, M., Hoshino, M. and Zelenyi, L. M. (2012). Particle acceleration in the magnetotail and aurora. *Space Sci. Rev.*, **173**, 49–102, doi:10.1007/s11214-012-9874-4.

Boardsen, S. A., Anderson, B. J., Acuña, M. H., Slavin, J. A., Korth, H. and Solomon, S. C. (2009a). Narrow-band ultra-low-frequency wave observations by MESSENGER during its January 2008 flyby through Mercury's magnetosphere. *Geophys. Res. Lett.*, **36**, L01104, doi:10.1029/2008GL036034.

Boardsen, S. A., Slavin, J. A., Anderson, B. J., Korth, H. and Solomon, S. C. (2009b). Comparison of ultra-low-frequency waves at Mercury under northward and southward IMF. *Geophys. Res. Lett.*, **36**, L18106, doi:10.1029/2009GL039525.

Boardsen, S. A., Sundberg, T., Slavin, J. A., Anderson, B. J., Korth, H., Solomon, S. C. and Blomberg, L. G. (2010). Observations of Kelvin–Helmholtz waves along the dusk-side boundary of Mercury's magnetosphere during MESSENGER's third flyby. *Geophys. Res. Lett.*, **37**, L12101, doi:10.1029/2010GL043606.

Boardsen, S. A., Slavin, J. A., Anderson, B. J., Korth, H., Schriver, D. and Solomon, S. C. (2012). Survey of coherent 1 Hz waves in Mercury's inner magnetosphere from MESSENGER observations. *J. Geophys. Res.*, **117**, A00M05, doi:10.1029/2012JA017822.

Boardsen, S. A., Kim, E.-H., Raines, J. M, Slavin, J. A., Gershman, D. J., Anderson, B. J., Korth, H., Sundberg, T., Schriver, D. and Trávníček, P. (2015). Interpreting ~1 Hz magnetic compressional waves in Mercury's inner magnetosphere in terms of propagating ion-Bernstein waves. *J. Geophys. Res. Space Physics*, **120**, 4213–4228, doi:10.1002/2014JA020910.

Borovsky, J. E., Nemzek, R. J. and Belian, R. D. (1993). The occurrence rate of magnetospheric-substorm onsets: Random and periodic substorms. *J. Geophys. Res.*, **98**, 3807–3813, doi:10.1029/92JA02556.

Christon, S., Daly, S., Eraker, J., Perkins, M., Simpson, J. and Tuzzolino, A. (1979). Electron calibration of instrumentation for low energy, high intensity particle measurements at Mercury. *J. Geophys. Res.*, **84**, 4277–4288, doi:10.1029/JA084iA08p04277.

Christon, S. P., Feynman, J. and Slavin, J. A. (1987). Substorm injection fronts: Similar magnetospheric phenomena at Earth and Mercury. In *Magnetotail Physics*, ed. A. T. Y. Lui. Baltimore, MD: Johns Hopkins University Press, pp. 393–402.

Cooling, B. M. A., Owen, C. J. and Schwartz, S. J. (2001). Role of the magnetosheath flow in determining the motion of open flux tubes. *J. Geophys. Res.*, **106**, 18,763–18,776, doi:10.1029/2000JA000455.

Coroniti, F. V. and Kennel, C. F. (1973). Can the ionosphere regulate magnetospheric convection? *J. Geophys. Res.*, **78**, 2837–2851, doi:10.1029/JA078i016p02837.

Cowley, S. W. H. and Owen, C. J. (1989). A simple illustrative model of open flux tube motion over the dayside magnetopause. *Planet. Space Sci.*, **39**, 1465–1475.

Delcourt, D. C. (2013). On the supply of heavy planetary material to the magnetotail of Mercury. *Ann. Geophys.*, **31**, 1673–1679, doi:10.5194/angeo-31-1673-2013, 2013.

Delcourt, D. C., Moore, T. E., Orsini, S., Millilo, A. and Sauvaud, J.-A. (2002). Centrifugal acceleration of ions near Mercury. *Geophys. Res. Lett.*, **29**, 1591, doi:10.1029/2001GL013829.

Delcourt, D. C., Grinald, S., Leblanc, F., Berthelier, J.-J., Millilo, A., Mura, A., Orsini, S. and Moore, T. E. (2003). A quantitative model of the planetary Na$^+$ contribution to Mercury's magnetosphere. *Ann. Geophys.*, **21**, 1723–1736, doi:10.5194/angeo-21-1723-2003.

Delcourt, D. C., Seki, K., Terada, N. and Miyoshi, Y. (2005). Electron dynamics during substorm dipolarization in Mercury's magnetosphere. *Ann. Geophys.*, **23**, 3389–3398, doi:10.5194/angeo-23-3389-2005.

Delcourt, D. C., Moore, T. E. and Fok, M.-C. H. (2010). Ion dynamics during compression of Mercury's magnetosphere. *Ann. Geophys.*, **28**, 1467–1474, doi:10.5194/angeo-28–1467-2010.

Delcourt, D. C., Seki, K., Terada, N. and Moore, T. E. (2012). Centrifugally stimulated exospheric ion escape at Mercury. *Geophys. Res. Lett.*, **39**, L22105, doi:10.1029/2012GL054085.

Dewey, R. M., Baker, D. N., Anderson, B. J., Benna, M., Johnson, C. L., Korth, H., Gershman, D. J., Ho, G. C., McClintock, W. E., Odstrcil, D., Philpott, L. C., Raines, J. M., Schriver, D., Slavin, J. A., Solomon, S. C. Winslow, R. M. and Zurbuchen, T. H. (2015). WSA-ENLIL model with the Cone extension. *J. Geophys. Res. Space Physics*, **120**, 5667–5685, doi:10.1002/2015JA021194.

DiBraccio, G. A., Slavin, J. A., Boardsen, S. A., Anderson, B. J., Korth, H., Zurbuchen, T. H., Raines, J. M., Baker, D. N., McNutt, R. L., Jr. and Solomon, S. C. (2013). MESSENGER observations of magnetopause structure and dynamics at Mercury. *J. Geophys. Res. Space Physics*, **118**, 997–1008, doi:10.1002/jgra.50123.

DiBraccio, G. A., Slavin, J. A., Imber, S. M., Gershman, D. J., Raines, J. M., Jackman, C. J., Boardsen, S. A., Anderson, B. J., Korth, H., Zurbuchen, T. H., McNutt, R. L., Jr. and Solomon, S. C. (2015a). MESSENGER observations of flux ropes in Mercury's magnetotail. *Planet. Space Sci.*, **115**, 77–89, doi:10.1016/j.pss.2014.12.016.

DiBraccio, G. A., Slavin, J. A., Raines, J. M., Gershman, D. J., Tracy, P. J., Boardsen, J. A., Zurbuchen, T. H., Anderson, B. J., Korth, H., McNutt, R. L., Jr. and Solomon, S. C. (2015b). First observations of Mercury's plasma mantle by MESSENGER. *Geophys. Res. Lett.*, **42**, 9666–9675, doi:10.1002/2015GL065805.

Domingue, D. L., Chapman, C. R., Killen, R. M., Zurbuchen, T. H., Gilbert, J. A., Sarantos, M., Benna, M., Slavin, J. A., Schriver, D., Trávníček, P. M., Orlando, T. M., Sprague, A. L., Blewett, D. T., Gillis-Davis J. J., Feldman, W. C., Lawrence, D. J., Ho, G. C., Ebel, D. S., Nittler, L. R., Vilas, F., Pieters, C. M., Solomon, S. C., Johnson, C. L., Winslow, R. M., Helbert, J., Peplowski, P. N., Weider, S. Z., Mouawad, N., Izenberg, N. R. and McClintock, W. E. (2014). Mercury's weather-beaten surface: Understanding Mercury in the context of lunar and asteroidal space weathering studies. *Space Sci. Rev.*, **181**, 121–214, doi:10.1007/s11214-014–0039-5.

Drake, J. F., Swisdak, M., Che, H. and Shay, M. A. (2006). Electron acceleration from contracting magnetic islands during reconnection. *Nature*, **443**, 553–556, doi:10.1038/nature05116.

Dungey, J. W. (1961). Interplanetary magnetic field and the auroral zones. *Phys. Rev. Lett.*, **6**, 47.

Eastwood, J. P., Phan, T. D., Øieroset, M., Shay, M. A., Malakit, K., Swisdak, M., Drake, J. F. and Masters, A. (2013). Influence of asymmetries and guide fields on the magnetic diffusion region in collisionless space plasmas. *Plasma Phys. Control. Fusion*, **55**, 124001, doi:10.1088/0741–3335/55/12/124001.

Eraker, J. H. and Simpson, J. A. (1986). Acceleration of charged particles in Mercury's magnetosphere. *J. Geophys. Res.*, **91**, 9973–9993, doi:10.1029/JA091iA09p09973.

Fairfield, D. H. and Behannon K. W. (1976). Bow shock and magnetosheath waves at Mercury. *J. Geophys. Res.*, **81**, 3897–3906, doi:10.1029/JA081i022p03897.

Farrugia, C. J., Erkaev, N. V., Biernat, H. K. and Burlaga, L. F. (1995). Anomalous magnetosheath properties during Earth passage of an interplanetary magnetic cloud. *J. Geophys. Res.*, **10**, 19,245–19,257, doi:10.1029/95JA0108.

Fear, R. C., Milan, S. E., Fazakerley, A. N., Owen, C. J., Asikainen, T., Taylor, M. G. G. T., Lucek, E. A., Rème, H., Dandouras, I. and Daly, P. W. (2007). Motion of flux transfer events: A test of the Cooling model. *Ann. Geophys.*, **25**, 1669–1690, doi:10.5194/angeo-25–1669-2007.

Fujimoto, M., Baumjohann, W., Kabin, K., Nakamura, R., Slavin, J. A., Terada, N. and Zelenyi, L. (2007). Hermean magnetosphere–solar wind interaction. *Space Sci. Rev.*, **132**, 529–550, doi:10.1007/s11214-007–9245-8.

Fuselier, S. A. and Lewis, W. S. (2011). Properties of near-Earth magnetic reconnection from in-situ observations. *Space Sci. Rev.*, **160**, 95–121, doi:10.1007/s11214-011–9820-x.

Gershman, D. J., Zurbuchen, T. H., Fisk, L. A., Gilbert, J. A., Raines, J. M., Anderson, B. J., Smith, C. W., Korth, H. and Solomon, S. C. (2012). Solar wind alpha particles and heavy ions in the inner heliosphere. *J. Geophys. Res.*, **117**, A00M02, doi:10.1029/2012JA017829.

Gershman, D. J., Slavin, J. A., Raines, J. M., Zurbuchen, T. H. Anderson, B. J., Korth, H., Baker, D. N. and Solomon, S. C. (2013). Magnetic flux pileup and plasma depletion in Mercury's subsolar magnetosheath. *J. Geophys. Res. Space Physics*, **118**, 7181–7199, doi:10.1002/2013JA019244.

Gershman, D. J., Slavin, J. A., Raines, J. M., Zurbuchen, T. H., Anderson, B. J., Korth, H., Baker, D. N. and Solomon, S. C. (2014). Ion kinetic properties in Mercury's pre-midnight plasma sheet. *Geophys. Res. Lett.*, **41**, 5740–5747, doi:10.1002/2014GL060468.

Gershman, D. J., Raines, J. M., Slavin, J. A., Zurbuchen, T. H., Sundberg, T., Boardsen, S. A., Anderson, B. J., Korth, H. and Solomon, S. C. (2015a). MESSENGER observations of multi-scale Kelvin–Helmholtz vortices at Mercury. *J. Geophys. Res. Space Physics*, **120**, 4354–4368, doi:10.1002/2014JA020903.

Gershman, D. J., Raines, J. M., Slavin, J. A., Zurbuchen, T. H., Anderson, B. J., Korth, H., Ho, G. C., Boardsen, S. A., Cassidy, T. A., Walsh, B. M. and Solomon, S. C. (2015b). MESSENGER observations of solar energetic electrons within Mercury's magnetosphere. *J. Geophys. Res. Space Physics*, **120**, 8559–8571, doi:10.1002/2015JA021610.

Gershman, D. J., Dorelli, J. C., DiBraccio, G. A., Raines, J. M., Slavin, J. A., Poh, G. and Zurbuchen, T. H. (2016). Ion-scale structure in Mercury's magnetopause reconnection diffusion region. *Geophys. Res. Lett.*, **43**, 5935–5942, doi:10.1002/2016GL069163.

Gingell, P. W., Sundberg, T. and Burgess D. (2015). The impact of a hot sodium ion population on the growth of the Kelvin–Helmholtz instability in Mercury's magnetotail. *J. Geophys. Res. Space Physics*, **120**, 5432–5442, doi:10.1002/2015JA021433.

Glassmeier, K.-H. (2000). Currents in Mercury's magnetosphere. In *Magnetospheric Current Systems*, ed. S. Ohtani, R. Fujii, M. Hesse and R. L. Lysak, Geophysical Monograph 118. Washington, DC: American Geophysical Union, pp. 371–380.

Glassmeier, K.-H. and Espley, J. (2006). ULF waves in planetary magnetospheres. In *Magnetospheric ULF Waves: Synthesis and New Directions*, ed. K. Takahashi, P. J. Chi, R. E. Denton and

R. L. Lysak, Geophysical Monograph 169. Washington, DC: American Geophysical Union, pp. 341–359, doi:10.1029/169GM22.

Glassmeier, K.-H., Grosser, J., Auster, U., Constantinescu, D., Narita, Y. and Stellmach, S. (2007). Electromagnetic induction effects and dynamo action in the Hermean system. *Space Sci. Rev.*, **132**, 511–527, doi:10.1007/s11214-007-9244-9.

Goldsten, J. O., Rhodes, E. A., Boynton, W. V., Feldman, W. C., Lawrence, D. J., Trombka, J. I., Smith, D. M., Evans, L. G., White, J., Madden, N. W., Berg, P. C., Murphy, G. A., Gurnee, R. S., Strohbehn, K., Williams, B. D., Schaefer, E. D., Monaco, C. A., Cork, C. P., Eckels, J. D., Miller, W. O., Burks, M. T., Hagler, L. B., DeTeresa, S. J. and Witte, M. C. (2007). The MESSENGER Gamma-Ray and Neutron Spectrometer. *Space Sci. Rev.*, **131**, 339–391, doi:10.1007/s11214-007-9262-7.

Hasegawa, H., Wang, J., Dunlop, M. W., Pu, Z. Y., Zhang, Q.-H., Lavraud, B., Taylor, M. G. G. T., Constantinescu, O. D., Berchem, J., Angelopoulos, V., McFadden, J. P., Frey, H. U., Panov, E. V., Volwerk, M. and Bogdanova, Y. V. (2010). Evidence for a flux transfer event generated by multiple X-line reconnection at the magnetopause. *Geophys. Res. Lett.*, **37**, L16101, doi:10.1029/2010GL044219.

Hesse, M. and Kivelson, M. G. (1998). The formation and structure of flux ropes in the magnetotail. In *New Perspectives on the Earth's Magnetotail*, ed. A. Nishida, D. N. Baker and S. W. H. Cowley. Geophysical Monograph 105 Washington, DC: American Geophysical Union, pp. 139–151, doi:10.1029/GM105p0139.

Heyner, D., Nabert, C., Liebert, E. and Glassmeier, K.-H. (2016). Concerning reconnection-induction balance at the magnetopause of Mercury. *J. Geophys. Res. Space Physics*, **121**, 2935–2961, doi:10.1002/2015JA021484.

Hill, T. W., Dessler, A. J. and Wolf, R. A. (1976). Mercury and Mars: The role of ionospheric conductivity in the acceleration of magnetospheric particles. *Geophys. Res. Lett.*, **3**, 429–432.

Ho, G. C., Krimigis, S. M., Gold, R. E., Baker, D. N., Slavin, J. A., Anderson, B. J., Korth, H., Starr, R. D., Lawrence, D. J., McNutt, R. L., Jr. and Solomon, S. C. (2011). MESSENGER observations of transient bursts of energetic electrons in Mercury's magnetosphere. *Science*, **333**, 1865–1868, doi:10.1126/science.1211141.

Ho, G. C., Krimigis, S. M., Gold, R. E., Baker, D. N., Anderson, B. J., Korth, H., Slavin, J. A., McNutt, R. L., Jr., Winslow, R. M. and Solomon, S. C. (2012). Spatial distribution and spectral characteristics of energetic electrons in Mercury's magnetosphere. *J. Geophys. Res.*, **117**, A00M04, doi:10.1029/2012JA017983.

Ho, G. C., Starr, R. D., Krimigis, S. M., Vandegriff, J. D., Baker, D. N., Gold, R. E., Anderson, B. J., Korth, H., Schriver, D., McNutt, R. L., Jr. and Solomon, S. C. (2016). MESSENGER observations of suprathermal electrons in Mercury's magnetosphere. *Geophys. Res. Lett.*, **43**, 550–555, doi:10.1002/2015GL066850.

Hones, E. W., Jr. (1977). Substorm processes in the magnetotail: Comments on 'On hot tenuous plasmas, fireballs, and boundary layers in the Earth's magnetotail' by L. A. Frank, K. L. Ackerson and R. P. Lepping, *J. Geophys. Res.*, **82**, 5633–5640, doi:10.1029/JA082i035p05633.

Hones, E. W., Jr., Baker, D. N., Bame, S. J., Feldman, W. C., Gosling, J. T., McComas, D. J., Zwickl, R. D., Slavin, J. A., Smith, E. J. and Tsurutani, B. T. (1984). Structure of the magnetotail at 220 R_E and its response to geomagnetic activity. *Geophys. Res. Lett.*, **11**, 5–7, doi:10.1029/GL011i001p00005.

Hood, L. and Schubert, G. (1979). Inhibition of solar wind impingement on Mercury by planetary induction currents. *J. Geophys. Res.*, **84**, 2641–2647, doi:10.1029/JA084iA06p02641.

Hoshino, M. (2005). Electron surfing acceleration in magnetic reconnection. *J. Geophys. Res.*, **110**, A10215, doi:10.1029/2005JA011229.

Huang, C.-S., DeJong, A. D. and Cai, X. (2009). Magnetic flux in the magnetotail and polar cap during sawteeth, isolated substorms, and steady magnetospheric convection events. *J. Geophys. Res.*, **114**, A07202, doi:10.1029/2009JA014232.

Ieda, A., Machida, S., Mukai, T., Saito, Y., Yamamoto, T., Nishida, A., Terasawa, T. and Kokubun, S. (1998). Statistical analysis of the plasmoid evolution with Geotail observations. *J. Geophys. Res.*, **103**, 4453–4465, doi:10.1029/97JA03240.

Iijima, T. and Potemra, T. A. (1976). The amplitude distribution of field-aligned currents at northern high latitudes observed by Triad. *J. Geophys. Res.*, **81**, 2165–2174.

Imber, S. M., Slavin, J. A., Auster, U. and Angelopoulos, V. (2011). A THEMIS survey of flux ropes and traveling compression regions: Location of the near-Earth reconnection site during solar minimum. *J. Geophys. Res.*, **116**, A02201, doi:10.1029/2010JA016026.

Imber, S. M., Slavin, J. A., Boardsen, S. A., Anderson, B. J., Korth, H., McNutt, R. L., Jr. and Solomon, S. C. (2014). MESSENGER observations of large dayside flux transfer events: Do they drive Mercury's substorm cycle? *J. Geophys. Res. Space Physics*, **119**, 5613–5623, doi:10.1002/2014JA019884.

Ip, W.-H. (1987). Dynamics of electrons and heavy ions in Mercury's magnetosphere. *Icarus*, **71**, 441–447, doi:10.1016/0019-1035(87)90039-X.

Janhunen, P. and Kallio, E. (2004). Surface conductivity of Mercury provides current closure and may affect magnetospheric symmetry. *Ann. Geophys.*, **22**, 1829–1830, doi:10.5194/angeo-22-1829-2004.

Jia, X., Kivelson, M. G., Khurana, K. K., Linker, J. A. and Walker, R. J. (2010). Dynamics of Ganymede's magnetopause: Intermittent reconnection under steady external conditions. *J. Geophys. Res.*, **115**, A12202, doi:10.1029/2010JA015771.

Jia, X., Daldorff, L. K. S., Gombosi, T. I., van der Holst, B., Slavin, J. A. and Toth, G. (2015). Global MHD simulations of Mercury's magnetosphere with coupled planetary interior: Induction effect of the planetary conducting core on the global interaction. *J. Geophys. Res. Space Physics*, **120**, 4763–4775, doi:10.1002/2015JA021143.

Johnson, C. L., Purucker, M. E., Korth, H., Anderson, B. J., Winslow, R. M., Al Asad, M. M. H., Slavin, J. A., Alexeev, I. I., Phillips, R. J., Zuber, M. T. and Solomon, S. C. (2012). MESSENGER observations of Mercury's magnetic field structure. *J. Geophys. Res.*, **117**, E00L14, doi:10.1029/2012JE004217.

Johnson, C. L., Philpott, L. C., Anderson, B. J., Korth, H., Hauck, S. A., II, Heyner, D., Phillips, R. J., Winslow, R. M. and Solomon, S. C. (2016). MESSENGER observations of induced magnetic fields in Mercury's core. *Geophys. Res. Lett.*, **43**, 2436–2444, doi:10.1002/2015GL067370.

Kabin, K., Gombosi, T. I., DeZeeuw, D. L. and Powell, K. G. (2000). Interaction of Mercury with the solar wind. *Icarus*, **84**, 397–406, doi:10.1006/icar.1999.6252.

Kan, J. R., Zhu, L. and Akasofu, S.-I. (1988). A theory of substorms: Onset and subsidence. *J. Geophys. Res.*, **93**, 5624–5640, doi:10.1029/JA093iA06p05624.

Kepko, L., Glassmeier, K.-H., Slavin, J. A. and Sundberg, T. (2015). Substorm current wedge at Earth and Mercury. In *Magnetotails in the Solar System*, ed. A. Keiling, C. M. Jackman and P. A. Delamere. Hoboken, NJ: John Wiley & Sons, pp. 361–372, doi:10.1002/9781118842324.ch21.

Kidder, A., Winglee, R. M. and Harnett, E. M. (2008). Erosion of the dayside magnetosphere at Mercury in association with ion outflows and flux rope generation. *J. Geophys. Res.*, **113**, A09223, doi:10.1029/2008JA013038.

Klimas, A. (2015). New expression for collisionless magnetic reconnection rate. *Phys. Plasmas*, **22**, 04290, doi:10.1063/1.4917068.

Korth, H., Anderson, B. J., Zurbuchen, T. H., Slavin, J. A., Perri, S., Boardsen, S. A., Baker, D. N., Solomon, S. C. and McNutt, R. L., Jr. (2011a). The interplanetary magnetic field environment at Mercury's orbit. *Planet. Space Sci.*, **59**, 2075–2085, doi:10.1016/j.pss.2010.10.014.

Korth, H., Anderson, B. J., Raines, J. M., Slavin, J. A., Zurbuchen, T. H., Johnson, C. L., Purucker, M. E., Winslow, R. M., Solomon, S. C. and McNutt, R. L., Jr. (2011b). Plasma pressure in Mercury's equatorial magnetosphere derived from MESSENGER Magnetometer observations. *Geophys. Res. Lett.*, **38**, L22201, doi:10.1029/2011GL049451.

Korth, H., Anderson, B. J., Gershman, D. J., Raines, J. M., Slavin, J. A., Zurbuchen, T. H., Solomon, S. C. and McNutt, R. L., Jr. (2014). Plasma distribution in Mercury's magnetosphere derived from MESSENGER Magnetometer and Fast Imaging Plasma Spectrometer observations. *J. Geophys. Res. Space Physics*, **119**, 2917–2932, doi:10.1002/2013JA019567.

Krauss-Varban, D. and Omidi, N. (1991). Structure of medium Mach number quasi-parallel shocks: Upstream and downstream waves. *J. Geophys. Res.*, **96**, 17,715–17,731, doi:10.1029/91JA01545.

Kuo, H., Le, G. and Russell, C. T. (1995). Statistical studies of flux transfer events. *J. Geophys. Res.*, **100**, 3513–3519, doi:10.1029/94JA02498.

Kuznetsova, M. M., Hesse, M., Rastaetter, L., Taktakishvili, A., Toth, G., De Zeeuw, D. L., Ridley, A. and Gombosi, T. I. (2007). Multiscale modeling of magnetospheric reconnection. *J. Geophys. Res.*, **112**, A10210, doi:10.1029/2007JA012316.

Lavraud, B. and Borovsky, J. E. (2008). Altered solar wind–magnetosphere interaction at low Mach numbers: Coronal mass ejections. *J. Geophys. Res.*, **113**, A00B08, doi:10.1029/2008JA013192.

Lawrence, D. J., Anderson, B. J., Baker, D. N., Feldman, W. C., Ho, G. C., Korth, H., McNutt, R. L., Jr., Peplowski, P. N., Solomon, S. C., Starr, R. D., Vandegriff, J. D. and Winslow, R. M. (2015). Comprehensive survey of energetic electron events in Mercury's magnetosphere with data from the MESSENGER Gamma-Ray and Neutron Spectrometer. *J. Geophys. Res. Space Physics*, **120**, 2851–2876, doi:10.1002/2014JA020792.

Le, G., Chi, P. J., Blanco-Cano, X., Boardsen, S. A., Slavin, J. A., Anderson, B. J. and Korth, H. (2013). Upstream ultra-low frequency waves in Mercury's foreshock region: MESSENGER magnetic field measurements. *J. Geophys. Res. Space Physics*, **118**, 2809–2823, doi:10.1002/jgra.50342.

Lee, L. C. and Fu, Z. F. (1985). A theory of magnetic flux transfer at the Earth's magnetopause. *Geophys. Res. Lett.*, **12**, 105–108.

Lepping, R. P., Fairfield, D. H., Jones, J., Frank, L. A., Paterson, W. R., Kokubun, S. and Yamamoto, T. (1995). Cross-tail magnetic flux ropes as observed by the Geotail spacecraft. *Geophys. Res. Lett.*, **22**, 1193–1196, doi:10.1029/94GL01114.

Lepping, R. P., Slavin, J. A., Hesse, M., Jones, J. A. and Szabo, A. (1996). Analysis of magnetotail flux ropes with strong core fields: ISEE 3 observations. *J. Geomag. Geoelectr.*, **48**, 589–601, doi:10.5636/jgg.48.589.

Liljeblad, E., Sundberg, T., Karlsson, T. and Kullen, A. (2015a). Statistical investigation of Kelvin–Helmholtz waves at the magnetopause of Mercury. *J. Geophys. Res. Space Physics*, **119**, 9670–9683, doi:10.1002/2014JA020614.

Liljeblad, E., Karlsson, T., Raines, J. M., Slavin, J. A., Kullen, A., Sundberg, T. and Zurbuchen, T. H. (2015b). MESSENGER observations of the dayside low-latitude boundary layer in Mercury's magnetosphere. *J. Geophys. Res. Space Physics*, **120**, 8387–8400, doi:10.1002/2015JA021662.

Lindsay, S. T., James, M. K., Bunce, E. J., Imber, S. M., Korth, H., Martindale, A. and Yeoman, T. K. (2016). MESSENGER X-ray observations of magnetosphere–surface interaction on the nightside of Mercury. *Planet. Space Sci.*, **125**, 72–79, doi:10.1016/j.pss.2016.03.005i.

Liu, J., Angelopoulos, V., Sibeck, D., Phan, T., Pu, Z. Y., McFadden, J., Glassmeier, K. H. and Auster, H. U. (2008). THEMIS observations of the dayside traveling compression region and flows surrounding flux transfer events. *Geophys. Res. Lett.*, **35**, L17S07, doi:10.1029/2008GL033673.

Lyatsky, W. Khazanov, G. V. and Slavin, J. A. (2010). Alfven wave reflection model of field-aligned currents at Mercury. *Icarus*, **209**, 40–45.

Marsch, E., Mühlhäuser, K.-H., Rosenbauer, H., Schwenn, R. and Neubauer, F. M. (1982). Solar wind helium ions: Observations of the Helios solar probes between 0.3 and 1 AU. *J. Geophys. Res.*, **87**, 35–51, doi:10.1029/JA087iA01p00035.

Massetti, S., Orsini, S., Milillo, A., Mura, A., deAngelis, E., Lammer, H. and Wurz, P. (2003). Mapping of the cusp plasma precipitation on the surface of Mercury. *Icarus*, **166**, 229–237, doi:10.1016/j.icarus.2003.08.005.

Masters, A., Eastwood, J. P., Swisdak, M., Thomsen, M. F., Russell, C. T., Sergis, N. F., Crary, J., Dougherty, M. K., Coates, A. J., Crary, J. and Krimigis, S. M. (2012). The importance of plasma β conditions for magnetic reconnection at Saturn's magnetopause. *Geophys. Res. Lett.*, **39**, L08103, doi:10.1029/2012GL051372.

McPherron, R. L. (1991). Physical processes producing magnetospheric substorms and magnetic storms. In *Geomagnetism*, ed. J. Jacobs. San Diego, CA: Academic Press, pp. 593–739.

McPherron, R. L., Parks, G. K., Colburn, D. S. and Montgomery, M. D. (1973). Satellite studies of magnetospheric substorms on August 15, 1968: 2. Solar wind and outer magnetosphere. *J. Geophys. Res.*, **78**, 3054–3061, doi:10.1029/JA078i016p03054.

Middleton, H., Slavin, J. A., Raines, J. M., Jia, X., Anderson, B., Mays, M. L. and Zurbuchen, T. H. (2014). Mercury's disappearing dayside magnetosphere events (MESSENGER): Evidence for severe dayside erosion and/or compression. Presented at 2014 Fall Meeting, American Geophysical Union, abstract P21C-3919. San Francisco, CA, 15–19 December.

Milan, S. E., Cowley, S. W. H., Lester, M., Wright, D. M., Slavin, J. A., Fillingim, M., Carlson, C. W. and Singer, H. J. (2004). Response of the magnetotail to changes in the open flux content of the magnetosphere. *J. Geophys. Res.*, **109**, A04220, doi:10.1029/2003JA010350.

Moldwin, M. B. and Hughes, W. J. (1992). On the formation and evolution of plasmoids: A survey of ISEE 3 Geotail data. *J. Geophys. Res.*, **97**, 19,259–19,282, doi:10.1029/92JA01598.

Mozer, F. S. and Retinò, A. (2007). Quantitative estimates of magnetic field reconnection properties from electric and magnetic field measurements. *J. Geophys. Res.*, **112**, A10206, doi:10.1029/2007JA012406.

Muller, J., Simon, S., Wang, Y.-C., Motschmann, U., Heyner, D., Schüle, J., Ip, W.-H., Keindienst, G. and Pringle, G. J. (2012). Origin of Mercury's double magnetopause: 3D hybrid simulation with A.I.K.E.F. *Icarus*, **218**, 666–687, doi:10.1016/j.icarus.2011.12.028.

Nagai, T., Fujimoto, M., Saito, Y., Machida, S., Terasawa, T., Nakamura, R., Yamamoto, T. T., Mukai, T., Nishida, A. and Kokubun, S. (1998). Structure and dynamics of magnetic reconnection for substorm onsets with Geotail observations. *J. Geophys. Res.*, **103**, 4419–4440, doi:10.1029/97JA02190.

Nagano, H. (1979). Effect of finite ion Larmor radius on the Kelvin–Helmholtz instability of the magnetopause. *Planet. Space Sci.*, **27**, 881–884, doi:10.1016/0032-0633(79)90013-8.

Nakamura, T., Hasegawa, H. and Shinohara, I. (2010). Kinetic effects on the Kelvin–Helmholtz instability in ion-to-magnetohydrodynamic scale transverse velocity shear layers: Particle simulations. *Phys. Plasmas*, **17**, 042119, doi:10.1063/1.3385445.

Ness, N. F. (1979). The magnetic field of Mercury. *Phys. Earth Planet. Inter.*, **20**, 209–217.

Ness, N. F., Behannon, K. W., Lepping, R. P., Whang, Y. C. and Schatten, K. H. (1974). Magnetic field observations near Mercury: Preliminary results from Mariner 10. *Science*, **185**, 151–185.

Odstrcil, D., Riley, P. and Zhao, X. P. (2004). Numerical simulation of the 12 May 1997 interplanetary CME event. *J. Geophys. Res.*, **109**, A02116, doi:10.1029/2003JA010135.

Ohtani, S., Shay, M. A. and Mukai, T. (2004). Temporal structure of the fast convective flow in the plasma sheet: Comparison between observations and two-fluid simulations. *J. Geophys. Res.*, **109**, A03210, doi:10.1029/2003JA010002.

Øieroset, M., Lin, R. P., Phan, T. D., Larson, D. E. and Bale, S. D. (2002). Evidence for electron acceleration up to ~300 keV in the magnetic reconnection diffusion region of Earth's magnetotail. *Phys. Rev. Lett.*, **89**, 195001, doi:10.1103/PhysRevLett.89.195001.

Paral, J. and Rankin, R. (2013). Dawn–dusk asymmetry in the Kelvin–Helmholtz instability at Mercury. *Nature Commun.*, **4**, 1645, doi:10.1038/ncomms2676.

Paschmann, G., Papamastorakis, I., Baumjohann, W., Carlson, B. U., Lühr, H., Sckope, N. W. and Sonnerup, O. (1986). The magnetopause for large magnetic shear: AMPTE/IRM observations. *J. Geophys. Res.*, **91**, 11,099–11,115, doi:10.1029/JA091iA10p11099.

Pfleger, M., Lichtenegger, H. I. M., Wurtz, P., Lammer, H., Kallio, E., Alho, M., Mura, A., McKenna-Lawlor, S. and Martin-Fernandez, J. A. (2015). 3D-modeling of Mercury's solar wind sputtered surface-exosphere environment. *Planet. Space Sci.*, **115**, 90–101, doi:10.1016/j.pss.2015.04.016.

Poh, G. K., Slavin, J. A., Jia, X., DiBraccio, G. A., Raines, J. M., Imber, S. M., Gershman, D. J., Sun, W., Anderson, B. J., Korth, H., Zurbuchen, T. H., McNutt, R. L., Jr. and Solomon, S. C. (2016). MESSENGER observations of cusp plasma filaments at Mercury. *J. Geophys. Res. Space Physics*, **120**, 8260–8285, doi:10.1002/2016JA022552.

Raeder, J. (2006). Flux transfer events: 1. Generation mechanism for strong southward IMF. *Ann. Geophys.*, **24**, 381–392, doi:10.5194/angeo-24-381-2006.

Raeder, J., Zhu, P., Ge, Y. and Siscoe, G. (2010). Open Geospace General Circulation Model simulation of a substorm: Axial tail instability and ballooning mode preceding substorm onset. *J. Geophys. Res.*, **115**, A00I16, doi:10.1029/2010JA015876.

Raines, J. M., Slavin, J. A., Zurbuchen, T. H. Gloeckler, G., Anderson, B. J., Baker, D. N., Korth, H., Krimigis, S. M. and McNutt, R. L., Jr. (2011). MESSENGER observations of the plasma environment near Mercury. *Planet. Space Sci.*, **59**, 2004–2015, doi:10.1016/j/pss.2011.02.004.

Raines, J. M., Gershman, D. J., Zurbuchen, T. H., Sarantos, M., Slavin, J. A., Gilbert, J. A., Korth, H., Anderson, B. J., Gloeckler, G., Krimigis, S. M., Baker, D. N., McNutt, R. L., Jr. and Solomon, S. C. (2013). Distribution and compositional variations of plasma ions in Mercury's space environment: The first three Mercury years of MESSENGER observations. *J. Geophys. Res. Space Physics*, **118**, 1604–1619, doi:10.1029/2012JA018073.

Raines, J. M., Gershman, D. J., Slavin, J. A., Zurbuchen, T. H., Korth, H., Anderson, B. J. and Solomon, S. C. (2014). Structure and dynamics of Mercury's magnetospheric cusp: MESSENGER measurements of protons and planetary ions. *J. Geophys. Res. Space Physics*, **119**, 6587–6602, doi:10.1002/2014JA020120.

Raines, J. M., DiBraccio, G. A., Cassidy, T. A., Delcourt, D. C., Fujimoto, M., Jia, X., Mangano, V., Milillo, A., Sarantos, M., Slavin, J. A. and Wurz, P. (2015). Plasma sources in planetary magnetospheres: Mercury. *Space Sci. Rev.*, **192**, 91–144, doi:10.1007/s11214-015-0193-4.

Richardson, I. G., Owen, C. J. and Slavin, J. A. (1996). Energetic electron bursts in the deep geomagnetic tail observed by ISEE-3: Association with substorms and magnetotail structures. *J. Geomagnet. Geoelectr.*, **5–6** (48), 657–673.

Rijnbeek, R. P., Cowley, S. W. H., Southwood, D. J. and Russell, C. T. (1984). A survey of dayside flux transfer events observed by ISEE-1 and ISEE-2 magnetometers. *J. Geophys. Res.*, **89**, 786–800, doi:10.1029/JA89iA02p00786.

Russell, C. T. (1989). ULF waves in the Mercury magnetosphere. *Geophys. Res. Lett.*, **16**, 1253–1256, doi:10.1029/GL016i011p01253.

Russell, C. T. and Elphic, R. C. (1978). Initial ISEE magnetometer results: Magnetopause observations. *Space Sci. Res.*, **22**, 681–715, doi:10.1007/BF00212619.

Russell, C. T. and Walker, R. J. (1985). Flux transfer events at Mercury. *J. Geophys. Res.*, **90**, 11067–11074, doi:10.1029/JA090iA11p11067.

Russell, C. T., Baker, D. N. and Slavin, J. A. (1988). The magnetosphere of Mercury. In *Mercury*, ed. F. Vilas, C. R. Chapman and M. S. Matthews. Tucson, AZ: University of Arizona Press, pp. 514–561.

Sanny, J., Beck, C. and Sibeck, D. G. (1998). A statistical study of the magnetic signatures of FTEs near the dayside magnetopause. *J. Geophys. Res.*, **103**, 4683–4692, doi:10.1029/97JA03246.

Sarantos, M. and Slavin, J. A. (2009). On the possible formation of Alfvén wings at Mercury during encounters with coronal mass ejections. *Geophys. Res. Lett.*, **36**, L04107, doi:10.1029/2008GL036747.

Sarantos, M., Killen, R. M. and Kim, D. (2007). Predicting the long-term solar wind ion-sputtering source at Mercury. *Planet. Space Sci.*, **55**, 1584–1595, doi:10.1016/j.pss.2006.10.011.

Sarantos, M., Slavin, J. A., Benna, M., Boardsen, S. A., Killen, R. M., Schriver, D. and Trávníček, P. (2009). Sodium ion pickup observed above the magnetopause during MESSENGER's first Mercury flyby: Constraints on neutral exospheric models. *Geophys. Res. Lett.*, **36**, L04106, doi:10.1029/2008GL036207.

Sarris, E. T., Krimigis, S. M. and Armstrong, T. P. (1976). Observations of magnetospheric bursts of high-energy protons and electrons at ~35 R_E with Imp 7. *J. Geophys. Res.*, **81**, 2341–2355, doi:10.1029/JA081i013p02341.

Schindler, K. (1974). A theory of the substorm mechanism. *J. Geophys. Res.*, **79**, 2803–2810, doi:10.1029/JA079i019p02803.

Schriver, D., Trávníček, P., Ashour-Abdalla, M., Richard, R. L., Hellinger, P., Slavin, J. A., Anderson, B. J., Baker, D. N., Benna, M., Boardsen, S. A., Gold, R. E., Ho, G. C., Korth, H., Krimigis, S. M., McClintock, W. E., McLain, J. L. Orlando, T. M., Sarantos, M., Sprague, A. L. and Starr, R. D. (2011a). Electron transport and precipitation at Mercury during the MESSENGER flybys: Implications for electron-stimulated desorption. *Planet. Space Sci.*, **59**, 2026–2036, doi:10.1016/j.pss.2011.03.008.

Schriver, D., Trávníček, P. M., Anderson, B. J., Ashour-Abdalla, M., Baker, D. N., Benna, M., Boardsen, S. A., Gold, R. E.,

Hellinger, P., Ho, G. C., Korth, H., Krimigis, S. M., McNutt, R. L., Jr., Raines, J. M., Richard, R. L., Slavin, J. A., Solomon, S. C., Starr, R. D. and Zurbuchen, T. H. (2011b). Quasi-trapped ion and electron populations at Mercury. *Geophys. Res. Lett.*, **38**, L23103, doi:10.1029/2011GL049629.

Scurry, L., Russell, C. T. and Gosling, J. T. (1994). Geomagnetic activity and the beta dependence of the dayside reconnection rate. *J. Geophys. Res.*, **99**, 14,811–14,814, doi:10.1029/94JA00794.

Seki, K., Nagy, A., Jackman, C. M., Crary, F., Fontaine, D., Zarka, P., Wurz, P., Milillo, A., Slavin, J. A., Delcourt, D. C., Wiltberger, M., Ilie, R., Jia, X., Ledvina, S. A., Liemohn, M. W. and Schunk, R. W. (2015). A review of general physical and chemical processes related to plasma sources and losses for solar system magnetospheres. *Space Sci. Rev.*, **192**, 27–89, doi:10.1007/s11214-015-0170-y.

Simpson, J. A., Eraker, J. H., Lamport, J. E. and Walpole, P. H. (1974). Electrons and protons accelerated in Mercury's magnetosphere. *Science*, **185**, 160–166.

Siscoe, G. and Christopher, L. (1975). Variations in the solar wind stand-off distance at Mercury. *Geophys. Res. Lett.*, **2**, 158–160, doi:10.1029/GL002i004p00158.

Siscoe, G. L., Ness, N. F. and Yeates, C. M. (1975). Substorms on Mercury? *J. Geophys. Res.*, **80**, 4359–4363, doi:10.1029/JA080i031p04359.

Sitnov, M. I., Swisdak, M. and Divin, A. V. (2009). Dipolarization fronts as a signature of transient reconnection in the magnetotail. *J. Geophys. Res.*, **114**, A04202, doi:10.1029/2008JA013980.

Slavin, J. A. (2004). Mercury's magnetosphere. *Adv. Space Res.*, **33**, 1587–1872, doi:10.1016/j.asr.2003.02.019.

Slavin, J. A. and Holzer, R. E. (1979). The effect of erosion on the solar wind stand-off distance at Mercury. *J. Geophys. Res.*, **84**, 2076–2082, doi:10.1029/JA084iA05p02076.

Slavin, J. A. and Holzer, R. E. (1981). Solar wind flow about the terrestrial planets, 1. Modeling bow shock position and shape. *J. Geophys. Res.*, **86**, 11,401–11,418, doi:10.1029/JA086iA13p11401.

Slavin, J. A., Smith, E. J., Tsurutani, B. T., Sibeck, D. G., Singer, H. J., Baker, D. N., Gosling, J. T., Hones, E. W. and Scarf, F. L. (1984). Substorm associated traveling compression regions in the distant tail: ISEE-3 Geotail observations. *Geophys. Res. Lett.*, **11**, 657–660, doi:10.1029/GL011i007p00657.

Slavin, J. A., Smith, E. J., Sibeck, D. G., Baker, D. N., Zwickl, R. D. and Akasofu, S.-I. (1985). An ISEE-3 study of average and substorm conditions in the distant magnetotail. *J. Geophys. Res.*, **90**, 10,875–10,895, doi:10.1029/JA090iA11p10875.

Slavin, J. A., Lepping, R. P. and Baker, D. N. (1990). IMP-8 observations of traveling compression regions: New evidence for near-Earth plasmoids and neutral lines. *Geophys. Res. Lett.*, **17**, 913–916, doi:10.1029/GL017i007p00913.

Slavin, J. A., Smith, M. F., Mazur, E. L., Baker, D. N., Iyemori, T., Singer, H. J. and Greenstadt, E. W. (1992). ISEE 3 plasmoid and TCR observations during an extended interval of substorm activity. *Geophys. Res. Lett.*, **19**, 825–828, doi:10.1029/92GL00394.

Slavin, J. A., Smith, M. F., Mazur, E. L., Baker, D. N., Hones, E. W., Jr., Iyemori, T. and Greenstadt, E. W. (1993). ISEE 3 observations of traveling compression regions in the Earth's magnetotail. *J. Geophys. Res.*, **98**, 15,425–15,446, doi:10.1029/93JA01467.

Slavin, J. A., Fairfield, D. H., Lepping, R. P., Hesse, M., Ieda, A., Tanskanen, E., Østgaard, N., Mukai, T., Nagai, T., Singer, H. J. and Sutcliffe, P. R. (2002). Simultaneous observations of earthward flow bursts and plasmoid ejection during magnetospheric substorms. *J. Geophys. Res.*, **107**, SMP 13-1–SMP 13-23, doi:10.1029/2000JA003501.

Slavin, J. A., Lepping, R. P., Gjerloev, J., Fairfield, D. H., Hesse, M., Owen, C. J., Moldwin, M. B., Nagai, T., Ieda, A. and Mukai, T. (2003). Geotail observations of magnetic flux ropes in the plasma sheet. *J. Geophys. Res.*, **108**, 1015, doi:10.1029/2002JA009557.

Slavin, J. A., Tanskanen, E., Hesse, M., Owen, C. J., Dunlop, M. W., Imber, S., Lucek, E. A., Balogh, A. and Glassmeier, K.-H. (2005). Cluster observations of traveling compression regions in the near-tail. *J. Geophys. Res.*, **110**, A06207, doi:10.1029/2004JA010878.

Slavin, J. A., Krimigis, S. M., Acuña, M. H., Anderson, B. J., Baker, D. N., Koehn, P. L., Korth, H., Krimigis, S. M., Livi, S., Mauk, B. H., Solomon, S. C. and Zurbuchen, T. H. (2007). MESSENGER at Mercury: Exploring the magnetosphere. *Space Sci. Rev.*, **131**, 133–160, doi:10.1007/s11214-007-9154-x.

Slavin, J. A., Acuña, M. H., Anderson, B. J., Baker, D. N., Benna, M., Gloeckler, G., Gold, R. E., Ho, G. C., Killen, R. M., Korth, H., Krimigis, S. M., McNutt, R. L., Jr., Nittler, L. R., Raines, J. M., Schriver, D., Solomon, S. C., Starr, R. D., Trávníček, P. and Zurbuchen, T. H. (2008). Mercury's magnetosphere after MESSENGER's first flyby. *Science*, **321**, 85–89, doi:10.1126/science.1159040.

Slavin, J. A., Acuña, M. H., Anderson, B. J., Baker, D. N., Benna, M., Boardsen, S. A., Gloeckler, G., Gold, R. E., Ho, G. C., Korth, H., Krimigis, S. M., McNutt, R. L., Jr., Raines, J. M., Sarantos, M., Schriver, D., Solomon, S. C., Trávníček, P. and Zurbuchen, T. H. (2009a). MESSENGER observations of magnetic reconnection in Mercury's magnetosphere. *Science*, **324**, 606–610, doi:10.1126/science.1172011.

Slavin, J. A., Anderson, B. J., Zurbuchen, T. H., Baker, D. N., Krimigis, S. M., Acuña, M. H., Benna, M., Boardsen, S. A., Gloeckler, G., Gold, R. E., Ho, G. C., Korth, H., McNutt, R. L., Jr., Raines, J. M., Sarantos, M., Schriver, D., Solomon, S. C. and Trávníček, P. (2009b). MESSENGER observations of Mercury's magnetosphere during northward IMF. *Geophys. Res. Lett.*, **36**, L02101, doi:10.1029/2008GL036158.

Slavin, J. A., Anderson, B. J., Baker, D. N., Benna, M., Boardsen, S. A., Gloeckler, G., Gold, R. E., Ho, G. C., Korth, H., Krimigis, S. M., McNutt, R. L., Jr., Nittler, L. R., Raines, J. M., Sarantos, M., Schriver, D., Solomon, S. C., Starr, R. D., Trávníček, P. M. and Zurbuchen, T. H. (2010a). MESSENGER observations of extreme loading and unloading of Mercury's magnetic tail. *Science*, **329**, 665–668, doi:10.1126/science.1188067.

Slavin, J. A., Lepping, R. P., Wu, C.-C., Anderson, B. J., Baker, D. N., Benna, M., Boardsen, S. A., Killen, R. M., Korth, H., Krimigis, S. M., McClintock, W. E., McNutt, R. L., Jr., Sarantos, M., Schriver, D., Solomon, S. C., Trávníček, P. and Zurbuchen, T. H. (2010b). MESSENGER observations of large flux transfer events at Mercury. *Geophys. Res. Lett.*, **37**, L02105, doi:10.1029/2009GL041485.

Slavin, J. A., Anderson, B. J., Baker, D. N., Benna, M., Boardsen, S. A., Gold, R. E., Ho, G. C., Imber, S. M., Korth, H., Krimigis, S. M., McNutt, R. L., Jr., Raines, J. M., Sarantos, M., Schriver, D., Solomon, S. C., Trávníček, P. and Zurbuchen, T. H. (2012a). MESSENGER and Mariner 10 flyby observations of magnetotail structure and dynamics at Mercury. *J. Geophys. Res.*, **117**, A01215, doi:10.1029/2011JA016900.

Slavin, J. A., Imber, S. M., Boardsen, S. A., DiBraccio, G. A., Sundberg, T., Sarantos, M., Nieves-Chinchilla, T., Szabo, A., Anderson, B. J., Korth, H., Zurbuchen, T. H., Raines, J. M., Johnson, C. L., Winslow, R. M. Killen, R. M., McNutt, R. L., Jr. and Solomon, S. C. (2012b). MESSENGER observations of a flux-transfer-event shower at Mercury. *J. Geophys. Res.*, **117**, A00M06, doi:10.1029/2012JA017926.

Slavin, J. A., DiBraccio, G. A., Gershman, D. J., Ho, G., Imber, S. M., Poh, G. K., Raines, J. M., Zurbuchen, T. H., Jia, X., Baker, D. N., Glassmeier, K.-H., Livi, S. A., Boardsen, S. A., Cassidy, T. A., Sarantos, M., Sundberg, T., Masters, A., Johnson, C. L., Winslow, R. M., Anderson, B. J., Korth, H., McNutt, R. L., Jr. and Solomon, S. C. (2014). MESSENGER observations of Mercury's dayside magnetosphere under extreme solar wind conditions. *J. Geophys. Res. Space Physics*, **119**, 8087–8116, doi:10.1002/2014JA020319.

Smith, D. E., Zuber, M. T., Phillips, R. J., Solomon, S. C., Hauck, S. A., II, Lemoine, F. G., Mazarico, E., Neumann, G. A., Peale, S. J., Margot, J.-L., Johnson, C. L., Torrence, M. H., Perry, M. E., Rowlands, D. D., Goossens, S., Head, J. W. and Taylor, A. H. (2012). Gravity field and internal structure of Mercury from MESSENGER. *Science*, **336**, 214–217.

Solomon, S. C., McNutt, R. L., Jr., Gold, R. E. and Domingue, D. L. (2007). MESSENGER mission overview. *Space Sci. Rev.*, **131**, 3–39, doi:10.1007/s11214-007-9247-6.

Sonnerup, B. U. Ö. (1974). Magnetopause reconnection rate. *J. Geophys. Res.*, **79**, 1546–1549, doi:10.1029/JA079i010p01546.

Sonnerup, B. U. Ö, Papmastorakis, I., Paschmann, G. and Lühr, H., (1990). The magnetopause for large magnetic shear: Analysis of convection electric fields from AMPTE/IRM. *J. Geophys. Res.*, **95**, 10,541–10,557, doi:10.1029/JA095iA07p10541.

Spreiter, J. R., Summers, A. L. and Alksne, A. Y. (1966). Hydromagnetic flow around the magnetosphere. *Planet. Space Sci.*, **14**, 223–253, doi:10.1016/0032-0633(66)90124-3.

Starr, R. D., Schriver, D., Nittler, L. R., Weider, S. Z., Byrne, P. K., Ho, G. C., Rhodes, E. A., Schlemm, C. E., II, Solomon, S. C. and Trávníček, P. M. (2012). MESSENGER detection of electron-induced X-ray fluorescence from Mercury's surface. *J. Geophys. Res.*, **117**, E00L02, doi:10.1029/2012JE004118.

Suess, S. T. and Goldstein, B. E. (1979). Compression of the Hermaean magnetosphere by the solar wind. *J. Geophys. Res.*, **84**, 3306–3312, doi:10.1029/JA084iA07p03306.

Sun, W.-J., Slavin, J. A., Fu, S., Raines, J. M., Zong, Q.-G., Imber, S. M., Shi, Q., Yao, Z., Poh, G. K., Gershman, D. J., Pu, Z., Sundberg, T., Anderson, B. J., Korth, H. and Baker, D. N. (2015a). MESSENGER observations of magnetospheric substorm activity in Mercury's near magnetotail. *Geophys. Res. Lett.*, **42**, 3692–3699, doi:10.1002/2015GL064052.

Sun, W.-J., Slavin, J. A., Fu, S., Raines, J. M., Sundberg, T., Zong, Q.-G., Jia, X., Shi, Q., Shen, X., Poh, G., Pu, Z. and Zurbuchen, T. H. (2015b). MESSENGER observations of Alfvénic and compressional waves during Mercury's substorms. *Geophys. Res. Lett.*, **42**, 6189–6198, doi:10.1002/2015GL065452.

Sundberg, T. and Slavin, J. A. (2015). Mercury's magnetotail. In *Magnetotails in the Solar System*, ed. A. Keiling, C. M. Jackman and P. A. Delamere. Hoboken, NJ: John Wiley & Sons, pp. 23–42, doi:10.1002/9781118842324.ch2.

Sundberg, T., Boardsen, S. A., Slavin, J. A., Blomberg, L. G. and Korth, H. (2010). The Kelvin–Helmholtz instability at Mercury: An assessment. *Planet. Space Sci.*, **58**, 1434–1441, doi:10.1016/j.pss.2010.06.008.

Sundberg, T., Boardsen, S. A., Slavin, J. A., Anderson, B. J., Korth, H., Zurbuchen, T. H., Raines, J. M. and Solomon, S. C. (2012a). MESSENGER orbital observations of large-amplitude Kelvin–Helmholtz waves at Mercury's magnetopause. *J. Geophys. Res.*, **117**, A04216, doi:10.1029/2011JA017268.

Sundberg, T., Slavin, J. M., Boardsen, S. A., Anderson, B. J., Korth, H., Ho, G. C., Schriver, D., Uritsky, V. M., Zurbuchen, T. H., Raines, J. M., Baker, D. N., Krimigis, S. M., McNutt, R. L., Jr. and Solomon, S. C. (2012b). MESSENGER observations of dipolarization events in Mercury's magnetotail. *J. Geophys. Res.*, **117**, A00M03, doi:10.1029/2012JA017756.

Sundberg, T., Boardsen, S. A., Slavin, J. A., Uritsky, V. M., Anderson, B. J., Korth, H., Gershman, D. J., Raines, J. M., Zurbuchen, T. H. and Solomon, S. C. (2013). Cyclic reformation of a quasi-parallel bow shock at Mercury: MESSENGER observations. *J. Geophys. Res. Space Physics*, **118**, 6457–6464, doi:10.1002/jgra.50602.

Sundberg, T., Boardsen, S. A., Burgess, D. and Slavin, J. A. (2015). Coherent wave activity in Mercury's magnetosheath. *J. Geophys. Res. Space Physics*, **120**, 7342–7356, doi:10.1002/2015JA021499.

Taguchi, S., Slavin, J. A., Kiyohara, M., Nose, M., Reeves, G. and Lepping, R. P. (1998). Temporal relationship between mid-tail TCRs and substorm onset: Evidence for NENL formation in the late growth phase. *J. Geophys. Res.*, **103**, 26,607–26,612, doi:10.1029/98JA02617.

Tanskanen, E. I. (2009). A comprehensive high-throughput analysis of substorms observed by IMAGE magnetometer network: Years 1993–2003 examined. *J. Geophys. Res.*, **114**, A05204, doi:10.1029/2008JA013682.

Toth, G. and Odstrcil, D. (1996). Comparison of some flux corrected transport and total variation diminishing numerical schemes for hydrodynamic and magnetohydrodynamic problems. *J. Comput. Phys.*, **128**, 82–100.

Trávníček, P. M., Hellinger, P., Schriver, D., Herčík, D., Slavin, J. A. and Anderson, B. J. (2009). Kinetic instabilities in Mercury's magnetosphere: Three-dimensional simulation results. *Geophys. Res. Lett.*, **36**, L07104, doi:10.1029/2008GL036630.

Trávníček, P. M., Schriver, D., Hellinger, P., Herčík, D., Anderson, B. J., Sarantos, M. and Slavin, J. A. (2010). Mercury's magnetosphere–solar wind interaction for northward and southward interplanetary magnetic field: Hybrid simulations. *Icarus*, **209**, 11–22, doi:10.1016/j.icarus.2010.01.008.

Trenchi, L., Marcucci, M. F., Rème, H., Carr, C. M. and Cao, J. B. (2011). TC-1 observations of a flux rope: Generation by multiple X line reconnection. *J. Geophys. Res.*, **116**, A05202, doi:10.1029/2010JA015986.

Uritsky, V. M., Slavin, J. A., Khazanov, G. V., Donovan, E. F., Boardsen, S. A., Anderson, B. J. and Korth, H. (2011). Kinetic-scale magnetic turbulence and finite Larmor radius effects at Mercury. *J. Geophys. Res.*, **116**, A09236, doi:10.1029/2011JA016744.

Uritsky, V. M., Slavin, J. A., Boardsen, S. A., Sundberg, T., Raines, J. M., Gershman, D. J., Collinson, G., Sibeck, D., Khazanov, G. V., Anderson, B. J. and Korth, H. (2014). Active current sheets and candidate hot flow anomalies upstream of Mercury's bow shock. *J. Geophys. Res. Space Physics*, **119**, 853–876, doi:10.1002/2013JA019052.

Varela, J., Pantellini, F. and Moncuquet, M. (2015). The effect of interplanetary magnetic field orientation on the solar wind flux impacting Mercury's surface. *Planet. Space Sci.*, **119**, 264–269, doi:10.1016/l.pss.2015.10.004.

Vasyliunas, V. M. (2004). Comparative magnetospheres: Lessons from Earth. *Adv. Space Res.*, **33**, 2113–2120, doi:10.1016/j.asr.2003.04.051.

Walsh, B. M., Ryou, A. S., Sibeck, D. G. and Alexeev, I. I. (2013). Energetic particle dynamics in Mercury's magnetosphere. *J. Geophys. Res. Space Physics*, **118**, 1992–1999, doi:10.1002/jgra.50266.

Wang, Y. C., Motschmann, U., Müller, J. and Ip, W. H. (2010). A hybrid simulation of Mercury's magnetosphere for the MESSENGER encounters in year 2008. *Icarus*, **209**, 46–52.

Winglee, R. M., Harnett, E. M. and Kidder, A. (2009). Relative timing of substorm processes as derived from multifluid/multiscale simulations: Internally driven substorms. *J. Geophys. Res.*, **114**, A09213, doi:10.1029/2008JA013750.

Winslow, R. M., Johnson, C. L., Anderson, B. J., Korth, H., Slavin, J. A., Purucker, M. E. and Solomon, S. C. (2012). Observations of Mercury's northern cusp region with MESSENGER's Magnetometer. *Geophys. Res. Lett.*, **39**, L08112, doi:10.1029/2012GL051472.

Winslow, R. M., Anderson, B. J., Johnson, C. L., Slavin, J. A., Korth, H., Purucker, M. E., Baker, D. N. and Solomon, S. C. (2013). Mercury's magnetopause and bow shock from MESSENGER observations. *J. Geophys. Res. Space Physics*, **118**, 2213–2227, doi:10.1002/jgra.50237.

Zhong, J., Wan, W. X., Slavin, J. A., Wei, Y., Lin, R. L., Chai, L. H., Raines, J. M., Rong, Z. J. and Han, X. H. (2015a). Mercury's three-dimensional asymmetric magnetopause. *J. Geophys. Res. Space Physics*, **120**, 7658–7671, doi:10.1002/2015JA021425.

Zhong, J., Wan, W. X., Wei, Y., Slavin, J. A., Raines, J. M., Rong, Z. J., Chai, L. H. and Han, X. H. (2015b). Compressibility of Mercury's dayside magnetosphere. *Geophys. Res. Lett.*, **42**, 10,135–10,139, doi:10.1002/2015GL067063.

Zong, Q.-G., Fritz, T. A., Spence, H., Zhang, H., Huang, Z. Y., Pu, Z. Y., Glassmeier, K.-H., Korth, A., Daly, P. W., Balogh, A. and Reme, H. (2005). Plasmoid in the high latitude boundary/cusp region observed by Cluster. *Geophys. Res. Lett.*, **32**, L01101, doi:10.1029/2004GL020960.

Zurbuchen, T. H., Raines, J. M., Gloeckler, G., Krimigis, S. M., Slavin, J. A., Koehn, P. L., Killen, R. M., Sprague, A. L., McNutt, R. L., Jr. and Solomon, S. C. (2008). MESSENGER observations of the compositions of Mercury's ionized exosphere and plasma environment. *Science*, **321**, 90–92, doi:10.1126/science.1159314.

Zurbuchen, T. H., Raines, J. M., Slavin, J. A., Gershman, D. J., Gilbert, J. A., Gloeckler, G., Anderson, B. J., Banker, D. N., Korth, H., Krimigis, S. M., Sarantos, M., Schriver, D., McNutt, R. L., Jr. and Solomon, S. C. (2011). MESSENGER observations of the spatial distribution of planetary ions near Mercury. *Science*, **333**, 1862, doi:10.1126/science.1211302.

Zwan, B. J. and Wolf, R. A. (1976). Depletion of solar wind plasma near a planetary boundary. *J. Geophys. Res.*, **81**, 1636–1648, doi:10.1029/JA081i010p01636.

18

The Elusive Origin of Mercury

DENTON S. EBEL AND SARAH T. STEWART

18.1 INTRODUCTION

The MESSENGER mission to Mercury was driven by several scientific goals. A principal question was "What planetary formational processes led to Mercury's high ratio of metal to silicate?" (Solomon et al., 2001, 2007; Chapter 1). This ratio is the prime anomaly for Mercury relative to other planetary bodies, but by no means the only one. In this chapter, we review Mercury's anomalous chemistry in the context of astrophysical disk processes and planet formation as currently understood. We examine previous and current hypotheses for Mercury's origin in light of the new data from MESSENGER. Finally, we discuss the potential for Mercury-like planets in the rapidly emerging catalog of extrasolar planetary systems. A discussion of the evolution of Mercury as a planet, focusing on processes occurring after many or all of its anomalous properties were established, is given in Chapter 19.

18.2 HOW ANOMALOUS IS MERCURY?

By far the smallest terrestrial planet, Mercury is about half the mass of Mars. Mercury is also locked in a 3:2 spin–orbit resonance with the Sun. These factors may have affected Mercury's collisional environment (Chapter 9). Here, we focus on data returned by the MESSENGER mission that may provide insights into Mercury's formation. A surprising result from the MESSENGER mission is that Mercury's large ratio of metal to silicate is not accompanied by extreme depletions in volatile elements compared with the other terrestrial planets. The planet's mantle is much more chemically reduced than Earth's mantle, and also compared with carbonaceous chondrite meteorites, the usual solar system standard. These MESSENGER findings have stimulated many experimental studies bearing on the possible S and Si abundances in Mercury's large core. We begin by summarizing the evidence returned by the MESSENGER mission in the context of chemical anomalies.

18.2.1 Density

It has long been known that Mercury's zero-pressure, or uncompressed, density is significantly greater than that of Earth – 5000 versus 4400 kg/m^3 according to Urey (1950), and 5300 versus 4100 kg/m^3 according to Mahoney (2014) – and greater than the grain density of most meteorite parent bodies (Figure 18.1). It can be assumed that Mercury's density is attributable to a high abundance of iron and heavy, siderophile (iron-loving) elements, relative to silicon, magnesium, and other lithophile (rock-loving) elements. Upon planetary differentiation, the siderophile elements accompany iron that sinks to form the planetary core, and the lithophile elements form the mantle (Righter, 2003). Given that Mercury differentiated in the same manner as other planets, it is, therefore, the core/mantle mass ratio that is anomalous. The metal/silicate or Fe/Si ratios of primitive meteorites may be considered reference points for processes that led to Mercury's high density (Chapter 2).

In 1988, Chapman noted that the origin of Mercury was "the least well-understood topic" in the compendium of Mercury knowledge (Chapman, 1988). He divided solutions to the origin of Mercury's metal-rich composition into two categories. Either the composition reflects primordial radial gradients of "orderly" chemical fractionation and dynamic processes, or it results from "chaotic" evolutionary processes that catastrophically overprinted primordial signatures (Wetherill, 1994). Indeed, the relative roles of such processes have implications for the origin of all the planets in this and other planetary systems.

The CI chondrite meteorites contain most non-volatile elements in ratios (relative to Si or Mg) nearly identical to those measured by spectroscopy in the solar photosphere, which is thought to represent the bulk composition of the solar system (Lodders, 2003; Lodders et al., 2009; Sneden et al., 2009). Approximately chondritic ratios of the major rock-forming elements (Si, Mg, Fe, Ca, Al, and Ni) characterize Venus, Earth, and Mars (Righter et al., 2006). A rough calculation shows that if Mercury were once a differentiated planet with near-chondritic bulk chemistry it must have lost a substantial mass of silicate mantle to yield its present density. The thickness of a shell of such additional silicate would be ~1480 km to yield a present Mercury radius of 2440 km with 74% of its mass in a metallic, Si-free core. An initially chondritic Mercury would have been about 530 km larger in radius than Mars (3389.5 km mean radius) prior to stripping of ~64% of its mass, all from its mantle. Alternatively, Mercury was never chondritic in major elements and owes its anomalous density to as yet unknown nebular fractionation processes.

18.2.2 FeO-Free Silicates

The second major chemical anomaly for Mercury is the planet's extremely reduced nature. Here, we compare Mercury to the oxidation/reduction, or redox, state of other solar system bodies and materials. Earth-based astronomical measurements of

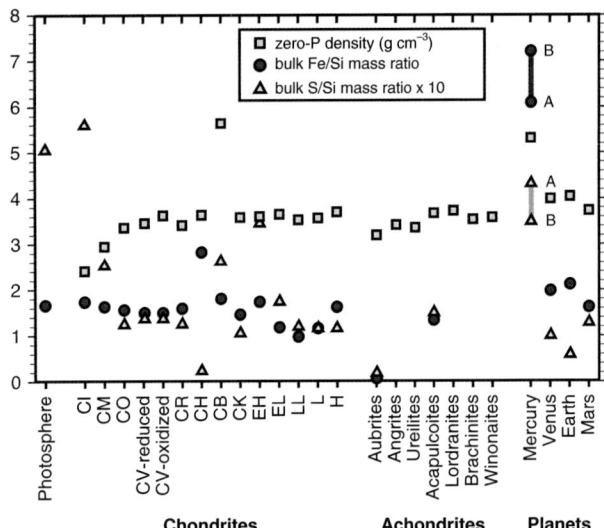

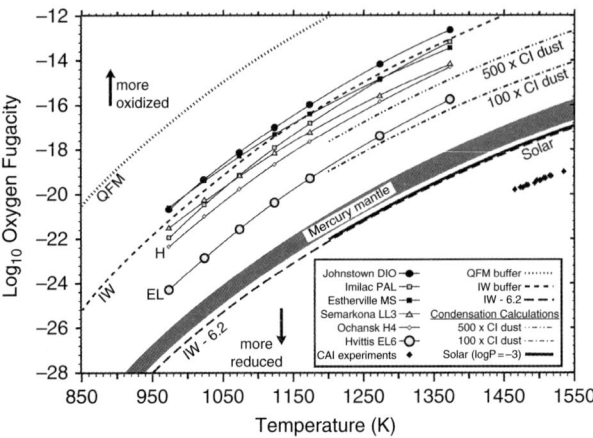

Figure 18.1. Mean grain densities of chondrites and achondrites and zero-pressure densities of planets (squares), bulk Fe/Si mass ratios (circles), and bulk S/Si mass ratios (multiplied by 10, triangles). Grain density is the bulk material density in the absence of porosity. Mercury values are calculated as described in the text, showing ranges (vertical bars) for Fe/Si and S/Si for cases A and B, as marked. Density data are from Macke (2012, meteorites) and Lodders and Fegley (1998, planets). Fe/Si data are from Wasson and Kallemeyn (1988): CM, CO, CV, EH, EL, LL, L, H; Jarosewich (1990): CB; Lodders (2003): solar photosphere; Lodders et al. (2009): CI; Lodders and Fegley (1998): CR, CH, CK, Earth, aubrites, acapulcoites, Venus, Mars.

Figure 18.2. Intrinsic oxygen fugacity (fO_2) for selected meteorites (Brett and Sato, 1984: LL3, H4, EL6 chondrites; Hewins and Ulmer, 1984: mesosiderites MS, diogenites DIO; Righter et al., 1990: pallasites PAL) and during condensation of possible early disk vapors (solar composition and solar composition enriched in CI dust by factors of 500 and 1000), all at 100 Pa (10^{-3} bar) total pressure (Ebel and Grossman, 2000). Buffer curves are for quartz–fayalite–magnetite (QFM), iron–wüstite (Fe–FeO, IW), and IW−6.2, which coincides with the calculated solar condensation curve. Condensation curves are nearly independent of the total pressure of the calculation (Ebel and Grossman, 2000). Results from the low-fO_2 experiments of Beckett (1986) are denoted by solid diamonds. The band shows the range (IW−5.4 ± 0.4) of fO_2 estimated by Namur et al. (2016) for mantle source regions of >75% of mercurian lavas, on the basis of experiments.

Mercury's spectral reflectance have long indicated a very low FeO content in surface silicates (Vilas, 1988; Sprague and Roush, 1998; Robinson and Taylor, 2001; Warell and Blewett, 2004). Microwave emissivity measurements (Mitchell and de Pater, 1994) precluded the possibility of widespread reduction to iron nanophases by space weathering (Domingue et al., 2014). The MESSENGER mission has confirmed the extreme depletion of Fe in any form on Mercury's surface and also in material excavated from depth (Murchie et al., 2015; Chapter 8). Indeed, the common crustal minerals olivine, pyroxene, and feldspar lack sufficient Fe^{2+} to yield spectral features, indicating that Fe is present in reduced form as metal or sulfides (Klima et al., 2013; Izenberg et al., 2014; Chapter 8). Calcium in most terrestrial and extraterrestrial rocks forms oxides, but a positive correlation between Ca and S in the measurements by MESSENGER's X-Ray Spectrometer (XRS) suggests the occurrence of CaS (oldhamite) on Mercury's surface (Weider et al., 2012).

The chemical availability of oxygen, known as the oxygen fugacity $f(O_2)$, is commonly described on a logarithmic scale relative to the $f(O_2)$ of a standard equilibrium reaction buffer, e.g., between iron metal and wüstite (~FeO), which is denoted by IW (Figure 18.2). As $f(O_2)$ decreases in an otherwise closed chemical system, more Si can dissolve into molten or solid Fe-rich metal, and more S can dissolve in silicate melts. From a variety of criteria based primarily on experimental metal–silicate partitioning and the MESSENGER XRS measurements of surface S and Fe abundances (Nittler et al., 2011), McCubbin et al. (2012) determined that the oxygen fugacity of Mercury's interior lies between a low of IW−6.3 and a high of IW−2.6, that is, $10^{-2.6}$ to $10^{-6.3}$ below IW. The high end of this range is an extremely generous upper limit (cf. Zolotov et al., 2013). The results from MESSENGER also stimulated other experiments that led to similar findings. Namur et al. (2016) parameterized S and metal solubility in magma compositions representative of mercurian lavas, and given mantle/core equilibrium they calculated $f(O_2)$ = IW−5.4 ± 0.4 ($10^{-5.0}$ to $10^{-5.8}$ below IW; Figure 8.2). Earth, on the other hand, was not as reduced during its accretion (e.g., Frost et al., 2008), and modern mid-ocean ridge basalts record upper mantle $f(O_2)$ at 10^2 above IW (IW+2), near the quartz–fayalite–magnetite (QFM) buffer (Cottrell and Kelley, 2011; Figure 8.2). The mantle source of venusian lavas is inferred to have fO_2 similar to Earth's upper mantle (Wadhwa, 2008).

The known or inferred $f(O_2)$ values for a broad range of solar system materials are summarized in Figure 18.2. Mercury is the most reduced planet and more reduced than all measured early solar system materials except, possibly, the enstatite chondrite and enstatite achondrite meteorites (aubrites) and a small class of Ca-, Al-rich inclusions (CAIs) in some chondritic meteorites (Figure 18.2; Beckett, 1986). Recent geochemical thermodynamic modeling suggests that, as Earth and Mars accreted, their oxidation states progressively increased (Righter et al., 2008., 2016; Wood et al., 2009; Badro et al., 2015; Rubie et al., 2015). Earth's lower mantle, Mars, and Vesta (represented by diogenite meteorites) all record mantle $f(O_2)$ near the iron–wüstite buffer curve (Figure 18.2) (Ghosal et al., 1998; Wadhwa, 2001; Frost et al., 2008; Szymanski et al., 2010; Tuff et al., 2013). The FeO content of silicates and the Si content of metals are proxies for the $f(O_2)$ during their formation. The silicates in pallasite meteorites record $f(O_2)$ similar to that in Earth's lower mantle (Figure 18.2), with olivine, $(Mg,Fe)_2SiO_4$, ranging from 10 to 20 mol% Fe_2SiO_4 (11–19 wt% FeO) (Righter et al., 1990) and very low Si in metal. The most Si-rich iron meteorite, Horse Creek, contains 2.5 wt% Si in metal (Buchwald, 1975). Thus, the meteorites that represent cores or lower mantles of early differentiated planetesimals all record higher $f(O_2)$ than Mercury.

Intrinsic fugacities, inferred from thermodynamic analyses of measured mineral compositions, have been calculated for some chondrite classes. The intrinsic $f(O_2)$ measured for the ordinary chondrites least equilibrated on their parent bodies (LL3, H4) range as low as IW−1; however, the equilibrated enstatite chondrite Hvittis (EL6) reaches IW−3 (Figure 18.2; Brett and Sato, 1984). The reduced, metal-rich CH and CB chondrites contain primarily <5 wt% FeO in olivine (Weisberg et al., 2001), and the CH chondrites contain metal grains, a very few of which reach 8 wt% Si (Weisberg et al., 1988). The enstatite achondrites (aubrites) are similar to the enstatite chondrites in their degree of reduction; for example, Mount Egerton contains 2.1 wt% Si in kamacite metal (Wasson and Wai, 1970). A new subgroup of metal-rich chondrites has recently been described, with ~22 vol.% metal and nearly FeO-free silicates (Weisberg et al., 2015). The equilibrated enstatite chondrites contain ubiquitous Si-bearing metal (~3.2 wt% in EH4–5, ~1.6 wt% in EL6), no olivine, and <0.9 wt% FeO in enstatite (Keil, 1968; Weisberg and Kimura, 2012, their Figure 5). The higher Si content of metal in equilibrated EH chondrites indicates that they are more chemically reduced than the EL6 chondrites. Of all the rocky bodies in the solar system, only some enstatite chondrites and Mercury approach the low $f(O_2)$ of a vapor of solar composition that condenses solids along a buffer curve of IW−6.2 (Ebel and Grossman, 2000).

Vigarano-like carbonaceous (CV) chondrites contain once-melted (igneous) CAIs that crystallized Al-, Ti-rich calcic pyroxene (fassaite) containing Ti^{3+} (Simon et al., 2007). Experiments by Beckett (1986) constrained the stability conditions of such pyroxenes in CAI-like melts to $\log_{10} f(O_2)$ ~ IW−8 at temperature T in the range 1470 K < T < 1540 K (Figure 18.2). The substantial Ti^{3+} in these pyroxenes indicates highly reduced crystallization conditions in their parental melts, which also record the oldest radiometric ages of all solar system materials (Russell et al., 2006). These CAIs are thought to have formed near the Sun, in highly reducing environments, and then been dynamically transported outward to accrete with more oxidized chondritic material (e.g., Krot et al., 2009; Brownlee, 2014). The host CV chondrites for these rare pyroxene-rich igneous CAIs are younger and not especially reduced, and it is only the CAIs in these meteorites that record such low f(O_2) values. These CAIs offer tantalizing evidence that local regions in the earliest nebula were highly reduced.

In summary, there exist reservoirs of undifferentiated early solar system material that record highly reduced conditions during formation and accretion of solids in the early nebula. A preponderance of evidence indicates that these materials formed by gradual, orderly processes. The vast majority of sampled materials, however, record much more oxidizing conditions of formation, accretion, and differentiation than does Mercury.

18.2.3 Elemental Abundances

The primary constraints on Mercury's formation come from compositional information (Lewis, 1988; Boynton et al., 2007; Chapter 2) and inferred internal structure (Chapter 4). Before MESSENGER, our best estimate of Mercury's bulk composition was largely based on a theory of planet formation in which volatility varied with radial distance from the early Sun (Morgan and Anders, 1980; Lodders and Fegley, 1998). MESSENGER has changed the equations markedly. While the MESSENGER mission has established tighter bounds on Mercury's reduced mantle, it has also shown that Mercury is not anomalous in several important ways. The less volatile lithophile elements Si, Ca, Al, and Mg are all broadly chondritic in the crust (Weider et al., 2015; Chapter 2). Crustal values must be extrapolated to a bulk planetary composition, as is done to establish model bulk compositions for the other terrestrial planets.

The elemental ratios in Figures 18.1 and 18.3 are derived from the mean mantle source compositions inferred for the source regions of magmas that solidified to form the northern smooth plains (NSP) and intercrater plains and heavily cratered terrain (IcP-HCT) and given in Chapter 2 (Table 2.2). Two cases are plotted. Case A is consistent with Chapter 2, adding 7 wt% S and 2 wt% C (Section 2.6; Namur et al., 2016) to a 340-km-thick mantle shell of the mean NSP plus IcP-HCT "mass balance" composition (Table 2.2), with an FeS lower mantle shell of 80 km thickness and a 2020-km-radius core with 5% Ni, 1.5% S, 0.5% C, 8 wt% Si, and the remainder Fe (Table 2.3; Hauck et al., 2013; Chapter 4). Case B replaces all but 0.01 km of the FeS shell with the same mantle as A and sequesters 3 wt% S and only 4 wt% Si in the core. Case B brings the elemental Si/Mg and S/Mg ratios to near EH chondrite levels and also decreases Fe/Mg. This calculation may be within the uncertainty regarding (1) the existence of an FeS shell, (2) Mercury's internal thermal profile, and (3) the olivine/pyroxene ratio of the mantle source, assumed to be near the low end of the range for lherzolites in Chapter 2. The calculation reveals another significant compositional anomaly, the high planetary Si/Mg mass ratio caused even by a conservative inference of only 4 wt% Si in the core. However, current estimates of the light elements in the core are based on single-stage core formation models under the assumption of thermodynamic equilibrium between core and mantle. More complete core formation models should reexamine the implications of a high bulk Si/Mg ratio for Mercury.

Earth, most primitive meteorites, and probably Venus, if not Mars, are all depleted in sulfur and other volatile elements relative to the chondritic (CI) reference composition. The average surface S/Si mass ratio of Mercury (0.092 ± 0.015, Evans et al., 2012) is comparable to or higher than bulk Earth estimates. The volatile element depletion of the bulk Earth (dashed line in Figure 18.4) is known from the lithophile element abundances in mantle-derived rocks. Earth's bulk planetary S/Mg atomic ratio is inferred to be ~7% of chondritic (relative atomic abundance of 1 in Figure 18.4), with most S thought to be in the core, which is why S plots so far below the dashed line. By contrast, experimental and thermodynamic constraints (Namur et al., 2016) indicate that Mercury's bulk S/Mg mass ratio

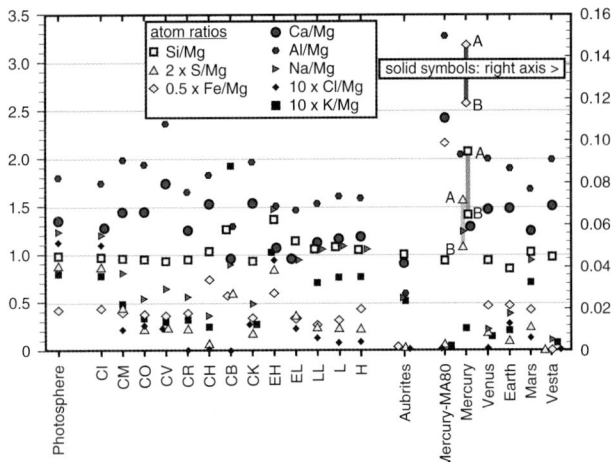
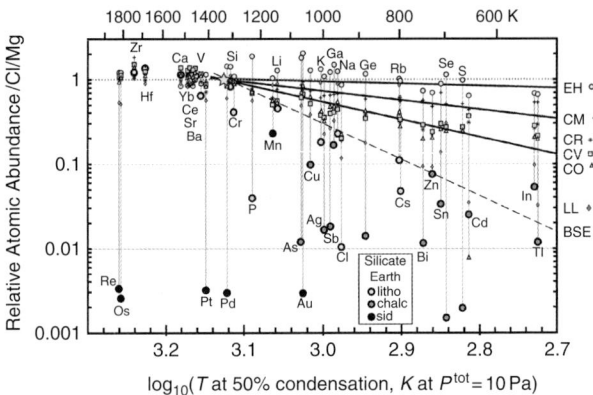

Figure 18.3. Mass ratios of lithophile elements in the solar photosphere, meteorites, and selected planets. Mercury values are calculated as described in the text, showing ranges (vertical bars) for Fe/Mg, S/Mg, and Si/Mg for cases A and B, as marked. "Mercury-MA80" and Venus values are from Morgan and Anders (1980). Earth and Mars are from Lodders and Fegley (1998; after Kargel and Lewis, 1993; and Lodders and Fegley, 1997; respectively). Other data sources are as in Figure 18.1. Si/Mg, 2 × S/Mg, and 0.5 × Fe/Mg (open symbols) refer to the left-hand axis. No Cl/Mg ratios for bulk CB, aubrites, or Mercury were calculated.

Figure 18.4. Volatile depletion of solar system reservoirs. Bulk silicate Earth (BSE, large circles), LL (diamonds), CO (triangles), CV (squares), CR (asterisks), CM (crosses), and EH (small circles) chondrite atomic abundances of selected elements, normalized to Mg (star) and CI chondrite (dotted line at a relative atomic abundance of 1) are plotted against the temperature $T_{50\%}$ at which they are 50% condensed from a vapor of solar composition at 10 Pa (10^{-4} bar) total pressure (Lodders, 2003). Earth data are from McDonough (2014); CI chondrite from Lodders et al. (2009); CM, CO, CV, EH, LL from Wasson and Kallemeyn (1988); and CR from Lodders and Fegley (1998). Trend lines for bulk silicate Earth (dashed), CV, CM, and EH are estimated. In accord with Goldschmidt's (1937) geochemical classification, bulk silicate Earth's lithophile, chalcophile, and siderophile elements are distinguished by shading.

(Figure 18.3) is nearly identical to that of the solar photosphere, the most primitive chondrites (CI), and also EH chondrites.

Mercury's crustal K/Th and K/U ratios are slightly higher than those of Earth and Mars (Peplowski et al., 2011, 2012; Figure 2.3). The ratios of the moderately volatile ($T_{50\%}$ ~ 1000 K, see Figure 18.4) element K to the highly refractory ($T_{50\%}$ > 1600 K) elements Th or U reflect differences in the abundances of moderately volatile elements between planetary bodies. These elements are all highly incompatible in crystals that remain behind when partial melts rise to planetary surfaces, so their ratios are preserved in volcanic rocks that result from mantle melting processes. However, if some U and Th were incorporated into the core, which is possible under highly reduced conditions, the total mercurian inventory of K may be lower than that inferred from surface measurements of K/U and K/Th, under the assumption of chondritic mantle K, U, and Th (Malavergne et al., 2010; McCubbin et al., 2012). Furthermore, the crustal Cl/K ratio is close to chondritic (Evans et al., 2015; Chapter 2), as is the crustal Na/Si ratio (Evans et al., 2012; Peplowski et al., 2014). Overall, Cl and Na appear to be present in near-chondritic abundances. Mercury's abundances of moderately volatile K, Na, and Cl relative to refractory Mg are inferred to be most similar to those measured for the EH enstatite chondrites (Figure 18.4).

In summary, Mercury is anomalously enriched in Fe, S, and possibly also Si, relative to the other terrestrial planets (Figure 18.3). Its apparent bulk Na and possibly Cl compositions are also enriched above chondritic, whereas K is depleted, but less so than for Earth (Figure 18.3). A model for Mercury's origin must elevate Si-rich Fe metal compared with Mg-silicates and retain volatile S and Na at near-chondritic values relative to Mg, while maintaining abundance ratios of K, Ca, and Al that are similar to those for the other terrestrial planets. A reasonable inference from these observations is that the formation processes that led to the depletion of moderately volatile elements in planets compared with chondrites were decoupled from the origin of the large metal fraction of Mercury.

18.2.4 Surface Reflectance

The surface reflectance of airless bodies depends upon chemical composition and regolith maturity. Mercury has a lower global reflectance than the Moon, but matures at a rate that is more rapid by as much as a factor of 4 (Robinson et al., 2008; Braden and Robinson, 2013; Chapter 8). Comparison of immature surfaces indicates that the difference in reflectance is likely due primarily to differences in the compositions of those surfaces. Darkening agents on the Moon include Fe- and Ti-bearing phases, but Mercury's surface is depleted in both elements. Low-reflectance material (LRM) on Mercury is up to 30% lower in reflectance than the global mean, but LRM is not enriched in either Fe or Ti, and Fe concentration does not correlate with reflectance (Weider et al., 2012, 2014; Murchie et al., 2015). The LRM likely represents excavated mid to lower crustal material (Ernst et al., 2010), so the darkening agent may represent a major component of the silicate portion of Mercury.

Gamma-Ray Spectrometer (GRS) measurements are consistent with 0–4.1 wt% C on Mercury's surface (Peplowski et al., 2015, 2016; Chapter 2). High thermal-neutron count rates measured by the MESSENGER Neutron Spectrometer (NS) correlate with LRM, consistent with LRM C abundances 1–3 wt% greater than in surrounding higher-reflectance material (Peplowski et al.,

2016). Experiments on materials analogous to Mercury's mantle indicate that the only major mineral that would be buoyant in a Mercury magma ocean would be graphite (Vander Kaaden and McCubbin, 2015). Peplowski et al. (2016) inferred that the LRM in particular, and the volcanic upper crust generally, samples remnants of an early graphite-rich flotation crust subsequently mixed and modified by impacts and magmatic intrusions and later excavated by large craters and/or assimilated into later volcanic magmas (Chapter 6). Attribution of the low reflectance of Mercury to elevated elemental carbon is consistent with the high S/Si ratio in surface materials. While graphite is a common mineral in enstatite chondrites, those meteorites contain only ~10% of the C measured in CI chondrites.

18.2.5 Summary

The anomalously large Si-bearing Fe core, oxidation state, volatile enrichment, and reflectance of Mercury all demand explanation, but they are also clues to Mercury's formation. The Si enrichment of the core follows from the observed oxidation state, which probably also controls S distribution in the planet. In the meteorite record there exist rare materials that are similarly iron-rich, or similarly reduced, but not both. The EH enstatite chondrites are reduced, enriched in Si relative to Mg, Al, and Ca, and similarly volatile-rich. Although the EH chondrites contain Cl and K in high abundance, inferences from MESSENGER data for Mercury may represent lower bounds, subject to further experimental partitioning studies (McCubbin et al., 2012). The low reflectance of Mercury's surface may be closely related to its reduced chemistry if the planet formed with a substantial carbon content, as suggested by MESSENGER measurements (Peplowski et al., 2016).

18.3 PLANET FORMATION IN DISKS

18.3.1 Theoretical Considerations

The eighteenth century concept of the solar nebula (Kant, 1755) has evolved into modern astrophysical disk theory that treats three main stages of planet formation: (1) mineral dust concentrates in the midplane of the solar nebula and accretes to form multi-kilometer-sized planetesimals, (2) the largest planetesimals grow in annuli by runaway and preferential accretion to the largest bodies (oligarchic growth) to form Moon- to Mars-mass planetary embryos, and (3) the final terrestrial planets form by energetic, stochastic collisions between embryos driven by gravitational interactions (Safronov, 1972; Morbidelli et al., 2012). Stage 1 is poorly understood and is thought to be rapid (~10^5 yr), stage 2 (~10^6 yr) forms embryos with characteristic spacing, with gas dissipating in a few million years, and stage 3 (~10^8 yr) establishes the final radius, mass, and composition of each of the terrestrial planets through violent collisions between planetesimals and embryos from disparate solar radii (Chambers, 2004, 2009a). The challenge lies in understanding the details of how these processes occurred.

The cornerstone for chemical models for planet formation is the chondritic model for condensates in the solar nebula. The variations in the volatile compositions of the classes of chondritic meteorites are thought to reflect differences in the pressure–temperature conditions of equilibration of component solids with the gas in the solar nebula, overprinted with variable abundances of volatile ices (Figure 18.4; Davis, 2006). How these conditions varied over time with distance from the Sun and height in the disk is not known, but the conditions are probably recorded in the oldest, most volatile-rich meteorites, from bodies that did not differentiate into cores and mantles (Alexander et al., 2001). These meteorites come from asteroids and were accreted from high-temperature, 100–1000-micrometer-size chondrules and Ca-, Al-rich inclusions that were once free-floating nebular solids. In the chondrites least altered by water and heat on their parent bodies, these objects are surrounded by a fine-grained matrix containing presolar interstellar grains, organic matter, amorphous particles, and other materials (Alexander et al., 2007). The physical origin of meteoritic assemblages of chondrules and matrix materials remains elusive (e.g., Ebel et al., 2016). Although separation of metal and silicate among different components is observable at the millimeter scale, and there is significant variation in the metal content of chondrite groups, only one group (CH chondrites) has a metal/silicate ratio similar to that of Mercury (Figure 18.3). Understanding the origin of Mercury is significantly handicapped by the lack of certainty about the formation of terrestrial planets in general. The physics of growth from dust to Mars-mass bodies is particularly poorly constrained. As a result, the chemical evolution of the precursor materials of planets cannot be robustly predicted, and certainly not as a function of time and distance from the Sun.

In the past several years, new ideas have challenged traditional models of the orderly growth of planets. For example, planetesimals may rapidly grow into embryos by so-called "pebble accretion." In this model, the accretion efficiency of centimeter- to sub-meter-sized "pebbles" is greatly enhanced by Stokes drag in the atmospheres around growing embryos (Lambrechts and Johansen, 2012; Johansen et al., 2014). Calculations of embryo growth by pebble accretion have successfully led to systems that resemble the solar system's outer planets (e.g., Chambers, 2014; Levison et al., 2015a) and the terrestrial planets (Levison et al., 2015b). However, neither the physical origin nor the chemical nature of pebbles is constrained, because the models require only that pebbles are objects with a favorable Stokes number, such that the gas-drag stopping time is comparable to the time it takes for the pebble to cross the embryo's region of gravitational influence (Hill radius). In most models, pebbles are larger than the mostly sub-millimeter-sized chondrules found in meteorites (Friedrich et al., 2015), which would be strongly coupled to the gas. Thus, the relationship between pebbles and meteorites is currently unknown, and the chemical relationship between meteorites and growing embryos in this model is unexplored.

Central to the accretion of the terrestrial planets are the motions of the giant planets. The Nice model describing the outward migration of the giant planets from an earlier, more compact configuration is now a widely accepted basis for scenarios describing the early history of our solar system (e.g., Tsiganis et al., 2005; Levison et al., 2007; Morbidelli et al., 2007; Batygin and Brown, 2010). The possibility of inward and outward migration, as proposed in the Grand Tack model

(Walsh et al., 2011), is still under scrutiny. Competing models, such as pebble accretion, offer alternative solutions to the low mass of Mars (Levison et al., 2015b). Giant planet migration excites the orbits of bodies in the terrestrial zone, increasing the distribution of collisional energy and radial mixing of materials. Importantly, the probability of collisions of different energy depends on the overall context for terrestrial planet formation (e.g., Carter et al., 2015). For example, more destructive collisions are probable when the motion of a giant planet excites the orbits of bodies in the terrestrial region. By comparison, collisions generally lead to partial accretion during most of planet formation under standard models.

Most physical models of planet formation have not been well coupled to chemical models. However, recent research provides constraints on the chemical evolution of planets during the late stages of planet formation. Abundant evidence, including Earth's chondritic mantle ^{107}Ag/^{109}Ag ratio (Schönbächler et al., 2010), correlated volatile–refractory element pair variations (Wänke, 1981), and young U–Pb ages of old feldspars (Albarède, 2009), suggests that volatiles must have been delivered to Earth late or after the origin of the Moon. Other recent studies, however, have argued that Earth must have accreted water and other volatiles earlier in its growth history (Halliday, 2013; Dauphas and Morbidelli, 2014). Without a better understanding of volatile incorporation into all planets, it is difficult to assess whether the highly reduced state of Mercury falls in the expected range of final planets or if it requires an anomalous event that altered the chemical evolution of the planet.

Perhaps the only commonalities across all terrestrial planet formation models are the initial existence of a dust-rich nebula and a terminal period of giant impacts among planetary embryos. During the period of giant impacts, substantial chemical modifications of planets are possible (Wetherill, 1994; Asphaug, 2010; Stewart and Leinhardt, 2012; Asphaug and Reufer, 2014), including highly energetic collisions that could remove mantle material. The likelihood of mantle-stripping event(s) on a proto-Mercury can be assessed only in the context of specific models for the general process of terrestrial planet formation. However, dynamical studies of orbital evolution and merging of bodies leading to terrestrial planet formation (N-body simulations) universally exclude the inner region of the disk where Mercury orbits today because calculating the direct gravitational forces between all protoplanets on short orbital timescales is computationally expensive and difficult to parallelize.

Now that MESSENGER has provided new and important chemical observations, Mercury can be used in future studies as a powerful constraint and test of different models for terrestrial planet formation.

18.3.2 Observations

Protoplanetary disks are rotationally supported structures of gas and dust around young stars (Williams and Cieza, 2011; Armitage, 2011). Disks are observed around many T-Tauri type stars, actively accreting low-mass (<$3M_{sun}$, where M_{sun} is the solar mass) pre-main-sequence stars like the young Sun prior to initiation of nuclear fusion (McClure et al., 2013).

Statistical analyses of populations indicate that protoplanetary disks persist only for a few million years (Haisch et al., 2001). Understanding the mechanism of the observed rapid spindown of low-mass stars to order 10% of the break-up velocity while they are actively but episodically accreting mass is a difficult astrophysical problem, since angular momentum transfer into the disk must be rapid (Hartmann, 2009). High accretion rates are associated with strong outflows (Reipurth and Bally, 2001) and X-ray emission (Feigelson, 2010). Strong and variable magnetic fields in such high-energy environments complicate magnetohydrodynamic (MHD) modeling (McNally et al., 2013). Yet it is during this stage that planetesimals grow, with chemical compositions that may be strongly affected by the local physical environment. If Mercury's anomalous composition results from early chemical–physical processes at solar distances less than 0.5 AU, understanding those processes presents an extreme challenge for both astronomical observation and astrophysical MHD models.

The conditions in the innermost regions of planetary systems have been further complicated by observations of extrasolar planetary systems. In some exosystems, the inner regions contain substantially more mass than our solar system (Barnes et al., 2008; Fabrycky et al., 2014). For example, Kepler-11 has six known planets less than 5 Earth radii in size all orbiting inside 0.5 AU (Figure 18.5; Lissauer et al., 2011). Closely packed inner planetary systems are not universal, however, and other systems are more similar to our own, with only one planet

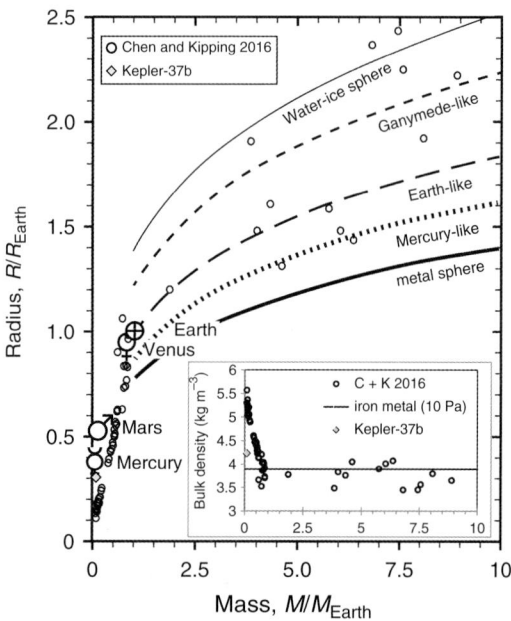

Figure 18.5. Mass–radius relations of selected low-mass (mass M_p < 10 M_{Earth}), small (radius R/R_{Earth} < 2.5) exoplanets (Chen and Kipping, 2016, their Appendix I) compared with the terrestrial planets. Inset shows calculated bulk densities versus M_p/M_{Earth} for the same data set. Curves are for Earth-like, Mercury-like, and Ganymede-like planets, and for a metal sphere and water-ice sphere (after Wagner et al., 2011). Values for Kepler-37b are from Barclay et al. (2013), using their relation (M_p/M_{Earth} = (R_p/R_{Earth})$^{2.05}$ and inferred radius. The tight, nearly linear behavior of the relations for the smallest exoplanets is a direct result of model assumptions.

having an orbit less than 200 days (Fang and Margot, 2012). The bulk densities of close-in exoplanets are difficult to measure (Chen and Kipping, 2016). So far, observations are consistent with many rocky exoplanets having Earth-like metal core fractions (Dressing and Charbonneau, 2015; Zeng et al., 2016). Mass–radius relations for close-in exoplanets (Figure 18.5; Chen and Kipping, 2016) yield bulk densities exceeding that of Mercury. Future observations should enable better calculations of masses and radii for planets very close to their host stars. At this time, the abundance of dense, inner members of stable multi-planet systems (e.g., around Kepler-37, Barclay et al., 2013) suggests that Mercury analogs are not rare.

18.4 "CHAOTIC" MODELS FOR MERCURY'S ORIGIN

18.4.1 Modeling Collisions

Several proposals for Mercury's origin are variations on the removal or separation of material by collisions. Because the possible scales of impacts vary from numerous small impactors to a single giant impact event, the possible time frame for the origin of Mercury's mass anomaly could span the period from embryo growth to the final stages of planet formation. Until recently, numerical N-body simulations of terrestrial planet formation were based on the assumption that collisions between bodies resulted in perfect merging (e.g., Kokubo and Ida, 1996, 1998; Raymond et al., 2009). As a result, these studies could not directly address the question of removal of silicate material by energetic collisions.

Major developments have improved numerical models of compositional evolution during planetary accretion. First, new analytic formulations have simplified the calculation of collision outcomes (Genda et al., 2011; Leinhardt and Stewart, 2011; Leinhardt et al., 2015). These formulations have been partially implemented into N-body simulations (Chambers, 2013; Bonsor et al., 2015; Carter et al., 2015; Dwyer et al., 2015; Leinhardt et al., 2015; Quintana et al., 2016). In general, the number of collision fragments is limited in order to keep the calculation tractable, a restriction that prevents detailed investigation of processes that involve the smallest bodies. Only one of these studies explicitly tracked the evolving metal mass fraction of the bodies and debris (Carter et al., 2015), and most studies have attempted to investigate compositional variations by post-processing the simulation data, although this approach is not robust (Bonsor et al., 2015). Carter et al. (2015) focused on growing planetary embryos; this stage did produce diversity in the metal/silicate mass ratios of embryos and planetesimals, with greater variations in the migrating giant planet Grand Tack scenario. However, embryos with metal mass fractions as high as Mercury were not produced in the limited number of simulations.

Second, N-body simulations with perfect merging after collision have been combined with metal–silicate equilibration models to predict the evolution of core and mantle compositions during accretion (e.g., Rubie et al., 2015). Such studies can address the redox state of the mantle of a growing planet subject to assumptions about the initial composition and chemistry of the embryos and planetesimals and how the initial ice/rock ratio varies through the solar system.

The evolution of the redox state of accreting planets is very much a topic of debate. In the approach of Rubie et al. (2015), variations in oxidation state are primarily controlled by the addition of water ice. In alternative approaches that are not fully coupled to accretion simulations (e.g., Wood et al., 2006; Badro et al., 2015), the mantle oxidation state evolves primarily from changes in the pressure–temperature conditions on the planet rather than from changes in accreting material. The development of linked accretion and compositional evolution models may be able to address the origin of Mercury's reduced mantle.

Third, our understanding of individual giant impacts is largely guided by studies of the origin of the Moon. Over the past several years, the giant impact hypothesis for lunar origin has been scrutinized (Asphaug, 2014; Melosh, 2014) because of the conflict between the predicted composition of the Moon and observations that the isotopic compositions of Earth and the Moon are very similar (Burkhardt, 2014). The canonical model (Canup and Asphaug, 2001; Canup, 2004, 2008) predicts that most of the lunar material would be derived from the impactor, which is expected to have isotopic signatures different from those of the proto-Earth. This conflict motivated studies of new styles of giant impacts that predict similar fractions of impactor material in Earth and the Moon (Canup, 2012; Ćuk and Stewart, 2012); however, these proposed solutions did not fully resolve the similarities and differences in the chemical makeup of Earth and the Moon (Elkins-Tanton, 2013; Asphaug, 2014; Melosh, 2014). The latest developments in lunar origin studies are predictions of the details of the chemical composition of the Moon due to its formation in a circumterrestrial disk (Canup et al., 2015; Lock et al., 2016). These studies find that the depletion of volatile elements in the Moon is due to incomplete condensation in the circumterrestrial disk rather than thermally driven loss by the energy of the giant impact. This result has direct implications for proposed collisional models for the origin of Mercury.

18.4.2 A Giant Impact

A longstanding proposal for the origin of Mercury's large core is the stripping of most of the planet's mantle by one or more giant impacts (Smith, 1979; Benz et al., 1988; Cameron et al., 1988). This scenario typically involves a differentiated proto-Mercury with more than twice the present mass that is impacted by a smaller differentiated body with sufficient energy to disperse a portion of the silicate mantles of the two bodies. If the colliding bodies had initial core mass fractions similar to that of Earth, a single impact must have gravitationally dispersed about half the total mass, primarily the silicates, to achieve Mercury's current core mass fraction (Benz et al., 1988, 2007). This class of collisions, called "catastrophic," is achieved by impact energies equivalent to a few to several times the gravitational binding energy of the total mass (Leinhardt and Stewart, 2011). Thus, most of the impact energy is converted into thermal energy, and a substantial portion of the planet and ejected fragments are transiently vaporized (Benz et al., 2007).

In a catastrophic collision between differentiated bodies, most of the ejecta are derived from the silicate layers of the colliding bodies, and most of the metal cores merge to form the new core of the largest post-collision body (Marcus et al., 2009, 2010b). The total mass of the largest remnant is determined by the self-gravity of a transiently decompressed, hot cloud of (generally segregated) metal and silicate debris, and the timescale of re-accretion into a compressed planetary structure is several hours. At that point, the planet would have a gravitationally separated molten core and silicate mantle. The mantle would have gained sufficient entropy to be a liquid to supercritical fluid at the highest pressures and pure vapor at the lowest pressures. At the lowest pressures, metal may be miscible in the silicate fluid. Mercury's thermal state would be analogous to that of Earth after the Moon-forming giant impact (e.g., Canup, 2008; Nakajima and Stevenson, 2015).

While the dynamics of catastrophic impacts are reasonably well understood, the potential for chemical changes to the colliding bodies is not. For example, the extent of metal–silicate re-equilibration is not understood in giant impacts because numerical simulations cannot model the small-length-scale processes that would enable greater mixing and chemical equilibration. Incomplete metal–silicate equilibration during accretion is inferred for Earth on the basis of the ^{182}W content of the mantle (Kleine et al., 2004; Rudge et al., 2010; Rizo et al., 2016). From dynamical considerations, portions of the cores of the impacting objects directly merge during giant impacts (Marcus et al., 2009), and only a portion of the silicate mantle is equilibrated with each small impactor (Dahl and Stevenson, 2010; Morishima et al., 2013). Thus, the effects of a giant impact on the redox state of Mercury's mantle and core, and their degree of subsequent equilibration, remain unquantified.

One of the major questions about the proposed giant impact origin for Mercury is its effect on the volatile content of the planet. When MESSENGER determined that the K/Th and K/U ratios of Mercury are similar to those of the other terrestrial planets, the giant impact hypothesis was rejected as inconsistent with a high-temperature event (Peplowski et al., 2011). At the time, there were no quantitative models available to address how moderately volatile elements would be affected by a giant impact. As a result, the observed depletion in volatile and moderately volatile elements in the Moon (Ringwood and Kesson, 1977) guided inferences about the outcome of other giant impact events.

However, all models of planet formation predict that the terrestrial planets (with the possible exception of Mars) experienced a stochastic number of giant impact events during accretion. Thus, if giant impacts substantially depleted moderately volatile element abundances on growing planets by removal, one would expect a different magnitude of depletion on each planet. The similar K/Th and K/U ratios for terrestrial planets, for which the impact histories must vary greatly, imply that giant impacts are not primarily responsible for the magnitude of moderately volatile element depletion in planets compared with CI chondrites (Figure 8.4), although detailed chemical calculations have not been accomplished to address this process. Giant impacts primarily remove bulk mantle silicates without chemical fractionation. Vaporized mantle would remain gravitationally bound and would recondense upon planetary cooling (Stewart et al., 2013, 2016).

This inference is supported by recent calculations demonstrating that the depletion of K and Na on the Moon compared with Earth is a result of incomplete condensation in the lunar disk (Canup et al., 2015; Lock et al., 2016). Hence, the abundances of moderately volatile elements on the Moon are different from those of the planets because of the Moon's origin in a circumplanetary disk (Stewart et al., 2016). Instead, Earth's moderately volatile element abundances are a better analog to a post-impact Mercury.

After a giant impact, a planet is a mixture of the compositions of the colliding bodies, with the portion from each body strongly dependent on the impact geometry and velocity (e.g., Canup, 2004, 2012; Ćuk and Stewart, 2012; Reufer et al., 2012). Because the post-impact planet is defined by self-gravitationally bound material, thermally driven escape is limited to the high-velocity tail of the Boltzmann distribution and the small mass fraction in the collisionless outer regions of the atmosphere. Radiative cooling quickly leads to re-condensation of silicates and sulfides from the vapor atmosphere, and thermally driven escape is limited (Stewart et al., 2016). At present, a giant impact origin for Mercury does not have a predicted chemical signature, other than the motivating observation of an enhanced metal core fraction, with which to test the hypothesis.

Even in the absence of a chemical test for the giant impact hypothesis, such an origin for Mercury is difficult. The principal problem with a single impact hypothesis is the re-accretion of debris. The orbits of gravitationally ejected material would intersect Mercury's orbit. Numerical simulations that track debris production and re-accretion find that most of the debris is quickly re-accreted, which limits the variation in core fraction of growing embryos (Carter et al., 2015).

Benz et al. (1988, 2007) proposed that the ejecta could be separated from Mercury if the ejected particles were sufficiently small to evolve quickly under non-gravitational forces such as Poynting–Robertson drag. Benz et al. (2007) calculated that most of the ejecta would be shock-heated to vapor and recondense as ~centimeter-sized particles. They calculated the orbital evolution of the ejecta and found that about one-third was re-accreted to Mercury in the first 2 Myr. This timescale is comparable to the half-life of centimeter-sized particles undergoing Poynting–Robertson drag into the Sun. Benz et al. (2007) proposed that Mercury could be formed by a "super-catastrophic" impact onto proto-Mercury, followed by partial re-accretion of the ejecta and substantial loss of ejecta to the Sun.

Gladman and Coffey (2009) also considered the dynamics of ejecta from a giant impact onto Mercury. They found that the debris field would be optically thick, which limits the role of Poynting–Robertson drag. In addition, the debris would collide with itself, further increasing the optical thickness and likelihood of re-accretion onto Mercury.

Changes in the chemical composition of the ejecta have not been modeled in detail. If the ejecta cool sufficiently quickly that silicates and sulfides recondense faster than the timescale of dynamical separation of condensates and gas, then the composition of re-accreting ejecta may be similar to that of the original colliding planets. As mentioned above, the dynamical evolution and re-accretion of condensed ejecta have timescales of millions of years (Benz et al., 2007; Gladman and Coffey, 2009). Adiabatic expansion and radiative cooling would lead to re-condensation of silicates and sulfides on much shorter timescales (e.g., days to years). Whereas highly volatile elements may remain in the gas phase in orbit around the Sun, the condensed ejecta should have compositions similar to the silicate layers of the source planets.

Ejecta may also be accreted onto other planets or planetary embryos. If some other nearby planet had a larger mass than Mercury, a majority of ejecta could be accreted to that body. Such a scenario has motivated a variation of the giant impact hypothesis for Mercury in which proto-Mercury is the smaller body in a so-called "hit-and-run" event.

18.4.3 A Hit-and-Run Impact

Agnor and Asphaug (2004) first realized that a large fraction of collisions between similarly sized bodies do not lead to mergers. During the giant impact stage (3), about one-third of collisions between planetary embryos are hit-and-run events where the two bodies obliquely collide and then gravitationally separate (Stewart and Leinhardt, 2012). In most cases, the two bodies collide and merge in a subsequent encounter (Kokubo and Genda, 2010; Carter et al., 2015). However, it is possible for a planetary embryo to be scattered to the inner edge of the terrestrial accretion zone and survive to the end of accretion (Hansen, 2009).

In a hit-and-run encounter, the smaller body may be catastrophically disrupted (Asphaug et al., 2006; Asphaug, 2010; Leinhardt and Stewart, 2011). Sarid et al. (2014) and Asphaug and Reufer (2014) proposed that proto-Mercury was the smaller body. Stripping the mantle of the smaller body requires a less energetic collision – or multiple low-energy collisions – than in the standard single giant impact scenario. However, the arguments presented above regarding the lack of chemical changes (e.g., volatile content and redox state) to the silicate mantle would also apply to the hit-and-run scenario regardless of the intensity of heating and vaporization.

In the hit-and-run variation, the ejecta would primarily be accreted onto the larger-mass body, perhaps a proto-Venus planetary embryo. Multiple hit-and-run events are possible between a pair of embryos, so this variation does not require that the enhanced metal mass fraction of Mercury be achieved in a single event. At this time, hit-and-run scenarios have not been tested in full numerical simulations of terrestrial planet formation, let alone with chemistry, so the likelihood of a Mercury-like outcome is not known. The key issue for this hypothesis is preventing the smaller proto-Mercury from ultimately being accreted onto the larger body.

The amount of mass lost from the smaller body is extremely sensitive to the impact geometry and velocity. At this time, the numerical simulations of individual hit-and-run encounters have not yet been distilled into a simple analytic formula, which is needed for future investigation of this scenario in N-body simulations of planet formation.

18.4.4 Collisional Erosion

Another variation on collisional stripping of Mercury's mantle is the accumulated effects of many small high-velocity impactors (Vityazev et al., 1988; Svetsov, 2011). In these scenarios, the impact velocities must exceed ~25 km/s in order for each collision to eject a total mass greater than that of the impacting body (Svetsov, 2011). For example, in order to erode a proto-Mercury with an Earth-like core fraction to the present core mass fraction requires more than a Mercury mass of planetesimals impacting at ~30 km/s (Svetsov, 2011). During planet formation, the typical collision velocities are a factor of 1 to 3 greater than the escape velocity of the largest bodies (e.g., O'Brien et al., 2006; Raymond et al., 2009). Thus typical Mars-mass embryos would not experience primarily erosive planetesimal bombardment. Indeed, most planetesimal collisions result in net accretion unless a migrating giant planet dynamically excites the system (Carter et al., 2015). Even if such a dynamically excited bombardment occurred, the re-accretion of ejecta remains an issue unless ejecta could be preferentially accreted onto other protoplanets. In addition, heavy bombardment of Mercury would also affect the composition of the other inner planets.

Investigations of collisional erosion of Mercury illustrate the general problem with invoking collisional erosion of planetary crusts to account for geochemical observations, as first proposed by O'Neill and Palme (2008). The required high-velocity bombardment is not predicted in current models of planet formation. In a giant impact, portions of both mantle and crust are ejected rather than all of the crust. Numerical simulations that include probable collision outcomes indicate that small-scale debris is continuously recycled as it is ejected and then re-accreted during the growth of planetary embryos (Bonsor et al., 2015; Carter et al., 2015). Thus, collisions are not expected to permanently remove incompatible elements that are concentrated in planetary crusts. Indeed, Mercury does not appear to have experienced such removal.

Bulk ejection of silicate layers, and their likely re-accretion, would not chemically fractionate the silicate component of the planet. Thus, the observed chondritic Cl/K ratio (Evans et al., 2015) on Mercury does not provide a major constraint on the impact history of the planet. Venus, Earth, and the Moon have lower bulk Cl/K ratios than Mercury, Mars, and chondrites (see Evans et al., 2015). The origin of these differences among bodies is not likely to be a result of collisional erosion of planetary crusts. Indeed, the role of collisional erosion in the inner solar system was originally motivated by Earth's ^{142}Nd abundances (Boyet and Carlson, 2005). These data have recently been reinterpreted in such a way that no geochemical support for collisional erosion remains (Bouvier and Boyet, 2016; Burkhardt et al., 2016).

18.5 "ORDERLY" PROCESSES FOR MERCURY'S ORIGIN

The arguments for a "chaotic" stage of planet formation (stage 3) after the formation of embryos (stage 2) are strong. Such embryos would reflect the chemical conditions and dynamical processes in narrow radial annuli or feeding zones. The chemical signatures of planetesimals and embryos established during stage-2 formation would be largely scrambled during stage 3, remaining only as cryptic heterogeneities such as perhaps the oxygen isotopic differences among Earth + Moon, Mars, and Vesta (Clayton and Mayeda, 1996). However, Mercury could represent an embryo, a remnant of stage 2, and record in its chemical composition extreme processes occurring uniquely in the innermost solar nebula. Several such extreme processes have been put forward to explain Mercury's anomalous density. Here we consider each in the light of MESSENGER results.

18.5.1 "Old School" Condensation

Following Urey (1950), Lewis (1973) stated that "the present quantitative theory for the composition and volatile content of solar system bodies attributes both to one parameter alone, formation temperature" (cf. Lewis, 1972, 1988). In this scenario, a hot disk of dust and gas, with dust concentrated in the midplane and innermost few astronomical units, cools and condenses solids slowly over time from the outer material toward the center (Larimer and Anders, 1967; Grossman and Larimer, 1974; Ebel, 2006), and those solids accrete onto the planets as we see them today, with very limited mixing. The best twentieth-century estimates for the bulk compositions of Venus and Mercury were derived with this assumption (Morgan and Anders, 1980; Lodders and Fegley, 1998). Modern theories of disk structure and evolution have gone well beyond such a simple picture, which is inconsistent with both the observed architectures of exoplanetary systems and the volatile enrichment of Mercury's silicate fraction discovered by MESSENGER. Furthermore, as discussed in detail by Weidenschilling (1978), a temperature at which the Fe/Si ratio of solid condensates matches Mercury's Fe/Si ratio is possible, but removal of nebular gas must occur within a very narrow cooling window of 10–50 K, before significant Mg-silicate condensation (Ebel, 2006, Plates 7 and 10).

Cameron (1985) suggested that the early Sun was sufficiently hot that the mantle of differentiating Mercury evaporated and was stripped by the strong solar wind. Fegley and Cameron (1987) calculated that vaporization of ~70–80% of the silicate from an initially chondritic mantle would be required to match Mercury's uncompressed density. This combined thermochemical/dynamical model predicts strong depletion in moderately volatile elements (K, Cl, S, Na) and enrichment in refractory lithophile elements (Ca, Al, Ti, Th). Both SiO_2 and FeO become depleted in the mantle, and U would deplete relative to Th due to its propensity to form the gaseous oxide. This model is not supported by the MESSENGER results.

Morgan and Anders (1980) superposed physical processes on the original nebular condensates, such as preferential settling, size sorting, or ferromagnetic attraction, as well as thermal processing (melting, evaporation). Condensation arguments and observed K and U abundances were used to derive planetary compositions (Figure 18.3); however, they chose to use lunar values in the absence of data from Mercury. Going further, they assumed, "tenuously!" in their words, a mantle FeO abundance trend for the terrestrial planets that increases monotonically with solar distance, with Mercury pyroxenes at 5.5 wt% FeO. Results from MESSENGER invalidate such assumptions.

18.5.2 Metal/Silicate Fractionation

Chondritic meteorites represent the oldest, undifferentiated materials formed in the solar system and are thought to be the precursors of the terrestrial planets. Metal abundance is one of the major components describing the variation among chondritic meteorites (Grossman, 1996). There are numerous hints in primitive chondrites of metal fractionation from silicates early in solar system history. Rare, metal-rich chondrites may record metal–silicate fractionation in some part of the early nebula. Iron-rich asteroids also suggest metal–silicate fractionation by dynamical processes.

18.5.2.1 Metal-Rich Chondrites

The metal-rich Bencubbin-like (CB) and CH carbonaceous chondrites (Weisberg et al., 2001) have many primitive characteristics and a higher grain density, due to an abundance of metal grains, than uncompressed Mercury (Figure 18.1). Chondrules and metal grains in these chondrites have been attributed to planetesimal impact up to 5 Myr after CAI formation, the usually accepted "origin" time of the solar system (Krot et al., 2005; Fedkin et al., 2015), although some of their compositional characteristics are difficult to reconcile with such an origin (e.g., their CAIs, Weisberg et al., 2016). Collisions among bodies of ordinary chondrite composition in a disk from which most of the H_2-rich gas has been removed will recondense solids from oxidizing vapors, yielding FeO-rich olivine. Carbonaceous chondrite vapor is even more oxidizing (Ebel, 2001), so persistence of H_2-rich gas in the solar nebula is required to make these metal-rich meteorites in late-stage impacts. Persistence of a gas is also required to prevent fragmentation during chondrite accretion. These metal-rich chondrites, however, are not nearly as reduced as the enstatite chondrites, which contain CaS and MgS. If the CB chondrites formed late in solar system history, their influence on planetary embryo compositions is likely to have been minimal.

18.5.2.2 Asteroid Cores

A large number of magmatic iron meteorite parent bodies are inferred from the meteorite record, but complementary planetesimal mantles are missing. Differential comminution by impacts, followed by Poynting–Robertson drag over long time periods, is the most likely explanation for loss of silicates, primarily to the Sun (Burbine et al., 1996). W–Hf isotopic chronometry shows that the differentiation of many parent bodies occurred within 1 Myr of CAI formation (Yang et al., 2007; Burkhardt et al., 2008; Kleine et al., 2009). Density determinations appear to require that many iron-rich asteroids are bound, macroscopic, metallic rubble (e.g., 216 Kleopatra, Ostro et al., 2000; Consolmagno and Britt, 2004; Carry, 2012). The behaviors of big metal chunks and small, brittle mantle silicate fragments in the disk before and after gas dispersal would differ (Weidenschilling, 1978; Section 18.5.2.3). However, the iron-rich asteroids represented in the meteorite record do not offer themselves as potential sources for Mercury's excess metal, since they are, with few exceptions, lacking in the alloyed metallic Si typical of highly reduced bodies (Section 18.2.2). The oxidation states of iron meteorites are quite different from that inferred for Mercury on the basis of MESSENGER observations and experiments (Section 18.2).

18.5.2.3 Photophoresis and Dynamical Fractionation

The photophoretic effect has been explored to explain the wide variation in chondrite meteorite metal contents (Wurm et al., 2013) and size sorting of particles (Loesche et al., 2016). Photophoresis occurs when non-isotropic (e.g., Sun-derived) thermal radiation produces thermal gradients in millimeter-sized particles that drive those particles toward colder regions. This force is proportional to the temperature gradient in each particle (Krauss and Wurm, 2005). Differences in radial velocity relative to the Sun could separate metal, rock, and dust aggregate particles because metal has much higher thermal conductivity, so metal grains lack temperature gradients. Thus, photophoresis could have affected early collisional dust evolution and accretion (Loesche et al., 2016). If such effects were widespread in a particular annulus of the disk, resulting planetesimals could have inherited the effects of photophoresis, including metal/silicate fractionation (Cuzzi et al., 2008). This effect is expected to be strongest closer to the Sun.

Opacity is expected to be high in the midplane of a planet-forming disk, inhibiting the photophoretic effect. For photophoresis to work, a sufficient fraction of Mercury precursor material must see sufficient sunlight, and rocky grains must be forced outward, leaving an anomalously metal-rich feeding zone for Mercury. However, inner disk "walls" are observed in transition disks, which are stellar objects with substantially cleared inner disks, perhaps the result of planet formation (Espaillat et al., 2014), suggesting that accretion zones closest to stars could be affected by photophoretic separation of grains on the basis of their physical properties, e.g., size and porosity (McNally and McClure, 2017). Testing this hypothesis requires a model that addresses the chemistry of small solids in the inner disk, subject to photophoresis, and their accretion onto planetary precursors.

An "aerodynamic fractionation" model was proposed by Weidenschilling (1978), who recognized that more effective removal of silicates relative to metal from a chondritic feeding zone could account for Mercury's anomalous density. In the inner disk, small grains were entrained in the accreting gas, whereas larger grains (>1 m) in Keplerian orbits experienced a headwind due to the radiation pressure experienced by the gas. This gas drag caused meter-sized grains to drift sunward faster than the gas. Larger and/or denser boulders experienced slower orbital decay. Weidenschilling proposed that precursor multi-meter-size solids (boulders) with differing Fe/Si ratios could dynamically interact in such a way that silicates were preferentially removed into the Sun by gas drag.

Hubbard (2014) explored a "magnetic erosion" model that combined the magnetic attraction of metallic grains and the differential comminution of silicates and metal by collision. He called on particular conditions at the outer edge of the inner disk to enhance magnetization of metal-rich grains, thus enhancing both the collisional removal of non-magnetic silicate and the rapid collisional growth of large metal grains to sizes beyond the meter barrier. The magnetic field requirement restricts this mechanism to the inner disk.

All three of these models are consistent with Mercury's anomalous density and Earth-like mantle abundances, and they all enrich Mercury in metal that is commonly associated with sulfide, consistent with Mercury's likely S enrichment. None of these models, however, addresses the extremely reduced nature of Mercury. They all call on special astrophysical conditions in the innermost nebula, and none has been explored with fully three-dimensional chemical–physical models. If precursor metal is assumed to carry sulfur, then Mg-silicates and refractory elements might be assumed to be in large grains (chondrules, CAIs), whereas K, Na, and Cl would have concentrated in smaller grains, the matrix mineral dust of chondrites. MESSENGER results for K indicate that Mercury's volatility curve is not steeper than Earth's (Figure 18.4). The fates of the volatiles in these scenarios are difficult to predict.

Several recent astrophysical models have yielded results consistent with Mercury's metal enrichment and reduced chemistry for disks around stars similar to the Sun. Pasek et al. (2005) used a two-dimensional steady-state α-disk solution to yield time–temperature–pressure histories in the disk midplane at various solar radii along with chemistry codes to compute condensation fronts. With a model for diffusive transport driven by condensation fronts in the inner disk, they then calculated condensation histories at progressively more O-depleted inner radii. They found that the effects of water depletion on nebular S speciation formed reduced enstatite-chondrite-like rocks at Mercury-like astrocentric radii.

Moriarty et al. (2014) applied a disk model (Chambers, 2009b) to calculate disk temperature, pressure, and density over time. They then calculated equilibrium chemistry at steps in time and radius, removing gas and dust into planetesimals growing at a prescribed rate and decoupled from the gas. They also accounted for the effects of radial gas movement (with perfectly coupled dust) on the chemical inventory. Finally, their planetesimals were input into an N-body simulation of late-stage planet formation. Their model produced C-rich, short-period planetesimals around stars with C/O ratios slightly above the solar ratio.

Pignatale et al. (2016) used a disk model (D'Alessio et al., 1999) to prescribe temperature and pressure in a two-dimensional disk and then calculated condensation under each of those conditions. They then applied dust-settling and radial-migration models to calculate redistributions of material. Their model produced sulfide- and enstatite-rich zones within 1 AU of the young Sun. Although feedback between chemistry and dynamics is highly limited or missing from efforts to date to couple chemistry and dynamics, model results consistently predict reduced, metal-enriched inner disks, the probable feeding zone for planet Mercury.

18.5.2.4 C-Rich Condensation

The gross separation of inner volatile-poor planets and outer volatile-rich planets immediately suggests a nebular "snow line," a complex time-dependent surface inside of which water remained in the vapor phase. In the presence of free oxygen, a similar "C-line" would have marked the locus inside of which graphite would have oxidized to CO and/or CO_2. The time–temperature and oxygen distribution histories of the inner disk are not known, nor is the time-dependent accretion flux of interstellar material onto the innermost disk well quantified. The balance between C and O as a function of time and radius in the terrestrial planet-forming region may, therefore, have been quite heterogeneous. Ebel and Alexander (2011) investigated the consequences of carbon enrichment and oxygen depletion on the stability of minerals in a cooling H_2-rich vapor. They noted that the most abundant interplanetary dust particles (IDPs) in the present solar system are 50–1000-μm sized, anhydrous, porous, chondritic "C-IDPs." The C-IDPs are aggregates containing highly primitive sub-micrometer silicates, metal, sulfide and presolar grains all attached together by poorly graphitized carbon (Messenger et al., 2003; Busemann et al., 2009; Bradley, 2014). Their original C content has been diminished by pre-capture stratospheric entry.

Ebel and Alexander (2011) explored the consequences of equilibrium condensation in systems enriched in a C-enriched, O-depleted analog C-IDP dust. At high (1000×) enrichments in such a dust (relative to H_2), condensates at 1650 K (at 10 Pa total pressure) have atomic Fe/Si reaching 50% of the estimate for bulk Mercury, because Si remains in the vapor to low temperatures (<1000 K) as $SiS_{(gas)}$. Sulfur behaves as a more refractory element in this reducing system than in systems enriched in chondritic dust (Ebel and Grossman, 2000), stable minerals are FeO-poor, and the minerals stable at high temperature include CaS and MgS. These results reproduce mineralogical and petrological characteristics of enstatite chondrites. These calculations suggest that enrichment in anhydrous chondritic dust, with variation in O/C ratios by O depletion, can produce environments in which highly reduced, high Fe/Si condensate assemblages are stable above 1200 K. If such condensates were isolated from the gas phase, they could have formed enstatite chondrite parent planetesimals, which could then have accumulated to form Mercury-like planets.

In follow-up work, Ebel and Sack (2013) computed the stability of djerfisherite, $K_6(Fe,Ni,Cu)_{25}S_{26}Cl$, in similar C-IDP dust-enriched systems. Comparison with djerfisherite occurrences in EH3 enstatite chondrites allowed them to conclude that both K and Cl, as well as S, can behave as refractory elements under the hypothesized nebular conditions.

This line of reasoning relies on the assumption that C-rich silicate dust could reach the inner solar system well inside the evaporation radius of more oxidized material. There are other hints at such chemical gradients, but all are cryptic. Initial enrichment of Mercury in elemental carbon would provide both a light element in the core (in addition to S and Si) and an explanation for Mercury's low spectral reflectance (Vander Kaaden and McCubbin, 2015; Peplowski et al., 2016).

18.6 BEARING ON EXTRASOLAR PLANETS

We are entering the third decade of a revolution in the observation and theoretical understanding of extrasolar planets. One of the critical questions regards the relationship between observed high-mass inner planets, "super-Earths," and the terrestrial planets in our own solar system (Batalha et al., 2011; Morbidelli and Raymond, 2016). Wagner et al. (2011; cf. Dressing and Charbonneau, 2015; Zeng et al., 2016) calculated radii R/R_E and masses M/M_E for hypothetical, fully differentiated, thermally equilibrated Earth-like (32.5 wt% iron core, 67.5% silicate mantle), Mercury-like (70% core, 30% silicate), and Ganymede-like (6.5% core, 48.5% silicate, 45% water ice) planets (Figure 18.5), where M_E is Earth's mass and R_E is Earth's radius. These curves are illustrated in Figure 18.5 and compared with inferred properties of a subset of planets discovered up to early 2016.

The difficulty of estimating exoplanet mass and size is illustrated by the fact that some exoplanets plot below the low-mass extension of the metal sphere mass–radius curve of Wagner et al. (2011) in Figure 18.5. The observational bias and the large uncertainties in data even for transiting planets make comparisons with our terrestrial planets premature. Marcus et al. (2010a) estimated the maximum amount of mantle material that could be removed from a planet by a single giant impact. This minimum radius is widely used as a lower limit on the radius of rocky exoplanets because pure-iron planets are unlikely. Planets with densities indicative of high metal core fractions are at the lower edge of the range of planet sizes and orbital periods explored by the Kepler mission (Borucki, 2016), but the abundance of Mercury analogs in the existing data (e.g., Barclay et al., 2013; Chen and Kipping, 2016) suggests that they are a natural outcome of dynamical–cosmochemical processes in protoplanetary disks.

18.7 SUMMARY

18.7.1 "Chaotic" Models

The origin of Mercury remains elusive. The MESSENGER mission has provided strong constraints on potential models, but models to date lack the detail to be tested against the observations. Here, we summarize the major models and their current challenges. Three dynamical-impact-related or "chaotic" models (Section 18.2.1) have been considered:

- A single giant impact onto proto-Mercury: The energy required for a single event falls in the expected range of planet formation. The primary challenge lies in the expected re-accretion of debris. The orbits of the ejected silicates must be dynamically separated from the post-collision Mercury.
- A hit-and-run with proto-Mercury as the smaller body: This type of encounter is expected to be common during stage 3 of planet formation. Preferential accretion of debris onto the larger body is possible but has not yet been investigated quantitatively. The model demands that the post-impact Mercury should not be accreted onto the larger body, which may require special circumstances such as dynamical ejection out of the zone of planet formation.
- Multiple impact scenarios: Current models for terrestrial planet formation do not predict a bombardment of small

planetesimals with sufficient energy to remove Mercury's mantle if it began with a chondritic bulk composition. A mixture of small and large impacts may produce a range of final core mass fractions, but the limited simulations conducted so far have not produced core enhancements as large as observed on Mercury.

All such models of the inner solar system have fallen short of directly addressing the origin of Mercury. Future detailed calculations of terrestrial planet formation should be able to address the plausible range of final core mass fractions during collisional accretion and evolution of planets. Combined with chemical models of the growing planet, such studies should also be able to address the origin of Mercury's extremely reduced mantle and range of possible core compositions, constrained by experimental petrology. At present, there are no obvious geochemical tests of the collisional origin hypotheses. Current geochemical observations from all the terrestrial planets and physical studies of collisional accretion do not support earlier suggestions that collisions should preferentially remove moderately volatile elements or incompatible elements.

18.7.2 "Orderly" Models

"Orderly" processes (Section 18.2.1) perhaps fare better than "chaotic processes" in approaching explanations for Mercury's origin. MESSENGER findings definitively refute previous models in which only the most refractory elements condense and accrete to form Mercury from a vapor of solar composition in the hot innermost part of the protoplanetary disk (e.g., Section 18.2.3; Morgan and Anders, 1980). MESSENGER results also refute early hypotheses of direct evaporation of silicates from proto-Mercury by an energetic young Sun. Such a process would rapidly deplete Mercury in many moderately volatile elements, including Si and Mg. The Si-poor metal in most metal-rich chondritic or iron-rich asteroidal material and the measurable FeO in silicates in most chondrites eliminate their potential role as major precursors to highly reduced Mercury. Only a very tiny fraction of solar system materials in the meteorite record, the EH chondrites, have experienced reducing conditions similar to Mercury's.

Dynamical models in the context of a disk in which composition, temperature, and density vary smoothly with heliocentric radius offer another route to chemical fractionation close to the Sun. Although photophoresis models are not sufficiently mature to produce Mercury's density and redox anomalies, they do separate metal from silicate, particularly in the innermost disk. Aerodynamic fractionation of Fe-rich "boulders" from smaller, brittle silicates and preferential removal of non-magnetic silicates by "magnetic erosion" are both consistent with Mercury's anomalous density, but neither model appears to address the highly reduced chemistry of Mercury.

The possibility that the inner region of the disk was carbon-rich allows consideration of processes that could yield Mercury-producing scenarios. Condensation in C-rich, O-poor systems has been explored with very conservative parameters, yielding enstatite chondrite condensates (Ebel and Alexander, 2011). This parameter space might yield condensates with high Fe/Si at high temperatures. There is strong evidence for the volatility-controlled accretion of solids onto various chondrites and the planets. A scenario of high C/O condensation would require a dynamical argument for accretion of high Fe/Si solids at the high temperatures at which Si remains in the vapor phase. A successful model describing the physical chemistry of the inner disk must address radiative transfer, chemistry–opacity feedbacks, and active mass transfer among vapor, liquid, and solid phases.

18.8 CONCLUSION

MESSENGER confirmed the anomalous bulk Fe/Mg ratio and anomalously reduced nature of Mercury and demonstrated that the abundances of moderately volatile elements are consistent across the terrestrial planets and asteroids. These findings indicate that these bodies record disk processing rather than special circumstances of individual planets. Models that focus on terrestrial planet formation do not commonly produce Mercury-like planets. However, the EH chondrites offer a way forward in understanding how gradual processes in the innermost disk might have produced a small, reduced, Fe- and probably Si-rich planet.

It is plausible that the present state of Mercury is the sum of several processes at work during planet accretion. None of the proposed formation models can explain all of the observations, and all suffer from our lack of information about the chemical, thermal, and dynamical conditions at the inner edge of the protoplanetary disk. Increasingly detailed astronomical observations of protoplanetary disks and other planetary systems will aid development of more detailed planet formation theories. Such theories must be informed by better coupling of physical and geochemical processes in models of inner disk processes. Whether Mercury requires a special circumstance to account for its unique properties remains to be answered.

ACKNOWLEDGEMENTS

The authors thank Erik Asphaug, Larry Nittler, and Sean Solomon for comprehensive reviews. Denton Ebel thanks the Arthur Ross Foundation and the American Museum of Natural History for support of his participation on the MESSENGER Science Team. This research has made use of NASA's Astrophysics Data System.

REFERENCES

Agnor, C. and Asphaug, E. (2004). Accretion efficiency during planetary collisions. *Astrophys. J. Lett.*, **613**, L157–L160.

Albarède, F. (2009). Volatile accretion history of the terrestrial planets and dynamic implications. *Nature*, **461**, 1227–1233.

Alexander, C. M. O'D., Boss, A. P. and Carlson, R. W. (2001). The early evolution of the inner solar system: A meteoritic perspective. *Science*, **293**, 64–68.

Alexander, C. M. O'D., Boss, A. P., Keller, L. P., Nuth, J. A. and Weinberger, A. (2007). Astronomical and meteoritic evidence for thermal processing of interstellar dust in protoplanetary disks. In *Protostars and Planets V*, ed. B. Reipurth, D. Jewitt and K. Keil. Tucson, AZ: University of Arizona Press, pp. 801–813.

Armitage, P. J. (2011). Dynamics of protoplanetary disks. *Annu. Rev. Astron. Astrophys.*, **49**, 195–236.

Asphaug, E. (2010). Similar-sized collisions and the diversity of planets. *Chemie der Erde-Geochemistry*, **70**, 199–219.

Asphaug, E. (2014). Impact origin of the Moon? *Annu. Rev. Earth Planet. Sci.*, **42**, 551–578.

Asphaug, E. and Reufer, A. (2014). Mercury and other iron-rich planetary bodies as relics of inefficient accretion. *Nature Geosci.*, **7**, 564–568. doi:10.1038/ngeo2189.

Asphaug, E., Agnor, C. B. and Williams, Q. (2006). Hit-and-run planetary collisions. *Nature*, **439**, 155–160.

Badro, J., Brodholt, J. P., Siebert, J. and Ryerson, F. J. (2015). Core formation and core composition from coupled geochemical and geophysical constraints. *Proc. Natl. Acad. Sci.*, **112**, 12,310–12,314.

Barclay, T., Rowe, J. F., Lissauer, J. J., Huber, D., Fressin, F., Howell, S. B., Bryson, S. T., Chaplin, W. J., Désert, J.-M., Lopez, E. D., Marcy, G. W., Mullaly, F., Ragozzine, D., Torres, G., Adams, E. R., Agol, E., Barrado, D., Basu, S., Bedding, T. R., Buchhave, L. A., Charbonneau, D., Christiansen, J. L., Christensen-Dalsgaard, J., Ciardi, D., Cochran, W. D., Dupree, A. K., Elsworth, Y., Everett, M., Fischer, D. A., Ford, E. B., Fortney, J. J., Geary, J. C., Haas, M. R., Handberg, R., Hekker, S., Henze, C. E., Horch, E., Howard, A. W., Hunter, R. C., Isaacson, H., Jenkins, J. M., Karoff, C., Kawaler, S. D., Kjeldsen, H., Klaus, T. C., Latham, D. W., Li, J., Lillo-Box, J., Lund, M. N., Lundkvist, M., Metcalfe, T. S., Miglio, A., Morris, R. L., Quintana, E. V., Stello, D., Smith, J. C., Still, M. and Thompson, S. E. (2013). A sub-Mercury-sized exoplanet. *Nature*, **494**, 452–454.

Barnes, R., Gozdziewski, K. and Raymond, S. N. (2008). The successful prediction of the extrasolar planet HD 74156d. *Astrophys. J.*, **680**, L57–L60.

Batalha, N. M., Borucki, W. J., Bryson, S. T., Buchhave, L. A., Caldwell, D. A., Christensen-Dalsgaard J., Ciardi, D., Dunham, E. W., Fressin, F., Gautier, T. N., III, Gilliland, R. L., Haas, M. R., Howell, S. B., Jenkins, J. M., Kjeldsen, H., Koch, D. G., Latham, D. W., Lissauer, J. J., Marcy, G. W., Rowe, J. F., Sasselov, D. D., Seager, S., Steffen, J. H., Torres, G., Basri, G. S., Brown, T. M., Charbonneau, D., Christiansen, J., Clarke, B., Cochran, W. D., Dupree, A., Fabrycky, D. C., Fischer, D., Ford, E. B., Fortney, J., Girouard, F. R., Holman, M. J., Johnson, J., Isaacson, H., Klaus, T. C., Machalek, P., Moorehead, A. V., Morehead, R. C., Ragozzine, D., Tenenbaum, P., Twicken, J., Quinn, S., VanCleve, J., Walkowicz, L. M., Welsh, W. F., Devore, E. and Gould, A. (2011). Kepler's first rocky planet: Kepler-10b. *Astrophys. J.*, **729**, 27–48, doi:10.1088/0004-637X/729/1/27.

Batygin, K. and Brown, M. E. (2010). Early dynamical evolution of the solar system: Pinning down the initial conditions of the Nice model. *Astrophys. J.*, **716**, 1323–1331.

Beckett, J. R. (1986). The origin of calcium-, aluminum-rich inclusions from carbonaceous chondrites: An experimental study. Ph.D. thesis, University of Chicago, Chicago, IL.

Benz, W., Slattery, W. L. and Cameron, A. G. W. (1988). Collisional stripping of Mercury's mantle. *Icarus*, **74**, 516–528.

Benz, W., Anic, A., Horner, J. and Whitby, J. A. (2007). The origin of Mercury. *Space Sci. Rev.*, **132**, 189–202.

Bonsor, A., Leinhardt, Z. M., Carter, P. J., Elliott, T., Walter, M. J. and Stewart, S. T. (2015). A collisional origin to Earth's non-chondritic composition? *Icarus*, **247**, 291–300.

Borucki, W. J. (2016). KEPLER mission: Development and overview. *Rep. Prog. Phys.*, **79**, 036901.

Bouvier, A. and Boyet, M. (2016). Primitive Solar System materials and Earth share a common initial ^{142}Nd abundance. *Nature*, **537**, 399–402.

Boyet, M. and Carlson, R. W. (2005). ^{142}Nd evidence for early (>4.53 Ga) global differentiation of the silicate Earth. *Science*, **309**, 576–581.

Boynton, W. V., Sprague, A. L., Solomon, S. C., Starr, R. D., Evans, L. G., Feldman, W. C., Trombka, J. I. and Rhodes, E. A. (2007). MESSENGER and the chemistry of Mercury's surface. *Space Sci. Rev.*, **131**, 85–104.

Braden, S. E. and Robinson, M. S. (2013). Relative rates of optical maturation of regolith on Mercury and the Moon. *J. Geophys. Res. Planets*, **118**, 1903–1914, doi:10.1002/jgre.20143

Bradley, J. P. (2014). Early solar nebula grains – interplanetary dust particles. In *Meteorites and Cosmochemical Processes*, ed. A. M. Davis, *Treatise on Geochemistry*, 2nd edn, Vol. 1, ed. H. D. Holland and K. Turekian. Amsterdam, Oxford: Elsevier, pp. 287–308.

Brett, R. and Sato, M. (1984). Intrinsic oxygen fugacity measurements on seven chondrites, a pallasite, and a tektite and the redox state of meteorite parent bodies. *Geochim. Cosmochim. Acta*, **48**, 111–120.

Brownlee, D. (2014). The Stardust mission: Analyzing samples from the edge of the solar system. *Annu. Rev. Earth Planet. Sci.*, **42**, 179–205.

Buchwald, V. F. (1975). *Handbook of Iron Meteorites, Their History, Distribution, Composition and Structure*. Berkeley, CA: University of California Press, 1426 pp.

Burbine, T. H., Meibom, A. and Binzel, R. P. (1996). Mantle material in the main belt: Battered to bits? *Meteorit. Planet. Sci.*, **31**, 607–620.

Burkhardt, C. (2014). Isotopic composition of the Moon and the lunar isotopic crisis. In *Encyclopedia of Lunar Science*. Springer (online), doi:10.1007/SpringerReference_440362.

Burkhardt, C., Kleine, T., Bourdon, B., Palme, H., Zipfel, J., Friedrich, J. and Ebel, D. S. (2008). Hf-W systematics of Ca-Al-rich inclusions from carbonaceous chondrites: Dating the age of the solar system and core formation in asteroids. *Geochim. Cosmochim. Acta*, **72**, 6177–6197.

Burkhardt C., Borg, L. E., Brennecka, G. A., Shollenberger, Q. R., Dauphas, N. and Kleine T. (2016). A nucleosynthetic origin for the Earth's anomalous ^{142}Nd composition. *Nature*, **537**, 394–398.

Busemann, H., Nguyen, A. N., Cody, G. D., Hoppe, P., Kilcoyne, A. L. D., Stroud, R. M., Zega, T. J. and Nittler, L. R. (2009). Ultra-primitive interplanetary dust particles from the comet 26P/Grigg-Skjellerup dust stream collection. *Earth Planet. Sci. Lett.*, **288**, 44–57.

Cameron, A. G. W. (1985). The partial volatilization of Mercury. *Icarus*, **64**, 285–294.

Cameron, A. G. W., Fegley, B., Benz, W. and Slattery, W. L. (1988). The strange density of Mercury: Theoretical considerations. In *Mercury*, ed. F. Vilas, C. R. Chapman and M. S. Matthews. Tucson, AZ: University of Arizona Press, pp. 692–708.

Canup, R. M. (2004). Dynamics of lunar formation. *Annu. Rev. Astron. Astrophys.*, **42**, 441–475.

Canup, R. M. (2008). Accretion of the Earth. *Phil. Trans. Roy. Soc. London A*, **366**, 4061–4075.

Canup, R. M. (2012). Forming a Moon with an Earth-like composition via a giant impact. *Science*, **338**, 1052–1055.

Canup, R. M. and Asphaug, E. (2001). Origin of the Moon in a giant impact near the end of the Earth's formation. *Nature*, **412**, 708–712.

Canup, R. M., Visscher, C., Salmon, J. and Fegley, B., Jr. (2015). Lunar volatile depletion due to incomplete accretion within an impact-generated disk. *Nature Geosci.*, **8**, 918–921, doi:10.1038/ngeo2574.

Carry, B. (2012). Density of asteroids. *Planet. Space Sci.*, **73**, 98–118.

Carter, P. J., Leinhardt, Z. M., Elliott, T., Walter, M. J. and Stewart, S. T. (2015). Compositional evolution during rocky protoplanet accretion. *Astrophys. J.*, **813**, 72–91.

Chambers, J. E. (2004). Planetary accretion in the inner Solar System. *Earth Planet. Sci. Lett.*, **223**, 241–252.

Chambers, J. E. (2009a). Planetary migration: What does it mean for planet formation? *Annu. Rev. Earth Planet. Sci.*, **37**, 321–344.

Chambers, J. E. (2009b). An analytical model for the evolution of a viscous, irradiated disk. *Astrophys. J.*, **705**, 1206–1214.

Chambers, J. E. (2013). Late-stage planetary accretion including hit-and-run collisions and fragmentation. *Icarus*, **224**, 43–56.

Chambers, J. E. (2014). Giant planet formation with pebble accretion. *Icarus*, **233**, 83–100.

Chapman, C. R. (1988). Mercury: Introduction to an end-member planet. In *Mercury*, ed. F. Vilas, C. R. Chapman and M. S. Matthews. Tucson, AZ: University of Arizona Press, pp. 1–23.

Chen, J. and Kipping, D. (2016). Probabilistic forecasting of the masses and radii of other worlds. *Astrophys. J.*, **834**, 17–30.

Clayton, R. N. and Mayeda, T. K. (1996). Oxygen isotope studies of achondrites. *Geochim. Cosmochim. Acta*, **60**, 1999–2017.

Consolmagno, G. J. and Britt, D. T. (2004). Meteoritical evidence and constraints on asteroid impacts and disruption. *Planet. Space Sci.*, **52**, 1119–1128.

Cottrell, E. and Kelley, K. A. (2011). The oxidation state of Fe in MORB glasses and the oxygen fugacity of the upper mantle. *Earth Planet. Sci. Lett.*, **305**, 270–282.

Ćuk, M. and Stewart, S. T. (2012). Making the Moon from a fast-spinning Earth: A giant impact followed by resonant despinning. *Science*, **338**, 1047–1052, doi:10.1126/science.1225542.

Cuzzi, J. N., Hogan, R. C. and Shariff, K. (2008). Toward planetesimals: Dense chondrule clumps in the protoplanetary nebula. *Astrophys. J.*, **687**, 1432–1447.

Dahl, T. W. and Stevenson, D. J. (2010). Turbulent mixing of metal and silicate during planet accretion – And interpretation of the Hf–W chronometer. *Earth Planet. Sci. Lett.*, **295**, 177–186.

D'Alessio, P., Calvet, N., Hartmann, L., Lizano, S. and Cantó, J. (1999). Accretion disks around young objects. II. Tests of well-mixed models with ISM dust. *Astrophys. J.*, **527**, 893–909.

Dauphas, N. and Morbidelli, A. (2014). Geochemical and planetary dynamical views on the origin of Earth's atmosphere and oceans. In *The Atmosphere – History*, ed. J. Farquhar, *Treatise on Geochemistry*, 2nd edn, Vol. 6, ed. H. D. Holland and K. Turekian. Amsterdam, Oxford: Elsevier, pp. 1–35.

Davis, A. M. (2006). Volatile evolution and loss. In *Meteorites and the Early Solar System II*, ed. D. Lauretta and H. Y. McSween, Jr. Tucson, AZ: University of Arizona Press, pp. 295–307.

Domingue, D. L., Chapman, C. R., Killen, R. M., Zurbuchen, T. H., Gilbert, J. A., Sarantos, M., Benna, M., Slavin, J. A., Schriver, D., Trávníček, P. M., Orlando, T. M., Sprague, A. L., Blewett, D. T., Gillis-Davis, J. J., Feldman, W. C., Lawrence, D. J., Ho, G. C., Ebel, D. S., Nittler, L. R., Vilas, F., Pieters, C. M., Solomon, S. C., Johnson, C. L., Winslow, R. M., Helbert, J., Peplowski, P. N., Weider, S. Z., Mouawad, N., Izenberg, N. R. and McClintock, W. E. (2014). Mercury's weather-beaten surface: Understanding Mercury in the context of lunar and asteroidal space weathering studies. *Space Sci. Rev.*, **181**, 121–214, doi:10.1007/s11214-014-0039-5.

Dressing, C. D. and Charbonneau, D. (2015). The occurrence of potentially habitable planets orbiting M dwarfs estimated from the full Kepler dataset and an empirical measurement of the detection sensitivity. *Astrophys. J.*, **807**, 45–68.

Dwyer, C. A., Nimmo, F. and Chambers, J. E. (2015). Bulk chemical and Hf-W isotopic consequences of incomplete accretion during planet formation. *Icarus*, **245**, 145–152.

Ebel, D. S. (2001). Vapor/liquid/solid equilibria when chondrites collide. *Meteorit. Planet. Sci. Suppl.*, **36**, A52–A53.

Ebel, D. S. (2006). Condensation of rocky material in astrophysical environments. In *Meteorites and the Early Solar System II*, ed. D. Lauretta and H. Y. McSween, Jr. Tucson, AZ: University of Arizona Press, pp. 253–277.

Ebel, D. S. and Alexander, C. M. O'D. (2011). Equilibrium condensation from chondritic porous IDP enriched vapor: Implications for Mercury and enstatite chondrite origins. *Planet. Space. Sci.*, **59**, 1888–1894.

Ebel, D. S. and Grossman, L. (2000). Condensation in dust-enriched systems. *Geochim. Cosmochim. Acta*, **64**, 339–366.

Ebel, D. S. and Sack, R. O. (2013). Djerfisherite: Nebular source of refractory potassium. *Contrib. Mineral. Petrol.*, **166**, 923–934.

Ebel, D. S., Brunner, C., Leftwich, K., Erb, I., Lu, M., Konrad, K., Rodriguez, H., Friedrich, J. M. and Weisberg, M. K. (2016). Abundance, composition and size of inclusions and matrix in CV and CO chondrites. *Geochim. Cosmochim. Acta*, **172**, 322–356, doi:10.1016/j.gca.2015.10.007.

Elkins-Tanton, L. T. (2013). Planetary science: Occam's origin of the Moon. *Nature Geosci.*, **6**, 996–998.

Ernst, C. M., Murchie, S. L., Barnouin, O. S., Robinson, M. L., Denevi, B. W., Blewett, D. T., Head, J. W., Izenberg, N. R., Solomon, S. C. and Roberts, J. H. (2010). Exposure of spectrally distinct material by impact craters on Mercury: Implications for global stratigraphy. *Icarus*, **209**, 210–223, doi:10.1016/j.icarus.2010.05.022.

Espaillat, D., Muzerolle, J., Najita, J., Andrews, S., Zhu, Z., Calvet, N., Kraus, S., Hashimoto, J., Kraus, A. and D'Alessio, P. (2014). An observational perspective of transitional disks. In *Protostars and Planets VI*, ed. H. Beuther, R. Klessen, C. Dullemond and Th. Henning. Tucson, AZ: University of Arizona Press, pp. 497–520.

Evans, L. G., Peplowski, P. N., Rhodes, E. A., Lawrence, D. J., McCoy, T. J., Nittler, L. R., Solomon, S. C., Sprague, A. L., Stockstill-Cahill, K. R., Starr, R. D., Weider, S. Z., Boynton, W. F., Hamara, D. K. and Goldsten, J. O. (2012). Major-element abundances on the surface of Mercury: Results from the MESSENGER Gamma-Ray Spectrometer. *J. Geophys. Res.*, **117**, E00L07, doi:10.1029/2012JE004178.

Evans, L. G., Peplowski, P. N., McCubbin, F. M., McCoy, T. J., Nittler, L. R., Zolotov, M. Yu., Ebel, D. S., Lawrence, D. J., Starr, R. D., Weider, S. Z. and Solomon, S. C. (2015). Chlorine on the surface of Mercury: MESSENGER gamma-ray measurements and implications for the planet's formation and evolution. *Icarus*, **257**, 417–427.

Fabrycky, D. C., Lissauer, J. J., Ragozzine, D., Fowe, J. F., Steffen, J. H., Agol, E., Barclay, T., Batalha, N., Borucki, W. and Ciardi, D. R. (2014). Architecture of Kepler's multi-transiting systems. II. New investigations with twice as many candidates. *Astrophys. J.*, **790**, 146–157.

Fang, J. and Margot, J. L. (2012). Architecture of planetary systems based on Kepler data: Number of planets and coplanarity. *Astrophys. J.*, **761**, 92–105.

Fedkin, A. V., Grossman, L., Humayun, M., Simon, S. B. and Campbell, A. J. (2015). Condensates from vapor made by impacts between metal-, silicate-rich bodies: Comparison with metal and chondrule in CB chondrites. *Geochim. Cosmochim. Acta*, **164**, 236–261.

Fegley, B., Jr. and Cameron, A. G. W. (1987). A vaporization model for iron/silicate fractionation in the Mercury protoplanet. *Earth Planet. Sci. Lett.*, **82**, 207–222.

Feigelson, E. D. (2010). X-ray insights into star and planet formation. *Proc. Natl. Acad. Sci.*, **107**, 7153–7157.

Friedrich, J. M., Weisberg, M. K., Ebel, D. S., Biltz, A. E., Corbett, B. M., Iotzov, I. V., Khan, W. S. and Wolman, M. D. (2015). Chondrule size and density in all meteorite groups: A compilation and evaluation of current knowledge. *Chemie der Erde*, **75**, 419–443, doi:10.1016/j.chemer.2014.08.003.

Frost, D. J., Mann, U., Asahara, Y. and Rubie, D. C. (2008). The redox state of the mantle during and just after core formation. *Phil. Trans. Roy. Soc. London A*, **366**, 4315–4337.

Genda, H., Kokubo, E. and Ida, S. (2011). Merging criteria for giant impacts of protoplanets. *Astrophys. J.*, **744**, 137–144.

Ghosal, S., Sack, R. O., Ghiorso, M. S. and Lipschutz, M. E. (1998). Evidence for a reduced, Fe-depleted martian mantle source region of shergottites. *Contrib. Mineral. Petrol.*, **130**, 346–357.

Gladman, B. and Coffey, J. (2009). Mercurian impact ejecta: Meteorites and mantle. *Meteorit. Planet. Sci.*, **44**, 285–291.

Goldschmidt, V. M. (1937). The principles of distribution of chemical elements in minerals and rocks. The seventh Hugo Müller Lecture, delivered before the Chemical Society on March 17th, 1937. *J. Chem. Soc.*, **1937**, 655–673.

Grossman, J. N. (1996). Chemical fractionations of chondrites: Signatures of events before chondrule formation. In *Chondrules and the Protoplanetary Disk*, ed. R. H. Hewins, R. H. Jones and E. R. D. Scott. New York: Cambridge University Press, pp. 243–253.

Grossman, L. and Larimer, J. W. (1974). Early chemical history of the solar system. *Rev. Geophys. Space Phys.*, **12**, 71–101.

Haisch, K. E., Lada, E. A. and Lada, C. J. (2001). Circumstellar disks in the IC 348 cluster. *Astrophys. J.*, **121**, 2065–2074.

Halliday, A. N. (2013). The origins of volatiles in the terrestrial planets. *Geochim. Cosmochim. Acta*, **105**, 146–171.

Hansen, B. M. S. (2009). Formation of the terrestrial planets from a narrow annulus. *Astrophys. J.*, **703**, 1131–1140.

Hartmann, L. (2009). The star-jet-disk system and angular momentum transfer. In *Protostellar Jets in Context*, ed. K. R. Tsinganos and M. Stute. Berlin: Springer, pp. 23–32.

Hauck S. A., II, Margot, J.-L., Solomon, S. C., Phillips, R. J., Johnson, C. L., Lemoine, F. G., Mazarico, E., McCoy, T. J., Padovan, S., Peale, S. J., Perry, M. E., Smith, D. E. and Zuber, M. T. (2013). The curious case of Mercury's internal structure. *J. Geophys. Res. Planets*, **118**, 1204–1220.

Hewins, R. H. and Ulmer, G. C. (1984). Intrinsic oxygen fugacities of diogenites and mesosiderite clasts. *Geochim. Cosmochim. Acta*, **48**, 1555–1560.

Hubbard, A. (2014). Explaining Mercury's density through magnetic erosion. *Icarus*, **241**, 329–335.

Izenberg, N. R., Klima, R. L., Murchie, S. L., Blewett, D. T., Holsclaw, G. M., McClintock, W. E., Malaret, E., Mauceri, C., Vilas, F., Sprague, A. L., Helbert, J., Domingue, D. L., Head, J. W., III, Goudge, T. A., Solomon, S. C., Hibbitts, C. A. and Dyar, M. D. (2014). The low-iron, reduced surface of Mercury as seen in spectral reflectance by MESSENGER. *Icarus*, **228**, 364–374.

Jarosewich, E. (1990). Chemical analyses of meteorites: A compilation of stony and iron meteorite analyses. *Meteoritics*, **25**, 323–337.

Johansen, A., Blum, J., Tanaka, H., Ormel, C., Bizzarro, M. and Rickman, H. (2014). The multifaceted planetesimal formation process. In *Protostars and Planets VI*, ed. H. Beuther, R. Klessen, C. Dullemond and Th. Henning. Tucson, AZ: University of Arizona Press, pp. 547–570.

Kant, I. (1755). Universal Natural History and Theory of the Heavens. In *Kant's Critical Religion* (2000), translated by S. Palmquist. Aldershot: Ashgate, 320 pp.

Kargel, J. S. and Lewis, J. S. (1993). The composition and early evolution of Earth. *Icarus*, **105**, 1–25.

Keil, K. (1968). Mineralogical and chemical relationships among enstatite chondrites. *J. Geophys. Res.*, **73**, 6945–6976.

Kleine, T., Mezger, K., Palme, H. and Münker, C. (2004). The W isotope evolution of the bulk silicate Earth: Constraints on the timing and mechanisms of core formation and accretion. *Earth Planet. Sci. Lett.*, **228**, 109–123.

Kleine, T., Touboul, M., Bourdon, B., Nimmo, F., Mezger, K., Palme, H., Jacobsen, S. B., Yin, Q.-Z. and Halliday, A. N. (2009). Hf-W chronology of the accretion and early evolution of asteroids and terrestrial planets. *Geochim. Cosmochim. Acta*, **73**, 5150–5188.

Klima, R. L., Izenberg, N. R., Murchie, S., Meyer, H. M., Stockstill-Cahill, K. R., Blewett, D. T., D'Amore, M., Denevi, B. W., Ernst, C. M., Helbert, J., McCoy, T., Sprague, A. L., Vilas, F., Weider, S. Z. and Solomon, S. C. (2013). Constraining the ferrous iron content of silicate minerals in Mercury's crust. *Lunar Planet. Sci.*, **44**, abstract 1602.

Kokubo, E. and Genda, H. (2010). Formation of terrestrial planets from protoplanets under a realistic accretion condition. *Astrophys. J. Lett.*, **714**, L21–L25.

Kokubo, E. and Ida, S. (1996). On runaway growth of planetesimals. *Icarus*, **123**, 180–191.

Kokubo, E. and Ida, S. (1998). Oligarchic growth of protoplanets. *Icarus*, **131**, 171–178.

Krauss, O. and Wurm, G. (2005). Photophoresis and the pile-up of dust in young circumstellar disks. *Astrophys. J.*, **630**, 1088–1092.

Krot, A. N., Amelin, Y., Cassen, P. and Meibom, A. (2005). Young chondrules in CB chondrites from a giant impact in the early Solar System. *Nature*, **436**, 989–992.

Krot, A. N., Amelin, Y., Bland, P., Ciesla, F. J., Connelly, H. J., Jr., Davis, A. M., Huss, G. R., Hutcheon, I. D., Makide, K., Nagashima, K., Nyquist, L. E., Russell, S. S., Scott, E. R. D., Thrane, K., Yurimoto, H. and Yin, Q.-Z. (2009). Origin and chronology of chondritic components: A review. *Geochim. Cosmochim. Acta*, **73**, 4963–4997.

Lambrechts, M. and Johansen, A. (2012). Rapid growth of gas-giant cores by pebble accretion. *Astron. Astrophys.*, 544, A32–A45.

Larimer, J. W. and Anders, E. (1967). Chemical fractionation in meteorites: II. Abundance patterns and their interpretation. *Geochim. Cosmochim. Acta*, **31**, 1239–1270.

Leinhardt, Z. M. and Stewart, S. T. (2011). Collisions between gravity-dominated bodies. I. Outcome regimes and scaling laws. *Astrophys. J.*, **745**, 79–106.

Leinhardt, Z. M., Dobinson, J., Carter, P. J. and Lines, S. (2015). Numerically predicted indirect signatures of terrestrial planet formation. *Astrophys. J.*, **806**, 23–32.

Levison, H. F., Morbidelli, A., Gomes, R. and Backman, D. (2007). Planet migration in planetesimal disks. In *Protostars and Planets V*, ed. B. Reipurth, D. Jewitt and K. Keil. Tucson, AZ: University of Arizona Press, pp. 669–684.

Levison, H. F., Kretke, K. A. and Duncan, M. J. (2015a). Growing the gas-giant planets by the gradual accumulation of pebbles. *Nature*, **524**, 322–324.

Levison, H. F., Kretke, K. A., Walsh, K. J. and Bottke W. F. (2015b). Growing the terrrestrial planets from the gradual accumulation of submeter-sized objects. *Proc. Natl. Acad. Sci.*, **112**, 14,180–14,185.

Lewis, J. S. (1972). Metal/silicate fractionation in the solar system. *Earth Planet. Sci. Lett.*, **15**, 286–290.

Lewis, J. S. (1973). Chemistry of the planets. *Annu. Rev. Phys. Chem.*, **24**, 339–352.

Lewis, J. S. (1988). Origin and composition of Mercury. In *Mercury*, ed. F. Vilas, C. R. Chapman and M. S. Matthews. Tucson, AZ: University of Arizona Press, pp. 651–666.

Lissauer, J. J., Fabrycky, D. C., Ford, E. B., Borucki, W. J., Fressin, F., Marcy, G. W., Rorosz, J. A., Rowe, J. F., Torres, G., Welsh, W. F., Batalha, N. M., Bryson, S. T., Buchhave, L. A., Caldwell, D. A., Cartre, J. A., Charbonneau, D., Christiansen, J. L., Cochran, W. D., Desert, J.-M., Dunham, E. W., Fanelli, M. N., Fortney, J. J., Gautier, T. N., III, Geary, J. C., Gilliland, R. L., Haas, M. R., Hall, J. R., Holman, M. J., Coch, D. G., Latham, D. W., Lopez, E., McCauliff, S., Miller, N., Morehead, R. C., Quintana, E. V., Ragozzine, D., Sasselov, D., Short, D. R. and Stefffen, J. H. (2011). A closely packed system of low-mass, low-density planets transiting Kepler-11. *Nature*, **470**, 53–58.

Lock S. J., Stewart, S. T., Petaev, M. I., Leinhardt, Z. M., Mace, M., Jacobsen, S. B. and Ćuk, M. (2016). A new model for lunar origin: Equilibration with Earth beyond the hot spin stability limit. *Lunar Planet. Sci.*, **47**, abstract 2881.

Lodders, K. (2003). Solar system abundances and condensation temperatures of the elements. *Astrophys. J.*, **591**, 1220–1247.

Lodders, K. and Fegley, B., Jr. (1997). An oxygen isotope model for the composition of Mars. *Icarus*, **126**, 373–394.

Lodders, K. and Fegley, B., Jr. (1998). *The Planetary Scientist's Companion*. New York: Oxford University Press.

Lodders, K., Palme, H. and Gail, H. P. (2009). Abundances of the elements in the solar system. In *Landolt-Börnstein, New Series*, Vol. VI/4B, ed. J. E. Trümper. Berlin, Heidelberg, New York: Springer-Verlag, pp. 560–630.

Loesche, C., Wurm, G., Kelling, T., Teiser, J. and Ebel, D. S. (2016). The motion of chondrules and other particles in a protoplanetary disk with temperature fluctuations. *Mon. Not. Roy. Astron. Soc.*, **463**, 4167–4174.

Macke, R. J. (2012). Survey of meteorite physical properties: Density, porosity and magnetic susceptibility. Ph.D. thesis, University of Central Florida, Orlando, FL.

Mahoney, T. J. (2014). *Mercury, A Compendium of the Astronomical Lexicon, Part A: Gazetteer and Atlas of Astronomy, Vol. I: The Terrestrial Planets, Part 1.* New York: Springer; doi:10.1007/978-1-4614-7951-2.

Malavergne, V., Toplis, M. J., Berthet, S. and Jones J. (2010). Highly reducing conditions during core formation on Mercury: Implications for internal structure and the origin of a magnetic field. *Icarus*, **206**, 199–209, doi:10.1016/j.icarus.2009.09.001.

Marcus, R. A., Stewart, S. T., Sasselov, D. and Hernquist, L. (2009). Collisional stripping and disruption of super-Earths. *Astrophys. J. Lett.*, **700**, L118–L122.

Marcus, R. A., Sasselov, D., Hernquist, L. and Stewart, S. T. (2010a). Minimum radii of super-Earths: Constraints from giant impacts. *Astrophys. J. Lett.*, **712**, L73–L76.

Marcus, R. A., Sasselov, D., Stewart, S. T. and Hernquist, L. (2010b). Water/icy super-Earths: Giant impacts and maximum water content. *Astrophys. J. Lett.*, **719**, L45–L49.

McClure, M. K., D'Alessio, P., Calvet, N., Espaillat, C., Hartmann, L., Sargent, B., Watson, D. M., Ingleby, L. and Hernández, J. (2013). Curved walls: Grain growth, settling, and composition patterns in T Tauri disk dust sublimation fronts. *Astrophys. J.*, **775**, 114–124.

McCubbin, F. M., Riner, M. A., Vander Kaaden, K. E. and Burkemper L. K. (2012). Is Mercury a volatile-rich planet? *Geophys. Res. Lett.*, **39**, L09202, doi:10.1029/2012GL051711.

McDonough W. F. (2014). Compositional model for the Earth's core. In *The Mantle and Core*, ed. R. W. Carlson, *Treatise on Geochemistry*, 2nd edn, Vol. 3, ed. H. D. Holland and K. Turekian. Amsterdam, Oxford: Elsevier, pp. 559–577.

McNally, C. P. and McClure, M. K. (2017). Photophoretic levitation and trapping of dust in the inner regions of protoplanetary disks. *Astrophys. J.*, **834**, 48–60.

McNally, C. P., Hubbard, A., Mac Low, M.-M., Ebel, D. S. and D'Alessio, P. (2013). Mineral processing by short circuits in protoplanetary disks. *Astrophys. J.*, **767**, L2–L7.

Melosh, H. J. (2014), New approaches to the Moon's isotopic crisis. *Phil. Trans. Roy. Soc. London A*, **372**, 20130168.

Messenger, S., Keller, L. P., Stadermann, F. J., Walker, R. M. and Zinner, E. (2003). Samples of stars beyond the Solar System: Silicate grains in interplanetary dust. *Science*, **300**, 105–108.

Mitchell, D. and de Pater, I. (1994). Microwave imaging of Mercury's thermal emission at wavelengths from 0.3 to 20.5 cm. *Icarus*, **110**, 2–32.

Morbidelli, A. and Raymond, S. N. (2016). Challenges in planet formation. *J. Geophys. Res. Planets*, **121**, 1962–1980, doi:10.1002/2016JE005088.

Morbidelli, A., Tsiganis, K., Crida, A., Levison, H. F. and Gomes, R. (2007). Dynamics of the giant planets of the Solar System in the gaseous protoplanetary disk and their relationship to the current orbital architecture. *Astron. J.*, **134**, 1790–1798.

Morbidelli, A., Lunine, J. I., O'Brien, D. P., Raymond, S. N. and Walsh, K.J. (2012). Building terrestrial planets. *Annu. Rev. Earth Planet. Sci.*, **40**, 251–275.

Morgan, J. W. and Anders, E. (1980). Chemical composition of Earth, Venus, and Mercury. *Proc. Natl. Acad. Sci.*, **77**, 6973–6977.

Moriarty, J., Madhusudhan, N. and Fischer, D. (2014) Chemistry in an evolving protoplanetary disk: Effects on terrestrial planet composition. *Astrophys. J.*, **787**, 81–91.

Morishima, R., Golabek, G. J. and Samuel, H. (2013). *N*-body simulations of oligarchic growth of Mars: Implications for Hf–W chronology. *Earth Planet. Sci. Lett.*, **366**, 6–16.

Murchie, S. L., Klima, R. L., Denevi, B. W., Ernst, C. M., Keller, M. R., Domingue, D. L., Blewett, D. T., Chabot, N. L., Hash, C. D., Malaret, E., Izenberg, N. R., Vilas, F., Nittler, L. R., Gillis-Davis, J. J., Head, J. W. and Solomon, S. C. (2015). Orbital multispectral mapping of Mercury with the MESSENGER Mercury Dual Imaging System: Evidence for the origins of plains units and low-reflectance material. *Icarus*, **254**, 287–305.

Nakajima, M. and Stevenson, D. J. (2015). Melting and mixing states of the Earth's mantle after the Moon-forming impact. *Earth Planet. Sci. Lett.*, **427**, 286–295.

Namur, O., Charlier, B., Holtz, F., Cartier, C. and McCammon, C. (2016). Sulfur solubility in reduced mafic silicate melts: Implications for the speciation and distribution of sulfur on Mercury. *Earth Planet. Sci. Lett.*, **448**, 102–114.

Nittler, L. R., Starr, R. D., Weider, S. Z., McCoy, T. J., Boynton, W. V., Ebel, D. S., Ernst, C. M., Evans, L. G., Goldsten, J. O., Hamara, D. K., Lawrence, D. J., McNutt, R. L., Jr., Schlemm, C. E., II, Solomon, S. C. and Sprague, A. L. (2011). The major-element composition of Mercury's surface from MESSENGER X-ray spectrometry. *Science*, **333**, 1847–1850.

O'Brien, D. P., Morbidelli A. and Levison, H. F. (2006). Terrestrial planet formation with strong dynamical friction. *Icarus*, **184**, 39–58.

O'Neill, H. St. C. and Palme H. (2008). Collisional erosion and the non-chondritic composition of the terrestrial planets. *Phil. Trans. Roy. Soc. London A*, **366**, 4205–4238.

Ostro, S. J., Hudson, R. S., Nolan, M. C., Margo, J-L., Scheeres, D. J., Campbell, D. B., Magri, C., Giorgini, J. D. and Yeomans, D. K. (2000). Radar observations of asteroid 216 Kleopatra. *Science*, **288**, 836–839. doi:10.1126/science.288.5467.836.

Pasek, M. A., Milsom, J. A., Ciesla, F. J., Lauretta, D. S., Sharp C. M. and Lunine, J. I. (2005). Sulfur chemistry with time-varying oxygen abundance during Solar System formation. *Icarus*, **175**, 1–14.

Peplowski, P. N., Evans, L. G., Hauck, S. A., II, McCoy, T. J., Boynton, W. V., Gillis-Davis, J.-J., Ebel, D. S., Goldsten, J. O., Hamara, D. K., Lawrence, D. J., McNutt, R. L., Jr., Nittler, L. R., Solomon, S. C., Rhodes, E. A., Sprague, A. L., Starr, R. D. and Stockstill-Cahill, K. R. (2011). Radioactive elements on Mercury's surface from MESSENGER: Implications for the planet's formation and evolution. *Science*, **333**, 1850–1852.

Peplowski, P. N., Lawrence, D. J., Rhodes, E. A., Sprague, A. L., McCoy, T. J., Denevi, B. W., Evans, L. G., Head, J. W., Nittler, L. R., Solomon, S. C., Stockstill-Cahill, K. R. and Weider, S. Z. (2012). Variations in the abundances of potassium and thorium on the surface of Mercury: Results from the MESSENGER Gamma-Ray Spectrometer. *J. Geophys. Res.*, **117**, E00L04, doi:10.1029/2012JE004141.

Peplowski, P. N., Evans, L. G., Stockstill-Cahill, K. R., Lawrence, D. J., Goldsten, J. O., McCoy, T. J., Nittler, L. R., Solomon, S. C., Sprague, A. L., Starr, R. D. and Weider, S. Z. (2014). Enhanced sodium abundance in Mercury's north polar region revealed by the MESSENGER Gamma-Ray Spectrometer. *Icarus*, **228**, 86–95.

Peplowski, P. N., Lawrence, D. J., Evans, L. G., Klima, R. L., Blewett, D. T., Goldsten, J. O., Murchie, S. L., McCoy, T. J., Nittler, L. R., Solomon, S. C., Starr, R. D. and Weider, S. Z. (2015). Constraints on the abundance of carbon in near-surface materials on Mercury: Results from the MESSENGER gamma-ray spectrometer. *Planet. Space Sci.*, **108**, 98–107.

Peplowski, P. N., Klima, R. L., Lawrence, D. J., Ernst, C. M., Denevi, B. W., Frank, E. A., Goldsten, J. O., Murchie, S. L., Nittler, L. R. and Solomon, S. C. (2016). Remote sensing evidence for an ancient carbon-bearing crust on Mercury, *Nature Geosci.*, **9**, 273–276, doi:10.1038/ngeo2669.

Pignatale, F. C., Liffman, K., Maddison, S. T. and Brooks, G. (2016). 2D condensation model for the inner Solar Nebula: An enstatite-rich environment. *Mon. Not. Roy. Astron. Soc.*, **457**, 1359–1370.

Quintana, E. V., Barclay, T., Borucki, W., Rowe, J. F. and Chambers, J. E. (2016). Giant impacts on Earth-like worlds. *Astrophys. J.*, **821**, 126–139.

Raymond, S. N., O'Brien, D. P., Morbidelli, A. and Kaib, N. A. (2009). Building the terrestrial planets: Constrained accretion in the inner Solar System. *Icarus*, **203**, 644–662.

Reipurth, B. and Bally, J. (2001). Herbig-Haro flows: Probes of early stellar evolution. *Annu. Rev. Astron. Astrophys.*, **49**, 195–236.

Reufer, A., Meier, M. M. M., Benz, W. and Weiler, R. (2012). A hit-and-run giant impact scenario. *Icarus*, **221**, 296–299.

Righter, K. (2003). Metal-silicate partitioning of siderophile elements and core formation in the early Earth. *Annu. Rev. Earth Planet. Sci.*, **31**, 135–174.

Righter, K., Arculus, R. J., Delano, J. W. and Paslick C. (1990). Electrochemical measurements and thermodynamic calculations of redox equilibria in pallasite meteorites: Implications for the eucrite parent body. *Geochim. Cosmochim. Acta*, **54**, 1803–1815.

Righter, K., Drake, M. J. and Scott, E. (2006). Compositional relationships between meteorites and terrestrial planets. In *Meteorites and the Early Solar System II*, ed. D. Lauretta and H. Y. McSween, Jr. Tucson, AZ: University of Arizona Press, pp. 803–828.

Righter, K., Humayun, M. and Danielson, L. (2008). Partitioning of palladium at high pressures and temperatures during core formation. *Nature Geosci.*, **1**, 321–323, doi:10.1038/ngeo180.

Righter, K., Sutton, S. R., Danielson, L., Pando, K. and Newville, M. (2016). Redox variations in the inner solar system with new constraints from vanadium XANES in spinels. *Amer. Mineral.*, **101**, 1928–1942.

Ringwood, A. E. and Kesson, S. E. (1977). Basaltic magmatism and the bulk composition of the Moon. *Moon*, **16**, 425–464.

Rizo, H., Walker, R. J., Carlson, R. W., Horan, M. F., Mukhopadhyay, S., Manthos, V., Francis, D. and Jackson, M. G. (2016). Preservation of Earth-forming events in the tungsten isotopic composition of modern flood basalts. *Science*, **352**, 809–812.

Robinson, M. S. and Taylor, G. J. (2001). Ferrous oxide in Mercury's crust and mantle. *Meteorit. Planet. Sci.*, **36**, 841–847.

Robinson, M. S., Murchie, S. L., Blewett, D. T., Domingue, D. L., Hawkins, S. E., III, Head, J. W., Holsclaw, G. M., McClintock, W. E., McCoy, T. J., McNutt, R. L., Jr., Prockter, L. M., Solomon, S. C. and Watters, T. R. (2008). Reflectance and color variations on Mercury: Regolith processes and compositional heterogeneity. *Science*, **321**, 66–69, doi:10.1126/science.1160080.

Rubie, D. C., Jacobson, S. A., Morbidelli, A., O'Brien, D. P., Young, E. D., de Vries, J., Nimmo, F., Palme, H. and Frost, D. J. (2015). Accretion and differentiation of the terrestrial planets with implications for the compositions of early-formed Solar System bodies and accretion of water. *Icarus*, **248**, 89–108.

Rudge, J. F., Kleine, T. and Bourdon, B. (2010). Broad bounds on Earth's accretion and core formation constrained by geochemical models. *Nature Geosci.*, **3**, 439–443.

Russell, S. S., Hartmann, L., Cuzzi, J. N., Krot, A. N., Gounelle, M. and Weidenschilling, S. J. (2006). Timescales of the solar protoplanetary disk. In *Meteorites and the Early Solar System II*, ed. D. Lauretta and H. Y. McSween, Jr. Tucson, AZ: University of Arizona Press, pp. 233–251.

Safronov, V. S. (1972). *Evolution of the Protoplanetary Cloud and Formation of the Earth and Planets*. Tech. Transl. F-677. Washington, DC: NASA.

Sarid, G., Stewart, S. T. and Leinhardt, Z. M. (2014). Mercury, the impactor. *Lunar Planet. Sci.*, **46**, abstract 2723.

Schönbächler, M., Carlson, R. W., Horan M. F., Mock T. D. and Hauri, E. H. (2010). Heterogeneous accretion and the moderately volatile element budget of Earth. *Science*, **328**, 884–887.

Simon, S. B., Sutton, S. R. and Grossman, L. (2007). Valence of titanium and vanadium in pyroxene in refractory inclusion interiors and rims. *Geochim. Cosmochim. Acta*, **71**, 3098–3118.

Smith, J. V. (1979). Mineralogy of the planets: A voyage in space and time. *Mineral. Mag.*, **43**, 1–89.

Sneden, S., Lawler, J. E., Cowan, J. J., Ivans, I. I. and Den Hartog, E. A. (2009). *Astrophys. J. Suppl. Ser.*, **182**, 80–96.

Solomon, S. C., McNutt, R. L., Jr., Gold, R. E., Acuña, M. H., Baker, D. N., Boynton, W. V., Chapman, C. R., Cheng, A. F., Gloeckler, G., Head, J. W., III, Krimigis, S. M., McClintock, W. E., Murchie, S. L., Peale, S. J., Phillips, R. J., Robinson, M. S., Slavin, J. A., Smith, D. E., Strom, R. G., Trombka, J. I. and Zuber, M T. (2001). The MESSENGER mission to Mercury: Scientific objectives and implementation. *Planet. Space Sci.*, **49**, 1445–1465.

Solomon, S. C., McNutt, R. L., Jr., Gold, R. E. and Domingue, D. L. (2007). MESSENGER mission overview. *Space Sci. Rev.*, **131**, 3–39.

Sprague, A. L. and Roush, T. L. (1998). Comparison of laboratory emission spectra with Mercury telescopic data. *Icarus*, **133**, 174–183.

Stewart, S. T. and Leinhardt, Z. M. (2012). Collisions between gravity-dominated bodies. II. The diversity of impact outcomes during the end stage of planet formation. *Astrophys. J.*, **751**, 32–49.

Stewart, S. T., Leinhardt, Z. M. and Humayun, M. (2013). Giant impacts, volatile loss, and the K/Th ratios on the Moon, Earth, and Mercury. *Lunar Planet. Sci.*, **44**, abstract 2306.

Stewart, S. T., Lock, S. J., Petaev, M. I., Jacobsen, S. B., Sarid, G., Leinhardt, Z. M., Mukhopadhyay, S. and Humayun, M. (2016). Mercury impact origin hypothesis survives the volatile crisis: Implications for terrestrial planet formation. *Lunar Planet. Sci.*, **47**, abstract 2954.

Svetsov, V. (2011). Cratering erosion of planetary embryos. *Icarus*, 214, 316–326.

Szymanski, A., Brenker, F. E., Palme, H. and El Goresy, A. (2010). High oxidation state during formation of Martian nakhlites. *Meteorit. Planet. Sci.*, **45**, 21–31.

Tsiganis, K., Gomes, R., Morbidelli, A. and Levison, H. F. (2005). Origin of the orbital architecture of the giant planets of the Solar System. *Nature*, **435**, 459–461.

Tuff, J., Wade, J. and Wood, B. J. (2013). Volcanism on Mars controlled by early oxidation of the upper mantle. *Nature*, **498**, 342–345.

Urey, H. (1950). The origin and development of the Earth and other terrestrial planets. *Geochim. Cosmochim. Acta*, **1**, 209–277.

Vander Kaaden, K. E. and McCubbin, F. M. (2015). Exotic crust formation on Mercury: Consequences of a shallow, FeO-poor mantle. *J. Geophys. Res. Planets*, **120**, 195–209, doi:10.1002/2014je004733.

Vilas, F. (1988). Surface composition of Mercury from reflectance spectrophotometry. In *Mercury*, ed. F. Vilas, C. R. Chapman and M. S. Matthews. Tucson, AZ: University of Arizona Press, pp. 59–76.

Vityazev, A. V., Pechernikova, G. V. and Safronov, V. S. (1988). Formation of Mercury and removal of its silicate shell. In *Mercury*, ed. F. Vilas, C. Chapman and M. S. Matthews. Tucson, AZ: University of Arizona Press, pp. 667–669.

Wadhwa, M. (2001). Redox state of Mars' upper mantle and crust from Eu anomalies in shergottite pyroxenes. *Science*, **291**, 1527–1530.

Wadhwa, M. (2008). Redox conditions on small bodies, the Moon and Mars. *Rev. Mineral. Geochem.*, **68**, 493–510.

Wagner, F. W., Sohl, F., Hussmann, H., Grott, M. and Rauer, H. (2011). Interior structure models of solid exoplanets using material laws in the infinite pressure limit. *Icarus*, **214**, 366–376.

Walsh, K. J., Morbidelli, A., Raymond, S. N., O'Brien, D. P. and Mandell, A. M. (2011). A low mass for Mars from Jupiter's early gas-driven migration. *Nature*, **475**, 206–209, doi:10.1038/nature10201.

Wänke, H. (1981) Constitution of terrestrial planets. *Phil. Trans. Roy. Soc. London A*, **303**, 287–302.

Warell, J. and Blewett, D. T. (2004). Properties of the Hermean regolith: V. New optical reflectance spectra, comparison with lunar anorthosites, and mineralogical modelling. *Icarus*, **168**, 257–276.

Wasson, J. T. and Kallemeyn, G. W. (1988). Composition of chondrites. *Phil. Trans. Roy. Soc. London A*, **325**, 535–544.

Wasson, J. T. and Wai, C. M. (1970). Composition of the metal, schreibersite and perryite of enstatite achondrites and the origin of enstatite chondrites and achondrites. *Geochim. Cosmochim. Acta*, **34**, 169–184.

Weidenschilling, S. J. (1978). Iron/silicate fractionation and the origin of Mercury. *Icarus*, **35**, 99–111.

Weider, S. Z., Nittler, L. R., Starr, R. D., McCoy, T. J., Stockstill-Cahill, K. R., Byrne, P. K., Denevi, B. W., Head, J. W. and Solomon, S. C. (2012). Chemical heterogeneity on Mercury's surface revealed by the MESSENGER X-Ray Spectrometer. *J. Geophys. Res.*, **117**, E00L05, doi:10.1029/2012JE004153.

Weider, S. Z., Nittler, L. R., Starr, R. D., McCoy, T. J. and Solomon, S. C. (2014). Variations in the abundance of iron on Mercury's surface from MESSENGER X-Ray Spectrometer observations. *Icarus*, **235**, 170–186.

Weider, S. Z., Nittler, L. R., Starr, R. D., Crapster-Pregont, E. J., Peplowski, P. N., Denevi, B. W., Head, J. W., Byrne, P. K., Hauck, S. A., II, Ebel, D. S. and Solomon, S. C. (2015). Evidence for geochemical terranes on Mercury: Global mapping of major elements with MESSENGER's X-Ray Spectrometer. *Earth Planet. Sci. Lett.*, **416**, 109–120.

Weisberg, M. K. and Kimura, M. (2012). The unequilibrated enstatite chondrites. *Chemie der Erde*, **72**, 101–115.

Weisberg, M. K., Prinz, M. and Nehru, C. E. (1988). Petrology of ALH85085: A chondrite with unique characteristics. *Earth Planet. Sci. Lett.*, **91**, 19–32.

Weisberg, M. K., Prinz, M., Clayton, R. N., Mayeda, T. K., Sugiura, N., Zashu, S. and Ebihara, M. (2001). A new metal-rich chondrite grouplet. *Meteorit. Planet. Sci.*, **36**, 3401–3418.

Weisberg, M. K., Ebel, D. S., Nakashima, D., Kita, N. T. and Humayun, M. (2015). Petrology and geochemistry of chondrules and metal in NWA 5492 and GRO 95551: A new type of metal-rich chondrite. *Geochim. Cosmochim. Acta*, **167**, 269–285.

Weisberg, M. K., Bigolski, J., Ebel, D. S. and Walker, D. (2016). Calcium-aluminum-rich (CAI) and sodium-aluminum-rich (NAI) inclusions in the PAT 91546 CH chondrite. *Lunar Planet. Sci.*, **47**, abstract 2152.

Wetherill, G. W. (1994). Provenance of the terrestrial planets. *Geochim. Cosmochim. Acta*, **58**, 4513–4520.

Williams, J. P. and Cieza, L. A. (2011). Protoplanetary disks and their evolution. *Annu. Rev. Astron. Astrophys.*, **49**, 67–118.

Wood, B. J., Walter, M. J. and Wade, J. (2006). Accretion of the Earth and segregation of its core. *Nature*, **441**, 825–833.

Wood, B. J., Wade, J. and Kilburn, M. (2009). Core formation and the oxidation state of the Earth: Additional constraints from Nb, V and Cr partitioning. *Geochim. Cosmochim. Acta*, **72**, 1415–1426.

Wurm, G., Trieloff, M. and Rauer, H. (2013). Photophoretic separation of metals and silicates: The formation of Mercury like planets and metal depletion in chondrites. *Astrophys. J.*, **769**, 78–85.

Yang J., Goldstein J. I. and Scott E. R. D. (2007). Iron meteorite evidence for early formation and catastrophic disruption of protoplanets. *Nature*, **446**, 888–891.

Zeng, L., Sasselov, D. and Jacobsen, S. (2016). Mass–radius relation for rocky planets based on PREM. *Astrophys. J.*, **819**, 127–131.

Zolotov, M. Yu., Sprague, A. L., Hauck, S. A., II, Nittler, L. R., Solomon, S. C. and Weider, S. Z. (2013). The redox state, FeO content, and origin of sulfur-rich magmas on Mercury. *J. Geophys. Res. Planets*, **118**, 138–146, doi:10.1029/2012JE004274.

19

Mercury's Global Evolution

STEVEN A. HAUCK, II, MATTHIAS GROTT, PAUL K. BYRNE, BRETT W. DENEVI, SABINE STANLEY,
AND TIMOTHY J. MCCOY

19.1 INTRODUCTION

From formation to quiescence, the history of a planet is the consequence of an intricate set of relationships between processes that both shape the surface and operate through the entirety of the planet (Kaula, 1975). MESSENGER, which completed the first orbital investigation of Mercury in April 2015 (Chapter 1), has revealed that planet to be as rich an example of that intricacy as any of the major bodies of the inner solar system. Mercury has long been known as a planet of enigmas, from its 3:2 spin–orbit resonance with the Sun, to its global contraction, to its unexpected magnetic field (Solomon, 2003). Now, MESSENGER has unveiled the majority of the planet that was previously unseen (Chapters 6, 9–13), characterized the large-scale chemical composition and heterogeneity of the surface (Chapters 2, 7–8), determined Mercury's shape, gravity, and rotational state (Chapters 3–4), and revealed unknown structure and ancient activity of the magnetic field (Chapter 5).

The broad set of observations of Mercury's surface and interior by MESSENGER places fundamental constraints on the processes governing the planet's evolution. Although few of these observations individually lead to unique conclusions about the history of the innermost planet, taken as a whole, and in combination with an understanding of the processes that operate on and within planets in general, they provide an important picture of how Mercury evolved. At its most basic level, a planet seen today is the consequence of how material and heat have been transported on and to its surface and within the interior. Mercury's early history was marked by both intense bombardment and widespread volcanism (Chapters 6, 9, 11). Generally overprinting this record of crustal growth and reworking is a global set of tectonic features, predominantly shortening in nature and indicative of substantial contraction of Mercury, formed largely since the end of the period of heaviest bombardment of the planet (Chapter 10). MESSENGER's observations of remanent crustal magnetism during its final year in orbit revealed that Mercury possessed an internal magnetic field early in the planet's history (Chapter 5). This result indicates that within the first several hundred million years of Mercury's history, the deep interior where the magnetic field was generated was vigorously active. Each of these findings is set against the backdrop of a geochemically diverse and quite surprising surface and, by inference, interior composition (Chapters 2, 7). Indeed, MESSENGER found Mercury to be the most chemically reduced terrestrial planet on the basis of its low surface abundance of iron and relatively large surface abundance of sulfur (Nittler et al., 2011). Furthermore, MESSENGER observations showed the planet to be unexpectedly volatile rich, including considerable abundances of the heat-producing elements potassium, thorium, and uranium (Peplowski et al., 2011). The chemically reduced interior has major implications for the composition of Mercury's core, its structure, and how the magnetic field is generated, as does the newly constrained understanding of the abundance of heat-producing elements, which control the rate at which the planet cooled and its ability to generate magma.

In order to better understand how Mercury evolved over the past 4.5 billion years we synthesize observations by MESSENGER that elucidate the primary processes that have governed its history. We begin by outlining results from MESSENGER that clarify both how the crust of the planet formed and the history of the crust and lithosphere, including constraints from observations of surface geochemistry, the record of volcanism and tectonics, and the structure of the crust. Then we focus on observations that provide information on the state, structure, and behavior of the deeper interior. In tandem, we investigate the thermochemical evolution of the interior of Mercury subject to the constraints provided by MESSENGER's observations. Finally, we discuss the implications of these results for the history of the planet and outline prospects for future progress on understanding how the whole of Mercury has evolved.

19.2 EARLIEST HISTORY OF THE CRUST

19.2.1 Geological Constraints

The geologic record of Mercury's earliest crust – the outermost, petrologically distinct layer of the silicate portion of the planet derived from melting of the mantle (e.g., Brown and Elkins-Tanton, 2009; Namur et al., 2016; Namur and Charlier, 2017; Chapter 3) – is largely obscured by resurfacing by both impacts and volcanism (e.g., Trask and Guest, 1975; Spudis and Guest, 1988; Strom and Neukum, 1988; Denevi et al., 2009). Indeed, the most heavily cratered terrain has been estimated to have an age of 4.0–4.1 Gyr (Marchi et al., 2013). However, despite the fact that there are no areas of the crust that can be quantifiably ascribed to the first ~500 Myr of Mercury's history, important clues to the nature and origin of the crust are found in several areas that appear to have undergone only minimal resurfacing as well as in material exposed from depth by large impact events (Chapter 6).

Spectral units termed low-reflectance material (LRM) (Robinson et al., 2008; Denevi et al., 2009; Murchie et al., 2015; Klima et al., 2016) appear to be one key to our understanding of Mercury's crust. With a reflectance of just 4–5% at 550 nm (Chapter 8), LRM is ~30% darker than Mercury's average surface and is found concentrated in the ejecta of large impact craters (Denevi et al., 2009; Ernst et al., 2010; Klima et al., 2016). The reflectance and spectral properties of the LRM are consistent with the deposits having a graphite component (Murchie et al., 2015). Furthermore, increases in thermal neutron count rates associated with LRM deposits suggest a carbon abundance that is 1–3 wt% higher than that of surrounding terrain (Peplowski et al., 2015a, 2016). These observations are consistent with the hypothesis that Mercury developed a carbon-rich flotation crust due to buoyancy of graphite in an early magma ocean (Vander Kaaden and McCubbin, 2015).

Any early crust, particularly one as thin as a graphite-rich crust might have been, was surely disrupted heavily by impacts, modified by magmatic intrusions, and buried by volcanic deposits. Therefore, the modern distribution of this primordial material on the surface is limited, as it has been substantially mixed and diluted with other materials. By this reasoning, LRM is the material with the greatest concentration of carbon in a C-rich crust (Peplowski et al., 2015a). The depth of origin of LRM, calculated from the excavation depth of impact craters, is often several to tens of kilometers (Denevi et al., 2009; Ernst et al., 2010, 2015; Peplowski et al., 2015a). These depth estimates provide lower bounds to the depth of burial by impact and volcanic deposits subsequent to the formation of the original flotation crust. In some of the most heavily cratered terrains, the overall surface is relatively low in reflectance and all impact craters in the region expose LRM, suggesting that these regions may have experienced less resurfacing than average (Chapter 6). However, in other large regions, no LRM is found in any crater smaller than ~150 km in diameter, suggesting burial by at least 8 km of volcanic material (Chapter 6). Rivera-Valentin and Barr (2014) explored impact redistribution models for an impactor population consistent with Mercury's cratering record and found that the LRM is consistent with a darkening agent approximately 30 km deep, which would be within the lowermost crust or upper mantle (James et al., 2015; Padovan et al., 2015). Concentration of a darkening agent, such as graphite, from a layer deep within the crust may also imply that volcanism was substantial and occurred with a flux much greater than that of impact redistribution of upper crustal material in the period before the onset of the late heavy bombardment (LHB). Otherwise, the darkening agent would have been efficiently mixed throughout the crust and unlikely to display variations associated with exhumation from depth.

19.2.2 Geochemical State of the Crust and Mantle

The composition and chemical diversity of the surface of Mercury provide important insights into the nature, origin, and evolution of the crust and mantle. Given that Mercury is strongly differentiated with an uncommonly low silicate-to-metal ratio (Chapters 2, 4), understanding the mechanisms that may be responsible for Mercury's crustal formation has been a long-standing question. At their most basic, models for the formation of the crust include partial melting of an undifferentiated, chondritic-like mantle; formation as the uppermost layer of a solidifying magma ocean; or products of remelting of a magma ocean.

Geochemical observations and the relative ages of the surface units of Mercury argue against an undifferentiated mantle as the source region for melts erupted onto the surface. Melting of enstatite chondrites has been investigated experimentally and modeled from phase equilibria to understand both the origin of the highly reduced aubrite parent body (McCoy et al., 1999) and Mercury (Burbine et al., 2002; Malavergne et al., 2010). An undifferentiated chondritic mantle would produce sodium-rich melts at low degrees of partial melting, consistent with the composition of the northern smooth plains (NSP) (Vander Kaaden and McCubbin, 2016). However, the high Mg/Si and low Al/Si ratios observed for Mercury's average surface composition require relatively high degrees of partial melting (Burbine et al., 2002; Nittler et al., 2011). Further, the formation of the high-sodium flood basalts of the NSP relatively late in the history of Mercury would require a fertile mantle source that had not experienced earlier partial melting. Finally, the highly differentiated nature of Mercury, including the presence of a large core, argues against preservation of a wholly undifferentiated mantle.

A widely accepted model of the mantle and crust suggests that Mercury once had a magma ocean responsible for an initial stage of silicate differentiation. Prior to MESSENGER's orbit insertion and the early geochemical measurements of the surface of Mercury, the nature of the crust and the bulk composition of the surface and planet were poorly constrained, although the surface was known to be FeO-poor and the bulk composition of the planet to be rich in iron metal (Taylor and Scott, 2003, and references therein). This uncertainty led to a range of magma ocean models producing either a plagioclase flotation crust or a low-FeO magmatic crust, depending on the bulk composition of the magma ocean (Brown and Elkins-Tanton, 2009; Riner et al., 2009). Some of these petrologic models produce gravitationally unstable mantles that would experience overturn, similar to that posited for the lunar mantle.

With the realization that the crust of Mercury is neither a plagioclase-rich flotation crust nor chemically homogeneous, models emerged that considered a magma ocean with subsequent remelting (Charlier et al., 2013; Vander Kaaden and McCubbin, 2015, 2016). Charlier et al. (2013) suggested that compositional heterogeneity observed during early MESSENGER orbital observations could have been the result of melting of different layers within the mantle during convection and adiabatic pressure-release melting, even in the absence of mantle overturn. Vander Kaaden and McCubbin (2015) strengthened the argument against a significant primary flotation crust experimentally by demonstrating that graphite is the only phase that would be buoyant in a Mercury magma ocean. The equivalent thickness of such a graphite layer is directly dependent on the concentration of carbon in the silicate portion of the planet. Should Mercury have a bulk silicate carbon content similar to those of Earth, Mars, or the Moon, that layer might be up to ~100-m thick. However, if Mercury's carbon content is more similar to that of chondritic materials, a graphite crust could range in thickness from as little as 100 m to more than 10 km, with the largest values for bulk silicate compositions similar to carbonaceous chondrites (Vander Kaaden and McCubbin, 2015). These authors further noted that, unlike the case in many other planetary bodies, partial melts derived from mantle melting on Mercury are buoyant throughout the mantle and would rise to the surface without stalling at some neutral buoyancy depth. Thus, the crust of Mercury is likely composed of an impact-gardened mixture of primary crust formed during a magma ocean stage and subsequent volcanic deposits. Vander Kaaden and McCubbin (2016) further refined this idea by noting that a crystallizing magma ocean without buoyant silicate phases would concentrate incompatible elements, including volatiles, near the surface of the planet. Thus, remelting of shallow cumulates can produce volatile-rich compositions, like the NSP, even at high degrees of partial melting.

19.3 HISTORY OF THE CRUST AND LITHOSPHERE

Geological observations provide compelling evidence that Mercury's crust is largely volcanic in origin and has experienced widespread tectonic deformation. The accumulated, observable history of Mercury's crust and lithosphere contains fundamental clues to the processes that shaped the surface of the planet, and, importantly, the time progression of these processes. Whereas the earliest history of the planet may have included a magma ocean and the generation of a thin and rather exotic flotation crust, it is the subsequent history that is more discernable. MESSENGER's collected geophysical, geological, and geochemical observations of Mercury provide important insights into both the planet's integrated history and many discrete events, of variable duration, that reflect its evolutionary path.

19.3.1 Crustal Thickness

In addition to the geochemical and geological markers of crustal formation, Mercury's gravity field and topography provide important clues to the nature and formation of the crust (Perry et al., 2015; Tosi et al., 2015; Chapter 3). Mercury's crust is the product of the combined processes of crystallization of any magma ocean and upward transport of mantle partial melts integrated over the course of the planet's history. Therefore, knowledge of the thickness of the crust is a crucial indicator of the efficiency and pattern of igneous differentiation of the planet, which in turn depend strongly on Mercury's internal activity.

Orbital observations of Mercury's gravity field by MESSENGER provided the first detailed measurements of its mass distribution. MESSENGER's eccentric orbit (Chapter 1), with the periapsis located at a high northern latitude, resulted in gravity field measurements that have the highest spatial resolution in the north and that resolve only much longer wavelengths in the southern hemisphere (Smith et al., 2012; Mazarico et al., 2014; Verma and Margot, 2016). Focusing on the higher-resolution information in the northern hemisphere, several estimates of the thickness of the crust have been calculated (Smith et al., 2012; James et al., 2015; Padovan et al., 2015). The most recent models place the average crustal thickness of the northern hemisphere at 35 ± 18 km on the basis of geoid-to-topography ratios (GTRs) (Padovan et al., 2015) and place a minimum on the average thickness of 38 km with a model that accounts for both crustal and mantle sources of compensation (James et al., 2015). Density differences between the crust and mantle are a major source of uncertainty in crustal thickness models. Padovan et al. (2015) considered a range of crustal densities from 2700 to 3100 kg m^{-3}, with the upper bound consistent with grain densities they inferred from MESSENGER elemental compositions, the lower bound the result of including 12% porosity throughout the crust, as has been inferred for the Moon (Wieczorek et al., 2013), and a mantle density of 3300 kg m^{-3}. This range overlaps independent estimates of the grain densities calculated from experimental determinations of the modal mineralogy consistent with the range of surface compositions across Mercury (Namur and Charlier, 2017). Similarly, the inversion approach of James et al. (2015) was for a nominal crustal density of 3200 kg m^{-3} and a mantle density of 3400 kg m^{-3}. Generally speaking, the small difference in grain density between the crust and mantle, approximately 200 kg m^{-3}, is a reflection of the inferred low iron content of Mercury's silicate layers. This density difference is also important for crustal flow models, as the driving stress for any topographic relaxation via lower-crustal flow scales directly with the density contrast (e.g., Nimmo and Stevenson, 2001), so a small density contrast implies less lower-crustal flow. Potentially of greater importance is that the inferred crustal thickness values when compared with the thickness of the mantle imply that Mercury has experienced the most efficient extraction of crust among the terrestrial bodies. Indeed, Mercury's crust represents approximately 10% of all silicate material on the planet (James et al., 2015; Padovan et al., 2015). Such efficient extraction is likely the result of relatively high degrees of partial melting, consistent with geochemical observations of the surface and inferences for the interior (Chapters 2, 7).

Compared with Mercury's global shape as derived from laser altimetry and radio occultation measurements, the geoid has a spectral power of only ~1% that of the shape at spherical harmonic degree and order 2, which indicates that topographic variations on Mercury at the longest wavelengths are largely isostatically compensated (Perry et al., 2015). Should the variations at degree and order 2 be compensated just by variations in the thickness of the crust, this difference would imply a ~24 km pole-to-equator change in crustal thickness. However, other mechanisms such as variations in density due to temperature or composition may contribute to the compensation, potentially reducing any long-wavelength crustal thickness variation (Perry et al., 2015; Tosi et al., 2015; Chapter 3). Regardless, a substantial latitudinal variation in the crustal thickness of Mercury would be an

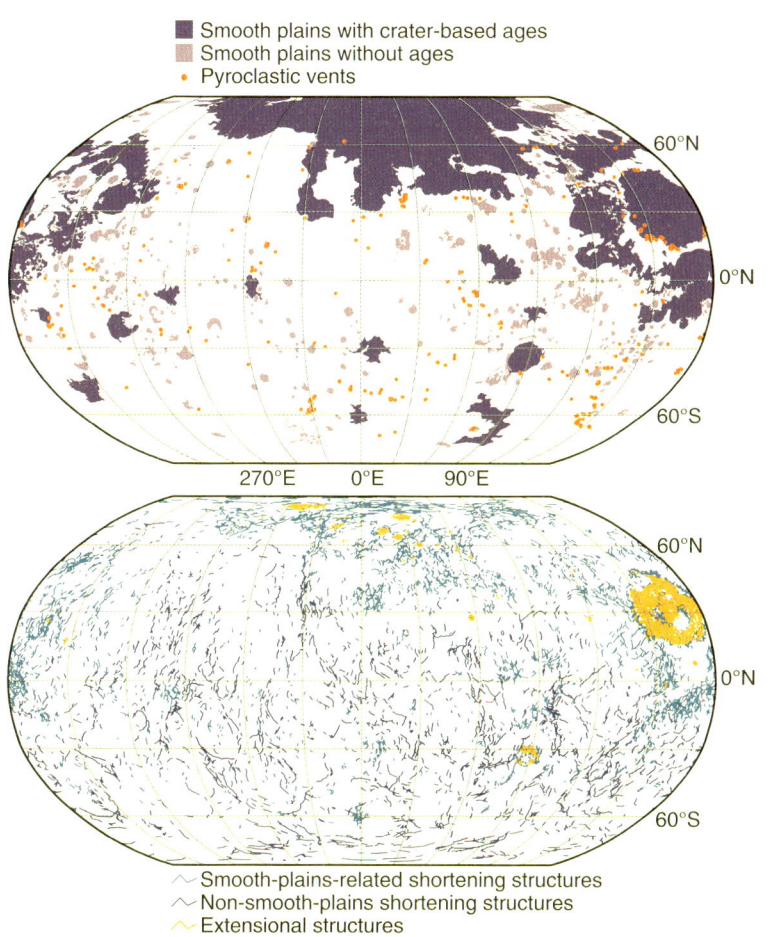

Figure 19.1. Overview of major geological features on Mercury. Top: Smooth plains are in purple; the darker units have estimated ages whereas the lighter-shade units are too small for reliable crater-based ages. Mapped units are from Denevi et al. (2013a) and Byrne et al. (2016). Locations of pyroclastic vents are from the compilation of Thomas et al. (2014). Bottom: Compilation of tectonic structures, with shortening structures outlined in shades of blue and extensional structures in orange. Structures in light blue are associated with smooth plains units as outlined in the top map in purple. The shortening structures are from Byrne et al. (2014), and the extensional structures are from Klimczak et al. (2012), Ferrari et al. (2014), and Chapter 10.

important, if as yet poorly understood, constraint on crustal production (Chapter 3).

19.3.2 Surface History

One of the more direct measures of the evolution of a planet's crust is the geological history of its surface. To first order, Mercury's surface can be classified into units of either smooth plains or intercrater plains (Chapter 6). The former type of unit is texturally smooth and relatively sparsely cratered, displays sharp boundaries with adjacent regions, and is level to gently sloped over baselines of ~100–200 km (Trask and Guest, 1975; Denevi et al., 2013a; Chapter 6). These smooth plains units occupy about 27% of the planet's surface (Figure 19.1) and are predominantly located in the northern hemisphere in the NSP and within and adjacent to the Caloris basin. The remainder of the surface is largely dominated by intercrater plains, which are characterized by gently rolling terrain with gradational boundaries and a greater density of secondary craters 5–10 km in diameter than on smooth plains (Trask and Guest, 1975; Denevi et al., 2013a). The intercrater plains are situated between individual large (>30 km) craters and clusters of such craters, which generally superpose the plains and are the source of the secondary craters. As the density of superposed impact craters appears to be the main distinction between the varieties of plains (Byrne et al., 2016), their main difference likely reflects a range in age rather than specific lithological or rheological differences (Murray et al., 1975; Strom, 1977; Spudis and Guest, 1988; Denevi et al., 2009; Whitten et al., 2014). Little evidence remains of an older, more heavily cratered surface apart from several regions that have undergone only partial resurfacing or portions of basin massifs that predate the intercrater plains (Chapter 6).

Observations of Mercury have established that the planet has been heavily shaped by volcanic activity. For example, the majority of smooth plains units are interpreted as effusive volcanic deposits, on the basis of their distinct unit boundaries, embayment relations with surrounding topography, the presence of buried "ghost craters" within these units, spectral differences from neighboring terrain, and deposits located far from any large basins (Murray et al., 1974, 1975; Strom et al., 1975; Spudis and Guest, 1988; Robinson and Lucey, 1997; Head et al., 2008, 2011; Murchie et al., 2008; Robinson et al., 2008; Denevi et al., 2009, 2013a; Chapter 11). A number of other volcanic landforms formed by effusive activity have also been reported across the planet, including a small shield volcano, lobate flow margins, and lava-sculpted valles (Head et al., 2008,

2011; Byrne et al., 2013; Hurwitz et al., 2013). Landforms attributed to explosive volcanism (e.g., Kerber et al., 2009; Thomas et al., 2014), often in close spatial proximity to smooth plains, have also been identified.

The major smooth plains deposits on Mercury have crater densities that vary by up to a factor of 5 for craters larger than 10 km. However, because of the inferred rapid decline in cratering during their formation, their derived model ages are the same, within statistical error, for any of the published model production function (PF) chronologies for Mercury (Strom and Neukum, 1988; Neukum et al., 2001; Marchi et al., 2009; Le Feuvre and Wieczorek, 2011), though differences among the model chronologies are greater for lower-density (younger) deposits (Chapter 9). Crater size–frequency analyses have shown that the NSP, the single largest smooth plains deposit on the planet (Chapter 6), as well as the plains interior to the Caloris and the Rembrandt impact basins, were emplaced around 3.8–3.7 Ga (Fassett et al., 2009; Head et al., 2011; Strom et al., 2011; Denevi et al., 2013a; Ferrari et al., 2014; Ostrach et al., 2015; Chapter 9). The areal densities of impact craters for two additional large smooth plains deposits on Mercury, those near the Faulkner crater and the Rachmaninoff basin, are comparable to the densities for the NSP and Caloris interior plains (Fassett et al., 2009; Denevi et al., 2013a; Whitten et al., 2014; Ostrach et al., 2015), implying that these other units are similar in age. Crater density measurements for several additional, smaller smooth plains deposits yield ages of ~3.8–3.5 Ga for these sites (Byrne et al., 2016) with the crater model production function of Le Feuvre and Wieczorek (2011). Only one definitively volcanic smooth plains deposit has been identified on the planet with a substantially younger age than those above. Situated within the inner peak ring of the Rachmaninoff impact basin, this deposit is considerably smaller than other plains units for which ages have been determined (Prockter et al., 2010; Marchi et al., 2011). The distribution of the model ages of smooth plains units (in particular the units shaded dark purple in Figure 19.1), which are stratigraphically the youngest effusive volcanic features on Mercury, suggest therefore that flood volcanism was largely completed by ~3.5 Ga (Byrne et al., 2016).

Similar to the smooth plains, intercrater plains units have a range of crater areal densities, the lowest values of which overlap the highest corresponding values for smooth plains units (Whitten et al., 2014; Byrne et al., 2016). The model ages of the intercrater plains are ~4.1–3.9 Ga (e.g., Whitten et al., 2014; Chapter 9). Notably, nowhere on Mercury is as heavily cratered as the lunar highlands (Strom, 1977; Strom et al., 2008; Fassett et al., 2011; Marchi et al., 2011), and the most heavily cratered regions on Mercury have been dated at just 4.1–4.0 Ga (Marchi et al., 2013) with the chronology of Marchi et al. (2009). These model age results suggest that little remains of the geologic record of the earliest ~500 Myr of Mercury's surface history (Chapters 6, 9).

The origin of Mercury's intercrater plains is less certain than that of the smooth plains, but they may also be dominantly products of volcanism. The main line of evidence lies in their age: model ages of 4.1–4.0 Ga imply major resurfacing of the earliest crust, with volcanism being a likely major cause (Head et al., 2011; Denevi et al., 2013a; Whitten et al., 2014; Chapters 6, 11). For example, Whitten et al. (2014) showed that cratering of smooth plains, particularly by secondaries from nearby primary craters, renders those smooth deposits texturally similar to intercrater plains. Large regions within the intercrater plains have also been interpreted as volcanic in origin on the basis of a substantial deficit of the most degraded class of craters, as well as stratigraphic and color relationships that are analogous to volcanic smooth plains deposits (Denevi et al., 2013b; Chapter 6). Although discrete volcanic landforms may not have survived the history of impact bombardment of Mercury prior to the emplacement of the smooth plains, thermochemical evolution models of the planet imply that voluminous and widespread effusive volcanic activity operated for at least the planet's first half-billion years (Michel et al., 2013; Tosi et al., 2013). If so, then the intercrater plains we observe today are likely just older smooth plains deposits (e.g., Strom, 1977; Spudis and Guest, 1988; Denevi et al., 2009; Whitten et al., 2014). This inference is consistent with the observed compositional heterogeneity on Mercury, where differences in composition do not always follow morphologic boundaries, and where smooth and intercrater plains can share similar compositions (Weider et al., 2015).

The cessation of large-scale effusive volcanism on Mercury, as seen in the smooth plains and the older intercrater plains, effectively heralded the end of the crust-building phase of Mercury's evolution, but volcanic activity in some form continued thereafter. For example, the identification of irregular pits across Mercury, often characterized by a lack of a raised rim, scalloped edges, and diffuse-edged deposits with a distinct reddish color, provides evidence for explosive volcanism having occurred on the planet (Head et al., 2008; Murchie et al., 2008; Kerber et al., 2009; Chapter 11). Some of these pyroclastic deposits may be as young as ~1 Ga (Thomas et al., 2014). Many of Mercury's explosive volcanic landforms and deposits are spatially associated with areas of pre-existing crustal weaknesses, including the surface breaks of thrust faults underlying lobate scarps and within the heavily fractured central peaks and peak rings of craters (Figure 19.1) (Kerber et al., 2011; Thomas et al., 2014; Chapter 10). Additionally, the areal extents of pyroclastic deposits are far less than those of effusive volcanic deposits. Although widely distributed, the role of explosive volcanism in the building and resurfacing of Mercury's crust was negligible compared with the contribution from effusive volcanism.

The history of Mercury's surface is recorded as much in its tectonic landforms as in its volcanic ones. Indeed, the surface of Mercury is replete with tectonic features, including landforms termed "wrinkle ridges" and "lobate scarps" (see the bottom panel of Figure 19.1), interpreted to have accommodated crustal shortening in response to global contraction (Strom et al., 1975). The number and structural relief of this ensemble of landforms correspond to a decrease in planetary radius of at least 5 to 7 km (Byrne et al., 2014; Chapter 10). These figures are in stark contrast with earlier estimates from more limited Mariner 10 data and early flyby data from MESSENGER, which suggested that perhaps no more than 2 km of contraction was likely (Strom et al., 1975; Watters et al., 1998, 2009). Importantly, crater and thrust fault superposition relations indicate that global contraction was underway by around the time that widespread effusive volcanism came

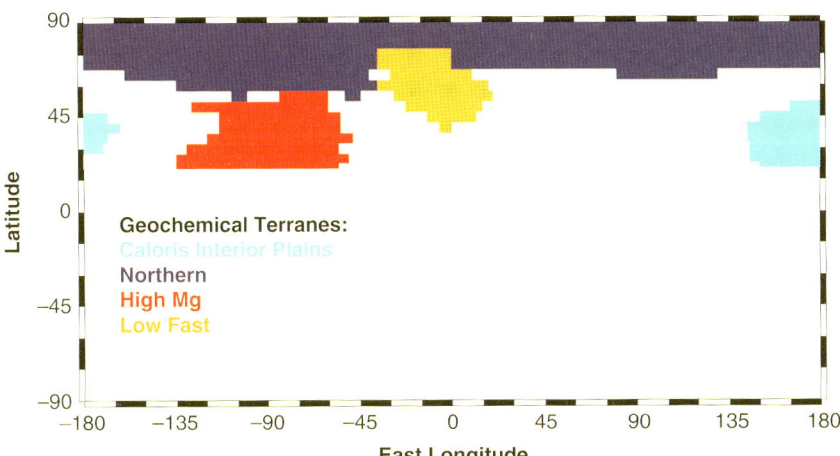

Figure 19.2. Overview of geochemically distinct terranes on Mercury (Chapters 2, 7). From Patrick Peplowski.

to an end (Banks et al., 2015; Byrne et al., 2016). Observations of craters that formed during the Calorian system (Spudis and Guest, 1988; Chapter 9) and superpose scarps show that shortening of Mercury's surface on at least a regional scale had begun at some time before ~3.6 Ga (Banks et al., 2015). Further, the discovery with MESSENGER low-altitude image data of a population of lobate scarps at least an order of magnitude smaller than previously recognized (Watters et al., 2015b), and the stratigraphic relationships between such scarps and impact craters with a range of degradation states, is suggestive that tectonic accommodation of global contraction persisted over most of Mercury's history (Banks et al., 2015).

Observations made with MESSENGER data have helped characterize the resurfacing mechanisms and history of the innermost planet. Voluminous magma genesis within Mercury's interior likely resulted in globally extensive effusive volcanism that persisted for at least several hundred million years. This volcanic activity, together with an increase in the impact flux at the start of the LHB, has obscured the geological record of the first ~500 Myr of Mercury's surface history. With a reduction in magma genesis as a result of secular cooling and with the horizontal compressive state in Mercury's lithosphere resulting from global contraction, widespread effusive volcanism began to wane, with eruptive volumes decreasing with time, before ultimately ending by about 3.5 Ga. Explosive volcanism endured for far longer, but the vast majority of Mercury's crust was in place prior to 4 Ga, and smooth plains formation constituted the tapering end of the planet's crust-building phase.

19.3.3 Chemical and Petrological Constraints on Crustal Formation

Observations by MESSENGER's suite of geochemical sensors have provided important insight into the composition of the planet, the makeup of the crust, and how it formed (Chapters 2, 7). In particular, the X-Ray Spectrometer (XRS), Gamma-Ray Spectrometer (GRS), and Neutron Spectrometer (NS) provided spatially resolved surface abundances of U, K, and Th, as well as Si-normalized elemental abundances for Na, Mg, Al, S, Cl, Ca, Ti, Cr, Mn, Fe, and O. On a global scale, XRS measurements (Nittler et al., 2011) indicate that the surface of Mercury exhibits a high Mg/Si ratio (0.33–0.67), which is intermediate between those of terrestrial oceanic and lunar mare basalts and highly magnesian komatiites. Mercury's surface also exhibits lower Al/Si and Ca/Si ratios than typical terrestrial or lunar basalts. Most surprising, high S/Si ratios (0.05–0.15) suggest abundances of the moderately volatile element S up to ~4 wt%. Observations from the GRS further argue against a volatile-depleted composition for Mercury. For example, Mercury's K/Th ratio is comparable with that of other terrestrial planets and is much higher than observed in the volatile-depleted lunar crust (Peplowski et al., 2011). Moreover, large ratios of Na/Si (0.12) and Cl/Si (0.0057) are also observed (Evans et al., 2012, 2015). Together, these observations suggest a magnesium-rich, iron-poor crust formed under chemically reducing conditions, yet not depleted in volatiles as had been predicted for an iron-rich planet so close to the Sun (e.g., Taylor and Scott, 2003).

The surface of Mercury exhibits considerable chemical and, therefore by extension, mineralogical diversity. This diversity is best documented in the northern hemisphere, where high-spatial-resolution measurements allow us to distinguish discrete geochemical terranes (Figure 19.2; Chapters 2, 7). These include the Northern Geochemical Terrane, the Caloris Interior Plains Terrane, the High-Magnesium Terrane, and the "Low-Fast" Terrane (so named because it has a low count rate for fast neutrons). Among these terranes, the Northern Geochemical Terrane and the Low-Fast Terrane are present largely, though not exclusively, within the NSP. The Caloris Interior Plains Terrane corresponds spatially to the boundaries of the smooth plains within the Caloris impact basin. In contrast, the High-Magnesium Terrane is geochemically coherent in a number of features but exhibits no clear correlation with spectral or morphometric features across the entirety of the region. However, while the crustal thickness within the majority of the High-Magnesium Terrane is similar to the average for the northern hemisphere, the northern and eastern boundaries are approximately coincident with areas that transition from average to thicker-than-average crust (Chapter 3). In contrast to the well-resolved XRS measurements in the northern hemisphere, the large XRS footprints in the southern hemisphere yield only a single hemispheric average composition.

Chemical compositions derived from the four distinct, northern hemisphere geochemical terranes range in composition from basaltic andesite to trachyte on the basis of their total alkali content (Na and K) compared with silica, but ultimately all share a boninite classification due to their high MgO (> 8 wt%) and low TiO_2 (<0.5 wt%) concentrations. A common feature of all these geochemical terranes is that each has high volatile element concentrations, with Na ranging from 2.6 to 5.7 wt% and S from 1.8 to 2.9 wt%. Considerable geochemical differences do exist among the terranes, particularly with respect to Na, Mg, Al, and Fe, all of which differ by factors of 1.8 or greater among the terranes. The Low-Fast Terrane is most similar to the average surface composition for the planet. However, it is geochemically distinct from the Northern Geochemical Terrane, with the two terranes combined occupying much of the NSP. Mineralogically, these terranes share the common feature of being unusually rich in normative plagioclase (37–58 wt%) (see Chapter 7). If classified as plutonic igneous rocks, these terranes would include norite, anorthositic norite, and anorthositic gabbro, reflecting differences in plagioclase abundance and the ratio of high-calcium to low-calcium pyroxene. The High-Magnesium Terrane is distinctive in its unusually high concentration of normative olivine (31 wt%).

19.3.4 Evolution of the Lithosphere

The mechanical behavior of the lithosphere – the outer portion of the planet that behaves as a mechanically strong layer and includes portions of the crust and possibly mantle – provides key insights into the history of planetary stresses and temperatures. This behavior is recorded in both the tectonic landforms that the lithosphere hosts (e.g., Chapter 10) and in the flexural response to loads inferred from gravity and topography data (e.g., Chapter 3). For example, elastic dislocation modeling of topographic profiles derived from Mariner 10 stereophotogrammetric data of select lobate scarp features indicates that the underlying faults penetrate, and thus the lithosphere deformed in a brittle fashion, to depths of 25–30 km, at the time of faulting (e.g., Watters and Nimmo, 2010). A comprehensive assessment of the spatial variation in lithospheric thickness from this or similar techniques, however, has yet to be completed. Interestingly, this estimate of the depth to the brittle–ductile transition from lobate scarp fault depths is consistent with models of lithospheric strength for time periods prior to ~3.5 Ga (Williams et al., 2011).

At regional and more local scales, stresses in the lithosphere can be modified by loads produced by volcanism and impact basin formation and evolution (e.g., Kennedy et al., 2008; Freed et al., 2009; Blair et al., 2013). Nonetheless, the history of the state of stress in Mercury's lithosphere has been dominated by some combination of two independent, globally acting processes: despinning from a likely early, rapid rate of rotation, and global changes in planetary radius arising from internal temperature changes (and cooling in particular) (Chapter 10). During planetary spindown, an equatorial bulge supported by the planet's lithosphere would have relaxed (Melosh, 1977), forming a global set of near-surface joints with no preferred orientation at the poles but with an increasingly prominent east–west fabric toward the equator, under the assumption of a globally uniform lithospheric thickness (Klimczak, 2015). Similarly, also under the assumption of globally uniform lithospheric properties, a reduction in planetary volume from cooling of the core and mantle and from mineralogical phase changes (e.g., core crystallization) would yield a stress state in which horizontal compressive stresses exceed vertical stresses and under which a global set of thrust faults with no preferred orientation would develop (Melosh and McKinnon, 1988). However, spatial variations in lithospheric thickness, such as those imparted by long-lived latitudinal and longitudinal differences in surface temperature (Williams et al., 2011; Tosi et al., 2015), may further have influenced how tectonic deformation on Mercury was exhibited (Beuthe, 2010). Furthermore, the combination of stresses from both despinning and global contraction may have had a substantial influence on the stress state and the style of brittle tectonic deformation at the surface of the planet (Melosh and Dzurisin, 1978; Dombard and Hauck, 2008; Beuthe, 2010). Reorientation as a result of true polar wander, for example, driven by the formation of a large load such as the Caloris basin (Matsuyama and Nimmo, 2009), could also have altered the prevailing stress state.

Global mapping of Mercury's tectonic landforms from MESSENGER image data (Byrne et al., 2014; Watters et al., 2015a) has not revealed, to first order, any evidence of planet-wide, organized patterns of tectonic landforms predicted by earlier studies of despinning (e.g., Melosh, 1977; Melosh and McKinnon, 1988; Matsuyama and Nimmo, 2009; Beuthe, 2010). Given Mercury's near-zero obliquity, solar illumination at the equator is always due east or due west, which facilitates the identification of tectonic landforms that strike roughly north–south (Byrne et al., 2014) more easily than those trending east–west at low to mid latitudes. Landforms between 60°S and 60°N, in particular, show a predominantly north–south orientation. Landforms north of 60°N show some clustering at southwest–northeast trends but are not as strongly oriented as those at mid to low latitudes; landforms south of 60°S show no preferred orientations (Byrne et al., 2014). Nonetheless, when the effect of solar azimuthal illumination is considered, a general north–south trend for mid- to low-latitude landforms remains (Watters et al., 2015a). Under the assumption that currently published tectonic maps are generally complete and that no bias in lighting geometry has obscured substantial ~east–west-trending landforms yet to be identified (Chapter 10), the history of stress within Mercury's lithosphere must be reconciled with these observations.

The lack of opening-mode fractures on Mercury, aside from those identified within volcanically flooded impact features (Freed et al., 2012; Klimczak et al., 2012; Chapter 10), indicates that no direct evidence of tidal despinning alone remains. Given that tidal despinning likely occurred geologically rapidly – although the timing of this process remains to be characterized fully – it is perhaps no surprise that such evidence is missing from the geological record, especially given that the oldest terrain on Mercury is ~4.1 Ga (Section 19.3.2). On the other hand, shortening structures on the innermost planet do not form a globally random pattern, the expected result of global contraction alone.

It may be, then, that the most straightforward interpretation of the global pattern of orientations of Mercury's tectonic landforms represents some combination of despinning and global contraction (Klimczak, 2015; Chapter 10). Thrust faults developing in such a stress state would have developed with preferred north–south orientations near the equator (Klimczak, 2015), though the expected pattern near the poles could be oriented either somewhat randomly (Klimczak, 2015), depending on rock strength, or in a more organized pattern if latitudinal variations in lithospheric thickness were substantial (Beuthe, 2010). Nevertheless, this general pattern in the mid latitudes with a differing orientation at high latitudes is similar to that observed from global mapping (Byrne et al., 2014). As such, it indicates either that tidal despinning temporally overlapped with global contraction or that despinning imparted some fabric to Mercury's lithosphere that survived until the onset of, and influenced the tectonic deformation from, global contraction. The lack of a clear signature of reorientation stresses such as would be reflected in an orientation of the lobate scarp structures (Matsuyama and Nimmo, 2009) suggests that true polar wander was not a major component of the processes that drove tectonic deformation. Thus, the relative timing of global contraction and despinning, as well as the effects of spatial variations in lithospheric thickness (e.g., Beuthe, 2010) when considered with possible values for the degree of lithospheric fracturing (e.g., Klimczak, 2015), are important questions that remain outstanding.

19.4 KNOWLEDGE OF THE INTERIOR

19.4.1 Constraints on Core Composition

Observations of low ferrous iron concentrations and larger-than-expected sulfur abundances on Mercury's surface indicate that the planet's surface and, by extension, interior are strongly chemically reduced (Nittler et al., 2011; Weider et al., 2012; Chapter 2). Inferred oxygen fugacities range between 3 and 7.3 $\log_{10}$ units below the iron–wüstite (IW) buffer, with a consistent overlap between published estimates of IW–4.5 to IW–6.3 (Malavergne et al., 2010; McCubbin et al., 2012; Zolotov et al., 2013). Under these highly reducing conditions, elements that are normally lithophile and incorporated into silicates and oxides can instead have chalcophile or siderophile behavior, combining to form sulfides or metallic phases. This behavior is observed in aubrite meteorites, which formed at similarly reducing conditions and contain a host of exotic sulfides, metals, carbides, phosphides, and nitrides (Keil, 1989). For the surface of Mercury, correlated S/Si and Ca/Si abundances have been invoked to postulate the presence of oldhamite (CaS) (Nittler et al., 2011), although more recent studies (Stockstill-Cahill et al., 2012; Vander Kaaden and McCubbin, 2016; Vander Kaaden et al., 2017) favor complex sulfides of Fe, Mg, Ti, Cr, and Mn, as well as Ca.

Although Fe and Ni are expected to be primary components of Mercury's core, the highly reducing conditions inferred from surface materials, if indicative of conditions in Mercury's interior, will also have led to the incorporation of light elements into the core, most notably Si and S. As conditions become more reducing, Si becomes more soluble in the metal phase (e.g., Berthet et al., 2009; Malavergne et al., 2010; Chabot et al., 2014). A similar trend is also observed as temperatures increase (McCoy et al., 1999). During planetary differentiation, Si can be incorporated into the metallic phase and thus into the core of a planetary body. In practice, the segregated core material under reducing conditions contains S as well, so that the Fe–S–Si system rather than the binary Fe–Si system governs phase relations. Within the Fe–S–Si system (Raghavan, 1988), liquid immiscibility can occur, producing separate S-rich and Si-rich metallic liquids. Chabot et al. (2014) noted this behavior in experiments in which Fe–S–Si liquids occurred in equilibrium with silicate melts for which the S concentration was comparable with that observed on the surface of Mercury (1–4 wt% S). These authors noted that, whereas the co-existing Fe–S–Si melts (and, by extension, the core of Mercury) can readily contain both S and Si, changing oxygen fugacity may result in either high-Si, low-S or high-S, low-Si melts, either of which could satisfy the constraint imposed by the presence of S-rich silicate melts on the surface of Mercury.

Additional, less well-constrained light elements that might be incorporated into the core are C and P. The inference that graphite could both crystallize from and occur as a flotation product in an early Mercury magma ocean (Vander Kaaden and McCubbin, 2015) suggests that Mercury's core might be saturated in carbon if the mantle and core were in equilibrium, as carbon tends to be siderophile at reducing conditions. Although likely a minor constituent in Mercury's core, it may have a substantial effect on its melting behavior (Deng et al., 2013a; Martin et al., 2014, 2015). Phosphorus also behaves as a siderophile under reducing conditions, forming Fe,Ni-phosphides. No measurement of phosphorus on the surface of Mercury yet exists, although its incorporation into the core in minor concentrations would be expected, potentially resulting in a complex behavior of the core governed by the Fe–S–Si–C–P system.

19.4.2 Internal Structure

Planetary evolution is intimately intertwined with the distribution of materials within the interior of the body. The processes of metal–silicate differentiation, core crystallization, mantle convection, and magmatism tend to result in a layered compositional and density structure within the interior of a planet. Such a layered structure is typically comprised of one or more Fe-rich layers in the planet's core, as well as one or more silicate mantle layers, all topped by a silicate crust. The thickness, material properties, and heat-producing element content of each of these layers controls how the planet generates and loses heat, generates magma, and produces a magnetic field – although neither of the latter two is guaranteed.

Mercury's large bulk density of ~5430 kg m^{-3} has long been understood to imply that the planet has an unusually large metal-to-silicate ratio (Siegfried and Solomon, 1974; Solomon, 1976). Consequently, Mercury has a relatively large metallic core with a comparatively thin layer of overlying silicate material. Mariner 10's discovery of Mercury's magnetic field (Ness et al., 1975) suggested the possibility that the core could be partially molten. The presence of a liquid layer within the core was subsequently confirmed by Earth-based radar observations of the libration and orientation of Mercury (Margot et al., 2007; Chapter 4). Pre-MESSENGER studies also used the long-wavelength gravity field to estimate the thickness of the planet's crust (Anderson et al., 1996) at ~100–300 km. Such a large crustal thickness was exceptionally surprising because it represented one-sixth to one-half of the estimated silicate content of the planet, far in excess of that for any other known planetary body. Constrained only by the radius and bulk density of the planet, the relative size of Mercury's core, and whether it contained a solid inner core, remained similarly uncertain. However, in models of Mercury's internal evolution it was commonly assumed that the silicate portion of the planet was ~600 km thick, with the remainder of the interior composed of an Fe-rich core (e.g., Schubert et al., 1988; Hauck et al., 2004; Redmond and King, 2007; Grott et al., 2011).

Measurements of Mercury's gravity field by the MESSENGER spacecraft have led to greatly improved estimates of the planet's internal structure (Smith et al., 2012; Hauck et al., 2013; Rivoldini and Van Hoolst, 2013; Chapter 4). That Mercury occupies a Cassini state, wherein the rotation axis is approximately perpendicular to the plane of its orbit about the Sun and the spin and precession rates of the planet are equal, presents an opportunity to estimate the planet's structure. Indeed, a procedure was developed to determine the normalized polar moment of inertia and the fraction of that moment contributed by the outermost solid portion of the planet (e.g., Peale, 1988; Peale et al., 2002) as a result of Mercury's special rotation state. The background and details of this experiment and its interpretation were discussed at length in Chapter 4. The fundamental result is that through measurement of just four quantities – the polar and equatorial oblateness of the gravity field expressed as the second-degree spherical harmonic coefficients C_{20} and C_{22}, the amplitude of the physical libration, and the obliquity of the planet – it is possible to resolve two measures of the radial density distribution of the planet (Peale, 1988; Peale et al., 2002; Margot et al., 2007; Margot et al., 2012; Chapter 4). These quantities are the normalized polar moment of inertia C/MR^2 and the fraction of the polar moment of inertia contributed by the solid shell of the planet that overlies the liquid outer core C_{m+cr}/C.

The MESSENGER-derived moment of inertia values and the bulk density of the planet have been used to constrain the relative thicknesses of the silicate mantle and metallic core, and their respective densities, in suites of models of varying complexity (Margot et al., 2012; Hauck et al., 2013; Rivoldini and Van Hoolst, 2013; Dumberry and Rivoldini, 2015; Chapter 4). Early estimates for the average densities of the outermost solid layer of the planet and the metallic core of 3380 ± 200 kg m^{-3} and 6980 ± 280 kg m^{-3}, where the boundary between these two layers is ~420 ± 30 km below the planet's surface (Hauck et al., 2013), are consistent with the most recent estimates of Mercury's gravity field and rotational parameters (Mazarico et al., 2014), because of the similarity to previous estimates of these parameters (Margot et al., 2012).

Detailed models of Mercury's interior have been designed to resolve additional layers within the interior. For example, as discussed in Section 19.3.1, detailed analyses of gravity and topography data returned by MESSENGER have led to improved estimates of the thickness of the silicate crust of 35 ± 18 km or >38 km, depending on the method employed (James et al., 2015; Padovan et al., 2015; Chapter 3). That the average density of the outermost solid shell of the planet is greater than expected for iron-poor silicate materials, together with estimates of the composition of Mercury's core inferred from the strongly chemically reducing conditions discovered at the surface, has led to the consideration of a solid iron sulfide layer at the top of the core (Smith et al., 2012; Hauck et al., 2013; Padovan et al., 2014; Chapter 4). Given that both Si and S should have partitioned into the core (Chapter 2; Section 19.4.1), at the modest pressures prevalent at the top of the core, melting Fe–S–Si can yield two immiscible liquids (one Fe–S rich and the other Fe–Si rich) over a broad range of bulk compositions (Morard and Katsura, 2010). This behavior would lead to segregation of the S-bearing liquids to the shallowest portions of the liquid core, including the core–mantle boundary. Recent metal–silicate partitioning experiments at 100 kPa (1 bar) pressure, however, suggest that the range of potential core S and Si contents consistent with the surface S content may not lead to core compositions that permit immiscibility and compositional segregation (Chabot et al., 2014) (see also Chapter 4). Additional experimental work at higher pressures and varying silicate compositions are necessary to fully test the importance of liquid immiscibility in Mercury's core and the possibility of a solid FeS layer. However, measurement of induced magnetic fields at Mercury has led to estimates of the depth to the top of the core (Johnson et al., 2016; Chapter 5) that are consistent with internal structure models. Taken together, the consistency between the internal structure models that give an estimate of the depth to the top of the liquid outer core, and the induced magnetic field analyses that yield the depth to the top of an electrically conducting layer, indicates that any FeS layer, if present, is limited in thickness.

Similarly, an Fe-rich solid inner core may also be present, though constraints on its size are sparse. Internal structure models consistent with the gravity field and rotational state of Mercury are generally limited in their ability to resolve the inner core (Chapter 4), though there does appear to be a slight tendency toward relatively modest inner core radii (Hauck et al., 2013; Dumberry and Rivoldini, 2015), perhaps smaller than half the total core radius. Recent work on the dynamic coupling of the rotation of the inner core to the outer, librating solid shell of the planet indicates that, for inner cores larger than ~30% of the radius of the planet, it is necessary to know the size of the inner core in addition to the gravity field and rotation data in order to infer the moments of inertia of the planet (Peale et al., 2016; Chapter 4). Although at present it is not possible to determine independently the size of the inner core, models with inner cores larger than 30% of the radius of the planet tend to have silicate layer densities less than the densities of magnesian olivine and pyroxene, the likely dominant constituents of Mercury's mantle. Thus, Mercury's inner core, if present, is unlikely to have a radius more than 30% of the planet's radius.

19.4.3 Magnetic Field

Mercury's magnetic field observations demonstrate that a global-scale field is presently being generated by a core dynamo

(Chapter 5). Initial data from Mariner 10, along with the more recent MESSENGER mission measurements, show that Mercury's dynamo-generated field is relatively weak and dominated by an axially aligned dipole. The dipole dominance of the field suggests, at first glance, that Mercury's dynamo may be quite Earth-like in its morphology, although a suite of characteristics of Mercury's field suggest that it has distinctive properties.

The weak intensity of the field challenges our understanding of how Mercury's magnetic field is generated. Both energy- and force-balance arguments suggest that Mercury's observed magnetic field should be at least two orders of magnitude stronger than the field measured by Mariner 10 and MESSENGER. Although the dipole is the largest harmonic in the field, the quadrupole component is relatively large, at approximately 40% of the dipole strength. This quadrupolar component – equivalent to an offset of the dipole from Mercury's center – is larger than observed for other planets with dipole-dominated fields. Indeed, it is larger than those of other planets even when corrections are made for the relatively shorter distance from the surface to the core–mantle boundary (CMB) at Mercury, with a proportionately smaller attenuation of the quadrupole component with distance from the dynamo region. The multipolar terms beyond the quadrupole, though, are quite small. Furthermore, a property that has not received much attention to date is the axisymmetry of the dipole and quadrupole components. With the possible exception of Saturn, no other planet has a field as axisymmetric as Mercury. The combination of these three characteristics requires alterations to dynamo scenarios previously proposed for Mercury.

The weakness of Mercury's field was the first puzzle to be confronted, and several solutions have been suggested (e.g., Wicht and Heyner, 2014). For example, numerical dynamo models with very large inner cores (Stanley et al., 2005) or with very small inner cores (Heimpel et al., 2005) could produce relatively weak fields. However, current compositional, thermal, and structural models for Mercury's core suggest that the inner core is unlikely to be sufficiently large to satisfy the large inner core models, even if the size of the inner core is weakly constrained at best (see Section 19.4.2). Another explanation for the relative weakness of Mercury's field is that the outer portion of the core may be stably stratified, an idea consistent with the small magnitudes of the terms beyond the quadrupole in the field's multipolar expansion. This stratification could be thermal (the result of subadiabatic heat flux at the CMB) or compositional (due to light element segregation) in origin. Such a stably stratified layer may attenuate the field intensity observed at the surface (Christensen, 2006; Christensen and Wicht, 2008), although double-diffusive convection in the stable layer may hinder the attenuation (Manglik et al., 2010). A third suggestion is that feedback between currents generated in Mercury's magnetosphere and those in Mercury's core may result in a weak-field state (Glassmeier et al., 2007; Heyner et al., 2011). A fourth possibility is that, if S is the principal light element in the core, temperatures may drop below the melting temperature near the top and the middle of the core in regions often termed Fe-snow zones, where Fe would crystallize and then sink through the core; this situation contrasts with that of Earth, where crystallization first occurs at the center of the planet (Chen et al., 2008). A proposed consequence of such top-down crystallization in Mercury's core is that there could be two separate regions of dynamo generation and that the dipole components oppose each other, yielding a weak net external field (Vilim et al., 2010).

Although these scenarios offer promising avenues for understanding the weakness of Mercury's field, they must also explain the other characteristics of the magnetic field observed by MESSENGER. None of these proposed mechanisms, by themselves, have yet been shown to lead naturally to magnetic fields with large quadrupole components and very axisymmetric fields. The combination of a large quadrupole component and an axisymmetric dipole component is particularly challenging because dynamo theory demonstrates that when a fluid velocity mode excites the generation of the axial quadrupole component, it will also excite the non-axisymmetric dipole component (Bullard and Gellman, 1954). Special circumstances may therefore apply in order to dampen only one of these magnetic modes.

Two recent studies have had some success in this vein. Cao et al. (2014) imposed a north–south symmetric thermal perturbation at the CMB in a numerical dynamo model (resulting in higher heat flux at the CMB equator: see Figure 19.3) along with volumetric heat sources throughout the core. Their model matched the dipole–quadrupole dominance and axisymmetry in Mercury observations, but it did not reproduce the relatively low strength of Mercury's field. The likelihood that such a thermal perturbation is present at Mercury's CMB is also unclear. In contrast, a numerical dynamo model by Tian et al. (2015) instead imposed a north–south antisymmetric thermal perturbation (i.e., of spherical harmonic degree 1) at the CMB (Figure 19.3), resulting in higher heat flux in the northern hemisphere. In addition, a thin, stably stratified layer was imposed at

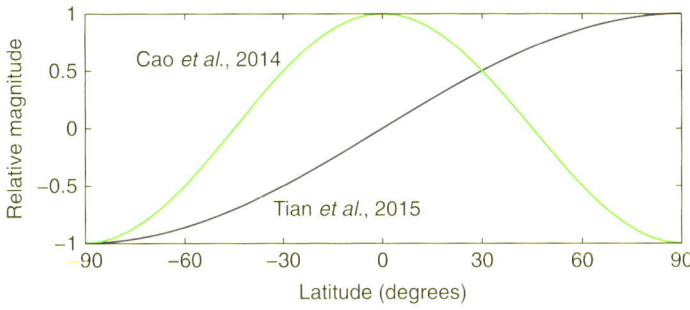

Figure 19.3. Relative variations in the imposed heat flux along the core–mantle boundary in MESSENGER-era dynamo models. The work of Cao et al. (2014) invoked a core heat flux that is higher at, and symmetric about, the equator, whereas Tian et al. (2015) assumed a core heat flux that is greater in the northern hemisphere than in the southern hemisphere.

the top of the core in this model. This combination of properties resulted in a magnetic field that reproduced the dipole–quadrupole dominance, the axisymmetry, and the weakness of Mercury's field. The north–south antisymmetric thermal perturbation in this model was justified on the basis of the concentration of smooth plains in Mercury's northern hemisphere (Head et al., 2011).

Recent work by Philpott et al. (2014) also suggested that there has been little to no secular variation in the large-scale magnetic field components between the time of the Mariner 10 flybys (1974–1975) and the four years that MESSSENGER was in orbit about Mercury. A study by Stanley and Bloxham (2016) of the saturnian dynamo suggests that if Mercury possesses a stably stratified layer at the top of the core, and if the magnetic field is very axisymmetric, then very slow secular variation of the field is a natural result. This correspondence between slow secular variation and a stably stratified layer may help to explain the lack of observed secular variation in Mercury's magnetic field.

19.4.4 Core Properties

The relative dominance of Mercury's core as a fraction of the planet's mass and volume (Chapter 4) underscores the influence of the core in the planet's overall evolution. The basic properties of the core, and particularly its thermodynamic attributes, are critical for understanding how it has evolved. Siegfried and Solomon (1974) utilized a thermal conduction model for heat transport through the planet, in concert with knowledge of the thermodynamic properties of iron for the core, to investigate the thermal history and core crystallization of Mercury. More recent approaches have generally considered heat transport through the mantle via convection and various alloys of iron and sulfur for the core (e.g., Schubert et al., 1988; Hauck et al., 2004; Grott et al., 2011; Tosi et al., 2013).

Over the past two decades, knowledge of the behavior of a variety of potential core-forming materials has grown considerably. The pressures within Mercury's core, ~5–40 GPa (Hauck et al., 2013), are directly accessible in laboratory experiments. Of particular interest are the temperature and pressure dependencies of the properties of iron alloys, including the thermal conductivity, thermal expansivity, and melting behavior. It is well-known that Fe–S alloys have the peculiar behavior that their eutectic melting temperature decreases with increasing pressure (e.g., Fei et al., 1997, 2000; Li et al., 2001; Chudinovskikh and Boehler, 2007; Stewart et al., 2007; Chen et al., 2008) up to 14 GPa, with shifts in the eutectic composition toward more Fe-rich compositions at pressures up to at least 40 GPa (Stewart et al., 2007).

Iron–silicon alloys, which may be present in the core as a consequence of Mercury's chemically reduced conditions (see Section 19.4.1), behave differently from alloys of iron and sulfur. The primary distinctions are that the presence of silicon results in a smaller melting point depression than with S and the Fe–Si alloys show a strong solid solution (Kuwayama and Hirose, 2004), particularly when compared with the limited solubility of S in solid Fe, even at high pressure (Li et al., 2001). Furthermore, temperature differences between the liquidus and solidus are <50 K and the compositional differences between coexisting liquid and solid are <2 wt% Si on the Fe side of the eutectic in this system at 21 GPa (Kuwayama and Hirose, 2004). In contrast to the Fe–S system, it appears that the eutectic temperature increases with pressure (Kuwayama and Hirose, 2004; Morard et al., 2011; Fischer et al., 2013) up to at least 50 GPa, and the Si content of the eutectic composition increases with pressure up to at least 21 GPa (Kuwayama and Hirose, 2004).

The more likely situation is that Mercury's core contains multiple alloying elements, particularly S and Si (e.g., Chapter 2; Section 19.4.1) because of a broad trade-off from S- to Si-bearing Fe alloys as a function of decreasing oxygen fugacity, with mixtures of the two quite likely (Malavergne et al., 2010; Hauck et al., 2013; Chabot et al., 2014). Depending on the composition of the core, liquid immiscibility at pressures less than 12 GPa is possible, with the result that S- and Si-rich liquids would separate due to their differential buoyancy and would lead to a S-rich liquid at the top of the core (Section 19.4.2) – though recent experiments suggest that a single miscible Fe–S–Si liquid is more likely (Chabot et al., 2014).

In addition to the melting behavior of Fe-rich alloys likely to be present in Mercury's core, other thermodynamic properties of these materials are critical to the planet's evolution. Among the most relevant is the thermal expansivity of these alloys, which is a controlling parameter in both the temperature gradient and in the amount the planet expands or contracts with temperature changes (Siegfried and Solomon, 1974; Schubert et al., 1988; Hauck et al., 2004, 2013; Williams, 2009; Grott et al., 2011; Tosi et al., 2013; Jing et al., 2014). Measurement of the thermal expansivity of Fe alloys is challenging, particularly at high pressure and for liquids. The majority of pre-MESSENGER-era models of Mercury's interior were predicated on a constant value for thermal expansivity in the core (e.g., Siegfried and Solomon, 1974; Schubert et al., 1988; Hauck et al., 2004; Grott et al., 2011), consistent with expectations for Earth's core. However, the thermal expansivity is clearly a function of pressure and composition (e.g., Williams, 2009), and recent work on both the internal structure (Smith et al., 2012; Hauck et al., 2013; Rivoldini and Van Hoolst, 2013; Chapter 4) and the contraction of Mercury (Tosi et al., 2013) has aimed to accommodate this variation. Recent experimental measurements of the density and sound velocity of Fe alloys at high pressure and temperature have altered our understanding of the variation in core thermal gradients and the potential for contraction in these relevant systems, particularly for Fe–S alloys (Jing et al., 2014). The primary consequence of these new data and models is the potential for steeper adiabatic thermal gradients and larger amounts of thermal contraction than previously appreciated.

The thermal conductivity of Fe alloys is particularly important for understanding both core heat transfer and the evolution of Mercury's magnetic field. Recent experimental and numerical estimates of the thermal conductivity of Fe and Fe alloys at the conditions of Earth's core are greater than previous canonical values by a factor of 2–3 (e.g., de Koker et al., 2012; Pozzo et al., 2012; Gomi et al., 2013), leading to questions regarding the relative role of thermal buoyancy in driving Earth's dynamo. However, molecular dynamics calculations by Zhang et al. (2015) are more consistent with lower values of the conductivity, as are recent direct measurements of the thermal conductivity of solid iron at high pressure (Konôpková et al., 2016). At Mercury's core conditions, experimental work also appears to suggest a larger thermal conductivity in pure Fe (Deng et al., 2013b) than is typically found in models of the planet's core heat transport and magnetic field generation. Moreover, Si is known to substantially decrease the thermal conductivity in Fe alloys (Seagle et al., 2013), perhaps reducing even the larger estimates of thermal conductivity toward the values assumed for Mercury's core in previous models of the planet's interior. Ultimately, the thermal conductivity of Mercury's core depends on pressure, temperature, and composition and plays a major role in the thermal gradient and the longevity and pervasiveness of core convection necessary for driving the magnetic field.

19.5 THERMOCHEMICAL MODELS OF INTERIOR EVOLUTION

The volcanic and tectonic evolution of Mercury as recorded on its surface is closely connected to the amount of heat in, and the transfer of that heat from, the planet's interior. Therefore, understanding Mercury's geologic evolution also requires an understanding of the processes acting in the deep interior. Thermal or thermochemical evolution models are usually employed to shed light on the working of a planet's interior heat engine (Solomon, 1977; Stevenson et al., 1983; Schubert et al., 1988; Hauck et al., 2004; Redmond and King, 2007; King, 2008; Grott et al., 2011; Michel et al., 2013; Tosi et al., 2013). In order to understand the evolution of the entire planet we utilize models of planetary evolution that incorporate crucial thermal, chemical, magmatic, and tectonic processes constrained by MESSENGER observations.

19.5.1 Modeling Approaches

The most straightforward models of the internal evolution of planets consider the global energy balance for the mantle and core via the relations

$$\rho_m c_m V_m \frac{dT_m}{dt} = -q_m A_m + Q_m V_m,$$

$$\rho_c c_c V_c \frac{dT_c}{dt} = -q_c A_c + Q_c V_c,$$

where ρ is density, c is heat capacity, V is volume, T is temperature, A is surface area, and the subscripts m and c refer to the mantle and core, respectively. Q is the rate of heat released in the interior per unit volume by the decay of the long-lived radioactive elements ^{40}K, ^{232}Th, and ^{238}U, and q_c and q_m are the heat flow out of the core and the mantle, respectively. A model based on the above energy balance is generally sufficient to quantify the amount of secular cooling of the planet, and parameterizations for the heat flow values q_m and q_c derived from scaling relations between key dimensionless numbers are usually employed to describe the heat transport from the mantle to the surface (e.g., Moresi and Solomatov, 1995; Grasset and Parmentier, 1998; Reese et al., 1998), where heat is ultimately radiated to space.

Perhaps the most important parameter for governing how heat moves through a planetary mantle is the solid-state viscosity. The mantle viscosity has a strong dependence on temperature such that the cool, outermost portion of the planet is rigid, yet at temperatures several hundred to a thousand degrees hotter, mantle material behaves like a slow-moving fluid. A consequence of this behavior for most planets is that an immobile upper layer called a stagnant lid rapidly develops. This situation is in contrast to the mechanism of plate tectonics operating on Earth. The slow diffusion of heat from the interior through the thick stagnant lid is considerably less efficient than the advectively dominated heat transport from lithospheric recycling in plate tectonics. As a result, cooling is slow in most planets, and the interior is kept warm over extended periods of time.

Planetary thermal evolution calculated via the parameterized energy balance approach can then be combined with a model of mantle melting behavior to quantify the amount of melt generated in the interior (e.g., Hauck et al., 2004; Grott et al., 2011). With limited melting experiments tailored to the low-Fe and highly reducing conditions in Mercury's mantle, the well-characterized solidus of terrestrial KLB-1 peridotite has often been used as a proxy for the mantle solidus (e.g., Herzberg et al., 2000). In these one-dimensional parameterized mantle convection models, melt is then generated whenever the mantle temperature exceeds the model solidus and is assumed to be extracted instantaneously, whereas the melt region is replenished with undepleted material on a timescale associated with the mantle convection speed. A schematic view of the relevant temperatures in the interior and the generation of partial melt is shown in Figure 19.4.

One of the major constraints on models of the thermochemical evolution of Mercury is the amount by which the planet has radially contracted as documented by its surface tectonic landforms (Chapter 10; Section 19.3.2). Three global processes contribute to radial contraction, and the magnitude of each can be estimated once the thermochemical evolution of mantle and core has been calculated (Hauck et al., 2004; Grott et al., 2011; Tosi et al., 2013). Cooling causes the mantle and core to contract, resulting in a contribution ΔR_{th} to the change in planetary radius. Phase changes in the core and mantle can result in changes in the specific volume of their constituent materials, which further contribute to a change in the radius of the planet. Usually, consideration of phase changes is restricted to partial melting of the mantle and freezing of an inner core. The products (i.e., crust and the mantle residuum) of mantle differentiation have a larger volume than the primordial mantle, resulting in a net expansion, ΔR_{md}, of the planet (Kirk and Stevenson, 1989). In contrast, solidification of the solid inner core results in a decrease in volume and a hence a radial contraction, ΔR_{ic} (Solomon, 1976). However, with Si present in the core, the density difference between the liquid and solid phase will be small as a result of the nearly similar compositions of the two phases. In total, the radius change of the planet can be expressed as the sum of the individual contributions:

$$\Delta R_p = \Delta R_{th} + \Delta R_{md} + \Delta R_{ic},$$

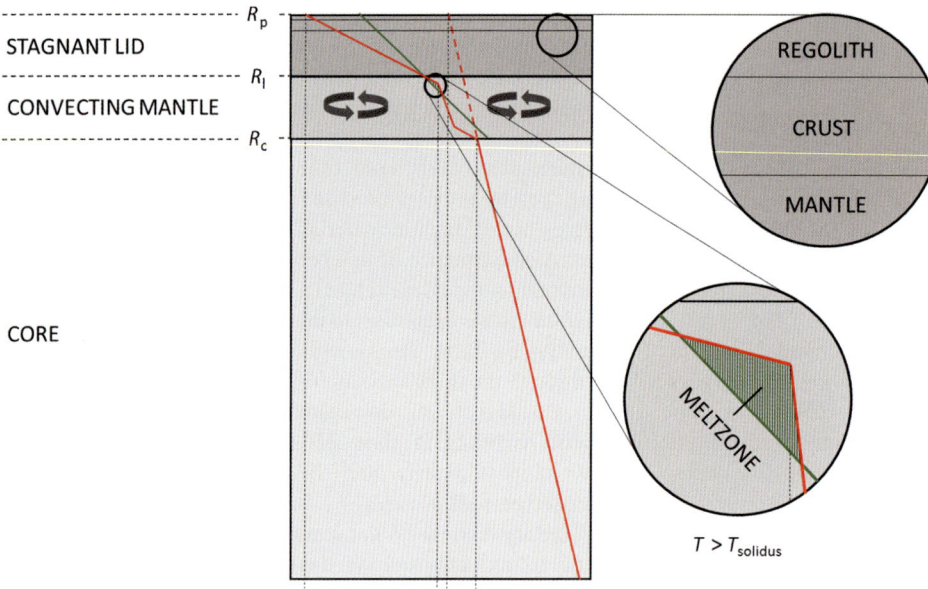

Figure 19.4. Schematic diagram of the reservoirs considered in parameterized thermal evolution models, including the stagnant lid, the convecting mantle, and the core. The planetary radius R_p, stagnant lid radius R_l, and core radius R_c are indicated. Temperatures shown are the surface temperature T_s, the upper mantle temperature T_m, the mantle potential temperature T_p, and the core temperature T_c, with temperature increasing to the right. A nominal temperature profile is shown in red, a nominal mantle solidus temperature is shown in green, and the melt zone in which the local temperature exceeds the solidus is indicated by the filled area in the lower inset.

where the details of the amount of radius change depend on the assumed chemical composition of mantle and core as well as on the associated thermal expansion coefficients (e.g., Grott et al., 2011; Tosi et al., 2013).

Whereas one-dimensional models have been shown to be sufficient to study the global evolution in terms of secular cooling, crustal production, and planetary contraction (Tosi et al., 2013), more complex two- and three-dimensional models are necessary to understand the planform of mantle convection, the efficiency of mantle mixing, and the persistence of mantle convection to the present (Redmond and King, 2007; King, 2008; Michel et al., 2013; Tosi et al., 2013). Instead of parameterizing the heat flow into and out of the mantle, such models involve self-consistent solutions to the equations of mass, energy, and momentum transport in the mantle and directly calculate convective velocities and the temperature distribution in the interior. The chemical composition of the mantle is often tracked with particle tracers (Plesa et al., 2013; Tosi et al., 2013), and the resulting buoyancy is included in the momentum conservation equation.

The increase in model detail of two- and three-dimensional simulations comes at the price of higher computational cost, and running a large number of Monte-Carlo-style simulations, as is increasingly common with one-dimensional models, becomes prohibitively expensive for fully dynamical models. Instead, the parameter space is usually sampled with a few representative models. Depending on the aim of the investigation, model complexity can be reduced by considering two-dimensional models (Redmond and King, 2007; Michel et al., 2013), or by disregarding crustal production (Redmond and King, 2007; King, 2008) or mantle mixing of melt residuum (Redmond and King, 2007; King, 2008; Michel et al., 2013).

The other major constraint on the internal evolution of Mercury is its internally generated magnetic field. A magnetic field generated by a core dynamo requires fluid motions within the electrically conductive fluid portion of the core. A commonly employed minimum, though not necessarily sufficient, requirement for dynamo generation is that if the motions are the result of thermal convection then the core heat flux must exceed the amount of heat that can be transported by thermal conduction along the adiabatic thermal gradient that convection imparts. Energy for driving convective motions also may be derived from compositional buoyancy, such as is generated by the expulsion of a relatively light-element-rich fluid upon crystallization of the core. It has also been suggested (e.g., Christensen, 2006) that Mercury's core may not be entirely convecting and could instead have a stable layer at its top, which would account for its relatively weak magnetic field (see Section 19.4.3). In both of these latter cases, the heat flux at the CMB flows from the core to the mantle but may be less than can be conducted along the adiabatic thermal gradient, with convection restricted to deeper portions of the core.

19.5.2 Persistence of Mantle Convection

Mercury's large core and relatively thin silicate shell raise important questions about how the planet has cooled through its history, in particular the role of mantle convection. These questions are important because upwelling mantle is generally a critical ingredient in magma generation, and convection leads to larger rates of cooling, which help drive the fluid flow in the core necessary for magnetic field generation. Solid-state convection within a layer in a planetary body depends on several material properties as well as the temperature contrast across the layer and particularly the thickness of that layer. For bottom-heated convection, the vigor of mantle convection is described by the non-dimensional mantle Rayleigh number:

$$Ra = \frac{\rho g \alpha \Delta T D^3}{\kappa \eta},$$

where ρ is density, g is gravitational acceleration, α is thermal expansivity, ΔT is the temperature difference across the convecting layer, κ is thermal diffusivity, η is mantle viscosity, and $D = R_p - R_c$ is the thickness of the convecting layer, where R_p is the planetary radius and R_c is the core radius. Convection requires that Ra be larger than some critical value, and above that value the vigor of convection increases with Ra. Therefore, mantle convection is more difficult in a thin mantle than in a thicker one because of the cubic dependence on layer thickness. As a consequence, mantle convection in Mercury is expected to be sluggish compared with convection in planets with thicker mantles. Furthermore, mantle cooling can result in a transition from convection to thermal conduction in the mantle if the Rayleigh number falls below its critical value. The cessation of mantle convection would result in the end of pressure-release melting during convective ascent and most likely the end of global-scale volcanism, though local volcanism could continue, such as is observed in impact basins and sites of small-scale explosive volcanism (Chapter 11).

As a result of Mercury's thin mantle, predictions of rather modest internal heat production, and the strong temperature dependence of the viscosity of mantle rocks, questions were raised in the pre-MESSENGER era about the persistence of mantle convection to the present. Some studies with one-dimensional parameterized convection models found that, although convection was important for much of the planet's history, it may have ceased before the present (Hauck et al., 2004). However, other work with two- and three-dimensional fluid dynamic models (Breuer et al., 2007; Redmond and King, 2007), as well as studies with one-dimensional models that considered the insulating capacity of the near-surface regolith and crust (Grott et al., 2011), generally found that mantle convection persisted throughout the planet's history.

MESSENGER's observations of Mercury have substantially improved our understanding of the planet's interior and so have helped refine many of the assumptions and boundary conditions required for models of its internal evolution. The most important of these constraints are the improved knowledge of radiogenic heat production (Peplowski et al., 2011, 2012) and the thickness of the outer solid shell of the planet (Chapter 4). Typically, earlier work was based on the assumption that the core–mantle boundary was ~600 km deep, compared with the ~420 km determined by MESSENGER (Hauck et al., 2013).

That data from MESSENGER indicate Mercury's silicate shell is nearly one-third thinner than in previous models has led to a reevaluation of whether mantle convection continued throughout the planet's history. Furthermore, although the precise partitioning of heat-producing elements between the near surface (where they have been measured) and the interior is only weakly constrained, the relative amounts of U, Th, and K, as well as their surface abundances, provide important (and previously unavailable) constraints. Indeed, the finding of surprisingly abundant K (Peplowski et al., 2011, 2012) is important for quantifying Mercury's internal evolution because of the strong heat output of ^{40}K coupled with its relatively shorter half-life than the long-lived isotopes of U and Th. Taking these new data into account, Michel et al. (2013) reevaluated the issue of convection within Mercury's mantle utilizing two-dimensional axisymmetric, spherical shell fluid dynamic calculations. They found that, for a broad range of conditions of mantle heat production, mantle viscosity, and initial internal temperatures, cessation of mantle convection within the past several billion years is common in models with silicate layers less than ~440 km thick. These results are consistent with those of Tosi et al. (2013), who evaluated the internal evolution of Mercury in one-, two-, and three-dimensional models of mantle convection additionally constrained by ~3 km of global radial contraction, as had been inferred from mapping ~21% of Mercury's surface by Di Achille et al. (2012). In the models of Tosi et al. (2013), cessation of mantle convection within the past 1–1.5 Gyr was the norm. However, the 5–7 km of radial contraction inferred from more recent global mapping (Byrne et al., 2014) warrants additional thermal evolution calculations, because a larger total cumulative contraction may require higher rates of cooling, which may be more consistent with prolonged mantle convection than with thermal conduction only. As a result of its small obliquity, its proximity to the Sun, and its 3:2 spin–orbit resonance (Chapter 4), Mercury has large spatial variations in surface temperature. By including the latitudinal variation in temperature, Michel et al. (2013) found that cessation of mantle convection may be delayed by a few hundred million years relative to typical models with a spatially constant surface temperature.

With a Monte Carlo approach and the inferred magmatic evolution and global contraction as model constraints, Tosi et al. (2013) determined the times at which convection stopped in one-dimensional models of Mercury's thermochemical evolution. An update of their calculations, taking into account the larger amount of global contraction and the observation that Mercury had an ancient magnetic field as well as a modern one, is shown in Figure 19.5. In all, about 40% of the models consistent with the presently available constraints are found to convect to the present. This outcome is a direct consequence of the more recent estimate of global contraction, which allows for Mercury to have experienced more efficient mantle cooling than in the models of Tosi et al. (2013), which permitted only 3 km of contraction and had vanishingly few outcomes in which mantle convection operated at present.

19.5.3 Internal Evolution Models Consistent with Observational Constraints

Our current understanding of the timing of major processes in Mercury's evolution, as described in the preceding sections, is summarized in Figure 19.6. Evidence from the first ~500 Myr is limited mainly as a result of resurfacing by intercrater plains formation and impact cratering before and during the late heavy bombardment. An internally generated magnetic field was active prior to 3.9–3.7 Ga (Johnson et al., 2015; Chapter 5) and is active today (Ness et al., 1976; Ness, 1979; Anderson et al., 2011, 2012), implying a cooling core within which either thermal or chemical convection operated during each era. The magnetic field history between ~3.8 Ga and the present is currently unknown, and either a continuously operating core dynamo or an early shutdown of the dynamo followed by a later reinitialization is a plausible scenario (Chapter 5). Effusive volcanism was widespread early in Mercury's recorded history (e.g., Marchi et al., 2013), and the areal extent of volcanism waned rapidly from the LHB until perhaps ~3.5 Ga, after which effusive activity largely ended, with the exception of local

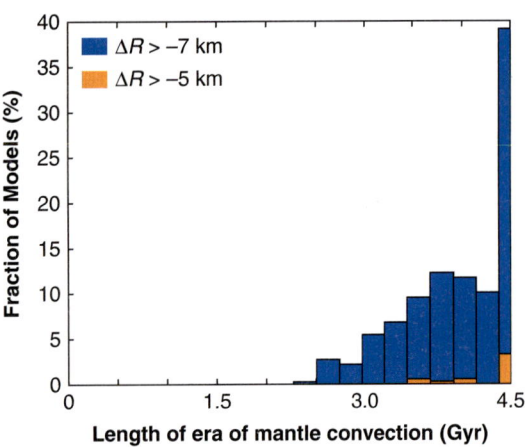

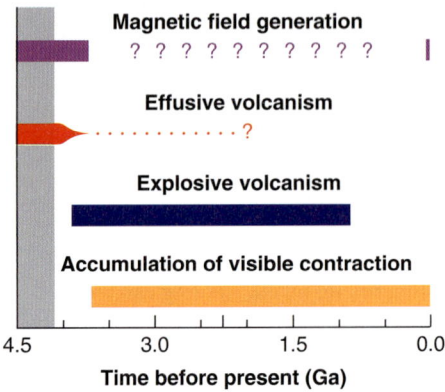

Figure 19.5. Results of Monte Carlo simulations of Mercury's thermal evolution for the duration of mantle convection. A total of 351 (blue) out of 2000 models from the simulations are consistent with the constraints posed by Mercury's magmatic evolution, global contraction, and magnetic field generation. The histogram shows the fraction of models in which mantle convection stopped at a given time. About 40% of the successful models convect to the present. Models shown in blue predict a reduction in planetary radius of between 5 and 7 km. This result should be compared with the models in orange, in which global contraction less than 5 km occurs but which otherwise satisfy the constraints, indicating the sensitivity of the inference on the longevity of mantle convection to the total observed radial contraction. Because of the uncertainty in core composition (Section 19.4.4), contraction from inner core growth is neglected in these calculations. Note that the convention for global contraction here is a negative change in radius.

Figure 19.6. Schematic timeline of major processes in Mercury's evolution. Evidence of the planet's history during the first ~500 Myr has been erased by effusive volcanism and impact bombardment, as indicated by the gray shading.

activity within younger impact basins (Prockter et al., 2010; Denevi et al., 2013a; Byrne et al., 2016; Chapter 11). Explosive volcanism continued for a longer time period than did widespread plains volcanism (Kerber et al., 2009; Thomas et al., 2014). The global contraction accumulated on shortening tectonic landforms records planetary cooling from the end of the LHB to the present (Chapter 10).

Following the approach of Tosi et al. (2013), the range of models that satisfy the following major constraints can be determined. Successful models must (1) produce at least 5 km of crust by partial melting of the mantle, which is a minimal requirement for producing the intercrater and smooth plains, (2) show 5–7 km of global contraction following the end of the late heavy bombardment, and (3) exhibit heat flow from the core that would permit, though not require, the generation of a magnetic field. It is worth noting that the choice of the thickness of extracted crust has little influence on the results, as long as some crust is produced. Furthermore, the requirement on heat flow from the core serves to reject those models that would preclude a thermally driven core dynamo during the earliest evolution, but it is not particularly restrictive later in the planet's history, as core heat flux is generally small after 4 Gyr of evolution.

Models that satisfy all of these constraints show some common trends. Slow cooling of the planet is required, and model mantle reference viscosities at 1600 K range from 10^{20} to 10^{22} Pa s. Additionally, most models also show an early phase of mantle heating, whereas the core cools monotonically throughout evolution. Also, although up to 100 km of crust can be produced, most models produce less than 75 km. Furthermore, surface heat flow declines from about 30 mW/m² at the beginning of evolution to ~10 mW/m² today, consistent with an estimate derived from tectonic modeling (Egea-González et al., 2012). Finally, and most interestingly, the ratio of the concentration of heat-producing elements in the crust to that in the primordial mantle is found to be between 2 and 4.5, which is similar to the results obtained by Tosi et al. (2013) for a core radius of 1940 km. On the other hand, the initial mantle temperature in the models is poorly constrained and can range from 1600 to 1900 K, similar to the range in initial core temperature.

A typical thermochemical evolution model that satisfies the above constraints is shown in Figure 19.7, where the core and mantle temperature; the core, mantle, and surface heat flow; the radius change from thermal contraction and mantle differentiation; and the crustal thickness, stagnant lid thickness, and extent of the partial melt zone are shown as functions of time. In this model, the initial mantle temperature is 1700 K, the initial core temperature is 1875 K, the crustal thermal conductivity is 2.5 W m^{-1} K^{-1}, a poorly conducting regolith layer, 5 km thick and with a thermal conductivity of 0.2 W m^{-1} K^{-1} is included, and the mantle viscosity is $10^{20.5}$ Pa s. With surface abundances of radiogenic elements of 1288 ppm ^{40}K, 155 ppb ^{232}Th, and 90 ppb ^{238}U (Peplowski et al., 2012) and a crustal enrichment factor of 3.5, this typical model has bulk silicate concentrations of heat-producing elements of 368 ppm ^{40}K, 44 ppb ^{232}Th, and 25 ppb ^{238}U, similar to values for Earth and Mars. Following the late heavy bombardment (i.e., at ~3.8 Ga), the model monotonically cools at a rate of 40 K Gyr^{-1}, with the core and mantle cooling at the same rate. Global crustal production ceases around 2.5 Ga (though is largely complete nearly 1 Gyr earlier), and a total of 25 km of crust is produced, resulting in a final crustal thickness of 30 km. Total radial contraction is just short of 7 km, with continuous accumulation of contraction following the late heavy bombardment. It is worth noting that care must be taken when interpreting the timing of crustal production from such one-dimensional

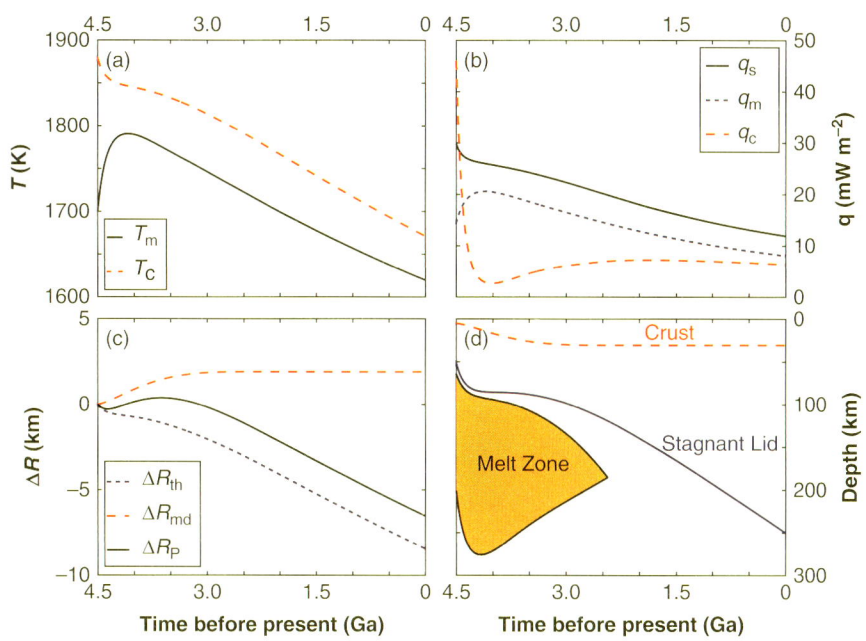

Figure 19.7. Representative thermochemical evolution model for Mercury with parameters as discussed in the text. (a) Evolution of mantle temperature T_m and CMB temperature T_c; (b) evolution of the surface heat flux q_s, mantle heat flux q_m, and core heat flux q_c; (c) evolution of the planetary radius change from thermal expansion and contraction of the mantle and core, R_{th}, from mantle differentiation, R_{md}, and from the sum of the two contributions, R_P; (d) evolution of the thickness of the secondary crust, of the stagnant lid, and of the region in which partial melting occurs.

models, as this timing can differ considerably from that determined with fully dynamical two- or three-dimensional models, which generally have crustal production concentrated earlier in the planet's evolution but result in similar total crustal thickness values (e.g., Tosi et al., 2013).

Given the uncertainties associated with the state and composition of Mercury's core, the model shown in Figure 19.7 focuses on the most robust aspects of the core and considers only thermal contraction of the core; it does not take into account contraction by core solidification. Although, for a given amount of inner core growth, this solidification could be a major contribution to planetary contraction for an Fe–FeS core composition (e.g., Solomon, 1976; Schubert et al., 1988; Knibbe and van Westrenen, 2015), it would be less so if Si were the major alloying light element in the core (Fei et al., 2011) as the density difference between solid and liquid would be smaller because of the very small difference in Si content between solid and liquid (Kuwayama and Hirose, 2004). However, the melting behavior of core material is an important factor in core contraction arising from crystallization: S-bearing cores would experience less inner core growth due to their stronger melting point depression relative to Si-bearing alloys. The true contribution of core freezing to global contraction will likely fall between these two limiting cases, but this effect is difficult to quantify without further data on the equation of state of the Fe–S–Si system. More importantly, it is clear that there is little room for a large contribution to the observed global contraction from core crystallization. The solidification of a large volume fraction of the core would lead to significantly more total contraction than that from thermal contraction alone, e.g., crystallization of >2.5% the volume of the core (equivalent to an inner core >30% of the radius of the core) would lead to at least 2 km of additional contraction (Grott et al., 2011). Thus, the contribution of core crystallization is likely limited, as fewer models would be permitted because they would exceed the 7 km of radial contraction accommodated by tectonic deformation and even the 9 km inferred for total planetary contraction that includes the elastic accommodation of radial contraction prior to the formation of major faults (Chapter 10). This result implies that either core solidification was close to complete by the end of the late heavy bombardment, or that only a small inner core started freezing in the recent past. Because of indications that the inner core is likely to be small (Chapter 4), the latter scenario is more likely.

A three-dimensional view of the thermal evolution of a model with the same properties as discussed above is shown in Figure 19.8. Additionally, the surface temperature variation imposed by Mercury's 3:2 spin–orbit resonance is taken into account (Chapter 4). The model is similar to that presented by Tosi et al. (2015), in which chemical composition is tracked with a particle tracer technique (Plesa et al., 2013), and uses the same initial conditions as the model shown in Figure 19.7. Figure 19.8a shows the variation in the average annual surface temperature, which ranges from 260 K to 430 K between the poles and the equatorial regions. The mantle convection pattern shown in Figure 19.8b reflects this type of temperature distribution, with downwellings (blue) more focused near the polar regions. As a result of the small thickness of Mercury's mantle, the convective pattern shows only small-scale up- and downwellings, and the more linear structures found in earlier simulations of mantle convection with a mantle thickness of 600 km (King, 2008) are not reproduced. Toward the end of the model run, mantle convection ceases, resulting in a conductive temperature profile in the mantle (Figure 19.8 c). In this model, modern mantle temperatures reflect the forcing imposed by the insolation pattern. However, it should be noted that it takes a few hundred million years for the perturbation from insolation to diffuse to any meaningful depth. Therefore, the full extent of the temperature forcing will be reflected in the deep interior only if the 3:2 spin–orbit resonance has been stable for an extended period of time (Correia and Laskar, 2004; Noyelles et al., 2014).

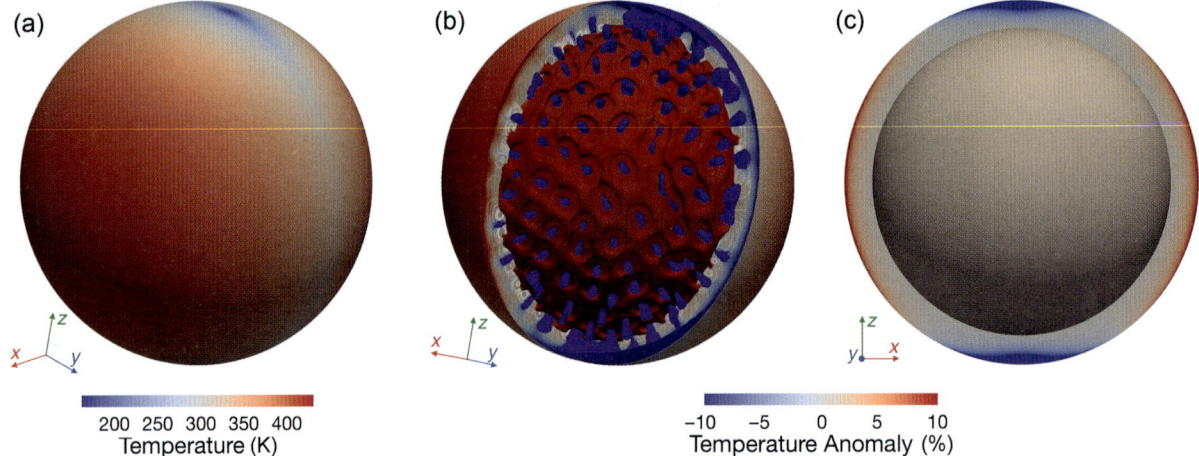

Figure 19.8. (a) Distribution of Mercury's average near-surface temperature according to the model of Vasavada et al. (1999). Hot equatorial poles are located at 0° and 180°E longitude, whereas cold poles are located at ±90°E. (b) Interior temperature anomalies after 1 Gyr of evolution when the mantle was still convecting. The color scale refers to the two mantle slices passing through the 0° and 90° meridional planes (the x–z and y–z planes, respectively), on top of which streamlines are plotted. Within the sector of mantle between the two meridional planes, blue volumes mark the locations of downwelling flow in which temperatures are 4–5% colder than average at that depth, and the red surface is the shallowest surface of the volume of upwelling material in which temperatures are 1–2% hotter than average at that depth. (c) Interior temperature anomalies at present after the mantle transitioned to a conductive state, shown on the 0° meridional plane (x–z). Figure courtesy of Nicola Tosi.

Although the general picture of Mercury's thermochemical evolution is consistent with the constraints provided by MESSENGER observations, details of the models may change as more data are analyzed and further data are eventually provided by new missions such as BepiColombo (Chapter 20). In particular, the amount of radial contraction documented in shortening tectonic structures has been continuously refined (Strom et al., 1975; Watters et al., 2009; Di Achille et al., 2012; Byrne et al., 2014), resulting in less stringent constraints on Mercury's thermal evolution. Current best estimates for the total radial contraction accumulated by brittle structures since the late heavy bombardment range from 5 to 7 km (Byrne et al., 2014) but may be as large as ~9 km when elastic deformation is considered, or less than 5 km if the dip angles of the thrust faults are uniformly and surprisingly steep (Chapter 10). Importantly, larger values (>7 km) of contraction would allow for lower mantle viscosities and thus more efficient mantle convection. Alternatively, such greater contraction could also allow for a larger contribution of core solidification to the total contraction of Mercury, depending on core composition, or more likely some combination of increased cooling and core solidification.

19.5.4 Other Factors Influencing Mercury's Thermochemical Evolution

One of the factors not considered in the above models is the potential presence of heat-producing elements in Mercury's core. At the low oxygen fugacities inferred from the high S abundance and low FeO content in Mercury's crust (Zolotov et al., 2013), lithophile elements such as K, Th, and U can become more siderophile (Malavergne et al., 2010). McCubbin et al. (2012) estimated that up to 10% of the total inventory of U and potentially Th could have partitioned into the core, thus providing an additional heat source that could slow global contraction. However, the differences in global contraction between models with and without heat-producing elements in the core have been found to be minor (Tosi et al., 2013), as the total inventory of heat-producing elements in the interior is only weakly affected. Partitioning of U and Th into the core tends to increase the heat flux out of the core and can extend the period during which a thermal-buoyancy-generated dynamo can operate by as much as 100 Myr.

In addition to the production of partial melt in the interior, Mercury's surface compressive stress state has likely been an important factor controlling effusive volcanism. On a contracting planet such as Mercury, extrusive volcanism may be substantially inhibited as magma pathways to the surface are shut off by maximum compressive stresses in the horizontal direction (Chapter 11). Therefore, the longevity of volcanism as observed on the surface may not be a direct indicator of the timing of melt production in the deep interior. On the other hand, local factors such as variations in the thickness of an insulating crust and/or regolith layer, which would have a lower thermal conductivity than the mantle (Section 19.5.3), largely due to higher porosities (Schumacher and Breuer, 2006), are usually not fully taken into account in thermochemical evolution models. Therefore, local volcanism may be ongoing even if global models, particularly one-dimensional models, do not predict the production of partial melt at a given time.

Another energy source not treated in the above discussion is impact heating, which would be expected to contribute to the global energy balance mainly during the early phases of Mercury's evolution. Impact heating associated with the formation of the Caloris impact basin was modeled by Roberts and Barnouin (2012), who showed that impact heat can alter mantle dynamics. In addition to the production of melt at the impact site itself, partial melting may be induced even far from the impact. Thus, the smooth plains within and adjacent to the Caloris basin could be at least in part the consequence of the impact itself, the heat for which was stored in the mantle over an extended period of time. On the other hand, the influence of isolated impacts on the global evolution of the planet is relatively small (Roberts and Barnouin, 2012), and the conclusions drawn from the simpler models discussed above remain essentially unchanged.

19.5.5 Core Evolution

MESSENGER's unveiling of Mercury's internal structure and the geometry and history of its internal magnetic field underscore the important role of the metallic core on the planet's evolution. Taken in concert with the growing understanding of the properties of materials at the conditions of Mercury's core (Section 19.4.4), which indicate the potential importance of zones of top-down crystallization and liquid–liquid immiscibility, it is clear that core evolution in Mercury differed from that in Earth. Ultimately, models of core evolution on Mercury must account for the planet's magnetic field structure and history (Chapter 5), match the internal structure (Chapter 4), and be consistent with the magnitude of the planet's contraction (Chapter 10).

The driving mechanisms of core evolution are cooling and the chemical differentiation that results from crystallization as the core cools below its melting temperature. The rate of core cooling depends strongly on how the mantle is cooling, as all of the heat from the core must pass through the mantle on its way to the planet's surface. Early in the planet's history, core cooling may have been relatively rapid (Figure 19.7), especially if the planet was hot, because high internal temperatures would reduce the viscosity of the mantle and make it easier to remove heat quickly by convection. Of course, just as the cooling of the mantle slows as its initial store of heat of formation is lost and heat production follows the decay of radioactive elements, the cooling of the core slows as well. The rate of cooling of the core is important because a source of convection is necessary to drive the motions in the electrically conductive liquid metal that generate the magnetic field. A minimum condition for thermal convection throughout the entire core is that the heat flux through the CMB must exceed that which can be conducted along the adiabat. Given a thermal conductivity of 40 W m^{-1} K^{-1}, previous workers (Hauck et al., 2004; Tosi et al., 2013) found the minimum core heat flux for thermal convection to be in the range of 12–19 mW m^{-2} for a range of possible thermal expansivity values. Such core heat fluxes were exceeded only early in Mercury's history. The more recent, higher estimates of the thermal conductivity of pure iron at pressures near that of Mercury's CMB (Deng et al., 2013b) of 40–120 W m^{-1} K^{-1} could increase this minimum heat flux by up to a factor of 3. Such high thermal conductivities would limit thermally driven core convection to a very short time period following planet formation. However, the presence of light alloying elements tends to decrease the thermal conductivity; for example, as little as 9 wt% Si reduces the thermal conductivity of the Fe alloy to 41–60 W m^{-1} K^{-1} (Seagle et al., 2013) at Earth's core conditions. As Mercury's core likely hosts considerable abundances of light elements (Section 19.4.1), the earlier value adopted for thermal conductivity may not be far off, though the uncertainty may be considerable.

Although it is possible that Mercury's early magnetic field (Chapter 5) was driven by thermal convection, the present-day field is likely dominated by flows driven by compositional buoyancy. The simplest mechanism for generating compositional buoyancy is crystallization of a core alloy in a situation where the compositional difference between the precipitating solid and residual liquid is large, such as has been previously described in the Fe–S system. Sulfur-bearing systems are the best-studied analog for Mercury because of the broad literature on Fe–S melting and because S has such a large melting-point depression even at high pressure (e.g., Fei et al., 1997). The consequence of the decreasing melting temperatures and eutectic S contents with increasing pressure (Section 19.4.4) is that, if the core is composed of an Fe–S alloy, then it is likely that the crystallization of core material at these pressures began at the top, rather than the bottom, of the core (Hauck et al., 2006; Stewart et al., 2007; Chen et al., 2008; Williams, 2009). An interesting consequence of the combination of the shifts in eutectic temperature and compositions, which vary with pressure, is that two radially separated regions of the core may experience such top-down crystallization, also termed Fe snow (Chen et al., 2008).

Both pre- and post-MESSENGER models (Chen et al., 2008; Dumberry and Rivoldini, 2015) of an Fe–S core indicate multiple modes of crystallization, including bottom-up (like Earth) and top-down (Fe snow). In such a system, at low S contents of ~5 wt% or less and with small inner cores, Dumberry and Rivoldini (2015) found that bottom-up crystallization would be expected. However, those workers did not model the non-ideal mixing behavior observed at 14 GPa in the Fe–S system (Chen et al., 2008), which essentially requires a zone of Fe snow between 10 and 14 GPa at even very small S contents because the decrease in melting temperature is so large. With larger S contents or with larger inner core sizes, various top-down crystallization regimes are possible, whether there is a layer of crystallizing material overlying a layer in which the Fe snow re-melts, whether the crystallizing material simply falls to the top of the growing inner core (Hauck et al., 2006; Dumberry and Rivoldini, 2015), or whether there is a second layer of top-down crystallization (Chen et al., 2008).

Top-down crystallization is a consequence of a situation in which the melting temperature increases as a function of depth more slowly than the actual temperature (Hauck et al., 2006; Williams, 2009). In the Fe–S system there is a marked decrease in the eutectic melting temperature with increasing pressure, as well as a reduction in the S content of the eutectic with increasing pressure, both of which lead to melting temperatures decreasing with depth for a wide range of bulk compositions. Measurements of the density and sound velocity of Fe–S liquids at high pressure also indicate that S tends to result in larger adiabatic temperature gradients relative to pure Fe liquids, enhancing this effect and extending to even lower S contents (Jing et al., 2014). As a result of the small melting-point reduction in Fe–S alloy cores with low abundances of S, such systems tend to have large inner cores, which in turn tends to concentrate S in the outer core because of the low solubility of S in solid Fe. As a consequence, Fe–S-dominated cores are likely to have experienced Fe snow regardless of their composition. However, such large inner cores are not favored in structural models constrained by Mercury's rotational dynamics (Chapter 4).

Even though our understanding of the evolutionary paths of Mercury's core under scenarios in which S is the sole light element is becoming more mature, it is also clear that other light elements in addition to, or instead of, S are likely to be present in the core (Section 19.4.1; Chapter 2). As noted above, carbon is generally a siderophile element, but it has been suggested that C is present as graphite in the mantle and that graphite may have formed an early flotation crust on the planet (Vander Kaaden and McCubbin, 2015), an idea that is consistent with spectral reflectance and neutron spectroscopy observations of the surface (Murchie et al., 2015; Peplowski et al., 2015a, 2016). Consequently, if the core and mantle formed in equilibrium then the core may be saturated in C, although the total amount would be small as the maximum solubility of C in Fe is ~4 wt% and that value decreases with increasing pressure (Lord et al., 2009). This value would be larger if Fe_3C were present, but the density and compressibility of C-bearing alloys are such that it would be difficult for C to be the sole light element in Mercury's core. However, the consequences of even some C being present might be important. For example, the decreasing amount of C in eutectic melts with increasing pressure in the $Fe-Fe_3C$ system is consistent with top-down crystallization, even without S.

In contrast, the presence of silicon, which is likely because of the planet's strongly reducing conditions (see Sections 19.4.1 and 19.4.4), has rather different implications for the evolution of the core. The melting behavior of Si-bearing Fe alloys at conditions appropriate to Mercury is more poorly known than for alloys with S or even C. The phase diagram of Fe-FeSi at 21 GPa determined experimentally by Kuwayama and Hirose (2004) is instructive, as they found that the eutectic point is at both a higher temperature and a larger Si abundance than at 100 kPa (1 bar). They also found, as noted above, that the difference in composition between the coexisting solid and liquid phases at temperatures between the solidus and liquidus on the Fe side of the eutectic is very small: there is a maximum of ~2 wt% Si between the solid and liquid phases. An important consequence of this behavior is an Earth-like bottom-up crystallization of the core, but with residual liquids left by crystallizing of Fe–Si core material that would be only marginally less dense than surrounding material, limiting the buoyancy available to drive convection were the core sufficiently chemically reduced that Si were the only light alloying element present.

Perhaps most critical to understanding the evolution of Mercury's core is the behavior of Fe alloys with combinations of S, Si, and possibly C. Despite the fact that the thermodynamic properties of multi-component Fe alloys are less well known than for the binary systems, the data that are available suggest interesting evolutionary paths for Mercury's core. For example, liquid immiscibility, such as displayed in both Fe–S–C (e.g., Dasgupta et al., 2009) and Fe–S–Si liquids (Section 19.4.1), has potential consequences for compositional segregation within the outer core. Fe–S–C immiscibility would have an influence within only a relatively thin layer near Mercury's CMB because immiscible behavior occurs only at pressures less than 6 GPa (Dasgupta et al., 2009), which is close to the possible CMB pressure (Chapter 4). However, immiscibility in the Fe–S–Si system would extend deeper within Mercury's outer core (Section 19.4.4). Such segregation, if present, likely developed early in the planet's history during metal–silicate differentiation and immediately thereafter. For bulk core compositions near the miscibility limit, however, there is a possibility that the progressive crystallization of an Fe–Si-rich solid and resultant increase in concentration of S in the liquid would drive Mercury's core into a liquid immiscibility state later in its evolution. For this situation to occur, however, relatively large inner core growth would be required to substantially change the outer core composition, an outcome that is inconsistent with models of Mercury's thermal contraction discussed above and estimates of the planet's internal structure (Chapter 4).

A relative lack of experimental data limits firm conclusions about the crystallization behavior in an Fe–S–Si core. Recent experimental results on the Fe–S–Si–C system (Martin et al., 2015) indicate eutectic melting temperatures similar to those of the Fe–S–C system at ~4–15 GPa, with minimal pressure dependence of the eutectic. Top-down crystallization would be favored in that system. However, data on the pressure dependence of melting in the Fe–S–Si system are not available at present. While the melting behavior of the Fe–S and Fe–S–Si–C systems suggest that top-down crystallization is likely, the Fe–Si system appears more consistent with a bottom-up crystallization sequence; whether the effects of alloying with S or Si would dominate that behavior is unclear without further data. Determination of melting behavior in the Fe–S–Si system, and of the thermodynamic properties that control the adiabatic temperature gradient, are crucially needed in order to understand more fully the crystallization of Mercury's core.

19.6 DISCUSSION

MESSENGER observations have substantially altered our understanding of how Mercury has evolved over its history, but several crucial questions remain open. In particular, we are at a relatively early stage in understanding the connection between the dynamics of the mantle and the production of the crust and the generation of the magnetic field. We next discuss these issues in more detail, focusing on open questions that may be addressed through a combination of analysis of MESSENGER data, modeling, and the acquisition of new observations from BepiColombo and other future missions.

19.6.1 Crustal Production and Mantle Dynamics

Global crustal production through time is a primary indicator of the evolution of a planet – that of its crust and of the interior from which the crust was derived. For planets without crustal recycling, the crust represents a nearly complete time history of intrusive and extrusive volcanism. This history, even when known only to first order, places powerful constraints on our understanding of the evolution of the interior (e.g., Hauck and Phillips, 2002). On Mercury, the clearest constraints on crustal formation are that the best estimate of its average thickness is approximately 35 km (James et al., 2015; Padovan et al., 2015; Chapter 3) and that the tail end of the era of effusive volcanism postdates the Caloris impact by perhaps a few hundred million years at most (Byrne et al., 2016; Chapters 6, 11). Intercrater plains, also interpreted to be dominantly volcanic in origin, are more areally extensive than the smooth plains and in places are as old as 4.1–4.0 Ga (Chapter 6). The first ~500 Myr of Mercury's surface history is also veiled by the overprinting of the late heavy bombardment. Regardless, MESSENGER observations have revealed that Mercury has experienced the most efficient and complete differentiation of mantle and crust among the terrestrial planets, and that this crust was largely built up by successive episodes of effusive volcanism that were likely largely complete within the first 1 Gyr of planet history. Given that Mercury has such a thin mantle, prone to relatively sluggish mantle flow and even the cessation of mantle convection entirely, it is remarkable that generation of the crust could have been so efficient and rapid – particularly in light of the idea that crustal products of a magma ocean may have been only meters thick (e.g., Vander Kaaden and McCubbin, 2015), leaving virtually all of the crust to be produced by serial magmatism. However, because of the low FeO content and modest pressures in Mercury's mantle, the partial melts produced throughout the mantle would be exceptionally buoyant and less susceptible to stalling during ascent (Vander Kaaden and McCubbin, 2015), perhaps facilitating such efficient crustal formation.

The heterogeneity of Mercury's crustal production as observed in its geochemical diversity (e.g., Weider et al., 2015; Chapters 2, 7), and the spatial distribution of smooth plains volcanism, also provide important clues to the history and dynamics of the interior. Indeed, observations by MESSENGER's suite of geochemical sensors indicate both a range of crustal compositions that point to partial melting from multiple sources (Charlier et al., 2013), and a spatial heterogeneity in compositions that does not always follow geomorphological unit boundaries (Peplowski et al., 2015b; Weider et al., 2015). Interestingly, in a manner similar to the Moon's spatial dichotomy in mare volcanism between its near and far sides, and the asymmetric concentration of volcanism on Mars near the Tharsis province, there is a distinctive spatial difference in the abundance of smooth plains units between Mercury's northern and southern hemispheres (Chapters 6, 11). The largest expanses of smooth plains reside at high northern latitudes and within and around the Caloris basin, which is also located in the northern hemisphere. Smaller areas of smooth plains are found generally in proximity to impact basins, with little difference in areal coverage between the hemispheres (Chapter 6). Consequently, the processes responsible for the formation of smooth plains in the Caloris region and the northern volcanic plains may be different from those that yielded the isolated, small smooth plains units distributed more evenly throughout the northern and southern hemispheres. Any hemispherical differences in the earlier volcanic activity that produced the intercrater plains are not clear at this time, though some regions also appear to be associated with impact basins (Denevi et al., 2013b). Although MESSENGER provided global geochemical coverage of Mercury, the spacecraft's highly eccentric orbit and high northern periapsis resulted in measurements only at low spatial resolution in the southern hemisphere. That these measurements cannot resolve distinct geochemical terranes in the southern hemisphere limits our understanding of the global evolution of Mercury. The planned orbit for the Mercury Planetary Orbiter on the BepiColombo mission (Chapter 20) will yield higher-resolution southern hemisphere data and may resolve additional geochemical terranes analogous to those observed by MESSENGER in the northern hemisphere.

These heterogeneities in the geochemical and volcanic character of the surface are largely connected to the thermal and chemical properties of the mantle. Mercury's thin mantle yielded a style of mantle convection that was both relatively sluggish and strongly spatially restricted, because the size of individual convective cells would have been on the order of the thickness of the mantle itself. Thus, the large expanses of volcanism in the northern hemisphere require conditions that either permit extraordinarily voluminous magma production from spatially restricted upwellings or conditions that substantially altered the mantle flow dynamics from that expected on the basis of Mercury's mantle thickness. One such mechanism for altering those dynamics is a large impact, such as that which formed the Caloris basin. Indeed, the large thermal perturbation imparted by shock heating from the Caloris impact event may have led to substantial heating of the shallow mantle beneath the impact, but it might also have enhanced some nearby, preexisting mantle upwellings that generated magma distal from the impact site (Roberts and Barnouin, 2012). Such a mechanism could have been a major contributor to the generation of the Caloris-centric volcanism, but the northern volcanic plains do not appear to host such a large impact capable of triggering such volcanism, even though Caloris and the northern volcanic plains have indistinguishable crater size–frequency distributions and thus ages (e.g., Ostrach et al., 2015). On the other hand, both the broad geochemical heterogeneity across the surface and the smaller, more distributed areas of smooth plains on Mercury could be direct consequences of the small, spatially restricted upwellings and inefficient mixing in a mantle of small thickness. This fluid dynamic behavior of the mantle could act to preserve large-scale geochemical heterogeneities, yet also focus volcanism in locally restricted areas. An important question regarding the era of dwindling effusive volcanism is the relative importance of the pattern of convection (e.g., small yet relatively abundant upwellings) to the total cooling of Mercury that led to a strongly compressive stress state, one that tended to favor intrusive over extrusive volcanic activity.

19.6.2 Evolution of the Core and Magnetic Field

The operation of an internally generated planetary magnetic field is a fundamental indicator of the dynamical behavior of the deep interior of a planet. MESSENGER observations of Mercury's magnetic field have provided important constraints on the character of field generation at present as well as early in the planet's history. Orbital measurements of the geochemical character of the surface materials, as well as gravity and rotational state determinations by MESSENGER, also help to characterize the core. However, these new observations raise a number of interesting questions about the behavior of the interior over the history of the planet. In particular, the mechanism of magnetic field generation may require a number of special conditions in order to produce a weak, axisymmetric field with a large dipole offset. Further, the magnetic field, with remanent crustal magnetism indicating an ancient field in addition to the modern field, places limits on the rate of cooling over the planet's history.

Although explaining Mercury's weak magnetic field has long been a challenge (e.g., Heimpel et al., 2005; Stanley et al., 2005; Christensen, 2006), it is the combination of the weakness of the field and its axial alignment and asymmetry about the equator that makes understanding the dynamo mechanism even more intriguing. A common thread in many models of Mercury's magnetic field is the presence of a layer stable against convection (e.g., Christensen, 2006; Vilim et al., 2010; Tian et al., 2015). If such a layer is present, most likely at the top of the fluid core, then the heat flux out of the core may be less than what can be conducted along the adiabatic temperature gradient. In addition, compositional stratification may also be present. As discussed above, it is quite likely that there is a thermal component to the stability of such a layer, as thermal history calculations generally predict a subadiabatic heat flux at present. Furthermore, many potential core alloy compositions favor top-down crystallization regimes that lead to compositionally stratified layers. Thus, it seems likely that Mercury's core contains a stable layer that plays a role in the strength and geometry of the planet's magnetic field.

Yet the presence of a stable layer alone appears insufficient for explaining the strength and geometry of Mercury's magnetic field. To that end, recent models have included additional heterogeneity capable of further influencing magnetic field character (e.g., Figure 19.3). In particular, both Cao et al. (2014) and Tian et al. (2015) imposed laterally variable heat flux conditions at the CMB. Cao et al. (2014) utilized a heat flux pattern symmetric about the equator similar to the latitudinal variation in surface temperature consistent with Mercury's small axial tilt. Should the mantle be in a conductive, rather than convective state, then surface temperature variations at the surface may also have a signature at the CMB if enough time has passed since the end of the convective motions. Cao et al. (2014) investigated models with the highest or lowest heat flow at the equator, and they found that models with higher heat flow near the equator were better able to stabilize fields with geometries similar to those observed by MESSENGER. However, the mechanism for inducing larger heat fluxes along the equator, rather than at the poles, is unclear. Diffusion of surface temperatures to the CMB would result in relatively lower mantle temperatures near the poles, and therefore larger temperature differences and heat fluxes across the CMB there, rather than at the equator. As demonstrated in Figure 19.8, the limited thickness of the mantle seems to preclude long-wavelength convective patterns, so a degree-2 style of mantle convection appears unlikely as well. Therefore, some other mechanism for inducing a symmetric equator-to-pole variation in heat flux appears necessary for this mode of dynamo generation to operate.

Alternatively, Tian et al. (2015) imposed an asymmetric heat flux boundary condition along the CMB, with a higher heat flux out of the core near the north pole of Mercury (Figure 19.3). Those authors postulated that the higher heat flux there is a remnant of the magmatism that produced the NSP. As discussed in the previous section, there is a notable spatial dichotomy in the distribution of the youngest smooth plains on Mercury, with the largest expanses in the northern hemisphere (e.g., Ostrach et al., 2015; Chapters 6, 11). However, as those volcanic deposits were emplaced at 3.8–3.7 Ga, the thermal conditions that generated them are likely no longer present. Furthermore, smaller though still extensive ($>10^5$ km^2 area) (Byrne et al., 2016) smooth plains units, the youngest effusive volcanic deposits on Mercury, are relatively well distributed between the northern and southern hemispheres, exclusive of the NSP and the plains associated with Caloris. Thus, it is worth considering whether the mechanisms for the large volcanic deposits and the smaller, more evenly distributed smooth plains deposits are the same (including whether some of the smaller deposits are even volcanic). Whereas the relatively larger concentrations of K at high northern latitudes on Mercury (Peplowski et al., 2012; Chapter 7) might argue for a mantle source more enriched in heat-producing elements, such enhanced heat production would in fact lead to smaller temperature contrasts and a lower heat flux across the CMB. Interestingly, the K enhancement at high northern latitudes does not respect the morphologic boundaries of the northern plains, nor are the lavas in Caloris so enriched. However, if the generation of the NSP substantially depleted the mantle at high northern latitudes of heat-producing elements compared with the rest of the planet, then core heat fluxes might be somewhat higher there due to the cooler mantle temperatures. The relatively limited amount of lateral mixing of the mantle expected under low-Rayleigh-number convection, coupled with the small scale of convection, could act to preserve such heterogeneity.

It is worth noting that MESSENGER gravity and topography data indicate that the domical rise within the northern volcanic plains is substantively compensated within ~100 km of the CMB (James et al., 2015). James et al. (2015) investigated a variety of mechanisms for the source of this compensation, including relief along a compositional interface (e.g., between the silicate mantle and a possible solid FeS layer at the top of the core) as well as other density variations. Variations in the thickness of an FeS layer would also result in changes in the thermal conductivity profile above the liquid core, leading to lateral differences in heat flux. A variety of compositions or viscosities at that depth may also have induced additional thermal heterogeneity, though the impact of such variations relative to the remainder of the planet remains to be investigated.

It is clear that heterogeneity within Mercury's mantle may influence the mechanisms by which the planet's magnetic field is generated, though more work – and the need for further observations – remains. Indeed, any geochemical and petrologic heterogeneity (Chapters 2, 7) inherited from Mercury's earliest history may have substantially influenced the planet's history; yet, as less is known about the geochemical and geophysical character of the entire southern hemisphere than the north, we have much more to learn about the distribution of any heterogeneous properties of the interior.

Mercury's internal structure and chemical makeup strongly influence the manner by which the planet's core, and therefore its magnetic field, has evolved. The discovery of Mercury's remanent crustal magnetism (Johnson et al., 2015; Chapter 5) in crust that was last emplaced before ~3.7 Ga raises the question of how a planet cooling as modestly as suggested by its record of global contraction could have hosted both a relatively protracted period of early magnetic field generation and a modern field. A purely thermally generated dynamo that spans both time periods is unlikely, as the thermal history models indicate that core heat flux dropped below the critical value for convection early in the planet's history and remains so. Indeed, early-onset thermal dynamos would tend to be short-lived, as evidenced by Figure 19.7 and previous modeling efforts (Hauck et al., 2004; Grott et al., 2011; Tosi et al., 2013). Although much shorter than the upper bound of ~800 Myr implied by the surface age of the crust in areas of remanent magnetism, such shorter-duration dynamos are potentially consistent with observations, as the column of crust hosting the remanence may predate the surface age. Models with longer-lived supercritical core heat fluxes are also possible. Under that scenario, the simplest explanation for the modern magnetic field is that it restarted comparatively recently as a result of the onset of core crystallization and perhaps even inner core growth. Alternatively, core crystallization that operated throughout the past 3.7 Gyr would account for both the ancient and modern fields. This mechanism is possible, yet would likely result in solidification of a substantial fraction of the core and greater contraction of the planet than has been documented so far. A large inner core does not appear to be compatible with the planet's internal structure (Hauck et al., 2013; Dumberry and Rivoldini, 2015; Peale et al., 2016) nor with magnetic field generation, as compositional gradients imposed by top-down crystallization, coupled with a large inner core, may serve to stabilize the entire core against convection (Dumberry and Rivoldini, 2015; Rückriemen et al., 2015). Thus, a full understanding of the operation and evolution of Mercury's magnetic field depends on characterizing the age distribution of remanent crustal magnetism and understanding how core evolution, including the effects of core chemistry, was coupled to mantle convection and cooling through time.

19.7 CONCLUSIONS

MESSENGER has been instrumental in unveiling key elements of the global evolution of Mercury. From firmly establishing the occurrence of volcanism and its distribution in space and time, to substantively resolving the long-standing paradox between predicted and observed values for Mercury's global contraction and cooling, MESSENGER has brought new insight to fundamental questions about the planet that stood for nearly four decades. In turn, and as with all new missions of discovery, MESSENGER has raised new questions about how Mercury has operated over its history. With Mercury's remarkably thin mantle, which is incapable of significantly homogenizing its chemical character by mantle convection, it is clear that chemical heterogeneity has played an important role in the planet's history. The weak, axially aligned, and northward offset geometry of the internally generated magnetic field may be a distinct manifestation of internal heterogeneity. However, it is the discovery of Mercury's ancient magnetic field, recorded in the crustal rocks, that may hold some of the deepest clues to the planet's internal evolution.

REFERENCES

Anderson, B. J., Johnson, C. L., Korth, H., Purucker, M. E., Winslow, R. M., Slavin, J. A., Solomon, S. C., McNutt, R. L., Jr., Raines, J. M. and Zurbuchen, T. H. (2011). The global magnetic field of Mercury from MESSENGER orbital observations. *Science*, **333**, 1859–1862, doi:10.1126/science.1211001.

Anderson, B. J., Johnson, C. L., Korth, H., Winslow, R. M., Borovsky, J. E., Purucker, M. E., Slavin, J. A., Solomon, S. C., Zuber, M. T. and McNutt, R. L., Jr. (2012). Low-degree structure in Mercury's planetary magnetic field. *J. Geophys. Res.*, **117**, E00L12, doi:10.1029/2012je004159.

Anderson, J. D., Jurgens, R. F., Lau, E. L., Slade, M. A., III and Schubert, G. (1996). Shape and orientation of Mercury from radar ranging data. *Icarus*, **124**, 690–697, doi:10.1006/icar.1996.0242.

Banks, M. E., Xiao, Z., Watters, T. R., Strom, R. G., Braden, S. E., Chapman, C. R., Solomon, S. C., Klimczak, C. and Byrne, P. K. (2015). Duration of activity on lobate-scarp thrust faults on Mercury. *J. Geophys. Res. Planets*, **120**, 1751–1762, doi:10.1002/2015je004828.

Berthet, S., Malavergne, V. and Righter, K. (2009). Melting of the Indarch meteorite (EH4 chondrite) at 1 GPa and variable oxygen fugacity: Implications for early planetary differentiation processes. *Geochim. Cosmochim. Acta*, **73**, 6402–6420, doi:10.1016/j.gca.2009.07.030.

Beuthe, M. (2010). East–west faults due to planetary contraction. *Icarus*, **209**, 795–817, doi:10.1016/j.icarus.2010.04.019.

Blair, D. M., Freed, A. M., Byrne, P. K., Klimczak, C., Prockter, L. M., Ernst, C. M., Solomon, S. C., Melosh, H. J. and Zuber, M. T. (2013). The origin of graben and ridges in Rachmaninoff, Raditladi, and Mozart basins, Mercury. *J. Geophys. Res. Planets*, **118**, 47–58, doi:10.1029/2012JE004198.

Breuer, D., Hauck, S. A., II, Buske, M., Pauer, M. and Spohn, T. (2007). Interior evolution of Mercury. *Space Sci. Rev.*, **132**, 229–260, doi:10.1007/s11214-007-9228-9.

Brown, S. M. and Elkins-Tanton, L. T. (2009). Compositions of Mercury's earliest crust from magma ocean models. *Earth Planet. Sci. Lett.*, **286**, 446–455, doi:10.1016/j.epsl.2009.07.010.

Bullard, E. and Gellman, H. (1954). Homogeneous dynamos and terrestrial magnetism. *Phil. Trans. Roy. Soc. London A*, **247**, 213–278, doi:10.1098/rsta.1954.0018.

Burbine, T. H., McCoy, T. J., Nittler, L. R., Benedix, G. K., Cloutis, E. A. and Dickinson, T. L. (2002). Spectra of extremely reduced assemblages: Implications for Mercury. *Meteorit. Planet. Sci.*, **37**, 1233–1244, doi:10.1111/j.1945-5100.2002.tb00892.x.

Byrne, P. K., Klimczak, C., Williams, D. A., Hurwitz, D. M., Solomon, S. C., Head, J. W., Preusker, F. and Oberst, J. (2013). An assemblage of lava flow features on Mercury. *J. Geophys. Res. Planets*, **118**, 1303–1322, doi:10.1002/jgre.20052.

Byrne, P. K., Klimczak, C., Şengör, A. M. C., Solomon, S. C., Watters, T. R. and Hauck, S. A., II (2014). Mercury's global contraction much greater than earlier estimates. *Nature Geosci.*, **7**, 301–307, doi:10.1038/ngeo2097.

Byrne, P. K., Ostrach, L. R., Fassett, C. I., Chapman, C. R., Denevi, B. W., Evans, A. J., Klimczak, C., Banks, M. E., Head, J. W. and Solomon, S. C. (2016). Widespread effusive volcanism on Mercury likely ended by about 3.5 Ga. *Geophys. Res. Lett.*, **43**, 7408–7416, doi:10.1002/2016GL069412.

Cao, H., Aurnou, J. M., Wicht, J., Dietrich, W., Soderlund, K. M. and Russell, C. T. (2014). A dynamo explanation for Mercury's anomalous magnetic field. *Geophys. Res. Lett.*, **41**, 4127–4134, doi:10.1002/2014gl060196.

Chabot, N. L., Wollack, E. A., Klima, R. L. and Minitti, M. E. (2014). Experimental constraints on Mercury's core composition. *Earth Planet. Sci. Lett.*, **390**, 199–208, doi:10.1016/j.epsl.2014.01.004.

Charlier, B., Grove, T. L. and Zuber, M. T. (2013). Phase equilibria of ultramafic compositions on Mercury and the origin of the compositional dichotomy. *Earth Planet. Sci. Lett.*, **363**, 50–60, doi:10.1016/j.epsl.2012.12.021.

Chen, B., Li, J. and Hauck, S. A., II (2008). Non-ideal liquidus curve in the Fe–S system and Mercury's snowing core. *Geophys. Res. Lett.*, **35**, L07201, doi:10.1029/2008gl033311.

Christensen, U. R. (2006). A deep dynamo generating Mercury's magnetic field. *Nature*, **444**, 1056–1058, doi:10.1038/nature05342.

Christensen, U. R. and Wicht, J. (2008). Models of magnetic field generation in partly stable planetary cores: Applications to Mercury and Saturn. *Icarus*, **196**, 16–34, doi:10.1016/j.icarus.2008.02.013.

Chudinovskikh, L. and Boehler, R. (2007). Eutectic melting in the system Fe–S to 44 GPa. *Earth Planet. Sci. Lett.*, **257**, 97–103, doi:10.1016/j.epsl.2007.02.024.

Correia, A. C. M. and Laskar, J. (2004). Mercury's capture into the 3/2 spin-orbit resonance as a result of its chaotic dynamics. *Nature*, **429**, 848–850, doi:10.1038/nature02609.

Dasgupta, R., Buono, A., Whelan, G. and Walker, D. (2009). High-pressure melting relations in Fe–C–S systems: Implications for formation, evolution, and structure of metallic cores in planetary bodies. *Geochim. Cosmochim. Acta*, **73**, 6678–6691, doi:10.1016/j.gca.2009.08.001.

de Koker, N., Steinle-Neumann, G. and Vlcek, V. (2012). Electrical resistivity and thermal conductivity of liquid Fe alloys at high P and T, and heat flux in Earth's core. *Proc. Natl. Acad. Sci.*, **109**, 4070–4073, doi:10.1073/pnas.1111841109.

Denevi, B. W., Robinson, M. S., Solomon, S. C., Murchie, S. L., Blewett, D. T., Domingue, D. L., McCoy, T. J., Ernst, C. M., Head, J. W., Watters, T. R. and Chabot, N. L. (2009). The evolution of Mercury's crust: A global perspective from MESSENGER. *Science*, **324**, 613–618, doi:10.1126/science.1172226.

Denevi, B. W., Ernst, C. M., Meyer, H. M., Robinson, M. S., Murchie, S. L., Whitten, J. L., Head, J. W., Watters, T. R., Solomon, S. C., Ostrach, L. R., Chapman, C. R., Byrne, P. K., Klimczak, C. and Peplowski, P. N. (2013a). The distribution and origin of smooth plains on Mercury. *J. Geophys. Res. Planets*, **118**, 891–907, doi:10.1002/jgre.20075.

Denevi, B. W., Ernst, C. M., Whitten, J. L., Head, J. W., Murchie, S. L., Watters, T. R., Byrne, P. K., Blewett, D. T., Solomon, S. C. and Fassett, C. I. (2013b). The volcanic origin of a region of intercrater plains on Mercury. *Lunar Planet. Sci.*, **44**, abstract 1218.

Deng, L., Fei, Y., Liu, X., Gong, Z. and Shahar, A. (2013a). Effect of carbon, sulfur and silicon on iron melting at high pressure: Implications for composition and evolution of the planetary terrestrial cores. *Geochim. Cosmochim. Acta*, **114**, 220–233, doi:10.1016/j.gca.2013.01.023.

Deng, L., Seagle, C., Fei, Y. and Shahar, A. (2013b). High pressure and temperature electrical resistivity of iron and implications for planetary cores. *Geophys. Res. Lett.*, **40**, 33–37, doi:10.1029/2012GL054347.

Di Achille, G., Popa, C., Massironi, M., Epifani, E. M., Zusi, M., Cremonese, G. and Palumbo, P. (2012). Mercury's radius change estimates revisited using MESSENGER data. *Icarus*, **221**, 456–460.

Dombard, A. and Hauck, S. A., II (2008). Despinning plus global contraction and the orientation of lobate scarps on Mercury: Predictions for MESSENGER. *Icarus*, **198**, 274–276, doi:10.1016/j.icarus.2008.06.008.

Dumberry, M. and Rivoldini, A. (2015). Mercury's inner core size and core-crystallization regime. *Icarus*, **248**, 254–268, doi:10.1016/j.icarus.2014.10.038.

Egea-González, I., Ruiz, J., Fernández, C., Williams, J.-P., Márquez, Á. and Lara, L. M. (2012). Depth of faulting and ancient heat flows in the Kuiper region of Mercury from lobate scarp topography. *Planet. Space Sci.*, **60**, 193–198, doi:10.1016/j.pss.2011.08.003.

Ernst, C. M., Murchie, S. L., Barnouin, O. S., Robinson, M. S., Denevi, B. W., Blewett, D. T., Head, J. W., Izenberg, N. R., Solomon, S. C. and Roberts, J. H. (2010). Exposure of spectrally distinct material by impact craters on Mercury: Implications for global stratigraphy. *Icarus*, **209**, 210–223, doi:10.1016/j.icarus.2010.05.022.

Ernst, C. M., Denevi, B. W., Barnouin, O. S., Klimczak, C., Chabot, N. L., Head, J. W., Murchie, S. L., Neumann, G. A., Prockter, L. M., Robinson, M. S., Solomon, S. C. and Watters, T. R. (2015). Stratigraphy of the Caloris basin, Mercury: Implications for volcanic history and basin impact melt. *Icarus*, **250**, 413–429, doi:10.1016/j.icarus.2014.11.003.

Evans, L. G., Peplowski, P. N., Rhodes, E. A., Lawrence, D. J., McCoy, T. J., Nittler, L. R., Solomon, S. C., Sprague, A. L., Stockstill-Cahill, K. R., Starr, R. D., Weider, S. Z., Boynton, W. V., Hamara, D. K. and Goldsten, J. O. (2012). Major-element abundances on the surface of Mercury: Results from the MESSENGER Gamma-Ray Spectrometer. *J. Geophys. Res.*, **117**, E00L07, doi:10.1029/2012je004178.

Evans, L. G., Peplowski, P. N., McCubbin, F. M., McCoy, T. J., Nittler, L. R., Zolotov, M. Yu., Ebel, D. S., Lawrence, D. J., Starr, R. D., Weider, S. Z. and Solomon, S. C. (2015). Chlorine on the surface of Mercury: MESSENGER gamma-ray measurements and implications for the planet's formation and evolution. *Icarus*, **257**, 417–427, doi:10.1016/j.icarus.2015.04.039.

Fassett, C. I., Head, J. W., Blewett, D. T., Chapman, C. R., Dickson, J. L., Murchie, S. L., Solomon, S. C. and Watters, T. R. (2009). Caloris impact basin: Exterior geomorphology, stratigraphy, morphometry, radial sculpture, and smooth plains deposits. *Earth Planet. Sci. Lett.*, **285**, 297–308, doi:10.1016/j.epsl.2009.05.022.

Fassett, C. I., Kadish, S. J., Head, J. W., Solomon, S. C. and Strom, R. G. (2011). The global population of large craters on Mercury and comparison with the Moon. *Geophys. Res. Lett.*, **38**, L10202, doi:10.1029/2011gl047294.

Fei, Y., Bertka, C. M. and Finger, L. W. (1997). High-pressure iron sulfur compound, Fe_3S_2, and melting relations in the Fe–FeS system. *Science*, **275**, 1621–1623, doi:10.1126/science.275.5306.1621.

Fei, Y., Li, J., Bertka, C. M. and Prewitt, C. T. (2000). Structure type and bulk modulus of Fe_3S, a new iron-sulfur compound. *Amer. Mineral.*, **85**, 1830–1833, doi:10.2138/am-2000-11-1229.

Fei, Y., Hillgren, V. J., Shahar, A. and Solomon, S. C. (2011). On the silicon content of Mercury's core and implications for core mineralogy, structure, and density. *Lunar Planet. Sci.*, **42**, abstract 1949.

Ferrari, S., Massironi, M., Marchi, S., Byrne, P. K., Klimczak, C., Martellato, E. and Cremonese, G. (2014). Age relationships of the Rembrandt basin and Enterprise Rupes, Mercury. In *Volcanism and Tectonism Across the Solar System*, ed. T. Platz, M. Massironi, P. K. Byrne and H. Hiesinger, Special Publication 401. London: Geological Society, pp. 159–172, doi:10.1144/SP401.20.

Fischer, R. A., Campbell, A. J., Reaman, D. M., Miller, N. A., Heinz, D. L., Dera, P. and Prakapenka, V. B. (2013). Phase relations in the Fe–FeSi system at high pressures and temperatures. *Earth Planet. Sci. Lett.*, **373**, 54–64, doi:10.1016/j.epsl.2013.04.035.

Freed, A. M., Solomon, S. C., Watters, T. R., Phillips, R. J. and Zuber, M. T. (2009). Could Pantheon Fossae be the result of the

Apollodorus crater-forming impact within the Caloris basin, Mercury? *Earth Planet. Sci. Lett.*, **285**, 320–327, doi:10.1016/j.epsl.2009.02.038.

Freed, A. M., Blair, D. M., Watters, T. R., Klimczak, C., Byrne, P. K., Solomon, S. C., Zuber, M. T. and Melosh, H. J. (2012). On the origin of graben and ridges within and near volcanically buried craters and basins in Mercury's northern plains. *J. Geophys. Res.*, **117**, E00L06, doi:10.1029/2012je004119.

Glassmeier, K.-H., Auster, H.-U. and Motschmann, U. (2007). A feedback dynamo generating Mercury's magnetic field. *Geophys. Res. Lett.*, **34**, L22201, doi:10.1029/2007gl031662.

Gomi, H., Ohta, K., Hirose, K., Labrosse, S., Caracas, R., Verstraete, M. J. and Hernlund, J. W. (2013). The high conductivity of iron and thermal evolution of the Earth's core. *Phys. Earth Planet. Inter.*, **224**, 88–103, doi:10.1016/j.pepi.2013.07.010.

Grasset, O. and Parmentier, E. M. (1998). Thermal convection in a volumetrically heated, infinite Prandtl number fluid with strongly temperature-dependent viscosity: Implications for planetary thermal evolution. *J. Geophys. Res.*, **103**, 18,171–18,181, doi:10.1029/98JB01492.

Grott, M., Breuer, D. and Laneuville, M. (2011). Thermo-chemical evolution and global contraction of Mercury. *Earth Planet. Sci. Lett.*, **307**, 135–146, doi:10.1016/j.epsl.2011.04.040.

Hauck, S. A., II and Phillips, R. J. (2002). Thermal and crustal evolution of Mars. *J. Geophys. Res.*, **107**, 5052, doi:10.1029/2001JE001801.

Hauck, S. A., II, Dombard, A. J., Phillips, R. J. and Solomon, S. C. (2004). Internal and tectonic evolution of Mercury. *Earth Planet. Sci. Lett.*, **222**, 713–728, doi:10.1016/j.epsl.2004.03.037.

Hauck, S. A., II, Aurnou, J. M. and Dombard, A. J. (2006). Sulfur's impact on core evolution and magnetic field generation on Ganymede. *J. Geophys. Res.*, **111**, E09008, doi:10.1029/2005je002557.

Hauck, S. A., II, Margot, J.-L., Solomon, S. C., Phillips, R. J., Johnson, C. L., Lemoine, F. G., Mazarico, E., McCoy, T. J., Padovan, S., Peale, S. J., Perry, M. E., Smith, D. E. and Zuber, M. T. (2013). The curious case of Mercury's internal structure. *J. Geophys. Res. Planets*, **118**, 1204–1220, doi:10.1002/jgre.20091.

Head, J. W., Murchie, S. L., Prockter, L. M., Robinson, M. S., Solomon, S. C., Strom, R. G., Chapman, C. R., Watters, T. R., McClintock, W. E., Blewett, D. T. and Gillis-Davis, J. J. (2008). Volcanism on Mercury: Evidence from the first MESSENGER Flyby. *Science*, **321**, 69–72, doi:10.1126/science.1159256.

Head, J. W., Chapman, C. R., Strom, R. G., Fassett, C. I., Denevi, B. W., Blewett, D. T., Ernst, C. M., Watters, T. R., Solomon, S. C., Murchie, S. L., Prockter, L. M., Chabot, N. L., Gillis-Davis, J. J., Whitten, J. L., Goudge, T. A., Baker, D. M. H., Hurwitz, D. M., Ostrach, L. R., Xiao, Z., Merline, W. J., Kerber, L., Dickson, J. L., Oberst, J., Byrne, P. K., Klimczak, C. and Nittler, L. R. (2011). Flood volcanism in the northern high latitudes of Mercury revealed by MESSENGER. *Science*, **333**, 1853–1856, doi:10.1126/science.1211997.

Heimpel, M. H., Aurnou, J. M., Al-Shamali, F. M. and Gomez Perez, N. (2005). A numerical study of dynamo action as a function of spherical shell geometry. *Earth Planet. Sci. Lett.*, **236**, 542–557, doi:10.1016/j.epsl.2005.04.032.

Herzberg, C. T., Raterron, P. and Zhang, J. (2000). New experimental observations on the anhydrous solidus for peridotite KLB-1. *Geophys. Geochem. Geosyst.*, **1**, doi:10.1029/2000GC000089.

Heyner, D., Wicht, J., Gómez-Pérez, N., Schmitt, D., Auster, H.-U. and Glassmeier, K.-H. (2011). Evidence from numerical experiments for a feedback dynamo generating Mercury's magnetic field. *Science*, **334**, 1690–1693, doi:10.1126/science.1207290.

Hurwitz, D. M., Head, J. W., Byrne, P. K., Xiao, Z., Solomon, S. C., Zuber, M. T., Smith, D. E. and Neumann, G. A. (2013). Investigating the origin of candidate lava channels on Mercury with MESSENGER data: Theory and observations. *J. Geophys. Res. Planets*, **118**, 471–486, doi:10.1029/2012je004103.

James, P. B., Zuber, M. T., Phillips, R. J. and Solomon, S. C. (2015). Support of long-wavelength topography on Mercury inferred from MESSENGER measurements of gravity and topography. *J. Geophys. Res. Planets*, **120**, 287–310, doi:10.1002/2014je004713.

Jing, Z., Wang, Y., Kono, Y., Yu, T., Sakamaki, T., Park, C., Rivers, M. L., Sutton, S. R. and Shen, G. (2014). Sound velocity of Fe–S liquids at high pressure: Implications for the Moon's molten outer core. *Earth Planet. Sci. Lett.*, **396**, 78–87, doi:10.1016/j.epsl.2014.04.015.

Johnson, C. L., Phillips, R. J., Purucker, M. E., Anderson, B. J., Byrne, P. K., Denevi, B. W., Feinberg, J. M., Hauck, S. A., Head, J. W., Korth, H., James, P. B., Mazarico, E., Neumann, G. A., Philpott, L. C., Siegler, M. A., Tsyganenko, N. A. and Solomon, S. C. (2015). Low-altitude magnetic field measurements by MESSENGER reveal Mercury's ancient crustal field. *Science*, **348**, 892–895, doi:10.1126/science.aaa8720.

Johnson, C. L., Philpott, L. C., Anderson, B. J., Korth, H., Hauck, S. A., II, Heyner, D., Phillips, R. J., Winslow R. M. and Solomon S. C. (2016). MESSENGER observations of induced magnetic fields in Mercury's core, *Geophys. Res. Lett.*, **43**, 2436–2444, doi:10.1002/2015GL067370.

Kaula, W. M. (1975). The seven ages of a planet. *Icarus*, **26**, 1–15, doi:10.1016/0019-1035(75)90138-4.

Keil, K. (1989). Enstatite meteorites and their parent bodies. *Meteoritics*, **24**, 195–208.

Kennedy, P. J., Freed, A. M. and Solomon, S. C. (2008). Mechanisms of faulting in and around Caloris basin, Mercury. *J. Geophys. Res.*, **113**, E08004, doi:10.1029/2007JE002992.

Kerber, L., Head, J. W., Solomon, S. C., Murchie, S. L., Blewett, D. T. and Wilson, L. (2009). Explosive volcanic eruptions on Mercury: Eruption conditions, magma volatile content, and implications for interior volatile abundances. *Earth Planet. Sci. Lett.*, **285**, 263–271, doi:10.1016/j.epsl.2009.04.037.

Kerber, L., Head, J. W., Blewett, D. T., Solomon, S. C., Wilson, L., Murchie, S. L., Robinson, M. S., Denevi, B. W. and Domingue, D. L. (2011). The global distribution of pyroclastic deposits on Mercury: The view from MESSENGER flybys 1–3. *Planet. Space Sci.*, **59**, 1895–1909, doi:10.1016/j.pss.2011.03.020.

King, S. D. (2008). Pattern of lobate scarps on Mercury's surface reproduced by a model of mantle convection. *Nature Geosci.*, **1**, 229–232, doi:10.1038/ngeo152.

Kirk, R. L. and Stevenson, D. J. (1989). The competition between thermal contraction and differentiation in the stress history of the Moon. *J. Geophys. Res.*, **94**, 12133–12144, doi:10.1029/JB094iB09p12133.

Klima, R. L., Blewett, D. T., Denevi, B. W. Ernst, C. M., Frank, E. A., Head, J. W., III, Izenberg, N. R., Murchie, S. L., Nittler, L. R., Peplowski, P. N. and Solomon, S. C. (2016). Global distribution and spectral properties of low-reflectance material on Mercury. *Lunar Planet. Sci.*, **47**, abstract 1195.

Klimczak, C. (2015). Limits on the brittle strength of planetary lithospheres undergoing global contraction. *J. Geophys. Res. Planets*, **120**, 2135–2151, doi:10.1002/2015je004851.

Klimczak, C., Watters, T. R., Ernst, C. M., Freed, A. M., Byrne, P. K., Solomon, S. C., Blair, D. M. and Head, J. W. (2012). Deformation associated with ghost craters and basins in volcanic smooth plains on Mercury: Strain analysis and implications for plains evolution. *J. Geophys. Res.*, **117**, E00L03, doi:10.1029/2012je004100.

Knibbe, J. S. and van Westrenen, W. (2015). The interior configuration of planet Mercury constrained by moment of inertia and planetary contraction. *J. Geophys. Res. Planets*, **120**, 1904–1923, doi:10.1002/2015JE004908.

Konôpková, Z., McWilliams, R. S., Gómez-Pérez, N. and Goncharov, A. F. (2016). Direct measurement of thermal conductivity in solid iron at planetary core conditions. *Nature*, **534**, 99–101, doi:10.1038/nature18009.

Kuwayama, Y. and Hirose, K. (2004). Phase relations in the system Fe–FeSi at 21 GPa. *Amer. Mineral.*, **89**, 273–276, doi:10.2138/am-2004-2-303.

Le Feuvre, M. and Wieczorek, M. A. (2011). Nonuniform cratering of the Moon and a revised crater chronology of the inner Solar System. *Icarus*, **214**, 1–20, doi:10.1016/j.icarus.2011.03.010.

Li, J., Fei, Y., Mao, H. K., Hirose, K. and Shieh, S. R. (2001). Sulfur in the Earth's inner core. *Earth Planet. Sci. Lett.*, **193**, 509–514, doi:10.1016/S0012-821X(01)00521-0.

Lord, O. T., Walter, M. J., Dasgupta, R., Walker, D. and Clark, S. M. (2009). Melting in the Fe–C system to 70 GPa. *Earth Planet. Sci. Lett.*, **284**, 157–167, doi:10.1016/j.epsl.2009.04.017.

Malavergne, V., Toplis, M. J., Berthet, S. and Jones, J. (2010). Highly reducing conditions during core formation on Mercury: Implications for internal structure and the origin of a magnetic field. *Icarus*, **206**, 199–209, doi:10.1016/j.icarus.2009.09.001.

Manglik, A., Wicht, J. and Christensen, U. R. (2010). A dynamo model with double diffusive convection for Mercury's core. *Earth Planet. Sci. Lett.*, **289**, 619–628, doi:10.1016/j.epsl.2009.12.007.

Marchi, S., Mottola, S., Cremonese, G., Massironi, M. and Martellato, E. (2009). A new chronology for the Moon and Mercury. *Astron. J.*, **137**, 4936–4948, doi:10.1088/0004-6256/137/6/4936.

Marchi, S., Massironi, M., Cremonese, G., Martellato, E., Giacomini, L. and Prockter, L. (2011). The effects of the target material properties and layering on the crater chronology: The case of Raditladi and Rachmaninoff basins on Mercury. *Planet. Space Sci.*, **59**, 1968–1980, doi:10.1016/j.pss.2011.06.007.

Marchi, S., Chapman, C. R., Fassett, C. I., Head, J. W., Bottke, W. F. and Strom, R. G. (2013). Global resurfacing of Mercury 4.0–4.1 billion years ago by heavy bombardment and volcanism. *Nature*, **499**, 59–61, doi:10.1038/nature12280.

Margot, J. L., Peale, S. J., Jurgens, R. F., Slade, M. A. and Holin, I. V. (2007). Large longitude libration of Mercury reveals a molten core. *Science*, **316**, 710–714, doi:10.1126/science.1140514.

Margot, J.-L., Peale, S. J., Solomon, S. C., Hauck, S. A., II, Ghigo, F. D., Jurgens, R. F., Padovan, S. and Campbell, D. B. (2012). Mercury's moment of inertia from spin and gravity data. *J. Geophys. Res.*, **117**, E00L09, doi:10.1029/2012JE004161.

Martin, A. M., Van Orman, J., Hauck, S. A., II, Chen, B., II, Sun, N., Moore, R. D. and Han, J. (2014). In situ determination of the eutectic melting temperature of Fe–FeS–Fe$_3$C between 4.5 and 24.5 GPa and implications for Mercury's core. *Lunar Planet. Sci.*, **45**, abstract 2854.

Martin, A. M., Van Orman, J., Hauck, S. A., II, Sun, N., Yu, T. and Wang, Y. (2015). Role of sulfur, silicon and carbon on the crystallization processes in Mercury's core inferred from in-situ melting experiments between 4.5 and 15.5 GPa. *Lunar Planet. Sci.*, **46**, abstract 2627.

Matsuyama, I. and Nimmo, F. (2009). Gravity and tectonic patterns of Mercury: Effect of tidal deformation, spin-orbit resonance, non-zero eccentricity, despinning, and reorientation. *J. Geophys. Res.*, **114**, E01010, doi:10.1029/2008je003252.

Mazarico, E., Genova, A., Goossens, S., Lemoine, F. G., Neumann, G. A., Zuber, M. T., Smith, D. E. and Solomon, S. C. (2014). The gravity field, orientation, and ephemeris of Mercury from MESSENGER observations after three years in orbit. *J. Geophys. Res. Planets*, **119**, 2417–2436, doi:10.1002/2014je004675.

McCoy, T. J., Dickinson, T. L. and Lofgren, G. E. (1999). Partial melting of the Indarch (EH4) meteorite: A textural, chemical, and phase relations view of melting and melt migration. *Meteorit. Planet. Sci.*, **34**, 735–746, doi:10.1111/j.1945-5100.1999.tb01386.x.

McCubbin, F. M., Riner, M. A., Vander Kaaden, K. E. and Burkemper, L. K. (2012). Is Mercury a volatile-rich planet? *Geophys. Res. Lett.*, **39**, L09202, doi:10.1029/2012gl051711.

Melosh, H. J. (1977). Global tectonics of a despun planet. *Icarus*, **31**, 221–243, doi:10.1016/0019-1035(77)90035-5.

Melosh, H. J. and Dzurisin, D. (1978). Mercurian global tectonics: A consequence of tidal despinning? *Icarus*, **35**, 227–236, doi:10.1016/0019-1035(78)90007-6.

Melosh, H. J. and McKinnon, W. B. (1988). The tectonics of Mercury. In *Mercury*, ed. F. Vilas, C. R. Chapman and M. S. Matthews. Tucson, AZ: University of Arizona Press, pp. 374–400.

Michel, N. C., Hauck, S. A., II, Solomon, S. C., Phillips, R. J., Roberts, J. H. and Zuber, M. T. (2013). Thermal evolution of Mercury as constrained by MESSENGER observations. *J. Geophys. Res. Planets*, **118**, 1033–1044, doi:10.1002/jgre.20049.

Morard, G. and Katsura, T. (2010). Pressure–temperature cartography of Fe–S–Si immiscible system. *Geochim. Cosmochim. Acta*, **74**, 3659–3667, doi:10.1016/j.gca.2010.03.025.

Morard, G., Andrault, D., Guignot, N., Siebert, J., Garbarino, G. and Antonangeli, D. (2011). Melting of Fe–Ni–Si and Fe–Ni–S alloys at megabar pressures: Implications for the core–mantle boundary temperature. *Phys. Chem. Minerals*, **38**, 767–776, doi:10.1007/s00269-011-0449-9.

Moresi, L. N. and Solomatov, V. S. (1995). Numerical investigation of 2D convection with extremely large viscosity variations. *Phys. Fluids*, **7**, 2154–2162, doi:10.1063/1.868465.

Murchie, S. L., Watters, T. R., Robinson, M. S., Head, J. W., Strom, R. G., Chapman, C. R., Solomon, S. C., McClintock, W. E., Prockter, L. M., Domingue, D. L. and Blewett, D. T. (2008). Geology of the Caloris basin, Mercury: A view from MESSENGER. *Science*, **321**, 73–76, doi:10.1126/science.1159261.

Murchie, S. L., Klima, R. L., Denevi, B. W., Ernst, C. M., Keller, M. R., Domingue, D. L., Blewett, D. T., Chabot, N. L., Hash, C. D., Malaret, E., Izenberg, N. R., Vilas, F., Nittler, L. R., Gillis-Davis, J. J., Head, J. W. and Solomon, S. C. (2015). Orbital multispectral mapping of Mercury with the MESSENGER Mercury Dual Imaging System: Evidence for the origins of plains units and low-reflectance material. *Icarus*, **254**, 287–305, doi:10.1016/j.icarus.2015.03.027.

Murray, B. C., Belton, M. J. S., Danielson, G. E., Davies, M. E., Gault, D. E., Hapke, B., O'Leary, B., Strom, R. G., Suomi, V. and Trask, N. (1974). Mercury's surface: Preliminary description and interpretation from Mariner 10 pictures. *Science*, **185**, 169–179.

Murray, B. C., Strom, R. G., Trask, N. J. and Gault, D. E. (1975). Surface history of Mercury: Implications for terrestrial planets. *J. Geophys. Res.*, **80**, 2508–2514, doi:10.1029/JB080i017p02508.

Namur, O. and Charlier, B. (2017). Silicate mineralogy at the surface of Mercury. *Nature Geosci.*, **10**, 9–13, doi:10.1038/ngeo2860.

Namur, O., Collinet, M., Charlier, B., Grove T. L., Holtz F. and McCammon, C. (2016). Melting processes and mantle sources of lavas on Mercury. *Earth Planet. Sci. Lett.*, **439**, 117–128, doi:10.1016/j.epsl.2016.01.030.

Ness, N. F. (1979). The magnetic field of Mercury. *Phys. Earth Planet. Inter.*, **20**, 209–217, doi:10.1016/0031-9201(79)90044-X.

Ness, N. F., Behannon, K. W., Lepping, R. P. and Whang, Y. C. (1975). The magnetic field of Mercury. I. *J. Geophys. Res.*, **80**, 2708–2716, doi:10.1029/JA080i019p02708.

Ness, N. F., Behannon, K. W., Lepping, R. P. and Whang, Y. C. (1976). Observations of Mercury's magnetic field. *Icarus*, **28**, 479–488, doi:10.1016/0019-1035(76)90121-4.

Neukum, G., Ivanov, B. A. and Hartmann, W. K. (2001). Cratering records in the inner solar system in relation to the lunar reference system. *Space Sci. Rev.*, **96**, 55–86.

Nimmo, F. and Stevenson, D. J. (2001). Estimates of Martian crustal thickness from viscous relaxation of topography. *J. Geophys. Res.*, **106**, 5085–5098, doi:10.1029/2000JE001331.

Nittler, L. R., Starr, R. D., Weider, S. Z., McCoy, T. J., Boynton, W. V., Ebel, D. S., Ernst, C. M., Evans, L. G., Goldsten, J. O., Hamara, D. K., Lawrence, D. J., McNutt, R. L., Jr., Schlemm, C. E., II, Solomon, S. C. and Sprague, A. L. (2011). The major-element composition of Mercury's surface from MESSENGER X-ray spectrometry. *Science*, **333**, 1847–1850, doi:10.1126/science.1211567.

Noyelles, B., Frouard, J., Makarov, V. V. and Efroimsky, M. (2014). Spin-orbit evolution of Mercury revisited. *Icarus*, **241**, 26–44, doi:10.1016/j.icarus.2014.05.045.

Ostrach, L. R., Robinson, M. S., Whitten, J. L., Fassett, C. I., Strom, R. G., Head, J. W. and Solomon, S. C. (2015). Extent, age, and resurfacing history of the northern smooth plains on Mercury from MESSENGER observations. *Icarus*, **250**, 602–622, doi:10.1016/j.icarus.2014.11.010.

Padovan, S., Margot, J.-L., Hauck, S. A., II, Moore, W. B. and Solomon, S. C. (2014). The tides of Mercury and possible implications for its interior structure. *J. Geophys. Res. Planets*, **119**, 850–866, doi:10.1002/2013je004459.

Padovan, S., Wieczorek, M. A., Margot, J.-L., Tosi, N. and Solomon, S. C. (2015). Thickness of the crust of Mercury from geoid-to-topography ratios. *Geophys. Res. Lett.*, **42**, 1029–1038, doi:10.1002/2014gl062487.

Peale, S. J. (1988). The rotational dynamics of Mercury and the state of its core. In *Mercury*, ed. F. Vilas, C. R. Chapman and M. S. Matthews. Tucson, AZ: University of Arizona Press, pp. 461–493.

Peale, S. J., Phillips, R. J., Solomon, S. C., Smith, D. E. and Zuber, M. T. (2002). A procedure for determining the nature of Mercury's core. *Meteorit. Planet. Sci.*, **37**, 1269–1283, doi:10.1111/j.1945-5100.2002.tb00895.x.

Peale, S. J., Margot, J.-L., Hauck, S. A., II and Solomon, S. C. (2016). Consequences of a solid inner core on Mercury's spin configuration. *Icarus*, **264**, 443–455, doi:10.1016/j.icarus.2015.09.024.

Peplowski, P. N., Evans, L. G., Hauck, S. A., II, McCoy, T. J., Boynton, W. V., Gillis-Davis, J. J., Ebel, D. S., Goldsten, J. O., Hamara, D. K., Lawrence, D. J., McNutt, R. L., Jr., Nittler, L. R., Solomon, S. C., Rhodes, E. A., Sprague, A. L., Starr, R. D. and Stockstill-Cahill, K. R. (2011). Radioactive elements on Mercury's surface from MESSENGER: Implications for the planet's formation and evolution. *Science*, **333**, 1850–1852, doi:10.1126/science.1211576.

Peplowski, P. N., Lawrence, D. J., Rhodes, E. A., Sprague, A. L., McCoy, T. J., Denevi, B. W., Evans, L. G., Head, J. W., Nittler, L. R., Solomon, S. C., Stockstill-Cahill, K. R. and Weider, S. Z. (2012). Variations in the abundances of potassium and thorium on the surface of Mercury: Results from the MESSENGER Gamma-Ray Spectrometer. *J. Geophys. Res.*, **117**, E00L04, doi:10.1029/2012JE004141.

Peplowski, P. N., Lawrence, D. J., Evans, L. G., Klima, R. L., Blewett, D. T., Goldsten, J. O., Murchie, S. L., McCoy, T. J., Nittler, L. R., Solomon, S. C., Starr, R. D. and Weider, S. Z. (2015a). Constraints on the abundance of carbon in near-surface materials on Mercury: Results from the MESSENGER Gamma-Ray Spectrometer. *Planet. Space Sci.*, **108**, 98–107, doi:10.1016/j.pss.2015.01.008.

Peplowski, P. N., Lawrence, D. J., Feldman, W. C., Goldsten, J. O., Bazell, D., Evans, L. G., Head, J. W., Nittler, L. R., Solomon, S. C. and Weider, S. Z. (2015b). Geochemical terranes of Mercury's northern hemisphere as revealed by MESSENGER neutron measurements. *Icarus*, **253**, 346–363, doi:10.1016/j.icarus.2015.02.002.

Peplowski, P. N., Klima, R. L., Lawrence, D. J., Ernst, C. M., Denevi, B. W., Frank, E. A., Goldsten, J. O., Murchie, S. L., Nittler, L. R. and Solomon, S. C. (2016), Remote sensing evidence for an ancient carbon-bearing crust on Mercury, *Nature Geosci.*, **9**, 273–276, doi:10.1038/ngeo2669.

Perry, M. E., Neumann, G. A., Phillips, R. J., Barnouin, O. S., Ernst, C. M., Kahan, D. S., Solomon, S. C., Zuber, M. T., Smith, D. E., Hauck, S. A., II, Peale, S. J., Margot, J.-L., Mazarico, E., Johnson, C. L., Gaskell, R. W., Roberts, J. H., McNutt, R. L., Jr. and Oberst, J. (2015). The low-degree shape of Mercury. *Geophys. Res. Lett.*, **42**, 6951–6958, doi:10.1002/2015gl065101.

Philpott, L. C., Johnson, C. L., Winslow, R. M., Anderson, B. J., Korth, H., Purucker, M. E. and Solomon, S. C. (2014). Constraints on the secular variation of Mercury's magnetic field from the combined analysis of MESSENGER and Mariner 10 data. *Geophys. Res. Lett.*, **41**, 6627–6634, doi:10.1002/2014gl061401.

Plesa, A. C., Tosi, N. and Hüttig, C. (2013). Thermo-chemical convection in planetary mantles: Advection methods and magma ocean overturn simulations. In *Integrated Information and Computing Systems for Natural, Spatial, and Social Sciences*, ed. R. Claus-Peter. Hershey, PA: IGI Global, pp. 302–323, doi:10.4018/978-1-4666-2190-9.ch015.

Pozzo, M., Davies, C., Gubbins, D. and Alfe, D. (2012). Thermal and electrical conductivity of iron at Earth's core conditions. *Nature*, **485**, 355–358, doi:10.1038/nature11031.

Prockter, L. M., Ernst, C. M., Denevi, B. W., Chapman, C. R., Head, J. W., Fassett, C. I., Merline, W. J., Solomon, S. C., Watters, T. R., Strom, R. G., Cremonese, G., Marchi, S. and Massironi, M. (2010). Evidence for young volcanism on Mercury from the third MESSENGER flyby. *Science*, **329**, 668–671, doi:10.1126/science.1188186.

Raghavan, V. (1988). *Phase Diagrams of Ternary Iron Alloys, Part 2: Ternary Systems Containing Iron and Sulphur*. Calcutta: Indian Institute of Metals.

Redmond, H. L. and King, S. D. (2007). Does mantle convection currently exist on Mercury? *Phys. Earth Planet. Inter.*, **164**, 221–231, doi:10.1016/j.pepi.2007.07.004.

Reese, C. C., Solomatov, V. S. and Moresi, L. N. (1998). Heat transport efficiency for stagnant lid convection with dislocation viscosity: Application to Venus and Mars. *J. Geophys. Res.*, **103**, 13643–13658, doi:10.1029/98JE01047.

Riner, M. A., Lucey, P. G., Desch, S. J. and McCubbin, F. M. (2009). Nature of opaque components on Mercury: Insights into a Mercurian magma ocean. *Geophys. Res. Lett.*, **36**, L02201, doi:10.1029/2008GL036128.

Rivera-Valentin, E. G. and Barr, A. C. (2014). Impact-induced compositional variations on Mercury. *Earth Planet. Sci. Lett.*, **391**, 234–242, doi:10.1016/j.epsl.2014.02.003.

Rivoldini, A. and Van Hoolst, T. (2013). The interior structure of Mercury constrained by the low-degree gravity field and the rotation of Mercury. *Earth Planet. Sci. Lett.*, **377–378**, 62–72, doi:10.1016/j.epsl.2013.07.021.

Roberts, J. H. and Barnouin, O. S. (2012). The effect of the Caloris impact on the mantle dynamics and volcanism of Mercury. *J. Geophys. Res.*, **117**, E02007, doi:10.1029/2011JE003876.

Robinson, M. S. and Lucey, P. G. (1997). Recalibrated Mariner 10 color mosaics: Implications for Mercurian volcanism. *Science*, **275**, 197–200, doi:10.1126/science.275.5297.197.

Robinson, M. S., Murchie, S. L., Blewett, D. T., Domingue, D. L., Hawkins, S. E., Head, J. W., Holsclaw, G. M., McClintock, W. E., McCoy, T. J., McNutt, R. L., Prockter, L. M., Solomon, S. C. and Watters, T. R. (2008). Reflectance and color variations on Mercury: Regolith processes and compositional heterogeneity. *Science*, **321**, 66–69, doi:10.1126/science.1160080.

Rückriemen, T., Breuer, D. and Spohn, T. (2015). The Fe snow regime in Ganymede's core: A deep-seated dynamo below a stable snow zone. *J. Geophys. Res. Planets*, **120**, 1095–1118, doi:10.1002/2014JE004781.

Schubert, G., Ross, M. N., Stevenson, D. J. and Spohn, T. (1988). Mercury's thermal history and the generation of its magnetic field. In *Mercury*, ed. F. Vilas, C. R. Chapman and M. S. Matthews. Tucson, AZ: University of Arizona Press, pp. 429–460.

Schumacher, S. and Breuer, D. (2006). Influence of a variable thermal conductivity on the thermochemical evolution of Mars. *J. Geophys. Res.*, **111**, E02006, doi:10.1029/2005JE002429.

Seagle, C. T., Cottrell, E., Fei, Y., Hummer, D. R. and Prakapenka, V. B. (2013). Electrical and thermal transport properties of iron and iron–silicon alloy at high pressure. *Geophys. Res. Lett.*, **40**, 5377–5381, doi:10.1002/2013gl057930.

Siegfried, R. W., II and Solomon, S. C. (1974). Mercury: Internal structure and thermal evolution. *Icarus*, **23**, 192–205, doi:10.1016/0019-1035(74)90005-0.

Smith, D. E., Zuber, M. T., Lemoine, F. G., Solomon, S. C., Hauck, S. A., II, Lemoine, F. G., Mazarico, E., Phillips, R. J., Neumann, G. A., Peale, S. J., Margot, J.-L., Johnson, C. L., Torrence, M. H., Perry, M. E., Rowlands, D. D., Goossens, S., Head, J. W. and Taylor, A. H. (2012). Gravity field and internal structure of Mercury from MESSENGER. *Science*, **336**, 214–217, doi:10.1126/science.1218809.

Solomon, S. C. (1976). Some aspects of core formation in Mercury. *Icarus*, **28**, 509–521, doi:10.1016/0019-1035(76)90124-X.

Solomon, S. C. (1977). The relationship between crustal tectonics and internal evolution in the Moon and Mercury. *Phys. Earth Planet. Inter.*, **15**, 135–145, doi:10.1016/0031-9201(77)90026-7.

Solomon, S. C. (2003). Mercury: The enigmatic innermost planet. *Earth Planet. Sci. Lett.*, **216**, 441–455, doi:10.1016/s0012-821x(03)00546-6.

Spudis, P. D. and Guest, J. E. (1988). Stratigraphy and geologic history of Mercury. In *Mercury*, ed. F. Vilas, C. R. Chapman and M. S. Matthews. Tucson, AZ: University of Arizona Press, pp. 118–164.

Stanley, S. and Bloxham, J. (2016). On the secular variation of Saturn's magnetic field. *Phys. Earth Planet. Inter.*, **250**, 31–34, doi:10.1016/j.pepi.2015.11.002.

Stanley, S., Bloxham, J., Hutchison, W. and Zuber, M. (2005). Thin shell dynamo models consistent with Mercury's weak observed magnetic field. *Earth Planet. Sci. Lett.*, **234**, 27–38, doi:10.1016/j.epsl.2005.02.040.

Stevenson, D. J., Spohn, T. and Schubert, G. (1983). Magnetism and thermal evolution of the terrestrial planets. *Icarus*, **54**, 466–489, doi:10.1016/0019-1035(83)90241-5.

Stewart, A. J., Schmidt, M. W., van Westrenen, W. and Liebske, C. (2007). Mars: A new core-crystallization regime. *Science*, **316**, 1323–1325, doi:10.1126/science.1140549.

Stockstill-Cahill, K. R., McCoy, T. J., Nittler, L. R., Weider, S. Z. and Hauck, S. A., II (2012). Magnesium-rich crustal compositions on Mercury: Implications for magmatism from petrologic modeling. *J. Geophys. Res.*, **117**, E00L15, doi:10.1029/2012JE004140.

Strom, R. G. (1977). Origin and relative age of lunar and Mercurian intercrater plains. *Phys. Earth Planet. Inter.*, **15**, 156–172, doi:10.1016/0031-9201(77)90028-0.

Strom, R. G. and Neukum, G. (1988). The cratering record on Mercury and the origin of impacting objects. In *Mercury*, ed. F. Vilas, C. R. Chapman and M. S. Matthews. Tucson, AZ: University of Arizona Press, pp. 336–373.

Strom, R. G., Trask, N. J. and Guest, J. E. (1975). Tectonism and volcanism on Mercury. *J. Geophys. Res.*, **80**, 2478–2507, doi:10.1029/JB080i017p02478.

Strom, R. G., Chapman, C. R., Merline, W. J., Solomon, S. C. and Head, J. W. (2008). Mercury cratering record viewed from MESSENGER's first flyby. *Science*, **321**, 79–81, doi:10.1126/science.1159317.

Strom, R. G., Banks, M. E., Chapman, C. R., Fassett, C. I., Forde, J. A., Head, J. W., III, Merline, W. J., Prockter, L. M. and Solomon, S. C. (2011). Mercury crater statistics from MESSENGER flybys: Implications for stratigraphy and resurfacing history. *Planet. Space Sci.*, **59**, 1960–1967, doi:10.1016/j.pss.2011.03.018.

Taylor, G. J. and Scott, E. R. D. (2003). Mercury. In *Meteorites, Comets and Planets*, ed. A. M. Davis, *Treatise on Geochemistry*, Vol. 1, ed. H. D. Holland and K. K. Turekian. Oxford: Pergamon, pp. 477–485, doi:10.1016/B0-08-043751-6/01071-9.

Thomas, R. J., Rothery, D. A., Conway, S. J. and Anand, M. (2014). Mechanisms of explosive volcanism on Mercury: Implications from its global distribution and morphology. *J. Geophys. Res. Planets*, **119**, 2239–2254, doi:10.1002/2014je004692.

Tian, Z., Zuber, M. T. and Stanley, S. (2015). Magnetic field modeling for Mercury using dynamo models with a stable layer and laterally variable heat flux. *Icarus*, **260**, 263–268, doi:10.1016/j.icarus.2015.07.019.

Tosi, N., Grott, M., Plesa, A. C. and Breuer, D. (2013). Thermochemical evolution of Mercury's interior. *J. Geophys. Res. Planets*, **118**, 2474–2487, doi:10.1002/jgre.20168.

Tosi, N., Čadek, O., Běhounková, M., Káňová, M., Plesa, A. C., Grott, M., Breuer, D., Padovan, S. and Wieczorek, M. A. (2015). Mercury's low-degree geoid and topography controlled by insolation-driven elastic deformation. *Geophys. Res. Lett.*, **42**, 7327–7335, doi:10.1002/2015gl065314.

Trask, N. J. and Guest, J. E. (1975). Preliminary geologic terrain map of Mercury. *J. Geophys. Res.*, **80**, 2461–2477, doi:10.1029/JB080i017p02461.

Vander Kaaden, K. E. and McCubbin, F. M. (2015). Exotic crust formation on Mercury: Consequences of a shallow, FeO-poor mantle. *J. Geophys. Res. Planets*, **120**, 195–209, doi:10.1002/2014je004733.

Vander Kaaden, K. E. and McCubbin, F. M. (2016). The origin of boninites on Mercury: An experimental study of the northern volcanic plains lavas. *Geochim. Cosmochim. Acta*, **173**, 246–263, doi:10.1016/j.gca.2015.10.016.

Vander Kaaden, K. E., McCubbin, F. M., Nittler, L. R., Peplowski, P. N., Weider, S. Z., Frank, E. A. and McCoy, T. J. (2017). Geochemistry, mineralogy, and petrology of boninitic and komatiitic rocks on the mercurian surface: Insights into the mercurian mantle. *Icarus*, **285**, 155–168, doi:10.1016/j.icarus.2016.11.041.

Vasavada, A. R., Paige, D. A. and Wood, S. E. (1999). Near-surface temperatures on Mercury and the Moon and the stability of polar ice deposits. *Icarus*, **141**, 179–193, doi:10.1006/icar.1999.6175.

Verma, A. and J.-L. Margot (2016). Mercury's gravity, tides, and spin from MESSENGER radio science data, *J. Geophys. Res. Planets*, **121**, 1627–1640, doi:10.1002/2016JE005037.

Vilim, R., Stanley, S. and Hauck, S. A., II (2010). Iron snow zones as a mechanism for generating Mercury's weak observed magnetic field. *J. Geophys. Res.*, **115**, E11003, doi:10.1029/2009JE003528.

Watters, T. R. and Nimmo, F. (2010). Tectonism on Mercury. In *Planetary Tectonics*, ed. T. R. Watters and R. A. Schultz. Cambridge: Cambridge University Press, pp. 15–80.

Watters, T. R., Robinson, M. S. and Cook, A. C. (1998). Topography of lobate scarps on Mercury: New constraints on the planet's contraction. *Geology*, **26**, 991–994.

Watters, T. R., Solomon, S. C., Robinson, M. S., Head, J. W., André, S. L., Hauck, S. A., II and Murchie, S. L. (2009). The tectonics of Mercury: The view after MESSENGER's first flyby. *Earth Planet. Sci. Lett.*, **285**, 283–296, doi:10.1016/j.epsl.2009.01.025.

Watters, T. R., Selvans, M. M., Banks, M. E., Hauck, S. A., II, Becker, K. J. and Robinson, M. S. (2015a). Distribution of large-scale contractional tectonic landforms on Mercury: Implications for the origin of global stresses. *Geophys. Res. Lett.*, **42**, 3755–3763, doi:10.1002/2015gl063570.

Watters, T. R., Solomon, S. C., Daud, K., Banks, M. E., Selvans, M. M., Robinson, M. S., Murchie, S. L., Chabot, N. L., Denevi, B. W., Ernst, C. M., Chapman, C. R., Fassett, C. I., Klimczak, C., Byrne, P. K. and Blewett, D. T. (2015b). Small thrust fault scarps on Mercury revealed in low-alitude MESSENGER images. *Lunar Planet. Sci.*, **46**, abstract 2240.

Weider, S. Z., Nittler, L. R., Starr, R. D., McCoy, T. J., Stockstill-Cahill, K. R., Byrne, P. K., Denevi, B. W., Head, J. W. and Solomon, S. C. (2012). Chemical heterogeneity on Mercury's surface revealed by the MESSENGER X-Ray Spectrometer. *J. Geophys. Res.*, **117**, E00L05, doi:10.1029/2012je004153.

Weider, S. Z., Nittler, L. R., Starr, R. D., Crapster-Pregont, E. J., Peplowski, P. N., Denevi, B. W., Head, J. W., Byrne, P. K., Hauck, S. A., II, Ebel, D. S. and Solomon, S. C. (2015). Evidence for geochemical terranes on Mercury: Global mapping of major elements with MESSENGER's X-Ray Spectrometer. *Earth Planet. Sci. Lett.*, **416**, 109–120, doi:10.1016/j.epsl.2015.01.023.

Whitten, J. L., Head, J. W., Denevi, B. W. and Solomon, S. C. (2014). Intercrater plains on Mercury: Insights into unit definition, characterization, and origin from MESSENGER datasets. *Icarus*, **241**, 97–113, doi:10.1016/j.icarus.2014.06.013.

Wicht, J. and Heyner, D. (2014). Mercury's magnetic field in the MESSENGER era. In *Planetary Geodesy and Remote Sensing*, ed. S. Jin. New York: CRC Press, pp. 223–262.

Wieczorek, M. A., Neumann, G. A., Nimmo, F., Kiefer, W. S., Taylor, G. J., Melosh, H. J., Phillips, R. J., Solomon, S. C., Andrews-Hanna, J. C., Asmar, S. W., Konopliv, A. S., Lemoine, F. G., Smith, D. E., Watkins, M. M., Williams, J. G. and Zuber, M. T. (2013). The crust of the Moon as seen by GRAIL. *Science*, **339**, 671–675, doi:10.1126/science.1231530.

Williams, J.-P., Ruiz, J., Rosenburg, M. A., Aharonson, O. and Phillips, R. J. (2011). Insolation driven variations of Mercury's lithospheric strength. *J. Geophys. Res.*, **116**, E01008, doi:10.1029/2010JE003655.

Williams, Q. (2009). Bottom-up versus top-down solidification of the cores of small solar system bodies: Constraints on paradoxical cores. *Earth Planet. Sci. Lett.*, **284**, 564–569, doi:10.1016/j.epsl.2009.05.019.

Zhang, P., Cohen, R. E. and Haule, K. (2015). Effects of electron correlations on transport properties of iron at Earth's core conditions. *Nature*, **517**, 605–607, doi:10.1038/nature14090.

Zolotov, M. Yu., Sprague, A. L., Hauck, S. A., II, Nittler, L. R., Solomon, S. C. and Weider, S. Z. (2013). The redox state, FeO content, and origin of sulfur-rich magmas on Mercury. *J. Geophys. Res. Planets*, **118**, 138–146, doi:10.1029/2012je004274.

20

Future Missions: Mercury after MESSENGER

RALPH L. MCNUTT, JR., JOHANNES BENKHOFF, MASAKI FUJIMOTO, AND BRIAN J. ANDERSON

20.1 INTRODUCTION

Mercury has now been explored by two spacecraft, Mariner 10, which flew by the planet three times in 1974–1975, and MErcury Surface, Space ENvironment, GEochemistry, and Ranging (MESSENGER), which in 2015 completed four years of orbital observations, the results of which are discussed in detail throughout this book. In this chapter we consider Mercury exploration going forward in the broad context of the history of Mercury exploration and its link to scientific policy and progress in space technology. The MESSENGER scientific and technical achievements are briefly reviewed in light of the consensus science policies for Mercury to set the stage for an overview of the next mission to Mercury, the dual-spacecraft BepiColombo mission, developed jointly by the European Space Agency (ESA) and the Japan Aerospace Exploration Agency (JAXA) and scheduled for launch in October 2018. The prospects for more ambitious missions to land on the surface or even return samples from the planet are then discussed, highlighting the key science questions that only such missions could address and considering their feasibility given the broad trends of spaceflight technology.

20.2 MERCURY AS A PLANETARY EXPLORATION TARGET

We begin by considering the history of US science policy with respect to Mercury exploration. The initial assessments of priorities for solar system exploration were made in the early 1960s (e.g., Space Science Board, 1962). Lunar and planetary priorities were focused on the Moon and the needs of the manned Apollo Moon-landing program. With respect to planetary targets, Mars had already been called out as a priority, because of its perceived potential for exobiological activity and for its possibility for eventual human exploration. Mercury was not mentioned explicitly in the report of the Space Science Board (1962) of the National Research Council:

> Lunar specialists emphasized that certain kinds of scientific data about the Moon must be obtained relatively early, not because they are of greater scientific importance but because they are required for the proper execution of the Apollo mission. A comprehensive recommendation cutting across all other interests is that, in the early exploration of Mars, biological and biochemical studies must have the right-of-way until we find either that there is no life on Mars or that the risk of irrevocably destroying it by the introduction of terrestrial organisms is negligibly small. These examples do not by any means exhaust the list of specific priorities; nevertheless the subject as a whole is still largely open.

By the time of the next comprehensive review of the future of space research by the Space Science Board (1966), Mars had become the primary scientific priority in planetary research, but a priority scheme for all of the planets was also laid out. Mercury was ranked sixth in importance, just after "comets and asteroids" and before Pluto (seventh) and dust (eighth and last). Advances in our scientific knowledge of Mercury were viewed as following from future Earth-based observations, a "close fly-by mission," a lander, and a geodetic satellite:

> Mercury is an object unique in the solar system, and should be included in a comprehensive program of planetary exploration. On the one hand, it can be classed as a terrestrial planet; on the other, it can be classed with the major satellites such as the Moon, Io, Triton, etc. From either point of view, comparative study of surfaces, atmospheres, and interiors is relevant to the more general problem of the evolution and origin of the planets and the solar system as a whole.

It is worth noting that it was at this time that the Space Science Board advocated the use of large launch vehicles, the Saturn IB and Saturn V, both being developed to support the Apollo lunar landing program, to support large scientific payloads to the surface of Mars. Indeed, "In light of the excellent progress on Saturn V, we recommend that the Office of Space Science and Applications (OSSA) and the Office of Manned Space Flight (OMSF) jointly undertake a study of the early use of Saturn V for exploration of the planets with special emphasis on a Martian capsule landing in the early 1970's" (Space Science Board, 1966). The contemplated large surface laboratory mission to Mars was part of the first "Voyager" missions[1] (Cortright, 1968), but this mission as well as potentially larger missions were subsequently ruled out due to cost realities (Ezell and Ezell, 1984). The proposal of a biological laboratory to the martian surface was down-scoped and eventually flown on the less capable Titan Centaur in the form of the independent launches of Viking 1 and 2 landers and orbiters.

---

[1] Although initially billed as "a program of planetary exploration to be carried out with automated spacecraft during the 1970's," with "Mars and Venus will be the primary objects for investigation during that time period," the only real mission planed was a dual launch with a Saturn V, and with a projected cost of $2.2 billion in April 1967, the program was deleted from the 1968 budget.

In 1968 the Space Science Board established a special panel to examine the comprehensive scientific exploration of the solar system. Of the principal recommendations (Space Science Board, 1968), item 3 (d) was "We accord next priorities (in descending order) to a Mariner-class Venus-Mercury fly-by in 1973 or 1975, a multiple dropsonde mission to Venus in 1975, and a major lander on Mars, perhaps in 1975 (page 6)." (In this context, "dropsonde" denoted a battery-powered payload that would return data during its descent through the venusian atmosphere and possibly from the surface.) That Mercury was locked into a 3:2 spin–orbit resonance rather than synchronous rotation had been demonstrated only three years earlier by Earth-based radar observations. It had also only just been realized that a gravity assist at Venus could enable a mission to the innermost planet with a far less powerful launch vehicle than would be required for a spacecraft capable of deceleration at Mercury after a direct trajectory from Earth ("Exploration of Mercury would not otherwise be possible without employing a very much larger booster.") The study noted that such a mission could be carried out in 1973 or 1975, and that the next opportunity would not be until the following decade.

The presence of an atmosphere on Mercury was considered doubtful but unknown. Earth-orbital telescopic observations were also considered problematic because of thermal constraints driven by Mercury's small elongation. (A footnote says with respect to high-resolution telescopic observations: "The planet Mercury is excluded because of the difficulties in arriving at a thermal design for an Earth-orbital telescope pointing within 20° of the Sun. These difficulties are unlikely to be solved in the next few years." Indeed, this is still the case as Mercury is excluded from Hubble Space Telescope observations because of its proximity to the Sun.) The report therefore rated atmospheric study of Mercury rather low: "Atmospheric investigations of Mercury should not command a high priority in the near future. Fly-by missions to the planet may take place for other reasons, however. If so, radio-occultation and flourescence [sic] measurements would be of interest."

The 1968 report had indeed laid some interesting groundwork for future observations ("In the absence of other information we may take this figure of 1 km as the upper limit of resolution needed to detect past volcanic activity on Mercury or Mars. Other processes require even better resolution." "It is also important to attempt at least an exploratory, preliminary examination of all the terrestrial planets. In practical terms, this means that we place considerable value upon a photographic reconnaissance of Mercury and radar examination of Venus."). The summary of knowledge was given as follows:

> Mercury is the smallest and most dense of the terrestrial planets. It is also the closest to the Sun. Albedo and radar reflection suggest a surface resembling the lunar maria. Only indistinct surface markings are visible from Earth.
>
> The great range in temperature between subsolar and midnight positions, the large solar radiation flux, and the probable lack of an atmosphere must all influence the nature of the surface in a major way. Yet the remoteness of Mercury from Earth has served to lessen its attractiveness to those interested in planetary exploration. Recently, however, an upsurge in interest has been sparked by the realization that in 1973 or 1975 a spacecraft launched by a modest-sized vehicle can take advantage of the Venus gravitational field to gain acceleration and thereby enter a trajectory that will send it past Mercury.

With respect to a "Venus-Mercury flyby" mission, the 1968 report continued:

> The Venus swing-by mission proposed for 1973 provides the first opportunity to examine Mercury. As a prime objective, we recommend photographing the planet with a resolution of about 2 km. Similar photography of Venus during the fly-by portion of the orbit may reveal cloud patterns indicative of the atmospheric circulation system. Additional experiments for the Mercury encounter should include a magnetometer to determine if Mercury has a magnetic field and some form of emission line photometer to determine if Mercury has an atmosphere. If the trajectory permits occultation, a second test of the existence of an atmosphere can be achieved from the S-band radio links.
>
> If imagery of Mercury at 2-km resolution is indeed obtainable from a Pioneer class Venus-Mercury fly-by, it becomes a most significant experiment from the viewpoint of planetary surfaces. The Sun may have a profound effect on its nearest neighbor so that an unusual balance of internal and external activity is evidenced by surface topography. The best way to underscore the scientific value of a limited imagery mission for Mercury is to recall the major changes in our thinking about Mars produced by the 4-km resolution Mariner 4 pictures.
>
> The short lead time, the low cost of a Pioneer mission, and the value to our national prestige of a planetary first are strong arguments supporting the basic scientific value of the mission.

Further, the 1968 perspective already included the notions of orbiters and landers:

> From the point of view of planetary dynamics, Mercury is perhaps the most important object in the solar system. Being closest to the Sun, it is the most sensitive detector of departures from the laws proposed to account for planetary orbital motions. Its spin is also unusual, being coupled to its orbital motion in a three-halves resonance state. In view of its unusually high density, the interior of Mercury is also of special interest.
>
> A space-probe fly-by of Mercury could provide important information on both its dynamics and its interior. From photographs we may obtain the precise orientation of Mercury. Combined with similar pictures from later fly-bys, the vital knowledge of the direction of Mercury's spin axis and the fractional difference in its equatorial moments of inertia can be determined. Search for magnetic field strengths and determination of the electromagnetic radiation from Mercury's surface, as well as photographs of the surface, will provide important data on its interior structure. The radius, mass, and hence, density, and the orbit can also be refined from the fly-by data. An orbiter is required to

determine the detailed gravitational field of Mercury and a lander to study the interior by monitoring seismic activity.

From this early vantage point, perhaps the most interesting comment with respect to Mercury is actually aimed at the future exploration of Mars:

> With respect to its interior and dynamical properties, Mars is of no greater intrinsic interest than Venus or Mercury, perhaps less. However, most of the appropriate measurements are much more easily made for Mars; we address ourselves to the determination of the dynamics and interior of Mars with special cognizance of this fact and of the unique biological and geological interest in the planet.

The science questions raised and means of addressing them, notably by orbiters and landers all were – and remain – relevant to Mercury. The significant difference between obtaining new scientific knowledge at Mercury and Mars from the perspective of the late 1960s was, of course, the difference in difficulty to reach these different planets.

The 1968 study cemented the ground for the subsequent Mariner 10 mission to Mercury, making use of a Venus gravity assist to enable the mission on an available launch vehicle (the Atlas Centaur) and using resonant flybys, following the first Mercury encounter on 29 March 1974, to make two more visits to the planet (21 September 1974 and 16 March 1975). Owing to the novel Venus flyby trajectory and motivated by interest in conducting reconnaissance in the near term, the Mariner Venus–Mercury mission was developed and renamed Mariner 10 following its successful launch. The observations of Mercury during the three Mariner 10 flybys provided tantalizing clues that greatly increased interest in the planet.

The Mariner 10 mission was extremely successful (Balogh et al., 2007a), but it was also clear that substantial further progress in the scientific exploration of Mercury required an orbiter. Orbiting Mercury, however, was recognized as far more difficult than "simple" flybys because of the need to insert a spacecraft into orbit in Mercury's relatively weak gravity field deep within that of the Sun, while also dealing with the harsh thermal environment close to the Sun and near Mercury's hot dayside surface (Balogh et al., 2007b). The next study assessing solar system exploration priorities was a NASA internal study conducted by a Terrestrial Bodies Science Working Group chartered in 1976 (Toksöz et al., 1977). This study advocated a Mercury polar orbiter "with a subsatellite and/or surface lander" having a launch in (calendar year) 1986 and operations in 1988 (the same year as a "Mars Surface Sample Return (MSSR) with mobility"). Low-altitude (≤500 km) circular orbits of the planet enabled by "low-thrust propulsion system such as solar sail or ion drive to meet the science objectives" were advocated. The proposed science was similar to that recommended a year later, but the NASA study also provided details for scientific instruments, which perhaps not surprisingly targeted observational objectives very similar to those achieved by the MESSENGER payload (Solomon et al., 2007; Chapter 1) and closely aligned with the instrumentation to be flown on BepiColombo (discussed below).

The most comprehensive study articulating NASA's overall strategy for solar system exploration in the late 1970s was that of the Committee on Planetary and Lunar Exploration, or COMPLEX (1978). The basic scheme advocated by COMPLEX was to follow flyby reconnaissance with exploration by orbiters and entry probes (for planets with atmospheres) and then with intensive study, e.g., by soft landers conducting in situ investigations, and this strategy has held well. As technologies for remote automation and mobility have advanced, the use of surface rovers has been favored, when practical, over stationary landers for many targets.

In the COMPLEX report, low-thrust solar electric propulsion (SEP) was advocated for providing the means to place a spacecraft into orbit around Mercury:

> Advances in the exploration of Mercury, however, are predicated on observations and measurements from a circular orbit of the planet, which the present U.S. launch capability cannot currently provide. *Steps should be made to prepare for the investigation of Mercury after definition of an adequate propulsion capability and in advance of availability of the system.*

Sample return in general was advocated as an extremely important part of the intensive study portfolio of approaches. COMPLEX did not discuss such a mission for Mercury, no doubt due to the difficulties at that time just of achieving an orbital mission. "In general, sample return from any extraterrestrial body cannot be considered as a viable mission option ab initio." Careful deliberations to establish the scientific need and context were required to establish the specific targets. For the period under consideration, primary objectives for Mercury included determining the chemical composition of the surface; ascertaining the structure and state of the interior; and extending coverage and resolution of orbital imaging. All of these objectives were viewed as achievable with an orbiter. Exploration of the magnetosphere and magnetic field, measuring the planetary heat flow, and providing global gravity and topography of the planet were all relegated to "secondary planetary objectives." The committee cautioned that in the absence of new propulsion technology "undertaking the next investigations of Mercury must remain indeterminate" and that, even with the advent of such new propulsive means, the exploration of Mercury should "*not inhibit or detrimentally affect the primary emphasis on the triad Earth-Mars-Venus*" [emphasis in original].

20.3 BREAKTHROUGH FOR MERCURY ORBITAL MISSIONS

In the absence of in-space propulsion breakthroughs, the problem of placing a spacecraft in orbit about Mercury remained a major impediment to efforts to motivate an orbital mission to the planet. Direct trajectories from Earth required too large an excess velocity at Mercury approach for chemical propulsion systems to compensate. The approach speed at Mercury would be ~9 km/s, and a braking burn that imparted a velocity change (ΔV) of at least ~7.5 km/s would be needed to enter orbit. In the face of this technical obstacle, investigation of Mercury from orbit was simply not feasible.

20.3.1 Multiple Gravity-Assist Solution

It was the 1985 discovery by Chen-wan Yen of multiple Venus and Mercury gravity assist combinations that, at the expense of longer mission flight times, enabled Mercury approach speeds within the capabilities of chemical propulsion (Yen, 1985, 1989; Stern and Vilas, 1988; Belcher et al., 1991; Balogh et al., 2007b). From that time on, all serious proposals for orbital missions to Mercury relied at some level on multiple gravity assists at Venus and Mercury to enable orbit insertion. This was the case for multiple proposals in both Europe and the United States, including several mission proposals to the NASA Discovery Program and the MESSENGER mission as implemented. The ESA–JAXA BepiColombo mission is no exception. With its diverse payload and dual-spacecraft implementation, it uses both planetary gravity assists and solar electric propulsion to reach Mercury orbit with a large payload mass.

Following the discovery of this multiple planetary gravity assist technique to enable a Mercury orbiter, NASA chartered a Mercury Orbiter Science Working Team (MeO SWT) to consider an orbital mission to Mercury (Belcher et al., 1991). The approach envisioned two identical spacecraft with 11 instruments (Table 20.1), each focused on a combination of magnetospheric and planetary science objectives and employing a single Titan IV–Centaur launch. The "Dual Orbiter" concept of the MeO SWT represented a nearly optimal mission concept for Mercury exploration, but the projected high cost relative to NASA's established priority of Mercury relative to Mars and Venus (COMPLEX, 1978) precluded its implementation. Even so, the MeO SWT concept provided the justification for Mercury mission proposals to the NASA Discovery Program in the 1990s (Nelson et al., 1995; McNutt et al., 2006).

Table 20.1. *Notional instruments for each of the MeO SWT Dual Orbiters (Belcher et al., 1991).*

Item	Instrument	Mass (kg)	Power (W)
1	DC Electric Field Analyzer	18.2	7.0
2	Energetic Particle Detector	15.0	15.0
3	Fast Electron Analyzer	4.0	5.0
4	Fast Ion Analyzer	4.0	5.0
5	Gamma/X-Ray Spectrometer	17.0	14.3
6	Ion Composition Plasma Analyzer	10.0	12.0
7	Solar Wind Analyzer	10.0	10.0
8	Line-Scan Imaging (and Thermoelectric Cooler)	5.1	11.0
9	Magnetometer	5.3	5.3
10	Radio/Plasma Wave Analyzer	7.2	7.2
11	Solar Neutron Analyzer	10.0	10.0
Total		105.8	101.8

20.3.2 A Focused Mercury Orbital Mission: MESSENGER

On 29 January 1996, Charles Elachi, Director of the Jet Propulsion Laboratory (JPL), issued a "Dear Colleague" letter to assemble a roadmap development team to prepare input for a "Mission to the Solar System" roadmap. A community meeting at Caltech on 5–6 March 1996 began the process that led to the final Roadmap report (Gulkis et al., 1998). The roadmap included Mercury Orbiter and Mercury Magnetospheric Multi-Satellites missions. At the same time, a workshop was set up to begin deliberations on a roadmap focused on NASA's Office of Space Science (OSS) Solar Connections theme, with its meeting scheduled at the Johns Hopkins University Applied Physics Laboratory (APL) on 10–12 April 1996. By 29 March 1996, the program for the workshop had been set to begin with four talks on potential future missions to Mercury and on Mercury magnetospheric science on the first day (10 April 1996).

In March 1996 discussions and initial analysis at APL focused on how the eight key science questions for a Mercury orbiter mission suggested some years earlier (Stern and Vilas, 1988) might be addressed within the Discovery Program constraints to identify a feasible suite of payload instruments and corresponding measurement requirements. Preliminary analysis was made of an "Orbiting Mercury Observatory" (OMO), and on 2 April 1996 Sean Solomon, then the Director of the Department of Terrestrial Magnetism of the Carnegie Institution of Washington, agreed to lead the effort.

The need for substantial propulsion to achieve the required ΔV to enter orbit about Mercury, together with a "focused" (i.e., low-mass) payload to allow for as high a mass ratio of fuel to payload as possible, were the immediate concerns of the core team of scientists, engineers, and managers assembled to develop the mission concept. Initial estimates were made of component masses. Combinations of gravity assist, chemical propulsion, and SEP were all considered, along with assessments of the technological readiness of the various approaches. Mission tour, science questions, payload, and instrument requirements were all discussed. An initial 11-instrument payload, with a "floor" of six instruments, was identified, and an initial science team was assembled. The mission acronym "MESSENGER" was adopted in May 1996. Use of an SEP system was ruled out because of the lack of technological maturity, and the use of an all-chemical system with "staging" at Venus was considered as well in July 1996 but was rejected as not viable. The review cycle of the initial MESSENGER proposal yielded critical guidance for work on key technology issues, chief among which were the development and qualification of a solar array design that could withstand Mercury's thermal environment. The early decisions on science and basic approach held as the mission analyses continued through a second, and successful, proposal round in 1998–1999 (McNutt et al., 2006).

The MESSENGER mission, spacecraft, and payload have been described in some detail, both in their initially accepted configurations (Gold et al., 2001; Solomon et al., 2001; Santo et al., 2001) and in the configuration as built and flown (Anderson et al., 2007; Andrews et al., 2007; Cavanaugh et al., 2007; Goldsten et al., 2007; Hawkins et al., 2007; Leary et al., 2007; McAdams et al., 2007; McClintock and Lankton, 2007; Schlemm et al., 2007; Solomon et al., 2007; Srinivasan et al., 2007) as

Figure 20.1. The last image downlinked from MESSENGER prior to impact on Mercury. Central latitude and longitude are 72.0°N and 223.8°E. The image was acquired on 30 April 2015 at 11:07:43 UTC during orbit 4104 (the last complete orbit of Mercury by the spacecraft). The area imaged is about 1 km across, and the resolution is 2.1 m/pixel.

summarized also in Chapter 1. Moreover, the progress of the mission has been chronicled from its beginning through launch (McNutt et al., 2006), the Venus flybys (McNutt et al., 2008), the first Mercury flybys (McNutt et al., 2010), its primary orbital mission (Bedini et al., 2012; McNutt et al., 2014), and its first extended mission (XM1) from 18 March 2012 to 17 March 2013 (McNutt et al., 2012). In addition, more than 400 articles have been published by the MESSENGER team in the scientific and engineering literature. A second extended mission (XM2) was proposed in February 2013 to last through a planned impact of MESSENGER onto Mercury in March 2015 after propellant for orbit corrections was exhausted. A final extension (XM2′) was proposed on 7 November 2014 and accepted 12 days later to take full advantage of the very low periapsis passages of the final orbits of the mission, allowing for unprecedented, and not likely to be repeated, close-up measurements of the surface and near-surface environment (Chapters 1 and 5).

Out of propellant to raise its periapsis altitude further, MESSENGER impacted Mercury's surface at an estimated spacecraft event time of 3:26 pm Eastern Daylight Time (EDT) (3:34 pm EDT ground receipt time) on 30 April 2015. The impact occurred 3922 days after launch from Cape Canaveral Air Force Station and 1504 days following Mercury orbit insertion, after starting orbit 4105 about the innermost planet of the solar system. The last image downlinked from the spacecraft is shown in Figure 20.1.

20.3.3 Pointing the Way to Deeper Discoveries

The progression in capability from Mariner 10 to MESSENGER follows the paradigm of planetary exploration outlined above of initial reconnaissance followed by more in-depth exploration. Mariner 10 was a low-budget mission relative to other options available, and this choice limited the scope of the scientific payload. Nonetheless, imaging of almost half the planet, discovery of the magnetic field and magnetosphere, discovery of energetic particles, and detection of the planetary exosphere provided many of the insights required to optimize the MESSENGER payload, both in function and in capabilities, given the corresponding limitations on budget and mass for a mission in NASA's Discovery Program. Primary among these limitations were the inability to include several candidate instruments, such as a plasma wave detector, a thermal emission spectrometer, and electric field probes, and the need to choose one hemisphere for a more detailed investigation than the other, in order to accommodate thermal design constraints. With body-fixed instruments and only moderate downlink rates, pointing scenarios and data return were judged to be adequate for an initial orbital survey but less than ideal for some science questions that were deferred to future missions.

The concept for the ideal orbital mission to Mercury articulated by the MeO SWT was never implemented by NASA because of the higher priority of other exploration targets. However, the priority for a dual-orbiter mission to Mercury was higher in Europe and Japan, enabling implementation of the BepiColombo mission as described in detail below. There is a striking correspondence between the MeO SWT strawman payload (Table 20.1) and that of the BepiColombo spacecraft (see Section 20.5). A dual-spacecraft mission with a more extensive payload, however, comes at substantially higher cost than MESSENGER. The total cost for MESSENGER through all mission extensions was ~$500 million in real-year dollars, compared with the projected costs for the BepiColombo mission, which are currently ~$1.3 billion for ESA and $140 million for JAXA. Although not initially planned in such a way (Balogh et al., 2007b), the sequenced approach of the NASA Discovery-class mission MESSENGER arriving at Mercury nearly 15 years earlier than BepiColombo yielded many benefits from a scientific viewpoint. MESSENGER provided answers to the key science questions that it targeted, but it also turned up a number of unexpected results that raise new questions to be addressed by BepiColombo. Moreover, the vast amount of new information about Mercury from the MESSENGER mission is helping the BepiColombo mission with targeted investigations and mission operations planning.

20.4 SCIENTIFIC AND TECHNICAL ADVANCES OF MESSENGER: FACILITATING FUTURE MERCURY EXPLORATION

The MESSENGER mission led to important achievements in both science and engineering that profoundly advanced our understanding of Mercury and point to new directions for future exploration. The scientific advances are summarized by mission phase in Chapter 1 and covered in detail in the other chapters of this volume. Although MESSENGER answered the scientific questions that framed its primary and extended missions, this deeper understanding has led to new mysteries. As the first extended mission built on the results from the primary mission, and the second extension on those of the first, the BepiColombo mission to Mercury will build on the results from MESSENGER. In addition to the scientific direction and focus that MESSENGER provides, there were a number of engineering achievements that are relevant to future exploration. To set the context for the description of BepiColombo, it is useful to review the MESSENGER results and achievements from the perspective of the technical opportunities they enable and new questions they raise.

MESSENGER and BepiColombo have been and remain complementary missions (McNutt et al., 2004). Aside from repeating some measurements more than a decade later, thereby providing a longer temporal baseline for a variety of measurements and observations, BepiColombo will collect a number of high-resolution observations and make many entirely new observations. Compared with MESSENGER, the southern hemisphere will be observed from markedly lower altitudes, and the two BepiColombo spacecraft will perform simultaneous measurements of the magnetic field and its dynamic response to changes in solar activity from two spacecraft in different orbits. BepiColombo contains payload instruments not included on MESSENGER, such as a thermal infrared spectrometer, a full complement of plasma physics instrumentation, and a triple-band, radio-science instrument with an on-board accelerometer to obtain high-precision measurements to determine the orbit and test gravitational theory. Indeed, that the BepiColombo mission, nearly ready for launch, is well suited to resolve many of the new mysteries revealed by the discoveries from the MESSENGER mission makes this an especially exciting time for Mercury science.

20.4.1 Surface Composition

Elemental composition mapping from MESSENGER was largely limited to the northern hemisphere, and most maps are at low (hundreds of kilometers) spatial resolution because of limited count rates in the measuring instruments and the eccentricity of MESSENGER's orbit (Chapter 2). Acquiring compositional maps at higher spatial resolution is key to understanding processes responsible for different geologic units and other surface features discovered with MESSENGER imaging (Chapters 6, 7, 8, and 11). The compositions of the hollows (Chapter 12) and polar deposits (Chapter 13) are particularly interesting and could not be well resolved in the MESSENGER data. Resolving compositional variations may also allow determination of unique signatures that might establish whether there are any samples from Mercury in meteorite collections on Earth. More globally, the composition derived from MESSENGER revealed a low oxygen-to-silicon ratio of surface materials, a characteristic not understood; the anomaly could be a surface processing effect or may reflect the bulk properties of the surface materials (Chapter 7). MESSENGER showed, too, that carbon in the form of graphite is likely the dominant darkening agent on Mercury's surface (Chapters 7 and 8), but improved measurements of the surface carbon abundance and its variation with geochemical terrane and morphological and spectral unit would be valuable. The nature of the high-magnesium region (Chapters 2 and 7), mechanisms of source processes for neutral and ionized species in the exosphere and magnetosphere (Chapters 14, 15, 16, and 17), and details of the space weathering of Mercury's surface (Chapter 8) are also open issues, as are the identification and roles of various minor exospheric species (Chapters 14 and 15).

20.4.2 Interior Structure

The radio science and altimetry data from MESSENGER have yielded significant new data and constraints on the geophysical properties of Mercury that in turn raise many new questions (Chapters 3 and 4). MESSENGER observations of degree-2 coefficients in the spherical harmonic expansion of Mercury's gravity field, C_{20} and C_{22}, together with Earth-based radar observations of the libration amplitude and obliquity, have been used to determine the planet's interior structure (Chapter 4). These results constrain the core radius to be 2000–2050 km and the outer silicate shell thickness to be ~400 km. However, owing to MESSENGER's eccentric orbit, the gravity field in the northern hemisphere was determined to much higher resolution than that in the southern hemisphere (Chapter 3), so determination of the distribution of mascons and other gravity anomalies in the south must await comparable coverage in the southern hemisphere. Tracking of the BepiColombo spacecraft will help to refine the gravity field of Mercury and will yield refinements to the size and physical state of its core. The mission will provide additional constraints on models of the planet's internal structure and test theories of gravity with unprecedented accuracy, e.g., by obtaining full global coverage of topography because of the more nearly circular orbit of the BepiColombo Mercury Planetary Orbiter (see Section 20.5.4) and the shift of its periapsis over the course of the mission. In addition, the equatorial asymmetry in the global magnetic field identified by MESSENGER, the crustal magnetic field of Mercury, discovered during MESSENGER's second extended mission, and the nature of how field-aligned currents at Mercury close through the planet's interior (Chapter 5) bear further investigation with lower altitude observations in the southern hemisphere and simultaneous observations at high altitudes, in the solar wind, and near the planet.

20.4.3 Volcanic Processes

Not only did the MESSENGER observations provide definitive confirmation that volcanism played a substantial role in Mercury's surface evolution, but they revealed that multiple volcanic processes have been active (Chapters 6 and 11). The initial characterization of the diverse forms of volcanism on Mercury motivates many new questions. The relationship between volcanism and mantle processes, the variation of magma composition with time, the hemispheric difference in the areal density of impact basins, and the hemispheric asymmetry in the distribution of smooth plains remain to be fully explained (Chapters 6 and 11). Specifically, it is not known how pyroclastic and effusive eruptions are related. Understanding relationships between volcanism and global contraction may allow greater understanding of the planet's thermal evolution.

Further insights into these questions can be attained with high-resolution, global measurements from BepiColombo instrumentation. BepiColombo can contribute to the search for additional basins and improve our understanding of spatial and temporal variations in volcanic eruptive style. BepiColombo high-resolution imaging and topography can be helpful, particularly in the southern hemisphere where imaging resolution by MESSENGER was poorer than in the north. Measurements to determine relations among fault activity, plains emplacement, and craters of differing ages can also be provided by the higher-resolution compositional, altimetric, gravity, and spectroscopic observations from BepiColombo.

20.4.4 Mineralogy

One of the chief results from MESSENGER is that Mercury's surface has been subjected to intense space weathering, which has resulted in substantial muting of spectroscopic absorption signatures from near-ultraviolet to near-infrared wavelengths characteristic of different minerals (Chapters 6 and 7). The muted spectral signatures are in part the result of the low iron abundance in Mercury's silicate fraction (Chapter 2) and in part the result of higher rates of space weathering on Mercury than on the Moon or other airless bodies (Domingue et al., 2014; Chapter 8). Characterizing the surface mineralogy of Mercury has therefore proved to be more challenging than anticipated, and the mineralogy of Mercury's surface has largely been inferred to date from elemental composition and petrological considerations (Chapter 7).

Broad spectral variations between geologic units are present, however, and raise key questions that the instrumentation carried by BepiColombo can help answer. Spectral units are distinguishable on the basis of color and overall reflectance (Chapter 8), and crater ejecta, hollows, and volcanic deposits often differ in spectral character from their surroundings (Chapters 9, 11, and 12). Understanding the physical processes that produced these differences requires characterization of the mineralogical variations at scales as small as individual geologic features, which proved a challenge to MESSENGER measurements. BepiColombo will make more sensitive spectral observations and measurements over a greater range of wavelengths (e.g., mid-infrared) than MESSENGER and is therefore likely to improve our understanding of surface mineralogy and its variation among spectral and geological units.

20.4.5 Polar Deposits

MESSENGER confirmed earlier suggestions from Earth-based radar imaging that Mercury's polar deposits consist predominantly of water ice and discovered that most of these deposits are overlain by a layer of dark volatile material several tens of centimeters in thickness (Chapter 13). The formation dynamics and sources of the polar deposits are not known; nor is the composition of the dark surficial layer, the nature of its interactions with the underlying water ice, or the extent to which the trapping of water ice and other frozen volatiles is sporadic or ongoing. Moreover, MESSENGER showed that Mercury's crustal material is generally richer in volatiles than previously thought, and the compositions are not consistent with most scenarios previously advanced for planet formation (Chapters 18 and 19).

The orbits and instrumentation of BepiColombo should allow major advances in understanding the physical processes and geological implications of the polar deposits. Most obviously, the BepiColombo Mercury Planetary Orbiter (see Section 20.5.4) will enable study of the characteristics of volatile emplacement in the south polar region of Mercury as well as the north polar region. The expanded instrumentation on BepiColombo will enable combined observations to investigate the chemical and physical characteristics of some of these deposits.

20.4.6 Hollows

The discovery of hollows – rimless depressions commonly associated with bright haloes – on Mercury's surface was a complete surprise (Chapter 12). Hollow formation is thought to involve a volatile wasting process that may have occurred via sublimation, space weathering, pyroclastic volcanism, or outgassing. Understanding the physical processes that produce the hollows and linking this information to the geologic processes that bring volatile material to the surface requires compositional and mineralogical characterization on spatial scales at least as small as clusters of hollows, and the BepiColombo instruments are likely to provide invaluable observations to constrain the physical processes and implications of these remarkable formations.

20.4.7 Neutral Particle Environment: Exosphere and Dust

MESSENGER was not instrumented to survey the dust environment of Mercury, although the altitude distribution and temporal variation of the density of neutral atoms in the exosphere point to micrometeoroid impacts as an important source process (Chapters 14 and 15). Hence, direct measurements of the dust environment as planned for BepiColombo should confirm these inferences of the dynamic sources for the exosphere. The discovery of hollows and confirmation of water-ice deposits in the polar regions by MESSENGER suggest that sensitive neutral and ion instruments might be able to probe material sputtered from these regions, and the higher-sensitivity neutral particle spectrometers carried by BepiColombo may allow specific inferences on the material in hollows and polar deposits.

20.4.8 Magnetosphere

The extensive survey of the magnetic and ion plasma dynamics of Mercury's magnetosphere conducted by MESSENGER revealed a remarkably dynamic and active space environment with substantial acceleration and precipitation to the surface of energetic particles (Chapters 16 and 17). These results make it even more imperative to measure the thermal electron population, plasma convection dynamics, and electromagnetic wave environment, which MESSENGER was not instrumented to resolve but which are key targets of BepiColombo instrumentation. Moreover, the remarkable dynamics of the magnetosphere raises fundamental questions regarding internal magnetospheric processes versus those driven directly by the interaction with the solar wind, a topic that will be better addressed with the two-point magnetic field observations that are planned with BepiColombo.

Characterizing the magnetosphere is of primary importance to deriving the structure of Mercury's internally generated magnetic field (Chapters 6 and 16). There are several ways in which BepiColombo can advance quantitative determination of the magnetospheric magnetic field. Because the lower-altitude (<800 km) Mercury Planetary Orbiter will provide the first survey within the magnetosphere in the southern hemisphere, it will yield improved estimates of the planetary field and its north–south asymmetry. By assessing the field-aligned currents in the southern hemisphere, BepiColombo may also refine knowledge of Mercury's electrical conductivity structure via quantification of the intensity, distribution, and closure path of these currents. In addition, since BepiColombo will arrive at Mercury more than 10 years after the end of MESSENGER's observations, it can further constrain any secular variation in Mercury's field (e.g., Philpott et al., 2014). Simultaneous magnetic field measurements by the Mercury Planetary and Magnetospheric Orbiters (see below) should allow parameterization of the magnetospheric response to solar wind forcing (e.g., variations in interplanetary magnetic field magnitude and direction), which in turn may yield key improvements in specifications for the inner edge of the tail current and the current sheet thickness. The additional plasma, electric field, and wave instrumentation will enable a thorough specification of the convection flows and greatly increase understanding of magnetotail dynamics, acceleration mechanisms of energetic electron bursts, and the relationship between convection and particle precipitation to the surface (Chapter 17).

20.4.9 Engineering and Technical Advances

In addition to the scientific results from MESSENGER that have prompted the concentration of subsequent exploration on key unresolved questions, the engineering solutions implemented in the course of the MESSENGER mission provide solid bases for future missions. The MESSENGER mission was the first to use solar radiation pressure to achieve gradual thrusting during passive cruise operations (O'Shaughnessy et al., 2014). This "solar sailing" conserved precious propellant and was a major factor that enabled the second extended mission at Mercury and particularly the low-altitude campaign over the final months of operations (Chapter 1). Future missions can be expected to adopt similar strategies in the inner solar system, where the solar intensity is sufficiently high that this modest force is large enough over the long cruise periods required for the multiple flyby trajectories to Mercury to reduce the propulsion budgets for these missions.

The spacecraft operations and instrument commanding needed to accomplish the complex observations for MESSENGER required novel approaches to identify and select observation opportunities within operational constraints and automatically generate spacecraft attitude control and instrument operations commands (Choo et al., 2014; Chapter 1). Similar operational challenges are presented for other orbital missions that require complex observation plans. Whether similar processes will prove necessary for the next Mercury orbiter mission remains to be determined, as the orbital operations planning for BepiColombo is in its early phases.

The harsh thermal environment at Mercury, the result of both intense sunlight and radiant heating from the dayside planetary surface, required novel techniques to shield, dissipate, reflect, or withstand the severe environment. The solar cell and reflector (mirror) bonding techniques developed for MESSENGER have provided solid engineering bases for designing solar cell and mirror arrays for subsequent missions to the inner solar system, such as the Parker Solar Probe (Fox et al., 2016). The ceramic cloth sunshade (Leary et al., 2007) on MESSENGER demonstrated the success of refractory materials to shield and re-radiate thermal energy to protect the spacecraft structure, subsystem components, and electronics. MESSENGER was also the first interplanetary mission to use a phased-array antenna for communications, a decision that eliminated the need for a gimbaled antenna structure and reduced the spacecraft pointing for communications operations (Leary et al., 2007; Srinivasan et al., 2007). The successful demonstration of these technologies has facilitated subsequent designs for spacecraft facing harsh environments, whether three-axis stabilized or spinning platforms.

Finally, by executing the first successful orbit injection, navigation in orbit, and orbit maintenance at Mercury, MESSENGER provided valuable guidance for BepiColombo by quantifying the gravity field of Mercury as well as the effects of radiation pressure and solar and Mercury gravity on orbit stability (McAdams et al., 2015). MESSENGER also introduced the use of propulsion system pressurant (gaseous helium) as propellant to delay final spacecraft impact (Kirk et al., 2015). The collective MESSENGER experience has enhanced confidence in mission planning for BepiColombo.

20.5 BEPICOLOMBO: THE NEXT STEP IN MERCURY EXPLORATION

20.5.1 Historical Background

The joint ESA–JAXA BepiColombo mission was named in honor of Giuseppe (Bepi) Colombo (1920–1984), best known in the planetary science community for his work on the planet Mercury (Colombo, 1965; Anderson et al., 1987) and specifically for motivating calculations made by J. Beerer at the Jet Propulsion Laboratory on how to place a spacecraft into a resonant orbit with Mercury with multiple flybys (Beerer, 1970), a substantial contribution to the success of the Mariner 10 mission. Colombo also explained Mercury's spin–orbit resonance, showing that the planet rotates three times for every two orbits around the Sun (Colombo and Shapiro, 1966; Ward et al., 1976).

In May 1993, a mission to Mercury was proposed to the European Space Agency in response to a "Call for Ideas." The mission was selected as a cornerstone candidate in the Horizons 2000 scientific programme of the Agency in 1996.

On 15 October 2000, ESA's Science Programme Committee (SPC) approved BepiColombo as ESA's fifth cornerstone mission, with launch in 2009–2010 and consisting of three major elements, the Mercury Planetary Orbiter (MPO), the Mercury Magnetospheric Orbiter (MMO), and a Mercury Surface Element (MSE). At that time, the mission scenario involved separate launches of the MPO and MMO on two Soyuz–Fregat vehicles within the same launch window. A third launch vehicle was required for the MSE. A severe reduction of the science budget after the Ministerial Conference in November 2001 caused the MSE to be dropped from the mission baseline.

Between 1 October 2002 and 30 June 2003, BepiColombo went through a reassessment process with the aim of maximizing the scientific performance through optimization of the payload complement, while attempting to reduce costs and programmatic risk. The preferred mission scenario that emerged from the reassessment was to launch the MPO and MMO together on a single launch vehicle (Soyuz–Fregat 2-1B, which was changed later to an Ariane 5 as a result of growth in the required launch mass). To achieve the above goal with adequate resource margins, the MPO payload resources, particularly mass, had to be substantially reduced while ensuring that the mission scientific objectives were retained. This combination was achieved by defining, from the analysis of the science objectives, the corresponding payload complement and resulting instrument requirements. An optimized reference payload suite was defined by which instruments share common functions and resources, an action that led to an enhanced science performance at significantly lower cost. ESA's Solar System Working Group (SSWG) at its 112th meeting recognized that a traditional full announcement-of-opportunity process would not in these circumstances be possible, and the Executive proposed a different approach to payload procurement. Subsequently, on 6 November 2003, ESA's Science Programme Committee (SPC) approved the BepiColombo mission with the MPO–MMO complement as a part of their reconstructed Cosmic Vision Programme. The payload selection procedure for the MPO payload as outlined at the 105th meeting of the ESA Science Programme Committee on 6 November 2003 [reference document ESA/SPC 2003(41)] was unanimously approved.

On the JAXA side, a Mercury Exploration Working Group (MEWG) was formed in June 1997 under the Steering Committee for Space Science (SCSS) in the former Institute of Space and Astronautical Science (ISAS) to investigate a mission to Mercury. The MEWG published the Japanese plan, which was based on a spinning Mercury orbiter with chemical propulsion and multiple Venus and Mercury flybys. The possibility of collaboration with the ESA BepiColombo mission was discussed at the time of an Inter-Agency Consultative Group (ICAG) meeting in November 1999 and set out in a letter from the Director-General of ISAS to the Directorate of the Science Programme of ESA dated 31 July 2000. With the approval of BepiColombo as the fifth cornerstone mission of ESA, the MEWG was re-formed for the investigation of the MMO for the BepiColombo mission. The International Mercury Exploration Mission in the framework of the BepiColombo program was approved by the SCSS of ISAS in January 2002, followed by the formal approval by the Space Activities Commission in June 2003. The Japan Aerospace Exploration Agency (JAXA) was formed in October 2003 as a merger of ISAS, the National Space Development Agency (NASDA), and the National Aerospace Laboratory (NAL).

From the beginning, there was sustained cooperation and scientific exchange between the MESSENGER and BepiColombo teams. As a result, the design, science focus, ground calibration efforts, and operations planning for BepiColombo have been adjusted to yield an overall science return that leverages the technical and scientific results from MESSENGER to optimize the focus and discoveries to be achieved by BepiColombo.

20.5.2 Overview

The BepiColombo mission implementation, illustrated in Figure 20.2, consists of two spacecraft to be placed into orbit around Mercury. The MPO is three-axis-stabilized and nadir pointing, and the MMO is a spinning spacecraft. Both spacecraft are carried by the Mercury Transfer Module (MTM), which hosts the SEP to execute the mission trajectory through orbit capture at Mercury after launch. Communication and navigation are handled via the MPO. Since the the cruise configuration of BepiColombo is incompatible with the thermal design of the spinning MMO, the Magnetospheric Orbiter Sunshield and Interface (MOSIF) protects the MMO spacecraft from the Sun during cruise. The MPO and MMO during integration and testing are shown in Figure 20.3. The entire assembly will be launched on an Ariane 5 rocket in October 2018; there is a backup launch opportunity in April 2019.

The MPO and its suite of scientific instruments will focus on global characterization of Mercury through the investigation of the planet's interior, surface, exosphere, and magnetic field. In addition, it will conduct a high-precision test of Einstein's theory of general relativity. The MMO carries five instruments to study the environment around the planet, including the planet's exosphere and magnetosphere and their interaction with the solar wind and the planet, and will characterize the interplanetary dust environment at Mercury.

Up to arrival at Mercury, the two spacecraft will be mounted on top of the MTM, which provides the thrusting for targeting planetary flybys during cruise, leading to orbit capture at Mercury. The MTM propulsion is achieved by an SEP system. Upon arrival in December 2025 after a cruise phase of about seven years, the MTM will be jettisoned, and chemical propulsion on the MPO will be used to inject both spacecraft into their dedicated polar orbits. The MMO will be released first, after which an additional thrust phase will insert the MPO into its final orbit. The choice of the orbits is mainly a compromise between the ambitious science objectives and the thermal load on the spacecraft. The science objectives are best met with global, high-resolution coverage, implying a polar orbit at low altitude. However, the closer the spacecraft will be over Mercury's surface, the larger the received thermal flux from the planet's dayside, which will come in addition to the Sun's flux of up to a factor of 11 greater than that at Earth. As a result, the

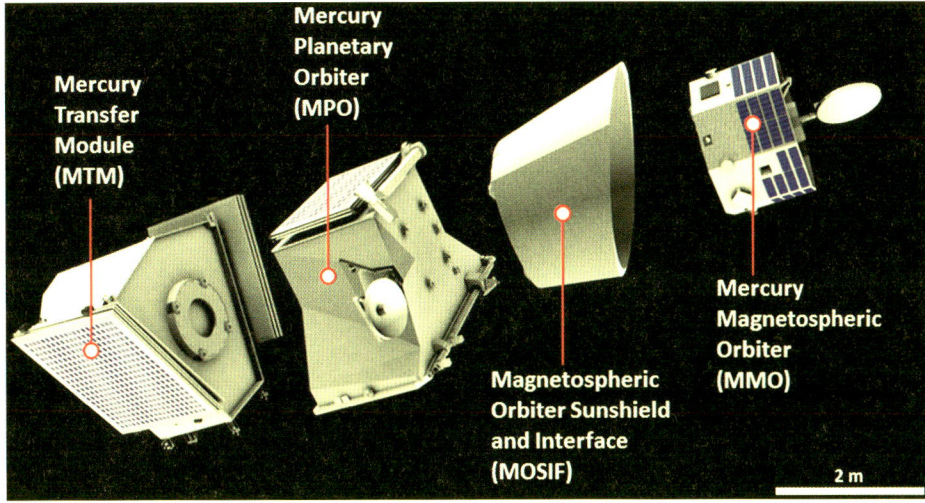

Figure 20.2. BepiColombo spacecraft elements. All four elements are launched together on an Ariane 5 rocket. The MTM provides solar-electric propulsion for the cruise phase and is jettisoned prior to orbit insertion. Thereafter the MPO provides chemical propulsion to lower the orbit. The MMO and the MOSIF separate from the MPO before the MPO completes its orbit adjustment maneuvers to its final orbit. The MOSIF is jettisoned after MMO deployment. The scale shown is approximate for dimensions projected onto the plane of this perspective view. The MPO overall dimensions are 3.9 × 2.2 × 1.7 m, and the MMO diameter is 1.8 m.

Figure 20.3. (Left) The MPO undergoing radio testing. (Right) The MMO spacecraft at the Institute of Space and Astronautical Science prior to shipment.

MPO orbit is as low as feasible within the thermal constraints.

For the MPO, a polar orbit at 480 × 1500-km altitude with a 2.3-h period was selected with its apoapsis at the equator on the dayside when Mercury is at perihelion to obtain full high-resolution mapping coverage of the planet, one of the main scientific objectives of the mission. At this subsolar point the thermal load on the spacecraft is at its maximum. Half a Mercury year later, at aphelion, the subsolar point occurs when the spacecraft is at its minimum distance from the planet. For the MMO, a highly eccentric orbit at 590 × 11,640-km altitude with a 9.3-h period, co-planar with the MPO orbit, was selected to allow mapping of the magnetic field and study of the magnetosphere, including the bow shock, magnetotail, and magnetopause. The scientific instrument payloads for the MPO and MMO are listed in Table 20.2 and Table 20.3, respectively. The combined instrument complement is more comprehensive than that of MESSENGER and tailored to the scientific objectives addressed from the two different orbits.

With two spacecraft, the MPO optimized for remote sensing and the MMO designed for in situ magnetospheric measurements, and with magnetometers on each, many of the viewing

Table 20.2. *Instruments on the BepiColombo Mercury Planetary Orbiter.*

Instrument	Observational objective	Principal investigator (country)
BELA: BepiColombo Laser Altimeter	Characterize the topography and surface morphology of Mercury.	Hauke Hussmann (Germany) Nicolas Thomas (Switzerland)
MORE: Mercury Orbiter Radio Science Experiment	Determine Mercury's gravity field as well as the size and physical state of its core.	Luciano Iess (Italy)
ISA: Italian Spring Accelerometer	Study Mercury's interior structure and test Einstein's theory of relativity.	Valerio Lafolla (Italy)
MPO-MAG: Mercury Magnetometer	Describe Mercury's magnetic field and its source.	Karl-Heinz Glassmeier (Germany)
MERTIS: Mercury Thermal Infrared Spectrometer	Determine Mercury's mineralogical composition and obtain a global map of the surface temperature.	Harald Hiesinger (Germany)
MGNS: Mercury Gamma-ray and Neutron Spectrometer	Determine the elemental composition of Mercury's surface distribution of volatiles in the polar areas.	Igor Mitrofanov (Russia)
MIXS: Mercury Imaging X-ray Spectrometer	Obtain a global map of the surface atomic composition.	Emma Bunce (United Kingdom)
PHEBUS: Probing of Hermean Exosphere by Ultraviolet Spectroscopy	Characterize the composition and dynamics of Mercury's exosphere.	Eric Quremerais (France)
SERENA: Search for Exosphere Refilling and Emitted Neutral Abundances	Study the interactions among the surface, exosphere, magnetosphere, and solar wind.	Stefano Orsini (Italy)
SIMBIO-SYS: Spectrometer and Imagers for MPO BepiColombo-Integrated Observatory SYStem	Provide global, high-resolution, and infrared imaging of the surface.	Gabriele Cremonese (Italy)
SIXS: Solar Intensity X-ray Spectrometer	Perform measurements of solar X-rays and energetic particles at high time resolution.	Juhani Huovelin (Finland)

Table 20.3. *Instruments on the BepiColombo Mercury Magnetospheric Orbiter.*

Instrument	Observational Objective	Principal Investigator (Country)
MMO-MAG: Mercury Magnetometer	Provide a detailed description of Mercury's magnetosphere and of its interaction with the planetary magnetic field and the solar wind.	Wolfgang Baumjohann (Austria)
MPPE: Mercury Plasma Particle Experiment	Study low- and high-energy particles in the magnetosphere.	Yoshifumi Saito (Japan)
PWI: Plasma Wave Instrument	Make a detailed analysis of the structure and dynamics of the magnetosphere.	Yasumasa Kasaba (Japan)
MSASI: Mercury Sodium Atmospheric Spectral Imager	Measure the abundance, distribution, and dynamics of sodium in Mercury's exosphere.	Ichiro Yoshikawa (Japan)
MDM: Mercury Dust Monitor	Study the distribution of interplanetary dust in the orbit of Mercury.	Masanori Kobayashi (Japan)

conflicts encountered by MESSENGER can be avoided while also distinguishing spatial and temporal effects in the magnetosphere (as had been envisioned in the MeO SWT study of 1991). After NASA's Mariner 10 and MESSENGER missions, BepiColombo will address a comprehensive set of scientific questions to gain further knowledge about the planet, its evolution, and its surrounding environment, providing important new constraints on the origin and formation of the terrestrial planets.

The BepiColombo mission will follow up and complement the work of MESSENGER. It will address the many new questions raised by the MESSENGER results. The comprehensive set of instruments on the MPO in a close polar orbit around the planet will allow global, high-resolution coverage. The MMO, with its more than 15 different sensors on the five instruments, will allow a thorough monitoring of processes and dynamics in the near and far environment of Mercury. In particular,

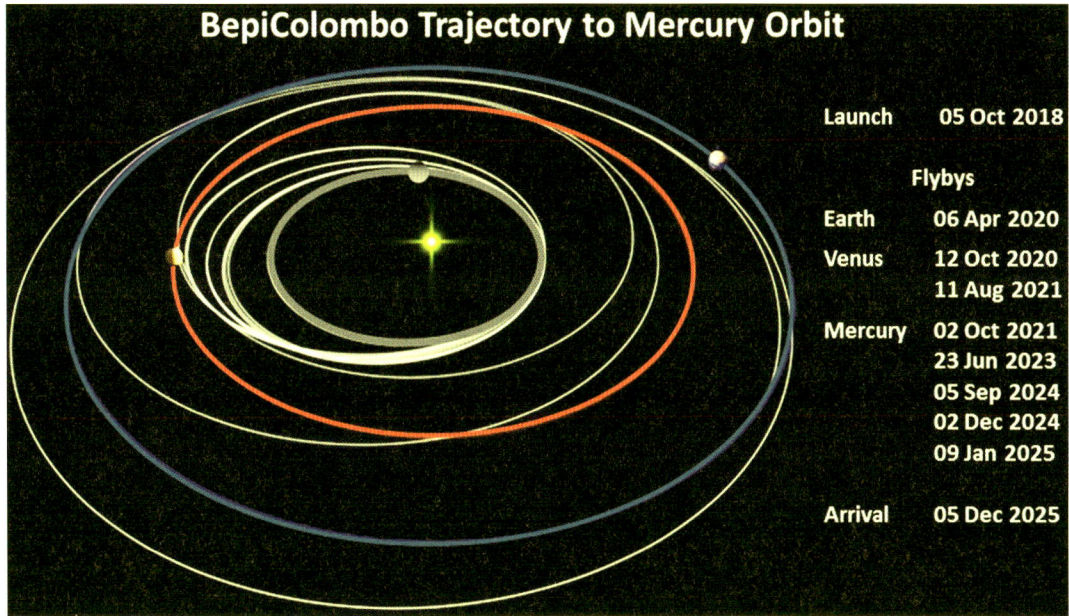

Figure 20.4. Cruise phase of BepiColombo from Earth to Mercury. The various gravity assists are listed at right for a nominal launch date of 5 October 2018. The orbits of Earth, Venus, and Mercury are shown in blue, red, and gray, respectively.

BepiColombo will provide a rare opportunity to collect two-point measurements in a planetary environment. It is foreseen that the orbits of MPO and MMO will allow close encounters of the two spacecraft throughout the mission and will enable study of the planet and its surroundings in an optimal way. This geometry will allow acquisition of scientifically valuable information in a setting in which both spatial and temporal scales can be very short.

20.5.3 The BepiColombo Mission Scenario

BepiColombo will launch on an Ariane 5 from Europe's spaceport in Kourou, French Guiana. It will use the gravity of Earth, Venus, and Mercury in combination with the thrust provided by solar-electric propulsion to reach Mercury. The two orbiters will voyage to Mercury together as a single composite spacecraft, with the MTM providing power. In addition, the MOSIF solar shield will protect the MMO from solar radiation during the cruise phase. The trajectory is shown in Figure 20.4 for an October 2018 launch date, a 7.2-year cruise, and Mercury orbit capture in December 2025. The mission trajectory includes two consecutive Venus flybys (300 days apart), which reduce the perihelion to nearly that of Mercury. A sequence of six Mercury flybys passing through 3:2, 4:3, 5:4, and 1:1 resonances with Mercury's orbit period and a 180° singular transfer (Langevin, 2000) lowers the spacecraft velocity relative to the planet to 1.85 km/s. Five final thrust arcs further reduce the relative velocity such that the spacecraft will be weakly captured by Mercury's gravity on 5 December 2025 even if no orbit insertion maneuver takes place.

Near the Earth–Sun orbit distance, ~1 AU, power will be available from the MTM solar arrays for only one thruster to be operated at a thrust level between 100 and 130 mN. Thrust arcs inside the orbit of Venus can be carried out with up to 290 mN thrust (two thrusters in simultaneous operation), using the increased power from the MTM solar arrays closer to the Sun. The total velocity increment to be provided by electric propulsion is 5.4 km/s for the baseline October 2018 launch and 5.8 km/s for the 2019 backup launch.

One advantage of the long thrust trajectories is that they allow flexibility in the trajectory design within the boundaries of launch vehicle capability, spacecraft mass, and flight duration, at very low propellant cost. The SEP system uses QinetiQ T6 gridded Kaufmann ion thrusters with xenon propellant at a specific impulse of almost 4300 s (Lewis et al., 2015). The xenon tanks are sized to carry ~580 kg of xenon. The mission requires the delivery of 2250 kg of mass for the MPO and MMO in the initial Mercury orbit. The lift-off mass will be 4.2 metric tons (including a launch vehicle adapter), of which approximately 32% is propellant. After arriving in polar orbit around the planet, the MTM will be separated. When the MMO's desired elliptical polar orbit is reached (taking it from 590 to 11,640 km above the planet's surface), the MMO will be released and the MOSIF jettisoned. An additional thrust phase will insert the MPO into its final orbit, lowering the apoapsis to about 1500 km and rotating the line of apsides to achieve its operational eccentric orbit ranging between 480 and 1500 km above the planet's surface.

BepiColombo will be the first mission to employ gravity capture at a planet. This weak gravity capture strategy has significant advantages over the more usual large propulsive maneuver approach (Jehn et al., 2004). It reduces the fuel needed for injection and eliminates the necessity of an immediate orbit insertion burn, which can be a single-point failure in traditional hyperbolic approaches. Significantly, it also provides several recovery opportunities in case of failed insertion (Novara, 2002) and increases the flexibility of choosing appropriate sets of arrival conditions. In particular, a Mercury true anomaly at arrival that is >60° away from the thermally critical perihelion conditions is allowed. The final Mercury orbit insertion maneuvers will be performed by chemical propulsion engines integrated into the MPO. The MPO chemical propulsion system will provide the required thrust for a firm Mercury capture and orbit injection, using a set of four 22 N thrusters. The velocity changes needed to insert the MMO and MPO into their nominal orbits are 325 m/s and 620 m/s, respectively. The arrival conditions are constrained such that the MPO and MMO operational orbits will have their lines of apsides close to the Mercury equator and their periapses on the antisolar side at Mercury perihelion, which yields a more benign thermal environment than other geometries. The polar orbits are optimal to obtaining global coverage of the measurements.

The baselined lifetime of the MPO and MMO in Mercury orbit is one Earth year (about four Mercury years, or two Mercury solar days). A mission extension by another Earth year is optional and has been factored into the consumables and expected radiation damage. Orbit maintenance is not required over the planned operational lifetimes of the MPO and MMO.

20.5.4 Mercury Planetary Orbiter

The BepiColombo MPO accommodates 11 scientific instruments (Table 20.2) and has a box-like shape with a size of 3.7 × 2.2 × 1.7 m (Figures 20.2 and 20.3). The entire MPO totals up to 1230 kg of dry mass, including 85 kg of science payload. The structure allows most units and payloads to be mounted on the MPO outer face, ensuring good accessibility during integration. The primary structure carries a thin cage frame to which the high-temperature multi-layer insulation (MLI) is fixed. In the center of the MPO are two tanks that carry the chemical propellant. The MPO is designed to take scientific measurements in all parts of the orbit throughout the Mercury year, implying that most of the apertures of the remote sensing instruments are continuously nadir pointing. As a consequence, five out of six spacecraft faces may be illuminated by the Sun at some point. This total leaves only one spacecraft side for a radiator to dump excess heat into space and to avoid solar exposure of the radiator. A further consequence is that a spacecraft flip-over maneuver is needed twice per Mercury year.

One of the biggest challenges in exploring Mercury is the planet's thermal environment. Not only is the intensity of the solar radiation up to 11 times higher than at Earth, up to 14,000 W/m^2, but the dayside of Mercury also reflects about 600 W/m^2 of sunlight and emits infrared radiation of as much as ~5400 W/m^2 at the MPO orbital altitude. This environment imposes strong requirements on the spacecraft design, particularly elements that are exposed to the Sun and Mercury, such as the solar array, mechanisms, antennas, multi-layer insulation, thermal coatings, and radiator. The development of these elements and the solar electric propulsion system have been the main cost drivers for the MPO. In addition, some late design updates to cope with the harsh environment are responsible for an increase of the overall spacecraft mass by more than a metric ton compared with the initial plans when the mission was adopted.

The MPO design incorporates numerous features to deal with the harsh thermal environment. The outer MLI surface of the MPO has a low solar absorptivity to reflect most of the sunlight. Nevertheless, it heats up to more than 360°C, which prevents the utilization of standard MLI, so a ceramic fabric with titanium layers is used. Low absorptivity is also adopted for the high-gain antenna coating, which, because of its position, is fully exposed to the Sun and the reflected light and infrared radiation from the planet. In addition, the radio science experiment requires a very stable antenna, which is constructed largely of titanium. Although the radiator is not exposed to direct sunlight, it will receive intense reflected light and heat from Mercury. To minimize the influence of this heat flux on the radiator, highly reflective fins (polished and geometrically reflecting outwards) have been mounted to it at an appropriate angle, to minimize absorption of heat radiated from Mercury while allowing radiation to deep space. Inside the spacecraft, temperatures are kept within the standard range (0–40°C), and for specific instruments interface temperatures below −10°C are provided. Thermal transport from spacecraft components to the radiators is accomplished passively or via heat pipes when necessary. The design yields an even temperature distribution within the spacecraft.

The average power demand of the MPO in Mercury orbit, when conducting scientific measurements, will be approximately 1300 W, which will be provided by a solar array. The solar array must continuously track the Sun in order to keep the array temperature below 200°C. This requirement will be met by choosing solar incidence angles of up to 80° that yield sufficient power but minimize heating of the solar array.

Communication with Earth is ensured via a high-gain antenna, a medium-gain antenna, and two low-gain antennas. The medium-gain antenna is mounted on a boom and provides global coverage, ensuring that contact is maximized with respect to spacecraft attitude and Earth position. The high-gain antenna provides a link with a high data rate for science data transmission and is the basis for the MORE radio science experiment. This link is achieved using X-band uplink for commanding, both Ka and X-band for data downlink, and both Ka and X-band uplink and downlink for radio science. Over one Earth year in Mercury orbit, about 1550 Gbit will be downlinked to Earth. Precise range and range rate (Doppler) measurements are enabled by the dual-frequency uplink and downlink, allowing accurate orbit determination.

Spacecraft attitude control is provided by a set of four reaction wheels and small thrusters for angular momentum management. Three star trackers, Sun sensors, and a high-precision gyroscope package are employed as sensors for attitude control. The combination of the star trackers and gyroscopes ensure a precise attitude determination (a few arcseconds), required by several experiments. The Sun sensors are important to minimize the solar irradiation of sensitive surfaces in case of an anomaly that entails a loss of attitude. The control concept ensures that in no case will a sensitive surface be exposed to the Sun for longer than 85 s.

The MPO propulsion system consists of four redundant 22 N thrusters pointing from the nadir face of the spacecraft. The thrusters are employed in orbital maneuvers only until final orbit acquisition, after which they will be rendered passive.

For reaction wheel desaturation and attitude control, four redundant 10 N thrusters are mounted on the radiator. The two propellant tanks in the center of the spacecraft contain pure hydrazine as fuel and nitrogen tetroxide (NTO) as oxidizer. In this dual-mode system the 22 N thrusters are operated in a bipropellant mode using hydrazine and NTO to yield a high specific impulse, whereas the 10 N thrusters operate just with hydrazine, which minimizes contamination by the exhaust plume.

20.5.5 Instruments on the Mercury Planetary Orbiter

20.5.5.1 BELA

The BepiColombo Laser Altimeter (BELA) (Gunderson and Thomas, 2010) will characterize and measure the figure, topography, and surface morphology of Mercury. It will provide absolute topographic height and position with respect to a Mercury-centered coordinate system. This information will be used to create a digital terrain model that allows quantitative exploration of the planet's geology and tectonics. In synergy with the stereo camera, BELA will improve knowledge of Mercury's geology, geomorphology, tectonics, volcanism, and the evolution of the planet. BELA uses a classic approach to laser altimetry. A Nd:YAG laser produces a 50 mJ pulse at 1064-nm wavelength with a duration of 5–8 ns. This beam produces a 60-µrad footprint (20–50 m) on the surface of the planet.

The laser light is reflected from the surface, received with a telescope, and fed into the pulse discrimination electronics, which determines the time of flight, the integrated pulse intensity, and the pulse width. Onboard data compression and data storage are essential. BELA requires significant baffling and thermal control and can operate over the dayside and nightside hemispheres, allowing optimum data acquisition. Performance estimates show that data return is expected at altitudes up to 1050 km above the surface with very low probabilities of false detections. Samples will be acquired every 250 m along ground tracks that will be separated by 25 km at the equator. The experiment will provide return pulse intensity and width information, allowing an assessment of surface reflectance and roughness.

20.5.5.2 ISA

The Italian Spring Accelerometer (ISA) (Iafolla et al., 2010) is a three-axis, high-sensitivity accelerometer devoted to measurement of the acceleration related to the non-gravitational perturbations (NGPs) acting on the surface of the MPO spacecraft. The three sensitive elements are based on a mechanical harmonic oscillator with a resonance frequency of 3.5 Hz. The weakest accelerations that ISA is able to measure cause a displacement of the proof masses of about 2×10^{-11} m. These displacements are detected by means of a capacitive pickup system in a bridge configuration. The NGPs in Mercury orbit are mainly due to the incoming solar visible radiation and visible and infrared radiation from the planet. The scientific objectives of ISA are strongly related to the BepiColombo Radio Science Experiment (RSE) of MORE (Section 20.5.5.7). ISA's key role is to remove the NGPs from the list of unknowns in the MPO equation of motion in the precise orbit determination (POD) that is the core of the RSE. As a "byproduct," ISA will deliver a characterization of the MPO dynamics at low frequencies. The ISA accelerometer is able to measure variable accelerations in the frequency band 3×10^{-5} to 0.1 Hz with a sensitivity of 10^{-8} m/s^2 Hz$^{0.5}$.

20.5.5.3 MPO-MAG

The primary objective of the MPO Magnetometer (MPO-MAG) (Glassmeier et al., 2010) is to collect magnetic field measurements to describe Mercury's planetary magnetic field and its source in great detail. These measurements will improve understanding of the origin, evolution, and current state of the planetary interior. The requirement is to determine all the terms associated with the internal field up to octupole components with high accuracy, using accurate magnetic field measurements on the low portions of the MPO orbit. This campaign will be supported by similar measurements to be made by MMO (Section 20.5.6), to distinguish the effects of the magnetospheric currents on the MPO measurements and to use the MMO measurements directly to augment the database for the determination of the internal field.

The secondary objectives of MPO-MAG are related to the interaction of the solar wind with Mercury's magnetic field and the planet itself. This interaction leads to the formation of highly dynamic global magnetospheric current systems. In particular, measurements close to the planet will allow a determination of the conditions for access of the solar wind to the planetary surface and an assessment of the role and importance of different current systems, including subsurface induction currents sensitive to the conductivity of the interior. These objectives will again be assisted by the planned close association with the magnetic field investigation on the MMO.

The MPO-MAG experiment consists of a dual fluxgate magnetometer system that will measure vector magnetic fields from direct current (DC) to 128 Hz frequency within ±2048 nT with a digital resolution better than 60 pT. In order to determine and remove magnetic contamination (alternating current and DC) from the spacecraft, MPO-MAG consists of two sensors, an inboard sensor and an outboard sensor, mounted on a 2.8-m-long boom and separated by 50 cm.

20.5.5.4 MERTIS

The goal of the Mercury Thermal Infrared Spectrometer (MERTIS) instrument (Hiesinger and Helbert, 2010) is to provide detailed information about the mineralogical composition of Mercury's surface material by globally mapping spectral emittance at high spectral resolution. MERTIS will cover a wavelength range from 7 to 14 µm with a spectral resolution of 90 nm, although some binning will be needed to improve the signal-to-noise ratio for identifying subtle spectral features at lower spectral resolution. This resolution will allow the detection and identification of the characteristic features of surface minerals in this spectral region, such as the Christiansen frequencies, Reststrahlen bands, and transparency features. In addition, MERTIS will be able to measure thermophysical properties of the surface, such as thermal inertia and surface texture. MERTIS is an infrared imaging spectrometer and will make use of micro-bolometer technology for which no cooling is required. MERTIS will globally map the planet with a spatial resolution of 500 m.

20.5.5.5 MGNS

The scientific goal of the Mercury Gamma-ray and Neutron Spectrometer (MGNS) (Mitrofanov et al., 2010) is to measure the elemental surface and subsurface composition for distinguishable regions over the entire surface of Mercury by measuring (a) the nuclear lines of major elements in Mercury surface material (Na, Fe, Ti, Al, Mg, Si, Ca, O, K, U, and Th), (b) the leakage flux of neutrons, and (c) the lines of naturally radioactive elements, including U, Th, and K. It will also determine the regional distribution of volatile deposits in permanently shadowed polar areas of Mercury and provide a map of column density of these deposits with an accuracy of 0.1 g cm^{-2} and a surface resolution of about 400 km.

20.5.5.6 MIXS

The Mercury Imaging X-ray Spectrometer (MIXS) (Fraser et al., 2010) will measure the planetary X-ray flux from Mercury, stimulated by high-energy solar X-rays and charged particle interactions with the surface of the planet. This information, in combination with simultaneous measurements of the solar X-ray flux with the Solar Intensity X-ray and particle Spectrometer (Section 20.5.5.11) (Huovelin et al., 2010), will allow measurements at high spectral and spatial resolution of the planetary surface composition. In order to achieve its science objectives, MIXS consists of two channels: MIXS-C, a collimator providing efficient flux collection over a broad range of energies with a wide field of view for global planetary mapping, and MIXS-T, an imaging telescope with a narrow field of view for high-resolution measurements of the surface.

By use of the X-ray fluorescence technique, MIXS will provide a global view of Mercury's surface composition in both hemispheres. It is expected that MIXS will obtain global elemental abundance maps of key rock-forming elements (e.g., Na, Mg, Al, Si, S, K, Ca, Ti, and Fe) to an accuracy of 10–20%. MIXS will also obtain high-spatial-resolution measurements of the distribution of chemical elements on the local scale. These measurements will enable the composition determination of small surface features (down to the few-kilometer scale) during periods of high solar activity. In addition, the MIXS data set will allow the study of complex interactions among Mercury's surface, its local environment, and the solar wind, via remote sensing of the energetic electrons and particles that are known to precipitate to the surface and produce X-ray emission (Starr et al., 2012).

20.5.5.7 MORE

The Mercury Orbiter Radio Science Experiment (MORE) addresses scientific goals in geodesy, geophysics, and fundamental physics. It will help to determine the gravity field of Mercury as well as the size and physical state of its core. It will provide crucial experimental constraints to model the planet's internal structure and test theories of gravity with unprecedented accuracy. MORE will also measure the gravitational oblateness of the Sun and test and characterize the most advanced interplanetary tracking system ever built. Finally, it will assess the performance of the novel tracking system for precise orbit determination and space navigation. These scientific goals will be achieved by means of several data types, generated by MORE itself at the ground station, other onboard instruments (BELA, ISA, and SIMBIO-SYS), and the onboard attitude determination and control system. MORE will also contribute to the determination of Mercury's obliquity (i.e., the angle that the spin axis makes to the normal to the orbital plane) and the amplitude of its 88-day physical libration in longitude. These two quantities, together with the coefficients of the second-degree harmonics of the gravity field, will more precisely constrain the outer radius of the planet's molten core.

20.5.5.8 PHEBUS

The Probing of Hermean Exosphere by Ultraviolet Spectroscopy (PHEBUS) experiment (Chassefière et al., 2010) is an ultraviolet spectrometer devoted to the characterization (structure, composition, and dynamics) of Mercury's exosphere and to the understanding of the coupled surface–exosphere–magnetosphere system. The spectral range of PHEBUS spans the major resonance lines of most detected or expected species. One of the key objectives is to produce an average exosphere, i.e., the altitude profile of density for key atmospheric species at different distances from the Sun, and to quantify further north–south and east–west asymmetries (Sprague et al., 1997; Potter et al., 1999). An aim of the experiment is to produce such maps every one-eighth of a Mercury year, that is, on a timescale of 10 Earth days. Mercury's exosphere is expected to vary rapidly in response to solar wind variations, and therefore it is important to provide partial maps of the exosphere on timescales of less than a few hours. The polar orbit of the spacecraft will allow the exosphere to be monitored at all latitudes but only within a narrow longitudinal region along the orbit, with the restriction that only regions of the exosphere illuminated by the Sun may be observed.

The spectral range of PHEBUS is covered by three instruments: an extreme ultraviolet detector (EUV), a far-ultraviolet detector (FUV), and a near-ultraviolet spectrometer (NUV). The EUV detector covers emission lines at 25–155 nm, and the FUV detector does the same at 145–315 nm. The FUV detector is protected by a vacuum cover consisting of a sealed MgF$_2$ window that is transparent above 115 nm, allowing FUV to cover the wavelength range 145–422 nm. The EUV detector must respond to wavelengths shorter than the MgF$_2$ window cutoff, so it has a cover that is opened in flight. The NUV spectrometer monitors the spectrum out to >425 nm to observe exospheric Ca and K emissions at 404 and 422 nm, respectively. The wavelength resolution of all observations will be better than 1 nm. A vertical scanning range covers the altitude range 0–1500 km, with a vertical resolution of about 20 km.

20.5.5.9 SERENA

The Search for Exosphere Refilling and Emitted Neutral Abundances (SERENA) experiment (Orsini et al., 2010) will provide information about the global surface–exosphere–magnetosphere system and its interaction with the solar wind.

The experiment consists of four sensors that can be operated individually: (1) Emitted Low-Energy Neutral Atoms (ELENA) is a 4° × 76° one-dimensional imager (the spacecraft track will provide the second dimension) of energetic neutral particles emitted from the surface of Mercury (with energies from about 50 eV up to 5 keV); (2) Strofio is a mass spectrograph that determines particle mass per charge (mass resolution $m/\Delta m \geq 60$) by a time-of-flight (TOF) technique; (3) Miniature Ion Precipitation Analyser (MIPA) is an ion mass analyzer with a hemispheric field of view for the energy range 10 eV – 15 keV and a time cadence of 22 s per full distribution function to measure ions that precipitate toward the surface; and (4) Planetary Ion Camera (PICAM) is an ion mass spectrometer (mass resolution $m/\Delta m$ about 50) with a field of view of 0.4π sr, in the energy range from the spacecraft potential up to ~3 keV with 32 energy channels.

SERENA will measure in situ both neutral species and ions and, in addition to investigating the planetary response to external forcing, it will complement the MMO for magnetospheric dynamics investigations. The key scientific objectives of SERENA include the identification and localization of source and sink processes of neutral and charged particles as well as estimates of their relative efficiencies. The latter depend on surface composition and external forcing such as solar irradiance, plasma, precipitation, or micrometeoroid impact, and both spatial and temporal variability are expected. Measurement objectives further include the composition and altitude profile of neutral particles and ions in the exosphere for all species, including their energy spectra and spatial distributions. The dynamics of the neutral and ionized exosphere, e.g., circulation from day to night and active to inactive regions, will be investigated as well as atmosphere–magnetosphere exchange and transport processes.

20.5.5.10 SIMBIO-SYS

The Spectrometer and Imagers for MPO BepiColombo-Integrated Observatory SYStem (SIMBIO-SYS) instrument suite is an integrated package for imaging and spectroscopic investigation of the surface of Mercury (Flamini et al., 2010). The science goals of SIMBIO-SYS are to examine the surface geology (stratigraphy, geomorphology), volcanism (lava plain emplacement, volcano identification), global tectonics (structural geology, mechanical properties of the lithosphere), surface age (crater population and morphometry, degradation processes), surface composition (maturity and crustal differentiation, weathering, rock-forming mineral abundance determination), and geophysics (libration measurements, internal planet dynamics) of Mercury. It incorporates capabilities to perform global mapping at medium spatial resolution in stereo and color imaging using two pan-chromatic and four broad-band filters, respectively, as well as high-spatial-resolution imaging with pan-chromatic and three broadband filters and imaging spectroscopy at visible to near-infrared wavelengths.

The instrument suite consists of three units: (a) The Stereo Channel (STC) will provide global color coverage of the surface in full stereo at 60 m/pixel resolution with the aim of defining the main geological units, large-scale tectonic features, impact crater population, and volcanic edifices. The STC design, composed of two "sub-channels" that utilize the same detector and based on a push-frame acquisition mode, yields good stereo performance with general compactness, saving mass, volume, and power resources (Cremonese et al., 2009; Da Deppo et al., 2010). (b) The High spatial Resolution Imaging Channel (HRIC) will characterize special surface targets with high-resolution images at ground pixel sizes of about 6 m/pixel from 480-km altitude in four different bands. (c) The Visible Infrared Hyperspectral Imager Channel (VIHI) is a hyperspectral imager in the visible to near-infrared wavelength range (400–2000 nm) that will map the planet to provide the global mineralogical composition of the surface at a spectral resolution of 6.25 nm and at 500 m/pixel size and will give coverage of selected areas with a resolution as good as 125 m (Capaccioni et al., 2010).

20.5.5.11 SIXS

The Solar Intensity X-ray Spectrometer (SIXS) experiment (Huovelin et al., 2010) will monitor solar X-rays (SIXS-X) and energetic particles (SIXS-P). The X-ray data are required for a fluorescence analysis of MIXS spectra. Because the intensity and energy spectrum of both X-rays and energetic particles emitted by the Sun are highly variable, simultaneous operation of SIXS and MIXS is a strong requirement. Scientific objectives for SIXS-X are to monitor the solar X-ray corona and solar flares and to determine their temporal variability and spectral classification. Therefore, SIXS-X needs a clear view to the Sun as continuously as possible, whatever the spacecraft attitude. The sensor contains three detectors, each having about a 100°-wide field of view and covering a spectral range of 1–20 keV with about 300-eV resolution. SIXS-P will monitor solar energetic electron and proton fluxes and their variations. The key scientific objective is to study the interaction of this radiation with Mercury's exosphere, magnetosphere, and surface.

SIXS is a key supporting instrument to MIXS, but SIXS can also be operated independently of MIXS (whereas the opposite is not true), as its observations will be desirable for other investigations for which the measurements of solar X-rays and energetic particles are important or necessary inputs. These investigations include exospheric studies with SERENA and PHEBUS on MPO and most studies with the MMO payload. In addition, X-ray observations by SIXS of the side of the Sun not visible to instruments near Earth can be useful to space weather studies at Earth.

20.5.6 Mercury Magnetospheric Orbiter

The BepiColombo MMO will be a spin-stabilized spacecraft once it has separated from the MPO following Mercury orbit insertion. The MMO is optimized for in situ measurements of plasma and electromagnetic fields and waves in orbit about Mercury. The nominal spin rate is 15 rpm (or a spin period of 4 s) to meet the scientific requirements. The spin axis is pointed nearly perpendicular to the Mercury orbital plane. The total MMO mass is 255 kg, including 45 kg for the science payload and N_2 gas for attitude control after separation. The MMO main structure consists of two decks (upper and lower), a central cylinder (thrust tube), and four bulkheads. Normal to its short dimension, the spacecraft has an octagonal shape, which can be surrounded by a 1.8-m-diameter circle. The height of the main body is 1.1 m, and the octagonal structure is divided into three parts: upper, middle, and lower sections. The external surface of the upper section is covered in solar cells and optical solar reflectors (OSRs) in a 50:50 ratio, and OSRs are put on the internal surface of the upper section to reduce the cell temperature. The middle section has OSRs on the exterior and MLI on the interior, whereas the lower section is entirely covered with OSRs to reflect the direct solar flux. The instruments are located on the upper and lower decks, which are separated by 37 cm. Most of the scientific instruments (e.g., particle sensors) are mounted on the upper side of the bottom deck, whereas the four deployment units of the electric probe antennas for the Plasma Wave Instrument (PWI) are installed on the lower side of the bottom deck.

During the interplanetary cruise phase, the MMO is shielded and thus cannot produce its own power. Therefore, the BepiColombo MPO provides heater power and energy for regular status checks. After MMO separation, the solar cells experience wide temperature variations because of the large range in Mercury's distance from the Sun. All external surfaces have high electrical conductivity to keep the surface at the same electric potential with respect to the environment, which is essential to measurements of DC electric fields and low-energy electrons.

The MMO is controlled by a combination of passive and active thermal design techniques to maintain the onboard equipment and spacecraft structure within the proper temperature range during all mission phases. The passive control elements are the OSRs, a thermal shield, paints, films, and MLI blankets. The internal surfaces of the upper and lower deck have high-emissivity surfaces (black paint) to equalize internal temperature. The external surface of the upper deck is covered by MLI for insulation from the external thermal environment, whereas the external surface of the lower deck is covered by OSRs to give low absorptivity and high emissivity. The octagonal structure (substrate) is insulated from the upper and lower decks with thermal standoffs. Most of the internal components have a surface with high emissivity (black paint) to equalize the internal temperature. The temperatures of the batteries are controlled independently with the aid of radiators and heaters, which are installed on a battery panel. This panel is attached to the bottom of the central cylinder and insulated from the main structure by MLI and thermal standoffs. The radiator has a surface with high emissivity (OSRs). The antenna despun motor (ADM) and gaseous nitrogen (GN2) tank are covered with MLI for insulation from the external thermal environment. The high-gain antenna (HGA) disks are painted white.

The 80-cm-diameter HGA is used for the high-rate X-band telemetry (TLM) command (CM) and ranging link, with the use of a 20-W power amplifier. The MMO HGA is pointed toward Earth with the ADM and antenna pointing mechanism (APM) to control elevation angle to between −90° and +15°, depending on the positions of Mercury and Earth. A medium-gain antenna (MGA) is accommodated for emergency TLM/CM link. The MGA is installed on the lower surface of the MMO and will be deployed after MMO separation. In orbit about Mercury, the MMO telemetry rate will change as a function of the distance from Earth. The average data rate of the HGA is 16 kbps, which translates into 40 MB/day given a 6-h pass in view of a radio antenna on Earth.

A pair of Sun sensors on one side panel will measure the spin-stabilized MMO spacecraft attitude, and a star scanner is attached to the bottom surface. The attitude is controlled by the propulsion system with a cold gas jet. A nutation damper installed inside the central cylinder is used for passive nutation damping. The MMO propulsion system uses a cold-gas jet system, since only attitude control capability is required (i.e., no orbit control function is needed). The system consists of one propellant tank, six 0.2-N nitrogen-gas jet thrusters, valves, piping, and thermal control equipment (heaters and sensors). The four tangential thrusters for roll control are all on side panels, whereas the two axial thrusters are mounted at the bottom of the spacecraft body. The GN2 tank consists of titanium alloy liner and carbon fiber shell. The tank volume is 14.7 liter, and the maximum designed pressure (MDP) is 27.6 MPa. About 4 kg of GN2 will be loaded at the launch site, including residual propellant of 0.25 kg. The downstream part of the propulsion system consists of a fully redundant system.

20.5.7 MMO Sunshield and Interface Structure

The MOSIF provides the interface structure between the MPO and the MMO and protects the MMO from the full intensity of the Sun until the probe's separation, at which time the spacecraft will have reached its operational orbit. The sunshield is a metal truss structure covered with MLI with appropriate thermal finishes inside and out to ensure suitable temperatures for the MMO. The conical shape of the MOSIF – with an opening angle of about 16° – is needed to allow for the lateral velocity and wobble of the MMO generated during its spin-up at separation.

20.5.8 Instruments on the Mercury Magnetospheric Orbiter

20.5.8.1 MDM

The main objective of the Mercury Dust Monitor (MDM) (Nogami et al., 2010) is to measure the dust environment at Mercury's region of the solar system (0.31–0.47 AU). The impact of micrometeoroids may provide an important source process for the planet's exosphere. At Mercury's orbit, the main dust components are Keplerian dust particles and beta-meteoroid particles. The Keplerian dust particles are interplanetary dust particles (IDPs) that originate from asteroids or comets and gradually decrease their solar-centric distance by the Poynting–Robertson effect, whereas beta-meteoroids are dust particles that are on unbound orbits from the direction of the Sun. The MDM system is composed of a 64-cm^2 piezo-electric lead zirconate titanate (PZT) sensor unit (MDM-S) attached to the outside of the side panel and the electronics unit (MDM-E) installed inside the spacecraft. This instrument can detect impact momentum, crude direction, and the number density of dust particles in the local environment. The viewing direction covers nearly a hemisphere. The PZT is a very simple device that can withstand high temperatures (about 230°C) and does not need any bias voltage or high voltage for operation.

20.5.8.2 MMO-MAG

The primary objective of the MMO Magnetometer (MMO-MAG) (Baumjohann et al., 2010) is to collect magnetic field measurements to study the variability of Mercury's magnetosphere and probe the planetary interior. The MMO-MAG consists of fluxgate magnetometers, an outer sensor (MGO-O) that is a so-called digital type and an inner sensor (MGF-I) that is a traditional analog type. Both have their own electronics boards, and both are mounted on a 4.4-m-long boom. The outer sensor is mounted on the tip of the boom, and the inner sensor is mounted 1.6 m from the boom tip. The instrument is designed to measure magnetic fields with an accuracy of about 10 pT, a dynamic range of ±2048 nT, and a sampling rate of up to 128 Hz. The sampling rate of the data transmitted to Earth is flexible and adapted to study each particular process in the different regions of observation.

20.5.8.3 MPPE

The magnetic field of Mercury, although weak, is sufficiently strong to stand off the solar wind much of the time and to form a magnetosphere. The Mercury Plasma Particle Experiment (MPPE) is designed to investigate the plasma and particle environment around the planet (Saito et al., 2010). MPPE is a comprehensive instrument suite for measurements of plasma, high-energy particles, and energetic neutral atoms. It consists of seven sensors: two Mercury Electron Analyzers (MEA1 and MEA2) (Sauvaud et al., 2010), the Mercury Ion Analyzer (MIA) (Miyake et al., 2009), the Mercury mass Spectrum Analyzer (MSA) (Delcourt et al., 2009), the High Energy Particle instrument for electrons (HEP-ele), the High Energy Particle instrument for ions (HEP-ion), and the Energetic Neutrals Analyzer (ENA). The first six sensors perform in situ observations and cover the range of particle species and energy of interest from the perspective of space plasma physics. The MEA will provide fast electron measurements at Mercury's orbit, MIA will provide precise measurements of both solar wind ions and Mercury magnetospheric ions, MSA will provide plasma composition information with high mass resolution, HEP-ele will measure the energy and angular distributions of electrons in the energy range 30–700 keV, and HEP-ion will provide the energy or velocity distribution of ions in the energy range 30–1500 keV. The ENA will image Mercury's magnetosphere in energetic neutrals created via charge-exchange on the Mercury exosphere and map the plasma precipitation via backscattered neutral particles to investigate the global dynamics of the Mercury magnetosphere and exosphere–magnetosphere interactions.

20.5.8.4 MSASI

Direct exposure of Mercury's rocky surface to the space environment gives the planet distinct characteristics in its atmospheric composition. Its tenuous atmosphere is known to have a substantial sodium component. The Mercury Sodium Atmospheric Spectral Imager (MSASI) (Yoshikawa et al., 2010) is a high-dispersion visible spectrometer working in the spectral range around the wavelength of the sodium D2 emission (589 nm). A Fabry–Pérot etalon is used to achieve a compact design. A one-degree-of-freedom scanning mirror is employed to obtain full-disk images of the planet. A radiation-tolerant complementary metal-oxide semiconductor (CMOS) device with an image intensifier is used as a photon detector. The Fabry–Pérot interferometer comprises two parallel, flat, transparent plates coated with a film of high reflectivity. Its principal advantage is that its throughput is much higher than that of a prism or grating spectrometer. The MSASI will be the first use of such a device for a planetary mission. The combination of Fabry–Pérot etalon and filter accommodates the mass and power limitations on the instrument and provides high sensitivity (16 counts per 2 ms sampling time per bin per 10 kR, achieving a signal-to-noise ratio of 4) and spectral resolution (0.009 nm or better).

20.5.8.5 PWI

Plasma wave and radio wave receivers provide rich information regarding a plasma environment, but no such device flew on either the Mariner 10 or MESSENGER spacecraft. The Plasma Wave Instrument (PWI) on MMO (Kasaba et al., 2010) was designed and developed in a collaboration between Japanese and European scientists. The PWI will provide the first electric field, plasma wave, and radio wave data from the Mercury plasma environment. It will give important information regarding energy exchange processes in the small magnetosphere, where the role of microphysics is particularly important for global dynamics. The PWI consists of three sets of receivers connected to two sets of electric field sensors and two magnetic field sensors. The receivers include an Electric Field Detector (EFD), WaveForm Capture (WFC), and Onboard Frequency Analyzer (OFA), together denoted as EWO for EFD–WFC–OFA; a Spectroscopie des Ondes Radio et du Bruit Electrostatique Thermique (SORBET), which translates to "spectroscopy of radio waves and thermal electrostatic noise;" and the Active Measurement of Mercury's Plasma (AM2P). The electric field sensors are the Mercury Electric Field In-Situ TOol (MEFISTO) and Wire-Probe anTenna (WPT). The magnetic field sensors are the Low-Frequency Search Coil (LF-SC) and Dual-Band Search Coil (DB-SC). The PWI will observe both waveforms and spectra in the frequency range from DC to 10 MHz for the electric field and from 0.1 Hz to 640 kHz for the magnetic field.

20.5.9 Mercury Transfer Module

The Mercury Transfer Module provides the acceleration and braking required during interplanetary cruise to reach eventual capture by Mercury and the large amount of power required by the solar electric propulsion system. The MTM also constitutes the bottom element in the overall spacecraft structure. The MTM is equipped with a bipropellant propulsion system of 10-N thrusters that are used for attitude control activities during cruise. The bipropellant system is also able to provide navigation ΔV maneuvers during cruise. By far the major part of the ΔV required during the cruise trajectory is delivered by the SEP system, using its four 145-mN ion thrusters, which are initially operated singly and later in pairs. The MTM solar arrays use the same high-temperature technologies as developed for the MPO and are rotated away from the Sun for temperature control. At their peak output, the MTM solar array delivers 14 kW, of which 10.6 kW is required by the SEP system.

The MTM structure is based on a carbon-fiber-reinforced plastic (CFRP) conical primary structure interfacing with the launch vehicle adapter and the MPO. The mechanical interfaces to MPO are characterized by cup-cone separation systems for in-flight separation 6 years after launch, and they provide the primary load path through the Mercury composite spacecraft (MCS) structure at launch.

20.6 INTENSIVE INVESTIGATIONS: MERCURY LANDER/ROVER

20.6.1 Overview

To make further substantive steps forward in our scientific understanding of Mercury, the next mission to Mercury after BepiColombo should be a lander, one that functions as long as possible on the planet's surface. A lander would provide ground truth on geochemical remote sensing and perform geophysical and chemical measurements. There are several scientifically compelling landing sites that could be selected on the basis of our current understanding. For example, a lander in a permanently shadowed crater near one of the poles could provide answers to questions about the composition and physical characteristics of the volatiles in Mercury's polar deposits, including the dark material that covers the water ice in most shadowed regions. A landing site near one of the hollows could address the nature of the host material for these features, the volatile material lost during their formation, and their relation to source processes for the exosphere and ions in the magnetosphere. A landing site on volcanic plains or a pyroclastic deposit could ascertain the composition of volcanic material on Mercury. These three types of landing sites are most likely mutually exclusive, but the general questions of planetary heat flow, seismicity, the abundance of graphite as a darkening agent in the crust, and the effects of space weathering could be addressable at any site.

As demonstrated by landers and rovers on Mars, the mobility provided by a rover is a distinct advantage to sample better the vicinity of a landing site uncontaminated by the propulsive plume from the landing system. The trade-off of rover complexity and mass versus lander simplicity and limitations would, of course, be subject to detailed scrutiny in the trade-off against transport capabilities and scientific priorities. Thermal constraints would likely lead to a requirement to land at night, soon after "sundown" before the site had cooled to its lowest temperatures (depending upon the thermal inertia of surface layers at the site). A radioisotope power supply, such as the Multi-Mission Radioisotope Thermoelectric Generator (MMRTG) (Hammel et al., 2013) powering the Curiosity rover on Mars (Grotzinger et al., 2012), would be required. Artificial illumination of the work site would be needed, although some illumination, albeit at a low level, might be provided by emission from sodium atoms in the exosphere or by indirect lighting from illuminated portions of topography in view from the landing site.

Whether long-lived solar-powered landers could be implemented in sufficiently favorable thermal locations near Mercury's poles could be considered as well. Given the large thermal variations even in the polar environment (Paige et al., 2013; Chapter 13), such an approach would require extremely accurate autonomous descent and landing. One could also consider the implementation of a short-lived lander, such as is under study for Europa (Hand et al., 2017). Given the same types of mass constraints that would be encountered for a Europa lander (due to the large amounts of propellant required), a similar lifetime of 20 days or less would be probable for such an implementation.

To place a lander or rover on the innermost planet, we may well be at a similar crossroads as faced by those considering early landers on Mars. The early Voyager Program considered the use of the Saturn V launch vehicle to land large spacecraft on Mars (Cortright, 1968), i.e., exploiting large launch vehicles primarily designed to support an expanded human presence in space. Saturn V costs and miniaturization of robotic systems led to the implementation of Mars landers and rovers that could be delivered with smaller launch systems. The much longer flight times to Mercury, coupled with the planet's depth in the Sun's gravity well, once again pushes the implementation of current state-of-the-art landers and rovers, e.g., those recently and currently operating on Mars, to extremely large launch vehicles, at this time the Space Launch System (SLS) Block 1B (McNutt and Vernon, 2016), now under development to provide human access with the Orion spacecraft to cislunar space, and, potentially, beyond.

20.6.2 Decadal Survey

At the time of the last decadal survey for planetary science (Committee on the Planetary Science Decadal Survey, 2011), MESSENGER was en route to Mercury orbit insertion, although the first flybys of Mercury since those of Mariner 10 had been completed. The orbital phase of MESSENGER's primary mission remained in the future. BepiColombo was also well into development, but its launch was still several years away. Hence, what was known scientifically about Mercury as described in the decadal survey was at a relatively primitive state compared with what is now known. At the time, measurements of Mercury's forced libration via Earth-based radar had demonstrated that Mercury has a liquid outer core, but quantification of the extent of the core had to await the better definition of Mercury's gravitational field, possible only after orbital observations by MESSENGER (Chapters 4 and 19). Of the three main goals for inner planet research in the decadal survey, two are relevant to Mercury, namely "Understand the origin and diversity of terrestrial planets" and "Understand how the evolution of terrestrial planets enables and limits the origin and evolution of life." The third: "Understand the processes that control climate on Earth-like planets" does not apply to Mercury.

The first goal of the decadal survey mentioned above is largely achieved for Mercury by the findings documented in this book, with more to come from BepiColombo. The second goal is also relevant to Mercury, although less so: "The Moon and Mercury are unlikely to harbor life, but they provide critical records of processes and information about the early solar system when life emerged on Earth." The objectives under this goal include (Committee on the Planetary Science Decadal Survey, 2011):

- Understand the composition and distribution of volatile chemical compounds;
- Understand the effects of internal planetary processes on life and habitability; and
- Understand the effects of processes external to a planet on life and habitability.

MESSENGER results that relate to the first objective concern volatile elements and compounds and their distribution and physical properties, especially the northern polar deposits, where water ice has been confirmed and organic materials have been suggested to make up the dark layer that covers and insulates the water ice in most areas of permanent shadow (Chapter 13). Confirmation and detailed mapping of the corresponding features in the southern polar regions, along with an improved understanding of their nature, are expected outcomes from BepiColombo with its closer view of the southern hemisphere.

With respect to the second objective (Committee on the Planetary Science Decadal Survey, 2011):

Despite the dearth of spacecraft missions to explore the inner planets in the past decade, there have been several important discoveries about internal processes. Recent flybys of Mercury by MESSENGER have confirmed the dipole field measured by Mariner 10. Flyby data also confirm that Mercury's plains are volcanic and show that some are far younger than previously had been proposed. Further improvements in our knowledge of Mercury's internal structure and geologic history are expected after MESSENGER enters its mapping orbit in 2011.

Those "further improvements" have indeed materialized during MESSENGER's orbital mission (Chapters 3, 4, 6–13, 18, and 19).

With respect to the third objective, the issues noted in the decadal survey center on volatile influx and escape as well as transport on atmosphere-free bodies, both with (Mercury) and without (the Moon) a global magnetic field. The hollows and the possible presence of organic compounds cold trapped with water ice near the poles of Mercury are both relevant to this objective and were unknown prior to MESSENGER's orbital mission. Both of these MESSENGER discoveries feed into the identified "Future Directions for Investigations and Measurements" and will have a role in discussions for the next planetary decadal survey.

Mercury has been and remains an important part of the study of the inner planets, both in its own right and as it relates to Earth and the origin of life in our solar system. As with all solar system exploration, the advancement of scientific goals at Mercury is a multi-decadal process. As noted in the decadal survey (Committee on the Planetary Science Decadal Survey, 2011):

A series of National Research Council (NRC) reports, culminating in the 2003 planetary science decadal survey, affirm that the exploration of Mercury is central to the scientific understanding of the solar system. The successful achievement of science objectives of the NASA MESSENGER and the European Space Agency-Japan Aerospace Exploration Agency (ESA-JAXA) BepiColombo missions remains a high priority. Given all the advances that will likely come from MESSENGER and BepiColombo, as well as ongoing technology and capability enhancement work, the high priority of Mercury landed science could be revisited at the earliest opportunity in the mid to late years of this decade.

With respect to further exploration, in Box 5.1 "Planetary Roadmaps" the decadal survey notes (Committee on the Planetary Science Decadal Survey, 2011):

For Mercury, the current MESSENGER mission will provide a wealth of new information that could further redefine our understanding of the planet and modify priorities for future missions. The planned European Space Agency (ESA) BepiColombo mission will augment those data and fill important data gaps. Given these missions, the next logical step for the exploration of Mercury would be a landed mission to perform in situ investigations, such as those delineated in the committee's study of a Mercury lander concept (Appendixes D and G). Additional Discovery missions and ground-based observations (e.g., at the Arecibo Observatory in Puerto Rico and the National Radio Astronomy Observatory in Green Bank, West Virginia) will be important in addressing data gaps not filled by current and planned missions. Later Mercury missions would likely include the establishment of a geophysical network and sample return.

The possibility of follow-up NASA Discovery Program missions to Mercury is also noted, as well as landed seismic networks (Table 11.2 of Committee on the Planetary Science Decadal Survey, 2011).

Given the richness and diversity of data and knowledge returned by the MESSENGER mission from Mercury, along with the exploding knowledge of extrasolar planets, many of which orbit their primary star much closer than Mercury orbits our own Sun, it is clear that Mercury is a much higher-priority interest in planetary system studies than had been thought by many at the conclusion of the Mariner 10 flybys. With the upcoming BepiColombo mission, it is also apparent that Mercury science will have a significant role to play in the next planetary decadal survey. In addition to the questions raised by MESSENGER observations of hollows and polar materials noted above, MESSENGER measurements of the magnetic and gravity fields of Mercury at low altitudes near the end of the mission have also posed new intriguing questions for further study (Chapters 3 and 5).

Not all additional, in-depth studies would necessarily require taking the arduous path to Mercury's location. For example, further studies of Mercury's atmosphere and magnetosphere by remote sensing from Earth orbital, Lagrange point, or even lunar observatories that have low solar elongation capability, perhaps carried out in conjunction with the BepiColombo operations at Mercury, could significantly add to our understanding of Mercury.

Figure 20.5. Concept of a Mercury Lander during local night when the surface temperature is low. The dominant illumination comes from indirect lighting from topographical features in sunlight in view from the landing site. Emissions from exospheric Na (yellow D-line) are shown as faintly visible since the brightness of Na emissions at Mercury is comparable to bright aurorae at Earth (Cassidy et al., 2016). Figure credit: Johns Hopkins University Applied Physics Laboratory.

20.6.3 An Implementation Example

A Mercury Lander was studied for technical implementation (Hauck et al., 2010) during the course of the last planetary decadal survey (Committee on the Planetary Science Decadal Survey, 2011). This mission concept was one of 12 studied but not selected by the committee for analysis with the Aerospace Corporation's cost and technical evaluation methodology, and a notional lander is shown in Figure 20.5. The science objectives for a lander on Mercury, as defined prior to MESSENGER's orbital mission, included (Hauck et al., 2010):

- Characterize major and minor elements of the chemical composition of Mercury's surface.
- Characterize the mineralogy and structural state of the materials at Mercury's surface.
- Investigate the magnitude and time dependence of Mercury's magnetic field, for at least one location on the surface.
- Characterize geologic activity (e.g., volcanism, tectonism, impact cratering) at scales ranging from regional to local.
- Determine the rotational state of Mercury.

Given the results of MESSENGER, and those anticipated from BepiColombo, the most significant modification would be landing-site selection – possibly concentrating on cold-trapped material in a permanently shadowed crater, as illustrated in Figure 20.5; an example of hollows; or a volcanic deposit well characterized by orbital remote sensing.

A potential payload was posed but not analyzed; the focus was placed on "getting there" with a simple lander. Approaches with both ballistic trajectories and chemical propulsion – like MESSENGER – and solar electric propulsion – similar to BepiColombo – were considered, but the former (ballistic) approach was estimated to be lower in cost. In any case, a ~5-year interplanetary cruise time with multiple planetary gravity assists was envisioned. With a reduced 21-kg payload mass and a landed dry mass of 289 kg, including a 30% reserve margin, the launch mass was 4630 kg (including margins) on an Atlas V 551 launch vehicle.

We can compare these characteristics with those of the Curiosity rover currently operating on Mars: 75 kg of instrumentation and a landed mass of 899.2 kg powered with ~110 W of electricity from an MMRTG (Grotzinger et al., 2012). The launch mass was 3893 kg, including a 539-kg fueled stage for cruise and a 2401-kg entry, descent, and landing (EDL) system (aeroshell plus fueled descent stage). So for a Mercury lander about one-third the mass of the Curiosity rover, the launch mass is about 800 kg higher, but this figure is still with a ~5-year interplanetary cruise duration. By using an SLS Block 1B with upper (solid) stages, a direct landing with such a robotic probe on Mercury may be possible, but further study is required.

20.7 MERCURY SAMPLE RETURN?

While not an "end game" in planetary exploration, planetary sample returns continue to have the potential for major advances in knowledge of solar system bodies (COMPLEX, 1978). The only real change in viewpoint over the last 40 years regarding such missions has been how technically challenging – and expensive – they actually are. The main challenges of returning a sample from the surface of Mercury are those that have challenged all missions to the innermost planet of the solar system: the location of Mercury deep in the gravity well of the Sun and the associated extreme thermal and radiation environment.

As with architectures for sample return missions to Mars, the most tenable robotic approach will be transport of the means of sample collection to the surface of the planet, and, either on that or a subsequent mission, transport to the surface of an ascent vehicle capable of delivering the sample in its sealed transport canister to planetary orbit to await orbital rendezvous, transfer to a return vehicle, and subsequent return of the sample to Earth. In the same way that emplacement of a lander or rover onto the surface of Mercury in a relatively short time will likely require high-capability transport, e.g., an SLS-class chemical rocket, transport of a Mercury Ascent Vehicle (MeAV) to the surface and the subsequent orbital pickup and transport to Earth will likely require two more robotic missions, both of which will also rely on large chemical rockets [compare with the current, notional Mars Sample Return (MSR) architecture (Committee on the Planetary Science Decadal Survey, 2011)].

Once a robotic MSR has been accomplished successfully (and ahead of a human mission to Mars' surface), some fortuitous similarities between the sizes and surface gravitational acceleration of Mars and Mercury suggest that some commonalities in required flight hardware could be exploited.

Operating near Mercury provides thermal challenges not present for a Mars mission, but a simple thermal-shield approach, such as implemented on the Mariner 10, MESSENGER, and BepiColombo missions, can deal with heating by the Sun. In addition, for the surface part of the system, the problems of landing and operating on the nightside of the planet or a permanently shadowed region will need to have already been solved for a precursor lander or rover prior to sample return. Operations at the far higher surface temperatures on the sunlit side of the planet would be significantly more challenging technically and are not a pressing science requirement (at least given our current state of knowledge of Mercury).

Following Earth and Venus, Mars and Mercury have the two highest surface escape velocities of any of the solid bodies in the solar system, 5.02 km/s and 4.25 km/s, respectively. Although Mercury is smaller than Mars (0.382 versus 0.532 Earth radii, respectively), and less massive (0.055 versus 0.107 Earth masses, respectively), the higher density of Mercury yields a surface gravitational acceleration on that planet nearly equal to that on Mars, 0.38 that at the surface of Earth. Hence, the design of a Mars Ascent Vehicle (MAV) for a sample return will require about the same acceleration, and thus thrust, for a similar-mass MeAV, although the MAV's resultant parking orbit will reside deeper in Mars' gravity well than will the MeAV in Mercury's. The nominally desired 520-km altitude orbit at Mars (Ross et al., 2012) can be translated into an approximate circular orbit at Mercury on the basis of energy considerations.

For any planet, the change in energy per mass in going from the planetary surface (radius R) to a circular orbit of altitude h can be estimated as

$$\frac{\Delta E}{m} = -\frac{1}{2}\frac{GM}{(R+h)} + \frac{GM}{R} = \frac{GM}{R}\left(1 - \frac{1}{2}\frac{1}{(1+h/R)}\right)$$
$$= \frac{1}{2}v_{esc}^2\left(1 - \frac{1}{2}\frac{1}{(1+h/R)}\right),$$

where m and M are the mass of the spacecraft and planet, respectively, G is the gravitational constant, and v_{esc} is the escape velocity. If we assume for simplicity that the MAV and MeAV vehicles have the same mass (there are trade-offs, of course; the MeAV must deal with a more challenging thermal environment, and the MAV must deal with the atmosphere of Mars), we can equate the energy per mass at each planet that the two vehicles can provide and solve for a corresponding circular orbit altitude at Mercury. For h_{Mars} = 520 km, R_{Mars} = 3394 km, and $R_{Mercury}$ = 2439 km, with the ratio of the square of the escape speeds $v_{esc,Mercury}^2/v_{esc,Mars}^2$ = 0.717, we obtain $h_{Mercury}$ = 3380 km.

That is, by using a suitably modified MAV flight system, a Mercury surface sample of the same size could be put into a far higher orbit at Mercury to await pickup. The thermal design would have to be developed, but the relatively high orbit would help to minimize the thermal input from the planet itself, leaving a sunward-pointed thermal shield as the major hardware requirement. The stability of a high circular orbit would need to be assessed, as this would drive the duration of the period over which the sample could be parked in Mercury orbit before retrieval by the Earth-return system.

Even more than for a Mercury lander or rover, the use of high-performance launch vehicles for a Mercury Sample Return mission is compelling. Although low-thrust, in-space propulsion systems, notably solar sails and SEP, have been discussed for some time for use on Mercury orbital missions (e.g., Friedlander and Feingold, 1977), only SEP has been demonstrated to date, e.g., on the Deep Space 1 (Rayman and Lehman, 1997; Boice et al., 2000; Rayman and Varghese, 2001) and Dawn (Russell et al., 2004) missions. Solar sails have also been advocated specifically as enabling for Mercury Sample Return (Hughes and McInnes, 2002), but these remain at a very low technology readiness level (TRL) and have been flagged as a low developmental priority for NASA (Steering Committee for NASA Technology Roadmaps, 2012).

BepiColombo is an SEP mission but still retains the need for multiple planetary gravity assists, leading to a lengthy transit time to Mercury. Whether future SEP systems can be made more mass efficient to enable faster transit times with very large, SLS-class launch vehicles is an open question. Until that technical and cost issue is settled, the best approach to a sample return mission to Mercury will remain open.

A Mercury sample return, built on large launch systems and a successful Mars sample return architecture, is not without challenges but appears technically feasible. Unlike the case at Mars, there is no current advocacy for a human mission to Mercury, and a robotic approach is required. Such a mission, if targeted to a landing site on one of Mercury's polar deposits, could provide definitive identification of all aspects of the dark surficial materials that insulate water ice in many of Mercury's permanently shadowed craters. Such identification could well lead to a paradigm shift in our understanding of the transport of organic materials from the outer solar system, the rate and timing of that activity, and its potential connection to the origin(s) of life on Earth.

20.8 A POSTSCRIPT

Exploration is never over, nor should it be. New scientific results lead to new insights and new questions. How the current era of solar system exploration will play out is a work in progress and will remain so for many decades to come. In any event, it will only be with a lander on Mercury – and perhaps a returned sample – that some of our current questions about the innermost planet can be definitively answered. The curious 3:2 spin–orbit resonance of Mercury (Goldreich and Peale, 1966; Correia and Laskar, 2004) will not, however, be one of those questions. The origin of this resonance is a different class of problem – and one relating to the origin of the solar system itself – which may be "solved" only with new insight gained from extrasolar planet research (Brown et al., 2014). Additional sample returns following a first will remain even more problematic than for Mars because of the technical difficulties. This challenge will place even more pressure on sampling the "right" spot the first time (COMPLEX, 1978), which has been a topic of extended discussion for Mars (e.g., COMPLEX, 1996), a situation further complicated at Mars with the possibility of planetary back-contamination (Nealson et al., 1997). From the current vantage point, it is clear that human missions to Mercury would be so technically challenging that even their serious consideration would be warranted only by some truly unforeseen development in the future. Hence, robotic sampling of Mercury's surface – with or without returned samples – will most likely be the extent of the scientific exploration of Mercury as far as one can foresee. That said, there is still much to do.

REFERENCES

Anderson, B. J., Acuña, M. H., Lohr, D. A., Scheifele, J., Raval, A., Korth, H. and Slavin, J. A. (2007). The Magnetometer instrument on MESSENGER. *Space Sci. Rev.*, **131**, 417–450.

Anderson, D. J., Colombo, G., Esposito, P. B., Lau, E. L. and Trager, G. B. (1987). The mass, gravity field, and ephemeris of Mercury, *Icarus*, **71**, 337–349, doi:10.1016/0019-1035(87)90033-9.

Andrews, G. B., Zurbuchen, T. H., Mauk, B. H., Malcom, H., Fisk, L. A., Gloeckler, G., Ho, G. C., Kelley, J. S., Koehn, P. L., Lefevere, T. W., Livi, S. S., Lundgren, R. A. and Raines, J. M. (2007). The Energetic Particle and Plasma Spectrometer instrument on the MESSENGER spacecraft. *Space Sci. Rev.*, **131**, 523–556.

Balogh, A., Ksanfomality, L. and von Steiger, R. (2007a). Introduction. *Space Sci. Rev.*, **132**, 183–187, doi:10.1007/s11214-007-9293-0.

Balogh, A., Grard, R., Solomon, S. C., Schulz, R., Langevin, Y., Kasaba, Y. and Fujimoto, M. (2007b). Missions to Mercury. *Space Sci. Rev.*, **132**, 611–645, doi:10.1007/s11214-007-9212-4.

Baumjohann, W., Matsuoka, A., Magnes, W., Glassmeier, K.-H., Nakamura, R., Biernat, H., Delva, M., Schwingenschuh, K., Zhang, T., Auster, H.-U., Fornacon, K.-H., Richter, I., Balogh, A., Cargill, P., Carr, C., Dougherty, M., Horbury, T. S., Lucek, E. A., Tohyama, F., Takahashi, T., Tanaka, M., Nagai, T., Tsunakawa, H., Matsushima, M., Kawano, H., Yoshikawa, A., Shibuya, H., Nakagawa, T., Hoshino, M., Tanaka, Y., Kataoka, R., Anderson, B. J., Russell, C. T., Motschmann, U. and Shinohara, M. (2010). Magnetic field investigation of Mercury's magnetosphere and the inner heliosphere by MMO/MGF. *Planet. Space Sci.*, **58**, 279–286, doi:10.1016/j.pss.2008.05.019.

Bedini, P. D., Solomon, S. C., Finnegan, E. J., Calloway, A. B., Ensor, S. L., McNutt, R. L., Jr., Anderson, B. J. and Prockter, L. M. (2012). MESSENGER at Mercury: A mid-term report. *Acta Astronautica*, **81**, 369–379, doi:10.1016/j.actaastro.2012.07.011.

Beerer, J. (1970). Historical account of return trajectory. Interoffice Memorandum, Beerer to Gordon, 16 July 1970. Pasadena, CA: Jet Propulsion Laboratory.

Belcher, J. W., Slavin, J. A., Armstrong, T. P., Farquhar, R. W., Akasofu, S. I., Baker, D. N., Cattell, C. A., Cheng, A. F., Chupp, E. L., Clark, P. E., Davies, M. E., Hones, E. W., Kurth, W. S., Maezawa, J. K., Mariani, F., Marsch, E., Parks, G. K., Shelley, E. G., Siscoe, G. L., Smith, E. J., Strom, R., Trombka, J. I., Williams, D. J. and Yen, C. (1991). *Mercury Orbiter: Report of the Science Working Team*. NASA Technical Memorandum 4255. Greenbelt, MD: NASA Goddard Space Flight Center, 134 pp.

Boice, D. C., Soderblom, L. A., Britt, D. T., Brown, R. H., Sandel, B. R., Yelle, R. V., Buratti, B. J., Hicks, M. D., Nelson, R. M., Rayman, M. D., Oberst, J. and Thomas, N. (2000). The Deep Space 1 encounter with comet 19p/Borrelly. *Earth Moon Planets*, **89**, 301–324.

Brown, S. P., Mead, A. J., Forgan, D. H., Raven, J. A. and Cockell, C. S. (2014). Photosynthetic potential of planets in 3:2 spin–orbit resonances. *Int. J. Astrobiology*, **13**, 279–289, doi:10.1017/S1473550414000068.

Capaccioni, F., Sanctis, M. C. D., Filacchione, G., Piccioni, G., Ammannito, E., Tommasi, L., Veltroni, I. F., Cosi, M., Debei, S., Calamai, L. and Flamini, E. (2010). VIS-NIR imaging spectroscopy of Mercury's surface: SIMBIO-SYS/VIHI experiment onboard the BepiColombo mission. *IEEE Trans. Geosci. Remote Sensing*, **48**, 3932–3940, doi:10.1109/TGRS.2010.2051676.

Cassidy, T. A., McClintock, W. E., Killen, R. M., Sarantos, M., Merkel, A. W., Vervack, R. J., Jr. and Burger, M. H. (2016). A cold-pole enhancement in Mercury's sodium exosphere. *Geophys. Res. Lett.*, **43**, 11,121–11,128, doi:10.1002/2016GL071071.

Cavanaugh, J. F., Smith, J. C., Sun, X., Bartels, A. E., Ramos-Izquierdo, L., Krebs, D. J., McGarry, J. F., Trunzo, R., Novo-Gradac, A. M., Britt, J. L., Karsh, J., Katz, R. B., Lukemire, A. T., Szymkiewicz, R., Berry, D. L., Swinski, J. P., Neumann, G. A., Zuber, M. T. and Smith, D. E. (2007). The Mercury Laser Altimeter instrument for the MESSENGER mission. *Space Sci. Rev.*, **131**, 451–479.

Chassefière, E., Maria, J. L., Goutail, J. P., Quémerais, E., Leblanc, F., Okano, S., Yoshikawa, I., Korablev, O., Gnedykh, V., Naletto, G., Nicolosi, P., Pelizzo, M. G., Correia, J. J., Gallet, S., Hourtoule, C., Mine, P. O., Montaron, C., Rouanet, N., Rigal, J. B., Muramaki, G., Yoshioka, K., Kozlov, O., Kottsov, V., Moisseev, P., Semena, N., Bertaux, J. L., Capria, M. T., Clarke, J., Cremonese, G., Delcourt, D., Doressoundiram, A., Erard, S., Gladstone, R., Grande, M., Hunten, D., Ip, W., Izmodenov, V., Jambon, A., Johnson, R., Kallio, E., Killen, R., Lallement, R., Luhmann, J., Mendillo, M., Milillo, A., Palme, H., Potter, A., Sasaki, S., Slater, D., Sprague, A., Stern, A. and Yan, N. (2010). PHEBUS: A double ultraviolet spectrometer to observe Mercury's exosphere. *Planet. Space Sci.*, **58**, 201–223, doi:10.1016/j.pss.2008.05.018.

Choo, T. H., Murchie, S. L., Bedini, P. D., Steele, R. J., Skura, J. P., Nguyen, L., Nair, H., Lucks, M., Berman, A. F., McGovern, J. A. and Turner, F. S. (2014). SciBox, an end-to-end automated science planning and commanding system. *Acta Astronautica*, **93**, 490–496.

Colombo, G. (1965). Rotational period of the planet Mercury. *Nature*, **208**, 575, doi:10.1038/208575a0.

Colombo, G. and Shapiro, I. I. (1966). The rotation of the planet Mercury. *Astrophys. J.*, **145**, 295–307, doi:10.1086/148762.

Committee on the Planetary Science Decadal Survey (2011). *Vision and Voyages for Planetary Science in the Decade 2013–2022*. Washington, DC: National Academies Press, 382 pp.

COMPLEX (Committee on Planetary and Lunar Exploration) (1978). *Strategy for Exploration of the Inner Planets: 1977–1987*. Washington, DC: National Academies Press, 53 pp.

COMPLEX (Committee on Planetary and Lunar Exploration) (1996). *On NASA Mars Sample-Return Mission Options*, Letter Report. Washington, DC: National Academies Press, 19 pp.

Correia, A. C. M. and Laskar, J. (2004). Mercury's capture into the 3/2 spin-orbit resonance as a result of its chaotic dynamics. *Nature*, **429**, 848–850.

Cortright, E. M. (1968). The Voyager Program. *American Astronautical Society Science and Technology Series*, **16**, 65–92.

Cremonese, G., Fantinel, D., Giro, E., Capria, M. T., Deppo, V. D., Naletto, G., Forlani, G., Massironi, M., Giacomini, L., Sgavetti, M., Simioni, E., Bettanini, C., Debei, S., Zaccariotto, M., Borin, P., Marinangeli, L. and Flamini, E. (2009). The stereo camera on the BepiColombo ESA/JAXA mission: A novel approach. In *Advances in Geosciences*, ed. A. Bhardwaj. Singapore: World Scientific Publishing, pp. 305–322.

Da Deppo, V., Naletto, G., Cremonese, G. and Calamai, L. (2010). Optical design of the single-detector planetary stereo camera for the BepiColombo European Space Agency mission to Mercury. *Appl. Optics*, **49**, 2910–2919, doi:10.1364/AO.49.002910.

Delcourt, D., Saito, Y., Illiano, J. M., Krupp, N., Berthelier, J. J., Fontaine, D., Fraenz, M., Leblanc, F., Fischer, H., Yokota, S.,

Michalik, H., Godefroy, M., Saint-Jacques, E., Techer, J. D., Fiethe, B., Covinhes, J., Gastou, J. and Attia, D. (2009). The mass spectrum analyzer (MSA) onboard BEPI COLOMBO MMO: Scientific objectives and prototype results. *Adv. Space Res.*, **43**, 869–874, doi:10.1016/j.asr.2008.12.002.

Domingue, D. L., Chapman, C. R., Killen, R. M., Zurbuchen, T. H., Gilbert, J. A., Sarantos, M., Benna, M., Slavin, J. A., Schriver, D., Trávníček, P. M., Orlando, T. M., Sprague, A. L., Blewett, D. T., Gillis-Davis, J. J., Feldman, W. C., Lawrence, D. J., Ho, G. C., Ebel, D. S., Nittler, L. R., Vilas, F., Pieters, C. M., Solomon, S. C., Johnson, C. L., Winslow, R. M., Helbert, J., Peplowski, P. N., Weider, S. Z., Mouawad, N., Izenberg N. R. and McClintock, W. E. (2014). Mercury's weather-beaten surface: Understanding Mercury in the context of lunar and asteroidal space weathering studies. *Space Sci. Rev.*, **181**, 121–214.

Ezell, E. C. and Ezell, L. N. (1984). Voyager: Perils of advanced planning, 1960–1967. In *On Mars: Exploration of the Red Planet 1958–1978*, Chapter 4, Special Publication SP-4212. Washington, DC: National Aeronautics and Space Administration, pp. 83–119.

Flamini, E., Capaccioni, F., Colangeli, L., Cremonese, G., Doressoundiram, A., Josset, J. L., Langevin, Y., Debei, S., Capria, M. T., De Sanctis, M. C., Marinangeli, L., Massironi, M., Mazzotta Epifani, E., Naletto, G., Palumbo, P., Eng, P., Roig, J. F., Caporali, A., Da Deppo, V., Erard, S., Federico, C., Forni, O., Sgavetti, M., Filacchione, G., Giacomini, L., Marra, G., Martellato, E., Zusi, M., Cosi, M., Bettanini, C., Calamai, L., Zaccariotto, M., Tommasi, L., Dami, M., Ficai Veltroni, J., Poulet, F. and Hello, Y. (2010). SIMBIO-SYS: The spectrometer and imagers integrated observatory system for the BepiColombo planetary orbiter. *Planet. Space Sci.*, **58**, 125–143, doi:10.1016/j.pss.2009.06.017.

Fox, N. J., Velli, M. C., Bale, S. D., Decker, R., Driesman, A., Howard, R. A., Kasper, J. C., Kinnison, J., Kusterer, M., Lario, D., Lockwood, M. K., McComas, D. J., Raouafi, N. E. and Szabo, A. (2016). The Solar Probe Plus mission: Humanity's first visit to our star. *Space Sci. Rev.*, **204**, 7–48.

Fraser, G. W., Carpenter, J. D., Rothery, D. A., Pearson, J. F., Martindale, A., Huovelin, J., Treis, J., Anand, M., Anttila, M., Ashcroft, M., Benkoff, J., Bland, P., Bowyer, A., Bradley, A., Bridges, J., Brown, C., Bulloch, C., Bunce, E. J., Christensen, U., Evans, M., Fairbend, R., Feasey, M., Giannini, F., Hermann, S., Hesse, M., Hilchenbach, M., Jorden, T., Joy, K., Kaipiainen, M., Kitchingman, I., Lechner, P., Lutz, G., Malkki, A., Muinonen, K., Näränen, J., Portin, P., Prydderch, M., Juan, J. S., Sclater, E., Schyns, E., Stevenson, T. J., Strüder, L., Syrjasuo, M., Talboys, D., Thomas, P., Whitford, C. and Whitehead, S. (2010). The mercury imaging X-ray spectrometer (MIXS) on bepicolombo. *Planet. Space Sci.*, **58**, 79–95, doi:10.1016/j.pss.2009.05.004.

Friedlander, A. L. and Feingold, H. (1977). *Mercury Orbiter Transport Study*. Report SAI 1-120-580-176, NASA-CR-158658. Schaumburg, IL: Science Applications, Incorporated, 128 pp.

Glassmeier, K.-H., Auster, H.-U., Heyner, D., Okrafka, K., Carr, C., Berghofer, G., Anderson, B. J., Balogh, A., Baumjohann, W., Cargill, P., Christensen, U., Delva, M., Dougherty, M., Fornaçon, K.-H., Horbury, T. S., Lucek, E. A., Magnes, W., Mandea, M., Matsuoka, A., Matsushima, M., Motschmann, U., Nakamura, R., Narita, Y., O'Brien, H., Richter, I., Schwingenschuh, K., Shibuya, H., Slavin, J. A., Sotin, C., Stoll, B., Tsunakawa, H., Vennerstrom, S., Vogt, J. and Zhang, T. (2010). The fluxgate magnetometer of the BepiColombo Mercury Planetary Orbiter. *Planet. Space Sci.*, **58**, 287–299, doi:10.1016/j.pss.2008.06.018.

Gold, R. E., Solomon, S. C., McNutt, R. L., Jr., Santo, A. G., Abshire, J. B., Acuña, M. H., Afzal, R. S., Anderson, B. J., Andrews, G. B., Bedini, P. D., Cain, J., Cheng, A. F., Evans, L. G., Feldman, W. C., Follas, R. B., Gloeckler, G., Goldsten, J. O., Hawkins, S. E., III, Izenberg, N. R., Jaskulek, S. E., Ketchum, E. A., Lankton, M. R., Lohr, D. A., Mauk, B. H., McClintock, W. E., Murchie, S. L., Schlemm, C. E., II, Smith, D. E., Starr, R. D. and Zurbuchen, T. H. (2001). The MESSENGER mission to Mercury: Scientific payload. *Planet. Space Sci.*, **49**, 1467–1479.

Goldreich, P. and Peale, S. J. (1966). Spin-orbit coupling in the solar system, *Astron. J.*, **71**, 425–438.

Goldsten, J. O., Rhodes, E. A., Boynton, W. V., Feldman, W. C., Lawrence, D. J., Trombka, J. I., Smith, D. M., Evans, L. G., White, J., Madden, N. W., Berg, P. C., Murphy, G. A., Gurnee, R. S., Strohbehn, K., Williams, B. D., Schaefer, E. D., Monaco, C. A., Cork, C. P., Del Eckels, J., Miller, W. O., Burks, M. T., Hagler, L. B., Deteresa, S. J. and Witte, M. C (2007). The MESSENGER Gamma-Ray and Neutron Spectrometer. *Space Sci. Rev.*, **131**, 339–391.

Grotzinger, J. P., Crisp, J., Vasavada, A. R., Anderson, R. C., Baker, C. J., Barry, R., Blake, D. F., Conrad, P., Edgett, K. S., Ferdowski, B., Gellert, R., Gilbert, J. B., Golombek, M., Gómez-Elvira, J., Hassler, D. M., Jandura, L., Litvak, M., Mahaffy, P., Maki, J., Meyer, M., Malin, M. C., Mitrofanov, I., Simmonds, J. J., Vaniman, D., Welch, R. V. and Wiens, R. C. (2012). Mars Science Laboratory mission and science investigation. *Space Sci. Rev.*, **170**, 5–56, doi:10.1007/s11214-012-9892-2.

Gulkis, S., Stetson, D. S. and Stofan, E. R. (1998). *Mission to the Solar System: Exploration and Discovery – A Mission and Technology Roadmap*. Publication 97–12. Pasadena, CA: Jet Propulsion Laboratory, 65 pp.

Gunderson, K. and Thomas, N. (2010). BELA receiver performance modeling over the BepiColombo mission lifetime. *Planet. Space Sci.*, **58**, 309–318, doi:10.1016/j.pss.2009.08.006.

Hammel, T. E., Bennett, R., Sievers, R. K., Keyser, S., Otting, W. and Gard, L. (2013). Multi-Mission Radioisotope Thermoelectric Generator (MMRTG) performance data and application to life modeling. 11th International Energy Conversion Engineering Conference, American Institute of Aeronautics and Astronautics, 8 pp., doi:10.2514/6.2013-3925.

Hand, K. P., Murray, A. E., Garvin, J. B., Brinckerhoff, W. B., Christner, B. C., Edgett, K. S., Ehlmann, B. L., German, C. R., Hayes, A. G., Hoehler, T. M., Horst, S. M., Lunine, J. I., Nealson, K. H., Paranicas, C., Schmidt, B. E., Smith, D. E., Rhoden, A. R., Russell, M. J., Templeton, A. S., Willis, P. A., Yingst, R. A., Phillips, C. B., Cable, M. L., Craft, K. L., Hofmann, A. E., Nordheim, T. A., Pappalardo, R. P. and the Project Engineering Team (2017). *Report of the Europa Lander Science Definition Team*. Pasadena, CA: Jet Propulsion Laboratory, 264 pp.

Hauck, S. A., II, Eng, D. A., Treiman, A., Tahu, G., Lindstrom, K., Blewett, D., Seifert, H., Stambaugh, K., Chavers, G., Oleson, S., Mcguire, M., Guo, Y., Dankanich, J., Dong, C., Burke, L., Wolfarth, L., Hahn, M., Drexler, J., Holdridge, M., Cockrell, J., Miller, T., Trinh, H., Fittje, J., Verhey, T., Gyekenyesi, J., Ercol, J., Abel, E., Colozza, T., Sequeira, B., Warner, J., Fraeman, M., Williams, G., Schmitz, P., Lowery, E., Landis, G., Hojniki, J., Adams, D., Martini, M., Williams, S. and Drexler, J. (2010). *Mercury Lander Mission Concept Study*. Laurel, MD: The Johns Hopkins University Applied Physics Laboratory, 132 pp.

Hawkins, S. E., III, Boldt, J. D., Darlington, E. H., Espiritu, R., Gold, R. E., Gotwols, B., Grey, M. P., Hash, C. D., Hayes, J. R., Jaskulek, S. E., Kardian, C. J., Keller, M. R., Malaret, E. R., Murchie, S. L., Murphy, P. K., Peacock, K., Prockter, L. M.,

Reiter, R. A., Robinson, M. S., Schaefer, E. D., Shelton, R. G., Sterner, R. E., Taylor, H. W., Watters, T. R. and Williams, B. D. (2007). The Mercury Dual Imaging System on the MESSENGER spacecraft. *Space Sci. Rev.*, **131**, 247–338.

Hiesinger, H., Helbert, J. and MERTIS Co-I Team (2010). The Mercury Radiometer and Thermal Infrared Spectrometer (MERTIS) for the BepiColombo mission. *Planet. Space Sci.*, **58**, 144–165, doi:10.1016/j.pss.2008.09.019.

Hughes, G. and McInnes, C. (2002). Mercury sample return missions using solar sail propulsion. 34th Scientific Assembly of the Committee on Space Research and 53rd International Astronautical Congress of the International Astronautical Federation, World Space Congress 2002, paper IAC-02-W.2.08. Houston, TX, 11 pp.

Huovelin, J., Vainio, R., Andersson, H., Valtonen, E., Alha, L., Mälkki, A. Grande, M., Fraser, G. W., Kato, M., Koskinen, H., Muinonen, K., Näränen, J., Schmidt, W., Syrjäsuo, M., Anttila, M., Vihavainen, T., Kiuru, E., Roos, M., Peltonen, J., Lehti, J., Talvioja, M., Portin, P. and Prydderch, M. (2010). Solar Intensity X-ray and particle Spectrometer (SIXS). *Planet. Space Sci.*, **58**, 96–107, doi:10.1016/j.pss.2008.11.007.

Iafolla, V., Fiorenza, E., Lefevre, C., Morbidini, A., Nozzoli, S., Peron, R., Persichini, M., Reale, A. and Santoli, F. (2010). Italian Spring Accelerometer (ISA): A fundamental support to BepiColombo radio science experiments. *Planet. Space Sci.*, **58**, 300–308, doi:10.1016/j.pss.2009.04.005.

Jehn, R., Campagnola, S., Garcia, D. and Kemble, S. (2004), Low-thrust approach and gravitational capture at Mercury. 18th International Symposium on Space Flight Dynamics. Munich, Germany, 6 pp.

Kasaba, Y., Bougeret, J.-L., Blomberg, L. G., Kojima, H. Yagitani, S., Moncuquet, M., Trotignon, J.-G., Chanteur, G., Kumamoto, A., Kasahara, Y., Lichtenberger, J., Omura, Y., Ishisaka, K. and Matsumoto, H. (2010). The Plasma Wave Investigation (PWI) onboard the BepiColombo/MMO: First measurement of electric fields, electromagnetic waves, and radio waves around Mercury. *Planet. Space Sci.*, **58**, 238–278, doi:10.1016/j.pss.2008.07.017.

Kirk, M. N., Flanigan, S. H., O'Shaughnessy, D. J., Bushman, S. S. and Rosendall, P. E. (2015). MESSENGER maneuver performance during the low-altitude hover camaign. Astrodynamics Specialist Conference, American Astronautical Society, paper AAS 15–652. Vail, CO, 19 pp.

Langevin, Y. (2000). Chemical and solar electric propulsion options for a cornerstone mission to Mercury. *Acta Astronautica*, **47**, 443–452, doi:10.1016/S0094-5765(00)00084-9.

Leary, J. C., Conde, R. F., Dakermanji, G., Engelbrecht, C. S., Ercol, C. J., Fielhauer, K. B., Grant, D. G., Hartka, T. J., Hill, T. A., Jaskulek, S. E., Mirantes, M. A., Mosher, L. E., Paul, M. V., Persons, D. F., Rodberg, E. H., Srinivasan, D. K., Vaughan, R. M. and Wiley, S. R. (2007). The MESSENGER spacecraft. *Space Sci. Rev.*, **131**, 187–217.

Lewis, R. A., Pérez Luna, J., Coombs, N. and Guarducci, F. (2015). Qualification of the T6 thruster for BepiColombo. Joint Conference of the 30th International Symposium on Space Technology and Science, 34th International Electric Propulsion Conference, and 6th Nano-satellite Symposium. Hyogo-Kobe, Japan, 10 pp.

McAdams, J. V., Farquhar, R. W., Taylor, A. H. and Williams, B. G. (2007). MESSENGER mission design and navigation. *Space Sci. Rev.*, **131**, 219–246.

McAdams, J. V., Bryan, C. G., Bushman, S. S., Calloway, A. B., Carranza, E., Flanigan, S. H., Kirk, M. N., Korth, H., Moessner, D. P., O'Shaughnessy, D. J. and Williams, K. E. (2015). Engineering MESSENGER's grand finale at Mercury – The low-altitude hover campaign. Astrodynamics Specialist Conference, American Astronautical Society, paper AAS 15–634. Vail, CO, 20 pp.

McClintock, W. E. and Lankton, M. R. (2007). The Mercury Atmospheric and Surface Composition Spectrometer for the MESSENGER mission. *Space Sci. Rev.*, **131**, 481–521.

McNutt, R. L., Jr. and Vernon, S. R. (2016). Enabling solar system science with the Space Launch System (SLS). 67th International Astronautical Congress, International Astronautical Federation. Guadalajara, Mexico, 7 pp.

McNutt, R. L., Jr., Solomon, S. C., Grard, R., Novara, M. and Mukai, T. (2004). An international program for Mercury exploration: Synergy of MESSENGER and BepiColombo. *Adv. Space Res.*, **33**, 2126–2132, doi:10.1016/s0273-1177(03)00439-3.

McNutt, R. L., Jr., Solomon, S. C., Gold, R. E., Leary, J. C. and the MESSENGER Team (2006). The MESSENGER mission to Mercury: Development history and early mission status, *Adv. Space Res.*, **38**, 564–571, doi:10.1016/j.asr.2005.05.044.

McNutt, R. L., Jr., Solomon, S. C., Grant, D. G., Finnegan, E. J. and Bedini, P. D. (2008). The MESSENGER mission to Mercury: Status after the Venus flybys. *Acta Astronautica*, **63**, 68–73.

McNutt, R. L., Jr., Solomon, S. C., Bedini, P. D., Finnegan, E. J. and Grant, D. G. (2010). The MESSENGER mission: Results from the first two Mercury flybys. *Acta Astronautica*, **67**, 681–687, doi:10.1016/j.actaastro.2010.05.020.

McNutt, R. L., Jr., Solomon, S. C., Nittler, L. R., Bedini, P. D., Finnegan, E. J., Winters, H. L. and Grant, D. G. (2012). The MESSENGER mission continues: Transition to the extended mission. International Astronautical Congress. Naples, Italy, 15 pp.

McNutt, R. L., Jr., Solomon, S. C., Bedini, P. D., Anderson, B. J., Blewett, D. T., Evans, L. G., Gold, R. E., Krimigis, S. M., Murchie, S. L., Nittler, L. R., Phillips, R. J., Prockter, L. M., Slavin, J. A., Zuber, M. T., Finnegan, E. J. and Grant, D. G. (2014). MESSENGER at Mercury: Early orbital operations. *Acta Astronautica*, **93**, 509–515, doi:10.1016/j.actaastro.2012.08.012.

Mitrofanov, I. G., Kozyrev, A. S., Konovalov, A., Litvak, M. L., Malakhov, A. A., Mokrousov, M. I., Sanin, A. B., Tret'yakov, V. I., Vostrukhin, A. V., Bobrovnitskij, Y. I., Tomilina, T. M., Gurvits, L. and Owens, A. (2010). The Mercury Gamma and Neutron Spectrometer (MGNS) on board the Planetary Orbiter of the BepiColombo mission. *Planet. Space Sci.*, **58**, 116–124, doi:10.1016/j.pss.2009.01.005.

Miyake, W., Saito, Y., Harada, M., Saito, M., Hasegawa, H., Ieda, A., Machida, S., Nagai, T., Nagatsuma, T., Seki, K., Shinohara, I. and Terasawa, T. (2009). Mercury Ion Analyzer (MIA) onboard Mercury Magnetospheric Orbiter: MMO. *Adv. Space Res.*, **43**, 1986–1992, doi:10.1016/j.asr.2009.03.011.

Nealson, K. H., Carr, M. H., Clark, B. C., Doolittle, R. F., Jakosky, B. M., Korwek, E. L., Pace, N. R., Poindexter, J. S., Race, M. S., Reysenbach, A.-L., Schopf, J. W. and Stevens, T. O. (1997). *Mars Sample Return: Issues and Recommendations*. Washington, DC: National Academies Press, 57 pp.

Nelson, R. M., Horn, L. J., Weiss, J. R. and Smythe, W. D. (1995). Hermes Global Orbiter: A Discovery mission in gestation. *Acta Astronautica*, **35**, Suppl. 1, 387–395, doi:10.1016/0094-5765(94)00204-5.

Nogami, K., Fujii, M., Ohashi, H., Miyachi, T., Sasaki, S., Hasegawa, S., Yano, H., Shibata, H., Iwai, T., Minami, S., Takechi, S., Grün, E. and Srama, R. (2010). Development of the Mercury dust monitor (MDM) onboard the BepiColombo mission. *Planet. Space Sci.*, **58**, 108–115, doi:10.1016/j.pss.2008.08.016.

Novara, M. (2002). The BepiColombo ESA cornerstone mission to Mercury. *Acta Astronautica*, **51**, 387–395, doi:10.1016/S0094-5765(02)00065-6.

Orsini, S., Livi, S., Torkar, K., Barabash, S., Milillo, A., Wurz, P., Di Lellis, A. M. and Kallio, E. (2010). SERENA: A suite of four instruments (ELENA, STROFIO, PICAM and MIPA) on board BepiColombo-MPO for particle detection in the Hermean environment. *Planet. Space Sci.*, **58**, 166–181, doi:10.1016/j.pss.2008.09.012.

O'Shaughnessy, D. J., McAdams, J. V., Bedini, P. D., Calloway, A. B., Williams, K. E. and Page, B. R. (2014). MESSENGER's use of solar sailing for cost and risk reduction. *Acta Astronautica*, **93**, 483–489.

Paige, D. A., Siegler, M. A., Harmon, J. K., Neumann, G. A., Mazarico, E. M., Smith, D. E., Zuber, M. T., Harju, E., Delitsky, M. L. and Solomon, S. C. (2013). Thermal stability of volatiles in the north polar region of Mercury. *Science*, **339**, 300–303.

Philpott, L. C., Johnson, C. L., Winslow, R. M., Anderson, B. J., Korth, H., Purucker, M. E. and Solomon, S. C. (2014), Constraints on the secular variation of Mercury's magnetic field from the combined analysis of MESSENGER and Mariner 10 data. *Geophys. Res. Lett.*, **41**, 6627–6634, doi:10.1002/2014GL061401.

Potter, A. E., Killen, R. M. and Morgan, T. H. (1999). Rapid changes in the sodium exosphere of Mercury. *Planet. Space Sci.*, **47**, 1441–1448, doi:10.1016/S0032-0633(99)00070-7.

Rayman, M. D. and Lehman, D. H. (1997). Deep Space One: NASA's first deep-space technology validation mission. *Acta Astronautica*, **41**, 289–299.

Rayman, M. D. and Varghese, P. (2001). The Deep Space 1 extended mission. *Acta Astronautica*, **48**, 693–705.

Ross, D., Russell, J. and Sutter, B. (2012). Mars Ascent Vehicle (MAV): Designing for high heritage and low risk. 2012 IEEE Aerospace Conference. Big Sky, MT, 6 pp., doi:10.1109/AERO.2012.6187296.

Russell, C. T., Coradini, A., Christensen, U., De Sanctis, M. C., Feldman, W. C., Jaumann, R., Keller, H. U., Konopliv, A. S., McCord, T. B., McFadden, L. A., McSween, H. Y., Mottola, S., Neukum, G., Pieters, C. M., Prettyman, T. H., Raymond, C. A., Smith, D. E., Sykes, M. V., Williams, B. G., Wise, J. and Zuber, M. T. (2004). Dawn: A journey in space and time. *Planet. Space Sci.*, **52**, 465–489, doi:10.1016/j.pss.2003.06.013.

Saito, Y., Sauvaud, J. A., Hirahara, M., Barabash, S., Delcourt, D., Takashima, T. and Asamura, K. (2010). Scientific objectives and instrumentation of Mercury Plasma Particle Experiment (MPPE) onboard MMO. *Planet. Space Sci.*, **58**, 182–200, doi:10.1016/j.pss.2008.06.003.

Santo, A. G., Gold, R. E., McNutt, R. L., Jr., Solomon, S. C., Ercol, C. J., Farquhar, R. W., Hartka, T. J., Jenkins, J. E., McAdams, J. V., Mosher, L. E., Persons, D. F., Artis, D. A., Bokulic, R. S., Conde, R. F., Dakermanji, G., Goss, M. E., Jr., Haley, D. R., Heeres, K. J., Maurer, R. H., Moore, R. C., Rodberg, E. H., Stern, T. G., Wiley, S. R., Williams, B. G., Yen, C. L. and Peterson, M. R. (2001). The MESSENGER mission to Mercury: Spacecraft and mission design. *Planet. Space Sci.*, **49**, 1481–1500.

Sauvaud, J. A., Fedorov, A., Aoustin, C., Seran, H. C., Le Comte, E., Petiot, M., Rouzaud, J., Saito, Y., Dandouras, J., Jacquey, C., Louarn, P., Mazelle, C. and Médale, J. L. (2010). The Mercury Electron Analyzers for the Bepi Colombo mission. *Adv. Space Res.*, **46**, 1139–1148, doi:10.1016/j.asr.2010.05.022.

Schlemm, C. E., II, Starr, R. D., Ho, G. C., Bechtold, K. E., Hamilton, S. A., Boldt, J. D., Boynton, W. V., Bradley, W., Fraeman, M. E., Gold, R. E., Goldsten, J. O., Hayes, J. R., Jaskulek, S. E., Rossano, E., Rumpf, R. A., Schaefer, E. D., Strohbehn, K., Shelton, R. G., Thompson, R. E., Trombka, J. I. and Williams, B. D. (2007). The X-Ray Spectrometer on the MESSENGER spacecraft. *Space Sci. Rev.*, **131**, 393–415.

Solomon, S. C., McNutt, R. L., Jr., Gold, R. E., Acuña, M. H., Baker, D. N., Boynton, W. V., Chapman, C. R., Cheng, A. F., Gloeckler, G., Head, J. W., Krimigis, S. M., McClintock, W. E., Murchie, S. L., Peale, S. J., Phillips, R. J., Robinson, M. S., Slavin, J. A., Smith, D. E., Strom, R. G., Trombka, J. I. and Zuber, M. T. (2001). The MESSENGER mission to Mercury: Scientific objectives and implementation. *Planet. Space Sci.*, **49**, 1445–1465.

Solomon, S. C., McNutt, R. L., Jr., Gold, R. E. and Domingue, D. L. (2007). MESSENGER mission overview. *Space Sci. Rev.*, **131**, 3–39, doi:10.1007/s11214-007-9247-6.

Space Science Board (1962). *A Review of Space Research: The Report of the Summer Study Conducted under the Auspices of the National Academy of Sciences at the State University of Iowa, Iowa City, Iowa, June 17-August 10, 1962*. Washington, DC: National Academies Press, 604 pp.

Space Science Board (1966). *Space Research: Directions for the Future*. Washington, DC: National Academies Press, 653 pp.

Space Science Board (1968). *Planetary Exploration: 1968–1975*. Washington, DC: National Academies Press, 30 pp.

Sprague, A. L., Kozlowski, R. W. H., Hunten, D. M., Schneider, N. M., Domingue, D. L., Wells, W. K., Schmitt, W. and Fink, U. (1997). Distribution and abundance of sodium in Mercury's atmosphere, 1985–1988. *Icarus*, **129**, 506–527, doi:10.1006/icar.1997.5784.

Srinivasan, D. K., Perry, M. E., Fielhauer, K. B., Smith, D. E. and Zuber, M. T. (2007). The radio frequency subsystem and radio science on the MESSENGER mission. *Space Sci. Rev.*, **131**, 557–571.

Starr, R. D., Schriver, D., Nittler, L. R., Weider, S. Z., Byrne, P. K., Ho, G. C., Rhodes, E. A., Schlemm, C. E., II, Solomon, S. C. and Trávníček, P. M. (2012). MESSENGER detection of electron-induced X-ray fluorescence from Mercury's surface. *J. Geophys. Res.*, **117**, E00L02, doi:10.1029/2012je004118.

Steering Committee for NASA Technology Roadmaps (2012). *NASA Space Technology Roadmaps and Priorities – Restoring NASA's Technological Edge and Paving the Way for a New Era in Space*. Washington, DC: National Academies Press, 122 pp., doi:10.17226/13354.

Stern, S. A. and Vilas, F. (1988). Future observations of and missions to Mercury. In *Mercury*, ed. F. Vilas, C. R. Chapman and M. S. Matthews. Tucson, AZ: University of Arizona Press, pp. 24–36.

Toksöz, M. N., Malin, M. C., Albee, A. L., Brandt, J. C., Briggs, G. A., Chapman, C. R., Coroniti, F. V., Duke, M. B., Fanale, F. P., Flinn, E. A., Haskin, L. A., Hayes, J. M., Johnson, T. V., Kaula, W. M., Masursky, H., McCord, T. B., Prinn, R., Schopf, J. W., Sonett, C. P., Stewart, A. I., Trombka, J. I., Wood, J. A. and Young, R. E. (1977). *Report of the Terrestrial Bodies Science Working Group*. Report NASA-CR-155189, Publication 77–51. Pasadena, CA: Jet Propulsion Laboratory, 33 pp.

Ward, W. R., Colombo, G. and Franklin, F. A. (1976). Secular resonance, solar spin down, and the orbit of Mercury. *Icarus*, **28**, 441–452, doi:10.1016/0019-1035(76)90117-2.

Yen, C.-W. (1985). Ballistic Mercury orbiter mission via Venus and Mercury gravity assists. Astrodynamics Specialist Conference, paper AAS-85-346. Vail, CO, 15 pp.

Yen, C.-W. (1989). Ballistic Mercury orbiter mission via Venus and Mercury gravity assists. *J. Astron. Sci.*, **37**, 417–432.

Yoshikawa, I., Korablev, O., Kameda, S., Rees, D., Nozawa, H., Okano, S., Gnedykh, V., Kottsov, V., Yoshioka, K., Murakami, G., Ezawa, F. and Cremonese, G. (2010). The Mercury sodium atmospheric spectral imager for the MMO spacecraft of Bepi-Colombo. *Planet. Space Sci.*, **58**, 224–237, doi:10.1016/j.pss.2008.07.008.

INDEX

253 Mathilde, 196
2P/Encke, 392
4 Vesta, 195, 196, 350
433 Eros, 195, 196, 339

activation energy, 409, 412
adiabat, 38
adiabatic decompression melting, 38, 60, 168, 186
adiabatic gradient, 96
admittance, 64, 65, 74, 271
aerodynamic fractionation, 507, 509
Airy isostasy, 64
Al. *See* aluminum
Al exosphere. *See* aluminum exosphere
albedo, 192, 198
   compared with other bodies, 196
Alfvén Mach number, 430, 433, 442, 463
aluminum, 36, 38, 147, 177, 178–184, 185, 186, 209, 210
aluminum exosphere, 371, 399–400, 403, 423–424
   ground-based observations, 423
andesite, 179, 182, 183
Andrade creep function, 100
Andrade rheological model, 100
anorthosite, 30, 210
anticline, 70, 251
Apollo program, 544
apparent depth of compensation, 74
Arecibo Observatory, 346, 347, 348
Ariane 5, 552, 553, 555
asteroid impacts, 217, 225, 232, 347, 365
asteroids, 30, 40, 191, 195, 225, 233, 235, 365, 506–507, 509
   E-belt, 239
   main belt, 233, 234, 236
   near-Earth, 233, 237, 238, 239
   size–frequency distribution, 233, 239
Atlas Centaur, 546
aubrites, 40, 498, 523
average radius, 88

basal décollement, 255
basalt, 36, 60, 61, 145, 206, 210, 261, 262, 263, 413
basin tectonics, 268–270
BCFDs. *See* bright crater-floor deposits
BDT. *See* brittle–ductile transition
bencubbinites, 39, 43, 184, 185, 499, 501, 506
bending moment, 65
bending stress, 64

BepiColombo, 46, 109, 134, 136, 138, 279, 314, 315, 366, 403, 463, 487, 488, 535, 544, 546, 547, 548–562, 563, 564, 565
BELA. *See* BepiColombo: BepiColombo Laser Altimeter
BepiColombo Laser Altimeter, 554, 557, 558
   gravity assists, 555
   gyroscope, 556
HGA. *See* BepiColombo: high-gain antenna
high-gain antenna, 556, 560
ISA. *See* BepiColombo: Italian Spring Accelerometer
Italian Spring Accelerometer, 549, 554, 557, 558
Magnetospheric Orbiter Sunshield and Interface, 552, 553, 555, 560
MDM. *See* BepiColombo: Mercury Dust Monitor
Mercury Dust Monitor, 554, 560–561
Mercury flybys, 555
Mercury Gamma-ray and Neutron Spectrometer, 554, 558
Mercury Imaging X-ray Spectrometer, 558
Mercury Magnetospheric Orbiter, 552, 553, 554, 555, 556, 557, 559–561
Mercury Orbiter Radio Science Experiment, 554, 556–558
Mercury Planetary Orbiter, 366, 549, 550, 551, 552, 553, 554, 555, 556–559, 560, 562
Mercury Plasma Particle Experiment, 554, 561
Mercury Sodium Atmospheric Spectral Imager, 554, 561
Mercury Thermal Infrared Spectrometer, 366, 554, 557–558
Mercury Transfer Module, 552, 553, 555, 561–562
MERTIS. *See* BepiColombo: Mercury Thermal Infrared Spectrometer
MGNS. *See* BepiColombo: Mercury Gamma-ray and Neutron Spectrometer
MMO. *See* BepiColombo: Mercury Magnetospheric Orbiter
MMO Magnetometer, 554, 561
MMO-MAG. *See* BepiColombo: MMO Magnetometer
MORE. *See* BepiColombo: Mercury Orbiter Radio Science Experiment
MOSIF. *See* BepiColombo: Magnetospheric Orbiter Sunshield and Interface
MPO. *See* BepiColombo: Mercury Planetary Orbiter
MPO Magnetometer, 554, 557
MPO periapsis, 549
MPO-MAG. *See* BepiColombo: MMO Magnetometer
MPPE. *See* BepiColombo: Mercury Plasma Particle Experiment
MSASI. *See* BepiColombo: Mercury Sodium Atmospheric Spectral Imager
MTM. *See* BepiColombo: Mercury Transfer Module
MTM thrusters, 561
perihelion, 555
PHEBUS. *See* BepiColombo: Probing of Hermean Exosphere by Ultraviolet Spectroscopy
Plasma Wave Instrument, 556, 559, 561

Probing of Hermean Exosphere by Ultraviolet Spectroscopy, 554, 558, 559
propellant, 556, 560
PWI. *See* BepiColombo: Plasma Wave Instrument
reaction wheels, 556
Search for Exosphere Refilling and Emitted Neutral Abundances, 554, 558–559
SERENA. *See* BepiColombo: Search for Exosphere Refilling and Emitted Neutral Abundances
SIMBIO-SYS. *See* BepiColombo: Spectrometers and Imagers for MPO BepiColombo-Integrated Observatory SYStem
SIXS. *See* BepiColombo: Solar Intensity X-ray Spectrometer
solar arrays, 556, 562
Solar Intensity X-ray Spectrometer, 554, 559
Spectrometers and Imagers for MPO BepiColombo-Integrated Observatory SYStem, 554, 558, 559
star trackers, 556
thrusters, 556, 560
Venus flybys, 555
biannual average temperature, 355
biannual maximum surface temperature, 355
bipropellant, 561
Birch–Murnaghan equation of state, 96
bombardment history, 238, 241
boninite, 36, 145, 180, 181, 182, 183, 206, 307, 522
Bouguer correction, 61
Bouguer disturbance, 61
boundary-normal coordinates, 467
bow shock, 430, 431, 432–434, 441–442, 443, 455, 463, 466, 470, 474, 481, 487
  quasi-parallel, 481, 482
  standoff distance, 442
bright crater-floor deposits, 324–326
brittle deformation, 261–263
brittle–ductile transition, 62, 263
broad rises, 17
bulk composition, 43–45, 497, 499, 517
bulk density, 88, 92, 95
Byerlee's law, 262

C. *See* carbon
Ca. *See* calcium
Ca exosphere. *See* calcium exosphere
calcium, 33, 36, 38, 147, 177, 178–184, 185, 186, 206, 209, 210, 336–337, 339, 498
calcium exosphere, 371, 390–392, 402, 419–421
  dawn enhancement, 392, 402, 419, 420
  ground-based observations, 390, 419, 421
  MESSENGER observations, 390–392
  seasonal variation, 392, 410, 419
  source, 410, 417
  tail, 392, 419
  temperature, 390, 392
calcium sulfide, 208, 337, 339, 340, 341
calderas, 299
Callisto, 338
Calorian period, 157, 159, 166, 169, 238, 272, 310, 312
Calorian System, 157, 159, 166–167
Caloris exterior plains. *See* circum-Caloris plains
Caloris interior plains, 17, 136, 150, 152, 153, 200, 205, 206, 210, 222, 223, 229, 237, 238, 520
canali, 296
Cape Canaveral Air Force Station, 548

carbon, 34, 37, 180, 184, 185, 191, 206–207, 209, 336–337, 339, 340, 499, 500, 508
carbonaceous chondrites, 168, 497, 501
Cassini state, 86–87, 90, 103, 524
cavi, 326
cavus, 326
CB chondrites. *See* bencubbinites
center of figure, 56
center of mass, 56
CH chondrites. *See* bencubbinites
CH_4. *See* methane
Chamberlain model, 372, 418, 421
Chapman–Ferraro currents. *See* magnetopause: current
charged particle environment, 18, 430–455, 461–488
chemical convection, 114
chemical sputtering, 379
chlorine, 33, 34, 177, 179, 500, 505
Christiansen frequencies, 557
chromium, 36, 177, 179, 182, 183, 184
CIPW norm, 177–178, 180, 182, 183
circum-Caloris plains, 78, 136, 150, 152, 169, 201, 205, 221, 222, 224, 229, 232, 233, 235, 236, 237, 238, 254, 269, 292, 294, 298, 303, 310
Cl. *See* chlorine
closest-approach altitude, 116, 117
CMB. *See* core–mantle boundary
CME. *See* coronal mass ejection
cold poles, 74, 349
cold traps, 346
collisional erosion, 505
collisions, 501, 503, 505, 506
Colombo, Giuseppe (Bepi), 551
comet impacts, 217, 225, 347, 365
comets, 225, 235
Committee on Planetary and Lunar Exploration, 546, 547, 564, 565
COMPLEX. *See* Committee on Planetary and Lunar Exploration
complex craters. *See* craters: complex
compositional buoyancy, 533
condensation, 44, 501, 506, 508
convecting mantle, 65
convection, 115
core, 3, 14–15, 114, 115, 117, 124, 129, 130, 133, 250, 472, 526–527, 549, 554, 558
  convection, 114, 533–534
  cooling, 533
  density, 93, 94, 97
  dynamo, 135, 138, 524, 528
  dynamo field, 114, 116, 117, 120, 124, 126, 130, 132, 133, 134, 135, 138
  dynamo models, 125, 138
  evolution, 533–534, 535–537
  field structure, 117, 126, 135
  induced fields, 126–129
  induction, 126
  radius, 93, 96, 98, 99, 104, 108, 115, 116, 128, 129, 138, 549
core composition, 39–43, 499, 523
  carbon, 41, 523, 534
  hydrogen, 40–41
  iron, 94, 115
  light elements, 115
  oxygen, 41
  phosphorus, 40, 523
  potassium, uranium, thorium, 43

core composition (cont.)
   silicon, 41, 42, 94, 523, 534
   sulfur, 41, 42, 94, 523, 533
core–mantle boundary, 38, 61, 97, 114, 115, 116, 126, 129, 131, 133, 134, 135, 136, 270, 525
coronal mass ejection, 463, 466, 472, 474, 487
correlation spectra, 65
Cosmic Vision Programme, 552
Coulomb criterion, 262
Cr. See chromium
crater degradation, 225, 227, 232, 235
   classification system, 154–157
   state, 154–157, 311
crater ejecta, 221, 230, 231, 235, 550
   blanket, 224, 225, 226, 232, 234
   deposits, 223, 224, 225, 226, 229, 230
   morphology, 227–229
   rays, 168, 198, 224, 225, 227, 229, 230, 332
crater obliteration, 234
crater rays. See crater ejecta: rays
crater retention age, 520
crater saturation, 231, 234, 235, 236, 238
cratered plains structures, 253
crater-related structures, 253
craters
   areal density, 145–151, 156, 157, 159, 161, 167, 168, 169, 222, 223, 224, 231, 232, 234, 236, 237, 238
   buried. See ghost craters
   central peak, 218, 219, 220, 221, 224
   class 1, 232
   complex, 218, 219, 220, 221, 225, 226, 227, 229, 241, 348, 349
   crater chains, 223, 230, 301, 311
   depth, 226
   depth-to-diameter ratio, 219, 226–227, 229
   elliptical, 229
   floor area, 219
   floor uplift, 220
   polygonal, 229
   population 1, 232, 233, 239
   population 2, 232, 233, 239
   primary, 225
   secondary, 146, 159, 217, 224, 225, 226, 227, 228, 229, 230, 232, 234, 235, 236, 238, 301, 337
   SFD. See craters: size–frequency distribution
   simple, 220, 226, 227, 346, 348, 349, 355
   size–frequency distribution, 150, 156, 225, 226, 231, 232, 233, 234, 235, 236, 237, 238, 239, 292, 309
   tectonics, 226
   volcanic fill, 219, 221, 226, 229
crater-to-basin transition, 218–220
creep strength, 63
creep stress, 78
C-rich condensation, 507, 509
CrMB. See crust–mantle boundary
cross-polar-cap potential, 470
crust, 60, 79, 114, 117, 129, 130
crustal density, 93, 104, 518
crustal electrical conductivity, 138
crustal formation, 168–170, 517, 534
crustal magnetic fields, 23, 115, 116, 117, 120, 126, 130–132, 138, 549
crustal magnetization, 115, 117, 130, 133, 136–138, 536
crustal stratigraphy, 305
crustal structure, 206, 210
crustal thickness, 14, 60–62, 66, 67, 79, 93, 136, 518–519, 524

crust–mantle boundary, 60, 65, 66, 79, 258, 263
crust–mantle density contrast, 61
Curie temperature, 137
Curiosity rover, 562, 564
currents
   Birkeland, 22, 114, 117, 119, 126, 129, 134, 138, 431, 444–446, 447, 448, 455, 478, 549, 551
   cross-tail, 431, 435, 443–444, 446–447, 452, 453, 475
   magnetopause. See magnetopause: current
   magnetotail. See magnetotail: current
cusp. See magnetospheric cusp
cusp filaments, 465, 466, 469–470, 487

dark spots, 326, 330, 334
dayside boundary layer, 440–441, 448
DC. See degree of compensation
Deep Space Network, 56
deformational history, 272–276
degree of compensation, 77
degree-2 geoid, 70–79
degree-2 shape, 70–79
Deimos, 196
DEM. See digital elevation model
depth extent of faulting, 65, 78
desorption, 412, 413, 425
diamagnetic depressions, 437, 438
digital elevation model, 55, 90, 250, 288, 355, 557
digital terrain model. See digital elevation model
dike propagation, 270
dikes, 221, 301, 313
diopside, 183, 185
dipolarization, 476, 479
   events, 477, 478
   front, 476, 478, 486, 487
dipole
   axial, 118, 128, 134
   azimuth, 121
   equivalent source, 120, 131, 133, 134
   field, 117, 119, 120, 121, 125, 129, 137
   induction, 129
   moment, 115, 120, 121, 123, 124, 125, 128, 430, 431, 446, 447, 453
   offset, 123, 125, 431, 432, 438, 444, 446, 447, 453, 454–455
   origin, 116
   planetary, 121
   planetocentric, 118
   structure, 120
   tilt, 120, 121, 123, 125, 443
direct trajectory, 545
displacement–length scaling, 250, 264
dissociative ionization, 421, 424
djerfisherite, 508
Doppler, 556
downlink, 548, 556
DSN. See Deep Space Network
DTM. See digital terrain model
ductile deformation, 263
ductile strength, 263
Dungey cycle, 444, 452, 463, 464, 475, 478
dynamic compliance, 100
dynamic pressure, 66
dynamic recrystallization, 263
dynamic viscosity, 63, 100

dynamo, 108, 114, 115, 133, 135, 136, 137
    field, 115, 136, 137, 138
    models, 115, 133, 135, 138, 525
    origin, 120
    self-sustaining, 114
    thermoelectric, 116

early bombardment, 217, 234
early crust, 160–163, 168–170, 205, 516–518
East Kaibab monocline, 251
effective elastic lithosphere, 63
effusive volcanism, 166, 221, 225, 232, 236, 309–311, 519
EID. *See* electron impact dissociation
elastic lithosphere, 62, 66, 76, 78
    thickness, 62, 65, 79
elastic–plastic lithosphere, 69
electrical conductance, 129
electrical conductivity, 114, 115, 117, 126, 129, 130, 136, 138
electrical conductivity structure, 126–130
electromagnetic wave. *See* plasma wave
electron impact dissociation, 420
electron reflectometry, 124
electron-stimulated desorption, 410, 411, 420, 421
elemental abundance maps, 32, 34, 164, 178
elemental abundances. *See* surface composition
emergence angle, 192, 197
Encke dust stream, 410, 421, 422, 424
energetic electron bursts, 451, 476, 484–487
energetic electrons, 14, 18, 22, 451–452, 461, 486, 551, 558, 559
energetic neutral atoms, 559, 561
Energetic Particle and Plasma Spectrometer, 8, 372
energetic particle bursts. *See* energetic electron bursts
Energetic Particle Spectrometer, 8, 451, 461, 476, 484
energetic particles, 548, 550, 554, 561
enstatite, 38, 94, 97, 182, 186
enstatite chondrites, 36, 39, 43, 136, 137, 178, 184, 498, 499, 501, 506, 508, 509
EOS. *See* equation of state
epithermal neutrons, 31, 32, 350, 351, 352
EPPS. *See* Energetic Particle and Plasma Spectrometer
EPS. *See* Energetic Particle Spectrometer
equation of state, 94, 95, 531
equator
    geographic, 118, 119, 122, 123, 125, 133
    magnetic, 121, 122, 123, 125, 126, 431, 436, 443, 447, 448, 450, 451, 452, 454, 455
    magnetic dip, 122, 123
ER. *See* electron reflectometry
ESA. *See* European Space Agency
escarpments, 249
ESD. *See* electron-stimulated desorption
Europa, 338
Europe spaceport, 555
European Space Agency, 544, 547, 551–552, 563
exobase, 372
exosphere, 3–4, 15, 17–18, 30, 371–403, 407–425, 461, 548, 549, 550, 552, 554, 558, 559, 560, 561, 562
    loss processes, 407, 413–415, 421
    observational techniques, 372–378
    source processes, 407–413, 416, 417, 419, 424
    tail, 413, 414
exosphere species, 378–402
    aluminum. *See* aluminum exosphere
    calcium. *See* calcium exosphere

helium, 371, 398
hydrogen. *See* hydrogen exosphere
ionized calcium, 371, 401, 403, 423–424
iron, lithium, silicon, 402
magnesium. *See* magnesium exosphere
manganese, 371, 401, 403, 423–424
oxygen, 371, 399, 403, 424
potassium. *See* potassium exosphere
sodium. *See* sodium exosphere
explosive volcanism, 13, 15, 167, 277, 297–299, 311, 312, 520
external fields, 116, 117, 119, 120, 121, 123, 134
extrasolar planets, 502, 508, 509, 563, 565
extreme ultraviolet, 558

failure criteria, 262
far ultraviolet, 364, 558
Fast Imaging Plasma Spectrometer, 8, 372, 436, 437, 438–440, 441, 449, 455, 461, 469, 470, 472, 474, 475, 476
fast neutrons, 31, 32, 36, 176, 182, 183, 350, 351, 521
fault dip angle, 251, 252, 265
fault displacement-gradient fold, 251
fault heave, 265
fault throw, 265
fault-bend fold, 251
fault-propagation fold, 251
Fe. *See* iron
feldspar, 33, 181, 182, 210, 498
Fe–Ni. *See* iron–nickel
FeO. *See* ferrous iron
ferrous iron, 30, 39, 145, 191, 193, 195, 196, 209, 210, 337, 498, 508
FeS. *See* iron sulfide
FeS layer. *See* iron sulfide layer
field-aligned currents. *See* currents: Birkeland
FIPS. *See* Fast Imaging Plasma Spectrometer
flexural stress, 62
flood basalts, 188, 279, 281, 289, 292, 293, 294, 295, 297, 299, 305, 314, 517
flood volcanism, 34, 180, 182, 184, 276–277, 290, 293, 296, 314, 520
floor-fractured craters, 300
flotation crust, 60, 163, 168, 184, 188, 206, 207, 210, 305, 501, 517
fluidized impact ejecta, 287, 303
flux rope, 443, 454, 468, 469–470, 475, 479–480, 486, 487
flux transfer event, 466, 469–470, 474, 479, 487
fO_2. *See* oxygen fugacity
fold-and-thrust belt, 17, 69–70, 256
formation, 2, 13, 497–509
forsterite, 38, 39, 97, 186
fossil bulge, 71
fractionation, 497
free precession, 87
free-air gravity anomaly, 58
frictional resistance, 262
FTB. *See* fold-and-thrust belt
FTE. *See* flux transfer event

gabbro, 180, 182, 183, 184
galactic cosmic rays, 31, 350, 363
Gamma-Ray and Neutron Spectrometer, 6–7, 30, 94, 176, 191, 288, 350, 401, 461, 476, 486
Gamma-Ray Spectrometer, 6, 30, 31, 36, 37, 147, 179, 201, 336, 339, 451, 454, 486, 500
    anticoincidence shield, 31, 147, 486
Ganymede, 338
gas–surface interaction, 412–413, 417, 418

general relativity, 552
geochemical modeling, 37–39, 201, 206, 207, 209, 498
geochemical terranes, 22, 33, 36, 176–188, 521
    Caloris Interior Plains Terrane, 178, 184, 185, 186, 188
    High-Magnesium Terrane, 178, 183, 185, 188
    Low-Fast Terrane, 178, 183–184, 186
    Northern Terrane, 178, 182, 184, 186
geodesy, 85
GEODYN software, 57
geoid, 52, 58, 70
    equatorial ellipticity, 73
    polar flattening, 71, 73
geoid-to-topography ratio, 62, 518
geological history, 2–3, 13–14, 250, 519–521
geomorphology, 206, 338, 557
ghost craters, 149, 229, 268, 287, 292, 309, 519
giant impact, 41, 44, 503–505, 508
giant planet migration, 233, 234, 238
global contraction, 3, 17, 21, 69, 107, 167, 249, 255, 264–266, 276–277, 278, 311, 313, 520, 522, 527, 549
global evolution, 516–537
Goldstone Deep Space Communications Complex, 90, 346, 347
graben, 221, 222, 223, 224, 226, 229, 230, 249, 251, 258, 268, 301
grain density, 94, 104
Grand Tack model, 501
graphite, 37, 60, 153, 163, 167, 168, 180, 182, 184, 188, 191, 200, 206–207, 209, 210, 305, 336–337, 339, 340, 341, 501, 517
gravitational constant, 88
gravitational parameter, 88
gravitational potential, 56
gravitational torques, 86
gravity anomaly, 58, 59, 68, 271
gravity assist, 545, 546, 547, 564, 565
gravity disturbance, 58
    range, 58
gravity field, 56–60, 79, 87, 518
    degree strength, 58, 59, 64
    equatorial ellipticity, 56, 71
    polar flattening, 56, 71
    resolution, 58
gravity–shape correlation, 64, 65
GRNS. *See* Gamma-Ray and Neutron Spectrometer
GRS. *See* Gamma-Ray Spectrometer
GTR. *See* geoid-to-topography ratio
gyroradius, 441, 452, 453, 454

H. *See* hydrogen
H exosphere. *See* hydrogen exosphere
hanging wall, 251
harzburgite, 184, 185, 186, 188
He. *See* helium
heat pipes, 556
heat production rate, 101
heat-producing elements, 249, 266, 527, 529, 532
heavily cratered terrain, 37, 39, 40, 101, 104, 131, 161, 180, 182, 226, 227, 232, 233, 234, 235, 236, 237, 238, 253, 302, 303, 311, 336, 499, 516, 517
high-admittance shape, 66, 74
high-energy particles. *See* energetic particles
highly siderophile elements, 217
high-Mg region, 33, 34, 38, 100, 147, 164, 183, 205, 210, 218, 308, 549
high-reflectance red material, 163–164, 166, 168
high-reflectance red plains, 145, 163, 166, 180, 182, 184, 196, 200, 201, 203, 205, 206, 222, 223, 224, 225, 330

high-relief ridge, 251
high-speed stream, 472
high-terrain-bounding structures, 253
hit-and-run collision, 41, 44, 217, 505, 508
Hoek–Brown criterion, 262
hollows, 14, 15, 17, 19, 35, 37, 167, 199, 200, 201, 204, 210, 217, 222, 224, 225, 226, 311, 324–341, 549, 550, 562, 563, 564
    color and spectral properties, 335–336
    composition, 336–337, 339
    definition, 326
    evidence for young ages, 332–334
    formation, 207–208, 224, 338–341
    formation rate, 334
    geographic distribution and compositional affinities, 328–332
    geological setting, 326
    planetary analogs, 338–341
    sizes, shapes, depths, 326–327
    texture, 335
homologous temperature, 102
horizontal shortening, 249
hot poles, 62, 74, 349
hot-pole longitudes, 328
HRP. *See* high-reflectance red plains
Hubble Space Telescope, 545
hydrazine, 557
hydrogen exosphere, 371, 396–398, 403
    temperature, 396
hydrostatic equilibrium, 71, 73

ICAG. *See* Inter-Agency Consultative Group
icy surfaces, 338
illumination bias, 256, 522
ilmenite, 31, 192, 200, 337
IMF. *See* interplanetary magnetic field
immiscible liquids, 524
impact basins, 70, 147, 151, 152, 154–160, 185, 217–225, 268–270, 303
    distribution, 218
    ejecta, 222, 223, 225
    multi-ring basins, 218, 220, 221, 241
    peak-ring basins, 219, 220, 221, 222, 223, 224, 225, 241
    protobasins, 220, 221, 226
    rim-crest diameter, 219, 221
    tectonics, 222, 225
    volcanic fill, 223, 224
impact cratering, 217–241
    cratering rate, 229
    impact velocity, 217, 219, 227, 232, 235, 237, 238
    PF. *See* impact cratering: production function
    production function, 232, 234, 235, 238, 239
    target properties, 217, 218, 226, 227, 229, 234, 236, 237, 239
impact gardening, 358
impact heating, 532
impact melt, 156, 157, 167, 199, 218, 219, 220, 221, 222, 223, 224, 225, 230–231, 340
impact vaporization, 384, 385, 390, 392, 396, 403, 407, 409–410, 411, 413, 416, 417, 418, 421, 422, 423, 424
impactor size–frequency distribution, 233
IMPs. *See* Moon: irregular mare patches
in situ modulus of deformation, 261
incidence angle, 192, 197
induced magnetic fields, 22, 524
induced magnetization, 136
induction, 115, 121, 136, 472, 473, 487

induction currents, 464, 557
infrared spectrometer, 549
injections
    energetic electron, 454
    plasma, 443
inner core, 94, 95, 98, 103–104, 114, 524
    density, 103
    growth, 115
    radius, 104, 133
insolation models, 348, 355
instantaneous Laplace plane, 87
Institute of Space and Astronautical Science, 552, 553
Inter-Agency Consultative Group, 552
intercrater plains, 78, 132, 133, 137, 145–148, 157–168, 200, 205, 232, 233, 234, 236, 287, 301–305, 310, 315, 336, 499, 519
    crater density, 145–148
    origin, 147, 169
    reflectance, 147
    roughness, 146
    thickness, 163
interior outgassing, 364
intermediate plains, 163, 200, 201, 203, 205, 206, 209
intermediate plains stratigraphic unit, 150, 151
internal fields, 126, 130
internal structure, 15, 52–109, 114, 117, 499, 523–524
International Mercury Exploration Mission, 552
interplanetary dust, 347, 392, 410, 421, 422, 508, 552, 554, 560
interplanetary magnetic field, 114, 430, 432, 435, 437, 441, 442, 443, 444, 452, 455, 463, 469, 472, 474, 551
interplanetary shock, 464
intrusive magmatism, 298
ion sputtering, 379, 390, 470
ion-enhanced diffusion, 379
IP. See intermediate plains
iron, 31, 36, 94, 147, 169, 177, 178–184, 192, 200, 201, 205, 207, 209, 210, 497, 498
    alloys, 136
    minerals, 137
    multidomain, 137
    partitioning, 136
iron meteorites, 40, 499, 506
iron snow, 115, 136, 525, 533
iron sulfide, 136, 206, 207, 339, 499
iron sulfide layer, 68, 97, 128, 524
iron–nickel, 40
iron–wüstite buffer, 36, 177
ISAS. See Institute of Space and Astronautical Science
isostasy, 52, 62, 73
isostatic compensation, 79, 257, 518
isothermal bulk modulus, 96

Japan Aerospace Exploration Agency, 544, 547, 548, 551, 552, 563
Jet Propulsion Laboratory, 547, 551
joints, 259
JPL. See Jet Propulsion Laboratory

K. See potassium
K exosphere. See potassium exosphere
Ka-band, 556
Kaula rule, 57
Kelvin–Helmholtz
    instability, 441, 449
    waves, 441, 471–472, 487, 488
KH. See Kelvin–Helmholtz

kinetic escape, 415, 416
kīpukas, 296
komatiite, 36, 60, 145, 206, 307, 340
KREEP, 33, 35, 166
KT14. See magnetic field model: KT14
Kuiperian period, 157, 167, 230, 238, 239, 240, 241, 273, 310, 332
Kuiperian System, 157, 159, 167, 332

lag deposit, 339, 340, 341, 346, 355, 359
Laplace plane, 86
large igneous province, 289, 293
laser altimetry, 53–54
late heavy bombardment, 225, 232, 233, 234, 235, 236, 237, 238, 239, 240, 241, 253, 312, 314, 517
lava flow fronts, 293
LBP. See low-reflectance blue plains
Legendre polynomial, 53
LHB. See late heavy bombardment
lherzolite, 186, 188, 499
libration, 549, 558, 559
    88-day, 86
    forced, 562
libration amplitude, 86, 89, 91
limb profile, 55, 375, 385
limb scans, 375, 393, 416, 419, 420, 422
    dayside, 416, 418, 419, 421, 424
    south pole, 416
Liouville's theorem, 421
liquid immiscibility, 95
liquidus temperature, 177, 180, 181, 182, 183, 184, 185, 186
lithosphere, 62–63, 72, 260–264, 522
lithospheric folding, 69, 271
lithospheric loading, 249, 269
lithospheric state of stress, 262
lithospheric strength envelope, 262
lobate scarp, 62, 65, 217, 221, 223, 251, 252, 253, 311, 520, 521, 522, 523
long-term librations, 91
long-wavelength shape, 67, 77
long-wavelength topographic undulations, 68, 223, 259–260, 271, 275–276
low-latitude boundary layer, 471
low-reflectance blue plains, 200, 201, 203, 205, 206, 330, 339, 340
low-reflectance material, 14, 23, 37, 60, 145, 150, 161, 164, 166, 168–170, 185, 200, 203, 205, 206, 222, 224, 225, 305, 329, 330, 336, 339, 340, 500, 517
    composition and formation, 206–207
    spectral heterogeneity, 205
LRM. See low-reflectance material
Lunar Crater Observation and Sensing Satellite, 363
Lunar Reconnaissance Orbiter Camera, 146
Lyman alpha photodissociation, 362

Mach number, 466
MAG. See Magnetometer
magma, 33, 34, 38, 166–170, 184, 186, 209–210, 276–278, 295, 299–301, 307–308, 312–314, 337, 340, 364–365, 498–499, 517, 521, 528, 532, 535, 549
magma ascent, 312–314
magma buoyancy, 60, 518
magma conduits, 277
magma ocean, 37, 60, 163, 168, 184–185, 186, 188, 207, 305, 517
magmatic volatiles, 34, 208–210, 308, 332, 340, 341

magmatism, 137
magnesium, 33, 36, 38, 147, 164, 177, 178–184, 186, 204, 206, 209, 210, 336–337, 339, 499
magnesium exosphere, 371, 392–396, 402, 421–423
   morning enhancement, 394, 396, 402
   tail, 421, 422
   temperature, 393
magnesium sulfide, 208, 336, 337, 339, 340, 341
magnetic carriers, 136
magnetic diffusion time, 135
magnetic dipole, 3, 14
magnetic erosion, 507, 509
magnetic field, 3, 14, 108, 430–455, 461, 524–526, 545, 546, 548, 549, 551, 552, 553, 554, 561, 563, 564
   global, 114, 116, 124, 135
   planetary, 554, 557
   spherical harmonic descriptions, 164
magnetic field lines, 114
magnetic field model, 438, 446–448
   KT14, 446, 447–448
   paraboloid, 438, 446–448, 449, 455
   scaled Earth, 455
magnetic field modeling, 120, 132
magnetic quadrupole, 525
magnetic reconnection, 14, 22, 117, 126, 430, 432, 435, 438, 443, 444, 447, 449, 452, 453, 454, 463–464, 465, 468, 469, 470, 472–474, 475, 476–480, 486, 487–488
   magnetotail, 453
   rate, 443, 464, 466, 467, 468
   X-lines, 435, 453, 469, 473, 475, 476, 479, 480, 486
magnetic remanence, 114
magnetic residuals, 121, 123, 126, 127, 444, 445, 447, 448
magnetization, 114, 115, 116, 120
Magnetometer, 7, 117, 118, 123, 124, 126, 128, 402, 433, 436, 438, 439, 441, 445, 446, 461, 545, 553, 557, 561
magnetopause, 114, 117, 118, 119, 121, 122, 125, 126, 128, 133, 430, 431, 432–434, 437, 438, 440, 441, 442, 443, 444, 447, 448, 449, 452, 453, 454, 455, 463–464, 466, 467, 470, 471, 472, 553
   current, 114, 116, 119, 120, 432, 438, 444, 446, 447, 469, 487
   field, 119, 123, 124, 128, 129
   flaring, 435
   shape, 119, 121, 446, 447
   standoff distance, 433, 447
   subsolar, 126, 128, 473
   subsolar distance, 116
magnetosheath, 430, 432, 437, 438, 441, 442, 461, 463, 464, 469, 470, 471, 472, 474, 481, 482, 487
   magnetic field, 432, 438, 441, 443
   plasma, 437, 438, 452
   plasma beta, 430, 442
   thermal pressure, 435
   thickness, 433
magnetosphere, 22–23, 114, 116, 117, 118, 119, 120, 121, 122, 123, 124, 126, 128, 133, 138, 430–455, 461–488, 546, 548, 549, 550–551, 552, 553, 554, 558, 559, 561, 562, 563
   disappearing dayside, 474
magnetospheric activity index, 120
magnetospheric convection, 487
magnetospheric current systems, 431, 446, 447, 455
magnetospheric cusp, 124, 126, 127, 133, 134, 437, 438–441, 443, 444, 448, 450, 452, 453, 455, 464, 466, 470, 487
magnetospheric polar cap, 438, 439, 440, 452, 455
magnetotail, 118, 121, 122, 430, 435, 439, 444, 446, 447, 452, 453, 454, 463, 464, 466, 469, 470, 475, 476–480, 482, 486, 487, 551, 553
   current, 119, 120, 443, 444, 447
   field, 123, 124
   flaring, 433
   loading–unloading, 464, 475, 479–480, 487
   lobes, 430, 453
   structure, 434, 435
   X-line, 435, 453
Malaita anticlines, 257
manganese, 36, 177, 179, 182, 183, 184
manganese sulfide, 208
Mansurian period, 157, 159, 167, 230, 238, 239, 240, 273, 310
Mansurian System, 157, 159, 166–167
mantle, 60, 128, 549
   conductivity, 129, 136, 138
   magnetization, 136
mantle composition, 37–39, 185–186
mantle convection, 60, 69, 76, 105, 186, 188, 249, 270–271, 528–529
mantle overturn, 60
mantle partial melting, 37–39, 60, 61, 62, 78, 79, 94, 102, 164, 166, 169, 170, 185, 186, 187, 188, 305, 517, 518, 527, 530, 531, 535
mantle solidus, 527
mantle stripping, 44, 497, 502, 503–505
Mariner 10, 1, 30, 53, 57, 88, 108, 115, 116, 117, 119, 125, 135, 144, 152, 191, 192, 217, 222, 226, 227, 229, 230, 240, 249–250, 258, 287, 288, 301, 324, 346, 371, 374, 375, 396, 398, 399, 544, 546, 548, 551, 554, 561, 562, 563, 564
   Mercury flybys, 115, 116, 117, 125
   Venus flyby, 546
Mars, 3, 33, 34, 44, 116, 168, 169, 217, 219, 226, 227, 228, 232, 233, 235, 236, 238, 257, 296, 301, 338, 346, 498, 535, 544, 545, 546, 547, 562, 564, 565
   cratering record, 232
   craters, 226
   degree-2 geoid, 71
   degree-2 shape, 71
   impactor velocity, 232
   moment of inertia, 71
   Olympus Mons, 269
   rampart craters, 229
   Tharsis rise, 71, 271
Mars lander missions, 544, 545, 562
mascons, 14, 56, 58, 69, 549
MASCS. See Mercury Atmospheric and Surface Composition Spectrometer
mass, 88, 92
maximum faulting depth, 78
Maxwell rheological model, 100
Maxwell rheology, 63
Maxwell time, 63, 100
Maxwell–Boltzmann distribution, 408, 410
MBAs. See asteroids: main belt
MBF. See Mercury body-fixed coordinates
MDIS. See Mercury Dual Imaging System
mean radius, 53
mechanical equilibrium, 64
mechanical erosion, 296
mechanical lithosphere, 63
   thickness, 63
megaregolith, 163, 225, 227, 237
membrane stress, 62, 64
MeO SWT. See Mercury Orbiter Science Working Team
Mercury Atmospheric and Surface Composition Spectrometer, 8, 191, 335, 371, 375, 385, 407, 420

Mercury body-fixed coordinates, 116, 117, 120, 122, 123, 126, 127, 130, 134, 135
Mercury Dual Imaging System, 6, 55, 60, 144, 191, 193–195, 220, 227, 236, 240, 250, 288, 324, 335, 347
Mercury Exploration Working Group, 552
Mercury lander mission, 315, 544, 545, 546, 562, 564, 565
Mercury Laser Altimeter, 7–8, 53, 90, 134, 219, 223, 226, 227, 250, 288, 348, 352
Mercury Orbiter Science Working Team, 547, 548, 554
Mercury sample return mission, 563, 564–565
Mercury solar magnetospheric coordinates, 117, 118, 432, 433, 435, 436, 446, 447, 453, 466
Mercury solar orbital coordinates, 116, 117, 121, 122, 124, 126, 432, 433, 438, 439, 444, 447, 448, 470
MESSENGER mission
  Earth flyby, 11, 53
  first extended mission, 15–18
  first extended mission objectives, 15–16
  first extended mission project requirements, 16
  key scientific questions, 2–4
  low-altitude campaigns, 19
  Mercury flybys, 12, 54, 116, 117, 120, 193, 198, 324, 371, 376, 381, 384, 391, 392, 548
  Mercury orbit insertion, 551
  orbit-correction maneuvers, 131
  primary mission, 11–12, 53
  project requirements, 4
  science data acquisition planning, 8–11
  science observation performance, 10–11
  science planning, 10
  scientific objectives, 4
  second extended mission, 18–23
  second extended mission objectives, 19
  second extended mission project requirements, 19
  solar radiation pressure, 551
  solar sailing, 551
  Venus flybys, 11, 548
MESSENGER spacecraft, 4–5
  altitude, 119, 120, 121, 127
  apoapsis altitude, 118
  attitude control, 5
  initial orbit, 8, 12
  launch, 1, 11, 117, 548
  launch vehicle, 4
  Mercury orbit insertion, 1, 12
  mission operations constraints, 9
  orbit, 53, 350
  orbit design, 8
  orbit period, 8, 12, 16
  orbit-correction maneuver, 8, 12, 16, 19
  payload, 5–8
  periapsis altitude, 12, 16, 19, 53, 117, 118, 120, 121, 127, 130, 131, 133, 258, 548
  periapsis latitude, 12, 16, 19, 53, 57
  propellant, 548, 551
  science observation constraints, 9
  solar arrays, 5, 547
  sunshade, 5, 551
  surface impact, 19, 548
  telecommunications system, 5
metal/silicate fractionation, 506–508
metal/silicate ratio, 2, 13, 217, 523
metasomatism, 315
meteoroid impacts, 407, 409, 417, 424

methane, 339
MEWG. *See* Mercury Exploration Working Group
Mg. *See* magnesium
Mg exosphere. *See* magnesium exosphere
micrometeoroid impacts, 365, 407, 409, 410, 413, 417, 419, 420, 421, 424, 550, 559, 560
microphase iron, 205, 337
mineralogical composition, 554, 557, 559
mineralogy, 178–184, 185–186, 307, 550, 564
MLA. *See* Mercury Laser Altimeter
Mn. *See* manganese
MnS. *See* manganese sulfide
model production function, 310, 520
molecular dissociation, 419, 420, 421, 423, 424
moment of inertia. *See* polar moment of inertia
moment of inertia of core, 87
moment of inertia of mantle and crust, 87, 92, 95
monocline, 251
Monte Carlo modeling, 416, 419, 422
Moon, 33, 34, 37, 146, 152, 168, 169, 191, 192, 195, 196, 210, 217, 225, 226, 227, 229, 232, 234, 326, 332, 500, 504, 535
  composition, 237
  crater areal density, 169, 238
  crater size–frequency distribution, 235, 236
  cratering rate, 238, 241
  cratering record, 232
  craters, 221, 225, 226, 234, 241
  degree-2 geoid, 71
  highlands, 147, 148, 160–163, 192, 204, 234, 238
  Imbrium basin, 156, 222
  impact basins, 217, 218, 219
  impact flux, 217, 235
  impact melt, 231
  impactor velocity, 232, 237
  impactors, 240
  irregular mare patches, 337
  lunar light plains, 287
  lunar polar deposits, 363–364
  Mare Australe, 152
  mare basalt, 35, 152, 153, 176, 192, 210, 231, 237, 287, 308, 338, 521
  maria, 2, 152, 192, 251, 545
  non-hydrostatic shape, 72
  Orientale basin, 147, 156, 161, 222
  origin, 217
  peak-ring basins, 219, 220
  resurfacing rate, 235
  South Pole–Aitken basin, 217
  surface chronology, 237, 238, 239, 241
  Timocharis crater, 228
  Tsiolkovsky basin, 228
MPF. *See* model production function
MSM. *See* Mercury solar magnetospheric coordinates
MSO. *See* Mercury solar orbital coordinates
multiple saturation point, 38, 186

Na. *See* sodium
Na exosphere. *See* sodium exosphere
NAC. *See* narrow-angle camera
nanophase iron, 205, 337
narrow-angle camera, 6, 55, 220, 250, 288, 324
NASA Discovery Program, 547, 563
National Research Council, 544, 563
NE Rachmaninoff pyroclastic deposit, 35, 199, 209, 337, 340
near-infrared, 550, 559

near-surface thermal models, 355
near-ultraviolet, 550, 558
Neutron Spectrometer, 6, 30, 31, 36, 37, 60, 185, 191, 207, 209, 336, 337, 350, 451, 486, 500
neutron spectroscopy, 31–32, 350
NGP. See non-gravitational perturbation
Ni. See nickel
Nice model, 234, 501
nickel, 39
non-gravitational perturbation, 557
non-hydrostatic shape, 76, 79
non-volatile elements, 37
norite, 36, 180, 182, 184
normative mineralogy, 177, 183, 184
north polar region, 347, 352
northern rise, 55, 67–68, 259, 271, 276, 536
northern smooth plains, 17, 33, 36, 38, 55, 62, 67, 78, 100, 131, 132, 137, 150, 152, 155, 166, 180, 184, 188, 200, 205, 210, 227, 228, 229, 232, 233, 236, 237, 238, 239, 240, 253, 254, 255, 268, 269, 270, 271, 275, 292, 293, 294, 295, 296, 298, 299, 301, 303, 309, 310, 499, 517, 520
northern terrane, 33, 36
NS. See Neutron Spectrometer
NSP. See northern smooth plains
nuclear spallation reactions, 350

O. See oxygen
obliquity, 63, 76, 79, 86, 89, 90–91, 524, 549, 558
OCM. See MESSENGER spacecraft: orbit-correction maneuver
Odin-type plains, 292
offset of centers, 56
oldhamite, 38, 498
olivine, 100, 104, 180, 182, 183, 185, 204, 210, 307, 498
OMCT. See oxygen–metal charge transfer
opening-mode fractures, 259
optical maturity, 193, 204, 332
orbit capture, 552
orbit determination, 56, 556, 557, 558
orbit precession period, 86
orbit precession rate, 87
orbital eccentricity, 75, 77, 86, 137
orbital inclination, 86
orbital precession, 86, 87
orbit-plane normal, 86
ordinary chondrites, 195, 339, 341, 499, 506
Orion spacecraft, 562
orthopyroxene, 94, 104
outer core, 86, 91, 95, 114, 115, 136, 523, 562
outer core radius, 94, 97, 104
outflow channel systems, 296
oxygen, 36, 37, 177, 178–184
oxygen fugacity, 36, 38, 41, 42, 43, 44, 94, 97, 176, 177, 178, 206, 308, 498–499, 523
oxygen–metal charge transfer, 191, 195, 200, 203, 205, 210, 299, 336

P. See core composition: phosphorus
parameterized convection models, 527
partial melting, 38, 60, 184, 185, 186, 188
PDL. See plasma depletion layer
pebble accretion, 501
perihelion, 553, 556
permanently shadowed regions, 32, 35, 346, 347, 562, 564, 565
petrologic modeling, 36, 39, 177–178, 337

phase angle, 192, 197, 198, 335
phase-ratio analysis, 197, 198–200, 335, 341
Phobos, 196
photodesorption, 379
photodissociation, 392
photoionization, 413, 414–415, 420, 421, 422
  lifetime, 414, 415, 417, 419, 420, 422, 423
  rate, 414, 415
photometric model, 192, 197–200
photon-stimulated desorption, 390, 407, 408, 409, 410–411, 412, 413, 416, 417, 418, 419
photophoresis, 507, 509
physical librations, 86, 108, 524
pickup ions, 441, 449
Pioneer missions, 545
pit complexes, 222
pitch angle, 124
pit-floor craters, 299, 300
plagioclase, 180, 181, 182, 183, 184, 185, 186, 205, 307, 522
planet formation theory, 501–502, 507
planetary accretion, 217, 233, 498, 501, 505, 509
planetary embryos, 501, 505
planetary magnetic fields, 114, 116
planetary reorientation, 271–272
plasma, 119, 123, 124, 131, 133, 135, 430, 431, 432, 436, 445, 447, 448–455, 549, 550, 559, 561
  beta, 430, 436, 442, 443, 467, 487
  composition, 431, 449–452
  convection, 550
  cusp, 438, 439
  density, 441, 442
  electrons, 476
  flow, 441, 442
  heating, 452–454
  losses, 454–455
  precipitation, 14, 18
  pressure, 126, 437, 438
  properties, 436
  solar wind, 440
  sources, 436, 448–449
  temperature, 437
  thermal pressure, 430, 436
  transport, 438, 440, 452–454
plasma depletion layer, 441–443, 467, 487
plasma sheet, 126, 430, 434–437, 443, 444, 448, 450, 452–453, 455, 466, 475, 476, 487
plasma wave, 548, 550, 551, 559, 561
  ion-Bernstein, 483
plasmoid, 435, 454
Poisson's ratio, 62, 261
polar deposits, 3, 15, 22, 346–366, 549, 550, 558, 562, 563, 565
  boundaries, 359
  epithermal neutron flux, 350
  high-reflectance surfaces, 354
  illumination conditions, 347–350
  imaging, 357–360
  insulating layer, 346, 352, 355, 359
  low-reflectance surfaces, 353, 355, 359
  low-temperature silicates, 347, 350, 361
  neutron spectroscopy, 350–352
  organic compounds, 355, 357
  relative age, 359
  sulfur, 347, 350, 361
  surface reflectance, 352–354, 358

thermal models, 355
thickness, 350, 358, 363
water ice fraction, 363
polar moment of inertia, 57, 86, 87, 92, 95, 103, 524
ponded lavas, 289
potassium, 31, 33, 36, 166, 177, 178, 500, 505, 507
potassium exosphere, 371, 389–390, 403, 411
ground-based observations, 425
Poynting–Robertson effect, 560
Pratt isostasy, 64
precipitation, 454, 455
ion, 438, 439, 440, 455
Preliminary Reference Mercury Model, 85, 105, 107
pressure-release melting, 78
pre-Tolstojan period, 166, 272, 310
pre-Tolstojan System, 157–159, 160–166, 169
primary crust, 60, 170, 305, 518
principal axes, 75
principal component analysis, 203, 325, 336
principal moments of inertia, 86
principal-axis coordinate system, 52, 71
PRMM, 105. See Preliminary Reference Mercury Model
propulsion
chemical, 546, 547, 552, 553, 556, 564
solar electric, 546, 547, 552, 555, 556, 561, 562, 564, 565
proton reflectometry, 124–125, 134, 135
protoplanetary disks, 502, 508, 509
PSD. See photon-stimulated desorption
PSRs. See permanently shadowed regions
pyroclastic deposits, 14, 15, 23, 167, 199, 200, 208, 210, 297, 299, 300, 308, 311, 331, 336, 339, 340, 520, 562
pyroclastic volcanism, 34, 36, 37, 167, 208–210, 337, 549, 550
pyroxene, 94, 102, 104, 177, 180, 182, 183, 184, 185, 186, 188, 192, 204, 210, 307, 498
pyrrhotite, 136, 137

quality factor, 100
quartz, 182, 183

radar backscatter, 346, 363
radar circular polarization ratio, 346, 363
radar-bright deposits. See polar deposits
radiation pressure, 372, 407, 413–414, 415, 416, 417, 420, 421, 422
radio frequency, 53
radio occultations, 54–55
radio science, 8, 54, 90, 549, 556
radius change, 264–266
ram point, 425
Rayleigh number, 105, 528
red unit, 200, 201, 203, 205, 208–210
spectral heterogeneity, 205
reducing conditions, 36, 38, 41, 43, 45, 46, 94, 176, 184, 206, 209, 337, 497–499, 507, 509, 521, 523
reference ellipsoid, 58
reference sphere, 52, 58
regolith, 162, 166, 168, 197, 225, 226, 229, 235, 237, 324, 352
conductance, 129, 130
conductivity, 129
grain size, 198
porosity, 198
scattering properties, 198
thickness, 130
regolith gardening, 347

relative plot, 231, 232, 233, 234, 235
remanence
crustal magnetization, 115, 137
magnetic field, 115, 137
magnetization, 115, 117, 136, 137
shock magnetization, 115
thermal magnetization, 115, 136
viscous magnetization, 115, 136
remanent. See remanence
Reststrahlen bands, 557
resurfacing, 145–148, 163–166
impact, 147
volcanic, 147, 164
reverse fault, 251
RF. See radio frequency
rheology, 62–63, 73, 99–100
ridges, 249
rotational bulge, 71
rover mission, 546, 564, 565
R-plot. See relative plot

S. See sulfur
sample return mission, 546
satellites of Mercury, 240–241
satellite-to-satellite tracking, 56
saturation equilibrium, 161, 234
Saturn V, 544, 562
scarp retreat, 340, 341
SciBox, 9–10
secondary craters. See craters: secondary
secondary crust, 60, 170, 305
second-degree gravity coefficients, 56, 73, 87, 88
secular resonance, 234
secular variation, 117, 125–126, 133, 134, 135, 138, 526
self-gravitation, 71
SEP. See solar electric propulsion
SFD. See craters: size–frequency distribution
SH. See spherical harmonics
shape, 52, 53–56, 79
compensation, 52, 63–64
dynamic range, 55
equatorial ellipticity, 55
equatorial shape, 54
polar flattening, 55
shape ellipsoid, 70
shear modulus, 63, 261
shield volcano, 297
shortening strain, 265
Si. See silicon
silicon, 36, 38, 147, 178–184, 209, 499
sills, 301, 313
simple craters. See craters: simple
simple-to-complex crater transition, 226, 227
single-scattering albedo, 198
sinuous rilles, 296
SLS. See Space Launch System
smelting, 37
smooth plains, 2, 131, 132, 133, 149–154, 193, 200, 218, 221, 222, 223, 224, 226, 227, 229, 232, 235, 238–239, 241, 287, 309, 519
color, 152–153
crater density, 151
distribution, 152
formation, 223

smooth plains (cont.)
  tectonic deformation, 149
  thickness, 150
smooth plains structures, 253
snow line, 507
sodium, 33, 38, 177, 178–184, 186, 500
sodium exosphere, 371, 379–389, 402, 407, 408, 410, 412, 413, 415–419, 424, 425
  ground-based observations, 379–384, 416, 419, 424
  MESSENGER observations, 384–389
  seasonal variation, 416, 417
  spatial distribution, 379, 384–389
  tail, 371, 383, 411, 413, 415, 417, 418
  temperature, 385
  thermalization, 385
  velocity distribution, 381
sodium ions, 466, 472, 475, 487
sodium tail. See sodium exosphere: tail
solar cycle maximum, 18
solar cycle minimum, 463
solar energetic particles, 469
solar heating, 332, 339, 341
solar oblateness, 558
solar wind, 114, 116, 117, 119, 126, 430, 434, 435, 441, 442, 444, 452, 463, 464, 549, 551, 552, 554, 557, 558, 561
  dynamic pressure, 464, 467, 472–474
  environment, 117, 126, 431, 432, 437, 441, 442, 448, 450, 455
  fast, 431, 432, 448
  flow, 430
  ions, 561
  Mach number, 432, 433, 442
  plasma, 436, 437, 439, 448, 449, 450, 453
  plasma beta, 442
  ram pressure, 114, 117, 126, 138, 430, 432, 434, 435, 442
  slow, 431, 432
  speed, 117, 436, 437
  velocity, 122
solar wind–surface interactions, 364
solid basal layer, 95
SOR. See spin–orbit resonance
source-free region, 118
south polar region, 347, 366
southern smooth plains, 240
Soyuz–Fregat, 552
space environment, 550, 561
Space Launch System, 562
Space Science Board, 544–546
space weathering, 19, 126, 159, 168, 191, 205, 210, 332, 336, 340, 438, 455, 464, 549, 550, 562
spectral absorption, 200, 210
  0.6-μm feature, 200, 203, 205, 208, 210
  1-μm crystal-field absorption, 191, 192, 193, 200–201
spectral properties, 200–205
  relationship to composition, 210
spectral reflectance, 37, 145, 191–210, 324, 335, 337, 340, 500, 508
spectral slope, 192, 193, 199, 200, 201, 204, 324, 336, 337
spectral units, 200–201
  spatial distribution, 205
spectral variability, 200–205
spectrally red pitted ground, 332, 335, 339
spherical harmonic coefficients, 53, 60, 524
spherical harmonic expansion, 120, 121, 123, 124, 125, 126, 128, 129, 132, 134, 135, 138
spherical harmonics, 52, 87

spin axis, 86
spin axis orientation, 90
spin precession period, 86
spin rate, 91–92
spin–orbit resonance, 63, 72, 76, 79, 86, 107, 266, 267, 347, 551, 565
sputtering, 408, 409, 413, 415, 417, 419, 422, 423, 424
  chemical, 407, 412
  ion, 407, 421
  models, 421
stagnant lid, 169, 527
standardized reflectance, 192, 196, 198, 201
Steering Committee for Space Science, 552
stereo imaging, 55
stereophotogrammetry, 250, 288
Stokes flow, 65
stratigraphic column, 206
strike-slip deformation, 255
sub-isostatic state, 64
sublimation, 338, 341, 550
sublimation degradation, 338
substorm, 464, 470, 475, 477–478, 479, 480, 486, 487
sulfides, 167, 177, 180, 182, 183, 206, 209, 210, 307, 337, 339, 340
sulfur, 31, 33, 34, 36, 38, 94, 177, 178–184, 185, 206, 208, 209, 210, 336–337, 339, 340, 498, 499
super-isostatic state, 64
surface chronology, 236–239, 240
  absolute, 225, 237–239, 240, 241
  oldest terrains, 238
  relative, 236–237
  smooth plains, 238–239
surface composition, 13, 17, 22, 23, 32–37, 145, 153, 176–188, 206, 308, 315, 336, 499–500, 517, 521–522, 558, 559
surface reflectance, 168, 180, 352
surface roughness, 198
surface temperature, 63, 74, 75, 115, 137, 264, 346, 522, 529
synchronous rotation, 267
syntaxis, 257

talus, 333, 341
TAS. See total alkalis versus silica diagram
TCR. See traveling compression region
tectonic grid, 255
tectonics, 3, 17, 21–22, 218, 222, 241, 249–279, 520, 557, 559, 564
  extensional structures, 221, 222, 223, 258–260, 275
  shortening structures, 221, 222, 250–258, 272–274
tephra, 297
Th. See thorium
thermal conductivity, 101, 526
thermal contraction, 269, 526, 530, 531, 534
thermal desorption, 381, 388, 403, 407–409
thermal erosion, 296
thermal evolution. See thermal history
thermal expansion coefficient, 96, 526
thermal fracturing, 341
thermal history, 63, 76, 105, 114, 115, 138, 249, 264, 276, 311–312, 527–534
thermal lithosphere, 63
thermal neutrons, 31–32, 36, 60, 145, 207, 209, 299, 305, 350, 500, 517
thermochemical convection, 114
thermochemical evolution models, 527–534
thermoelastic strain, 77
thermoelastic stress, 77, 78, 79
thorium, 31, 33, 36, 500
thrust duplex, 251

thrust fault, 251, 293
thrusters, 556
   ion, 555, 562
Ti. *See* titanium
tidal bulge, 71
tidal despinning, 69, 77, 107, 255, 266, 278, 522
tidal forcing, 89, 100
tidal potential, 99
   Love number, 57, 89, 99
tidal response, 99–103
tilted crater floors, 68, 223, 259, 275
time-stratigraphic system, 154, 157–168
titanium, 31, 36, 177, 179, 182, 183, 184, 206
Tolstojan period, 166, 272, 310
Tolstojan System, 157–159, 163–166
topography, 13, 52, 224
total alkalis versus silica diagram, 178, 180, 181, 182, 183
total macroscopic neutron absorption cross section, 177
trachyandesite, 145
trachyte, 181
traveling compression region, 469, 470, 479, 480
Triton, 338
TRM. *See* remanence: thermal magnetization
troilite. *See* iron sulfide
true polar wander, 522

U. *See* uranium
Ultraviolet and Visible Spectrometer, 8, 191, 193–195, 336, 375, 377, 390, 392, 396, 401, 407, 408, 411, 415, 416, 417, 419, 421, 422, 423, 424, 425
uncompressed density, 497
unrelaxed rigidity, 100
uplink, 556
uranium, 36, 500
UVVS. *See* Ultraviolet and Visible Spectrometer

valles, 260, 276, 295
velocity change, 546, 556
velocity distribution, 372
   exosphere, 407, 409, 410, 411, 413, 415, 422, 423
   meteoroid, 409
Venus, 168, 170, 219, 232, 256, 270, 296, 498, 547, 555, 565
vergence, 251
Very Large Array, 346
Viking landers, 544
VIRS. *See* Visible and Infrared Spectrograph
viscosity, 63, 100, 101, 177, 180, 181, 182, 183, 184, 307, 527
Visible and Infrared Spectrograph, 8, 191, 193–195, 335
volatile deposits, 353, 558

volatile elements, 13, 15, 32–36, 362, 499, 504, 509, 522, 563
volatile loss, 332, 338–341
volatile organic compounds, 15, 22, 355, 356, 357, 359, 361, 362, 365, 563
volatile phases, 207–208, 338–341
volatile wasting, 550
volatiles
   crustal, 550
   frozen, 550, 554, 562
volcanic history, 238–239, 308–312
volcanic landforms, 153–154
volcanic pits, 337
volcanic vents, 225, 226, 297, 311
volcanism, 3, 13, 17, 105, 115, 147, 168, 218, 221, 222, 224, 231, 233, 241, 276–277, 287–316, 339, 519, 549–550, 557, 559, 564
   early, 241
Voyager Program, 544, 562
VRM. *See* remanence: viscous magnetization
vulcanoids, 225, 240–241

WAC. *See* wide-angle camera
water ice, 3, 15, 22, 346, 348, 351, 361, 550, 562, 563, 565
   organic synthesis within, 363
   source, 361–366
   stability, 346
   subsurface, 355
   surface exposures, 354, 355, 357
   thermal sublimation, 346, 362
wehrlite, 184, 185, 186
Weibull distribution, 409, 410, 412, 418
wide-angle camera, 6, 55, 144, 191, 193–195, 220, 240, 250, 288, 324, 336, 347, 357
wrinkle ridge, 221, 223, 224, 229, 230, 251, 311

X-band, 556, 560
X-ray fluorescence, 31, 558
X-Ray Spectrometer, 7, 30, 36, 37, 94, 147, 164, 176, 179, 191, 201, 207, 209, 288, 336, 339, 401, 451, 461, 476, 498
XRS. *See* X-Ray Spectrometer

Yakima fold belt, 257
Yarkovsky effect, 233, 240
yield strength envelope, 62
yield stress, 78
Young's modulus, 62, 261
YSE. *See* yield strength envelope

Σ_a. *See* total macroscopic neutron absorption cross section

INDEX OF PLACE NAMES

Abedin basin, 221, 228, 275
Ahmad Baba basin, 303, 304
Ailey crater, 167, 168, 332
Akutagawa crater, 207
Alver crater, 310
Amaral crater, 160, 230
Andal–Coleridge basin, 218
Aneirin basin, 159, 309
Angkor Vallis, 295
Antoniadi Dorsum, 252, 256, 257
Apārangi Planitia, 150, 151, 152, 272
Apollodorus crater, 221, 260
Atget crater, 261, 276

Balanchine crater, 324, 332, 333, 334
Balzac crater, 160, 324, 335
Barma crater, 151, 309
Bartók crater, 160
Bashō crater, 160, 205, 208, 332
Beagle Rupes, 251, 252, 255
Bechet crater, 353
Beckett crater, 300
Beethoven basin, 151, 152, 159, 164, 197, 218, 277, 294, 295, 298, 303, 309, 310, 312
Belgica Rupes, 255
Benoit crater, 300
Blossom Rupes, 255, 272
Borealis basin, 55, 70, 207, 218
Borealis Planitia, 152, 253
Boznańska crater, 326, 328, 329
Budh basin, 55, 70, 218
Budh Planitia, 152, 155, 292
Burke crater, 348, 349

Cahokia Vallis, 295
Calder crater, 151
Calder–Hodgkins basin, 159, 164, 165
Caloris basin, 17, 34, 36, 55, 65, 66, 67, 68–69, 70, 131, 132, 136, 137, 138, 145, 150, 151, 152, 154, 155, 156, 158, 159, 161, 162, 166, 182, 185, 188, 199, 205, 217, 218, 220, 221–223, 237, 238, 241, 249, 252, 253, 254, 258, 259, 260, 261, 265, 268, 269, 270, 271, 275, 276, 277, 292, 296, 298, 301, 309, 310, 312, 326, 337, 519, 521, 522, 532, 535
    Nervo Formation, 156
    Odin Formation, 156
    Van Eyck Formation, 156, 159
Caloris Montes, 224
Caloris Planitia, 151, 152, 153, 155, 157, 249, 260, 277, 292, 294, 295, 297, 298, 299, 303, 309, 310, 312

Calvino crater, 204, 206
Calypso Rupes, 274
Caral Vallis, 295
Carnegie Rupes, 59, 131, 252, 273
Catuilla Planum, 152
Chao Meng-Fu crater, 55, 346, 347, 348
Chesterton crater, 349, 354, 355, 359
Copland crater, 328
Cunningham crater, 228, 332, 335

Dali basin, 291
de Graft crater, 160, 324, 326
Debussy crater, 160, 168, 230, 231
Degas crater, 160, 167, 168, 259, 260, 275, 301, 326, 332
Derzhavin crater, 207
Derzhavin–Sor Juana basin, 218
Desprez crater, 349, 350
Discovery Rupes, 65, 78, 263
Disney crater, 310
Dominici crater, 200, 326, 332, 336
Dostoevskij basin, 159, 218
Duccio crater, 251, 252

Eastman crater, 290
Egonu crater, 349, 357
Eminescu basin, 199, 324, 326, 335, 341
Endeavour Rupes, 256, 257
Ensor crater, 349, 360
Enterprise Rupes, 70, 252, 254, 255, 268, 269, 272
Erté crater, 160

Faulkner crater, 205, 309, 520
Firdousi crater, 203
Fonteyn crater, 220
Fuller crater, 260, 349, 353, 354, 355, 359, 360
Futabatei crater, 160

Gaudí crater, 303, 304
Gibran basin, 151, 302
Glinka crater, 277
Goethe basin, 55, 70, 150, 159, 218, 229, 260, 294

Hawthorne–Riemenschneider basin, 218
Hodgkins crater, 151
Hokusai crater, 145, 160, 167, 168, 228, 229, 275, 301, 311, 366
Homer basin, 159, 218
Hopper crater, 200, 324, 325, 326, 336

Jokai crater, 151

Kandinsky crater, 349, 354, 355, 359
Kertész crater, 324, 326, 340
Kofi basin, 295
Kuiper crater, 145, 159, 160, 166, 167, 168, 230, 324, 332
Kunisada basin, 290
Kuniyoshi crater, 277, 311
Kyosai crater, 326

La Dauphine Rupes, 255
Laxness crater, 349, 353, 354, 360
Lennon–Picasso basin, 159, 165
Lermontov basin, 324, 331, 339
Lugus Planitia, 151, 152

Machaut crater, 231
Mansur crater, 159
Martial crater, 273
Matabei crater, 197
Matisse–Repin basin, 218, 253
Mearcair Planitia, 152, 155
Monk crater, 349, 357
Mozart basin, 155, 221, 222, 224–225, 258, 268, 269, 275, 301

Nabokov crater, 163, 164, 165, 203

Odin Planitia, 55, 70, 152, 155, 156
Otaared Planitia, 152, 165

Paestum Vallis, 295
Pantheon Fossae, 221, 223, 258, 260, 269, 270, 275, 299, 301
Papsukkal Planitia, 151, 152, 153
Paramour Rupes, 255
Picasso crater, 165, 203
Praxiteles basin, 311, 339
Prokofiev crater, 55, 349, 350, 353, 354, 355, 357, 358, 359, 362
Pushkin crater, 151

Qiu Ying crater, 353

Rachmaninoff basin, 35, 36, 55, 62, 70, 145, 154, 167, 199, 201, 205, 221, 222, 223–224, 225, 239, 240, 258, 269, 275, 299, 301, 310, 326, 328, 337, 339, 520
Raditladi basin, 55, 70, 155, 220, 221, 222, 224, 225, 239, 240, 258, 269, 275, 301, 324, 326, 331, 333, 334
Raphael basin, 159, 218
Rembrandt basin, 17, 70, 145, 150, 151, 152, 153, 157, 159, 161, 197, 203, 205, 218, 221, 223, 224, 240, 252, 258, 268, 269, 270, 272, 277, 293, 294, 295, 298, 299, 301, 303, 309, 310, 312
Rudaki crater, 201, 287, 309, 310
Rustaveli basin, 331

Sanai basin, 159
Sander crater, 324, 326, 340
Santa Maria Rupes, 263
Sapkota crater, 348, 349
Scarlatti basin, 339
Seuss crater, 160, 326
Shakespeare basin, 131, 133, 159, 218
Sholem Aleichem basin, 151, 207, 333
Sibelius crater, 231
Sihtu Planitia, 151, 152, 164
Simonides crater, 151
Sobkou basin, 55, 70, 131, 218
Sobkou Planitia, 145, 152, 155, 159, 164, 205, 304
Sor Juana crater, 55, 70, 207
Spitteler crater, 160
Stevenson crater, 230
Stieglitz crater, 228, 303, 304
Stilbon Planitia, 152
Strindberg basin, 131, 133, 303, 304
Suisei Planitia, 131, 132, 133, 136, 152, 155
Sveinsdóttir crater, 229, 251, 252

Theophanes crater, 324, 335
Thoreau crater, 151
Timgad Vallis, 295, 297
Tir Planitia, 70, 151, 152, 155, 292
Tolkien crater, 349, 354, 355, 359
Tolstoj basin, 145, 151, 152, 153, 158, 159, 203, 205, 218, 224, 240, 277, 294, 295, 298, 299, 303, 309, 310, 312
Tryggvadóttir crater, 349, 354, 355, 359
Turgenev crater, 131, 133
Turms Planitia, 151, 152
Tyagaraja crater, 14, 160, 201, 203, 324, 331, 335, 339

Utaridi crater, 151
Utaridi Planitia, 152

Van Eyck Formation, 296
Velázquez basin, 220
Victoria Rupes, 131, 256, 257, 311
Vincente crater, 151
Vincente–Yakovlev basin, 218
Vivaldi basin, 163, 164, 256, 326
Vyāsa basin, 55, 70, 159, 218

Warhol crater, 326, 327
Waters crater, 199

Xiao Zhao crater, 326, 327, 332

Zeami crater, 324, 335

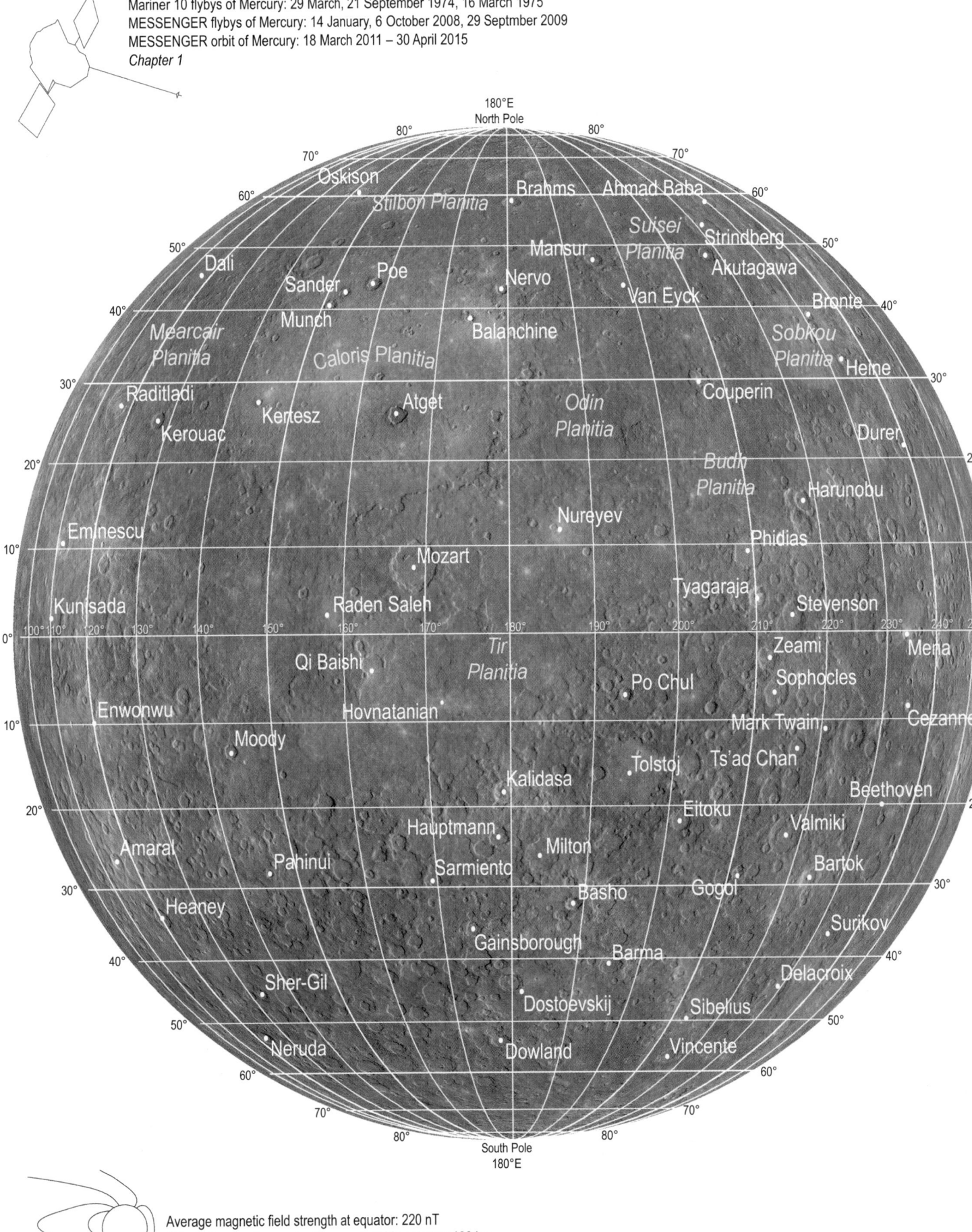

Mariner 10 flybys of Mercury: 29 March, 21 September 1974, 16 March 1975
MESSENGER flybys of Mercury: 14 January, 6 October 2008, 29 Septmber 2009
MESSENGER orbit of Mercury: 18 March 2011 – 30 April 2015
Chapter 1

Average magnetic field strength at equator: 220 nT
Offset of magnetic equator from geographic equator: 480 km
Chapter 5